The Latin rules of grammar pertain to singular and plural forms of scientific names.

	Gender		
	Feminine	Masculine	Neuter
Singular	-a	-us	-um
Plural	-ae	-i	-a
Examples	alga, algae	fungus, fungi	bacterium, bacteria

a-, an- absence, lack. Examples: abiotic, in the absence of life; anaerobic, in the absence of air.

-able able to, capable of. Example: viable, having the ability to live or exist.

actino- ray. Example: actinomycetes, bacteria that form star-shaped (with rays) colonies.

aer- air. Examples: aerobic, in the presence of air; aerate, to add air.

albo- white. Example: *Streptomyces albus* produces white colonies.

ameb- change. Example: ameboid, movement involving changing shapes.

amphi- around. Example: amphitrichous, tufts of flagella at both ends of a cell.

amyl- starch. Example: amylase, an enzyme that degrades starch.

ana- up. Example: anabolism, building up.

ant-, anti- opposed to, preventing. Example: antimicrobial, a substance that prevents microbial growth.

archae- ancient. Example: archaeobacteria, "ancient" bacteria, thought to be like the first form of life.

asco- bag. Example: ascus, a baglike structure holding spores.

aur- gold. Example: *Staphylococcus aureus,* gold-pigmented colonies.

aut-, auto- self. Example: autotroph, self-feeder.

bacillo- a little stick. Example: bacillus, rod-shaped.

basid- base, pedestal. Example: basidium, a cell that bears spores.

bdell- leech. Example: *Bdellovibrio,* a predatory bacterium.

bio- life. Example: biology, the study of life and living organisms.

blast- bud. Example: blastospore, spores formed by budding.

bovi- cattle. Example: *Mycobacterium bovis,* a bacterium found in cattle.

brevi- short. Example: *Lactobacillus brevis,* a bacterium with short cells.

butyr- butter. Example: butyric acid, formed in butter, responsible for rancid odor.

campylo- curved. Example: *Campylobacter,* curved rod.

carcin- cancer. Example: carcinogen, a cancer-causing agent.

caseo- cheese. Example: caseous, cheeselike.

caul- a stalk. Example: *Caulobacter,* appendaged or stalked bacteria.

cerato- horn. Example: keratin, the horny substance making up skin and nails.

chlamydo- covering. Example: chlamydospores, spores formed inside hypha.

chloro- green. Example: chlorophyll, green-pigmented molecule.

chrom- color. Examples: chromosome, readily stained structure; metachromatic, intracellular colored granules.

chryso- golden. Example: *Streptomyces chryseus,* golden colonies.

-cide killing. Example: bactericide, an agent that kills bacteria.

cili- eyelash. Example: cilia, a hairlike organelle.

cleisto- closed. Example: cleistothecium, completely closed ascus.

co-, con- together. Example: concentric, having a common center, together in the center.

cocci- a berry. Example: coccus, a spherical cell.

coeno- shared. Example: coenocyte, a cell with many nuclei not separated by septa.

col-, colo- colon. Examples: colon, large intestine; *Escherichia coli,* a bacterium found in the large intestine.

conidio- dust. Example, conidia, spores developed at the end of aerial hypha, never enclosed.

coryne- club. Example: *Corynebacterium,* club-shaped cells.

-cul small form. Example: particle, a small part.

-cut the skin. Example: Firmicutes, bacteria with a firm cell wall, gram-positive.

cyano- blue. Example: cyanobacteria, blue-green pigmented organisms.

cyst- bladder. Example: cystitis, inflammation of the urinary bladder.

cyt- cell. Example: cytology, the study of cells.

de- undoing, reversal, loss, removal. Example: deactivation, becoming inactive.

di-, diplo- twice, double. Example: diphlococci, pairs of cocci.

dia- through, between. Example: diaphragm, the wall through or between two areas.

dys- difficult, faulty, painful. Example: dysfunction, disturbed function.

ec-, ex-, ecto- out, outside, away from. Example: excrete, to remove materials from the body.

en-, em- in, inside. Example: encysted, enclosed in a cyst.

entero– intestine. Example: *Enterobacter,* a bacterium found in the intestine.

eo– dawn, early. Example: *Eobacterium,* a 3.4-billion-year-old fossilized bacterium.

epi– upon, over. Example: epidemic, number of cases of a disease over the normally expected number.

erythro– red. Example: erythema, redness of the skin.

eu– well, proper. Example: eukaryote, a proper cell.

exo– outside, outer layer. Example: exogenous, from outside the body.

extra– outside, beyond. Example: extracellular, outside the cells of an organism.

firmi– strong. Example: *Bacillus firmus* forms resistant endospores.

flagell– a whip. Example: flagellum, a projection from a cell; in eukaryotic cells, it pulls cells in a whiplike fashion.

flav– yellow. Example: *Flavobacterium* cells produce yellow pigment.

fruct– fruit. Example: fructose, fruit sugar.

–fy to make. Example: magnify, to make larger.

galacto– milk. Example: galactose, monosaccharide from milk sugar.

gamet– to marry. Example: gamete, a reproductive cell.

gastr– stomach. Example: gastritis, inflammation of the stomach.

gel– to stiffen. Example: gel, a solidified colloid.

–gen an agent that initiates. Example: pathogen, any agent that produces disease.

–genesis formation. Example: pathogenesis, production of disease.

germ, germin– bud. Example: germ, part of an organisms capable of developing.

–gony reproduction. Example: schizogony, multiple fission producing many new cells.

gracili– thin. Example: *Aquaspirillum gracile,* a thin cell.

halo– salt. Example: halophile, an organism that can live in high salt concentrations.

haplo– one, single. Example: haploid, half the number of chromosomes or one set.

hema–, hemato–, hemo– blood. Example: *Haemophilus,* a bacterium that requires nutrients from red blood cells.

hepat– liver. Example: hepatitis, inflammation of the liver.

herpes creeping. Example: herpes, or shingles, lesions appear to creep along the skin.

hetero– different, other. Example: heterotroph, obtains organic nutrients from other organisms; other feeder.

hist– tissue. Example: histology, the study of tissues.

hom–, homo– same. Example: homofermenter, an organism that produces only lactic acid from fermentation of a carbohydrate.

hydr–, hydro– water. Example: dehydration, loss of body water.

hyper– excess. Example: hypertonic, having a greater osmotic pressure in comparison with another.

hypo– below, deficient. Example: hypotonic, having a lesser osmotic pressure in comparison with another.

im– not, in. Example: impermeable, not permitting passage.

inter– between. Example: intercellular, between the cells.

intra– within, inside. Example: intracellular, inside the cell.

io– violet. Example: iodine, a chemical element that produces a violet vapor.

iso– equal, same. Example: isotonic, having the same osmotic pressure when compared with another.

–itis inflammation of. Example: colitis, inflammation of the large intestine.

–karyo, –caryo a nut. Example: eukaryote, a cell with a membrane-enclosed nucleus.

kin– movement. Example: streptokinase, an enzyme that lyses or moves fibrin.

lacti– milk. Example: lactose, the sugar in milk.

lepis– scaly. Example: leprosy, disease characterized by skin lesions.

lepto– thin. Example: *Leptospira,* thin spirochete.

leuko– whiteness. Example: leukocyte, a white blood cell.

lip–, lipo– fat, lipid. Example: lipase, an enzyme that breaks down fats.

–logy the study of. Example: pathology, the study of changes in structure and function brought on by disease.

lopho– tuft. Example: lophotrichous, having a group of flagella on one side of a cell.

luc–, luci– light. Example: luciferin, a substance in certain organisms that emits light when acted upon by the enzyme luciferase.

lute–, luteo– yellow. Example: *Micrococcus luteus,* yellow colonies.

–lysis loosening, to break down. Example: hydrolysis, chemical decomposition of a compound into other compounds as a result of taking up water.

macro– largeness. Example: macromolecules, large molecules.

mendosi– faculty. Example: mendosicutes, archaeobacteria lacking peptidoglycan.

meningo– membrane. Example: meningitis, inflammation of the membranes of the brain.

meso– middle. Example: mesophile, an organism whose optimum temperature is in the middle range.

The Microbiology Place

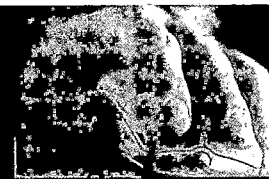

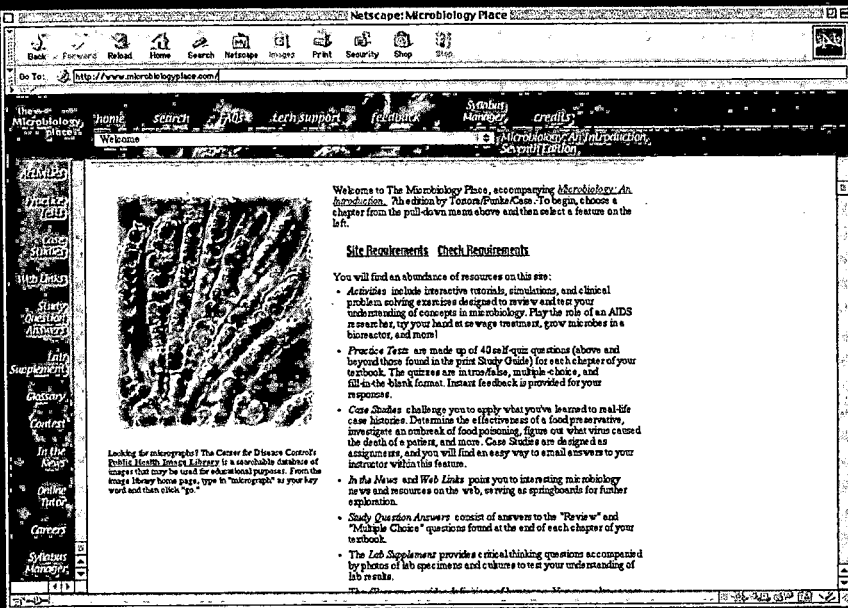

Log on.

Tune in.

Succeed.

To help you succeed in microbiology, your professor has arranged for you to enjoy access to great media resources, including two interactive CD-ROMs and **The Microbiology Place** web site. You will find that these resources that accompany your textbook will enhance your course materials.

Here's your personal ticket to success:

How to log on to www.microbiologyplace.com:

1. Go to www.microbiologyplace.com.
2. Click "Register Here."
3. Enter your pre-assigned Access Code exactly as it appears below and click "Submit."
4. Complete online registration form to create your own personal User ID and Password.
5. Once your User ID and Password are confirmed by email, go back to www.microbiologyplace.com, type in your new User ID and Password, and click "Enter."

Your Access Code is:

WSTM-CROOK-CREPY-TIGON-DAUNT-ROSES

Detach this card and keep it handy. It's your ticket to valuable information.

Important: Please read the License Agreement, located on the launch screen before using The Microbiology Place or the Microbiology Interactive Student Tutorial CD-ROM. By using the web site, you indicate that you have read, understood, and accepted the terms of this agreement.

Got technical questions?
For technical support, please visit www.aw.com/techsupport, send an email to online.support@pearsoned.com (for web site questions) or send an email to media.support@pearsoned.com (for CD-ROM questions) with a detailed description of your computer system and the technical problem. You can also call our tech support hotline at 1-800-677-6337 Monday-Friday, 8 a.m. to 5 p.m. CST.

What your system needs to use these media resources:
WINDOWS
166 MHz Intel Pentium processor or greater
Windows 95, 98 NT4, 2000
32 MB RAM installed
800 X 600 screen resolution
4x CD-ROM drive
Browser: Internet Explorer 5.0 or Netscape Communicator 4.7
Plug-Ins: Shockwave Player 8, Flash Player 4, Chime

MACINTOSH
120 MHz PowerPC
OS 8.1 or higher
32 MB RAM available
800 x 600 screen resolution, thousands of colors
4x CD-ROM Drive
Browser: Internet Explorer 5.0 or Netscape Communicator 4.7
Plug-Ins: Shockwave Player 8, Flash Player 4, Chime

Your User ID

Your Password

Microbiology

AN INTRODUCTION

Media Update

Gerard J. Tortora

Bergen Community College

Berdell R. Funke

North Dakota State University

Christine L. Case

Skyline College

Benjamin
Cummings

San Francisco   Boston   New York
Cape Town   Hong Kong   London   Madrid   Mexico City
Montreal   Munich   Paris   Singapore   Sydney   Tokyo   Toronto

Publisher: Daryl Fox

Sponsoring Editor: Amy Folsom

Assistant Editor: Erin Joyce

Managing Editor: Wendy Earl

Production Editor: Sharon Montooth

Art and Design Director: Bradley Burch

Art and Photo Supervisor: Donna M. Kalal

Photo Editor: Stuart Kenter

Text and Cover Designer: Carolyn Deacy

Copy Editor: Sally Peyrefitte

Proofreader: Martha Ghent

Illustrations: Precision Graphics

Compositor: GTS Graphics

Manufacturing Supervisor: Stacey Weinberger

Marketing Manager: Lauren Harp

Cover Photograph: Susan M. Barns, Los Alamos National Laboratory

Credits appear following the Glossary.

Library of Congress Cataloging-in-Publication Data

Tortora, Gerard J.
   Microbiology: an introduction media update/Gerard J. Tortora, Berdell R. Funke, Christine L. Case.
      p. cm.
   This is an update of the 7th ed.
   Includes bibliographical references and index.
   ISBN 0-8053-7602-X (pbk.)
      I. Microbiology. I. Funke, Berdell R. II. Case, Christine L., 1948– . III. Title.

QR41.2. T67 2002
579—dc21

2001037298

ISBN 0-8053-7602-X

10 9 8 7 6 5 4 3 2 1-VHP-05 04 03 02 01

www.aw.com/bc

About the Authors

Gerard J. Tortora Jerry Tortora is a professor of biology and teaches microbiology and human anatomy and physiology at Bergen Community College in Paramus, New Jersey. He received his M.A. in biology from Montclair State College in 1965. He belongs to numerous biology/microbiology organizations, such as the American Society of Microbiology (ASM), Human Anatomy and Physiology Society (HAPS), American Association for the Advancement of Science (AAAS), National Education Association (NEA), New Jersey Education Association (NJEA), and the Metropolitan Association of College and University Biologists (MACUB). Jerry is the author of a number of biological science textbooks. In 1995, he was selected as one of the finest faculty scholars at Bergen Community College and was named Distinguished Faculty Scholar. In 1996, Jerry received a National Institute for Staff and Organizational Development (NISOD) excellence award from the University of Texas and was selected to represent Bergen Community College in a campaign to increase awareness of the contributions of community colleges to higher education.

Berdell R. Funke Bert Funke received his Ph.D., M.S., and B.S. in microbiology from Kansas State University. He has spent his professional years as a professor of microbiology at North Dakota State University. He is teaching introductory microbiology, including laboratory sections, general microbiology, food microbiology, soil microbiology, and clinical parasitology. He has also taught pathogenic microbiology. As a research scientist in the Experiment Station at North Dakota State, he has published numerous papers in soil microbiology and food microbiology.

Christine L. Case Chris Case is a registered microbiologist and a professor of microbiology at Skyline College in San Bruno, California, where she has taught for the past 30 years. She received her Ed.D. in curriculum and instruction from Nova Southeastern University and her M. A. in microbiology from San Francisco State University. She was Director for the Society for Industrial Microbiology (SIM) and is an active member of the ASM, SIM, the Northern California Association of Microbiologists (NCASM), Northern California SIM, and the Sigma Xi Research Society. She received the NCASM outstanding educator award. In addition to teaching, Chris contributes regularly to the professional literature, develops innovative educational methodologies, and maintains a personal and professional commitment to conservation and the importance of science in society. Chris is also an avid photographer, and many of her photographs appear in this book.

Preface

The seventh edition of *Microbiology: An Introduction* continues to be a comprehensive text for students in a wide variety of programs. Among these are allied health sciences, biological sciences, environmental studies, animal science, forestry, agriculture, home economics, and liberal arts. It is a beginning text, assuming no previous study of biology or chemistry.

During the 18 years since the publication of the first edition, this book has been used at more than 1000 colleges and universities by over 800,000 students, making it the best selling introductory text around the world. We have been gratified to hear from instructors and students alike that the book has become a favorite among their textbooks—a learning tool that is both effective and enjoyable.

We have retained in this new edition the features that made the previous editions so popular. These include:

- **An appropriate balance between microbiological fundamentals and applications, and between medical applications and other applied areas of microbiology.** As in previous editions, basic microbiological principles are given greater emphasis than applications, and health-related applications are emphasized. Applications are integrated throughout the text, and considerable attention is devoted to microorganisms in habitats outside the human body. We hope students will gain an appreciation for the fascinating diversity of microbial life, the central roles of microorganisms in nature, and the importance of microorganisms in our daily lives.

- **Straightforward presentation of complex topics.** Each section of the text has been revised with the student in mind, to maintain the clarity of explanation for which our book has become known.

- **The integrated learning objectives and end-of-chapter questions help students check their understanding of key chapter concepts and learn critical problem-solving skills needed in clinical and industrial situations.**

- **Applications and discovery-oriented boxes focus on modern, practical uses of microbiology and biotechnology and emphasize the process of scientific discovery.** The applications boxes show real people doing science to provide students with examples of career opportunities in microbiology.

Features of the Seventh Edition

Our primary goal for the seventh edition of *Microbiology: An Introduction* was to again update the book throughout, making certain it reflects the important new discoveries and reclassification of microbes during the past few years. This updating is evident in such topics as emerging infectious diseases, the phylogenetic classification of prokaryotes in accordance with the forthcoming 2nd edition of *Bergey's Manual of Systematic Bacteriology*, phylogenetic classification of fungi and protozoa, chromosome mapping and genomics, DNA chips, trichothecene mycotoxins, acute-phase proteins, vaccine safety, new groups of antibiotics, revised discussion of bioremediation and biofilms, new advances in microscopy, diagnostic immunology, AIDS, and industrial uses of microbes. At the same time, we wanted to maintain a manageable length with an emphasis on clear coverage of the fundamental principles of microbiology. Every page and every figure of the book was scrutinized with these goals in mind. The narration has been updated throughout and every element of the art program has been revised, refined, and updated.

The following list highlights the major changes in this edition:

- **NEW In-text instructions for using technology.** Within the Study Outlines at the end of each chapter, icons for the Student Tutorial Version 2.0 direct students to the appropriate topic, subtopic and interactive activity on the CD. In addition, new end-of-chapter "Learning with Technology" sections give students specific instructions for completing relevant CD-ROM and website exercises and quizzes for each chapter. Students will now find it easier to make the best use of their two free CD-ROMs (the Microbiology Interactive Student Tutorial Version 2.0 and VirtualUnknown™ Microbiology) as well as The Microbiology Place website.

- **NEW The Microbiology Place Web site** (www.microbiologyplace.com) This new, in-depth, subscription-based web site contains exciting and highly interactive microbiology study aids and teaching resources. The site includes tutorial animations, simulations, chapter quizzes, articles, web links, case studies, a syllabus manager, answers to the end-of-chapter study questions, and a Microbe of the Month contest. A FREE 12-month subscription is included with each new student copy of the text.

- **Updated Applications boxes.** To stay abreast of new developments in the field, most boxes have either been updated or are new to the Seventh Edition. *Applications of Microbiology* boxes focus on modern, practical uses of microbiology and biotechnology. *Microbiology in the News* boxes make sense of stories in today's headlines. *Morbidity and Mortality Weekly Report* boxes review the latest epidemiological cases from the CDC. New *Clinical Problem Solving* boxes use case histories adapted from recent issues of *Morbidity and Mortality Weekly Report* to encourage critical thinking in the examination of a clinical problem.

- **An exceptional art program includes many new micrographs and figures.** Well-developed, full-color illustrations throughout the book support and enhance the text. Key concepts and questions appear in figure legends to encourage students to think critically about illustrations.

Key features of the illustration program include:

Consistent use of symbols and colors. Molecules such as phosphate groups (ⓟ) and ATP (ATP) are the same color and shape throughout the book, enabling students to progress from familiar parts of illustrated processes to unfamiliar ones with confidence.

Orientation diagrams. These miniature versions of overview illustrations, with appropriate parts highlighted, have been retained and expanded.

Step-by-step descriptions. Step-by-step descriptions, in both the narration and illustrations, walk students through important microbiological processes and help them visualize the order of events. Color-coded numbers link text and figure legends to corresponding art.

Micrograph icons. Icons appear throughout the book to identify the types of microscopes used in the micrographs. Red **SEM** (scanning electron micrographs) icons and **TEM** (transmission electron micrographs) icons indicate the use of colorized micrographs. The **LM** icon indicates light micrographs.

Approximately 350 key illustrations are included in the transparency acetate package, with additional selections provided in the Digital Library Art CD-ROM. See "Supplementary Materials for the Instructor."

Chapter Highlights of the Seventh Edition

Every chapter in this edition has been thoroughly revised, and all the data in the text, tables, and figures have been updated through January 2000 where possible. The main changes for each chapter are summarized below.

Part One: Fundamentals of Microbiology

- **Chapter 1, The Microbial World and You** Living organisms are classified into Bacteria, Archaea, and Eukarya domains. Emerging infectious disease discussion includes bovine spongiform encephalopathy, *E. coli* O157:H7, Ebola hemorrhagic fever, *Hantavirus* pulmonary syndrome, and cryptosporidiosis. Topics are reorganized and several new photos are included.

- **Chapter 2, Chemical Principles,** includes an enhanced discussion of enzymes.

- **Chapter 3, Observing Microorganisms Through a Microscope,** has revised illustrations and an atomic-force micrograph.

- **Chapter 4, Functional Anatomy of Prokaryotic and Eukaryotic Cells,** includes new illustrations of eukaryotic cells and a revised discussion of eukaryotic cells to include peroxisomes and centrosomes.

- **Chapter 5, Microbial Metabolism,** includes revised inset diagrams of metabolic pathways.

- **Chapter 6, Microbial Growth** Hyperthermophiles are included with a new diagram illustrating the relationships between hydrothermal vent worms and their symbiotic bacteria.

- **Chapter 7, The Control of Microbial Growth,** includes several new and updated discussions including: biocides, plasma gas sterilization, bisphenols (triclosan), conditions that influence microbial control, and relative susceptibility of microbes to biocides. A new *Clinical Problem Solving* box shows the link between resistance to chemical disinfectants and hospital-acquired infections.

- **Chapter 8, Microbial Genetics,** has been carefully revised to include 5'–3' orientation of DNA and RNA. Miniature orientation diagrams have been added to figures to show the relationship of specific processes to vertical and horizontal gene transmission. Chromosome mapping and genomics are described, and DNA chips are introduced.

- **Chapter 9, Biotechnology and Recombinant DNA,** has been revised to emphasize microbiological techniques such as selection and mutation. Antisense DNA technology is introduced, and the use of biotechnology to make blue jeans is described in a new box.

Part Two:
A Survey of the Microbial World

- **Chapter 10, Classification of Microorganisms** Phylogenetic classification of prokaryotic organisms is updated to reflect application of rRNA sequencing and *Bergey's Manual of Systematic Bacteriology,* Second Edition. Figures of domains include taxa that students will encounter in Chapters 11 and 12.

- **Chapter 11, The Prokaryotes: Domains Bacteria and Archaea,** has been completely revised to reflect the phylogenetic classification of prokaryotes, in accordance with *Bergey's Manual of Systematic Bacteriology,* Second Edition. Learning objectives encourage students to develop dichotomous keys to practice differentiating between bacteria and to learn characteristics of selected bacteria.

- **Chapter 12, The Eukaryotes: Fungi, Algae, Protozoa, and Helminths** The phylogenetic classification of fungi and protozoa is updated. An emerging infectious disease is introduced in a new MMWR box on *Pfiesteria* and possible estuarine-associated syndrome.

- **Chapter 13, Viruses, Viroids, and Prions** A figure showing how prions can be infectious is added. The MMWR box on AIDS risk to health care workers is updated.

Part Three:
Interaction Between Microbe and Host

- **Chapter 14, Principles of Disease and Epidemiology** The pioneering epidemiologic works of Snow, Semmelweis, and Nightingale are included. A new *Clinical Problem Solving* box describes an actual nosocomial outbreak of streptocococcal toxic-shock syndrome. Morbidity data on nosocomial infections, AIDS, Lyme disease, and tuberculosis are updated. There is a revised discussion of nosocomial infections and incidence and prevalence of disease.

- **Chapter 15, Microbial Mechanisms of Pathogenicity** Trichothecene mycotoxins are included.

- **Chapter 16, Nonspecific Defenses of the Host,** includes a new box that describes nitric oxide activation in macrophages. The role of normal microbiota in nonspecific resistance is discussed, and acute-phase proteins are included.

- **Chapter 17, Specific Defenses of the Host: The Immune Response,** contains an expanded discussion of colostrum as a factor in immunity. There is an updated and revised discussion of apoptosis, or programmed cell death. Figure 17.18, which describes the dual nature of the immune system, has been extensively revised.

- **Chapter 18, Practical Applications of Immunology,** includes a new discussion on vaccine safety. There is also expanded coverage of the development of new and novel vaccines, better administration methods for vaccines, and ways to increase the effectiveness of present vaccines, such as the use of adjuvants. New disease diagnostic methods are also included.

- **Chapter 19, Disorders Associated with the Immune System** The section on AIDS has been thoroughly updated, and now includes coverage of the theoretical origins of the virus and the beginning of the epidemic, the theoretical concepts of the pathology of the disease, a discussion of new tests for HIV infection and for the detection of the virus in donated blood, and an updated discussion of the progress of the infection worldwide. The MMWR box on AIDS drugs is also updated.

- **Chapter 20, Antimicrobial Drugs** The new groups of antibiotics, streptogramins, is included, and there is a revised discussion of many commonly used antibiotics. The topic of rising resistance to antibiotics, antibiotic safety, and the search for entirely new antibiotics is extensively updated and revised.

Part Four:
Microorganisms and Human Disease

- **Chapter 21, Microbial Diseases of the Skin and Eyes,** contains a revised discussion of acne and its treatment; new artwork illustrates the latency of certain herpes viruses; the childhood disease roseola has been added; and the discussion of eye infections caused by the use of contact lenses has been updated and revised. A new *Clinical Problem Solving* box describes an emerging necrotizing mycobacterial disease.

- **Chapter 22, Microbial Diseases of the Nervous System** A new *Clinical Problem Solving* box describes an actual case of rabies and a new figure illustrates the pathogenesis of rabies. Recent data comparing the incidence of meningitis attributed to different organisms is included. In the discussion of leprosy, the terms now used by the World Health Organization to describe the stages have been introduced. A discussion of the outbreak of the arbovirus disease, West Nile Virus, in New York, has been added. The discussion of prion-causing diseases has been updated and expanded.

- **Chapter 23, Microbial Diseases of the Cardiovascular and Lymphatic Systems** An entirely new discussion of the disease leishmaniasis has been introduced. The discussion of brucellosis has been completely revised. The discussion of anthrax has been revised with an emphasis on its possible use as an agent of biological terrorism. The discussions about cat-scratch disease and schistosomiasis have been updated and revised. A new *Clinical Problem Solving* box describes the re-emergence of malaria in the United States.

- **Chapter 24, Microbial Disease of the Respiratory System** The discussion of otitis media has been updated and expanded; the disease caused by respiratory syncytial virus extensively updated and revised; the potential for new vaccines for influenza is described. The new drugs now available for treatment of influenza are discussed. A new *Clinical Problem Solving* box follows diagnosis of an actual case of streptococcal pneumonia.

- **Chapter 25, Microbial Diseases of the Digestive System** Methods for detecting the serious pathogen *Escherichia coli* 0157:H7 are discussed. The discussion of Hepatitis C is expanded. A new *Clinical Problem Solving* box describes a multistate outbreak of shigellosis from one source.

- **Chapter 26, Microbial Diseases of the Urinary and Reproductive Systems** Diagnostic methods for syphilis have been updated and revised. The discussion of genital herpes has been updated and expanded. A new *Clinical Problem Solving* box describes a urinary tract infection among athletes.

Part Five:
Environmental and Applied Microbiology

- **Chapter 27, Environmental Microbiology** The discussion of acid deposition (acid rain) has been revised, and there is an expanded discussion of bioremediation and biofilms. There is also more coverage of the forms of life that are not dependent upon energy inputs from sunshine, with a special emphasis on microbial life found deep in the rocks of the earth's mantle.

- **Chapter 28, Applied and Industrial Microbiology,** contains an expanded discussion of irradiation of food. There is also a revised discussion of microbiological processes in (copper) ore extraction.

Course Sequences

We have organized the book in what we think is a useful fashion, while recognizing that the material might be effectively presented in a number of other sequences. For those who wish to use a different order, we have made each chapter as independent as possible and have included numerous cross-references. Fundamentals of microbiology are included in Chapters 1 through 9, and other chapters can be used to illustrate and emphasize topics for specific courses. The Instructor's Guide provides detailed guidelines for organizing the material in several other ways. The various diseases are organized into chapters according to the host organ system most affected. A taxonomic guide to diseases covered in the text is included in Appendix F.

A Comprehensive Teaching and Learning Package

The supplementary materials developed by Benjamin Cummings for the Seventh Edition of Microbiology are more innovative and thorough than ever before—in print, on CD-ROM, and on the Web—for instructor and student alike.

Supplementary Materials for the Student

- **REVISED Microbiology Interactive Student Tutorial CD-ROM Version 2.0** Included with every text, this CD includes tutorials, interactive activities, and simulations covering microbiology's toughest topics: viruses, microbial growth, genetics, biotechnology, microbial interactions, metabolism, biodiversity, and immunology. The new, easy-to-use interface emphasizes the interrelationships of micro-biology concepts and provides a direct link to The Microbiology Place Web site. Version 2.0 features five new interactive activities: Using PCR to identify microbes in environmental samples, Nitrogen cycle, Gene cloning practice, Virus identification, Host and pathogen responses during infection, and three new simulations: The effect of sewage treatment on wa-terways, Industrial fermentations in a bioreactor, and Enzyme inhibitors drugs. It also includes a complete glossary with pronunciations and quizzes for each chapter. FREE with text.

- **NEW The Microbiology Place Web site** (www.microbiologyplace.com) FREE 12-month subscription with every student copy of the book. (See "Features of the Seventh Edition" for more details.)

- **NEW VirtualUnknown™ Microbiology CD-ROM** This complete lab simulation program teaches students how to identify unknown bacteria, as well as reinforces proper laboratory techniques. It contains more than 70 bacterial species and more than 40 laboratory tests. VirtualUnknown™ exercises are integrated into appropriate chapters in the "Learning with Technology" sections of the text. An overview of the program is included in Appendix G; instructions for use are included in the liner notes accompanying the CD-ROM.

- **REVISED Student Study Guide** by Berdell Funke (0-8053-7585-6) Revised for the new edi-tion, the study guide includes concise explanations of key concepts, definitions of important terms, art la-beling exercises, critical thinking problems, and a variety of self-test questions with answers.

- **NEW** *Microbiology: A Photographic Atlas for the Laboratory* by Steve K. Alexander and Dennis Strete (0-8053-2732-0) Tailored for the introductory mi-crobiology laboratory, this atlas contains approxi-mately 400 high-quality, color photographs that demonstrate the results of laboratory procedures and the morphology of important microorganisms.

- **REVISED** *Laboratory Experiments in Microbiology, Sixth Edition* by Ted Johnson and Christine Case (0-8053-7589-9) Includes 57 classroom-tested ex-periments that meet ASM recommendations for the laboratory core curriculum considered essential to teach in every introductory microbiology laboratory. These experiments reinforce lecture concepts in me-tabolism, growth, genetics, and immunology and also promote critical thinking, laboratory thinking skills, and good laboratory practices. A **Preparation Guide** (0-8053-7586-4) is available to instructors to aid in the preparation of lab sessions.

- *Microbiology Coloring Book* by I. Edward Alcamo and Lawrence Elson (0-06-041925-3) This unique study tool contains 105 plates, including text and il-lustrations to color, which enhance understanding and comprehension of important microbiological concepts.

Supplementary Materials for the Instructor

All supplemental teaching materials are available to qual-ified adopting instructors. Please contact your local Ben-jamin Cummings sales consultant or call our Customer Service at 781-944-3700 x5100. Some package items may not be available to adopters outside of the United States.

- **REVISED Instructor's Guide and Test Bank** (0-8053-7584-8) Written and revised by Christine L. Case, the Instructor's Guide contains teaching tips, alternative course outlines, ideas for using special text features, answers to text questions, and suggestions on how to incorporate the use of technology into the classroom. The Test Bank features over 1200 multiple-choice questions with answers as well as 3–5 essay questions per chapter.

- **UPDATED Computerized Test Bank CD-ROM** (Test Gen 3.0 0-8053-7588-0 Windows/Macintosh)

Easy-to-use testing program allows you to view and edit electronic questions from the Test Bank, create multiple tests, and print them in a variety of formats.

- **EXPANDED Transparency Acetates** (0-8053-7591-0) Features approximately 350 key illustrations from the text, 50% more than in the previous edition package.

- **NEW Digital Library: Instructor's Presentation CD-ROM** (0-8053-7545-7) Includes all illustrations from the text and several tables which can be edited and customized for lecture presentations. Images can be easily exported into PowerPoint or course Web pages in three formats: images with labels and leaders, images with leaders only, and images with no labels or leaders. Cross platform CD-ROM.

- **NEW Course Management Systems, including WebCT, Blackboard, and eCollege** Useful for on-line course management and distance learning courses, content selected from The Microbiology Place and the entire computerized Test Bank is available in the three leading course management systems. Please contact your local Benjamin Cummings sales consultant for more details.

Acknowledgements

In the preparation of this textbook, we have benefited from the guidance and advice of a large number of microbiology instructors across the country. The reviewers and focus group participants listed on this page provided constructive criticism and valuable suggestions at various stages of the revision. We gratefully acknowledge our debt to these individuals. We also thank Ken Kubo of American River College for his help with the Digital Library Art CD-ROM and Rick Corbett of Midlands Technical College for his help in choosing the acetate package. We offer special thanks to Laura Bonazzoli for her developmental edit of the sixth edition. Her thoughtful suggestions helped us refine and improve the text in many ways.

We also thank the staff of Benjamin/Cummings for their dedication to excellence. Amy Folsom, our project editor, successfully kept us all focused on where we wanted this revision to go. Sally Peyrefitte's careful attention to continuity and detail in her copyedit of both text and art served to keep concepts and information clear throughout. Sharon Montooth and Wendy Earl expertly guided the text through the production process. Donna Kalal effectively managed the large art program. The photo editor, Stuart Kenter, made sure we had clear and striking images throughout the book. Carolyn Deacy created the interior design and the beautiful cover. GTS Graphics did their usual outstanding job moving this book quickly and beautifully through composition; the skilled team was led by Sandie Sigrist. Stacey Weinberger guided the book through the manufacturing process. In addition, Erin Joyce did an excellent job preparing the manuscript and managing the extensive supplements package. Larry Olsen expertly guided the production of the supplements.

We would all like to acknowledge our spouses and families, who have provided invaluable support throughout the writing process.

Finally, we have an enduring appreciation for our students, whose comments and suggestions provide insight and remind us of their needs. This text is for them.

Gerard J. Tortora
Berdell R. Funke
Christine L. Case

Seventh Edition Reviewers and Focus Group Participants

Virgil Bain, Indian Hills Community College
Susan Cockayne, Brigham Young University
Judy Dacus, Louisiana State University
Judith Davis, Florida Community College at Jacksonville
Axel Duwe, Diablo Valley College
Greg Garman, Centralia College
Charles Goldberg, Borough of Manhattan Community College
Diane Hilker, Mercer County Community College
Jerry Hinkley, College of Lake County
Carol Lauzon, California State University, Hayward
Francis Maxin, Community College of Allegheny County
Donald McGary, Kennesaw State University
Mohan Menon, DeAnza Community College
Remo Morelli, San Francisco State University
Brina Nathanson, Borough of Manhattan Community College
Philip Penner, Borough of Manhattan Community College
Holly Pinkart, Central Washington University
Lisa Shimeld, Crafton Hills College
Richard Shippee, Vincennes University
Doris Spangord, DeAnza Community College
Keith Sternes, Sul Ross State University
Mary Talbot, San Jose State University
Carole Toebe, City College of San Francisco

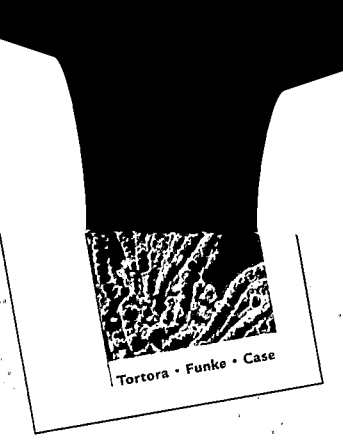

Tortora · Funke · Case

Your textbook includes several built-in tools that can help you study more effectively. Each major section begins with a learning objective, and each chapter ends with an outline and study questions. All of these tools are designed to help you focus on the key concepts in the chapter.

The new end-of-chapter "Learning with Technology" sections provide specific instructions for completing relevant CD-ROM and Web site exercises for each chapter.

Learning with Technology

MP = The Microbiology Place website   **ST** = Student Tutorial CD-ROM   **BID** = Bacteria ID CD-ROM

MP   Don't forget to go to The Microbiology Place website (http://www.microbiologyplace.com) to take the practice tests, explore the interactive activity and case study, and check out the news articles and web links for this chapter.

ST   Remember there is also a quiz for this chapter on the Microbiology Interactive Student Tutorial CD-ROM.

BID   Your neighbor calls you to look at his strawberry plants because they appear to be diseased. You observe that the plants' vascular tissue is filled with a sticky gel. You quickly realize that this gel may be useful as a thickener and emulsifier in foods and other products so you culture the bacterium from the strawberry plants. What is the organism? (Set 3, #8)

Microbiology Interactive Student Tutorial CD-ROM

As you review the end-of-chapter Study Outlines, you will notice that many chapters contain ST icons reminding you to use your **Microbiology Interactive Student Tutorial CD-ROM Version 2.0** to review important concepts in the chapter. These direct you to the appropriate topic, subtopic, and interactive activity or simulation to explore. This cross-platform CD-ROM is fun and easy to use!

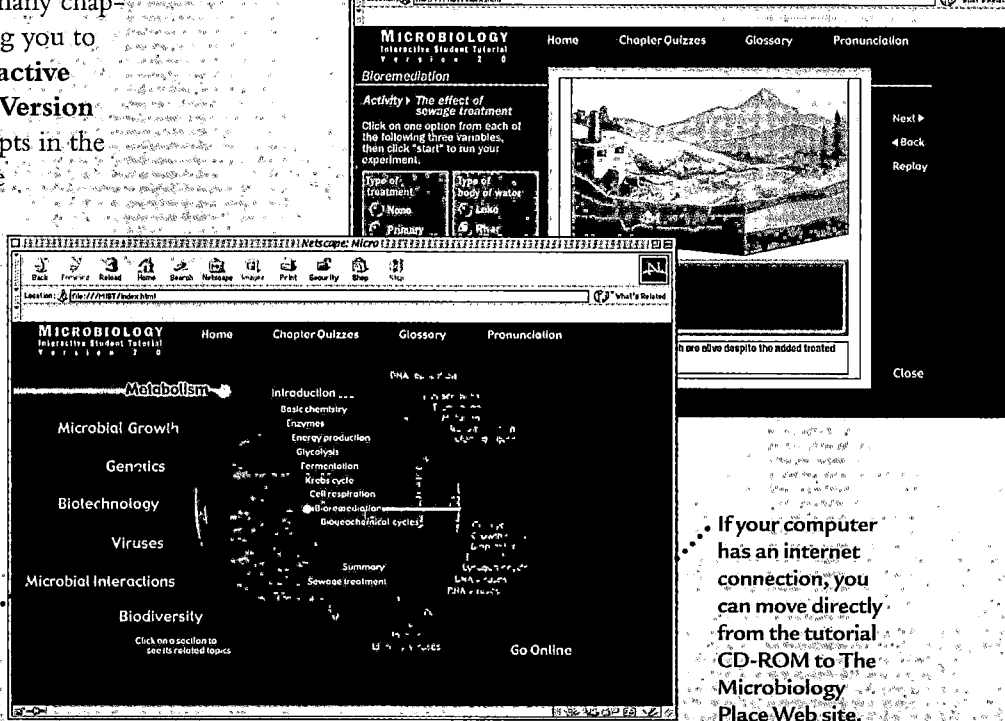

If your computer has an internet connection, you can move directly from the tutorial CD-ROM to The Microbiology Place Web site.

The CD-ROM also includes chapter quizzes, a glossary of terms, and a pronunciation guide.

The Microbiology Place Web site

If you have purchased a new copy of this book, the first page will contain a password for a 12-month subscription to **The Microbiology Place** Web site. The site provides exciting and highly interactive tools to help you prepare for exams, including practice tests, tutorial animations, simulations, case studies, and articles.

Subscribers are also invited to participate in the Microbe of the Month contest.

www.microbiologyplace.com

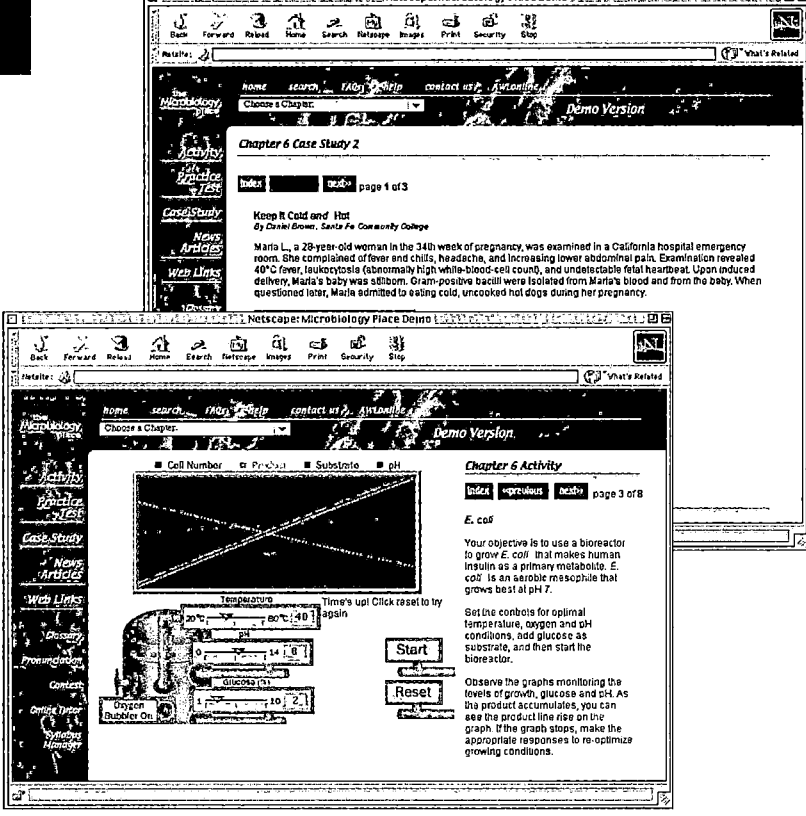

VirtualUnknown™ Microbiology CD-ROM

VirtualUnknown™ Microbiology assigns you an unknown microbe at the beginning of each session—it's then up to you to decide which tests to perform in order to identify your mystery microbe! There are interesting practice problems for you to solve using this CD-ROM in the "Learning with Technology" sections of selected chapters.

Detailed instructions for using **VirtualUnknown™ Microbiology** are found in the Tutorial section of the CD-ROM, on the CD-ROM liner notes, and in Appendix G.

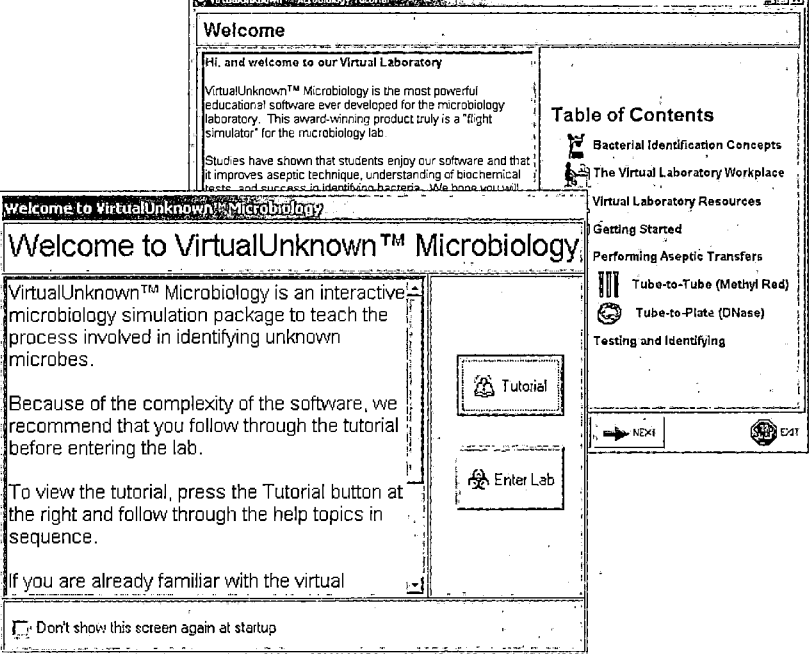

Brief Contents

Contents

CHAPTER 3

Observing Microorganisms Through a Microscope 56

CHAPTER 4

Functional Anatomy of Prokaryotic and Eukaryotic Cells 77

CHAPTER 5

Microbial Metabolism 113

CHAPTER 6

Microbial Growth 156

CHAPTER 7

The Control of Microbial Growth 184

PART TWO
A Survey of the Microbial World

PART THREE

Interaction Between Microbe and Host

CHAPTER 14

Principles of Disease
and Epidemiology 406

CHAPTER 15

Microbial Mechanisms
of Pathogenicity 435

CHAPTER 16

Nonspecific Defenses of the Host 454

CHAPTER 17

Specific Defenses of the Host: The Immune Response 476

CHAPTER 18

Practical Applications of Immunology 500

CHAPTER 19

Disorders Associated with the Immune System 520

CHAPTER 20

Antimicrobial Drugs 549

PART FOUR

Microorganisms and Human Disease

CHAPTER 24

Microbial Diseases of the Respiratory System 656

CHAPTER 25

Microbial Diseases of the Digestive System 685

CHAPTER 26

Microbial Diseases of the Urinary and Reproductive Systems 720

PART FIVE

Environmental and Applied Microbiology

CHAPTER 27

Environmental Microbiology 742

CHAPTER 28

Applied and Industrial Microbiology 771

part one

Fundamentals of Microbiology

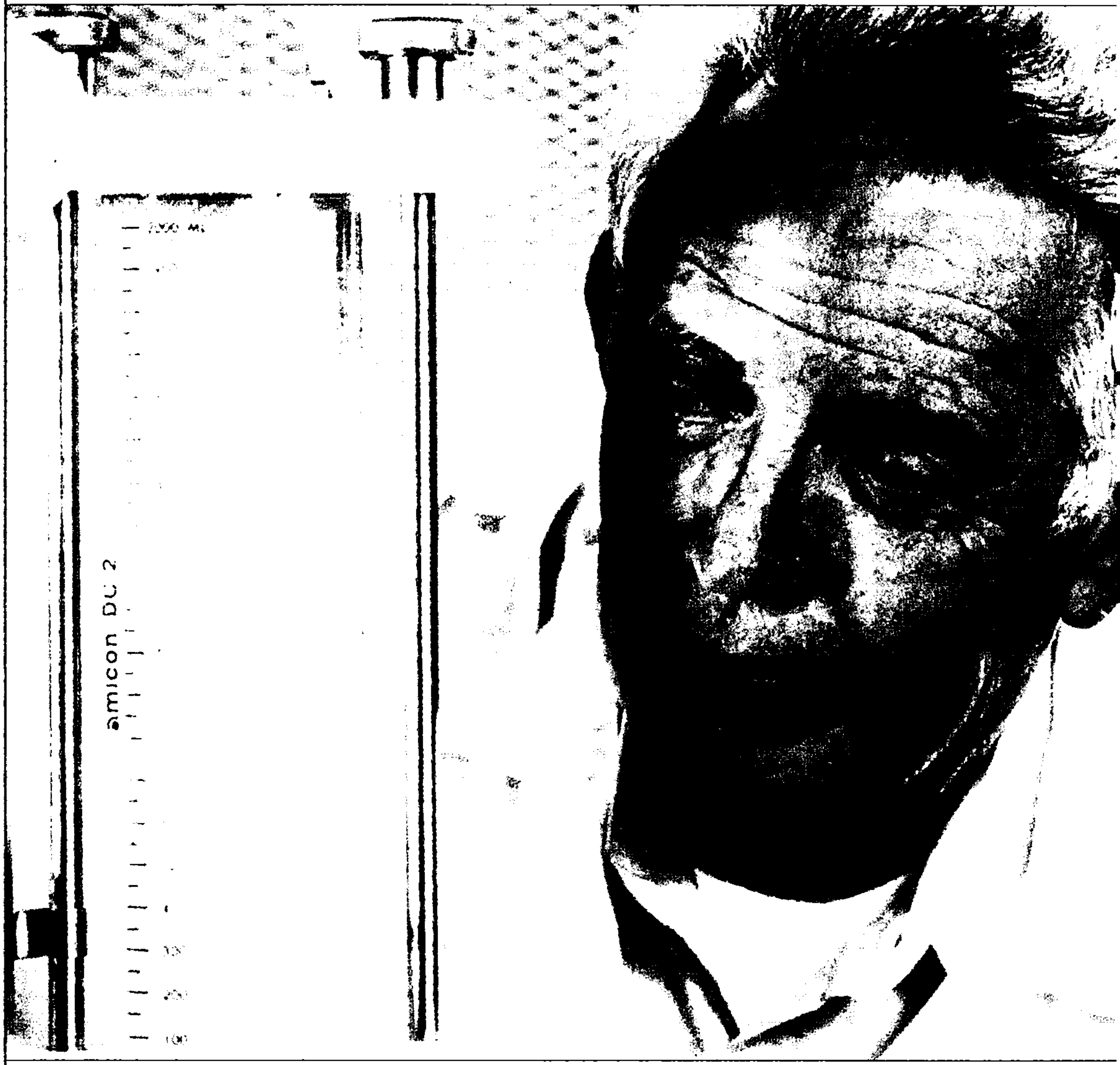

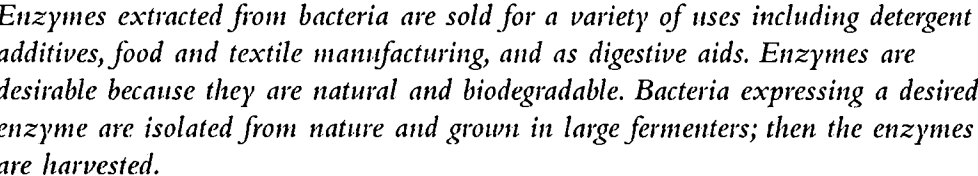

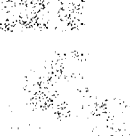

Enzymes extracted from bacteria are sold for a variety of uses including detergent additives, food and textile manufacturing, and as digestive aids. Enzymes are desirable because they are natural and biodegradable. Bacteria expressing a desired enzyme are isolated from nature and grown in large fermenters; then the enzymes are harvested.

chapter 1

The Microbial World and You

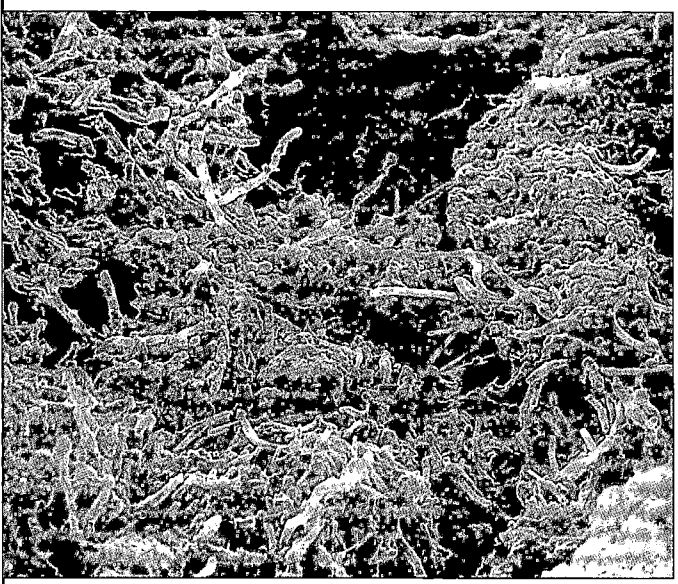

Normal microbiota of the human mouth. Normal microbiota in and on the human body can provide protection against harmful microorganisms.

The overall theme of this textbook is the relationship between microbes and our lives. This relationship involves not only the familiar harmful effects of certain microorganisms, such as causing disease and spoiling food, but also their many beneficial effects. In this chapter we will introduce you to some of the many ways microbes affect our lives. They have been fruitful subjects of study for a number of years, as you will see in our short history of microbiology. We then introduce the incredible diversity of microorganisms. The chapter next discusses the ecological importance of microbes, which help maintain balance in the environment by recycling chemical elements such as carbon and nitrogen between the soil and the atmosphere. We will also examine how microbes are used in commercial and industrial applications to produce foods, chemicals, and drugs (such as penicillin) and to treat sewage, control pests, and clean up pollutants. Finally, we will discuss microbes as the cause of diseases such as AIDS, mad cow disease, diarrhea, and hemorrhagic fever.

Microbes in Our Lives

Learning Objective

■ *List several ways in which microbes affect our lives.*

For many people, the words *germ* and *microbe* bring to mind a group of tiny creatures that do not quite fit into any of the categories in that old question, "Is it animal, vegetable, or mineral?" Microbes, also called microorganisms, are minute living things that individually are too small to be seen with the unaided eye. The group includes bacteria (Chapter 11), fungi (yeasts and molds), protozoa, and microscopic algae (Chapter 12). It also includes viruses, those noncellular entities sometimes regarded as being at the border between life and nonlife (Chapter 13). You will be introduced to each of these groups of microbes shortly.

We tend to associate these small organisms only with major diseases such as AIDS, uncomfortable infections, or such common inconveniences as spoiled food. However, the majority of microorganisms make crucial contributions to the welfare of the world's inhabitants by helping to maintain the balance of living organisms and chemicals in our environment. Marine and freshwater microorganisms form the basis of the food chain in oceans, lakes, and rivers. Soil microbes help break down wastes and incorporate nitrogen gas from the air into organic compounds, thereby recycling chemical elements in the soil, water, and air. Certain microbes play important roles in photosynthesis, a food- and oxygen-generating process that is critical to life on Earth. Humans and many other animals depend on the microbes in their intestines for digestion and the synthesis of some vitamins that their bodies require, including some B vitamins for metabolism and vitamin K for blood clotting.

Microorganisms also have many commercial applications. They are used in the synthesis of such chemical products as acetone, organic acids, enzymes, alcohols, and many drugs. The process by which microbes produce acetone and butanol was discovered in 1914 by Chaim Weizmann, a Russian-born chemist working in England. When World War I broke out in August of that year, the production of acetone was very important for making cordite (gunpowder) for use in the manufacture of munitions. Weizmann's discovery played a significant role in determining the outcome of the war.

The food industry also uses microbes in producing vinegar, sauerkraut, pickles, alcoholic beverages, green olives, soy sauce, buttermilk, cheese, yogurt, and bread (see the box on page 5). In addition, enzymes from microbes can now be manipulated such that the microbes produce substances they normally do not synthesize. These substances include cellulose, digestive aids, and drain cleaner, plus important therapeutic substances such as insulin.

Though only a minority of microorganisms are **pathogenic** (disease-producing), practical knowledge of microbes is necessary for medicine and the related health sciences. For example, hospital workers must be able to protect patients from common microbes that are normally harmless but pose a threat to the sick and injured.

Today we understand that microorganisms are found almost everywhere. Yet not long ago, before the invention of the microscope, microbes were unknown to scientists. Thousands of people died in devastating epidemics, the causes of which were not understood. Food spoilage often could not be controlled, and entire families died because vaccinations and antibiotics were not available to fight infections.

We can get an idea of how our current concepts of microbiology developed by looking at a few of the historic milestones in microbiology that have changed our lives. Before doing this, however, we will first take a look at the major groups of microbes and how they are named and classified.

Naming and Classifying Microorganisms

Learning Objective

- *Recognize the system of scientific nomenclature that uses two names: a genus and a specific epithet.*

Nomenclature

The system of nomenclature (naming) for organisms in use today was established in 1735 by Carolus Linnaeus. Scientific names are latinized because Latin was the language traditionally used by scholars. Scientific nomenclature assigns each organism two names—the **genus** (plural: **genera**) is the first name and is always capitalized; the **specific epithet (species** name) follows and is not capitalized. The organism is referred to by both the genus and the specific epithet, and both names are underlined or italicized. By custom, after a scientific name has been mentioned once, it can be abbreviated with the initial of the genus followed by the specific epithet.

Scientific names can, among other things, describe an organism, honor a researcher, or identify the habitat of a species. For example, consider *Staphylococcus aureus* (staf-i-lō-kok'kus ô'rē-us), a bacterium commonly found on human skin. *Staphylo-* describes the clustered arrangement of the cells; *coccus* indicates that they are shaped like spheres. The specific epithet, *aureus,* is Latin for golden, the color of many colonies of this bacterium. The genus of the bacterium *Escherichia coli* (esh-er-i'kē-ä kō'lī or kō'lē) is named for a scientist, Theodor Escherich, whereas its specific epithet, *coli,* reminds us that *E. coli* live in the colon, or large intestine.

Types of Microorganisms

Learning Objective

- *Differentiate among the major characteristics of each group of microorganisms.*

Bacteria

Bacteria (singular: *bacterium*) are relatively simple, single-celled (unicellular) organisms whose genetic material is not enclosed in a special nuclear membrane (Figure 1.1a). For this reason, bacterial cells are called **prokaryotes** (prō-kar'-ē-ōts), from Greek words meaning prenucleus. Prokaryotes include the bacteria and the archaea.

Bacterial cells generally appear in one of several shapes. *Bacillus* (rodlike), *coccus* (spherical or ovoid), and *spiral* (corkscrew or curved) are among the most common shapes, but some bacteria are star-shaped or square

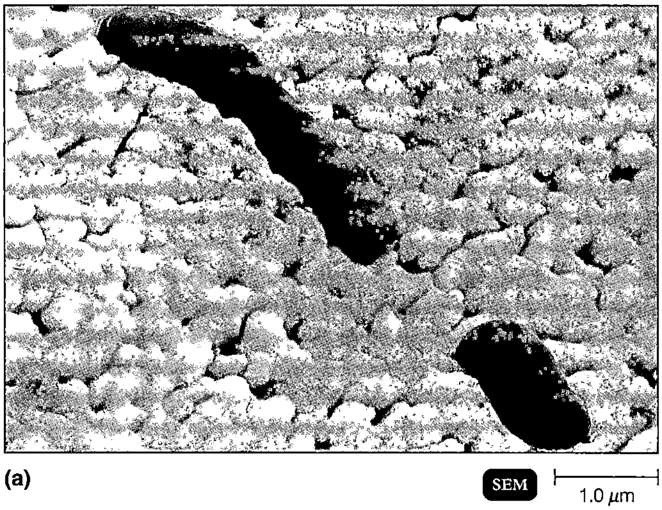

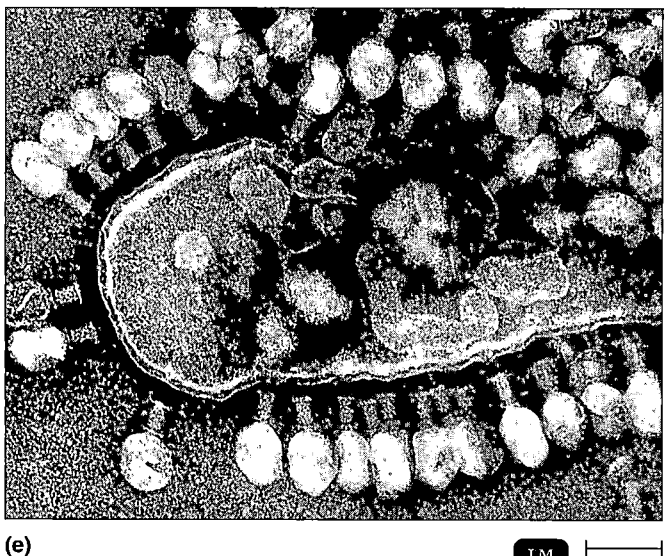

(a) SEM ⊢────⊣ 1.0 μm

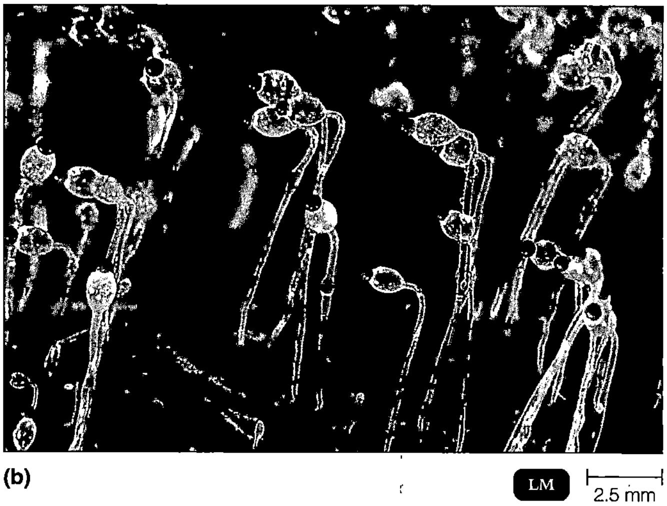

(b) LM ⊢────⊣ 2.5 mm

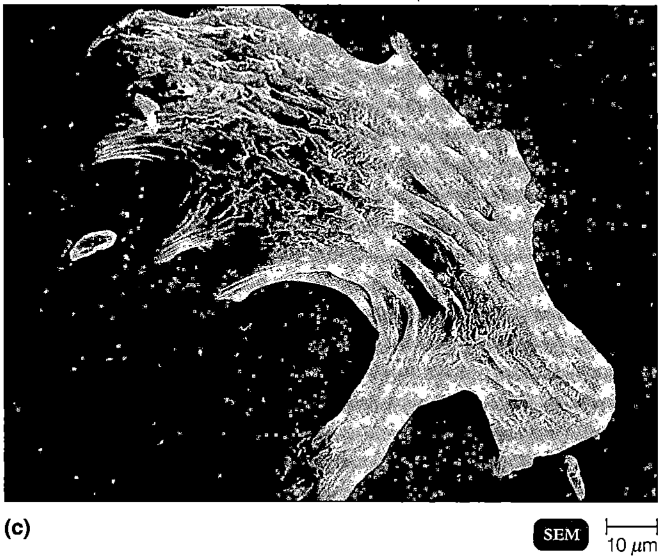

(c) SEM ⊢────⊣ 10 μm

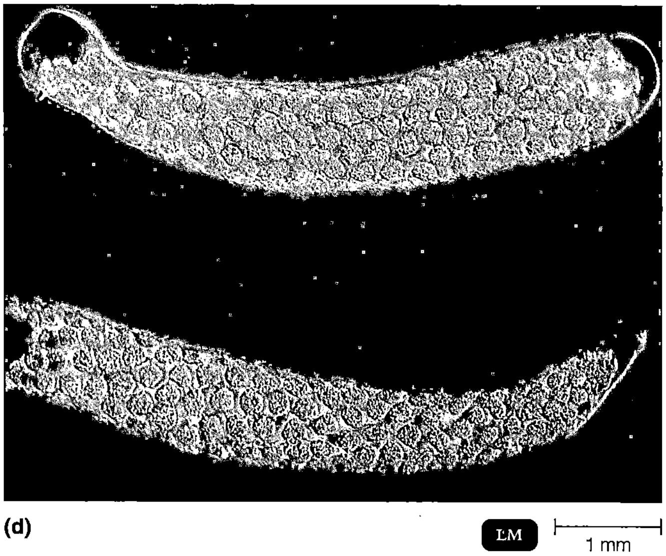

(d) LM ⊢────⊣ 1 mm

FIGURE 1.1 Types of microorganisms. (a) The rod-shaped bacterium *Haemophilus influenzae,* one of the bacterial causes of pneumonia. **(b)** *Pilobolus,* a mold, which is a type of fungus, with its sporangia oriented toward the light. When released, the sporangia will stick to grasses and be eaten by grazing animals. **(c)** An amoeba, a protozoan, approaching food. **(d)** The alga *Characiosiphon.* **(e)** Several T4 bacteriophages (viruses that attack bacteria) attached to *E. coli.*

■ The classification of organisms on the basis of cellular structure includes Bacteria, Archaea, and Eukarya (Protists, Fungi, Plants, and Animals).

(e) LM ⊢────⊣ 100 nm

What Makes Sourdough Bread Different?

Imagine being a miner during the California gold rush. You've just made bread dough from your last supplies of flour and salt when someone yells, "Gold!" Temporarily forgetting your hunger, you run off to the gold fields. Many hours later you return. The dough has been rising longer than usual, but you are too cold, tired, and hungry to start a new batch. Later you find that your bread tastes different from previous batches; it is slightly sour. During the gold rush, miners baked so many sour loaves that they were nicknamed "sourdoughs."

Conventional bread is made from flour, water, sugar, salt, shortening, and a living microbe, yeast. The yeast belongs to the Kingdom Fungi and is named *Saccharomyces cerevisiae*. When flour is mixed with water, an enzyme in the flour breaks its starch into two sugars, maltose and glucose. After the ingredients for the bread are mixed, the yeast metabolizes the sugars and produces alcohol (ethanol) and carbon dioxide as waste products. This metabolic process is called *fermentation*. The dough rises as carbon dioxide bubbles get trapped in the sticky matrix. The alcohol, which evaporates during baking, and the carbon dioxide gas form spaces that remain in the bread.

Originally, breads were leavened by wild yeast from the air, which had been trapped in the dough. Later, bakers kept a starter culture of yeast—dough from the last batch of bread—to leaven each new batch of dough. Sourdough bread is made with a special sourdough starter culture that is added to flour, water, and salt. Perhaps the most famous sourdough bread made today comes from San Francisco, where a handful of bakeries have continuously cultivated their

starters for more than 100 years and have meticulously maintained the starters to keep out unwanted microbes that can produce different and undesirable flavors. After bakeries in other areas made several unsuccessful attempts to match the unique flavor of San Francisco sourdough, rumors attributed the taste to a unique local climate or contamination from bakery walls. Ted F. Sugihara and Leo Kline from the United States Department of Agriculture (USDA) set out to debunk these rumors and to determine the microbiological basis for the bread's different taste so that it could be made in other areas.

The USDA workers found that sourdough is eight to ten times more acidic than conventional bread because of the presence of lactic and acetic acids. These acids account for the sour flavor of the bread. The workers isolated

and identified the yeast in the starter as *Saccharomyces exiguus*, a unique yeast that does not ferment maltose and thrives in the acidic environment of this dough. The sourdough question was not answered, however, because the yeast did not produce the acids and did not use maltose. Sugihara and Kline searched the starter for a second agent capable of fermenting maltose and producing the acids (see the figure). The bacterium they isolated, so carefully guarded all those years, was classified into the genus *Lactobacillus*. Many members of this genus are used in dairy fermentations and are found naturally on humans and other mammals. Analyses of cell structure and genetic composition showed that the sourdough bacterium is genetically different from other previously characterized lactobacilli. It has been given the name *Lactobacillus sanfrancisco*.

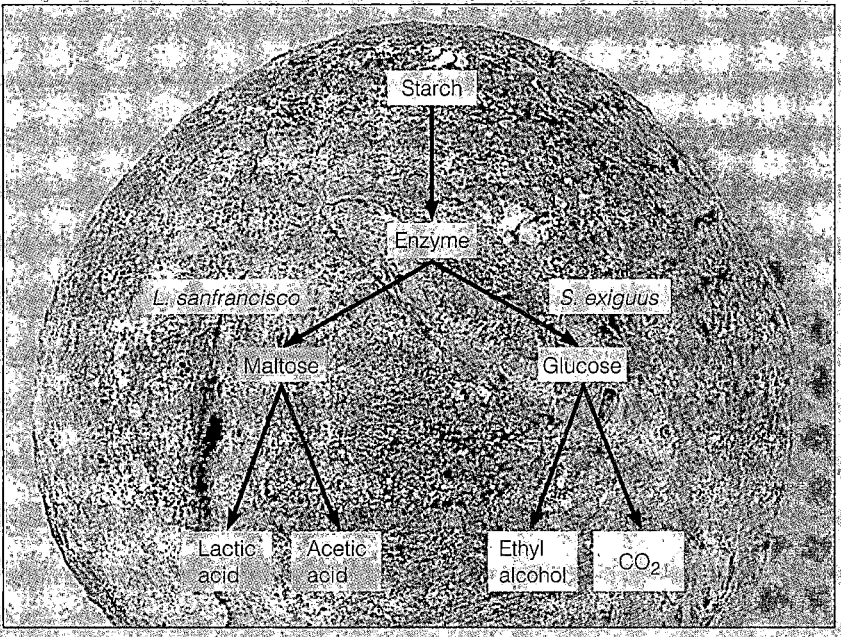

(see Figures 4.1–4.4 on pages 78–80). Individual bacteria may form pairs, chains, clusters, or other groupings; such formations are usually characteristic of a particular genus or species of bacteria.

Bacteria are enclosed in cell walls that are largely composed of a substance called *peptidoglycan*. (By contrast, cellulose is the main substance of plant and algal cell walls.) Bacteria generally reproduce by dividing into two equal daughter cells; this process is called *binary fission*. For nutrition, most bacteria use organic chemicals, which in nature can be derived from either dead or living organisms. Some bacteria can manufacture their own food by photosynthesis, and some can derive nutrition from inorganic substances. Many bacteria can "swim" by using moving appendages called *flagella*. (For a complete discussion of bacteria, see Chapter 11.)

Archaea

Like bacteria, **Archaea** (är′kē-ä) consist of prokaryotic cells but, if they have cell walls, the walls lack peptidoglycan. Archaea, often found in extreme environments, are divided into three main groups. The *methanogens* produce methane as a waste product from respiration. The *extreme halophiles* live in extremely salty environments such as the Great Salt Lake and the Dead Sea. The *extreme thermophiles* live in hot sulfurous water such as hot springs at Yellowstone National Park.

Fungi

Fungi (singular: *fungus*) are **eukaryotes** (yū-kar′-ē-ōts), organisms whose cells have a distinct nucleus containing the cell's genetic material (DNA), surrounded by a special envelope called the nuclear membrane. Organisms in the Kingdom Fungi may be unicellular or multicellular (see Chapter 12, page 331). Large multicellular fungi, such as mushrooms, may look somewhat like plants, but they cannot carry out photosynthesis, as most plants can. True fungi have cell walls composed primarily of a substance called *chitin*. The unicellular forms of fungi, *yeasts,* are oval microorganisms that are larger than bacteria. The most typical fungi are *molds* (Figure 1.1b). Molds form visible masses called *mycelia,* which are composed of long filaments *(hyphae)* that branch and intertwine. The cottony growths sometimes found on bread and fruit are mold mycelia. Fungi can reproduce sexually or asexually. They obtain nourishment by absorbing solutions of organic material from their environment—whether soil, seawater, fresh water, or an animal or plant host.

Protozoa

Protozoa (singular: *protozoan*) are unicellular, eukaryotic microbes (see Chapter 12, page 349). Protozoa move by pseudopods, flagella, or cilia. Amoebas (Figure 1.1c) move by using extensions of their cytoplasm called *pseudopods* (false feet). Other protozoa have long *flagella* or numerous shorter appendages for locomotion called *cilia.* Protozoa have a variety of shapes and live either as free entities or as parasites (organisms that derive nutrients from living hosts) that absorb or ingest organic compounds from their environment. Protozoa can reproduce sexually or asexually.

Algae

Algae (singular: *alga*) are photosynthetic eukaryotes with a wide variety of shapes and both sexual and asexual reproductive forms (Figure 1.1d). The algae of interest to microbiologists are usually unicellular (see Chapter 12, page 344). The cell walls of many algae, like those of plants, are composed of *cellulose*. Algae are abundant in fresh and salt water, in soil, and in association with plants. As photosynthesizers, algae need light and air for food production and growth, but they do not generally require organic compounds from the environment. As a result of photosynthesis, algae produce oxygen and carbohydrates, which are utilized by other organisms, including animals. Thus, they play an important role in the balance of nature.

Viruses

Viruses (Figure 1.1e) are very different from the other microbial groups mentioned here. They are so small that most can be seen only with an electron microscope, and they are acellular (not cellular). Structurally very simple, a virus particle contains a core made of only one type of nucleic acid, either DNA or RNA. This core is surrounded by a protein coat. Sometimes the coat is encased by an additional layer, a lipid membrane called an envelope. All living cells have RNA *and* DNA, can carry out chemical reactions, and can reproduce as self-sufficient units. Viruses can reproduce only by using the cellular machinery of other organisms. Thus, all viruses are parasites of other forms of life. (Viruses will be discussed in detail in Chapter 13.)

Multicellular Animal Parasites

Although multicellular animal parasites are not strictly microorganisms, they are of medical importance and therefore will be discussed in this text. The two major groups of parasitic worms are the flatworms and the roundworms, collectively called **helminths** (see Chapter 12, page 356). During some stages of their life cycle, helminths are microscopic in size. Laboratory identification of these organisms includes many of the same techniques used for the identification of microbes.

Classification of Microorganisms

Before the existence of microbes was known, all organisms were grouped into either the animal kingdom or the plant kingdom. When microscopic organisms with characteristics of animals or plants were discovered late in the seventeenth century, a new system of classification was needed. Still, biologists could not agree on the criteria for classifying the new organisms they were seeing until the late 1960s.

In 1978, Carl Woese devised a system of classification based on the cellular organization of organisms. It groups all organisms in three domains as follows:

- Bacteria (cell walls contain peptidoglycan)
- Archaea (cell walls, if present, lack peptidoglycan)
- Eukarya, which includes the following:
  Protists (slime molds, protozoa, and algae)
  Fungi (unicellular yeasts, multicellular molds, and mushrooms)
  Plants (includes mosses, ferns, conifers, and flowering plants)
  Animals (includes sponges, worms, insects, and vertebrates)

Classification will be discussed in more detail in Part Two (Chapters 10–13).

A Brief History of Microbiology

The science of microbiology dates back only a few hundred years, yet the recent discovery of *Mycobacterium tuberculosis* (mī-kō-bak-ti′rē-um) DNA in 3000-year-old Egyptian mummies reminds us that microorganisms have been around for much longer. While we know relatively little about what earlier people thought about the causes, transmission, and treatment of disease, the history of the past few hundred years is better known. Let's look now at some key developments in microbiology that have helped the field progress to its current high-technology state.

The First Observations

Learning Objective

- *Explain the importance of observations made by Hooke and van Leeuwenhoek.*

One of the most important discoveries in the history of biology occurred in 1665 with the help of a relatively crude microscope. An Englishman, Robert Hooke, reported to the world that life's smallest structural units were "little boxes," or "cells," as he called them. Using his improved version of a compound microscope (one that

uses two sets of lenses), Hooke was able to see individual cells. Hooke's discovery marked the beginning of the **cell theory**—the theory that *all living things are composed of cells.* Subsequent investigations into the structure and functions of cells were based on this theory.

Though Hooke's microscope was capable of showing cells, he lacked the staining techniques that would have allowed him to see microbes clearly. The Dutch merchant and amateur scientist Antoni van Leeuwenhoek was probably the first to actually observe live microorganisms through magnifying lenses. Between 1673 and 1723, he wrote a series of letters to the Royal Society of London describing the "animalcules" he saw through his simple, single-lens microscope. Van Leeuwenhoek's detailed drawings of "animalcules" in rainwater, in liquid in which peppercorns had soaked, and in material scraped from his teeth have since been identified as representations of bacteria and protozoa (Figure 1.2).

The Debate Over Spontaneous Generation

Learning Objectives

- *Compare the theories of spontaneous generation and biogenesis.*
- *Identify the contributions to microbiology made by Needham, Spallanzani, Virchow, and Pasteur.*

After van Leeuwenhoek discovered the previously "invisible" world of microorganisms, the scientific community of the time became interested in the origins of these tiny living things. Until the second half of the nineteenth century, many scientists and philosophers believed that some forms of life could arise spontaneously from nonliving matter; they called this hypothetical process **spontaneous generation.** Not much more than 100 years ago, people commonly believed that toads, snakes, and mice could be born of moist soil; that flies could emerge from manure; and that maggots, the larvae of flies, could arise from decaying corpses.

Evidence Pro and Con

A strong opponent of spontaneous generation, the Italian physician Francesco Redi, set out in 1668 (even before van Leeuwenhoek's discovery of microscopic life) to demonstrate that maggots do not arise spontaneously from decaying meat. Redi filled three jars with decaying meat and sealed them tightly. Then he arranged three other jars similarly but left them open. Maggots appeared in the open vessels after flies entered the jars and laid their eggs, but the sealed containers showed no signs of maggots. Still, Redi's antagonists were not convinced; they

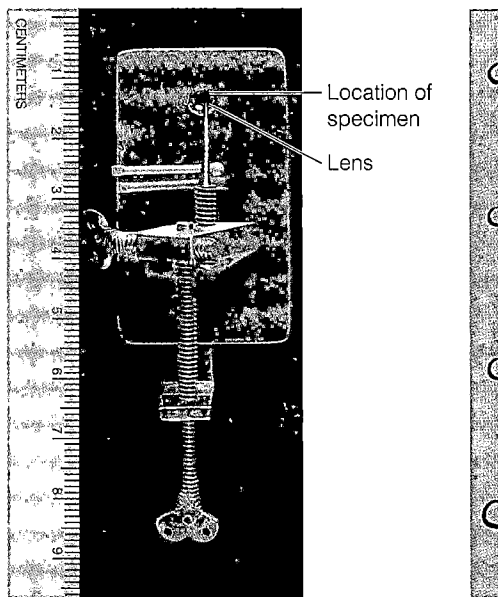
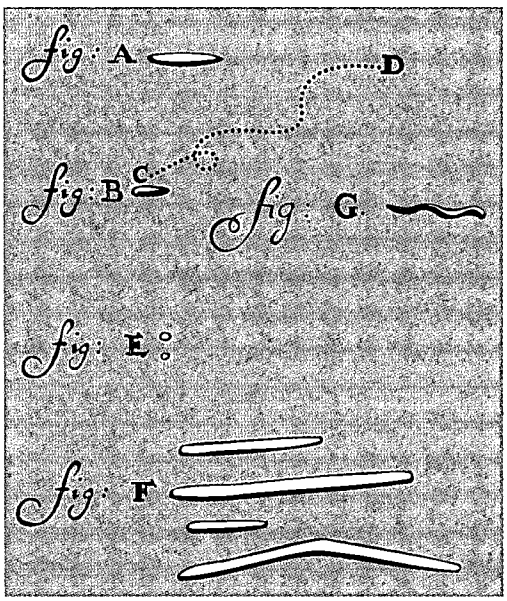

(a) Microscope replica

(b) Drawings of bacteria

FIGURE 1.2 Antoni van Leeuwenhoek's microscopic observations. (a) A replica of a simple microscope made by van Leeuwenhoek to observe living organisms too small to be seen with the unaided eye. The specimen was placed on the tip of the adjustable point and viewed from the other side through the tiny, nearly spherical lens. The highest magnification possible with his microscopes was about 300×(times). **(b)** Some of van Leeuwenhoek's drawings of bacteria, made in 1683. The letters represent various shapes of bacteria. C–D represents a path of motion he observed.

> Van Leeuwenhoek was the first person to view microorganisms that we now call bacteria and protozoa.

claimed that fresh air was needed for spontaneous generation. So Redi set up a second experiment, in which three jars were covered with a fine net instead of being sealed. No larvae appeared in the gauze-covered jars, even though air was present. Maggots appeared only if flies were allowed to leave their eggs on the meat.

Redi's results were a serious blow to the long-held belief that large forms of life could arise from nonlife. However, many scientists still believed that small organisms, such as van Leeuwenhoek's "animalcules," were simple enough to be generated from nonliving materials.

The case for spontaneous generation of microorganisms seemed to be strengthened in 1745 when John Needham, an Englishman, found that even after he heated nutrient fluids (chicken broth and corn broth) before pouring them into covered flasks, the cooled solutions were soon teeming with microorganisms. Needham claimed that microbes developed spontaneously from the fluids. Twenty years later, Lazzaro Spallanzani, an Italian scientist, suggested that microorganisms from the air probably had entered Needham's solutions after they were boiled. Spallanzani showed that nutrient fluids heated *after* being sealed in a flask did not develop micro-

bial growth. Needham responded by claiming the "vital force" necessary for spontaneous generation had been destroyed by the heat and was kept out of the flasks by the seals.

This intangible "vital force" was given all the more credence shortly after Spallanzani's experiment, when Laurent Lavoisier showed the importance of oxygen to life. Spallanzani's observations were criticized on the grounds that there was not enough oxygen in the sealed flasks to support microbial life.

The Theory of Biogenesis

The issue was still unresolved in 1858, when the German scientist Rudolf Virchow challenged spontaneous generation with the concept of **biogenesis,** the claim that living cells can arise only from preexisting living cells. Arguments about spontaneous generation continued until 1861, when the issue was resolved by the French scientist Louis Pasteur.

With a series of ingenious and persuasive experiments, Pasteur demonstrated that microorganisms are present in the air and can contaminate sterile solutions, but air itself does not create microbes. He filled several

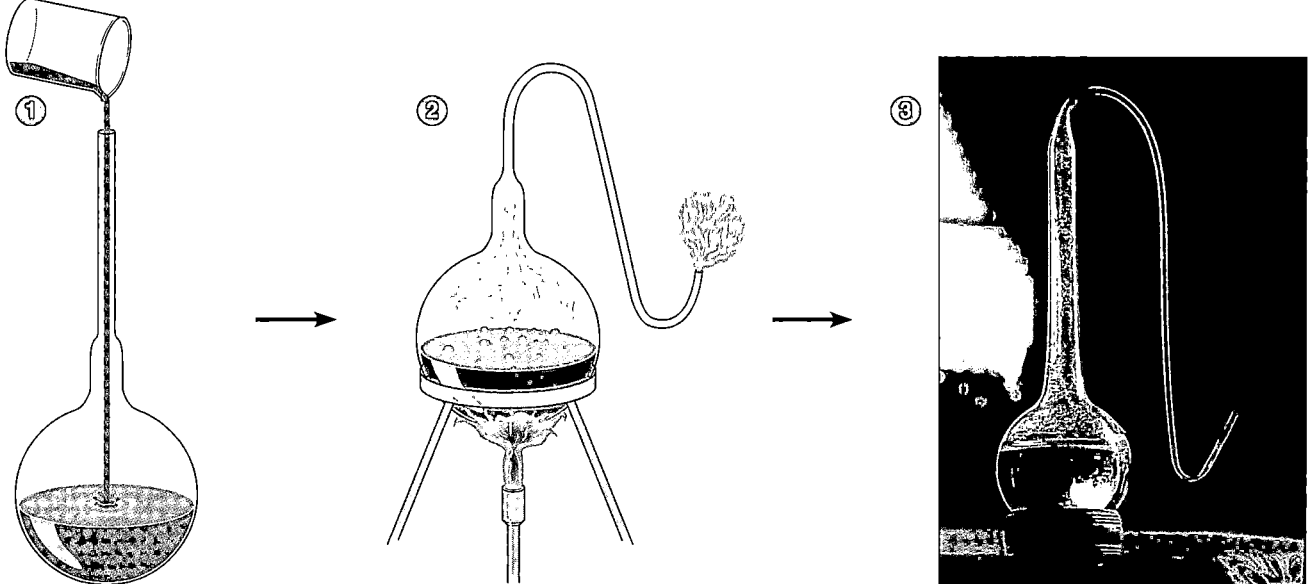

FIGURE 1.3 Pasteur's experiment disproving the theory of spontaneous generation.
① Pasteur first poured beef broth into a long-necked flask. ② Next he heated the neck of the flask and bent it into an S-shaped curve; then he boiled the broth for several minutes. ③ Microorganisms did not appear in the cooled solution, even after long periods, as you can see in this recent photograph of an actual flask used by Pasteur in a similar experiment.

■ Pasteur's discoveries form the basis for aseptic techniques, procedures that prevent contamination by unwanted organisms.

short-necked flasks with beef broth and then boiled their contents. Some were then left open and allowed to cool. In a few days, these flasks were found to be contaminated with microbes. The other flasks, sealed after boiling, were free of microorganisms. From these results, Pasteur reasoned that microbes in the air were the agents responsible for contaminating nonliving matter such as the broths in Needham's flasks.

Pasteur next placed broth in open-ended long-necked flasks and bent the necks into S-shaped curves (Figure 1.3). The contents of these flasks were then boiled and cooled. The broth in the flasks did not decay and showed no signs of life, even after months. Pasteur's unique design allowed air to pass into the flask, but the curved neck trapped any airborne microorganisms that might contaminate the broth. (Some of these original vessels, which were later sealed, are on display at the Pasteur Institute in Paris. Like the flask shown in Figure 1.3, they still show no sign of contamination more than 100 years later.)

Pasteur showed that microorganisms can be present in nonliving matter—on solids, in liquids, and in the air. Furthermore, he demonstrated conclusively that microbial life can be destroyed by heat and that methods can be devised to block the access of airborne microorganisms to nutrient

environments. These discoveries form the basis of **aseptic techniques,** techniques that prevent contamination by unwanted microorganisms, which are now the standard practice in laboratory and many medical procedures. Modern aseptic techniques are among the first and most important things that a beginning microbiologist learns.

Pasteur's work provided evidence that microorganisms cannot originate from mystical forces present in nonliving materials. Rather, any appearance of "spontaneous" life in nonliving solutions can be attributed to microorganisms that were already present in the air or in the fluids themselves. Scientists now believe that a form of spontaneous generation probably did occur on the primitive Earth when life first began, but they agree that this does not happen under today's environmental conditions.

The Golden Age of Microbiology

Learning Objectives

■ Identify the importance of Koch's postulates.
■ Explain how Pasteur's work influenced Lister and Koch.
■ Identify the importance of Jenner's work.

For about 60 years, beginning with the work of Pasteur, there was an explosion of discoveries in microbiology.

The period from 1857 to 1914 has been appropriately named the Golden Age of Microbiology. During this period, rapid advances, spearheaded mainly by Pasteur and Robert Koch, led to the establishment of microbiology as a science. Discoveries during these years included both the agents of many diseases and the role of immunity in the prevention and cure of disease. During this productive period, microbiologists studied the chemical activities of microorganisms, improved the techniques for performing microscopy and culturing microorganisms, and developed vaccines and surgical techniques. Some of the major events that occurred during the Golden Age of Microbiology are listed in Figure 1.4.

Fermentation and Pasteurization

One of the key steps that established the relationship between microorganisms and disease occurred when a group of French merchants asked Pasteur to find out why wine and beer soured. They hoped to develop a method that would prevent spoilage when those beverages were shipped long distances. At the time, many scientists believed that air converted the sugars in these fluids into alcohol. Pasteur found instead that microorganisms called yeasts convert the sugars to alcohol in the absence of air. This process, called **fermentation** (see Chapter 5, page 133), is used to make wine and beer. Souring and spoilage are caused by different microorganisms called bacteria. In the presence of air, bacteria change the alcohol in the beverage into vinegar (acetic acid).

Pasteur's solution to the spoilage problem was to heat the beer and wine just enough to kill most of the bacteria that caused the spoilage; the process, called **pasteurization,** is now commonly used to kill spoilage and potentially harmful bacteria in milk as well as in some alcoholic drinks. Showing the connection between spoilage of food and microorganisms was a major step toward establishing the relationship between disease and microbes.

The Germ Theory of Disease

As we have seen, the fact that many kinds of diseases are related to microorganisms was unknown until relatively recently. Before the time of Pasteur, effective treatments for many diseases were discovered by trial and error, but the causes of the diseases were unknown.

The realization that yeasts play a crucial role in fermentation was the first link between the activity of a microorganism and physical and chemical changes in organic materials. This discovery alerted scientists to the possibility that microorganisms might have similar relationships with plants and animals—specifically, that microorganisms might cause disease. This idea was known as the **germ theory of disease.**

The germ theory was a difficult concept for many people to accept at that time because for centuries disease was believed to be punishment for an individual's crimes or misdeeds. When the inhabitants of an entire village became ill, people often blamed the disease on demons appearing as foul odors from sewage or on poisonous vapors from swamps. Most people born in Pasteur's time found it inconceivable that "invisible" microbes could travel through the air to infect plants and animals, or remain on clothing and bedding to be transmitted from one person to another. But gradually scientists accumulated the information needed to support the new germ theory.

In 1865, Pasteur was called upon to help fight silkworm disease, which was ruining the silk industry throughout Europe. Years earlier, in 1835, Agostino Bassi, an amateur microscopist, had proved that another silkworm disease was caused by a fungus. Using data provided by Bassi, Pasteur found that the more recent infection was caused by a protozoan, and he developed a method for recognizing afflicted silkworm moths.

In the 1860s, Joseph Lister, an English surgeon, applied the germ theory to medical procedures. Lister was aware that in the 1840s, the Hungarian physician Ignaz Semmelweis had demonstrated that physicians, who at the time did not disinfect their hands, routinely transmitted infections (puerperal, or childbirth, fever) from one obstetrical patient to another. Lister had also heard of Pasteur's work connecting microbes to animal diseases. Disinfectants were not used at the time, but Lister knew that phenol (carbolic acid) kills bacteria, so he began treating surgical wounds with a phenol solution. The practice so reduced the incidence of infections and deaths that other surgeons quickly adopted it. Lister's technique was one of the earliest medical attempts to control infections caused by microorganisms. In fact, his findings proved that microorganisms cause surgical wound infections.

The first proof that bacteria actually cause disease came from Robert Koch in 1876. Koch, a German physician, was Pasteur's young rival in the race to discover the cause of anthrax, a disease that was destroying cattle and sheep in Europe. Koch discovered rod-shaped bacteria now known as *Bacillus anthracis* (bä-sil′lus an-thrā′sis) in the blood of cattle that had died of anthrax. He cultured the bacteria on nutrients and then injected samples of the culture into healthy animals. When these animals became sick and died, Koch isolated the bacteria in their blood and compared them with the bacteria originally isolated. He found that the two sets of blood cultures contained the same bacteria.

Koch thus established a sequence of experimental steps for directly relating a specific microbe to a specific

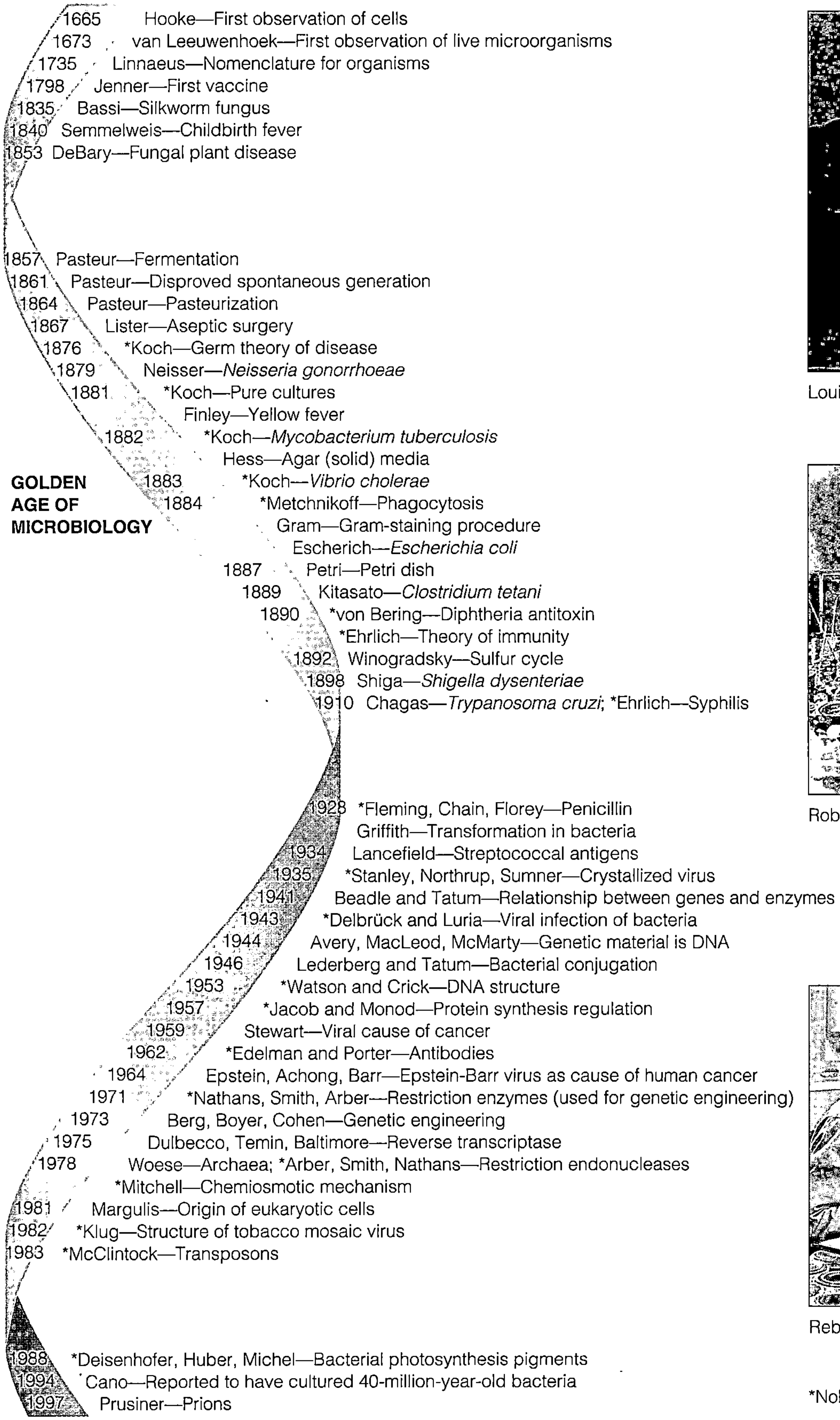

1665   Hooke—First observation of cells
1673 · van Leeuwenhoek—First observation of live microorganisms
1735 · Linnaeus—Nomenclature for organisms
1798 · Jenner—First vaccine
1835· Bassi—Silkworm fungus
1840 Semmelweis—Childbirth fever
1853 DeBary—Fungal plant disease

1857 Pasteur—Fermentation
1861 Pasteur—Disproved spontaneous generation
1864 Pasteur—Pasteurization
1867   Lister—Aseptic surgery
1876   *Koch—Germ theory of disease
1879   Neisser—*Neisseria gonorrhoeae*
1881   *Koch—Pure cultures
         Finley—Yellow fever
1882   *Koch—*Mycobacterium tuberculosis*
          Hess—Agar (solid) media

GOLDEN   1883   *Koch—*Vibrio cholerae*
AGE OF   1884   *Metchnikoff—Phagocytosis
MICROBIOLOGY   Gram—Gram-staining procedure
                   Escherich—*Escherichia coli*
         1887   Petri—Petri dish
         1889   Kitasato—*Clostridium tetani*
         1890   *von Bering—Diphtheria antitoxin
                 *Ehrlich—Theory of immunity
              1892 Winogradsky—Sulfur cycle
              1898 Shiga—*Shigella dysenteriae*
              1910 Chagas—*Trypanosoma cruzi*; *Ehrlich—Syphilis

Louis Pasteur (1822–1895)

Robert Koch (1843–1910)

1928 *Fleming, Chain, Florey—Penicillin
      Griffith—Transformation in bacteria
1934 Lancefield—Streptococcal antigens
1935 *Stanley, Northrup, Sumner—Crystallized virus
1941   Beadle and Tatum—Relationship between genes and enzymes
1943   *Delbrück and Luria—Viral infection of bacteria
1944   Avery, MacLeod, McMarty—Genetic material is DNA
1946   Lederberg and Tatum—Bacterial conjugation
1953   *Watson and Crick—DNA structure
1957   *Jacob and Monod—Protein synthesis regulation
1959   Stewart—Viral cause of cancer
1962   *Edelman and Porter—Antibodies
1964   Epstein, Achong, Barr—Epstein-Barr virus as cause of human cancer
1971   *Nathans, Smith, Arber—Restriction enzymes (used for genetic engineering)
1973   Berg, Boyer, Cohen—Genetic engineering
1975   Dulbecco, Temin, Baltimore—Reverse transcriptase
1978   Woese—Archaea; *Arber, Smith, Nathans—Restriction endonucleases
        *Mitchell—Chemiosmotic mechanism
1981 · Margulis—Origin of eukaryotic cells
1982 *Klug—Structure of tobacco mosaic virus
1983 *McClintock—Transposons

Rebecca C. Lancefield (1895–1981)

1988 *Deisenhofer, Huber, Michel—Bacterial photosynthesis pigments
1994 ·Cano—Reported to have cultured 40-million-year-old bacteria
1997 Prusiner—Prions

*Nobel laureate

FIGURE 1.4 Milestones in microbiology, highlighting those that occurred during the Golden Age of Microbiology.

■ The Golden Age of Microbiology is so named because during this period, numerous discoveries led to the establishment of microbiology as a science.

disease. These steps are known today as **Koch's postulates** (see Figure 14.3 on page 411). During the past 100 years, these same criteria have been invaluable in investigations proving that specific microorganisms cause many diseases. Koch's postulates, their limitations, and their application to disease will be discussed in greater detail in Chapter 14.

Vaccination

Often a treatment or preventive procedure is developed before scientists know why it works. The smallpox vaccine is an example of this. On May 4, 1796, almost 70 years before Koch established that a specific microorganism causes anthrax, Edward Jenner, a young British physician, embarked on an experiment to find a way to protect people from smallpox.

Smallpox epidemics were greatly feared. The disease periodically swept through Europe, killing thousands, and it wiped out 90% of the Native Americans on the East Coast when European settlers first brought the infection to the New World.

When a young milkmaid informed Jenner that she couldn't get smallpox because she already had been sick from cowpox—a much milder disease—he decided to put the girl's story to the test. First Jenner collected scrapings from cowpox blisters. Then he inoculated a healthy 8-year-old volunteer with the cowpox material by scratching the person's arm with a pox-contaminated needle. The scratch turned into a raised bump. In a few days, the volunteer became mildly sick but recovered and never again contracted either cowpox or smallpox. The process was called *vaccination,* from the Latin word *vacca,* meaning cow. Pasteur gave it this name in honor of Jenner's work. The protection from disease provided by vaccination (or by recovery from the disease itself) is called **immunity.** We will discuss the mechanisms of immunity in Chapter 17.

Years after Jenner's experiment, in about 1880, Pasteur discovered why vaccinations work. He found that the bacterium that causes fowl cholera lost its ability to cause disease (lost its *virulence,* or became *avirulent*) after it was grown in the laboratory for long periods. However, it—and other microorganisms with decreased virulence—was able to induce immunity against subsequent infections by its virulent counterparts. The discovery of this phenomenon provided a clue to Jenner's successful experiment with cowpox. Both cowpox and smallpox are caused by viruses. Even though cowpox virus is not a laboratory-produced derivative of smallpox virus, it is so closely related to the smallpox virus that it can induce immunity to both viruses. Pasteur used the term *vaccine* for cultures of avirulent microorganisms used for preventive inoculation.

Jenner's experiment was the first time in a Western culture that a living viral agent—the cowpox virus—was used to produce immunity. Physicians in China had immunized patients by removing scales from drying pustules of a person suffering from a mild case of smallpox, grinding the scales to a fine powder, and inserting the powder into the nose of the person to be protected.

Some vaccines are still produced from avirulent microbial strains that stimulate immunity to the related virulent strain. Other vaccines are made from killed virulent microbes, from isolated components of virulent microorganisms, or by genetic engineering techniques.

The Birth of Modern Chemotherapy: Dreams of a "Magic Bullet"

Learning Objective

- *Identify the contributions to microbiology made by Ehrlich, Fleming, and Dubos.*

After the relationship between microorganisms and disease was established, medical microbiologists next focused on the search for substances that could destroy pathogenic microorganisms without damaging the infected animal or human. Treatment of disease by using chemical substances is called **chemotherapy.** (The term also commonly refers to chemical treatment of noninfectious diseases, such as cancer.) Chemotherapeutic agents prepared from chemicals in the laboratory are called **synthetic drugs.** Chemicals produced naturally by bacteria and fungi to act against other microorganisms are called **antibiotics.** The success of chemotherapy is based on the fact that some chemicals are more poisonous to microorganisms than to the hosts infected by the microbes. Antimicrobial therapy will be discussed in further detail in Chapter 20.

The First Synthetic Drugs

Paul Ehrlich, a German physician, was the imaginative thinker who fired the first shot in the chemotherapy revolution. As a medical student, Ehrlich speculated about a "magic bullet" that could hunt down and destroy a pathogen without harming the infected host. Ehrlich launched a search for such a bullet. In 1910, after testing hundreds of substances, he found a chemotherapeutic agent called *salvarsan,* an arsenic derivative effective against syphilis. The agent was named salvarsan because it was considered to offer salvation from syphilis and it contained arsenic. Before this discovery, the only known chemical in Europe's medical arsenal was an extract from the bark of a South American tree, *quinine,* which had been used by Spanish conquistadors to treat malaria.

By the late 1930s, researchers had developed several other synthetic drugs that could destroy microorganisms. Most of these drugs were derivatives of dyes. This is because the dyes synthesized and manufactured for fabrics were routinely tested for antimicrobial qualities by microbiologists looking for a "magic bullet." In addition, *sulfonamides* (sulfa drugs) were synthesized at about the same time.

A Fortunate Accident—Antibiotics

In contrast to the sulfa drugs, which were deliberately developed from a series of industrial chemicals, the first antibiotic was discovered by accident. Alexander Fleming, a Scottish physician and bacteriologist, almost tossed out some culture plates that had been contaminated by mold. Fortunately, he took a second look at the curious pattern of growth on the contaminated plates. There was a clear area around the mold where the bacterial culture had stopped growing (see Figure 20.1 on page 550). Fleming was looking at a mold that could inhibit the growth of a bacterium. The mold was later identified as *Penicillium notatum* (pen-i-sil'lē-um nō-tä'tum), and in 1928 Fleming named the mold's active inhibitor *penicillin*. Thus, penicillin is an antibiotic produced by a fungus. The enormous usefulness of penicillin was not apparent until the 1940s, when it was finally tested clinically and mass-produced.

Interest in penicillin came after 1939, when René Dubos, a French microbiologist, discovered two antibiotics called *gramicidin* and *tyrocidine.* Both were produced by a bacterium, *Bacillus brevis,* cultured from soil.

Since the early discoveries of antibiotics, thousands of others have been discovered. Unfortunately, antibiotics and other chemotherapeutic drugs are not without problems. Many antimicrobial chemicals are too toxic to humans for practical use; they kill the pathogenic microbes, but they also damage the infected host. For reasons we will discuss later, toxicity to humans is a particular problem in the development of drugs for treating viral diseases. Because viral growth is so closely linked with the life processes of normal host cells, there are very few successful antiviral drugs.

Another major problem associated with antimicrobial drugs is the emergence and spread of new varieties of microorganisms that are resistant to antibiotics. Over the years, more and more microbes have developed resistance to antibiotics that at one time were very effective against them. Drug resistance results from an adaptive response of microbes that enables them to tolerate a certain amount of an antibiotic that would normally inhibit them (see the box in Chapter 28). These changes might include the production by microbes of chemicals (enzymes) that inactivate antibiotics, changes in the surface of a microbe that prevent an antibiotic from attaching to it, and prevention of an antibiotic from entering the microbe.

The recent appearance of vancomycin-resistant *Staphylococcus aureus* and *Enterococcus faecalis* has alarmed health care professionals because some previously treatable bacterial infections may soon be impossible to treat with antibiotics.

Modern Developments in Microbiology

Learning Objectives

- *Define bacteriology, mycology, parasitology, immunology, and virology.*
- *Explain the importance of recombinant DNA technology.*

The quest to solve drug resistance, identify viruses, and develop vaccines requires sophisticated research techniques and correlated studies that were never dreamed of in the days of Koch and Pasteur.

The groundwork laid during the Golden Age of Microbiology provided the basis for several monumental achievements during the twentieth century (Table 1.1). New branches of microbiology were developed, including immunology and virology. Most recently, the development of a set of new methods called recombinant DNA technology has revolutionized research and practical applications in all areas of microbiology.

Bacteriology, Mycology, and Parasitology

Bacteriology, the study of bacteria, began with van Leeuwenhoek's first examination of tooth scrapings. New pathogenic bacteria are still discovered regularly. Many bacteriologists, like their predecessor Pasteur, look at the roles of bacteria in food and the environment. One intriguing discovery came in 1997, when Heide Schulz discovered a bacterium large enough to be seen with the unaided eye (0.2 mm wide). This bacterium, which she named *Thiomargarita namibiensis* (thī'o-mä-gär-e-tä na'mib-e-en-sis), lives in the mud on the African coast. *Thiomargarita* is unusual because of its size and its ecological niche. The bacterium consumes hydrogen sulfide, which would be toxic to mud-dwelling animals (Figure 11.26 on page 327).

Mycology, the study of fungi, includes medical, agricultural, and ecological branches. Recall that Bassi's work leading up to the germ theory of disease was on a fungal pathogen. Fungal infection rates have been rising during the past decade, accounting for 10% of hospital-acquired infections. Climatic and environmental changes (severe drought) are thought to account for the tenfold increase in *Coccidioides immitis* (kok-sid-ē-oi'dēz im'mitis) infections in California. New techniques for diagnosing and treating fungal infections are currently being investigated.

table 1.1	Selected Nobel Prizes Awarded for Research in Microbiology			
Nobel Laureates	Year of Presentation	Country of Birth	Contribution
Emil A. von Behring	1901	Germany	Developed a diphtheria antitoxin.
Robert Koch	1905	Germany	Cultured tuberculosis bacteria.
Paul Ehrlich	1908	Germany	Developed theories on immunity.
Elie Metchnikoff	1908	Russia	Described phagocytosis, the intake of solid materials by cells.
Alexander Fleming, Ernst Chain, and Howard Florey	1945	Scotland England	Discovered penicillin.
Selman A. Waksman	1952	Ukraine	Discovered streptomycin.
Hans A. Krebs	1953	Germany	Discovered chemical steps of the Krebs cycle in carbohydrate metabolism.
John F. Enders, Thomas H. Weller, and Frederick C. Robbins	1954	United States	Cultured poliovirus in cell cultures.
Joshua Lederberg, George Beadle, and Edward Tatum	1958	United States	Described genetic control of biochemical reactions.
James D. Watson, Frances H. C. Crick, and Maurice A. F. Wilkins	1962	United States England New Zealand	Identified the physical structure of DNA.
François Jacob, Jacques Monod, and André Lwoff	1965	France	Described how protein synthesis is regulated in bacteria.
Robert Holley, Har Gobind Khorana, and Marshall W. Nirenberg	1968	United States India United States	Discovered the genetic code for amino acids.
Max Delbrück, Alfred D. Hershey, and Salvador E. Luria	1969	Germany United States Italy	Described the mechanism of viral infection of bacterial cells.
Gerald M. Edelman and Rodney R. Porter	1972	United States England	Described the nature and structure of antibodies.
Renato Dulbecco, Howard Temin, and David Baltimore	1975	United States	Discovered reverse transcriptase and described how RNA viruses could cause cancer.
Daniel Nathans, Hamilton Smith, and Werner Arber	1978	United States United States Switzerland	Described the action of restriction enzymes (now used in genetic engineering).

Parasitology is the study of protozoa and parasitic worms. Because many parasitic worms are large enough to be seen with the unaided eye, they have been known by people for thousands of years. One hypothesis is that the medical symbol the caduceus represents the removal of parasitic guinea worms (Figure 1.5 on page 16). New parasitic diseases in humans are being discovered as laborers become exposed while clearing rain forests. Previ-ously unknown parasitic diseases are also being found in patients whose immune systems have been suppressed by organ transplants, cancer chemotherapy, and AIDS.

Bacteriology, mycology, and parasitology are currently going through a "golden age of classification." Recent advances in **genomics,** the study of all of an organism's genes, have allowed scientists to classify bacteria and fungi according to their relationships with other bacteria,

Nobel Laureates	Year of Presentation	Country of Birth	Contribution
Peter Mitchell	1978	England	Described the chemiosmotic mechanism for ATP synthesis.
Paul Berg	1980	United States	Performed experiments in gene splicing (genetic engineering).
Walter Gilbert	1980	United States	Discovered a method of DNA sequencing.
Aaron Klug	1982	South Africa	Described the structure of tobacco mosaic virus (TMV).
Barbara McClintock	1983	United States	Discovered transposons (small segments of DNA that can move from one region of a DNA molecule to another).
César Milstein, Georges J.F. Köhler, and Niels Kai Jerne	1984	Argentina Germany Denmark	Developed a technique for producing monoclonal antibodies (single pure antibodies).
Susumu Tonegawa	1987	Japan	Described the genetics of antibody production.
Johann Deisenhofer, Robert Huber, and Hartmut Michel	1988	Germany	Described the structure of bacterial photosynthetic pigments.
J. Michael Bishop and Harold E. Varmus	1989	United States	Discovered cancer-causing genes called oncogenes.
Joseph E. Murray and E. Donnall Thomas	1990	United States	Performed the first successful organ transplants by using immunosuppressive agents.
Edmond H. Fisher and Edwin G. Krebs	1992	United States	Discovered protein kinases, enzymes that regulate cell growth.
Richard J. Roberts and Phillip A. Sharp	1993	Great Britain United States	Discovered that genes can be separated on different segments of DNA.
Kary B. Mullis	1993	United States	Discovered the polymerase chain reaction to amplify (copy) DNA.
Michael Smith	1993	Canada	Discovered a procedure to modify DNA to make new proteins.
Alfred Gilman and Martin Rodbell	1994	United States	Identified G proteins, which translate and integrate external signals that affect a wide range of biological activities.
Peter C. Doherty and Rolk M. Zinkernagel	1996	Australia Switzerland	Discovered how cytotoxic T cells recognize virus-infected cells prior to destroying them.
Stanley B. Prusiner	1997	United States	Discovered and named proteinaceous infectious particles (prions) and demonstrated a relationship between prions and deadly neurological diseases in humans and animals.

fungi, and protozoa. Previously these microorganisms were classified according to a limited number of visible characteristics.

Immunology

Immunology, the study of immunity, actually dates back in Western culture to Jenner's first vaccine in 1796. Since then, knowledge about the immune system has accumu-

lated steadily and has expanded rapidly during the twentieth century. Vaccines are now available for numerous diseases, including measles, rubella (German measles), mumps, chickenpox, pneumonia, tetanus, tuberculosis, influenza, whooping cough, polio, hepatitis B, and Lyme disease. The smallpox vaccine was so effective that the disease has been eliminated. Public health officials estimate that polio will be eradicated within a few years because of

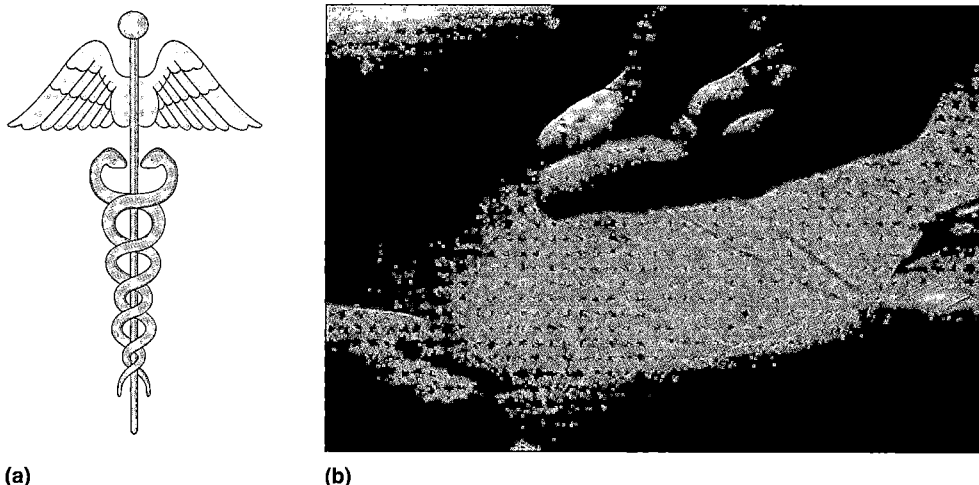

(a) **(b)**

FIGURE 1.5 Parasitology: the study of protozoa and parasitic worms. (a) The caduceus, the symbol of the medical profession, may have been designed after the procedure for removing parasitic guinea worms. **(b)** A doctor removes a guinea worm
(Dracunculus medininsis) from the subcutaneous tissue of a patient.

■ How do bacteriology, mycology, and parasitology differ?

the polio vaccine. In 1960, interferons, substances generated by the body's own immune system, were discovered.
Interferons inhibit replication of viruses and have triggered considerable research related to the treatment of viral diseases and cancer. One of today's biggest challenges
for immunologists is learning how the immune system
might be stimulated to ward off the virus responsible for
AIDS, a disease that destroys the immune system.

A major advance in immunology occurred in 1933,
when Rebecca Lancefield proposed that streptococci be
classified according to serotypes (variants within a
species) based on certain components in the cell walls of
the bacteria. Streptococci are responsible for a variety of
diseases, such as sore throat (strep throat), streptococcal
toxic shock, and septicemia (blood poisoning).

Virology

The study of viruses, **virology,** actually originated during the Golden Age of Microbiology. In 1892, Dmitri
Iwanowski reported that the organism that caused mosaic disease of tobacco was so small that it passed through
filters fine enough to stop all known bacteria. At the
time, Iwanowski was not aware that the organism in
question was a virus in the sense that we now understand
the term. In 1935, Wendell Stanley demonstrated that
the organism, called tobacco mosaic virus (TMV), was
fundamentally different from other microbes and so simple and homogeneous that it could be crystallized like a
chemical compound. Stanley's work facilitated the study
of viral structure and chemistry. Since the development

of the electron microscope in the 1940s, microbiologists
have been able to observe the structure of viruses in detail, and today much is known about their structure and
activity.

Recombinant DNA Technology

Microorganisms can now be genetically engineered to
manufacture large amounts of human hormones and
other urgently needed medical substances. In the late
1960s, Paul Berg showed that fragments of human or animal DNA that code for important proteins (genes) can
be attached to bacterial DNA. The resulting hybrid was
the first example of **recombinant DNA.** When recombinant DNA is inserted into bacteria (and other microbes), it can be used to make large quantities of the
desired protein. The technology that developed from this
technique is called **recombinant DNA technology,** or
genetic engineering, and it had its origins in two related fields. The first, **microbial genetics,** studies the
mechanisms by which microorganisms inherit traits. The
second, **molecular biology,** specifically studies how genetic information is carried in molecules of DNA and
how DNA directs the synthesis of proteins.

Although molecular biology encompasses all organisms, much of our knowledge of how genes determine
specific traits has been revealed through experiments
with bacteria. Until the 1930s, all genetic research was
based on the study of plant and animal cells. But in the
1940s, scientists turned to unicellular organisms, primarily bacteria, which have several advantages for genetic and

biochemical research. For one thing, bacteria are less complex than plants and animals. For another, the life cycles of many bacteria require less than an hour, so scientists can cultivate very large numbers of individuals for study in a relatively short time.

Once science turned to the study of unicellular life, progress in genetics began to occur rapidly. In 1941, George W. Beadle and Edward L. Tatum demonstrated the relationship between genes and enzymes. DNA was established as the hereditary material in 1944 by Oswald Avery, Colin MacLeod, and Maclyn McCarty. In 1946, Joshua Lederberg and Edward L. Tatum discovered that genetic material could be transferred from one bacterium to another by a process called conjugation. Then, in 1953, James Watson and Francis Crick proposed a model for the structure and replication of DNA. The early 1960s witnessed a further explosion of discoveries relating to the way DNA controls protein synthesis. In 1961, François Jacob and Jacques Monod discovered messenger RNA (ribonucleic acid), a chemical involved in protein synthesis, and later they made the first major discoveries about the regulation of gene function in bacteria. During the same period, scientists were able to break the genetic code and thus understand the biochemical signals that DNA transmits.

Microbes and Human Welfare

Learning Objective

■ *List at least four beneficial activities of microorganisms.*

As mentioned earlier, only a minority of all microorganisms are pathogenic. Microbes that cause food spoilage, such as soft spots on fruits and vegetables, decomposition of meats, and rancidity of fats and oils, are also a minority. The vast majority of microbes benefit humans, other animals, and plants in many ways. The following sections outline some of these beneficial activities. In later chapters, we will discuss these activities in greater detail.

Recycling Vital Elements

Discoveries made by two microbiologists in the 1880s have formed the basis for today's understanding of the biochemical cycles that support life on Earth. Martinus Beijerinck and Sergei Winogradsky were the first to show how bacteria help recycle vital elements between the soil and the atmosphere. **Microbial ecology,** the study of the relationship between microorganisms and their environment, originated with the work of Beijerinck and Winogradsky. Today, microbial ecology has branched out and includes the study of how microbial populations interact with plants and animals in various environments. Among the concerns of microbial ecologists are water pollution and toxic chemicals in the environment.

The chemical elements carbon, nitrogen, oxygen, sulfur, and phosphorus are essential for life and abundant, but not necessarily in forms that organisms can use. Microorganisms are mostly responsible for converting these elements into forms that can be used by plants and animals. Microorganisms, mostly bacteria and fungi, play a key role in returning carbon dioxide to the atmosphere when they decompose organic wastes and dead plants and animals. Algae, cyanobacteria, and higher plants use the carbon dioxide during photosynthesis to produce carbohydrates for animals, fungi, and bacteria. Nitrogen is abundant in the atmosphere but must be made into ammonia by bacteria so that nitrogen is available for plants and animals.

Sewage Treatment: Using Microbes to Recycle Water

With our growing awareness of the need to preserve the environment, we are conscious of our responsibility to recycle precious water and prevent the pollution of rivers and oceans. One major pollutant is sewage, which consists of human excrement, waste water, industrial wastes, and surface runoff. Sewage is about 99.9% water, with a few hundredths of 1% suspended solids. The remainder is a variety of dissolved materials.

Sewage treatment plants remove the undesirable materials and harmful microorganisms. Treatments combine various physical and chemical processes with the action of beneficial microbes. Large solids such as paper, wood, glass, gravel, and plastic are removed from sewage; left behind are liquid and organic materials that bacteria convert into such by-products as carbon dioxide, nitrates, phosphates, sulfates, ammonia, hydrogen sulfide, and methane. (We will discuss sewage treatment in detail in Chapter 27.)

Bioremediation: Using Microbes to Clean Up Pollutants

In 1988, scientists began using microbes to clean up pollutants and toxic wastes produced by various industrial processes. For example, some bacteria can actually turn pollutants into energy sources that they consume; others produce enzymes that break down toxins into less harmful substances. By using bacteria in these ways—a process known as **bioremediation**—toxins can be removed from underground wells, chemical spills, toxic waste sites, and oil spills, such as the *Exxon Valdez* disaster of 1989 (see the box in Chapter 2, page 35). In addition, bacterial

table 1.2   Representative Products Produced Through Recombinant DNA Techniques

Product	Description
Monoclonal antibodies	Used to diagnose cancer and assist in AIDS research.
Orthoclone®	Used to suppress the immune system in transplant recipients.
Vaccines	Vaccines against hepatitis B and influenza have already been developed. Research is under way to develop vaccines against AIDS, herpes, and malaria.
Enzymes	
Antitrypsin	Assists emphysema patients.
PEG-SOD	Reduces damage to nervous tissue following trauma.
rhDNase	Helps patients with cystic fibrosis.
Other Proteins	
Insulin	Used to lower blood sugar levels in diabetics.
Human growth hormone	Required for growth during childhood.
Interferons	Antiviral, and possibly anticancer, substances.
Interleukins	Regulate the immune functions of white blood cells.
Factor VIII	A clotting protein missing in the blood of most hemophiliacs.
Tissue plasminogen activator	A substance used to dissolve blood clots.
Hemoglobin A	A type of artificial blood.
Erythropoietin	Used to stimulate red blood cell production.
Relaxin	Used to assist childbirth.
Beta-endorphin	A pain suppressant.

enzymes are used in drain cleaners to remove clogs without adding harmful chemicals to the environment. In some cases, microorganisms indigenous to the environment are used; in others, genetically engineered microbes are used. Among the most commonly used bioremedial microbes are certain species of bacteria of the genera *Pseudomonas* and *Bacillus. Bacillus* enzymes are also used in household detergents to remove spots from clothing.

Insect Pest Control by Microorganisms

Besides spreading diseases, insects can cause devastating crop damage. Insect pest control is therefore important for both agriculture and the prevention of human disease.

The bacterium *Bacillus thuringiensis* (bä-sil'lus thür-in-jē-en'sis) has been used extensively in the United States to control such pests as alfalfa caterpillars, bollworms, corn borers, cabbageworms, tobacco budworms, and fruit tree leaf rollers. It is incorporated into a dusting powder that is applied to the crops these insects eat, or is commercially available in a form that is genetically engineered into plants. The bacteria produce protein crystals that are toxic to the digestive systems of the insects.

By using microbial rather than chemical insect control, farmers can avoid harming the environment. Many chemical insecticides, such as DDT, remain in the soil as toxic pollutants and are eventually incorporated into the food chain.

Modern Biotechnology and Genetic Engineering

Learning Objective

- List two examples of biotechnology that use genetic engineering and two examples that do not.

Earlier, we touched on the commercial use of microorganisms to produce some common foods and chemicals. Such practical applications of microbiology are called **biotechnology.** Although biotechnology has been used in some form for centuries, techniques have become much more sophisticated in the past few decades. In the last several years biotechnology has undergone a revolution through the advent of genetic engineering, the use of recombinant DNA technology to expand the potential of bacteria, viruses, and yeast cells and other fungi as miniature biochemical factories. Cultured plant and animal cells, as well as intact plants and animals, are also used as recombinant cells and organisms.

The applications of genetic engineering are increasing with each passing year. Recombinant DNA techniques have been used thus far to produce a number of natural proteins, vaccines, and enzymes. Such substances have great potential for medical use; some of them are described in Table 1.2.

A very exciting and important outcome of recombinant DNA techniques is **gene therapy**—inserting a missing gene or replacing a defective one in human cells. This technique uses a harmless virus to carry the missing or new gene into certain host cells, where the gene is picked up and inserted into the appropriate chromosome. Since 1990, gene therapy has been used to treat patients with ADA (adenosine deaminase) deficiency, a cause of severe combined immunodeficiency disease (SCID), in which cells of the immune system are inactive or missing; Duchenne's muscular dystrophy, a muscle-destroying disease; cystic fibrosis, a disease of the secreting portions of the respiratory passages, pancreas, salivary glands, and sweat glands; and LDL-receptor (low-density lipoprotein) deficiency, a condition in which LDL receptors are defective and LDL cannot enter cells. The LDL remains in the blood in high concentrations and increases the risk of atherosclerosis and coronary disease because it leads to fatty plaque formation in blood vessels. Results are still being evaluated. Certain genetic diseases may also be treatable by gene therapy in the future, including hemophilia, an inability of the blood to clot normally; diabetes, elevated blood sugar levels; sickle-cell disease, an abnormal kind of hemoglobin; and one type of hypercholesterolemia, high blood cholesterol.

Beyond medical applications, recombinant DNA techniques have also been applied to agriculture. For example, genetically altered strains of bacteria have been developed to protect fruit against frost damage, and bacteria are being engineered to control insects that damage crops. Bacteria have also been used to improve the appearance, flavor, and shelf life of fruits and vegetables. Potential agricultural uses of genetic engineering include drought resistance, resistance to insects and microbial diseases, and increased temperature tolerance in crops.

Microbes and Human Disease

Learning Objectives

- *Define normal microbiota and resistance.*
- *Define and describe several infectious diseases.*
- *Define emerging infectious diseases.*

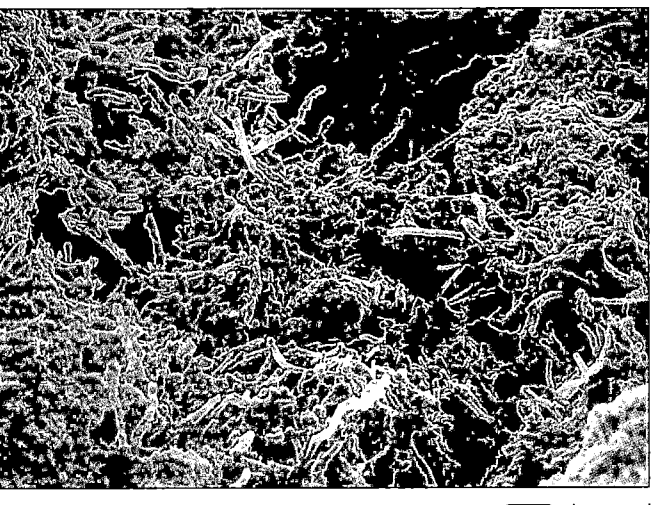

SEM |⊢———⊣| 3 μm

FIGURE 1.6 Several types of bacteria found as part of the normal microbiota inside the human mouth.

■ Normal microbiota generally do not harm us and, in some cases, actually benefit us.

Normal Microbiota

We all live in a world filled with microbes, from birth until death, and we all have a variety of microorganisms on and inside our bodies. These microorganisms make up our **normal microbiota** or *flora* (Figure 1.6). Although the normal microbiota do us no harm, and in some cases actually benefit us, under some circumstances they can make us sick or infect people we contact.

When is a microbe a welcome part of a healthy human, and when is it a harbinger of disease? The distinction between health and disease is in large part a balance between the natural defenses of the body and the disease-producing properties of microorganisms. Whether our bodies overcome the offensive tactics of a particular microbe depends on our **resistance**—the ability to ward off diseases. Important resistance is provided by the barrier of the skin, mucous membranes, cilia, stomach acid, and antimicrobial chemicals such as interferons. Microbes can be destroyed by white blood cells, the inflammatory response, fever, and by specific responses of our immune systems. Sometimes, when our natural defenses are not strong enough to overcome an invader, they have to be supplemented by antibiotics or other drugs.

Infectious Diseases

An **infectious disease** is one in which pathogens invade a susceptible host, such as a human or an animal. In the process, the pathogen carries out at least part of its life cycle inside the host, and as a result, disease frequently occurs. By the end of World War II, many people believed that infectious diseases were under control. They thought malaria would be eradicated through the use of the insecticide DDT to kill mosquitoes, that a vaccine would prevent diphtheria, and that improved sanitation measures would help prevent cholera transmission. Malaria is far from eliminated. Since 1986, local outbreaks have been identified in New Jersey, California, Florida, New York, and Texas, and the disease infects 300 million people worldwide. In 1994, diphtheria appeared in the United States, brought by travelers from the newly independent states of the former Soviet Union, which were experiencing a massive diphtheria epidemic that was brought under control in 1998.

Emerging Infectious Diseases

These incidents point to the fact that infectious diseases are not only not disappearing, but seem to be reemerging and increasing. In addition, a number of new diseases— **emerging infectious diseases (EIDs)**—have cropped up in recent years. These are diseases that are new or changing, and are increasing or have the potential to increase in incidence in the near future. Some of the factors that have contributed to the emergence of EIDs are evolutionary changes in existing organisms; the spread of known diseases to new geographic regions or populations; and increased human exposure to new, unusual infectious agents in areas that are undergoing ecologic changes such as deforestation and construction. An increasing number of incidents in recent years highlights the extent of the problem.

In 1996, countries worldwide were refusing to import beef from the United Kingdom, and the U.K. had to kill hundreds of thousands of cattle born after 1988 because of an epidemic of **bovine spongiform encephalopathy (BSE)** or **mad cow disease.** BSE first came to the attention of microbiologists in 1986 as one of a handful of diseases caused by an infectious protein called a *prion*. Studies suggest that the source of disease was cattle feed prepared from sheep infected with their own version of the disease. Cattle are herbivores (plant-eaters), but their growth and health are improved by adding protein to their feed. **Creutzfeldt-Jakob disease (CJD)** is a human disease also caused by a prion. The incidence of CJD in the U. K. is similar to the incidence in other countries. However, by 2000, the U. K. reported 46 human cases of CJD caused by a new variant related to the bovine disease (see Chapter 22, page 619).

Escherichia coli is a normal inhabitant of the large intestine of vertebrates, including humans, and its presence is beneficial because it helps produce certain vitamins and breaks down otherwise undigestible foodstuffs (see Chapter 25). However, a strain called *E. coli* O157:H7 causes bloody diarrhea when it grows in the intestines. This strain was first recognized in 1982 and since then has emerged as a public health problem. It is now one of the leading causes of diarrhea worldwide. In 1996, some 9000 people in Japan became ill, and 7 died, as a result of infection by *E. coli* O157:H7. The recent outbreaks of *E. coli* O157:H7 in the United States, associated with contamination of undercooked meat and unpasteurized beverages, have made public health officials aware that new methods of testing for bacteria in food must be developed.

In 1995, so-called flesh-eating bacteria were reported on the front pages of major newspapers. The bacteria are more correctly named invasive group A *Streptococcus,* or IGAS. There has been a trend toward increasing rates of IGAS in the United States, Scandinavia, England, and Wales. The reasons for the recent increase are unclear (see Chapter 21, page 586).

In 1995, a hospital laboratory technician in Congo (Zaire) who had fever and bloody diarrhea underwent surgery for a suspected perforated bowel. Subsequent to surgery, he started hemorrhaging, and his blood began clotting in his blood vessels. A few days later, health care workers in the hospital where he was staying developed similar symptoms. One of them was transferred to a hospital in a different city; personnel in the second hospital who cared for this patient also developed symptoms. By the time the epidemic was over, 315 people had contracted **Ebola hemorrhagic fever (EHF),** and over 75% of them died. The epidemic was controlled when microbiologists instituted training on the use of protective equipment and educational measures in the community. Human-to-human transmission occurs when there is close personal contact with infectious blood or other body fluids or tissue (see Chapter 23 on page 641).

Microbiologists first isolated Ebola viruses from humans during earlier outbreaks in Congo in 1976. (The virus is named after Congo's Ebola River.) In 1994, a single case of infection from a newly described Ebola virus occurred in Côte d'Ivoire. In 1989 and 1996, outbreaks among monkeys imported into the United States from the Philippines were caused by another Ebola virus but were not associated with human disease. Microbiologists have been studying many animals but have not yet discovered the natural reservoir (source) of EHF virus.

Hantavirus **pulmonary syndrome** caught the attention of the public in 1993 when two people in the same household became ill and died within five days of each other. Their illnesses began with a fever and cough and rapidly progressed to respiratory failure. Within a month, 23 additional cases, including 10 deaths, were re-

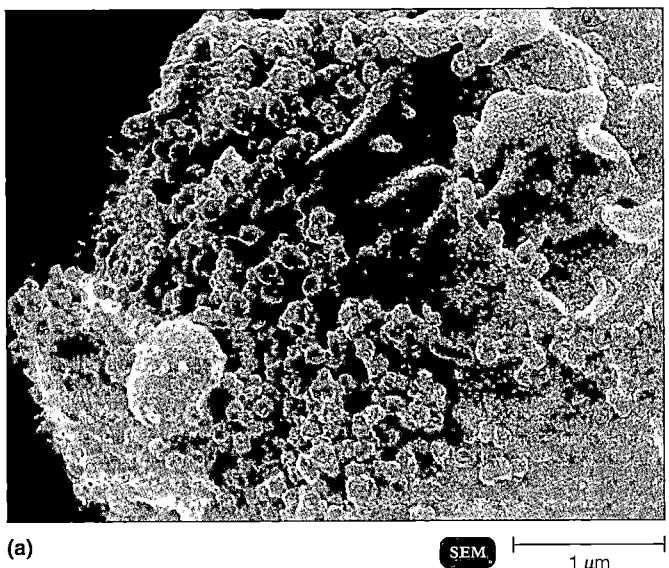

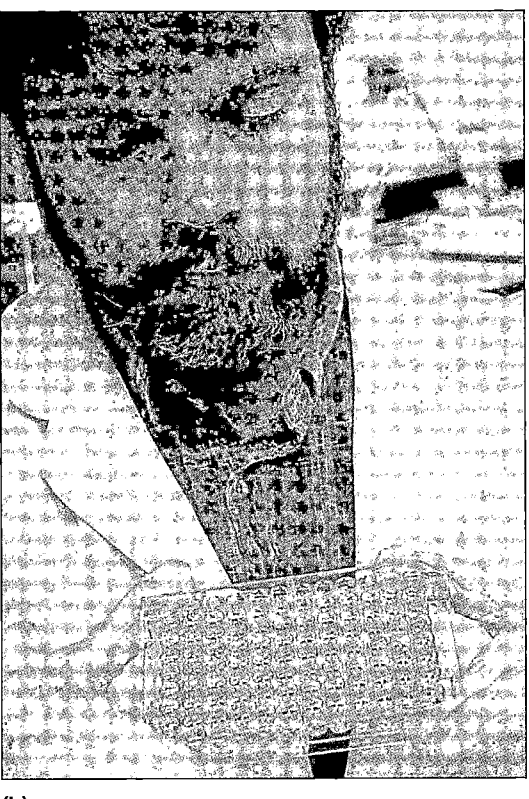

FIGURE 1.7 Human immunodeficiency virus (HIV). (a) This micrograph shows HIVs (small red spheres), the virus that causes AIDS, budding from one of its target cells, a white blood cell called a CD4 lymphocyte. **(b)** A laboratory technician performing the standard enzyme-linked immunosorbent assay (ELISA) that is used to indirectly detect the presence of HIV in the body.

(b)

ported in the Four Corners area of the American Southwest. Worried tourists began canceling vacations to the area, and residents wondered who would get the disease next. Using techniques that only became available in the 1990s, microbiologists determined that the cause was a newly recognized *Hantavirus* called the Sin Nombre virus, which is carried by deer mice. (The virus is named after the Hantaan River in Korea, where the virus was first discovered.) Researchers developed a test to rapidly identify the virus and made recommendations to help people reduce their risk of exposure to potentially infected rodents. It is probably not a new virus. Comparisons of *Hantavirus* genes suggest that the virus probably came to North America with the first rats from the Old World (see Chapter 23).

Also in 1993, an outbreak of **cryptosporidiosis** transmitted through the public water supply in Milwaukee, Wisconsin, resulted in diarrheal illness in an estimated 403,000 persons. The microorganism responsible for this outbreak was the protozoan *Cryptosporidium* (krip-tō-spô-ri′dē-um). First reported as a cause of human disease in 1976, it is responsible for up to 30% of the diarrheal illness in developing countries. In the United States, transmission has occurred via drinking water, swimming pools, and contaminated hospital supplies.

AIDS (acquired immunodeficiency syndrome) first came to public attention in 1981 with reports from Los Angeles that a few young homosexual men had died

of a previously rare type of pneumonia known as *Pneumocystis* (nü-mō-sis′tis) pneumonia. These men had experienced a severe weakening of the immune system, which normally fights infectious diseases. Soon these cases were correlated with an unusual number of occurrences of a rare form of cancer, Kaposi's sarcoma, among young homosexual men. Similar increases in such rare diseases were found among hemophiliacs and intravenous drug users.

By the end of 1999, nearly 700,000 people in the United States had been diagnosed as having AIDS, and more than 60% of them had died as a result of the disease. A great many more people had tested positive for the presence of the AIDS virus in their blood. As of 1999, health officials estimated that 600,000–900,000 Americans carried the virus. In 1999, the World Health Organization (WHO) estimated that over 47 million people worldwide had become infected with the AIDS virus.

Researchers quickly discovered that the cause of AIDS was a previously unknown virus (Figure 1.7). The virus, now called **human immunodeficiency virus (HIV),** destroys certain white blood cells of the immune system called CD4 lymphocytes, one of the cells of the body's defense. Sickness and death result from microorganisms or cancerous cells that might otherwise have been defeated by the body's natural defenses. So far, the disease has been inevitably fatal once symptoms develop.

By studying disease patterns, medical researchers found that HIV could be spread through sexual intercourse, by contaminated needles, from infected mothers to their fetuses before birth, and by blood transfusions—in short, by the transmission of body fluids from one person to another. Since 1985, blood used for transfusions has been carefully checked for the presence of HIV, and it is now quite unlikely that the virus can be spread by this means.

Since 1994, new treatments have extended the lifespan of people with AIDS; however, approximately 40,000 new cases occur annually in the United States. The majority of individuals with AIDS are in the sexually active age group, and because heterosexual partners of AIDS sufferers are at high risk of infection, public health officials are concerned that even more women and minorities will contract AIDS. In 1997, AIDS diagnoses began increasing among women and minorities. Among the AIDS cases reported for individuals 13 to 24 years of age, 44% were females and 63% were African American.

In the months and years to come, microbiological techniques will continue to be applied to help scientists learn more about the structure of the deadly HIV, how it is transmitted, how it grows in cells and causes disease, how drugs can be directed against it, and whether an effective vaccine can be developed. Public health officials have also focused on prevention through education.

AIDS poses one of this century's most formidable health threats, but it is not the first serious epidemic of a sexually transmitted disease. Syphilis was also once a fatal epidemic disease. As recently as 1941, syphilis caused an estimated 14,000 deaths per year in the United States. With few drugs available for treatment and no vaccines to prevent it, efforts to control the disease focused mainly on restraining sexual behavior and on the use of condoms. The eventual development of drugs to treat syphilis contributed significantly to preventing the spread of the disease. According to the Centers for Disease Control and Prevention (CDC), reported cases of syphilis dropped from a record high of 575,000 in 1943 to 6277 cases in 1999.

Just as microbiological techniques helped researchers in the fight against syphilis and smallpox, they will help scientists discover the causes of new emerging infectious diseases in the twenty-first century. Undoubtedly there will be new diseases. In 1996, a protozoan and a worm that cause disease in humans were discovered. Ebola virus and *Hantavirus* are examples of viruses that may be changing their abilities to infect different host species. Emerging infectious diseases will be discussed further in Chapter 14 on page 423.

Infectious diseases may reemerge because of the development of resistance to antibiotics. The breakdown of public health measures for previously controlled infections has resulted in unexpected cases of tuberculosis, whooping cough, and diphtheria (see Chapter 24).

★   ★   ★

The diseases we have mentioned are caused by viruses, bacteria, protozoa, and prions—types of microorganisms. This book introduces you to the enormous variety of microscopic organisms. It shows you how microbiologists use specific techniques and procedures to study the microbes that cause such diseases as AIDS and diarrhea—and diseases that have yet to be discovered. You will also learn about the body's responses to microbial infection and the ways certain drugs combat microbial diseases. Finally, you will learn about the many beneficial roles that microbes play in the world around us.

Study Outline

MICROBES IN OUR LIVES (pp. 2–3)

1. Living things too small to be seen with the unaided eye are called microorganisms.
2. Microorganisms are important in the maintenance of an ecological balance on Earth.
3. Some microorganisms live in humans and other animals and are needed to maintain the animal's health.
4. Some microorganisms are used to produce foods and chemicals.
5. Some microorganisms cause disease.

NAMING AND CLASSIFYING MICROORGANISMS (pp. 3–7)

1. In a nomenclature system designed by Carolus Linnaeus (1735), each living organism is assigned two names.

2. The two names consist of a genus and a specific epithet, both of which are underlined or italicized.

Types of Microorganisms (pp. 3–6)

Bacteria (p. 3)

1. Bacteria are unicellular organisms. Because they have no nucleus, the cells are described as prokaryotic.
2. The three major basic shapes of bacteria are bacillus, coccus, and spiral.
3. Most bacteria have a peptidoglycan cell wall; they divide by binary fission, and they may possess flagella.
4. Bacteria can use a wide range of chemical substances for their nutrition.

Archaea (p. 6)

1. Archaea consist of prokaryotic cells; they lack peptidoglycan in their cell walls.

2. Archaea include methanogens, extreme halophiles, and extreme thermophiles.

Fungi (p. 6)

1. Fungi (mushrooms, molds, and yeasts) have eukaryotic cells (with a true nucleus). Most fungi are multicellular.

2. Fungi obtain nutrients by absorbing organic material from their environment.

Protozoa (p. 6)

1. Protozoa are unicellular eukaryotes.

2. Protozoa obtain nourishment by absorption or ingestion through specialized structures.

Algae (p. 6)

1. Algae are unicellular or multicellular eukaryotes that obtain nourishment by photosynthesis.

2. Algae produce oxygen and carbohydrates that are used by other organisms.

Viruses (p. 6)

1. Viruses are noncellular entities that are parasites of cells.

2. Viruses consist of a nucleic acid core (DNA or RNA) surrounded by a protein coat. An envelope may surround the coat.

Multicellular Animal Parasites (p. 6)

1. The principal groups of multicellular animal parasites are flatworms and roundworms, collectively called helminths.

2. The microscopic stages in the life cycle of helminths are identified by traditional microbiological procedures.

Classification of Microorganisms (pp. 6–7)

1. All organisms are classified into Bacteria, Archaea, and Eukarya. Eukarya include protists, fungi, plants, and animals.

A BRIEF HISTORY OF MICROBIOLOGY (pp. 7–17)

The First Observations (p. 7)

1. Robert Hooke observed that plant material was composed of "little boxes"; he introduced the term *cell* (1665).

2. Hooke's observations laid the groundwork for development of the cell theory, the concept that all living things are composed of cells.

3. Antoni van Leeuwenhoek, using a simple microscope, was the first to observe microorganisms (1673).

The Debate Over Spontaneous Generation (pp. 7–9)

1. Until the mid-1880s, many people believed in spontaneous generation, the idea that living organisms could arise from nonliving matter.

2. Francesco Redi demonstrated that maggots appear on decaying meat only when flies are able to lay eggs on the meat (1668).

3. John Needham claimed that microorganisms could arise spontaneously from heated nutrient broth (1745).

4. Lazzaro Spallanzani repeated Needham's experiments and suggested that Needham's results were due to microorganisms in the air entering his broth (1765).

5. Rudolf Virchow introduced the concept of biogenesis: Living cells can arise only from preexisting cells (1858).

6. Louis Pasteur demonstrated that microorganisms are in the air everywhere and offered proof of biogenesis (1861).

7. Pasteur's discoveries led to the development of aseptic techniques used in laboratory and medical procedures to prevent contamination by microorganisms in the air.

The Golden Age of Microbiology (pp. 9–12)

1. Rapid advances in the science of microbiology were made between 1857 and 1914.

Fermentation and Pasteurization (p. 10)

1. Pasteur found that yeast ferment sugars to alcohol and that bacteria can oxidize the alcohol to acetic acid.

2. A heating process called pasteurization is used to kill bacteria in some alcoholic beverages and milk.

The Germ Theory of Disease (pp. 10–12)

1. Agostino Bassi (1835) and Pasteur (1865) showed a causal relationship between microorganisms and disease.

2. Joseph Lister introduced the use of a disinfectant to clean surgical wounds in order to control infections in humans (1860s).

3. Robert Koch proved that microorganisms cause disease. He used a sequence of procedures called Koch's postulates (1876), which are used today to prove that a particular microorganism causes a particular disease.

Vaccination (p. 12)

1. In a vaccination, immunity (resistance to a particular disease) is conferred by inoculation with a vaccine.

2. In 1798, Edward Jenner demonstrated that inoculation with cowpox material provides humans with immunity to smallpox.

3. About 1880, Pasteur discovered that avirulent bacteria could be used as a vaccine for fowl cholera; he coined the word *vaccine.*

4. Modern vaccines are prepared from living avirulent microorganisms or killed pathogens, from isolated components of pathogens, and by recombinant DNA techniques.

The Birth of Modern Chemotherapy: Dreams of a "Magic Bullet" (pp. 12–13)

1. Chemotherapy is the chemical treatment of a disease.
2. Two types of chemotherapeutic agents are synthetic drugs (chemically prepared in the laboratory) and antibiotics (substances produced naturally by bacteria and fungi to inhibit the growth of other microorganisms).
3. Paul Ehrlich introduced an arsenic-containing chemical called salvarsan to treat syphilis (1910).
4. Alexander Fleming observed that the mold (fungus) *Penicillium* inhibited the growth of a bacterial culture. He named the active ingredient penicillin (1928).
5. Penicillin has been used clinically as an antibiotic since the 1940s.
6. In 1939, René Dubos discovered two antibiotics produced by the bacterium *Bacillus*.
7. Researchers are tackling the problem of drug-resistant microbes.

Modern Developments in Microbiology (pp. 13–17)

1. Bacteriology is the study of bacteria, mycology is the study of fungi, and parasitology is the study of parasitic protozoa and worms.
2. Microbiologists are using genomics, the study of all of an organism's genes, to classify bacteria, fungi, and protozoa.
3. The study of AIDS, analysis of the action of interferons, and the development of new vaccines are among the current research interests in immunology.
4. New techniques in molecular biology and electron microscopy have provided tools for advancement of our knowledge of virology.
5. The development of recombinant DNA technology has helped advance all areas of microbiology.

MICROBES AND HUMAN WELFARE (pp. 17–19)

1. Microorganisms degrade dead plants and animals and recycle chemical elements to be used by living plants and animals.
2. Bacteria are used to decompose organic matter in sewage.
3. Bioremediation processes use bacteria to clean up toxic wastes.
4. Bacteria that cause diseases in insects are being used as biological controls of insect pests. Biological controls are specific for the pest and do not harm the environment.
5. Using microbes to make products such as foods and chemicals is called biotechnology.
6. Using recombinant DNA, bacteria can produce important substances such as proteins, vaccines, and enzymes.
7. In gene therapy, viruses are used to carry replacements for defective or missing genes into human cells.
8. Genetically engineered bacteria are used in agriculture to protect plants from frost and insects and to improve the shelf life of produce.

MICROBES AND HUMAN DISEASE (pp. 19–22)

1. Everyone has microorganisms in and on the body; these make up the normal microbiota or flora.
2. The disease-producing properties of a species of microbe and the host's resistance are important factors in determining whether a person will contract a disease.
3. An infectious disease is one in which pathogens invade a susceptible host.
4. An emerging infectious disease (EID) is a new or changing disease showing an increase in incidence in the recent past or a potential to increase in the near future.

Study Questions

REVIEW

1. How did the idea of spontaneous generation come about?
2. Some proponents of spontaneous generation believed that air is necessary for life. They thought that Spallanzani did not really disprove spontaneous generation because he hermetically sealed his flasks to keep air out. How did Pasteur's experiments address the air question without allowing the microbes in the air to ruin his experiment?
3. Briefly state the role played by microorganisms in each of the following:
   a. Biological control of pests
   b. Recycling of elements
   c. Normal microbiota
   d. Sewage treatment
   e. Human insulin production
   f. Vaccine production
4. Into which field of microbiology would the following scientists best fit?

Researcher Who	Field
___ Studies biodegradation of toxic wastes	(a) Biotechnology
___ Studies the causative agent of *Hantavirus* pulmonary syndrome	(b) Immunology
	(c) Microbial ecology
___ Studies the production of human proteins by bacteria	(d) Microbial genetics
	(e) Microbial physiology
___ Studies the symptoms of AIDS	(f) Molecular biology
___ Studies the production of toxin by *E. coli*	(g) Mycology
___ Studies the life cycle of *Cryptosporidium*	(h) Virology
___ Develops gene therapy for a disease	
___ Studies the fungus *Candida albicans*	

5. Match the following people to their contribution toward the advancement of microbiology.

___ Avery, MacLeod, and McCarty	(a) Developed vaccine against smallpox
___ Beadle and Tatum	(b) Discovered antibiotics produced by bacteria
___ Berg	(c) Discovered how DNA controls protein synthesis in a cell
___ Dubos	
___ Ehrlich	
___ Fleming	(d) Discovered penicillin
___ Hooke	(e) Discovered that DNA can be transferred from one bacterium to another
___ Iwanowski	
___ Jacob and Monod	
___ Jenner	(f) Disproved spontaneous generation
___ Koch	
___ Lancefield	(g) First to characterize a virus
___ Lederberg and Tatum	(h) First to use disinfectants in surgical procedures
___ Lister	(i) First to observe bacteria
___ Pasteur	(j) First to observe cells in plant material and name them
___ Stanley	
___ van Leeuwenhoek	(k) Observed that viruses are filterable
___ Virchow	
___ Weizmann	(l) Proved that DNA is the hereditary material

(m) Proved that microorganisms can cause disease

(n) Said living cells arise from preexisting living cells

(o) Showed that genes code for enzymes

(p) Spliced animal DNA to bacterial DNA

(q) Used bacteria to produce acetone

(r) Used the first synthetic chemotherapeutic agent

(s) Proposed a classification system for streptococci based on antigens in their cell walls

6. The genus name of a bacterium is "erwinia" and the specific epithet is "carotovora." Write the scientific name of this organism correctly. Using this name as an example, explain how scientific names are chosen.

7. Match the following microorganisms to their descriptions.

___ Archaea	(a) Not composed of cells
___ Algae	(b) Cell wall made of chitin
___ Bacteria	(c) Cell wall made of peptidoglycan
___ Fungi	(d) Cell wall made of cellulose; photosynthetic
___ Helminths	
___ Protozoa	(e) Complex cell structure lacking a cell wall
___ Viruses	(f) Multicellular animals
	(g) Prokaryote without peptidoglycan cell wall

8. It is possible to purchase the following microorganisms in a retail store. Provide a reason for buying each.
   a. *Bacillus thuringiensis*
   b. *Saccharomyces*

MULTIPLE CHOICE

1. Which of the following is a scientific name?
   a. *Mycobacterium tuberculosis*
   b. Tubercle bacillus

2. Which of the following is *not* a characteristic of bacteria?
   a. are prokaryotic
   b. have peptidoglycan cell walls
   c. have the same shape
   d. grow by binary fission
   e. have the ability to move

3. Which of the following is the most important element of Koch's germ theory of disease? The animal shows disease symptoms when
   a. the animal has been in contact with a sick animal.
   b. the animal has a lowered resistance.
   c. a microorganism is observed in the animal.
   d. a microorganism is inoculated into the animal.
   e. microorganisms can be cultured from the animal.

4. Recombinant DNA is
   a. DNA in bacteria.
   b. the study of how genes work.
   c. the DNA resulting when genes of two different organisms are mixed.
   d. the use of bacteria in the production of foods.
   e. the production of proteins by genes.

5. Which of the following statements is the best definition of biogenesis?
   a. Nonliving matter gives rise to living organisms.
   b. Living cells can only arise from preexisting cells.
   c. A vital force is necessary for life.
   d. Air is necessary for living organisms.
   e. Microorganisms can be generated from nonliving matter.

6. Which of the following is a beneficial activity of microorganisms?
   a. Some microorganisms are used as food for humans.
   b. Some microorganisms use carbon dioxide.
   c. Some microorganisms provide nitrogen for plant growth.
   d. Some microorganisms are used in sewage treatment processes.
   e. all of the above

7. It has been said that bacteria are essential for the existence of life on Earth. Which of the following would be the essential function performed by bacteria?
   a. control insect populations
   b. directly provide food for humans
   c. decompose organic material and recycle elements
   d. cause disease
   e. produce human growth hormones such as insulin

8. Which of the following is an example of bioremediation?
   a. application of oil-degrading bacteria to an oil spill
   b. application of bacteria to a crop to prevent frost damage

c. fixation of gaseous nitrogen into usable nitrogen

d. production by bacteria of a human protein such as interferon

e. all of the above

9. Spallanzani's conclusion about spontaneous generation was challenged because Lavoisier had just shown that oxygen was the vital component of air. Which of the following statements is true?

a. All life requires air.

b. Only disease-causing organisms require air.

c. Some microbes do not require air.

d. Pasteur kept air out of his biogenesis experiments.

e. Lavoisier was mistaken.

10. Which of the following statements about *E. coli* is *not* true?

a. *E. coli* was the first disease-causing bacterium identified by Koch.

b. *E. coli* is part of the normal microbiota of humans.

c. *E. coli* is beneficial in human intestines.

d. A disease-causing strain of *E. coli* causes bloody diarrhea.

e. none of the above

CRITICAL THINKING

1. How did the theory of biogenesis lead the way for the germ theory of disease?

2. Even though the germ theory of disease was not demonstrated until 1876, why did Semmelweis (1840) and Lister (1867) argue for the use of aseptic techniques?

3. Find at least three supermarket products made by microorganisms. (*Hint:* The label will state the scientific name of the organism or include the word *culture, fermented,* or *brewed.*)

4. People believed all microbial diseases would be controlled during the twentieth century. List three reasons why we are identifying new diseases now.

CLINICAL APPLICATIONS

1. The prevalence of arthritis in the United States is 1 in 100,000 children. However, 1 in 10 children in Lyme, Connecticut, developed arthritis between June and September in 1973. Allen Steere, a rheumatologist at Yale University, investigated the cases in Lyme and found that 25% of the patients remembered having a skin rash during their arthritic episode and that the disease was treatable with penicillin. Steere concluded that this was a new infectious disease and did not have an environmental, genetic, or immunologic cause.

a. What was the factor that caused Steere to reach his conclusion?

b. What is the disease?

c. Why was the disease more prevalent between June and September?

2. In 1864, Lister observed that patients recovered completely from simple fractures, but compound fractures had "disastrous consequences." He knew that the application of phenol (carbolic acid) to fields in the town of Carlisle prevented cattle disease. In 1864, Lister treated compound fractures with phenol, and his patients recovered without complications. How was Lister influenced by Pasteur's work? Why was Koch's work still needed?

Learning with Technology

MP = **The Microbiology Place website**   **ST** = **Student Tutorial CD-ROM**   **VU** = **VirtualUnknown CD-ROM**

MP   Don't forget to go to The Microbiology Place website (http://www.microbiologyplace.com) to take the practice tests, explore the interactive activity and case study, and check out the news articles and web links for this chapter.

ST   Remember there is also a quiz for this chapter on the Microbiology Interactive Student Tutorial CD-ROM.

VU   Follow the installation instructions on the liner notes accompanying the VirtualUnknown™ Microbiology CD-ROM included with this textbook. Enter the Virtual Lab, click the arrow next to the Session field, click Textbook Exercises, and select Chapter 1. Read the Case Study and then proceed.

1. Select Tryptone Broth from the Media list. You will be prompted to label the tube. Type "tryptone" in the field provided in the Medium Label box.

2. Observe the culture tubes that appear. Think about the culture tubes and the Petri dishes you may have used in your previous classes. Illustrate the route air must take to enter (1) a tube of culture medium, and (2) a Petri dish of agar medium.

3. Compare your drawings with the route air must take to enter the swan-neck flask shown in Figure 1.3 on page 9 of the textbook. Why are these designs effective in preventing microbial contamination?

4. Compare the turbidity (cloudiness) of the tube of inoculum at left with the sterile tube of tryptone broth at right. Closeups of the two tubes can be obtained by using the magnifying tool available from the tool bar above the Virtual Lab. What is responsible for the turbidity?

5. Dispose of the tubes by clicking and dragging them to the white bucket at the right of your screen. Do not record any results when prompted. Exit the Virtual Lab.

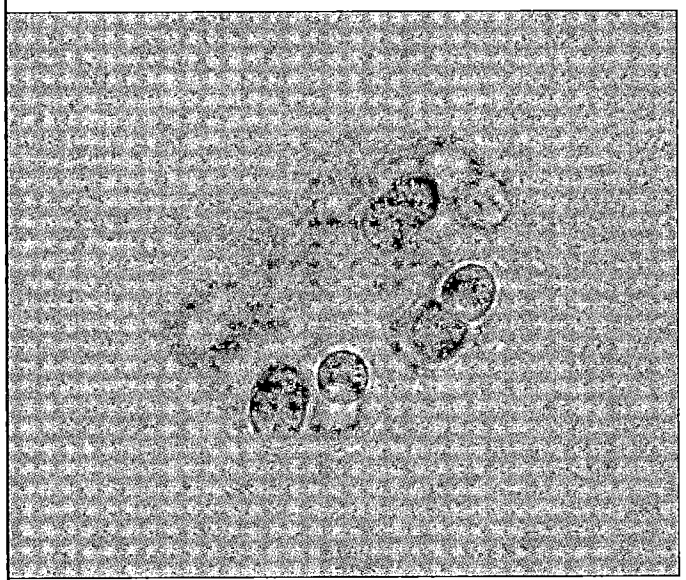

Cyanobacteria. Light energy drives the oxidation reactions in these photosynthetic bacteria. All organisms use oxidation reactions to get energy.

We can see a tree rot and smell milk going sour, but we might not realize what is happening on a microscopic level. In both cases, microbes are conducting chemical operations. The tree rots when microorganisms decompose the wood. Milk turns sour from the production of lactic acid by bacteria. Most of the activities of microorganisms are the result of a series of chemical reactions.

Like all organisms, microorganisms must use nutrients to make chemical building blocks that are used for growth and for all the other functions essential to life. For most microorganisms, synthesizing these building blocks requires them to break down nutrient substances and use the energy released to assemble the resulting molecular fragments into new substances. These chemical reactions take place minute by minute in countless microenvironments.

All matter—whether air, rock, or a living organism—is made up of small units called **atoms.** Atoms interact with each other in certain combinations to form **molecules.** Living cells are made up of molecules, some of which are very complex. The science of the interaction between atoms and molecules is called **chemistry.**

The chemistry of microbes is one of the most important concerns of microbiologists. Knowledge of chemistry is essential to understanding the roles of microorganisms in nature, how they cause disease, how methods for diagnosing disease are developed, how the body's defenses combat infection, and how antibiotics and vaccines are produced to combat the harmful effects of microbes. To understand the changes that occur in microorganisms and the changes microbes make in the world around us, we need to know how molecules are formed and how they interact.

The Structure of Atoms

Learning Objective

■ *Describe the structure of an atom and its relation to the chemical properties of elements.*

Atoms are the smallest units of matter that enter into chemical reactions. Every atom has a centrally located **nucleus** and particles called **electrons** that move around the nucleus in patterns known as electronic configurations (Figure 2.1). The nuclei of most atoms are stable—that is, they do not change spontaneously—and nuclei do not participate in chemical reactions. The nucleus is made up of positively (+) charged particles called **protons** and uncharged (neutral) particles called **neutrons.** The nucleus, therefore, bears a net positive charge. Neutrons and protons have approximately the same weight, which is about 1840 times that of an electron. The charge on electrons is negative (−), and in all atoms the number of electrons is equal to the number of protons. Because the total positive charge of the nucleus equals the total negative charge of the electrons, each atom is electrically neutral.

The number of protons in an atomic nucleus ranges from one (in a hydrogen atom) to more than 100 (in the largest atoms known). Atoms are often listed by their **atomic number,** the number of protons in the nucleus. The total number of protons and neutrons in an atom is its approximate **atomic weight.**

Chemical Elements

All atoms with the same number of protons behave the same way chemically and are classified as the same **chemical element.** Each element has its own name and a one- or two-letter symbol, usually derived from the English or Latin name for the element. For example, the symbol for the element hydrogen is H, and the symbol for carbon is C. The symbol for sodium is Na—the first two letters of its Latin name, *natrium*—to distinguish it from nitrogen, N, and from sulfur, S. There are 92 naturally occurring elements. However, only about 26 elements are commonly found in living things. Table 2.1 lists some of the chemical elements found in living organisms, including their atomic numbers and weights. The elements most abundant in living matter are hydrogen, carbon, nitrogen, and oxygen.

Most elements have several **isotopes**—atoms with different numbers of neutrons in their nuclei. All isotopes of an element have the same number of protons in their nuclei, but their atomic weights differ because of the difference in the number of neutrons. For example, in a natural sample of oxygen, all the atoms will contain eight protons. However, 99.76% of the atoms will have eight neutrons, 0.04% will contain nine neutrons, and the remaining 0.2% will contain ten neutrons. Therefore, the three isotopes composing a natural sample of oxygen will have atomic weights of 16, 17, and 18, although all will have the atomic number 8. Atomic numbers are written as a subscript to the left of an element's chemical symbol. Atomic weights are written as a superscript above the atomic number. Thus, natural oxygen isotopes are represented as $_8^{16}O$, $_8^{17}O$, and $_8^{18}O$. Isotopes of certain elements are extremely useful in biological research, medical diagnosis, the treatment of some disorders, and in some forms of sterilization.

Electronic Configurations

In an atom, electrons are arranged in **electron shells,** which are regions corresponding to different **energy levels.** The arrangement is called an **electronic configuration.** Shells are layered outward from the nucleus, and each shell can hold a characteristic maximum number of electrons—two electrons in the innermost shell (lowest energy level), eight electrons in the second shell, and eight electrons in the third shell, if it is the atom's outermost (valence) shell. The fourth, fifth, and sixth electron shells can each accommodate 18 electrons, although there are some exceptions to this generalization. Table 2.2 shows the electronic configurations for atoms of some elements found in living organisms.

There is a tendency for the outermost shell to be filled with the maximum number of electrons. An atom can give up, accept, or share electrons with other atoms to fill this shell. The chemical properties of atoms are largely a function of the number of electrons in the outermost electron shell. When its outer shell is filled, the atom is

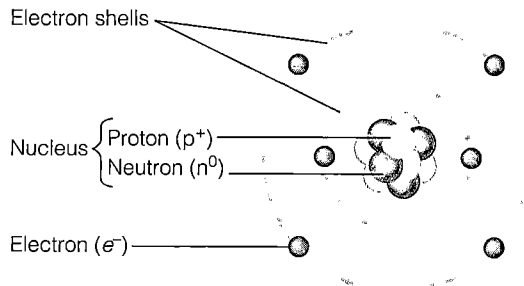

FIGURE 2.1 The structure of an atom. In this simplified diagram of a carbon atom, note the central location of the nucleus. The nucleus contains six neutrons and six protons, although not all are visible in this view. The six electrons move about the nucleus in regions called electron shells, shown here as circles.

■ What is the atomic number of this atom?

table 2.1	The Elements of Life*		
Element	**Symbol**	**Atomic Number**	**Approximate Atomic Weight**
Hydrogen	H	1	1
Carbon	C	6	12
Nitrogen	N	7	14
Oxygen	O	8	16
Sodium	Na	11	23
Magnesium	Mg	12	24
Phosphorus	P	15	31
Sulfur	S	16	32
Chlorine	Cl	17	35
Potassium	K	19	39
Calcium	Ca	20	40
Iron	Fe	26	56
Iodine	I	53	127

*Hydrogen, carbon, nitrogen, and oxygen are the most abundant chemical elements in living organisms.

chemically stable, or inert: It does not tend to react with other atoms. Helium (atomic number 2) and neon (atomic number 10) are examples of atoms of inert gases that have filled outer shells.

When an atom's outer electron shell is only partially filled, the atom is chemically unstable. Such an atom reacts with other atoms, and this reaction depends, in part, on the degree to which the outer energy levels are filled. Notice the number of electrons in the outer energy levels of the atoms in Table 2.2. We will see later how the number correlates with the chemical reactivity of the elements.

How Atoms Form Molecules: Chemical Bonds

Learning Objective

■ *Define ionic bond, covalent bond, hydrogen bond, molecular weight, and mole.*

When the outermost energy level of an atom is not completely filled by electrons, you can think of it as having either unfilled spaces or extra electrons in that energy level, depending on whether it is easier for the atom to gain or lose electrons. For example, an atom of oxygen, with two

electrons in the first energy level and six in the second, has two unfilled spaces in the second electron shell; an atom of magnesium has two extra electrons in its outermost shell. The most chemically stable configuration for any atom is to have its outermost shell filled, as do the inert gases. Therefore, for these two atoms to attain that state, oxygen must gain two electrons, and magnesium must lose two electrons. All atoms tend to combine so that the extra electrons in the outermost shell of one atom fill the spaces of the outermost shell of the other atom; for example, oxygen and magnesium combine so that the outermost shell of each atom has the full complement of eight electrons.

The **valence,** or combining capacity, of an atom is the number of extra or missing electrons in its outermost electron shell. For example, hydrogen has a valence of 1 (one unfilled space, or one extra electron), oxygen has a valence of 2 (two unfilled spaces), carbon has a valence of 4 (four unfilled spaces, or four extra electrons), and magnesium has a valence of 2 (two extra electrons).

Basically, atoms achieve the full complement of electrons in their outermost energy shells by combining to form molecules, which are made up of atoms of one or more elements. A molecule that contains at least two different kinds of atoms, such as H_2O, the water molecule, is called a **compound.** In H_2O, the subscript 2 indicates that there are two atoms of hydrogen; the absence of a subscript indicates that there is only one atom of oxygen. Molecules hold together because the valence electrons of the combining atoms form attractive forces, called **chemical bonds,** between the atomic nuclei. Therefore, valence may also be viewed as the bonding capacity of an element. Because energy is required for chemical bond formation, each chemical bond possesses a certain amount of potential chemical energy.

In general, atoms form bonds in one of two ways: by either gaining or losing electrons from their outer electron shell, or by sharing outer electrons. When atoms have gained or lost outer electrons, the chemical bond is called an ionic bond. When outer electrons are shared, the bond is called a covalent bond. Although we will discuss ionic and covalent bonds separately, the kinds of bonds actually found in molecules do not belong entirely to either category. Instead, bonds range from the highly ionic to the highly covalent.

Ionic Bonds

Atoms are electrically neutral when the number of positive charges (protons) equals the number of negative charges (electrons). But when an isolated atom gains or loses electrons, this balance is upset. If the atom gains

table 2.2	Electronic Configurations for the Atoms of Some Elements Found in Living Organisms						
Element	First Electron Shell	Second Electron Shell	Third Electron Shell	Diagram	Number of Valence (Outermost) Shell Electrons	Number of Unfilled Spaces	Maximum Number of Bonds Formed
Hydrogen	1	—	—	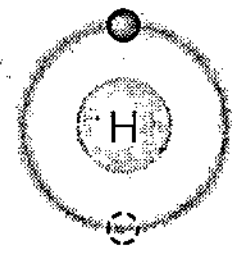	1	1	1
Carbon	2	4	—	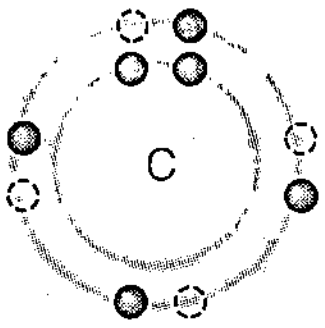	4	4	4
Nitrogen	2	5	—	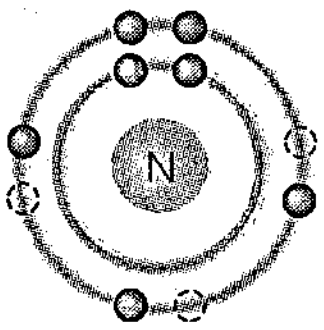	5	3	3
Oxygen	2	6	—	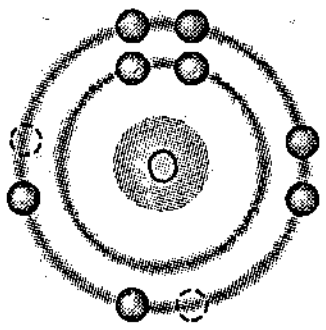	6	2	2
Magnesium	2	8	2	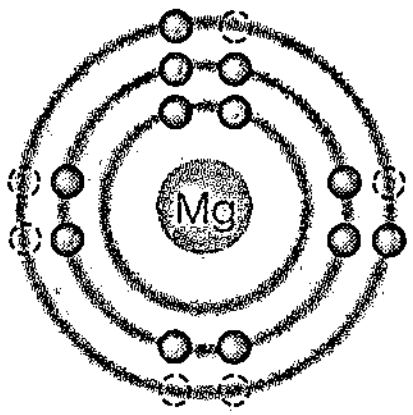	2	6	2
Phosphorus	2	8	5	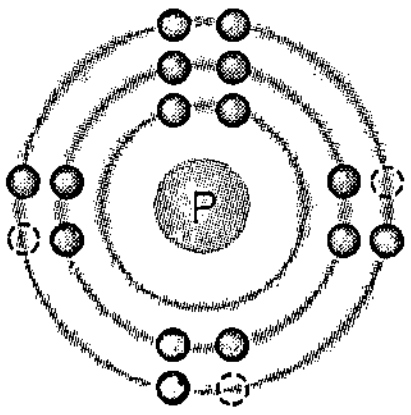	5	3	5
Sulfur	2	8	6	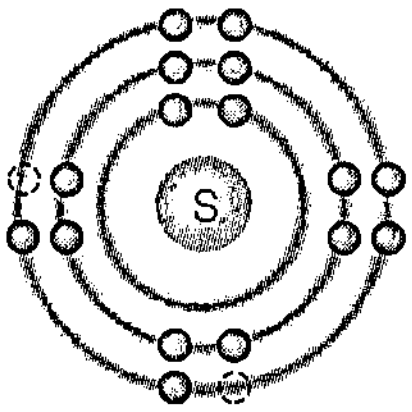	6	2	2

Key:
- electron
- unfilled space
- atomic nucleus

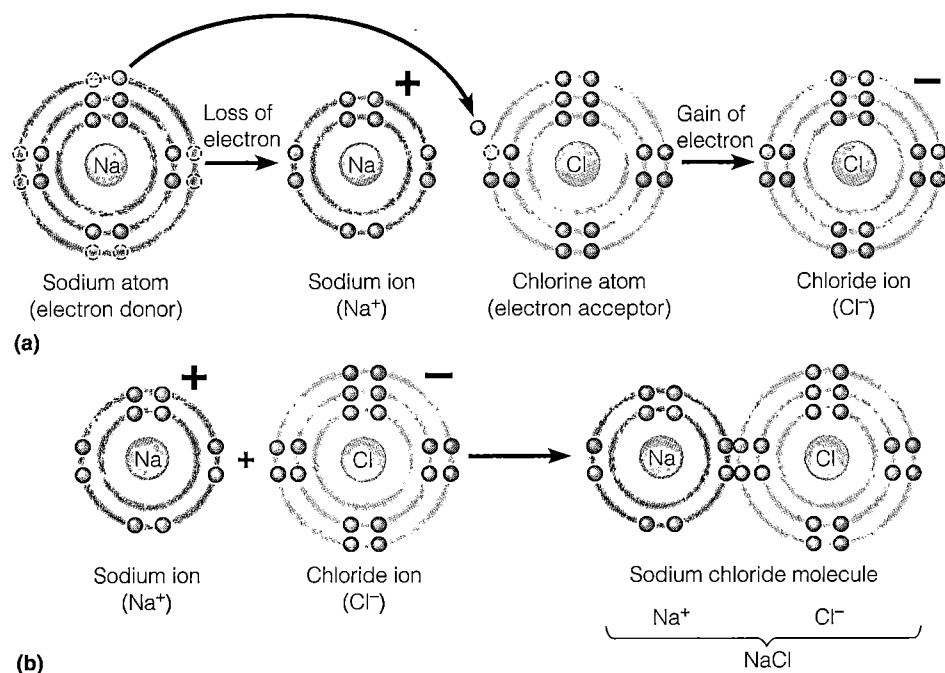

Sodium atom
(electron donor)

Sodium ion
(Na⁺)

Chlorine atom
(electron acceptor)

Chloride ion
(Cl⁻)

(a)

Sodium ion
(Na⁺)

Chloride ion
(Cl⁻)

Sodium chloride molecule

Na⁺   Cl⁻

NaCl

(b)

**FIGURE 2.2 Ionic bond formation.
(a)** A sodium atom (Na), left, loses one electron to an electron acceptor and forms a sodium ion (Na^+). A chlorine atom (Cl), right, accepts one electron from an electron donor to become a chloride ion (Cl^-). **(b)** The sodium and chloride ions are attracted because of their opposite charges and are held together by an ionic bond to form a molecule of sodium chloride.

■ An ionic bond is a chemical attraction that holds together ions with different charges.

electrons, it acquires an overall negative charge; if the atom loses electrons, it acquires an overall positive charge. Such a negatively or positively charged atom (or group of atoms) is called an **ion.**

Consider the following examples. Sodium (Na) has 11 protons and 11 electrons, with one electron in its outer electron shell. Sodium tends to lose the single outer electron; it is an *electron donor* (Figure 2.2a). When sodium donates an electron to another atom, it is left with 11 protons and only 10 electrons and so has an overall charge of +1. This positively charged sodium atom is called a sodium ion and is written as Na^+. Chlorine (Cl) has a total of 17 electrons, seven of them in the outer electron shell. Because this outer shell can hold eight electrons, chlorine tends to pick up an electron that has been lost by another atom; it is an *electron acceptor* (see Figure 2.2a). By accepting an electron, chlorine totals 18 electrons. However, it still has only 17 protons in its nucleus. The chloride ion therefore has a charge of −1 and is written as Cl^-.

The opposite charges of the sodium ion (Na^+) and chloride ion (Cl^-) attract each other. The attraction, an ionic bond, holds the two atoms together, and a molecule is formed (Figure 2.2b). The formation of this molecule, called sodium chloride (NaCl) or table salt, is a common example of ionic bonding. Thus, an **ionic bond** is an attraction between ions of opposite charge that holds them together to form a stable molecule. Put another way, an ionic bond is an attraction between atoms in which one atom loses electrons and another atom gains electrons. Strong ionic bonds, such as those that hold Na^+ and Cl^-

together in salt crystals, have limited importance in living cells. But the weaker ionic bonds formed in aqueous (water) solutions are important in biochemical reactions in microbes and other organisms. For example, weaker ionic bonds assume a role in certain antigen–antibody reactions—that is, reactions in which molecules produced by the immune system (antibodies) combine with foreign substances (antigens) to combat infection.

In general, an atom whose outer electron shell is less than half-filled will lose electrons and form positively charged ions, called **cations.** Examples of cations are the potassium ion (K^+), calcium ion (Ca^{2+}), and sodium ion (Na^+). When an atom's outer electron shell is more than half-filled, the atom will gain electrons and form negatively charged ions, called **anions.** Examples are the iodide ion (I^-), chloride ion (Cl^-), and sulfide ion (S^{2-}).

Covalent Bonds

A **covalent bond** is a chemical bond formed by two atoms sharing one or more pairs of electrons. Covalent bonds are stronger and far more common in organisms than are true ionic bonds. In the hydrogen molecule, H_2, two hydrogen atoms share a pair of electrons. Each hydrogen atom has its own electron plus one electron from the other atom (Figure 2.3a). The shared pair of electrons actually orbits the nuclei of both atoms. Therefore, the outer electron shells of both atoms are filled. When only one pair of electrons is shared between atoms, a *single covalent bond* is formed. For simplicity, a single covalent bond is

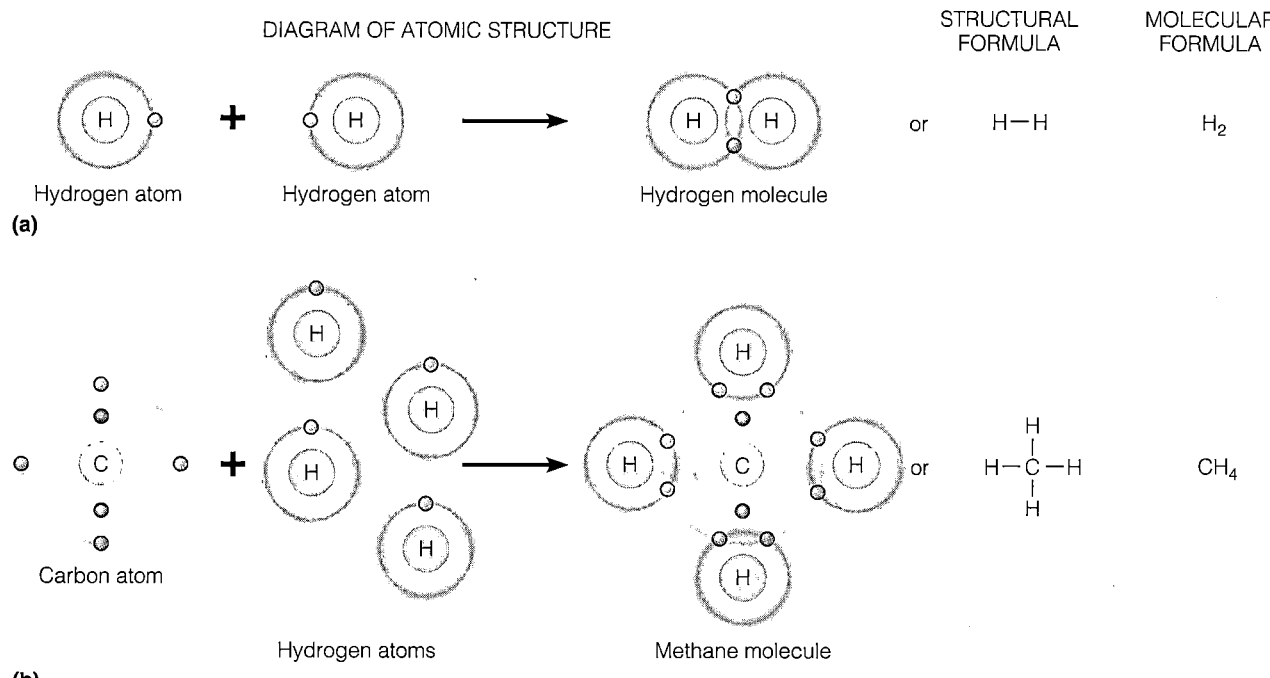

DIAGRAM OF ATOMIC STRUCTURE   STRUCTURAL   MOLECULAR
                                                         FORMULA   FORMULA

Hydrogen atom   Hydrogen atom   Hydrogen molecule   or   H—H H_2

(a)

Carbon atom

Hydrogen atoms   Methane molecule   or   H—C—H (with H above and below) CH_4

(b)

FIGURE 2.3 Covalent bond formation. (a) A single covalent bond between two hydrogen atoms. **(b)** Single covalent bonds between four hydrogen atoms and a carbon atom, forming a methane molecule. On the right are simpler ways to represent molecules. In structural formulas, each covalent bond is written as a straight line between the symbols for two atoms. In molecular formulas, the number of atoms in each molecule is noted by subscripts.

▨ In a covalent bond, two atoms share one, two, or three pairs of valance electrons.

expressed as a single line between the atoms (H—H). When two pairs of electrons are shared between atoms, a *double covalent bond* is formed, expressed as two single lines (=). A *triple covalent bond,* expressed as three single lines (≡), occurs when three pairs of electrons are shared.

The principles of covalent bonding that apply to atoms of the same element also apply to atoms of different elements. Methane (CH_4) is an example of covalent bonding between atoms of different elements (Figure 2.3b). The outer electron shell of the carbon atom can hold eight electrons but has only four; each hydrogen atom can hold two electrons but has only one. Consequently, in the methane molecule the carbon atom gains four hydrogen electrons to complete its outer shell, and each hydrogen atom completes its pair by sharing one electron from the carbon atom. Each outer electron of the carbon atom orbits both the carbon nucleus and a hydrogen nucleus. Each hydrogen electron orbits both its own nucleus and the carbon nucleus.

Elements such as hydrogen and carbon, whose outer electron shells are half-filled, form covalent bonds quite

easily. In fact, in living organisms, carbon almost always forms covalent bonds; it almost never becomes an ion. *Remember:* Covalent bonds are formed by the *sharing* of electrons between atoms. Ionic bonds are formed by *attraction* between atoms that have lost or gained electrons and are therefore positively or negatively charged.

Hydrogen Bonds

Another chemical bond of special importance to all organisms is the **hydrogen bond,** in which a hydrogen atom that is covalently bonded to one oxygen or nitrogen atom is attracted to another oxygen or nitrogen atom. Such bonds are weak and do not bind atoms into molecules. However, they do serve as bridges between different molecules or between various portions of the same molecule.

When hydrogen combines with atoms of oxygen or nitrogen, the relatively large nucleus of these larger atoms attracts the hydrogen electron more strongly than does the small hydrogen nucleus. Thus, in a molecule of water

(H$_2$O), all the electrons tend to be closer to the oxygen nucleus than to the hydrogen nuclei. The oxygen portion of the molecule thus has a slightly negative charge, and the hydrogen portion of the molecule has a slightly positive charge (Figure 2.4a). When the positively charged end of one molecule is attracted to the negatively charged end of another molecule, a hydrogen bond is formed (Figure 2.4b). This attraction can also occur between hydrogen and other atoms of the same molecule, especially in large molecules. Oxygen and nitrogen are the elements most frequently involved in hydrogen bonding.

Hydrogen bonds are considerably weaker than either ionic or covalent bonds; they have only about 5% of the strength of covalent bonds. Consequently, hydrogen bonds are formed and broken relatively easily. This property accounts for the temporary bonding that occurs between certain atoms of large and complex molecules, such as proteins and nucleic acids. Even though hydrogen bonds are relatively weak, large molecules containing several hundred of these bonds have considerable strength and stability.

Molecular Weight and Moles

You have seen that bond formation results in the creation of molecules. Molecules are often discussed in terms of units of measure called molecular weight and moles. The **molecular weight** of a molecule is the sum of the atomic weights of all its atoms. To relate the molecular level to the laboratory level, we use a unit called the mole. One **mole** of a substance is its molecular weight expressed in grams. For example, 1 mole of water weighs 18 grams because the molecular weight of H$_2$O is 18 [(2 × 1) + 16].

Chemical Reactions

Learning Objectives

- *Diagram three basic types of chemical reactions.*
- *Identify the role of enzymes in chemical reactions.*

As we said earlier, **chemical reactions** involve the making or breaking of bonds between atoms. After a chemical reaction, the total number of atoms remains the same, but there are new molecules with new properties because the atoms have been rearranged.

Energy in Chemical Reactions

Some change of energy occurs whenever bonds between atoms are formed or broken during chemical reactions.

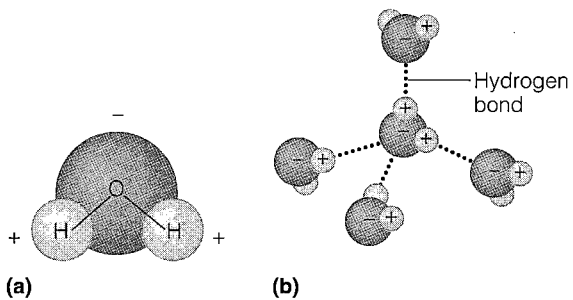

(a)   **(b)**

FIGURE 2.4 Hydrogen bond formation in water. (a) In a water molecule, the electrons of the hydrogen atoms are strongly attracted to the oxygen atom. Therefore, the part of the water molecule containing the oxygen atom has a slightly negative charge, and the part containing hydrogen atoms has a slightly positive charge. **(b)** In a hydrogen bond between water molecules, the hydrogen of one water molecule is attracted to the oxygen of another water molecule. Many water molecules may be attracted to each other by hydrogen bonds (black dots).

How do hydrogen bonds affect the temperature of water?

This energy is called **chemical energy.** When a chemical bond is formed, energy is required. Such a chemical reaction that absorbs more energy than it releases is called an **endergonic reaction** (*end* = within), meaning that energy is directed inward. When a bond is broken, energy is released. A chemical reaction that releases more energy than it absorbs is called an **exergonic reaction** (*ex* = out), meaning that energy is directed outward.

In this section we will look at three basic types of chemical reactions common to all living cells. By becoming familiar with these reactions, you will be able to understand the specific chemical reactions we will discuss later, particularly in Chapter 5.

Synthesis Reactions

When two or more atoms, ions, or molecules combine to form new and larger molecules, the reaction is called a **synthesis reaction.** To synthesize means to put together, and a synthesis reaction *forms new bonds.* Synthesis reactions can be expressed in the following way:

$$\underset{\substack{\text{Atom, ion,}\\\text{or molecule A}}}{A} + \underset{\substack{\text{Atom, ion,}\\\text{or molecule B}}}{B} \xrightarrow{\substack{\text{Combine}\\\text{to form}}} \underset{\substack{\text{New molecule}\\\text{AB}}}{AB}$$

The combining substances, A and B, are called the *reactants;* the substance formed by the combination, AB, is the *product.* The arrow indicates the direction in which the reaction proceeds.

FIGURE 2.5   Energy requirements of a chemical reaction. This graph shows the progress of the reaction AB → A + B both without (black line) and with (magenta line) an enzyme. The presence of an enzyme lowers the activation energy of the reaction (see arrows). Thus, more molecules of reactant AB are converted to products A and B because more molecules of reactant AB possess the activation energy needed for the reaction.

■ How do enzymes speed up chemical reactions?

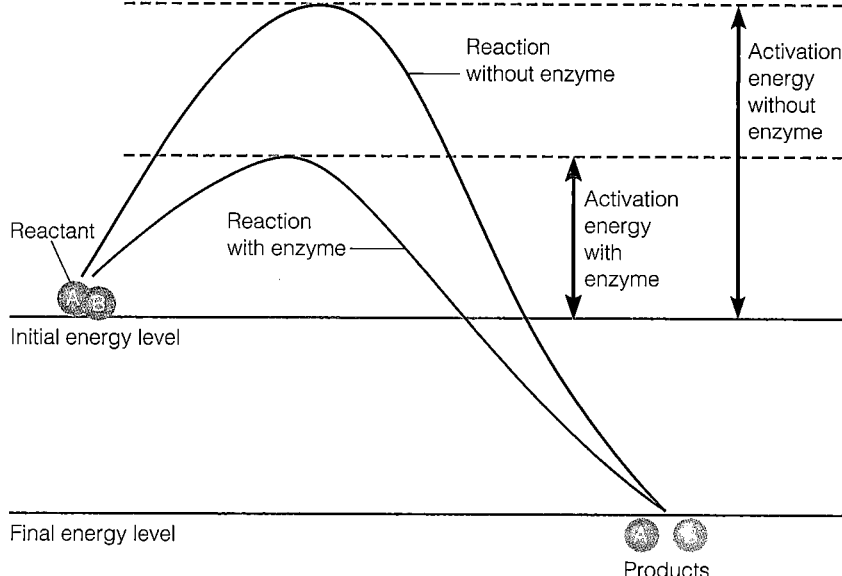

chemical reaction is its **activation energy,** which is the amount of energy needed to disrupt the stable electronic configuration of any specific molecule so that the electrons can be rearranged.

The **reaction rate**—the frequency of collisions containing sufficient energy to bring about a reaction—depends on the number of reactant molecules at or above the activation energy level. One way to increase the reaction rate of a substance is to raise its temperature. By causing the molecules to move faster, heat increases both the frequency of collisions and the number of molecules that attain activation energy. The number of collisions also increases when pressure is increased or when the reactants are more concentrated (because the distance between molecules is thereby decreased). In living systems, enzymes increase the reaction rate without raising the temperature.

Enzymes and Chemical Reactions

Substances that can speed up a chemical reaction without being permanently altered themselves are called **catalysts.** In living cells, **enzymes** serve as biological catalysts. As catalysts, enzymes are specific. Each acts on a specific substance, called the enzyme's **substrate** (or substrates, when there are two or more reactants), and each catalyzes

only one reaction. For example, sucrose (table sugar) is the substrate of the enzyme sucrase, which catalyzes the hydrolysis of sucrose to glucose and fructose.

As catalysts, enzymes typically accelerate chemical reactions. The three-dimensional enzyme molecule has an *active site,* a region that will interact with a specific chemical substance (see Figure 5.3 on page 117). The chemical substance (reactant) an enzyme acts on is referred to as the enzyme's substrate.

The enzyme orients the substrate into a position that increases the probability of a reaction. The **enzyme–substrate complex** formed by the temporary binding of enzyme and reactants enables the collisions to be more effective and lowers the activation energy of the reaction (Figure 2.5). The enzyme therefore speeds up the reaction by increasing the number of AB molecules that attain sufficient activation energy to react.

An enzyme's ability to accelerate a reaction without the need for an increase in temperature is crucial to living systems because a significant temperature increase would destroy cellular proteins. The crucial function of enzymes, therefore, is to speed up biochemical reactions at a temperature that is compatible with the normal functioning of the cell.

IMPORTANT BIOLOGICAL MOLECULES

Biologists and chemists divide compounds into two principal classes: inorganic and organic. **Inorganic compounds** are defined as molecules, usually small and structurally simple, that typically lack carbon and in which ionic bonds may play an important role. Inorganic

compounds include water, oxygen, carbon dioxide, and many salts, acids, and bases.

Organic compounds always contain carbon and hydrogen and are typically structurally complex. Carbon is a unique element because it has four electrons in its outer shell and four unfilled spaces. It can combine with a variety of atoms, including other carbon atoms, to form straight or branched chains and rings. Carbon chains form the basis of many organic compounds in living cells, including sugars, amino acids, and vitamins. Organic compounds are held together mostly or entirely by covalent bonds. Some organic molecules, such as polysaccharides, proteins, and nucleic acids, are very large and usually contain thousands of atoms. Such giant molecules are called *macromolecules*. In the following section we will discuss inorganic and organic compounds that are essential for cells.

Inorganic Compounds

Water

Learning Objective

■ *List several properties of water that are important to living systems.*

All living organisms require a wide variety of inorganic compounds for growth, repair, maintenance, and reproduction. Water is one of the most important, as well as one of the most abundant, of these compounds, and it is particularly vital to microorganisms. Outside the cell, nutrients are dissolved in water, which facilitates their passage through cell membranes. And inside the cell, water is the medium for most chemical reactions. In fact, water is by far the most abundant component of almost all living cells. Water makes up at least 5–95% of every cell, the average being between 65% and 75%. Simply stated, no organism can survive without water.

Water has structural and chemical properties that make it particularly suitable for its role in living cells. As we discussed, the total charge on the water molecule is neutral, but the oxygen region of the molecule has a slightly negative charge and the hydrogen region has a slightly positive charge (see Figure 2.4a on page 33). Any molecule having such an unequal distribution of charges is called a **polar molecule.** The polar nature of water gives it four characteristics that make it a useful medium for living cells.

First, every water molecule is capable of forming four hydrogen bonds with nearby water molecules (see Figure 2.4b). This property results in a strong attraction between water molecules. Because of this strong attraction, a great deal of heat is required to separate water molecules from each other to form water vapor; thus, water has a relatively high boiling point (100°C). Because water has such a high boiling point, it exists in the liquid state on most of the Earth's surface. Furthermore, the hydro-

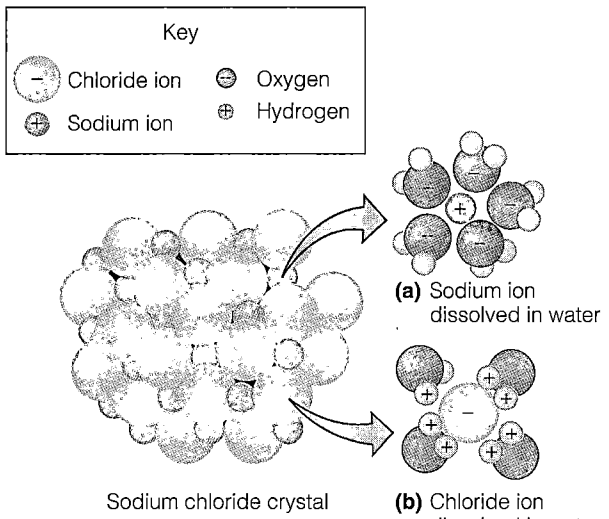

(a) Sodium ion dissolved in water

Sodium chloride crystal   **(b)** Chloride ion dissolved in water

FIGURE 2.6 How water acts as a solvent for sodium chloride (NaCl). (a) The positively charged sodium ion (Na^+) is attracted to the negative part of the water molecule. **(b)** The negatively charged chloride ion (Cl^-) is attracted to the positive part of the water molecule. In the presence of water molecules, the bonds between the Na^+ and Cl^- are disrupted, and the NaCl dissolves in the water.

■ **Ionization is the separation of molecules into ions.**

gen bonding between water molecules affects the density of water, depending on whether it occurs as ice or a liquid. For example, the hydrogen bonds in the crystalline structure of water (ice) make ice take up more space. As a result, ice has fewer molecules than an equal volume of liquid water. This makes its crystalline structure less dense than liquid water. For this reason, ice floats and can serve as an insulating layer on the surfaces of lakes and streams that harbor living organisms.

Second, the polarity of water makes it an excellent dissolving medium or **solvent.** Many polar substances undergo **dissociation,** or separation, into individual molecules in water—that is, they dissolve—because the negative part of the water molecules is attracted to the positive part of the molecules in the **solute,** or dissolving substance, and the positive part of the water molecules is attracted to the negative part of the solute molecules. Substances (such as salts) that are composed of atoms (or groups of atoms) held together by ionic bonds tend to dissociate into separate cations and anions in water. Thus, the polarity of water allows molecules of many different substances to separate and become surrounded by water molecules (Figure 2.6).

Third, polarity accounts for water's characteristic role as a reactant or product in many chemical reactions. Its polarity facilitates the splitting and rejoining of hydrogen ions

FIGURE 2.7 Acids, bases, and salts. (a) In water, hydrochloric acid (HCl) dissociates into H$^+$ and Cl$^-$. **(b)** Sodium hydroxide (NaOH), a base, dissociates into OH$^-$ and Na$^+$ in water. **(c)** In water, table salt (NaCl) dissociates into positive ions (Na$^+$) and negative ions (Cl$^-$), neither of which are H$^+$ or OH$^-$.

■ How do acids and bases differ?

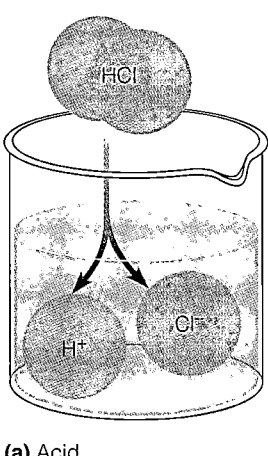

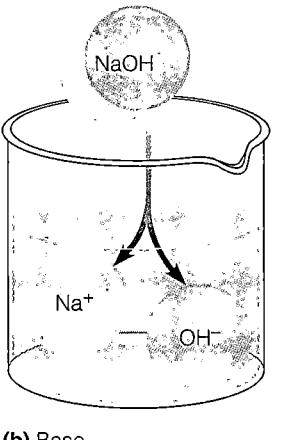

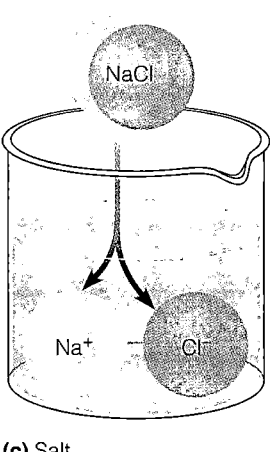

(a) Acid   (b) Base   (c) Salt

(H$^+$) and hydroxide ions (OH$^-$). Water is a key reactant in the digestive processes of organisms, whereby larger molecules are broken down into smaller ones. Water molecules are also involved in synthetic reactions; water is an important source of the hydrogen and oxygen that are incorporated into numerous organic compounds in living cells.

Finally, the relatively strong hydrogen bonding between water molecules (see Figure 2.4b) makes water an excellent temperature buffer. A given quantity of water requires a great gain of heat to increase its temperature and a great loss of heat to decrease its temperature, compared with many other substances. Normally, heat absorption by molecules increases their kinetic energy and thus increases their rate of motion and their reactivity. In water, however, heat absorption first breaks hydrogen bonds rather than increasing the rate of motion. Therefore, much more heat must be applied to raise the temperature of water than to raise the temperature of a non–hydrogen-bonded liquid. The reverse is true as water cools. Thus, water more easily maintains a constant temperature than other solvents and tends to protect a cell from fluctuations in environmental temperatures.

Acids, Bases, and Salts

Learning Objective

■ *Define acid, base, salt, and pH.*

As we saw in Figure 2.6, when inorganic salts such as sodium chloride (NaCl) are dissolved in water, they undergo **ionization** or *dissociation;* that is, they break apart into ions. Substances called acids and bases show similar behavior.

An **acid** can be defined as a substance that dissociates into one or more hydrogen ions (H$^+$) and one or more negative ions (anions). Thus, an acid can also be defined as a proton (H$^+$) donor. A **base** dissociates into one or more

positive ions (cations) plus one or more negatively charged hydroxide ions (OH$^-$) that can accept, or combine with, protons. Thus, sodium hydroxide (NaOH) is a base because it dissociates to release OH$^-$, which have a strong attraction for protons and are among the most important proton acceptors. A **salt** is a substance that dissociates in water into cations and anions, neither of which is H$^+$ or OH$^-$. Figure 2.7 shows common examples of each type of compound and how they dissociate in water.

Acid–Base Balance

An organism must maintain a fairly constant balance of acids and bases to remain healthy. In the aqueous environment within organisms, acids dissociate into hydrogen ions (H$^+$) and anions. Bases, on the other hand, dissociate into hydroxide ions (OH$^-$) and cations. The more hydrogen ions that are free in a solution, the more acidic the solution is. Conversely, the more hydroxide ions that are free in a solution, the more basic, or alkaline, it is.

Biochemical reactions—that is, chemical reactions in living systems—are extremely sensitive to even small changes in the acidity or alkalinity of the environments in which they occur. In fact, H$^+$ and OH$^-$ are involved in almost all biochemical processes, and the functions of a cell are modified greatly by any deviation from its narrow band of normal H$^+$ and OH$^-$ concentrations. For this reason, the acids and bases that are continually formed in an organism must be kept in balance.

It is convenient to express the amount of H$^+$ in a solution by a logarithmic **pH** scale, which ranges from 0 to 14 (Figure 2.8). The term *pH* means potential of hydrogen. On a logarithmic scale, a change of one whole number represents a tenfold change from the previous concentration. Thus, a solution of pH 1 has ten times more hydrogen ions than a solution of pH 2 and has 100 times more hydrogen ions than a solution of pH 3.

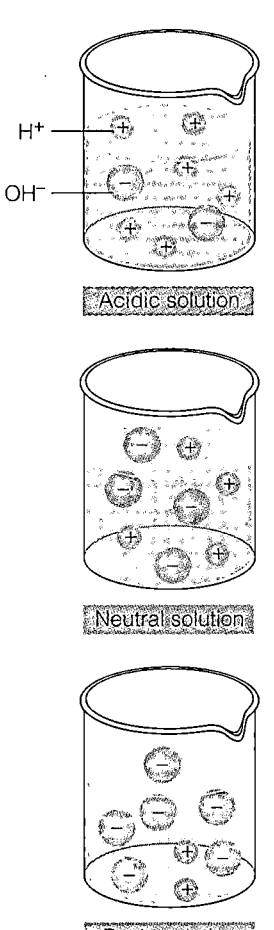

Acidic solution

Neutral solution

Basic solution

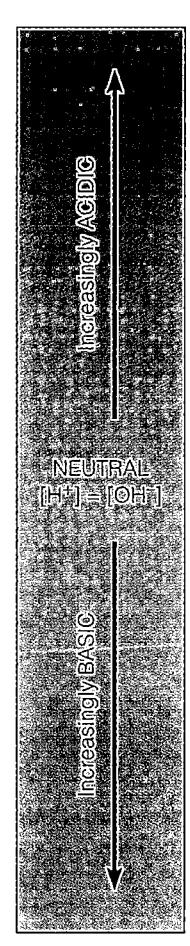

Increasingly ACIDIC

NEUTRAL
[H⁺] = [OH⁻]

Increasingly BASIC

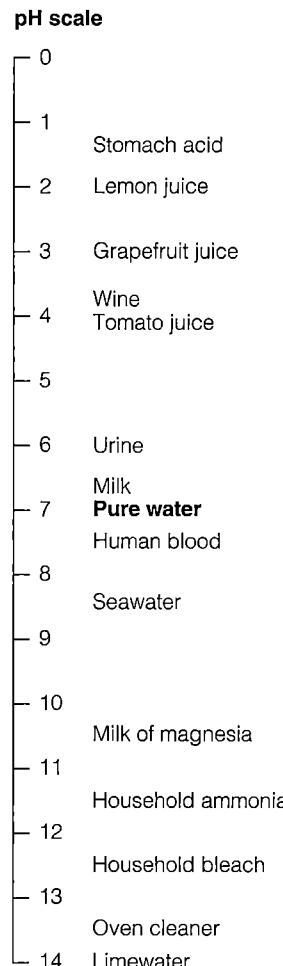

pH scale

pH	
0	
1	Stomach acid
2	Lemon juice
3	Grapefruit juice
4	Wine / Tomato juice
5	
6	Urine
7	Milk / **Pure water** / Human blood
8	Seawater
9	
10	
11	Milk of magnesia
12	Household ammonia
13	Household bleach / Oven cleaner
14	Limewater

FIGURE 2.8 The pH scale. As pH values decrease from 14 to 0, the H⁺ concentration increases. Thus, the lower the pH, the more acidic the solution; the higher the pH, the more basic the solution. If the pH value of a solution is below 7, the solution is acidic; if the pH is above 7, the solution is basic (alkaline). The approximate pH values of some human body fluids and common substances are shown next to the pH scale.

■ *At what pH is the concentration of H⁺ and OH⁻ equal?*

A solution's pH is calculated as $-\log_{10}[H^+]$, the negative logarithm to the base 10 of the hydrogen ion concentration (denoted by brackets), determined in moles per liter $[H^+]$. For example, if the H^+ concentration of a solution is 1.0×10^{-4} moles/liter, or 10^{-4}, its pH equals $-\log_{10}10^{-4} = -(-4) = 4$; this is about the pH value of wine (see Appendix D). The pH values of some human body fluids and other common substances are also shown in Figure 2.8. In the laboratory, you will usually measure the pH of a solution with a pH meter or with chemical test papers.

Acidic solutions contain more H^+ than OH^- and have a pH lower than 7. If a solution has more OH^- than H^+, it is a basic, or alkaline, solution. In pure water, a small percentage of the molecules are dissociated into H^+ and OH^-, so it has a pH of 7. Because the concentrations of H^+ and OH^- are equal, this pH is said to be the pH of a neutral solution.

Keep in mind that the pH of a solution can be changed. We can increase its acidity by adding substances that will increase the concentration of hydrogen ions. As

a living organism takes up nutrients, carries out chemical reactions, and excretes wastes, its balance of acids and bases tends to change, and the pH fluctuates. Fortunately, organisms possess natural pH **buffers,** compounds that help keep the pH from changing drastically. But the pH in our environment's water and soil can be altered by waste products from organisms, pollutants from industry, or fertilizers used in agricultural fields or gardens. When bacteria are grown in a laboratory medium, they excrete waste products such as acids that can alter the pH of the medium. If this effect were to continue, the medium would become acidic enough to inhibit bacterial enzymes and cause the death of the bacteria. To prevent this problem, pH buffers are added to the culture medium. One very effective pH buffer for some culture media uses a mixture of K_2HPO_4 and KH_2PO_4 (see Table 6.3 on page 166).

Different microbes function best within different pH ranges, but most organisms grow best in environments with a pH value between 6.5 and 8.5. Among microbes, fungi are best able to tolerate acidic conditions, whereas the prokaryotes called cyanobacteria tend to do well in alkaline habitats. *Propionibacterium acnes* (prō-pē-on-ē-bakti′rē-um ak′nēz), a bacterium that contributes to acne, has as its natural environment human skin, which tends to be slightly acidic, with a pH of about 4. *Thiobacillus ferrooxidans* (thī-ō-bä-sil′lus fer-rō-oks′i-danz) is a bacterium that grows on elemental sulfur and produces sulfuric acid (H_2SO_4). Its pH range for optimum growth is from 1 to 3.5. The sulfuric acid produced by this bacterium in mine water is important in dissolving uranium and copper from low-grade ore (see the box in Chapter 28 on page 786).

Organic Compounds

Learning Objectives

- *Distinguish between organic and inorganic compounds.*
- *Identify the building blocks of carbohydrates, simple lipids, phospholipids, proteins, and nucleic acids.*

Inorganic compounds, excluding water, constitute about 1–1.5% of living cells. These relatively simple components, whose molecules have only a few atoms, cannot be used by cells to perform complicated biological functions. Organic molecules, whose carbon atoms can combine in an enormous variety of ways with other carbon atoms and with atoms of other elements, are relatively complex and thus are capable of more complicated biological functions.

Structure and Chemistry

In the formation of organic molecules, carbon's four outer electrons can participate in up to four covalent bonds, and carbon atoms can bond to each other to form straight-chain, branched-chain, or ring structures.

In addition to carbon, the most common elements in organic compounds are hydrogen (which can form one bond), oxygen (two bonds), and nitrogen (three bonds). Sulfur (two bonds) and phosphorus (five bonds) appear less often. Other elements are found, but only in a relatively few organic compounds. The elements that are most abundant in living organisms are the same as those that are most abundant in organic compounds (see Table 2.1, page 29).

The chain of carbon atoms in an organic molecule is called the **carbon skeleton;** a huge number of combinations is possible for carbon skeletons. Most of these carbons are bonded to hydrogen atoms. The bonding of other elements with carbon and hydrogen forms characteristic **functional groups,** specific groups of atoms that are most commonly involved in chemical reactions and

table 2.3	*Representative Functional Groups and the Compounds in Which They Are Found*	
Structure	**Name of Group**	**Class of Compounds**
R—O—H	Hydroxyl	Alcohol
R—C(=O)H	Carbonyl (terminal)*	Aldehyde
R—C(=O)—R	Carbonyl (internal)*	Ketone
R—CH(H)—NH₂	Amino	Amine
R—C(=O)—O—R′	Ester	Ester
R—CH(H)—O—CH(H)—R′	Ether	Ether
R—CH(H)—SH	Sulfhydryl	Sulfhydryl
R—C(=O)—OH	Carboxyl	Organic acid

*A "terminal" carbonyl group is at an end of a molecule. In contrast, an "internal" carbonyl is at least one C atom removed from either end.

are responsible for most of the characteristic chemical properties and many of the physical properties of a particular organic compound (Table 2.3).

Different functional groups confer different properties on organic molecules. For example, the hydroxyl group of alcohols is hydrophilic (water-loving) and thus attracts water molecules to it. This attraction helps dissolve organic molecules containing hydroxyl groups. Since the carboxyl group is a source of hydrogen ions, molecules containing it have acidic properties. Amino groups, by contrast, function as bases because they readily accept hydrogen ions. The sulfhydryl group helps stabilize the intricate structure of many proteins.

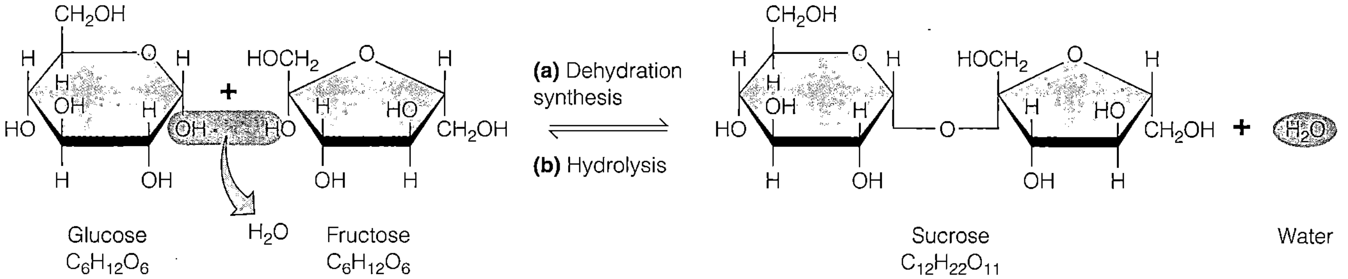

FIGURE 2.9 Dehydration synthesis and hydrolysis. (a) In dehydration synthesis (left to right), the monosaccharides glucose and fructose combine to form a molecule of the disaccharide sucrose. A molecule of water is released in the reaction. **(b)** In hydrolysis (right to left), the sucrose molecule breaks down into the smaller molecules glucose and fructose. For the hydrolysis reaction to proceed, water must be added to the sucrose.

■ What is the difference between a polymer and a monomer?

Functional groups help us classify organic compounds. For example, the —OH group is present in each of the following molecules:

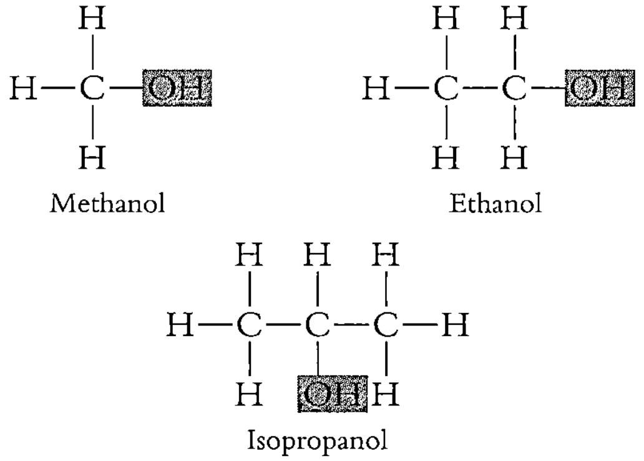

Because the characteristic reactivity of the molecules is based on the —OH group, they are grouped together in a class called alcohols. The —OH group is called the *hydroxyl group* and is not to be confused with the *hydroxide ion* (OH^-) of bases. The hydroxyl group of alcohols does not ionize at neutral pH; it is covalently bonded to a carbon atom.

When a class of compounds is characterized by a certain functional group, the letter **R** can be used to stand for the remainder of the molecule. For example, alcohols in general may be written R—OH.

Frequently, more than one functional group is found in a single molecule. For example, an amino acid molecule contains both amino and carboxyl groups. The amino acid glycine has the following structure:

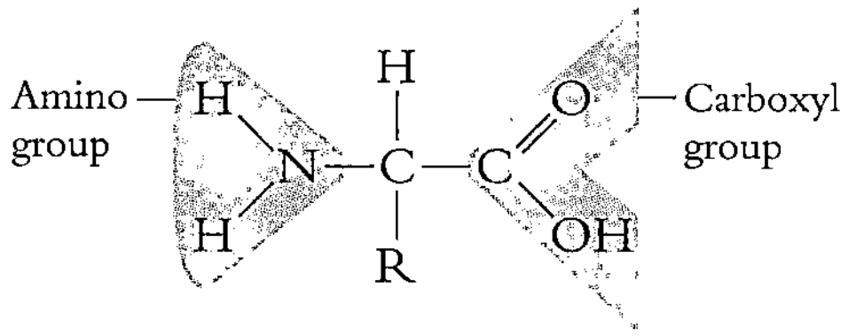

Most of the organic compounds found in living organisms are quite complex; a large number of carbon atoms form the skeleton, and many functional groups are attached. In organic molecules, it is important that each of the four bonds of carbon is satisfied (attached to another atom) and that each of the attaching atoms has its characteristic number of bonds satisfied. Because of this, such molecules are chemically stable.

Small organic molecules can be combined into very large molecules called **macromolecules** (*macro* = large). Macromolecules are usually **polymers** (*poly* = many; *mers* = parts), large molecules formed by covalent bonding of many repeating small molecules called **monomers** (*mono* = one). When two monomers join together, the reaction usually involves the elimination of a hydrogen atom from one monomer and a hydroxyl group from the other; the hydrogen atom and the hydroxyl group combine to produce water:

This type of exchange reaction is called **dehydration synthesis** (*de* = from; *hydra* = water), or a **condensation reaction,** because a molecule of water is released (Figure 2.9a). Such macromolecules as carbohydrates, lipids, proteins, and nucleic acids are assembled in the cell, essentially by dehydration synthesis. However, other molecules must also participate to provide energy for bond formation. ATP, the cell's chief energy provider, is discussed at the end of this chapter.

Carbohydrates

The **carbohydrates** are a large and diverse group of organic compounds that includes sugars and starches. Carbohydrates perform a number of major functions in living systems. For instance, one type of sugar (deoxyribose) is a building block of deoxyribonucleic acid (DNA), the molecule that carries hereditary information. Other sugars help form the cell walls of bacterial cells. Simple carbohydrates are used in the synthesis of amino acids and fats or fatlike substances, which are used to build structures and provide an emergency source of energy. Macromolecular carbohydrates function as food reserves. The principal function of carbohydrates, however, is to fuel cell activities with a ready source of energy.

Carbohydrates are made up of carbon, hydrogen, and oxygen atoms. The ratio of hydrogen to oxygen atoms is always 2:1 in simple carbohydrates. This ratio can be seen in the formulas for the carbohydrates ribose ($C_5H_{10}O_5$), glucose ($C_6H_{12}O_6$), and sucrose ($C_{12}H_{22}O_{11}$). Although there are exceptions, the general formula for carbohydrates is $(CH_2O)_n$, where n indicates that there are three or more CH_2O units. Carbohydrates can be classified into three major groups on the basis of size: monosaccharides, disaccharides, and polysaccharides.

Monosaccharides

Simple sugars are called **monosaccharides** (*sacchar* = sugar); each molecule contains from three to seven carbon atoms. The number of carbon atoms in the molecule of a simple sugar is indicated by the prefix in its name. For example, simple sugars with three carbons are called trioses. There are also tetroses (four-carbon sugars), pentoses (five-carbon sugars), hexoses (six-carbon sugars), and heptoses (seven-carbon sugars). Pentoses and hexoses are extremely important to living organisms. Deoxyribose is a pentose found in DNA. Glucose, a common hexose, is the main energy-supplying molecule of living cells.

Disaccharides

Disaccharides (*di* = two) are formed when two monosaccharides bond in a dehydration synthesis reaction.* For example, molecules of two monosaccharides, glucose and fructose, combine to form a molecule of the disaccharide sucrose (table sugar) and a molecule of water (see Figure 2.9a). Similarly, the dehydration synthesis of the monosaccharides glucose and galactose forms the disaccharide lactose (milk sugar).

It may seem odd that glucose and fructose have the same chemical formula (see Figure 2.9), even though they are different monosaccharides. The positions of the

oxygens and carbons differ in the two different molecules, and consequently the molecules have different physical and chemical properties. Two molecules with the same chemical formula but different structures and properties are called **isomers** (*iso* = same).

Disaccharides can be broken down into smaller, simpler molecules when water is added. This chemical reaction, the reverse of dehydration synthesis, is called **hydrolysis** (*hydro* = water; *lysis* = to loosen) (see Figure 2.9b). A molecule of sucrose, for example, may be hydrolyzed (digested) into its components of glucose and fructose by reacting with the H^+ and OH^- of water.

As you will see in Chapter 4, the cell walls of bacterial cells are composed of disaccharides and proteins that together form a substance called peptidoglycan.

Polysaccharides

Carbohydrates in the third major group, the **polysaccharides,** consist of tens or hundreds of monosaccharides joined through dehydration synthesis. Polysaccharides often have side chains branching off the main structure and are classified as macromolecules. Like disaccharides, polysaccharides can be split apart into their constituent sugars through hydrolysis. Unlike monosaccharides and disaccharides, however, they usually lack the characteristic sweetness of sugars such as fructose and sucrose and usually are not soluble in water.

One important polysaccharide is *glycogen,* which is composed of glucose subunits and is synthesized as a storage material by animals and some bacteria. *Cellulose,* another important glucose polymer, is the main component of the cell walls of plants and most algae. Although cellulose is the most abundant carbohydrate on Earth, it can be digested by only a few organisms that have the appropriate enzyme. The polysaccharide *dextran,* which is produced as a sugary slime by certain bacteria, is used in a blood plasma substitute. *Starch* is a polymer of glucose produced by plants and used as food by humans.

These four polysaccharides are all composed of glucose subunits, but the glucose molecules are bonded together differently in each case. Many animals, including humans, produce enzymes called *amylases* that can break the bonds between the glucose molecules in starch. However, this enzyme cannot break the bonds in cellulose. Bacteria and fungi that produce enzymes called *cellulases* can digest cellulose. Cellulases from the fungus *Trichoderma* (trik'ō-dėr-mä) are used for a variety of industrial purposes. One of the more unusual uses is producing stone-washed denim. Because washing the fabric with rocks damages washing machines, cellulase is used to digest, and therefore soften, the cotton. (See the box in Chapter 9 on page 251.)

---

* Carbohydrates composed of two to about 20 monosaccharides are called **oligosaccharides** (*oligo* = few). Disaccharides are the most common oligosaccharides.

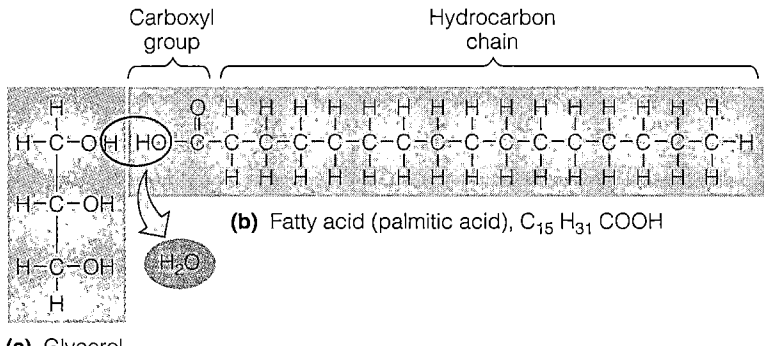

(a) Glycerol

(b) Fatty acid (palmitic acid), $C_{15}H_{31}COOH$

Carboxyl group

Hydrocarbon chain

FIGURE 2.10 Structural formulas of simple lipids.
(a) Glycerol. **(b)** Palmitic acid, a fatty acid. **(c)** The chemical combination of a molecule of glycerol and three fatty acid molecules (palmitic, stearic, and oleic in this example) forms one molecule of fat (triglyceride) and three molecules of water in a dehydration synthesis reaction. The bond between glycerol and each fatty acid is called an ester linkage. The addition of three water molecules to a fat forms glycerol and three fatty acid molecules in a hydrolysis reaction.

■ **How do saturated and unsaturated fatty acids differ?**

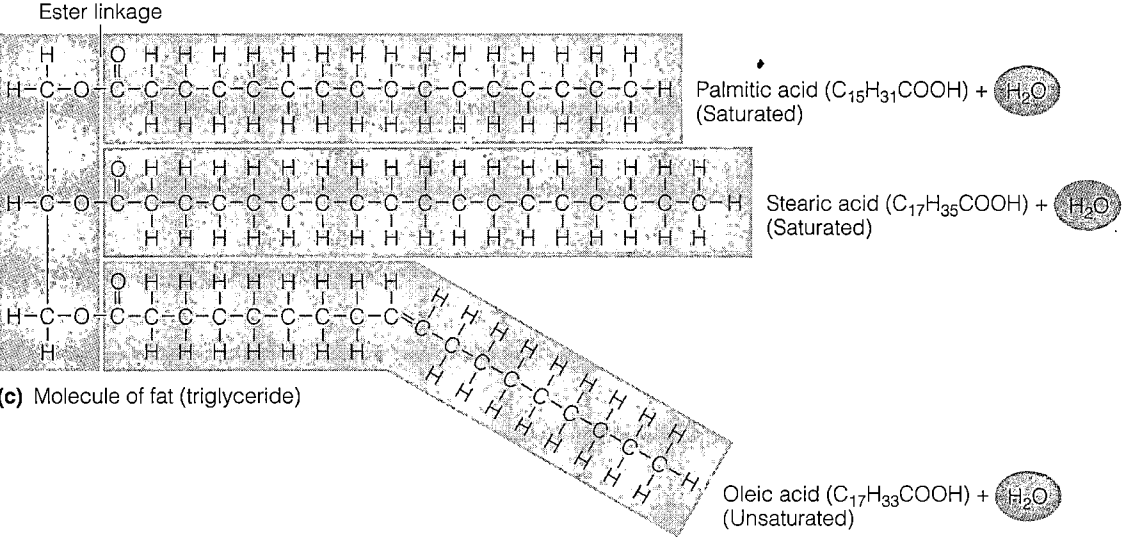

Ester linkage

(c) Molecule of fat (triglyceride)

Palmitic acid ($C_{15}H_{31}COOH$) + H_2O
(Saturated)

Stearic acid ($C_{17}H_{35}COOH$) + H_2O
(Saturated)

Oleic acid ($C_{17}H_{33}COOH$) + H_2O
(Unsaturated)

Lipids

If lipids were suddenly to disappear from the Earth, all living cells would collapse in a pool of fluid, because lipids are essential to the structure and function of membranes that separate living cells from their environment. **Lipids** (*lip* = fat) are a second major group of organic compounds found in living matter. Like carbohydrates, they are composed of atoms of carbon, hydrogen, and oxygen, but lipids lack the 2:1 ratio between hydrogen and oxygen atoms. Even though lipids are a very diverse group of compounds, they share one common characteristic: They are *nonpolar* molecules; unlike water, they do not have a positive and a negative end (pole). Therefore, most lipids are insoluble in water but dissolve readily in nonpolar solvents, such as ether and chloroform. Lipids function in energy storage and provide part of the structure of membranes and cell walls.

Simple Lipids

Simple lipids, called *fats* or *triglycerides,* contain an alcohol called *glycerol* and a group of compounds known as *fatty acids.* Glycerol molecules have three carbon atoms to

which are attached three hydroxyl (—OH) groups (Figure 2.10a). Fatty acids consist of long hydrocarbon chains (composed only of carbon and hydrogen atoms) ending in a carboxyl (—COOH, organic acid) group (Figure 2.10b). Most common fatty acids contain an even number of carbon atoms.

A fat molecule is formed when a molecule of glycerol combines with one to three fatty acid molecules to form a monoglyceride, diglyceride, or triglyceride (Figure 2.10c). In the reaction, one to three molecules of water are formed (dehydration), depending on the number of fatty acid molecules reacting. The chemical bond formed where the water molecule is removed is called an *ester linkage*. In the reverse reaction, hydrolysis, a fat molecule is broken down into its component fatty acid and glycerol molecules.

Because the fatty acids that form lipids have different structures, there is a wide variety of lipids. For example, three molecules of fatty acid A might combine with a glycerol molecule. Or one molecule each of fatty acids A, B, and C might unite with a glycerol molecule (see Figure 2.10c).

The primary function of lipids is the formation of plasma membranes that enclose cells. A plasma membrane

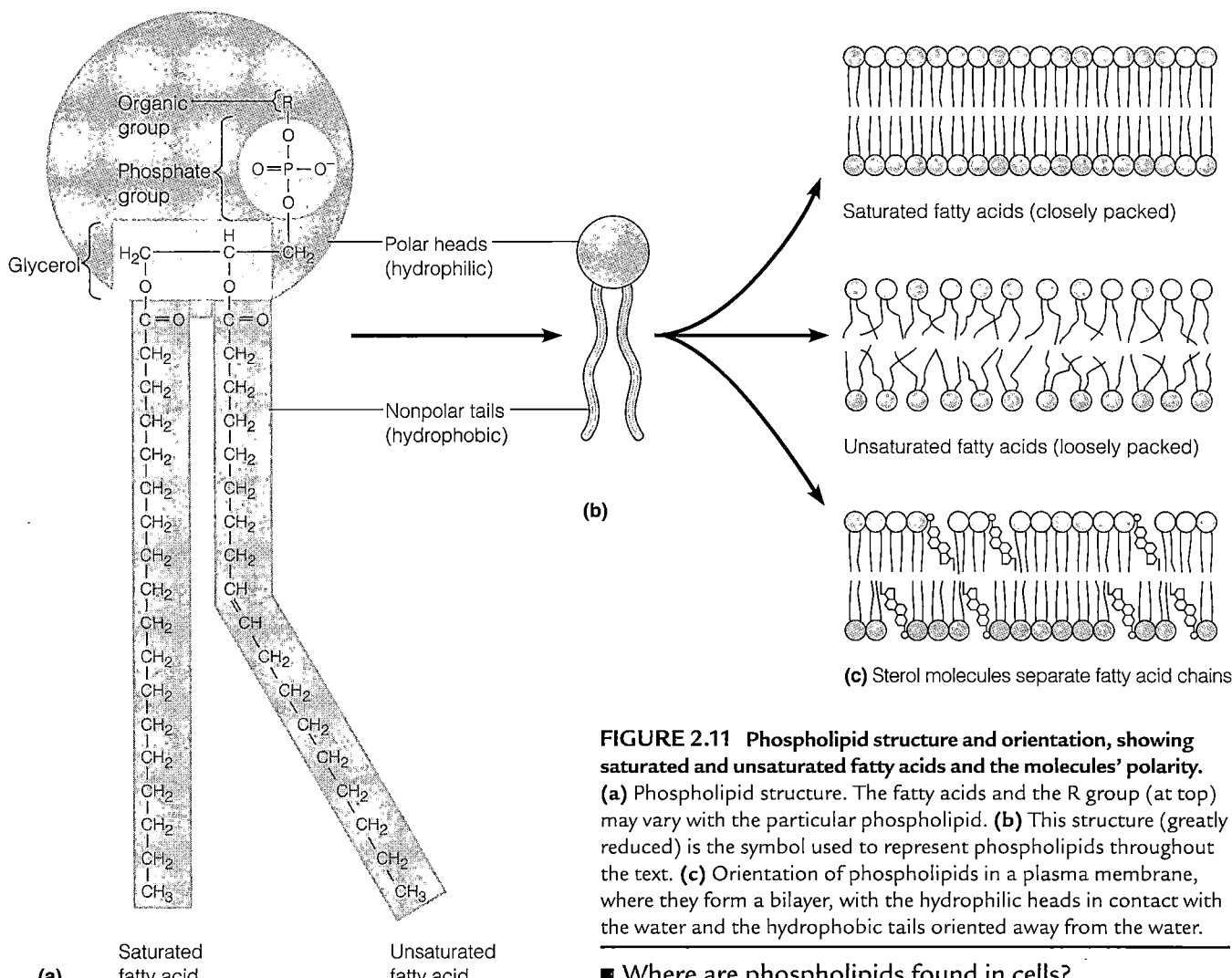

Glycerol

Polar heads (hydrophilic)

Nonpolar tails (hydrophobic)

(b)

Saturated fatty acids (closely packed)

Unsaturated fatty acids (loosely packed)

(c) Sterol molecules separate fatty acid chains

Saturated fatty acid   Unsaturated fatty acid

(a)

FIGURE 2.11 Phospholipid structure and orientation, showing saturated and unsaturated fatty acids and the molecules' polarity.
(a) Phospholipid structure. The fatty acids and the R group (at top) may vary with the particular phospholipid. **(b)** This structure (greatly reduced) is the symbol used to represent phospholipids throughout the text. **(c)** Orientation of phospholipids in a plasma membrane, where they form a bilayer, with the hydrophilic heads in contact with the water and the hydrophobic tails oriented away from the water.

■ **Where are phospholipids found in cells?**

supports the cell and allows nutrients and wastes to pass in and out; therefore, the lipids must maintain the same viscosity, regardless of the temperature of the surroundings. The membrane must be about as viscous as olive oil, without getting too fluid when warmed or too thick when cooled. As everyone who has ever cooked a meal knows, animal fats (such as butter) are usually solid at room temperature, whereas vegetable oils are usually liquid at room temperature. The difference in their respective melting points is caused by the degrees of saturation of the fatty acid chains. A fatty acid is said to be *saturated* when it has no double bonds; then the carbon skeleton contains the maximum number of hydrogen atoms (see Figures 2.10c and 2.11a). Saturated chains become solid more easily because they are relatively straight and are thus able to pack together more closely than unsaturated chains. The double bonds of *unsaturated* chains create kinks in the chain, which keep the chains apart from one another (Figure 2.11b).

Complex Lipids

Complex lipids contain such elements as phosphorus, nitrogen, and sulfur, in addition to the carbon, hydrogen, and oxygen found in simple lipids. The complex lipids called *phospholipids* are made up of glycerol, two fatty acids, and, in place of a third fatty acid, a phosphate group bonded to one of several organic groups (Figure 2.11a). Phospholipids are the lipids that build membranes; they are essential to a cell's survival. Phospholipids have polar as well as nonpolar regions (Figure 2.11a and b; see also Figure 4.13 on page 91). When placed in water, phospholipid molecules twist themselves in such a way that all polar (hydrophilic) portions will orient themselves toward the polar water molecules, with which they then form hydrogen bonds. (Recall that *hydrophilic* means water-loving.) This forms the basic structure of a plasma membrane (Figure 2.11c). Polar portions consist of a phosphate group and glycerol. In contrast to the polar regions, all nonpolar

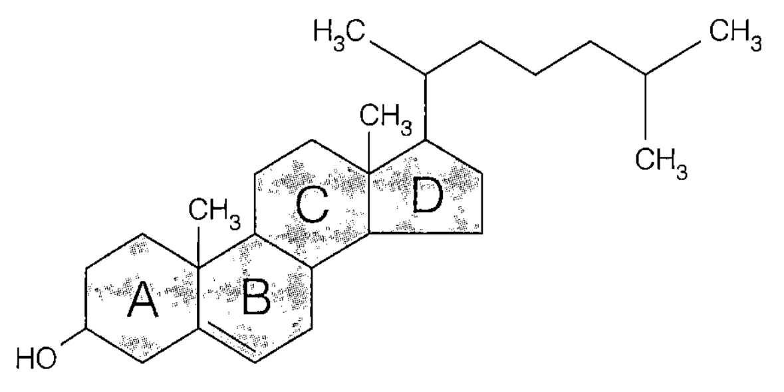

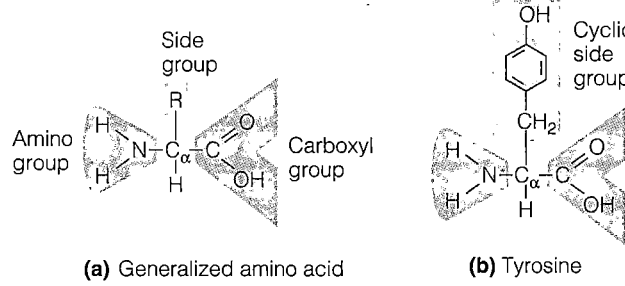

FIGURE 2.12 Cholesterol, a steroid. Note the four "fused" carbon rings (labeled A–D), which are characteristic of steroid molecules. The hydrogen atoms attached to the carbons at the corners of the rings have been omitted. The —OH group (colored red) makes this molecule a sterol.

■ Where are sterols found in cells?

FIGURE 2.13 Amino acid structure. (a) The general structural formula for an amino acid. The alpha-carbon (C_α) is shown in the center. Different amino acids have different R groups, also called side groups. **(b)** Structural formula for the amino acid tyrosine, which has a cyclic side group.

■ Each amino acid has a unique R group.

(hydrophobic) parts of the phospholipid make contact only with the nonpolar portions of neighboring molecules. (*Hydrophobic* means water-fearing.) Nonpolar portions consist of fatty acids. This characteristic behavior makes phospholipids particularly suitable to being a major component of the membranes that enclose cells. Phospholipids enable the membrane to act as a barrier that separates the contents of the cell from the water-based environment in which it lives.

Some complex lipids are useful in identifying certain bacteria. For example, the cell wall of *Mycobacterium tuberculosis* (mī-kō-bak-ti'rē-um tü-bér-kū-lō'sis), the bacterium that causes tuberculosis, is distinguished by its lipid-rich content. The cell wall contains complex lipids such as waxes and glycolipids (lipids with carbohydrates attached) that give the bacterium distinctive staining characteristics. Cell walls rich in such complex lipids are characteristic of all members of the genus *Mycobacterium*.

Steroids

Steroids are structurally very different from lipids. Figure 2.12 shows the structure of the steroid cholesterol, with the four interconnected carbon rings that are characteristic of steroids. When an —OH group is attached to one of the rings, the steroid is called a *sterol* (an alcohol). Sterols are important constituents of the plasma membranes of animal cells and of one group of bacteria (mycoplasmas), and they are also found in fungi and plants. The sterols separate the fatty acid chains and thus prevent the packing that would harden the plasma membrane at low temperatures (see Figure 2.11c).

Proteins

Proteins are organic molecules that contain carbon, hydrogen, oxygen, and nitrogen. Some also contain sulfur. If you were to separate and weigh all the groups of organic compounds in a living cell, the proteins would tip the scale. Hundreds of different proteins can be found in any single cell, and together they make up 50% or more of a cell's dry weight.

Proteins are essential ingredients in all aspects of cell structure and function. We have already mentioned enzymes, the proteins that catalyze biochemical reactions. But proteins have other functions as well. *Transporter proteins* help transport certain chemicals into and out of cells. Other proteins, such as the *bacteriocins* produced by many bacteria, kill other bacteria. Certain toxins, called exotoxins, produced by some disease-causing microorganisms are also proteins. Some proteins play a role in the contraction of animal muscle cells and the movement of microbial and other types of cells. Other proteins are integral parts of cell structures such as walls, membranes, and cytoplasmic components. Still others, such as the hormones of certain organisms, have regulatory functions. As we will see in Chapter 17, proteins called antibodies play a role in vertebrate immune systems.

Amino Acids

Just as monosaccharides are the building blocks of larger carbohydrate molecules, and fatty acids and glycerol are the building blocks of fats, **amino acids** are the building blocks of proteins. Amino acids contain at least one carboxyl (—COOH) group and one amino (—NH₂) group attached to the same carbon atom, called an alpha-carbon (written C_α) (Figure 2.13a). Such amino acids are called *alpha-amino acids*. Also attached to the alpha-carbon is a side group (R group), which is the amino acid's distinguishing feature. The side group can be a hydrogen atom, an unbranched or branched chain of atoms, or a

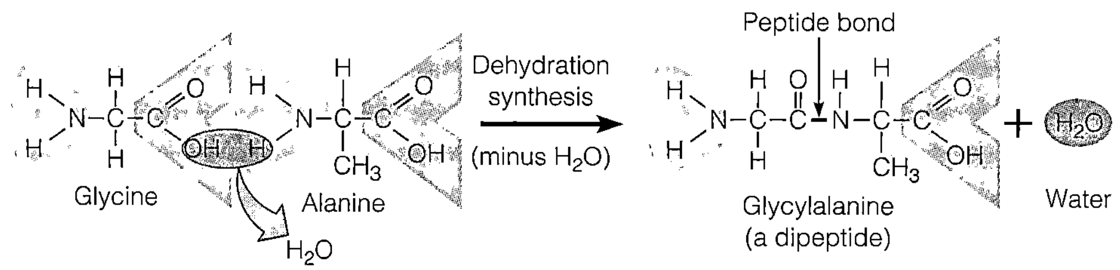

FIGURE 2.15 Peptide bond formation by dehydration synthesis. The amino acids glycine and alanine combine to form a dipeptide. The newly formed bond between the carbon atom of glycine and the nitrogen atom of alanine is called a peptide bond.

Glycine | Alanine | Dehydration synthesis (minus H₂O) | Peptide bond | Glycylalanine (a dipeptide) | Water | H₂O

■ Amino acids are the building blocks of proteins.

incorrect amino acid in a blood protein can produce the deformed hemoglobin molecule characteristic of sickle-cell disease. But proteins do not exist as long, straight chains. Each polypeptide chain folds and coils in specific ways into a relatively compact structure with a characteristic three-dimensional shape.

A protein's *secondary structure* is the localized, repetitious twisting or folding of the polypeptide chain. This aspect of a protein's shape results from hydrogen bonds joining the atoms of peptide bonds at different locations along the polypeptide chain. The two types of secondary protein structures are clockwise spirals called helices (singular: helix) and pleated sheets, which form from roughly parallel portions of the chain (Figure 2.16b). Both structures are held together by hydrogen bonds between oxygen or nitrogen atoms that are part of the polypeptide's backbone.

Tertiary structure refers to the overall three-dimensional structure of a polypeptide chain (Figure 2.16c). The folding is not repetitive or predictable, as in secondary structure. Whereas secondary structure involves hydrogen bonding between atoms of the amino and carboxyl groups involved in the peptide bonds, tertiary structure involves several interactions between various amino acid side groups in the polypeptide chain. For example, amino acids with nonpolar (hydrophobic) side groups usually interact at the core of the protein, out of contact with water. This *hydrophobic interaction* helps contribute to tertiary structure. Hydrogen bonds between side groups, and ionic bonds between oppositely charged side groups, also contribute to tertiary structure. Proteins that contain the amino acid cysteine form strong covalent bonds called *disulfide bridges.* These bridges form when two cysteine molecules are brought close together by the folding of the protein. Cysteine molecules contain sulfhydryl groups (—SH), and the sulfur of one cysteine molecule bonds to the sulfur on another, forming (by the removal of hydrogen atoms) a disulfide bridge (S—S) that holds parts of the protein together.

Some proteins have a *quaternary structure,* which consists of an aggregation of two or more individual polypeptide chains (subunits) that operate as a single functional

unit. Figure 2.16d shows a hypothetical protein consisting of two polypeptide chains. More commonly, proteins have two or more kinds of polypeptide subunits. The bonds that hold a quaternary structure together are basically the same as those that maintain tertiary structure. The overall shape of a protein may be globular (compact and roughly spherical) or fibrous (threadlike).

If a protein encounters a hostile environment in terms of temperature, pH, or salt concentrations, it may unravel and lose its characteristic shape. This process is called **denaturation** (see Figure 5.5 on page 119). As a result of denaturation, the protein is no longer functional. This process will be discussed in more detail in Chapter 5 with regard to denaturation of enzymes.

The proteins we have been discussing are *simple proteins,* which contain only amino acids. *Conjugated proteins* are combinations of amino acids with other organic or inorganic components. Conjugated proteins are named by their non–amino acid component. Thus, glycoproteins contain sugars, nucleoproteins contain nucleic acids, metalloproteins contain metal atoms, lipoproteins contain lipids, and phosphoproteins contain phosphate groups. Phosphoproteins are important regulators of activity in eukaryotic cells. Bacterial synthesis of phosphoproteins may be important for the survival of bacteria such as *Legionella pneumophila* that grow inside host cells.

Nucleic Acids

In 1944, three American microbiologists—Oswald Avery, Colin MacLeod, and Maclyn McCarty—discovered that a substance called **deoxyribonucleic acid (DNA)** is the substance of which genes are made. Nine years later, James Watson and Francis Crick, working with molecular models and X-ray information supplied by Maurice Wilkins and Rosalind Franklin, identified the physical structure of DNA. In addition, Crick suggested a mechanism for DNA replication and how it works as the hereditary material. DNA, and another substance called **ribonucleic acid (RNA),** are together referred to as **nucleic acids** because they were first discovered in the nuclei of cells. Just as amino acids are the structural units of proteins, nucleotides are the structural units of nucleic acids.

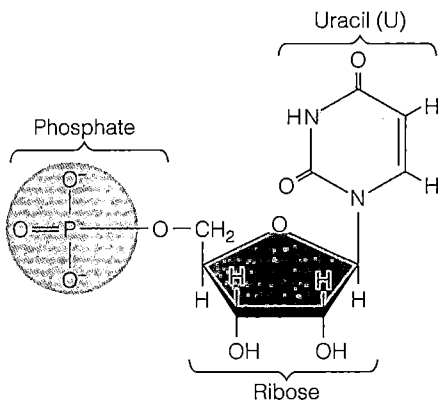

Uracil (U)

Phosphate

Ribose

FIGURE 2.18   A uracil nucleotide of RNA.

■ How do DNA and RNA differ in structure?

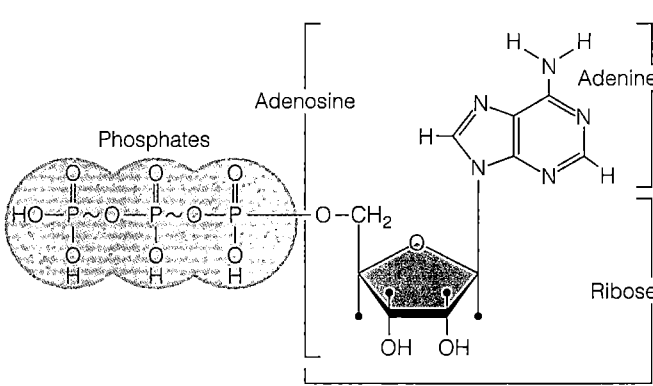

Adenosine

Phosphates

Adenine

Ribose

OH   OH

FIGURE 2.19   The structure of ATP. High-energy phosphate bonds are indicated by wavy lines. When ATP breaks down to ADP and inorganic phosphate, a large amount of chemical energy is released for use in other chemical reactions.

■ How is ATP similar to a nucleotide in RNA? In DNA?

ure 8.3 to see how nucleotides are bonded.) The nitrogen-containing bases make up the rungs of the ladder. Note that the purine A is always paired with the pyrimidine T and that the purine G is always paired with the pyrimidine C. The bases are held together by hydrogen bonds; A and T are held by two hydrogen bonds, and G and C are held by three.

The order in which the nitrogen base pairs occur along the backbone is extremely specific and in fact contains the genetic instructions for the organism. Nucleotides form genes, and a single DNA molecule may contain thousands of genes. Genes determine all hereditary traits, and they control all the activities that take place within cells.

One very important consequence of nitrogen-containing base pairing is that if the sequence of bases of one strand is known, then the sequence of the other strand is also known. For example, if one strand has the sequence . . . ATGC . . . , then the other strand has the sequence . . . TACG. . . . Because the sequence of bases of one strand is determined by the sequence of bases of the other, the bases are said to be *complementary*. The actual transfer of information becomes possible because of DNA's unique structure and will be discussed further in Chapter 8.

RNA

RNA, the second principal kind of nucleic acid, differs from DNA in several respects. Whereas DNA is double-stranded, RNA is usually single-stranded. The five-carbon sugar in the RNA nucleotide is ribose, which has one more oxygen atom than deoxyribose. Also, one of RNA's bases is uracil (U) instead of thymine (Figure 2.18). The other three bases (A, G, C) are the same as DNA. Three major kinds of RNA have been identified in cells. These are referred to as **messenger RNA (mRNA), ribosomal RNA (rRNA),** and **transfer RNA (tRNA).** As we will see in Chapter 8, each type of RNA has a specific role in protein synthesis.

Adenosine Triphosphate (ATP)

Learning Objective

■ Describe the role of ATP in cellular activities.

Adenosine triphosphate (ATP) is the principal energy-carrying molecule of all cells and is indispensable to the life of the cell. It stores the chemical energy released by some chemical reactions, and it provides the energy for reactions that require energy. ATP consists of an adenosine unit, composed of adenine and ribose, with three phosphate groups (abbreviated ⓟ) attached (Figure 2.19). In other words, it is an adenine nucleotide (also called adenosine monophosphate, or AMP) with two extra phosphate groups. ATP is called a high-energy molecule because it releases a large amount of usable energy when it is hydrolyzed to become **adenosine diphosphate (ADP).** This reaction can be represented as follows:

Adenosine—ⓟ—ⓟ—ⓟ + H_2O ⇌

Adenosine
triphosphate

Water

Adenosone—ⓟ—ⓟ + ⓟ$_i$ + Energy

Adenosine
diphosphate

Inorganic
phosphate

A cell's supply of ATP at any particular time is limited. Whenever the supply needs replenishing, the reaction goes in the reverse direction; the addition of a phosphate group to ADP and the input of energy produces more ATP. The energy required to attach the terminal phosphate group to ADP is supplied by the cell's various decomposition reactions, particularly the decomposition of glucose. ATP can be stored in every cell, where its potential energy is not released until needed.

Study Outline   [ST] Student Tutorial CD-ROM

INTRODUCTION (p. 27)

1. The science of the interaction between atoms and molecules is called chemistry.
2. The metabolic activities of microorganisms involve complex chemical reactions.
3. Nutrients are broken down by microbes to obtain energy and to make new cells.

THE STRUCTURE OF ATOMS (pp. 28–29)

[ST] *To review, go to Metabolism: Basic Chemistry: Define Atomic Structure*

1. Atoms are the smallest units of chemical elements that enter into chemical reactions.
2. Atoms consist of a nucleus, which contains protons and neutrons, and electrons that move around the nucleus.
3. The atomic number is the number of protons in the nucleus; the total number of protons and neutrons is the atomic weight.

Chemical Elements (p. 28)

1. Atoms with the same number of protons and the same chemical behavior are classified as the same chemical element.
2. Chemical elements are designated by abbreviations called chemical symbols.
3. About 26 elements are commonly found in living cells.
4. Atoms that have the same atomic number (are of the same element) but different atomic weights are called isotopes.

Electronic Configurations (pp. 28–29)

1. In an atom, electrons are arranged around the nucleus in electron shells.
2. Each shell can hold a characteristic maximum number of electrons.
3. The chemical properties of an atom are largely due to the number of electrons in its outermost shell.

HOW ATOMS FORM MOLECULES: CHEMICAL BONDS (pp. 29–33)

[ST] *Metabolism: Basic Chemistry: Ionic and Covalent Bonding*

1. Molecules are made up of two or more atoms; molecules consisting of at least two different kinds of atoms are called compounds.
2. Atoms form molecules in order to fill their outermost electron shells.
3. Attractive forces that bind the atomic nuclei of two atoms together are called chemical bonds.
4. The combining capacity of an atom—the number of chemical bonds the atom can form with other atoms—is its valence.

Ionic Bonds (pp. 29–31)

1. A positively or negatively charged atom or group of atoms is called an ion.
2. A chemical attraction between ions of opposite charge is called an ionic bond.
3. To form an ionic bond, one ion is an electron donor, and the other ion is an electron acceptor.

Covalent Bonds (pp. 31–32)

1. In a covalent bond, atoms share pairs of electrons.
2. Covalent bonds are stronger than ionic bonds and are far more common in organisms.

Hydrogen Bonds (pp. 32–33)

[ST] *Metabolism: Basic Chemistry: Hydrogen Bonds*

1. A hydrogen bond exists when a hydrogen atom covalently bonded to one oxygen or nitrogen atom is attracted to another oxygen or nitrogen atom.
2. Hydrogen bonds form weak links between different molecules or between parts of the same large molecule.

Molecular Weight and Moles (p. 33)

1. The molecular weight is the sum of the atomic weights of all the atoms in a molecule.
2. A mole of an atom, ion, or molecule is equal to its atomic or molecular weight expressed in grams.

CHEMICAL REACTIONS (pp. 33–36)

1. Chemical reactions are the making or breaking of chemical bonds between atoms.
2. A change of energy occurs during chemical reactions.
3. Endergonic reactions require energy; exergonic reactions release energy.
4. In a synthesis reaction, atoms, ions, or molecules are combined to form a larger molecule.
5. In a decomposition reaction, a larger molecule is broken down into its component molecules, ions, or atoms.
6. In an exchange reaction, two molecules are decomposed, and their subunits are used to synthesize two new molecules.
7. The products of reversible reactions can readily revert to form the original reactants.
8. For a chemical reaction to take place, the reactants must collide with each other.
9. The minimum collision energy that can produce a chemical reaction is called its activation energy.
10. Specialized proteins called enzymes accelerate chemical reactions in living systems by lowering the activation energy.

IMPORTANT BIOLOGICAL MOLECULES (pp. 36–51)

INORGANIC COMPOUNDS (pp. 37–40)

1. Inorganic compounds are usually small, ionically bonded molecules.
2. Water and many common acids, bases, and salts are examples of inorganic compounds.

Water (pp. 37–38)

1. Water is the most abundant substance in cells.
2. Because water is a polar molecule, it is an excellent solvent.
3. Water is a reactant in many of the decomposition reactions of digestion.
4. Water is an excellent temperature buffer.

Acids, Bases, and Salts (p. 38)

1. An acid dissociates into H^+ and anions.
2. A base dissociates into OH^- and cations.
3. A salt dissociates into negative and positive ions, neither of which is H^+ or OH^-.

Acid–Base Balance (pp. 38–40)

1. The term pH refers to the concentration of H^+ in a solution.
2. A solution of pH 7 is neutral; a pH value below 7 indicates acidity; pH above 7 indicates alkalinity.
3. A pH buffer, which stabilizes the pH inside a cell, can be used in culture media.

ORGANIC COMPOUNDS (pp. 40–51)

1. Organic compounds always contain carbon and hydrogen.
2. Carbon atoms form up to four bonds with other atoms.
3. Organic compounds are mostly or entirely covalently bonded, and many of them are large molecules.

Structure and Chemistry (pp. 40–41)

1. A chain of carbon atoms forms a carbon skeleton.
2. Functional groups of atoms are responsible for most of the properties of organic molecules.
3. The letter R may be used to denote the remainder of an organic molecule.
4. Frequently encountered classes of molecules are R—OH (alcohols), R—COOH (organic acids), and H_2N—R—COOH (amino acids).
5. Small organic molecules may combine into very large molecules called macromolecules.
6. Monomers usually bond together by dehydration synthesis, or condensation reactions, that form water and a polymer.
7. Organic molecules may be broken down by hydrolysis, a reaction involving the splitting of water molecules.

Carbohydrates (pp. 41–42)

1. Carbohydrates are compounds consisting of atoms of carbon, hydrogen, and oxygen, with hydrogen and oxygen in a 2:1 ratio.
2. Carbohydrates include sugars and starches.
3. Carbohydrates can be classified as monosaccharides, disaccharides, and polysaccharides.
4. Monosaccharides contain from three to seven carbon atoms.
5. Isomers are two molecules with the same chemical formula but different structures and properties—for example, glucose ($C_6H_{12}O_6$) and fructose ($C_6H_{12}O_6$).
6. Monosaccharides may form disaccharides and polysaccharides by dehydration synthesis.

Lipids (pp. 43–45)

1. Lipids are a diverse group of compounds distinguished by their insolubility in water.
2. Simple lipids (fats) consist of a molecule of glycerol and three molecules of fatty acids.
3. A saturated lipid has no double bonds between carbon atoms in the fatty acids; an unsaturated lipid has one or more double bonds. Saturated lipids have higher melting points than unsaturated lipids.
4. Phospholipids are complex lipids consisting of glycerol, two fatty acids, and a phosphate group.
5. Steroids have carbon ring structures; sterols have a functional hydroxyl group.

Proteins (pp. 45–48)

1. Amino acids are the building blocks of proteins.
2. Amino acids consist of carbon, hydrogen, oxygen, nitrogen, and sometimes sulfur.
3. Twenty amino acids occur naturally.
4. By linking amino acids, peptide bonds (formed by dehydration synthesis) allow the formation of polypeptide chains.
5. Proteins have four levels of structure: primary (sequence of amino acids), secondary (helices or pleats), tertiary (overall three-dimensional structure of a polypeptide), and quaternary (two or more polypeptide chains).
6. Conjugated proteins consist of amino acids combined with other organic or inorganic compounds.

Nucleic Acids (pp. 48–51)

1. Nucleic acids—DNA and RNA—are macromolecules consisting of repeating nucleotides.
2. A nucleotide is composed of a pentose, a phosphate group, and a nitrogen-containing base. A nucleoside is composed of a pentose and a nitrogen-containing base.
3. A DNA nucleotide consists of deoxyribose (a pentose) and one of the following nitrogen-containing bases: thymine or cytosine (pyrimidines) or adenine or guanine (purines).

4. DNA consists of two strands of nucleotides wound in a double helix. The strands are held together by hydrogen bonds between purine and pyrimidine nucleotides: AT and GC.

5. Genes consist of sequences of nucleotides.

6. An RNA nucleotide consists of ribose (a pentose) and one of the following nitrogenous bases: cytosine, guanine, adenine, or uracil.

Adenosine Triphosphate (ATP) (p. 51)

1. ATP stores chemical energy for various cellular activities.

2. When the bond to ATP's terminal phosphate group is hydrolyzed, energy is released.

3. The energy from decomposition reactions is used to regenerate ATP from ADP and inorganic phosphate.

Study Questions

REVIEW

1. What is a chemical element?

2. Diagram the electronic configuration of a carbon atom.

3. How does $^{14}_6C$ differ from $^{12}_6C$?

4. What type of bonding exists between water molecules?

5. What type of bonds holds the following atoms together?
   a. Li^+ and Cl^- in LiCl
   b. Carbon and oxygen atoms in CO_2
   c. Oxygen atoms in O_2
   d. A hydrogen atom of one nucleotide to a nitrogen or oxygen atom of another nucleotide in:

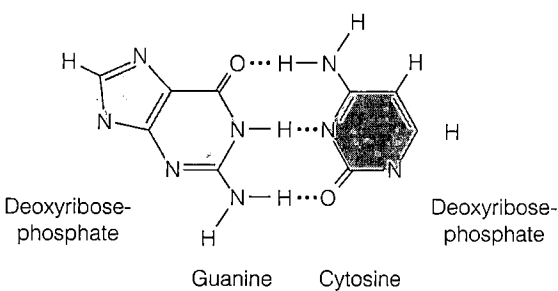

Deoxyribose-phosphate   Deoxyribose-phosphate

Guanine   Cytosine

6. Vinegar, pH 3, is how many times more acidic than pure water, pH 7?

7. Calculate the molecular weight of 1 mole of $C_6H_{12}O_6$.

8. Classify the following types of chemical reactions.
   a. Glucose + fructose → sucrose + H_2O
   b. Lactose → glucose + galactose
   c. $NH_4Cl + H_2O → NH_4OH + HCl$
   d. ATP → ADP + $\textcircled{P}_i$

9. Bacteria use the enzyme urease to obtain nitrogen in a form they can use from urea in the following reaction:

$$CO(NH_2)_2 + H_2O \longrightarrow 2NH_3 + CO_2$$
   Urea   Ammonia   Carbon
                                   dioxide

   What purpose does the enzyme serve in this reaction? What type of reaction is this?

10. Classify the following as subunits of either a carbohydrate, lipid, protein, or nucleic acid.
   a. $CH_3-(CH_2)_7-CH=CH-(CH_2)_7-COOH$
      Oleic acid

b.
$$\begin{array}{c} NH_2 \\ | \\ H-C-COOH \\ | \\ CH_2 \\ | \\ OH \end{array}$$
   Serine

   c. $C_6H_{12}O_6$
   d. Thymine nucleotide

11. Add the appropriate functional group(s) to this ethyl group to produce each of the following compounds: ethanol, acetic acid, acetaldehyde, ethanolamine, diethyl ether.

$$\begin{array}{cc} H & H \\ | & | \\ H-C-C- \\ | & | \\ H & H \end{array}$$

12. The artificial sweetener aspartame, or NutraSweet®, is made by joining aspartic acid to methlyated phenylalanine, as shown below.

a. What types of molecules are aspartic acid and phenylalanine?
b. What direction is the hydrolysis reaction (left to right or right to left)?
c. What direction is the dehydration synthesis reaction?
d. Circle the atoms involved in the formation of water.
e. Identify the peptide bond.

13. The energy-carrying property of the ATP molecule is due to energy dynamics that favor the breaking of bonds between _____. What type of bonds are these?

14. The following diagram shows a protein molecule. Indicate the regions of primary, secondary, and tertiary structure. Does this protein have quaternary structure?

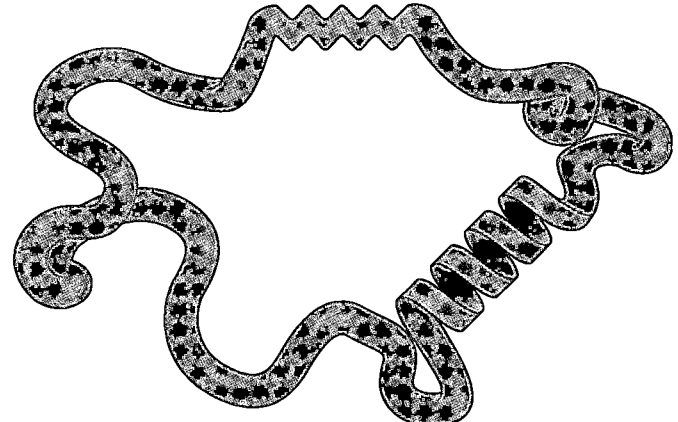

15. Draw a simple lipid, and show how it could be modified to a phospholipid.

MULTIPLE CHOICE

Radioisotopes are frequently used to label molecules in a cell. The fate of atoms and molecules in a cell can then be followed. This process is the basis for questions 1–3.

1. Assume *E. coli* are grown in a nutrient medium containing the radioisotope ^{16}N. After a 48-hour incubation period, the ^{16}N would most likely be found in the *E. coli*'s
a. carbohydrates.   c. proteins.   e. none of the above
b. lipids.   d. water.

2. If *Pseudomonas* bacteria are supplied with radioactively labeled cytosine, after a 24-hour incubation period this cytosine would most likely be found in the cells'
a. carbohydrates.   c. lipids.   e. proteins.
b. DNA.   d. water.

3. If *E. coli* were grown in a medium containing the radioactive isotope ^{32}P, the ^{32}P would be found in all of the following molecules of the cell *except*
a. ATP.   c. DNA.   e. none of the above
b. carbohydrates.   d. plasma membrane.

4. A carbonated drink, pH 3, is _____ times more acid than distilled water.
a. 4   c. 100   e. 10,000
b. 10   d. 1000

5. The best definition of ATP is that it is:
a. a molecule stored for food use.
b. a molecule that supplies energy to do work.
c. a molecule stored for an energy reserve.
d. a molecule used as a source of phosphate.

6. Which of the following is an organic molecule?
a. H_2O (water)   d. FeO (iron oxide)
b. O_2 (oxygen)   e. $F_2C=CF_2$ (Teflon)
c. $C_{18}H_{29}SO_3$ (Styrofoam)

Classify the molecules shown in questions 7–10 using the following choices. The dissociation products of the molecules are shown to help you.
a. acid   b. base   c. salt

7. $HNO_3 \rightarrow H^+ + NO_3^-$ **9.** $NaOH \rightarrow Na^+ + OH^-$

8. $H_2SO_4 \rightarrow 2H^+ + SO_4^{2-}$ **10.** $MgSO_4 \rightarrow Mg^{2+} + SO_4^{2-}$

CRITICAL THINKING

1. When you blow bubbles into a glass of water, the following reactions take place:

$$H_2O + CO_2 \xrightarrow{A} H_2CO_3 \xrightarrow{B} H^+ + HCO_3^-$$

a. What type of reaction is *A*?
b. What does reaction *B* tell you about the type of molecule H_2CO_3 is?

2. What are the common structural characteristics of ATP and DNA molecules?

3. What happens to the relative amount of unsaturated lipids in its plasma membrane when *E. coli* grown at 25°C are then grown at 37°C?

4. Giraffes, termites, and koalas eat only plant matter. Since animals cannot digest cellulose, how do you suppose these animals get nutrition from the leaves and wood they eat?

APPLICATIONS

1. Laundry detergents and drain cleaners contain enzymes from *Bacillus subtilis*. What is the value of these enzymes in clothes washing and drain cleaning?

2. *Thiobacillus ferrooxidans* was responsible for destroying buildings in the Midwest by causing changes in the Earth. The original rock, which contained lime ($CaCO_3$) and pyrite (FeS_2), expanded as bacterial metabolism caused gypsum ($CaSO_4$) crystals to form. How did *T. ferrooxidans* bring about the change from lime to gypsum?

3. When growing in an animal, *Bacillus anthracis* produces a capsule that is resistant to phagocytosis. The capsule is composed of D-glutamic acid. Why is this capsule resistant to digestion by the host's phagocytes? (Phagocytes are white blood cells that engulf bacteria.)

4. The antibiotic amphotericin B causes leaks in cells by combining with sterols in the plasma membrane. Would you expect to use amphotericin B against a bacterial infection? A fungal infection? Offer a reason why amphotericin B has severe side effects in humans.

Learning with Technology

MP = **The Microbiology Place website**   **ST** = **Student Tutorial CD-ROM**

MP   Don't forget to go to The Microbiology Place website (http://www.microbiologyplace.com) to take the practice tests, explore the interactive activity and case study, and check out the news articles and web links for this chapter.

ST   Remember there is also a quiz for this chapter on the Microbiology Interactive Student Tutorial CD-ROM.

chapter 3

Observing Microorganisms Through a Microscope

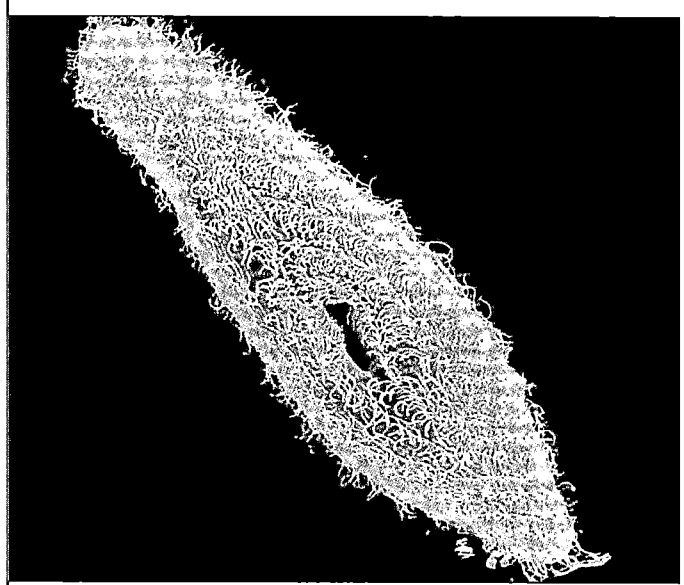

Paramecium caudatum. *Surface structures including cilia and the oral groove of this* Paramecium *can be seen with a scanning electron microscope.*

Microorganisms are much too small to be seen with the unaided eye. In order to make them visible, they must be observed with a microscope. The word microscope is derived from the Latin word *micro,* which means small, and the Greek word *skopos,* to look at. Modern microbiologists have access to microscopes that produce, with great clarity, magnifications that range anywhere from ten to thousands of times greater than those of van Leeuwenhoek's single lens (see Figure 1.2a on page 8). Part of this chapter will describe how different types of microscopes function, and why one type of microscope might be used in preference to another.

Some microbes are more readily visible than others because of their larger size or more easily observable features. Many microbes, however, must undergo several staining procedures before their cell walls, membranes, and other structures lose their colorless natural state. The last part of this chapter will explain some of the more commonly used methods of preparing specimens for examination through a light microscope.

You may be wondering how we are going to sort out, count, and measure the specimens we will be studying. This chapter opens with a discussion of how to use the metric system for measuring microbes.

Units of Measurement

Learning Objective

- *List the metric units of measurement used for microorganisms, and their metric equivalents.*

Because microorganisms and their component parts are so very small, they are measured in units that are unfamiliar to many of us in everyday life. When measuring microorganisms, we use the metric system. The standard unit of length in the metric system is the meter (m). A major advantage of the metric system is that the units are related to each other by factors of 10. Thus, 1 m equals 10 decimeters (dm) or 100 centimeters (cm) or 1000 millimeters (mm). Units in the U.S. system of measure do not have the advantage of easy conversion by a single factor of 10. For example, we use 3 feet or 36 inches to equal 1 yard.

table 3.1	Metric Units of Length and U.S. Equivalents		
Metric Unit	**Meaning of Prefix**	**Metric Equivalent**	**U.S. Equivalent**
1 kilometer (km)	*kilo* = 1000	1000 m = 10^3 m	3280.84 ft or 0.62 mi; 1 mi = 1.61 km
1 meter (m)		Standard unit of length	39.37 in or 3.28 ft or 1.09 yd
1 decimeter (dm)	*deci* = 1/10	0.1 m = 10^{-1} m	3.94 in
1 centimeter (cm)	*centi* = 1/100	0.01 m = 10^{-2} m	0.394 in; 1 in = 2.54 cm
1 millimeter (mm)	*milli* = 1/1000	0.001 m = 10^{-3} m	
1 micrometer (μm)	*micro* = 1/1,000,000	0.000001 m = 10^{-6} m	
1 nanometer (nm)	*nano* = 1/1,000,000,000	0.000000001 m = 10^{-9} m	

Microorganisms and their structural components are measured in even smaller units, such as micrometers and nanometers. A **micrometer (μm)** is equal to 0.000001 m (10^{-6} m). The prefix *micro* indicates that the unit following it should be divided by 1 million, or 10^6 (see the Exponential Notation section in Appendix E). A **nanometer (nm)** is equal to 0.000000001 m (10^{-9} m). Angstrom (Å) was previously used for 10^{-10} m or 0.1 nm.

Table 3.1 presents the basic metric units of length and some of their U.S. equivalents. In Table 3.1, you can compare the microscopic units of measurement with the commonly known macroscopic units of measurement, such as centimeters, meters, and kilometers. If you look ahead to Figure 3.2 (page 59), you will see the relative sizes of various organisms on the metric scale.

Microscopy: The Instruments

The simple microscope used by van Leeuwenhoek in the seventeenth century had only one lens and was similar to a magnifying glass. However, van Leeuwenhoek was the best lens grinder in the world in his day. His lenses were ground with such precision that a single lens could magnify a microbe 300×. His simple microscopes enabled him to be the first person to see bacteria.

Contemporaries of van Leeuwenhoek, such as Robert Hooke, built compound microscopes, which have multiple lenses. In fact, a Dutch spectacles maker, Zaccharias Janssen, is credited with making the first compound microscope around 1600. However, these early compound microscopes were of poor quality and could not be used to see bacteria. It was not until about 1830 that a significantly better microscope was developed by Joseph Jackson Lister (the father of Joseph Lister). Various improvements to Lister's microscope resulted in the development of the modern compound microscope, the kind used in microbiology laboratories today. Micro-

scopic studies of live specimens have revealed dramatic interactions between microbes (see the box on page 60).

Light Microscopy

Learning Objectives

- Diagram the path of light through a compound microscope.
- Define total magnification and resolution.
- Identify a use for darkfield, phase-contrast, DIC, fluorescence, and confocal microscopy, and compare each with brightfield illumination.

Light microscopy refers to the use of any kind of microscope that uses visible light to observe specimens. Here we examine several types of light microscopy.

Compound Light Microscopy

A modern **compound light microscope (LM)** has a series of lenses and uses visible light as its source of illumination (Figure 3.1a). With a compound light microscope, we can examine very small specimens as well as some of their fine detail. A series of finely ground lenses (Figure 3.1b) forms a clearly focused image that is many times larger than the specimen itself. This magnification is achieved when light rays from an **illuminator,** the light source, are passed through a **condenser,** which has lenses that direct the light rays through the specimen. From here, light rays pass into the **objective lenses,** the lenses closest to the specimen. The image of the specimen is magnified again by the **ocular lens,** or *eyepiece.*

We can calculate the **total magnification** of a specimen by multiplying the objective lens magnification (power) by the ocular lens magnification (power). Most microscopes used in microbiology have several objective lenses, including 10× (low power), 40× (high power), and 100× (oil immersion, which is described shortly).

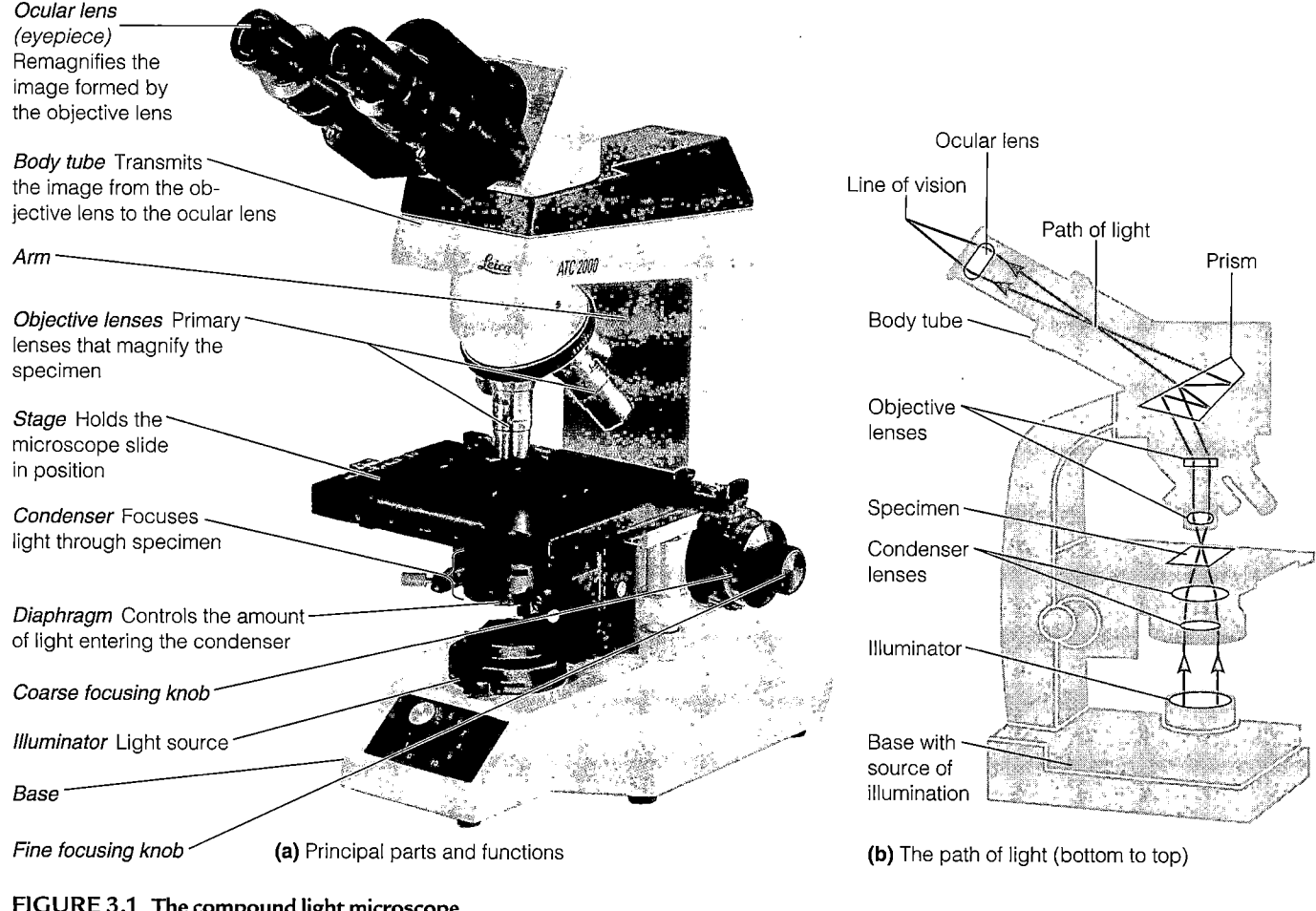

Ocular lens (eyepiece) Remagnifies the image formed by the objective lens

Body tube Transmits the image from the objective lens to the ocular lens

Arm

Objective lenses Primary lenses that magnify the specimen

Stage Holds the microscope slide in position

Condenser Focuses light through specimen

Diaphragm Controls the amount of light entering the condenser

Coarse focusing knob

Illuminator Light source

Base

Fine focusing knob

(a) Principal parts and functions

Ocular lens

Line of vision

Path of light

Prism

Body tube

Objective lenses

Specimen

Condenser lenses

Illuminator

Base with source of illumination

(b) The path of light (bottom to top)

FIGURE 3.1 The compound light microscope.

■ How is the total magnification of a compound light microscope calculated?

Most oculars magnify specimens by a factor of 10. Multiplying the magnification of a specific objective lens with that of the ocular, we see that the total magnifications would be 100× for low power, 400× for high power, and 1000× for oil immersion. Some compound light microscopes can achieve a magnification of 2000× with the oil immersion lens.

Resolution (also called *resolving power*) is the ability of the lenses to distinguish fine detail and structure. Specifically, it refers to the ability of the lenses to distinguish between two points a specified distance apart. For example, if a microscope has a resolving power of 0.4 nm, it can distinguish between two points if they are at least 0.4 nm apart. A general principle of microscopy is that the shorter the wavelength of light used in the instrument, the greater the resolution. The white light used in a compound light microscope has a relatively long wavelength and cannot resolve structures smaller than about 0.2 μm. This fact and other practical considerations limit

the magnification achieved by even the best compound light microscopes to about 2000×. By comparison, van Leeuwenhoek's microscopes had a resolution of 1 μm.

Figure 3.2 shows various specimens that can be resolved by the human eye, light microscope, and electron microscope.

To obtain a clear, finely detailed image under a compound light microscope, specimens must be made to contrast sharply with their *medium* (substance through which light passes). To attain such contrast, we must change the refractive index of specimens from that of their medium. The **refractive index** is a measure of the light-bending ability of a medium. We change the refractive index of specimens by staining them, a procedure we will discuss shortly. Light rays move in a straight line through a single medium. After staining, when light rays pass through the two materials (the specimen and its medium) with different refractive indexes, the rays change direction (refract) from a straight path by bending or changing an angle at

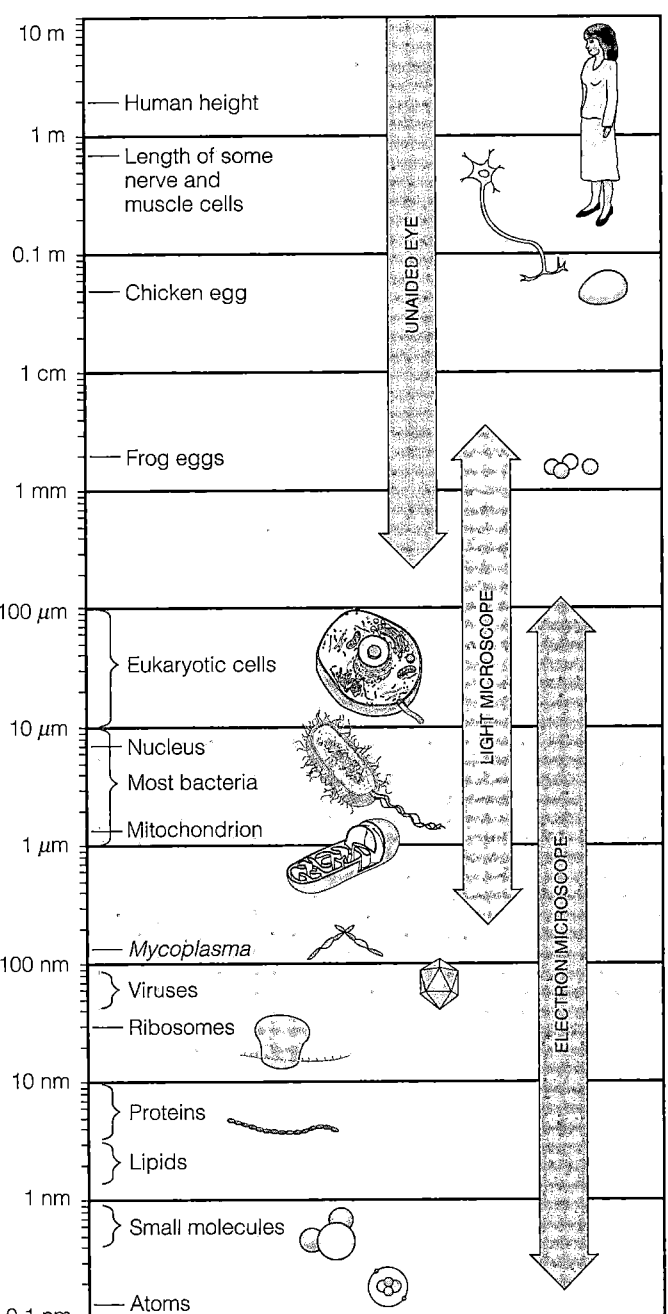

FIGURE 3.2 Relationships between the sizes of various specimens and the resolution of the human eye, light microscope, and electron microscope. The area in yellow shows the size range of most of the organisms we will be studying in this book.

■ **Resolution is the ability of lenses to distinguish between two points a specified distance apart.**

the boundary between the materials and increase the image's contrast between the specimen and the medium. As the light rays travel away from the specimen, they spread out and enter the objective lens, and the image is thereby magnified.

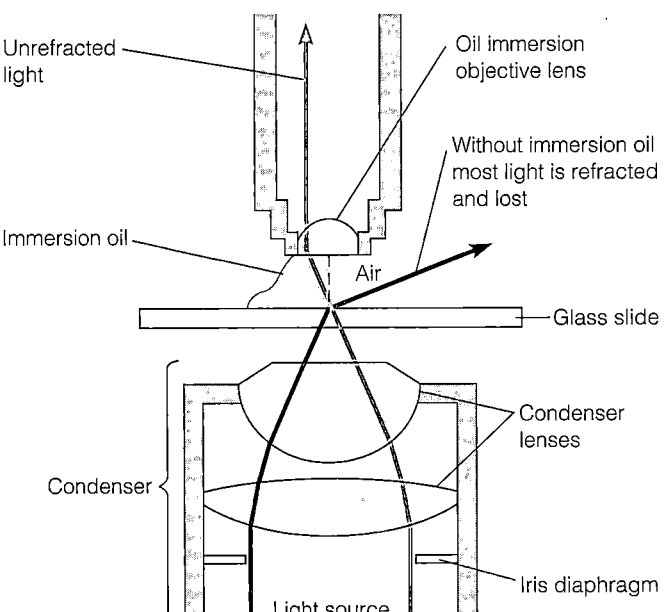

FIGURE 3.3 Refraction in the compound microscope using an oil immersion objective lens. Because the refractive indexes of the glass microscope slide and immersion oil are the same, the light rays do not refract when passing from one to the other when an oil immersion objective lens is used. This method produces images with better resolution at magnifications greater than 900×.

To achieve high magnification (100×) with good resolution, the objective lens must be small. Although we want light traveling through the specimen and medium to refract differently, we do not want to lose light rays after they have passed through the stained specimen. To preserve the direction of light rays at the highest magnification, immersion oil is placed between the glass slide and the oil immersion objective lens (Figure 3.3). The immersion oil has the same refractive index as glass, so the oil becomes part of the optics of the glass of the microscope. Unless immersion oil is used, light rays are refracted as they enter the air from the slide, and the objective lens would have to be increased in diameter to capture most of them. The oil has the same effect as increasing the objective lens diameter; therefore, it improves the resolving power of the lenses. If oil is not used with an oil immersion objective lens, the image becomes fuzzy, with poor resolution.

Under usual operating conditions, the field of vision in a compound light microscope is brightly illuminated. By focusing the light, the condenser produces a **bright-field illumination** (Figure 3.4a on page 61).

It is not always desirable to stain a specimen. However, an unstained cell has little contrast with its surroundings and is therefore difficult to see. Unstained cells are more easily observed with the modified compound microscopes described in the next section.

Even through a microscope, appearances can be deceptive, and first impressions can be quite misleading. For example, at first glance, the bacterium *Bdellovibrio* looks like a perfectly ordinary 1-µm-long curved, gram-negative rod. *Bdellovibrio* cells exhibit the normal runs and tumbles of flagellated bacteria, although they move about ten times faster than other bacteria of the same size. However, more patient observation, or use of a microscope equipped with a video camera, reveals that *Bdellovibrio*'s behavior is unlike that of any other known bacterium.

A unique microbial drama occurs when *Bdellovibrio* cells are placed in a suspension with other bacteria, such as *E. coli*. Apparently drawn by chemical attractants, *Bdellovibrio* speeds toward and rams the larger, slower *E. coli*. The two cells become attached and rotate rapidly. As the two bacteria whirl about, the *Bdellovibrio* paralyzes the *E. coli* by an unknown mechanism that probably involves one of *Bdellovibrio*'s protein products. *Bdellovibrio* kills *E. coli* without breaking down its contents.

Next, in preparation for moving into its newly acquired home, *Bdellovibrio* adds structural components to the outer membrane of *E. coli*. This activity, which stabilizes the outer membrane, is another ability unique to *Bdellovibrio*. After it penetrates *E. coli*'s cell wall and detaches its flagellum, *Bdellovibrio* snuggles into the periplasm (the space between the plasma membrane and the cell wall) and prepares for a feast.

Macromolecules leak from *E. coli*'s cytoplasm into the periplasm. *Bdellovibrio* digests the macromolecules and grows into a long, coiled filament. In 2–3 hours, the cytoplasmic contents of the host (including its DNA) are digested, and the *Bdellovibrio* filament fragments into as many as 20 individual

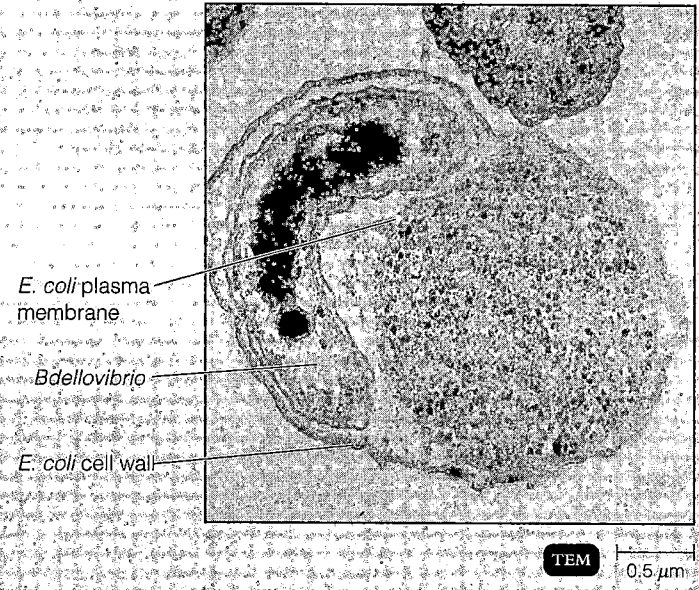

E. coli plasma membrane

Bdellovibrio

E. coli cell wall

TEM   0.5 µm

A *Bdellovibrio* bacterium has penetrated the cell wall of an *E. coli* cell. The *Bdellovibrio*, having killed the prey cell, uses the *E. coli* cell's contents for its elongation and multiplication.

curved rods. Each rod develops a flagellum and is released when the host cell is broken open by a *Bdellovibrio* enzyme. Within 4 hours, new *Bdellovibrio* cells scurry away to feed on other bacteria.

Bdellovibrio was once regarded as simply a parasite. However, thorough studies have shown that *Bdellovibrio* does not merely siphon off its host's resources and energy; it actually digests the cell's contents as food, so *Bdellovibrio* is now considered to be a bacterial predator. The ecological importance of *Bdellovibrio* is that it, along with protozoa and bacteriophages (viruses that infect bacteria), helps control the population growth of soil bacteria.

In addition, now that its habits are known, *Bdellovibrio* may have a new career as a bacterial exterminator. Researchers led by Richard Whiting at the U.S. Department of Agriculture

(USDA) are investigating the use of *Bdellovibrio* to control gram-negative foodborne pathogens such as *E. coli* O157:H7 and *Salmonella*. Although these bacteria are destroyed by cooking, foods that contain raw eggs or unpasteurized apple juice are a problem. It isn't easy to kill *E. coli* on meat and fruit without heating. Even washing in organic acids (pH 3.6) does not kill *E. coli* O157:H7. However, *Bdellovibrio* can kill even high concentrations of *E. coli* at 20°C and pH 5.6–8.6. In addition, researchers hope that *Bdellovibrio* can be introduced into chickens via food or water to kill salmonellae.

As harmful as they are to other bacteria, *Bdellovibrio* cells are safe for humans. Because *Bdellovibrio* cells prey on gram-negative bacteria, the USDA researchers hope to use *Bdellovibrio* for the biological control of pathogenic and spoilage organisms in food.

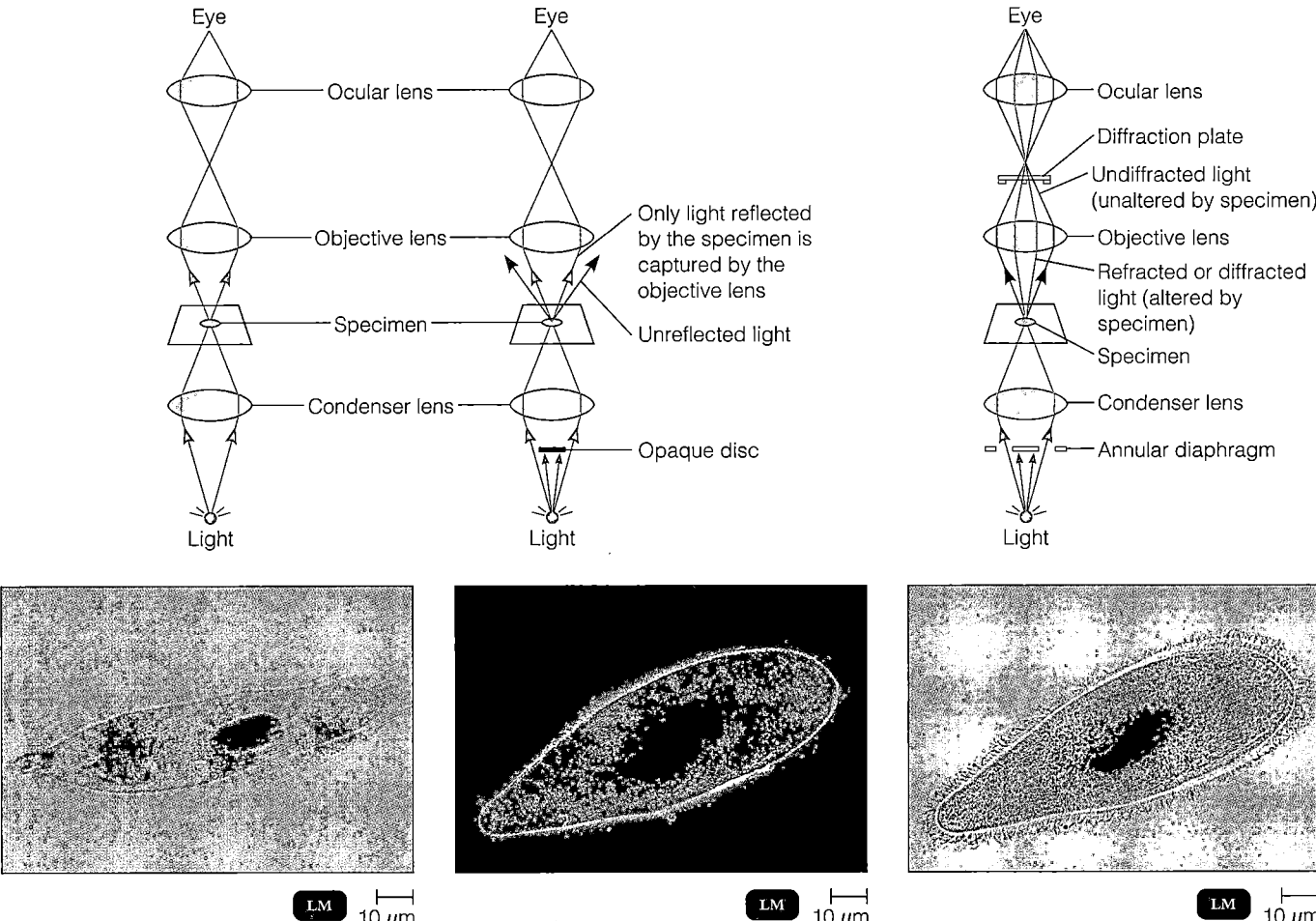

(a) Brightfield. (Top) The path of light in brightfield microscopy, the type of illumination produced by regular compound light microscopes. (Bottom) Brightfield illumination shows internal structures and the outline of the transparent sheath (external covering).

(b) Darkfield. (Top) The darkfield microscope uses a special condenser with an opaque disc that eliminates all light in the center of the beam. The only light that reaches the specimen comes in at an angle; thus, only light reflected by the specimen (blue rays) reaches the objective lens. (Bottom) Against the black background seen with darkfield microscopy, edges of the cell are bright, some internal structures seem to sparkle, and the sheath is almost visible.

(c) Phase-contrast. (Top) In phase-contrast microscopy, the specimen is illuminated by light passing through an annular diaphragm. Direct light rays (unaltered by the specimen) travel a different path than light rays that are reflected or diffracted as they pass through the specimen. These two sets of rays are combined at the eye. Reflected or diffracted light rays are indicated in blue; direct rays are red. (Bottom) Phase-contrast microscopy shows greater differentiation of internal structures and also shows the sheath wall.

FIGURE 3.4 Brightfield, darkfield, and phase-contrast microscopy. The illustrations show the contrasting light pathways of each of these types of microscopy. The photographs compare the same *Paramecium* specimen using these different microscopy techniques.

■ What are the advantages of brightfield, darkfield, and phase-contrast microscopy?

Darkfield Microscopy

A **darkfield microscope** is used for examining live microorganisms that either are invisible in the ordinary light microscope, cannot be stained by standard methods, or are so distorted by staining that their characteristics then cannot be identified. Instead of the normal condenser, a darkfield microscope uses a darkfield condenser that contains an opaque disc. The disc blocks light that would enter the objective lens directly. Only light that is reflected off (turned away from) the specimen enters the objective lens. Because there is no direct background light, the specimen appears light against a black background—the

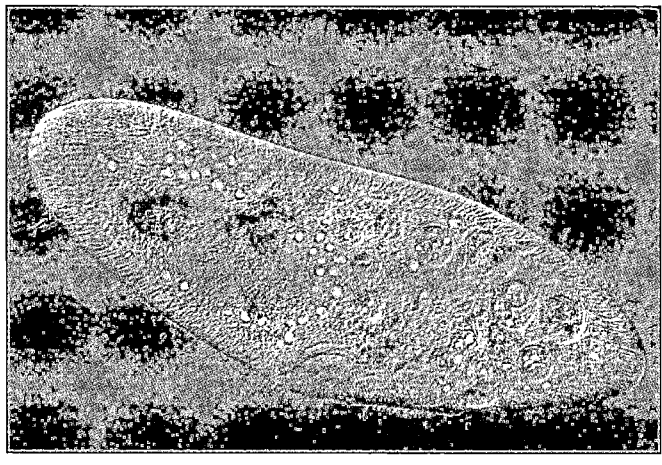

LM |⎯⎯| 10 μm

FIGURE 3.5 Differential interference contrast (DIC) microscopy. Like phase-contrast, DIC uses differences in refractive indexes to produce an image, in this case of the protozoan *Paramecium*. The colors in the image are produced by prisms that split the two light beams used in this process.

■ **Why is the resolution of a DIC microscope higher than that of a phase-contrast microscope?**

dark field (Figure 3.4b). This technique is frequently used to examine unstained microorganisms suspended in liquid. One use for darkfield microscopy is the examination of very thin spirochetes, such as *Treponema pallidum* (tre-pō-nē′mä pal′li-dum), the causative agent of syphilis.

Phase-Contrast Microscopy

Another way to observe microorganisms is with a **phase-contrast microscope.** Phase-contrast microscopy is especially useful because it permits detailed examination of internal structures in *living* microorganisms. In addition, it is not necessary to fix (attach the microbes to the microscope slide) or stain the specimen—procedures that could distort or kill the microorganisms.

The principle of phase-contrast microscopy is based on the wave nature of light rays, and the fact that light rays can be *in phase* (their peaks and valleys match) or *out of phase*. If the wave peak of light rays from one source coincides with the wave peak of light rays from another source, the rays interact to produce *reinforcement* (relative brightness). However, if the wave peak from one light source coincides with the wave trough from another light source, the rays interact to produce *interference* (relative darkness). In a phase-contrast microscope, one set of light rays comes directly from the light source. The other set comes from light that is reflected or diffracted from a particular structure in the specimen. (*Diffraction* is the scat-

tering of light rays as they "touch" a specimen's edge. The diffracted rays are bent away from the parallel light rays that pass farther from the specimen.) When the two sets of light rays—direct rays and reflected or diffracted rays—are brought together, they form an image of the specimen on the ocular lens, containing areas that are relatively light (in phase), through shades of gray, to black (out of phase; Figure 3.4c). In phase-contrast microscopy, the internal structures of a cell become more sharply defined.

Differential Interference Contrast (DIC) Microscopy

Differential interference contrast (DIC) microscopy is similar to phase-contrast microscopy in that it uses differences in refractive indexes. However, a DIC microscope uses two beams of light instead of one. In addition, prisms split each light beam, adding contrasting colors to the specimen. Therefore, the resolution of a DIC microscope is higher than that of a standard phase-contrast microscope. Also, the image is brightly colored and appears nearly three-dimensional (Figure 3.5).

Fluorescence Microscopy

Fluorescence microscopy takes advantage of **fluorescence,** the ability of substances to absorb short wavelengths of light (ultraviolet) and give off light at a longer wavelength (visible). Some organisms fluoresce naturally under ultraviolet light; if the specimen to be viewed does not naturally fluoresce, it is stained with one of a group of fluorescent dyes called *fluorochromes*. When microorganisms stained with a fluorochrome are examined under a fluorescence microscope with an ultraviolet or near-ultraviolet light source, they appear as luminescent, bright objects against a dark background.

Fluorochromes have special attractions for different microorganisms. For example, the fluorochrome auramine O, which glows yellow when exposed to ultraviolet light, is strongly absorbed by *Mycobacterium tuberculosis*, the bacterium that causes tuberculosis. When the dye is applied to a sample of material suspected of containing the bacterium, the bacterium can be detected by the appearance of bright yellow organisms against a dark background (Table 3.2, pages 64–65). *Bacillus anthracis*, the causative agent of anthrax, appears apple green when stained with another fluorochrome, fluorescein isothiocyanate (FITC).

The principal use of fluorescence microscopy is a diagnostic technique called the **fluorescent-antibody (FA) technique,** or **immunofluorescence. Antibodies** are natural defense molecules that are produced by humans and many animals in reaction to a foreign sub-

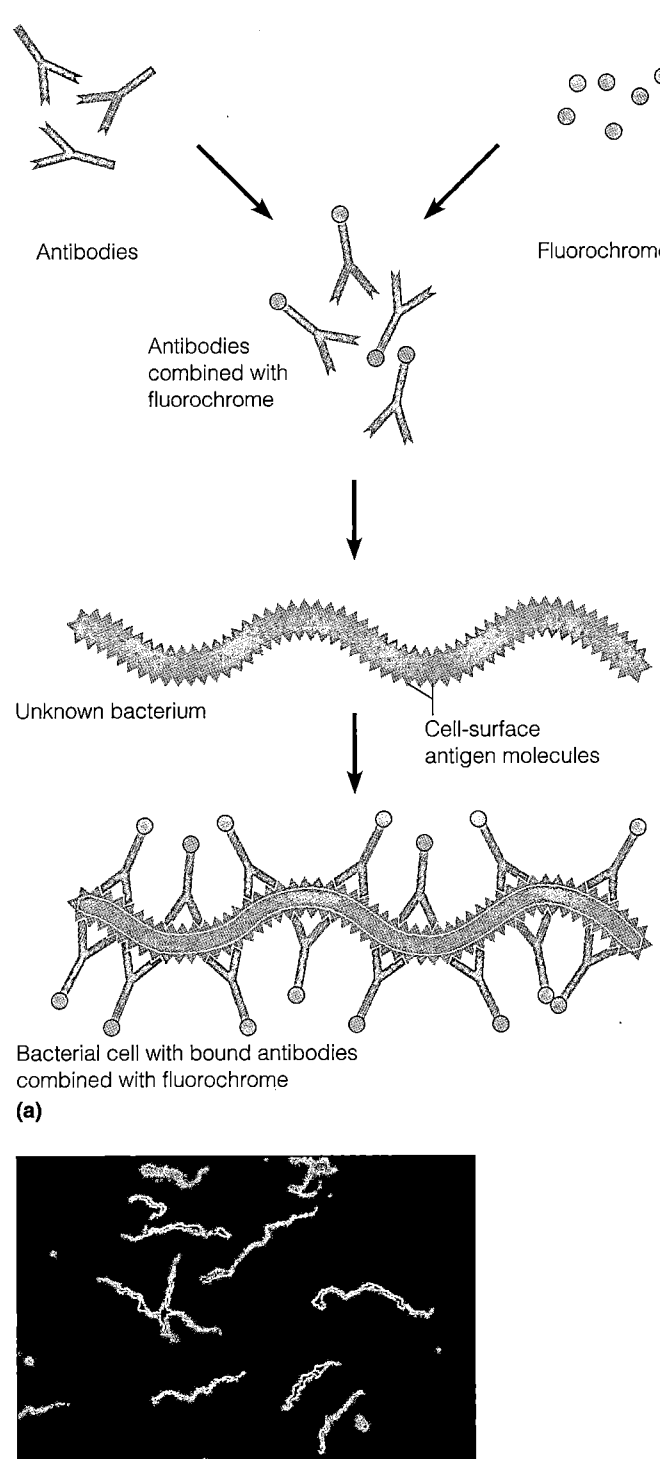

Antibodies

Fluorochrome

Antibodies combined with fluorochrome

Unknown bacterium

Cell-surface antigen molecules

Bacterial cell with bound antibodies combined with fluorochrome

(a)

(b)   LM   10 μm

FIGURE 3.6 The principle of immunofluorescence. (a) A type of fluorochrome is combined with antibodies against a specific type of bacterium. When the preparation is added to bacterial cells on a microscope slide, the antibodies attach to the bacterial cells, and the cells fluoresce when illuminated with ultraviolet light. **(b)** In the fluorescent treponemal antibody absorption (FTA-ABS) test for syphilis shown here, *Treponema pallidum* shows up as light regions against a darker background.

■ **Why won't other bacteria fluoresce in the FTA-ABS test?**

moved from the serum of the animal. Next, as shown in Figure 3.6a, a fluorochrome is chemically combined with the antibodies. These fluorescent antibodies are then added to a microscope slide containing an unknown bacterium. If this unknown bacterium is the same bacterium that was injected into the animal, the fluorescent antibodies bind to antigens on the surface of the bacterium, causing it to fluoresce.

This technique can detect bacteria or other pathogenic microorganisms, even within cells, tissues, or other clinical specimens (Figure 3.6b). Of paramount importance, it can be used to identify a microbe in minutes. Immunofluorescence is especially useful in diagnosing syphilis and rabies. We will say more about antigen–antibody reactions and immunofluorescence in Chapter 18.

Confocal Microscopy

A fairly recent development in light microscopy is known as **confocal microscopy.** Like fluorescent microscopy, specimens are stained with fluorochromes so they will emit, or return, light. In confocal microscopy, one plane of a small region of a specimen is illuminated with a laser, which passes the returned light through an aperture aligned with the illuminated region. Each plane corresponds to an image of a fine slice that has been physically cut from a specimen. Successive planes and regions are illuminated until the entire specimen has been scanned. Because confocal microscopy uses a pinhole aperture, it eliminates the blurring that occurs with other microscopes. As a result, exceptionally clear two-dimensional images can be obtained, with improved resolution of up to 40% over that of other microscopes.

Most confocal microscopes are used in conjunction with computers to construct three-dimensional images. The scanned planes of a specimen, which resemble a stack of images, are converted to a digital form that can be used by a computer to construct a three-dimensional representation. The reconstructed images can be rotated and viewed in any orientation. Using this technique,

stance, or *antigen.* Fluorescent antibodies for a particular antigen are obtained as follows: An animal is injected with a specific antigen, such as a bacterium; the animal then begins to produce specific antibodies against that antigen. After a sufficient time, the antibodies are re-

table 3.2   A Summary of Various Types of Microscopes

Microscope Type	Distinguishing Features	Typical Image	Principal Uses
Light			
Brightfield	Uses visible light as a source of illumination; cannot resolve structures smaller than about 0.2 μm; specimen appears against a bright background. Inexpensive and easy to use.	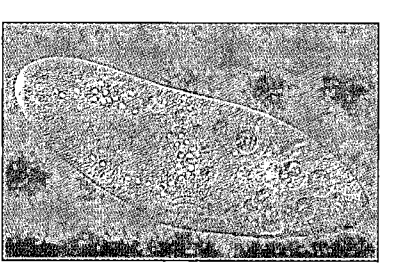 LM ⊢⊣ 10 μm	To observe various stained specimens and to count microbes; does not resolve very small specimens, such as viruses.
Darkfield	Uses a special condenser with an opaque disc that blocks light from entering the objective lens directly; light reflected by specimen enters the objective lens, and the specimen appears light against a black background.	LM ⊢⊣ 10 μm	To examine living microorganisms that are invisible in brightfield microscopy, do not stain easily, or are distorted by staining; frequently used to detect *Treponema pallidum* in the diagnosis of syphilis.
Phase-contrast	Uses a special condenser containing an annular (ring-shaped) diaphragm. The diaphragm allows direct light to pass through the condenser, focusing light on the specimen and a diffraction plate in the objective lens. Direct and reflected or diffracted light rays are brought together to produce the image. No staining required.	LM ⊢⊣ 10 μm	To facilitate detailed examination of the internal structures of living specimens.
Differential interference contrast (DIC)	Like phase-contrast, uses differences in refractive indexes to produce images. Uses two beams of light separated by prisms; the specimen appears colored as a result of the prism effect. No staining required.	LM ⊢⊣ 10 μm	To provide three-dimensional images.
Fluorescence	Uses an ultraviolet or near-ultraviolet source of illumination that causes fluorescent microbes (yellow-colored) in a specimen to emit light.	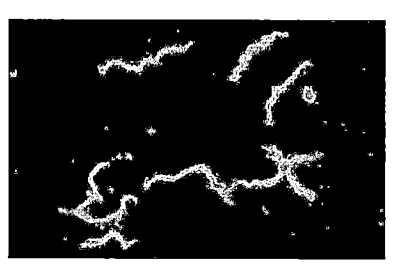 LM ⊢――⊣ 10 μm	For fluorescent-antibody techniques (immunofluorescence) to rapidly detect and identify microbes in tissues or clinical specimens.

table 3.2	(continued)		
Microscope Type	**Distinguishing Features**	**Typical Image**	**Principal Uses**
Confocal	Uses laser light to illuminate one plane of a specimen at a time.	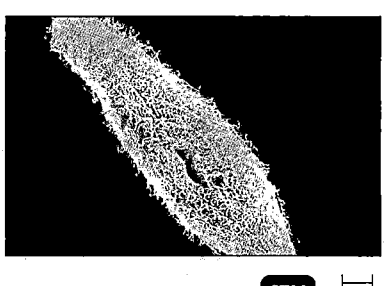 LM ⊢—⊣ 6 μm	To obtain two- and three-dimensional images of cells for biomedical applications.
Electron Transmission	Uses a beam of electrons instead of light; because of the shorter wavelength of electrons, structures smaller than 0.2 μm can be resolved. The image produced is two-dimensional.	TEM ⊢—⊣ 10 μm	To examine viruses or the internal ultrastructure in thin sections of cells (usually magnified 10,000–100,000×).
Scanning	Uses a beam of electrons instead of light; because of the shorter wavelength of electrons, structures smaller than 0.2 μm can be resolved. The image produced appears three-dimensional.	SEM ⊢—⊣ 10 μm	To study the surface features of cells and viruses (usually magnified 1000–10,000×).
Scanned-probe Scanning tunneling	Uses a thin metal probe that scans a specimen and produces an image revealing the bumps and depressions of the atoms on the surface of the specimen. Resolving power is much greater than that of an electron microscope. No special preparation required.	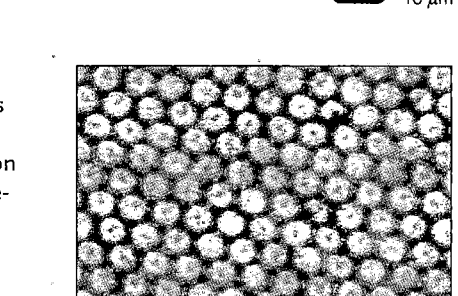 SEM ⊢—⊣ 20 nm	Provides very detailed views of molecules inside cells.
Atomic force	Uses a metal-and-diamond probe gently forced down along the surface of the specimen. Produces a three-dimensional image. No special preparation required.	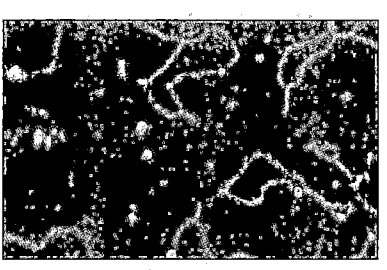 SEM ⊢—⊣ 15 nm	Provides images of biological molecules in nearly atomic detail and molecular processes.

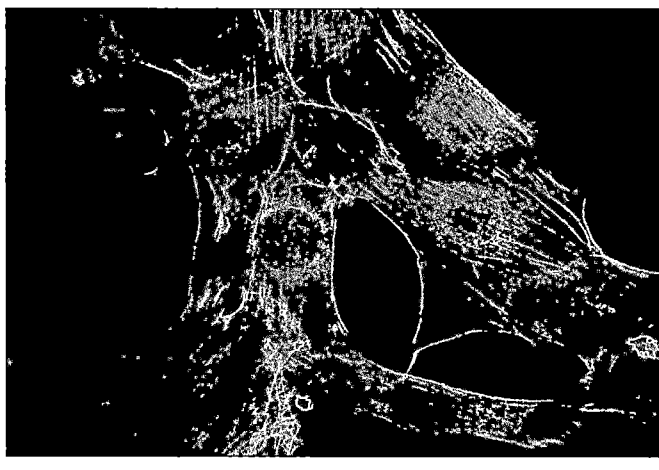

FIGURE 3.7 Confocal microscopy. Confocal microscopy produces three-dimensional images and can be used to look inside cells.

three-dimensional images of entire cells and cellular components have been obtained (Figure 3.7). In addition, confocal microscopy can be used to evaluate cellular physiology by monitoring the distributions and concentrations of substances such as ATP and calcium ions.

Electron Microscopy

Learning Objectives

- Explain how electron microscopy differs from light microscopy.
- Identify one use for the TEM, SEM, and scanned-probe microscopes.

Objects smaller than about 0.2 μm, such as viruses or the internal structures of cells, must be examined with an **electron microscope.** In electron microscopy, a beam of electrons is used instead of light. Free electrons travel in waves. The resolving power of the electron microscope is far greater than that of the other microscopes described here so far. The better resolution of electron microscopes is due to the shorter wavelengths of electrons; the wavelengths of electrons are about 100,000 times smaller than the wavelengths of visible light. Thus, electron microscopes are used to examine structures too small to be resolved with light microscopes. Images produced by electron microscopes are always black and white, but they may be colored artificially to accentuate certain details.

Instead of using glass lenses, an electron microscope uses electromagnetic lenses to focus a beam of electrons onto a specimen. There are two types of electron microscopes: the transmission electron microscope and the scanning electron microscope.

Transmission Electron Microscopy

In the **transmission electron microscope (TEM),** a finely focused beam of electrons from an electron gun passes through a specially prepared, ultrathin section of the specimen (Figure 3.8a). The beam is focused on a small area of the specimen by an electromagnetic condenser lens that performs roughly the same function as the condenser of a light microscope—directing the beam of electrons in a straight line to illuminate the specimen.

Electron microscopes use electromagnetic lenses to control illumination, focus, and magnification. Instead of being placed on a glass slide, as in light microscopes, the specimen is usually placed on a copper mesh grid. The beam of electrons passes through the specimen and then through an electromagnetic objective lens, which magnifies the image. Finally, the electrons are focused by an electromagnetic projector lens, rather than by an ocular lens as in a light microscope, onto a fluorescent screen or photographic plate. The final image, called a *transmission electron micrograph,* appears as many light and dark areas, depending on the number of electrons absorbed by different areas of the specimen.

In practice, the transmission electron microscope can resolve objects as close together as 2.5 nm, and objects are generally magnified 10,000–100,000×. Because most microscopic specimens are so thin, the contrast between their ultrastructures and the background is weak. Contrast can be greatly enhanced by using a "stain" that absorbs electrons and produces a darker image in the stained region. Salts of various heavy metals, such as lead, osmium, tungsten, and uranium, are commonly used as stains. These metals can be fixed onto the specimen *(positive staining)* or used to increase the electron opacity of the surrounding field *(negative staining).* Negative staining is useful for the study of the very smallest specimens, such as virus particles, bacterial flagella, and protein molecules.

In addition to positive and negative staining, a microbe can be viewed by a technique called *shadow casting.* In this procedure, a heavy metal such as platinum or gold is sprayed at an angle of about 45° so that it strikes the microbe from only one side. The metal piles up on one side of the specimen, and the uncoated area on the opposite side of the specimen leaves a clear area behind it as a shadow. This gives a three-dimensional effect to the specimen and provides a general idea of the size and shape of the specimen (see Figure 4.6a on page 82).

Transmission electron microscopy has high resolution and is extremely valuable for examining different layers of specimens. However, it does have certain disadvantages. Because electrons have limited penetrating power, only a

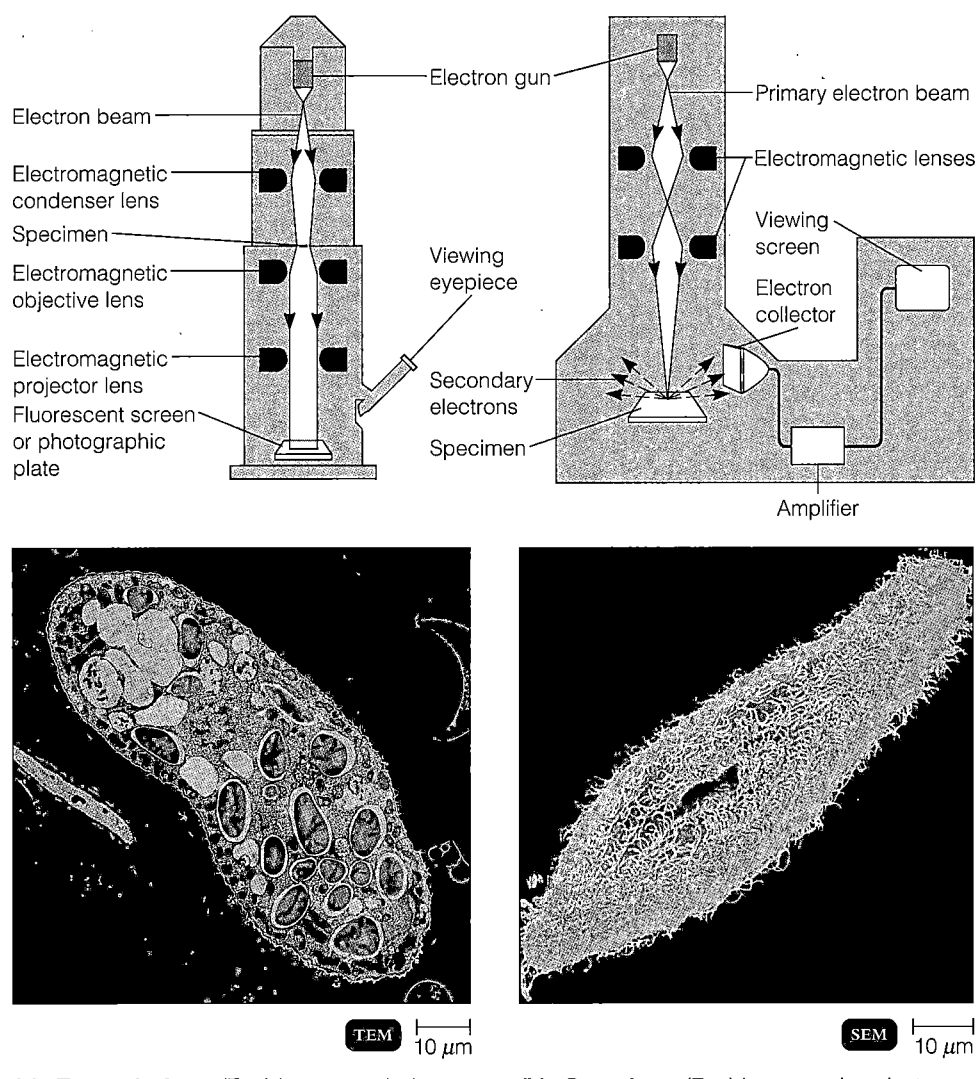

(a) Transmission. (Top) In a transmission electron microscope, electrons pass through the specimen and are scattered. Magnetic lenses focus the image onto a fluorescent screen or photographic plate. (Bottom) This colorized transmission electron micrograph (TEM) shows a thin slice of a *Paramecium*. In this type of microscopy, the internal structures present in the slice can be seen.

(b) Scanning. (Top) In a scanning electron microscope, primary electrons sweep across the specimen and knock electrons from its surface. These secondary electrons are picked up by a collector, amplified, and transmitted onto a viewing screen or photographic plate. (Bottom) In this colorized scanning electron micrograph (SEM), the surface structures of a *Paramecium* can be seen. Note the three-dimensional appearance of this cell, in contrast to that of the cell in part a.

FIGURE 3.8   Transmission and scanning electron microscopy. The illustrations show the pathways of electron beams used to create images of the specimens. The photographs show a *Paramecium* viewed with both of these types of electron microscopes.

▨ How do TEM and SEM images of the same organism differ?

very thin section of a specimen (about 100 nm) can be studied effectively. Thus, the specimen has no three-dimensional aspect. In addition, specimens must be fixed, dehydrated, and viewed under a high vacuum to prevent electron scattering. These treatments not only kill the specimen but also cause some shrinkage and distortion, sometimes to the extent that there may appear to be additional structures in a prepared cell. Structures that appear as a result of the method of preparation are called *artifacts*.

(a)
20 nm

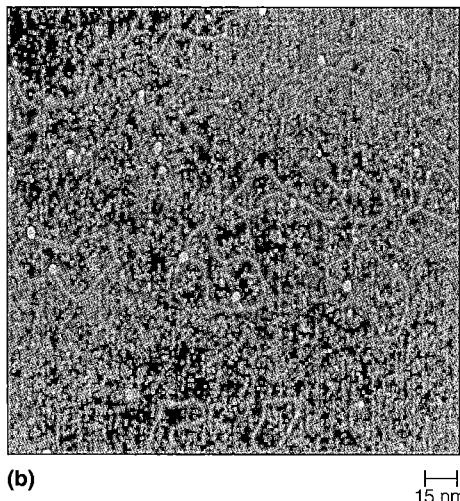

(b)
15 nm

FIGURE 3.9 Scanned-probe microscopy. **(a)** Scanning tunneling microscopy (STM) of the surface of the gram-positive bacterium Deinococcus radiodurans. Unlike other gram-positive bacteria, Deinococcus has an outer membrane. **(b)** Atomic force microscopy (AFM) of plasmid DNA being digested by a restriction enzyme.

Scanning Electron Microscopy

The **scanning electron microscope (SEM)** overcomes the problem of sectioning associated with a transmission electron microscope. A scanning electron microscope provides striking three-dimensional views of specimens (Figure 3.8b). In scanning electron microscopy, an electron gun produces a finely focused beam of electrons called the primary electron beam. These electrons pass through electromagnetic lenses and are directed over the surface of the specimen. The primary electron beam knocks electrons out of the surface of the specimen, and the secondary electrons thus produced are transmitted to an electron collector, amplified, and used to produce an image on a viewing screen or photographic plate. The image is called a *scanning electron micrograph*. This microscope is especially useful in studying the surface structures of intact cells and viruses. In practice, it can resolve objects as close together as 20 nm, and objects are generally magnified 1000–10,000×.

Scanned-Probe Microscopy

Since the early 1980s, several new types of microscopes, called **scanned-probe microscopes,** have been developed. They use various kinds of probes to examine the surface of a specimen at very close range, and they do so without modifying the specimen or exposing it to damaging, high-energy radiation. Such microscopes can be used to map atomic and molecular shapes, to characterize

magnetic and chemical properties, and to determine temperature variations inside cells. Among the new scanned-probe microscopes are the scanning tunneling microscope and the atomic force microscope, discussed next.

Scanning Tunneling Microscopy

Scanning tunneling microscopy (STM) uses a thin metal (tungsten) probe that scans a specimen and produces an image revealing the bumps and depressions of the atoms on the surface of the specimen (Figure 3.9a). The resolving power of an STM is much greater than that of an electron microscope; it can resolve features that are only about 1/100 the size of an atom. Moreover, special preparation of the specimen for observation is not needed. STMs are used to provide incredibly detailed views of molecules such as DNA.

Atomic Force Microscopy

In **atomic force microscopy (AFM),** a metal-and-diamond probe is gently forced down onto a specimen. As the probe moves along the surface of the specimen, its movements are recorded and a three-dimensional image is produced (Figure 3.9b). As with STM, AFM does not require special specimen preparation. AFM is used to image both biological substances (in nearly atomic detail) and molecular processes (such as the assembly of fibrin, a component of a blood clot).

The various types of microscopy described above are summarized in Table 3.2 (pages 64–65).

Preparation of Specimens for Light Microscopy

Learning Objectives

- *Differentiate between an acidic dye and a basic dye.*
- *Compare simple, differential, and special stains.*
- *List the steps in preparing a Gram stain, and describe the appearance of gram-positive and gram-negative cells after each step.*
- *Compare and contrast the Gram stain and the acid-fast stain.*

Because most microorganisms appear almost colorless when viewed through a standard light microscope, we often must prepare them for observation. One of the ways this can be done is by staining (coloring). Next we will discuss several different staining procedures.

Preparing Smears for Staining

Most initial observations of microorganisms are made with stained preparations. **Staining** simply means coloring the microorganisms with a dye that emphasizes certain structures. Before the microorganisms can be stained, however, they must be **fixed** (attached) to the microscope slide. Fixing simultaneously kills the microorganisms and attaches them to the slide. It also preserves various parts of microbes in their natural state with only minimal distortion.

When a specimen is to be fixed, a thin film of material containing the microorganisms is spread over the surface of the slide. This film, called a **smear,** is allowed to air dry. In most staining procedures the slide is then fixed by passing it through the flame of a Bunsen burner several times, smear side up, or by covering the slide with methyl alcohol for 1 minute. Air drying and flaming fix the microorganisms to the slide. Stain is applied and then washed off with water; then the slide is blotted with absorbent paper. Without fixing, the stain might wash the microbes off the slide. The stained microorganisms are now ready for microscopic examination.

Stains are salts composed of a positive and a negative ion, one of which is colored and is known as the *chromophore.* The color of so-called **basic dyes** is in the positive ion; in **acidic dyes,** it is in the negative ion. Bacteria are slightly negatively charged at pH 7. Thus, the colored positive ion in a basic dye is attracted to the negatively charged bacterial cell. Basic dyes, which include crystal violet, methylene blue, malachite green, and safranin, are more commonly used than acidic dyes. Acidic dyes are not attracted to most types of bacteria because the dye's negative ions are repelled by the negatively charged bacterial surface, so the stain colors the background instead. Preparing colorless bacteria against a colored background is called **negative staining.** It is valuable in the observation of overall cell shapes, sizes, and capsules because the cells are made highly visible against a contrasting dark background (see Figure 3.12a on page 72). Distortions of cell size and shape are minimized because heat fixing is not necessary and the cells do not pick up the stain. Examples of acidic dyes are eosin, acid fuchsin, and nigrosin.

To apply acidic or basic dyes, microbiologists use three kinds of staining techniques: simple, differential, and special.

Simple Stains

A **simple stain** is an aqueous or alcohol solution of a single basic dye. Although different dyes bind specifically to different parts of cells, the primary purpose of a simple stain is to highlight the entire microorganism so that cellular shapes and basic structures are visible. The stain is applied to the fixed smear for a certain length of time and then washed off, and the slide is dried and examined. Occasionally, a chemical is added to the solution to intensify the stain; such an additive is called a **mordant.** One function of a mordant is to increase the affinity of a stain for a biological specimen; another is to coat a structure (such as a flagellum) to make it thicker and easier to see after it is stained with a dye. Some of the simple stains commonly used in the laboratory are methylene blue, carbolfuchsin, crystal violet, and safranin.

Differential Stains

Unlike simple stains, **differential stains** react differently with different kinds of bacteria and thus can be used to distinguish among them. The differential stains most frequently used for bacteria are the Gram stain and the acid-fast stain.

Gram Stain

The **Gram stain** was developed in 1884 by the Danish bacteriologist Hans Christian Gram. It is one of the most useful staining procedures because it classifies bacteria into two large groups: gram-positive and gram-negative.

In this procedure (Figure 3.10a), ① a heat-fixed smear is covered with a basic purple dye, usually crystal violet. Because the purple stain imparts its color to all cells, it is referred to as a **primary stain.** ② After a short time, the purple dye is washed off, and the smear is covered with iodine, a mordant. When the iodine is washed off, both gram-positive and gram-negative bacteria appear dark

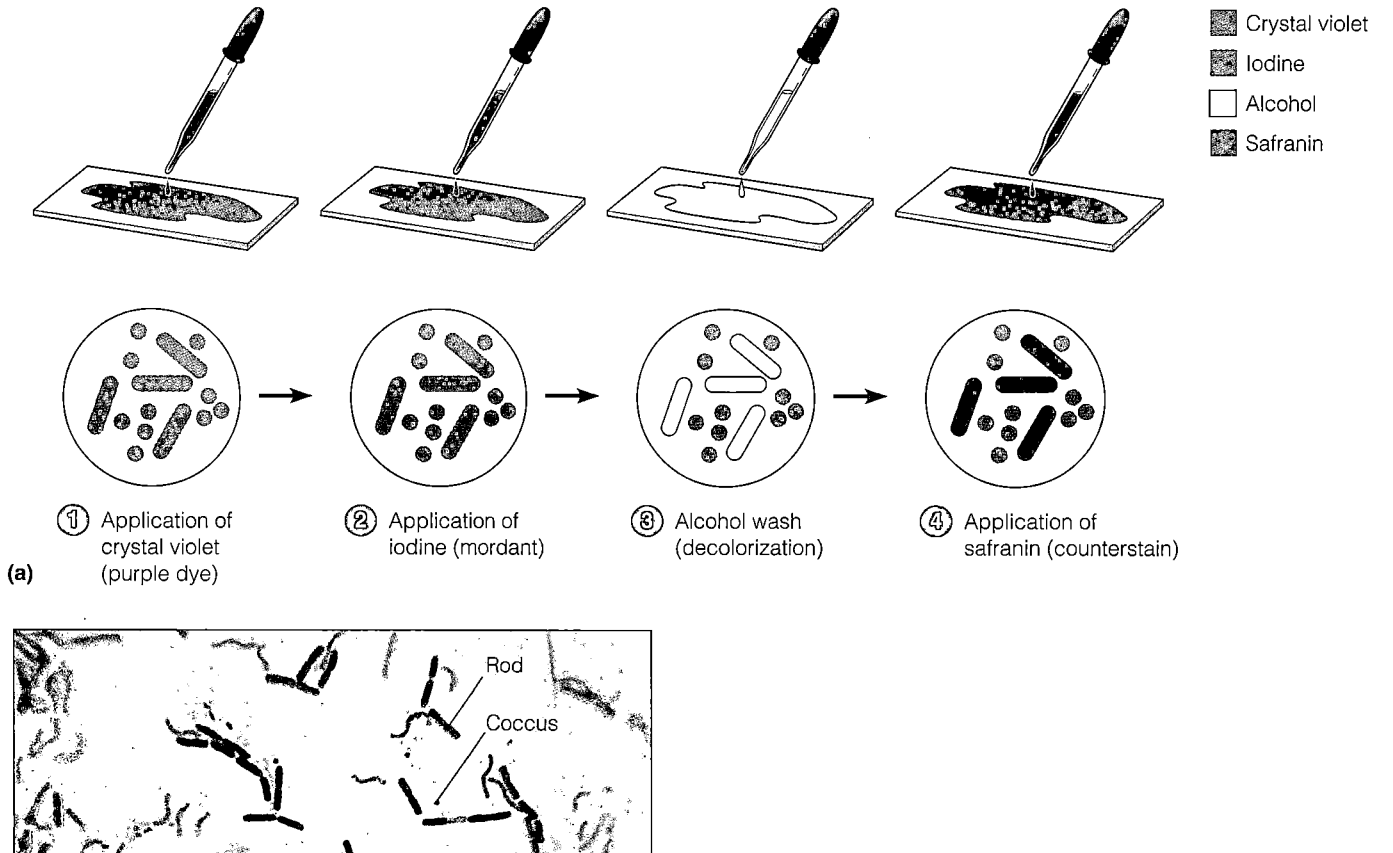

Crystal violet
Iodine
Alcohol
Safranin

① Application of crystal violet (purple dye)   ② Application of iodine (mordant)   ③ Alcohol wash (decolorization)   ④ Application of safranin (counterstain)

(a)

(b)

Rod
Coccus
Vibrio
LM ⊢—⊣ 5 μm

FIGURE 3.10 Gram staining. (a) Procedure. **(b)** Micrograph of gram-stained bacteria. The rods and cocci (purple) are gram-positive, and the vibrios (pink) are gram-negative.

■ **How can the Gram reaction be useful in prescribing antibiotic treatment?**

violet or purple. ③ Next, the slide is washed with alcohol or an alcohol-acetone solution. This solution is a **decolorizing agent,** which removes the purple from the cells of some species but not from others. ④ The alcohol is rinsed off, and the slide is then stained with safranin, a basic red dye. The smear is washed again, blotted dry, and examined microscopically.

The purple dye and the iodine combine in the cytoplasm of each bacterium and color it dark violet or purple. Bacteria that retain this color after the alcohol has attempted to decolorize them are classified as **gram-positive;** bacteria that lose the dark violet or purple color after decolorization are classified as **gram-negative** (Figure 3.10b). Because gram-negative bacteria are colorless after the alcohol wash, they are no longer visible. This is why the basic dye safranin is applied; it turns the gram-negative bacteria pink. Stains such as safranin that have a contrasting

color to the primary stain are called **counterstains.** Because gram-positive bacteria retain the original purple stain, they are not affected by the safranin counterstain.

As you will see in Chapter 4, different kinds of bacteria react differently to the Gram stain, because structural differences in their cell walls affect the retention or escape of a combination of crystal violet and iodine, called the crystal violet-iodine (CV–I) complex. Among other differences, gram-positive bacteria have a thicker peptidoglycan (disaccharides and amino acids) cell wall than gram-negative bacteria. In addition, gram-negative bacteria contain a layer of lipopolysaccharide (lipids and polysaccharides) as part of their cell wall (see Figure 4.12 on page 87). When applied to both gram-positive and gram-negative cells, crystal violet and then iodine readily enter the cells. Inside the cells, the crystal violet and iodine combine to form CV–I. This complex is larger than the crystal violet molecule that entered the cells, and, because of its size, it cannot be washed out of the intact peptidoglycan layer of gram-positive cells by alcohol. Consequently, gram-positive cells retain the color of the crystal violet dye. In gram-negative cells, however, the alcohol wash disrupts the outer lipopolysaccharide layer, and the CV–I complex is washed out through the thin layer of peptidoglycan. As a result, gram-negative cells are colorless until counterstained with safranin, after which they are pink.

In summary, gram-positive cells retain the dye and remain purple. Gram-negative cells do not retain the dye; they are colorless until counterstained with a red dye, after which they appear pink.

The Gram method is one of the most important staining techniques in medical microbiology. But Gram staining results are not universally applicable because some bacterial cells stain poorly or not at all. The Gram reaction is most consistent when it is used on young, growing bacteria.

The Gram reaction of a bacterium can provide valuable information for the treatment of disease. Gram-positive bacteria tend to be killed easily by penicillins and cephalosporins. Gram-negative bacteria are generally more resistant because the antibiotics cannot penetrate the lipopolysaccharide layer. Some resistance to these antibiotics among both gram-positive and gram-negative bacteria is due to bacterial inactivation of the antibiotics.

Acid-Fast Stain

Another important differential stain (one that differentiates bacteria into distinctive groups) is the **acid-fast stain,** which binds strongly only to bacteria that have a waxy material in their cell walls. Microbiologists use this stain to identify all bacteria in the genus *Mycobacterium,* including the two important pathogens *Mycobacterium tuberculosis,* the causative agent of tuberculosis, and *Mycobacterium leprae* (lep′rī), the causative agent of leprosy. This stain is also used to identify the pathogenic strains of the genus *Nocardia* (nō-kär′dē-ä).

In the acid-fast staining procedure, the red dye carbolfuchsin is applied to a fixed smear, and the slide is gently heated for several minutes. (Heating enhances penetration and retention of the dye.) Then the slide is cooled and washed with water. The smear is next treated with acid-alcohol, a decolorizer, which removes the red stain from bacteria that are not acid-fast. The acid-fast microorganisms retain the red color because the carbolfuchsin is more soluble in the cell wall lipids than in the acid-alcohol (Figure 3.11). In non–acid-fast bacteria, whose cell walls lack the lipid components, the carbolfuchsin is rapidly removed during decolorization, leaving

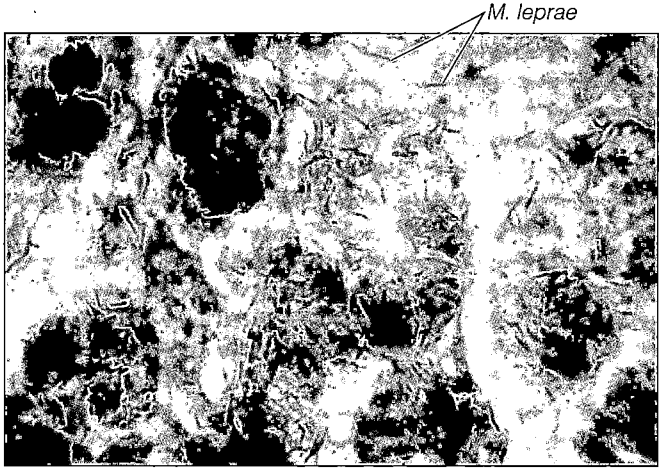

FIGURE 3.11 **Acid-fast bacteria.** The *Mycobacterium leprae* bacteria that have infected this tissue have been stained red with an acid-fast stain.

■ **What two diseases can be diagnosed using the acid-fast stain?**

the cells colorless. The smear is then stained with a methylene blue counterstain. Non–acid-fast cells appear blue after application of the counterstain.

Special Stains

Learning Objective

■ *Explain why each of the following is used: capsule stain, endospore stain, flagellar stain.*

Special stains are used to color and isolate specific parts of microorganisms, such as endospores and flagella, and to reveal the presence of capsules.

Negative Staining for Capsules

Many microorganisms contain a gelatinous covering called a **capsule,** which we will discuss in our examination of the prokaryotic cell in Chapter 4. In medical microbiology, demonstrating the presence of a capsule is a means of determining the organism's **virulence,** the degree to which a pathogen can cause disease.

Capsule staining is more difficult than other types of staining procedures because capsular materials are soluble in water and may be dislodged or removed during rigorous washing. To demonstrate the presence of capsules, a microbiologist can mix the bacteria in a solution containing a fine colloidal suspension of colored particles (usually India ink or nigrosin) to provide a dark background and then stain the bacteria with a simple stain, such as safranin

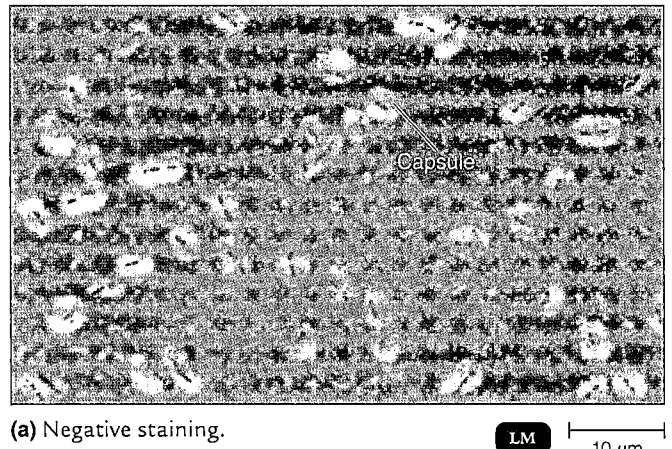

(a) Negative staining.

LM ├──────┤
10 μm

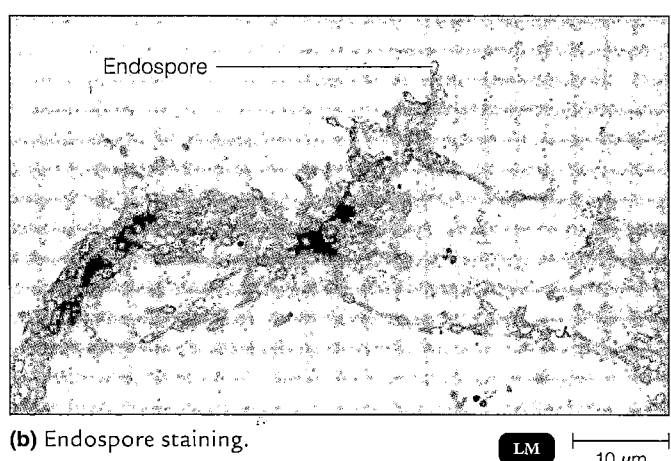

(b) Endospore staining.

LM ├──────┤
10 μm

FIGURE 3.12 Special staining. (a) Capsule staining provides a contrasting background, so the capsules of these bacteria, *Klebsiella pneumoniae,* show up as light areas surrounding the stained cells. **(b)** Endospores are seen as green ovals in these rod-shaped cells of the bacterium *Bacillus cereus,* using the Schaeffer-Fulton endospore stain. **(c)** Flagella are shown as wavy extensions from the ends of these cells of the bacterium *Spirillum volutans.* In relation to the body of the cell, the flagella are much thicker than normal because layers of the stain have accumulated from treatment of the specimen with a mordant.

■ **Of what value are capsules, endospores, and flagella to bacteria?**

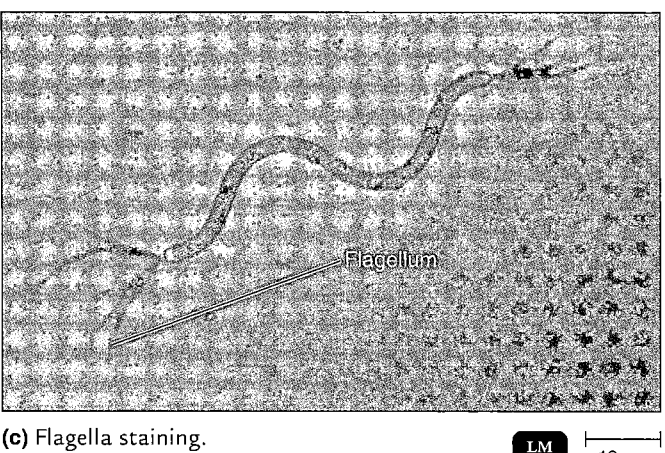

(c) Flagella staining.

LM ├──────┤
10 μm

(Figure 3.12a). Because of their chemical composition, capsules do not accept most biological dyes, such as safranin, and thus appear as halos surrounding each stained bacterial cell.

Endospore (Spore) Staining

An **endospore** is a special resistant, dormant structure formed within a cell that protects a bacterium from adverse environmental conditions. Although endospores are relatively uncommon in bacterial cells, they can be formed by a few genera of bacteria. Endospores cannot be stained by ordinary methods, such as simple staining and Gram staining, because the dyes do not penetrate the wall of the endospore.

The most commonly used endospore stain is the *Schaeffer-Fulton endospore stain* (Figure 3.12b). Malachite green, the primary stain, is applied to a heat-fixed smear and heated to steaming for about 5 minutes. The heat helps the stain penetrate the endospore wall. Then the preparation is washed for about 30 seconds with water to remove the malachite green from all of the cells' parts except the endospores. Next, safranin, a counterstain, is applied to the

smear to stain portions of the cell other than endospores. In a properly prepared smear, the endospores appear green within red or pink cells. Because endospores are highly refractive, they can be detected under the light microscope when unstained, but they cannot be differentiated from inclusions of stored material without a special stain.

Flagella Staining

Bacterial **flagella** (singular: *flagellum*) are structures of locomotion too small to be seen with a light microscope without staining. A tedious and delicate staining procedure uses a mordant and the stain carbolfuchsin to build up the diameters of the flagella until they become visible under the light microscope (Figure 3.12c). Microbiologists use the number and arrangement of flagella as diagnostic aids.

★ ★ ★

A summary of stains is presented in Table 3.3. In the next chapter we will take a closer look at the structure of microbes and how they protect, nourish, and reproduce themselves.

table 3.3	A Summary of Various Stains and Their Uses
Stain	**Principal Uses**
Simple (methylene blue, carbolfuchsin, crystal violet, safranin)	Aqueous or alcohol solution of a single basic dye. (Sometimes a mordant is added to intensify the stain.) Used to highlight microorganisms to determine cellular shapes and arrangements.
Differential	React differently with different kinds of bacteria in order to distinguish among them.
Gram	Classifies bacteria into two large groups: gram-positive and gram-negative. Gram-positive bacteria retain the crystal violet stain and appear purple. Gram-negative bacteria do not retain the crystal violet stain and remain colorless until counterstained with safranin and then appear pink.
Acid-fast	Used to distinguish *Mycobacterium* species and some species of *Nocardia*. Acid-fast bacteria, once stained with carbolfuchsin and treated with acid-alcohol, remain red because they retain the carbolfuchsin stain. Non–acid-fast bacteria, when stained and treated the same way and then stained with methylene blue, appear blue because they lose the carbolfuchsin stain and are then able to accept the methylene blue stain.
Special	Used to color and isolate various structures, such as capsules, endospores, and flagella; sometimes used as a diagnostic aid.
Negative	Used to demonstrate the presence of capsules. Because capsules do not accept most stains, the capsules appear as unstained halos around bacterial cells and stand out against a dark background.
Endospore	Used to detect the presence of endospores in bacteria. When malachite green is applied to a heat-fixed smear of bacterial cells, the stain penetrates the endospores and stains them green. When safranin (red) is then applied, it stains the remainder of the cells red or pink.
Flagella	Used to demonstrate the presence of flagella. A mordant is used to build up the diameters of flagella until they become visible microscopically when stained with carbolfuchsin.

Study Outline

UNITS OF MEASUREMENT (pp. 56–57)

1. The standard unit of length is the meter (m).
2. Microorganisms are measured in micrometers, μm (10^{-6} m), and in nanometers, nm (10^{-9} m).

MICROSCOPY: THE INSTRUMENTS (pp. 57–68)

1. A simple microscope consists of one lens; a compound microscope has multiple lenses.

Light Microscopy (pp. 57–66)

Compound Light Microscopy (pp. 57–60)

1. The most common microscope used in microbiology is the compound light microscope (LM).
2. The total magnification of an object is calculated by multiplying the magnification of the objective lens by the magnification of the ocular lens.
3. The compound light microscope uses visible light.
4. The maximum resolution, or resolving power (the ability to distinguish between two points) of a compound light microscope is 0.2 μm; maximum magnification is 2000×.

5. Specimens are stained to increase the difference between the refractive indexes of the specimen and the medium.
6. Immersion oil is used with the oil immersion lens to reduce light loss between the slide and the lens.
7. Brightfield illumination is used for stained smears.
8. Unstained cells are more productively observed using darkfield, phase-contrast, or DIC microscopy.

Darkfield Microscopy (pp. 61–62)

1. The darkfield microscope shows a light silhouette of an organism against a dark background.
2. It is most useful for detecting the presence of extremely small organisms.

Phase-Contrast Microscopy (p. 62)

1. A phase-contrast microscope brings direct and reflected or diffracted light rays together (in phase) to form an image of the specimen on the ocular lens.
2. It allows the detailed observation of living organisms.

Differential Interference Contrast (DIC) Microscopy (p. 62)

1. The DIC microscope provides a colored, three-dimensional image of the object being observed.
2. It allows the detailed observations of living cells.

Fluorescence Microscopy (pp. 62–63)

1. In fluorescence microscopy, specimens are first stained with fluorochromes and then viewed through a compound microscope by using an ultraviolet light source.

2. The microorganisms appear as bright objects against a dark background.

3. Fluorescence microscopy is used primarily in a diagnostic procedure called fluorescent-antibody (FA) technique, or immunofluorescence.

Confocal Microscopy (pp. 63–66)

1. In confocal microscopy, a specimen is stained with a fluorescent dye and illuminated one plane at a time.

2. Using a computer to process the images, two-dimensional and three-dimensional images of cells can be produced.

Electron Microscopy (pp. 66–68)

1. A beam of electrons, instead of light, is used with an electron microscope.

2. Electromagnets, instead of glass lenses, control focus, illumination, and magnification.

3. Thin sections of organisms can be seen in an electron micrograph produced using a transmission electron microscope (TEM). Magnification: 10,000–100,000×. Resolving power: 2.5 nm.

4. Three-dimensional views of the surfaces of whole microorganisms can be obtained with a scanning electron microscope (SEM). Magnification: 1000–10,000×. Resolving power: 20 nm.

Scanned-Probe Microscopy (p. 68)

1. Scanning tunneling microscopy (STM) and atomic force microscopy (AFM) produce three-dimensional images of the surface of a molecule.

PREPARATION OF SPECIMENS FOR LIGHT MICROSCOPY (pp. 69–73)

Preparing Smears for Staining (p. 69)

1. Staining means coloring a microorganism with a dye to make some structures more visible.

2. Fixing uses heat or alcohol to kill and attach microorganisms to a slide.

3. A smear is a thin film of material used for microscopic examination.

4. Bacteria are negatively charged, and the colored positive ion of a basic dye will stain bacterial cells.

5. The colored negative ion of an acidic dye will stain the background of a bacterial smear; a negative stain is produced.

Simple Stains (p. 69)

1. A simple stain is an aqueous or alcohol solution of a single basic dye.

2. It is used to make cellular shapes and arrangements visible.

3. A mordant may be used to improve bonding between the stain and the specimen.

Differential Stains (pp. 69–71)

1. Differential stains, such as the Gram stain and acid-fast stain, differentiate bacteria according to their reactions to the stains.

2. The Gram stain procedure uses a purple stain (crystal violet), iodine as a mordant, an alcohol decolorizer, and a red counterstain.

3. Gram-positive bacteria retain the purple stain after the decolorization step; gram-negative bacteria do not and thus appear pink from the counterstain.

4. Acid-fast microbes, such as members of the genera *Mycobacterium* and *Nocardia,* retain carbolfuchsin after acid-alcohol decolorization and appear red; non–acid-fast microbes take up the methylene blue counterstain and appear blue.

Special Stains (pp. 71–73)

1. The endospore stain and flagella stain are special stains that color only certain parts of bacteria.

2. Negative staining is used to make microbial capsules visible.

Study Questions

REVIEW

1. Fill in the following blanks:

   $1 \ \mu m =$ _____ m

   1 _____ $= 10^{-9}$ m

   $1 \ \mu m =$ _____ nm

2. Label the parts of the compound light microscope in the following figure:

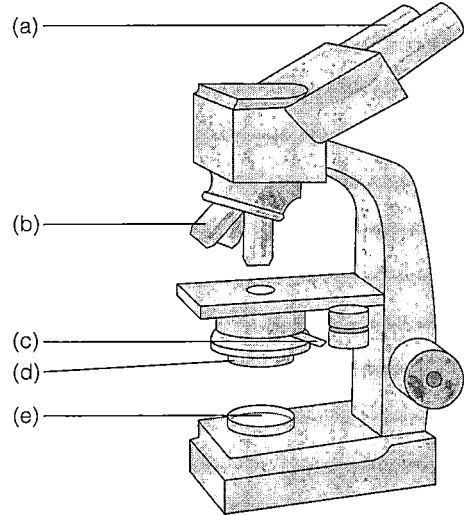

(a) _____

(b) _____

(c) _____

(d) _____

(e) _____

a. _____   d. _____

b. _____   e. _____

c. _____

3. Calculate the total magnification of the nucleus of a cell being observed through a compound light microscope with a 10× ocular lens and an oil immersion lens.

4. Which type of microscope would be best to use to observe each of the following?

   a. a stained bacterial smear

   b. unstained bacterial cells when the cells are small and no detail is needed

   c. unstained live tissue when it is desirable to see some intracellular detail

   d. a sample that emits light when illuminated with ultraviolet light

   e. intracellular detail of a cell that is 1 μm long

   f. unstained live cells in which intracellular structures are shown in color

5. An electron microscope differs from a light microscope in that _____ focused by _____ is used instead of light, and the image is viewed not through the ocular lenses but on _____.

6. The maximum magnification of a compound microscope is _____; that of an electron microscope, _____. The maximum resolution of a compound microscope is _____; that of an electron microscope, _____. One advantage of a scanning electron microscope over a transmission electron microscope is _____.

7. Why do basic dyes stain bacterial cells? Why don't acidic dyes stain bacterial cells?

8. When is it most appropriate to use each of the following?

   a. a simple stain   c. a negative stain

   b. a differential stain   d. a flagella stain

9. Why is a mordant used in the Gram stain? In the flagella stain?

10. What is the purpose of a counterstain in the acid-fast stain?

11. What is the purpose of a decolorizer in the Gram stain? In the acid-fast stain?

12. Choose from the following terms to fill in the blanks: counterstain, decolorizer, mordant, primary stain. In the endospore stain, safranin is the _____. In the Gram stain, safranin is the _____.

13. Fill in the following table regarding the Gram stain:

	Appearance After This Step of	
Steps	Gram-Positive Cells	Gram-Negative Cells
Crystal violet		
Iodine		
Alcohol-acetone		
Safranin		

MULTIPLE CHOICE

1. You are trying to identify an unknown gram–negative bacterium. Which of the following stains is *not* necessary?

   a. negative stain   d. endospore stain

   b. acid-fast stain   e. both b and d

   c. flagella stain

2. Assume you stain *Bacillus* by applying malachite green with heat and then counterstain with safranin. Through the microscope, the green structures are

   a. cell walls.   d. flagella.

   b. capsules.   e. impossible to identify.

   c. endospores.

3. Carbolfuchsin can be used as a simple stain and a negative stain. As a simple stain, the pH is

   a. 2.

   b. higher than the negative stain.

   c. lower than the negative stain.

   d. the same as the negative stain.

4. Looking at the cell of a photosynthetic microorganism, you observe that the chloroplasts are green in brightfield microscopy and red in fluorescence microscopy. You conclude that

   a. chlorophyll is fluorescent.

   b. the magnification has distorted the image.

   c. you're not looking at the same structure in both microscopes.

   d. the stain masked the green color.

   e. none of the above

5. Which of the following is *not* a functionally analogous pair of stains?
   a. nigrosin and malachite green
   b. crystal violet and carbolfuchsin
   c. safranin and methylene blue
   d. ethanol-acetone and acid-alcohol
   e. none of the above

6. Which of the following pairs is mismatched?
   a. capsule—negative stain
   b. cell arrangement—simple stain
   c. cell size—negative stain
   d. gram stain—bacterial identification
   e. none of the above

7. Assume you stain *Clostridium* by applying a basic stain, carbolfuchsin, with heat, decolorizing with acid-alcohol, and counterstaining with an acid stain, nigrosin. Through the microscope, the endospores are ____1____ and the cells are stained ____2____.
   a. 1—red; 2—black
   b. 1—black; 2—colorless
   c. 1—colorless; 2—black
   d. 1—red; 2—colorless
   e. 1—black; 2—red

8. Assume that you are viewing a Gram-stained field of red cocci and blue baccili through the microscope. You can safely conclude that you have
   a. made a mistake in staining.
   b. two different species.
   c. old bacterial cells.
   d. young bacterial cells.
   e. none of the above

9. In 1996, scientists described a new tapeworm parasite that has killed at least one person. The initial examination of the patient's abdominal mass was most likely made using
   a. brightfield microscopy.
   b. darkfield microscopy.
   c. electron microscopy.
   d. phase-contrast microscopy.
   e. fluorescence microscopy.

10. Which of the following is *not* a modification of a compound light microscope?
    a. brightfield microscopy
    b. darkfield microscopy
    c. electron microscopy
    d. phase-contrast microscopy
    e. fluorescence microscopy

CRITICAL THINKING

1. In a Gram stain, one step could be omitted and still allow differentiation between gram-positive and gram-negative cells. What is that one step?

2. Using a good compound light microscope with a resolving power of 0.3 μm, a 10× ocular lens, and a 100× oil immersion lens, would you be able to discern two objects separated by 3 μm? 0.3 μm? 300 nm?

3. Why isn't the Gram stain used on acid-fast bacteria? If you did Gram stain acid-fast bacteria, what would their Gram reaction be? What is the Gram reaction of non–acid-fast bacteria?

4. Endospores can be seen as refractile structures in unstained cells and as colorless areas in Gram-stained cells. Why is it necessary to do an endospore stain to verify the presence of endospores?

CLINICAL APPLICATIONS

1. In 1882, German bacteriologist Paul Erhlich described a method for staining *Mycobacterium* and noted, "It may be that all disinfecting agents which are acidic will be without effect on this [tubercle] bacillus, and one will have to be limited to alkaline agents." How did he reach this conclusion without testing disinfectants?

2. Laboratory diagnosis of *Neisseria gonorrhoeae* infection is based on microscopic examination of Gram-stained pus. Identify the bacteria in this light micrograph. What is the disease?

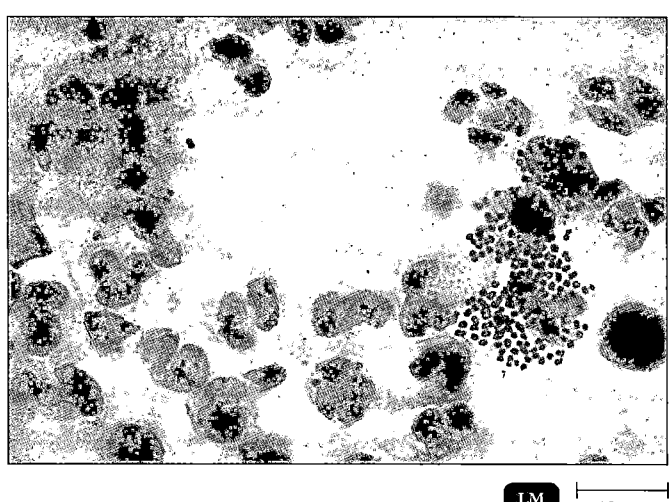

LM ├────────┤
10 μm

3. Assume that you are viewing a Gram-stained sample of vaginal discharge. Large (10 μm) red cells are coated with small (0.5 μm × 1.5 μm) blue cells on their surfaces. What is the most likely explanation for the red and blue cells?

4. A sputum sample from Calle, a 30-year-old Asian elephant, was smeared onto a slide and air dried. The smear was fixed, covered with carbolfuchsin, and heated for 5 minutes. After washing with water, acid-alcohol was placed on the smear for 30 seconds. Finally, the smear was stained with methylene blue for 30 seconds, washed with water, and dried. On examination at 1000×, the zoo veterinarian saw red rods on the slide. What infections do the results suggest? (Calle was treated and recovered.)

Learning with Technology

MP = **The Microbiology Place website**   **ST** = **Student Tutorial CD-ROM**

MP   Don't forget to go to The Microbiology Place website (http://www.microbiologyplace.com) to take the practice tests, explore the interactive activity and case study, and check out the news articles and web links for this chapter.

ST   Remember there is also a quiz for this chapter on the Microbiology Interactive Student Tutorial CD-ROM.

chapter 4

Functional Anatomy of Prokaryotic and Eukaryotic Cells

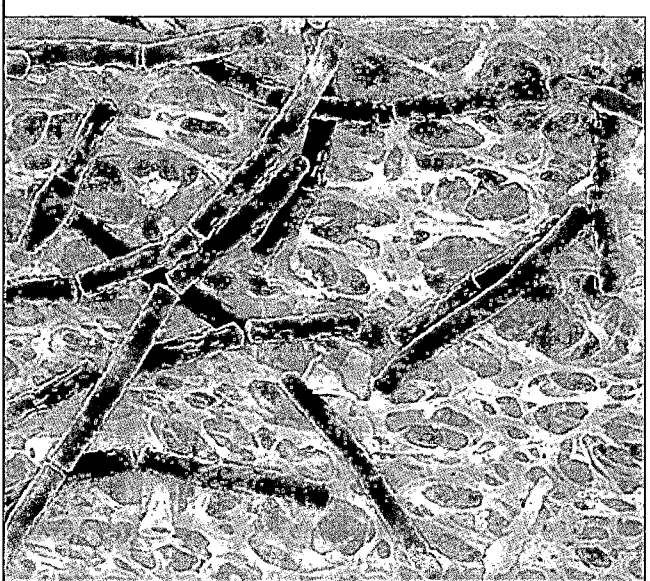

Bacillus. *Cells of the genus* Bacillus *frequently form long chains.*

Despite their complexity and variety, all living cells can be classified into two groups, prokaryotes and eukaryotes, based on their ultrastructure as seen with the electron microscope. Plants and animals are entirely composed of eukaryotic cells. In the microbial world, Bacteria and Archaea are prokaryotes. Other cellular microbes—fungi (yeasts and molds), protozoa, and algae—are eukaryotes.

Viruses, as noncellular elements, do not fit into any organizational scheme of living cells. They are genetic particles that replicate but are unable to perform the usual chemical activities of living cells. Viral structure and activity will be discussed in Chapter 13. In this chapter we will concentrate on describing prokaryotic and eukaryotic cells.

Comparing Prokaryotic and Eukaryotic Cells: An Overview

Learning Objective

- *Compare and contrast the overall cell structure of prokaryotes and eukaryotes.*

Prokaryotes and eukaryotes are chemically similar, in the sense that they both contain nucleic acids, proteins, lipids, and carbohydrates. They use the same kinds of chemical reactions to metabolize food, build proteins, and store energy. It is primarily the structure of cell walls and membranes, and the absence of *organelles* (specialized cellular structures that have specific functions), that distinguish prokaryotes from eukaryotes.

The chief distinguishing characteristics of **prokaryotes** (from the Greek words meaning prenucleus) are as follows:

1. Their DNA (genetic material) is not enclosed within a membrane and is one circular chromosome.
2. Their DNA is not associated with histones (special chromosomal proteins found in eukaryotes); other proteins are associated with the DNA.
3. They lack membrane-enclosed organelles.
4. Their cell walls almost always contain the complex polysaccharide peptidoglycan.
5. They usually divide by **binary fission.** During this process, the DNA is copied and the cell splits into two cells. Binary fission involves fewer structures and processes than eukaryotic cell division.

Eukaryotes (from the Greek words meaning true nucleus) have the following distinguishing characteristics:

1. Their DNA is found in the cell's nucleus, which is separated from the cytoplasm by a nuclear membrane, and the DNA is found in multiple chromosomes.

2. Their DNA is consistently associated with chromosomal proteins called histones and with nonhistones.

3. They have a number of membrane-enclosed organelles, including mitochondria, endoplasmic reticulum, Golgi complex, lysosomes, and sometimes chloroplasts.

4. Their cell walls, when present, are chemically simple.

5. They usually divide by mitosis, in which chromosomes replicate and an identical set is distributed into each of two nuclei. This process is guided by the mitotic spindle, a football-shaped assembly of microtubules (described later). Division of the cytoplasm and other organelles follows so that the two cells produced are identical to each other.

Additional differences between prokaryotic and eukaryotic cells are listed in Table 4.2 on page 99. Next we describe, in detail, the parts of the prokaryotic cell.

THE PROKARYOTIC CELL

The members of the prokaryotic world make up a vast heterogeneous group of very small unicellular organisms. Prokaryotes include bacteria and archaea. The majority of prokaryotes, including the photosynthesizing cyanobacteria, are included in the bacteria. Although bacteria and archaea look similar, they are different in chemical composition, as will be described later. The thousands of species of bacteria are differentiated by many factors, including morphology (shape), chemical composition (often detected by staining reactions), nutritional requirements, biochemical activities, and source of energy (sunlight or chemicals). For a discussion of how bacteria interact, see the box on page 81.

The Size, Shape, and Arrangement of Bacterial Cells

Learning Objective

- *Identify the three basic shapes of bacteria.*

There are a great many sizes and shapes among bacteria. Most bacteria range from 0.2–2.0 μm in diameter and from 2–8 μm in length. They have a few basic shapes: spherical **coccus** (plural: *cocci,* meaning berries), rod-shaped **bacillus** (plural: *bacilli,* meaning little staffs), and **spiral.**

Cocci are usually round but can be oval, elongated, or flattened on one side. When cocci divide to reproduce, the cells can remain attached to one another. Cocci that remain in pairs after dividing are called **diplococci;** those that divide and remain attached in chainlike patterns are called **streptococci** (Figure 4.1a). Those that divide in two planes and remain in groups of four are known as **tetrads** (Figure 4.1b). Those that divide in

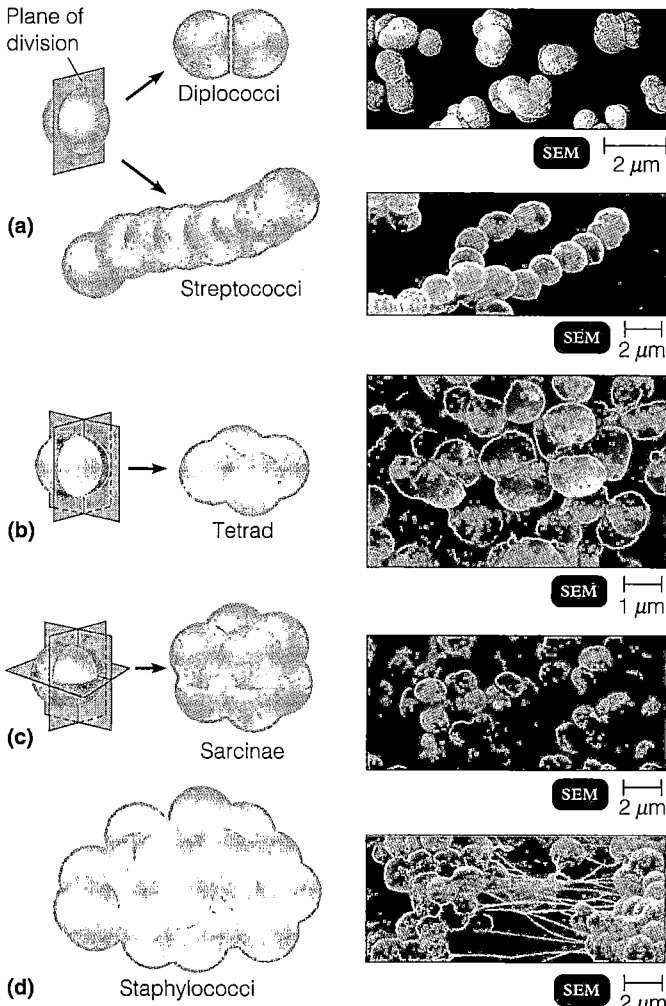

FIGURE 4.1 Arrangements of cocci. (a) Division in one plane produces diplococci and streptococci. **(b)** Division in two planes produces tetrads. **(c)** Division in three planes produces sarcinae, and **(d)** Division in multiple planes produces staphylococci.

■ **The planes in which the cell divides determine the arrangement of cells.**

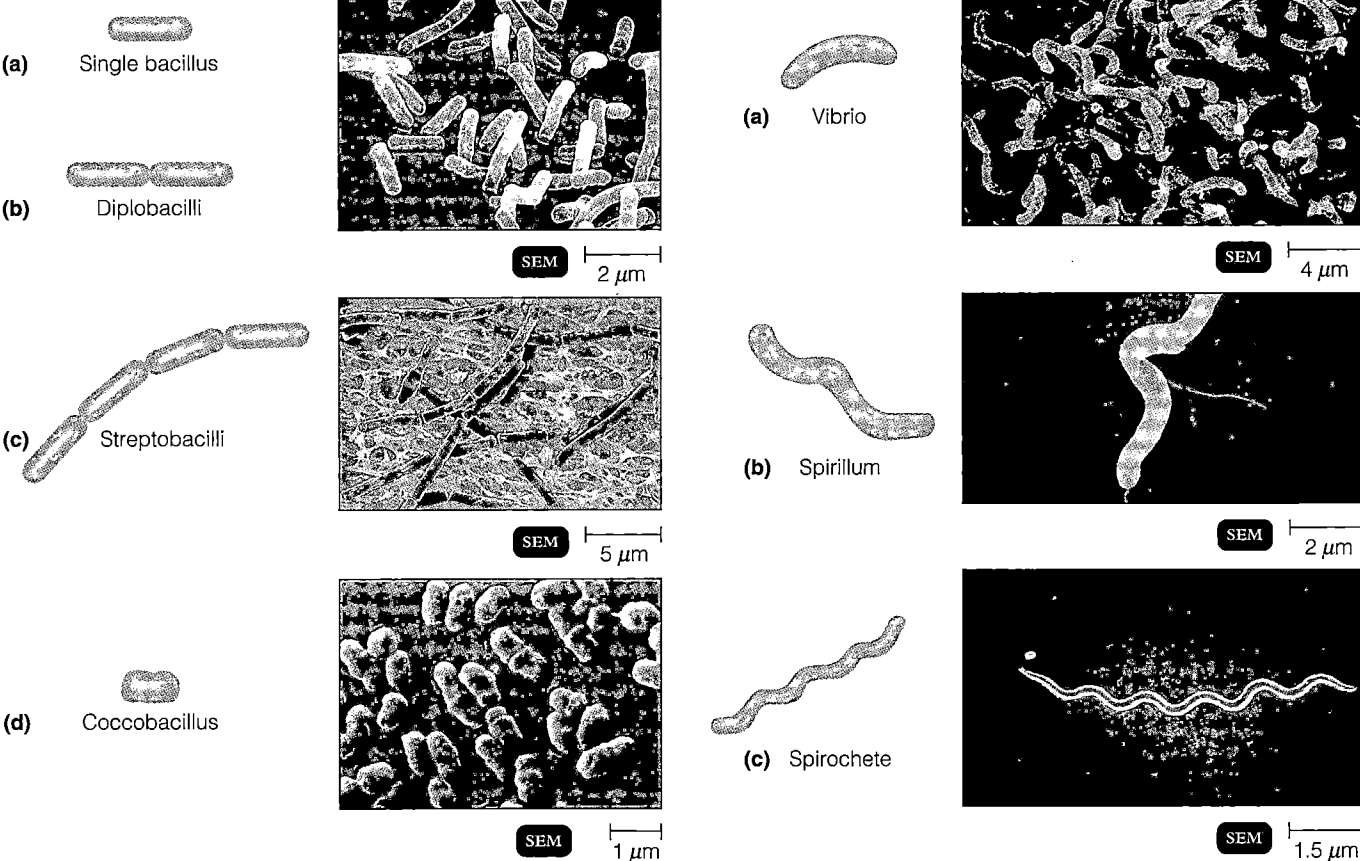

FIGURE 4.2 Bacilli. (a) Single bacilli. **(b)** Diplobacilli. In the top micrograph, a few joined pairs of bacilli serve as examples of diplobacilli. **(c)** Streptobacilli. **(d)** Coccobacilli.

■ **Why don't bacilli form tetrads or clusters?**

FIGURE 4.3 Spiral bacteria. (a) Vibrios, **(b)** Spirilla, and **(c)** Spirochetes.

■ **Spiral bacteria have one or more twists.**

three planes and remain attached in cubelike groups of eight are called **sarcinae** (Figure 4.1c). Those that divide in multiple planes and form grapelike clusters or broad sheets are called **staphylococci** (Figure 4.1d). These group characteristics are frequently helpful in the identification of certain cocci.

Bacilli divide only across their short axis, so there are fewer groupings of bacilli than of cocci. Most bacilli appear as single rods (Figure 4.2a). **Diplobacilli** appear in pairs after division (Figure 4.2b), and **streptobacilli** occur in chains (Figure 4.2c). Some bacilli look like straws. Others have tapered ends, like cigars. Still others are oval and look so much like cocci that they are called **coccobacilli** (Figure 4.2d).

"Bacillus" has two meanings in microbiology. As we have just used it, bacillus refers to a bacterial shape. When capitalized and italicized, it refers to a specific genus. For example, the bacterium *Bacillus anthracis* is the causative agent of anthrax.

Spiral bacteria have one or more twists; they are never straight. Bacteria that look like curved rods are called **vibrios** (Figure 4.3a). Others, called **spirilla,** have a helical shape, like a corkscrew, and fairly rigid bodies (Figure 4.3b). Yet another group of spirals are helical and flexible; they are called **spirochetes** (Figure 4.3c). Unlike the spirilla, which use whiplike external appendages called flagella to move, spirochetes move by means of axial filaments, which resemble flagella but are contained within a flexible external sheath.

In addition to the three basic shapes, there are star-shaped cells (genus *Stella*); rectangular, flat cells (halophilic archaea) of the genus *Haloarcula* (Figure 4.4); and triangular cells.

The shape of a bacterium is determined by heredity. Genetically, most bacteria are **monomorphic;** that is, they maintain a single shape. However, a number of environmental conditions can alter that shape. If the shape is altered, identification becomes difficult. Moreover, some

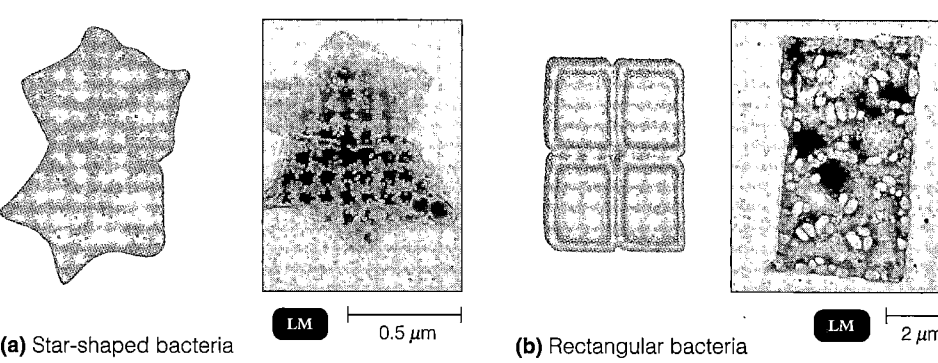

(a) Star-shaped bacteria

(b) Rectangular bacteria

FIGURE 4.4 Star-shaped and rectangular prokaryotes.
(a) *Stella* (star-shaped).
(b) *Haloarcula,* a type of halophilic archaea (rectangular cells).

bacteria, such as *Rhizobium* (rī-zō′bē-um) and *Corynebacterium* (kô-rī-nē-bak-ti′rē-um), are genetically **pleomorphic,** which means they can have many shapes, not just one.

The structure of a typical prokaryotic cell is shown in Figure 4.5. We will discuss its components according to the following organization: (1) structures external to the cell wall, (2) the cell wall itself, and (3) structures internal to the cell wall.

Structures External to the Cell Wall

Learning Objective

■ *Describe the structure and function of the glycocalyx, flagella, axial filaments, fimbriae, and pili.*

Among the structures external to the prokaryotic cell wall are the glycocalyx, flagella, axial filaments, fimbriae, and pili.

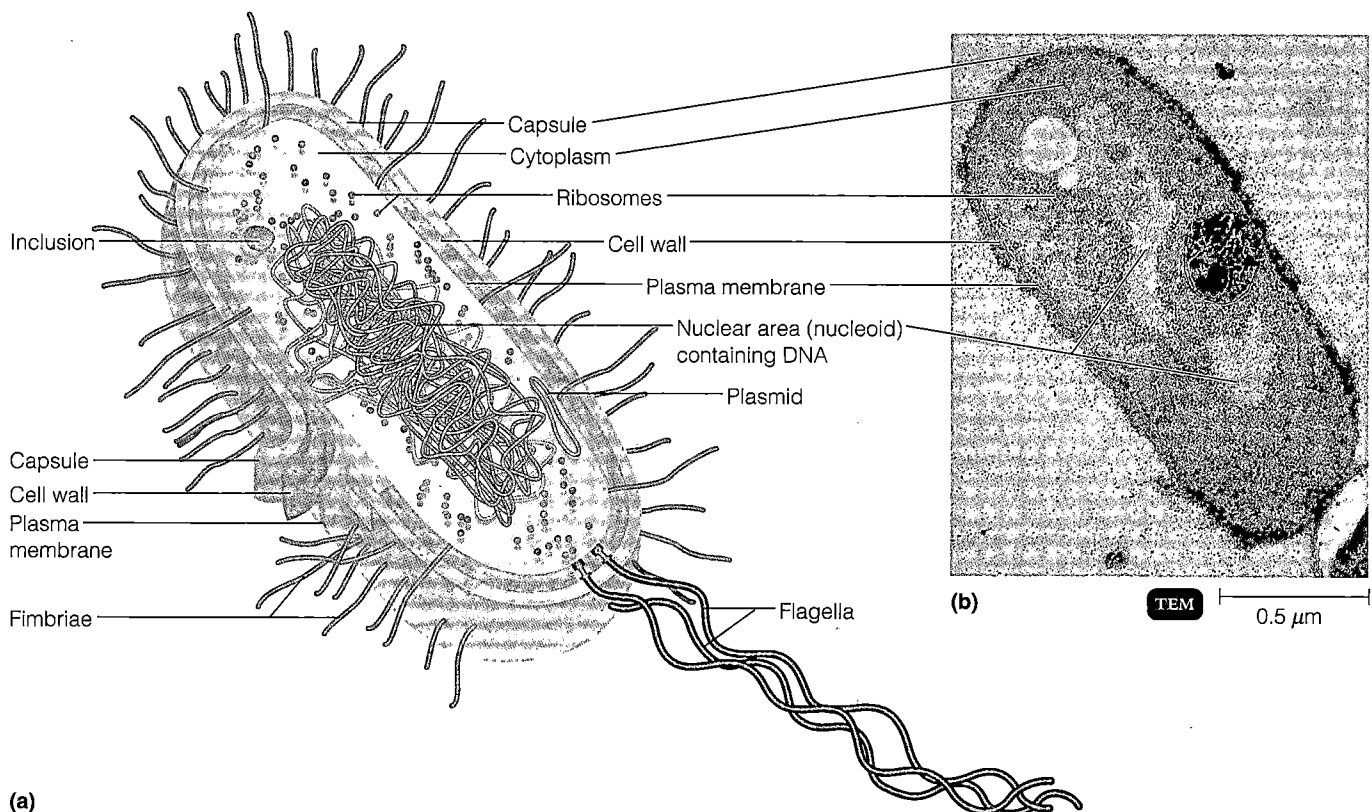

(a)

FIGURE 4.5 A prokaryotic cell showing typical structures. Both the drawing **(a)** and micrograph **(b)** show a bacterium lengthwise to reveal the internal composition.

■ Prokaryotic cells are distinguished from eukaryotic cells primarily by the structure of the cell wall and membranes and the absence of organelles.

Are Bacteria Multicellular?

The idea that bacteria are unicellular has been challenged by a number of biologists who claim that bacterial cells do not act as unicellular organisms when they are growing in a colony. Instead, the cells interact and exhibit multicellular organization. That is, cells in a colony are not identical; they differentiate, resulting in some cells having different structures and functions from their neighbor cells. These researchers cite a number of examples that are leading them to believe bacteria have multicellular organization.

STAPHYLOCOCCUS

Researchers at the University of Montana found that the source of recurring septicemia in a patient was a colony of *Staphylococcus aureus* growing on his pacemaker. The colony was resistant to penicillin because it was protected by a slime layer. However, individual cells that broke off from the colony were sensitive to penicillin. Cells in the colony were acting differently than cells growing independently.

BACILLUS

Bacillus cells that inherit certain genes form the highly organized chains shown in Figure a. After cell division, cell sepa-

ration is inhibited, so each daughter cell remains attached at the division septum. The resulting chain of cells twists and bends until it forms a helically twisted fiber. Researchers at the University of Arizona and Cambridge University speculate that the folded structure is more organized than a moving chain and therefore more stable.

PROTEUS

One look at a swarming *Proteus* colony suggests that all the cells in the colony are not alike (Figure b). The cells differentiate to produce normal, 2-μm-long proteal cells with 6–10 flagella and 40-μm swarmer cells with thousands of flagella. The swarmer cells move to the periphery of the colony and for several hours cause expansion of the colony by moving out over the agar. Then they undifferentiate back to normal proteal cells, and new swarmer cells form. Individual swarmer cells, removed from the colony, are nonmotile, indicating the cells in the colony are communicating to cause the colony to expand.

MYXOBACTERIA

Myxobacteria are found in decaying organic material and fresh water throughout the world. Although they

are bacteria, many myxobacteria never exist as individual cells. *Myxococcus xanthus* appears to hunt in packs. In its natural aqueous habitat, *M. xanthus* cells form spherical colonies that surround prey bacteria, where they can secrete digestive enzymes and absorb the nutrients. On solid substrates, other myxobacterial cells glide over a solid surface, leaving slime trails that are followed by other cells. When food is scarce, the cells aggregate to form a mass. Cells within the mass differentiate into a fruiting body consisting of a slime stalk and clusters of spores, as shown in Figure c.

★ ★ ★

It is clear that a complex series of events occurs between bacterial cells, perhaps mediated by extracellular signal molecules like those in multicellular eukaryotes. Do bacteria form multicellular colonies, or do individual cells communicate to act together? Either way, it may be time to reevaluate the 100-year-old view that bacteria are *simply* unicellular organisms.

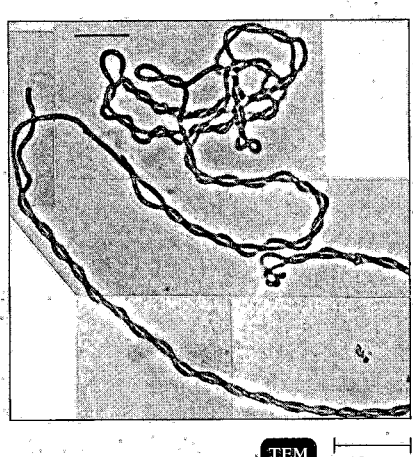

(a) A double-stranded helix formed by *Bacillus subtilis*.

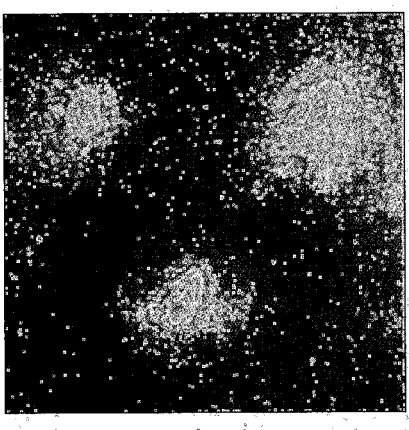

(b) A swarming colony of *Proteus mirabilis*.

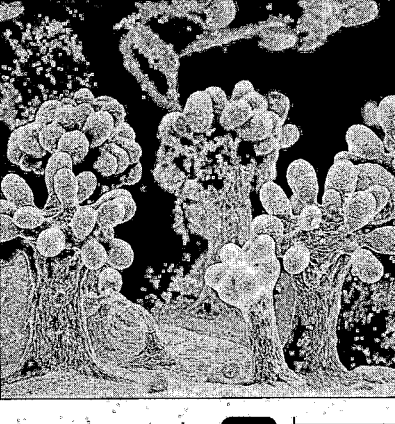

(c) A fruiting body of a myxobacterium.

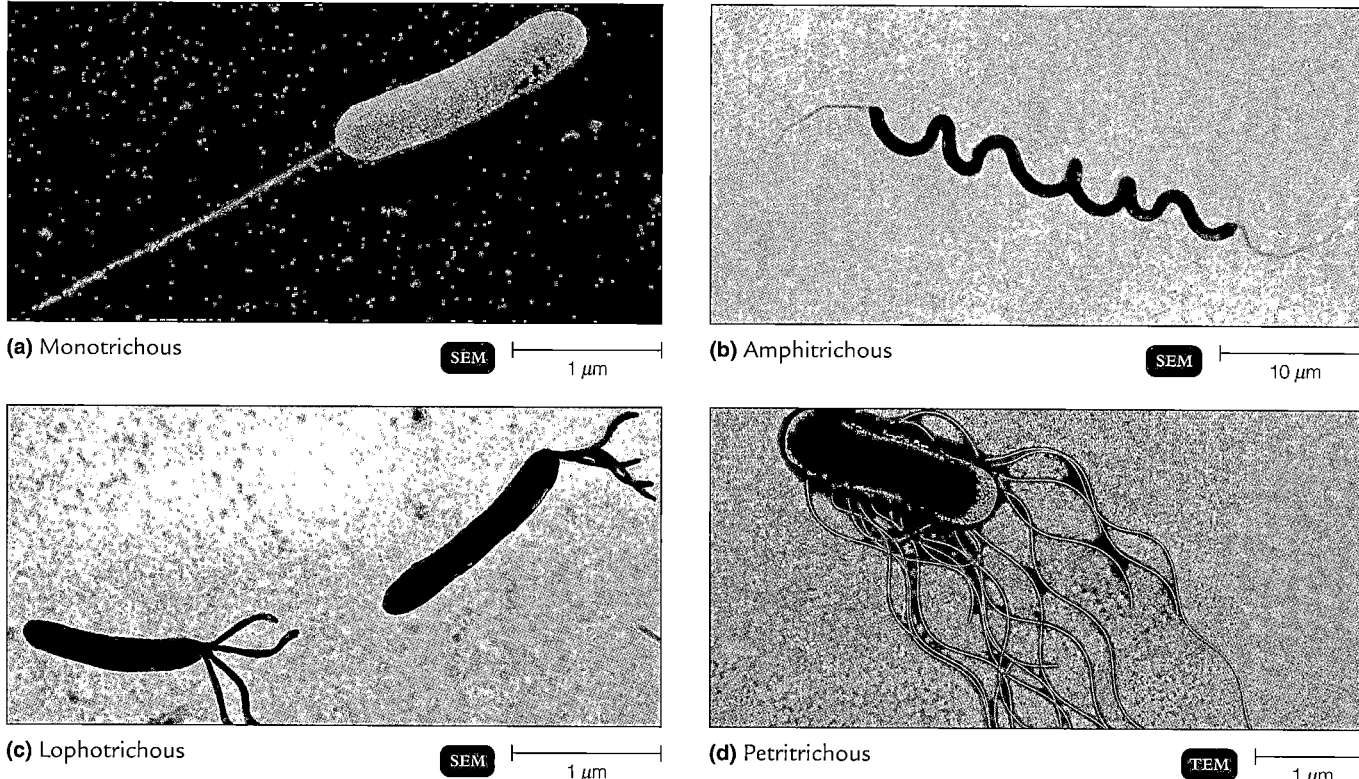

(a) Monotrichous   SEM   1 μm

(b) Amphitrichous   SEM   10 μm

(c) Lophotrichous   SEM   1 μm

(d) Petritrichous   TEM   1 μm

FIGURE 4.6   Four basic arrangements of bacterial flagella.

Glycocalyx

Glycocalyx (meaning sugar coat) is the general term used for substances that surround cells. The bacterial glycocalyx is a viscous (sticky), gelatinous polymer that is external to the cell wall and composed of polysaccharide, polypeptide, or both. Its chemical composition varies widely with the species. For the most part, it is made inside the cell and excreted to the cell surface. If the substance is organized and is firmly attached to the cell wall, the glycocalyx is described as a **capsule.**

The presence of a capsule can be determined by using negative staining, described in Chapter 3 (see Figure 3.12a on page 72). If the substance is unorganized and only loosely attached to the cell wall, the glycocalyx is described as a **slime layer.**

In certain species, capsules are important in contributing to bacterial virulence (the degree to which a pathogen causes disease). Capsules often protect pathogenic bacteria from phagocytosis by the cells of the host. For example, *Bacillus anthracis,* which produces a capsule of D-glutamic acid. (Recall from Chapter 2 that the D forms of amino acids are unusual.) Because only encapsulated *B. anthracis* causes anthrax, it is speculated that the capsule may prevent its being destroyed by phagocytosis.

Another example involves *Streptococcus pneumoniae*

(strep-tō-kok'kus nü-mō'nē-ī) causes pneumonia only when the cells are protected by a polysaccharide capsule. Unencapsulated *S. pneumoniae* cells cannot cause pneumonia and are readily phagocytized. The polysaccharide capsule of *Klebsiella* (kleb-sē-el'lä) also prevents phagocytosis and allows the bacterium to adhere to and colonize the respiratory tract. A glycocalyx made of sugars is called an **extracellular polysaccharide (EPS).** The EPS enables a bacterium to attach to various surfaces in its natural environment in order to survive. Through attachment, bacteria can grow on diverse surfaces such as rocks in fast-moving streams, plant roots, human teeth, medical implants, water pipes, and even other bacteria. *Streptococcus mutans* (mū'tans), an important cause of dental caries, attaches itself to the surface of teeth by a glycocalyx. *S. mutans* may use its capsule as a source of nutrition by breaking it down and utilizing the sugars when energy stores are low. A glycocalyx can protect a cell against dehydration. Also, its viscosity may inhibit the movement of nutrients out of the cell.

Flagella

Some prokaryotic cells have **flagella** (singular: *flagellum,* meaning whip), which are long filamentous appendages that propel bacteria (see Figure 4.6).

Bacterial cells have four arrangements of flagella (Figure 4.6): **monotrichous** (a single polar flagellum), **amphitrichous** (a tuft of flagella at each end of the cell), **lophotrichous** (two or more flagella at one pole of the cell), and **peritrichous** (flagella distributed over the entire cell).

A flagellum has three basic parts (Figure 4.7). The long outermost region, the *filament,* is constant in diameter and contains the globular (roughly spherical) protein *flagellin* arranged in several chains that intertwine and form a helix around a hollow core. Certain pathogenic bacteria can be identified by their flagellar proteins. In most bacteria, filaments are not covered by a membrane or sheath, as in eukaryotic cells. The filament is attached to a slightly wider *hook,* consisting of a different protein. The third portion of a flagellum is the *basal body,* which anchors the flagellum to the cell wall and plasma membrane.

The basal body is composed of a small central rod inserted into a series of rings. Gram-negative bacteria contain two pairs of rings; the outer pair of rings is anchored to various portions of the cell wall, and the inner pair of rings is anchored to the plasma membrane. In gram-positive bacteria, only the inner pair is present. As you will see later, the flagella (and cilia) of eukaryotic cells are more complex than those of prokaryotic cells.

Bacteria with flagella are motile; that is, they have the ability to move on their own. Each prokaryotic flagellum is a semirigid, helical structure that moves the cell by rotating from the basal body. The rotation of a flagellum is either clockwise or counterclockwise around its long axis. (Eukaryotic flagella, by contrast, undulate in a wavelike motion.) The movement of a prokaryotic flagellum results from rotation of its basal body and is similar to the movement of the shaft of an electric motor. As the flagella rotate, they form a bundle that pushes against the surrounding liquid and propels the bacterium. Although the mechanochemical basis for this biological "motor" is not completely understood, we know that it depends on the cell's continuous generation of energy.

Bacterial cells can alter the speed and direction of rotation of flagella and thus are capable of various patterns of **motility,** the ability of an organism to move by itself. When a bacterium moves in one direction for a length of time, the movement is called a "run" or "swim." "Runs" are interrupted by periodic, abrupt, random changes in direction called "tumbles." Then, a "run" resumes. "Tumbles" are caused by a reversal of flagellar rotation (Figure 4.8a). Some species of bacteria endowed with many flagella—*Proteus* (prō'tē-us), for example (Figure 4.8b on page 84 and the box on page 81)—can "swarm," or show rapid wavelike movement across a solid culture medium.

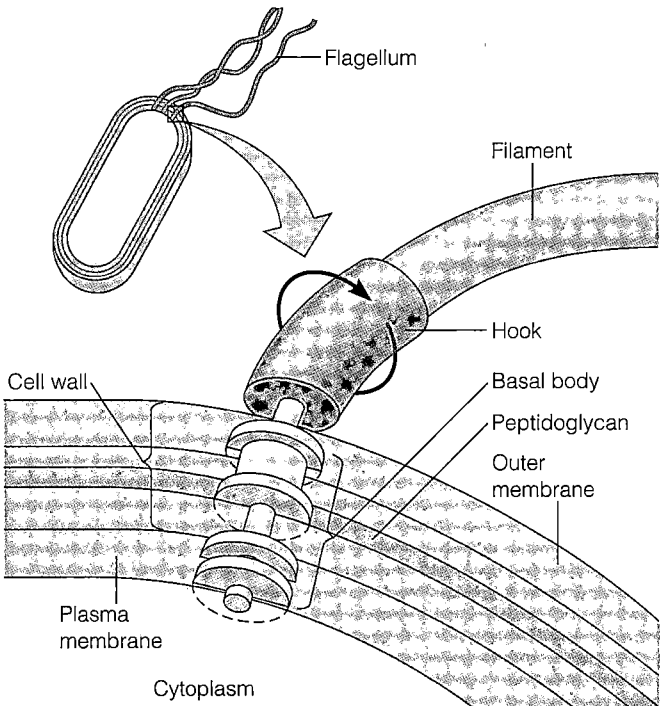

FIGURE 4.7 The structure of a prokaryotic flagellum. The parts and attachment of a flagellum of a gram-negative bacterium are shown in this highly schematic diagram.

■ What are the three basic parts of a flagellum?

One advantage of motility is that it enables a bacterium to move toward a favorable environment or away from an adverse one. The movement of a bacterium toward or away from a particular stimulus is called **taxis.** Such stimuli include chemicals **(chemotaxis)** and light **(phototaxis).** Motile bacteria contain receptors in various locations, such as in or just under the cell wall. These receptors pick up chemical stimuli, such as oxygen, ribose, and galactose. In response to the stimuli, information is passed to the flagella. If the chemotactic signal is positive, called an *attractant,* the bacteria move toward the stimulus with many runs and few tumbles. If the chemotactic signal is negative, called a *repellent,* the frequency of tumbles increases as the bacteria move away from the stimulus.

The flagellar protein called H antigen is useful for distinguishing among **serovars,** or variations within a species, of gram-negative bacteria. For example, there are at least 50 different H antigens for *E. coli.* Those serovars identified as *E. coli* O157:H7 are associated with foodborne epidemics (see Chapter 1, page 20).

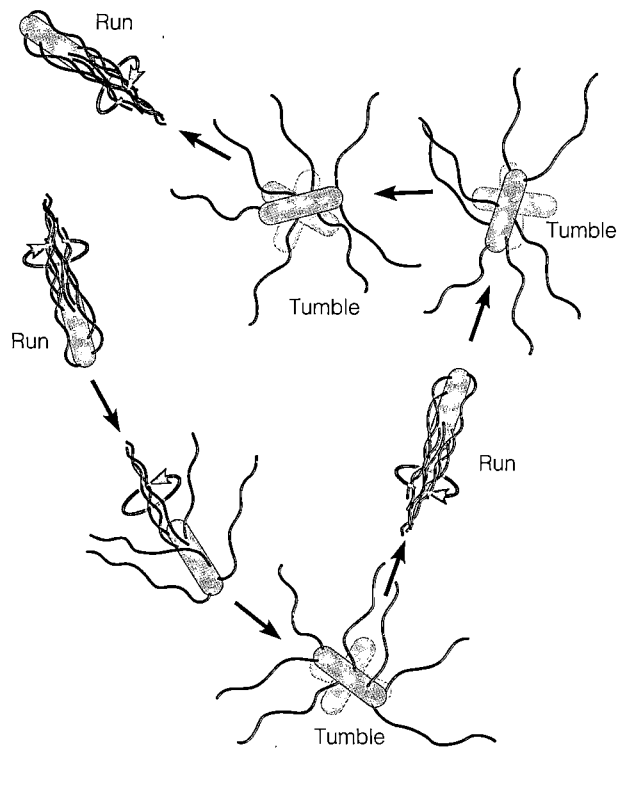

(a)

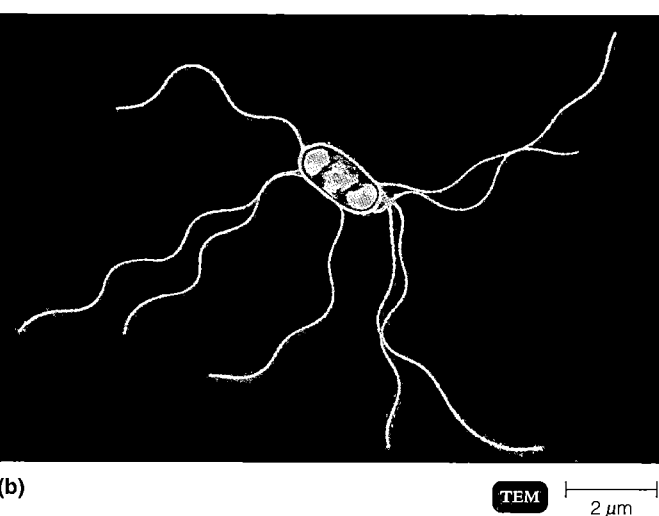

(b)

TEM ⊢———⊣ 2 μm

FIGURE 4.8 Flagella and bacterial motility. (a) A bacterium running and tumbling. Notice that the direction of flagellar rotation determines which of these movements occurs. (b) *Proteus* cells in the swarming stage may have more than 1000 peritrichous flagella.

■ Motility is the ability of microbes to move on their own.

Axial Filaments

Spirochetes are a group of bacteria that have unique structure and motility. One of the best-known spirochetes is *Treponema pallidum,* the causative agent of syphilis. Another spirochete is *Borrelia burgdorferi,* the causative agent of Lyme disease. Spirochetes move by means of **axial filaments,** or **endoflagella,** bundles of fibrils that arise at the ends of the cell beneath an outer sheath and spiral around the cell (Figure 4.9).

Axial filaments, which are anchored at one end of the spirochete, have a structure similar to that of flagella. The rotation of the filaments produces a movement of the outer sheath that propels the spirochetes in a spiral motion. This type of movement is similar to the way a corkscrew moves through a cork. This corkscrew motion probably enables a bacterium such as *T. pallidum* to move effectively through body fluids.

Fimbriae and Pili

Many gram-negative bacteria contain hairlike appendages that are shorter, straighter, and thinner than flagella and are used for attachment rather than for motility.

These structures, which consist of a protein called *pilin* arranged helically around a central core, are divided into two types, fimbriae and pili, having very different functions. (Some microbiologists use the two terms interchangeably to refer to all such structures, but we distinguish between them.)

Fimbriae (singular: *fimbria*) can occur at the poles of the bacterial cell, or they can be evenly distributed over the entire surface of the cell. They can number anywhere from a few to several hundred per cell (Figure 4.10). Like the glycocalyx, fimbriae enable a cell to adhere to surfaces, including the surfaces of other cells. For example, fimbriae attached to the bacterium *Neisseria gonorrhoeae* (nī-se'rē-ä go-nôr-rē'ī), the causative agent of gonorrhea, help the microbe colonize mucous membranes. Once colonization occurs, the bacteria can cause disease. When fimbriae are absent (because of genetic mutation), colonization cannot happen, and no disease ensues.

Pili (singular: *pilus*) are usually longer than fimbriae and number only one or two per cell. Pili join bacterial cells in preparation for the transfer of DNA from one cell to another. For this reason, they are sometimes also called **sex pili** (see Chapter 8).

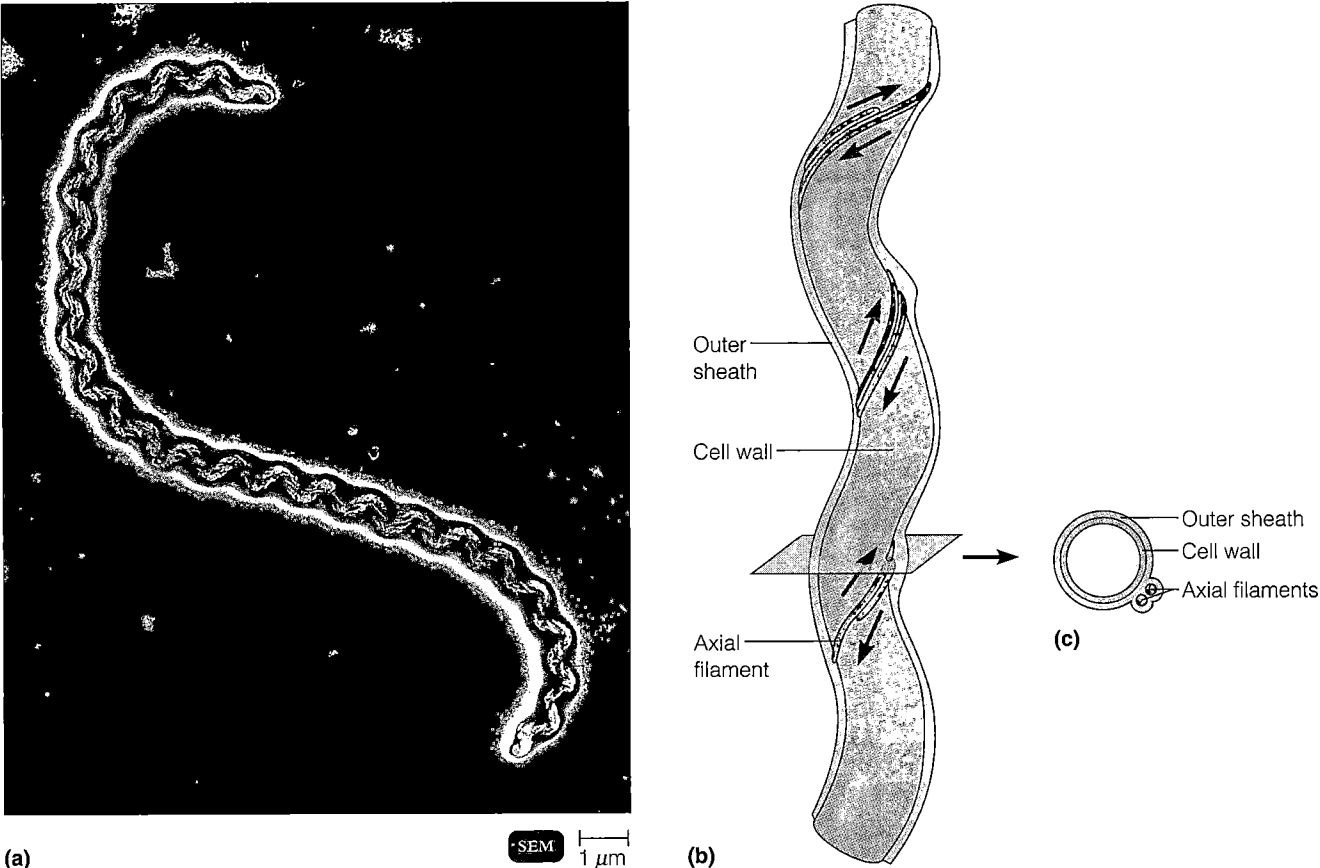

(a)

(c)

Outer sheath
Cell wall
Axial filaments

Outer sheath
Cell wall
Axial filament

(b)

FIGURE 4.9 Axial filaments. (a) A photomicrograph of the spirochete *Leptospira*, showing an axial filament. **(b)** A diagram of axial filaments wrapping around part of a spirochete. **(c)** A cross-sectional diagram of the spirochete, showing the position of axial filaments.

■ How do spirochetes and spirilla differ?

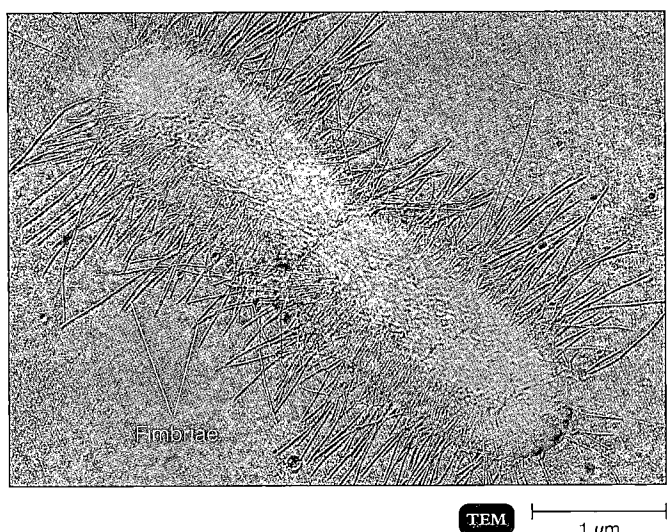

Fimbriae

FIGURE 4.10 Fimbriae. The fimbriae seem to bristle from this *E. coli* cell, which is beginning to divide.

■ Fimbriae enable bacterial cells to adhere to surfaces.

The Cell Wall

Learning Objectives

- *Compare and contrast the cell walls of gram-positive bacteria, gram-negative bacteria, archaea, and mycoplasmas.*
- *Differentiate between protoplast and spheroplast.*

The **cell wall** of the bacterial cell is a complex, semi-rigid structure responsible for the shape of the cell. The cell wall surrounds the underlying, fragile plasma (cytoplasmic) membrane and protects it and the interior of the cell from adverse changes in the outside environment (see Figure 4.5). Almost all prokaryotes have cell walls.

The major function of the cell wall is to prevent bacterial cells from rupturing when the water pressure inside the cell is greater than that outside the cell. It also helps maintain the shape of a bacterium and serves as a point of anchorage for flagella. As the volume of a bacterial cell

N-acetylglucosamine   N-acetylmuramic
(NAG)   acid (NAM)

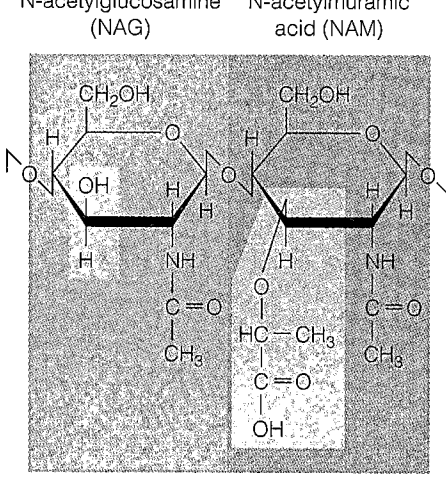

**FIGURE 4.11 N-acetylglucosamine (NAG) and N-acetylmu-
ramic acid (NAM) joined as in peptidoglycan.** The light purple
areas show the differences between the two molecules. The linkage between them is called a β-1,4 linkage.

increases, its plasma membrane and cell wall extend as
needed. Clinically, the cell wall is important because it
contributes to the ability of some species to cause disease
and is the site of action of some antibiotics. In addition,
the chemical composition of the cell wall is used to differentiate major types of bacteria.

Although the cells of some eukaryotes, including
plants, algae, and fungi, have cell walls, their walls differ
chemically from those of prokaryotes, are simpler in
structure, and are less rigid.

Composition and Characteristics

The bacterial cell wall is composed of a macromolecular
network called **peptidoglycan** (also known as *murein*),
which is present either alone or in combination with
other substances. Peptidoglycan consists of a repeating
disaccharide attached by polypeptides to form a lattice
that surrounds and protects the entire cell. The disaccharide portion is made up of monosaccharides called
N-acetylglucosamine (NAG) and N-acetylmuramic acid
(NAM) (from *murus*, meaning wall), which are related to
glucose. The structural formulas for NAG and NAM are
shown in Figure 4.11.

The various components of peptidoglycan are assembled in the cell wall (Figure 4.12a). Alternating NAM and
NAG molecules are linked in rows of 10 to 65 sugars to
form a carbohydrate "backbone" (the glycan portion of
peptidoglycan). Adjacent rows are linked by **polypeptides** (the peptide portion of peptidoglycan). Although
the structure of the polypeptide link varies, it always in-

cludes *tetrapeptide side chains,* which consist of four amino
acids attached to NAMs in the backbone. The amino
acids occur in an alternating pattern of D and L forms (see
Figure 2.14 on page 47). This is unique because the
amino acids found in other proteins are L forms. Parallel
tetrapeptide side chains may be directly bonded to each
other or linked by a *peptide cross-bridge,* consisting of a
short chain of amino acids.

Penicillin interferes with the final linking of the peptidoglycan rows by peptide cross-bridges (Figure 4.12a).
As a result, the cell wall is greatly weakened and the cell
undergoes **lysis,** destruction caused by rupture of the
plasma membrane and the loss of cytoplasm.

Gram-Positive Cell Walls

In most gram-positive bacteria, the cell wall consists of
many layers of peptidoglycan, forming a thick, rigid
structure (Figure 4.12b). By contrast, gram-negative cell
walls contain only a thin layer of peptidoglycan.

In addition, the cell walls of gram-positive bacteria
contain *teichoic acids,* which consist primarily of an alcohol (such as glycerol or ribitol) and phosphate. There are
two classes of teichoic acids: *lipoteichoic acid,* which spans
the peptidoglycan layer and is linked to the plasma membrane, and *wall teichoic acid,* which is linked to the peptidoglycan layer. Because of their negative charge (from the
phosphate groups), teichoic acids may bind and regulate
the movement of cations (positive ions) into and out of
the cell. They may also assume a role in cell growth, preventing extensive wall breakdown and possible cell lysis.
Finally, teichoic acids provide much of the wall's antigenic specificity and thus make it possible to identify bacteria by certain laboratory tests (see Chapter 10).

The cell walls of gram-positive streptococci are covered with various polysaccharides that allow them to be
grouped into medically significant types. The cell walls of
acid-fast bacteria, such as *Mycobacterium,* consist of as
much as 60% mycolic acid, a waxy lipid, whereas the rest
is peptidoglycan. These bacteria can be stained with the
Gram stain and are considered gram-positive.

Gram-Negative Cell Walls

The cell walls of gram-negative bacteria consist of one or
a very few layers of peptidoglycan and an outer membrane (Figure 4.12c). The peptidoglycan is bonded to
lipoproteins (lipids covalently linked to proteins) in the
outer membrane and is in the *periplasm,* a fluid-filled
space between the outer membrane and the plasma
membrane. The periplasm contains a high concentration
of degradative enzymes and transport proteins. Gram-
negative cell walls do not contain teichoic acids. Because

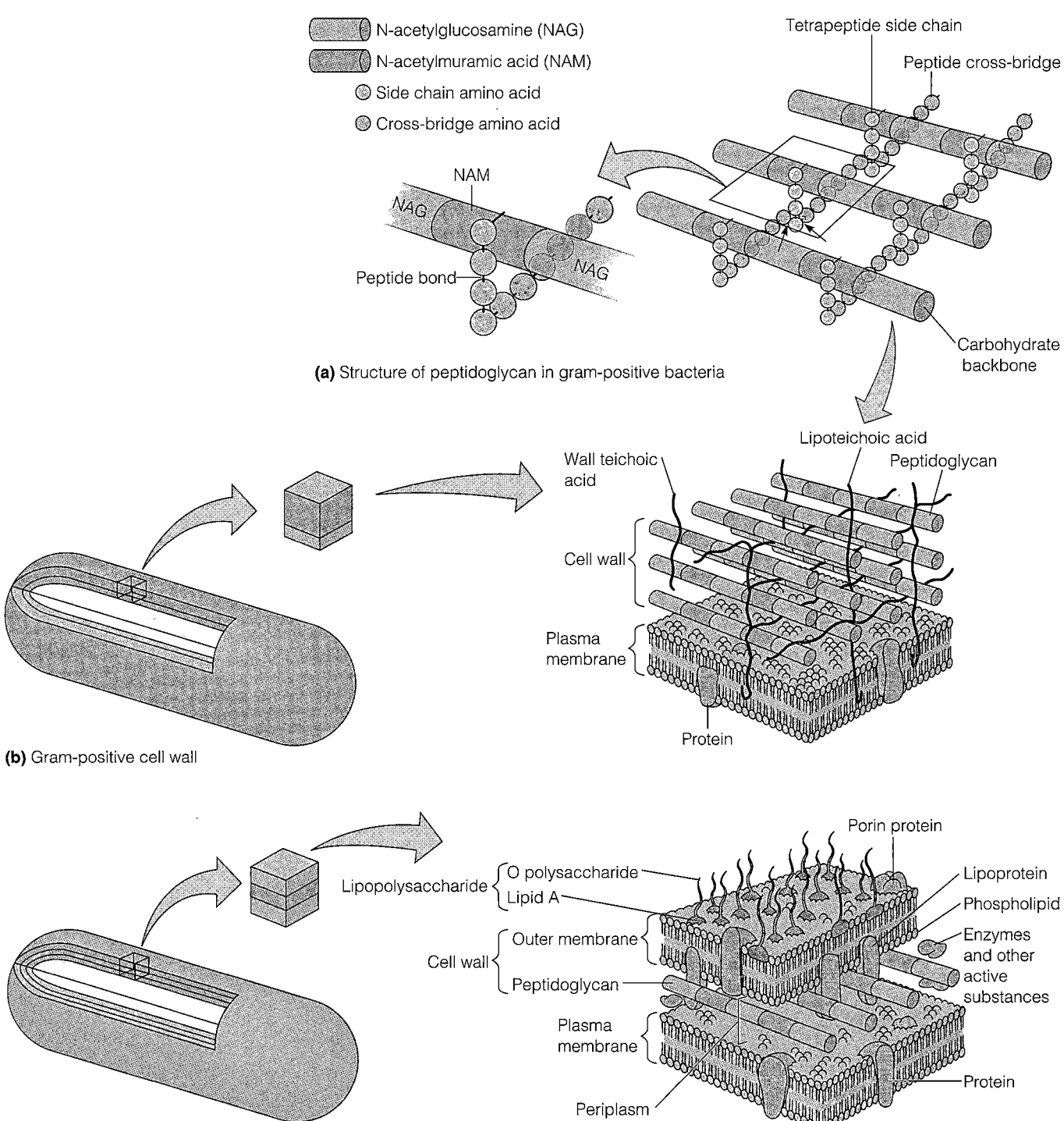

FIGURE 4.12 Bacterial cell walls. (a) The structure of peptidoglycan in gram-positive bacteria. Together the carbohydrate backbone (glycan portion) and tetrapeptide side chains (peptide portion) make up peptidoglycan. The frequency of peptide cross-bridges and the number of amino acids in these bridges vary with the species of bacterium. The small arrows indicate where penicillin interferes with the linkage of peptidoglycan rows by peptide cross-bridges. **(b)** A gram-positive cell wall. **(c)** A gram-negative cell wall.

▓ **What are the major structural differences between gram-positive and gram-negative cell walls?**

the cell walls of gram-negative bacteria contain only a small amount of peptidoglycan, they are more susceptible to mechanical breakage.

The *outer membrane* of the gram-negative cell consists of lipopolysaccharides (LPS), lipoproteins, and phospholipids (see Figure 4.12c). The outer membrane has several specialized functions. Its strong negative charge is an important factor in evading phagocytosis and the actions of complement (lyses cells and promotes phagocytosis), two components of the defenses of the host (discussed in detail in Chapter 16). The outer membrane also provides a barrier to certain antibiotics (for example, penicillin), digestive enzymes such as lysozyme, detergents, heavy metals, bile salts, and certain dyes. The box in Chapter 7 (page 207) illustrates the importance of the outer membrane in health care.

However, the outer membrane does not provide a barrier to all substances in the environment because nutrients must pass through to sustain the metabolism of the cell. Part of the permeability of the outer membrane is due to proteins in the membrane, called **porins,** that form channels. Porins permit the passage of molecules such as nucleotides, disaccharides, peptides, amino acids, vitamin B_{12}, and iron.

The LPS component of the outer membrane provides two important characteristics of gram-negative bacteria. First, the polysaccharide portion is composed of sugars, called O *polysaccharides,* that function as antigens and are useful for distinguishing species of gram-negative bacteria. For example, the foodborne pathogen *E. coli* O157:H7 is distinguished from other serovars by certain laboratory tests that test for the specific antigens. This role is comparable to that of teichoic acids in gram-positive cells. Second, the lipid portion of the lipopolysaccharide, called *lipid A,* is referred to as *endotoxin* and is toxic when in the host's bloodstream or gastrointestinal tract. It causes fever and shock. The nature and importance of endotoxins and other bacterial toxins will be discussed in Chapter 15.

Cell Walls and the Gram Stain Mechanism

Now that you have studied the Gram stain (in Chapter 3, page 69) and the chemistry of the bacterial cell wall (in the previous section), it is easier to understand the mechanism of the Gram stain. The mechanism is based on differences in the structure of the cell walls of gram-positive and gram-negative bacteria and how each reacts to the various reagents (substances used for producing a chemical reaction). Crystal violet, the primary stain, stains both gram-positive and gram-negative cells purple because the dye enters the cytoplasm of both types of cells. When iodine (the mordant) is applied, it forms large crystals with the dye that are too large to escape through the cell wall. The ap-

plication of alcohol dehydrates the peptidoglycan of gram-positive cells to make it more impermeable to the crystal violet-iodine. The effect on gram-negative cells is quite different; alcohol dissolves the outer membrane of gram-negative cells and even leaves small holes in the thin peptidoglycan layer through which crystal violet-iodine diffuse. Because gram-negative bacteria are colorless after the alcohol wash, the addition of safranin (the counterstain) turns the cells pink.

In any population of cells, some gram-positive cells will give a gram-negative response. These cells are usually dead. However, there are a few gram-positive genera that show an increasing number of gram-negative cells as the culture ages. *Bacillus, Clostridium,* and *Mycobacterium* are the most notable of these, and are often described as *gram-variable.*

A comparison of some of the characteristics of gram-positive and gram-negative bacteria is presented in Table 4.1.

Atypical Cell Walls

Among prokaryotes, certain types of cells have no walls or have very little wall material. These include members of the genus *Mycoplasma* (mī-kō-plaz′mä) and related organisms. Mycoplasmas are the smallest known bacteria that can grow and reproduce outside living host cells. Because of their size and because they have no cell walls, they pass through most bacterial filters and were first mistaken for viruses. Their plasma membranes are unique among bacteria in having lipids called *sterols,* which are thought to help protect them from lysis (rupture).

Archaea may lack walls or may have unusual walls composed of polysaccharides and proteins but not peptidoglycan. These walls do, however, contain a substance similar to peptidoglycan called *pseudomurein.* Pseudomurein contains N-acetyltalosaminuronic acid instead of NAM and lacks the D amino acids found in bacterial cell walls. A representative of the archaea is described in the box in Chapter 5 (page 144).

Damage to the Cell Wall

Chemicals that damage bacterial cell walls, or interfere with their synthesis, often do not harm the cells of an animal host because the bacterial cell wall is made of chemicals unlike those in eukaryotic cells. Thus, cell wall synthesis is the target for some antimicrobial drugs. One way the cell wall can be damaged is by exposure to the digestive enzyme *lysozyme.* This enzyme occurs naturally in some eukaryotic cells and is a constituent of tears, mucus, and saliva. Lysozyme is particularly active on the major cell wall components of most gram-positive bacteria, making them vulnerable to lysis. Lysozyme catalyzes hy-

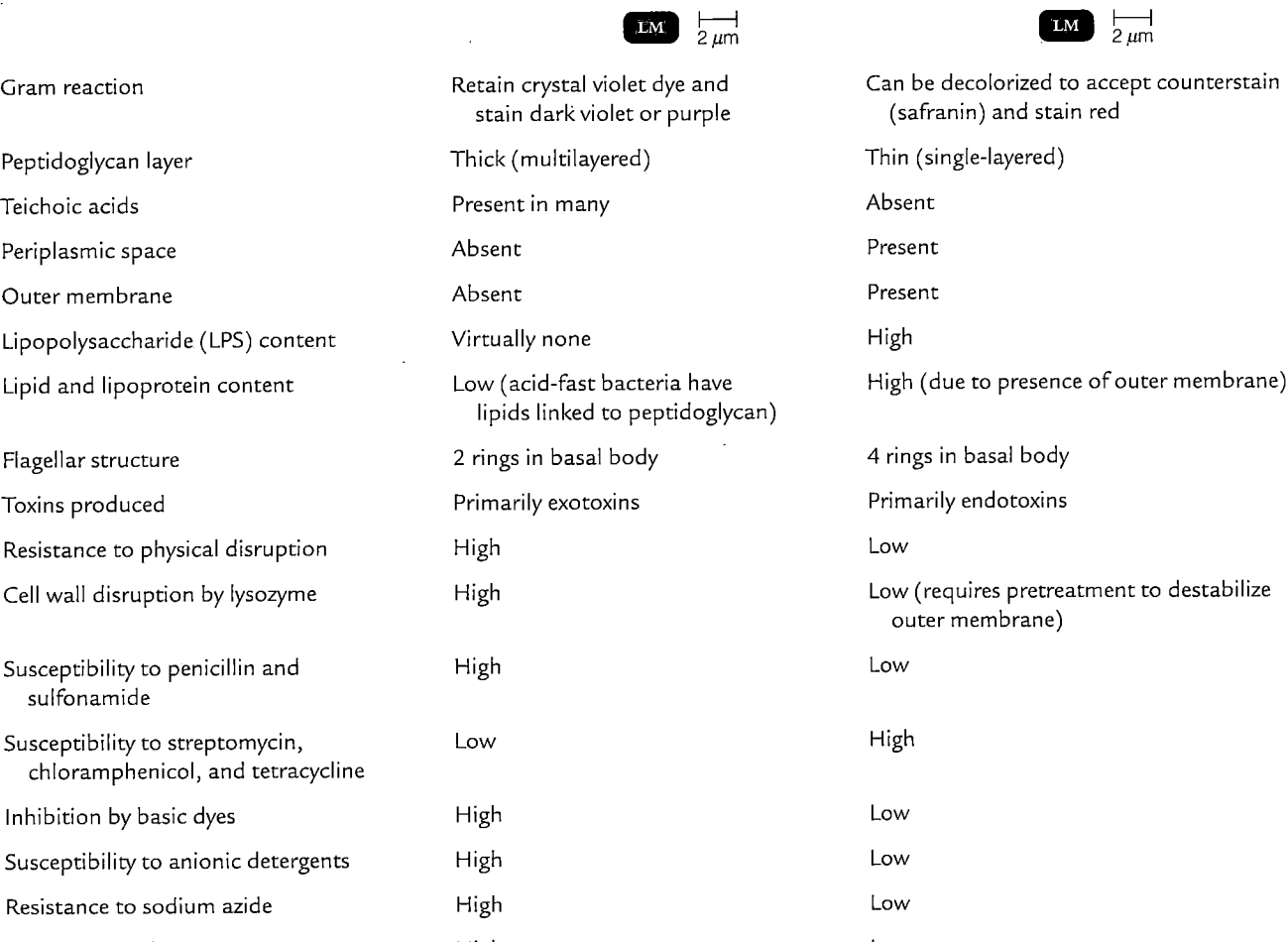

table 4.1 **Some Comparative Characteristics of Gram-Positive and Gram-Negative Bacteria**

Characteristic	Gram-Positive	Gram-Negative
Gram reaction	Retain crystal violet dye and stain dark violet or purple	Can be decolorized to accept counterstain (safranin) and stain red
Peptidoglycan layer	Thick (multilayered)	Thin (single-layered)
Teichoic acids	Present in many	Absent
Periplasmic space	Absent	Present
Outer membrane	Absent	Present
Lipopolysaccharide (LPS) content	Virtually none	High
Lipid and lipoprotein content	Low (acid-fast bacteria have lipids linked to peptidoglycan)	High (due to presence of outer membrane)
Flagellar structure	2 rings in basal body	4 rings in basal body
Toxins produced	Primarily exotoxins	Primarily endotoxins
Resistance to physical disruption	High	Low
Cell wall disruption by lysozyme	High	Low (requires pretreatment to destabilize outer membrane)
Susceptibility to penicillin and sulfonamide	High	Low
Susceptibility to streptomycin, chloramphenicol, and tetracycline	Low	High
Inhibition by basic dyes	High	Low
Susceptibility to anionic detergents	High	Low
Resistance to sodium azide	High	Low
Resistance to drying	High	Low

drolysis of the bonds between the sugars in the repeating disaccharide "backbone" of peptidoglycan. This act is analogous to cutting the steel supports of a bridge with a cutting torch: The gram-positive cell wall is almost completely destroyed by lysozyme. The cellular contents that remain surrounded by the plasma membrane may remain intact if lysis does not occur; this wall-less cell is termed a **protoplast.** Typically, a protoplast is spherical and is still capable of carrying on metabolism.

When lysozyme is applied to gram-negative cells, usually the wall is not destroyed to the same extent as in gram-positive cells; some of the outer membrane also re-

mains. In this case, the cellular contents, plasma membrane, and remaining outer wall layer are called a **spheroplast,** also a spherical structure. For lysozyme to exert its effect on gram-negative cells, the cells are first treated with ethylenediaminetetraacetic acid (EDTA). EDTA weakens ionic bonds in the outer membrane and thereby damages it, giving the lysozyme access to the peptidoglycan layer.

Protoplasts and spheroplasts burst in pure water or very dilute salt or sugar solutions because the water molecules from the surrounding fluid rapidly move into and enlarge the cell, which has a much lower internal concentration of water. This rupturing is called **osmotic lysis** and will be discussed in detail shortly.

As noted earlier, certain antibiotics, such as penicillin, destroy bacteria by interfering with the formation of the peptide cross-bridges of peptidoglycan, thus preventing the formation of a functional cell wall. Most gram-negative bacteria are not as susceptible to penicillin as gram-positive bacteria are because the outer membrane of gram-negative bacteria forms a barrier that inhibits the entry of this and other substances, and gram-negative bacteria have fewer peptide cross-bridges. However, gram-negative bacteria are quite susceptible to some β-lactam antibiotics that penetrate the outer membrane better than penicillin. Antibiotics will be discussed in more detail in Chapter 20.

Structures Internal to the Cell Wall

Thus far, we have discussed the prokaryotic cell wall and structures external to it. We will now look inside the prokaryotic cell and discuss the structures and functions of the plasma membrane and components within the cytoplasm of the cell.

The Plasma (Cytoplasmic) Membrane

Learning Objectives

- *Describe the structure, chemistry, and functions of the prokaryotic plasma membrane.*
- *Define simple diffusion, facilitated diffusion, osmosis, active transport, and group translocation.*

The **plasma (cytoplasmic) membrane** (or *inner membrane*) is a thin structure lying inside the cell wall and enclosing the cytoplasm of the cell (see Figure 4.5). The plasma membrane of prokaryotes consists primarily of phospholipids (see Figure 2.11 on page 44), which are the most abundant chemicals in the membrane, and proteins. Eukaryotic plasma membranes also contain carbohydrates and sterols, such as cholesterol. Because they lack sterols, prokaryotic plasma membranes are less rigid than eukaryotic membranes. One exception is the wall-less prokaryote *Mycoplasma,* which contains membrane sterols.

Structure

In electron micrographs, prokaryotic and eukaryotic plasma membranes (and the outer membranes of gram-negative bacteria) look like two-layered structures; there are two dark lines with a light space between the lines (Figure 4.13a). The phospholipid molecules are arranged in two parallel rows, called a *phospholipid bilayer* (Figure 4.13b). As introduced in Chapter 2, each phospholipid molecule contains a polar head, composed of a phosphate group and glycerol that is hydrophilic (water-loving) and soluble in water, and nonpolar tails, composed of fatty acids that are hydrophobic (water-fearing) and insoluble in water (Figure 4.13c). The polar heads are on the two surfaces of the phospholipid bilayer, and the nonpolar tails are in the interior of the bilayer.

The protein molecules in the membrane can be arranged in a variety of ways. Some, called *peripheral proteins,* are easily removed from the membrane by mild treatments and lie at the inner or outer surface of the membrane. They may function as enzymes that catalyze chemical reactions, as a "scaffold" for support, and as mediators of changes in membrane shape during movement. Other proteins, called *integral proteins,* can be removed from the membrane only after disrupting the bilayer (by using detergents, for example). Some integral proteins are believed to penetrate the membrane completely and are called transmembrane proteins. Some integral proteins are channels that have a pore or hole through which substances enter and exit the cell.

Studies have demonstrated that the phospholipid and protein molecules in membranes are not static but move quite freely within the membrane surface. This movement is most probably associated with the many functions performed by the plasma membrane. Because the fatty acid tails cling together, phospholipids in the presence of water form a self-sealing bilayer, with the result that breaks and tears in the membrane will heal themselves. The membrane must be about as viscous as olive oil to allow membrane proteins to move freely enough to perform their functions without destroying the structure of the membrane. This dynamic arrangement of phospholipids and proteins is referred to as the **fluid mosaic model.**

Functions

The most important function of the plasma membrane is to serve as a selective barrier through which materials enter and exit the cell. In this function, plasma membranes have **selective permeability** (sometimes called *semipermeability*). This term indicates that certain molecules and ions pass through the membrane, but others are prevented from passing through it. The permeability of the membrane depends on several factors. Large molecules (such as proteins) cannot pass through the plasma membrane, possibly because these molecules are larger than the pores in integral proteins that function as channels. But smaller molecules (such as water, oxygen, carbon dioxide, and some simple sugars) usually pass through easily. Ions penetrate the membrane very slowly. Substances that dissolve easily in lipids (such as oxygen, carbon dioxide, and non-

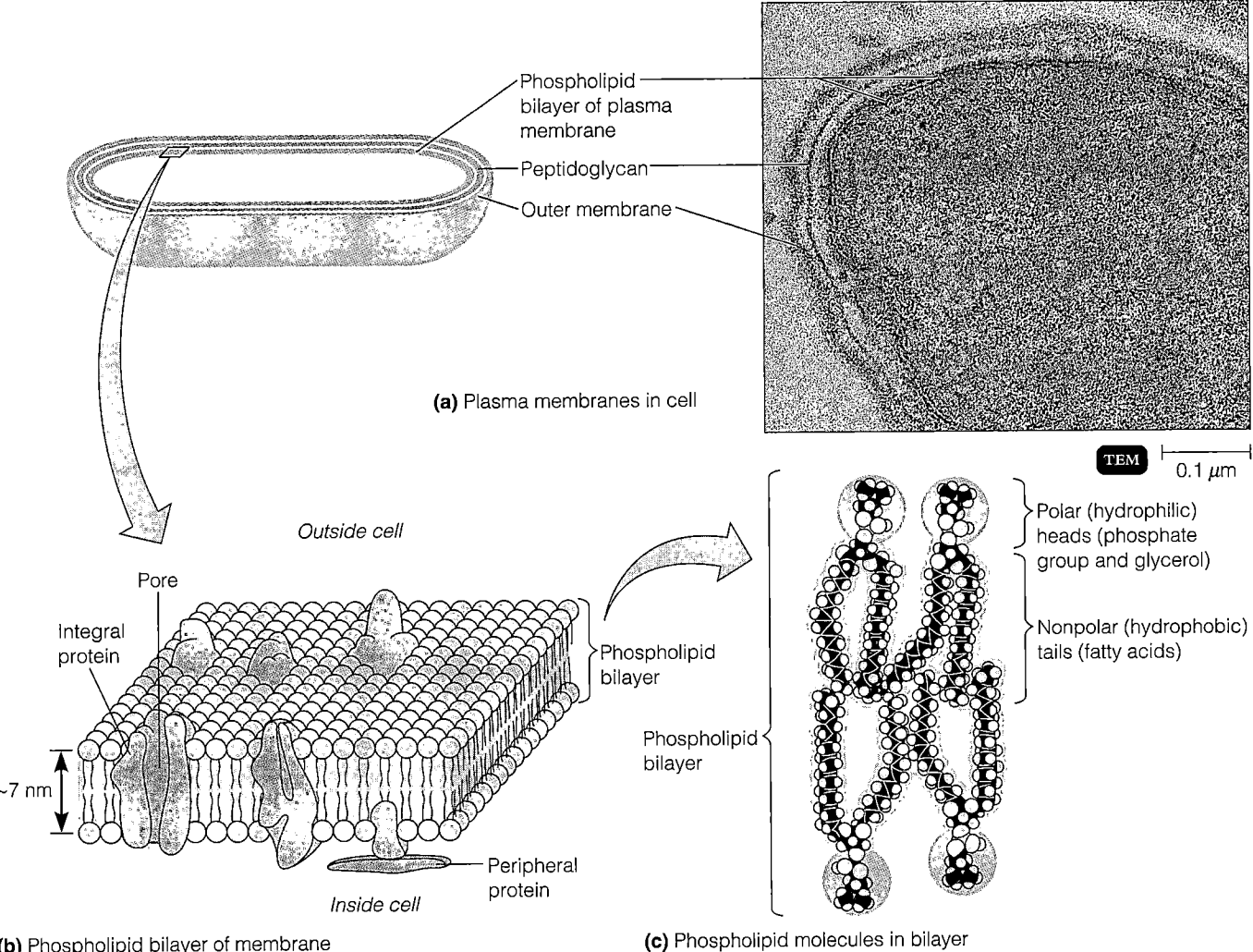

(a) Plasma membranes in cell

TEM |⎯⎯⎯⎯|
     0.1 μm

Outside cell

Pore

Integral protein

~7 nm

Inside cell

Peripheral protein

(b) Phospholipid bilayer of membrane

Phospholipid bilayer of plasma membrane

Peptidoglycan

Outer membrane

Phospholipid bilayer

Phospholipid bilayer

Polar (hydrophilic) heads (phosphate group and glycerol)

Nonpolar (hydrophobic) tails (fatty acids)

(c) Phospholipid molecules in bilayer

FIGURE 4.13 Plasma membrane. (a) A diagram and micrograph showing the phospholipid bilayer forming the inner plasma membrane of the gram-negative bacterium *Aquaspirillum serpens*. Layers of the cell wall, including the outer membrane, can be seen outside the inner membrane. **(b)** A portion of the inner membrane showing the phospholipid bilayer and proteins. The outer membrane of gram-negative bacteria is also a phospholipid bilayer. **(c)** Space-filling models of several molecules as they are arranged in the phospholipid bilayer.

■ **What is the action of polymyxins on plasma membranes?**

polar organic molecules) enter and exit more easily than other substances because the membrane consists mostly of phospholipids. The movement of materials across plasma membranes also depends on transporter molecules, which will be described shortly.

Plasma membranes are also important to the breakdown of nutrients and the production of energy. The plasma membranes of bacteria contain enzymes capable of catalyzing the chemical reactions that break down nutrients and produce ATP. In some bacteria, pigments and enzymes involved in photosynthesis are found in infoldings of the

plasma membrane that extend into the cytoplasm. These membranous structures are called **chromatophores** or **thylakoids** (Figure 4.14).

When viewed with an electron microscope, bacterial plasma membranes often appear to contain one or more large, irregular folds called **mesosomes.** Many functions have been proposed for mesosomes. However, it is now known that they are artifacts, not true cell structures. Mesosomes are believed to be folds in the plasma membrane that develop by the process used for preparing specimens for electron microscopy.

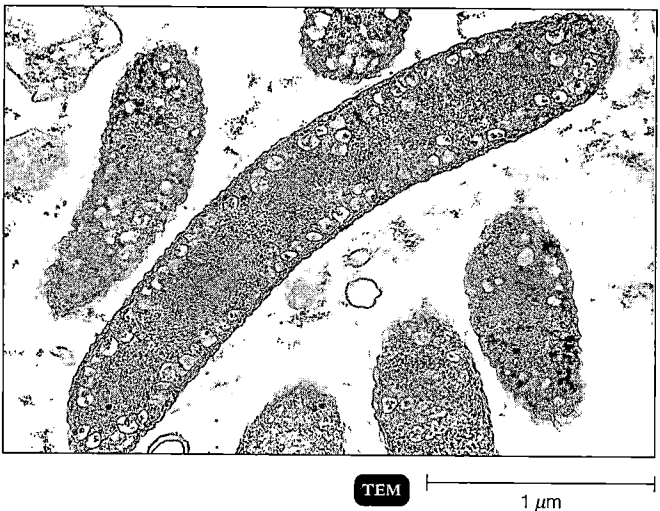

FIGURE 4.14 **Chromatophores.** In this micrograph of *Rhodospirillum rubrum,* a purple (nonsulfur) bacterium, the chromatophores are visible.

■ Chromatophores are sites of photosynthesis.

Destruction of the Plasma Membrane by Antimicrobial Agents

Because the plasma membrane is vital to the bacterial cell, it is not surprising that several antimicrobial agents exert their effects at this site. In addition to the chemicals that damage the cell wall and thereby indirectly expose the membrane to injury, many compounds specifically damage plasma membranes. These compounds include certain alcohols and quaternary ammonium compounds, which are used as disinfectants. By disrupting the membrane's phospholipids, a group of antibiotics known as the *polymyxins* cause leakage of intracellular contents and subsequent cell death. This mechanism will be discussed in Chapter 20.

The Movement of Materials Across Membranes

Materials move across plasma membranes of both prokaryotic and eukaryotic cells by two kinds of processes: passive and active. In *passive processes,* substances cross the membrane from an area of high concentration to an area of low concentration (move with the concentration gradient, or difference), without any expenditure of energy (ATP) by the cell. In *active processes,* the cell must use energy (ATP) to move substances from areas of low concentration to areas of high concentration (against the concentration gradient).

Passive Processes Passive processes include simple diffusion, facilitated diffusion, and osmosis.

Simple diffusion is the net (overall) movement of molecules or ions from an area of high concentration to

an area of low concentration (Figure 4.15). The movement continues until the molecules or ions are evenly distributed. The point of even distribution is called *equilibrium.* Cells rely on simple diffusion to transport certain small molecules, such as oxygen and carbon dioxide, across their cell membranes.

In **facilitated diffusion,** the substance (glucose, for example) to be transported combines with a plasma membrane protein called a *transporter.* In one proposed mechanism for facilitated diffusion, transporters (sometimes called *permeases*) bind a substance on one side of the membrane and, by changing shape, move it to the other side of the membrane, where it is released (Figure 4.16). Facilitated diffusion is similar to simple diffusion in that the cell does not need to expend energy because the substance moves from a high to a low concentration. The process differs from simple diffusion in its use of transporters.

In some cases, molecules that bacteria need are too large to be transported into the cells by these methods. Most bacteria, however, produce enzymes that can break down large molecules into simpler ones (such as proteins into amino acids, or polysaccharides into simple sugars). Such enzymes, which are released by the bacteria into the surrounding medium, are appropriately called *extracellular enzymes.* Once the enzymes degrade the large molecules, the subunits move into the cell with the help of transporters. For example, specific carriers retrieve DNA bases, such as the purine guanine, from extracellular media and bring them into the cell's cytoplasm.

Osmosis is the net movement of solvent molecules across a selectively permeable membrane from an area with a high concentration of solvent molecules to an area of low concentration. In living systems, the chief solvent is water.

Osmosis may be demonstrated with the apparatus shown in Figure 4.17 on page 94. A sack constructed from cellophane, which is a selectively permeable membrane, is filled with a solution of 20% sucrose (table sugar). The cellophane sack is placed into a beaker containing distilled water. Initially, the concentrations of water on either side of the membrane are different. Because of the sucrose molecules, the concentration of water is lower inside the cellophane sack. Therefore, water moves from the beaker (where its concentration is higher) into the cellophane sack (where its concentration is lower).

There is no movement of sugar out of the cellophane sack into the beaker, however, because the cellophane is impermeable to molecules of sugar—the sugar molecules are too large to go through the pores of the membrane. As water moves into the cellophane sack, the sugar solution becomes increasingly dilute, and, because the cellophane sack has expanded to its limit as a result of an

(a)

(b)

FIGURE 4.15 The principle of simple diffusion. (a) After a dye pellet is put into a beaker of water, the molecules of dye in the pellet diffuse into the water from an area of high dye concentration to areas of low dye concentration. **(b)** The dye potassium permanganate in the process of diffusing.

▨ In passive processes such as simple diffusion, no energy (ATP) is expended by a cell.

increased volume of water, water begins to move up the glass tube. In time, the water that has accumulated in the cellophane sack and the glass tube exerts a downward pressure that forces water molecules out of the cellophane sack and back into the beaker. This movement of water through a selectively permeable membrane produces a pressure called osmotic pressure. **Osmotic pressure** is the pressure required to prevent the movement of pure water (water with no solutes) into a solution containing some solutes. In other words, osmotic pressure is the pressure needed to stop the flow of water across the selectively permeable membrane (cellophane). When water molecules leave and enter the cellophane sack at the same rate, equilibrium is reached.

A bacterial cell may be subjected to any of three kinds of osmotic solutions: isotonic, hypotonic, or hypertonic. An **isotonic (isosmotic) solution** is a medium in which the overall concentration of solutes equals that found inside a cell (*iso* means equal). Water leaves and enters the cell at the same rate (no net change); the cell's contents are in equilibrium with the solution outside the cell wall (Figure 4.17c).

Earlier we mentioned that lysozyme and certain antibiotics damage bacterial cell walls, causing the cells to rupture, or lyse. Such rupturing occurs because bacterial cytoplasm usually contains such a high concentration of solutes that, when the wall is weakened or removed, additional water enters the cell by osmosis. The damaged (or removed) cell wall cannot constrain the swelling of the cytoplasmic membrane, and the membrane bursts. This is an example of osmotic lysis caused by immersion in a hypotonic solution. A **hypotonic (hypoosmotic) solution** outside the cell is a medium whose concentration of solutes is lower than that inside the cell (*hypo* means under or less). Most bacteria live in hypotonic solutions, and swelling is contained by the cell wall. Cells with weak cell walls, such as gram-negative bacteria, may burst or undergo osmotic lysis as a result of excessive water intake (Figure 4.17d).

A **hypertonic (hyperosmotic) solution** is a medium having a higher concentration of solutes than inside the cell has (*hyper* means above or more). Most bacterial cells placed in a hypertonic solution shrink and collapse or plasmolyze because water leaves the cells by osmosis (Figure 4.17e). Keep in mind that the terms *isotonic, hypotonic,* and *hypertonic*

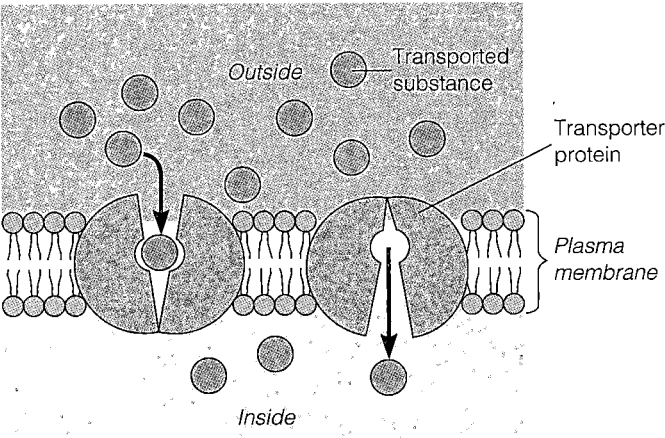

FIGURE 4.16 Facilitated diffusion. Transporter proteins in the plasma membrane transport molecules across the membrane from an area of high concentration to one of low concentration (with the concentration gradient). The transporter undergoes a change in shape to transport the substance. The process does not require ATP.

▨ How does simple diffusion differ from facilitated diffusion?

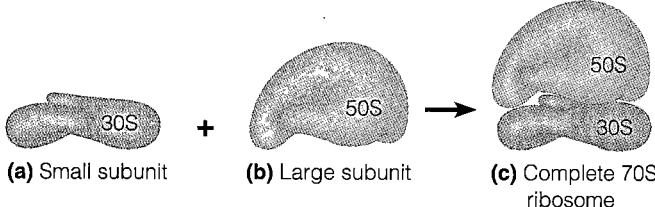

(a) Small subunit   **(b)** Large subunit   **(c)** Complete 70S ribosome

FIGURE 4.18 The prokaryotic ribosome. (a) A small 30S subunit and **(b)** a large 50S subunit make up **(c)** the complete 70S prokaryotic ribosome.

■ **What is the importance of the differences between prokaryotic and eukaryotic ribosomes with regard to antibiotic therapy?**

during ultra–high-speed centrifugation. Sedimentation rate is a function of the size, weight, and shape of a particle. The subunits of a 70S ribosome are a small 30S subunit containing one molecule of rRNA and a larger 50S subunit containing two molecules of rRNA.

Several antibiotics work by inhibiting protein synthesis on prokaryotic ribosomes. Antibiotics such as streptomycin and gentamicin attach to the 30S subunit and interfere with protein synthesis. Other antibiotics, such as erythromycin and chloramphenicol, interfere with protein synthesis by attaching to the 50S subunit. Because of differences in prokaryotic and eukaryotic ribosomes, the microbial cell can be killed by the antibiotic while the eukaryotic host cell remains unaffected.

Inclusions

Within the cytoplasm of prokaryotic cells are several kinds of reserve deposits, known as **inclusions.** Cells may accumulate certain nutrients when they are plentiful and use them when the environment is deficient. Evidence suggests that macromolecules concentrated in inclusions avoid the increase in osmotic pressure that would result if the molecules were dispersed in the cytoplasm. Some inclusions are common to a wide variety of bacteria, whereas others are limited to a small number of species and therefore serve as a basis for identification.

Metachromatic Granules

Metachromatic granules are large inclusions that take their name from the fact that they sometimes stain red with certain blue dyes such as methylene blue. Collectively they are known as **volutin.** Volutin represents a reserve of inorganic phosphate (polyphosphate) that can be used in the synthesis of ATP. It is generally formed by cells that grow in phosphate-rich environments. Metachromatic granules are found in algae, fungi, and protozoa, as well as in bacteria. These granules are characteristic of *Corynebacterium diphtheriae* (kô-rī-nē-bak-tī'rē-um dif-thi'rē-ĭ), the causative agent of diphtheria; thus, they have diagnostic significance.

Polysaccharide Granules

Inclusions known as **polysaccharide granules** typically consist of glycogen and starch, and their presence can be demonstrated when iodine is applied to the cells. In the presence of iodine, glycogen granules appear reddish brown and starch granules appear blue.

Lipid Inclusions

Lipid inclusions appear in various species of *Mycobacterium, Bacillus, Azotobacter* (ä-zō-tō-bak'tėr), *Spirillum* (spī-ril'lum), and other genera. A common lipid-storage material, one unique to bacteria, is the polymer *poly-β-hydroxybutyric acid.* Lipid inclusions are revealed by staining cells with fat-soluble dyes, such as Sudan dyes.

Sulfur Granules

Certain bacteria—for example, the "sulfur bacteria" that belong to the genus *Thiobacillus*—derive energy by oxidizing sulfur and sulfur-containing compounds. These bacteria may deposit **sulfur granules** in the cell, where they serve as an energy reserve.

Carboxysomes

Carboxysomes are inclusions that contain the enzyme ribulose 1,5-diphosphate carboxylase. Bacteria that use carbon dioxide as their sole source of carbon require this enzyme for carbon dioxide fixation during photosynthesis. Among the bacteria containing carboxysomes are nitrifying bacteria, cyanobacteria, and thiobacilli.

Gas Vacuoles

Hollow cavities found in many aquatic prokaryotes, including cyanobacteria, anoxygenic photosynthetic bacteria, and halobacteria are called **gas vacuoles.** Each vacuole consists of rows of several individual *gas vesicles,* which are hollow cylinders covered by protein. Gas vacuoles maintain buoyancy so that the cells can remain at the depth in the water appropriate for them to receive sufficient amounts of oxygen, light, and nutrients.

Magnetosomes

Magnetosomes are inclusions of iron oxide (Fe_3O_4), formed by several gram-negative bacteria such as *Aquaspirillum magnetotacticum,* that act like magnets (Figure 4.19).

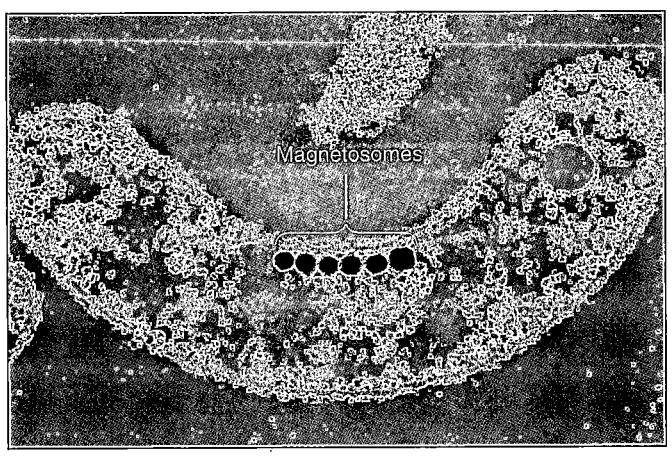

FIGURE 4.19 Magnetosomes. This micrograph of *Aquaspirillum magnetotacticum* shows a chain of magnetosomes. The outer membrane of the gram-negative wall is also visible.

■ **Magnetosomes are iron-oxide inclusions, formed by some gram-negative bacteria, that act like magnets.**

Bacteria may use magnetosomes to move downward until they reach a suitable attachment site. In vitro, magnetosomes can decompose hydrogen peroxide, which forms in cells in the presence of oxygen. Researchers speculate that magnetosomes may protect the cell against hydrogen peroxide accumulation. Industrial microbiologists are developing culture methods to obtain large quantities of magnetite from bacteria to use in the production of magnetic tapes for sound and data recording.

Endospores

Learning Objective

■ *Describe the functions of endospores, sporulation, and endospore germination.*

When essential nutrients are depleted, certain gram-positive bacteria, such as those of the genera *Clostridium* and *Bacillus,* form specialized "resting" cells called **endospores** (Figure 4.20). Unique to bacteria, endospores are highly durable dehydrated cells with thick walls and additional layers. They are formed internal to the bacterial cell membrane.

When released into the environment, they can survive extreme heat, lack of water, and exposure to many toxic chemicals and radiation. For example, 7500-year-old endospores of *Thermoactinomyces vulgaris* (thèr-mō-ak-tin-ō-mī'sēs vul-ga'ris) from the freezing muds of Elk Lake in Minnesota have germinated when rewarmed and placed in a nutrient medium, and 25-million-year-old en-

dospores trapped in amber are reported to have germinated when placed in nutrient media. Although true endospores are found in gram-positive bacteria, one gram-negative species, *Coxiella burnetii* (käks-ē-el'lä bèr-ne'tē-ē), the cause of Q fever, forms endosporelike structures that resist heat and chemicals and can be stained with endospore stains.

The process of endospore formation within a vegetative (parent) cell takes several hours and is known as **sporulation** or **sporogenesis** (see Figure 4.20a). It is not clear what nutrients actually trigger this process. In the first observable stage of sporulation, a newly replicated bacterial chromosome and a small portion of cytoplasm are isolated by an ingrowth of the plasma membrane called a *spore septum.* The spore septum becomes a double-layered membrane that surrounds the chromosome and cytoplasm. This structure, entirely enclosed within the original cell, is called a *forespore.* Thick layers of peptidoglycan are laid down between the two membrane layers. Then a thick *spore coat* of protein forms around the outside membrane. This coat is responsible for the resistance of endospores to many harsh chemicals.

The diameter of the endospore may be the same as, smaller than, or larger than the diameter of the vegetative cell. Depending on the species, the endospore might be located *terminally* (at one end), *subterminally* (near one end; Figure 4.20b), or *centrally* inside the vegetative cell. When the endospore matures, the vegetative cell wall ruptures (lyses), killing the cell, and the endospore is freed.

Most of the water present in the forespore cytoplasm is eliminated by the time sporulation is complete, and endospores do not carry out metabolic reactions. The highly dehydrated endospore core contains only DNA, small amounts of RNA, ribosomes, enzymes, and a few important small molecules. The latter include a strikingly large amount of an organic acid called *dipicolinic acid* (found in the cytoplasm), which is accompanied by a large number of calcium ions. These cellular components are essential for resuming metabolism later.

Endospores can remain dormant for thousands of years. An endospore returns to its vegetative state by a process called **germination.** Germination is triggered by physical or chemical damage to the endospore's coat. The endospore's enzymes then break down the extra layers surrounding the endospore, water enters, and metabolism resumes. Because one vegetative cell forms a single endospore, which, after germination, remains one cell, sporulation in bacteria is *not* a means of reproduction. This process does not increase the number of cells.

Endospores are important from a clinical viewpoint and in the food industry because they are resistant to processes that normally kill vegetative cells. Such

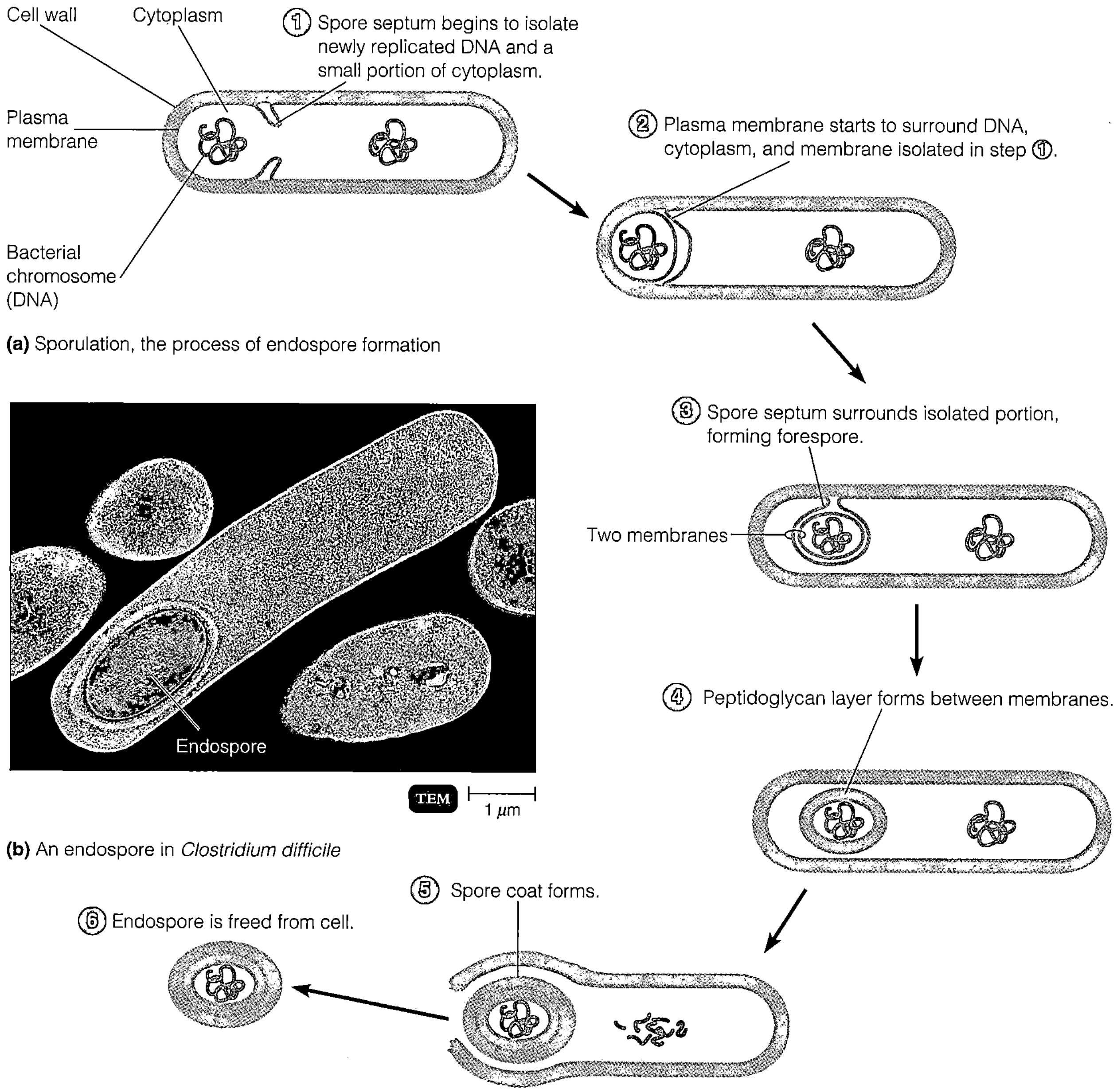

Cell wall   Cytoplasm   ① Spore septum begins to isolate
                                    newly replicated DNA and a
                                    small portion of cytoplasm.

Plasma
membrane

② Plasma membrane starts to surround DNA,
   cytoplasm, and membrane isolated in step ①.

Bacterial
chromosome
(DNA)

(a) Sporulation, the process of endospore formation

③ Spore septum surrounds isolated portion,
   forming forespore.

Two membranes

④ Peptidoglycan layer forms between membranes.

Endospore

TEM ├─────┤
     1 μm

(b) An endospore in *Clostridium difficile*

⑤ Spore coat forms.

⑥ Endospore is freed from cell.

FIGURE 4.20   Formation of endospores by sporulation.

■ Under what conditions are endospores formed by bacteria?

processes include heating, freezing, desiccation, use of chemicals, and radiation. Whereas most vegetative cells are killed by temperatures above 70°C, endospores can survive in boiling water for several hours or more. Endospores of thermophilic (heat-loving) bacteria can survive in boiling water for 19 hours. Endospore-forming bacteria are a problem in the food industry because they are likely to survive underprocessing, and if conditions for growth occur, some species produce toxins and disease. Special methods for controlling organisms that produce endospores are discussed in Chapter 7.

As noted in Chapter 3, endospores are difficult to stain for detection. Thus, a specially prepared stain must be used along with heat. (The Schaeffer-Fulton endospore stain is commonly used.)

★ ★ ★

Having examined the functional anatomy of the prokaryotic cell, we will now look at the functional anatomy of the eukaryotic cell.

table 4.2	*Principal Differences Between Prokaryotic and Eukaryotic Cells*	
Characteristic	**Prokaryotic**	**Eukaryotic**

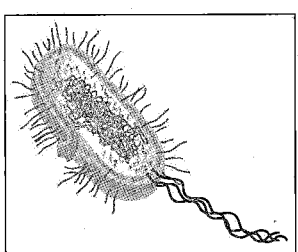 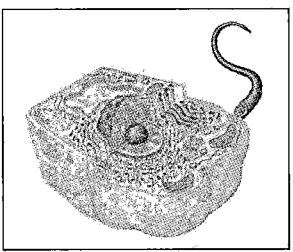

Size of cell	Typically 0.2–2.0 μm in diameter	Typically 10–100 μm in diameter
Nucleus	No nuclear membrane or nucleoli	True nucleus, consisting of nuclear membrane and nucleoli
Membrane-enclosed organelles	Absent	Present; examples include lysosomes, Golgi complex, endoplasmic reticulum, mitochondria, and chloroplasts
Flagella	Consist of two protein building blocks	Complex; consist of multiple microtubules
Glycocalyx	Present as a capsule or slime layer	Present in some cells that lack a cell wall
Cell wall	Usually present; chemically complex (typical bacterial cell wall includes peptidoglycan)	When present, chemically simple
Plasma membrane	No carbohydrates and generally lacks sterols	Sterols and carbohydrates that serve as receptors present
Cytoplasm	No cytoskeleton or cytoplasmic streaming	Cytoskeleton; cytoplasmic streaming
Ribosomes	Smaller size (70S)	Larger size (80S); smaller size (70S) in organelles
Chromosome (DNA) arrangement	Single circular chromosome; lacks histones	Multiple linear chromosomes with histones
Cell division	Binary fission	Mitosis
Sexual reproduction	No meiosis; transfer of DNA fragments only	Involves meiosis

THE EUKARYOTIC CELL

As mentioned earlier, eukaryotic organisms include algae, protozoa, fungi, higher plants, and animals. The eukaryotic cell is typically larger and structurally more complex than the prokaryotic cell (Figure 4.21). By comparing the structure of the prokaryotic cell in Figure 4.5 with that of the eukaryotic cell, the differences between the two types of cells become apparent. The principal differences between prokaryotic and eukaryotic cells are summarized in Table 4.2.

The following discussion of eukaryotic cells will parallel our discussion of prokaryotic cells by starting with structures that extend to the outside of the cell.

Flagella and Cilia

Learning Objective

■ *Differentiate between prokaryotic and eukaryotic flagella.*

Many types of eukaryotic cells have projections that are used for cellular locomotion or for moving substances along the surface of the cell. These projections contain cytoplasm and are enclosed by the plasma membrane. If the projections are few and are long in relation to the size of the cell, they are called **flagella.** If the projections are numerous and short, resembling hairs, they are called **cilia** (singular: *cilium*).

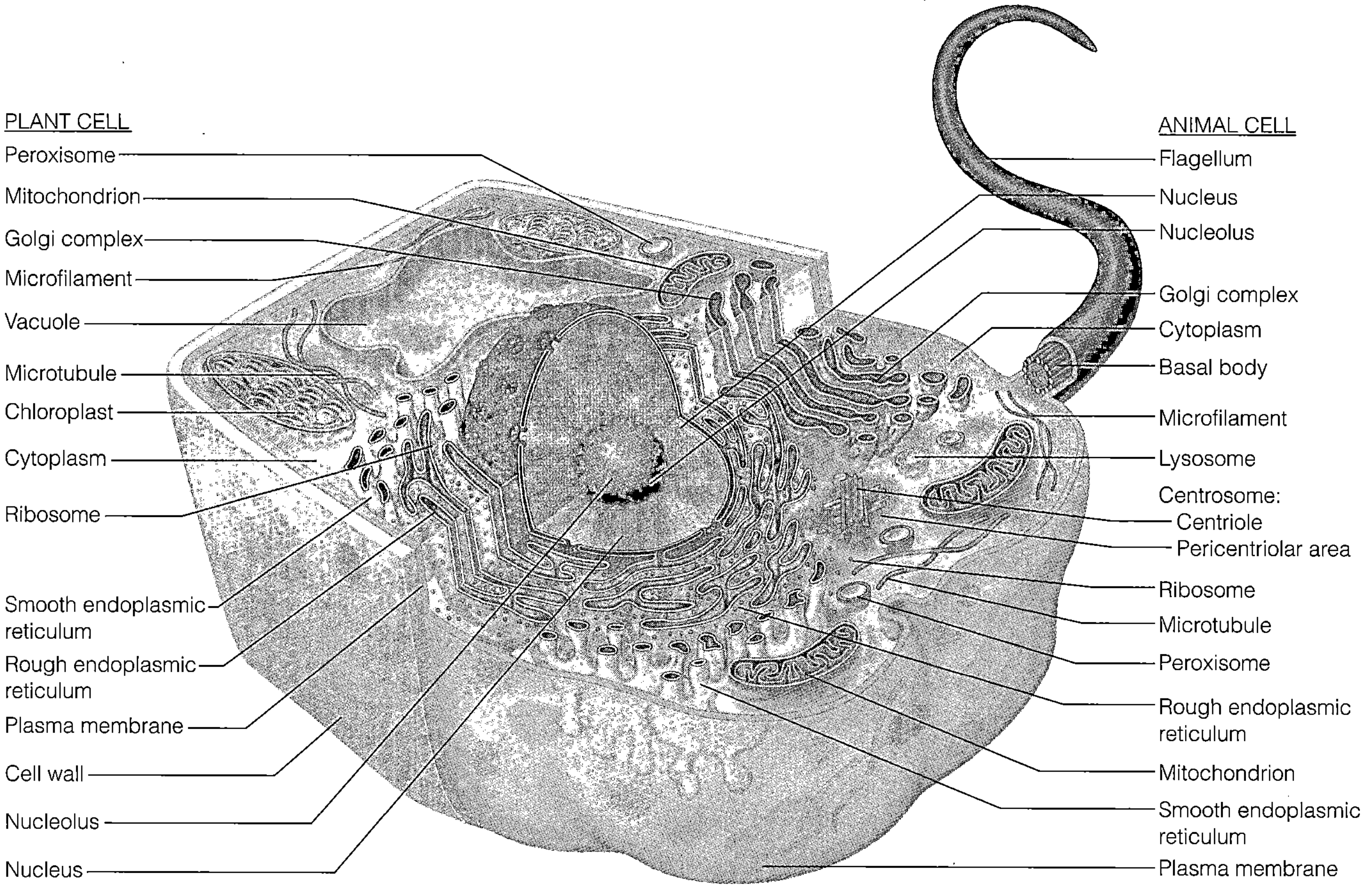

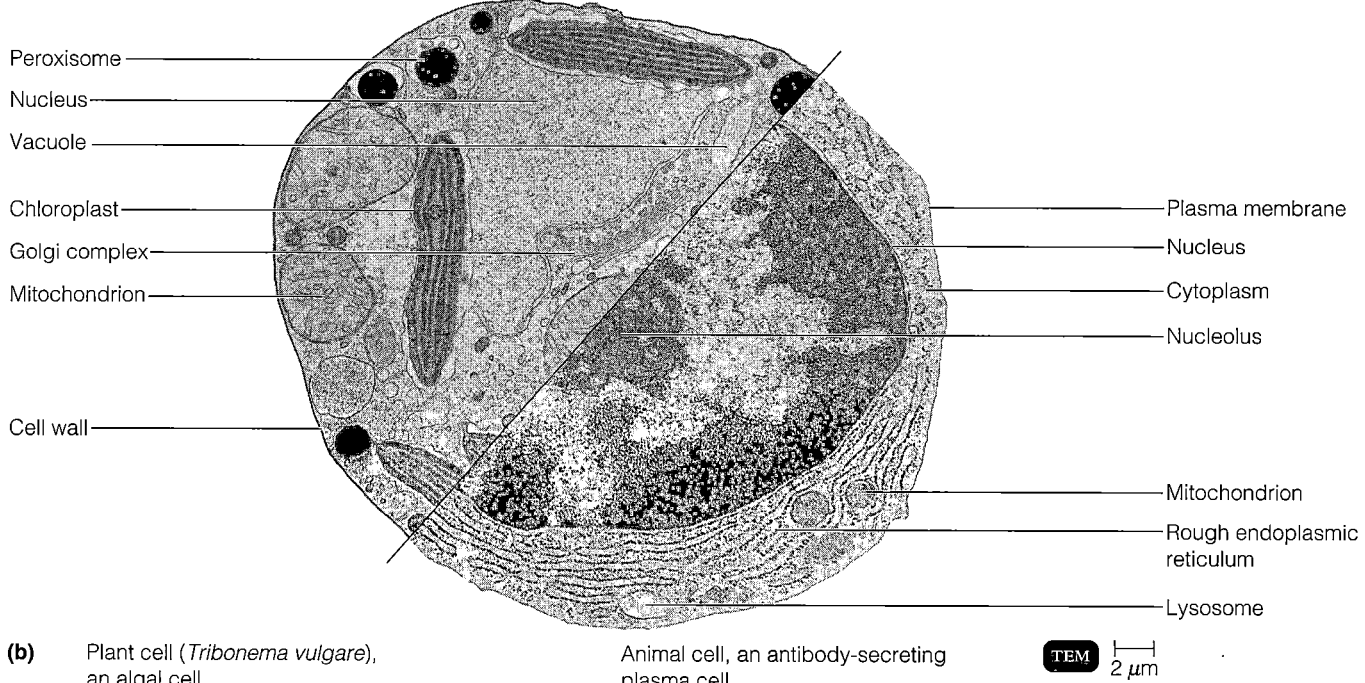

PLANT CELL
- Peroxisome
- Mitochondrion
- Golgi complex
- Microfilament
- Vacuole
- Microtubule
- Chloroplast
- Cytoplasm
- Ribosome
- Smooth endoplasmic reticulum
- Rough endoplasmic reticulum
- Plasma membrane
- Cell wall
- Nucleolus
- Nucleus

ANIMAL CELL
- Flagellum
- Nucleus
- Nucleolus
- Golgi complex
- Cytoplasm
- Basal body
- Microfilament
- Lysosome
- Centrosome:
  - Centriole
  - Pericentriolar area
- Ribosome
- Microtubule
- Peroxisome
- Rough endoplasmic reticulum
- Mitochondrion
- Smooth endoplasmic reticulum
- Plasma membrane

(a) Highly schematic diagram of a composite eukaryotic cell, half plant and half animal

- Peroxisome
- Nucleus
- Vacuole
- Chloroplast
- Golgi complex
- Mitochondrion
- Cell wall

- Plasma membrane
- Nucleus
- Cytoplasm
- Nucleolus
- Mitochondrion
- Rough endoplasmic reticulum
- Lysosome

(b) Plant cell (*Tribonema vulgare*), an algal cell   Animal cell, an antibody-secreting plasma cell   TEM   ⊢—⊣ 2 μm

FIGURE 4.21 Eukaryotic cells showing typical structures.

▓ What kingdoms contain eukaryotic organisms?

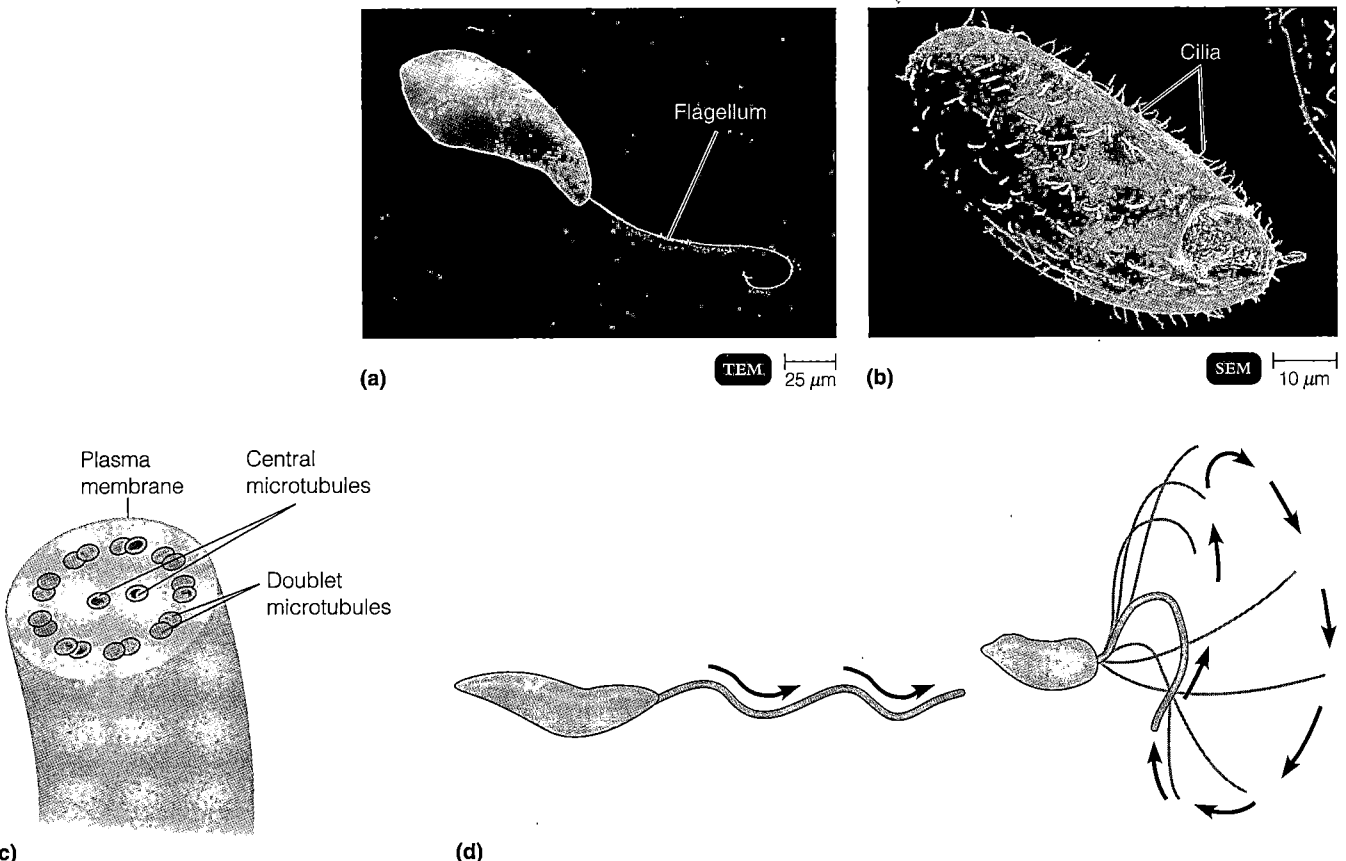

FIGURE 4.22 Eukaryotic flagella and cilia. (a) A micrograph of *Euglena*, a chlorophyll-containing alga, with its flagellum. **(b)** A micrograph of *Tetrahymena*, a common freshwater protozoan, with cilia. **(c)** The internal structure of a flagellum (or cilium), showing the 9 + 2 arrangement of microtubules. **(d)** The pattern of movement of a eukaryotic flagellum.

■ How do prokaryotic and eukaryotic flagella differ?

Euglenoid algae use a flagellum for locomotion, whereas protozoa, such as *Tetrahymena* (tet-rä-hī'me-nä), use cilia for locomotion (Figure 4.22a and b). Both flagella and cilia are anchored to the plasma membrane by a basal body, and both consist of nine pairs of microtubules (doublets) arranged in a ring, plus another two microtubules in the center of the ring, an arrangement called a *9 + 2 array* (Figure 4.22c). **Microtubules** are long, hollow tubes made up of a protein called tubulin. A prokaryotic flagellum rotates, but a eukaryotic flagellum moves in a wavelike manner (Figure 4.22d). To help keep foreign material out of the lungs, ciliated cells of the human respiratory system move the material along the surface of the cells in the bronchial tubes and trachea toward the throat and mouth (see Figure 16.4 on page 457).

The Cell Wall and Glycocalyx

Learning Objective

■ *Compare and contrast prokaryotic and eukaryotic cell walls and glycocalyxes.*

Most eukaryotic cells have cell walls, although they are generally much simpler than those of prokaryotic cells. Many algae have cell walls consisting of the polysaccharide *cellulose* (as do all plants); other chemicals may be present as well. Cell walls of some fungi also contain cellulose, but in most fungi the principal structural component of the cell wall is the polysaccharide *chitin*, a polymer of N-acetylglucosamine (NAG) units. (Chitin is also the main structural component of the exoskeleton of crustaceans and insects.) The cell walls of yeasts contain the polysaccharides *glucan* and *mannan*. In eukaryotes that lack a cell wall, the plasma membrane may be the outer covering; however, cells that have direct contact with the environment may have coatings outside the plasma membrane. Protozoa do not have a typical cell wall; instead, they have a flexible outer covering called a *pellicle.*

In other eukaryotic cells, including animal cells, the plasma membrane is covered by a **glycocalyx,** a layer of material containing substantial amounts of sticky carbohydrates. Some of these carbohydrates are covalently bonded to proteins and lipids in the plasma membrane, forming glycoproteins and glycolipids that anchor the glycocalyx to the cell. The glycocalyx strengthens the cell surface, helps attach cells together, and may contribute to cell-cell recognition.

Eukaryotic cells do not contain peptidoglycan, the framework of the prokaryotic cell wall. This is significant medically because antibiotics, such as penicillins and cephalosporins, act against peptidoglycan and therefore do not affect human eukaryotic cells.

The Plasma (Cytoplasmic) Membrane

Learning Objective

- *Compare and contrast prokaryotic and eukaryotic plasma membranes.*

The **plasma membrane** of eukaryotic and prokaryotic cells is very similar in function and basic structure. There are, however, differences in the types of proteins found in the membranes. Eukaryotic membranes also contain carbohydrates, which serve as receptor sites that assume a role in such functions as cell-cell recognition. These carbohydrates also provide attachment sites for bacteria. Eukaryotic plasma membranes also contain *sterols,* complex lipids not found in prokaryotic plasma membranes (with the exception of *Mycoplasma* cells). Sterols seem to be associated with the ability of the membranes to resist lysis resulting from increased osmotic pressure.

Substances can cross eukaryotic and prokaryotic plasma membranes by simple diffusion, facilitated diffusion, osmosis, or active transport. Group translocation does not occur in eukaryotic cells. However, eukaryotic cells can use a mechanism called **endocytosis.** This occurs when a segment of the plasma membrane surrounds a particle or large molecule, encloses it, and brings it into the cell. Endocytosis is one of the ways viruses can enter animal cells.

Two very important types of endocytosis are phagocytosis and pinocytosis. During *phagocytosis,* cellular projections called pseudopods engulf particles and bring them into the cell. Phagocytosis is used by white blood cells to destroy bacteria and foreign substances (see Figure 16.8 and further discussion in Chapter 16). In *pinocytosis,* the plasma membrane folds inward, bringing extracellular fluid into the cell, along with whatever substances are dissolved in the fluid.

Cytoplasm

Learning Objective

- *Compare and contrast prokaryotic and eukaryotic cytoplasms.*

The **cytoplasm** of eukaryotic cells encompasses the substance inside the plasma membrane and outside the nucleus (see Figure 4.21). The cytoplasm is the substance in which various cellular components are found. (The term **cytosol** refers to the fluid portion of cytoplasm.) A major difference between eukaryotic and prokaryotic cytoplasm is that eukaryotic cytoplasm has a complex internal structure, consisting of exceedingly small rods called *microfilaments* and *intermediate filaments* and cylinders called *microtubules.* Together, they form the **cytoskeleton.** The cytoskeleton provides support and shape and assists in transporting substances through the cell (and even in moving the entire cell, as in phagocytosis). The movement of eukaryotic cytoplasm from one part of the cell to another, which helps distribute nutrients and move the cell over a surface, is called **cytoplasmic streaming.** Another difference between prokaryotic and eukaryotic cytoplasm is that many of the important enzymes found in the cytoplasmic fluid of prokaryotes are sequestered in the organelles of eukaryotes.

Organelles

Learning Objectives

- *Define organelle.*
- *Describe the functions of the nucleus, endoplasmic reticulum, ribosomes, Golgi complex, lysosomes, vacuoles, mitochondria, chloroplasts, peroxisomes, and centrosomes.*

Organelles are structures with specific shapes and specialized functions and are characteristic of eukaryotic cells. They include the nucleus, endoplasmic reticulum, ribosome, Golgi complex, lysosomes, vacuoles, mitochondria, chloroplasts, peroxisomes, and centrosomes. In addition, eukaryotic cells contain ribosomes that are larger and denser than prokaryotic ribosomes.

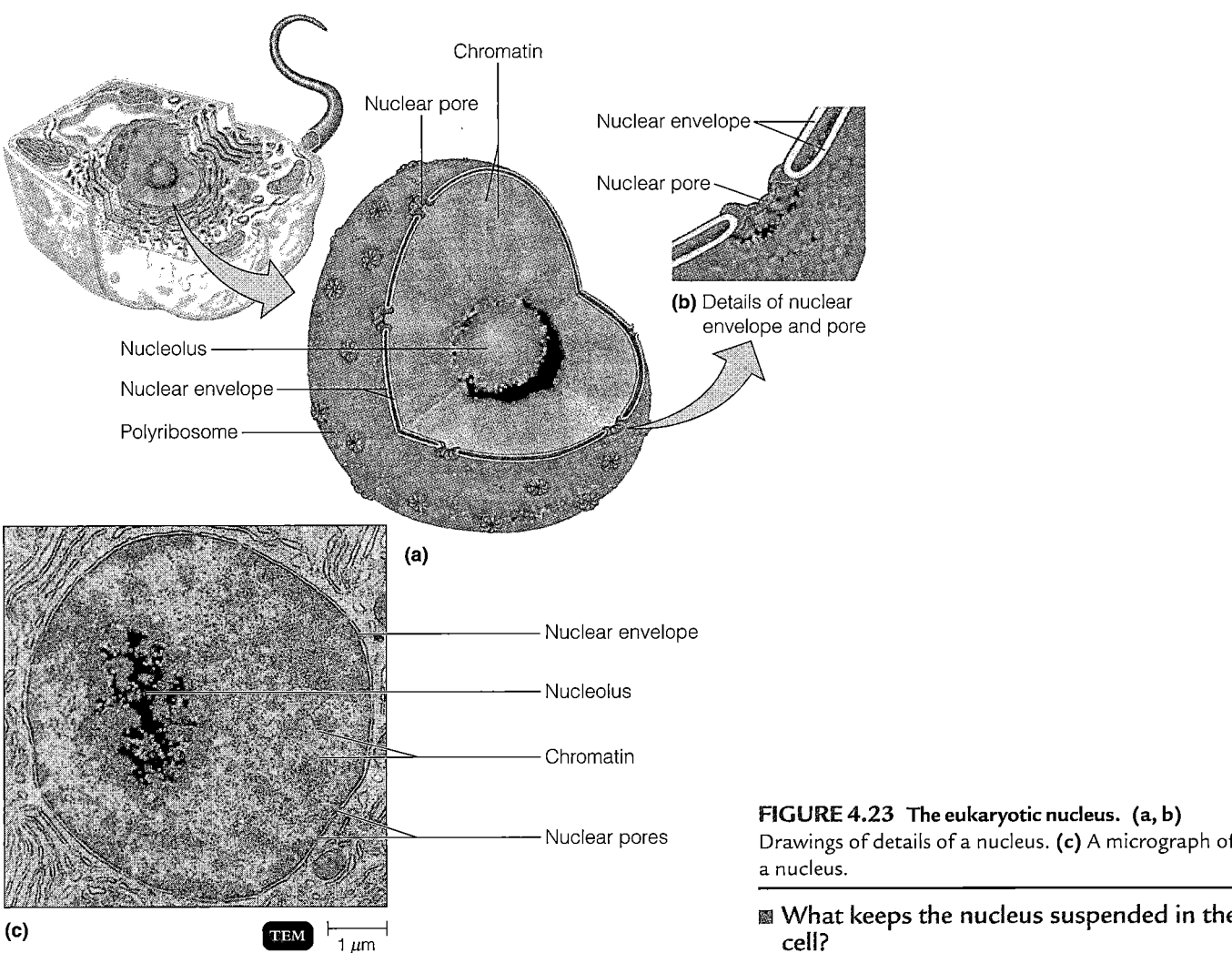

(b) Details of nuclear envelope and pore

(a)

(c) TEM 1 μm

FIGURE 4.23 The eukaryotic nucleus. (a, b)
Drawings of details of a nucleus. **(c)** A micrograph of
a nucleus.

▨ **What keeps the nucleus suspended in the cell?**

The Nucleus

The most characteristic eukaryotic organelle is the nucleus (see Figure 4.21). The **nucleus** (Figure 4.23) is usually spherical or oval, is frequently the largest structure in the cell, and contains almost all of the cell's hereditary information (DNA). Some DNA is also found in mitochondria and in the chloroplasts of photosynthetic organisms.

The nucleus is surrounded by a double membrane called the **nuclear envelope.** Both membranes resemble the plasma membrane in structure. Tiny channels in the membrane called **nuclear pores** allow the nucleus to communicate with the cytoplasm (see Figure 4.23). Nuclear pores control the movement of substances between the nucleus and cytoplasm. Within the nuclear envelope are one or more spherical bodies called **nucleoli** (singular: *nucleolus*). Nucleoli are actually condensed regions of chromosomes where ribosomal RNA is being synthesized. Ribosomal RNA is an essential component of ribosomes (described shortly).

The nucleus also contains most of the cell's DNA, which is combined with several proteins, including some basic proteins called **histones** and nonhistones. The combination of about 165 base pairs of DNA and 9 molecules of histones is referred to as a *nucleosome*. When the cell is not reproducing, the DNA and its associated proteins appear as a threadlike mass called **chromatin.** During nuclear division, the chromatin coils into shorter and thicker rodlike bodies called **chromosomes.** Prokaryotic chromosomes do not undergo this process, do not have histones, and are not enclosed in a nuclear envelope.

Eukaryotic cells require two elaborate mechanisms: mitosis and meiosis to segregate chromosomes prior to cell division. Neither process occurs in prokaryotic cells.

Endoplasmic Reticulum

Within the cytoplasm of eukaryotic cells is the **endoplasmic reticulum** or **ER**, an extensive network of flattened membranous sacs or tubules called **cisterns**

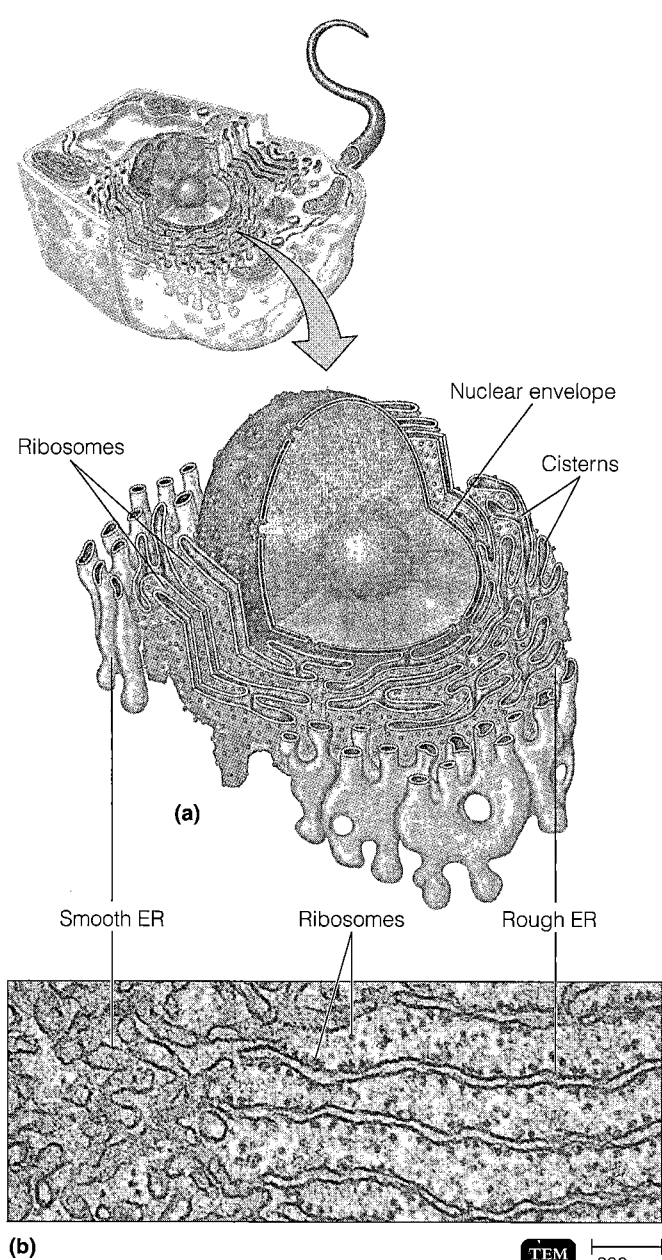

FIGURE 4.24 Rough endoplasmic reticulum and ribosomes. (a) A drawing of details of endoplasmic reticulum. (b) A micrograph of endoplasmic reticulum and ribosomes.

🔳 **What is the difference between rough ER and smooth ER?**

(Figure 4.24). The ER network is continuous with the nuclear envelope (see Figure 4.21a).

Most eukaryotic cells contain two distinct, but interrelated, forms of ER that differ in structure and function. The membrane of **rough ER** is continuous with the nuclear membrane and usually unfolds into a series of flattened sacs. The outer surface of rough ER is studded with ribosomes, the sites of protein synthesis. Proteins synthe-

sized by ribosomes attached to rough ER enter cisterns within the ER for processing and sorting. In some cases, enzymes within the cisterns attach the proteins to carbohydrates to form glycoproteins. In other cases, enzymes attach the proteins to phospholipids, also synthesized by rough ER. These molecules may be incorporated into organelle membranes or the plasma membrane. Thus, rough ER is a factory for synthesizing secretory proteins and membrane molecules.

Smooth ER extends from the rough ER to form a network of membrane tubules (Figure 4.24). Unlike rough ER, smooth ER does not have ribosomes on the outer surface of its membrane. However, smooth ER contains unique enzymes that make it functionally more diverse than rough ER. Although it does not synthesize proteins, smooth ER does synthesize phospholipids, as does rough ER. Smooth ER also synthesizes fats and steroids, such as estrogens and testosterone. In liver cells, enzymes of the smooth ER help release glucose into the bloodstream and inactivate or detoxify drugs and other potentially harmful substances (for example, alcohol). In muscle cells, calcium ions released from the sarcoplasmic reticulum, a form of smooth ER, trigger the contraction process.

Ribosomes

Attached to the outer surface of rough ER are **ribosomes** (see Figure 4.24), which are also found free in the cytoplasm. As in prokaryotes, ribosomes are the sites of protein synthesis in the cell.

The ribosomes of eukaryotic ER and cytoplasm are somewhat larger and denser than those of prokaryotic cells. These eukaryotic ribosomes are 80S ribosomes, each of which consists of a large 60S subunit containing three molecules of rRNA and a smaller 40S subunit with one molecule of rRNA. The subunits are made separately in the nucleolus and, once produced, exit the nucleus and join together in the cytosol. Chloroplasts and mitochondria contain 70S ribosomes, which may indicate their evolution from prokaryotes. This theory is discussed later in this chapter. The role of ribosomes in protein synthesis will be discussed in more detail in Chapter 8.

Some ribosomes, called *free ribosomes,* are unattached to any structure in the cytoplasm. Primarily, free ribosomes synthesize proteins used *inside* the cell. Other ribosomes, called *membrane-bound ribosomes,* attach to the nuclear membrane and the endoplasmic reticulum. These ribosomes synthesize proteins destined for insertion in the plasma membrane or for export from the cell. Ribosomes are also located within mitochondria, where they synthesize mitochondrial proteins. Sometimes 10–20 ri-

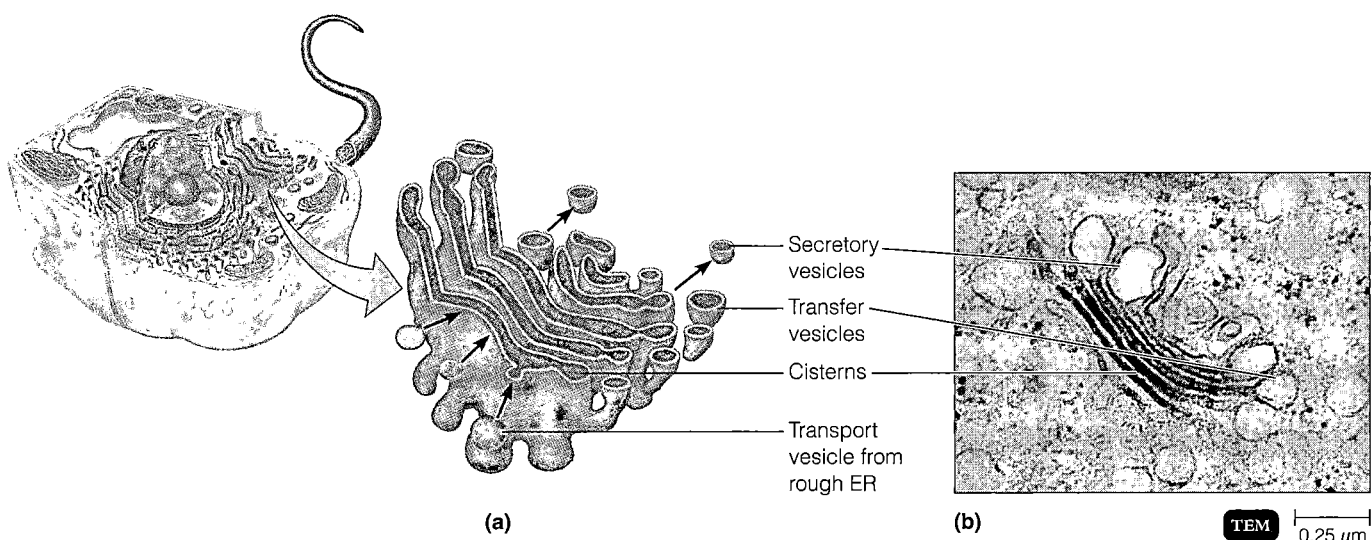

(a)   **(b)**   TEM ⊢———⊣
                                                                                         0.25 μm

FIGURE 4.25 Golgi complex. (a) A drawing of details of a Golgi complex. **(b)** A micrograph of a Golgi complex.

■ The Golgi complex modifies, sorts, and packages proteins received from the ER; discharges proteins via exocytosis; replaces portions of the plasma membrane; and forms lysosomes.

bosomes join together in a stringlike arrangement called a *polyribosome.*

Golgi Complex

Most of the proteins synthesized by ribosomes attached to rough ER are ultimately transported to other regions of the cell. The first step in the transport pathway is through an organelle called the **Golgi complex.** It consists of 3–20 flattened membranous sacs with bulging edges, called **cisterns,** that resemble a stack of pita bread (Figure 4.25). The cisterns are often curved, giving the Golgi complex a cuplike shape.

Proteins synthesized by ribosomes on the rough ER are surrounded by a portion of the ER membrane, which eventually buds from the membrane surface to form a **transport vesicle.** The transport vesicle fuses with a cistern of the Golgi complex, releasing proteins into the cistern. The proteins are modified and move from one cistern to another via **transfer vesicles** that bud from the cisterns' edges. Enzymes in the cisterns modify the proteins to form glycoproteins, glycolipids, and lipoproteins. Some of the processed proteins leave the cisterns in **secretory vesicles,** which detach from the cistern and deliver the proteins to the plasma membrane, where they are discharged by exocytosis. Other processed proteins leave the cisterns in vesicles that deliver their contents to the plasma membrane for incorporation into the membrane. Finally, some processed proteins leave the cisterns in vesicles that are called **storage vesicles.** The major

storage vesicle is a lysosome, whose structure and functions are discussed next.

Lysosomes

Lysosomes are formed from Golgi complexes and look like membrane-enclosed spheres. Unlike mitochondria, lysosomes have only a single membrane and lack internal structure (see Figure 4.21). But they contain as many as 40 different kinds of powerful digestive enzymes capable of breaking down various molecules. Moreover, these enzymes can also digest bacteria that enter the cell. Human white blood cells, which use phagocytosis to ingest bacteria, contain large numbers of lysosomes.

Vacuoles

A **vacuole** (see Figure 4.21) is a space or cavity in the cytoplasm of a cell that is enclosed by a membrane called a *tonoplast.* In plant cells, vacuoles may occupy 5–90% of the cell volume, depending on the type of cell. Vacuoles are derived from the Golgi complex and have several diverse functions. Some vacuoles serve as temporary storage organelles for substances such as proteins, sugars, organic acids, and inorganic ions. Other vacuoles form during endocytosis to help bring food into the cell. Many plant cells also store metabolic wastes and poisons that would otherwise be injurious if they accumulated in the cytoplasm. Finally, vacuoles may take up water, enabling plant cells to increase in size and also providing rigidity to leaves and stems.

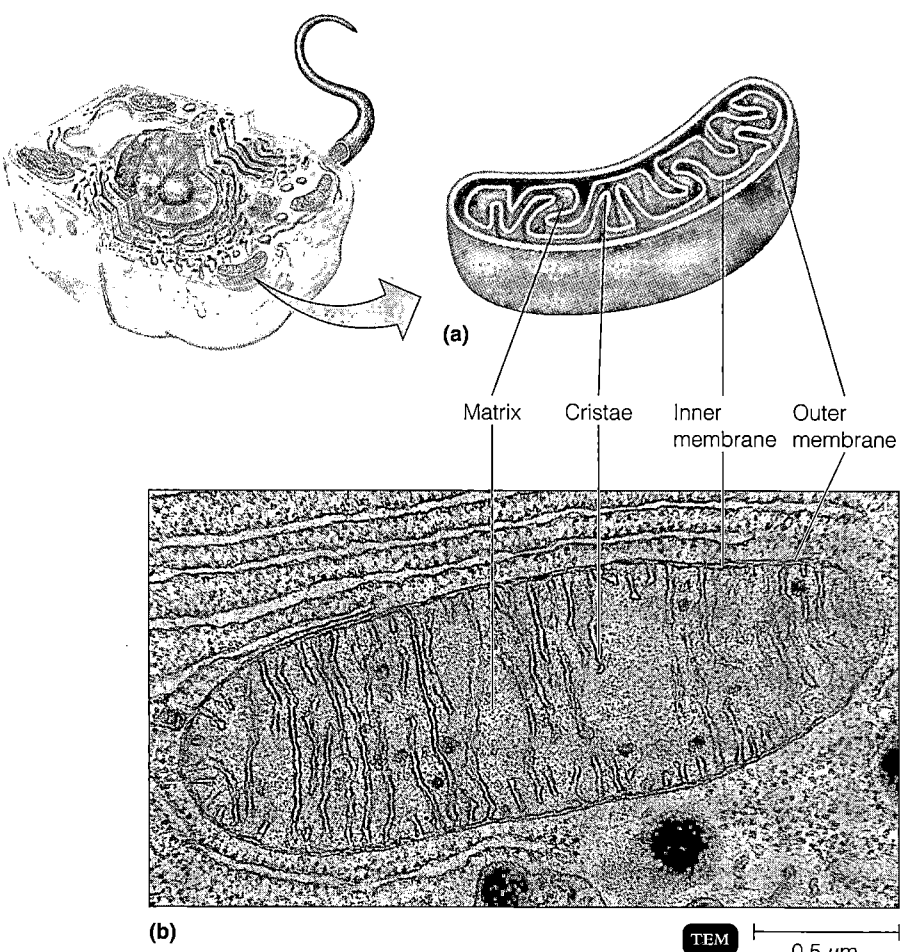

FIGURE 4.26 Mitochondria.
(a) A drawing of details of a mitochondrion. **(b)** A micrograph of a
mitochondrion from a rat pancreas cell.

▩ **How are mitochondria similar
to prokaryotic cells?**

Matrix   Cristae   Inner   Outer
membrane membrane

(b)   TEM   ⊢————————⊣
0.5 μm

Mitochondria

Spherical or rod-shaped organelles called **mitochondria**
(singular: *mitochondrion*) appear throughout the cytoplasm
of most eukaryotic cells (see Figure 4.21). The number of
mitochondria per cell is highly variable. For example, the
protozoan *Giardia* has none, while liver cells contain
1000–2000 per cell. A mitochondrion consists of a double membrane similar in structure to the plasma membrane (Figure 4.26). The outer mitochondrial membrane
is smooth, but the inner mitochondrial membrane is
arranged in a series of folds called **cristae** (singular: *crista*).
The center of the mitochondrion is a semifluid substance
called the **matrix.** Because of the nature and arrangement of the cristae, the inner membrane provides an
enormous surface area on which chemical reactions can
occur. Some proteins that function in cellular respiration,
including the enzyme that makes ATP, are located on the
cristae of the inner mitochondrial membrane, and many
of the metabolic steps involved in cellular respiration are
concentrated in the matrix (see Chapter 5). Mitochon-

dria are often called the "powerhouses of the cell" because of their central role in ATP production.

Mitochondria contain 70S ribosomes and some DNA
of their own, as well as the machinery necessary to replicate, transcribe, and translate the information encoded by
their DNA. In addition, mitochondria can reproduce
more or less on their own by growing and dividing in two.

Chloroplasts

Algae and green plants contain a unique organelle called a
chloroplast (Figure 4.27), a membrane-enclosed structure that contains both the pigment chlorophyll and the
enzymes required for the light-gathering phases of photosynthesis (see Chapter 5). The chlorophyll is contained in
flattened membrane sacs called **thylakoids;** stacks of thylakoids are called *grana* (singular: *granum*) (Figure 4.27).

Like mitochondria, chloroplasts contain 70S ribosomes, DNA, and enzymes involved in protein synthesis.
They are capable of multiplying on their own within the
cell. The way both chloroplasts and mitochondria multi-

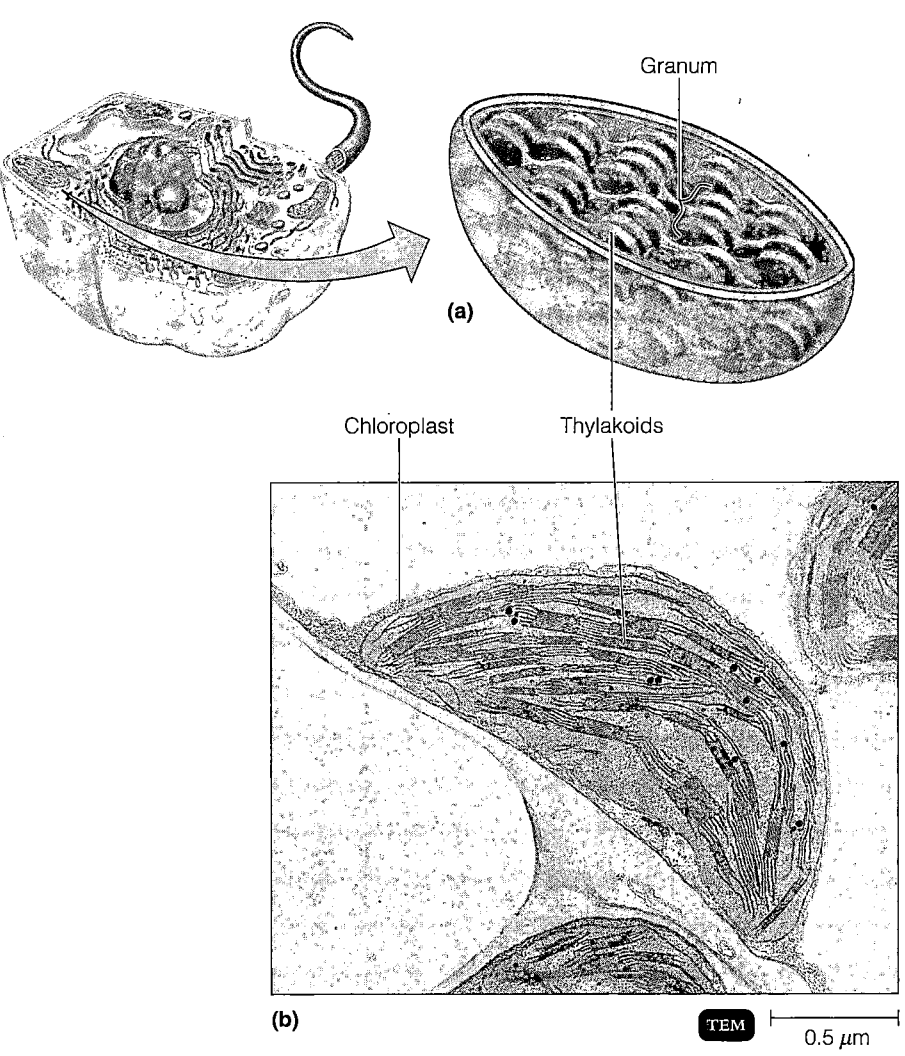

Granum

Chloroplast   Thylakoids

(a)

(b)   TEM |———————|
0.5 μm

FIGURE 4.27 Chloroplasts. Photosynthesis occurs in chloroplasts; the light-trapping pigments are located on the thylakoids. **(a)** A drawing of details of a chloroplast, showing grana. **(b)** A micrograph of chloroplasts in a plant cell.

■ **What are the similarities between chloroplasts and prokaryotic cells?**

ply—by increasing in size and then dividing in two—is strikingly reminiscent of bacterial multiplication.

Peroxisomes

Organelles similar in structure to lysosomes, but smaller, are called **peroxisomes** (see Figure 4.21). Although peroxisomes were once thought to form by budding off the ER, it is now generally agreed that they form by the division of preexisting peroxisomes.

Peroxisomes contain one or more enzymes that can oxidize various organic substances. For example, substances such as amino acids and fatty acids are oxidized in peroxisomes as part of normal metabolism. In addition, enzymes in peroxisomes oxidize toxic substances, such as alcohol. A by-product of the oxidation reactions is hydrogen peroxide (H_2O_2), a potentially toxic compound. However, peroxisomes also contain the enzyme *catalase,* which decomposes H_2O_2 (see Chapter 6, page 163). Be-

cause the generation and degradation of H_2O_2 occurs within the same organelle, peroxisomes protect other parts of the cell from the toxic effects of H_2O_2.

Centrosome

The **centrosome,** located near the nucleus, consists of two components: the pericentriolar area and centrioles (see Figure 4.21). The *pericentriolar area* is a region of the cytosol composed of a dense network of small protein fibers. This area is the organizing center for the mitotic spindle, which plays a critical role in cell division, and for microtubule formation in nondividing cells. Within the pericentriolar area is a pair of cylindrical structures called *centrioles,* each of which is composed of nine clusters of three microtubules (triplets) arranged in a circular pattern, an arrangement called a *9 + 0 array.* The 9 refers to the nine clusters of microtubules, and the 0 refers to the absence of microtubules in the center. The long axis of one centriole is at a right angle to the long axis of the other. Centrioles play a role in the formation or regeneration of cilia and flagella.

The Evolution of Eukaryotes

Learning Objective

■ Discuss evidence that supports the endosymbiotic theory of eukaryotic evolution.

The **endosymbiotic theory** states that the forerunners of eukaryotic cells were associations of small, symbiotic (living together) prokaryotic cells existing within larger prokaryotic cells. This theory focuses primarily on the origin of mitochondria and chloroplasts (which multiply in a way similar to that of bacteria). Good evidence indicates that the ancestors of mitochondria were probably oxygen-requiring heterotrophic bacteria. (*Heterotrophs* obtain food molecules by eating other organisms or their by-products; see Chapter 5). Chloroplasts are believed to be descendants of photosynthetic prokaryotes. It has been suggested that oxygen-requiring heterotrophs and photosynthetic prokaryotes both gained entry into a large prokaryotic cell as undigested prey or internal parasites. The nuclear envelope and other internal membranes may have evolved from infoldings of the plasma membrane (see Figure 10.3).

Our next concern is to examine microbial metabolism. In Chapter 5, you will learn about the importance of enzymes to microorganisms, and the ways microbes produce and use energy.

Study Outline

COMPARING PROKARYOTIC AND EUKARYOTIC CELLS: AN OVERVIEW (pp. 77-78)

1. Prokaryotic and eukaryotic cells are similar in their chemical composition and chemical reactions.
2. Prokaryotic cells lack membrane-enclosed organelles (including a nucleus).
3. Peptidoglycan is found in prokaryotic cell walls but not in eukaryotic cell walls.
4. Eukaryotic cells have a membrane-bound nucleus and other organelles.

THE PROKARYOTIC CELL (pp. 78-98)

1. Bacteria are unicellular, and most of them multiply by binary fission.
2. Bacterial species are differentiated by morphology, chemical composition, nutritional requirements, biochemical activities, and source of energy.

THE SIZE, SHAPE, AND ARRANGEMENT OF BACTERIAL CELLS (pp. 78-80)

1. Most bacteria are 0.2–2.0 μm in diameter and 2–8 μm in length.
2. The three basic bacterial shapes are coccus (spherical), bacillus (rod-shaped), and spiral (twisted).
3. Pleomorphic bacteria can assume several shapes.

STRUCTURES EXTERNAL TO THE CELL WALL (pp. 80-85)

Glycocalyx (p. 82)

1. The glycocalyx (capsule, slime layer, or extracellular polysaccharide) is a gelatinous polysaccharide and/or polypeptide covering.
2. Capsules may protect pathogens from phagocytosis.
3. Capsules enable adherence to surfaces, prevent desiccation, and may provide nutrients.

Flagella (pp. 82-83)

1. Flagella are relatively long filamentous appendages consisting of a filament, hook, and basal body.
2. Prokaryotic flagella rotate to push the cell.
3. Motile bacteria exhibit taxis; positive taxis is movement toward an attractant, and negative taxis is movement away from a repellent.
4. Flagellar (H) protein functions as an antigen.

Axial Filaments (p. 84)

1. Spiral cells that move by means of an axial filament (endoflagellum) are called spirochetes.
2. Axial filaments are similar to flagella, except that they wrap around the cell.

Fimbriae and Pili (pp. 84-85)

1. Fimbriae and pili are short, thin appendages.
2. Fimbriae help cells adhere to surfaces.
3. Pili join cells for the transfer of DNA from one cell to another.

THE CELL WALL (pp. 85–90)

Composition and Characteristics (p. 86)

1. The cell wall surrounds the plasma membrane and protects the cell from changes in water pressure.
2. The bacterial cell wall consists of peptidoglycan, a polymer consisting of NAG and NAM and short chains of amino acids.
3. Penicillin interferes with peptidoglycan synthesis.
4. Gram-positive cell walls consist of many layers of peptidoglycan and also contain teichoic acids.
5. Gram-negative bacteria have a lipopolysaccharide-lipoprotein-phospholipid outer membrane surrounding a thin peptidoglycan layer.
6. The outer membrane protects the cell from phagocytosis and from penicillin, lysozyme, and other chemicals.
7. Porins are proteins that permit small molecules to pass through the outer membrane; specific channel proteins allow other molecules to move through the outer membrane.
8. The lipopolysaccharide component of the outer membrane consists of sugars (O polysaccharides) that function as antigens and lipid A, which is an endotoxin.

Cell Walls and the Gram Stain Mechanism (p. 88)

1. The crystal violet–iodine complex combines with peptidoglycan.
2. The decolorizer removes the lipid outer membrane of gram-negative bacteria and washes out the crystal violet.

Atypical Cell Walls (p. 88)

1. *Mycoplasma* is a bacterial genus that naturally lacks cell walls.
2. Archaea have pseudomurein; they lack peptidoglycan.

Damage to the Cell Wall (pp. 88–90)

1. In the presence of lysozyme, gram-positive cell walls are destroyed, and the remaining cellular contents are referred to as a protoplast.
2. In the presence of lysozyme, gram-negative cell walls are not completely destroyed, and the remaining cellular contents are referred to as a spheroplast.
3. Protoplasts and spheroplasts are subject to osmotic lysis.
4. Antibiotics such as penicillin interfere with cell wall synthesis.

STRUCTURES INTERNAL TO THE CELL WALL (pp. 90–98)

The Plasma (Cytoplasmic) Membrane (pp. 90–95)

1. The plasma membrane encloses the cytoplasm and is a phospholipid bilayer with peripheral and integral proteins (the fluid mosaic model).
2. The plasma membrane is selectively permeable.

3. Plasma membranes carry enzymes for metabolic reactions, such as nutrient breakdown, energy production, and photosynthesis.
4. Mesosomes, irregular infoldings of the plasma membrane, are artifacts, not true cell structures.
5. Plasma membranes can be destroyed by alcohols and polymyxins.

The Movement of Materials Across Membranes (pp. 92–95)

1. Movement across the membrane may be by passive processes, in which materials move from areas of higher to lower concentration, and no energy is expended by the cell.
2. In simple diffusion, molecules and ions move until equilibrium is reached.
3. In facilitated diffusion, substances are transported by transporter proteins across membranes from areas of high to low concentration.
4. Osmosis is the movement of water from areas of high to low concentration across a selectively semipermeable membrane until equilibrium is reached.
5. In active transport, materials move from areas of low to high concentration by transporter proteins, and the cell must expend energy.
6. In group translocation, energy is expended to modify chemicals and transport them across the membrane.

Cytoplasm (p. 95)

1. Cytoplasm is the fluid component inside the plasma membrane.
2. The cytoplasm is mostly water, with inorganic and organic molecules, DNA, ribosomes, and inclusions.

The Nuclear Area (p. 95)

1. The nuclear area contains the DNA of the bacterial chromosome.
2. Bacteria can also contain plasmids, which are circular, extrachromosomal DNA molecules.

Ribosomes (pp. 95–96)

1. The cytoplasm of a prokaryote contains numerous 70S ribosomes; ribosomes consist of rRNA and protein.
2. Protein synthesis occurs at ribosomes; it can be inhibited by certain antibiotics.

Inclusions (pp. 96–97)

1. Inclusions are reserve deposits found in prokaryotic and eukaryotic cells.
2. Among the inclusions found in bacteria are metachromatic granules (inorganic phosphate), polysaccharide granules (usually glycogen or starch), lipid inclusions, sulfur granules, carboxysomes (ribulose 1,5-diphosphate carboxylase), magnetosomes (Fe_3O_4), and gas vacuoles.

7. Which of the following pairs is mismatched?
   a. glycocalyx—adherence
   b. pili—reproduction
   c. membrane—DNA synthesis
   d. cell wall—protection
   e. plasma membrane—transport

8. Which of the following pairs is mismatched?
   a. metachromatic granules—stored phosphates
   b. polysaccharide granules—stored starch
   c. lipid inclusions—poly-β-hydroxybutyric acid
   d. sulfur granules—energy reserve
   e. ribosomes—protein storage

9. You have isolated a motile, gram-positive cell with no visible nucleus. You can assume this cell has
   a. ribosomes.   **d.** a Golgi complex.
   b. mitochondria.   **e.** all of the above
   c. an endoplasmic reticulum.

10. The antibiotic amphotericin B disrupts plasma membranes by combining with sterols; it will affect all of the following cells *except*
   a. animal cells.   **d.** *Mycoplasma* cells
   b. bacterial cells.   **e.** plant cells.
   c. fungal cells.

CRITICAL THINKING

1. Why can prokaryotic cells be smaller than eukaryotic cells and still carry on all the functions of life?

2. The smallest eukaryotic cell is the motile alga *Micromonas*. What is the minimum number of organelles this alga must have?

3. Two types of prokaryotic cells have been distinguished: bacteria and archaea. How do these cells differ from each other? How are they similar?

4. In 1985, a 0.5-mm cell was discovered in surgeonfish and named *Epulopiscium fishelsoni* (see Figure 11.19 on page 322). It was presumed to be a protozoan. In 1993, researchers determined that *Epulopiscium* was actually a gram-positive bacterium. What do you suppose caused the initial identification of this organism as a protozoan? What evidence would change the classification to bacterium?

5. When *E. coli* cells are exposed to a hypertonic solution, the bacteria produce a transporter protein that can move K^+ (potassium ions) into the cell. Of what value is the active transport of K^+, which requires ATP?

CLINICAL APPLICATIONS

1. A child with a bloodborne *Neisseria* infection was treated with gentamicin. After treatment, *Neisseria* could not be cultured from her blood, indicating that the bacteria were killed. However, her symptoms became worse. Annually, nearly half of similar patients die. Explain why antibiotic treatment made her symptoms increase.

2. *Clostridium botulinum* is a strict anaerobe; that is, it is killed by the molecular oxygen (O_2) present in air. Humans can die of botulism from eating foods in which *C. botulinum* is growing. How does this bacterium survive on plants picked for human consumption? Why are home-canned foods most often the source of botulism?

3. Within a three-day period at a large hospital, five patients undergoing hemodialysis developed fever and chills. *Pseudomonas aeruginosa* and *Klebsiella pneumoniae* were isolated from three of the patients. *P. aeruginosa, K. pneumoniae,* and *Enterobacter agglomerans* were isolated from the dialysis system. Why do all three bacteria cause similar symptoms?

4. A South San Francisco child enjoyed bath time at his home because of the colorful orange and red water. The water did not have this rusty color at its source, and the water department could not culture the *Thiobacillus* bacteria responsible for the rusty color from the source. How were the bacteria getting into the household water? What bacterial structures make this possible?

5. Live cultures of *Bacillus thuringiensis* (Dipel®) and *B. subtilis* (Kodiak®) are sold for agricultural use. What bacterial structures make it possible to package and sell these bacteria? For what purpose is each product used? (*Hint:* Refer to Chapter 11.)

Learning with Technology

VU = VirtualUnknown CD-ROM

VU   Enter the Virtual Lab, click the arrow next to the Session field, click Textbook Exercises, and select Chapter 4. Read the Case Study and then proceed.

1. Display the Gram stain for this unknown organism by selecting Gram Stain from the View menu in the Main Menu bar. Predict the cell wall structure represented here, based on Figure 4.12 on page 87 of the textbook.

2. Provide the Gram reaction, morphology, and arrangement for the bacteria found in the Gram stain. Which dye from the Gram-stain procedure is responsible for the color of the cells?

3. Indicate which of the following cellular features would be found in an organism exhibiting this Gram reaction.
   a) LPS layer present
   b) peptidoglycan present
   c) teichoic acids present
   d) periplasmic space present
   e) flagella with only two rings in basal body
   f) endotoxin present
   g) resistance to salts and drying
   h) resistance to penicillins

4. Display the motility video clip by selecting Motility from the View menu in the Main Menu bar. Play the motility video clip. Does the video clip demonstrate true motility or false motility?

5. Figure 4.7 on page 83 of the textbook shows the anchoring of flagella within the Gram-negative cell wall. Draw your prediction of how flagella are anchored in Gram-positive cell walls.

6. Close the Gram Stain window and exit the Virtual Lab.

chapter 5

Microbial Metabolism

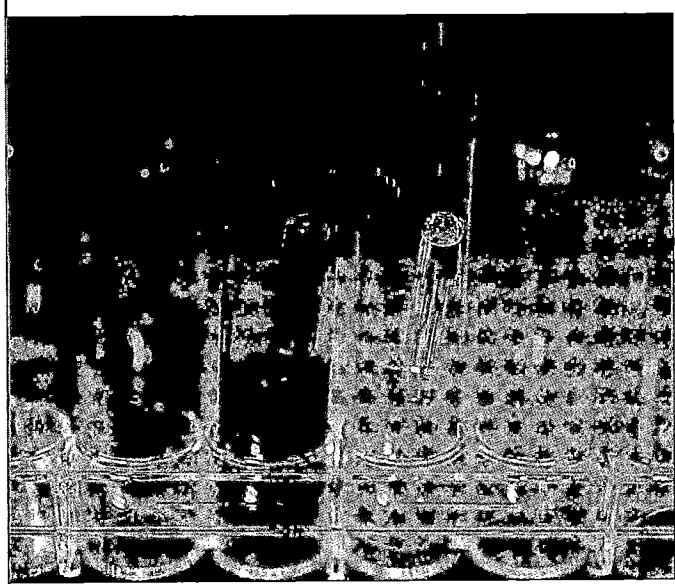

Fermentation test. Biochemical tests such as fermentation are used to study bacterial metabolism and to understand the role of bacteria in the environment.

Now that you are familiar with the structure of prokaryotic cells, we can discuss the activities that enable these microbes to thrive. The life-support processes of even the most structurally simple organism involve a large number of complex biochemical reactions. Most, although not all, of the biochemical processes of bacteria also occur in eukaryotic microbes and in the cells of multicellular organisms, including humans. However, the reactions that are unique to bacteria are fascinating because they allow microorganisms to do things we cannot do. For example, some bacteria can live on cellulose, while others can live on petroleum. Through their metabolism, bacteria recycle elements after other organisms have used them. Chemoautotrophs can live on diets of such inorganic substances as carbon dioxide, iron, sulfur, hydrogen gas, and ammonia.

This chapter examines some representative chemical reactions that either produce energy (the catabolic reactions) or use energy (the anabolic reactions) in microorganisms. We will also look at how these various reactions are integrated within the cell.

Catabolic and Anabolic Reactions

Learning Objectives

- Define *metabolism*, and describe the fundamental differences between anabolism and catabolism.
- Identify the role of *ATP* as an intermediate between catabolism and anabolism.

We use the term **metabolism** to refer to the sum of all chemical reactions within a living organism. Because chemical reactions either release or require energy, metabolism can be viewed as an energy-balancing act. Accordingly, metabolism can be divided into two classes of chemical reactions: those that release energy and those that require energy.

In living cells, the chemical reactions that release energy are generally the ones involved in **catabolism**, the breakdown of complex organic compounds into simpler ones. These reactions are called *catabolic*, or *degradative*, reactions. Catabolic reactions are generally hydrolytic reactions (reactions that use water and in which chemical bonds are broken), and they are exergonic (produce more

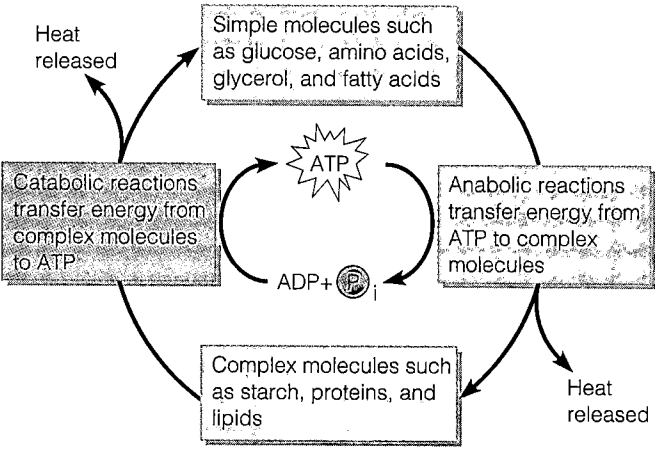

FIGURE 5.1 The role of ATP in coupling anabolic and catabolic reactions. When complex molecules are split apart (catabolism), some of the energy is transferred to and trapped in ATP, and the rest is given off as heat. When simple molecules are combined to form complex molecules (anabolism), ATP provides the energy for synthesis, and again some energy is given off as heat.

■ **The coupling of anabolic and catabolic reactions is achieved through ATP.**

energy than they consume). An example of catabolism occurs when cells break down sugars into carbon dioxide and water.

The energy-requiring reactions are mostly involved in **anabolism,** the building of complex organic molecules from simpler ones. These reactions are called *anabolic,* or *biosynthetic,* reactions. Anabolic processes often involve dehydration synthesis reactions (reactions that release water), and they are endergonic (consume more energy than they produce). Examples of anabolic processes are the formation of proteins from amino acids, nucleic acids from nucleotides, and polysaccharides from simple sugars. These biosynthetic reactions generate the materials for cell growth.

Catabolic reactions furnish the energy needed to drive anabolic reactions. This coupling of energy-requiring and energy-releasing reactions is made possible through the molecule adenosine triphosphate (ATP). (You can review its structure in Figure 2.19 on page 51.) ATP stores energy derived from catabolic reactions and releases it later to drive anabolic reactions and perform other cellular work. Recall from Chapter 2 that a molecule of ATP consists of an adenine, a ribose, and three phosphate groups. When the terminal phosphate group is split from ATP, adenosine diphosphate (ADP) is formed, and energy is released to drive anabolic reactions. Using ⓟ to represent a phosphate group, we write this reaction as:

$$ATP \longrightarrow ADP + ⓟ_i + energy$$

Then, the energy from catabolic reactions is used to combine ADP and a ⓟ to resynthesize ATP:

$$ADP + ⓟ_i + energy \longrightarrow ATP$$

Thus, anabolic reactions are coupled to ATP breakdown, and catabolic reactions are coupled to ATP synthesis. This concept of coupled reactions is very important; you will see why by the end of this chapter. For now, you should know that the chemical composition of a living cell is constantly changing: Some molecules are broken down while others are being synthesized. This balanced flow of chemicals and energy maintains the life of a cell.

The role of ATP in coupling anabolic and catabolic reactions is shown in Figure 5.1. Only part of the energy released in catabolism is actually available for cellular functions, for part of the energy is lost to the environment as heat. Because the cell must use energy to maintain life, it has a continuous need for new external sources of energy.

Before we discuss how cells produce energy, let's first consider the principal properties of a group of proteins involved in almost all biologically important chemical reactions. These proteins, the enzymes, were described briefly in Chapter 2. It is important to understand that a cell's **metabolic pathways** (sequences of chemical reactions) are determined by its enzymes, which are in turn determined by the cell's genetic makeup.

Enzymes

Learning Objectives

- *Identify the components of an enzyme.*
- *Describe the mechanism of enzymatic action.*
- *List the factors that influence enzymatic activity.*
- *Define ribozyme.*

We indicated in Chapter 2 that chemical reactions occur when chemical bonds are formed or broken. In order for reactions to take place, atoms, ions, or molecules must collide. Whether a collision produces a reaction depends on the speed of the particles, the activation energy, and the specific configuration of the particles. Paradoxically, the inherent physiological temperature and pressure of organisms are too low for chemical reactions to occur quickly enough to maintain the life of the organism. Raising the temperature and pressure and the number of reacting molecules would increase the frequency of collisions and the rate of chemical reactions. However, such changes could also damage or kill the organism. The solution to this problem is enzymes (see Figure 2.5 on page 36), which can speed up chemical reactions in several

table 5.1   Enzyme Classification Based on Type of Chemical Reaction Catalyzed

Class	Type of Chemical Reaction Catalyzed	Examples
Oxidoreductase	Oxidation-reduction in which oxygen and hydrogen are gained or lost	Cytochrome oxidase, lactate dehydrogenase
Transferase	Transfer of functional groups, such as an amino group, acetyl group, or phosphate group	Acetate kinase, alanine deaminase
Hydrolase	Hydrolysis (addition of water)	Lipase, sucrase
Lyase	Removal of groups of atoms without hydrolysis	Oxalate decarboxylase, isocitrate lyase
Isomerase	Rearrangement of atoms within a molecule	Glucose-phosphate isomerase, alanine racemase
Ligase	Joining of two molecules (using energy usually derived from the breakdown of ATP)	Acetyl-CoA synthetase, DNA ligase

ways. For example, an enzyme may bring two reactant molecules close together and may properly orient them to react. Whatever the method, the result is that the enzyme lowers the activation energy for the reaction without increasing the temperature or pressure inside the cell.

In living cells, enzymes serve as biological catalysts. As catalysts, enzymes are specific. Each acts on a specific substance (substrate), and each catalyzes only one reaction. For example, sucrose (table sugar) is the substrate of the enzyme sucrase, which catalyzes the hydrolysis of sucrose to glucose and fructose.

The specificity of enzymes is made possible by their structures. Enzymes are generally large globular proteins that range in molecular weight from about 10,000 to several million. Each of the thousands of known enzymes has a characteristic three-dimensional shape with a specific surface configuration as a result of its primary, secondary, and tertiary structures (see Figure 2.16 on page 49). The unique configuration of each enzyme enables it to "find" the correct substrate from among the large number of diverse molecules in the cell.

Enzymes are extremely efficient. Under optimum conditions, they can catalyze reactions at rates 10^8–10^{10} times (up to 10 billion times) higher than those of comparable reactions without enzymes. The **turnover number** (maximum number of substrate molecules an enzyme molecule converts to product each second) is generally between 1 and 10,000 and can be as high as 500,000. For example, the enzyme DNA polymerase I, which participates in the synthesis of DNA, has a turnover number of 15, whereas the enzyme lactate dehydrogenase, which removes hydrogen atoms from lactic acid, has a turnover number of 1000.

Many enzymes exist in the cell in both active and inactive forms. The rate at which enzymes switch between these two forms is determined by the cellular environment.

Naming Enzymes

The names of enzymes usually end in *-ase*. All enzymes can be grouped into six classes, according to the type of chemical reaction they catalyze (Table 5.1). Enzymes within each of the major classes are named according to the more specific types of reactions they assist. For example, the class called oxidoreductases is involved with oxidation-reduction reactions (described shortly). Enzymes in the oxidoreductase class that remove hydrogen from a substrate are called *dehydrogenases;* those that add molecular oxygen (O_2) are called *oxidases*. As you will see later, dehydrogenase and oxidase enzymes have even more specific names, such as lactate dehydrogenase and cytochrome oxidase, depending on the specific substrates on which they act.

Enzyme Components

Although some enzymes consist entirely of proteins, most consist of both a protein portion called an **apoenzyme** and a nonprotein component called a **cofactor.** Ions of iron, zinc, magnesium, or calcium are examples of cofactors. If the cofactor is an organic molecule, it is called a **coenzyme.** Apoenzymes are inactive by themselves; they must be activated by cofactors. Together, the apoenzyme and cofactor form a **holoenzyme,** or whole, active enzyme (Figure 5.2). If the cofactor is removed, the apoenzyme will not function.

FIGURE 5.2 Components of a holoenzyme. Many enzymes require both an apoenzyme (protein portion) and a cofactor (nonprotein portion) to become active. The cofactor can be a metal ion, or if it is an organic molecule, it is called a coenzyme (as shown here). The apoenzyme and cofactor together make up the holoenzyme, or whole enzyme. The substrate is the reactant acted upon by the enzyme.

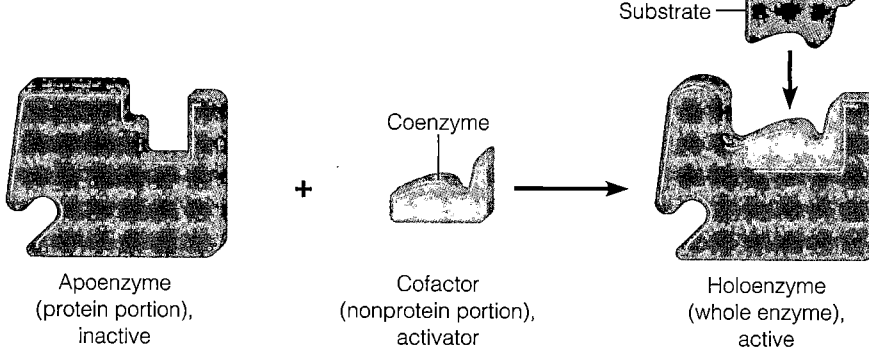

Apoenzyme (protein portion), inactive

Coenzyme

Cofactor (nonprotein portion), activator

Substrate

Holoenzyme (whole enzyme), active

■ What is the function of enzymes in living organisms?

Coenzymes may assist the enzyme by accepting atoms removed from the substrate or by donating atoms required by the substrate. Some coenzymes act as electron carriers, removing electrons from the substrate and donating them to other molecules in subsequent reactions. Many coenzymes are derived from vitamins (Table 5.2). Two of the most important coenzymes in cellular metabolism are **nicotinamide adenine dinucleotide (NAD$^+$)** and **nicotinamide adenine dinucleotide phosphate (NADP$^+$).** Both compounds contain derivatives of the B vitamin nicotinic acid (niacin), and both function as electron carriers. Whereas NAD$^+$ is primarily involved in catabolic (energy-yielding) reactions, NADP$^+$ is primarily involved in anabolic (energy-requiring) reactions. The flavin coenzymes, such as **flavin mononucleotide (FMN)** and **flavin adenine dinucleotide (FAD),** contain derivatives of the B vitamin riboflavin and are also electron carriers. Another important coenzyme, **coenzyme A (CoA),** contains a derivative of pantothenic acid, another B vitamin. This coenzyme plays an important role in the synthesis and breakdown of fats and in a series of oxidizing reactions called the Krebs cycle. We will come across all of these coenzymes in our discussion of metabolism later in the chapter.

As noted earlier, some cofactors are metal ions, including iron, copper, magnesium, manganese, zinc, calcium, and cobalt. Such cofactors may help catalyze a reaction by forming a bridge between the enzyme and a substrate. For example, magnesium (Mg^{2+}) is required by many phosphorylating enzymes (enzymes that transfer a phosphate group from ATP to another substrate). The Mg^{2+} can form a link between the enzyme and the ATP molecule. Most trace elements required by living cells are probably used in some such way to activate cellular enzymes.

The Mechanism of Enzymatic Action

Although scientists do not completely understand how enzymes lower the activation energy of chemical reactions, the general sequence of events in enzyme action is as follows (Figure 5.3a):

① The surface of the substrate contacts a specific region of the surface of the enzyme molecule called the **active site.**

② A temporary intermediate compound forms, called an **enzyme–substrate complex.**

③ The substrate molecule is transformed by the rearrangement of existing atoms, the breakdown of the substrate molecule, or in combination with another substrate molecule.

④ The transformed substrate molecules—the products of the reaction—are released from the enzyme molecule because they no longer fit in the active site of the enzyme.

⑤ The unchanged enzyme is now free to react with other substrate molecules.

As a result of these events, an enzyme speeds up a chemical reaction.

As mentioned earlier, enzymes have *specificity* for particular substrates. For example, a specific enzyme may be able to hydrolyze a peptide bond only between two specific amino acids. Other enzymes can hydrolyze starch but not cellulose; even though both starch and cellulose are polysaccharides composed of glucose subunits, the orientations of the subunits in the two polysaccharides differ. Enzymes have this specificity because the three-dimensional shape of the active site fits the substrate somewhat as a lock fits with its key (Figure 5.3b). However, the active site and substrate are flexible, and they change shape somewhat as they meet to fit together more tightly. The substrate is usually much smaller than the

table 5.2	Selected Vitamins and Their Coenzymatic Functions
Vitamin	**Function**
Vitamin B$_1$ (thiamine)	Part of coenzyme cocarboxylase; has many functions, including the metabolism of pyruvic acid
Vitamin B$_2$ (riboflavin)	Coenzyme in flavoproteins; active in electron transfers
Niacin (nicotinic acid)	Part of NAD molecule; active in electron transfers
Vitamin B$_6$ (pyridoxine)	Coenzyme in amino acid metabolism
Vitamin B$_{12}$ (cyanocobalamin)	Coenzyme (methyl cyanocobalamide) involved in the transfer of methyl groups; active in amino acid metabolism
Pantothenic acid	Part of coenzyme A molecule; involved in the metabolism of pyruvic acid and lipids
Biotin	Involved in carbon dioxide fixation reactions and fatty acid synthesis
Folic acid	Coenzyme used in the synthesis of purines and pyrimidines
Vitamin E	Needed for cellular and macromolecular syntheses
Vitamin K	Coenzyme used in electron transport (naphthoquinones and quinones)

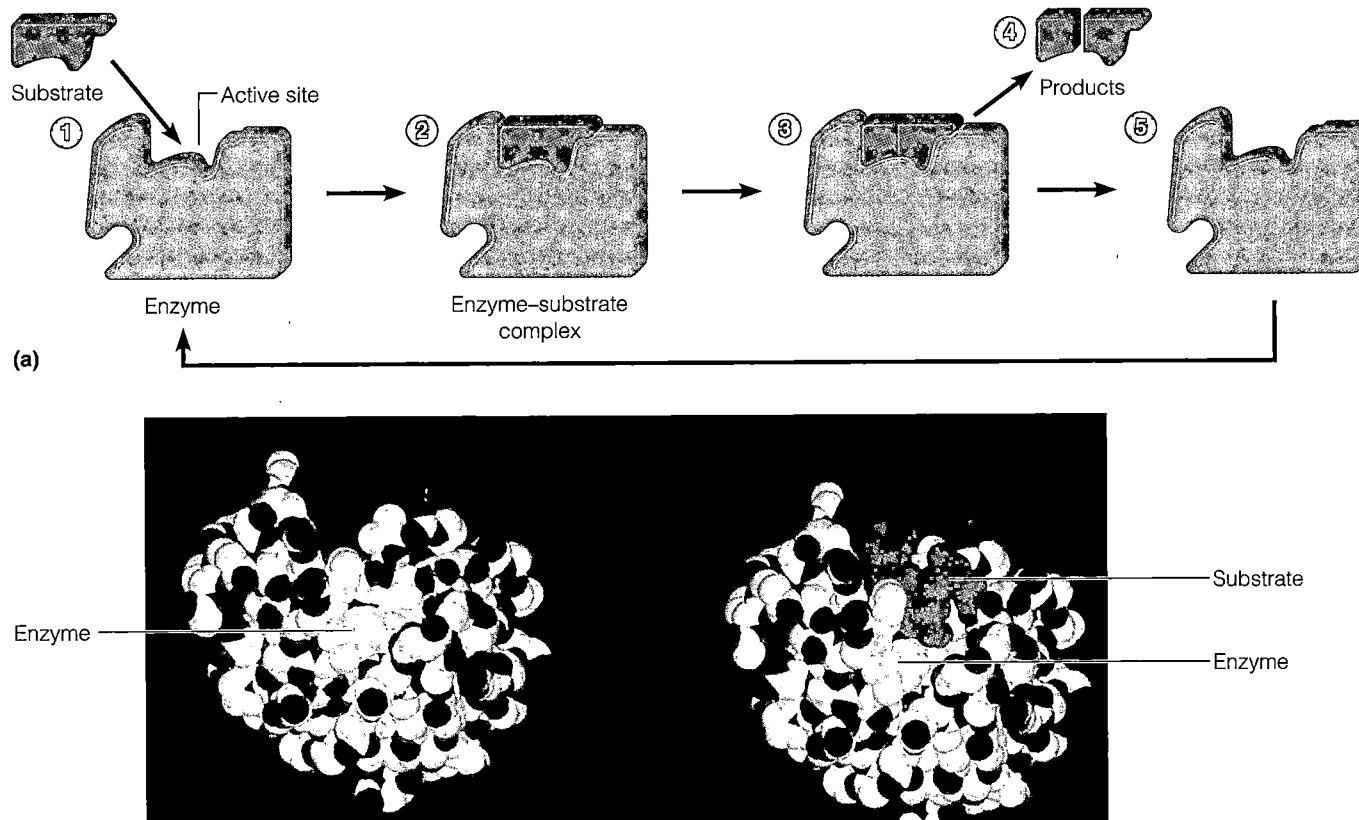

FIGURE 5.3   The mechanism of enzymatic action. (a) ① The substrate contacts the active site on the enzyme to form ② an enzyme–substrate complex. ③ The substrate is then transformed into products, ④ the products are released, and ⑤ the enzyme is recovered unchanged. In the example shown, the transformation into products involves a breakdown of the substrate into two products. Other transformations, however, may occur. **(b)** Left: A molecular model of the enzyme in step ① of part a. The active site of the enzyme can be seen here as a groove on the surface of the protein. Right: As the enzyme and substrate meet in step ② of part a, they change shape slightly to fit together more tightly.

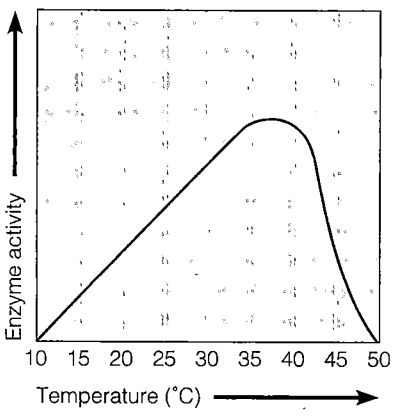

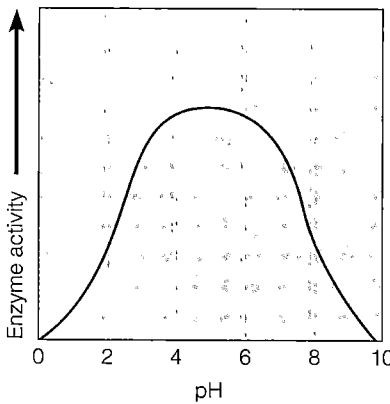

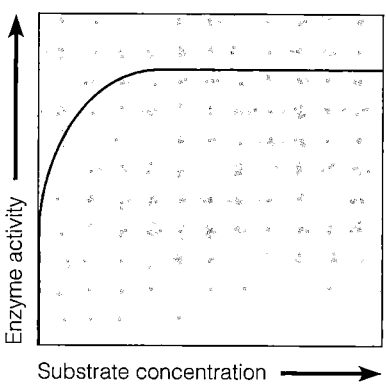

(a) Temperature. The enzymatic activity (rate of reaction catalyzed by the enzyme) increases with increasing temperature until the enzyme, a protein, is denatured by heat and inactivated. At this point, the reaction rate falls steeply.

(b) pH. The enzyme illustrated is most active at about pH 5.0.

(c) Substrate concentration. With increasing concentration of substrate molecules, the rate of reaction increases until the active sites on all the enzyme molecules are filled, at which point the maximum rate of reaction is reached.

FIGURE 5.4 Factors that influence enzymatic activity, plotted for a hypothetical enzyme.

■ How will this enzyme act at 25°C? At 45°C? At pH 7?

enzyme, and relatively few of the enzyme's amino acids make up the active site.

A certain compound can be a substrate for several different enzymes that catalyze different reactions, so the fate of a compound depends on the enzyme that acts upon it. Glucose 6-phosphate, a molecule important in cell metabolism, can be acted upon by at least four different enzymes, and each reaction will yield a different product.

Factors Influencing Enzymatic Activity

Enzymes are subject to various cellular controls. Two primary types are the control of enzyme *synthesis* (see Chapter 8) and the control of enzyme *activity* (how much enzyme is present versus how active it is).

Several factors influence the activity of an enzyme. Among the more important are temperature, pH, substrate concentration, and the presence or absence of inhibitors.

Temperature

The rate of most chemical reactions increases as the temperature increases. Molecules move more slowly at lower temperatures than at higher temperatures and so may not have enough energy to cause a chemical reaction. For enzymatic reactions, however, elevation beyond a certain temperature (the optimal temperature) drastically reduces the rate of reaction (Figure 5.4a). The optimal temperature for most disease-producing bacteria in the human body is between 35°C and 40°C. The reduced rate of reaction be-

yond the optimal temperature is due to the enzyme's **denaturation,** the loss of its characteristic three-dimensional structure (tertiary configuration) (Figure 5.5). Denaturation of a protein involves the breakage of hydrogen bonds and other noncovalent bonds; a common example is the transformation of uncooked egg white (a protein called albumin) to a hardened state by heat.

Denaturation of an enzyme changes the arrangement of the amino acids in the active site, altering its shape and causing the enzyme to lose its catalytic ability. In some cases, denaturation is partially or fully reversible. However, if denaturation continues until the enzyme has lost its solubility and coagulates, the enzyme cannot regain its original properties. Enzymes can also be denatured by concentrated acids, bases, heavy-metal ions (such as lead, arsenic, or mercury), alcohol, and ultraviolet radiation.

pH

Most enzymes have an optimum pH at which their activity is characteristically maximal. Above or below this pH value, enzyme activity, and therefore the reaction rate, decline (Figure 5.4b). When the H^+ concentration (pH) in the medium is changed, the protein's three-dimensional structure is altered. Extreme changes in pH can cause denaturation.

Substrate Concentration

There is a maximum rate at which a certain amount of enzyme can catalyze a specific reaction. Only when the

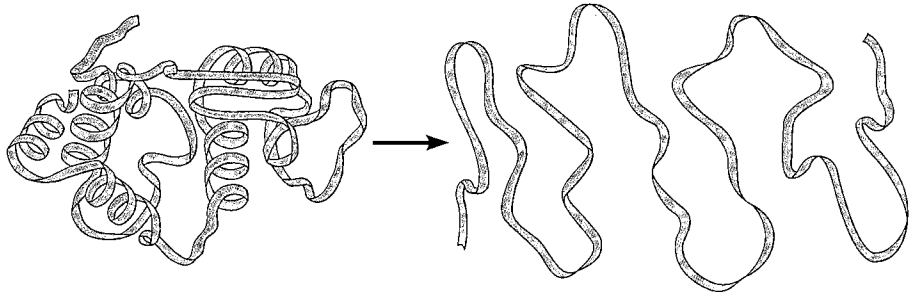

Active (functional) protein   Denatured protein

FIGURE 5.5 Denaturation of a protein. Breakage of the noncovalent bonds (such as hydrogen bonds) that hold the active protein in its three-dimensional shape renders the denatured protein nonfunctional.

■ **What factors may cause denaturation?**

concentration of substrate(s) is extremely high can this maximum rate be attained. Under conditions of high substrate concentration, the enzyme is said to be in **saturation;** that is, its active site is always occupied by substrate or product molecules. In this condition, a further increase in substrate concentration will not affect the reaction rate because all active sites are already in use (Figure 5.4c). Under normal cellular conditions, enzymes are not saturated with substrate(s). At any given time, many of the enzyme molecules are inactive for lack of substrate; thus, the rate of reaction is likely to be influenced by the substrate concentration.

Inhibitors

An effective way to control the growth of bacteria is to control their enzymes. Certain poisons, such as cyanide, arsenic, and mercury, combine with enzymes and prevent them from functioning. As a result, the cells stop functioning and die.

Enzyme inhibitors are classified as either competitive or noncompetitive inhibitors (Figure 5.6). **Competitive inhibitors** fill the active site of an enzyme and compete with the normal substrate for the active site. A competitive inhibitor can do this because its shape and chemical

structure are similar to those of the normal substrate (Figure 5.6b). However, unlike the substrate, it does not undergo any reaction to form products. Some competitive inhibitors bind irreversibly to amino acids in the active site, preventing any further interactions with the substrate. Others bind reversibly, alternately occupying and leaving the active site; these slow the enzyme's interaction with the substrate. Reversible competitive inhibition can be overcome by increasing the substrate concentration. As active sites become available, more substrate molecules than competitive inhibitor molecules are available to attach to the active sites of enzymes.

One good example of a competitive inhibitor is sulfanilamide (a sulfa drug), which inhibits the enzyme whose normal substrate is para-aminobenzoic acid (PABA):

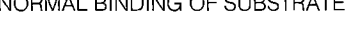

Sulfanilamide   PABA

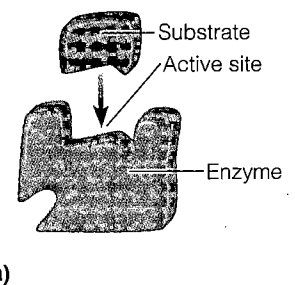
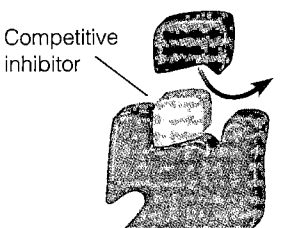
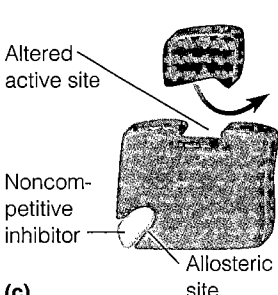

NORMAL BINDING OF SUBSTRATE   ACTION OF ENZYME INHIBITORS

(a)   (b)   (c)

FIGURE 5.6 Enzyme inhibitors. (a) An uninhibited enzyme and its normal substrate. **(b)** A competitive inhibitor. **(c)** One type of noncompetitive inhibitor, causing allosteric inhibition.

■ **Competitive inhibitors fill the active site of an enzyme and compete with the normal substrate for the active site.**

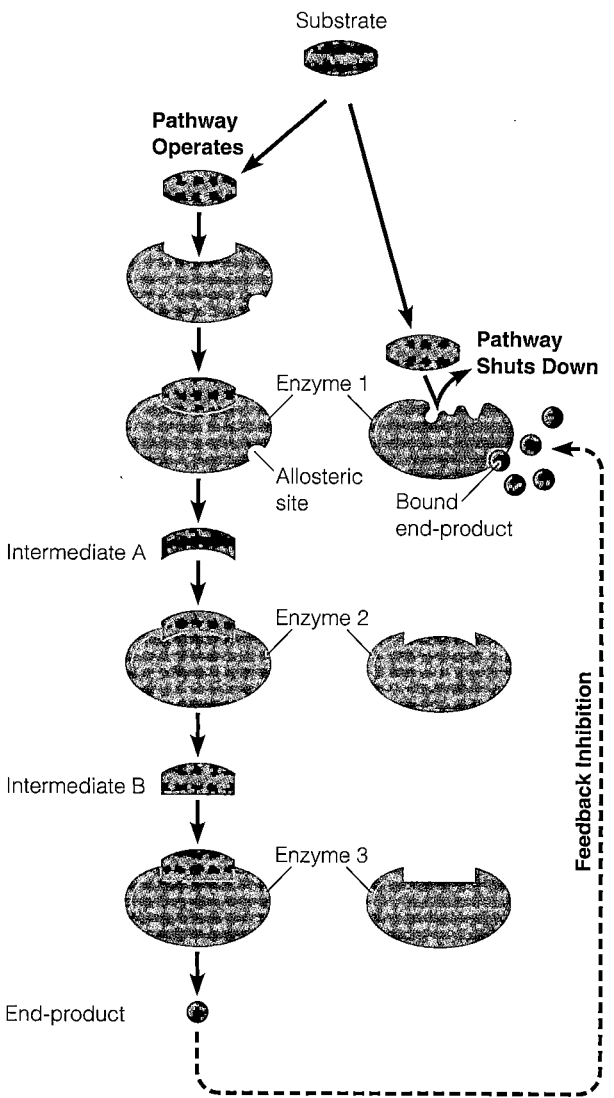

FIGURE 5.7 Feedback inhibition.

■ In feedback inhibition, a series of enzymes makes an end-product that inhibits the first enzyme in the series, thus shutting down the entire pathway when sufficient end-product has been made.

PABA is an essential nutrient used by many bacteria in the synthesis of folic acid, a vitamin that functions as a coenzyme. When sulfanilamide is administered to bacteria, the enzyme that normally converts PABA to folic acid combines instead with the sulfanilamide. Folic acid is not synthesized, and the bacteria cannot grow. Because human cells do not use PABA to make their folic acid, sulfanilamide kills bacteria but does not harm human cells.

Noncompetitive inhibitors do not compete with the substrate for the enzyme's active site; instead, they interact with another part of the enzyme (Figure 5.6c). In this process, called **allosteric** ("other space") **inhibition,** the inhibitor binds to a site on the enzyme other than the substrate's binding site, called the **allosteric site.** This binding

causes the active site to change its shape, making it nonfunctional. As a result, the enzyme's activity is reduced. This effect can be either reversible or irreversible, depending on whether or not the active site can return to its original shape. In some cases, allosteric interactions can activate an enzyme rather than inhibit it. Another type of noncompetitive inhibition can operate on enzymes that require metal ions for their activity. Certain chemicals can bind or tie up the metal ion activators and thus prevent an enzymatic reaction. Cyanide can bind the iron in iron-containing enzymes, and fluoride can bind calcium or magnesium. Substances such as cyanide and fluoride are sometimes called *enzyme poisons* because they permanently inactivate enzymes. In low concentrations, fluoride kills bacteria in the mouth that can contribute to tooth decay.

Feedback Inhibition

Allosteric inhibitors play a role in a kind of biochemical control called **feedback inhibition,** or **end-product inhibition.** This control mechanism stops the cell from wasting chemical resources by making more of a substance than it needs. In some metabolic reactions, several steps are required for the synthesis of a particular chemical compound, called the *end-product*. The process is similar to an assembly line, with each step catalyzed by a separate enzyme (Figure 5.7). In many anabolic pathways, the final product can allosterically inhibit the activity of one of the enzymes earlier in the pathway. This phenomenon is feedback inhibition.

Feedback inhibition generally acts on the first enzyme in a metabolic pathway (similar to shutting down an assembly line by stopping the first worker). Because the enzyme is inhibited, the product of the first enzymatic reaction in the pathway is not synthesized. Because that unsynthesized product would normally be the substrate for the second enzyme in the pathway, the second reaction stops immediately as well. Thus, even though only the first enzyme in the pathway is inhibited, the entire pathway shuts down and no new end-product is formed. By inhibiting the first enzyme in the pathway, the cell also keeps metabolic intermediates from accumulating. As the existing end-product is used up by the cell, the first enzyme's allosteric site will more often remain unbound, and the pathway will resume activity.

The bacterium *E. coli* can be used to demonstrate feedback inhibition in the synthesis of the amino acid isoleucine, which is required for the cell's growth. In this metabolic pathway, five steps are taken to enzymatically convert the amino acid threonine to isoleucine. If isoleucine is added to the growth medium for *E. coli*, it inhibits the first enzyme in the pathway, and the bacteria stop synthesizing isoleucine. This condition is maintained until the supply of isoleucine is depleted. This type of feedback inhibition is also involved in regulating the cells' production of other amino acids, as well as vitamins, purines, and pyrimidines.

Ribozymes

Prior to 1982, it was believed that only protein molecules had enzymatic activity. Researchers working on microbes discovered a unique type of RNA called a **ribozyme.** Like protein enzymes, ribozymes function as catalysts, have active sites that bind to substrates, and are not used up in a chemical reaction. Ribozymes specifically act on strands of RNA by removing sections and splicing together the remaining pieces. In this respect, ribozymes are more restricted than protein enzymes in terms of the diversity of substrates with which they interact.

Energy Production

Learning Objectives

- *Explain what is meant by oxidation-reduction.*
- *List and provide examples of three types of phosphorylation reactions that generate ATP.*

Nutrient molecules, like all molecules, have energy associated with the electrons that form bonds between their atoms. When it is spread throughout the molecule, this energy is difficult for the cell to use. Various reactions in catabolic pathways, however, concentrate the energy into the bonds of ATP, which serves as a convenient energy carrier. ATP is generally referred to as having "high-energy" bonds. Actually, a better term is probably unstable bonds. Although the amount of energy in these bonds is not exceptionally large, it can be released quickly and easily. In a sense, ATP is similar to a highly flammable liquid such as kerosene. Although a large log might eventually burn to produce more heat than a cup of kerosene, the kerosene is easier to ignite and provides heat more quickly and conveniently. In a similar way, the "high-energy" unstable bonds of ATP provide the cell with readily available energy for anabolic reactions.

Before discussing the catabolic pathways, we will consider two general aspects of energy production: the concept of oxidation-reduction and the mechanisms of ATP generation.

Oxidation-Reduction Reactions

Oxidation is the removal of electrons (e^-) from an atom or molecule, a reaction that often produces energy. Figure 5.8 shows an example of an oxidation in which molecule

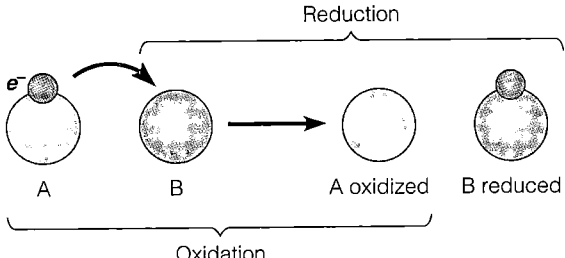

FIGURE 5.8 Oxidation-reduction. An electron is transferred from molecule A to molecule B. In the process, molecule A is oxidized and molecule B is reduced.

- Oxidation is the removal of electrons, and reduction is the addition of electrons.

A loses an electron to molecule B. Molecule A has undergone oxidation (meaning that it has lost one or more electrons), whereas molecule B has undergone **reduction** (meaning that it has gained one or more electrons).* Oxidation and reduction reactions are always coupled; in other words, each time one substance is oxidized, another is simultaneously reduced. The pairing of these reactions is called **oxidation-reduction** or a **redox reaction.**

In many cellular oxidations, electrons and protons (hydrogen ions, H^+) are removed at the same time; this is equivalent to the removal of hydrogen atoms, because a hydrogen atom is made up of one proton and one electron (see Table 2.2, page 30). Because most biological oxidations involve the loss of hydrogen atoms, they are also called **dehydrogenation** reactions. Figure 5.9 shows an example of a biological oxidation. An organic molecule is oxidized by the loss of two hydrogen atoms, and a molecule of NAD^+ is reduced. Recall from our earlier discussion of coenzymes that NAD^+ assists enzymes by accepting hydrogen atoms removed from the substrate, in this case the organic molecule. As shown in Figure 5.9 on page 123, NAD^+ accepts two electrons and one proton. One proton (H^+) is left over and is released into the surrounding medium. The reduced coenzyme, NADH, contains more energy than NAD^+. This energy can be used to generate ATP in later reactions.

An important point to remember about biological oxidation-reduction reactions is that cells use them in catabolism to extract energy from nutrient molecules. Cells take nutrients, some of which serve as energy sources, and degrade them from highly reduced compounds (with

---

*The terms do not seem logical until one considers the history of the discovery of these reactions. When mercury is roasted, it gains weight as mercuric oxide is formed; this was called *oxidation.* Later it was determined that the mercury actually *lost* electrons. Reduction, therefore, is the *addition* of electrons.

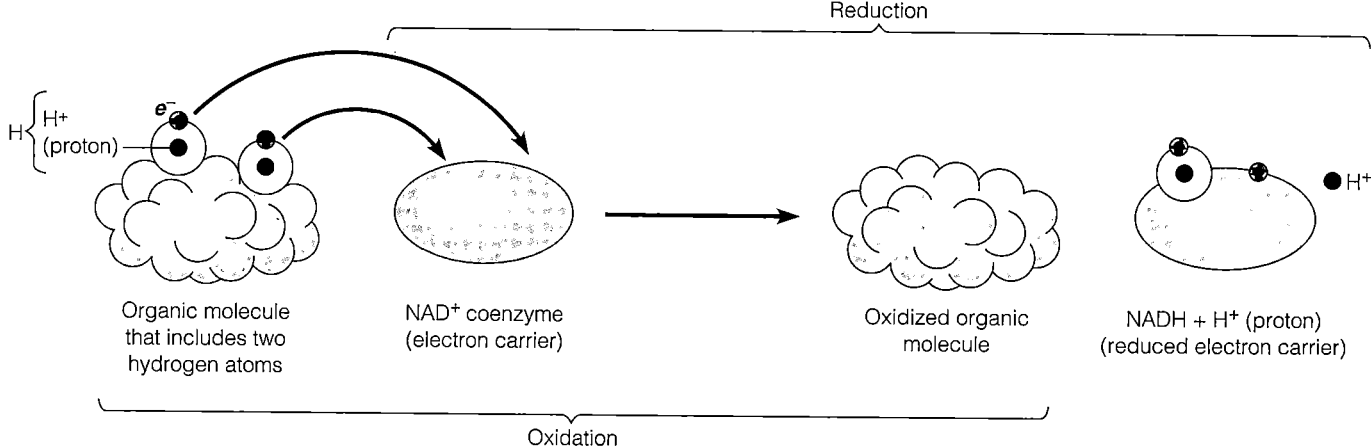

FIGURE 5.9 **Representative biological oxidation.** Two electrons and two protons (altogether equivalent to two hydrogen atoms) are transferred from an organic substrate molecule to a coenzyme, NAD^+. NAD^+ actually receives one hydrogen atom and one electron, and one proton is released into the medium. NAD^+ is reduced to NADH, which is a more energy-rich molecule.

■ Organisms use oxidation-reduction reactions in catabolism to extract energy from nutrient molecules, such as glucose.

many hydrogen atoms) to highly oxidized compounds. For example, when a cell oxidizes a molecule of glucose ($C_6H_{12}O_6$) to CO_2 and H_2O, the energy in the glucose molecule is removed in a stepwise manner and ultimately is trapped by ATP, which can then serve as an energy source for energy-requiring reactions. Compounds such as glucose that have many hydrogen atoms are highly reduced compounds, containing a large amount of potential energy. Thus, glucose is a valuable nutrient for organisms.

The Generation of ATP

Much of the energy released during oxidation-reduction reactions is trapped within the cell by the formation of ATP. Specifically, a phosphate group, Ⓟ, is added to ADP with the input of energy to form ATP:

ADP
$$\overbrace{\text{Adenosine}—Ⓟ\text{~}Ⓟ}+\text{Energy}+Ⓟ\rightarrow$$
$$\underbrace{\text{Adenosine}—Ⓟ\text{~}Ⓟ\text{~}Ⓟ}$$
ATP

The symbol ~ designates a "high-energy" bond—that is, one that can readily be broken to release usable energy. The high-energy bond that attaches the third Ⓟ in a sense contains the energy stored in this reaction. When this Ⓟ is removed, usable energy is released. The addition of Ⓟ to a chemical compound is called **phosphorylation.** Organisms use three mechanisms of phosphorylation to generate ATP from ADP.

Substrate-Level Phosphorylation

In **substrate-level phosphorylation,** ATP is generated when a high-energy Ⓟ is directly transferred from a phosphorylated compound (a substrate) to ADP. Generally, the Ⓟ has acquired its energy during an earlier reaction in which the substrate itself was oxidized. The following example shows only the carbon skeleton and the Ⓟ of a typical substrate:

$$C—C—C \text{~} Ⓟ + ADP \longrightarrow C—C—C + ATP$$

Oxidative Phosphorylation

In **oxidative phosphorylation,** electrons are transferred from organic compounds to one group of electron carriers (usually to NAD^+ and FAD). Then, the electrons are passed through a series of different electron carriers to molecules of oxygen (O_2) or other inorganic molecules. This process occurs in the plasma membrane of prokaryotes and in the inner mitochondrial membrane of eukaryotes. The sequence of electron carriers used in oxidative phosphorylation is called an **electron transport chain (system)** (see Figure 5.13). The transfer of electrons from one electron carrier to the next releases energy, some of which is used to generate ATP from ADP through a process called *chemiosmosis,* to be described shortly.

Photophosphorylation

The third mechanism of phosphorylation, **photophosphorylation,** occurs only in photosynthetic cells, which contain light-trapping pigments such as chlorophylls. In photosynthesis, organic molecules, especially sugars, are synthesized with the energy of light from the energy-poor building blocks carbon dioxide and water. Photophosphorylation starts this process by converting light energy to the chemical energy of ATP and NADPH, which, in turn, are used to synthesize organic molecules. As in oxidative phosphorylation, an electron transport chain is involved.

Metabolic Pathways of Energy Production

Learning Objective

■ *Explain the overall function of metabolic pathways.*

Organisms release and store energy from organic molecules by a series of controlled reactions rather than in a single burst. If the energy were released all at once, as a large amount of heat, it could not be readily used to drive chemical reactions and would, in fact, damage the cell. To extract energy from organic compounds and store it in chemical form, organisms pass electrons from one compound to another through a series of oxidation-reduction reactions.

As noted earlier, a sequence of enzymatically catalyzed chemical reactions occurring in a cell is called a metabolic pathway. Below is a hypothetical pathway that converts starting material A to end-product F in a series of five steps:

The first step is the conversion of molecule A to molecule B. The curved arrow indicates that the reduction of coenzyme NAD^+ to NADH is coupled to that reaction; the electrons and protons come from molecule A. Similarly, the two arrows in ③ show a coupling of two reactions. As C is converted to D, ADP is converted to ATP; the energy needed comes from C as it transforms into D. The reaction converting D to E is readily reversible, as indicated by the double arrow. In the fifth step, the curved arrow leading from O_2 indicates that O_2 is a reactant in the reaction. The curved arrows leading to CO_2 and H_2O indicate that these substances are secondary products produced in the reaction, in addition to F, the end-product that (presumably) interests us the most. Secondary prod-

ucts such as CO_2 and H_2O shown here are sometimes called "by-products" or "waste products." Keep in mind that almost every reaction in a metabolic pathway is catalyzed by a specific enzyme; sometimes the name of the enzyme is printed near the arrow.

Carbohydrate Catabolism

Learning Objectives

■ *Describe the chemical reactions of glycolysis.*
■ *Explain the products of the Krebs cycle.*
■ *Describe the chemiosmotic model for ATP generation.*

Most microorganisms oxidize carbohydrates as their primary source of cellular energy. **Carbohydrate catabolism,** the breakdown of carbohydrate molecules to produce energy, is therefore of great importance in cell metabolism. Glucose is the most common carbohydrate energy source used by cells. Microorganisms can also catabolize various lipids and proteins for energy production, as you will see later.

To produce energy from glucose, microorganisms use two general processes: *cellular respiration* and *fermentation.* (In discussing cellular respiration, we frequently refer to the process simply as respiration, but it should not be confused with breathing.) Both processes usually start with the same first step, glycolysis, but follow different subsequent pathways (Figure 5.10). Before examining the details of glycolysis, respiration, and fermentation, we will first look at a general overview of the processes.

As shown in Figure 5.10, the respiration of glucose typically occurs in three principal stages: glycolysis, the Krebs cycle, and the electron transport chain (system).

① Glycolysis is the oxidation of glucose to pyruvic acid with the production of some ATP and energy-containing NADH.

② The Krebs cycle is the oxidation of acetyl (a derivative of pyruvic acid) to carbon dioxide, with the production of some ATP, energy-containing NADH, and another reduced electron carrier, $FADH_2$.

③ In the electron transport chain (system), NADH and $FADH_2$ are oxidized, contributing the electrons they have carried from the substrates to a "cascade" of oxidation-reduction reactions involving a series of additional electron carriers. Energy from these reactions is used to generate a considerable amount of ATP. In respiration, most of the ATP is generated in the third step.

Because respiration involves a long series of oxidation-reduction reactions, the entire process can be thought of as involving a flow of electrons from the energy-rich glucose

FIGURE 5.10 An overview of respiration and fermentation.
① Glycolysis produces ATP and reduces NAD^+ to NADH while oxidizing glucose to pyruvic acid. In respiration, the pyruvic acid is converted into the first reactant in ② the Krebs cycle, which produces ATP and reduces NAD^+ (and another electron carrier called $FADH_2$) while giving off CO_2. The NADH from both processes carries electrons to ③ the electron transport chain, in which their energy is used to produce a great deal of ATP. In fermentation, the pyruvic acid and the electrons carried by NADH from glycolysis are incorporated into fermentation end-products. A small version of this figure will be included in figures throughout the chapter to indicate the relationships of different reactions to the overall processes.

■ **What is the basic difference between respiration and fermentation?**

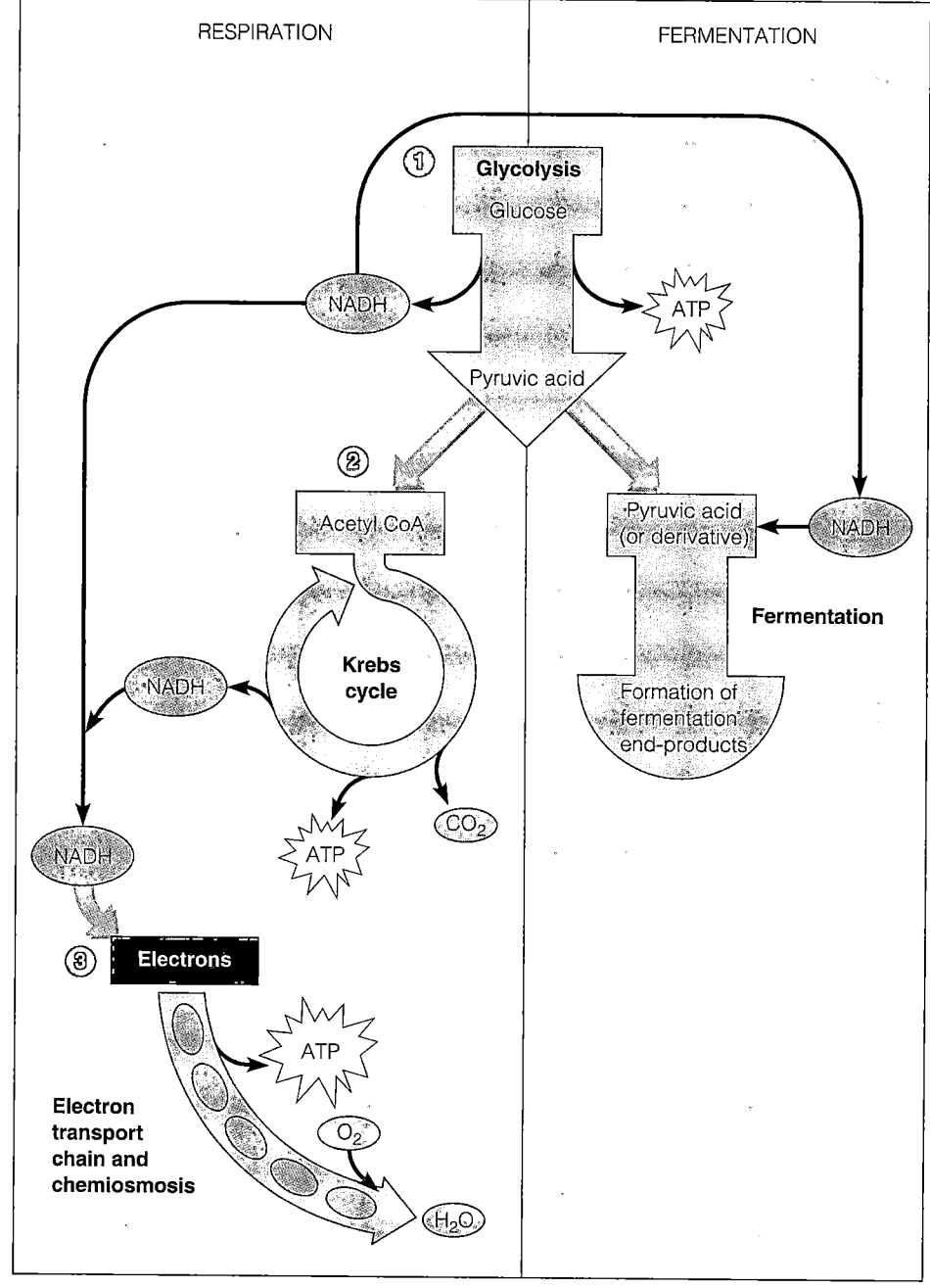

molecule to the relatively energy-poor CO_2 and H_2O molecules. The coupling of ATP production to this flow is somewhat analogous to the production of electrical power by using energy from a flowing stream. Carrying the analogy further, you could imagine a stream flowing down a gentle slope during glycolysis and the Krebs cycle, supplying energy to turn two old-fashioned waterwheels. Then the stream rushes down a steep slope in the electron transport chain, supplying energy for a large modern power plant. In a similar way, glycolysis and the Krebs cycle generate a small amount of ATP and also supply the electrons

that generate a great deal of ATP at the electron transport chain stage.

Typically, the initial stage of fermentation is also glycolysis (see Figure 5.10). However, once glycolysis has taken place, the pyruvic acid is converted into one or more different products, depending on the type of cell. These products might include alcohol (ethanol) and lactic acid. Unlike respiration, there is no Krebs cycle or electron transport chain in fermentation. Accordingly, the ATP yield, which comes only from glycolysis, is much lower.

Glycolysis

Glycolysis, the oxidation of glucose to pyruvic acid, is usually the first stage in carbohydrate catabolism. Most microorganisms use this pathway; in fact, it occurs in most living cells.

Glycolysis is also called the *Embden-Meyerhof pathway*. The word *glycolysis* means splitting of sugar, and this is exactly what happens. The enzymes of glycolysis catalyze the splitting of glucose, a six-carbon sugar, into two three-carbon sugars. These sugars are then oxidized, releasing energy, and their atoms are rearranged to form two molecules of pyruvic acid. During glycolysis NAD$^+$ is reduced to NADH, and there is a net production of two ATP molecules by substrate-level phosphorylation. Glycolysis does not require oxygen; it can occur whether oxygen is present or not. This pathway is a series of ten chemical reactions, each catalyzed by a different enzyme. The steps are outlined in Figure 5.11; see also Appendix C for a more detailed representation of glycolysis.

To summarize the process, glycolysis consists of two basic stages, a preparatory stage and an energy-conserving stage:

1. First, in the preparatory stage (steps ①–④ in Figure 5.11), two molecules of ATP are used as a six-carbon glucose molecule is phosphorylated, restructured, and split into two three-carbon compounds: glyceraldehyde 3-phosphate (GP) and dihydroxyacetone phosphate (DHAP). ⑤ DHAP is readily converted to GP. (The reverse reaction may also occur.) The conversion of DHAP into GP means that from this point on in glycolysis, two molecules of GP are fed into the remaining chemical reactions.

2. In the energy-conserving stage (steps ⑥–⑩ in Figure 5.11), the two three-carbon molecules are oxidized in several steps to two molecules of pyruvic acid. In these reactions, two molecules of NAD$^+$ are reduced to NADH, and four molecules of ATP are formed by substrate-level phosphorylation.

Because two molecules of ATP were needed to get glycolysis started and four molecules of ATP are generated by the process, *there is a net gain of two molecules of ATP for each molecule of glucose that is oxidized.*

Alternatives to Glycolysis

Many bacteria have another pathway in addition to glycolysis for the oxidation of glucose. The most common alternative is the pentose phosphate pathway; another alternative is the Entner-Doudoroff pathway.

The Pentose Phosphate Pathway

The **pentose phosphate pathway** (or *hexose monophosphate shunt*) operates simultaneously with glycolysis and provides a means for the breakdown of five-carbon sugars (pentoses) as well as glucose (see the detailed figure in Appendix C). A key feature of this pathway is that it produces important intermediate pentoses used in the synthesis of (1) nucleic acids, (2) glucose from carbon dioxide in photosynthesis, and (3) certain amino acids. The pathway is an important producer of the reduced coenzyme NADPH from NADP$^+$. The pentose phosphate pathway yields a net gain of only one molecule of ATP for each molecule of glucose oxidized. Bacteria that use the pentose phosphate pathway include *Bacillus subtilis* (sub′til-us), *E. coli, Leuconostoc mesenteroides* (lü-kō-nos′tok mes-en-ter-oi′dēz), and *Enterococcus faecalis* (fē-kāl′is).

The Entner-Doudoroff Pathway

From each molecule of glucose, the **Entner-Doudoroff pathway** produces two molecules of NADPH and one molecule of ATP for use in cellular biosynthetic reactions (see Appendix C for a more detailed representation). Bacteria that have the enzymes for the Entner-Doudoroff pathway can metabolize glucose without either glycolysis or the pentose phosphate pathway. The Entner-Doudoroff pathway is found in some gram-negative bacteria, including *Rhizobium, Pseudomonas* (sū-dō-mō′nas), and *Agrobacterium* (ag-rō-bak-ti′rē-um); it is generally not found among gram-positive bacteria. Tests for the ability to oxidize glucose by this pathway are sometimes used to identify *Pseudomonas* in the clinical laboratory.

Cellular Respiration

Learning Objective

- *Compare and contrast aerobic and anaerobic respiration.*

After glucose has been broken down to pyruvic acid, the pyruvic acid can be channeled into the next step of either fermentation (described later) or cellular respiration (see Figure 5.10). **Cellular respiration,** or simply **respiration,** is defined as an ATP-generating process in which molecules are oxidized and the final electron acceptor is (almost always) an inorganic molecule. An essential feature of respiration is the operation of an electron transport chain.

There are two types of respiration, depending on whether an organism is an **aerobe,** which uses oxygen, or an **anaerobe,** which does not use oxygen and may even be killed by it. In **aerobic respiration,** the final

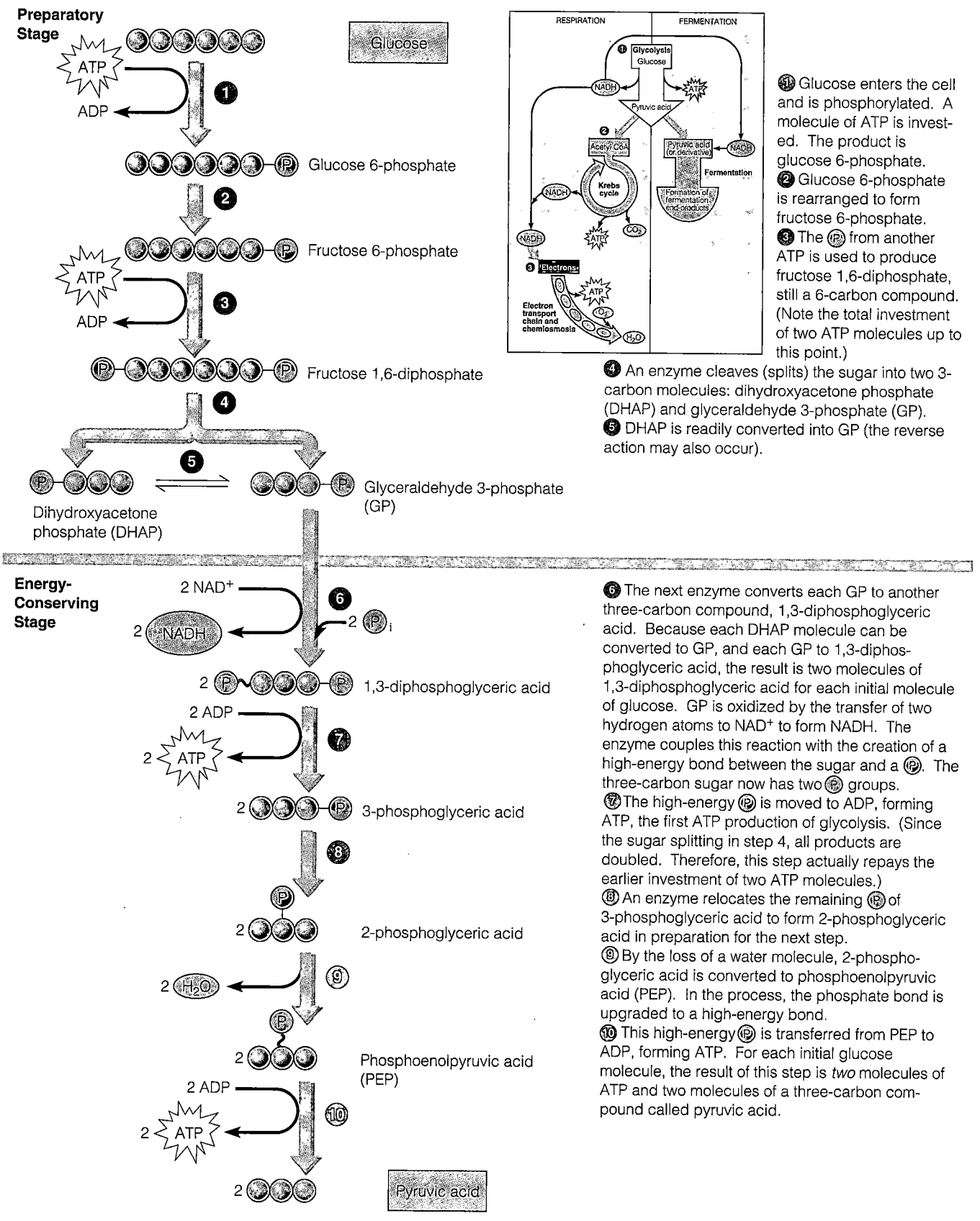

Preparatory Stage

Glucose

Glucose 6-phosphate

Fructose 6-phosphate

Fructose 1,6-diphosphate

Dihydroxyacetone phosphate (DHAP)

Glyceraldehyde 3-phosphate (GP)

Energy-Conserving Stage

2 NAD+

2 NADH

2 P_i

1,3-diphosphoglyceric acid

2 ADP

2 ATP

3-phosphoglyceric acid

2-phosphoglyceric acid

2 H_2O

Phosphoenolpyruvic acid (PEP)

2 ADP

2 ATP

Pyruvic acid

❶ Glucose enters the cell and is phosphorylated. A molecule of ATP is invested. The product is glucose 6-phosphate.
❷ Glucose 6-phosphate is rearranged to form fructose 6-phosphate.
❸ The ℗ from another ATP is used to produce fructose 1,6-diphosphate, still a 6-carbon compound. (Note the total investment of two ATP molecules up to this point.)
❹ An enzyme cleaves (splits) the sugar into two 3-carbon molecules: dihydroxyacetone phosphate (DHAP) and glyceraldehyde 3-phosphate (GP).
❺ DHAP is readily converted into GP (the reverse action may also occur).

❻ The next enzyme converts each GP to another three-carbon compound, 1,3-diphosphoglyceric acid. Because each DHAP molecule can be converted to GP, and each GP to 1,3-diphosphoglyceric acid, the result is two molecules of 1,3-diphosphoglyceric acid for each initial molecule of glucose. GP is oxidized by the transfer of two hydrogen atoms to NAD+ to form NADH. The enzyme couples this reaction with the creation of a high-energy bond between the sugar and a ℗. The three-carbon sugar now has two ℗ groups.
❼ The high-energy ℗ is moved to ADP, forming ATP, the first ATP production of glycolysis. (Since the sugar splitting in step 4, all products are doubled. Therefore, this step actually repays the earlier investment of two ATP molecules.)
❽ An enzyme relocates the remaining ℗ of 3-phosphoglyceric acid to form 2-phosphoglyceric acid in preparation for the next step.
❾ By the loss of a water molecule, 2-phosphoglyceric acid is converted to phosphoenolpyruvic acid (PEP). In the process, the phosphate bond is upgraded to a high-energy bond.
❿ This high-energy ℗ is transferred from PEP to ADP, forming ATP. For each initial glucose molecule, the result of this step is *two* molecules of ATP and two molecules of a three-carbon compound called pyruvic acid.

FIGURE 5.11 An outline of the reactions of glycolysis. The inset indicates the relationship of glycolysis to the overall processes of respiration and fermentation. A more detailed version of glycolysis is presented in Appendix C.

■ Glycolysis is the oxidation of glucose to pyruvic acid with the production of ATP and NADH.

electron acceptor is O_2; in **anaerobic respiration,** the final electron acceptor is an inorganic molecule other than molecular oxygen or, rarely, an organic molecule. First we will describe respiration as it typically occurs in an aerobic cell.

Aerobic Respiration

The Krebs Cycle The **Krebs cycle,** also called the *tricarboxylic acid (TCA) cycle* or *citric acid cycle,* is a series of biochemical reactions in which the large amount of potential chemical energy stored in acetyl CoA is released step by step (see Figure 5.10). In this cycle, a series of oxidations and reductions transfer that potential energy, in the form of electrons, to electron carrier coenzymes, chiefly NAD^+. The pyruvic acid derivatives are oxidized; the coenzymes are reduced.

Pyruvic acid, the product of glycolysis, cannot enter the Krebs cycle directly. In a preparatory step, it must lose one molecule of CO_2 and become a two-carbon compound (Figure 5.12, at top). This process is called **decarboxylation.** The two-carbon compound, called an *acetyl group,* attaches to coenzyme A through a high-energy bond; the resulting complex is known as *acetyl coenzyme A (acetyl CoA).* During this reaction, pyruvic acid is also oxidized and NAD^+ is reduced to NADH.

Remember that the oxidation of one glucose molecule produces two molecules of pyruvic acid, so for each molecule of glucose, two molecules of CO_2 are released in this preparatory step, two molecules of NADH are produced, and two molecules of acetyl CoA are formed. Once the pyruvic acid has undergone decarboxylation and its derivative (the acetyl group) has attached to CoA, the resulting acetyl CoA is ready to enter the Krebs cycle.

As acetyl CoA enters the Krebs cycle, CoA detaches from the acetyl group. The two-carbon acetyl group combines with a four-carbon compound called oxaloacetic acid to form the six-carbon citric acid. This synthesis reaction requires energy, which is provided by the cleavage of the high-energy bond between the acetyl group and CoA. The formation of citric acid is thus the first step in the Krebs cycle. The major chemical reactions of this cycle are outlined in Figure 5.12; a more detailed representation of the Krebs cycle is provided in Appendix C. Keep in mind that each reaction is catalyzed by a specific enzyme.

The chemical reactions of the Krebs cycle fall into several general categories; one of these is decarboxylation. For example, in step ③ isocitric acid, a six-carbon compound, is decarboxylated to the five-carbon compound called α-ketoglutaric acid. Another decarboxylation takes place in step ④. Because one decarboxylation has taken place in the preparatory step and two in the Krebs cycle, all three carbon atoms in pyruvic acid are

eventually released as CO_2 by the Krebs cycle. This represents the conversion to CO_2 of all six carbon atoms contained in the original glucose molecule.

Another general category of Krebs cycle chemical reactions is oxidation-reduction. For example, in step ③, two hydrogen atoms are lost during the conversion of the six-carbon isocitric acid to a five-carbon compound. In other words, the six-carbon compound is oxidized. Hydrogen atoms are also released in the Krebs cycle in steps ④, ⑥, and ⑧ and are picked up by the coenzymes NAD^+ and FAD. Because NAD^+ picks up two electrons but only one additional proton, its reduced form is represented as NADH; however, FAD picks up two complete hydrogen atoms and is reduced to $FADH_2$.

If we look at the Krebs cycle as a whole, we see that for every two molecules of acetyl CoA that enter the cycle, four molecules of CO_2 are liberated by decarboxylation, six molecules of NADH and two molecules of $FADH_2$ are produced by oxidation-reduction reactions, and two molecules of ATP are generated by substrate-level phosphorylation. Many of the intermediates in the Krebs cycle also play a role in other pathways, especially in amino acid biosynthesis (discussed later in the chapter).

The CO_2 produced in the Krebs cycle is ultimately liberated into the atmosphere as a gaseous by-product of aerobic respiration. (Humans produce CO_2 from the Krebs cycle in most cells of the body and discharge it through the lungs during exhalation.) The reduced coenzymes NADH and $FADH_2$ are the most important products of the Krebs cycle because they contain most of the energy originally stored in glucose. During the next phase of respiration, a series of reductions indirectly transfers the energy stored in those coenzymes to ATP. These reactions are collectively called the electron transport chain.

The Electron Transport Chain (System) An **electron transport chain (system)** consists of a sequence of carrier molecules that are capable of oxidation and reduction. As electrons are passed through the chain, there is a stepwise release of energy, which is used to drive the chemiosmotic generation of ATP, to be described shortly. The final oxidation is irreversible. In eukaryotic cells, the electron transport chain is contained in the inner membrane of mitochondria; in prokaryotic cells, it is found in the plasma membrane.

There are three classes of carrier molecules in electron transport chains. The first are **flavoproteins.** These proteins contain flavin, a coenzyme derived from riboflavin (vitamin B_2), and are capable of performing alternating oxidations and reductions. One important flavin coenzyme is flavin mononucleotide (FMN). The second class

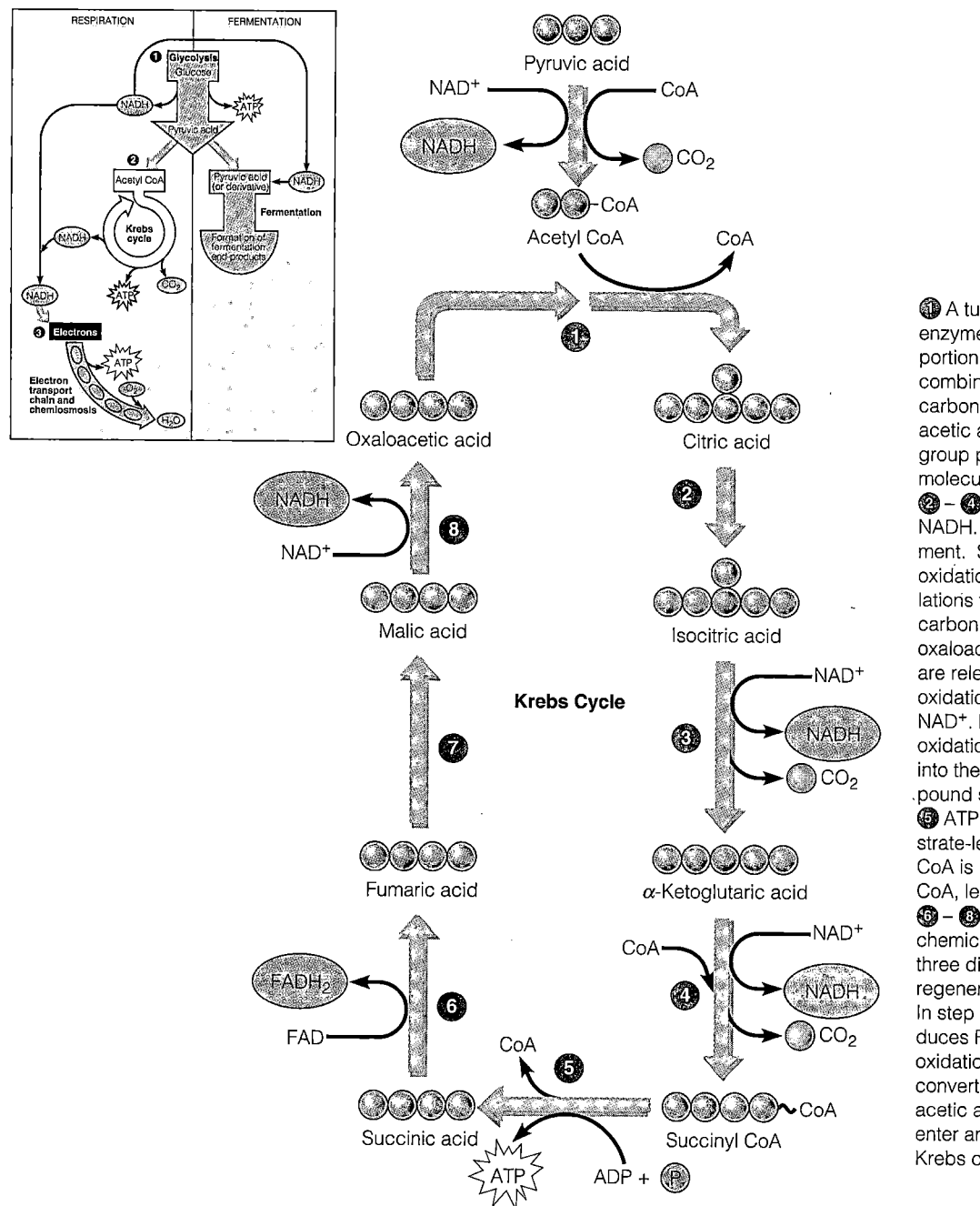

A turn of the cycle begins as enzymes strip off the CoA portion from acetyl CoA and combine the remaining two-carbon acetyl group with oxalo-acetic acid. Adding the acetyl group produces the six-carbon molecule citric acid.

② – ④ Oxidations generate NADH. Step 2 is a rearrangement. Steps 3 and 4 combine oxidations and decarboxylations to dispose of the two carbon atoms that came from oxaloacetic acid. The carbons are released as CO_2, and the oxidations generate NADH from NAD^+. During the second oxidation (step 4), CoA is added into the cycle, forming the compound succinyl CoA.

⑤ ATP is produced by substrate-level phosphorylation. CoA is removed from succinyl CoA, leaving succinic acid.

⑥ – ⑧ Enzymes rearrange chemical bonds, producing three different molecules before regenerating oxaloacetic acid. In step 6, an oxidation produces $FADH_2$. In step 8, a final oxidation generates NADH and converts malic acid to oxaloacetic acid, which is ready to enter another round of the Krebs cycle. See Appendix C.

FIGURE 5.12 The Krebs cycle. The inset indicates the relationship of the Krebs cycle to the overall process of respiration.

■ The Krebs cycle is the oxidation of acetyl groups to CO_2 with the production of ATP, NADH, and $FADH_2$.

of carrier molecules are **cytochromes,** proteins with an iron-containing group (heme) capable of existing alternately as a reduced form (Fe^{2+}) and an oxidized form (Fe^{3+}). The cytochromes involved in electron transport chains include cytochrome b (cyt b), cytochrome c_1 (cyt c_1), cytochrome c (cyt c), cytochrome a (cyt a), and cytochrome a_3 (cyt a_3). The third class is known as

ubiquinones, or **coenzyme Q,** symbolized Q; these are small nonprotein carriers.

The electron transport chains of bacteria are somewhat diverse, in that the particular carriers used by a bacterium and the order in which they function may differ from those of other bacteria and from those of eukaryotic mitochondrial systems. Even a single bacterium may have

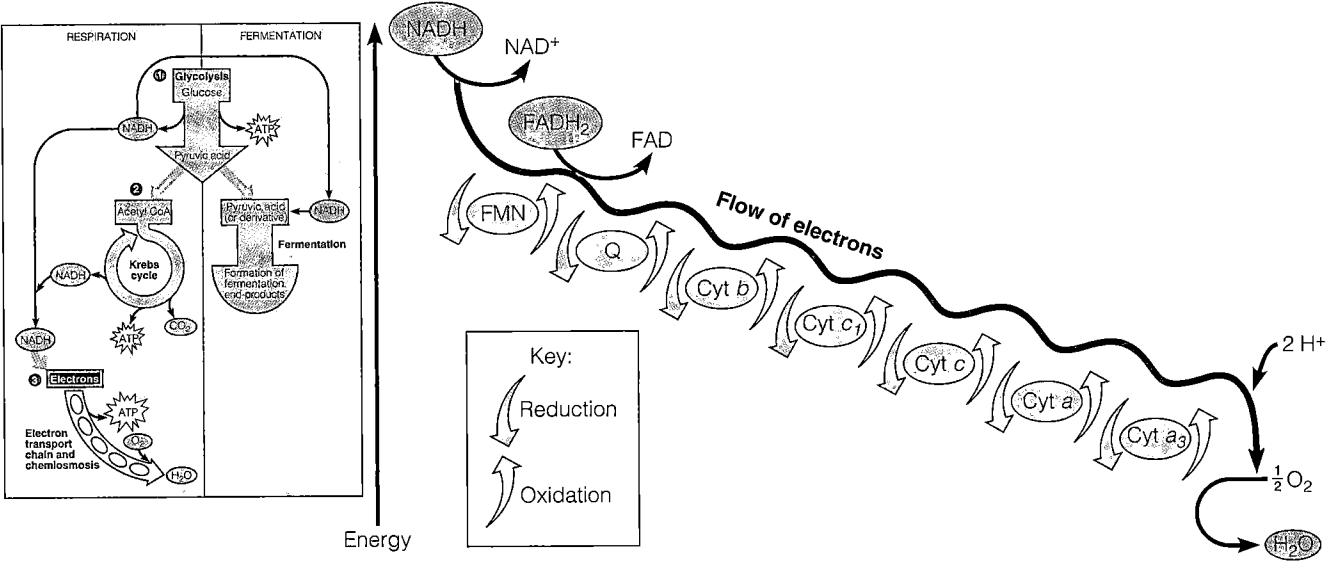

FIGURE 5.13 An electron transport chain (system). The inset indicates the relationship of the electron transport chain to the overall process of respiration. In the mitochondrial electron transport chain shown, the electrons pass along the chain in a gradual and stepwise fashion, so energy is released in manageable quantities. To learn where ATP is formed, see Figure 5.15.

■ In the electron transport chain, NADH and FADH$_2$ are oxidized, and electron carriers are oxidized and reduced to produce ATP.

several types of electron transport chains. However, keep in mind that all electron transport chains achieve the same basic goal, that of releasing energy as electrons are transferred from higher-energy compounds to lower-energy compounds. Much is known about the electron transport chain in the mitochondria of eukaryotic cells, so this is the chain we will describe.

The first step in the mitochondrial electron transport chain involves the transfer of high-energy electrons from NADH to FMN, the first carrier in the chain (Figure 5.13). This transfer actually involves the passage of a hydrogen atom with two electrons to FMN, which then picks up an additional H$^+$ from the surrounding aqueous medium. As a result of the first transfer, NADH is oxidized to NAD$^+$, and FMN is reduced to FMNH$_2$. In the second step in the electron transport chain, FMNH$_2$ passes 2H$^+$ to the other side of the mitochondrial membrane (see Figure 5.15) and passes two electrons to Q. As a result, FMNH$_2$ is oxidized to FMN. Q also picks up an additional 2H$^+$ from the surrounding aqueous medium and releases it on the other side of the membrane.

The next part of the electron transport chain involves the cytochromes. Electrons are passed successively from Q to cyt b, cyt c_1, cyt c, cyt a, and cyt a_3. Each cytochrome in the chain is reduced as it picks up electrons and is oxidized as it gives up electrons. The last cytochrome, cyt a_3, passes its electrons to molecular oxygen (O$_2$), which becomes negatively charged and then picks up protons from the surrounding medium to form H$_2$O.

Notice that Figure 5.13 shows FADH$_2$, which is derived from the Krebs cycle, as another source of electrons. However, FADH$_2$ adds its electrons to the electron transport chain at a lower level than NADH. Because of this, the electron transport chain produces about one-third less energy for ATP generation when FADH$_2$ donates electrons than when NADH is involved.

An important feature of the electron transport chain is the presence of some carriers, such as FMN and Q, that accept and release protons as well as electrons, and other carriers, such as cytochromes, that transfer electrons only. Electron flow down the chain is accompanied at several points by the active transport (pumping) of protons from the matrix side of the inner mitochondrial membrane to the opposite side of the membrane. The result is a buildup of protons on one side of the membrane. Just as water behind a dam stores energy that can be used to generate electricity, this buildup of protons provides energy for the generation of ATP by the chemiosmotic mechanism.

The Chemiosmotic Mechanism of ATP Generation The mechanism of ATP synthesis using the electron transport chain is called **chemiosmosis.** To understand

Source	ATP Yield (Method)

Glycolysis

1. Oxidation of glucose to pyruvic acid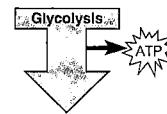

2. Production of 2 NADH

2 ATP (substrate-level phosphorylation)

6 ATP (oxidative phosphorylation in electron transport chain)

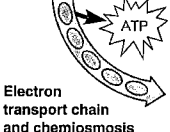

Electron transport chain and chemiosmosis

Preparatory Step

1. Formation of acetyl CoA produces 2 NADH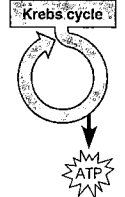

6 ATP (oxidative phosphorylation in electron transport chain)

Krebs Cycle

1. Oxidation of succinyl CoA to succinic acid

2 GTP (equivalent of ATP; substrate-level phosphorylation)

2. Production of 6 NADH

18 ATP (oxidative phosphorylation in electron transport chain)

3. Production of 2 FADH

4 ATP (oxidative phosphorylation in electron transport chain)

Total: 38 ATP

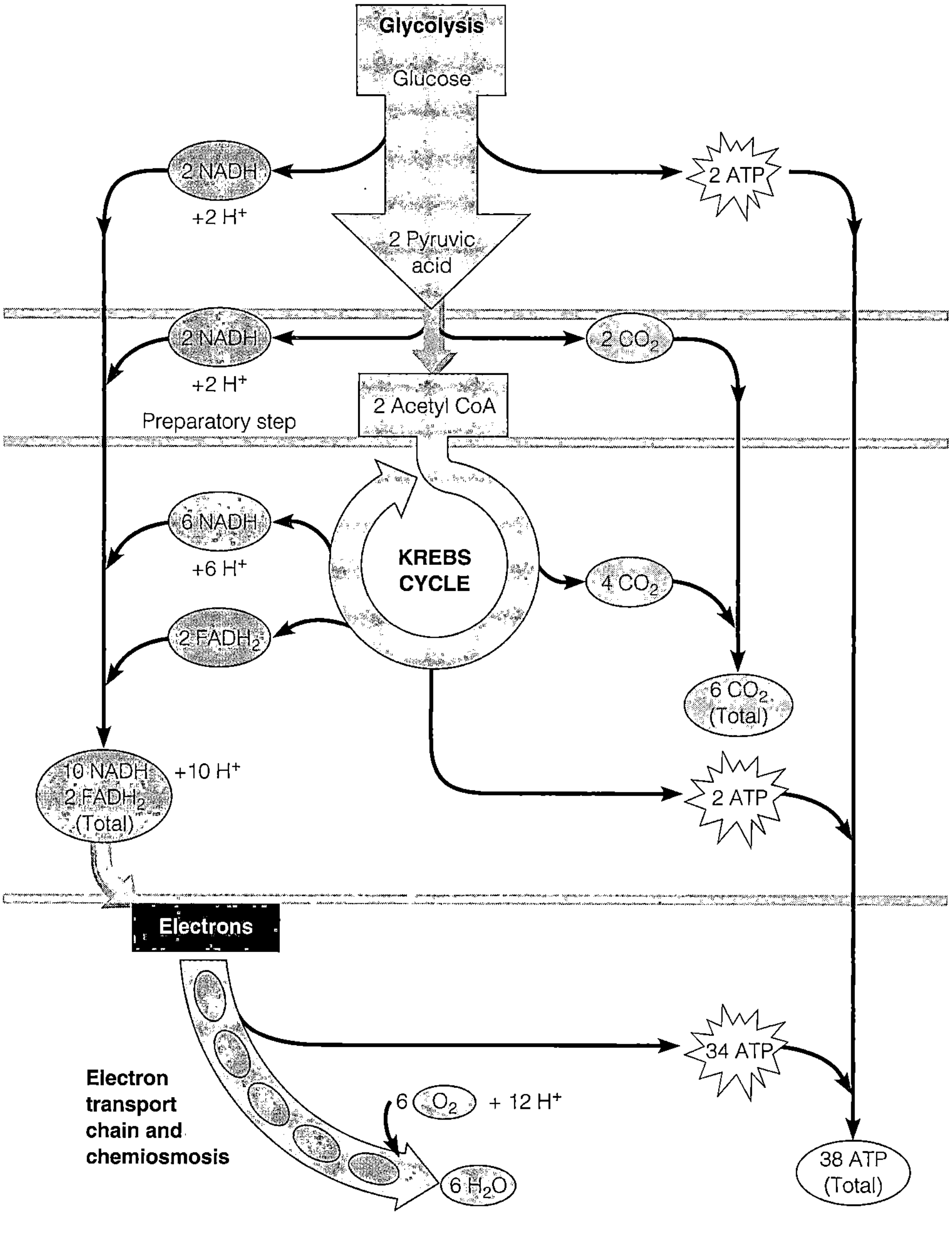

FIGURE 5.16 A summary of aerobic respiration in prokaryotes. Glucose is broken down completely to carbon dioxide and water, and ATP is generated. This process has three major phases: glycolysis, the Krebs cycle, and the electron transport chain. The preparatory step is between glycolysis and the Krebs cycle. The key event in aerobic respiration is that electrons are picked up from intermediates of glycolysis and the Krebs cycle by NAD$^+$ or FAD and are carried by NADH or FADH$_2$ to the electron transport chain. NADH is also produced in the conversion of pyruvic acid to acetyl CoA. Most of the ATP generated by aerobic respiration is made by the chemiosmotic mechanism during the electron transport chain phase; this is called oxidative phosphorylation.

■ How do aerobic and anaerobic respiration differ?

APPLICATIONS OF MICROBIOLOGY

What Is Fermentation?

To many people, fermentation simply means the production of alcohol. Grains and fruits are fermented to produce beer and wine. If a food soured, you might say it was "off" or fermented. Here are some definitions of fermentation. They range from informal, general usage to more scientific definitions. Fermentation is:

1. Any spoilage of food by microorganisms (general use).

2. Any process that produces alcoholic beverages or acidic dairy products (general use).

3. Any large-scale microbial process occurring with or without air (common definition used in industry).

4. Any energy-releasing metabolic process that takes place only under anaerobic conditions (becoming more scientific).

5. Any metabolic process that releases energy from a sugar or other organic molecule, does not require oxygen or an electron transport system, and uses an organic molecule as the final electron acceptor. (This is the definition we use in this book.)

Anaerobic Respiration

In anaerobic respiration, the final electron acceptor is an inorganic substance other than oxygen (O_2). Some bacteria, such as *Pseudomonas* and *Bacillus*, can use a nitrate ion (NO_3^-) as a final electron acceptor; the nitrate ion is reduced to a nitrite ion (NO_2^-), nitrous oxide (N_2O), or nitrogen gas (N_2). Other bacteria, such as *Desulfovibrio* (dē-sul-fō-vib'rē-ō), use sulfate (SO_4^{2-}) as the final electron acceptor to form hydrogen sulfide (H_2S). Still other bacteria use carbonate (CO_3^{2-}) to form methane (CH_4). Anaerobic respiration by bacteria using nitrate and sulfate as final acceptors is essential for the nitrogen and sulfur cycles that occur in nature. The amount of ATP generated in anaerobic respiration varies with the organism and the pathway. Because only part of the Krebs cycle operates under anaerobic conditions, and since not all the carriers in the electron transport chain participate in anaerobic respiration, the ATP yield is never as high as in aerobic respiration. Accordingly, anaerobes tend to grow more slowly than aerobes.

Fermentation

Learning Objective

- *Describe the chemical reactions of, and list some products of, fermentation.*

After glucose has been broken down into pyruvic acid, the pyruvic acid can be completely broken down in respiration, as previously described, or it can be converted to an organic product in fermentation (see Figure 5.10). **Fermentation** can be defined in several ways (see the box above), but we define it here as a process that:

1. releases energy from sugars or other organic molecules, such as amino acids, organic acids, purines, and pyrimidines;

2. does not require oxygen (but sometimes can occur in its presence);

3. does not require use of the Krebs cycle or an electron transport chain;

4. uses an organic molecule as the final electron acceptor;

5. produces only small amounts of ATP (only one or two ATP molecules for each molecule of starting material) because much of the original energy in glucose remains in the chemical bonds of the organic end-products, such as lactic acid or ethanol.

During fermentation, electrons are transferred (along with protons) from reduced coenzymes (NADH, NADPH) to pyruvic acid or its derivatives (Figure 5.17a). Those final electron acceptors are reduced to the end-products shown in Figure 5.17b. In the process, NAD^+ and $NADP^+$ are regenerated and can enter another round of glycolysis. An essential function of the second stage of fermentation is to ensure a steady supply of NAD^+ and $NADP^+$ so that glycolysis can continue. In fermentation, ATP is generated only during glycolysis.

Various microorganisms can ferment various substrates; the end-products depend on the particular microorganism, the substrate, and the enzymes that are present and active. Chemical analyses of these end-products are useful in identifying microorganisms. We next consider two of the more important processes: lactic acid fermentation and alcohol fermentation.

Lactic Acid Fermentation

During glycolysis, which is the first phase of **lactic acid fermentation,** a molecule of glucose is oxidized to two molecules of pyruvic acid (Figure 5.18; see also Figure 5.11). This oxidation generates the energy that is used to

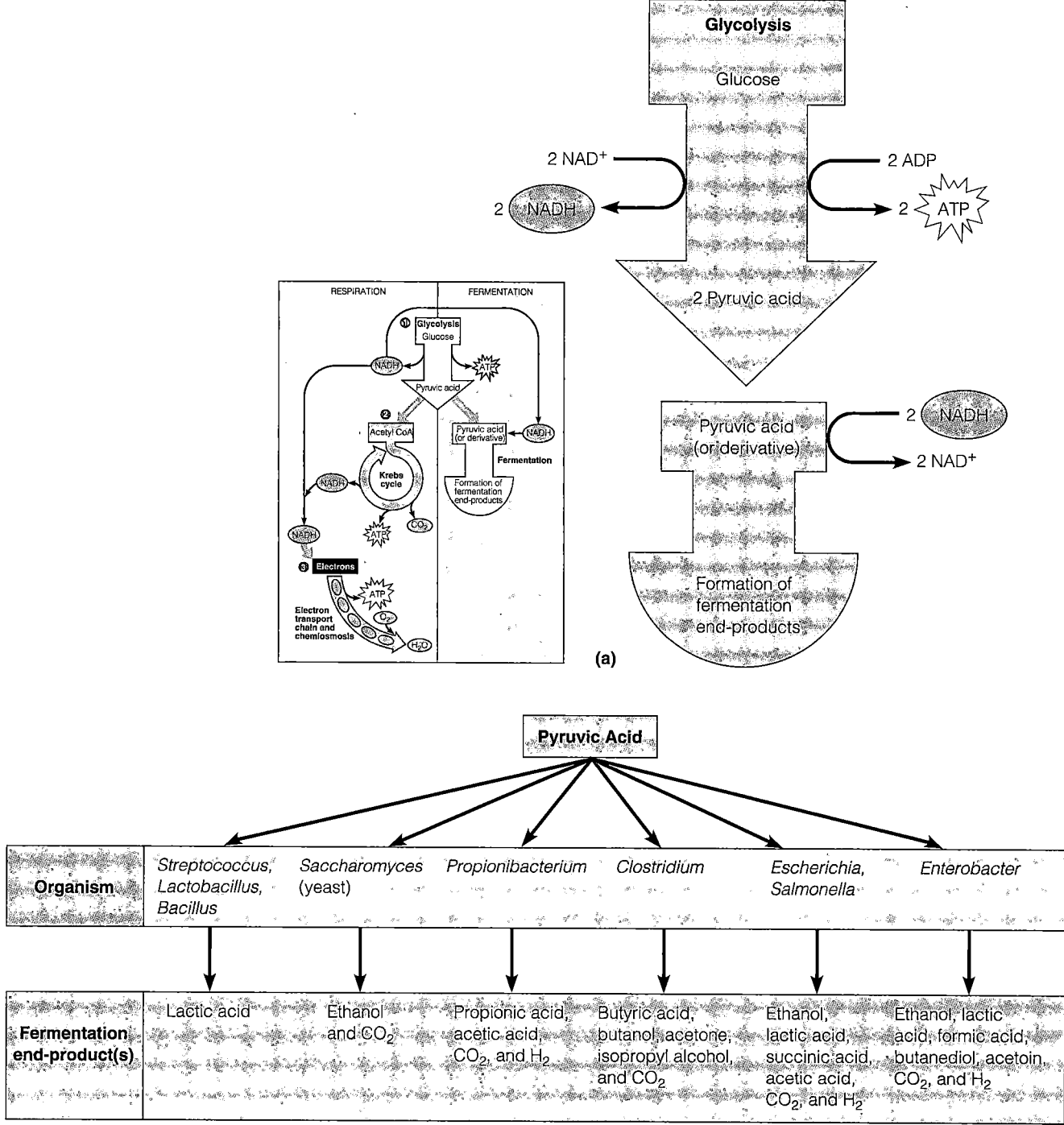

(a)

(b)

FIGURE 5.17 **Fermentation.** The inset indicates the relationship of fermentation to the overall energy-producing processes. **(a)** An overview of fermentation. The first step is glycolysis, the conversion of glucose to pyruvic acid. In the second step, the reduced coenzymes from glycolysis or its alternatives (NADH, NADPH) donate their electrons and hydrogen ions to pyruvic acid or a derivative to form a fermentation end-product. **(b)** End-products of various microbial fermentations.

■ In fermentation, ATP is generated only during glycolysis.

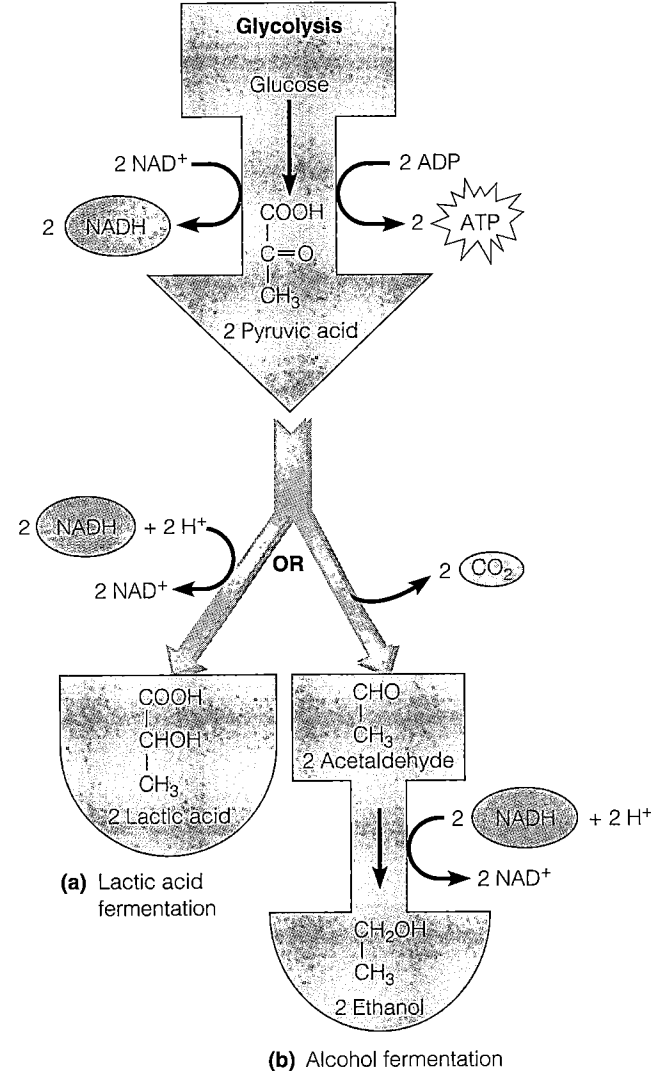

FIGURE 5.18 Types of fermentation.

What is the difference between homolactic and heterolactic fermentation?

form the two molecules of ATP. In the next step, the two molecules of pyruvic acid are reduced by two molecules of NADH to form two molecules of lactic acid (Figure 5.18a). Because lactic acid is the end-product of the reaction, it undergoes no further oxidation, and most of the energy produced by the reaction remains stored in the lactic acid. Thus, this fermentation yields only a small amount of energy.

Two important genera of lactic acid bacteria are *Streptococcus* and *Lactobacillus* (lak-tō-bä-sil'lus). Because these microbes produce only lactic acid, they are referred to as **homolactic** (or *homofermentative*). Lactic acid fermentation can result in food spoilage. However, the process can also produce yogurt from milk, sauerkraut from fresh cabbage, and pickles from cucumbers.

Alcohol Fermentation

Alcohol fermentation also begins with the glycolysis of a molecule of glucose to yield two molecules of pyruvic acid and two molecules of ATP. In the next reaction, the two molecules of pyruvic acid are converted to two molecules of acetaldehyde and two molecules of CO_2 (Figure 5.18b). The two molecules of acetaldehyde are next reduced by two molecules of NADH to form two molecules of ethanol. Again, alcohol fermentation is a low-energy-yield process because most of the energy contained in the initial glucose molecule remains in the ethanol, the end-product.

Alcohol fermentation is carried out by a number of bacteria and yeasts. The ethanol and carbon dioxide produced by the yeast *Saccharomyces* (sak-ä-rō-mī'sēs) are waste products for yeast cells but are useful to humans. Ethanol made by yeasts is the alcohol in alcoholic beverages, and carbon dioxide made by yeasts causes bread dough to rise (see the box in Chapter 1, page 5).

Organisms that produce lactic acid as well as other acids or alcohols are known as **heterolactic** (or *heterofermentative*) and often use the pentose phosphate pathway.

Table 5.4 lists some of the various microbial fermentations used by industry to convert inexpensive raw materials into useful end-products. A summary comparison of aerobic respiration, anaerobic respiration, and fermentation is given in Table 5.5.

Lipid and Protein Catabolism

Learning Objective

■ *Describe how lipids and proteins undergo catabolism.*

Our discussion of energy production has emphasized the oxidation of glucose, the main energy-supplying carbohydrate. However, microbes also oxidize lipids and proteins, and the oxidations of all these nutrients are related.

Recall that fats are lipids consisting of fatty acids and glycerol. Microbes produce extracellular enzymes called *lipases* that break fats down into their fatty acid and glycerol components. Each component is then metabolized separately (Figure 5.19). The Krebs cycle functions in the oxidation of glycerol and fatty acids. Many bacteria that hydrolyze fatty acids can use the same enzymes to degrade petroleum products. Although these bacteria are a nuisance when they grow in a fuel storage tank, they are beneficial when they grow in oil spills. Beta oxidation (the oxidation of fatty acids) of petroleum is discussed in the box in Chapter 2 (page 35).

Proteins are too large to pass unaided through plasma membranes. Microbes produce extracellular *proteases* and

table 5.4	Some Industrial Uses for Different Types of Fermentations		
Fermentation End-Product(s)	**Industrial or Commercial Use**	**Starting Material**	**Microorganism**
Ethanol	Beer	Malt extract	*Saccharomyces cerevisiae* (yeast, a fungus)
	Wine	Grape or other fruit juices	*Saccharomyces cerevisiae* var. *ellipsoideus*
	Fuel	Agricultural wastes	*Saccharomyces cerevisiae*
Acetic acid	Vinegar	Ethanol	*Acetobacter* (bacterium)
Lactic acid	Cheese, yogurt	Milk	*Lactobacillus, Streptococcus* (bacteria)
	Rye bread	Grain, sugar	*Lactobacillus bulgaricus* (bacterium)
	Sauerkraut	Cabbage	*Lactobacillus plantarum* (bacterium)
	Summer sausage	Meat	*Pediococcus* (bacterium)
Propionic acid and carbon dioxide	Swiss cheese	Lactic acid	*Propionibacterium freudenreichii* (bacterium)
Acetone and butanol	Pharmaceutical, industrial uses	Molasses	*Clostridium acetobutylicum* (bacterium)
Glycerol	Pharmaceutical, industrial uses	Molasses	*Saccharomyces cerevisiae*
Citric acid	Flavoring	Molasses	*Aspergillus* (fungus)
Methane	Fuel	Acetic acid	*Methanosarcina* (bacterium)
Sorbose	Vitamin C (ascorbic acid)	Sorbitol	*Acetobacter*

table 5.5	Aerobic Respiration, Anaerobic Respiration, and Fermentation Compared			
Energy-Producing Process	**Growth Conditions**	**Final Hydrogen (Electron) Acceptor**	**Type of Phosphorylation Used to Generate ATP**	**ATP Molecules Produced per Glucose Molecule**
Aerobic respiration	Aerobic	Molecular oxygen (O_2)	Substrate-level and oxidative	36 or 38*
Anaerobic respiration	Anaerobic	Usually an inorganic substance (such as NO_3^-, SO_4^{2-}, or CO_3^{2-}), but not molecular oxygen (O_2)	Substrate-level and oxidative	Variable (fewer than 38 but more than 2)
Fermentation	Aerobic or anaerobic	An organic molecule	Substrate-level	2

*In prokaryotic aerobic respiration, 38 ATP molecules are produced; in eukaryotic aerobic respiration, 36 ATP molecules are produced.

peptidases, enzymes that break down proteins into their component amino acids, which can cross the membranes. However, before amino acids can be catabolized, they must be enzymatically converted to other substances that can enter the Krebs cycle. In one such conversion, called **deamination,** the amino group of an amino acid is removed and converted to an ammonium ion (NH_4^+),

which can be excreted from the cell. The remaining organic acid can enter the Krebs cycle. Other conversions involve **decarboxylation** (the removal of —COO) and **dehydrogenation.**

A summary of the interrelationships of carbohydrate, lipid, and protein catabolism is shown in Figure 5.20 on page 138.

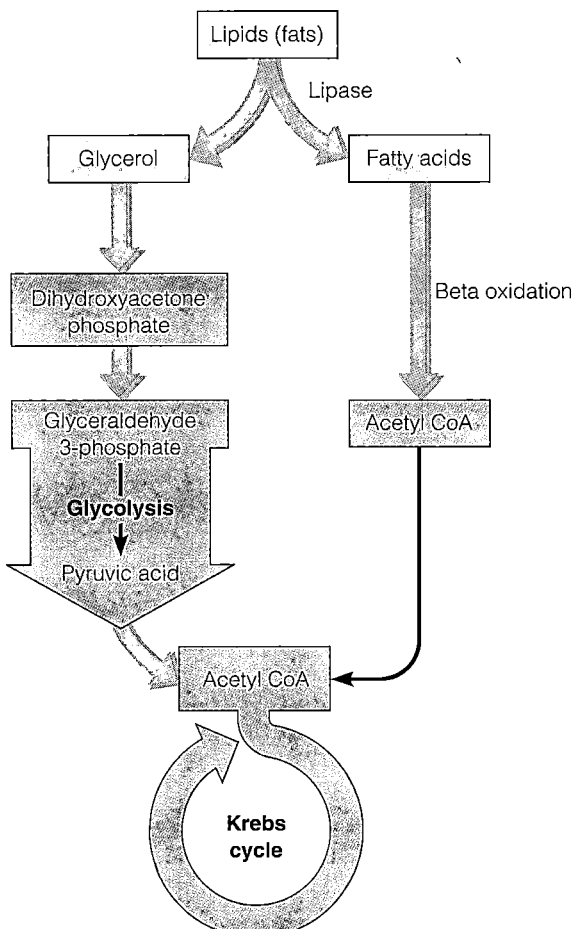

FIGURE 5.19 Lipid catabolism. Glycerol is converted into di-hydroxyacetone phosphate (DHAP) and catabolized via glycolysis and the Krebs cycle. Fatty acids undergo beta oxidation, in which carbon fragments are split off two at a time to form acetyl CoA, which is catabolized via the Krebs cycle.

■ Glycerol and fatty acids are metabolized along different metabolic pathways.

Biochemical Tests and Bacterial Identification

Learning Objective

■ *Describe two examples of the use of biochemical tests to identify bacteria in the laboratory.*

Biochemical testing is frequently used to identify bacteria and yeasts because different species produce different enzymes. Such biochemical tests are designed to detect the presence of enzymes. One type of biochemical test is the detection of amino acid catabolizing enzymes involved in decarboxylation and dehydrogenation (discussed on page 136; Figure 5.21 on page 139). Another is a **fermenta-**

tion test. The test medium contains protein, a single carbohydrate, a pH indicator, and a fermentation tube to capture gas (Figure 5.22 on page 139). Bacteria inoculated into the tube can use the protein or carbohydrate as a carbon and energy source. If they catabolize the carbohydrate and produce acid, the pH indicator changes color. Some organisms produce gas as well as acid from carbohydrate catabolism. The presence of a bubble in the Durham tube indicates gas formation (Figure 5.22b–d on page 139). Another example of the use of biochemical tests is shown in Figure 10.8 on page 288.

Note that in some instances, the waste products of one microorganism can be used as a carbon and energy source by another species. *Acetobacter* (ä-sē-tō-bak'ter) bacteria oxidize ethanol made by yeast. *Propionibacterium* (prō-pē-on-ē-bak-ti're-um) can use lactic acid produced by other bacteria. Propionibacteria convert lactic acid to pyruvic acid in preparation for the Krebs cycle. During the Krebs cycle, propionic acid and CO_2 are made. The holes in Swiss cheese are formed by the accumulation of the CO_2 gas.

Photosynthesis

Learning Objectives

■ *Compare and contrast cyclic and noncyclic photophosphorylation.*

■ *Compare and contrast the light and dark reactions of photosynthesis.*

■ *Compare and contrast oxidative phosphorylation and photophosphorylation.*

In all of the metabolic pathways just discussed, organisms obtain energy for cellular work by oxidizing organic compounds. But where do organisms obtain these organic compounds? Some, including animals and many microbes, feed on matter produced by other organisms. For example, bacteria may catabolize compounds from dead plants and animals or may obtain nourishment from a living host.

Other organisms synthesize complex organic compounds from simple inorganic substances. The major mechanism for such synthesis is a process called **photosynthesis,** which is used by plants and many microbes. Essentially, photosynthesis is the conversion of light energy from the sun into chemical energy. The chemical energy is then used to convert CO_2 from the atmosphere to more reduced carbon compounds, primarily sugars. The word *photosynthesis* summarizes the process: *photo* means light, and *synthesis* refers to the assembly of organic compounds. This synthesis of sugars by using carbon atoms from CO_2 gas is also called **carbon fixation.** Continuation of life as

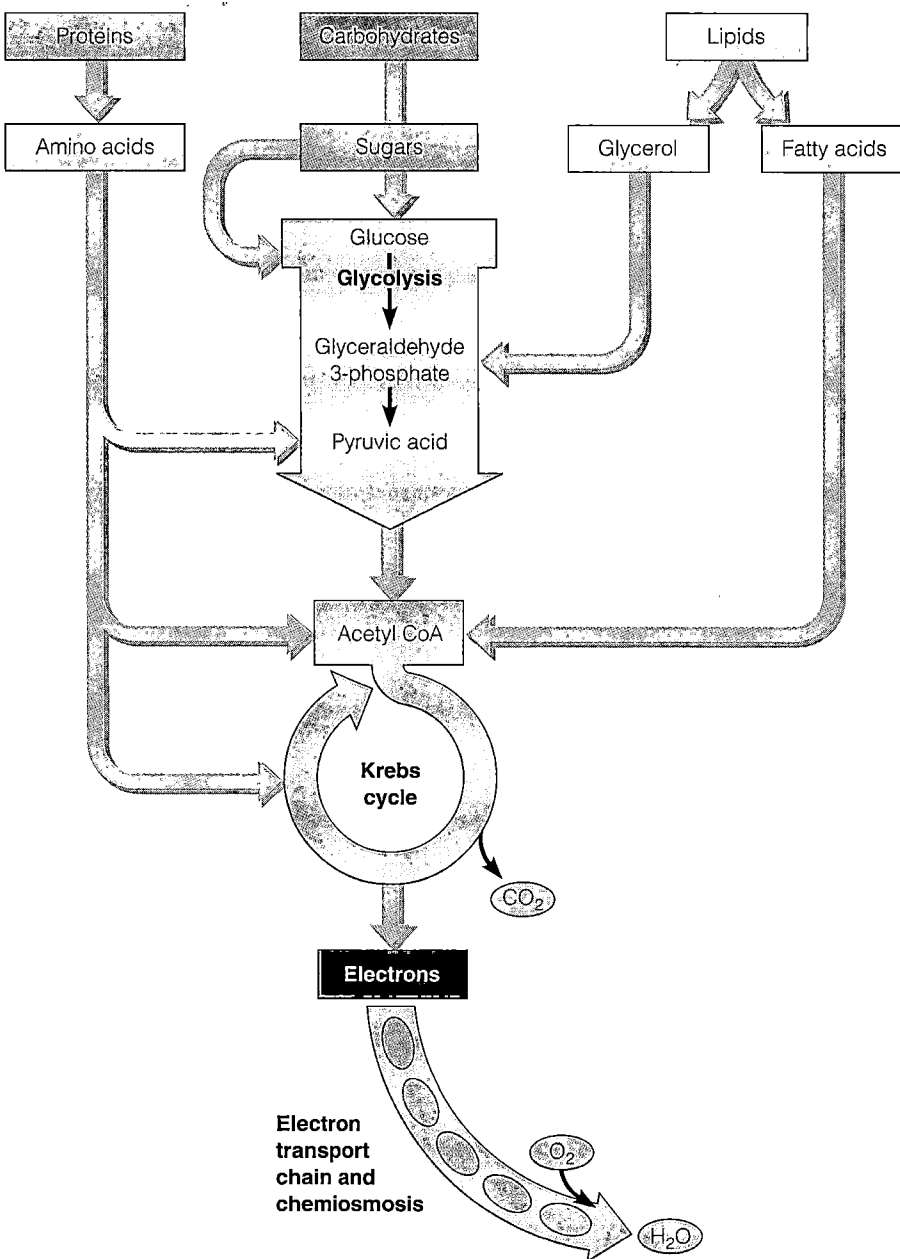

FIGURE 5.20 Catabolism of various organic food molecules. Proteins, carbohydrates, and lipids can all be sources of electrons and protons for respiration. These food molecules enter glycolysis or the Krebs cycle at various points.

■ Glycolysis and the Krebs cycle are catabolic funnels through which high-energy electrons from all kinds of organic molecules flow on their energy-releasing pathways.

we know it on Earth depends on the recycling of carbon in this way. Cyanobacteria, algae, and green plants all contribute to this vital recycling with photosynthesis.

Photosynthesis can be summarized as follows:

$$6 \ CO_2 + 12 \ H_2O + \text{Light energy} \longrightarrow$$
$$C_6H_{12}O_6 + 6 \ O_2 + 6 \ H_2O$$

In the course of photosynthesis, electrons are taken from the hydrogen atoms of water, an energy-poor molecule, and incorporated into sugar, an energy-rich molecule. The energy boost is supplied by light energy, although indirectly.

Photosynthesis takes place in two stages. In the first stage, called the **light (light-dependent) reactions,**

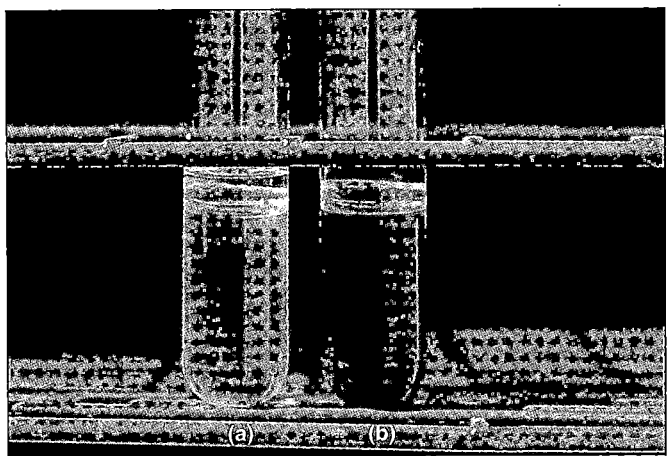

FIGURE 5.21 Detecting amino acid catabolizing enzymes in the lab. Bacteria are inoculated in tubes containing glucose, a pH indicator, and a specific amino acid. **(a)** The pH indicator turns to yellow when bacteria produce acid from glucose. **(b)** Alkaline products from decarboxylation turn the indicator to purple.

■ What is decarboxylation?

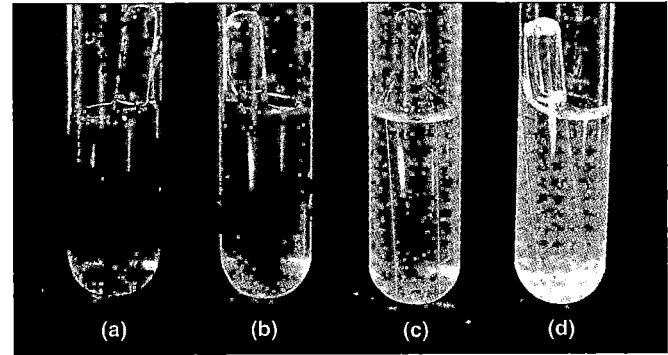

FIGURE 5.22 A fermentation test. (a) An uninoculated fermentation tube containing the carbohydrate mannitol. **(b)** *Staphylococcus epidermidis* grew on the protein but did not use the carbohydrate. This organism is described as mannitol −. **(c)** *Staphylococcus aureus* produced acid but not gas. This species is mannitol+. **(d)** *Escherichia coli* is also mannitol+ and produced acid and gas from mannitol. The gas is trapped in the inverted tube.

■ What is the importance of biochemical testing?

light energy is used to convert ADP and ℗ to ATP. In addition, in the predominant form of the light reactions, the electron carrier NADP is reduced to NADPH. The coenzyme NADPH, like NADH, is an energy-rich carrier of electrons. In the second stage, the **dark (light-independent) reactions,** these electrons are used along with energy from ATP to reduce CO_2 to sugar.

The Light Reactions: Photophosphorylation

Photophosphorylation is one of the three ways ATP is formed, and it occurs only in photosynthetic cells. In this mechanism, light energy is absorbed by chlorophyll molecules in the photosynthetic cell, exciting some of the molecules' electrons. The chlorophyll principally used by green plants, algae, and cyanobacteria is *chlorophyll a.* It is located in the membranous thylakoids of chloroplasts in algae and green plants (see Figure 4.27 on page 107) and in the thylakoids found in the photosynthetic structures of cyanobacteria. Other bacteria use *bacteriochlorophylls.*

The excited electrons jump from the chlorophyll to the first of a series of carrier molecules, an electron transport chain similar to that used in respiration. As electrons are passed along the series of carriers, protons are pumped across the membrane, and ADP is converted to ATP by chemiosmosis. In **cyclic photophosphorylation** the electron eventually returns to chlorophyll (Figure 5.23a). In **noncyclic photophosphorylation,** which is the more common process, the electrons released from chlorophyll do not return to chlorophyll but become in-

corporated into NADPH (Figure 5.23b). The electrons lost from chlorophyll are replaced by electrons from a reducing substance other than H_2O or another oxidizable compound, such as hydrogen sulfide (H_2S). To summarize: The products of noncyclic photophosphorylation are ATP (formed by chemiosmosis using energy released in an electron transport chain), O_2 (from water molecules), and NADPH (in which the hydrogen electrons and protons were derived ultimately from water).

The Dark Reactions: The Calvin-Benson Cycle

The dark (light-independent) reactions are so named because no light is directly required for them to occur. They include a complex cyclic pathway called the **Calvin-Benson cycle,** in which CO_2 is "fixed"—that is, used to synthesize sugars (Figure 5.24 on page 141; see also Appendix C).

A Summary of Energy Production Mechanisms

Learning Objective

■ *Write a sentence to summarize energy production in cells.*

In the living world, energy passes from one organism to another in the form of the potential energy contained in the bonds of chemical compounds. Organisms obtain the

FIGURE 5.23 Photophosphorylation.
(a) In cyclic photophosphorylation, electrons released from chlorophyll by light return to chlorophyll after passage along the electron transport chain. The energy from electron transfer is converted to ATP. **(b)** In noncyclic photophosphorylation, electrons released from chlorophyll are replaced by electrons from water. The chlorophyll electrons are passed along the electron transport chain to the electron acceptor $NADP^+$. $NADP^+$ combines with electrons and with hydrogen ions from water, forming NADPH.

■ How are oxidative phosphorylation and photophosphorylation similar?

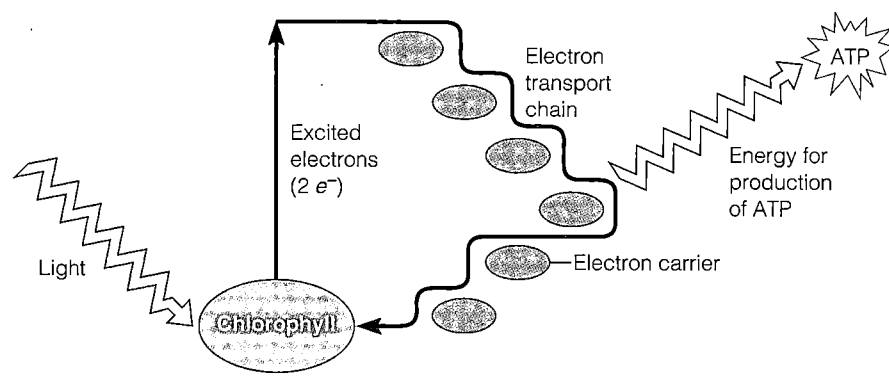

(a) Cyclic photophosphorylation

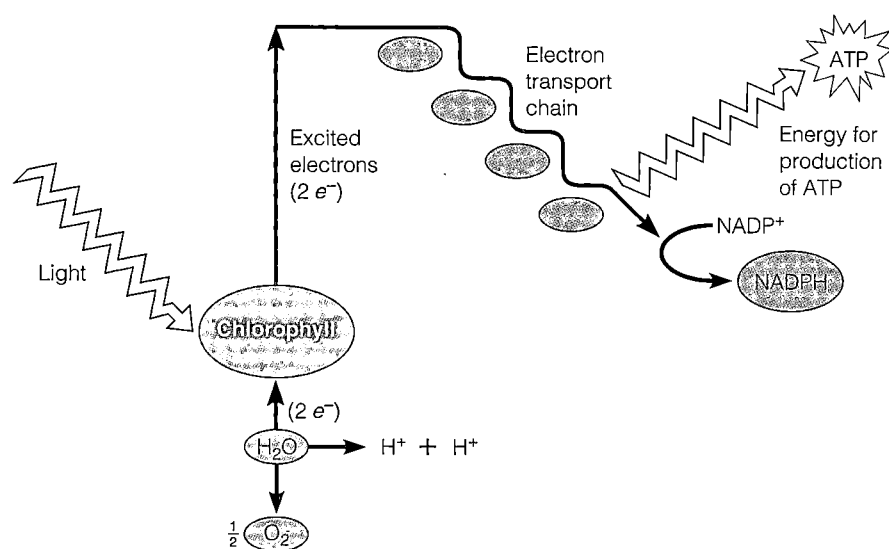

(b) Noncyclic photophosphorylation

energy from oxidation reactions. To obtain energy in a usable form, a cell must have an electron (or hydrogen) donor, which serves as an initial energy source within the cell. Electron donors can be as diverse as photosynthetic pigments, glucose or other organic compounds, elemental sulfur, ammonia, or hydrogen gas (Figure 5.25 on page 142). Next, electrons removed from the chemical energy sources are transferred to electron carriers, such as the coenzymes NAD^+, $NADP^+$, and FAD (see Figure 5.25). This transfer is an oxidation-reduction reaction; the initial energy source is oxidized as this first electron carrier is reduced. During this phase, some ATP is produced. In the third stage, electrons are transferred from electron carriers to their final electron acceptors in further oxidation-reduction reactions (see Figure 5.25), producing more ATP.

In aerobic respiration, oxygen (O_2) serves as the final electron acceptor. In anaerobic respiration, inorganic substances other than oxygen, such as nitrate ions (NO_3^-) or sulfate ions (SO_4^{2-}), serve as the final electron acceptors. In fermentation, organic compounds serve as the final

electron acceptors. In aerobic and anaerobic respiration, a series of electron carriers called an electron transport chain releases energy that is used by the mechanism of chemiosmosis to synthesize ATP. Regardless of their energy sources, all organisms use similar oxidation-reduction reactions to transfer electrons, and similar mechanisms to use the energy released to produce ATP.

Metabolic Diversity Among Organisms

Learning Objective

■ *Categorize the various nutritional patterns among organisms according to carbon source and mechanisms of carbohydrate catabolism and ATP generation.*

We have looked in detail at some of the energy-generating metabolic pathways that are used by animals and plants, as well as by many microbes. Microbes are distinguished by

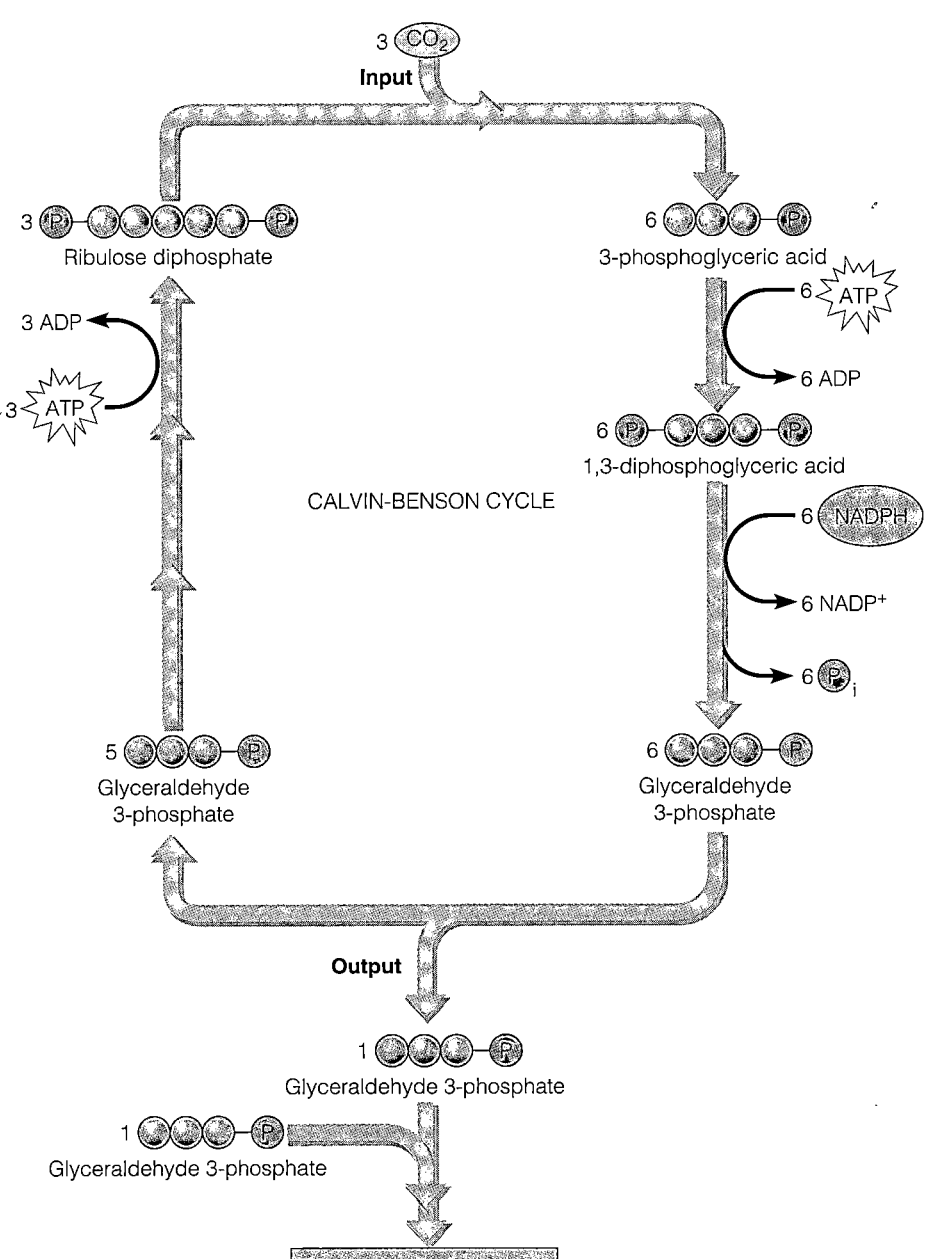

3 CO_2

Input

3 P-$OOOOO$-P
Ribulose diphosphate

6 OOO-P
3-phosphoglyceric acid

3 ADP

6 ATP

·3 ATP

6 ADP

6 P-OOO-P
1,3-diphosphoglyceric acid

CALVIN-BENSON CYCLE

6 NADPH

6 NADP+

6 P_i

5 OOO-P
Glyceraldehyde
3-phosphate

6 OOO-P
Glyceraldehyde
3-phosphate

Output

1 OOO-P
Glyceraldehyde 3-phosphate

1 OOO-P
Glyceraldehyde 3-phosphate

Glucose and other sugars

FIGURE 5.24 A simplified version of the Calvin-Benson cycle. This diagram shows three turns of the cycle, in which three molecules of CO_2 are fixed and one molecule of glyceraldehyde 3-phosphate is produced and leaves the cycle. Two molecules of glyceraldehyde 3-phosphate are needed to make one molecule of glucose. Therefore, the cycle must turn six times for each glucose molecule produced, requiring a total investment of 6 molecules of CO_2, 18 molecules of ATP, and 12 molecules of NADPH. A more detailed version of this cycle is presented in Appendix C.

■ In the Calvin-Benson cycle, CO_2 is used to synthesize sugars.

their great metabolic diversity, however, and some can sustain themselves on inorganic substances by using pathways that are unavailable to either plants or animals. All organisms, including microbes, can be classified metabolically according to their *nutritional pattern*—their source of energy and their source of carbon.

First considering the energy source, we can generally classify organisms as phototrophs or chemotrophs. **Phototrophs** use light as their primary energy source, whereas **chemotrophs** depend on oxidation-reduction reactions of inorganic or organic compounds for energy. For their principal carbon source, **autotrophs** (self-

feeders) use carbon dioxide, and **heterotrophs** (feeders on others) require an organic carbon source. Autotrophs are also referred to as *lithotrophs* (rock eating), and heterotrophs are also referred to as *organotrophs*.

If we combine the energy and carbon sources, we derive the following nutritional classifications for organisms: *photoautotrophs, photoheterotrophs, chemoautotrophs,* and *chemoheterotrophs* (Figure 5.26 on page 143). Almost all of the medically important microorganisms discussed in this book are chemoheterotrophs. Typically, infectious organisms catabolize substances obtained from the host.

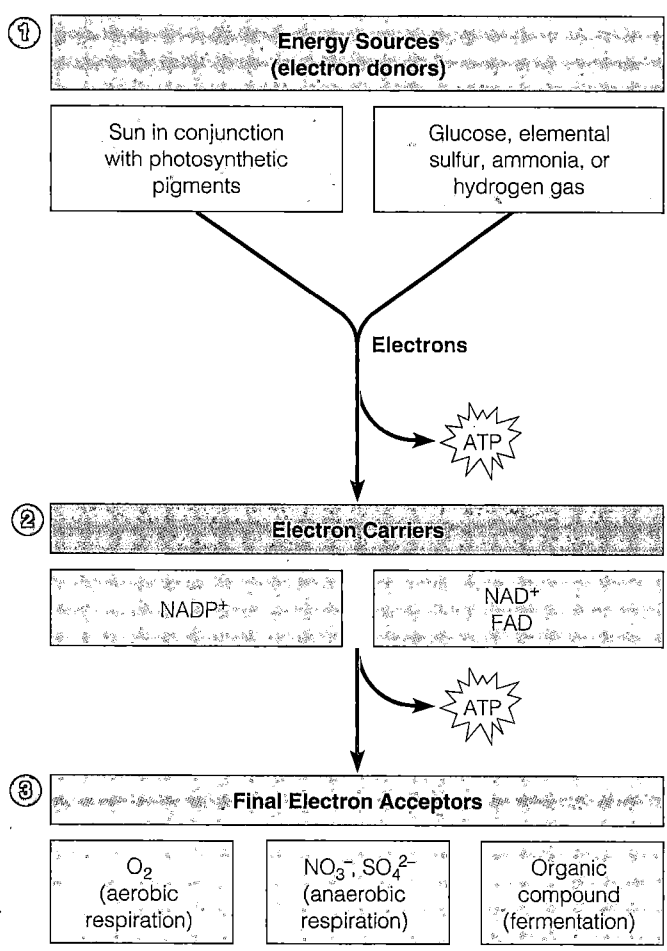

FIGURE 5.25 Requirements of ATP production. The production of ATP requires ① an energy source (electron donor), ② the transfer of electrons to an electron carrier during an oxidation-reduction reaction, and ③ the transfer of electrons to a final electron acceptor.

Photoautotrophs

Photoautotrophs use light as a source of energy and carbon dioxide as their chief source of carbon. They include photosynthetic bacteria (green sulfur and purple sulfur bacteria and cyanobacteria), algae, and green plants. In the photosynthetic reactions of cyanobacteria, algae, and green plants, the hydrogen atoms of water are used to reduce carbon dioxide, and oxygen gas is given off. Because this photosynthetic process produces O_2, it is sometimes called **oxygenic.**

In addition to the cyanobacteria (see Figure 11.20 on page 323), there are several other families of photosynthetic prokaryotes. Each is classified according to the way it reduces CO_2. These bacteria cannot use H_2O to reduce CO_2 and cannot carry on photosynthesis when oxygen is present (they must have an anaerobic environment).

Consequently, their photosynthetic process does not produce O_2 and is called **anoxygenic.** The anoxygenic photoautotrophs are the green sulfur and purple sulfur bacteria. The **green sulfur bacteria,** such as *Chlorobium* (klô-rō'bē-um), use sulfur (S), sulfur compounds (such as hydrogen sulfide, H_2S), or hydrogen gas (H_2) to reduce carbon dioxide and form organic compounds. Applying the energy from light and the appropriate enzymes, these bacteria oxidize sulfide (S^{2-}) or sulfur (S) to sulfate (SO_4^{2-}), or hydrogen gas to water (H_2O). The **purple sulfur bacteria,** such as *Chromatium* (krō-mā'tē-um), also use sulfur, sulfur compounds, or hydrogen gas to reduce carbon dioxide. They are distinguished from the green sulfur bacteria by their type of chlorophyll, location of stored sulfur, and ribosomal RNA.

The chlorophylls used by these photosynthetic bacteria are called *bacteriochlorophylls,* and they absorb light at longer wavelengths than that absorbed by chlorophyll *a*. Bacteriochlorophylls of green sulfur bacteria are found in vesicles called *chlorosomes* (or *chlorobium vesicles*) underlying and attached to the plasma membrane. In the purple sulfur bacteria, the bacteriochlorophylls are located in invaginations of the plasma membrane (*intracytoplasmic membranes*).

Several characteristics that distinguish eukaryotic photosynthesis from prokaryotic photosynthesis are presented in Table 5.6. See the box on page 144 for a discussion of an exceptional photosynthetic system that exists in *Halobacterium.* The system does not use chlorophyll.

Photoheterotrophs

Photoheterotrophs use light as a source of energy but cannot convert carbon dioxide to sugar; rather, they use organic compounds, such as alcohols, fatty acids, other organic acids, and carbohydrates, as sources of carbon. They are anoxygenic. The **green nonsulfur bacteria,** such as *Chloroflexus* (klô-rō-flex'us), and **purple nonsulfur bacteria,** such as *Rhodopseudomonas* (rō-dō-sū-dō-mō'nas), are photoheterotrophs.

Chemoautotrophs

Chemoautotrophs use the electrons from reduced inorganic compounds as a source of energy, and they use CO_2 as their principal source of carbon (see Figure 27.1, the carbon cycle). Inorganic sources of energy for these organisms include hydrogen sulfide (H_2S) for *Beggiatoa* (bej-jē-ä-tō'ä); elemental sulfur (S) for *Thiobacillus thiooxidans;* ammonia (NH_3) for *Nitrosomonas* (nī-trō-sō-mō'näs); nitrite ions (NO_2^-) for *Nitrobacter* (nī-trō-bak'tėr); hydrogen gas (H_2) for *Hydrogenomonas* (hī-drō-je-nō-mō'näs); and ferrous iron (Fe^{2+}) for *Thiobacillus ferrooxidans.* The

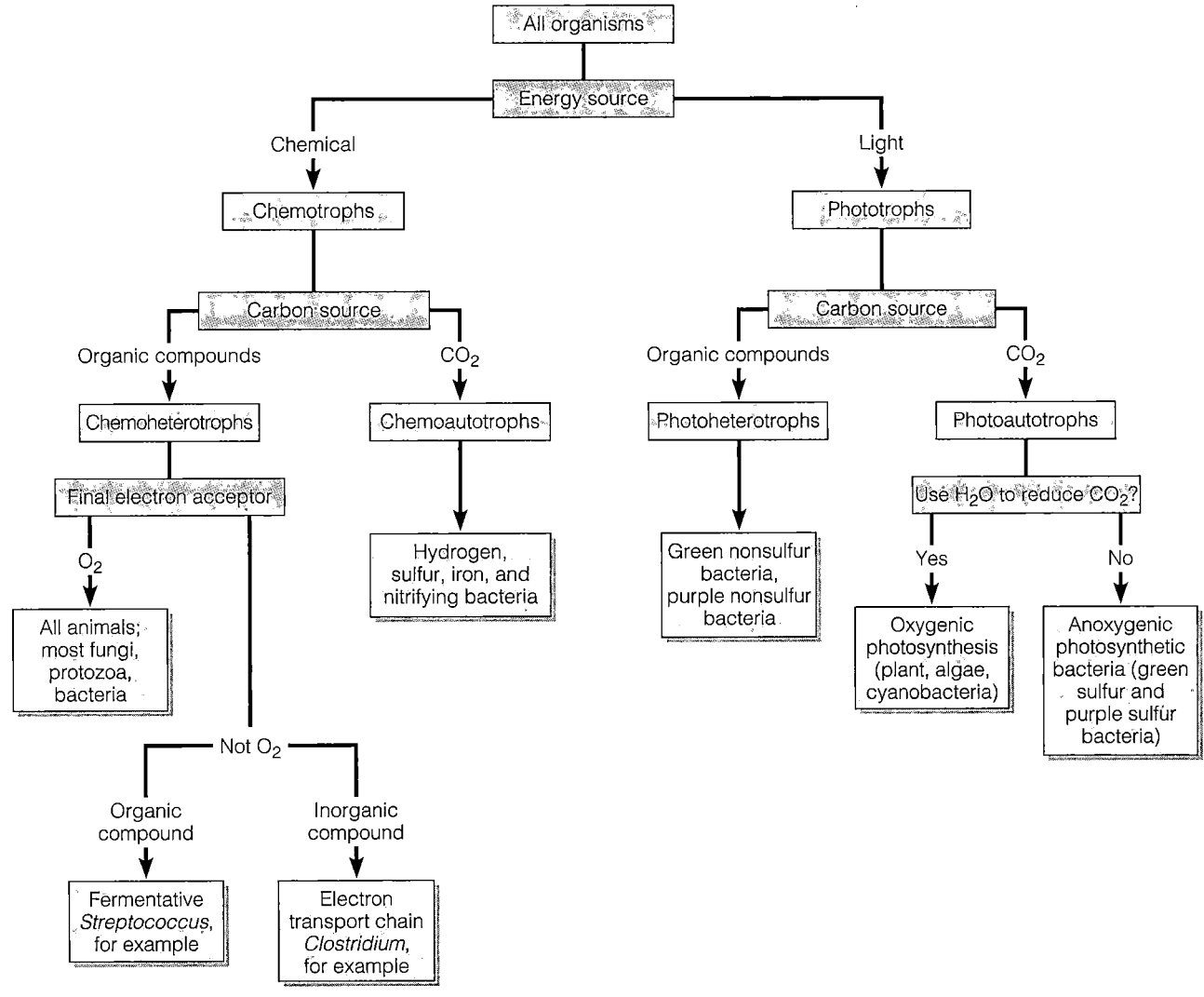

FIGURE 5.26 A nutritional classification of organisms.

■ What is the basic difference between chemotrophs and phototrophs?

table 5.6	Photosynthesis Compared in Selected Eukaryotes and Prokaryotes			
Characteristic	**Eukaryotes**		**Prokaryotes**	
			Green Sulfur Bacteria	**Purple Sulfur Bacteria**
	Algae, Plants	**Cyanobacteria**		
Substance that reduces CO_2	H atoms of H_2O	H atoms of H_2O	Sulfur, sulfur compounds, H_2 gas	Sulfur, sulfur compounds, H_2 gas
Oxygen production	Oxygenic	Oxygenic (and anoxygenic)	Anoxygenic	Anoxygenic
Light-trapping pigment	Chlorophyll a	Chlorophyll a	Bacteriochlorophyll a	Bacteriochlorophyll a or b
Site of photosynthesis	Chloroplasts with thylakoids	Thylakoids	Chlorosomes	Intracytoplasmic membrane
Environment	Aerobic	Aerobic (and anaerobic)	Anaerobic	Anaerobic

An interesting member of the Archaea, *Halobacterium*, lives where very little else can grow. This bacterium is found in salt lakes, salt licks on ranches, salt flats, or any other environment with a concentration of salt that is five to seven times that of the ocean. Halobacteria are easy to detect because they turn their environment purple (see the photograph). These bacteria cannot ferment carbohydrates and do not contain chlorophyll, so it was assumed that all their energy comes from oxidative phosphorylation. The exciting discovery of a new system of photophosphorylation arose through the study of the plasma membrane of *Halobacterium halobium*.

Researchers found that the plasma membrane of *H. halobium* fragments into two fractions (red and purple) when the cell is broken down and its components are sorted. The red fraction, which constitutes most of the membrane, contains cytochromes, flavoproteins, and other parts of the electron transport chain, which carries out oxidative phosphorylation. The purple fraction is more interesting. This purple membrane occurs in distinct patches of hexagonal lattices within the plasma membrane. The purple color comes from a protein that makes up 75% of the purple membrane. This protein is similar to the retinal pigment in the rod cells of the human eye, rhodopsin, so the protein was named bacteriorhodopsin. At the time it was discovered, its function was not known.

Further studies showed that *H. halobium* can grow in the presence of either light or oxygen but cannot grow when neither is present.

This unexpected result suggested that *Halobacterium* can obtain energy by using either of two systems, one that operates in the presence of oxygen (oxidative phosphorylation) and one that operates in the presence of light (some

Halobacterium (purple color) grows in the high-salt concentration of solar evaporation ponds used for manufacturing salt around San Francisco Bay.

kind of photophosphorylation). The rate of ATP synthesis by *H. halobium* was found to be highest when the cells receive light that is between 550 and 600 nm in wavelength; this range exactly corresponds to the absorption spectrum of bacteriorhodopsin.

Researchers hypothesized that bacteriorhodopsin, like chlorophyll-containing systems, acts as a proton pump to create a proton gradient across a cell membrane; in this case, the gradient is created across the purple membrane. The proton gradient can do cellular work—can drive the synthesis of ATP or transport solutes.

HALOBACTERIUM REPLACES THE SILICON CHIP

At Syracuse University's Center of Molecular Electronics, Robert Birge grows *Halobacterium* for five days in 5-liter batches and then extracts bacteriorhodopsin from the cells for a novel

use. Birge has developed a computer chip made of a thin layer of bacteriorhodopsin.

Conventional computers store information on thin wafers of silicon. Computers process information by "reading" a series of zeros and ones produced as electrons flow through switches etched in the silicon. Electrons passing through a switch represent a one; a switch that stops the electron flow represents a zero. However, silicon can't hold enough information or process information fast enough for such applications as artificial intelligence or robot vision.

In contrast, the bacteriorhodopsin chip will be able to store more information than a silicon chip and process the information faster, more like a human brain. The bacteriorhodopsin chip works with light, which of course moves at the speed of light, much faster than the flow of electrons. Green light causes the protein to fold; a folded protein is read as a one, whereas an unfolded protein represents a zero. Laser light is used to "see" the configuration of the protein.

At present, the protein chip needs to be stored at −4°C to maintain its structure, but Birge and his coworkers are hopeful they will solve this problem. Russian scientists have made a protein processor for military radar, and the U.S. military is apparently using the protein chips in their combat planes. If such a plane crashes, the cooling system will go off, thus destroying the chip and keeping classified information from being stolen. Eventually, these smaller, faster, and higher-capacity chips will probably make it possible to develop computers that perform functions closer to human intelligence, such as acting as eyes for blind people.

energy derived from the oxidation of these inorganic compounds is eventually stored in ATP, which is produced by oxidative phosphorylation.

Chemoheterotrophs

When we discuss photoautotrophs, photoheterotrophs, and chemoautotrophs, it is easy to categorize the energy source and carbon source because they occur as separate entities. However, in chemoheterotrophs, the distinction is not as clear because the energy source and carbon source are usually the same organic compound—glucose, for example. **Chemoheterotrophs** specifically use the electrons from hydrogen atoms in organic compounds as their energy source.

Heterotrophs are further classified according to their source of organic molecules. **Saprophytes** live on dead organic matter, and **parasites** derive nutrients from a living host. Most bacteria, and all fungi, protozoa, and animals, are chemoheterotrophs.

Bacteria and fungi can use a wide variety of organic compounds for carbon and energy sources. This is why they can live in diverse environments. Understanding microbial diversity is scientifically interesting and economically important. In some situations microbial growth is undesirable, such as when rubber-degrading bacteria destroy a gasket or shoe sole. However, these same bacteria might be beneficial if they decomposed discarded rubber products such as tires. *Rhodococcus erythropolis* (rō-dō-kok′kus er-i-throp′ō-lis) is widely distributed in soil and can cause disease in humans and other animals. This same species is able to replace sulfur atoms in petroleum with atoms of oxygen. A Texas company is currently using *R. erythropolis* to produce desulfurized oil.

★   ★   ★

We will next consider how cells use ATP pathways for the synthesis of organic compounds such as carbohydrates, lipids, proteins, and nucleic acids.

Metabolic Pathways of Energy Use

Learning Objective

■ *Describe the major types of anabolism and their relationship to catabolism.*

Up to now we have been considering energy production. Through the oxidation of organic molecules, organisms produce energy by aerobic respiration, anaerobic respiration, and fermentation. Much of this energy is given off as heat. The complete metabolic oxidation of glucose to carbon dioxide and water is considered a very efficient process, but about 45% of the energy of glucose is lost as heat. Cells use the remaining energy, which is trapped in the bonds of ATP, in a variety of ways. Microbes use ATP to provide energy for the transport of substances across plasma membranes—the process called active transport that we discussed in Chapter 4. Microbes also use some of their energy for flagellar motion (also discussed in Chapter 4). Most of the ATP, however, is used in the production of new cellular components. This production is a continuous process in cells, and, in general, is faster in prokaryotic cells than in eukaryotic cells.

Autotrophs build their organic compounds by fixing carbon dioxide in the Calvin–Benson cycle (see Figure 5.24). This requires both energy (ATP) and electrons (from the oxidation of NADPH). Heterotrophs, by contrast, must have a ready source of organic compounds for biosynthesis—the production of needed cellular components, usually from simpler molecules. The cells use these compounds as both the carbon source and the energy source. We will next consider the biosynthesis of a few representative classes of biological molecules: carbohydrates, lipids, amino acids, purines, and pyrimidines. As we do so, keep in mind that synthesis reactions require a net input of energy.

Polysaccharide Biosynthesis

Microorganisms synthesize sugars and polysaccharides. The carbon atoms required to synthesize glucose are derived from the intermediates produced during processes such as glycolysis and the Krebs cycle and from lipids or amino acids. After synthesizing glucose (or other simple sugars), bacteria may assemble it into more complex polysaccharides such as glycogen. For bacteria to build glucose into glycogen, glucose units must be phosphorylated and linked. The product of glucose phosphorylation is glucose 6-phosphate. Such a process involves the expenditure of energy, usually in the form of ATP. In order for bacteria to synthesize glycogen, a molecule of ATP is added to glucose 6-phosphate to form *adenosine diphosphoglucose (ADPG)* (Figure 5.27). Once ADPG is synthesized, it is linked with similar units to form glycogen.

Using a nucleotide called uridine triphosphate (UTP) as a source of energy and glucose 6-phosphate, animals synthesize glycogen (and many other carbohydrates) from *uridine diphosphoglucose, UDPG* (see Figure 5.27). A compound related to UDPG, called *UDP-N-acetylglucosamine (UDPNAc),* is a key starting material in the

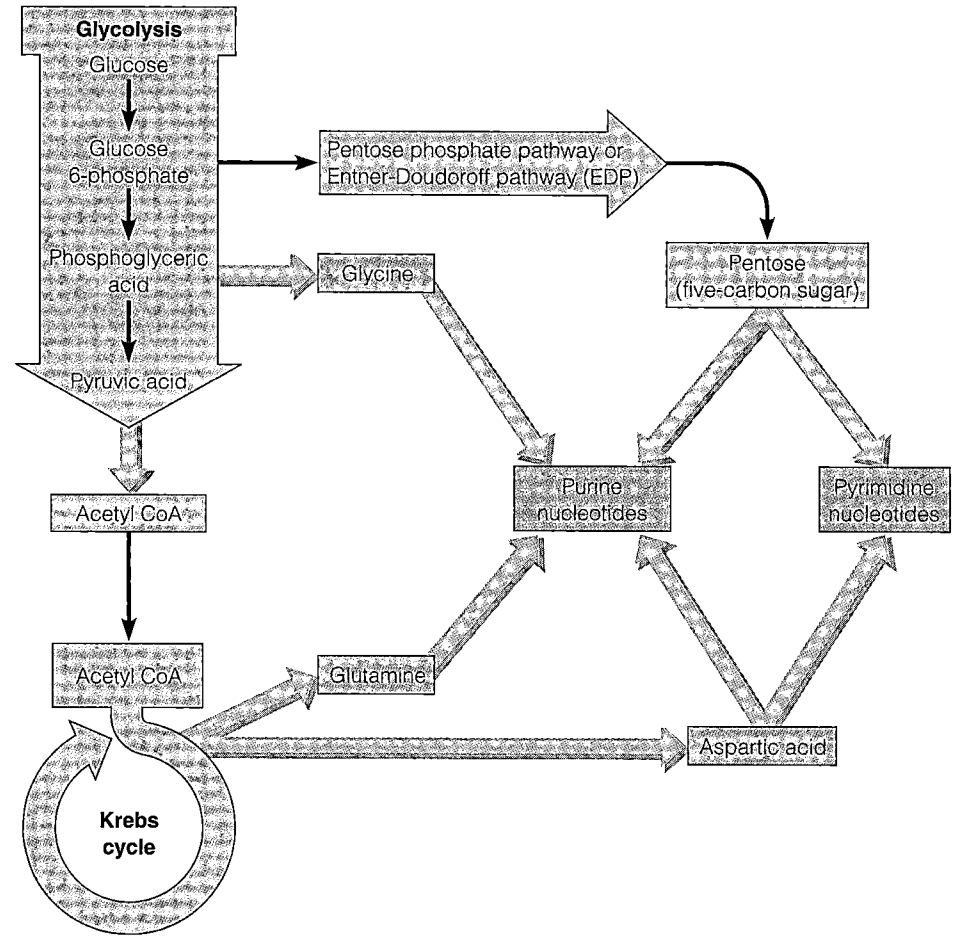

FIGURE 5.30 The biosynthesis of purine and pyrimidine nucleotides.

■ What are the functions of nucleotides in a cell?

energy is used for biosynthesis. With such a variety of activity, you might imagine that anabolic and catabolic reactions occur independently of each other in space and time. Actually, anabolic and catabolic reactions are joined through a group of common intermediates (identified as key intermediates in Figure 5.31). Both anabolic and catabolic reactions also share some metabolic pathways, such as the Krebs cycle. For example, reactions in the Krebs cycle not only participate in the oxidation of glucose but also produce intermediates that can be converted to amino acids. Metabolic pathways that function in both anabolism and catabolism are called **amphibolic pathways,** meaning that they are dual-purpose.

Amphibolic pathways bridge the reactions that lead to the breakdown and synthesis of carbohydrates, lipids, proteins, and nucleotides. Such pathways enable simultaneous reactions to occur in which the breakdown product formed in one reaction is used in another reaction to synthesize a different compound, and vice versa.

Because various intermediates are common to both anabolic and catabolic reactions, mechanisms exist that regulate synthesis and breakdown pathways and allow these reactions to occur simultaneously. One such mechanism involves the use of different coenzymes for opposite pathways. For example, NAD^+ is involved in catabolic reactions, whereas $NADP^+$ is involved in anabolic reactions. Enzymes can also coordinate anabolic and catabolic reactions by accelerating or inhibiting the rates of biochemical reactions.

The energy stores of a cell can also affect the rates of biochemical reactions. For example, if ATP begins to accumulate, an enzyme shuts down glycolysis; this control helps to synchronize the rates of glycolysis and the Krebs cycle. Thus, if citric acid consumption increases, either because of a demand for more ATP or because anabolic pathways are draining off intermediates of the citric acid cycle, glycolysis accelerates and meets the demand.

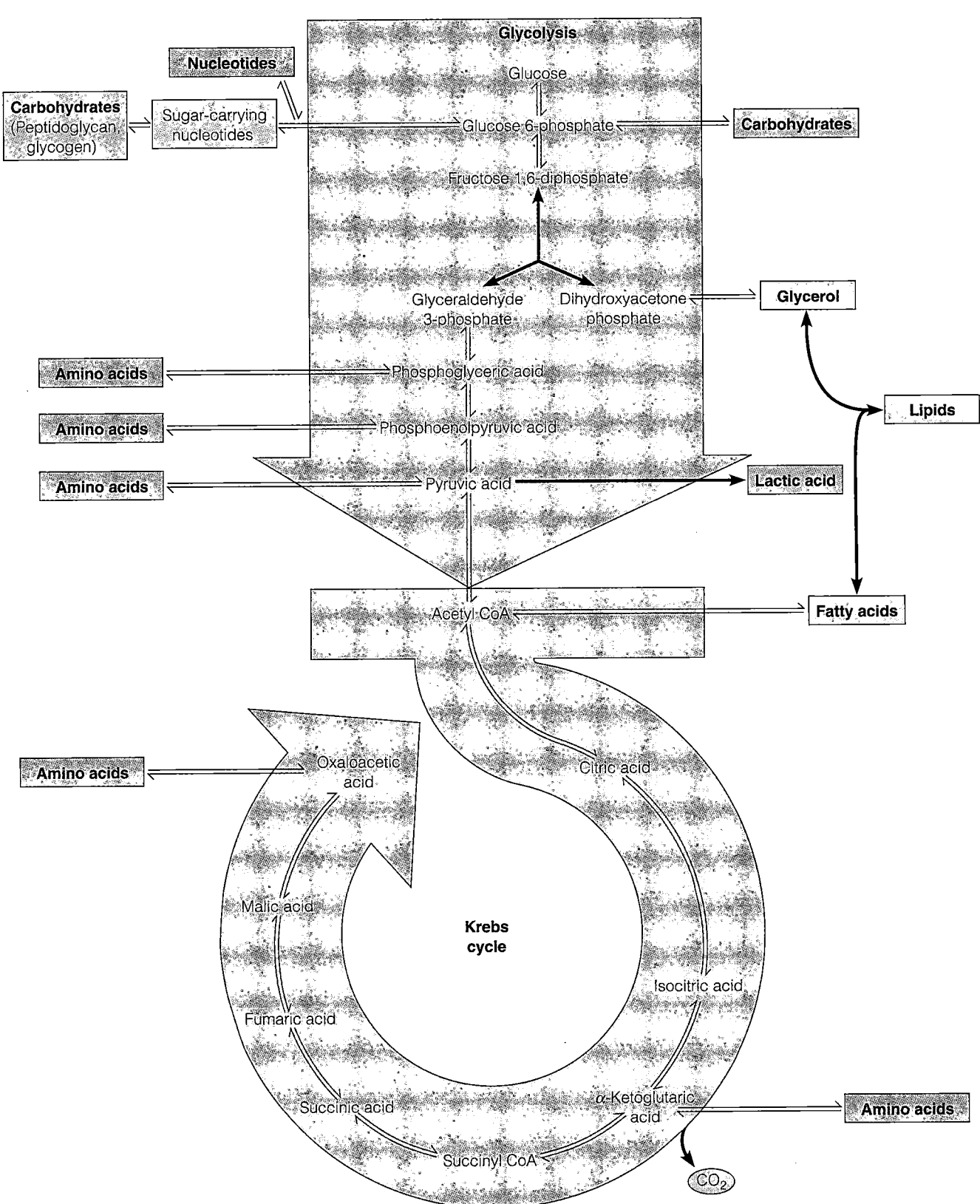

FIGURE 5.31   The integration of metabolism. Key intermediates are shown. Although not indicated in the figure, amino acids and ribose are used in the synthesis of purine and pyrimidine nucleotides (see Figure 5.30). The double arrows indicate amphibolic pathways.

■ What is an amphibolic pathway?

Study Outline ▧ Student Tutorial CD-ROM

CATABOLIC AND ANABOLIC REACTIONS (pp. 113–114)

1. The sum of all chemical reactions within a living organism is known as metabolism.

2. Catabolism refers to chemical reactions that result in the breakdown of more complex organic molecules into simpler substances. Catabolic reactions usually release energy.

3. Anabolism refers to chemical reactions in which simpler substances are combined to form more complex molecules. Anabolic reactions usually require energy.

4. The energy of catabolic reactions is used to drive anabolic reactions.

5. The energy for chemical reactions is stored in ATP.

ENZYMES (pp. 114–121)

▧ To review, go to Metabolism: Energy Production: Oxidation-Reduction

1. Enzymes are proteins, produced by living cells, that catalyze chemical reactions by lowering the activation energy.

2. Enzymes are generally globular proteins with characteristic three-dimensional shapes.

3. Enzymes are efficient, can operate at relatively low temperatures, and are subject to various cellular controls.

Naming Enzymes (p. 115)

1. Enzyme names usually end in -ase.

2. The six classes of enzymes are defined on the basis of the types of reactions they catalyze.

Enzyme Components (pp. 115–116)

1. Most enzymes are holoenzymes, consisting of a protein portion (apoenzyme) and a nonprotein portion (cofactor).

2. The cofactor can be a metal ion (iron, copper, magnesium, manganese, zinc, calcium, or cobalt) or a complex organic molecule known as a coenzyme (NAD^+, $NADP^+$, FMN, FAD, or coenzyme A).

The Mechanism of Enzymatic Action (pp. 116–118)

▧ Metabolism: Enzyme: Enzymatic Action

1. When an enzyme and substrate combine, the substrate is transformed, and the enzyme is recovered.

2. Enzymes are characterized by specificity, which is a function of their active sites.

Factors Influencing Enzymatic Activity (pp. 118–120)

▧ Metabolism: Enzyme: Enzymatic Activity and Enzyme Inhibitors

1. At high temperatures, enzymes undergo denaturation and lose their catalytic properties; at low temperatures, the reaction rate decreases.

2. The pH at which enzymatic activity is maximal is known as the optimum pH.

3. Within limits, enzymatic activity increases as substrate concentration increases.

4. Competitive inhibitors compete with the normal substrate for the active site of the enzyme. Noncompetitive inhibitors act on other parts of the apoenzyme or on the cofactor and decrease the enzyme's ability to combine with the normal substrate.

Feedback Inhibition (pp. 120–121)

1. Feedback inhibition occurs when the end-product of a metabolic pathway inhibits an enzyme's activity near the start of the pathway.

Ribozymes (p. 121)

1. Ribozymes are enzymatic RNA molecules that cut and splice RNA in eukaryotic cells.

ENERGY PRODUCTION (pp. 121–123)

▧ Metabolism: Energy Production: Oxidation-Reduction

Oxidation-Reduction Reactions (pp. 121–122)

1. Oxidation is the removal of one or more electrons from a substrate. Protons (H^+) are often removed with the electrons.

2. Reduction of a substrate refers to its gain of one or more electrons.

3. Each time a substance is oxidized, another is simultaneously reduced.

4. NAD^+ is the oxidized form; NADH is the reduced form.

5. Glucose is a reduced molecule; energy is released during a cell's oxidation of glucose.

The Generation of ATP (pp. 122–123)

1. Energy released during certain metabolic reactions can be trapped to form ATP from ADP and ℗ (phosphate). Addition of a ℗ to a molecule is called phosphorylation.

2. During substrate-level phosphorylation, a high-energy ℗ from an intermediate in catabolism is added to ADP.

3. During oxidative phosphorylation, energy is released as electrons are passed to a series of electron acceptors (an electron transport chain) and finally to O_2 or another inorganic compound.

4. During photophosphorylation, energy from light is trapped by chlorophyll, and electrons are passed through a series of electron acceptors. The electron transfer releases energy used for the synthesis of ATP.

Metabolic Pathways of Energy Production (p. 123)

1. A series of enzymatically catalyzed chemical reactions called metabolic pathways store energy in and release energy from organic molecules.

CARBOHYDRATE CATABOLISM (pp. 123–135)

1. Most of a cell's energy is produced from the oxidation of carbohydrates.

2. Glucose is the most commonly used carbohydrate.

3. The two major types of glucose catabolism are respiration, in which glucose is completely broken down, and fermentation, in which it is partially broken down.

Glycolysis (p. 125)

SIT *Metabolism: Glycolysis*

1. The most common pathway for the oxidation of glucose is glycolysis. Pyruvic acid is the end-product.

2. Two ATP and two NADH molecules are produced from one glucose molecule.

Alternatives to Glycolysis (p. 125)

1. The pentose phosphate pathway is used to metabolize five-carbon sugars; one ATP and 12 NADPH molecules are produced from one glucose molecule.

2. The Entner-Doudoroff pathway yields one ATP and two NADPH molecules from one glucose molecule.

Cellular Respiration (pp. 125–133)

1. During respiration, organic molecules are oxidized. Energy is generated from the electron transport chain.

2. In aerobic respiration, O_2 functions as the final electron acceptor.

3. In anaerobic respiration, the final electron acceptor is an inorganic molecule other than O_2.

Aerobic Respiration (pp. 127–132)

The Krebs Cycle (p. 127)

SIT *Metabolism: Krebs Cycle*

1. Decarboxylation of pyruvic acid produces one CO_2 molecule and one acetyl group.

2. Two-carbon acetyl groups are oxidized in the Krebs cycle. Electrons are picked up by NAD^+ and FAD for the electron transport chain.

3. From one molecule of glucose, oxidation produces six molecules of NADH, two molecules of $FADH_2$, and two molecules of ATP.

4. Decarboxylation produces six molecules of CO_2.

The Electron Transport Chain (System) (pp. 127–129)

1. Electrons are brought to the electron transport chain by NADH.

2. The electron transport chain consists of carriers, including flavoproteins, cytochromes, and ubiquinones.

The Chemiosmotic Mechanism of ATP Generation (pp. 129–130)

SIT *Metabolism: Cell Respiration: Chemiosmotic Model*

1. Protons being pumped across the membrane generate a proton motive force as electrons move through a series of acceptors or carriers.

2. Energy produced from movement of the protons back across the membrane is used by ATP synthase to make ATP from ADP and ⓟ.

3. In eukaryotes, electron carriers are located in the inner mitochondrial membrane; in prokaryotes, electron carriers are in the plasma membrane.

A Summary of Aerobic Respiration (pp. 130–132)

SIT *Metabolism: Cell Respiration: Aerobic and anaerobic Respiration*

1. In aerobic prokaryotes, 38 ATP molecules can be produced from complete oxidation of a glucose molecule in glycolysis, the Krebs cycle, and the electron transport chain.

2. In eukaryotes, 36 ATP molecules are produced from complete oxidation of a glucose molecule.

Anaerobic Respiration (p. 133)

1. The final electron acceptors in anaerobic respiration include NO_3^-, SO_4^{2-}, and CO_3^{2-}.

2. The total ATP yield is less than in aerobic respiration because only part of the Krebs cycle operates under anaerobic conditions.

Fermentation (pp. 133–135)

SIT *Metabolism: Fermentation*

1. Fermentation releases energy from sugars or other organic molecules by oxidation.

2. O_2 is not required in fermentation.

3. Two ATP molecules are produced by substrate-level phosphorylation.

4. Electrons removed from the substrate reduce NAD^+.

5. The final electron acceptor is an organic molecule.

6. In lactic acid fermentation, pyruvic acid is reduced by NADH to lactic acid.

7. In alcohol fermentation, acetaldehyde is reduced by NADH to produce ethanol.

8. Heterolactic fermenters can use the pentose phosphate pathway to produce lactic acid and ethanol.

LIPID AND PROTEIN CATABOLISM (pp. 135–137)

1. Lipases hydrolyze lipids into glycerol and fatty acids.

2. Fatty acids and other hydrocarbons are catabolized by beta oxidation.

3. Catabolic products can be further broken down in glycolysis and the Krebs cycle.

4. Before amino acids can be catabolized, they must be converted to various substances that enter the Krebs cycle.

5. Transamination, decarboxylation, and dehydrogenation reactions convert the amino acids to be catabolized.

BIOCHEMICAL TESTS AND BACTERIAL IDENTIFICATION (p. 137)

1. Bacteria and yeast can be identified by detecting action of their enzymes.

2. Fermentation tests are used to determine whether an organism can ferment a carbohydrate to produce acid and gas.

PHOTOSYNTHESIS (pp. 137–139)

1. Photosynthesis is the conversion of light energy from the sun into chemical energy; the chemical energy is used for carbon fixation.

The Light Reactions: Photophosphorylation (p. 139)

1. Chlorophyll *a* is used by green plants, algae, and cyanobacteria; it is found in thylakoid membranes.
2. Electrons from chlorophyll pass through an electron transport chain, from which ATP is produced by chemiosmosis.
3. In cyclic photophosphorylation, the electrons return to the chlorophyll.
4. In noncyclic photophosphorylation, the electrons are used to reduce $NADP^+$, and electrons are returned to chlorophyll from H_2O or H_2S.
5. When H_2O is oxidized by green plants, algae, and cyanobacteria, O_2 is produced.

The Dark Reactions: The Calvin-Benson Cycle (p. 139)

1. CO_2 is used to synthesize sugars in the Calvin-Benson cycle.

A SUMMARY OF ENERGY PRODUCTION MECHANISMS (pp. 139–140)

1. Sunlight is converted to chemical energy in oxidation-reduction reactions carried on by phototrophs. Chemotrophs can use this chemical energy.
2. In oxidation-reduction reactions, energy is derived from the transfer of electrons.
3. To produce energy, a cell needs an electron donor (organic or inorganic), a system of electron carriers, and a final electron acceptor (organic or inorganic).

METABOLIC DIVERSITY AMONG ORGANISMS (pp. 140–145)

1. Photoautotrophs obtain energy by photophosphorylation and fix carbon from CO_2 via the Calvin-Benson cycle to synthesize organic compounds.
2. Cyanobacteria are oxygenic phototrophs. Green sulfur bacteria and purple sulfur bacteria are anoxygenic phototrophs.

3. Photoheterotrophs use light as an energy source and an organic compound for their carbon source or electron donor.
4. Chemoautotrophs use inorganic compounds as their energy source and carbon dioxide as their carbon source.
5. Chemoheterotrophs use complex organic molecules as their carbon and energy sources.

METABOLIC PATHWAYS OF ENERGY USE (pp. 145–147)

Polysaccharide Biosynthesis (pp. 145–146)

1. Glycogen is formed from ADPG.
2. UDPNAc is the starting material for the biosynthesis of peptidoglycan.

Lipid Biosynthesis (p. 146)

1. Lipids are synthesized from fatty acids and glycerol.
2. Glycerol is derived from dihydroxyacetone phosphate, and fatty acids are built from acetyl CoA.

Amino Acid and Protein Biosynthesis (pp. 146–147)

1. Amino acids are required for protein biosynthesis.
2. All amino acids can be synthesized either directly or indirectly from intermediates of carbohydrate metabolism, particularly from the Krebs cycle.

Purine and Pyrimidine Biosynthesis (p. 147)

1. The sugars composing nucleotides are derived from either the pentose phosphate pathway or the Entner-Doudoroff pathway.
2. Carbon and nitrogen atoms from certain amino acids form the backbones of the purines and pyrimidines.

THE INTEGRATION OF METABOLISM (pp. 147–149)

1. Anabolic and catabolic reactions are integrated through a group of common intermediates.
2. Such integrated metabolic pathways are referred to as amphibolic pathways.

Study Questions

REVIEW

1. Define metabolism.
2. Distinguish between catabolism and anabolism. How are these processes related?
3. Using the diagrams to the right, show:
   a. where the substrate will bind.
   b. where the competitive inhibitor will bind.
   c. where the noncompetitive inhibitor will bind.

d. which of the four elements could be the inhibitor in feedback inhibition.

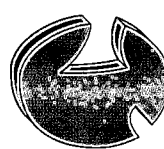
Enzyme

Substrate

Competitive inhibitor

Noncompetitive inhibitor

4. What will the effect of the reactions in question 3 be?

5. Name the pathways diagrammed in a, b, and c.

6. Show where glycerol is catabolized and where fatty acids are catabolized.

7. Show where the amino acid glutamic acid is catabolized:

$$HOOC-CH_2-CH_2-\underset{\underset{NH_2}{|}}{\overset{\overset{H}{|}}{C}}-COOH$$

8. Show how these pathways are related.

9. Where is ATP required in pathways a and b?

10. Where is CO_2 released in pathways b and c?

11. Show where a long-chain hydrocarbon such as petroleum is catabolized.

12. Where is NADH (or $FADH_2$ or NADPH) used and produced in these pathways?

13. Identify four places where anabolic and catabolic pathways are integrated.

14. Why are most enzymes active at one particular temperature? Why are enzymes less active below this temperature? What happens above this temperature?

15. List four compounds that can be made from pyruvic acid by an organism that uses fermentation only.

16. Fill in the following table with the carbon source and energy source of each type of organism.

Organism	Carbon Source	Energy Source
Photoautotroph		
Photoheterotroph		
Chemoautotroph		
Chemoheterotroph		

17. There are three mechanisms for the phosphorylation of ADP to produce ATP. Write the name of the mechanism that describes each of the reactions in the following table.

ATP Generated by	Reaction
_____	An electron, liberated from chlorophyll by light, is passed down an electron transport chain.
_____	Cytochrome c passes two electrons to cytochrome a.
_____	$\underset{\underset{COOH}{|}}{\overset{\overset{CH_2}{\|}}{C}}-O\sim\text{℗}$ → $\underset{\underset{COOH}{|}}{\overset{\overset{CH_3}{|}}{C}}=O$
	Phosphoenolpyruvic acid Pyruvic acid

Use the following diagrams for questions 5–13.

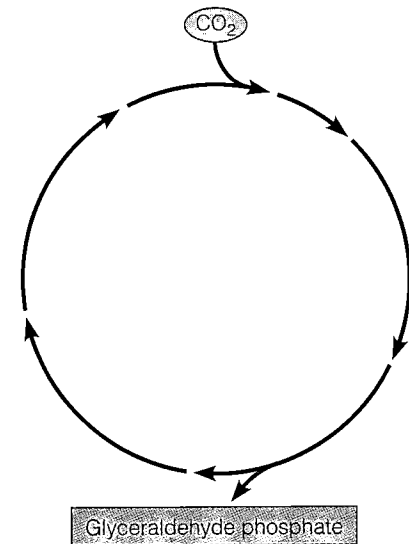

(a)

Glucose
$(C-C-C-C-C-C)$

Glyceraldehyde 3-phosphate $(C-C-C-\text{℗})$ ⇌ Dihydroxyacetone phosphate $(C-C-C-\text{℗})$

Two molecules of pyruvic acid (two $C-C-C$)

(b)

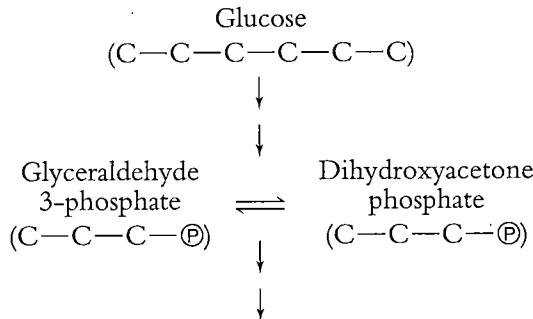

(c)

18. Define oxidation–reduction, and differentiate between the following terms:
    a. aerobic and anaerobic respiration
    b. respiration and fermentation
    c. cyclic and noncyclic photophosphorylation

19. The pentose phosphate pathway produces only one ATP. List four advantages of this pathway for the cell.

20. All of the energy-producing biochemical reactions that occur in cells, such as photophosphorylation and glycolysis, are _____ reactions.

21. Explain how ATP is a key intermediate in metabolism.

22. An enzyme and substrate are combined. The rate of reaction begins as shown in the following graph. To complete the graph, show the effect of increasing substrate concentration on a constant enzyme concentration. Show the effect of increasing temperature.

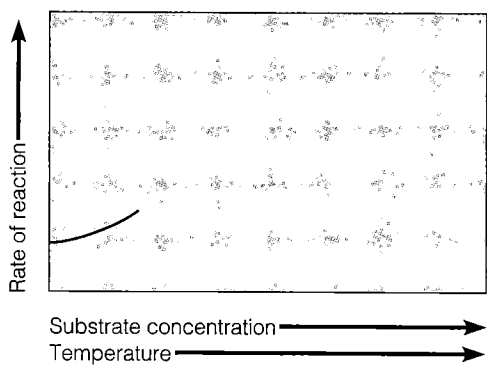

Substrate concentration ⟶
Temperature ⟶

MULTIPLE CHOICE

1. Which substance in the following reaction is being reduced?

$$\underset{\text{Acetaldehyde}}{\overset{\displaystyle H}{\underset{\displaystyle CH_3}{C\!\!=\!\!O}}} + NADH + H^+ \quad \rightarrow \quad \underset{\text{Ethanol}}{\overset{\displaystyle H}{H\!-\!\underset{\displaystyle CH_3}{C}\!-\!OH}} + NAD^+$$

    a. Acetaldehyde
    b. NADH
    c. Ethanol
    d. NAD$^+$

2. Which of the following reactions produces the most molecules of ATP during aerobic metabolism?
    a. glucose → glucose-6-P
    b. phosphoenolpyruvic acid → pyruvic acid
    c. glucose → pyruvic acid
    d. acetyl CoA → CO_2 + H_2O
    e. succinic acid → fumaric acid

3. Which of the following processes does not generate ATP?
    a. photophosphorylation
    b. the Calvin-Benson cycle
    c. oxidative phosphorylation
    d. substrate-level phosphorylation
    e. none of the above

4. Which of the following compounds has the greatest amount of energy for a cell?
    a. CO_2
    b. ATP
    c. glucose
    d. O_2
    e. lactic acid

5. Which of the following is the best definition of the Krebs cycle?
    a. the oxidation of pyruvic acid
    b. the way cells produce CO_2
    c. a series of chemical reactions in which NADH is produced from the oxidation of pyruvic acid
    d. a method of producing ATP by phosphorylating ADP
    e. a series of chemical reactions in which ATP is produced from the oxidation of pyruvic acid.

6. Which of the following is the best definition of respiration?
    a. a sequence of carrier molecules with O_2 as the final electron acceptor
    b. a sequence of carrier molecules with an inorganic molecule as the final electron acceptor
    c. a method of generating ATP
    d. the complete oxidation of glucose to CO_2 and H_2O
    e. a series of reactions in which pyruvic acid is oxidized to CO_2 and H_2O

Use the following choices to answer questions 7–10.
    a. *E. coli* growing in glucose broth at 35°C with O_2 for 5 days
    b. *E. coli* growing in glucose broth at 35°C without O_2 for 5 days
    c. both a and b
    d. neither a nor b

7. Which culture produces the most lactic acid?

8. Which culture produces the most ATP?

9. Which culture uses NAD$^+$?

10. Which culture uses the most glucose?

CRITICAL THINKING

1. Write your own definition of the chemiosmotic mechanism of ATP generation. On Figure 5.15, mark the following using the appropriate letter:
    a. the acidic side of the membrane
    b. the side with a positive electrical charge
    c. potential energy
    d. kinetic energy

2. Explain why, even under ideal conditions, *Streptococcus* grows slowly.

3. Why must NADH be reoxidized? How does this happen in an organism that uses respiration? Fermentation?

4. The following graph shows the normal rate of reaction of an enzyme and its substrate (black) and the rate when an excess of competitive inhibitor is present (red). Explain why the graph appears as it does.

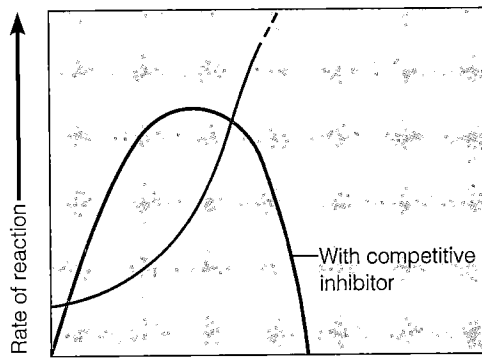

5. Compare and contrast carbohydrate catabolism and energy production in the following bacteria:
   a. *Pseudomonas,* an aerobic chemoheterotroph
   b. *Spirulina,* an oxygenic photoautotroph
   c. *Ectothiorhodospira,* an anoxygenic photoautotroph

6. How much ATP could be obtained from the complete oxidation of one molecule of glucose? From one molecule of butterfat containing one glycerol and three 12-carbon chains?

7. The chemoautotroph *Thiobacillus* can obtain energy from the oxidation of arsenic ($As^{3+} \rightarrow As^{5+}$). How does this reaction provide energy? How can this bacterium be put to use by humans?

CLINICAL APPLICATIONS

1. *Haemophilus influenzae* requires hemin (X factor) to synthesize cytochromes and NAD^+ (V factor) from other cells. For what does it use these two growth factors? What diseases does *H. influenzae* cause?

2. The drug HIVID®, also called ddC, inhibits DNA synthesis. It is used to treat HIV infection and AIDS. Compare the following illustration to Figure 2.17 on page 50. How does this drug work?

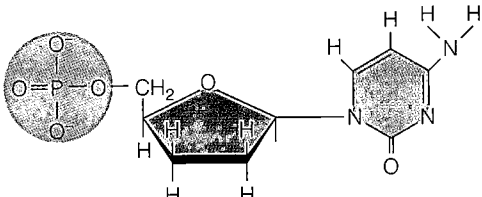

3. The bacterial enzyme streptokinase is used to digest fibrin (blood clots) in patients with atherosclerosis. Why doesn't injection of streptokinase cause a streptococcal infection? How do we know the streptokinase will digest fibrin only and not good tissues?

Learning with Technology

VU = VirtualUnknown CD-ROM

VU   *You must work through the VirtualUnknown Tutorial at least once before completing the VirtualUnknown exercises for this and all subsequent chapters.* You may also reference the liner notes accompanying your CD for instructions on how to perform laboratory tests.

Enter the Virtual Lab, click the arrow next to the Session field, click Textbook Exercises, and select Chapter 5a. Read the Case Study. This organism is *Escherichia coli.*

1. Complete the following tests, recording the results observed.
   - Gram stain
   - Acid from glucose
   - Gas from glucose
   - Lactose fermentation test
   - Sucrose fermentation test
   - Indole production
   - Acetoin production
   - Citrate utilization
   - Hydrogen sulfide (H^2S) production
   - Urea hydrolysis
   - Lysine decarboxylase activity
   - Growth in potassium cyanide
   - Malonate utilization
   - Nitrate reduction test

   After these tests are completed, print out the Virtual Laboratory Report.

2. Repeat the above assignment for each of the following organisms from the Textbook Exercises list:
   - Chapter 5b: *Enterobacter aerogenes*
   - Chapter 5c: *Salmonella typhi*
   - Chapter 5d: *Proteus vulgaris*

   Be sure to print out a Virtual Laboratory Report for each of these organisms.

3. Create a table that compares and contrasts the results of the four bacteria for this collection of tests. Use the table, the Help files in VirtualUnknown™ Microbiology, and the information in Chapter 5 to answer the following questions:

   a) All four bacteria were able to metabolize glucose, but not all produced identical results. Consult Figure 5.18 on page 135 of the textbook to provide an explanation for the inability of some bacteria to produce gas from glucose. Predict the metabolic products for the four bacteria you have just identified.

   b) If all four bacteria were able to metabolize glucose, explain why they were not all able to metabolize lactose and sucrose.

   c) How do microbes use different approaches to introduce the amino acids lysine and phenylalanine into the pathways for energy production?

   d) Create a dichotomous tree that contains the minimum tests necessary (from those included in your table) to clearly distinguish between these four bacteria. An example of the format for a dichotomous tree can be found in Figure 10.8 on page 288 of the textbook.

chapter 6

Microbial Growth

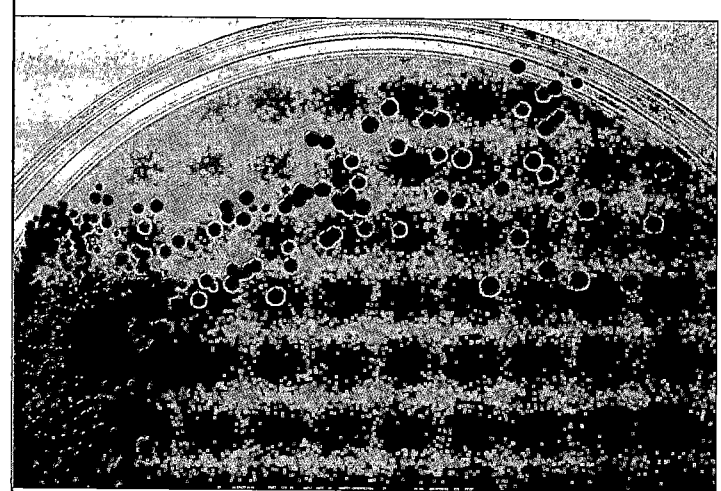

Staphylococcus aureus.
*Colonies of coagulase-positive S.
aureus can be identified by their
black colonies on tellurite-glycine
medium.*

When we talk about microbial growth, we are really referring to the *number* of cells, not the *size* of the cells. Microbes that are "growing" are increasing in number, accumulating into *colonies* (groups of cells large enough to be seen without a microscope) of hundreds of thousands of cells, or *populations* of billions of cells. For the most part, we are not concerned with the growth of individual cells. Although individual cells approximately double in size during their lifetime, this change is not very significant compared with the size increases observed during the lifetime of plants and animals.

Microbial populations can become incredibly large in a very short time, as we will see later in this chapter. By understanding the conditions necessary for microbial growth, we can determine how to control the growth of microbes that cause diseases and food spoilage. We can also learn how to encourage the growth of helpful microbes and those we wish to study. The manager of a sewage treatment plant and the manager of a brewery may not seem to have much in common, but both would be interested in promoting rapid microbial activity.

In this chapter we will examine the physical and chemical requirements for microbial growth, the various kinds of culture media, bacterial division, the phases of microbial growth, and the methods of measuring microbial growth. In later chapters we will see how unwanted microbial growth can be controlled.

The Requirements for Growth

The requirements for microbial growth can be divided into two main categories: physical and chemical. Physical aspects include temperature, pH, and osmotic pressure. Chemical requirements include sources of carbon, nitrogen, sulfur, phosphorus, trace elements, oxygen, and organic growth factors.

Physical Requirements

Learning Objectives

- *Classify microbes into five groups on the basis of preferred temperature range.*
- *Identify how and why the pH of culture media is controlled.*
- *Explain the importance of osmotic pressure to microbial growth.*

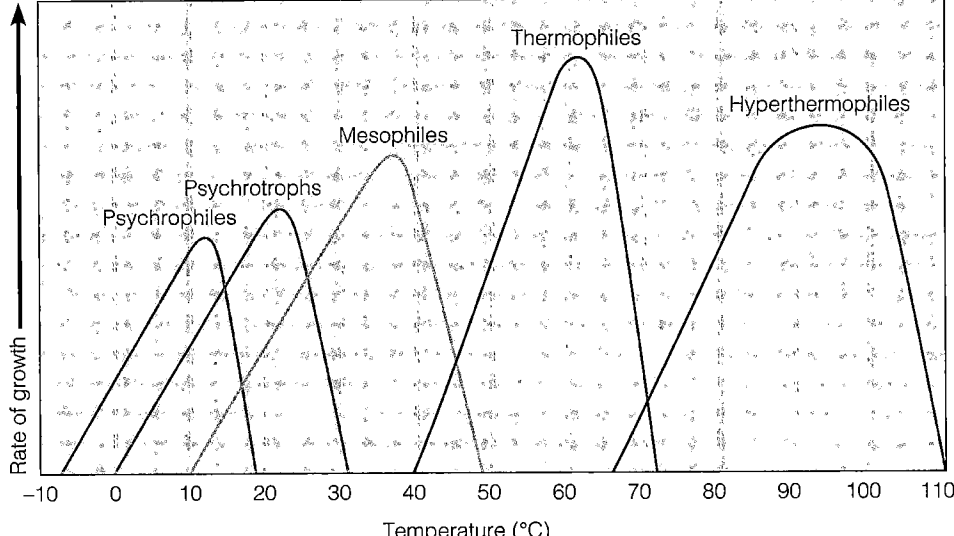

FIGURE 6.1 Typical growth rates of different types of microorganisms in response to temperature. Optimum growth (fastest reproduction) is represented by the peak of the curve. Notice that the reproductive rate drops off very quickly at temperatures only a little above the optimum. At either extreme of the temperature range, the reproductive rate is much lower than the rate at the optimum temperature.

▨ Why is it difficult to define psychrophile, mesophile, and thermophile?

Temperature

Most microorganisms grow well at the temperatures favored by humans. However, certain bacteria are capable of growing at extremes of temperature that would certainly hinder the survival of almost all eukaryotic organisms.

Microorganisms are classified into three primary groups on the basis of their preferred range of temperature: **psychrophiles** (cold-loving microbes), **mesophiles** (moderate-temperature–loving microbes), and **thermophiles** (heat-loving microbes). Most bacteria grow only within a limited range of temperatures, and their maximum and minimum growth temperatures are only about 30°C apart. They grow poorly at the temperature extremes within their range.

Each bacterial species grows at particular minimum, optimum, and maximum temperatures. The **minimum growth temperature** is the lowest temperature at which the species will grow. The **optimum growth temperature** is the temperature at which the species grows best. The **maximum growth temperature** is the highest temperature at which growth is possible. By graphing the growth response over a temperature range, we can see that the optimum growth temperature is usually near the top of the range; above that temperature the rate of growth drops off rapidly (Figure 6.1). This presumably happens because the high temperature has inactivated necessary enzymatic systems of the cell.

The ranges and maximum growth temperatures that define bacteria as psychrophiles, mesophiles, or thermophiles are not rigidly defined. Psychrophiles, for example, were originally considered simply to be organisms capable of growing at 0°C. However, there seem to be two fairly distinct groups capable of growth at that temperature. One group, composed of psychrophiles in the

strictest sense, can grow at 0°C but has an optimum growth temperature of about 15°C. Most of these organisms are so sensitive to higher temperatures that they will not even grow in a reasonably warm room (25°C). Found mostly in the oceans' depths or in certain polar regions, such organisms seldom cause problems in food preservation. The other group that can grow at 0°C has higher optimum temperatures, usually 20–30°C and cannot grow above about 40°C. Organisms of this type are much more common than psychrophiles and are the most likely to be encountered in low-temperature food spoilage because they grow fairly well at refrigerator temperatures. We will use the term **psychrotrophs** for the group of microorganisms responsible for such spoilage.

Refrigeration is the most common method of preserving household food supplies. It is based on the principle that microbial reproductive rates decrease at low temperatures. Although microbes usually survive even subfreezing temperatures (they might become entirely dormant), they gradually decline in number. Some species decline faster than others. Psychrotrophs actually do not grow well at low temperatures, except in comparison with other organisms, but given time they are able to slowly degrade food. Such spoilage might take the form of mold mycelium, slime on food surfaces, or off-tastes or off-colors in foods. The temperature inside a properly set refrigerator will greatly slow the growth of most spoilage organisms and will entirely prevent the growth of all but a few pathogenic bacteria. Figure 6.2 illustrates the importance of low temperatures for preventing the growth of spoilage and disease organisms. When large amounts of food must be refrigerated, it is important to keep in mind the slow cooling rate of a large quantity of warm food (Figure 6.3).

FIGURE 6.2 Food spoilage temperatures. The principle of refrigeration is based on the fact that low temperatures decrease microbial reproductive rates.

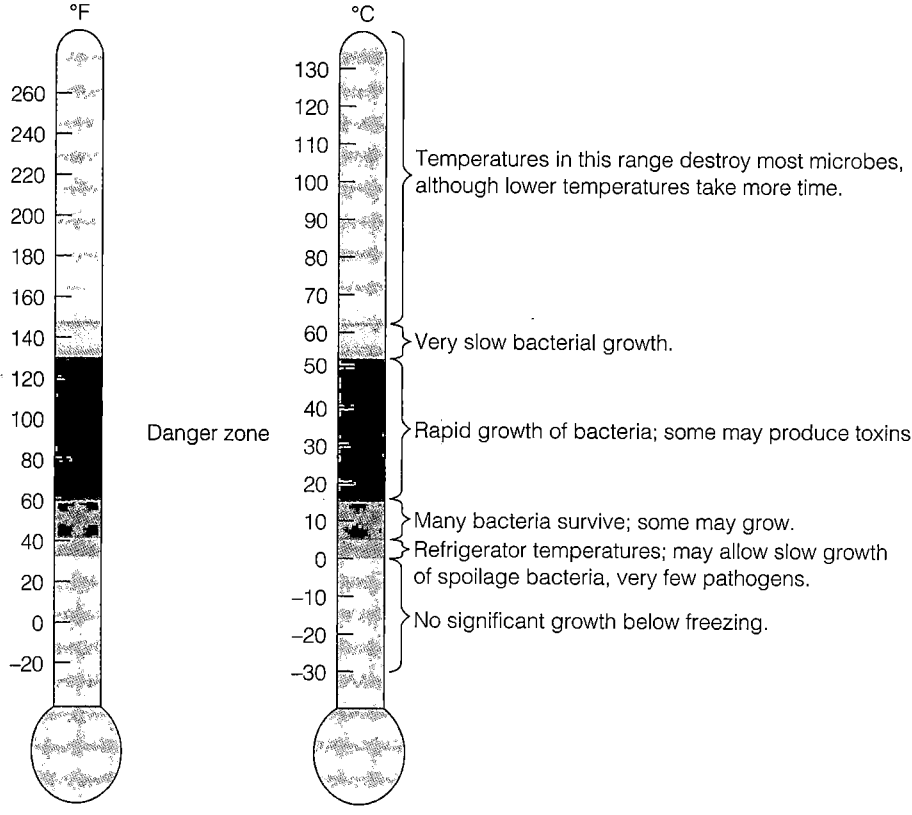

Temperatures in this range destroy most microbes, although lower temperatures take more time.

Very slow bacterial growth.

Danger zone

Rapid growth of bacteria; some may produce toxins.

Many bacteria survive; some may grow.

Refrigerator temperatures; may allow slow growth of spoilage bacteria, very few pathogens.

No significant growth below freezing.

FIGURE 6.3 The effect of the amount of food on its cooling rate in a refrigerator and its chance of spoilage. Notice that in this example the pan of rice with a depth of 5 cm (2 in) cooled through the incubation temperature of the *Bacillus cereus* in about 1 hour, whereas the pan of rice with a depth of 15 cm (6 in) remained in this temperature range for about 5 hours.

■ The slow cooling of large amounts of food is an essential consideration for people who prepare food for large groups.

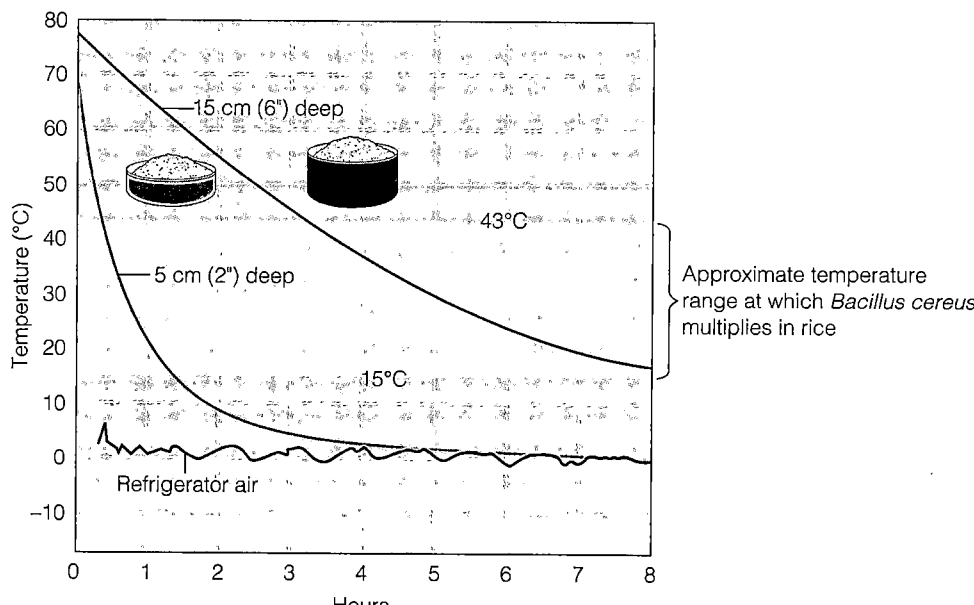

Approximate temperature range at which *Bacillus cereus* multiplies in rice

Mesophiles, with an optimum growth temperature of 25–40°C, are the most common type of microbe. Organisms that have adapted to live in the bodies of animals usually have an optimum temperature close to that of their hosts. The optimum temperature for many pathogenic bacteria is about 37°C, and incubators for clinical cultures are usually set at about this temperature. The mesophiles include most of the common spoilage and disease organisms.

Thermophiles are microorganisms capable of growth at high temperatures. Many of these organisms have an optimum growth temperature of 50–60°C, about the temperature of water from a hot water tap. Such temperatures can also be reached in sunlit soil and in thermal

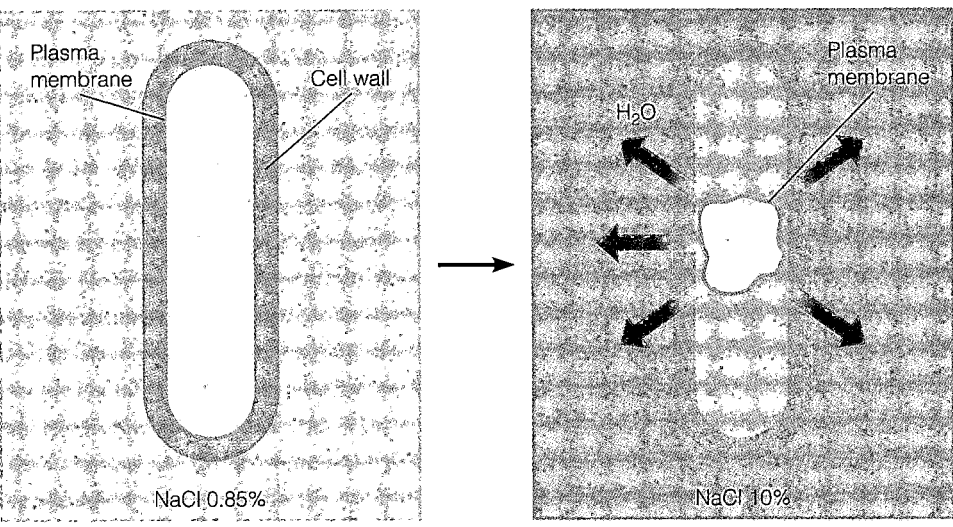

Normal cell in isotonic solution   Plasmolyzed cell in hypertonic solution

FIGURE 6.4 **Plasmolysis.**
(Left): A normal cell in which the osmotic pressure in the cell is equivalent to 0.85% sodium chloride (NaCl). (Right): If the concentration of solutes is higher in the surrounding medium than in the cell (hypertonic), water tends to leave the cell.

■ The addition of salts to a solution increases osmotic pressure and is used to preserve food.

waters such as hot springs. Remarkably, many thermophiles cannot grow at temperatures below about 45°C. Endospores formed by thermophilic bacteria are unusually heat resistant and may survive the usual heat treatment given canned goods. Although elevated storage temperatures may cause surviving endospores to germinate and grow, thereby spoiling the food, these thermophilic bacteria are not considered a public health problem. Thermophiles are important in organic compost piles (see Chapter 27, page 753), in which the temperature can rise rapidly to 50–60°C.

Some microbes, members of the Archaea, have an optimum growth temperature of 80°C or higher. These bacteria are called **hyperthermophiles,** or sometimes **extreme thermophiles.** Most of these organisms live in hot springs associated with volcanic activity; sulfur is usually important in their metabolic activity. The known record for bacterial growth at high temperatures is about 110°C near deep-sea hydrothermal vents. The immense pressure in the ocean depths prevents water from boiling even at temperatures well above 100°C.

pH

Recall from Chapter 2 (page 38), that pH refers to the acidity or alkalinity of a solution. Most bacteria grow best in a narrow pH range near neutrality, between pH 6.5 and 7.5. Very few bacteria grow at an acidic pH below about pH 4. This is why a number of foods, such as sauerkraut, pickles, and many cheeses, are preserved from spoilage by acids produced by bacterial fermentation. Nonetheless, some bacteria, called **acidophiles,** are remarkably tolerant of acidity. One type of chemoautotrophic bacteria, which is found in the drainage water from coal mines and oxidizes sulfur to form sulfuric acid,

can survive at a pH value of 1 (see Chapter 28, page 784). Molds and yeasts will grow over a greater pH range than bacteria will, but the optimum pH of molds and yeasts is generally below that of bacteria, usually about pH 5–6. Alkalinity also inhibits microbial growth but is rarely used to preserve foods.

When bacteria are cultured in the laboratory, they often produce acids that eventually interfere with their own growth. To neutralize the acids and maintain the proper pH, chemical buffers are included in the growth medium. The peptones and amino acids in some media act as buffers, and many media also contain phosphate salts. Phosphate salts have the advantage of exhibiting their buffering effect in the pH growth range of most bacteria. They are also nontoxic; in fact, they provide phosphorus, an essential nutrient.

Osmotic Pressure

Microorganisms obtain almost all their nutrients in solution from the surrounding water. Thus, they require water for growth and are made up of 80–90% water. High osmotic pressures have the effect of removing necessary water from a cell. When a microbial cell is in a solution that has a higher concentration of solutes than in the cell (hypertonic), the cellular water passes out through the plasma membrane to the high solute concentration. (See the discussion of osmosis in Chapter 4, page 92, and review Figure 4.17 for the three types of solution environments a cell may encounter.) This osmotic loss of water causes **plasmolysis,** or shrinkage of the cell's plasma membrane (Figure 6.4).

The importance of this phenomenon is that the growth of the cell is inhibited as the plasma membrane pulls away from the cell wall. Thus, the addition of salts

APPLICATIONS OF MICROBIOLOGY

Studying Hydrothermal Bacteria

Until humans explored the deep ocean floor, scientists believed that only a few forms of life could survive in that high-pressure, completely dark, oxygen-poor environment. Then, in 1977, the first manned vehicle capable of penetrating to the bottom of the deepest oceans carried two scientists 2600 meters below the surface at the Galápagos Rift, about 350 km northeast of the Galápagos Islands. There, amid the vast expanse of barren basalt rocks, the scientists found unexpectedly rich oases of life, including mollusks, crustaceans, and worms (see the photograph). How do such creatures survive in these harsh conditions? Many bacterial samples have been taken to the surface and are being studied.

ECOSYSTEM OF THE HYDROTHERMAL VENTS

Life at the surface of the world's oceans depends on photosynthetic organisms, such as bacteria and algae, which harness the sun's energy to fix carbon dioxide (CO_2) to make carbohydrates. At the deep ocean floor, where no light penetrates, photosynthesis is not possible. The scientists found that the primary producers at the ocean floor are chemoautotrophic bacteria. Using chemical energy from hydrogen sulfide (H_2S) as a source of energy to fix CO_2, the chemoautotrophs create an environment that supports higher life forms.

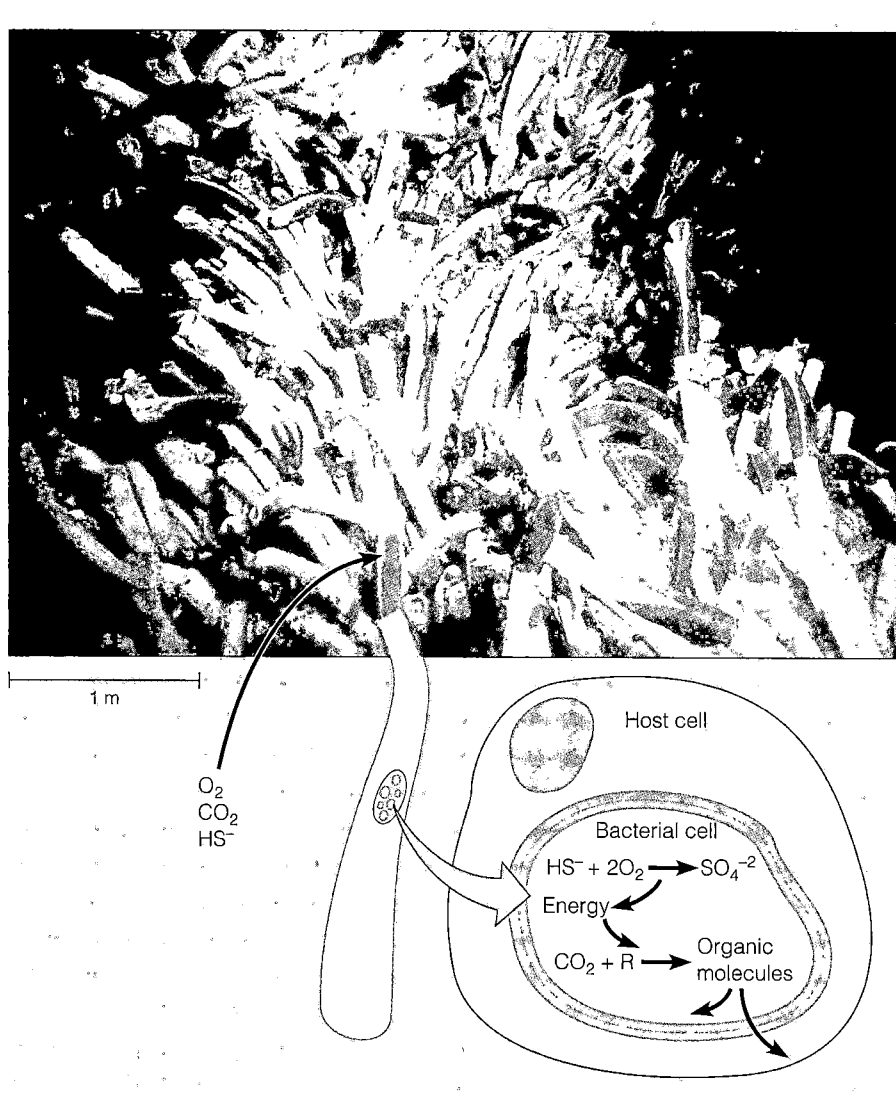

1 m

O_2
CO_2
HS^-

Host cell

Bacterial cell

$HS^- + 2O_2 \longrightarrow SO_4^{-2}$

Energy

$CO_2 + R \longrightarrow$ Organic molecules

Tubeworms and mussels at the Galápagos vent.

(or other solutes) to a solution, and the resulting increase in osmotic pressure, can be used to preserve foods. Salted fish, honey, and sweetened condensed milk are preserved largely by this mechanism; the high salt or sugar concentrations draw water out of any microbial cells that are present and thus prevent their growth. These effects of osmotic pressure are roughly related to the number of dissolved molecules and ions in a volume of solution.

Some organisms, called **extreme halophiles**, have adapted so well to high salt concentrations that they actually require them for growth. In this case, they may be termed **obligate halophiles.** Organisms from such saline waters as the Dead Sea often require nearly 30% salt, and the inoculating loop (a device for handling bacteria in the laboratory) used to transfer them must first be dipped into a saturated salt solution. More common are **facultative halophiles,** which do not require high salt concentrations but are able to grow at salt concentrations up to 2%, a concentration that inhibits the growth of many other organisms. A few species of facultative halophiles can even tolerate 15% salt. (A halophile is described in the box in Chapter 5, page 144.)

Studying Hydrothermal Bacteria (continued)

Hydrothermal vents in the seafloor supply the H_2S and CO_2. As superheated water from within the Earth rises through fractures in the Earth's crust called vents, it reacts with surrounding rock and dissolves metal ions, sulfides, and CO_2. The ecosystems of these vents depend on an abundance of sulfur compounds in the hot water. The concentration of sulfide (S^{2-}) is three times greater than the concentration of molecular oxygen in and around the vents. Such high sulfide concentrations are toxic to many organisms, but not to the creatures that inhabit this exotic environment.

Mats of bacteria grow along the sides of the vents, where temperatures exceed 100°C. These are the highest temperatures that any organism is known to tolerate. Above the vent, where temperatures are about 30°C, the concentration of bacteria is about four

times greater than that in water farther from the vents, and the growth rate of bacteria is equal to that found in productive, sunlit coastal waters. The bacteria in and around the vents create an environment in which bacteria are the producers supporting consumers (such as clams, mussels, and worms) and decomposers.

BIOTECHNOLOGICAL BENEFITS FROM HYDROTHERMAL VENTS

Researchers at Oak Ridge National Laboratory in Tennessee have identified two archaea living near deep-sea vents that hold promise for a renewable energy source. *Thermoplasma acidophilus* and *Pyrococcus furiosus* can produce the fuel, hydrogen gas, and an extracellular polysaccharide from glucose. Since glucose can be produced by bacteria and plants, it is a renewable resource. An additional benefit is that the polysaccha-

ride is produced in large quantities and may have industrial applications—for example, as a thickener in foods or other products.

DNA polymerases (enzymes that synthesize specific polymers) isolated from two archaea living near deep-sea vents are being used in the polymerase chain reaction (PCR), a technique for making many copies of DNA. In PCR, single-stranded DNA is made by heating a chromosome fragment to 98°C and cooling it so that DNA polymerase can copy each strand. DNA polymerase from *Thermococcus litoralis*, called Vent $_R$, and from *Pyrococcus*, called Deep Vent$_R$, are not denatured at 98°C. These enzymes can be used in automatic thermalcyclers to repeat the heating and cooling cycles, allowing many copies of DNA to be made easily and quickly. Vent$_R$ and Deep Vent$_R$ add bases to DNA at a rate of about 1000 bases per minute.

Most microorganisms, however, must be grown in a medium that is nearly all water. For example, the concentration of agar (a complex polysaccharide isolated from marine algae) used to solidify microbial growth media is usually about 1.5%. If markedly higher concentrations are used, the growth of some bacteria can be inhibited by the increased osmotic pressure.

If the osmotic pressure is unusually low (hypotonic), such as in distilled water, for example, water tends to enter the cell rather than leave it. Some microbes that have a relatively weak cell wall may be lysed by such treatment.

Chemical Requirements

Learning Objectives

- *Provide a use for each of the four elements (carbon, nitrogen, sulfur, and phosphorus) needed in large amounts for microbial growth.*
- *Explain how microbes are classified on the basis of oxygen requirements.*
- *Identify ways in which aerobes avoid damage by toxic forms of oxygen.*

Carbon

Besides water, one of the most important requirements for microbial growth is carbon. Carbon is the structural backbone of living matter; it is needed for all the organic compounds that make up a living cell. Half the dry weight of a typical bacterial cell is carbon. Chemoheterotrophs get most of their carbon from the source of their energy—organic materials such as proteins, carbohydrates, and lipids. Chemoautotrophs and photoautotrophs derive their carbon from carbon dioxide.

Nitrogen, Sulfur, and Phosphorus

In addition to carbon, other elements are needed by microorganisms for the synthesis of cellular material. For example, protein synthesis requires considerable amounts of nitrogen as well as some sulfur. The syntheses of DNA and RNA also require nitrogen and some phosphorus, as does the synthesis of ATP, the molecule so important for the storage and transfer of chemical energy within the cell. Nitrogen makes up about 14% of the dry weight of a bacterial cell, and sulfur and phosphorus together constitute about another 4%.

4. The **hydroxyl radical** (OH·) is another intermediate form of oxygen and probably the most reactive. It is formed in the cellular cytoplasm by ionizing radiation. Most aerobic respiration produces some hydroxyl radicals.

Obligate anaerobes usually produce neither superoxide dismutase nor catalase. Because aerobic conditions probably lead to an accumulation of superoxide free radicals in their cytoplasm, obligate anaerobes are extremely sensitive to oxygen.

Aerotolerant anaerobes (Table 6.1d) cannot use oxygen for growth, but they tolerate it fairly well. On the surface of a solid medium, they will grow without the special techniques (discussed later) required for obligate anaerobes. Many of the aerotolerant bacteria characteristically ferment carbohydrates to lactic acid. As lactic acid accumulates, it inhibits the growth of aerobic competitors and establishes a favorable ecological niche for lactic acid producers. A common example of lactic acid–producing aerotolerant anaerobes is the lactobacilli used in the production of many acidic fermented foods, such as pickles and cheese. In the laboratory, they are handled and grown much like any other bacteria, but they make no use of the oxygen in the air. These bacteria can tolerate oxygen because they possess SOD or an equivalent system that neutralizes the toxic forms of oxygen previously discussed.

A few bacteria are **microaerophiles** (Table 6.1e). They are aerobic; they do require oxygen. However, they grow only in oxygen concentrations lower than those in air. In a test tube of solid nutrient medium, they grow only at a depth where small amounts of oxygen have diffused into the medium; they do not grow at the oxygen-rich surface or below the narrow zone of adequate oxygen. This limited tolerance is probably due to their sensitivity to superoxide free radicals and peroxides, which they produce in lethal concentrations under oxygen-rich conditions.

Organic Growth Factors

Essential organic compounds an organism is unable to synthesize are known as **organic growth factors;** they must be directly obtained from the environment. One group of organic growth factors for humans is vitamins. Most vitamins function as coenzymes, the organic cofactors required by certain enzymes in order to function. Many bacteria can synthesize all their own vitamins and are not dependent on outside sources. However, some bacteria lack the enzymes needed for the synthesis of certain vitamins, and for them those vitamins are organic growth factors. Other organic growth factors required by some bacteria are amino acids, purines, and pyrimidines.

Culture Media

Learning Objectives

- *Distinguishing between chemically defined and complex media.*
- *Justify the use of each of the following: anaerobic techniques, living host cells, candle jars, selective and differential media, enrichment medium.*

A nutrient material prepared for the growth of microorganisms in a laboratory is called a **culture medium.** Some bacteria can grow well on just about any culture medium; others require special media, and still others cannot grow on any nonliving medium yet developed. When microbes are introduced into a culture medium to initiate growth, they are called an **inoculum.** The microbes that grow and multiply in or on a culture medium are referred to as a **culture.**

Suppose we want to grow a culture of a certain microorganism, perhaps the microbes from a particular clinical specimen. What criteria must the culture medium meet? First, it must contain the right nutrients for the specific microorganism we want to grow. It should also contain sufficient moisture, a properly adjusted pH, and a suitable level of oxygen, perhaps none at all. The medium must initially be **sterile**—that is, it must initially contain no living microorganisms—so that the culture will contain only the microbes (and their offspring) we add to the medium. Finally, the growing culture should be incubated at the proper temperature.

A wide variety of media are available for the growth of microorganisms in the laboratory. Most of these media, which are available from commercial sources, have premixed components and require only the addition of water and then sterilization. Media are constantly being developed or revised for use in the isolation and identification of bacteria that are of interest to researchers in such fields as food, water, and clinical microbiology.

When it is desirable to grow bacteria on a solid medium, a solidifying agent such as agar is added to the medium. A complex polysaccharide derived from a marine alga, **agar** has long been used as a thickener in foods such as jellies and ice cream.

Agar has some very important properties that make it valuable to microbiology, and no satisfactory substitute has ever been found. Few microbes can degrade agar, so it remains solid. Also, agar liquifies at about 100°C (the boiling point of water) and at sea level remains liquid until the temperature drops to about 40°C. For laboratory use, agar is held in water baths at about 50°C. At this temperature, it does not injure most bacteria when it is

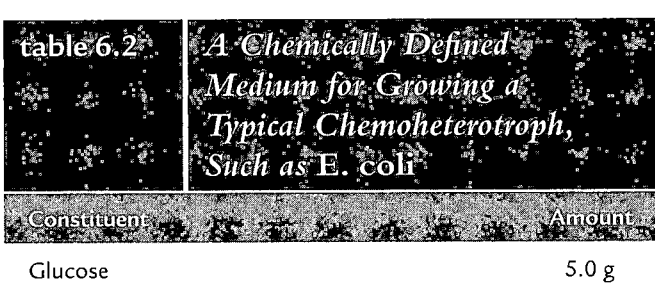

Constituent	Amount
Glucose	5.0 g
Ammonium phosphate, monobasic ($NH_4H_2PO_4$)	1.0 g
Sodium chloride (NaCl)	5.0 g
Magnesium sulfate ($MgSO_4 \cdot 7H_2O$)	0.2 g
Potassium phosphate, dibasic (K_2HPO_4)	1.0 g
Water	1 liter

poured over them (as shown in Figure 6.16a). Once the agar has solidified, it can be incubated at temperatures approaching 100°C before it again liquifies; this property is particularly useful when thermophilic bacteria are being grown.

Agar media are usually contained in test tubes or Petri dishes. The test tubes are called *slants* when they are allowed to solidify with the tube held at an angle so that a large surface area for growth is available. When the agar solidifies in a vertical tube, it is called a *deep*. Petri dishes, named for their inventor, are shallow dishes with a lid that nests over the bottom to prevent contamination; when filled, they are called Petri (or culture) plates.

Chemically Defined Media

To support microbial growth, a medium must provide an energy source, as well as sources of carbon, nitrogen, sulfur, phosphorus, and any organic growth factors the organism is unable to synthesize. A **chemically defined medium** is one whose exact chemical composition is known. For a chemoheterotroph, the chemically defined medium must contain organic growth factors that serve as a source of carbon and energy. For example, as shown in Table 6.2, glucose is included in the medium for growing the chemoheterotroph *E. coli*.

As Table 6.3 shows, many organic growth factors must be provided in the chemically defined medium used to cultivate a species of *Neisseria*. Organisms that require many growth factors are described as "fastidious." Organisms of this type, such as *Lactobacillus,* are sometimes used in tests that determine the concentration of a particular vitamin in a substance. To perform such a *microbiological assay,* a growth medium is prepared that contains all the

growth requirements of the bacterium except the vitamin being assayed. Then the medium, test substance, and bacterium are combined, and the growth of bacteria is measured. This bacterial growth, which is reflected by the amount of lactic acid produced, will be proportional to the amount of vitamin in the test substance. The more lactic acid, the more the *Lactobacillus* cells have been able to grow, so the more vitamin is present.

Complex Media

Chemically defined media are usually reserved for laboratory experimental work or for the growth of autotrophic bacteria. Most heterotrophic bacteria and fungi, such as you would work with in an introductory lab course, are routinely grown on **complex media,** made up of nutrients such as extracts from yeasts, meat, or plants, or digests of proteins from these and other sources. The exact chemical composition varies slightly from batch to batch. Table 6.4 shows one widely used recipe.

In complex media, the energy, carbon, nitrogen, and sulfur requirements of the growing microorganisms are primarily provided by protein. Protein is a large, relatively insoluble molecule that a minority of microorganisms can utilize directly, but a partial digestion by acids or enzymes reduces protein to shorter chains of amino acids called *peptones.* These small, soluble fragments can be digested by the bacteria.

Vitamins and other organic growth factors are provided by meat extracts or yeast extracts. The soluble vitamins and minerals from the meats or yeasts are dissolved in the extracting water, which is then evaporated so that these factors are concentrated. (These extracts also supplement the organic nitrogen and carbon compounds.) Yeast extracts are particularly rich in the B vitamins. If a complex medium is in liquid form, it is called **nutrient broth.** When agar is added, it is called **nutrient agar.** (This terminology can be confusing; just remember that agar itself is not a nutrient.)

Anaerobic Growth Media and Methods

The cultivation of anaerobic bacteria poses a special problem. Because anaerobes might be killed by exposure to oxygen, special media called **reducing media** must be used. These media contain ingredients, such as sodium thioglycolate, that chemically combine with dissolved oxygen and deplete the oxygen in the culture medium. To routinely grow and maintain pure cultures of obligate anaerobes, microbiologists use reducing media stored in ordinary, tightly capped test tubes. These media are heated shortly before use, to drive off absorbed oxygen.

table 6.3	A Chemically Defined Medium for Growing a Fastidious Chemoheterotrophic Bacterium, Such as Neisseria gonorrhoeae		
Constituent	**Amount**	**Constituent**	**Amount**
Carbon and energy sources		Amino acids	
Glucose	9.1 g	Cysteine	1.5 g
Starch	9.1 g	Arginine, proline (each)	0.3 g
Sodium acetate	1.8 g	Glutamic acid, methionine (each)	0.2 g
Sodium citrate	1.4 g	Asparagine, isoleucine, serine (each)	0.2 g
Oxaloacetate	0.3 g	Cystine	0.06 g
Salts		Organic growth factors	
Potassium phosphate, dibasic (K_2HPO_4)	12.7 g	Calcium pantothenate	0.02 g
Sodium chloride (NaCl)	6.4 g	Thiamine	0.02 g
Potassium phosphate, monobasic (KH_2PO_4)	5.5 g	Nicotinamide adenine dinucleotide	0.01 g
Sodium bicarbonate ($NaHCO_3$)	1.2 g	Uracil	0.006 g
Potassium sulfate (K_2SO_4)	1.1 g	Biotin	0.005 g
Sodium sulfate (Na_2SO_4)	0.9 g	Hypoxanthine	0.003 g
Magnesium chloride ($MgCl_2$)	0.5 g	Reducing agent	
Ammonium chloride (NH_4Cl)	0.4 g	Sodium thioglycolate	0.00003 g
Potassium chloride (KCl)	0.4 g	Water	1 liter
Calcium chloride ($CaCl_2$)	0.006 g		
Ferric nitrate [$Fe(NO_3)_3$]	0.006 g		

SOURCE: R. M. Atlas, *Handbook of Microbiological Media*, Ann Arbor, MI: CRC Press, 1993.

table 6.4	Composition of Nutrient Agar, a Complex Medium for the Growth of Heterotrophic Bacteria
Constituent	**Amount**
Peptone (partially digested protein)	5.0 g
Beef extract	3.0 g
Sodium chloride	8.0 g
Agar	15.0 g
Water	1 liter

When the culture must be grown in Petri plates to observe individual colonies, special anaerobic jars are used (Figure 6.5). The culture plates are placed in the jar, and oxygen is removed by the following process: A packet of chemicals (sodium bicarbonate and sodium borohy-dride) in the jar is moistened with a few milliliters of water, and the jar is sealed. Hydrogen and carbon dioxide are produced by the reaction of the chemicals with the water. A palladium catalyst in the jar combines the oxygen in the jar with the hydrogen produced by the chemical reaction, and water is formed. As a result, the oxygen quickly disappears. Moreover, the carbon dioxide that is produced aids the growth of many anaerobic bacteria.

Researchers regularly working with anaerobes use transparent anaerobic chambers equipped with air locks and filled with inert gases (Figure 6.6). Technicians can manipulate the equipment by inserting their hands into airtight rubber gloves called glove ports, which are fitted to the wall of the chamber.

Special Culture Techniques

Some bacteria have never been successfully grown on artificial laboratory media. *Mycobacterium leprae,* the leprosy bacillus, is now usually grown in armadillos, which have a relatively low body temperature that matches the require-

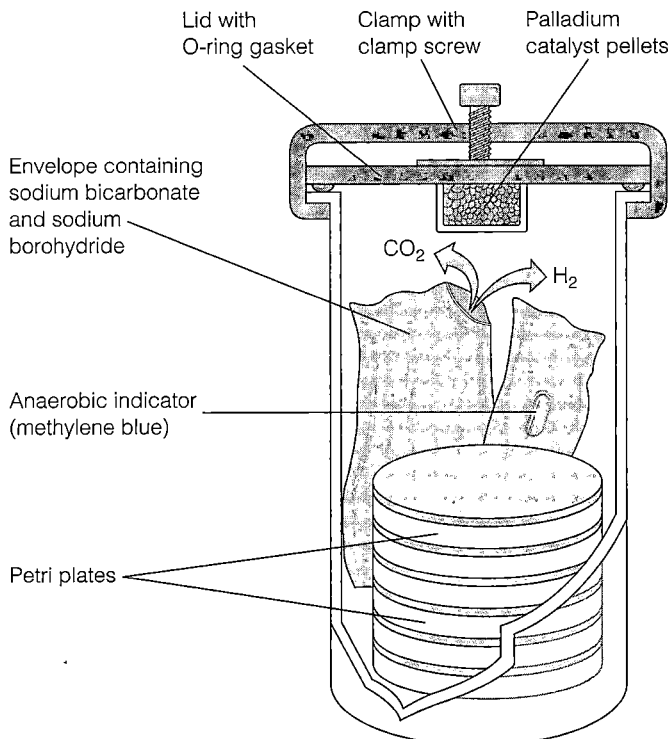

Lid with O-ring gasket — Clamp with clamp screw — Palladium catalyst pellets

Envelope containing sodium bicarbonate and sodium borohydride

CO_2 — H_2

Anaerobic indicator (methylene blue)

Petri plates

FIGURE 6.5 A jar for cultivating anaerobic bacteria on Petri plates. When water is mixed with the chemical packet containing sodium bicarbonate and sodium borohydride, hydrogen and carbon dioxide are generated. Reacting on the surface of a palladium catalyst in a screened reaction chamber, which may also be incorporated into the chemical packet, the hydrogen and atmospheric oxygen in the jar combine to form water. The oxygen is thus removed. Also in the jar is an anaerobic indicator, containing methylene blue, which is blue when oxidized (as shown here) and turns colorless when the oxygen is removed.

▨ **The CO_2 produced in the jar aids the growth of many anaerobic bacteria.**

ments of the microbe. Another example is the syphilis spirochete, although certain nonpathogenic strains of this microbe have been grown on laboratory media. With few exceptions, the obligate intracellular bacteria, such as the rickettsias and the chlamydias, do not grow on artificial media. Like viruses, they can reproduce only in a living host cell.

Many clinical laboratories have special *carbon dioxide incubators* in which to grow aerobic bacteria that require concentrations of CO_2 higher or lower than that found in the atmosphere. Desired CO_2 levels are maintained by electronic controls. High CO_2 levels are also obtained with simple *candle jars* (Figure 6.7a). Cultures are placed in a large sealed jar containing a lighted candle, which consumes oxygen. The candle stops burning when the air in the jar has a lowered concentration of oxygen (but one still adequate for the growth of aerobic bacteria). An ele-

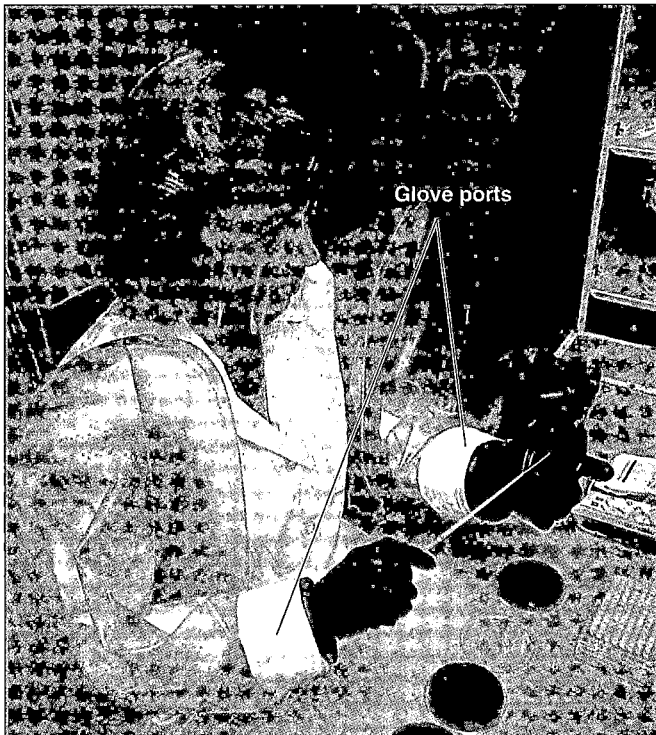

FIGURE 6.6 An anaerobic chamber. The technician is holding a Petri plate and an inoculating loop that are inside an anaerobic chamber filled with an inert, oxygen-free gas. Organisms and materials enter and leave through an air lock, and equipment is manipulated by means of glove ports.

vated concentration of CO_2 is also present. Microbes that grow better at high CO_2 concentrations are called **capnophiles.** The low-oxygen, high-CO_2 conditions resemble those found in the intestinal tract, respiratory tract, and other body tissues where pathogenic bacteria grow.

Candle jars are still used occasionally, but more often commercially available chemical packets are used to generate carbon dioxide atmospheres in containers (Figure 6.7b). When only one or two Petri plates of cultures are to be incubated, clinical laboratory investigators often use small plastic bags with self-contained chemical gas generators that are activated by crushing the packet or moistening it with a few milliliters of water. These packets are sometimes specially designed to provide precise concentrations of carbon dioxide (usually higher than can be obtained in candle jars) and oxygen, for culturing organisms such as the microaerophilic *Campylobacter* bacteria.

Selective and Differential Media

In clinical and public health microbiology, it is frequently necessary to detect the presence of specific microorganisms associated with disease or poor sanitation. For this task, selective and differential media are used.

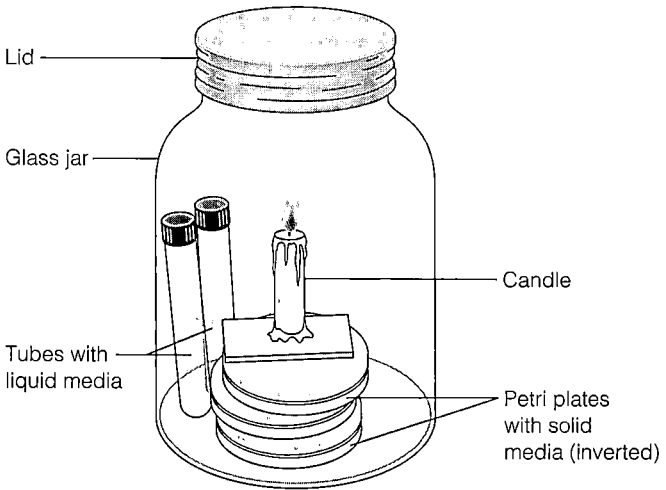

(a) Candle jar

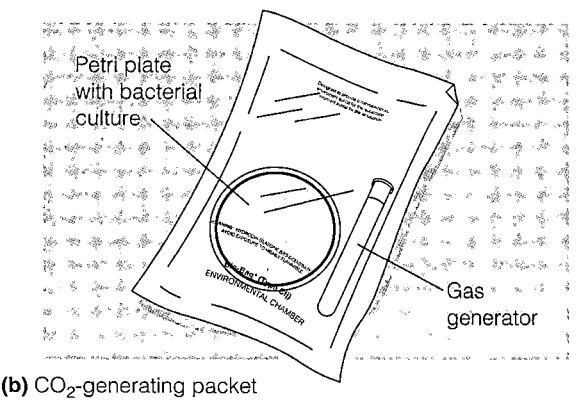

(b) CO₂-generating packet

FIGURE 6.7   Equipment for producing CO₂-rich environments.
(a) Plates and tubes inoculated with, for example, *Neisseria meningitidis* are placed in a jar with a lighted candle, and the jar is sealed. This will provide a CO₂ atmosphere of approximately 3%. **(b)** The packet consists of a bag containing a Petri plate and a CO₂ gas generator. The gas generator is crushed to mix the chemicals it contains and start the reaction that produces CO₂. This gas reduces the oxygen concentration in the bag to about 5% and provides a CO₂ concentration of about 10%.

■ Microbes that grow better at high CO₂ concentrations are called capnophiles.

Selective media are designed to suppress the growth of unwanted bacteria and encourage the growth of the desired microbes. For example, bismuth sulfite agar is one medium used to isolate the typhoid bacterium, the gram-negative *Salmonella typhi* (tī fē), from feces. Bismuth sulfite inhibits gram-positive bacteria and most gram-negative intestinal bacteria (other than *S. typhi,*) as well. Sabouraud's dextrose agar, which has a pH of 5.6, is used to isolate fungi that outgrow most bacteria at this pH. Dyes such as brilliant green selectively inhibit gram-positive bacteria, and this dye is the basis of

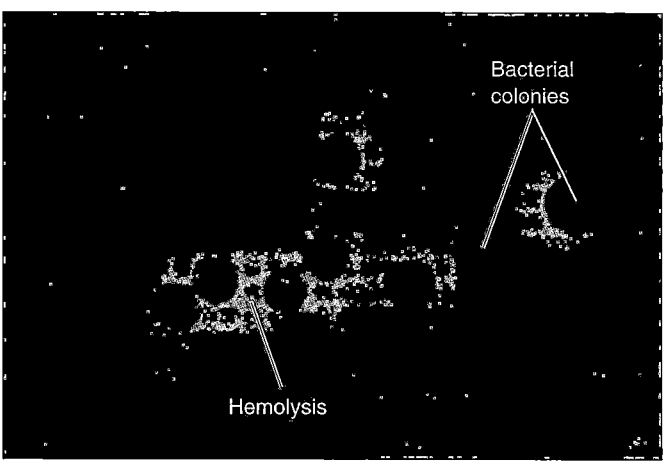

FIGURE 6.8   Blood agar, a differential medium containing red blood cells. The bacteria have lysed the red blood cells (hemolysis), causing the clear areas around the colonies.

■ Differential media are designed to distinguish colonies of the desired organism from other colonies growing on the same plate.

a medium called brilliant green agar that is used to isolate the gram-negative *Salmonella*.

Differential media make it easier to distinguish colonies of the desired organism from other colonies growing on the same plate. Similarly, pure cultures of microorganisms have identifiable reactions with differential media in tubes or plates. Blood agar (which contains red blood cells) is a medium that microbiologists often use to identify bacterial species that destroy red blood cells. These species, such as *Streptococcus pyogenes* (pī-äj'en-ēz), the bacterium that causes strep throat, show a clear ring around their colonies where they have lysed the surrounding blood cells (Figure 6.8).

Sometimes, selective and differential characteristics are combined in a single medium. Suppose we want to isolate the common bacterium *Staphylococcus aureus,* found in the nasal passages. This organism has a tolerance for high concentrations of sodium chloride; it can also ferment the carbohydrate mannitol to form acid. Mannitol salt agar contains 7.5% sodium chloride, which will discourage the growth of competing organisms and thus *select for* (favor the growth of) *S. aureus.* This salty medium also contains a pH indicator that changes color if the mannitol in the medium is fermented to acid; the mannitol-fermenting colonies of *S. aureus* are thus *differentiated from* colonies of bacteria that do not ferment mannitol. Bacteria that grow at the high salt concentration *and* ferment mannitol to acid can be readily identified by the color change. These are probably colonies of *S. aureus,* and their identification can be confirmed by additional tests.

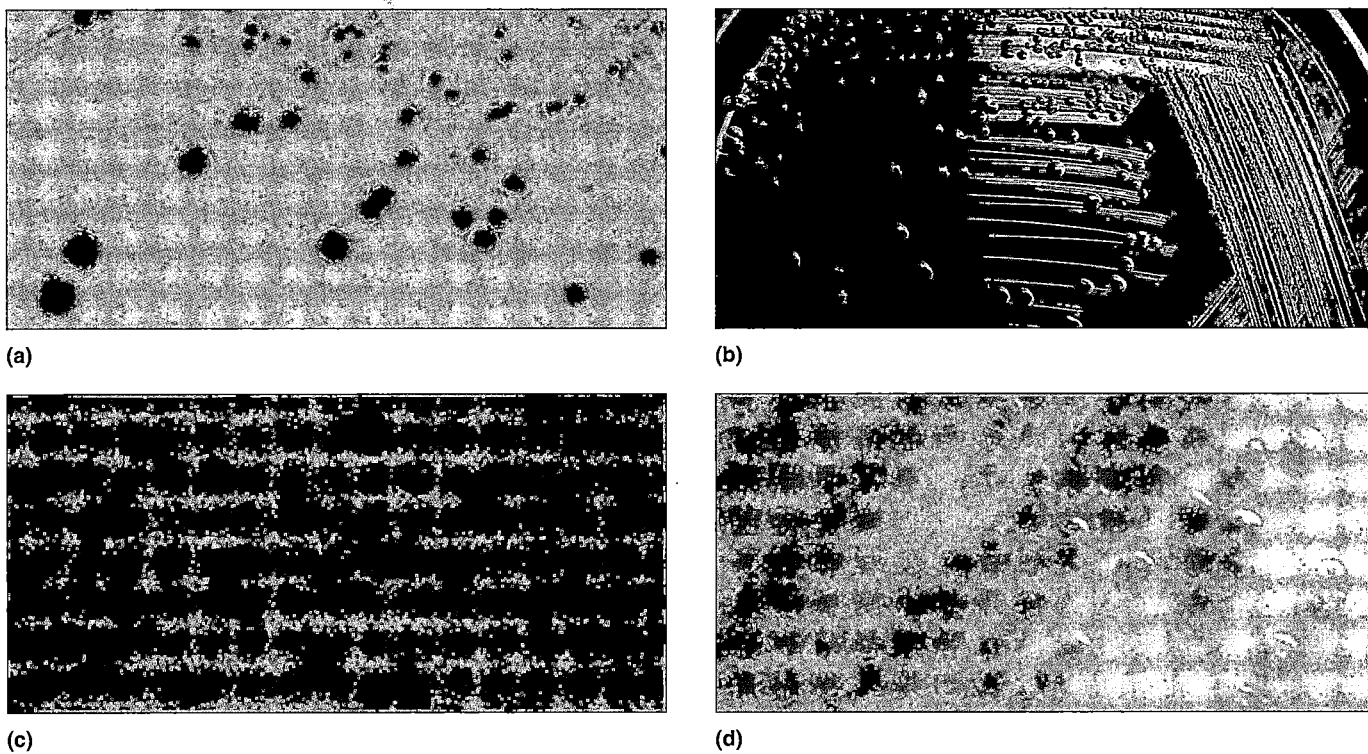

FIGURE 6.9 Bacterial colonies on several differential media.
(a) *Staphylococcus aureus* on tellurite-glycine medium.
(b) *Escherichia coli* on eosin methylene blue (EMB) medium. The black-centered colonies are surrounded by a characteristic metallic green sheen.
(c) *Enterobacter aerogenes* on EMB medium showing characteristic dark-centered colonies.
(d) On pseudomonas agar P (PSP) medium, *Pseudomonas aeruginosa* produces a blue-green water-soluble pigment.

■ Which media pictured here are both selective and differential?

Another medium that is both selective and differential is MacConkey agar. This medium contains bile salts and crystal violet, which inhibit the growth of gram-positive bacteria. Because this medium also contains lactose, gram-negative bacteria that can grow on lactose can be differentiated from similar bacteria that cannot. The ability to distinguish between lactose fermenters (red or pink colonies) and nonfermenters (colorless colonies) is useful in distinguishing between the pathogenic *Salmonella* bacteria and other related bacteria. Figure 6.9 shows the appearance of bacterial colonies on several differential media.

Enrichment Culture

Because bacteria present in small numbers can be missed, especially if other bacteria are present in much larger numbers, it is sometimes necessary to use an **enrichment culture.** This is often the case for soil or fecal samples. The

medium (enrichment medium) for an enrichment culture is usually liquid and provides nutrients and environmental conditions that favor the growth of a particular microbe but not others. In this sense, it is also a selective medium, but it is designed to increase very small numbers of the desired type of organism to detectable levels.

Suppose we want to isolate from a soil sample a microbe that can grow on phenol and is present in much smaller numbers than other species. If the soil sample is placed in a liquid enrichment medium in which phenol is the only source of carbon and energy, microbes unable to metabolize phenol will not grow. The culture medium is allowed to incubate for a few days, and then a small amount of it is transferred into another flask of the same medium. After a series of such transfers, the surviving population will consist of bacteria capable of metabolizing phenol. The bacteria are given time to grow in the medium between transfers; this is the enrichment stage. See the box in Chapter 11 on page 328. Any nutrients in

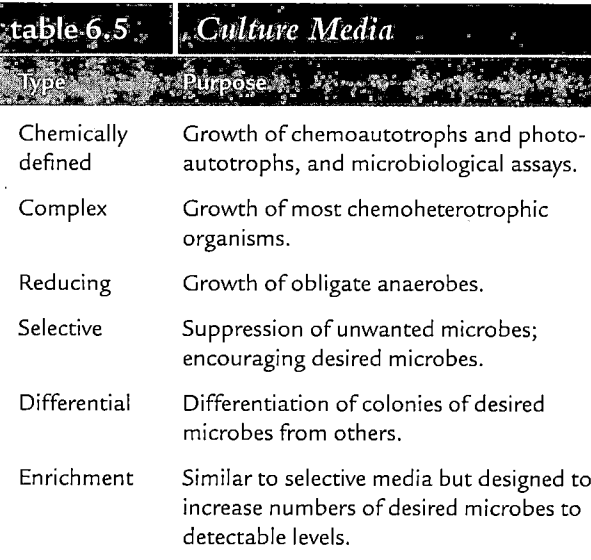

table 6.5	Culture Media	
Type	**Purpose**	
Chemically defined	Growth of chemoautotrophs and photo-autotrophs, and microbiological assays.	
Complex	Growth of most chemoheterotrophic organisms.	
Reducing	Growth of obligate anaerobes.	
Selective	Suppression of unwanted microbes; encouraging desired microbes.	
Differential	Differentiation of colonies of desired microbes from others.	
Enrichment	Similar to selective media but designed to increase numbers of desired microbes to detectable levels.	

the original inoculum are rapidly diluted out with the successive transfers. When the last dilution is streaked onto a solid medium of the same composition, only those colonies of organisms capable of using phenol should grow. A remarkable aspect of this particular technique is that phenol is normally lethal to most bacteria.

★ ★ ★

Table 6.5 summarizes the purposes of the main types of culture media.

Obtaining Pure Cultures

Learning Objectives

■ Define colony.
■ Describe how pure cultures can be isolated by using the streak plate method.

Most infectious materials, such as pus, sputum, and urine, contain several different kinds of bacteria; so do samples of soil, water, or food. If these materials are plated out onto the surface of a solid medium, colonies will form that are exact copies of the original organism. A visible **colony** theoretically arises from a single spore or vegetative cell or from a group of the same microorganisms attached to one another in clumps or chains. Microbial colonies often have a distinctive appearance that distinguishes one microbe from another (see Figure 6.9). The bacteria must be distributed widely enough so that the colonies are visibly separated from each other.

Most bacteriological work requires pure cultures, or clones, of bacteria. The isolation method most commonly

used to get pure cultures is the **streak plate method** (Figure 6.10). A sterile inoculating loop is dipped into a mixed culture that contains more than one type of microbe and is streaked in a pattern over the surface of the nutrient medium. As the pattern is traced, bacteria are rubbed off the loop onto the medium. The last cells to be rubbed off the loop are far enough apart to grow into isolated colonies. These colonies can be picked up with an inoculating loop and transferred to a test tube of nutrient medium to form a pure culture containing only one type of bacterium.

The streak plate method works well when the organism to be isolated is present in large numbers relative to the total population. However, when the microbe to be isolated is present only in very small numbers, its numbers must be greatly increased by selective enrichment before it can be isolated with the streak plate method.

Preserving Bacterial Cultures

Learning Objective

■ Explain how microorganisms are preserved by deep-freezing and lyophilization (freeze-drying).

Refrigeration can be used for the short-term storage of bacterial cultures. Two common methods of preserving microbial cultures for long periods are deep-freezing and lyophilization. **Deep-freezing** is a process in which a pure culture of microbes is placed in a suspending liquid and quick-frozen at temperatures ranging from $-50°$ to $-95°C$. The culture can usually be thawed and cultured even several years later. During **lyophilization (freeze-drying)**, a suspension of microbes is quickly frozen at temperatures ranging from $-54°$ to $-72°C$, and the water is removed by a high vacuum (sublimation). While under vacuum, the container is sealed by a high-temperature torch. The remaining powderlike residue that contains the surviving microbes can be stored for years. The organisms can be revived at any time by hydration with a suitable liquid nutrient medium.

The Growth of Bacterial Cultures

Learning Objective

■ Define bacterial growth, including binary fission.

Being able to graphically represent the enormous populations resulting from the growth of bacterial cultures is an

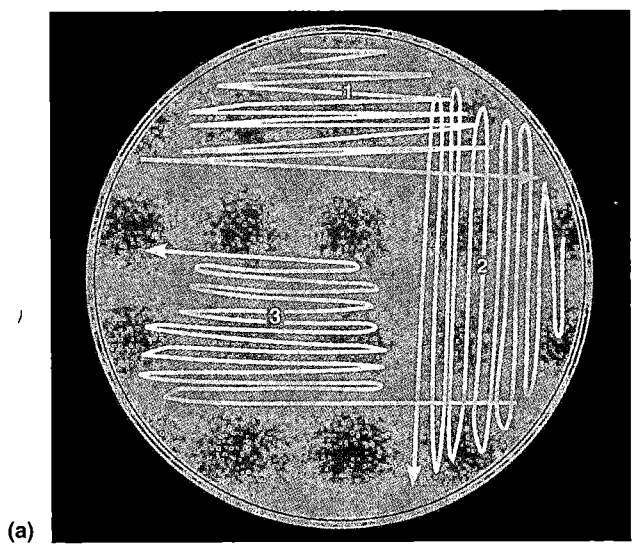

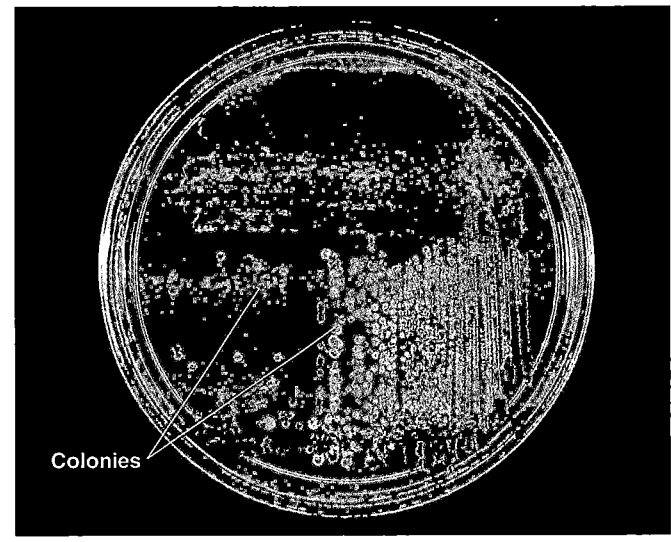

FIGURE 6.10 The streak plate method for isolating pure bacterial cultures. (a) The direction of streaking is indicated by arrows. Streak series 1 is made from the original bacterial culture. The inoculating loop is sterilized following each streak series. In series 2 and 3, the loop picks up bacteria from the previous series, diluting the number of cells each time. There are numerous variants of such patterns. **(b)** In series 3 of this example, notice that well-isolated colonies of bacteria of two different types have been obtained.

■ Streak plates are the most common technique used to obtain pure cultures of microorganisms.

essential part of microbiology. It is also necessary to be able to determine microbial numbers, either directly, by counting, or indirectly, by measuring their metabolic activity.

Bacterial Division

As we mentioned at the beginning of the chapter, bacterial growth refers to an increase in bacterial numbers, not an increase in the size of the individual cells. Bacteria normally reproduce by **binary fission** (Figure 6.11).

A few bacterial species reproduce by **budding;** they form a small initial outgrowth (a bud) that enlarges until its size approaches that of the parent cell, and then it separates. Some filamentous bacteria (certain actinomycetes) reproduce by producing chains of conidiospores carried externally at the tips of the filaments. A few filamentous species simply fragment, and the fragments initiate the growth of new cells.

Generation Time

For purposes of calculating the generation time of bacteria, we will consider only reproduction by binary fission, which is by far the most common method. As you can see

in Figure 6.12, one cell's division produces two cells, two cells' divisions produce four cells, and so on. When the arithmetic number of cells in each generation is expressed as a power of 2, the exponent tells the number of doublings (generations) that have occurred.

The time required for a cell to divide (and its population to double) is called the **generation time.** It varies considerably among organisms and with environmental conditions such as temperature. Most bacteria have a generation time of 1–3 hours; others require more than 24 hours per generation. (The math required to calculate generation times is presented in Appendix D.) If binary fission continues unchecked, an enormous number of cells will be produced. If a doubling occurred every 20 minutes—which is the case for *E. coli* under favorable conditions—after 20 generations a single initial cell would increase to over 1 million cells. This would require a little less than 7 hours. In 30 generations, or 10 hours, the population would be 1 billion, and in 24 hours it would be a number trailed by 21 zeros. It is difficult to graph population changes of such enormous magnitude by using arithmetic numbers. This is why logarithmic scales are generally used to graph bacterial

FIGURE 6.11 Binary fission in bacteria. (a) A diagram of the sequence of cell division. **(b)** A thin section of a cell of *Bacillus licheniformis* starting to divide.

■ As a result of binary fission, two individual cells are formed, each identical to the parent cell.

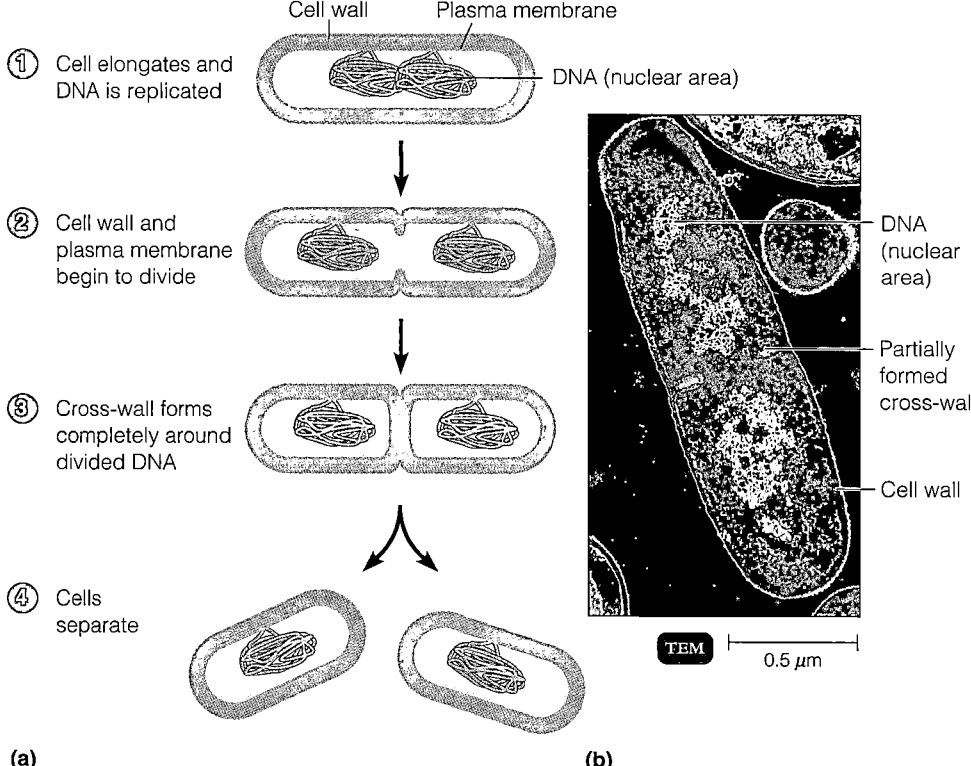

① Cell elongates and DNA is replicated

Cell wall   Plasma membrane

DNA (nuclear area)

② Cell wall and plasma membrane begin to divide

③ Cross-wall forms completely around divided DNA

④ Cells separate

DNA (nuclear area)

Partially formed cross-wall

Cell wall

TEM   0.5 μm

(a)   (b)

FIGURE 6.12 Cell division. When the arithmetic number of cells in each generation is expressed as a power of 2, the exponent tells the number of doublings (generations) that have occurred.

■ If a single bacterium reproduced every 20 minutes, how many would there be in 2 hours?

Arithmetic Numbers of Cells	Numbers Expressed as a Power of 2	Visual Representation of Numbers
1	2^0	
2	2^1	
4	2^2	
8	2^3	
16	2^4	
32	2^5	

Generation Number	Arithmetic Number of Cells	Log_{10} of Arithmetic Number of Cells
0	1	0
5 (2^5) =	32	1.51
10 (2^{10}) =	1,024	3.01
15 (2^{15}) =	32,768	4.52
16 (2^{16}) =	65,536	4.82
17 (2^{17}) =	131,072	5.12
18 (2^{18}) =	262,144	5.42
19 (2^{19}) =	524,288	5.72
20 (2^{20}) =	1,048,576	6.02

growth. Understanding logarithmic representations of bacterial populations requires some use of mathematics and is necessary for anyone studying microbiology. (See Appendix D.)

Logarithmic Representation of Bacterial Populations

To illustrate the difference between logarithmic and arithmetic graphing of bacterial populations, let's express 20 bacterial generations both logarithmically and arithmetically. In five generations (2^5), there would be 32 cells; in ten generations (2^{10}), there would be 1024 cells, and so on. (If your calculator has a y^x key and a log key, you can duplicate the calculations in the third column.)

In Figure 6.13, notice that the arithmetically plotted line (solid) does not clearly show the population changes in the early stages of the growth curve at this scale. In fact, the first ten generations do not appear to leave the base-

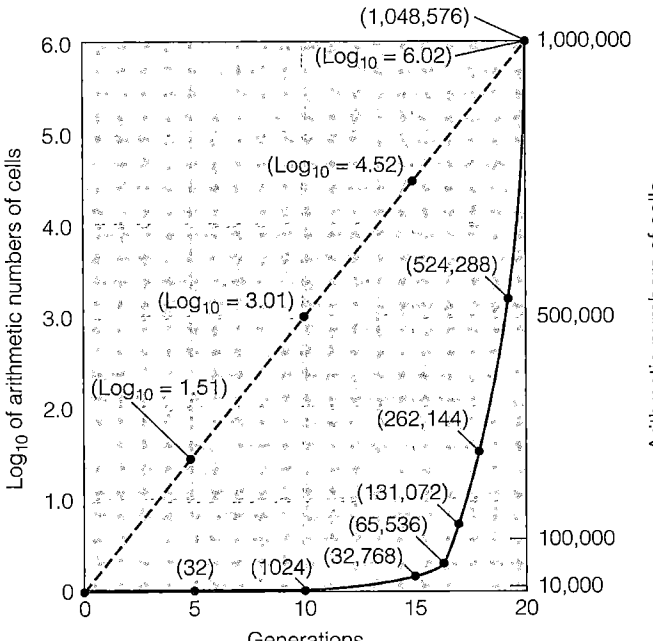

FIGURE 6.13   A growth curve for an exponentially increasing population, plotted logarithmically (dashed line) and arithmetically (solid line).

■ If the arithmetic line were plotted for two more generations, would it still be on the page?

line. Furthermore, another one or two arithmetic generations graphed to the same scale would greatly increase the height of the graph and take the line off the page.

The dashed line in Figure 6.13 shows how these plotting problems can be avoided by graphing the $\log_{10}$ of the population numbers. The $\log_{10}$ of the population is plotted at 5, 10, 15, and 20 generations. Notice that a straight line is formed and that a thousand times this population (1,000,000,000, or $\log_{10}$ 9.0) could be accommodated in relatively little extra space. However, this advantage is obtained at the cost of distorting our "common sense" perception of the actual situation. We are not accustomed to thinking in logarithmic relationships, but it is necessary for a proper understanding of graphs of microbial populations.

Phases of Growth

Learning Objective

■ *Compare the phases of microbial growth, and describe their relation to generation time.*

When a few bacteria are inoculated into a liquid growth medium and the population is counted at intervals, it is possible to plot a **bacterial growth curve** that shows the growth of cells over time (Figure 6.14). There are four basic phases of growth: the lag, log, stationary, and death phases.

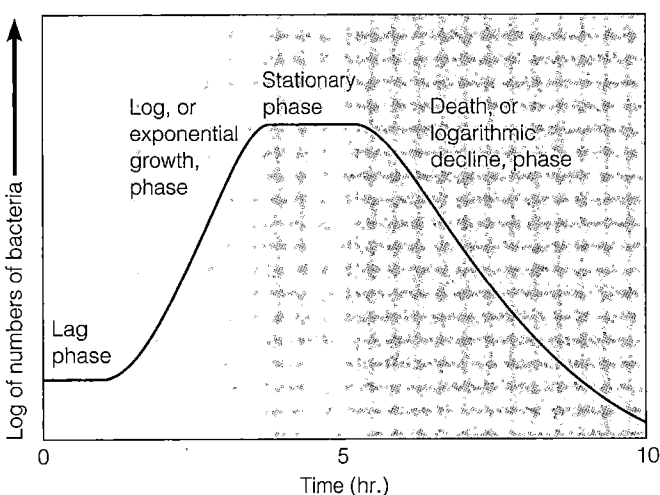

FIGURE 6.14   A bacterial growth curve, showing the four typical phases of growth.

■ A bacterial growth curve shows the growth of a population over time.

The Lag Phase

For a while, the number of cells changes very little because the cells do not immediately reproduce in a new medium. This period of little or no cell division is called the **lag phase,** and it can last for 1 hour or several days. During this time, however, the cells are not dormant. The microbial population is undergoing a period of intense metabolic activity involving, in particular, DNA and enzyme synthesis. (The situation is analogous to a factory newly equipped to produce automobiles; there is considerable tooling-up activity but no immediate increase in the automobile population.)

The Log Phase

Eventually, the cells begin to divide and enter a period of growth, or logarithmic increase, called the **log phase,** or **exponential growth phase.** Cellular reproduction is most active during this period, and generation time reaches a constant minimum. Because the generation time is constant, a logarithmic plot of growth during the log phase is a straight line. The log phase is the time when cells are most active metabolically and is preferred for industrial purposes where, for example, a product needs to be produced efficiently.

However, during their log phase of growth, microorganisms are particularly sensitive to adverse conditions. Radiation and many antimicrobial drugs—for example, the antibiotic penicillin—exert their effect by interfering with some important step in the growth process and are therefore most harmful to cells during this phase.

The Stationary Phase

If exponential growth continues unchecked, startlingly large numbers of cells could arise. For example, a single bacterium (at a weight of 9.5×10^{-13} g per cell) dividing every 20 minutes for only 25.5 hours can theoretically produce a population equivalent in weight to that of an 80,000-ton aircraft carrier. In reality, this does not happen. Eventually, the growth rate slows, the number of microbial deaths balances the number of new cells, and the population stabilizes. The metabolic activities of individual surviving cells also slow at this stage. This period of equilibrium is called the **stationary phase.**

What causes exponential growth to stop is not always clear. The exhaustion of nutrients, accumulation of waste products, and harmful changes in pH may all play a role. In a specialized apparatus called a *chemostat,* a population can be kept in the exponential growth phase indefinitely by draining off spent medium and adding fresh medium. This type of *continuous* culture is used in industrial fermentations (Chapter 28, page 779).

The Death Phase

The number of deaths eventually exceeds the number of new cells formed, and the population enters the **death phase,** or **logarithmic decline phase.** This phase continues until the population is diminished to a tiny fraction of the number of cells in the previous phase, or the population dies out entirely. Some species pass through the entire series of phases in only a few days; others retain some surviving cells almost indefinitely. Microbial death will be discussed further in Chapter 7.

Direct Measurement of Microbial Growth

Learning Objective

■ *Explain four direct methods of measuring cell growth.*

The growth of microbial populations can be measured in a number of ways. Some methods measure cell numbers; other methods measure the population's total mass, which is often directly proportional to cell numbers. Population numbers are usually recorded as the number of cells in a milliliter of liquid or in a gram of solid material. Because bacterial populations are usually very large, most methods of counting them are based on direct or indirect counts of very small samples; calculations then determine the size of the total population. Assume, for example, that a millionth of a milliliter (10^{-6} ml) of sour milk is found to contain 70 bacterial cells. Then there must be 70 times 1 million, or 70 million, cells per milliliter.

However, it is not practical to measure out a millionth of a milliliter of liquid or a millionth of a gram of food. Therefore, the procedure is done indirectly, in a series of dilutions. For example, if we add 1 ml of milk to 99 ml of water, each milliliter of this dilution now has one-hundredth as many bacteria as each milliliter of the original sample had. By making a series of such dilutions, we can readily estimate the number of bacteria in our original sample. To count microbial populations in foods (such as hamburger), an homogenate of one part food to nine parts water is finely ground in a food blender. Samples of this initial one-tenth dilution can then be transferred with a pipette for further dilutions or cell counts.

Plate Counts

The most frequently used method of measuring bacterial populations is the **plate count.** An important advantage of this method is that it measures the number of viable cells. One disadvantage may be that it takes some time, usually 24 hours or more, for visible colonies to form. This can be a serious problem in some applications, such as quality control of milk, when it is not possible to hold a particular lot for this amount of time.

Plate counts assume that each live bacterium grows and divides to produce a single colony. This is not always true because bacteria frequently grow linked in chains or as clumps (see Figure 4.1 on page 78). Therefore, a colony often results, not from a single bacterium, but from short segments of a chain or from a bacterial clump. To reflect this reality, plate counts are often reported as **colony-forming units (CFU).**

When a plate count is performed, it is important that only a limited number of colonies develop in the plate. When too many colonies are present, some cells are overcrowded and do not develop; these conditions cause inaccuracies in the count. Generally, only plates with 25–250 colonies are counted. To ensure that some colony counts will be within this range, the original inoculum is diluted several times in a process called **serial dilution** (Figure 6.15).

Serial Dilutions Let's say, for example, that a milk sample has 10,000 bacteria per milliliter. If 1 ml of this sample were plated out, there would theoretically be 10,000 colonies formed in the Petri plate of medium. Obviously, this would not produce a countable plate. If 1 ml of this sample were transferred to a tube containing 9 ml of sterile water, each milliliter of fluid in this tube would now contain 1000 bacteria. If 1 ml of this sample were inoculated into a Petri plate, there would still be too many potential colonies to count on a plate. Therefore, another

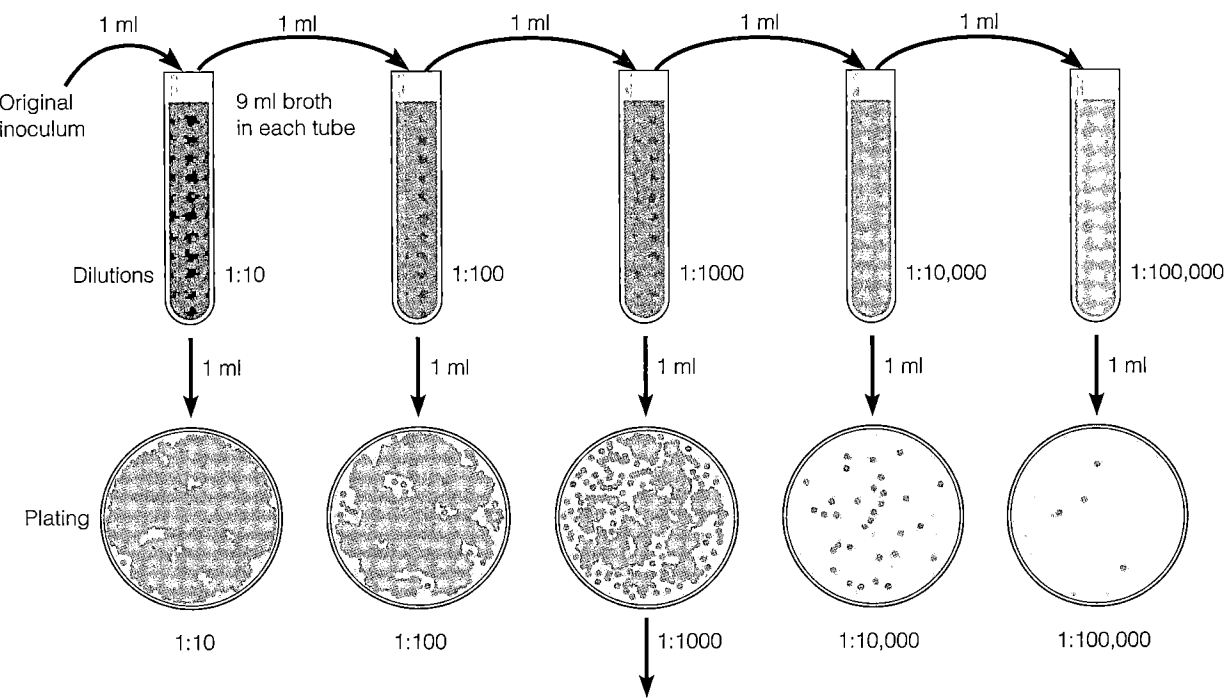

Calculation: Number of colonies on plate × reciprocal of dilution of sample = number of bacteria/ml
(For example, if 32 colonies are on a plate of $^1/_{10,000}$ dilution, then the count is 32 × 10,000 = 320,000/ml in sample.)

FIGURE 6.15 Plate counts and serial dilutions. In serial dilutions, the original inoculum is diluted in a series of dilution tubes. In our example, each succeeding dilution tube will have only one-tenth the number of microbial cells as the preceding tube. Then samples of the dilution are used to inoculate Petri plates, on which colonies grow and can be counted. This count is then used to estimate the number of bacteria in the original sample.

■ The most frequently used method of measuring bacterial populations is the plate count.

serial dilution could be made. One milliliter containing 1000 bacteria would be transferred to a second tube of 9 ml of water. Each milliliter of this tube would now contain only 100 bacteria, and if 1 ml of the contents of this tube were plated out, potentially 100 colonies would be formed—an easily countable number. Learning how to do serial dilutions is an important part of certain experiments in microbiology laboratory classes.

Pour Plates and Spread Plates A plate count is done by either the pour plate method or the spread plate method. The **pour plate method** follows the procedure shown in Figure 6.16a. Either 1.0 ml or 0.1 ml of dilutions of the bacterial suspension is introduced into a Petri dish. The nutrient medium, in which the agar is kept liquid by holding it in a water bath at about 50°C, is poured over the sample, which is then mixed into the medium by gentle agitation of the plate. When the agar solidifies, the plate is incubated. With the pour plate technique,

colonies will grow within the nutrient agar (from cells suspended in the nutrient medium as the agar solidifies) as well as on the surface of the agar plate.

This technique has some drawbacks because some relatively heat-sensitive microorganisms may be damaged by the melted agar and will therefore be unable to form colonies. Also, when certain differential media are used, the distinctive appearance of the colony on the surface is essential for diagnostic purposes. Colonies that form beneath the surface of a pour plate are not satisfactory for such tests. To avoid these problems, the **spread plate method** is frequently used instead (Figure 6.16b). A 0.1-ml inoculum is added to the surface of a prepoured, solidified agar medium. The inoculum is then spread uniformly over the surface of the medium with a specially shaped, sterilized glass rod. This method positions all the colonies on the surface and avoids contact of the cells with melted agar.

FIGURE 6.16 Methods of preparing plates for plate counts.

▣ What are the advantages of each method?

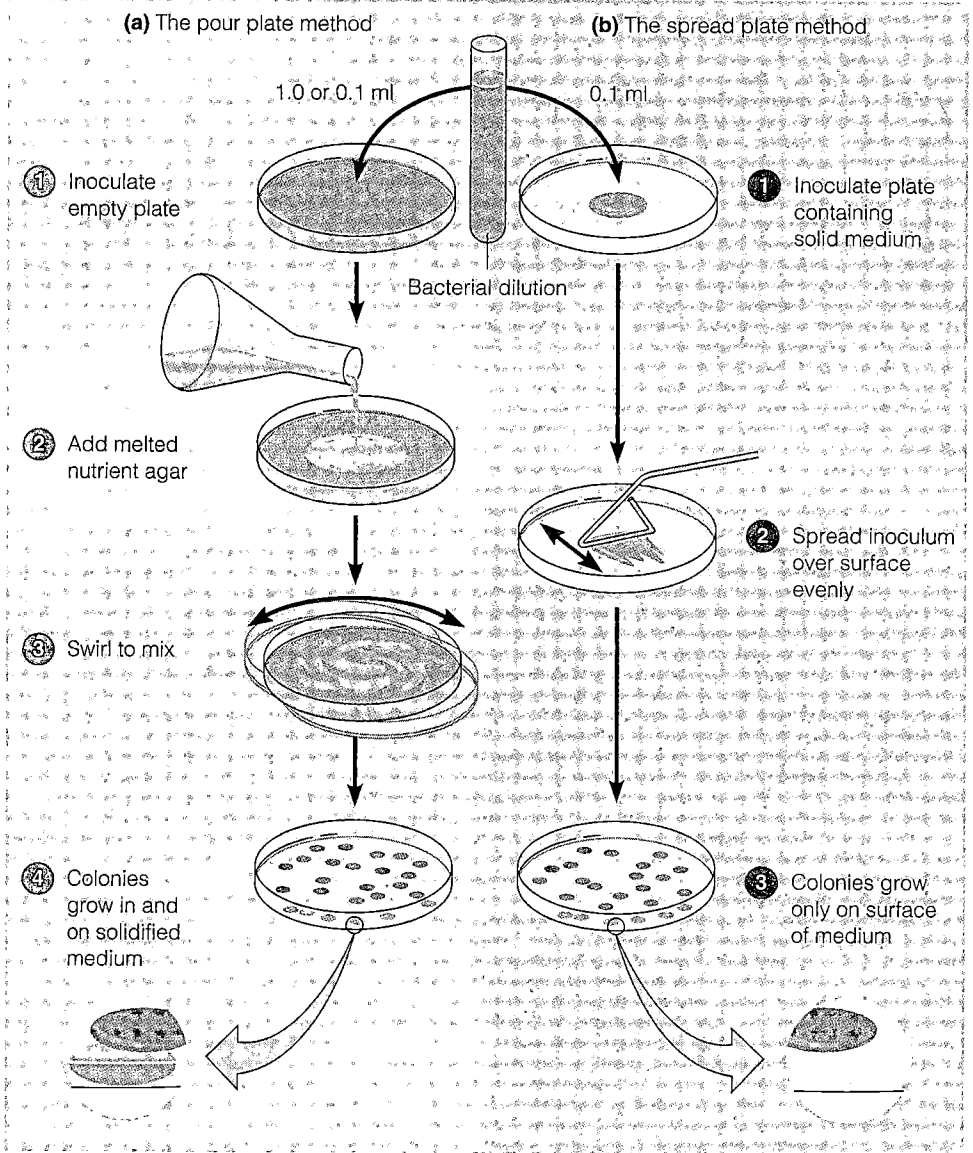

(a) The pour plate method

(b) The spread plate method

1.0 or 0.1 ml

0.1 ml

① Inoculate empty plate

① Inoculate plate containing solid medium

Bacterial dilution

② Add melted nutrient agar

② Spread inoculum over surface evenly

③ Swirl to mix

④ Colonies grow in and on solidified medium

③ Colonies grow only on surface of medium

Filtration

When the quantity of bacteria is very small, as in lakes or relatively pure streams, bacteria can be counted by **filtration** methods (Figure 6.17). In this technique, at least 100 ml of water are passed through a thin membrane filter whose pores are too small to allow bacteria to pass. Thus, the bacteria are filtered out and retained on the surface of the filter. This filter is then transferred to a Petri dish containing a pad soaked in liquid nutrient medium, where colonies arise from the bacteria on the filter's surface. This method is applied frequently to detection and enumeration of coliform bacteria, which are indicators of fecal pollution of food or water (see Chapter 27, page 758). The colonies formed by these bacteria are distinctive when a differential nutrient medium is used. (The colonies shown in Figures 6.9b and c are examples of coliforms.)

The Most Probable Number Method

Another method for determining the number of bacteria in a sample is the **most probable number (MPN) method,** illustrated in Figure 6.18. This statistical estimating technique is based on the fact that the greater the number of bacteria in a sample, the more dilution is needed to reduce the density to the point at which no bacteria are left to grow in the tubes in a dilution series. The MPN method is most useful when the microbes being counted will not grow on solid media (such as the chemoautotrophic nitrifying bacteria). It is also useful when the growth of bacteria in a liquid differential medium is used to identify the microbes (such as coliform bacteria, which selectively ferment lactose to acid, in water testing). The MPN is only a statement that there is a 95% chance that the bacterial population falls within

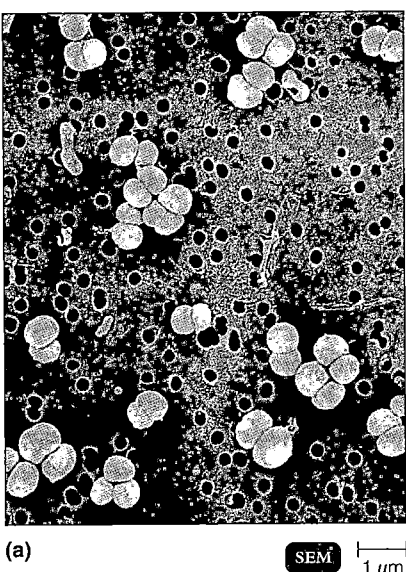

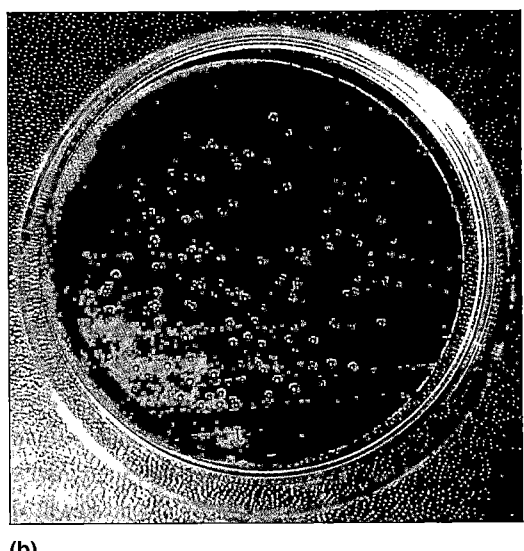

(a)   SEM ⊢——⊣ 1 μm   (b)

FIGURE 6.17 Counting bacteria by filtration. (a) The bacteria in 100 ml of water were sieved out onto the surface of a membrane filter. (b) Such a filter, with the bacteria much more widely spaced, was placed on a pad saturated with liquid Endo medium, which is selective for gram-negative bacteria. The individual bacteria grew into visible colonies. One hundred twenty-four colonies are visible, so we would record 124 bacteria per 100 ml of water sample.

■ **Bacteria can be counted by filtration when their concentration is very small.**

Combination of Positives	MPN Index/ 100 ml	95% Confidence Limits Lower	95% Confidence Limits Upper
4-2-0	22	9	56
4-2-1	26	12	65
4-3-0	27	12	67
4-3-1	33	15	77
4-4-0	34	16	80
5-0-0	23	9	86
5-0-1	30	10	110
5-0-2	40	20	140
5-1-0	30	10	120
5-1-1	50	20	150
5-1-2	60	30	180
5-2-0	50	20	170
5-2-1	70	30	210
5-2-2	90	40	250
5-3-0	80	30	250
5-3-1	110	40	300
5-3-2	140	60	360

Volume of Inoculum for Each Set of Five Tubes

Tubes of Nutrient Medium (Sets of Five Tubes)	Number of Positive Tubes in Set
10 ml	5
1 ml	3
0.1 ml	1

FIGURE 6.18 **The most probable number (MPN) method.** In this example, there are three sets of tubes and five tubes in each set. Each tube in the first set of five tubes receives 10 ml of the inoculum, such as a sample of water. Each tube in the second set of five tubes receives 1 ml of the sample, and the third set, 0.1 ml each. There were enough bacteria in the sample so that all five tubes in the first set showed bacterial growth and were recorded as positive. In the second set, which received only one-tenth as much inoculum, only three tubes were positive. In the third set, which received one-hundredth as much inoculum, only one tube was positive. MPN tables enable us to calculate for a sample the microbial numbers that are statistically likely to lead to such a result. The number of positive tubes is recorded for each set: In the shaded example, 5, 3, and 1. If we look up this combination in an MPN table, we find that the MPN index per 100 ml is 110. Statistically, this means that 95% of the water samples that give this result contain 40–300 bacteria, with 110 being the most frequent number.

■ **Under what circumstances is the MPN method used to determine the number of bacteria in a sample?**

FIGURE 6.19 Direct microscopic count of bacteria with a Petroff-Hausser cell counter. The average number of cells within a large square multiplied by a factor of 1,250,000 gives the number of bacteria per milliliter.

■ Direct microscopic counts are useful for certain applications but have a disadvantage of requiring rather high microbial populations for them to be countable.

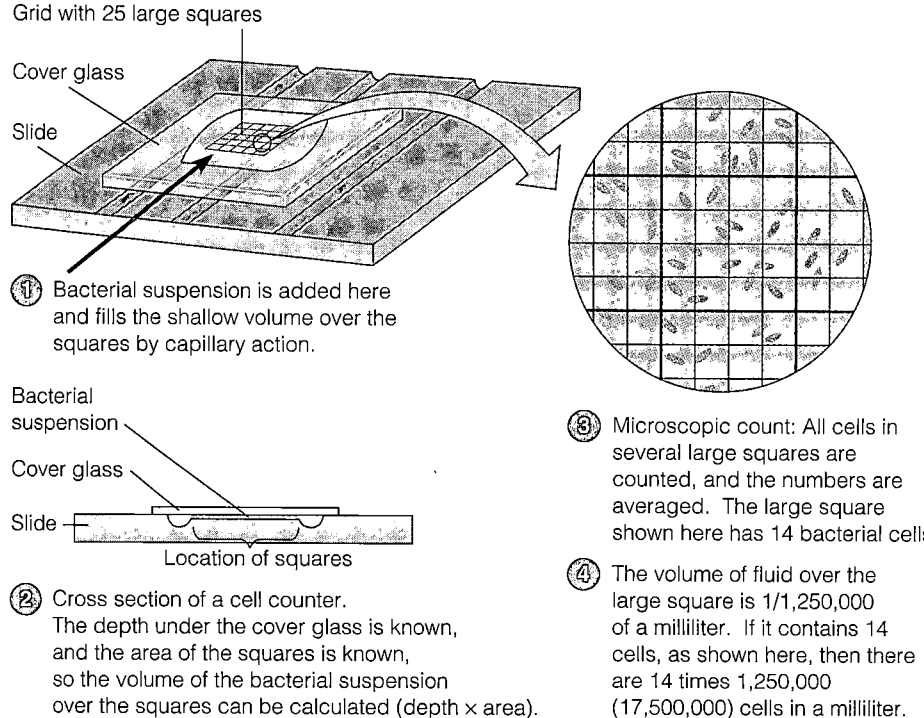

Grid with 25 large squares

Cover glass

Slide

① Bacterial suspension is added here and fills the shallow volume over the squares by capillary action.

Bacterial suspension

Cover glass

Slide

Location of squares

② Cross section of a cell counter. The depth under the cover glass is known, and the area of the squares is known, so the volume of the bacterial suspension over the squares can be calculated (depth × area).

③ Microscopic count: All cells in several large squares are counted, and the numbers are averaged. The large square shown here has 14 bacterial cells.

④ The volume of fluid over the large square is 1/1,250,000 of a milliliter. If it contains 14 cells, as shown here, then there are 14 times 1,250,000 (17,500,000) cells in a milliliter.

---

a certain range and that the MPN is statistically the most probable number.

Direct Microscopic Count

In the method known as the **direct microscopic count,** a measured volume of a bacterial suspension is placed within a defined area on a microscope slide. In the *Breed count method* which is used to count the number of bacteria in milk, for example, a 0.01-ml sample is spread over a marked square centimeter of slide, stain is added so that the bacteria can be seen, and the sample is viewed under the oil immersion objective lens. The area of the viewing field of this objective can be determined. Once the number of bacteria has been counted in several different fields, the average number of bacteria per viewing field can be calculated. From these data, the number of bacteria in the square centimeter over which the sample was spread can also be calculated. Because this area on the slide contained 0.01 ml of sample, the number of bacteria in each milliliter of the suspension is the number of bacteria in the sample times 100.

A specially designed slide called a *Petroff-Hausser cell counter* is also used in direct microscopic counts (Figure 6.19). ① In the center of the microscope slide the surface is inscribed with a grid of squares of known area. Above this grid is a space of known volume formed because it is slightly lower than the portions of the slide that support the cover glass. ② The volume over the grid is filled with the microbial suspension. ③–④ The average number of

bacteria in each of a series of these squares is calculated and then multiplied by the factor that produces the count per milliliter.

Motile bacteria are difficult to count by this method, and, as happens with other microscopic methods, dead cells are about as likely to be counted as live ones. In addition to these disadvantages, a rather high concentration of cells is required to be countable—about 10 million bacteria per milliliter. The chief advantage of microscopic counts is that no incubation time is required, and they are usually reserved for applications in which time is the primary consideration. This advantage also holds for *electronic cell counters,* sometimes known as *Coulter counters,* which automatically count the number of cells in a measured volume of liquid. These instruments are used in some research laboratories and hospitals.

Estimating Bacterial Numbers by Indirect Methods

Learning Objectives
- *Differentiate between direct and indirect methods of measuring cell growth.*
- *Explain three indirect methods of measuring cell growth.*

It is not always necessary to count microbial cells to estimate their numbers. In science and industry, microbial numbers and activity are determined by some of the following indirect means as well.

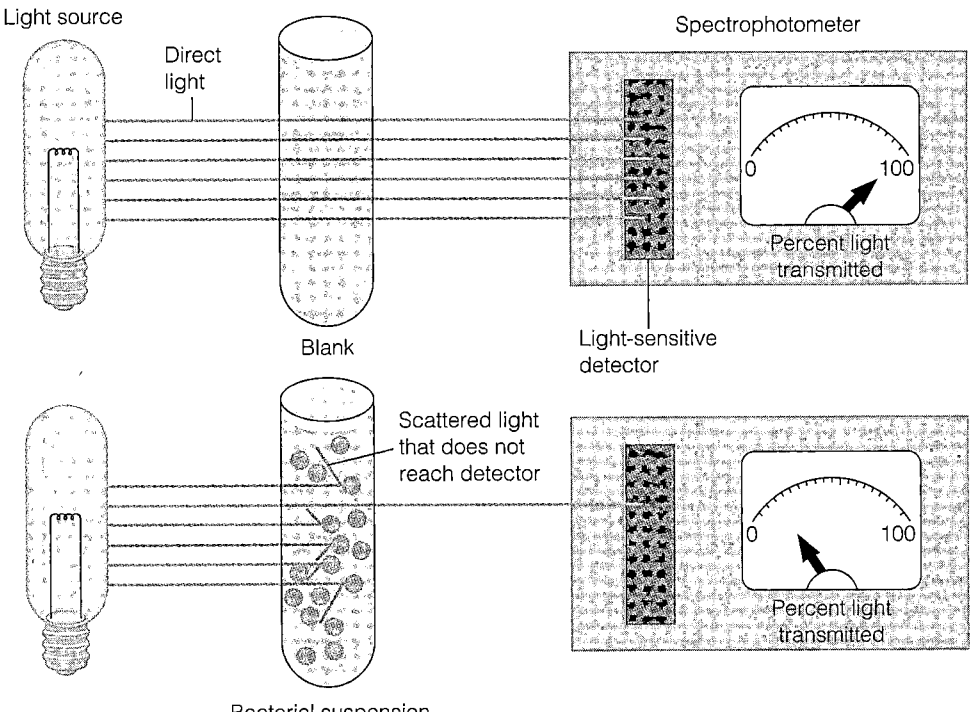

Light source
Direct light
Spectrophotometer
Blank
Light-sensitive detector
0 100
Percent light transmitted

Scattered light that does not reach detector
0 100
Percent light transmitted

Bacterial suspension

FIGURE 6.20 Turbidity estimation of bacterial numbers. The amount of light striking the light-sensitive detector on the spectrophotometer is inversely proportional to the number of bacteria under standardized conditions. The less light transmitted, the more bacteria in the sample.

■ Is turbidity a direct or an indirect method of measuring bacterial growth?

Turbidity

For some types of experimental work, estimating *turbidity* is a practical way of monitoring bacterial growth. As bacteria multiply in a liquid medium, the medium becomes turbid, or cloudy with cells.

The instrument used to measure turbidity is a *spectrophotometer* (or colorimeter). In the spectrophotometer, a beam of light is transmitted through a bacterial suspension to a light-sensitive detector (Figure 6.20). As bacterial numbers increase, less light will reach the detector. This change of light will register on the instrument's scale as the *percentage of transmission*. Also printed on the instrument's scale is a logarithmic expression called the *absorbance* (sometimes called *optical density*, or *OD*); a value derived from the percentage of transmission may also be reported. The absorbance is used to plot bacterial growth. When the bacteria are in logarithmic growth or decline, a graph of absorbance versus time will form an approximately straight line. If absorbance readings are matched with plate counts of the same culture, this correlation can be used in future estimations of bacterial numbers obtained by measuring turbidity.

More than a million cells per milliliter must be present for the first traces of turbidity to be visible. About 10 million to 100 million cells per milliliter are needed to make a suspension turbid enough to be read on a spectrophotometer. Therefore, turbidity is not a useful measure of contamination of liquids by relatively small numbers of bacteria.

Metabolic Activity

Another indirect way to estimate bacterial numbers is to measure a population's *metabolic activity*. This method assumes that the amount of a certain metabolic product, such as acid or CO_2, is in direct proportion to the number of bacteria present. An example of a practical application of a metabolic test is the microbiological assay in which acid production is used to determine amounts of vitamins.

Dry Weight

For filamentous organisms, such as molds, the usual measuring methods are less satisfactory. A plate count would not measure this increase in filamentous mass. In plate counts of molds, the number of asexual spores is counted instead, but often this is not a good measure of growth. One of the better ways to measure the growth of filamentous organisms is by *dry weight*. In this procedure, the fungus is removed from the growth medium, filtered to remove extraneous material, placed in a weighing bottle, and dried in a desiccator. For bacteria, the same basic procedure is followed. The bacteria are usually removed from the culture medium by centrifugation.

★   ★   ★

You now have a basic understanding of the requirements for, and measurements of, microbial growth. In Chapter 7, we will look at how this growth is controlled in laboratories, hospitals, industry, and our homes.

Study Outline $\boxed{\text{ST}}$ Student Tutorial CD-ROM

THE REQUIREMENTS FOR GROWTH (pp. 156–164)

1. The growth of a population is an increase in the number of cells.

2. The requirements for microbial growth are both physical and chemical.

Physical Requirements (pp. 156–160)

1. On the basis of preferred temperature ranges, microbes are classified as psychrophiles (cold-loving), mesophiles (moderate-temperature–loving), and thermophiles (heat-loving).

2. The minimum growth temperature is the lowest temperature at which a species will grow, the optimum growth temperature is the temperature at which it grows best, and the maximum growth temperature is the highest temperature at which growth is possible.

3. Most bacteria grow best at a pH value between 6.5 and 7.5.

4. In a hypertonic solution, most microbes undergo plasmolysis; halophiles can tolerate high salt concentrations.

Chemical Requirements (pp. 160–164)

1. All organisms require a carbon source; chemoheterotrophs use an organic molecule, and autotrophs typically use carbon dioxide.

2. Nitrogen is needed for protein and nucleic acid synthesis. Nitrogen can be obtained from the decomposition of proteins or from NH_4^+ or NO_3^-; a few bacteria are capable of nitrogen (N_2) fixation.

3. On the basis of oxygen requirements, organisms are classified as obligate aerobes, facultative anaerobes, obligate anaerobes, aerotolerant anaerobes, and microaerophiles.

4. Aerobes, facultative anaerobes, and aerotolerant anaerobes must have the enzymes superoxide dismutase ($2\ O_2^- + 2\ H^+ \longrightarrow O_2 + H_2O_2$) and either catalase ($2\ H_2O_2 \longrightarrow 2\ H_2O + O_2$) or peroxidase ($H_2O_2 + 2\ H^+ \longrightarrow 2\ H_2O$).

5. Other chemicals required for microbial growth include sulfur, phosphorus, trace elements, and, for some microorganisms, organic growth factors.

CULTURE MEDIA (pp. 164–170)

1. A culture medium is any material prepared for the growth of bacteria in a laboratory.

2. Microbes that grow and multiply in or on a culture medium are known as a culture.

3. Agar is a common solidifying agent for a culture medium.

Chemically Defined Media (p. 165)

1. A chemically defined medium is one in which the exact chemical composition is known.

Complex Media (p. 165)

1. A complex medium is one in which the exact chemical composition varies slightly from batch to batch.

Anaerobic Growth Media and Methods (pp. 165–166)

1. Reducing media chemically remove molecular oxygen (O_2) that might interfere with the growth of anaerobes.

2. Petri plates can be incubated in an anaerobic jar or anaerobic chamber.

Special Culture Techniques (pp. 166–167)

1. Some parasitic and fastidious bacteria must be cultured in living animals or in cell cultures.

2. CO_2 incubators or candle jars are used to grow bacteria requiring an increased CO_2 concentration.

Selective and Differential Media (pp. 167–169)

1. By inhibiting unwanted organisms with salts, dyes, or other chemicals, selective media allow growth of only the desired microbes.

2. Differential media are used to distinguish among different organisms.

Enrichment Culture (pp. 169–170)

1. An enrichment culture is used to encourage the growth of a particular microorganism in a mixed culture.

OBTAINING PURE CULTURES (p. 170)

1. A colony is a visible mass of microbial cells that theoretically arose from one cell.

2. Pure cultures are usually obtained by the streak plate method.

PRESERVING BACTERIAL CULTURES (p. 170)

1. Microbes can be preserved for long periods of time by deep-freezing or lyophilization (freeze-drying).

THE GROWTH OF BACTERIAL CULTURES (pp. 170–179)

$\boxed{\text{ST}}$ *To review: go to Microbial Growth: Cell Cycles*

Bacterial Division (p. 171)

1. The normal reproductive method of bacteria is binary fission, in which a single cell divides into two identical cells.

2. Some bacteria reproduce by budding, aerial spore formation, or fragmentation.

Generation Time (pp. 171–172)

1. The time required for a cell to divide or a population to double is known as the generation time.

Logarithmic Representation of Bacterial Populations (pp. 172–173)

    SI *Microbial Growth: Growth Curve*

1. Bacterial division occurs according to a logarithmic progression (two cells, four cells, eight cells, etc.).

Phases of Growth (pp. 173–174)

1. During the lag phase, there is little or no change in the number of cells, but metabolic activity is high.

2. During the log phase, the bacteria multiply at the fastest rate possible under the conditions provided.

3. During the stationary phase, there is an equilibrium between cell division and death.

4. During the death phase, the number of deaths exceeds the number of new cells formed.

Direct Measurement of Microbial Growth (pp. 174–178)

1. A standard plate count reflects the number of viable microbes and assumes that each bacterium grows into a single colony;

plate counts are reported as number of colony-forming units (CFU).

2. A plate count may be done by either the pour plate method or the spread plate method.

3. In filtration, bacteria are retained on the surface of a membrane filter and then transferred to a culture medium to grow and subsequently be counted.

4. The most probable number (MPN) method can be used for microbes that will grow in a liquid medium; it is a statistical estimation.

5. In a direct microscopic count, the microbes in a measured volume of a bacterial suspension are counted with the use of a specially designed slide.

Estimating Bacterial Numbers by Indirect Methods (pp. 178–179)

1. A spectrophotometer is used to determine turbidity by measuring the amount of light that passes through a suspension of cells.

2. An indirect way of estimating bacterial numbers is measuring the metabolic activity of the population (for example, acid production or oxygen consumption).

3. For filamentous organisms such as fungi, measuring dry weight is a convenient method of growth measurement.

Study Questions

REVIEW

1. Describe binary fission.

2. Draw a typical bacterial growth curve. Label and define each of the four phases.

3. Macronutrients (needed in relatively large amounts) are often listed as CHONPS. What does each of these letters indicate, and why are they needed by the cell?

4. Most bacteria grow best at pH _____.

5. Why can high concentrations of salt or sugar be used to preserve food?

6. Define and explain the importance of each of the following:
     a. catalase   **d.** superoxide free radical
     b. hydrogen peroxide   **e.** superoxide dismutase
     c. peroxidase

7. *Clostridium* can be cultured in an anaerobic incubator or in the presence of atmospheric oxygen if thioglycolate is added to the nutrient broth. Compare these two techniques. Using terms from question 6, explain why elaborate culture techniques are used for *Clostridium*.

8. Eight methods of measuring microbial growth were explained in this chapter. Categorize each as either a direct or an indirect method.

9. By deep-freezing, bacteria can be stored without harm for extended periods. Why do refrigeration and freezing preserve foods?

10. A pastry chef accidentally inoculated a cream pie with six *S. aureus* cells. If *S. aureus* has a generation time of 60 minutes, how many cells would be in the cream pie after 7 hours?

11. Nitrogen and phosphorus added to beaches following an oil spill encourage the growth of natural oil-degrading bacteria. Explain why the bacteria do not grow if nitrogen and phosphorus are not added.

12. Differentiate between complex and chemically defined media.

13. Draw the following growth curves for *E. coli*, starting with 100 cells with a generation time of 30 minutes at 35°C.
    a. The cells are incubated for 5 hours at 35°C.
    b. After 5 hours, the temperature is changed to 20°C for 2 hours.
    c. After 5 hours at 35°C, the temperature is changed to 5°C for 2 hours followed by 35°C for 5 hours.

MULTIPLE CHOICE

Use the following information to answer questions 1 and 2. Two culture media were inoculated with four different bacteria. After incubation, the following results were obtained:

Organism	Medium 1	Medium 2
Escherichia coli	Red colonies	No growth
Staphylococcus aureus	No growth	Growth
S. epidermidis	No growth	Growth
Salmonella enteritidis	Colorless colonies	No growth

1. Medium 1 is
   a. selective.
   b. differential.
   c. both selective and differential.

2. Medium 2 is
   a. selective.
   b. differential.
   c. both selective and differential.

Use the following graph to answer questions 3 and 4.

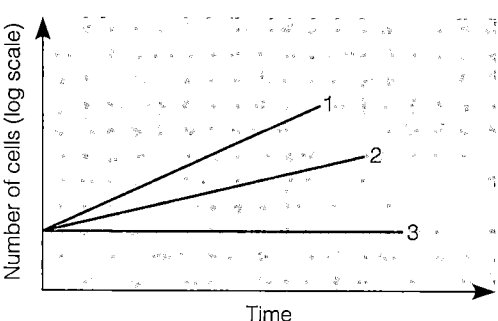

3. Which of the lines best depicts the log phase of a thermophile incubated at room temperature?

4. Which of the lines best depicts the log phase of *Listeria monocytogenes* growing in a human?

5. Assume you inoculated 100 facultatively anaerobic cells onto nutrient agar and incubated the plate aerobically. You then inoculated 100 cells of the same species onto nutrient agar and incubated the second plate anaerobically. After incubation for 24 hours, you should have
   a. more colonies on the aerobic plate.
   b. more colonies on the anaerobic plate.
   c. the same number of colonies on both plates.

6. The term *trace elements* refers to
   a. the elements CHONPS.
   b. vitamins.
   c. nitrogen, phosphorus, and sulfur.
   d. small mineral requirements.
   e. toxic substances.

7. Which one of the following temperatures would most likely kill a mesophile?
   a. −50°C   c. 9°C   e. 60°C
   b. 0°C   d. 37°C

8. All of the following are true about agar *except:*
   a. It is a source of nutrients in culture media.
   b. It is a polysaccharide.
   c. It liquifies at 100°C.
   d. It solidifies at approximately 40°C.
   e. It is metabolized by few bacteria.

9. Which of the following types of media would *not* be used to culture aerobes?
   a. selective media
   b. reducing media
   c. enrichment media
   d. differential media
   e. complex media

10. An organism that has peroxidase and superoxide dismutase but lacks catalase is most likely an
    a. aerobe.
    b. aerotolerant anaerobe.
    c. obligate anaerobe.

CRITICAL THINKING

1. *E. coli* was incubated with aeration in a nutrient medium containing two carbon sources, and the following growth curve was made from this culture.
   a. Explain what happened at the time marked *x*.
   b. Which substrate provided "better" growth conditions for the bacteria? How can you tell?

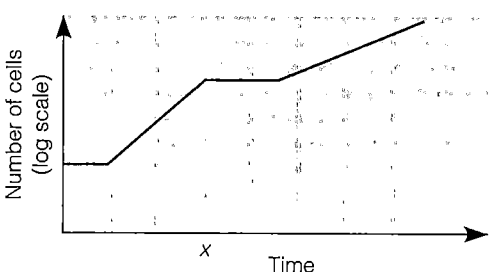

2. *Clostridium* and *Streptococcus* are both catalase-negative. *Streptococcus* grows by fermentation. Why is *Clostridium* killed by oxygen, whereas *Streptococcus* is not?

3. Most laboratory media contain a fermentable carbohydrate and peptone because the majority of bacteria require carbon, nitrogen, and energy sources in these forms. How are these three needs met by glucose–minimal salts medium? (*Hint:* See Table 6.2.)

4. Flask A contains yeast cells in glucose–minimal salts broth incubated at 30°C with aeration. Flask B contains yeast cells in glucose–minimal salt broth incubated at 30°C in an anaerobic jar. The yeasts are facultative anaerobes.
   a. Which culture produced more ATP?
   b. Which culture produced more alcohol?
   c. Which culture had the shorter generation time?
   d. Which culture had the greater cell mass?
   e. Which culture had the higher absorbance?

CLINICAL APPLICATIONS

1. Assume that, after washing your hands, you leave ten bacterial cells on a new bar of soap. You then decide to do a plate count of the soap after it was left in the soap dish for 24 hours. You dilute 1 g of the soap 1:10^6 and plate it on standard plate count agar. After 24 hours of incubation, there are 168 colonies. How many bacteria were on the soap? How did they get there?

2. Heat lamps are commonly used to maintain foods at about 50°C for as long as 12 hours in cafeteria serving lines. The following experiment was conducted to determine whether this practice poses a potential health hazard.

   Beef cubes were surface-inoculated with 500,000 bacterial cells and incubated at 43–53°C to establish temperature limits for bacterial growth. The following results were obtained from standard plate counts performed on beef cubes at 6 and 12 hours after inoculation:

		Bacteria/Gram of Beef After	
	Temp. (°C)	6 hr	12 hr
S. aureus	43	140,000,000	740,000,000
	51	810,000	59,000
	53	650	300
S. typhimurium	43	3,200,000	10,000,000
	51	950,000	83,000
	53	1,200	300
C. perfringens	43	1,200,000	3,600,000
	51	120,000	3,800
	53	300	300

   Draw the growth curves for each organism. What holding temperature would you recommend? Assuming that cooking kills bacteria in foods, how could these bacteria contaminate the cooked foods? What disease does each organism cause? (*Hint:* See Chapter 25.)

3. The number of bacteria in saliva samples was determined by collecting the saliva, making serial dilutions, and inoculating nutrient agar by the pour plate method. The plates were incubated aerobically for 48 hours at 37°C.

	Bacteria/ml Saliva	
	Before Using Mouthwash	After Using Mouthwash
Product 1	13.1 × 10^6	10.9 × 10^6
Product 2	11.7 × 10^6	14.2 × 10^5
Product 3	9.3 × 10^5	7.7 × 10^5

   What can you conclude from these data? Did all the bacteria present in each saliva sample grow?

Learning with Technology

MP = The Microbiology Place website   **ST** = Student Tutorial CD-ROM   **VU** = VirtualUnknown CD-ROM

MP   Don't forget to go to The Microbiology Place website (http://www.microbiologyplace.com) to take the practice tests, explore the interactive activity and case study, and check out the news articles and web links for this chapter.

ST   Remember there is also a quiz for this chapter on the Microbiology Interactive Student Tutorial CD-ROM.

VU   Enter the Virtual Lab, click the arrow next to the Session field, click Textbook Exercises, and select Chapter 6. Read the Case Study. Use your textbook and the reference resource books provided in the software to complete the questions below.

1. For each of the following media, describe a) whether the medium is complex or synthetic, and b) whether the medium is selective, differential, or both.
   Malonate broth
   Tryptone broth
   KCN broth
   Nutrient gelatin
   OF glucose broth
   DNase agar
   Ornithine decarboxylase broth

2. Name the a) carbon source and b) nitrogen source for each of the following media:
   Phenol red adonitol broth
   KCN broth
   Phenylalanine deaminase agar
   Simmons' citrate agar
   Spirit blue agar
   Nutrient gelatin

3. Both Christiansen's urea broth and phenol red glucose broth contain phenol red as the pH indicator, yet, the test for urease is positive when phenol red turns hot pink and the test for glucose fermentation is positive when phenol red turns yellow. Explain.

4. Complete the OF glucose tests for your unknown bacterium by inoculating, incubating, and interpreting the results using OF glucose broth and OF glucose broth with oil overlay. Use these results to classify your unknown as an aerobe, anaerobe, facultative anaerobe, or microaerophil. Predict (and draw) the expected appearance of fluid thioglycollate medium containing a culture of your unknown.

5. Select tryptone broth from the Media list. Note that the inoculum at left and the sterile medium initially provided at right have very different appearances (NOTE: You may view closeups of the tubes by using the magnifying tool in the tool bar above the Virtual Lab.). Use the information provided on page 179 of the textbook to explain the difference in appearance.

The Control of Microbial Growth

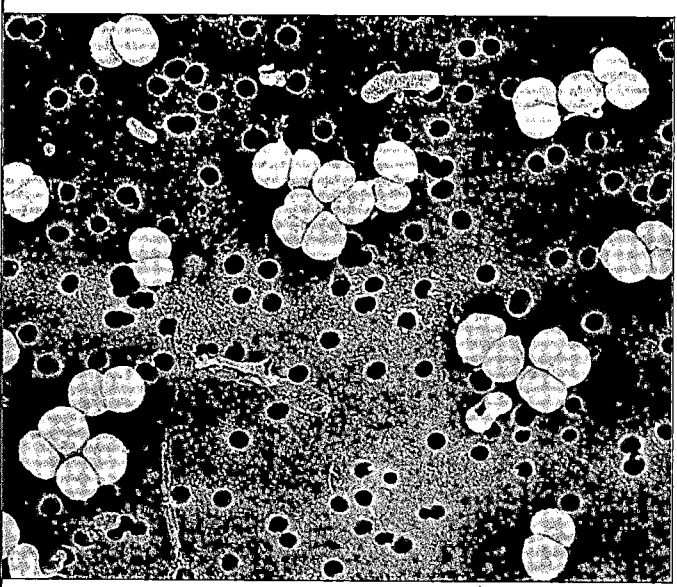

Bacteria trapped in a membrane filter. Filtration can be used to remove microorganisms from water and solutions.

The scientific control of microbial growth began only about 100 years ago. Recall from Chapter 1 that Pasteur's work on microorganisms led scientists to believe that microbes were a possible cause of disease. In the mid-1800s, the Hungarian physician Ignatz Semmelweis and English physician Joseph Lister used this thinking to develop some of the first microbial control practices for medical procedures. These practices included hand washing with microbe-killing chloride of lime and **aseptic surgery** techniques to prevent microbial contamination of surgical wounds. Until that time, hospital-acquired infections, or *nosocomial infections,* were the cause of death in at least 10% of surgical cases, and deaths of delivering mothers were as high as 25%. Ignorance of microbes was such that, during the American Civil War, a surgeon might have cleaned his scalpel on his bootsole between incisions.

Over the last century, scientists have continued to develop a variety of physical methods and chemical agents to control microbial growth. Physical methods include the use of heat, filtration, low temperatures, desiccation, osmotic pressure, and radiation. Chemical agents include several groups of substances that destroy microbes or limit microbial growth on body surfaces and on inanimate objects. In Chapter 20 we will discuss methods for the control of microbes once infection has occurred, mainly antibiotic chemotherapy.

The Terminology of Microbial Control

Learning Objective

- *Define the following key terms related to microbial control: sterilization, disinfection, antisepsis, degerming, sanitization, biocide, germicide, bacteriostasis, and asepsis.*

A word frequently used, and misused, in discussing the control of microbial growth is sterilization. Strictly speaking, **sterilization** is the destruction of *all forms* of microbial life, including endospores, which are the most resistant form. Heating is the most common method used for sterilization.

One would think that canned food in the supermarket is completely sterile. In reality, the heat treatment re-

table 7.1	Terminology Relating to the Control of Microbial Growth	
	Definition	**Comments**
Sterilization	Destruction of all forms of microbial life, including endospores.	Usually done by steam under pressure or a sterilizing gas such as ethylene oxide.
Commercial Sterilization	Sufficient heat treatment to kill endospores of *Clostridium botulinum* in canned food.	More-resistant endospores of thermophilic bacteria may survive, but will not germinate and grow under normal storage conditions.
Disinfection	Destruction of vegetative pathogens.	May make use of physical or chemical methods.
Antisepsis	Destruction of vegetative pathogens on living tissue.	Treatment is almost always by chemical antimicrobials.
Degerming	Removal of microbes from a limited area, such as the skin around an injection site.	Mostly a mechanical removal by an alcohol-soaked swab.
Sanitization	Treatment intended to lower microbial counts on eating and drinking utensils to safe public health levels.	May be done with high-temperature washing or by dipping into a chemical disinfectant.

quired to ensure absolute sterility would unnecessarily degrade the quality of the food. Instead, food is subjected only to enough heat to destroy the endospores of *Clostridium botulinum,* which can produce a deadly toxin. This limited heat treatment is termed **commercial sterilization.** The endospores of a number of thermophilic bacteria, capable of causing food spoilage but not human disease, are considerably more resistant to heat than *C. botulinum.* If present, they will survive, but their survival is usually of no practical consequence; they will not grow at normal food storage temperatures. If canned foods in a supermarket were incubated at temperatures in the growth range of these thermophiles (above about 45°C), significant food spoilage would occur.

Complete sterilization is often not required in other settings. For example, the body's normal defenses can cope with a few microbes entering a surgical wound. A drinking glass or a fork in a restaurant requires only enough microbial control to prevent the transmission of possibly pathogenic microbes from one person to another.

Control directed at destroying harmful microorganisms is called **disinfection.** It usually refers to the destruction of vegetative (non–endospore-forming) pathogens, which is not the same thing as complete sterility. Disinfection might make use of chemicals, ultraviolet radiation, boiling water, or steam. In practice, the term is most commonly applied to the use of a chemical (a *disinfectant*) to treat an inert surface or substance. When this treatment is directed at living tissue, it is called **antisepsis,** and the chemical is then called an *antiseptic.* There-

fore, in practice the same chemical might be called a disinfectant for one use and an antiseptic for another. Of course, many chemicals suitable for swabbing a tabletop would be too harsh to use on living tissue.

There are modifications of disinfection and antisepsis. For example, when someone is about to receive an injection, the skin is swabbed with alcohol—the process of **degerming** (or *degermation*), which mostly results in the mechanical removal, rather than the killing, of most of the microbes in a limited area. Restaurant glassware, china, and tableware are subjected to **sanitization,** which is intended to lower microbial counts to safe public health levels and minimize the chances of disease transmission from one user to another. This is usually accomplished by high-temperature washing or, in the case of glassware in a bar, washing in a sink followed by a dip in a chemical disinfectant.

Table 7.1 summarizes the terminology relating to the control of microbial growth.

Names of treatments that cause the outright death of microbes have the suffix *-cide,* meaning kill. A **biocide,** or **germicide,** kills microorganisms (usually with certain exceptions, such as endospores); a *fungicide* kills fungi; a *virucide* inactivates viruses; and so on. Other treatments only inhibit the growth and multiplication of bacteria; their names have the suffix *-stat* or *-stasis,* meaning to stop or to steady, as in **bacteriostasis.** Once a bacteriostatic agent is removed, growth might resume.

Sepsis, from the Greek for decay or putrid, indicates bacterial contamination, as in septic tanks for sewage treatment. (The term is also used to describe a disease

table 7.2	Microbial Death Rate: An Example	
Time (min)	Deaths per Minute	Number of Survivors
0	0	1,000,000
1	900,000	100,000
2	90,000	10,000
3	9000	1000
4	900	100
5	90	10
6	9	1

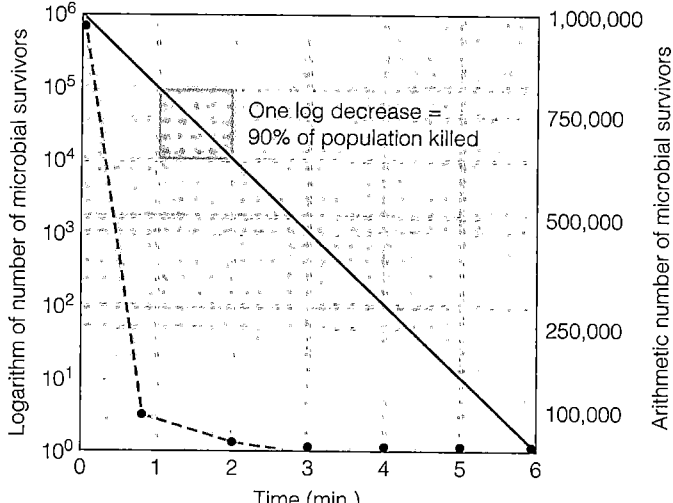

(a)

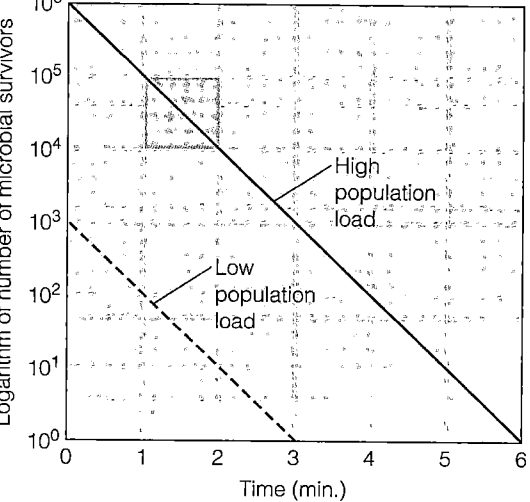

(b)

FIGURE 7.1 A microbial death curve. (a) The curve is plotted logarithmically (solid line) and arithmetically (broken line). In this case, the cells are dying at a rate of 90% each minute. **(b)** The effect of high or low initial load of microbes. If the rate of killing is the same, it will take longer to kill all members of a larger population than a smaller one. This is true for both heat and chemical treatments.

■ **If the graph in part a reflects the experience with a vegetative bacterium, what would the logarithmic curve look like for the endospores of the same bacterium? Less steep? Steeper?**

condition; see Chapter 23 on page 626.) *Aseptic* means that an object or area is free of pathogens. Recall from Chapter 1 that **asepsis** is the absence of significant contamination. Aseptic techniques are important in surgery to minimize contamination from the instruments, operating personnel, and the patient. Aseptic packaging is used in the food-processing industry to make packages in a sterile environment that are then filled with presterilized food.

The Rate of Microbial Death

Learning Objective

■ *Describe the patterns of microbial death caused by treatments with microbial control agents.*

When bacterial populations are heated or treated with antimicrobial chemicals, they usually die at a constant rate. For example, suppose a population of 1 million microbes has been treated for 1 minute, and 90% of the population has died. We are now left with 100,000 microbes. If the population is treated for another minute, 90% of *those* microbes die, and we are left with 10,000 survivors. In other words, for each minute the treatment is applied, 90% of the remaining population is killed (Table 7.2). If the death curve is plotted logarithmically, the death rate is constant, as shown by the straight line in Figure 7.1a.

Several factors influence the effectiveness of antimicrobial treatments:

■ *The number of microbes.* The more microbes there are to begin with, the longer it takes to eliminate the entire population (Figure 7.1b).

■ *Environmental influences.* The presence of organic matter often inhibits the action of chemical antimicrobials. In hospitals, the presence of organic matter

in blood, vomit, or feces influences the selection of disinfectants. Microbes in surface biofilms, shown in Figure 27.10, are difficult for biocides to reach effectively. Because their activity is due to temperature-dependent chemical reactions, disinfectants work somewhat better under warm conditions. Directions on disinfectant containers frequently specify the use of a warm solution.

The nature of the suspending medium is also a factor in heat treatment. Fats and proteins are especially protective, and a medium rich in these substances protects microbes, which will then have a higher survival rate. Heat is also measurably more effective under acidic conditions.

■ *Time of exposure.* Chemical antimicrobials often require extended exposure for more-resistant microbes or endospores to be affected. In heat treatments, a longer exposure can compensate for a lower temperature, a phenomenon of particular importance to pasteurization of dairy products. The effects of irradiation on microbes, other factors being equal, are also very dependent upon time of exposure.

■ *Microbial characteristics.* The concluding section of this chapter discusses how microbial characteristics affect chemical and physical control methods.

Actions of Microbial Control Agents

Learning Objective

■ *Describe the effects of microbial control agents on cellular structures.*

In this section, we examine the ways various agents actually kill or inhibit microbes.

Alteration of Membrane Permeability

A microorganism's plasma membrane, located just inside the cell wall, is the target of many microbial control agents. This membrane actively regulates the passage of nutrients into the cell and the elimination of wastes from the cell. Damage to the lipids or proteins of the plasma membrane by antimicrobial agents causes cellular contents to leak into the surrounding medium and interferes with the growth of the cell.

Damage to Proteins and Nucleic Acids

Bacteria are sometimes thought of as "little bags of enzymes." Enzymes, which are primarily protein, are vital to all cellular activities. Recall that the functional properties of proteins are the result of their three-dimensional shape (see Figure 2.16 on page 49). This shape is maintained by chemical bonds that link adjoining portions of the amino acid chain as it folds back and forth upon itself. Some of those bonds are hydrogen bonds, which are susceptible to

breakage by heat or certain chemicals; breakage results in denaturation of the protein. Covalent bonds, which are stronger, are also subject to attack. For example, disulfide bridges, which play an important role in protein structure by joining amino acids with exposed sulfhydryl (—SH) groups, can be broken by certain chemicals or sufficient heat.

The nucleic acids DNA and RNA are the carriers of the cell's genetic information. Damage to these nucleic acids by heat, radiation, or chemicals is frequently lethal to the cell; the cell can no longer replicate, nor can it carry out normal metabolic functions such as the synthesis of enzymes.

Physical Methods of Microbial Control

Learning Objectives

■ *Compare the effectiveness of moist heat (boiling, autoclaving, pasteurization) and dry heat.*
■ *Describe how filtration, low temperatures, desiccation, and osmotic pressure suppress microbial growth.*
■ *Explain how radiation kills cells.*

As early as the Stone Age, it is likely that humans were already using some physical methods of microbial control to preserve foods. Drying (desiccation) and salting (osmotic pressure) were probably among the earliest techniques.

When selecting methods of microbial control, consideration must be given to effects on things besides the microbes. For example, certain vitamins or antibiotics in a solution might be inactivated by heat. Many laboratory or hospital materials, such as rubber and latex tubing, are damaged by repeated heating. There are also economic considerations; for example, it may be less expensive to use presterilized, disposable plasticware than to repeatedly reuse and resterilize glassware.

Heat

A visit to any supermarket will demonstrate that heat-preserved canned goods represent one of the most common methods of food preservation. Laboratory media and glassware, and hospital instruments, are also usually sterilized by heat. Heat appears to kill microorganisms by denaturing their enzymes; the resultant changes to the three-dimensional shapes of these proteins inactivate them (see Figure 5.5 on page 119).

Heat resistance varies among different microbes; these differences can be expressed through the concept

FIGURE 7.2 An autoclave. The entering steam forces the air out of the bottom (blue arrows). The automatic ejector valve remains open as long as an air-steam mixture is passing out of the waste line. When all the air has been ejected, the higher temperature of the pure steam closes the valve, and the pressure in the chamber increases.

■ Autoclaving is the preferred method of sterilization, provided the material to be sterilized will not be damaged by heat or moisture.

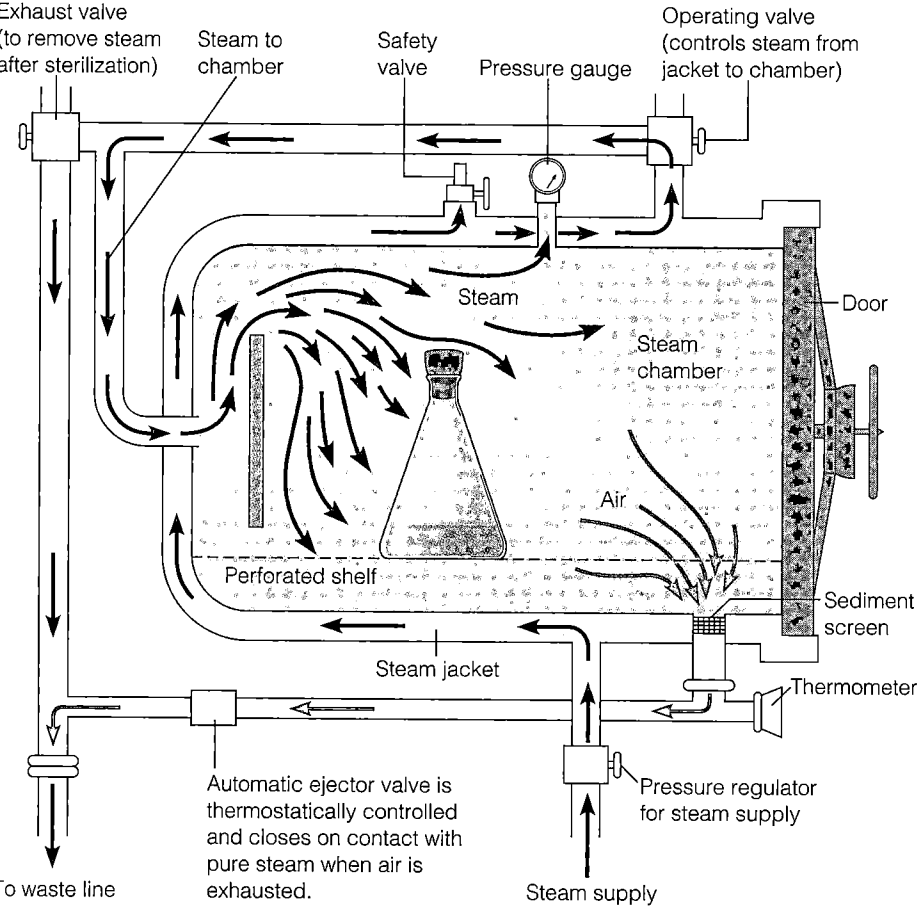

Exhaust valve (to remove steam after sterilization)   Steam to chamber   Safety valve   Pressure gauge   Operating valve (controls steam from jacket to chamber)

Steam   Steam chamber   Door   Air   Perforated shelf   Sediment screen   Steam jacket   Thermometer

To waste line   Automatic ejector valve is thermostatically controlled and closes on contact with pure steam when air is exhausted.   Pressure regulator for steam supply   Steam supply

of thermal death point. **Thermal death point (TDP)** is the lowest temperature at which all the microorganisms in a liquid suspension will be killed in 10 minutes.

Another factor to be considered in sterilization is the length of time required. This is expressed as **thermal death time (TDT)**, the minimal length of time for all bacteria in a liquid culture to be killed at a given temperature. Both TDP and TDT are useful guidelines that indicate the severity of treatment required to kill a given population of bacteria.

Decimal reduction time (DRT, or *D value*) is a third concept related to bacterial heat resistance. DRT is the time, in minutes, in which 90% of a population of bacteria at a given temperature will be killed (in Table 7.2 and Figure 7.1a, DRT is 1 minute). DRT is especially useful in the canning industry.

The heat used in sterilization can be applied in the form of moist heat or dry heat.

Moist Heat

Moist heat kills microorganisms primarily by the coagulation of proteins, which is caused by breakage of the hydrogen bonds that hold the proteins in their three-dimensional structure. This coagulation process is familiar to anyone who has watched an egg white frying. Protein coagulation, or denaturation, occurs more quickly in the presence of water.

One type of moist heat sterilization is boiling, which kills vegetative forms of bacterial pathogens, almost all viruses, and fungi and their spores within about 10 minutes, usually much faster. Free-flowing (unpressurized) steam is equivalent in temperature to boiling water. Endospores and some viruses, however, are not destroyed this quickly. Some hepatitis viruses, for example, can survive up to 30 minutes of boiling, and some bacterial endospores have resisted boiling for more than 20 hours. Boiling is therefore not always a reliable sterilization procedure. However, brief boiling, even at high altitudes, will kill most pathogens. The use of boiling to sanitize baby bottles is a familiar example.

Reliable sterilization with moist heat requires temperatures above that of boiling water. These high temperatures are most commonly achieved by steam under pressure in an **autoclave** (Figure 7.2). Autoclaving is the preferred method of sterilization, unless the material to be sterilized can be damaged by heat or moisture.

table 7.3	The Relationship Between the Pressure and Temperature of Steam at Sea Level*	
Pressure (psi in excess of atmospheric pressure)		**Temperature (°C)**
0 psi		100
5 psi		110
10 psi		116
15 psi		121
20 psi		126
30 psi		135

*At higher altitudes, where the atmosphere is thinner, the pressure indicated on the gauge would be higher than shown here. For example, in Denver, the pressure shown would be higher than 15 psi at an autoclave temperature of 121°C.

table 7.4	The Effect of Container Size on Autoclave Sterilization Times for Liquid Solutions*	
Container Size	**Liquid Volume**	**Sterilization Time (min)**
Test tube: 18 × 150 mm	10 ml	15
Erlenmeyer flask: 125 ml	95 ml	15
Erlenmeyer flask: 2000 ml	1500 ml	30
Fermentation bottle: 9000 ml	6750 ml	70

*Sterilization times in the autoclave include the time for the contents of the containers to reach sterilization temperatures. For smaller containers, this is only 5 min or less; but for a 9000-ml bottle, it might be as much as 70 min. A container is usually not filled past 75% of its capacity.

The higher the pressure in the autoclave, the higher the temperature. For example, when free-flowing steam at a temperature of 100°C is placed under a pressure of 1 atmosphere above sea level pressure—that is, about 15 pounds of pressure per square inch (psi)—the temperature rises to 121°C. Increasing the pressure to 20 psi raises the temperature to 126°C. The relationship between temperature and pressure is shown in Table 7.3.

Sterilization in an autoclave is most effective when the organisms are either contacted by the steam directly or are contained in a small volume of aqueous (primarily water) liquid. Under these conditions, steam at a pressure of about 15 psi (121°C) will kill *all* organisms and their endospores in about 15 minutes.

Autoclaving is used to sterilize culture media, instruments, dressings, intravenous equipment, applicators, solutions, syringes, transfusion equipment, and numerous other items that can withstand high temperatures and pressures. Large industrial autoclaves are called *retorts*, but the same principle applies for the common household pressure cooker used in the home canning of foods.

Heat requires extra time to reach the center of solid materials, such as canned meats, because such materials do not develop the efficient heat-distributing convection currents that occur in liquids. Heating large containers also requires extra time. Table 7.4 shows the different time requirements for sterilizing liquids in various container sizes. Unlike sterilizing aqueous solutions, in order to sterilize the surface of a solid, steam must actually contact it. To sterilize dry glassware, bandages, and the like, care must be taken to ensure that steam contacts all surfaces.

For example, aluminum foil is impervious to steam and should not be used to wrap dry materials that are to be sterilized; paper should be used instead. Care should also be taken to avoid trapping air in the bottom of a dry container because trapped air will not be replaced by steam, which is lighter than air. The trapped air is the equivalent of a small hot-air oven, which, as we will see shortly, requires a higher temperature and longer time to sterilize materials. Containers that can trap air should be placed in a tipped position so that the steam will force out the air. Products that do not permit penetration by moisture, such as mineral oil or petroleum jelly, are not sterilized by the same methods that would sterilize aqueous solutions.

Several commercially available methods can indicate whether sterilization has been achieved by heat treatment. Some of these are chemical reactions in which an indicator changes color when the proper times and temperatures have been reached (Figure 7.3). In some designs, the word "sterile" or "autoclaved" appears on wrappings or tapes. In another method, a pellet contained within a glass vial melts. A widely used test consists of preparations of specified species of bacterial endospores impregnated into paper strips. After autoclaving, these can then be aseptically inoculated into culture media. Growth in the culture media indicates survival of the endospores and therefore inadequate processing. Other designs use endospore suspensions that can be released, after heating, into a surrounding culture medium within the same vial.

Steam under pressure fails to sterilize when the air is not completely exhausted. This is usually due to the

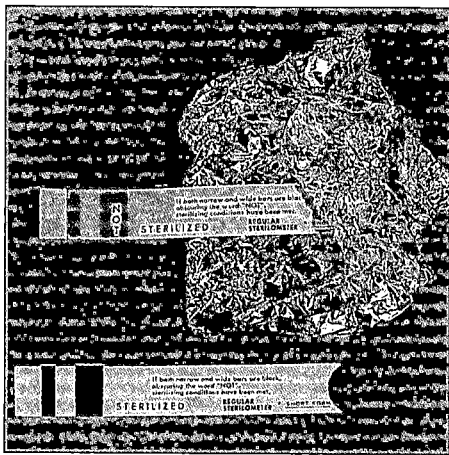

FIGURE 7.3 Examples of sterilization indicators. The strips indicate if the item has been properly heat sterilized; the word NOT appears if heating has been inadequate. In the illustration, the indicator that was wrapped with aluminum foil was not sterilized because steam couldn't penetrate the foil.

premature closing of the autoclave's automatic ejector valve (see Figure 7.2). The principles of heat sterilization have a direct bearing on home canning. As anyone familiar with home canning knows, the steam must flow vigorously out of the valve in the lid for several minutes to carry with it all the air before the pressure cooker is sealed. If the air is not completely exhausted, the container will not reach the temperature expected for a given pressure. Because of the possibility of botulism, a kind of food poisoning resulting from improper canning methods (see Chapter 22, page 608), people involved in home canning should obtain reliable directions and follow them exactly.

Pasteurization

Recall from Chapter 1 that in the early days of microbiology, Louis Pasteur found a practical method of preventing the spoilage of beer and wine. Pasteur used mild heating, which was sufficient to kill the organisms that caused the particular spoilage problem without seriously damaging the taste of the product. The same principle was later applied to milk to produce what we now call pasteurized milk. The intent of pasteurization of milk was to eliminate pathogenic microbes. It also lowers microbial numbers, which prolongs milk's good quality under refrigeration. Many relatively heat-resistant (**thermoduric**) bacteria survive pasteurization, but these are unlikely to cause disease or cause refrigerated milk to spoil.

Products other than milk, such as ice cream, yogurt, and beer, all have individual pasteurization times and temperatures, which often differ considerably. Reasons for variation include less efficient heating in more viscous foods, and the protective effects of fats. The dairy industry routinely uses a test to determine whether products have been pasteurized: the *phosphatase test* (phosphatase is an enzyme naturally present in milk). If the product has been pasteurized, phosphatase will have been inactivated.

In the classic pasteurization treatment of milk, the milk was exposed to a temperature of about 63°C for 30 minutes. Most milk pasteurization today uses higher temperatures, at least 72°C, but for only 15 seconds. This treatment, known as **high-temperature short-time (HTST) pasteurization,** is applied as the milk flows continuously past a heat exchanger. In addition to killing pathogens, HTST pasteurization lowers total bacterial counts, so the milk keeps well under refrigeration.

Milk can also be sterilized—something quite different from pasteurization—by **ultra-high-temperature (UHT) treatments** so that it can be stored without refrigeration. This is more useful in parts of the world where refrigeration facilities are not always available. In the United States, sterilization is sometimes used on the small containers of coffee creamers found in restaurants. To avoid giving the milk a cooked taste, a UHT system is used in which the liquid milk never touches a surface hotter than the milk itself while being heated by steam. The milk falls in a thin film through a chamber of superheated steam and reaches 140°C in less than a second. It is held for 3 seconds in a holding tube and then cooled in a vacuum chamber, where the steam flashes off. With this process, in less than 5 seconds the milk temperature rises from 74°C to 140°C and drops back to 74°C.

The heat treatments we have just discussed illustrate the concept of **equivalent treatments:** As the temperature is increased, much less time is needed to kill the same number of microbes. For example, the destruction of highly resistant endospores might take 70 minutes at 115°C, whereas only 7 minutes would be needed at 125°C. Both treatments yield the same result. The concept of equivalent treatments also explains why classic pasteurization at 63°C for 30 minutes, HTST treatment at 72°C for 15 seconds, and UHT treatment at 140°C for less than a second can have similar effects.

Dry Heat Sterilization

Dry heat kills by oxidation effects. A simple analogy is the slow charring of paper in a heated oven, even when the temperature remains below the ignition point of paper. One of the simplest methods of dry heat sterilization is direct **flaming.** You will use this procedure many times in the microbiology laboratory when you sterilize inoculating loops. To effectively sterilize the inoculating loop, you heat the wire to a red glow. A similar principle is used in *incineration,* an effective way to sterilize and dispose of contaminated paper cups, bags, and dressings.

Another form of dry heat sterilization is **hot-air sterilization.** Items to be sterilized by this procedure are placed in an oven. Generally, a temperature of about 170°C maintained for nearly 2 hours ensures sterilization. The longer period and higher temperature (relative to moist heat) are required because the heat in water is more readily transferred to a cool body than is the heat in air. For example, imagine the different effects of immersing your hand in boiling water at 100°C (212°F) and of holding it in a hot-air oven at the same temperature for the same amount of time.

Filtration

Recall from Chapter 6 that *filtration* is the passage of a liquid or gas through a screenlike material with pores small enough to retain microorganisms (often the same apparatus used for counting; see Figure 6.17 on page 177). A vacuum that is created in the receiving flask helps gravity pull the liquid through the filter. Filtration is used to sterilize heat-sensitive materials, such as some culture media, enzymes, vaccines, and antibiotic solutions.

Some operating theaters and rooms occupied by burn patients receive filtered air to lower the numbers of airborne microbes. **High-efficiency particulate air (HEPA) filters** remove almost all microorganisms larger than about 0.3 μm in diameter.

In the early days of microbiology, hollow candle-shaped filters of unglazed porcelain were used to filter liquids. The long and indirect passageways through the walls of the filter adsorbed the bacteria. Unseen pathogens that passed through the filters (causing such diseases as rabies) were called *filterable viruses.*

In recent years, **membrane filters,** composed of such substances as cellulose esters or plastic polymers, have become popular for industrial and laboratory use (Figure 7.4). These filters are only 0.1 mm thick. The pores of membrane filters include, for example, 0.22-μm and 0.45-μm sizes, which are intended for bacteria. Some very flexible bacteria, such as spirochetes, or the wall-less mycoplasma, will sometimes pass through such filters, however. Filters are available with pores as small as 0.01 μm, a size that will retain viruses and even some large protein molecules.

Low Temperatures

The effect of low temperatures on microorganisms depends on the particular microbe and the intensity of the application. For example, at temperatures of ordinary refrigerators (0–7°C), the metabolic rate of most microbes is so reduced that they cannot reproduce or synthesize

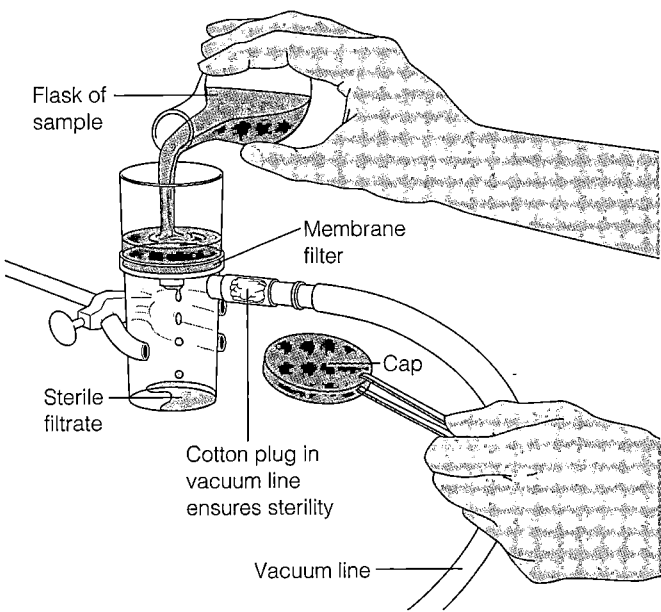

FIGURE 7.4 Filter sterilization with a disposable, presterilized plastic unit. The sample is placed into the upper chamber and forced through the membrane filter by a vacuum in the lower chamber. Pores in the membrane filter are smaller than the bacteria, so bacteria are retained on the filter. The sterilized sample can then be decanted from the lower chamber. Similar equipment with removable filter disks is used to count bacteria in samples (see Figure 6.17).

Labels in figure: Flask of sample; Membrane filter; Cap; Sterile filtrate; Cotton plug in vacuum line ensures sterility; Vacuum line

■ **Filters are available with pores small enough to retain viruses.**

toxins. In other words, ordinary refrigeration has a bacteriostatic effect. Yet psychrotrophs do grow slowly at refrigerator temperatures and will alter the appearance and taste of foods after a time. For example, one microbe reproducing only three times a day would reach a population of more than 2 million within a week. Pathogenic bacteria generally will not grow at refrigerator temperatures, but for at least one important exception, see the discussion of listeriosis in Chapter 22 on page 606.

Surprisingly, some bacteria can grow at temperatures several degrees below freezing. Rapidly attained subfreezing temperatures tend to render microbes dormant but do not necessarily kill them. Slow freezing is more harmful to bacteria; the ice crystals that form and grow disrupt the cellular and molecular structure of the bacteria. Once frozen, one-third of the population of some vegetative bacteria might survive a year, whereas other species might have very few survivors after this time. Many parasites, such as the roundworms that cause trichinosis, are killed by several days of freezing temperatures. Some important temperatures associated with microorganisms and food spoilage are shown in Figure 6.2 on page 158.

Desiccation

In the absence of water, a condition known as **desiccation**, microorganisms cannot grow or reproduce but can remain viable for years. Then, when water is made available to them, they can resume their growth and division. This ability is used in the laboratory when microbes are preserved by lyophilization, or freeze-drying, a process described in Chapter 6 on page 170. Certain foods are also freeze-dried (for example, coffee and some fruit additives for dry cereals).

The resistance of vegetative cells to desiccation varies with the species and the organism's environment. For example, the gonorrhea bacterium can withstand dryness for only about an hour, but the tuberculosis bacterium can remain viable for months. Viruses are generally resistant to desiccation, but they are not as resistant as bacterial endospores, some of which have survived for centuries. This ability of certain dried microbes and endospores to remain viable is important in a hospital setting. Dust, clothing, bedding, and dressings might contain infectious microbes in dried mucus, urine, pus, and feces.

Osmotic Pressure

The use of high concentrations of salts and sugars to preserve food is based on the effects of *osmotic pressure*. High concentrations of these substances create a hypertonic environment that causes water to leave the microbial cell (see Figure 6.4 on page 159). This process resembles preservation by desiccation, in that both methods deny the cell the moisture it needs for growth. The principle of osmotic pressure is used in the preservation of foods. For example, concentrated salt solutions are used to cure meats, and thick sugar solutions are used to preserve fruits.

As a general rule, molds and yeasts are much more capable than bacteria of growing in materials with low moisture or high osmotic pressures. This property of molds, sometimes combined with their ability to grow under acidic conditions, is the reason fruits and grains are spoiled by molds rather than by bacteria. It is also part of the reason molds are able to form mildew on a damp wall or a shower curtain.

Radiation

Radiation has various effects on cells, depending on its wavelength, intensity, and duration. Radiation that kills microorganisms (sterilizing radiation) is of two types: ionizing and nonionizing.

Ionizing radiation—gamma rays, X rays, or high-energy electron beams—has a wavelength shorter than that of nonionizing radiation, less than about 1 nm. Therefore, it carries much more energy (Figure 7.5).

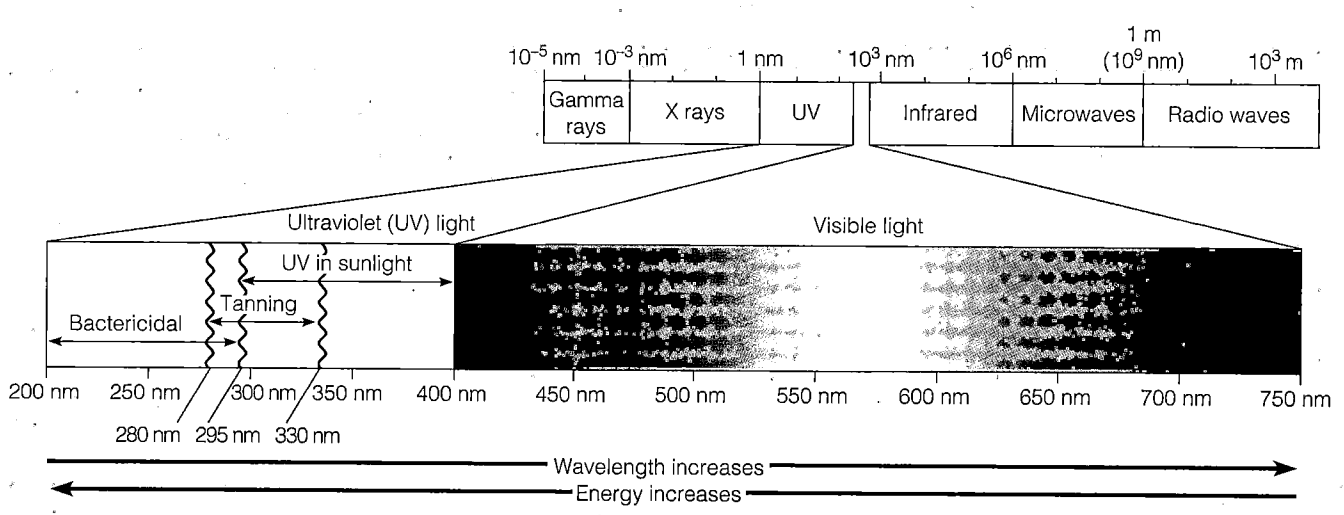

FIGURE 7.5 The radiant energy spectrum. Visible light and other forms of radiant energy radiate through space as waves of various lengths. Ionizing radiation, such as gamma rays and X rays, has a wavelength shorter than 1 nm. Nonionizing radiation, such as ultraviolet (UV) light, has a wavelength between 1 nm and about 380 nm, where the visible spectrum begins.

▥ What effect might increased UV radiation (due to decrease in the ozone layer) have on the Earth's ecosystems?

Gamma rays are emitted by certain radioactive elements such as cobalt, and electron beams are produced by accelerating electrons to high energies in special machines. *X rays,* which are produced by machines in a manner similar to the production of electron beams, are similar in nature to gamma rays. Gamma rays penetrate deeply but may require hours to sterilize large masses; *high-energy electron beams* have much lower penetrating power but usually require only a few seconds of exposure. The principal effect of ionizing radiation is the ionization of water, which forms highly reactive hydroxyl radicals (see the discussion of toxic forms of oxygen in Chapter 6, pages 162–163). These radicals react with organic cellular components, especially DNA.

The so-called target theory of damage by radiation supposes that ionizing particles, or packets of energy, pass through or close to vital portions of the cell; these constitute "hits." One, or a few, hits may only cause nonlethal mutations, some of them conceivably useful. More hits are likely to cause sufficient mutations to kill the microbe.

The food industry has recently renewed its interest in the use of radiation for food preservation (discussed more fully in Chapter 28 on page 774). Low-level ionizing radiation, used for years in many countries, has been approved in the United States for processing spices and certain meats and vegetables. Ionizing radiation, especially high-energy electron beams, is used for the sterilization of pharmaceuticals and disposable dental and medical supplies, such as plastic syringes, surgical gloves, suturing materials, and catheters.

Nonionizing radiation has a wavelength longer than that of ionizing radiation, usually greater than about 1 nm. The best example of nonionizing radiation is ultraviolet (UV) light. UV light damages the DNA of exposed cells by causing bonds to form between adjacent thymines in DNA chains (see Figure 8.20). These *thymine dimers* inhibit correct replication of the DNA during reproduction of the cell. The UV wavelengths most effective for killing microorganisms are about 260 nm; these wavelengths are specifically absorbed by cellular DNA. UV radiation is also used to control microbes in the air. A UV or "germicidal" lamp is commonly found in hospital rooms, nurseries, operating rooms, and cafeterias. UV light is also used to disinfect vaccines and other medical products. A major disadvantage of UV light as a disinfectant is that the radiation is not very penetrating, so the organisms to be killed must be directly exposed to the rays. Organisms protected by solids and such coverings as paper, glass, and textiles are not affected. Another potential problem is that UV light can damage human eyes, and prolonged exposure can cause burns and skin cancer in humans.

Sunlight contains some UV radiation, but the shorter wavelengths—those most effective against bacteria—are screened out by the ozone layer of the atmosphere. The antimicrobial effect of sunlight is due almost entirely to the formation of singlet oxygen in the cytoplasm (see Chapter 6 on page 162). Many pigments produced by bacteria provide protection from sunlight.

Microwaves do not have much direct effect on microorganisms, and bacteria can readily be isolated from the interior of recently operated microwave ovens. Moisture-containing foods are heated by microwave action, and the heat will kill most vegetative pathogens. Solid foods heat unevenly because of the uneven distribution of moisture. For this reason, pork cooked in a microwave oven has been responsible for outbreaks of trichinosis.

★ ★ ★

Table 7.5 summarizes the physical methods of microbial control.

Chemical Methods of Microbial Control

Chemical agents are used to control the growth of microbes on both living tissue and inanimate objects. Unfortunately, few chemical agents achieve sterility; most of them merely reduce microbial populations to safe levels or remove vegetative forms of pathogens from objects. A common problem in disinfection is the selection of an agent. No single disinfectant is appropriate for all circumstances.

Principles of Effective Disinfection

Learning Objective

■ *List the factors related to effective disinfection.*

By reading the label, we can learn a great deal about a disinfectant's properties. The label will usually indicate what groups of organisms the disinfectant will be effective against. Remember that the concentration of a disinfectant affects its action, so it should always be diluted exactly as specified by the manufacturer.

Also consider the nature of the material being disinfected. For example, are organic materials present that might interfere with the action of the disinfectant? Similarly, the pH of the medium often has a great effect on a disinfectant's activity.

Another very important consideration is whether the disinfectant will easily make contact with the microbes. An area might need to be scrubbed and rinsed before the

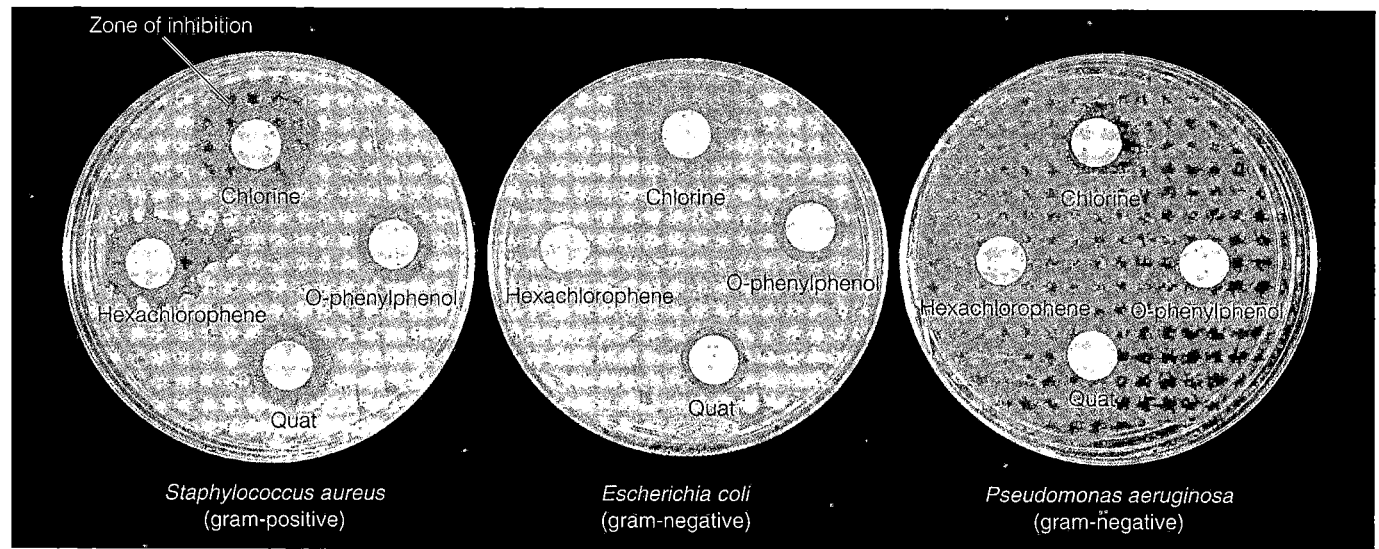

Zone of inhibition

| Staphylococcus aureus (gram-positive) | Escherichia coli (gram-negative) | Pseudomonas aeruginosa (gram-negative) |

FIGURE 7.6 Evaluation of disinfectants by the disk-diffusion method. In this experiment, paper disks are soaked in a solution of disinfectant and placed on the surface of a nutrient medium on which a culture of test bacteria has been spread to produce uniform growth. At the top of each plate, the tests show that chlorine (as sodium hypochlorite) was effective against all the test bacteria but was more effective against gram-positive bacteria. At the bottom row of each plate, the tests show that the quaternary ammonium compound ("quat") was also more effective against the gram-positive bacteria, but it did not affect the pseudomonads at all. At the left side of each plate, the tests show that hexachlorophene was effective against gram-positive bacteria only. At the right sides, O-phenylphenol was ineffective against pseudomonads but was almost equally effective against the gram-positive bacteria and the gram-negative bacteria. All four chemicals worked against the gram-positive test bacteria, but only one of the four chemicals affected pseudomonads.

■ **Which group of bacteria are most resistant to the disinfectants tested?**

particularly susceptible to hexachlorophene, so it is often used to control such infections in nurseries. However, excessive use of this bisphenol, such as bathing infants with it several times a day, can lead to neurological damage.

Another widely used bisphenol is *triclosan* (Figure 7.7d), an ingredient in antibacterial soaps and at least one toothpaste. Triclosan has even been incorporated into kitchen cutting boards and the handles of knives and other plastic kitchenware. Its use is now so widespread that resistant bacteria have been reported, and concerns about its effect on microbes' resistance to certain antibiotics have been raised. Triclosan's mode of action is unknown, but its primary effects appear to be on the plasma membrane. It has a broad spectrum of activity against bacteria, especially gram–positives, and fungi.

Biguanides

Chlorhexidine is a member of the **biguanide** group with a broad spectrum of activity. It is frequently used for microbial control on skin and mucous membranes. Combined with a detergent or alcohol, chlorhexidine is also

used for surgical hand scrubs and preoperative skin preparation in patients. In such applications, its strong affinity for binding to the skin or mucous membranes is an advantage, as is its low toxicity. However, contact with the eyes can cause damage. Its killing effect is related to the injury it causes to the plasma membrane. It is biocidal against most vegetative bacteria and fungi. Mycobacteria are relatively resistant, and endospores and protozoan cysts are not affected. The only viruses affected are certain enveloped (lipophilic) types (see Chapter 13).

Halogens

The **halogens**, particularly iodine and chlorine, are effective antimicrobial agents, both alone and as constituents of inorganic or organic compounds. *Iodine* (I_2) is one of the oldest and most effective antiseptics. It is effective against all kinds of bacteria, many endospores, various fungi, and some viruses. One proposed mechanism for the activity of iodine is that it combines with certain amino acids of enzymes and other cellular proteins. The exact mode of action is not known.

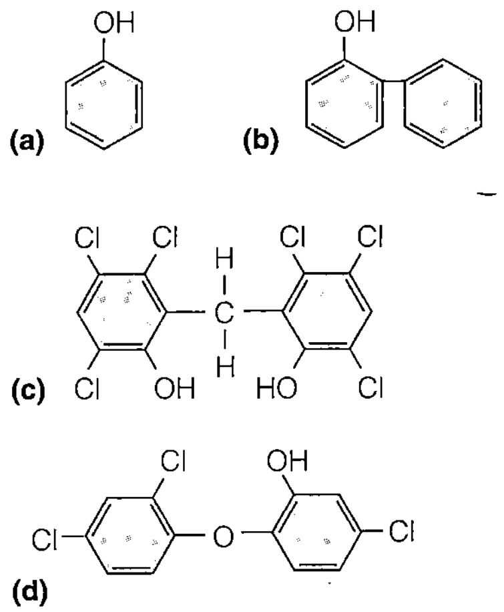

FIGURE 7.7 The structure of phenolics and bisphenols.
(a) Phenol. (b) O-phenylphenol. (c) Hexachlorophene (a bis-phenol). (d) Triclosan (a bisphenol).

■ At concentrations above 10%, phenol has a significant antibacterial effect.

Iodine is available as a **tincture**—that is, in solution in aqueous alcohol—and as an iodophor. An **iodophor** is a combination of iodine and an organic molecule, from which the iodine is released slowly. Iodophors have the antimicrobial activity of iodine, but they do not stain and are less irritating. The most common commercial preparations are Betadine® and Isodine®. These are *povidone-iodines;* povidone is a surface-active iodophor that improves the wetting action and serves as a reservoir of free iodine. Iodines are used mainly for skin disinfection and wound treatment. Many campers are familiar with iodine for water treatment. To treat water, iodine tablets are added or the water can be passed through iodine-treated resin filters.

Chlorine (Cl_2), as a gas or in combination with other chemicals, is another widely used disinfectant. Its germicidal action is caused by the hypochlorous acid (HOCl) that forms when chlorine is added to water:

(1)
$$Cl_2 \; + \; H_2O \; \rightleftharpoons \; H^+ \; + \; Cl^- \; + \; HOCl$$

| Chlorine | Water | Hydrogen ion | Chloride ion | Hypochlorous acid |

(2)
$$HOCl \; \rightleftharpoons \; H^+ \; + \; OCl^-$$

| Hypochlorous acid | | Hydrogen ion | Hypochlorite ion |

Exactly how hypochlorous acid exerts its killing power is not known. It is a strong oxidizing agent that prevents much of the cellular enzyme system from functioning. Hypochlorous acid is the most effective form of chlorine because it is neutral in electrical charge and diffuses as

rapidly as water through the cell wall. Because of its negative charge, the hypochlorite ion (OCl^-) cannot enter the cell freely.

A liquid form of compressed chlorine gas is used extensively for disinfecting municipal drinking water, water in swimming pools, and sewage. Several compounds of chlorine are also effective disinfectants. For example, solutions of *calcium hypochlorite* [$Ca(OCl)_2$] are used to disinfect dairy equipment and restaurant eating utensils. This compound, once called chloride of lime, was used as early as 1825, long before the concept of a germ theory for disease, to soak hospital dressings in Paris hospitals. It was also the disinfectant used in the 1840s by Semmelweis to control hospital infections during childbirth, as mentioned in Chapter 1, page 10. Another chlorine compound, *sodium hypochlorite* (NaOCl; see Figure 7.6), is used as a household disinfectant and bleach (Clorox®), and as a disinfectant in dairies, food-processing establishments, and hemodialysis systems. When the quality of drinking water is in question, household bleach can provide a rough equivalent of municipal chlorination. After two drops of bleach are added to a liter of water (four drops if the water is cloudy) and the mixture has sat for 30 minutes, the water is considered safe for drinking under emergency conditions. U.S. military forces in the field are issued a tablet (Chlor-Floc®) that contains *sodium dichloroisocyanurate,* a form of chlorine combined with an agent that flocculates (coagulates) suspended materials in a water sample, causing them to settle out, thus clarifying it.

Another group of chlorine compounds, the *chloramines,* consist of chlorine and ammonia. They are used as disinfectants, antiseptics, or sanitizing agents. Chloramines are very stable compounds that release chlorine over long periods. They are relatively effective in organic matter, but they have the disadvantages of acting more slowly and being less effective purifiers than many other chlorine compounds. Chloramines are used to sanitize glassware and eating utensils and to treat dairy and food-manufacturing equipment. Ammonia is usually mixed with chlorine in municipal water-treatment systems to form chloramines. The chloramines control taste and odor problems caused by the reaction of chlorine with other nitrogenous compounds in the water. Because chloramines are less effective as germicides, sufficient chlorine must be added to ensure a residual of chlorine in the form of HOCl. (Chloramines are toxic to aquarium fish, but pet shops sell chemicals to neutralize them.)

Alcohols

Alcohols effectively kill bacteria and fungi but not endospores and nonenveloped viruses. The mechanism of action of alcohol is usually protein denaturation, but

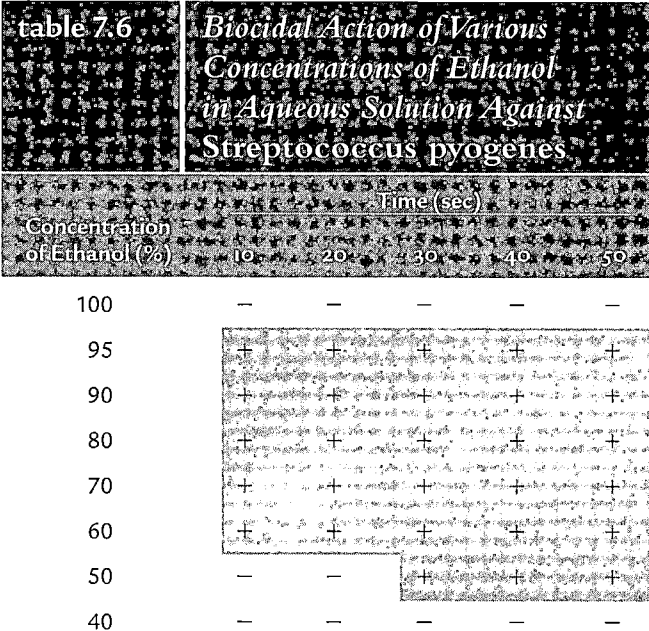

table 7.6 *Biocidal Action of Various Concentrations of Ethanol in Aqueous Solution Against Streptococcus pyogenes*

Concentration of Ethanol (%)	Time (sec)				
	10	20	30	40	50
100	−	−	−	−	−
95	+	+	+	+	+
90	+	+	+	+	+
80	+	+	+	+	+
70	+	+	+	+	+
60	+	+	+	+	+
50	−	−	+	+	+
40	−	−	−	−	−

NOTES: A minus sign indicates no biocidal action (bacterial growth); a plus sign indicates biocidal action (no bacterial growth). The highlighted area represents bacteria killed by biocidal action.

alcohol can also disrupt membranes and dissolve many lipids, including the lipid component of enveloped viruses. Alcohols have the advantage of acting and then evaporating rapidly and leaving no residue. When the skin is swabbed (degermed) before an injection, most of the microbial control activity comes from simply wiping away dirt and microorganisms, along with skin oils. However, alcohols are unsatisfactory antiseptics when applied to wounds. They cause coagulation of a layer of protein under which bacteria continue to grow.

Two of the most commonly used alcohols are ethanol and isopropanol. The recommended optimum concentration of *ethanol* is 70%, but concentrations between 60% and 95% seem to kill as fast (Table 7.6). Pure ethanol is less effective than aqueous solutions (ethanol mixed with water) because denaturation requires water. *Isopropanol,* often sold as rubbing alcohol, is slightly superior to ethanol as an antiseptic and disinfectant. Moreover, it is less volatile, less expensive, and more easily obtained than ethanol.

Ethanol and isopropanol are often used to enhance the effectiveness of other chemical agents. For example, an aqueous solution of Zephiran® (described on page 199) kills about 40% of the population of a test organism in two minutes, whereas a tincture of Zephiran kills about 85% in the same period. To compare the effectiveness of tinctures and aqueous solutions, see Figure 7.10 on page 200.

Heavy Metals and Their Compounds

Several heavy metals can be biocidal or antiseptic, including silver, mercury, and copper. The ability of very small amounts of heavy metals, especially silver and copper, to exert antimicrobial activity is referred to as **oligodynamic action** (*oligo* means few). This action can be seen when we place a coin or other clean piece of metal containing silver or copper on a culture on an inoculated Petri plate. Extremely small amounts of metal diffuse from the coin and inhibit the growth of bacteria for some distance around the coin (Figure 7.8). This effect is produced by the action of heavy metal ions on microbes. When the metal ions combine with the sulfhydryl groups on cellular proteins, denaturation results.

Silver is used as an antiseptic in a 1% *silver nitrate* solution. At one time, many states required that the eyes of newborns be treated with a few drops of silver nitrate to guard against an infection of the eyes called gonorrheal ophthalmia neonatorum, which the infants might have contracted as they passed through the birth canal. In recent years, antibiotics have replaced silver nitrate for this purpose.

Recently, there has been renewed interest in silver as an antimicrobial agent. Silver-impregnated dressings that slowly release silver ions have proven especially useful when antibiotic-resistant bacteria are a problem. A combination of silver and the drug sulfadiazine, *silver-sulfadiazine,* is the most common formulation. It is available as a topical cream for use on burns. Silver can also be incorporated into indwelling catheters, which are a common source of hospital infections, and in wound dressings.

Inorganic mercury compounds, such as *mercuric chloride,* have a long history of use as disinfectants. They have a very broad spectrum of activity; their effect is primarily bacteriostatic. However, their use is now limited because of their toxicity, corrosiveness, and ineffectiveness in organic matter. At present, the primary use of mercurials is to control mildew in paints. In addition, mercurochrome antiseptic, an organic mercury compound, is often found in home medicine chests.

Copper in the form of *copper sulfate* is used chiefly to destroy green algae (algicide) that grow in reservoirs, swimming pools, and fish tanks. If the water does not contain excessive organic matter, copper sulfate is effective in concentrations of one part per million of water. To prevent mildew, copper compounds such as copper 8-hydroxyquinoline are sometimes included in paint.

Another metal used as an antimicrobial is zinc. The effect of trace amounts of zinc can be seen on roofs of buildings down-slope from galvanized (zinc-coated) fittings. The roof is lighter-colored where biological growth

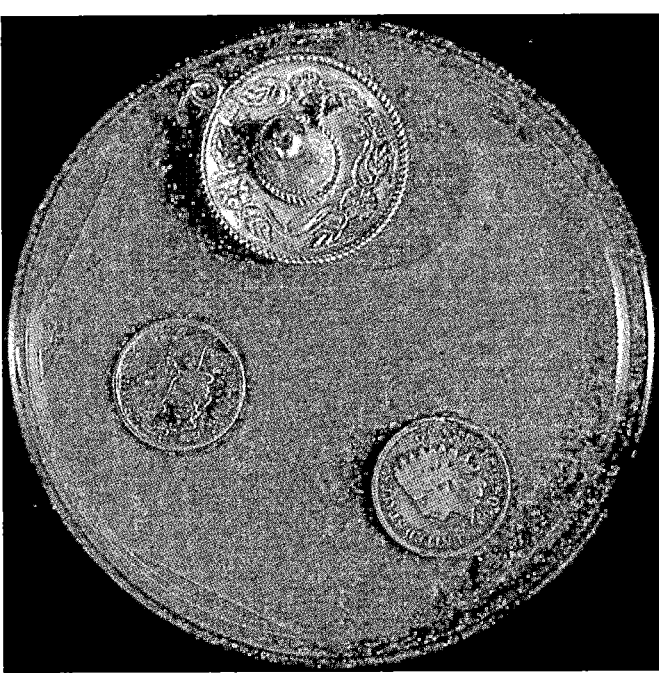

FIGURE 7.8 **Oligodynamic action of heavy metals.** Clear zones where bacterial growth has been inhibited are seen around the sombrero charm (pushed aside), the dime, and the penny. The charm and the dime contain silver; the penny contains copper.

■ **The coins used in this demonstration were minted many years ago; why were current coins not used?**

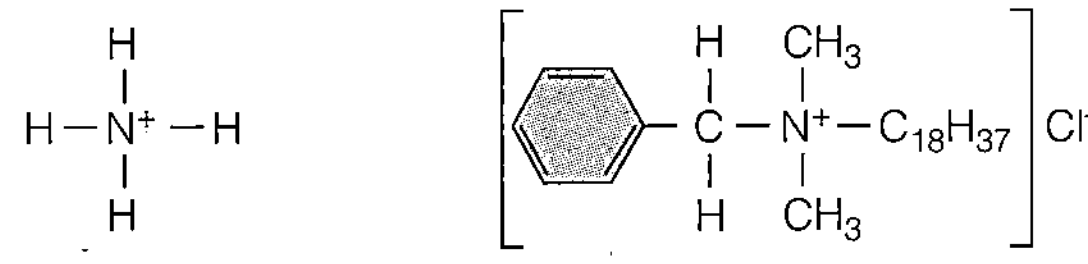

is impeded. Copper- and zinc-treated shingles are available. *Zinc chloride* is a common ingredient in mouthwashes, and *zinc oxide* is probably the most widely used antifungal in paints, mainly because it is often part of the pigment formulation.

Surface-Active Agents

Surface-active agents, or **surfactants,** can decrease surface tension among molecules of a liquid. Such agents include soaps and detergents. Soap has little value as an antiseptic, but it does have an important function in the mechanical removal of microbes through scrubbing. The skin normally contains dead cells, dust, dried sweat, microbes, and oily secretions from oil glands. Soap breaks the oily film into tiny droplets, a process called *emulsification,* and the water and soap together lift up the emulsified oil and debris and float them away as the lather is washed off. In this sense, soaps are good degerming agents. Many so-called deodorant soaps contain *triclocarban.* This compound is particularly active against gram-positive bacteria.

Acid-anionic surface-active sanitizers are very important in the cleaning of dairy utensils and equipment. Their sanitizing ability is related to the negatively charged portion

FIGURE 7.9 **The ammonium ion and a quaternary ammonium compound, benzalkonium chloride (Zephiran®).** Notice how other groups replace the hydrogens of the ammonium ion.

■ **Quaternary ammonium compounds are strongly bactericidal against gram-positive bacteria, but less effective against gram-negative bacteria.**

(anion) of the molecule, which reacts with the plasma membrane. These sanitizers, which act on a wide spectrum of microbes, including troublesome thermoduric bacteria, are nontoxic, noncorrosive, and fast acting.

Quaternary Ammonium Compounds (Quats)

The most widely used surface-active agents are the cationic detergents, especially the **quaternary ammonium compounds (quats).** Their cleansing ability is related to the positively charged portion—the cation—of the molecule. Their name is derived from the fact that they are modifications of the four-valence ammonium ion, NH_4^+ (Figure 7.9). Quaternary ammonium compounds are strongly bactericidal against gram-positive bacteria and somewhat less active against gram-negative bacteria (Figure 7.6).

Quats are also fungicidal, amoebicidal, and virucidal against enveloped viruses. They do not kill endospores or mycobacteria. Their chemical mode of action is unknown, but they probably affect the plasma membrane. They change the cell's permeability and cause the loss of essential cytoplasmic constituents, such as potassium.

Two popular quats are Zephiran, a brand name of *benzalkonium chloride* (see Figure 7.9), and Cepacol®, a brand name of *cetylpyridinium chloride.* They are strongly antimicrobial, colorless, odorless, tasteless, stable, easily diluted, and nontoxic, except at high concentrations. If your mouthwash bottle fills with foam when shaken, the mouthwash probably contains a quat. However, organic matter interferes with their activity, and they are rapidly neutralized by soaps and anionic detergents.

Anyone associated with medical applications of quats should remember that certain bacteria, such as some species of *Pseudomonas,* not only survive in quaternary ammonium compounds but actively grow in them. This resistance occurs not only to the disinfectant solution but also to moistened gauze and bandages, whose fibers tend to neutralize the quats. (See the box on page 201.)

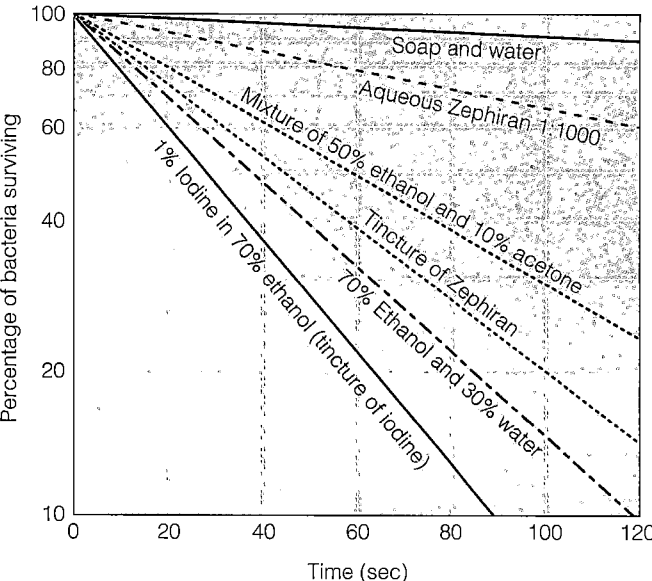

FIGURE 7.10 A comparison of the effectiveness of various antiseptics. The steeper the downward slope of the killing curve of the antiseptic, the more effective it is. A 1% iodine in 70% ethanol solution is the most effective; soap and water are the least effective. Notice that a tincture of Zephiran® is more effective than an aqueous solution of the same antiseptic.

***

▨ **Why is the tincture of Zephiran® more effective than the aqueous solution?**

***

Before we move on to the next group of chemical agents, refer to Figure 7.10, which compares the effectiveness of some of the antiseptics we have discussed so far.

Chemical Food Preservatives

Chemical preservatives are frequently added to foods to retard spoilage. *Sulfur dioxide* (SO_2) has long been used as a disinfectant, especially in wine-making. Homer mentioned its use in the *Odyssey*, written nearly 2800 years ago. Among the more common additives are sodium benzoate, sorbic acid, and calcium propionate. These chemicals are simple organic acids, or salts of organic acids, which the body readily metabolizes and which are generally judged to be safe in foods. *Sorbic acid,* or its more soluble salt *potassium sorbate,* and *sodium benzoate* prevent molds from growing in certain acidic foods, such as cheese and soft drinks. Such foods, usually with a pH of 5.5 or lower, are most susceptible to mold-type spoilage. *Calcium propionate,* an effective fungistat used in bread, prevents the growth of surface molds and the *Bacillus* bacterium that causes ropy bread. These organic acids inhibit mold growth, not by affecting the pH but by interfering with the mold's metabolism or the integrity of the plasma membrane.

Sodium nitrate and *sodium nitrite* are added to many meat products, such as ham, bacon, hot dogs, and sausage. The active ingredient is sodium nitrite, which certain bacteria in the meats can also produce from sodium nitrate. These bacteria use nitrate as a substitute for oxygen under anaerobic conditions. The nitrite has two main functions: to preserve the pleasing red color of the meat by reacting with blood components in the meat, and to prevent the germination and growth of any botulism endospores that might be present. There has been some concern that the reaction of nitrites with amino acids can form certain carcinogenic products known as **nitrosamines,** and the amount of nitrites added to foods has generally been reduced recently for this reason. However, the use of nitrites continues because of their established value in preventing botulism. Because nitrosamines are formed in the body from other sources, the added risk posed by a limited use of nitrates and nitrites in meats is lower than was once thought.

Antibiotics

The antimicrobials discussed in this chapter are not useful for ingestion or injection to treat disease. Antibiotics are used for this purpose. The use of antibiotics is highly restricted; however, at least two have considerable use in food preservation. Neither is of value for clinical purposes. *Nisin,* which is often added to cheese to inhibit the growth of certain endospore-forming spoilage bacteria, is an example of a bacteriocin, a protein that is produced by one bacterium and inhibits another (see Chapter 8, page 240). Nisin is present naturally in small amounts in many dairy products. It is tasteless, readily digested, and nontoxic. *Natamycin* (pimaricin) is an antifungal antibiotic approved for use in foods, mostly cheese.

Aldehydes

Aldehydes are among the most effective antimicrobials. Two examples are formaldehyde and glutaraldehyde. They inactivate proteins by forming covalent cross-links with several organic functional groups on proteins ($—NH_2$, $—OH$, $—COOH$, and $—SH$). *Formaldehyde gas* is an excellent disinfectant. However, it is more commonly available as *formalin,* a 37% aqueous solution of formaldehyde gas. Formalin was once used extensively to preserve biological specimens and inactivate bacteria and viruses in vaccines.

Glutaraldehyde is a chemical relative of formaldehyde that is less irritating and more effective than

CHAPTER 7 The Control of Microbial Growth   **201**

formaldehyde. Glutaraldehyde is used to disinfect hospital instruments, including respiratory-therapy equipment. When used in a 2% solution (Cidex™), it is bactericidal, tuberculocidal, and virucidal in 10 minutes and sporicidal in 3–10 hours. Glutaraldehyde is one of the few liquid chemical disinfectants that can be considered a sterilizing agent. However, 30 minutes is often considered the maximum time allowed for a sporicide to act, which is a criterion glutaraldehyde cannot meet. Both glutaraldehyde and formalin are used by morticians for embalming.

Gaseous Chemosterilizers

Gaseous chemosterilizers are chemicals that sterilize in a closed chamber (similar to an autoclave). A gas suitable for this method is *ethylene oxide*:

$$H_2C\!-\!CH_2$$
$$\diagdown \diagup$$
$$O$$

Its activity depends on the denaturation of proteins: The proteins' labile hydrogens, such as $-SH$, $-COOH$, or $-OH$, are replaced by alkyl groups *(alkylation)*, such as

—CH_2CH_2OH. Ethylene oxide kills all microbes and endospores but requires a lengthy exposure period of 4–18 hours. It is toxic and explosive in its pure form, so it is usually mixed with a nonflammable gas, such as carbon dioxide or nitrogen. One of its principal advantages is that it is highly penetrating, so much so that ethylene oxide was chosen to sterilize spacecraft sent to land on the moon and Mars. Using heat to sterilize the electronic gear on these vehicles was not practical.

Because of their ability to sterilize without heat, gases like ethylene oxide are also widely used on medical supplies and equipment. Many large hospitals have ethylene oxide chambers, some large enough to sterilize mattresses, as part of their sterilizing equipment. Propylene oxide and beta-propiolactone are also used for gaseous sterilization:

$$H_3C-CH-CH_2 \qquad H_2C-CH_2$$

Propylene oxide   Beta-propiolactone

A disadvantage of all these gases is that they are suspected carcinogens, especially beta-propiolactone. For this reason, there has been concern about the exposure of hospital workers to ethylene oxide from such sterilizers. Because of this hazard, these gases may eventually be replaced by *plasma gas sterilization*. This makes use of vapors of hydrogen peroxide (discussed shortly) subjected to radio frequencies or microwave radiation to produce reactive free radicals. No by-products toxic to humans are produced, and it is an effective sterilant.

Peroxygens (Oxidizing Agents)

Peroxygens exert antimicrobial activity by oxidizing cellular components of the treated microbes. Examples are ozone, hydrogen peroxide, and peracetic acid. *Ozone* (O_3) is a highly reactive form of oxygen that is generated by passing oxygen through high-voltage electrical discharges. It is responsible for the air's rather fresh odor after a lightning storm, in the vicinity of electrical sparking, or around an ultraviolet light. Ozone is often used to supplement chlorine in the disinfection of water, because it helps neutralize tastes and odors. Although ozone is a more effective killing agent, its residual activity is difficult to maintain in water, and it is more expensive than chlorine.

Hydrogen peroxide is an antiseptic found in many household medicine cabinets and in hospital supply rooms. It is not a good antiseptic for open wounds; in fact, it may slow wound healing. It is quickly broken down to water and gaseous oxygen by the action of the enzyme catalase, which is present in human cells (see Chapter 6, page 163). However, hydrogen peroxide does effectively disinfect inanimate objects, an application in which it is even sporicidal, especially at elevated temperatures. On a nonliving surface, the normally protective enzymes of aerobic bacteria and facultative anaerobes are overwhelmed by the high concentrations of peroxide used. Because of these factors, the food industry is increasing its use of hydrogen peroxide for aseptic packaging (see Chapter 28 on page 773). The packaging materials pass through a hot solution of the chemical before being assembled into a container. In addition, many wearers of contact lenses are familiar with disinfection by hydrogen peroxide. After disinfection, a platinum catalyst in the lens-disinfecting kit destroys residual hydrogen peroxide so that it does not persist on the lens, where it might be an irritant.

Oxidizing agents are useful for irrigating deep wounds, where the oxygen released makes an environment that inhibits the growth of anaerobic bacteria.

Benzoyl peroxide is another compound useful for treating wounds infected by anaerobic pathogens, but it is probably more familiar as the main ingredient in over-the-counter medications for acne, which is caused by a type of anaerobic bacterium infecting hair follicles.

Peracetic acid is one of the most effective liquid chemical sporicides available and is considered a sterilant. It is generally effective on endospores and viruses within 30 minutes, and kills vegetative bacteria and fungi in less than 5 minutes. Peracetic acid has many applications in the disinfection of food-processing and medical equipment because it leaves no toxic residues and is minimally affected by the presence of organic matter.

Microbial Characteristics and Microbial Control

Learning Objective

■ *Explain how the control of microbial growth is affected by the type of microbe.*

Many biocides tend to be more effective against gram-positive bacteria, as a group, than against gram-negative bacteria. This is illustrated in Figure 7.11, which presents a simplified hierarchy of relative resistance of major microbial groups to biocides. A principal factor in this relative resistance to biocides is the external lipopolysaccharide layer of gram-negative bacteria. Within gram-negative bacteria, members of the genera *Pseudomonas* and *Burkholderia* are of special interest.

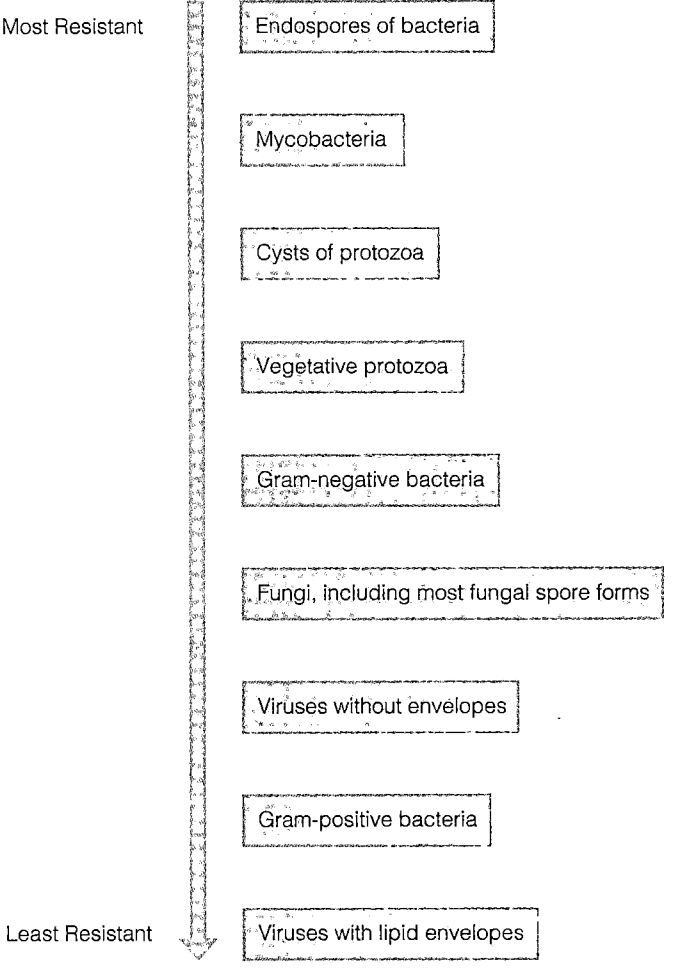

Most Resistant — Endospores of bacteria

Mycobacteria

Cysts of protozoa

Vegetative protozoa

Gram-negative bacteria

Fungi, including most fungal spore forms

Viruses without envelopes

Gram-positive bacteria

Least Resistant — Viruses with lipid envelopes

FIGURE 7.11 Decreasing order of resistance of microorganisms to chemical biocides.

table 7.7	The Effectiveness of Chemical Antimicrobials Against Endospores and Mycobacteria	
Chemical Agent	Endospores	Mycobacteria
Mercury	No activity	No activity
Phenolics	Poor	Good
Bisphenols	No activity	No activity
Quaternary ammonium compounds	No activity	No activity
Chlorines	Fair	Fair
Iodine	Poor	Good
Alcohols	Poor	Good
Glutaraldehyde	Fair	Good
Chlorhexidine	No activity	Fair

These closely related bacteria are unusually resistant to biocides (Figure 7.6) and will even grow actively in some disinfectants and antiseptics, most notably the quaternary ammonium compounds. (See the box on page 201). Later, in Chapter 20, you will see that these bacteria are also resistant to many antibiotics. This resistance to chemical antimicrobials is mostly related to the characteristics of their *porins* (structural openings in the wall of gram-negative bacteria, see Figure 4.12c on page 87). Porins are highly selective of molecules that they permit to enter the cell.

The mycobacteria are another group of non-endospore-forming bacteria that exhibit greater than normal resistance to chemical biocides. This group includes *Mycobacterium tuberculosis,* the pathogen that causes tuberculosis. The cell wall of this organism and other members of this genus have a waxy, lipid-rich component. Instruction labels on disinfectants often state whether they are tuberculocidal, indicating if they are effective against mycobacteria. Special tuberculocidal tests have been developed to evaluate the effectiveness of biocides against this bacterial group.

Bacteria endospores are affected by relatively few biocides. (The activity of the major chemical antimicrobial groups against mycobacteria and endospores is summarized in Table 7.7.) The cysts and oocysts of protozoa are also relatively resistant to chemical disinfection.

Viruses are not especially resistant to biocides, but a distinction must be made between those that possess a lipid-containing envelope, and those that do not. Antimicrobials that are lipid-soluble are more likely to be effective against enveloped viruses. If so, this will usually be indicated on the label by a statement that they are effective against lipophilic viruses. Non-enveloped viruses, with only a protein coat, are more resistant—fewer biocides are active against them.

In summary, it is important to remember that microbial control methods, especially biocides, are not uniformly effective against all microbes.

★ ★ ★

A summary of chemical agents used to control microbial growth is presented in Table 7.8.

The compounds discussed in this chapter are not generally useful in the treatment of diseases. Because antibiotics are used in chemotherapy, antibiotics and the pathogens against which they are active will be discussed together, in Chapter 20.

table 7.8	Chemical Agents Used to Control Microbial Growth		
Chemical Agent	Mechanism of Action	Preferred Use	Comment
Phenol and Phenolics			
1. Phenol	Disruption of plasma membrane, denaturation of enzymes.	Rarely used, except as a standard of comparison.	Seldom used as a disinfectant or antiseptic because of its irritating qualities and disagreeable odor.
2. Phenolics	Disruption of plasma membrane, denaturation of enzymes.	Environmental surfaces, instruments, skin surfaces, and mucous membranes.	Derivatives of phenol that are reactive even in the presence of organic material; O-phenylphenol is an example.
3. Bisphenols	Probably disruption of plasma membrane.	Disinfectant hand soaps and skin lotions.	Triclosan is an especially common example of a bisphenol. Broad spectrum, but most effective against gram-positives.
Biguanides (Chlorhexidine)	Disruption of plasma membrane.	Skin disinfection, especially for surgical scrubs.	Bactericidal to gram-positives and gram-negatives; nontoxic, persistent.
Halogens	Iodine inhibits protein function and is a strong oxidizing agent; chlorine forms the strong oxidizing agent hypochlorous acid, which alters cellular components.	Iodine is an effective antiseptic available as a tincture and an iodophor; chlorine gas is used to disinfect water; chlorine compounds are used to disinfect dairy equipment, eating utensils, household items, and glassware.	Iodine and chlorine may act alone or as components of inorganic and organic compounds.
Alcohols	Protein denaturation and lipid dissolution.	Thermometers and other instruments; in swabbing the skin with alcohol before an injection, most of the disinfecting action probably comes from a simple wiping away (degerming) of dirt and some microbes.	Bactericidal and fungicidal, but not effective against endospores or nonenveloped viruses; commonly used alcohols are ethanol and isopropanol.
Heavy Metals and Their Compounds	Denaturation of enzymes and other essential proteins.	Silver nitrate may be used to prevent gonorrheal opthalmia neonatorum; mercurochrome disinfects skin and mucous membranes; copper sulfate is an algicide.	Heavy metals such as silver and mercury are biocidal.
Surface-Active Agents			
1. Soaps and acid-anionic detergents	Mechanical removal of microbes through scrubbing.	Skin degerming and removal of debris.	Many antibacterial soaps contain antimicrobials.
2. Acid-anionic detergents	Not certain; may involve enzyme inactivation or disruption.	Sanitizers in dairy and food-processing industries.	Wide spectrum of activity; nontoxic, noncorrosive, fast-acting.
3. Cationic detergents (quaternary ammonium compounds)	Enzyme inhibition, protein denaturation, and disruption of plasma membranes.	Antiseptic for skin, instruments, utensils, rubber goods.	Bactericidal, bacteriostatic, fungicidal, and virucidal against enveloped viruses; examples of quats are Zephiran® and Cepacol®.

table 7.8 (continued)			
Chemical Agent	**Mechanism of Action**	**Preferred Use**	**Comment**
Organic Acids	Metabolic inhibition, mostly affecting molds; action not related to their acidity.	Sorbic acid and benzoic acid effective at low pH; parabens much used in cosmetics, shampoos; calcium propionate used in bread; all are mainly antifungals.	Widely used to control molds and some bacteria in foods and cosmetics.
Aldehydes	Protein denaturation.	Glutaraldehyde (Cidex™) is less irritating than formaldehyde and is used for disinfection of medical equipment.	Very effective antimicrobials.
Gaseous Sterilants	Protein denaturation.	Excellent sterilizing agent, especially for objects that would be damaged by heat.	Ethylene oxide is the most commonly used.
Peroxygens (Oxidizing Agents)	Oxidation.	Contaminated surfaces; some deep wounds, in which they are very effective against oxygen-sensitive anaerobes.	Ozone is widely used as a supplement for chlorination; hydrogen peroxide is a poor antiseptic but a good disinfectant. Peracetic acid is especially effective.

Study Outline

THE TERMINOLOGY OF MICROBIAL CONTROL (pp. 184–186)

1. The control of microbial growth can prevent infections and food spoilage.
2. Sterilization is the process of destroying all microbial life on an object.
3. Commercial sterilization is heat treatment of canned foods to destroy C. botulinum endospores.
4. Disinfection is the process of reducing or inhibiting microbial growth on a nonliving surface.
5. Antisepsis is the process of reducing or inhibiting microorganisms on living tissue.
6. The suffix -cide means to kill; the suffix -stat means to inhibit.
7. Sepsis is bacterial contamination.

THE RATE OF MICROBIAL DEATH (pp. 186–187)

1. Bacterial populations subjected to heat or antimicrobial chemicals usually die at a constant rate.
2. Such a death curve, when plotted logarithmically, shows this constant death rate as a straight line.
3. The time it takes to kill a microbial population is proportional to the number of microbes.
4. Microbial species and life cycle phases (e.g., endospores) have different susceptibilities to physical and chemical controls.

5. Organic matter may interfere with heat treatments and chemical control agents.
6. Longer exposure to lower heat can produce the same effect as shorter time at higher heat.

ACTIONS OF MICROBIAL CONTROL AGENTS (p. 187)

Alteration of Membrane Permeability (p. 187)

1. The susceptibility of the plasma membrane is due to its lipid and protein components.
2. Certain chemical control agents damage the plasma membrane by altering its permeability.

Damage to Proteins and Nucleic Acids (p. 187)

1. Some microbial control agents damage cellular proteins by breaking hydrogen bonds and covalent bonds.
2. Other agents interfere with DNA and RNA replication and protein synthesis.

PHYSICAL METHODS OF MICROBIAL CONTROL (pp. 187–193)

Heat (pp. 187–191)

1. Heat is frequently used to eliminate microorganisms.
2. Moist heat kills microbes by denaturing enzymes.

3. Thermal death point (TDP) is the lowest temperature at which all the microbes in a liquid culture will be killed in 10 minutes.

4. Thermal death time (TDT) is the length of time required to kill all bacteria in a liquid culture at a given temperature.

5. Decimal reduction time (DRT) is the length of time in which 90% of a bacterial population will be killed at a given temperature.

6. Boiling (100°C) kills many vegetative cells and viruses within 10 minutes.

7. Autoclaving (steam under pressure) is the most effective method of moist heat sterilization. The steam must directly contact the material to be sterilized.

8. In HTST pasteurization, a high temperature is used for a short time (72°C for 15 seconds) to destroy pathogens without altering the flavor of the food. Ultra-high-temperature (UHT) treatment (140°C for 3 seconds) is used to sterilize dairy products.

9. Methods of dry heat sterilization include direct flaming, incineration, and hot-air sterilization. Dry heat kills by oxidation.

10. Different methods that produce the same effect (reduction in microbial growth) are called equivalent treatments.

Filtration (p. 191)

1. Filtration is the passage of a liquid or gas through a filter with pores small enough to retain microbes.

2. Microbes can be removed from air by high-efficiency particulate air filters.

3. Membrane filters composed of nitrocellulose or cellulose acetate are commonly used to filter out bacteria, viruses, and even large proteins.

Low Temperatures (p. 191)

1. The effectiveness of low temperatures depends on the particular microorganism and the intensity of the application.

2. Most microorganisms do not reproduce at ordinary refrigerator temperatures (0–7°C).

3. Many microbes survive (but do not grow) at the subzero temperatures used to store foods.

Desiccation (p. 192)

1. In the absence of water, microorganisms cannot grow but can remain viable.

2. Viruses and endospores can resist desiccation.

Osmotic Pressure (p. 192)

1. Microorganisms in high concentrations of salts and sugars undergo plasmolysis.

2. Molds and yeasts are more capable than bacteria of growing in materials with low moisture or high osmotic pressure.

Radiation (pp. 192–193)

1. The effects of radiation depend on its wavelength, intensity, and duration.

2. Ionizing radiation (gamma rays, X rays, and high-energy electron beams) has a high degree of penetration and exerts its effect primarily by ionizing water and forming highly reactive hydroxyl radicals.

3. Ultraviolet (UV) radiation, a form of nonionizing radiation, has a low degree of penetration and causes cell damage by making thymine dimers in DNA that interfere with DNA replication; the most effective germicidal wavelength is 260 nm.

4. Microwaves can kill microbes indirectly as materials get hot.

CHEMICAL METHODS OF MICROBIAL CONTROL (pp. 193–203)

1. Chemical agents are used on living tissue (as antiseptics) and on inanimate objects (as disinfectants).

2. Few chemical agents achieve sterility.

Principles of Effective Disinfection (pp. 193–195)

1. Careful attention should be paid to the properties and concentration of the disinfectant to be used.

2. The presence of organic matter, degree of contact with microorganisms, and temperature should also be considered.

Evaluating a Disinfectant (p. 195)

1. In the use-dilution test, bacterial (*S. choleraesuis, S. aureus,* and *P. aeruginosa*) survival in the manufacturer's recommended dilution of a disinfectant is determined.

2. Viruses, endospore-forming bacteria, mycobacteria, and fungi can also be used in the use-dilution test.

3. In the disk-diffusion method, a disk of filter paper is soaked with a chemical and placed on an inoculated agar plate; a clear zone of inhibition indicates effectiveness.

Types of Disinfectants (pp. 196–203)

Phenol and Phenolics (p. 196)

1. Phenolics exert their action by injuring plasma membranes.

Bisphenols (pp. 196–197)

1. Bisphenols such as triclosan (over the counter) and hexachlorophene (prescription) are widely used in household products.

Biguanides (p. 197)

1. Chlorhexidine damages plasma membranes of vegetative cells.

Halogens (pp. 197–198)

1. Some halogens (iodine and chlorine) are used alone or as components of inorganic or organic solutions.

2. Iodine may combine with certain amino acids to inactivate enzymes and other cellular proteins.

3. Iodine is available as a tincture (in solution with alcohol) or as an iodophor (combined with an organic molecule).

4. The germicidal action of chlorine is based on the formation of hypochlorous acid when chlorine is added to water.

5. Chlorine is used as a disinfectant in gaseous form (Cl₂) or in the form of a compound, such as calcium hypochlorite, sodium hypochlorite, sodium dichloroisocyanurate, and chloramines.

Alcohols (p. 198)

1. Alcohols exert their action by denaturing proteins and dissolving lipids.

2. In tinctures, they enhance the effectiveness of other antimicrobial chemicals.

3. Aqueous ethanol (60–95%) and isopropanol are used as disinfectants.

Heavy Metals and Their Compounds (pp. 198–199)

1. Silver, mercury, copper, and zinc are used as germicidals.

2. They exert their antimicrobial action through oligodynamic action. When heavy metal ions combine with sulfhydryl (—SH) groups, proteins are denatured.

Surface-Active Agents (p. 199)

1. Surface-active agents decrease the surface tension among molecules of a liquid; soaps and detergents are examples.

2. Soaps have limited germicidal action but assist in the removal of microorganisms through scrubbing.

3. Acid-anionic detergents are used to clean dairy equipment.

Quaternary Ammonium Compounds (Quats) (pp. 199–200)

1. Quats are cationic detergents attached to NH_4^+.

2. By disrupting plasma membranes, they allow cytoplasmic constituents to leak out of the cell.

3. Quats are most effective against gram-positive bacteria.

Chemical Food Preservatives (p. 200)

1. SO₂, sorbic acid, benzoic acid, and propionic acid inhibit fungal metabolism and are used as food preservatives.

2. Nitrate and nitrite salts prevent germination of *Clostridium botulinum* endospores in meats.

Antibiotics (p. 200)

1. Nisin and natamycin are antibiotics used to preserve foods, especially cheese.

Aldehydes (p. 201)

1. Aldehydes such as formaldehyde and glutaraldehyde exert their antimicrobial effect by inactivating proteins.

2. They are among the most effective chemical disinfectants.

Gaseous Chemosterilizers (p. 202)

1. Ethylene oxide is the gas most frequently used for sterilization.

2. It penetrates most materials and kills all microorganisms by protein denaturation.

Peroxygens (Oxidizing Agents) (pp. 202–203)

1. Ozone, peroxide, and peracetic acid are used as antimicrobial agents.

2. They exert their effect by oxidizing molecules inside cells.

MICROBIAL CHARACTERISTICS AND MICROBIAL CONTROL (pp. 203–205)

1. Gram-negative bacteria are generally more resistant than gram-positive bacteria to disinfectants and antiseptics.

2. Mycobacteria, endospores, and protozoan cysts and oocysts are very resistant to disinfectants and antiseptics.

3. Nonenveloped viruses are generally more resistant than enveloped viruses to disinfectants and antiseptics.

Study Questions

REVIEW

1. Name the cause of cell death resulting from damage to each of the following:
  a. cell wall   **c.** proteins
  b. plasma membrane   **d.** nucleic acids

2. The thermal death time for a suspension of *B. subtilis* endospores is 30 minutes in dry heat and less than 10 minutes in an autoclave. Which type of heat is more effective? Why?

3. If pasteurization does not achieve sterilization, why is food treated by pasteurization?

4. Thermal death point is not considered an accurate measure of the effectiveness of heat sterilization. List three factors that can alter thermal death point.

5. The antimicrobial effect of gamma radiation is due to _____. The antimicrobial effect of ultraviolet radiation is due to _____.

6. A bacterial culture was in log phase in the following figure. At time *x*, an antibacterial compound was added to the culture. Which line indicates addition of a bactericidal compound? A bacteriostatic compound? How can you tell? Explain why the viable count does not immediately drop to zero at *x*.

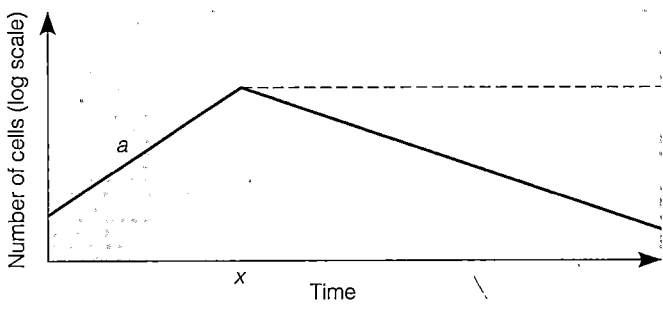

7. Fill in the following table:

Method of Sterilization	Temp.	Time	Type of Heat	Pref'd Use	Type of Action
Autoclaving					
Hot Air					
Pasteurization					

8. How do the examples in question 7 illustrate the concept of equivalent treatments?

9. How do salts and sugars preserve foods? Why are these considered physical rather than chemical methods of microbial control? Name one food that is preserved with sugar and one preserved with salt. How do you account for the occasional growth of *Penicillium* mold in jelly, which is 50% sucrose?

10. List five factors to consider before selecting a disinfectant.

11. Give the method of action and at least one standard use of each of the following types of disinfectants:
   a. phenolics
   b. iodine
   c. chlorine
   d. alcohol
   e. heavy metals
   f. aldehydes
   g. ethylene oxide
   h. oxidizing agents

12. The use–dilution values for two disinfectants tested under the same conditions are: Disinfectant A—1:2; Disinfectant B—1:10,000. If both disinfectants are designed for the same purpose, which would you select?

13. A large hospital washes burn patients in a stainless steel tub. After each patient, the tub is cleaned with a quat. It was noticed that 14 of 20 burn patients acquired *Pseudomonas* infections after being bathed. Provide an explanation for this high rate of infection.

MULTIPLE CHOICE

1. Which of the following does *not* kill endospores?
   a. autoclaving
   b. incineration
   c. hot-air sterilization
   d. pasteurization
   e. none of the above

2. Which of the following is most effective for sterilizing mattresses and plastic Petri dishes?
   a. chlorine
   b. ethylene oxide
   c. glutaraldehyde
   d. autoclaving
   e. nonionizing radiation

3. Which of these disinfectants does *not* act by disrupting the plasma membrane?
   a. phenolics
   b. phenol
   c. quaternary ammonium compounds
   d. halogens
   e. biguanides

4. Which of the following *cannot* be used to sterilize a heat-labile solution stored in a plastic container?
   a. gamma radiation
   b. ethylene oxide
   c. nonionizing radiation
   d. autoclaving
   e. short-wavelength radiation

5. Which of the following is *not* a characteristic of quaternary ammonium compounds?
   a. bactericidal against gram-positive bacteria
   b. sporicidal
   c. amoebicidal
   d. fungicidal
   e. kills enveloped viruses

6. A classmate is trying to determine how a disinfectant might kill cells. You observed that when he spilled the disinfectant in your reduced litmus milk, the litmus turned blue again. You suggest to your classmate that
   a. the disinfectant might inhibit cell wall synthesis.
   b. the disinfectant might oxidize molecules.
   c. the disinfectant might inhibit protein synthesis.
   d. the disinfectant might denature proteins.
   e. he take his work away from yours.

7. Which of the following is most likely to be bactericidal?
   a. membrane filtration
   b. ionizing radiation
   c. freeze-drying
   d. deep-freezing
   e. all of the above

8. Which of the following is used to control microbial growth in foods?
   a. organic acids
   b. alcohols
   c. aldehydes
   d. heavy metals
   e. all of the above

Use the following information to answer questions 9 and 10. The data were obtained from a use-dilution test comparing four disinfectants against *Salmonella choleraesuis*.

	Bacterial Growth After Exposure to			
Dilution	Disinfectant A	Disinfectant B	Disinfectant C	Disinfectant D
1:2	−	+	−	−
1:4	−	+	−	+
1:8	−	+	+	+
1:16	+	+	+	+

9. Which disinfectant is the most effective?

10. Which disinfectant(s) is (are) bactericidal?
   a. A, B, C, and D
   b. A, C, and D
   c. A only
   d. B only
   e. none of the above

CRITICAL THINKING

1. The disk–diffusion method was used to evaluate three disinfectants. The results were as follows:

Disinfectant	Zone of Inhibition
X	0 mm
Y	5 mm
Z	10 mm

   a. Which disinfectant was the most effective against the organism?
   b. Can you determine whether compound Y was bactericidal or bacteriostatic?

2. Why is each of the following bacteria often resistant to disinfectants?
   a. *Mycobacterium*
   b. *Pseudomonas*
   c. *Bacillus*

3. A use-dilution test was used to evaluate two disinfectants against *Salmonella choleraesuis*. The results were as follows:

Time of Exposure (min)	Bacterial Growth After Exposures		
	Disinfectant A	Disinfectant B Diluted with Distilled Water	Disinfectant B Diluted with Tap Water
10	+	−	+
20	+	−	−
30	−	−	−

   a. Which disinfectant was the most effective?
   b. Which disinfectant should be used against *Staphylococcus*?

4. To determine the lethal action of microwave radiation, two 10^5 suspensions of *E. coli* were prepared. One cell suspension was exposed to microwave radiation while wet, whereas the other was lyophilized (freeze–dried) and then exposed to radiation. The results are shown in the following figure. Dashed lines indicate the temperature of the samples. What is the most likely method of lethal action of microwave radiation? How do you suppose these data might differ for *Clostridium*?

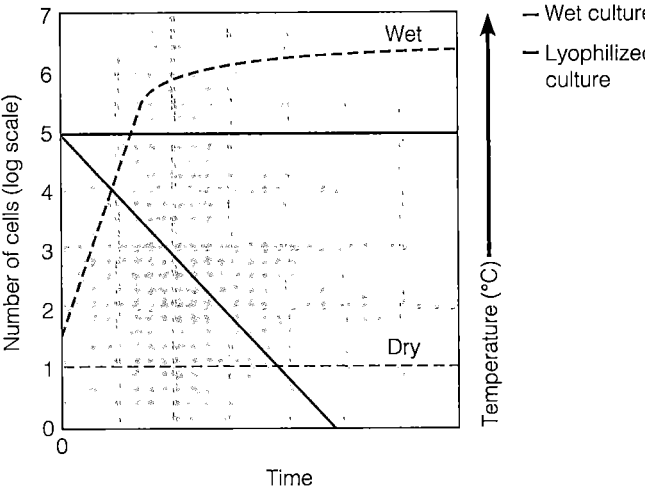

CLINICAL APPLICATIONS

1. *Entamoeba histolytica* and *Giardia lamblia* were isolated from the stool sample of a 45-year-old man, and *Shigella sonnei* was isolated from the stool sample of an 18-year-old woman. Both patients experienced diarrhea and severe abdominal cramps, and prior to onset of digestive symptoms, both had been treated by the same chiropractor. The chiropractor had administered colonic irrigations (enemas) to these patients.

The device used for this treatment was a gravity-dependent apparatus using 12 liters of tap water. There were no check valves to prevent backflow, so all parts of the apparatus could have become contaminated with feces during each colonic treatment. The chiropractor provided colonic treatment to four or five patients per day. Between patients, the adaptor piece that is inserted into the rectum was placed in a "hot-water sterilizer."

What two errors were made by the chiropractor?

2. Between March 9 and April 12, five chronic peritoneal dialysis patients at one hospital became infected with *Pseudomonas aeruginosa*. Four patients developed peritonitis (inflammation of the abdominal cavity), and one developed a skin infection at the catheter insertion site. All patients with peritonitis had low-grade fever, cloudy peritoneal fluid, and abdominal pain. All patients had permanent indwelling peritoneal catheters, which the nurse wiped with gauze soaked with an iodophor solution each time the catheter was connected to or disconnected from the machine tubing. Aliquots of the iodophor were transferred from stock bottles to small in-use bottles. Cultures from the dialysate concentrate and the internal areas of the dialysis machines were negative; iodophor from a small in-use plastic container yielded a pure culture of *P. aeruginosa*.

What improper technique led to this infection?

3. Eleven patients received injections of methylprednisolone and lidocaine to relieve the pain and inflammation of arthritis at the same orthopedic surgery office. All of them developed septic arthritis caused by *Serratia marcescens*. Unopened bottles of methylprednisolone from the same lot numbers tested sterile; the methylprednisolone was preserved with a quat. Cotton balls were used to wipe multiple-use injection vials before the medication was drawn into a disposable syringe. The site of injection on each patient was also wiped with a cotton ball. The cotton balls were soaked in benzalkonium chloride, and fresh cotton balls were added as the jar was emptied. Opened methylprednisolone containers and the jar of cotton balls contained *S. marcescens*.

How was the infection transmitted? What part of the routine procedure caused the contamination?

Learning with Technology

MP = The Microbiology Place website   **ST** = Student Tutorial CD-ROM

MP   Don't forget to go to The Microbiology Place website (http://www.microbiologyplace.com) to take the practice tests, explore the interactive activity and case study, and check out the news articles and web links for this chapter.

ST   Remember there is also a quiz for this chapter on the Microbiology Interactive Student Tutorial CD-ROM.

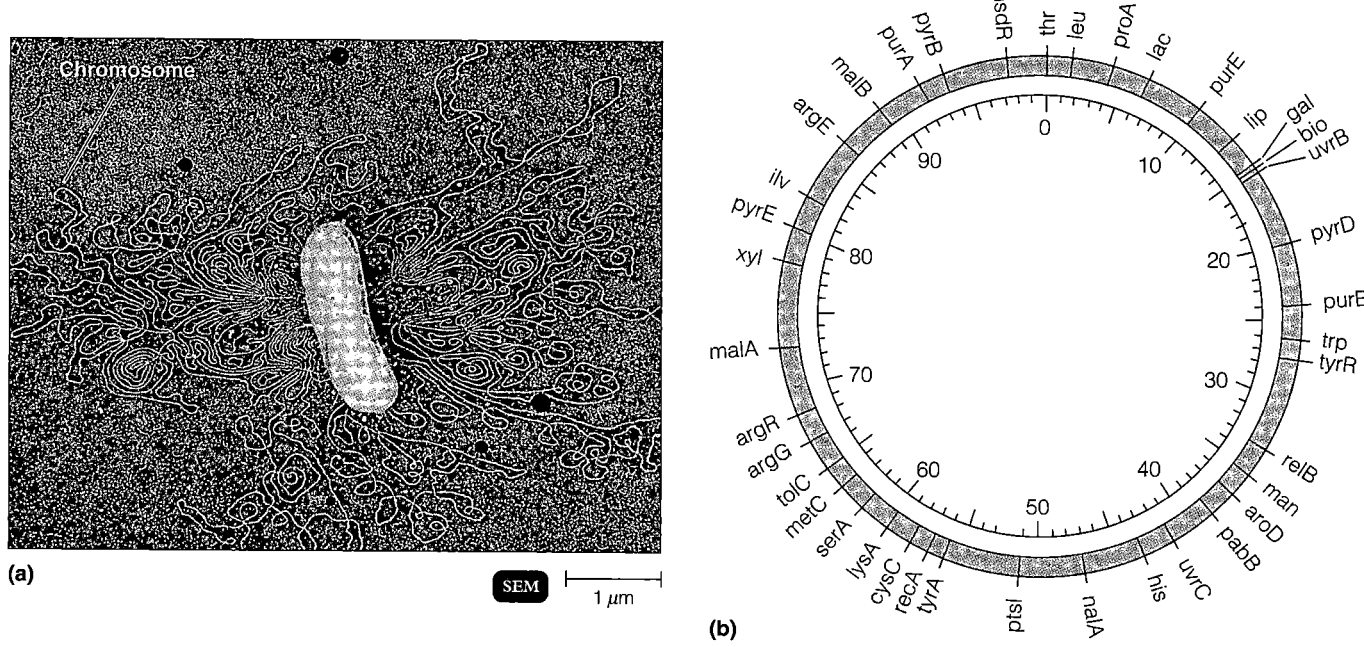

FIGURE 8.1 Chromosomes. (a) A prokaryotic chromosome. The tangled mass and looping strands of DNA emerging from this disrupted *E. coli* cell are part of its single chromosome. (b) A genetic map of the chromosome of *E. coli*. The units are minutes based on the length of time it takes to transfer the genes during mating between two cells.

DNA and Chromosomes

Bacteria typically have a single circular chromosome consisting of a single circular molecule of DNA with associated proteins. The chromosome is looped and folded (Figure 8.1a) and attached at one or several points to the plasma membrane. The DNA of *E. coli*, the most-studied bacterial species, has about 4 million base pairs and is about 1 mm long—1000 times longer than the entire cell. However, because DNA is very thin and tightly packed inside the cell, this twisted, coiled macromolecule takes up only about 10% of the cell's volume.

The location of genes on a bacterial chromosome is determined by experiments on the transfer of genes from one cell to another. These processes will be discussed later in this chapter. The bacterial chromosome map that results is marked in minutes corresponding to when the genes are transferred from a donor cell to a recipient cell (Figure 8.1b).

In recent years, the complete base sequences of several bacterial chromosomes have been determined. The sequencing and molecular characterization of genomes is called **genomics.** The most important genetic questions being addressed concern how cells use the genetic information available to them.

The Flow of Genetic Information

DNA replication makes possible the flow of genetic information from one generation to the next. As shown in Figure 8.2, the DNA of a cell replicates before cell division, so that each daughter cell receives a chromosome identical to the parent's. Within each metabolizing cell, the genetic information contained in DNA also flows in another way: It is transcribed into mRNA and then translated into protein. We describe the processes of transcription and translation later in this chapter.

DNA Replication

Learning Objective

- Describe the process of DNA replication.

In DNA replication, one "parental" double-stranded DNA molecule is converted to two identical "daughter" molecules. The complementary structure of the nitrogenous base sequences in the DNA molecule is the key to understanding DNA replication. Because the bases along the two strands of double-helical DNA are complementary, one strand can act as a template for the production of the other strand (Figure 8.3 on page 214).

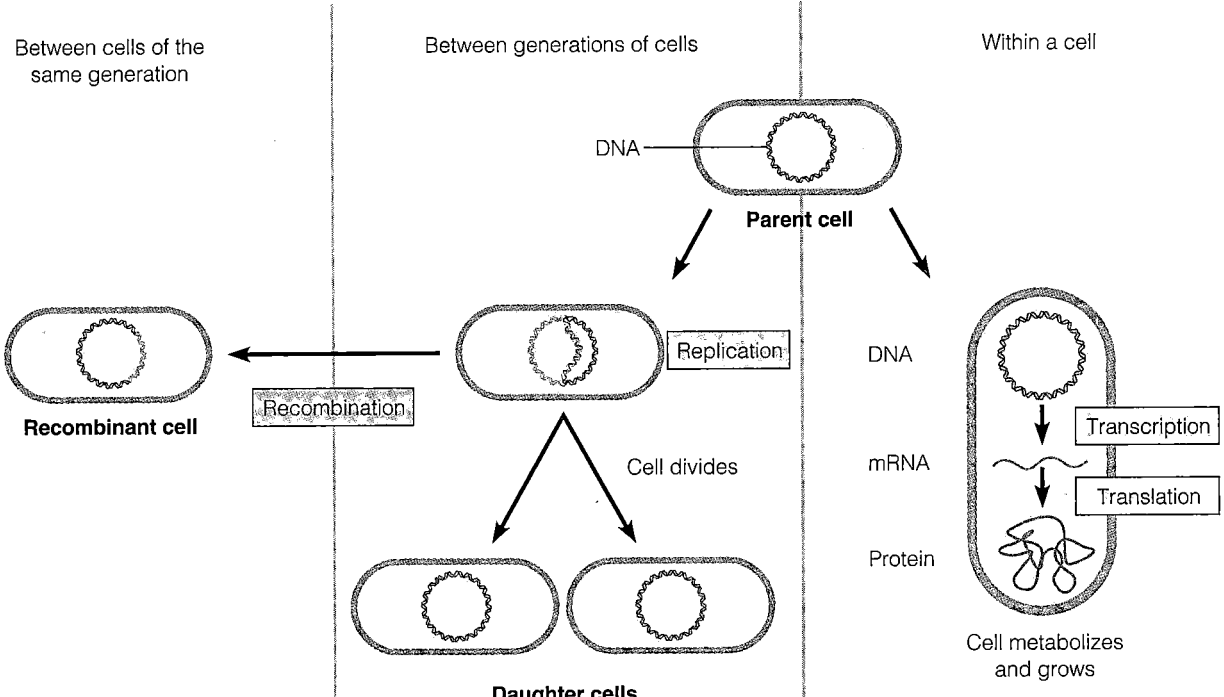

Between cells of the
same generation

Between generations of cells

Within a cell

DNA

Parent cell

Replication

DNA

Recombination

Recombinant cell

Cell divides

mRNA

Transcription

Translation

Protein

Daughter cells

Cell metabolizes
and grows

FIGURE 8.2 An overview of the flow of genetic information. Genetic information
can be transferred between generations of cells, through DNA replication. Occasionally, genetic information can be transferred between cells of the same generation,
through recombination. Genetic information is also used within a cell to produce the
proteins the cell needs to function, through transcription and translation. The cell
represented here is a bacterium with a single circular chromosome. A small version of
this figure will be included in figures throughout this chapter to indicate the relationships of different processes.

■ **All of these processes can occur at the same time in a bacterial cell.
Which process results in reproduction?**

DNA replication requires the presence of complex cellular proteins that direct a particular sequence of events. When replication begins, the two strands of parental DNA are unwound and separated from each other in one small DNA segment after another. Free nucleotides present in the cytoplasm of the cell are matched up to the exposed bases of the single-stranded parental DNA. Where thymine is present on the original strand, only adenine can fit into place on the new strand; where guanine is present on the original strand, only cytosine can fit into place, and so on. Any bases that are improperly base-paired are removed and replaced by replication enzymes. Once aligned, the newly added nucleotide is joined to the growing DNA strand by an enzyme called **DNA polymerase.** Then the parental DNA is unwound a bit further to allow the addition of the next nucleotides. The point at which replication occurs is called the *replication fork.*

As the replication fork moves along the parental DNA, each of the unwound single strands combines with new nucleotides. The original strand and this newly synthesized daughter strand then rewind. Because each new double-stranded DNA molecule contains one original, conserved strand and one new strand, the process of replication is referred to as **semiconservative replication.**

Before looking at DNA replication in more detail, let's take a closer look at the structure of DNA (see Figure 2.17 on page 50). It is important to understand the concept that the paired DNA strands are oriented in opposite directions relative to each other. Notice in Figure 2.17 that the carbon atoms of the sugar component of each nucleotide are numbered 1′ (pronounced "one prime") to 5′. In order for the paired bases to be next to each other, the sugar components in one strand are upside-down relative to the other. The end with the hydroxyl attached to the 3′ carbon is called the 3′ end of the DNA strand; the end having a phosphate attached to the 5′ carbon is called the 5′ end. The way in which the two strands fit together dictates that the 5′ → 3′ direction of one strand runs

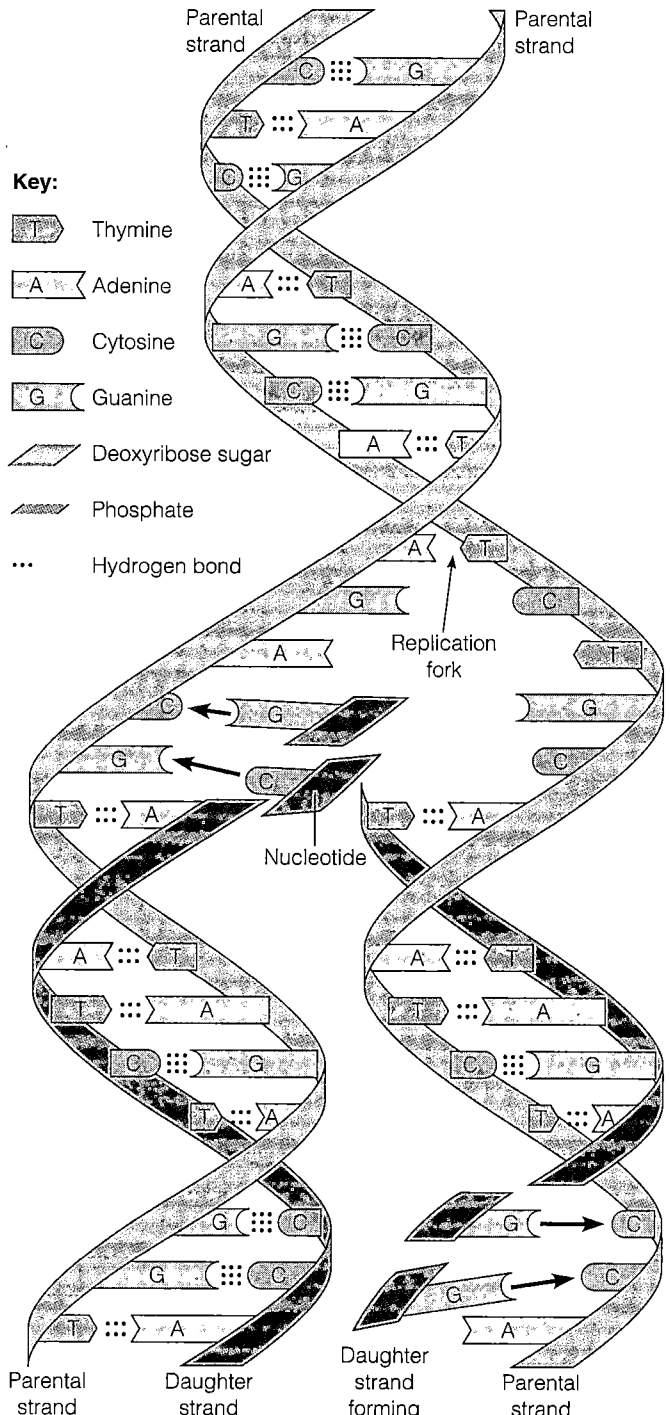

Key:

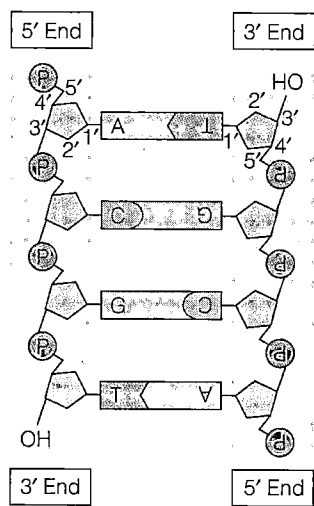

FIGURE 8.4 The two strands of DNA are antiparallel. The sugar phosphate backbone of one strand is upside-down relative to the backbone of the other strand. Turn the book upside-down to demonstrate this.

FIGURE 8.3 DNA replication. The double helix of the parental DNA separates as weak hydrogen bonds between the nucleotides on opposite strands break in response to the action of replication enzymes. Next, hydrogen bonds form between new complementary nucleotides and each strand of the parental template to form new base pairs. Enzymes catalyze the formation of sugar-phosphate bonds between sequential nucleotides on each resulting daughter strand.

■ What is meant by semiconservative replication?

counter to the 5' → 3' direction of the other strand (Figure 8.4). This structure of DNA affects the replication process because DNA polymerases can add new nucleotides to the 3' end only. Therefore, as the replication fork moves along the parental DNA, the two new strands must grow in different directions.

DNA replication requires a great deal of energy. The energy is supplied from the nucleotides, which are actually nucleoside triphosphates. You already know about ATP; the only difference between ATP and the adenine nucleotide in DNA is the sugar component. Deoxyribose is the sugar in the nucleosides used to synthesize DNA, and nucleoside triphosphates with ribose are used to synthesize RNA. Two phosphate groups are removed to add nucleotide to a growing strand of DNA; hydrolysis of the nucleoside is exergonic and provides energy to make the new bonds in the DNA strand (Figure 8.5).

Figure 8.6 on page 216 provides more detail about the many steps that go into this complex process. ①–② Once the parental DNA is unwound and stabilized, the replication fork forms at a fixed site called the origin of replication. ③ One new DNA strand, called the **leading strand,** is synthesized continuously as the DNA polymerase moves toward the replication fork making DNA in the 5' → 3' direction. ④ Remember that DNA polymerase can only add new nucleotides to the 3' end, so a short piece of RNA called an **RNA primer** is needed to start synthesis. DNA polymerase can then add nucleotides

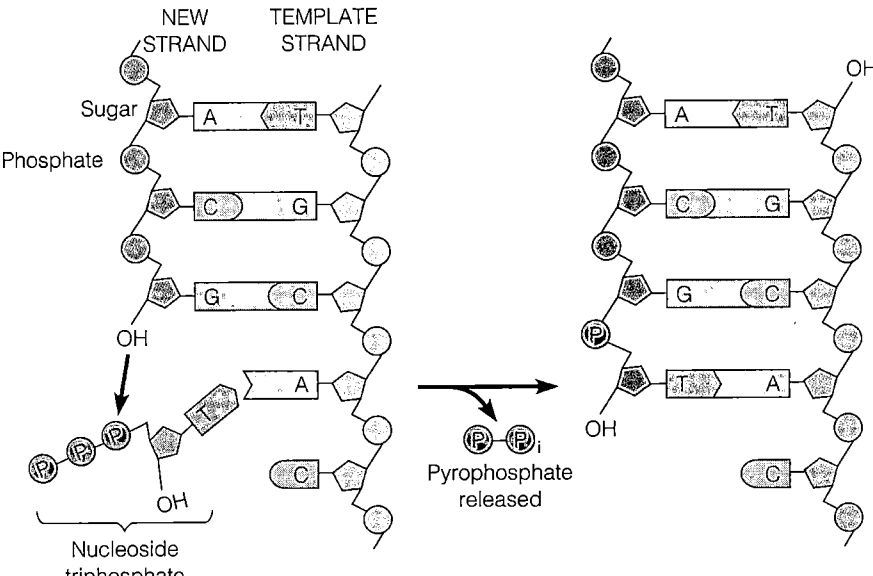

FIGURE 8.5 Adding a nucleotide to DNA. When a nucleoside triphosphate bonds to the sugar in a growing DNA strand, it loses two phosphates. Hydrolysis of the phosphate bonds provides the energy for the reaction.

to the 3' end of the RNA. Consequently, the **lagging strand** of new DNA is synthesized in pieces consisting of about 1000 nucleotides as the DNA polymerase moves away from the replication fork. ⑤ DNA polymerase removes the RNA primer, and ⑥ the enzyme **DNA ligase** joins the newly made DNA fragments.

DNA replication of some bacteria, such as *E. coli,* goes *bidirectionally* around the chromosome (Figure 8.7). Two replication forks move in opposite directions away from the origin of replication. Because the bacterial chromosome is a closed loop, the replication forks eventually meet when replication is completed. Much evidence shows an association between the bacterial plasma membrane and the origin of replication. After duplication, if each copy of the origin binds to the membrane at opposite poles, then each daughter cell would receive one copy of the DNA molecule—that is, one complete chromosome.

DNA replication is an amazingly accurate process. Typically, mistakes are made at a rate of only 1 in every 10^{10} bases incorporated. Such accuracy is largely due to the *proofreading* capability of DNA polymerase. As each new base is added, the enzyme evaluates whether it forms the proper complementary base-pairing structure. If not, the enzyme excises the improper base and replaces it with the correct one. In this way, DNA replication can be performed very accurately, allowing each daughter chromosome to be virtually identical to the parental DNA.

RNA and Protein Synthesis

Learning Objective

■ *Describe protein synthesis, including transcription, RNA processing, and translation.*

How is the information in DNA used to make the proteins that control cell activities? In the process of *transcription,* genetic information in DNA is copied, or transcribed, into a complementary base sequence of RNA. The cell then uses the information encoded in this RNA to synthesize specific proteins through the process of *translation.* We now take a closer look at these two processes as they occur in a bacterial cell.

Transcription

Transcription is the synthesis of a complementary strand of RNA from a DNA template. As mentioned earlier, there are three kinds of RNA in bacterial cells: messenger RNA, ribosomal RNA, and transfer RNA. Ribosomal RNA forms an integral part of ribosomes, the cellular machinery for protein synthesis. Transfer RNA is also involved in protein synthesis, as we will see. **Messenger RNA (mRNA)** carries the coded information for making specific proteins from DNA to ribosomes, where proteins are synthesized.

During transcription, a strand of mRNA is synthesized using a specific gene—a portion of the cell's

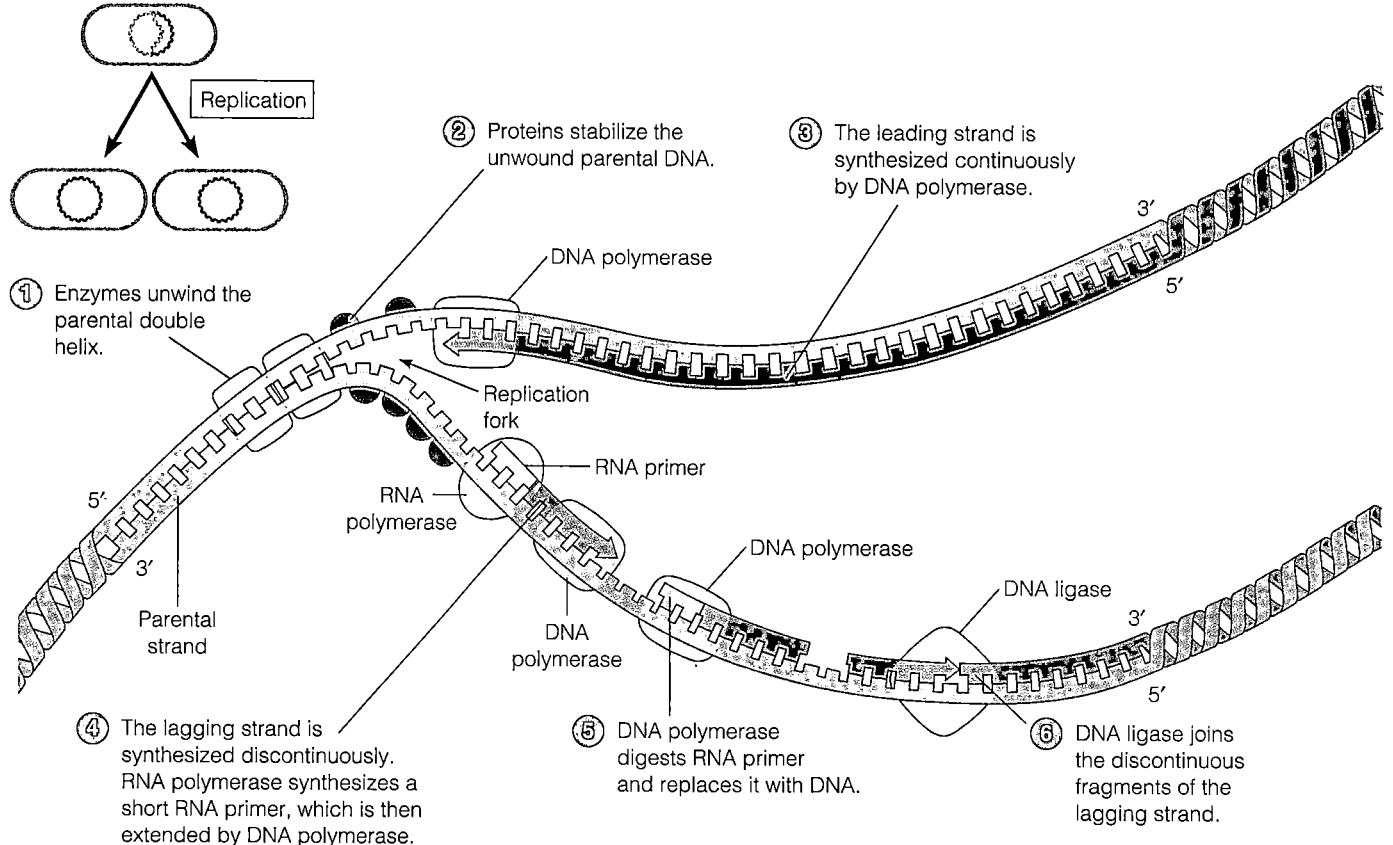

① Enzymes unwind the
parental double
helix.

② Proteins stabilize the
unwound parental DNA.

③ The leading strand is
synthesized continuously
by DNA polymerase.

DNA polymerase

Replication
fork

RNA primer

RNA
polymerase

DNA polymerase

DNA ligase

Parental
strand

DNA
polymerase

④ The lagging strand is
synthesized discontinuously.
RNA polymerase synthesizes a
short RNA primer, which is then
extended by DNA polymerase.

⑤ DNA polymerase
digests RNA primer
and replaces it with DNA.

⑥ DNA ligase joins
the discontinuous
fragments of the
lagging strand.

FIGURE 8.6 A summary of events at the DNA replication fork. Enzymes at the replication fork unwind the parental double helix. DNA polymerase synthesizes a continuous strand of new DNA, using one of the parental strands as a template. DNA polymerase also uses the other parental strand as a template, but because the orientation of the sugars is opposite, RNA polymerase starts the synthesis by adding a short stretch of RNA called an RNA primer. The DNA polymerase digests away the RNA as it makes a small piece of DNA. The small units are subsequently joined by DNA ligase.

■ Why is one strand of DNA synthesized discontinuously?

DNA—as a template. In other words, the genetic information stored in the sequence of nitrogenous bases of DNA is rewritten so that the same information appears in the base sequence of mRNA. As in DNA replication, a G in the DNA template dictates a C in the mRNA being made, a C in the DNA template dictates a G in the mRNA, and a T in the DNA template dictates an A in the mRNA. However, an A in the DNA template dictates a uracil (U) in the mRNA because RNA contains U instead of T. (U has a chemical structure slightly different from T, but it base-pairs in the same way.) If, for example, the template portion of DNA has the base sequence ATGCAT, the newly synthesized mRNA strand will have the complementary base sequence UACGUA.

The process of transcription requires both an enzyme called *RNA polymerase* and a supply of RNA nucleotides (Figure 8.8). Transcription begins when ① RNA poly-

merase binds to the DNA at a site called the **promoter.** Only one of the two DNA strands serves as the template for RNA synthesis for a given gene. Like DNA, RNA is synthesized in the 5′ → 3′ direction. ② RNA polymerase assembles free nucleotides into a new chain, using complementary base pairing as a guide. ③ As the new RNA chain grows, RNA polymerase moves along the DNA. ④ RNA synthesis continues until RNA polymerase reaches a site on the DNA called the **terminator.** ⑤ When this happens, RNA polymerase and the newly formed, single-stranded mRNA are released from the DNA.

The process of transcription allows the cell to produce short-term copies of genes that can be used as the direct source of information for protein synthesis. Messenger RNA acts as an intermediate between the permanent storage form, DNA, and the process that uses the information, translation.

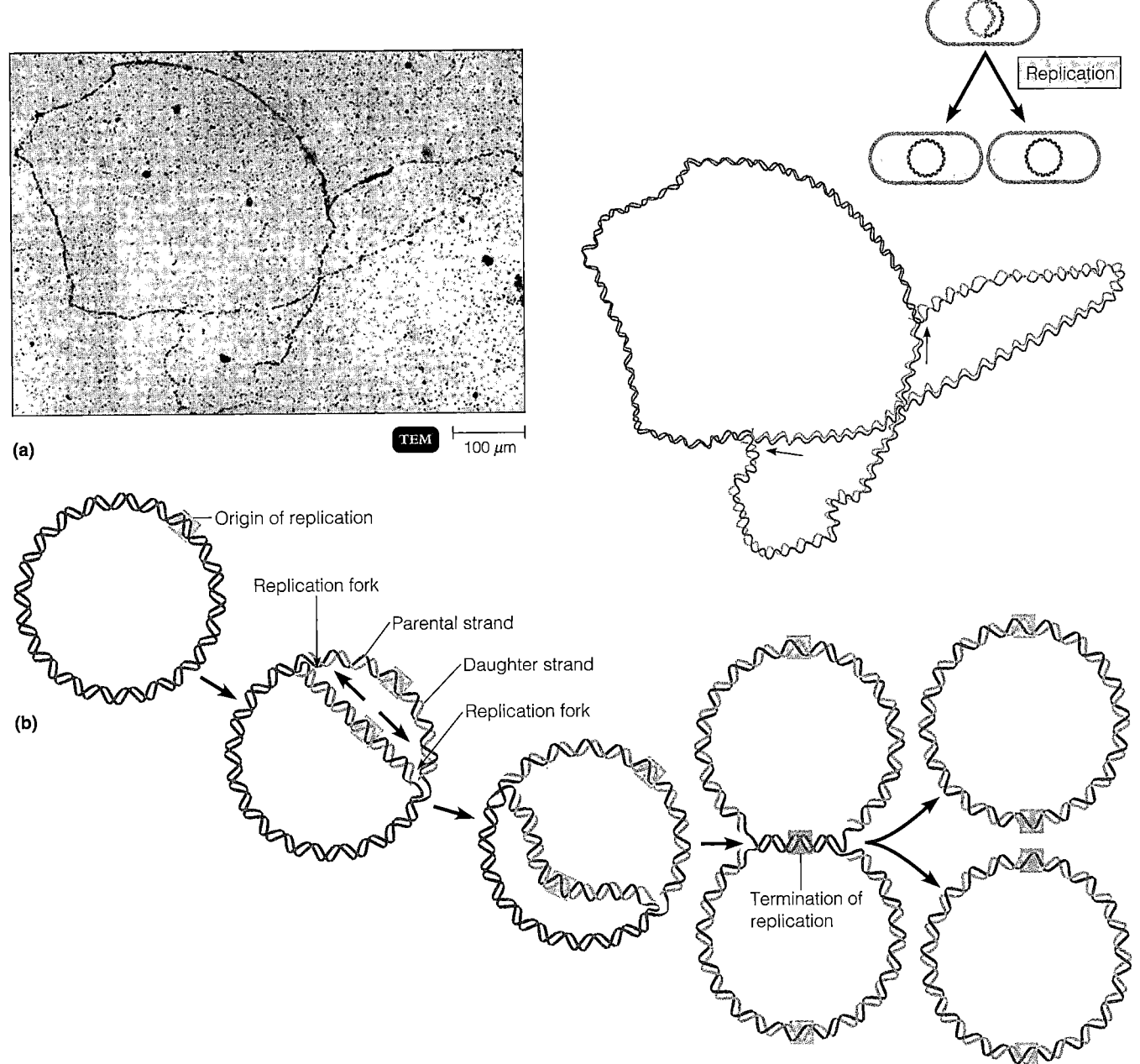

(a)

(b)

Origin of replication

Replication fork

Parental strand

Daughter strand

Replication fork

Termination of replication

Replication

TEM | 100 μm

FIGURE 8.7 Replication of bacterial DNA. (a) An *E. coli* chromosome in the process of replicating. (In the corresponding diagram at right, the arrows point to the two replication forks.) The chromosome is about one-third replicated. Notice that one of the new helixes is crossed over the other one. **(b)** A diagram of the bidirectional replication of a circular bacterial DNA molecule. The new strands are shown in blue.

■ What is the origin of replication?

Translation

We have seen how the genetic information in DNA is transferred to mRNA during transcription. Now we will see how mRNA serves as the source of information for the synthesis of proteins. Protein synthesis is called **translation** because it involves decoding the "language" of nucleic acids and converting that information into the "language" of proteins.

The language of mRNA is in the form of **codons,** groups of three nucleotides, such as AUG, GGC, or AAA. The sequence of codons on an mRNA molecule determines the sequence of amino acids that will be in the

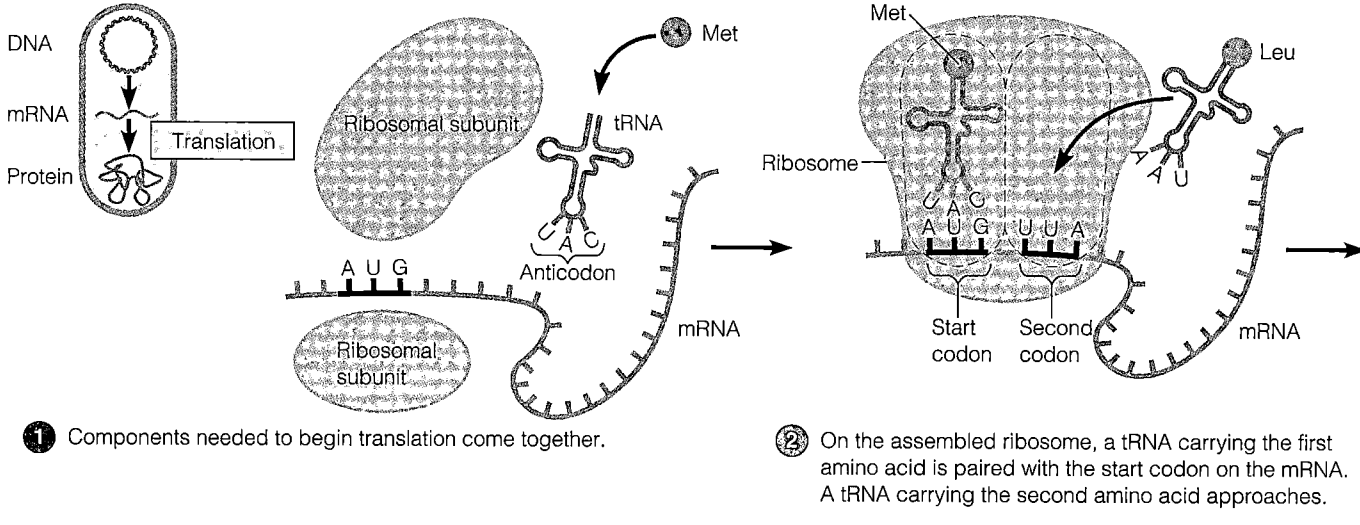

① Components needed to begin translation come together.

② On the assembled ribosome, a tRNA carrying the first amino acid is paired with the start codon on the mRNA. A tRNA carrying the second amino acid approaches.

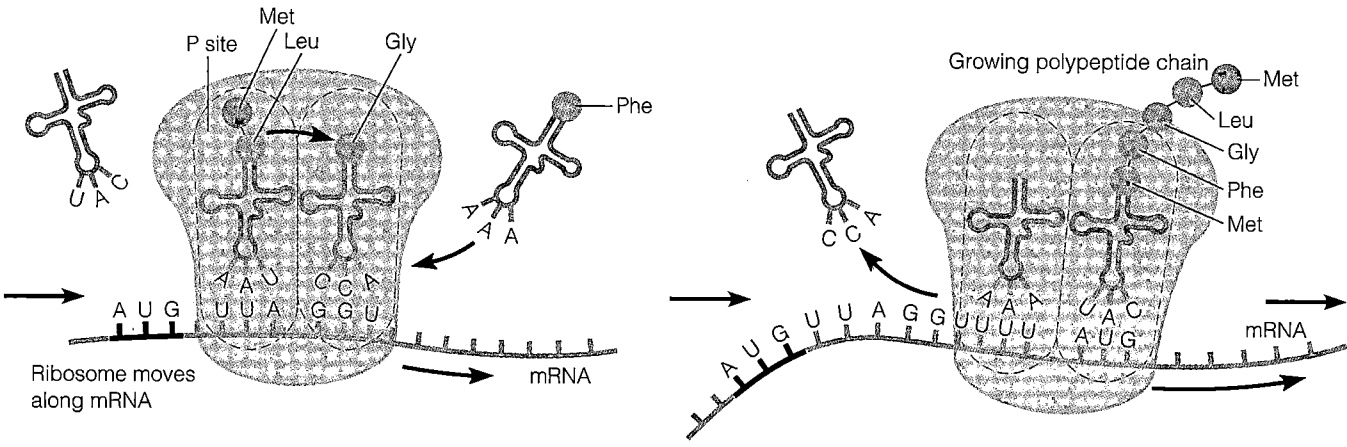

⑤ The ribosome moves along the mRNA until the second tRNA is in the P site, and the process continues.

⑥ The ribosome continues to move along the mRNA, and new amino acids are added to the polypeptide.

FIGURE 8.10 The process of translation. The overall goal of translation is to produce proteins using mRNAs as the source of biological information. The complex cycle of events illustrated here shows the primary role of tRNA and ribosomes in the decoding of this information. The ribosome acts as the site where the mRNA-encoded information is decoded, as well as the site where individual amino acids are connected into polypeptide chains. The tRNA molecules act as the actual "translators"—one end of each tRNA recognizes a specific mRNA codon, while the other end carries the amino acid coded for by that codon.

■ Translation is the process by which the base sequence of mRNA determines the amino acid sequence of a protein.

★ ★ ★

To summarize, genes are the units of biological information encoded by the sequence of nucleotide bases in DNA. A gene is expressed, or turned into a product within the cell, through the processes of transcription and translation. The genetic information carried in DNA is transferred to a temporary mRNA molecule by transcription. Then, during translation, the mRNA directs the assembly of amino acids into a polypeptide chain: mRNA attaches to a ribosome, tRNAs deliver the amino acids to the ribosome as directed by the mRNA codon

sequence, and the ribosome assembles the amino acids into the chain that will be the newly synthesized protein.

The Regulation of Bacterial Gene Expression

Learning Objective

■ *Explain the regulation of gene expression in bacteria by induction, repression, and catabolic repression.*

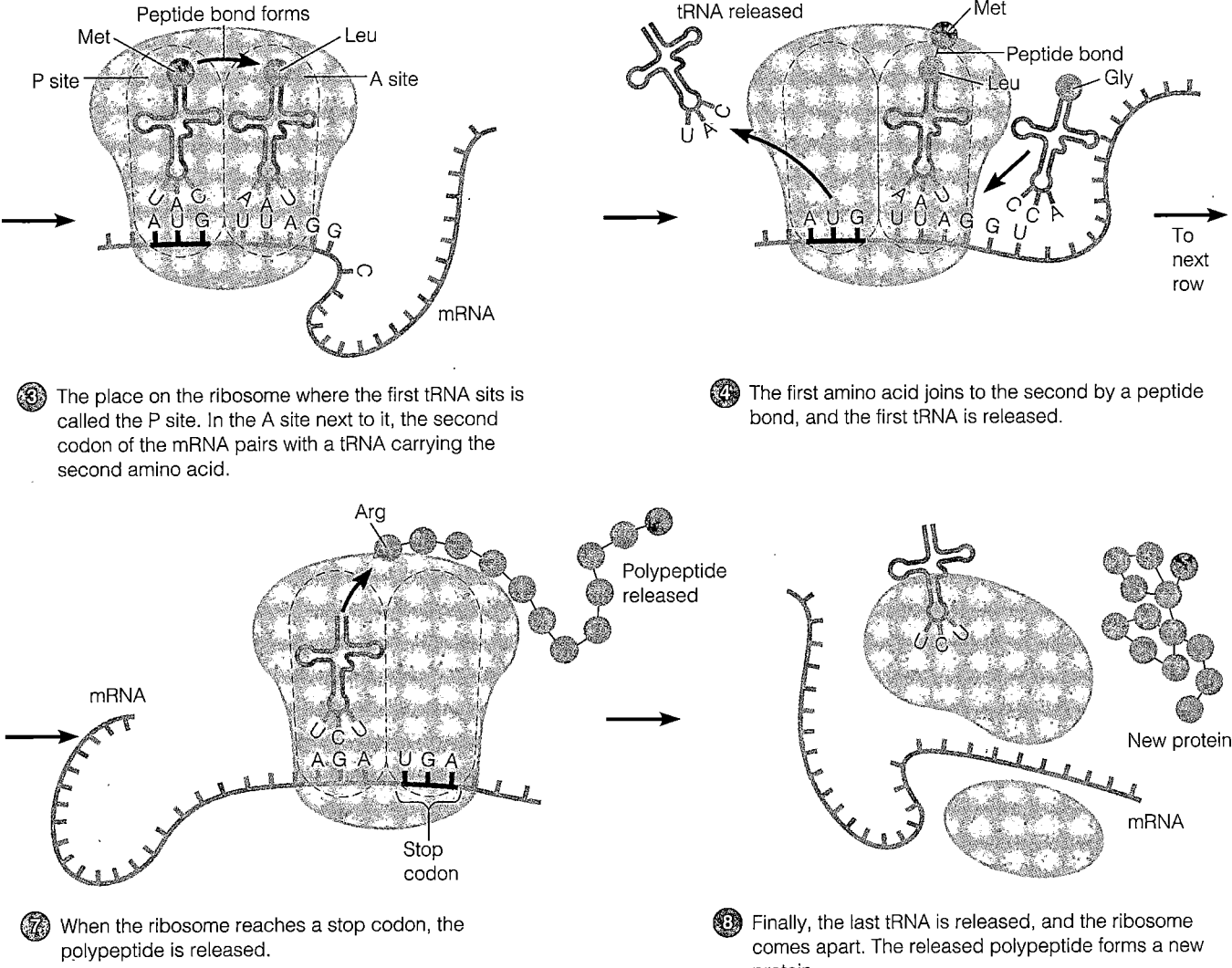

3 The place on the ribosome where the first tRNA sits is called the P site. In the A site next to it, the second codon of the mRNA pairs with a tRNA carrying the second amino acid.

4 The first amino acid joins to the second by a peptide bond, and the first tRNA is released.

7 When the ribosome reaches a stop codon, the polypeptide is released.

8 Finally, the last tRNA is released, and the ribosome comes apart. The released polypeptide forms a new protein.

FIGURE 8.10 The process of translation (continued).

A cell's genetic machinery and its metabolic machinery are integrated and interdependent. Recall from Chapter 5 that the bacterial cell carries out an enormous number of metabolic reactions. The common feature of all metabolic reactions is that they are catalyzed by enzymes. Also recall from Chapter 5 (page 120) that feedback inhibition stops a cell from performing unneeded chemical reactions. Feedback inhibition stops enzymes that have already been synthesized. We will now look at mechanisms to prevent synthesis of enzymes that are not needed.

We have seen that genes, through transcription and translation, direct the synthesis of proteins, many of which serve as enzymes—the very enzymes used for cellular metabolism. Because protein synthesis requires a tremendous expenditure of energy, the regulation of protein synthesis is important to the cell's energy economy. The cell conserves energy by making only those proteins needed at a particular time. We will next look at how chemical reactions are regulated by controlling the synthesis of the enzymes.

Many genes, perhaps 60–80%, are not regulated but are instead *constitutive,* meaning that their products are constantly produced at a fixed rate. Usually these genes, which are effectively turned on all the time, code for enzymes that the cell needs in fairly large amounts for its major life processes; the enzymes of glycolysis are examples. The production of other enzymes is regulated so that they are present only when needed. *Trypanosoma,* the protozoan parasite that causes African sleeping sickness, has hundreds of genes coding for surface glycoproteins. Each protozoan cell turns on only one glycoprotein gene. As the host's immune system kills parasites with one type of surface molecule, parasites with different surface glycoproteins can continue to grow.

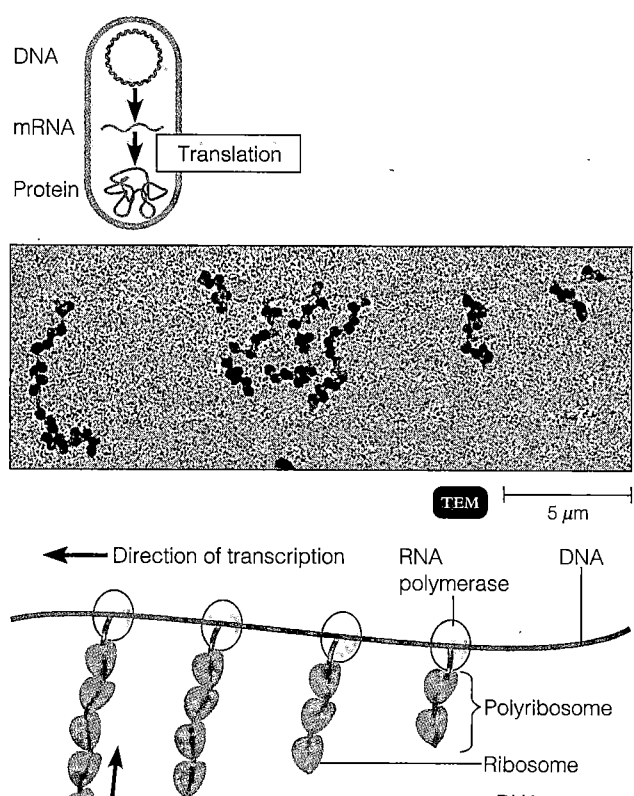

Direction of transcription   RNA polymerase   DNA

Polyribosome

Ribosome

mRNA

Direction of translation

FIGURE 8.11 Simultaneous transcription and translation in bacteria. The micrograph and diagram show these processes in a single bacterial gene. Many molecules of mRNA are being synthesized simultaneously. The longest mRNA molecules were the first to be transcribed at the promoter. Notice the ribosomes attached to the newly forming mRNA. The newly synthesized polypeptides are not shown. Reprinted with permission from O. L. Miller, Jr., B. A. Hamkalo, and C. A. Thomas, Jr., *Science* 169 (24 July 1970): 392. © 1970 American Association for the Advancement of Science.

■ Why can translation begin before transcription is complete in prokaryotes but not in eukaryotes?

Repression and Induction

Two genetic control mechanisms known as repression and induction regulate the transcription of mRNA and consequently the synthesis of enzymes from them. These mechanisms control the formation and amounts of enzymes in the cell, not the activities of the enzymes.

Repression

The regulatory mechanism that inhibits gene expression and decreases the synthesis of enzymes is called **repression.** Repression is usually a response to the overabundance of an end-product of a metabolic pathway; it causes a decrease in the rate of synthesis of the enzymes leading to the formation of that product. Repression is mediated by regulatory proteins called **repressors,** which block the ability of RNA polymerase to initiate transcription from the repressed genes (Figure 8.13).

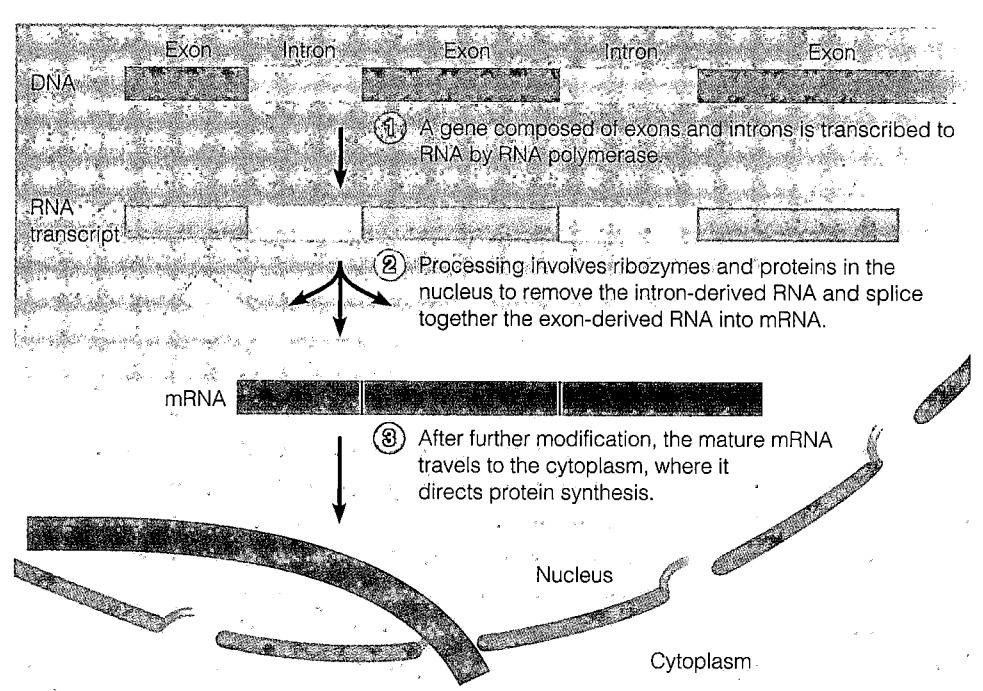

① A gene composed of exons and introns is transcribed to RNA by RNA polymerase.

② Processing involves ribozymes and proteins in the nucleus to remove the intron-derived RNA and splice together the exon-derived RNA into mRNA.

③ After further modification, the mature mRNA travels to the cytoplasm, where it directs protein synthesis.

Nucleus

Cytoplasm

FIGURE 8.12 RNA processing in eukaryotic cells.

■ Why can't the RNA transcript be used for translation?

Induction

The process that turns on the transcription of a gene or genes is **induction.** A substance that acts to induce transcription of a gene is called an **inducer,** and enzymes that are synthesized in the presence of inducers are *inducible enzymes.* The genes required for lactose metabolism in *E. coli* are a well-known example of an inducible system. One of these genes codes for the enzyme β-galactosidase, which splits the substrate lactose into two simple sugars, glucose and galactose. (β refers to the type of linkage that joins the glucose and galactose.) If *E. coli* is placed into a medium in which no lactose is present, the organisms contain almost no β-galactosidase; however, when lactose is added to the medium, the bacterial cells produce a large quantity of the enzyme. Lactose is converted in the cell to the related compound allolactose, which is the inducer for these genes; the presence of lactose thus indirectly induces the cells to synthesize more enzyme. This response, which is under genetic control, is termed **enzyme induction.**

The Operon Model of Gene Expression

Details of the control of gene expression by induction and repression are described by the operon model. François Jacob and Jacques Monod formulated this general model in 1961 to account for the regulation of protein synthesis. They based their model on studies of the induction of the enzymes of lactose catabolism in *E. coli.* In addition to β-galactosidase, these enzymes include permease, which is involved in the transport of lactose into the cell, and transacetylase, which metabolizes certain disaccharides other than lactose.

The genes for the three enzymes involved in lactose uptake and utilization are next to each other on the bacterial chromosome and are regulated together (Figure 8.14a). These genes, which determine the structures of proteins, are called **structural genes** to distinguish them from an adjoining control region on the DNA. When lactose is introduced into the culture medium, the *lac* structural genes are all transcribed and translated rapidly and simultaneously. We will now see how this regulation occurs.

In the control region of the *lac* operon are two relatively short segments of DNA. One, the *promoter,* is the region of DNA where RNA polymerase initiates transcription. The other is the **operator,** which is like a traffic light that acts as a go or stop signal for transcription of the structural genes. A set of operator and promoter sites, and the structural genes they control, are what define an **operon;** thus, the combination of the three *lac* structural genes and the adjoining control regions is called the *lac* operon.

Without repressor:

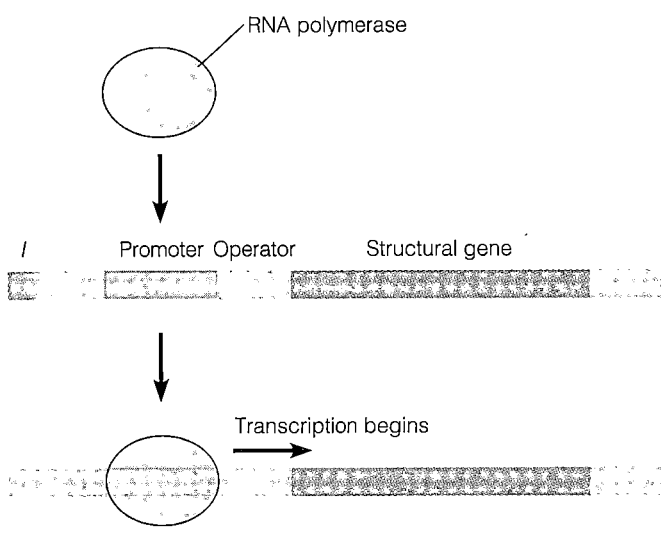

With repressor:

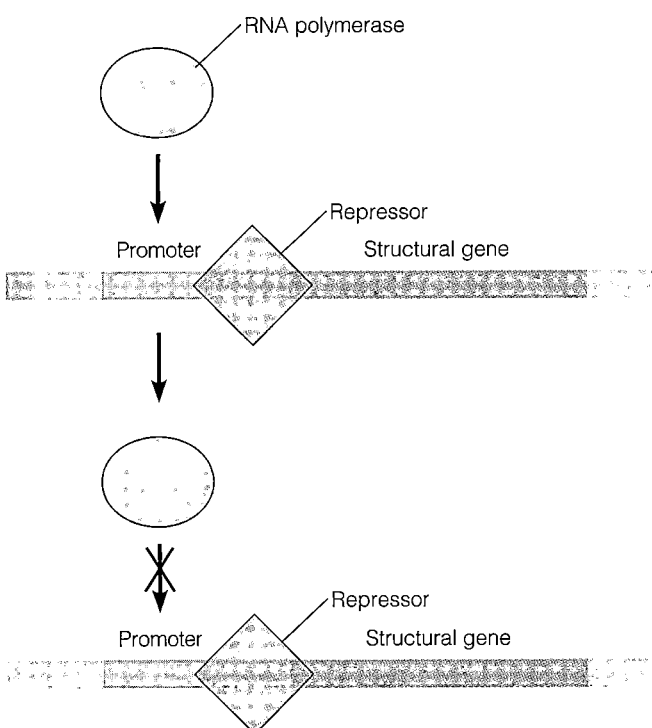

FIGURE 8.13 A model for repression. The repressor protein is encoded by the *I* gene. When a repressor protein is present, it either blocks RNA polymerase from binding to the promoter, or it blocks its progress along the DNA. In either case, the repressor effectively stops transcription of the gene.

■ Repression is a regulatory mechanism that inhibits gene expression and decreases enzyme synthesis.

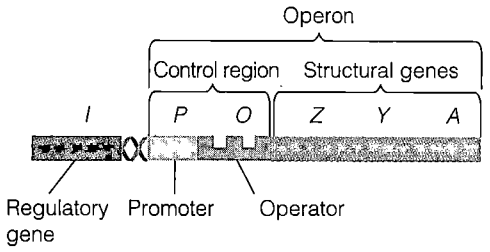

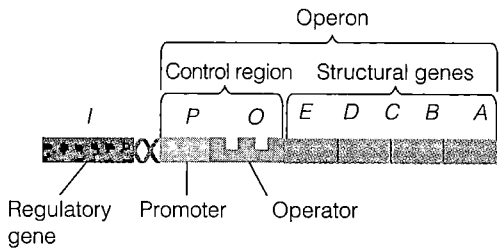

1 **Structure of the operon.** The operon consists of the promoter (*P*), and operator (*O*) sites, and structural genes which code for the protein. The operon is regulated by the product of the regulatory gene (*I*).

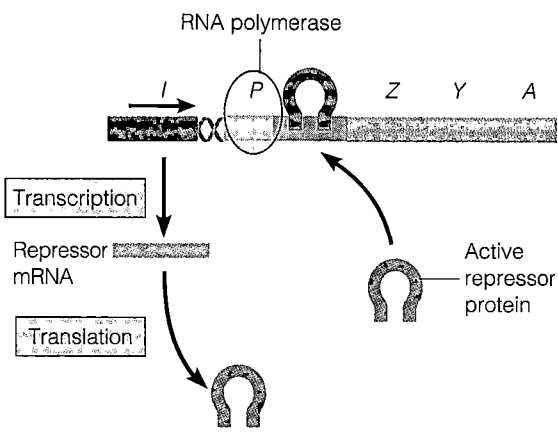

2 **Repressor active, operon off.** The repressor protein binds with the operator, preventing transcription from the operon.

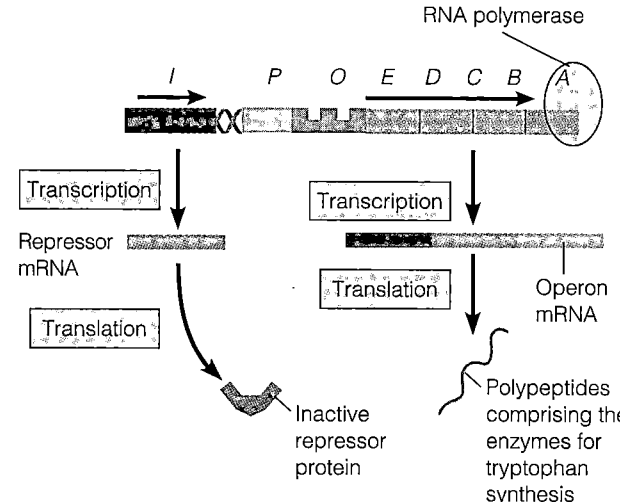

2 **Repressor inactive, operon on.** The repressor is inactive and transcription and translation proceed leading to the synthesis of tryptophan.

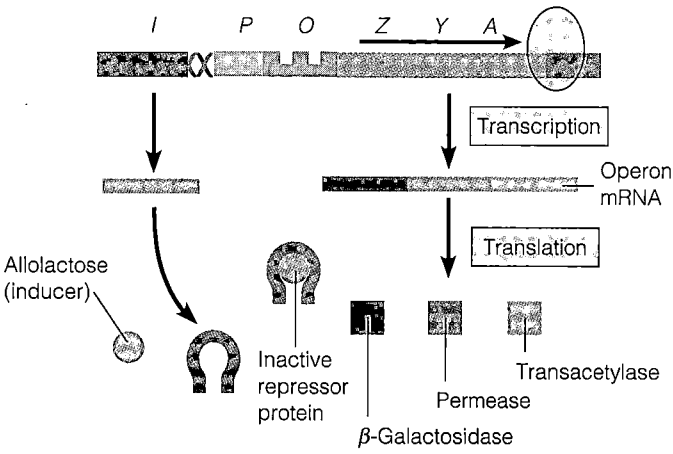

3 **Repressor inactive, operon on.** When the inducer allolactose binds to the repressor protein, the inactivated repressor can no longer block transcription. The structural genes are transcribed, ultimately resulting in the production of the enzymes needed for lactose catabolism.

(a) An inducible operon

3 **Repressor active, operon off.** When the corepressor tryptophan binds to the repressor protein, the activated repressor binds with the operator, preventing transcription from the operon.

(b) A repressible operon

FIGURE 8.14 **The operon: regulation of gene expression.** **(a)** Lactose is digested by a catabolic pathway catalyzed by inducible enzymes. **(b)** Tryptophan is an amino acid produced by an anabolic pathway catalyzed by repressible enzymes.

■ How does a repressible enzyme differ from an inducible enzyme?

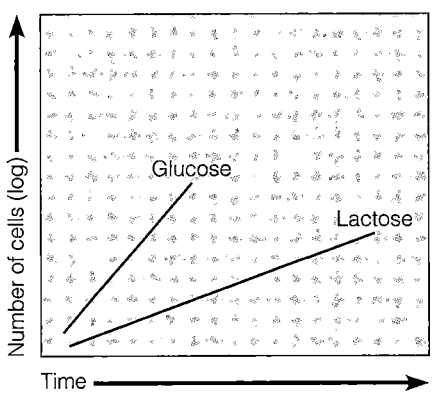

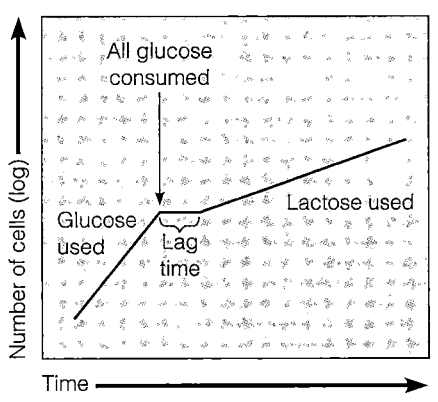

(a) Growth on glucose or lactose alone

(b) Growth on glucose and lactose combined

FIGURE 8.15 The growth rate of *E. coli* on glucose and lactose. The steeper the straight line, the faster the growth. **(a)** Bacteria growing on glucose as the sole carbon source grow faster than on lactose. **(b)** Bacteria growing in a medium containing glucose and lactose first consume the glucose, and then, after a short lag time, the lactose. During the lag time, intracellular cyclic AMP increases, the *lac* operon is transcribed, more lactose is transported into the cell, and β-galactosidase is synthesized to break down lactose.

■ Will transcription of the *lac* operon occur in the presence of lactose and glucose? In the presence of lactose and the absence of glucose? In the presence of glucose and the absence of lactose?

Near the *lac* operon on the bacterial DNA is a regulatory gene called the *I* gene, which codes for a repressor protein. ❷ When lactose is absent, the repressor protein binds tightly to the operator site. This binding prevents RNA polymerase from transcribing the adjacent structural genes; consequently, no mRNA is made and no enzymes are synthesized. ❸ But when lactose is present, some of it is transported into the cells and converted into the inducer allolactose. The inducer binds to the repressor protein and alters it so it cannot bind to the operator site. In the absence of an operator-bound repressor protein, RNA polymerase can transcribe the structural genes into mRNA, which is then translated into enzymes. This is why, in the presence of lactose, enzymes are produced. Lactose is said to induce enzyme synthesis, and the *lac* operon is called an inducible operon.

In repressible operons, the structural genes are transcribed until they are turned off or *repressed* (Figure 8.14b). ❶ The genes for the enzymes involved in the synthesis of tryptophan are regulated in this manner. ❷ The structural genes are transcribed and translated leading to tryptophan synthesis. ❸ When excess tryptophan is present, the tryptophan acts as a **corepressor** binding to the repressor protein. The repressor protein cannot bind to the operator, stopping further tryptophan synthesis.

Regulation of the lactose operon also depends on the level of glucose in the medium, which in turn controls the intracellular level of the small molecule **cyclic AMP (cAMP),** a substance derived from ATP that serves as a cellular alarm signal. Enzymes that metabolize glucose are constitutive, and cells grow at their maximal rate with glucose as their carbon source because they can use it most efficiently (Figure 8.15). When glucose is no longer available, cyclic AMP (cAMP) accumulates in the cell. The cAMP binds to the allosteric site of *cAMP receptor protein (CRP).* CRP then binds to the *lac* promoter which initiates transcription by making it easier for RNA polymerase to bind to the promoter. Thus transcription of the *lac* operon requires both the presence of lactose and the absence of glucose (see Figure 8.14a).

Cyclic AMP is an example of an *alarmone,* a chemical alarm signal the cell uses to respond to environmental or nutritional stress. (In this case, the stress is the lack of glucose.) The same mechanism involving cAMP allows the cell to grow on other sugars. Inhibition of the metabolism of alternative carbon sources by glucose is termed **catabolic repression** (or the *glucose effect*). When glucose is available, the level of cAMP in the cell is low, and consequently CRP is not bound.

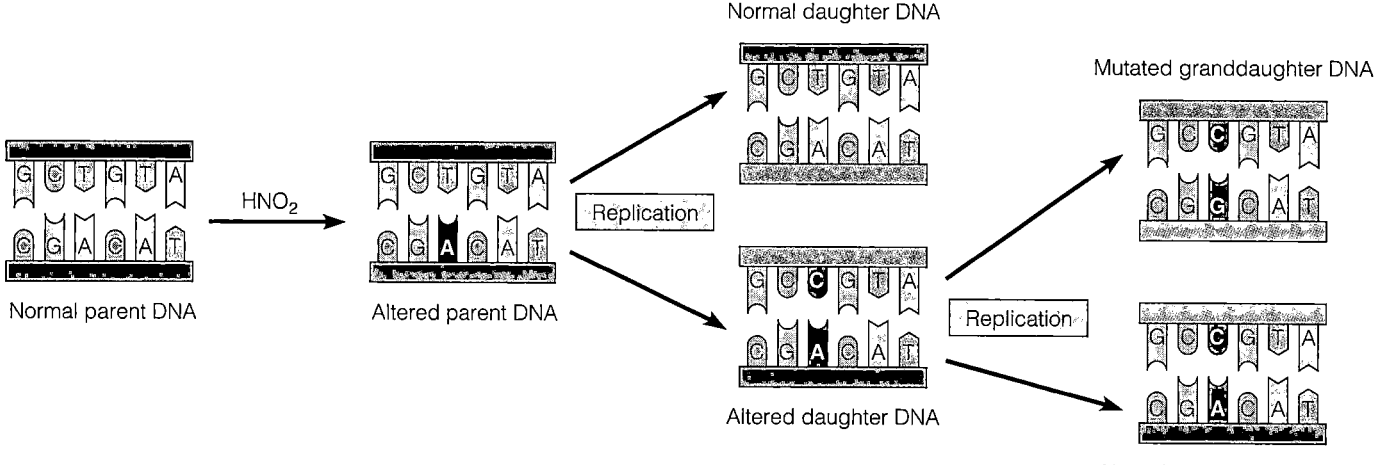

FIGURE 8.18 Nitrous acid (HNO₂) as a mutagen. The nitrous acid alters an adenine in such a way that it pairs with cytosine instead of thymine.

■ A mutagen is any environmental agent that can directly or indirectly bring about a mutation.

only a fragment is synthesized. A base substitution resulting in a nonsense codon is thus called a **nonsense mutation** (Figure 8.17c).

Besides base-pair mutations, there are also changes in DNA called **frameshift mutations,** in which one or a few nucleotide pairs are deleted or inserted in the DNA (Figure 8.17d). This mutation can shift the "translational reading frame"—that is, the three-by-three grouping of nucleotides recognized as codons by the tRNAs during translation. For example, deleting one nucleotide pair in the middle of a gene causes changes in many amino acids downstream from the site of the original mutation. Frameshift mutations almost always result in a long stretch of altered amino acids and the production of an inactive protein from the mutated gene. In most cases, a nonsense codon will eventually be encountered and thereby terminate translation.

Occasionally, mutations occur where significant numbers of bases are added to (inserted into) a gene. Huntington's disease, for example, is a progressive neurological disorder caused by extra bases inserted into a particular gene. The reason these insertions occur in this particular gene is still being studied.

Base substitutions and frameshift mutations may occur spontaneously because of occasional mistakes made during DNA replication. These **spontaneous mutations** apparently occur in the absence of any mutation-causing agents. Agents in the environment, such as certain chemicals and radiation, that directly or indirectly bring about mutations are called **mutagens.** Almost any agent that can chemically or physically react with DNA can potentially cause

mutations. A wide variety of chemicals, many of which are common in nature or in households, are known to be mutagens. Many forms of radiation, including X rays and ultraviolet light, are also mutagenic, as discussed shortly.

In the microbial world, certain mutations result in resistance to antibiotics or altered pathogenicity. A mutation in a gene encoding the outer membrane may increase pathogenicity; for example, *Salmonella typhimurium* with an altered outer membrane can survive in phagocytes. A mutation in a capsule-encoding gene may result in decreased pathogenicity because phagocytes can destroy the bacteria, as in the cases of *Streptococcus pneumoniae, Haemophilus influenzae,* and *Neisseria meningitidis.*

Mutagens

Learning Objectives

■ *Define mutagen.*
■ *Describe two ways mutations can be repaired.*

Chemical Mutagens

One of the many chemicals known to be a mutagen is nitrous acid. Figure 8.18 shows how exposure of DNA to nitrous acid can convert the base adenine (A) to a form that no longer pairs with thymine (T) but instead pairs with cytosine (C). When DNA containing such modified adenines replicates, one daughter DNA molecule will have a base-pair sequence different from that of the parent DNA. Eventually, some AT base pairs of the parent will have been changed to GC base pairs in a granddaughter cell. Nitrous

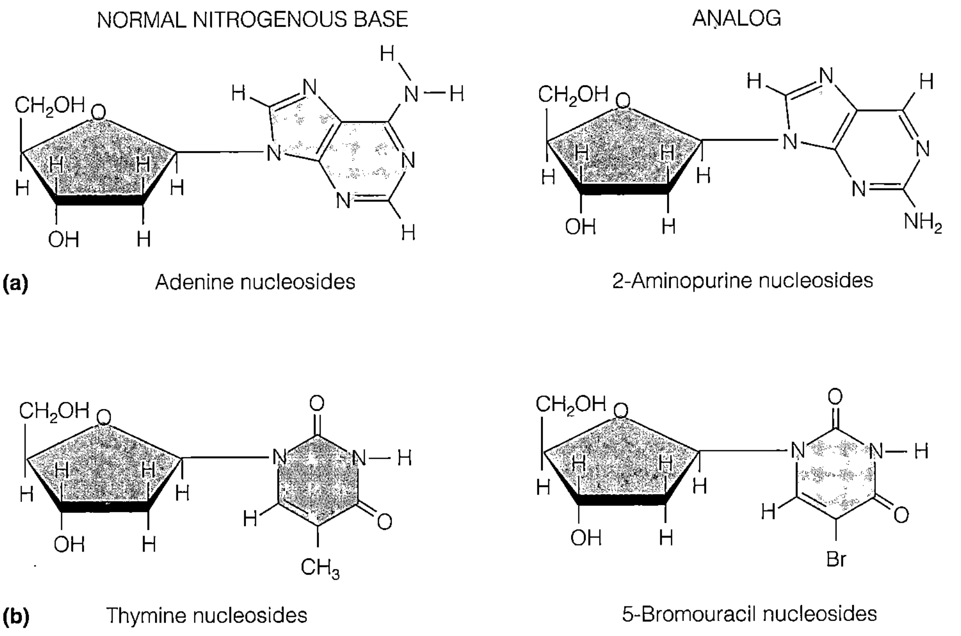

NORMAL NITROGENOUS BASE

ANALOG

(a)   Adenine nucleosides

2-Aminopurine nucleosides

(b)   Thymine nucleosides

5-Bromouracil nucleosides

FIGURE 8.19 Nucleoside analogs and the nitrogenous bases they replace. **(a)** Adenine and 2-aminopurine nucleosides. **(b)** Thymine and 5-bromouracil nucleosides.

■ What is a nucleoside analog? .

acid makes a specific base-pair change in DNA. Like all mutagens, it alters DNA at random locations.

Another type of chemical mutagen is the **nucleoside analog.** These molecules are structurally similar to normal nitrogenous bases, but they have slightly altered base-pairing properties. Examples are 2-aminopurine and 5-bromouracil. The 2-aminopurine molecule is incorporated into DNA in place of adenine but can sometimes pair with cytosine (Figure 8.19a). The 5-bromouracil molecule is incorporated into DNA in place of thymine (T) but often pairs with guanine (Figure 8.19b). When nucleoside analogs are given to growing cells, the analogs are randomly incorporated into cellular DNA in place of the normal bases. Then, during DNA replication, the analogs cause mistakes in base pairing. The incorrectly paired bases will be copied during subsequent replication of the DNA, resulting in base-pair substitutions in the progeny cells. Some antiviral and antitumor drugs are nucleoside analogs, including AZT (azidothymidine), one of the primary drugs used to treat HIV infection.

Still other chemical mutagens cause small deletions or insertions, which can result in frameshifts. For instance, under certain conditions, benzpyrene, which is present in smoke and soot, is an effective *frameshift mutagen*. And aflatoxin—produced by *Aspergillus flavus* (a-spèr-jil′lus flā′vus), a mold that grows on peanuts and grain—is a frameshift mutagen, as are the acridine dyes used experimentally against herpesvirus infections. Frameshift mutagens usually have the right size and chemical properties to slip between the stacked base pairs of the DNA double helix. They may work by slightly offsetting the two strands

of DNA, leaving a gap or bulge in one strand or the other. When the staggered DNA strands are copied during DNA synthesis, one or more base pairs can be inserted or deleted in the new double-stranded DNA. Interestingly, frameshift mutagens are often potent carcinogens.

Radiation

X rays and gamma rays are forms of radiation that are potent mutagens because of their ability to ionize atoms and molecules. The penetrating rays of ionizing radiation cause electrons to pop out of their usual shells (see Chapter 2). These electrons bombard other molecules and cause more damage, and many of the resulting ions and free radicals (molecular fragments with unpaired electrons) are very reactive. Some of these ions can combine with bases in DNA, resulting in errors in DNA replication and repair that produce mutations. An even more serious outcome is the breakage of covalent bonds in the sugar-phosphate backbone of DNA, which causes physical breaks in chromosomes.

Another form of mutagenic radiation is ultraviolet (UV) light, a nonionizing component of ordinary sunlight. However, the most mutagenic component of UV light (wavelength 260 nm) is screened out by the ozone layer of the atmosphere. The most important effect of direct UV light on DNA is the formation of harmful covalent bonds between certain bases. Adjacent thymines in a DNA strand can cross-link to form thymine dimers. Such dimers, unless repaired, may cause serious damage or death to the cell because it cannot properly transcribe or replicate such DNA.

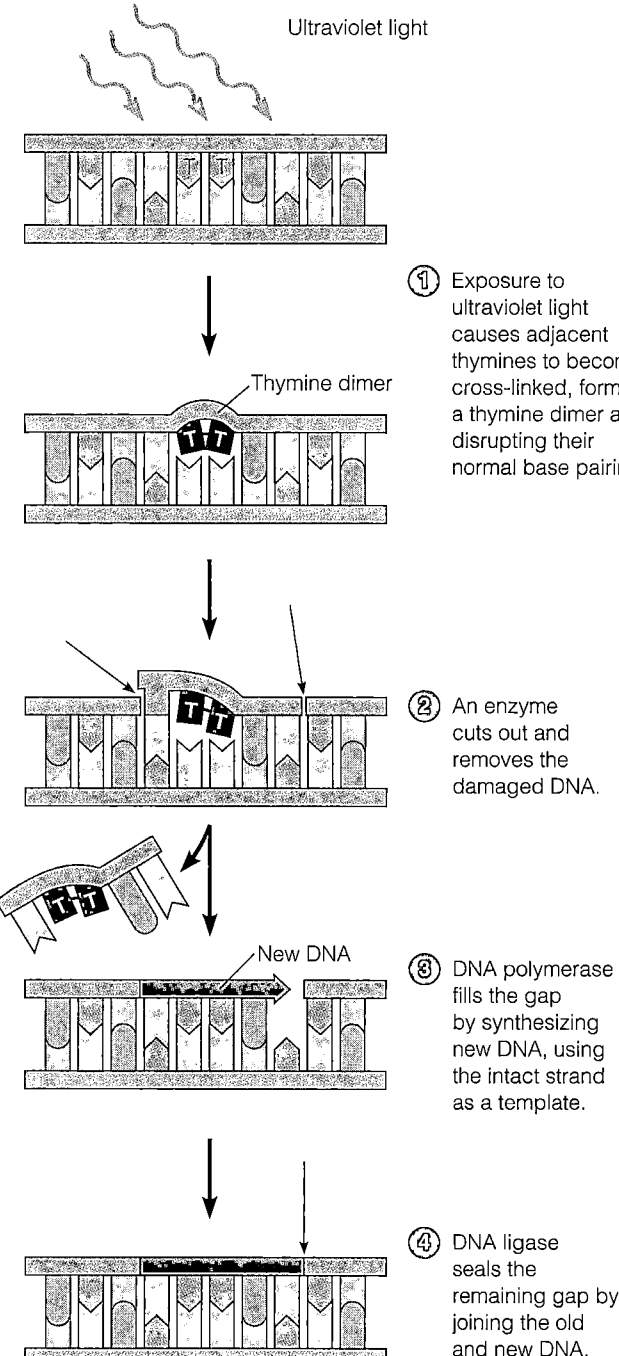

Ultraviolet light

Thymine dimer

① Exposure to ultraviolet light causes adjacent thymines to become cross-linked, forming a thymine dimer and disrupting their normal base pairing.

② An enzyme cuts out and removes the damaged DNA.

New DNA

③ DNA polymerase fills the gap by synthesizing new DNA, using the intact strand as a template.

④ DNA ligase seals the remaining gap by joining the old and new DNA.

FIGURE 8.20 The creation and repair of a thymine dimer caused by ultraviolet light. After exposure to UV light, adjacent thymines can become cross-linked, forming a thymine dimer. In the absence of visible light, the nucleotide excision repair mechanism is used in a cell to repair the damage.

■ Bacteria and other organisms have enzymes that can repair radiation damage.

Bacteria and other organisms have enzymes that can repair UV-induced damage. **Light-repair enzymes,** called *photolyases,* use visible light energy to separate the dimer back to the original two thymines. **Nucleotide excision repair,** shown in Figure 8.20, is not restricted to UV-induced damage; it can repair mutations from other causes as well. Enzymes cut out the distorted cross-linked thymines by opening a wide gap. They then fill in the gap with newly synthesized DNA that is complementary to the undamaged strand. By this means, the original base-pair sequence is restored. The last step is the covalent sealing of the DNA backbone by the enzyme DNA ligase. Occasionally, such a repair process makes an error, and the original base-pair sequence is not properly restored. The result of this error is a mutation.

Exposure to UV light in humans, such as by excessive suntanning, causes a large number of thymine dimers in skin cells. Unrepaired dimers may result in skin cancers. Humans with xeroderma pigmentosum, an inherited condition that results in increased sensitivity to UV light, have a defect in nucleotide excision repair; consequently, they have an increased risk of skin cancer.

The Frequency of Mutation

Learning Objective

■ Describe the effect of mutagens on the mutation rate.

The **mutation rate** is the probability that a gene will mutate when a cell divides. The rate is usually stated as a power of 10, and because mutations are very rare, the exponent is always a negative number. For example, if there is one chance in 10,000 that a gene will mutate when the cell divides, the mutation rate is 1/10,000, which is expressed as 10^{-4}. Spontaneous mistakes in DNA replication occur at a very low rate, perhaps only once in 10^9 replicated base pairs (a mutation rate of 10^{-9}). Because the average gene has about 10^3 base pairs, the spontaneous rate of mutation is about once in 10^6 (a million) replicated genes.

Mutations usually occur more or less randomly along a chromosome. The occurrence of random mutations at low frequency is an essential aspect of the adaptation of species to their environment, for evolution requires that genetic diversity be generated randomly and at a low rate. For example, in a bacterial population of significant size—say, greater than 10^7 cells—a few new mutant cells will always be produced in every generation. Most mutations are either harmful and likely to be removed from the gene pool when the individual cell dies, or are neutral. However, a few mutations may be beneficial. For example, a mutation that confers antibiotic resistance is beneficial if a population of bacteria is regularly exposed to antibiotics.

Once such a trait has appeared through mutation, cells carrying the mutated gene are more likely than other cells to survive and reproduce. Soon most of the cells in the population will have the gene; an evolutionary change will have occurred, although on a small scale.

A mutagen usually increases the spontaneous rate of mutation, which is about once in 10^6 replicated genes, by a factor of 10–1000 times. In other words, in the presence of a mutagen, the normal rate of 10^{-6} mutations per replicated gene becomes a rate of 10^{-5} to 10^{-3} per replicated gene. Mutagens are used experimentally to enhance the production of mutant cells for research on the genetic properties of microorganisms, and for commercial purposes.

Identifying Mutants

Learning Objective

■ *Outline the methods of direct and indirect selection of mutants.*

Mutants can be detected by selecting or testing for an altered phenotype. Whether or not a mutagen is used, mutant cells with specific mutations are always rare compared with other cells in the population. The problem is detecting such a rare event.

Experiments are usually performed with bacteria because they reproduce rapidly, so large numbers of organisms (more than 10^9 per milliliter of nutrient broth) can easily be used. Furthermore, because bacteria generally have only one copy of each gene per cell, the effects of a mutated gene are not masked by the presence of a normal version of the gene, as in many eukaryotic organisms.

Positive (direct) selection involves the detection of mutant cells by rejection of the unmutated parent cells. For example, suppose we were trying to find mutant bacteria that are resistant to penicillin. When the bacterial cells are plated on a medium containing penicillin, the mutant can be identified directly. The few cells in the population that are resistant (mutants) will grow and form colonies, whereas the normal, penicillin-sensitive parental cells cannot grow.

To identify mutations in other kinds of genes, **negative (indirect) selection** can be used. This process selects a cell that cannot perform a certain function, using the technique of **replica plating.** For example, suppose we wanted to use replica plating to identify a bacterial cell that has lost the ability to synthesize the amino acid histidine (Figure 8.21). First, about 100 bacterial cells are inoculated onto an agar plate. This plate, called the master plate, contains a medium with histidine on which all cells will grow. After several hours of incubation, each cell reproduces to form a colony. Then a pad of sterile material, such as latex,

filter paper, or velvet, is pressed over the master plate, and some of the cells from each colony adhere to the velvet. Next, the velvet is pressed down onto two (or more) sterile plates. One plate contains a medium with histidine, and one contains a medium without histidine on which the original, nonmutant bacteria can grow. Any colony that grows on the medium with histidine on the master plate but that cannot synthesize its own histidine will not be able to grow on the medium without histidine. The mutant colony can then be identified on the master plate. Of course, because mutants are so rare (even those induced by mutagens), many plates must be screened with this technique to isolate a specific mutant.

Replica plating is a very effective means of isolating mutants that require one or more new growth factors. Any mutant microorganism having a nutritional requirement that is absent in the parent is known as an **auxotroph.** For example, an auxotroph may lack an enzyme needed to synthesize a particular amino acid and will therefore require that amino acid as a growth factor in its nutrient medium.

Identifying Chemical Carcinogens

Learning Objective

■ *Identify the purpose of and outline the procedure for the Ames test.*

Many known mutagens have been found to be **carcinogens,** substances that cause cancer in animals, including humans. In recent years, chemicals in the environment, the workplace, and the diet have been implicated as causes of cancer in humans. The usual subjects of tests to determine potential carcinogens are animals, and the testing procedures are time-consuming and expensive. Now there are faster and less expensive procedures for the preliminary screening of potential carcinogens. One of these, called the **Ames test,** uses bacteria as carcinogen indicators.

The Ames test is based on the observation that exposure of mutant bacteria to mutagenic substances may cause new mutations that reverse the effect (the change in phenotype) of the original mutation. These are called back-mutations, or *reversions.* Specifically, the test measures the reversion of histidine auxotrophs of *Salmonella* (his$^-$ cells, mutants that have lost the ability to synthesize histidine) to histidine-synthesizing cells (his$^+$) after treatment with a mutagen (Figure 8.22). Bacteria are incubated in both the presence and absence of the substance being tested. Because many chemicals must be activated (transformed chemically into forms that are chemically reactive) by animal enzymes for mutagenic or carcinogenic activity to appear, the chemical to be tested

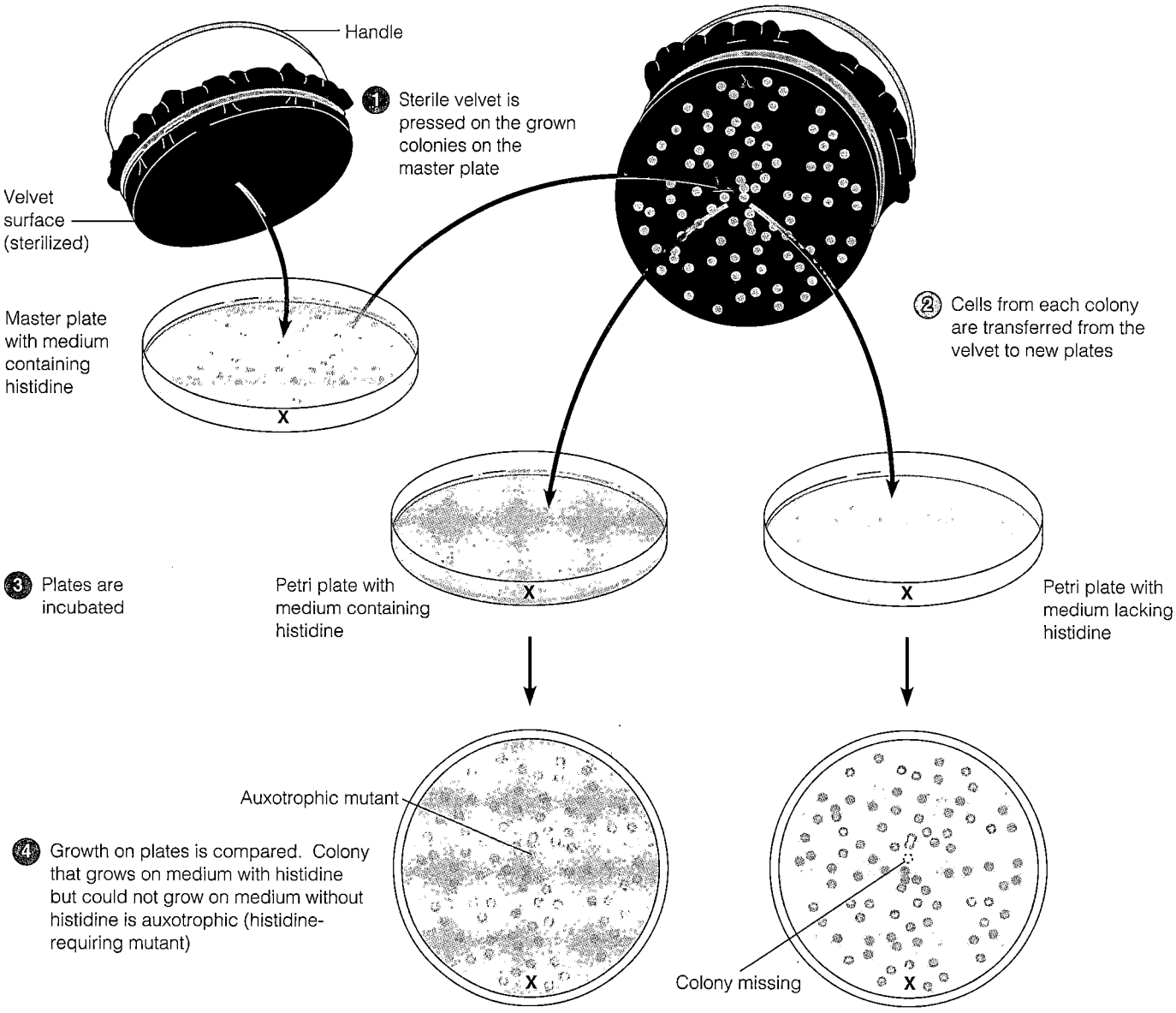

Handle

1 Sterile velvet is pressed on the grown colonies on the master plate

Velvet surface (sterilized)

Master plate with medium containing histidine

X

2 Cells from each colony are transferred from the velvet to new plates

3 Plates are incubated

Petri plate with medium containing histidine

X

X

Petri plate with medium lacking histidine

4 Growth on plates is compared. Colony that grows on medium with histidine but could not grow on medium without histidine is auxotrophic (histidine-requiring mutant)

Auxotrophic mutant

X

Colony missing

X

FIGURE 8.21 Replica plating. In this example, the auxotrophic mutant cannot synthesize histidine. The plates must be carefully marked (with an X here) to maintain orientation so that colony positions are known in relation to the original master plate.

■ Replica plating is used to identify auxotrophic mutants, bacteria that have lost the ability to synthesize an essential nutrient.

and the mutant bacteria are incubated together with rat liver extract, a rich source of activation enzymes. If the substance being tested is mutagenic, it will cause the reversion of his$^-$ bacteria to his$^+$ bacteria at a rate higher than the spontaneous reversion rate. The number of observed revertants provides an indication of the degree to which a substance is mutagenic and therefore possibly carcinogenic.

The test can be used in many ways. Several potential mutagens can be qualitatively tested by spotting the indi-vidual chemicals on small paper disks on a single plate inoculated with bacteria. In addition, mixtures such as wine, blood, smoke condensates, and extracts of foods can also be tested to see if they contain mutagenic substances. For an example of the use of the Ames test in exploring the role of bacteria in cancer, see the box on page 234.

About 90% of the substances found by the Ames test to be mutagenic have also been shown to be carcinogenic in animals. By the same token, the more mutagenic substances have generally been found to be more carcinogenic.

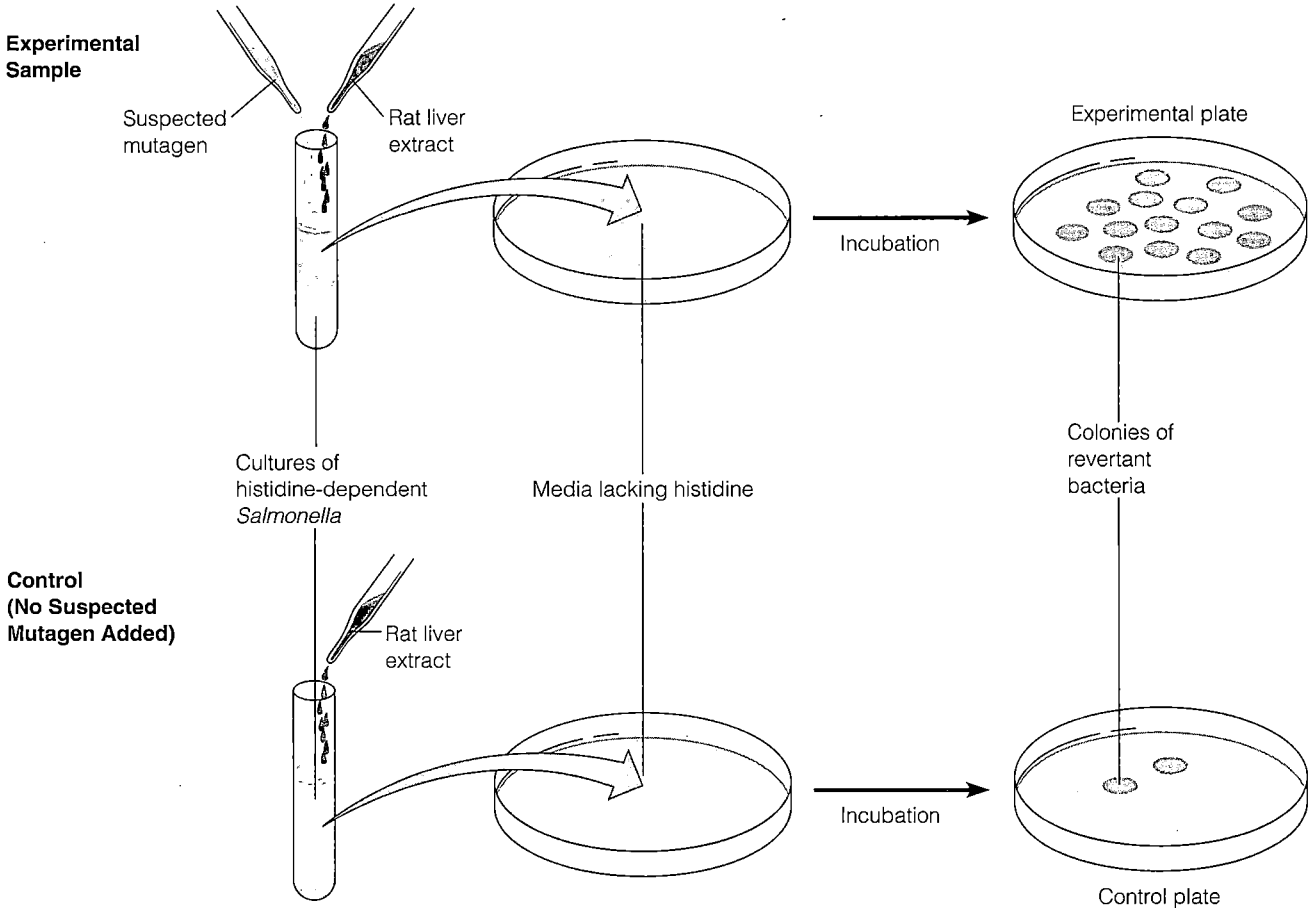

Experimental Sample

Suspected mutagen

Rat liver extract

Cultures of histidine-dependent *Salmonella*

Media lacking histidine

Incubation

Experimental plate

Colonies of revertant bacteria

Control (No Suspected Mutagen Added)

Rat liver extract

Incubation

Control plate

① Two cultures are prepared of *Salmonella* bacteria that have lost the ability to synthesize histidine (histidine-dependent).

② The suspected mutagen is added to the experimental sample only; rat liver extract (an activator) is added to both samples.

③ Each sample is poured onto a plate of medium lacking histidine. The plates are then incubated at 37°C for two days. Only bacteria whose histidine-dependent phenotype has mutated back (reverted) to histidine-synthesizing will grow into colonies.

④ The numbers of colonies on the experimental and control plates are compared. The control plate may show a few spontaneous histidine-synthesizing revertants. The test plates will show an increase in the number of histidine-synthesizing revertants if the test chemical is indeed a mutagen and potential carcinogen. The higher the concentration of mutagen used, the more revertant colonies will result.

FIGURE 8.22 The Ames test.

■ The Ames test is used to screen mutagens suspected of causing cancer in humans.

Genetic Transfer and Recombination

Learning Objectives

■ *Compare the mechanisms of genetic recombination in bacteria.*

■ *Differentiate between horizontal and vertical gene transfer.*

Genetic recombination refers to the exchange of genes between two DNA molecules to form new combinations of genes on a chromosome. Figure 8.23 shows one type of genetic recombination occurring between two pieces of DNA, which we will call A and B and regard as chromosomes for the sake of simplicity. If these two chromosomes break and rejoin as shown—a process called **crossing over**—some of the genes carried by

MICROBIOLOGY IN THE NEWS

The Role of Bacteria in Cancer

In 1996, the International Agency for Research on Cancer classified the bacterium *Helicobacter pylori* as a carcinogen. No specific mechanism was proposed to explain the relationship between *H. pylori* and gastric cancer. Researchers have known since the early 1970s that there is a relationship between diet and certain types of cancer. However, it may not be the diet itself that causes cancer but the interaction between diet and normal microbiota, the hundreds of species of microorganisms that flourish in the average person.

The importance of microorganisms in cancer inductions was first noticed when rats were fed cycasin, a substance that occurs naturally in the nut of the cycad plant. The bacterial enzyme β-glucosidase converts cycasin to methylazoxymethanol, which caused cancer in the rats. Germ-free rats were unaffected by the cycasin, but rats with the usual intestinal microbiota developed colon cancer.

To determine the role of intestinal bacteria in converting chemicals to carcinogens, Elena McCoy and her coworkers at New York Medical College used the Ames test to compare the mutagenic abilities of 2-aminofluorene when activated by liver cell enzymes, intestinal cell enzymes alone, *Bacteroides fragilis* enzymes alone, and intestinal cell enzymes plus *B. fragilis* enzymes. They found that liver enzymes produced the highest conversion of this chemical to a mutagenic form. Intestinal enzymes alone and bacterial enzymes alone produced some mutagenic conversion. And a combination of intestinal and bacterial enzymes had an effect almost as great as that produced by liver enzymes.

Compounds formed on the surface of fried or grilled meat and fish have been associated with an increased risk of colon cancer. One of these compounds, called IQ, is converted to a chemical called OHIQ, by intestinal bacteria. Researchers at the American Health Foundation in New York found that OHIQ is a mutagen in the Ames test, and adding liver enzymes increased its mutagenicity.

To determine the role of bacteria in bladder cancer, researchers at the Kyushu Medical Center in Japan used the Ames test to evaluate the mutagenic abilities of human urine activated by bacteria. The bacteria were isolated from patients with urinary tract infections. The bacterial isolates were tested for their ability to reduce nitrate ion (NO_3^-) to nitrite ion (NO_2^-). The mutagenic activity seen with the addition of nitrite ion indicates that nitrite is mutagenic (see the table below). Their results suggest that a chemical reaching the bladder can be activated to a carcinogen by bacterial enzymes.

Variations in diet produce little change in the kinds of bacteria that live in the intestines, but they produce dramatic changes in the metabolic activity of the bacteria. It may be the bacterial enzymes that are important. Some examples are bacterial β-glucuronidase, which retoxifies carcinogens that were previously detoxified by the liver; bacterial azoreductase, which can produce carcinogens from certain food dyes; and bacterial nitroreductase, which can produce carcinogenic nitrosamines from amino acids and nitrates. The ammonia produced by *H. pylori*'s urease may promote cell division.

Test Substance	Relative Amounts of Bacterial Growth in Ames Test
Normal human urine	–
Urine + NO_3^--reducing bacteria	+++
Urine + non-NO_3^--reducing bacteria	+
Urine + NO_2^-	+++

these chromosomes are shuffled. The original chromosomes have recombined, so that each now carries a portion of the other's genes.

If A and B represent DNA from different individuals, how are they brought close enough together to recombine? In eukaryotes, genetic recombination is an ordered process that usually occurs as part of the sexual cycle of the organism. Recombination generally takes place during the formation of reproductive cells, such that these cells contain recombinant DNA. In bacteria, genetic recombination can happen in a number of ways, which we will discuss in the following sections.

Like mutation, genetic recombination contributes to a population's genetic diversity, which is the source of variation in evolution. In highly evolved organisms such as present-day microbes, recombination is more likely than mutation to be beneficial because recombination will less likely destroy a gene's function and may bring together combinations of genes that enable the organism to carry out a valuable new function.

The major protein that constitutes the flagella of *Salmonella* is also one of the primary proteins that causes our immune systems to respond. However, these bacteria have the capability of producing two different flagellar

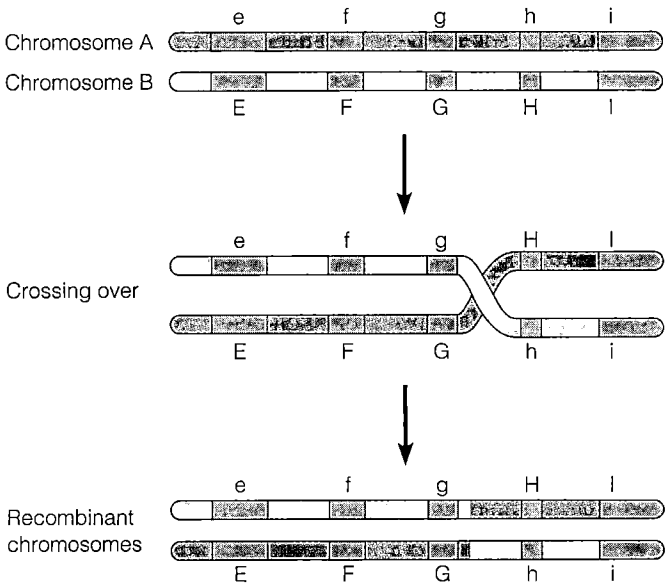

Chromosome A
Chromosome B

Crossing over

Recombinant
chromosomes

FIGURE 8.23 Genetic recombination by crossing over between two related chromosomes. Chromosomes A and B each carry one version of genes E through I. The chromosomes cross over by breaking and rejoining, sometimes in more than one location. The result is two recombinant chromosomes, each of which carries genes originating from both chromosomes.

■ Genetic recombination is the exchange of genes between two DNA molecules to form new combinations of genes.

proteins. As our immune system mounts a response against those cells containing one form of the flagellar protein, those organisms producing the second are not affected. Which flagellar protein is produced is determined by a recombination event that apparently occurs somewhat randomly within the chromosomal DNA. Thus, by altering the flagellar protein produced, *Salmonella* can better avoid the defenses of the host.

Vertical gene transfer occurs when genes are passed from an organism to its offspring. Plants and animals transmit their genes by vertical transmission. Bacteria can pass their genes not only to their offspring, but also laterally, to other microbes of the same generation. This is known as **horizontal gene transfer.** Horizontal gene transfer between bacteria occurs in several ways. In all of the mechanisms, the transfer involves a **donor cell** that gives a portion of its total DNA to a **recipient cell.** Once transferred, part of the donor's DNA is usually incorporated into the recipient's DNA; the remainder is degraded by cellular enzymes. The recipient cell that incorporates donor DNA into its own DNA is called a *recombinant.* The transfer of genetic material between bacteria is by no means a frequent event; it may occur in

only 1% or less of an entire population. Let's examine in detail the specific types of genetic transfer.

Transformation in Bacteria

During the process of **transformation,** genes are transferred from one bacterium to another as "naked" DNA in solution. This process was first demonstrated over 70 years ago, although it was not understood at the time. Not only did transformation show that genetic material could be transferred from one bacterial cell to another, but study of this phenomenon eventually led to the conclusion that DNA is the genetic material. The initial experiment on transformation was performed by Frederick Griffith in England in 1928 while he was working with two strains of *Streptococcus pneumoniae.* One, a virulent (pathogenic) strain, has a polysaccharide capsule that prevents phagocytosis. The bacteria grow and cause pneumonia. The other, an avirulent strain, lacks the capsule and does not cause disease.

Griffith was interested in determining whether injections of heat-killed bacteria of the encapsulated strain could be used to vaccinate mice against pneumonia. As he expected, injections of living encapsulated bacteria killed the mouse (Figure 8.24a); injections of live nonencapsulated bacteria (Figure 8.24b) or dead encapsulated bacteria did not kill the mouse (Figure 8.24c). However, when the dead encapsulated bacteria were mixed with live nonencapsulated bacteria and injected into the mice, many of the mice died. In the blood of the dead mice, Griffith found living, encapsulated bacteria. Hereditary material (genes) from the dead bacteria had entered the live cells and changed them genetically so that their progeny were encapsulated and therefore virulent (Figure 8.24d).

Subsequent investigations based on Griffith's research revealed that bacterial transformation could be carried out without mice. A broth was inoculated with live nonencapsulated bacteria. Dead encapsulated bacteria were then added to the broth. After incubation, the culture was found to contain living bacteria that were encapsulated and virulent. The nonencapsulated bacteria had been transformed; they had acquired a new hereditary trait by incorporating genes from the killed encapsulated bacteria.

The next step was to extract various chemical components from the killed cells to determine which component caused the transformation. These crucial experiments were performed in the United States by Oswald T. Avery and his associates Colin M. MacLeod and Maclyn McCarty. After years of research, they announced in 1944 that the component responsible for transforming harmless *S. pneumoniae* into virulent strains was DNA.

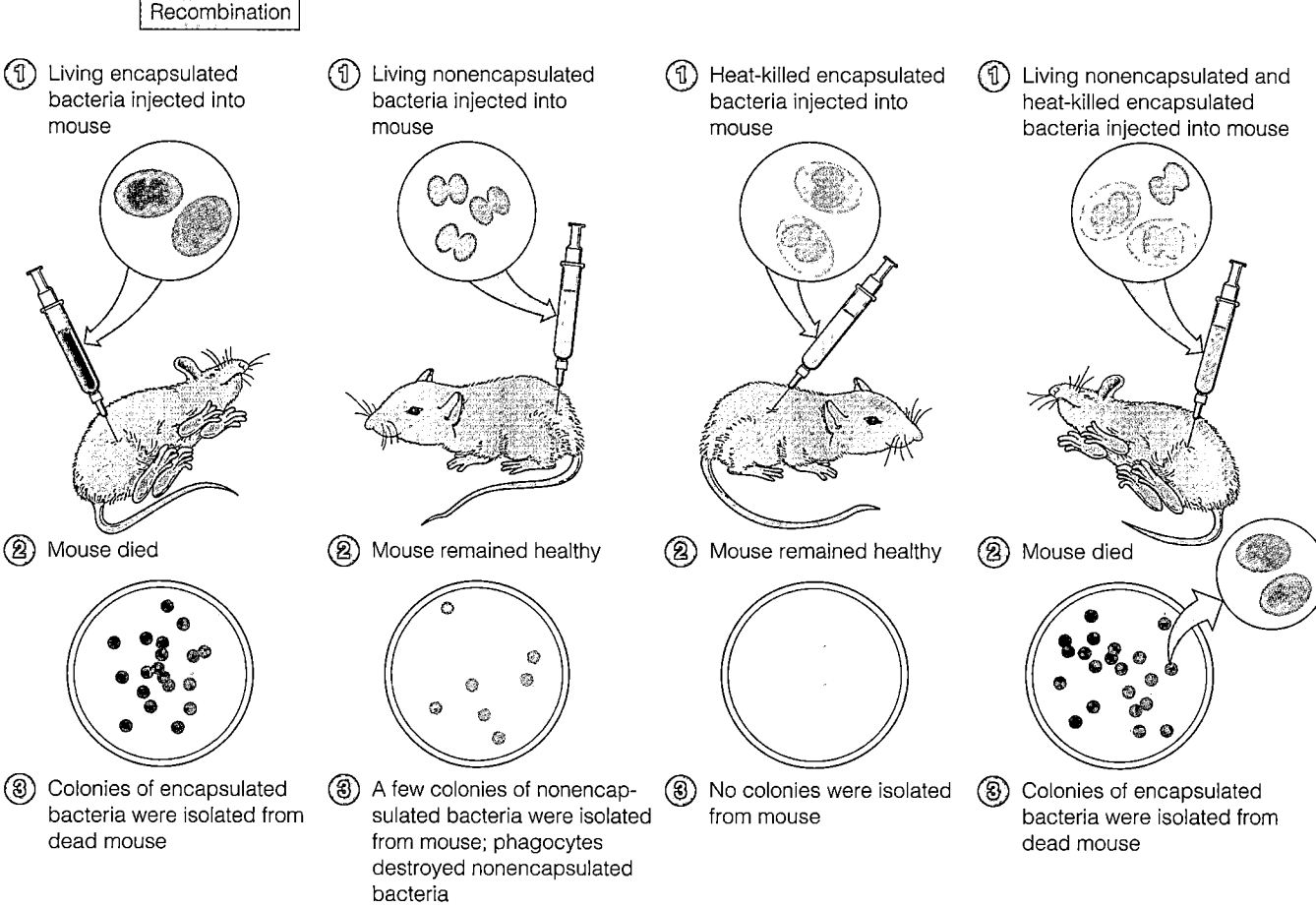

FIGURE 8.24 Griffith's experiment demonstrating genetic transformation.
(a) Living encapsulated bacteria caused disease and death when injected into a mouse.
(b) Living nonencapsulated bacteria are readily destroyed by the phagocytic defenses
of the host, so the mouse remained healthy after injection. **(c)** After being killed by
heat, encapsulated bacteria lost the ability to cause disease. **(d)** However, the combination of living nonencapsulated bacteria and heat-killed encapsulated bacteria
(neither of which alone cause disease) did cause disease. Somehow, the live nonencapsulated bacteria were transformed by the dead encapsulated bacteria so that they
acquired the ability to form a capsule and therefore cause disease. Subsequent experiments proved the transforming factor to be DNA.

■ Why did encapsulated bacteria kill the mouse while nonencapsulated bacteria did not? What killed the mouse in (d)?

Their results provided one of the conclusive indications
that DNA was indeed the carrier of genetic information.

Since the time of Griffith's experiment, considerable information has been gathered about transformation. In nature, some bacteria, perhaps after death and cell lysis, release
their DNA into the environment. Other bacteria can then
encounter the DNA and, depending on the particular
species and growth conditions, take up fragments of DNA
and integrate them into their own chromosomes by recombination. A recipient cell with this new combination of
genes is a kind of hybrid, or recombinant cell (Figure 8.25).
All the descendants of such a recombinant cell will be identical to it. Transformation occurs naturally among very few
genera of bacteria, including *Bacillus, Haemophilus* (hē-
mä'fi-lus), *Neisseria, Acinetobacter* (a-si-ne'tō-bak-tėr), and
certain strains of the genera *Streptococcus* and *Staphylococcus*.

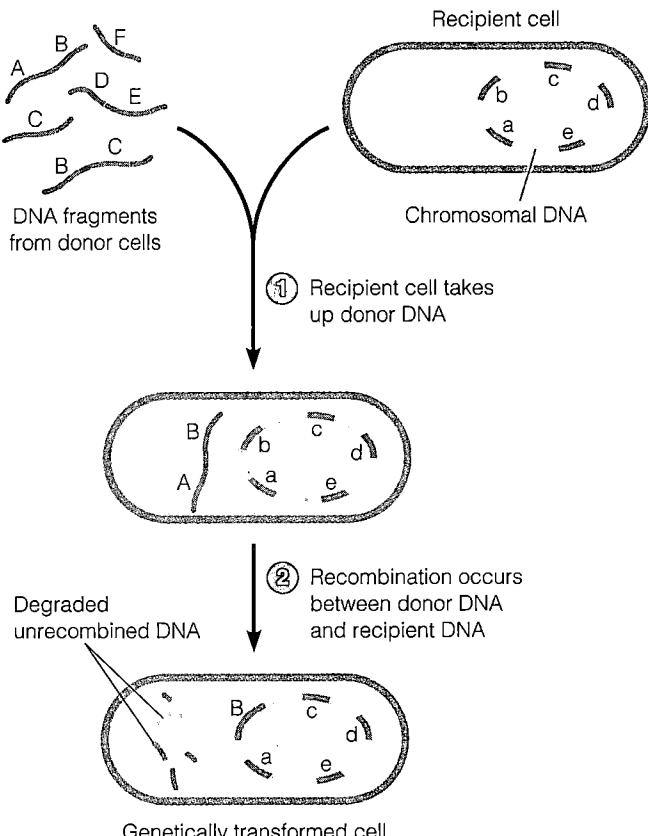

DNA fragments
from donor cells

Recipient cell

Chromosomal DNA

① Recipient cell takes
up donor DNA

Degraded
unrecombined DNA

② Recombination occurs
between donor DNA
and recipient DNA

Genetically transformed cell

FIGURE 8.25   The mechanism of genetic transformation in bacteria.

■ A recipient cell with a new combination of genes is a hybrid or recombinant cell.

Transformation works best when the donor and recipient cells are very closely related. Even though only a small portion of a cell's DNA is transferred to the recipient, the molecule that must pass through the recipient cell wall and membrane is still very large. When a recipient cell is in a physiological state in which it can take up the donor DNA, it is said to be competent. **Competence** results from alterations in the cell wall that make it permeable to large DNA molecules.

The well-understood and widely used bacterium *E. coli* is not naturally competent for transformation. However, a simple laboratory treatment enables *E. coli* to readily take up DNA. The discovery of this treatment has enabled researchers to use *E. coli* for genetic engineering, discussed in Chapter 9.

Conjugation in Bacteria

Another mechanism by which genetic material is transferred from one bacterium to another is known as **conjugation.** Conjugation is mediated by one kind of

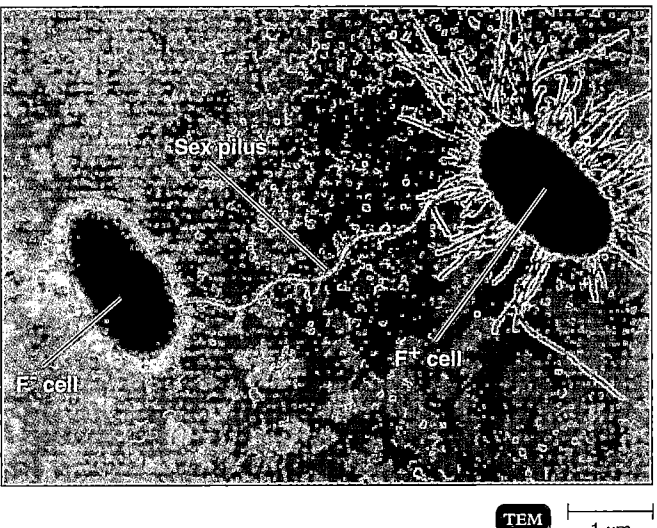

FIGURE 8.26   Bacterial conjugation. The sex pilus connecting these cells undergoing conjugation allows the transfer of genetic information. At the actual time of genetic exchange, the cells' connecting bridge contracts and the cells are much closer together. Notice that one cell has numerous fimbriae.

■ In gram-negative bacteria, the plasmid carries genes that code for the synthesis of sex pili.

plasmid, a circular piece of DNA that replicates independently from the cell's chromosome (discussed later in this chapter). However, plasmids differ from bacterial chromosomes in that the genes they carry are usually not essential for the growth of the cell under normal conditions. The plasmids responsible for conjugation are transmissible between cells during conjugation.

Conjugation differs from transformation in two major ways. First, conjugation requires direct cell-to-cell contact. Second, the conjugating cells must generally be of opposite mating type; donor cells must carry the plasmid, and recipient cells usually do not. In gram-negative bacteria, the plasmid carries genes that code for the synthesis of *sex pili,* projections from the donor's cell surface that contact the recipient and help bring the two cells into direct contact. Gram-positive bacterial cells produce sticky surface molecules that cause cells to come into direct contact with each other. In the process of conjugation, the plasmid is replicated during the transfer of a single-stranded copy of the plasmid DNA to the recipient, where the complementary strand is synthesized.

Because most experimental work on conjugation has been done with *E. coli,* we will describe the process in this organism. In *E. coli,* the **F factor (fertility factor)** was the first plasmid observed to be transferred between cells during conjugation (Figure 8.26). Donors carrying F factors (F$^+$ cells) transfer the plasmid to recipients (F$^-$

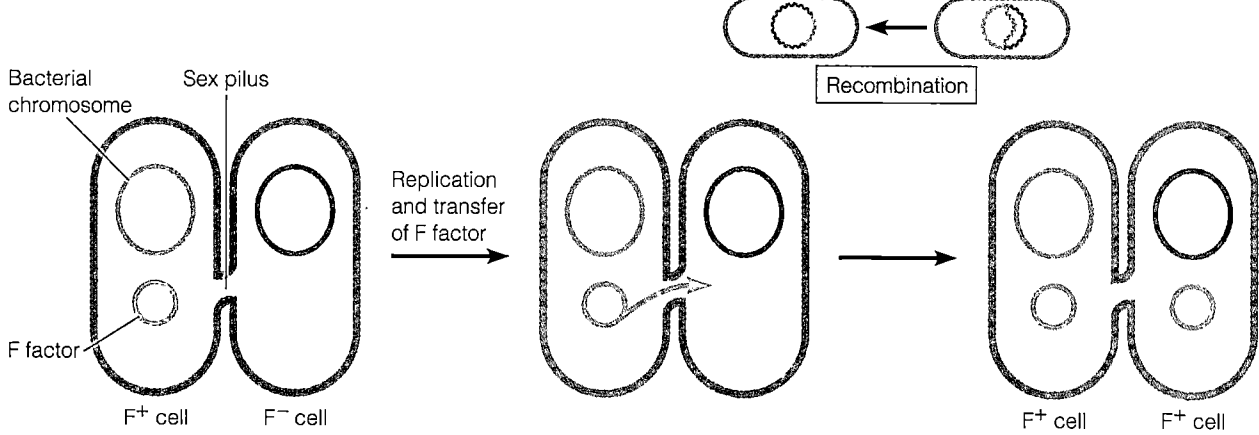

(a) When an F factor (a plasmid) is transferred from a donor (F⁺) to a recipient (F⁻), the F⁻ cell is converted into an F⁺ cell.

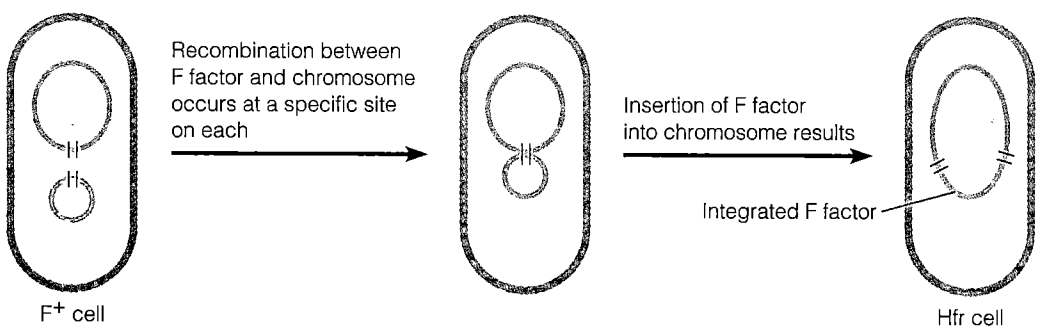

(b) When an F factor becomes integrated into the chromosome of an F⁺ cell, it makes the cell a high frequency of recombination (Hfr) cell.

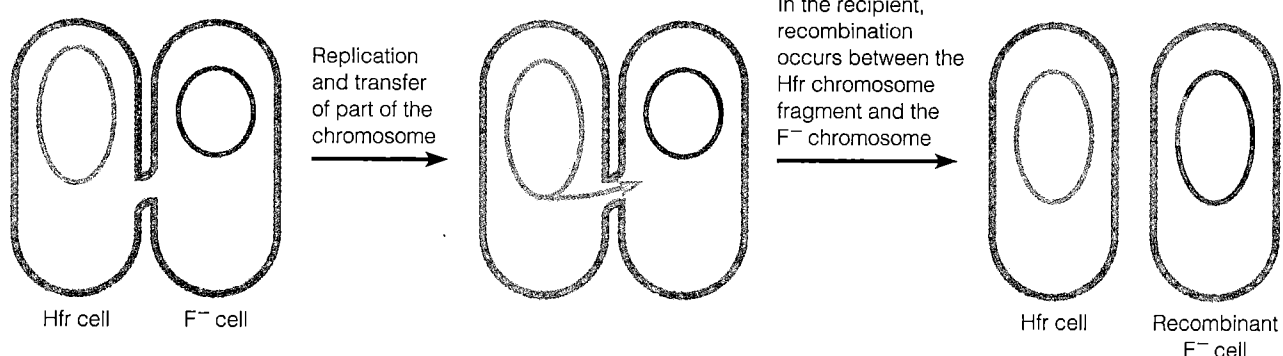

(c) When an Hfr donor passes a portion of its chromosome into an F⁻ recipient, a recombinant F⁻ cell results.

FIGURE 8.27   Conjugation in *E. coli*.

■ How does conjugation differ from transformation?

cells), which become F⁺ cells as a result (Figure 8.27a). In some cells carrying F factors, the factor integrates into the chromosome, converting the F⁺ cell to an **Hfr cell** (high frequency of recombination) (Figure 8.27b). When conjugation occurs between an Hfr cell and an F⁻ cell,

the Hfr cell's chromosome (with its integrated F factor) replicates, and a parental strand of the chromosome is transferred to the recipient cell (Figure 8.26c). Replication of the Hfr chromosome begins in the middle of the integrated F factor, and a small piece of the F factor leads

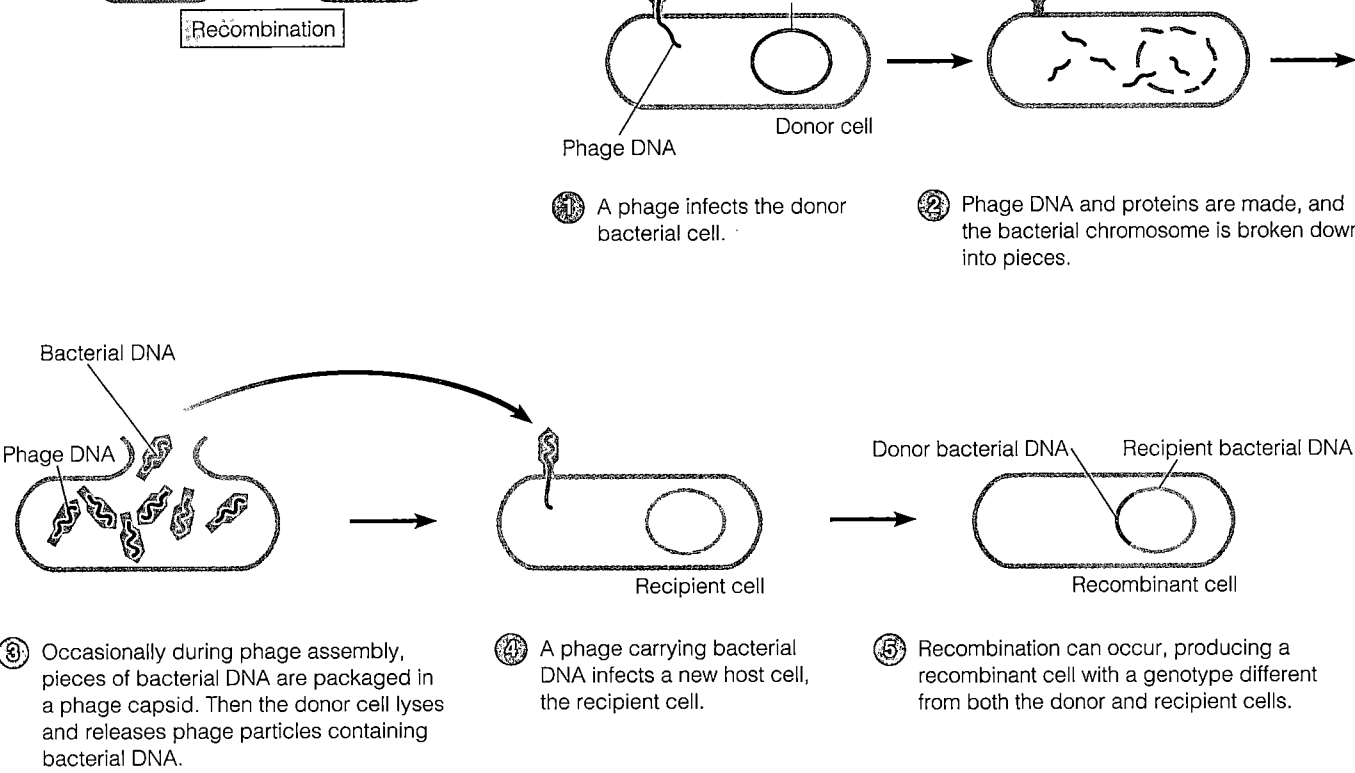

① A phage infects the donor bacterial cell.

② Phage DNA and proteins are made, and the bacterial chromosome is broken down into pieces.

③ Occasionally during phage assembly, pieces of bacterial DNA are packaged in a phage capsid. Then the donor cell lyses and releases phage particles containing bacterial DNA.

④ A phage carrying bacterial DNA infects a new host cell, the recipient cell.

⑤ Recombination can occur, producing a recombinant cell with a genotype different from both the donor and recipient cells.

FIGURE 8.28 Transduction by a bacteriophage. Shown here is generalized transduction, in which any bacterial DNA can be transferred from one cell to another.

■ What is transduction?

the chromosomal genes into the F^- cell. Usually, the chromosome breaks before it is completely transferred. Once within the recipient cell, donor DNA can recombine with the recipient's DNA. (Donor DNA that is not integrated is degraded.) Therefore, by conjugation with an Hfr cell, an F^- cell may acquire new versions of chromosomal genes (just as in transformation). However, it remains an F^- cell because it did not receive a complete F factor during conjugation.

Conjugation is used to map the location of genes on a bacterial chromosome (see Figure 8.1b). Notice that genes for the synthesis of threonine *(thr)* and leucine *(leu)* are first, reading clockwise from 0. Their locations were determined by conjugation experiments. Assume that conjugation is allowed for only 1 minute between an Hfr strain that is his^+, pro^+, thr^+ and leu^+, and an F^- strain that is his^-, pro^-, thr^-, and leu^-. If the F^- acquired the ability to synthesize threonine, then the *thr* gene is located early in the chromosome, between 0 and 1 minute. If after 2 minutes the F^- cell now becomes thr^+ and leu^+, the order of these two genes on the chromosome must be *thr, leu.*

Transduction in Bacteria

A third mechanism of genetic transfer between bacteria is **transduction.** In this process, bacterial DNA is transferred from a donor cell to a recipient cell inside a virus that infects bacteria, called a **bacteriophage,** or **phage.** (Phages will be discussed further in Chapter 13.)

To understand how transduction works, we will consider the life cycle of one type of transducing phage of *E. coli;* this phage carries out **generalized transduction** (Figure 8.28). **①** In the process of infection, the phage attaches to the donor bacterial cell wall and injects its DNA into the bacterium. **②** The phage DNA acts as a template for the synthesis of new phage DNA and also directs the synthesis of phage protein coats. During phage development inside the infected cell, the bacterial chromosome is broken apart by phage enzymes, and **③** at least some pieces of bacterial DNA are mistakenly packaged inside phage protein coats. (Even plasmid DNA or DNA of another virus that is inside the cell can be given phage protein coats.) The resulting phage particles then carry bacterial DNA instead of phage DNA. **④** When the released phage particles later infect a new population of bacteria, bacterial genes will be transferred to the newly infected recipient cells at low frequency. ❺ Transduction of cellular DNA by a virus can lead to recombination between the DNA of the donor host cell and the DNA of the recipient host cell. The process of generalized transduction is typical of bacteriophages such as phage P1 of *E. coli* and phage P22 of *Salmonella*.

All genes contained within a bacterium infected by a generalized transducing phage are equally likely to be packaged in a phage coat and transferred. In another type of transduction, called **specialized transduction**, only certain bacterial genes are transferred. In one type of specialized transduction, the phage codes for certain toxins produced by their bacterial hosts, such as *Corynebacterium diphtheriae* for diphtheria toxin, *Streptococcus pyogenes* for erythrogenic toxin, and *E. coli* O157:H7 for shigalike toxin that causes bloody diarrhea. Specialized transduction will be discussed in Chapter 13 on page 385. In addition to mutation, transformation, and conjugation, transduction is another way bacteria acquire new genotypes.

Plasmids and Transposons

Learning Objective

■ *Describe the functions of plasmids and transposons.*

Plasmids and transposons are genetic elements that provide additional mechanisms for genetic change. They occur in both prokaryotic and eukaryotic organisms, but this discussion focuses on their role in genetic change in prokaryotes.

Plasmids

As mentioned earlier, plasmids are self-replicating, gene-containing circular pieces of DNA about 1–5% the size of the bacterial chromosome (Figure 8.29a). They are found mainly in bacteria but also in some eukaryotic microorganisms, such as *Saccharomyces cerevisiae*. The F factor is a **conjugative plasmid** that carries genes for sex pili and for the transfer of the plasmid to another cell. Although plasmids are usually dispensable, under certain conditions genes carried by plasmids can be crucial to the survival and growth of the cell. For example, **dissimilation plasmids** code for enzymes that trigger the catabolism of certain unusual sugars and hydrocarbons. Some species of *Pseudomonas* can actually use such exotic substances as toluene, camphor, and hydrocarbons of petroleum as primary carbon and energy sources because they have catabolic enzymes encoded by genes carried on plasmids. Such specialized capabilities permit the survival

of those microorganisms in very diverse and challenging environments. Because of their ability to degrade and detoxify a variety of unusual compounds, many of them are being investigated for possible use in the cleanup of environmental wastes.

Other plasmids code for proteins that enhance the pathogenicity of a bacterium. The strain of *E. coli* that causes infant diarrhea and traveler's diarrhea carries plasmids that code for toxin production and for bacterial attachment to intestinal cells. Without these plasmids, *E. coli* is a harmless resident of the large intestine; with them, it is pathogenic. Other plasmid-encoded toxins include the exfoliative toxin of *Staphylococcus aureus, Clostridium tetani* neurotoxin, and toxins of *Bacillus anthracis*. Still other plasmids contain genes for the synthesis of **bacteriocins,** toxic proteins that kill other bacteria. These plasmids have been found in many bacterial genera, and they are useful markers for the identification of certain bacteria in clinical laboratories.

R factors (resistance factors) are plasmids that have significant medical importance. They were first discovered in Japan in the late 1950s after several dysentery epidemics. In some of these epidemics, the infectious agent was resistant to the usual antibiotic. Following isolation, the pathogen was also found to be resistant to a number of different antibiotics. In addition, other normal bacteria from the patients (such as *E. coli*) proved to be resistant as well. Researchers soon discovered that resistance was acquired by these bacteria through the spread of genes from one organism to another. The plasmids that mediated this transfer are R factors.

R factors carry genes that confer upon their host cell resistance to antibiotics, heavy metals, or cellular toxins. Many R factors contain two groups of genes. One group is called the **resistance transfer factor (RTF)** and includes genes for plasmid replication and conjugation. The other group, the **r-determinant,** has the resistance genes; it codes for the production of enzymes that inactivate certain drugs or toxic substances (Figure 8.29b). Different R factors, when present in the same cell, can recombine to produce R factors with new combinations of genes in their r-determinants.

In some cases, the accumulation of resistance genes within a single plasmid is quite remarkable. For example, Figure 8.29b shows a genetic map of resistance plasmid R100. Carried on this plasmid are resistance genes for sulfonamides, streptomycin, chloramphenicol, and tetracycline, as well as genes for resistance to mercury. This particular plasmid can be transferred between a number of enteric species, including *Escherichia, Klebsiella,* and *Salmonella*.

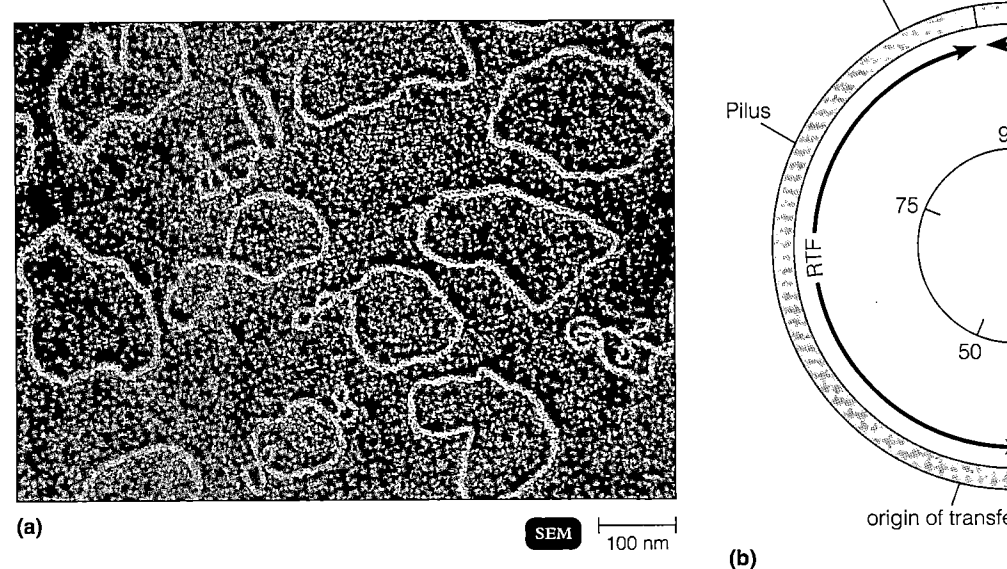

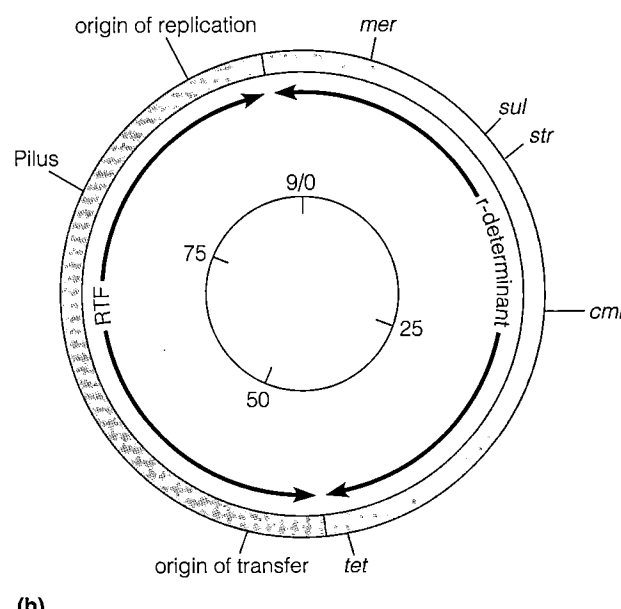

(a)

SEM | 100 nm |

(b)

FIGURE 8.29 R factor, a type of plasmid. (a) A plasmid isolated from the bacterium
Bacteroides fragilis that encodes resistance to the antibiotic clindamycin. (b) A diagram
of an R factor, which has two parts: The RTF contains genes needed for plasmid replication and transfer of the plasmid by conjugation; the r-determinant carries genes for
resistance to four different antibiotics.

■ **Why are R factors important in the treatment of infectious diseases?**

R factors present very serious problems for the treatment of infectious diseases with antibiotics. The widespread use of antibiotics in medicine and agriculture
(many types of animal feed contain antibiotics) has led to
the preferential survival (selection) of bacteria that have
R factors, so populations of resistant bacteria grow larger
and larger. The transfer of resistance between bacterial
cells of a population, and even between bacteria of different genera, also contributes to the problem. The ability to sexually reproduce with members of its own
species defines a eukaryotic species. However, a bacterial
species can conjugate and transfer plasmids to other
species. *Neisseria* may have acquired its penicillinase-
producing plasmid from *Streptococcus*, and *Agrobacterium*
can transfer plasmids to plant cells (see Figure 9.17 on
page 267). Nonconjugative plasmids may be transferred
from one cell to another by inserting themselves into a
conjugative plasmid or a chromosome. Insertion is made
possible by an insertion sequence, which will be discussed shortly.

Plasmids are an important tool for genetic engineering, discussed in Chapter 9 on page 252.

Transposons

Transposons are small segments of DNA that can move
(be "transposed") from one region of a DNA molecule to
another. These pieces of DNA are 700–40,000 base pairs
long.

In the 1950s, American geneticist Barbara McClintock discovered transposons in corn, but they occur in all
organisms and have been studied most thoroughly in microorganisms. They may move from one site to another
site on the same chromosome, or to another chromosome or plasmid. As you might imagine, the frequent
movement of transposons could wreak havoc inside a
cell. For example, as transposons move about on chromosomes, they may insert themselves *within* genes, inactivating them. Fortunately, transposition occurs relatively
rarely. The frequency of transposition is comparable to
the spontaneous mutation rate that occurs in bacteria—
that is, from 10^{-5} to 10^{-7} per generation.

All transposons contain the information for their own
transposition. As shown in Figure 8.30a, the simplest
transposons, also called **insertion sequences (IS),** contain only a gene that codes for an enzyme (*transposase,*

Repression and Induction (pp. 222-223)

1. Repression controls the synthesis of one or several (repressible) enzymes.

2. When cells are exposed to a particular end-product, the synthesis of enzymes related to that product decreases.

3. In the presence of certain chemicals (inducers), cells synthesize more enzymes. This process is called induction.

4. An example of induction is the production of β-galactosidase by E. coli in the presence of lactose, lactose can then be metabolized.

The Operon Model of Gene Expression (pp. 223-225)

1. The formation of enzymes is determined by structural genes.

2. In bacteria, a group of coordinately regulated structural genes with related metabolic functions, plus the promoter and operator sites that control their transcription, are called an operon.

3. In the operon model for an inducible system, a regulatory gene codes for the repressor protein.

4. When the inducer is absent, the repressor binds to the operator, and no mRNA is synthesized.

5. When the inducer is present, it binds to the repressor so that it cannot bind to the operator; thus, mRNA is made, and enzyme synthesis is induced.

6. In repressible systems, the repressor requires a corepressor in order to bind to the operator site; thus, the corepressor controls enzyme synthesis.

7. Transcription of structural genes for catabolic enzymes (such as β-galactosidase) is induced by the absence of glucose. Cyclic AMP and CRP must bind to a promoter in the presence of an alternative carbohydrate.

8. The presence of glucose inhibits the metabolism of alternative carbon sources by catabolic repression.

MUTATION: CHANGE IN THE GENETIC MATERIAL (pp. 226-233)

[ST] *Genetics: Mutation*

1. A mutation is a change in the nitrogenous base sequence of DNA; that change causes a change in the product coded for by the mutated gene.

2. Many mutations are neutral, some are disadvantageous, and others are beneficial.

Types of Mutations (pp. 227-228)

1. A base substitution occurs when one base pair in DNA is replaced with a different base pair.

2. Alterations in DNA can result in missense mutations (which cause amino acid substitutions) or nonsense mutations (which create stop codons).

3. In a frameshift mutation, one or a few base pairs are deleted or added to DNA.

4. Mutagens are agents in the environment that cause permanent changes in DNA.

5. Spontaneous mutations occur without the presence of any mutagen.

Mutagens (pp. 228-230)

1. Chemical mutagens include base-pair mutagens (for example, nitrous acid), nucleoside analogs (e.g., 2-aminopurine and 5-bromouracil), and frameshift mutagens (e.g., benzpyrene).

2. Ionizing radiation causes the formation of ions and free radicals that react with DNA; base substitutions or breakage of the sugar-phosphate backbone results.

3. Ultraviolet (UV) radiation is nonionizing; it causes bonding between adjacent thymines.

4. Damage to DNA caused by UV radiation can be repaired by enzymes that cut out and replace the damaged portion of DNA.

5. Light-repair enzymes repair thymine dimers in the presence of visible light.

The Frequency of Mutation (pp. 230-231)

1. Mutation rate is the probability that a gene will mutate when a cell divides; the rate is expressed as 10 to a negative power.

2. Mutations usually occur randomly along a chromosome.

3. A low rate of spontaneous mutations is beneficial in providing the genetic diversity needed for evolution.

Identifying Mutants (p. 231)

1. Mutants can be detected by selecting or testing for an altered phenotype.

2. Positive selection involves the selection of mutant cells and the rejection of nonmutated cells.

3. Replica plating is used for negative selection—to detect, for example, auxotrophs that have nutritional requirements not possessed by the parent (nonmutated) cell.

Identifying Chemical Carcinogens (pp. 231-233)

1. The Ames test is a relatively inexpensive and rapid test for identifying possible chemical carcinogens.

2. The test assumes that a mutant cell can revert to a normal cell in the presence of a mutagen, and that many mutagens are carcinogens.

3. Histidine auxotrophs of Salmonella are exposed to an enzymatically treated potential carcinogen, and reversions to the nonmutant state are selected.

GENETIC TRANSFER AND RECOMBINATION (p. 233-242)

[ST] *Genetics: Recombination*

1. Genetic recombination, the rearrangement of genes from separate groups of genes, usually involves DNA from different organisms; it contributes to genetic diversity.

2. In crossing over, genes from two chromosomes are recombined into one chromosome containing some genes from each original chromosome.

3. Vertical gene transfer occurs during reproduction when genes are passed from an organism to its offspring.

4. Horizontal gene transfer in bacteria involves a portion of the cell's DNA being transferred from donor to recipient.

5. When some of the donor's DNA has been integrated into the recipient's DNA, the resultant cell is called a recombinant.

Transformation in Bacteria (pp. 234–237)

1. During this process, genes are transferred from one bacterium to another as "naked" DNA in solution.

2. This process was first demonstrated in *Streptococcus pneumoniae,* and occurs naturally among a few genera of bacteria.

Conjugation in Bacteria (pp. 237–239)

1. This process requires contact between living cells.

2. One type of genetic donor cell is an F^+; recipient cells are F^-. F cells contain plasmids called F factors; these are transferred to the F^- cells during conjugation.

3. When the plasmid becomes incorporated into the chromosome, the cell is called an Hfr (high frequency of recombination) cell.

4. During conjugation, an Hfr cell can transfer chromosomal DNA to an F^- cell. Usually, the Hfr chromosome breaks before it is fully transferred.

Transduction in Bacteria (pp. 239–240)

1. In this process, DNA is passed from one bacterium to another in a bacteriophage and is then incorporated into the recipient's DNA.

2. In generalized transduction, any bacterial genes can be transferred.

Plasmids and Transposons (pp. 240–242)

1. Plasmids are self-replicating circular molecules of DNA carrying genes that are not usually essential for the cell's survival.

2. There are several types of plasmids, including conjugative plasmids, dissimilation plasmids, plasmids carrying genes for toxins or bacteriocins, and resistance factors.

3. Transposons are small segments of DNA that can move from one region to another region of the same chromosome, or to a different chromosome or a plasmid.

4. Transposons are found in the main chromosomes of organisms, in plasmids, and in the genetic material of viruses. They vary from simple (insertion sequences) to complex.

5. Complex transposons can carry any type of gene, including antibiotic-resistance genes, and are thus a natural mechanism for moving genes from one chromosome to another.

GENES AND EVOLUTION (p. 242)

1. Diversity is the precondition for evolution.

2. Genetic mutation and recombination provide a diversity of organisms, and the process of natural selection allows the growth of those best adapted to a given environment.

Study Questions

REVIEW

1. Briefly describe the components of DNA, and explain its functional relationship to RNA and protein.

2. Draw a diagram showing a portion of a chromosome undergoing replication.
   a. Identify the replication fork.
   b. What is the role of DNA polymerase? Of RNA polymerase?
   c. How does this process represent semiconservative replication?

3. The following is a code for a strand of DNA.

```
DNA 3'     A T A T _ _ _ T T T _ _ _ _ _ _ _ _ _
           1 2 3 4 5 6 7 8 9 10 11 12 13 14 15 16 17 18 19
mRNA                       C G U         U G A
tRNA                             U G G
Amino acid        Met   ____  ____  ____  ____
ATAT = Promoter sequence
```

a. Using the genetic code provided in Figure 8.9, fill in the blanks to complete the segment of DNA shown.
b. Fill in the blanks to complete the sequence of amino acids coded for by this strand of DNA.
c. Write the code for the complementary strand of DNA completed in part a.
d. What would be the effect if C was substituted for T at base 10?
e. What would be the effect if A was substituted for G at base 11?
f. What would be the effect if G was substituted for T at base 14?
g. What would be the effect if C was inserted between bases 9 and 10?
h. How would UV radiation affect this strand of DNA?
i. Identify a nonsense sequence in this strand of DNA.

4. Describe translation, and be sure to include the following terms: ribosome, rRNA, amino acid activation, tRNA, anticodon, and codon.

5. Explain how you would find an antibiotic-resistant mutant by direct selection and how you would find an antibiotic-sensitive mutant by indirect selection.

6. Match the following examples of mutagens.

   _____ A mutagen that is incorporated into DNA in place of a normal base

   _____ A mutagen that causes the formation of highly reactive ions

   _____ A mutagen that alters adenine so that it base-pairs with cytosine

   _____ A mutagen that causes insertions

   _____ A mutagen that causes the formation of pyrimidine dimers

   a. Frameshift mutagen
   b. Nucleoside analog
   c. Base-pair mutagen
   d. Ionizing radiation
   e. Nonionizing radiation

7. Describe the principle of the Ames test for identifying chemical carcinogens.

8. Define plasmids, and explain the relationship between F factors and conjugation.

9. Use the following metabolic pathway to answer the questions that follow it.

   enzyme *a*   enzyme *b*
   Substrate *A* ⟶ Intermediate *B* ⟶ End-product *C*

   a. If enzyme *a* is inducible and is not being synthesized at present, a _____ protein must be bound tightly to the _____ site. When the inducer is present, it will bind to the _____ so that _____ can occur.

   b. If enzyme *a* is repressible, end-product *C*, called a _____, causes the _____ to bind to the _____. What causes derepression?

   c. If enzyme *a* is constitutive, what effect, if any, will the presence of *A* or *C* have on it?

10. Identify three ways of preventing mistakes in DNA.

11. Define the following terms:
    a. genotype
    b. phenotype
    c. recombination

12. Which sequence is the best target for damage by UV radiation: AGGCAA, CTTTGA, or GUAAAU? Why aren't all bacteria killed when they are exposed to sunlight?

13. You are provided with cultures with the following characteristics:

    Culture 1: F$^+$, genotype A^+ B^+ C^+
    Culture 2: F$^-$, genotype A^- B^- C^-

    a. Indicate the possible genotypes of a recombinant cell resulting from the conjugation of cultures 1 and 2.

    b. Indicate the possible genotypes of a recombinant cell resulting from conjugation of the two cultures after the F$^+$ has become an Hfr-cell.

14. Why are semiconservative replication and degeneracy of the genetic code advantageous to the survival of species?

15. Why are mutation and recombination important in the process of natural selection and the evolution of organisms?

MULTIPLE CHOICE

Match the following terms to the definitions in questions 1 and 2.
   a. conjugation
   b. transcription
   c. transduction
   d. transformation
   e. translation

1. The transfer of DNA from a donor to a recipient cell by a bacteriophage.

2. The transfer of DNA from a donor to a recipient as naked DNA in solution.

3. Feedback inhibition differs from repression because feedback inhibition
   a. is less precise.
   b. is slower acting.
   c. stops the action of preexisting enzymes.
   d. stops the synthesis of new enzymes.
   e. all of the above

4. Bacteria can acquire antibiotic resistance by
   a. mutation.
   b. insertion of transposons.
   c. acquiring plasmids.
   d. all of the above
   e. none of the above

5. Suppose you inoculate three flasks of minimal salts broth with *E. coli*. Flask A contains glucose. Flask B contains glucose and lactose. Flask C contains lactose. After a few hours of incubation, you test the flasks for the presence of β-galactosidase. Which flask(s) do you predict will have this enzyme?
   a. A
   b. B
   c. C
   d. A and B
   e. B and C

6. Plasmids differ from transposons because plasmids
   a. become inserted into chromosomes.
   b. are self-replicated outside the chromosome.
   c. move from chromosome to chromosome.
   d. carry genes for antibiotic resistance.
   e. none of the above

Use the following choices to answer questions 7 and 8.
   a. catabolic repression
   b. DNA polymerase
   c. induction
   d. repression
   e. translation

7. The mechanism by which the presence of glucose inhibits the *lac* operon.

8. The mechanism by which lactose controls the *lac* operon.

9. Two daughter cells are most likely to inherit which one of the following from the parent cell?
   a. a change in a nucleotide in mRNA
   b. a change in a nucleotide in tRNA
   c. a change in a nucleotide in rRNA
   d. a change in a nucleotide in DNA
   e. a change in a protein

10. Which of the following is *not* a method of horizontal gene transfer?
    a. binary fission
    b. conjugation
    c. integration of a transposon
    d. transduction
    e. transformation

CRITICAL THINKING

1. Nucleoside analogs and ionizing radiation are used in the treatment of cancer. These mutagens can cause cancer, so how do you suppose they are used to treat the disease?

2. Replication of the *E. coli* chromosome takes 40–45 minutes, but the organism has a generation time of 26 minutes. How does the cell have time to make complete chromosomes for each daughter cell? For each granddaughter cell?

3. *Pseudomonas* has a plasmid containing the *mer* operon, which includes the gene for mercuric reductase. This enzyme catalyzes the reduction of the mercuric ion Hg^{2+} to the uncharged form of mercury, Hg^0. Hg^{2+} is quite toxic to cells; Hg^0 is not.
   a. What do you suppose is the inducer for this operon?
   b. The protein encoded by one of the *mer* genes binds Hg^{2+} in the periplasm and brings it into the cell. Why would a cell bring in a toxin?
   c. What is the value of the *mer* operon to *Pseudomonas*?

CLINICAL APPLICATIONS

1. Chloroquine, erythromycin, and acyclovir are used to treat microbial infections. Chloroquine acts by fitting between base pairs in the DNA molecule. Erythromycin binds in front of the A site on the 50S subunit of a ribosome. Acyclovir is a guanine analog.
   a. What steps in protein synthesis are inhibited by each drug?
   b. Which drug is more effective against bacteria? Why?
   c. Which drug is more effective against viruses? Why?
   d. Which drugs will have effects on the host's cells? Why?
   e. Use the index to identify the disease for which chloroquine is primarily used. Why is it more effective than erythromycin for treating this disease?
   f. Use the index to identify the disease for which acyclovir is primarily used. Why is it more effective than erythromycin for treating this disease?

2. HIV, the virus that causes AIDS, was isolated from three individuals, and the amino acid sequences for the viral coat were determined. Of the amino acid sequences shown below, which two of the viruses are most closely related? How can these amino acid sequences be used to identify the source of a virus?

Patient	Viral Amino Acid Sequence
A	Asn Gln Thr Ala Ala Ser Lys Asn Ile Asp Ala Glu Leu
B	. Asn Leu His Ser Asp Lys Ile Asn Ile Ile Leu Gln Leu
C	Asn Gln Thr Ala Asp Ser Ile Val Ile Asp Ala Cys Leu

3. Human herpesvirus-8 (HHV-8) is common in certain parts of Africa, the Middle East, and the Mediterranean, but rare elsewhere except in AIDS patients. Genetic analyses indicate that the African strain is not changing, whereas the Western strain is accumulating changes. Using the portions of the HHV-8 genomes (shown below) that encode one of the viral proteins, how similar are these two viruses? What mechanism can account for the changes? What disease does HHV-8 cause?

   Western   3' ATGGAGTTCTTCTGGACAAGA
   African   3' ATAAACTTTTTCTTGACAACG

Learning with Technology

MP = The Microbiology Place website   **ST** = Student Tutorial CD-ROM

MP   Don't forget to go to The Microbiology Place website (http://www.microbiologyplace.com) to take the practice tests, explore the interactive activity and case study, and check out the news articles and web links for this chapter.

ST   Remember there is also a quiz for this chapter on the Microbiology Interactive Student Tutorial CD-ROM.

chapter 9

Biotechnology and Recombinant DNA

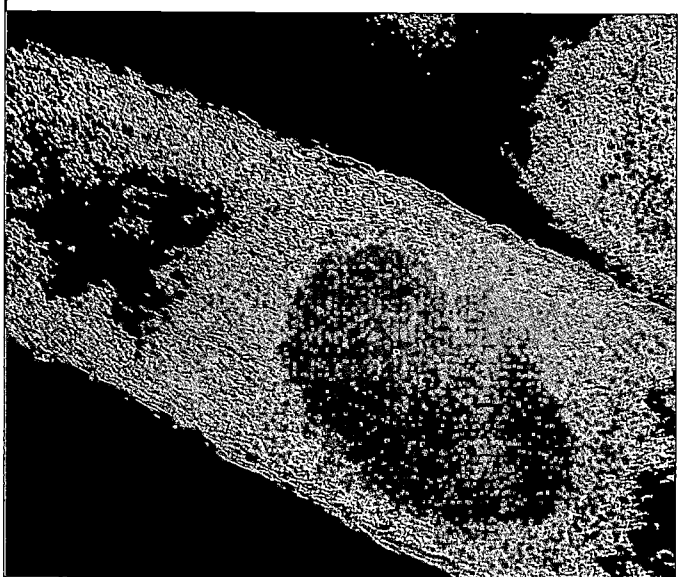

Escherichia coli. *This bacterium has been genetically engineered to produce a human protein, gamma interferon. Unlike human cells, it doesn't secrete the protein so the wells will be lysed to harvest the protein.*

For thousands of years, people have been consuming foods that are produced by the action of microorganisms. Bread, chocolate, and soy sauce are some of the best known examples. But it was only just over 100 years ago that scientists showed that microorganisms are responsible for these products. This knowledge opened the way for using microorganisms to produce other important products. Since World War I, microbes have been used to produce a variety of chemicals, such as ethyl alcohol, acetone, and citric acid. Since World War II, microorganisms have been grown on a large scale to produce antibiotics. More recently, microbes and their enzymes are replacing a variety of chemical processes involved in manufacturing such products as paper, textiles, and fructose. Using microbes or their enzymes instead of chemical syntheses offers several advantages: microbes may use inexpensive, abundant raw materials, such as starch; microbes work at normal temperatures and pressure, thereby avoiding the need for expensive and dangerous pressurized systems; and microbes don't produce toxic, hard-to-treat wastes. Currently, the search for microbes that produce new antibiotics and useful enzymes continues to broaden and expand (see the box on page 251).

The processes used for large-scale manufacture of microbes and their products will be discussed in Chapter 28. In this chapter you will learn the tools and techniques that are used to research and develop a product.

Introduction to Biotechnology

Learning Objective

- *Compare and contrast biotechnology, genetic engineering, and recombinant DNA technology.*

All the products just discussed are the result of **biotechnology,** the use of microorganisms, cells, or cell components to make a product. Microbes have been used in the commercial production of foods, vaccines, antibiotics, and vitamins for years. Bacteria are also used in mining to extract valuable elements from ore (see Figure 28.13 on page 785). Additionally, animal cells have been used to produce viral vaccines since the 1950s. Until the 1980s, products made by living cells were all naturally made by the cells; the role of scientists was to find the appropriate cell and develop a method for large-scale cultivation of the cells.

Now, microorganisms as well as entire plants are being used as "factories" to produce chemicals that the organisms don't naturally make. The latter is made possible by inserting genes into cells by a process called **genetic engineering.** The development of genetic engineering is expanding the practical applications of biotechnology almost beyond imagination.

Recombinant DNA Technology

Recall from Chapter 8 that recombination of DNA occurs naturally in microbes. In the 1970s and 1980s, scientists developed artificial techniques for making recombinant DNA. These techniques are called **recombinant DNA technology.**

A gene from a vertebrate animal, including humans, can be inserted into the DNA of a bacterium, or a gene from a virus into a yeast. In many cases, the recipient can then be made to express the gene, which may code for a commercially useful product. Thus, bacteria with genes for human insulin are now being used to produce insulin for treating diabetes, and a vaccine for hepatitis B is being made by yeast carrying a gene for part of the hepatitis virus (the yeast produces a viral coat protein). Scientists hope that such an approach may prove useful in producing vaccines against other infectious agents, eliminating the need to use whole organisms, as in conventional vaccines.

The new recombinant DNA techniques can also be used to make thousands of copies of the same DNA molecule—to *amplify* DNA, thus generating sufficient DNA for various kinds of experimentation and analysis. This technique has practical application for identifying microbes, such as viruses, that can't be cultured.

An Overview of Recombinant DNA Procedures

Learning Objective

■ *Identify the roles of a clone and a vector in genetic engineering.*

Figure 9.1 presents an overview of some of the procedures typically used in genetic engineering, along with some promising applications. ❶–❷ After several preparatory steps, ❸ the gene of interest is inserted into the vector DNA in vitro. In this example, the vector is a plasmid. The DNA molecule chosen as a vector must be a self-replicating type, such as a plasmid or a viral genome. ❹ Next, this recombinant vector DNA is taken up by a cell such as a bacterium, where it can multiply. ❺ The cell containing the recombinant vector is then grown in culture to form a **clone** of many genetically identical cells, each of which

carries a copy of the vector. This cell clone therefore contains many copies of the gene of interest. This is why DNA vectors are often called *gene-cloning vectors,* or simply *cloning vectors.* (In addition to referring to a culture of identical cells, the word *clone* is also routinely used as a verb, to describe the entire process, as in "to clone a gene.")

The final step varies according to whether the gene itself or the product of the gene is of interest. ❻ From the cell clone, the genetic engineer may isolate ("harvest") large quantities of the gene of interest, which may then be used for a variety of purposes. The gene may even be inserted into another vector for introduction into another kind of cell (such as a plant or animal cell). Alternatively, ❸ if the gene of interest is expressed (transcribed and translated) in the cell clone, ❼ its protein product can be harvested and used for a variety of purposes.

The advantages of genetic engineering for obtaining such proteins is illustrated by one of its early successes, human growth hormone (hGH). Some individuals do not produce adequate amounts of hGH, and their growth is stunted. In the past, hGH needed to correct this deficiency had to be obtained from human pituitary glands at autopsy. (hGH from other animals is not effective in humans.) This practice was not only expensive but also dangerous because on several occasions neurological diseases were transmitted with the hormone. Human growth hormone produced by genetically engineered *E. coli* is a pure and cost-effective product. Recombinant DNA techniques also result in faster production of the hormone than traditional methods might allow.

Tools of Biotechnology

Learning Objective

■ *Compare selection and mutation.*

Research scientists and technicians isolate bacteria and fungi from natural environments such as soil and water to find or *select* the organisms that produce a desired product. The selected organism can be mutated to make more product or to make a better product.

Selection

In nature, organisms with characteristics that enhance survival are more likely to survive and reproduce than are variants that lack the desirable traits. This is called *natural selection.* Humans use **artificial selection** to select desirable breeds of animals or strains of plants to cultivate. As microbiologists learned how to isolate and grow microorganisms in pure culture, they were able to

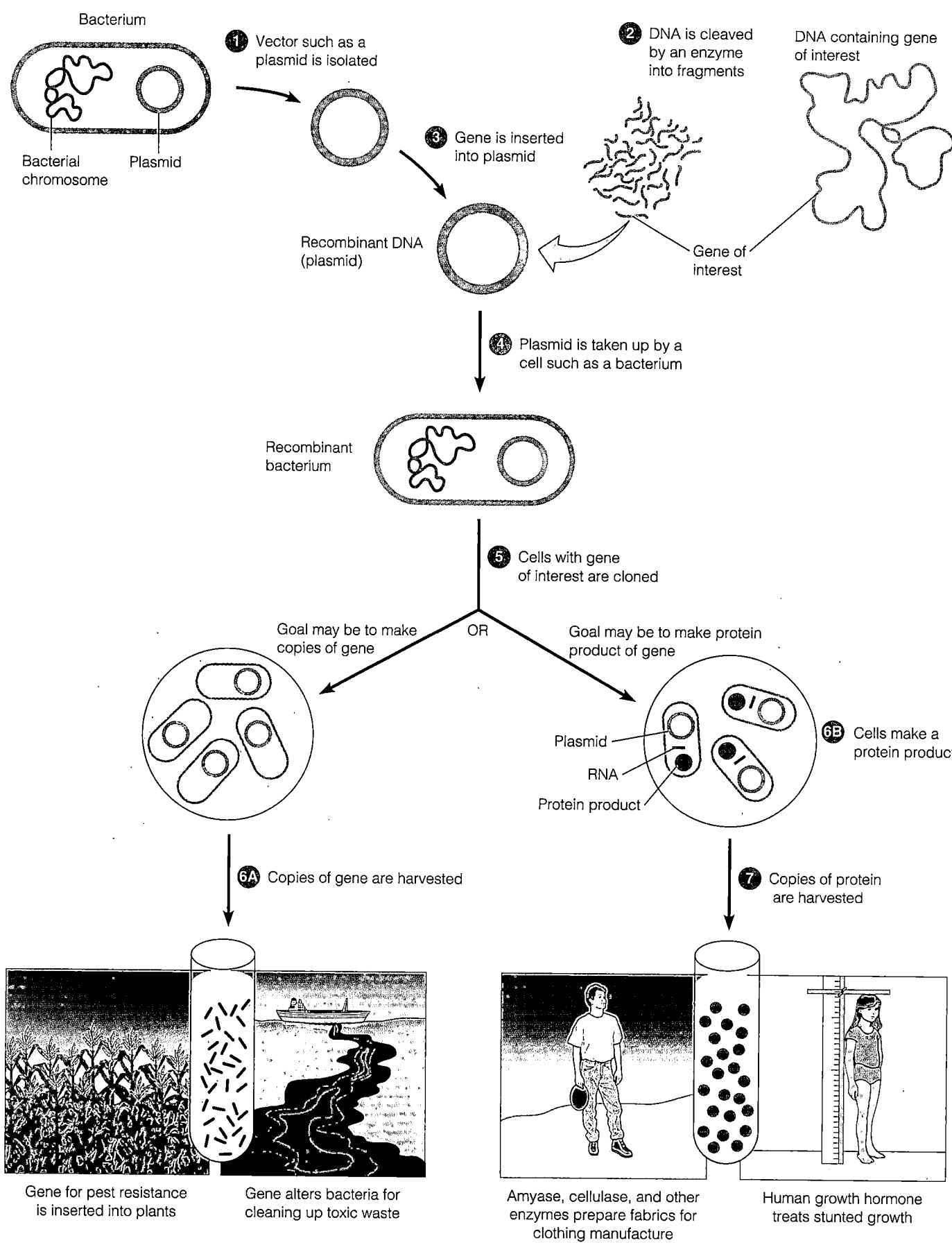

FIGURE 9.1 A typical genetic engineering procedure, with examples of applications.

■ What is genetic engineering?

APPLICATIONS OF MICROBIOLOGY

Designer Jeans

Denim blue jeans have become increasingly popular ever since they were first made for California gold miners in 1873 by Levi Strauss and Jacob Davis. Now, many companies manufacture jeans for sale around the world. Over the years, very little has changed in the denim-making process.

The denim for jeans is made from cotton. Cotton plants require water, costly fertilizer, and potentially harmful pesticides. The availability of cotton is dependent on weather and the plants' resistance to disease. After harvesting, cotton is made into yarn that is bleached with chlorine. Gaseous chlorine is quite hazardous but the safer liquid form (hypochlorite) is more expensive. After weaving yarn into cloth, the cloth is treated with starch to hold its shape, and dyed with indigo. The original indigo was made by fermentation of the indigo plant by *Clostridium*. However, this process was replaced over a century ago by a chemical process that requires mutagenic chemicals and a very high pH.

Now, companies that manufacture blue jeans and other products are returning to microbiology to develop environmentally sound production methods that minimize toxic wastes and the costs associated with treating toxic wastes. Moreover, microbiological methods can reduce their production costs and provide abundant, renewable raw materials.

In the 1980s, a softer denim, called "stone-washed," was introduced. The fabric is not really washed with rocks. Enzymes, called cellulases, from *Trichoderma* fungus are used to digest some of

the cellulose in the cotton, hence, softening it. Other microbial enzymes may be able to make the entire product safely. Thousands of enzymes have been identified in microbes and because enzymes are proteins they are readily degraded for removal from wastewater. Enzymes catalyze reactions in seconds rather than the days or weeks required of chemical methods and they usually operate at safe temperatures and pHs as opposed to high temperatures and extreme pHs of chemical reactions.

Bleaching Peroxide is a safer bleaching agent than chlorine and can be easily removed from fabric and wastewater by enzymes. Researchers at Novo Nordisk Biotech cloned a mushroom peroxidase gene in yeast and grew the yeast in washing machine conditions. The yeast that survived the washing machine were selected as the peroxidase producers.

Cotton Even in the days of Pasteur, scientists knew that *Acetobacter xylinum* could weave a film of cellulose over fermented wine and rotten fruit. In nature, *A. xylinum* make narrow, twisted bands of cellulose which may serve to help the bacteria bind to their substrate. Cellulose is made when glucose units are added to simple chains in the outer membrane of the bacterial cell. The cellulose microfibrils are extruded through pores in the outer membrane and bundles of microfibrils then twist into ribbons. Researchers now find that certain chemical dyes cause the bacteria to produce broad ribbons of cellulose that accumulate in culture. Each bacterium can produce 3 to 5 mm of cellulose a

day and ribbon-making ability is passed on every time the bacterium divides. So a colony of *A. xylinum* might discharge large quantities of cellulose thread within hours. This cellulose may augment conventional agricultural cotton or provide an alternative source that does not require large tracts of land and is not dependent on weather.

Indigo The current chemical synthesis of indigo requires a high pH and produces waste that explodes in contact with air. This will change if biotechnology companies have their way. A California biotechnology company, Genencor, is giving samples of microbially produced indigo to blue-jean makers. In the Genencor labs, *E. coli* turned blue when researchers put the gene for conversion of indole to indigo from a soil bacterium, *Pseudomonas putida*, into the *E. coli*. Then, using site-directed mutagenesis and modifications of the fermentation medium, Genencor scientists increased *E. coli*'s production of indigo. Another company, W. R. Grace has developed a mutated fungal culture that also produces indigo.

Plastic Microbes can even make plastic zippers and packaging material for the jeans. Over 25 bacteria make polyhydoxyalkanoate (PHA) inclusion granules as a nutrient reserve. PHAs are similar to common plastics and, because they are made by bacteria, they are also readily degraded by many bacteria. PHAs could provide a biodegradable alternative to conventional plastic which is made from petroleum.

select the ones that could accomplish the desired objective, such as brewing beer more efficiently, for example, or producing a new antibiotic. Over 2000 strains of antibiotic-producing bacteria have been discovered by testing soil bacteria and selecting the strains that produce an antibiotic. The box in Chapter 11 on page 328 describes the selection of a bacterium that converts a waste product into a valuable product.

Mutation

As we saw in Chapter 8, mutations are responsible for much of the diversity of life. A bacterium with a mutation that confers resistance to an antibiotic will survive and reproduce in the presence of that antibiotic. Biologists working with antibiotic-producing microbes discovered that they could create new strains by exposing microbes to mutagens. After random mutations were created in penicillin-producing *Penicillium* by exposing fungal cultures to radiation, the highest-yielding variant among the survivors was selected for another exposure to a mutagen. Using mutations, biologists increased the amount of penicillin produced by the fungus over 1000 times.

Screening each mutant for penicillin production is a tedious process. **Site-directed mutagenesis** can be used to make a specific change in a gene. Suppose you determine that changing one amino acid will make a laundry enzyme work better in cold water. Using the genetic code (see Figure 8.9 on page 219), you could produce the sequence of DNA that encodes that amino acid and insert it into the gene for that enzyme using the techniques described next.

The science of genetic engineering has advanced to such a degree that many routine cloning procedures are performed using prepackaged materials and procedures that are very much like cookbook recipes. Genetic engineers have a grab-bag of methods from which to choose, depending on the ultimate application of their experiments. Next we describe some of the most important tools and techniques, and later we will consider some specific applications.

Restriction Enzymes

Learning Objective

- *Define restriction enzymes, and outline how they are used to make recombinant DNA.*

Genetic engineering has its technical roots in the discovery of **restriction enzymes,** a special class of DNA-cutting enzymes that exist in many bacteria. First isolated in 1970, restriction enzymes in nature had actually been observed earlier, when certain bacteriophages were found to have a restricted host range. If these phages were used to infect bacteria other than their usual hosts, restriction enzymes in the new host destroyed almost all the phage DNA. Restriction enzymes protect a bacterial cell by hydrolyzing phage DNA. The bacterial DNA is protected from digestion because the cell **methylates** (adds methyl groups to) some of the cytosines in its DNA. The purified forms of these bacterial enzymes are used by today's genetic engineers.

What is important for genetic engineering is that a restriction enzyme recognizes and cuts, or digests, only one particular sequence of nucleotide bases in DNA, and it cuts this sequence in the same way each time. Typical restriction enzymes used in cloning experiments recognize four-, six-, or eight-base sequences. Many restriction enzymes make staggered cuts in the two strands of a DNA molecule—cuts that are not di-

rectly opposite each other (Figure 9.2). ❿ Notice in this figure that whereas the blue base sequences on the two strands are the same, they run in opposite directions. ❷ Staggered cuts leave stretches of single-stranded DNA at the ends of the DNA fragments. These are called *sticky ends* because they can "stick" to complementary stretches of single-stranded DNA by base pairing. Hundreds of restriction enzymes are known, each producing DNA fragments with characteristic ends. ❸ If two fragments of DNA from different sources have been produced by the action of the same restriction enzyme, the two pieces will have identical sets of sticky ends and can be spliced (recombined) in vitro. ❹ The sticky ends first join spontaneously by hydrogen bonding (base pairing) in either a linear or a circular form. ❺ Then the enzyme DNA ligase is used to covalently link the backbones of the DNA pieces, producing a recombinant DNA molecule.

Vectors

Learning Objectives

- *List the four properties of vectors.*
- *Describe the use of plasmid and viral vectors.*

A great variety of different types of DNA molecules can serve as vectors, provided that they have certain properties. The most important property is self-replication; once in a cell, a vector must be capable of replicating. Any DNA that is cloned in the vector will be replicated in the process. Thus, vectors serve as vehicles for the replication of desired DNA sequences.

Vectors also need to be of a size that allows them to be manipulated outside the cell during recombinant DNA procedures. Smaller vectors are more easily manipulated than larger DNA molecules, which tend to be more fragile. Preservation is another important property of vectors. The circular form of DNA molecules is important in protecting the DNA of the vector from destruction by the recipient of the vector. Notice in Figure 9.3 on page 254 that the DNA of a plasmid is circular. Another preservation mechanism occurs when the DNA of a virus inserts itself quickly into the chromosome of the host (see Chapter 13, page 384).

When it is necessary to retrieve cells containing the vector, a marker gene contained within the vector can often help make selection easy. Common selectable marker genes are for antibiotic resistance or for an enzyme that carries out an easily identified reaction.

Plasmids are one of the primary vectors in use, particularly variants of R factor plasmids. Plasmid DNA can be cut with the same restriction enzymes as the DNA to be

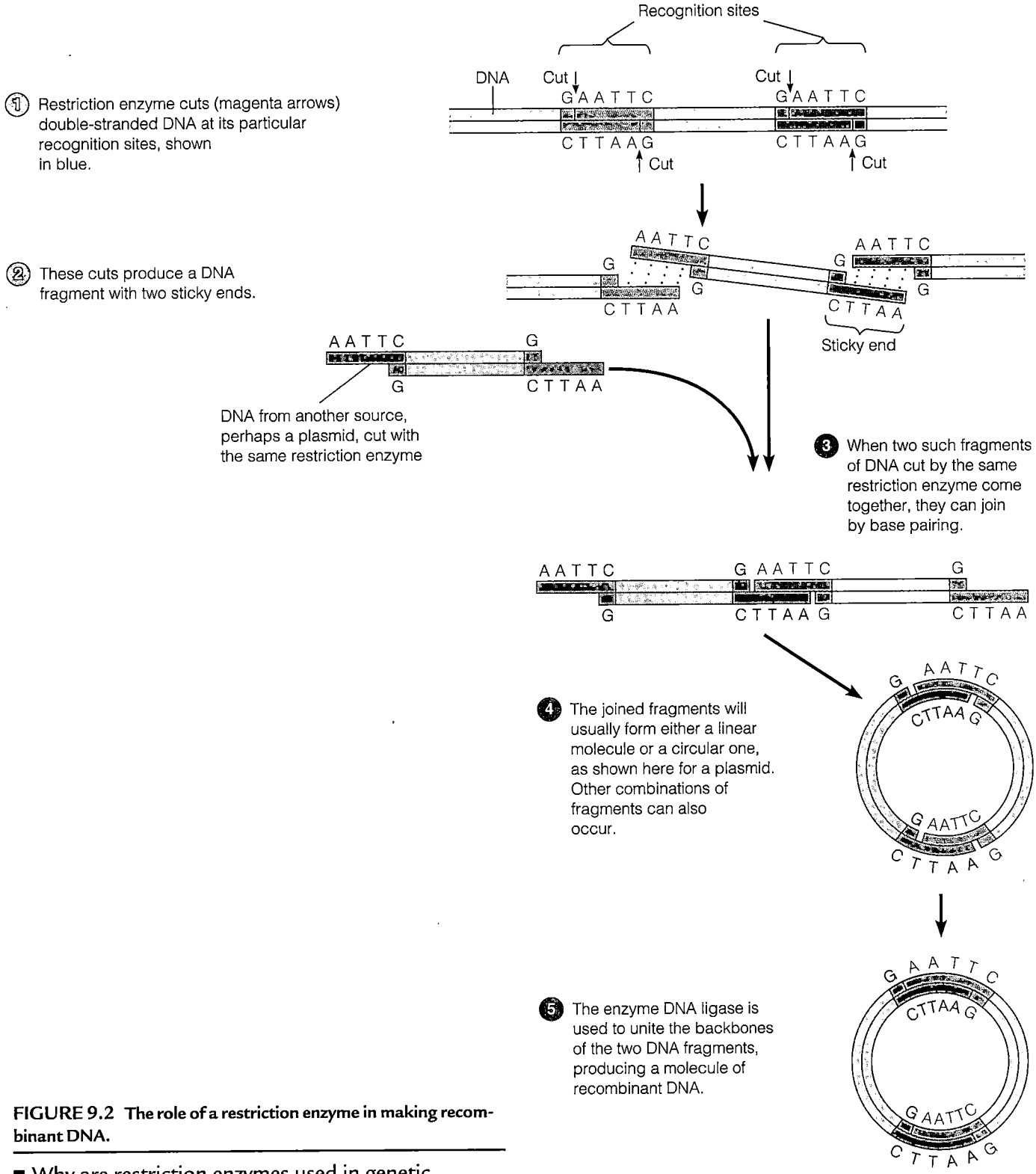

1. Restriction enzyme cuts (magenta arrows) double-stranded DNA at its particular recognition sites, shown in blue.

2. These cuts produce a DNA fragment with two sticky ends.

DNA from another source, perhaps a plasmid, cut with the same restriction enzyme

3. When two such fragments of DNA cut by the same restriction enzyme come together, they can join by base pairing.

4. The joined fragments will usually form either a linear molecule or a circular one, as shown here for a plasmid. Other combinations of fragments can also occur.

5. The enzyme DNA ligase is used to unite the backbones of the two DNA fragments, producing a molecule of recombinant DNA.

Recombinant DNA

FIGURE 9.2   The role of a restriction enzyme in making recombinant DNA.

■ Why are restriction enzymes used in genetic engineering?

cloned, so that all pieces of the DNA will have the same sticky ends. When the pieces are mixed, the DNA to be cloned will become inserted into the plasmid (see Figure 9.3). Note that other possible combinations of fragments can occur as well, including the plasmid re-forming a circle with no DNA inserted.

Some plasmids are capable of existing in several different species. They are called **shuttle vectors** and can be

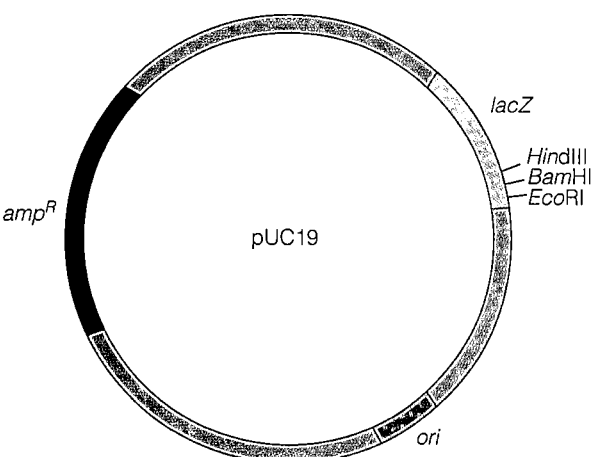

FIGURE 9.3 A plasmid used for cloning. pUC19 is a plasmid vector used for cloning in the bacterium *E. coli*. An origin of replication *(ori)* allows the plasmid to be self-replicating. Two genes, one encoding resistance to the antibiotic ampicillin *(amp)* and one encoding the enzyme β-galactosidase *(lacZ)*, serve as marker genes. Foreign DNA can be inserted at the restriction enzyme sites.

■ What is a vector in genetic engineering?

used to move cloned DNA sequences among organisms, such as among bacterial, yeast, and mammalian cells, or among bacterial, fungal, and plant cells. Shuttle vectors can be very useful in the process of genetically modifying multicellular organisms—for example, by trying to insert herbicide resistance genes into plants.

A different kind of vector is viral DNA. This type of vector can usually accept much larger pieces of foreign DNA than plasmids can. After the DNA has been inserted into the viral vector, it can be cloned in the virus's host cells. The choice of a suitable vector depends on many factors, including the organism that will receive the new gene and the size of the DNA to be cloned. Retroviruses, adenoviruses, and herpesviruses are being used to insert corrective genes into human cells that have defective genes. Gene therapy is discussed on page 264.

Polymerase Chain Reaction

Learning Objective

■ *Outline the steps in PCR, and provide an example of its use.*

The **polymerase chain reaction (PCR)** is a technique by which small samples of DNA can be quickly amplified, that is, increased to quantities that are large enough for analysis.

Starting with just one gene-sized piece of DNA, PCR can be used to make literally billions of copies in

only a few hours. The PCR process is shown in Figure 9.4. ① Each strand of the target DNA will serve as a template for DNA synthesis. ② To this DNA is added a supply of the four nucleotides (for assembly into new DNA) and the enzyme for catalyzing the synthesis, DNA polymerase (see Chapter 8, page 212). Short pieces of nucleic acid called primers are also added to help start the reaction. The primers are complementary to the ends of the target DNA and ③ will hybridize to the fragments to be amplified. ④ Then, the polymerase synthesizes new complementary strands. ⑤ After each cycle of synthesis, the DNA is heated to convert all the new DNA into single strands. Each newly synthesized DNA strand serves in turn as a template for more new DNA. As a result, the process proceeds exponentially. All of the necessary reagents are added to a tube, which is placed in a *thermalcycler.* The thermalcycler can be set for the desired temperatures, times, and number of cycles. Use of an automated thermalcycler is made possible by the use of DNA polymerase taken from a thermophilic bacterium such as *Thermus aquaticus;* the enzyme from such organisms can survive the heating phase without being destroyed (see the box in Chapter 6, page 162). Thirty cycles, completed in just a few hours, will increase the amount of target DNA by more than a billion times.

Note that PCR can only be used to amplify relatively small, specific sequences of DNA as determined by the choice of primers. It cannot be used to amplify an entire genome.

PCR can be applied to any situation that requires the amplification of DNA. Especially noteworthy are diagnostic tests that use PCR to detect the presence of infectious agents in situations in which they would otherwise be undetectable.

Techniques of Genetic Engineering

Inserting Foreign DNA into Cells

Learning Objective

■ *Describe five ways of getting DNA into a cell.*

Recombinant DNA procedures require that DNA molecules be manipulated outside the cell and then returned to living cells. There are several ways to introduce DNA into cells. The choice of method is usually determined by the type of vector and host cell being used.

In nature, plasmids are usually transferred between closely related microbes by cell-to-cell contact, such as in conjugation. In genetic engineering, a plasmid must be inserted into a cell by **transformation,** a procedure

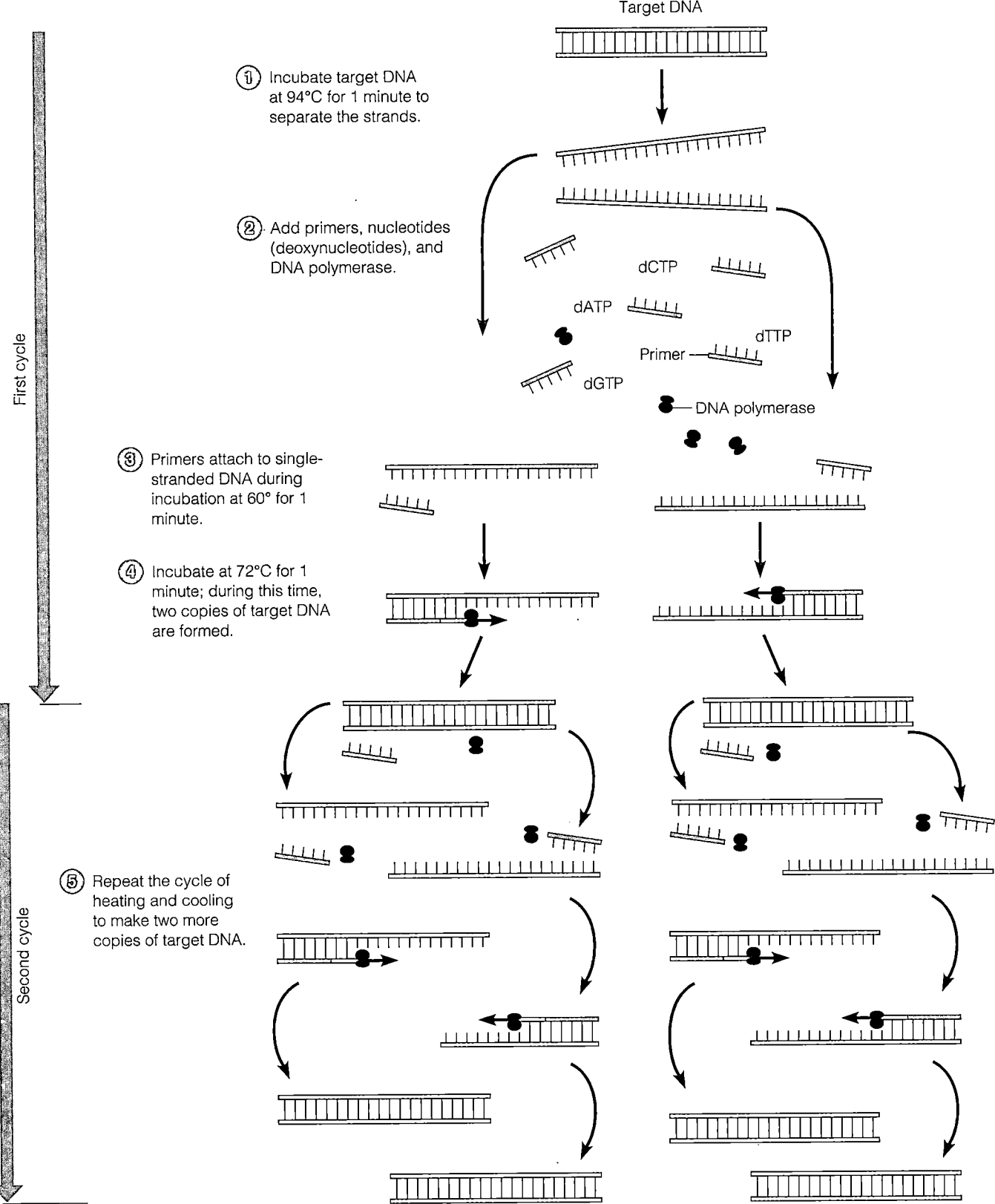

FIGURE 9.4 The polymerase chain reaction.

■ The polymerase chain reaction is used to increase samples of DNA to quantities large enough for analysis.

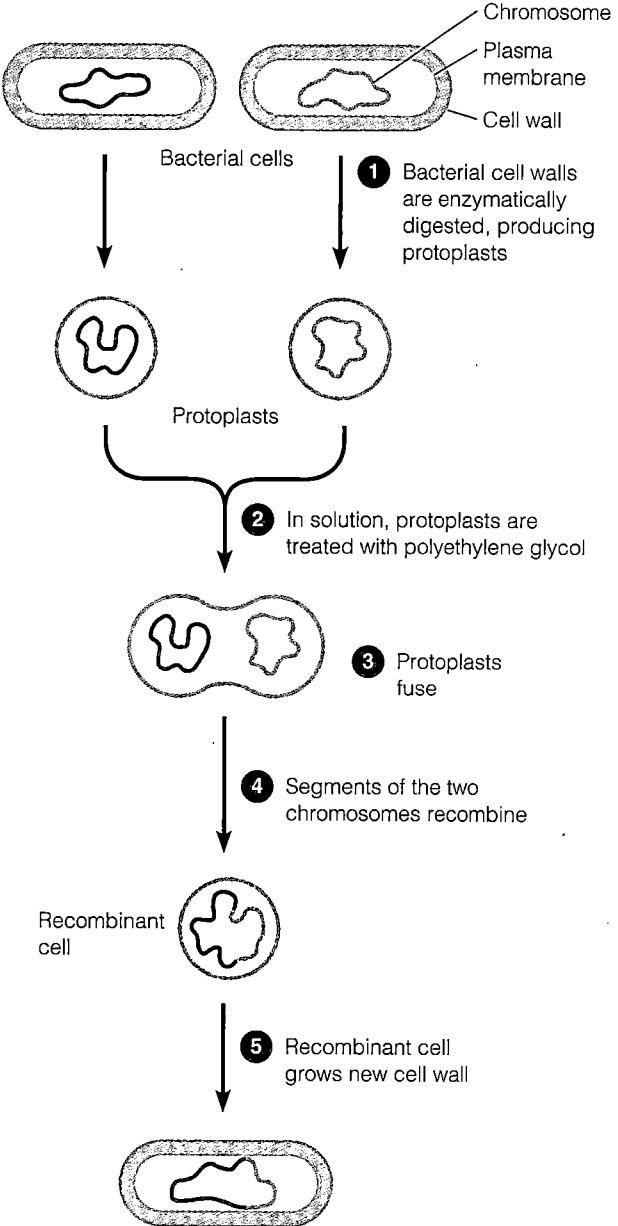

Chromosome

Plasma membrane

Cell wall

Bacterial cells

1 Bacterial cell walls are enzymatically digested, producing protoplasts

Protoplasts

2 In solution, protoplasts are treated with polyethylene glycol

3 Protoplasts fuse

4 Segments of the two chromosomes recombine

Recombinant cell

5 Recombinant cell grows new cell wall

(a) Process of protoplast fusion

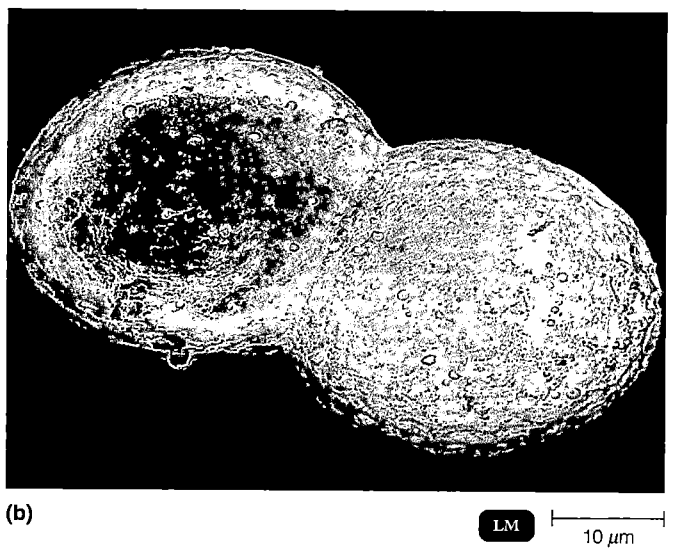

(b)

LM |————————|
      10 μm

FIGURE 9.5 Protoplast fusion. (a) A diagram of protoplast fusion with bacterial cells. **(b)** Two protoplast leaf cells from a tobacco plant are shown fusing. Removal of the cell wall left only the delicate plasma membrane to bind the cell contents together, allowing the exchange of DNA.

■ Protoplasts are produced by enzymatically removing the cell wall.

during which cells can take up DNA from the surrounding environment (see Chapter 8, page 234). Many cell types, including *E. coli*, yeast, and mammalian cells, do not naturally transform; however, simple chemical treatments can make all of these cell types *competent*, or able to take up external DNA. For *E. coli*, the procedure

for making cells competent is to soak them in a solution of calcium chloride for a brief period. Following this treatment, the now-competent cells are mixed with the cloned DNA and given a mild heat shock. Some of these cells will then take up the DNA.

There are other ways to transfer DNA to cells. A process called **electroporation** uses an electrical current to form microscopic pores in the membranes of cells; the DNA then enters the cells through the pores. Electroporation is generally applicable to all cells; those with cell walls often must be converted to protoplasts first (see Chapter 4, page 88). **Protoplasts** are produced by enzymatically removing the cell wall, thereby allowing more direct access to the plasma membrane.

The process of **protoplast fusion** also takes advantage of the properties of protoplasts. Protoplasts in solution will fuse at a low but significant rate; the addition of polyethylene glycol increases the frequency of fusion (Figure 9.5a). In the new hybrid cell, the DNA derived from the two "parent" cells may undergo natural recombination. This method is especially valuable in the genetic manipulation of plant cells (Figure 9.5b).

A remarkable way of introducing foreign DNA into plant cells is to literally shoot it directly through the thick cellulose walls using a gene gun (Figure 9.6). Microscopic particles of tungsten or gold are coated with DNA and propelled by a burst of helium through the plant cell walls. Some of the cells express the introduced DNA as if it were their own.

DNA can be introduced directly into an animal cell by **microinjection.** This technique requires the use of a glass micropipette with a diameter that is much smaller than the cell. The micropipette punctures the plasma membrane, and DNA can be injected through it (Figure 9.7).

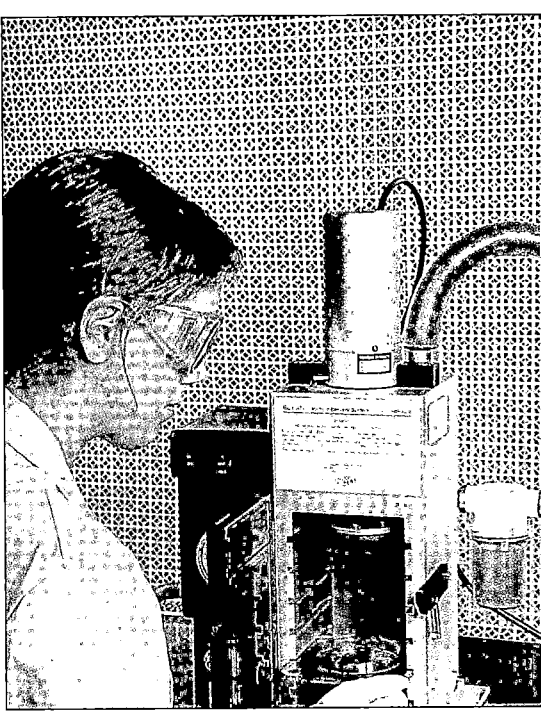

FIGURE 9.6 A gene gun, which can be used to insert DNA-coated "bullets" into a cell.

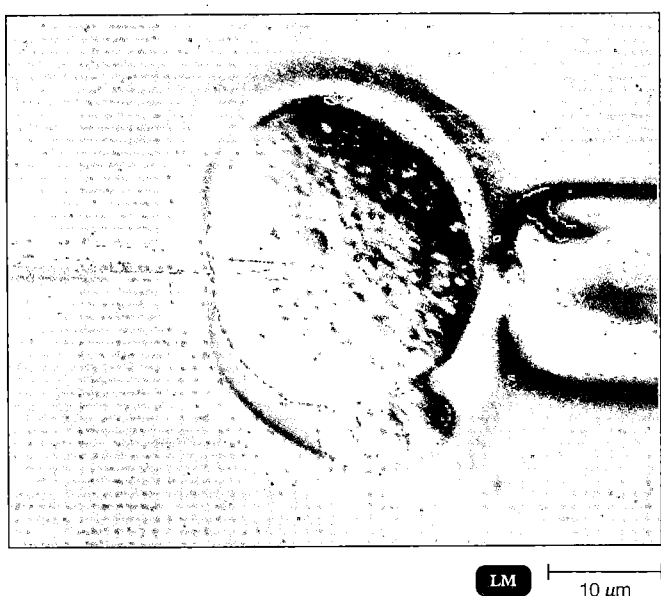

FIGURE 9.7 **The microinjection of foreign DNA into a fertilized mouse egg.** The egg is first immobilized by applying mild suction to the large, blunt, holding pipette (right). Several hundred copies of the gene of interest are then injected into the nucleus of the cell through the tiny end of the micropipette (left).

■ Microinjection is the introduction of DNA into a cell using a glass micropipette that is smaller in diameter than the cell.

Thus, there is a great variety of different restriction enzymes, vectors, and methods of inserting DNA into cells. But foreign DNA will survive only if it is either present on a self-replicating vector or incorporated into one of the cell's chromosomes by recombination.

Obtaining DNA

We have seen how genes can be cloned into vectors by using restriction enzymes, and how genes can be transformed or transferred into a variety of cell types. But how do genetic engineers obtain the genes they are interested in? There are two main sources of genes: (1) gene libraries containing either natural copies of genes or cDNA copies of genes made from mRNA, and (2) synthetic DNA.

Gene Libraries

Learning Objective

■ *Describe how a gene library is made.*

Isolating specific genes as individual pieces of DNA is seldom practical. Therefore, researchers interested in genes from a particular organism start by extracting the organism's DNA, which can be obtained from cells of any organism, whether plant, animal, or microbe, by lysing the cells and precipitating the DNA. This process results in a DNA mass that includes the organism's entire genome.

After the DNA is digested by restriction enzymes, the restriction fragments are then spliced into plasmid or phage vectors, and the recombinant vectors are introduced into bacterial cells. The goal is to make a collection of clones large enough to ensure that at least one clone exists for every gene in the organism. This collection of clones containing different DNA fragments is called a **gene library**; each "book" is a bacterial or phage strain that contains a fragment of the genome (Figure 9.8). Such libraries are essential for maintaining and retrieving DNA clones; they can even be purchased commercially.

Cloning genes from eukaryotic organisms presents a specific problem. Genes of eukaryotic cells generally contain both **exons,** stretches of DNA that code for protein, and **introns,** intervening stretches of DNA that do not code for protein. When the RNA transcript of such a gene is converted to mRNA, the introns are removed by a process called **splicing.** In cloning genes of eukaryotic cells, it is desirable to use a version of the gene that lacks introns because a gene that includes introns may be too large to work with easily. In addition, if such a gene is put into a bacterial cell, the bacterium will not usually be able to remove the introns from the RNA transcript, and therefore it will not be able to make the correct protein product. However, an artificial gene that contains only

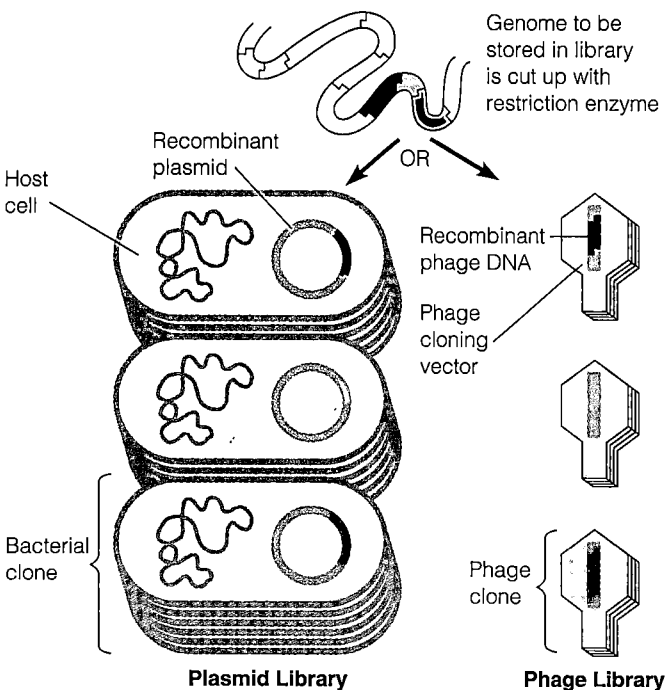

FIGURE 9.8 Gene libraries. Each fragment of DNA, containing about one gene, is carried by a vector, either a plasmid within a bacterial cell or a phage.

■ A gene library is a collection of a large number of DNA fragments from a genome.

exons can be produced by using an enzyme called **reverse transcriptase** to synthesize **complementary DNA (cDNA)** from an mRNA template (Figure 9.9). This synthesis is the reverse of the normal DNA-to-RNA transcription process. A DNA copy of mRNA is produced by reverse transcriptase. Following this, the mRNA is enzymatically digested away. DNA polymerase then synthesizes a complementary strand of DNA, creating a double-stranded piece of DNA containing the information from the mRNA. Molecules of cDNA produced from a mixture of all the mRNAs from a tissue or cell type can then be cloned to form a cDNA library.

The cDNA method is the most common method of obtaining eukaryotic genes. A difficulty with this method is that long molecules of mRNA may not be completely reverse-transcribed into DNA; the reverse transcription often aborts, forming only parts of the desired gene.

Synthetic DNA

Learning Objective

■ *Differentiate cDNA from synthetic DNA.*

Under certain circumstances, genes can be made in vitro with the help of DNA synthesis machines (Figure 9.10).

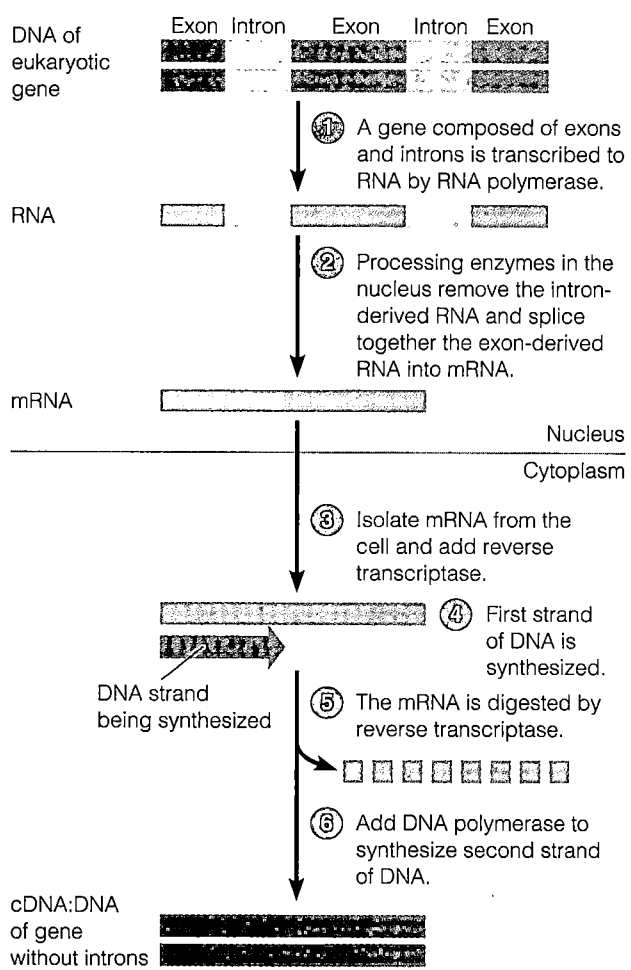

FIGURE 9.9 Making complementary DNA (cDNA) for a eukaryotic gene. Reverse transcriptase catalyzes the synthesis of double-stranded DNA from an RNA template.

■ How does reverse transcriptase differ from DNA polymerase?

A keyboard on the machine is used to enter the desired sequence of nucleotides, much as letters are entered into a word processor to compose a sentence. A microprocessor controls the synthesis of the DNA from stored supplies of nucleotides and the other necessary reagents. A chain of over 120 nucleotides can be synthesized by this method. Unless the gene is very small, at least several chains must be synthesized separately and linked together to form an entire gene.

The difficulty of this approach, of course, is that the sequence of the gene must be known before it can be synthesized. If the gene has not already been isolated, then the only way to predict the DNA sequence is by knowing the amino acid sequence of the protein product of the gene. If this amino acid sequence is known, in principle one can work backward through the genetic code to obtain the DNA sequence. Unfortunately, the

FIGURE 9.10 A DNA synthesis machine. Short sequences of DNA can be synthesized by instruments such as this one.

■ **What are some of the disadvantages of a DNA synthesis machine?**

degeneracy of the code prevents an unambiguous determination; thus, if the protein contains a leucine, for example, which of the six codons for leucine is the one in the gene?

For these reasons, it is rare to clone a gene by synthesizing it directly, although some commercial products, such as insulin, interferon, and somatostatin are produced from chemically synthesized genes. Desired restriction sites were added to the synthetic genes so that the genes could be inserted into plasmid vectors for cloning in *E. coli*. Synthetic DNA plays a much more useful role in selection procedures, as we will see.

Selecting a Clone

Learning Objective

■ *Explain how each of the following is used to locate a clone: antibiotic-resistance genes, DNA probes, gene product.*

In cloning, it is necessary to select the particular cell that contains the specific gene of interest. This is difficult, because out of millions of cells, only a very few cells might contain the desired gene. Here we will examine a typical screening procedure known as *blue-white screening,* from the color of the bacterial colonies formed at the end of the screening process.

The plasmid vector used contains a gene *(amp^R)* coding for resistance to the antibiotic ampicillin. The host bacterium will not be able to grow on the test medium, which contains ampicillin, unless the vector has transferred the ampicillin-resistance gene. The plasmid vector also contains a second gene, this one for the enzyme β-galactosidase *(lacZ)*. Notice in Figure 9.3 that there are several sites in *lacZ* that can be cut by restriction enzymes.

The procedure is shown in Figure 9.11. The two genes, called marker genes, are used so that the insertion of plasmid DNA into the host bacterium can be determined. ① The plasmid vector and foreign DNA are digested with the same restriction enzyme. ② The foreign DNA will insert into the gene for β-galactosidase. Therefore, the bacterium receiving the plasmid vector will not produce the enzyme β-galactosidase if foreign DNA has been inserted into the plasmid. ③ The recombinant plasmid is introduced into a culture of ampicillin-sensitive bacteria by transformation.

④ In the blue-white screening procedure, a library of bacteria is cultured in a medium called X-gal. X-gal contains two essential components other than those necessary to support normal bacterial growth. One is the antibiotic ampicillin, which prevents the growth of any bacterium that has not successfully received the ampicillin-resistance gene from the plasmid. The other, called X-gal, is a substrate for β-galactosidase. ⑤ Only bacteria that picked up the plasmid will grow—because they are now ampicillin resistant. Bacteria that picked up the recombinant plasmid—in which the new gene was inserted into the *lacZ* gene—will not hydrolyze lactose and will produce white colonies. If a bacterium received the original plasmid containing the intact *lacZ* gene, the cells will hydrolyze X-gal to produce a blue-colored compound; the colony will be blue.

What remains to be done can still be difficult. The above procedure has isolated white colonies known to contain foreign DNA, but it is still not known whether this is the desired fragment of foreign DNA. A second procedure is needed to identify these bacteria. If the foreign DNA in the plasmid codes for the production of an identifiable product, the bacterial isolate only needs to be grown in culture and tested. However, in some cases the gene itself must be identified in the host bacterium.

Colony hybridization is a common method of identifying cells that carry a specific cloned gene. **DNA probes,** short segments of single-stranded DNA that are complementary to the desired gene, are synthesized. If the DNA probe finds a match, it will adhere to the target gene. The DNA probe is labeled with a radioactive

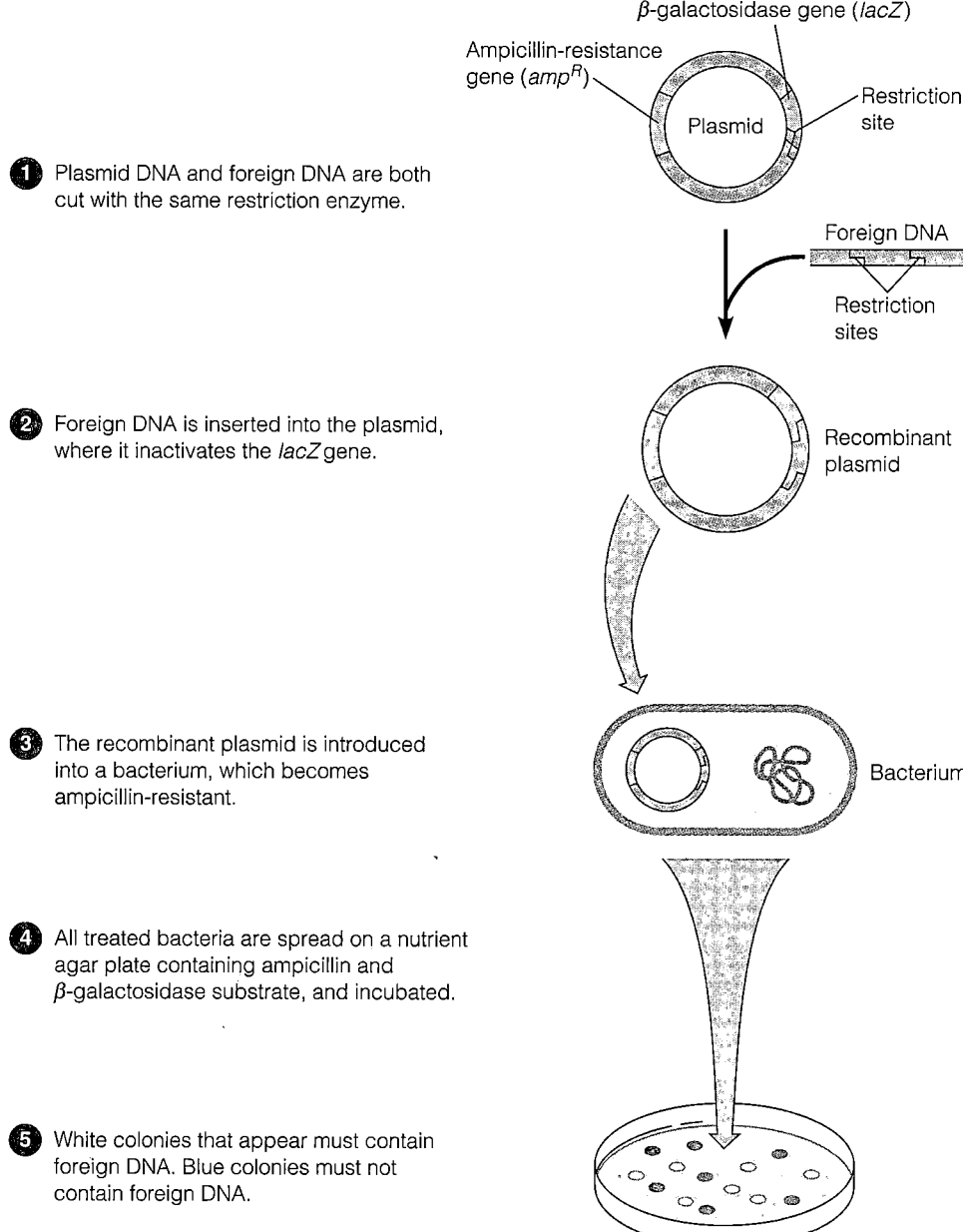

1 Plasmid DNA and foreign DNA are both cut with the same restriction enzyme.

2 Foreign DNA is inserted into the plasmid, where it inactivates the *lacZ* gene.

3 The recombinant plasmid is introduced into a bacterium, which becomes ampicillin-resistant.

4 All treated bacteria are spread on a nutrient agar plate containing ampicillin and β-galactosidase substrate, and incubated.

5 White colonies that appear must contain foreign DNA. Blue colonies must not contain foreign DNA.

FIGURE 9.11   One method of selecting recombinant bacteria.

■ Why are some colonies blue and others white?

element or fluorescent dye so its presence can be determined. A typical colony hybridization experiment is shown in Figure 9.12.

Making a Gene Product

Learning Objective

■ List one advantage of engineering each of the following: E. coli, Saccharomyces cerevisiae, *mammalian cells, plant cells.*

We have just seen that one way of identifying cells carrying a particular gene is by testing for the gene product. Such products are themselves, of course, a frequent objective of genetic engineering. Most of the earliest work in genetic engineering used *E. coli* to synthesize the gene products. *E. coli* is easily grown, and researchers are very familiar with this bacterium and its genetics. For example, some inducible promoters, such as that of the *lac* operon, have been cloned, and cloned genes can be attached to such promoters. The synthesis of great amounts of the

Master plate with colonies of bacteria containing cloned segments of foreign genes.

Nitrocellulose filter

① Make replica of master plate on nitrocellulose filter.

② Treat filter with detergent (SDS) to lyse bacteria.

Strands of bacterial DNA

③ Treat filter with sodium hydroxide (NaOH) to separate DNA into single strands.

Radioactively labeled probes

④ Add radioactively labeled probes.

Bound DNA probe

Gene of interest

Single-stranded DNA

⑤ Probe will hybridize with desired gene from bacterial cells.

Developed film

⑥ Wash filter to remove unbound probe and expose filter to X-ray film.

Colonies containing genes of interest

Replica plate

⑦ Developed film is compared with replica of master plate to identify colonies containing gene of interest.

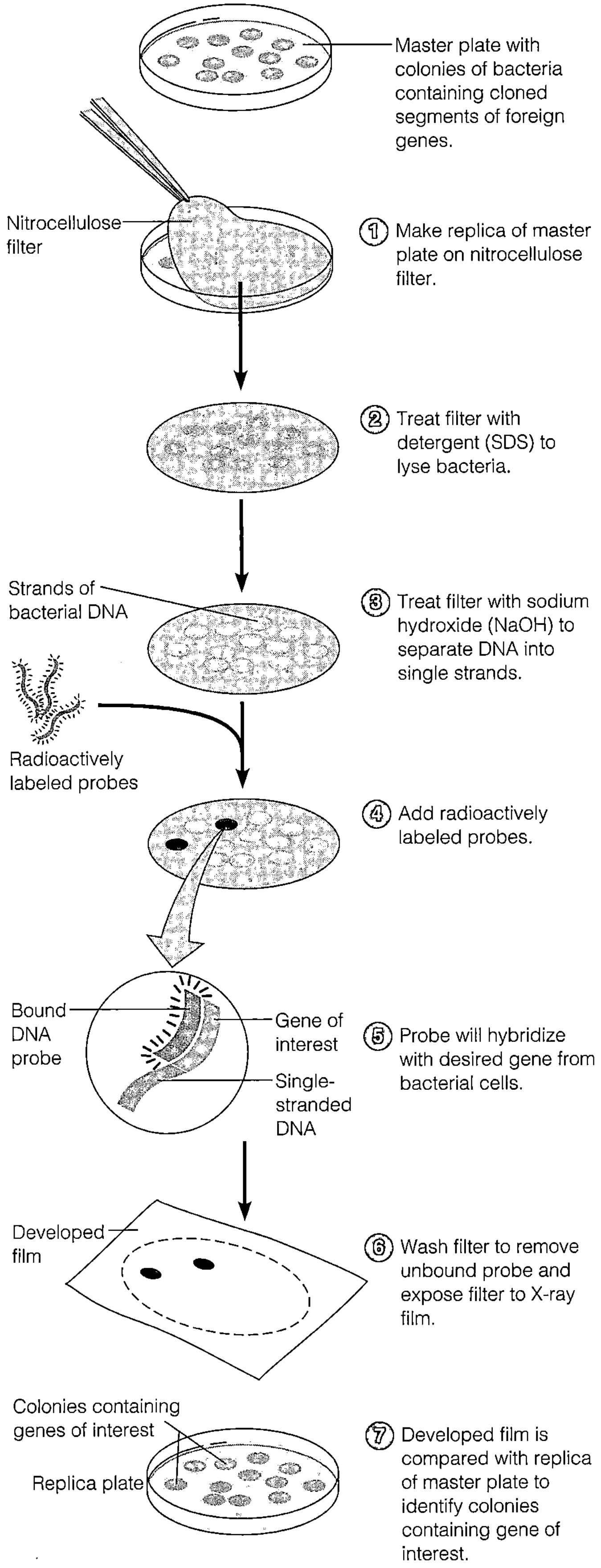

FIGURE 9.12 Colony hybridization: using a DNA probe to identify a cloned gene of interest.

■ A probe is a single-stranded DNA with a base sequence complementary to that of the gene of interest, that will hybridize with the gene from bacterial cells.

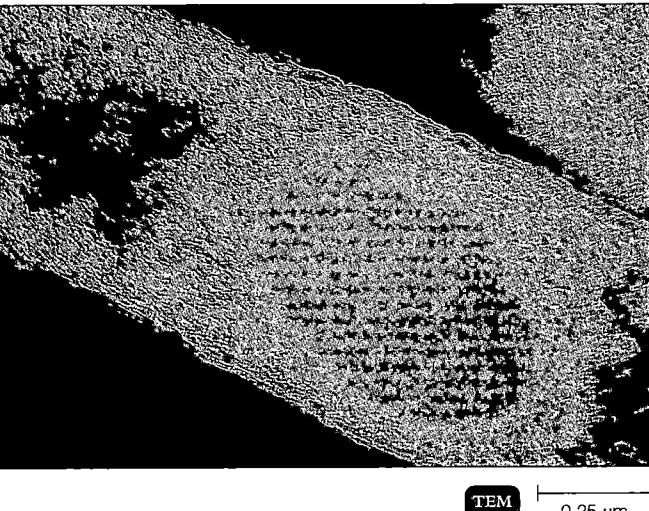

TEM ⊢0.25 μm⊣

FIGURE 9.13 *E. coli* genetically engineered to produce gamma interferon, a human protein that promotes an immune response. The product, visible here as an orange mass, can be released by lysis of the cell.

■ What is one advantage of using *E. coli* for genetic engineering? One disadvantage?

cloned gene product can then be directed by the addition of an inducer. Such a method has been used to produce gamma interferon in *E. coli* (Figure 9.13). However, *E. coli* also has several disadvantages. Like other gram-negative bacteria, it produces endotoxins as part of the outer layer of its cell wall. Because endotoxins cause fever and shock in animals, their accidental presence in products intended for human use would be a serious problem.

Another disadvantage of *E. coli* is that it does not usually secrete protein products. To obtain a product, cells must usually be broken open and the product purified from the resulting "soup" of cell components. Recovering the product from such a mixture is expensive when done on an industrial scale. It is more economical to have an organism secrete the product so that it can be recovered continuously from the growth medium. One approach has been to link the product to a natural *E. coli* protein that the bacterium does secrete. However, gram-positive bacteria, such as *Bacillus subtilis*, are more likely to secrete their products and are often preferred industrially for that reason.

Another microbe being used as a vehicle for expressing genetically engineered genes is baker's yeast, *Saccharomyces cerevisiae*. Its genome is only about four times larger than that of *E. coli* and is probably the best understood eukaryotic genome. Yeasts may carry plasmids, and the plasmids are easily transferred into yeast cells after their cell walls have been removed. As eukaryotic cells, yeasts may be more successful in expressing foreign eukaryotic genes than bacteria. Furthermore, yeasts are likely to continuously secrete the product. Because of all these factors, yeasts have become the eukaryotic workhorse of biotechnology.

Mammalian cells in culture, even human cells, can be used much like bacteria to produce genetically engineered products. Scientists have developed effective methods of growing certain mammalian cells in culture as hosts for growing viruses (see Chapter 13, page 380). In genetic engineering, mammalian cells are often the best suited to making protein products for medical use; these products include hormones, cytokines (which regulate cells of the immune system), and interferon (a natural antiviral substance also used to treat some cancers). Using mammalian cells to make foreign gene products on an industrial scale often requires a preliminary step of cloning the gene in bacteria. Consider the example of colony-stimulating factor (CSF). A protein produced naturally in tiny amounts by white blood cells, CSF is valuable because it stimulates the growth of certain cells that protect against infection. To produce huge amounts of CSF industrially, the gene is first inserted into a plasmid, and bacteria are used to make multiple copies of the plasmid (see Figure 9.1). The recombinant plasmids are inserted into mammalian cells that are grown in bottles.

Plant cells can also be grown in culture, altered by recombinant DNA techniques, and then used to generate genetically engineered plants. Such plants may prove useful as sources of valuable products, such as plant alkaloids (the painkiller codeine, for example), the isoprenoids that are the basis of synthetic rubber, and melanin (the animal skin pigment) for use in sunscreens. (We will return to the topic of genetic engineering of plants later in the chapter, on page 266.)

Applications of Genetic Engineering

Learning Objective

- List at least five applications of genetic engineering.

We have now described the entire sequence of events in cloning a gene. As indicated earlier, such cloned genes can be applied in a variety of ways. One is to produce useful substances more efficiently and less expensively (see the box on page 251). Another is to obtain information from the cloned DNA that is useful for either basic research or medical applications. A third is to use cloned genes to alter the characteristics of cells or organisms. The box in Chapter 28 (page 786) describes the use of recombinant cells to detect pollutants.

Therapeutic Applications

An extremely valuable pharmaceutical product is the hormone insulin, a small protein produced by the pancreas that controls the body's uptake of glucose from blood. For many years, people with insulin-dependent diabetes have controlled their disease by injecting insulin obtained from the pancreases of slaughtered animals. Obtaining this insulin is an expensive process, and the insulin from animals is not as effective as human insulin.

Because of the value of human insulin and the small size of the protein, the production of human insulin by recombinant DNA techniques was an early goal for the pharmaceutical industry. To produce the hormone, synthetic genes were first constructed for each of the two short polypeptide chains that make up the insulin molecule. The small size of these chains—only 21 and 30 amino acids long—made it possible to use synthetic genes. Following the procedure described earlier, each of the two synthetic genes was inserted into a plasmid vector and linked to the end of a gene coding for the bacterial enzyme β-galactosidase, so that the insulin polypeptide was coproduced with the enzyme. Two different *E. coli* bacterial cultures were used, one to produce each of the insulin polypeptide chains. The polypeptides were then recovered from the bacteria, separated from the β-galactosidase, and chemically joined to make human insulin. This accomplishment was one of the early commercial successes of genetic engineering, and it illustrates a number of the principles and procedures discussed in this chapter.

Another human hormone that is now being produced commercially by genetic engineering of *E. coli* is somatostatin. At one time 500,000 sheep brains were needed to produce 5 mg of animal somatostatin for experimental purposes. By contrast, only 8 liters of a genetically engineered bacterial culture are now required to obtain the equivalent amount of the human hormone.

Tissue-plasminogen activator (t-PA) and streptokinase dissolve blood clots. Consequently, they accelerate recovery from heart attack and help prevent recurrences, particularly if they are administered very soon after an initial attack. Tissue-plasminogen activator, a product of

table 9.1	Some Pharmaceutical Products of Genetic Engineering
Product	**Comments**
Tissue plasminogen activator (Activase®)	Dissolves the fibrin of blood clots; therapy for heart attacks; produced by mammalian cell culture.
Erythropoietin (EPO)	Treatment of anemia; produced by mammalian cell culture.
Human insulin	Therapy for diabetes; better tolerated than insulin extracted from animals; produced by *Escherichia coli*.
Interleukin-2 (IL-2)	Possible treatment for cancer; stimulates the immune system; produced by *E. coli*.
Alpha-interferon	Possible cancer and viral disease therapy; produced by *E. coli* and *Saccharomyces cerevisiae* (yeast).
Gamma-interferon	Treatment of chronic granulomatous disease; produced by *E. coli*.
Tumor necrosis factor (TNF)	Causes disintegration of tumor cells; produced by *E. coli*.
Human growth hormone (hGH)	Corrects growth deficiencies in children; produced by *E. coli*.
Epidermal growth factor (EGF)	Heals wounds, burns, ulcers; produced by *E. coli*.
Prourokinase	Anticoagulant; therapy for heart attacks; produced by *E. coli* and yeast.
Factor VIII	Treatment of hemophilia; improves clotting; produced by mammalian cell culture.
Colony-stimulating factor (CSF)	Counteracts effects of chemotherapy; improves resistance to infectious disease such as AIDS; treatment of leukemia; produced by *E. coli* and *S. cerevisiae*.
Superoxide dismutase (SOD)	Minimizes damage caused by oxygen free radicals when blood is resupplied to oxygen-deprived tissues; produced by *S. cerevisiae* and *Pichia pastoris* (yeast).
Monoclonal antibodies	Possible therapy for cancer and transplant rejection; used in diagnostic tests; produced by mammalian cell culture (from fusion of cancer cell and antibody-producing cell).
Hepatitis B vaccine	Produced by *S. cerevisiae* that carries hepatitis-virus gene on a plasmid.
Bone morphogenic proteins	Induces new bone formation; useful in healing fractures and reconstructive surgery; produced by mammalian cell culture.
Taxol	Plant product used for treatment for ovarian cancer; produced in *E. coli*.
Orthoclone®	Monoclonal antibody used in transplant patients to help suppress the immune system, reducing the chance of tissue rejection; produced by mouse cells.
Relaxin	Used to ease childbirth; produced by *E. coli*.
Pulmozyme® (rhDNase)	Enzyme used to break down mucous secretions in cystic fibrosis patients; produced by mammalian cell culture.

genetic engineering, is produced by mammalian cell culture; streptokinase, an older product, is a natural product of bacteria.

Subunit vaccines, consisting only of a protein portion of a pathogen, are being made by genetically engineering yeasts. Subunit vaccines have been produced for a number of diseases, notably hepatitis B. One of the advantages of a subunit vaccine is that there is no chance of becoming infected from the vaccine. The protein is harvested from engineered cells and purified for use as a vaccine. Animal viruses such as vaccinia virus can be engineered to carry a gene for another microbe's surface protein. When injected, the virus acts as a vaccine against the other microbe. Table 9.1 lists some other important genetically engineered products used in medical therapy.

The importance of recombinant DNA technology to medical research cannot be emphasized enough. For example, the cloning of HIV genes has formed the basis of current subunit vaccine testing (see the box in Chapter 19, p. 543). Moreover, the recently developed *protease inhibitors,* drugs that prevent a certain stage of HIV development, resulted from a detailed study of the HIV protease gene.

The rapid advances in genetic engineering of different organisms have resulted in a wide variety of medically related products and procedures. Artificial blood for use in transfusions can now be prepared using human hemoglobin produced in genetically engineered pigs. Sheep have also been genetically engineered to produce a number of drugs in their milk. This procedure has no apparent effect upon the sheep, and it provides a ready source of raw material for the engineered product that does not require sacrificing animals.

Gene therapy may eventually provide cures for some genetic diseases. It is possible to imagine removing some cells from a person and transforming them with a normal gene to replace a defective or mutated gene. When these cells are returned to the person, they should function normally. For example, a rare and deadly mutation can result in adenosine deaminase deficiency, a disease that leaves the individual without a functional immune system. Recent gene therapy experiments concerning this disease have had positive results. The first gene therapy to treat hemophilia in humans was done in 1999. An attenuated retrovirus was used as the vector. The number of gene therapy trials will increase as technical improvements are made and initial attempts are successful. However, there is a great deal of preliminary work to do, and cures may not be possible for all genetic diseases. Antisense DNA (see page 267) introduced into cells is also being explored to treat hepatitis, cancers, and one type of coronary artery disease.

Scientific Applications

Learning Objectives

- Diagram the Southern blotting procedure, and provide an example of its use.
- Diagram DNA fingerprinting, and provide an example of its use.

Recombinant DNA technology can be used to make products, but this is not its only important application. Because of its ability to produce many copies of DNA, it can serve as a sort of DNA "printing press." Once a large amount of a particular piece of DNA is available, various analytic techniques, discussed in this section, can be used to "read" the information contained in the DNA.

What kind of information can be obtained from cloned DNA? One kind is provided by the process of **DNA sequencing**—the determination of the exact sequence of nucleotide bases in DNA. Recent automation of this process enables a researcher to determine the sequence of over 1000 bases per day. The sequences of en-

tire viral genomes are now relatively easy to obtain. The genomes of *Sacchromyces cerevisiae* yeast, *E. coli,* and several other microbes have been mapped. A massive project to sequence all the DNA of a human, called the Human Genome Project, is under way (see page 269).

An example of the use of human DNA sequencing is the identification and cloning of the mutant gene that causes cystic fibrosis (CF). CF is characterized by the oversecretion of mucus, leading to blocked respiratory passageways. The sequence of the mutated gene can be used as a diagnostic tool in a hybridization technique called **Southern blotting** (Figure 9.14) (named for Ed Southern, who developed the technique in 1975).

In this technique, ① human DNA is first digested with a restriction enzyme, yielding thousands of fragments of various sizes. The different fragments are then separated by **gel electrophoresis.** ② The fragments are put in a well at one end of a layer of agarose gel. Then an electrical current is passed through the gel. While the charge is applied, the different-size pieces of DNA migrate through the gel at different rates. The fragments are called **RFLPs,** for restriction fragment length polymorphisms. ③–④ The separated fragments are transferred onto a filter by blotting. ⑤ The fragments on the filter are then exposed to a radioactive probe made from the cloned gene of interest, in this case the CF gene. The probe will hybridize to this mutant gene but not to the normal gene. ⑥ Fragments to which the probe bind are identified by exposing the filter to X-ray film. With this method, any person's DNA can be tested for the presence of the mutated gene.

This process, called **genetic screening,** can now be used to screen for several hundred genetic diseases. Such screening procedures can be performed on prospective parents and also on fetal tissue. Two of the more commonly screened genes are those associated with inherited forms of breast cancer and the gene responsible for Huntington's disease.

Several important diagnostic tools are now available as a result of genetic engineering, many of which rely on the technique of hybridization. Recall that this procedure enables the identification of a particular DNA sequence among many others. This is exactly what is necessary in most diagnostic situations—the identification of a particular pathogen among many others. Viral infections can often be diagnosed through hybridization: A probe specific for a particular viral gene is synthesized and used in Southern blotting procedures (described in the next section) to determine whether that virus is present in the blood or tissue of a patient. Genetic engineering has also been important in the development of ELISA diagnostic tools (see Chapter 18, page 511).

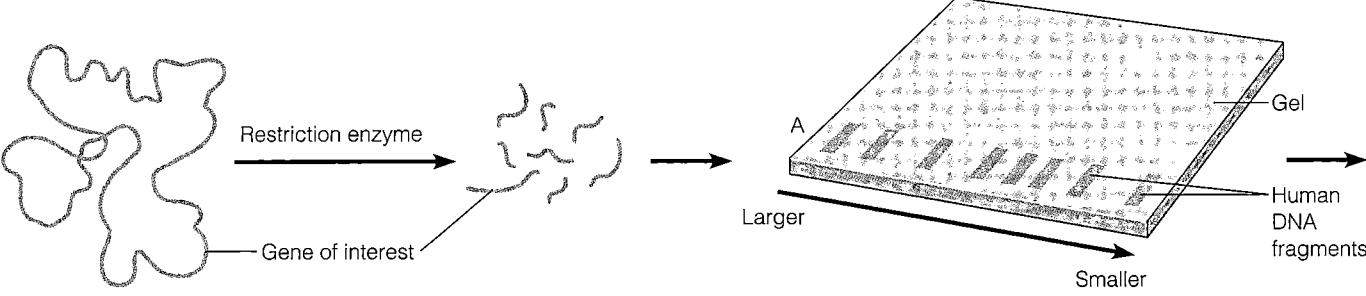

(1) DNA containing the gene of interest is extracted from human cells and cut into fragments by restriction enzymes.

(2) The fragments are separated according to size by gel electrophoresis. Each band consists of many copies of a particular DNA fragment. The bands are invisible but can be made visible by a dye that fluoresces under ultraviolet light.

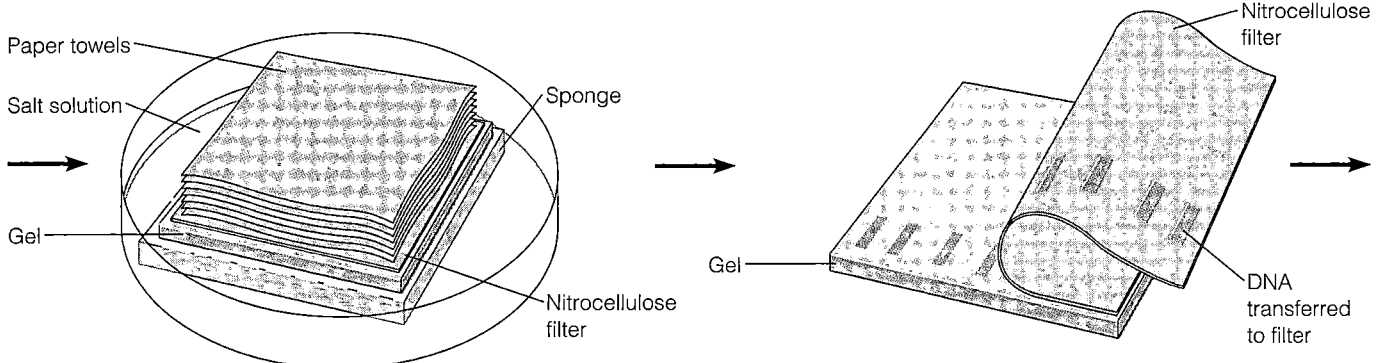

(3) The DNA bands are transferred to a nitrocellulose filter by blotting. The solution passes through the gel and filter to the paper towels.

(4) This produces a nitrocellulose filter with DNA fragments positioned exactly as on the gel.

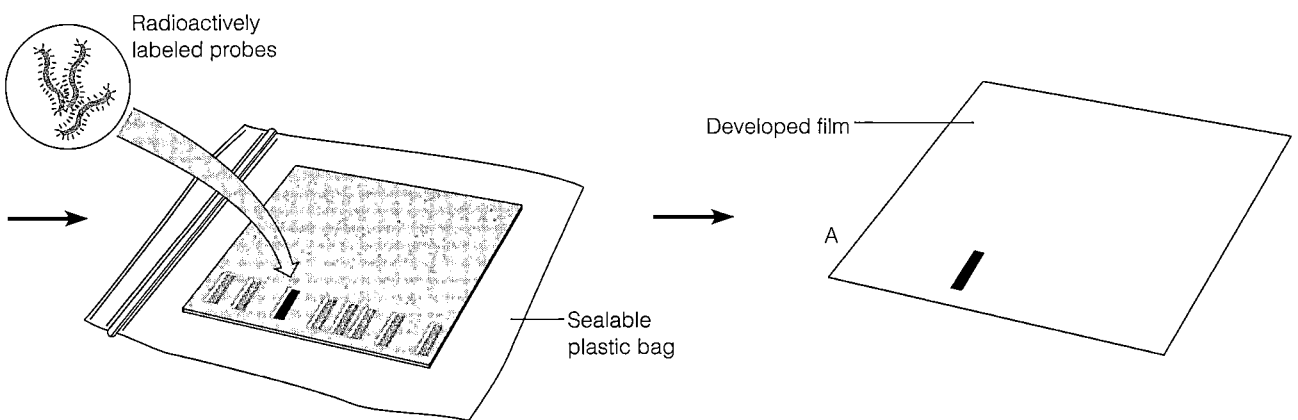

(5) The filter is exposed to a radioactively labeled probe for a specific gene. The probe will base-pair (hybridize) with a short sequence present on the gene.

(6) The filter is then exposed to X-ray film. The fragment containing the gene of interest is identified by a band on the developed film.

FIGURE 9.14 Southern blotting.

■ What is the purpose of Southern blotting?

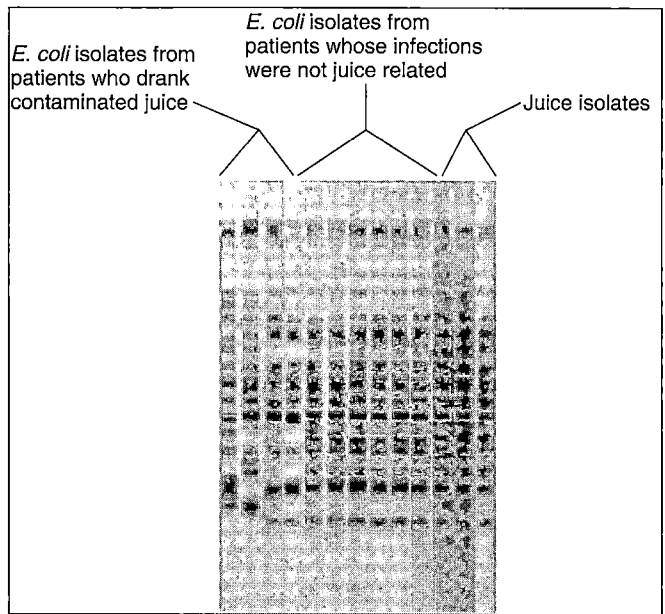

FIGURE 9.15 DNA fingerprints used to track an infectious disease. This figure shows the relationship among DNA patterns of bacterial isolates from an outbreak of *Escherichia coli* O157:H7. The isolates from apple juice are identical to the patterns of isolates from patients who drank the contaminated juice but different from those from patients whose infections were not juice-related. (CDC)

FIGURE 9.16 Crown gall disease on a tomato plant. The tumor-like growth is stimulated by a gene on the Ti plasmid carried by the bacterium *Agrobacterium tumefaciens*, which has infected the plant.

■ **What are some of the agricultural applications of recombinant DNA technology?**

RFLPs are also used in a method of identification known as **DNA fingerprinting,** which can identify bacterial or viral pathogens (Figure 9.15). DNA fingerprinting is also used in forensic medicine to determine paternity or to prove that blood on a murder suspect's clothes came from the murder victim. Southern blotting requires substantial amounts of DNA. As mentioned earlier, small samples of DNA can be quickly amplified, or increased to quantities large enough for analysis, by PCR.

DNA can often be extracted from preserved and fossilized materials, including mummies and extinct plants and animals. Although such material is very rare, and usually partially degraded, PCR enables researchers to study this genetic material that no longer exists in its natural form. The study of unusual organisms has also led to advances in basic taxonomy; this will be discussed in Chapter 10.

DNA probes such as the ones used to screen gene libraries are promising tools for the rapid identification of microorganisms. For use in medical diagnosis, these probes are derived from the DNA of a pathogenic microbe and are labeled (with a radioactive tag, for example). The probe then serves a diagnostic function by combining with the DNA of the pathogen to reveal its location in body tissue (or perhaps its presence in food). Probes are also being used in nonmedical aspects of microbiology—for example, for locating and identifying specific microbes in soil. We will discuss PCR and DNA probes further in Chapter 10.

Agricultural Applications

Learning Objective

■ *Outline genetic engineering with* Agrobacterium.

The process of selecting for genetically desirable plants has always been a time-consuming one. Performing conventional plant crosses is laborious and involves waiting for the planted seed to germinate and for the plant to mature. Plant breeding has been revolutionized by the use of plant cells grown in culture. Clones of plant cells, including cells that have been genetically altered by recombinant DNA techniques, can be grown in large numbers. These cells can then be induced to regenerate whole plants, from which seeds can be harvested.

Recombinant DNA can be introduced into plant cells in several ways. Previously we mentioned protoplast fusion and the use of DNA-coated "bullets." The most elegant method, however, makes use of a plasmid called the **Ti plasmid** (Ti stands for tumor–inducing), which occurs naturally in the bacterium *Agrobacterium tumefaciens.* This bacterium infects certain plants, in which the Ti plasmid causes the formation of a tumorlike growth called a crown gall (Figure 9.16). A part of the Ti plasmid, called T-DNA, integrates into the genome of the infected plant. The T-DNA stimulates local cellular growth (the crown gall) and simultaneously causes the production of

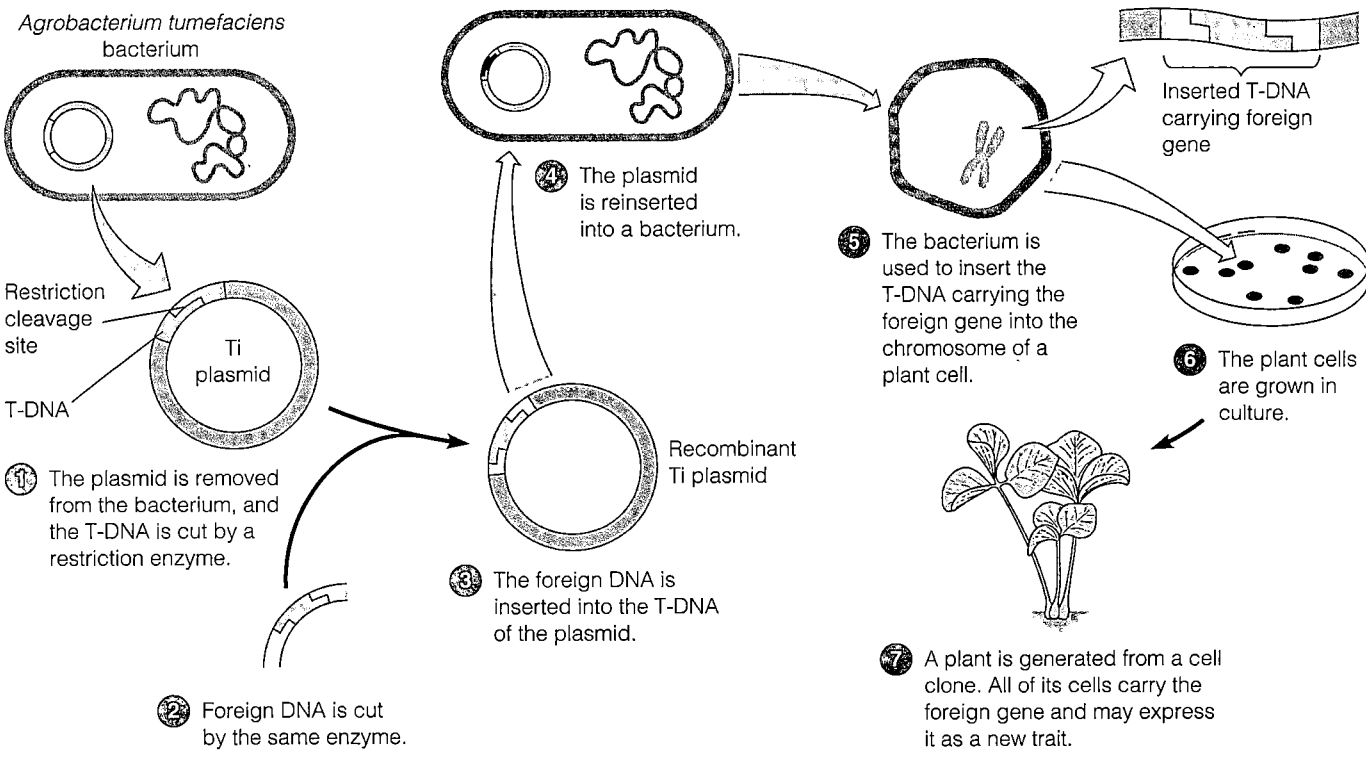

Agrobacterium tumefaciens
bacterium

Restriction
cleavage
site

Ti
plasmid

T-DNA

1 The plasmid is removed
from the bacterium, and
the T-DNA is cut by a
restriction enzyme.

2 Foreign DNA is cut
by the same enzyme.

3 The foreign DNA is
inserted into the T-DNA
of the plasmid.

Recombinant
Ti plasmid

4 The plasmid
is reinserted
into a bacterium.

5 The bacterium is
used to insert the
T-DNA carrying the
foreign gene into the
chromosome of a
plant cell.

Inserted T-DNA
carrying foreign
gene

6 The plant cells
are grown in
culture.

7 A plant is generated from a cell
clone. All of its cells carry the
foreign gene and may express
it as a new trait.

FIGURE 9.17 Using the Ti plasmid as a vector for genetic engineering in plants.

■ Why is the Ti plasmid important to biotechnology?

certain products used by the bacteria as a source of nutritional carbon and nitrogen.

For plant scientists, the attraction of the Ti plasmid is that it provides a vehicle for introducing genetically engineered DNA into a plant (Figure 9.17). A scientist can insert foreign genes into the T-DNA, put the recombinant plasmid back into the *Agrobacterium* cell, and use the bacterium to insert the recombinant Ti plasmid into a plant cell. The plant cell with the foreign gene can then be used to generate a new plant. With luck, the new plant will express the foreign gene. Unfortunately, *Agrobacterium* does not naturally infect grasses, so it cannot be used to improve grains such as wheat, rice, or corn.

Noteworthy accomplishments of this approach are the introduction into plants of resistance to the herbicide glyphosate, and a *Bacillus thuringiensis*–derived insecticidal toxin (Bt). Normally, the herbicide kills both weeds and useful plants by inhibiting an enzyme necessary for making certain essential amino acids. *Salmonella* bacteria happen to have this enzyme, and some salmonellae have a mutant enzyme that is resistant to the herbicide. When the DNA for this enzyme is introduced into a crop plant, the crop becomes resistant to the herbicide, which then kills only the weeds. There is now a variety of plants in

which different herbicide and pesticide resistances have been engineered. Resistance to drought, viral infection, and several other environmental stresses has also been engineered into crop plants. *Bacillus thuringiensis* bacteria are pathogenic to some insects because they produce a protein called Bt toxin that interferes with the insect digestive tract. The Bt gene has been inserted into a variety of crop plants, including cotton and potatoes, so insects that eat the plants will be killed.

Another example involves MacGregor tomatoes, which stay firm after harvest because the gene for polygalacturonase (PG), the enzyme that breaks down pectin, is suppressed. The suppression was accomplished by **antisense DNA technology.** First, a length of DNA complementary to the PG mRNA is synthesized. This antisense DNA is taken up by the cell and binds to the mRNA to inhibit translation. The DNA-RNA hybrid is broken down by the cell's enzymes, freeing the antisense DNA to disable another mRNA.

Perhaps the most exciting potential use of genetic engineering in plants concerns nitrogen fixation, the ability to convert the nitrogen gas in the air to compounds that living cells can use (see Chapter 27). The availability of such nitrogen-containing nutrients is usually the main

table 9.2	Some Agriculturally Important Products of Genetic Engineering
Product	**Comments**
Agricultural Products	
Pseudomonas syringae, ice-minus bacterium	Lacks normal protein product that initiates undesirable ice formation on plants.
Pseudomonas fluorescens bacterium	Has toxin-producing gene from insect pathogen *Bacillus thuringiensis;* toxin kills root-eating insects that ingest bacteria.
Rhizobium meliloti bacterium	Modified for enhanced nitrogen fixation.
Round-Up® (glyphosate)-resistant crops	Plants have bacterial gene; allows use of herbicide on weeds without damaging crops.
Bt cotton and Bt corn	Plants have toxin-producing gene from *B. thuringiensis;* toxin kills insects that eat plants.
MacGregor tomatoes	Gene for pectin degradation is removed so fruits have longer shelf life.
Animal Husbandry Products	
Porcine growth hormone (PGH)	Improves weight gain in swine; produced by *E. coli.*
Bovine growth hormone (BGH)	Improves weight gain and milk production in cattle; produced by *E. coli.*
Transgenic animals	Genetically alters animals to produce medically useful products in their milk.
Other Food Production Products	
Rennin	Causes formation of milk curds for dairy products; produced by *Aspergillus niger.*
Cellulase	Enzymes that degrade cellulose to make animal feedstocks; produced by *E. coli.*

factor limiting crop growth. But in nature, only certain bacteria have genes for carrying out this process. Some plants, such as alfalfa, benefit from a symbiotic relationship with these microbes. Species of the symbiotic bacterium *Rhizobium* have already been genetically engineered for enhanced nitrogen fixation. In the future, *Rhizobium* strains may be designed that can colonize such crop plants as corn and wheat, perhaps eliminating their requirement for nitrogen fertilizer. The ultimate goal would be to introduce functioning nitrogen-fixation genes directly into the plants. Although this goal cannot be achieved with our current knowledge, work toward it will continue because of its potential for dramatically increasing the world's food supply.

An example of a genetically engineered bacterium now in agricultural use is *Pseudomonas fluorescens* that has been engineered to produce a toxin normally produced by *Bacillus thuringiensis*. This toxin kills certain plant pathogens, such as the European corn borer. The genetically altered *Pseudomonas*, which produces much more toxin than *B. thuringiensis*, can be added to plant seeds and in time will enter the vascular system of the growing plant. Its toxin is ingested by the feeding borer larvae and kills them (but is harmless to humans and other warm-blooded animals).

Animal husbandry has also benefited from genetic engineering. We have seen how one of the early commercial products of genetic engineering was human growth

hormone. By similar methods it is possible to manufacture bovine growth hormone (BGH). When BGH is injected into beef cattle, it increases their weight gain; in dairy cows, it also causes a 10% increase in milk production. Such procedures have met with resistance from consumers, especially in Europe, primarily as a result of as-yet unsubstantiated fears that some of the BGH would be present in the milk or meat of these cattle and might be harmful to humans.

Table 9.2 lists these and several other genetically engineered products used in agriculture and animal husbandry.

Safety Issues and the Ethics of Genetic Engineering

Learning Objectives

- List the advantages of, and problems associated with, the use of genetic engineering techniques.
- Discuss some possible results of sequencing the human genome.

There will always be concern about the safety of any new technology, and genetic engineering and biotechnology are certainly no exceptions. One reason for this concern is that it is nearly impossible to prove that something is entirely safe under all conceivable conditions. People worry that the same techniques that can alter a microbe or plant to make them useful to humans could also inadvertently make them pathogenic to humans or otherwise dangerous to living organisms, or could create an ecological nightmare. Therefore, laboratories engaged in recombinant DNA research must meet rigorous standards of control to avoid either accidentally releasing genetically engineered organisms into the environment or exposing humans to any risk of infection. To reduce risk further, microbiologists engaged in genetic engineering often delete from the microbes' genomes certain genes that are essential for growth in environments outside the laboratory. Finally, recombinant DNA–carrying microbes intended for use in the environment (in agriculture, for example) may be engineered to contain "suicide genes"—genes that eventually turn on to produce a toxin that kills the microbes, thus ensuring that they will not survive in the environment for very long after they have accomplished their task.

The safety issues in agricultural biotechnology are similar to those concerning chemical pesticides: toxicity to humans and to nonpest species. Although not shown to be harmful, genetically engineered foods have not been popular with consumers. In 1999, researchers in Ohio noticed that humans may develop allergies to *Bacillus thuringiensis* (Bt) toxin after working in fields sprayed with the insecticide. And an Iowa study showed that the caterpillar stage of Monarch butterflies could be killed by ingesting windblown Bt-carrying pollen that landed on milkweed, the caterpillars' normal food. Crop plants can be genetically engineered for herbicide resistance so that fields can be sprayed to eliminate weeds without killing the desired crop. However, if the engineered plants pollinate related weed species, weeds could become resistant to herbicides, making it more difficult to control unwanted plants. An unanswered question is whether releasing genetically engineered organisms will alter evolution as genes move to wild species.

These developing technologies also raise a variety of moral and ethical issues. If genetic screening for diseases becomes routine, who should have access to this information? Should employers or insurance companies have the right to know the results of such tests? Restricting access to such information will be very difficult, which raises questions concerning the right to privacy. How can we be assured that such information will not be used to discriminate against certain groups?

Applications of genetic screening techniques are not limited to adults. The ability to diagnose a genetic disease in a fetus adds even more controversy to the abortion debate. Genetic counseling, which provides advice and counseling to prospective parents with family histories of genetic disease, is becoming more important in considerations about whether or not to have children. As more is learned about genetic causes for various diseases, such as cancer or Huntington's disease, reproductive choices might become more difficult for some families.

What extra burdens does genetic engineering place on our already overtaxed health care system? Genetic screening and gene therapy are expensive procedures, and we need to consider how they will be delivered to the public as the technology develops. Will a sufficient number of genetic counselors be available to work with people? Will expensive medical cures and treatments be available only to those who can afford them?

There are probably just as many harmful applications of a new technology as there are helpful ones. It is particularly easy to imagine genetic engineering being used to develop new and powerful biological weapons. In addition, because such research efforts are performed under top-secret conditions, it is virtually impossible for the general public to learn of them.

Perhaps more than most new technologies, genetic engineering holds the promise of affecting human life in previously unimaginable ways. It is important that society and individuals be given every opportunity to understand the potential impact of these new developments.

Like the invention of the microscope, the development of recombinant DNA techniques is causing profound changes in science, agriculture, and human health care. With this technology only slightly more than 30 years old, it is difficult to predict exactly what changes will occur. However, it is likely that within another 30 years, many of the treatments and diagnostic methods discussed in this book will have been replaced by far more powerful techniques based on the unprecedented ability to manipulate DNA precisely.

One monumental project involving much of the new technology is the previously mentioned Human Genome Project. The goal of this project is to map the 70,000 or so genes in human DNA and to sequence the entire genome, approximately 3 billion nucleotide pairs. The technical aspects of the project have been compared to reconstructing the contents of several sets of shredded encyclopedias. By 2000, virtually all (3 billion nucleotide pairs) had been sequenced, and chromosome 22 was completely mapped. Even before it is completed, however, the Human Genome Project is of immense value to our understanding of biology. It will also eventually be of great medical benefit, especially for the diagnosis of, and possibly the repair of, genetic diseases.

Study Outline
[ST] Student Tutorial CD-ROM

INTRODUCTION TO BIOTECHNOLOGY
(pp. 248–249)

1. Biotechnology is the use of microorganisms, cells, or cell components to make a product.

Recombinant DNA Technology (p. 249)

1. Closely related organisms can exchange genes in natural recombination.
2. Genes can be transferred among unrelated species via laboratory manipulation, called genetic engineering.
3. Recombinant DNA is DNA that has been artificially manipulated to combine genes from two different sources.

An Overview of Recombinant DNA Technologies (p. 249)

1. A desired gene is inserted into a DNA vector, such as a plasmid or a viral genome.
2. The vector inserts the DNA into a new cell, which is grown to form a clone.
3. Large quantities of the gene product can be harvested from the clone.

TOOLS OF BIOTECHNOLOGY (pp. 249–254)

Selection (pp. 249–251)

1. Microbes with desirable traits are selected for culturing by artificial selection.

Mutation (pp. 251–252)

1. Mutagens are used to cause mutations that might result in a microbe with desirable traits.
2. Site-directed mutagenesis is used to change a specific codon in a gene.

Restriction Enzymes (p. 252)

[ST] To review, go to Biotechnology: Cloning a Gene: Cloning Practice

1. Prepackaged kits are available for many genetic-engineering techniques.
2. A restriction enzyme recognizes and cuts only one particular nucleotide sequence in DNA.
3. Some restriction enzymes produce sticky ends, short stretches of single-stranded DNA at the ends of the DNA fragments.
4. Fragments of DNA produced by the same restriction enzyme will spontaneously join by hydrogen bonding. DNA ligase can covalently link the DNA backbones.

Vectors (pp. 252–254)

1. Shuttle vectors are plasmids that can exist in several different species.

2. A plasmid containing a new gene can be inserted into a cell by transformation.
3. A virus containing a new gene can insert the gene into a cell.

Polymerase Chain Reaction (p. 254)

[ST] Biotechnology: PCR: Clone DNA

1. The polymerase chain reaction (PCR) is used to make multiple copies of a desired piece of DNA enzymatically.
2. PCR can be used to increase the amounts of DNA in samples to detectable levels. This may allow sequencing of genes, the diagnosis of genetic diseases, or the detection of viruses.

TECHNIQUES OF GENETIC ENGINEERING (pp. 254–262)

Inserting Foreign DNA into Cells (pp. 254–257)

[ST] Biotechnology: Cloning a Gene: Genetically Engineer

1. Cells can take up naked DNA by transformation. Chemical treatments are used to make cells that are not naturally transformed competent to take up DNA.
2. Pores made in protoplasts and animal cells by electric current in the process of electroporation can provide entrance for new pieces of DNA.
3. Protoplast fusion is the joining of cells whose cell walls have been removed.
4. Foreign DNA can be introduced into plant cells by shooting DNA-coated particles into the cells.
5. Foreign DNA can be injected into animal cells by using a fine glass micropipette.

Obtaining DNA (pp. 257–259)

1. Gene libraries can be made by cutting up an entire genome with restriction enzymes and inserting the fragments into bacterial plasmids or phages.
2. cDNA made from mRNA by reverse transcription can be cloned in gene libraries.
3. Synthetic DNA can be made in vitro by a DNA synthesis machine.

Selecting a Clone (pp. 259–260)

1. Antibiotic-resistance markers on plasmid vectors are used to identify cells containing the engineered vector by direct selection.
2. In blue-white screening, the vector contains the genes for amp^R and β-galactosidase.
3. The desired gene is inserted into the β-galactosidase gene site, destroying the gene.
4. Clones containing the recombinant vector will be resistant to ampicillin and unable to hydrolyze X-gal (white colonies). Clones containing the vector without the new gene will be blue. Clones lacking the vector will not grow.

5. Clones containing foreign DNA can be tested for the desired gene product.

6. A short piece of labeled DNA called a DNA probe can be used to identify clones carrying the desired gene.

Making a Gene Product (pp. 260–262)

1. *E. coli* is used to produce proteins by genetic engineering because it is easily grown and its genomics are well understood.

2. Efforts must be made to ensure that *E. coli*'s endotoxin does not contaminate a product intended for human use.

3. To recover the product, *E. coli* must be lysed or the gene must be linked to a gene that produces a naturally secreted protein.

4. Yeasts can be genetically engineered and are likely to continuously secrete a gene product.

5. Mammalian cells can be engineered to produce proteins such as hormones for medical use.

6. Plant cells can be engineered and used to produce plants with new properties.

APPLICATIONS OF GENETIC ENGINEERING (pp. 262–268)

1. Cloned DNA is used to produce products, study the cloned DNA, and alter the phenotype of an organism.

Therapeutic Applications (pp. 262–264)

1. Synthetic genes linked to the β-galactosidase gene *(lacZ)* in a plasmid vector were inserted into *E. coli*, allowing *E. coli* to produce and secrete the two polypeptides used to make human insulin.

2. Cells can be engineered to produce a pathogen's surface protein, which can be used as a subunit vaccine.

3. Animal viruses can be engineered to carry a gene for a pathogen's surface protein. When the virus is used as a vaccine, the host develops an immunity to the pathogen.

4. Gene therapy can be used to cure genetic diseases by replacing the defective or missing gene.

Scientific Applications (pp. 264–266)

1. Recombinant DNA techniques can be used to increase understanding of DNA, for genetic fingerprinting, and for gene therapy.

2. DNA sequencing machines are used to determine the nucleotide base sequence in a gene.

3. Southern blotting can be used to locate a gene in a cell.

4. Genetic screening uses Southern blotting to look for mutations responsible for inherited diseases in humans.

5. Southern blotting is used in DNA fingerprinting to identify bacterial or viral pathogens.

6. DNA probes can be used to quickly identify a pathogen in body tissue or food.

Agricultural Applications (pp. 266–268)

1. Cells from plants with desirable characteristics can be cloned to produce many identical cells. These cells can then be used to produce whole plants from which seeds can be harvested.

2. Plant cells can be engineered by using the Ti plasmid vector. The tumor-producing T genes are replaced with desired genes, and the recombinant DNA is inserted into *Agrobacterium*. The bacterium naturally transforms its plant hosts.

3. Genes for glyphosate resistance, BT toxin, and pectinase suppression have been engineered into crop plants.

4. *Rhizobium* has been engineered for enhanced nitrogen fixation.

5. *Pseudomonas* has been engineered to produce *Bacillus thuringiensis* toxin against insects.

6. Bovine growth hormone is being produced by *E. coli*.

SAFETY ISSUES AND THE ETHICS OF GENETIC ENGINEERING (pp. 268–269)

1. Strict safety standards are used to avoid the accidental release of genetically engineered microorganisms.

2. Some microbes used in genetic engineering have been altered so that they cannot survive outside the laboratory.

3. Microorganisms intended for use in the environment may be engineered to contain suicide genes so that the organisms do not persist in the environment.

4. Genetic technology raises ethical questions such as: Should employers and insurance companies have access to a person's genetic records? Will some people be targeted for either breeding or sterilization? Will genetic counseling be available to everyone?

5. Genetically-engineered crops must be safe for consumption and for release in the environment.

6. Genetic engineering techniques are being used to map the human genome through the Human Genome Project.

7. This will provide tools for diagnosis and possibly the repair of genetic diseases.

Study Questions

REVIEW

1. Differentiate recombinant DNA from genetic engineering.

2. Compare and contrast the following terms:
   a. cDNA and gene
   b. restriction fragment and gene
   c. DNA probe and gene
   d. DNA polymerase and DNA ligase
   e. recombinant DNA and cDNA

3. How is each of the following used in genetic engineering?
   a. plasmid
   b. viral genome
   c. antibiotic-resistance genes
   d. restriction enzyme

4. Differentiate between a gene library and synthetic DNA.

5. Differentiate among the following terms. Which one is "hit and miss"—that is, does not add a specific gene to a cell?
   a. protoplast fusion
   b. gene gun
   c. microinjection
   d. electroporation

6. Some commonly used restriction enzymes are listed in the following table. The cutting site is indicated by ↓. Indicate which enzymes produce sticky ends. Of what value are sticky ends in making recombinant DNA?

Enzyme	Bacterial Source	Recognition Sequence
BamHI	Bacillus amyloliquefaciens	G↓G A T C C G C T A G↑G
EcoRI	Escherichia coli	G↓A A T T C C T T A A↑G
HaeIII	Haemophilus aegyptius	G G↓C C C C↑G G
HindIII	Haemophilus influenzae	A↓A G C T T T T C G A↑A

7. Suppose you want multiple copies of a gene you have synthesized. How would you obtain the necessary copies by cloning? By PCR?

8. Describe a genetic engineering experiment in two or three sentences. Use the following terms: intron, exon, DNA, mRNA, cDNA, RNA polymerase, reverse transcriptase.

9. List at least two examples of the use of genetic engineering in medicine and in agriculture.

10. You are attempting to insert a gene for saltwater tolerance into a plant by using the Ti plasmid. In addition to the desired gene, you add a gene for tetracycline resistance *(tet^R)* to the plasmid. What is the purpose of the *tet^R* gene?

MULTIPLE CHOICE

1. Restriction enzymes were first discovered with the observation that
   a. DNA is restricted to the nucleus.
   b. phage DNA is destroyed in a host cell.
   c. foreign DNA is kept out of a cell.
   d. foreign DNA is restricted to the cytoplasm.
   e. all of the above

2. The DNA probe, 3'GGCTTA, will hybridize with DNA containing
   a. 5'CCGUUA.   d. 3'CCGAAT.
   b. 5'CCGAAT.   e. 3'GGCAAU.
   c. 5'GGCTTA.

3. Which of the following is the fourth basic step of genetic engineering?
   a. transformation   d. restriction-enzyme digestion of gene
   b. ligation
   c. plasmid cleavage   e. isolation of gene

4. The following steps are used to make cDNA. What is the second step?
   a. reverse transcription   c. transcription
   b. RNA processing to remove introns   d. translation

5. If you put a gene in a virus, the next step in genetic engineering would be
   a. insertion of a plasmid.   d. PCR.
   b. transformation.   e. Southern blotting.
   c. transduction.

6. You have a small gene that you want replicated by PCR. You add radioactively labeled nucleotides to the PCR thermalcycler. After three replication cycles, what percentage of the DNA single-strands are radioactively labeled?
   a. 0%
   b. 12.5%
   c. 50%
   d. 87.5%
   e. 100%

Match the following choices to the statements in questions 7 through 10.
   a. antisense
   b. clone
   c. library
   d. Southern blot
   e. vector

7. Pieces of human DNA stored in yeast cells.

8. A population of cells carrying a desired plasmid.

9. Self-replicating DNA for transmitting a gene from one organism to another.

10. A gene that hybridizes with mRNA.

CRITICAL THINKING

1. Using the following map of plasmid pMICRO, give the number of restriction fragments that would result from digesting pMICRO with *EcoRI*, *HindIII*, and both enzymes together. Which enzyme makes the smallest fragment containing the tetracycline-resistance gene?

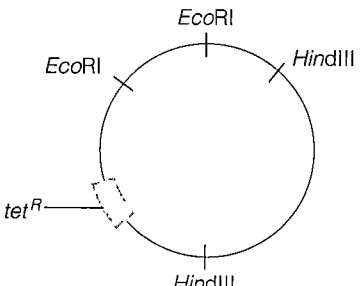

2. Design an experiment using vaccinia virus to make a vaccine against the AIDS virus (HIV).

3. Why did the use of DNA polymerase from the bacterium *Thermus aquaticus* allow researchers to add the necessary reagents to tubes in a preprogrammed heating block?

4. The following picture shows bacterial colonies growing on X-gal plus ampicillin in a blue-white screening test. Which colonies have the recombinant plasmid? The small satellite colonies do not have the plasmid. Why did they start growing on the medium 48 hours after the larger colonies?

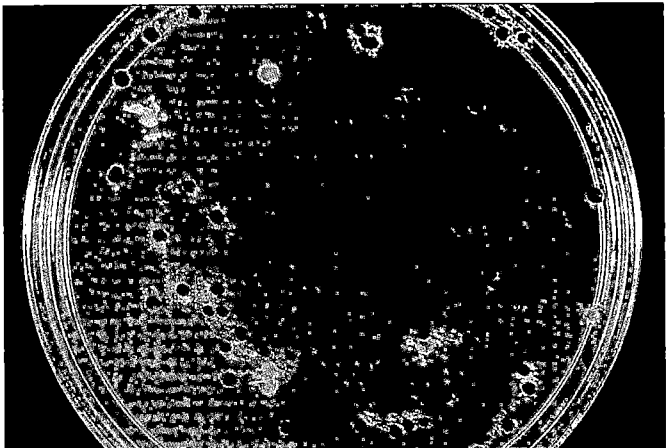

CLINICAL APPLICATIONS

1. PCR has been used to examine oysters for the presence of *Vibrio cholerae*. Oysters from different areas were homogenized, and DNA was extracted from the homogenates. The DNA was digested by the restriction enzyme *Hinc*II. A primer for the hemolysin gene of *V. cholerae* was used for the PCR reaction. After PCR, each sample was electrophoresed and stained with a probe for the hemolysin gene. Which of the oyster samples were (was) positive for *V. cholerae*? How can you tell? Why look for *V. cholerae* in oysters? What is the advantage of PCR over conventional biochemical tests to identify the bacteria?

Sample

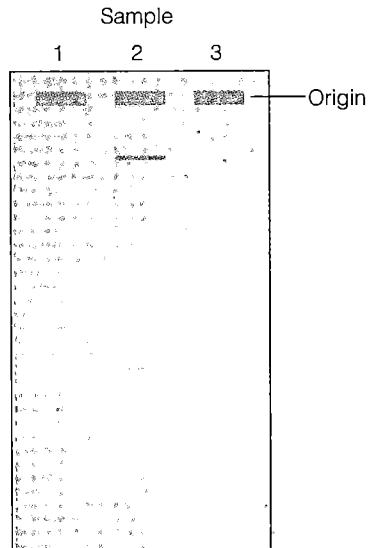

2. Using the restriction enzyme *Eco*RI, the following gel electrophoresis patterns were obtained from digests of various DNA molecules from a transformation experiment. Can you conclude from these data that transformation occurred? Explain why or why not.

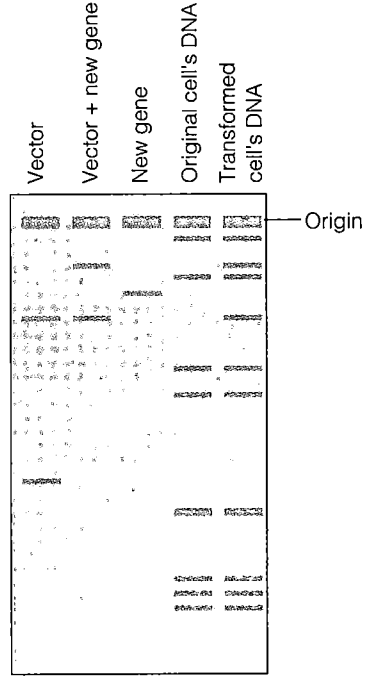

Learning with Technology

MP = The Microbiology Place website **ST** = Student Tutorial CD-ROM **VU** = VirtualUnknown CD-ROM

MP Don't forget to go to The Microbiology Place website (http://www.microbiologyplace.com) to take the practice tests, explore the interactive activity and case study, and check out the news articles and web links for this chapter.

ST Remember there is also a quiz for this chapter on the Microbiology Interactive Student Tutorial CD-ROM.

VU Enter the Virtual Lab, click the arrow next to the Session field, click Textbook Exercises, and select Chapter 9. Read the Case Study. Use this unknown for your work on the following problem:

You have been hired by a biotech firm to find new insecticides that are more environmentally friendly. You believe there are safe and effective ways to genetically engineer natural flora of pest insects to produce a recently discovered insecticidal protein. Always looking for potential hosts for the toxic gene, you have called your old classmates. One, an environmental microbiologist working for NASA, sends you this unknown as a candidate. Use the media and the tests in VirtualUnknown™ Microbiology to identify this unknown organism. Would you suggest that this microbe has economic potential for genetic engineering to control pest insects?

A Survey of the Microbial World

Microbiologists examine a DNA sequencing gel. The dark bands are pieces of DNA which migrate at different rates in response to an electrical current and are made visible by staining. DNA fingerprints like this are used to classify newly discovered organisms and to identify the causative agent of an infectious disease.

chapter 10

Classification of Microorganisms

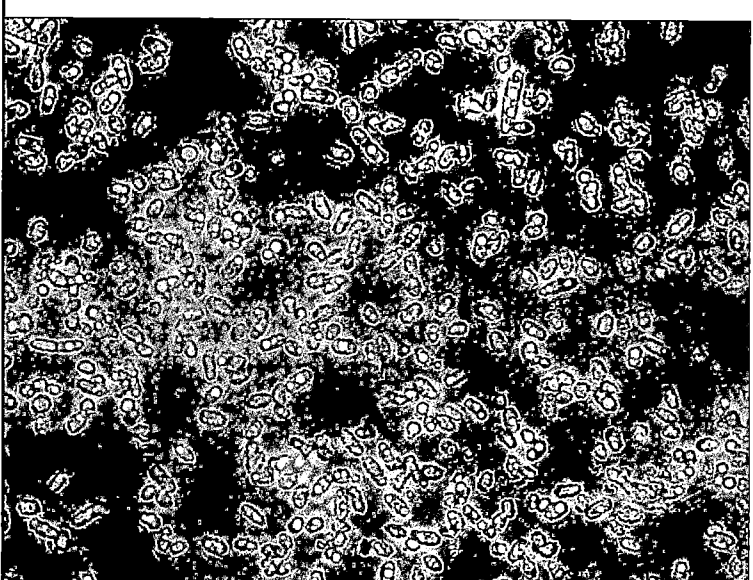

Chromatium. *Purple sulfur bacteria use hydrogen sulfide as an electron donor and store globules of sulfur in their cells.*

The science of classification, especially the classification of living forms, is called **taxonomy** (from the Greek for orderly arrangement). The objective of taxonomy is to classify living organisms—that is, to establish the relationships between one group of organisms and another and to differentiate between them. There may be as many as 30 million different living organisms, but fewer than 10% have been discovered, much less classified and identified.

A taxonomic system enables us to identify a previously unknown organism and then group or classify it with other organisms that have similar characteristics. For example, in 1985, investigators began looking into the cause of a diarrheal illness affecting people in Peru and Nepal. Microscopic studies, including electron micrographs, indicated that the cells excreted by patients were ovoid prokaryotic cells. In 1993, Ynes Ortega, after observing the unknown organism's life cycle, conclusively identified it as a new species of protozoan (a eukaryote), *Cyclospora cayetanensis* (sī-klō-spô-rä kī'ē-tan-en-sis), which now has been identified as the cause of disease worldwide.

Taxonomy also provides a common reference for identifying organisms already classified. For example, when a bacterium suspected of causing a specific disease is isolated from a patient, characteristics of that isolate are matched to lists of characteristics of previously classified bacteria to identify the isolate. After the bacterium has been identified, drugs can be selected that affect it. Finally, taxonomy is a basic and necessary tool for scientists, providing a universal language of communication.

Modern taxonomy is an exciting and dynamic field. New techniques in molecular biology and genetics are providing new insights into classification and evolution. In this chapter, you will learn the various classification systems, the different criteria used for classification, and tests that are used to identify microorganisms that have already been classified.

The Study of Phylogenetic Relationships

Learning Objectives

- *Define taxonomy, taxon, and phylogeny.*
- *Discuss the limitations of a two-kingdom classification system.*

There is tremendous unity and diversity in life. To date, biologists have identified more than 1.5 million different organisms. Among these different organisms are many similarities. For example, all organisms are composed of cells surrounded by a plasma membrane, use ATP for energy, and store their genetic information in DNA. These similarities are the result of evolution, or descent from a common ancestor. In 1859, the English naturalist Charles Darwin proposed that natural selection was responsible for the similarities as well as the differences among organisms. The differences can be attributed to the survival of organisms with traits best suited to a particular environment.

To facilitate research, scholarship, and communication, we arrange organisms into taxonomic categories, or **taxa** (singular: *taxon*), to show degrees of similarities among organisms. **Systematics,** or **phylogeny,** is the study of the evolutionary history of a group of organisms. The hierarchy of taxa reveals evolutionary or *phylogenetic* relationships.

From the time of Aristotle, living organisms were categorized in just two ways, as either plants or animals. As the biological sciences developed, however, biologists began looking for a *natural* classification system—one that groups organisms based on ancestral relationships and allows us to see the order in life. In 1857, Carl von Nägeli, a contemporary of Pasteur, proposed that bacteria and fungi be placed in the plant kingdom. In 1866, Ernst Haeckel proposed the Kingdom Protista, to include bacteria: protozoa, algae, and fungi. Because of disagreements over the definition of protists, for the next 100 years biologists continued to follow von Nägeli's placement of bacteria and fungi in the plant kingdom. It is ironic that recent DNA sequencing places fungi closer to animals than plants. Fungi were placed in their own kingdom in 1959.

With the advent of electron microscopy, the physical differences between cells became apparent. The term *prokaryote* was introduced in 1937 by Edward Chatton to distinguish cells having no nucleus from the nucleated cells of plants and animals. In 1961, Roger Stanier provided the current definition of prokaryotes: cells in which the nuclear material (nucleoplasm) is not surrounded by a nuclear membrane. In 1968, Robert G.E. Murray proposed the Kingdom Prokaryotae.

In 1969, Robert H. Whittaker founded the five-kingdom system in which prokaryotes were placed in the Kingdom Prokaryotae, or Monera, and eukaryotes comprised the other four kingdoms. The Kingdom Prokaryotae had been based on microscopic observations. Subsequently, new techniques in molecular biology revealed that there are actually two types of prokaryotic cells and one type of eukaryotic cell.

The Three Domains

Learning Objectives

- *Discuss the advantages of the three-domain system.*
- *List the characteristics of the Bacteria, Archaea, and Eukarya domains.*

The discovery of three cell types was based on the observations that ribosomes are not the same in all cells (see Chapter 4, page 95). Ribosomes provide a method of comparing cells because ribosomes are present in all cells. Comparing the sequences of nucleotides in ribosomal RNA (rRNA) from different kinds of cells shows that there are three distinctly different cell groups: the eukaryotes and two different types of prokaryotes—the bacteria and the archaea.

In 1978, Carl R. Woese proposed elevating the three cell types to a level above kingdom, called domain. Woese believed that the archaea and the bacteria, although similar in appearance, should form their own separate domains on the evolutionary tree (Figure 10.1). In this widely accepted scheme, animals, plants, fungi, and protists are kingdoms in the Domain **Eukarya.** Organisms are classified by cell type in the three domain systems. In addition to differences in rRNA, the three domains differ in membrane lipid structure, transfer RNA molecules, and sensitivity to antibiotics (Table 10.1 on page 279).

The Domain **Bacteria** includes all of the pathogenic prokaryotes as well as many of the nonpathogenic prokaryotes found in soil and water. The photoautotrophic prokaryotes are also in this domain. The Domain **Archaea** includes prokaryotes that do not have peptidoglycan in their cell walls. They often live in extreme environments, and they carry out unusual metabolic processes. Archaea include three major groups:

1. The methanogens, strict anaerobes that produce methane (CH_4) from carbon dioxide and hydrogen.

2. Extreme halophiles, which require high concentrations of salt for survival.

3. Hyperthermophiles, which normally grow in hot, acidic environments.

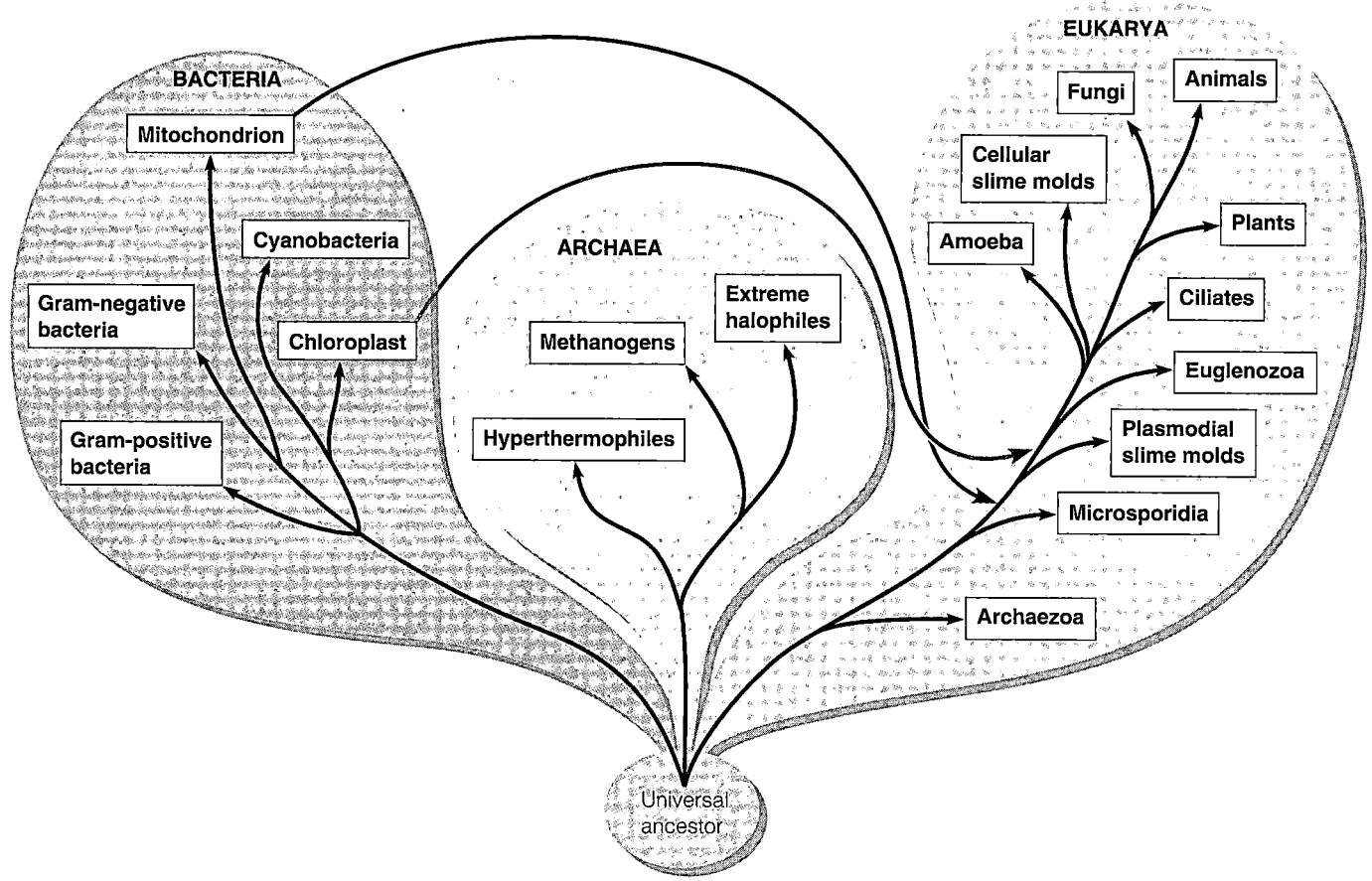

FIGURE 10.1 The three-domain system. The system recognizes the three types of cells. The Domain Eukarya includes the Kingdoms Fungi, Plantae, and Animalia. It also includes the protists discussed on page 284.

■ What characteristics are used to differentiate the three domains?

The evolutionary relationship of the three domains is the subject of current research by Woese and others. Originally, archaea were thought to be the most primitive group, whereas bacteria were assumed to be more closely related to eukaryotes. However, studies of rRNA indicate that a universal ancestor split into three lineages. That split led to the Archaea, the Bacteria, and what eventually became the nucleoplasm of the eukaryotes. The oldest known fossils are the remains of prokaryotes that lived more than 3.5 billion years ago. Eukaryotic cells evolved more recently, about 1.4 billion years ago. According to the Endosymbiotic Theory, eukaryotic cells evolved from prokaryotic cells living inside one another, as endosymbionts (see Chapter 4, page 108). In fact, the similarities between prokaryotic cells and eukaryotic organelles provide striking evidence for this endosymbiotic relationship (Table 10.2).

The original nucleoplasmic cell was prokaryotic. However, infoldings in its plasma membrane may have surrounded the nuclear region to produce a true nucleus (Figure 10.2). That cell provided the original host in which endosymbiotic bacteria developed into organelles. An example of a modern prokaryote living in a eukaryotic cell is shown in Figure 10.3. The cyanobacteriumlike cell and the eukaryotic host require each other for survival.

In sequencing the genome of a prokaryote called *Thermotoga maritima,* microbiologist Karen Nelson has discovered that this species has genes similar to both Bacteria and Archaea. Her findings suggest that *Thermotoga* is one of the earliest cells, arising before the Bacteria and Archaea split apart. For this reason, *Thermotoga* is referred to as one of the "Deeply Branching Genera."

Taxonomy provides tools for clarifying the evolution of organisms, as well as their interrelationships. New organisms are being discovered every day, and taxonomists continue to search for a natural classification system that reflects phylogenetic relationships.

table 10.1 | Some Characteristics of Archaea, Bacteria, and Eukaryotes

	Archaea	Bacteria	Eukaryotes
	Methanosarcina SEM ⊢10 μm⊣	*E. coli* SEM ⊢1 μm⊣	*Amoeba* SEM ⊢1 μm⊣
Cell Type	Prokaryotic	Prokaryotic	Eukaryotic
Cell Wall	Varies in composition; contains no peptidoglycan	Contains peptidoglycan	Varies in composition; contains carbohydrates
Membrane Lipids	Composed of branched carbon chains attached to glycerol by ether linkage	Composed of straight carbon chains attached to glycerol by ester linkage	Composed of straight carbon chains attached to glycerol by ester linkage
Start Signal for Protein Synthesis	Methionine	Formylmethionine	Methionine
Antibiotic Sensitivity	No	Yes	No
rRNA Loop*	Lacking	Present	Lacking
Common Arm of tRNA**	Lacking	Present	Present

*Binds to ribosomal protein; found in all bacteria.
**A sequence of bases on tRNA found in all eukaryotes and bacteria: guanine-thymine-pseudouridine-cytosine-guanine.

table 10.2 | Prokaryotic Cells and Eukaryotic Organelles Compared

	Prokaryotic Cell	Eukaryotic Cell	Eukaryotic Organelles (Mitochondria and Chloroplasts)
DNA	Circular	Linear	Circular
Histones	No	Yes	No
Ribosomes	70S	80S	70S
Growth	Binary fission	Mitosis	Binary fission

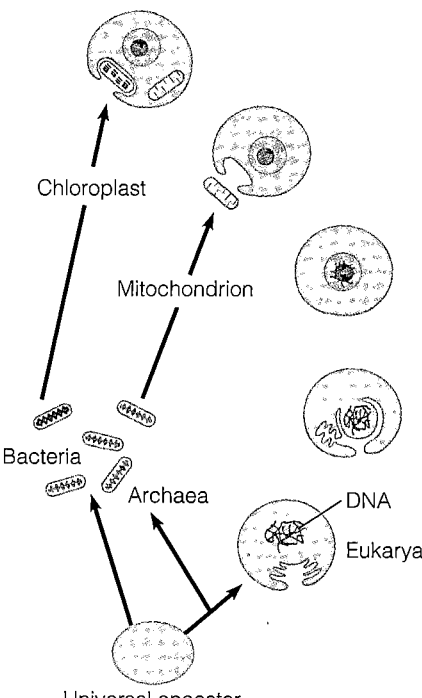

FIGURE 10.2 A model of the origin of eukaryotes. Invagination of the plasma membrane may have formed the nuclear envelope and endoplasmic reticulum. Similarities, including rRNA sequences, indicate that endosymbiotic prokaryotes gave rise to mitochondria and chloroplasts.

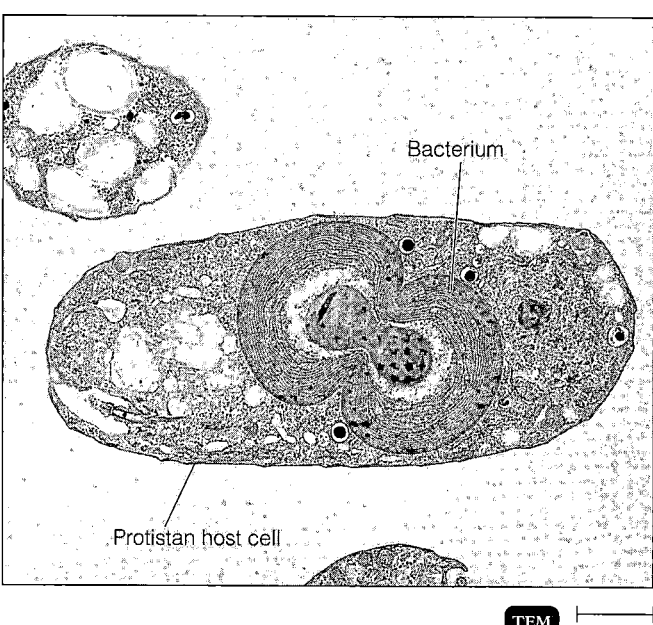

FIGURE 10.3 _Cyanophora paradoxa._ This organism, in which the eukaryotic host and the bacterium require each other for survival, provides a modern example of how eukaryotic cells might have evolved.

■ Eukaryotic organelles evolved from prokaryotic cells living inside host prokaryotic cells.

A Phylogenetic Hierarchy

In a phylogenetic hierarchy, grouping organisms according to common properties implies that a group of organisms evolved from a common ancestor; each species retains some of the characteristics of the ancestor. Some of the information used to classify and determine phylogenetic relationships in higher organisms comes from fossils. Bones, shells, or stems that contain mineral matter or have left imprints in rock that was once mud are examples of fossils.

The structures of most microorganisms are not readily fossilized. Some exceptions are the following:

■ A marine protist whose fossilized colonies form the White Cliffs of Dover, England.

■ Stromatolites, the fossilized remains of microbial communities that flourished between 0.5 and 2 billion years ago.

■ Fossilized cyanobacteria found in rocks in Western Australia that are 3.0–3.5 billion years old. These are the oldest known fossils. Some fossils of prokaryotes are shown in Figure 10.4.

Because fossil evidence is not available for most prokaryotes, their phylogeny must be based on other ev-

idence. But in one notable exception, scientists isolated living bacteria and yeast 25–40 million years old. In 1995, the American microbiologist Raul Cano and his colleagues reported growing _Bacillus sphaericus_ and other as-yet-unidentified microorganisms that had survived embedded in amber (fossilized plant resin) for millions of years. If confirmed, this discovery should provide more information about the evolution of microorganisms.

Conclusions from rRNA sequencing and DNA hybridization studies (discussed on page 295) of selected orders and families of eukaryotes are in agreement with the fossil records. This has encouraged workers to use DNA hybridization and rRNA sequencing to gain an understanding of the evolutionary relationships among prokaryotic groups.

Classification of Organisms

Learning Objective

■ _Differentiate among eukaryotic, prokaryotic, and viral species._

Living organisms are grouped according to similar characteristics (classification), and each organism is assigned a

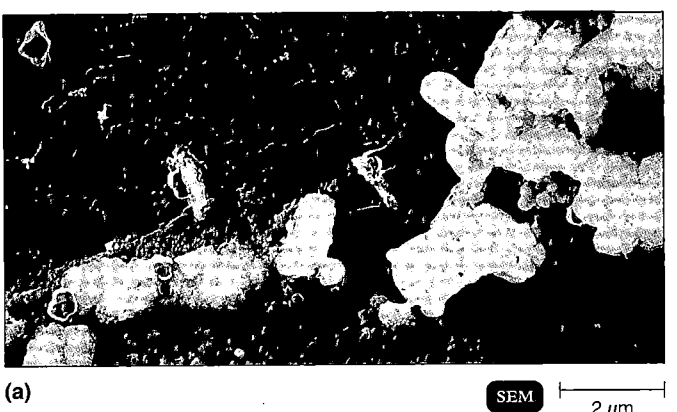
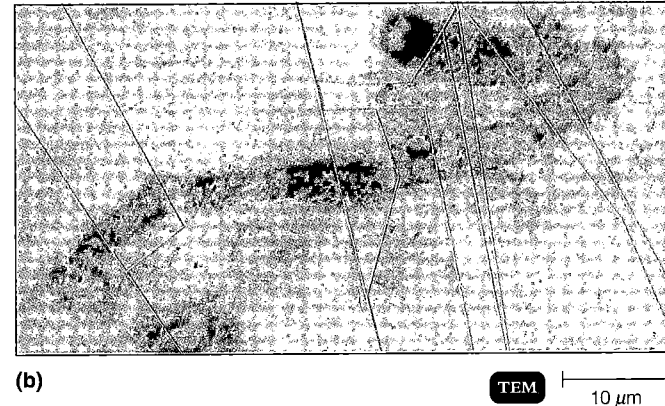

(a)   SEM ⊢————⊣ 2 μm   (b)   TEM ⊢————⊣ 10 μm

FIGURE 10.4 Fossilized prokaryotes. (a) Coccoid cyanobacteria from the Late Precambrian (ca. 850 million years ago) of central Australia. **(b)** Filamentous prokaryotes from the Early Precambrian (ca. 3.5 billion years ago) of western Australia.

■ **Because fossil evidence is not available for most prokaryotes, their phylogeny must be based on other kinds of evidence.**

unique scientific name. The rules for classifying and naming, which are used by biologists worldwide, are discussed next.

Scientific Nomenclature

Learning Objective

■ *Explain why scientific names are used.*

In a world inhabited by millions of living organisms, biologists must be sure they know exactly which organism is being discussed. We cannot use common names, because the same name is often used for many different organisms in different locales. For example, there are two different plants with the common name Spanish moss, and neither one is actually a moss. Three different animals are referred to as a gopher. Because common names are rarely specific and can often be misleading, a system of scientific names, referred to as *scientific nomenclature,* was developed in the eighteenth century.

Recall from Chapter 1 (page 3) that every organism is assigned two names, or a binomial. These names are the **genus** name and **specific epithet (species),** and both names are printed underlined or italicized. The genus name is always capitalized and is always a noun. The species name is lowercase and is usually an adjective. Because this system gives two names to each organism, the system is called **binomial nomenclature.**

Let's consider some examples. Our own genus and specific epithet are *Homo sapiens* (hō'mō sā'pē-ens). The noun, or genus, means man; the adjective, or specific epithet, means wise. A mold that contaminates bread is called

Rhizopus nigricans (rī'zō-pŭs nī'gri-kans). *Rhizo-* (root) describes rootlike structures on the fungus; *nigr-* (black) identifies the color of its spore sacs. Table 10.3 contains more examples.

Binomials are used by scientists worldwide, regardless of their native language, which enables them to share knowledge efficiently and accurately. Several scientific entities are responsible for establishing rules governing the naming of organisms. Rules for assigning names for protozoa and parasitic worms are published in the *International Code of Zoological Nomenclature.* Rules for assigning names for fungi and algae are published in the *International Code of Botanical Nomenclature.* Rules for naming newly classified bacteria and for assigning bacteria to taxa are established by the International Committee on Systematic Bacteriology and are published in the *Bacteriological Code.* Descriptions of bacteria and evidence for their classifications are published in the *International Journal of Systematic Bacteriology* before being incorporated into a reference called *Bergey's Manual.* According to the *Bacteriological Code,* scientific names are to be taken from Latin (a genus name can be taken from Greek) or latinized by the addition of the appropriate suffix. Suffixes for order and family are *-ales* and *-aceae,* respectively.

As new laboratory techniques make more detailed characterizations of bacteria possible, two genera may be reclassified as a single genus, or a genus may be divided into two or more genera. For example, the genera "Diplococcus" and *Streptococcus* were combined in 1974; the only diplococcal species is now called *Streptococcus pneumoniae.* (The *Bacteriological Code* states that the older genus name should be retained in such cases.) In 1984, DNA hybridization studies

table 10.3 *Making Scientific Names Familiar*

Use the word roots guide inside the book's front and back covers to find out what the name means. The name will not seem so strange if you translate it. When you encounter a new name, practice saying it out loud. The exact pronunciation is not as important as the familiarity you will gain. Guidelines for pronunciation are given in the back of this book.

Following are some examples of microbial names you may encounter in the popular press as well as in the lab.

	Pronunciation	Source of Genus Name	Source of Specific Epithet
Klebsiella pneumoniae (bacterium)	kleb-sē-el'lä nü-mō'nē-ī	Honors bacteriologist Edwin Klebs	The disease it causes
Salmonella typhimurium (bacterium)	sal-mōn-el'lä tī-fi-mür'ē-um	Honors public health microbiologist Daniel Salmon	Causes stupor *(typh-)* in mice *(muri-)*
Streptococcus pyogenes (bacterium)	strep-tō-kok'kus pī-äj'en-ēz	Appearance of cells in chains *(strepto-)*	Pus *(pyo-)* forming
Saccharomyces cerevisiae (yeast)	sak-ä-rō-mī'ses se-ri-vis'ē-ī	Fungus *(-myces)* that uses sugar *(saccharo-)*	Makes beer *(cerevisia)*
Penicillium notatum (fungus)	pen-i-sil'lē-um nō-tä'tum	Tuftlike or paintbrush *(penicill-)* appearance microscopically	Spores easily spread in air *(notus,* wind)
Trypanosoma cruzi (protozoan)	tri-pa-nō-sō'mä krūz'ē	Corkscrew *(trypano-,* borer; *soma-,* body)	Honors epidemiologist Oswaldo Cruz

indicated that "Streptococcus faecalis" and "Streptococcus faecium" were only distantly related to the other streptococcal species; consequently, a new genus called *Enterococcus* was created, and these species were renamed *E. faecalis* and *E. faecium* because the rules require that the original specific epithets be retained.

Making the transition to a new name can be confusing. The old name is often written in parentheses. For example, a physician looking for information on the cause of a patient's fever and eye irritation (cat-scratch disease) would find the bacterial name *Bartonella* (Rochalimaea) *henselae* (bär tō-nel'lä rō-chä-lē'mä-ä hen'sel-ī).

Obtaining the name of the organism is important in order to know what treatment to use; antifungal drugs will not work against bacteria, and antibacterial drugs will not work against viruses.

The Taxonomic Hierarchy

Learning Objective

- *List the major taxa.*

All organisms can be grouped into a series of subdivisions that make up the taxonomic hierarchy. A **eukaryotic species** is a group of closely related organisms that breed among themselves. (Bacterial species will be discussed shortly.) A genus consists of species that differ from each other in certain ways but are related by descent. For example, *Quercus* (kwer'kus), the genus name for oak, consists of all types of oak trees (white oak, red oak, bur oak, velvet oak, and so on). Even though each species of oak differs from every other species, they are all related genetically. Just as a number of species make up a genus, related genera make up a **family.** A group of similar families constitutes an **order,** and a group of similar orders makes up a **class.** Related classes, in turn, make up a **phylum.** (In botany, the comparable term *division* is used.) Thus, a particular organism (or species) has a genus name and specific epithet and belongs to a family, order, class, and phylum or division. All phyla or divisions that are related to each other make up a **kingdom,** and related kingdoms are grouped into a **domain.** (See Figure 10.5.)

Classification of Prokaryotes

The taxonomic classification scheme for prokaryotes is found in *Bergey's Manual of Systematic Bacteriology,* 2nd edition (see Appendix A). The first volume is due to be published in 2000, with the remaining four volumes to follow over the next few years. The contents of each volume are

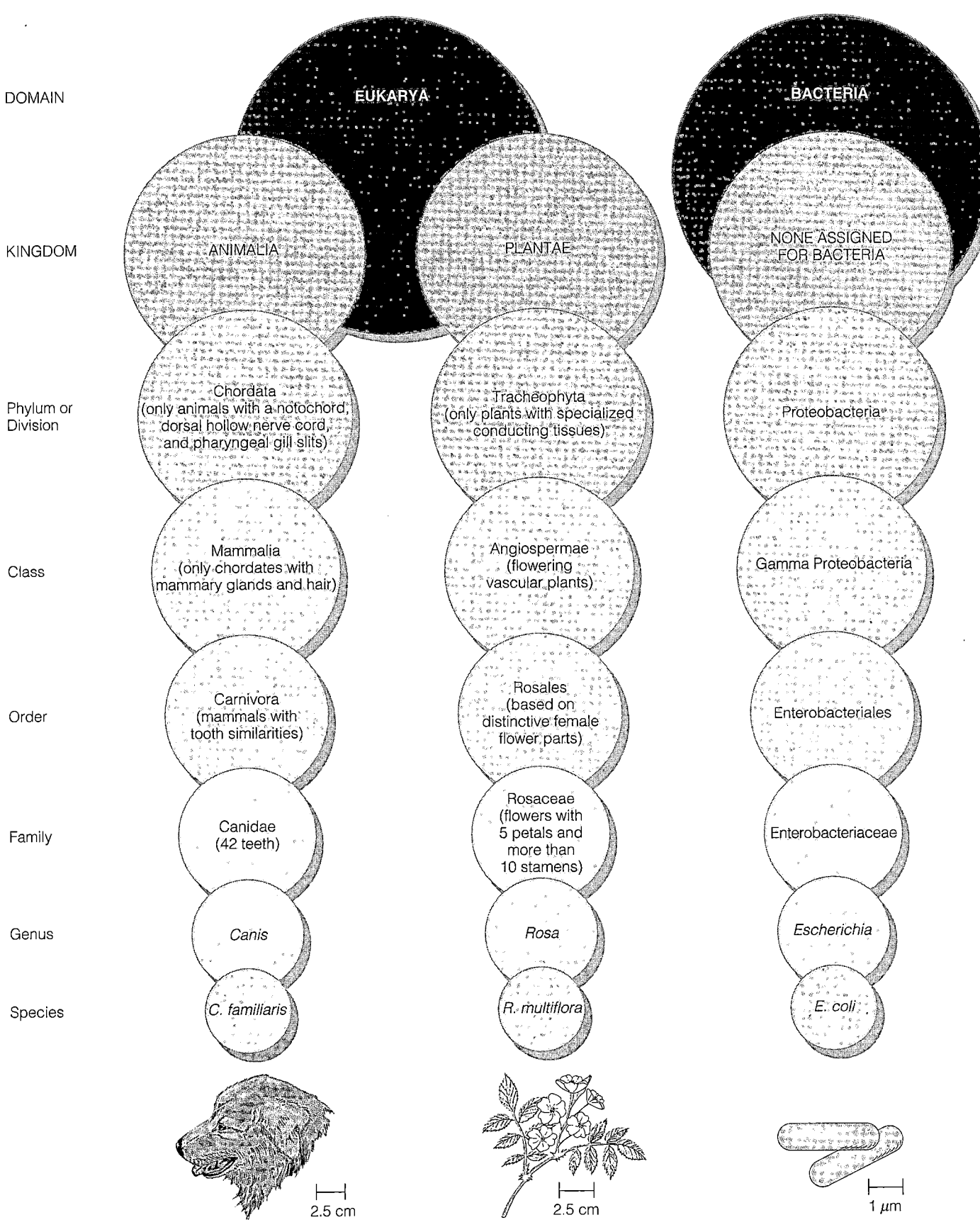

DOMAIN

EUKARYA

BACTERIA

KINGDOM

ANIMALIA

PLANTAE

NONE ASSIGNED
FOR BACTERIA

**Phylum or
Division**

Chordata
(only animals with a notochord,
dorsal hollow nerve cord,
and pharyngeal gill slits)

Tracheophyta
(only plants with specialized
conducting tissues)

Proteobacteria

Class

Mammalia
(only chordates with
mammary glands and hair)

Angiospermae
(flowering
vascular plants)

Gamma Proteobacteria

Order

Carnivora
(mammals with
tooth similarities)

Rosales
(based on
distinctive female
flower parts)

Enterobacteriales

Family

Canidae
(42 teeth)

Rosaceae
(flowers with
5 petals and
more than
10 stamens)

Enterobacteriaceae

Genus

Canis

Rosa

Escherichia

Species

C. familiaris

R. multiflora

E. coli

2.5 cm

2.5 cm

1 μm

FIGURE 10.5 The taxonomic hierarchy.

What is the biological definition of a family?

table 10.4	Classification of Prokaryotes (Bergey's Manual of Systematic Bacteriology, 2nd edition)	
Volume	**Contents**	**Notes**
1	Archaea, cyanobacteria, phototrophs, and deeply branching genera	Includes the Domain Archaea and some gram-negative Bacteria
2	Proteobacteria	Phylum Proteobacteria; these are related gram-negative bacteria
3	Low G+C gram-positives	Phylum Firmicutes (gram-positive cell wall) Phylum Mycoplasmas (wall-less)
4	High G+C gram-positives	Includes actinomyces
5	Planctomyces, spirochetes, sphingobacteria, bacteroids, and fusiforms	These are distinct phyla of bacteria, each with a unique rRNA sequence

shown in Table 10.4. In *Bergey's Manual,* prokaryotes are divided into two domains: Bacteria and Archaea. Each domain is divided into phyla. Remember, the classification is based on similarities in nucleotide sequences in rRNA. Classes are divided into orders; orders, into families; families, into genera; and genera, into species.

A prokaryotic species is defined somewhat differently than a eukaryotic species, which is a group of closely related organisms that can interbreed. Unlike reproduction in eukaryotic organisms, cell division in bacteria is not directly tied to sexual conjugation, which is infrequent and does not always need to be species-specific. A **prokaryotic species,** therefore, is defined simply as a population of cells with similar characteristics. (The types of characteristics will be discussed later in the chapter.) The members of a bacterial species are essentially indistinguishable from each other but are distinguishable from members of other species, usually on the basis of several features. In some cases, pure cultures of the same species are not identical in all ways. Each such group is called a **strain,** which is a collection of cells derived from a single cell. Strains are identified by numbers, letters, or names that follow the specific epithet.

Bergey's Manual provides a reference for identifying bacteria in the laboratory, as well as a classification scheme for bacteria. One scheme for the evolutionary relationships of bacteria is shown in Figure 10.6. Characteristics used to classify and identify bacteria are discussed in Chapter 11.

Classification of Eukaryotes

Learning Objective

■ List the major characteristics used to differentiate the four kingdoms in the Eukarya.

Some kingdoms in the Domain Eukarya are shown in Figure 10.1 (page 278).

Since 1969, simple eukaryotic organisms, mostly unicellular, have been grouped as the Kingdom **Protista,** a catchall kingdom for a variety of organisms. Historically, eukaryotic organisms that didn't fit into other kingdoms were placed in the Protista. Ribosomal RNA sequencing is making it possible to divide protists into groups based on their descent from common ancestors. Many biologists propose dividing Protista into five new kingdoms. For convenience, we will continue to use the term *protist* to refer to unicellular eukaryotes and their close relatives. These organisms will be discussed in Chapter 12.

Fungi, plants, and animals make up the three kingdoms of more complex eukaryotic organisms, most of which are multicellular.

The Kingdom **Fungi** includes the unicellular yeasts, multicellular molds, and macroscopic species such as mushrooms. To obtain raw materials for vital functions, a fungus absorbs dissolved organic matter through its plasma membrane. The cells of a multicellular fungus are commonly joined to form thin tubes called *hyphae.* The hyphae are divided into multinucleated units by cross-walls that have holes, so that cytoplasm can flow between the cell-like units. Most fungi lack flagella. Fungi develop from spores or from fragments of hyphae.

The Kingdom **Plantae** (plants) includes some algae and all mosses, ferns, conifers, and flowering plants. All members of this kingdom are multicellular. To obtain energy, a plant uses photosynthesis, the process that converts carbon dioxide and water into organic molecules used by the cell.

The kingdom of multicellular organisms called **Animalia** (animals) includes sponges, various worms, insects, and animals with backbones (vertebrates). Animals obtain nutrients and energy by ingesting organic matter through a mouth of some kind.

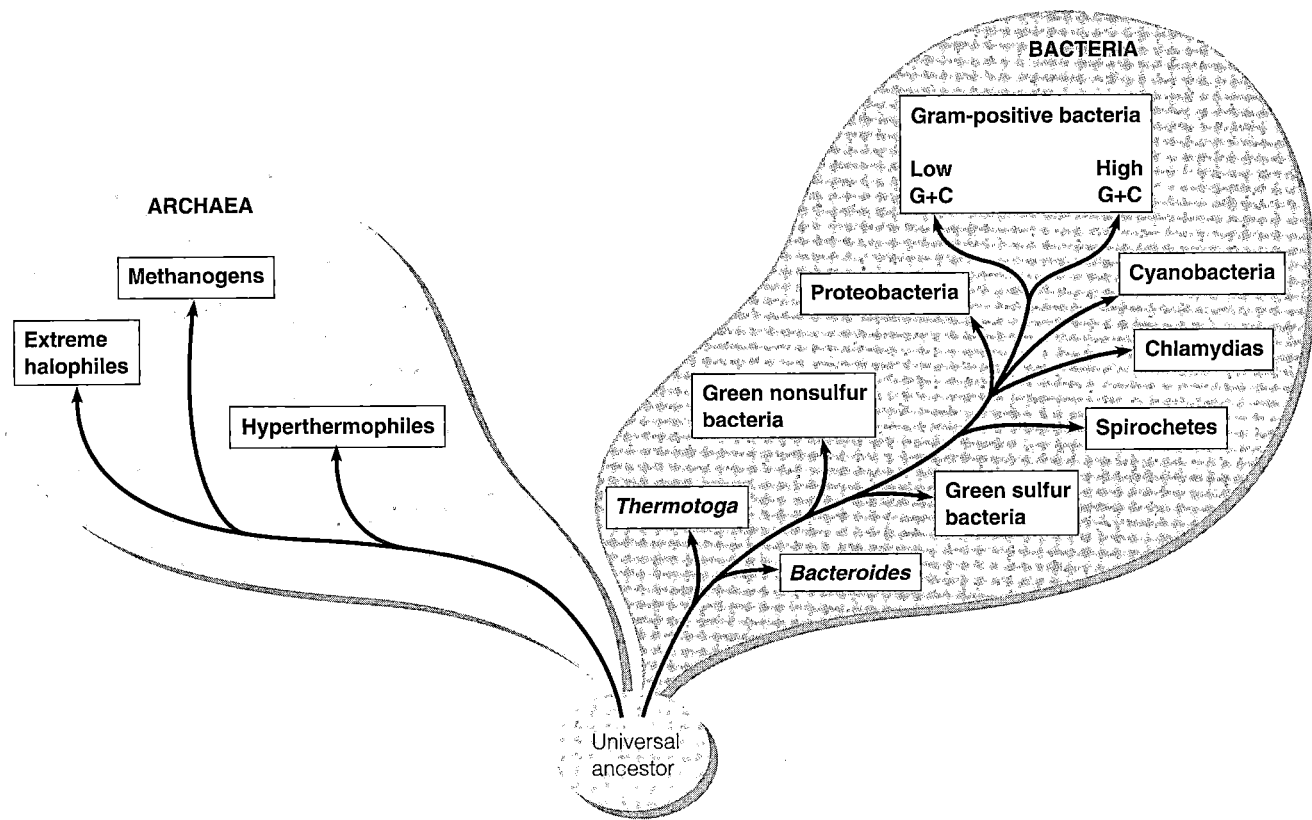

FIGURE 10.6 Phylogenetic relationships of prokaryotes. Arrows indicate major lines of descent of bacterial groups. Selected phyla are indicated by the white boxes.

■ Members of which phylum can be identified by Gram staining?

Classification of Viruses

Viruses are not composed of cells, and they use the anabolic machinery within living host cells to multiply. Viruses are not classified as part of any of the three domains because they don't have ribosomes. A viral genome can direct biosynthesis inside a host cell, and some viral genomes can become incorporated into the host genome. The ecological niche of a virus is its specific host cell, so viruses may be more closely related to their hosts than to other viruses. In 1991, the International Committee on Taxonomy of Viruses defined a **viral species** as a population of viruses with similar characteristics (including morphology, genes, and enzymes) that occupies a particular ecological niche.

Viruses are obligatory intracellular parasites, so they must have evolved after a suitable host cell had evolved. There are two hypotheses on the origin of viruses: (1) They arose from independently replicating strands of nucleic acids (such as plasmids), and (2) they developed from degenerative cells that, through many generations, gradu-

ally lost the ability to survive independently but could survive when associated with another cell. Viruses will be discussed in Chapter 13.

Methods of Classifying and Identifying Microorganisms

Learning Objectives

■ Compare and contrast classification and identification.
■ Explain the purpose of Bergey's Manual.

A classification scheme provides a list of characteristics and a means for comparison to aid in the identification of an organism. Once an organism is identified, it can be placed into a previously devised classification scheme. Microorganisms are *identified* for practical purposes—for example, to determine an appropriate treatment for an infection. They are not necessarily identified by the same techniques by which they are classified. Most identification procedures are easily performed in a laboratory and use as few procedures or tests as possible. Protozoa, parasitic worms, and fungi can usually be identified microscopically. Most prokaryotic organisms do not have distinguishing morphological features or even much variation in size and shape. Consequently, microbiologists have developed a variety of methods to test metabolic reactions and other characteristics to identify prokaryotes.

Bergey's Manual of Determinative Bacteriology has been a widely used reference since the first edition was published in 1923. *Bergey's Manual of Determinative Bacteriology* (9th ed., 1994) does not classify bacteria according to evolutionary relatedness but provides identification (determinative) schemes based on such criteria as cell wall composition, morphology, differential staining, oxygen requirements, and biochemical testing.* The majority of Bacteria and Archaea have not been cultured, and scientists estimate that only 1% of these microbes have been discovered.

Medical microbiology (the branch of microbiology dealing with human pathogens) has dominated the interest in microbes, and this interest is reflected in many identification schemes. However, to put the pathogenic properties of bacteria in perspective, of the more than 2600 species listed in the *Approved Lists of Bacterial Names*, fewer than 10% are human pathogens.

We next discuss several criteria and methods for the classification of microorganisms and the routine identification of some of those organisms. In addition to properties of the organism itself, the source and habitat of a bacterial isolate is considered as part of the classification and identification processes.

Morphological Characteristics

Morphological (structural) characteristics have helped taxonomists classify organisms for 200 years. Higher organisms are frequently classified according to observed anatomical detail. But many microorganisms look too similar to be classified by their structures. Through a microscope, organisms that might differ in metabolic or physiological properties may look alike. Literally hundreds of bacterial species are small rods or small cocci.

Larger size and the presence of intracellular structures does not always mean easy classification, however. *Pneumocystis* (nü-mō-sis'tis) pneumonia is the most common

opportunistic infection in immunosuppressed individuals and is a significant cause of death in AIDS patients. The causative agent of this infection, *Pneumocystis carinii* (kär-i'-nē-ē), was not considered a human pathogen until the 1970s. *Pneumocystis* lacks structures that can be easily used for identification (see Figure 24.18), and its taxonomic position has been uncertain since its discovery in 1909. Although it was provisionally classified as a protozoan, recent studies comparing its rRNA sequence with those of other protozoa, *Euglena*, cellular slime molds, plants, mammals, and fungi have shown that it might actually be a member of the Kingdom Fungi. Researchers have not been able to culture *Pneumocystis*, but they have developed some useful treatments for *Pneumocystis* pneumonia. Perhaps as researchers take into account this organism's relatedness to fungi, appropriate culture methods and treatments will result.

Cell morphology tells us little about phylogenetic relationships. However, morphological characteristics are still useful in identifying bacteria. For example, differences in such structures as endospores or flagella can be helpful.

Differential Staining

Learning Objective

■ *Describe how staining and biochemical tests are used to identify bacteria.*

Recall from Chapter 3 that one of the first steps in identifying bacteria is differential staining. Most bacteria are either gram-positive or gram-negative. Other differential stains, such as the acid-fast stain, can be useful for a more limited group of microorganisms. Recall that these stains are based on the chemical composition of cell walls and therefore are not useful in identifying either the wall-less bacteria or the archaea with unusual walls. Microscopic examination of a Gram stain or an acid-fast stain is used to obtain information quickly in the clinical environment. A physician can sometimes get sufficient information from a technician's lab report to begin appropriate treatment (Figure 10.7 and the box in Chapter 21, page 584).

Biochemical Tests

Enzymatic activities are widely used to differentiate bacteria. Even closely related bacteria can usually be separated into distinct species by subjecting them to biochemical tests, such as one to determine their ability to ferment an assortment of selected carbohydrates. For one example of the use of biochemical tests to identify bacteria (in this instance, in marine mammals),

---

*Both *Bergey's Manual of Systematic Bacteriology* (see page 282) and *Bergey's Manual of Determinative Biology* are referred to simply as *Bergey's Manual*; the complete titles are used when the information under discussion is found in one but not the other, for example, an identification table.

MICROBIOLOGY REQUISITION	Date:	Time:	Slip prepared by:
Lab: Date, time received:	Physician name:	Collected by:	Patient ID#:

⌐DO NOT WRITE BELOW THIS LINE⌐	**USE SEPARATE SLIP FOR EACH REQUEST**		

GRAM STAIN REPORT		**SOURCE OF SPECIMEN**	**TEST(S) REQUESTED**	
☐ GRAM POS. COCCI, GROUPS ☐ GRAM POS. COCCI, PAIRS/CHAIN ☐ GRAM POS. RODS ☒ GRAM NEG. COCCI ☐ GRAM NEG. RODS ☐ GRAM NEG. COCCOBACILLI ☐ YEAST ☐ OTHER	☐ NO GROWTH ☐ NO GROWTH IN ___DAYS ☐ MIXED MICROBIOTA ☐ SPECIMEN IMPROPERLY COLLECTED OR TRANSPORTED ☐ ___DIFFERENT TYPES OF ORGANISMS ☐ NEGATIVE FOR *SALMONELLA, SHIGELLA,* AND *CAMPYLOBACTER* ☐ NO OVA, CYSTS, OR PARASITES SEEN ☒ OXIDASE-POSITIVE GRAM-NEGATIVE DIPLOCOCCI ☐ PRESUMPTIVE BETA STREP GROUP A BY BACITRACIN	☐ BLOOD ☐ CEREBROSPINAL FLUID ☐ FLUID (Specify Source) _____ ☐ THROAT ☐ SPUTUM, expectorated ☐ OTHER Respiratory (Describe) _____ ☐ URINE, Clean Catch Midstream ☐ URINE, Indwelling Catheter ☐ URINE, Straight Catheter ☐ URINE, Entire First Morning ☐ URINE, Other (Describe) _____ ☐ STOOL ☒ GU (Specify Source) _____ ☐ ABSCESS (Specify Source) _____ ☐ TISSUE (Specify Source) _____ ☐ ULCER (Specify Source) _____ ☐ WOUND (Specify Source) _____ ☐ STERILIZER TEST	**Bacterial** ☐ **Routine culture;** Gram stain, anaerobic culture, susceptibility testing. Throats done for Gp A Strep. ☐ *Legionella* culture ☐ *Rochalimaea* ☐ Blood Culture **Other Non-Routine Cultures** ☐ *E. coli* 0157:H7 ☐ *Vibrio* ☐ *Yersinia* ☒ *H. ducreyi* ☐ *B. pertussis* ☐ Other _____ **Screening Cultures** ☒ Gonococci ☐ Group B Strep ☐ Group A Strep ☐ Other _____ ☐ **ACID-FAST BACILLI**	☐ **FUNGAL** **VIRAL** ☐ Routine culture ☐ Herpes simplex ☐ Direct FA for _____ **PARASITOLOGY** ☐ Exam for intestinal ova and parasites ☐ *Giardia* immunoassay ☐ *Cryptosporidium* ☐ Pinworm prep ☐ Blood parasites ☐ Filaria concentration ☐ *Trichomonas* ☐ Other _____ **TOXIN ASSAY** ☐ *Clostridium difficile* **DIRECT (Antigen Detection)** ☐ Cryptococcal antigen-CSF only ☐ Bacterial antigens (Specify) _____ **SPECIAL** ☐ Antimicrobial tests (MIC)

*filled out by one person   *filled out by different person

FIGURE 10.7 A clinical microbiology lab report form. In health care, morphology and differential staining are important in determining the proper treatment for microbial diseases. A clinician completes the form to identify the sample and specific tests. In this case, a genitourinary sample will be examined for sexually transmitted diseases. The lab technician reported the Gram stain and culture results. [Minimal inhibitory concentration (MIC) of antibiotics will be discussed in Chapter 20, page 567.]

■ A clinician requests information about patients with infectious diseases on a microbiology requisition.

see the box on page 292. Moreover, biochemical tests can provide insight into a species' niche in the ecosystem. For example, a bacterium that can fix nitrogen gas or oxidize elemental sulfur will provide important nutrients for plants and animals. This will be discussed in Chapter 27.

Enteric, gram–negative bacteria are a large heterogeneous group of microbes whose natural habitat is the intestinal tract of humans and other animals. This family contains several pathogens that cause diarrheal illness. A number of tests have been developed so that technicians can quickly identify the pathogen, then a clinician can

FIGURE 10.8 The use of metabolic characteristics to identify selected genera of enteric bacteria.

■ Even closely related bacteria can usually be separated into distinct species by biochemical testing.

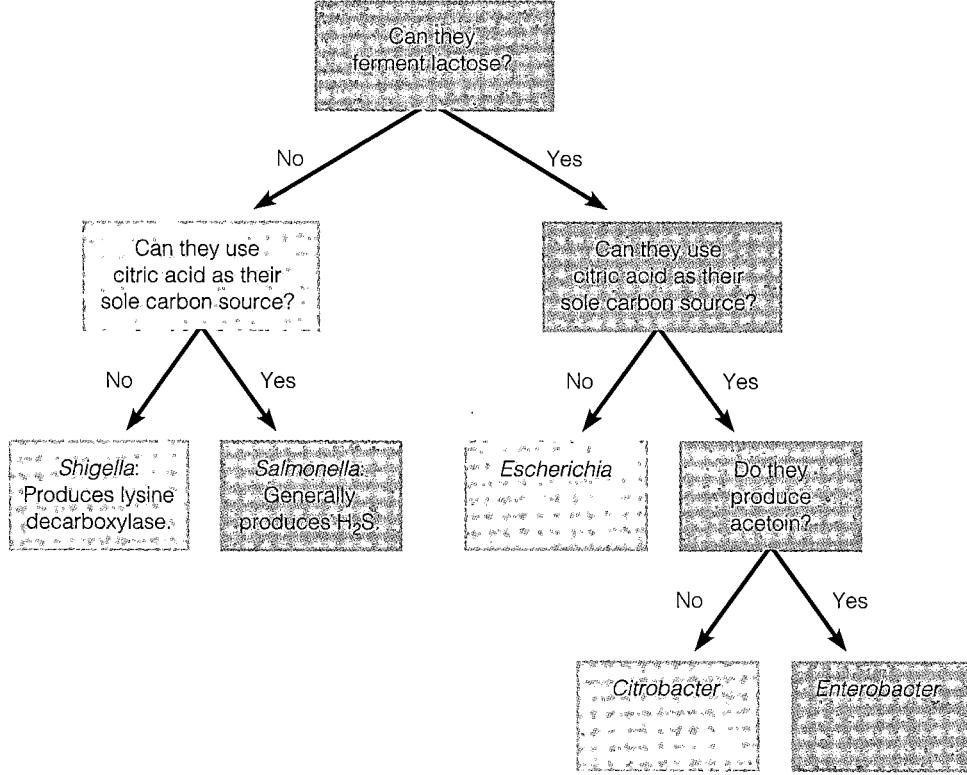

provide appropriate treatment, and epidemiologists can locate the source of an illness. All members of the family Enterobacteriaceae are oxidase-negative. Among the enteric bacteria are members of the genera *Escherichia, Enterobacter* (en-te-rō-bak'tĕr), *Shigella* (shi-gel'lä), *Citrobacter* (sit'rō-bak-tĕr), and *Salmonella. Escherichia, Enterobacter,* and *Citrobacter,* which ferment lactose to produce acid and gas, can be distinguished from *Salmonella* and *Shigella,* which do not. Further biochemical testing, as represented in Figure 10.8, can differentiate among the genera.

The time needed to identify bacteria can be reduced considerably by the use of selective and differential media or by rapid identification methods. Recall from Chapter 6 (page 167) that selective media contain ingredients that suppress the growth of competing organisms and encourage the growth of desired ones, and that differential media allow the desired organism to form a colony that is somehow distinctive.

Rapid identification tools are manufactured for groups of medically important bacteria, such as the enterics. Such tools are designed to perform several biochemical tests simultaneously and can identify bacteria within 4–24 hours. The results of each test are assigned a number (which varies with the tool being used) based on the relative reliability and importance of each test.

In the example shown in Figure 10.9, ① an unknown enteric bacterium is inoculated into a tube designed to perform 15 biochemical tests. ② After incubation, results in each compartment are observed. ③ Positive results are marked on the scoring form. Notice that each test is assigned a value; the number derived from scoring all the tests is called the I.D. value. Fermentation of glucose is important, and a positive reaction is valued at 2, compared with the production of acetoin (V-P test), which has no value. ④ A computerized interpretation of the simultaneous test results is essential and is provided by the manufacturer. In this example, the test results indicate that the bacterium is *Enterobacter cloacae,* with the typical test results. A limitation of biochemical testing is that mutations and plasmid acquisition can result in strains with different characteristics. Unless a large number of tests are used, an organism could be incorrectly identified.

Bergey's Manual does not evaluate the relative importance of each biochemical test and does not always describe strains. In the clinical diagnosis of a disease, a particular species and even a particular strain must be identified in order to proceed with proper treatment. To this end, specific series of biochemical tests have been developed for fast identification in hospital laboratories. Rapid biochemical systems have been developed for yeasts and other fungi, as well as bacteria.

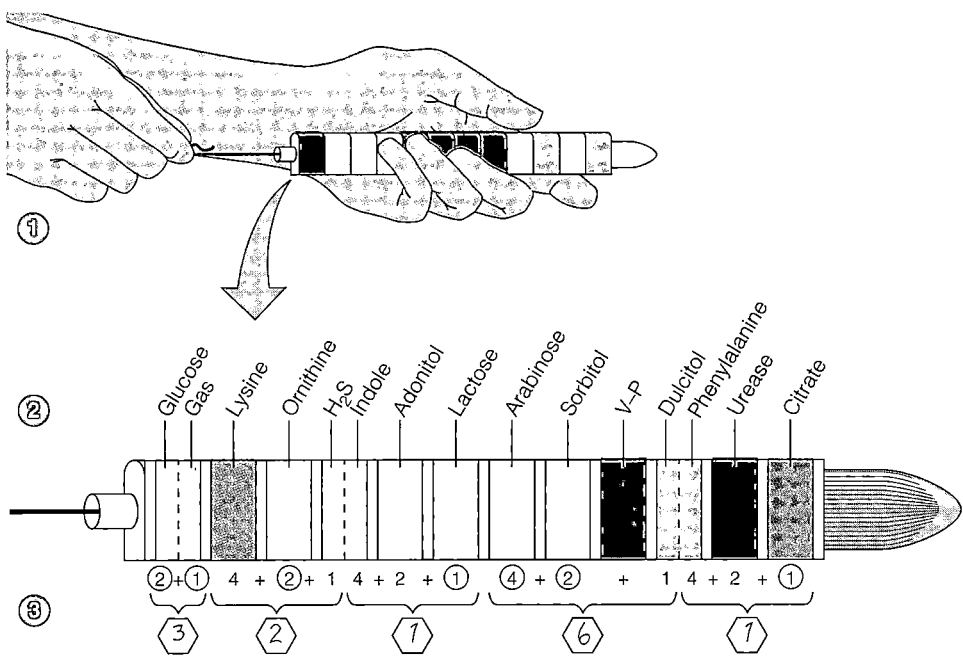

FIGURE 10.9 One type of rapid identification method for bacteria: Enterotube™ II from Becton Dickinson. ① One tube containing media for 15 tests is inoculated. ② After incubation, the tube is observed for results. ③ The value for each positive test is circled, and the numbers from each group of tests are added to give each portion of the I.D. value. ④ Comparing the resultant I.D. value with a computerized listing shows that the organism in the tube is *Enterobacter cloacae*. This example shows results for a typical strain of *E. cloacae*; however, other strains may produce different test results, which are listed in the Atypical Test Results column. The V-P test is used to confirm an identification.

■ In rapid identification methods, test results are weighted according to their relative importance.

ID Value	Organism	Atypical Test Results	Confirmatory Test
32143	*Enterobacter cloacae*	Sorbitol⁻	—
	Enterobacter sakazakii	Urea⁺	+
32161	*Enterobacter cloacae*	None	VP⁺
32162	*Enterobacter cloacae*	Citrate⁻	

Serology

Learning Objectives

■ *Differentiate Western blotting from Southern blotting.*
■ *Explain how serological tests and phage typing can be used to identify an unknown bacterium.*

Serology is the science that studies blood serum and immune responses that are evident in serum (see Chapter 18). Microorganisms are antigenic; that is, microorganisms that enter an animal's body stimulate it to form antibodies. Antibodies are proteins that circulate in the blood and combine in a highly specific way with the bacteria that caused their production. For example, the immune system of a rabbit injected with killed typhoid bacteria (antigens) responds by producing antibodies against typhoid bacteria. Solutions of such antibodies used in the identification of many medically important microorganisms are commercially available; such a solution is called an **antiserum** (plural: *antisera*). If an unknown bacterium is isolated from a patient, it can be tested against known antisera and often identified quickly.

In a procedure called a **slide agglutination test,** samples of an unknown bacterium are placed in a drop of saline on each of several slides. Then a different known antiserum is added to each sample. The bacteria agglutinate (clump) when mixed with antibodies that were produced in response to that species or strain of bacterium; a positive test is indicated by the presence of agglutination. Positive and negative slide agglutination tests are shown in Figure 10.10.

Serological testing can differentiate not only among microbial species but also among strains within species. As mentioned in Chapter 1, Rebecca Lancefield was able to classify serotypes of streptococci by studying serological reactions. She found that the different antigens in the cell walls of various serotypes of streptococci stimulate the formation of different antibodies. In contrast, because closely related bacteria also produce some of the same antigens, serological testing can be used to screen bacterial isolates for possible similarities. If an antiserum reacts with proteins from different bacterial species or strains, these bacteria can be tested further for relatedness.

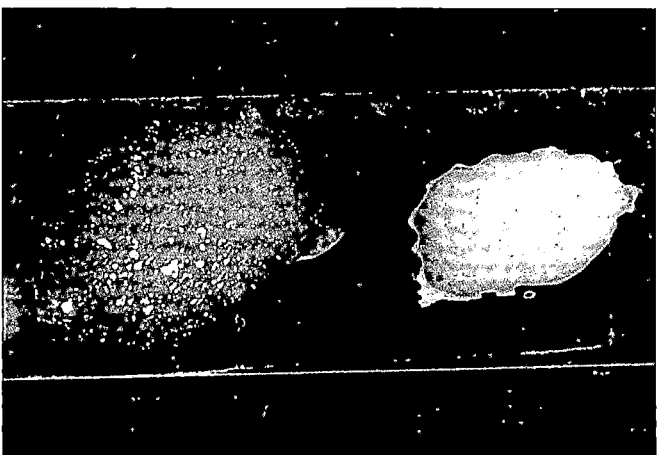

(a) Positive test   **(b)** Negative test

FIGURE 10.10 A slide agglutination test. (a) In a positive test, the grainy appearance is due to the clumping (agglutination) of the bacteria. **(b)** In a negative test, the bacteria are still evenly distributed in the saline and antiserum.

■ Agglutination results when the bacteria are mixed with antibodies that were produced in response to the same strain.

Serological testing was used to determine whether the increase in number of cases of necrotizing fasciitis in the United States and England since 1987 was due to a common source of the infections. No common source was located, but there has been an increase in two serotypes of *Streptococcus pyogenes* that have been dubbed the flesh-eating bacteria.

The **enzyme-linked immunosorbent assay (ELISA)** is widely used because it is fast and can be read by a computer scanner (Figure 10.11; see Chapter 18, page 511). In a direct ELISA, known antibodies are placed in the wells of a microplate, and an unknown type of bacterium is added to each well. A reaction between the known antibodies and the bacteria provides identification of the bacteria. An ELISA is used in AIDS testing to detect the presence of antibodies against human immunodeficiency virus (HIV), the virus that causes AIDS (see Figure 1.7a on page 21).

Another serological test, **Western blotting,** is used to identify bacterial antigens in a patient's serum (Figure 10.12). HIV infection is confirmed by Western blotting, and Lyme disease, caused by *Borrelia burgdorferi,* is often diagnosed by the Western blot (see Chapter 23, page 000). ① Proteins (including bacterial proteins) in the patient's serum are separated by a process called electrophoresis (Chapter 9, page 264). ② The proteins are then transferred to a filter by blotting. ③ Next, antibodies tagged with a dye are washed over the filter. If the specific antigen (in this case, *Borrelia* proteins) is present in the serum, the antibodies will combine with it and will be visible as a colored band on the filter.

Phage Typing

Like serological testing, phage typing looks for similarities among bacteria. Both techniques are useful in tracing the origin and course of a disease outbreak. **Phage typing** is a test for determining which phages a bacterium is sus-

(a)

(b)

FIGURE 10.11 Using a computer scanner to read the results of an ELISA test.
(a) A technician uses a micropipette to add samples to a microplate for an ELISA.
(b) ELISA results are then read by the computer scanner.

■ What are the similarities between the slide agglutination test and the ELISA test?

① If Lyme disease is suspected in a patient, a sample of the patient's serum is taken. Electrophoresis is used to separate proteins in the serum sample. Each band consists of many molecules of a particular protein (antigen). The bands are not visible at this point.

② The bands are transferred to a nitrocellulose filter by blotting.

③ The filter with proteins (antigens) is positioned exactly as it was on the gel. The filter is then washed with antibodies against a particular antigen. The antibodies are tagged with a dye so that they are visible when they combine with their specific antigen (shown here in purple).

④ The test is read. If the tagged antibodies stick to the filter, evidence of the presence of the microorganism in question—in this case, *Borrelia burgdorferi*—has been found in the patient's serum.

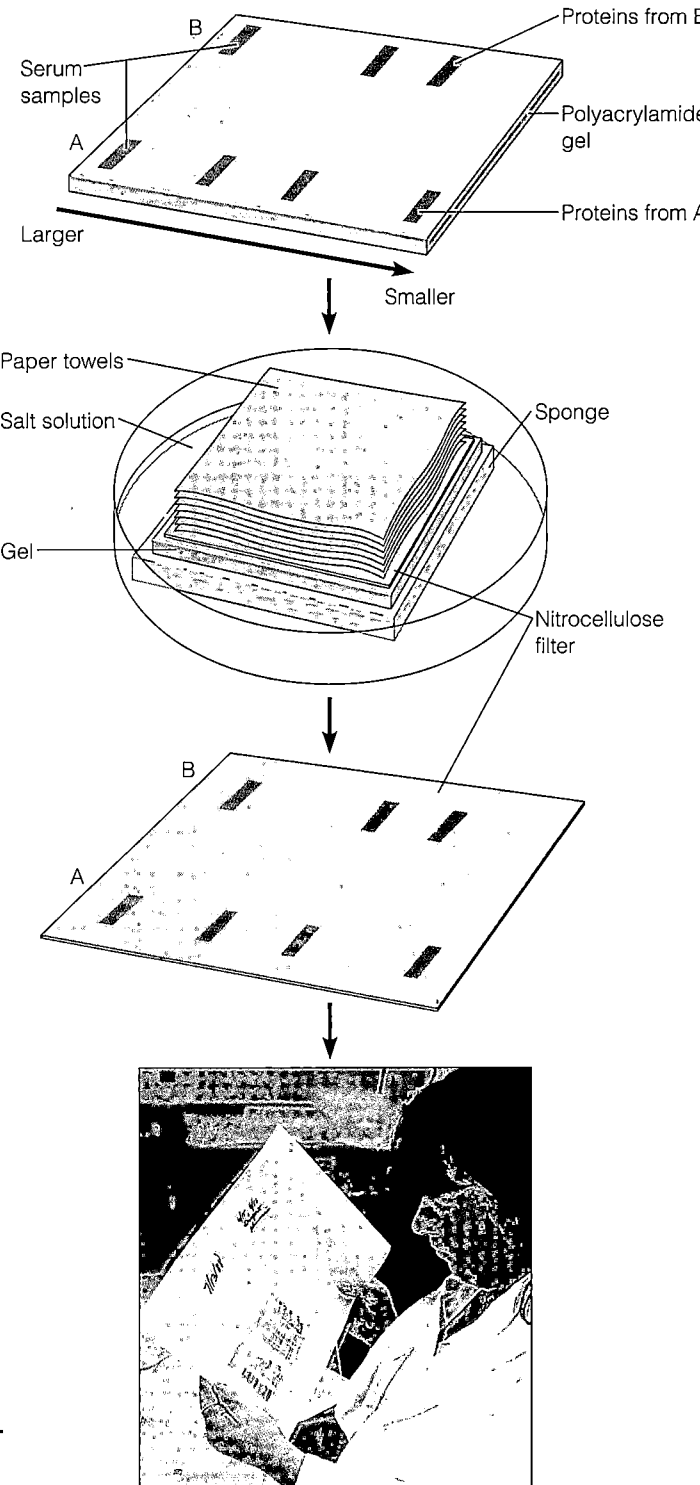

FIGURE 10.12 The Western blot. Proteins separated by electrophoresis can be detected by their reactions with antibodies.

■ Name two diseases that may be diagnosed by Western blotting.

ceptible to. Recall from Chapter 8 (page 239) that bacteriophages (phages) are bacterial viruses and that they usually cause lysis of the bacterial cells they infect. They are highly specialized, in that they usually infect only members of a particular species, or even particular strains within a species. One bacterial strain might be susceptible to two different phages, whereas another strain of the same species might be susceptible to those two phages plus a third phage. Bacteriophages will be discussed further in Chapter 13.

The sources of food-associated infections can be traced by phage typing. One version of this procedure

A record number of manatees died in Florida waters during 1996. Over half the deaths were due to pneumonia. In 1994, researchers at the Marine Mammal Center in California reported the presence of a skin disease in northern elephant seals. Microbiologists are trying to determine the causes of these diseases.

Another potentially alarming instance of unusual disease in marine mammals occurred among the dolphin population. During an average year, 10 or 12 dead dolphins are found washed ashore on beaches between New Jersey and Virginia. However, in one recent year the death toll climbed to more than 200, and scientists believe that hundreds more may have died offshore. These mortality figures raise concerns that entire populations of marine mammals may ultimately be destroyed. Moreover, alarmed swimmers have asked whether they are also in danger from marine mammal diseases. The U.S. Office of Naval Research is continuing to fund marine mammal research at colleges and universities to find the cause of the dolphins' deaths.

Large numbers of opportunistic pathogens, including 55 species of *Vib-*

rio, were found in the dolphins. These bacteria are a part of a dolphin's normal microbiota and the biota of coastal waters. It can cause disease only if the animals' immune system, their normal defense against infection, has been weakened. To find the ultimate cause of death, scientists must find out what has weakened the immune systems.

One possibility is chemical pollutants. Insecticides and polychlorinated biphenyls (PCBs) have been found in dolphins, manatees, and elephant seals. Daniel Martineau of Cornell University says that PCBs are strong immunosuppressants. Another possibility is a viral infection that affects the immune system. The cause of thousands of seal deaths in northern Europe in April 1988 is now known to be a previously undiscovered virus. Could a new virus be infecting other marine mammals as well?

INFORMATION IS SCARCE

Such questions are the concern of veterinary microbiology, which until recently has been a neglected branch of medical microbiology. Although the diseases of such animals as cattle, chick-

ens, and mink have been studied, partly because of their availability to researchers, the microbiology of wild animals, especially marine mammals, is a relatively newly emerging field. Gathering samples of animals that live in the open ocean and performing bacteriological analyses on them are very difficult. Currently, the animals being studied are those that live in captivity (see the photograph) and those that come onto the shore to breed, such as the northern fur sea lion.

The scientists are identifying bacteria in marine mammals by using conventional test batteries (see the figure) and genomic data on known species. The bacteria are compared with species described in Bergey's Manual to assign names or identify them. Perhaps new species of bacteria will be found in the marine mammals.

Veterinary microbiologists hope that increased study of the microbiology of wild animals, including marine mammals, will not only promote improved wildlife management but also provide models for the study of human diseases.

starts with a plate totally covered with bacteria growing on agar. A drop of each different phage type to be used in the test is then placed on the bacteria. Wherever the phages are able to infect and lyse the bacterial cells, clearings in the bacterial growth (called plaques) appear (Figure 10.13 on page 294). Such a test might show, for instance, that bacteria isolated from a surgical wound have the same pattern of phage sensitivity as those isolated from the operating surgeon or surgical nurses. This establishes that the surgeon or a nurse is the source of infection.

Fatty Acid Profiles

Bacteria synthesize a wide variety of fatty acids, and in general, these fatty acids are constant for a particular species. A commercial system has been designed to separate cellular fatty acids to compare them to fatty acid pro-

files of known organisms. Fatty acid profiles often need to be backed up by biochemical testing, and they can be used only for identification, not to determine phylogenetic relatedness.

Flow Cytometry

Flow cytometry can be used to identify bacteria in a sample without culturing the bacteria. In a *flow cytometer,* a moving fluid containing bacteria is forced through a small opening (see Figure 18.11). The simplest method detects the presence of bacteria by detecting the difference in electrical conductivity between cells and the surrounding medium. If the fluid passing through the opening is illuminated by a laser, the scattering of light provides information about the cell size, shape, density, and surface, which is analyzed by a computer. Fluores-

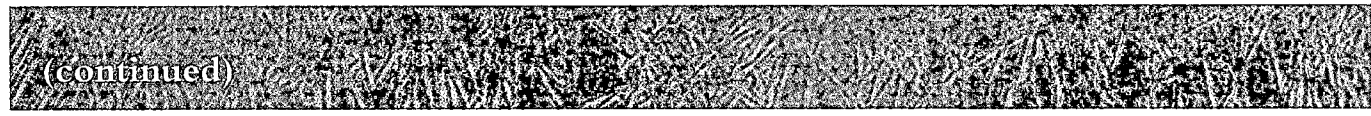

Marine mammal researchers examine a Pacific bottlenosed dolphin.

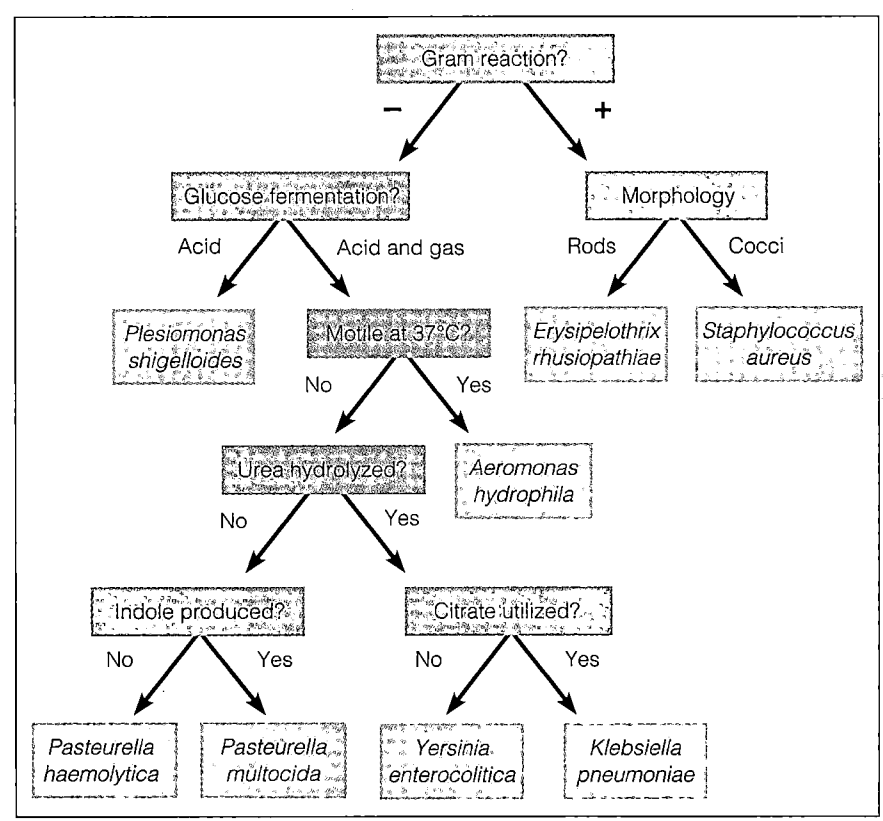

Biochemical tests used to identify selected species of human pathogens isolated from marine mammals.

cence can be used to detect naturally fluorescent cells, such as *Pseudomonas,* or cells tagged with fluorescent dyes.

Milk can be a vehicle for disease transmission. A proposed test that uses flow cytometry to detect *Listeria* in milk could save time because the bacteria would not need be cultured for identification. Antibodies against *Listeria* can be labeled with a fluorescent dye and added to the milk to be tested. The milk is passed through the flow cytometer, which records the light scattering of the antibody-labeled cells.

DNA Base Composition

Learning Objective

■ *Describe how a newly discovered microbe can be classified by: DNA base composition, rRNA sequencing, DNA fingerprinting, PCR, and nucleic acid hybridization.*

A classification technique that has come into wide use among taxonomists is the determination of an organism's **DNA base composition.** This base composition is usually expressed as the percentage of guanine plus cytosine (G + C). The base composition of a single species is theoretically a fixed property; thus, a comparison of the G + C content in different species can reveal the degree of species relatedness. As we saw in Chapter 8, each guanine (G) in DNA has a complementary cytosine (C). Similarly, each adenine (A) in the DNA has a complementary thymine (T). Therefore, the percentage of DNA bases that are GC pairs also tells us the percentage that are AT pairs (GC + AT = 100%). Two organisms that are closely related and hence have many identical or similar genes will have similar amounts of the various bases in their DNA. However, if there is a difference of more than 10% in their percentage of GC pairs (for example, if one bacterium's DNA contains 40% GC and another bacterium

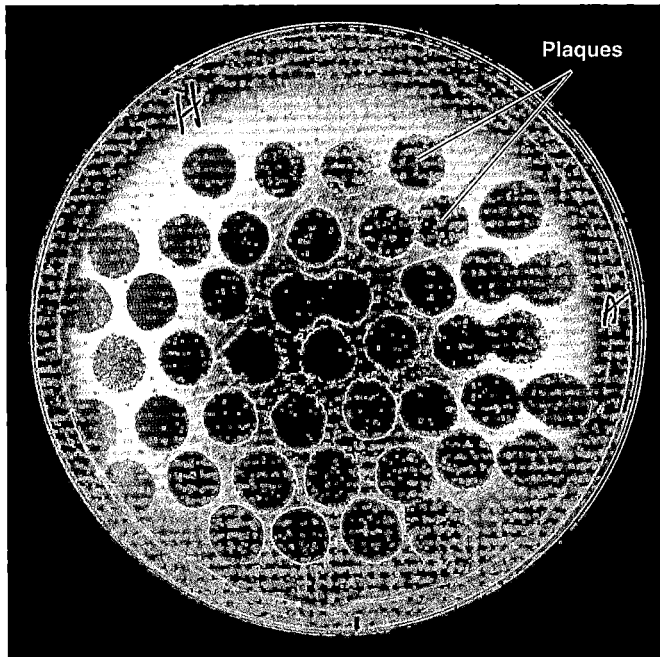

FIGURE 10.13 Phage typing of a strain of *Salmonella typhi*.
The tested strain was grown over the entire plate. Plaques, or areas of lysis, were produced by bacteriophages, indicating that the strain was sensitive to infection by these phages. Bacteria with this pattern of phage sensitivity are designated as *Salmonella typhi* phage type A.

■ What is being identified in phage typing?

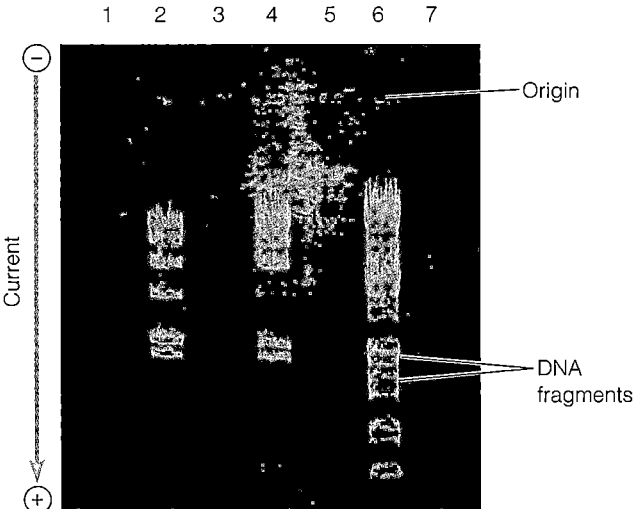

FIGURE 10.14 DNA fingerprints. Plasmids from five different bacteria were digested with the same restriction enzyme. Each digest was put in a different well (origin) in the agarose gel. (Wells 1 and 7 are empty.) An electrical current was then applied to the gel to separate the fragments by size and electrical charge. The DNA was made visible by staining with ethidium bromide, which fluoresces under ultraviolet light. Comparison of the lanes shows that none of the DNA samples (and therefore none of the bacteria) is identical.

■ DNA fingerprinting is used to determine the source of hospital-acquired infections.

has 60% GC), then these two organisms are probably not related. Of course, two organisms that have the same percentage of GC are not necessarily closely related; other supporting data are needed to draw conclusions about their phylogenetic relationship.

DNA Fingerprinting

Actually determining the entire sequence of bases in an organism's DNA is now possible with modern biochemical methods, but this is currently impractical for all but the smallest microorganisms (viruses) because of the great amount of time required. However, the use of restriction enzymes enables researchers to compare the base sequences of different organisms. Restriction enzymes cut a molecule of DNA everywhere a specific base sequence occurs, producing restriction fragments (as discussed in Chapter 9, page 252). For example, the enzyme *Eco*RI cuts DNA at the arrows in the sequence

$$...G^\downarrow A\ A\ T\ T\ C...$$
$$...C\ T\ T\ A\ A_\uparrow G...$$

In this technique, the DNA from two microorganisms is treated with the same restriction enzyme, and the restriction fragments produced are separated by electrophoresis on a thin layer of agar (see Chapter 9, page 264). A comparison of the number and sizes of restriction fragments that are produced from different organisms provides information about their genetic similarities and differences; the more similar the patterns, or *DNA fingerprints,* the more closely related the organisms are expected to be (Figure 10.14).

DNA fingerprinting is used to determine the source of hospital-acquired infections. In one hospital, patients undergoing coronary-bypass surgery developed infections caused by *Rhodococcus bronchialis* (rō-dō-kok′kus bron-kē′al-is). The DNA fingerprints of the patients' bacteria and the bacteria of one nurse were identical. The hospital was thus able to break the chain of transmission of this infection by encouraging this nurse to use aseptic technique. The use of DNA fingerprinting to locate the source of restaurant-associated diarrhea is described in the box in Chapter 25 on page 694.

Ribosomal RNA Sequencing

Ribosomal RNA (rRNA) sequencing is currently being used to determine the diversity of organisms and the phylogenetic relationships among them. There are several advantages to using rRNA. First, all cells contain ribosomes. Two closely related organisms will have fewer different bases in their rRNA than two organisms that are distantly related. Another advantage is that RNA genes have undergone few changes over time. The rRNA used most often is a component of the smaller portion of ribosomes. A third advantage of rRNA sequencing is that cells do not have to be cultured in the laboratory.

DNA in a soil or water sample can be amplified by the polymerase chain reaction (PCR) using an rRNA primer (see below). The rRNA gene can be sequenced to determine evolutionary relationships between organisms, especially microorganisms.

The Polymerase Chain Reaction

When a microorganism cannot be cultured by conventional methods, the causative agent of an infectious disease might not be recognized. However, a technique called the **polymerase chain reaction (PCR)** can be used to increase the amount of microbial DNA to levels that can be tested by gel electrophoresis (see Chapter 9, page 254).

In 1992, researchers used PCR to determine the causative agent of Whipple's disease, which was previously an unknown bacterium now named *Tropheryma whippelii* (trō-fer-ē-mä whip-pél'ē-ē). Whipple's disease was first described in 1907 by George Whipple as a gastrointestinal and nervous system disorder caused by an unknown bacillus. No one had been able to culture the bacterium to identify it, and thus there were no reliable methods of diagnosing and treating the disease.

In recent years, PCR has been used successfully in several discoveries that were not otherwise possible. For example, in 1992, Raul Cano used PCR to amplify DNA from *Bacillus* bacteria in amber that was 25–40 million years old. These primers were made from rRNA sequences in living *B. circulans* to amplify DNA coding for rRNA in the amber. These primers will cause amplification of DNA from other *Bacillus* species but do not cause amplification of DNA from other bacteria that might have been present, such as *Escherichia* or *Pseudomonas*. The DNA was sequenced after amplification. This information was used to determine the relationships between the ancient bacteria and modern bacteria.

In 1993, microbiologists identified a *Hantavirus* as the cause of an outbreak of hemorrhagic fever in the Ameri-can Southwest using PCR. The identification was made in record time—less than 2 weeks. PCR was used in 1994 to identify the causative agent of a new tickborne disease (human granulocytic ehrlichiosis) as the bacterium *Ehrlichia chaffeensis* (ėr'lik-ē-ä chaf'ē-en-sis). The use of PCR to identify rabies viruses is described in the box in Chapter 22 (page 617).

In 1996, Applied Biosystems introduced TaqMan™, a system that uses PCR to identify pathogenic *E. coli* in food and water. With this system, the newly amplified *E. coli* DNA fluoresces and can be detected using gel electrophoresis.

Nucleic Acid Hybridization

If a double-stranded molecule of DNA is subjected to heat, the complementary strands will separate as the hydrogen bonds between the bases break. If the single strands are then cooled slowly, they will reunite to form a double-stranded molecule identical to the original double strand. (This reunion occurs because the single strands have complementary sequences.) When this technique is applied to separated DNA strands from two different organisms, it is possible to determine the extent of similarity between the base sequences of the two organisms. This method is known as **nucleic acid hybridization**. The procedure assumes that if two species are similar or related, a major portion of their nucleic acid sequences will also be similar. The procedure measures the ability of DNA strands from one organism to hybridize (bind through complementary base pairing) with the DNA strands of another organism (Figure 10.15). The greater the degree of hybridization, the greater the degree of relatedness.

Recall from Chapter 8 that RNA is single-stranded and is transcribed from one strand of DNA; a particular strand of RNA, therefore, is complementary to the strand of DNA from which it was transcribed and will hybridize with that separated strand of DNA. DNA-RNA hybridization can thus be used to determine relatedness between DNA from one organism and RNA from another organism in the same way that DNA-DNA hybridization is used.

Nucleic acid hybridization can be used to identify unknown microorganisms by **Southern blotting** (see Figure 9.12). In addition, rapid identification methods using **DNA probes** are being developed. One method involves breaking DNA extracted from *Salmonella* into fragments with a restriction enzyme, then selecting a specific fragment as the probe for *Salmonella* (Figure 10.16). This fragment must be able to hybridize with the DNA of all *Salmonella* strains, but not with the DNA of closely

FIGURE 10.15 DNA-DNA hybridization. The greater the amount of pairing between DNA strands from different organisms (hybridization), the more closely the organisms are related.

■ What is the principle involved in DNA probes?

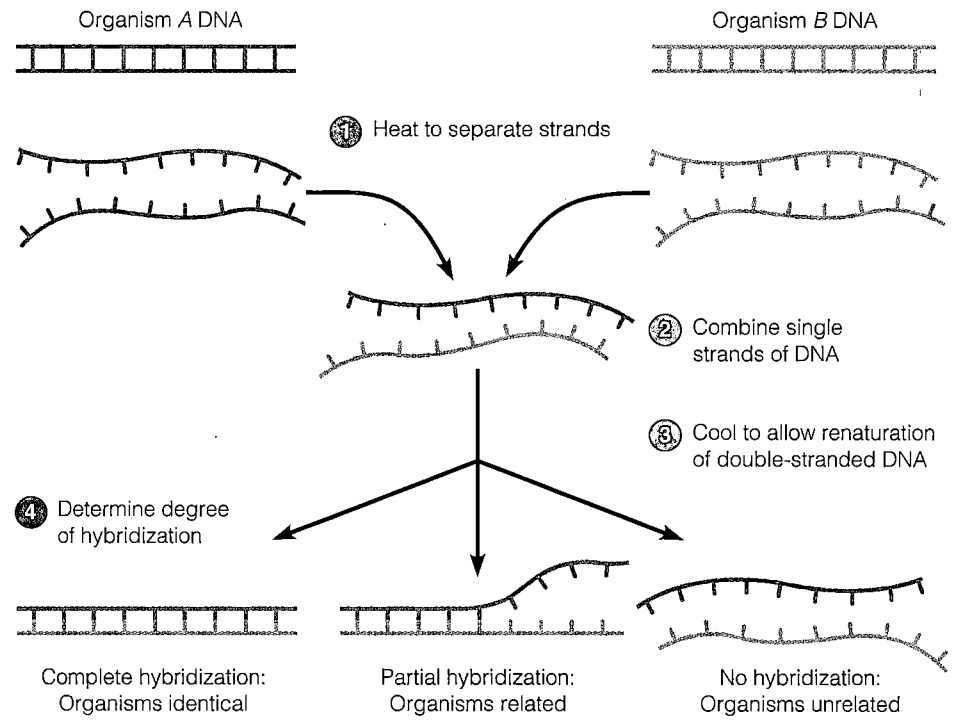

FIGURE 10.16 A DNA probe used to identify bacteria.

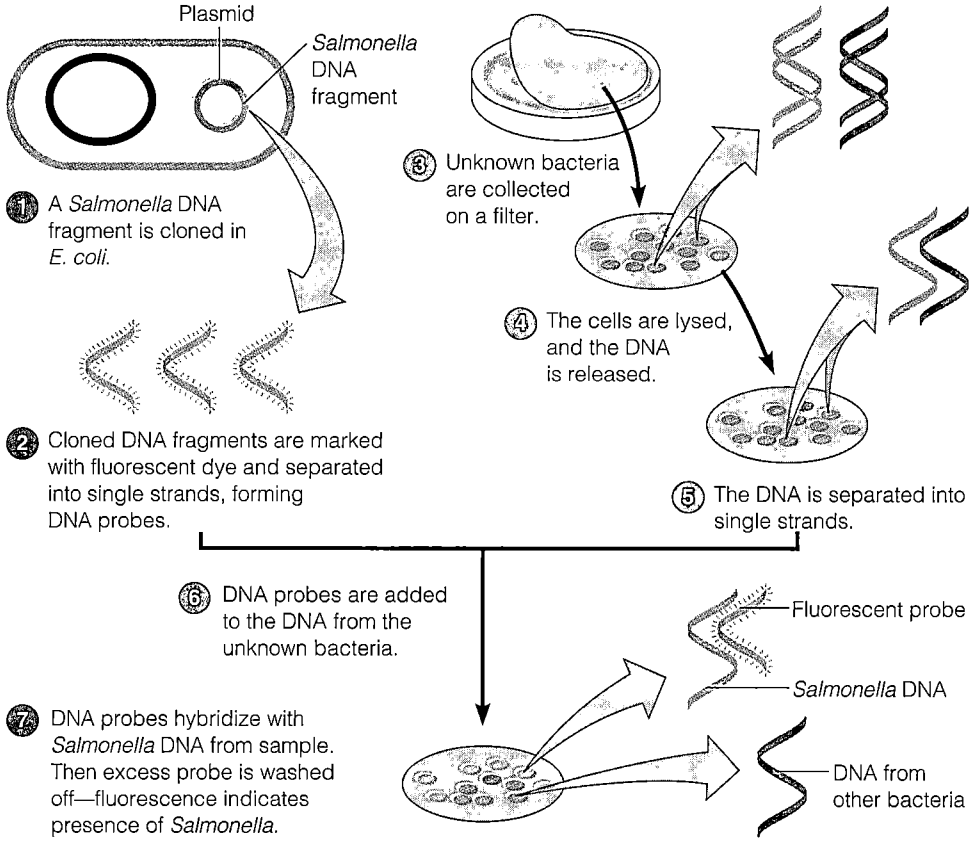

table 10.5	Taxonomic Criteria and Methods for Classifying and Identifying Bacteria	
	Used for	
Criterion or Method	Classification	Identification
Morphological characteristics	No (yes for cyanobacteria)	Yes
Differential staining	Yes (for cell wall type)	Yes
Biochemical testing	No	Yes
Serology	No	Yes
Phage typing	No	Yes
Fatty acid profiles	No	Yes
Flow cytometry	No	Yes
DNA base composition	Yes	No
DNA fingerprinting	No	Yes
rRNA sequencing	Yes	No
PCR	Yes	Yes
Nucleic acid hybridization	Yes	Yes (DNA probes, DNA chips)

related enteric bacteria. ① The chosen DNA fragment is cloned in a plasmid in *E. coli,* producing hundreds of specific *Salmonella* DNA fragments. ② These fragments are tagged with radioactive isotopes or a fluorescent dye and are separated into single strands of DNA. ③–⑥ The resulting DNA probes can then be mixed with single-stranded DNA prepared from a food sample suspected of containing *Salmonella*. ⑦ If *Salmonella* is present, the DNA probes will hybridize with the DNA of the *Salmonella,* and hybridization can be detected by the radioactivity or fluorescence of the probes.

An exciting new technology is the **DNA chip,** which will make it possible to quickly sequence entire genomes or detect a pathogen in a host by identifying a gene that is unique to that pathogen. The DNA chip is composed of DNA probes and a fluorescent dye indicator. A sample containing DNA from an unknown organism can be added to the chip. Hybridization between the probe DNA and DNA in the sample is detected by fluoresence.

Putting Classification Methods Together

Learning Objective

■ *Differentiate a dichotomous key from a cladogram.*

Morphological characteristics, differential staining, and biochemical testing were the only identification tools available just a few years ago. Technological advancements

are making it possible to use nucleic acid analysis techniques, once reserved for classification, for routine identification. (A practical example is described in the box in Chapter 22, page 617.) A summary of taxonomic criteria and methods is provided in Table 10.5.

Information about microbes obtained by these methods is used to identify and classify the organisms. Two methods of using the information are described below.

Dichotomous Keys

Dichotomous keys are widely used for identification. In a dichotomous key, identification is based on successive questions, and each question has two possible answers (dichotomous means cut in two). After answering one question, the investigator is directed to another question until an organism is identified. Although these keys often have little to do with phylogenetic relationships, they are invaluable for identification. For example, a dichotomous key for bacteria could begin with an easily determined characteristic, such as cell shape, and move on to the ability to ferment a sugar. Dichotomous keys are shown in Figure 10.8 and in the box on page 293.

Cladograms

Cladograms are maps that show evolutionary relationships among organisms (*clado-* means branch). Cladograms are shown in Figures 10.1 and 10.6. Each branch point on

FIGURE 10.17 Building a cladogram.

■ Why do *L. brevis* and *L. acidophilus* branch from the same node?

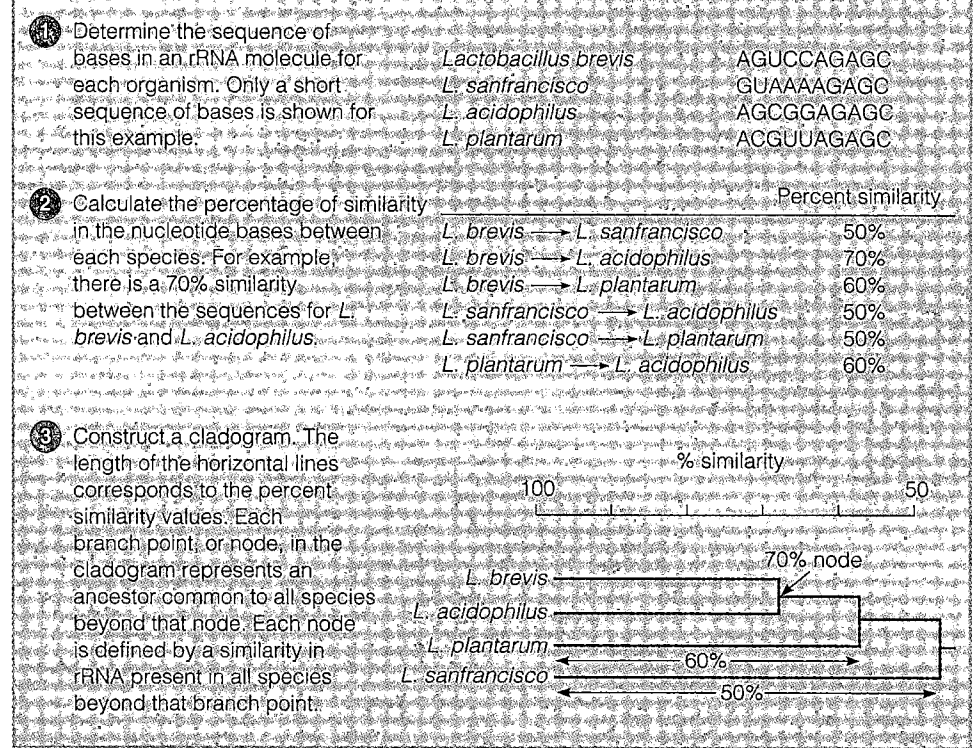

① Determine the sequence of bases in an rRNA molecule for each organism. Only a short sequence of bases is shown for this example.

Lactobacillus brevis	AGUCCAGAGC
L. sanfrancisco	GUAAAAGAGC
L. acidophilus	AGCGGAGAGC
L. plantarum	ACGUUAGAGC

② Calculate the percentage of similarity in the nucleotide bases between each species. For example, there is a 70% similarity between the sequences for *L. brevis* and *L. acidophilus*.

	Percent similarity
L. brevis → *L. sanfrancisco*	50%
L. brevis → *L. acidophilus*	70%
L. brevis → *L. plantarum*	60%
L. sanfrancisco → *L. acidophilus*	50%
L. sanfrancisco → *L. plantarum*	50%
L. plantarum → *L. acidophilus*	60%

③ Construct a cladogram. The length of the horizontal lines corresponds to the percent similarity values. Each branch point, or node, in the cladogram represents an ancestor common to all species beyond that node. Each node is defined by a similarity in rRNA present in all species beyond that branch point.

the cladogram is defined by a feature shared by various species on that branch. Historically, cladograms for vertebrates were made using fossil evidence; however, rRNA sequences are now being used to confirm assumptions based on fossils. As we said earlier, most microorganisms do not leave fossils; therefore, rRNA sequencing is primarily used to make cladograms for microorganisms. The small rRNA subunit used has 1500 bases, and computer programs do the calculations. The steps for constructing a cladogram are shown in Figure 10.17. ① Two rRNA sequences are aligned, and ② the percentage of similarity between the sequences is calculated. ③ Then the horizontal branches are drawn in a length proportional to the calculated percent similarity. All species beyond a node (branch point) have similar rRNA sequences, suggesting that they arose from an ancestor at that node.

Study Outline   Student Tutorial CD-ROM

INTRODUCTION (p. 276)

1. Taxonomy is the science of the classification of organisms, with the goal of showing relationships among organisms.

2. Taxonomy also provides a means of identifying organisms.

THE STUDY OF PHYLOGENETIC RELATIONSHIPS (pp. 277–280)

1. Phylogeny is the evolutionary history of a group of organisms.

2. The taxonomic hierarchy shows evolutionary, or phylogenetic, relationships among organisms.

3. Bacteria were separated into the Kingdom Prokaryotae in 1968.

4. Living organisms were divided into five kingdoms in 1969.

The Three Domains (pp. 277–280)

1. Living organisms are currently classified into three domains. A domain can be divided into kingdoms.

2. In this system, plants, animals, fungi, and protists belong to the Domain Eukarya.

3. Bacteria (with peptidoglycan) form a second domain.

4. Archaea (with unusual cell walls) are placed in the Domain Archaea.

A Phylogenetic Hierarchy (p. 280)

1. Organisms are grouped into taxa according to phylogenetic relationships (from a common ancestor).

2. Some of the information for eukaryotic relationships is obtained from the fossil record.

3. Prokaryotic relationships are determined by rRNA sequencing.

CLASSIFICATION OF ORGANISMS (pp. 280–285)

Scientific Nomenclature (pp. 281–282)

1. According to scientific nomenclature, each organism is assigned two names, or binomial: a genus and a specific epithet, or species.

2. Rules for the assignment of names to bacteria are established by the International Committee on Systematic Bacteriology.

3. Rules for naming fungi and algae are published in the *International Code of Botanical Nomenclature*.

4. Rules for naming protozoa are found in the *International Code of Zoological Nomenclature*.

The Taxonomic Hierarchy (p. 282)

1. A eukaryotic species is a group of organisms that interbreeds but does not breed with individuals of another species.

2. Similar species are grouped into a genus; similar genera are grouped into a family; families, into an order; orders, into a class; classes, into a division or phylum; phyla, into a kingdom; and kingdoms, into a domain.

Classification of Prokaryotes (pp. 282–284)

1. *Bergey's Manual of Systematic Bacteriology* is the standard reference on bacterial classification.

2. A group of bacteria derived from a single cell is called a strain.

3. Closely related strains constitute a bacterial species.

Classification of Eukaryotes (p. 284)

1. Eukaryotic organisms may be classified into the Kingdom Fungi, Plantae, or Animalia.

2. Protists are mostly unicellular organisms; these organisms are currently being assigned to kingdoms.

3. Fungi are absorptive chemoheterotrophs that develop from spores.

4. Multicellular photoautotrophs are placed in the Kingdom Plantae.

5. Multicellular ingestive heterotrophs are classified as Animalia.

Classification of Viruses (p. 285)

1. Viruses are not placed in a kingdom. They are not composed of cells and cannot grow without a host cell.

2. A viral species is a population of viruses with similar characteristics that occupies a particular ecological niche.

METHODS OF CLASSIFYING AND IDENTIFYING MICROORGANISMS (pp. 285–298)

1. *Bergey's Manual of Determinative Bacteriology* is the standard reference for laboratory identification of bacteria.

2. Morphological characteristics are useful in identifying microorganisms, especially when aided by differential staining techniques.

3. The presence of various enzymes, as determined by biochemical tests, is used in identifying microorganisms.

4. Serological tests, involving the reactions of microorganisms with specific antibodies, are useful in determining the identity of strains and species, as well as relationships among organisms. ELISA and Western blotting are examples of serological tests.

5. Phage typing is the identification of bacterial species and strains by determining their susceptibility to various phages.

6. Fatty acid profiles can be used to identify some organisms.

7. Flow cytometry measures physical and chemical characteristics of cells.

8. The percentage of GC base pairs in the nucleic acid of cells can be used in the classification of organisms.

9. The number and sizes of DNA fragments, or DNA fingerprints, produced by restriction enzymes are used to determine genetic similarities.

10. The sequence of bases in ribosomal RNA can be used in the classification of organisms.

11. The polymerase chain reaction (PCR) can be used to detect small amounts of microbial DNA in a sample.

    SM *To review, go to Biodiversity: PCR: Environmental Samples*

12. Single strands of DNA, or of DNA and RNA, from related organisms will hydrogen-bond to form a double-stranded molecule; this bonding is called nucleic acid hybridization.

13. Southern blotting and DNA probes are examples of hybridization techniques.

14. Dichotomous keys are used for the identification of organisms. Cladograms show phylogenetic relationships among organisms.

Study Questions

REVIEW

1. What is taxonomy?

2. Discuss the evidence that supports classifying organisms into three domains.

3. Explain why a 2-, 3-, or 5- kingdom system is no longer acceptable for classification.

4. List and define the three kingdoms of eukaryotic organisms.

5. Compare and contrast the following terms:
   a. archaea and bacteria
   b. bacteria and eukarya
   c. archaea and eukarya

6. What is binomial nomenclature?

7. Why is binomial nomenclature preferable to the use of common names?

8. Using *Escherichia coli* and *Entamoeba coli* as examples, explain why the genus name must always be written out the first time you use it in a report.

9. Put the following terms in the correct sequence, from the most general to the most specific: order, class, genus, domain, species, phylum, family.

10. Define the following terms: eukaryotic species, bacterial species, and viral species.

11. List the twelve taxonomic criteria and methods discussed in this chapter for the classification of microorganisms. Separate your list into those tests used primarily for taxonomic classification and those used primarily for identifying microorganisms already classified.

12. Higher organisms are arranged into taxonomic groups on the basis of evolutionary relationships. Why is this type of classification only now being developed for bacteria?

13. Can you tell which of the following organisms are most closely related? Are any two the same species? On what did you base your answer?

Characteristic	A	B	C	D
Morphology	Rod	Coccus	Rod	Rod
Gram reaction	+	−	−	+
Glucose utilization	Fermentative	Oxidative	Fermentative	Fermentative
Cytochrome oxidase	Present	Present	Absent	Absent
GC moles %	48–52	23–40	50–54	49–53

MULTIPLE CHOICE

1. *Bergey's Manual of Systematic Bacteriology* differs from *Bergey's Manual of Determinative Bacteriology* in that the former
   a. groups bacteria into species.
   b. groups bacteria according to phylogenetic relationships.
   c. groups bacteria according to pathogenic properties.
   d. groups bacteria into 19 species.
   e. all of the above

2. *Bacillus* and *Lactobacillus* are not in the same order. This indicates that which one of the following is *not* sufficient to assign an organism to a taxon?
   a. biochemical characteristics
   b. amino acid sequencing
   c. phage typing
   d. serology
   e. morphological characteristics

3. Which of the following is used to classify organisms into the Kingdom Fungi?
   a. ability to photosynthesize; possess a cell wall
   b. unicellular; possess cell wall; prokaryotic
   c. unicellular; lacking cell wall; eukaryotic
   d. absorptive; possess cell wall; eukaryotic
   e. ingestive; lacking cell wall; multicellular; prokaryotic

4. Which of the following is *not* true about scientific nomenclature?
   a. Each name is specific.
   b. Names vary with geographical location.
   c. The names are standardized.
   d. Each name consists of a genus and specific epithet.
   e. It was first designed by Linnaeus.

5. You could identify an unknown bacterium by all of the following *except*
   a. hybridizing a DNA probe from a known bacterium with the unknown's DNA.
   b. making a fatty acid profile of the unknown.
   c. specific antiserum agglutinating the unknown.
   d. ribosomal RNA sequencing.
   e. percentage of guanine + cytosine.

6. The wall-less mycoplasmas are considered to be related to gram-positive bacteria. Which of the following would provide the most compelling evidence for this?
   a. They share common rRNA sequences.
   b. Some gram-positive bacteria and some mycoplasmas produce catalase.
   c. Both groups are prokaryotic.
   d. Some gram-positive bacteria and some mycoplasmas have coccus-shaped cells.
   e. Both groups contain human pathogens.

Use the following choices to answer questions 7 and 8.
   a. Animalia
   b. Fungi
   c. Plantae
   d. Firmicutes (gram-positive bacteria)
   e. Proteobacteria (gram-negative bacteria)

7. Into which kingdom would you place a multicellular organism that has a mouth and lives inside the human liver?

8. Into which kingdom would you place a photosynthetic organism that lacks a nucleus and has a thin peptidoglycan wall surrounded by an outer membrane?

Use the following choices to answer questions 9 and 10.
   1. 9+2 flagella
   2. 70S ribosome
   3. fimbria
   4. nucleus
   5. peptidoglycan
   6. plasma membrane

9. Which is (are) found in all three domains?
   a. 2, 6
   b. 5
   c. 2, 4, 6
   d. 1, 3, 5
   e. all six

10. Which is (are) found *only* in prokaryotes?
   a. 1, 4, 6
   b. 3, 5
   c. 1, 2
   d. 4
   e. 2, 4, 5

CRITICAL THINKING

1. Here is some additional information on the organisms in review question 13:

Organisms	% DNA Hybridization
A and B	5–15
A and C	5–15
A and D	70–90
B and C	10–20
B and D	2–5

Which of these organisms are most closely related? Compare this answer with your response to review question 13.

2. The GC content of *Micrococcus* is 66–75 moles %, and of *Staphylococcus,* 30–40 moles %. According to this information, would you conclude that these two genera are closely related?

3. Describe the use of a DNA probe and PCR for:
   a. rapid identification of an unknown bacterium.
   b. determining which of a group of bacteria are most closely related.

4. SF medium is a selective medium, developed in the 1940s, to test for fecal contamination of milk and water. Only certain gram-positive cocci can grow in this medium. Why is it named SF? Using this medium, which genus will you culture? (*Hint:* Refer to Chapter 11.)

CLINICAL APPLICATIONS

1. A 55-year-old veterinarian was admitted to a hospital with a 2-day history of fever, chest pain, and cough. Gram-positive cocci were detected in his sputum, and he was treated for lobar pneumonia with penicillin. The next day, another Gram stain of his sputum revealed gram-negative rods, and he was switched to ampicillin and gentamicin. A sputum culture showed biochemically inactive gram-negative rods identified as *Enterobacter agglomerans.* After fluorescent-antibody staining and phage typing, *Yersinia pestis* was identified in the patient's sputum and blood, and chloramphenicol and tetracycline were administered. The patient died 3 days after admission to the hospital. Tetracycline was given to his 220 contacts (hospital personnel, family, and coworkers). What disease did the patient have? Discuss what went wrong in the diagnosis and how his death might have been prevented. Why were the 220 other people treated? (*Hint:* Refer to Chapter 23.)

2. A 6-year-old girl was admitted to a hospital with endocarditis. Blood cultures showed a gram-positive, aerobic rod identified by the hospital laboratory as *Corynebacterium xerosis.* The girl died after 6 weeks of treatment with intravenous penicillin and chloramphenicol. The bacterium was tested by another laboratory and identified as *C. diphtheriae.* The following test results were obtained by each laboratory:

	Hospital Lab	Other lab
Catalase	+	+
Nitrate reduction	+	+
Urea	−	−
Esculin hydrolysis	−	−
Glucose fermentation	+	+
Sucrose fermentation	−	+
Serological test for toxin production	Not done	+

Provide a possible explanation for the incorrect identification. What are the potential public health consequences of misidentifying *C. diphtheriae*? (*Hint:* Refer to Chapter 24.)

3. Use the information in the following table to construct a dichotomous key to these organisms. What is the purpose of a dichotomous key? Look up each genus in Chapter 11, and provide an example of why this organism is of interest to humans.

	Morphology	Gram Reaction	Acid from Glucose	Growth in Air (21% O_2)	Motile by Peritrichous Flagella	Presence of Cytochrome Oxidase	Produce Catalase
Staphylococcus aureus	Coccus	+	+	+	−	−	+
Streptococcus pyogenes	Coccus	+	+	+	−	−	−
Mycoplasma pneumoniae	Coccus	−	+	+ (Colonies < 1mm)	−	−	+
Clostridium botulinum	Rod	+	+	−	+	−	−
Escherichia coli	Rod	−	+	+	+	−	+
Pseudomonas aeruginosa	Rod	−	+	+	−	+	+
Campylobacter fetus	Vibrio	−	−	−	−	+	+
Listeria monocytogenes	Rod	+	+	+	+	−	+

4. Use this additional information to construct a cladogram for some of the organisms used in question 3. What is the purpose of a cladogram? How does your cladogram differ from a dichotomous key for these organisms?

	Similarity in rRNA Bases
P. aeruginosa—M. pneumoniae	52%
P. aeruginosa—C. botulinum	52%
P. aeruginosa—E. coli	79%
M. pneumoniae—C. botulinum	65%
M. pneumoniae—E. coli	52%
E. coli—C. botulinum	52%

Learning with Technology

MP = **The Microbiology Place website**   **ST** = **Student Tutorial CD-ROM**   **VU** = **VirtualUnknown CD-ROM**

MP   Don't forget to go to The Microbiology Place website (http://www.microbiologyplace.com) to take the practice tests, explore the interactive activity and case study, and check out the news articles and web links for this chapter.

ST   Remember there is also a quiz for this chapter on the Microbiology Interactive Student Tutorial CD-ROM.

VU   Enter the Virtual Lab, click the arrow next to the Session field, click Textbook Exercises, and select Chapter 10a.

1. Identify the unknown organism using the tests and media provided in the software. Print out your Virtual Laboratory Report when you are finished.

2. Create a blank copy of the microbiology lab report form illustrated in Figure 10.7 on page 287. Use the information obtained during your identification of the culture to complete the report form. What does the patient have?

Now select Chapter 10b.

1. Perform the tests listed in Figure 10.8 on page 288 and determine a preliminary identification.

2. Which tests shown in Figure 10.9 on page 289 can be accomplished using the software? Conduct further tests on this unknown and print out the Virtual Laboratory Report. Use the results to predict the appearance of an Enterotube™ II when inoculated with this organism.

chapter 11

The Prokaryotes: Domains Bacteria and Archaea

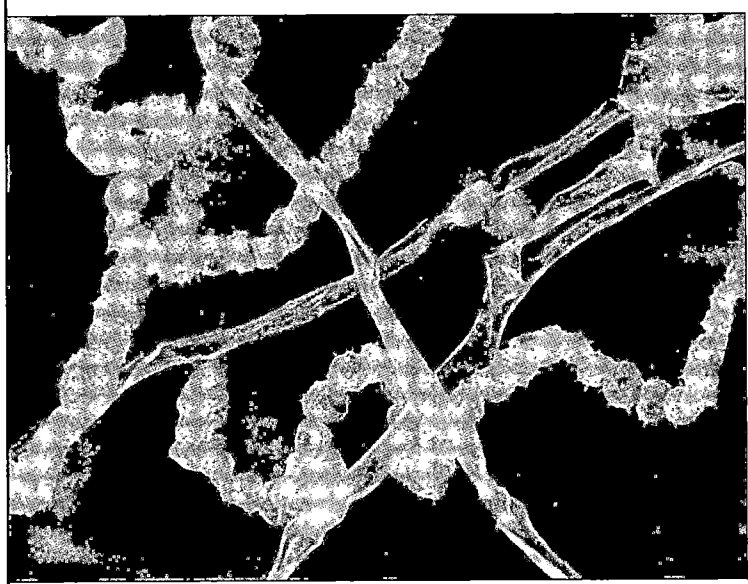

Streptomyces. *Streptomyces bacteria grow in filaments and make condiospores. These bacteria produce most of the antibiotics used in medicine.*

When biologists first encountered microscopic bacteria, they were puzzled as to how to classify them. Bacteria were clearly not animals or rooted plants. At first, bacteria were included with the fungi, many of which were also microscopic. Attempts to build a taxonomic system for bacteria based on the phylogenetic system developed for plants and animals, as discussed in Chapter 10, failed. Bacteria were grouped by morphology (rod, coccus, helix), staining reactions, presence of endospores, and other obvious features. Although this system had its practical uses, microbiologists were aware that it had many imperfections. For example, lactic acid bacteria were placed in widely separated groups distinguished by either rod-shaped morphology or spherical morphology. Subsequent studies of their ribosomal RNA (rRNA) showed that the two groups are actually quite closely related. In 2000, the first volume of a forthcoming five-volume second edition of *Bergey's Manual* was published. During the years since publication of the first edition, knowledge of bacteria at the molecular level has expanded to such a degree that it is possible to base this new edition on a phylogenetic system. One of the most important expressions of phylogenetic difference is rRNA. Above the species level, rRNA is especially useful. It is slow to change and performs the same functions in all organisms. It is now possible to begin constructing phylogenetic trees in which gene sequences can be compared. Organisms can then be grouped by a measure of differences in selected genes, sometimes expressed as evolutionary distance (see Figure 10.6, page 285). In general, there will be a tendency to greater variation as time allows differences in the genetic makeup to accumulate.

PROKARYOTIC GROUPS

In the second edition of *Bergey's Manual*, the prokaryotes are grouped into two domains, the Archaea and the Bacteria. Both domains consist of prokaryotic cells. All other organisms are assigned to the Domain Eukarya, which contains all the unicellular and multicellular eukaryotic organisms. The grouping into domains is shown in Figure 10.1 and Table 10.1, and the differences between the two prokaryotic domains are summarized in Figure 10.6. Each domain is divided into phyla, each phylum into classes, and so on. The phyla discussed in this chapter are summarized in Table 11.1. (Also see Appendix A.)

DOMAIN BACTERIA

Most of us have been conditioned to think of bacteria as invisible, potentially harmful little creatures. Actually, relatively few species of bacteria cause disease in humans, animals, plants, or any other organisms. Indeed, once you have completed a course in microbiology, you will realize that without bacteria, much of life as we know it would not be possible. In fact, all organisms made up of eukaryotic cells probably evolved from bacterialike organisms, which were some of the earliest forms of life. In this chapter, you will learn how bacterial groups are differentiated from each other and how important bacteria are in the world of microbiology. Our discussion emphasizes bacteria that are considered to be of practical importance, those important in medicine, or those that illustrate biologically unusual or interesting principles.

The Proteobacteria

The proteobacteria, which includes most of the gram-negative, chemoheterotrophic bacteria, are presumed to have arisen from a common photosynthetic ancestor. They are now the largest taxonomic group of bacteria. However, few are now photosynthetic; other metabolic and nutritional capacities have arisen to replace this characteristic. The phylogenetic relationship in these groups is based upon rRNA studies. The name **Proteobacteria** was taken from the mythological Greek god, Proteus, who could assume many shapes. Distinctive subgroups of the proteobacteria are designated by Greek letters.

The α (alpha) Proteobacteria

Learning Objective

The Learning Objectives in this chapter will help you to become familiar with these organisms and to look for similarities and differences between organisms. You will draw a dichotomous key to differentiate the bacteria described in each group. We'll draw the first one to get you started. Make a dichotomous key to distinguish among the α-proteobacteria described in this chapter.

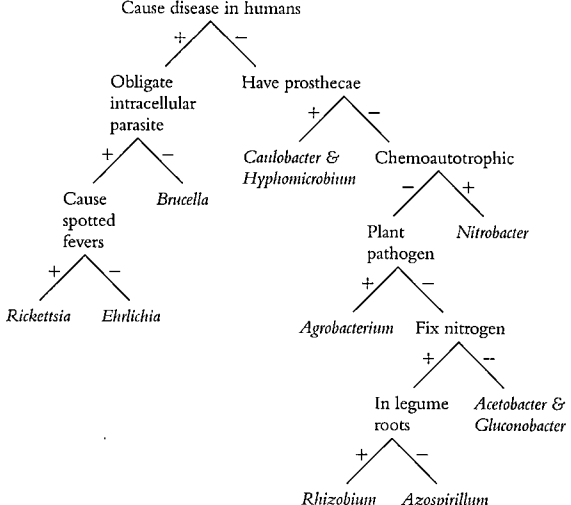

As a group, the α-proteobacteria includes most of the proteobacteria that are capable of growth at very low levels of nutrients. Some have unusual morphology, including protrusions such as stalks or buds known as **prosthecae**. The α-proteobacteria also include agriculturally important bacteria capable of inducing nitrogen fixation in symbiosis with plants, and several plant and human pathogens.

Azospirillum Agricultural microbiologists have been interested in members of the genus *Azospirillum* (ā-zō-spī'ril-lum), a soil bacterium that grows in close association with the roots of many plants, especially tropical grasses. It uses nutrients excreted by the plants and in return fixes nitrogen from the atmosphere. This form of nitrogen fixation is most significant in some tropical grasses and in sugar cane, although the organism can be isolated from the root system of many temperate-climate plants, such as corn. (The prefix *azo-* is frequently encountered in nitrogen-fixing genera of bacteria. It is derived from *a* (without) and *zo* (life), in reference to the early days of chemistry when oxygen was removed, by a burning candle, from an experimental atmosphere. Presumably, mostly nitrogen remained and mammalian life was found to be not possible in this atmosphere. Hence, nitrogen came to be associated with absence of life.)

table 11.1 *Selected Characteristics of Prokaryotes, from* Bergey's Manual of Systematic Bacteriology, Second Edition

Name of Phylum / Name of class	Important Genera	Gram Reaction	Special Features	See Also
Domain Bacteria				
Proteobacteria				
α-proteobacteria	Acetobacter	Negative	Grow at low nutrient levels; some with stalks	
	Agrobacterium			page 266 (Chapter 9)
	Azospirillum			
	Bradyrhizobiuim			page 748 (Chapter 27)
	Brucella			page 630 (Chapter 23)
	Caulobacter			
	Ehrlichia			page 635 (Chapter 23)
	Gluconobacter			
	Hyphomicrobrium			
	Nitrobacter			page 746 (Chapter 27)
	Rhizobium			page 748 (Chapter 27)
	Rickettsia			page 635 (Chapter 23)
β-proteobacteria	Bordetella	Negative	Found in soil, water; some are human pathogens; some are autotrophic	page 663 (Chapter 24)
	Burkholderia			page 201 (Chapter 7)
	Neisseria			page 605 (Chapter 22)
	Nitrosomonas			
	Sphaerotilus			
	Spirillum			
	Thiobacillus			page 784 (Chapter 28)
	Zooglea			
γ-proteobacteria	Azotobacter	Negative	Includes most of the common chemoheterotrophs, including the enterics	page 748 (Chapter 27)
	Azomonas			page 745 (Chapter 27)
	Beggiatoa			page 750 (Chapter 27)
	Francisella			page 629 (Chapter 23)
	Moraxella			page 671 (Chapter 24)
	Pseudomonas			page 586 (Chapter 21)
Order:				
Legionellales	Coxiella	Negative	Coxiella produces sporelike body	page 671 (Chapter 23)
	Legionella			page 669 (Chapter 24)
Vibrionales	Vibrio	Negative	Facultatively anaerobic curved rods; found in aquatic habitats	page 696 (Chapter 25)
Enterobacteriales	Enterobacter	Negative	Small rods; found in soil and in animal intestinal tracts; many important pathogens	
	Erwinia			
	Escherichia			page 697 (Chapter 25)
	Klebsiella			page 671 (Chapter 24)
	Proteus			
	Serratia			
	Shigella			page 691 (Chapter 25)
	Salmonella			page 692 (Chapter 25)
	Yersinia			page 633 (Chapter 23)
Pasteurellales	Haemophilus	Negative	Nonmotile; some important human pathogens	
	Pasteurella			page 632 (Chapter 23)
δ-proteobacteria	Bdellovibrio	Negative	Some are predators on other bacteria; some dissimilatory sulfur reducers	page 743 (Chapter 27)
	Desulfovibrio			page 756 (Chapter 27)
	Myxococcus			page 81 (Chapter 4)
ε-proteobacteria	Campylobacter	Negative	Helical or vibrioid	page 697 (Chapter 25)
	Helicobacter			page 697 (Chapter 25)

Name of Phylum Name of class	Important Genera	Gram Reaction	Special Features	See Also
Cyanobacteria	*Anabaena* *Gloeocapsa* *Spirulina*	Negative	Produce oxygen during photosynthesis; many species fix nitrogen	
Chlamydiae	*Chlamydia*	Negative	Intracellular parasites; life cycle includes elementary body and reticulate body	page 727 (Chapter 26)
Spirochetes	*Borrelia* *Leptospira* *Treponema*	Negative	Helical morphology; motility by axial filaments; several important pathogens	page 634 (Chapter 23) page 723 (Chapter 26) page 728 (Chapter 26)
Bacteroidetes	*Bacteroides*	Negative	Anaerobes; some in human intestinal tract	
	Cytophaga	Negative	Degrade cellulose; gliding motility	Chapter 27
Fusobacteria	*Fusobacterium*	Negative	Anaerobes; some in human mouth	
Firmicutes Order:				
Clostridiales	*Clostridium* *Epulopiscium* *Veillonella*	Positive	Found in soil and in animals; includes obligate anaerobes; Clostridia produce endospores	page 608 (Chapter 22)
Mycoplasmatales	*Mycoplasma* *Spiroplasma* *Ureaplasma*	Negative	Pleomorphic; lack cell walls; parasites of animals and plants	
Bacillales	*Bacillus*	Positive	Several produce endospores	page 631 (Chapter 23) page 699 (Chapter 25)
Lactobacillales	*Lactobacillus* *Lactococcus* *Listeria* *Staphylococcus* *Streptococcus*	Positive	Some aerotolerant anaerobes (lack cytochromes); lactic-acid bacteria; some important human pathogens	page 606 (Chapter 22) page 582 (Chapter 21) pages 585, 627 (Chapters 21, 23)
Actinobacteria	*Actinomyces* *Corynebacterium* *Frankia* *Gardnerella* *Mycobacterium* *Nocardia* *Propionibacterium* *Streptomyces*	Positive	Common in soil; some produce branching filaments with reproductive conidiospores	page 660 (Chapter 24) page 663 (Chapter 24)
Domain Archaea **Crenarchaeota**				
Hyperthermophiles	*Pyrodictium* *Sulfolobus*	Negative	High temperature; very acidic environments	page 16 (Chapter 1)
Euryarchaeota				
The Methanogens	*Methanobacterium*	Some positive; some negative	Anaerobic environments; useful CH_4 producers	
The Halobacteria	*Halobacterium* *Halococcus*	Negative or variable	High osmotic pressure environments	page 194 (Chapter 5)

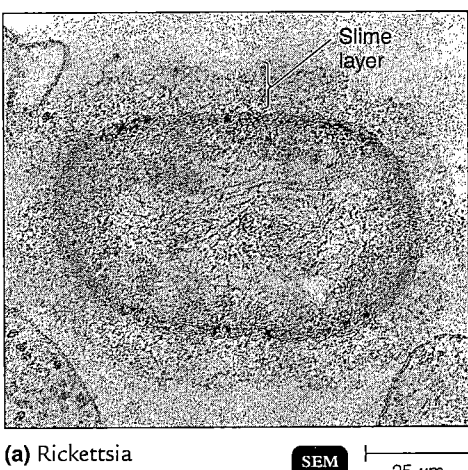

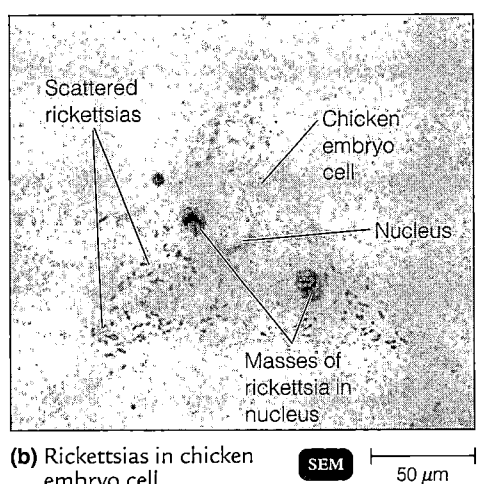

(a) Rickettsia — SEM ⊢ 25 μm ⊣

(b) Rickettsias in chicken embryo cell — SEM ⊢ 50 μm ⊣

FIGURE 11.1 Rickettsias. (a) A micrograph of *Rickettsia prowazekii,* which causes epidemic typhus. **(b)** The triangular body is a chicken embryo cell infected with intracellular rickettsias. Notice the scattered rickettsias in the cytoplasm and the two distinct compact masses of rickettsia in the oval nuclear area of the cell.

■ How are rickettsias transmitted from one host to another?

Acetobacter and *Gluconobacter* *Acetobacter* (ä′sē-tō-bak-tèr) and *Gluconobacter* (glü′kon-ō-bak-tèr) are industrially important aerobic organisms that convert ethanol into acetic acid (vinegar).

Rickettsia In the first edition of *Bergey's Manual,* the *Rickettsia, Coxiella,* and *Chlamydia* were grouped closely because they share the common characteristic of being obligate intracellular parasites—that is, they reproduce only within a mammalian cell. In the second edition they are now widely separated. A comparison of rickettsias, chlamydias, and viruses appears in Table 13.1, page 372.

The rickettsias are gram-negative rod-shaped bacteria or coccobacilli, (Figure 11.1a). One distinguishing feature of most rickettsias is that they are transmitted to humans by bites of insects and ticks, as are the *Coxiella* (discussed later with γ-proteobacteria). Rickettsia enter their host cell by inducing phagocytosis. They quickly enter the cytoplasm of the cell and begin reproducing by binary fission (Figure 11.1b). They can usually be cultivated artificially in cell culture or chick embryos (Chapter 13, page 380).

The rickettsias are responsible for a number of diseases known as the spotted fever group. These include epidemic typhus, caused by *Rickettsia prowazekii* (ri-ket′sē-ä prou-wä-ze′kē-ē) and transmitted by lice; endemic murine typhus, caused by *R. typhi* (tī′fē) and transmitted by rat fleas, and Rocky Mountain spotted fever, caused by *R. rickettsii* (ri-ket′sē-ē) and transmitted by ticks. In humans, rickettsial infections damage the permeability of blood capillaries, which results in a characteristic spotted rash.

Ehrlichia Ehrlichiae are gram-negative, rickettsialike bacteria that live obligately within white blood cells. *Ehrlichia* (èr′lik-ē-ä) species are transmitted by ticks to humans and cause ehrlichiosis, a sometimes fatal disease.

Caulobacter and Hyphomicrobium Both *Caulobacter* and *Hyphomicrobium* produce prominent prosthecae. Members of the genus *Caulobacter* (kô-lō-bak′tèr) are found in low-nutrient aquatic environments, such as lakes. They feature stalks that anchor the organs to surfaces (Figure 11.2). This arrangement increases their nutrient uptake because they are exposed to a continuously changing flow of water and because the stalk increases the surface-to-volume ratio of the cell. Also, if the surface to which they anchor is a living host, these bacteria can use the host's excretions as nutrients. When the nutrient concentration is exceptionally low, the size of the stalk increases, evidently to provide an even greater surface area for nutrient absorption.

Budding bacteria do not divide by binary fission into two nearly identical cells. The budding process resembles the asexual reproductive processes of many yeasts (Figure 12.3 on page 334). The parent cell retains its identity while the bud increases in size until it separates as a complete new cell. An example is the genus *Hyphomicrobium* (hī-fō-mī-krō′bē-um), as shown in Figure 11.3. These bacteria, like the caulobacteria, are found in low-nutrient aquatic environments and have even been found growing in laboratory water baths.

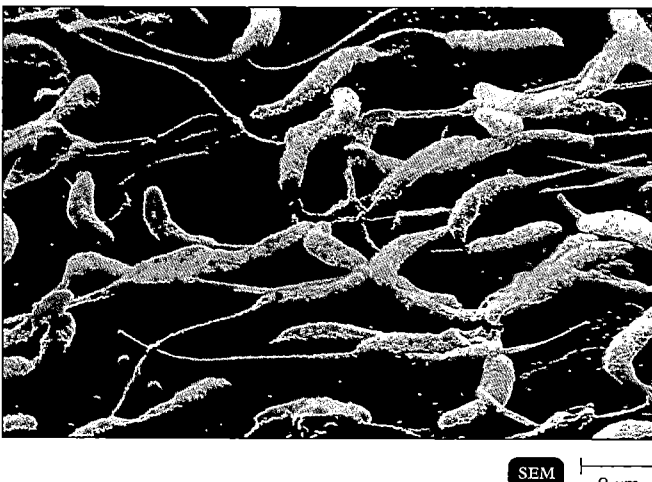

FIGURE 11.2 *Caulobacter.* Most of the *Caulobacter* in this photo are in the stalked stage.

■ Of what advantage are stalks?

Rhizobium and Agrobacterium The genus *Rhizobium* (rī-zō'bē-um) includes agriculturally important bacteria that specifically infect the roots of leguminous plants, such as beans, peas, or clover. The presence of the bacteria leads to formation of nodules in which the bacteria and plant form a symbiotic relationship, resulting in the fixation of nitrogen from the air for use by the plant (see Figure 27.4, page 749).

Like *Rhizobium*, the genus *Agrobacterium* has the ability to invade plants. However, they do not induce root nodules or fix nitrogen. Of particular interest is *Agrobacterium tumefaciens*. This is a plant pathogen that causes a disease called crown gall; the crown is the area of the plant where the roots and stem merge. The tumorlike gall is induced when *A. tumefaciens* inserts a plasmid containing bacterial genetic information into the plant's chromosomal DNA (see Figure 9.16, page 266). For this reason, microbial geneticists are very interested in this organism. Plasmids are the most common vector that genetic engineers use to carry new genes into a cell, and the thick wall of plants is especially difficult to penetrate (see Figure 9.17, page 267).

Brucella Brucella (brü'sel-la) is a small nonmotile coccobacillus. All species of *Brucella* are obligate parasites of mammals and cause the disease brucellosis. Of medical interest is the ability of *Brucella* to survive phagocytosis, an important element of our defense against bacteria (see Chapter 16, page 458).

Nitrobacter and Nitrosomonas These organisms, members of the nitrifying bacteria, are of great importance to the environment and to agriculture. They are chemoau-

FIGURE 11.3 *Hyphomicrobium,* a type of budding bacterium.

■ Most bacteria do not reproduce by budding; what method do they use?

totrophs capable of using inorganic chemicals as energy sources and carbon dioxide as the only source of carbon, from which they synthesize all of their complex chemical makeup. The energy sources of the genera *Nitrobacter* (nī-trō-bak'tér) and *Nitrosomonas* (nī-trō-sō-mō'nas) (this latter a member of the β-proteobacteria) are reduced nitrogenous compounds. *Nitrobacter* species oxidize ammonium (NH_4^+) to nitrite (NO_2^-), which is in turn oxidized by *Nitrosomonas* species to nitrates (NO_3^-) in the process of *nitrification*. Nitrate is important to agriculture; it is a nitrogen form that is highly mobile in soil and therefore likely to be encountered and used by plants.

The β (beta) Proteobacteria

Learning Objective

■ Make a dichotomous key to distinguish among the β-proteobacteria described in this chapter.

There is considerable overlap between the β-proteobacteria and the α-proteobacteria, for example among the nitrifying bacteria discussed earlier. The β-proteobacteria often use nutrient substances that diffuse away from areas of anaerobic decomposition of organic matter, such as hydrogen gas,

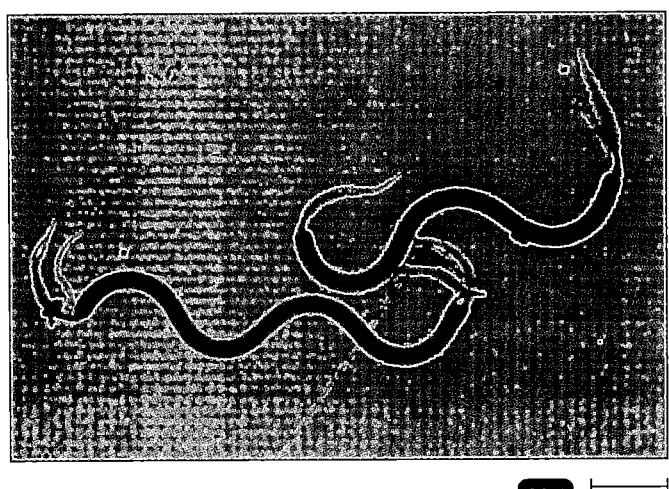

FIGURE 11.4 *Spirillum volutans.* This large helical bacterium is found in aquatic environments. Note the polar flagella.

■ **Is this bacterium motile? How can you tell?**

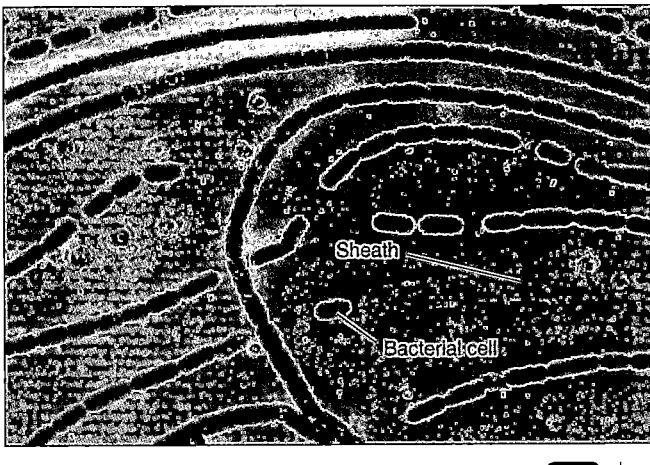

FIGURE 11.5 *Sphaerotilus natans.* This sheathed bacterium is found in dilute sewage and aquatic environments. It forms elongated sheaths in which the bacteria live. The bacteria have flagella (not visible here) and can eventually swim free of the sheath.

■ **How does the sheath help the cell?**

ammonia, and methane. Several important pathogenic bacteria are found in this group.

Thiobacillus *Thiobacillus* (thī-ō-bä-sil′lus) species and other sulfur-oxidizing bacteria are important in the sulfur cycle (Figure 27.6). These chemoautotrophic bacteria are capable of obtaining energy by oxidizing the reduced forms of sulfur, such as hydrogen sulfide (H_2S), or elemental sulfur (S^0), into sulfates (SO_4^{2-}).

Spirillum In the second edition of *Bergey's Manual,* the genus *Spirillum* (spī-ril′lum) is grouped in the same family as the thiobacilli. Their habitat is mainly fresh water. An important morphological difference from the helical spirochetes (discussed later) is that they are motile by conventional polar flagella, rather than axial filaments. The spirilla are relatively large, gram-negative, aerobic bacteria. *Spirillum volutans* (vō-lū-tans) is often used as a demonstration slide when microbiology students are first introduced to the operation of the microscope (Figure 11.4).

Sphaerotilus Sheathed bacteria, which include *Sphaerotilus natans* (sfe-rä′ti-lus na′tans), are found in freshwater and in sewage. These gram-negative bacteria with polar flagella form a hollow, filamentous sheath in which to live (Figure 11.5). Sheaths are protective and also aid in nutrient accumulation. *Sphaerotilus* probably contributes to bulking, an important problem in sewage treatment (see Chapter 27).

Burkholderia The genus *Burkholderia* has only recently been reclassified; formerly it was grouped with the genus *Pseudomonas,* which is now classified under the γ-proteobacteria. Like the pseudomonads, almost all

Burkholderia species are motile by a single polar flagellum or tuft of flagella. The best known species is the aerobic, gram-negative rod *Burkholderia cepacia* (berk′hōld-ér-ē-ä se-pā′se-ä). It has an extraordinary nutritional spectrum and is capable of degrading more than 100 different organic molecules. See the box in Chapter 7, page 201. This capability is often a factor in the contamination of equipment and drugs in hospitals; these bacteria may actually grow in disinfectant solutions. This bacterium is also a problem for cystic fibrosis patients, in whom it metabolizes accumulated respiratory secretions.

Bordetella Of special importance is the nonmotile, aerobic, gram-negative rod *Bordetella pertussis* (bór-de-tel′lä pér-tus′sis). This serious pathogen is the cause of pertussis, or whooping cough.

Neisseria Bacteria of the genus *Neisseria* (nī-se′rē-ä) are aerobic, gram-negative cocci that usually inhabit the mucous membranes of mammals. Pathogenic species include the genococcus bacterium *Neisseria gonorrhoeae* (go-nôr-rē′ī), the causative agent of gonorrhoea (Figure 11.6), and *N. meningitidis* (men-nin-ji′ti-dis), the agent of meningococcal meningitis.

Zoogloea The genus *Zoogloea* (zō′ō-glē-ä) is important in the context of aerobic sewage-treatment processes, such as the activated sludge system (Chapter 27). As they grow, *Zoogloea* bacteria form fluffy, slimy masses that are essential to the proper operation of such systems.

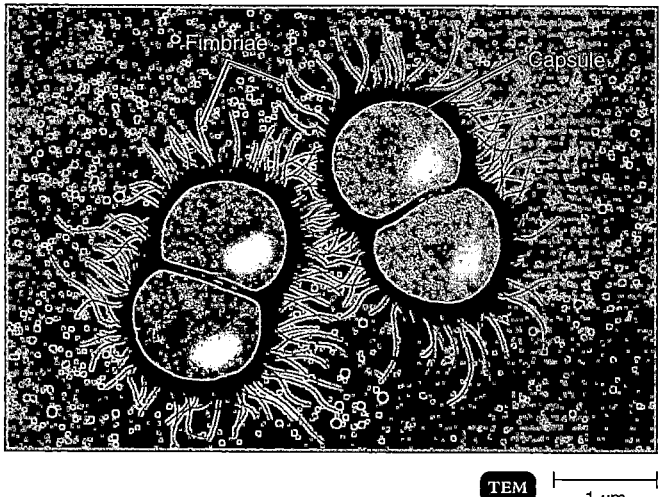

FIGURE 11.6 **The gram-negative coccus *Neisseria gonorrhoeae*.** Notice the paired arrangement (diplococci). The fimbriae enabled the organism to attach to mucous membranes and thus contribute to its pathogenicity. *N. gonorrhoeae* causes gonorrhea.

■ For what are fimbriae used?

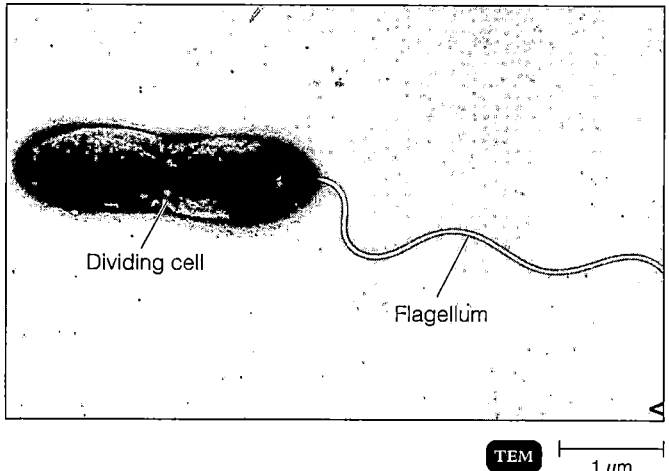

FIGURE 11.7 *Pesudomonas.* This photo of a *Pseudomonas* bacterium illustrates the polar flagellum that is characteristic of the genus.

■ Pseudomonads are gram-negative rods.

The γ (gamma) Proteobacteria

Learning Objective

■ *Make a dichotomous key to distinguish among the orders of γ-proteobacteria described in this chapter.*

The γ-proteobacteria constitute the largest subgroup of the proteobacteria and include a great variety of physiological types. One species that is used in industrial microbiology is described in the box on page 328.

Beggiatoa *Beggiatoa alba* (bej'jē-ä-tō-ä al'ba), the only species of this unusual genus, grows in aquatic sediments at the interface between the aerobic and anaerobic layers. Morphologically, it resembles certain filamentous cyanobacteria, but it is not photosynthetic. Motility is by gliding. Nutritionally it uses hydrogen sulfide (H_2S) as an energy source and accumulates internal granules of sulfur. In our discussion of the sulfur cycle in Chapter 27 on page 749, we show how this bacterium was a factor in the discovery of autotrophic metabolism.

Francisella *Francisella* (fran'sis-el'lä) is a genus of small, pleomorphic bacteria that grow only on complex media enriched with blood or tissue extracts. *Francisella tularensis* (tü'lär-en-sis) causes the disease tularemia.

Pseudomonadales

Members of the order Pseudomonadales are gram-negative aerobic rods or cocci. The most important genus in this group is *Pseudomonas.*

Pseudomonas A very important genus, *Pseudomonas* (sü-dō-mō'nas) consists of aerobic, gram-negative rods that are motile by polar flagella, either single or tufts (Figure 11.7). Many species of pseudomonads excrete extracellular, water-soluble pigments that diffuse into their media. One species, *Pseudomonas aeruginosa* (ä-rü-ji-nō'sä), produces a soluble, blue-green pigmentation. Under certain conditions, particularly in weakened hosts, this organism can infect the urinary tract, burns, and wounds, and can cause septicemia (blood infections), abscesses, and meningitis. Other pseudomonads produce soluble fluorescent pigments that glow when illuminated by ultraviolet light. One species, *P. syringae* (sèr'in-gī), is an occasional plant pathogen. (Some species of *Pseudomonas* have been transferred, based upon rRNA studies, to the genus *Burkholderia,* which was discussed previously with the α-proteobacteria.)

Pseudomonads are very common in soil and other natural environments. These bacteria are less efficient than some other heterotrophic bacteria in utilizing many of the common nutrients, but they compensate for this in other ways. For example, pseudomonads are capable of synthesizing an unusually large number of enzymes and can metabolize a wide variety of substrates. Therefore, they probably contribute significantly to the decomposition of chemicals, such as pesticides, that are added to soil. Many pseudomonads can grow at refrigerator temperatures. This characteristic, combined with an ability to utilize proteins and lipids, makes them an important contributor to food spoilage.

In hospitals and other places where pharmaceutical agents are prepared, the ability of pseudomonads to grow on minute traces of unusual carbon sources, such as soap residues or cap-liner adhesives found in a solution, has been unexpectedly troublesome. Pseudomonads are even capable of growth in some antiseptics, such as quaternary ammonium compounds. Their resistance to most antibiotics has also been a source of medical concern. This resistance is probably related to the characteristics of the cell wall porins, which control the entrance of molecules through the cell wall (see Chapter 4, page 88).

Although pseudomonads are classified as aerobic, some are capable of substituting nitrate for oxygen as a terminal electron acceptor. This process, anaerobic respiration, yields almost as much energy as aerobic respiration (see Chapter 5). In this way, pseudomonads cause important losses of valuable nitrogen in fertilizer and soil. Nitrate (NO_3^-) is the form of fertilizer nitrogen most easily used by plants. Under anaerobic conditions, as in waterlogged soil, pseudomonads eventually convert this valuable nitrate into nitrogen gas (N_2), which is lost to the atmosphere (see Chapter 27).

Azotobacter and Azomonas Some nitrogen-fixing bacteria, such as *Azotobacter* (ā-zō-tō-bak′tėr) and *Azomonas* (ā-zō-mō′nas), are free-living in soil. These large, ovoid, heavily capsulated bacteria are frequently used in laboratory demonstrations of nitrogen fixation. However, to fix agriculturally significant amounts of nitrogen, they would require energy sources, such as carbohydrates, that are in limited supply in soil.

Moraxella Members of the genus *Moraxella* (mô-raks-el′lä) are strictly aerobic coccobacilli—that is, intermediate in shape between cocci and rods. *Moraxella lacunata* (la-kü-nä′tä) is implicated in conjunctivitis, an inflammation of the conjunctiva, the membrane that covers the eye and lines the eyelids.

Legionellales

The genera *Legionella* and *Coxiella* are closely associated in the second edition of *Bergey's Manual,* where both are placed in the same order, Legionellales. Because the *Coxiella* share an intracellular lifestyle with the rickettsial bacteria, they were previously considered rickettsial in nature and grouped with them. *Legionella* bacteria have a conventional reproductive system and grow readily on suitable artificial media.

Legionella *Legionella* (lē-jä-nel′lä) was originally isolated during a search for the cause of an outbreak of pneumonia now known as legionellosis. The search was difficult

because these bacteria did not grow on the usual laboratory isolation media then available. After intensive effort, special media were developed that enabled researchers to isolate and culture the first *Legionella*. Microbes of this genus are now known to be relatively common in streams, and they colonize such habitats as warm-water supply lines in hospitals and water in the cooling towers of air conditioning systems. An ability to survive and reproduce within aquatic amoebae often makes them difficult to eradicate in water systems.

Coxiella *Coxiella burnetii* (käks-ē-el′lä bėr-ne′tē-ē), which causes Q fever, was formerly grouped with the rickettsia. Like them, *Coxiella* bacteria require a mammalian host cell in order to reproduce. Unlike rickettsias, *Coxiella* are not transmitted among humans by insect or tick bites. Although cattle ticks harbor the organism, it is most commonly transmitted by aerosols or contaminated milk. A sporogenic cycle has been reported for *C. burnetii* (see Figure 24.15b). This might explain the bacterium's relatively high resistance to the stresses of airborne transmission and heat treatment. Around 1950, the recommended temperature for pasteurization of milk was raised to ensure elimination of this pathogen.

Vibrionales

Members of this order are facultatively anaerobic gram-negative rods. Many are slightly curved. They are found mostly in aquatic habitats.

Vibrio Members of the genus *Vibrio* (vib′rē-ō) are rods that are often slightly curved (Figure 11.8). One important pathogen is *Vibrio cholerae* (kol′er-ī), the causative agent of cholera. The disease is characterized by a profuse and watery diarrhea. *V. parahaemolyticus* (pa-rä-hē-mō-li′ti-kus) causes a less serious form of gastroenteritis. Usually inhabiting coastal salt waters, it is transmitted to humans mostly by raw or undercooked shellfish.

Enterobacteriales

The members of the order Enterobacteriales are facultatively anaerobic, gram-negative rods that are, if motile, peritrichously flagellated. Morphologically, the rods are straight. They have simple nutritional requirements. This is an important bacterial group, often commonly called **enterics.** This reflects the fact that they inhabit the intestinal tracts of humans and other animals. Most enterics are active fermenters of glucose and other carbohydrates.

Because of the clinical importance of enterics, there are many techniques to isolate and identify them. An identification key to some enterics is shown in Figure 10.9, and a modern tool using 15 biochemical tests is

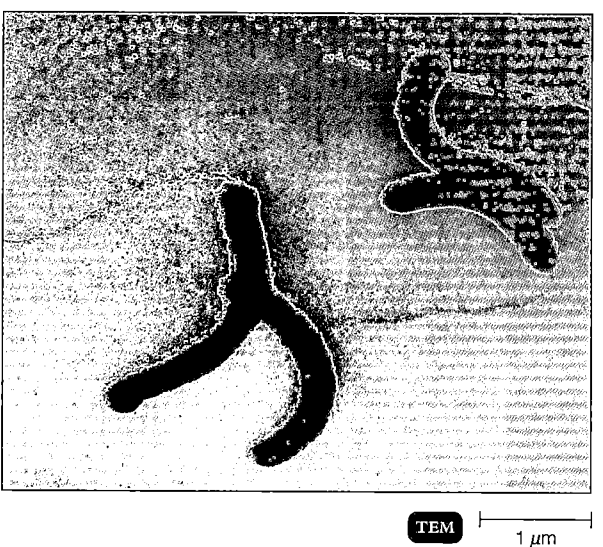

TEM ├──────────┤
1 μm

FIGURE 11.8 *Vibrio cholerae*. Notice the slight curvature of these rods, which is a characteristic of the genus.

■ **What disease does *Vibrio cholerae* cause?**

shown in Figure 10.9. Biochemical tests are especially important in clinical laboratory work and in food and water microbiology. Enterics can also be distinguished from each other according to antigens present on their surfaces, that is, by serology (see Chapter 18).

Enterics have fimbriae that help them adhere to surfaces or mucous membranes. Specialized sex pili aid in the exchange of genetic information between cells, which often includes antibiotic resistance (see Figures 8.26 and 8.27).

Enterics, like many bacteria, produce proteins called bacteriocins that cause the lysis of closely related species of bacteria. Bacteriocins may help maintain the ecological balance of various enterics in the intestines.

Escherichia The bacterium *Escherichia coli* is one of the most common inhabitants of the human intestinal tract and is probably the most familiar organism in microbiology. Recall from previous chapters that a great deal is known about the biochemistry and genetics of *E. coli*, and it continues to be an important tool for basic biological research—many researchers consider it almost a laboratory pet. Its presence in water or food is an indication of fecal contamination (see Chapter 27). *E. coli* is not usually pathogenic. However, it can be a cause of urinary tract infections, and certain strains produce enterotoxins that cause traveler's diarrhea and occasionally cause very serious foodborne disease (see *E. coli* O157:H7 in Chapter 25 on page 697).

Salmonella Almost all members of the genus *Salmonella* (sal-mön-el'lä) are potentially pathogenic. Accordingly, there are extensive biochemical and serological tests to

clinically isolate and identify salmonellae. Salmonellae are common inhabitants of the intestinal tracts of many animals, especially poultry and cattle. Under unsanitary conditions, they can contaminate food.

The nomenclature of the genus *Salmonella* is unusual. Instead of multiple species, the genus *Salmonella* can be considered for practical purposes a single species, *Salmonella enterica* (en-ter'i-kä), divided into more than 2300 **serovars,** as in serological varieties. The term **serotype** is often used to mean the same thing. When salmonellae are injected into appropriate animals, their flagella, capsules, and cell walls serve as *antigens* that cause the animals to form *antibodies* in their blood that are specific for each of these structures. Thus, serological means are used to differentiate the microorganisms. Serology is discussed more fully in Chapter 18, but for now it will be sufficient to state that it can be used to differentiate and identify bacteria.

Technically, then, a serovar such as *Salmonella typhimurium* (tī-fi-mür'ē-um), is not a species and therefore should be more properly written as "*Salmonella enterica* serovar Typhimurium." There are several variations on this theme, but little universal agreement. To avoid confusion, we will identify serovars of salmonellae as we would species, that is, *S. typhimurium*.

Specific antibodies, which are available commercially, can be used to differentiate *Salmonella* serovars by a system known as the Kauffmann–White scheme. This scheme designates an organism by numbers and letters that correspond to specific antigens on the organism's capsule, cell wall, and flagella, which are identified by the letters K, O, and H, respectively. For example, the antigenic formula for the bacterium *S. typhimurium* is O1,4,[5],12:H:i,1,2.* Many salmonellae are named only by their antigenic formulas. Serovars can be further differentiated by special biochemical or physiological properties into **biovars,** or **biotypes.**

A recent taxonomic arrangement based upon the latest molecular technology adds another species, *Salmonella bongori* (bon'gôr-ē). This is a resident of "cold-blooded" animals—it was originally isolated from a lizard in the town of Bongor in the African desert nation of Chad—and is rarely found in humans.

---

*The letters derive from the original German usage: K represents the German for capsule. (Salmonellae with capsules are identified serologically by a particular capsular antigen named Vi, for virulence.) Colonies that spread in a thin film over the agar surface were described by the German word for film, *hauch*. The motility needed to form a film implied the presence of flagella, and the letter H came to be assigned to the antigens of flagella. Nonmotile bacteria were described as *ohne hauch*, without film, and the O came to be assigned to the cell surface antigens. This terminology is also used in the naming of *E. coli* O157:H7 *Vibrio cholerae* O:1 and others.

Typhoid fever, caused by *Salmonella typhi,* is the most severe illness caused by any member of the genus *Salmonella.* A less severe gastrointestinal disease caused by other salmonellae is called salmonellosis. Salmonellosis is one of the most common forms of foodborne illness.

Shigella Species of *Shigella* (shi-gel'lä) are responsible for a disease called bacillary dysentery, or shigellosis. Unlike salmonellae, they are found only in humans. These organisms are second only to *E. coli* as a cause of traveler's diarrhea. Some strains of *Shigella* can cause life-threatening dysentery (see Chapter 25, page 691).

Klebsiella Members of the genus *Klebsiella* (kleb-sē-el'lä) are commonly found in soil or water. Many isolates are capable of fixing nitrogen from the atmosphere, which has been proposed as being a nutritional advantage in isolated populations with little protein nitrogen in their diet. The species *Klebsiella pneumoniae* (nü-mō'nē-ī) occasionally causes a serious form of pneumonia in humans.

Serratia *Serratia marcescens* (ser-rä'tē-ä mär-ses'sens) is distinguished by its production of red pigment. In hospital situations the organism can be found on catheters, in saline irrigation solutions, and in other supposedly sterile solutions. Such contamination is probably the cause of many urinary and respiratory tract infections in hospitals.

Proteus Colonies of *Proteus* (prō'tē-us) growing on agar exhibit a swarming type of growth, spreading across the agar plate and appearing as a series of concentric rings. See the photo in the box in Chapter 4, page 81. Their motility is by peritrichous flagella (Figure 11.9). They are implicated in many infections of the urinary tract, where the activity of their enzyme urease is a factor, and in wounds.

Yersinia *Yersinia pestis* (yèr-sin'ē-ä pes'tis) causes plague, the Black Death of medieval Europe. Urban rats in some parts of the world and ground squirrels in the American Southwest carry these bacteria. Fleas usually transmit the organisms among animals and to humans, although contact with respiratory droplets from infected animals and people can be involved in transmission.

Erwinia *Erwinia* (èr-wi'nē-ä) species are primarily plant pathogens; some cause plant soft-rot diseases. These species produce enzymes that hydrolyze the pectin between individual plant cells. This causes the plant cells to separate from each other, a disease that plant pathologists term plant rot.

Enterobacter Two *Enterobacter* (en-te-rō-bak'tèr) species, *E. cloacae* (klō-ā'kī), and *E. aerogenes* (ā-rä'jen-ēz) can cause urinary tract infections and hospital-acquired in-

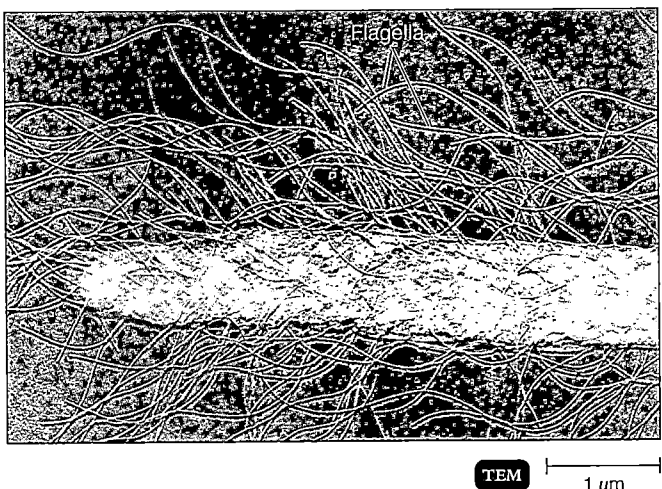

FIGURE 11.9 *Proteus mirabilis,* **a gram-negative rod with peritrichous flagella.**

■ How does the flagellar arrangement of this enteric bacterium differ from that of *Pseudomonas?*

fections. They are widely distributed in humans and animals, as well as in water, sewage, and soil.

Pasteurellales

The bacteria in this order are nonmotile; they are best known as human and animal pathogens.

Pasteurella This genus is primarily known as a pathogen of domestic animals. It causes septicemia in cattle, fowl cholera in chickens and other fowl, and pneumonia in several types of animals. The best-known species is *Pasteurella* (pas-tyèr-el'lä) *multocida* (mul-tō'si-dä), which can be transmitted to humans by dog and cat bites.

Haemophilus *Haemophilus* (hē-mä'fil-us) is a very important genus of pathogenic bacteria. These organisms commonly inhabit the mucous membranes of the upper respiratory tract, mouth, vagina, and intestinal tract. The best-known species that affects humans is *Haemophilus influenzae* (in-flü-en'zī), named long ago because of the erroneous belief that it was responsible for influenza.

The name *Haemophilus* is derived from the bacteria's requirement for blood in their culture medium (*hemo* = blood). They are unable to synthesize important parts of the cytochrome system needed for respiration, and they obtain these substances from the heme fraction, known as the **X factor,** of blood hemoglobin. The culture medium must also supply the cofactor nicotinamide adenine dinucleotide (NAD^+ or $NADP^+$), which is known as **V factor.** Clinical laboratories use tests for the requirement of X and V factors to identify isolates as *Haemophilus* species.

table 11.2	Selected Characteristics of Photosynthesizing Bacteria				
Common Name	Example	Phylum	Comments	Electron Donor for CO$_2$ Reduction	Oxygenic or Anoxygenic
Cyanobacteria	*Anabaena*	Cyanobacteria	Plantlike photosynthesis; some use bacterial photosynthesis under anaerobic conditions	Usually H$_2$O	Usually oxygenic
Green nonsulfur bacteria	*Chloroflexus*	Chloroflexi	Grow chemohetero-trophically in aerobic environments	Organic compounds	Anoxygenic
Green sulfur bacteria	*Chlorobium*	Chlorobi	Deposit sulfur granules outside cells	Usually H$_2$S	Anoxygenic
Purple nonsulfur bacteria	*Rhodospirillum*	Proteobacteria	Can grow chemohetero-trophically as well	Organic compounds	Anoxygenic
Purple sulfur bacteria	*Chromatium*	Proteobacteria	Deposit sulfur granules outside cells	Usually H$_2$S	Anoxygenic

Haemophilus influenzae is responsible for several important diseases. It has been a common cause of meningitis in young children and is a frequent cause of earaches. Other clinical conditions caused by *H. influenzae* include epiglotitis (a life-threatening condition in which the epiglottis becomes infected and inflamed), septic arthritis in children, bronchitis, and pneumonia. *Haemophilus ducreyi* (dü-krā'ē) is the cause of the sexually transmitted disease chancroid.

Purple and Green Photosynthetic Bacteria

Learning Objective

- *Compare and contrast purple and green photosynthetic bacteria with the cyanobacteria.*

The photosynthetic bacteria are taxonomically confusing. This physiological group includes purple sulfur bacteria and purple nonsulfur bacteria, and green sulfur and green nonsulfur bacteria. Photosynthetic bacteria are scattered throughout various subgroups (Table 11.2). The important purple sulfur bacteria are members of the γ-proteobacteria. The purple nonsulfur bacteria are found in the α-proteobacteria. The green sulfur and green nonsulfur bacteria are nonproteobacteria. For simplicity, we will discuss all of these at this point. These photosynthetic bacteria, which are not necessarily colored purple or green, are generally anaerobic. Their habitat is usually the deep sediments of lakes and ponds. Like plants, algae, and the cyanobacteria, purple and green bacteria carry out photosynthesis to make carbohydrates (CH$_2$O). Growing as they do in aquatic depths (Figure 27.11), these bacteria

possess chlorophyll that makes use of parts of the visible spectrum not intercepted by photosynthetic organisms located at higher levels. Also, unlike plantlike photosynthesis, the photosynthesis of purple or green bacteria is *anoxygenic*—it does not produce oxygen.

Plants, algae, and the cyanobacteria, which we will discuss shortly, produce oxygen (O$_2$) from water (H$_2$O) as they carry out photosynthesis:

$$(1)\ 2\ H_2O + CO_2 \xrightarrow{\text{light}} (CH_2O) + H_2O + O_2$$

The *purple sulfur* and *green sulfur bacteria* use reduced sulfur compounds, such as hydrogen sulfide (H$_2$S), instead of water, and they produce granules of sulfur (S^0) rather than oxygen, as follows:

$$(2)\ 2H_2S + CO_2 \xrightarrow{\text{light}} (CH_2O) + H_2O + 2S^0$$

Chromatium (krō-mā'tē-um), shown in Figure 11.10, is a representative genus. At one time, an important question in biology concerned the source of the oxygen produced by plant photosynthesis; was it from CO$_2$ or from H$_2$O? Until the introduction of radioisotope tracers, which traced the oxygen in water and carbon dioxide and finally settled the question, comparison of equations 1 and 2 was the best evidence that the oxygen source was from H$_2$O. It is important, also, to compare these two equations for an understanding of how reduced sulfur compounds, such as H$_2$S, can substitute for photosynthesis. Complex, interrelated life in lightless caves and deep, dark ocean depths often depends upon this energy source.

Other photoautotrophs, the *purple nonsulfur* and *green nonsulfur bacteria,* use organic compounds, such as acids

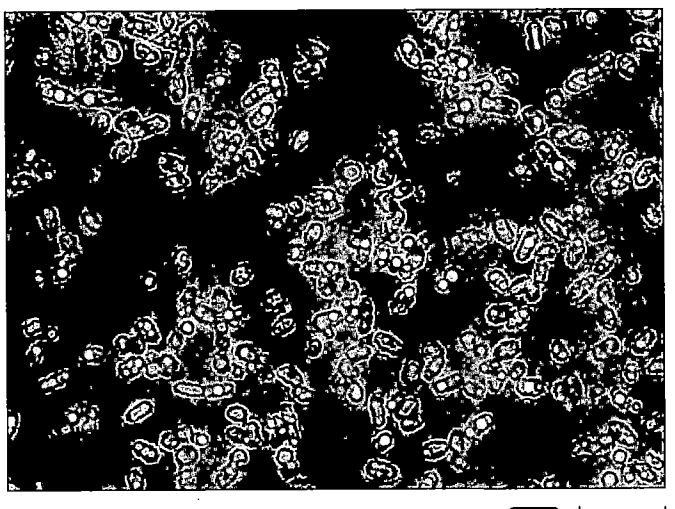

LM $\vdash$———$\dashv$ 10 μm

FIGURE 11.10 Purple sulfur bacteria. This photomicrograph of cells of the genus *Chromatium* shows the intracellular sulfur granules as multicolored refractile objects. The reason the sulfur accumulates can be surmised from inspection of equation 2 in the discussion.

■ **Purple sulfur bacteria are anoxygenic photo-autotrophs.**

and carbohydrates, for the photosynthetic reduction of carbon dioxide.

Morphologically, the photosynthetic bacteria are very diverse, with spirals, rods, cocci, and even budding forms.

The δ (delta) Proteobacteria

Learning Objective

■ *Make a dichotomous key to distinguish among the δ-proteobacteria described in this chapter.*

The δ-proteobacteria are distinctive in that they include some bacteria that are predators on other bacteria. Bacteria in this group are also important contributors to the sulfur cycle.

Bdellovibrio *Bdellovibrio* (del-lō-vib′rē-ō) is a particularly interesting genus. It attacks other gram-negative bacteria (*bdella* = leech). It attaches tightly, and after penetrating the outer layer of gram-negative bacteria, it reproduces within the periplasm. There, the cell elongates into a tight spiral, which then fragments almost simultaneously into several individual flagellated cells. The host cell then lyses, releasing them (see the box in Chapter 3, page 60).

Desulfovibrionales

Members of the order Desulfovibrionales are sulfur-reducing bacteria. They are obligately anaerobic bacteria that use oxidized forms of sulfur, such as sulfates (SO_4^{2-})

or elemental sulfur (S^0), rather than oxygen as electron acceptors. The product of this reduction is hydrogen sulfide (H_2S). (Because the H_2S is not assimilated as a nutrient, this type of metabolism is termed *dissimilatory*.) The activity of these bacteria releases millions of tons of H_2S into the atmosphere every year and plays a key part in the sulfur cycle (Figure 27.6). Sulfur-oxidizing bacteria such as *Beggiatoa* are able to use H_2S either as part of photosynthesis or as an autotrophic energy source.

Desulfovibrio The best studied sulfur-reducing genus is *Desulfovibrio* (dē′sul-fō-vib′rē-ō), which is found in anaerobic sediments, and also in the intestinal tracts of humans and animals. Sulfur-reducing and sulfate-reducing bacteria use organic compounds such as lactate, ethanol, or fatty acids as electron donors. This reduces sulfur or sulfate to H_2S. When H_2S reacts with iron it forms insoluble FeS, which is responsible for the black color of many sediments.

Myxococcales

In the first edition of *Bergey's Manual* the Myxococcales were classified among the fruiting and gliding bacteria. They illustrate the most complex life cycle of all bacteria, part of which is predatory upon other bacteria.

Myxococcus Vegetative cells of the myxobacteria (*myxo* = nasal mucus) move by gliding and leave behind a slime trail (Figure 11.11a). *Myxococcus xanthus* (micks-ō-kok′kus zan′thus) and *M. fulvus* (ful′vus) are well-studied representatives of the genus. As they move, their source of nutrition is the bacteria they encounter, enzymatically lyse, and digest. Large numbers of these gram-negative microbes eventually aggregate. Where the moving cells aggregate, they differentiate and form a macroscopic stalked fruiting body that contains resting cells called *myxospores* (Figure 11.11b). Differentiation is usually triggered by low nutrients. Under proper conditions, usually a change in nutrients, the myxospores germinate and form new vegetative gliding cells. You might note the resemblance to the life cycle of the eukaryotic cellular slime molds in Figure 12.21 on page 356.

The ε (epsilon) Proteobacteria

Learning Objective

■ *Make a dichotomous key to distinguish among the ε-proteobacteria described in this chapter.*

The ε-proteobacteria are slender gram-negative rods that are helical or vibrioid. Vibrioid is a term applied to helical bacteria that do not have a complete turn. We will discuss the two important genera, both of which are motile by means of flagella and are microaerophilic.

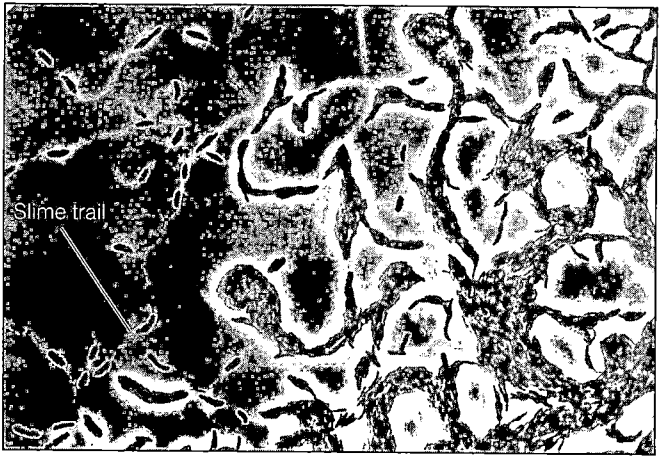

(a) Bacterial slime trails **LM** 10 μm

(b) Myxobacterium with sporangioles **LM** 100 μm

FIGURE 11.11 Myxococcales. (a) Slime trails of the gliding, fruiting myxobacterium *Myxococcus fulvus*. Under appropriate conditions, these bacteria aggregate and form a vertical stalk that bears a fruiting body similar to that shown in part b. (b) *Stigmatella aurantiaca*, another example of a gliding, fruiting bacterium. Numerous gliding vegetative cells have aggregated to form the stalk shown here, on which several sporangioles (fruiting bodies) have formed. Each sporangiole contains about 10,000 myxospores. When growth conditions improve, each of these myxospores can germinate into a new vegetative cell with gliding motility.

■ A few bacteria move by gliding over surfaces, usually leaving a trail.

Campylobacter Campylobacter are microaerophilic vibrios; each cell has one polar flagellum. One species of *Campylobacter*, *C. fetus* (kam'pi-lō-bak-tèr fē'tus), causes spontaneous abortion in domestic animals. Another species, *C. jejuni* (je-ju'ni), is a leading cause of outbreaks of foodborne intestinal disease.

Helicobacter Helicobacter are microaerophilic curved rods with multiple flagella. The species *Helicobacter pylori* (hē-lik-ō-bak-tèr pī-lôr'ē) has only relatively recently been identified as the most common cause of peptic ulcers in humans (Figure 11.12 and Figure 25.12).

The Nonproteobacteria Gram-Negative Bacteria

Learning Objective

■ *Make a dichotomous key to distinguish among the gram-negative nonproteobacteria described in this chapter.*

The nonproteobacteria gram-negative bacteria are not closely related to the gram-negative proteobacteria. This

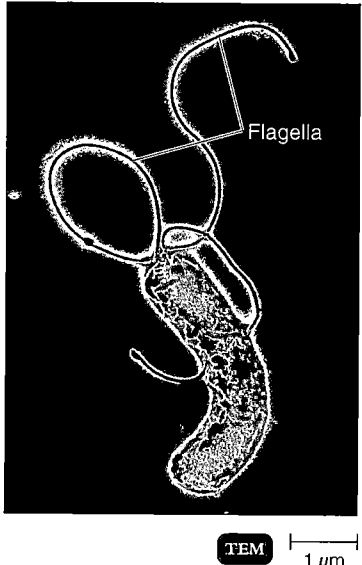

TEM 1 μm

FIGURE 11.12 *Helicobacter pylori,* **an example of a helical bacterium that does not make a complete twist.**

■ How do helical bacteria differ from spirochetes?

group includes a number of physiologically distinctive (photosynthetic) bacteria and morphologically distinctive bacteria, such as the spirochetes. The phylogenetic relationship in these groups is based upon rRNA studies.

Cyanobacteria

The cyanobacteria, named for their characteristic blue-green (cyan) pigmentation, were once called blue-green algae. Although they resemble the eukaryotic algae and often occupy the same environmental niches, this is a misnomer because they are bacteria; algae are not.

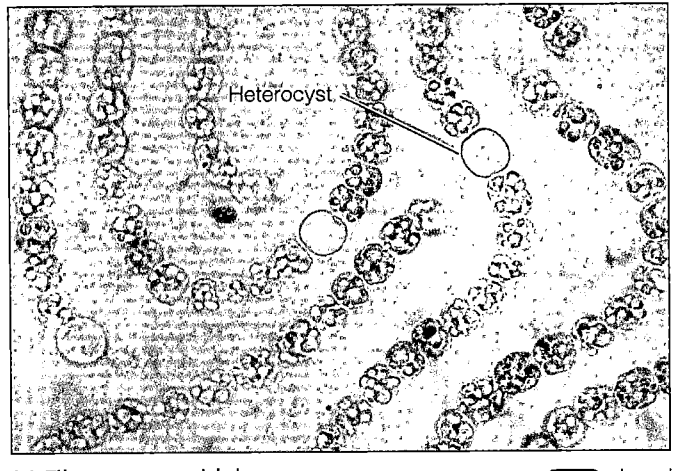

(a) Filamentous, with heterocysts   LM ⊢—⊣ 10 μm

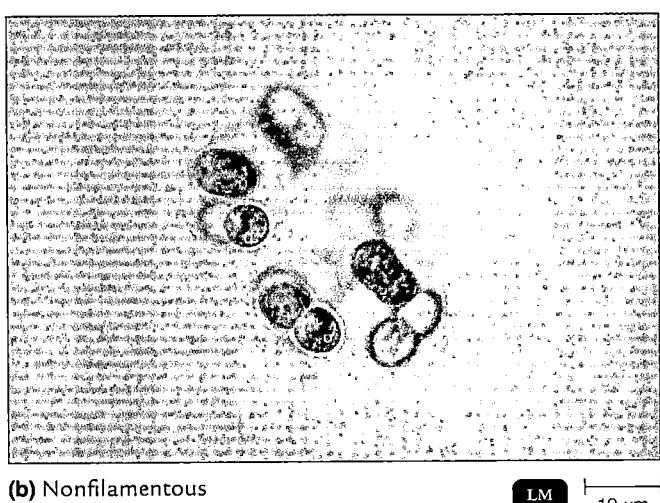

(b) Nonfilamentous   LM ⊢—⊣ 10 μm

FIGURE 11.13 Cyanobacteria (a) A filamentous cyanobacterium showing heterocysts, in which nitrogen-fixing activity is located. **(b)** A unicellular, nonfilamentous cyanobacterium, *Gleocapsa.* Groups of these cells, which divide by binary fission, are held together by the surrounding glycocalyx. **(c)** A branching, filamentous cyanobacterium.

■ How does the photosynthesis of the cyanobacteria differ from that of the purple sulfur bacteria?

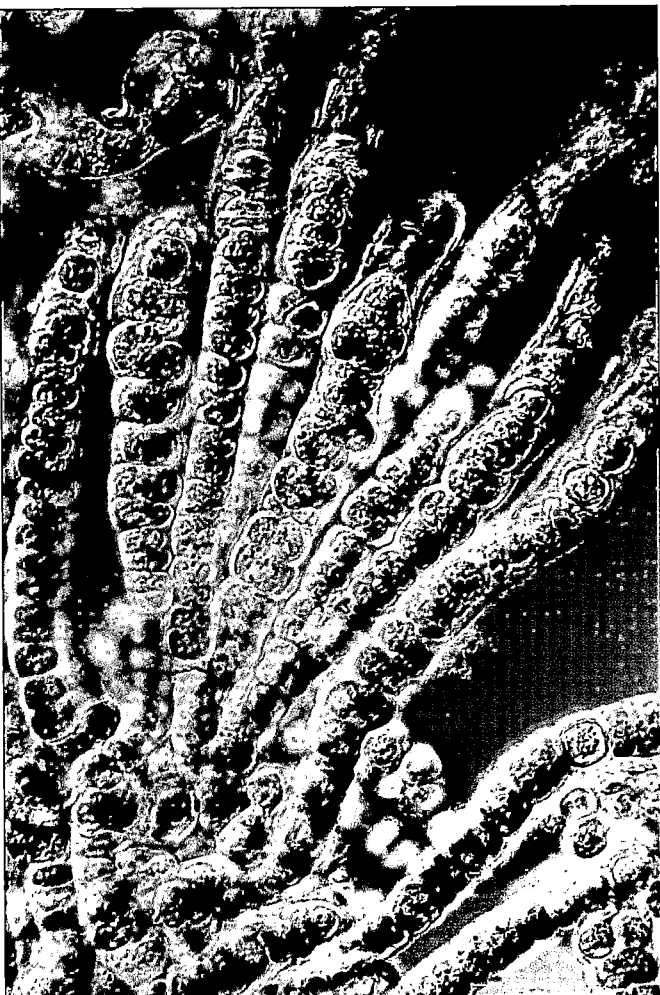

(c) Branching filamentous   LM ⊢—⊣ 10 μm

Cyanobacteria carry out oxygen-producing photosynthesis, much as plants and the eukaryotic algae do (see Chapter 12). Many cyanobacteria are capable of fixing nitrogen. Specialized cells called **heterocysts** contain enzymes that fix nitrogen gas (N_2) into ammonium (NH_4^+) for use by the growing cell (Figure 11.13a). *Gas vacuoles,* found in many species that grow in water, are a series of chambers or gas vesicles surrounded by a protein wall that is permeable to air but not to water. Gas vacuoles provide buoyancy that helps the cell float to favorable environments. Cyanobacteria that are motile move about by gliding.

Cyanobacteria are morphologically varied. They have unicellular forms that divide by simple binary fission (Figure 11.13b), colonial forms that divide by multiple fission, and filamentous forms that reproduce by fragmentation of the filaments. The filamentous forms usually exhibit some differentiation of cells, which are often bound together within an envelope or sheath (Figure 11.13c).

The oxygen-producing cyanobacteria played an important part in the development of life on Earth, which originally had essentially no free oxygen that would support life as we know it. Fossil evidence indicates that when cyanobacteria first appeared, the atmosphere contained only about 0.1% oxygen. When oxygen-producing eukaryotic plants first appeared, the concentration of oxygen was more than 10%. The increase presumably was a result of photosynthetic activity by cyanobacteria. The atmosphere we breathe today contains about 20% oxygen.

Cyanobacteria, especially those that fix nitrogen, are extremely important to the environment. They occupy

FIGURE 11.24 *Actinomyces.*
Notice the branched filamentous morphology.

■ **Why are these bacteria not classified as fungi?**

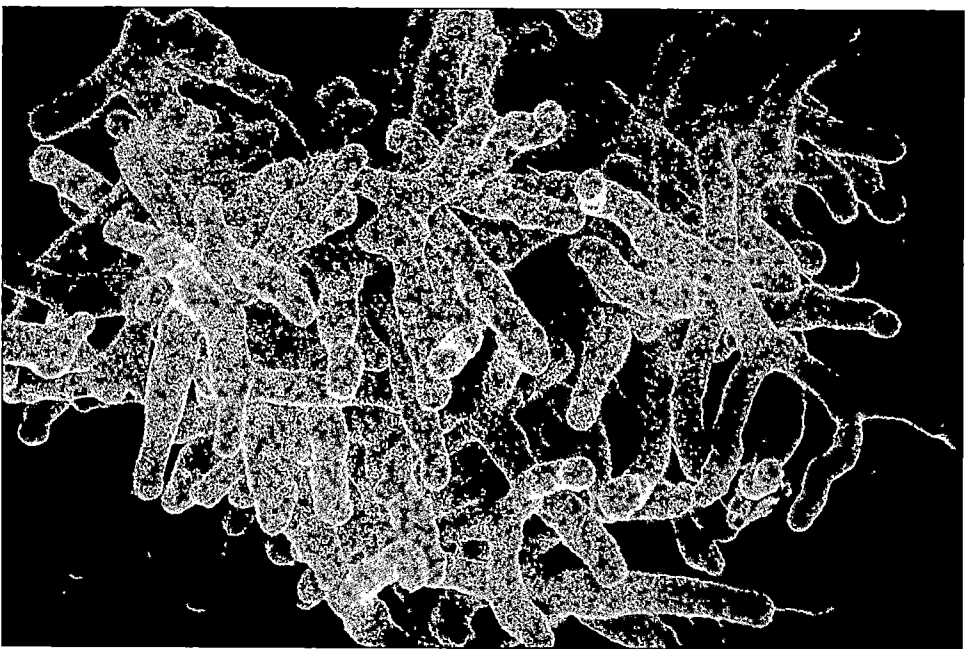

SEM ⊢—⊣ 1 μm

DOMAIN ARCHAEA

Learning Objective

■ *Name a habitat for each group of archaea.*

In the late 1970s, a distinctive type of prokaryotic cell was discovered. So different from what were known then as the Eubacteria, it was believed to constitute almost a third form of life. Most strikingly, their cell walls lacked the peptidoglycan common to most bacteria. It soon became clear that they also shared many rRNA sequences, and the sequences were different from those of the Domain Bacteria and the eukaryotic organisms. Analysis of the genomes of the archaea shows that although some of the genes are shared with conventional bacteria, more than half have never been seen before.

This exceptionally interesting group of prokaryotes is highly diverse. Most are of conventional morphology, that is, rods, cocci, and helixes, but some are of very unusual morphology, as illustrated in Figure 11.25. Some are gram-positive, others gram-negative; some may divide by binary fission, others by fragmentation or budding; a few lack cell walls. Organisms in this domain are physiologically diverse as well, ranging from aerobic, to facultatively anaerobic, to strictly anaerobic. Nutritionally, they include chemoautotrophs, photoautotrophs, and chemoheterotrophs. Of special interest to microbiologists is the

fact that the archaea are frequent inhabitants of exceptionally extreme environments of heat, cold, acidity, and pressure.

Prominent among the archaea are the extreme halophiles, bacteria that survive in very high concentrations of salt, such as the Great Salt Lake and solar evaporating ponds. Examples are *Halobacterium* (hā-lō-bak-ti'rē-um) (see the box in Chapter 5, page 144) and *Halococcus* (hā-lō-kok-kus), which live in high concentrations of sodium chloride (NaCl) and actually require such environmental conditions for growth.

Other archaea thrive in acidic, sulfur-rich hot springs. One such organism is *Sulfolobus* (sul'fō-lō-bus), which has a pH optimum of about 2 and a temperature optimum of more than 70°C. Archaea are also found in ocean depths near hydrothermal vents. Examples of hyperthermophiles, archaea that can thrive in extraordinarily high temperatures, are described in the box in Chapter 6, page 160.

The obligately anaerobic methane-producing members of the archaea, members of the *Methanobacterium* (meth-a-nō-bak-tér'ē-um), are of considerable economic importance. They are found in the human intestines and are used in sewage-treatment processes (see Chapter 27). These archaea derive energy from combining hydrogen (H_2) with carbon dioxide (CO_2) to form methane (CH_4). An essential part of treating sewage is encouraging the growth of these microbes in anaerobic digestion tanks to convert the sludge into CH_4.

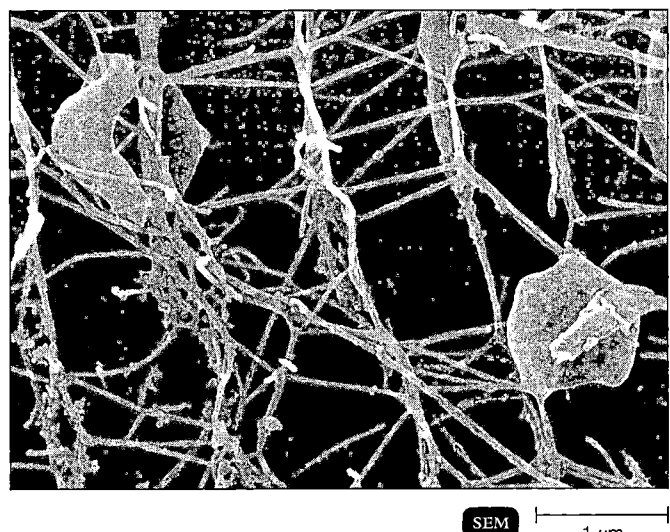

FIGURE 11.25 **Archaea.** *Pyridictum abyssi,* an unusual member of the archaea found growing in deep ocean sediment at a temperature of 110°C. The cells are disk-shaped with a network of tubules. Most archaea are more conventional in their morphology.

MICROBIAL DIVERSITY

Learning Objective

■ *List two factors that contribute to the limits of our knowledge of microbial diversity.*

Earlier in this chapter, we described the giant bacterium *Epulopiscium.* In 1999, another giant bacterium was discovered 100 meters deep in the sediments of the coastal waters off Namibia, on the southwestern coast of Africa. Named *Thiomargarita namibiensis* (thī'ō-mär-gär-ē-tà na'mi-bē-én-sis), meaning "sulfur pearl of Namibia," these spherical organisms are as large as 750 μm in diameter (Figure 11.26). This is a bit larger than the size of a period at the end of this sentence and even larger than *Epulopiscium.* As we have mentioned, a factor that limits the size of prokaryotic cells is that nutrients must enter the cytoplasm by simple diffusion. *T. namibiensis* minimizes this problem by resembling a fluid-filled balloon; the vacuole in the interior being surrounded by a relatively thin outer layer of cytoplasm. Its nutrient supply is essentially hydrogen sulfide, which is plentiful in the sediments in which it is normally found, and nitrate, which it must extract intermittently from nitrate-rich seawaters when storms stir the loose sediment. The cell's interior vacuole, which makes up about 98% of the bacterium's volume, serves as a storage space to hold the nitrate between recharging of its supply. The cell's energy is derived from the oxidation of hydrogen sulfide; the nitrate,

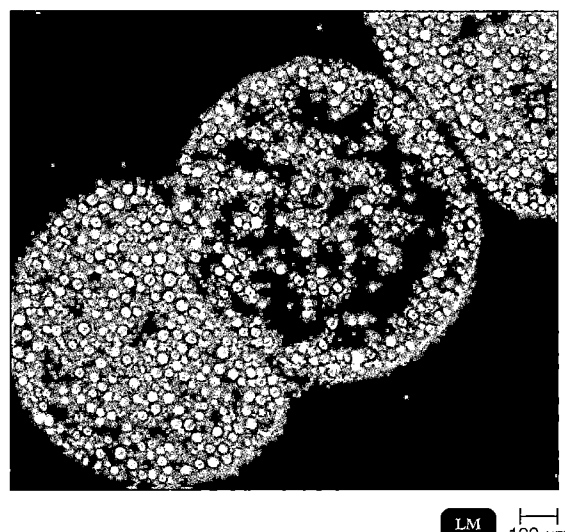

FIGURE 11.26 *Thiomargarita namibiensis.* *Thiomargarita namibiensis* gets its energy from reduced sulfur compounds such as hydrogen sulfide.

■ The interior of this giant bacterium is mostly a fluid-filled vacuole; thus, it can obtain nutrients by diffusion, like much smaller bacteria.

while a source of nutritional nitrogen, primarily serves as an electron acceptor in the absence of oxygen.

The discovery of uniquely large bacteria has raised the question of how large a prokaryotic cell can be and still absorb nutrients by diffusion. At the other extreme, reports of *nannobacteria* as small as 0.02–0.03 μm in deep rocks has raised another question: Is there a lower size limit for microorganisms? Some scientists have used theoretical considerations to calculate that a cell with a significant metabolism would have to have a diameter of at least 0.1 μm. Until now, microbiologists have described only about 5000 bacterial species, of which about 3000 are listed in *Bergey's Manual.* The true number may be in the millions. Many bacteria in soil or water, or elsewhere in nature, cannot be cultivated with the media and conditions normally used for bacterial growth. Moreover, some bacteria are part of complex food chains and can only grow in the presence of other microbes that supply specific growth requirements. Recently, researchers have been using the polymerase chain reaction (PCR) to make millions of copies of genes found at random in a soil sample. By comparing the genes found in many repetitions of this process, researchers can estimate the different bacterial species in such a sample. One report indicates that a single gram of soil may contain 10,000 or so bacterial types—about twice as many as have ever been described. In Chapter 27, which discusses environmental microbiology, you will encounter bacteria that live kilometers deep in rock, in boiling water, and under immense pressures deep in the oceans.

Xanthomonas campestris is a gram-negative rod that causes a disease called black rot in plants. After gaining access to a plant's vascular tissues, the bacteria use the glucose transported in those tissues to produce a sticky, gumlike substance. This substance builds up to form gumlike masses, which eventually block the plant's transport of nutrients. The gum that makes up these masses, xanthan, is composed of a high-molecular-weight polymer of mannose (see the photograph).

In contrast to its effects in plants, xanthan has no adverse effects when ingested by humans. Consequently, xanthan can be used as a thickener in foods such as dairy products and salad dressings, and in cosmetics such as cold creams and shampoos.

So when researchers at the U.S. Department of Agriculture (USDA) wanted to find some useful product that could be made out of whey, a liquid waste produced in abundance by the dairy industry, they thought of turning it into xanthan. However, because whey is mostly water and lactose, researchers had to figure out how to get *X. campestris* to produce xanthan using lactose rather than glucose.

One research team decided to engineer *X. campestris* to hydrolyze lactose more efficiently. They used an F^+ strain of *E. coli* that contains a plasmid with a *lac* operon. Organisms of this strain were incubated with *X. campestris,* and, as proven by later restriction enzyme digests, the plasmids containing the *lac* gene were transferred by conjugation.

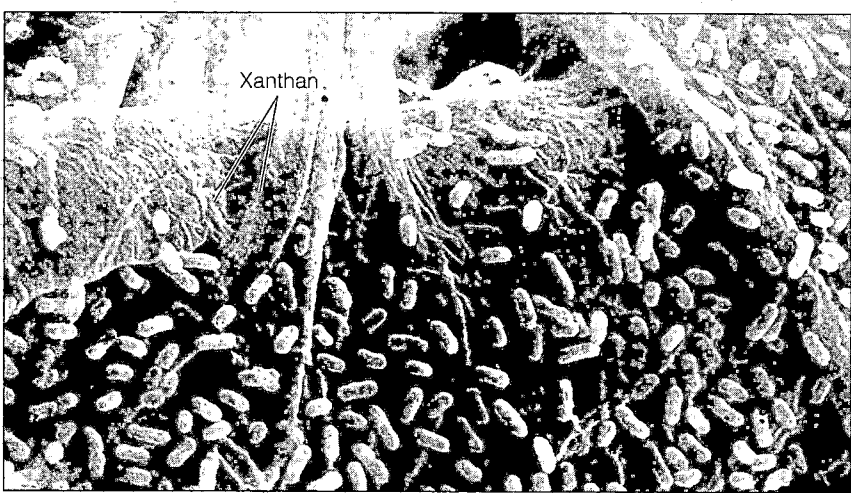

Xanthomonas campestris *producing gooey xanthan.* [SEM] ⊢———⊣ 2 μm

However, deletion mutations subsequently occurred in the *X. campestris* DNA that reduced their β-galactosidase (lactose-utilizing) ability.

Another research team working with the USDA at Stauffer Chemical Company used a simpler approach, which was based on satisfying only two requirements: that the bacteria grow on whey and make xanthan. First, they inoculated a whey medium with *X. campestris* and incubated it for 24 hours. Then they transferred an inoculum of this culture to a flask of lactose broth, to select a lactose-utilizing cell. The strain did not have to make xanthan from this broth; it only had to grow and use lactose.

A lactose-utilizing strain was isolated through serial transfers, selecting for the strain with the best ability to grow. After

incubation for 10 days, an inoculum was transferred to another flask of lactose broth, and the procedure was repeated two more times. When transferred to a flask of whey medium, the final lactose-utilizing bacteria grew in the whey, and the culture became extremely viscous—xanthan was being produced.

The process then had to be fine-tuned. The researchers addressed and solved problems such as how to sterilize the whey without destroying necessary ingredients, and how to handle the extremely viscous fermentations that resulted from the procedure. The final result was a process in which 40 g/L of whey powder is converted into 30 g/L of xanthan gum. A quick survey of labels in your neighborhood supermarket will demonstrate just how successful this project was.

Study Outline

INTRODUCTION (p. 303)

1. *Bergey's Manual* categorizes bacteria into taxa based on rRNA sequences.
2. *Bergey's Manual* lists identifying characteristics such as Gram stain reaction, cellular morphology, oxygen requirements, and nutritional properties.

PROKARYOTIC GROUPS (p. 304)

1. Prokaryotic organisms are classified into two domains: Archaea and Bacteria.

DOMAIN BACTERIA (pp. 304–326)

1. Bacteria are essential to life on Earth.

THE PROTEOBACTERIA (pp. 304–316)

1. Members of the phylum Proteobacteria are gram-negative.
2. Alpha-proteobacteria includes nitrogen-fixing bacteria, chemoautotrophs, and chemoheterotrophs.
3. The β-proteobacteria include chemoautotrophs and chemoheterotrophs.
4. Pseudomonadales, Legionellales, Vibrionales, Enterobacteriales, and Pasteurellales are classified as γ-proteobacteria.
5. Purple and green photosynthetic bacteria are photoautotrophs that use light energy and CO_2 and do not produce O_2.
6. *Myxococcus* and *Bdellovibrio* in the δ-proteobacteria prey on other bacteria.
7. Epsilon-proteobacteria include *Campylobacter* and *Helicobacter.*

THE NONPROTEOBACTERIA GRAM-NEGATIVE BACTERIA (pp. 316–320)

1. Several phyla of gram-negative bacteria are not related phylogenetically to the Proteobacteria.

2. Cyanobacteria are photoautotrophs that use light energy and CO_2 and do produce O_2.

3. Chemoheterotrophic examples are *Chlamydia,* spirochetes, *Bacteroides,* and *Fusobacterium.*

THE GRAM-POSITIVE BACTERIA (pp. 320–325)

1. In *Bergey's Manual,* gram-positive bacteria are divided into those that have low G + C ratio and those that have high G + C ratio.

2. Low G + C gram-positive bacteria include common soil bacteria, the lactic acid bacteria, and several human pathogens.

3. High G + C gram-positive bacteria include mycobacteria, corynebacteria, and actinomycetes.

DOMAIN ARCHAEA (p. 326)

1. Extreme halophiles, extreme thermophiles, and methanogens are included in the archaea.

MICROBIAL DIVERSITY (p. 327)

1. Few of the total number of different prokaryotes have been isolated and identified.

2. PCR can be used to uncover the presence of bacteria that can't be cultured in the laboratory.

Study Questions

REVIEW

1. The following outline is a key that can be used to identify medically important bacteria. Fill in a representative genus in the space provided.

Name of
Representative
Genus

I. Gram-positive
  A. Endospore-forming rod
    1. Obligate anaerobe   _____
    2. Not obligate anaerobe   _____
  B. Nonendospore-forming
    1. Cells are rods
      a. Produce conidiospores   _____
      b. Acid-fast   _____
    2. Cells are cocci
      a. Lack cytochrome system   _____
      b. Use aerobic respiration   _____
II. Gram-negative
  A. Cells are helical or curved
    1. Axial filament   _____
    2. No axial filament   _____
  B. Cells are rods
    1. Aerobic, nonfermenting   _____
    2. Facultatively anaerobic   _____
III. Lack cell walls   _____
IV. Obligate intracellular parasites
  A. Transmitted by ticks   _____
  B. Reticulate bodies in host cells   _____

2. Compare and contrast each of the following:
  a. Cyanobacteria and algae
  b. Actinomycetes and fungi
  c. *Bacillus* and *Lactobacillus*
  d. *Pseudomonas* and *Escherichia*
  e. *Leptospira* and *Spirillum*
  f. *Veillonella* and *Bacteroides*
  g. *Rickettsia* and *Chlamydia*
  h. *Ureaplasma* and *Mycoplasma*

3. Matching:
  I. Gram-positive
    A. Nitrogen-fixing ____
  II. Gram-negative
    A. Phototrophic
      1. Anoxygenic ____
      2. Oxygenic ____
    B. Chemoautotrophic
      1. Oxidize NO_2^- ____
      2. Reduce CO_2 to CH_4 ____
    C. Chemoheterotrophic
      1. Cells inside a sheath ____
      2. Form myxospores ____
      3. Reduce sulfate to H_2S
        a. Anaerobic ____
        b. Thermophilic ____
      4. Long filaments, found in sewage ____
      5. Form projections from the cell ____

  a. Cyanobacteria
  b. *Cytophaga*
  c. *Desulfovibrio*
  d. *Frankia*
  e. *Hyphomicrobium*
  f. Methanogens
  g. Myxobacteria
  h. *Nitrobacter*
  i. Purple bacteria
  j. *Sphaerotilus*
  k. *Sulfolobus*

MULTIPLE CHOICE

1. If you Gram-stained the bacteria that live in the human intestine, you would expect to find mostly
  a. gram-positive cocci.
  b. gram-negative rods.
  c. gram-positive, endospore-forming rods.
  d. gram-negative, nitrogen-fixing bacteria.
  e. all of the above

2. Which of the following does *not* belong with the others?
  a. Enterobacteriales
  b. Lactobacillales
  c. Legionellales
  d. Pasteurellales
  e. Vibrionales

3. Pathogenic bacteria can be
  a. motile.
  b. rods.
  c. cocci.
  d. anaerobic.
  e. all of the above

4. Which of the following is an intracellular parasite?
   a. *Rickettsia*   d. *Staphylococcus*
   b. *Mycobacterium*   e. *Streptococcus*
   c. *Bacillus*

5. Which of the following terms is the most specific?
   a. bacillus   d. endospore-forming rods
   b. *Bacillus*   and cocci
   c. gram-positive   e. anaerobic

6. Which of the following produce most of the antibiotics that are used to treat diseases?
   a. endospore-forming rods   c. *Streptomyces*
      and cocci   d. β-proteobacteria
   b. *Streptococcus*   e. fungi

7. Which of the following pairs is mismatched?
   a. Anaerobic endospore-forming gram-positive rods— *Clostridium*
   b. Facultatively anaerobic gram-negative rods—*Escherichia*
   c. Facultatively anaerobic gram-negative rods—*Shigella*
   d. Pleomorphic gram-positive rods—*Corynebacterium*
   e. Spirochete—*Helicobacter*

8. *Spirillum* is not classified as a spirochete because spirochetes
   a. do not cause disease.   d. are prokaryotes.
   b. possess axial filaments. e. none of the above
   c. possess flagella.

9. When *Legionella* was newly discovered, it was classified with the pseudomonads because
   a. it is a pathogen.   c. it is difficult to culture.
   b. it is an aerobic gram-   d. it is found in water.
      negative rod.   e. none of the above

10. Cyanobacteria differ from purple and green phototrophic bacteria because cyanobacteria
   a. produce oxygen during   d. have a membrane-enclosed
      photosynthesis.   nucleus.
   b. do not require light.   e. all of the above
   c. use H_2S as an electron
      donor.

CRITICAL THINKING

1. Place each section listed in Table 11.1 in the appropriate category:
   a. typical gram-positive cell wall
   b. typical gram-negative cell wall
   c. no peptidoglycan in cell wall
   d. no cell wall

2. To which of the following is the photosynthetic bacterium *Chromatium* most closely related? Briefly explain why.
   a. cyanobacteria   c. *Escherichia*
   b. *Chloroflexus*

3. Identify the genus that best fits each of the following descriptions:
   a. This organism can produce a fuel used for home heating and for generating electricity.
   b. This gram-positive genus presents the greatest source of bacterial damage to the beekeeping industry.
   c. This gram-positive rod is used in dairy fermentations.
   d. This γ-proteobacterial genus is well suited to degrade hydrocarbons in an oil spill.

CLINICAL APPLICATIONS

1. After contact with a patient's spinal fluid, a lab technician developed fever, nausea, and purple lesions on her neck and extremities. A throat culture grew gram-negative diplococci. What is the genus of the bacteria?

2. Between April 1 and May 15 of one year, 22 children in three states developed diarrhea, fever, and vomiting. The children had each received pet ducklings. Gram-negative, facultatively anaerobic bacteria were isolated from both the patients' and the ducks' feces; the bacteria were identified as serovar C2. What is the genus of these bacteria?

3. A woman complaining of lower abdominal pain with a temperature of 39°C gave birth soon after to a stillborn baby. Blood cultures from the infant revealed gram-positive rods. The woman had a history of eating unheated hot dogs during her pregnancy. Which organism is most likely involved?

Learning with Technology

MP = The Microbiology Place website   **ST** = Student Tutorial CD-ROM   **VU** = VirtualUnknown CD-ROM

MP   Don't forget to go to The Microbiology Place website (http://www.microbiologyplace.com) to take the practice tests, explore the interactive activity and case study, and check out the news articles and web links for this chapter.

ST   Remember there is also a quiz for this chapter on the Microbiology Interactive Student Tutorial CD-ROM.

VU   Enter the Virtual Lab, click the arrow next to the Session field, click Textbook Exercises, and select Chapter 11.

1. Consult Table 11.1 on page 306 of your textbook to determine which laboratory tests and key features would be helpful in placing an unknown bacterium in one of the phyla and orders. List those initial key tests and special features.

2. Read the Case Study carefully. Perform the initial key tests from your list. Based on the information provided in Table 11.1 and the results of the Virtual Laboratory tests you have performed, in which phylum and order should the unknown microbe be placed?

3. Complete additional tests in the Virtual Lab that would allow you to give a more informed placement of the organism in the taxonomic scheme. Do these results confirm your original prediction? If not, what is your modified prediction for its taxonomic placement?

4. Compare the survivors in the Identification Matrix with genera listed in Table 11.1. Did your predicted taxonomic placement match the final placement? Explain your reasoning for both the prediction and final placement.

5. If you have not done so, complete your identification of the unknown organism. Consult the Bacteria Reference resource book to provide an explanation for the presence of the identified bacterium in this habitat.

6. Print out your Virtual Laboratory Report for your instructor.

chapter 12

The Eukaryotes:

Fungi, Algae, Protozoa, and Helminths

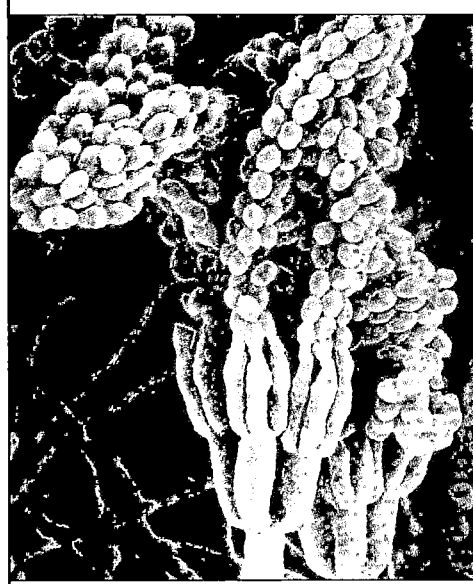

Penicillium. *Fungi such as Penicillium are important decomposers that recycle organic matter in the soil.*

Over half of the world's population is infected with eukaryotic pathogens. The World Health Organization (WHO) ranks six parasitic diseases among the top 20 microbial causes of death in the world. Every year, there are over 5 million cases each of malaria, schistosomiasis, amoebiasis, hookworm, African trypanosomiasis, and intestinal parasites reported in developing countries. Emerging eukaryotic pathogens in developed countries include *Pneumocystis,* the leading cause of death in AIDS patients; the protozoan *Cryp-* *tosporidium,* which caused disease in 400,000 people in Milwaukee in 1993; a new protozoan parasite *Cyclospora,* discovered in the United States in 1993; a new respiratory disease caused by the fungus *Stachybotrys,* which killed several infants in Cleveland; and new and increased poisonings due to algae.

In this chapter, we examine the eukaryotic microorganisms that affect humans: fungi, algae, protozoa, and parasitic helminths, or worms. (For a comparison of their characteristics, see Table 12.1.)

Fungi

Over the last 10 years, the incidence of serious fungal infections has been increasing. These infections are occurring as nosocomial infections and in people with compromised immune systems. In addition, thousands of fungal diseases afflict economically important plants, costing more than one billion dollars annually.

Fungi are also beneficial. They are important in the food chain because they decompose dead plant matter, thereby recycling vital elements. Through the use of extracellular enzymes such as cellulases, fungi are the primary decomposers of the hard parts of plants, which cannot be digested by animals. Nearly all plants depend on symbiotic fungi, known as **mycorrhizae,** which help their roots absorb minerals and water from the soil (see Chapter 27). Fungi are also valuable to animals. Fungi-farming ants cultivate fungi in order to break down cellulose and lignin from plants so the ants can then digest them. Fungi are used by humans for food (mushrooms) and to produce foods (bread and citric

table 12.1	Major Differences Among Eukaryotic Microorganisms: Fungi, Algae, Protozoa, and Helminths			
	Fungi	**Algae**	**Protozoa**	**Helminths**
Kingdom	Fungi	Protist	Protist	Animalia
Nutritional type	Chemoheterotroph	Photoautotroph	Chemoheterotroph	Chemoheterotroph
Multicellularity	All, except yeasts	Some	None	All
Cellular arrangement	Unicellular, filamentous, fleshy (such as mushrooms)	Unicellular, colonial, filamentous; tissues	Unicellular	Tissues and organs
Food acquisition method	Absorptive	Absorptive	Absorptive; ingestive (cytostome)	Ingestive (mouth); absorptive
Characteristic features	Sexual and asexual spores	Pigments	Motility; some form cysts	Many have elaborate life cycles, including egg, larva, and adult
Embryo formation	None	None	None	All

table 12.2	Selected Features of Fungi and Bacteria Compared	
	Fungi	**Bacteria**
Cell type	Eukaryotic	Prokaryotic
Cell membrane	Sterols present	Sterols absent, except in *Mycoplasma*
Cell wall	Glucans; mannans; chitin (no peptidoglycan)	Peptidoglycan
Spores	Produce a wide variety of sexual and asexual reproductive spores	Endospores (not for reproduction); some asexual reproductive spores
Metabolism	Limited to heterotrophic; aerobic, facultatively anaerobic	Heterotrophic, chemoautotrophic, photoautotrophic; aerobic, facultatively anaerobic, anaerobic
Sensitivity to antibiotics	Often sensitive to polyenes, imidazoles, and griseofulvin	Often sensitive to penicillins, tetracyclines, and aminoglycosides

After B. D. Davis et al., *Microbiology,* 4th Ed. Philadelphia: J. B. Lippincott, 1990, p. 746.

acid) and drugs (alcohol and penicillin). Of the more than 100,000 species of fungi, only about 100 are pathogenic to humans and animals.

The study of fungi is called **mycology.** We will first look at the structures that are the basis of fungal identification in a clinical laboratory. Then we will explore their life cycles, mainly because fungi are identified by the sexual stage of their life cycle. Recall from Chapter 10 that identification of a pathogen is often required in order to properly treat a disease and to prevent its spread.

We will also examine nutritional needs. All fungi are chemoheterotrophs, requiring organic compounds for energy and carbon. Fungi are aerobic or facultatively anaerobic; only a few anaerobic fungi are known.

Table 12.2 lists the basic differences between fungi and bacteria.

Characteristics of Fungi

Learning Objectives

- List the defining characteristics of fungi.
- Differentiate asexual from sexual reproduction, and describe each of these processes in fungi.

Yeast identification, like bacterial identification, uses biochemical tests. However, multicellular fungi are identified on the basis of physical appearance, including colony characteristics and reproductive spores.

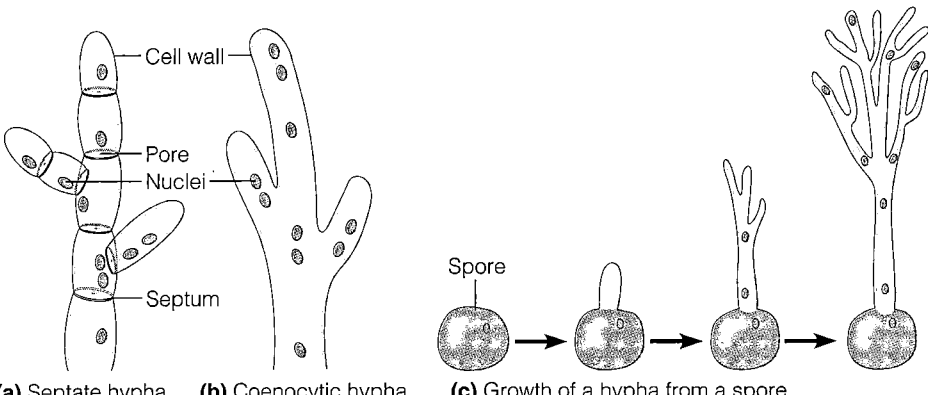

(a) Septate hypha   **(b)** Coenocytic hypha   **(c)** Growth of a hypha from a spore

FIGURE 12.1 Characteristics of fungal hyphae. (a) Septate hyphae have cross-walls, or septa, dividing the hyphae into cell-like units. **(b)** Coenocytic hyphae lack septa. **(c)** Hyphae grow by elongating at the tips.

■ What is a hypha? A mycelium?

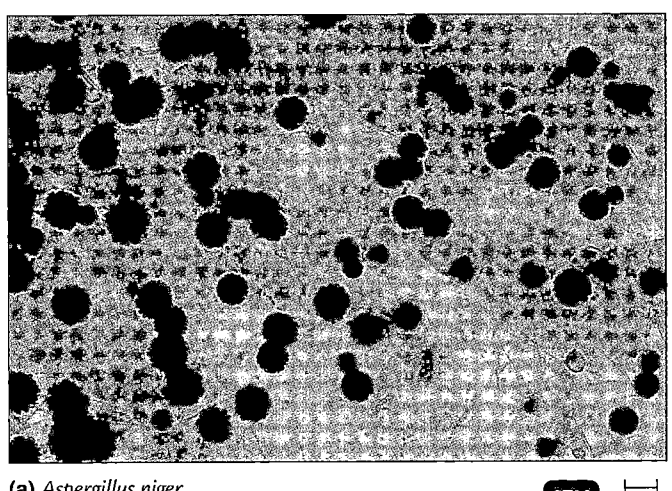

(a) *Aspergillus niger*   SEM   ⊢—⊣ 20 μm

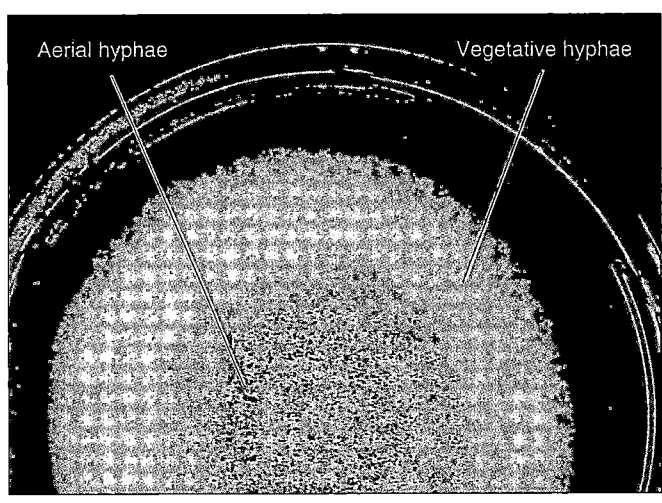

(b) *A. niger* on agar

FIGURE 12.2 Aerial and vegetative hyphae. (a) A photomicrograph of aerial hyphae, showing reproductive spores. **(b)** A colony of *Aspergillus niger* grown on a glucose agar plate, showing both vegetative and aerial hyphae.

■ How do fungal colonies differ from bacterial colonies?

Vegetative Structures

Fungal colonies are described as **vegetative** structures because they are composed of the cells involved in catabolism and growth.

Molds and Fleshy Fungi. The **thallus** (body) of a mold or fleshy fungus consists of long filaments of cells joined together; these filaments are called **hyphae** (singular: *hypha*). Hyphae can grow to immense proportions. The hyphae of a single fungus in Michigan extend across 40 acres and are estimated to weigh over 10 tons.

In most molds, the hyphae contain cross-walls called **septa** (singular: *septum*), which divide them into distinct, uninucleate (one-nucleus) cell-like units. These hyphae are called **septate hyphae** (Figure 12.1a). In a few classes of fungi, the hyphae contain no septa and appear as long, continuous cells with many nuclei. These are called **coenocytic hyphae** (Figure 12.1b). Even in

fungi with septate hyphae, there are usually openings in the septa that make the cytoplasm of adjacent "cells" continuous; these fungi are actually coenocytic organisms, too.

Hyphae grow by elongating at the tips (Figure 12.1c). Each part of a hypha is capable of growth, and when a fragment breaks off, it can elongate to form a new hypha. In the laboratory, fungi are usually grown from fragments obtained from a fungal thallus.

The portion of a hypha that obtains nutrients is called the *vegetative hypha;* the portion concerned with reproduction is the *reproductive* or *aerial hypha,* so named because it projects above the surface of the medium on which the fungus is growing. Aerial hyphae often bear reproductive spores (Figure 12.2a), discussed later. When environmental conditions are suitable, the hyphae grow to form a filamentous mass called a **mycelium,** which is visible to the unaided eye (Figure 12.2b).

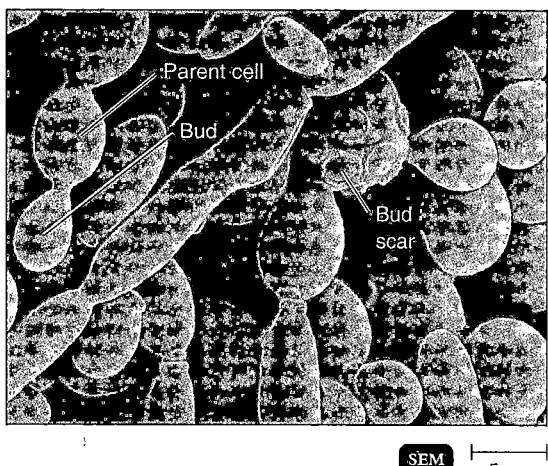

FIGURE 12.3 A budding yeast. A micrograph of *Saccharomyces cerevisiae* in various stages of budding.

■ Following mitosis, some yeast cells produce a small bud that grows into a full-size cell.

Yeasts. Yeasts are nonfilamentous, unicellular fungi that are typically spherical or oval. Like molds, yeasts are widely distributed in nature; they are frequently found as a white powdery coating on fruits and leaves. **Budding yeast,** such as *Saccharomyces* (sak-ä-rō-mī'sēs), divide unevenly.

In budding (Figure 12.3), the parent cell forms a protuberance (bud) on its outer surface. As the bud elongates, the parent cell's nucleus divides, and one nucleus migrates into the bud. Cell wall material is then laid down between the bud and parent cell, and the bud eventually breaks away.

One yeast cell can in time produce up to 24 daughter cells by budding. Some yeasts produce buds that fail to detach themselves; these buds form a short chain of cells called a **pseudohypha.** *Candida albicans* (kan'did-ä al'bi-kanz) attaches to human epithelial cells as a yeast but usually requires pseudohyphae to invade deeper tissues (see Figure 21.17a on page 594).

Fission yeasts, such as *Schizosaccharomyces* (skiz-ō-sak-ä-rō-mī'sēs), divide evenly to produce two new cells. During fission, the parent cell elongates, its nucleus divides, and two daughter cells are produced. Increases in the number of yeast cells on a solid medium produce a colony similar to a bacterial colony.

Yeasts are capable of facultative anaerobic growth. Yeasts can use oxygen or an organic compound as the final electron acceptor; this is a valuable attribute because it allows these fungi to survive in various environments. If given access to oxygen, yeasts perform aerobic respiration to metabolize carbohydrates to carbon dioxide and water; denied oxygen, they ferment carbohydrates and produce ethanol and carbon dioxide. This fermentation is used in

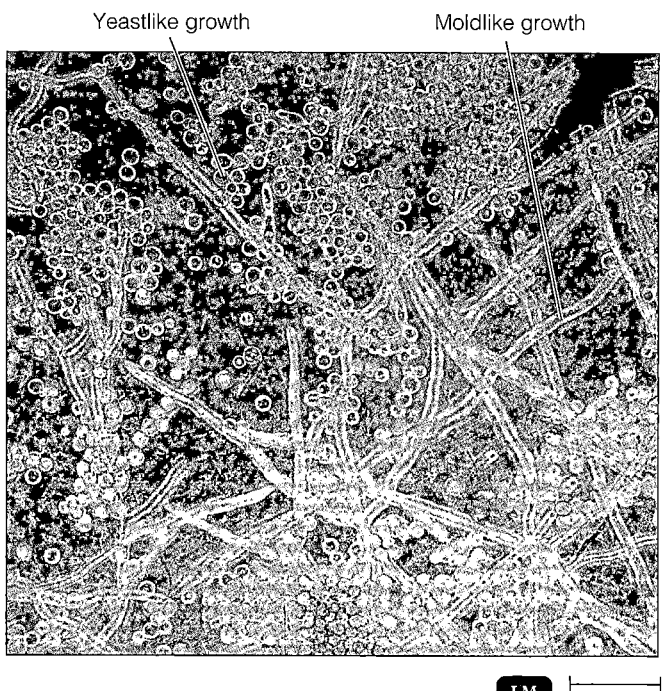

FIGURE 12.4 Fungal dimorphism. Dimorphism in the fungus *Mucor rouxii* depends on CO_2 concentration. On the agar surface, *Mucor* exhibits yeastlike growth, but in the agar it is moldlike.

■ What is fungal dimorphism?

the brewing, wine-making, and baking industries. Species of *Saccharomyces* produce ethanol in brewed beverages and carbon dioxide for leavening bread dough.

Dimorphic Fungi. Some fungi, most notably the pathogenic species, exhibit **dimorphism**—two forms of growth. Such fungi can grow either as a mold or as a yeast. The moldlike forms produce vegetative and aerial hyphae; the yeastlike forms reproduce by budding. Dimorphism in pathogenic fungi is temperature-dependent: At 37°C, the fungus is yeastlike, and at 25°C, it is moldlike. However, the appearance of the dimorphic (in this instance, nonpathogenic) fungus shown in Figure 12.4 changes with CO_2 concentration. (See Figure 24.17 on page 675.)

Life Cycle

Filamentous fungi can reproduce asexually by fragmentation of their hyphae. In addition, both sexual and asexual reproduction in fungi occurs by the formation of **spores.** In fact, fungi are usually identified by spore type.

Fungal spores, however, are quite different from bacterial endospores. Bacterial endospores allow a bacterial cell to survive adverse environmental conditions (see Chapter 4). A single vegetative bacterial cell forms one endospore, which eventually germinates to produce a single vegetative bacterial cell. This process is not reproduction because it does not increase the total number of bacterial cells. But after a mold forms a spore, the spore detaches from the parent and germinates into a new mold (see Figure 12.1c). Unlike the bacterial endospore, this is a true reproductive spore; a second organism grows from the spore. Although fungal spores can survive for extended periods in dry or hot environments, most do not exhibit the extreme tolerance and longevity of bacterial endospores.

Spores are formed from aerial hyphae in a number of different ways, depending on the species. Fungal spores can be either asexual or sexual. **Asexual spores** are formed by the hyphae of one organism. When these spores germinate, they become organisms that are genetically identical to the parent. **Sexual spores** result from the fusion of nuclei from two opposite mating strains of the same species of fungus. Fungi produce sexual spores less frequently than asexual spores. Organisms that grow from sexual spores will have genetic characteristics of both parental strains. Because spores are of considerable importance in the identification of fungi, we will next look at some of the various types of asexual and sexual spores.

Asexual Spores. Asexual spores are produced by an individual fungus through mitosis and subsequent cell division; there is no fusion of the nuclei of cells. Several types of asexual spores are produced by fungi. One type is a **conidium**, a unicellular or multicellular spore that is not enclosed in a sac (Figure 12.5a). Conidia are produced in a chain at the end of a **conidiophore.** Such spores are produced by *Aspergillus.* One type of conidium, an **arthrospore,** is formed by the fragmentation of a septate hypha into single, slightly thickened cells (see Figures 12.5b). One species that produces such spores is *Coccidioides immitis* (kok-sid-ē-oi′dēz im′mi-tis). Another type of conidiospore, **blastoconidia,** consist of buds coming off the parent cell (Figure 12.5c). Such spores are found in some yeasts, such as *Candida albicans* and *Cryptococcus.*

A second type of asexual spore is a **chlamydospore,** a thick-walled spore formed by rounding and enlargement within a hyphal segment (Figure 12.5d). A fungus that produces chlamydospores is the yeast *C. albicans.* The third type of asexual spore is a **sporangiospore,** formed within a **sporangium,** or sac, at the end of an aerial hypha called a **sporangiophore.** The sporangium can contain hundreds of sporangiospores (Figure 12.5e). Such spores are produced by *Rhizopus.*

Sexual Spores. A fungal sexual spore results from sexual reproduction, consisting of three phases:

1. **Plasmogamy.** A haploid nucleus of a donor cell (+) penetrates the cytoplasm of a recipient cell (−).

2. **Karyogamy.** The (+) and (−) nuclei fuse to form a diploid zygote nucleus.

3. **Meiosis.** The diploid nucleus gives rise to haploid nuclei (sexual spores), some of which may be genetic recombinants.

The sexual spores produced by fungi are the criterion used to classify the fungi into several divisions. In laboratory settings, most fungi exhibit only asexual spores. Consequently, clinical identification is based on microscopic examination of asexual spores.

Nutritional Adaptations

Fungi are generally adapted to environments that would be hostile to bacteria. Fungi are chemoheterotrophs, and, like bacteria, they absorb nutrients rather than ingesting them as animals do. However, fungi differ from bacteria in certain environmental requirements and in the following nutritional characteristics:

- Fungi usually grow better in an environment with a pH of about 5, which is too acidic for the growth of most common bacteria.
- Almost all molds are aerobic. Most yeasts are facultative anaerobes.
- Most fungi are more resistant to osmotic pressure than bacteria; most can therefore grow in relatively high sugar or salt concentrations.
- Fungi can grow on substances with a very low moisture content, generally too low to support the growth of bacteria.
- Fungi require somewhat less nitrogen than bacteria for an equivalent amount of growth.
- Fungi are often capable of metabolizing complex carbohydrates, such as lignin (a component of wood), that most bacteria cannot use for nutrients.

These characteristics enable fungi to grow on such unlikely substrates as bathroom walls, shoe leather, and discarded newspapers.

Medically Important Phyla of Fungi

Learning Objective

- *List the defining characteristics of the three phyla of fungi described in this chapter.*

This section provides an overview of medically important phyla of fungi. The actual diseases they cause will be studied in Part Four, Chapters 21–26. Note that not all fungi cause disease. Other phyla of fungi are not discussed here because they contain no pathogens.

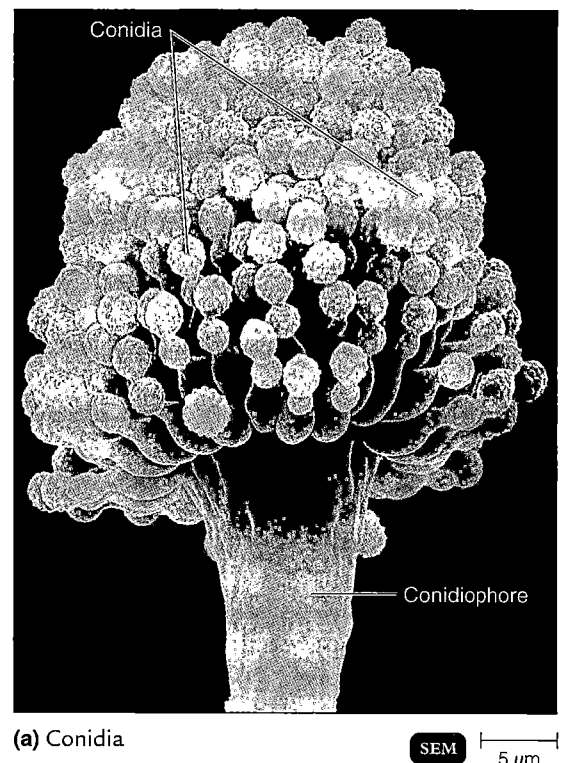

(a) Conidia

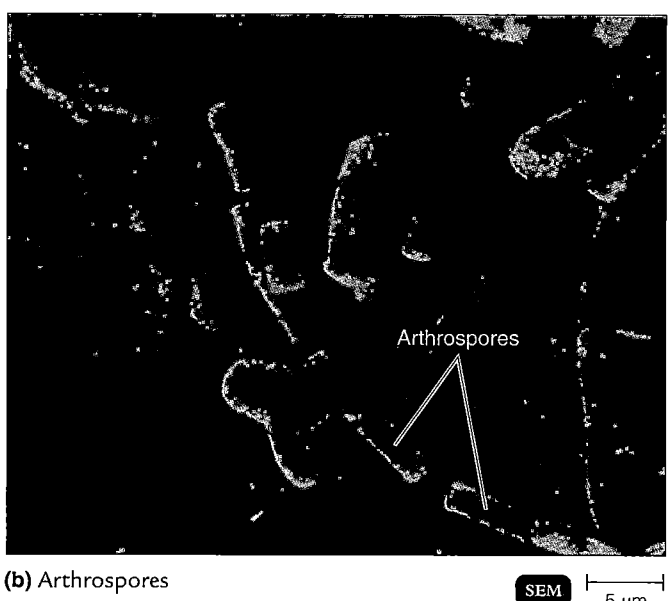

(b) Arthrospores

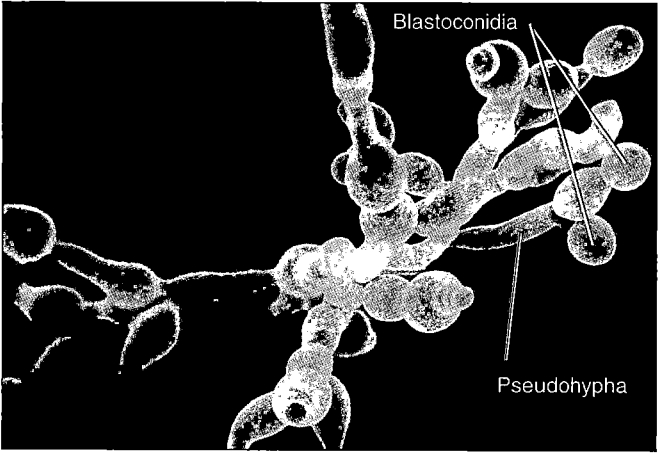

(c) Blastoconidia

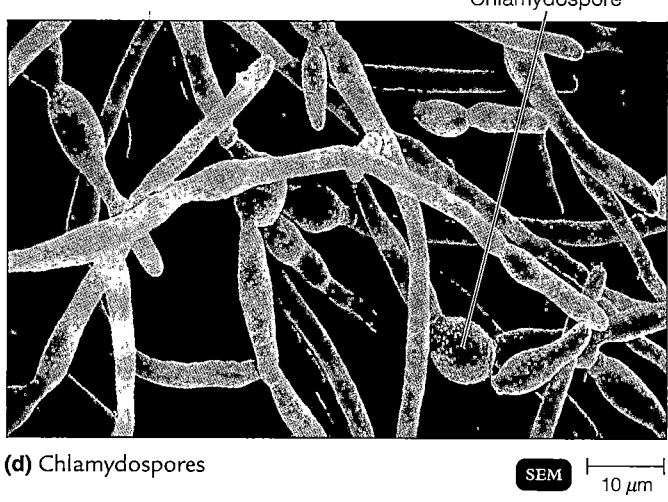

(d) Chlamydospores

FIGURE 12.5 Representative asexual spores. (a) Conidia are arranged in chains at the end of a conidiophore on this *Aspergillus flavus.* **(b)** Fragmentation of hyphae results in the formation of arthrospores in this *Coccidioides immitis.* **(c)** Blastoconidia are formed from the buds of a parent cell of *Candida albicans.*
(d) Chlamydospores are thick-walled cells within hyphae of this *C. albicans.* **(e)** Sporangiospores are formed within a sporangium (spore sac) of this *Rhizopus.*

■ **What are the green powdery structures on moldy food?**

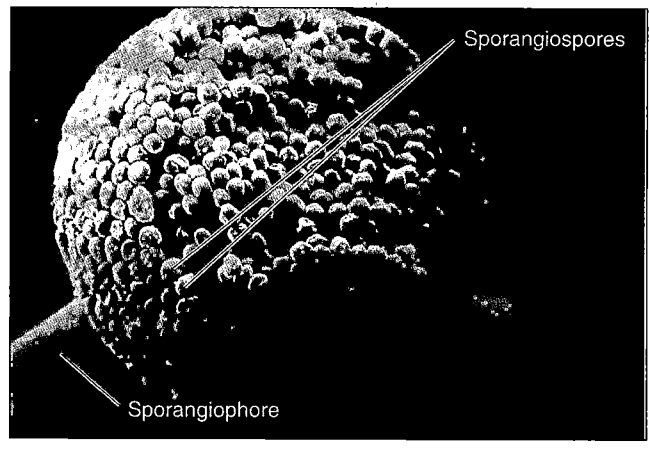

(e) Sporangiospores

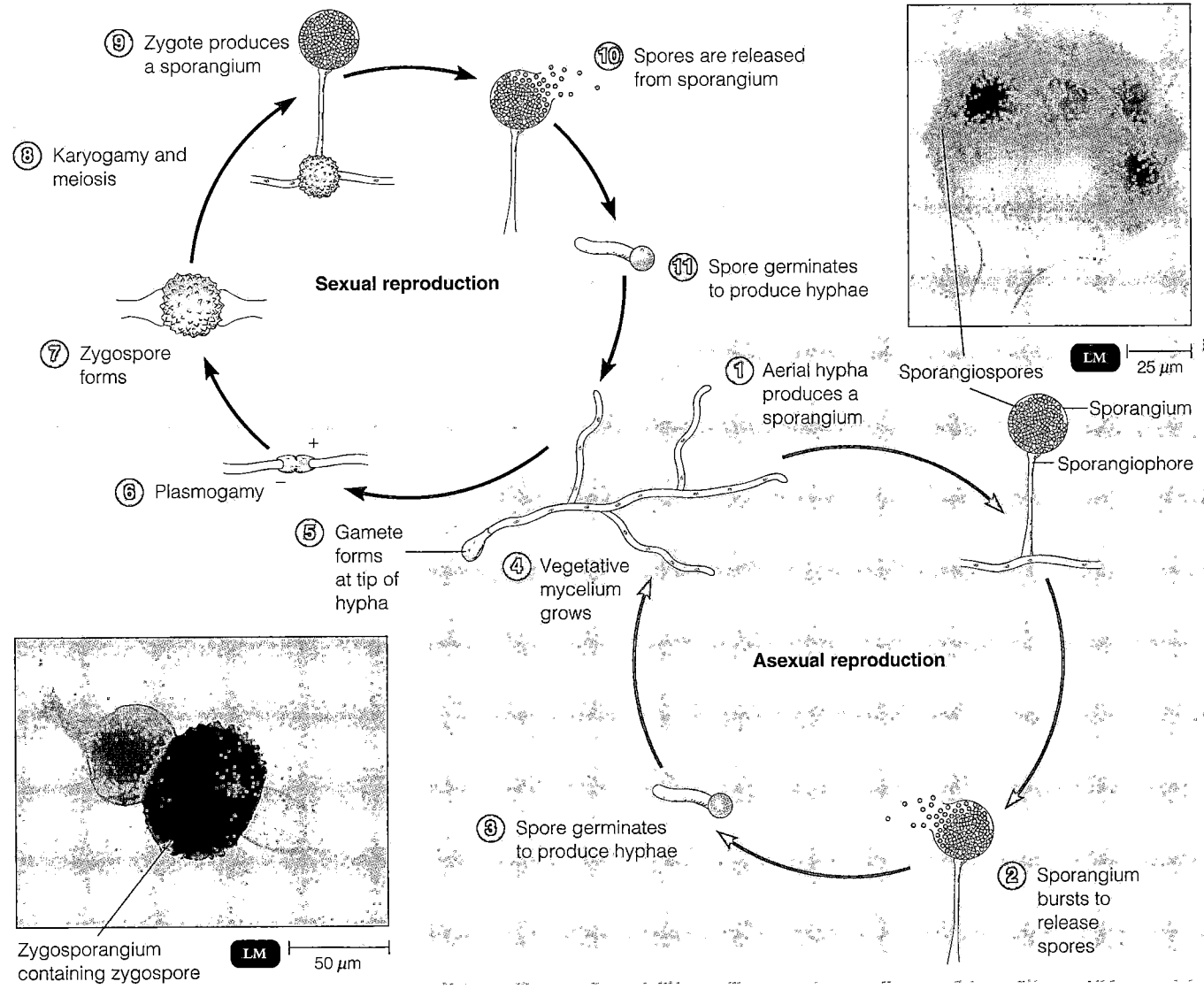

FIGURE 12.6 The life cycle of *Rhizopus*, a zygomycete. This fungus will reproduce asexually most of the time. Two opposite mating strains (designated + and −) are necessary for sexual reproduction.

■ **What is an opportunistic mycosis?**

The genera named in the following phyla include many that are readily found as contaminants in foods and in laboratory bacterial cultures. Although these genera are not all of primary medical importance, they are typical examples of their respective groups.

Zygomycota

The Zygomycota, or conjugation fungi, are saprophytic molds that have coenocytic hyphae. An example is *Rhizopus nigricans*, the common black bread mold. The asexual spores of *Rhizopus* are sporangiospores (Figure 12.6, upper right). The dark sporangiospores inside the spo-

rangium give *Rhizopus* its descriptive common name. When the sporangium breaks open, the sporangiospores are dispersed. If they fall on a suitable medium, they will germinate into a new mold thallus.

The sexual spores are zygospores. A **zygospore** is a large spore enclosed in a thick wall (Figure 12.6, lower left). This type of spore results from the fusion of the nuclei of two cells that are morphologically similar to each other.

Ascomycota

The Ascomycota, or sac fungi, include molds with septate hyphae and some yeasts. Their asexual spores are usually

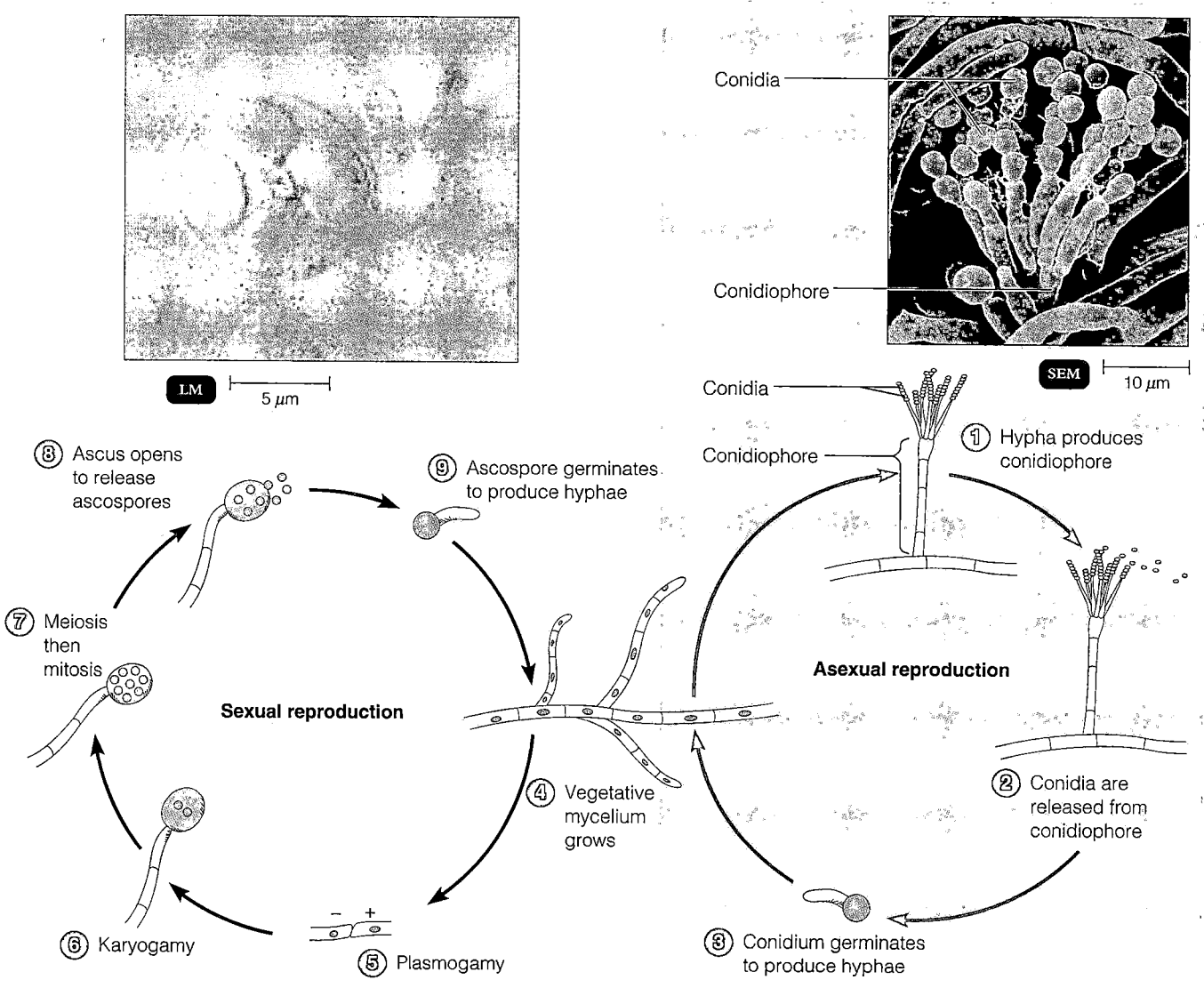

FIGURE 12.7 The life cycle of *Talaromyces,* an ascomycete. Occasionally, when two opposite mating cells from two different strains (+ and −) fuse, sexual reproduction occurs.

■ Name one ascomycete that can infect humans.

conidia produced in long chains from the conidiophore. The term *conidia* means dust, and these spores freely detach from the chain at the slightest disturbance and float in the air like dust.

An **ascospore** results from the fusion of the nuclei of two cells that can be either morphologically similar or dissimilar. These spores are produced in a saclike structure called an **ascus** (Figure 12.7, left). The members of this phylum are called sac fungi because of the ascus.

Basidiomycota

The Basidiomycota, or club fungi, also possess septate hyphae. This phylum includes fungi that produce mushrooms. **Basidiospores** are formed externally on a base pedestal called a **basidium** (Figure 12.8). (The common name of the fungus is derived from the shape of the basidium.) There are usually four basidiospores per basidium. Some of the basidiomycota produce asexual conidiospores. Representative basidiomycetes are shown in Figure 12.9, page 340. The fungi we have looked at thus far are **teleo-**

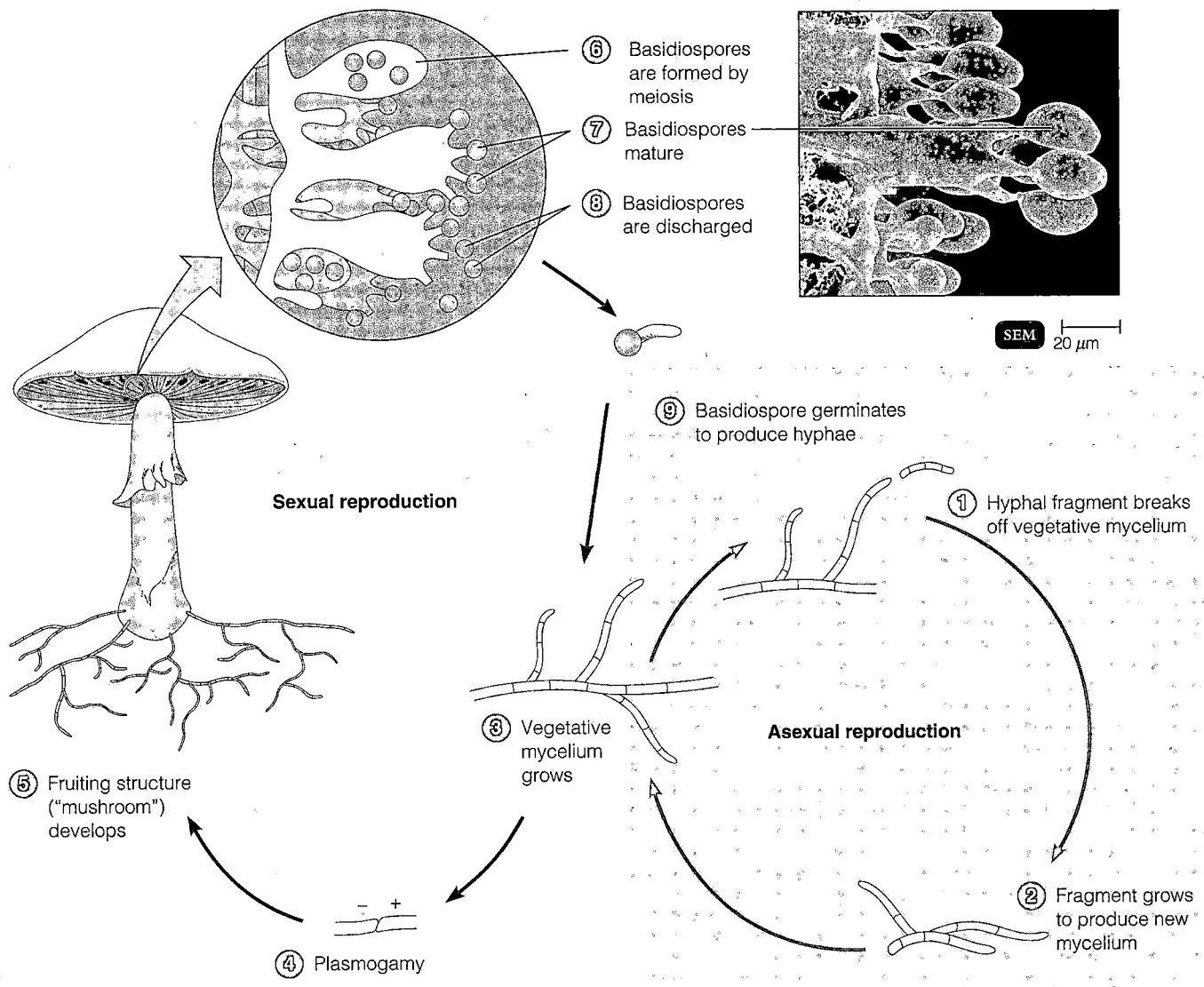

FIGURE 12.8 **A generalized life cycle of a basidiomycete.** Mushrooms appear after cells from two mating strains (+ and −) have fused.

■ On what basis are fungi classified into phyla?

morphs; that is, they produce both sexual and asexual spores. Some ascomycetes have lost the ability to reproduce sexually. These asexual fungi are called **anamorphs.** *Penicillium* is an example of an anamorph that arose from a mutation in a teleomorph. Historically, fungi whose sexual cycle had not been observed were put in a "holding category" called **Deuteromycota.** Now, mycologists are using rRNA sequencing to classify these organisms. Most of these previously unclassified deuteromycetes are anamorph phases of **Ascomycota,** and a few are basidiomycetes.

Table 12.3 on pages 342–343 lists some fungi that cause human diseases. Two generic names are given for some of the fungi because medically important fungi that are well known by their anamorph, or asexual, name are often referred to by that name.

Fungal Diseases

Any fungal infection is called a **mycosis.** Mycoses are generally chronic (long-lasting) infections because fungi

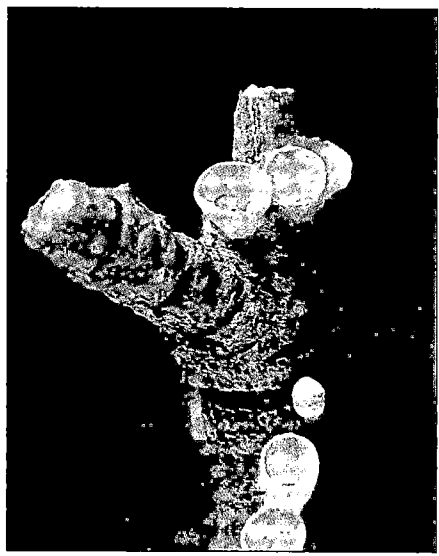

(a) *Crucibulum vulgare*

(b) *Amanita muscaria*

FIGURE 12.9 Representative basidiomycetes. (a) A bird's nest fungus (*Crucibulum vulgare*) growing on a twig. The basidiospores visible in one of the cups will pop out when the cup is hit by a raindrop. **(b)** *Amanita muscaria* grows in close association with plant roots (mycorrhiza) and produces a neurotoxin and a possible antitumor chemical.

■ What is the primary role of fungi in the ecosystem?

grow slowly. Mycoses are classified into five groups according to the degree of tissue involvement and mode of entry into the host: systemic, subcutaneous, cutaneous, superficial, or opportunistic. *Pneumocystis* is an opportunistic pathogen in individuals with compromised immune systems and is the leading cause of death in AIDS patients (see Figure 24.22 on page 677). It was first classified as a protozoan, but recent studies of its RNA indicate it is a unicellular ascomycete. In Chapter 10, we saw that fungi are related to animals. This fact makes fungal infections of humans and other animals often difficult to treat.

Systemic mycoses are fungal infections deep within the body. They are not restricted to any particular region of the body but can affect a number of tissues and organs. Systemic mycoses are usually caused by fungi that live in the soil. Inhalation of spores is the route of transmission; these infections typically begin in the lungs and then spread to other body tissues. They are not contagious from animal to human or from human to human. Two systemic mycoses, histoplasmosis and coccidioidomycosis, are discussed in Chapter 24.

Subcutaneous mycoses are fungal infections beneath the skin caused by saprophytic fungi that live in soil and on vegetation. Infection occurs by direct implantation of spores or mycelial fragments into a puncture wound in the skin.

Fungi that infect only the epidermis, hair, and nails are called **dermatophytes,** and their infections are called

dermatomycoses or **cutaneous mycoses** (see Figure 21.14). Dermatophytes secrete keratinase, an enzyme that degrades **keratin,** a protein found in hair, skin, and nails. Infection is transmitted from human to human or from animal to human by direct contact or by contact with infected hairs and epidermal cells (as from barber shop clippers or shower room floors).

The fungi that cause **superficial mycoses** are localized along hair shafts and in superficial (surface) epidermal cells. These infections are prevalent in tropical climates.

An **opportunistic pathogen** is generally harmless in its normal habitat but can become pathogenic in a host who is seriously debilitated or traumatized, who is under treatment with broad-spectrum antibiotics, or whose immune system is suppressed by drugs or by an immune disorder. AIDS patients are quite susceptible to opportunistic pathogens.

An example of an opportunistic pathogen is the fungus *Stachybotrys* (sta'ke-bo-tris), which normally grows on cellulose found in dead plants but in recent years has been found growing on water-damaged walls of homes. Its toxic spores can cause fatal pulmonary hemorrhage in infants. Mucormycosis is an opportunistic mycosis caused by *Rhizopus* and *Mucor* (mū-kôr); the infection occurs mostly in patients with diabetes mellitus, with leukemia, or undergoing treatment with immunosuppressive drugs. Aspergillosis is also an opportunistic mycosis; it is caused by *Aspergillus* (see Figure 12.2b). This disease occurs in people who have debilitating lung diseases or cancer and have inhaled *Aspergillus* spores. **Yeast infection,** or candidiasis, is most frequently caused by *Candida albicans* and may occur as vulvovaginal candidiasis or thrush, a mucocutaneous candidiasis. Candidiasis frequently occurs in newborns, in people with AIDS, and in people being treated with broad-spectrum antibiotics (see Figure 21.17 on page 594).

Some fungi cause disease by producing toxins. These toxins are discussed in Chapter 15.

Economic Effects of Fungi

Learning Objective

- *Identify two beneficial and two harmful effects of fungi.*

Fungi have been used in biotechnology for many years. *Aspergillus niger,* for example, has been used to produce citric acid for foods and beverages since 1914. The yeast *Saccharomyces cerevisiae* is used to make bread and wine. It is also genetically engineered to produce a variety of proteins, including hepatitis B vaccine. *Saccharomyces* and another yeast, *Torulopsis,* are used as protein supplements for humans and cattle. *Trichoderma* is used commercially to produce the enzyme cellulase, which is used to remove plant cell walls to produce a clear fruit juice. When the anticancer drug taxol, which is produced by yew trees, was discovered, there was concern that the yew forests of the U.S. Northwest coast would be decimated to harvest the drug. However, in 1993, Andrea and Donald Stierle saved the yews by discovering that the fungus *Taxomyces* also produces taxol.

Fungi are used as biological controls of pests. In 1990, the fungus *Entomorphaga* unexpectedly proliferated and killed gypsy moths that were destroying trees in the eastern United States. Scientists are investigating whether this fungus can be used in place of chemical insecticides. Annually, 25–50% of harvested fruits and vegetables are ruined by fungi. Chemical fungicides cannot be used to prevent this decay because of safety and environmental concerns. However, another fungus, *Candida oleophila,* can be and is used to prevent fungal growth on harvested fruits. This process of biocontrol works because the *C. oleophila* grows on the fruit surface before spoilage fungi grow.

In contrast to these beneficial effects, fungi can have undesirable effects for industry and agriculture because of their nutritional adaptations. As most of us have observed, mold spoilage of fruits, grains, and vegetables is relatively common, but bacterial spoilage of such foods is not. There is little moisture on the unbroken surfaces of such foods, and the interiors of fruits are too acidic for many bacteria to grow there. Jams and jellies also tend to be acidic, and they have a high osmotic pressure from the sugars they contain. These factors all discourage bacterial growth but readily support the growth of molds. A paraffin layer on top of a jar of homemade jelly helps deter mold growth because molds are aerobic and the paraffin layer keeps out the oxygen. However, fresh meats and certain other foods are such good substrates for bacterial growth that bacteria not only will outgrow molds but also will actively suppress mold growth in these foods.

In Ireland during the mid-1800s, 1 million people died when the country's potato crop failed. The fungus that caused the great potato blight, *Phytophthora infestans* (fī-tof'thô-rä in-fes'tans), was one of the first microorganisms to be associated with a disease. Today, *Phytophthora* infects soybeans, potatoes, and cocoa.

The spreading chestnut tree, of which Longfellow wrote, no longer grows in the United States except in a few widely isolated locations; a fungal blight killed virtually all of them. This blight was caused by the ascomycete *Cryphonectria parasitica* (kri-fō-nek'trē-ä par-ä-si'ti-kä), which was introduced from China around 1904. The fungus allows the tree roots to live and put forth shoots regularly, but then it kills the shoots just as regularly. *Cryphonectria*-resistant chestnuts are being developed. Another imported fungal plant disease is Dutch elm disease, caused by *Ceratocystis ulmi* (sē-rä-tō-sis'tis ul'mē). Carried from tree to tree by a bark beetle, the fungus blocks the afflicted tree's circulation. The disease has devastated the U.S. elm population.

Lichens

Learning Objectives

- *List the distinguishing characteristics of lichens, and describe their nutritional needs.*
- *Describe the roles of the fungus and the alga in a lichen.*

A **lichen** is a combination of a green alga (or a cyanobacterium) and a fungus. Lichens are placed in the Kingdom Fungi and are classified according to the fungal partner, most often an ascomycete. The two organisms exist in a *mutualistic* relationship, in which each partner benefits. The lichen is very different from either the alga or fungus growing alone, and if the partners are separated, the lichen no longer exists. Approximately 13,500 species of lichens occupy quite diverse habitats. Because they can inhabit areas in which neither fungi nor algae could survive alone, lichens are often the first life forms to colonize newly exposed soil or rock. Lichens secrete organic acids

table 12.3   Characteristics of Some Pathogenic Fungi

Phylum	Growth Characteristics	Asexual Spore Types	Human Pathogens
Zygomycota	Nonseptate hyphae	Sporangiospores	*Rhizopus* *Mucor*
Ascomycota	Dimorphic	Conidia	*Aspergillus* *Blastomyces** (*Ajellomyces***) *dermatitidis* *Histoplasma** (*Ajellomyces***) *capsulatum*
	Septate hyphae, strong affinity for keratin	Conidia Arthrospores	*Microsporum* *Trichophyton** (*Arthroderma***)
Anamorphs		Conidia	*Epidermophyton*
	Dimorphic	Conidia Arthrospores	*Sporothrix schenckii, Stachybotrys* *Coccidioides immitis*
	Yeastlike, pseudohyphae	Chlamydospores	*Candida albicans*
	Unknown	Unknown	*Pneumocystis*
Basidiomycota	Septate hyphae; includes rusts and smuts, and plant pathogens; yeastlike encapsulated cells	Conidia	*Cryptococcus neoformans** (*Filobasidiella*)**

*Anamorph name.
**Teleomorph name.

that chemically weather rock, and they accumulate nutrients needed for plant growth. Also found on trees, concrete structures, and rooftops, lichens are some of the slowest-growing organisms on Earth.

Lichens can be grouped into three morphologic categories (Figure 12.10a on page 344). *Crustose lichens* grow flush or encrusting onto the substratum, *foliose lichens* are more leaflike, and *fruticose lichens* have fingerlike projections. The lichen's thallus, or body, forms when fungal hyphae grow around algal cells to become the **medulla** (Figure 12.10b). Fungal hyphae project below the lichen body to form **rhizines,** or holdfasts. Fungal hyphae also form a **cortex,** or protective covering, over the algal layer and sometimes under it as well. After incorporation into a lichen thallus, the alga continues to grow, and the growing hyphae can incorporate new algal cells.

When the algal partner is cultured separately in vitro, about 1% of the carbohydrates produced during photosynthesis are released into the culture medium; however, when the alga is associated with a fungus, the algal plasma membrane is more permeable, and up to 60% of the products of photosynthesis are released to the fungus or are found as end-products of fungus metabolism. The fungus clearly benefits from this association. The alga,

while giving up valuable nutrients, is in turn compensated; it receives from the fungus both protection from desiccation (cortex) and attachment (holdfast).

Lichens had considerable economic importance in ancient Greece and other parts of Europe as dyes for clothing. Usnic acid from *Usnea* is used as an antimicrobial agent in China. Erythrolitmin, the dye used in litmus paper to indicate changes in pH, is extracted from a variety of lichens. Some lichens or their acids can cause allergic contact dermatitis in humans.

Populations of lichens readily incorporate cations (positively charged ions) into their thalli. Therefore, the concentrations and types of cations in the atmosphere can be determined by chemical analyses of lichen thalli. In addition, the presence or absence of species that are quite sensitive to pollutants can be used to ascertain air quality. A 1985 study in the Cuyahoga Valley in Ohio revealed that 81% of the 172 lichen species that were present in 1917 were gone. Because this area is severely affected by air pollution, the inference is that air pollutants, primarily sulfur dioxide (the major contributor to acid rain), caused the death of sensitive species.

Lichens are the major food for tundra herbivores such as caribou and reindeer. After the 1986 Chernobyl nu-

table 12.3   *(continued)*

Habitat	Type of Mycosis	Clinical Notes	Page Reference
Ubiquitous	Systemic	Opportunistic pathogen	678
Ubiquitous	Systemic	Opportunistic pathogen	678
Ubiquitous	Systemic	Opportunistic pathogen	678
Unknown	Systemic	Inhalation	678
Soil	Systemic	Inhalation	674
Soil, animals	Cutaneous	Tinea capitis (ringworm)	593
Soil, animals	Cutaneous	Tinea pedis (athlete's foot)	593
Soil, humans	Cutaneous	Tinea cruris (jock itch), tinea unguium (of fingernails or toenails)	593
Soil	Subcutaneous	Puncture wound	594
Soil	Systemic	Inhalation	675
Human normal microbiota	Cutaneous, systemic, mucocutaneous	Opportunistic pathogen	594
Ubiquitous	Systemic	Opportunistic pathogen	676
Soil, bird feces	Systemic	Inhalation	616

clear disaster, 70,000 reindeer in Lapland that had been raised for food had to be destroyed because of high levels of radiation. The lichens on which the reindeer fed had absorbed radioactive cesium-137, which had spread in the air.

Algae

Algae are familiar as the large brown kelp in coastal waters, the green scum in a puddle, and the green stains on soil or on rocks. A few algae are responsible for food poisonings. Some algae are unicellular; others form chains of cells (are filamentous); and a few have thalli.

Algae are mostly aquatic, although some are found in soil or on trees when sufficient moisture is available there. Unusual algal habitats include the hair of both the sedentary South American sloth and the polar bear. Water is necessary for physical support, reproduction, and the diffusion of nutrients. Generally, algae are found in cool temperate waters, although the large floating mats of the brown alga *Sargassum* (sär-gas'sum) are found in the subtropical Sargasso Sea. Some species of brown algae grow in antarctic waters.

Characteristics of Algae

Learning Objective

■ List the defining characteristics of algae.

Algae are relatively simple eukaryotic photoautotrophs that lack the tissues (roots, stem, and leaves) of plants. The identification of unicellular and filamentous algae requires microscopic examination. Most algae are found in the ocean. Their locations depend on the availability of appropriate nutrients, wavelengths of light, and surfaces on which to grow. Probable locations for representative algae are shown in Figure 12.11a.

Vegetative Structures

The body of a multicellular alga is called a thallus. Thalli of the larger multicellular algae, those commonly called seaweeds, consist of branched **holdfasts** (which anchor the alga to a rock), stemlike and often hollow **stipes,** and leaflike **blades** (Figure 12.11b). The cells covering the thallus can carry out photosynthesis. The thallus lacks the conductive tissue (xylem and phloem) characteristic of vascular plants; algae absorb nutrients from the water over their entire surface. The stipe is not lignified or woody, so

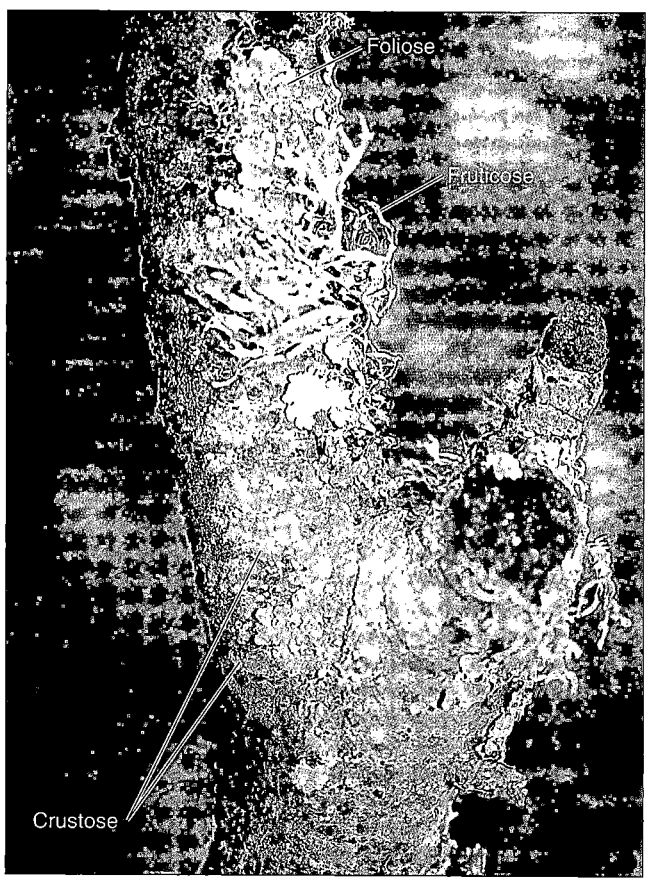

(a)

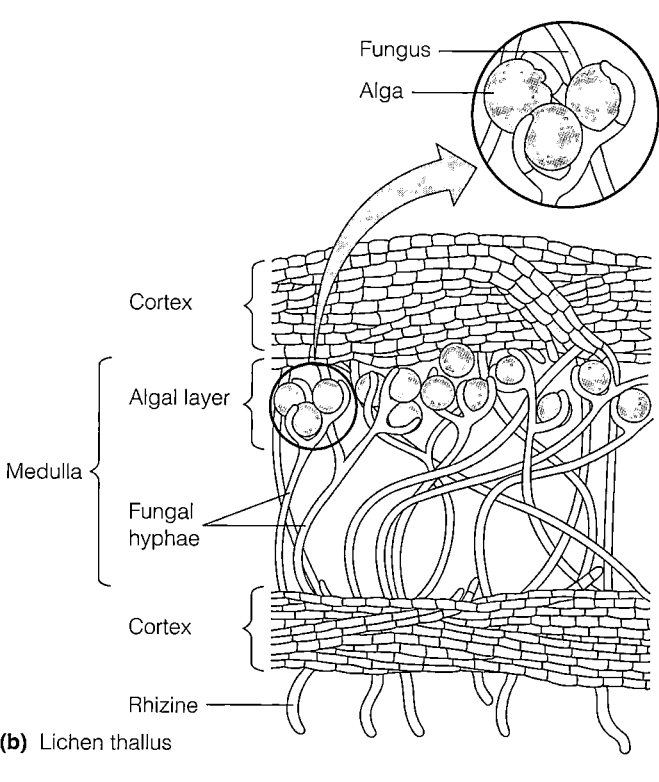

Fungus

Alga

Cortex

Algal layer

Medulla

Fungal hyphae

Cortex

Rhizine

(b) Lichen thallus

FIGURE 12.10 Lichens. (a) Three types of lichens. **(b)** The medulla is composed of fungal hyphae surrounding the algal layer. The protective cortex is a layer of fungal hyphae that covers the surface and sometimes the bottom of the lichen.

■ In what ways are lichens unique?

it does not offer the support of a plant's stem; instead, the surrounding water supports the algal thallus; some algae are also buoyed by a floating, gas-filled bladder called a *pneumatocyst*.

Life Cycle

All algae can reproduce asexually. Multicellular algae with thalli and filamentous forms can fragment; each piece is capable of forming a new thallus or filament. When a unicellular alga divides, its nucleus divides (mitosis), and the two nuclei move to opposite parts of the cell. The cell then divides into two complete cells (cytokinesis).

Sexual reproduction occurs in algae (Figure 12.12 on page 346). In some species, asexual reproduction may occur for several generations and then, under different conditions, the same species reproduce sexually. Other species alternate generations so that the offspring resulting from sexual reproduction reproduce asexually, and the next generation then reproduces sexually.

Nutrition

Algae are photoautotrophs and are therefore found throughout the photic (light) zone of bodies of water. Chlorophyll *a* (a light-trapping pigment) and accessory pigments involved in photosynthesis are responsible for the distinctive colors of many algae.

Algae are classified according to their structures, pigments, and other qualities (Table 12.4 on page 346). Following are descriptions of some divisions of algae.

Selected Phyla of Algae

Learning Objective

■ *List the outstanding characteristics of the five phyla of algae discussed in this chapter.*

The *brown algae,* or kelp, are macroscopic; some reach lengths of 50 m (see Figure 12.11b). Most brown algae are found in coastal waters. Brown algae have a phenom-

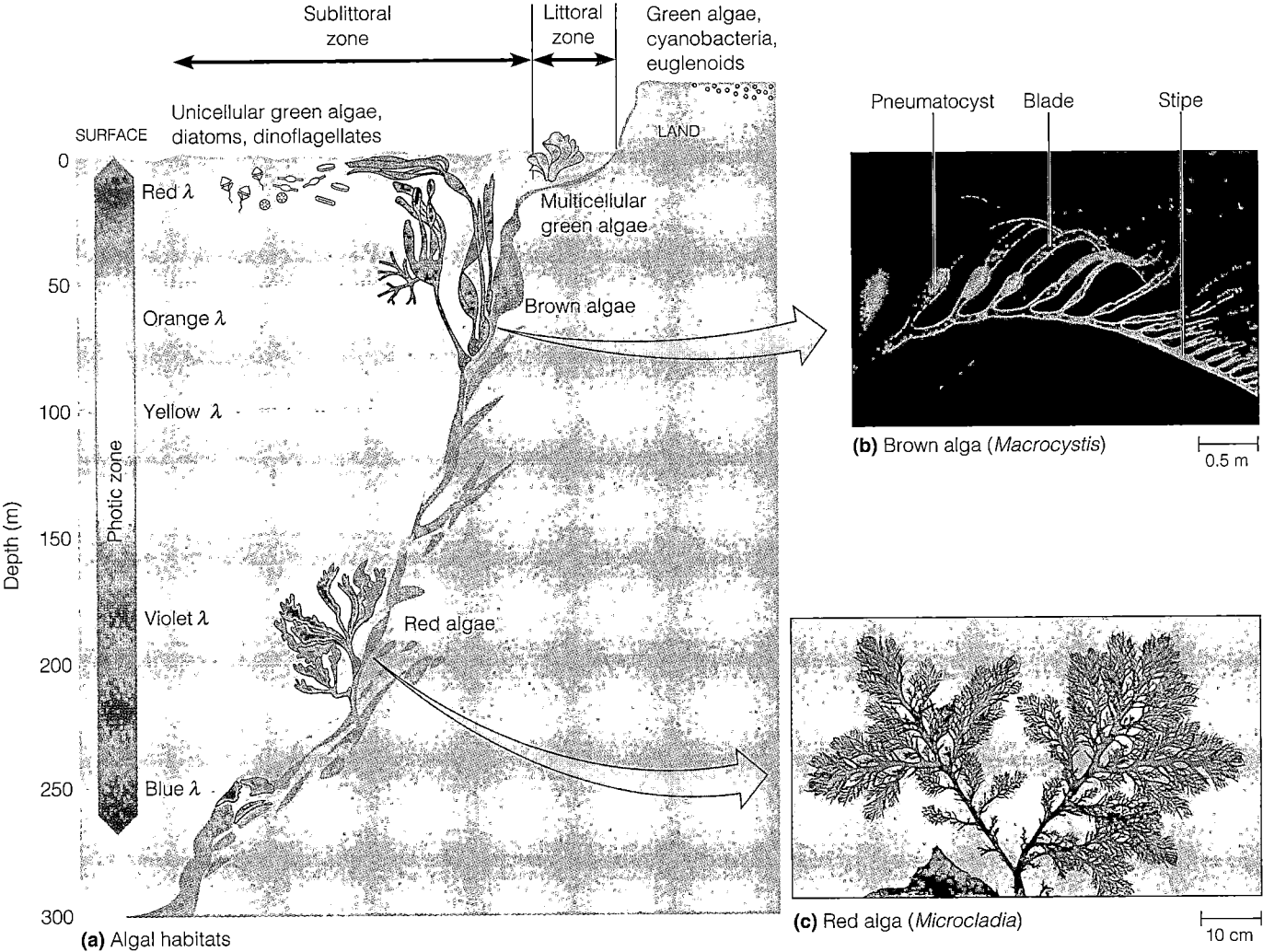

(a) Algal habitats

(b) Brown alga (*Macrocystis*) |——————| 0.5 m

(c) Red alga (*Microcladia*) |——————| 10 cm

FIGURE 12.11 Algae and their habitats. (a) Although unicellular and filamentous algae can be found on land, they frequently exist in marine and freshwater environments as plankton. Multicellular green, brown, and red algae require a suitable attachment site, adequate water for support, and light of the appropriate wavelengths. **(b)** *Macrocystis porifera*, a brown alga. The hollow stipe and gas-filled pneumatocysts hold the thallus upright to ensure that sufficient sunlight is received for growth. **(c)** *Microcladia*, a red alga. The delicately branched red algae get their color from phycobiliprotein accessory pigments.

■ **What red alga is toxic for humans?**

---

enal growth rate. Some grow at rates exceeding 20 cm per day and therefore can be harvested regularly. **Algin,** a thickener used in many foods (such as ice cream and cake decorations), is extracted from their cell walls. Algin is also used in the production of a wide variety of nonfood goods, including rubber tires and hand lotion. The brown alga *Laminaria japonica* is used to induce vaginal dilation before surgical entry into the uterus through the vagina.

Most *red algae* have delicately branched thalli and can live at greater ocean depths than other algae (Figure 12.11c). The thalli of a few red algae form crustlike coat-

ings on rocks and shells. The red pigments enable red algae to absorb the blue light that penetrates deepest into the ocean. The agar used in microbiological media is extracted from many red algae. Another gelatinous material, carrageenan, comes from a species of red algae commonly called Irish moss. Carrageenan and agar can be a thickening ingredient in evaporated milk, ice cream, and pharmaceutical agents. *Gracilaria* species, which grow in the Pacific Ocean, are used by humans for food. However, members of this genus can produce a lethal toxin.

M M W R

Fish Killer Emerges as Human Pathogen

The periodic proliferation of dinoflagellates has long been linked to human death. When the dinoflagellates' toxins become concentrated in animals that are eaten by humans and fish, deaths due to these dinoflagellates are not uncommon. However, the number of these algal blooms has increased during the last 50 years. Researchers suggest the increase is due to several factors:

- Increases in nitrogen and phosphorous from fertilizers and animal feces in runoff from fields.

- Ships' introducing algae into new waters with their ballast water

- Dams that affect the flow and nutrient levels of rivers, which in turn alters the chemistry of the river's estuary

- Global warming and increased ultraviolet radiation that may affect algal growth

In 1991, a billion fish died in North Carolina's Neuse Estuary, and in 1997, thousands of fish died in Chesapeake Bay estuaries. The 1997 fish kill was unusual because a woman became sick after only handling a fish from the poisoned waters. The CDC calls this possible estuary-associated syndrome (PEAS), and it can include memory loss, confusion, acute skin burning, headaches, skin rash, eye irritation, upper respiratory irritation, muscle cramps, and gastrointestinal complaints (e.g., nausea, vomiting, diarrhea, and abdominal cramps). Elevated liver enzyme levels in the blood of one patient also suggest liver damage.

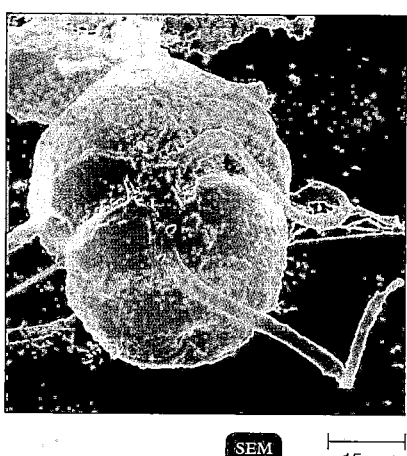

SEM |———————|
                15 μm

The culprit is a dinoflagellate, named *Pfiesteria* (see the figure), that was first discovered in 1988 in fish cultures at the North Carolina School of Veterinary Medicine by JoAnn Burkholder. Since then it has been implicated as the cause of half of the major fish kills from Delaware Bay to the Gulf Coast. The dinoflagellate was named in honor of the dinoflagellate biologist Lois Pfiester.

Most of the information on human disease was obtained from researchers working with toxic cultures of the dinoflagellates. They inadvertently discovered that *Pfiesteria* is unusual in that its toxic effects can be contracted by inhaling cultures containing 2000 or more cells per milliliter. It is unclear whether people exposed to *Pfiesteria* while swimming or boating are at risk for developing the illness. PEAS is not infectious and has not been associated with eating

fish or mollusks caught in waters where *Pfiesteria* has been found. CDC recommends avoiding areas with large numbers of diseased, dying, or dead fish.

The CDC is conducting multistate surveillance and epidemiologic studies for PEAS and has established that the following conditions may represent adverse consequences to *Pfiesteria* exposure.

1. 20% of a sample of 50 fish with *Pfiesteria* lesions or toxicity

2. A fish kill involving fish with *Pfiesteria* lesions

3. A fish kill involving fish without *Pfiesteria* lesions if there is no alternative reason for the fish kill

LIFE CYCLE

Pfiesteria has a complex life cycle that may include as many as 24 stages. Like other dinoflagellates, *Pfiesteria* can photosynthesize—but unlike other dinoflagellates, it must steal chloroplasts from algae. Without stolen chloroplasts, amoeboid and flagellated forms of *Pfiesteria* feed on bacteria, algae, and small animals in the water column. When large numbers of fish are in an area, their excreta causes *Pfiesteria* to become toxic. The excreted toxin causes skin ulcers, hemorrhage, and death to the fish. *Pfiesteria* then feed on the organic matter sloughed off from the dying fish. When food is depleted, the *Pfiesteria* encyst into a dormant stage.

SOURCE: Adapted from *MMWR* 48(18): 381–382 (5/14/99).

thrive in high concentrations of organic materials that exist in sewage or industrial wastes. When algae die, the decomposition of the large numbers of cells associated with an algal bloom depletes the level of dissolved oxygen in the water. (This phenomenon is discussed in Chapter 27.)

Much of the world's petroleum was formed from diatoms and other planktonic organisms that lived several

million years ago. When such organisms died and were buried by sediments, the organic molecules they contained did not decompose to be returned to the carbon cycle as CO_2. Heat and pressure resulting from the Earth's geologic movements altered the oil stored in the cells, as well as the cell membranes. Oxygen and other elements were eliminated, leaving a residue of hydrocarbons in the form of petroleum and natural gas deposits.

Many unicellular algae are symbionts in animals. The giant clam *Tridacna* (trī-dak'nä) has evolved special organs that host dinoflagellates. As the clam sits in shallow water, the algae proliferate in these organs when they are exposed to the sun. The algae release glycerol into the clam's bloodstream, thus supplying the clam's carbohydrate requirement. In addition, evidence suggests that the clam gets essential proteins by phagocytizing old algae.

Protozoa

Protozoa are unicellular, eukaryotic chemoheterotrophic organisms. Among the protozoa are many variations on this cell structure, as we shall see. Protozoa inhabit water and soil. The feeding and growing stage, or **trophozoite,** feeds upon bacteria and small particulate nutrients. Some protozoa are part of the normal microbiota of animals. Of the nearly 20,000 species of protozoa, relatively few cause disease.

Characteristics of Protozoa

Learning Objective

■ *List the defining characteristics of protozoa.*

The term *protozoan* means "first animal," which generally describes its animal-like nutrition. In addition to getting food, a protozoan must reproduce, and parasitic species must be able to get from one host to another.

Life Cycle

Protozoa reproduce asexually by fission, budding, or schizogony. **Schizogony** is multiple fission; the nucleus undergoes multiple divisions before the cell divides. After many nuclei are formed, a small portion of cytoplasm concentrates around each nucleus, and then the single cell separates into daughter cells.

Sexual reproduction has been observed in some protozoa. The ciliates, such as *Paramecium,* reproduce sexually by **conjugation** (Figure 12.15), which is very different from the bacterial process of the same name (see Figure 8.26 on page 237). During protozoan conjugation, two cells fuse, and a haploid nucleus (the micronucleus) from each cell migrates to the other cell. This haploid micronucleus fuses with the haploid macronucleus within the cell. The parent cells separate, each now a fertilized cell. When the cells later divide, they produce daughter cells with recombined DNA. Some protozoa produce **gametes (gametocytes),** haploid sex cells. During reproduction, two gametes fuse to form a diploid zygote.

Encystment. Under certain adverse conditions, some protozoa produce a protective capsule called a **cyst.** A

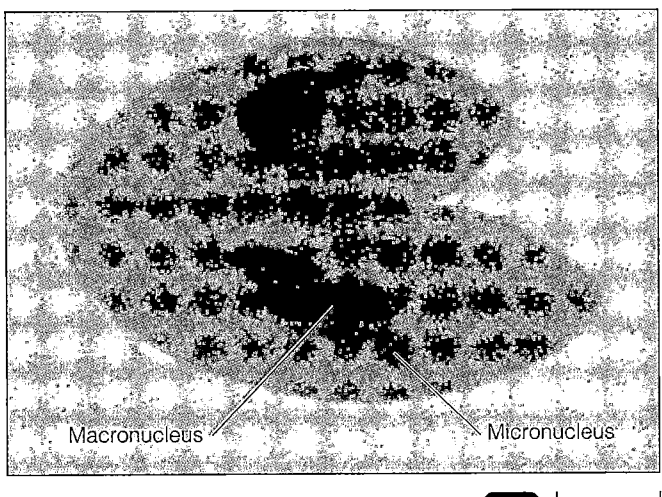

FIGURE 12.15 **Conjugation in the ciliate protozoan *Paramecium*.** Sexual reproduction in ciliates is by conjugation. Each cell has two nuclei: a micronucleus and a macronucleus. The micronucleus is haploid and is specialized for conjugation. One micronucleus from each cell will migrate to the other cell during conjugation. Both cells will then go on to produce two daughter cells with recombined DNA.

■ Some protozoa can reproduce sexually as well as asexually.

cyst permits the organism to survive when food, moisture, or oxygen are lacking, when temperatures are not suitable, or when toxic chemicals are present. A cyst also enables a parasitic species to survive outside a host. This is important because parasitic protozoa may have to be excreted from one host in order to get to a new host. The cyst form in members of the phylum Apicomplexa is called an **oocyst.** It is a reproductive structure in which new cells are produced asexually.

Nutrition

Protozoa are mostly aerobic heterotrophs, although many intestinal protozoa are capable of anaerobic growth. Two chlorophyll-containing groups, dinoflagellates and euglenoids, are often studied with algae.

All protozoa live in areas with a large supply of water. Some protozoa transport food across the plasma membrane. However, some have a protective covering, or *pellicle,* and thus require specialized structures to take in food. Ciliates take in food by waving their cilia toward a mouthlike opening called a **cytostome.** Amoebas engulf food by surrounding it with pseudopods and phagocytizing it. In all protozoa, digestion takes place in membrane-enclosed **vacuoles,** and waste may be eliminated through the plasma membrane or through a specialized **anal pore.**

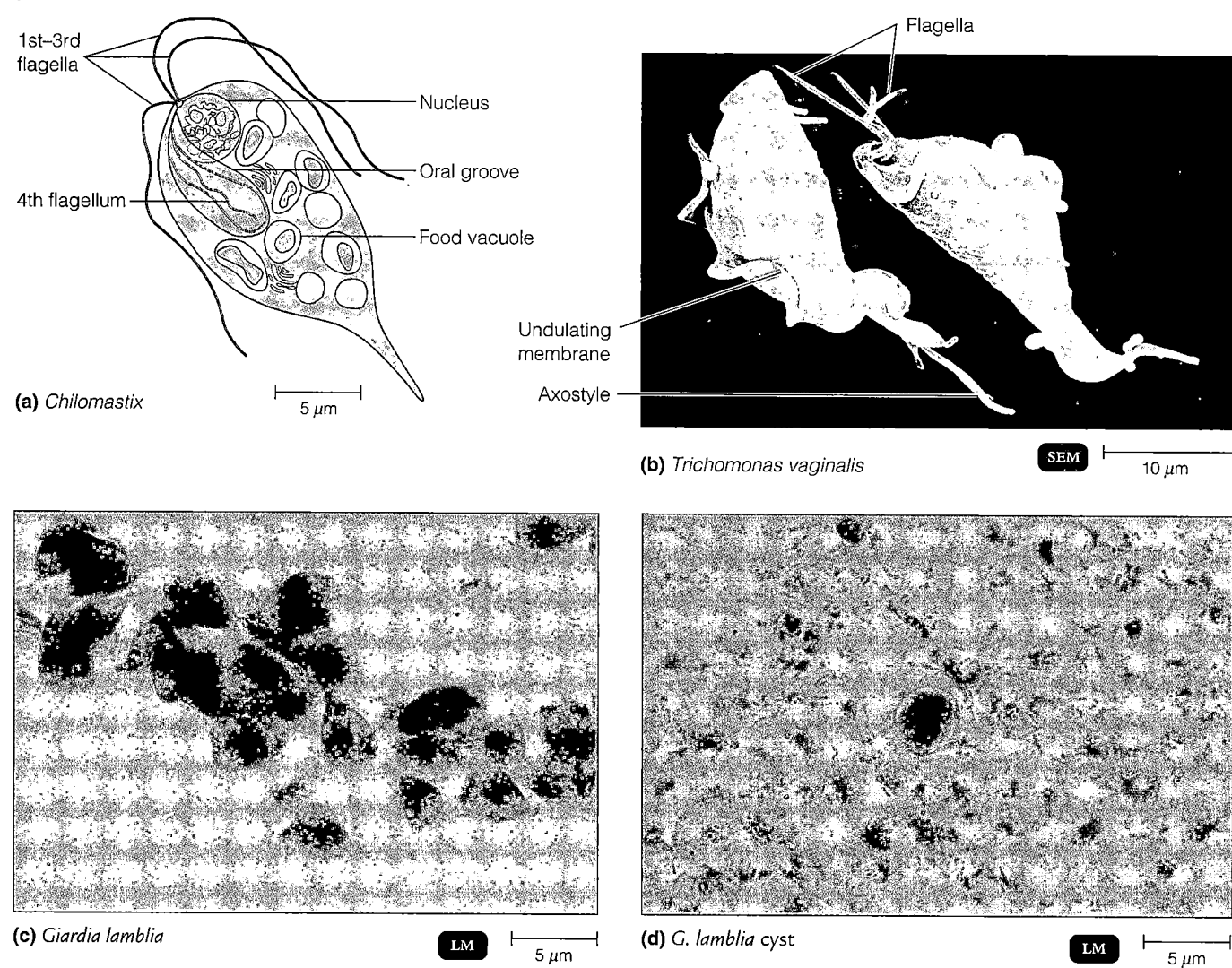

FIGURE 12.16 Archaezoa. (**a**) *Chilomastix.* This flagellate, found in the human intestine, may be mildly pathogenic. The cysts survive for months outside a human host. The fourth flagellum is used to move food into the oral groove, where food vacuoles are formed. (**b**) *Trichomonas vaginalis.* This flagellate causes urinary and genital tract infections. Notice the small undulating membrane. This flagellate does not have a cyst stage. (**c**) *Giardia lamblia.* The trophozoite of this intestinal parasite has eight flagella and two prominent nuclei, giving it a distinctive appearance. (**d**) The *G. lamblia* cyst provides protection from the environment before it is ingested by a new host.

■ How do archaezoans obtain energy without mitochondria?

Medically Important Phyla of Protozoa

Learning Objectives

■ *Describe the outstanding characteristics of the seven phyla of protozoa discussed in this chapter, and give an example of each.*

■ *Differentiate an intermediate host from a definitive host.*

The biology of protozoa is discussed in this chapter. Descriptions of diseases caused by protozoa are in Part Four.

Protozoa are a large and diverse group. Current schemes of classifying protozoan species into phyla are based on rRNA sequencing. Researchers have begun sorting out groups within the protists based on their evolutionary history; that is, members in a group derive from a single ancestor. At present, the following groups are phyla in the Kingdom Protista. As more information is obtained, some of these groups may be classified as kingdoms in the Domain Eukarya.

Archaezoa

The **Archaezoa** are eukaryotes that lack mitochondria. The absence of mitochondria suggests that they may have evolved before the endosymbiotic event that gave rise to mitochondria, or that they have lost the ability to make mitochondria. Many archaezoans live as symbionts in the digestive tracts of animals. Archaezoans are typically spindle-shaped, with flagella projecting from the front end (Figure 12.16a). Most have two or more flagella, which move in a whiplike manner that pulls the cells through their environment.

An example of an archaezoan that is a human parasite is *Trichomonas vaginalis* (trik-ō-mōn'as va-jin-al'is), shown in Figure 12.16b. Like some other flagellates, *T. vaginalis* has an **undulating membrane,** which consists of a membrane bordered by a flagellum. *T. vaginalis* does not have a cyst stage and must be transferred from host to host quickly before desiccation occurs. *T. vaginalis* is found in the vagina and in the male urinary tract. It is usually transmitted by sexual intercourse but can also be transmitted by toilet facilities or towels.

Another parasitic archaezoan is *Giardia lamblia* (jē-är'dē-ä lam'lē-ä) (see Figures 12.16c and 25.16 for the vegetative trophozoite and Figure 12.16d for the cyst stage). The parasite is found in the small intestine of humans and other mammals. It is excreted in the feces as a cyst and survives in the environment before being ingested by the next host. Diagnosis of giardiasis, the disease caused by *G. lamblia,* is often based on the identification of cysts in feces.

Microsporidia

Microsporidia, like the Archaezoa, are unusual eukaryotes because they lack mitochondria. Microsporidia do not have microtubules (see Chapter 4, page 100), and they are obligate intracellular parasites. Microsporidial protozoa have been reported since 1984 to be the cause of a number of human diseases, including chronic diarrhea and keratoconjunctivitis (inflammation of the conjunctiva near the cornea), most notably in AIDS patients.

Rhizopoda

The **rhizopoda** or amoebas move by extending blunt, lobelike projections of the cytoplasm called **pseudopods** (Figure 12.17a). Any number of pseudopods can flow from one side of the amoeba, and the rest of the cell will flow toward the pseudopods.

Entamoeba histolytica (en-tä-mē'bä his-tō-lī'ti-kä), the causative agent of amoebic dysentery, is the only pathogenic amoeba found in the human intestine (Figure 12.17b). The primary food of *E. histolytica* is red blood cells. *E. histolytica* is transmitted between humans through ingestion of the cysts that are excreted in the feces of the infected person. *Acanthamoeba* growing in water, including tap water, can infect the cornea and cause blindness (see Figure 21.20 on page 597).

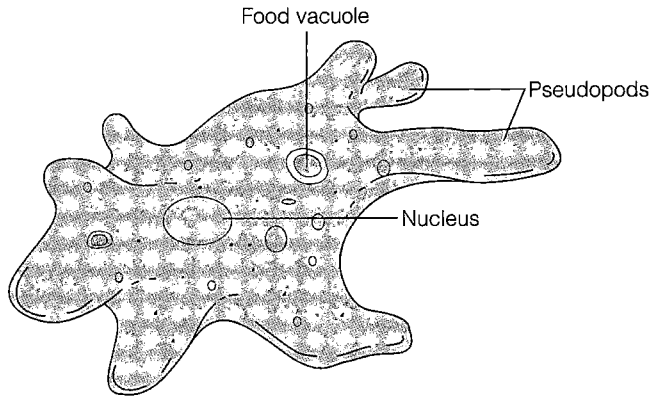

(a) *Amoeba proteus*

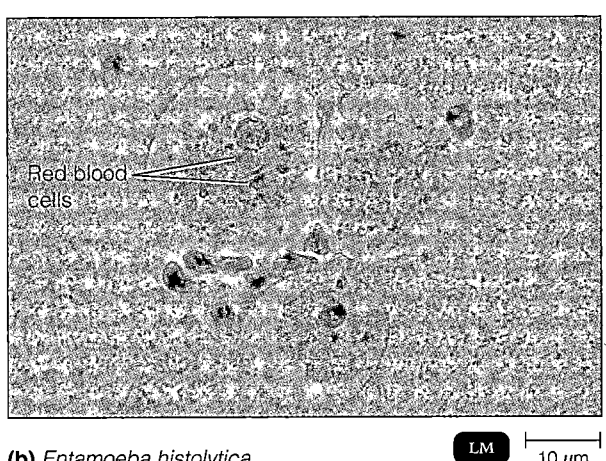

(b) *Entamoeba histolytica*   LM   10 μm

FIGURE 12.17   Rhizopoda. (a) To move and to engulf food, amoebas (such as this *Amoeba proteus*) extend cytoplasmic structures called pseudopods. Food vacuoles are created when pseudopods surround food and bring it into the cell. **(b)** *Entamoeba histolytica.* The presence of ingested red blood cells is diagnostic for *Entamoeba.*

■ **How do amoebic dysentery and bacillary dysentery differ?**

Apicomplexan

The **Apicomplexa** are not motile in their mature forms and are obligate intracellular parasites. Apicomplexans are characterized by the presence of a complex of special organelles at the apexes (tips) of their cells (hence the phylum name). The organelles in these apical complexes contain enzymes that penetrate the host's tissues.

Apicomplexans have a complex life cycle that involves transmission between several hosts. An example of an apicomplexan is *Plasmodium* (plaz-mō'dē-um), the causative agent of malaria. The complex life cycle makes it difficult to develop a vaccine against malaria (see the box in Chapter 18, page 505).

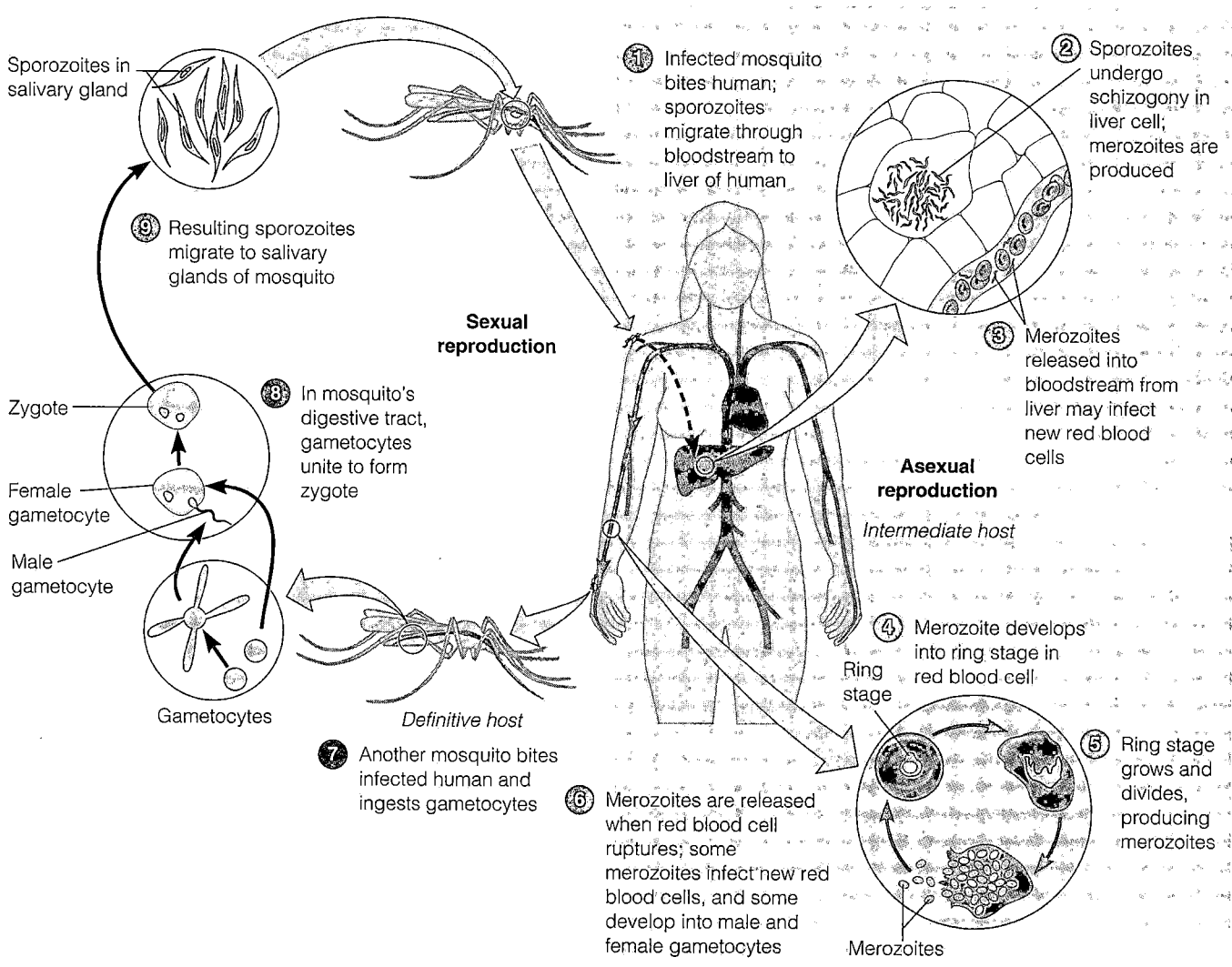

FIGURE 12.18 The life cycle of *Plasmodium vivax*, the apicomplexan that causes malaria. Asexual reproduction (schizogony) of the parasite takes place in the liver and in the red blood cells of a human host. Sexual reproduction occurs in the intestine of an *Anopheles* mosquito after the mosquito has ingested gametocytes.

■ What is the definitive host for *Plasmodium*?

Plasmodium grows by sexual reproduction in the *Anopheles* (an-of' el-ēz) mosquito (Figure 12.18). ① When an *Anopheles* carrying the infective stage of *Plasmodium*, called a **sporozoite**, bites a human, sporozoites can be injected into the human. The sporozoites are carried by the blood to the liver. ② They undergo schizogony in liver cells and produce thousands of progeny called **merozoites.** ③ Merozoites enter the bloodstream and infect red blood cells. ④ The young trophozoite looks like a ring in which the nucleus and cytoplasm are visible. This is called a **ring stage.** ⑤ The ring stage enlarges and divides repeatedly, and ⑥ the red blood cells eventually rupture and release more merozoites. Upon release of the merozoites, their waste products, which cause fever and chills,

are also released. Most of the merozoites infect new red blood cells and perpetuate their cycle of asexual reproduction. However, some develop into male and female sexual forms (gametocytes). Even though the gametocytes themselves cause no further damage, ⑦ they can be picked up by the bite of another *Anopheles* mosquito; they then enter the mosquito's intestine and begin their sexual cycle. ⑧ Here the male and female gametocytes unite to form a zygote. The zygote forms an oocyst, in which cell division occurs, and asexual sporozoites are formed. ⑨ When the oocyst ruptures, the sporozoites migrate to the salivary glands of the mosquito. They can then be injected into a new human host by the biting mosquito. The mosquito is the **definitive host** because it harbors the sexually reproducing stage of *Plasmodium*. The host in which the parasite undergoes asexual reproduction (in this case, the human) is the **intermediate host.**

Malaria is diagnosed in the laboratory by microscopic observation of thick blood smears for the presence of *Plasmodium* (see Figure 23.23 on page 645). A peculiar characteristic of malaria is that the interval between periods of fever caused by the release of merozoites is always the same for a given species of *Plasmodium* and is always a multiple of 24 hours. The reason and mechanism for such precision have intrigued scientists. After all, why should a parasite need a biological clock? Frank Hawking and his coworkers demonstrated that *Plasmodium*'s development is regulated by the host's body temperature, which normally fluctuates over a 24-hour period. The parasite's careful timing ensures that gametocytes are mature at night, when *Anopheles* mosquitoes are feeding, to facilitate transmission of the parasite to a new host.

Another apicomplexan parasite of red blood cells is *Babesia microti* (ba-bē-sē-ä mī-krō′tē). *Babesia* causes fever and anemia in immunosuppressed individuals. In the United States, it is transmitted by the tick *Ixodes scapularis.*

Toxoplasma gondii (toks-ō-plaz′mä gon′dē-ē) is another apicomplexan intracellular parasite of humans. The life cycle of this parasite involves domestic cats. The trophozoites, called **tachyzoites,** reproduce sexually and asexually in an infected cat, and **oocysts,** each containing eight sporozoites, are excreted with feces. If the oocysts are ingested by humans or other animals, the sporozoites emerge as trophozoites, which can reproduce in the tissues of the new host (see Figure 23.21, on page 643). *T. gondii* is dangerous to pregnant women, as it can cause congenital infections in utero. Tissue examination and observation of *T. gondii* are used for diagnosis. Antibodies may be detected by ELISA and by indirect fluorescent-antibody tests (see Chapter 18).

Cryptosporidium (krip-tō-spô-ri′dē-um) is a newly recognized parasite of humans. In AIDS patients and other immunosuppressed people, *Cryptosporidium* can cause respiratory and gallbladder infections and may be a major cause of death. The organism, which lives inside the cells lining the small intestine, can be transmitted to humans through the feces of cows, rodents, dogs, and cats. Waterborne and nosocomial infections have also been reported. Inside the host cell, each *Cryptosporidium* organism forms four oocysts (see Figure 25.18 on page 707), each containing four sporozoites. When the oocyst ruptures, sporozoites may infect new cells in the host or be released with the feces. The disease is diagnosed by acid-fast staining or fluorescent-antibody tests.

During the 1980s, epidemics of waterborne diarrhea were identified on every continent except Antarctica. The causative agent was misidentified as a cyanobacterium

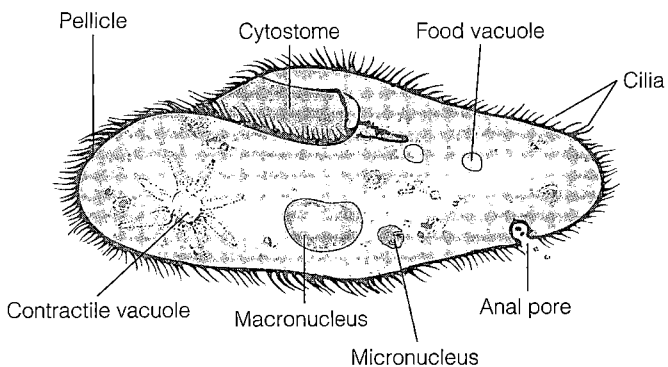

(a) *Paramecium*

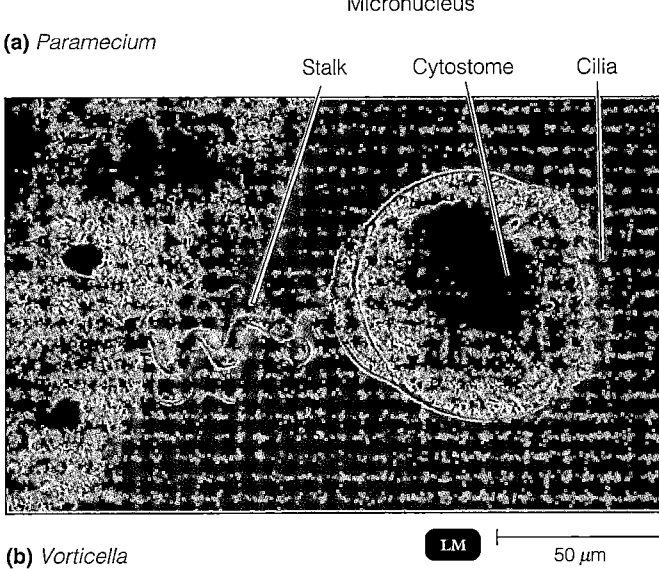

(b) *Vorticella* LM ├─────── 50 µm ───────┤

FIGURE 12.19 Ciliates. (a) *Paramecium* is covered with rows of cilia. It has specialized structures for ingestion (mouth), elimination of wastes (anal pore), and the regulation of osmotic pressure (contractile vacuoles). The macronucleus is involved with protein synthesis and other ongoing cellular activities. The micronucleus functions in sexual reproduction. **(b)** *Vorticella* attaches to objects in water by the base of its stalk. The springlike stalk can expand so *Vorticella* can feed in different areas. Cilia surround its cytostome.

■ **What ciliate can cause disease in humans?**

because the outbreaks occurred during warm months, and the disease agent looked like a prokaryotic cell. In 1993, the organism was identified as an apicomplexan similar to *Cryptosporiduim*. In 1996, the new parasite, named *Cyclospora cayetanensis* (sī-klō-spô-rä kī′ē-tan-en-sis), was responsible for 850 cases of diarrhea associated with raspberries in the United States and Canada.

Ciliophora

Members of the phylum **Ciliophora** or ciliates have cilia that are similar to but shorter than flagella. The cilia are arranged in precise rows on the cell (Figure 12.19). They are moved in unison to propel the cell through its environment and to bring food particles to the mouth.

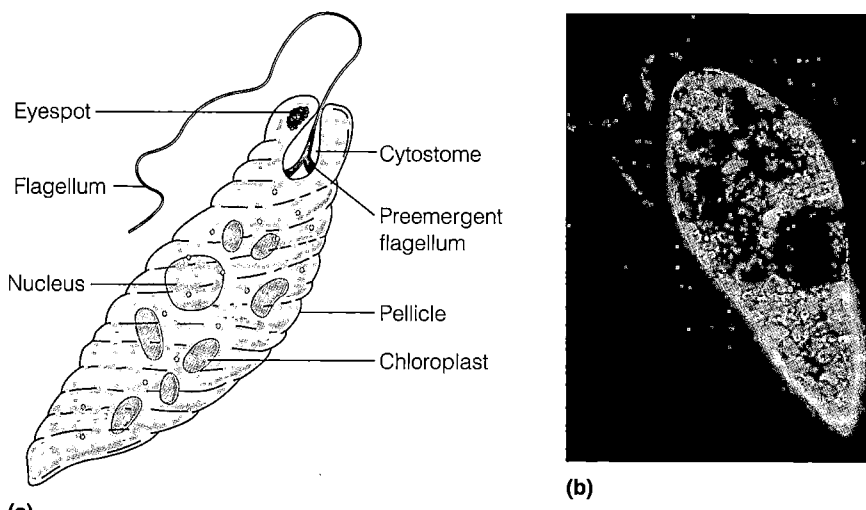

(b)

(a)

FIGURE 12.20 *Euglena.* Euglenoids are photoautotrophs. Semirigid rings supporting the pellicle allow *Euglena* to change shape.

■ Why is *Euglena* classified with the hemoflagellates?

The only ciliate that is a human parasite is *Balantidium coli* (bal-an-tid'ē-um kō'lī), the causative agent of a severe, though rare, type of dysentery. When cysts are ingested by the host, they enter the large intestine, into which the trophozoites are released. The trophozoites feed on bacteria and fecal debris as they multiply, and cysts are excreted with feces.

Ciliates, apicomplexans, and dinoflagellates (page 347) may be placed in their own phylum or kingdom, called **Alveolata** because they all have membrane-bound cavities (alveoli) under the cell surface and rRNA sequences in common.

Euglenozoa

Two groups of flagellated cells are included in the **Euglenozoa** based on common rRNA sequences, disk-shaped mitochondria, and absence of sexual reproduction.

Euglenoids are photoautotrophs (Figure 12.20). Euglenoids have a semirigid plasma membrane called a pellicle, and they move by means of a flagellum at the anterior end. Most euglenoids also have a red *eyespot* at the anterior end. This carotenoid-containing organelle senses light and directs the cell in the appropriate direction by using a *preemergent flagellum*. Some euglenoids are facultative chemoheterotrophs. In the dark, they ingest organic matter through a cytostome. Euglenoids are frequently studied with algae because they can photosynthesize.

The **hemoflagellates** (blood parasites) are transmitted by the bites of blood-feeding insects and are found in the circulatory system of the bitten host. To survive in this viscous fluid, hemoflagellates have long, slender bodies and an undulating membrane. The genus *Trypanosoma* (tripa'nō-sō-mä) includes the species that causes African sleeping sickness, *T. brucei gambiense* (brüs'ē gam-bē-ens'), which is transmitted by the tsetse fly. *T. cruzi* (kruz'ē; see Figure 23.20, on page 642), the causative agent of Chagas' disease, is transmitted by the "kissing bug," so named because it bites on the face (see Figure 12.31d). After entering the insect, the trypanosome rapidly multiplies by fission. If the insect then defecates while biting a human, it can release trypanosomes that can contaminate the bite wound.

Table 12.5 lists some typical parasitic protozoa and the diseases they cause.

Slime Molds

Learning Objective

■ Compare and contrast cellular slime molds and plasmodial slime molds.

Slime molds have both fungal and amoebal characteristics; they are probably more closely related to amoeba. There are two phyla of slime molds: cellular and plasmodial. **Cellular slime molds** are typical eukaryotic cells that resemble amoebas. In the life cycle of cellular slime molds (Figure 12.21 on page 356), ❶ the amoeboid cells live and grow by ingesting fungi and bacteria by phagocytosis. Cellular slime molds are of interest to biologists who study cellular migration and aggregation, because when conditions are unfavorable, ❷ –❸ large numbers of amoe-

table 12.5 *Some Representative Parasitic Protozoa*

Phylum	Human Pathogens	Distinguishing Features	Disease	Source of Human Infections	Figure Reference	Page Reference
Archaezoa	*Giardia lamblia*	Two nuclei, eight flagella	Giardial enteritis	Fecal contamination of drinking water	25.17	706
	Trichomonas vaginalis	No encysting stage	Urethritis, vaginitis	Contact with vaginal-urethral discharge	Table 26.1	732
Microsporidia	*Nosema*	Unknown	Diarrhea, kerato-conjunctivitis, conjunctivitis	Other animals	—	—
Rhizopoda	*Acanthamoeba*	Pseudopods	Keratitis	Water	21.20	567
	Entamoeba histolytica	Pseudopods	Amoebic dysentery	Fecal contamination of drinking water	25.19	708
Alveolata (Proposed) Apicomplexa	*Babesia microti*	Complex	Babesiosis	Domestic animals, ticks	—	—
	Cryptosporidium	Life cycles may require more than one host	Diarrhea	Humans, other animals, water	25.18	707
	Cyclospora	—	Diarrhea	Water	Table 25.2	712
	Isospora	—	Coccidiosis	Domestic animals	Table 19.5	540
	Plasmodium	—	Malaria	Bite of *Anopheles* mosquito	12.18 23.23	352 642
	Toxoplasma gondii	—	Toxoplasmosis	Cats, meat of other animals; congenital	23.21	642
Dinoflagellates	*Alexandrium, Pfiesteria*	Photosynthetic, see Table 12.4	Paralytic shellfish poisoning; ciguatera	Ingestion of dinoflagellates in mollusks, fish	12.14, 27.15	347, 758
Ciliophora	*Balantidium coli*	Only parasitic ciliate of humans	Balantidial dysentery	Fecal contamination of drinking water	—	—
Euglenozoa	*Leishmania*	Flagellated form in sand fly; ovoid form in vertebrate host	Leishmaniasis	Bite of sand fly (*Phlebotomus*)	23.24	645
	Naegleria fowleri	Flagellated and amoeboid forms	Meningo-cephalitis	Water in which people swim	22.16	618
	Trypanosoma cruzi	Undulating membrane	Chagas' disease	Bite of *Triatoma* (kissing bug)	23.20	641
	T. brucei gambiense, T.b. rhodesiense		African trypanosomiasis	Bite of tsetse fly	22.15	618

boid cells aggregate to form a single structure. This aggregation occurs because some individual amoebas produce the chemical cyclic AMP (cAMP), toward which the other amoebas migrate. ❹ The aggregated amoebas are enclosed in a slimy sheath called a *slug*. The slug migrates as a unit toward light. ❺ After a period of hours, the slug ceases to migrate and begins to form differentiated structures. ❻ Some of the amoeboid cells form a stalk; others swarm up the stalk to form a spore cap, and ❼ most of these differentiate into spores. ❽ When spores are released under favorable conditions, ❾ they germinate to form single amoebas.

In 1973, a Dallas resident discovered a pulsating red blob in his backyard. The news media claimed that a "new life form" had been found. For some people, the "creature" evoked spine-chilling recollections of an old science fiction movie. Before imaginations got carried away too far, biologists calmed everyone's worst fears (or highest

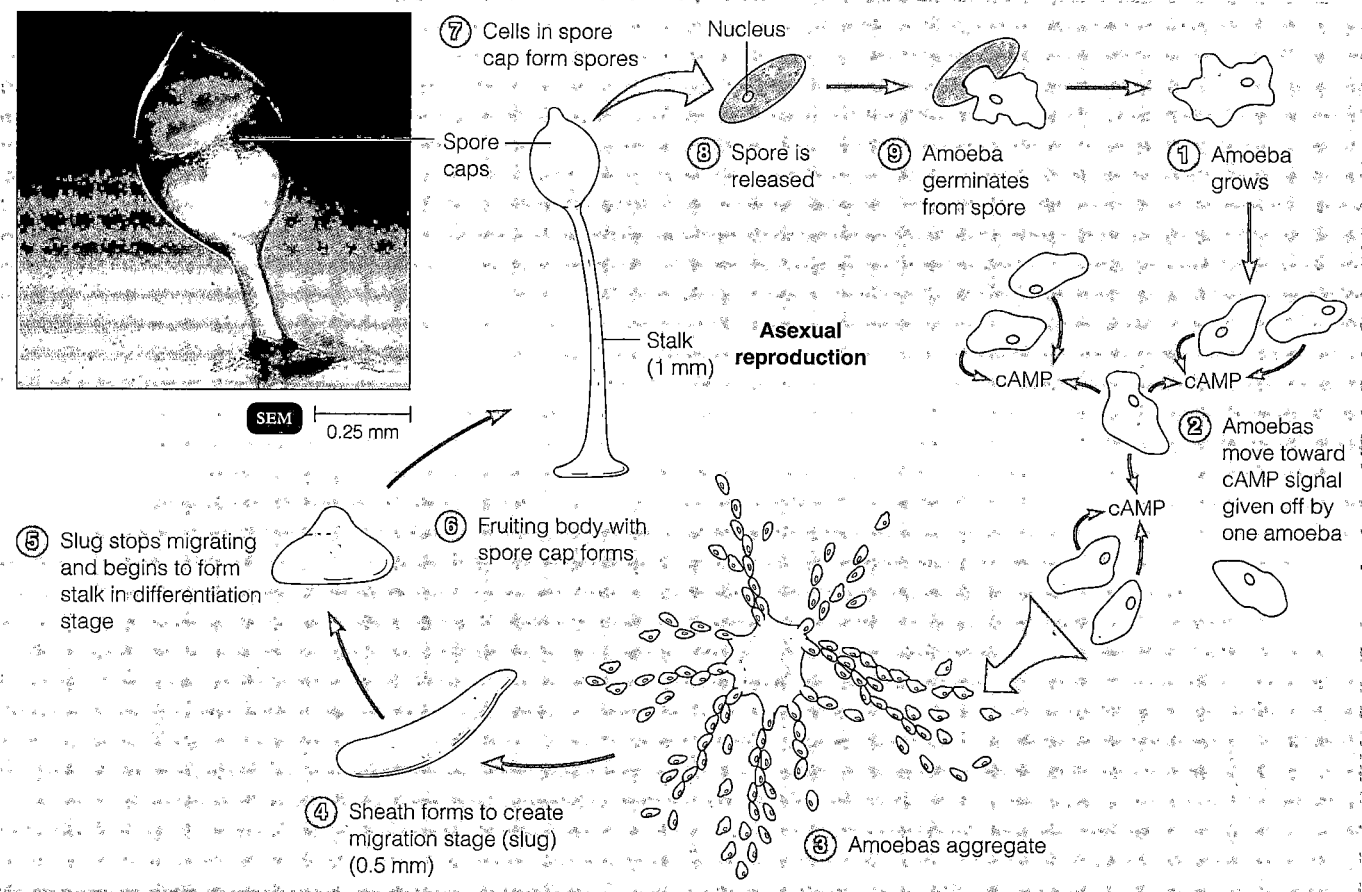

FIGURE 12.21 The generalized life cycle of a cellular slime mold. The micrograph shows a spore cap of *Dictyostelium*.

■ Slime molds have both fungal and protozoal characteristics.

hopes). The amorphous mass was merely a plasmodial slime mold, they explained. But its unusually large size— 46 cm in diameter—startled even scientists.

Plasmodial slime molds were first scientifically reported in 1729. They belong to a separate phylum. A plasmodial slime mold exists as a mass of protoplasm with many nuclei (it is multinucleated). This mass of protoplasm is called a **plasmodium** (Figure 12.22). ① The entire plasmodium moves as a giant amoeba; it engulfs organic debris and bacteria. Biologists have found that musclelike proteins forming microfilaments account for the movement of the plasmodium. ② When plasmodial slime molds are grown in laboratories, a phenomenon called **cytoplasmic streaming** is observed, during which the protoplasm within the plasmodium moves and changes both its speed and direction so that the oxygen and nutrients are evenly distributed.

The plasmodium continues to grow as long as there is enough food and moisture for it to thrive. ③ When either

is in short supply, the plasmodium separates into many groups of protoplasm; ④ each of these groups forms a stalked sporangium, ⑤ in which spores (a resistant, resting form of the slime mold) develop. ⑥ Nuclei within these spores undergo meiosis and form uninucleate haploid cells. ⑦ The spores are then released. ⑧ When conditions improve, these spores germinate, ⑨ fuse to form diploid cells, and ⑩ develop into a multinucleated plasmodium.

Helminths

There is a number of parasitic animals that spend part or all of their lives in humans. Most of these animals belong to two phyla: Platyhelminthes (flatworms) and Nematoda (roundworms). These worms are commonly called **helminths.** There are also free-living species in these phyla, but we will limit our discussion to the parasitic species. Diseases caused by parasitic worms are discussed in Part Four.

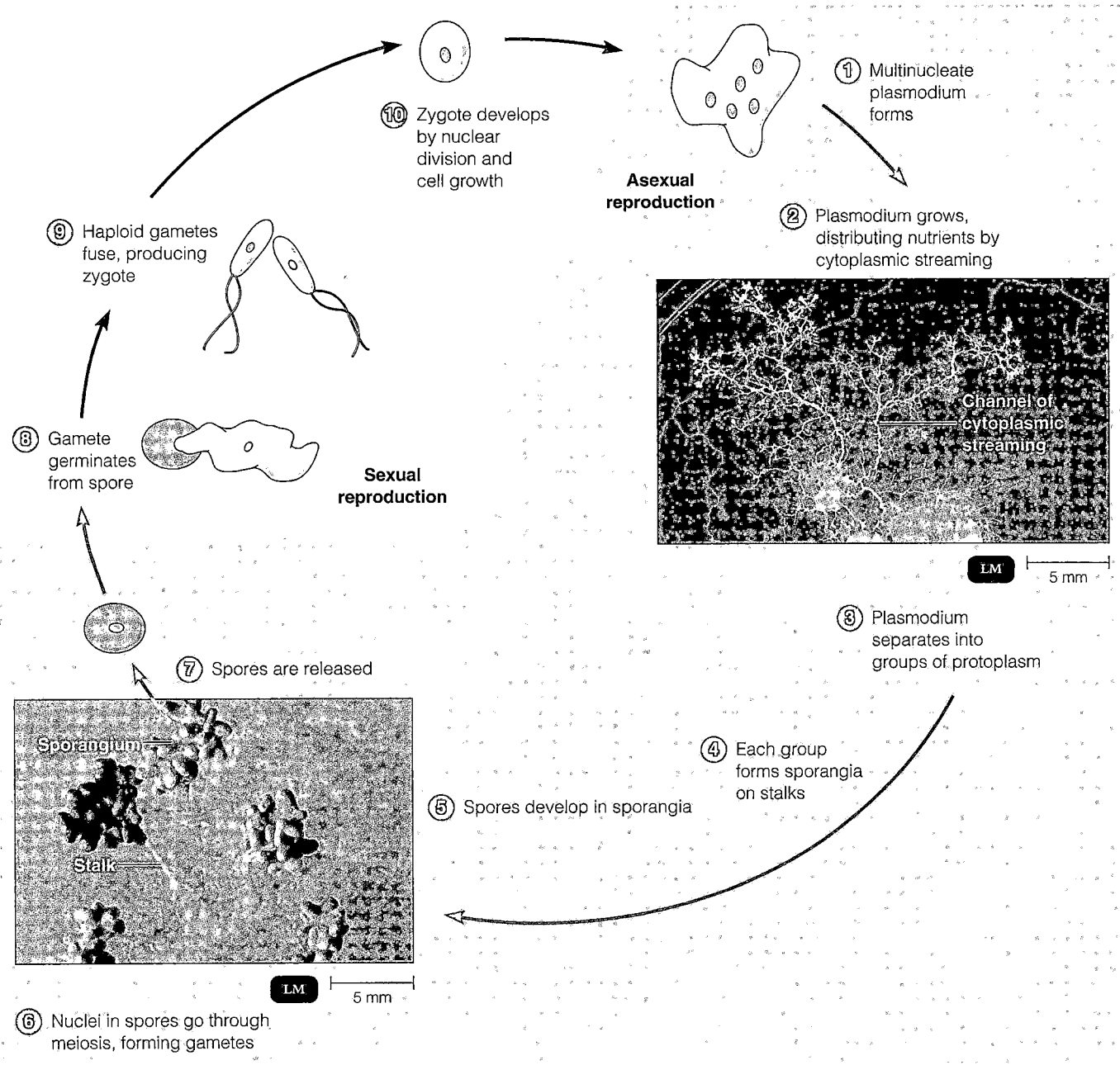

(10) Zygote develops by nuclear division and cell growth

(1) Multinucleate plasmodium forms

Asexual reproduction

(2) Plasmodium grows, distributing nutrients by cytoplasmic streaming

(9) Haploid gametes fuse, producing zygote

(8) Gamete germinates from spore

Sexual reproduction

Channel of cytoplasmic streaming

LM |—————| 5 mm

(3) Plasmodium separates into groups of protoplasm

(7) Spores are released

(4) Each group forms sporangia on stalks

(5) Spores develop in sporangia

Sporangium

Stalk

LM |—————| 5 mm

(6) Nuclei in spores go through meiosis, forming gametes

FIGURE 12.22 The life cycle of a plasmodial slime mold. Pictured in the photomicrographs is *Physarum*.

■ How do cellular and acellular slime molds differ?

Characteristics of Helminths

Learning Objectives

- *List the distinguishing characteristics of parasitic helminths.*
- *Provide a rationale for the elaborate life cycle of parasitic worms.*

Helminths are multicellular eukaryotic animals that generally possess digestive, circulatory, nervous, excretory, and reproductive systems. Parasitic helminths must be highly specialized to live inside their hosts. The following generalizations distinguish parasitic helminths from their free-living relatives:

1. They may *lack* a digestive system. They can absorb nutrients from the host's food, body fluids, and tissues.

2. Their nervous system is *reduced*. They do not need an extensive nervous system because they do not

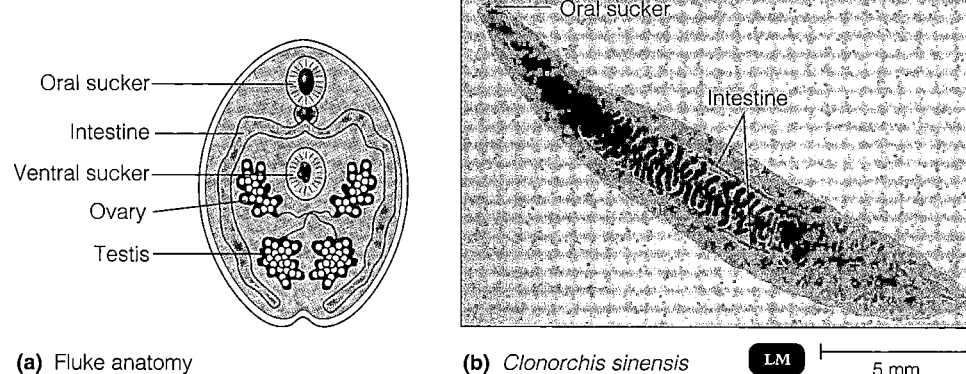

Oral sucker

Intestine

Ventral sucker

Ovary

Testis

(a) Fluke anatomy

(b) *Clonorchis sinensis*   LM   5 mm

FIGURE 12.23 Flukes. (a) General anatomy of an adult fluke, shown in cross section. The oral and ventral suckers attach the fluke to the host. The mouth is located in the center of the oral sucker. Flukes are hermaphroditic; each animal contains both testes and ovaries. **(b)** The Asian liver fluke *Clonorchis sinensis.* Heavy infestations may block bile ducts from the liver. Notice the incomplete digestive system.

■ **The flatworm digestive system has one opening that functions as the mouth and anus.**

have to search for food or respond much to their environment. The environment within a host is fairly constant.

3. Their means of locomotion is occasionally *reduced* or *completely lacking.* Because they are transferred from host to host, they do not need to search actively for a suitable habitat.

4. Their reproductive system is often complex; an individual produces large numbers of eggs, by which a suitable host is infected.

Life Cycle

The life cycle of parasitic helminths can be extremely complex, involving a succession of intermediate hosts for completion of each **larval** (developmental) stage of the parasite and a definitive host for the adult parasite.

Adult helminths may be **dioecious;** male reproductive organs are in one individual, and female reproductive organs are in another. In those species, reproduction occurs only when two adults of the opposite sex are in the same host.

Adult helminths may also be **monoecious** or **hermaphroditic**—one animal has both male and female reproductive organs. Two hermaphrodites may copulate and simultaneously fertilize each other. A few types of hermaphrodites fertilize themselves.

Platyhelminths

Learning Objectives

■ List the characteristics of the three groups of parasitic helminths, and give an example of each.

■ Describe a parasitic infection in which humans serve as a definitive host, as an intermediate host, and as both.

Members of the phylum Platyhelminthes, the **flatworms,** are flattened from front to back. The classes of parasitic flatworms include the trematodes and cestodes.

Trematodes

Trematodes, or **flukes,** often have flat, leaf-shaped bodies with a ventral sucker and an oral sucker (Figure 12.23). The suckers hold the organism in place. Flukes can also obtain food by absorbing it through their nonliving outer covering, called the **cuticle.** Flukes are given common names according to the tissue of the definitive host in which the adults live (for example, lung fluke, liver fluke, blood fluke). The Asian liver fluke *Clonorchis sinensis* (klonôr′kis si-nen′sis) is occasionally seen in immigrants in the United States, but it cannot be transmitted because its intermediate hosts are not in the United States.

To exemplify a fluke's life cycle, let's look at the lung fluke, *Paragonimus westermani* (păr-ä-gōn′e-mus we-sterma′nē), shown in Figure 12.24. The intermediate hosts for this fluke, and therefore the fluke itself, occur

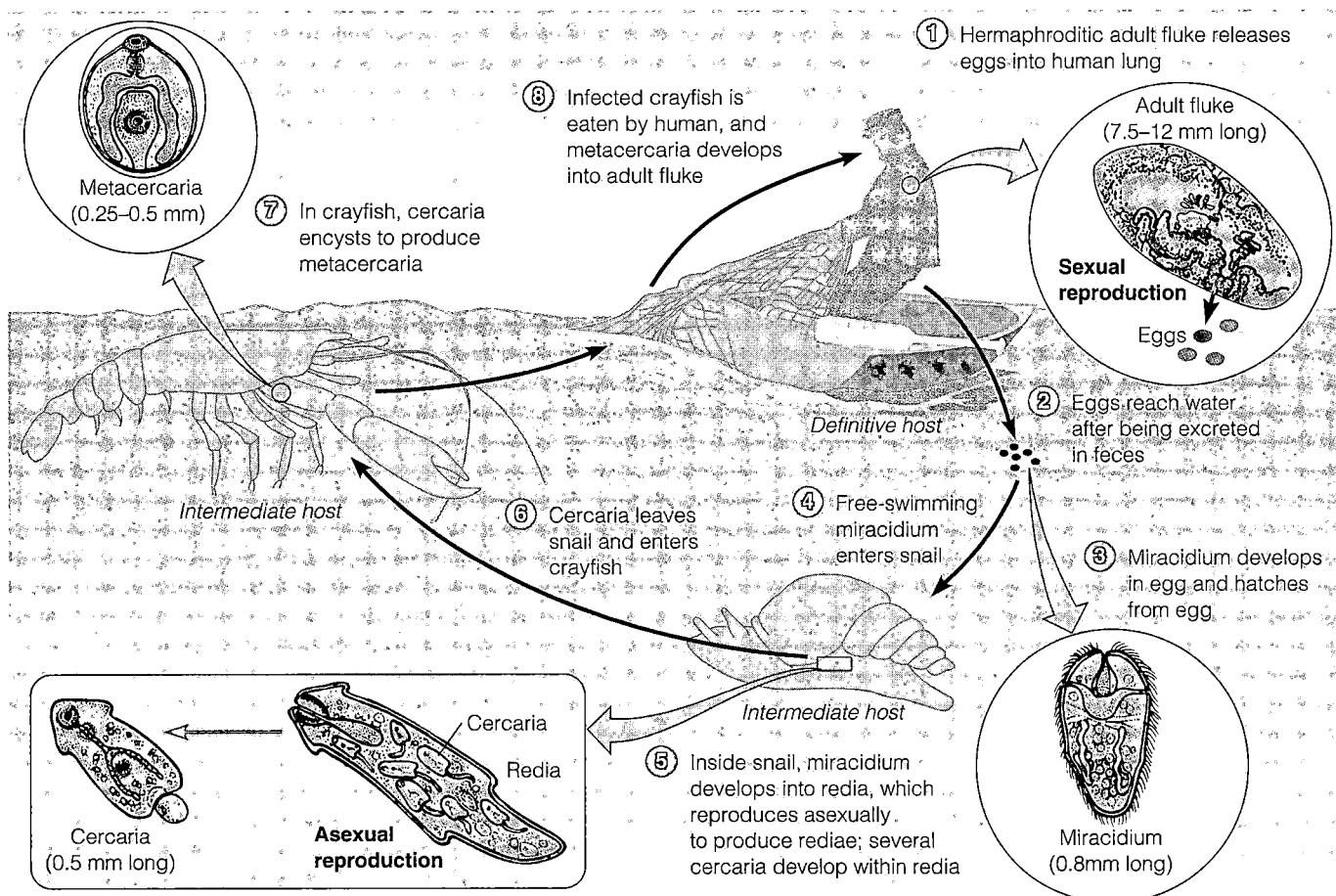

FIGURE 12.24 The life cycle of the lung fluke *Paragonimus westermani*. The trematode reproduces sexually in a human and asexually in a snail, its first intermediate host. Larvae encysted in the second intermediate host, a crayfish, infect humans when ingested. Also see the *Schistosoma* life cycle in Figure 23.25 (page 648).

■ **Parasites often have complex life cycles that ensure that their progeny get into a suitable host.**

throughout the world, including the United States and Canada. The adult lung fluke lives in the bronchioles of humans and other mammals, and is approximately 6 mm wide and 12 mm long.

⓵ The hermaphroditic adults liberate eggs into the bronchi. Because sputum that contains eggs is frequently swallowed, the eggs are usually excreted in feces of the definitive host. If the life cycle is to continue, ② the eggs must reach a body of water. ③ Inside the egg, a miracidial larva, or **miracidium,** then develops. When the egg hatches, ④ the larva enters a suitable snail. Only certain species of aquatic snails can be this first intermediate host. ⑤ Inside the snail, the miracidium develops into a **redia,** which then undergoes asexual reproduction and produces more rediae. Each redia produces **cercariae** that ⑥ bore out of the snail and penetrate the cuticle of a

crayfish. ⑦ The parasite encysts as a **metacercaria** in the muscles and other tissues of the crayfish. ⑧ When the crayfish is eaten by a human, the metacercaria is freed in the human's small intestine. It bores out and wanders around until it penetrates the lungs, enters the bronchioles, and develops into an adult lung fluke.

In a laboratory diagnosis, sputum and feces are examined microscopically for fluke eggs. Infection results from eating undercooked crayfish, and the disease can be prevented by thoroughly cooking crayfish.

The cercariae of the blood fluke *Schistosoma* (shis-tō-sō'ma) are not ingested. Instead, they burrow through the skin of the human host and enter the circulatory system. The adults are found in certain abdominal and pelvic veins. The disease schistosomiasis is a major world health problem; it will be discussed further in Chapter 23 (page 647).

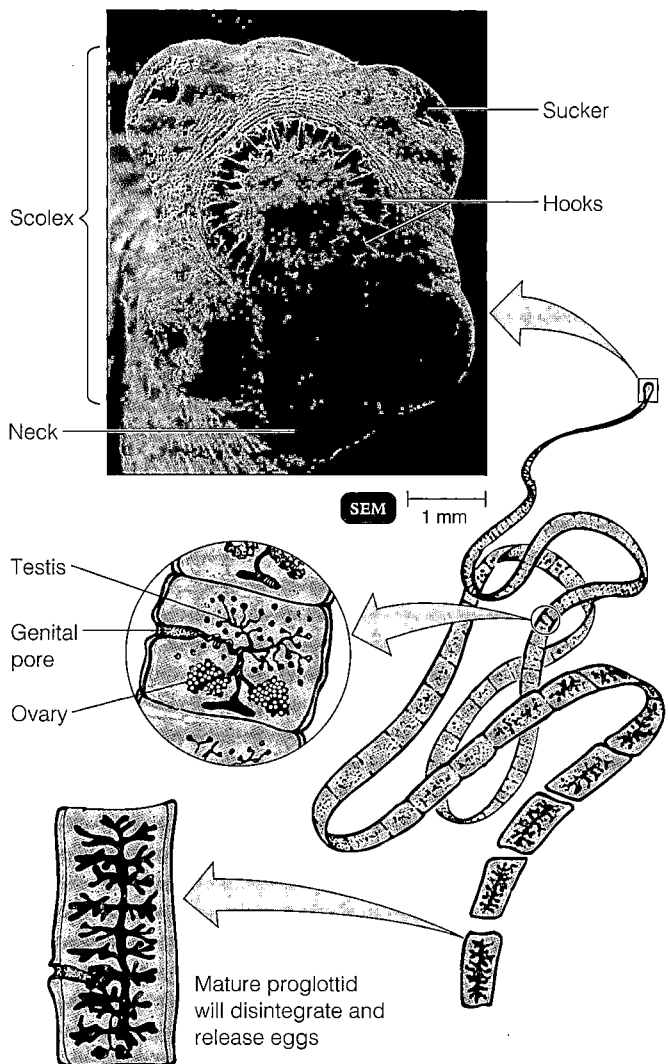

Scolex

Neck

SEM |—| 1 mm

Sucker

Hooks

Testis

Genital
pore

Ovary

Mature proglottid
will disintegrate and
release eggs

FIGURE 12.25 General anatomy of an adult tapeworm. The scolex, shown in the micrograph, consists of suckers and hooks that attach to the host's tissues. The body lengthens as new proglottids form at the neck. Each mature proglottid contains both testes and ovaries.

■ **What are the similarities between tapeworms and flukes?**

Cestodes

Cestodes, or **tapeworms,** are intestinal parasites. Their structure is shown in Figure 12.25. The head, or **scolex** (plural: *scoleces*), has suckers for attaching to the intestinal mucosa of the definitive host; some species also have small hooks for attachment. Tapeworms do not ingest the tissues of their hosts; in fact, they completely lack a digestive system. To obtain nutrients from the small intestine, they absorb food through their cuticle. The body consists of segments called **proglottids.** Proglottids are continually produced by the neck region of the scolex, as long as the scolex is attached and alive. Each mature proglottid

contains both male and female reproductive organs. The proglottids farthest away from the scolex are the mature ones containing eggs. Mature proglottids are essentially bags of eggs, each of which is infective to the proper intermediate host.

Humans as Definitive Hosts. The adults of *Taenia saginata* (te'nē-ä sa-ji-nä'tä), the beef tapeworm, live in humans and can reach a length of 6 m. The scolex is about 2 mm long and is followed by a thousand or more proglottids. The feces of an infected human contain mature proglottids, each of which contains thousands of eggs. As the proglottids wriggle away from the fecal material, they increase their chances of being ingested by an animal that is grazing. Upon ingestion by cattle, the larvae hatch from the eggs and bore through the intestinal wall. The larvae migrate to muscle (meat), in which they encyst as **cysticerci.** When the cysticerci are ingested by humans, all but the scolex is digested. The scolex anchors itself in the small intestine and begins producing proglottids.

Diagnosis of tapeworm infection in humans is based on the presence of mature proglottids and eggs in feces. Cysticerci can be seen macroscopically in meat; their presence is referred to as "measly beef." Inspecting beef that is intended for human consumption for "measly" appearance is one way to prevent infections by beef tapeworm. Another method of prevention is to avoid the use of untreated human sewage as fertilizer in grazing pastures.

Humans are the only known definitive host of the pork tapeworm, *Taenia solium.* Adult worms living in the human intestine produce eggs, which are passed out in feces. When eggs are eaten by pigs, the larval helminth encysts in the pig's muscles; humans become infected when they eat undercooked pork. The human–pig–human cycle of *T. solium* is common in Latin America, Asia, and Africa. In the United States, however, *T. solium* is virtually nonexistent in pigs; the parasite is transmitted from human to human. Eggs shed by one person and ingested by another person hatch, and the larvae encyst in the brain and other parts of the body, causing cysticercosis (see Figure 25.21 on page 709). The human hosting *T. solium*'s larvae is serving as an intermediate host. Approximately 7% of the few hundred cases reported in recent years were acquired by people who had never been outside the United States. They may have become infected through household contact with people who were born in or had traveled in other countries.

Humans as Intermediate Hosts. Humans are the intermediate hosts for *Echinococcus granulosus* (ē-kīn-ō-kok'kus gra-nū-lō'sus), shown in Figure 12.26. Dogs and coyotes are the definitive hosts for this minute (2–8 mm) tapeworm.

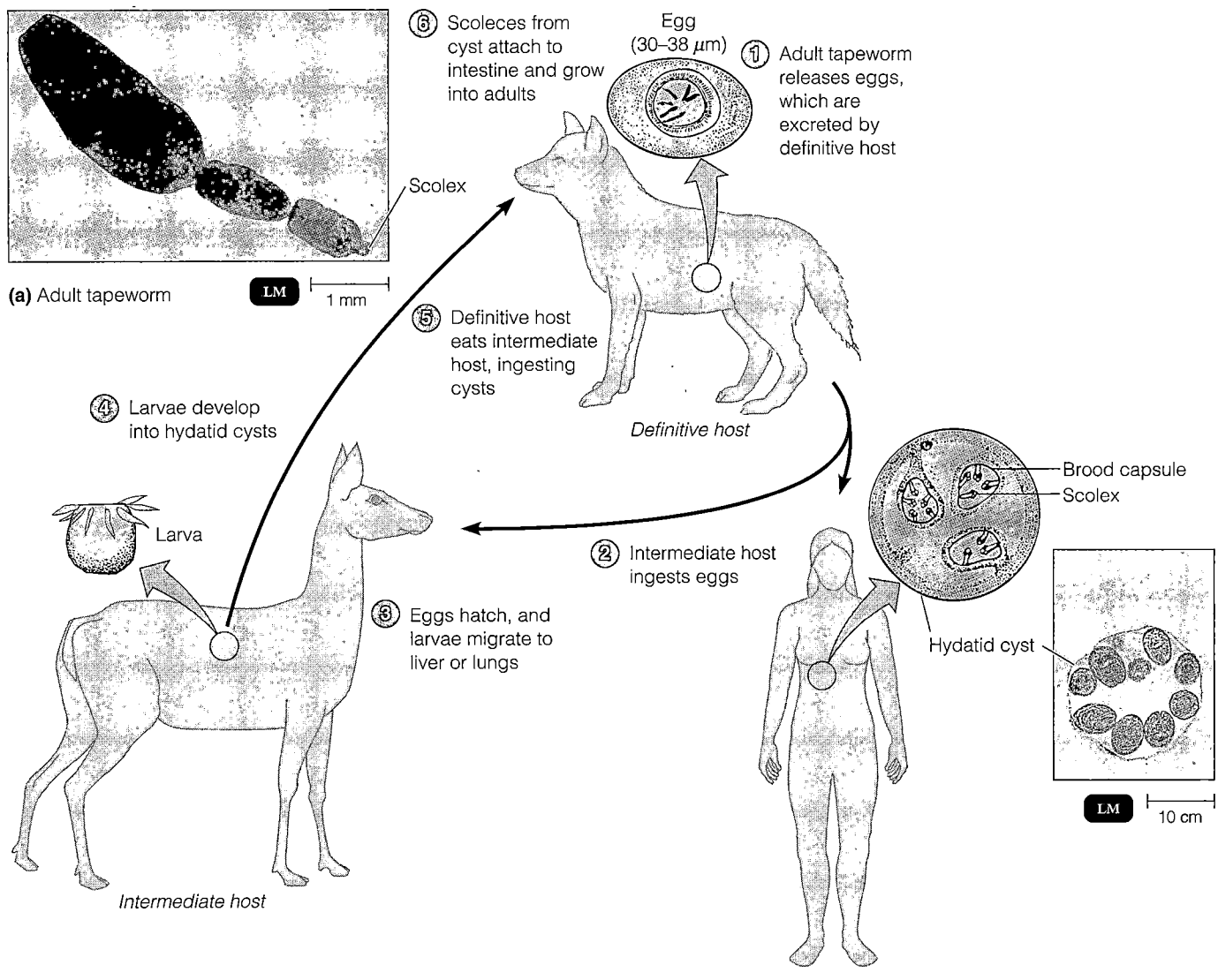

(a) Adult tapeworm

(b) Life cycle

FIGURE 12.26 The tapeworm *Echinococcus granulosus.* This tiny tapeworm is found in the intestines of dogs, wolves, and foxes. **(a)** The adult of the closely related *Echinococcus multilocularis.* **(b)** Life cycle. The photomicrograph shows a hydatid cyst. The parasite can complete its life cycle only if the cysts are ingested by a definitive host that eats the intermediate host.

▪ **A human serving as the intermediate host for a tapeworm is a dead end for the parasite, unless the human is eaten by an animal.**

① Eggs are excreted with feces and ② are ingested by deer, sheep, or humans. Humans can also become infected by contaminating their hands with dog feces or saliva from a dog that has licked itself. ③ The eggs hatch in the human's small intestine, and the larvae migrate to the liver or lungs. ④ The larva develops into a **hydatid cyst.** The cyst contains "brood capsules," from which thousands of scoleces might be produced. ⑤ Humans are a dead-end for the parasite, but in the wild, the cysts might be in a deer that is eaten by a wolf. ⑥ The scoleces would be able

to attach themselves in the wolf's intestine and produce proglottids. Diagnosis of hydatid cysts is frequently made only on autopsy, although X rays can detect the cysts (see Figure 25.22 on page 709).

Nematodes

Members of the Phylum Nematoda, the **roundworms,** are cylindrical and tapered at each end. Roundworms have a *complete* digestive system, consisting of a mouth, an

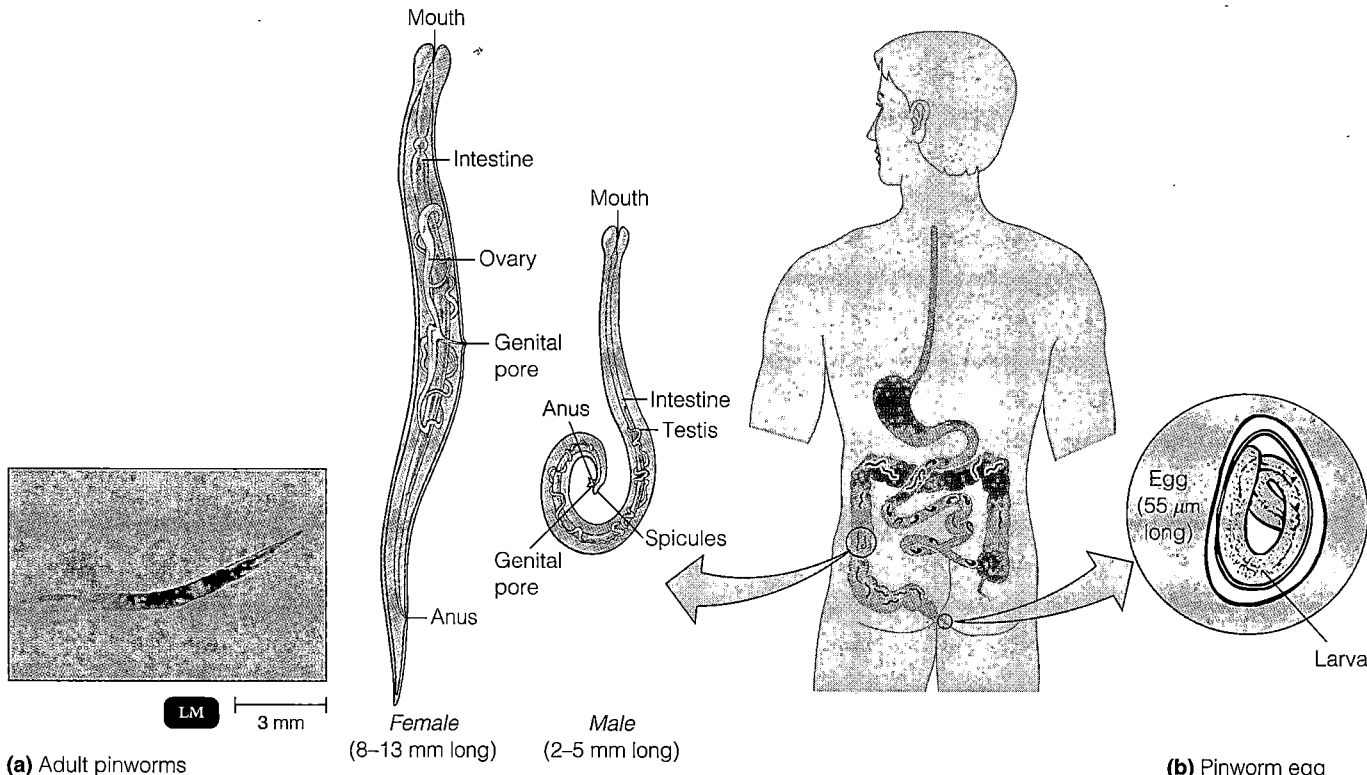

(a) Adult pinworms

Female
(8–13 mm long)

Male
(2–5 mm long)

(b) Pinworm egg

FIGURE 12.27 The pinworm *Enterobius vermicularis*. **(a)** Adult pinworms live in the large intestine of humans. Most roundworms are dioecious, and the female (left and photomicrograph) is often distinctly larger than the male (right). **(b)** Pinworm eggs are deposited by the female on the perianal skin at night.

■ Roundworms have a complete digestive system with a mouth, intestine, and anus.

intestine, and an anus. Most species are dioecious. Males are smaller than females and have one or two hardened **spicules** on their posterior ends. Spicules are used to guide sperm to the female's genital pore.

Some species of nematodes are free-living in soil and water, and others are parasites on plants and animals. Some nematodes pass their entire life cycle, from egg to mature adult, in a single host.

Nematode infections of humans can be divided into two categories: those in which the egg is infective, and those in which the larva is infective.

Eggs Infective for Humans

The pinworm *Enterobius vermicularis* (en-te-rō'bē-us ver-mi-kū-lar'is) spends its entire life in a human host (Figure 12.27). Adult pinworms are found in the large intestine. From there, the female pinworm migrates to the anus to deposit her eggs on the perianal skin. The eggs can be ingested by the host or by another person exposed through contaminated clothing or bedding. Pinworm infections

are diagnosed by the Graham sticky-tape method. A piece of transparent tape is placed on the perianal skin in such a way that the sticky side picks up eggs that were deposited earlier. The tape is then microscopically examined for the presence of eggs.

Ascaris lumbricoides (as'kar-is lum-bri-koi'dēz) is a large nematode (30 cm in length) (Figure 25.24 on page 710). It is dioecious with **sexual dimorphism;** that is, the male and female worms look distinctly different, the male being smaller with a curled tail. The adult *Ascaris* lives in the small intestines of humans and domestic animals (such as pigs and horses); it feeds primarily on semidigested food. Eggs, excreted with feces, can survive in the soil for long periods until accidentally ingested by another host. The eggs hatch in the small intestine of the host, mature in the lungs, and from there migrate to the intestines.

Diagnosis is frequently made when the adult worms are excreted with feces. Preventing infection in humans is managed by proper sanitary habits. The life cycle of *Ascaris* in pigs can be interrupted by keeping pigs in areas free of fecal material.

Larvae Infective for Humans

Adult hookworms, *Necator americanus* (ne-kā-tôr ä-me-ri-ka'nus) and Ancylostoma duodenale, live in the small intestine of humans (Figures 12.28 and 25.23); the eggs are excreted in feces. The larvae hatch in the soil, where they feed on bacteria. A larva enters its host by penetrating the host's skin. It then enters a blood or lymph vessel, which carries it to the lungs. It is coughed up in sputum, swallowed, and finally carried to the small intestine. Diagnosis is based on the presence of eggs in feces. People can avoid hookworm infections by wearing shoes.

Worldwide, *Trichinella spiralis* (trik-in-el'lä spī-ra'lis) infections, called trichinosis, are usually acquired by eating encysted larvae in poorly cooked pork. In the United States, trichinosis is more often acquired from eating game animals, such as bears. In the human digestive tract, the larvae are freed from the cysts. They mature into adults in the small intestine and sexually reproduce there. Eggs develop in the female, and she gives birth to live larvae. The larvae enter lymph and blood vessels in the intestines and migrate from there throughout the body. They encyst in muscles and other tissues and remain there until ingested by another host (see Figure 25.25 on page 711).

Diagnosis of trichinosis is made by microscopic examination for larvae in a muscle biopsy. Trichinosis can be prevented by thoroughly cooking meat prior to consumption.

Four genera of roundworms called anisakines, or wriggly worms, can be transmitted to humans from infected fish and squid. Anisakine larvae are in the fish's intestinal mesenteries and migrate to the muscle when the fish dies. Freezing or thorough cooking will kill the larvae.

Table 12.6 lists representative parasitic helminths of each phylum and class and the diseases they cause.

Arthropods as Vectors

Learning Objectives

- *Define arthropod vector.*
- *Differentiate between a tick and a mosquito, and name a disease transmitted by each.*

Arthropods are animals characterized by segmented bodies, hard external skeletons, and jointed legs. With nearly 1 million species, this is the largest phylum in the animal kingdom. Although not microbes themselves, we will briefly describe arthropods here because a few suck the blood of humans and other animals and can transmit microbial diseases while doing so. Arthropods that carry

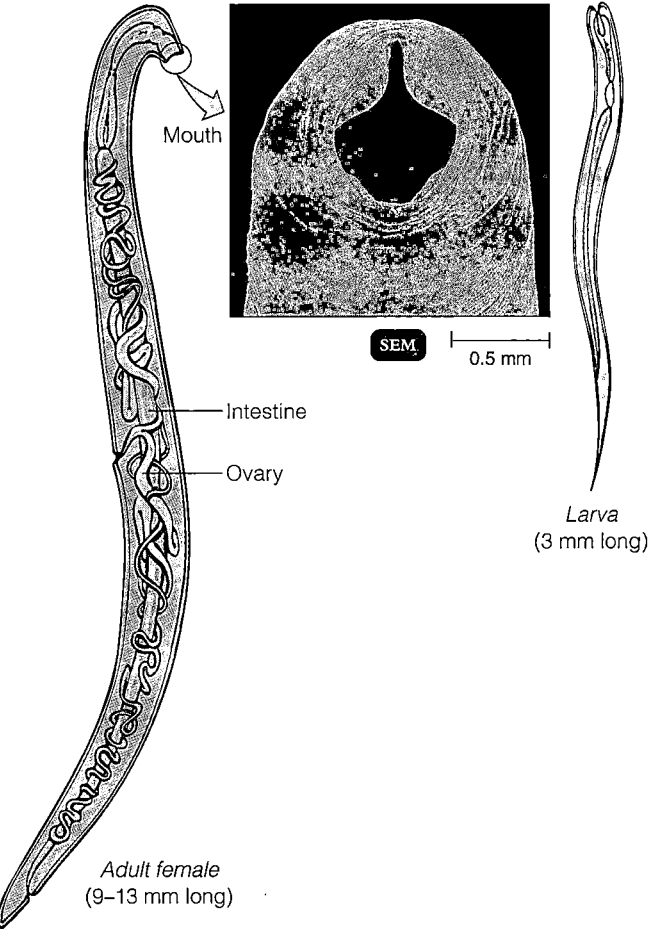

FIGURE 12.28 The hookworm *Necator americanus.* Adults are found in the small intestine of humans. The hooks located around the mouth are used for attaching to and removing food from the host tissue. The free-living larvae inhabit the soil and infect their definitive host (humans) by penetrating the skin.

■ How do roundworms and flatworms differ?

pathogenic microorganisms are called **vectors.** Scabies is a disease that is caused by an arthropod. Scabies is described in Chapter 21, page 595.

Representative classes of arthropods include the following:

- Arachnida (eight legs): spiders, mites, ticks.
- Crustacea (four antennae): crabs, crayfish.
- Insecta (six legs): bees, flies, lice.

Table 12.7 lists those arthropods that are important vectors, and Figures 12.29, 12.30, and 12.31 on pages 365–366 illustrate some of them. These insects and ticks reside on an animal only when they are feeding. An exception to this is the louse, which spends its entire life on its host and cannot survive for long away from a host.

table 12.6 — Representative Parasitic Helminths

Phylum	Class	Human Parasites	Intermediate Host	Definitive Host Site	Stage Passed to Humans; Method	Disease	Location in Humans	Figure Reference
Platyhelminthes	Trematodes	Paragonimus westermani	Freshwater snails and crayfish	Humans; lungs	Metacercaria in crayfish; ingested	Paragonimiasis (lung fluke)	Lungs	12.24
		Schistosoma	Freshwater snails	Humans	Cercariae; through skin	Schistosomiasis	Veins	23.25 23.26
	Cestodes	Taenia saginata	Cattle	Humans; small intestine	Cysticerci in beef; ingested	Tapeworm	Small intestine	—
		Taenia solium	Humans; pigs	Humans	Eggs; ingested	Neurocysticercosis	Brain; any tissue	25.21
		Echinococcus granulosus	Humans	Dogs and other animals; intestines	Eggs from other animals; ingested	Hydatidosis	Lungs, liver, brain	12.26 25.22
Nematoda		Ascaris lumbricoides	—	Humans; small intestine	Eggs; ingested	Ascariasis	Small intestine	25.24
		Enterobius vermicularis	—	Humans; large intestine	Eggs; ingested	Pinworm	Large intestine	12.27
		Necator americanus	—	Humans; small intestine	Larvae; through skin	Hookworm	Small intestine	12.28
		Ancylostoma duodenalis	—	Humans; small intestine	Larvae; through skin	Hookworm	Small Intestine	25.23
		Trichinella spiralis	—	Humans, swine, and other mammals; small intestine	Larvae; ingested	Trichinosis	Muscles	25.25
		Anisakines	Marine fish and squid	Marine mammals	Larvae in fish; ingested	Anisakiasis (sashimi worms)	Gastrointestinal tract	—

table 12.7 — Important Arthropod Vectors of Human Diseases

Class	Order	Vector	Disease	Figure Reference
Arachnida	Mites and ticks	Dermacentor (tick)	Rocky Mountain spotted fever	—
		Ixodes (tick)	Lyme disease, babesiosis, ehrlichiosis	12.30
		Ornithodorus (tick)	Relapsing fever	—
Insecta	Sucking lice	Pediculus (human louse)	Epidemic typhus	12.31a
	Fleas	Xenopsylla (rat flea)	Endemic murine typhus, plague	12.31b
	True flies	Chrysops (deer fly)	Tularemia	12.31c
		Aedes (mosquito)	Dengue fever, yellow fever	—
		Anopheles (mosquito)	Malaria	12.29
		Culex (mosquito)	Arboviral encephalitis	—
		Glossina (tsetse fly)	African trypanosomiasis	—
	True bugs	Triatoma (kissing bug)	Chagas' disease	12.31d

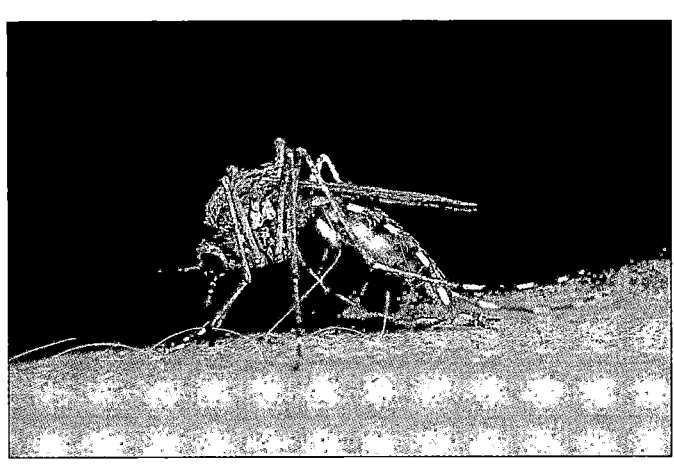

(a) Female mosquito

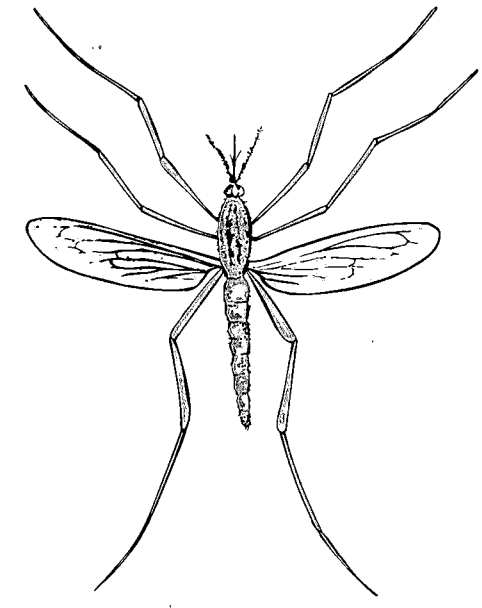

(b) *Anopheles*

FIGURE 12.29   Mosquitoes. **(a)** A female mosquito sucking blood from human skin. Mosquitoes transmit several diseases from person to person, including the viral diseases yellow fever and dengue fever. **(b)** The *Anopheles* mosquito transmits malaria.

■ **Mosquitoes are one family of flies.**

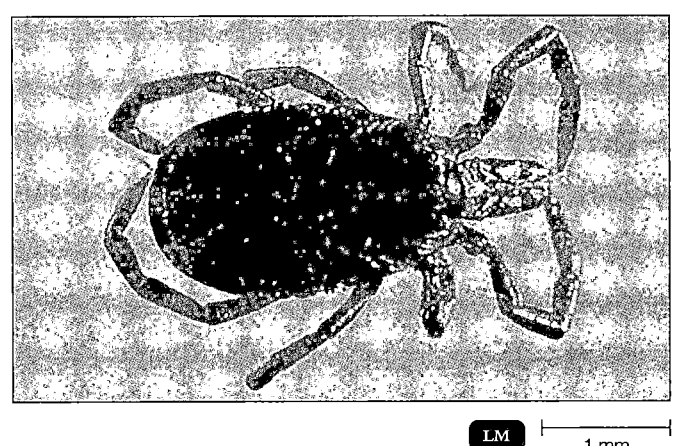

LM   1 mm

FIGURE 12.30   Ticks. *Ixodes pacificus* is the Lyme disease vector on the West Coast.

■ **Why aren't ticks classified as insects?**

Some vectors are just a mechanical means of transport for a pathogen. For example, houseflies lay their eggs on decaying organic matter such as feces. While doing so, a housefly can pick up a pathogen on its feet or body and transport the pathogen to our food.

Some parasites multiply in their vectors. When this happens, the parasites can accumulate in the vector's feces or saliva. Large numbers of parasites can then be deposited on or in the host while the vector is feeding there. The spirochete that causes Lyme disease is transmitted by ticks in this manner (see Chapter 23, page 634),

and the dengue fever virus is transmitted in the same way by mosquitoes (see Chapter 23, page 640).

As discussed earlier, *Plasmodium* is an example of a parasite that requires that its vector also be the definitive host. *Plasmodium* can sexually reproduce only in the gut of an *Anopheles* mosquito. *Plasmodium* is introduced into a human host with the mosquito's saliva, which acts as an anticoagulant to keep blood flowing.

To eliminate vectorborne diseases (such as African sleeping sickness), health workers focus on eradicating the vectors.

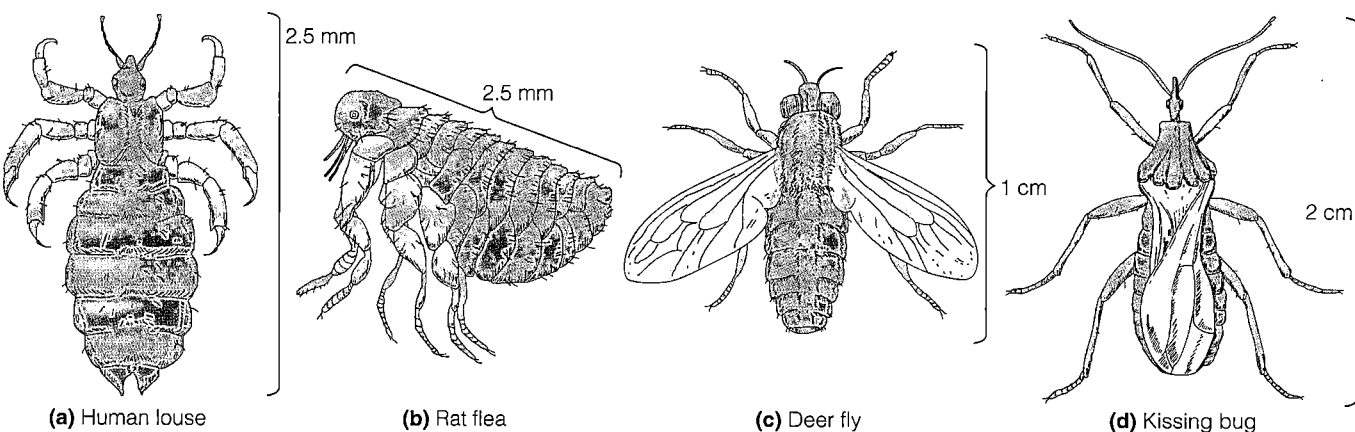

(a) Human louse   **(b)** Rat flea   **(c)** Deer fly   **(d)** Kissing bug

FIGURE 12.31 Arthropod vectors (above). (a) The human louse, *Pediculus.* **(b)** The rat flea, *Xenopsylla.* **(c)** The deer fly, *Chrysops.* **(d)** The kissing bug, *Triatoma.*

▨ Name one disease carried by each of these vectors.

Study Outline

FUNGI (pp. 331–342)

1. Mycology is the study of fungi.

2. The number of serious fungal infections is increasing.

3. Fungi are aerobic or facultatively anaerobic chemo-heterotrophs.

4. Most fungi are decomposers, and a few are parasites of plants and animals.

Characteristics of Fungi (pp. 332–335)

1. A fungal thallus consists of filaments of cells called hyphae; a mass of hyphae is called a mycelium.

2. Yeasts are unicellular fungi. To reproduce, fission yeasts divide symmetrically, whereas budding yeasts divide asymmetrically.

3. Buds that do not separate from the mother cell form pseudohyphae.

4. Pathogenic dimorphic fungi are yeastlike at 37°C and moldlike at 25°C.

5. These spores can be produced asexually: chlamydospores, sporangiospores, and conidiospores (including arthrospores and blastoconidia).

6. Fungi are classified according to the type of sexual spore that they form.

7. Sexual spores are usually produced in response to special circumstances, often changes in the environment.

8. Fungi can grow in acidic, low-moisture, aerobic environments.

9. They are able to metabolize complex carbohydrates.

Medically Important Phyla of Fungi (pp. 335–339)

1. The Zygomycota have coenocytic hyphae and produce sporangiospores and zygospores.

2. The Ascomycota have septate hyphae and produce ascospores and frequently conidiospores.

3. Basidiomycota have septate hyphae and produce basidiospores; some produce conidiospores.

4. Teleomorphic fungi produce sexual and asexual spores; anamorphic fungi produce asexual spores only.

Fungal Diseases (pp. 339–341)

1. Systemic mycoses are fungal infections deep within the body that affect many tissues and organs.

2. Subcutaneous mycoses are fungal infections beneath the skin.

3. Cutaneous mycoses affect keratin-containing tissues such as hair, nails, and skin.

4. Superficial mycoses are localized on hair shafts and superficial skin cells.

5. Opportunistic mycoses are caused by normal microbiota or fungi that are not usually pathogenic.

6. Opportunistic mycoses include mucormycosis, caused by some zygomycetes; aspergillosis, caused by *Aspergillus;* and candidiasis, caused by *Candida.*

7. Opportunistic mycoses can infect any tissues. However, they are usually systemic.

Economic Effects of Fungi (pp. 341–342)

1. *Saccharomyces* and *Trichoderma* are used in the production of foods.

2. Fungi are used for the biological control of pests.

3. Mold spoilage of fruits, grains, and vegetables is more common than bacterial spoilage of these products.

4. Many fungi cause diseases in plants (for example, in potatoes, chestnuts, and elms).

LICHENS (pp. 342-344)

1. A lichen is a mutualistic combination of an alga (or a cyanobacterium) and a fungus.

2. The alga photosynthesizes, providing carbohydrates for the lichen; the fungus provides a holdfast.

3. Lichens colonize habitats that are unsuitable for either the alga or the fungus alone.

4. Lichens may be classified on the basis of morphology as crustose, foliose, or fruticose.

5. Lichens are used for their pigments and as air quality indicators.

ALGAE (pp. 344-349)

1. Algae are unicellular, filamentous, or multicellular (thallic).

2. Most algae live in aquatic environments.

Characteristics of Algae (p. 344)

1. All algae are eukaryotic photoautotrophs.

2. The thallus (body) of multicellular algae usually consists of a stipe, a holdfast, and blades.

3. Algae reproduce asexually by cell division and fragmentation.

4. Many algae reproduce sexually.

5. Algae are photoautotrophs that produce oxygen.

6. Algae are classified according to their structures and pigments.

Selected Divisions of Algae (pp. 344-347)

1. Brown algae (kelp) may be harvested for algin.

2. Red algae grow deeper in the ocean than other algae because their red pigments can absorb the blue light that penetrates to deeper levels.

3. Green algae have cellulose and chlorophyll *a* and *b* and store starch.

4. Diatoms are unicellular and have pectin and silica cell walls; some produce a neurotoxin.

5. Dinoflagellates produce neurotoxins that cause paralytic shellfish poisoning and ciguatera.

Roles of Algae in Nature (pp. 347-349)

1. Algae are the primary producers in aquatic food chains.

2. Planktonic algae produce most of the molecular oxygen in the Earth's atmosphere.

3. Petroleum is the fossil remains of planktonic algae.

4. Unicellular algae are symbionts in such animals as *Tridacna*.

PROTOZOA (pp. 349-354)

1. Protozoa are unicellular, eukaryotic chemoheterotrophs.

2. Protozoa are found in soil and water and as normal microbiota in animals.

Characteristics of Protozoa (p. 349)

1. The vegetative form is called a trophozoite.

2. Asexual reproduction is by fission, budding, or schizogony.

3. Sexual reproduction is by conjugation.

4. During ciliate conjugation, two haploid nuclei fuse to produce a zygote.

5. Some protozoa can produce a cyst which provides protection during adverse environmental conditions.

6. Protozoa have complex cells with a pellicle, a cytostome, and an anal pore.

Medically Important Phyla of Protozoa (pp. 350-354)

1. Archaezoa lack mitochondria and have flagella; they include *Trichomonas* and *Giardia*.

2. Microsporidia lack mitochondria and microtubales; microsporans cause diarrhea in AIDS patients.

3. Rhizopoda are amoeba; they include *Entamoeba* and *Acanthamoeba*.

4. Apicomplexa have apical organelles for penetrating host tissue; they include *Plasmodium* and *Cryptosporidium*.

5. Ciliophora move by means of cilia; *Balantidium coli* is the human parasitic ciliate.

6. Euglenozoa move by means of flagella and lack sexual reproduction; they include *Trypanosoma*.

SLIME MOLDS (pp. 354-356)

1. Cellular slime molds resemble amoebas and ingest bacteria by phagocytosis.

2. Plasmodial slime molds consist of a multinucleated mass of protoplasm that engulfs organic debris and bacteria as it moves.

HELMINTHS (pp. 356-363)

1. Parasitic flatworms belong to the Phylum Platyhelminthes.

2. Parasitic roundworms belong to the Phylum Nematoda.

Characteristics of Helminths (pp. 357-358)

1. Helminths are multicellular animals; a few are parasites of humans.

2. The anatomy and life cycle of parasitic helminths are modified for parasitism.

3. The adult stage of a parasitic helminth is found in the definitive host.

4. Each larval stage of a parasitic helminth requires an intermediate host.

5. Helminths can be monoecious or dioecious.

Platyhelminths (pp. 358-361)

1. Flatworms are dorsoventrally flattened animals; parasitic flatworms may lack a digestive system.

2. Adult trematodes, or flukes, have an oral and ventral sucker with which they attach to host tissue.

3. Eggs of trematodes hatch into free-swimming miracidia that enter the first intermediate host; two generations of rediae develop in the first intermediate host; the rediae become cercariae that bore out of the first intermediate host and penetrate the second intermediate host; cercariae encyst as metacercariae in the second intermediate host; after they are ingested by the definitive host, the metacercariae develop into adults.

4. A cestode, or tapeworm, consists of a scolex (head) and proglottids.

5. Humans serve as the definitive host for the beef tapeworm, and cattle are the intermediate host.

6. Humans serve as the definitive host and can be an intermediate host for the pork tapeworm.

7. Humans serve as the intermediate host for *Echinococcus granulosus;* the definitive hosts are dogs, wolves, and foxes.

Nematodes (pp. 361–363)

1. Roundworms have a complete digestive system.

2. The nematodes that infect humans with their eggs are *Enterobius vermicularis* (pinworm) and *Ascaris lumbricoides.*

3. The nematodes that infect humans with their larvae are *Necator americanus* (hookworm), *Trichinella spiralis,* and anisakine worms.

ARTHROPODS AS VECTORS (pp. 363–365)

1. Jointed-legged animals, including ticks and insects, belong to the Phylum Arthropoda.

2. Arthropods that carry diseases are called vectors.

3. Elimination of vectorborne diseases is best done by the control or eradication of the vectors.

Study Questions

REVIEW

1. Contrast the mechanisms of conidiospore and ascospore formation by a fungus.

2. Fill in the following table.

Phylum	Spore Type(s)	
	Sexual	Asexual
Zygomycota		
Ascomycota		
Basidiomycota		

3. Following is a list of fungi, their methods of entry into the body, and sites of infections they cause. Categorize each type of mycosis as cutaneous, opportunistic, subcutaneous, superficial, or systemic.

Genus	Method of Entry	Site of Infection	Mycosis
Blastomyces	Inhalation	Lungs	
Sporothrix	Puncture	Ulcerative lesions	
Microsporum	Contact	Fingernails	
Trichosporon	Contact	Hair shafts	
Aspergillus	Inhalation	Lungs	

4. A mixed culture of *Escherichia coli* and *Penicillium chrysogenum* is inoculated onto the following culture media. On which medium would you expect each to grow? Why?
   a. 0.5% peptone in tap water
   b. 10% glucose in tap water

5. What is the role of the alga in a lichen? What is the role of the fungus?

6. Briefly discuss the importance of lichens in nature. Briefly discuss the importance of algae.

7. Complete the following table.

Division	Cell Wall Composition	Special Features/ Importance
Dinoflagellates		
Diatoms		
Red algae		
Brown algae		
Green algae		

Indicate which divisions consist primarily of unicellular forms. Which division could you include in the plant kingdom? Why?

8. Differentiate between cellular and plasmodial slime molds. How does each survive adverse environmental conditions?

9. Complete the following table.

Phylum	Method of Motility	One Human Parasite
Archaezoa		
Microsporidia		
Rhizopoda		
Apicomplexa		
Ciliophora		
Euglenozoa		

10. Why is it significant that *Trichomonas* does not have a cyst stage? Name a protozoan parasite that does have a cyst stage.

11. Recall the life cycle of *Plasmodium*. Where does asexual reproduction occur? Where does sexual reproduction occur? Identify the definitive host. Identify the vector.

12. By what means are helminthic parasites transmitted to humans?

13. To what phylum and class does this animal belong?

List two characteristics that put it in this phylum. Name the body parts. What is the name of the encysted larva of this animal?

14. Most nematodes are dioecious. What does this term mean? To what phylum do nematodes belong?

15. Vectors can be divided into three major types, according to the roles they play for the parasite. List the three types of vectors and a disease transmitted by each.

MULTIPLE CHOICE

1. How many phyla are represented in the following list of organisms: *Echinococcus, Cyclospora, Aspergillus, Taenia, Toxoplasma, Trichinella?*
   a. 1   **d.** 4
   b. 2   **e.** 5
   c. 3

Use the following choices to answer questions 2 and 3:
   (1) metacercaria   **(4)** miracidium
   (2) redia   **(5)** cercaria
   (3) adult

2. Put the above stages in order of development, beginning with the egg.
   a. 5, 4, 1, 2, 3
   b. 4, 2, 5, 1, 3
   c. 2, 5, 4, 3, 1
   d. 3, 4, 5, 1, 2
   e. 2, 4, 5, 1, 3

3. If a snail is the first intermediate host of a parasite with these stages, which stage would be found in the snail?
   a. 1   **d.** 4
   b. 2   **e.** 5
   c. 3

4. Which of the following statements about yeasts are true?
   (1) Yeasts are fungi.
   (2) Yeasts can form pseudohyphae.
   (3) Yeasts reproduce asexually by budding.
   (4) Yeasts are facultatively anaerobic.
   (5) All yeasts are pathogenic.
   (6) All yeasts are dimorphic.
   a. 1, 2, 3, 4
   b. 3, 4, 5, 6
   c. 2, 3, 4, 5
   d. 1, 3, 5, 6
   e. 2, 3, 4

5. Which of the following events follows cell fusion in an ascomycete?
   a. conidiophore formation   **d.** ascospore formation
   b. conidiospore germination **e.** conidiospore release
   c. ascus opening

6. The definitive host for *Plasmodium vivax* is
   a. human.   **c.** a sporocyte.
   b. *Anopheles.*   **d.** a gametocyte.

7. Fleas are the intermediate host for *Dipylidium caninum* tapeworm and dogs are the definitive host. Which stage of the parasite could be found in the flea?
   a. cysticerus larva   **c.** scolex
   b. proglottids   **d.** adult

Use the following choices to answer questions 8 and 9:
   a. Apicomplexa   **c.** Dinoflagellates
   b. Ciliophora   **d.** Microspora

8. These are obligate intracellular parasites that lack mitochondria.

9. These are nonmotile parasites with special organelles for penetrating host tissue.

10. These photosynthetic organisms can cause paralytic shellfish poisoning.

CRITICAL THINKING

1. A generalized life cycle of the liver fluke *Clonorchis sinensis* is shown below. Identify the intermediate host(s). Identify the definitive host(s). To what phylum and class does this animal belong?

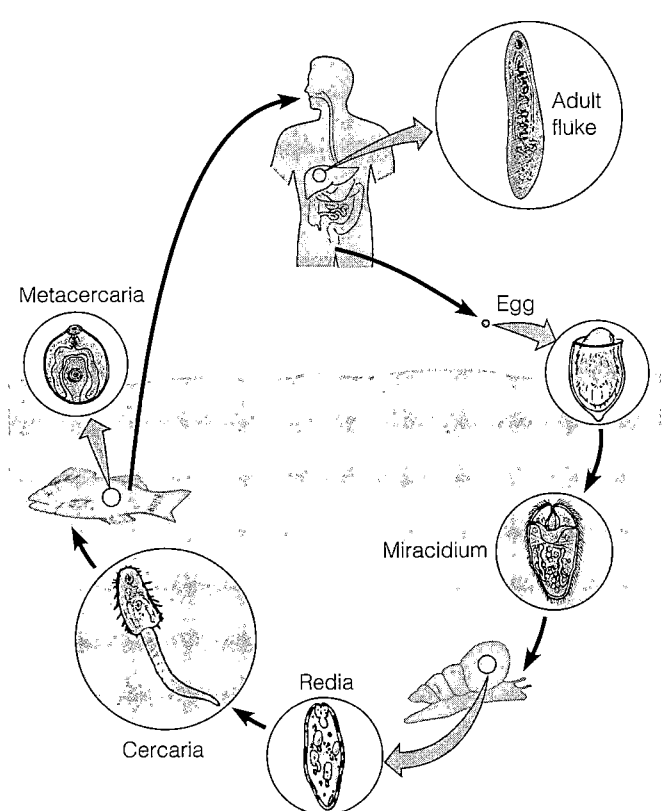

Adult fluke

Metacercaria

Egg

Miracidium

Redia

Cercaria

2. The size of a cell is limited by its surface-to-volume ratio; that is, if the volume becomes too great, internal heat cannot be dissipated, and nutrients and wastes cannot be efficiently transported. How do plasmodial slime molds manage to circumvent the surface-to-volume rule?

3. The life cycle of the fish tapeworm *Diphyllobothrium* is similar to that of *Taenia saginata,* except that the intermediate host is fish. Describe the life cycle and method of transmission to humans. Why are freshwater fish more likely to be a source of tapeworm infection than marine fish?

4. *Trypanosoma brucei gambiense* (part a) is the causative agent of African sleeping sickness. To what phylum does it belong? Part b shows a simplified life cycle for *T. b. gambiense.* Identify the host and vector of this parasite.

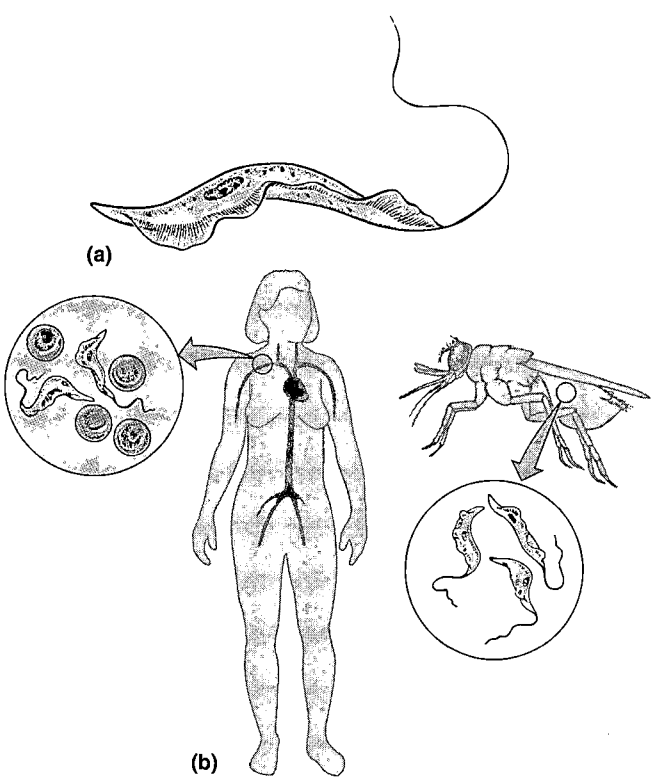

(a)

(b)

CLINICAL APPLICATIONS

1. A girl developed generalized seizures. A CAT scan revealed a single brain lesion consistent with a tumor. Biopsy of the lesion showed a cysticercus. The patient lived in South Carolina and had never traveled outside the state. What parasite caused her disease? How is this disease transmitted? How might it be prevented?

2. A California farmer developed a low-grade fever, myalgia, and cough. A chest X ray revealed an infiltrate in the lung. Microscopic examination of the sputum revealed round, budding cells. A sputum culture grew mycelia and arthrospores. What organism is most likely the cause of the symptoms? What is causing the man's disease? How is this disease transmitted? How might it be prevented?

3. A teenage male in California complained of remittent fever, chills, and headaches. A blood smear revealed ring-shaped cells in his white blood cells. He was successfully treated with primaquine and chloroquine. The patient lives near the San Luis Rey River and has no history of foreign travel, blood transfusion, or IV drug use. What is the disease? How was it acquired?

4. Seventeen patients in ten hospitals had cutaneous infections caused by *Rhizopus.* In all 17 patients, Elastoplast bandages were placed over sterile gauze pads to cover wounds. Fourteen of the patients had surgical wounds, two had venous line insertion sites, and one had a bite wound. Lesions present when the bandages were removed ranged from vesiculopustular eruptions to ulcerations and skin necrosis requiring debridement.
   a. How did the wounds most likely get contaminated?
   b. Why is a fungus a more likely contaminant than a bacterium in this instance?

5. In mid-December, a female who had been on prednisone and is an insulin-dependent diabetic fell and received an abrasion on the dorsal side of her right hand. She was placed on penicillin. By the end of January, the ulcer had not healed, and she was referred to a plastic surgeon. On January 30, a swab of the wound was cultured at 35°C on blood agar. On the same day, a smear was made for Gram staining. The Gram stain showed fungal elements. What would you do next? Brownish, waxy colonies grew on the blood agar. Slide cultures set up on February 1 and incubated at 25°C showed septate hyphae and single conidia. What caused the infection? What treatment do you recommend? (See Chapter 20.)

Learning with Technology

MP = **The Microbiology Place website**   **ST** = **Student Tutorial CD-ROM**

MP   Don't forget to go to The Microbiology Place website (http://www.microbiologyplace.com) to take the practice tests, explore the interactive activity and case study, and check out the news articles and web links for this chapter.

ST   Remember there is also a quiz for this chapter on the Microbiology Interactive Student Tutorial CD-ROM.

chapter 13

Viruses, Viroids, and Prions

Mastadenovirus. *This virus causes upper respiratory tract infections in humans.*

In 1886, the Dutch chemist Adolf Mayer showed that tobacco mosaic disease was transmissible from a diseased plant to a healthy plant. In 1892, in an experiment to isolate the cause of tobacco mosaic disease, the Russian bacteriologist Dmitri Iwanowski filtered the sap of diseased tobacco plants through a porcelain filter that was designed to retain bacteria. He expected to find the microbe trapped in the filter; instead, he found that the infectious agent had passed through the minute pores of the filter. When he injected healthy plants with the filtered fluid, they contracted tobacco mosaic disease.

Iwanowski still believed the infectious agent to be a bacterium that was small enough to pass through the filter. But later, other scientists, led by the Dutch botanist Martinus Beijerinck, observed that the behavior of this agent was different from that of bacteria. In the early 1900s, a distinction was finally made between bacteria and viruses, the filterable agents that cause tobacco mosaic disease and many other diseases. (*Virus* is the Latin word for poison.) In 1935, American chemist Wendell M. Stanley isolated the tobacco mosaic virus (genus *Tobamovirus*), making it possible for the first time to carry out chemical and structural studies on a purified virus. At about the same time, the invention of the electron microscope made it possible to see viruses for the first time.

We now know that viruses are found as parasites in all types of organisms. They can reproduce only within cells. Many human diseases are known to be caused by viruses; some diseases of agriculturally important animals and plants are also known to be caused by viruses. Viruses infect fungi, bacteria, and protists as well.

Advances in molecular biology techniques in the 1980s and 1990s have led to the recognition of new human viruses. These newly recognized viruses are not necessarily new but may be new to Western medicine. Human immunodeficiency virus (HIV) had a long biological history before it was recognized by Western medicine in 1983. The most common source of emerging viral diseases is wild and domestic animals. Although *Hantavirus* Sin Nombre had previously been isolated from rodents, the disease had not been reported in the United States until 1993, when many people in the American Southwest contracted the disease and several died. In 1999, West Nile Virus was determined to be the cause of an outbreak of encephalitis in New York. The source of the infections may have been migratory birds.

Public health officials are concerned that these emerging viruses will pose health risks and that the ease of world travel and changing environments will spread viruses into new areas.

General Characteristics of Viruses

Learning Objective

▪ *Differentiate a virus from a bacterium.*

The question of whether viruses are living organisms has an ambiguous answer. Life can be defined as a complex set of processes resulting from the actions of proteins specified by nucleic acids. The nucleic acids of living cells are in action all the time. Because viruses are inert outside living host cells, in this sense they are not considered to be living organisms. However, once viruses enter a host cell, the viral nucleic acids become active, and viral multiplication results. In this sense, viruses are alive when they multiply in the host cells they infect. From a clinical point of view, viruses can be considered alive because they cause infection and disease, just as pathogenic bacteria, fungi, and protozoa do. Depending on one's viewpoint, a virus may be regarded as an exceptionally complex aggregation of nonliving chemicals, or as an exceptionally simple living microorganism.

How, then, do we define a virus? Viruses were originally distinguished from other infectious agents because they are especially small (filterable) and because they are **obligatory intracellular parasites**—that is, they absolutely require living host cells in order to multiply. However, both of these properties are shared by certain small bacteria, such as some rickettsias. Viruses and bacteria are compared in Table 13.1.

The truly distinctive features of viruses are now known to relate to their simple structural organization and their mechanism of multiplication. Accordingly, **viruses** are entities that:

▪ Contain a single type of nucleic acid, either DNA or RNA.

▪ Contain a protein coat (sometimes itself enclosed by an envelope of lipids, proteins, and carbohydrates) that surrounds the nucleic acid.

▪ Multiply inside living cells by using the synthesizing machinery of the cell.

▪ Cause the synthesis of specialized structures that can transfer the viral nucleic acid to other cells.

Viruses have few or no enzymes of their own for metabolism; for example, they lack enzymes for protein synthesis and ATP generation. To multiply, viruses must take over the metabolic machinery of the host cell. This fact has considerable medical significance for the development of antiviral drugs, because most drugs that would interfere

table 13.1   Viruses and Bacteria Compared

| | Bacteria | | Viruses |
	Typical Bacteria	Rickettsias/ Chlamydias	
Intracellular parasite	No	Yes	Yes
Plasma membrane	Yes	Yes	No
Binary fission	Yes	Yes	No
Pass through bacteriological filters	No	No/Yes	Yes
Possess both DNA and RNA	Yes	Yes	No
ATP-generating metabolism	Yes	Yes/No	No
Ribosomes	Yes	Yes	No
Sensitive to antibiotics	Yes	Yes	No
Sensitive to interferon	No	No	Yes

with viral multiplication would also interfere with the functioning of the host cell and therefore are too toxic for clinical use. (Antiviral drugs are discussed in Chapter 20.)

Host Range

The **host range** of a virus is the spectrum of host cells the virus can infect. There are viruses that infect invertebrates, vertebrates, plants, protists, fungi, and bacteria. However, most viruses are able to infect specific types of cells of only one host species. In this chapter we are concerned mainly with viruses that infect either humans or bacteria. Viruses that infect bacteria are called **bacteriophages,** or **phages.**

The particular host range of a virus is determined by the virus's requirements for its specific attachment to the host cell and the availability within the potential host of cellular factors required for viral multiplication. For the virus to infect the host cell, the outer surface of the virus must chemically interact with specific receptor sites on the surface of the cell. The two complementary components are held together by weak bonds, such as hydrogen bonds. For some bacteriophages, the receptor site is part of the cell wall of the host; in other cases, it is part of the fimbriae or flagella. For animal viruses, the receptor sites are on the plasma membranes of the host cells.

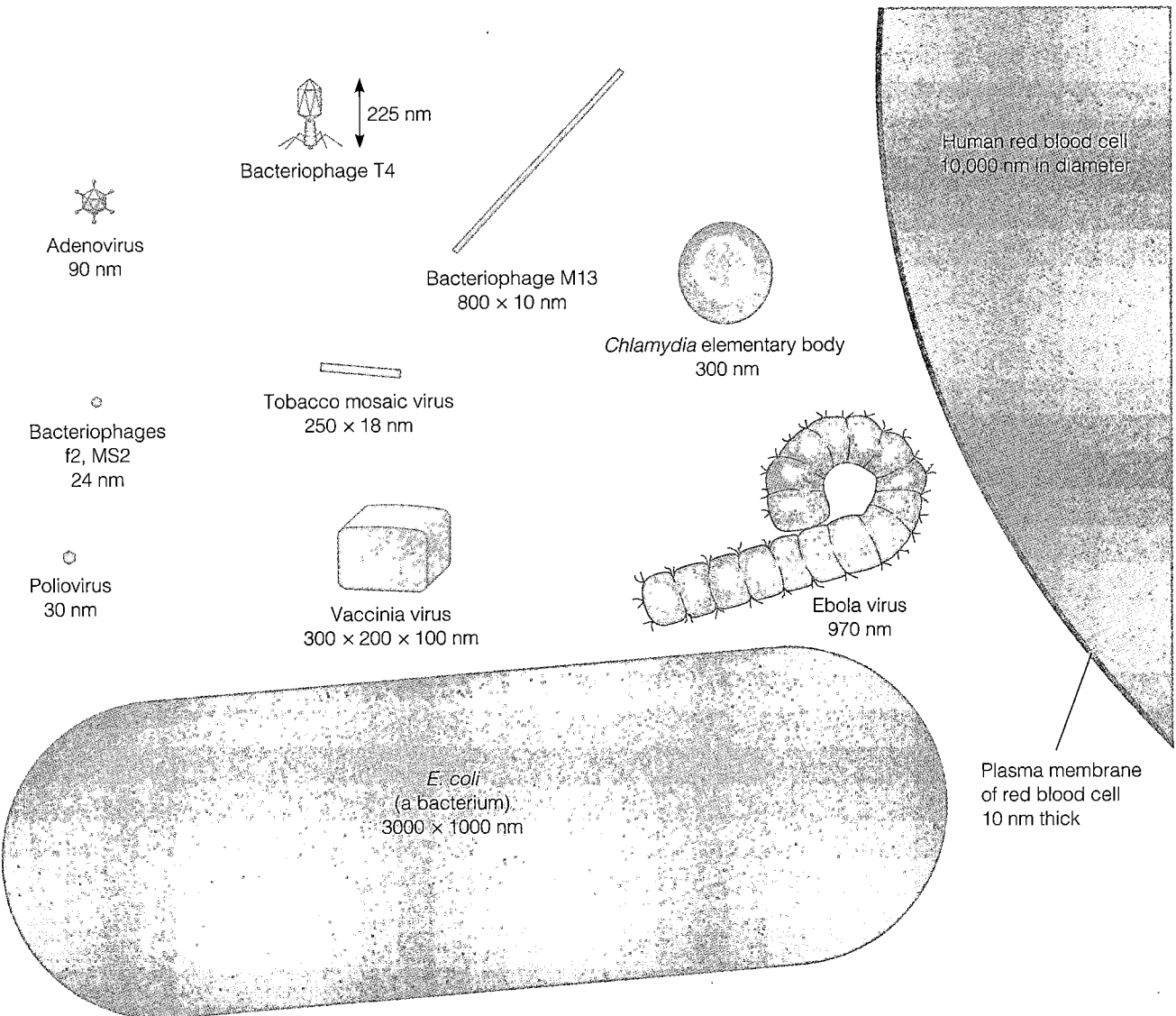

FIGURE 13.1 Virus sizes. The sizes of several viruses (teal) and bacteria (tan) are compared with a human red blood cell, shown to the right of the microbes. Dimensions are given in nanometers (nm) and are either diameters or length by width.

■ How do viruses differ from bacteria?

Viral Size

Viral sizes are determined with the aid of electron microscopy. Different viruses vary considerably in size. Although most are quite a bit smaller than bacteria, some of the larger viruses (such as the vaccinia virus) are about the same size as some very small bacteria (such as the mycoplasmas, rickettsias, and chlamydias). Viruses range from 20 to 14,000 nm in length. The comparative sizes of several viruses and bacteria are shown in Figure 13.1.

Viral Structure

Learning Objective

■ *Describe the chemical and physical structure of both an enveloped and a nonenveloped virus.*

A **virion** is a complete, fully developed, infectious viral particle composed of nucleic acid and surrounded by a protein coat that protects it from the environment and is a vehicle of transmission from one host cell to another. Viruses are classified by differences in the structures of these coats.

FIGURE 13.2 Morphology of a nonenveloped polyhedral virus.
(a) A diagram of a polyhedral (icosahedral) virus. (b) A micrograph of the adenovirus *Mastadenovirus*. Individual capsomeres in the protein coat are visible.

▨ A viral capsid is composed of capsomeres that often form an icosahedron.

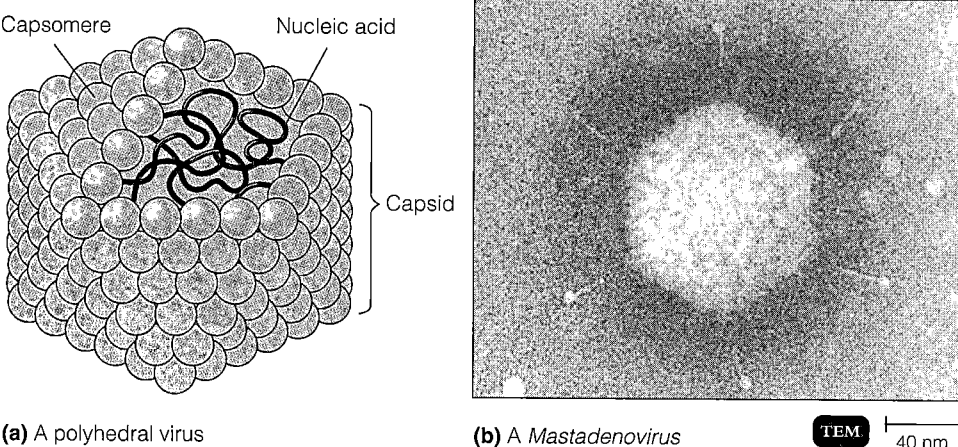

(a) A polyhedral virus

(b) A *Mastadenovirus*

TEM ⊢——⊣ 40 nm

Nucleic Acid

In contrast to prokaryotic and eukaryotic cells, in which DNA is always the primary genetic material (and RNA plays an auxiliary role), a virus can have either DNA or RNA, but never both. The nucleic acid of a virus can be single-stranded or double-stranded. Thus, there are viruses with the familiar double-stranded DNA, with single-stranded DNA, with double-stranded RNA, and with single-stranded RNA. Depending on the virus, the nucleic acid can be linear or circular. In some viruses (such as the influenza virus), the nucleic acid is in several separate segments.

The percentage of nucleic acid in relation to protein is about 1% for the *Influenzavirus* and about 50% for certain bacteriophages. The total amount of nucleic acid varies from a few thousand nucleotides (or pairs) to as many as 250,000 nucleotides. (*E. coli*'s chromosome consists of approximately 4 million nucleotide pairs.)

Capsid and Envelope

The nucleic acid of a virus is surrounded by a protein coat called the **capsid** (Figure 13.2a). The structure of the capsid is ultimately determined by the viral nucleic acid and accounts for most of the mass of a virus, especially of small ones. Each capsid is composed of protein subunits called **capsomeres.** In some viruses, the proteins composing the capsomeres are of a single type; in other viruses, several types of protein may be present. Individual capsomeres are often visible in electron micrographs (see Figure 13.2b for an example). The arrangement of capsomeres is characteristic of a particular type of virus.

In some viruses, the capsid is covered by an **envelope** (Figure 13.3a), which usually consists of some combination of lipids, proteins, and carbohydrates. Some animal viruses are released from the host cell by an extrusion process that coats the virus with a layer of the host cell's plasma membrane; that layer becomes the viral envelope. In many cases, the envelope contains proteins determined by the viral nucleic acid and materials derived from normal host cell components.

Depending on the virus, envelopes may or may not be covered by **spikes,** which are carbohydrate-protein complexes that project from the surface of the envelope. Some viruses attach to host cells by means of spikes. Spikes are such a reliable characteristic of some viruses that they can be used as a means of identification. The ability of certain viruses, such as the *Influenzavirus*, to clump red blood cells is associated with spikes. Such viruses bind to red blood cells and form bridges between them. The resulting clumping is called *hemagglutination* and is the basis for several useful laboratory tests.

Viruses whose capsids are not covered by an envelope are known as **nonenveloped viruses** (see Figure 13.2). The capsid of a nonenveloped virus protects the nucleic acid from nuclease enzymes in biological fluids and promotes the virus's attachment to susceptible host cells.

When the host has been infected by a virus, the host immune system is stimulated to produce antibodies (proteins that react with the surface proteins of the virus). This interaction between host antibodies and virus proteins should inactivate the virus and stop the infection. However, some viruses can escape antibodies because regions of the genes that code for these viruses' surface proteins are susceptible to mutations. The progeny of mutant viruses have altered surface proteins, such that the antibodies are not able to react with them. *Influenzavirus* frequently undergoes such changes in its spikes. This is why you can get influenza more than once. Although you may have produced antibodies to one *Influenzavirus*, the virus can mutate and infect you again.

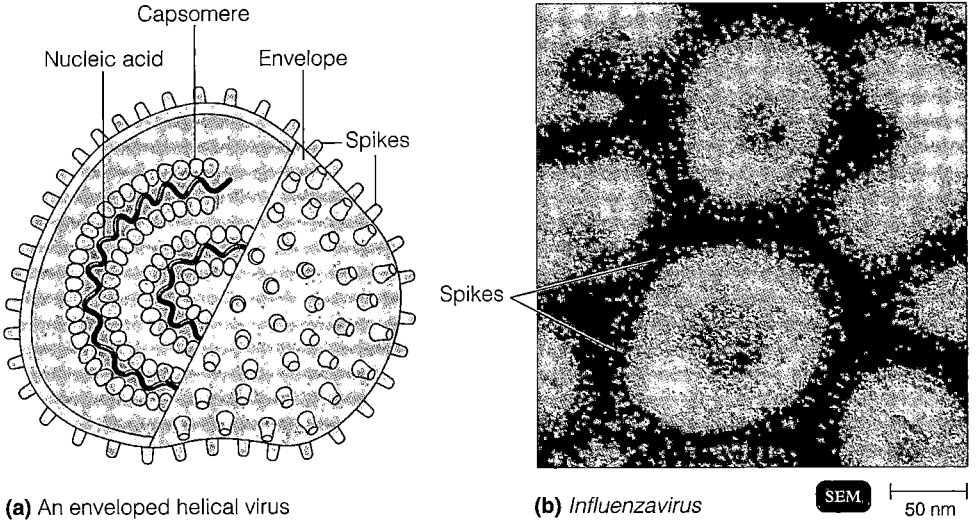

(a) An enveloped helical virus

(b) *Influenzavirus*   SEM |———| 50 nm

FIGURE 13.3 Morphology of an enveloped helical virus. (a) A diagram of an enveloped helical virus. **(b)** A micrograph of *Influenzavirus* A2. Notice the halo of spikes projecting from the outer surface of each envelope (see Chapter 24).

■ **What is the nucleic acid in a virus?**

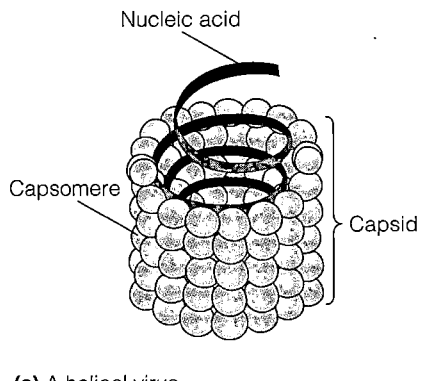

(a) A helical virus

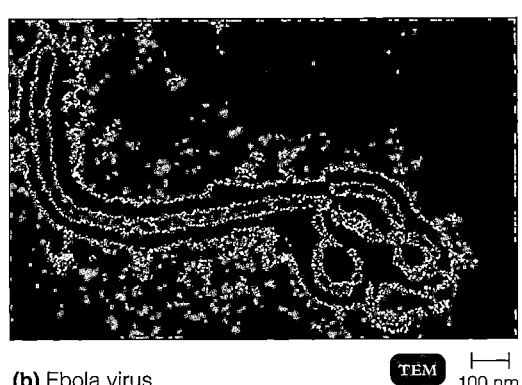

(b) Ebola virus   TEM |—| 100 nm

FIGURE 13.4 Morphology of a helical virus. (a) A diagram of a portion of a helical virus. Several rows of capsomeres have been removed to reveal the nucleic acid. **(b)** A micrograph of Ebola virus, a filovirus showing helical rods.

▣ **Helical viruses look like long or coiled threads.**

General Morphology

Viruses may be classified into several different morphological types on the basis of their capsid architecture. The structure of these capsids has been revealed by electron microscopy and a technique called X-ray crystallography.

Helical Viruses

Helical viruses resemble long rods that may be rigid or flexible. The viral nucleic acid is found within a hollow, cylindrical capsid that has a helical structure (Figure 13.4). The viruses that cause rabies and Ebola Hemorrhagic Fever are helical viruses.

Polyhedral Viruses

Many animal, plant, and bacterial viruses are polyhedral, or many-sided, viruses. The capsid of most polyhedral viruses is in the shape of an *icosahedron,* a regular polyhedron with 20 triangular faces and 12 corners (see Figure 13.2a). The capsomeres of each face form an equilateral triangle. An example of a polyhedral virus in the shape of an icosahedron is the adenovirus (shown in Figure 13.2b). Another icosahedral virus is the poliovirus.

Enveloped Viruses

As noted earlier, the capsid of some viruses is covered by an envelope. Enveloped viruses are roughly spherical. When helical or polyhedral viruses are enclosed by envelopes, they are called *enveloped helical* or *enveloped polyhedral viruses.* An example of an enveloped helical virus is the *Influenzavirus* (see Figure 13.3b). An example of an enveloped polyhedral (icosahedral) virus is the herpes simplex virus (genus *Simplexvirus*) (see Figure 13.14).

Complex Viruses

Some viruses, particularly bacterial viruses, have complicated structures and are called **complex viruses.**

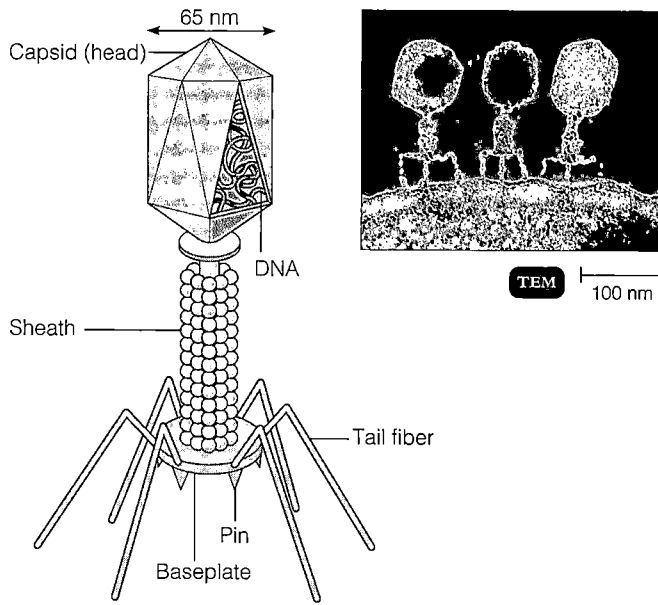

(a) A T-even bacteriophage

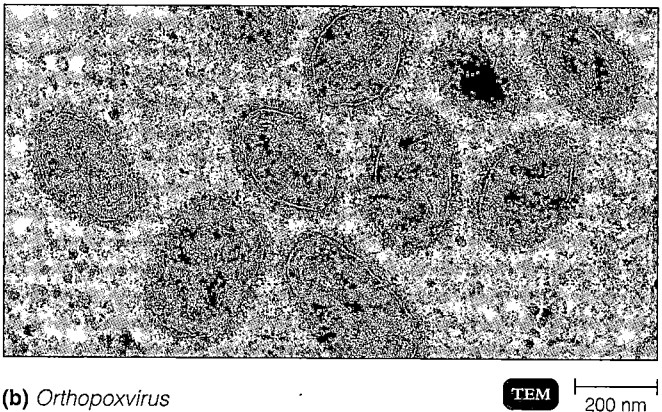

(b) *Orthopoxvirus*

FIGURE 13.5 Morphology of complex viruses. (a) A diagram and micrograph of a T-even bacteriophage. **(b)** A micrograph of variola virus, a species in the genus *Orthopoxvirus*, which causes smallpox.

■ **What is the chemical composition of a capsid?**

One example of a complex virus is a bacteriophage. Some bacteriophages have capsids to which additional structures are attached (Figure 13.5a). In this figure, notice that the capsid (head) is polyhedral and the tail sheath is helical. The head contains the nucleic acid. Later in the chapter we will discuss the functions of the other structures, such as the tail sheath, tail fibers, plate, and pin. Another example of complex viruses are poxviruses, which do not contain clearly identifiable capsids but have several coats around the nucleic acid (Figure 13.5b).

Taxonomy of Viruses

Learning Objectives
- *Define viral species.*
- *Give an example of a family, genus, and common name for a virus.*

Just as we need taxonomic categories of plants, animals, and bacteria, we need viral taxonomy to help us organize and understand newly discovered organisms. The oldest classification of viruses is based on symptomatology, such as for diseases that affect the respiratory system. This system was convenient but not scientifically acceptable because the same virus may cause more than one disease, depending on the tissue affected. In addition, this system artificially grouped viruses that do not infect humans.

Virologists began addressing the problem of viral taxonomy in 1966 with the formation of the International Committee on the Taxonomy of Viruses (ICTV). Since then, the ICTV has been grouping viruses into families based on (1) nucleic acid type, (2) strategy for replication, and (3) morphology. The suffix *-virus* is used for genus names; family names end in *-viridae;* and order names end in *-ales.* Two orders have been approved by the ICTV. Coronaviruses and equine arteritis virus are in the order Nidovirales. Viruses with one negative strand of RNA have been classified in the order Mononegavirales. In formal usage, the family and genus names are used in the following manner: Family Herpesviridae, genus *Simplexvirus,* human herpes virus 2.

A **viral species** is a group of viruses sharing the same genetic information and ecological niche. Specific epithets for viruses are not used. Thus, viral species are designated by descriptive common names, such as human immunodeficiency virus (HIV), with subspecies (if any) designated by a number (HIV-1). Table 13.2 presents a summary of the classification of viruses that infect humans.

The Isolation, Cultivation, and Identification of Viruses

The fact that viruses cannot multiply outside a living host cell complicates their detection, enumeration, and identification. Viruses must be provided with living cells instead of a fairly simple chemical medium. Living plants and animals are difficult and expensive to maintain, and pathogenic viruses that grow only in higher primates and human hosts cause additional complications. However, viruses that use bacterial cells as a host (bacteriophages)

table 13.2	*Families of Viruses That Affect Humans*		
Characteristics/ Dimensions	**Viral Family**	**Important Genera**	**Clinical or Special Features**
Single-stranded DNA nonenveloped 18–25 nm	Parvoviridae	*Dependovirus*	Depend on coinfection with adenoviruses; cause fetal death, gastroenteritis.
Double-stranded DNA nonenveloped 70–90 nm	Adenoviridae	*Mastadenovirus*	Medium-sized viruses that cause various respiratory infections in humans; some cause tumors in animals.
40–57 nm	Papovaviridae	*Papillomavirus* (human wart virus) *Polyomavirus*	Small viruses that induce tumors; the human wart virus (papilloma) and certain viruses that produce cancer in animals (polyoma and simian) belong to this family. Refer to Chapters 21 and 26.
Double-stranded DNA enveloped 200–350 nm	Poxviridae	*Orthopoxvirus* (vaccinia and smallpox viruses) *Molluscipoxvirus*	Very large, complex, brick-shaped viruses that cause diseases such as smallpox (variola), molluscum contagiosum (wartlike skin lesion), and cowpox. Refer to Chapter 21.
150–200 nm	Herpesviridae	*Simplexvirus* (HHV-1 and 2) *Varicellovirus* (HHV-3) *Lymphocryptovirus* (HHV-4) *Cytomegalovirus* (HHV-5) *Roseolovirus* (HHV-6) HHV-7 Kaposi's sarcoma (HHV-8)	Medium-sized viruses that cause various human diseases, such as fever blisters, chickenpox, shingles, and infectious mononucleosis; implicated in a type of human cancer called Burkitt's lymphoma. Refer to Chapters 21, 23, and 26.
42 nm	Hepadnaviridae	*Hepadnavirus* (hepatitis B virus)	After protein synthesis, hepatitis B virus uses reverse transcriptase to produce its DNA from mRNA; causes hepatitis B and liver tumors. Refer to Chapter 25.
Single-stranded RNA, + strand nonenveloped 28–30 nm	Picornaviridae	*Enterovirus* *Rhinovirus* (common cold virus) Hepatitis A virus	At least 70 human enteroviruses are known, including the polio-, coxsackie-, and echoviruses; more than 100 rhinoviruses exist and are the most common cause of colds. Refer to Chapters 22, 23, 24, and 25.
35–40 nm	Caliciviridae	Hepatitis E virus Norwalk agent	Includes causes of gastroenteritis and one cause of human hepatitis. Refer to Chapter 25.
Single-stranded RNA, + strand enveloped 60–70 nm	Togaviridae	*Alphavirus* *Rubivirus* (rubella virus)	Included are many viruses transmitted by arthropods (*Alphavirus*); diseases include eastern equine encephalitis (EEE) and western equine encephalitis (WEE). Rubella virus is transmitted by the respiratory route. Refer to Chapters 21, 22, and 23.

▶

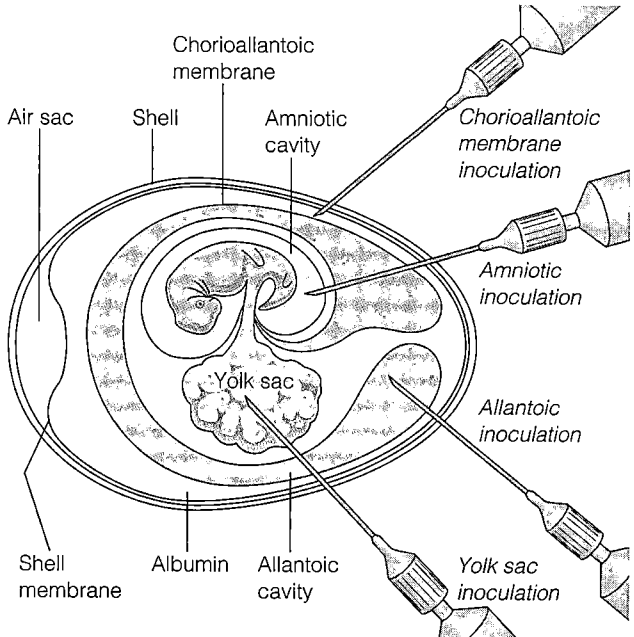

FIGURE 13.7 Inoculation of an embryonated egg. The injection site determines the membrane on which the viruses will grow.

■ Viruses must be grown in living host cells.

In Embryonated Eggs

If the virus will grow in an *embryonated egg,* this can be a fairly convenient and inexpensive form of host for many animal viruses. A hole is drilled in the shell of the embryonated egg, and a viral suspension or suspected virus-containing tissue is injected into the fluid of the egg. There are several membranes in an egg, and the virus is injected near the one most appropriate for its growth (Figure 13.7). Viral growth is signaled by the death of the embryo, by embryo cell damage, or by the formation of typical pocks or lesions on the egg membranes. This method was once the most widely used method of viral isolation and growth, and it is still used to grow viruses for some vaccines. For this reason, you may be asked if you are allergic to eggs before receiving a vaccination because egg proteins may be present in the viral vaccine preparations. (Allergic reactions will be discussed in Chapter 19.)

In Cell Cultures

Cell cultures have replaced embryonated eggs as the preferred type of growth medium for many viruses. Cell cultures consist of cells grown in culture media in the laboratory. Because these cultures are generally rather homogeneous collections of cells and can be propagated and handled much like bacterial cultures, they are more convenient to work with than whole animals or embryonated eggs.

Cell culture lines are started by treating a slice of animal tissue with enzymes that separate the individual cells (Figure 13.8). These cells are suspended in a solution that provides the osmotic pressure, nutrients, and growth factors needed for the cells to grow. Normal cells tend to adhere to the glass or plastic container and reproduce to form a monolayer. Viruses infecting such a monolayer sometimes cause the cells of the monolayer to deteriorate as they multiply. This cell deterioration is called **cytopathic effect (CPE),** illustrated in Figure 13.9. CPE can be detected and counted in much the same way as plaques caused by bacteriophages on a lawn of bacteria.

Viruses may be grown in primary or continuous cell lines. **Primary cell lines,** derived from tissue slices, tend to die out after only a few generations. Certain cell lines, called **diploid cell lines,** developed from human embryos can be maintained for about 100 generations and are widely used for culturing viruses that require a human host. Cell lines developed from embryonic human cells are used to culture rabies virus (genus *Lyssavirus*) for a rabies vaccine called human diploid culture vaccine (see Chapter 22).

When viruses are routinely grown in a laboratory, **continuous cell lines** are used. These are transformed (cancerous) cells that can be maintained through an indefinite number of generations, and they are sometimes called immortal cell lines (see the discussion of transformation later in the chapter). One of these, the HeLa cell line, was isolated from the cancer of a woman who died in 1951. After years of laboratory cultivation, many such cell lines have lost almost all the original characteristics of the cell, but these changes have not interfered with the use of the cells for viral propagation. In spite of the success of cell culture in viral isolation and growth, there are still some viruses that have never been successfully cultivated in cell culture.

The idea of cell culture dates back to the end of the nineteenth century, but it was not a practical laboratory technique until the development of antibiotics in the years following World War II. A major problem with cell culture is that the cell lines must be kept free of microbial contamination. The maintenance of cell culture lines requires trained technicians with considerable experience working on a full-time basis. Because of these difficulties, most hospital laboratories and many state health laboratories do not isolate and identify viruses in clinical work. Instead, the tissue or serum samples are sent to central laboratories that specialize in such work.

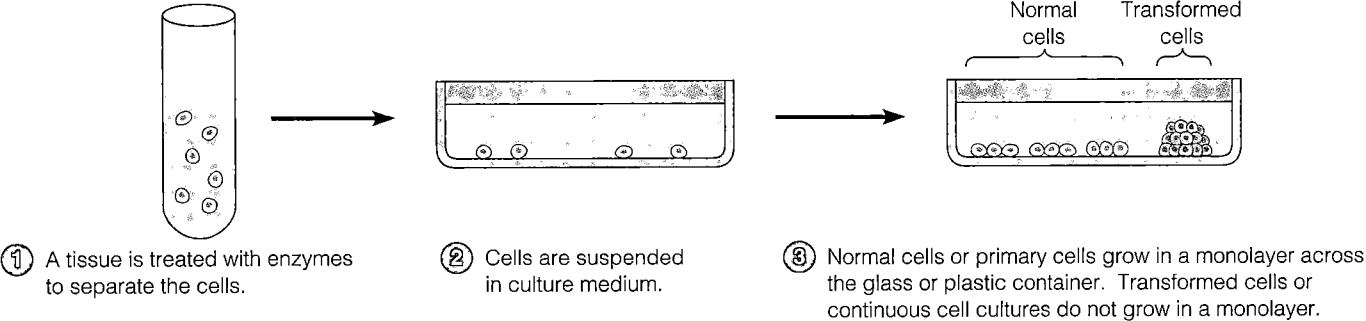

① A tissue is treated with enzymes to separate the cells.

② Cells are suspended in culture medium.

③ Normal cells or primary cells grow in a monolayer across the glass or plastic container. Transformed cells or continuous cell cultures do not grow in a monolayer.

FIGURE 13.8 Cell cultures. Transformed cells can be grown indefinitely in laboratory culture.

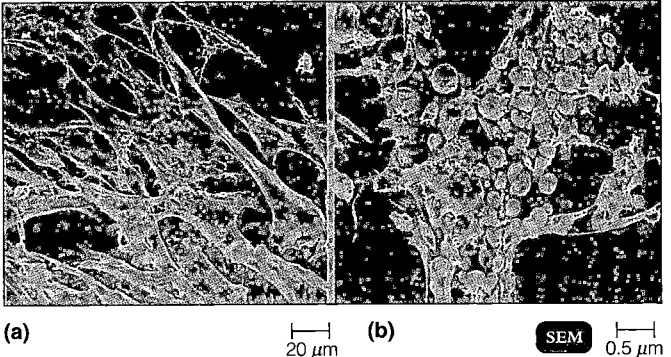

(a) |—| 20 μm **(b)** SEM |—| 0.5 μm

FIGURE 13.9 The cytopathic effect of viruses. (a) Uninfected mouse cells align next to each other forming a monolayer. **(b)** The same cells 24 hours after infection with vesicular stomatitis virus (VSV) (see Figure 13.18b). Notice the cells pile up and "round up."

■ How did VSV infection affect the cells?

Viral Identification

Learning Objective

■ *List three techniques used to identify viruses.*

The identification of viral isolates is not an easy task. For one thing, viruses cannot be seen at all without the use of an electron microscope. Serological methods, such as Western blotting, are the most commonly used means of identification (see Figure 10.12, page 291). In these tests, the virus is detected and identified by its reaction with antibodies. We will discuss antibodies in detail in Chapter 17 and a number of immunological tests for identifying viruses in Chapter 18. Observation of cytopathic effects, described in Chapter 15 on page 447, is also useful for the identification of a virus.

Virologists can identify and characterize viruses by using such modern molecular methods as restriction fragment length polymorphisms (RFLPs) and the polymerase chain reaction (PCR) (Chapter 9, pages 255 and 266). PCR was used to amplify RNA from birds and humans to identify the 1999 West Nile Virus outbreak in the United States.

Viral Multiplication

The nucleic acid in a virion contains only a few of the genes needed for the synthesis of new viruses. These include genes for the virion's structural components, such as the capsid proteins, and genes for a few of the enzymes used in the viral life cycle. These enzymes are synthesized and functional only when the virus is within the host cell. Viral enzymes are almost entirely concerned with replicating or processing viral nucleic acid. Enzymes needed for protein synthesis, ribosomes, tRNA, and energy production are supplied by the host cell and are used for synthesizing viral proteins, including viral enzymes. Although the smallest nonenveloped virions do not contain any preformed enzymes, the larger virions may contain one or a few enzymes, which usually function in helping the virus penetrate the host cell or replicate its own nucleic acid.

Thus, for a virus to multiply, it must invade a host cell and take over the host's metabolic machinery. A single virion can give rise to several or even thousands of similar viruses in a single host cell. This process can drastically change the host cell and can even cause its death.

Multiplication of Bacteriophages

Learning Objectives

■ *Describe the lytic cycle of T-even bacteriophages.*
■ *Describe the lysogenic cycle of bacteriophage lambda.*

Although the means by which a virus enters and exits a host cell may vary, the basic mechanism of viral multiplication is similar for all viruses. The best-understood viral life cycles are those of the bacteriophages. Phages can multiply by two alternative mechanisms: the lytic cycle or the lysogenic cycle. The **lytic cycle** ends with the lysis and death of the host cell, whereas the host cell remains alive in the **lysogenic cycle.** Because the *T-even bacteriophages* (T2, T4, and T6) have been studied most extensively, we will describe the multiplication of T-even bacteriophages in their host, *E. coli,* as an example of the lytic cycle.

T-Even Bacteriophages: The Lytic Cycle

The virions of T-even bacteriophages are large, complex, and nonenveloped, with a characteristic head-and-tail structure shown in Figures 13.5a and 13.10. The length of DNA contained in these bacteriophages is only about 6% of that contained in *E. coli,* yet the phage has enough DNA for over 100 genes. The multiplication cycle of these phages, like that of all viruses, occurs in five distinct stages: attachment, penetration, biosynthesis, maturation, and release (see Figure 13.10).

Attachment. ① After a chance collision between phage particles and bacteria, *attachment,* or *adsorption,* occurs. During this process, an attachment site on the virus attaches to a complementary receptor site on the bacterial cell. This attachment is a chemical interaction in which weak bonds are formed between the attachment and receptor sites. T-even bacteriophages use fibers at the end of the tail as attachment sites. The complementary receptor sites are on the bacterial cell wall.

Penetration. ② After attachment, the T-even bacteriophage injects its DNA (nucleic acid) into the bacterium. To do this, the bacteriophage's tail releases an enzyme, phage **lysozyme,** which breaks down a portion of the bacterial cell wall. During the process of *penetration,* the tail sheath of the phage contracts, and the tail core is driven through the cell wall. When the tip of the core reaches the plasma membrane, the DNA from the bacteriophage's head passes through the tail core, through the plasma membrane, and enters the bacterial cell. The capsid remains outside the bacterial cell. Therefore, the phage particle functions like a hypodermic syringe to inject its DNA into the bacterial cell.

Biosynthesis. ③ Once the bacteriophage DNA has reached the cytoplasm of the host cell, the biosynthesis of viral nucleic acid and protein occurs. Host protein synthesis is stopped by virus-induced degradation of the host DNA, viral proteins that interfere with transcription, or the repression of translation.

Initially, the phage uses the host cell's nucleotides and several of its enzymes to synthesize many copies of phage DNA. Soon after, the biosynthesis of viral proteins begins. Any RNA transcribed in the cell is mRNA transcribed from phage DNA for the biosynthesis of phage enzymes and capsid proteins. The host cell's ribosomes, enzymes, and amino acids are used for translation. Genetic controls regulate when different regions of phage DNA are transcribed into mRNA during the multiplication cycle. For example, early messages are translated into early phage proteins, the enzymes used in the synthesis of phage DNA. Also, late messages are translated into late phage proteins for the synthesis of capsid proteins.

For several minutes following infection, complete phages cannot be found in the host cell. Only separate components—DNA and protein—can be detected. The period during viral multiplication when complete, infective virions are not yet present is called the **eclipse period.**

Maturation. ④ In the next sequence of events, *maturation* occurs. In this process, bacteriophage DNA and capsids are assembled into complete virions. The viral components essentially assemble into a viral particle spontaneously, eliminating the need for many nonstructural genes and gene products. The phage heads and tails are separately assembled from protein subunits, and the head is filled with phage DNA and attached to the tail.

Release. ⑤ The final stage of viral multiplication is the *release* of virions from the host cell. The term **lysis** is generally used for this stage in the multiplication of T-even phages because in this case, the plasma membrane actually breaks open (lyses). Lysozyme, which is coded for by a phage gene, is synthesized within the cell. This enzyme causes the bacterial cell wall to break down, and the newly produced bacteriophages are released from the host cell. The released bacteriophages infect other susceptible cells in the vicinity, and the viral multiplication cycle is repeated within those cells.

A One-Step Growth Experiment

The time that elapses from phage attachment to release is known as **burst time** and averages 20–40 minutes. The number of newly synthesized phage particles released from a single cell is referred to as **burst size** and usually ranges from about 50 to 200.

The various stages involved in the multiplication of phages can be demonstrated experimentally in what is known as a *one-step growth experiment* (Figure 13.11). In this procedure, a phage suspension is diluted until a sample containing only a few phage particles is obtained. These particles are then introduced into a culture of host cells. Periodically, samples of phage particles are removed from the culture and inoculated onto a plate culture of susceptible host cells; the plaque method is used to determine the number of infective phage particles on this culture (see Figure 13.6). A few minutes after attachment, no infective particles are present. However, phage nucleic acid is found

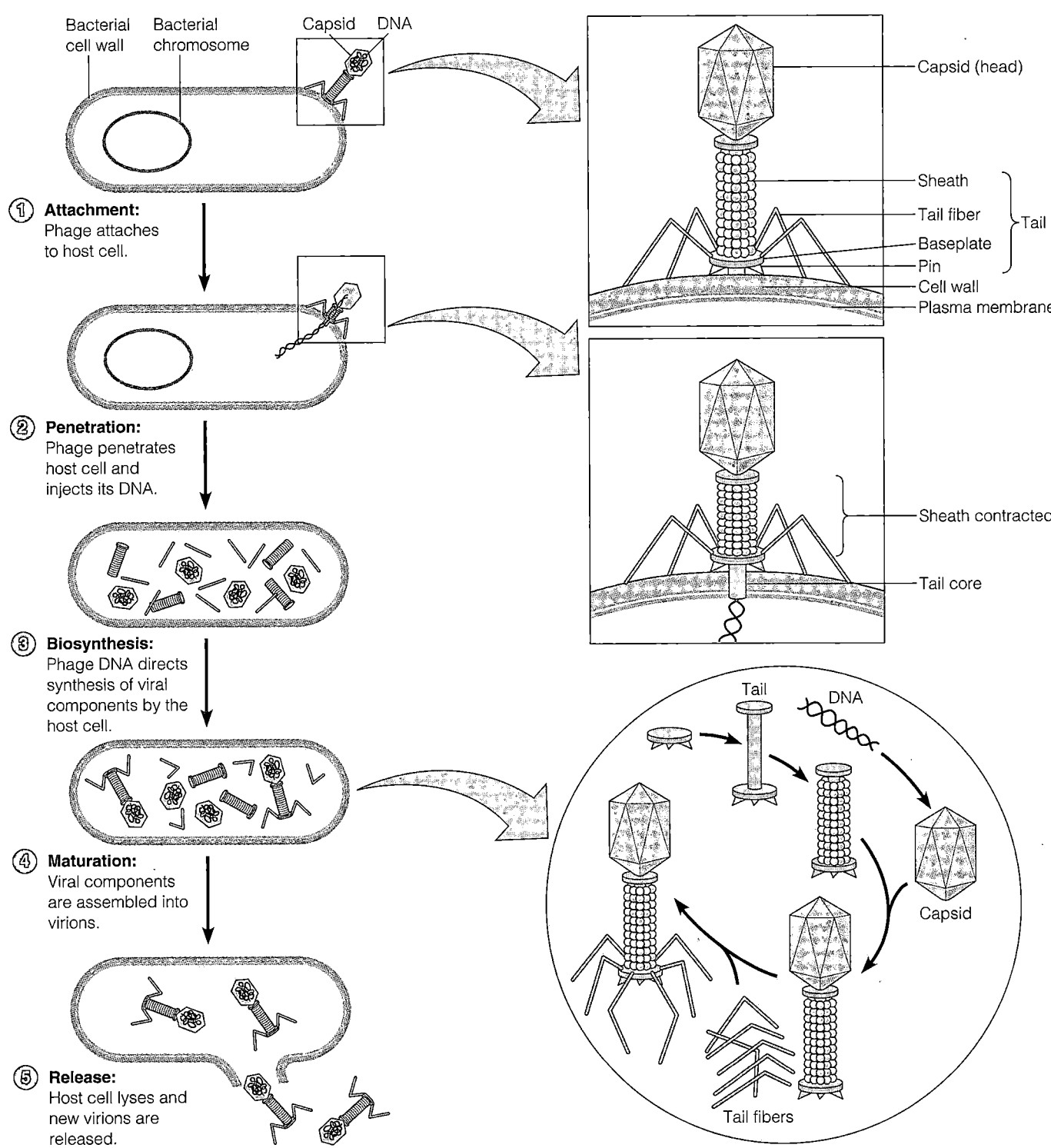

Bacterial cell wall

Bacterial chromosome

Capsid   DNA

① **Attachment:**
Phage attaches
to host cell.

② **Penetration:**
Phage penetrates
host cell and
injects its DNA.

③ **Biosynthesis:**
Phage DNA directs
synthesis of viral
components by the
host cell.

④ **Maturation:**
Viral components
are assembled into
virions.

⑤ **Release:**
Host cell lyses and
new virions are
released.

Capsid (head)

Sheath
Tail fiber } Tail
Baseplate
Pin
Cell wall
Plasma membrane

Sheath contracted

Tail core

Tail   DNA

Capsid

Tail fibers

FIGURE 13.10 The lytic cycle of a T-even bacteriophage.

▦ The lytic cycle results in lysis of the host cell and the release of new
phage particles.

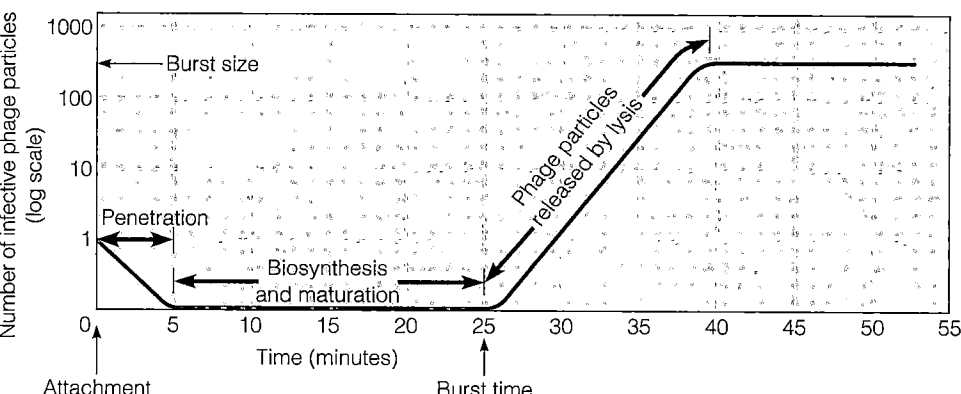

FIGURE 13.11 A bacterio-phage one-step growth curve. No new infective phage particles are found in a culture until after biosynthesis and maturation have taken place.

■ **What can be found in the cell during biosynthesis and maturation?**

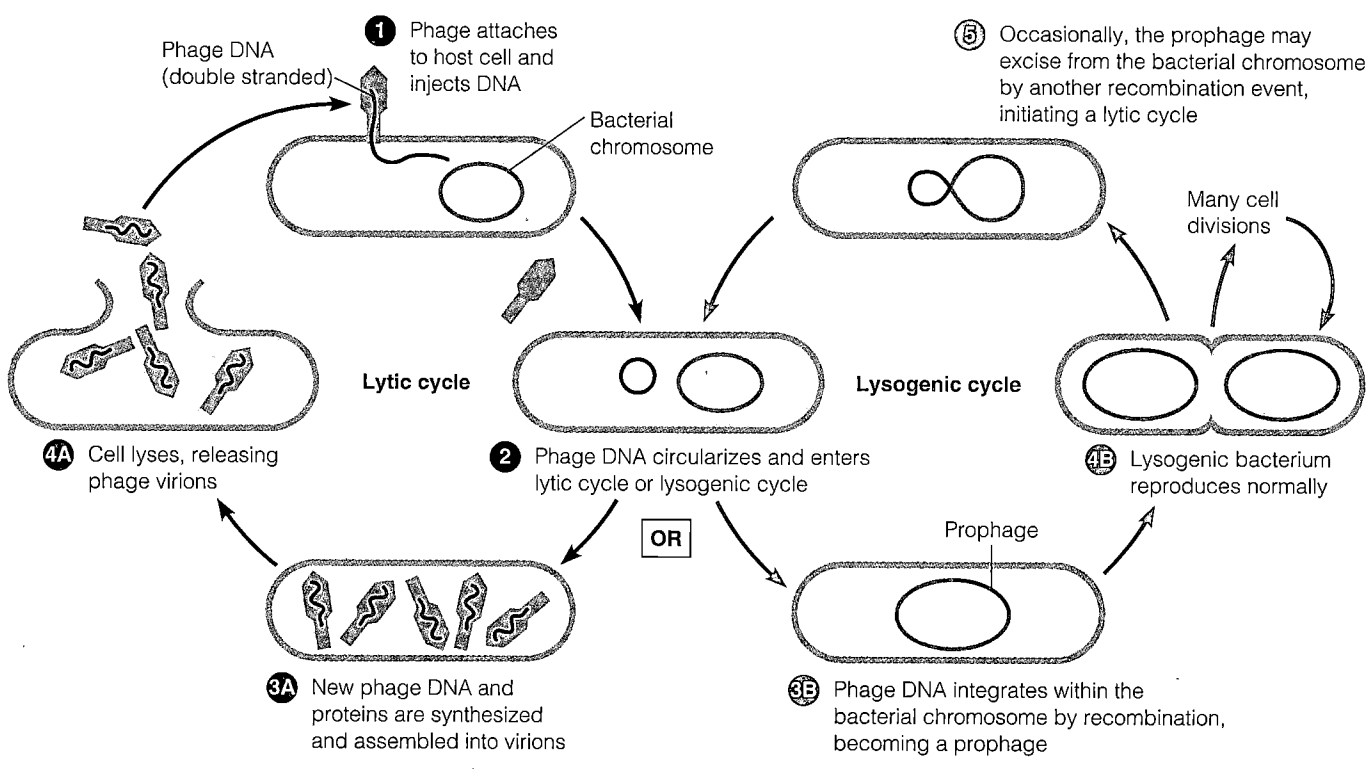

FIGURE 13.12 The lysogenic cycle of bacteriophage λ in E. coli.

■ **How does lysogeny differ from the lytic cycle?**

inside the infected cells, and capsid proteins will be synthesized. The time required for maturation is the interval between the appearance of phage nucleic acid and the synthesis of mature phages. After a few minutes, the number of infective phage particles found in subcultures begins to rise. The burst size is determined once the number of infective phage particles remains constant, indicating that no further phage multiplication will occur.

Bacteriophage Lambda: The Lysogenic Cycle

In contrast to T-even bacteriophages, some viruses do not cause lysis and death of the host cell when they multiply.

These *lysogenic phages* (also called *temperate phages*) may indeed proceed through a lytic cycle, but they are also capable of incorporating their DNA into the host cell's DNA to begin a lysogenic cycle. In **lysogeny,** the phage remains latent (inactive). The participating bacterial host cells are known as *lysogenic cells.*

We will use the bacteriophage λ (lambda), a well-studied lysogenic phage, as an example of the lysogenic cycle (Figure 13.12). ❶ Upon penetration into an *E. coli* cell, ❷ the originally linear phage DNA forms a circle. ❸Ⓐ This circle can multiply and be transcribed, ❹Ⓐ leading to the production of new phage and to cell lysis (the lytic cycle). ⑫ Alternatively, the circle can recombine with and become part of the circular bacterial DNA (the lysogenic cycle). The inserted phage DNA is now called a **prophage.** Most of the prophage genes are repressed by two repressor proteins that are the products of phage genes. These repressors stop transcription of all the other phage genes by binding to operators. Thus, the phage genes that would otherwise direct the synthesis and release of new virions are turned off, in much the same way that the genes of the *E. coli lac* operon are turned off by the *lac* repressor (Figure 8.14, page 224).

Every time the host cell's machinery replicates the bacterial chromosome, ⑬ it also replicates the prophage DNA. The prophage remains latent within the progeny cells. ⑭ However, a rare spontaneous event, or the action of UV light or certain chemicals, can lead to the excision (popping-out) of the phage DNA, and to initiation of the lytic cycle.

There are three important results of lysogeny. First, the lysogenic cells are immune to reinfection by the same phage. (However, the host cell is not immune to infection by other phage types.) The second result of lysogeny is **phage conversion;** that is, the host cell may exhibit new properties. For example, the bacterium *Corynebacterium diphtheriae,* which causes diphtheria, is a pathogen whose disease-producing properties are related to the synthesis of a toxin. The organism can produce toxin only when it carries a temperate phage, because the prophage carries the gene coding for the toxin. As another example, only streptococci carrying a temperate phage are capable of producing the toxin associated with scarlet fever. The toxin produced by *Clostridium botulinum,* which causes botulism, is encoded by a prophage gene, as is the choleratoxin produced by pathogenic strains of *Vibrio cholerae.*

The third result of lysogeny is that it makes **specialized transduction** possible. Recall from Chapter 8 that bacterial genes can be picked up in a phage coat and transferred to another bacterium in a process called generalized transduction (see Figure 8.28 on page 239). Any bacterial genes can be transferred by generalized transduction because the host chromosome is broken down into fragments, any of which can be packaged into a phage coat. In specialized transduction, however, only certain bacterial genes can be transferred.

Specialized transduction is mediated by a lysogenic phage, which packages bacterial DNA *along with* its own DNA in the same capsid. When a prophage is excised from the host chromosome, adjacent genes from either side may remain attached to the phage DNA. In Figure 13.13, bacteriophage λ has picked up the *gal* gene for galactose fermentation from its galactose-positive host.

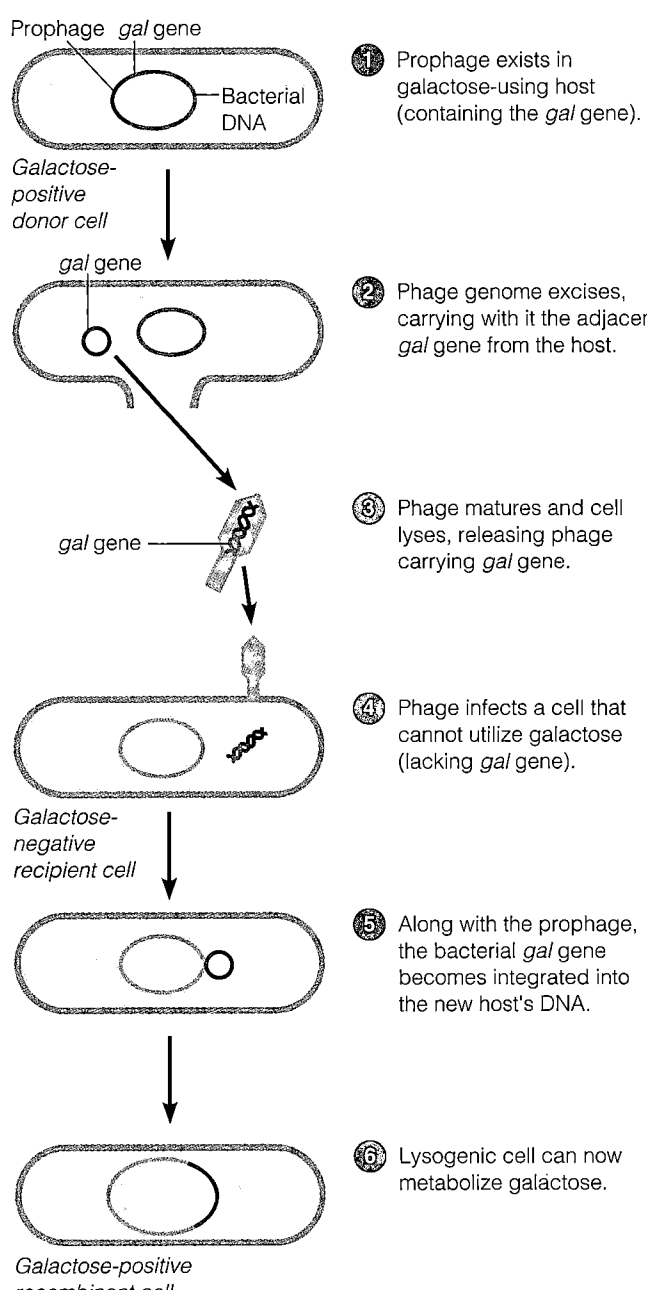

FIGURE 13.13 Specialized transduction. When a prophage is excised from its host chromosome, it can take with it a bit of the adjacent DNA from the bacterial chromosome.

■ How does specialized transduction differ from the lytic cycle?

The phage carries this gene to a galactose–negative cell, which then becomes galactose–positive.

Certain animal viruses can undergo processes very similar to lysogeny. Animal viruses that can remain latent in cells for long periods without multiplying or causing disease may become inserted in a host chromosome or

table 13.4	The Biosynthesis of DNA and RNA Viruses Compared	
Viral Nucleic Acid	**Virus Family**	**Special Features of Biosynthesis**
DNA, single-stranded	Adenoviridae	Cellular enzyme transcribes viral DNA in nucleus
DNA, double-stranded	Herpesviridae Papovaviridae	Cellular enzyme transcribes viral DNA in nucleus
	Poxviridae	Viral enzyme transcribes viral DNA in virion, in cytoplasm
DNA, reverse transcriptase	Hepadnaviridae	Cellular enzyme transcribes viral DNA in nucleus; reverse transcriptase copies mRNA to make viral DNA
RNA, + strand	Picornaviridae Togaviridae	Viral RNA functions as a template for synthesis of RNA polymerase which copies − strand RNA to make mRNA in cytoplasm
RNA, − strand	Rhabdoviridae	Viral enzyme copies viral RNA to make mRNA in cytoplasm
RNA, double-stranded	Reoviridae	Viral enzyme copies − strand RNA to make mRNA in cytoplasm
RNA, reverse transcriptase	Retroviridae	Viral enzyme copies viral RNA to make DNA in cytoplasm; DNA moves to nucleus

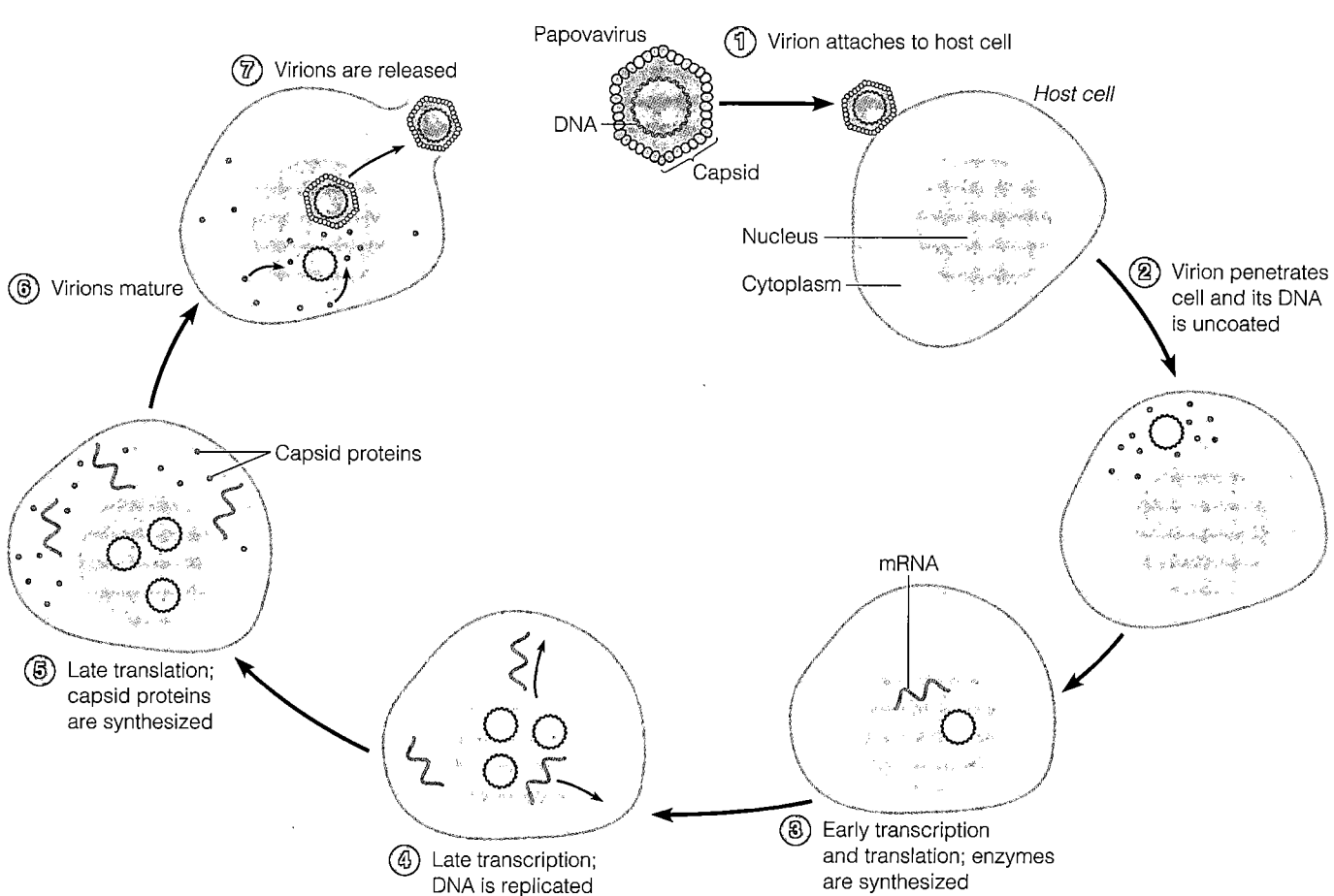

FIGURE 13.15   Multiplication of *Papovavirus*, a DNA-containing virus.

■ Why is mRNA made?

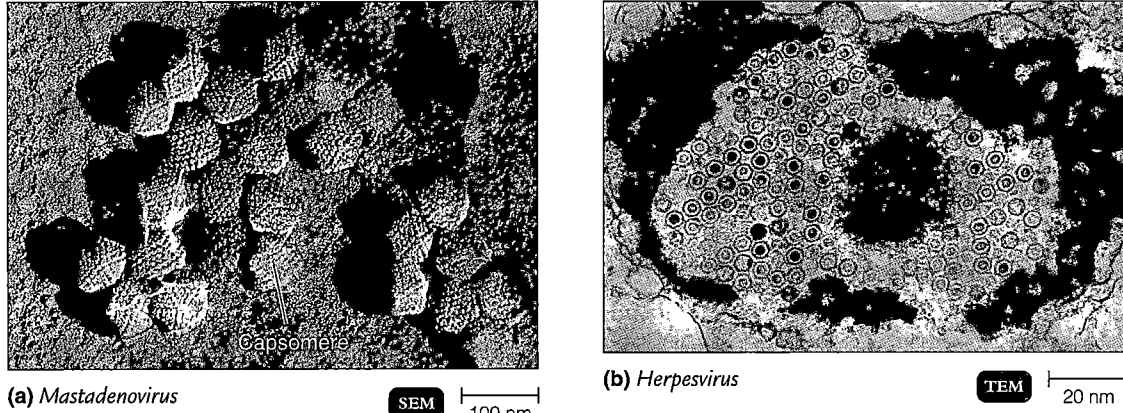

(a) *Mastadenovirus*   SEM ├─────┤ 100 nm

(b) *Herpesvirus*   TEM ├────┤ 20 nm

FIGURE 13.16 DNA-containing animal viruses. (a) Negatively stained adenovirus that have been concentrated in a centrifuge gradient. The individual capsomeres are clearly visible. **(b)** A negatively stained capsid of a herpesvirus, a member of the family Herpesviridae.

▨ What is the morphology of these viruses?

capsid proteins assemble to form complete viruses, which are then released ⑦ from the host cell.

Adenoviridae. Named after adenoids, from which they were first isolated, adenoviruses cause acute respiratory diseases—the common cold (Figure 13.16a).

Poxviridae. All diseases caused by poxviruses, including smallpox and cowpox, include skin lesions (see Figure ·13.6b). *Pox* refers to pus-filled lesions. Viral multiplication is started by viral transcriptase; the viral components are synthesized and assembled in the cytoplasm of the host cell.

Herpesviridae. Nearly 100 herpesviruses are known (Figure 13.16b). They are named after the spreading (*herpetic*) appearance of cold sores. Species of human herpesviruses (HHV) include HHV-1 and HHV-2, both in the genus *Simplexvirus,* which causes cold sores; HHV-3, genus *Varicellavirus,* which causes chickenpox; HHV-4, genus *Lympocryptovirus,* which causes infectious mononucleosis; HHV-5, genus *Cytomegalovirus,* which causes CMV inclusion disease; HHV-6, genus *Roseolovirus,* which causes roseola; HHV-7 which infects most infants causing measleslike rashes; and HHV-8 which causes Kaposi's sarcoma, primarily in AIDS patients.

Papovaviridae. Papovaviruses are named for *pa*pillomas (warts), *po*lyomas (tumors), and *va*cuolation (cytoplasmic vacuoles produced by some of these viruses). Warts are caused by members of the genus *Papillomavirus.* Some *Papillomavirus* species are capable of transforming cells and causing cancer. Viral DNA is replicated in the host cell's

nucleus along with host cell chromosomes. Host cells may proliferate, resulting in a tumor.

Hepadnaviridae. Hepadnaviridae are so named because they cause *hepa*titis and contain *DNA* (Figure 25.15, page 703). The only genus in this family causes hepatitis B. (Hepatitis A, C, D, E, F, and G viruses, although not related to each other, are RNA viruses. Hepatitis is discussed in Chapter 25.) Hepadnaviruses differ from other DNA viruses because they synthesize DNA by copying RNA, using viral reverse transcriptase. This enzyme is discussed later with the retroviruses, the only other family with reverse transcriptase.

The Biosynthesis of RNA Viruses

The multiplication of RNA viruses is essentially the same as that of DNA viruses, except that several different mechanisms of mRNA formation occur among different groups of RNA viruses (see Table 13.4). Although the details of these mechanisms are beyond the scope of this text, for comparative purposes we will trace the multiplication cycles of the four nucleic acid types of RNA viruses (three of which are shown in Figure 13.17). RNA viruses multiply in the host cell's cytoplasm. The major differences among the multiplication processes of these viruses lie in how mRNA and viral RNA are produced. Once viral RNA and viral proteins are synthesized, maturation occurs by similar means among all animal viruses, as will be discussed shortly.

Picornaviridae. Picornaviruses, such as poliovirus, are single-stranded RNA viruses. They are the smallest viruses; and the prefix *pico-* (small) plus *RNA* gives these

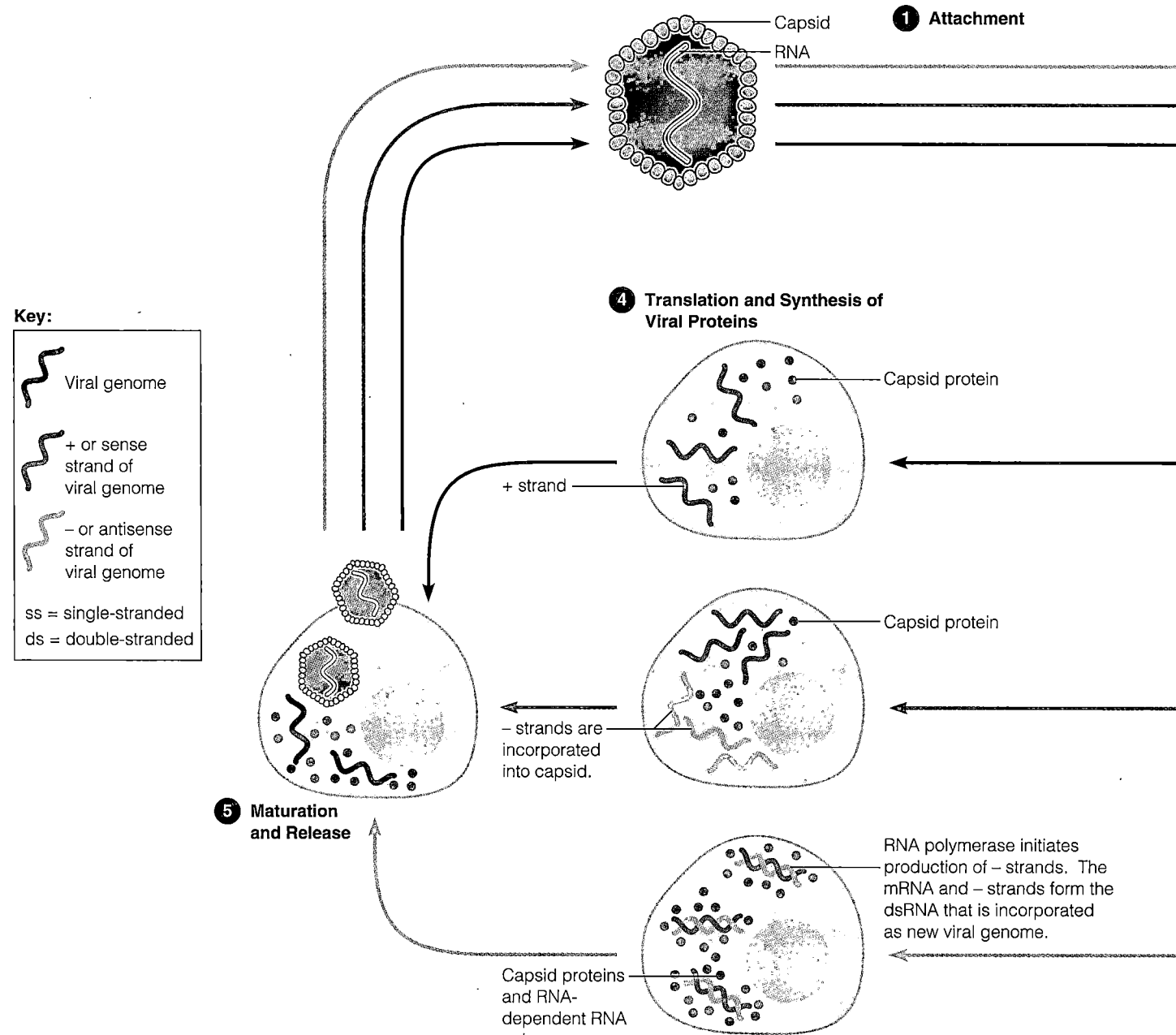

Key:

⌇ Viral genome

⌇ + or sense strand of viral genome

⌇ − or antisense strand of viral genome

ss = single-stranded
ds = double-stranded

① Attachment

Capsid
RNA

④ Translation and Synthesis of Viral Proteins

Capsid protein

+ strand

Capsid protein

− strands are incorporated into capsid.

⑤ Maturation and Release

RNA polymerase initiates production of − strands. The mRNA and − strands form the dsRNA that is incorporated as new viral genome.

Capsid proteins and RNA-dependent RNA polymerase

FIGURE 13.17 Pathways of multiplication used by various RNA-containing viruses.
(a) After uncoating, ssRNA viruses with a + strand genome are able to synthesize proteins directly from their + strand. Using the + strand as a template, they transcribe − strands to produce additional + strands to serve as mRNA and be incorporated into capsid proteins as the viral genome. **(b)** The ssRNA viruses with a − strand genome must transcribe a + strand to serve as mRNA before they begin synthesizing proteins. The mRNA transcribes additional − strands for incorporation into capsid protein. Both ssRNA and **(c)** dsRNA must use mRNA (+ strand) to code for proteins, including capsid proteins.

■ Why is − strand RNA made by picornaviruses and reoviruses? By rhabdoviruses?

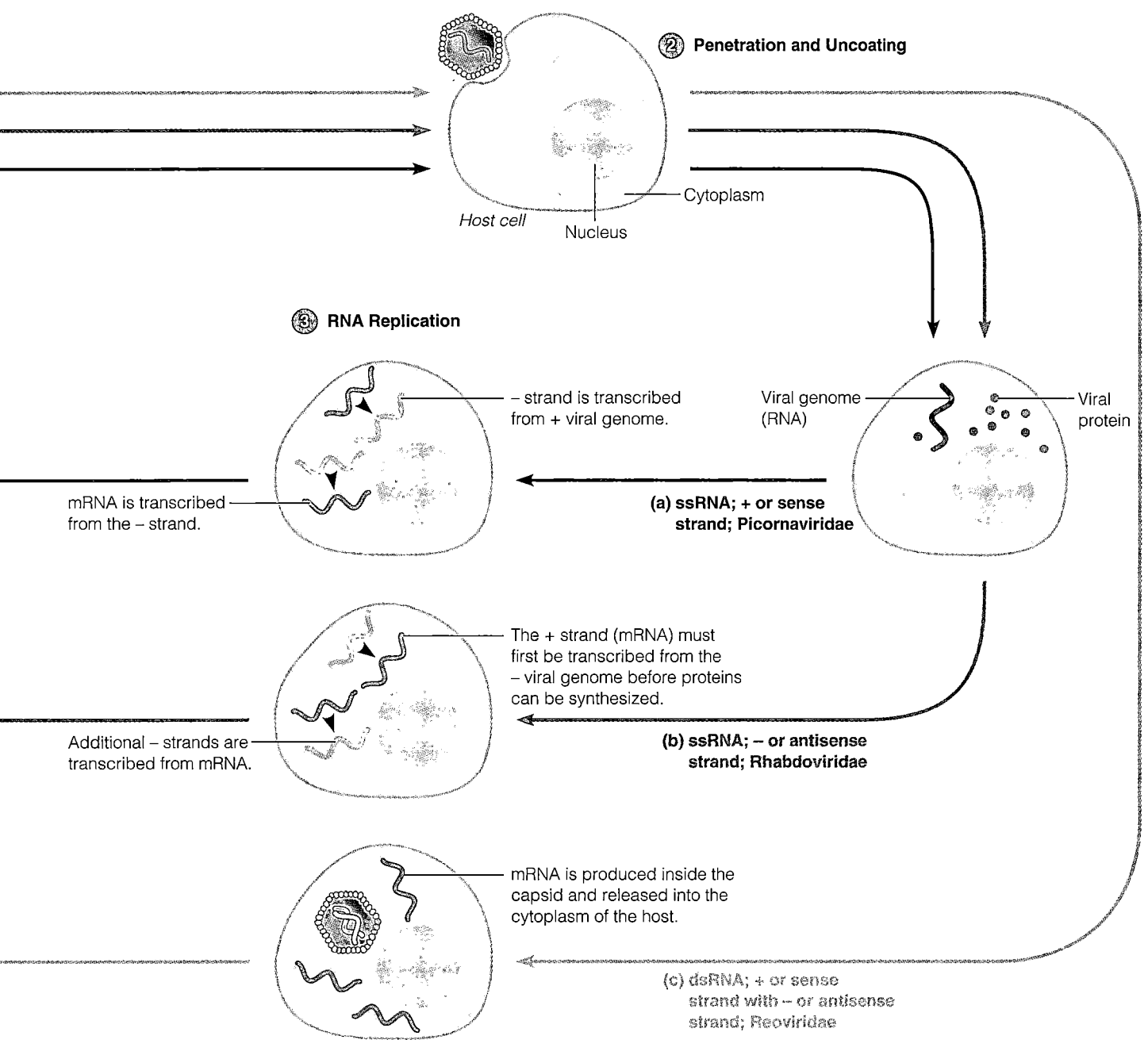

② Penetration and Uncoating

Cytoplasm

Host cell Nucleus

③ RNA Replication

– strand is transcribed
from + viral genome.

Viral genome
(RNA)

Viral
protein

mRNA is transcribed
from the – strand.

**(a) ssRNA; + or sense
strand; Picornaviridae**

The + strand (mRNA) must
first be transcribed from the
– viral genome before proteins
can be synthesized.

Additional – strands are
transcribed from mRNA.

**(b) ssRNA; – or antisense
strand; Rhabdoviridae**

mRNA is produced inside the
capsid and released into the
cytoplasm of the host.

**(c) dsRNA; + or sense
strand with – or antisense
strand; Reoviridae**

viruses their name. The RNA within the virion is identified as a **sense strand** (or **+ strand**), because it can act as mRNA. After attachment, penetration, and uncoating are completed, the single-stranded viral RNA (Figure 13.17a) is translated into two principal proteins, which inhibit the host cell's synthesis of RNA and protein and which form an enzyme called *RNA-dependent RNA polymerase*. This enzyme catalyzes the synthesis of another strand of RNA, which is complementary in base sequence to the original infecting strand. This new strand, called an **antisense strand** (or **– strand**), serves as a template to produce additional + strands. The + strands

may serve as mRNA for the translation of capsid proteins, may become incorporated into capsid proteins to form a new virus, or may serve as a template for continued RNA multiplication. Once viral RNA and viral protein are synthesized, maturation occurs.

Togaviridae. Togaviruses, which include arthropodborne alphaviruses or arboviruses (see Chapter 22), also contain a single + strand of RNA. Togaviruses are enveloped viruses; their name is from the Latin word for covering, *toga* (Figure 13.18a). Keep in mind that these are not the only enveloped viruses. After a – strand is

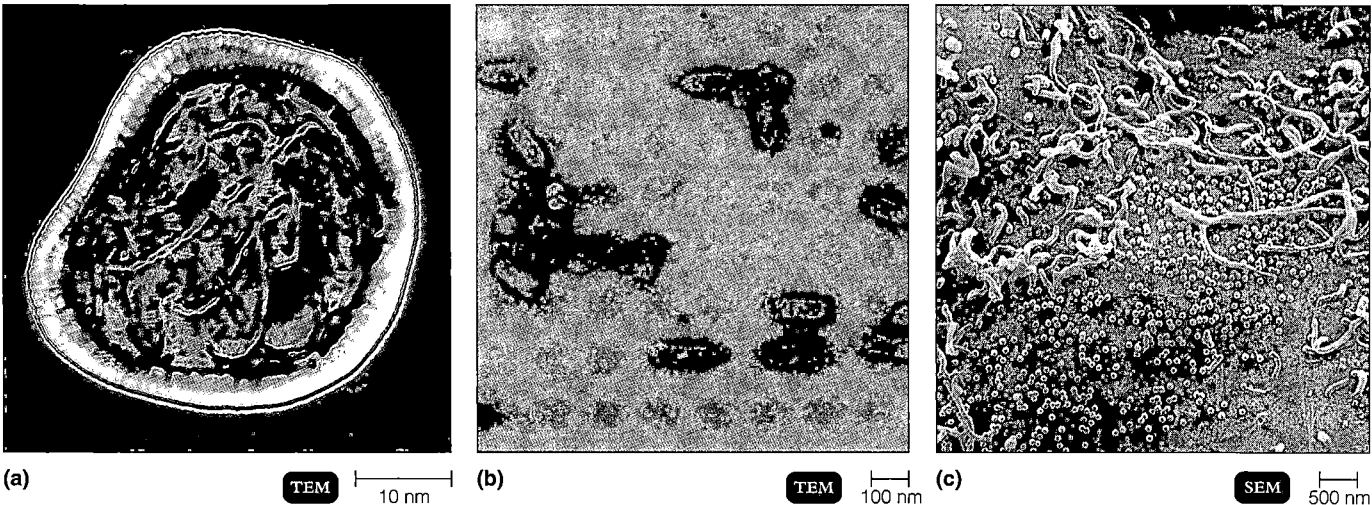

FIGURE 13.18 RNA-containing animal viruses. (a) A single rubella virus (*Rubivirus*), which belongs to the togavirus family. **(b)** Particles of vesicular stomatitis virus (*Vesiculovirus*), a member of the family Rhabdoviridae. **(c)** Spherical mouse mammary tumor virus (MMTV), in the family Retroviridae.

■ The RNA of + strand RNA viruses serves as mRNA; − strand RNA viruses must make + strand RNA to serve as mRNA.

made from the + strand, two types of mRNA are transcribed from the − strand. One type of mRNA is a short strand that codes for envelope proteins; the other, longer strand serves as mRNA for capsid proteins and can become incorporated into a capsid.

Rhabdoviridae. Rhabdoviruses, such as rabiesvirus (genus *Lyssavirus*), are usually bullet-shaped (Figure 13.18b). *Rhabdo* is from the Greek word for rod, which is not really an accurate description of their morphology. They contain a single − strand of RNA (Figure 13.17b). They also contain an RNA-dependent RNA polymerase that uses the − strand as a template from which to produce a + strand. The + strand serves as mRNA and as a template for synthesis of new viral RNA. (See Chapter 22, page 613.)

Reoviridae. Reoviruses are found in the respiratory and enteric (digestive) systems of humans. They were not associated with any diseases when first discovered so they were considered orphan viruses. Their name comes from the first letters of *r*espiratory, *e*nteric, and *o*rphan. Three serotypes are now known to cause respiratory tract and intestinal tract infections.

The capsid containing the double-stranded RNA is digested upon entering a host cell. Viral mRNA is produced in the cytoplasm, where it is used to synthesize more viral proteins (Figure 13.17c). One of the newly synthesized viral proteins acts as RNA-dependent RNA

polymerase to produce more − strands of RNA. The mRNA (+) and − strands form the double-stranded RNA that is then surrounded by capsid proteins.

Retroviridae. Many retroviruses infect vertebrates (Figure 13.18c). One genus of retrovirus, *Lentivirus*, includes the subspecies HIV-1 and HIV-2, which cause AIDS (see the box on page 395 and in Chapter 19, page 535). The retroviruses that cause cancer will be discussed later in this chapter.

The formation of mRNA and RNA for new retrovirus virions is shown in Figure 13.19. ① These viruses carry their own polymerase, which is an RNA-dependent DNA polymerase called **reverse transcriptase**, so named because it carries out a reaction (RNA → DNA) that is exactly the reverse of the familiar transcription of DNA → RNA. The name *retrovirus* is derived from the first letters of *re*verse *tr*anscriptase. ② Reverse transcriptase uses the RNA of the virus to synthesize a complementary strand of DNA, which in turn is replicated to form double-stranded DNA. This enzyme also degrades the original viral RNA.

The formation of complete viruses requires that DNA be transcribed back into the RNA that will serve as mRNA for viral protein synthesis and be incorporated into new virions. ③ However, before transcription can take place, the viral DNA must be integrated into the DNA of a host cell chromosome. In this integrated state, the viral DNA is called a **provirus.** Unlike a prophage,

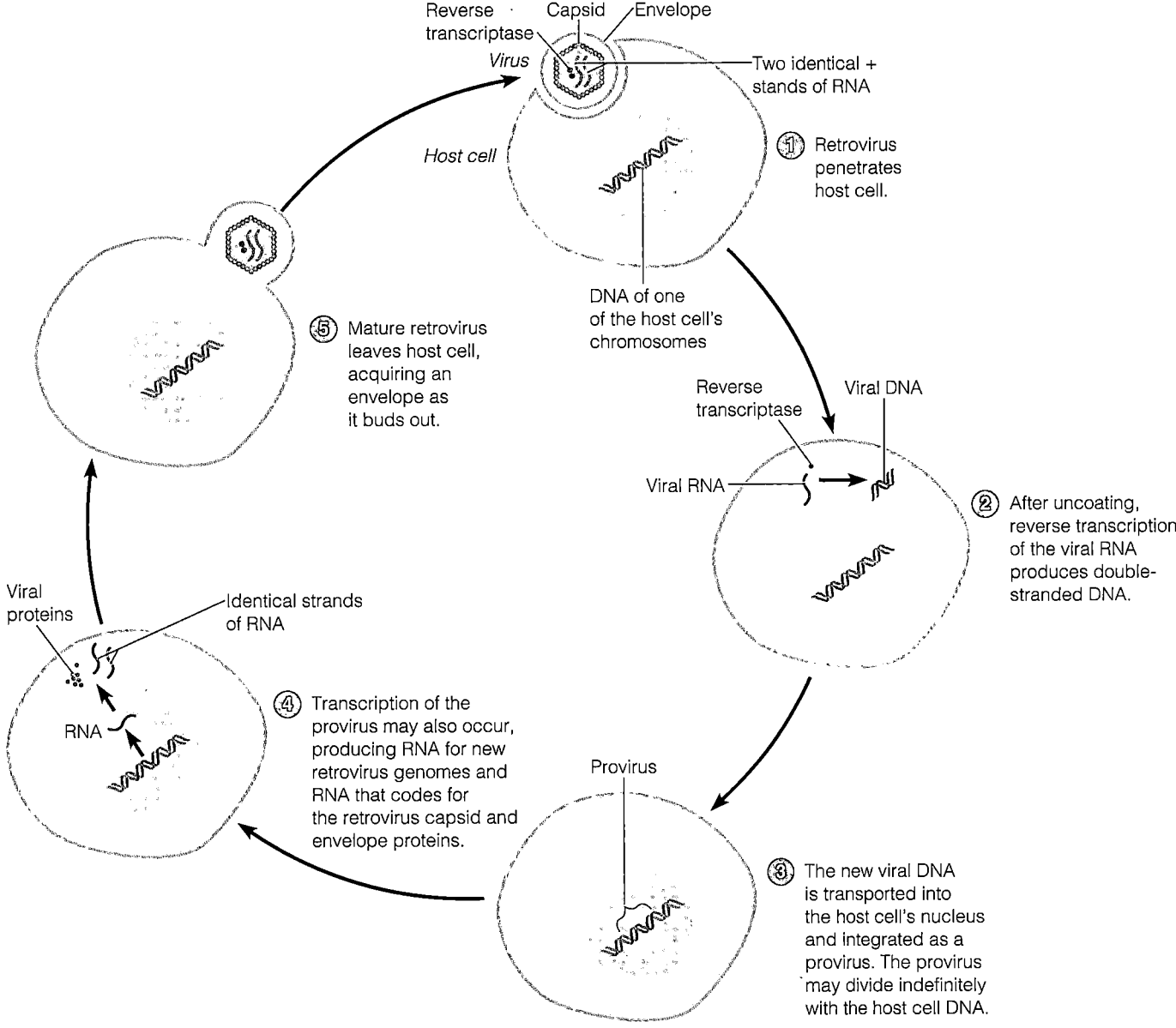

FIGURE 13.19 Multiplication and inheritance processes of the Retroviridae. A retrovirus may become a provirus that replicates in a latent state, and it may produce new retroviruses.

■ How does the biosynthesis of a retrovirus differ from that of other RNA viruses?

the provirus never comes out of the chromosome. As a provirus, HIV is protected from the host's immune system and antiviral drugs.

Once the provirus is integrated into the host cell's DNA, several things can happen. Sometimes the provirus simply remains in a latent state and replicates when the DNA of the host cell replicates. ④ In other cases, the provirus is expressed and ⑤ produces new viruses, which may infect adjacent cells. Mutagens such as gamma radia-

tion can induce expression of a provirus. The provirus can also convert the host cell into a tumor cell; possible mechanisms for this phenomenon will be discussed later.

Maturation and Release

The first step in viral maturation is the assembly of the protein capsid; this assembly is usually a spontaneous process. The capsids of many animal viruses are enclosed by an envelope consisting of protein, lipid, and

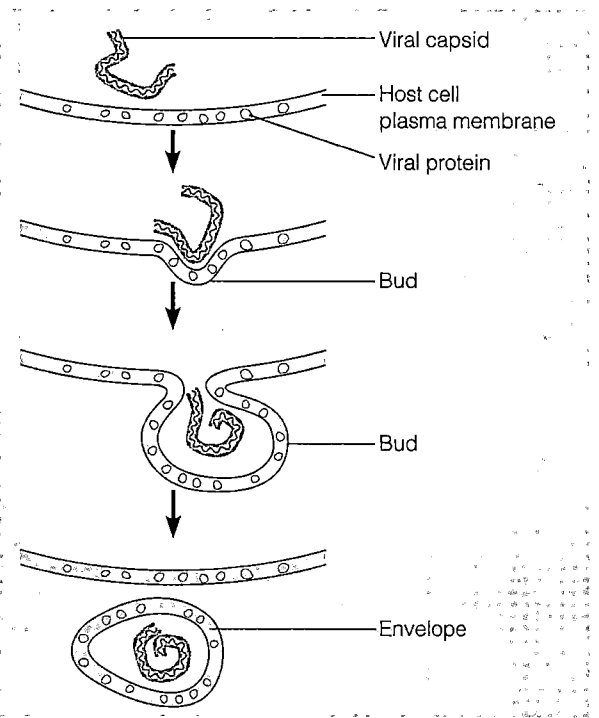

(a) Release by budding

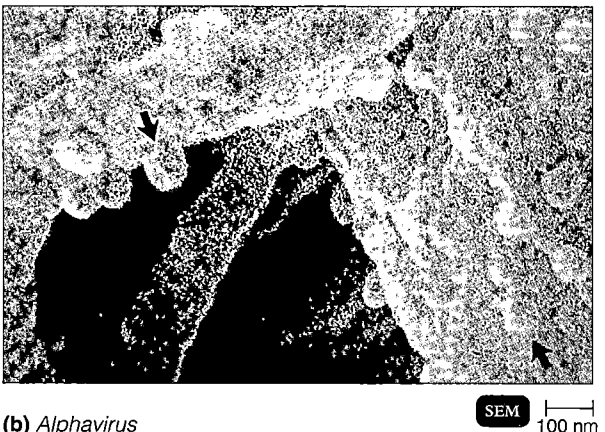

(b) *Alphavirus*   SEM ⊢————⊣ 100 nm

FIGURE 13.20 Budding of an enveloped virus. (a) A diagram of the budding process. **(b)** The small "bumps" seen on this freeze-fractured plasma membrane are Sindbis virus (*Alphavirus*) particles caught in the act of budding out from an infected cell.

■ Most enveloped viruses take part of the host's plasma membrane for their envelope.

carbohydrate, as noted earlier. Examples of such viruses include orthomyxoviruses and paramyxoviruses. The envelope protein is encoded by the viral genes and is incorporated into the plasma membrane of the host cell. The envelope lipid and carbohydrate are encoded

by host cell genes and are present in the plasma membrane. The envelope actually develops around the capsid by a process called **budding** (Figure 13.20).

After the sequence of attachment, penetration, uncoating, and biosynthesis of viral nucleic acid and protein, the assembled capsid containing nucleic acid pushes through the plasma membrane. As a result, a portion of the plasma membrane, now the envelope, adheres to the virus. This extrusion of a virus from a host cell is one method of release. Budding does not immediately kill the host cell, and in some cases the host cell survives.

Nonenveloped viruses are released through ruptures in the host cell plasma membrane. In contrast to budding, this type of release usually results in the death of the host cell.

Viruses and Cancer

Learning Objectives

■ *Define oncogene and transformed cell.*
■ *Discuss the relationship between DNA- and RNA-containing viruses and cancer.*

Several types of cancer are now known to be caused by viruses. Research on cancer-causing viruses has advanced our general understanding of cancer. Molecular biological research shows that the mechanisms of the diseases are similar, even when a virus does not cause the cancer.

The relationship between cancers and viruses was first demonstrated in 1908, when virologists Wilhelm Ellerman and Olaf Bang, working in Denmark, were trying to isolate the causative agent of chicken leukemia. They found that leukemia could be transferred to healthy chickens by cell-free filtrates that contained viruses. Three years later, F. Peyton Rous, working at the Rockefeller Institute in New York, found that a chicken **sarcoma** (cancer of connective tissue) can be similarly transmitted. Virus-induced **adenocarcinomas** (cancers of glandular epithelial tissue) in mice were discovered in 1936. At that time, it was clearly shown that mouse mammary gland tumors are transmitted from mother to offspring through the mother's milk. A human cancer-causing virus was discovered and isolated in 1972, by American bacteriologist Sarah Stewart.

The viral cause of cancer can often go unrecognized for several reasons. First, most of the particles of some viruses infect cells but do not induce cancer. Second, cancer might not develop until long after viral infection. Third, cancers do not seem to be contagious, as viral diseases usually are.

M M W R

AIDS: The Risk to Health Care Workers

With the advent of the AIDS epidemic, health care workers are understandably concerned about the risk of contracting AIDS after exposure to the body fluids of infected patients. However, when precautions are observed, the risk to workers is very small, even for those treating AIDS patients.

UNDERSTANDING THE RISK

The first protection for health care workers is a clear understanding of how AIDS can (and cannot) be transmitted in the course of their work. To date, direct inoculation of infected material is the only proven method of transmission in the health care environment. Infected materials that can transmit AIDS are blood, semen, vaginal secretions, and breast milk. The most common route of transmission is through accidental needle sticks. However, inoculation is also possible if infected material contacts mucous membranes or a break in the health care worker's skin. There is no evidence of HIV transmission by aerosols, the fecal-oral route, mouth-to-mouth or casual contact, or contact with environmental surfaces such as floors, walls, chairs, and toilets. Although HIV has been detected in fluids such as saliva, tears, cerebrospinal fluid, amniotic fluid, and urine, HIV is unlikely to be transmitted through exposure to these fluids.

The number of health care workers who have become infected, and the sources of their infections, are being carefully monitored by the Centers for Disease Control and Prevention (CDC). In the United States, by mid-1997, 52 health care workers who denied other risk factors and who had occupational exposure to HIV had contracted HIV infections, 45 of these following needlestick injuries. Five

cases had mucous membrane contact. Three cases were from exposure to concentrated virus in a laboratory.

The CDC estimates that the probability of infection following a needle-stick injury with HIV-infected blood is 3 out of 1000 exposures, or 0.3%. The probability of infection from mucous membrane exposure to blood is 0.1%. The probability of acquiring hepatitis B from a needlestick injury with blood containing HBV (hepatitis B virus) is 6–30%.

PRECAUTIONS

The CDC has developed the strategy of "universal precautions," which should be followed in *all* health care settings. These are described below.

Gloves. Disposable gloves should be used for direct exposure to infected blood, other body fluids, and tissues. Double-gloving is recommended during invasive surgical procedures. Personnel should not work when they have open skin lesions, dermatitis exuding fluids, or cutaneous wounds.

Gowns, Masks, and Goggles. Masks and protective eyewear are recommended when splashes are expected, such as during airway manipulation, endoscopy, and dental procedures, and in the laboratory.

Needles. To minimize the risk of needlesticks, needles should not be resheathed and should be put in a puncture-proof container for sterilization and disposal. More expensive, safer needle devices that minimize the risk of needlesticks are available.

Disinfection. Routine housekeeping in health care settings should include

washing floors, walls, and other areas not normally associated with disease transmissions with a 1:100 dilution of household bleach. A 1:10 dilution is recommended for disinfecting a spill.

PREVENTIVE TREATMENT AFTER EXPOSURE

Avoidance of exposure is the health care worker's first line of defense. Admittedly, however, accidental exposure cannot always be prevented. Studies indicate that exposed individuals can reduce their risk by the prophylactic (preventive) use of the antiviral drug, zidovudine (AZT).

THE RISK TO PATIENTS

Transmission of HBV from health care workers to patients during invasive dental procedures (tooth extractions) has been documented. The only documented cases of HIV transmission from a health care worker to patients involved a dentist in Florida and a surgeon in France. The risk of HIV transmission from an infected health care worker to patients is small and can be reduced with universal precautions.

Restrictions on the procedures that can be performed by health care workers with HIV infection have also been considered by the American Medical Association, the American Dental Association, the CDC, and other organizations. Their recommendations state that the risk of HIV transmission from health care workers to patients is greatest during invasive procedures, and that decisions regarding the restriction of patient care by infected workers who perform such procedures should be made on an individual basis.

SOURCE: Adapted from *MMWR* 38(S-2) (5/12/89); *MMWR* 47(RR-7)(5/15/98).

The Transformation of Normal Cells into Tumor Cells

Almost anything that can alter the genetic material of a eukaryotic cell has the potential to make a normal cell cancerous. These cancer-causing alterations to cellular DNA affect parts of the genome called **oncogenes.** Oncogenes were first identified in cancer-causing viruses and were thought to be a part of the normal viral genome. However, American microbiologists J. Michael Bishop and Harold E. Varmus received the 1989 Nobel Prize in Medicine for proving that the cancer-inducing genes carried by viruses are actually derived from animal cells. In 1976, Bishop and Varmus showed that the cancer-causing *src* gene in avian sarcoma viruses is derived from a normal part of chicken genes.

Oncogenes can be activated to abnormal functioning by a variety of agents, including mutagenic chemicals, high-energy radiation, and viruses. Viruses capable of inducing tumors in animals are called **oncogenic viruses,** or *oncoviruses.* Approximately 10% of cancers are known to be virus-induced.

Both DNA- and RNA-containing viruses are capable of inducing tumors in animals. When this occurs, the tumor cells undergo **transformation;** that is, they acquire properties that are distinct from the properties of uninfected cells or from infected cells that do not form tumors. An outstanding feature of all oncogenic viruses is that their genetic material integrates into the host cell's DNA and replicates along with the host cell's chromosome. This mechanism is similar to the phenomenon of lysogeny in bacteria, and it can alter the host cell's characteristics in the same way.

Normal animal cells in tissue culture move about randomly by amoeboid movement and divide repeatedly until they come into contact with one another. Then, both movement and cell division stop. This phenomenon is known as **contact inhibition.** Transformed cells in cell culture do not exhibit contact inhibition but instead form tumorlike cell masses. Transformed cells sometimes produce tumors when injected into susceptible animals.

After being transformed by viruses, many tumor cells contain a virus-specific antigen on their cell surface, called **tumor-specific transplantation antigen (TSTA),** or an antigen in their nucleus, called the **T antigen.** Transformed cells tend to be less round than normal cells, and they tend to exhibit certain chromosomal abnormalities, such as unusual numbers of chromosomes and fragmented chromosomes.

DNA Oncogenic Viruses

Oncogenic viruses are found within several families of DNA-containing viruses. These groups include the Adenoviridae, Herpesviridae, Poxviridae, Papovaviridae, and Hepadnaviridae. Among the papovaviruses, papillomaviruses cause uterine (cervical) cancer.

A genus of the Herpesviridae, *Lymphocryptovirus,* includes the Epstein–Barr virus (EB virus), which causes infectious mononucleosis and two human cancers, Burkitt's lymphoma and nasopharyngeal carcinoma. Burkitt's lymphoma is a rare cancer of the lymphatic system; it affects mostly children in certain parts of Africa. Nasopharyngeal carcinoma, a cancer of the nose and throat, is found worldwide. Some researchers have also suggested that EB virus is involved in Hodgkin's disease, a cancer of the lymphatic system.

About 90% of the U.S. population probably carry the latent stage of the EB virus in their lymphocytes but have no signs of disease. This latent stage is indicated by the presence of antibodies to the virus in blood serum. Although infection with this virus leads to mild symptoms in healthy children, it can cause infectious mononucleosis, mostly in teenagers.

EB virus was isolated from Burkitt's lymphoma cells in 1964 by Michael Epstein and Yvonne Barr. The proof that EB virus can cause cancer was accidentally demonstrated in 1985 when a 12-year-old boy known only as David received a bone marrow transplant. David was born without a functioning immune system and was kept from all microbes inside a sterile "bubble" his entire life; the news media called him "the bubble boy." (The lymphocytes that provide immunity are produced in bone marrow.) Several months after the transplant, he developed signs of infectious mononucleosis, and a few months after that, he died of cancer. An autopsy revealed that the virus had been unwittingly introduced into the boy with the bone marrow transplant. The case supported suspicions that EB virus can induce cancer in immunosuppressed individuals.

Another DNA virus that causes cancer is hepatitis B virus (HBV, genus *Hepadnavirus*). Many animal studies have been performed that have clearly indicated the causal role of HBV in liver cancer. In one human study, virtually all people with liver cancer had previous HBV infections. Moreover, results of this study indicated that the risk of getting liver cancer is over 98 times higher in people with prior HBV infection than in those with no prior infection.

RNA Oncogenic Viruses

Among the RNA viruses, only the oncoviruses in the family Retroviridae cause cancer. The human T-cell leukemia viruses (HTLV-1 and HTLV-2) are retroviruses that cause adult T-cell leukemia and lymphoma in humans. (T cells are a type of white blood cell involved in the immune response.)

Sarcoma viruses of cats, chickens, and rodents, and the mammary tumor viruses of mice, are also retroviruses. Another retrovirus, feline leukemia virus (FeLV), causes leukemia in cats and is transmissible among cats. There is a test to detect the virus in cat serum.

The ability of retroviruses to induce tumors is related to their production of a reverse transcriptase by the mechanism described earlier (see Figure 13.19). The provirus, which is the double-stranded DNA molecule synthesized from the viral RNA, becomes integrated into the host cell's DNA; new genetic material is thereby introduced into the host's genome, and this is the key reason retroviruses can contribute to cancer. Some retroviruses contain oncogenes; others contain promoters that turn on oncogenes or other cancer-causing factors.

Latent Viral Infections

Learning Objective

■ *Provide an example of a latent viral infection.*

A virus can remain in equilibrium with the host and not actually produce disease for a long period, often many years. All of the human herpesviruses can remain in host cells throughout the life of an individual. When herpesviruses are reactivated by immunosuppression (for example, AIDS), the resulting infection may be fatal. The classic example of such a **latent infection** in viruses is the infection of the skin by herpes simplex virus, which produces cold sores. This virus inhabits the host's nerve cells but causes no damage until it is activated by a stimulus such as fever or sunburn—hence the term *fever blister.*

In some individuals, viruses are produced but symptoms never appear. Even though a large percentage of the human population carries the herpes simplex virus, only 10–15% of people carrying the virus exhibit the disease.

The virus of some latent infections can exist in a lysogenic state within host cells.

The chickenpox virus (genus *Varicellovirus*) can also exist in a latent state. Chickenpox (varicella) is a skin disease that is usually acquired in childhood. The virus gains access to the skin via the blood. From the blood, some viruses may enter nerves, where they remain latent. Later, changes in the immune (T-cell) response can activate these latent viruses, causing shingles (zoster). The shingles rash appears on the skin along the nerve in which the virus was latent. Shingles occurs in 10–20% of people who have had chickenpox.

Persistent Viral Infections

Learning Objective

■ *Differentiate persistent viral infections from latent viral infections.*

The term *slow viral infection* was coined in the 1950s to refer to a disease process that occurs gradually over a long period and was thought to be caused by a virus. A better term to describe these prolonged diseases is **persistent viral infection.** Typically, persistent viral infections are fatal.

A number of persistent viral infections have in fact been shown to be caused by conventional viruses. For example, several years after causing measles, the measles virus can be responsible for a rare form of encephalitis called subacute sclerosing panencephalitis (SSPE). A persistent viral infection is apparently different from a latent viral infection in that, in most persistent viral infections, detectable infectious virus gradually builds up over a long period, rather than appearing suddenly.

Several examples of persistent viral infections caused by conventional viruses are listed in Table 13.5.

table 13.5	Examples of Persistent Viral Infections in Humans	
Disease	**Primary Effect**	**Causative Virus**
Subacute sclerosing panencephalitis (SSPE)	Mental deterioration	Measles virus (*Morbillivirus*)
Progressive encephalitis	Rapid mental deterioration	Rubella virus (*Rubivirus*)
Progressive multifocal leukoencephalopathy	Brain degeneration	Papovavirus (family Papovaviridae)
AIDS dementia complex	Brain degeneration	HIV (*Lentivirus*)
Persistent enterovirus infection	Mental deterioration associated with AIDS	Echoviruses

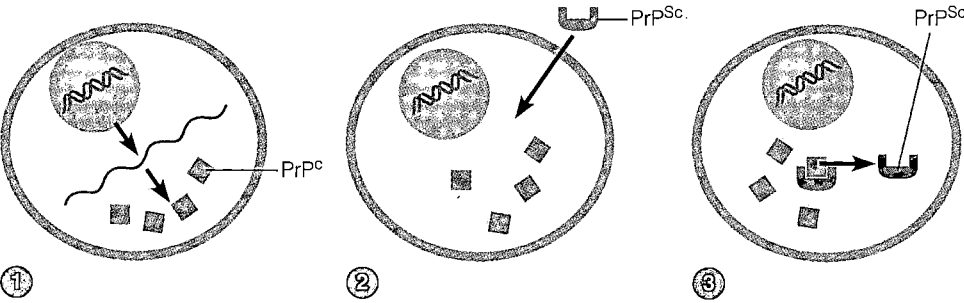

FIGURE 13.21 How a protein can be infectious. If an abnormal prion protein
(PrP^{Sc}) enters a cell, it changes a normal prion protein to PrP which now can change
another normal PrP, resulting in an accumulation of the abnormal PrP^{Sc}.

Prions

Learning Objective

■ *Discuss how a protein can be infectious.*

Other infectious diseases that have not been found to
have a viral cause might be caused by prions. In 1982,
American neurobiologist Stanley Prusiner proposed that
infectious proteins caused a neurological disease in sheep
called scrapie. The infectivity of scrapie-infected brain tissue
is reduced by treatment with proteases but not by
treatment with radiation, suggesting that the infectious
agent is pure protein. Prusiner coined the name **prion**
for *pro*teinaceous *in*fectious particle.

Nine animal diseases now fall into this category, including
the "mad cow disease" that emerged in cattle in
Great Britain in 1987. All nine are neurological diseases
called spongiform encephalopathies because large vacuoles
develop in the brain (Figure 22.17a, page 619). The
human diseases are kuru, Creutzfeldt-Jakob disease (CJD),
Gerstmann-Sträussler-Scheinker syndrome, and fatal familial
insomnia. (Neurological diseases are discussed in
Chapter 22.) These diseases run in families, which indicates
a possible genetic cause. However, they cannot be purely
inherited because mad cow disease arose from feeding
scrapie-infected sheep meat to cattle, and the new (bovine)
variant was transmitted to humans who ate undercooked
beef from infected cattle (see Chapter 1, page 20). Additionally,
CJD has been transmitted with transplanted nerve
tissue and contaminated surgical instruments.

One hypothesis for how an infectious agent can lack
nucleic acid is shown in Figure 13.21. ❶ The major portion
of the infectious agent is a protein (PrP, for prion
protein), whose gene is found in normal host DNA. The
PrP gene is located on chromosome 20 in humans. Normal
human PrP is called PrP^{C} (C = cellular), abnormal
form of PrP, designated PrP^{Sc}, is found in the brains of
animals with scrapie. ❷ The injection of PrP^{Sc} into experimental
animals causes disease in those animals. ❸
PrP^{Sc} may cause a change in PrP^{C}. When one PrP^{Sc} contacts
a normal PrP and causes it to refold into PrP^{Sc}, the
new PrP^{Sc} molecules attack other normal PrP^{C} molecules.

The actual cause of cell damage is not known. Fragments
of PrP^{Sc} molecules accumulate in the brain, forming
plaques. These plaques are used for postmortem
diagnosis, but they do not appear to be the cause of cell
damage.

Plant Viruses and Viroids

Learning Objectives

■ *Differentiate virus, viroid, and prion.*

■ *Name a virus that causes a plant disease.*

Plant viruses resemble animal viruses in many respects:
Plant viruses are morphologically similar to animal
viruses, and they have similar types of nucleic acid (Table
13.6). In fact, some plant viruses can multiply inside insect
cells. Plant viruses cause many diseases of economically
important crops, including beans (bean mosaic
virus), corn and sugarcane (wound tumor virus), and
potatoes (potato yellow dwarf virus). Viruses can cause
color change, deformed growth, wilting, and stunted
growth in their plant hosts. Some hosts, however, remain
symptomless and only serve as reservoirs of infection.

Plant cells are generally protected from disease by an
impermeable cell wall. Viruses must enter through
wounds or be assisted by other plant parasites, including
nematodes, fungi, and, most often, insects that suck the

table 13.6	*Classification of Some Major Plant Viruses*			
Characteristic	**Viral Family**	**Viral Genus or Unclassified Members**	**Morphology**	**Method of Transmission**
Double-stranded DNA, nonenveloped	Papovaviridae	Cauliflower mosaic virus		Aphids
Single-stranded RNA, + strand, nonenveloped	Picornaviridae	Bean mosaic virus		Pollen
	Tetraviridae	*Tobamovirus*		Wounds
Single-stranded RNA, strand, enveloped −	Rhabdoviridae	Potato yellow dwarf virus		Leafhoppers and aphids
Double-stranded RNA, nonenveloped	Reovirus	Wound tumor virus		Leafhoppers

plant's sap. Once one plant is infected, it can spread infection to other plants in its pollen and seeds.

In laboratories, plant viruses are cultured in protoplasts (plant cells with the cell walls removed) and in insect cell cultures.

Some plant diseases are caused by **viroids,** short pieces of naked RNA, only 300–400 nucleotides long, with no protein coat. The nucleotides are often internally paired, so the molecule has a closed, folded, three-dimensional structure that presumably helps protect it from attack by cellular enzymes. The RNA does not code for any proteins. Thus far, viroids have been conclusively identified as pathogens only of plants. Annually, infections by viroids, such as potato spindle tuber viroid, result in losses of millions of dollars from crop damage (Figure 13.22).

Current research on viroids has revealed similarities between the base sequences of viroids and introns. Recall from Chapter 8 (page 222) that introns are sequences of genetic material that do not code for polypeptides. This observation has led to the hypothesis that viroids evolved from introns, leading to speculation that future researchers may discover animal viroids.

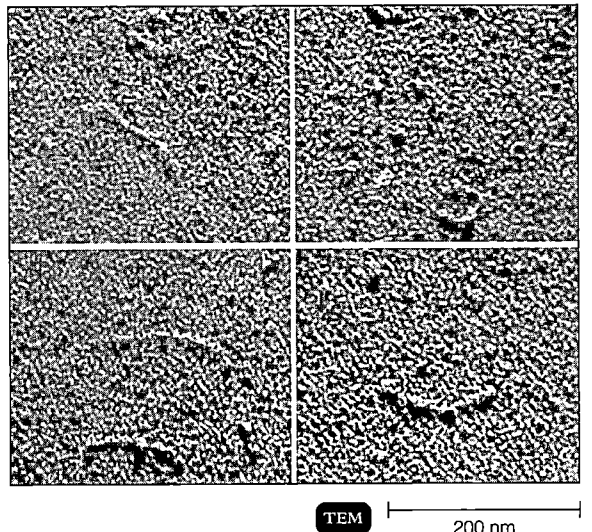

TEM ├──────── 200 nm ────────┤

FIGURE 13.22 Linear and circular potato spindle tuber viroid (PSTV).

■ A viroid is infectious RNA; a prion is infectious protein.

Study Outline   **STF** Student Tutorial CD-ROM

GENERAL CHARACTERISTICS OF VIRUSES (pp. 372–373)

1. Depending on one's viewpoint, viruses may be regarded as exceptionally complex aggregations of nonliving chemicals or as exceptionally simple living microbes.
2. Viruses contain a single type of nucleic acid (DNA or RNA) and a protein coat, sometimes enclosed by an envelope composed of lipids, proteins, and carbohydrates.
3. Viruses are obligatory intracellular parasites. They multiply by using the host cell's synthesizing machinery to cause the synthesis of specialized elements that can transfer the viral nucleic acid to other cells.

Host Range (p. 372)

1. Host range refers to the spectrum of host cells in which a virus can multiply.
2. Most viruses infect only specific types of cells in one host species.
3. Host range is determined by the specific attachment site on the host cell's surface and the availability of host cellular factors.

Viral Size (p. 373)

1. Viral size is ascertained by electron microscopy.
2. Viruses range from 20 to 14,000 nm in length.

VIRAL STRUCTURE (pp. 373–376)

1. A virion is a complete, fully developed viral particle composed of nucleic acid surrounded by a coat.

Nucleic Acid (p. 374)

1. Viruses contain either DNA or RNA, never both, and the nucleic acid may be single- or double-stranded, linear or circular, or divided into several separate molecules.
2. The proportion of nucleic acid in relation to protein in viruses ranges from about 1% to about 50%.

Capsid and Envelope (p. 374)

1. The protein coat surrounding the nucleic acid of a virus is called the capsid.
2. The capsid is composed of subunits, capsomeres, which can be a single type of protein or several types.
3. The capsid of some viruses is enclosed by an envelope consisting of lipids, proteins, and carbohydrates.
4. Some envelopes are covered with carbohydrate-protein complexes called spikes.

General Morphology (pp. 375–376)

1. Helical viruses (for example, Ebola virus) resemble long rods, and their capsids are hollow cylinders surrounding the nucleic acid.

2. Polyhedral viruses (for example, adenovirus) are many-sided. Usually the capsid is an icosahedron.
3. Enveloped viruses are covered by an envelope and are roughly spherical but highly pleomorphic. There are also enveloped helical viruses (for example, *Influenzavirus*) and enveloped polyhedral viruses (for example, *Simplexvirus*).
4. Complex viruses have complex structures. For example, many bacteriophages have a polyhedral capsid with a helical tail attached.

TAXONOMY OF VIRUSES (p. 376)

1. Classification of viruses is based on type of nucleic acid, strategy for replication, and morphology.
2. Virus family names end in *-viridae;* genus names end in *-virus.*
3. A viral species is a group of viruses sharing the same genetic information and ecological niche.

THE ISOLATION, CULTIVATION, AND IDENTIFICATION OF VIRUSES (pp. 376–381)

1. Viruses must be grown in living cells.
2. The easiest viruses to grow are bacteriophages.

Growing Bacteriophages in the Laboratory (p. 379)

1. The plaque method mixes bacteriophages with host bacteria and nutrient agar.
2. After several viral multiplication cycles, the bacteria in the area surrounding the original virus are destroyed; the area of lysis is called a plaque.
3. Each plaque originates with a single viral particle; the concentration of viruses is given as plaque-forming units.

Growing Animal Viruses in the Laboratory (pp. 379–380)

1. Cultivation of some animal viruses requires whole animals.
2. Simian AIDS and feline AIDS provide models for studying human AIDS.
3. Some animal viruses can be cultivated in embryonated eggs.
4. Cell cultures are cells growing in culture media in the laboratory.
5. Primary cell lines and embryonic diploid cell lines grow for a short time in vitro.
6. Continuous cell lines can be maintained in vitro indefinitely.
7. Viral growth can cause cytopathic effects in the cell culture.

Viral Identification (p. 381)

STF *To review, go to Genetics: PCR: Identify organisms*

1. Serological tests are used most often to identify viruses.
2. Viruses may be identified by RFLPs and PCR.

VIRAL MULTIPLICATION (pp. 381–394)

1. Viruses do not contain enzymes for energy production or protein synthesis.

2. For a virus to multiply, it must invade a host cell and direct the host's metabolic machinery to produce viral enzymes and components.

Multiplication of Bacteriophages (pp. 381–386)

1. During a lytic cycle, a phage causes the lysis and death of a host cell.

   [ST] *Viruses: Lytic cycle*

2. Some viruses can either cause lysis or have their DNA incorporated as a prophage into the DNA of the host cell. The latter situation is called lysogeny.

3. The T-even bacteriophages that infect *E. coli* have been studied extensively.

4. During the attachment phase of the lytic cycle, sites on the phage's tail fibers attach to complementary receptor sites on the bacterial cell.

5. In penetration, phage lysozyme opens a portion of the bacterial cell wall, the tail sheath contracts to force the tail core through the cell wall, and phage DNA enters the bacterial cell. The capsid remains outside.

6. In biosynthesis, transcription of phage DNA produces mRNA coding for proteins necessary for phage multiplication. Phage DNA is replicated, and capsid proteins are produced. During the eclipse period, separate phage DNA and protein can be found.

7. During maturation, phage DNA and capsids are assembled into complete viruses.

8. During release, phage lysozyme breaks down the bacterial cell wall, and the multiplied phages are released.

9. The time from phage adsorption to release is called burst time (20–40 minutes). Burst size, the number of newly synthesized phages produced from a single infected cell, ranges from 50 to 200.

10. During the lysogenic cycle, prophage genes are regulated by a repressor coded for by the prophage. The prophage is replicated each time the cell divides.

    [ST] *Viruses: Lysogenic cycle*

11. Exposure to certain mutagens can lead to excision of the prophage and initiation of the lytic cycle.

12. Because of lysogeny, lysogenic cells become immune to reinfection with the same phage and may undergo phage conversion.

13. A lysogenic phage can transfer bacterial genes from one cell to another through transduction. Any genes can be transferred in generalized transduction, and specific genes can be transferred in specialized transduction.

Multiplication of Animal Viruses (pp. 386–394)

1. Animal viruses attach to the plasma membrane of the host cell.

2. Penetration occurs by endocytosis or fusion.

3. Animal viruses are uncoated by viral or host cell enzymes.

4. The DNA of most DNA viruses is released into the nucleus of the host cell. Transcription of viral DNA and translation produce viral DNA and, later, capsid proteins. Capsid proteins are synthesized in the cytoplasm of the host cell.

   [ST] *Viruses: DNA viruses*

5. DNA viruses include members of the families Adenoviridae, Poxviridae, Herpesviridae, Papovaviridae, and Hepadnaviridae.

6. Multiplication of RNA viruses occurs in the cytoplasm of the host cell. RNA-dependent RNA polymerase synthesizes a double-stranded RNA.

   [ST] *Viruses: RNA viruses*

7. Picornaviridae + strand RNA acts as mRNA and directs the synthesis of RNA-dependent RNA polymerase.

8. Togaviridae + strand RNA acts as a template for RNA-dependent RNA polymerase, and mRNA is transcribed from a new − RNA strand.

9. Rhabdoviridae − strand RNA is a template for viral RNA-dependent RNA polymerase, which transcribes mRNA.

10. Reoviridae are digested in host cell cytoplasm to release mRNA for viral biosynthesis.

11. Retroviridae reverse transcriptase (RNA-dependent DNA polymerase) transcribes DNA from RNA.

12. After maturation, viruses are released. One method of release (and envelope formation) is budding. Naked viruses are released through ruptures in the host cell membrane.

    [ST] *Viruses: Virus Identification*

VIRUSES AND CANCER (pp. 394–397)

1. The earliest relationship between cancer and viruses was demonstrated in the early 1900s, when chicken leukemia and chicken sarcoma were transferred to healthy animals by cell-free filtrates.

The Transformation of Normal Cells into Tumor Cells (p. 396)

1. When activated, oncogenes transform normal cells into cancerous cells.

2. Viruses capable of producing tumors are called oncogenic viruses.

3. Several DNA viruses and retroviruses are oncogenic.

4. The genetic material of oncogenic viruses becomes integrated into the host cell's DNA.

5. Transformed cells lose contact inhibition, contain virus-specific antigens (TSTA and T antigen), exhibit chromosome abnormalities, and can produce tumors when injected into susceptible animals.

DNA Oncogenic Viruses (p. 396)

1. Oncogenic viruses are found among the Adenoviridae, Herpesviridae, Poxviridae, and Papovaviridae.

2. The EB virus, a herpesvirus, causes Burkitt's lymphoma and nasopharyngeal carcinoma. *Hepadnavirus* causes liver cancer.

RNA Oncogenic Viruses (pp. 396–397)

1. Among the RNA viruses, only retroviruses seem to be oncogenic.

2. HTLV-1 and HTLV-2 have been associated from human leukemia and lymphoma.

3. The virus's ability to produce tumors is related to the production of reverse transcriptase. The DNA synthesized from the viral RNA becomes incorporated as a provirus into the host cell's DNA.

4. A provirus can remain latent, can produce viruses, or can transform the host cell.

LATENT VIRAL INFECTIONS (p. 397)

1. A latent viral infection is one in which the virus remains in the host cell for long periods without producing an infection.

2. Examples are cold sores and shingles.

PERSISTENT VIRAL INFECTIONS (p. 397)

1. Persistent viral infections are disease processes that occur over a long period and are generally fatal.

2. Persistent viral infections are caused by conventional viruses; viruses accumulate over a long period.

PRIONS (p. 398)

1. Prions are infectious proteins first discovered in the 1980s.

2. Prion diseases, such as CJD and mad cow disease, all involve the degeneration of brain tissue.

3. Prion diseases are the result of an altered protein; the cause can be a mutation in the normal gene for PrP or contact with an altered protein (PrP^{Sc}).

PLANT VIRUSES AND VIROIDS (pp. 398–399)

1. Plant viruses must enter plant hosts through wounds or with invasive parasites, such as insects.

2. Some plant viruses also multiply in insect (vector) cells.

3. Viroids are infectious pieces of RNA that cause some plant diseases, such as potato spindle tuber viroid disease.

Study Questions

REVIEW

1. Viruses were first detected because they are filterable. What do we mean by the term *filterable,* and how could this property have helped researchers detect viruses before the invention of the electron microscope?

2. Why do we classify viruses as obligatory intracellular parasites?

3. List the four properties that define a virus. What is a virion?

4. Describe the four morphological classes of viruses, then diagram and give an example of each.

5. Describe how bacteriophages are detected and enumerated by the plaque method.

6. Why are continuous cell lines of more practical use than primary cell lines for culturing viruses? What is unique about continuous cell lines?

7. *Vibrio cholerae* produces toxin and is capable of causing cholera only when it is lysogenic. What does this mean?

8. Describe the principal events of attachment, penetration, uncoating, biosynthesis, maturation, and release of an enveloped DNA-containing virus.

9. Recall from Chapter 1 that Koch's postulates are used to determine the etiology of a disease. Why is it difficult to determine the etiology of
   a. a viral infection such as influenza?
   b. cancer?

10. Persistent viral infections such as _____ might be caused by _____ that are _____.

11. The DNA of DNA-containing oncogenic viruses can become integrated into the host DNA. When integrated, the DNA is called a _____. How does this process result in transformation of the cell? Describe the changes of transformation. How can an RNA-containing virus be oncogenic?

12. Contrast viroids and prions. Name a disease caused by each.

13. Plant viruses cannot penetrate intact plant cells because _____; therefore, they enter cells by _____. Plant viruses can be cultured in _____.

MULTIPLE CHOICE

1. Place the following in the most likely order for biosynthesis of a bacteriophage: (1) phage lysozyme; (2) mRNA; (3) DNA; (4) viral proteins; (5) DNA polymerase.
   a. 5, 4, 3, 2, 1
   b. 1, 2, 3, 4, 5
   c. 5, 3, 4, 2, 1
   d. 3, 5, 2, 4, 1
   e. 2, 5, 3, 4, 1

2. The molecule serving as mRNA can be incorporated in the newly synthesized virus capsids of all of the following *except*
   a. + strand RNA picornaviruses.
   b. + strand RNA togaviruses.
   c. − strand RNA rhabdoviruses.
   d. double-stranded RNA reoviruses.
   e. double-stranded DNA herpesviruses.

3. A virus with RNA-dependent RNA polymerase
   a. synthesizes DNA from an RNA template.
   b. synthesizes double-stranded RNA from an RNA template.
   c. synthesizes double-stranded RNA from a DNA template.
   d. transcribes mRNA from DNA.
   e. none of the above

4. Which of the following would be the first step in the biosynthesis of a virus with reverse transcriptase?
   a. A complementary strand of RNA must be synthesized.
   b. Double-stranded RNA must be synthesized.
   c. A complementary strand of DNA must be synthesized from an RNA template.
   d. A complementary strand of DNA must be synthesized from a DNA template.
   e. none of the above

5. An example of lysogeny in animals could be
   a. slow viral infections.
   b. latent viral infections.
   c. T-even bacteriophages.
   d. infections resulting in cell death.
   e. none of the above

6. The ability of a virus to infect an organism is regulated by
   a. the host species.
   b. the type of cells.
   c. the availability of an attachment site.
   d. cell factors necessary for viral replication.
   e. all of the above

7. Which of the following statements is *not* true?
   a. Viruses contain DNA or RNA.
   b. The nucleic acid of a virus is surrounded by a protein coat.
   c. Viruses multiply inside living cells using viral mRNA, tRNA, and ribosomes.
   d. Viruses cause the synthesis of specialized infectious elements.
   e. Viruses multiply inside living cells.

8. Place the following in the order in which they are found in a host cell: (1) capsid proteins; (2) infective phage particles; (3) phage nucleic acid.
   a. 1, 2, 3
   b. 3, 2, 1
   c. 2, 1, 3
   d. 3, 1, 2
   e. 1, 3, 2

9. Which of the following does *not* initiate DNA synthesis?
   a. A double-stranded DNA virus (Poxviridae)
   b. A DNA virus with reverse transcriptase (Hepadnaviridae)
   c. An RNA virus with reverse transcriptase (Retroviridae)
   d. A single-stranded RNA virus (Togaviridae)
   e. none of the above

10. A viral species is not defined on the basis of the disease symptoms it causes. The best example of this is
   a. polio.
   b. rabies.
   c. hepatitis.
   d. chickenpox and shingles.
   e. measles.

CRITICAL THINKING

1. Discuss the arguments for and against the classification of viruses as living organisms.

2. In some viruses, capsomeres function as enzymes as well as structural supports. Of what advantage is this to the virus?

3. Why was the discovery of simian AIDS and feline AIDS important?

4. Prophages and proviruses have been described as being similar to bacterial plasmids. What similar properties do they exhibit? How are they different?

CLINICAL APPLICATIONS

1. A 40-year-old male who was seropositive for HIV experienced abdominal pain, fatigue, and low-grade fever (38°C) for 2 weeks. A chest X-ray examination revealed lung infiltrates. Gram and acid-fast stains were negative. A viral culture revealed the cause of his symptoms: a large, enveloped polyhedral virus with double-stranded DNA. What is the disease? Which virus causes it? Why was a viral culture done after the Gram and acid-fast stain results were obtained?

2. A newborn female developed extensive vesicular and ulcerative lesions over her face and chest. What is the most likely cause of her symptoms? How would you determine the viral cause of this disease without doing a viral culture?

3. Thirty-two people in the same town reported to their physicians with fever (40°C), jaundice, and tender abdomen. All 32 had eaten an ice-slush beverage purchased from a local convenience store. Liver function tests were abnormal. Over the next several months, the symptoms subsided and liver function returned to normal. What is the disease? This disease could be caused by a member of the Picornaviridae, Hepadnaviridae, or Flaviviridae. Differentiate among these families by method of transmission, morphology, nucleic acid, and type of replication.

Learning with Technology

MP = The Microbiology Place website   **ST** = Student Tutorial CD-ROM

MP   Don't forget to go to The Microbiology Place website (http://www.microbiologyplace.com) to take the practice tests, explore the interactive activity and case study, and check out the news articles and web links for this chapter.

ST   Remember there is also a quiz for this chapter on the Microbiology Interactive Student Tutorial CD-ROM.

chapter 14

Principles of Disease and Epidemiology

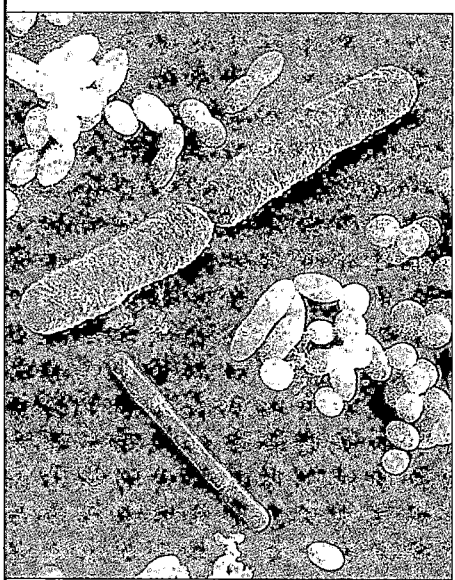

Intestinal bacteria. Millions of bacteria live in the human large intestine and are essential for good health.

Now that you have a basic understanding of the structures and functions of microorganisms and some idea of the variety of microorganisms that exist, we can consider how the human body and various microorganisms interact in terms of health and disease.

We all have our defense mechanisms to keep us healthy. For instance, unbroken skin and mucous membranes are effective barriers against microbial invasion. Within the body, certain cells and certain specialized proteins called antibodies can work together to destroy microbes. In spite of our defenses, however, we are still susceptible to **pathogens** (disease-causing microorganisms). Some bacteria can invade our tissues and resist our defenses by producing capsules or enzymes; other bacteria release toxins that can seriously affect our health. A rather delicate balance exists between our defenses and the pathogenic mechanisms of microorganisms. When our defenses resist these pathogenic capabilities, we maintain our health. But when the pathogen's capability overcomes our defenses, disease results. After the disease has become established, an infected person may recover completely, suffer temporary

or permanent damage, or die. The outcome depends on many factors.

Infectious disease mortality in the United States decreased markedly during most of the 20th century. However, between 1980 and 1992, the death rate from infectious disease increased 58% (including only those people for whom the primary cause of death was an infectious disease). The sharp increase in infectious disease deaths in 1918 and 1919 was caused by the influenza pandemic, which killed more than 20 million people worldwide and over 500,000 people in the United States. This episode illustrates the volatility of infectious disease death rates.

In Part Three we examine some of the principles of infection and disease, the mechanisms by which pathogens cause disease, the body's defenses against disease, and the ways that microbial diseases can be prevented by immunization and controlled by drugs. This first chapter discusses the general principles of disease, starting with a discussion of the meaning and scope of pathology. In the last section of this chapter, Epidemiology, you will learn how these principles are useful in studying and controlling disease.

Pathology, Infection, and Disease

Learning Objective

■ *Define pathology, etiology, infection, and disease.*

Pathology is the scientific study of disease (*pathos* = suffering; *logos* = science). Pathology is first concerned with the cause, or **etiology**, of disease. Second, it deals with **pathogenesis**, the manner in which a disease develops. Third, pathology is concerned with the structural and functional changes brought about by disease and with their final effects on the body.

Although the terms *infection* and *disease* are sometimes used interchangeably, they differ somewhat in meaning. **Infection** is the invasion or colonization of the body by pathogenic microorganisms; **disease** occurs when an infection results in any change from a state of health. Disease is an abnormal state in which part or all of the body is not properly adjusted or incapable of performing its normal functions. An infection may exist in the absence of detectable disease. For example, the body may be infected with the virus that causes AIDS, but there may be no symptoms of the disease.

The presence of a particular type of microorganism in a part of the body where it is not normally found is also called an infection—and may lead to disease. For example, although large numbers of *E. coli* are normally present in the healthy intestine, their infection of the urinary tract usually results in disease.

Few microorganisms are pathogenic. In fact, the presence of some microorganisms can even benefit the host. Therefore, before we discuss the role of microorganisms in causing disease, let's examine the relationship of the microorganisms to the healthy human body.

Normal Microbiota

Learning Objective

■ *Define normal and transient microbiota.*

Animals, including humans, are generally free of microbes in utero. At birth, however, normal and characteristic microbial populations begin to establish themselves. Just before a woman gives birth, lactobacilli in her vagina multiply rapidly. The newborn's first contact with microorganisms is usually with these lactobacilli, and they become the predominant organisms in the newborn's intestine. More microorganisms are introduced to the newborn's body from the environment when breathing begins and feeding starts. After birth, *E. coli* and other bacteria acquired from foods begin to inhabit the large intestine. These microorganisms remain there throughout life and, in response to altered environmental conditions, may increase or decrease in number and contribute to disease.

Many other usually harmless microorganisms establish themselves inside other parts of the normal adult body and on its surface. A typical human body contains 1×10^{13} body cells, yet harbors an estimated 1×10^{14} bacterial cells. This gives you an idea of the abundance of microorganisms that normally reside in the human body. The microorganisms that establish more or less permanent residence (colonize) but that do not produce disease under normal conditions are members of the body's **normal microbiota** or **normal flora** (Figure 14.1). Others, called **transient microbiota,** may be present for several days, weeks, or months and then disappear. Microorganisms are not found throughout the entire human body but are localized in certain regions, as shown in Figure 14.2.

The principal normal microbiota in different regions of the body and some distinctive features of each region are listed in Table 14.1 on page 409. Normal microbiota are also discussed more specifically in Part Four.

Relationships Between the Normal Microbiota and the Host

Learning Objective

■ *Compare commensalism, mutualism, and parasitism, and give an example of each.*

Once established, the normal microbiota can benefit the host by preventing the overgrowth of harmful microorganisms. This phenomenon is called **microbial antagonism.** Microbial antagonism involves competition among microbes. One consequence of this competition is that the normal microbiota protect the host against colonization by potentially pathogenic microbes by competing for nutrients, producing substances harmful to the invading microbes, and affecting conditions such as pH and available oxygen. When this balance between normal microbiota and pathogenic microbes is upset, disease can result. For example, the normal bacterial microbiota of the adult human vagina maintains a local pH of 3.5–4.5. The presence of normal microbiota inhibits the overgrowth of the yeast *Candida albicans,* which cannot grow under these conditions. If the bacterial population is eliminated by antibiotics, excessive douching, or deodorants, the pH of the vagina reverts to nearly neutral, and

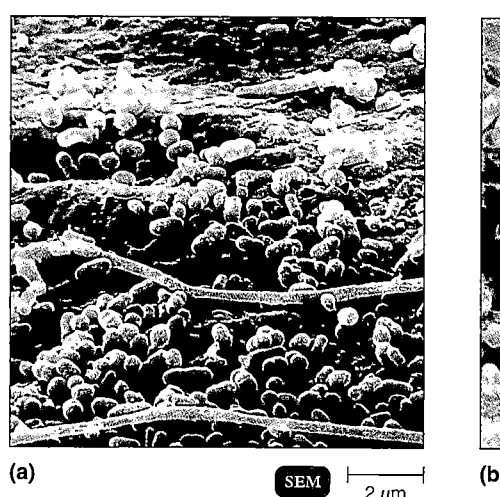 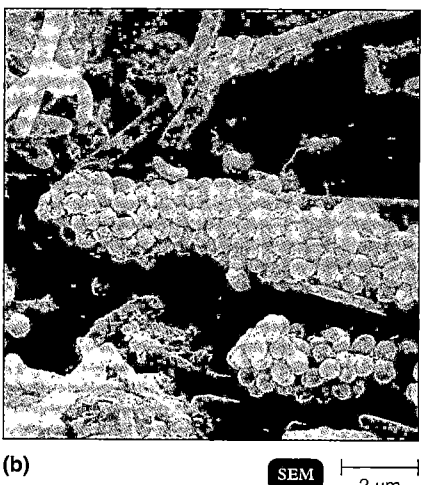 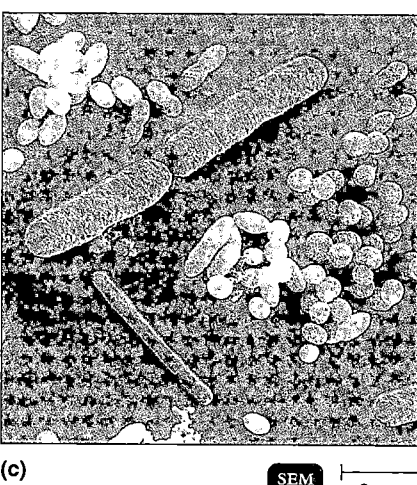

(a) SEM ⊢———⊣ 2 μm   (b) SEM ⊢———⊣ 2 μm   (c) SEM ⊢———⊣ 2 μm

FIGURE 14.1 Representative normal microbiota for different regions of the body.
(a) Bacteria on the surface of the skin. (b) Plaque on enamel near the gums. The bacteria that cause dental plaque, although part of the normal microbiota, will cause gum disease and caries (cavities) unless removed frequently. (c) Bacteria of the large intestine.

■ Microbes that establish more or less permanent residence but do not produce disease under normal conditions are referred to as normal microbiota.

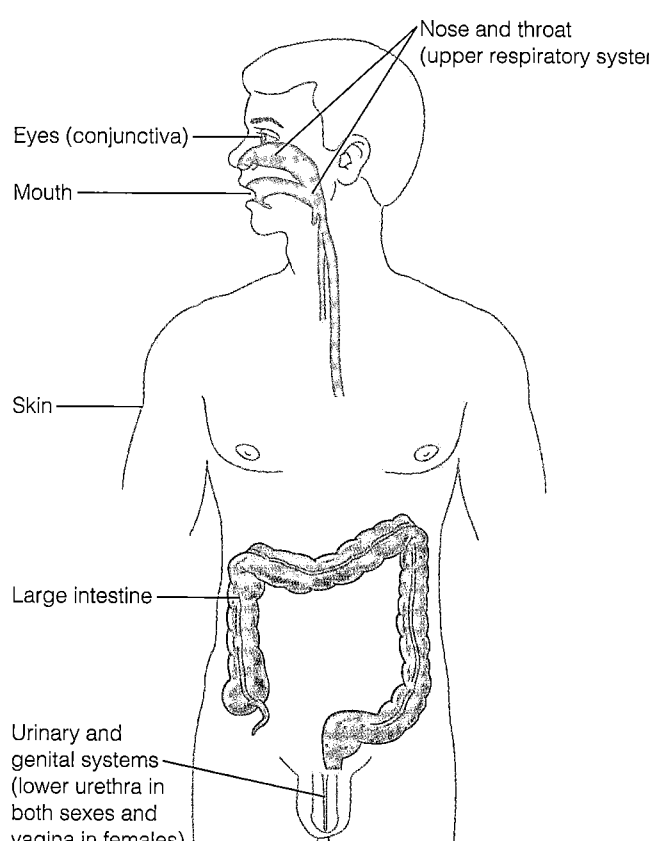

Nose and throat
(upper respiratory system)

Eyes (conjunctiva)

Mouth

Skin

Large intestine

Urinary and genital systems (lower urethra in both sexes and vagina in females)

FIGURE 14.2 Locations of normal microbiota on and in the human body.

■ Why is the normal microbiota essential for good health?

C. albicans can flourish and become the dominant microorganism there. This condition can lead to a form of vaginitis (vaginal infection).

Another example of microbial antagonism is seen in the mouth, where streptococci produce compounds that prevent the growth of most gram-positive and gram-negative cocci. And in the large intestine, *E. coli* cells produce *bacteriocins,* proteins that inhibit the growth of other bacteria of the same or closely related species, such as pathogenic *Salmonella* and *Shigella.* A bacterium that makes a particular bacteriocin is not killed by that bacteriocin, but may be killed by other ones. Bacteriocins are used in medical microbiology to help identify different strains of bacteria. Such identification helps determine whether several outbreaks of an infectious disease are caused by one or more strains of a bacterium.

A final example involves another bacterium, *Clostridium difficile,* also in the large intestine. The normal microbiota of the large intestine effectively inhibit *C. difficile,* possibly by making host receptors unavailable, competing for available nutrients, or producing bacteriocins. However, if the normal microbiota are eliminated, for example, by antibiotics, *C. difficile* can become a problem. This microbe is responsible for nearly all gastrointestinal infections that follow antibiotic therapy, from mild diarrhea to severe or even fatal colitis (inflammation of the colon). The microbe invades the intestinal lining and releases toxins that induce diarrhea, fever, and abdominal pain.

table 14.1 | Representative Members of the Normal Microbiota by Body Region*

Region	Principal Components	Comments
Skin	Propionibacterium acnes, Staphylococcus epidermidis, Staphylococcus aureus, Corynebacterium xerosis, Pityrosporum spp. (fungus), Candida spp. (fungus)	Most of the microbes in direct contact with skin do not become residents because secretions from sweat and oil glands have antimicrobial properties.
Eyes (conjunctiva)	S. epidermidis, S. aureus, diphtheroids	The conjunctiva, a continuation of the skin or mucous membrane, contains basically the same microbiota found on the skin.
Nose and throat (upper respiratory system)	S. aureus, S. epidermidis, and aerobic diphtheroids in the nose; S. epidermidis, S. aureus, diphtheroids, Streptococcus pneumoniae, Haemophilus, and Neisseria in the throat	Although some normal microbiota are potential pathogens, their ability to cause disease is reduced by microbial antagonism.
Mouth	Various species of Streptococcus, Lactobacillus, Actinomyces, Bacteroides, Fusobacterium, Treponema, Corynebacterium, and Candida (fungus)	Abundant moisture, warmth, and the constant presence of food make the mouth an ideal environment that supports very large and diverse microbial populations on the tongue, cheeks, teeth, and gums.
Large intestine	Bacteroides, Fusobacterium, Lactobacillus, Enterococcus, Bifidobacterium, Escherichia coli, Enterobacter, Citrobacter, Proteus, Klebsiella, Shigella, Candida (fungus)	The large intestine contains the largest numbers of resident microbiota in the body because of its available moisture and nutrients.
Urogenital system	Staphylococcus epidermidis, aerobic micrococci, Enterococcus, Lactobacillus, aerobic diphtheroids, Pseudomonas, Klebsiella, and Proteus in urethra; lactobacilli, aerobic diphtheroids, Streptococcus, Staphylococcus, Bacteroides, Clostridium, Candida albicans (fungus), and Trichomonas vaginalis (protozoan) in vagina	The lower urethra in both sexes has a resident population; the vagina has its acid-tolerant population of microbes because of the nature of its secretions.

*This table lists some microbiota that are not discussed in this chapter; they will be discussed in Part Four. Unless indicated, the organisms are bacteria.

The relationship between the normal microbiota and the host is called **symbiosis** (living together). In the symbiotic relationship called **commensalism,** one of the organisms is benefited and the other is unaffected. Many of the microorganisms that make up our normal microbiota are commensals; these include the corynebacteria that inhabit the surface of the eye and certain saprophytic mycobacteria that inhabit the ear and external genitals. These bacteria live on secretions and sloughed-off cells, and they bring no apparent benefit or harm to the host.

Mutualism is a type of symbiosis that benefits both organisms. For example, the large intestine contains bacteria, such as *E. coli,* that synthesize vitamin K and some B vitamins. These vitamins are absorbed into the bloodstream and distributed for use by body cells. In exchange, the large intestine provides nutrients for the bacteria so that they can survive. In still another kind of symbiosis, one organism is benefited at the expense of the other; this

relationship is called **parasitism.** Many disease-causing bacteria are parasites.

Opportunistic Microorganisms

Learning Objective

■ *Contrast normal and transient microbiota with opportunistic microorganisms.*

Although categorizing symbiotic relationships by type is convenient, keep in mind that under certain conditions the relationship can change. For example, given the proper circumstances, a mutualistic organism, such as *E. coli,* can become harmful. *E. coli* is generally harmless as long as it remains in the large intestine; but if it gains access to other body sites, such as the urinary bladder, lungs, spinal cord, or wounds, it may cause urinary tract infections, pulmonary infections, meningitis, or abscesses, respectively.

Microbes such as *E. coli* are called **opportunistic pathogens**. They ordinarily do not cause disease in their normal habitat in a healthy person but may do so in a different environment. For example, microbes that gain access through broken skin or mucous membranes can cause opportunistic infections. Or, if the host is already weakened or compromised by infection, microbes that are usually harmless can cause disease. AIDS is often accompanied by a common opportunistic infection, *Pneumocystis* pneumonia, caused by the opportunistic organism *Pneumocystis carinii* (see Figure 24.18). This secondary infection can develop in AIDS patients because their immune systems are suppressed. Before the AIDS epidemic, this type of pneumonia was rare.

In addition to the usual symbionts, many people carry other microorganisms that are generally regarded as pathogenic but that may not cause disease in those people. Among the pathogens that are frequently carried in healthy individuals are echoviruses (*echo* comes from *en*teric *cy*topathogenic *h*uman *o*rphan), which can cause intestinal diseases; and adenoviruses, which can cause respiratory diseases. *Neisseria meningitidis,* which often resides benignly in the respiratory tract, can cause meningitis, a disease that inflames the coverings of the brain and spinal cord. *Streptococcus pneumoniae,* a normal resident of the nose and throat, can cause a type of pneumonia.

Cooperation Among Microorganisms

In addition to competition among microbes in causing disease, there are situations in which cooperation among microbes is a factor in causing disease. One example of cooperation among microbes in the development of disease involves oral streptococci that colonize the teeth. Pathogens that cause periodontal disease and gingivitis have been found to have receptors for the streptococci, rather than for the teeth.

The Etiology of Infectious Diseases

Learning Objective

■ *List Koch's postulates.*

Some diseases—such as polio, Lyme disease, and tuberculosis—have a well-known etiology. Some have an etiology that is not completely understood—for example, the relationship between ulcers and *Helicobacter pylori*. For still others, such as Alzheimer's disease, the etiology is unknown. Of course, not all diseases are caused by microorganisms. For example, the disease hemophilia is an

inherited *(genetic) disease;* osteoarthritis and cirrhosis are considered *degenerative diseases.* There are several other categories of disease, but here we will discuss only *infectious diseases,* those caused by microorganisms. To see how microbiologists determine the etiology of an infectious disease, we will discuss in greater detail the work of Robert Koch, which was introduced in Chapter 1 on page 10.

Koch's Postulates

In the historical overview of microbiology presented in Chapter 1, we briefly discussed Koch's famous postulates. Recall that Koch was a German physician who played a major role in establishing that microorganisms cause specific diseases. In 1877, he published some early papers on anthrax, a disease of cattle that can also occur in humans. Koch demonstrated that certain bacteria, today known as *Bacillus anthracis,* were always present in the blood of animals that had the disease and were not present in healthy animals. He knew that the mere presence of the bacteria did not prove that they had caused the disease; the bacteria could have been there as a result of the disease. Thus, he experimented further.

He took a sample of blood from a diseased animal and injected it into a healthy one. The second animal developed the same disease and died. He repeated this procedure many times, always with the same results. (A key criterion in the validity of any scientific proof is that experimental results be repeatable.) Koch also cultivated the microorganism in fluids outside the animal's body, and he demonstrated that the bacterium would cause anthrax even after many culture transfers.

Koch showed that a specific infectious disease (anthrax) is caused by a specific microorganism (*B. anthracis*) that can be isolated and cultured on artificial media. He later used the same methods to show that the bacterium *Mycobacterium tuberculosis* is the causative agent of tuberculosis.

Koch's research provides a framework for the study of the etiology of any infectious disease. Today, we refer to Koch's experimental requirements as **Koch's postulates** (Figure 14.3). They are summarized as follows:

1. The same pathogen must be present in every case of the disease.

2. The pathogen must be isolated from the diseased host and grown in pure culture.

3. The pathogen from the pure culture must cause the disease when it is inoculated into a healthy, susceptible laboratory animal.

4. The pathogen must be isolated from the inoculated animal and must be shown to be the original organism.

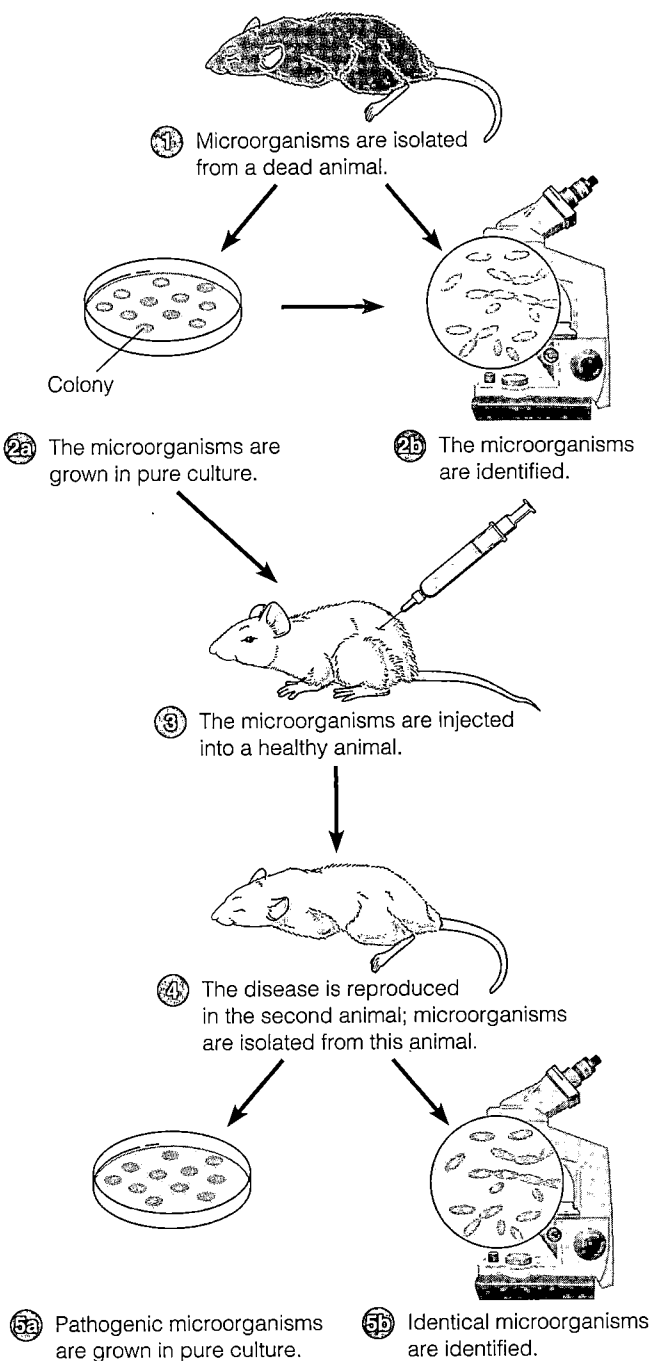

① Microorganisms are isolated from a dead animal.

Colony

②a The microorganisms are grown in pure culture.

②b The microorganisms are identified.

③ The microorganisms are injected into a healthy animal.

④ The disease is reproduced in the second animal; microorganisms are isolated from this animal.

⑤a Pathogenic microorganisms are grown in pure culture.

⑤b Identical microorganisms are identified.

FIGURE 14.3 Application of Koch's postulates.

■ Koch's postulates provide a framework for the study of the etiology of any infectious disease.

Exceptions to Koch's Postulates

Although Koch's postulates are useful in determining the causative agent of most bacterial diseases, there are some exceptions. For example, some microbes have unique culture requirements. The bacterium *Treponema pallidum* is known to cause syphilis, but virulent strains have never been cultured on artificial media. The causative agent of leprosy, *Mycobacterium leprae,* has also never been grown on artificial media. Moreover, many rickettsial and viral pathogens cannot be cultured on artificial media because they multiply only within cells.

The discovery of microorganisms that cannot grow on artificial media has necessitated some modifications of Koch's postulates and the use of alternative methods of culturing and detecting certain microbes. For example, when researchers looking for the microbial cause of legionellosis (Legionnaires' disease) were unable to isolate the microbe directly from a victim, they took the alternative step of inoculating a victim's lung tissue into guinea pigs. These guinea pigs developed the disease's pneumonialike symptoms, whereas guinea pigs inoculated with tissue from an unafflicted person did not. Then tissue samples from the diseased guinea pigs were cultured in yolk sacs of chick embryos, a method (see Figure 13.7, page 380) that reveals the growth of extremely small microbes. After the embryos were incubated, electron microscopy revealed rod-shaped bacteria in the chick embryos. Finally, modern immunological techniques (which will be discussed in Chapter 18) were used to show that the bacteria in the chick embryos were the same bacteria as those in the guinea pigs and in afflicted humans.

In a number of situations, a human host exhibits certain signs and symptoms that are associated only with a certain pathogen and its disease. For example, the pathogens responsible for diphtheria and tetanus cause distinguishing signs and symptoms that can be produced by no other microbe. They are unequivocally the only organisms that produce their respective diseases. But some infectious diseases are not as clear-cut and provide another exception to Koch's postulates. For example, nephritis (inflammation of the kidneys) can involve any of several different pathogens, all of which cause the same signs and symptoms (described shortly). Thus, it is often difficult to know which particular microorganism is causing a disease. Other infectious diseases that sometimes have poorly defined etiologies are pneumonia, meningitis, and peritonitis (inflammation of the peritoneum, the membrane that lines the abdomen and pelvis and covers the organs within them).

Still another exception to Koch's postulates results because some pathogens can cause several disease conditions. *Mycobacterium tuberculosis,* for example, is implicated in diseases of the lungs, skin, bones, and internal organs. *Streptococcus pyogenes* can cause sore throat, scarlet fever, skin infections (such as erysipelas), and osteomyelitis (inflammation of bone), among other diseases. When clinical signs and symptoms are used together with laboratory methods, these infections can usually be distinguished from infections of the same organs by other pathogens.

Ethical considerations may also impose an exception to Koch's postulates. For example, some agents that cause disease in humans have no other known host. An example is human immunodeficiency virus (HIV), the cause of AIDS. This poses the ethical question of whether humans can be intentionally inoculated with infectious agents. In 1721, King George I told condemned prisoners they could be inoculated with smallpox to test a smallpox vaccine (see Chapter 18). He promised their freedom if they lived. Human experiments with untreatable diseases are not acceptable today. Sometimes accidental inoculation does occur. A contaminated bone marrow transplant provided the third Koch's postulate to prove that a herpesvirus caused cancer.

Classifying Infectious Diseases

Learning Objective

■ *Differentiate a communicable from a noncommunicable disease.*

Every disease that affects the body alters body structures and functions in particular ways, and these alterations are usually indicated by several kinds of evidence. For example, the patient may experience certain **symptoms,** or changes in body function, such as pain and *malaise* (a vague feeling of body discomfort). These *subjective* changes are not apparent to an observer. The patient can also exhibit **signs,** which are *objective* changes the physician can observe and measure. Frequently evaluated signs include lesions (changes produced in tissues by disease), swelling, fever, and paralysis. A specific group of symptoms or signs may always accompany a particular disease; such a group is called a **syndrome.** The diagnosis of a disease is made by evaluation of the signs and symptoms, together with the results of laboratory tests.

Diseases are often classified in terms of how they behave within a host and within a given population. Any disease that spreads from one host to another, either directly or indirectly, is said to be a **communicable disease.** Chickenpox, measles, genital herpes, typhoid fever, and tuberculosis are examples. Chickenpox and measles are also examples of **contagious diseases,** that is, diseases that are *easily* spread from one person to another. A **noncommunicable disease** is not spread from one host to another. These diseases are caused by microorganisms that normally inhabit the body and only occasionally produce disease, or by microorganisms that reside outside the body and produce disease only when introduced into the body. An example is tetanus: *Clostridium tetani* produces disease only when it is introduced into the body via abrasions or wounds.

The Occurrence of a Disease

Learning Objective

■ *Categorize diseases according to frequency of occurrence.*

To understand the full scope of a disease, we should know something about its occurrence. The **incidence** of a disease is the number of people in a population who develop a disease during a particular time period. It is an indicator of the spread of the disease. The **prevalence** of a disease is the number of people in a population who develop a disease at a specified time, regardless of when it first appeared. Prevalence takes into account both old and new cases. It's an indicator of how seriously and how long a disease affects a population. For example, the incidence of AIDS in the United States in 1999 was 45,000, while the prevalence in that same year was estimated to be about 700,000. Knowing the incidence and the prevalence of a disease in different populations (for example, in populations representing different geographic regions or different racial groups) enables scientists to estimate the range of the disease's occurrence and its tendency to affect some groups of people more than others.

Frequency of occurrence is another criterion that is used in the classification of diseases. If a particular disease occurs only occasionally, it is called a **sporadic disease;** typhoid fever in the United States is such a disease. A disease constantly present in a population is called an **endemic disease;** an example of such a disease is the common cold. If many people in a given area acquire a certain disease in a relatively short period, it is called an **epidemic disease;** influenza is an example of a disease that often achieves epidemic status. Figure 14.4 shows the epidemic incidence of AIDS in the United States. Some authorities consider gonorrhea and certain other sexually transmitted diseases to be epidemic at this time as well (see Figures 26.5 and 26.6 on page 726). An epidemic disease that occurs worldwide is called a **pandemic disease.** We experience pandemics of influenza from time to time. Some authorities also consider AIDS to be pandemic.

The Severity or Duration of a Disease

Learning Objectives

■ *Categorize diseases according to severity.*
■ *Define herd immunity.*

Another useful way of defining the scope of a disease is in terms of its severity or duration. An **acute disease** is one that develops rapidly but lasts only a short time; a good example is influenza. A **chronic disease** develops more slowly, and the body's reactions may be less severe, but the

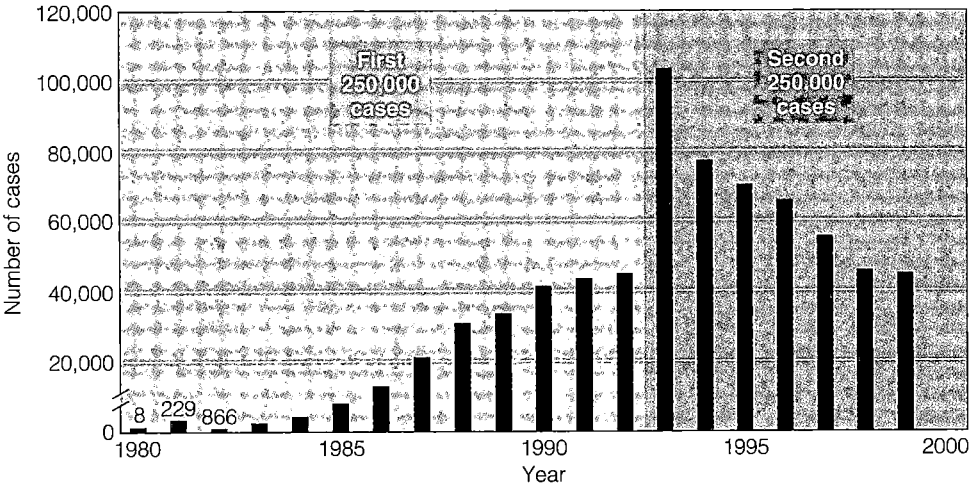

FIGURE 14.4 Reported AIDS cases in the United States. Notice that the first 250,000 cases occurred over a 12-year period, whereas the second 250,000 cases in this epidemic occurred in just 3 years. Much of the increase shown for 1993 is due to an expanded definition of AIDS cases adopted in that year. (Source: CDC.)

■ Distinguish among sporadic, endemic, epidemic, and pandemic diseases.

disease is likely to be continual or recurrent for long periods. Infectious mononucleosis, tuberculosis, and hepatitis B fall into this category. A disease that is intermediate between acute and chronic is described as a **subacute disease;** an example is subacute sclerosing panencephalitis, a rare brain disease characterized by diminished intellectual function and loss of nervous function. A **latent disease** is one in which the causative agent remains inactive for a time but then becomes active to produce symptoms of the disease; an example is shingles, one of the diseases caused by *Varicellovirus.*

The rate at which a disease, or an epidemic, spreads, and the number of individuals involved, are determined in part by the immunity of the population. Vaccination can provide long-lasting and sometimes lifelong protection of an individual against certain diseases. People who are immune to an infectious disease will not be carriers, thereby reducing the occurrence of the disease. Immune individuals act as a barrier to the spread of infectious agents. Even though a highly communicable disease may cause an epidemic, many nonimmune people will be protected because of the unlikelihood of their coming into contact with an infected person. A great advantage of vaccination is that enough individuals in a population will be protected from a disease to prevent its rapid spread to those in the population who are not vaccinated. When many immune people are present in a community, **herd immunity** exists.

The Extent of Host Involvement

Infections can also be classified according to the extent to which the host's body is affected. A **local infection** is one in which the invading microorganisms are limited to a relatively small area of the body. Some examples of local infections are boils and abscesses. In a **systemic (generalized) infection,** microorganisms or their products are spread throughout the body by the blood or lymph. Measles is an example of a systemic infection. Very often, agents of a local infection enter a blood or lymphatic vessel and spread to other specific parts of the body, where they are confined to specific areas of the body. This condition is called a **focal infection.** Focal infections can arise from infections in areas such as the teeth, tonsils, or sinuses.

The presence of bacteria in the blood is known as **bacteremia,** and if the bacteria actually multiply in the blood, the condition is called **septicemia. Toxemia** refers to the presence of toxins in blood (as occurs in tetanus), and **viremia** refers to the presence of viruses in blood.

The state of host resistance also determines the extent of infections. A **primary infection** is an acute infection that causes the initial illness. A **secondary infection** is one caused by an opportunistic pathogen after the primary infection has weakened the body's defenses. Secondary infections of the skin and respiratory tract are common and are sometimes more dangerous than the primary infections. *Pneumocystis* pneumonia as a consequence of AIDS is an example of a secondary infection;

streptococcal bronchopneumonia following influenza is an example of a secondary infection that is more serious than the primary infection. A **subclinical (inapparent) infection** is one that does not cause any noticeable illness. Poliovirus and hepatitis A virus, for example, can be carried by people who never develop the illness.

Patterns of Disease

A definite sequence of events usually occurs during infection and disease. As you will learn shortly, there must be a reservoir of infection as a source of pathogens for an infectious disease to occur. Next, the pathogen must be transmitted to a susceptible host by direct contact, by indirect contact, or by vectors. Transmission is followed by invasion, in which the microorganism enters the host and multiplies. Following invasion, the microorganism injures the host through a process called pathogenesis (discussed further in the next chapter). The extent of injury depends on the degree to which host cells are damaged, either directly or by toxins. Despite the effects of all these factors, the occurrence of disease ultimately depends on the resistance of the host to the activities of the pathogen.

Predisposing Factors

Learning Objective

■ *Identify four predisposing factors for disease.*

Certain predisposing factors also affect the occurrence of disease. A **predisposing factor** is one that makes the body more susceptible to a disease and may alter the course of the disease. Gender is sometimes a predisposing factor; for example, females have a higher incidence of urinary tract infections than males, whereas males have higher rates of pneumonia and meningitis. Other aspects of genetic background may play a role as well. For example, sickle-cell disease is a severe, life-threatening form of anemia that occurs when the genes for the disease are inherited from both parents. Individuals who carry only one sickle-cell gene have a condition called sickle-cell trait, and are normal unless specially tested. However, they are relatively resistant to the most serious form of malaria. The potential that individuals in a population might inherit a life-threatening disease is more than counterbalanced by protection from malaria among carriers of the gene for sickle-cell trait. Of course, in countries where malaria is not present, sickle-cell trait is an entirely negative condition.

Climate and weather seem to have some effect on the incidence of infectious diseases. In temperate regions, the incidence of respiratory diseases increases during the winter. This increase may be related to the fact that when

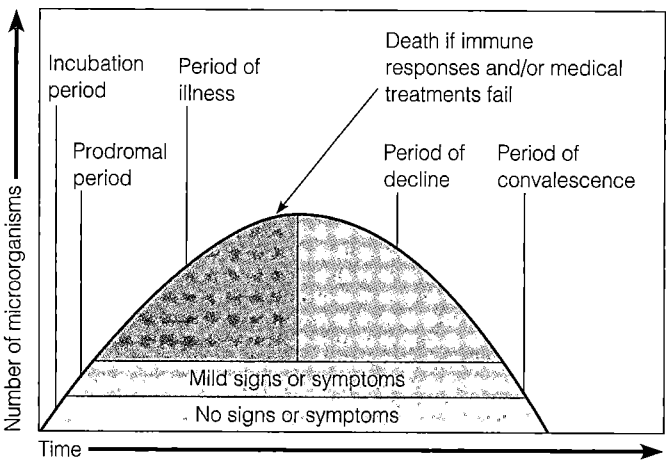

FIGURE 14.5 The stages of a disease.

■ During which period can a disease be transmitted?

people stay indoors and have closer contact with each other, the spread of respiratory pathogens is facilitated.

Other predisposing factors include inadequate nutrition, fatigue, age, environment, habits, lifestyle, occupation, preexisting illness, chemotherapy, and emotional disturbances. It is often difficult to know the exact relative importance of the various predisposing factors.

The Development of Disease

Learning Objective

■ *Put the following in proper sequence, according to the pattern of disease: period of decline, period of convalescence, period of illness, prodromal period, incubation period.*

Once a microorganism overcomes the defenses of the host, development of the disease follows a certain sequence that tends to be similar whether the disease is acute or chronic (Figure 14.5).

The Incubation Period

The **incubation period** is the time interval between the initial infection and the first appearance of any signs or symptoms. In some diseases, the incubation period is always the same; in others, it is quite variable. The time of incubation depends on the specific microorganism involved, its virulence (degree of pathogenicity), the number of infecting microorganisms, and the resistance of the host. (See Table 15.1 on page 438 for the incubation periods of a number of microbial diseases.)

The Prodromal Period

The **prodromal period** is a relatively short period that follows the period of incubation in some diseases. The prodromal period is characterized by early, mild symptoms of disease, such as general aches and malaise.

The Period of Illness

During the period of illness, the disease is most acute. The person exhibits overt signs and symptoms of disease, such as fever, chills, muscle pain (myalgia), sensitivity to light (photophobia), sore throat (pharyngitis), lymph node enlargement (lymphadenopathy), and gastrointestinal disturbances. During the period of illness, the number of white blood cells may increase or decrease. Generally, the patient's immune response and other defense mechanisms overcome the pathogen, and the period of illness ends. When the disease is not successfully overcome (or successfully treated), the patient dies during this period.

The Period of Decline

During the period of decline, the signs and symptoms subside. The fever decreases, and the feeling of malaise diminishes. During this phase, which may take from less than 24 hours to several days, the patient is vulnerable to secondary infections.

The Period of Convalescence

During the **period of convalescence,** the person regains strength and the body returns to its prediseased state. Recovery has occurred.

We all know that during the period of illness, people can serve as reservoirs of disease and can easily spread infections to other people. However, you should also know that people can spread infection during incubation and convalescence. This is especially true of diseases such as typhoid fever and cholera, in which the convalescing person carries the pathogenic microorganism for months or even years.

The Spread of Infection

Now that you have an understanding of normal microbiota, the etiology of infectious diseases, and the types of infectious diseases, we will examine the sources of pathogens and how diseases are transmitted.

Reservoirs of Infection

Learning Objectives

- *Define reservoir of infection.*
- *Contrast human, animal, and nonliving reservoirs, and give one example of each.*

For a disease to perpetuate itself, there must be a continual source of the disease organisms. This source can be either a living organism or an inanimate object that provides a pathogen with adequate conditions for survival and multiplication and an opportunity for transmission. Such a source is called a **reservoir of infection.** These reservoirs may be human, animal, or nonliving.

Human Reservoirs

The principal living reservoir of human disease is the human body itself. Many people harbor pathogens and transmit them directly ,or indirectly to others. People with signs and symptoms of a disease may transmit the disease; in addition, some people can harbor pathogens and transmit them to others without exhibiting any signs of illness. These people, called **carriers,** are important living reservoirs of infection. Some carriers have inapparent infections for which no signs or symptoms are ever exhibited. Other people, such as those with latent diseases, carry a disease during its symptom-free stages— during the incubation period (before symptoms appear) or during the convalescent period (recovery). Human carriers play an important role in the spread of such diseases as AIDS, diphtheria, typhoid fever, hepatitis, gonorrhea, amoebic dysentery, and streptococcal infections.

Animal Reservoirs

Both wild and domestic animals are living reservoirs of microorganisms that can cause human diseases. Diseases that occur primarily in wild and domestic animals and can be transmitted to humans are called **zoonoses** (singular: *zoonosis*). Rabies (found in bats, skunks, foxes, dogs, and cats), and Lyme disease (found in deer) are examples of zoonoses. Other representative zoonoses are presented in Table 14.2.

About 150 zoonoses are known. The transmission of zoonoses to humans can occur via one of many routes: by direct contact with infected animals; by direct contact with domestic pet waste (such as cleaning a litter box or bird cage); by contamination of food and water; by air from contaminated hides, fur, or feathers; by consuming infected animal products; or by insect vectors (insects that transmit pathogens).

Nonliving Reservoirs

The two major nonliving reservoirs of infectious disease are soil and water. Soil harbors such pathogens as fungi, which cause mycoses including ringworm and systemic infections; *Clostridium botulinum,* the bacterium that causes botulism; and *C. tetani,* the bacterium that causes tetanus. Because both species of clostridia are part of the normal intestinal microbiota of horses and cattle, the bacteria are found especially in soil where animal feces are used as fertilizer.

table 14.2	Selected Zoonoses			
Disease	**Causative Agent**	**Reservoir**	**Method of Transmission**	**Chapter Reference**
Viral				
Influenza (some types)	*Influenzavirus*	Swine, waterfowl	Direct contact	24
Rabies	*Lyssavirus*	Bats, skunks, foxes, dogs, cats	Direct contact (bite)	22
Western equine encephalitis	*Alphavirus*	Horses, birds	*Culex* mosquito bite	22
Hantavirus pulmonary syndrome (HPS)	*Hantavirus*	Rodents (primarily deer mice)	Direct contact with rodent saliva, feces, or urine	23
Bacterial				
Anthrax	*Bacillus anthracis*	Domestic livestock	Direct contact with contaminated hides or animals; air; food	23
Brucellosis	*Brucella* spp.	Domestic livestock	Direct contact with contaminated milk, meat, or animals	23
Bubonic plague	*Yersinia pestis*	Rodents	Flea bites	23
Cat-scratch disease	*Bartonella henselae*	Domestic cats	Direct contact	23
Leptospirosis	*Leptospira*	Wild mammals, domestic dogs and cats	Direct contact with urine, soil, water	26
Lyme disease	*Borrelia burgdorferi*	Field mice, deer	Tick bites	23
Psittacosis (ornithosis)	*Chlamydia psittaci*	Birds, especially parrots	Direct contact	24
Rocky Mountain spotted fever	*Rickettsia rickettsii*	Rodents	Tick bites	23
Salmonellosis	*Salmonella* spp.	Poultry, rats, turtles, reptiles	Ingestion of contaminated food and water and putting hands in mouth	25
Endemic typhus	*Rickettsia typhi*	Rodents	Flea bites	23
Fungal				
Ringworms	*Trichophyton Microsporum Epidermophyton*	Domestic mammals	Direct contact; fomites (nonliving objects)	21
Protozoan				
Malaria	*Plasmodium* spp.	Monkeys	*Anopheles* mosquito bite	23
Toxoplasmosis	*Toxoplasma gondii*	Cats and other mammals	Ingestion of contaminated meat or by direct contact with infected tissues or fecal matter	23
Helminthic				
Tapeworm (pork)	*Taenia solium*	Pigs	Ingestion of undercooked contaminated pork	25
Trichinosis	*Trichinella spiralis*	Pigs, bears	Ingestion of undercooked contaminated pork	25

(a)

(b)

FIGURE 14.6 Means of disease transmission. (a) Contact, **(b)** vehicles, and **(c)** vectors.

▣ What is a reservoir of infection? What is a disease vector?

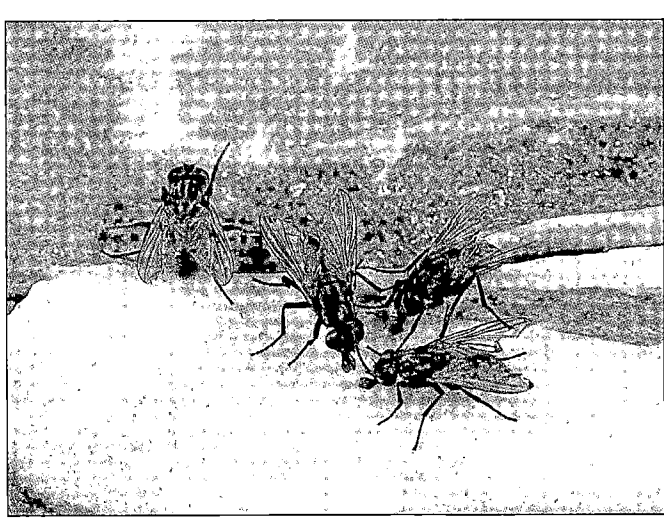

(c)

Water that has been contaminated by the feces of humans and other animals is a reservoir for several pathogens, notably those responsible for gastrointestinal diseases. These include *Vibrio cholerae,* which causes cholera, and *Salmonella typhi,* which causes typhoid fever. Other nonliving reservoirs include foods that are improperly prepared or stored. They may be sources of diseases such as trichinosis and salmonellosis.

The Transmission of Disease

Learning Objective

■ *Explain four methods of disease transmission.*

The causative agents of disease can be transmitted from the reservoir of infection to a susceptible host by three principal routes: contact, vehicles, and vectors (Figure 14.6).

Contact Transmission

Contact transmission is the spread of an agent of disease by direct contact, indirect contact, or droplet transmission. **Direct contact transmission,** also known as *person-to-person transmission,* is the direct transmission of an agent by physical contact between its source and a susceptible host; no intermediate object is involved (Figure 14.6a). The most common forms of direct contact transmission are touching, kissing, and sexual intercourse. Among the diseases that can be transmitted by direct contact are viral respiratory tract diseases (the common cold and influenza), staphylococcal infections, hepatitis A, measles, scarlet fever, and sexually transmitted diseases (syphilis, gonorrhea, and genital herpes). Direct contact is also one way of spreading AIDS and infectious mononucleosis. To guard against person-to-person transmission, health care workers use gloves and other protective measures (Figure 14.7). Potential pathogens can also be transmitted by direct contact from animals (or animal products) to humans. Examples are the pathogens causing rabies and anthrax.

Indirect contact transmission occurs when the agent of disease is transmitted from its reservoir to a susceptible host by means of a nonliving object. The general term for any nonliving object involved in the spread of an infection is a **fomite.** Examples of fomites are tissues, handkerchiefs, towels, bedding, diapers, drinking cups, eating utensils, toys, money, and thermometers. Contaminated syringes serve as fomites in the transmission of AIDS and hepatitis B. Other fomites may transmit diseases such as tetanus.

Nosocomial (Hospital-Acquired) Infections

Learning Objective

■ *Define nosocomial infections, and explain their importance.*

A **nosocomial infection** is one that does not show any evidence of being present or incubating at the time of admission to a hospital; it is acquired as a result of a hospital stay. (The word *nosocomial* is derived from the Greek word for hospital; the term also includes infections acquired in nursing homes and other health care facilities.) The Centers for Disease Control and Prevention (CDC) estimates that 5–15% of all hospital patients acquire some type of nosocomial infection. The work of pioneers in aseptic techniques such as Lister and Semmelweis (Chapter 1, page 10) decreased the rate of nosocomial infections considerably. However, despite modern advances in sterilization techniques and disposable materials, the rate of nosocomial infections has increased 36% during the last 20 years. In the United States, about 2 million people per year contract nosocomial infections, and nearly 90,000 die as a result.

Nosocomial infections result from the interaction of several factors: (1) microorganisms in the hospital environment, (2) the compromised (or weakened) status of the host, and (3) the chain of transmission in the hospital. Figure 14.9 illustrates that the presence of any one of these factors alone is generally not enough to cause infection; it is the interaction of all three factors that poses a significant risk of nosocomial infection.

Microorganisms in the Hospital

Although every effort is made to kill or check the growth of microorganisms in the hospital, the hospital environment is a major reservoir for a variety of pathogens. One reason is that certain normal microbiota of the human body are opportunistic and present a particularly strong danger to hospital patients. In fact, most of the microbes that cause nosocomial infections do not cause disease in healthy people but are pathogenic only for individuals whose defenses have been weakened by illness or therapy (see the boxes on page 423 and in Chapter 7, page 201).

In the 1940s and 1950s, most nosocomial infections were caused by gram-positive microbes. At one time, the gram-positive *Staphylococcus aureus* was the primary cause of nosocomial infections. In the 1970s, gram-negative rods, such as *E. coli* and *Pseudomonas aeruginosa,* were the

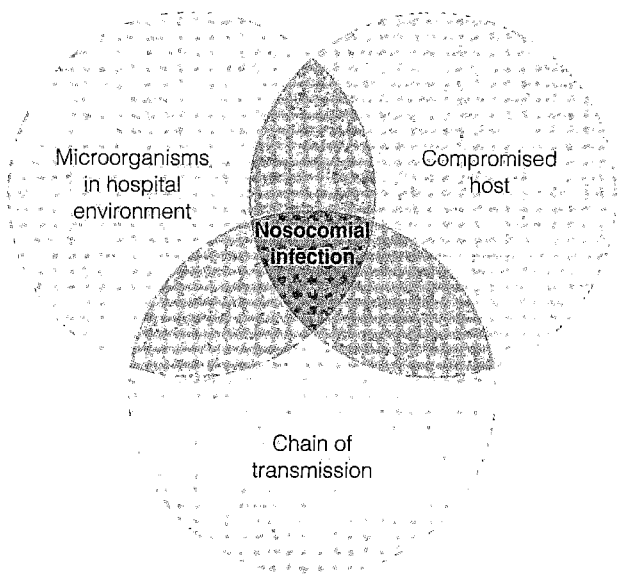

FIGURE 14.9 Nosocomial infections.

■ The interaction of microorganisms in the hospital environment, a compromised (weakened) host, and the chain of transmission in the hospital contributes to nosocomial infections.

most common causes of nosocomial infections. During the 1980s, antibiotic-resistant gram-positive bacteria, *Staphylococcus aureus,* coagulase-negative staphylococci, and *Enterococcus* spp., emerged as nosocomial pathogens. In the 1990s, these gram-positive bacteria accounted for 34% of nosocomial infections and four gram-negative pathogens accounted for 32%. The principal microorganisms involved in nosocomial infections are summarized in Table 14.4.

In addition to being opportunistic, some microorganisms in the hospital become resistant to antimicrobial drugs, which are commonly used there. For example, *P. aeruginosa* and other such gram-negative bacteria tend to be difficult to control with antibiotics because of their R factors, which carry genes that determine resistance to antibiotics (see Chapter 8, page 240). As the R factors recombine, new and multiple resistance factors are produced. These strains become part of the microbiota of patients and hospital personnel and become progressively more resistant to antibiotic therapy. In this way, people become part of the reservoir (and chain of transmission) for antibiotic-resistant strains of bacteria. Usually, if the host's resistance is high, the new strains are not much of a problem. But if disease, surgery, or trauma has weakened the host's defenses, secondary infections may be difficult to treat.

table 14.4	Microorganisms Involved in Most Nosocomial Infections		
Microorganism		**Percentage of Total Infections**	**Infections Caused**
Staphylococcus aureus, coagulase-negative staphylococci, enterococci		34%	Surgical wound infections, pneumonia, septicemia, urinary tract infections
Escherichia coli, Pseudomonas aeruginosa, Enterobacter spp., and *Klebsiella pneumoniae*		32%	Pneumonia and surgical wound infections
Clostridium difficile		17%	Causes nearly half of all nosocomial diarrhea
Fungi (mostly *Candida albicans*)		10%	Urinary tract infections and septicemia
Other gram-negative bacteria (*Acinetobacter, Citrobacter, Haemophilus*)		7%	Urinary tract infections and surgical wound infections

DATA SOURCE: CDC, National Nosocomial Infections Surveillance, 1990–1996.

The Compromised Host

Learning Objective

- *Define compromised host.*

A **compromised host** is one whose resistance to infection is impaired by disease, therapy, or burns. Two principal conditions can compromise the host: broken skin or mucous membranes, and a suppressed immune system.

As long as the skin and mucous membranes remain intact, they provide formidable physical barriers against most pathogens. Burns, surgical wounds, trauma (such as accidental wounds), injections, invasive diagnostic procedures, ventilators, intravenous therapy, and urinary catheters (used to drain urine) can all break the first line of defense and make a person more susceptible to disease in hospitals. Burn patients are especially susceptible to nosocomial infections because their skin is no longer an effective barrier to microorganisms.

The risk of infection is also related to other invasive procedures, such as administering anesthesia, which may alter breathing and contribute to pneumonia, and tracheotomy, in which an incision is made into the trachea to assist breathing. Patients who require invasive procedures usually have a serious underlying disease, which further increases susceptibility to infections. Invasive devices provide a pathway for microorganisms in the environment to enter the body; they also help transfer microbes from one part of the body to another. Pathogens can also proliferate on the devices themselves.

In healthy individuals, white blood cells called T lymphocytes (T cells) provide resistance to disease by killing pathogens directly, mobilizing phagocytes and other lymphocytes, and secreting chemicals that kill pathogens.

White blood cells called B lymphocytes (B cells), which develop into antibody-producing cells, also protect against infection. Antibodies provide immunity by such actions as neutralizing toxins, inhibiting the attachment of a pathogen to host cells, and helping to lyse pathogens. Drugs, radiation therapy, steroid therapy, burns, diabetes, leukemia, kidney disease, stress, and malnutrition can all adversely affect the actions of T and B lymphocytes and compromise the host. In addition, the AIDS virus destroys certain T cells.

A summary of the principal sites of nosocomial infections is presented in Figure 14.10 and Table 14.5.

The Chain of Transmission

Given the variety of pathogens (and potential pathogens) in the hospital and the compromised state of the host, routes of transmission are a constant concern. The principal routes

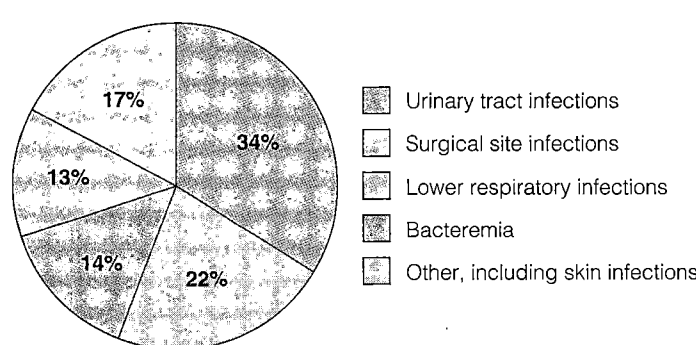

FIGURE 14.10 Relative frequency of nosocomial infections.

▨ Urinary tract infections account for most nosocomial infections.

table 14.5	Principal Sites of Nosocomial Infections
Type of Infection	**Comment**
Urinary tract infections	Most common, usually accounts for about 34% of all nosocomial infections. Typically related to urinary catheterization.
Surgical site infections	Ranks second in infection incidence (about 17%). An estimated 5–12% of all surgical patients develop postoperative infections; the percentage can reach 30% for certain surgeries, such as colon surgery and amputations.
Bacteremia	Bacteremias rank third in incidence and account for about 14% of nosocomial infections. Intravenous catheterization is implicated in nosocomial infections of the bloodstream, particularly infections caused by bacteria and fungi.
Lower respiratory infections	Nosocomial pneumonias account for about 13% and have high mortality rates (13–55%). Most of these pneumonias are related to respiratory devices that aid breathing or administer medications.
Cutaneous infections	Among the least common of all nosocomial infections. However, newborns have a high rate of susceptibility to skin and eye infections.

DATA SOURCE: CDC, National Nosocomial Infection Surveillance, May 1996.

of transmission of nosocomial infections are direct contact transmission from hospital staff to patient and from patient to patient, and indirect contact transmission through fomites and the hospital's ventilation system (airborne transmission).

Because hospital personnel are in direct contact with patients, they can often transmit disease. For example, a physician or nurse may transmit microbiota to a patient when changing a dressing, or a kitchen worker who carries *Salmonella* can contaminate a food supply.

Certain areas of a hospital are reserved for specialized care; these include the burn, hemodialysis, recovery, intensive care, and oncology units. Unfortunately, these units also group patients together and provide environments for the epidemic spread of nosocomial infections from patient to patient.

Many diagnostic and therapeutic hospital procedures provide a fomite route of transmission. The urinary catheter used to drain urine from the urinary bladder is a fomite in many nosocomial infections. Intravenous catheters, which pass through the skin and into a vein to provide fluids, nutrients, or medication, can also transmit nosocomial infections. Respiratory aids can introduce contaminated fluids into the lungs. Needles may introduce pathogens into muscle or blood, and surgical dressings can become contaminated and promote disease.

The Control of Nosocomial Infections

Learning Objectives

- List several methods of disease transmission in hospitals.
- Explain how nosocomial infections can be prevented.

Control measures aimed at preventing nosocomial infections vary from one institution to another, but certain procedures are generally implemented. It is important to reduce the number of pathogens to which patients are exposed by using aseptic techniques, handling contaminated materials carefully, insisting on frequent and thorough hand washing, educating staff members about basic infection control measures, and using isolation rooms and wards.

According to the CDC, hand washing is the single most important means of preventing the spread of infection. Nevertheless, in 1997, a CDC report revealed that in one long-term care facility, health care workers washed their hands before interacting with patients only 27% of the time. And in 1996, researchers identified hand washing rates as low as 31% in one emergency department.

In addition to hand washing, tubs used to bathe patients should be disinfected between uses so that bacteria from the previous patient will not contaminate the next one. Respirators and humidifiers provide both a suitable growth environment for some bacteria and a method of airborne transmission. These sources of nosocomial infections must be kept scrupulously clean and disinfected, and materials used for bandages and intubation (insertion of tubes into organs, such as the trachea) should be single-use disposable or sterilized before use. Packaging used to maintain sterility should be removed aseptically. Physicians can help improve patients' resistance to infection by prescribing antibiotics only when necessary, avoiding invasive procedures if possible, and minimizing the use of immunosuppressive drugs.

Accredited hospitals should have an infection control committee. Most hospitals have at least an infection control nurse or epidemiologist (an individual who studies disease in populations). The role of these staff members is to identify problem sources, such as antibiotic-resistant strains of bacteria and improper sterilization techniques. The infection control officer should make periodic ex-

An Outbreak of Streptococcal Toxic-Shock Syndrome

1. On December 23, a previously healthy 28-year-old woman underwent surgery to remove a parathyroid gland. On December 26, she developed kidney failure and died on December 29. On December 30, a previously healthy 56-year-old woman had her thyroid gland removed. She was discharged on December 31 and found dead later that day in her home. On December 30, a previously healthy 57-year-old woman also underwent a thyroidectomy; she was discharged the next day. On January 1, she was admitted to the intensive care unit with sepsis and kidney failure. She was discharged on February 4. Group A *Streptococcus* (GAS) was isolated from the neck wounds of all three patients. GAS is an unusual cause of surgical site or postpartum infections. The bacterium is isolated from <1% of surgical-site infections and 3% of infections after vaginal births.

   What organisms are most often involved in nosocomial infections? What other information do you need to proceed with your investigation?

2. Gram-negative bacteria account for over 40% of nosocomial infections. Review of the hospital's microbiol-

ogy records revealed no episodes of postoperative GAS during the six months immediately prior to the outbreak.

   What should you do next to determine how these people got sick?

3. Forty-one health care workers (HCWs) worked in the operating room and in the preoperative and postoperative areas on the days of surgery for these patients. Surgeon *A* was the only HCW who had contact in the operating room with all three patients. Surgeon *B* assisted with two patients and performed preoperative care on the third patient.

   Is it unusual that none of the HCWs had GAS infections? What samples should be cultured from these staff members?

4. GAS may be part of the normal or transient microbiota. The most common site of asymptomatic carriage among HCWs is the anus, but vaginal, skin, and pharyngeal carriage have also been recorded. Throat, rectal, and vaginal cultures were obtained from the 41 HCWs. All cultures were negative, except a throat culture from one orderly that grew GAS. Surgeon *A* received self-initiated penicillin on January

2, before adequate cultures were obtained.

   What could be done to determine whether the organisms cultured from the patients and orderly were the same?

5. Gene sequencing of the M-protein gene (called emm typing) showed the GAS isolated from all three patients were emm type 1. The orderly's GAS isolate was emm type STNS5.

   Is it possible to associate risk for illness with one staff member? What steps can be taken to prevent further infections from occurring? How is GAS transmitted?

6. Throat cultures from surgeon *A*'s household contacts were negative. Surgeons *A* and *B* were restricted from patient care until each had completed a 10-day course of antibiotics. No further postoperative GAS infection has occurred in this hospital. Most nosocomial transmission is traced to carriers involved in direct patient care. GAS carriers can shed the organism into the immediate environment despite proper gowning and gloving. The mode of transmission is presumed to be airborne.

SOURCE: Adapted from *MMWR* 48(8):163–166 (3/5/99).

---

aminations of hospital equipment to determine the extent of microbial contamination. Samples should be taken from tubing, catheters, respirator reservoirs, and other equipment.

Emerging Infectious Diseases

Learning Objective

■ *List several probable reasons for emerging infectious diseases, and name one example for each reason.*

As noted in Chapter 1, **emerging infectious diseases (EIDs)** are ones that are new or changing, showing an

increase in incidence in the recent past, or a potential to increase in the near future. An emerging disease can be caused by a virus, a bacterium, a fungus, a protozoan, or a helminth. Several criteria are used for identifying an EID. For example, some diseases present symptoms that are clearly distinctive from all other diseases. Some are recognized because improved diagnostic techniques allow the identification of a new pathogen. Others are identified when a local disease becomes widespread, a rare disease becomes common, a mild disease becomes more severe, or an increase in life span permits a slow disease to develop. Examples of emerging infectious diseases are listed in Table 14.6 and described in the boxes in Chapters 21 and 23.

table 14.6   Emerging Infectious Diseases

Microorganism	Year of Emergence	Disease Caused	Chapter Reference
Bacteria			
Borrelia burgdorferi	1975	Lyme disease	23
Legionella pneumophila	1976	Legionnaires' disease	24
Staphylococcus aureus	1978	Toxic shock syndrome	21
Escherichia coli O157:H7	1982	Hemorrhagic diarrhea	25
Bartonella henselae	1983	Cat-scratch disease	23
Ehrlichia chaffeenis	1986	Human monocytic ehrlichiosis (tickborne)	23
Vibrio cholerae O139	1992	New serovar of cholera, Asia	25
Corynebacterium diphtheriae	1994	Diphtheria epidemic, eastern Europe	24
Fungi			
Pneumocystis carinii	1981	Pneumonia in immunocompromised patients	24
Coccidioides immitis	1993	Coccidioidomycosis	24
Protozoa			
Cryptosporidium parvum	1976	Cryptosporidiosis	25
Cyclospora cayetanensis	1993	Severe diarrhea and wasting syndrome	25
Plasmodium spp.	1986	Malaria in United States; 40% of the world's population is at risk; insecticide-resistant mosquitoes and drug-resistant protozoa	23
Helminths			
Unidentified tapeworm	1996	Gastrointestinal pain	25
Viruses			
HIV	1983	AIDS	19
Dengue fever virus	1984	Dengue fever and dengue hemorrhagic fever, South and Central America and the Caribbean	23
Hepatitis C virus	1989	Hepatitis	25
Hepatitis E virus	1990	Hepatitis	25
Venezuelan hemorrhagic fever	1991	Hemorrhagic fever, South America	23
Hantavirus	1993	Hantavirus pulmonary syndrome	23
Hendra virus	1994	Encephalitis-like symptoms, Australia	22
Ebola virus	1995, 1979, 1975	Ebola hemorrhagic fever	23
Prions			
Bovine spongiform encephalitis agent	1996	Mad cow disease, Great Britain	22

A variety of factors contribute to the emergence of new infectious diseases:

■ A new serovar, such as *Vibrio cholerae* O139, may result from changes in or the evolution of existing microorganisms.

■ The widespread, and sometimes unwarranted, use of antibiotics and pesticides encourages the growth of more resistant populations of microbes and the insects (mosquitoes and lice) and ticks that carry them.

- Global warming may increase the distribution and survival of reservoirs and vectors, resulting in the introduction and dissemination of diseases such as malaria, and *Hantavirus* pulmonary syndrome disease.

- Known diseases, such as cholera, may spread to new geographic areas by modern transportation. This was less likely 100 years ago when travel took so long that infected travelers either died or recovered during passage.

- Previously unrecognized infections may appear in individuals living or working in regions undergoing ecological changes brought about by natural disaster, construction, wars, and expanding human settlement. In California, the incidence of coccidioidomycosis increased tenfold following the Northridge earthquake of 1989. Workers clearing South American forests are now contracting Venezuelan hemorrhagic fever.

- Even animal control measures may affect the incidence of a disease. The increase in Lyme disease in recent years could be due to rising deer populations resulting from the killing of deer predators.

- Failures in public health measures may be a contributing factor to the emergence of previously controlled infections. For example, the failure of adults to get a diphtheria booster vaccination led to a diphtheria epidemic in the newly independent republics of the former Soviet Union in the 1990s.

The CDC, the National Institutes of Health (NIH), and the World Health Organization (WHO) have developed plans to address issues relating to emerging infectious diseases. Their priorities include the following:

1. To detect, promptly investigate, and monitor emerging infectious pathogens, the diseases they cause, and factors that influence their emergence.

2. To expand basic and applied research on ecological and environmental factors, microbial changes and adaptations, and host interactions that influence EIDs.

3. To enhance the communication of public health information and the prompt implementation of prevention strategies regarding EIDs.

4. To establish plans to monitor and control EIDs worldwide.

The importance of emerging infectious diseases to the scientific community resulted in a new publication, *Emerging Infectious Diseases*, devoted exclusively to the topic. It was first published in January 1995. In addition, as a result of a global focus of attention, January 1996 was proclaimed Emerging Infection Month, in which 36 scientific journals worldwide documented the occurrence, causes, and consequences of emerging and reemerging infectious diseases.

Epidemiology

Learning Objective

- *Define epidemiology, and describe three types of epidemiologic investigations.*

In today's crowded, overpopulated world, in which frequent travel and the mass production and distribution of food and other goods are a way of life, diseases can spread rapidly. A contaminated food or water supply, for example, can affect many thousands of people very quickly. Identifying the causative agent is desirable so that a disease can be effectively controlled and treated. It is also desirable to understand the mode of transmission and geographical distribution of the disease. The science that studies when and where diseases occur and how they are transmitted in populations is called **epidemiology.**

Modern epidemiology began in the mid-1800s with three now-famous investigations. John Snow, a British physician, conducted a series of investigations related to outbreaks of cholera in London. As the cholera epidemic of 1848–1849 raged, Snow analyzed the death records attributed to cholera, gathered information about the victims, and interviewed survivors who lived in the neighborhood. Using the information he compiled, Snow made a map showing that most individuals who died of cholera drank or brought water from the Broad Street pump; those who used other pumps (or drank beer, like the workers at a nearby brewery) did not get cholera. He concluded that contaminated water from the Broad Street pump was the source of the epidemic. When the pump's handle was removed and people could no longer get water from this location, the number of cholera cases dropped significantly.

Between 1846 and 1848, Ignaz Semmelweis meticulously recorded the number of births and maternal deaths at Vienna General Hospital. The First Maternity Clinic had become a source of gossip throughout Vienna because the death rate due to puerperal sepsis ranged between 13% and 18%, four times that of the Second Maternity Clinic. Puerperal sepsis (childbirth fever) is a nosocomial infection that begins in the uterus as a result of childbirth or abortion. It is frequently caused by *Streptococcus pyogenes*. The infection progresses to the abdominal cavity (peritonitis) and in many cases to septicemia (proliferation of microbes in the blood). Wealthy women did not go to the clinic, and poor women had learned they had a better chance of surviving childbirth if they gave birth elsewhere before going to the hospital. Looking at his data, Semmelweis realized that wealthy women and the poor women who had given birth prior to entering the clinic were not examined by the medical students, who had spent their mornings dissecting cadavers. In May 1847, he ordered all medical students to wash their hands before entering the delivery room, and the mortality rate dropped to under 2%.

Florence Nightingale recorded statistics on epidemic typhus in the English civilian and military populations. In 1858, she published a thousand-page report using statistical comparisons to demonstrate that diseases, poor food, and unsanitary conditions were killing the soldiers. Her work resulted in reforms in the British Army and to her admission to the Statistical Society, their first female member.

These three careful analyses of where and when a disease occurs and how it is transmitted within a population constituted a new approach to medical research and demonstrated the importance of epidemiology. The works of Snow, Semmelweis, and Nightingale resulted in changes that lowered the incidence of diseases even though there was limited knowledge of the causes of infectious disease. Most physicians believed that the symptoms they saw were the causes of the disease, not the result of disease. Koch's work on the germ theory of disease was still 30 years in the future.

An epidemiologist not only determines the etiology of a disease but also identifies other possibly important factors and patterns concerning the people affected. An important part of the epidemiologist's work is assembling and analyzing such data as age, sex, occupation, personal habits, socioeconomic status, history of immunization, presence of any other diseases, and the common history of affected individuals (such as eating the same food or visiting the same doctor's office). Also important for the prevention of future outbreaks is knowledge of the site at which a susceptible host came into contact with the agent of infection. In addition, the epidemiologist considers the period during which the disease occurs, either on a seasonal basis (to indicate whether the disease is prevalent during the summer or winter) or on a yearly basis (to indicate the effects of immunization or an emerging or reemerging disease).

An epidemiologist is also concerned with various methods for controlling a disease. The strategies controlling diseases include the use of drugs (chemotherapy) and vaccines (immunization). Other methods include the control of human, animal, and nonliving reservoirs of infection, water treatment, proper sewage disposal (enteric diseases), cold storage, pasteurization, food inspection,

and adequate cooking (foodborne diseases), improved nutrition to bolster host defenses, changes in personal habits, and screening of transfused blood and transplanted organs.

Figure 14.11 contains graphs indicating the incidence of selected diseases. Such graphs provide information about whether disease outbreaks are sporadic or epidemic, and, if epidemic, how the disease might have spread. By establishing the frequency of a disease in a population and identifying the factors responsible for its transmission, an epidemiologist can provide physicians with information that is important in determining the prognosis and treatment of a disease. Epidemiologists also evaluate how effectively a disease is being controlled in a community—by a vaccination program, for example. Finally, epidemiologists can provide data to help in evaluating and planning overall health care for a community.

Epidemiologists use three basic types of investigations when analyzing the occurrence of a disease: descriptive, analytical, and experimental.

Descriptive Epidemiology

Descriptive epidemiology entails collecting all data that describe the occurrence of the disease under study. Relevant information usually includes information about the affected individuals and the place and period in which the disease occurred. Snow's search for the cause of the cholera outbreak in London is an example of descriptive epidemiology.

Such a study is generally *retrospective* (looking backward after the episode has ended). In other words, the epidemiologist backtracks to the cause and source of the disease (see the boxes in Chapters 23, 25, and 26). The search for the cause of toxic shock syndrome is an example of a fairly recent retrospective study. In the initial phase of an epidemiological study, retrospective studies are more common than *prospective* (looking forward) studies, in which an epidemiologist chooses a group of people who are free of a particular disease to study. The group's subsequent disease experiences are then recorded for a given period. Prospective studies were used to test the Salk polio vaccine in 1954 and 1955.

Analytical Epidemiology

Analytical epidemiology analyzes a particular disease to determine its probable cause. This study can be done in two ways. With the *case control method*, the epidemiologist looks for factors that might have preceded the disease. A group of people who have the disease is compared with another group of people who are free of the disease. For example, one group with meningitis and one without the disease might be matched by age, sex, socioeconomic status, and location. These statistics are compared to determine which of all the possible factors—genetic, environmental, nutritional, and so forth—might be responsible for the meningitis. Nightingale's work was an example of analytical epidemiology, in which she compared disease in soldiers and civilians. With the *cohort method*, the epidemiologist studies two populations: one that has had contact with the agent causing a disease and another that has not (both groups are called *cohort groups*). For example, a comparison of one group composed of people who have received blood transfusions and one composed of people who have not could reveal an association between blood transfusions and the incidence of hepatitis B virus.

Experimental Epidemiology

Experimental epidemiology begins with a hypothesis about a particular disease; experiments to test the hypothesis are then conducted with a group of people. One such hypothesis could be the assumed effectiveness of a drug. A group of infected individuals is selected and divided randomly so that some receive the drug and others receive a placebo, a substance that has no effect. If all other factors are kept constant between the two groups, and if those people who received the drug recover more rapidly than those who received the placebo, it can be concluded that the drug was the experimental factor (variable) that made the difference.

Case Reporting

We noted earlier in this chapter that establishing the chain of transmission for a disease is extremely important. Once known, the chain can be interrupted in order to slow down or stop the spread of the disease.

An effective way to establish the chain of transmission is *case reporting,* a procedure that requires health care workers to report specified diseases to local, state, and national health officials. Examples of such diseases are AIDS, measles, gonorrhea, tetanus, and typhoid fever. Case reporting provides epidemiologists with an approximation of the incidence and prevalence of a disease. This information helps officials decide whether or not to investigate a given disease.

Case reporting provided epidemiologists with valuable leads regarding the origin and spread of AIDS. In fact, one of the first clues about AIDS came from reports of young males with Kaposi's sarcoma, formerly a disease of older males. Using these reports, epidemiologists began various studies of the patients. If an epidemiological

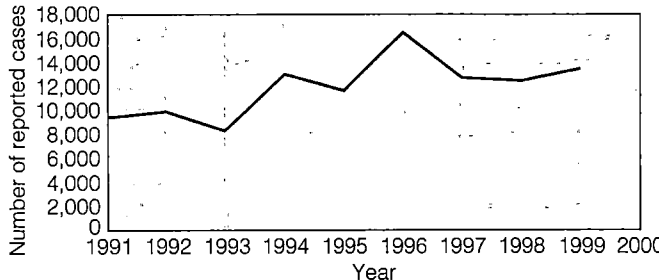

(a) Lyme disease cases, 1991-1999

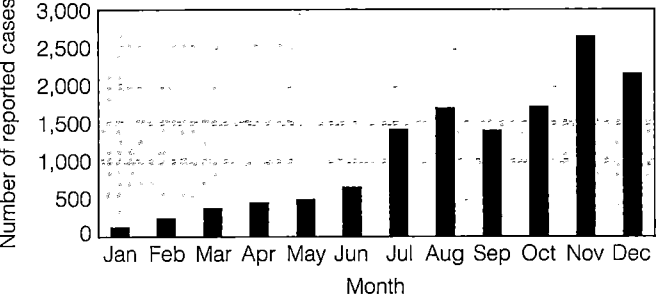

(b) Lyme disease by month, 1999

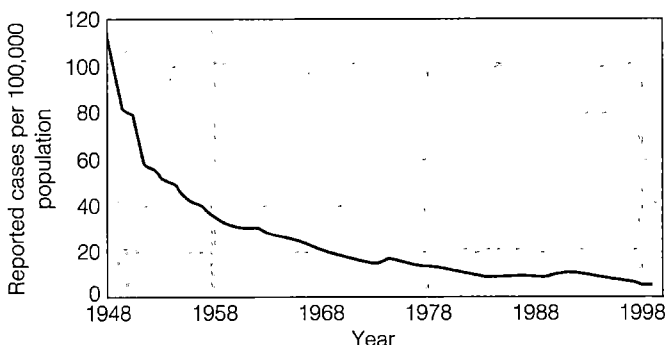

(c) Reported tuberculosis cases, 1948-1999

FIGURE 14.11 Epidemiological graphs. (a) A graph of Lyme disease cases, showing the annual occurrence of the disease during the covered period. (b) This bar graph represents Lyme disease from a different perspective, enabling epidemiologists to begin drawing some conclusions about the disease. Over a period of years, researchers saw that the disease was transmitted by a vector that was active during summer and fall. This graph records the number of cases per 100,000 people, rather than the total number of cases. (c) A graph of the incidence of tuberculosis shows a decrease in the rate of decline since 1958. Epidemiologists are investigating the reasons for the rate change. (Source: CDC.)

■ Graphs representing the incidence of diseases indicate whether disease outbreaks are sporadic or epidemic, how a disease might spread, and how effectively a disease is controlled.

table 14.7	Nationally Notifiable Diseases, 1999		
Acquired immunodeficiency disease (AIDS)	Malaria
Anthrax	Measles
Botulism	Meningococcal disease
Brucellosis	Mumps
Chancroid	Pertussis
Chlamydia trachomatis, genital infections	Plague
Cholera	Poliomyelitis, paralytic
Coccidioidomycosis	Psittacosis
Cryptosporidiosis	Rabies, animal
Diphtheria	Rabies, human
Encephalitis, California	Rocky Mountain spotted fever
Encephalitis, eastern equine	Rubella
Encephalitis, St. Louis	Rubella, congenital syndrome
Encephalitis, western equine	Salmonellosis
Escherichia coli O157:H7	Shigellosis
Gonorrhea	Streptococcal disease, invasive, group A
Haemophilus influenzae, invasive disease	*Streptococcus pneumoniae,* drug-resistant
Hansen's disease (Leprosy)	Streptococcal toxic shock syndrome
Hantavirus pulmonary syndrome	Syphilis
Hemolytic uremic syndrome, post-diarrheal	Syphilis, congenital
Hepatitis A	Tetanus
Hepatitis B	Toxic shock syndrome
Hepatitis C (non-A, non-B hepatitis)	Trichinosis
HIV infection, pediatric	Tuberculosis
Legionellosis	Typhoid fever
Lyme disease	Yellow fever

study shows that a large enough segment of the population is affected by a disease, an attempt is then made to isolate and identify its causative agent. Identification is accomplished by a number of different microbiological methods. Identification of the causative agent often provides valuable information regarding the reservoir for the disease.

Once the chain of transmission is discovered, it is possible to apply control measures to stop the disease from spreading. These might include elimination of the source of infection, isolation and segregation of infected people, the development of vaccines, and, as in the case of AIDS, education.

The Centers for Disease Control and Prevention (CDC)

Learning Objectives

- *Identify the function of the CDC.*
- *Define the following terms: morbidity, mortality, and notifiable disease.*

Epidemiology is a major concern of state and federal public health departments. The **Centers for Disease Control and Prevention (CDC),** a branch of the U.S. Public Health Service located in Atlanta, Georgia, is a central source of epidemiological information in the United States.

The CDC issues a publication called the *Morbidity and Mortality Weekly Report*. The *MMWR*, as it is called, is read by microbiologists, physicians, and other hospital and public health professionals. The *MMWR* contains data on **morbidity,** the incidence of specific notifiable diseases, and **mortality,** the number of deaths from these diseases. These data are usually organized by state. **Notifiable diseases,** shown in Table 14.7, are those for which physicians are required by law to report cases to the U.S. Public Health Service. As of January 1999, a total of 52 infectious diseases were reported at the national level. **Morbidity rate** is the number of people affected by a disease in a given period of time in relation to the total population. **Mortality rate** is the number of deaths resulting from a disease in a population in a given period of time in relation to the total population.

MMWR articles include reports of disease outbreaks, case histories of special interest, and summaries of the status of particular diseases during a recent period. These articles often include recommendations for procedures for diagnosis, immunization, and treatment. A number of graphs and other data in this textbook are from the *MMWR,* and the case history boxes are adapted from reports from this publication.

★ ★ ★

In the next chapter, we consider the mechanisms of pathogenicity. We will discuss in more detail the methods by which microorganisms enter the body and cause disease, the effects of disease on the body, and the means by which pathogens leave the body.

Study Outline

INTRODUCTION (p. 406)

1. Disease-causing microorganisms are called pathogens.

2. Pathogenic microorganisms have special properties that allow them to invade the human body or produce toxins.

3. When a microorganism overcomes the body's defenses, a state of disease results.

PATHOLOGY, INFECTION, AND DISEASE (p. 407)

1. Pathology is the scientific study of disease.

2. Pathology is concerned with the etiology (cause), pathogenesis (development), and effects of disease.

3. Infection is the invasion and growth of pathogens in the body.

4. A host is an organism that shelters and supports the growth of pathogens.

5. Disease is an abnormal state in which part or all of the body is not properly adjusted or is incapable of performing normal functions.

NORMAL MICROBIOTA (pp. 407–410)

1. Animals, including humans, are usually germ-free in utero.

2. Microorganisms begin colonization in and on the surface of the body soon after birth.

3. Microorganisms that establish permanent colonies inside or on the body without producing disease make up the normal microbiota.

4. Transient microbiota are microbes that are present for various periods and then disappear.

Relationships Between the Normal Microbiota and the Host (pp. 407–409)

1. The normal microbiota can prevent pathogens from causing an infection; this phenomenon is known as microbial antagonism.

2. Normal microbiota and the host exist in symbiosis (living together).

3. The three types of symbiosis are commensalism (one organism benefits and the other is unaffected), mutualism (both organisms benefit), and parasitism (one organism benefits and one is harmed).

Opportunistic Microorganisms (pp. 409–410)

1. Opportunistic pathogens do not cause disease under normal conditions but cause disease under special conditions.

Cooperation Among Microorganisms (p. 410)

1. In some situations, one microorganism makes it possible for another to cause a disease or produce more severe symptoms.

THE ETIOLOGY OF INFECTIOUS DISEASES (pp. 410–412)

Koch's Postulates (p. 410)

1. Koch's postulates are criteria for establishing that specific microbes cause specific diseases.

2. Koch's postulates have the following requirements: (a) the same pathogen must be present in every case of the disease; (b) the pathogen must be isolated in pure culture; (c) the pathogen isolated from pure culture must cause the same disease in a healthy, susceptible laboratory animal; and (d) the pathogen must be reisolated from the inoculated laboratory animal.

Exceptions to Koch's Postulates (pp. 411–412)

1. Koch's postulates are modified to establish etiologies of diseases caused by viruses and some bacteria, which cannot be grown on artificial media.
2. Some diseases, such as tetanus, have unequivocal signs and symptoms.
3. Some diseases, such as pneumonia and nephritis, may be caused by a variety of microbes.
4. Some pathogens, such as *S. pyogenes,* cause several different diseases.
5. Certain pathogens, such as HIV, cause disease in humans only.

CLASSIFYING INFECTIOUS DISEASES (pp. 412–414)

1. A patient may exhibit symptoms (subjective changes in body functions) and signs (measurable changes), which a physician uses to make a diagnosis (identification of the disease).
2. A specific group of symptoms or signs that always accompanies a specific disease is called a syndrome.
3. Communicable diseases are transmitted directly or indirectly from one host to another.
4. A contagious disease is one that is easily spread from one person to another.
5. Noncommunicable diseases are caused by microorganisms that normally grow outside the human body and are not transmitted from one host to another.

The Occurrence of a Disease (p. 412)

1. Disease occurrence is reported by incidence (number of people contracting the disease) and prevalence (number of cases at a particular time).
2. Diseases are classified by frequency of occurrence: sporadic, endemic, epidemic, and pandemic.

The Severity or Duration of a Disease (pp. 412–413)

1. The scope of a disease can be defined as acute, chronic, subacute, or latent.
2. Herd immunity is the presence of immunity to a disease in most of the population.

The Extent of Host Involvement (pp. 413–414)

1. A local infection affects a small area of the body; a systemic infection is spread throughout the body via the circulatory system.
2. A secondary infection can occur after the host is weakened from a primary infection.
3. An inapparent, or subclinical, infection does not cause any signs of disease in the host.

PATTERNS OF DISEASE (pp. 414–415)

Predisposing Factors (p. 414)

1. A predisposing factor is one that makes the body more susceptible to disease or alters the course of a disease.
2. Examples include gender, climate, age, fatigue, and inadequate nutrition.

The Development of Disease (pp. 414–415)

1. The incubation period is the time interval between the initial infection and the first appearance of signs and symptoms.
2. The prodromal period is characterized by the appearance of the first mild signs and symptoms.
3. During the period of illness, the disease is at its height, and all disease signs and symptoms are apparent.
4. During the period of decline, the signs and symptoms subside.
5. During the period of convalescence, the body returns to its prediseased state, and health is restored.

THE SPREAD OF INFECTION (pp. 415–419)

Reservoirs of Infection (pp. 415–417)

1. A continual source of infection is called a reservoir of infection.
2. People who have a disease or are carriers of pathogenic microorganisms are human reservoirs of infection.
3. Zoonoses are diseases that affect wild and domestic animals and can be transmitted to humans.
4. Some pathogenic microorganisms grow in nonliving reservoirs, such as soil and water.

The Transmission of Disease (pp. 417–419)

1. Transmission by direct contact involves close physical contact between the source of the disease and a susceptible host.
2. Transmission by fomites (inanimate objects) constitutes indirect contact.
3. Transmission via saliva or mucus in coughing or sneezing is called droplet transmission.
4. Transmission by a medium such as water, food, or air is called vehicle transmission.
5. Airborne transmission refers to pathogens carried on water droplets or dust for a distance greater than 1 meter.
6. Arthropod vectors carry pathogens from one host to another by both mechanical and biological transmission.

Portals of Exit (p. 419)

1. Just as pathogens have preferred portals of entry, they also have definite portals of exit.
2. Three common portals of exit are the respiratory tract via coughing or sneezing, the gastrointestinal tract via saliva or feces, and the urogenital tract via secretions from the vagina or penis.
3. Arthropods and syringes provide a portal of exit for microbes in blood.

NOSOCOMIAL (HOSPITAL-ACQUIRED) INFECTIONS (pp. 420–423)

1. A nosocomial infection is any infection that is acquired during the course of stay in a hospital, nursing home, or other health care facility.
2. About 5–15% of all hospitalized patients acquire nosocomial infections.

Microorganisms in the Hospital (p. 420)

1. Certain normal microbiota are often responsible for nosocomial infections when they are introduced into the body through such medical procedures as surgery and catheterization.
2. Opportunistic, drug-resistant gram-negative bacteria are the most frequent causes of nosocomial infections.

The Compromised Host (p. 421)

1. Patients with burns, surgical wounds, and suppressed immune systems are the most susceptible to nosocomial infections.

The Chain of Transmission (pp. 421–422)

1. Nosocomial infections are transmitted by direct contact between staff members and patients and between patients.
2. Fomites such as catheters, syringes, and respiratory devices can transmit nosocomial infections.

The Control of Nosocomial Infections (pp. 422–423)

1. Aseptic techniques can prevent nosocomial infections.
2. Hospital infection control staff members are responsible for overseeing the proper cleaning, storage, and handling of equipment and supplies.

EMERGING INFECTIOUS DISEASES (pp. 423–425)

1. New diseases and diseases with increasing incidences are called emerging infectious diseases (EIDs).
2. EIDs can result from the use of antibiotics and pesticides, climatic changes, travel, the lack of vaccinations, and improved case reporting.
3. The CDC, NIH, and WHO are responsible for surveillance and responses to emerging infectious diseases.

EPIDEMIOLOGY (pp. 425–429)

1. The science of epidemiology is the study of the transmission, incidence, and frequency of disease.
2. Modern epidemiology began in in the mid-1800s with the works of Snow, Semmelweis, and Nightingale.
3. Data about infected people are collected and analyzed in descriptive epidemiology.
4. In analytical epidemiology, a group of infected people is compared with an uninfected group.
5. Controlled experiments designed to test hypotheses are performed in experimental epidemiology.
6. Case reporting provides data on incidence and prevalence to local, state, and national health officials.
7. The Centers for Disease Control and Prevention (CDC) is the main source of epidemiologic information in the United States.
8. The CDC publishes the *Morbidity and Mortality Weekly Report* to provide information on morbidity (incidence) and mortality (deaths).

Study Questions

REVIEW

1. Differentiate the terms in each of the following pairs:
   a. etiology and pathogenesis
   b. infection and disease
   c. communicable disease and noncommunicable disease
2. What is meant by normal microbiota? How do they differ from transient microbiota?
3. Define symbiosis. Differentiate among commensalism, mutualism, and parasitism, and give an example of each.

4. What is a reservoir of infection? Match the following diseases with their reservoirs:
   ____ influenza   a. nonliving
   ____ rabies   b. human
   ____ botulism   c. animal
5. Describe how Koch's postulates establish the etiology of many infectious diseases. Why don't Koch's postulates apply to all infectious diseases?

6. Describe the various ways diseases can be transmitted in each of the following categories. Name one disease transmitted by each method.
   a. transmission by direct contact
   b. transmission by indirect contact
   c. transmission by arthropod vectors
   d. droplet transmission
   e. vehicle transmission
   f. airborne transmission

7. List four predisposing factors to disease.

8. Indicate whether each of the following conditions is typical of subacute, chronic, or acute infections.
   a. The patient experiences a rapid onset of malaise; symptoms last 5 days.
   b. The patient experiences cough and breathing difficulty for months.
   c. The patient has no apparent symptoms and is a known carrier.

9. Of all the hospital patients with infections, one-third do not enter the hospital with an infection. How do they acquire these infections? What is the method of transmission of these infections? What is the reservoir of infection?

10. Differentiate between endemic and epidemic states of infectious disease.

11. What is epidemiology? What is the role of the Centers for Disease Control and Prevention (CDC)?

12. Distinguish between symptoms and signs as signals of disease.

13. How can a local infection become a systemic infection?

14. Why are some organisms that constitute the normal microbiota described as commensals, whereas others are described as mutualistic?

15. Put the following in the correct order to describe the pattern of disease: period of convalescence, prodromal period, period of decline, incubation period, period of illness.

MULTIPLE CHOICE

1. The emergence of new infectious diseases is probably due to all of the following *except*
   a. the need of bacteria to cause disease.
   b. the ability of humans to travel by air.
   c. changing environments (e.g., flood, drought, pollution).
   d. a pathogen crossing the species barrier.
   e. the increasing human population.

2. All members of a group of ornithologists studying barn owls in the wild have had salmonellosis (*Salmonella* gastroenteritis). One birder is experiencing her third infection. What is the most likely source of their infections?
   a. The ornithologists are eating the same food.
   b. They are contaminating their hands while handling the owls and nests.
   c. One of the workers is a *Salmonella* carrier.
   d. Their drinking water is contaminated.

3. Which of the following statements is *not* true?
   a. *E. coli* never causes disease.
   b. *E. coli* provides vitamin K for its host.
   c. *E. coli* often exists in a mutual relationship with humans.
   d. *E. coli* gets nutrients from intestinal contents.

4. Which of the following is *not* one of Koch's postulates?
   a. The same pathogen must be present in every case of the disease.
   b. The pathogen must be isolated and grown in pure culture from the diseased host.
   c. The pathogen from pure culture must cause the disease when inoculated into a healthy, susceptible laboratory animal.
   d. The disease must be transmitted from a diseased animal to a healthy, susceptible animal by some form of contact.
   e. The pathogen must be isolated in pure culture from an experimentally infected lab animal.

Use the following information to answer questions 5–7.

On September 6, a 6-year old boy experienced fever, chills, and vomiting. On September 7, he was hospitalized with diarrhea and swollen lymph nodes under both arms. On September 3, the boy had been scratched and bitten by a cat. The cat was found dead on September 5, and *Yersinia pestis* was isolated from the cat. Chloramphenicol was administered to the boy from September 7, when *Y. pestis* was isolated from him. On September 17, the boy's temperature returned to normal, and on September 22, he was released from the hospital.

5. Identify the incubation period for this case of bubonic plague.
   a. September 3–5.
   b. September 3–6.
   c. September 6–7.
   d. September 6–17.

6. Identify the prodromal period for this disease.
   a. September 3–5.
   b. September 3–6.
   c. September 6–7.
   d. September 6–17.

7. Identify the crisis during this disease.
   a. September 6.
   b. September 7.
   c. September 6–17.
   d. September 17.

Use the following information to answer questions 8–10.

A Maryland woman was hospitalized with dehydration; *Vibrio cholerae* and *Plesiomonas shigelloides* were isolated from the patient. She had neither traveled outside the United States nor eaten raw shellfish during the preceding month. She had attended a party 2 days before her hospitalization. Two other people at the party had acute diarrheal illness and elevated levels of serum antibodies against *Vibrio*. Everyone at the party ate crabs and rice pudding with coconut milk. Crabs left over from this party were served at a second party. One of the 20 people at the second party had onset of mild diarrhea; specimens from 14 of these people were negative for vibriocidal antibodies.

8. This is an example of
   a. vehicle transmission.
   b. airborne transmission.
   c. transmission by fomites.
   d. direct contact transmission.
   e. nosocomial transmission.

9. The etiologic agent of the disease is
   a. *Plesiomonas shigelloides.*
   b. crabs.
   c. *Vibrio cholerae.*
   d. coconut milk.
   e. rice pudding.

10. The source of the disease was
   a. *Plesiomonas shigelloides.*
   b. crabs.
   c. *Vibrio cholerae.*
   d. coconut milk.
   e. rice pudding.

CRITICAL THINKING

1. Ten years before Robert Koch published his work on anthrax, Anton De Bary showed that potato blight was caused by the fungus *Phytophthora infestans.* Why do you suppose we use Koch's postulates instead of something called "De Bary's postulates"?

2. Florence Nightingale gathered the following data in 1855.

Population Sampled	Deaths from Contagious Diseases
Englishmen	0.2%
English soldiers (in England)	18.7%
English soldiers (in Crimean War)	42.7%
English soldiers (in Crimean War) after Nightingale's sanitary reforms	2.2%

Discuss how Nightingale used the three basic types of epidemiologic investigation. The contagious diseases were primarily cholera and typhus; how are these diseases transmitted and prevented?

3. Name the method of transmission of each of the following diseases:
   a. malaria
   b. tuberculosis
   c. nosocomial infections
   d. salmonellosis
   e. streptococcal pharyngitis
   f. mononucleosis
   g. measles
   h. hepatitis A
   i. tetanus
   j. hepatitis B
   k. chlamydial urethritis

4. The following graph shows the incidence of typhoid fever in the United States from 1954 to 1999. Mark the graph to show when this disease occurred sporadically and epidemically. What appears to be the endemic level? What would have to be shown to indicate a pandemic of this disease? How is typhoid fever transmitted?

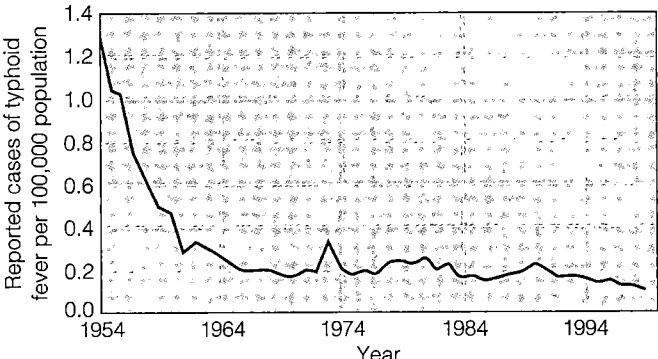

5. In 1991, a cholera epidemic caused by *Vibrio cholerae* O1 Inaba was discovered in Peru. During 1991 and 1992, 731,312 cholera cases were reported in 20 countries in the Western Hemisphere. The map below shows the location of cholera cases in the continental United States caused by *V. cholerae* O1 Inaba (gold), as well as where this strain has been found in coastal waters (purple).

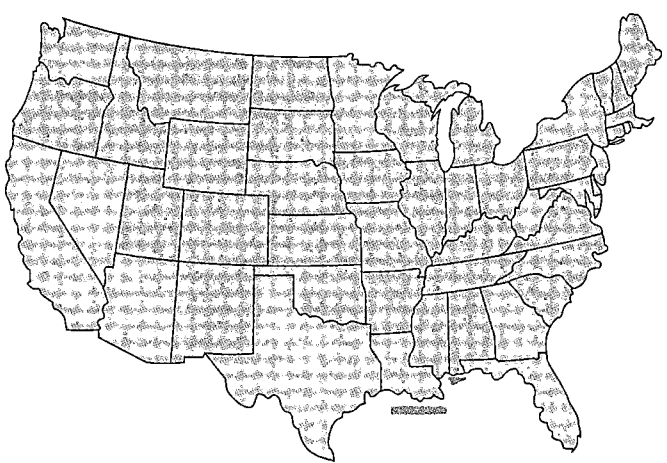

How is cholera transmitted? How did *V. cholerae* O1 Inaba get into the United States? Explain how eight people who did not travel outside the U.S. contracted *V. cholerae* O1 Inaba.

conjunctiva, a delicate membrane that covers the eyeballs and lines the eyelids. Most pathogens enter through the mucous membranes of the gastrointestinal and respiratory tracts.

The respiratory tract is the easiest and most frequently traveled portal of entry for infectious microorganisms. Microbes are inhaled into the nose or mouth in drops of moisture and dust particles. Diseases that are commonly contracted via the respiratory tract include the common cold, pneumonia, tuberculosis, influenza, measles, and smallpox.

Microorganisms can gain access to the gastrointestinal tract in food and water and via contaminated fingers. (See the box on page 437, which discusses how these and other human behaviors can, over time, influence the virulence of microorganisms.) Most microbes that enter the body in these ways are destroyed by hydrochloric acid (HCl) and enzymes in the stomach, or by bile and enzymes in the small intestine. Those that survive can cause disease. Microbes in the gastrointestinal tract can cause poliomyelitis, hepatitis A, typhoid fever, amoebic dysentery, giardiasis, shigellosis (bacillary dysentery), and cholera. These pathogens are then eliminated with feces and can be transmitted to other hosts via contaminated water, food, or fingers.

The genitourinary tract is a portal of entry for pathogens that are contracted sexually. Some microbes that cause sexually transmitted diseases (STDs) may penetrate an unbroken mucous membrane. Others require a cut or abrasion of some type. Examples of STDs are HIV infection, genital warts, chlamydia, herpes, syphilis, and gonorrhea.

Skin

The skin is one of the largest organs of the body in terms of surface area and is an important defense against disease. Unbroken skin is impenetrable by most microorganisms. Some microbes gain access to the body through openings in the skin, such as hair follicles and sweat gland ducts. In addition, larvae of the hookworm actually bore through intact skin, and some fungi grow on the keratin in skin or infect the skin itself.

The Parenteral Route

Other microorganisms gain access to the body when they are deposited directly into the tissues beneath the skin or into mucous membranes when these barriers are penetrated or injured. This route is called the **parenteral route.** Punctures, injections, bites, cuts, wounds, surgery, and splitting due to swelling or drying can all establish parenteral routes.

The Preferred Portal of Entry

Even after microorganisms have entered the body, they do not necessarily cause disease. The occurrence of disease depends on several factors, only one of which is the portal of entry. Many pathogens have a preferred portal of entry that is a prerequisite to their being able to cause disease. If they gain access to the body by another portal, disease might not occur. For example, the bacteria of typhoid fever, *Salmonella typhi,* produce all the signs and symptoms of the disease when swallowed (preferred route), but if the same bacteria are rubbed on the skin, no reaction (or only a slight inflammation) occurs. Streptococci that are inhaled (preferred route) can cause pneumonia; those that are swallowed generally do not produce signs or symptoms. Some pathogens, such as *Yersinia pestis,* the microorganism that causes plague, can initiate disease from more than one portal of entry. The preferred portals of entry for some common pathogens are listed in Table 15.1 on page 438.

Numbers of Invading Microbes

Learning Objective

- Define LD_{50} and ID_{50}.

If only a few microbes enter the body, they will probably be overcome by the host's defenses. However, if large numbers of microbes gain entry, the stage is probably set for disease. Thus, the likelihood of disease increases as the number of pathogens increases.

The virulence of a microbe or the potency of its toxin is often expressed as the **LD_{50}** (lethal dose for 50% of hosts), the number of microbes in a dose that will kill 50% of inoculated test animals. The dose required to produce a demonstrable infection in 50% of the test animals is called the **ID_{50}** (infectious dose for 50% of hosts). The 50 is not an absolute value. It is used to compare relative toxicities or to study experimental conditions. For example, the ID_{50} of *Vibrio cholerae* is 10^8 cells, but if stomach acid is neutralized with bicarbonate, the number of cells required to cause an infection decreases significantly.

Adherence

Learning Objective

- Using examples, explain how microbes adhere to host cells.

Once pathogens gain entry to a host, almost all of them have some means of attaching themselves to host tissues. For most pathogens, this attachment, called **adherence,** is a necessary step in pathogenicity. (Of course, non-

MICROBIOLOGY IN THE NEWS

How Human Behavior Influences the Evolution of Virulence in Microorganisms

According to human logic, a parasite that kills its host is harming itself. Thus, it seems reasonable that some parasites have evolved to a sort of commensalism with their human hosts. In this respect, tapeworms might be considered the "perfect parasite." Tapeworms cause no obvious symptoms in most patients, thus ensuring that their hosts will be able to move around and shed tapeworm eggs that will be ingested by intermediate hosts.

However, other diseases such as cholera, antibiotic-resistant tuberculosis, viral cancers, and syphilis sometimes do kill their hosts. Does it make sense for a parasite to kill its host? Remember that nature does not have a plan for evolution; the genetic variations that give rise to evolution are due to random mutations, not logic. However, according to natural selection, organisms best adapted to their environments will survive and reproduce. Coevolution between a parasite and its host seems to occur: The behavior of one influences that of the other.

Some parasites, such as the cholera bacterium, get transmitted before their host dies. *Vibrio cholerae* quickly induces diarrhea, threatening the host's life resulting from a loss of fluids and salts, but providing a way to transmit the parasite to another person through the ingestion of contaminated water. This cycle continues in countries where war and poverty have prevented the implementation of water purification systems.

However, the virulent form of cholera has been selected against in countries where water-treatment systems prevent the bacteria from getting to a new host. After water purification was instituted in India, the milder agent, *V. cholerae* el-tor, replaced the deadly *V. cholerae*.

When we interfere with a pathogen's survival by using antibiotics, we inadvertently select for antibiotic-resistant mutants. The rapid rise of multidrug-resistant *Mycobacterium tuberculosis*, documented in the popular and scientific media, demonstrates this. In an antibiotic-laden environment, the "fittest" bacterium is one that is resistant to antibiotics. Antibiotic resistance is selected for when chemotherapy is not completed and when antibiotics are not used correctly. For example, tuberculosis (TB) chemotherapy protocols are very long (up to 2 years). In the United States, only 75% of TB patients complete the prescribed chemotherapy. When a person doesn't take the complete regimen of the prescribed antibiotic, the level of the drug in the body decreases, and some resistant bacteria survive. These bacteria reproduce, and some of the new population are able to survive in even higher levels of antibiotics.

Comparisons of human T-cell leukemia virus type 1 (HTLV-1) infections in Japan and Jamaica also suggest that human sexual behavior influences the virulence of sexually transmitted diseases. HTLV-1 is a sexually transmit-

ted retrovirus that causes a type of adult leukemia. In Japan, the average age of onset of cancers caused by HTLV-1 is 60, whereas in Jamaica, onset occurs at about age 45. In Japan, barrier contraceptives are more widely used than birth control pills, so the virus needs a healthy host for a longer period of time. In Jamaica, barrier contraceptives are not widely used and HTLV-1 is more virulent, perhaps because it has a greater potential for sexual transmission.

Human behavior may also have influenced the evolution of syphilis. *Treponema pallidum* has a long incubation period and goes into noninfectious latent periods between the primary, secondary, and tertiary stages of the disease. Although it cannot be transmitted during the incubation and latent periods, these periods allow the bacteria to stay alive until the infected host changes sex partners. Partner changes might take several years in a monogamous society.

Paul Ewald of Amherst College reminds us that commensalism is not the inevitable end point of evolution. In order to control diseases, we must modify our behavior to make commensalism the most favorable outcome for the pathogen. Some examples of these modifications are ensuring completion of the prescribed antibiotic protocol for tuberculosis, and the use of condoms to prevent the transmission of sexually transmitted diseases.

---

pathogens also have structures for attachment.) The attachment between pathogen and host is accomplished by means of surface molecules on the pathogen called **adhesins** or **ligands** that bind specifically to complementary surface **receptors** on the cells of certain host tissues (Figure 15.1 on page 439). Adhesins may be located on a microbe's glycocalyx or on other microbial surface structures, such as fimbriae (see Chapter 4).

The majority of adhesins on the microorganisms studied so far are glycoproteins or lipoproteins. The receptors on host cells are typically sugars, such as mannose. Adhesins on different strains of the same species of pathogen can vary in structure. Different cells of the same host can also have different receptors that vary in structure. If adhesins, receptors, or both can be altered to interfere with adherence, infection can often be prevented (or at least controlled).

table 15.1 Portals of Entry for the Pathogens of Some Common Diseases

Portal of Entry	Pathogen	Disease	Incubation Period
Mucous Membranes			
Respiratory tract	Streptococcus pneumoniae	Pneumococcal pneumonia	Variable
	Mycobacterium tuberculosis[b]	Tuberculosis	Variable
	Bordetella pertussis	Whooping cough (pertussis)	12–20 days
	Influenza virus	Influenza	18–36 hours
	Measles virus (Morbillivirus)	Measles (rubeola)	11–14 days
	Rubella virus (Rubivirus)	German measles (rubella)	2–3 weeks
	Epstein-Barr virus (Lymphocryptovirus)	Infectious mononucleosis	2–6 weeks
	Varicella-zoster virus (Varicellavirus)	Chickenpox (varicella) (primary infection)	14–16 days
	Histoplasma capsulatum (fungus)	Histoplasmosis	5–18 days
Gastrointestinal tract	Shigella spp.	Bacillary dysentery (shigellosis)	1–2 days
	Brucella spp.	Brucellosis (undulant fever)	6–14 days
	Vibrio cholerae	Cholera	1–3 days
	Salmonella enterica,	Salmonellosis	7–22 hours
	Salmonella typhi	Typhoid fever	14 days
	Hepatitis A virus (Hepatovirus)	Hepatitis A	15–50 days
	Mumps virus (Rubulavirus)	Mumps	2–3 weeks
	Trichinella spiralis (helminth)	Trichinosis	2–28 days
Genitourinary tract	Neisseria gonorrhoeae	Gonorrhea	3–8 days
	Treponema pallidum	Syphilis	9–90 days
	Chlamydia trachomatis	Nongonococcal urethritis	1–3 weeks
	Herpes simplex virus type 2	Herpes virus infections	4–10 days
	Human immunodeficiency virus (HIV)[c]	AIDS	10 years
	Candida albicans (fungus)[c]	Candidiasis	2–5 days
Skin or Parenteral Route	Clostridium perfringens	Gas gangrene	1–5 days
	Clostridium tetani	Tetanus	3–21 days
	Rickettsia rickettsii	Rocky Mountain spotted fever	3–12 days
	Hepatitis B virus (Hepadnavirus)[b]	Hepatitis B	6 weeks–6 months
	Rabiesvirus (Lyssavirus)	Rabies	10 days–1 year
	Plasmodium spp. (protozoan)	Malaria	2 weeks

[a]All pathogens are bacteria, unless indicated otherwise. For viruses, the viral species and/or genus name is given.

[b]These pathogens can also cause disease after entering the body via the gastrointestinal tract.

[c]These pathogens can also cause disease after entering the body via the parenteral route.

The following examples illustrate the diversity of adhesins. *Streptococcus mutans,* a bacterium that plays a key role in tooth decay, attaches to the surface of teeth by its glycocalyx. An enzyme produced by *S. mutans,* called glucosyltransferase, converts glucose (derived from sucrose or table sugar) into a sticky polysaccharide called dextran, which forms the glycocalyx. *Actinomyces* cells have fimbriae that adhere to the glycocalyx of *S. mutans,* further contributing to plaque. Enteropathogenic strains of *E. coli*

(those responsible for gastrointestinal disease) have adhesins on fimbriae that adhere only to specific kinds of cells in certain regions of the small intestine. After adhering, *Shigella* and *E. coli* induce endocytosis as a vehicle to enter host cells and then multiply within them (see Figure 25.9 on page 693). *Treponema pallidum,* the causative agent of syphilis, uses its tapered end as a hook to attach to host cells. *Listeria monocytogenes* produces an adhesin for a specific receptor on host cells. *Neisseria gonorrhoeae,* the

Intestinal tissue   *E. coli*   Bacteria   Skin

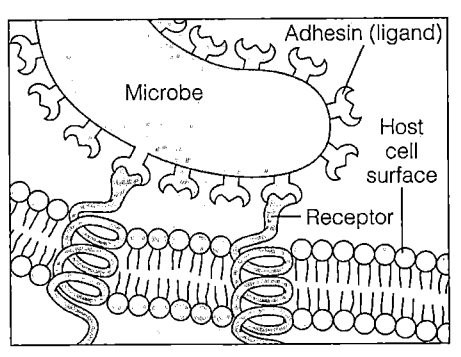

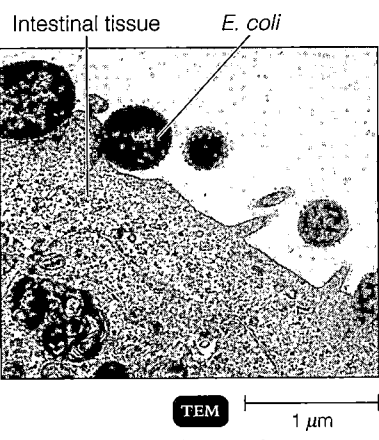

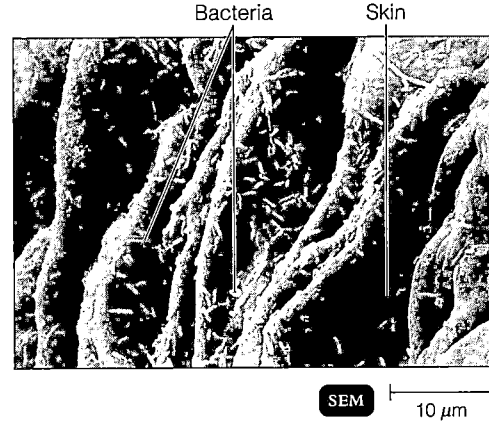

(a) Surface molecules on a pathogen, called adhesins or ligands, bind specifically to complementary surface receptors on cells of certain host tissues.

(b) Selective attachment of a pathogenic strain of *E. coli* to the intestinal tissue of a rabbit.

(c) Bacteria adhering to the skin of a salamander.

FIGURE 15.1 Adherence.

■ Most adhesins are glycoproteins or lipoproteins.

causative agent of gonorrhea, also has fimbriae containing adhesins, which in this case permit attachment to cells with appropriate receptors in the genitourinary tract, eyes, and pharynx. *Staphylococcus aureus*, which can cause skin infections, binds to skin by a mechanism of adherence that resembles viral attachment (see Chapter 13).

How Bacterial Pathogens Penetrate Host Defenses

Learning Objective

■ *Explain how capsules and cell wall components contribute to pathogenicity.*

Although some pathogens can cause damage on the surface of tissues, most must penetrate tissues to cause disease. Here we will consider several factors that contribute to the ability of bacteria to invade a host.

Capsules

Recall from Chapter 4 that some bacteria make glycocalyx material that forms capsules around their cell walls; this property increases the virulence of the species. The capsule resists the host's defenses by impairing phagocytosis, the process by which certain cells of the body engulf and destroy microbes (see Chapter 16, page 458). The chemical nature of the capsule appears to prevent the phagocytic cell from adhering to the bacterium. However, the human body can produce antibodies against the capsule, and when these antibodies are present on the

capsule surface, the encapsulated bacteria are easily destroyed by phagocytosis.

One bacterium that owes its virulence to the presence of a polysaccharide capsule is *Streptococcus pneumoniae*, the causative agent of pneumococcal pneumonia (see Figure 24.10). Some strains of this organism have capsules, and others do not. Strains with capsules are virulent, but strains without capsules are avirulent because they are susceptible to phagocytosis. Other bacteria that produce capsules related to virulence are *Klebsiella pneumoniae*, a causative agent of bacterial pneumonia; *Haemophilus influenzae*, a cause of pneumonia and meningitis in children; *Bacillus anthracis*, the cause of anthrax; and *Yersinia pestis*, the causative agent of bubonic plague. Keep in mind that capsules are not the only cause of virulence. Many nonpathogenic bacteria produce capsules, and the virulence of some pathogens is not related to the presence of a capsule.

Components of the Cell Wall

The cell walls of certain bacteria contain chemical substances that contribute to virulence. For example, *Streptococcus pyogenes* produces a heat-resistant and acid-resistant protein called **M protein** (see Figure 21.5 on page 585). This protein is found on both the cell surface and fimbriae. The M protein mediates attachment of the bacterium to epithelial cells of the host and helps the bacterium resist phagocytosis by white blood cells. The protein thereby increases the virulence of the microorganism. Immunity to *S. pyogenes* depends on the body's production of an antibody specific to M protein.

The waxes that make up the cell wall of *Mycobacterium tuberculosis* also increase virulence by resisting digestion by phagocytes. In fact, *M. tuberculosis* can even multiply inside phagocytes.

Enzymes

Learning Objective

- *Compare the effects of leukocidins, hemolysins, coagulases, kinases, hyaluronidase, and collagenase.*

The virulence of some bacteria is thought to be aided by the production of extracellular enzymes (exoenzymes) and related substances. These chemicals can break cells open, dissolve materials between cells, and form or dissolve blood clots, among other functions.

Substances called **leukocidins,** produced by some bacteria, can destroy neutrophils, leukocytes (white blood cells) that are very active in phagocytosis. Leukocidins are also active against macrophages (phagocytic cells) present in tissues. Among the bacteria that secrete leukocidins are staphylococci and streptococci. Leukocidins produced by streptococci degrade lysosomes within leukocytes, thereby causing the death of the white blood cell. Enzymes released from the lysosomes can damage other cellular structures and thus intensify streptococcal lesions. This type of damage to white blood cells decreases host resistance.

Hemolysins are bacterial enzymes that cause the lysis of erythrocytes (red blood cells). Bacteria produce hemolysins that differ in their ability to lyse different kinds of red blood cells (humans, sheep, and rabbits, for example) and the type of lysis they cause. Important producers of hemolysins are staphylococci; *Clostridium perfringens*, the most common causative agent of gas gangrene; and streptococci. *Streptolysins* are hemolysins produced by streptococci. One kind, called *streptolysin O (SLO)*, is so named because it is inactivated by atmospheric oxygen. Another kind is called *streptolysin S (SLS)* because it has an affinity for a protein (albumin) in blood serum. Both streptolysins can cause the lysis of not only red blood cells, but also white blood cells (whose function is to kill the streptococci) and other body cells.

Coagulases are bacterial enzymes that coagulate (clot) the fibrinogen in blood. Fibrinogen, a plasma protein produced by the liver, is converted by coagulases into fibrin, the threads that form a blood clot. The fibrin clot may protect the bacterium from phagocytosis and isolate it from other defenses of the host. Coagulases are produced by some members of the genus *Staphylococcus;* they may be involved in the walling-off process in boils produced by staphylococci. However, some staphylococci

that do not produce coagulases are still virulent. (Capsules may be more important to their virulence.)

Bacterial **kinases** are bacterial enzymes that break down fibrin and thus dissolve clots formed by the body to isolate the infection. One of the better-known kinases is *fibrinolysin (streptokinase)*, which is produced by such streptococci as *Streptococcus pyogenes*. Another kinase, *staphylokinase*, is produced by such staphylococci as *Staphylococcus aureus*. Injected directly into the blood, streptokinase has been used successfully to dissolve some types of blood clots in cases of heart attacks due to obstructed coronary arteries.

Hyaluronidase is another enzyme secreted by certain bacteria, such as streptococci. It hydrolyzes hyaluronic acid, a type of polysaccharide that holds together certain cells of the body, particularly cells in connective tissue. This digesting action is thought to be involved in the tissue blackening of infected wounds and to help the microorganism spread from its initial site of infection. Hyaluronidase is also produced by some clostridia that cause gas gangrene. For therapeutic use, hyaluronidase may be mixed with a drug to promote the spread of the drug through a body tissue.

Another enzyme, **collagenase,** produced by several species of *Clostridium,* facilitates the spread of gas gangrene. Collagenase breaks down the protein collagen, which forms the connective tissue of muscles and other body organs and tissues.

Other bacterial substances thought to contribute to virulence are *necrotizing factors*, which cause the death of body cells; *hypothermic factors*, which decrease body temperature; *lecithinase*, which destroys the plasma membrane, especially around red blood cells; *protease*, which breaks down proteins, especially in muscle tissue; and *siderophores*, which scavenge iron from the host's body fluids.

Penetration into the Host Cell Cytoskeleton

Learning Objective

- *Describe how bacteria use the host cell's cytoskeleton to enter the cell.*

As previously noted, microbes attach to host cells by adhesins. The interaction triggers signals in the host cell that activate factors that can result in the entrance of some bacteria. The actual mechanism is provided by the host cell cytoskeleton. Recall from Chapter 4 that eukaryotic cytoplasm has a complex internal structure consisting of protein filaments called microfilaments, intermediate filaments, and microtubules that comprise the cytoskeleton.

A major component of the cytoskeleton is a protein called actin, which is used by some microbes to penetrate host cells and by others to move through and between host cells.

Salmonella strains and *E. coli* make contact with the host cell plasma membrane. This leads to dramatic changes in the membrane at the point of contact. The microbes produce surface proteins called **invasins** that rearrange nearby actin filaments of the cytoskeleton. This causes a cytoplasmic structure to project from the host cell like a pedestal under the *Salmonella*. The structure supports the bacterial cells, then forms an actin basket that appears to cuddle the *Salmonella* and move it into the cell (Figure 15.2).

Once inside the host cell, certain bacteria such as *Shigella* species and *Listeria* species can actually use actin to propel themselves through the host cell cytoplasm and from one host cell to another. The condensation of actin on one end of the bacteria propels them through the cytoplasm. The bacteria also make contact with membrane junctions that form part of a transport network between host cells. The bacteria use a glycoprotein called cadherin, which bridges the junctions, to move from cell to cell.

The study of the numerous interactions between microbes and host cell cytoskeleton is a very intense area of investigation on virulence mechanisms.

How Bacterial Pathogens Damage Host Cells

When a microorganism invades a body tissue, it initially encounters phagocytes of the host. If the phagocytes are successful in destroying the invader, no further damage is done to the host. But if the pathogen overcomes the host defense, then the microorganism can damage host cells in three basic ways: (1) by causing direct damage in the immediate vicinity of the invasion; (2) by producing toxins, transported by blood and lymph, that damage sites far removed from the original site of invasion; and (3) by inducing hypersensitivity reactions. This latter mechanism is considered in detail in Chapter 19. For now, we will discuss only the first two mechanisms.

Direct Damage

Once a pathogen is attached to host cells, it can pass through them to invade other tissues. During this invasion, the pathogens metabolize and multiply to kill host cells. For example, some bacteria, such as *E. coli, Shigella, Salmonella,* and *Neisseria gonorrhoeae,* can induce host epithelial cells to engulf them by a process that resembles

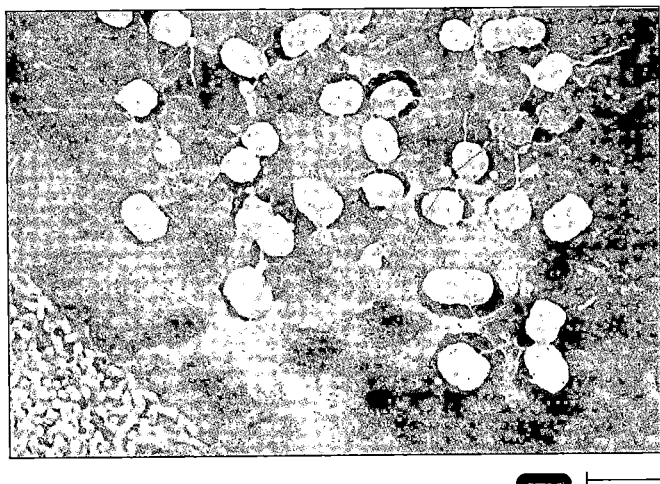

SEM 3 μm

FIGURE 15.2 *Salmonella* **entering epithelial cells.**

■ **Some microbes produce surface proteins called invasins that rearrange actin filaments of the cytoskeleton.**

phagocytosis. These pathogens can then be extruded from the host cells by a reverse phagocytosis process in order to enter other host cells. Some bacteria can also penetrate host cells by excreting enzymes and by their own motility; such penetration can itself damage the host cell. Most damage by bacteria, however, is done by toxins.

The Production of Toxins

Learning Objectives

■ *Contrast the nature and effects of exotoxins and endotoxins.*

■ *Outline the mechanisms of action of diphtheria toxin, botulinum toxin, tetanus toxin, cholera toxin, and lipid A.*

■ *Identify the importance of the LAL assay.*

Toxins are poisonous substances that are produced by certain microorganisms. They are often the primary factor contributing to the pathogenic properties of those microbes. The capacity of microorganisms to produce toxins is called **toxigenicity.** Toxins transported by the blood or lymph can cause serious, and sometimes fatal, effects. Some toxins produce fever, cardiovascular disturbances, diarrhea, and shock. Toxins can also inhibit protein synthesis, destroy blood cells and blood vessels, and disrupt the nervous system by causing spasms. Of the 220 or so known bacterial toxins, nearly 40% cause disease by damaging eukaryotic cell membranes. The term **toxemia** refers to the presence of toxins in the blood. Toxins are of two types: exotoxins and endotoxins.

FIGURE 15.3   Exotoxins and endotoxins.

■ What are the three principal types of exotoxins based on their mode of action?

Cell wall

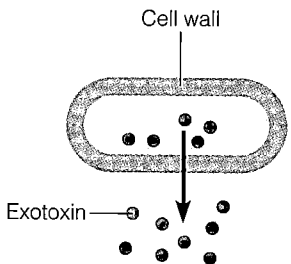

Exotoxin

Endotoxin

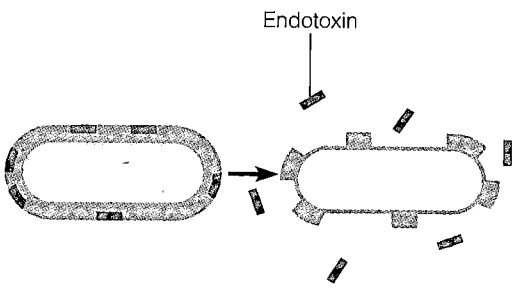

(a) Exotoxins are produced inside mostly gram-positive bacteria as part of their growth and metabolism. They are then released into the surrounding medium.

(b) Endotoxins are part of the outer portion of the cell wall (lipid A; see Figure 4.12c) of gram-negative bacteria. They are liberated when the bacteria die and the cell wall breaks apart.

Exotoxins

Exotoxins are produced inside some bacteria as part of their growth and metabolism and are released into the surrounding medium (Figure 15.3a). Exotoxins are proteins, and many are enzymes that catalyze only certain biochemical reactions. Most bacteria that produce exotoxins are gram-positive. The genes for most (perhaps all) exotoxins are carried on bacterial plasmids or phages. Because exotoxins are soluble in body fluids, they can easily diffuse into the blood and are rapidly transported throughout the body.

Exotoxins work by destroying particular parts of the host's cells or by inhibiting certain metabolic functions. They are highly specific in their effects on body tissues. Exotoxins may be grouped into three principal types, based on their mode of action: (1) **cytotoxins,** which kill host cells or affect their functions; (2) **neurotoxins,** which interfere with normal nerve impulse transmission; and (3) **enterotoxins,** which affect cells lining the gastrointestinal tract. Exotoxins are among the most lethal substances known. Only 1 mg of the botulinum exotoxin is enough to kill 1 million guinea pigs. Fortunately, only a few bacterial species produce such potent exotoxins.

Diseases caused by bacteria that produce exotoxins are often caused by minute amounts of exotoxins, not by the bacteria themselves. It is the exotoxins that produce the specific signs and symptoms of the disease. Thus, exotoxins are disease-specific. For example, the infection of a wound by *Clostridium tetani* need be no larger or more painful than a pin prick, yet the organisms in a wound that size can produce enough tetanus toxin to kill an unvaccinated human.

The body produces antibodies called **antitoxins** that provide immunity to exotoxins. When exotoxins are inactivated by heat or by formaldehyde, iodine, or other chemicals, they no longer cause the disease but can still stimulate the body to produce antitoxins. Such altered exotoxins are called *toxoids.* When toxoids are injected into the body as a vaccine, they stimulate antitoxin production so that immunity is produced. Diphtheria and tetanus can be prevented by toxoid vaccination.

Next we will briefly describe a few of the more notable exotoxins (antitoxins will be discussed further in Chapter 18).

Diphtheria Toxin   *Corynebacterium diphtheriae* produces the diphtheria toxin only when it is infected by a lysogenic phage carrying the *tox* gene. This cytotoxin inhibits protein synthesis in eukaryotic cells. The mechanism by which it does this is an excellent example of how an exotoxin interacts with host cells (Figure 15.4). ① Diphtheria toxin is a protein consisting of two different polypeptides, designated A (active) and B (binding). Although only polypeptide A causes symptoms in the host, polypeptide B is required by polypeptide A to be active. ② Polypeptide B binds to surface receptors on the host cell and causes the transport of the entire protein across the plasma membrane into the cell. ③ Once this is accomplished, polypeptide A inhibits protein synthesis within the target cell. (This model also applies to the cholera exotoxin, described on page 443.)

Erythrogenic Toxins   *Streptococcus pyogenes* has the genetic material to synthesize three types of cytotoxins, designated A, B, and C. These erythrogenic (*erythro* = red; *gen* = producing) toxins damage blood capillaries under the skin and produce a red skin rash. Scarlet fever, caused by *S. pyogenes* exotoxins, is named for this characteristic rash.

Botulinum Toxin   Botulinum toxin is produced by *Clostridium botulinum.* Although toxin production is associated with the germination of endospores and the

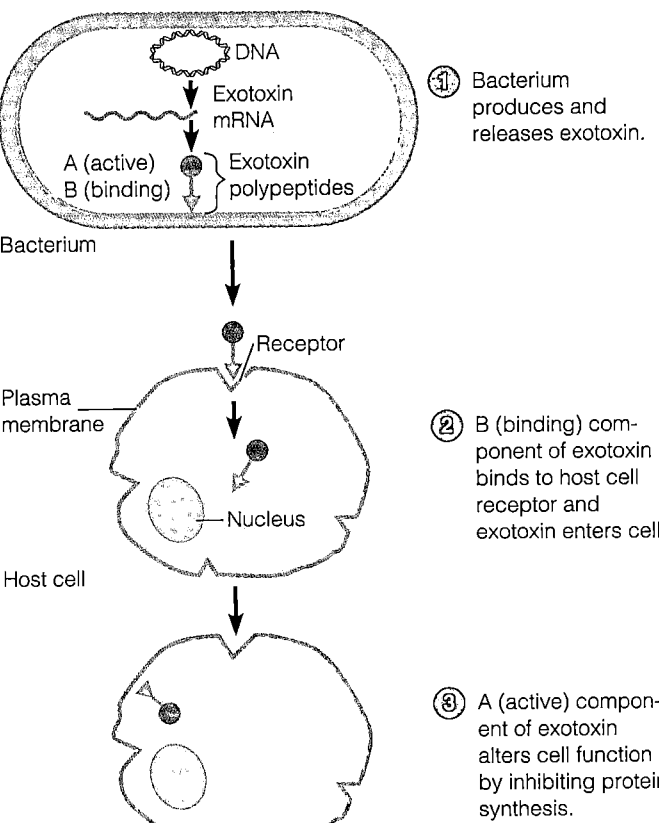

FIGURE 15.4 The action of an exotoxin. A proposed model for the mechanism of action of diphtheria toxin.

① Bacterium produces and releases exotoxin.

② B (binding) component of exotoxin binds to host cell receptor and exotoxin enters cell.

③ A (active) component of exotoxin alters cell function by inhibiting protein synthesis.

■ Diphtheria toxin consists of two different polypeptides: A (active) and B (binding); B is required to make A active.

growth of vegetative cells, little of the toxin appears in the medium until it is released by lysis late in growth. Botulinum toxin is a neurotoxin; it acts at the neuromuscular junction (the junction between nerve cell and muscle cell) and prevents the transmission of impulses from the nerve cell to the muscle. The toxin accomplishes this by binding to the nerve cell and inhibiting the release of a neurotransmitter called acetylcholine. As a result, botulinum toxin causes paralysis in which muscle tone is lacking (flaccid paralysis). *C. botulinum* produces several different types of botulinum toxin, and each possesses a different potency.

Tetanus Toxin *Clostridium tetani* produces tetanus neurotoxin, also known as tetanospasmin. This toxin reaches the central nervous system and binds to nerve cells that control the contraction of various skeletal muscles. These nerve cells normally send inhibiting impulses that prevent random contractions and terminate completed contrac-

tions. The binding of tetanospasmin blocks this relaxation pathway (see Chapter 22). The result is uncontrollable muscle contraction, producing the convulsive symptoms (spasmodic contractions) of tetanus, or "lockjaw."

Vibrio Enterotoxin *Vibrio cholerae* produces an enterotoxin called cholera toxin. Like diphtheria toxin, cholera toxin consists of two polypeptides, A (active) and B (binding). The B component binds to plasma membranes of epithelial cells lining the small intestine, and the A component induces the formation of cyclic AMP from ATP in the cytoplasm (see Chapter 8). As a result, epithelial cells discharge large amounts of fluids and electrolytes (ions). Normal muscular contractions are disturbed, leading to severe diarrhea that may be accompanied by vomiting. *Heat-labile enterotoxin* (so named because it is more sensitive to heat than are most toxins), produced by some strains of *E. coli*, has an action identical to that of cholera toxin.

Staphylococcal Enterotoxin *Staphylococcus aureus* produces an enterotoxin that affects the intestines in the same way as cholera toxin. A strain of *S. aureus* also produces enterotoxins that result in the symptoms associated with toxic shock syndrome (see Chapter 21). A summary of diseases produced by exotoxins is shown in Table 15.2.

Endotoxins

Endotoxins differ from exotoxins in several ways. Endotoxins are part of the outer portion of the cell wall of gram-negative bacteria (see Figure 15.3b). Recall from Chapter 4 that gram-negative bacteria have an outer membrane surrounding the peptidoglycan layer of the cell wall. This outer membrane consists of lipoproteins, phospholipids, and lipopolysaccharides (LPS) (see Figure 4.12c on page 87). The lipid portion of LPS, called **lipid A**, is the endotoxin. Thus, endotoxins are lipopolysaccharides, whereas exotoxins are proteins.

Endotoxins exert their effects when gram-negative bacteria die and their cell walls undergo lysis, thus liberating the endotoxin. Antibiotics used to treat diseases caused by gram-negative bacteria can lyse the bacterial cells; this reaction releases endotoxin and may lead to an immediate worsening of the symptoms, but the condition usually improves as the endotoxin breaks down. All endotoxins produce the same signs and symptoms, regardless of the species of microorganism, although not to the same degree. These include chills, fever, weakness, generalized aches, and, in some cases, shock and even death. Endotoxins can also induce miscarriage.

Another consequence of endotoxins is the activation of blood-clotting proteins, causing the formation of small

table 15.2	Diseases Caused by Exotoxins	
Disease	**Bacterium**	**Mechanism**
Botulism	*Clostridium botulinum*	Neurotoxin prevents the transmission of nerve impulses; flaccid paralysis results.
Tetanus	*Clostridium tetani*	Neurotoxin blocks nerve impulses to muscle relaxation pathway.
Gas gangrene and food poisoning	*Clostridium perfringens* and other species of *Clostridium*	One exotoxin (cytotoxin) causes massive red blood cell destruction (hemolysis); another exotoxin (enterotoxin) is related to food poisoning and causes diarrhea.
Diphtheria	*Corynebacterium diphtheriae*	Cytotoxin inhibits protein synthesis, especially in nerve, heart, and kidney cells.
Scalded skin syndrome, food poisoning, and toxic shock syndrome	*Staphylococcus aureus*	One exotoxin causes skin layers to separate and slough off (scalded skin); another exotoxin (enterotoxin) produces diarrhea and vomiting; still another exotoxin produces symptoms associated with TSS.
Cholera	*Vibrio cholerae*	Enterotoxin induces diarrhea.
Scarlet fever	*Streptococcus pyogenes*	Cytotoxins cause vasodilation that results in the characteristic rash.
Traveler's diarrhea	Enterotoxigenic *Escherichia coli* and *Shigella* spp.	Enterotoxin causes excessive secretion of ions and water; diarrhea results.

blood clots. These blood clots obstruct capillaries, and the resulting decreased blood supply induces the death of tissues. This condition is referred to as *disseminated intravascular clotting*.

The fever (pyrogenic response) caused by endotoxins is believed to occur as depicted in Figure 15.5. ① When gram-negative bacteria are ingested by phagocytes and ② degraded in vacuoles, the LPS of the bacterial cell wall are released. These endotoxins cause macrophages to produce small protein molecules called **interleukin-1 (IL-1)**, formerly called *endogenous pyrogen*, ③ which are carried via the blood to the hypothalamus, a temperature control center in the brain. ④ IL-1 induces the hypothalamus to release lipids called prostaglandins, which reset the thermostat in the hypothalamus at a higher temperature. The result is a fever. Bacterial cell death caused by lysis or antibiotics can also produce fever by this mechanism. Both aspirin and acetaminophen reduce fever by inhibiting the synthesis of prostaglandins. (The function of fever in the body is discussed in Chapter 16 on page 466.)

Shock refers to any life-threatening loss of blood pressure. Shock caused by gram-negative bacteria is called **septic shock** (or endotoxic shock). Like fever, the shock produced by endotoxins is related to the secretion of a substance by macrophages. Phagocytosis of gram-negative bacteria causes the phagocytes to secrete a polypeptide called **tumor necrosis factor (TNF)**, or *cachectin*. TNF binds to many tissues in the body and alters their metabo-

lism in a number of ways. One effect of TNF is damage to blood capillaries; their permeability is increased, and they lose large amounts of fluid. The result is a drop in blood pressure that results in shock. Low blood pressure has serious effects on the kidneys, lungs, and gastrointestinal tract. Another mechanism that causes septic shock is discussed in the box in Chapter 16 (page 463). In addition, the presence of gram-negative bacteria such as *Haemophilus influenzae* type b in cerebrospinal fluid causes the release of IL-1 and TNF. These, in turn, cause a weakening of the blood–brain barrier that normally protects the central nervous system from infection. The barrier is weakened to let phagocytes in, but this also lets more bacteria enter from the bloodstream. In the United States, 500,000 cases of septic shock occur each year. One-third of the patients die within a month, and nearly half die within six months.

Endotoxins do not promote the formation of effective antitoxins. Antibodies are produced, but they tend not to counter the effect of the toxin; sometimes, in fact, they actually enhance its effect.

Representative microorganisms that produce endotoxins are *Salmonella typhi* (the causative agent of typhoid fever), *Proteus* spp. (frequently the causative agents of urinary tract infections), and *Neisseria meningitidis* (the causative agent of meningococcal meningitis).

It is important to have a sensitive test to identify the presence of endotoxins in drugs, medical devices, and body fluids. Materials that have been sterilized may

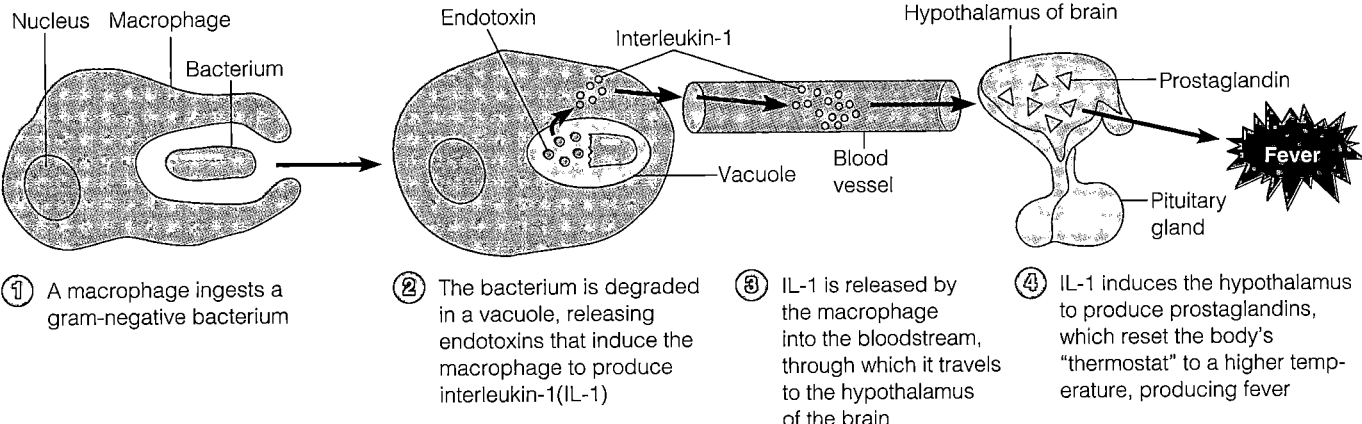

① A macrophage ingests a gram-negative bacterium

② The bacterium is degraded in a vacuole, releasing endotoxins that induce the macrophage to produce interleukin-1(IL-1)

③ IL-1 is released by the macrophage into the bloodstream, through which it travels to the hypothalamus of the brain

④ IL-1 induces the hypothalamus to produce prostaglandins, which reset the body's "thermostat" to a higher temp-erature, producing fever

FIGURE 15.5   Endotoxins and the pyrogenic response. The proposed mechanism by which endotoxins cause fever.

■ All endotoxins produce the same signs and symptoms, regardless of the species of microbe.

contain endotoxins, even though no bacteria can be cultured from them. One such laboratory test is called the *Limulus* **amoebocyte lysate (LAL) assay**, which can detect even minute amounts of endotoxin. The he-molymph (blood) of the Atlantic coast horseshoe crab, *Limulus polyphemus,* contains white blood cells called amoebocytes, which have large amounts of a protein (lysate) that causes clotting. In the presence of endo-toxin, amoebocytes in the crab hemolymph lyse and liberate their clotting protein. The resulting gel–clot (precipitate) is a positive test for the presence of endo-toxin. The degree of the reaction is measured by a spec-trophotometer, an instrument that measures degree of turbidity.

A comparison of exotoxins and endotoxins appears in Table 15.3.

Plasmids, Lysogeny, and Pathogenicity

Learning Objective

■ *Using examples, describe the roles of plasmids and lysogeny in pathogenicity.*

Recall from Chapters 4 (page 95) and 8 (page 240) that plasmids are small, circular DNA molecules that are not connected to the main bacterial chromosome and are ca-pable of independent replication. One group of plasmids, called R (resistance) factors, is responsible for the resis-tance of some microorganisms to antibiotics. In addition, a plasmid may carry the information that determines a microbe's pathogenicity. Examples of virulence factors

that are encoded by plasmid genes are tetanospasmin, heat-labile enterotoxin, and staphylococcal enterotoxin. Other examples are dextransucrase, an enzyme produced by *Streptococcus mutans* that is involved in tooth decay; ad-hesins and coagulase produced by *Staphylococcus aureus;* and a type of fimbria specific to enteropathogenic strains of *E. coli.*

In Chapter 13, we noted that some bacteriophages (viruses that infect bacteria) can incorporate their DNA into the bacterial chromosome, becoming a prophage, and thus remain latent and not cause lysis of the bac-terium. Such a state is called *lysogeny,* and cells containing a prophage are said to be lysogenic. One outcome of lysogeny is that the host bacterial cell and its progeny may exhibit new properties coded for by the bacteriophage DNA. Such a change in the characteristics of a microbe due to a prophage is called **lysogenic conversion.** As a result of lysogenic conversion, the bacterial cell is im-mune to infection by the same type of phage. In addition, lysogenic cells are of medical importance because some bacterial pathogenesis is caused by the prophages they contain.

Among the bacteriophage genes that contribute to pathogenicity are the genes for diphtheria toxin, eryth-rogenic toxins, staphylococcal enterotoxin and pyro-genic toxin, botulinum neurotoxin, and the capsule produced by *Streptococcus pneumoniae.* Pathogenic strains of *Vibrio cholerae* carry lysogenic phages. These phages can transmit the choleratoxin gene to nonpathogenic *V. cholerae* strains, increasing the number of pathogenic bacteria.

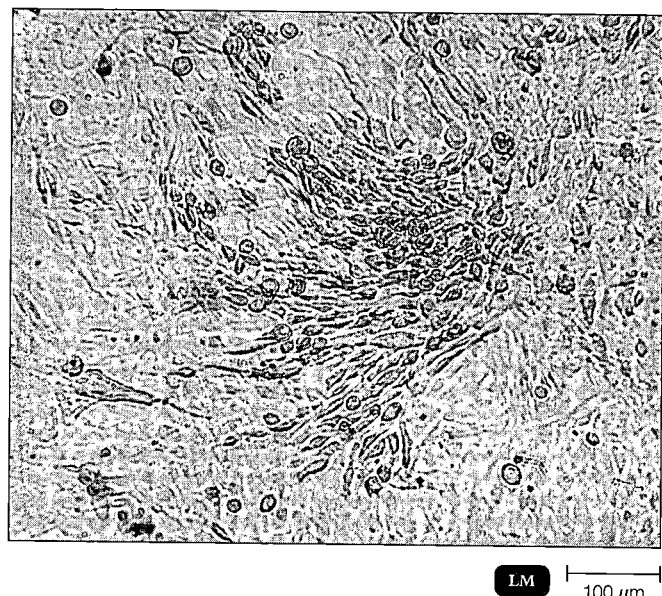

LM   100 μm

FIGURE 15.7 Transformed cells in culture. In the center of this photomicrograph is a cluster of chick embryo cells transformed by Rous sarcoma virus. Such a cluster results from the multiplication of a single cell infected with a transforming virus. Notice how the transformed cells appear dark, in contrast to the monolayer of light, flat, normal cells around them. This appearance is caused by their spindle shapes and their uninhibited growth due to the absence of contact inhibition.

■ What is contact inhibition?

measles, mumps, and the common cold (paramyxoviruses); some herpesviruses; and other viruses.

5. Some viral infections result in changes in the host cell's functions with no visible changes in the infected cells. For example, if a virus changes the production level of a hormone, the cells normally affected by that hormone will not function properly.

6. Some virus-infected cells produce substances called **interferons.** Viral infection induces cells to produce interferons, but the host cell's DNA actually codes for the interferon. This protects neighboring uninfected cells from viral infection. (Interferons will be discussed further in Chapter 16 on page 469.)

7. Many viral infections induce antigenic changes on the surface of the infected cells. These antigenic changes elicit a host antibody response against the infected cell, and thus they target the cell for destruction by the host's immune system.

8. Some viruses induce chromosomal changes in the host cell. For example, some viral infections result in chromosomal damage to the host cell, most often chromosomal breakage. Frequently oncogenes (cancer-causing genes) may be contributed or activated by a virus.

table 15.4	Cytopathic Effects of Selected Viruses
Virus (Genus)	**Cytopathic Effect**
Poliovirus (Enterovirus)	Cytocidal (cell death)
Papovavirus (family Papovaviridae)	Acidophilic inclusion bodies in nucleus
Adenovirus (Mastadenovirus)	Basophilic inclusion bodies in nucleus
Rhabdovirus (family Rhabdoviridae)	Acidophilic inclusion bodies in cytoplasm
Cytomegalovirus	Acidophilic inclusion bodies in nucleus and cytoplasm
Measles virus (Morbillivirus)	Cell fusion
Polyomavirus	Transformation
HIV (Lentivirus)	Destruction of T cells

9. Most normal cells cease growing in vitro when they come close to another cell, a phenomenon known as **contact inhibition.** Viruses capable of causing cancer *transform* host cells, as discussed in Chapter 13. Transformation results in an abnormal, spindle-shaped cell that does not recognize contact inhibition (Figure 15.7). The loss of contact inhibition results in unregulated cell growth.

Some representative viruses that cause cytopathic effects are presented in Table 15.4. In subsequent chapters, we will discuss the pathological properties of viruses in more detail.

Fungi, Protozoa, Helminths, and Algae

Learning Objective

■ *Discuss the causes of symptoms in fungal, protozoan, helminthic, and algal diseases.*

This section describes some general pathological effects of fungi, protozoa, helminths, and algae that cause human disease. Most specific diseases caused by fungi, protozoa, and helminths, along with the pathological properties of these organisms, are discussed in detail in Part Four (Chapters 21–26).

Fungi

Although fungi cause disease, they do not have a well-defined set of virulence factors. Some fungi have metabolic products that are toxic to human hosts. In such cases, however, the toxin is only an indirect cause of disease, as the fungus is already growing in or on the host. Chronic fungal infections, such as dermatomycoses, can also provoke an allergic response in the host.

Trichothecenes are fungal toxins that inhibit protein synthesis in eukaryotic cells. These toxins cause skin irritation, diarrhea, and fatal pulmonary hemorrhages. In recent years, trichothecenes produced by *Stachybotrys* (stak'e-bo-tris) growing on water-damaged walls have been responsible for the deaths of several infants.

There is evidence that some fungi do have virulence factors. Two fungi that can cause skin infections, *Candida albicans* and *Trichophyton* (trik-ō-fī'ton), secrete proteases. These enzymes may modify host cell membranes to allow attachment of the fungi. *Cryptococcus neoformans* (krip-tō-kok'kus nē-ō-fôr'manz) is a fungus that causes a type of meningitis; it produces a capsule that helps it resist phagocytosis.

The disease called ergotism, which was common in Europe during the Middle Ages, is caused by a toxin produced by an ascomycete plant pathogen, *Claviceps purpurea* (kla'vi-seps pùr-pū-rē'ä), that grows on grains. The toxin is contained in **sclerotia,** highly resistant portions of the mycelia of the fungus that can detach. The toxin itself, **ergot,** is an alkaloid that can cause hallucinations resembling those produced by LSD; in fact, ergot is a natural source of LSD. Ergot also constricts capillaries and can cause gangrene of the limbs by preventing proper blood circulation in the body. Although *C. purpurea* still occasionally occurs on grains, modern milling usually removes the sclerotia.

Several other toxins are produced by fungi that grow on grains or other plants. For example, peanut butter is occasionally recalled because of excessive amounts of **aflatoxin,** a toxin that has carcinogenic properties. Aflatoxin is produced by the growth of the mold *Aspergillus flavus.* When ingested, the toxin might be altered in a human body to a mutagenic compound.

A few mushrooms produce toxins called **mycotoxins** (toxins produced by fungi). Examples are **phalloidin** and **amanitin,** produced by *Amanita phalloides* (am-an-ī'ta fal-loi'dēz), commonly known as the death angel. These neurotoxins are so potent that ingestion of the *Amanita* mushroom may result in death.

Protozoa

The presence of protozoa and their waste products often produces disease symptoms in the host (see Table 12.6, page 364). Some protozoa, such as *Plasmodium,* the causative agent of malaria, invade host cells and reproduce within them, causing their rupture. *Toxoplasma* attaches to macrophages and gains entry by phagocytosis. The parasite prevents normal acidification and digestion; thus, it can grow in the phagocytic vacuole. Other protozoa, such as *Giardia lamblia,* the causative agent of giardiasis, attach to host cells and digest the cells and tissue fluids.

Some protozoa can evade host defenses and cause disease for very long periods of time. For example, *Giardia,* which causes diarrhea, and *Trypanosoma,* which causes African trypanosomiasis (sleeping sickness), both have a mechanism that enables them to stay one step ahead of the host's immune system. The immune system is alerted to recognize foreign substances called antigens; the presence of antigens causes the immune system to produce antibodies designed to destroy them (see Chapter 17). When *Trypanosoma* is introduced into the bloodstream by a tsetse fly, it produces and displays a specific antigen. In response, the body produces antibodies against that antigen. However, within 2 weeks, the microbe stops displaying the original antigen and instead produces and displays a different one. Thus, the original antibodies are no longer effective. Because the microbe can make up to 1000 different antigens, such an infection can last for decades.

Helminths

The presence of helminths also often produces disease symptoms in a host (see Table 12.6, page 364). Some of these organisms actually use host tissues for their own growth or produce large parasitic masses; the resulting cellular damage evokes the symptoms. An example is the roundworm *Wuchereria bancrofti* (vū-kèr-ār'ē-ä ban-krof'tē), the causative agent of elephantiasis. This parasite blocks lymphatic circulation, leading to an accumulation of lymph and eventually causing grotesque swelling of the legs and other body parts. Waste products of the metabolism of these parasites can also contribute to the symptoms of a disease.

Algae

A few species of algae produce neurotoxins. For example, some genera of dinoflagellates, such as *Alexandrium,* are important medically because they produce a neurotoxin called **saxitoxin.** Although mollusks that feed on the dinoflagellates that produce saxitoxin show no symptoms of disease, people who eat the mollusks develop symptoms similar to botulism called paralytic shellfish poisoning. Public health agencies frequently prohibit human consumption of mollusks during red tides.

★ ★ ★

In the next chapter, we will examine a group of nonspecific defenses of the host against disease. But before proceeding, examine Figure 15.8 carefully. It summarizes some key concepts of the microbial mechanisms of pathogenicity we have discussed in this chapter.

FIGURE 15.8 Microbial mechanisms of pathogenicity. A summary of how microorganisms cause disease.

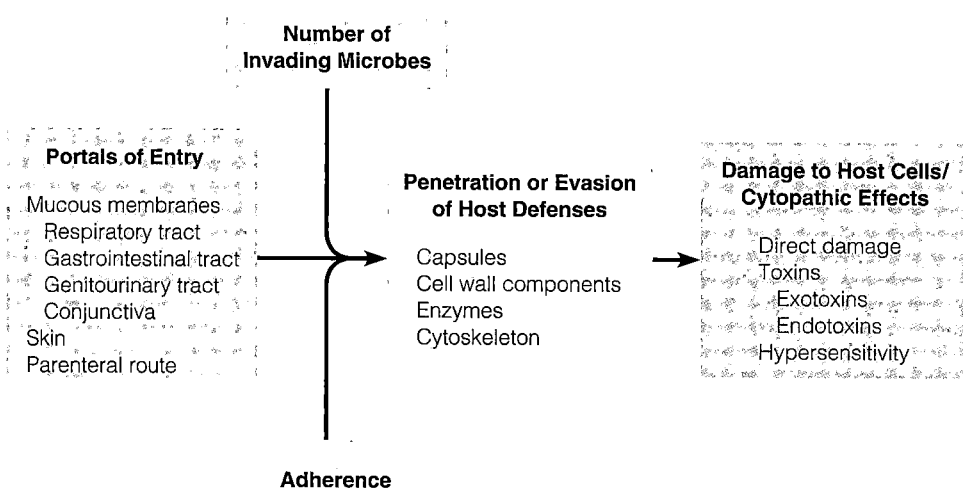

Study Outline

INTRODUCTION (p. 435)

1. Pathogenicity is the ability of a pathogen to produce a disease by overcoming the defenses of the host.
2. Virulence is the degree of pathogenicity.

HOW MICROORGANISMS ENTER A HOST (pp. 435–439)

1. The specific route by which a particular pathogen gains access to the body is called its portal of entry.

Portals of Entry (pp. 435–436)

1. Many microorganisms can penetrate mucous membranes of the conjunctiva and the respiratory, gastrointestinal, and genitourinary tracts.
2. Microorganisms that are inhaled with droplets of moisture and dust particles gain access to the respiratory tract.
3. The respiratory tract is the most common portal of entry.
4. Microorganisms that gain access via the genitourinary tract can enter the body through mucous membranes.
5. Microorganisms enter the gastrointestinal tract via food, water, and contaminated fingers.
6. Most microorganisms cannot penetrate intact skin; they enter hair follicles and sweat ducts.
7. Some fungi infect the skin itself.
8. Some microorganisms can gain access to tissues by inoculation through the skin and mucous membranes in bites, injections, and other wounds. This route of penetration is called the parenteral route.

The Preferred Portal of Entry (p. 436)

1. Many microorganisms can cause infections only when they gain access through their specific portal of entry.

Numbers of Invading Microbes (p. 436)

1. Virulence can be expressed as LD_{50} (lethal dose for 50% of the inoculated hosts) or ID_{50} (infectious dose for 50% of the inoculated hosts).

Adherence (pp. 436–439)

1. Surface projections on a pathogen called adhesins (ligands) adhere to complementary receptors on the host cells.
2. Adhesins can be glycoproteins or lipoproteins and are frequently associated with fimbriae.
3. Mannose is the most common receptor.

HOW BACTERIAL PATHOGENS PENETRATE HOST DEFENSES (pp. 439–441)

Capsules (p. 439)

1. Some pathogens have capsules that prevent them from being phagocytized.

Components of the Cell Wall (pp. 439–440)

1. Proteins in the cell wall can facilitate adherence or prevent a pathogen from being phagocytized.
2. Some microbes can reproduce inside phagocytes.

Enzymes (p. 440)

1. Leukocidins destroy neutrophils and macrophages.
2. Hemolysins lyse red blood cells.
3. Local infections can be protected in a fibrin clot caused by the bacterial enzyme coagulase.
4. Bacteria can spread from a focal infection by means of kinases (which destroy blood clots), hyaluronidase (which destroys a mucopolysaccharide that holds cells together), and collagenase (which hydrolyzes connective tissue collagen).

Penetration into the Host Cell Cytoskeleton
(pp. 440–441)

1. *Salmonella* bacteria produce invasins, proteins that cause the actin of the host cell's cytoskeleton to form a basket to carry the bacteria into the cell.

HOW BACTERIAL PATHOGENS DAMAGE HOST CELLS (pp. 441–445)

Direct Damage (p. 441)

1. Host cells can be destroyed when pathogens metabolize and multiply inside the host cells.

The Production of Toxins (pp. 441–445)

1. Poisonous substances produced by microorganisms are called toxins; toxemia refers to the presence of toxins in the blood. The ability to produce toxins is called toxigenicity.
2. Exotoxins are produced by bacteria and released into the surrounding medium. Exotoxins, not the bacteria, produce the disease symptoms.
3. Antibodies produced against exotoxins are called antitoxins.
4. Cytotoxins include diphtheria toxin (which inhibits protein synthesis) and erythrogenic toxins (which damage capillaries).
5. Neurotoxins include botulinum toxin (which prevents nerve transmission) and tetanus toxin (which prevents inhibitory nerve transmission).
6. *Vibrio cholerae* toxin and staphylococcal enterotoxin are enterotoxins, which induce fluid and electrolyte loss from host cells.
7. Endotoxins are lipopolysaccharides (LPS), the lipid A component of the cell wall of gram-negative bacteria.
8. Bacterial cell death, antibiotics, and antibodies may cause the release of endotoxins.
9. Endotoxins cause fever (by inducing the release of interleukin-1) and shock (because of a TNF-induced decrease in blood pressure).
10. Endotoxins allow bacteria to cross the blood–brain barrier.

11. The *Limulus* amoebocyte lysate (LAL) assay is used to detect endotoxins in drugs and on medical devices.

Plasmids, Lysogeny, and Pathogenicity (p. 445)

1. Plasmids may carry genes for antibiotic resistance, toxins, capsules, and fimbriae.
2. Lysogenic conversion can result in bacteria with virulence factors, such as toxins or capsules.

PATHOGENIC PROPERTIES OF NONBACTERIAL MICROORGANISMS (pp. 446–450)

Viruses (pp. 446–448)

1. Viruses avoid the host's immune response by growing inside cells.
2. Viruses gain access to host cells because they have attachment sites for receptors on the host cell.
3. Visible signs of viral infections are called cytopathic effects (CPE).
4. Some viruses cause cytocidal effects (cell death), and others cause noncytocidal effects (damage but not death).
5. Cytopathic effects include the stopping of mitosis, lysis, the formation of inclusion bodies, cell fusion, antigenic changes, chromosomal changes, and transformation.

Fungi, Protozoa, Helminths, and Algae (pp. 448–449)

1. Symptoms of fungal infections can be caused by capsules, toxins, and allergic responses.
2. Symptoms of protozoan and helminthic diseases can be caused by damage to host tissue or by the metabolic waste products of the parasite.
3. Some protozoa change their surface antigens while growing in a host so that the host's antibodies don't kill the protozoa.
4. Some algae produce neurotoxins that cause paralysis when ingested by humans.

Study Questions

REVIEW

1. List three portals of entry, and describe how microorganisms gain access through each.
2. Compare pathogenicity with virulence.
3. Explain how drugs that bind each of the following would affect pathogenicity:
   a. mannose on human cell membranes
   b. *Neisseria gonorrhoeae* fimbriae
   c. *Streptococcus pyogenes* M protein
4. Define cytopathic effects, and give five examples.
5. Compare and contrast the following aspects of endotoxins and exotoxins: bacterial source, chemistry, toxicity, and pharmacology. Give an example of each toxin.

6. How are capsules and cell wall components related to pathogenicity? Give specific examples.
7. Describe how hemolysins, leukocidins, coagulase, kinases, and hyaluronidase might contribute to pathogenicity.
8. Describe the factors contributing to the pathogenicity of fungi, protozoa, and helminths.
9. Which of the following genera is the most infectious?

Genus	ID_{50}	Genus	ID_{50}
Legionella	1 cell	*Shigella*	200 cells
Salmonella	10^5 cells	*Treponema*	52 cells

10. The LD_{50} of botulinum toxin is 0.000025 μg. The LD_{50} of *Salmonella* toxin is 200 μg. Which of these is the more potent toxin? How can you tell from the LD_{50} values?

11. Food poisoning can be divided into two categories: food infection and food intoxication. On the basis of toxin production by bacteria, explain the difference between these two categories.

12. How can viruses and protozoa avoid being killed by the host's immune response?

MULTIPLE CHOICE

1. The removal of plasmids reduces virulence in which of the following organisms?
   a. *Clostridium tetani*
   b. *Escherichia coli*
   c. *Staphylococcus aureus*
   d. *Streptococcus mutans*
   e. *Clostridium botulinum*

2. What is the LD_{50} for the bacterial toxin tested in the example below?

Dilution	No. of Animals Died	No. of Animals Survived
a. 6 μg/kg	0	6
b. 12.5 μg/kg	0	6
c. 25 μg/kg	3	3
d. 50 μg/kg	4	2
e. 100 μg/kg	6	0

3. Which of the following is *not* a portal of entry for pathogens?
   a. mucous membranes of the respiratory tract
   b. mucous membranes of the gastrointestinal tract
   c. skin
   d. blood
   e. parenteral route

4. All of the following can occur during bacterial infection. Which would prevent all of the others?
   a. vaccination against fimbriae
   b. phagocytosis
   c. inhibition of phagocytic digestion
   d. destruction of adhesins
   e. alteration of cytoskeleton

5. The ID_{50} for *Campylobacter* sp. is 500 cells; the ID_{50} for *Cryptosporidium* sp. is 100 cells. Which of the following statements is *not* true?
   a. Both microbes are pathogens.
   b. Both microbes produce infections in 50% of the inoculated hosts.
   c. *Cryptosporidium* is more virulent than *Campylobacter*.
   d. *Campylobacter* and *Cryptosporidium* are equally virulent; they cause infections in the same number of test animals.
   e. The severity of infections caused by *Campylobacter* and *Cryptosporidium* cannot be determined by the information provided.

6. An encapsulated bacterium can be virulent because the capsule
   a. resists phagocytosis.
   b. is an endotoxin.
   c. destroys host tissues.
   d. interferes with physiological processes.
   e. has no effect; since many pathogens do not have capsules, capsules do not contribute to virulence.

7. A drug that binds to mannose on human cells would prevent
   a. the entrance of *Vibrio* enterotoxin.
   b. the attachment of pathogenic *E. coli*.
   c. the action of botulinum toxin.
   d. streptococcal pneumonia.
   e. the action of diphtheria toxin.

8. The earliest smallpox vaccines were infected tissue rubbed into the skin of a healthy person. The recipient of such a vaccine usually developed a mild case of smallpox, recovered, and was immune thereafter. The most likely reason this vaccine did not kill more people is:
   a. Skin is the wrong portal of entry for smallpox.
   b. The vaccine consisted of a mild form of the virus.
   c. Smallpox is normally transmitted by skin-to-skin contact.
   d. Smallpox is a virus.
   e. The virus mutated.

9. Which of the following does *not* represent the same mechanism for avoiding host defenses as the others?
   a. Rabiesvirus attaches to the receptor for the neurotransmitter acetylcholine.
   b. *Salmonella* attaches to the receptor for epidermal growth factor.
   c. EB virus binds to the host receptor for C3.
   d. Surface protein genes in *Neisseria gonorrhoeae* mutate frequently.
   e. none of the above

10. Which of the following statements is true?
    a. The primary goal of a pathogen is to kill its host.
    b. Evolution selects for the most virulent pathogens.
    c. A successful pathogen doesn't kill its host before it is transmitted.
    d. A successful pathogen never kills its host.

CRITICAL THINKING

1. How can plasmids and lysogeny turn the normally harmless *E. coli* into a pathogen? Using the graph below, showing confirmed cases of enteropathogenic *E. coli*, what is the usual portal of entry for this bacterium?

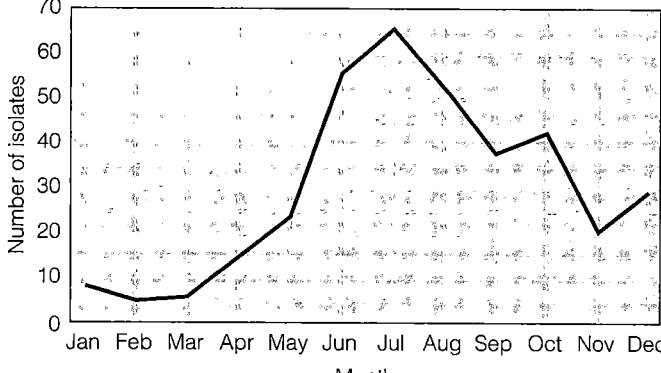

2. The cyanobacterium *Microcystis aeruginosa* produces a peptide that is toxic to humans. According to the graph below, when is this bacterium most toxic?

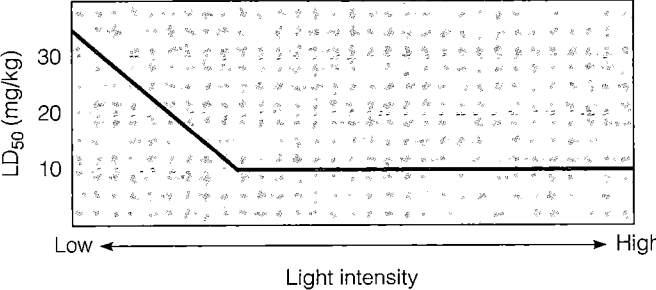

3. How do each of the following strategies contribute to the virulence of the pathogen? What disease does each organism cause?

Strategy	Pathogen
Changes its cell wall after entry into host	*Yersinia pestis*
Uses urea to produce ammonia	*Helicobacter pylori*
Causes host to make more receptors	Rhinovirus
Binds to site for epidermal growth factor on human cells	*Salmonella*

4. When injected into rats, the ID_{50} for *Salmonella typhimurium* is 10^6 cells. If sulfonamides are injected with the salmonellae, the ID_{50} is 35 cells. Explain the change in ID_{50} value.

CLINICAL APPLICATIONS

1. On July 8, a woman was given an antibiotic for presumptive sinusitis. However, her condition worsened, and she was unable to eat for 4 days because of severe pain and tightness of the jaw. On July 12, she was admitted to a hospital with severe facial spasms. She reported that on July 5 she had incurred a puncture wound at the base of her big toe; she cleaned the wound but did not seek medical attention. What caused her symptoms? Was her condition due to an infection or an intoxication? Can she transmit this condition to another person?

2. Explain whether each of the following examples is a food infection or intoxication. What is the probable etiological agent in each case?
   a. Eighty-two people who ate shrimp at a dinner in Port Allen, Louisiana, developed diarrhea, cramps, weakness, nausea, chills, headache, and fever from 4 hours to 2 days after eating.
   b. Two people in Vermont who ate barracuda caught in Florida developed malaise, nausea, blurred vision, breathing

difficulty, and numbness 3–6 hours after eating.

3. Washwater containing *Pseudomonas* was sterilized and used to wash cardiac catheters. Three patients undergoing cardiac catheterization developed fever, chills, and hypotension. The water and catheters were sterile. Why did the patients show these reactions?

4. Cancer patients undergoing chemotherapy are normally *more* susceptible to infections. However, a patient receiving an antitumor drug that inhibited cell division was resistant to *Salmonella*. Provide a possible mechanism for the resistance.

Learning with Technology

MP Don't forget to go to The Microbiology Place website (http://www.microbiologyplace.com) to take the practice tests, explore the interactive activity and case study, and check out the news articles and web links for this chapter.

STT Remember there is also a quiz for this chapter on the Microbiology Interactive Student Tutorial CD-ROM.

VU Enter the Virtual Lab, click the arrow next to the Session field, click Textbook Exercises, and select Chapter 15. Read the Case Study carefully, identify the unknown, and use what you learn to answer the following questions (consult your textbook for additional information):

1. Which of the following would describe the toxin most likely produced by this organism? (a) enterotoxin; (b) cytotoxin; (c) neurotoxin; (d) endotoxin

2. Which of the following would best predict symptomology observed in an infection caused by this organism? (a) red skin rash, sore throat, and fever; (b) blurred vision, vomiting, and watery diarrhea; (c) uncontrolled muscle contraction, fever, and convulsions; (d) fever, cramps, and diarrhea; (e) coughing, sneezing, and headache

3. Which of the following would be likely to be found in the microbe's cell wall? (circle all that apply): (a) M protein; (b) Lipid A; (c) Waxes; (d) Teichoic acids; (e) Mycolic acid

4. Which of the following describe the toxin produced by this organism? (circle all that apply): (a) heat stable; (b) can lead to septic shock; (c) impact frequently lessened by immunization; (d) low LD_{50}; (e) not released until cell death

5. What is the likely portal of entry in diseases caused by this organism?

chapter 16

Nonspecific Defenses of the Host

A phagocyte. Phagocytes provide protection against pathogens. This macrophage is engulfing bacteria which will prevent the bacteria from colonizing and causing disease.

rom our discussion to this point, you can see that pathogenic microorganisms are endowed with special properties that enable them to cause disease if given the right opportunity. If microorganisms never encountered resistance from the host, we would constantly be ill and would eventually die of various diseases. But in most cases, our body's defenses prevent this from happening. Some of these defenses are designed to keep out microorganisms altogether, other defenses remove the microorganisms if they do get in, and still others combat them if they remain inside. Our ability to ward off disease through our defenses is called **resistance.** Vulnerability or lack of resistance is known as **susceptibility.**

In discussing resistance, we will divide our body defenses into two general kinds: nonspecific and specific (Figure 16.1). **Nonspecific resistance** refers to defenses that protect us against any pathogen, regardless of species. These include a first line of defense (skin and mucous membranes) and a second line of defense (phagocytes, inflammation, fever, and antimicrobial substances). **Specific resistance,** or **immunity,** is a third line of defense the body provides against particular pathogens. Specific defenses are based on specialized cells of the immune system called lymphocytes (a type of white blood cell) and the production of specific proteins called antibodies. Chapter 17 deals with specific resistance. In this chapter, we consider nonspecific resistance.

Skin and Mucous Membranes

Learning Objectives

- Describe the role of the skin and mucous membranes in nonspecific resistance.
- Differentiate mechanical from chemical factors, and list five examples of each.
- Describe the role of normal microbiota in nonspecific resistance.

The skin and mucous membranes are the body's first line of defense against pathogens. This function results from both mechanical and chemical factors.

Mechanical Factors

The intact **skin** is one of the human body's largest organs in terms of surface area. It consists of two distinct portions: the dermis and the epidermis (Figure 16.2). The **dermis,**

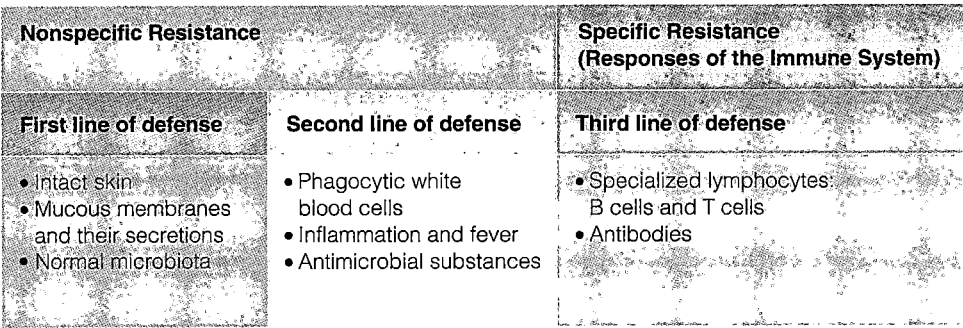

Nonspecific Resistance		Specific Resistance (Responses of the Immune System)
First line of defense	Second line of defense	Third line of defense
• Intact skin • Mucous membranes and their secretions • Normal microbiota	• Phagocytic white blood cells • Inflammation and fever • Antimicrobial substances	• Specialized lymphocytes: B cells and T cells • Antibodies

FIGURE 16.1 An overview of the body's defenses. Nonspecific resistance refers to defenses against any pathogen, regardless of species; specific resistance refers to defenses against a specific pathogen.

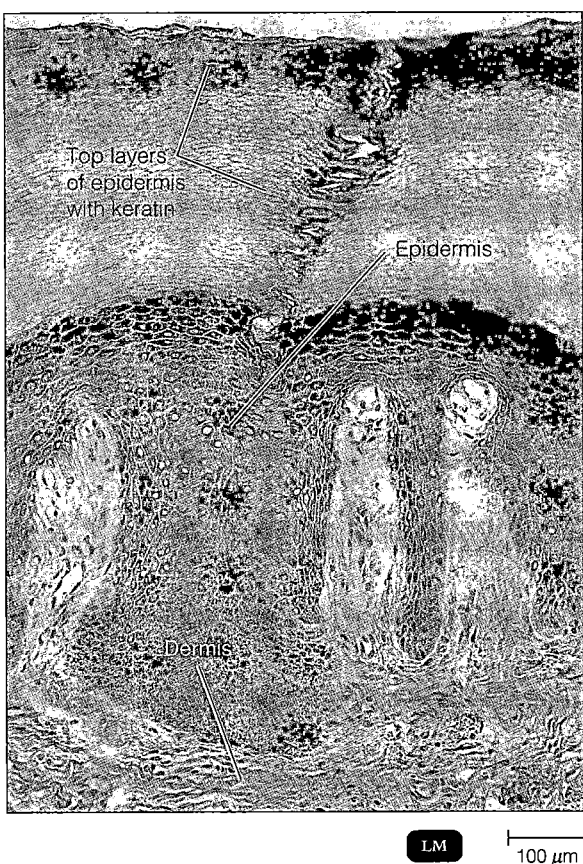

FIGURE 16.2 A section through human skin. The thin layers at the top of this photomicrograph contain keratin. These layers and the darker purple cells beneath them make up the epidermis. The lighter purple material near the bottom is the dermis.

■ **The intact skin and mucous membranes are components of the first line of defense against pathogens.**

the skin's inner, thicker portion, is composed of connective tissue. The **epidermis,** the outer, thinner portion, is in direct contact with the external environment. The epidermis consists of layers of continuous sheets of tightly packed epithelial cells with little or no material between the cells. Certain cells in the epidermis, called Langerhans cells, participate in immunity by assisting lymphocytes

called helper T (T$_H$) cells. Langerhans cells process and present antigens to helper T cells, which enhance antibody production, secrete substances that stimulate proliferation of T and B cells, and stimulate the inflammatory response. (Helper T cells will be discussed in Chapter 17.) The top layer of epidermal cells is dead and contains a protective protein called **keratin.** When the skin is moist, as in hot, humid climates, skin infections are quite common, especially fungal infections such as athlete's foot. These fungi hydrolyze keratin when water is available.

If we consider the closely packed cells, continuous layering, and presence of keratin, we can see why the intact skin provides such a formidable barrier to the entrance of microorganisms. The intact surface of healthy epidermis is rarely, if ever, penetrated by microorganisms. But when the epithelial surface is broken, a subcutaneous (below-the-skin) infection often develops. The bacteria most likely to cause infection are the staphylococci that normally inhabit the epidermis, hair follicles, and sweat and oil glands of the skin. Infections of the skin and underlying tissues frequently occur as a result of burns, cuts, stab wounds, or other conditions that break the skin.

Mucous membranes also consist of an epithelial layer and an underlying connective tissue layer. Although mucous membranes inhibit the entrance of many microorganisms, they offer less protection than the skin. Mucous membranes line the entire gastrointestinal, respiratory, and genitourinary tracts. The epithelial layer of a mucous membrane secretes a fluid (mucus), which prevents the tracts from drying out. Some pathogens that can thrive on the moist secretions of a mucous membrane are able to penetrate the membrane if the microorganism is present in sufficient numbers. *Treponema pallidum* is such a pathogen. This penetration may be facilitated by toxic substances produced by the microorganism, prior injury by viral infection, or mucosal irritation.

Besides the physical barrier presented by the skin and mucous membranes, several other mechanical factors help protect certain epithelial surfaces. One such mechanism that protects the eyes is the **lacrimal apparatus,** a

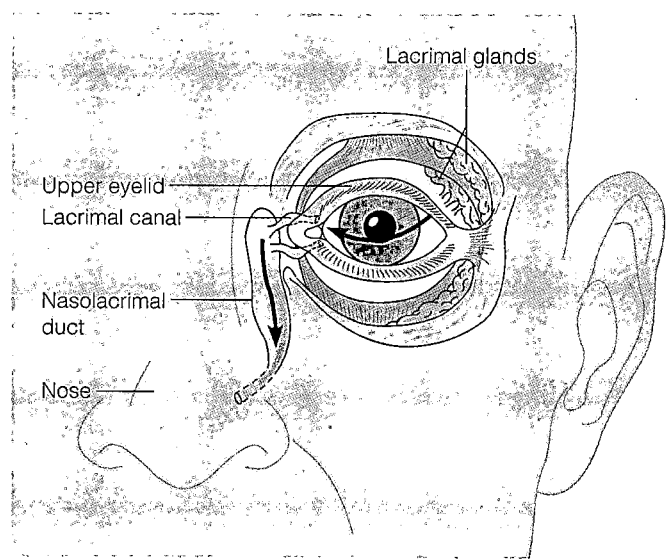

FIGURE 16.3 The lacrimal apparatus. The washing action of the tears is shown by the arrow passing over the surface of the eyeball. Tears produced by the lacrimal glands pass across the surface of the eyeball into two small holes that convey the tears into the lacrimal canals and the nasolacrimal duct.

■ The lacrimal apparatus protects the eye by continually producing tears that wash away microorganisms.

group of structures that manufactures and drains away tears (Figure 16.3). The lacrimal glands, located toward the upper, outermost portion of each eye socket, produce the tears and pass them under the upper eyelid. From here, tears pass toward the corner of the eye near the nose and into two small holes that lead through tubes (lacrimal canals) to the nose. The tears are spread over the surface of the eyeball by blinking. Normally, the tears evaporate or pass into the nose as fast as they are produced. This continual washing action helps keep microorganisms from settling on the surface of the eye. If an irritating substance or large numbers of microorganisms come in contact with the eye, the lacrimal glands start to secrete heavily, and the tears accumulate more rapidly than they can be carried away. This excessive production is a protective mechanism because the excess tears dilute and wash away the irritating substance or microorganisms.

In a cleansing action very similar to that of tears, **saliva** is produced by the salivary glands. Saliva helps dilute the numbers of microorganisms and wash them from both the surface of the teeth and the mucous membrane of the mouth. This helps prevent colonization by microbes.

The respiratory and gastrointestinal tracts have many mechanical forms of defense. **Mucus** is a slightly viscous (thick) secretion produced by goblet cells of a mucous membrane that traps many of the microorganisms that

enter the respiratory and gastrointestinal tracts. Goblet cells are modified epithelial cells that produce mucus. The mucous membrane of the nose also has mucus-coated hairs that filter inhaled air and trap microorganisms, dust, and pollutants. The cells of the mucous membrane of the lower respiratory tract are covered with cilia. By moving synchronously, these cilia propel inhaled dust and microorganisms that have become trapped in mucus upward toward the throat. This so-called **ciliary escalator** (Figure 16.4) keeps the mucus blanket moving toward the throat at a rate of 1–3 cm per hour; coughing and sneezing speed up the escalator. (Some substances in cigarette smoke are toxic to cilia and can seriously impair the functioning of the ciliary escalator.) Microorganisms are also prevented from entering the lower respiratory tract by a small lid of cartilage called the **epiglottis,** which covers the larynx (voicebox) during swallowing.

The cleansing of the urethra by the flow of **urine** is another mechanical factor that prevents microbial colonization in the genitourinary tract. **Vaginal secretions** likewise move microorganisms out of the female body.

Chemical Factors

Mechanical factors alone do not account for the high degree of resistance of skin and mucous membranes to microbial invasion. Certain chemical factors also play important roles.

Sebaceous (oil) glands of the skin produce an oily substance called **sebum** that prevents hair from drying and becoming brittle. Sebum also forms a protective film over the surface of the skin. One of the components of sebum is unsaturated fatty acids, which inhibit the growth of certain pathogenic bacteria and fungi. The low pH of the skin, between pH 3 and 5, is caused in part by the secretion of fatty acids and lactic acid. The skin's acidity probably discourages the growth of many other microorganisms.

Bacteria that live commensally on the skin decompose sloughed-off skin cells, and the resultant organic molecules and the end-products of their metabolism produce body odor. As we will see in Chapter 21, certain bacteria commonly found on the skin metabolize sebum, and this metabolism forms free fatty acids that cause the inflammatory response associated with acne. Isotretinoin (Accutane®), a derivative of vitamin A that prevents sebum formation, is a treatment for a very severe type of acne called cystic acne.

The sweat glands of the skin produce **perspiration,** which helps maintain body temperature, eliminate certain wastes, and flush microorganisms from the surface of the skin. Perspiration also contains **lysozyme,** an enzyme capable of breaking down cell walls of gram-positive bacteria and, to a lesser extent, gram-negative bacteria (see

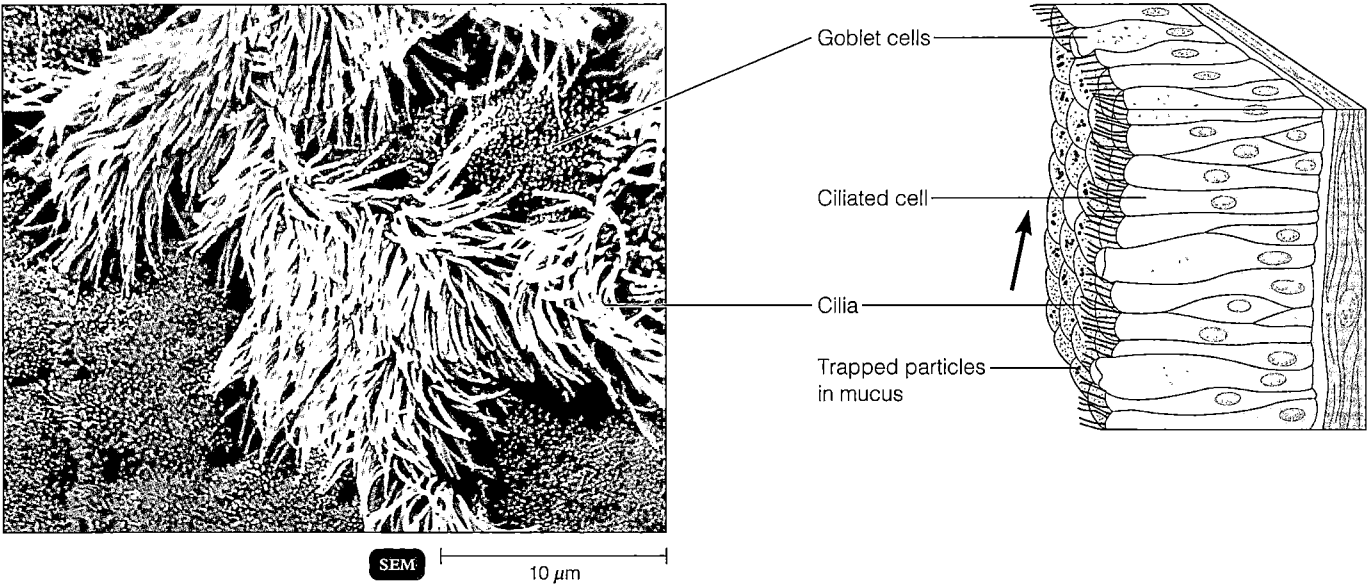

Goblet cells

Ciliated cell

Cilia

Trapped particles
in mucus

SEM ⊢————— 10 μm —————⊣

FIGURE 16.4 The ciliary escalator. Microorganisms in the lower respiratory tract
are trapped in mucus produced by goblet cells, then propelled upward by the synchro-
nized beating of cilia.

Figure 4.12 on page 87). Lysozyme is also found in tears,
saliva, nasal secretions, and tissue fluids, where it exhibits
its antimicrobial activity. Alexander Fleming was studying
lysozyme in 1929 when he accidentally discovered the antimicrobial effects of penicillin (see Figure 20.1).

Gastric juice is produced by the glands of the stomach. It is a mixture of hydrochloric acid, enzymes, and mucus. The very high acidity of gastric juice (pH 1.2–3.0) is
sufficient to preserve the usual sterility of the stomach.
This acidity destroys bacteria and most bacterial toxins,
except those of *Clostridium botulinum* and *Staphylococcus aureus.* However, many enteric pathogens are protected by
food particles and can enter the intestines via the gastrointestinal tract. The bacterium *Helicobacter pylori* neutralizes
stomach acid which allows the bacterium to grow in the
stomach. Its growth results in ulcers and gastritis.

Vaginal secretions are also slightly acidic, which discourages bacterial growth. Blood also contains antimicrobial chemicals. For example, iron-binding proteins, called
transferrins, inhibit bacterial growth by reducing the
amount of available iron. Iron is not only required for microbial growth (the synthesis of cytochromes and certain
enzymes), it also suppresses chemotaxis and phagocytosis.
Iron overload increases the risk of infection.

Normal Microbiota
and Nonspecific Resistance

In Chapter 14, several relationships between normal microbiota and host cells were described. Some of these re-

lationships assist in preventing the overgrowth of
pathogens and thus may be considered components of
nonspecific defense. For example, in microbial antagonism, the normal microbiota prevent pathogens from colonizing the host by competing with them for nutrients,
by producing substances that are harmful to the pathogens, and by altering conditions that affect the survival
of the pathogens, such as pH and oxygen availability. The
presence of normal microbiota in the vagina, for example, alters pH to prevent overpopulation by *Candida albicans,* a pathogenic yeast that causes vaginitis. In the large
intestine, *E. coli* produce bacteriocins that inhibit the
growth of *Salmonella* and *Shigella.*

In commensalism, one organism uses the body of a
larger organism as its physical environment and may
make use of the body to obtain nutrients. Thus in commensalism, one organism benefits while the other is unaffected. Most microbes that are part of the commensal
microbiota are found on the skin and in the gastrointestinal tract. The majority of such microbes are bacteria that
have highly specialized attachment mechanisms and precise environmental requirements for survival. Normally,
such microbes are harmless, but may cause disease if their
environmental conditions change. These opportunistic
pathogens include *E. coli* and *S. aureus.* In some cases,
however, commensals can benefit the host by preventing
more pathogenic microbes from colonizing, as occurs
with the intestinal microbiota; this is called *competititve exclusion.* Actually, commensalism may become mutualism,
in which both organisms benefit from the association.

table 16.1	Formed Elements in Blood		
Type of Cell		**Numbers per microliter (μl) or cubic mm (mm³)**	**Function**
Erythrocytes (red blood cells)		4.8–5.4 million	Transport of O_2 and CO_2
Leukocytes (white blood cells)		5000–10,000	
A. Granulocytes (stained) 1. Neutrophils (PMNs) (60–70% of leukocytes)			Phagocytosis
2. Basophils (0.5–1%)			Production of histamine
3. Eosinophils (2–4%)			Production of toxic proteins against certain parasites; some phagocytosis
B. Agranulocytes (stained) 1. Monocytes (3–8%)			Phagocytosis (when they mature into macrophages)
2. Lymphocytes (20–25%)			Antibody production (descendants of B lymphocytes); cell-mediated immunity (T lymphocytes)*
Platelets		150,000–400,000	Blood clotting

*Discussed in Chapter 17.

Phagocytosis

Learning Objective

- *Define phagocytosis and phagocyte.*

Phagocytosis (from Greek words meaning eat and cell) is the ingestion of a microorganism or any particulate matter by a cell. We have previously mentioned phagocytosis as the method of nutrition of certain protozoa. In this chapter, phagocytosis is discussed as a means by which cells in the human body counter infection—the second line of defense. The cells that perform this function are collectively called **phagocytes,** all of which are types of white blood cells or derivatives of white blood cells. Before we look at phagocytes specifically,

let's consider the various components of the blood more generally.

Formed Elements in Blood

Learning Objectives

- *Classify phagocytic cells, and describe the roles of granulocytes and monocytes.*
- *Define differential white blood cell count.*

Blood consists of a fluid called **plasma,** which contains **formed elements**—that is, cells and cell fragments (Table 16.1). Of the cells listed in Table 16.1, those that concern us at present are the **leukocytes,** or white blood cells (Figure 16.5).

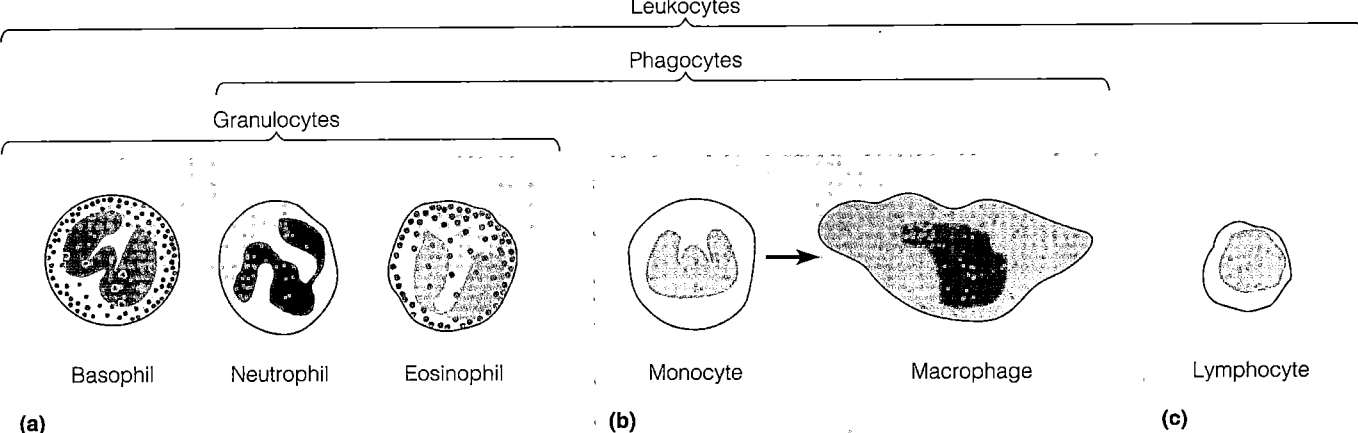

FIGURE 16.5 Principal kinds of leukocytes. **(a)** Granulocytes can be distinguished by staining. **(b)** Monocytes mature into actively phagocytic macrophages. Neutrophils, eosinophils, and macrophages are the phagocytic cells of nonspecific resistance. **(c)** Lymphocytes are important in specific resistance.

■ All phagocytes are leukocytes or cells derived from them.

During many kinds of infections, especially bacterial infections, the total number of white blood cells increases; this increase is called *leukocytosis*. During the active stage of infection, the leukocyte count might double, triple, or quadruple, depending on the severity of the infection. Diseases that might cause such an elevation in the leukocyte count include meningitis, infectious mononucleosis, appendicitis, pneumococcal pneumonia, and gonorrhea. Other diseases, such as salmonellosis and brucellosis, and some viral and rickettsial infections, may cause a *decrease* in the leukocyte count, called *leukopenia*. Leukopenia may be related to either impaired white blood cell production or the effect of increased sensitivity of white blood cell membranes to damage by complement, antimicrobial plasma proteins discussed later in the chapter. Leukocyte increase or decrease can be detected by a **differential white blood cell count,** which is a calculation of the percentage of each kind of white cell in a sample of 100 white blood cells. The percentages in a normal differential white blood cell count are shown in parentheses in the first column of Table 16.1.

Leukocytes are divided into two main categories based on their appearance under a light microscope: granulocytes and agranulocytes. **Granulocytes** owe their name to the presence of large granules in their cytoplasm that can be seen under a light microscope after staining. They are differentiated into three types on the basis of how the granules stain. The granules of **neutrophils** stain pale lilac with a mixture of acidic and basic dyes, those of **basophils** stain blue-purple with the basic dye methylene blue, and those of **eosinophils** stain red or orange with the acidic dye eosin.

Neutrophils are also commonly called *polymorphonuclear leukocytes (PMNs),* or *polymorphs.* (The term *polymorphonuclear* refers to the fact that the nuclei of neutrophils contain two to five lobes.) Neutrophils, which are highly phagocytic and motile, are active in the initial stages of an infection. They have the ability to leave the blood, enter an infected tissue, and destroy microbes and foreign particles.

The role of basophils is not clear. However, they release substances, such as histamine, that are important in inflammation and allergic responses.

Eosinophils are somewhat phagocytic and also have the ability to leave the blood. Their major function is to produce toxic proteins against certain parasites, such as helminths. Although eosinophils are physically too small to ingest and destroy helminths, they can attach to the outer surface of the parasites and discharge peroxide ions that destroy them. (See Figure 17.17 on page 495.) Their number increases significantly during certain parasitic worm infections and hypersensitivity (allergy) reactions.

Agranulocytes also have granules in their cytoplasm but they are not visible under the light microscope after staining. **Monocytes** are not actively phagocytic until they leave circulating blood, enter body tissues, and mature into **macrophages.** In fact, the maturation and proliferation of macrophages (along with lymphocytes) is one factor responsible for the swelling of lymph nodes during an infection. As blood and lymph that contain microorganisms pass through organs with macrophages, the microorganisms are removed by phagocytosis. Macrophages also dispose of worn out blood cells.

FIGURE 16.6 Components of the lymphatic system. Fluid circulating between tissue cells .(interstitial fluid) is picked up by lymphatic vessels. The fluid is then called lymph. Microorganisms can enter the lymphatic vessels. Macrophages in the lymph nodes remove microorganisms from the lymph. Lymph is transported into larger lymphatic vessels, to be collected in the right lymphatic duct and thoracic duct. Eventually, the lymph is returned to the blood at the subclavian veins, just before the blood enters the heart.

■ Why do lymph nodes swell during an infection?

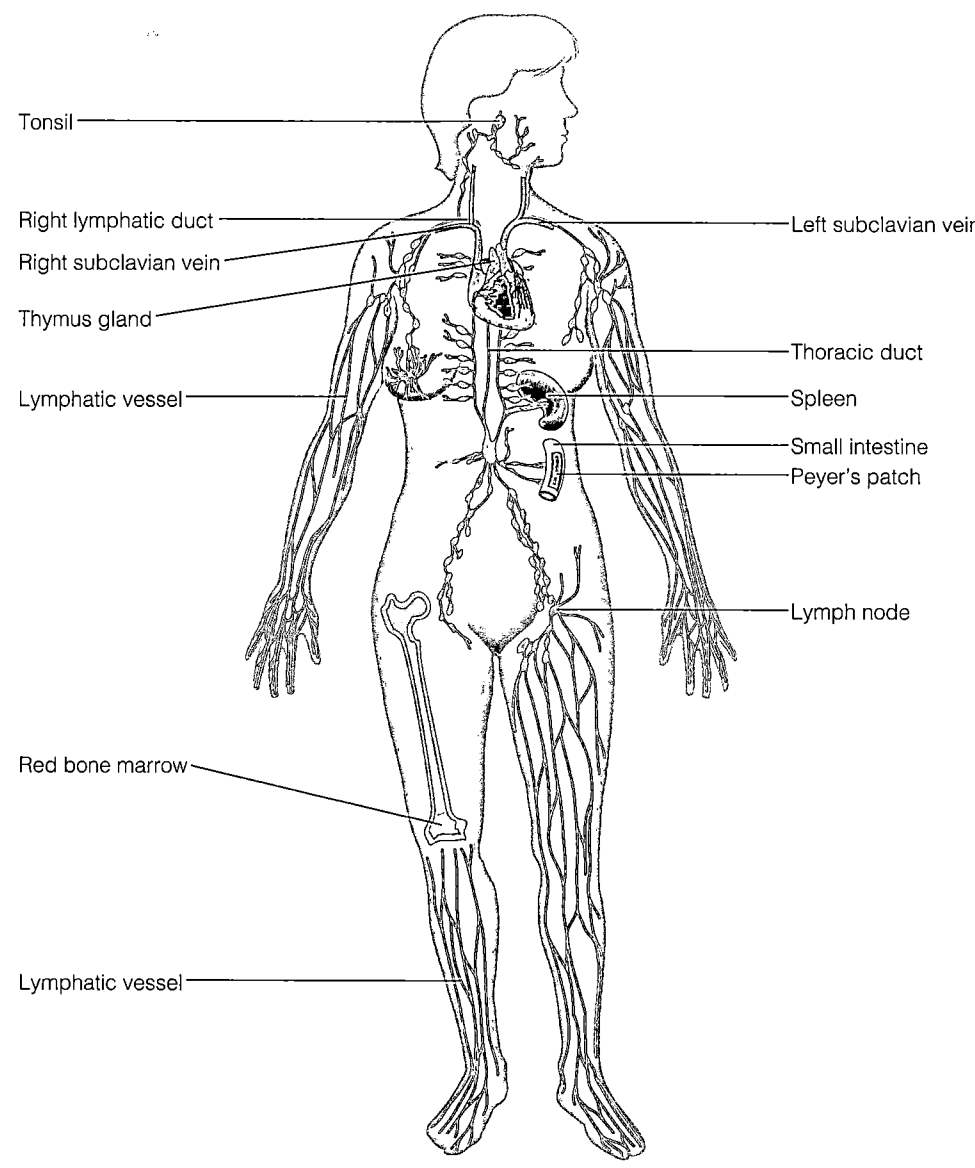

Tonsil

Right lymphatic duct

Right subclavian vein

Thymus gland

Lymphatic vessel

Red bone marrow

Lymphatic vessel

Left subclavian vein

Thoracic duct

Spleen

Small intestine

Peyer's patch

Lymph node

Lymphocytes are not phagocytic but play a key role in specific immunity (see Chapter 17). They occur in lymphoid tissues of the lymphatic system: in the tonsils, spleen, thymus gland, thoracic duct, red bone marrow, appendix, Peyer's patches of the small intestine, and lymph nodes in the respiratory, gastrointestinal, and reproductive tracts (Figure 16.6). They also circulate in the blood.

Phagocytes may be activated by components of bacteria such as lipid A or lipopolysaccharides (LPS). Among the more important activators are small protein hormones secreted by phagocytes and other cells involved in immunity called cytokines (see Chapter 17).

Actions of Phagocytic Cells

When an infection occurs, both granulocytes (especially neutrophils) and monocytes migrate to the infected area.

During this migration, monocytes enlarge and develop into actively phagocytic macrophages (Figure 16.7). Because these cells leave the blood and migrate through tissue to infected areas, they are called **wandering macrophages.** Some macrophages, called **fixed macrophages** or *histiocytes,* are located in certain tissues and organs of the body. Fixed macrophages are found in the liver (Kupffer's cells), lungs (alveolar macrophages), nervous system (microglial cells), bronchial tubes, spleen, lymph nodes, red bone marrow, and the peritoneal cavity surrounding abdominal organs. The various macrophages of the body constitute the **mononuclear phagocytic (reticuloendothelial) system.**

During the course of an infection, a shift occurs in the type of white blood cell that predominates in the bloodstream. Granulocytes, especially neutrophils, dominate dur-

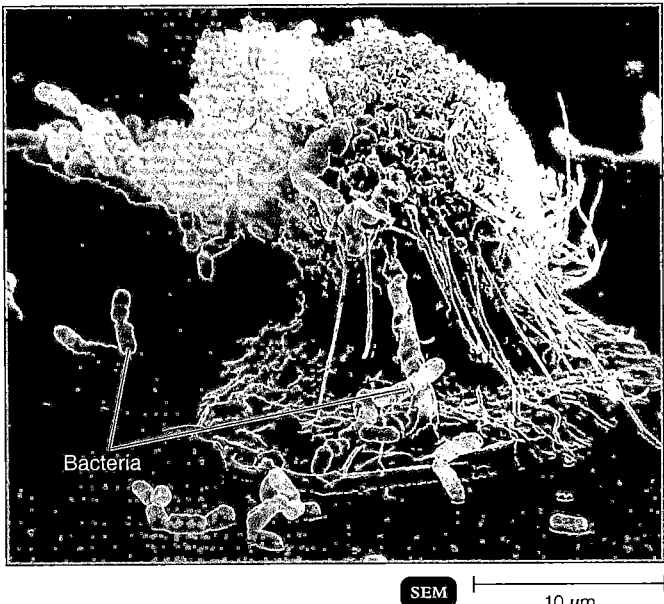

FIGURE 16.7 A macrophage engulfing rod-shaped bacteria.
Macrophages in the mononuclear phagocytic system remove microorganisms after the initial phase of infection.

table 16.2	Classification and Functions of Phagocytes	
Type of Phagocyte	Functional Cells	Functions
Granulocytes	Neutrophils and eosinophils	Phagocytic against microbes during the initial phase of infection
Agranulocytes	Mononuclear phagocytic system: wandering macrophages and fixed macrophages (histiocytes) developed from monocytes	Phagocytic against microbes as infection progresses and against worn-out blood cells as infection subsides; also involved in cell-mediated immunity (see Chapter 17)

ing the initial phase of bacterial infection, at which time they are actively phagocytic; this dominance is indicated by their increased number in a differential white blood cell count. However, as the infection progresses, the macrophages dominate; they scavenge and phagocytize remaining living bacteria and dead or dying bacteria. The increased number of monocytes (which develop into macrophages) is also reflected in a differential white blood cell count. In viral and fungal infections, macrophages predominate in all phases of defense.

Table 16.2 summarizes the different types of phagocytic cells and their functions.

The Mechanism of Phagocytosis

Learning Objective

■ *Describe the process of phagocytosis, and include the stages of adherence and ingestion.*

How does phagocytosis occur? For the convenience of study, we will divide phagocytosis into four main phases: chemotaxis, adherence, ingestion, and digestion (Figure 16.8).

Chemotaxis

① **Chemotaxis** is the chemical attraction of phagocytes to microorganisms. (The mechanism of chemotaxis is discussed in Chapter 4 on page 83.) Among the chemotactic chemicals that attract phagocytes are microbial

products, components of white blood cells and damaged tissue cells, and peptides derived from complement, a system of host defense discussed later in the chapter.

Adherence

As it pertains to phagocytosis, **adherence** is the attachment of the phagocyte's plasma membrane to the surface of the microorganism or other foreign material. In some instances, adherence occurs easily, and the microorganism is readily phagocytized. However, adherence can be inhibited by the presence of M protein or large capsules. As mentioned in Chapter 15 on page 439, the M protein of *Streptococcus pyogenes* inhibits the attachment of phagocytes to their surfaces and makes adherence more difficult. Organisms with large capsules include *Streptococcus pneumoniae* and *Haemophilus influenzae* type b. Heavily encapsulated microorganisms like these can be phagocytized only if the phagocyte traps the microorganism against a rough surface, such as a blood vessel, blood clot, or connective tissue fiber, from which the microbe cannot slide away.

Microorganisms can be more readily phagocytized if they are first coated with certain serum proteins that promote attachment of the microorganisms to the phagocyte. This coating process is called **opsonization.** The proteins that act as *opsonins* include some components of the complement system and antibody molecules (described later in this chapter and in Chapter 17).

Ingestion

② Following adherence, **ingestion** occurs. During this process, the plasma membrane of the phagocyte extends

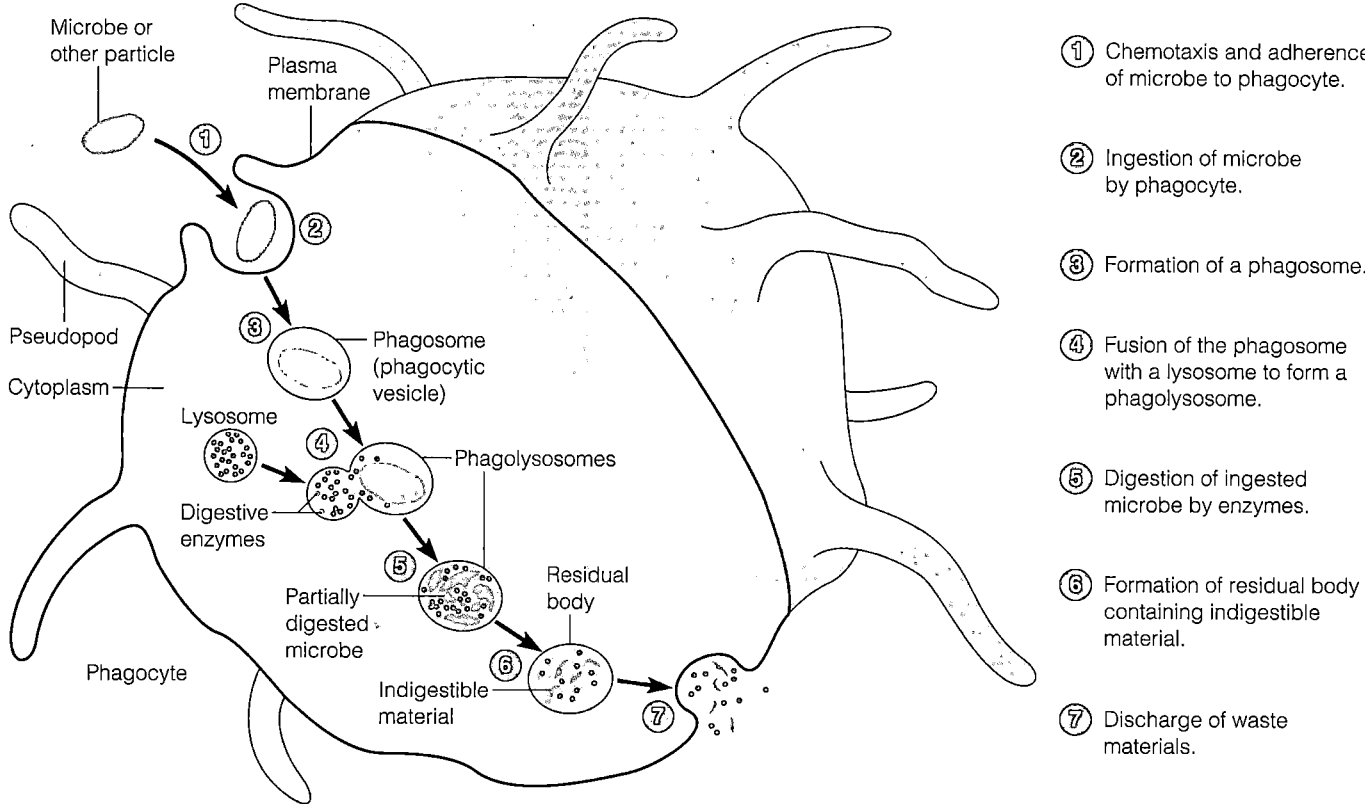

Microbe or other particle

Plasma membrane

Pseudopod

Cytoplasm

Lysosome

Digestive enzymes

Phagocyte

Phagosome (phagocytic vesicle)

Phagolysosomes

Partially digested microbe

Residual body

Indigestible material

① Chemotaxis and adherence of microbe to phagocyte.

② Ingestion of microbe by phagocyte.

③ Formation of a phagosome.

④ Fusion of the phagosome with a lysosome to form a phagolysosome.

⑤ Digestion of ingested microbe by enzymes.

⑥ Formation of residual body containing indigestible material.

⑦ Discharge of waste materials.

(a) Phases of phagocytosis

FIGURE 16.8 The mechanism of phagocytosis in a phagocyte.
The phases of phagocytosis are chemotaxis, adherence, ingestion, and digestion.

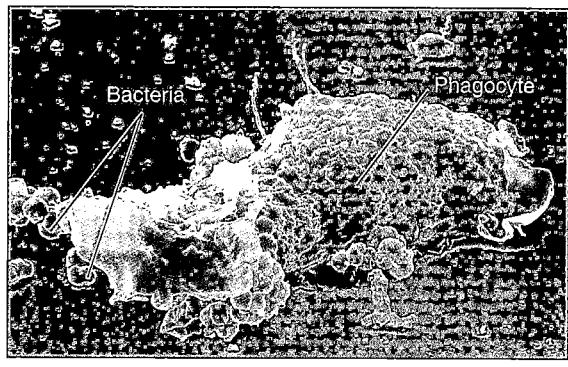

Bacteria   Phagocyte

(b) A phagocyte engulfing bacteria
(*Neisseria gonorrhoeae*)   SEM   ⊢——⊣ 4 μm

projections called **pseudopods** that engulf the microorganism. ③ Once the microorganism is surrounded, the pseudopods meet and fuse, surrounding the microorganism with a sac called a **phagosome** or *phagocytic vesicle*.

Digestion

In this phase of phagocytosis, the phagosome pinches off from the plasma membrane and enters the cytoplasm. Within the cytoplasm, it contacts lysosomes that contain digestive enzymes and bactericidal substances (see Chapter 4 on page 105). ④ Upon contact, the phagosome and lysosome membranes fuse to form a single, larger structure called a **phagolysosome.** ⑤ The contents of the phagolysosome take only 10–30 minutes to kill most types of bacteria.

Lysosomal enzymes that attack microbial cells directly include lysozyme, which hydrolyzes peptidoglycan in bacterial cell walls. A variety of other enzymes, such as lipases, proteases, ribonuclease, and deoxyribonuclease, hydrolyze other macromolecular components of microorganisms. The hydrolytic enzymes are most active at about pH 4, which is the phagolysosome's usual pH because of lactic acid produced by the phagocyte. Lysosomes also contain

enzymes that can produce toxic oxygen products such as superoxide radical (O_2^-), hydrogen peroxide (H_2O_2), singlet oxygen (O_2), and hydroxyl radical ($OH\cdot$) (see Chapter 6 on page 162). Other enzymes can make use of these toxic oxygen products in killing ingested microorganisms. For example, the enzyme myeloperoxidase converts chloride (Cl^-) ions and hydrogen peroxide into highly toxic hypochlorous acid (HOCl). The acid contains hypochlorous ions, which are found in household bleach and account for its antimicrobial activity (see Chapter 7 on page 197).

⑥ After enzymes have digested the contents of the phagolysosome brought into the cell by ingestion, the phagolysosome contains indigestible material and is

MICROBIOLOGY IN THE NEWS

Macrophages Say NO

In 1998, three scientists received the Nobel Prize for Physiology or Medicine for their discovery that nitric oxide (NO) is a powerful signaling molecule in the human body. NO is unusual for a chemical messenger in humans. Chemical messengers such as interferon or epinephrine are large organic molecules, but NO is composed of only two atoms, nitrogen and oxygen ($\cdot$N = O). Nitric oxide is not the same compound as the anesthetic nitrous oxide (N_2O), called laughing gas.

Robert Furchgott of the State University of New York Health Sciences Center, Ferid Murad of the University of Texas, Houston Health Science Center, and Louis Ignarro of the University of California, Los Angeles discovered that NO, released by endothelial cells lining blood vessels, is the molecule that relaxes the smooth muscle of the blood vessel to allow increased blood flow. Murad discovered that nitroglycerin dilates blood vessels because the nitroglycerin releases NO. Since its discovery, NO has been found to be involved

in a wide variety of processes from memory to immunity to impotence to blood pressure regulation.

High nitrate concentrations have been detected in the urine of people with certain types of infections. The reason is that macrophages also produce NO. The macrophage enzyme, NO synthase, produces NO from the amino acid arginine, and the NO is oxidized to nitrates and nitrites, which are excreted in urine.

Macrophages that are activated by γ interferon from T cells or bacterial endotoxin produce NO synthase, which makes NO. NO appears to kill microorganisms as well as tumor cells by inhibiting ATP production. Evidence for the role of NO is provided by experiments demonstrating that macrophages can't kill tumor cells or parasites if NO production is inhibited. Although NO is quickly destroyed after it is made, large amounts of NO will dilate blood vessels, which causes blood pressure to drop and the patient to go into shock. This is the reason gram-

negative bacteria cause septic shock. A future treatment for septic shock may involve inhibiting NO synthase.

NO SUPPRESSES AUTOIMMUNITY

Autoimmune diseases, such as lupus erythematosus and rheumatoid arthritis, in which a person makes antibodies against their own molecules, appear to be uncommon in less developed countries. Researchers in Australia suggest that the reason lies in the higher incidence of childhood infections in these countries. We now know that NO production is stimulated by infection and that NO can activate the suppressor cells that are responsible for preventing an immune response to self. Therefore, Bill Cowden at the Australian National University reasons that boosting NO production may prevent autoimmune diseases. He and his colleagues are developing a vaccine made from dead malarial parasites in the hope that the vaccine will prevent autoimmune reactions.

---

called a *residual body*. ⑦ This residual body then moves toward the cell boundary and discharges its wastes outside the cell. The box above describes another mechanism by which phagocytes can kill microorganisms and tumor cells.

Not all phagocytized microorganisms are killed by lysosomal enzymes. *Coxiella burnetii*, the causative agent of Q fever, actually requires the low pH inside a phagolysosome to replicate. The toxins of some microbes, such as toxin-producing staphylococci and the dental plaque bacterium *Actinobacillus*, can actually kill phagocytes. Some bacterial pathogens, such as *Listeria monocytogenes* (the causative agent of listeriosis) and *Shigella flexneri* (a causative agent of shigellosis) have enzymes that lyse phagolysosomes. They then use host cell cytoskeletal actin to propel themselves throughout the cytoplasm to obtain nutrients and even move into neighboring cells.

Other microorganisms such as HIV (the causative agent of AIDS), *Chlamydia, Mycobacterium tuberculosis, Leishmania* (a protozoan), and malarial parasites can evade

components of the immune system by entering phagocytes. In addition, these microbes can prevent both the fusion of a phagosome with a lysosome and the proper acidification of digestive enzymes. The microbes then multiply within the phagocyte, almost completely filling it. In most cases, the phagocyte dies and the microbes are released by autolysis to infect other cells. Still other microbes, such as the causative agents of tularemia and brucellosis, can remain dormant within phagocytes for months or years at a time.

    ★   ★   ★

In addition to providing nonspecific resistance for the host, phagocytosis plays a role in immunity. Macrophages help T and B lymphocytes perform vital immune functions. In Chapter 17, we will discuss in more detail how phagocytosis supports immunity.

In the next section, we will see how phagocytosis often occurs as part of another nonspecific mechanism of resistance: inflammation.

Inflammation

Learning Objective

■ *List the stages of inflammation.*

Damage to the body's tissues triggers a defensive response called **inflammation.** The damage can be caused by microbial infection, physical agents (such as heat, radiant energy, electricity, or sharp objects), or chemical agents (acids, bases, and gases). Inflammation is usually characterized by four signs and symptoms: *redness, pain, heat,* and *swelling.* Sometimes a fifth, *loss of function,* is present; its occurrence depends on the site and extent of damage.

In apparent contradiction to the signs and symptoms observed, however, the inflammatory response is beneficial. Inflammation has the following functions: (1) to destroy the injurious agent, if possible, and to remove it and its by-products from the body; (2) if destruction is not possible, to limit the effects on the body by confining or walling off the injurious agent and its by-products; and (3) to repair or replace tissue damaged by the injurious agent or its by-products.

During inflammation, concentrations of some proteins change. These proteins, whose concentration changes markedly during inflammation, are called **acute-phase** proteins. Acute-phase proteins include complement (pages 467–469), the cytokines (Chapter 17, pp. 487–489), and several specialized proteins such as fibrinogen for clotting, and kinins for vasodilation.

For purposes of our discussion, we will divide the process of inflammation into three stages: vasodilation and increased permeability of blood vessels, phagocyte migration and phagocytosis, and tissue repair.

Vasodilation and Increased Permeability of Blood Vessels

Learning Objective

■ *Describe the roles of vasodilation, kinins, prostaglandins, and leukotrienes in inflammation.*

Immediately following tissue damage, blood vessels dilate in the area of damage, and their permeability increases (Figure 16.9a and b). **Vasodilation** is an increase in the diameter of blood vessels. Vasodilation increases blood flow to the damaged area and is responsible for the redness (erythema) and heat associated with inflammation.

Increased permeability permits defensive substances normally retained in the blood to pass through the walls of the blood vessels and enter the injured area. The increase in permeability, which permits fluid to move from the blood into tissue spaces, is responsible for the **edema** (swelling) of inflammation. The pain of inflammation can be caused by nerve damage, irritation by toxins, or the pressure of edema.

❶ Vasodilation and the increase in permeability of blood vessels are caused by chemicals released by damaged cells in response to injury. One such substance is **histamine,** a chemical present in many cells of the body, especially in mast cells in connective tissue, circulating basophils, and blood platelets. Histamine is released in direct response to the injury of cells that contain it; it is also released in response to stimulation by certain components of the complement system (discussed later). Phagocytic granulocytes attracted to the site of injury can also produce chemicals that cause the release of histamine.

Kinins are another group of substances that cause vasodilation and increased permeability of blood vessels. Kinins are present in blood plasma, and once activated, they play a role in chemotaxis by attracting phagocytic granulocytes, chiefly neutrophils, to the injured area.

Prostaglandins, substances released by damaged cells, intensify the effects of histamine and kinins and help phagocytes move through capillary walls. **Leukotrienes** are substances produced by mast cells (cells especially numerous in the connective tissue of the skin and respiratory system, and in blood vessels) and basophils (one of the types of white blood cells). Leukotrienes cause increased permeability of blood vessels and help attach phagocytes to pathogens. Various components of the complement system stimulate the release of histamine, attract phagocytes, and promote phagocytosis.

Vasodilation and the increased permeability of blood vessels also help deliver clotting elements of blood into the injured area. ❷ The blood clots that form around the site of activity prevent the microbe (or its toxins) from spreading to other parts of the body. ❸ As a result, there may be a localized collection of **pus,** a mixture of dead cells and body fluids, in a cavity formed by the breakdown of body tissues. This focus of infection is called an **abscess.** Common abscesses include pustules and boils.

The next stage in inflammation involves the migration of phagocytes to the injured area.

Phagocyte Migration and Phagocytosis

Learning Objective

■ *Describe phagocyte migration.*

Generally, within an hour after the process of inflammation is initiated, phagocytes appear on the scene (Figure 16.9c). ❹ As the flow of blood gradually decreases,

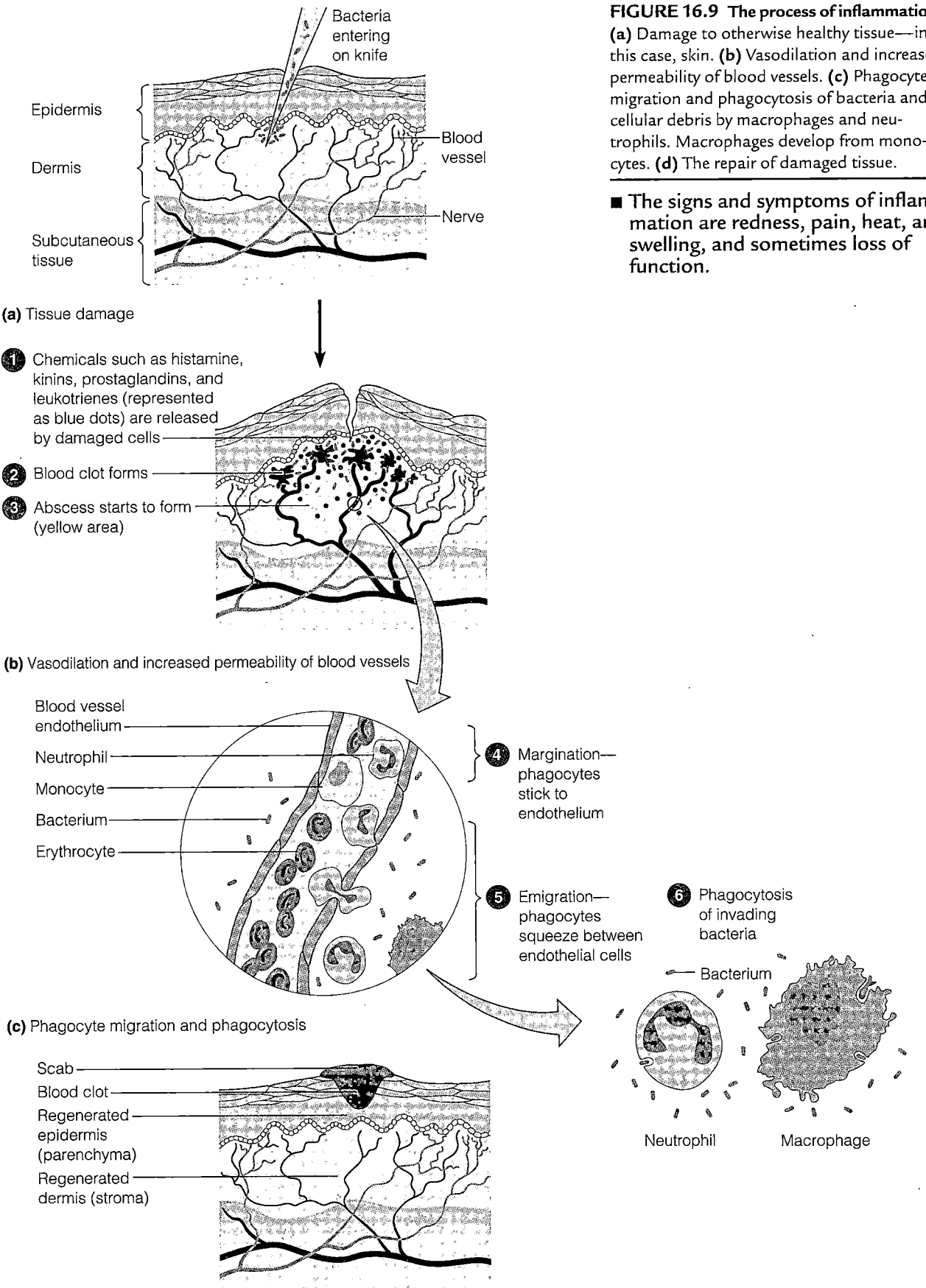

(a) Tissue damage

Epidermis

Dermis

Subcutaneous tissue

Bacteria entering on knife

Blood vessel

Nerve

1. Chemicals such as histamine, kinins, prostaglandins, and leukotrienes (represented as blue dots) are released by damaged cells

2. Blood clot forms

3. Abscess starts to form (yellow area)

(b) Vasodilation and increased permeability of blood vessels

Blood vessel endothelium

Neutrophil

Monocyte

Bacterium

Erythrocyte

4. Margination— phagocytes stick to endothelium

5. Emigration— phagocytes squeeze between endothelial cells

6. Phagocytosis of invading bacteria

Bacterium

Neutrophil   Macrophage

(c) Phagocyte migration and phagocytosis

Scab

Blood clot

Regenerated epidermis (parenchyma)

Regenerated dermis (stroma)

(d) Tissue repair

FIGURE 16.9 The process of inflammation.
(a) Damage to otherwise healthy tissue—in this case, skin. **(b)** Vasodilation and increased permeability of blood vessels. **(c)** Phagocyte migration and phagocytosis of bacteria and cellular debris by macrophages and neutrophils. Macrophages develop from monocytes. **(d)** The repair of damaged tissue.

■ The signs and symptoms of inflammation are redness, pain, heat, and swelling, and sometimes loss of function.

phagocytes (both neutrophils and monocytes) begin to stick to the inner surface of the endothelium (lining) of blood vessels. This sticking process is called **margination.** ⑤ Then the collected phagocytes begin to squeeze between the endothelial cells of the blood vessel to reach the damaged area. This migration, which resembles amoeboid movement, is called **emigration** (*diapedesis*); the migratory process can take as little as 2 minutes. ⑥ The phagocytes then begin to destroy invading microorganisms by phagocytosis.

As mentioned earlier, certain chemicals attract neutrophils to the site of injury (chemotaxis). These include chemicals produced by microorganisms and even other neutrophils; other chemicals are kinins, leukotrienes, and components of the complement system. The availability of a steady stream of neutrophils is ensured by the production and release of additional granulocytes from bone marrow.

As the inflammatory response continues, monocytes follow the granulocytes into the infected area. Once the monocytes are contained in the tissue, they undergo changes in biological properties and become wandering macrophages. The granulocytes predominate in the early stages of infection but tend to die off rapidly. Macrophages enter the picture during a later stage of the infection, once granulocytes have accomplished their function. They are several times more phagocytic than granulocytes and are large enough to phagocytize tissue that has been destroyed, granulocytes that have been destroyed, and invading microorganisms.

After granulocytes or macrophages engulf large numbers of microorganisms and damaged tissue, they themselves eventually die. As a result, pus forms, and its formation usually continues until the infection subsides. At times, the pus pushes to the surface of the body or into an internal cavity for dispersal. On other occasions the pus remains even after the infection is terminated. In this case, the pus is gradually destroyed over a period of days and is absorbed by the body.

As effective as phagocytosis is in contributing to nonspecific resistance, there are times when the mechanism becomes less functional in response to certain conditions. For example, some individuals are born with an inability to produce phagocytes. And with age, there is a progressive decline in the efficiency of phagocytosis. Recipients of heart or kidney transplants have impaired nonspecific defenses as a result of receiving drugs that prevent the rejection of the transplant. Radiation treatments can also depress nonspecific immune responses by damaging bone marrow. Even certain diseases such as AIDS and cancer can cause defective functioning of nonspecific defenses.

Tissue Repair

The final stage of inflammation is tissue repair, the process by which tissues replace dead or damaged cells (Figure 16.9d). Repair begins during the active phase of inflammation, but it cannot be completed until all harmful substances have been removed or neutralized at the site of injury. The ability of a tissue to regenerate, or repair itself, depends on the type of tissue. For example, skin has a high capacity for regeneration, whereas cardiac muscle tissue does not regenerate at all.

A tissue is repaired when its stroma or parenchyma produces new cells. The *stroma* is the supporting connective tissue, and the *parenchyma* is the functioning part of the tissue. For example, the capsule around the liver that encloses and protects it is part of the stroma because it is not involved in the functions of the liver; liver cells (hepatocytes) that perform the functions of the liver are part of the parenchyma. If only parenchymal cells are active in repair, a perfect or near-perfect reconstruction of the tissue occurs. A familiar example of perfect reconstruction is a minor skin cut, in which parenchymal cells are more active in repair. However, if repair cells of the stroma of the skin are more active, scar tissue is formed.

Fever

Learning Objective

■ *Describe the cause and effects of fever.*

Inflammation is a local response of the body to injury. There are also systemic, or overall, responses; one of the most important is **fever,** an abnormally high body temperature. The most frequent cause of fever is infection from bacteria (and their toxins) or viruses.

Body temperature is controlled by a part of the brain called the hypothalamus. The hypothalamus is sometimes called the body's thermostat, and it is normally set at 37°C (98.6°F). It is believed that certain substances affect the hypothalamus by setting it at a higher temperature. Recall from Chapter 15 that when phagocytes ingest gram-negative bacteria, the lipopolysaccharides (LPS) of the cell wall (endotoxins) are released, causing the phagocytes to release interleukin-1 (formerly called endogenous pyrogen). IL-1 causes the hypothalamus to release prostaglandins that reset the hypothalamic thermostat at a higher temperature, thereby causing fever (see Figure 15.5 on page 445).

Assume that the body is invaded by pathogens and that the thermostat setting is increased to 39°C (102.2°F). To adjust to the new thermostat setting, the body responds with blood vessel constriction, increased rate of metabolism, and **shivering,** all of which raise body temperature. Even though body temperature is climbing higher than normal, the skin remains cold, and shivering occurs. This condition, called a *chill,* is a definite sign that body temperature is rising. When body temperature reaches the setting of the thermostat, the chill disappears. But the body will continue to maintain its temperature at 39°C until the IL-1 is eliminated. The thermostat is then reset at 37°C. As the infection subsides, heat-losing mechanisms such as vasodilation and sweating go into operation. The skin becomes warm and the person begins to sweat. This phase of the fever, called the **crisis,** indicates that body temperature is falling.

Up to a certain point, fever is considered a defense against disease. Interleukin-1 helps step up the production of T lymphocytes. High body temperature intensifies the effect of interferons (antiviral proteins we will discuss shortly). They are believed to inhibit the growth of some microorganisms by decreasing the amount of iron available to them. Also, because the high temperature speeds up the body's reactions, it may help body tissues repair themselves more quickly.

Antimicrobial Substances

The body produces certain antimicrobial substances in addition to the chemical factors mentioned earlier. Among the most important of these are the proteins of the complement system and interferons.

The Complement System

Learning Objectives

- *List the components of the complement system.*
- *Describe two pathways of activating complement.*
- *Describe three consequences of complement activation.*

The **complement system** is a defensive system consisting of serum proteins that participate in lysis of foreign cells, inflammation, and phagocytosis. **Serum** is plasma minus its blood-clotting proteins. The complement system can be activated in either of two ways: (1) by an immune reaction of antibodies to antigens (described in Chapter 17) in the *classical pathway,* or (2) by the direct interaction of certain proteins with polysaccharides in the *alternative pathway.* Either way, the result is the same: the cleavage of a protein called C3 (C for complement), which triggers a sequence of subsequent events (Figure 16.10). These events, and the major components, are described next.

Complement is nonspecific: the same proteins can be activated in response to any foreign cell. However, in the classical pathway, complement assists, or complements, specific immunity.

Components

The complement system consists of a group of at least 20 interacting proteins found in normal serum. Proteins of the complement system make up about 5% of the serum proteins in vertebrates. The major components of the classical pathway are designated by a numbering system ranging from C1 through C9. The C1 protein also has three subcomponents: C1q, C1r, and C1s. The alternative pathway consists of proteins called factor B, factor D, and factor P (properdin), along with the C3 and C5 through C9 proteins active in the classical pathway.

Pathways of Activation

The proteins of the classical and alternative pathways act in an ordered sequence, or *cascade,* which, except for C4, follows the numerical designations. In a series of steps, each protein activates the next one in the series, usually by cleaving (splitting) it. The fragments of the cleaved proteins have new physiological or enzymatic functions. For example, one fragment might cause blood vessel dilation, whereas another fragment serves as part of the enzyme that cleaves the next protein in the series.

C3 plays a central role in the complement system. Its activation triggers several mechanisms that contribute to microbial destruction. As shown in Figure 16.10, either of two pathways can activate C3. The classical pathway is initiated by the binding of antibodies to antigen. The antigen could consist of bacteria or other cells. The classical pathway also involves the activation of C1, C2, and C4 proteins. The details of complement activation by this pathway are described below.

The alternative pathway does not involve antibodies; it is initiated by the interaction between certain polysaccharides and the proteins known as factors B, D, and P. Most of these polysaccharides are contained in the cell walls of certain bacteria and fungi, although they also include molecules on the surface of some foreign mammalian red blood cells. The alternative pathway is of particular importance in combating enteric gram-negative bacteria (many of which inhabit the intestines). The endotoxin (lipid A) triggers the alternative pathway. Note that this pathway does not involve C1, C2, or C4.

Consequences of Complement Activation

How does the complement system contribute to microbial destruction? Both the classical and the alternative

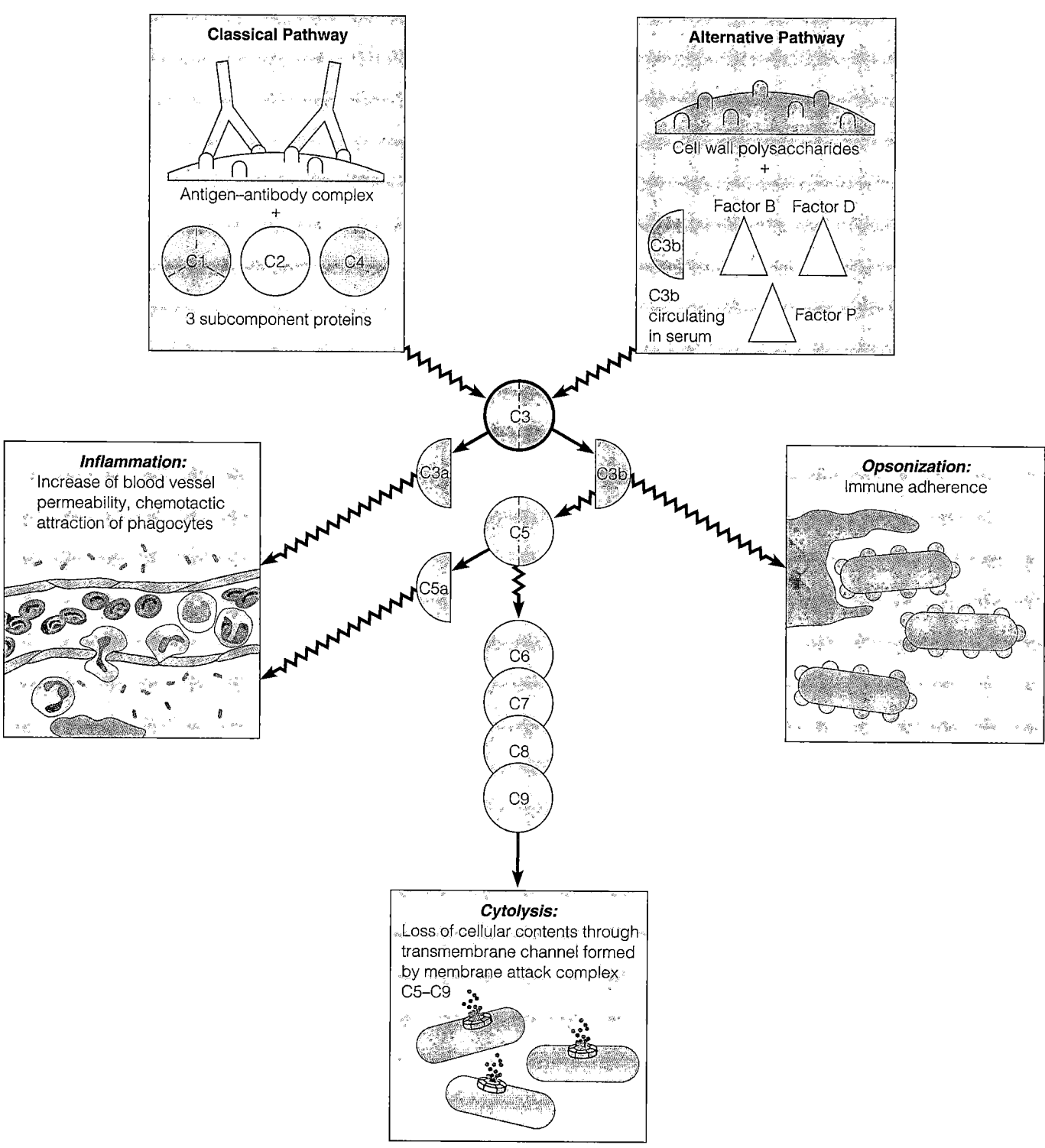

FIGURE 16.10 Complement activation via classical and alternative pathways.
Events in either the classical or the alternative pathway can initiate the cleavage of C3
into C3a and C3b. These fragments induce three kinds of consequences destructive to
microorganisms: cytolysis, inflammation, and opsonization.

■ Complement is a defensive system consisting of at least 20 interacting serum proteins.

pathways lead to the cleavage of C3 into two fragments, C3a and C3b. These fragments induce three processes that are destructive to microorganisms: cytolysis, inflammation, and opsonization.

Cytolysis The main function of the complement system is to destroy foreign cells by damaging their plasma membranes, causing the cellular contents to leak out. This process, called **cytolysis,** is accomplished as follows (Figure 16.11a). ① Once a pair of antibodies recognizes and attaches to the antigen, the complement protein C1 binds to two or more adjacent antibodies, thus activating C1. ② Next, activated C1 in turn activates C2 and C4. It does this by splitting the C2 and C4 proteins. C2 is split into fragments called C2a and C2b, and C4 into fragments called C4a and C4b. Then C2b and C4b combine to form another enzyme, which in turn activates C3 by splitting it into two fragments, C3a and C3b. ③ C3b initiates a sequence of reactions involving C5–C9, which are known collectively as the **membrane attack complex.** The activated components of these proteins, with C9 proteins possibly playing a key role, attack the invading cell's membrane and ④ produce circular lesions, called **transmembrane channels (membrane pores),** that lead to the loss of ions and eventual cytolysis.

The utilization of the complement components in this process is called **complement fixation;** it forms the basis of a clinical laboratory test that will be explained in Chapter 18 (see Figure 18.9 on page 511). Figure 16.11b illustrates the effects of complement on a microbial cell.

A number of intracellular pathogens secrete membrane attack complexes that lyse phagocyte cell membranes once inside the phagocyte. For example, *Trypanosoma cruzi* (the causative agent of American trypanosomiasis), and *Listeria monocytogenes* (the causative agent of listeriosis), produce membrane attack complexes that lyse phagolysosome membranes and release the microbes into the cytoplasm of the phagocyte, where they propagate. Later the microbes secrete more membrane attack complexes that lyse the plasma membrane and release the microbes from the phagocyte, resulting in lysis of the phagocyte and infection of neighboring cells by the microbe.

Inflammation C3a, a cleavage product from C3, and C5a, a cleavage product from C5, can contribute to the development of acute inflammation (Figure 16.12 on page 471). C3a and C5a bind to mast cells, basophils, and blood platelets to trigger the release of histamine, which increases blood vessel permeability. C5a also functions as a powerful chemotactic factor that attracts phagocytes to the site of complement fixation.

Opsonization When bound to the surface of a microorganism, C3b can interact with special receptors on phagocytes to promote phagocytosis (see Figure 16.10). As introduced earlier, this phenomenon is called *opsonization,* or immune adherence. In the process, C3b functions as an opsonin by coating the microorganism and promoting attachment of the phagocyte to the microbe.

Inactivation of Complement Once complement is activated, its destructive capabilities usually cease very quickly in order to minimize the destruction of host cells. This is accomplished by various regulatory proteins in the host's blood and on certain cells, such as blood cells. The proteins bring about the breakdown of activated complement and function as inhibitors and destructive enzymes.

Complement and Disease

In addition to its importance in defense, the complement system assumes a role in causing disease as a result of inherited deficiencies. For example, deficiencies of C1, C2, or C4 cause collagen vascular disorders that result in hypersensitivity (anaphylaxis); deficiency of C3, though rare, results in increased susceptibility to bacterial infections; and C5–C9 defects result in increased susceptibility to *Neisseria meningitidis* and *N. gonorrhoeae* infections.

Interferons

Learning Objectives

- *Define interferons.*
- *Compare and contrast the actions of α-IFN and β-IFN with γ-IFN.*

Because viruses depend on their host cells to provide many functions of viral multiplication, it is difficult to inhibit viral multiplication without affecting the host cell itself. One way the infected host counters viral infections is with interferons. **Interferons (IFNs)** are a class of similar antiviral proteins produced by certain animal cells after viral stimulation. One of the principal functions of interferons is to interfere with viral multiplication.

An interesting feature of interferons is that they are host-cell-specific but not virus-specific. Interferons produced by human cells protect human cells but produce little antiviral activity for cells of other species, such as mice or chickens. However, the interferons of a species are active against a number of different viruses.

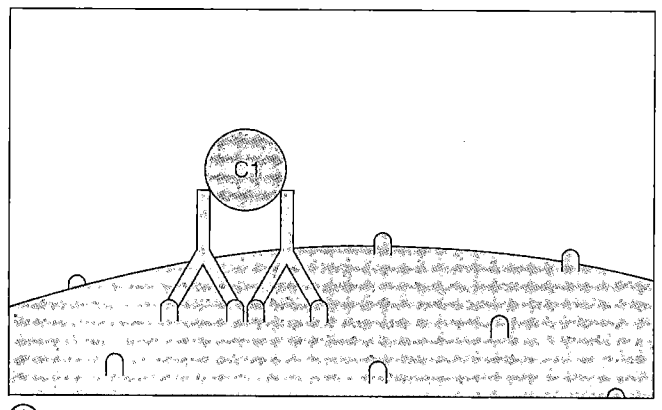

① Once antibodies recognize and attach to the antigen, complement protein C1 binds to two adjacent antibodies.

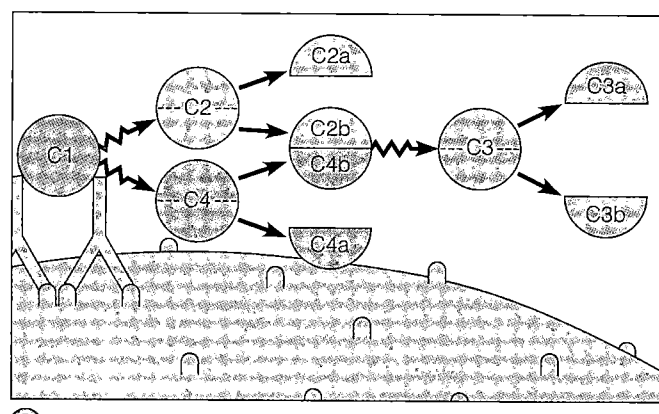

② C1 acts as an enzyme that splits the C2 and C4 proteins into fragments. Fragments C2b and C4b combine to form another enzyme, which splits C3 into two fragments. The active fragment is called C3b.

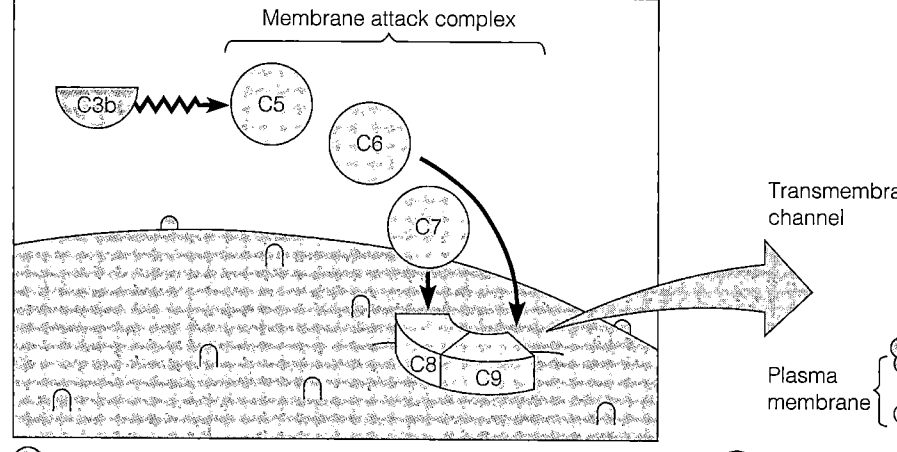

③ C3b initiates a series of reactions involving C5–C9, collectively called the membrane attack complex. This complex forms circular transmembrane channels (lesions) in the antigenic cell's membrane, with C9 proteins possibly playing a key role.

④ The result is leakage of the cell's contents—cytolysis, as shown in this enlarged view of a transmembrane channel.

(a)

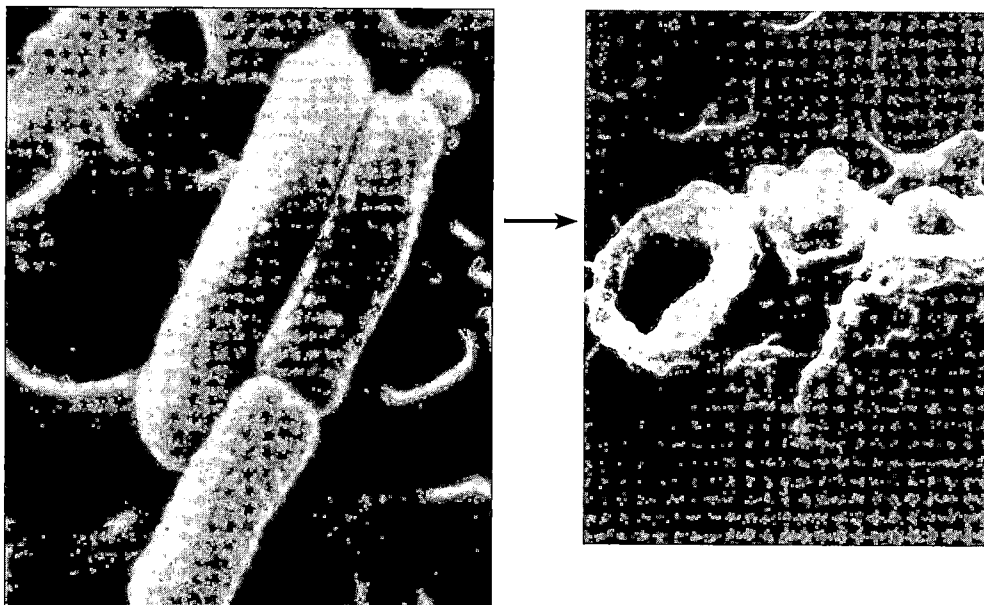

(b)

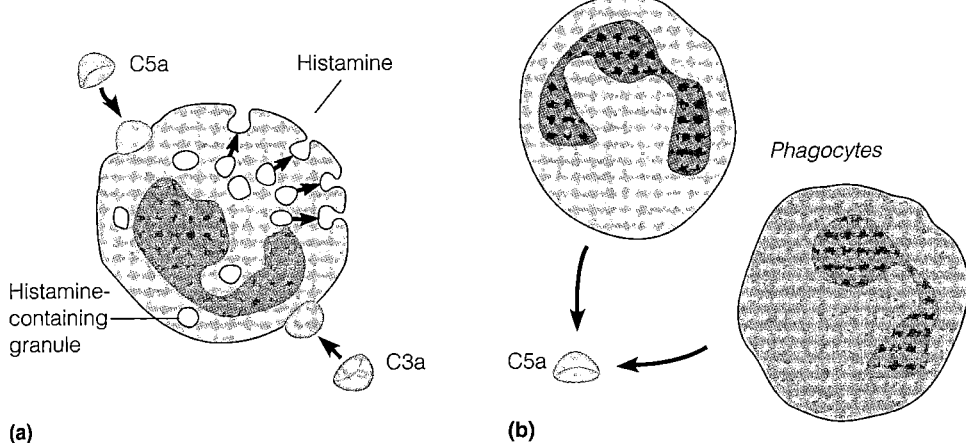

FIGURE 16.12 Inflammation stimulated by complement. (a) C3a and C5a bind to mast cells, basophils, and platelets to trigger the release of histamine, which increases blood vessel permeability. (b) C5a functions as a chemotactic factor that attracts phagocytes to the site of complement activation.

■ How is complement inactivated once it has completed its functions?

Just as different animal species produce different interferons, different types of cells in the same animal also produce different interferons. Human interferons are of three principal types: *alpha interferon* (α-IFN), *beta interferon* (β-IFN), and *gamma interferon* (γ-IFN). There are also various subtypes of interferons within each of the principal groups. In the human body, interferons are produced by fibroblasts in connective tissue and by lymphocytes and other leukocytes. Each of the three types of interferons produced by these cells can have a slightly different effect on the body.

All interferons are small proteins, with molecular weights between 15,000 and 30,000. They are quite stable at low pH and are fairly resistant to heat.

Both α-IFN and β-IFN are produced by virus-infected host cells only in very small quantities and dif-

FIGURE 16.11 Cytolysis caused by complement via the classical pathway. (a) A few key steps in the complex sequence that leads to the membrane attack complex forming a transmembrane channel in a plasma membrane. (b) Micrographs of a rod-shaped bacterium before cytolysis (left) and after cytolysis (right). Reprinted from Schreiber, R. D. et al. *"Bactericidal Activity of the Alternative Complement Pathway Generated from 11 Isolated Plasma Proteins." Journal of Experimental Medicine,* 149:870–882, 1979.

■ What is the relationship of complement in causing disease?

fuse to uninfected neighboring cells (Figure 16.13). They react with plasma or nuclear membrane receptors, inducing the uninfected cells to manufacture mRNA for the synthesis of **antiviral proteins (AVPs).** These proteins are enzymes that disrupt various stages of viral multiplication. For example, one AVP called *oligoadenylate synthetase,* degrades viral mRNA. Another, called *protein kinase,* inhibits protein synthesis.

Gamma-IFN is produced by lymphocytes; it causes neutrophils to kill bacteria. Neutrophils in individuals with an inherited condition called chronic granulomatous disease (CGD) do not kill bacteria. When these people take a recombinant γ-IFN, their neutrophils kill bacteria. Gamma-IFN is not a cure and must be taken for the life of the CGD individual.

The low concentrations at which interferons inhibit viral multiplication are nontoxic to uninfected cells. Because of their beneficial properties, they would seem to be ideal antiviral substances. But certain problems do exist. For one thing, interferons are effective for only short periods. They typically play a major role in infections that are acute and short-term, such as colds and influenza. Another problem is that they have no effect on viral multiplication in cells already infected.

The importance of interferons in protecting the body against viruses, as well as their potential as anticancer agents, has made their production in large quantities a top health priority. Several groups of scientists have successfully applied recombinant DNA technology in inducing certain species of bacteria to produce interferons. (This technique is described in Chapter 9.) The interferons produced with recombinant DNA techniques, called *recombinant interferons (rIFNs),* are important for two reasons: They are pure, and they are plentiful.

In clinical trials, IFNs have exhibited no effects against some types of tumors and only limited effects against others. Alpha-IFN (Intron A®) is approved in the United States for treating several virus-associated disorders. One is Kaposi's sarcoma, a cancer that often occurs in patients infected with HIV. Other approved uses for α-IFN include treating genital herpes caused by herpesvirus; hepatitis B and C, caused by the hepatitis B and C viruses; and hairy cell leukemia. Alpha-IFN also is being tested to see if it can slow the development of AIDS in HIV-infected people. A form of β-IFN (Betaseron®) slows the progression of multiple sclerosis (MS) and lessens the frequency and severity of MS attacks.

★ ★ ★

In Chapter 17, we will discuss specific immune responses. We will also look at the principal factors that contribute to immunity.

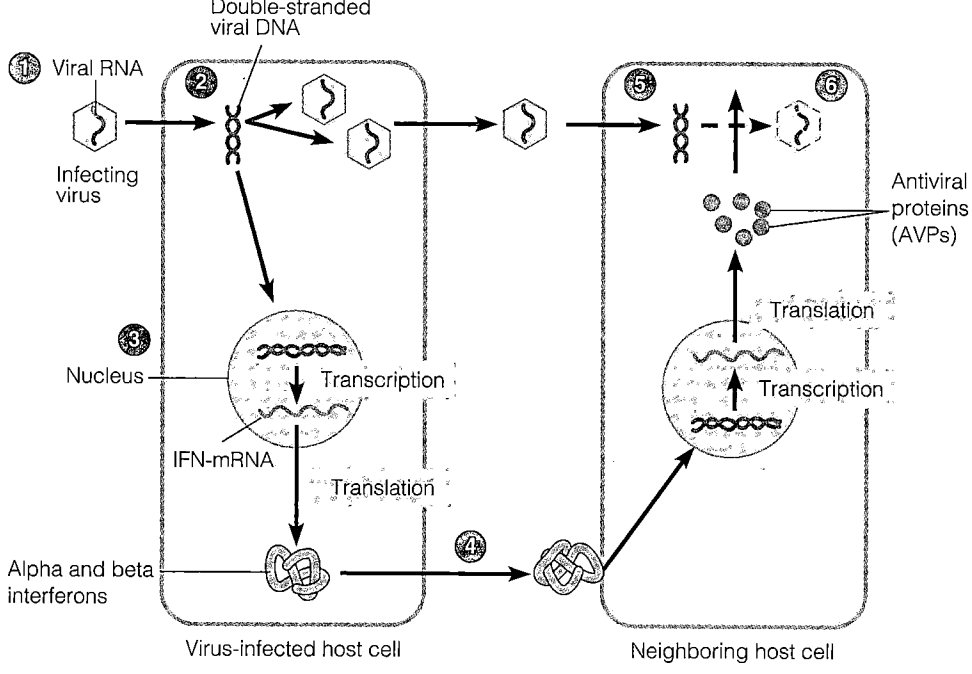

1. Viral RNA from an infecting virus enters the cell.
2. The infecting virus replicates into new viruses.
3. The infecting virus also induces the host cell to produce interferon on RNA (IFN-mRNA), which is translated into alpha and beta interferons.
4. Interferons released by the virus-infected host cell bind to plasma membrane or nuclear membrane receptors on uninfected neighboring host cells, inducing them to synthesize antiviral proteins (AVPs). These include oligoadenylate synthetase, and protein kinase.
5. New viruses released by the virus-infected host cell infect neighboring host cells.
6. AVPs degrade viral m-RNA and inhibit protein synthesis and thus interfere with viral replication.

FIGURE 16.13 Antiviral action of alpha and beta interferons (IFNs). Interferons are host-cell-specific but not virus-specific.

Study Outline   Student Tutorial CD-ROM

INTRODUCTION (p. 454)

To review, go to Microbial Interactions: Disease Host-pathogen responses

1. The ability to ward off disease through body defenses is called resistance.

2. Lack of resistance is called susceptibility.

3. Nonspecific resistance refers to all body defenses that protect the body against any kind of pathogen.

4. Specific resistance (immunity) refers to defenses (antibodies) against specific microorganisms.

SKIN AND MUCOUS MEMBRANES (pp. 454–457)

Mechanical Factors (pp. 454–456)

1. The structure of intact skin and the waterproof protein keratin provide resistance to microbial invasion.

2. Some pathogens, if present in large numbers, can penetrate mucous membranes.

3. The lacrimal apparatus protects the eyes from irritating substances and microorganisms.

4. Saliva washes microorganisms from teeth and gums.

5. Mucus traps many microorganisms that enter the respiratory and gastrointestinal tracts; in the lower respiratory tract, the ciliary escalator moves mucus up and out.

6. The flow of urine moves microorganisms out of the urinary tract, and vaginal secretions move microorganisms out of the vagina.

Chemical Factors (pp. 456–457)

1. Sebum contains unsaturated fatty acids, which inhibit the growth of pathogenic bacteria. Some bacteria commonly found on the skin can metabolize sebum and cause the inflammatory response associated with acne.

2. Perspiration washes microorganisms off the skin.

3. Lysozyme is found in tears, saliva, nasal secretions, and perspiration.

4. The high acidity (pH 1.2–3.0) of gastric juice prevents microbial growth in the stomach.

5. Normal microbiota prevent the growth of many pathogens.

Normal Microbiota and Nonspecific Resistance (p. 457)

1. Normal microbiota change the environment, which can prevent the growth of pathogens.

PHAGOCYTOSIS (pp. 458–463)

1. Phagocytosis is the ingestion of microorganisms or particulate matter by a cell.

2. Phagocytosis is performed by phagocytes, certain types of white blood cells or their derivatives.

Formed Elements in Blood (pp. 458–460)

1. Blood consists of plasma (fluid) and formed elements (cells and cell fragments).

2. Leukocytes (white blood cells) are divided into three categories: granulocytes (neutrophils, basophils, and eosinophils), lymphocytes, and monocytes.

3. During many infections, the number of leukocytes increases (leukocytosis); some infections are characterized by leukopenia (decrease in leukocytes).

4. Phagocytes are activated by bacterial components (for example, lipid A) and cytokines.

Actions of Phagocytic Cells (pp. 460–461)

1. Among the granulocytes, neutrophils are the most important phagocytes.

2. Enlarged monocytes become wandering macrophages and fixed macrophages.

3. Fixed macrophages are located in selected tissues and are part of the mononuclear phagocytic system.

4. Granulocytes predominate during the early stages of infection, whereas monocytes predominate as the infection subsides.

The Mechanism of Phagocytosis (pp. 461–463)

1. Chemotaxis is the process by which phagocytes are attracted to microorganisms.

2. The phagocyte then adheres to the microbial cells; adherence may be facilitated by opsonization—coating the microbe with serum proteins.

3. Pseudopods of phagocytes engulf the microorganism and enclose it in a phagocytic vesicle to complete ingestion.

4. Many phagocytized microorganisms are killed by lysosomal enzymes and oxidizing agents.

5. Some microbes are not killed by phagocytes and can even reproduce in phagocytes.

INFLAMMATION (pp. 464–466)

1. Inflammation is a bodily response to cell damage; it is characterized by redness, pain, heat, swelling, and sometimes the loss of function.

Vasodilation and Increased Permeability of Blood Vessels (p. 464)

1. The release of histamine, kinins, and prostaglandins causes vasodilation and increased permeability of blood vessels.

2. Blood clots can form around an abscess to prevent dissemination of the infection.

Phagocyte Migration and Phagocytosis (pp. 464–466)

1. Phagocytes have the ability to stick to the lining of the blood vessels (margination).

2. They also have the ability to squeeze through blood vessels (emigration).

3. Pus is the accumulation of damaged tissue and dead microbes, granulocytes, and macrophages.

Tissue Repair (p. 466)

1. A tissue is repaired when the stroma (supporting tissue) or parenchyma (functioning tissue) produces new cells.

2. Stromal repair by fibroblasts produces scar tissue.

FEVER (pp. 466–467)

1. Fever is an abnormally high body temperature produced in response to a bacterial or viral infection.

2. Bacterial endotoxins and interleukin-1 can induce fever.

3. A chill indicates a rising body temperature; crisis (sweating) indicates that the body's temperature is falling.

ANTIMICROBIAL SUBSTANCES
(PP. 467–471)

The Complement System (pp. 467–469)

1. The complement system consists of a group of serum proteins that activate one another to destroy invading microorganisms. Serum is the liquid remaining after blood plasma is clotted.

2. C1 binds to antigen–antibody complexes to eventually activate C3 protein. Factor B, factor D, factor P, and C3 bind to certain cell wall polysaccharides to activate C3b.

3. C3 activation can result in cell lysis, inflammation, and opsonization.

4. Complement is deactivated by host-regulatory proteins.

5. Complement deficiencies can result in an increased susceptibility to disease.

Interferons (pp. 469–471)

1. Interferons (IFNs) are antiviral proteins produced in response to viral infection.

2. There are three types of human interferon: α-IFN, β-IFN, and γ-IFN. Recombinant interferons have been produced.

3. The mode of action of α-IFN and β-IFN is to induce uninfected cells to produce antiviral proteins (AVPs) that prevent viral replication.

4. Interferons are host-cell-specific but not virus-specific.

5. Gamma-IFN activates neutrophils to kill bacteria.

Study Questions

REVIEW

1. Define the following terms:
   a. resistance
   b. susceptibility
   c. nonspecific resistance

2. Identify at least one mechanical and one chemical factor that prevent microbes from entering the body through each of the following:
   a. skin
   b. eyes
   c. digestive tract
   d. respiratory tract
   e. urinary tract
   f. reproductive tract

3. Describe the five different white blood cells, and name a function for each cell type.

4. Define phagocytosis.

5. Compare the structures and functions of granulocytes and monocytes in phagocytosis.

6. How do fixed and wandering macrophages differ?

7. Diagram the following processes that result in phagocytosis: margination, emigration, adherence, and ingestion.

8. Define inflammation, and list its characteristics.

9. Why is inflammation beneficial to the body?

10. How is fever related to nonspecific resistance?

11. What causes the periods of chill and crisis during fever?

12. What is complement? List the steps for complement fixation.

13. Summarize the major outcomes of complement activation.

14. How is the complement system activated by bacterial endotoxin in the bloodstream? How does endotoxic shock result in massive host cell destruction?

15. What are interferons? Discuss their roles in nonspecific resistance. Why do α-IFN and β-IFN share the same receptor on target cells yet γ-IFN has a different receptor?

MULTIPLE CHOICE

1. *Legionella* uses C3b receptors to enter monocytes. This
   a. prevents phagocytosis.
   b. degrades complement.
   c. inactivates complement.
   d. prevents inflammation.
   e. prevents cytolysis.

2. *Chlamydia* can prevent the formation of phagolysosomes, and therefore *Chlamydia* can
   a. avoid being phagocytized.
   b. avoid destruction by complement.
   c. prevent adherence.
   d. avoid being digested.
   e. none of the above

3. If the following are placed in the order of occurrence, which would be the *third* step?
   a. Emigration
   b. Digestion
   c. Formation of a phagosome
   d. Formation of a phagolysosome
   e. Margination

4. If the following are placed in the order of occurrence, which would be the *third* step?
   a. Activation of C5–C9
   b. Cell lysis
   c. Antigen–antibody reaction
   d. Activation of C3
   e. Activation of C2–C4

5. A human host can prevent a pathogen from getting enough iron by
   a. reducing dietary intake of iron.
   b. binding iron with transferrin.
   c. binding iron with hemoglobin.
   d. excreting excess iron.
   e. binding iron with siderophores.

6. A decrease in the production of C3 would result in
   a. increased susceptibility to infection.
   b. increased numbers of white blood cells.
   c. increased phagocytosis.
   d. activation of C5–C9.
   e. none of the above

7. In 1884, Elie Metchnikoff observed blood cells collected around a splinter inserted in a sea star embryo. This was the discovery of
   a. blood cells.
   b. sea stars.
   c. phagocytosis.
   d. immunity.
   e. none of the above

8. *Helicobacter pylori* uses the enzyme urease to counteract a chemical defense in the human organ in which it lives. This chemical defense is
   a. lysozyme.
   b. hydrochloric acid.
   c. superoxide radicals.
   d. sebum.
   e. complement.

9. Which of the following statements about α-IFN is *not* true?
   a. It interferes with viral replication.
   b. It is host specific.
   c. It is released by fibroblasts.
   d. It is virus-specific.
   e. It is released by lymphocytes.

10. Which of the following does not stimulate phagocytes?
   a. Cytokines
   b. γ-IFN
   c. C3b
   d. Lipid A
   e. Histamine

CRITICAL THINKING

1. Why do serum levels of iron increase during an infection? What can a bacterium do to respond to high levels of transferrin?

2. A variety of drugs with the ability to reduce inflammation are available. Comment on the danger of misuse of these anti-inflammatory drugs.

3. To be a successful parasite, a microbe must avoid destruction by complement. The following list provides examples of complement-evading techniques. For each microbe, identify the disease it causes, and describe how its strategy enables it to avoid destruction by complement.

Pathogen	Strategy
Group A streptococci	C3 does not bind to M protein
Haemophilus influenzae type b	Has a capsule
Pseudomonas aeruginosa	Sheds cell wall polysaccharides
Trypanosoma cruzi	Degrades C1

4. The list below identifies a virulence factor for a selected microorganism. Describe the effect of each factor listed. Name a disease caused by each organism.

Microorganism	Virulence Factor
Influenza virus	Causes release of lysosomal enzymes
Mycobacterium tuberculosis	Inhibits lysosome fusion
Toxoplasma gondii	Prevents phagosome acidification
Trichophyton	Secretes keratinase
Trypanosoma cruzi	Lyses phagosomal membrane

CLINICAL APPLICATIONS

1. People with *rhinovirus* infections of the nose and throat have an 80-fold increase in kinins and no increase in histamine. What do you expect for rhinoviral symptoms? What disease is caused by rhinoviruses?

2. A hematologist often performs a differential white blood cell count on a blood sample. Such a count determines the relative numbers of white blood cells. Why are these numbers important? What do you think a hematologist would find in a differential white blood cell count of a patient with mononucleosis? With neutropenia? With eosinophilia?

3. Leukocyte adherence deficiency (LAD) is an inherited disease resulting in the inability of neutrophils to recognize C3b-bound microorganisms. What are the most likely consequences of LAD?

4. The neutrophils of individuals with Chediak-Higashi syndrome (CHS) have fewer than normal chemotactic receptors and lysosomes that spontaneously rupture. What are the consequences of CHS?

Learning with Technology

[MP] **= The Microbiology Place website**　[ST] **= Student Tutorial CD-ROM**　[VU] **= VirtualUnknown CD-ROM**

[MP]　Don't forget to go to The Microbiology Place website (http://www.microbiologyplace.com) to take the practice tests, explore the interactive activity and case study, and check out the news articles and web links for this chapter.

[ST]　Remember there is also a quiz for this chapter on the Microbiology Interactive Student Tutorial CD-ROM.

[VU]　Enter the Virtual Lab, click the arrow next to the Session field, click Textbook Exercises, and select Chapter 16. Read the Case Study carefully, identify the unknown, and use what you learn to answer the following questions (consult your textbook for additional information):

1. List the nonspecific means by which humans fight off infections. Which of these would be involved in fighting the infection described in this case study?

2. List the nonspecific defenses the patient used against this microbe, and provide an explanation for how this microbe was able to defeat each defense to cause disease.

chapter 17

Specific Defenses of the Host:
The Immune Response

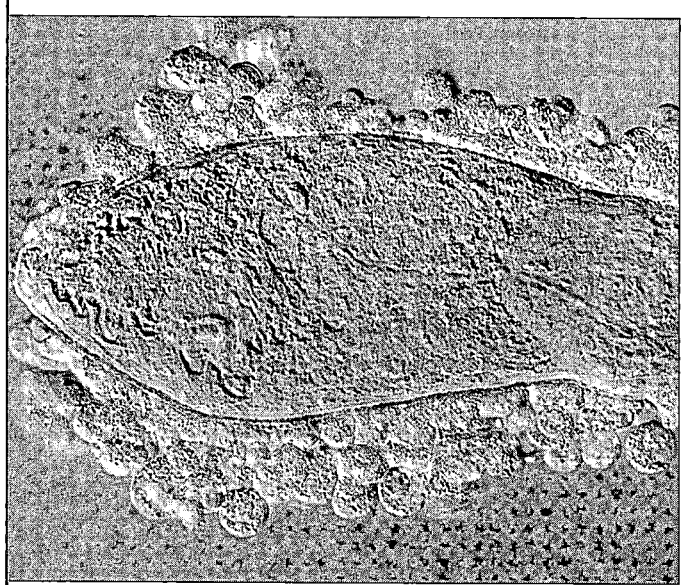

Eosinophils attacking a fluke larva. White blood cells such as these eosinophils can kill parasites by attaching to the parasite and secreting digestive enzymes.

n Chapter 16, we discussed nonspecific defenses of the host, including the skin and mucous membranes, phagocytosis, inflammation, and fever. In addition to nonspecific resistance, humans have **innate resistance** to certain illnesses. For example, all humans are resistant to many infectious diseases of other animals, such as canine distemper and hog cholera.

Even resistance to human diseases can vary from person to person. For example, the effect of measles is usually relatively mild for individuals of European ancestry, but the disease devastated the populations of Pacific Islanders when they were first exposed to it by European explorers. The reason lies in natural selection. For the Europeans, many generations of exposure to the measles virus presumably led to the selection of genes that conferred some resistance to the virus. An individual's resistance to disease also depends on other genetic factors, as well as on gender, age, nutritional status, and general health.

In this chapter, we will study immunity, another aspect of the body's defenses. (For a review of the body's defense systems, see Figure 16.1, page 455.)

Immunity

Learning Objectives

- *Differentiate between immunity and nonspecific resistance.*
- *Contrast the four types of acquired immunity.*

Immunity is a *specific* defensive response to an invasion by foreign organisms or other substances. The human immune system recognizes foreign substances as not belonging to the body, and it develops an immune response against them. Substances that provoke such a response are called *antigens.* This immune response involves the production of proteins called *antibodies* and specialized lymphocytes. (Antigens and antibodies are discussed in detail later in the chapter.) Both antibodies and specialized lymphocytes are specifically targeted for the antigens that cause their formation, and they can destroy or inactivate those antigens if they encounter them again.

Invading organisms might be pathogenic bacteria, viruses, fungi, protozoa, or helminths; other substances include pollen, insect venom, and transplanted tissue. Body cells that become cancerous are also recognized as foreign and may be eliminated. (However, if they survive to become established as solid tumors, the immune system is no longer effective against them.) We will see in this chapter and in Chapters 18 and 19 that the immune system is essential to survival, but that it can also cause harm when its efforts are misdirected.

Types of Acquired Immunity

Acquired immunity refers to the protection an animal develops against certain types of microbes or foreign substances. Acquired immunity is developed during an individual's lifetime. The various types of acquired immunity are summarized in Figure 17.1.

Immunity can be acquired either actively or passively. Immunity is acquired *actively* when a person is exposed to microorganisms or foreign substances and the immune system responds. Immunity is acquired *passively* when antibodies are transferred from one person to another. Passive immunity in the recipient lasts only as long as the antibodies are present—in most cases, a few weeks or months. Both actively acquired immunity and passively acquired immunity can be obtained by natural or artificial means.

Naturally Acquired Immunity

Naturally acquired active immunity is obtained when a person is exposed to antigens in the course of daily life. Once acquired, immunity is lifelong for some diseases, such as measles and chickenpox. For other diseases, especially intestinal diseases, the immunity may last for only a few years. *Subclinical infections* (those that produce no noticeable symptoms or signs of illness) can also confer immunity.

Naturally acquired passive immunity involves the natural transfer of antibodies from a mother to her infant. A pregnant woman's body passes some of her antibodies to her fetus across the placenta, a mechanism known as *transplacental transfer.* If the mother is immune to diphtheria, rubella, or polio, the newborn will be temporarily immune to these diseases as well. Certain antibodies are also passed from the mother to her nursing infant in breast milk, especially in the first secretions called *colostrum.* In the infant, immunity generally lasts only as long as the transmitted antibodies are present— usually a few weeks or months. These maternal antibodies are essential for providing immunity to the infant until its own immune system matures.

Colostrum is even more important to some other mammals. Calves, for example, do not have antibodies

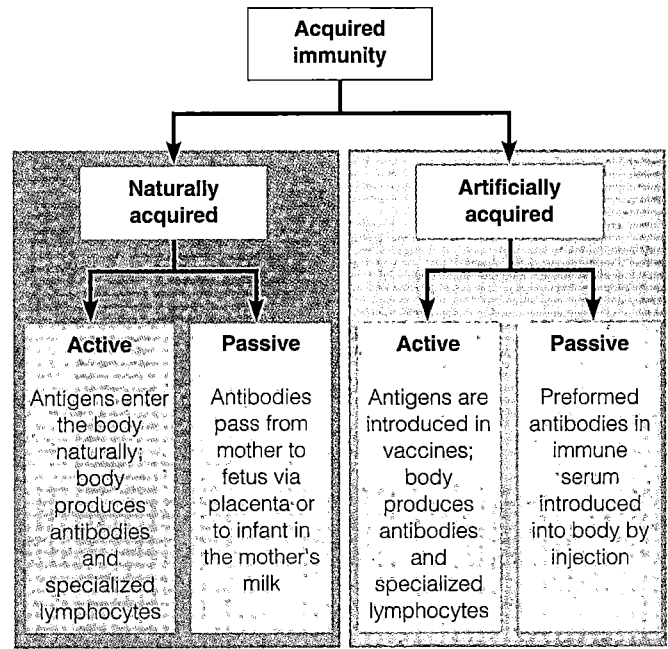

FIGURE 17.1 Types of acquired immunity.

■ Which type of immunity, active or passive, lasts longer?

that cross the placenta and rely on colostrum ingested during the first day of life for this type of immunity. Researchers often specify fetal calf serum for certain experimental or clinical uses because it does not contain maternal antibodies.

Artificially Acquired Immunity

Artificially acquired active immunity results from vaccination. **Vaccination,** also called **immunization,** introduces specially prepared antigens called **vaccines** into the body. Vaccines may be inactivated bacterial toxins (toxoids), killed microorganisms, living but attenuated (weakened) microorganisms, or parts of microorganisms such as capsules. These substances can no longer cause disease, but they can still stimulate an immune response, much as naturally acquired pathogens do. Vaccines are discussed in more detail in Chapter 18.

Artificially acquired passive immunity involves the introduction of antibodies (rather than antigens) into the body. These antibodies come from an animal or person who is already immune to the disease.

The antibodies are found in the serum (page 467) of the immune animal or individual. Because most of the antibodies remain in the serum, **antiserum** has become a generic term for blood-derived fluids containing antibodies. Hence, the study of reactions between antibodies and antigens is called **serology.** When serum is subjected

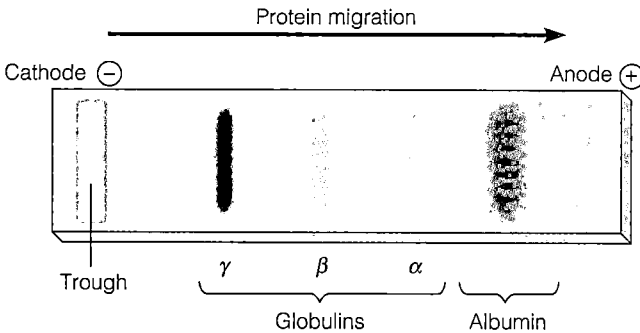

Cathode ⊖   Anode ⊕

Protein migration

Trough

γ   β   α

Globulins   Albumin

FIGURE 17.2 The separation of serum proteins by gel electrophoresis. In this procedure, serum is placed in a trough cut into a gel. In response to an electrical current, the negatively charged proteins of the serum migrate through the gel from the negatively charged end (cathode) to the positively charged end (anode).

■ Antibodies are concentrated mainly in the gamma (γ) fraction of globulins, hence the original term gamma globulin.

to an electrical current in the laboratory during gel electrophoresis (discussed in Chapter 9), the proteins within it move at different rates, as shown in Figure 17.2. The proteins separate into fractions, or **globulins,** simple proteins with certain solubility characteristics. The fractions are termed alpha, beta, and gamma globulins. The fraction that contains most of the antibodies that were present in the original sample is the gamma fraction. This antibody-rich serum component is therefore called **immune serum globulin,** or **gamma globulin.**

When immune serum globulin from an individual who is immune to a disease is injected into the body, it confers immediate protection against the disease. However, although artificially acquired passive immunity is immediate, it is short-lived because antibodies are degraded by the recipient. The half-life of an injected antibody (the time required for half of the antibodies to disappear) is typically about 3 weeks.

The Duality of the Immune System

Learning Objective

■ *Differentiate humoral (antibody-mediated) from cell-mediated immunity.*

The first Nobel Prize for Physiology or Medicine was awarded in 1901 to the German bacteriologist Emil von Behring for his discovery that immunity to diphtheria could be transferred from an animal immunized against diphtheria to another animal. The factors involved in this immunity were known as *humoral immunity,* because they were found in body fluids (formerly called humors in

medical parlance). These factors were identified much later, in the 1930s, as antibodies.

Even before von Behring's work, Russian biologist Elie Metchnikoff had shown the importance of phagocytic cells in the body's nonspecific defenses. He also observed that these cells were much more effective in immunized animals. (Metchnikoff received a Nobel Prize for his work in immunology in 1908.) Specific immunity of this type was called *cell-mediated immunity.* During the 1940s and 1950s, lymphocytes were identified as the responsible cells. In experiments similar to those of von Behring, it was shown that lymphocytes could transfer immunity to certain diseases, such as tuberculosis, between animals. As you will soon see, cell-mediated and humoral immunity are closely interrelated.

Humoral (Antibody-Mediated) Immunity

Humoral immunity, or **antibody-mediated immunity,** involves the production of antibodies that act against foreign organisms and substances. These antibodies are found in extracellular fluids, such as blood plasma, lymph, and mucus secretions. Cells called **B cells** (or B lymphocytes) are responsible for the production of antibodies. The humoral immune response defends primarily against bacteria, bacterial toxins, and viruses that are circulating freely in the body's fluids. It is also a factor in some reactions against transplanted tissue.

Cell-Mediated Immunity

Cell-mediated immunity involves specialized lymphocytes called **T cells** (or T lymphocytes) that act against foreign organisms or tissues. T cells also regulate the activation and proliferation of other immune system cells, such as macrophages. The cell-mediated immune response is most effective against bacteria and viruses located within phagocytic or infected host cells, and against fungi, protozoa, and helminths. This system is the primary responder to transplanted tissue, such as a foreign skin graft. Cell-mediated immunity mounts a response to reject the foreign tissue. It is also an important factor in our defense against cancer.

Antigens and Antibodies

Learning Objective

■ *Define antigen and hapten.*

Antigens and antibodies play key roles in the response of the immune system. Antigens (sometimes more descriptively called *immunogens*) provoke a highly specific immune response in an organism.

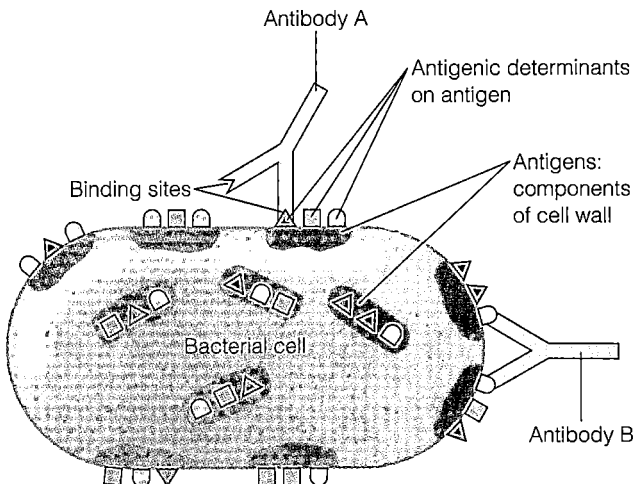

FIGURE 17.3 **Antigenic determinants (epitopes).** In this diagram, the antigenic molecules are components of the bacterial cell wall. Each antigen carries more than one antigenic determinant. Each antibody molecule has at least two binding sites that can attach to a specific kind of determinant site on an antigen. An antibody can also bind to identical determinants on two different cells at the same time (see Figure 18.4), which can cause neighboring cells to aggregate.

■ **Molecules that elicit an antibody response are called antigens.**

The Nature of Antigens

Normally, the immune system recognizes components of the body it protects as "self" and foreign matter as "nonself." This recognition is the reason people's defense systems do not usually produce antibodies against their own body tissues (although we will see in Chapter 19 that this sometimes does occur).

Most **antigens** are either proteins or large polysaccharides. Lipids and nucleic acids usually are antigenic only when combined with proteins and polysaccharides. Antigenic compounds are often components of invading microbes, such as the capsules, cell walls, flagella, fimbriae, and toxins of bacteria; the coats of viruses; or the surfaces of other types of microbes. Nonmicrobial antigens include pollen, egg white, blood cell surface molecules, serum proteins from other individuals or species, and surface molecules of transplanted tissues and organs.

Generally, antibodies recognize and interact with specific regions on antigens called **antigenic determinants,** or **epitopes** (Figure 17.3). The nature of this interaction depends on the size, shape, and chemical nature of the antigenic determinant, as well as the chemical structure of the binding site on the antibody molecule.

Most antigens have a molecular weight of 10,000 or higher. A foreign substance that has a low molecular weight is often not antigenic unless it is attached to a car-

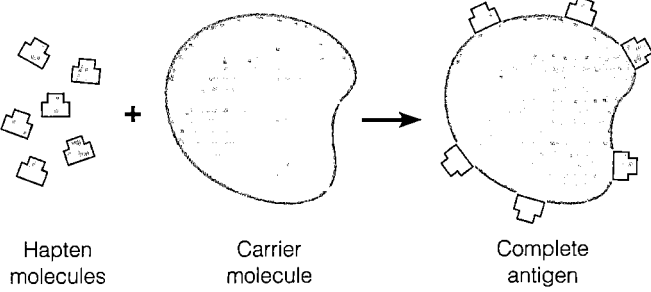

FIGURE 17.4 **Haptens.** A hapten is a molecule too small to stimulate antibody formation by itself. However, when the hapten is combined with a larger carrier molecule, usually a serum protein, the hapten and its carrier together function as an antigen and can stimulate an immune response.

■ **A hapten is a partial antigen.**

rier molecule. These small compounds are called **haptens** (from the Greek *hapto,* to grasp; Figure 17.4). Once an antibody against the hapten has been formed, the hapten will react with the antibody independent of the carrier molecule. Penicillin is a good example of a hapten. This drug is not antigenic by itself, but some people develop an allergic reaction to it. (Allergic reactions are a type of immune response.) In these people, when penicillin combines with serum proteins, the resulting combined molecule initiates an immune response.

The Nature of Antibodies

Learning Objectives

■ *Explain the function of antibodies, and describe their structural and chemical characteristics.*

■ *Name one function for each of the five classes of antibodies.*

Antibodies are proteins that are made in response to an antigen and can recognize and bind to that antigen. Antibodies can therefore help neutralize or destroy that antigen. Antibodies are highly specific in recognizing the antigen that stimulated their original formation. An antigen such as a bacterium or virus will probably have several antigenic determinant sites that cause the production of different antibodies.

Each antibody has at least two identical sites that bind to antigenic determinants. These sites are known as **antigen-binding sites.** The number of antigen-binding sites on an antibody is called the **valence** of that antibody. For example, most human antibodies have two binding sites; therefore, they are bivalent. Antibodies are members of a group of soluble proteins collectively known as **immunoglobulins (Igs).**

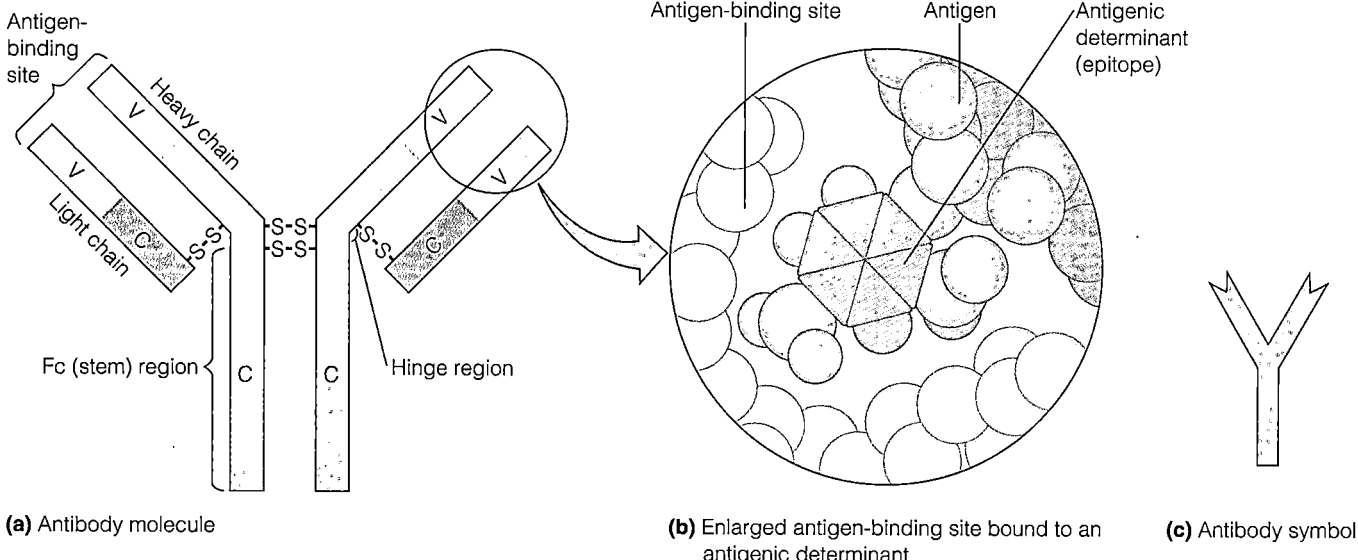

(a) Antibody molecule

(b) Enlarged antigen-binding site bound to an antigenic determinant

(c) Antibody symbol

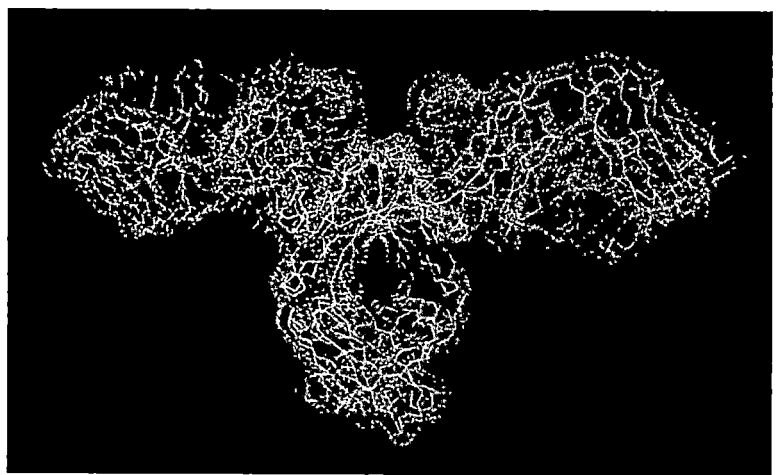

(d) Computer graphic model of an antibody molecule

FIGURE 17.5 The structure of a typical antibody molecule. (a) The Y-shaped molecule is composed of two light chains and two heavy chains linked by disulfide bridges (S—S). Most of the molecule is made up of constant regions (C), which are the same for all antibodies of the same class. The amino acid sequences of the variable regions (V), which form the two antigen-binding sites, differ from molecule to molecule. **(b)** One antigen-binding site is shown enlarged and bound to an antigenic determinant. **(c)** The symbol used to represent antibodies throughout the text. **(d)** A computer graphic of an antibody molecule.

■ **What is responsible for the specificity of each different antibody?**

Antibody Structure

Because a bivalent antibody has the simplest molecular structure, it is called a **monomer.** A typical antibody monomer has four protein chains: two identical *light (L) chains* and two identical *heavy (H) chains.* ("Light" and "heavy" refer to the relative molecular weights.) The chains are joined by disulfide links (Figure 2.16c on page 49) and other bonds to form a Y-shaped molecule (Figure 17.5). The Y-shaped molecule is flexible and can assume a T shape (notice the hinge region in Figure 17.5a).

The two sections located at the ends of the Y's arms are called *variable (V) regions.* The amino acid sequences, and therefore the three-dimensional structure, of these two variable regions are identical on any one antibody. Their structure reflects the nature of the antigen for which they are specific. These are the two antigen-binding sites found on each antibody monomer. The stem of the antibody monomer and the lower parts of the Y's arms are called *constant (C) regions.* There are five major types of C regions, which accounts for the five

table 17.1   *A Summary of Immunoglobulin Classes*

Characteristics	IgG	IgM	IgA	IgD	IgE
Structure	Monomer	Pentamer	Dimer (with secretory component)	Monomer	Monomer
Percentage of total serum antibody	80%	5–10%	10–15%*	0.2%	0.002%
Location	Blood, lymph, intestine	Blood, lymph, B-cell surface (as monomer)	Secretions (tears, saliva, mucus, intestine, milk), blood lymph	B-cell surface, blood, lymph	Bound to mast and basophil cells throughout body, blood
Molecular weight	150,000	970,000	405,000	175,000	190,000
Half-life in serum	23 days	5 days	6 days	3 days	2 days
Complement fixation	Yes	Yes	No**	No	No
Placental transfer	Yes	No	No	No	No
Known functions	Enhances phagocytosis; neutralizes toxins and viruses; protects fetus and newborn	Especially effective against microorganisms and agglutinating antigens; first antibodies produced in response to initial infection	Localized protection on mucosal surfaces	Serum function not known; presence on B cells functions in initiation of immune response	Allergic reactions; possibly lysis of parasitic worms

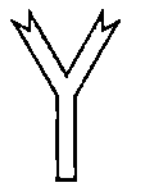

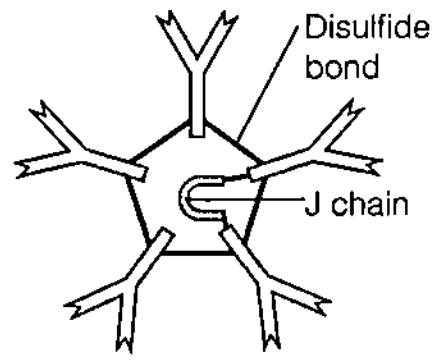

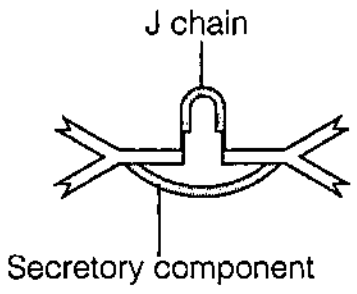

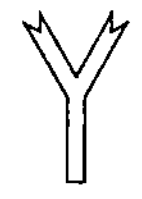

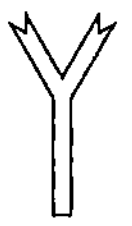

*Percentage in serum only; if mucous membranes and body secretions are included, percentage is much higher.

**May be yes via alternate pathway.

major classes of immunoglobulins, which will be described shortly.

The stem of the Y-shaped antibody monomer is called the *Fc region,* so named because in the early days of working out antibody structure, it was a fragment (F) that crystallized (c) in cold storage.

These Fc regions are often important in immunological reactions. If left exposed after both antigen–binding sites attach to an antigen such as a bacterium, the Fc regions of adjacent antibodies can bind complement. This leads to the destruction of the bacterium, as shown in Figure 16.11, page 470. Conversely, the Fc region may bind to a cell, leaving the antigen–binding sites of adjacent antibodies free to react with antigens, as shown in Figure 19.1, page 522.

Immunoglobulin Classes

The five classes of immunoglobulins (Igs) are designated IgG, IgM, IgA, IgD, and IgE. Each class plays a different role in the immune response. The structures of IgG, IgD, and IgE molecules resemble the structure shown in Figure 17.5a; molecules of IgA and IgM are usually two or five monomers, respectively, joined together by disulfide bonds. The structures and characteristics of the immunoglobulin classes are summarized in Table 17.1.

IgG IgG antibodies account for about 80% of all antibodies in serum. These monomer antibodies readily cross the walls of blood vessels and enter tissue fluids. Maternal IgG antibodies, for example, can cross the placenta and confer passive immunity to a fetus.

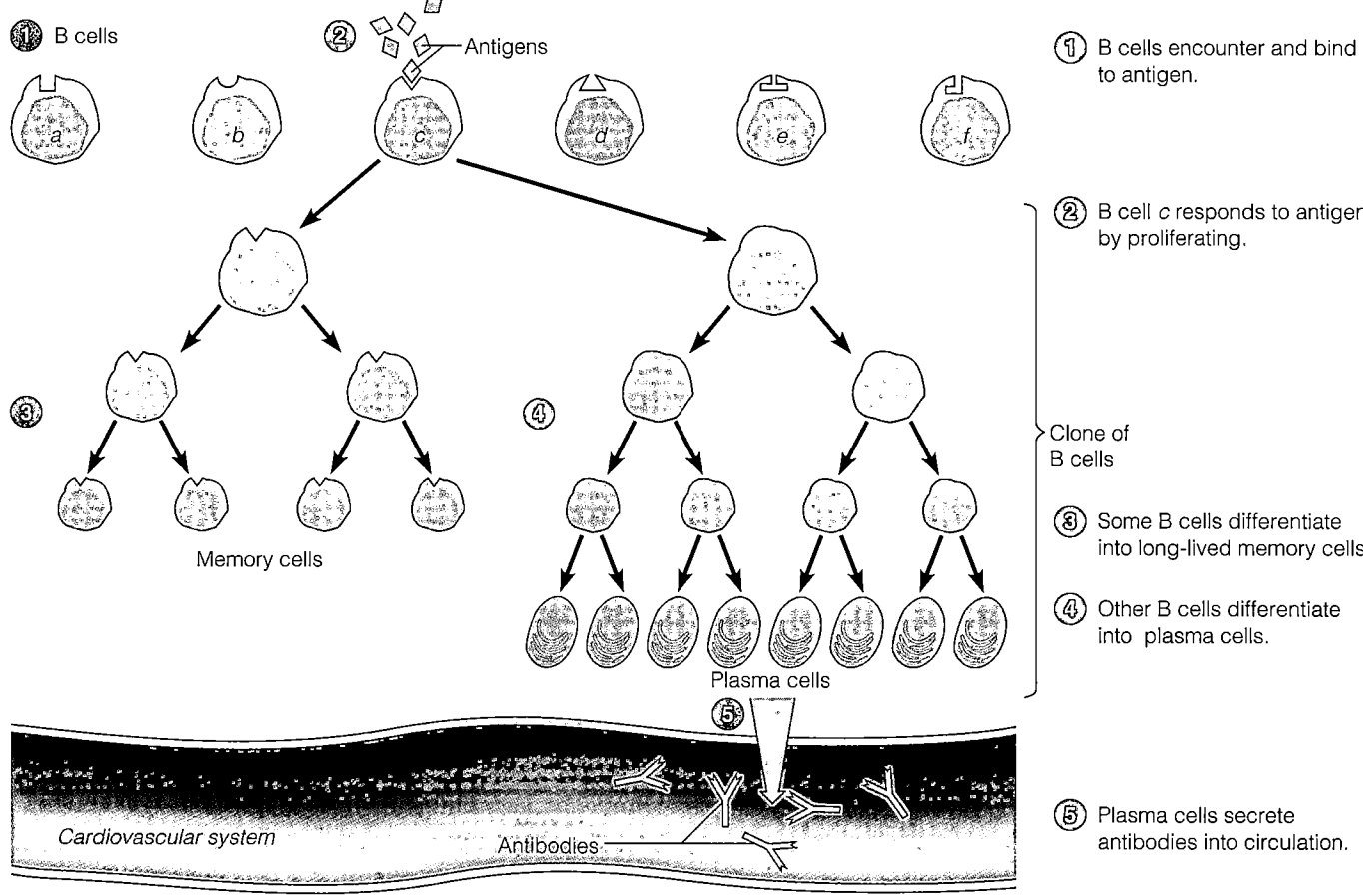

B cells

① B cells encounter and bind to antigen.

② B cell *c* responds to antigen by proliferating.

Clone of B cells

③ Some B cells differentiate into long-lived memory cells.

④ Other B cells differentiate into plasma cells.

Memory cells

Plasma cells

⑤ Plasma cells secrete antibodies into circulation.

Cardiovascular system

Antibodies

FIGURE 17.7 Clonal selection and differentiation of B cells. B cells can recognize an almost infinite number of antigens, but each particular cell recognizes only one type of antigen. An encounter with a particular antigen triggers the proliferation of a cell that is specific for that antigen (here, B cell *c*) into a clone of cells with the same specificity, hence the term clonal selection.

■ What is selected for in clonal selection?

a few days but can produce about 2000 antibody molecules per second.

Stimulation of a B cell by an antigen also results in the production of a population of cells called *memory cells,* which function in long-term immunity (discussed shortly).

An important question raised by the clonal selection theory is why the immune system does not react to the body's own cells and macromolecules, a phenomenon called **self-tolerance,** or the differentiation between *self* and *nonself.* The exact answer is still a mystery, but apparently the B and T cells that interact with self antigens are somehow destroyed during fetal development, most probably during passage through the thymus. This mechanism is called **clonal deletion.** A more recent proposal is that the body selectively learns to discriminate between the dangerous and the nondangerous rather than between self and nonself. Normal cell death, or apoptosis, leads to

toleration. Anything that violently kills cells signals danger and triggers an inflammatory response. For example, it is possible that the immune response that causes the rejection of an organ transplant is not against nonself, but is initiated by the cell damage associated with the transplant surgery.

Antigen—Antibody Binding and Its Results

Learning Objective

■ *Explain how an antibody reacts with an antigen, and identify the consequences of the reaction.*

When an antibody encounters an antigen for which it is specific, an **antigen—antibody complex** rapidly forms. An antibody binds to an antigen at the *antigen-binding site* (see Figure 17.5a, b). Sometimes, especially when large amounts of antigen are present, some antigens attach to

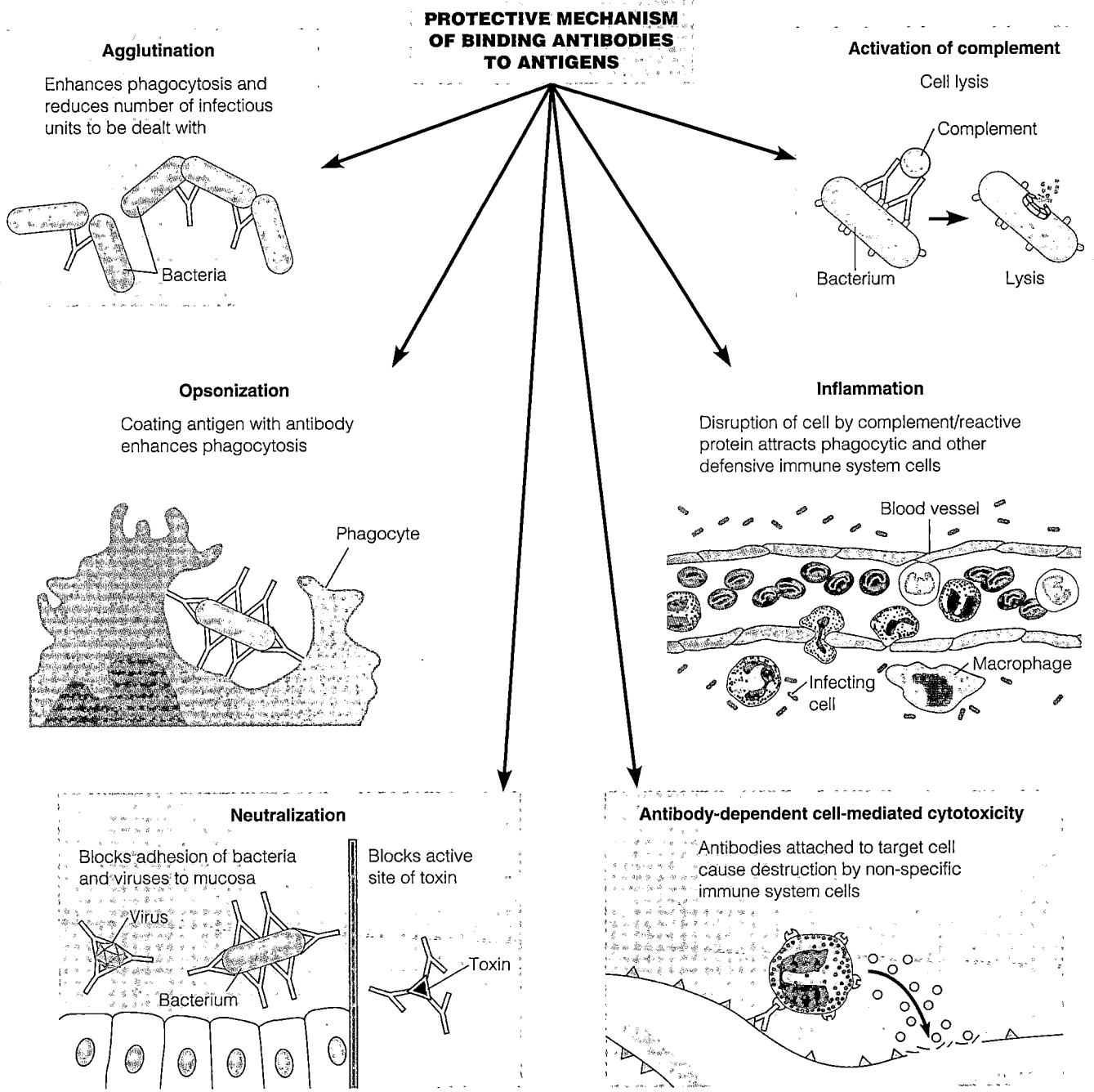

FIGURE 17.8 The results of antigen–antibody binding. The binding of antibodies to antigens to form antigen–antibody complexes tags foreign cells and molecules for destruction by phagocytes and complement.

■ What are some of the possible outcomes of an antigen–antibody reaction?

antigen-binding sites for which they are less than a perfect fit. Because of this, some antibodies are poorer matches for the antigen; these antibodies are said to have less *affinity*.

The binding of an antibody to an antigen basically protects the host by tagging foreign cells and molecules for destruction by phagocytes and complement. The antibody molecule itself is not damaging to the antigen. Foreign organisms and toxins are rendered harmless by only a few mechanisms, as summarized in Figure 17.8. These are agglutination, opsonization, neutralization, antibody-dependent cell-mediated cytotoxicity, and the

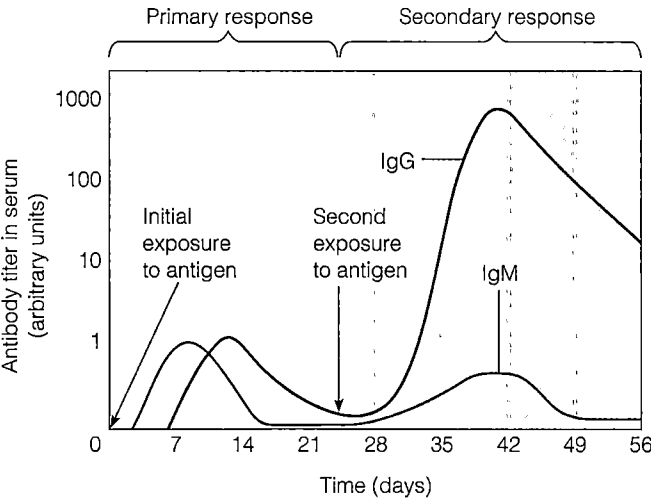

FIGURE 17.9 The primary and secondary immune responses to an antigen. IgM appears first in response to the initial exposure. IgG follows and provides longer-term immunity. The second exposure to the same antigen stimulates the memory cells formed at the time of initial exposure to rapidly produce a large amount of antibody. The antibodies produced in response to this second exposure are mostly IgG.

■ **Why do many diseases, such as measles, occur only once in a person yet others, such as colds, occur more than once?**

action of complement, resulting in cell lysis and inflammation (see Figure 16.11 on page 470).

In **agglutination,** antibodies cause antigens to clump together. For example, the two antigen-binding sites of an IgG antibody can combine with antigenic determinants on two different foreign cells, aggregating the cells into clumps that are more easily ingested by phagocytes. Because of its numerous binding sites, IgM is more effective at cross-linking and aggregating particulate antigens (see Figure 18.5 on page 508). IgG requires 100 to 1000 times as many molecules for the same results. (In Chapter 18, we will see how agglutination is important in the diagnosis of some diseases.)

In neutralization, IgG antibodies inactivate viruses by blocking their attachment to host cells, and they neutralize bacterial toxins by blocking their active sites. In opsonization, the antigen, such as a bacterium, is coated with antibodies that enhance its ingestion and lysis by phagocytic cells. Antibody-dependent cell-mediated cytotoxicity (see page 493 and Figure 17.17) resembles opsonization in that the target organism is coated with antibodies. However, destruction of the target cell is by nonspecific immune system cells that remain external to the target cell.

Finally, both IgG and IgM antibodies trigger the complement system, so named because it complements the action of antibodies. **Inflammation** is caused by in-

fection or tissue injury (see Figure 16.9). One aspect of inflammation is that it will often cause microbes in the inflamed area to become coated with certain reactive proteins. This leads to the attachment and activation of complement on the microbe's surface, which is then lysed. **Lysis** of the microbe attracts phagocytes and other defensive immune system cells to the area.

As we will see in Chapter 19, the action of antibodies can also damage the host. For example, immune complexes of antibody, antigen, and complement can damage host tissue. Antigens combining with IgE on mast cells can initiate allergic reactions, and antibodies can react with host cells and cause autoimmune disorders.

Immunological Memory

Learning Objective

■ *Distinguish a primary from a secondary immune response.*

The intensity of the humoral response is reflected by the **antibody titer,** the amount of antibody in the serum. After the initial contact with an antigen, the exposed person's serum contains no detectable antibodies for several days. Then there is a slow rise in antibody titer; first IgM antibodies are produced, followed by IgG (Figure 17.9). Finally, a gradual decline in antibody titer occurs. This pattern is characteristic of a **primary response** to an antigen.

The immune responses of the host intensify after a second exposure to an antigen. This **secondary response** is also called the **memory,** or **anamnestic, response.** As we saw in Figure 17.7, some activated B lymphocytes do not become antibody-producing plasma cells but persist as long-lived **memory cells.** Years (perhaps decades) later, if these cells are stimulated by the same antigen, they rapidly differentiate into antibody-producing plasma cells. This is the basis of the secondary immune response shown in Figure 17.9.

A similar response occurs with T cells, which, as we will see in Chapter 19, is necessary for establishing the lifelong memory for distinguishing between self and nonself.

Monoclonal Antibodies and Their Uses

Learning Objective

■ *Define monoclonal antibodies, and identify their advantage over conventional antibody production.*

We have been discussing antibodies primarily in the context of their protective effects against disease. Antibodies are also useful in clinical medicine, especially in diagnosing diseases, which will be emphasized in Chapter 18.

If we have a known antibody, it can be used, for example, to identify an unknown pathogen. Before monoclonal antibodies became available, antibodies were generally produced in animals by a procedure similar to natural infection. An antigen such as a toxin or microbial pathogen was injected into an animal. If large amounts were needed, a large animal was used. However, the amounts were still quite limited, and the serum contained many other antibodies produced by the animal. Ideally, if an antibody-producing B cell could be grown by standard cell-culture methods, it would produce the desired antibody in nearly unlimited amounts without contamination by other antibodies. Unfortunately, a B cell reproduces only a few times under such conditions.

However, in 1984, three immunologists shared the Nobel Prize for their discovery of a method to prolong the culture life of antibody-producing B cells. Niels Jerne, Georges Köhler, and César Milstein made this discovery in 1975. Scientists have long observed that antibody-producing B cells may become cancerous. In this case, their proliferation is unchecked and they are called *myelomas.* These cancerous B cells can be isolated and propagated indefinitely in cell culture. Cancer cells, in this sense, are "immortal." The breakthrough came in combining an "immortal" cancerous B cell with an antibody-producing normal B cell. When fused, this combination is termed a **hybridoma.**

When a hybridoma is grown in culture, its genetically identical cells continue to produce the type of antibody characteristic of the ancestral B cell. The importance of the technique is that clones of the antibody-secreting cells now can be maintained indefinitely in cell culture and can produce immense amounts of identical antibody molecules. Because all of these antibody molecules are produced by a single hybridoma clone, they are called **monoclonal antibodies** (Figure 17.10 on page 488).

Monoclonal antibodies are useful for three reasons: They are uniform, they are highly specific, and they can be readily produced in large quantities. Because of these qualities, monoclonal antibodies have assumed enormous importance as diagnostic tools. For instance, commercial kits use monoclonal antibodies to recognize chlamydial and streptococcal bacteria, and nonprescription pregnancy tests use monoclonal antibodies to indicate the presence of a hormone excreted only in the urine of a pregnant woman (see Figure 18.13 on page 515).

Monoclonal antibodies are also being used to overcome unwanted effects of the immune system, such as rejection of transplanted organs (see Chapter 19). In this case, antibodies are prepared that react with the T cells involved in rejecting the transplanted tissue; the antibodies suppress the T-cell activity.

The prospect of using monoclonal antibodies to treat cancer is also of great interest. One approach is to combine monoclonal antibodies specifically targeted against cancer cells with a toxin to make an **immunotoxin.** The idea is that immunotoxins could be carried throughout the body by the cardiovascular system to specifically attach to and kill cancer cells or to target the signals that stimulate the cells' growth.

The therapeutic use of monoclonal antibodies has been limited because these antibodies are currently produced by mouse cells. The immune systems of some people have reacted against the foreign mouse proteins. Monoclonal antibodies derived from human cells would probably provoke fewer reactions. Several approaches are being taken to solve this problem. One is to construct, by recombinant DNA methods, antibodies with variable regions derived from mouse cells and constant regions derived from human sources. These antibodies, which are more compatible with the human immune system, are called **chimeric monoclonal antibodies.** Genetic engineering might also be used to alter mouse antibodies so that they have characteristics that are "more human."

T Cells and Cell-Mediated Immunity

Learning Objective

- Identify at least one function of each of the following in cell-mediated immunity: cytokines, interleukins, interferons.

As briefly mentioned earlier, antigens that stimulate cell-mediated immunity are mostly intracellular in nature. This type of immunity also generally requires the continued presence of the antigen to maintain its effectiveness. Unlike humoral immunity, cell-mediated immunity is not transferred to the fetus via the placenta.

Cell-mediated immunity is based on the activity of certain specialized lymphocytes, primarily T cells.

Chemical Messengers of Immune Cells: Cytokines

As immunologists began working with cultures of immune system cells, they found that the fluids supporting these cultures contained a confusing array of soluble "factors"—chemical messengers—that regulated many other cells of the immune system. In effect, the cells communicated with each other, triggering their activity or causing them to differentiate. These factors were assigned names such as B-cell activating factor or B-cell differentiation factor. Only after decades of work, beginning in the 1950s, did the picture begin to be clarified.

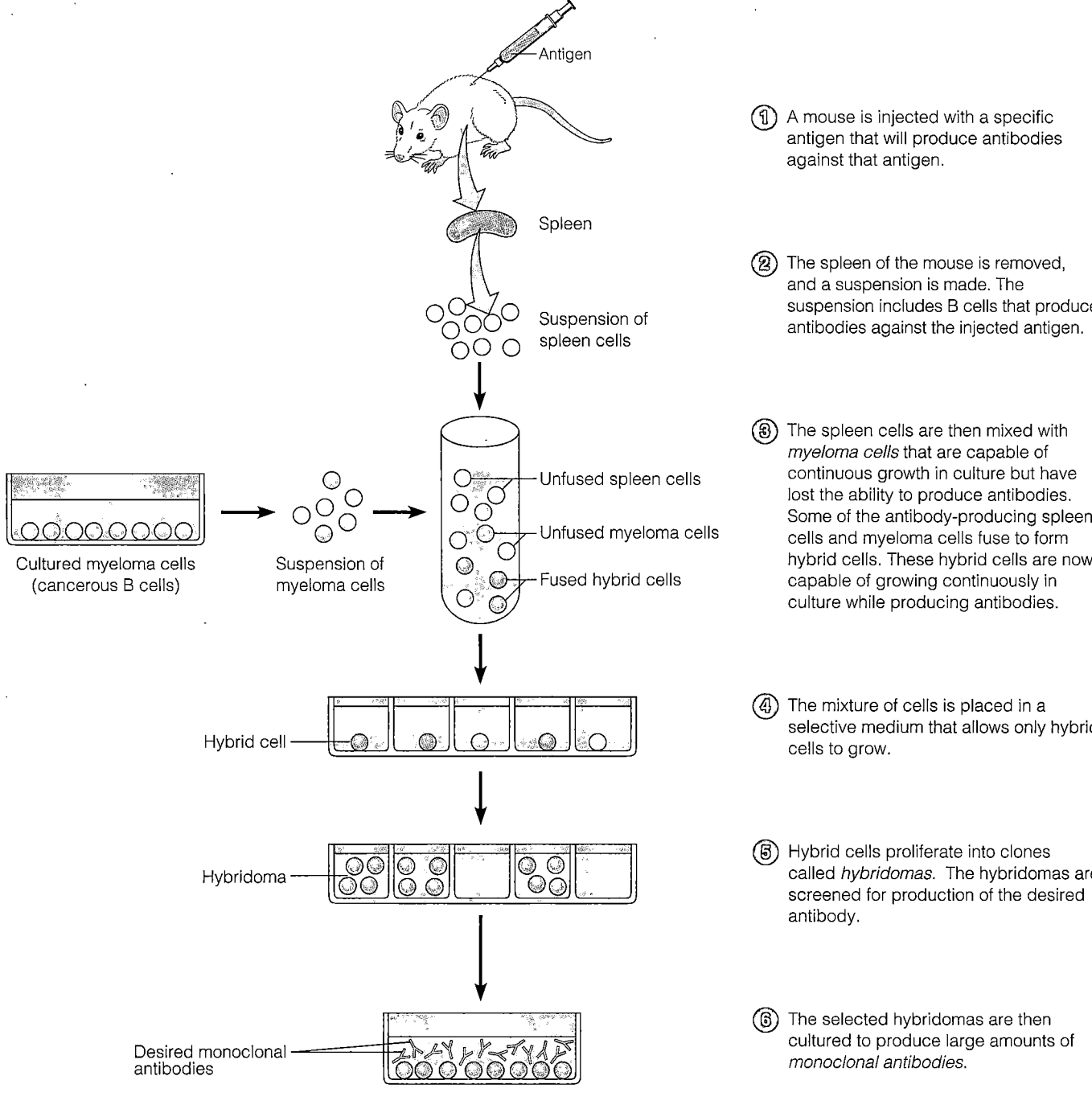

FIGURE 17.10   The production of monoclonal antibodies.

① A mouse is injected with a specific antigen that will produce antibodies against that antigen.

② The spleen of the mouse is removed, and a suspension is made. The suspension includes B cells that produce antibodies against the injected antigen.

③ The spleen cells are then mixed with *myeloma cells* that are capable of continuous growth in culture but have lost the ability to produce antibodies. Some of the antibody-producing spleen cells and myeloma cells fuse to form hybrid cells. These hybrid cells are now capable of growing continuously in culture while producing antibodies.

④ The mixture of cells is placed in a selective medium that allows only hybrid cells to grow.

⑤ Hybrid cells proliferate into clones called *hybridomas*. The hybridomas are screened for production of the desired antibody.

⑥ The selected hybridomas are then cultured to produce large amounts of *monoclonal antibodies*.

■ Monoclonal antibody production provides pure preparations of antibodies in large quantities.

Many of these variously named factors, now known as **cytokines,** were ultimately determined to be the same. There are now more than 60 known cytokines. The multiple activities of cytokines can be thought of much like words that have different meanings in different contexts; that is, cytokines might inhibit a response under one set of circumstances and stimulate a response under others.

Cytokines that serve as communicators between leukocytic (white cell) populations are now known as **interleukins** (between leukocytes). When enough information, including the amino acid sequence, is known, the cytokine is assigned an interleukin number by an international committee. At present, we know of 18 interleukins. For example, *interleukin-1 (IL-1)* is produced by every nucleated cell, especially macrophages, and affects an array of activities of practically all cell types. It has had at least eight descriptive names. *Interleukin-2 (IL-2)* is produced mostly by CD4 T$_H$ cells (which we will be discussing shortly), and is important because of its role in T-cell proliferation.

Also classed as cytokines are interferons, which help protect against the viral infection of cells. Tumor necrosis factor (TNF) is another important cytokine. Colony-stimulating factor (CSF) stimulates the growth of a number of infection-fighting cells.

One of the more recent members of the cytokine family are the **chemokines**—from chemotaxis; they induce the migration of leukocytes into infected areas. The best known of at least 30 identified chemokines is IL-8. Table 17.2 lists some of the better-known cytokines.

The role of cytokines in the stimulation of the immune system has suggested their use as therapeutic agents (see the box on page 490). Laboratory experiments have shown IL-1 to be of potential value in controlling several parasitic diseases, as well as inhibiting the blood flow that feeds animal tumors. The possible use of the immune system to control cancer is also focusing attention on cytokines. One approach is to genetically engineer tumor cells to make cytokines to stimulate cell-mediated immunity against the tumor.

Cellular Components of Immunity

Learning Objective

- *Describe at least one function for each of the following: T$_H$ cell, T$_C$ cell, T$_D$ cell, T$_S$ cell, APC, MHC, activated macrophage, NK cell.*

T cells are the key cellular component of immunity. Like B cells and all other cells involved in the immune response, T cells develop from stem cells in the bone marrow (see Figure 17.6). T cells are influenced by the thymus gland, where they differentiate into mature cells (the T in *T cell* is for thymus). Next they migrate to lymphoid organs, where they are apt to encounter antigens.

Most pathogens first enter the gastrointestinal tract or lungs, where they encounter a barrier of epithelial cells. Normally they can pass this barrier only by a scattered number of gateway cells called **M cells** (for membranous) that contain pockets filled with B cells, T cells, and macrophages. M cells are located directly over lymphoid structures where the antigens activate B cells. It is there that antibodies, mostly IgA essential for mucosal immunity, are formed and migrate to the mucosal lining.

Like the B cells of humoral immunity, T cells use antigen receptors to recognize and react to an antigen. Also,

table 17.2	A Summary of Some Important Cytokines
Cytokine	**Representative Activity**
Interleukin-1 (IL-1)	Stimulates T$_H$ cells in presence of antigens; chemically attracts phagocytes in inflammatory response
Interleukin-2 (IL-2)	Involved in proliferation of antigen-stimulated T$_H$ cells, proliferation and differentiation of B cells, and activation of T$_C$ cells and NK cells
Interleukin-8 (IL-8)	Chemoattractant for immune system cells and phagocytes to site of inflammation
Interleukin-12 (IL-12)	Mainly involved in differentiation of CD4-type T cells
Gamma-interferon (γ-IFN)	Inhibits intracellular viral replication; increases activity of macrophages against microbes and tumor cells
Tumor necrosis factor-beta (TNF-β)	Cytotoxic to tumor cells; enhances activity of phagocytic cells
Granulocyte-macrophage colony-stimulating factor (GM-CSF)	Stimulates formation of red and white blood cells from stem cells

like B cells, each T cell can react specifically with only a single antigen. As in humoral immunity, clonal selection is the mechanism by which T cells with particular antigen receptors are stimulated to be differentiated into the effector T cells that carry out cell-mediated immunity and proliferate. As with B cells, some of the *effector (antigen-stimulated) T cells* become memory cells, which provide the basis for a secondary immune response if they encounter the same antigen later. The body's ability to make new T cells decreases with age, beginning in late adolescence. Eventually, the T-cell-producing thymus gland becomes inactive, and bone marrow produces fewer B cells. As a result, the immune system is relatively weak in the elderly. However, sufficient T and B memory cells survive to make immunization effective for such diseases as influenza and pneumococcal pneumonia.

Types of T Cells

There seem to be four main functional types of T cells: helper T (T$_H$) cells, cytotoxic T (T$_C$) cells, delayed hypersensitivity T (T$_D$) cells, and perhaps, suppressor

Is IL-12 the Next "Magic Bullet"?

Worldwide, cancer and AIDS kill 10 million people annually. If laboratory tests are any indication of the future, the cytokine IL-12 (interleukin-12) could be the "magic bullet" against AIDS and many forms of cancer.

A dozen research teams in the United States and Italy are investigating the potential for IL-12 to treat disease. Since its discovery in the 1980s, IL-12 proved to be different from the other cytokines. For one thing, it is made from the products of two different genes instead of the usual one gene. IL-12 is released by B cells and macrophages in response to infections. It inhibits the humoral response and activates the cell-mediated response by activating T_H cells and recruiting NK cells. Gamma-interferon, γ-IFN, causes T_H cells and NK cells to produce more γ-IFN, which activates more T_H cells and NK cells.

Early studies have shown that mice infected with *Leishmania* and *Toxoplasma gondii* were cured by IL-12; and IL-12

may provide protection against mycobacteria. These microbes prevent phagocytes from digesting them; consequently, they can live in the phagocytes, protected from the host's immune response. IL-12 activated the phagocytes to kill the parasites.

IL-12 is known to inhibit about 20 kinds of tumors in mice by inhibiting blood vessel growth to tumors. The Genetics Institute of Massachusetts is currently conducting clinical trials to test the effectiveness of IL-12 in patients with advanced kidney cancer. (A safe dosage for IL-12 was determined in the Phase I trial.)

Researchers at Philadelphia's Children's Hospital have shown that HIV infection decreases the production of IL-12, which may make the patient more susceptible to opportunistic infections. When treated with IL-12, however, T_H cells taken from HIV-positive people responded to viruses, including HIV.

IL-12 could also provide the boost needed for DNA vaccines and gene therapy to be successful. Viruses are used as vectors in gene therapy to insert desirable genes into animal cells. Unfortunately, in this case, the introduction of a virus causes the recipient to produce antibodies, thereby preventing the administration of more than one dose. This is especially significant because many patients, such as those with cystic fibrosis, need more than one dose of the new genes. James Wilson of the University of Pennsylvania Medical Centers has observed that antibodies are not made against the gene-carrying virus if it is administered with IL-12.

Will IL-12 be a panacea? Further studies are needed to determine whether the exaggerated responses caused by IL-12 could have adverse effects, such as autoimmune disease.

T (T_S) cells. Mature T cells of these four types can be identified by their characteristic cell-surface molecules.

Another classification of T cells is based on the type of cell-surface receptor called CD, for clusters of differentiation. Two types of these receptors are CD4 and CD8. **CD4 cells** are primarily helper T cells, whereas **CD8 cells** include both cytotoxic and suppressor T cells. The cause of the pathogenicity of the AIDS virus was identified by using monoclonal antibodies to show that a population of CD4 cells was decimated by this infection. These CD4 cells are vital to a vigorous immune response.

Individual T cells interact specifically with only a single antigen, which must be displayed on a cell surface. These cells are known as **antigen-presenting cells (APCs).** The primary APCs are macrophages (see Figure 16.7) or **dendritic cells** (Figure 17.11). Dendritic cells have no other function. The APC ingests and processes the antigen; it then displays fragments of the antigen on the cell surface. A T cell will recognize an antigenic fragment on an APC only if it is in close association with certain cell-surface self molecules. These self molecules are components of the **major histocompatibility complex (MHC).** The particular MHC proteins a person carries

are unique to that individual. Thus, they serve as signals by which the immune system distinguishes self from nonself. The MHC in humans will be discussed in more detail in Chapter 19 in relation to autoimmune disease conditions.

Helper T cells Immune system cells known as **helper T (T_H) cells** play a central role in the immune response. With the aid of cytokines, they induce the formation of cytotoxic T cells and activate macrophages. T_H cells are especially prolific producers of cytokines. Once activated by an antigen, T_H cells can use these chemical messengers to influence the activity of other immune system cells. T_H cells are also essential to the formation of many antibodies by B cells, as will be seen later in the chapter.

In helper T cell activation (Figure 17.12), ① an antigen-presenting cell first encounters and processes an antigen. ② Binding of a T_H cell to an antigen-MHC complex on the APC stimulates the APC to secrete the cytokine IL-1. ③ The IL-1 in turn activates the T_H cell, which begins to synthesize interleukin-2 (IL-2). The IL-2 secreted by the T_H cell returns to receptors on the surface of the same cell, which then begins to proliferate and differentiate into mature T_H cells. Only T_H cells that have been stim-

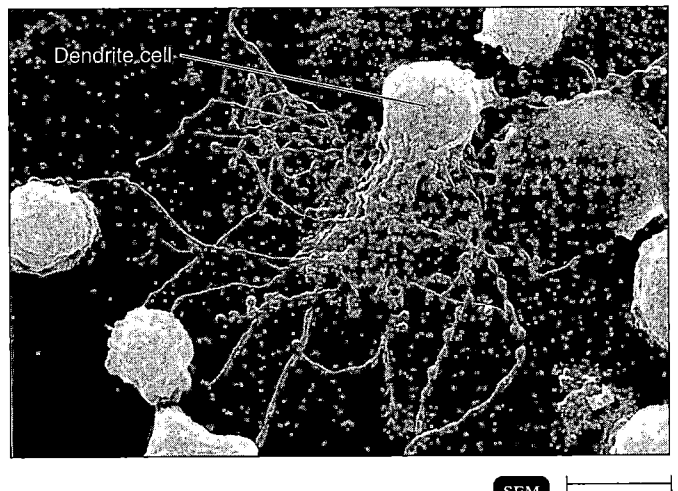

FIGURE 17.11 An antigen-presenting cell (APC). APCs are usually macrophages or dendritic cells that present antigen to helper T cells. This micrograph shows a dendritic cell exhibiting the "beaded" branches that suggested their name. The branches extend from the cell body. Dendritic cells are a type of leukocyte found in skin, airways, and lymphoid organs. Any antigen molecules being presented on the surface of the cells are too small to be seen here.

SOURCE: Microanatomy of lymphoid tissue during humoral immune responses: structure function relationships. A. K. Szakal et al., *Ann. Rev. Immunol.* 7:91–109 (1989).

■ What is the relationship between APCs and MHCs?

ulated by an antigen have receptors for IL-2, so although the cytokine IL-2 is nonspecific, the resulting T_H cells are specific to that antigen. ④ These activated T_H cells then begin to produce IL-2 and other cytokines. This IL-2 in turn stimulates other T_H cells specific to that antigen to proliferate and mature.

Cytotoxic T Cells Cytotoxic T (T_C) cells destroy target cells on contact. Because viruses, and some bacteria, reproduce within host cells, they cannot be attacked there by antibodies. The T_C cell attacks instead—for example, the virus-infected host cell shown in Figure 17.13. ① The T_C cell binds with the MHC-antigen complex on the cell's surface and then ② releases a protein called perforin. The perforin forms a pore in the membrane of the target cell, causing ③ lysis. T_C cells continue their activity as long as the stimulating antigen persists; when it disappears, they undergo apoptosis. (Chapter 19 discusses how cytotoxic T cells recognize and lyse cancerous cells.)

Delayed Hypersensitivity T Cells Delayed hypersensitivity T (T_D) cells are the cells for which cell-mediated immunity was originally named, because the transfer of these cells between animals was found to transfer immunity to tuberculosis. However, T_D cells are probably not a separate population, but mostly T_H cells and a few T_C cells that are involved in these types of immune

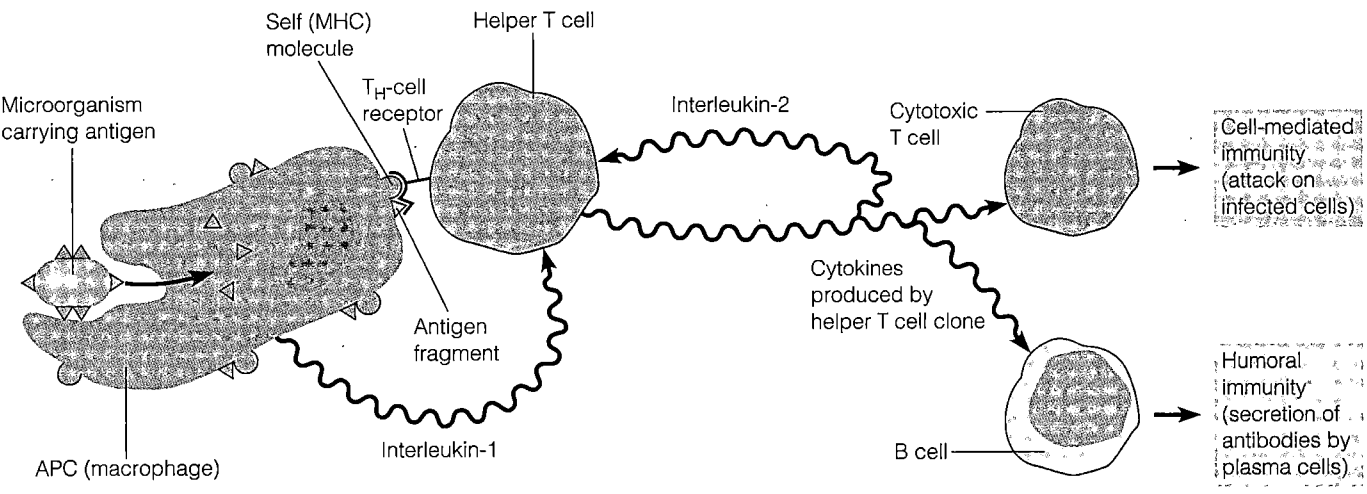

① An antigen-presenting cell (APC) encounters and processes an antigen, forming MHC-antigen complexes on its surface.

② A helper T (T_H) cell receptor binds to the complex, stimulating the APC to secrete interleukin-1.

③ This interleukin-1 stimulates the helper T cell to produce interleukin-2, which then stimulates that helper T cell to form a clone of helper T cells.

④ The cells of this clone in turn produce cytokines, stimulating cells of both immune systems.

FIGURE 17.12 The central role of helper T cells.

■ What is the function of T_H cells?

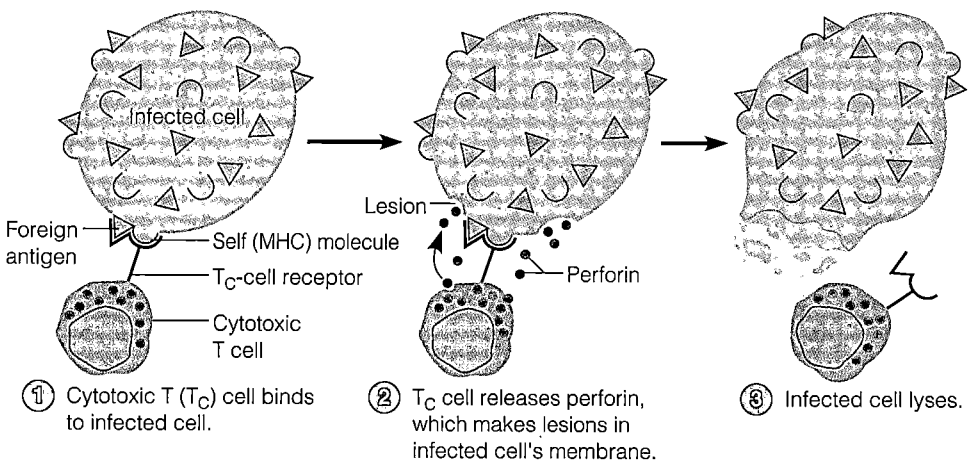

Foreign antigen — Self (MHC) molecule — T_C-cell receptor — Cytotoxic T cell

Lesion — Perforin

① Cytotoxic T (T_C) cell binds to infected cell.

② T_C cell releases perforin, which makes lesions in infected cell's membrane.

③ Infected cell lyses.

FIGURE 17.13 Cell-mediated cytotoxicity. A cytotoxic T (T_C) cell binds to the MHC-antigen complex on the surface of an infected cell. The T_C cell then discharges a protein called perforin, which lyses the infected cell.

■ What is the function of T_C cells?

functions. They are associated with certain allergic reactions, such as to poison ivy, and with the rejection of transplanted tissues (discussed fully in Chapter 19).

Suppressor T Cells Suppressor T (T_S) cells are not well understood. They are generally thought to be T cells that regulate the immune response by turning it off when an antigen is no longer present. Most immunologists consider it likely that T_S cells are not a distinct population but represent suppressive activity by the T_H and T_C cell populations.

Nonspecific Cellular Components

T cells, which are generally directed at specific antigens, are the primary warriors in cell-mediated defense. Other important elements are activated macrophages and natural killer cells, which are less specific in their activity.

Activated Macrophages Macrophages are phagocytic cells usually found in a resting state. Their phagocytic capabilities are greatly increased when they are stimulated to become **activated macrophages.** This stimulation is primarily by ingestion of antigenic material. However, cytokines from antigenically activated helper T cells may also activate macrophages. Activated macrophages are more effective, and their appearance becomes recognizably different as well; they are larger and become ruffled (Figure 17.14).

Activated macrophages are especially valued for their enhanced ability to eliminate certain virus-infected cells and pathogenic intracellular bacteria, such as the tubercle bacillus. Of great importance is their ability to attack and

destroy many cancerous cells. Furthermore, they function well as antigen-presenting cells.

Natural Killer Cells Certain lymphocytes called **natural killer (NK) cells** are capable of destroying other cells, especially virus-infected cells and tumor cells. They can also attack large parasites, as illustrated in Figure 17.17, page 495. In contrast to cytotoxic T cells, NK cells are not immunologically specific; that is, they do not need to be stimulated by an antigen. They are not phagocytic but must contact the target cell to lyse it.

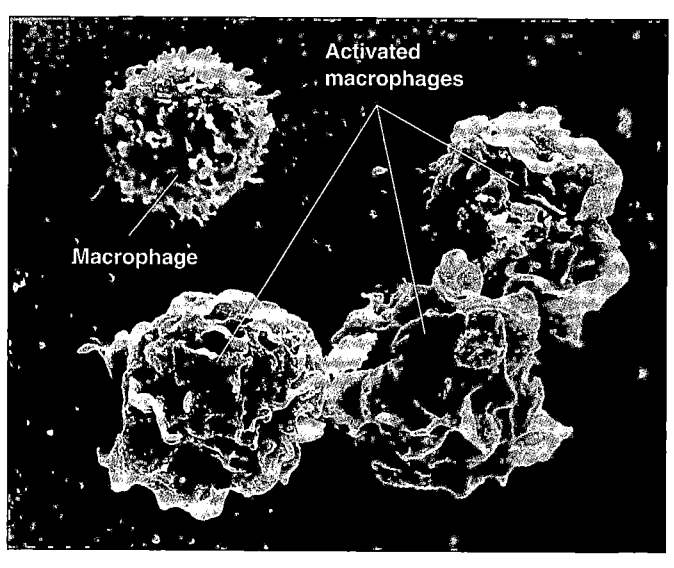

SEM |—————| 10 μm

FIGURE 17.14 Activated macrophages. When activated, macrophages became larger and ruffled.

■ How do macrophages become activated?

The functions of natural killer cells and the other principal cells involved in cell-mediated immunity are briefly summarized in Table 17.3.

The Interrelationship of Cell-Mediated and Humoral Immunity

Learning Objective

- *Compare and contrast cell-mediated and humoral immunity.*

Although cell-mediated and humoral immunity are considered separate branches of the specific immune system, they are closely interrelated. In our earlier discussion of the mechanisms by which humoral antibodies are protective (see Figure 17.8), we saw instances in which antibodies of the humoral immune system and cells of the cell-mediated immune system worked in concert. In fact, it is not even possible to describe the production of most antibodies—the core concept of humoral immunity—without introducing the involvement of helper T cells, which are central to the cell-mediated immune system.

The Production of Antibodies

Learning Objectives

- *Compare and contrast T-dependent antigens and T-independent antigens.*
- *Describe the role of antibodies and NK cells in antibody-dependent cell-mediated cytotoxicity.*

The production of antibodies against certain antigens requires the assistance of helper T cells; this type of antigen is therefore known as a **T-dependent antigen.** T-dependent antigens are mainly proteins, such as those found on viruses, bacteria, foreign red blood cells, and combinations of haptens and carrier molecules.

Figure 17.15 illustrates the process by which a T-dependent antigen stimulates the formation of an antibody. ① The antigen is ingested and processed by an APC. Fragments of the antigen are presented on the surface of the APC in close proximity to MHC molecules. ② A T_H cell specific for the antigen reacts with the MHC-antigen complex. (The need to recognize the complex rather than the antigen alone minimizes the possibility of producing antibodies against host tissues.) ③ The T_H cell reacting with the MHC-antigen complex then produces the cytokine IL-2 that influences a B cell to ④ differentiate into a plasma cell that begins to produce humoral antibodies. (Notice that, as shown in Figure 17.12, a similar mechanism activates T cells to become T_C cells.)

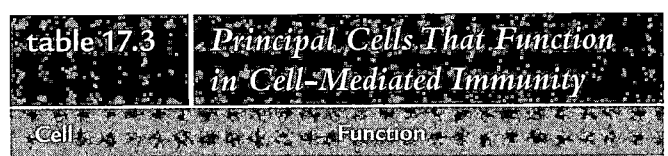

table 17.3	Principal Cells That Function in Cell-Mediated Immunity	
Cell	**Function**
Helper T (T_H) cell	Activates cytotoxic T cells and other helper T cells; necessary for B-cell activation by T-dependent antigens
Cytotoxic T (T_C) cell	Destroys target cells on contact
Delayed hypersensitivity T (T_D) cell	Provides protection against infectious agents; causes inflammation associated with allergic reactions and tissue transplant rejection
Suppressor T (T_S) cell	Regulates immune response and helps maintain tolerance
Activated macrophage	Enhanced phagocytic activity, attacks cancer cells
Natural killer (NK) cell	Attacks and destroys target cells; participates in antibody-dependent cell-mediated cytotoxicity

Antigens that can stimulate B cells directly, without the help of T cells, are called **T-independent antigens.** This type of antigen is usually composed of polysaccharides or lipopolysaccharides characterized by repeating subunits. Bacterial capsules are often good examples of T-independent antigens. The repeating subunits, as shown in Figure 17.16, can bind to multiple B-cell receptors, which is probably why they do not require T-cell assistance. T-independent antigens generally provoke a weaker immune response than T-dependent antigens, and the immune system of infants may not be stimulated by them until about age 2. These factors will be discussed further in the context of vaccines in Chapter 18.

Antibody-Dependent Cell-Mediated Cytotoxicity

With the help of antibodies produced by the humoral immune system, the cell-mediated immune system can stimulate NK cells and cells of the nonspecific defense system to kill targeted cells. In this way, an organism, such as a protozoan or helminth, that is too large to be phagocytized can be attacked by immune system cells. This process is referred to as **antibody-dependent cell-mediated cytotoxicity (ADCC).** In this process, ① the target cell must first be coated with antibodies, leaving their Fc (stem) regions pointing outward. ② In addition to NK

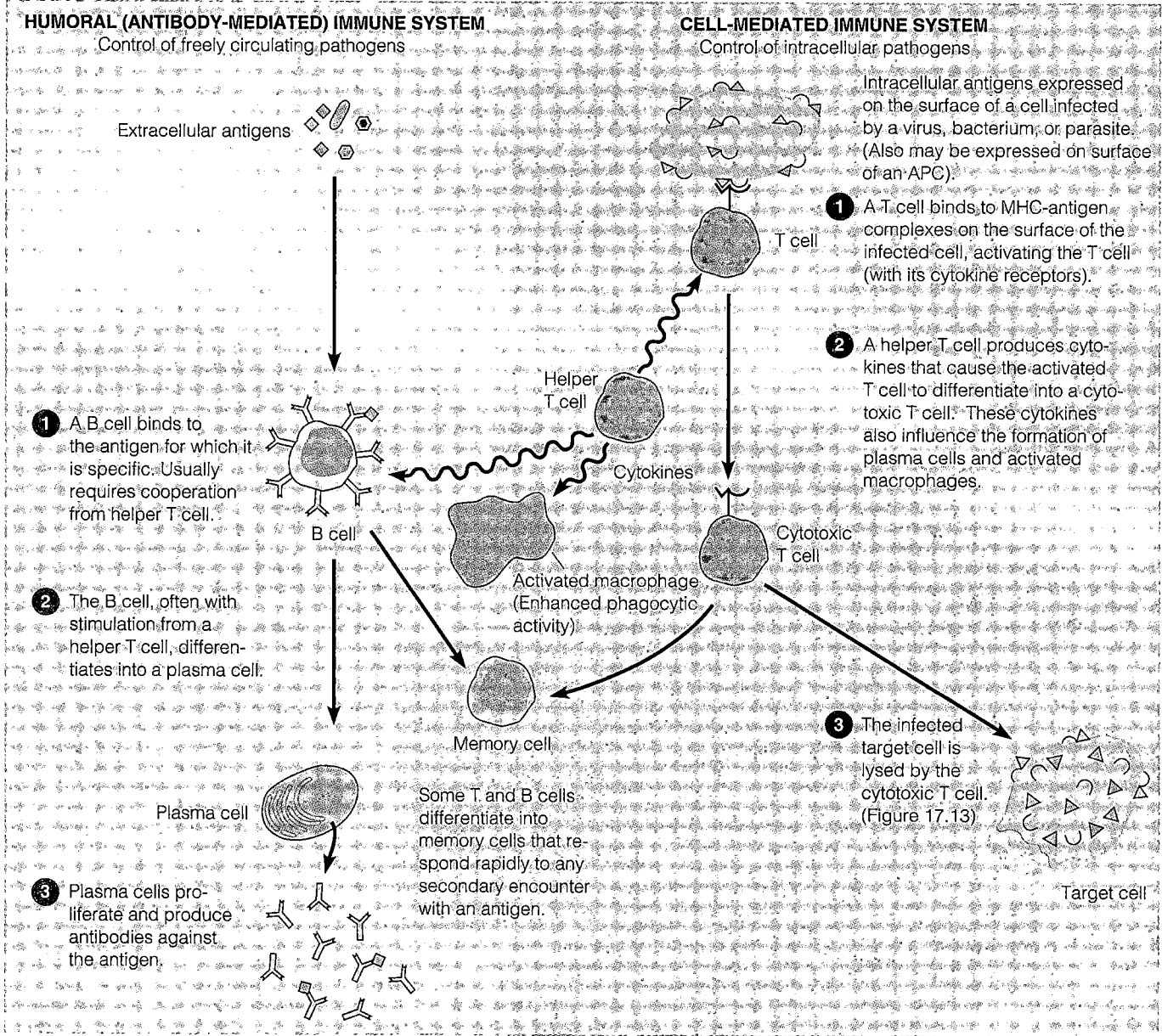

HUMORAL (ANTIBODY-MEDIATED) IMMUNE SYSTEM
Control of freely circulating pathogens

Extracellular antigens

1 A B cell binds to the antigen for which it is specific. Usually requires cooperation from helper T cell.

B cell

2 The B cell, often with stimulation from a helper T cell, differentiates into a plasma cell.

Plasma cell

3 Plasma cells proliferate and produce antibodies against the antigen.

Helper T cell

Cytokines

Activated macrophage (Enhanced phagocytic activity)

Memory cell

Some T and B cells differentiate into memory cells that respond rapidly to any secondary encounter with an antigen.

CELL-MEDIATED IMMUNE SYSTEM
Control of intracellular pathogens

Intracellular antigens expressed on the surface of a cell infected by a virus, bacterium, or parasite. (Also may be expressed on surface of an APC).

1 A T cell binds to MHC-antigen complexes on the surface of the infected cell, activating the T cell (with its cytokine receptors).

T cell

2 A helper T cell produces cytokines that cause the activated T cell to differentiate into a cytotoxic T cell. These cytokines also influence the formation of plasma cells and activated macrophages.

Cytotoxic T cell

3 The infected target cell is lysed by the cytotoxic T cell. (Figure 17.13)

Target cell

FIGURE 17.18 The duality of the immune system. Pathogens can come into contact with antibodies only when they circulate in the blood or spaces outside of cells. However, viruses and some bacterial pathogens and parasites reproduce only inside living cells, where circulating antibodies cannot reach them. Elimination of these intracellular pathogens requires cell-mediated immune responses that depend on T cells, especially cytotoxic T cells.

■ **When does a B cell require stimulation by a T_H cell?**

---

5. Serum containing antibodies is often called antiserum.

6. When serum is separated by gel electrophoresis, antibodies are found in the gamma fraction of the serum and are termed immune serum globulin, or gamma globulin.

The Duality of the Immune System (p. 478)

1. Humoral immunity is in body fluids.

2. Cell-mediated immunity is due to certain types of lymphocytes.

Humoral (Antibody-Mediated) Immunity (p. 478)

1. The humoral immune system involves antibodies produced by B cells in response to a specific antigen.

2. Antibodies primarily defend against bacteria, viruses, and toxins in blood plasma and lymph.

Cell-Mediated Immunity (p. 478)

1. The cell-mediated immune system depends on T cells and does not involve antibody production.

2. Cellular immunity is primarily a response to intracellular bacteria and viruses, multicellular parasites, transplanted tissue, and cancer cells.

ANTIGENS AND ANTIBODIES (pp. 478–482)

The Nature of Antigens (p. 479)

1. An antigen (or immunogen) is a chemical substance that causes the body to produce specific antibodies or sensitized T cells.

2. As a rule, antigens are foreign substances; they are not part of the body's chemistry.

3. Most antigens are components of invading microbes: proteins, nucleoproteins, lipoproteins, glycoproteins, or large polysaccharides with a molecular weight greater than 10,000.

4. Antibodies are formed against specific regions on the surface of an antigen called antigenic determinants.

5. Most antigens have many different determinants.

6. A hapten is a low-molecular-weight substance that cannot cause the formation of antibodies unless combined with a carrier molecule.

The Nature of Antibodies (pp. 479–482)

1. An antibody, or immunoglobulin, is a protein produced by B cells in response to the presence of an antigen and capable of combining specifically with that antigen.

2. An antibody has at least two identical antigen-binding (valence) sites.

Antibody Structure (pp. 479–481)

1. A single bivalent antibody unit is a monomer.

2. Most antibody monomers consist of four polypeptide chains. Two are heavy chains, and two are light chains.

3. Within each chain is a variable (V) region, where antigen binding occurs, and a constant (C) region, which serves as a basis for distinguishing the classes of antibodies.

4. An antibody monomer is Y- or T-shaped; the variable regions form the tips, and the constant regions form the base and Fc (stem) region.

5. The Fc region can attach to a host cell or complement.

Immunoglobulin Classes (pp. 481–482)

1. IgG antibodies are the most prevalent in serum; they provide naturally acquired passive immunity, neutralize bacterial toxins, participate in complement fixation, and enhance phagocytosis.

2. IgM antibodies consist of five monomers held by a joining chain; they are involved in agglutination and complement fixation.

3. Serum IgA antibodies are monomers; secretory IgA antibodies are dimers that protect mucosal surfaces from invasion by pathogens.

4. IgD antibodies are antigen receptors on B cells.

5. IgE antibodies bind to mast cells and basophils and are involved in allergic reactions.

B CELLS AND HUMORAL IMMUNITY (pp. 482–487)

1. Humoral immunity involves antibodies that are produced by B cells.

2. Bone marrow stem cells give rise to B cells.

3. Mature B cells migrate to lymphoid organs.

4. A mature B cell recognizes an antigen with antigen receptors.

Apoptosis (p. 483)

1. Lymphocytes that are not needed undergo apoptosis, or programmed cell death, and are destroyed by phagocytes.

Activation of Antibody-Producing Cells by Clonal Selection (pp. 483–484)

STI *To review, go to Microbial Interactions: Immunology: B cells*

1. According to the clonal selection theory, a B cell becomes activated when an antigen reacts with antigen receptors on its surface.

2. The activated B cell produces a clone of plasma cells and memory cells.

3. Plasma cells secrete antibodies. Memory cells recognize pathogens from previous encounters.

4. T cells and B cells that react with self antigens are destroyed during fetal development; this is called clonal deletion.

Antigen–Antibody Binding and Its Results (pp. 484–486)

1. An antigen binds to the antigen-binding site (variable region) of an antibody to form an antigen–antibody complex.

2. IgG antibodies inactivate viruses and also neutralize bacterial toxins.

3. Agglutination of cellular antigens occurs when an IgG or IgM antibody combines with two cells.

4. Antigen–antibody complexes involving IgG and IgM antibodies can fix complement, resulting in the lysis of a bacterial (antigenic) cell.

Immunological Memory (p. 486)

STI *Microbial Interactions: Immunology: Compare responses*

1. The amount of antibody in serum is called the antibody titer.

2. The response of the body to the first contact with an antigen is called the primary response. It is characterized by the appearance of IgM followed by IgG.

3. Subsequent contact with the same antigen results in a very high antibody titer and is called the secondary, anamnestic, or memory response. The antibodies are primarily IgG.

Monoclonal Antibodies and Their Uses (pp. 486–487)

1. Hybridomas are produced in the laboratory by fusing a cancerous cell with an antibody-secreting plasma cell.

2. A hybridoma cell culture produces large quantities of the plasma cell's antibody, called monoclonal antibodies.

3. Monoclonal antibodies are used in serologic identification tests, to prevent tissue rejections, and to make immunotoxins to treat cancer.

4. Immunotoxins can be made by combining a monoclonal antibody and a toxin; the toxin will then kill a specific antigen.

T CELLS AND CELL-MEDIATED IMMUNITY (PP. 487–493)

1. Cell-mediated immunity involves specialized lymphocytes, primarily T cells, that respond to intracellular antigens.

Chemical Messengers of Immune Cells: Cytokines (PP. 487–489)

1. Cells of the immune system communicate with each other by means of chemicals called cytokines.

2. Interleukins (IL) are cytokines that serve as communicators between leukocytes.

3. Interferons are cytokines that protect cells against viruses.

4. Chemokines cause leukocytes to move to the site of infection.

5. Cytokines may be useful in treating tumors.

Cellular Components of Immunity (PP. 489–493)

1. T cells are responsible for cell-mediated immunity.

2. After differentiation in the thymus gland, T cells migrate to lymphoid tissue.

3. T cells differentiate into effector T cells when they are stimulated by an antigen.

4. Some effector T cells become memory cells.

Types of T Cells (PP. 489–492)

1. T cells are classified according to their functions and cell-surface receptors called CDs.

2. The antigen must be processed by an antigen-presenting cell (APC) and positioned on the surface of the APC.

3. The major histocompatibility complex (MHC) consists of cell-surface proteins that are unique to each individual and provide self molecules.

4. A T cell recognizes antigens in association with MHC on an APC, causing the APC to release IL-1.

5. After binding to an APC, helper T (T_H) or CD4 cells secrete IL-2 to activate other T_H cells specific for that antigen.

6. Cytotoxic T (T_C) or CD8 cells release perforin to lyse cells carrying the target antigen and MHC.

7. Delayed hypersensitivity T (T_D) cells are associated with certain types of allergic reactions and transplant rejection.

8. Suppressor T (T_S) cells appear to regulate the immune response.

Nonspecific Cellular Components (PP. 492–493)

1. Macrophages that are stimulated by ingesting an antigen or by cytokines become activated to have enhanced phagocytic ability.

2. Natural killer (NK) cells lyse virus-infected and tumor cells. They are not T cells and are not antigenically specific.

THE INTERRELATIONSHIP OF CELL-MEDIATED AND HUMORAL IMMUNITY (PP. 493–495)

🖳 *Microbial Interactions: Immunology: T cell*

1. T_H cells activate B cells to produce antibodies against T-dependent antigens.

2. Antigens that directly activate B cells are called T-independent antigens.

3. In antibody-dependent cell-mediated cytotoxicity (ADCC), NK cells, macrophages, and other leukocytes lyse antibody-coated cells.

4. ADCC is useful against helminthic parasites.

Study Questions

REVIEW

1. Define immunity.

2. Contrast the terms in the following pairs:
   a. nonspecific resistance and immunity
   b. humoral and cell-mediated immunity
   c. active and passive immunity
   d. innate resistance and acquired immunity
   e. natural and artificial immunity
   f. T-dependent and T-independent antigens
   g. CD4 and CD8

3. Classify the following examples of immunity as naturally acquired active immunity, naturally acquired passive immunity, artificially acquired active immunity, or artificially acquired passive immunity:
   a. Immunity following the injection of diphtheria toxoid.

   b. Immunity following an infection.
   c. A newborn's immunity to yellow fever.
   d. Immunity following an injection of antirabies serum.

4. Explain what an antigen is. Distinguish an antigen from a hapten.

5. Explain what an antibody is by describing the characteristics of antibodies. Diagram the structure of a typical antibody; label the heavy chain, light chain, and constant, variable, and Fc regions.

6. Discuss the clonal selection mechanism.

7. By means of a diagram, explain the role of T cells and B cells in immunity.

8. Explain a function for the following types of cells: T_C, T_D, T_H, and T_S. What is a cytokine?

9. a. In the graph below, at time *A,* the host was injected with tetanus toxoid. At time *B,* the host was given a booster dose. Explain the meaning of the areas of the curve marked *a* and *b.*

b. Identify in the graph the antibody response of this same individual to exposure to a new antigen indicated at time *B.*

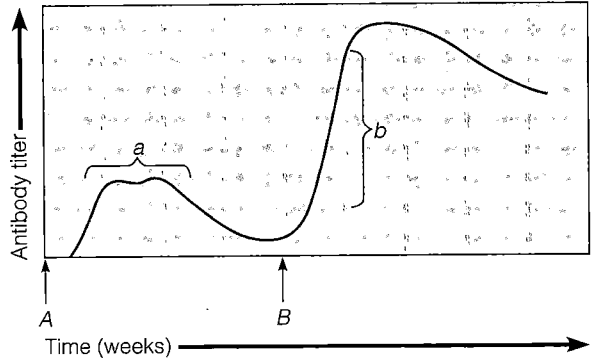

Time (weeks)

10. What effect does an antibody have on an antigen?

11. How does a T cell recognize an antigen?

12. What are natural killer cells?

13. How would each of the following prevent infection?
a. antibodies against *Neisseria gonorrhoeae* fimbriae
b. antibodies against host cell mannose

14. How are monoclonal antibodies produced?

15. Explain why a person who recovers from a disease can attend others with the disease without fear of contracting the disease.

16. Pooled human immune serum globulin is sometimes administered to a patient after exposure to hepatitis A. What is human immune serum globulin? What type of immunity might this confer on the patient?

MULTIPLE CHOICE

Match the following choices to questions 1–3:
a. innate resistance
b. naturally acquired active immunity
c. naturally acquired passive immunity
d. artificially acquired active immunity
e. artificially acquired passive immunity

1. The type of protection provided by the injection of a toxoid.

2. The type of protection provided by the injection of an antitoxin.

3. The type of protection resulting from recovery from an infection.

Match the following choices to the statements in questions 4–6:
a. IgA   **d.** IgG
b. IdD   **e.** IgM
c. IgE

4. Antibodies that protect the fetus and newborn.

5. The first antibodies synthesized; especially effective against microorganisms.

6. Antibodies that are bound to mast cells and involved in allergic reactions.

7. Put the following in the correct sequence to elicit an antibody response: (1) T_H cell recognizes B cell; (2) APC phagocytizes antigen; (3) antigen fragment goes to surface of APC; (4) T_H recognizes antigen digest and MHC; (5) B cell recognizes antigen.
a. 1, 2, 3, 4, 5   **d.** 2, 3, 4, 1, 5
b. 5, 4, 3, 2, 1   **e.** 4, 5, 3, 1, 2
c. 3, 4, 5, 1, 2

Use the following choices to answer questions 8 and 9:
a. antigen   **d.** IL-2
b. hapten   **e.** perforin
c. IL-1

8. Responsible for the differentiation of T and NK cells.

9. Activates a T cell to bind IL-2.

10. Patients with Chediak–Higashi syndrome suffer from various types of cancer. These patients are most likely lacking which of the following:
a. T_D cells   **d.** NK cells
b. T_H cells   **e.** T_S cells
c. B cells

CRITICAL THINKING

1. Injections of T_C cells completely removed all hepatitis B viruses from infected mice, but they killed only 5% of the infected liver cells. Explain how T_C cells cured the mice.

2. Provide an explanation for the following:
a. IL-2 has been used to treat pancreatic cancer.
b. IL-2 would exacerbate autoimmune diseases.

3. A positive tuberculin skin test shows cell-mediated immunity to *Mycobacterium tuberculosis.* How could a person acquire this immunity?

CLINICAL APPLICATIONS

1. A woman had life-threatening salmonellosis that was successfully treated with monoclonal antibodies. Why did this treatment work, when antibiotics and her own immune system failed?

2. A patient with AIDS has a low T_H/T_S cell ratio. What clinical manifestations would this cause?

3. A patient with chronic diarrhea was found to lack IgA in his secretions, although he had a normal level of serum IgA. What was this patient found to be unable to produce?

4. Newborns (under 1 year) who contract dengue fever have a higher chance of dying from it if their mothers had dengue fever prior to pregnancy. Explain why.

Learning with Technology

chapter 18

Practical Applications of Immunology

Immunofluorescence. These streptococci fluoresce under UV light because dye-tagged antibodies are attached to them.

n Chapter 17, we learned the basics of the immune system in which the body recognizes foreign microbes, toxins, or tissues. In response, it forms antibodies and activates other immune system cells that are programmed to recognize and neutralize or destroy this foreign material if it encounters it again. This specific immunity is, of course, an important element of our defenses against foreign pathogens.

In this chapter, we will discuss some useful tools that have been developed from knowledge of the basics of the immune system. Vaccines were briefly mentioned in the previous chapter; in this chapter we will expand our discussion of this important field of immunology. The diagnosis of disease frequently depends on tests that make use of the specificity of the immune system. Antibodies, especially monoclonal antibodies, are of special value in many of these diagnostic tests.

Vaccines

Learning Objective

- *Define vaccine.*

Long before the invention of vaccines, it was known that people who recovered from certain diseases, such as smallpox, were immune to the disease thereafter. As discussed in Chapter 1 on page 12, Chinese physicians may have been the first to try to exploit this phenomenon to prevent disease when they had children inhale dried smallpox scabs.

In 1717, Mary Montagu reported from her travels there that in Turkey, an "old woman comes with a nutshell full of the matter of the best sort of smallpox and asks what veins you please to have opened, and puts into the vein as much venom as can lie upon the head of her needle." This practice usually led to a week of mild illness, and the person was subsequently protected from smallpox. Called **variolation,** this procedure became commonplace in England. Unfortunately, it was known in some cases to backfire and kill the recipient. In eighteenth-century England, the mortality rate associated with variolation was about 1%, still a significant improvement over the 50% mortality rate that could be expected from smallpox.

One person who received this treatment, at the age of 8, was Edward Jenner. Later in life, as a physician, Jenner was intrigued by a dairymaid's assertion that she had no fear of smallpox because she had already had cowpox. Cowpox is a mild disease that causes lesions on cows' udders; dairymaids' hands often became infected during milking. Motivated by his childhood memory of variolation, Jenner began a series of experiments in 1798 in which he deliberately inoculated people with cowpox in an attempt to prevent smallpox. To honor Jenner's work, the term *vaccination* (from *vacca* = cow) was coined by Louis Pasteur. A **vaccine** is a suspension of organisms or fractions of organisms that is used to induce immunity. Two centuries later, smallpox has been eliminated worldwide by vaccination, and two other viral diseases, measles and polio, are also targeted for elimination.

Principles and Effects of Vaccination

Learning Objective

- *Explain why vaccination works.*

We now know that Jenner's inoculations worked because the cowpox virus, which is not a serious pathogen, is closely related to the smallpox virus. The injection, by skin scratches, provoked a primary immune response in the recipients, leading to the formation of antibodies and long-term memory cells. Later, when the recipient encountered the smallpox virus, the memory cells were stimulated, producing a rapid, intense secondary immune response (see Figure 17.9 on page 486). This response mimics the immunity gained by recovering from the dis-

ease. The cowpox vaccine was soon replaced by a vaccinia virus vaccine. The vaccinia virus also confers immunity to smallpox. Although, strangely, little is known with certainty about the origin of this important virus, it is probably a hybrid that arose long ago from an accidental mixing of cowpox and smallpox viruses. The development of vaccines based on the model of the smallpox vaccine is the single most important application of immunology.

Many communicable diseases can be controlled by behavioral and environmental methods. For example, proper sanitation can prevent the spread of cholera, and the use of condoms can slow the spread of sexually transmitted diseases. If prevention fails, bacterial diseases can often be treated with antibiotics. Viral diseases, however, are not readily treated once contracted. Therefore, vaccination is often the only feasible method of controlling viral disease. Controlling a disease does not necessarily require that everyone be immune to it. If most of the population is immune, called *herd immunity*, outbreaks are limited to sporadic cases because there are not enough susceptible individuals to support the spread of epidemics.

The principal vaccines used to prevent bacterial and viral diseases in the United States are listed in Tables 18.1 and 18.2. Recommendations for childhood immunizations against some of these diseases are given in Table 18.3. American travelers who might be exposed to cholera, yellow fever, or other diseases not endemic in the United States can obtain current inoculation recommendations from the U.S. Public Health Service and local public health agencies.

table 18.1   Principal Vaccines Used in the United States to Prevent Bacterial Diseases in Humans

Disease	Vaccine	Recommendation	Booster
Diphtheria	Purified diphtheria toxoid	See Table 18.3	Every 10 years for adults
Meningococcal meningitis	Purified polysaccharide from *Neisseria meningitidis*	For people with substantial risk of infection	Need not established
Pertussis (whooping cough)	Killed whole or acellular fragments of *Bordetella pertussis*	Children prior to school age; see Table 18.3	For high-risk adults
Pneumococcal pneumonia	Purified polysaccharide from *Streptococcus pneumoniae*	For adults with certain chronic diseases; people over 65	Normally not recommended
Tetanus	Purified tetanus toxoid	See Table 18.3	Every 10 years for adults
Haemophilus influenzae b meningitis	Polysaccharide from *Haemophilus influenzae* b conjugated with protein to enhance effectiveness	Children prior to school age; see Table 18.3	None recommended

table 18.2 *Principal Vaccines Used in the United States to Prevent Viral Diseases in Humans*

Disease	Vaccine	Recommendation	Booster
Influenza	Killed virus	For chronically ill people, especially with respiratory diseases, or for healthy people over 65	Annual
Measles	Attenuated virus	For infants age 15 months	See Table 18.3
Mumps	Attenuated virus	For infants age 15 months	(Duration of immunity not known)
Rubella	Attenuated virus	For infants age 15 months; for females of childbearing age who are not pregnant	(Duration of immunity not known)
Chickenpox	Attenuated virus	For infants age 12 months	(Duration of immunity not known)
Poliomyelitis	Attenuated or killed virus (enhanced potency type)	For children, see Table 18.3; for adults, as risk to exposure warrants	(Duration of immunity not known)
Rabies	Killed virus	For field biologists in contact with wildlife in endemic areas; for veterinarians; for people exposed to rabies virus by bites	Every 2 years
Hepatitis B	Antigenic fragments of virus	For children, see Table 18.3; for adults, especially health care workers, homosexual males, injecting drug users, heterosexual people with multiple partners, and household contacts of hepatitis B carriers	Duration of protection at least 7 years; need for boosters uncertain
Hepatitis A	Inactivated virus	Mostly for travel to endemic areas and protecting contacts during outbreaks	Duration of protection estimated at about 10 years

table 18.3 *Schedule of Childhood Immunizations*

Vaccine	Birth	1 mo	2 mos	4 mos	6 mos	12 mos	15 mos	18 mos	4–6 yrs	11–12 yrs	14–16 yrs
Hepatitis B (Hep B)	Hep B (if mother is HBV positive)	Hep B		Hep B						Hep B	
Diphtheria, Tetanus, Pertussis*			DTaP	DTaP	DTaP		DTaP		DTaP	Td	
H. influenzae type b (Hib)			Hib	Hib	Hib	Hib					
Polio**			IPV	IPV	IPV				IPV		
Measles, Mumps, Rubella (MMR)						MMR			MMR	MMR	
Varicella (Var)						Var				Var	

Vaccines are listed under routinely recommended ages. Bars indicate range of recommended ages for immunization. Any dose not given at the recommended age should be given as a "catch-up" immunization at any subsequent visit when indicated and feasible. Light Bars indicate "catch-up" vaccinations.

*DTaP (diphtheria and tetanus and acellular pertussis vaccine) is the preferred vaccine for all doses. (Td vaccines contain only tetanus and diphtheria toxoids.) For adolescents and adults only.

**IPV = inactivated polio vaccine.

***A vaccine for Hepatitis A is recommended in certain states or regions at ages 24 months through 12 years.
SOURCE: CDC

Experience has shown that vaccines against enteric bacterial pathogens, such as those causing cholera and typhoid, are not nearly as effective or long-lived as those against viral diseases such as measles and smallpox.

Types of Vaccines and Their Characteristics

Learning Objectives

- *Differentiate between the following, and provide an example of each: attenuated, inactivated, toxoid, subunit, and conjugated vaccines.*
- *Contrast subunit vaccines and nucleic acid vaccines.*

There are now several basic types of vaccine. Some of the newer vaccines take full advantage of knowledge and technology developed in recent years.

Attenuated whole-agent vaccines use living, but attenuated (weakened), microbes. Live vaccines more closely mimic an actual infection. Lifelong immunity, especially with viruses, is often achieved without booster immunizations, and an effectiveness rate of 95% is not unusual. This long-term effectiveness probably occurs because the attenuated viruses replicate in the body, increasing the original dose and acting as a series of secondary (booster) immunizations.

Examples of attenuated vaccines are the Sabin polio vaccine and those used against measles, mumps, and rubella (MMR). The widely used vaccine against the tuberculosis bacillus and certain of the newly introduced, orally administered typhoid vaccines contain attenuated bacteria. Attenuated microbes are usually derived from mutations accumulated during long-term culture. A danger of such vaccines is that the live microbes can back-mutate to a virulent form (discussed later in the chapter). Attenuated vaccines are not recommended for people whose immune systems are compromised. If available, inactivated vaccines are substituted.

Inactivated whole-agent vaccines use microbes that have been killed, usually by formalin or phenol. Inactivated virus vaccines used in humans include those against rabies (animals sometimes receive a live vaccine considered too hazardous for humans), influenza (Figure 18.1), and polio (the Salk polio vaccine). Inactivated bacterial vaccines include those for pneumococcal pneumonia and cholera. Several long-used inactivated vaccines that are being replaced for most uses by newer, more effective types are those for pertussis (whooping cough) and typhoid.

Toxoids, which are inactivated toxins, are vaccines directed at the toxins produced by a pathogen. The tetanus and diphtheria toxoids have long been part of the

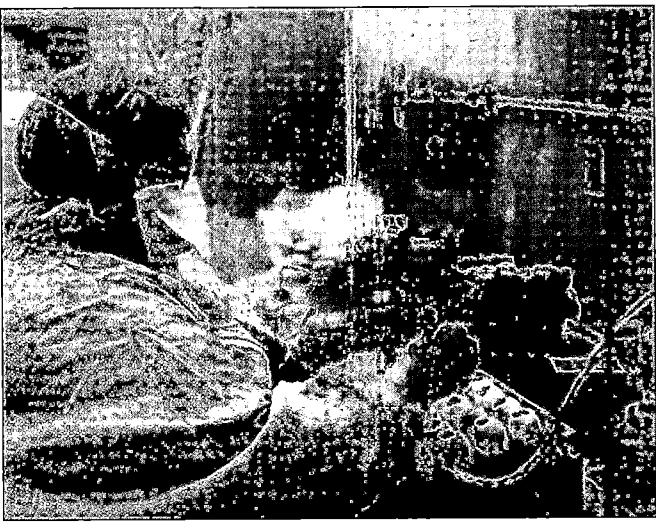

FIGURE 18.1 Influenza viruses are grown in embryonated eggs. (See Figure 13.7 on page 380.) The viruses will be inactivated to make a vaccine.

- Could this method of virus cultivation be a problem for persons allergic to eggs?

standard childhood inoculation series. They require a series of injections for full immunity, followed by boosters every 10 years. Many older adults have not received boosters; they are likely to have low levels of protection.

Subunit vaccines use only those antigenic fragments of a microorganism that best stimulate an immune response. Subunit vaccines that are produced by genetic engineering techniques, meaning that other microbes are programmed to produce the desired antigenic fraction, are called **recombinant vaccines.** For example, the vaccine against the hepatitis B virus consists of a portion of the viral protein coat that is produced by a genetically engineered yeast.

Subunit vaccines are inherently safer because they cannot reproduce in the recipient. They also contain little or no extraneous material and therefore tend to produce fewer adverse effects. Similarly, it is possible to separate the fractions of a disrupted bacterial cell, retaining the desired antigenic fractions. The newer **acellular vaccines** for pertussis use this approach.

Conjugated vaccines have been developed in recent years to deal with the poor immune response of children to vaccines based on capsular polysaccharides. As shown in Figure 17.16 (page 494), polysaccharides are T-independent antigens; children's immune systems do not respond well to these antigens until the age of 15–24 months. Therefore, the polysaccharides are combined with proteins such as diphtheria toxoid; this approach has led to the very successful vaccine for *Haemophilus influenzae* type b, which gives significant protection even at 2 months.

Nucleic-acid vaccines, or DNA vaccines, are one of the newest and most promising types of vaccine, although at this time they have not yet resulted in any vaccine for humans. Experiments with animals show that plasmids of "naked" DNA injected into muscle results in the production of the protein encoded in the DNA. (The "gene gun" method for injecting nucleic acids into plant cells is described in Chapter 9, Figure 9.6 on page 257.) These proteins persist and stimulate an immune response.

A problem with this type of vaccine is that the DNA remains effective only until it is degraded. Indications are that RNA, which could replicate in the recipient, might be a more effective agent.

The Development of New Vaccines

Learning Objectives

- Compare and contrast the production of whole-agent vaccines, recombinant vaccines, and DNA vaccines.
- Define adjuvant.

An effective vaccine is the most desirable method of disease control. It prevents the targeted disease from ever occurring at all in an individual, and it is generally the most economical. This is especially important in developing parts of the world. A "dream" vaccine would be swallowed instead of injected. It also would give lifelong immunity with a single dose, remain stable without refrigeration, and be affordable. This dream is currently far from realization, however. See the box on the next page.

Although interest in vaccine development declined with the introduction of antibiotics, it has intensified in recent years. Fear of litigation contributed to the decrease in the development of new vaccines in the United States. However, passage of the National Childhood Vaccine Injury Act in 1986, which limits vaccine-makers' liability, helped reverse this trend.

Historically, vaccines could be developed only by growing the pathogen in usefully large amounts. The early successful viral vaccines were developed by animal cultivation. The vaccinia virus for smallpox was grown on the shaved bellies of calves. The rabies virus used by Pasteur in his vaccine over 100 years ago was grown in the central nervous system of rabbits.

The introduction of vaccines against polio, measles, mumps, and a number of other viral diseases that would not grow in anything but a living human awaited the development of cell culture techniques. Cell cultures from human sources, or more often from animals such as monkeys that are closely related to humans, enabled growth of these viruses on a large scale. A convenient animal that will grow many viruses is the chick embryo (see Figure 13.7

on page 380), and viruses for several vaccines (influenza, for example) are grown this way (Figure 18.1). Interestingly, the first vaccine against hepatitis B virus used viral antigens extracted from the blood of chronically infected humans because no other source was available.

Recombinant vaccines and DNA vaccines do not need a cell or animal host to grow the vaccine's microbe. This avoids a major problem with certain viruses that so far have not been grown in cell culture—hepatitis B, for example.

The so-called golden age of immunology occurred from about 1870 to 1910, when most of the basic elements of immunology were discovered and several important vaccines were developed. We may soon be entering another golden age in which new technologies are brought to bear on emerging infectious diseases and problems arising from the decreasing effectiveness of antibiotics. It is remarkable that there are no useful vaccines against chlamydias, fungi, protozoa, or helminthic parasites of humans. Moreover, vaccines for some diseases, such as cholera and tuberculosis, are not reliably protective. At present, vaccines for at least 75 diseases are under development, ranging from those for prominent deadly diseases such as AIDS and malaria to such commonplace conditions as earaches. But we will probably find that the easy vaccines have already been made.

Infectious diseases are not the only possible target of vaccines. Researchers are investigating vaccines' potential for treating and preventing cocaine addiction and cancer and for contraception.

Work is underway to improve effectiveness of antigens that may not be very effective when injected by themselves. For example, chemicals added for this purpose, called **adjuvants,** greatly improve the effectiveness of many antigens. Only alum has been approved as an adjuvant for human use; others might be found. Injecting antigens attached to microscopic particles has been shown experimentally to improve their effectiveness.

For all of us who wish for a replacement for the needle as a delivery system for vaccines, help may be on the way. Perhaps you have already received injection by high-pressure "guns." Experiments are also underway to test delivery by intranasal aerosols, skin patches, and even antigens produced in foods, such as bananas, that vaccinate on ingestion.

Safety of Vaccines

We have seen how variation, the first attempt to provide immunity to smallpox, sometimes caused the disease it was intended to prevent. At the time, however, the risk was considered very worthwhile. As you will see later in

Vaccines were responsible for eradicating smallpox and hold promise for eradicating polio and measles within the next 5 to 10 years. Research in development of, and use of, vaccines over the next decade are expected to increase. Vaccines are under development for contraception and cancer, and even a vaccine for cocaine addiction has been proposed. It appears that there is no absolute limit to the number of vaccines that can be given to a person. All standard vaccines recommended for children and adults can be given to the same person, at separate anatomical sites, on the same day or weeks apart, with no risk of interference of one vaccine by another and no potential adverse effects. A few exceptions exist; for example, live vaccines should not be given to immunosuppressed or pregnant individuals, and live bacterial vaccines should not be given within 24 hours after administering antibiotics.

So why not use more vaccines? For example, the vaccines against rabies, botulism, plague, and yellow fever are not routinely used. In the United States in 1999, there were 0 cases of rabies, 8 cases of plague, and 58 of botulism; yellow fever has not occurred in the United States since 1924. One might argue that these cases of plague and botulism, few as there were, would have been prevented by vaccination. However, the plague vaccine affords only limited protection, and society must

balance the cost of manufacturing and distributing a vaccine against the overall benefit. The *primary* purpose of vaccination against an infectious disease is to provide herd immunity against it; if herd immunity already exists in a society, it is not usually necessary to undertake efforts at widespread vaccination of all individuals. Why don't we vaccinate against tuberculosis? Like the plague vaccine, the TB vaccine affords variable protection. This presents a risk; if people assume they are immune to a disease, they will not practice precautions.

Even in the United States, where the vaccination of children against diseases should be routine, all children do not get vaccinated. In 1993, the U.S. government initiated the Childhood Immunization Initiative (CII) to increase vaccination coverage levels to at least 90% among 2-year-old children by 1996 (for hepatitis B, the objective was set for 1998). By 1997, childhood immunization rates for the complete series of vaccinations were at an all-time high of 78%; however, 1 million children under age 2 still had not received all their immunizations.

While efforts to vaccinate children in the United States are increasing, the World Health Organization is working to extend protection against diseases preventable with vaccine to children of all nations. Measles still accounts for 10% of global mortality among children aged less than 5 years (approximately

1 million deaths annually). New vaccines are needed against diseases that affect millions of people in developing nations. Unfortunately, these diseases are not usually candidates for vaccine development. The affected populations cannot afford even minimal costs. Moreover, if electricity and therefore refrigeration are not available, many vaccines cannot be made available.

Recent attempts to develop a malaria vaccine provide an example of the clinical and technical difficulties involved in developing a vaccine. Some 300–500 million new cases of malaria, with 2 million deaths, occur every year throughout the world. Preliminary human trials of a vaccine against a malaria sporozoite antigen showed that human volunteers did develop antibodies against the antigen. However, the antibodies must kill the sporozoites within 30 minutes after the mosquito bites its victim, before the infective sporozoites enter the protection of the liver cells. A separate vaccine is needed to provoke an immune response against the later stage merozoites, which destroy red blood cells and cause the typical symptoms of malaria.

The CII has established ambitious goals for enhancing present vaccines and for developing new ones. The ultimate vaccine is an affordable, heat-stable, orally administered, multiple-antigen, single immunization given at birth.

this text, even today the oral polio vaccine may cause the disease on rare occasions. In 1999, a vaccine to prevent infant diarrhea caused by rotaviruses was withdrawn from the market because several recipients developed life-threatening intestinal obstruction. But public reaction to such risks has changed; today, reports or even rumors of harmful effects often lead people to avoid certain vaccines for themselves or their children, even though the risk of adverse effects from contracting the disease is much greater than the risks from the vaccine. In Chapter 24, you will read about such a situation, and its results, in the discussion of pertussis, or whooping cough. More re-

cently, a report in a medical journal in Great Britain suggested a connection between a vaccine combination that is almost universally administered and autism. In the same issue of that journal medical authorities discounted the association; nevertheless, intense concern in Britain and the United States led some parents to insist that the vaccines be administered individually rather than in their usual combination. Isolated instances, or anecdotal evidence, are not sufficient to prove a connection; after all, if almost the entire population of children receives a certain vaccine, there are certain to be cases of autism, neurological problems, or other conditions in that population.

Such connections must be proved by statistical analysis, a procedure that most lay persons do not understand very well and that takes considerable time and resources. There are other examples: the alleged association between Agent Orange (a defoliant used in Vietnam) and birth defects and cancer, and the systemic illnesses in women who had received silicone breast implants. Statistical analysis does not support any connection, but the public at large often remains unconvinced; it is always difficult for researchers to prove a negative. Although it is impossible to completely eliminate risk for any medical procedure, including vaccination, work will continue to make vaccines that cause fewer side effects and are more economical, more effective, and easier to administer. In a sense, vaccines are a victim of their own success. Few parents today have ever experienced the terror of epidemics of paralytic polio, helpless to know if their child would be the next victim. Few have ever seen a case of measles or diphtheria. Despite the potential risks, vaccines will still remain the safest and most effective means available to control infectious diseases.

Diagnostic Immunology

Learning Objective

- *Explain how antibodies are used to diagnose diseases.*

Throughout most of history, diagnosing a disease was essentially a matter of observing a patient's signs and symptoms. The writings of ancient and medieval physicians left descriptions of many diseases that are recognizable even today.

Knowledge of the high specificity of the immune system soon suggested that this might be used in diagnosing diseases. In fact, an accidental observation in the course of research into vaccines against tuberculosis led to one of the first diagnostic tests for an infectious disease. More than 100 years ago, Robert Koch was trying to develop a vaccine against tuberculosis. He observed that when guinea pigs with the disease were injected with a suspension of *Mycobacterium tuberculosis*, the site of the injection became red and slightly swollen a day or two later. You may recognize this symptom as the positive result of the test to determine whether a person has been infected by the tuberculosis pathogen. This test is widely used today; almost all of us have been subjected to the skin test for tuberculosis (Figure 24.11). Koch, of course, had no idea of the mechanism of cell-mediated immunity that caused this phenomenon, nor did he know of the existence of antibodies.

Since the time of Robert Koch, immunology has given us many other invaluable diagnostic tools, most of

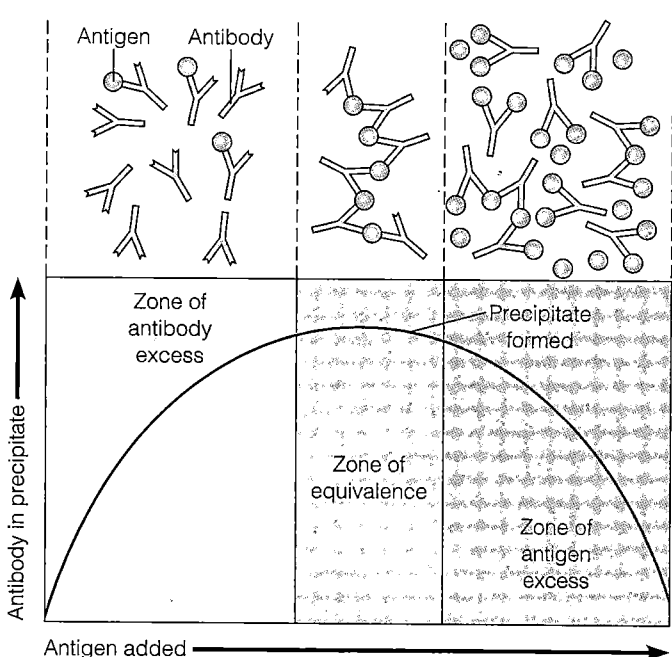

FIGURE 18.2 A precipitation curve. The curve is based on the ratio of antigen to antibody. The maximum amount of precipitate forms in the zone of equivalence, where the ratio is roughly equivalent.

■ Precipitation reactions involve soluble antigens.

which are based on interactions of humoral antibodies with antigens. In this section we will discuss several of the most important techniques used to detect antigens and antibodies.

One problem our diagnostic tools must overcome is that antibodies cannot be seen directly. Even at magnifications of well over 100,000X, they appear only as fuzzy, ill-defined particles. Therefore, we must establish their presence indirectly through a variety of reactions. In most instances the presence of antibodies is of primary interest, but the procedures can be reversed so that known antibodies can be used to detect antigens.

Precipitation Reactions

Learning Objective

- *Explain how precipitation reactions and immunodiffusion tests work.*

Precipitation reactions involve the reaction of soluble antigens with IgG or IgM antibodies to form large, interlocking molecular aggregates called lattices.

Precipitation reactions occur in two distinct stages. First, the antigens and antibodies rapidly form small antigen–antibody complexes (see Chapter 17). This in-

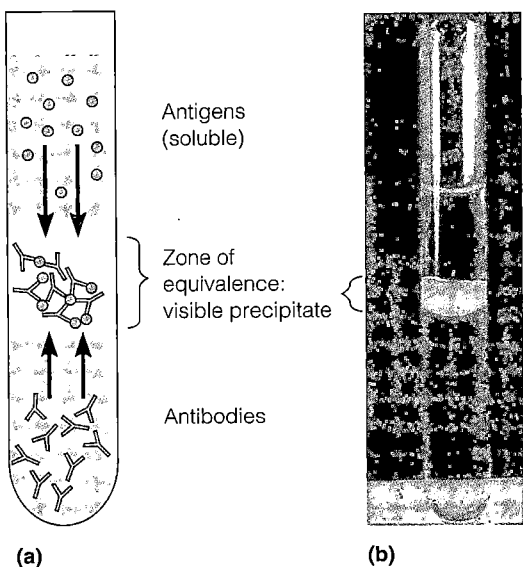

(a) **(b)**

FIGURE 18.3 The precipitin ring test. (a) This drawing shows the diffusion of antigens and antibodies toward each other in a small-diameter test tube. Where they reach equal proportions, in the zone of equivalence, a visible line or ring of precipitate is formed. **(b)** A photograph of a precipitin ring.

■ **What causes the visible line?**

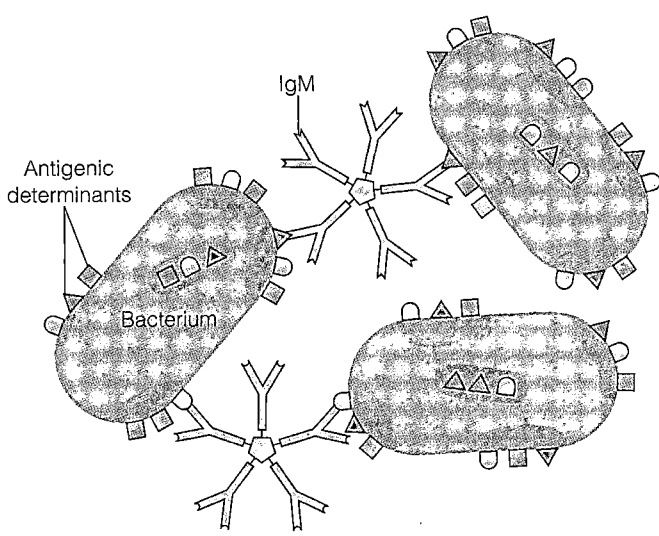

FIGURE 18.4 An agglutination reaction. When antibodies react with antigenic determinant sites on antigens carried on neighboring cells, such as these bacteria (or red blood cells), the particulate antigens (cells) agglutinate. IgM, the most efficient immunoglobulin for agglutination, is shown here, but IgG also participates in agglutination reactions.

■ Agglutination reactions involve particulate antigens.

teraction occurs within seconds and is followed by a slower reaction, which may take minutes to hours, in which the antigen–antibody complexes form lattices that precipitate from solution. Precipitation reactions normally occur only when the ratio of antigen to antibody is optimal. Figure 18.2 shows that no visible precipitate forms when either component is in excess. The optimal ratio is produced when separate solutions of antigen and antibody are placed adjacent to each other and allowed to diffuse together. In a **precipitin ring test** (Figure 18.3), a cloudy line of precipitation (ring) appears in the area in which the optimal ratio has been reached (the *zone of equivalence*).

Immunodiffusion tests are precipitation reactions carried out in an agar gel medium, on either a Petri plate or a microscope slide. A line of visible precipitate develops between the wells at the point where the optimal antigen–antibody ratio is reached.

Other tests use electrophoresis to speed up the movement of antigen and antibody in a gel, sometimes in less than an hour, with this method. The techniques of immunodiffusion and electrophoresis can be combined in a procedure called **immunoelectrophoresis.** The procedure is used in research to separate proteins in human serum and is the basis of certain diagnostic tests. It is an essential part of the Western blot test used in AIDS testing (see Figure 10.12 on page 291).

Agglutination Reactions

Learning Objectives

- *Differentiate direct from indirect agglutination tests.*
- *Differentiate agglutination from precipitation tests.*
- *Define hemagglutination.*

Whereas precipitation reactions involve *soluble* antigens, agglutination reactions involve either *particulate* antigens (particles such as cells that carry antigenic molecules) or soluble antigens adhering to particles. These antigens can be linked together by antibodies to form visible aggregates, a reaction called **agglutination** (Figure 18.4). Agglutination reactions are very sensitive, relatively easy to read (see Figure 10.10 on page 290), and available in great variety. Agglutination tests are classified as either direct or indirect.

Direct Agglutination Tests

Direct agglutination tests detect antibodies against relatively large cellular antigens, such as those on red blood cells, bacteria, and fungi. At one time they were carried out in a series of test tubes, but now they are usually done in plastic *microtiter plates,* which have many shallow wells that take the place of the individual test tubes. The amount of particulate antigen in each well is the

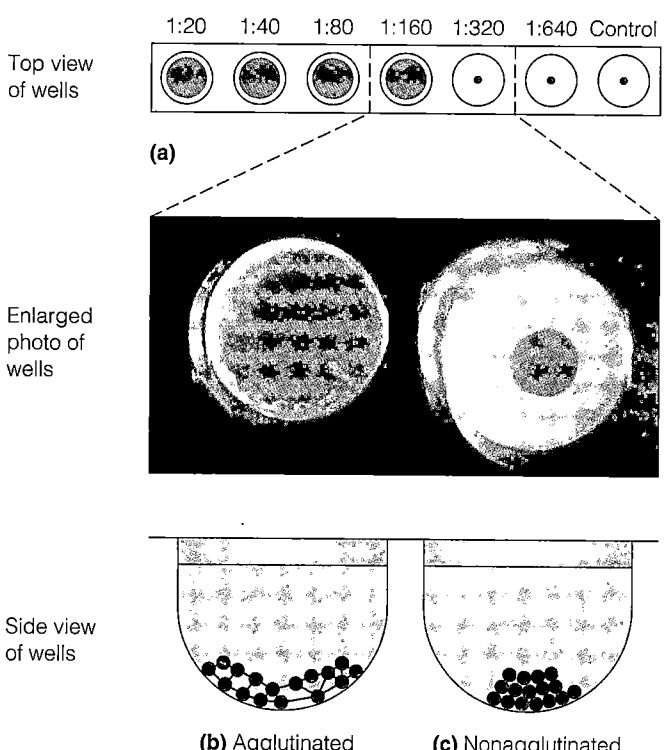

Top view of wells

1:20   1:40   1:80   1:160   1:320   1:640   Control

(a)

Enlarged photo of wells

Side view of wells

(b) Agglutinated   (c) Nonagglutinated

(a) Each well in this microtiter plate contains, from left to right, only half the concentration of serum that is contained in the preceding well. Each well contains the same concentration of particulate antigens, in this instance red blood cells.

(b) In a positive (agglutinated) reaction, sufficient antibodies are present in the serum to link the antigens together, forming a mat of antigen–antibody complexes on the bottom of the well.

(c) In a negative (nonagglutinated) reaction, not enough antibodies are present to cause the linking of antigens. The particulate antigens roll down the sloping sides of the well, forming a pellet at the bottom. In this example, the antibody titer is 160 because the well with a 1:160 concentration is the most dilute concentration that produces a positive reaction.

FIGURE 18.5   Measuring antibody titer with the direct agglutination test.

■ What is meant by the term *antibody titer?*

same, but the amount of serum containing antibodies is diluted so that each successive well has half the antibodies of the previous well. These tests are used, for example, to test for brucellosis and to separate *Salmonella* isolates into serovars, types defined by serological means.

Clearly, the more antibody we start with, the more dilutions it will take to lower the amount to the point where there is not enough antibody for the antigen to react with. This is the measure of **titer,** or concentration of serum antibody (Figure 18.5). For infectious diseases in general, the higher the serum antibody titer, the greater the immunity to the disease. However, the titer alone is of limited use in diagnosing an existing illness. There is no way to know whether the measured antibodies were generated in response to the immediate infection or to an earlier illness. For diagnostic purposes, *a rise in titer* is significant; that is, the titer is higher later in the course of the disease than at its outset. Also, if it can be demonstrated that the person's blood had no antibody titer before the illness but has a significant titer while the disease is progressing, this change, called **seroconversion,** is also diagnostic. This situation is frequently encountered with HIV infections.

Some diagnostic tests specifically identify IgM antibodies. As discussed in Chapter 17, short-lived IgM is more likely to reflect a response to a current disease condition.

Indirect (Passive) Agglutination Tests

Antibodies against soluble antigens can be detected by agglutination tests if the antigens are adsorbed onto particles such as bentonite clay, or most often, minute latex spheres, each about one-tenth of the diameter of a bacterium. Such tests, known as *latex agglutination tests,* are commonly used for the rapid detection of serum antibodies against many bacterial and viral diseases. In such **indirect (passive) agglutination tests,** the antibody reacts with the soluble antigen adhering to the particles (Figure 18.6). The particles then agglutinate with one another, much as particles do in the direct agglutination tests. The same principle can be applied in reverse by using particles coated with antibodies to detect the antigens against which they are specific. This approach is especially common in tests for the streptococci that cause sore throats. A diagnosis can be completed in about 10 minutes.

Hemagglutination

When agglutination reactions involve the clumping of red blood cells, the reaction is called **hemagglutination.** These reactions, which involve red blood cell surface antigens and their complementary antibodies, are used routinely in blood typing (see Table 19.2 on page 525) and in the diagnosis of infectious mononucleosis.

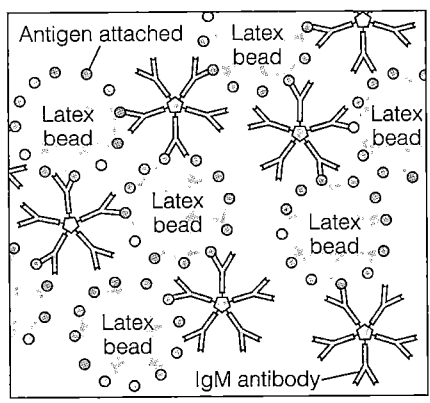

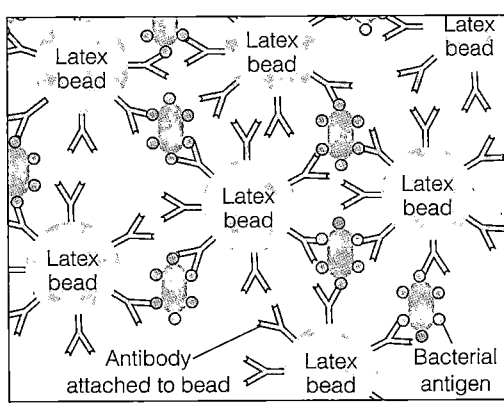

(a) Reaction in a positive indirect test for antibodies

(b) Reaction in a positive indirect test for antigens

FIGURE 18.6 Reactions in indirect agglutination tests. These tests are performed using antigens or antibodies coated onto particles such as minute latex spheres. **(a)** When particles are coated with antigens, agglutination indicates the presence of antibodies, such as the IgM shown here. **(b)** When particles are coated with monoclonal antibodies, agglutination indicates the presence of antigens.

▣ **Differentiate direct from indirect agglutination tests.**

Certain viruses, such as those causing mumps, measles, and influenza, have the ability to agglutinate red blood cells without an antigen–antibody reaction; this process is called **viral hemagglutination** (Figure 18.7). This type of hemagglutination can be inhibited by antibodies that neutralize the agglutinating virus. Diagnostic tests based on such neutralization reactions are discussed in the next section.

Neutralization Reactions

Learning Objectives

- *Explain how a neutralization test works.*
- *Differentiate precipitation from neutralization tests.*

Neutralization is an antigen–antibody reaction in which the harmful effects of a bacterial exotoxin or a virus are blocked by specific antibodies. These reactions were first described in 1890, when investigators observed that immune serum could neutralize the toxic substances produced by the diphtheria pathogen, *Corynebacterium diphtheriae*. Such a neutralizing substance, which is called an antitoxin, is a specific antibody produced by a host as it responds to a bacterial exotoxin or its corresponding toxoid (inactivated toxin). The antitoxin combines with the exotoxin to neutralize it (Figure 18.8a). Antitoxins produced in an animal

can be injected into humans to provide passive immunity against a toxin. Antitoxins from horses are routinely used to prevent or treat diphtheria and botulism; tetanus antitoxin is usually of human origin.

These therapeutic uses of neutralization reactions have led to their use as diagnostic tests. Viruses that exhibit their cytopathic (cell-damaging) effects in cell culture or embryonated eggs can be used to detect the presence of neutralizing viral antibodies (see Chapter 15 on page 447). If the serum to be tested contains antibodies against the particular virus, the antibodies will prevent that virus from infecting cells in the cell culture or eggs, and no cytopathic effects will be seen. Such tests, known as in vitro neutralization tests, can thus be used to both identify a virus and ascertain the viral antibody titer. In vitro neutralization tests are comparatively complex to carry out and are becoming less common in modern clinical laboratories.

A more frequently used neutralization test is the **viral hemagglutination inhibition test.** This test is used in the diagnosis of influenza, measles, mumps, and a number of other infections caused by viruses that can agglutinate red blood cells. If a person's serum contains antibodies against these viruses, these antibodies will react with the viruses and neutralize them (Figure 18.8b). For example, if hemagglutination occurs in a mixture of measles virus and red blood cells but does not occur when the patient's

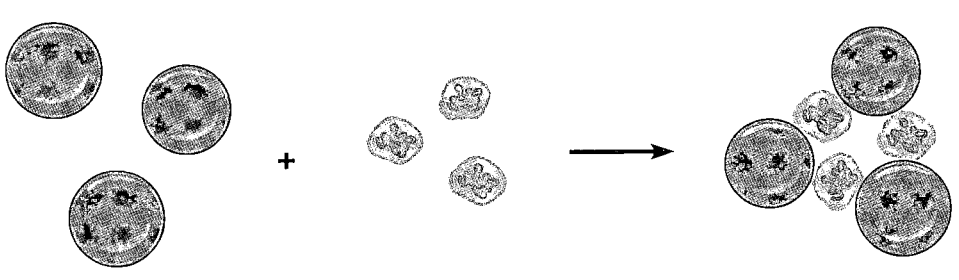

Red blood cells   Viruses   Hemagglutination

FIGURE 18.7 Viral hemagglutination. Viral hemagglutination is not an antigen–antibody reaction.

▣ **Some viral diseases can be diagnosed by the virus's ability to agglutinate red blood cells.**

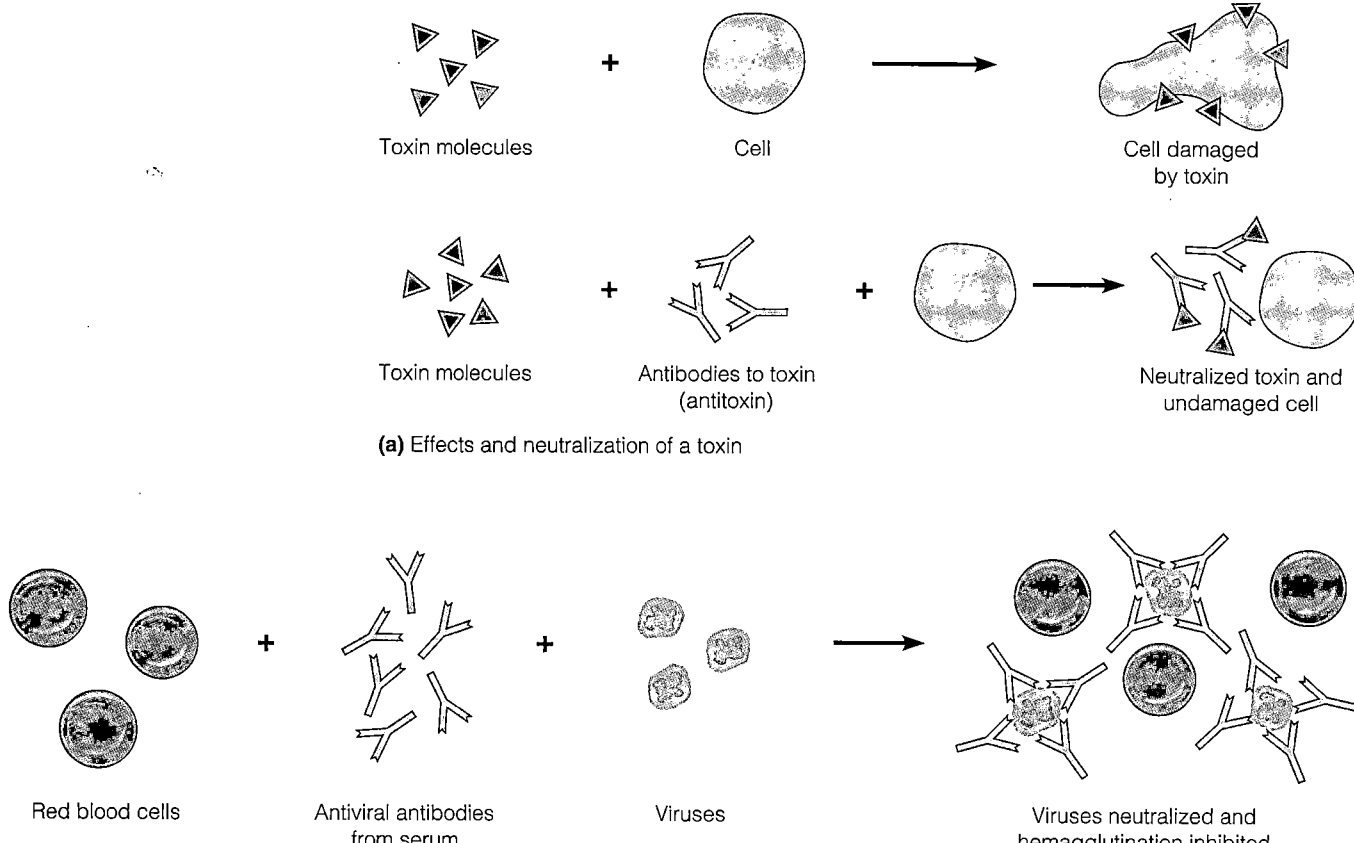

(a) Effects and neutralization of a toxin

Toxin molecules | Cell | Cell damaged by toxin

Toxin molecules | Antibodies to toxin (antitoxin) | Neutralized toxin and undamaged cell

Red blood cells | Antiviral antibodies from serum | Viruses | Viruses neutralized and hemagglutination inhibited

(b) Neutralization of viruses in positive hemagglutination inhibition test

FIGURE 18.8 Reactions in neutralization tests. (a) The effects of a toxin on a susceptible cell and neutralization of the toxin by antitoxin. **(b)** In the viral hemagglutination inhibition test, serum that may contain antibodies to a virus is mixed with red blood cells and the virus. If antibodies are present, as shown here, they neutralize the virus and inhibit hemagglutination.

■ Why does hemagglutination indicate a patient does not have a specific disease?

serum is added to the mixture, this result indicates that the serum contains antibodies that have bound to and neutralized the measles virus.

Complement-Fixation Reactions

Learning Objective

■ *Explain the basis for the complement-fixation test.*

In Chapter 16 (pages 467–469) we discussed a group of serum proteins collectively called complement. During most antigen–antibody reactions, complement binds to the antigen–antibody complex and is used up, or fixed. This process of **complement fixation** can be used to detect very small amounts of antibody. Antibodies that do not produce a visible reaction, such as precipitation or agglutination, can be demonstrated by the fixing of complement

during the antigen–antibody reaction. Complement fixation was once used in the diagnosis of syphilis (Wasserman test) and is still used to diagnose certain viral, fungal, and rickettsial diseases. Execution of the complement-fixation test requires great care and good controls, one reason the trend is to replace it with newer, simpler tests. The test is performed in two stages: complement fixation and indicator (Figure 18.9).

Fluorescent-Antibody Techniques

Learning Objective

■ *Compare and contrast direct and indirect fluorescent-antibody tests.*

Fluorescent-antibody (FA) techniques can identify microorganisms in clinical specimens and can detect the

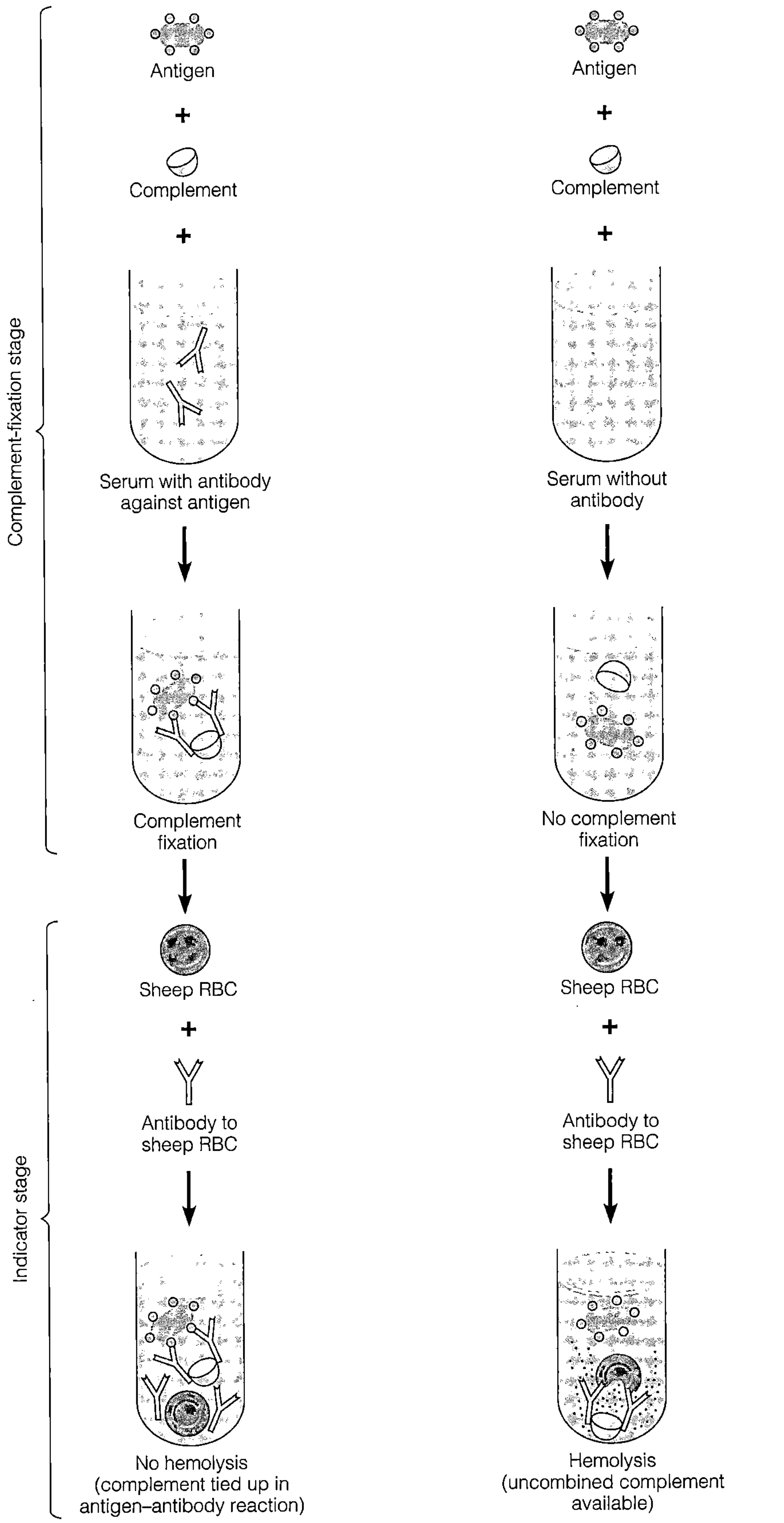

Complement-fixation stage

Antigen

+

Complement

+

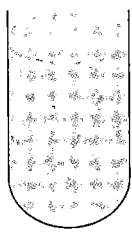

Serum with antibody
against antigen

↓

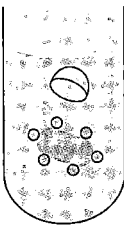

Complement
fixation

↓

Indicator stage

Sheep RBC

+

Antibody to
sheep RBC

↓

No hemolysis
(complement tied up in
antigen–antibody reaction)

(a) Positive test. All available
complement is fixed by the
antigen–antibody reaction; no
hemolysis occurs, so the test is
positive for the presence of antibodies.

Antigen

+

Complement

+

Serum without
antibody

↓

No complement
fixation

↓

Sheep RBC

+

Antibody to
sheep RBC

↓

Hemolysis
(uncombined complement
available)

(b) Negative test. No antigen–antibody
reaction occurs. The complement
remains, and the red blood cells are
lysed in the indicator stage, so the
test is negative.

FIGURE 18.9 The complement-fixation test. This test is used to indicate the presence of antibodies to a known antigen. Complement will combine (be fixed) with an antibody that is reacting with an antigen. If all the complement is fixed in the complement-fixation stage, then none will remain to cause hemolysis of the red blood cells in the indicator stage.

■ Why does red blood cell lysis indicate that the patient does not have a specific disease?

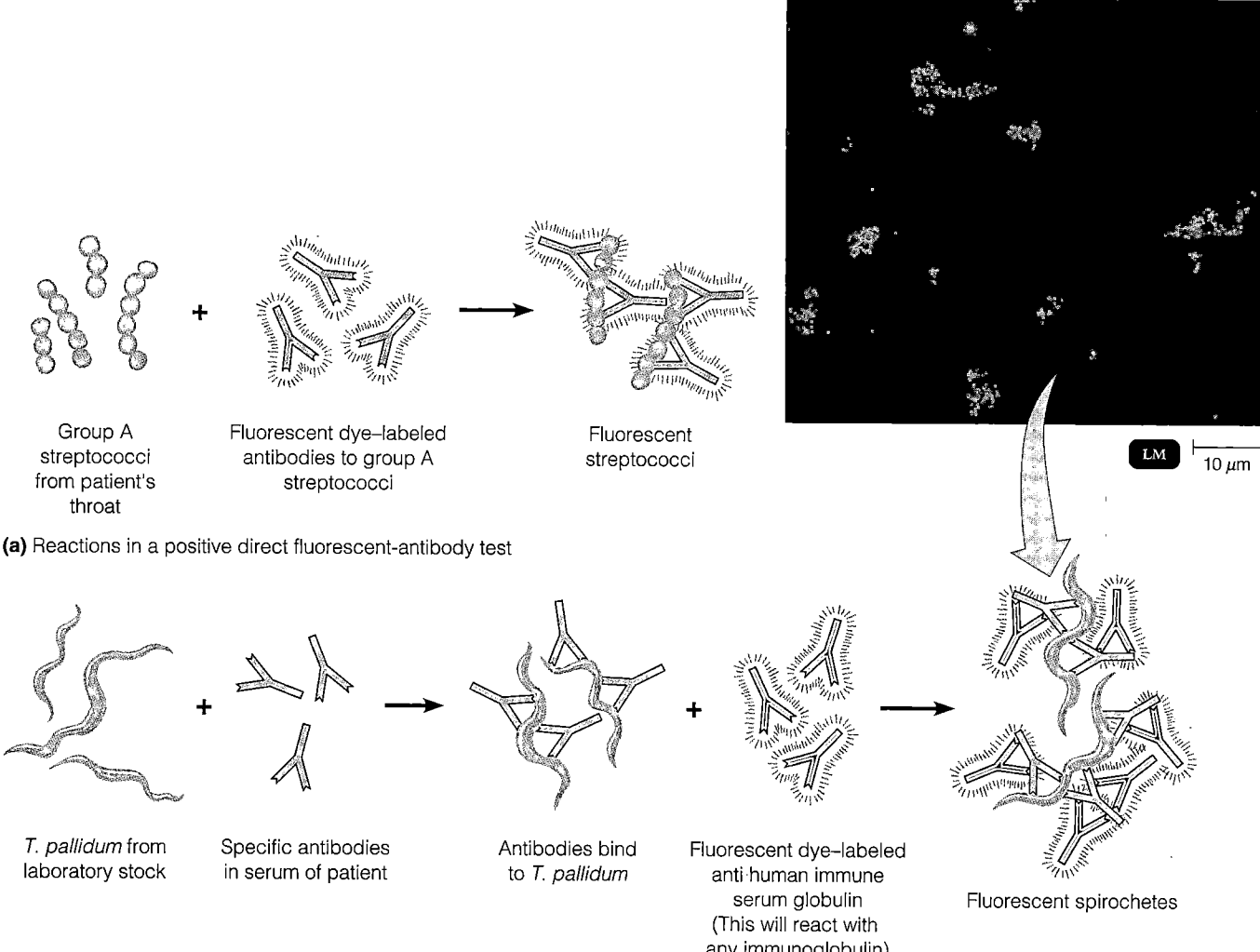

Group A
streptococci
from patient's
throat

Fluorescent dye–labeled
antibodies to group A
streptococci

Fluorescent
streptococci

(a) Reactions in a positive direct fluorescent-antibody test

T. pallidum from
laboratory stock

Specific antibodies
in serum of patient

Antibodies bind
to *T. pallidum*

Fluorescent dye–labeled
anti·human immune
serum globulin
(This will react with
any immunoglobulin)

Fluorescent spirochetes

(b).Reactions in a positive indirect fluorescent-antibody test

FIGURE 18.10 Fluorescent-antibody (FA) techniques. (a) A direct FA test to identify group A streptococci. **(b)** In an indirect FA test such as that used in the diagnosis of syphilis, the fluorescent dye is attached to antihuman immune serum globulin, which reacts with any human immunoglobulin (such as the *Treponema pallidum*–specific antibody) that has previously reacted with the antigen. The reaction is viewed through a fluorescence microscope (see the photomicrograph), and the antigen with which the dye-tagged antibody has reacted—in this case, *T. pallidum*—fluoresces (glows) in the ultraviolet illumination.

■ A direct FA test detects the presence of antigen in the patient. An indirect FA test detects antibodies in the patient.

presence of a specific antibody in serum (Figure 18.10). These techniques combine fluorescent dyes such as fluorescein isothiocyanate (FITC) with antibodies to make them fluoresce when exposed to ultraviolet light (see Figure 3.6 on page 63). These procedures are quick, sensitive, and very specific; the FA test for rabies can be performed in a few hours and has an accuracy rate close to 100%.

Fluorescent-antibody tests are of two types, direct and indirect. **Direct FA tests** are usually used to identify a microorganism in a clinical specimen (Figure 18.10a). During

this procedure, the specimen containing the antigen to be identified is fixed onto a slide. Fluorescein-labeled antibodies are then added, and the slide is incubated briefly. Next the slide is washed to remove any antibody not bound to antigen and is then examined under the fluorescence microscope for yellow-green fluorescence.

Indirect FA tests are used to detect the presence of a specific antibody in serum following exposure to a microorganism (Figure 18.10b). They are often more sensitive than direct tests. During this procedure, a known antigen is

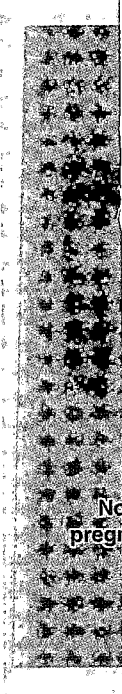

fixed onto a slide. The test serum is then added, and, if antibody that is specific to that microbe is present, it reacts with the antigen to form a bound complex. In order for the antigen–antibody complex to be seen, fluorescein-labeled **antihuman immune serum globulin (anti-HISG),** an antibody that reacts specifically with *any* human antibody, is added to the slide. Anti–HISG will be present only if the specific antibody has reacted with its antigen and is therefore present as well. After the slide has been incubated and washed (to remove unbound antibody), it is examined under a fluorescence microscope. If the known antigen fixed to the slide appears fluorescent, the antibody specific to the test antigen is present.

An especially interesting adaptation of fluorescent antibodies is the **fluorescence-activated cell sorter (FACS).** In Chapter 17, we learned that T cells carry antigenically specific receptors such as CD4 and CD8 on their surface, and these are characteristic of certain groups of T cells. The depletion of CD4 T cells is used to follow the progression of AIDS; their populations can be determined with a FACS.

The FACS is a modification of a *flow cytometer,* in which a suspension of cells leaves a nozzle as droplets containing no more than one cell each (see page 292). A laser beam strikes each cell-containing droplet and is then received by a detector that determines certain characteristics such as size (Figure 18.11). If the cells carry FA markers to identify them as CD4 or CD8 T cells, the detector can measure this fluorescence. As the laser beam detects a cell of a preselected size or fluorescence, an electrical charge, either positive or negative, can be imparted to it. As the charged droplet falls between electrically charged plates, it is attracted to one receiving tube or another, effectively separating cells of different types. Millions of cells can be separated in an hour with this process, all under sterile conditions, which allows them to be used in experimental work.

Enzyme-Linked Immunosorbent Assay (ELISA)

Learning Objective

- *Explain how direct and indirect ELISA tests work.*

The **enzyme-linked immunosorbent assay (ELISA)** is the most widely used of a group of tests known as *enzyme immunoassay (EIA).* There are two basic methods. The *direct ELISA* detects antigens, and the *indirect ELISA* detects antibodies. A microtiter plate with numerous shallow wells is used in both procedures (see Figure 10.11a). Variations of the test exist; for example, the reagents can

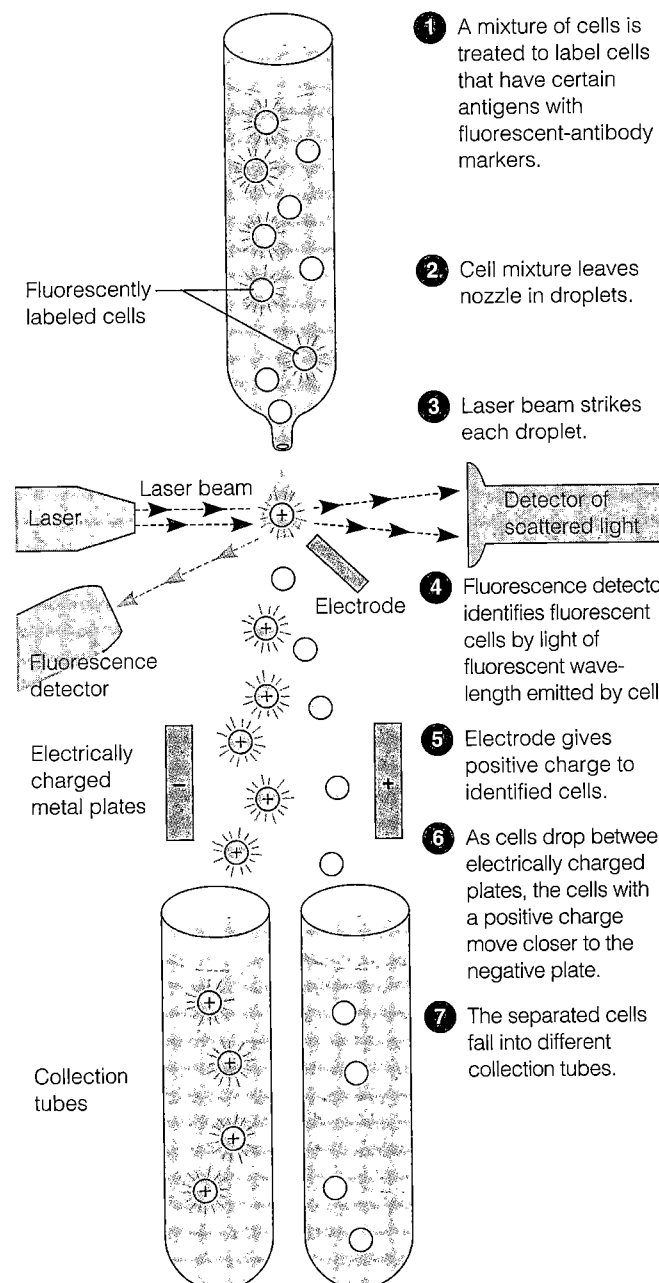

FIGURE 18.11 The fluorescence-activated cell sorter (FACS). This technique can be used to separate different classes of T cells. A fluorescence-labeled antibody reacts with, for example, the CD4 receptor on a T cell.

■ **The FACS is used to count T cells to follow progress of HIV infection.**

① A mixture of cells is treated to label cells that have certain antigens with fluorescent-antibody markers.

② Cell mixture leaves nozzle in droplets.

③ Laser beam strikes each droplet.

④ Fluorescence detector identifies fluorescent cells by light of fluorescent wavelength emitted by cell.

⑤ Electrode gives positive charge to identified cells.

⑥ As cells drop between electrically charged plates, the cells with a positive charge move closer to the negative plate.

⑦ The separated cells fall into different collection tubes.

be bound to tiny latex particles rather than to the surfaces of the microtiter plates. ELISA procedures are popular primarily because they require little interpretive skill to read; the results tend to be clearly positive or clearly negative.

FIGURE 18
home pregn
that is excre
contains two
capture mor
antibodies a
cific for the
urine that co
by the captu
create a visib
change in th

■ What is

Direct ELIS

The direct
page 514. I
antigen to
wells of the
ing uniden
antigen rea
the well, th
will be reta
bound anti
gen is ther
wall of the
the antigen
will be for
molecules.
added antib
ish peroxid

(Figure 18.12b). To see whether a serum sample contains antibodies against this antigen, ❷ the antiserum is added to the well. If the serum contains antibody specific to the antigen, the antibody will bind to the adsorbed antigen. All unreacted antiserum is washed from the well. Anti-HISG is then allowed to react with the antigen–antibody complex. ❸ The anti HISG, which has been linked with an enzyme, reacts with the antibodies that are bound to the antigens in the well. ❹ Finally, all unbound anti-HISG is rinsed away, and the correct substrate for the enzyme is added. A colored enzymatic reaction occurs in the wells in which bound antigen has combined with antibody in the serum sample. This procedure resembles that of the indirect FA test, except that the anti-HISG of the indirect assay is linked with an enzyme rather than with a fluorescing dye.

Many ELISA tests are available for clinical use in the form of commercially prepared kits. Procedures are often highly automated, with the results read by a scanner and printed out by computer (see Figure 10.11). Some tests based on this principle are also available for use by the public; one example is a commonly available home pregnancy test (Figure 18.13). ELISA tests are also used to screen blood for antibodies to HIV.

Radioimmunoassay

In ELISA-type tests, it is possible to use a radioactive tag, usually an isotope of iodine, rather than an enzyme. However, using radioactive isotopes carries several serious disadvantages: The isotopes have a relatively short shelf life, and special precautions are needed to handle radioactive materials and dispose of radioactive wastes. In fact, it was these considerations that led to the development of enzyme-based tags.

American biologist Rosalyn Yalow received the Nobel Prize in 1977 for a type of radioimmunoassay that today is still widely used in chemistry and toxicology but has little use in clinical microbiology. However, some clinical laboratories still use this assay method in testing for certain hepatitis-causing viruses. These tests are not directly comparable to the ELISA tests diagrammed in Figure 18.12. Their principle involves competitive binding of radioactively tagged antigen and untagged antigen to an antibody. The two types of antigen compete for binding sites on the antibody. The more untagged antigen in the sample, the more of the tagged antigen will be displaced from the binding sites on the antibody. The more displaced, tagged antigen that is detected, the more untagged antigen was originally present.

The Future of Diagnostic Immunology

Learning Objective

■ *Explain the importance of monoclonal antibodies.*

The introduction of monoclonal antibodies has revolutionized diagnostic immunology by making available large, economical amounts of specific antibodies. This has led to many newer diagnostic tests that are more sensitive, specific, rapid, and simpler to use. For example, tests to diagnose sexually transmitted chlamydial infections and certain protozoan-caused intestinal parasitic diseases are coming into common use. These tests had previously required relatively difficult cultural or microscopic methods for diagnosis. At the same time, the use of many of the classic serological tests, such as complement–fixation tests, is declining. Most newer tests will require less human judgment to read, and they often require fewer highly trained personnel.

The use of certain *nonimmunological* tests, such as the PCR and DNA probes that were discussed in Chapter 10, is increasing. Some of these tests will become automated to a significant degree. For example, a DNA chip containing over 50,000 DNA probes for genetic information expected in possible pathogens can be exposed to a test sample. This chip is scanned and its data automatically analyzed. PCR tests are also becoming highly automated. Federal approval has already been granted to instruments that measure quantities of several viral pathogens. At the other end of the technical spectrum, efforts are underway to provide low-cost diagnostic methods for developing nations, where the annual medical expenditure per person is only a few dollars. Interest in less invasive sampling methods, such as urinalysis and oral swabs, is also increasing.

The tests described in this chapter are most often used to detect existing disease. In the future, diagnostic testing will probably also be directed at preventing disease. Food and water, for example, are frequently contaminated with pathogens. Controlling diseases spread by ingesting them is a significant public health problem. At present, detecting and identifying such pathogens usually require that the microorganism be selectively isolated and cultivated before diagnostic tests can be applied. This time-consuming process poses significant problems when the source of infection is a perishable consumer product. Much current research in diagnostic methods is now aimed at enabling health workers to detect and identify pathogenic organisms directly, and quickly, in a food sample without the need for isolation and cultivation.

(a) A p

FIGUR

wells of

■ Diff

Study Outline

VACCINES (pp. 500-506)

1. Edward Jenner developed the modern practice of vaccination when he inoculated people with cowpox virus to protect them against smallpox.

Principles and Effects of Vaccination (pp. 501-503)

1. Herd immunity results when most of a population is immune to a disease.

Types of Vaccines and Their Characteristics (pp. 503-504)

1. Attenuated whole-agent vaccines consist of attenuated (weakened) microorganisms; attenuated virus vaccines generally provide lifelong immunity.

2. Inactivated whole-agent vaccines consist of killed bacteria or viruses.

3. Toxoids are inactivated toxins.

4. Subunit vaccines consist of antigenic fragments of a microorganism; these include recombinant vaccines and acellular vaccines.

5. Conjugated vaccines combine the desired antigen with a protein that boosts the immune response.

6. Nucleic acid vaccines, or DNA vaccines, are being developed. These cause the recipient to make the antigenic protein associated with class I MHC.

The Development of New Vaccines (p. 504)

1. Viruses for vaccines may be grown in animals, cell cultures, or chick embryos.

2. Recombinant vaccines and nucleic acid vaccines do not need to be grown in cells or animals.

3. Adjuvants improve the effectiveness of some antigens.

Safety of Vaccines (pp. 504-506)

1. Vaccines are the safest and most effective means of controlling infectious diseases.

DIAGNOSTIC IMMUNOLOGY (pp. 506-516)

1. Many tests based on the interactions of antibodies and antigens have been developed to determine the presence of antibodies or antigens in a patient.

Precipitation Reactions (pp. 506-507)

1. The interaction of soluble antigens with IgG or IgM antibodies leads to precipitation reactions.

2. Precipitation reactions depend on the formation of lattices and occur best when antigen and antibody are present in optimal proportions. Excesses of either component decrease lattice formation and subsequent precipitation.

3. The precipitin ring test is performed in a small tube.

4. Immunodiffusion procedures are precipitation reactions carried out in an agar gel medium.

5. Immunoelectrophoresis combines electrophoresis with immunodiffusion for the analysis of serum proteins.

Agglutination Reactions (pp. 507-509)

1. The interaction of particulate antigens (cells that carry antigens) with antibodies leads to agglutination reactions.

2. Diseases may be diagnosed by combining the patient's serum with a known antigen.

3. Diseases can be diagnosed by a rising titer or seroconversion (from no antibodies to the presence of antibodies).

4. Direct agglutination reactions can be used to determine antibody titer.

5. Antibodies cause visible agglutination of soluble antigens affixed to latex spheres in indirect or passive agglutination tests.

6. Hemagglutination reactions involve agglutination reactions using red blood cells. Hemagglutination reactions are used in blood typing, the diagnosis of certain diseases, and the identification of viruses.

Neutralization Reactions (pp. 509-510)

1. In neutralization reactions, the harmful effects of a bacterial exotoxin or virus are eliminated by a specific antibody.

2. An antitoxin is an antibody produced in response to a bacterial exotoxin or a toxoid that neutralizes the exotoxin.

3. In a virus neutralization test, the presence of antibodies against a virus can be detected by the antibodies' ability to prevent cytopathic effects of viruses in cell cultures.

4. Antibodies against certain viruses can be detected by their ability to interfere with viral hemagglutination in viral hemagglutination inhibition tests.

Complement-Fixation Reactions (p. 510)

1. Complement-fixation reactions are serological tests based on the depletion of a fixed amount of complement in the presence of an antigen–antibody reaction.

Fluorescent-Antibody Techniques (pp. 510-511)

1. Fluorescent-antibody techniques use antibodies labeled with fluorescent dyes.

2. Direct fluorescent-antibody tests are used to identify specific microorganisms.

3. Indirect fluorescent-antibody tests are used to demonstrate the presence of antibody in serum.

4. A fluorescence-activated cell sorter can be used to detect and count cells labeled with fluorescent antibodies.

Enzyme-Linked Immunosorbent Assay (ELISA) (pp. 511–513)

1. ELISA techniques use antibodies linked to an enzyme, such as horseradish peroxidase or alkaline phosphatase.

2. Antigen–antibody reactions are detected by enzyme activity. If the indicator enzyme is present in the test well, an antigen–antibody reaction has occurred.

3. The direct ELISA is used to detect antigens against a specific antibody bound in a test well.

4. The indirect ELISA is used to detect antibodies against an antigen bound in a test well.

Radioimmunoassay (pp. 513–515)

1. In RIA, antibodies against the compound are combined with a radioactively labeled antigen and a sample containing an unknown amount of antigen.

2. Analysis of radioactivity in the resulting antigen–antibody complexes indicates the amount of antigen in the sample.

The Future of Diagnostic Immunology (pp. 515–516)

1. The use of monoclonal antibodies will continue to make new diagnostic tests possible.

Study Questions

REVIEW

1. Classify by type the following vaccines. Which could cause the disease it is supposed to prevent?
   a. attenuated measles virus
   b. dead *Rickettsia prowazekii*
   c. *Vibrio cholerae* toxoid
   d. hepatitis B antigen produced in yeast cells
   e. purified polysaccharides from *Streptococcus pyogenes*
   f. *Haemophilus influenzae* polysaccharide bound to diphtheria toxoid
   g. a plasmid containing genes for influenza A protein

2. Explain the effects of excess antigen and antibody on the precipitation reaction. How is the precipitin ring test different from an immunodiffusion test?

3. How does the antigen in an agglutination reaction differ from that in a precipitation reaction?

4. Define the following terms, and give an example of how each reaction is used diagnostically:
   a. viral hemagglutination
   b. hemagglutination inhibition
   c. passive agglutination

5. Explain the fluorescent-antibody techniques.

6. Describe the direct and indirect FA tests in the following situations. Sketch how the reactions might appear in these tests if the antigens and antibodies could be seen.
   a. Rabies can be diagnosed postmortem by mixing fluorescent-labeled antibodies with brain tissue.
   b. Syphilis can be diagnosed by adding the patient's serum to a slide fixed with *Treponema pallidum*. Antihuman immune serum globulin tagged with a fluorescent dye is added.

7. Explain the ELISA techniques.

8. Describe the direct and indirect ELISA tests in the following situations:
   a. Respiratory secretions to detect respiratory syncytial virus.
   b. Blood to detect human immunodeficiency virus antibodies.

   Which of these tests provides definitive proof of disease?

9. Match the following serological tests to the descriptions.
   ___ Precipitation
   ___ Immuno-electrophoresis
   ___ Agglutination
   ___ Radioimmuno-assay
   ___ Complement fixation
   ___ Neutralization
   ___ ELISA

   a. Occurs with particulate antigens
   b. Uses an enzyme for the indicator
   c. Uses red blood cells for the indicator
   d. Uses antihuman immune serum globulin
   e. Occurs with a free soluble antigen
   f. Used to determine the presence of antitoxin
   g. Uses radioactive isotopes

10. Match each of the following tests to its positive reaction.
    ___ Agglutination
    ___ Complement fixation
    ___ ELISA
    ___ FA test
    ___ Neutralization
    ___ Precipitation

    a. Peroxidase activity
    b. Harmful effects of agents not seen
    c. No hemolysis
    d. Cloudy white line
    e. Cell clumping
    f. Fluorescence

MULTIPLE CHOICE

Use the following choices to answer questions 1 and 2:
   a. hemolysis
   b. hemagglutination
   c. hemagglutination-inhibition
   d. no hemolysis
   e. precipitin ring forms

1. Patient's serum, influenza virus, sheep red blood cells, and antisheep red blood cells are mixed in a tube. What happens if the patient has antibodies against influenza?

2. Patient's serum, *Chlamydia,* guinea pig complement, sheep red blood cells, and antisheep red blood cells are mixed in a tube. What happens if the patient has antibodies against *Chlamydia?*

3. The examples in questions 1 and 2 are
   a. direct tests.
   b. indirect tests.
Use the following choices to answer questions 4 and 5:
   a. anti-*Brucella*
   b. *Brucella*
   c. substrate for the enzyme
4. Which is the third step in a direct ELISA test?
5. Which item is from the patient in an indirect ELISA test?
6. In an immunodiffusion test, a strip of filter paper containing diphtheria antitoxin is placed on a solid culture medium. Then bacteria are streaked perpendicular to the filter paper. If the bacteria are toxigenic,
   a. the filter paper will turn red.
   b. a line of antigen–antibody precipitate will form.
   c. the cells will lyse.
   d. the cells will fluoresce.
   e. none of the above
Use the following choices to answer questions 7–9.
   a. direct fluorescent antibody
   b. indirect fluorescent antibody
   c. rabies immune globulin
   d. killed rabies virus
   e. none of the above
7. Treatment given to a person bitten by a rabid bat.
8. Test used to identify rabies virus in the brain of a dog.
9. Test used to detect the presence of antibodies in a patient's serum.
10. In an agglutination test, eight serial dilutions to determine antibody titer were set up: Tube #1 contained a 1:2 dilution; tube #2, a 1:4, and so on. If tube #5 is the last tube showing agglutination, what is the antibody titer?
   a. 5
   b. 1:5
   c. 32
   d. 1:32

CRITICAL THINKING

1. What problems are associated with the use of attenuated whole-agent vaccines?
2. The World Health Organization has announced the complete eradication of smallpox and is working toward the eradication of measles and polio. Why would vaccination be more likely to eradicate a viral disease than a bacterial disease?
3. Many of the serological tests require a supply of antibodies against pathogens. For example, to test for *Salmonella,* anti-*Salmonella* antibodies are mixed with the unknown bacterium. How are these antibodies obtained?
4. A test for antibodies against *Treponema pallidum* uses an antigen called cardiolipin and the patient's serum (suspected of having antibodies). Why do the antibodies react with cardiolipin? What is the disease?

CLINICAL APPLICATIONS

1. Which of the following is proof of a disease state? Why doesn't the other situation confirm a disease state? What is the disease?
   a. *Mycobacterium tuberculosis* is isolated from a patient.
   b. Antibodies against *M. tuberculosis* are found in a patient.
2. Streptococcal erythrogenic toxin is injected into a person's skin in the Dick test. What results are expected if a person has antibodies against this toxin? What type of immunological reaction is this? What is the disease?
3. The following data were obtained from FA tests for anti-*Legionella* in four people. What conclusions can you draw? What is the disease?

| | Antibody Titer | | | |
	Day 1	Day 7	Day 14	Day 21
Patient A	128	256	512	1024
Patient B	0	0	0	0
Patient C	256	256	256	256
Patient D	0	0	128	512

4. Extracting urine from soil and measuring the amount of the hormone estrone can be used to determine pregnancy in free-roaming feral horses without the stress of capturing or immobilizing the horses. The soil is combined with radioactively labeled estrone and with antibodies to the estrone. The radioactive estrone and the estrone will compete for antibody. Antigen–antibody complexes are then isolated and analyzed for their radioactivity. What does it mean if the antigen–antibody complexes have a high level of radioactive antigen? A low level?

Learning with Technology

chapter 19

Disorders Associated with the Immune System

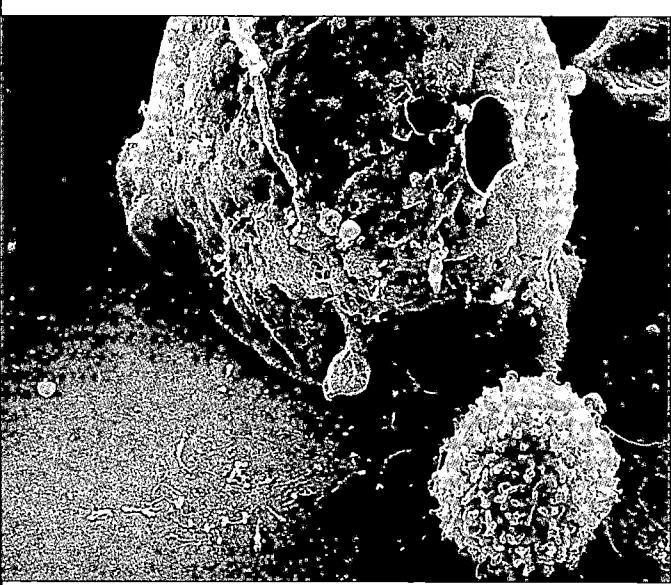

Cytotoxic T cells destroy cancer cells. Enzymes from the small cytotoxic T cell have made a hole in the larger cancer cell.

In this chapter, we will see that not all immune system responses produce a desirable result, such as immunity to disease. Occasionally the reactions of the immune system are harmful. A familiar example is hay fever, which results from repeated exposure to plant pollen. Most of us also know that a blood transfusion will be rejected if the blood of the donor and the blood of the recipient are not compatible, and that rejection is also a potential problem with transplanted organs. The rejection of transfusions or transplants occurs because the immune system identifies the donated cells as nonself and attacks them. One's own tissue may also be mistakenly attacked by the immune system, causing diseases we classify as autoimmune.

We may also be harmed if the immune system fails to function properly. The obvious example is that we may be the victim of some pathogen if a vaccine, or recovery from a pre-

vious illness, has failed to provide immunity to it. Cancer can be another, less apparent, indication of a failure of the immune system.

Certain antigens provoke such a drastic immune response that they are termed **superantigens.** Prominent among them are the staphylococcal enterotoxins that cause food poisoning and the toxin that causes toxic shock syndrome (see Chapter 15). These toxins are relatively nonspecific antigens and indiscriminately activate many T cell receptors at once, resulting in a damaging reaction.

Some people are born with a defective immune system, and in all of us, the effectiveness of our immune system declines with age. In addition, many people have their immune systems deliberately crippled (immunosuppressed) to minimize the potential for the rejection of transplanted organs, and others suffer from the effects of HIV, which attacks the immune system.

table 19.1	Types of Hypersensitivity		
Type of Reaction	**Time Before Clinical Signs**	**Characteristics**	**Examples**
Type I (anaphylactic)	<30 min	IgE binds to mast cells or basophils; causes degranulation of mast cell or basophil and release of reactive substances such as histamine	Anaphylactic shock from drug injections and insect venom; common allergic conditions, such as hay fever, asthma
Type II (cytotoxic)	5–12 hours	Antigen causes formation of IgM and IgG antibodies that bind to target cell; when combined with action of complement, destroys target cell	Transfusion reactions, Rh incompatibility
Type III (immune complex)	3–8 hours	Antibodies and antigens form complexes that cause damaging inflammation	Arthus reactions, serum sickness
Type IV (cell-mediated or delayed-type)	24–48 hours	Antigens cause formation of T_C that kill target cells.	Rejection of transplanted tissues; contact dermatitis, such as poison ivy; certain chronic diseases, such as tuberculosis

Hypersensitivity

Learning Objective

- *Define hypersensitivity.*

The term **hypersensitivity** refers to an antigenic response beyond that considered normal; the term **allergy** is more familiar and is essentially synonymous. Hypersensitivity responses occur in individuals who have been *sensitized* by previous exposure to an antigen, which in this context is sometimes called an **allergen.** Once sensitized, when an individual is exposed to that antigen again, his or her immune system reacts to it in a damaging manner. The four principal types of hypersensitivity reactions, summarized in Table 19.1, are anaphylactic, cytotoxic, immune complex, and cell-mediated, or delayed-type, reactions.

Type I (Anaphylactic) Reactions

Learning Objectives

- *Describe the mechanism of anaphylaxis.*
- *Compare and contrast systemic and localized anaphylaxis.*
- *Explain how allergy skin tests work.*
- *Define desensitization and blocking antibody.*

Type I, or anaphylactic, reactions often occur within 2 to 30 minutes after a person sensitized to an antigen is reexposed to that antigen. Anaphylaxis means "the opposite of protected," from the prefix *ana-,* against, and the Greek

phylaxis, protection. **Anaphylaxis** is an inclusive term for the reactions caused when certain antigens combine with IgE antibodies. Anaphylactic responses can be *systemic reactions,* which produce shock and breathing difficulties and are sometimes fatal, or *localized reactions,* which include common allergic conditions such as hay fever, asthma, and hives (slightly raised, often itchy and reddened areas of the skin).

The IgE antibodies produced in response to an antigen, such as insect venom or plant pollen, bind to the surfaces of cells such as mast cells and basophils. These two cell types are similar in morphology and in their contribution to allergic reactions. **Mast cells** are especially prevalent in the connective tissue of the skin and respiratory tract and in surrounding blood vessels. The name is from the German word *mastzellen,* meaning "well fed"; they are packed with granules that at one time were mistakenly thought to have been ingested. See Figure 19.1. **Basophils** circulate in the bloodstream, where they constitute fewer than 1% of the leukocytes. Both are packed with granules containing a variety of chemicals called *mediators.*

Mast cells and basophils can have as many as 500,000 sites for IgE attachment. The Fc (stem) region of an IgE antibody (see Figure 17.5) can attach to one of these specific receptor sites on such a cell, leaving two antigen-binding sites free. Of course, the attached IgE monomers will not all be specific for the same antigen. But when an antigen such as plant pollen encounters two adjacent antibodies of the same appropriate specificity, it can bind to

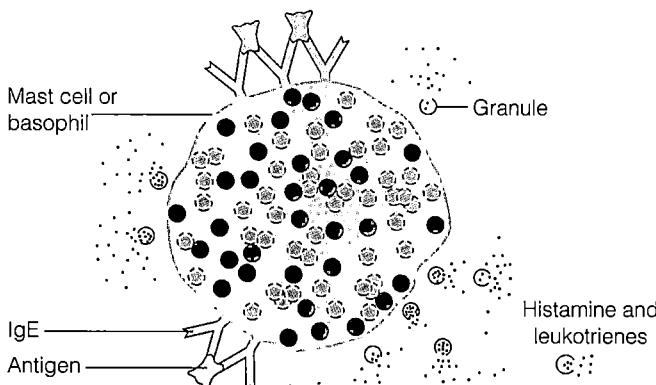

Mast cell or basophil

Granule

IgE

Antigen

Histamine and leukotrienes

(a) IgE antibodies, produced in response to an antigen, coat mast cells and basophils. When an antigen bridges the gap between two adjacent antibody molecules of the same specificity, the cell undergoes degranulation and releases mediators such as histamine and leukotrienes.

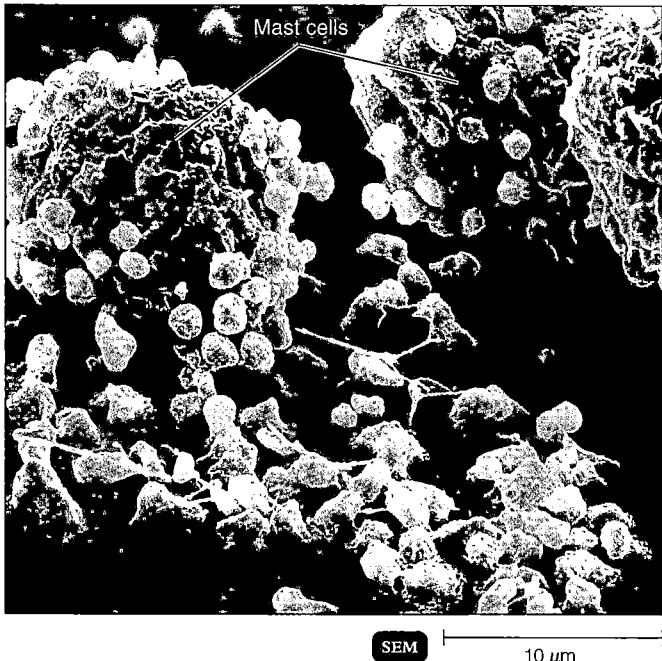

Mast cells

SEM   10 μm

(b) A degranulated mast cell that has reacted with an antigen and released granules of histamine and other reactive mediators.

FIGURE 19.1 **The mechanism of anaphylaxis.**

■ IgE and an antigen interact to cause common allergic reactions.

one antigen–binding site on each antibody, bridging the space between them. This bridge triggers the mast cell or basophil to undergo **degranulation,** which releases the granules inside these cells and also the mediators they contain (Figure 19.1).

These mediators cause the unpleasant and damaging effects of an allergic reaction. The best-known mediator

is **histamine.** The release of histamine increases the permeability and dilation of blood capillaries, resulting in edema (swelling) and erythema (redness). Other effects include increased mucus secretion (a runny nose, for example) and smooth muscle contraction, which in the respiratory bronchi results in breathing difficulty.

Other mediators include **leukotrienes** of various types and **prostaglandins.** These mediators are not preformed and stored in the granules but are synthesized by the antigen–triggered cell. Because leukotrienes tend to cause prolonged contractions of certain smooth muscles, their action contributes to the spasms of the bronchial tubes that occur during asthmatic attacks. Prostaglandins affect smooth muscles of the respiratory system and cause increased mucus secretion.

Collectively, all these mediators serve as chemotactic agents that, in a few hours, attract neutrophils and eosinophils to the site of the degranulated cell. They then activate various factors that cause inflammatory symptoms, such as dilation of the capillaries, swelling, increased secretion of mucus, and involuntary contractions of smooth muscles.

Systemic Anaphylaxis

At the turn of the century, two French biologists studied the responses of dogs to the venom of stinging jellyfish. Large doses of venom usually killed the dogs, but sometimes a few survived the injections. These surviving dogs were used for repeat experiments with the venom, and the results were surprising. Even a very tiny dose of the venom, one that should have been almost harmless, killed the dogs. They suffered difficulty in respiration, entered shock as their cardiovascular systems collapsed, and quickly died. This phenomenon was called *anaphylactic shock.*

Systemic anaphylaxis (or *anaphylactic shock*) can result when an individual sensitized to an antigen is exposed to it again. Injected antigens are more likely to cause a dramatic response than antigens introduced via other portals of entry. The release of mediators causes peripheral blood vessels throughout the body to dilate, resulting in a drop in blood pressure (shock). This reaction can be fatal within a few minutes. There is very little time to act once someone develops systemic anaphylaxis. Treatment usually involves self-administration with a preloaded syringe of epinephrine, a drug that constricts blood vessels and raises the blood pressure. In the United States, 50–60 people die each year from anaphylactic shock caused by insect stings.

You may be acquainted with someone who reacts to penicillin in this way. In these individuals, the penicillin, which is a hapten (it cannot induce antibody formation

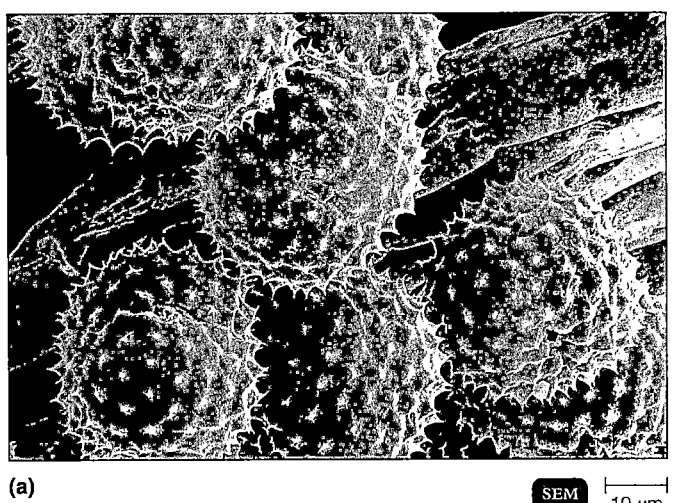

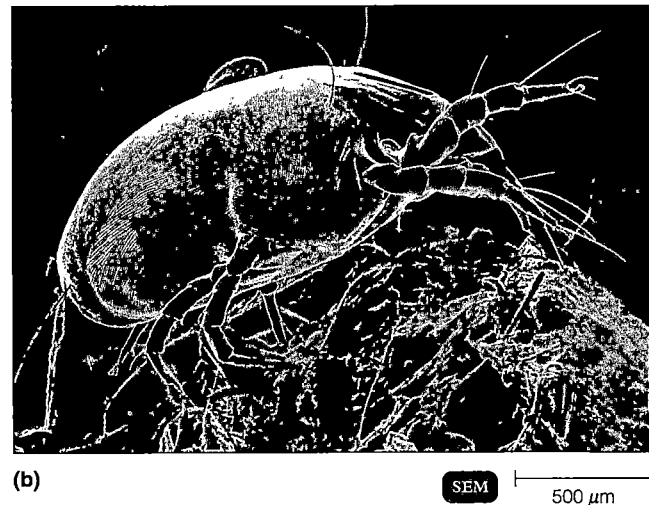

(a) SEM |—————| 10 μm   (b) SEM |—————| 500 μm

FIGURE 19.2 Localized anaphylaxis. (a) A micrograph of pollen grains. **(b)** A micrograph of a house dust mite.

■ Localized anaphylaxis is usually associated with antigens that are ingested or inhaled.

by itself; see Chapter 17), combines with a carrier serum protein. Only then is penicillin immunogenic. Penicillin allergy probably occurs in about 2% of the population. There is no completely reliable skin test for sensitivity to penicillin.

Localized Anaphylaxis

Whereas sensitization to injected antigens is a common cause of systemic anaphylaxis, **localized anaphylaxis** is usually associated with antigens that are ingested (foods) or inhaled (pollen)(Figure 19.2a). The symptoms depend primarily on the route by which the antigen enters the body.

In allergies involving the upper respiratory system, such as hay fever (allergic rhinitis), sensitization usually involves mast cells in the mucous membranes of the upper respiratory tract. Reexposure to the airborne antigen, which might be a common environmental material such as plant pollen, fungal spores, feces of house dust mites (Figure 19.2b), or animal dander.* The typical symptoms are itchy and teary eyes, congested nasal passages, coughing, and sneezing. Antihistamine drugs, which compete for histamine receptor sites, are often used to treat these symptoms.

Asthma is an allergic reaction that mainly affects the lower respiratory system. Symptoms such as wheezing

and shortness of breath are caused by the constriction of smooth muscles in the bronchial tubes.

For unknown reasons, asthma is becoming a near epidemic, affecting about 10% of children in Western society, although they often outgrow it in time. It is speculated that lack of childhood exposure in the developed world to many childhood infections, the so-called hygiene hypothesis, is a factor in the increase in the incidence of asthma. Mental or emotional stress can also be a contributing factor in precipitating an attack. Symptoms of asthma are usually controlled by aerosol inhalants, which, unfortunately, may be difficult for very small children to use.

Antigens that enter the body via the gastrointestinal tract can also sensitize an individual. Many of us may know someone who is allergic to a particular food. Many so-called food allergies may not be related to hypersensitivity at all and are more accurately described as *food intolerances.* For example, many people are unable to digest the lactose in milk because they lack the enzyme that breaks down this disaccharide milk sugar. The lactose enters the intestine, where it osmotically retains fluid, causing diarrhea.

Gastrointestinal upset is a common symptom of food allergies, but it can also result from many other factors. Hives are more characteristic of a true food allergy, and ingestion of the antigen may result in systemic anaphylaxis. Death has even resulted when a person sensitive to fish ate french fries that had been prepared in oil previously used to fry fish. Skin tests are not reliable indicators for the diagnosis of food-related allergies, and completely controlled tests for hypersensitivity to ingested foods are very difficult to perform. Only eight foods are responsible for

---

*Dander is a general term for microscopic particles from the fur or skin of animals. A cat, for example, carries about 100 mg of dander on its coat and sheds about 0.1 mg a day. This accumulates in upholstery and carpeting. People with allergies to mice, gerbils, and similar small animals are often actually allergic to components of urine accumulating in cages.

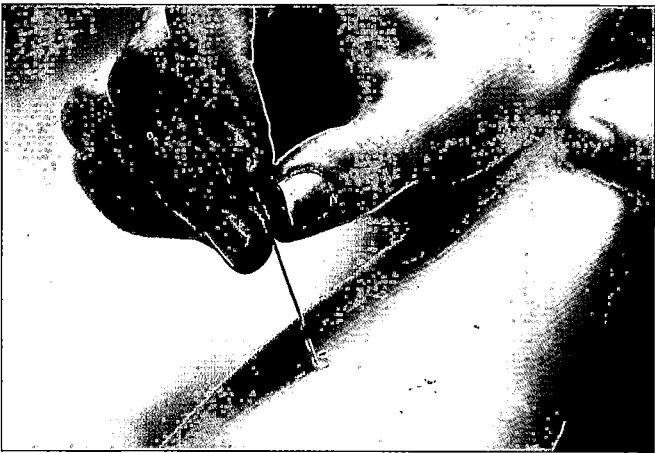

FIGURE 19.3 A skin test to identify allergens. Drops of fluid containing test substances are placed on the skin. A light scratch is made with a needle to allow the substances to penetrate the skin. Reddening and swelling at the site identify the substance as a probable cause of an allergic reaction.

■ In skin tests, sensitivity to a substance is indicated by a rapid inflammation to the inoculated antigen.

97% of food-related allergies: eggs, peanuts, tree-grown nuts, milk, soy, fish, wheat, and peas. Most children exhibiting allergies to milk, egg, wheat, and soy develop tolerance as they age, but reactions to peanuts, tree nuts, and seafood tend to persist. Sulfites, to which many people have an allergy, are a frequent problem. Their use is widespread in foods and beverages, and although food labels should indicate their presence, they may be difficult to avoid in practice.

Prevention of Anaphylactic Reactions

Avoiding contact with the sensitizing antigen is the most obvious way to prevent allergic reactions. Unfortunately, avoidance is not always possible. (Although commercial foods are required to list all ingredients, errors and oversights sometimes occur.) Some allergic individuals may never know exactly what antigen they are sensitive to. In other cases, skin tests might be of use in diagnosis (Figure 19.3). These tests involve inoculating small amounts of the suspected antigen just beneath the epidermis of the skin. Sensitivity to the antigen is indicated by a rapid inflammatory reaction that produces redness, swelling, and itching at the inoculation site. This small affected area is called a *wheal*.

Once the responsible antigen has been identified, the person can either try to avoid contact with it or undergo **desensitization.** This procedure consists of a series of dosages of the antigen carefully injected beneath the skin.

The objective is to cause the production of IgG rather than IgE antibodies in the hope that the circulating IgG antibodies will act as *blocking antibodies* to intercept and neutralize the antigens before they can react with cell-bound IgE. Desensitization is not a routinely successful procedure, but it is effective in 65–75% of individuals whose allergies are induced by inhaled antigens, and in a reported 97% for insect venom.

Type II (Cytotoxic) Reactions

Learning Objectives

■ *Describe the mechanism of cytotoxic reactions and how drugs can induce them.*

■ *Describe the basis of the ABO and Rh blood group systems.*

■ *Explain the relationship between blood groups, blood transfusions, and hemolytic disease of the newborn.*

Type II (cytotoxic) reactions generally involve the activation of complement by the combination of IgG or IgM antibodies with an antigenic cell. This activation stimulates complement to lyse the affected cell, which might be either a foreign cell or a host cell that carries a foreign antigenic determinant (such as a drug) on its surface. Additional cellular damage may be caused within 5 to 8 hours by the action of macrophages and other cells that attack antibody-coated cells.

The most familiar cytotoxic hypersensitivity reactions are *transfusion reactions,* in which red blood cells are destroyed as a result of reacting with circulating antibodies. These involve blood group systems that include the ABO and Rh antigens.

The ABO Blood Group System

In 1901, Karl Landsteiner discovered that human blood could be grouped into four principal types, which were designated A, B, AB, and O. This method of classification is called the **ABO blood group system.** Since then, other blood group systems, such as the Lewis system and the MN system, have been discovered, but our discussion will be limited to two of the best known, the ABO and the Rh systems. The main features of the ABO blood group system are summarized in Table 19.2.

A person's ABO blood type depends on the presence or absence of carbohydrate antigens located on the cell membranes of red blood cells (RBCs). Cells of blood type O lack both A and B antigens. Table 19.2 shows that the plasma of individuals with a given blood type, such as A, have antibodies against the alternative blood type, anti-B antibody. These antibodies arise in response to bacteria that have antigenic determinants very similar to blood

table 19.2	The ABO Blood Group System					Frequency (% U.S. Population)		
Blood Group	Erythrocyte or Red Blood Cell Antigens	Illustration	Plasma Antibodies	Blood That Can Be Received		White	Black	Asian
AB	A and B		Neither anti-A nor anti-B antibodies	A, B, AB, O Universal recipient		3	4	5
B	B		Anti-A	B, O		9	20	27
A	A		Anti-B	A, O		41	27	28
O	None		Anti-A and Anti-B	O Universal donor		47	49	40

group antigens. Individuals with type AB cells have plasma with no antibodies to either A or B antigens and therefore can receive cells with either A or B blood without a reaction. Type O individuals have antibodies against both A and B antigens. Lacking antigens, their blood cells can be transfused without difficulty to others, but they can receive transfusions of type O blood only.

When a transfusion is incompatible, as when type B blood is transfused into a person with type A blood, the antigens on the type B blood cells will react with anti-B antibodies in the recipient's serum. This antigen–antibody reaction activates complement, which in turn causes lysis of the donor's RBCs as they enter the recipient's system.

The Rh Blood Group System

In the 1930s, researchers discovered the presence of a different surface antigen on human red blood cells. Soon after they injected rabbits with RBCs from rhesus monkeys, the rabbit serum contained antibodies that were directed against the monkey blood cells but that would also agglutinate some human RBCs. This indicated that a common antigen was present on both human and monkey red blood cells. The antigen was named the **Rh factor** (Rh for Rhesus monkey). The roughly 85% of the population whose cells possess this antigen are called Rh^+; those lacking this RBC antigen (about 15%) are Rh^-. Antibodies that react with the Rh antigen do not occur naturally in the serum of Rh^- individuals, but exposure to this antigen can sensitize their immune systems to produce anti-Rh antibodies.

Blood Transfusions and Rh Incompatibility If blood from an Rh^+ donor is given to an Rh^- recipient, the donor's RBCs stimulate the production of anti-Rh antibodies in the recipient. If the recipient then receives Rh^+ RBCs in a subsequent transfusion, a rapid, serious hemolytic reaction will develop.

Hemolytic Disease of the Newborn Blood transfusions are not the only way in which an Rh^- person can become sensitized to Rh^+ blood. When an Rh^- woman and an Rh^+ man produce a child, there is a 50% chance that the child will be Rh^+ (Figure 19.4). If the child is Rh^+, the Rh^- mother can become sensitized to this antigen during birth, when the placental membranes tear and the fetal Rh^+ RBCs enter the maternal circulation, causing the mother's body to produce anti-Rh antibodies of the IgG type. If the fetus in a subsequent pregnancy is Rh^+, her anti-Rh antibodies will cross the placenta and destroy the fetal RBCs. The fetal body responds to this immune attack by producing large numbers of immature RBCs called erythroblasts. Thus, the term *erythroblastosis fetalis* was once used to describe what is now called **hemolytic disease of the newborn (HDNB)**. Before the birth of a fetus with this condition, the maternal

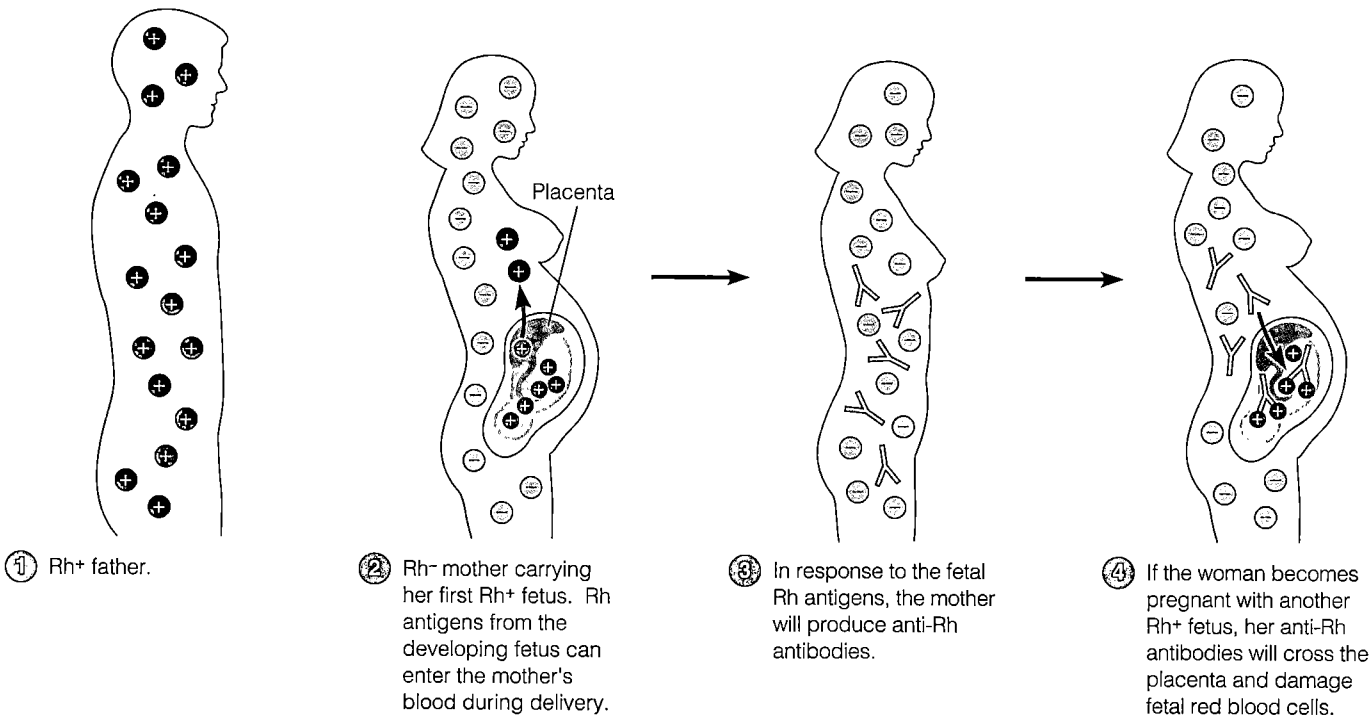

① Rh+ father.

② Rh⁻ mother carrying her first Rh+ fetus. Rh antigens from the developing fetus can enter the mother's blood during delivery.

③ In response to the fetal Rh antigens, the mother will produce anti-Rh antibodies.

④ If the woman becomes pregnant with another Rh+ fetus, her anti-Rh antibodies will cross the placenta and damage fetal red blood cells.

FIGURE 19.4 Hemolytic disease of the newborn.

■ Hemolytic disease of the newborn is caused by a mother's antibodies to the fetus crossing the placenta.

circulation removes most of the toxic by-products of fetal RBC disintegration. After birth, however, the fetal blood is no longer purified by the mother, and the newborn develops jaundice and severe anemia.

HDNB is usually prevented today by passive immunization of the Rh⁻ mother at the time of delivery of any Rh⁺ infant with anti-Rh antibodies, which are available commercially (Rhogam®). These anti-Rh antibodies combine with any fetal Rh⁺ RBCs that have entered the mother's circulation, so it is much less likely that she will become sensitized to the Rh antigen. If the disease is not prevented, the newborn's Rh⁺ blood, contaminated with maternal antibodies, may have to be replaced by transfusion of uncontaminated blood.

Drug-Induced Cytotoxic Reactions

Blood platelets (thrombocytes) are minute cell-like bodies that are destroyed by drug-induced cytotoxic reactions in the disease called **thrombocytopenic purpura.** The drug molecules are usually haptens because they are too small to be antigenic by themselves, but in the situation illustrated in Figure 19.5, a platelet has become coated with molecules of a drug (quinine is a familiar example), and the combination is antigenic. Both antibody and complement are needed for lysis of the platelet. Be-

cause platelets are necessary for blood clotting, their loss results in hemorrhages that appear on the skin as purple spots (purpura).

Drugs may bind similarly to white or RBCs, causing local hemorrhaging and yielding symptoms described as "blueberry muffin" skin mottling. Immune-caused destruction of granulocytic white cells is called **agranulocytosis,** and it affects the body's phagocytic defenses. When RBCs are destroyed in the same manner, the condition is termed **hemolytic anemia.**

Type III (Immune Complex) Reactions

Learning Objective

■ Describe the mechanism of immune complex reactions.

Type III reactions involve antibodies against soluble antigens circulating in the serum. (In contrast, type II immune reactions are directed against antigens located on cell or tissue surfaces.) The antigen–antibody complexes are deposited in organs and cause inflammatory damage.

Immune complexes form only when certain ratios of antigen and antibody occur. The antibodies involved are usually IgG. A significant excess of antibody leads to the formation of complement-fixing complexes that are rapidly

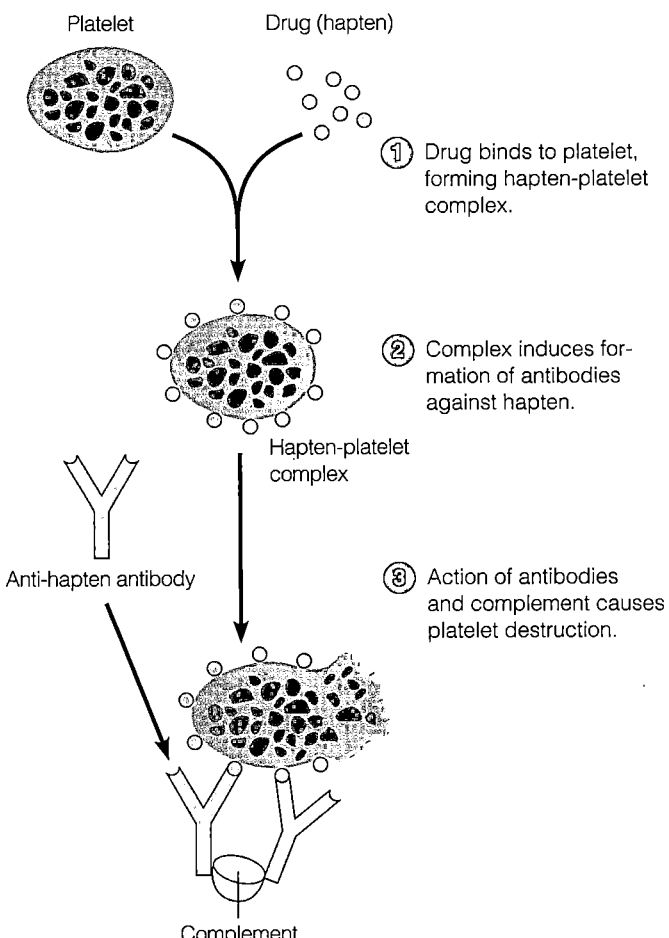

FIGURE 19.5 Drug-induced thrombocytopenic purpura. Molecules of a drug such as quinine accumulate on the surface of a platelet and stimulate an immune response that destroys the platelet.

■ **What actually destroys the platelets in thrombocytopenic purpura?**

removed from the body by phagocytosis. When there is a significant excess of antigen, soluble complexes form that do not fix complement and do not cause inflammation. However, when a certain antigen–antibody ratio exists, usually with a slight excess of antigen, the soluble complexes that form are small and escape phagocytosis.

Figure 19.6 illustrates the consequences. These complexes circulate in the blood, pass between endothelial cells of the blood vessels, and become trapped in the basement membrane beneath the cells. In this location, they may activate complement and cause a transient inflammatory reaction: attracting neutrophils that release enzymes. Repeated introduction of the same antigen can lead to more serious inflammatory reactions, causing damage to the basement membrane's endothelial cells within 2 to 8 hours.

Glomerulonephritis is an immune complex condition that causes inflammatory damage to the kidney glomeruli, which are sites of blood filtration.

Type IV (Cell-Mediated) Reactions

Learning Objective

■ *Describe the mechanism of cell-mediated reactions, and name two examples.*

Up to this point we have discussed humoral immune responses involving IgE, IgG, or IgM. Type IV reactions involve cell-mediated immune responses and are caused mainly by T cells. Instead of occurring within a few minutes or hours after a sensitized individual is again exposed to an antigen, these delayed reactions are not apparent for a day or more. A major factor in the delay is

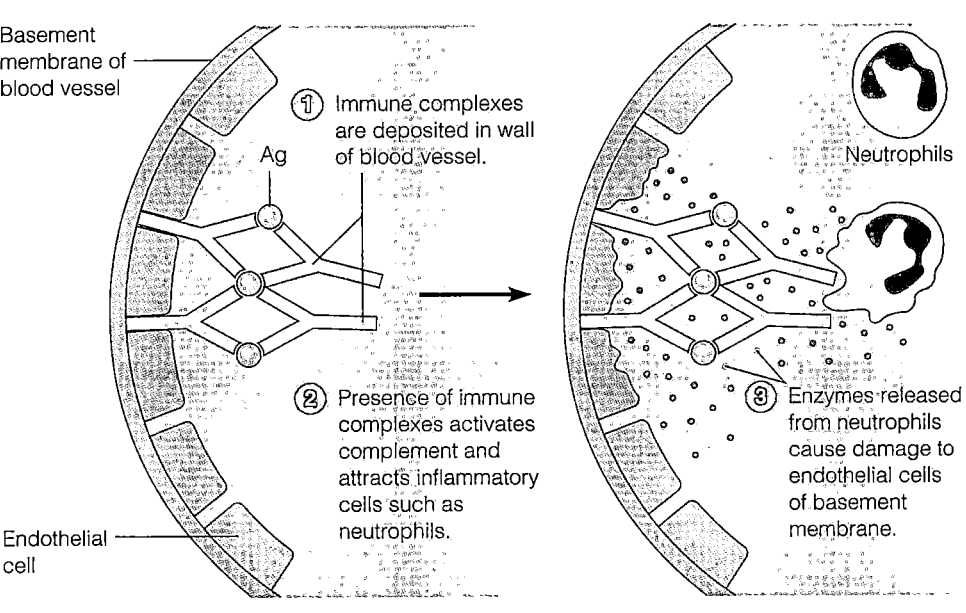

FIGURE 19.6 Immune complex–mediated hypersensitivity. ① Immune complexes are deposited on the basement membrane of the wall of a blood vessel, where they ② activate complement and attract inflammatory cells such as neutrophils to the site. ③ The neutrophils discharge enzymes as they react with the immune complexes, resulting in damage to tissue cells.

■ **Name one immune complex disease.**

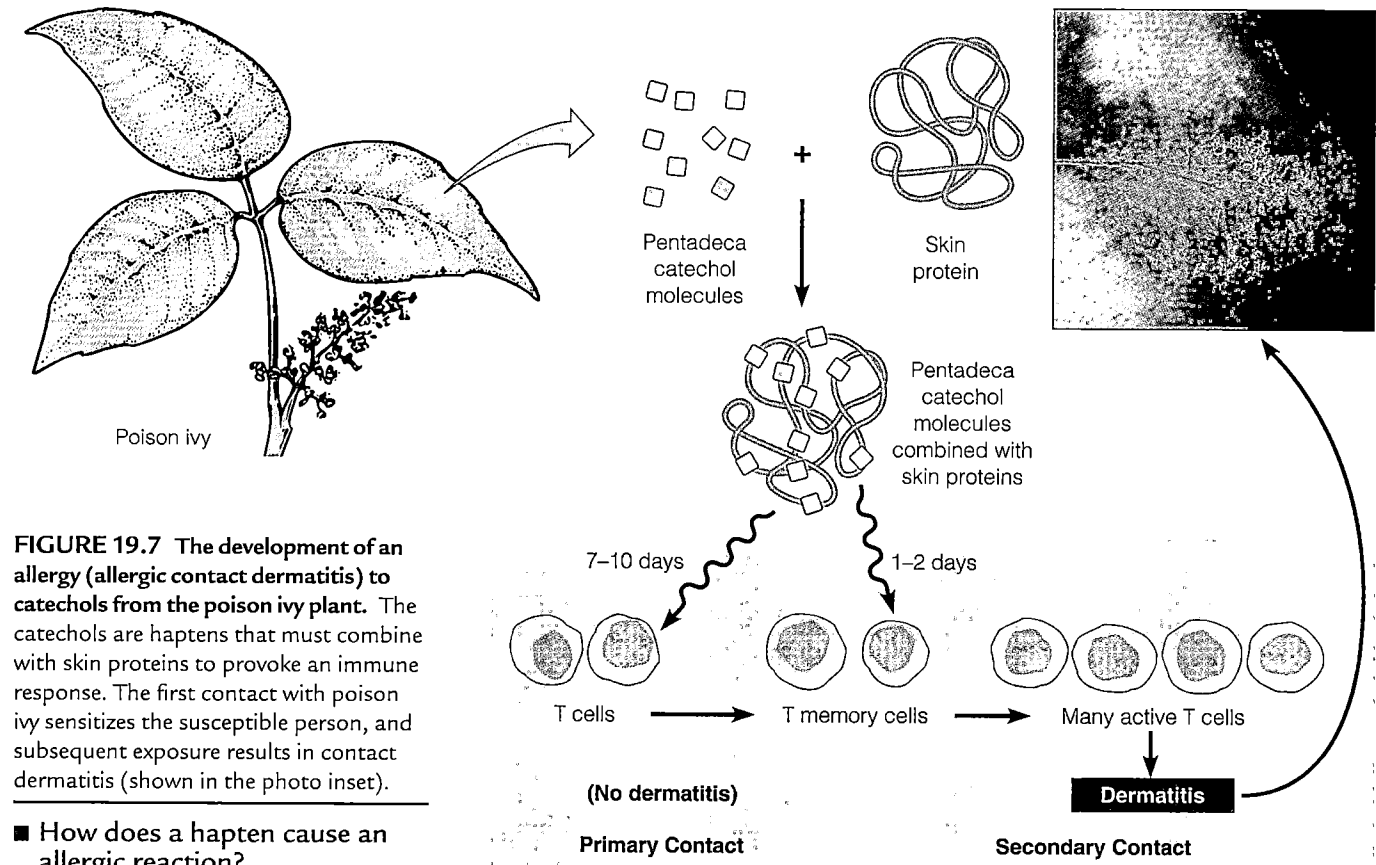

FIGURE 19.7 The development of an allergy (allergic contact dermatitis) to catechols from the poison ivy plant. The catechols are haptens that must combine with skin proteins to provoke an immune response. The first contact with poison ivy sensitizes the susceptible person, and subsequent exposure results in contact dermatitis (shown in the photo inset).

■ How does a hapten cause an allergic reaction?

the time required for the participating T cells and macrophages to migrate to and accumulate near the foreign antigens. The T cells involved in **delayed-type hypersensitivity** reactions are primarily T_D cells. In some types of hypersensitivities resulting in tissue damage, T_C cells may also participate.

Causes of Type IV Reactions

Sensitization for type IV hypersensitivity reactions occurs when certain foreign antigens, particularly those that bind to tissue cells, are phagocytized by macrophages and then presented to receptors on the T-cell surface. Contact between the antigenic determinant sites and the appropriate T cell causes the T cell to proliferate into mature differentiated T cells and memory cells.

When a person sensitized in this way is reexposed to the same antigen, a cell-mediated hypersensitivity reaction might result. Memory cells from the initial exposure activate T cells, which release destructive cytokines in their interaction with the target antigen. In addition, some cytokines contribute to the inflammatory reaction to the foreign antigen by attracting macrophages to the site and activating them (see Chapter 17, Figures 17.14 and 17.18 on pages 492 and 496).

Cell-Mediated Hypersensitivity Reactions of the Skin

We have seen that hypersensitivity symptoms are frequently displayed on the skin. One cell-mediated hypersensitivity reaction that involves the skin is the familiar skin test for tuberculosis. Because *Mycobacterium tuberculosis* is often located within macrophages, this organism can stimulate a cell-mediated immune response. As a screening test, protein components of the bacteria are injected into the skin. If the recipient has (or has had) a previous infection by tuberculosis bacteria, an inflammatory reaction to the injection of these antigens will appear on the skin in 1–2 days (Figure 24.11); this interval is typical of delayed hypersensitivity reactions.

Allergic contact dermatitis, another common manifestation of type IV hypersensitivity, is usually caused by haptens that combine with proteins in the skin of some people (particularly the amino acid lysine) to produce an immune response. Reactions to poison ivy (Figure 19.7), cosmetics, and the metals in jewelry (especially nickel) are familiar examples of these allergies.

The increasing exposure to latex in condoms, in certain catheters, and in gloves used by health care workers has led to a greater awareness of hypersensitivity to latex. Death from anaphylactic shock can also occur. During a recent 4-year period, 15 fatalities among patients were caused by exposure to latex tubing used for enemas or, during abdominal surgery, to the latex gloves of the surgeon. Among physicians and nurses, 3% or more have reported this type of hypersensitivity to surgical gloves. Delayed-type hypersensitivity reactions are not to latex itself, but to chemicals used in its manufacture. The result is a contact dermatitis such as that shown in Figure 19.8.

The powder in latex gloves is another source of problems; it absorbs allergens, becomes airborne, and can be inhaled. Latex paint, however, does not pose a threat of hypersensitivity reactions. Despite its name, latex paint contains no natural latex, but only synthetic nonallergenic chemical polymers.

The identity of the environmental factor causing the dermatitis can usually be determined by a *patch test*. Samples of suspected materials are taped to the skin; after 48 hours, the area is examined for inflammation.

Autoimmune Diseases

Learning Objectives

■ *Describe a mechanism for self-tolerance.*
■ *Give an example of type II, type III, and type IV autoimmune diseases.*

When the action of the immune system is in response to self-antigens and causes damage to one's own organs, the result is an **autoimmune disease.**

Autoimmune diseases occur when there is a loss of **self-tolerance,** the immune system's ability to discriminate self from nonself. In the generally accepted model by which T cells become capable of distinguishing self from nonself, the cells acquire this ability during their passage through the thymus. As we saw in Chapter 17 (page 484), any T cells that will target host cells are either eliminated by *clonal deletion* or inactivated during this period. This requirement makes it unlikely that the T cell will attack its own tissue cells.

In autoimmune diseases, the loss of self-tolerance leads to the production of antibodies or a response by sensitized T cells against a person's own tissue antigens. Autoimmune reactions, and the diseases they cause, can be cytotoxic, immune complex, or cell-mediated in nature.

Type I Autoimmunity

Type I autoimmunity involves antibodies that attack self. These antibodies may be made in response to an infectious agent such as a virus but sequence similarities between viral and self-proteins may result in the antibodies

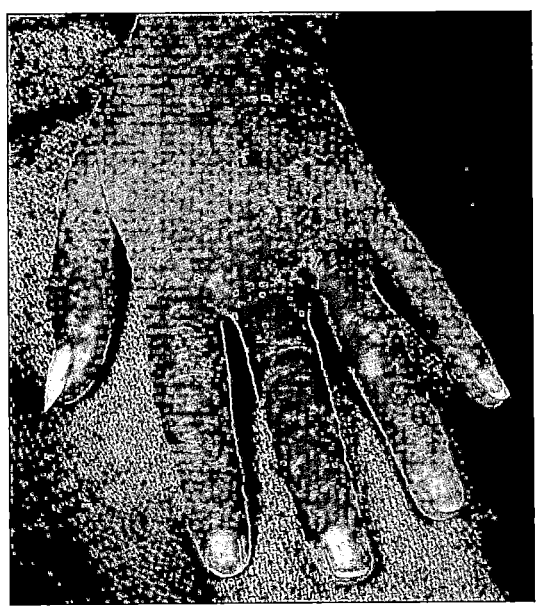

FIGURE 19.8 Allergic contact dermatitis. This person's hand exhibits a severe case of delayed-type contact dermatitis from wearing latex surgical gloves.

■ **What is allergic contact dermatitis?**

attacking self cells. Hepatitis C virus may be responsible for autoimmune hepatitis by this mechanism.

Type II (Cytotoxic) Autoimmune Reactions

Graves' disease and myasthenia gravis are two examples of disorders caused by type II autoimmune reactions. Both diseases involve antibody reactions to cell-surface antigens, although there is no cytotoxic destruction of the cells.

Graves' disease is caused by antibodies called longacting thyroid stimulators. These antibodies attach to receptors on thyroid gland cells that are the normal target cells of the thyroid-stimulating hormone produced by the pituitary gland. The result is that the thyroid gland is stimulated to produce increased amounts of thyroid hormones and becomes greatly enlarged. The most striking signs of the disease are goiter, which results from the enlarged thyroid, and markedly bulging, staring eyes.

Myasthenia gravis is a disease in which muscles become progressively weaker. It is caused by antibodies that coat the acetylcholine receptors at the junctions at which nerve impulses reach the muscles. Eventually, the muscles controlling the diaphragm and the rib cage may fail to receive the necessary nerve signals, and respiratory arrest and death result.

table 19.3	Diseases Related to Specific HLAs	
Disease	**Increased Risk of Occurrence with Specific HLA***	**Description**
Inflammatory Diseases		
Multiple sclerosis	5 times	Progressive inflammatory disease affecting nervous system
Rheumatic fever	4–5 times	Cross-reaction with antibodies against streptococcal antigen
Endocrine Diseases		
Addison's disease	4–10 times	Deficiency in production of hormones by adrenal gland
Graves' disease	10–12 times	Antibodies attached to certain receptors in the thyroid gland cause it to enlarge and produce excessive hormones
Malignant Disease		
Hodgkin's disease	1.4–1.8 times	Cancer of lymph nodes

*Compared to the general population.

Type III (Immune Complex) Autoimmune Reactions

Systemic lupus erythematosus is a systemic autoimmune disease, involving immune complex reactions, that mainly affects women. The etiology of the disease is not completely understood, but afflicted individuals produce antibodies directed at components of their own cells, including DNA, which is probably released during the normal breakdown of tissues, especially the skin. The most damaging effects of the disease result from deposits of immune complexes in the kidney glomeruli.

Crippling **rheumatoid arthritis** is a disease in which immune complexes of IgM, IgG, and complement are deposited in the joints. In fact, immune complexes called *rheumatoid factors* may be formed by IgM binding to the Fc region of normal IgG. These factors are found in 70% of individuals suffering from rheumatoid arthritis. The chronic inflammation caused by this deposition eventually leads to severe damage to the cartilage and bone of the joint.

Type IV (Cell-Mediated) Autoimmune Reactions

Hashimoto's thyroiditis is a result of the destruction of the thyroid gland, primarily by T cells of the cell-mediated immune system. It is a fairly common disorder and is often found in related family members. **Insulin-dependent diabetes mellitus** is a familiar condition caused by immunological destruction of insulin-secreting cells of the pancreas. T cells are clearly implicated in this disease; animals that are genetically likely to develop diabetes fail to do so when their thymus is removed in infancy.

Reactions Related to the Human Leukocyte Antigen (HLA) Complex

Learning Objective

- *Define HLA complex, and explain its importance in disease susceptibility and tissue transplants.*

The inherited genetic characteristics of individuals are expressed not only in the color of their eyes and the curl of their hair but also in the composition of the self molecules on their cell surfaces. Some of these are called **histocompatibility antigens.** The genes controlling the production of the most important of these self molecules are known as the **major histocompatibility complex (MHC).** In humans, these genes are called the **human leukocyte antigen (HLA) complex.** We encountered these self molecules in Chapter 17, where we saw that most antigens can stimulate an immune reaction only if they are associated with an MHC molecule.

A process called *HLA typing* is used to identify and compare HLAs. Certain HLAs are related to an increased susceptibility to specific diseases; one medical application of HLA typing is to identify such susceptibility. A few of these relationships are summarized in Table 19.3.

Another important medical application of HLA typing is in transplant surgery, in which the donor and the recipient must be matched by *tissue typing.* The serological technique shown in Figure 19.9 is the one most often used. In serological tissue typing, the laboratory uses standardized antisera or monoclonal antibodies that are specific for particular HLAs. ① Lymphocytes from the

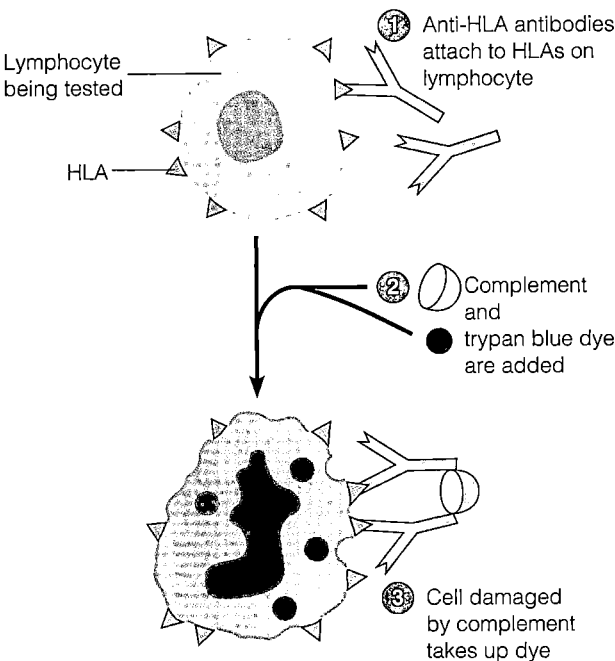

Lymphocyte being tested

HLA

① Anti-HLA antibodies attach to HLAs on lymphocyte

② Complement and trypan blue dye are added

③ Cell damaged by complement takes up dye

FIGURE 19.9 Tissue typing, a serological method. Lymphocytes from the person being tested are incubated with laboratory test stocks of anti-HLA antibodies specific for a particular HLA. If the antibodies react with the antigens on a lymphocyte, then complement damages the lymphocyte and dye can enter the cell. Such a positive test result indicates that the person has the particular HLA being tested for.

■ **Why is tissue typing done?**

person being tested are incubated with a selected specific antiserum. ② Complement and a dye, such as trypan blue, are then added. ③ If antibodies in the antiserum have reacted specifically with the lymphocyte, the cell is damaged. The damaged cell will take up the dye (undamaged cells will not), thus indicating that the lymphocyte possesses a certain antigen. This method is simple and rapid.

A promising new technique for analyzing HLA is the use of *PCR*, the *polymerase chain reaction,* to amplify the cell's DNA (see Figure 9.4 on page ·255). If this is done for both donor and recipient, a match between donor DNA and recipient DNA can then be made. Such a match should result in a much higher success rate in transplant surgery.

Two main classes of HLAs are important in such matches: *class I* (HLA-A, -B, and -C) and *class II* (HLA-DR, -DP, and -DQ). The genes that control the production of these self molecules are located close together on chromosome 6. Matching donors and recipients for class I self molecules, which are on all cells in the body, has long been a standard procedure. These self molecules stimulate a strong immune response by the antibodies and

T_C cells mediating transplant rejection. However, matching for class II self molecules, found primarily on the surfaces of specialized immune system cells, might well be more important, especially for tissue from an unrelated person. If the donor and the recipient do not share these self molecules, the transplant will probably be rejected. The donor and the recipient must also be of the same ABO blood type.

Other factors may be involved in the success of a transplant, however. In Chapter 17, we briefly introduced a hypothesis that the body's reaction to transplanted foreign tissue may be a response to surgery-damaged cells. In other words, tissue rejection may result from a learned reaction to the danger signal posed by damaged cells, rather than a learned reaction to nonself.

Reactions to Transplantation

Learning Objectives

■ *Explain how the rejection of a transplant occurs.*
■ *Define privileged site.*
■ *Define autograft, isograft, allograft, and xenograft.*
■ *Explain how graft-versus-host disease occurs.*

In sixteenth-century Italy, crimes were often punished by cutting off the offender's nose. A surgeon of the time, in his attempts to repair this mutilation, observed that if skin was taken from the patient, it healed properly, but if it was taken from another person, it did not. He called this a manifestation of "the force and power of individuality."

We now know the principles behind this phenomenon. Transplants recognized as nonself are rejected—attacked by T cells that directly lyse the grafted cells, by macrophages activated by T cells, and, in certain cases, by antibodies, which activate the complement system and injure blood vessels supplying the transplanted tissue. However, transplants that are not rejected can add many healthy years to a person's life.

Since the first kidney transplant was performed in 1954, this particular type of transplant has become a nearly routine medical procedure. Other types of transplants that are now feasible include bone marrow, thymus glands, heart, liver, and cornea. Tissues and organs for transplant are usually taken from recently deceased individuals, although one of a pair of organs, such as a kidney, occasionally comes from a living donor.

Privileged Sites and Privileged Tissue

Some transplants or grafts do not stimulate an immune response. A transplanted cornea, for example, is rarely rejected, mainly because antibodies usually do not circulate into that portion of the eye, which is therefore considered an immunologically **privileged site.** (However, rejections do occur, especially when the cornea has developed many blood vessels from corneal infections or damage.) The brain is also an immunologically privileged site, probably because it does not have lymphatic vessels and because the walls of the blood vessels in the brain differ from blood vessel walls elsewhere in the body (the blood–brain barrier is discussed in Chapter 22). Someday it may even be possible to graft foreign nerves to replace damaged nerves in the brain and spinal cord. Encouraging results have been obtained with these types of grafts in experiments with rats.

It is possible to transplant **privileged tissue** that does not stimulate an immune rejection. An example is replacing a person's damaged heart valve with a valve from a pig's heart. However, privileged sites and tissues are more the exception than the rule.

How animals tolerate pregnancy without rejecting the fetus is only partially understood. The uterus is not a privileged site; yet during pregnancy, the tissues of two genetically different individuals are in direct contact. One recently proposed theory to explain this phenomenon is that the placenta contains an enzyme that destroys tryptophan, a nutrient required by T cells that might otherwise attack the fetus.

Grafts

When one's own tissue is grafted to another part of the body, as is done in burn treatment or in plastic surgery, the graft is not rejected. Recent technology has made it possible to use a few cells of a burn patient's uninjured skin to culture extensive sheets of new skin. This new skin is an example of an **autograft.** Identical twins have the same genetic makeup; therefore, skin or organs such as kidneys may be transplanted between them without provoking an immune response. Such a transplant is called an **isograft.**

Most transplants, however, are made between people who are not identical twins, and these transplants do trigger an immune response. Attempts are made to match the HLAs of the donor and recipient as closely as possible so that the chances of rejection are reduced. Because HLAs of close relatives are most likely to match, blood relatives, especially siblings, are the preferred donors. Grafts between people who are not identical twins are called **allografts.**

Because of the shortage of available organs, it would be helpful if **xenografts,** or *xenotransplants,* transplanted tissue or organs from animals, could be more successfully performed on humans. However, the body tends to mount an especially severe immune assault on such transplants. Unsatisfactory attempts have been made to use organs from baboons and other nonhuman primates. Research interest is high in genetically engineering pigs—an animal in plentiful supply, of the right size, and one that generates relatively little public sympathy—to make acceptable donors of organs. The primary concern about xenografts is the possibility of transferring harmful animal viruses.

Preliminary research is under way that may eventually allow some bones and organs to be grown from the host's own tissue cells.

In order to be successful, xenografts must overcome **hyperacute rejection,** caused by the development in early infancy of antibodies against all distantly related animals such as pigs. With the aid of complement, these antibodies attack the transplanted animal tissue and destroy it within an hour. Hyperacute rejection occurs in human-to-human transplants only when antibodies have been preformed because of previous transfusions, transplantations, or pregnancies.

Bone Marrow Transplants

Transplants of bone marrow are frequently in the news. The recipients are usually individuals who lack the capacity to produce B cells and T cells vital for immunity, or who are suffering from leukemia. Recall from Chapter 17 that bone marrow stem cells give rise to red blood cells and immune system lymphocytes. The goal of bone marrow transplants is to enable the recipient to produce such vital cells. However, such transplants can result in **graft-versus-host (GVH) disease.** The transplanted bone marrow contains immunocompetent cells that mount primarily a cell-mediated immune response against the tissue into which they have been transplanted. Because the recipients lack effective immunity, GVH disease is a serious complication and can even be fatal.

An extremely promising technique for avoiding this problem is the use of *umbilical cord blood* instead of bone marrow. This blood is harvested from the placenta and umbilical cords of newborns, material that would otherwise be discarded. It is very rich in the stem cells (Figure 17.6, page 483) found in bone marrow. Not only do these cells proliferate into the variety of cells required by the recipient; but because stem cells from this source are younger and less mature, the "matching" requirements are also less stringent than with bone marrow. As a result, GVH disease is less likely to occur.

Immunosuppression

Learning Objective

■ *Explain how rejection of a transplant is prevented.*

To keep the problem of transplant rejection in perspective, it is useful to remember that the immune system is simply doing its job and has no way of recognizing that its attack against the transplant is not helpful. In an attempt to prevent rejection, the recipient of an allograft usually receives treatment to suppress this normal immune response against the graft.

In transplantation surgery, it is generally desirable to suppress cell-mediated immunity, the most important factor in transplant rejection. If humoral (antibody-based) immunity, is not suppressed, much of the ability to resist microbial infection will remain. In 1976, the drug *cyclosporine* was isolated from a mold. (Curiously, this fungus has a sexual cycle in a dung beetle that stimulates the beetle to climb high on vegetation and die, ensuring more efficient airborne distribution of fungal spores.)

The successful transplantation of organs such as hearts and livers generally dates from the discovery of cyclosporine. The secretion of interleukin-2 (IL-2) is suppressed by cyclosporine, disrupting cell-mediated immunity by cytotoxic T cells. Following the success of this drug, other immunosuppressant drugs soon followed. *Tacrolimus* (FK506) has a mechanism similar to cyclosporine and is a frequent alternative, although both have many serious side effects. Neither cyclosporine nor tacrolimus has much effect on antibody production by the humoral immune system.

Some newer drugs, such as *sirolimus* (rapamycin) are among those that inhibit both cell-mediated and humoral immunity. This can be an advantage if chronic or hyperacute rejection by antibodies is a consideration. Sirolimus blocks the action of IL-2. Drugs such as *mycophenolate mofetil* inhibit the proliferation of T cells and B cells. In a similar approach, in 1998 the FDA approved two chimeric monoclonal antibodies (see page 487), *basiliximab* and *daclizumab* that block IL-2. Immunosuppressive drugs are usually administered in combinations.

Several other drugs are being studied, all of which will probably improve the already impressive success rate of transplantation medicine.

Immune Deficiencies

Learning Objective

■ *Compare and contrast congenital and acquired immune deficiencies.*

The absence of a sufficient immune response is called an **immune deficiency.** Immune deficiencies can be either congenital or acquired.

Congenital Immune Deficiencies

Some people are born with a defective immune system. Defects in, or the absence of, a number of inherited genes

can result in **congenital immune deficiencies.** For example, individuals with a certain recessive trait may not have a thymus gland and therefore lack cell-mediated immunity. A different recessive trait causes lower numbers of B cells, affecting humoral immunity.

Acquired Immune Deficiencies

A variety of drugs, cancers, or infectious agents can result in **acquired immune deficiencies.** For example, Hodgkin's disease (a type of cancer) lowers the cell-mediated response. Many viruses are capable of infecting and killing lymphocytes, lowering the immune response. Removal of the spleen decreases humoral immunity. Table 19.4 summarizes several of the better known immune deficiency conditions, including AIDS.

The Immune System and Cancer

Learning Objective

■ *Describe the immune responses to cancer and how cells evade immune responses.*

Like an infectious disease, cancer represents a failure of the body's defenses, including the immune system. One of the most promising avenues for effective cancer therapy makes use of immunological techniques.

One long-held hypothesis is that cancer cells arise by mutation or virus-induced changes, and that the immune system constantly patrols and eliminates these cells before they become established tumors. This process is called *immunological surveillance,* and it has even been suggested that the cell-mediated immune system may have arisen primarily for this purpose. This concept is supported by the observation that cancers occur most often in older adults, whose immune systems are becoming less efficient, or in the very young, whose immune systems may not have developed fully or properly. Also, individuals who are immunosuppressed by either natural or artificial means are more susceptible to cancer, although these are usually restricted to cancers of the immune system. However, there are many observations that cannot be explained by the surveillance hypothesis.

A cell becomes cancerous when it undergoes transformation and begins to proliferate without control (see Chapter 13, page 396). Because of transformation, the surfaces of tumor cells may acquire tumor-associated antigens that mark them as nonself to the immune system. Activated T$_C$ cells respond to these nonself antigens by attaching to and lysing the cancerous cells that carry them (Figure 19.10 on page 535).

table 19.4 | *Immune Deficiencies*

Disease	Cells Affected	Comments
Acquired immunodeficiency syndrome (AIDS)	T cells (virus destroys T_H [CD4] cells)	Allows cancer and bacterial, viral, fungal, and protozoan diseases; caused by HIV infection
Selective IgA immunodeficiency	B, T cells	Affects about 1 in 700, causing frequent mucosal infections; specific cause uncertain
Common variable hypogammaglobulinema	B, T cells (decreased immunoglobulins)	Frequent viral and bacterial infections; second most common immune deficiency, affecting about 1 in 70,000; inherited
Reticular dysgenesis	B, T, and stem cells (a combined immunodeficiency; deficiencies in B and T cells and neutrophils)	Usually fatal in early infancy; very rare; inherited; bone marrow transplant a possible treatment
Severe combined immunodeficiency	B, T, and stem cells (deficiency of both B and T cells)	Affects about 1 in 100,000; allows severe infections, inherited; treated with bone marrow, fetal thymus transplants; gene therapy treatment is promising
Thymic aplasia (DiGeorge syndrome)	T cells (defective thymus causes deficiency of T cells)	Absence of cell-mediated immunity; usually fatal in infancy from *Pneumocystis* pneumonia or viral or fungal infections; due to failure to develop in embryo
Wiskott-Aldrich syndrome	B, T cells (few platelets in blood, abnormal T cells)	Frequent infections by viruses, fungi, protozoa; eczema, defective blood clotting; usually causes death in childhood; inherited on X chromosome
X-linked infantile (Bruton's) agammaglobulinemia	B cells (decreased immunoglobulins)	Frequent extracellular bacterial infections; affects about 1 in 200,000; the first immunodeficiency disorder recognized (1952); inherited on X chromosome

Cancer sometimes occurs even in people with a presumably healthy immune system. Initially, the cancer is an individual cell that arises from mutations, perhaps resulting from exposure to chemicals or radiation. These changes usually require multiple "hits" (constituting damage to genes, and often separated by considerable time), to complete the change from normal to cancerous. Viral infection can also change a normal cell to a cancerous cell.

Individual cancer cells that arise are attacked by the immune system, much as it responds to transplanted foreign tissue. Once the cancerous cell attaches to tissue and becomes vascularized (connected with the body's blood supply), it develops rapidly into a tumor that becomes resistant to immune rejection. At present, a number of mechanisms have been identified. Tumor cells often lack surface molecules necessary to activate T_C cells, or lack parts of their structure that in other ways stimulate the immune system.

Tumor cells sometimes actively suppress the immune system; they produce factors that reduce the effectiveness of T_C cells, and some cancer cells induce immune cells to undergo apoptosis. Once established, tumor cells may even reproduce so rapidly that they exceed the capacity of an effective immune response.

It is encouraging to know that occasionally some malignancies spontaneously disappear, presumably because of a successful immune response. The phenomenon of cancer's resistance to the immune system is of interest because a better understanding could lead to ways to defeat this resistance, as will be discussed in the next section on immunotherapy.

Immunotherapy

Learning Objective

- *Define immunotherapy, and give two examples.*

At the turn of the twentieth century, William B. Coley, a physician at a New York City hospital, observed that if cancer patients contracted typhoid fever, their cancers often diminished noticeably. Following this lead, Coley made vaccines of killed gram-negative bacteria, called Coley's toxins, to simulate a bacterial infection. Some of this work was very promising, but its results were inconsistent, and advances in surgery and radiation treatment caused it to be nearly forgotten.

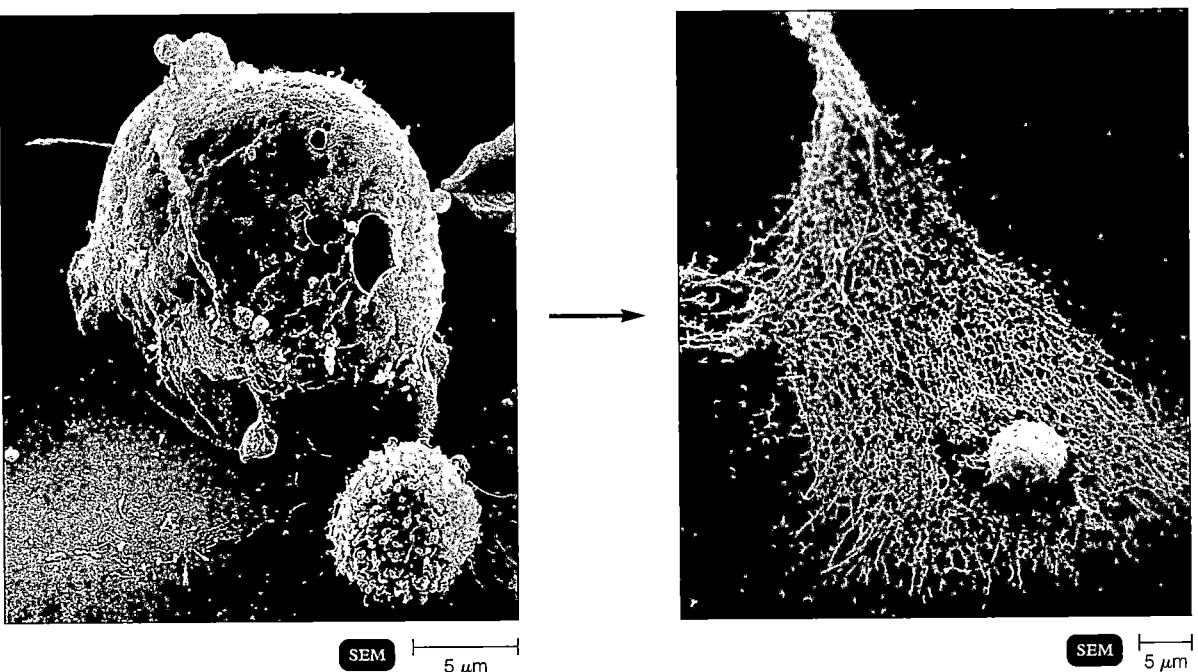

FIGURE 19.10 The interaction between a cytotoxic T (T_C) cell and a cancer cell. The small T_C cell (left panel) has already made a perforation in the cancer cell. At right, only the skeleton of the cancer cell remains.

■ T_C cells can lyse cancer cells. How do they do this? (*Hint:* see Figure 17.13.)

We now know that endotoxins from such bacteria are powerful stimulants for the production of tumor necrosis factor (TNF) by macrophages. TNF is a small protein that interferes with the blood supply of cancers in animals. There is now also much interest in cancer therapy with TNF and other cytokines such as interleukin-2 (see Chapter 17, page 488) and interferons (see Chapter 16, page 469).

The treatment of cancer by immunological means is termed **immunotherapy**, an approach that will probably be used in the future. One promising technique involves immunotoxins. Recall from Chapter 17 that an immunotoxin is a combination of a monoclonal antibody and a toxic agent, such as the drug ricin or a radioactive compound. The monoclonal antibody is directed against a particular type of tumor-associated antigen. The antibody selectively locates the cancer cell, and the attached toxic agent destroys the cell but causes little or no damage to healthy tissue. One such monoclonal antibody, *Herceptin,* is currently seeing limited use in certain applications in the treatment of breast cancer. It blocks a growth factor receptor on the tumor cell.

A promising new target for therapeutic antibodies is to neutralize cytokines that furnish the blood supply for tumor cell proliferation. This therapy would prevent the formation of connective tissue that makes up most of the mass of a tumor; thus, the cancerous cells would not develop into a tumor.

A vaccine against cancers also remains a possibility. For many years, for example, there has been an effective vaccine for preventing Marek's disease, a type of cancer in poultry. A cancer vaccine will require isolation and cultivation of suitable tumor antigens. This would be only the first step in a lengthy and expensive process of developing a practical vaccine against human cancers.

Acquired Immunodeficiency Syndrome (AIDS)

In 1981, a cluster of cases of *Pneumocystis* pneumonia (see Chapter 24) appeared in the Los Angeles area. This extremely rare disease usually only occurred in immunosuppressed individuals. Investigators soon correlated the appearance of this disease with an unusual incidence of a rare form of cancer of the skin and blood vessels called Kaposi's sarcoma. The people affected were all young homosexual men, and all showed a loss of immune function. By 1983, the pathogen causing the loss of immune function had been identified as a virus that selectively infects helper T cells. This virus is now known as human immunodeficiency virus (HIV).

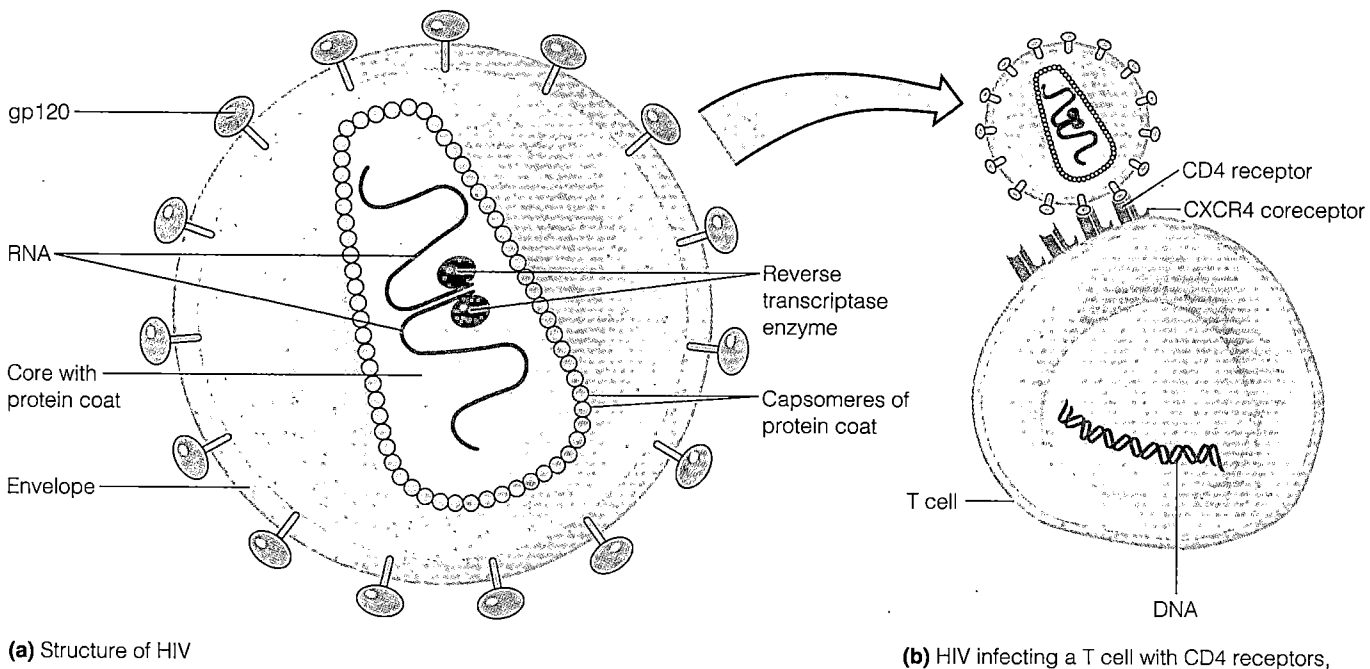

gp120

RNA

Core with protein coat

Envelope

Reverse transcriptase enzyme

Capsomeres of protein coat

(a) Structure of HIV

CD4 receptor

CXCR4 coreceptor

T cell

DNA

(b) HIV infecting a T cell with CD4 receptors, and CXCR4 coreceptors which are distributed over the surface of the cell

FIGURE 19.11   HIV structure and attachment to receptors on target T cell.

■ Why does HIV preferentially infect CD4 cells?

The Origin of AIDS

Learning Objective

■ *Give two examples of how emerging infectious diseases arise.*

HIV is now believed to have arisen by the mutation of a virus that had been endemic in some areas of central Africa for many years. Researchers have speculated that a relatively benign virus infecting monkeys and chimpanzees entered the human population when the animals were skinned and butchered for food. Mathematical models of the supposed evolution of HIV, by Bette Korber of the Los Alamos National Laboratory, calculate that the virus probably made the transition to humans around 1930. The disease may have smoldered with little notice as long as transmission was limited to small villages. The virus could not have killed or incapacitated its hosts quickly; otherwise, it could not have been maintained in the village population. With the sudden end of European colonialism, the social structure of sub-Saharan Africa was disrupted. The population became urbanized; the developments that result from urbanization, such as an increase in prostitution and the growth of highway transportation, are believed to be responsible for the spread of the disease. The earliest documented case of AIDS is from a patient in Leopoldville, Belgian Congo (now Kinshasha,

capital of the Democratic Republic of the Congo). This man died in 1959; preserved samples of his blood contain antibodies to HIV. In the Western world, the first confirmed case of AIDS was the death of a Norwegian sailor in 1976, who probably was infected in 1961 or 1962 by contacts in western Africa.

HIV Infection

Learning Objectives

■ *Explain the attachment of HIV to a host cell.*
■ *List two ways in which HIV avoids the host's antibodies.*
■ *Describe the stages of HIV infection.*
■ *Describe the effects of HIV infection on the immune system.*

One of the most common misconceptions is that HIV infection is synonymous with AIDS. But AIDS denotes only the final stage of a long infection.

The Structure of HIV

HIV, of the genus Lentivirus, is a retrovirus (see Figure 13.19 on page 393). It has two identical strands of RNA, the enzyme reverse transcriptase, and an envelope of phospholipid (Figure 19.11a). The envelope has spikes termed gp120 (the notation for a glycoprotein with a molecular weight of 120,000).

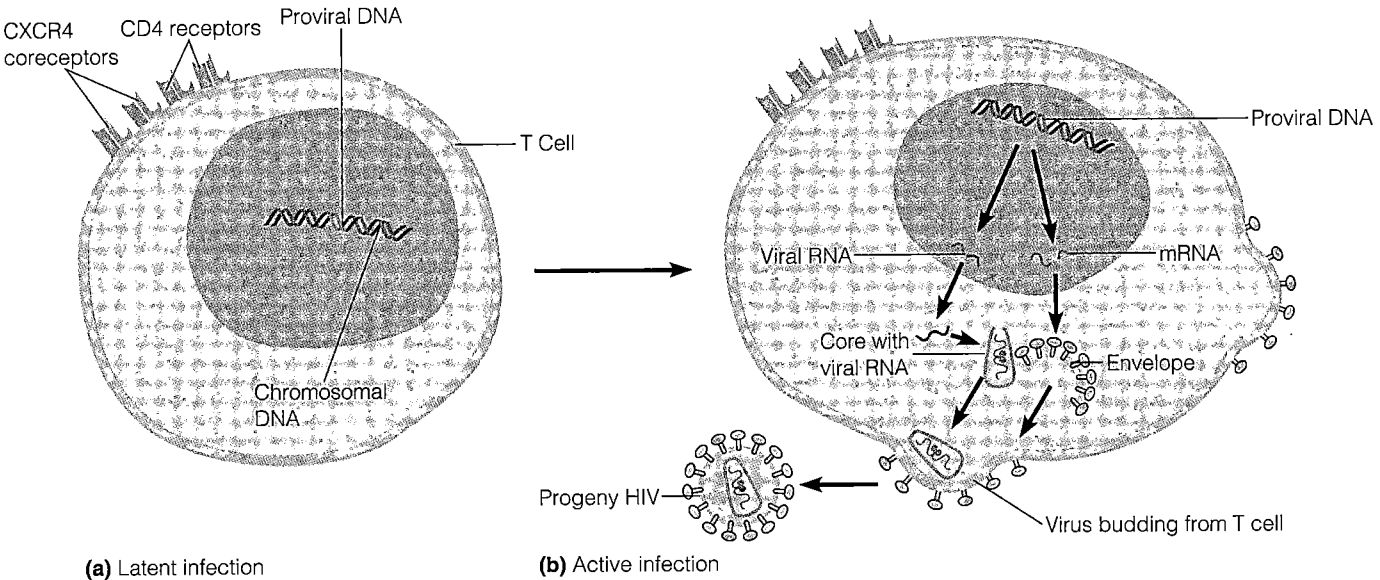

FIGURE 19.12   Latent and active HIV infection in CD4 T cells. (a) In latent infections, no viral genes are being transcribed and no new viruses are being formed. **(b)** In active infections, proviral DNA controls the synthesis of new viruses, which bud from the host cell.

■ **CD4 T cells may be latently or actively infected with HIV.**

The Infectiveness and Pathogenicity of HIV

The spikes enable the virus to attach to the CD4 receptor on the host cells (Figure 19.11b). CD4 receptors are found on helper T cells, macrophages, and dendritic cells—the main targets of HIV infection. A CD4 receptor by itself is not sufficient for HIV infection. Certain coreceptors, which are receptors for chemokines, are also required. The two best known chemokine coreceptors, originally called fusins, are named CCR5 and CXCR4. This intimidating nomenclature indicates the beginning amino acid sequence of these proteins. The term CCR5 indicates that the beginning sequence consists of cysteines, thus CC. If some other amino acid is located between the first two cysteines, this is shown by changing to CXC. The letter R is a convention for the balance of the chemical molecule. Simplistically, CCR5 is most important for the infection of macrophages and appears in the earlier stages of the infection. CXCR4 is involved mainly in the infection of T cells and predominates in later stages.

Attachment of the virus is followed by entry into the host cell (Figure 19.11b). In the host cell, viral RNA is released and transcribed into DNA by the enzyme reverse transcriptase. This viral DNA then becomes integrated into the chromosomal DNA of the host cell. The DNA may control the production of an active infection in which new viruses bud from the host cell, as shown in Figure 19.12b.

Alternatively, this integrated DNA may not produce new HIV but remains hidden in the host cell's chromosome as a *provirus* (Figures 19.12a and 19.13a). As a provirus, it is not detected by the immune system. HIV produced by a host cell is not necessarily released from the cell but may remain as *latent virions* in vacuoles within the cell, as shown in Figure 19.13b. This ability of the virus to remain as a provirus or latent virus within host cells is an important reason why anti-HIV antibodies developed by infected individuals fail to inhibit progression of the infection. Another way HIV evades the immune system is by cell-cell fusion, by which the virus moves from an infected cell to an adjacent uninfected cell.

The virus also evades immune defenses by undergoing rapid antigenic changes. Retroviruses, with the reverse transcriptase enzyme step, have a high mutation rate compared to DNA viruses. They also lack the corrective "proofreading" capacity of DNA viruses. As a result, a mutation is probably introduced at every position in the HIV genome many times each day in an infected person. This may amount to an accumulation of 1 million variants of the virus in an asymptomatic person and 100 million variants during the final stages of the infection. These dramatic numbers illustrate the potential problems of drug resistance and obstacles to the development of vaccines and diagnostic tests.

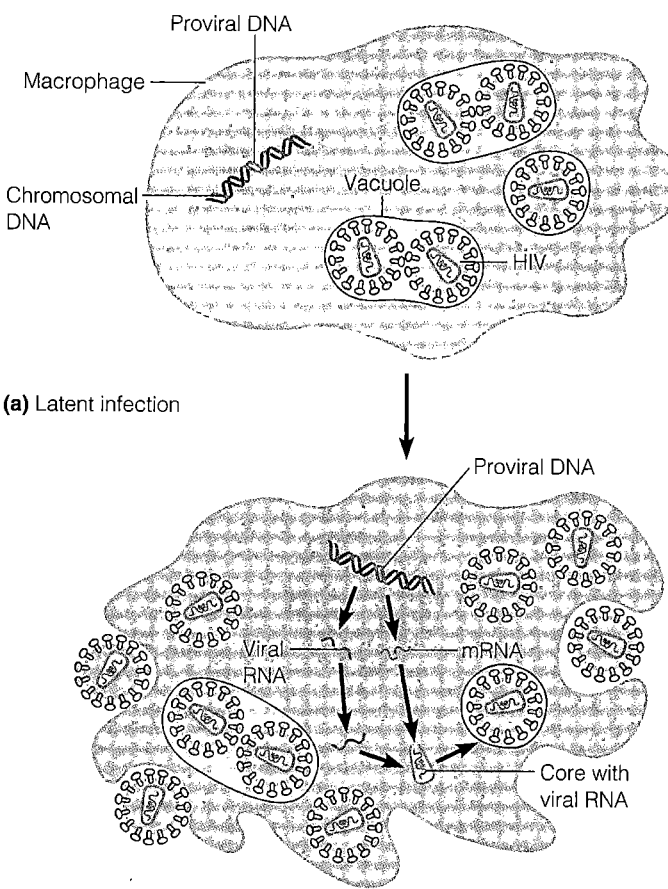

(a) Latent infection

(b) Active infection

FIGURE 19.13 Latent and active HIV infection in macrophages.
(a) In latent infections, the newly made viruses remain in vacuoles. They are also present in the form of a provirus in the cell's chromosome. If the macrophage (not the infecting virus) is activated, these viruses will be released. **(b)** In active infections, new viruses are being produced; some of these remain in vacuoles, while others are released. Also, notice the proviruses latent in the chromosome.

■ **How does an active infection differ from a latent infection?**

Clades (Subtypes) of HIV

Worldwide, HIV is beginning to separate into distinctive groups called *clades* (Greek for branches), or subtypes. Currently, *HIV-1,* the most common major type of HIV, has 11 such clades. Viruses may vary by 15–20% within a clade; between clades, they may vary by 30% or more. A second major HIV type, *HIV-2,* is found primarily in western Africa and is rare in the United States. The progression from infection to AIDS is much longer with HIV-2.

The Stages of HIV Infection

The CDC classification divides the progress of HIV infection in adults into three clinical stages, or categories (Figure 19.14):

1. *Category A.* At this stage, the infection may be asymptomatic or cause persistent lymphadenopathy (swollen lymph nodes).

2. *Category B.* This stage is characterized by persistent infections by the yeast *Candida albicans,* which can appear in the mouth, throat, or vagina. Other conditions may include shingles, persistent diarrhea and fever, whitish patches on the oral mucosa (hairy leukoplakia), and certain cancerous or precancerous conditions of the cervix.

3. *Category C.* This stage is clinical AIDS. Important AIDS indicator conditions are *Candida albicans* infections of the esophagus, bronchi, and lungs; cytomegalovirus eye infections; tuberculosis; *Pneumocystis* pneumonia; toxoplasmosis of the brain; and Kaposi's sarcoma (probably caused by Human herpesvirus 8).

The CDC also classifies the progress of HIV infections based on T-cell populations. The purpose is primarily to furnish guidance for treatment, such as when to administer certain drugs. The normal population of a healthy individual is 800–1000 CD4 T cells/mm^3. A count below 200/mm^3 is considered diagnostic for AIDS, regardless of the clinical category observed.

The progression from initial HIV infection to AIDS typically takes about 10 years in adults. Cellular warfare on an immense scale occurs during this time. At least 100 billion HIVs are generated every day, each with a remarkably short half-life of about 6 hours. These viruses must be cleared by the body's defenses, which include antibodies, cytotoxic T cells, and macrophages. Most HIVs are produced by infected CD4 T cells, which survive for only about 2 days (T cells normally live for several years). Every day, an average of about 2 billion CD4 T cells are produced to compensate for losses. Over time, however, there is a daily net loss of at least 20 million CD4 T cells, one of the main markers for the progression of HIV infection. The most recent studies show that the decrease in CD4 T cells is not due entirely to direct viral destruction of the cells; rather, it is caused primarily by shortened life of the cells and the body's failure to compensate by increasing production of replacement T cells. Suppression of viral numbers by chemotherapy apparently removes this inhibition to production of new T cells.

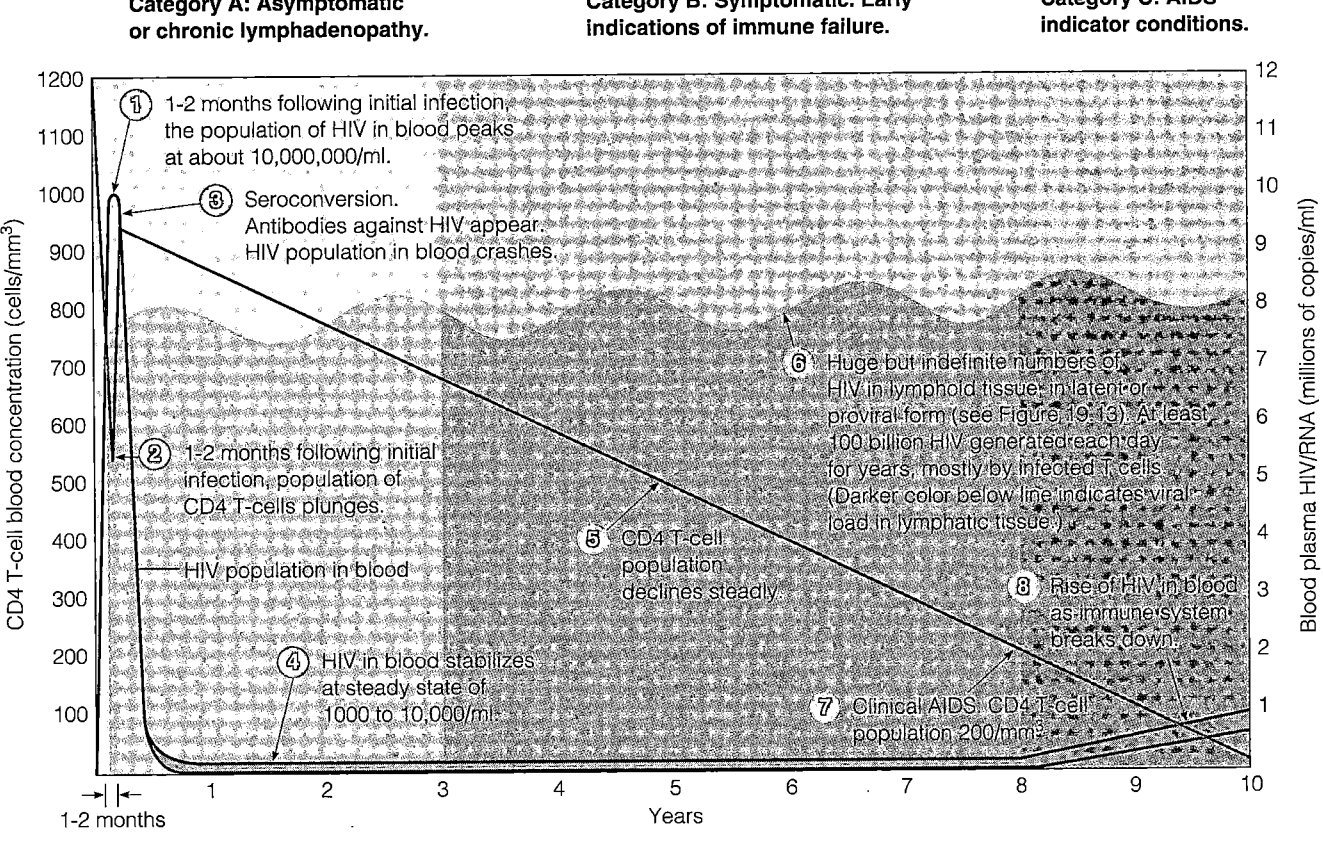

FIGURE 19.14 The progression of HIV infection.

■ What causes the HIV population to drop in the first few months following infection?

Survival with HIV Infection

HIV infection devastates the immune system, which is then unable to respond effectively to pathogens. The diseases or conditions most commonly associated with HIV infection and AIDS are summarized in Table 19.5. Success in treating these conditions has extended the lives of many HIV-infected people.

As the historical record of the AIDS epidemic lengthens, it has become apparent that not all HIV-positive individuals progress inexorably to AIDS and death. A significant group—about 5% of all those infected—are *long-term nonprogressors*. These infected individuals remain free of AIDS and even symptoms; and their CD4 T-cell counts remain stable. Survival exceeding 25 years is predicted. In some cases, the virus seems less virulent; but in most, the immune system, especially T_C cells, is apparently more effective.

The age of the infected person can also be an important factor. Older adults are less able to replace antiviral T-cell populations. Infants born to HIV-positive mothers

are not always infected—in fact, only a minority are. The rate of disease progression in affected infants is directly related to the severity of the disease in the mother. Infants most seriously infected survive less than 18 months.

Another surprising aspect of the AIDS epidemic is that certain people are subjected to multiple HIV exposures and yet never become infected at all. The evidence is that their CD4 T cells are innately resistant.

Diagnostic Methods

Learning Objective

■ Describe how HIV infection is diagnosed.

Generally speaking, it is simpler and less expensive to detect antibodies against HIV than to detect the virus itself. Therefore, most tests for HIV infection have been developed to detect HIV antibodies. The most commonly used screening tests are versions of the ELISA test (see Figure 18.12 on page 514). Positive ELISA tests are confirmed with the Western blot test (see Figure 10.12

table 19.5 Some Common Diseases Associated with AIDS

Pathogen or Disease	Disease Description
Protozoa	
Cryptosporidium parvum	Persistent diarrhea
Toxoplasma gondii	Encephalitis
Isospora belli	Gastroenteritis
Viruses	
Cytomegalovirus	Fever, encephalitis, blindness
Herpes simplex virus	Vesicles of skin and mucous membranes
Varicella-zoster virus	Shingles
Bacteria	
Mycobacterium tuberculosis	Tuberculosis
M. avium-intracellulare	May infect many organs; gastroenteritis and other highly variable symptoms
Fungi	
Pneumocystis carinii	Life-threatening pneumonia
Histoplasma capsulatum	Disseminated infection
Cryptococcus neoformans	Disseminated, but especially meningitis
C. albicans	Overgrowth on oral and vaginal mucous membranes (Category B stage of HIV infection)
C. albicans	Overgrowth in esophagus, lungs (Category C stage of HIV infection)
Cancers or Precancerous Conditions	
Kaposi's sarcoma	Cancer of skin and blood vessels (probably caused by HHV-8)
Hairy leukoplakia	Whitish patches on mucous membranes; commonly considered precancerous
Cervical dysplasia	Abnormal cervical growth

on page 291). A problem with antibody-type testing is the window of time between infection and the appearance of detectable antibodies, or **seroconversion.** Because of this delay, it is possible to transplant tissue or blood that contains HIV but that tests negative for anti-HIV antibodies. This window has been narrowed to about 25 days. Tests that detect HIV itself, nucleic acid tests, are now available that can identify as few as 10 HIV. The tests are significantly more costly and require 48–72 hours to complete.

Notice in Figure 19.14 that an early burst of HIV numbers precedes seroconversion. There are several methods of detecting and quantifying HIV. Viral RNA is detected after amplifying certain fragments of the viral genome using PCR. Currently, PCR is used to determine infections in newborns of HIV-infected mothers (in cases where maternal antibodies interfere with other tests), and certain specialized situations such as research that requires knowledge of the viral load. The Red Cross is now introducing PCR testing for HIV at all their regional centers. Pooled samples of 120 donations will be

tested. If HIV is detected, each individual sample will be tested until the infected sample is found. It is hoped that this will close the window of detection.

Even with the best available tests, however, there is always a small but real risk of contracting HIV from transfusions or transplants. Also, it is possible that current tests cannot detect all the myriad variants of HIV, especially if subtypes not normally present in this country are introduced unexpectedly. Tests for RNA of HIV sample only the virions circulating in the blood. This appears, however, to accurately reflect the much larger amounts of virus in the cells of the lymphatic system, which are constantly released into the peripheral circulation.

HIV Transmission

Learning Objective

- List the routes of HIV transmission.

The transmission of HIV requires the transfer of, or direct contact with, infected body fluids. The most important of

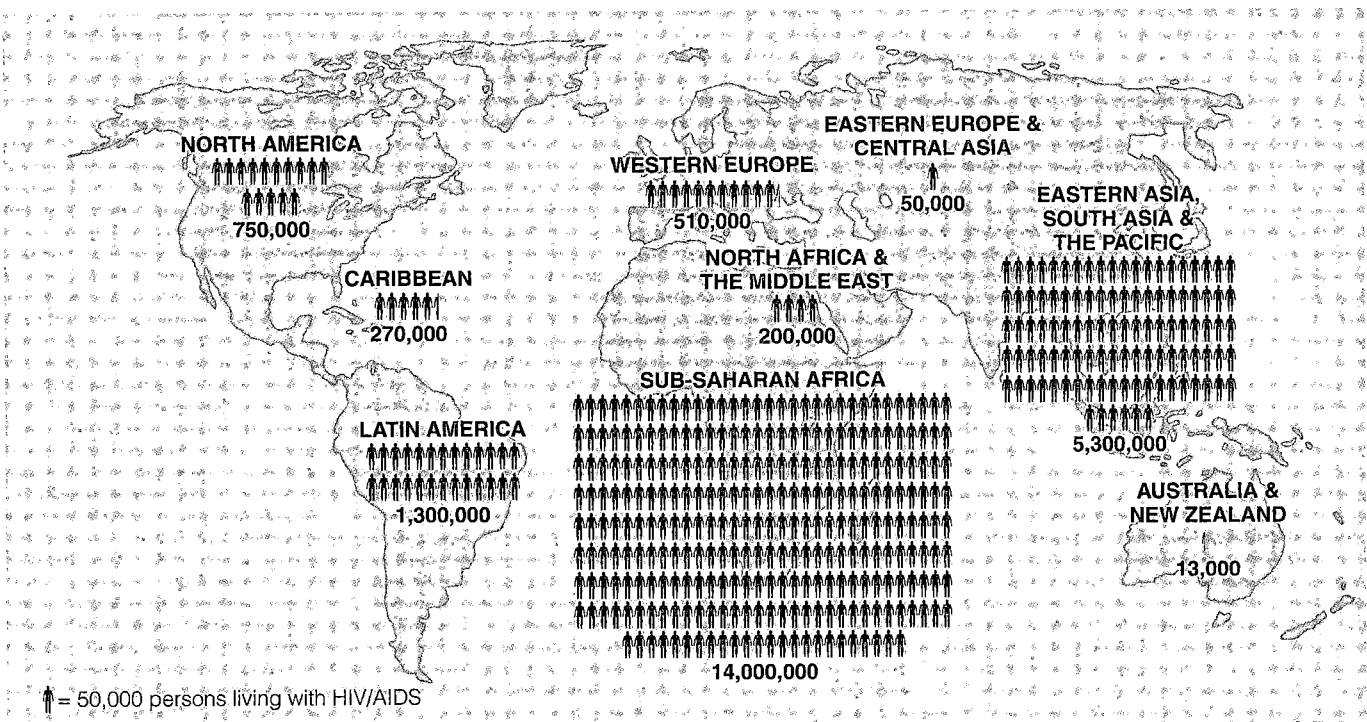

FIGURE 19.15   Distribution of HIV infection and AIDS in regions of the world. Each figure represents 50,000 persons living with HIV infection or AIDS. Worldwide, 43% of people living with HIV/AIDS are women.
SOURCE: Reprinted with permission from *Confronting AIDS: Public Priorities in a Global Epidemic.* Map produced from UNAIDS data by the Map Design Unit of the World Bank.

■ Where do you think the most accurate figures would be available?

these is blood, which contains 1000–100,000 infective viruses per milliliter, and semen, which contains about 10–50 viruses per milliliter. The viruses are often located within cells in these fluids, especially in macrophages. HIV can survive more than 1.5 days inside a cell but only about 6 hours outside a cell.

Routes of HIV transmission include intimate sexual contact, breast milk, transplacental infection of a fetus, blood-contaminated needles, organ transplants, artificial insemination, and blood transfusion. Probably the most dangerous form of sexual contact is anal-receptive intercourse. Vaginal intercourse is much more likely to transmit HIV from male to female than vice versa, and transmission either way is much greater when genital lesions are present. Although rare, transmission can occur by oral-genital contact.

HIV is not transmitted by insects or casual contact such as hugging or sharing household items. Saliva generally contains less than 1 virus per milliliter, and kissing is not known to transmit the virus. In developed countries, transmission by transfusion is unlikely because blood is tested for HIV antibodies. However, there will always be a slight risk, as discussed earlier.

AIDS Worldwide

Learning Objective

■ *Identify geographic patterns of HIV transmission.*

Only about two decades after AIDS was first recognized, HIV infection has become a global pandemic. An estimated 2.5 million AIDS deaths have occurred worldwide, and over 47 million adults, children, and infants are infected with HIV. It is the leading cause of death in sub-Saharan Africa and also in many of the largest cities in the United States and Europe (Figure 19.15). Worldwide, probably 16,000 new HIV infections occur each day.

Three basic epidemiological patterns have emerged:

■ In the United States, Canada, western Europe, Australia, northern Africa, and certain parts of South America, HIV has primarily affected injecting drug users (IDUs) and homosexual and bisexual males. In western Europe and North America, the incidence of heterosexual spread is increasing rapidly. It is expected that women will soon be infected at the same rate as men.

FIGURE 19.16 Modes of HIV transmission. (a) In most of the world, transmission is primarily by heterosexual sex. **(b)** In the United States, this form of transmission is much lower, but growing rapidly. Transmission in western Europe is similar to that in the United States.

■ **Worldwide, the primary mode of HIV transmission is heterosexual sex.**

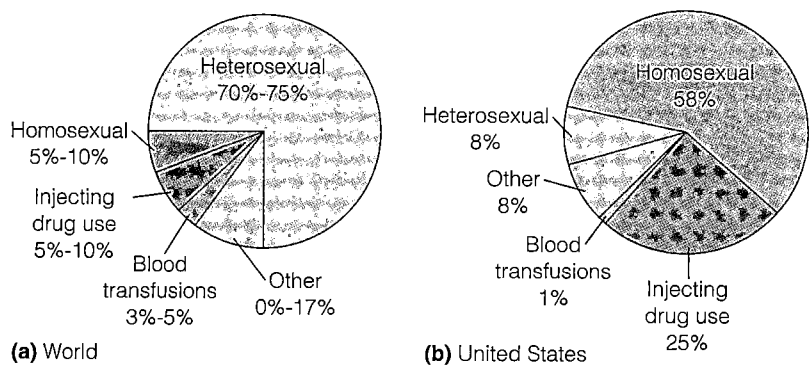

(a) World

(b) United States

■ In sub-Saharan Africa, HIV transmission has been almost entirely from heterosexual contact, and males and females are affected in about equal numbers. About 3% of the population is HIV-positive: in certain urban areas, the prevalence is as high as 30%.

■ The infection has been introduced more recently in eastern Europe, the Middle East, and, most importantly, in Southeast Asia. The World Health Organization (WHO) believes that Asia's epidemic may ultimately dwarf all others in scope and impact. In these areas, the epidemic involves IDUs, commercial sex workers, and young heterosexual males.

The HIV clades in Asia may be better adapted to heterosexual transmission through mucous membranes. In India, the WHO currently estimates that there are as many as 3.5 million infected adults. Female sex workers in urban areas are infected at rates sometimes approaching 50%, and the disease is increasingly appearing in rural villages. India, with its population of about 1 billion, may soon have more AIDS cases than any other country. HIV-infection statistics in China are murky, but the WHO estimates about 400,000.

Figure 19.16 compares modes of HIV transmission in the United States with other parts of the world.

The Prevention and Treatment of AIDS

Learning Objective

■ List the current methods of preventing and treating HIV infection.

At present, for most of the world the only practical means of control is by minimizing transmission. This requires education programs to promote the use of condoms, as well as discouraging sexual promiscuity. Attempts to prevent the use of contaminated needles among IDUs is also important. To be effective, educational programs often require fundamental social changes that are not easy to accomplish, but they have slowed the infection rate in some areas.

HIV Vaccines

An inexpensive vaccine is probably the only realistic way to effectively control this epidemic. The development of such a vaccine has been prevented by a number of obstacles, however. One is the lack of suitable animal hosts that develop recognizable AIDS. Much current work makes use of monkeys and the simian immunodeficiency virus (SIV), which has many similarities to HIV (see Chapter 13, page 379). The rapid mutation rate of HIV makes it difficult to develop a vaccine that is effective against all forms of the virus.

Another obstacle is the variety of routes by which HIV can be transmitted. An effective vaccine would have to protect against transmission by different mucosal routes, which is proving to be an elusive goal in SIV tests that have been made in monkeys, and against both virus-containing cells and free viruses. Affordability in poorer parts of the world, and legal liability, are additional concerns. A vaccine for AIDS will be a much more difficult challenge to develop than any existing vaccine, and the prospects are uncertain.

Chemotherapy

Much progress has been made in the use of chemotherapy to inhibit HIV infections. The first target of anti-HIV drugs was the enzyme reverse transcriptase, an enzyme not present in human cells. Several drugs, such as *zidovudine* (a drug known by the acronym AZT and occasionally, ZDV), *didanosine* (ddI), and *zalcitabine* (ddC), are analogs of nucleosides that terminate the synthesis of viral DNA by competitive inhibition (see Chapter 5, page 119).

Other drugs that inhibit reverse transcription are not analogs of nucleic acids; these non-nucleoside inhibitors of reverse transcriptase include *efavirenze* and *nevirapine*.

Another promising target is the enzyme protease. This enzyme cuts proteins into pieces that are then reassembled into the coat of new HIV particles. Protease inhibitors such as *idinavir, saquinavir,* and *ritonavir* are now in clinical use. Probably the next groups of anti-HIV drugs

MICROBIOLOGY IN THE NEWS

New Weapons Against AIDS

The ability of HIV to destroy, evade, and even hide inside cells of the human immune system makes it extremely difficult to prevent by vaccination and to treat. Because the virus is so unusual and so deadly, it has stimulated intense research into both conventional and extremely unconventional approaches to vaccination and chemotherapy.

VACCINES

Experimental vaccines are being made with gp120 and other HIV envelope glycoproteins. It is thought that antibodies against envelope proteins would not only attack free viruses, but also block the gp120 on the surfaces of infected T cells, thus preventing cell fusion and cell-to-cell transmission. Human trials of these and other vaccines are under way. Initially, these human tests can determine only whether the vaccines produce an immune response and whether they are safe. Determining whether they actually prevent AIDS is much more difficult.

Even if an uninfected volunteer became infected with HIV early in these tests, researchers don't expect to know whether the vaccines actually prevent AIDS for 3–5 years because AIDS develops so slowly.

VaxGen Company of California is beginning clinical trials of an AIDS vaccine. Three thousand uninfected people who have an HIV-positive partner will be vaccinated, and 2000 will receive placebo injections. All 5000 people will be checked for infection after 3 years.

HIV usually enters the body through mucous membranes; therefore, topically applied anti-HIV chemicals may prevent infection. The AIDS Vaccine Evaluation Group is considering oral, vaginal, and rectal administration of vaccines to boost production of IgA antibodies against HIV.

Although several clinical trials of gp120 vaccines are ongoing, researchers are not optimistic about the results. The vaccines induce antibodies against

gp120; however, the gp120 antigen is hidden on the surface of the virus and not readily accessible to the antibodies. Additionally, recent trials of an attenuated simian immunodeficiency virus vaccine in monkeys were fatal to three-fourths of the test monkeys. The attenuated virus was able to reproduce and was still virulent.

IMMUNOTHERAPY

Is is known that α-IFN (interferon) causes cells to produce antiviral proteins. In 1996, Interferon Sciences, Inc., started clinical trials of α-IFN to treat AIDS patients.

In a test conducted at the National Institute of Allergy and Infectious Diseases, HIV-infected people with moderate immune suppression were given IL-2 over 8 months. The IL-2 caused a 50% increase in the number of CD4 cells in the test group. In 1999, clinical trials of IL-2 began in the United States and several other countries.

▶

will include inhibitors of the enzyme integrase. This enzyme incorporates viral DNA into the DNA of the host cell. There are several other points at which the production of HIV can be selectively interrupted. Currently, early trials are under way to evaluate drugs classified as entry, or fusion, inhibitors. These are designed to prevent HIV from penetrating into the cell in which it replicates.

Although experience has shown that HIV rapidly develops resistance to any single drug, the multiple administration of "cocktails" of at least three different reverse transcriptase or protease inhibitors has demonstrated encouraging results. The number of HIVs in circulation has been reduced to fewer than can be detected. Although this is not necessarily the same as eradication, especially considering the viruses and proviruses "hidden" in lymphoid tissue. However, perhaps the viral load can be controlled sufficiently to allow the immune system to eliminate the infection. The sooner chemotherapy is initiated, the greater the likelihood of success—at least, through buying more years by slowing the progression to AIDS.

The long-term results of chemotherapy are still uncertain. One clearly successful application of chemotherapy,

however, has been to reduce the chance of HIV transmission from an infected mother to her newborn. The administration of even AZT alone sharply reduces the incidence. Unfortunately, the cost of these drugs is a serious problem in prosperous countries, and absolutely prohibitive throughout much of the world. (For more information about new ways to treat AIDS, see the box above.)

The AIDS Epidemic and the Importance of Scientific Research

The AIDS epidemic gives clear evidence of the value of basic scientific research. Without the advances in molecular biology of the past few decades, we would have been unable even to identify the causative agent of AIDS. We would not have been able to develop the tests for screening donated blood, to identify points in the viral life cycle for which selectively toxic drugs could be developed, or even to monitor the course of the infection. In the lifetime of most of us, we will have the opportunity to witness medical history being made as the struggle with this deadly and elusive virus continues.

MICROBIOLOGY IN THE NEWS

New Weapons Against AIDS (continued)

GENE THERAPY

In 1996, the company Cell Genesys began clinical trials using gene therapy to treat AIDS. The trials begin with the collection of the patient's own T cells. The T cells are genetically engineered to recognize and destroy HIV-infected cells; then the cells are returned to the patient.

MUCOSAL PROTECTION

The identification of compounds that afford protection at mucous membranes could lead to the development of vaginal gels to prevent HIV infection.

CORECEPTOR INHIBITORS

Currently, three classes of drugs are used to treat HIV infection: (1) nucleoside reverse transcriptase inhibitors, (2) nonnucleoside reverse transcriptase inhibitors, and (3) protease inhibitors. Researchers at Trimeris Company in North Carolina are investigating a fourth class: coreceptor inhibitors. Their new drug, T-20, prevents HIV from entering host cells by interfering with fusion of viral gp41 with the host cell. T-20 is currently undergoing clinical trials.

MICROBIAL ANTI-HIV TREATMENTS

Sharon Hill of the University of Pittsburgh has identified two lactic-acid-producing bacteria, *Lactobacillus crispatus* and *L. jensenii*, in the vaginal microbiota that produce bacteriocins that inhibit HIV. Studies have shown a correlation between the lack of these bacteria and transmission of HIV by sexual intercourse.

Yale Medical School researchers are experimenting with a virus that kills HIV-infected cells in cell cultures. They have genetically engineered a cow virus, vesicular stomatitis virus (VSV), to display CD4 and coreceptors on their surfaces. The VSVs attach to HIV-infected cells that display gp120 on their surfaces. The VSV then enters and kills the HIV-infected cell but does not harm uninfected cells.

Research into new treatments and vaccines against HIV continues to be intense. However, it will take years of continued research and testing to discover effective drugs and vaccines.

Study Outline

INTRODUCTION (p. 520)

1. Hay fever, transplant rejection, and autoimmunity are examples of harmful immune reactions.
2. Infection and immunosuppression are examples of failure of the immune system.
3. Superantigens activate many T-cell receptors, resulting in the release of excessive amounts of cytokines that can cause adverse host responses.

HYPERSENSITIVITY (pp. 521–529)

1. Hypersensitivity reactions represent immunological responses to an antigen (allergen) that lead to tissue damage rather than immunity.
2. Hypersensitivity reactions occur when a person has been sensitized to an antigen.
3. Hypersensitivity reactions can be divided into four classes: types I, II, and III are immediate reactions based on humoral immunity, and type IV is a delayed reaction based on cell-mediated immunity.

Type I (Anaphylactic) Reactions (pp. 521–524)

1. Anaphylactic reactions involve the production of IgE antibodies that bind to mast cells and basophils to sensitize the host.
2. The binding of two adjacent IgE antibodies to an antigen causes the target cell to release chemical mediators, such as histamine, leukotrienes, and prostaglandins, which cause the observed allergic reactions.
3. Systemic anaphylaxis may develop in minutes after injection or ingestion of the antigen; this may result in circulatory collapse and death.
4. Localized anaphylaxis is exemplified by hives, hay fever, and asthma.
5. Skin testing is useful in determining sensitivity to an antigen.
6. Desensitization to an antigen can be achieved by repeated injections of the antigen, which leads to the formation of blocking (IgG) antibodies.

Type II (Cytotoxic) Reactions (pp. 524–526)

1. Type II reactions are mediated by IgG or IgM antibodies and complement.
2. The antibodies are directed toward foreign cells or host cells. Complement fixation may result in cell lysis. Macrophages and other cells may also damage the antibody-coated cells.

The ABO Blood Group System (pp. 524–525)

1. Human blood may be grouped into four principal types, designated A, B, AB, and O.

2. The presence or absence of two carbohydrate antigens designated A and B on the surface of the red blood cell determines a person's blood type.

3. Naturally occurring antibodies are present or absent in serum against the opposite AB antigen.

4. Incompatible blood transfusions lead to the complement-mediated lysis of the donor red blood cells.

The Rh Blood Group System (pp. 525–526)

1. Approximately 85% of the human population possesses another blood group antigen, designated the Rh antigen; these individuals are designated Rh^+.

2. The absence of this antigen in certain individuals (Rh^-) can lead to sensitization upon exposure to it.

3. An Rh^+ person can receive Rh^+ or Rh^- blood transfusions.

4. When an Rh^- person receives Rh^+ blood, that person will produce anti-Rh antibodies.

5. Subsequent exposure to Rh^+ cells will result in a rapid, serious hemolytic reaction.

6. An Rh^- mother carrying an Rh^+ fetus will produce anti-Rh antibodies.

7. Subsequent pregnancies involving Rh incompatibility may result in hemolytic disease of the newborn.

8. The disease may be prevented by passive immunization of the mother with anti-Rh antibodies.

Drug-Induced Cytotoxic Reactions (p. 526)

1. In the disease thrombocytopenic purpura, platelets are destroyed by antibodies and complement.

2. Agranulocytosis and hemolytic anemia result from antibodies against one's own blood cells coated with drug molecules.

Type III (Immune Complex) Reactions (pp. 526–527)

1. Immune complex diseases occur when IgG antibodies and soluble antigen form small complexes that lodge in the basement membranes of cells.

2. Subsequent complement fixation results in inflammation.

3. Glomerulonephritis is an immune complex disease.

Type IV (Cell-Mediated) Reactions (pp. 527–529)

1. Delayed-type hypersensitivity reactions are due primarily to T_D cell proliferation.

2. Sensitized T cells secrete cytokines in response to the appropriate antigen.

3. Cytokines attract and activate macrophages and initiate tissue damage.

4. The tuberculin skin test and allergic contact dermatitis are examples of delayed hypersensitivities.

AUTOIMMUNE DISEASES (pp. 529–530)

1. Autoimmunity results from a loss of self-tolerance.

2. Self-tolerance occurs during fetal development; T cells that will target host cells are eliminated (clonal deletion) or inactivated.

3. Type I autoimmunity may be due to antibodies against infectious agents.

4. Graves' disease and myasthenia gravis are type II autoimmune reactions in which antibodies react to cell-surface antigens.

5. Systemic lupus erythematosus and rheumatoid arthritis are type III autoimmune reactions in which the deposition of immune complexes results in tissue damage.

6. Hashimoto's disease and insulin-dependent diabetes mellitus are type IV autoimmune reactions mediated by T cells.

REACTIONS RELATED TO THE HUMAN LEUKOCYTE ANTIGEN (HLA) COMPLEX (pp. 530–533)

1. Histocompatibility self molecules located on cell surfaces express genetic differences among individuals; these antigens are coded for by MHC or HLA gene complexes.

2. To prevent the rejection of transplants, HLA and ABO blood group antigens of the donor and recipient are matched as closely as possible.

Reactions to Transplantation (pp. 531–533)

1. Transplants recognized as foreign antigens may be lysed by T cells and attacked by macrophages and complement-fixing antibodies.

2. Transplantation to a privileged site (such as the cornea) or of a privileged tissue (such as pig heart valves) does not cause an immune response.

3. Four types of transplants have been defined on the basis of genetic relationships between the donor and the recipient: autografts, isografts, allografts, and xenografts.

4. Xenografts are subject to hyperacute rejection.

5. Bone marrow (with immunocompetent cells) can cause graft-versus-host disease.

6. Successful transplant surgery often requires immunosuppressant drugs to prevent an immune response to the transplanted tissue.

IMMUNE DEFICIENCIES (p. 533)

1. Immune deficiencies can be congenital or acquired.

2. Congenital immune deficiencies are due to defective or absent genes.

3. A variety of drugs, cancers, and infectious diseases can cause acquired immune deficiencies.

CRITICAL THINKING

1. When and how does our immune system discriminate between self- and nonself antigens?

2. The first preparations used for artificially acquired passive immunity were antibodies in horse serum. A complication that resulted from the therapeutic use of horse serum was immune complex disease. Why did this occur?

3. Do people with AIDS make antibodies? If so, why are they said to have an immune deficiency?

4. What are the methods of action of anti-AIDS drugs?

CLINICAL APPLICATIONS

1. Fungal infections such as athlete's foot are chronic. These fungi degrade skin keratin but are not invasive and do not produce toxins. Why do you suppose that many of the symptoms of a fungal infection are due to hypersensitivity to the fungus?

2. After working in a mushroom farm for several months, a worker develops these symptoms: hives, edema, and swelling lymph nodes.
   a. What do these symptoms indicate?
   b. What mediators cause these symptoms?
   c. How may sensitivity to a particular antigen be determined?
   d. Other employees do not appear to have any immunological reactions. What could explain this?
   (*Hint:* The allergen is conidiospores from molds growing in the mushroom farm.)

3. Physicians administering live, attenuated mumps and measles vaccines prepared in chick embryos are instructed to have epinephrine available. Epinephrine will not treat these viral infections. What is the purpose of keeping this drug on hand?

4. A woman with blood type A+ once received a transfusion of AB+ blood. When she carried a type B+ fetus, the fetus developed hemolytic disease of the newborn. Explain why this fetus developed this condition when another type B+ fetus in a type A+ mother was normal.

Learning with Technology

MP = The Microbiology Place website   **ST** = Student Tutorial CD-ROM   **VU** = VirtualUnkown CD-ROM

MP   Don't forget to go to The Microbiology Place website (http://www.microbiologyplace.com) to take the practice tests, explore the interactive activity and case study, and check out the news articles and web links for this chapter.

ST   Remember there is also a quiz for this chapter on the Microbiology Interactive Student Tutorial CD-ROM.

VU   Enter the Virtual Lab, click the arrow next to the Session field, click Textbook Exercises, and select Chapter 19. Read the Case Study carefully, identify the unknown, and use what you learn to answer the following questions (consult your textbook for additional information).

1. The microbe responsible for this illness is seldom a pathogen in healthy individuals. Explain how it is normally kept in check by a healthy immune system.

2. What types of genetic immunodeficiency would also render an individual susceptible to this organism?

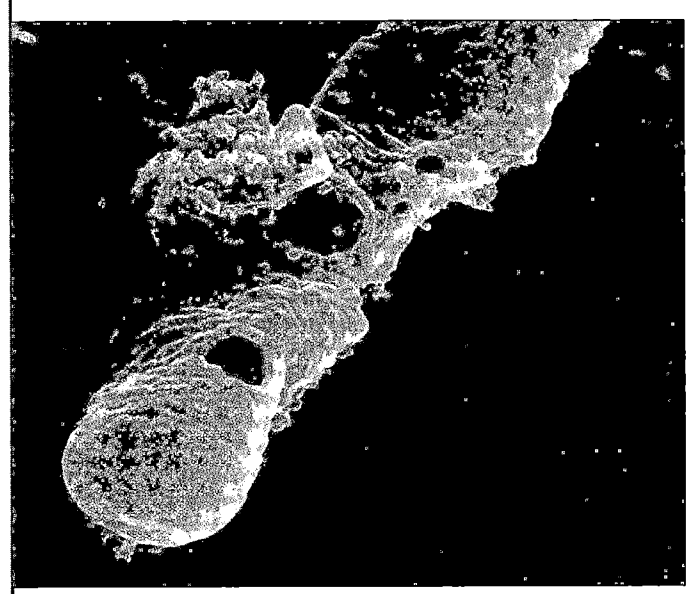

This bacterium is lysing because an antibiotic has prevented cell wall synthesis.

When the body's normal defenses cannot prevent or overcome a disease, it is often treated with **chemotherapy.** In this chapter we focus on antimicrobial drugs, the class of chemotherapeutic agents used to treat infectious diseases. Like the disinfectants discussed in Chapter 7, **antimicrobial drugs** act by interfering with the growth of microorganisms. Unlike disinfectants, however, they must often act *within* the host. Therefore, their effects on the cells and tissues of the host are important. The ideal antimicrobial drug kills the harmful microorganism without damaging the host; this is the principle of **selective toxicity.**

The History of Chemotherapy

Learning Objectives

- *Identify the contributions of Paul Ehrlich and Alexander Fleming to chemotherapy.*
- *Name the microbes that produce most antibiotics.*

The birth of modern chemotherapy is credited to the efforts of Paul Ehrlich in Germany during the early part of the twentieth century. While attempting to stain bacteria without staining the surrounding tissue, he speculated about some "magic bullet" that would selectively find and destroy pathogens but not harm the host. This idea provided the basis for chemotherapy, a term he coined.

In 1928, Alexander Fleming observed that the growth of the bacterium *Staphylococcus aureus* was inhibited in the area surrounding the colony of a mold that had contaminated a Petri plate (Figure 20.1). The mold was identified as *Penicillium notatum,* and its active compound, which was isolated a short time later, was named penicillin. Similar inhibitory reactions between colonies on solid media are commonly observed in microbiology, and the mechanism of inhibition is called *antibiosis.* From this word comes the term **antibiotic,** a substance produced by microorganisms that in small amounts inhibits another microorganism. Therefore, the wholly synthetic sulfa drugs, for example, technically are not antibiotics. However, this distinction is often ignored in practice.

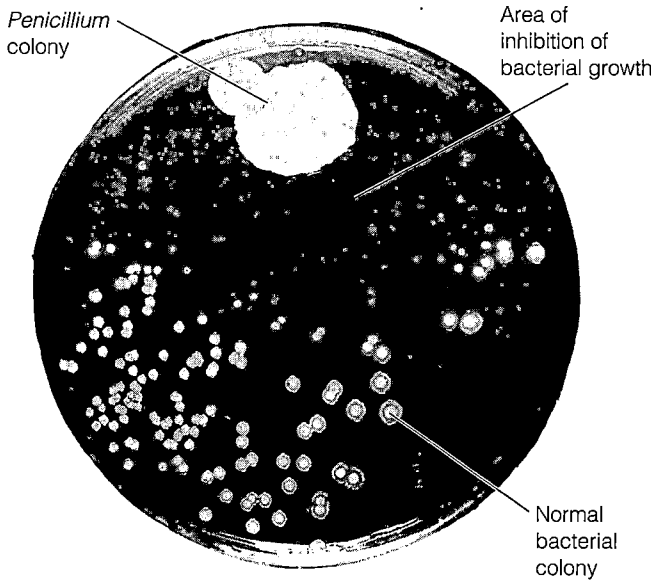

FIGURE 20.1 The discovery of penicillin. Alexander Fleming took this photograph in 1928. The colony of *Penicillium* mold accidentally contaminated the plate and is inhibiting nearby bacterial growth.

■ **New antibiotic-producing organisms are discovered by looking for areas of inhibition around colonies.**

table 20.1	Representative Sources of Antibiotics
Microorganism	**Antibiotic**
Gram-Positive Rods	
Bacillus subtilis	Bacitracin
Bacillus polymyxa	Polymyxin
Actinomycetes	
Streptomyces nodosus	Amphotericin B
Streptomyces venezuelae	Chloramphenicol
Streptomyces aureofaciens	Chlortetracycline and tetracycline
Streptomyces erythraeus	Erythromycin
Streptomyces fradiae	Neomycin
Streptomyces griseus	Streptomycin
Micromonospora purpurea	Gentamicin
Fungi	
Cephalosporium spp.	Cephalothin
Penicillium griseofulvum	Griseofulvin
Penicillium notatum	Penicillin

In 1940, a group of scientists at Oxford University headed by Howard Florey and Ernst Chain succeeded in the first clinical trials of penicillin. Intensive research in the United States then led to the isolation of especially productive *Penicillium* strains for use in the mass production of the antibiotic. (The most famous of these high-producing strains was originally isolated from a cantaloupe bought at a market in Peoria, Illinois.)

Antibiotics are actually rather easy to discover, but few are of medical or commercial value. Some are used commercially other than for treating disease—for example, as a supplement in animal feed (see the box on page 562). Many antibiotics are toxic to humans or lack any advantage over antibiotics already in use.

More than half of our antibiotics are produced by species of *Streptomyces*, filamentous bacteria that commonly inhabit soil. A few antibiotics are produced by bacteria of the genus *Bacillus*, and others are produced by molds, mostly of the genera *Penicillium* and *Cephalosporium* (sef-ä-lō-spô'rē-um). See Table 20.1 for the sources of many antibiotics in use today—a surprisingly limited group of organisms. It is especially interesting to note that practically all antibiotic-producing microbes have some sort of sporulation process.

The Spectrum of Antimicrobial Activity

Learning Objectives

■ Describe the problems of chemotherapy for viral, fungal, protozoan, and helminthic infections.

■ Define the following terms: spectrum of activity, broad-spectrum antibiotic, superinfection.

It is comparatively easy to find or develop drugs that are effective against prokaryotic cells and that do not affect the eukaryotic cells of humans. These two cell types differ substantially in many ways, such as in the presence or absence of cell walls, the fine structure of their ribosomes, and details of their metabolism. Thus, selective toxicity has numerous targets. The problem is more difficult when the pathogen is a eukaryotic cell, such as a fungus, protozoan, or helminth. At the cellular level, these organisms resemble the human cell much more closely than a bacterial cell does. We will see that our arsenal against these types of pathogens is much more limited than our arsenal of antibacterial drugs. Viral infections are particularly difficult to treat because the pathogen is within the human host's cells, and the genetic information of the virus is directing

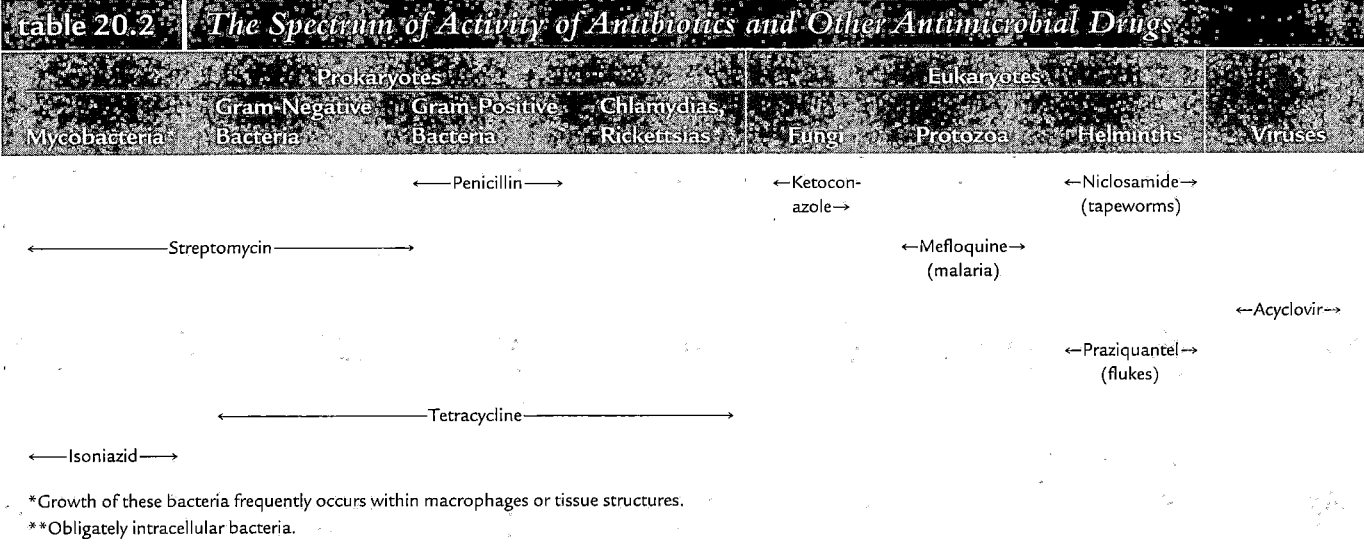

table 20.2	The Spectrum of Activity of Antibiotics and Other Antimicrobial Drugs

Prokaryotes				Eukaryotes			
Mycobacteria*	Gram-Negative Bacteria	Gram-Positive Bacteria	Chlamydias, Rickettsias**	Fungi	Protozoa	Helminths	Viruses

←——Penicillin——→

←—Ketocon-
  azole—→

←—Niclosamide—→
  (tapeworms)

←——————Streptomycin——————→

←—Mefloquine—→
  (malaria)

←—Acyclovir—→

←—Praziquantel—→
  (flukes)

←————————Tetracycline————————→

←——Isoniazid——→

*Growth of these bacteria frequently occurs within macrophages or tissue structures.
**Obligately intracellular bacteria.

the human cell to make viruses rather than to synthesize normal cellular materials.

Some drugs have a narrow **spectrum of microbial activity,** or range of different microbial types they affect. Penicillin, for example, affects gram-positive bacteria but very few gram-negative bacteria. Antibiotics that affect a broad range of gram-positive or gram-negative bacteria are therefore called **broad-spectrum antibiotics.**

A primary factor involved in the selective toxicity of antibacterial action lies in the lipopolysaccharide outer layer of gram-negative bacteria and the porins that form water-filled channels across this layer (see Figure 4.12c on page 87). Drugs that pass through the porin channels must be relatively small and preferably hydrophilic. Drugs that are lipophilic (having an affinity for lipids) or especially large do not enter gram-negative bacteria readily.

Table 20.2 summarizes the spectrum of activity of a number of chemotherapeutic drugs. Because the identity of the pathogen is not always immediately known, a broad-spectrum drug would seem to have an advantage in treating a disease by saving valuable time. The disadvantage is that many normal microbiota of the host are destroyed by broad-spectrum drugs. The normal microbiota ordinarily compete with and check the growth of pathogens or other microbes. If certain organisms in the normal microbiota are not destroyed by the antibiotic and their competitors are destroyed, the survivors may flourish and become opportunistic pathogens. An example that sometimes occurs is overgrowth by the yeastlike fungus *Candida albicans,* which is not sensitive to bacterial antibiotics. This overgrowth is called a **superinfection,** a term that is also applied to growth of a target pathogen that has developed resistance to the antibiotic. In this sit-

uation, such an antibiotic-resistant strain replaces the original sensitive strain, and the infection continues.

The Action of Antimicrobial Drugs

Learning Objective

- *Identify five modes of action of antimicrobial drugs.*

Antimicrobial drugs are either **bactericidal** (they kill microbes directly) or **bacteriostatic** (they prevent microbes from growing). In bacteriostasis, the host's own defenses, such as phagocytosis and antibody production, usually destroy the microorganisms. The major modes of action are summarized in Figure 20.2.

The Inhibition of Cell Wall Synthesis

Recall from Chapter 4 that the cell wall of a bacterium consists of a macromolecular network called peptidoglycan. Peptidoglycan is found only in bacterial cell walls. Penicillin and certain other antibiotics prevent the synthesis of intact peptidoglycan; consequently, the cell wall is greatly weakened, and the cell undergoes lysis (Figure 20.3). Because penicillin targets the synthesis process, only actively growing cells are affected by these antibiotics. And, because human cells do not have peptidoglycan cell walls, penicillin has very little toxicity for host cells.

The Inhibition of Protein Synthesis

Because protein synthesis is a common feature of all cells, whether prokaryotic or eukaryotic, it would seem

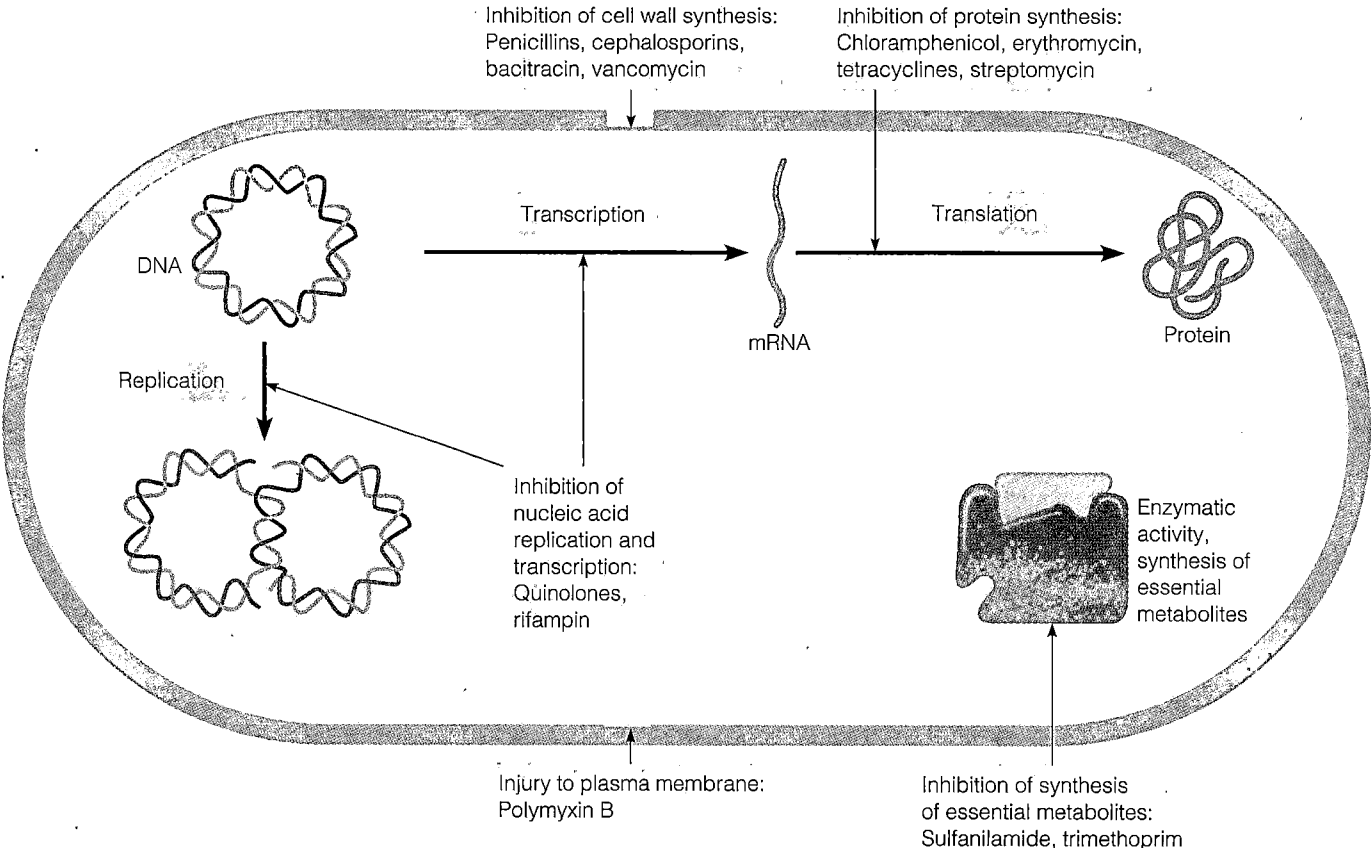

Inhibition of cell wall synthesis:
Penicillins, cephalosporins,
bacitracin, vancomycin

Inhibition of protein synthesis:
Chloramphenicol, erythromycin,
tetracyclines, streptomycin

DNA

Transcription

Translation

mRNA

Protein

Replication

Inhibition of
nucleic acid
replication and
transcription:
Quinolones,
rifampin

Enzymatic
activity,
synthesis of
essential
metabolites

Injury to plasma membrane:
Polymyxin B

Inhibition of synthesis
of essential metabolites:
Sulfanilamide, trimethoprim

FIGURE 20.2 A summary of the major modes of action of antimicrobial drugs. This illustration shows these actions as they might affect a highly diagrammatic representation of a bacterial cell.

■ Which modes of action are effective against prokaryotic cells? Against eukaryotic cells?

an unlikely target for selective toxicity. One notable difference between prokaryotes and eukaryotes, however, is the structure of their ribosomes. As discussed in Chapter 4 on page 95, eukaryotic cells have 80S ribosomes; prokaryotic cells have 70S ribosomes. (The 70S ribosome is made up of a 50S and a 30S unit.) The difference in ribosomal structure accounts for the selective toxicity of antibiotics that affect protein synthesis. However, mitochondria (important eukaryotic organelles) also contain 70S ribosomes similar to those of bacteria. Antibiotics targeting the 70S ribosomes can therefore have adverse effects on the cells of the host. Among the antibiotics that interfere with protein synthesis are chloramphenicol, erythromycin, streptomycin, and the tetracyclines (Figure 20.4).

Reacting with the 50S portion of the 70S prokaryotic ribosome, chloramphenicol inhibits the formation of peptide bonds in the growing polypeptide chain. Erythromycin also reacts with the 50S portion of the 70S prokaryotic ribosome. Most drugs that inhibit protein synthesis have a broad spectrum of activity; erythromycin is an exception.

Because it does not penetrate the gram-negative cell wall, it affects mostly gram-positive bacteria.

Some other antibiotics react with the 30S portion of the 70S prokaryotic ribosome. The tetracyclines interfere with the attachment of the tRNA carrying the amino acids to the ribosome, preventing the addition of amino acids to the growing polypeptide chain. Tetracyclines do not interfere with mammalian ribosomes because they do not penetrate very well into intact mammalian cells. However, at least small amounts are able to enter the host cell, as is apparent from the fact that the intracellular pathogenic rickettsias and chlamydias are sensitive to tetracyclines. The selective toxicity of the drug in this case is due to a greater sensitivity of the bacteria at the ribosomal level.

Aminoglycoside antibiotics, such as streptomycin and gentamicin, interfere with the initial steps of protein synthesis by changing the shape of the 30S portion of the 70S prokaryotic ribosome. This interference causes the genetic code on the mRNA to be read incorrectly.

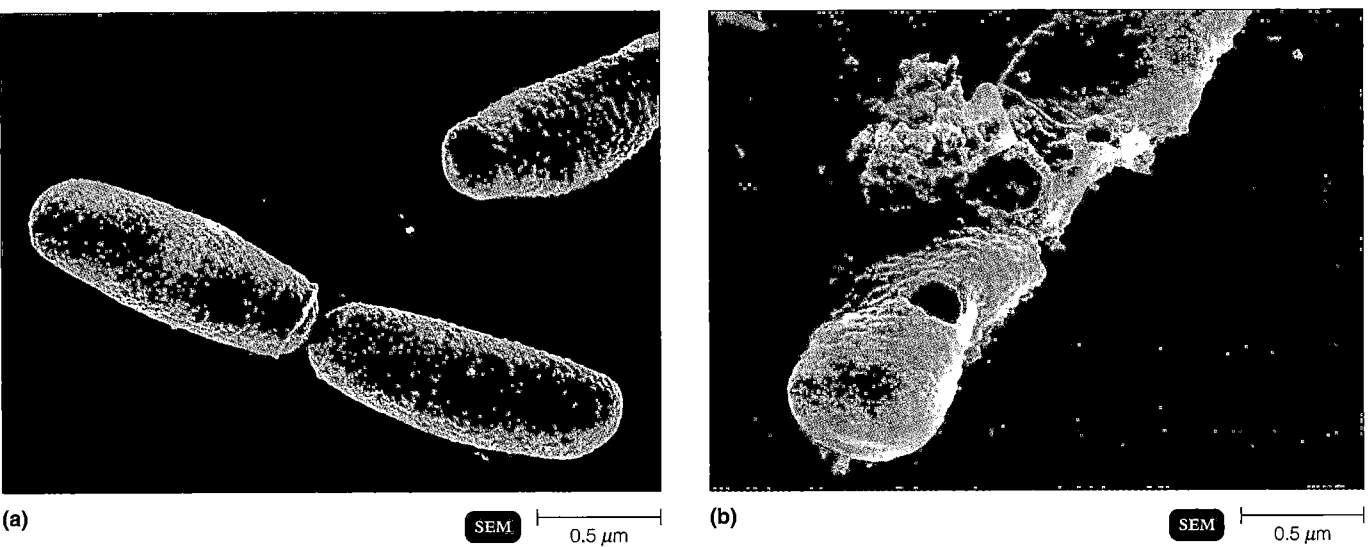

(a) SEM |———————| 0.5 μm   **(b)** SEM |———————| 0.5 μm

FIGURE 20.3 The inhibition of bacterial cell synthesis by penicillin. (a) Rod-shaped bacterium before penicillin. **(b)** The bacterial cell is lysing as penicillin weakens the cell wall.

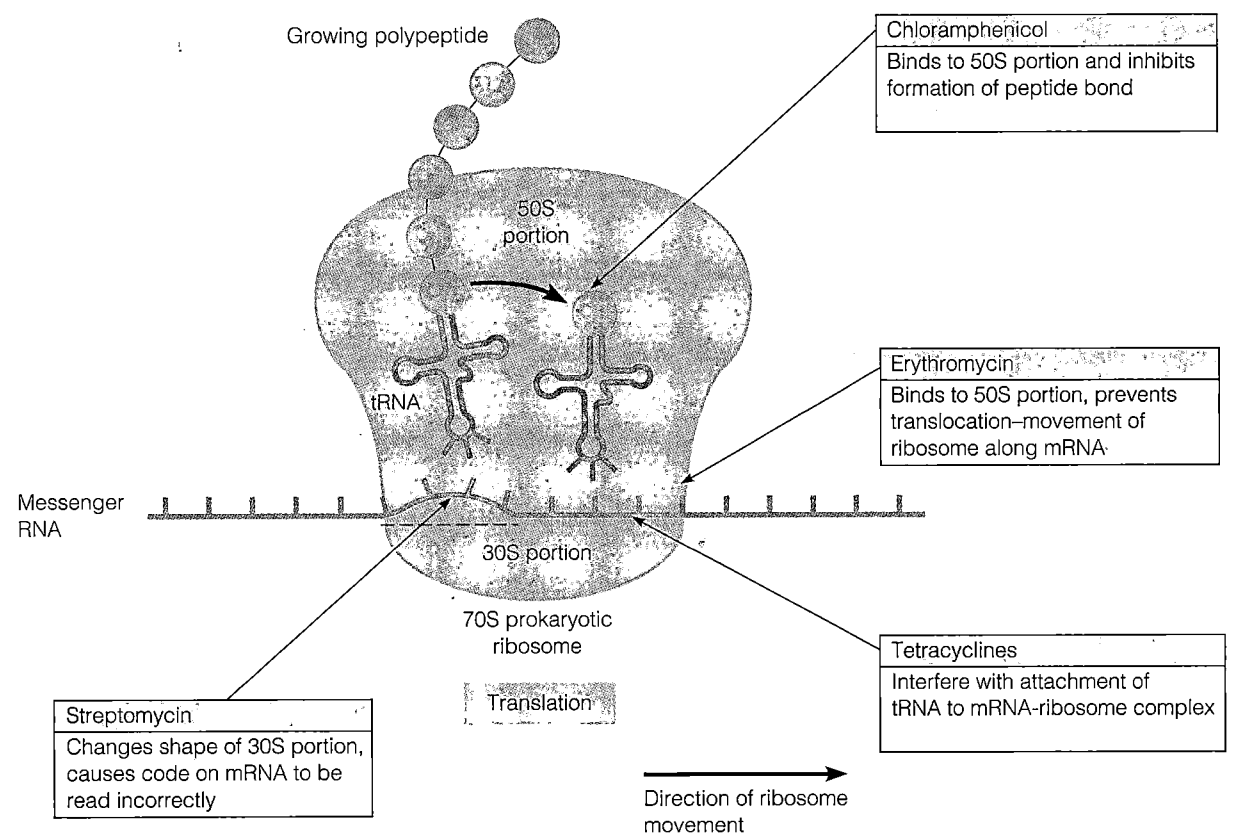

Growing polypeptide

Chloramphenicol
Binds to 50S portion and inhibits formation of peptide bond

50S portion

tRNA

Erythromycin
Binds to 50S portion, prevents translocation–movement of ribosome along mRNA

Messenger RNA

30S portion

70S prokaryotic ribosome

Streptomycin
Changes shape of 30S portion, causes code on mRNA to be read incorrectly

Translation

Tetracyclines
Interfere with attachment of tRNA to mRNA-ribosome complex

Direction of ribosome movement

FIGURE 20.4 The inhibition of protein synthesis by antibiotics. The black arrows indicate the specific points at which chloramphenicol, erythromycin, the tetracyclines, and streptomycin exert their activities.

■ **Why do antibiotics that inhibit protein synthesis affect bacteria and not human cells?**

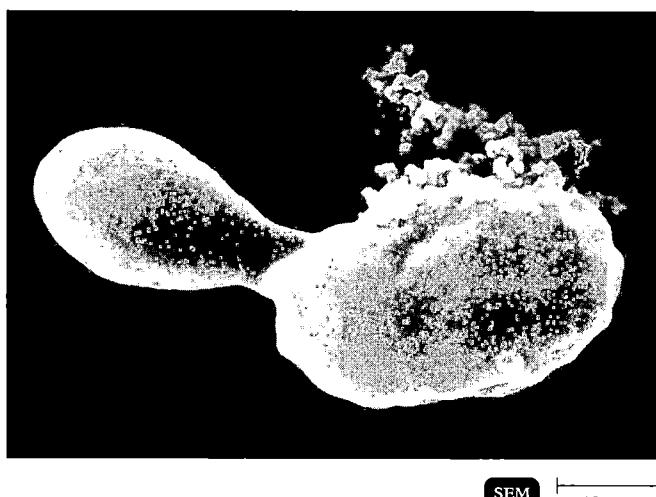

FIGURE 20.5 **Injury to the plasma membrane of a yeast cell caused by an antifungal drug.** The cell releases its cytoplasmic contents as the plasma membrane is disrupted by the antifungal drug miconazole.

☒ **Many antifungal drugs combine with sterols in the plasma membrane.**

Injury to the Plasma Membrane

Certain antibiotics, especially polypeptide antibiotics, bring about changes in the permeability of the plasma membrane; these changes result in the loss of important metabolites from the microbial cell. For example, polymyxin B causes disruption of the plasma membrane by attaching to the phospholipids of the membrane.

Some antifungal drugs, such as amphotericin B, miconazole, and ketoconazole, are effective against a considerable range of fungal diseases. Such drugs combine with sterols in the fungal plasma membrane to disrupt the membrane (Figure 20.5). Because bacterial plasma membranes generally lack sterols, these antibiotics do not act on bacteria. However, the plasma membranes of animal cells do contain sterols, and amphotericin B and ketoconazole can be toxic to the host. Fortunately, animal cell membranes have mostly *cholesterol,* and fungal cells have mostly *ergosterol,* against which the drug is most effective, so that the balance of the toxicity is tilted against the fungus.

The Inhibition of Nucleic Acid Synthesis

A number of antibiotics interfere with the processes of DNA replication and transcription in microorganisms. Some drugs with this mode of action have an extremely limited usefulness because they interfere with mammalian DNA and RNA as well. Others, such as rifampin and the quinolones, are more widely used in chemotherapy because they are more selectively toxic.

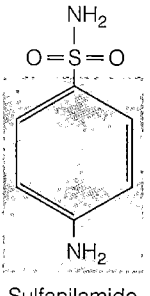
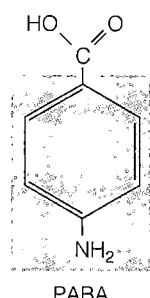

Sulfanilamide   PABA

FIGURE 20.6 **A comparison of the structures of PABA and sulfanilamide, a sulfa drug that inhibits the synthesis of essential metabolites.** Sulfanilamide's structural resemblance to PABA allows it to act as a competitive inhibitor, stopping the synthesis of folic acid, a compound necessary for microorganisms to grow.

☒ **Why are sulfonamides specific for bacteria?**

Inhibiting the Synthesis of Essential Metabolites

In Chapter 5, we mentioned that a particular enzymatic activity of a microorganism can be *competitively inhibited* by a substance (antimetabolite) that closely resembles the normal substrate for the enzyme (see Figure 5.6 on page 119). An example of competitive inhibition is the relationship between the antimetabolite sulfanilamide (a sulfa drug) and **para-aminobenzoic acid (PABA)** (Figure 20.6). In many microorganisms, PABA is the substrate for an enzymatic reaction leading to the synthesis of folic acid, a vitamin that functions as a coenzyme for the synthesis of the purine and pyrimidine bases of nucleic acids and many amino acids. In the presence of sulfanilamide, the enzyme that normally converts PABA to folic acid combines with the drug instead of with PABA. This combination prevents folic acid synthesis and stops the growth of the microorganism. Because humans do not produce folic acid from PABA (they obtain it as a vitamin in ingested foods), sulfanilamide exhibits selective toxicity—it affects microorganisms that synthesize their own folic acid but does not harm the human host. Other chemotherapeutic agents that act as antimetabolites are the sulfones and trimethoprim.

A Survey of Commonly Used Antimicrobial Drugs

Learning Objective

■ *Explain why the drugs described in this section are specific for bacteria.*

Tables 20.3 and 20.4 summarize the commonly used antimicrobial drugs.

table 20.3   *Antibacterial Drugs*

Drugs by Mode of Action	Comments
Inhibitors of Cell Wall Synthesis	
Natural penicillins	
Penicillin G	Gram-positive bacteria, injected. Bactericidal.
Penicillin V	Gram-positive bacteria, taken orally. Bactericidal.
Semisynthetic penicillins	
Ampicillin	Broad spectrum. Bactericidal.
Methicillin	Resistant to penicillinase. Bactericidal.
Monobactams	
Aztreonam	Gram-negative bacteria, including *Pseudomonas* spp. Bactericidal.
Cephalosporins	Resistant to penicillinase; gram-negative bacteria with a broad spectrum of activity. Bactericidal.
Carbapenems (imipenem)	β-lactam type, very broad spectrum. Bactericidal.
Bacitracin	Polypeptide type, gram-positive bacteria, topical use. Bactericidal.
Vancomycin	Glycopeptide type, penicillin-resistant gram-positive bacteria. Bactericidal.
Isoniazid (INH)	Mycobacteria (tuberculosis); inhibits synthesis of mycolic acid component of mycobacterial cell wall. Bacteriostatic.
Ethambutol	Mycobacteria (tuberculosis); inhibits incorporation of mycolic acid into mycobacterial cell wall. Bacteriostatic.
Inhibitors of Protein Synthesis	
Aminoglycosides	
Streptomycin	Broad spectrum, including mycobacteria. Bactericidal.
Neomycin	Topical use, broad spectrum. Bactericidal.
Gentamicin	Broad spectrum, including *Pseudomonas* spp. Bactericidal.
Tetracyclines	
Tetracycline, oxytetracycline, chlortetracycline	Broad spectrum, including chlamydias and rickettsias; animal feed additives. Bacteriostatic.
Chloramphenicol	Broad spectrum, potentially toxic. Bacteriostatic.
Macrolides	
Erythromycin	Alternative to penicillin. Bacteriostatic.
Streptogramins	
Quinupristin and dalfopristin (Synercid)	Inhibit protein synthesis by attaching to ribosomes.
Injury to the Plasma Membrane	
Polymyxin B	Topical use, gram-negative bacteria, including *Pseudomonas* spp. Bactericidal.
Inhibitors of Nucleic Acid Synthesis	
Rifamycins	
Rifampin	Inhibits synthesis of mRNA; treatment of tuberculosis. Bactericidal.
Quinolones and fluoroquinolones	
Nalidixic acid, nofloxacin, ciprofloxacin	Inhibits DNA synthesis; broad spectrum; urinary tract infections. Bactericidal.
Competitive Inhibitors of the Synthesis of Essential Metabolites	
Sulfonamides	
Trimethoprim-sulfamethoxazole	Broad spectrum; combination is widely used. Bacteriostatic.

| table 20.4 | Antifungal, Antiviral, Antiprotozoan, and Antihelminthic Drugs. |
	Mode of Action	Comments
Antifungal Drugs		
Polyenes		
Amphotericin B	Injury to plasma membrane	Systemic fungal infections. Fungicidal.
Azoles	Inhibition of plasma membrane synthesis.	Clotrimazole, miconazole: Topical use for fungal infections of skin and mucous membrane. Possibly fungicidal. Ketoconazole: Topical use for fungal infections of skin; orally for systemic fungal infections. Fungistatic.
Griseofulvin	Inhibition of mitotic microtubules.	Fungal infections of the skin. Fungistatic.
Tolnaftate	Unknown.	Athlete's foot. Fungicidal.
Flucytosine	Inhibits DNA and RNA synthesis.	Systemic fungal infections. Fungicidal.
Antiviral Drugs		
Amantadine	Blocks entry or uncoating.	Influenza A prevention.
Nucleoside analogs		
Acyclovir, ribavirin, ganciclovir, trifluoridine	Inhibit DNA or RNA synthesis.	Herpesviruses.
Zidovudine, didanosine, zalcitabine.	Inhibit DNA synthesis by reverse transcriptase.	HIV infection.
Enzyme inhibitors		
Indinavir, saquinavir	Inhibit viral protease.	HIV infection.
Interferons		
α-interferon	Inhibits spread of virus to new cells.	Viral hepatitis.
Antiprotozoan Drugs		
Chloroquine	Inhibits DNA synthesis.	Malaria; effective against red blood cell stage only.
Diiodohydroxyquin	Unknown.	Amoebic infections. Amoebicidal.
Metronidazole	Damages DNA.	Protozoans, *Entamoeba, Trichomonas;* anaerobic bacteria. Amoebicidal and trichomonicidal.
Antihelminthic Drugs		
Niclosamide	Prevents ATP generation in mitochondria.	Tapeworm infections. Kills tapeworms.
Praziquantel	Alters permeability of plasma membranes.	Tapeworm and fluke infestations. Kills flatworms.
Pyantel pamoate	Neuromuscular block.	Intestinal roundworms. Kills roundworms.

Antibacterial Antibiotics: Inhibitors of Cell Wall Synthesis

Learning Objectives

- List the advantages of each of the following over penicillin: semisynthetic penicillins, cephalosporins, and vancomycin.
- Explain why INH and ethambutol are antimycobacterial agents.

Penicillin

The term **penicillin** refers to a group of over 50 chemically related antibiotics (Figure 20.7). All penicillins have a common core structure containing a β-lactam ring called the nucleus. Penicillin molecules are differentiated by the chemical side chains attached to their nuclei. Penicillins can be produced either naturally or semisynthetically. Penicillins prevent the cross-linking of

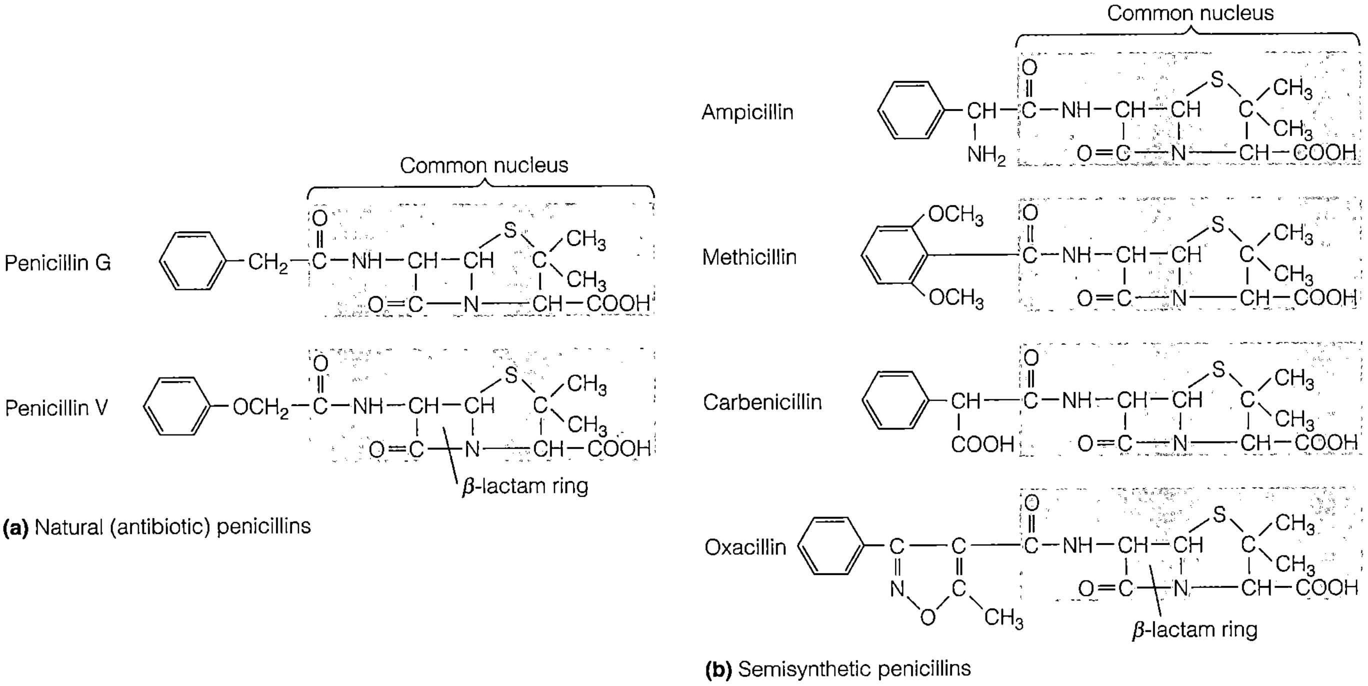

Common nucleus

Ampicillin

Methicillin

Carbenicillin

Oxacillin

β-lactam ring

(b) Semisynthetic penicillins

Common nucleus

Penicillin G

Penicillin V

β-lactam ring

(a) Natural (antibiotic) penicillins

FIGURE 20.7 **The structure of penicillins, antibacterial antibiotics.** The portion that all penicillins have in common—which contains the β-lactam ring—is shaded in purple. The unshaded portions represent the side chains that distinguish one penicillin from another.

■ Semisynthetic penicillins include the β-lactam ring made by *Penicillium* with various side chains to extend the spectrum of activity or resist penicillinase.

the peptidoglycans, which interferes with the final stages of the construction of the cell wall; see Figure 4.12a on page 87.

Natural Penicillins Penicillin extracted from cultures of the mold *Penicillium* exists in several closely related forms. These are the so-called **natural penicillins.** The prototype compound of all the penicillins is *penicillin G.* It has a narrow but useful spectrum of activity and is often the drug of choice against most staphylococci, streptococci, and several spirochetes. When injected intramuscularly, penicillin G is rapidly excreted from the body in 3–6 hours (Figure 20.8). When taken orally, the acidity of the digestive fluids in the stomach diminishes its concentration. *Procaine penicillin,* a combination of the drugs procaine and penicillin G, is retained at detectable concentrations for up to 24 hours; concentration peaks at about 4 hours. Still longer retention times can be achieved with *benzathine penicillin,* a combination of benzathine and penicillin G. Although retention times of as long as 4 months can be obtained, the concentration of the drug is so low that the organisms must be very sensitive to it. Penicillin V, which is stable in

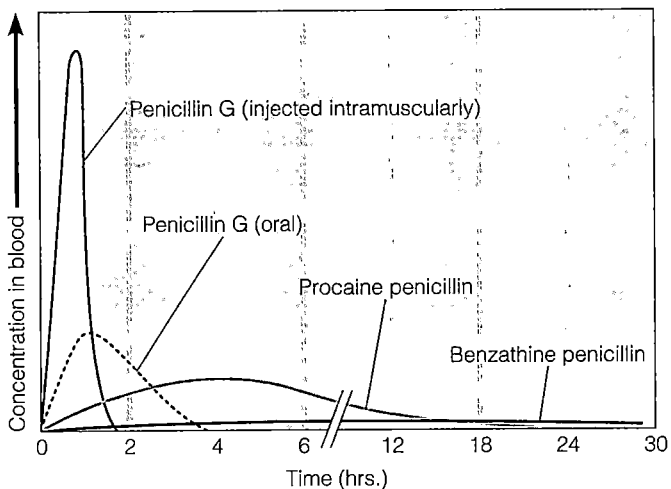

FIGURE 20.8 **Retention of penicillin G.** Penicillin G is normally injected (solid red line); when administered by this route, the drug is present in high concentrations in the blood but disappears quickly. Taken orally (dotted red line), penicillin G is destroyed by stomach acids and is not very effective. It is possible to improve retention of penicillin G by combining it with compounds such as procaine and benzathine (blue and black lines). However, the blood concentration reached is low, and the target bacterium must be extremely sensitive to the antibiotic.

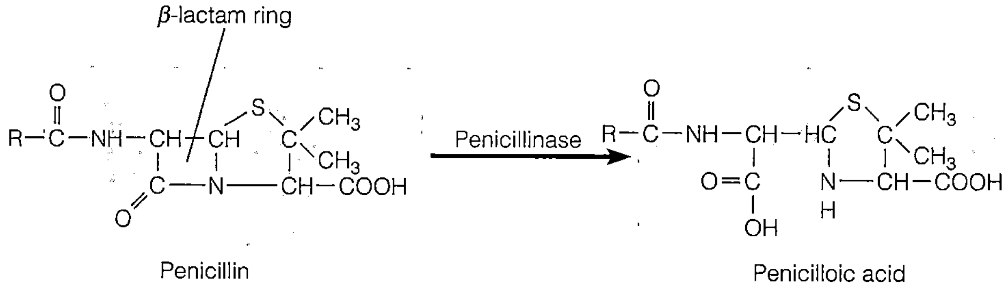

FIGURE 20.9 The effect of penicillinase on penicillins. Bacterial production of this enzyme, which is shown breaking the β-lactam ring, is by far the most common form of resistance to penicillins. R is an abbreviation for the chemical side groups that differentiate similar or otherwise-identical compounds.

stomach acids and is best taken orally, and penicillin G are the natural penicillins most often used.

Natural penicillins have some disadvantages. Chief among them are their narrow spectrum of activity and their susceptibility to penicillinases. *Penicillinases* are enzymes produced by many bacteria, especially *Staphylococcus* species, that cleave the β-lactam ring of the penicillin molecule (Figure 20.9). Because of this, penicillinases are sometimes called *β-lactamases.*

Semisynthetic Penicillins A large number of **semisynthetic penicillins** have been developed in attempts to overcome the disadvantages of natural penicillins (see Figure 20.7b). Scientists develop these penicillins in either of two ways. First, they can stop synthesis of the molecule by *Penicillium* and obtain only the common penicillin nucleus for use. Second, they can remove the side chains from the completed natural molecules and then chemically add other side chains that, among other things, make them more resistant to penicillinase. Thus the term semisynthetic: Part of the penicillin is produced by the mold, and part is added synthetically. *Methicillin,* designed to evade penicillinase-type resistance, was the first semisynthetic penicillin used widely. Many bacteria developed resistance even to it, so others such as *oxacillin* have been developed to replace it.

To overcome the problem of the narrow spectrum of activity of natural penicillins, broader-spectrum semisynthetic penicillins have been developed. These new penicillins are effective against many gram-negative bacteria as well as gram-positive ones, although they are not resistant to penicillinases. The first such penicillins were the aminopenicillins, such as *ampicillin* and *amoxicillin*. When bacterial resistance to these became more common, the carboxypenicillins were developed. Members of this group, such as *carbenicillin* and *ticarcillin,* have even greater activity against gram-negative bacteria and have the special advantage of activity against *Pseudomonas aeruginosa.* Among the more recent additions to the penicillin family are the ureidopenicillins, such as *mezlocillin* and *azlocillin.* These broader-spectrum penicillins are modifi-

cations of the structure of ampicillin. The search for even more effective modifications of penicillin continues.

A different approach to the proliferation of penicillinase is to combine penicillins with *potassium clavulanate (clavulanic acid),* a product of a streptomycete. Potassium clavulanate is a noncompetitive inhibitor of penicillinase with essentially no antimicrobial activity of its own. It has been combined with some new broader-spectrum penicillins, such as amoxicillin (the combination is called by its trade name Augmentin®). It is worth mentioning that the same antibiotic might be sold worldwide under numerous different trade names.

Monobactams

Another method of avoiding the effects of penicillinase is shown by *aztreonam,* which is the first member of a new class of antibiotics. It is a synthetic antibiotic that has only a single ring and is therefore known as a **monobactam,** rather than the conventional β-lactam double ring. Aztreonam's spectrum of activity is remarkable for a penicillin-related compound—this antibiotic affects only certain gram-negative bacteria, including pseudomonads and *E. coli.* Of special interest is the unusually low toxicity of this monobactam.

Cephalosporins

In structure, the nuclei of **cephalosporins** resemble those of penicillin (Figure 20.10). Cephalosporins inhibit cell wall synthesis in essentially the same way as do penicillins. However, cephalosporins differ from penicillin in that they are resistant to penicillinases and are effective against more gram-negative organisms than the natural penicillins. However, the cephalosporins are susceptible to a separate group of β-lactamases.

The number of second-, third-, and even fourth-generation cephalosporins has proliferated in recent years; there are now more than 70 versions. Each generation tends to be more effective against gram-negatives and has a broader spectrum of activity than the previous generation. Some characteristic cephalosporins are *cephalothin, cefamandole,* and *cefotaxime.*

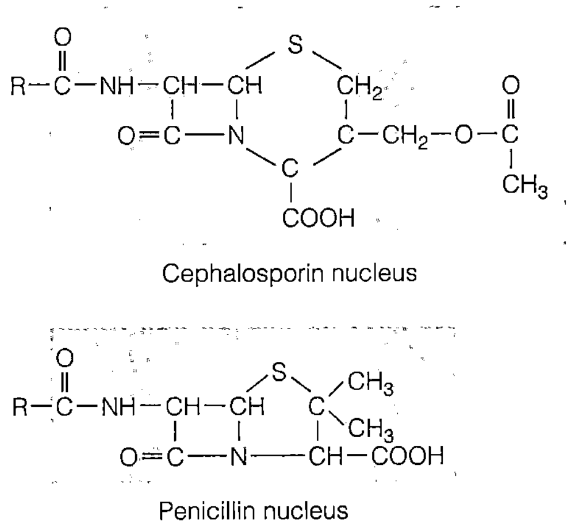

Cephalosporin nucleus

Penicillin nucleus

FIGURE 20.10 The nuclear structures of cephalosporin and penicillin compared.

■ **Why are cephalosporins resistant to penicillinase?**

Oral administration is preferred by patients, and several newer cephalosporins of this type, including *cefpodoxime* and *cefixime,* have been approved.

Carbapenems

One class of antibiotics, the **carbapenems,** is remarkable for its extremely broad spectrum of activity. Carbapenems work by inhibiting cell wall synthesis, and they represent another modification of the β-lactam structure. The first of this group is Primaxin®, a combination of a β-lactam antibiotic (*imipenem*) and *cilastatin sodium.* The cilastatin sodium prevents degradation of the combination in the kidneys. Tests have demonstrated that Primaxin is active against 98% of all organisms isolated from hospital patients.

Bacitracin

Bacitracin (the name is derived from its source, a *Bacillus* isolated from a wound on a girl named Tracy) is a polypeptide antibiotic effective primarily against gram-positive bacteria, such as staphylococci and streptococci. Bacitracin inhibits the synthesis of cell walls at an earlier stage than penicillins and cephalosporins. It interferes with the synthesis of the linear strands of the peptidoglycans. (See Figure 4.12a on page 87.) Its use is restricted to topical application for superficial infections.

Vancomycin

Vancomycin is one of a small group of glycopeptide antibiotics. It is a relatively toxic drug that must be administered carefully and has a very narrow spectrum of activity that is based on inhibiting cell wall synthesis.

Staphylococcus aureus, a serious and very common pathogen, has become increasingly resistant to all types of penicillin, including methicillin. For treatment of *S. aureus* infections vancomycin has been considered the last line of antibiotic defense. The recent appearance of strains of *S. aureus* and some other important pathogens, such as certain enterococci (see the box on page 562), that are resistant to vancomycin, is considered a medical emergency. In response to this situation, in 1999 the FDA accelerated the approval of Synercid®, a combination of two cyclic peptides produced by *Streptomyces* spp., *quinupristin* and *dalfopristin* which are distantly related to the macrolides. Synercid acts on the bacterial ribosome, disrupting protein synthesis. It is the first of a unique group of antibiotics, the **streptogramins,** which consist of two bacteriostatic agents that are bactericidal in combination. Although Synercid is an effective substitute for vancomycin, its expense and high incidence of adverse side effects limits its use.

Isoniazid (INH)

Isoniazid (INH) is a very effective synthetic antimicrobial drug against *Mycobacterium tuberculosis.* The primary effect of INH is to inhibit the synthesis of mycolic acids, which are components of cell walls only of the mycobacteria. It has little effect on nonmycobacteria. When used to treat tuberculosis, INH is usually administered simultaneously with other drugs, such as rifampin or ethambutol. This minimizes development of drug resistance. Because the tubercle bacillus is usually found only within macrophages or walled off in tissue, any antitubercular drug must be able to penetrate into such sites.

Ethambutol

Ethambutol is effective only against mycobacteria. The drug apparently inhibits incorporation of mycolic acid into the cell wall. It is a comparatively weak antitubercular drug; its principal use is as the secondary drug to avoid resistance problems.

Inhibitors of Protein Synthesis

Learning Objective

■ *Describe how each of the following inhibits protein synthesis: aminoglycosides, tetracyclines, chloramphenicol, macrolides.*

Aminoglycosides

Aminoglycosides are a group of antibiotics in which amino sugars are linked by glycoside bonds. With a broad spectrum of activity, they were among the first antibiotics to have significant activity against gram-negative bacteria.

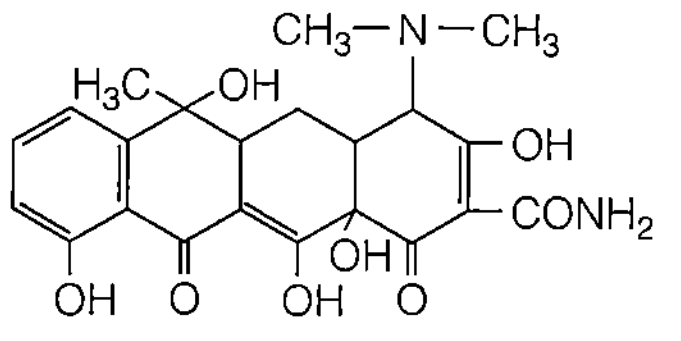

Tetracycline

FIGURE 20.11 The structure of the antibacterial antibiotic tetracycline. Other tetracycline-type antibiotics share the four-cyclic-ring structure of tetracycline and closely resemble it.

■ Tetracylines inhibit protein synthesis.

Probably the best-known aminoglycoside is *streptomycin,* which was discovered in 1944. Streptomycin is still used as an alternative drug in the treatment of tuberculosis, but rapid development of resistance and serious toxic effects have diminished its usefulness.

Aminoglycosides are bactericidal and inhibit protein synthesis. They can affect hearing by causing permanent damage to the auditory nerve, and damage to the kidneys has also been reported. Because of this, their use has been declining. *Neomycin* is present in many nonprescription topical preparations. *Gentamicin,* spelled with an "i" to reflect its source, the filamentous bacterium, *Micromonospora,* is especially useful against *Pseudomonas* infections. Pseudomonads are a major problem for persons suffering from cystic fibrosis. The aminoglycoside *tobramycin* is administered in an aerosol to aid in the control of infections that occur in patients with cystic fibrosis.

Tetracyclines

Tetracyclines are a group of closely related broad spectrum antibiotics, produced by *Streptomyces* spp., that inhibit protein synthesis. They are not only effective against gram-positive and gram-negative bacteria, but also penetrate body tissues well and are especially valuable against the intracellular rickettsias and chlamydias. Three of the more commonly used tetracyclines are *oxytetracycline* (Terramycin®), *chlortetracycline* (Aureomycin®), and tetracycline itself (Figure 20.11). Some semisynthetic tetracyclines, such as *doxycycline* and *minocycline,* are available. They have the advantage of longer retention in the body.

Tetracyclines are used to treat many urinary tract infections, mycoplasmal pneumonia, and chlamydial and rickettsial infections. They are also frequently used as alternative drugs for such diseases as syphilis and gonorrhea. Tetracyclines often suppress the normal intestinal microbiota because of their broad spectrum, causing gastrointestinal upsets and often leading to superinfections, particularly by the fungus *Candida albicans.* They are not advised for children, who might experience a brownish

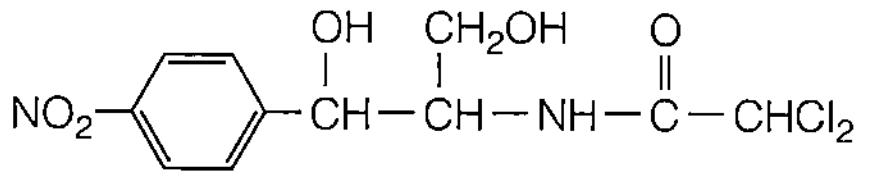

Chloramphenicol

FIGURE 20.12 The structure of the antibacterial antibiotic chloramphenicol. Notice the simple structure, which makes synthesizing this drug less expensive than isolating it from *Streptomyces.*

■ What effect does the binding of chloramphenicol to the 50S portion of the ribosomes have on a cell?

discoloration of the teeth, or for pregnant women, in whom they might cause liver damage. Tetracyclines are among the most common antibiotics added to animal feeds, where their use results in significantly faster weight gain; however, some human health problems can also result (see the box on page 562).

Chloramphenicol

Chloramphenicol is a broad-spectrum bacteriostatic antibiotic that interferes with protein synthesis on the 50S ribosome. Because of its relatively simple structure (Figure 20.12), it is less expensive for the pharmaceutical industry to synthesize it chemically than to isolate it from *Streptomyces.* Its relatively small molecular size promotes its diffusion into areas of the body that are normally inaccessible to many other drugs. However, chloramphenicol has serious adverse effects; most important is the suppression of bone marrow activity. This suppression affects the formation of blood cells. In about 1 in 40,000 users, the drug appears to cause aplastic anemia, a potentially fatal condition; the normal rate for this condition is only about 1 in 500,000 individuals. Physicians are advised not to use the drug for trivial conditions or ones for which suitable alternatives are available.

Macrolides

Macrolides are a group of antibiotics named for the presence of a macrocyclic lactone ring. The best-known macrolide in clinical use is *erythromycin* (Figure 20.13). Its mode of action is the inhibition of protein synthesis. However, erythromycin is not able to penetrate the cell walls of most gram-negative bacilli. Its spectrum of activity is therefore similar to that of penicillin G, and it is a frequent alternative drug to penicillin. Because it can be administered orally, an orange-flavored preparation of erythromycin is a frequent penicillin substitute for the treatment of streptococcal and staphylococcal infections in children. Erythromycin is the drug of choice for the treatment of legionellosis, mycoplasmal pneumonia, and several other infections.

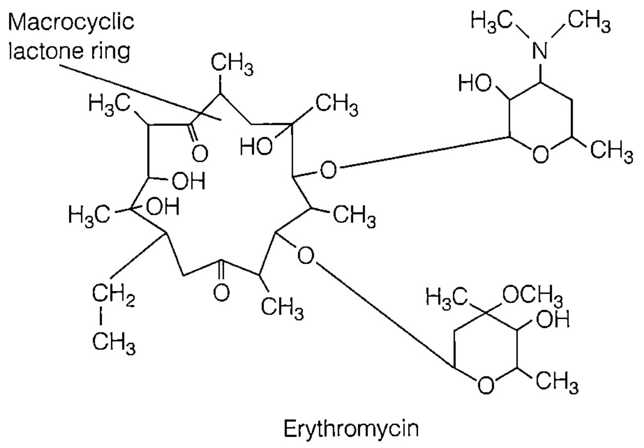

Macrocyclic
lactone ring

Erythromycin

FIGURE 20.13 The structure of the antibacterial antibiotic erythromycin, a representative macrolide. All macrolides have the macrocyclic lactone ring shown here.

■ Macrolides inhibit protein synthesis.

Injury to the Plasma Membrane

Learning Objective

■ *Compare the mode of action of polymyxin B, bacitracin, and neomycin.*

Polymyxin B is a bactericidal antibiotic effective against gram-negative bacteria. For many years, it was one of very few drugs used against infections by gram-negative *Pseudomonas*. The mode of action of polymyxin B is to injure plasma membranes. Polymyxin B is seldom used today except in the topical treatment of superficial infections.

Both *bacitracin* and *polymyxin B* are available in nonprescription antiseptic ointments, in which they are usually combined with *neomycin,* a broad-spectrum aminoglycoside. In a rare exception to the rule, these antibiotics do not require a prescription.

Inhibitors of Nucleic Acid (DNA/RNA) Synthesis

Learning Objective

■ *Describe how rifamycins and quinolones kill bacteria.*

Rifamycins

The best known derivative of the **rifamycin** family of antibiotics is *rifampin*. These drugs are structurally related to the macrolides and inhibit the synthesis of mRNA. By far the most important use of rifampin is against mycobacteria in the treatment of tuberculosis and leprosy. A valuable characteristic of rifampin is its ability to pene-

trate tissues and reach therapeutic levels in cerebrospinal fluid and abscesses. This characteristic is probably an important factor in its antitubercular activity, because the tuberculosis pathogen is usually located inside tissues or macrophages. An unusual side effect of rifampin is the appearance of orange-red urine, feces, saliva, sweat, and even tears.

Quinolones and Fluoroquinolones

In the early 1960s, the synthetic drug *nalidixic acid* was developed—the first of the **quinolone** group of antimicrobials. It exerted a unique bactericidal effect by selectively inhibiting an enzyme (DNA gyrase) needed for the replication of DNA. Although nalidixic acid found only limited use (its only application being for urinary tract infections), it led to the development in the 1980s of a prolific group of new antibiotics, the **fluoroquinolones.** *Norfloxacin* and *ciprofloxacin* are used most. Fluoroquinolones are frequently prescribed because they have a broad spectrum of activity, penetrate tissues well, and are safe for adults. However, their use in children, adolescents, and pregnant women is limited because they may adversely affect the development of cartilage.

Competitive Inhibitors of the Synthesis of Essential Metabolites

Learning Objective

■ *Describe how sulfa drugs inhibit microbial growth.*

Sulfonamides

As noted earlier, **sulfonamides,** or **sulfa drugs,** were among the first synthetic antimicrobial drugs used to treat microbial diseases. Antibiotics have diminished the importance of sulfa drugs in chemotherapy. But they continue to be used to treat certain urinary tract infections and have other specialized uses, as in the combination drug *silver sulfadiazine,* used to control infections in burn patients. Sulfonamides are bacteriostatic; as mentioned, their action is due to their structural similarity to para-aminobenzoic acid (PABA) (see Figure 20.6).

Probably the most widely used sulfa drug today is a combination of *trimethoprim* and *sulfamethoxazole* (TMP-SMZ). This combination is an excellent example of drug **synergism.** When used in combination, only 10% of the concentration is needed, compared to when each drug is used alone. The combination also has a broader spectrum of action and greatly reduces the emergence of resistant strains. (Synergism is discussed more fully later in the chapter; see Figure 20.23.)

MICROBIOLOGY IN THE NEWS

Antibiotics in Animal Feed Linked to Human Disease

Over 40 years ago, livestock growers began using antibiotics in the feed of closely penned animals that were being fattened for market. The drugs helped reduce the number of bacterial infections and controlled their spread in such ripe conditions. They also unexpectedly accelerated the animals' growth, an effect that led to wider and wider use. The growth-promoting effects are thought to be due to suppression of intestinal bacteria, such as *Clostridium* spp., that produce excessive gas and toxins, which may retard the animals' growth. Today, more than half the antibiotics used worldwide are given to farm animals to promote weight gain and treat infections.

Meat and milk that reach the consumer's table are not heavily laden with antibiotics because the U.S. Food and Drug Administration (FDA) has established limits for antibiotic residues in edible tissues. But the constant presence of antibiotics in these animals puts selective pressure on their normal microbiota, as well as on their pathogens, to develop drug resistance.

SALMONELLA

Salmonella can be transferred from animals to humans in meat or milk. The first known case of salmonellosis occurred in Germany in 1888, when 50 people became ill after eating ground beef from one cow. The use of antibiotics in animal feed preferentially allows the growth of strains of bacteria that are resistant to drugs commonly used to treat human infections. In the 1980s, researchers proved that antibiotic-resistant bacteria are being transferred from animals to humans directly in meat or milk. In 1984, Scott Holmberg and his associates at the CDC traced an antibiotic-resistant *Salmonella newport* infection from South Dakota beef cattle to 18 people in four states. Antibiotic-resistant *S. typhimurium* originating from one Illinois dairy infected 16,000 people in seven states in 1985. Reports from other countries have shown similar patterns of resistance. Before 1984 in Europe, *S. dublin* was sensitive to antibiotics. Since 1984, isolates

from cattle and from children with salmonellosis are resistant to chloramphenicol and streptomycin.

In 1985, nearly 1000 cases of infection by *S. newport* resistant to multiple antibiotics were reported in Los Angeles County, California. A radioactive DNA probe was used to identify an identical plasmid carrying multiple antibiotic resistance genes in 99% of the *S. newport* isolates. A plasmid profile analysis was used to track the bacterium from the sick people back to a slaughterhouse, to a meat-deboning plant, and finally to three farms.

VANCOMYCIN-RESISTANT ENTEROCOCCI

Vancomycin-resistant *Enterococcus* spp. (VRE) were first isolated in France in 1986, and were found in the United States in 1989. Vancomycin and another glycopeptide, avoparcin, are widely used in animal feed in Europe. Veterinary researchers in Denmark cultured vancomycin-resistant enterococci from horses, pigs, and poultry in the eight European countries that use avoparcin in animal feed. They did not find VRE in livestock in Sweden or the United States, which do not use avoparcin. Their conclusion is that use of avoparcin has created a reservoir for VRE in food animals. VRE are frequently present in food produced in European countries, and human VRE carriers are not uncommon. This suggests that VRE can be ingested from food in Europe. The researchers hypothesize that the high incidence of VRE nosocomial infections in the United States is due to (1) importation of the bacteria in travelers and imported food; (2) transmission by hospital personnel; and (3) high use of antibiotics in hospitals that selects for resistant microbes.

In 1996, veterinary use of avoparcin was banned in Germany. After the ban, VRE-positive samples were reduced from 100% to 25%, and the human carrier rate decreased from 12% to 3%. In 1997, all European Union countries banned the use of avoparcin.

MICROBIAL ALTERNATIVES TO ANTIBIOTICS

The U.S. Department of Agriculture (USDA) estimates that bacteria account for 3.6–7.1 million foodborne illnesses each year. Among the leading causes are *Salmonella*, *Campylobacter*, and *Escherichia coli* O157:H7, which colonize the intestines of farm animals and subsequently contaminate meat products during processing. *E. coli* O157:H7 colonize most cattle herds at one time or another, and 25% of raw chickens carry *Salmonella*. These bacteria can cause disease in humans when the food is undercooked and improperly stored. Because the pathogens may be present in low numbers of carcasses, screening is not practical. A variety of approaches may be necessary to reduce the possibility of illness: (1) prevent colonization in the animals at the farm, (2) reduce fecal contamination of meat during processing at the slaughterhouse, and (3) use proper storage and cooking methods.

Because these human pathogens are common in other animals, veterinary microbiologists are considering ways to prevent colonization. Competitive exclusion (CE), the use of bacteria to prevent the colonization of undesirable microorganisms, is being investigated to eliminate undesirable bacteria from the animals without using antibiotics.

Desirable bacteria can be included with the animals' feed. When ingested, these bacteria will colonize the digestive system and compete for nutrients, compete for attachment sites, and produce antibacterial compounds. Current research has shown that a mixture of three strains of bacteria can inhibit *Campylobacter* from colonizing in poultry. These three bacteria only grow on a component of mucus, so they will not cause infections in the chicken. These bacteria produce metabolites that inhibit *Campylobacter*.

Veterinary microbiologists are trying to identify possible CE bacteria against *E. coli*. The first step is to look at cattle herds that do not have *E. coli*. The microbiota in their digestive tract must then be tested to determine whether some of the bacteria can prevent the colonization of *E. coli*.

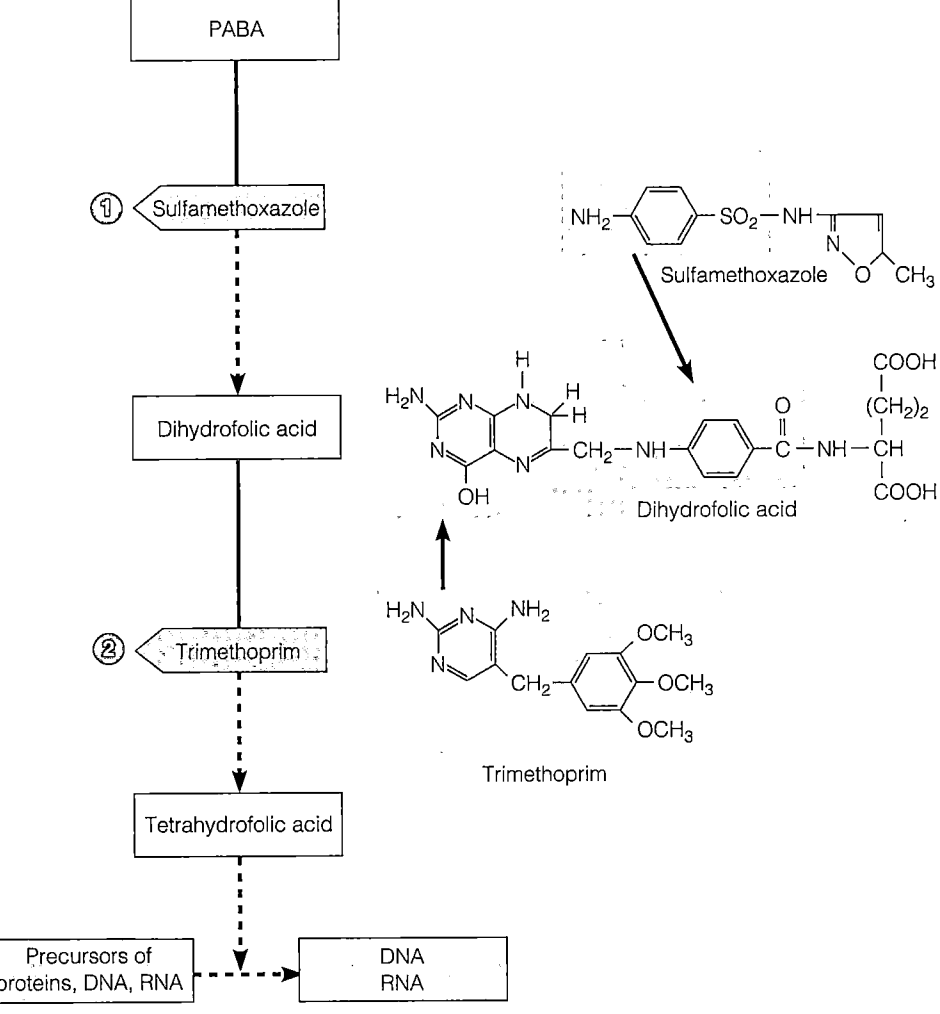

FIGURE 20.14 Actions of the antibacterial synthetics trimethoprim and sulfamethoxazole. TMP-SMZ works by inhibiting different stages in the synthesis of precursors of DNA, RNA, and proteins. Together the drugs are synergistic. ① Sulfamethoxazole, a sulfonamide that is a structural analog of PABA, competitively inhibits the synthesis of dihydrofolic acid from PABA. ② Then trimethoprim, a structural analog of a portion of dihydrofolic acid, competitively inhibits the synthesis of tetrahydrofolic acid.

■ In some instances, two drugs act synergistically to increase the effectiveness of each and decrease the potential for resistance.

Figure 20.14 illustrates how the two drugs interfere with different steps of a metabolic sequence leading to the synthesis of precursors of proteins, DNA, and RNA. ① Sulfamethoxazole inhibits the synthesis of dihydrofolic acid from PABA, so that the amount of dihydrofolic acid made is much less than usual. ② Then trimethoprim (which is not classified as a sulfa drug) inhibits the conversion of any dihydrofolic acid present into tetrahydrofolic acid. TMP-SMZ has a broad spectrum of activity, except against pseudomonads, and can be taken orally. It is even used for the control of *Pneumocystis* pneumonia, which is caused by a fungus and often occurs as a complication of AIDS. TMP-SMZ is very effective in penetrating the brain and cerebrospinal fluid.

Antifungal Drugs

Learning Objective

■ *Explain the modes of action of currently used antifungal drugs.*

Eukaryotes, such as fungi, use the same mechanisms to synthesize proteins and nucleic acids as higher animals. Therefore it is more difficult to find a point of selective toxicity in eukaryotes than in prokaryotes. Moreover, fungal infections are becoming more frequent because of their role as opportunistic infections in immunosuppressed individuals, especially those with AIDS.

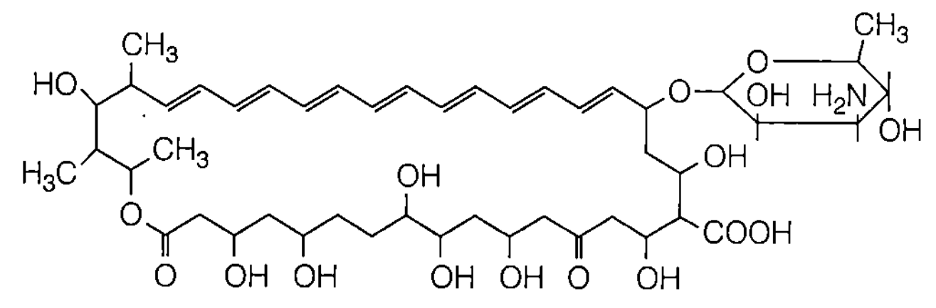

Amphotericin B

FIGURE 20.15 The structure of the antifungal drug amphotericin B, representative of the polyenes.

■ **Why do polyenes injure fungal plasma membranes and not bacterial membranes?**

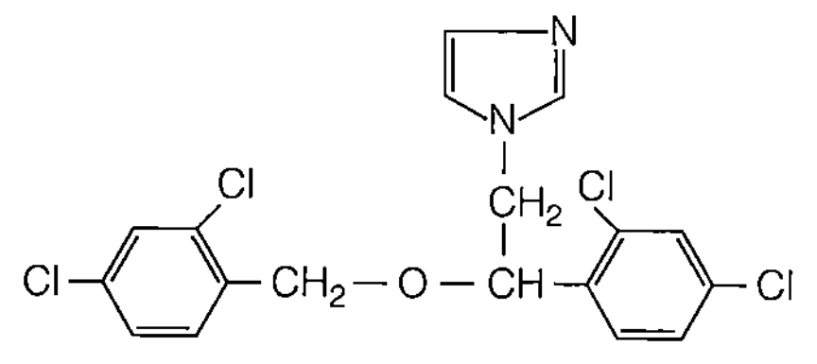

Polyenes

Amphotericin B is the most commonly used member of the antifungal **polyene antibiotics** (Figure 20.15). Produced by *Streptomyces* species of soil bacteria, these antibiotics combine with the sterols in fungal plasma membranes, making the membranes excessively permeable and killing the cell. Bacterial plasma membranes (except those of mycoplasmas) do not contain sterols, and the drugs therefore do not affect bacteria. For many years, amphotericin B has been a mainstay of clinical treatment for systemic fungal diseases, such as histoplasmosis, coccidioidomycosis, and blastomycosis. The drug's toxicity, particularly to the kidneys, is a strongly limiting factor in these uses. Administering the drug encapsulated in lipids (liposomes) appears to minimize toxicity.

Azoles

Imidazole and **triazole** are antifungal drugs that primarily interfere with sterol synthesis in fungi, although they probably have other antimetabolic effects. Imidazoles such as *clotrimazole* and *miconazole* (Figure 20.16) are generally used topically, and they are sold without a prescription for the treatment of cutaneous mycoses such as athlete's foot and vaginal yeast infections. When taken orally, the imidazole *ketoconazole* is effective and is a less toxic alternative to amphotericin B for many systemic fungal infections, although occasional liver damage has been reported. Ketoconazole ointments are used to treat dermatomycoses. Ketoconazole also has an unusually broad spectrum of activity.

Systemic fungal infections are often treated by the triazoles *fluconazole* and *itraconazole*. They have essentially the same mode of action as the other azole family of antifungals but are less toxic than other systemic antifungals.

Miconazole

FIGURE 20.16 The structure of the antifungal drug miconazole, representative of the imidazoles.

■ **Imidazoles injure fungal plasma membranes.**

Griseofulvin

Griseofulvin is an antibiotic produced by a species of *Penicillium*. It has the interesting property of being active against superficial dermatophytic fungal infections of the hair (tinea capitis, or ringworm) and nails, even though its route of administration is oral. The drug apparently binds selectively to the keratin found in the skin, hair follicles, and nails. Its mode of action is primarily to block microtubule assembly, which interferes with mitosis and thereby inhibits fungal reproduction.

Other Antifungal Drugs

Tolnaftate is a common alternative to miconazole as a topical agent for the treatment of athlete's foot. Its mechanism of action is not known. *Undecylenic acid* is a fatty acid that has antifungal activity against athlete's foot, although it is not as effective as tolnaftate or the imidazoles.

The antifungal drug *flucytosine,* an antimetabolite of the base cytosine, interferes with DNA and RNA synthesis. It is preferentially taken up by fungal cells. Its spectrum of activity is limited to only a few systemic fungal infections. Kidney and bone marrow toxicity limit its use. *Pentamidine isethionate* is used in the treatment of *Pneumocystis* pneumonia, a common complication in immunocompromised individuals, particularly AIDS patients. The drug's mode of action is unknown, but it appears to bind DNA.

Antiviral Drugs

Learning Objective

■ *Explain the modes of action of currently used antiviral drugs.*

In developed parts of the world, it is estimated that at least 60% of infectious illnesses are caused by viruses, and about 15% by bacteria. Every year, at least 90% of the U.S. population suffers from a viral disease. Yet relatively few antiviral drugs have been approved in the United States, and they are effective against only an extremely limited group of diseases. Many of the most recently developed antiviral drugs are directed at HIV. Targets for antiviral drugs are various points in viral reproduction (see Figures 13.16, 13.17, and 13.19), such as attachment to host cell, penetration, uncoating, DNA or RNA syntheses or reproduction, or assembly of the new virion.

Nucleoside and Nucleotide Analogs

Several important antiviral drugs are analogs of nucleosides and nucleotides (page 50). Among the nucleoside analogs, *acyclovir* is the one more widely used (Figure 20.17). While best known for treating genital herpes, it is generally useful for most herpesvirus infections, especially in immunosuppressed individuals. The antiviral drugs *famciclovir,* which can be taken orally, and *ganciclovir* are derivatives of acyclovir and have a similar mode of action. *Trifluridine,* used as a topical ointment to treat acyclovir-resistant herpes keratitis (an eye infection), is an analog of the thymine-containing nucleoside thymidine. *Ribavirin* resembles the nucleoside guanine and interferes with viral replication.

A drug well known in chemotherapy for HIV infections is *zidovudine* (AZT), a nucleoside analog, which blocks the synthesis of DNA from RNA by the enzyme reverse transcriptase. The stimulus of research into anti-HIV drugs has led to the development of other analogs of nucleosides that also block activity of reverse transcriptase; most of those are less toxic than zidovudine. These drugs include *didanosine* and *zalcitabine*.

Other Enzyme Inhibitors

Another approach to controlling HIV infections is to inhibit the enzymes that control the last stage of viral reproduction. When the host cell, at the direction of the infecting HIV, makes a new virus, it must begin by cutting up large proteins with protease enzymes. These fragments are used to assemble new viruses. The protease inhibitors are analogs of amino acid sequences in the large proteins and competitively interfere with the action of the proteases. The protease inhibitors *indinavir* and *saquinavir* have proved especially effective when combined with inhibitors of reverse transcriptase.

Two inhibitors of the enzyme neuraminidase have recently been introduced for treatment of influenza. These are *zanamivir* (Relenza®) and *oseltamivir phosphate* (Tamiflu®).

Interferons

Cells infected by a virus often produce interferon, which inhibits further spread of the infection. Interferons are classified as cytokines, discussed in Chapter 17. *Alpha-interferon* (see Chapter 16 on page 469) is currently the drug of choice for viral hepatitis infections.

Antiprotozoan and Antihelminthic Drugs

Learning Objective

- *Explain the modes of action of currently used antiprotozoan and antihelminthic drugs.*

For hundreds of years, quinine from the Peruvian cinchona tree was the only drug known to be effective for the treatment of a parasitic infection (malaria). It was first introduced into Europe in the early 1600s and was known as "Jesuit's powder." There are now many antiprotozoan and antihelminthic drugs, although many of them are still considered experimental. This does not preclude their use, however, by qualified physicians. The CDC provides several of them on request when they are not available commercially.

Antiprotozoan Drugs

Quinine is still used to control the protozoan disease malaria, but synthetic derivatives, such as *chloroquine,* have largely replaced it. For preventing malaria in areas where the disease has developed resistance to chloroquine, the new drug *mefloquine* is recommended. *Quinacrine* is the drug of choice for treating the protozoan disease giardiasis. *Diiodohydroxyquin (iodoquinol)* is an important drug prescribed for several intestinal amoebic diseases, but its dosage must be carefully controlled to avoid optic nerve damage. Its mode of action is unknown.

Metronidazole (Flagyl®) is one of the most widely used antiprotozoan drugs. It is unique in that it acts not only against parasitic protozoa but also against obligately anaerobic bacteria. For example, as an antiprotozoan agent, it is the drug of choice for vaginitis caused by *Trichomonas vaginalis.* It is also used in the treatment of giardiasis and amoebic dysentery. The mode of action is to interfere with anaerobic metabolism, which incidentally these protozoans share with certain obligately anaerobic bacteria, such as *Clostridium.*

Antihelminthic Drugs

With the increased popularity of sushi, a Japanese specialty often made with raw fish, the CDC began to notice an increased incidence of tapeworm infestations. To estimate the incidence, the CDC documents requests for *niclosamide,* which is the usual first choice in treatment. The drug is effective because it inhibits ATP production under aerobic conditions. *Praziquantel* is about equally effective for the treatment of tapeworms; it kills worms by altering the permeability of their plasma membranes. Praziquantel has a broad spectrum of activity and is highly recommended for treating several fluke-caused

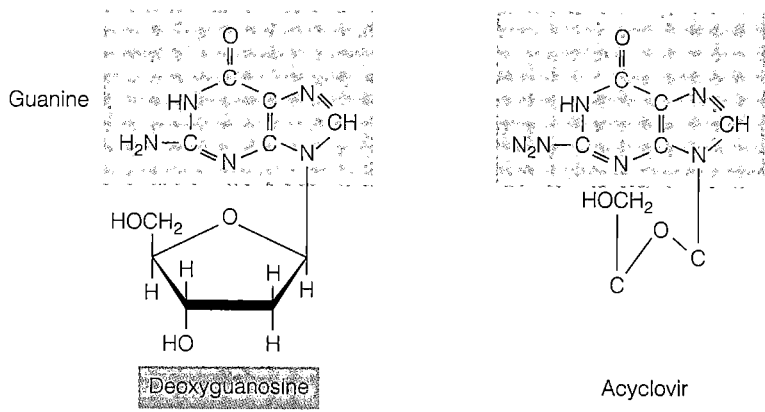

Guanine

Deoxyguanosine

Acyclovir

(a) Structural resemblance between acyclovir and guanine-containing nucleoside

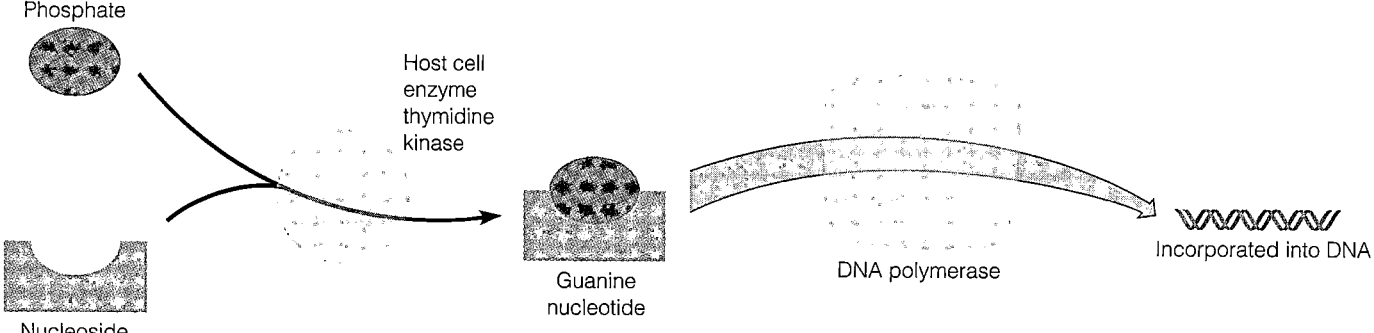

Phosphate

Host cell
enzyme
thymidine
kinase

Nucleoside

Guanine
nucleotide

DNA polymerase

Incorporated into DNA

(b) Synthesis of normal viral DNA guanine nucleotide

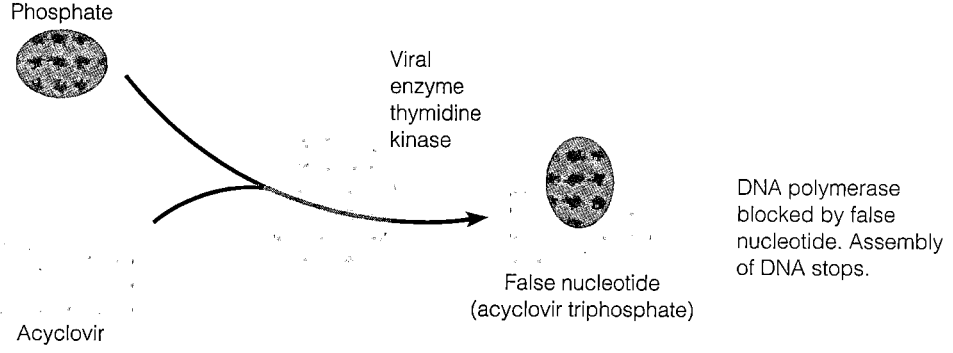

Phosphate

Viral
enzyme
thymidine
kinase

Acyclovir

False nucleotide
(acyclovir triphosphate)

DNA polymerase
blocked by false
nucleotide. Assembly
of DNA stops.

(c) Synthesis of false viral DNA nucleotide with acyclovir

FIGURE 20.17 The structure and function of the antiviral drug acyclovir. (a) Acyclovir structurally resembles the nucleoside 2'-deoxyguanosine. **(b)** The viral enzyme thymidine kinase combines phosphates with such nucleosides to form nucleotides, which are then incorporated into DNA. **(c)** Because acyclovir is a structural analog of the nucleoside, the enzyme uses it to assemble a false nucleotide that cannot be incorporated into DNA. The selective toxicity of acyclovir is based upon the fact that the phosphorylated form of the drug (acyclovir triphosphate) is a very strong inhibitor of DNA polymerase. The *viral enzyme* thymidine kinase is much more effective in forming acyclovir triphosphate than is the *host-cell* thymidine kinase.

■ Why are viral infections generally difficult to treat with chemotherapeutic agents?

diseases, especially schistosomiasis. It causes the helminths to undergo muscular spasms and apparently makes them susceptible to attack by the immune system. Apparently, its action exposes surface antigens, which antibodies can then reach.

Mebendazole is an antihelminthic drug frequently used in the treatment of several of the most common intestinal helminthic infections: ascariasis (*Ascaris lumbricoides*), pinworms (*Enterobius vermicularis*), and whipworms (*Trichuris trichiura;* trik′ĕr-is trik-ē-yėr′a). The mode of action is disruption of the microtubules in the cytoplasm, which indirectly reduces the worm's motility. *Pyantel pamoate* is used in the treatment of pinworms, hookworms, and ascariasis. It paralyzes the organisms, which soon exit the host's body. *Ivermectin,* widely used in the livestock industry, is very effective against several human nematodes, including those of ascariasis and whipworm. It acts by paralyzing the nematode.

Tests to Guide Chemotherapy

Learning Objective

- Describe two tests for microbial susceptibility to chemotherapeutic agents.

Different microbial species and strains have different degrees of susceptibility to different chemotherapeutic agents. Moreover, the susceptibility of a microorganism can change with time, even during therapy with a specific drug. Thus, a physician must know the sensitivities of the pathogen before treatment can be started. However, physicians often cannot wait for sensitivity tests and must begin treatment based on their "best guess" estimation of the most likely pathogen causing the illness.

Several tests can be used to indicate which chemotherapeutic agent is most likely to combat a specific pathogen. However, if the organisms have been identified—for example, as *Pseudomonas aeruginosa,* β-hemolytic streptococci, or gonococci—certain drugs can be selected without specific testing for susceptibility. Tests are necessary only when susceptibility is not predictable, or when antibiotic resistance problems develop.

The Diffusion Methods

Probably the most widely used, although not necessarily the best, method of testing is the **disk-diffusion method,** also known as the *Kirby–Bauer test* (Figure 20.18). A Petri plate containing an agar medium is inoculated ("seeded") uniformly over its entire surface with a standardized amount of a test organism. Next, filter paper disks impregnated with known concentrations of

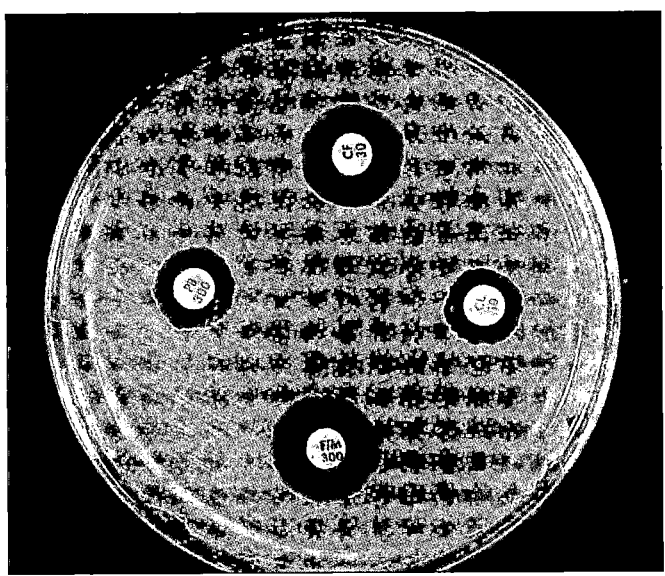

FIGURE 20.18 The disk-diffusion method for determining the activity of antimicrobials. Each disk contains a different chemotherapeutic agent, which diffuses into the surrounding agar. The clear zones indicate inhibition of growth of the microorganism swabbed onto the plate surface.

■ **Which agent is the most effective against the bacterium being tested?**

chemotherapeutic agents are placed on the solidified agar surface. During incubation, the chemotherapeutic agents diffuse from the disks into the agar. The farther the agent diffuses from the disk, the lower its concentration. If the chemotherapeutic agent is effective, a **zone of inhibition** forms around the disk after a standardized incubation. The diameter of the zone can be measured; in general, the larger the zone, the more sensitive the microbe is to the antibiotic. The zone diameter is compared to a standard table for that drug and concentration, and the organism is reported as *sensitive, intermediate,* or *resistant.* For a drug with poor solubility, however, the zone of inhibition indicating that the microbe is sensitive will usually be smaller than for another drug that is more soluble and has diffused more widely. Results obtained by the disk-diffusion method are often inadequate for many clinical purposes. However, the test is simple and inexpensive and is most often used when more sophisticated laboratory facilities are not available.

A more advanced diffusion method, the **E test,** enables a lab technician to estimate the **minimal inhibitory concentration (MIC),** the lowest antibiotic concentration that prevents visible bacterial growth. A plastic-coated strip contains a gradient of antibiotic concentrations, and the MIC can be read from a scale printed on the strip (Figure 20.19).

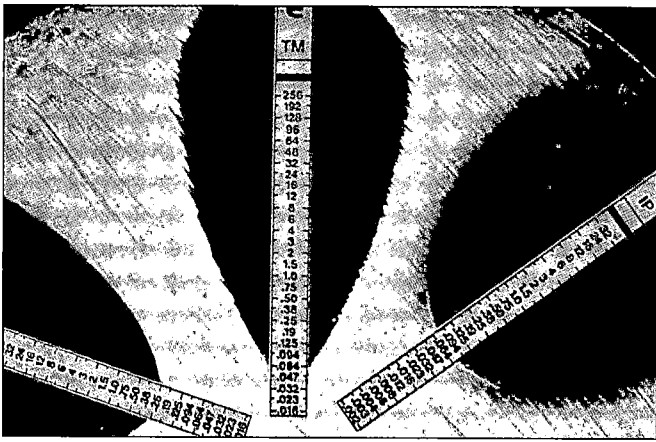

FIGURE 20.19 The E test, a diffusion method that determines antibiotic sensitivity and estimates minimal inhibitory concentration (MIC). The plastic strip, which is placed on an agar surface inoculated with test bacteria, contains an increasing gradient of the antibiotic.

■ **What other factors must be considered before treatment is prescribed?**

Broth Dilution Tests

A weakness of the diffusion method is that it does not determine whether a drug is bactericidal and not just bacteriostatic. A **broth dilution test** is often useful in determining the MIC and the **minimal bactericidal concentration (MBC)** of an antimicrobial drug. The MIC is determined by making a sequence of decreasing concentrations of the drug in a broth, which is then inoculated with the test bacteria (Figure 20.20). The wells that do not show growth (higher concentration than the MIC) can be cultured in broth free of the drug. If growth occurs in this broth, the drug was not bactericidal, and the MBC can be determined. Determining the MIC and MBC is important because it avoids the excessive or erroneous use of expensive antibiotics and minimizes the chance of toxic reactions that larger-than-necessary doses might cause.

Dilution tests are often highly automated. The drugs are purchased already diluted into broth in wells formed in a plastic tray. A suspension of the test organism is prepared and inoculated into all the wells simultaneously by a special inoculating device. After incubation, the turbidity may be read visually, although clinical laboratories with high workloads may read the trays with special scanners that enter the data into a computer that provides a printout of the MIC.

Other tests are also useful for the clinician; a determination of the microbe's ability to produce β-lactamase is one example. One popular, rapid method makes use of a

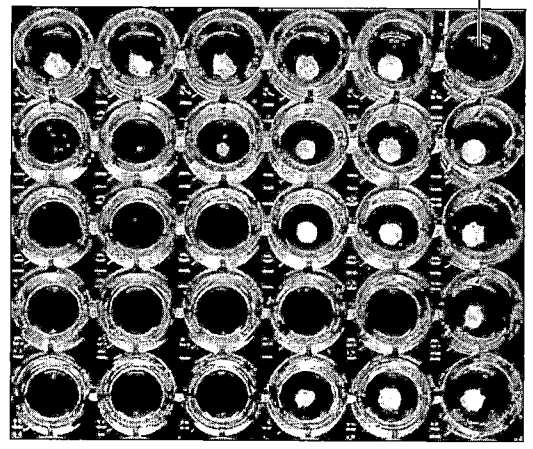

Broth only (negative control)

Doxycycline

Sulfameth-oxazole

Ethambutol

Streptomycin

Kanamycin

Decreasing concentration of drug ⟶

FIGURE 20.20 A microdilution (MIC) plate. Growth is shown by a white dot in the bottom of the well. In this test, doxycycline had no effect at all. Kanamycin and ethambutol were about equally effective. Streptomycin was effective at all concentrations. A trailing endpoint, as observed here with sulfamethoxazole, is read where there is an 80% reduction in growth. The negative control well contained broth only (no antimicrobial drugs and no inoculum); the positive control well (not shown) contained no drugs but was inoculated.

■ **The MIC is the lowest concentration that inhibits microbial growth.**

cephalosporin that changes color when its β-lactam ring is opened. In addition, a measurement of the *serum concentration* of an antimicrobial is especially important when toxic drugs are used. These assays tend to vary with the drug and may not always be suitable for smaller laboratories.

The Effectiveness of Chemotherapeutic Agents

Drug Resistance

Learning Objective

■ *Describe the mechanisms of drug resistance.*

Bacteria become resistant to chemotherapeutic agents by four major mechanisms: (1) destruction or inactivation of the drug (by β-lactamase, for example); (2) prevention of penetration to the target site within the microbe (a frequent mechanism for tetracycline resistance); (3) alteration of the drug's target sites (for example, a single amino acid change in the ribosome may be enough to make the microbe resistant to certain macrolide antibiotics); and (4) rapid efflux (ejection), which pumps the drug out of the cell before it can become effective.

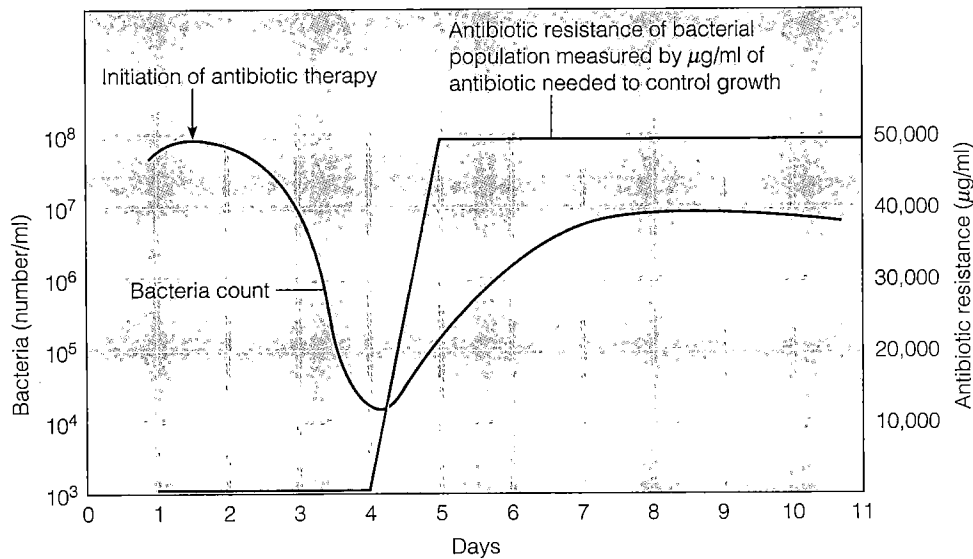

FIGURE 20.21 The development of an antibiotic-resistant mutant during antibiotic therapy. The patient, suffering from a chronic kidney infection caused by a gram-negative bacterium, was treated with streptomycin. Antibiotic sensitivity is expressed as the MIC of the antibiotic in micrograms per milliliter.
SOURCE: From *Biology of Microorganisms* 7th Ed. by T. D. Brock, M.T. Madigan, J. M. Martinko, and J. Parker, p. 410 © 1993. Adapted by permission of Prentice-Hall Inc., Upper Saddle River, New Jersey.

■ **Sensitive bacteria are killed, and resistant mutants are able to grow in the presence of an antibiotic.**

Variations on these mechanisms also occur. For example, a microbe could become resistant to trimethoprim by synthesizing very large amounts of the enzyme against which the drug is targeted. Conversely, polyene antibiotics can become less effective when resistant organisms produce smaller amounts of the sterols against which the drug is effective. Of particular concern is the possibility that such *resistant mutants* will increasingly replace the susceptible normal populations (see the box on page 562). Figure 20.21 shows how rapidly bacterial numbers increase during infection as resistance develops.

Hereditary drug resistance is often carried by plasmids, or by small segments of DNA called transposons that can jump from one piece of DNA to another (Chapter 8 on page 240). Some plasmids, including those called resistance (R) factors, can be transferred between bacterial cells in a population and between different but closely related bacterial populations (see Figure 8.27 on page 238). R factors often contain genes for resistance to several antibiotics.

Antibiotics have been a much misused product, nowhere more so than in the less developed areas of the world. Well-trained personnel are scarce, especially in rural areas, which is perhaps one reason why antibiotics can al-

most universally be purchased without prescription in these countries. A survey in rural Bangladesh, for example, showed that only 8% of antibiotics had been prescribed by a physician. In much of the world, antibiotics are sold to treat headaches and for other inappropriate uses (Figure 20.22). Even when the use of antibiotics is appropriate, dose regimens are usually shorter than needed to eradicate the infection. This encourages the survival of resistant strains of bacteria. Outdated and adulterated, and therefore weakened, antibiotics are also common. These problems have no easy solutions. Poverty is endemic and is complicated by political turmoil that disrupts already inadequate medical services.

The developed world is also contributing to the rise of antibiotic resistance. The CDC estimates that in the United States, 30% of antibiotic prescriptions for ear infections, 100% of prescriptions for the common cold, and 50% of prescriptions for sore throats were unnecessary. At least half of the antibiotics produced in the United States are used to promote animal growth, a practice many people feel should be curtailed. Patients also contribute to survival of antibiotic resistant microbes when they fail to finish the full regimen of the prescription or use leftover antibiotics prescribed for someone else.

FIGURE 20.22 Antibiotics are sold without prescriptions in much of the world. This frequently leads to development of resistant strains of pathogenic bacteria.

Strains of bacteria that are resistant to antibiotics are particularly common among people who work in hospitals, where antibiotics are in constant use. Drugs should not be used indiscriminately, and the concentrations administered should always be of optimum strength to minimize the chances of survival of resistant mutants. Many hospitals have special monitoring committees to review the use of antibiotics for effectiveness and cost.

Another way to minimize the development of resistant strains is to administer two or more drugs simultaneously. If a strain is resistant to one of the drugs, the other might be effective against it. The probability that the organism will acquire resistance to both drugs is roughly the product of the two individual probabilities. However, multidrug-resistant tuberculosis is an emerging problem (see the box in Chapter 15, page 437).

Bacteria that drugs are unable to control are the so-called *superbugs*.

It is hoped that if antibiotic use is controlled, the relative numbers of susceptible bacteria will rise. It is believed that they will reproduce in greater numbers than will the resistant bacteria because they do not have to expend energy to maintain the resistance genes.

There is concern that the use of antimicrobials (which are not antibiotics) such as triclosan (see page 196) in soaps and other products may promote the survival and growth of plasmid-carrying bacteria. These bacteria may also carry plasmids for antibiotic resistance.

Antibiotic Safety

In our discussions of antibiotics, we have occasionally mentioned side effects. These may be potentially serious, such as liver or kidney damage or hearing impairment. Administration of almost any drug involves an assessment of risks against benefits; this is called the *therapeutic index*. Sometimes, the use of another drug can cause toxic effects that do not occur when the drug is taken alone. One drug may also neutralize the intended effects of the other. For example, a few antibiotics have been reported to neutralize the effectiveness of contraceptive pills. However, the association between contraceptive failure and the use of those drugs has not been confirmed. Also, some individuals may have hypersensitivity reactions, for example, to penicillins.

A pregnant woman should take only those antibiotics that are classified by the U. S. Food and Drug Administration as presenting no evidence of risk to the fetus.

Effects of Combinations of Drugs

Learning Objective

■ *Compare and contrast synergism and antagonism.*

The chemotherapeutic effect of two drugs given simultaneously is sometimes greater than the effect of either given alone; (Figure 20.23). This phenomenon, called synergism, was introduced earlier. For example, in the treatment of bacterial endocarditis, penicillin and streptomycin are much more effective when taken together than when either drug is taken alone. Damage to bacterial cell walls by penicillin makes it easier for streptomycin to enter.

Other combinations of drugs can show **antagonism.** For example, the simultaneous use of penicillin and tetracycline is often less effective than when either drug is used alone. By stopping the growth of the bacteria, the bacteriostatic drug tetracycline interferes with the action of penicillin, which requires bacterial growth.

The Future of Chemotherapeutic Agents

Learning Objective

■ *Identify three areas of research on new chemotherapeutic agents.*

Antibiotics are clearly one of the greatest triumphs of medical science, but they rely upon a limited range of targets (Figure 20.2). Microbes are now developing resistance to these modes of action. Unfortunately, the problem of antibiotic resistance has now reached a crisis

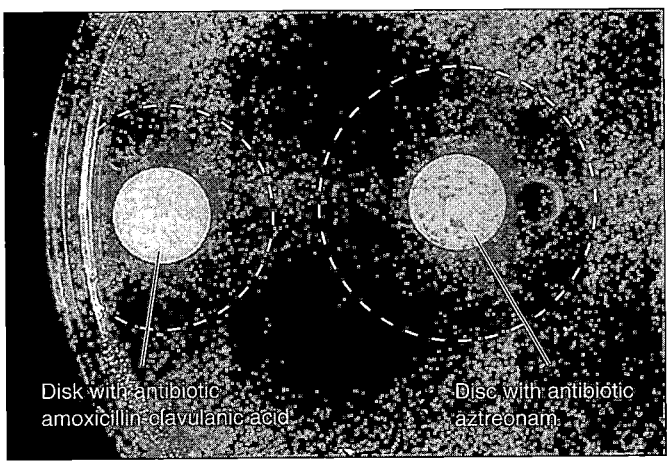

FIGURE 20.23 An example of synergism between two different antibiotics. The photograph shows the surface of a Petri plate seeded with bacteria. The paper disc at the left contains the antibiotic amoxicillin plus clavulanic acid. The disc on the right contains the antibiotic aztreonam. The dashed circles drawn over the photo show the clear areas surrounding each disc where bacterial growth would have been inhibited if there had been no synergy. The additional clear area between these two areas and outside the drawn circles, illustrates inhibition of bacterial growth through the effects of synergy.

point. Dealing with this crisis will require both intensified educational campaigns to promote the wise use of existing antibiotics and redoubled efforts toward the design of new ones.

A first step will be to modify existing drugs to extend their spectrum and prevent their destruction by bacterial enzymes of resistance. This is the approach that has been used in the penicillin group. Research is under way to prevent the resistance strategy of rapid antibiotic efflux. Investigators are also working to identify new targets for antimicrobial activity other than those summarized in Figure 20.2.

Much of this research will be based on our increasing knowledge of the bacterial genome and reevaluations of drugs used for veterinary or other uses and on screening of existing chemicals of all types.

Most of our present antibiotics are the products of other microorganisms. Interest is now also focusing on antibiotics produced by plants and animals that often exhibit extraordinary resistance to microbial infections. There is great interest concerning a class of drugs, the *antimicrobial peptides,* that are based on host–defense biology. These seem to be used by animals to defend against microbial infections, perhaps as a primitive form of immune system. They typically use a unique mechanism of action that disrupts the permeability of the microbial membrane without affecting those of the host's cells. Examples are *magainin* from the skin of certain frogs; *squalamine,* an easily synthesized peptide found in the tissue of the spiny dogfish shark; and *protegrins,* isolated from pigs. Apparently, human skin and epithelial cells of the tongue and trachea also are sources of antimicrobial peptides. *Nisin,* which we have mentioned for its use as a food preservative, resembles magainin in its mode of action on membranes. It is the subject of considerable research to see if nisin derivatives might replace vancomycin for use against gram-positive pathogens.

Insects lack a mammalian-type immune system and produce an array of antimicrobial peptides to defend against infecting bacteria and fungi. One such compound is *cecropin,* which is produced by certain moths.

Another promising new approach is the use of short synthetic strands of DNA, called antisense and triplex agents. The principle involved is to identify sites on the DNA and RNA of the pathogen that are responsible for the pathogenic effects. Segments of DNA are then synthesized that will selectively recognize and bind to the target site. If the DNA segment binds to the double strand of DNA, a **triplex agent,** or to the messenger RNA that encodes the target protein will be synthesized, an **antisense agent,** no pathogenic protein will be made. This approach has the great advantage of preventing the production of a pathogenic protein, rather than attempting to selectively neutralize it once it is made. The main problem encountered has been poor penetration into the target bacterium.

These techniques are still in the investigative stages. Their primary potential targets are viral infections and cancer. There is special interest in such antiviral drugs, as well as in antifungal and antiparasitic drugs, because our arsenal in these categories is especially limited.

★ ★ ★

The past few chapters have shown how profoundly science has changed the effects of infectious disease on human mortality and life span. At the turn of the twentieth century, the most common causes of death were infectious diseases. Most of these, including tuberculosis, typhoid fever, and diphtheria, were caused by bacteria. At the beginning of the twenty-first century, viral diseases such as influenza, viral pneumonias, and AIDS, are the only infectious diseases among the 10 leading causes of death in the United States. These facts bear testimony to the effectiveness of sanitation, vaccines, and (as discussed in this chapter) the discovery and use of antibiotics. In Chapters 21–26, we will see that the struggle against infectious disease continues.

Study Outline

INTRODUCTION (p. 549)

1. An antimicrobial drug is a chemical substance that destroys pathogenic microorganisms with minimal damage to host tissues.

2. Chemotherapeutic agents include chemicals that combat disease in the body.

THE HISTORY OF CHEMOTHERAPY (pp. 549–550)

1. Paul Ehrlich developed the concept of chemotherapy to treat microbial diseases; he predicted the development of chemotherapeutic agents, which would kill pathogens without harming the host.

2. Sulfa drugs came into prominence in the late 1930s.

3. Alexander Fleming discovered the first antibiotic, penicillin, in 1929; its first clinical trials were done in 1940.

THE SPECTRUM OF ANTIMICROBIAL ACTIVITY (pp. 550–551)

1. Antibacterial drugs affect many targets in a prokaryotic cell.

2. Fungal, protozoan, and helminthic infections are more difficult to treat because these organisms have eukaryotic cells.

3. Narrow-spectrum drugs affect only a select group of microbes—gram-positive cells, for example; broad-spectrum drugs affect a large number of microbes.

4. Small, hydrophilic drugs can affect gram-negative cells.

5. Antimicrobial agents should not cause excessive harm to normal microbiota.

6. Superinfections occur when a pathogen develops resistance to the drug being used or when normally resistant microbiota multiply excessively.

THE ACTION OF ANTIMICROBIAL DRUGS (pp. 551–554)

1. General action is either by directly killing microorganisms (bactericidal) or by inhibiting their growth (bacteriostatic).

2. Some agents, such as penicillin, inhibit cell wall synthesis in bacteria.

3. Other agents, such as chloramphenicol, erythromycin, tetracyclines, and streptomycin, inhibit protein synthesis by acting on 70S ribosomes.

4. Agents such as polymyxin B cause injury to plasma membranes.

5. Rifampin and the quinolones inhibit nucleic acid synthesis.

6. Agents such as sulfanilamide act as antimetabolites by competitively inhibiting enzyme activity.

A SURVEY OF COMMONLY USED ANTIMICROBIAL DRUGS (pp. 554–567)

Antibacterial Antibiotics: Inhibitors of Cell Wall Synthesis (pp. 556–559)

1. All penicillins contain a β-lactam ring.

2. Natural penicillins produced by *Penicillium* are effective against gram-positive cocci and spirochetes.

3. Penicillinases (β-lactamases) are bacterial enzymes that destroy natural penicillins.

4. Semisynthetic penicillins are made in the laboratory by adding different side chains onto the β-lactam ring made by the fungus.

5. Semisynthetic penicillins are resistant to penicillinases and have a broader spectrum of activity than natural penicillins.

6. The monobactam aztreonam affects only gram-negative bacteria.

7. Cephalosporins inhibit cell wall synthesis and are used against penicillin-resistant strains.

8. Carbapenems are broad-spectrum antibiotics that inhibit cell wall synthesis.

9. Polypeptides such as bacitracin are applied topically to treat superficial infections.

10. Bacitracin inhibits cell wall synthesis primarily in gram-positive bacteria.

11. Vancomycin inhibits cell wall synthesis and may be used to kill penicillinase-producing staphylococci. Streptogramins are bactericidal agents that inhibit protein synthesis and may be used to kill vancomycin-resistant bacteria.

12. Isoniazid (INH) inhibits mycolic acid synthesis in mycobacteria. INH is administered with rifampin or ethambutol to treat tuberculosis.

13. The antimetabolite ethambutol is used with other drugs to treat tuberculosis.

Inhibitors of Protein Synthesis (pp. 559–560)

1. Aminoglycosides, tetracyclines, chloramphenicol, and macrolides inhibit protein synthesis at 70S ribosomes.

Injury to the Plasma Membrane (p. 560)

1. Polymyxin B and bacitracin cause damage to plasma membranes.

Inhibitors of Nucleic Acid (DNA/RNA) Synthesis (p. 561)

1. Rifamycin inhibits mRNA synthesis; it is used to treat tuberculosis.

2. Quinolones and fluoroquinolones inhibit DNA gyrase for treatment of urinary tract infections.

Competitive Inhibitors of the Synthesis of Essential Metabolites (pp. 561–563)

1. Sulfonamides competitively inhibit folic acid synthesis.

2. TMP-SMZ competitively inhibits dihydrofolic acid synthesis.

Antifungal Drugs (pp. 563–564)

1. Polyenes, such as nystatin and amphotericin B, combine with plasma membrane sterols and are fungicidal.

2. Azoles interfere with sterol synthesis and are used to treat cutaneous and systemic mycoses.

3. Griseofulvin interferes with eukaryotic cell division and is used primarily to treat skin infections caused by fungi.

4. The antifungal agent flucytosine is an antimetabolite of cytosine.

Antiviral Drugs (pp. 564–565)

1. Nucleoside and nucleotide analogs, such as acyclovir, AZT, ddI, and ddC, inhibit DNA or RNA synthesis.

2. Protease inhibitors, such as indinavir and saquinavir, block activity of an HIV enzyme essential for assembly of a new viral coat.

3. Alpha-interferons inhibit the spread of viruses to new cells.

Antiprotozoan and Antihelminthic Drugs (pp. 565–567)

1. Chloroquine, quinacrine, diiodohydroxyquin, pentamidine, and metronidazole are used to treat protozoan infections.

2. Antihelminthic drugs include niclosamide, mebendazole, praziquantel, and piperazine.

3. Mebendazole disrupts microtubules; pyantel pamoate paralyzes intestinal roundworms.

TESTS TO GUIDE CHEMOTHERAPY (pp. 567–568)

1. These tests are used to determine which chemotherapeutic agent is most likely to combat a specific pathogen.

2. These tests are used when susceptibility cannot be predicted or when drug resistance arises.

The Diffusion Methods (p. 567)

1. In this test, also known as the Kirby-Bauer test, a bacterial culture is inoculated on an agar medium, and filter paper disks impregnated with chemotherapeutic agents are overlaid on the culture.

2. After incubation, the absence of microbial growth around a disk is called a zone of inhibition.

3. The diameter of the zone of inhibition, when compared with a standardized reference table, is used to determine whether the organism is sensitive, intermediate, or resistant to the drug.

4. MIC is the lowest concentration of drug capable of preventing microbial growth; MIC can be estimated using the E test.

Broth Dilution Tests (p. 568)

1. In a broth dilution test, the microorganism is grown in liquid media containing different concentrations of a chemotherapeutic agent.

2. The lowest concentration of a chemotherapeutic agent that kills bacteria is called the minimum bactericidal concentration (MBC).

THE EFFECTIVENESS OF CHEMOTHERAPEUTIC AGENTS (pp. 568–571)

Drug Resistance (pp. 568–570)

1. Resistance may be due to enzymatic destruction of a drug, prevention of penetration of the drug to its target site, or cellular or metabolic changes at target sites.

2. Hereditary drug resistance (R) factors are carried by plasmids and transposons.

3. Resistance can be minimized by the discriminating use of drugs in appropriate concentrations and dosages.

Antibiotic Safety (p. 570)

1. The risk (e.g., side-effects) versus the benefit (e.g., curing an infection) must be evaluated prior to use of antibiotics.

Effects of Combinations of Drugs (p. 570)

1. Some combinations of drugs are synergistic; they are more effective when taken together.

2. Some combinations of drugs are antagonistic; when taken together, both drugs become less effective than when taken alone.

The Future of Chemotherapeutic Agents (pp. 570–571)

1. Many bacterial diseases, previously treatable with antibiotics, have become resistant to antibiotics.

2. Chemicals produced by plants and animals are providing new antimicrobial agents, including antimicrobial peptides.

3. New antimicrobial drugs include DNA that is complementary to specific genes in a pathogen; the DNA will bind to the pathogen's DNA or mRNA and inhibit protein synthesis.

Study Questions

REVIEW

1. Fill in the following table:

Antimicrobial Agent	Synthetic or Antibiotic	Method of Action	Principal Use
Isoniazid			
Sulfonamides			
Ethambutol			
Trimethoprim			
Fluoroquinolones			
Penicillin, natural			
Penicillin, semisynthetic			
Cephalosporins			
Carbapenems			
Aminoglycosides			
Tetracyclines			
Chloramphenicol			
Macrolides			
Polypeptides			
Vancomycin			
Rifamycins			
Polyenes			
Griseofulvin			
Amantadine			
Zidovudine			
Niclosamide			

2. Define chemotherapeutic agent. Distinguish a synthetic drug from an antibiotic.

3. Discuss the contributions to chemotherapy made by Ehrlich and Fleming.

4. List and explain five criteria used to identify an effective antimicrobial agent.

5. What similar problems are encountered with antiviral, antifungal, antiprotozoan, and antihelminthic drugs?

6. Identify three modes of action of antiviral drugs. Give an example of a currently used antiviral drug for each mode of action.

7. Compare and contrast the broth dilution and disk-diffusion tests. Identify at least one advantage of each.

8. Describe the disk-diffusion test for microbial susceptibility. What information can you obtain from this test?

9. Define drug resistance. How is it produced? What measures can be taken to minimize drug resistance?

10. List the advantages of using two chemotherapeutic agents simultaneously to treat a disease. What problem can be encountered using two drugs?

11. Why does a cell die from the following antimicrobial actions?
   a. Colistimethate binds to phospholipids.
   b. Kanamycin binds to 70S ribosomes.

12. How is translation inhibited by each of the following?
   a. chloroamphenicol
   b. erythromycin
   c. tetracycline
   d. streptomycin

13. Dideoxyinosine (ddI) is an antimetabolite of guanine. The -OH is missing from carbon 3 in ddI. How does ddI inhibit DNA synthesis?

MULTIPLE CHOICE

1. Which of the following pairs is mismatched?
   a. antihelminthic—inhibition of oxidative phosphorylation
   b. antihelminthic—inhibition of cell wall synthesis
   c. antifungal—injury to plasma membrane
   d. antifungal—inhibition of mitosis
   e. antiviral—inhibition of DNA synthesis

2. All of the following are modes of action of antiviral drugs except
   a. inhibition of protein synthesis at 70S ribosomes.
   b. inhibition of DNA synthesis.
   c. inhibition of RNA synthesis.
   d. inhibition of uncoating.
   e. none of the above

3. Which of the following modes of action would not be fungicidal?
   a. inhibition of peptidoglycan synthesis
   b. inhibition of mitosis
   c. injury to the plasma membrane
   d. inhibition of nucleic acid synthesis
   e. none of the above

4. An antimicrobial agent should meet all of the following criteria except
   a. selective toxicity.
   b. the production of hypersensitivities.
   c. a narrow spectrum of activity.
   d. no production of drug resistance.
   e. none of the above

5. The most selective antimicrobial activity would be exhibited by a drug that
   a. inhibits cell wall synthesis.
   b. inhibits protein synthesis.
   c. injures the plasma membrane.
   d. inhibits nucleic acid synthesis.
   e. all of the above

6. Antibiotics that inhibit translation have side effects
   a. because all cells have proteins.
   b. only in the few cells that make proteins.
   c. because eukaryotic cells have 80S ribosomes.
   d. at the 70S ribosomes in eukaryotic cells.
   e. none of the above

7. Which of the following will *not* affect eukaryotic cells?
   a. inhibition of the mitotic spindle
   b. binding with sterols
   c. binding to 80S ribosomes
   d. binding to DNA
   e. none of the above

8. Cell membrane damage causes death because
   a. the cell undergoes osmotic lysis.
   b. cell contents leak out.
   c. the cell plasmolyzes.
   d. the cell lacks a wall.
   e. none of the above

9. A drug that intercalates into DNA has the following effects. Which one leads to the others?
   a. It disrupts transcription.
   b. It disrupts translation.
   c. It interferes with DNA replication.
   d. It causes mutations.
   e. It alters proteins.

10. Chloramphenicol binds to the 50S portion of a ribosome, which will interfere with
    a. transcription in prokaryotic cells.
    b. transcription in eukaryotic cells.
    c. translation in prokaryotic cells.
    d. translation in eukaryotic cells.
    e. DNA synthesis.

CRITICAL THINKING

1. Which of the following can affect human cells? Explain why or why not.
   a. penicillin
   b. indinavir
   c. erythromycin
   d. polymyxin

2. Why is idoxuridine effective if host cells also contain DNA?

3. Some bacteria become resistant to tetracycline because they don't make porins. Why can a porin-deficient mutant be detected by its inability to grow on a medium containing a single carbon source such as succinic acid?

4. The following data were obtained from a disk-diffusion test.

Antibiotic	Zone of Inhibition
A	15 mm
B	0 mm
C	7 mm
D	15 mm

   a. Which antibiotic was most effective against the bacteria being tested?
   b. Which antibiotic would you recommend for treatment of a disease caused by this bacterium?
   c. Was antibiotic A bactericidal or bacteriostatic? How can you tell?

5. Why do you suppose *Streptomyces griseus* produces an enzyme that inactivates streptomycin? Why is this enzyme produced early in metabolism?

6. The following results were obtained from a broth dilution test for microbial susceptibility.

Antibiotic Concentration	Growth	Growth in Subculture
200 μg	−	−
100 μg	−	−
50 μg	−	+
25 μg	+	+

   a. The MIC of this antibiotic is _____.
   b. The MBC of this antibiotic is _____.

CLINICAL APPLICATIONS

1. Penicillin and streptomycin can be used together under certain circumstances, but penicillin and tetracycline cannot be used together. Offer an explanation for this.

2. A patient with a urinary bladder infection took nalidixic acid, but her condition did not improve. Explain why her infection disappeared when she switched to a sulfonamide.

3. A patient with streptococcal sore throat takes penicillin for 2 days of a prescribed 10-day regimen. Because he feels better, he then saves the remaining penicillin for some other time. After 3 more days, he suffers a relapse of the sore throat. Discuss the probable cause of the relapse.

Learning with Technology

MP = **The Microbiology Place website**   **ST** = **Student Tutorial CD-ROM**   **VU** = **VirtualUnknown CD-ROM**

MP   Don't forget to go to The Microbiology Place website (http://www.microbiologyplace.com) to take the practice tests, explore the interactive activity and case study, and check out the news articles and web links for this chapter.

ST   Remember there is also a quiz for this chapter on the Microbiology Interactive Student Tutorial CD-ROM.

VU   Enter the Virtual Lab, click the arrow next to the Session field, click Textbook Exercises, and select Chapter 20. Read the Case Study carefully, identify the unknown, and use what you learn to answer the following questions (consult your textbook for additional information).

1. Table 20.3 on page 555 lists various antibacterial drugs. List the classes of antibiotics, and for each state whether it might be effective for treating this illness. If effective, how are the microbes defeated? If not effective, explain why.

2. Explain what special problems are encountered when using antibiotics to fight infections of the central nervous system.

Microoorganisms and Human Disease

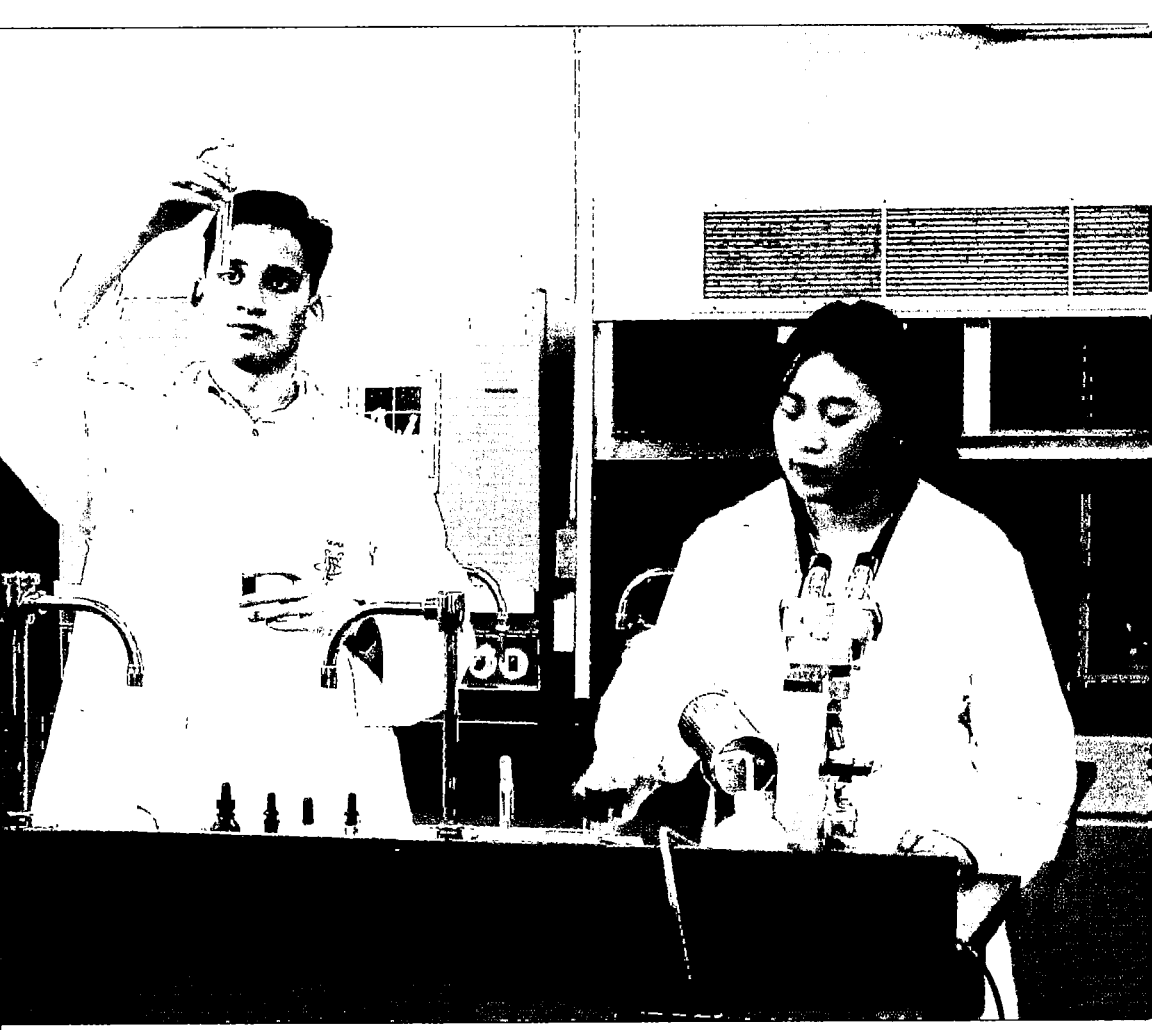

Technicians in a microbiology laboratory use powerful yet simple techniques to identify the microorganisms for illnesses. Although the microbiologist uses several modern, time-saving machines, much of their work involves plating and growing cultures. Here, microbiology students are learning and using the techniques used in a microbiology laboratory to identify microorganisms.

chapter 21

Microbial Diseases of the Skin and Eyes

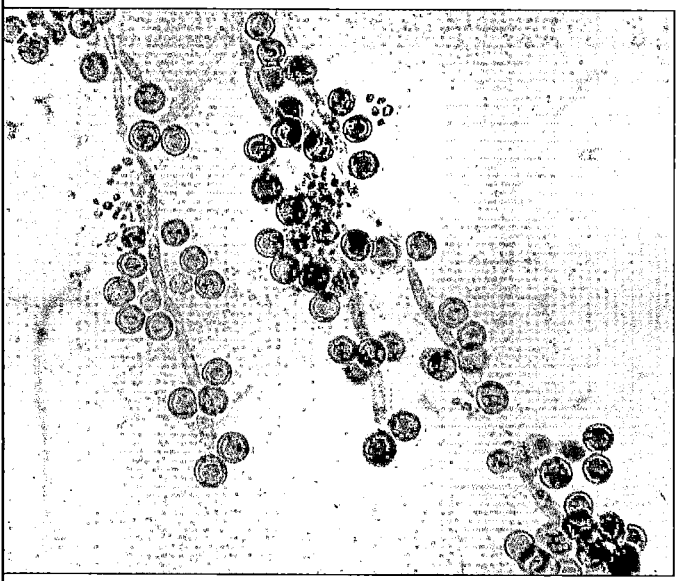

Candida albicans. *This yeast causes infections of the skin and mucous membranes and is a common cause of nosocomial bloodstream infections.*

T he skin, which covers and protects the body, is the body's first line of defense against pathogens. The skin is an inhospitable place for most microorganisms because the secretions of the skin are acidic and most of the skin contains little moisture. Moreover, much of the skin is exposed to radiation, which discourages microbial life. Some parts of the body, however, such as the armpit, have enough moisture to support relatively large bacterial populations. Other regions, such as the scalp, support rather small numbers of microorganisms. The skin is a physical as well as an ecological barrier, and it is almost impossible for pathogens to penetrate it. However, some can enter through skin breaks that are not readily apparent, and the larval forms of a few parasites can penetrate intact skin.

Structure and Function of the Skin

Learning Objective

- *Describe the structure of the skin and mucous membranes and the ways pathogens can invade the skin.*

The skin of an average adult occupies a surface area of about 1.9 m² and varies in thickness from 0.05 to 3.0 mm. As we mentioned in Chapter 16, skin consists of two principal parts, the epidermis and the dermis (Figure 21.1). The **epidermis** is the thin outer portion, composed of several layers of epithelial cells. The outermost layer of the epidermis, the *stratum corneum,* consists of dead cells that contain a waterproofing protein called **keratin.** The epidermis, when unbroken, is an effective physical barrier against microorganisms.

The **dermis** is the inner, relatively thick portion of skin, composed mainly of connective tissue. The hair follicles, sweat gland ducts, and oil gland ducts in the dermis provide passageways through which microorganisms can enter the skin and penetrate deeper tissues.

Perspiration provides moisture and some nutrients for microbial growth. However, it contains salt, which inhibits many microorganisms, and the enzyme lysozyme, which is capable of breaking down the cell walls of certain bacteria.

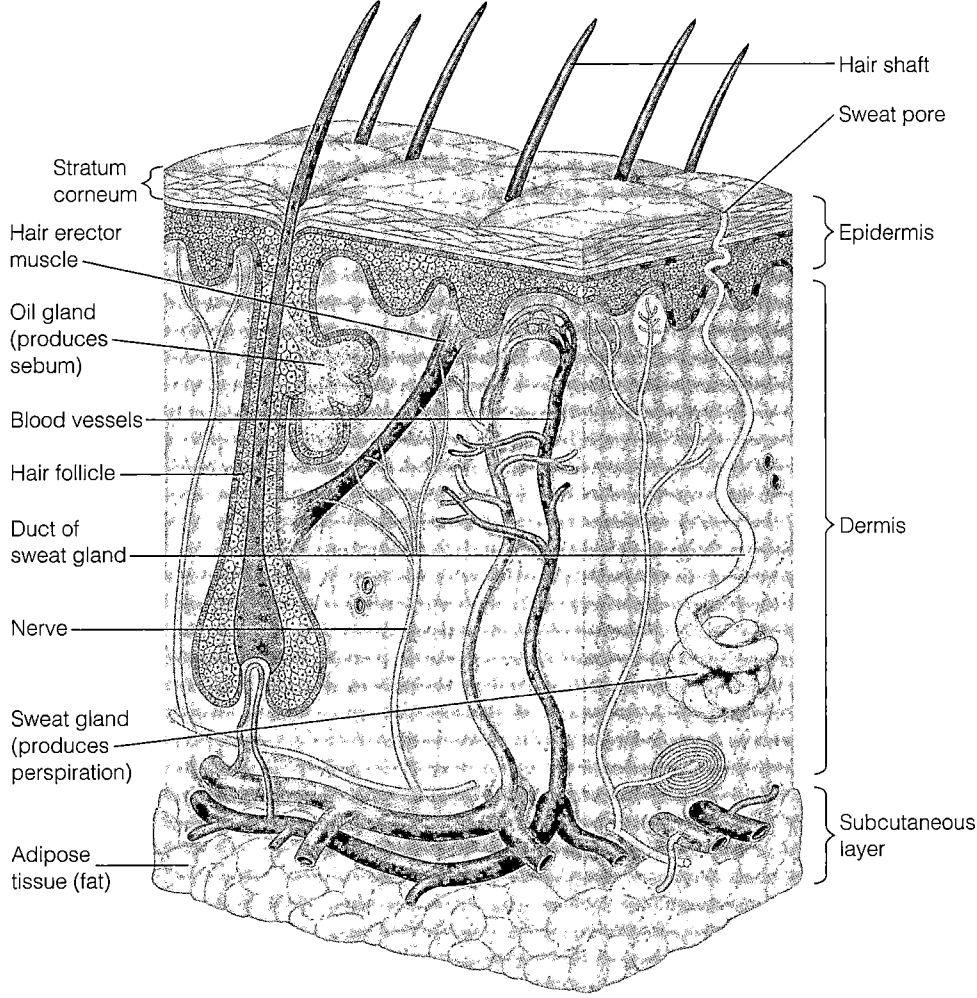

Stratum corneum

Hair erector muscle

Oil gland (produces sebum)

Blood vessels

Hair follicle

Duct of sweat gland

Nerve

Sweat gland (produces perspiration)

Adipose tissue (fat)

Hair shaft

Sweat pore

Epidermis

Dermis

Subcutaneous layer

FIGURE 21.1   The structure of human skin. Notice the passageways between the hair follicle and hair shaft through which microbes can penetrate the deeper tissues. They can also enter the skin through sweat pores.

■ The skin is a large, complex organ that protects the body.

Sebum, secreted by oil glands, is a mixture of lipids (unsaturated fatty acids), proteins, and salts that prevents skin and hair from drying out. Although the fatty acids inhibit the growth of certain pathogens, sebum, like perspiration, is also nutritive for many microorganisms.

Mucous Membranes

In the linings of body cavities, such as those associated with the gastrointestinal, respiratory, urinary, and genital tracts, the outer protective barrier differs from the skin. It consists of sheets of tightly packed *epithelial cells.* These cells are attached at their bases to a layer of extracellular material called the *basement membrane.* Many of these cells secrete mucus—hence the name **mucous membrane,** or **mucosa.** Other mucosal cells have cilia, and in the respiratory system, the mucous layer traps particles, including microorganisms, which the cilia sweep upward out of the body (see Figure 16.4 page 457). Mucous membranes are often acidic, which tends to limit their

microbial populations. Also, the membranes of the eyes are mechanically washed by tears, and the lysozyme in tears destroys the cell walls of certain bacteria. Mucous membranes are often folded to maximize surface area; the total surface area in an average human is about 400 m^2, much more than the surface area of the skin.

Normal Microbiota of the Skin

Learning Objective

■ *Provide examples of normal skin microbiota, and state the general locations and ecological roles of its members.*

Although the skin is generally inhospitable to most microorganisms, it supports the growth of certain microbes, which are established as part of the normal microbiota. On superficial skin surfaces, certain aerobic bacteria produce fatty acids from sebum. These acids inhibit many microbes and allow better-adapted bacteria to flourish.

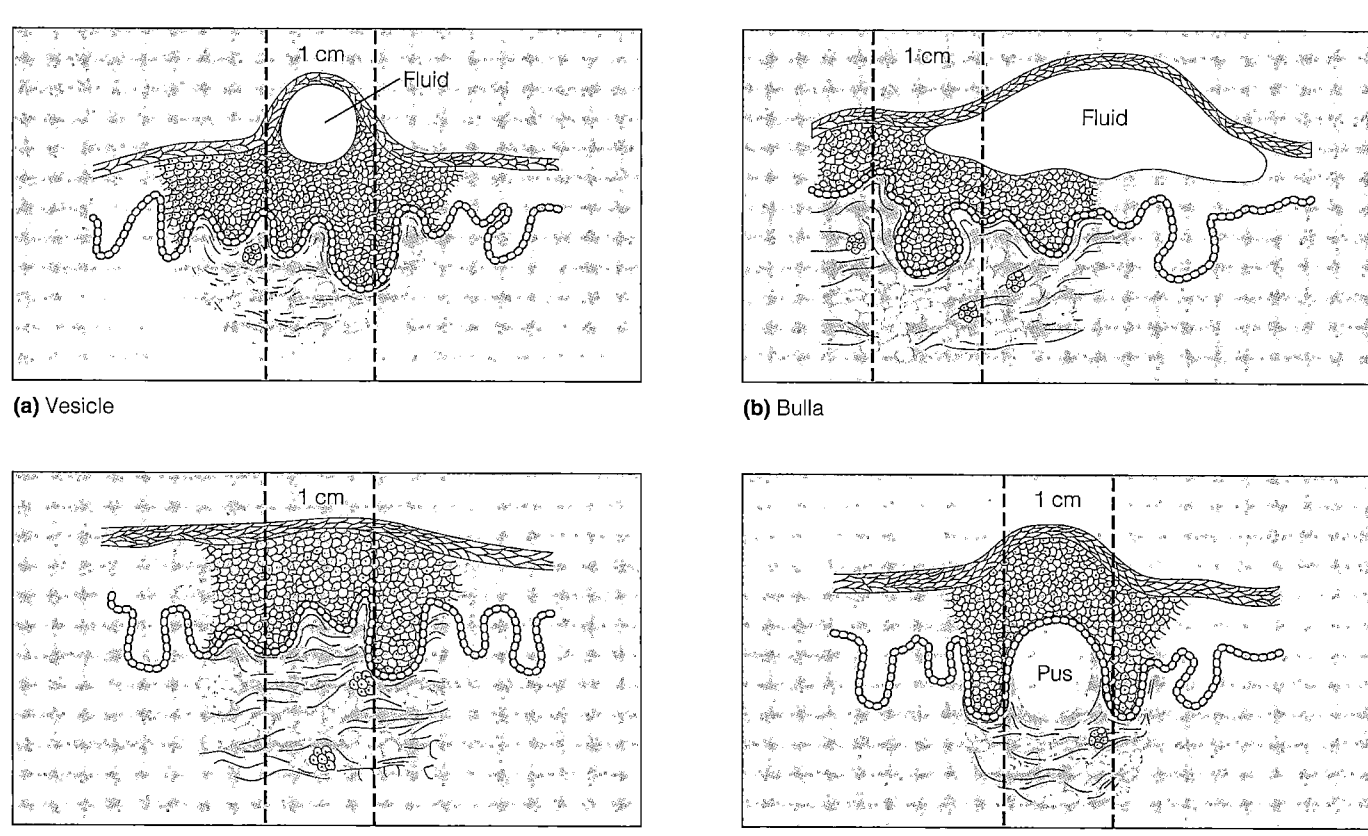

(a) Vesicle

(b) Bulla

(c) Macule

(d) Pustule (papule)

FIGURE 21.2 Skin lesions. (a) Vesicles are small, fluid-filled lesions. **(b)** Bullae are larger fluid-filled lesions. **(c)** Macules are flat lesions that are often reddish. **(d)** Papules are raised lesions; when they contain pus, as shown here, they are called pustules.

🔲 Microorganisms that enter the body through the respiratory tract often manifest themselves by skin lesions.

Microorganisms that find the skin a satisfactory environment are resistant to drying and to relatively high salt concentrations. The skin's normal microbiota contain relatively large numbers of gram-positive bacteria, such as staphylococci and micrococci. Some of these are capable of growth at sodium chloride (table salt) concentrations of 7.5% or more. Scanning electron micrographs show that bacteria on the skin tend to be grouped into small clumps (see Figure 14.1a, page 408). Vigorous washing can reduce their numbers but will not eliminate them. Microorganisms remaining in hair follicles and sweat glands after washing will soon reestablish the normal populations. Areas of the body with more moisture, such as the armpits and between the legs, have higher populations of microbes. These metabolize secretions from the sweat glands and are the main contributors to body odor.

Also part of the skin's normal microbiota are gram-positive pleomorphic rods called *diphtheroids*. Some diphtheroids, such as *Propionibacterium acnes,* are typically anaerobic and inhabit hair follicles. Their growth is sup-

ported by secretions from the oil glands (sebum), which, as we will see, makes them a factor in acne. These bacteria produce propionic acid, which helps maintain the low pH of skin, generally between 3 and 5. Other diphtheroids, such as *Corynebacterium xerosis* (ze-rō'sis), are aerobic and occupy the skin surface. A yeast, *Pityrosporum ovale* (pit-i-ros'pô-rum ō-vä'lē) is capable of growing on oily skin secretions and is thought to be responsible for the scaling skin condition known as dandruff. Shampoos for treating dandruff contain the antibiotic ketoconazole or zinc pyrithione or selenium sulfide. All are active against this yeast.

Microbial Diseases of the Skin

Rashes and lesions on the skin do not necessarily indicate an infection of the skin; in fact, many systemic diseases affecting internal organs are manifested in skin lesions. Variations in these lesions are often useful in describing the symptoms of the disease. For example, small, fluid-filled lesions are **vesicles** (Figure 21.2a). Vesicles larger than

table 21.1   Microbial Diseases of the Skin

	Pathogen	Characteristics	Treatment
Bacterial Diseases			
Impetigo	*Staphylococcus aureus;* occasionally, *Streptococcus pyogenes*	Superficial skin infection; isolated pustules	Penicillin (for *Streptococcus* infections only)
Folliculitis	*Staphylococcus aureus*	Infection of hair follicle	Drain pus; penicillin
Toxic shock syndrome	*Staphylococcus aureus*	Fever, rash, shock	Penicillin
Necrotizing fasciitis	*Streptococcus pyogenes*	Extensive tissue destruction	Surgical removal of tissue; penicillin
Erysipelas	*Streptococcus pyogenes*	Reddish patches on skin; often with high fever	Penicillin
Pseudomonas dermatitis	*Pseudomonas aeruginosa*	Superficial rash	Usually self-limiting
Otitis externa	*Pseudomonas aeruginosa*	Superficial infection of external ear canal	Fluoroquinolones
Acne	*Propionibacterium acnes*	Inflammatory lesions originating with accumulations of sebum that rupture a hair follicle	Benzoyl peroxide, isotretinoin, azelaic acid
Viral Diseases			
Warts	*Papillomavirus* spp.	A horny projection of the skin formed by proliferation of cells	May be removed by liquid nitrogen cryotherapy, electrodesiccation, acids, or lasers
Smallpox (variola)	Smallpox (variola) virus	Pustules that may be nearly confluent on skin; systemic viral infection affects many internal organs	None
Chickenpox (varicella)	Varicella-zoster virus	Vesicles in most cases confined to face, throat, and lower back	Acyclovir for immunocompromised patients
Shingles (herpes-zoster)	Varicella-zoster virus	Vesicles similar to chickenpox; typically on one side of waist, face and scalp, or upper chest	Acyclovir for immunocompromised patients
Herpes simplex	Herpes simplex virus type 1	Most commonly as cold sores—vesicles around mouth; can also affect other areas of skin and mucous membranes	Acyclovir may modify symptoms
Measles (rubeola)	Measles virus	Skin rash of reddish macules first appearing on face and spreading to trunk and extremities	None
Rubella (German measles)	Rubella virus	Mild disease with a rash resembling measles, but less extensive and disappears in 3 days or less	None

about 1 cm in diameter are termed **bullae** (Figure 21.2b). Flat, reddened lesions are known as **macules** (Figure 21.2c). Raised lesions are called **papules,** or **pustules** when they contain pus (Figure 21.2d). Although the focus of infection is often elsewhere in the body, it is convenient to classify these diseases by the organ most obviously affected: the skin.

Table 21.1 summarizes the most important diseases associated with the skin. An emerging skin disease is described in the box on page 584.

table 21.1	Microbial Diseases of the Skin (continued)		
Pathogen		**Characteristics**	**Treatment**
Fifth disease (erythema infectiosum)	Human parvovirus B19	Mild disease with a facial rash	None
Roseola	Human herpesvirus 6	Childhood disease; high fever followed by body rash	None
Fungal Diseases			
Ringworm (tinea)	Microsporum, Trichophyton, Epidermophyton spp.	Skin lesions of highly varied appearance; on scalp may cause local loss of hair	Griseofulvin (orally), miconazole, clotrimazole (topically)
Sporotrichosis	Sporothrix schenckii	Ulcer at site of infection spreading into nearby lymphatic vessels	Potassium iodide solution (orally)
Candidiasis	Candida albicans	Symptoms vary with infection site; usually affects mucous membranes or moist areas of skin	Miconazole, clotrimazole, (topically)
Parasitic Infestation			
Scabies	Sarcoptes scabiei (mite)	Papules due to hypersensitivity reaction to mites	Gamma benzene hexachloride, permethrin (topically)

Bacterial Diseases of the Skin

Learning Objectives

- Differentiate staphylococci from streptococci, and name several skin infections caused by each.
- List the causative agent, mode of transmission, and clinical symptoms of Pseudomonas dermatitis, otitis externa, and acne.

Two genera of bacteria, Staphylococcus and Streptococcus, are frequent causes of skin-related diseases and merit special discussion. We will also discuss these bacteria in subsequent chapters in relation to other organs and conditions. Superficial staphylococcal and streptococcal infections of the skin are very common. The bacteria frequently come into contact with the skin and have adapted fairly well to the physiological conditions there. Both genera also produce invasive enzymes and damaging toxins that contribute to the disease process.

Staphylococcal Skin Infections

Staphylococci are spherical gram-positive bacteria that form irregular clusters like grapes (see Figure 11.22, page 324). This characteristic occurs because the cells divide at random points about their circumference, and the daughter cells do not completely separate from each other (see Figure 4.1d on page 78). For almost all clinical purposes, these bacteria can be divided into those that produce **coagulase,** an enzyme that coagulates (clots) fibrin in

blood, and those that do not (coagulase-positive and coagulase-negative strains respectively). Coagulase-negative strains are very common on the skin, where they may represent 90% of the normal microbiota. They are generally pathogenic only when the skin barrier is broken or is invaded by medical procedures, such as the insertion and removal of catheters into veins. On the surface of the catheter (Figure 21.3), they are surrounded by a slime layer of capsular material that protects them from desiccation and disinfectants.

At one time, coagulase-negative staphylococci were considered to be all one species, Staphylococcus epidermidis. They have since been subdivided into several species, and the name S. epidermidis is applied now only to the predominant species from human skin.

Staphylococcus aureus is the most pathogenic of the staphylococci. Typically, it forms golden-yellow colonies. Almost all pathogenic strains of S. aureus are coagulase-positive. Fibrin clots may protect the microorganisms from phagocytosis and isolate them from other defenses of the host. There is a high correlation between the bacterium's ability to form coagulase and its production of damaging toxins, several of which may injure tissues. Staphylococci may also produce leukocidin, which destroys phagocytic leukocytes, and exfoliative toxin, which is responsible for scalded skin syndrome (discussed shortly). Some staphylococcal toxins, enterotoxins, affect the gastrointestinal tract and will be discussed in Chapter 25 with diseases of the digestive system.

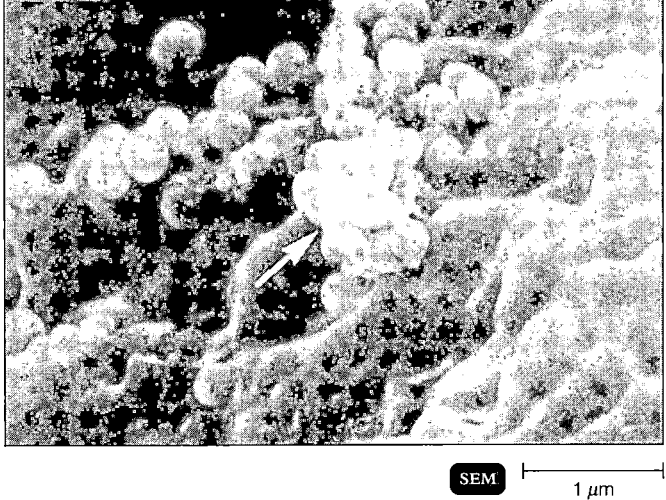

FIGURE 21.3 **Coagulase-negative staphylococci.** These slime-producing bacteria are shown here adhering to the surface of a plastic catheter.

■ **Microbes that are introduced into the body by catheters can cause disease.**

S. aureus is often a problem in the hospital environment. Because *S. aureus* is carried by patients, hospital staff members, and visitors, the danger of infection to surgical wounds and other breaks in the skin is great. And such infections are difficult to treat because *S. aureus* in the hospital environment is exposed to so many antibiotics that it quickly becomes resistant to them. At one time, this organism was almost uniformly extremely susceptible to penicillin, but now only about 10% of *S. aureus* strains are sensitive. You may recall the discussion of vancomycin-resistant *S. aureus* in Chapter 20, considered the last resort in combating penicillin-resistant strains.

The nasal passages provide an especially favorable environment for *S. aureus*, which is often present there in large numbers. In fact, its presence on unbroken skin is often the result of transport from the nasal passages. *S. aureus* often enters the body through a natural opening in the skin barrier, the hair follicle's passage through the epidermal layer. Infections of hair follicles, or **folliculitis,** often occur as pimples. The infected follicle of an eyelash is called a **sty.** A more serious hair follicle infection is the **furuncle (boil),** which is a type of **abscess,** a localized region of pus surrounded by inflamed tissue. Antibiotics do not penetrate well into abscesses, which are difficult to treat. Draining pus from the abscess is frequently a preliminary step to successful treatment.

When the body fails to wall off a furuncle, neighboring tissue can be progressively invaded. The extensive damage is called a **carbuncle,** a hard, round deep inflammation of tissue under the skin. At this stage of infection,

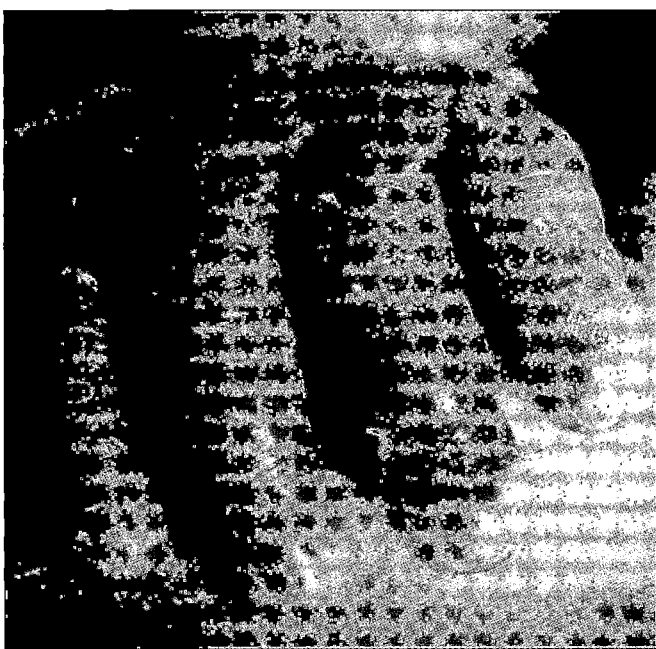

FIGURE 21.4 **Lesions of scalded skin syndrome.** Some staphylococci produce a toxin that causes the skin to peel off in sheets, as on the hand of this infant. It is especially likely to occur in children under age 2.

■ **A lysogenic phage causes *S. aureus* to produce scalded skin toxin.**

the patient usually exhibits the symptoms of generalized illness with fever.

Staphylococci are the primary cause of a very troublesome problem in hospital nurseries, **impetigo of the newborn.** Symptoms of this disease are thin-walled vesicles on the skin that rupture and later crust over. To prevent outbreaks, which can reach epidemic proportions, hexachlorophene-containing skin lotions are commonly prescribed (see Chapter 7).

Staphylococcal infections always carry the risk that the underlying tissue will become infected or that the infection will enter the bloodstream. The circulation of toxins, such as those produced by staphylococci, is called **toxemia.** One such toxin, which is produced by staphylococci lysogenized by certain phage types, causes **scalded skin syndrome.** (See Chapters 13 and 15 for a discussion of lysogeny and phage conversion.) This condition is first apparent as a lesion around the nose and mouth, which rapidly develops into a bright red area and spreads. Within 48 hours, the skin of the affected areas peels off in sheets when it is touched (Figure 21.4). Scalded skin syndrome is frequently observed in children under age 2, especially in newborns, as a complication of staphylococcal infections. These patients are seriously ill, and vigorous antibiotic therapy is required.

CLINICAL PROBLEM SOLVING

An Emerging Necrotizing Disease

You will see questions as you read through this problem. The questions are those that clinicians ask themselves as they proceed through a diagnosis and treatment. Try to answer each question before going on to the next one.

1. A 20-year-old man presented at a clinic in Melbourne, Australia, with a 10 cm ulcer and cellulitis (inflammation of tissue) (see the figure). The man reported feeling a subcutaneous nodule several weeks earlier. The nodule soon disappeared, however, and the area then became red and itchy. Over the following week, the site of the nodule hardened, and the skin ulcer appeared. He became concerned because the ulcer didn't heal and the surrounding skin was sloughing off. He had no recent animal bite or traumatic injury.
What diagnostic test would be done quickly?

2. A swab from the ulcer was used to prepare smears for staining and microscopic examination. Acid-fast staining of one smear revealed red-stained small rods.
What genus does this indicate? What treatment can you suggest?

3. *Mycobacterium* spp. are acid-fast rods. The antituberculosis drugs rifampin and ethambutol were administered. The bacteria grew on culture media and in cell culture at 30–33°C, incubated for 12 weeks.
If chemotherapy fails, can you suggest an alternative treatment?

4. Medical researchers have suggested that use of a device to increase skin temperature to 40°C might inhibit

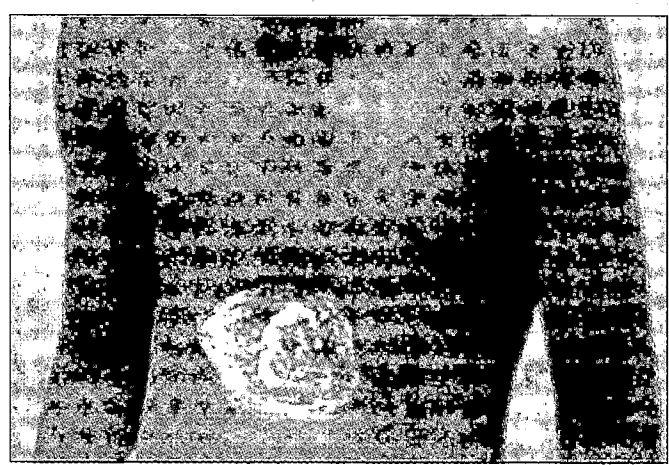

growth of this bacterium. The necrosis extended beyond the area in which the bacteria were growing.
What does the extended area of necrosis indicate?

5. The extensive damage suggests that the bacterium is producing an exotoxin. A study of clinic records revealed that 28 people had been seen for nodules, skin ulcers, or scarring from ulcers.
What type of information do you need to look for?

There is no indication of human-to-human transmission or arthropod transmission. All of the victims lived near a golf course. The bacterium's DNA could be found in soil and water by polymerase chain reaction (PCR), although the bacterium could not be cultured from environmental samples.

The disease described is called Buruli ulcer, named after Buruli, Uganda in Africa. The disease and its etiological agent, *Mycobacterium ulcerans*, were first

described in 1948 in Australia; since then, Buruli ulcer has appeared in 25 countries on five continents: Africa, Asia, Australia, South America, and North America. The areas in which the disease occurs are tropical, and all cases are associated with slow-moving or stagnant water. Buruli ulcer is an emerging disease that is expected to become more prevalent than leprosy, which is also caused by a *Mycobacterium*. The clinical stages of the disease are (1) subcutaneous nodule, (2) inflammation of subcutaneous fat, (3) cellulitis, and (4) scarring. Buruli ulcer is rarely fatal, but the extensive tissue destruction is frequently disfiguring and can result in impaired limb function or loss of the affected limb. The disease is relatively easy to diagnose; however, there is no effective treatment. In 1997, the World Health Organization formed the Global Buruli Ulcer Initiative to investigate and find a method of eradicating the disease.

Scalded skin syndrome is also characteristic of the late stages of **toxic shock syndrome (TSS).** In this potentially life-threatening condition, fever, vomiting, and a sunburnlike rash are followed by shock. TSS originally became known as a result of staphylococcal growth associated with the use of a new type of highly absorbent vaginal tampon; the correlation is especially high for cases in which the tampons remain in place too long. Staphylococcal toxin (mainly toxic shock syndrome toxin-1, or TSST-1) enters the bloodstream from the bacterial growth site in and around the tampon, and its circulation causes the symptoms.

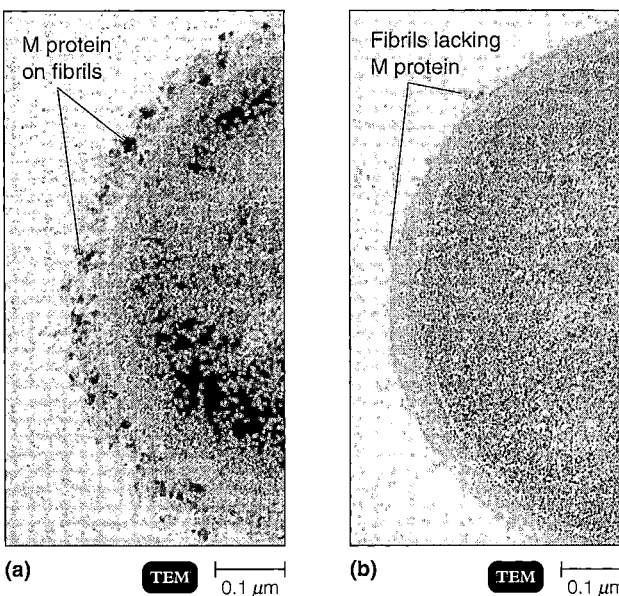

(a) `TEM` |————| 0.1 μm   **(b)** `TEM` |————| 0.1 μm

FIGURE 21.5 The M protein of group A β-hemolytic strepto-cocci. (a) Part of a cell that carries the M protein on a fuzzy layer of surface fibrils. **(b)** Part of a cell that lacks the M protein.

◘ The M protein is important to the pathogenicity of certain streptococci, especially in providing protection from phagocytosis.

Today only little more than half of the cases of TSS are associated with menstruation. Nonmenstrual TSS occurs from staphylococcal infections that follow nasal surgery in which absorbent packing is used, after surgical incisions, and in women who have just given birth.

Streptococcal Skin Infections

Streptococci are gram-positive spherical bacteria. Unlike staphylococci, streptococcal cells usually grow in chains (see Figure 11.21, page 324). Prior to division, the individual cocci elongate on the axis of the chain, and then the cells divide (see Figure 4.1a on page 78). Streptococci cause a wide range of disease conditions beyond those covered in this chapter, including meningitis, pneumonia, sore throats, otitis media, endocarditis, puerperal fever, and even dental caries.

As streptococci grow, they secrete toxins and enzymes, virulence factors that vary with the different streptococcal species. Among these toxins are *hemolysins,* which lyse red blood cells. Depending on the hemolysin they produce, streptococci are categorized as α-hemolytic, β-hemolytic, and γ-hemolytic (actually nonhemolytic) streptococci (see Figure 6.8 on page 168). Hemolysins can lyse not only red blood cells, but almost any type of cell. It is uncertain, though, just what part they play in streptococcal pathogenicity.

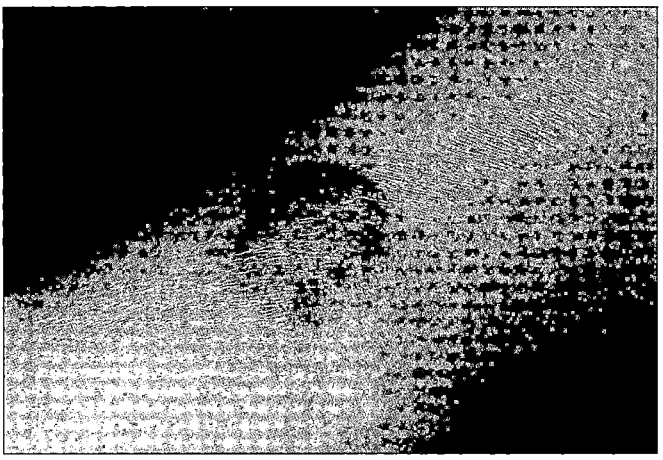

FIGURE 21.6 Lesions of erysipelas, caused by group A β-hemolytic streptococcal toxins.

■ Toxins produced by bacteria can cause reddening of the skin.

Beta-hemolytic streptococci are often associated with human disease. This group is further differentiated into serological groups, designated A through T, according to antigenic carbohydrates in their cell walls. The group A β-hemolytic streptococci are the most important. This group of pathogenic streptococci can be further subdivided into over 80 immunological types according to the antigenic properties of the M protein found in some strains (Figure 21.5). This protein is external to the cell wall on a fuzzy layer of fibrils. The M protein has antiphagocytic properties that greatly contribute to the strain's pathogenicity. It also appears to aid the bacteria in adhering to and colonizing mucous membranes.

The terms group A β-hemolytic streptococci and *Streptococcus pyogenes* are synonymous. When *S. pyogenes* infects the dermal layer of the skin, it causes a serious disease, **erysipelas.** In this disease, the skin erupts into reddish patches with raised margins (Figure 21.6). It can progress to local tissue destruction, and even enter the bloodstream, causing septicemia (page 626). The infection usually appears first on the face and often has been preceded by a streptococcal sore throat. High fever is common. Fortunately, *S. pyogenes* has remained sensitive to β-lactam-type antibiotics.

S. pyogenes, like the staphylococci, can cause the local infection **impetigo.** This is most common in toddlers and children of grade-school age. Streptococcal impetigo is characterized by isolated pustules that become crusted and rupture (Figure 21.7). The disease is spread mostly by contact, and the bacteria penetrate the skin through some minor abrasion or insect bite. Staphylococci are often

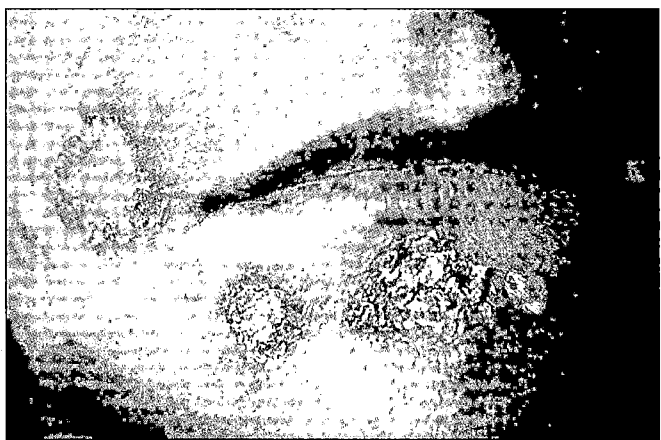

FIGURE 21.7 Lesions of impetigo. This disease is characterized by isolated pustules that become crusted. Impetigo in older children, as depicted here, is usually caused by streptococci.

■ **What other bacterial species also cause impetigo?**

FIGURE 21.8 Necrotizing fasciitis due to group A streptococcus.

■ **Differentiate group A β-hemolytic streptococci from invasive group A streptococci.**

found in this type of impetigo, and medical opinions differ as to whether they are the primary cause or only secondary invaders.

Streptococcal skin infections are generally localized, but if the bacteria reach deeper tissue, they can be highly destructive. They produce substances that promote the rapid spread of infection. Among these are *streptokinases* (enzymes that dissolve blood clots), *hyaluronidase* (an enzyme that dissolves hyaluronic acid, the cementing substance of connective tissue), *deoxyribonucleases* (enzymes that degrade DNA), and several *proteases*. Also produced are *erythrogenic toxins* (responsible for the red rash of scarlet fever). Some 15,000 cases of invasive group A streptococcal infection, caused by the so-called flesh-eating bacteria, occur each year in the United States (Figure 21.8). The infection may destroy tissue as rapidly as a surgeon can cut it out, and mortality rates can exceed 40%. The streptococci attack solid tissue (*cellulitis*), muscle (*myositis*), or the muscle covering (*necrotizing fasciitis*). An important factor seems to be an *exotoxin,* exotoxin A, which acts as a superantigen, causing the immune system to contribute to the damage (see Chapter 19).

Infections by Pseudomonads

Pseudomonads are aerobic gram-negative rods that are widespread in soil and water. Capable of surviving in any moist environment, they can grow on traces of even unusual organic matter, such as soap films or cap liner adhesives, and are resistant to many antibiotics and disinfectants. The most prominent species is *Pseudomonas aeruginosa,* which is considered a model of an opportunistic pathogen (see pages 310-311).

Pseudomonads frequently cause outbreaks of *Pseudomonas* **dermatitis.** This is a self-limiting rash of about two weeks' duration, often associated with swimming pools and pool-type saunas and hot tubs. When many people use these facilities, the alkalinity rises and the chlorines become less effective; at the same time, the concentration of nutrients that support the growth of pseudomonads increases. Hot water causes hair follicles to open wider, facilitating the entry of bacteria. Competition swimmers are often troubled with **otitis externa,** or swimmer's ear, a painful pseudomonad infection of the external ear canal leading to the eardrum.

P. aeruginosa produces several exotoxins that account for much of its pathogenicity. It also has an endotoxin. Except for superficial skin infections and otitis externa, infection by *P. aeruginosa* is rare in healthy people. However, it can cause opportunistic respiratory infections in people already compromised by immunological deficiencies (natural or drug-induced) or by chronic pulmonary disease, especially cystic fibrosis. (Respiratory infections are discussed in Chapter 24.)

P. aeruginosa is also a very common and serious opportunistic pathogen in burn patients, particularly those with second- and third-degree burns. Infection may produce blue-green pus, whose color is caused by the bacterial pigment **pyocyanin.** Of concern in many hospitals is the ease with which *P. aeruginosa* grows in flower vases, mop water, and even dilute disinfectants.

The relative resistance to antibiotics that characterizes pseudomonads is still a problem (see Chapter 20). However, in recent years, several new antibiotics have been developed, and chemotherapy to treat these infections is not as restricted as it once was. The fluoroquinolones and the newer, antipseudomonal β-lactam antibiotics are the usual drugs of choice. Silver sulfadiazine is very useful in the treatment of burn infections by *P. aeruginosa*.

Acne

Acne is probably the most common skin disease in humans, affecting an estimated 17 million people in the United States. More than 85% of all teenagers have the problem to some degree. Many suffer from an unusually severe acne, called **cystic acne.** In these cases, acne progresses to form inflamed pustules (cysts) and subsequent pitted scarring of the face and upper body. Although the occurrence of acne decreases after the teenage years, the scarring from these severe cases often remains.

Acne begins when channels for the passage of sebum to the skin surface are blocked. As sebum accumulates, whiteheads form; if the blockage breaks through the skin, lesions known as blackheads form. Bacteria, especially *Propionibacterium acnes*, an anaerobic diphtheroid commonly found on the skin, become involved at this stage. *P. acnes* has a nutritional requirement for glycerol in sebum; in metabolizing the sebum, it forms free fatty acids that cause an inflammatory response. Neutrophils that secrete enzymes which damage the wall of the hair follicle are attracted. This inflammation leads to the appearance of pustules and subsequent acne scars. Cosmetics, especially oil-based cosmetics, frequently aggravate the condition, but diet—including the consumption of chocolate—has been demonstrated to have no significant effect on the disease.

The familiar nonprescription acne treatments containing benzoyl peroxide are effective against bacteria, especially *P. acnes,* and also cause drying that helps loosen plugged follicles. Benzoyl peroxide is also available as a gel, benzamycin, where it is combined with the antibiotic erythromycin; the combination is more effective than either compound used alone. A recently introduced cream preparation containing azelaic acid is also effective against *P. acnes.* It reduces the formation of blackheads, but in dark-skinned users it may cause unwanted lightening of skin pigmentation. People with severe, cystic acne should consult a dermatologist for treatment rather than rely on over-the-counter remedies. Antibiotics, administered orally or topically, are often part of a dermatologist's treatment of acne.

Tretinoin (Retin-A®) is effective in preventing and clearing the lesions of acne. Because it inactivates benzoyl peroxide, the two preparations must be applied at different times of the day. However, tretinoin is synergistic when used in combination with azelaic acid. The most important development in treating cystic acne is isotretinoin (Accutane®). Taken orally, this derivative of vitamin A inhibits sebum formation, and dramatic improvement follows. Isotretinoin is not recommended to treat mild cases of acne because of its adverse effects. It is especially important to know that it is *teratogenic*—will cause damage to the fetus—if taken by a pregnant woman for even a few days.

Viral Diseases of the Skin

Learning Objective

- *List the causative agent, mode of transmission, and clinical symptoms of these skin infections: warts, smallpox, chickenpox, shingles, cold sores, measles, rubella, fifth disease, and roseola.*

Many viral diseases, although systemic in nature and transmitted by respiratory or other routes, are most apparent by their effects on the skin.

Warts

Warts, or papillomas, are generally benign skin growths caused by viruses. It was long known that warts can be transmitted from one person to another by contact, even sexually, but it was not until 1949 that viruses were identified in wart tissues. More than 50 types of papillomavirus are now known to cause different kinds of warts, often with greatly varying appearances.

After infection, there is an incubation period of several weeks before the warts appear. The most common medical treatments for warts are to apply extremely cold liquid nitrogen (cryotherapy), dry them with an electrical current (electrodesiccation), and burn them with acids. Topical application of prescription drugs such as podophyllum and podofilox is often effective. Warts that do not respond to any other treatments can be treated with injected interferon or lasers. The use of lasers results in a virus-laden aerosol. Physicians using lasers to remove warts have contracted warts themselves, especially in their nostrils.

The incidence of genital warts, discussed separately in Chapter 26, has reached epidemic proportions. Although warts are not a form of cancer, some skin and cervical cancers are associated with papillomaviruses.

Smallpox (Variola)

During the Middle Ages, an estimated 80% of the population of Europe contracted **smallpox** at some time during their lives. Those who recovered from the disease retained disfiguring scars. The disease, introduced by American colonists, was even more devastating to Native Americans who had had no previous exposure and thus little resistance.

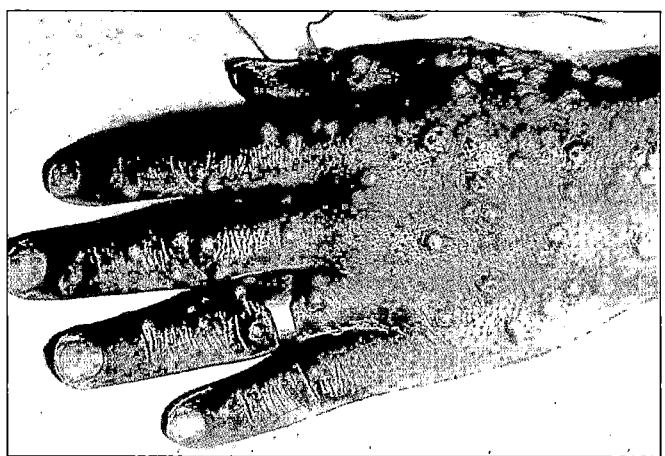

FIGURE 21.9 Smallpox lesions. In some severe cases, the lesions nearly run together.

■ Smallpox was eradicated in 1977.

Smallpox is caused by a poxvirus known as the smallpox (variola) virus. There are two basic forms of this disease: **variola major,** with a mortality rate of 20% or higher, and **variola minor,** with a mortality rate of less than 1%. (Variola minor first appeared around 1900.)

Transmitted first by the respiratory route, the viruses infect many internal organs before their eventual movement into the bloodstream leads to infection of the skin and the production of more recognizable symptoms. The growth of the virus in the epidermal layers of the skin causes lesions (Figure 21.9).

Smallpox was the first disease to which immunity was artificially induced (see Chapters 1 and 18) and the first to be eradicated from the human population. The last victim of a natural case of smallpox is believed to be an individual who recovered from variola minor in 1977 in Somalia. The eradication of smallpox was possible because an effective vaccine was developed, and because there are no animal host reservoirs for the disease. A concerted worldwide vaccination effort was coordinated by the World Health Organization.

In recent years, the smallpox virus collections in laboratories have been the most likely sources of new infections. The risk of such infection is not merely a hypothetical concern, as there have already been several laboratory-associated infections, one of which was fatal. Today, only two sites are known to maintain the smallpox virus, one in the United States and one in Russia. Dates for the destruction of these collections have been set and then postponed.

Chickenpox (Varicella) and Shingles (Herpes Zoster)

Chickenpox (varicella) is a relatively mild childhood disease. After gonorrhea and genital infections by chlamydia, it is the third most common reportable infectious disease in the United States. Surveys indicate that 95% of the population of the United States has been infected. The mortality rate from chickenpox is very low; even so, about 100 deaths occur annually, usually from complications such as encephalitis (infection of the brain) or pneumonia. Because the disease tends to be more serious if contracted in later life, almost half of these deaths occur in adults.

Chickenpox (Figure 21.10a) is the result of an initial infection with the herpesvirus varicella-zoster. (The official, but less used, name is human herpesvirus 3 [see Chapter 13].) We will encounter other instances in which both the vernacular and the official names of herpesviruses will be stated. The disease is acquired by entry of the virus into the respiratory system, and the infection localizes in skin cells after about 2 weeks. The infected skin is vesicular for 3–4 days. During that time, the vesicles fill with pus, rupture, and form a scab before healing. Lesions are mostly confined to the face, throat, and lower back but can also occur on the chest and shoulders. Chickenpox in adults, which does not occur frequently because the high incidence in childhood grants immunity to most individuals, is a more severe disease with a significant mortality rate. If varicella infection occurs during early pregnancy, serious fetal damage may occur in about 2% of cases.

Reye's syndrome is an occasional severe complication of chickenpox, influenza, and some other viral diseases. A few days after the initial infection has receded, the patient persistently vomits and exhibits signs of brain dysfunction, such as extreme drowsiness or combative behavior. Coma and death can follow. At one time, the death rate of reported cases approached 90%, but this rate has been declining with improved care and is now 30% or lower when the disease is recognized and treated in time. Survivors may show neurological damage, especially if very young. Reye's syndrome affects children and teenagers almost exclusively. The use of aspirin to lower fevers in chickenpox and influenza increases the chances of acquiring Reye's syndrome.

Like all herpesviruses, a characteristic of varicella-zoster virus is its ability to remain latent within the body. Following a primary infection, the virus enters the peripheral nerves and moves to a central nerve ganglion (a group of nerve cells lying outside the central nervous system), where it persists as viral DNA (see Figure 21.10a). Humoral antibodies cannot penetrate into the nerve cell,

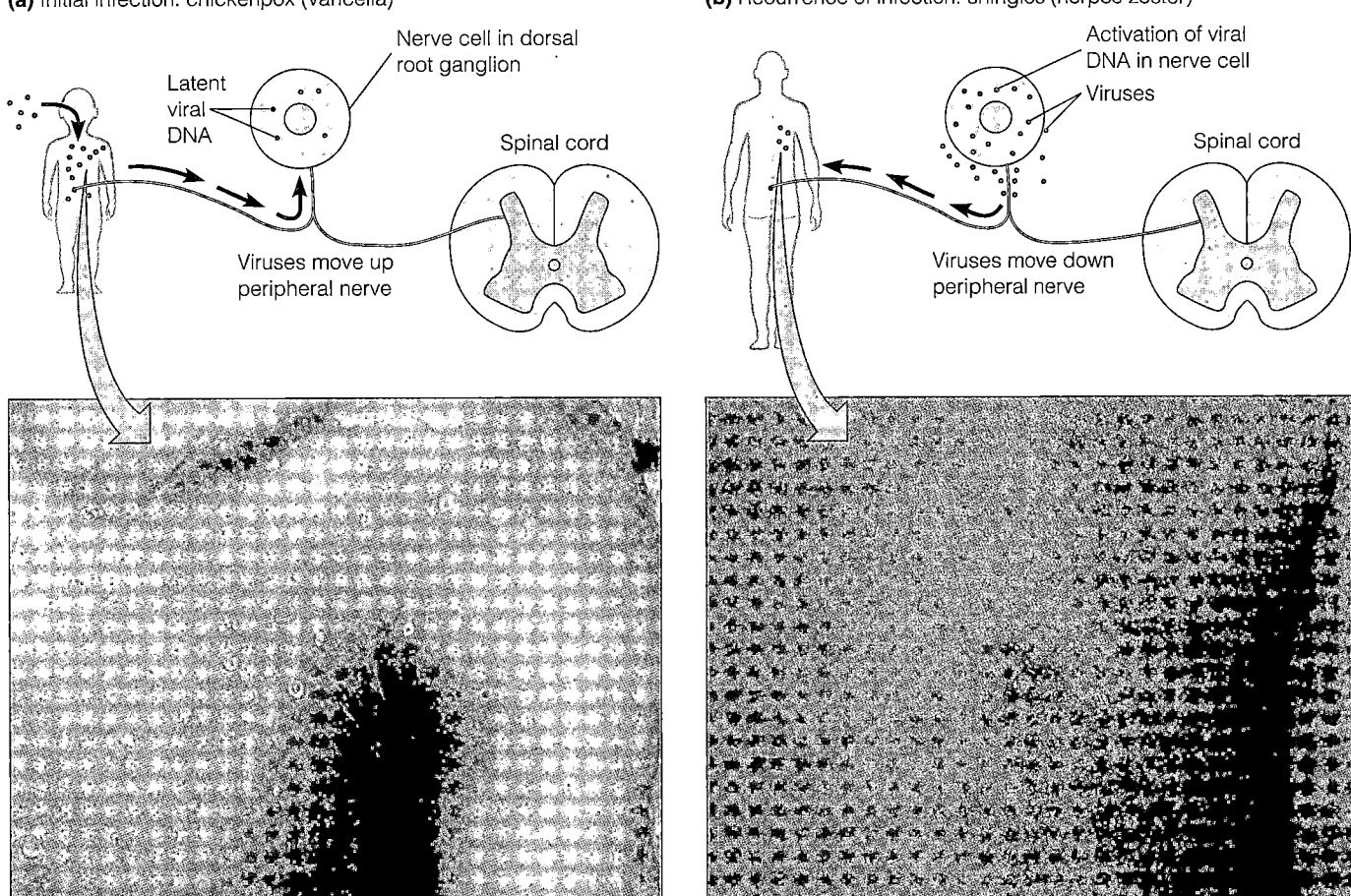

(a) Initial infection: chickenpox (varicella)

Nerve cell in dorsal root ganglion

Latent viral DNA

Spinal cord

Viruses move up peripheral nerve

(b) Recurrence of infection: shingles (herpes zoster)

Activation of viral DNA in nerve cell

Viruses

Spinal cord

Viruses move down peripheral nerve

FIGURE 21.10 Chickenpox and shingles, caused by the herpesvirus varicella-zoster.
(a) Initial infection with the virus, usually during childhood, causes chickenpox, characterized by the formation of the lesions shown here. The virus then moves to a dorsal root ganglion, near the spine, where it remains latent indefinitely. **(b)** Later, usually in adulthood, immune system depression or stress can trigger a reactivation of the virus, causing shingles. The typical lesions shown here are on a patient's back.

◾ **Viral diseases of the skin sometimes become latent in the nervous system and can reappear years later in a different form.**

and because no viral antigens are expressed on the surface of the nerve cell, cytotoxic T cells are not activated. Therefore, neither arm of the specific immune system disturbs the latent virus.

Latent varicella–zoster virus is located in the dorsal root ganglion near the spine. Later, perhaps as long as decades later, the virus may be reactivated (Figure 21.10b). The trigger can be stress or simply the lower immune competence associated with aging. The virions produced by the reactivated DNA move along the peripheral nerves to the cutaneous sensory nerves of the skin, where they cause a new outbreak of the virus in the form of **shingles** (herpes zoster).

In shingles, vesicles similar to those of chickenpox occur but are localized in distinctive areas. Typically, they are distributed about the waist, although facial shingles and infections of the upper chest and back also occur (see Figure 21.10b). The infection follows the distribution of the affected cutaneous sensory nerves and is usually limited to one side of the body at a time because these nerves are unilateral. Occasionally, such nerve infections can result in nerve damage that impairs vision or even causes paralysis. Severe pain is also frequently reported.

Shingles is simply a different expression of the virus that causes chickenpox; it expresses differently because the patient, having had chickenpox, now has partial immunity to the virus. Exposing children to shingles has led to their contracting chickenpox. Shingles seldom occurs in people under age 20, and by far the highest incidence is among older adults.

FIGURE 21.11   Cold sores, or fever blisters, caused by herpes simplex virus.

■ **Why can cold sores reappear, and why do they recur in the same place?**

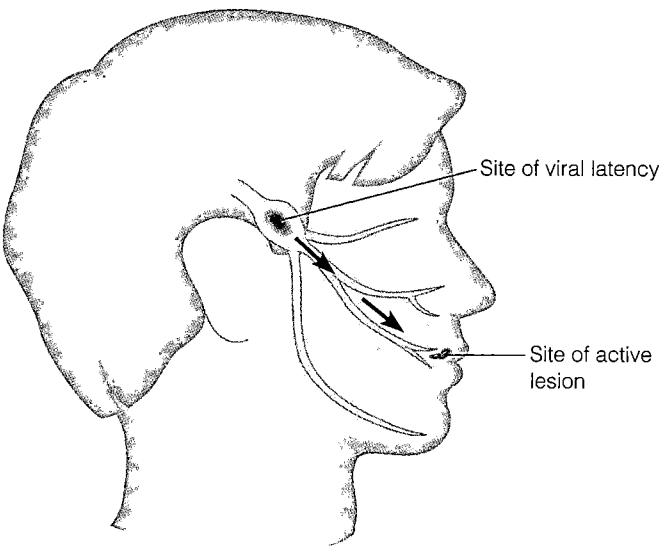

FIGURE 21.12   Site of latency of human herpesvirus type 1 in the trigeminal nerve ganglia.

■ **Why do you think this nerve system is termed trigeminal?**

Immunocompromised patients are in serious danger from infection by varicella–zoster virus; multiple organs become infected, and a mortality rate of 17% is common. In such cases, the antiviral drug acyclovir has proven effective. A live, attenuated varicella–zoster vaccine was licensed for use in 1995. Experience since that time indicates that the vaccine provides good long-term protection.

Herpes Simplex

Herpes simplex viruses (HSV) can be separated into two identifiable groups, HSV-1 and HSV-2. The name herpes simplex virus, used here, is the common or vernacular name. The official names are human herpesvirus 1 and 2. HSV-1 is transmitted primarily by oral or respiratory routes, and infection usually occurs in infancy. Serological surveys show that about 90% of the U.S. population has been infected. Frequently, this infection is subclinical, but many cases develop lesions known as **cold sores** or **fever blisters** (Figure 21.11). Usually occurring in the oral mucous membrane, these lesions heal as the infection subsides. However, HSV-1 usually remains latent in the trigeminal nerve ganglia communicating between the face and the central nervous system (Figure 21.12). Recurrences can be triggered by events such as excessive exposure to UV radiation from the sun, emotional upsets or the hormonal changes associated with menstruation. An important complication of these infections is herpetic keratitis, in which the cornea of the eye becomes infected. This condition is discussed in more detail later in this chapter.

HSV-1 infection can be transmitted by skin contact among wrestlers; this is colorfully termed **herpes gladiatorum.** Incidence as high as 3% has been reported among high school wrestlers.

A very similar virus, herpes simplex virus type 2 (HSV-2), is transmitted primarily by sexual contact. It is the usual cause of genital herpes (see Chapter 26). HSV-2 is differentiated from HSV-1 by its antigenic makeup and by its effect on cells in tissue culture. It is latent in the sacral nerve ganglia found near the base of the spine, a different location than that of HSV-1.

Very rarely, either type of the herpes simplex virus may spread to the brain, causing **herpes encephalitis.** Infections by HSV-2 are more serious, with a fatality rate as high as 70% if untreated. Only about 10% of survivors can expect to lead healthy lives. When administered promptly, acyclovir often cures such encephalitis. Even so, the mortality rate in certain outbreaks was still 28%, and only 38% of the survivors escaped serious neurological damage.

Measles (Rubeola)

Measles is an extremely contagious viral disease that is spread by the respiratory route. Because a person with measles is infectious before symptoms appear, quarantine is not an effective measure of prevention.

Humans are the only reservoir for measles; therefore, it can potentially be eradicated, much as smallpox was. Since the licensing of a vaccine in 1963, the number of

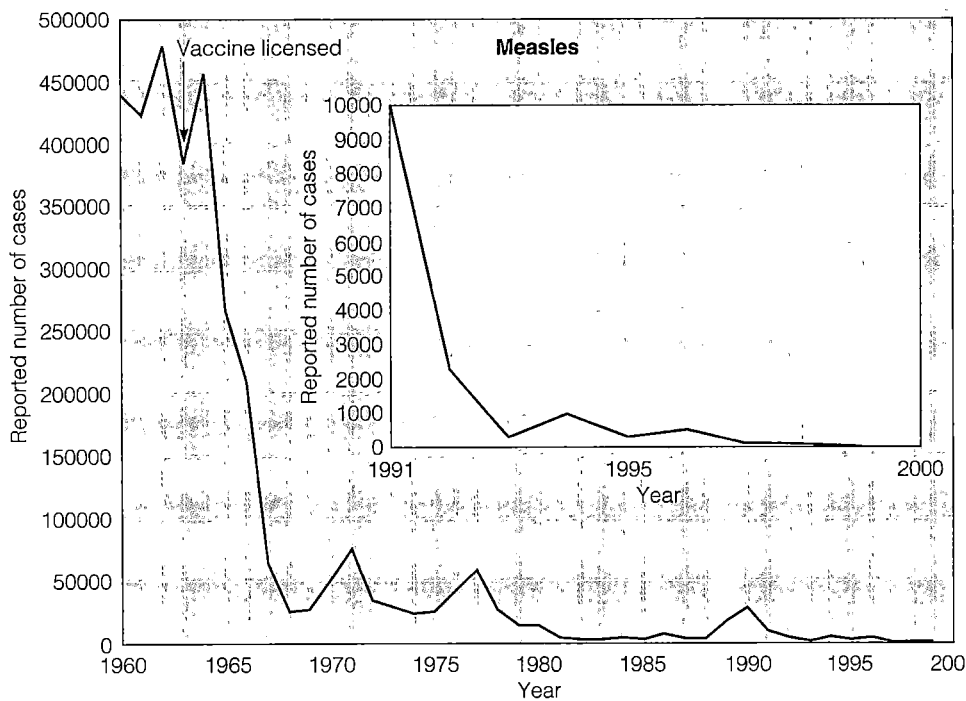

FIGURE 21.13 Reported numbers of measles cases in the United States, 1960–1999. Notice the sharp decline in cases after introduction of the measles vaccine in 1963. Inset: Measles is at its lowest reported level. However, previous periods of low activity have been followed by resurgences.
[SOURCES: CDC, *Summary of Notifiable Diseases 1998, MMWR 47(53) (12/31/99); MMWR 48(51)(1/7/00).*]

■ These graphs make dramatically clear the effectiveness of vaccination in controlling a disease.

measles cases in the United States has declined from more than a reported 400,000 cases per year (5 million were estimated; Figure 21.13). Currently, little more than 100 cases are reported each year and progress toward the goal of eliminating measles in the Western Hemisphere is good.

Despite laws requiring immunization for school entry, the immunization rate is low in some densely populated inner-city groups, and about 40% of measles cases occur in preschool children in these groups. A series of measles outbreaks occurred in such groups around 1990 (see Figure 21.13). In addition, although the vaccine is about 95% effective, cases continue to occur among those who do not develop or retain good immunity. Some of these infections are caused by contact with infected people who come from outside the United States.

An unexpected result of the measles vaccine is that many cases of measles today occur in children under 1 year of age. Measles is especially hazardous to infants; they are more likely to have serious complications. In prevaccination days, measles was rare at this age because infants were protected by maternal antibodies derived from their mothers' recovery from the disease. Unfortunately, antibodies made in response to the vaccine are not as effective in providing protection as are antibodies made in response to the disease. Because the vaccine is not effective when administered in early infancy, the initial vaccination is not made before 12 months. Therefore, the child is vulnerable for a significant time.

The development of rubeola is similar to that of smallpox and chickenpox. Infection begins in the upper respiratory system. After an incubation period of 10–12 days, symptoms develop resembling those of a common cold. Soon, a macular rash appears, beginning on the face and spreading to the trunk and extremities (Figure 21.14). Lesions of the oral cavity include *Koplik's spots*, tiny red patches with central white specks, on the oral mucosa opposite the molars. The presence of Koplik's spots is a diagnostic indicator of the disease.

Measles is an extremely dangerous disease, especially in very young and very old people. It is frequently complicated by middle ear infections or pneumonia caused by the virus itself or by a secondary bacterial infection. Encephalitis strikes approximately 1 in 1000 measles victims; its survivors are often left with permanent brain damage. As many as 1 in 3000 cases is fatal, mostly in infants. The virulence of the virus seems to vary with different epidemic outbreaks. A rare complication of measles (about 1 in 1,000,000 cases) is subacute sclerosing panencephalitis. Occurring mostly in males, it appears about 1–10 years after recovery from measles. Severe neurological symptoms result in death within a few years.

Rubella

Rubella, or German measles, is a much milder viral disease than rubeola (measles) and often goes undetected. A macular rash of small red spots and a light fever are the

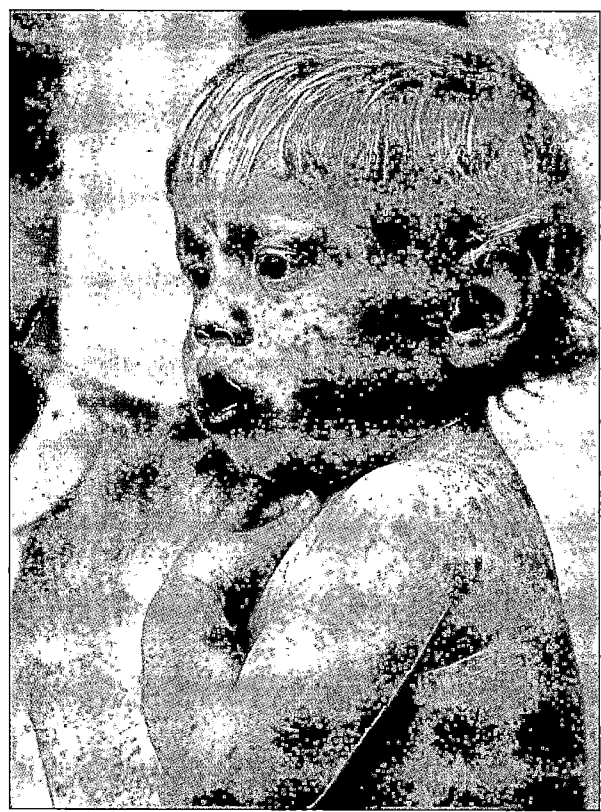

FIGURE 21.14 The rash of small raised spots typical of measles (rubeola). The rash usually begins on the face and spreads to the trunk and extremities.

■ **Why is it possible to eradicate measles?**

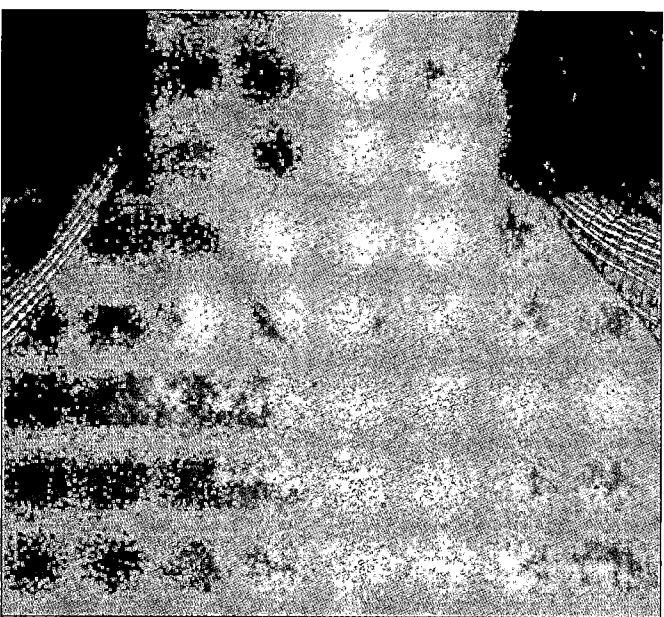

FIGURE 21.15 The rash of red spots characteristic of rubella. The spots are not raised above the surrounding skin.

■ **What is congenital rubella syndrome?**

usual symptoms (Figure 21.15). Complications are rare, especially in children, but encephalitis occurs in about 1 case in 6000, mostly in adults. Transmission is by the respiratory route, and an incubation time of 2–3 weeks is the norm. Recovery from clinical or subclinical cases appears to give a firm immunity.

The seriousness of rubella was not appreciated until 1941, when the association was made between certain severe birth defects and maternal infection during the first trimester (3 months) of pregnancy, a condition called **congenital rubella syndrome.** If a pregnant woman contracts the disease during this time, there is about a 35% incidence of serious fetal damage, including deafness, eye cataracts, heart defects, mental retardation, and death. Some 15% of babies with congenital rubella syndrome die during their first year. The last major epidemic of rubella in the United States was during 1964 and 1965. About 20,000 severely impaired children were born during this epidemic.

It is therefore important to identify women of childbearing age who are not immune to rubella. In some states, the blood test required for a marriage license includes a test for rubella antibodies. Serum antibody can be assayed by a number of commercially available laboratory tests. Accurate diagnosis of immune status always requires such tests; histories alone are unreliable.

In addition to this surveillance, a rubella vaccine was introduced in 1969; in 1979, a more effective version, which produces fewer adverse effects, replaced it. Follow-up studies indicate that more than 90% of vaccinated individuals are protected for at least 15 years. Because of these preventive measures, fewer than 10 cases of congenital rubella syndrome are now reported every year.

The vaccine is not recommended for pregnant women. However, in hundreds of cases in which women were vaccinated 3 months before or 3 months after their presumed date of conception, no case of congenital rubella syndrome defects has occurred. Individuals with an impaired immune system should not receive live vaccine of any disease.

Other Viral Rashes

Fifth Disease (Erythema Infectiosum) Parents with young children are often baffled by a diagnosis of fifth disease, which they have never heard of before. The name derives from a 1905 list of skin rash diseases: measles, scarlet fever, rubella, Filatow-Dukes disease (a mild form of scarlet fever), and the fifth disease on the list. **Fifth dis-**

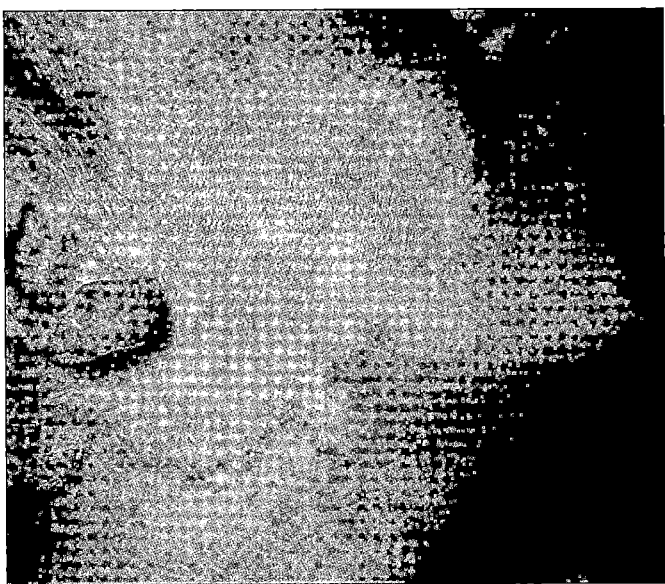

(a) Ringworm

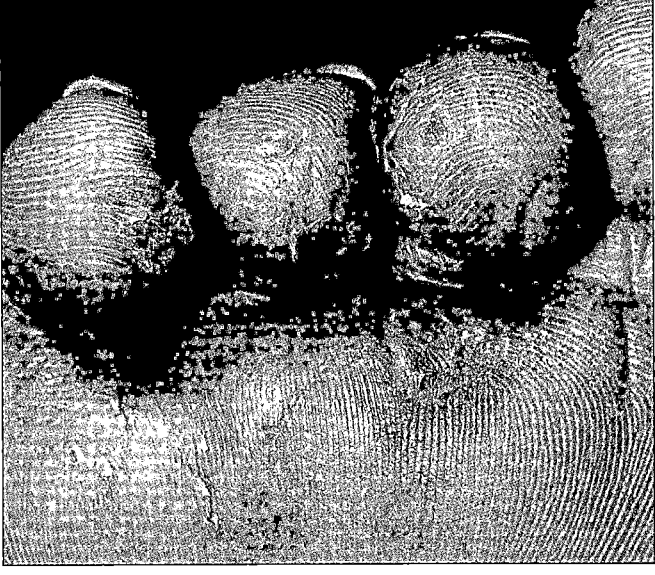

(b) Athlete's foot

FIGURE 21.16 Dermatomycoses.

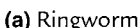

■ Dermatophytes possess the enzyme keratinase.

ease, or **erythema infectiosum,** produces no symptoms at all in about 20% of those infected by the virus (human parvovirus B19, first identified in 1989). Symptoms are similar to a mild case of influenza, but there is a distinctive "slapped-cheek" facial rash that slowly fades. In adults who missed an immunizing infection in childhood, the disease may cause anemia, an episode of arthritis, or, rarely, miscarriage.

Roseola Roseola is a mild childhood disease that is very common. The child has a high fever for a few days, which is followed by a rash over much of the body lasting for a day or two. Recovery leads to immunity. The pathogen is a herpesvirus, human herpesvirus 6 (HHV-6). The virus was first identified in 1988 and is present in the saliva of 85% of adults.

Fungal Diseases of the Skin

Learning Objectives

■ *Differentiate cutaneous from subcutaneous mycoses, and provide an example of each.*
■ *List the causative agent of and predisposing factors for candidiasis.*

The skin is most susceptible to microorganisms that can resist high osmotic pressure and low moisture. It is not surprising, therefore, that fungi cause a number of skin disorders. Any fungal infection of the body is called a **mycosis.**

Cutaneous Mycoses

Fungi that colonize the hair, nails, and the outer layer (stratum corneum) of the epidermis (see Figure 21.1) are called **dermatophytes;** they grow on the keratin present in those locations. Termed **dermatomycoses,** these fungal infections are more informally known as *tineas* or *ringworm.* **Tinea capitis,** or ringworm of the scalp, is fairly common among elementary school children and can result in bald patches. This characteristic led to the adoption by the Romans of the name *tinea,* Latin for clothes moth, because the infection resembles the holes left by the wormlike larvae of the moth in wool clothing. The infections tend to expand circularly, hence the term *ringworm* (Figure 21.16a). The infection is usually transmitted by contact with fomites. Dogs and cats are also frequently infected with fungi that cause ringworm in children. Ringworm of the groin, or jock itch, is known as **tinea cruris,** and ringworm of the feet, or athlete's foot, is known as **tinea pedis** (Figure 21.16b). The moisture in such areas favors fungal infections.

Three genera of fungi are involved in cutaneous mycosis. *Trichophyton* (trik-ō-fī'ton) can infect hair, skin, or nails; *Microsporum* (mī-krō-spô'rum) usually involves only the hair or skin; *Epidermophyton* (ep-i-dèr-mō-fī'ton) affects only the skin and nails. The topical drugs available without prescription for tinea infections include miconazole and clotrimazole. When hair is involved, topical treatment is not very effective. An oral antibiotic, griseofulvin, is often useful in such infections because it can

The most effective methods for disinfecting contact lenses involve the application of heat; lenses that cannot be heated can be disinfected with hydrogen peroxide, which is then neutralized.

Bacterial Diseases of the Eye

The bacterial microorganisms most commonly associated with the eye usually originate from the skin and upper respiratory tract.

Neonatal Gonorrheal Ophthalmia

Neonatal gonorrheal ophthalmia is a serious form of conjunctivitis caused by *Neisseria gonorrhoeae* (the cause of gonorrhea). Large amounts of pus are formed; if treatment is delayed, ulceration of the cornea will usually result. The disease is acquired as the infant passes through the birth canal, and infection carries a high risk of blindness. Early in the twentieth century, legislation required that the eyes of all newborn infants be treated with a 1% solution of silver nitrate, which proved to be a very effective treatment in preventing this eye infection. Between 1906 and 1959, the percentage of admissions to schools for the blind that could be attributed to neonatal ophthalmia declined from 24% to only 0.3%. Silver nitrate has been almost entirely replaced by antibiotics because of frequent coinfections by gonococci and sexually transmitted chlamydias, and silver nitrate is not effective against chlamydias. In parts of the world where the cost of antibiotics is prohibitive, a dilute solution of povidone-iodine has proven effective.

Inclusion Conjunctivitis

Chlamydial conjunctivitis, or **inclusion conjunctivitis,** is quite common today. It is caused by *Chlamydia trachomatis,* a bacterium that grows only as an obligate intracellular parasite. In infants, who acquire it in the birth canal, the condition tends to resolve spontaneously in a few weeks or months, but in rare cases it can lead to scarring of the cornea, much as in trachoma (discussed below). Chlamydial conjunctivitis also appears to spread in the unchlorinated waters of swimming pools; in this context, it is called swimming pool conjunctivitis. Tetracycline applied as an ophthalmic ointment is an effective treatment.

Trachoma

A serious infection caused by *C. trachomatis* is **trachoma,** the greatest single infectious cause of blindness in the world today. In the arid parts of Africa and Asia, almost all children are infected early in their lives. Worldwide, there are probably 500 million active cases and 7 million

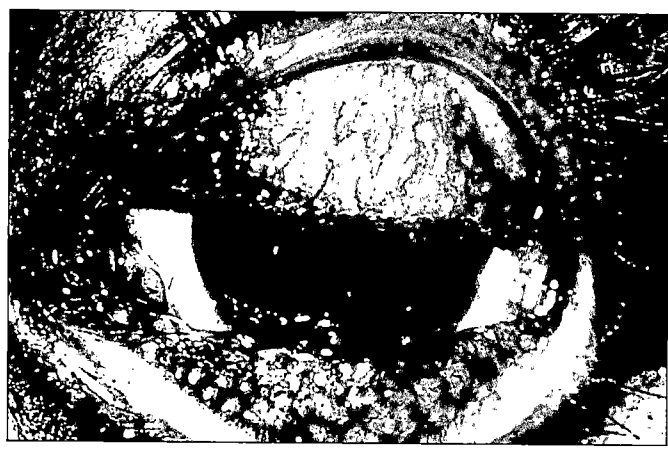

FIGURE 21.19 Trachoma lesions. The eyelids have been pulled back to show the inflammatory nodules on the conjunctival membrane in contact with the cornea, which is damaged by the abrasion and is susceptible to secondary infections.

■ **What organism causes trachoma?**

blinded victims. Trachoma also occurs occasionally in the southwestern United States, especially among Native Americans. The disease is transmitted largely by hand contact or by sharing such personal objects as towels. Flies may also carry the bacteria.

The disease is a conjunctivitis that eventually leads to permanent scarring. Blindness is a result of long-term mechanical abrasion of the cornea by these scars and by turned-in eyelashes (Figure 21.19). Simple surgery can prevent eyelashes from rubbing the eyes. Secondary infections by other bacterial pathogens are also a factor. Antibiotic ointments, especially tetracycline, are useful in treatment. Partial immunity is generated by recovery. The disease can be controlled through sanitary practices and health education.

Other Infectious Diseases of the Eye

Learning Objective

■ *List the causative agent, mode of transmission, and clinical symptoms of these eye infections: herpetic keratitis, Acanthamoeba keratitis.*

Microorganisms such as viruses and protozoa can also cause eye diseases. The diseases discussed here are characterized by inflammation of the cornea, which is called *keratitis.*

Herpetic Keratitis

Herpetic keratitis is caused by the same herpes simplex type 1 virus that causes cold sores and is latent in the trigeminal nerves (Figure 21.12). The disease is an infection of the cornea, often resulting in deep ulcers, that may be the most common cause of infectious blindness in the United States. The drug trifluridine is often an effective treatment.

Acanthamoeba Keratitis

The first case of *Acanthamoeba* (a-kan-thä-mē'bä) **keratitis** was reported in 1973 in a Texas rancher. Since then, well over 100 cases have been diagnosed in the United States. This amoeba has been found in fresh water, tap water, hot tubs, and soil. Most recent cases have been associated with the wearing of contact lenses. Contributing factors are inadequate, unsanitary, or faulty disinfecting procedures (only heat will reliably kill the cysts), homemade saline solutions, and wearing the contact lenses overnight and while swimming.

In its early stages, the infection consists of only a mild inflammation, but later stages are often accompanied by severe pain. Damage is often so severe as to require a corneal transplant, or even removal of the eye. Diagnosis is confirmed by the presence of trophozoites (Figure 21.20) and cysts in stained scrapings of the cornea.

The diseases of the eye are summarized in Table 21.2.

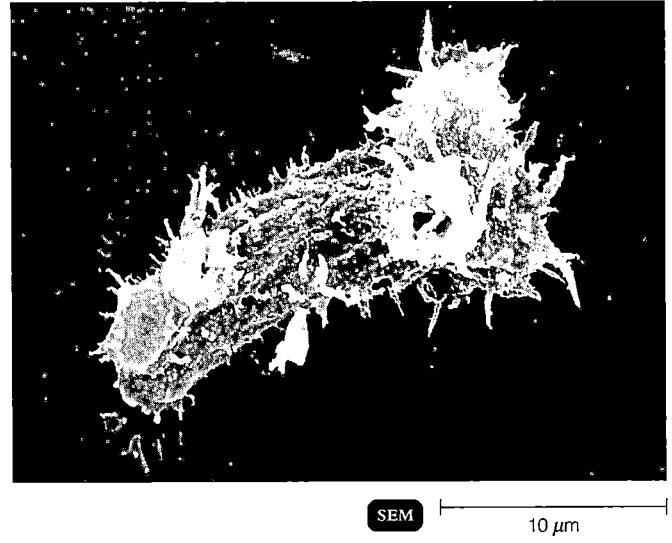

SEM ⊢————— 10 μm —————⊣

FIGURE 21.20 The trophozoite of an *Acanthamoeba*, the cause of *Acanthamoeba* keratitis. In water it assumes a flagellated, swimming morphology.

■ This protozoan is found in water supplies and can result in severe damage to contact lens wearers.

table 21.2	Microbial Diseases Associated with the Eye		
	Causative Agent	**Characteristics**	**Treatment**
Bacterial Diseases			
Neonatal gonorrheal ophthalmia	*Neisseria gonorrhoeae*	Acute infection with much pus formation; if treatment is delayed, ulcers form on cornea	Silver nitrate, tetracycline, or erythromycin for prevention
Inclusion conjunctivitis	*Chlamydia trachomatis*	Swelling of eyelid; mucus and pus formation	Tetracycline
Trachoma	*Chlamydia trachomatis*	Conjunctivitis causes scarring of eyelid that mechanically damages cornea, often causing secondary infections	Tetracycline
Viral Disease			
Herpetic keratitis	Herpes simplex type 1 virus	May progress to corneal ulcers and severe damage	Trifluridine may be effective
Protozoan Disease			
Acanthamoeba keratitis	*Acanthamoeba* spp.	Frequently results in severe eye damage	Topical propamidine isethionate or miconazole; corneal transplant or eye removal surgery may be required

Study Outline

INTRODUCTION (p. 578)

1. The skin is a physical and chemical barrier against microorganisms.

2. Moist areas of the skin (such as the armpit) support larger populations of bacteria than dry areas (such as the scalp).

STRUCTURE AND FUNCTION OF THE SKIN (pp. 578–579)

1. The outer portion of the skin, called the epidermis, contains keratin, a waterproof coating.

2. The inner portion of the skin, the dermis, contains hair follicles, sweat ducts, and oil glands that provide passageways for microorganisms.

3. Sebum and perspiration are secretions of the skin that can inhibit the growth of microorganisms.

4. Sebum and perspiration provide nutrients for some microorganisms.

5. Body cavities are lined with epithelial cells. When these cells secrete mucus, they constitute the mucous membrane.

NORMAL MICROBIOTA OF THE SKIN (pp. 579–580)

1. Microorganisms that live on skin are resistant to desiccation and high concentrations of salt.

2. Gram-positive cocci predominate on the skin.

3. The normal skin microbiota are not completely removed by washing.

4. Members of the genus *Propionibacterium* metabolize oil from the oil glands and colonize hair follicles.

5. *Pityrosporum ovale* yeast grows on oily secretions and may be the cause of dandruff.

MICROBIAL DISEASES OF THE SKIN (pp. 580–595)

1. Vesicles are small fluid-filled lesions; bullae are vesicles larger than 1 cm; macules are flat, reddened lesions; papules are raised lesions; and pustules are raised lesions containing pus.

Bacterial Diseases of the Skin (pp. 582–587)

Staphylococcal Skin Infections (pp. 582–585)

1. Staphylococci are gram-positive bacteria that often grow in clusters.

2. The majority of skin microbiota consist of coagulase-negative *S. epidermidis*.

3. Almost all pathogenic strains of *S. aureus* produce coagulase.

4. Pathogenic *S. aureus* can produce enterotoxins, leukocidins, and exfoliative toxin.

5. Many strains of *S. aureus* produce penicillinase; these are treated with vancomycin.

6. Localized infections (sties, pimples, and carbuncles) result from *S. aureus* entering openings in the skin.

7. Impetigo of the newborn is a highly contagious superficial skin infection caused by *S. aureus*.

8. Toxemia occurs when toxins enter the bloodstream; staphylococcal toxemias include scalded skin syndrome and toxic shock syndrome.

Streptococcal Skin Infections (pp. 585–586)

1. Streptococci are gram-positive cocci that often grow in chains.

2. Streptococci are classified according to their hemolytic enzymes and cell wall antigens.

3. Group A β-hemolytic streptococci (including *S. pyogenes*) are the pathogens most important to humans.

4. Group A β-hemolytic streptococci produce a number of virulence factors: M protein, erythrogenic toxin, deoxyribonuclease, streptokinases, and hyaluronidase.

5. Erysipelas (reddish patches) and impetigo (isolated pustules) are skin infections caused by *S. pyogenes*.

6. Invasive group A β-hemolytic streptococci cause severe and rapid tissue destruction.

Infections by Pseudomonads (pp. 586–587)

1. Pseudomonads are gram-negative rods. They are aerobes found primarily in soil and water that are resistant to many disinfectants and antibiotics.

2. *Pseudomonas aeruginosa* is the most prominent species; it produces an endotoxin and several exotoxins.

3. Diseases caused by *P. aeruginosa* include otitis externa, respiratory infections, burn infections, and dermatitis.

4. Infections have a characteristic blue-green pus caused by the pigment pyocyanin.

5. Fluoroquinolones are useful in treating *P. aeruginosa* infections.

Acne (p. 587)

1. *Propionibacterium acnes* can metabolize sebum trapped in hair follicles.

2. Metabolic end-products (fatty acids) cause an inflammatory response known as acne.

3. Tretinoin, benzoyl peroxide, erythromycin, and Accutane® are used to treat acne.

Viral Diseases of the Skin (pp. 587–593)

Warts (p. 587)

1. Papillomaviruses cause skin cells to proliferate and produce a benign growth called a wart or papilloma.

2. Warts are spread by direct contact.

3. Warts may regress spontaneously or be removed chemically or physically.

Smallpox (Variola) (pp. 587–588)

1. Variola virus causes two types of skin infections: variola major and variola minor.

2. Smallpox is transmitted by the respiratory route, and the virus is moved to the skin via the bloodstream.

3. The only host for smallpox is humans.

4. Smallpox has been eradicated as a result of a vaccination effort by the WHO.

Chickenpox (Varicella) and Shingles (Herpes Zoster) (pp. 588–590)

1. Varicella-zoster virus is transmitted by the respiratory route and is localized in skin cells, causing a vesicular rash.

2. Complications of chickenpox include encephalitis and Reye's syndrome.

3. After chickenpox, the virus can remain latent in nerve cells and subsequently activate as shingles.

4. Shingles (herpes zoster) is characterized by a vesicular rash along the affected cutaneous sensory nerves.

5. The virus can be treated with acyclovir. An attenuated live vaccine is available.

Herpes Simplex (p. 590)

1. Herpes simplex infection of mucosal cells results in cold sores and occasionally encephalitis.

2. The virus remains latent in nerve cells, and cold sores can recur when the virus is activated.

3. HSV-1 is transmitted primarily by oral and respiratory routes.

4. Herpes encephalitis occurs when herpes simplex viruses infect the brain.

5. Acyclovir has proven successful in treating herpes encephalitis.

Measles (Rubeola) (pp. 590–591)

1. Measles is caused by measles virus and transmitted by the respiratory route.

2. Vaccination provides effective long-term immunity.

3. After the virus has incubated in the upper respiratory tract, macular lesions appear on the skin, and Koplik's spots appear on the oral mucosa.

4. Complications of measles include middle ear infections, pneumonia, encephalitis, and secondary bacterial infections.

Rubella (pp. 591–592)

1. The rubella virus is transmitted by the respiratory route.

2. A red rash and light fever might occur in an infected individual; the disease can be asymptomatic.

3. Congenital rubella syndrome can affect a fetus when a woman contracts rubella during the first trimester of her pregnancy.

4. Damage from congenital rubella syndrome includes stillbirth, deafness, eye cataracts, heart defects, and mental retardation.

5. Vaccination with live rubella virus provides immunity of unknown duration.

Other Viral Rashes (pp. 592–593)

1. Human parvovirus B19 causes Fifth Disease and HHV-6 causes Roseola.

Fungal Diseases of the Skin (pp. 593–595)

Cutaneous Mycoses (pp. 593–594)

1. Fungi that colonize the outer layer of the epidermis cause dermatomycoses.

2. *Microsporum, Trichophyton,* and *Epidermophyton* cause dermatomycoses called ringworm, or tinea.

3. These fungi grow on keratin-containing epidermis, such as hair, skin, and nails.

4. Ringworm and athlete's foot are usually treated with topical antifungal chemicals.

5. Diagnosis is based on the microscopic examination of skin scrapings or fungal culture.

Subcutaneous Mycoses (p. 594)

1. Sporotrichosis results from a soil fungus that penetrates the skin through a wound.

2. The fungi grow and produce subcutaneous nodules along the lymphatic vessels.

Candidiasis (pp. 594–595)

1. *Candida albicans* causes infections of mucous membranes and is a common cause of thrush (in oral mucosa) and vaginitis.

2. *C. albicans* is an opportunistic pathogen that may proliferate when the normal bacterial microbiota are suppressed.

3. Topical antifungal chemicals may be used to treat candidiasis.

Parasitic Infestation of the Skin (p. 595)

1. Scabies is caused by a mite burrowing and laying eggs in the skin.

2. Topical application of gamma benzene hexachloride is used to treat scabies.

MICROBIAL DISEASES OF THE EYE (pp. 595–597)

1. The mucous membrane lining the eyelid and covering the eyeball is the conjunctiva.

Inflammation of the Eye Membranes: Conjunctivitis (pp. 595–596)

1. Conjunctivitis is caused by several bacteria and can be transmitted by improperly disinfected contact lenses.

Bacterial Diseases of the Eye (p. 596)

1. Bacterial microbiota of the eye usually originate from the skin and upper respiratory tract.

2. Neonatal gonorrheal ophthalmia is caused by the transmission of *Neisseria gonorrhoeae* from an infected mother to an infant during its passage through the birth canal.

3. All newborn infants are treated with an antibiotic to prevent the growth of *Neisseria* and *Chlamydia* infection.

4. Inclusion conjunctivitis is an infection of the conjunctiva caused by *Chlamydia trachomatis*. It is transmitted to infants during birth and is transmitted in unchlorinated swimming water.

5. In trachoma, which is caused by *C. trachomatis*, scar tissue forms on the cornea.

6. Trachoma is transmitted by hands, fomites, and perhaps flies.

Other Infectious Diseases of the Eye (pp. 596–597)

1. Inflammation of the cornea is called keratitis.

2. Herpetic keratitis causes corneal ulcers. The etiology is HSV-1 that invades the central nervous system and can recur.

3. Trifluridine is an effective treatment for herpes keratitis.

4. *Acanthamoeba* protozoa, transmitted via water, can cause a serious form of keratitis.

Study Questions

REVIEW

1. Discuss the usual mode of entry of bacteria into the skin. Compare bacterial skin infections with those caused by fungi and viruses with respect to mode of entry.

2. What bacteria are identified by a positive coagulase test? What bacteria are characterized as group A β-hemolytic?

3. Compare and contrast impetigo and erysipelas.

4. Complete the table of epidemiology below:

Disease	Etiologic Agent	Clinical Symptoms	Mode of Transmission
Acne			
Pimples			
Warts			
Chickenpox			
Fever blisters			
Measles			
Rubella			

5. How do sporotrichosis and athlete's foot differ? In what ways are they similar? How is each disease treated?

6. (a) Differentiate between conjunctivitis and keratitis.
   (b) Select a bacterial and viral eye infection, and discuss the epidemiology of each.

7. Describe the symptoms, etiologic agent, and treatment of candidiasis. How is candidiasis contracted?

8. Why does the blood test required for a marriage license include a test for antibodies against rubella for women?

9. Identify the diseases based on the symptoms in the chart below:

Symptoms	Disease
Koplik's spots	
Macular rash	
Vesicular rash	
Small, spotted rash	
Recurrent "blisters" on oral mucosa	
Corneal ulcer and swelling of lymph nodes	

10. What complications can occur from HSV-1 infections?

11. What is in the MMR vaccine?

12. Explain the relationship between shingles and chickenpox.

13. Why are the eyes of all newborn infants washed with an antiseptic or antibiotic?

14. What is the leading infectious cause of blindness in the world?

15. What is scabies? How is it treated?

MULTIPLE CHOICE

Use the following information to answer questions 1 and 2. A 6-year-old girl was taken to the physician for evaluation of a slowly growing bump on the back of her head. The bump was a raised, scaling lesion 4 cm in diameter. A fungal culture of material from the lesion was positive for a fungus with numerous conidia.

1. The girl's disease was
   a. rubella.
   b. candidiasis.
   c. dermatomycosis.
   d. a cold sore.
   e. none of the above

2. Besides the scalp, this disease can occur on all of the following *except*
   a. feet.
   b. nails.
   c. the groin.
   d. subcutaneous tissue.
   e. none of the above

Use the following information to answer questions 3 and 4. A 12-year-old boy had a fever, rash, headaches, sore throat, and cough. He also had a macular rash on his trunk, face, and arms. A throat culture was negative for *Streptococcus pyogenes*.

3. The boy most likely had
   a. streptococcal sore throat.
   b. measles.
   c. rubella.
   d. smallpox.
   e. none of the above

4. All of the following are complications of this disease *except*
   a. middle ear infections.
   b. pneumonia.
   c. birth defects.
   d. encephalitis.
   e. none of the above

5. A patient has conjunctivitis. If you isolated *Pseudomonas* from the patient's mascara, you would most likely conclude all of the following *except* that
   a. the mascara was the source of the infection.
   b. *Pseudomonas* is causing the infection.
   c. *Pseudomonas* has been growing in the mascara.
   d. the mascara was contaminated by the manufacturer.
   e. none of the above

6. You microscopically examine scrapings from a case of *Acanthamoeba* keratitis. You expect to see
   a. nothing.
   b. viruses.
   c. gram-positive cocci.
   d. eukaryotic cells.
   e. gram-negative cocci.

Use the following choices to answer questions 7 through 9.
   a. *Pseudomonas*
   b. *S. aureus*
   c. scabies
   d. *Sporothrix*
   e. virus

7. Nothing is seen in microscopic examination of a scraping from the patient's rash.

8. Microscopic examination of the patient's ulcer reveals ovoid cells.

9. Microscopic examination of scrapings from the patient's rash shows gram-negative rods.

10. Penicillin would be most effective against
    a. *Chlamydia*.
    b. *Pseudomonas*.
    c. *Candida*.
    d. *Streptococcus*.
    e. human herpesviruses.

CRITICAL THINKING

1. A laboratory test used to determine the identity of *Staphylococcus aureus* is its growth on mannitol salt agar. The medium contains 7.5% sodium chloride (NaCl). Why is it considered a selective medium for *S. aureus*?

2. Is it necessary to treat a patient for warts? Explain briefly.

3. Analyses of nine conjunctivitis cases provided the data in the chart below. How were these infections transmitted? How could they be prevented?

No.	Etiology	Isolated from Eye Cosmetics or Contact Lenses
5	S. epidermidis	+
1	Acanthamoeba	+
1	Candida	+
1	P. aeruginosa	+
1	S. aureus	+

4. What factors made the eradication of smallpox possible? What other diseases meet these criteria?

CLINICAL APPLICATIONS

1. A hospitalized patient recovering from surgery develops an infection that has blue-green pus and a grapelike odor. What is the probable etiology? How might the patient have acquired this infection?

2. A 12-year-old diabetic girl using continuous subcutaneous insulin infusion to manage her diabetes developed a fever (39.4°C), low blood pressure, abdominal pain, and erythroderma. She was supposed to change the needle-insertion site every 3 days after cleaning the skin with an iodine solution. Frequently she did not change the insertion site more often than every 10 days. Blood culture was negative, and abscesses at insertion sites were not cultured. What is the probable cause of her symptoms?

3. A teenage male with confirmed influenza was hospitalized when he developed respiratory distress. He had a fever, rash, and low blood pressure. *S. aureus* was isolated from his respiratory secretions. Discuss the relationship between his symptoms and the etiological agent.

Learning with Technology

MP = The Microbiology Place website   **ST** = Student Tutorial CD-ROM   **BID** = Bacteria ID CD-ROM

MP   Don't forget to go to The Microbiology Place website (http://www.microbiologyplace.com) to take the practice tests, explore the interactive activity and case study, and check out the news articles and web links for this chapter.

ST   Remember there is also a quiz for this chapter on the Microbiology Interactive Student Tutorial CD-ROM.

BID   1. An 8-year-old male went to the pediatrician complaining of pain in his right hip. He had no previous injury to the hip or leg. His temperature was 38°C. A bone scan revealed damage to the ball of the femur. Bacterial cultures of blood and fluid from the hip joint were positive for the same organism. What is the organism? (Set 2, #16)

2. In lab, you put a strand of your hair on a nutrient agar plate. After incubation for 4 days, this organism grows. What is the organism? (Set 2, #5)

chapter 22

Microbial Diseases of the Nervous System

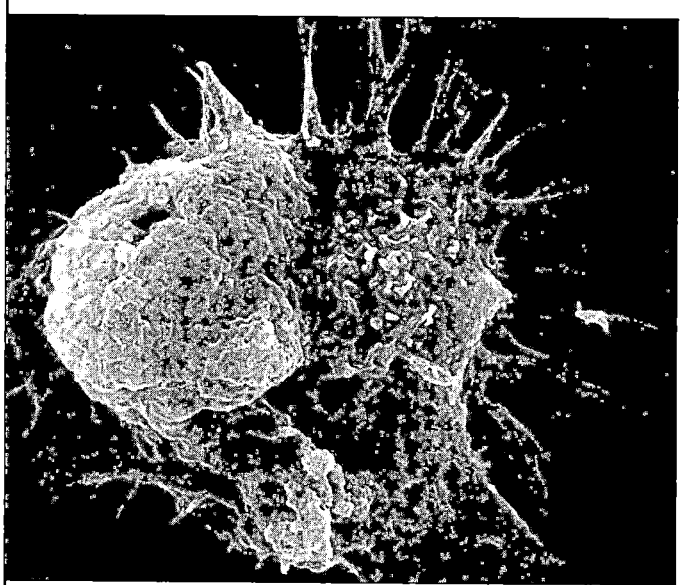

Naegleria fowleri. *Infections by this normally free-living amoeba are associated with swimming. The amoeba crosses nasal mucosa to enter the central nervous system.*

Some of the most devastating infectious diseases affect the nervous system, especially the brain and spinal cord. Because of the crucial importance of these control systems, they are strongly protected from acci-dent and infection by bone and other structures. Pathogens capable of causing diseases of the nervous system often have virulence characteristics of a special nature to penetrate these defenses.

Structure and Function of the Nervous System

Learning Objectives

- *Define central nervous system and blood-brain barrier.*
- *Differentiate meningitis from encephalitis.*

The human nervous system is organized into two divisions: the central nervous system and the peripheral nervous system (Figure 22.1). The **central nervous system (CNS)** consists of the brain and the spinal cord. As the control center for the entire body, the CNS picks up sensory information from the environment, interprets the information, and sends impulses that coordinate the body's activities. The **peripheral nervous system (PNS)** consists of all the nerves that branch off from the brain and spinal cord. These peripheral nerves are the lines of communication between the central nervous system, the various parts of the body, and the external environment.

Both the brain and the spinal cord are covered and protected by three continuous membranes called *meninges* (Figure 22.2 on page 604). These are the outermost *dura mater,* the middle *arachnoid,* and the innermost *pia mater.* Between the pia mater and arachnoid membranes is a space called the *subarachnoid space,* in which an adult has 100–160 ml of *cerebrospinal fluid (CSF)* circulating. Because CSF has low levels of complement or circulating antibodies and few phagocytic cells, bacteria can multiply in it with few checks.

An important feature of the brain is the **blood-brain barrier.** Certain capillaries permit some substances to pass from the blood into the brain but restrict others. These capillaries are less permeable than others within the body and are therefore more selective in passing materials.

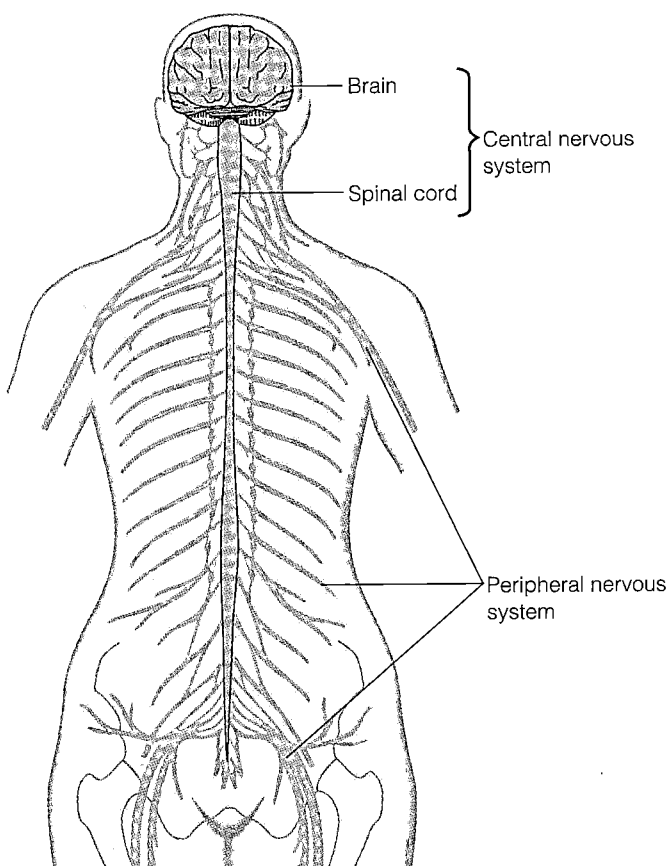

FIGURE 22.1 The human nervous system. This view shows the central and peripheral nervous systems.

■ **Encephalitis is an infection of the brain; meningitis is an infection of the meninges.**

Drugs cannot cross the blood-brain barrier unless they are lipid-soluble. (Glucose and many amino acids are not lipid-soluble, but they can cross the barrier because special transport systems exist for them.) The lipid-soluble antibiotic chloramphenicol enters the brain readily. Penicillin is only slightly lipid-soluble, but if it is taken in very large doses, enough may cross the barrier to be effective. Inflammations of the brain tend to alter the blood-brain barrier in such a way as to allow antibiotics to cross that would not be able to cross if there were no infection.

Even though the central nervous system has considerable protection, it can still be invaded by microorganisms in several ways. For example, microorganisms can gain access through trauma, such as a skull or backbone fracture, or through a medical procedure such as a spinal tap, in which a needle is inserted into the subarachnoid space of the spinal meninges to obtain a sample of cerebrospinal fluid for diagnosis. Some microorganisms can also move along peripheral nerves. But probably the most common

routes of CNS invasion are the bloodstream and lymphatic system (see Chapter 23), when inflammation alters permeability of the blood-brain barrier.

An inflammation of the meninges is called **meningitis.** An inflammation of the brain itself is called **encephalitis.**

Bacterial Diseases of the Nervous System

Microbial infections of the central nervous system are infrequent but often have serious consequences.

Bacterial Meningitis

Learning Objectives

■ *Discuss the epidemiology of meningitis caused by* Haemophilus influenzae, Neisseria meningitidis, Streptococcus pneumoniae, *and* Listeria monocytogenes.

■ *Explain how bacterial meningitis is diagnosed and treated.*

The initial symptoms of meningitis are not especially alarming: fever, headache, and a stiff neck. Nausea and vomiting often follow. Eventually, meningitis may progress to convulsions and coma. The mortality rate varies with the pathogen but is generally high for an infectious disease today. Many people who survive an attack suffer some degree of neurological damage.

Meningitis can be caused by different types of pathogens, including viruses, bacteria, fungi, and protozoa. Viral meningitis, mostly caused by echoviruses, is probably much more common than bacterial meningitis but tends to be a mild disease.

Only three bacterial species cause more than 70% of the meningitis cases and 70% of the related deaths. These are the gram-positive diplococcus *Streptococcus pneumoniae,* and the gram-negative bacteria *Haemophilus influenzae* and *Neisseria meningitidis* (Table 22.1). All three possess a capsule that protects them from phagocytosis as they replicate rapidly in the bloodstream, from which they might enter the cerebrospinal fluid. Death from bacterial meningitis often occurs very quickly, probably from shock and inflammation caused by the release of endotoxins of the gram-negative pathogens or the release of cell wall fragments (peptidoglycans and teichoic acids) of gram-positive bacteria.

Nearly 50 other species of bacteria have been reported to be opportunistic pathogens that occasionally cause meningitis. Especially important are *Listeria monocytogenes,* streptococci of group B, staphylococci, and

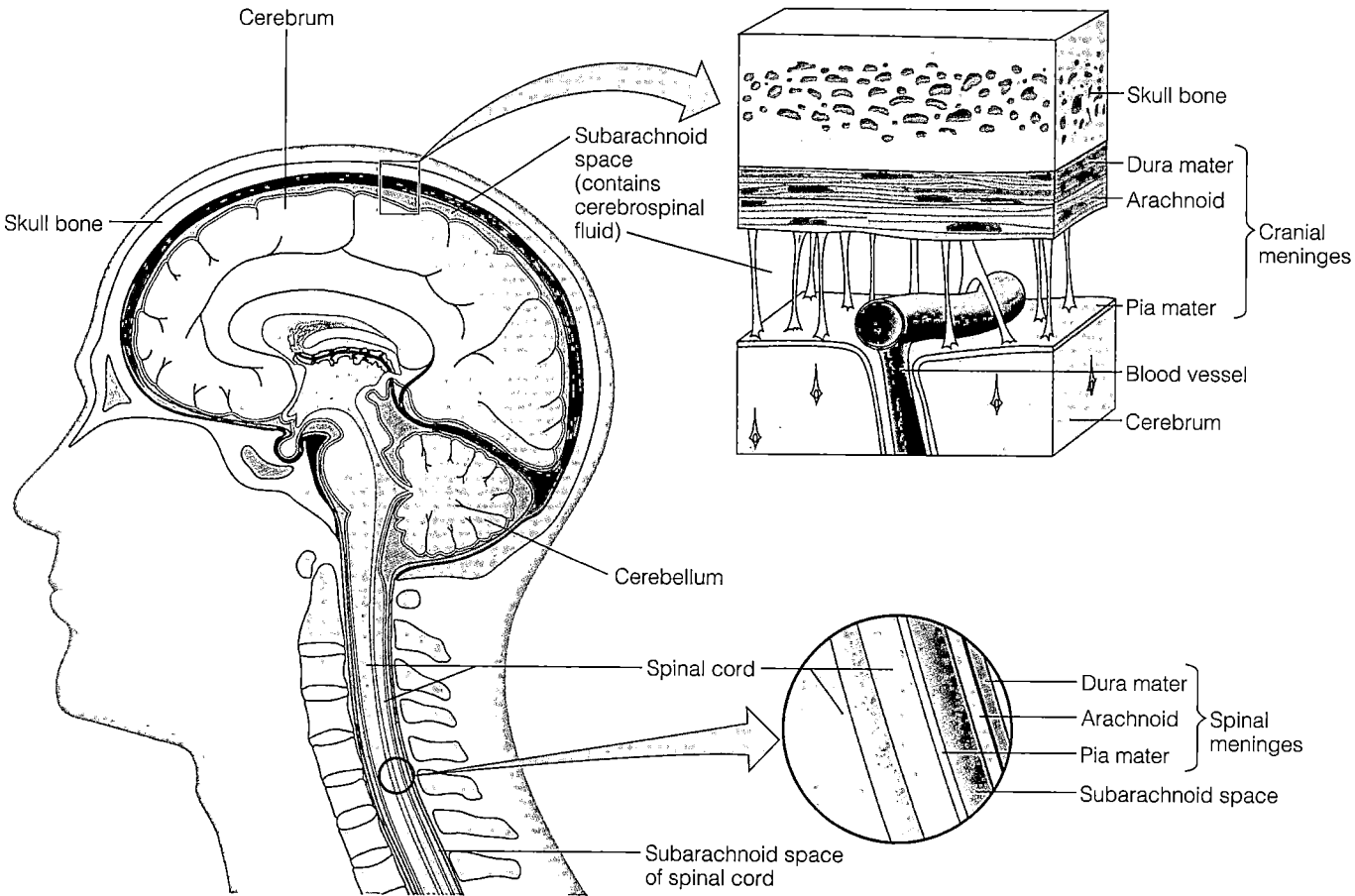

FIGURE 22.2 The meninges and cerebrospinal fluid. The meninges, whether cranial or spinal, consist of three layers: dura mater, arachnoid, and pia mater. Between the arachnoid and the pia mater is the subarachnoid space, in which cerebrospinal fluid circulates. Notice that the CSF is vulnerable to contamination by microbes carried in the blood that are able to penetrate the blood-brain barrier at the walls of the blood vessels.

■ Cerebrospinal fluid is vulnerable to contamination by microbes that cross the blood-brain barrier.

table 22.1	Relative Incidence of Bacterial Meningitis in United States and Canada	
Bacterium	**Percentage of Cases**	**Fatality Rate**
Streptococcus pneumoniae	30–50	19–46
Neisseria meningitidis	15–40	3–17
Haemophilus influenzae	2–7	3–11

Other bacteria causing meningitis account for 6–8% of cases.

SOURCE: Adapted from E.J. Phillips and A.E. Simor, "Bacterial Meningitis in Children and Adults." *Postgraduate Medicine* 103 (3):104 (1998).

certain gram-negative bacteria. Some of these principally affect newborns (*L. monocytogenes* and the streptococci), and others mainly result from infections related to surgical operations (staphylococci and the gram-negative bacteria).

Haemophilus influenzae Meningitis

Haemophilus influenzae is an aerobic, gram-negative bacterium that is a common member of the normal throat microbiota. In addition to causing meningitis, it is the cause of pneumonia, epiglottitis, and otitis media (earaches). The carbohydrate capsule of the bacterium is important to its pathogenicity, especially those bacteria with capsular antigens of type b. Medically, the bacterium is often referred to by the acronym *Hib*.

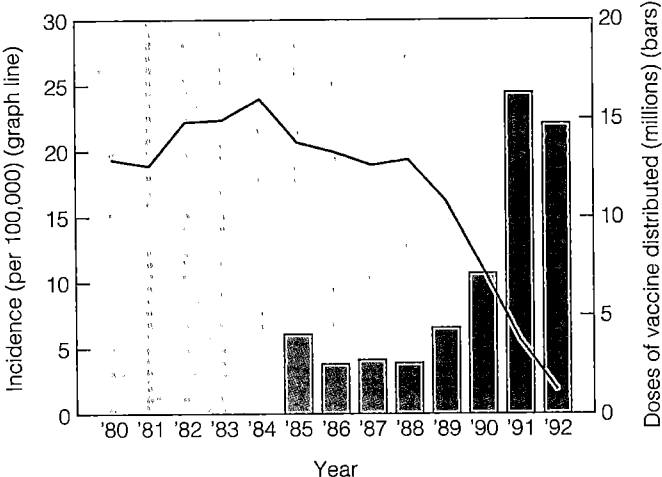

FIGURE 22.3 Incidence of H. influenzae meningitis in the United States. The graph shows the decline in cases of meningitis caused by *H. influeuzae* in children under 5 years as the incidence of vaccine administration increased.

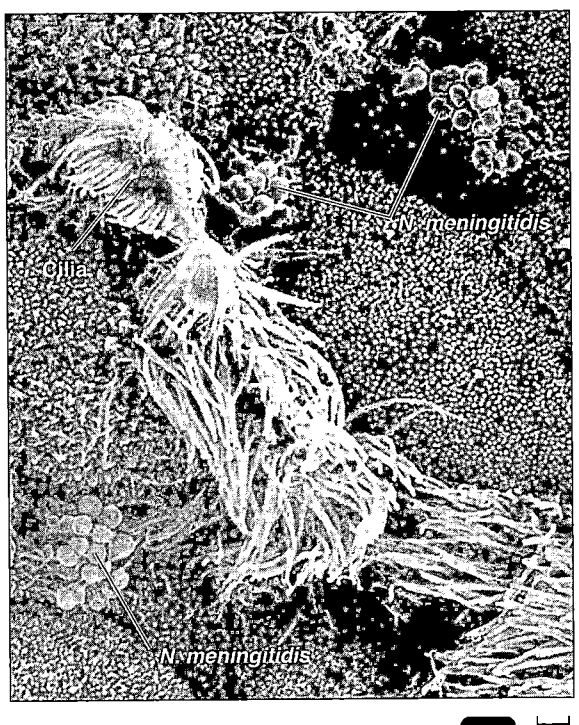

FIGURE 22.4 Neisseria meningitis. This scanning electron micrograph shows *Neisseria meningitidis* in clusters attached to cells on the mucous membrane in the pharynx.

The name *Haemophilus influenzae* was given because the microorganism was erroneously thought to be the causative agent of the influenza pandemics of 1890 and World War I. *H. influenzae* was probably only a secondary invader during those virus-caused pandemics. *Haemophilus* refers to the need the microorganism has for factors in blood for growth (*hemo* = blood; *philus* = loving).

Hib-caused meningitis occurs mostly in children under age 4, especially at about 6 months, when antibody protection provided by the mother weakens. The incidence is decreasing because of the Hib vaccine, which was introduced in 1988. The recommended vaccination series begins at 2 months, and effective immunity is achieved at about 6 months. The incidence of meningitis in infants and children age 5 or younger has declined by 99% and is targeted for elimination in the United States (Figure 22.3). The incidence in older children and adults remains little changed. Historically, *H. influenzae* meningitis accounted for most of the cases of reported bacterial meningitis (45%), with a mortality rate of about 6%.

Neisseria Meningitis (Meningococcal Meningitis)

Meningococcal meningitis is caused by *Neisseria meningitidis*. The meningococcus is an aerobic, gram-negative bacterium. Like Hib and the pneumococcus, it is frequently present in the nose and throat without causing disease symptoms (Figure 22.4). These carriers are a reservoir of infection. Meningococcal meningitis typically begins with a throat infection, leading to bacteremia and eventually meningitis. It usually occurs in children under 2 years of age, mostly after the maternal immunity weakens at about 6 months. Most symptoms

are caused by endotoxins. In some patients the bacteria begin to proliferate in the bloodstream, resulting in gram-negative sepsis, a condition discussed in Chapter 23, page 626. This can lead to extensive tissue destruction and even the need to amputate some of the affected extremities. Death can occur in a few hours. However, antibiotic therapy has helped reduce the mortality rate, which was once as high as 80%.

The meningococcus occurs in three major serotypes: A, B, and C. Group A causes widespread epidemics in Africa, China, and the Middle East, but generally does not occur in the United States, where group C predominates. Epidemics are unique to meningococci; other bacterial meningitis cases tend to occur sporadically. Epidemics often coincide with dry seasons, when the nasal mucous membranes are less resistant to bacterial invasion.

A vaccine is used routinely in the U. S. military, but, like other polysaccharide capsular vaccines, it is not effective in very young children.

Streptococcus pneumoniae Meningitis (Pneumococcal Meningitis)

Streptococcus pneumoniae, like *H. influenzae*, is a common inhabitant of the nasopharyngeal region. The pneumococcus is one of the most significant bacterial pathogens in the United States, and its capsule is the most important element in its pathogenicity. In addition to about 3000 annual cases of meningitis, it causes 500,000 cases of pneumonia and millions of painful earaches every year. About half of the cases of pneumococcal meningitis occur among children between the ages of 1 month and 4 years. The mortality rate is very high.

Hospitalized older adults are another susceptible group. For them, some protection is available from the vaccine for pneumococcal pneumonia. A conjugated vaccine, modeled after the Hib vaccine, has been introduced. It is recommended for infants under the age of two; the first dose can be given as early as 6 weeks of age. Antibiotic resistance is an increasing threat with the pneumococcal diseases.

Diagnosis and Treatment of the Most Common Types of Bacterial Meningitis

Prompt treatment of any type of bacterial meningitis is essential, and chemotherapy of suspected cases is usually initiated before identification of the pathogen is complete. Broad-spectrum third-generation cephalosporins are usually the first choice of antibiotics. As soon as identification is confirmed, or perhaps when antibiotic sensitivity has been determined from cultures, the antibiotic treatment may be changed.

A diagnosis of bacterial meningitis requires a sample of cerebrospinal fluid obtained by a spinal tap. A simple Gram stain is often useful; it will frequently determine the identity of the pathogen with considerable reliability. Cultures are also made from the fluid. For this purpose, prompt and careful handling is required because many of the likely pathogens are very sensitive and will not survive much storage time or even changes in temperature. The most frequently used type of serological tests performed on CSF are latex agglutination tests. Results are available within about 20 minutes. However, a negative result does not eliminate the possibility of less common bacterial pathogens or nonbacterial causes.

Listeriosis

Listeria monocytogenes is a gram-positive rod known to cause stillbirth and neurological disease in animals long before it was recognized as causing human disease. Excreted in animal feces, it is widely distributed in soil and water. The name is derived from the proliferation of monocytes (a type of leukocyte) found in some animals infected by it. The disease **listeriosis** is usually a mild, often symptomless disease in healthy adult humans, but recovering or apparently healthy individuals often shed the pathogen indefinitely in their feces. *L. monocytogenes* is in-

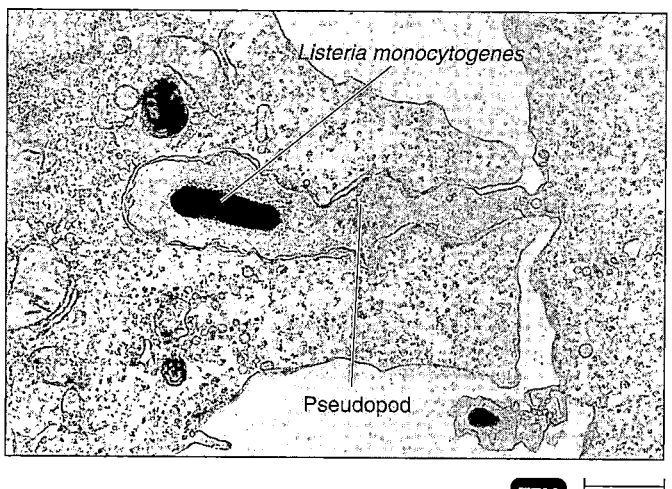

FIGURE 22.5 Cell-to-cell spread of *Listeria monocytogenes*, the cause of listeriosis. Notice that the bacterium has caused the macrophage on the right, in which it resided, to form a pseudopod that is now engulfed by the macrophage on the left. The pseudopod will soon be pinched off and the microbe transferred to the macrophage on the left.

■ How is listeriosis contracted?

gested by phagocytic cells but is not destroyed; it even proliferates within them. It also has the unusual capability of moving directly from one macrophage to an adjacent one (Figure 22.5).

As a pathogen, *L. monocytogenes* has two principal characteristics: It mainly affects adults who are immunosuppressed, pregnant, or have cancer; and it has a special affinity for growth in the central nervous system and the placenta, which provides the fetus with nutrients from the mother. Growth in the CNS is usually expressed as meningitis. When it infects a pregnant woman, the microbe's ability to spread from cell to cell is probably a factor in its ability to cross the barrier of the placenta and infect the fetus. The result is high rates of spontaneous abortions and stillbirths. A surviving newborn may be ill with septicemia and meningitis. The infant mortality rate associated with this type of infection is about 60%.

In human outbreaks, the organism is mostly foodborne. It is frequently isolated from a wide variety of foods; dairy products have been involved in several outbreaks. *L. monocytogenes* is one of the few pathogens capable of growth at refrigerator temperatures, which can lead to an increase in its numbers during a food's shelf life.

Efforts to improve the methods of detecting *L. monocytogenes* in foods are ongoing. Considerable progress has been made with selective growth media and rapid biochemical tests. However, eventually DNA probes and serological tests using monoclonal antibodies are ex-

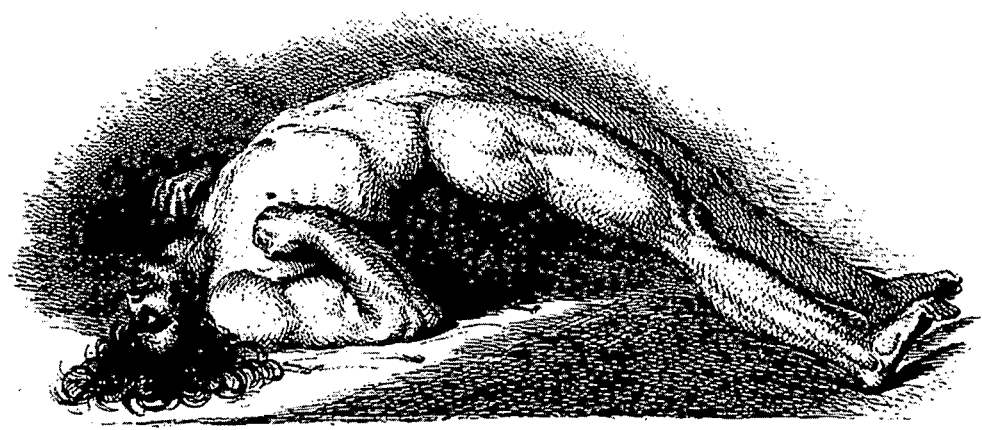

FIGURE 22.6 An advanced case of tetanus. A drawing of a British soldier during the Napoleonic wars. These spasms, known as opisthotonos, can actually result in a fractured spine. (Drawing by Charles Bell of the Royal College of Surgeons, Edinburgh.)

■ Tetanus toxin (tetanospasmin) is a neurotoxin that acts on nerves, resulting in the inhibition of muscle relaxation.

pected to be the most satisfactory (see Chapter 10). Diagnosis in humans depends on isolation and culturing of the pathogen, usually from blood or cerebrospinal fluid. Penicillin G is the antibiotic of choice for treatment.

Tetanus

Learning Objective

■ *Discuss the epidemiology of tetanus, including mode of transmission, etiology, disease symptoms, and preventive measures.*

The causative agent of **tetanus,** *Clostridium tetani,* is an obligately anaerobic, endospore-forming, gram-positive rod. It is especially common in soil contaminated with animal fecal wastes.

The symptoms of tetanus are caused by an extremely potent neurotoxin, *tetanospasmin,* that is released upon death and lysis of the growing bacteria (see Chapter 15). It enters the CNS via peripheral nerves or the blood. An amount of tetanospasmin weighing as much as the ink in one period on this page could kill 30 people. The bacteria themselves do not spread from the infection site, and there is no inflammation.

In a muscle's normal operation, a nerve impulse initiates contraction of the muscle. At the same time, an opposing muscle receives a signal to relax so as not to oppose the contraction. The tetanus neurotoxin blocks the relaxation pathway so that both opposing sets of muscles contract, resulting in the characteristic muscle spasms. The muscles of the jaw are affected early in the disease, preventing the mouth from opening, a condition known as *lockjaw.* In extreme cases, spasms of the back

muscles cause the head and heels to bow backward, a condition called *opisthotonos* (Figure 22.6). Gradually, other skeletal muscles become affected, including those involved in swallowing. Death results from spasms of the respiratory muscles.

Because the microbe is an obligate anaerobe, the wound by which it enters the body must provide anaerobic growth conditions—for example, improperly cleaned deep wounds such as those caused by rusty (and therefore presumably dirt-contaminated) nails. Injecting drug users are at high risk; sanitation during injection is not a priority; also, the drugs are often contaminated. However, many cases of tetanus arise from trivial injuries, such as sitting on a tack, that are considered too minor to bring to the attention of a physician.

Effective vaccines for tetanus have been available since the 1940s. But vaccination was not always as common as it is today, where it is part of the standard DTP childhood vaccine. Currently, about 96% of 6-year-olds in the United States have good immunity, but only about 30% of people age 70 do. The tetanus vaccine is a *toxoid,* an inactivated toxin that stimulates the formation of antibodies that neutralize the toxin produced by the bacteria. A booster is required every 10 years to maintain good immunity, but many people do not obtain these vaccinations. Serological surveys show that at least 50% of the U.S. population does not have adequate protection. In fact, 70% of U.S. tetanus cases occur in individuals over age 50. Some were never immunized at all, and others had lost effective antibody levels over time.

Even so, immunization has made tetanus in the United States a rare disease—typically, fewer than 50 cases a year. In 1903, 406 people died of fireworks-related tetanus injuries alone. (Fireworks explosions drive soil particles deep into human tissue.) Worldwide, there are an estimated 1 million cases annually. At least half occur in newborns. In many parts of the world, the severed umbilical cords of infants are dressed with materials such as soil, clay, and even cow dung. Estimates are that the mortality rate from tetanus is about 50% in developing areas; in the United States, it is about 25%.

When a wound is severe enough to need a physician's attention, the doctor must decide whether it is necessary to provide protection against tetanus. Usually there is not enough time to administer toxoid to produce antibodies and block the progression of the infection, even if given as a booster to a patient who has been immunized. However, temporary immunity can be conferred by tetanus immune globulin (TIG), prepared from the antibody-containing serum of immunized humans. (Prior to World War I, long before tetanus toxoid became available, similar preparations of preformed antibodies called antisera were used. Made by inoculating horses, antisera were very effective in lowering the incidence of tetanus in injured people.)

A physician's decision for treatment largely depends on the extent of the deep injuries and the immunization history of the patient, who may not be certain. Minor injuries with little tissue damage might not be considered a threat of tetanus. People with extensive injuries who have previously had three or more doses or had one within the past 10 years would be considered protected, requiring no action. For extensive wounds in patients with unknown or low immunity, TIG would be given to provide temporary protection. In addition, the first of a toxoid series would be administered to provide more permanent immunity. When TIG and toxoid are both injected, different sites must be used, to avoid neutralization of the toxoid by the TIG. Adults receive a Td vaccine that also boosts immunity to diphtheria. To minimize the production of more toxin, damaged tissue that provides growth conditions for the pathogen should be removed, a procedure called **debridement,** and antibiotics should be administered. However, once the toxin has attached to the nerves, such therapy is of little use.

Botulism

Learning Objective

■ *State the causative agent, symptoms, suspect foods, and treatment for botulism.*

Botulism, a form of food poisoning, is caused by *Clostridium botulinum,* an obligately anaerobic, endospore-forming gram-positive rod found in soil and many fresh-

water sediments. Ingesting the endospores usually does no harm, as will be explained shortly. However, in anaerobic environments, such as sealed cans, the microorganism produces an exotoxin that animal assays show to be the most potent of all natural toxins. This neurotoxin is highly specific for the synaptic end of the nerve, where it blocks the release of acetylcholine, a chemical necessary for transmission of nerve impulses across synapses.

Individuals suffering from botulism undergo a progressive *flaccid paralysis* for 1–10 days and may die from respiratory and cardiac failure. Nausea, but no fever, may precede the neurological symptoms. The initial neurological symptoms vary, but nearly all sufferers have double or blurred vision. Other symptoms include difficulty swallowing and general weakness. Incubation time varies, but symptoms typically appear within a day or two. As with tetanus, recovery from the disease does not confer immunity because the toxin is usually not present in amounts large enough to be effectively immunogenic.

Botulism was first described as a clinical disease in the early 1800s, when it was known as the sausage disease (*botulus* is the Latin word for sausage). Blood sausage was made by filling a pig stomach with blood and ground meats, tying shut all the openings, boiling it for a short time, and smoking it over a wood fire. The sausage was then stored at room temperature. This attempt at food preservation included most of the requirements for an outbreak of botulism. It killed competing bacteria but allowed the more heat-stable *C. botulinum* endospores to survive, and it provided anaerobic conditions and an incubation period for toxin production.

The botulinal toxin will be destroyed by most ordinary cooking methods that bring the food to a boil. Sausage rarely causes botulism today, largely because of the addition of nitrites. Nitrites prevent *C. botulinum* growth following germination of the endospores.

Botulinal toxin is not formed in acidic foods (below pH 4.7). Such foods as tomatoes can therefore be safely preserved without the use of a pressure cooker. There have been cases of botulism from acidic foods that normally would not have supported the growth of the botulism organisms. Most of these episodes are related to mold growth, which metabolized enough acid to allow the initiation of growth of *C. botulinum.*

Botulinal Types

There are several serological types of the botulinal toxin produced by different strains of the pathogen. These differ considerably in their virulence and other factors.

Type A toxin is probably the most virulent. Deaths have resulted from type A toxin when the food was only tasted but not swallowed. It is even possible to absorb lethal doses through skin breaks while handling laboratory samples. In untreated cases, the mortality rate is 60–70%. The type A endospore is the most heat-resistant of all *C. botulinum* strains. In the United States, it is found mainly in California, Washington, Colorado, Oregon, and New Mexico. The type A organism is usually proteolytic (the breakdown of proteins by clostridia releases amines with unpleasant odors), but obvious spoilage odor is not always apparent in low-protein foods, such as corn and beans.

Type B toxin is responsible for most European outbreaks of botulism and is the most common type in the eastern United States. The mortality rate in cases without treatment is about 25%. Type B botulism organisms occur in both proteolytic and nonproteolytic strains.

Type E toxin is produced by botulism organisms that are often found in marine or lake sediments. Therefore, outbreaks often involve seafood and are especially common in the Pacific Northwest, Alaska, and the Great Lakes area. The endospore of type E botulism is less heat-resistant than that of other strains and is usually destroyed by boiling. Type E is nonproteolytic, so the chance of detecting spoilage by odor in high-protein foods such as fish is minimal. The pathogen is also capable of producing toxin at refrigerator temperatures and requires less strictly anaerobic conditions for growth.

Incidence and Treatment of Botulism

Botulism is not a common disease. Only a few cases are reported each year, but outbreaks from restaurants occasionally involve 20–30 cases. About half the cases are type A, and types B and E account about equally for the balance. Native Americans in Alaska probably have the highest rate of botulism in the world, mostly of type E. The problem arises from food preparation methods that reflect a cultural tradition of avoiding the use of scarce fuels for heating or cooking. For example, one food involved in Alaskan outbreaks of botulism is *muktuk*. Muktuk is prepared by slicing the flippers of seals or whales into strips and then drying them for a few days. To tenderize them, they are stored anaerobically in a container of seal oil for several weeks until they approach putrefaction. The 40% mortality rate for type E botulism observed in recent years among Alaskan natives reflects the difficulty in getting prompt treatment for isolated ethnic groups.

Botulism organisms do not seem to be able to compete successfully with the normal intestinal microbiota, so the production of toxin by ingested bacteria almost never causes botulism in adults. However, the intestinal microbiota of infants is not well established, and they may suffer from **infant botulism.** An estimated 250 cases oc-

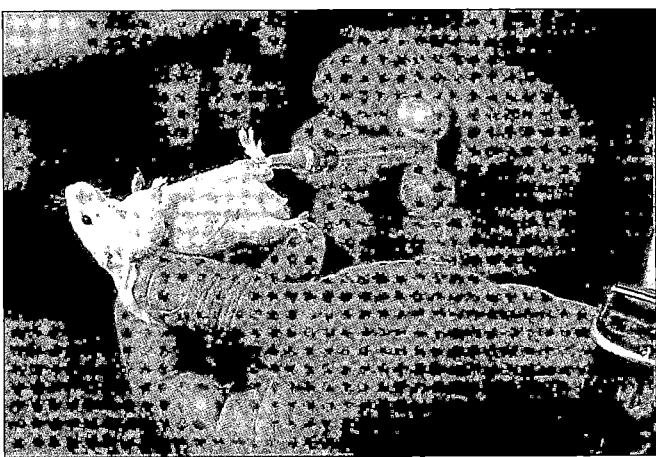

FIGURE 22.7 Diagnosis of botulism by the identification of botulinal toxin type. To determine whether botulinal toxin is present, mice are injected with the liquid portion of food extracts or cell-free cul! :es. If the mice die within 72 hours, this is evidence of the pres(:::· of toxin. To determine the specific type of toxin, groups of i ·i : are passively immunized with antisera specific for *C. botulinu* :· :·e A, B, or E. If one group of mice receiving a specific antito:.: lives and the other mice die, the type of toxin in the food or cul! : has been identified.

What are the symptoms of botulism?

cur in th.: United States annually, far more than any other form of : :tulism. Although infants have ample opportunity to in est soil and other materials contaminated with the endo: ·ores of the organism, many reported cases have been asso iated with honey. Endospores of *C. botulinum* are recov:.red with some frequency from honey, and a lethal dose may be as few as 2000 bacteria. The recommendation is not to feed honey to infants under 1 year of age; there is no problem with older children or adults who have normal intestinal microbiota.

The treatment of botulism relies heavily on supportive care. Recovery requires that the nerve endings regenerate; it therefore proceeds slowly. Extended respiratory assistance may be needed, and some neurological impairment ma:' persist for months. Antibiotics are of almost no use bec:ause the toxin is preformed. Antitoxins aimed at the neutralization of A, B, and E toxins are available and are usually administered together. This trivalent antitoxin will not ':ct the toxin already attached to the nerve endings :: l is probably more effective on type E than on types A a: l B.

Botu: u is diagnosed by the inoculation of mice with sam :ies from patient serum, stool, or vomitus specimens (F; ure 22.7). Different sets of mice are immunized with typ: A, B, or E antitoxin. All the mice are then inoculated with the test toxin; if, for example, those protected wi h type A antitoxin are the only survivors, then

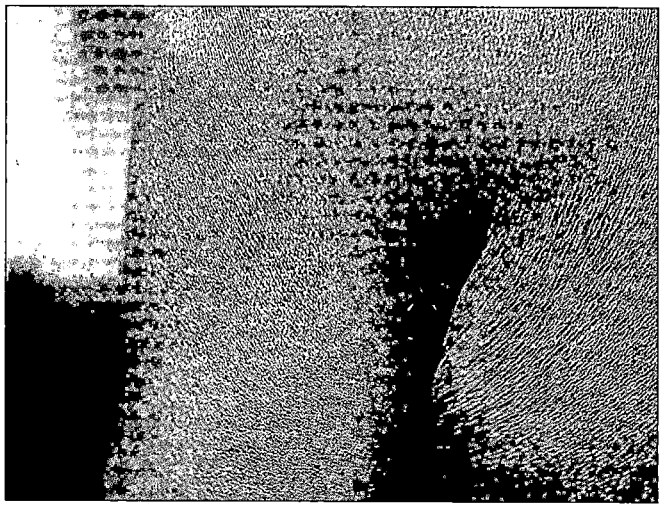

(a) Tuberculoid (neural) leprosy

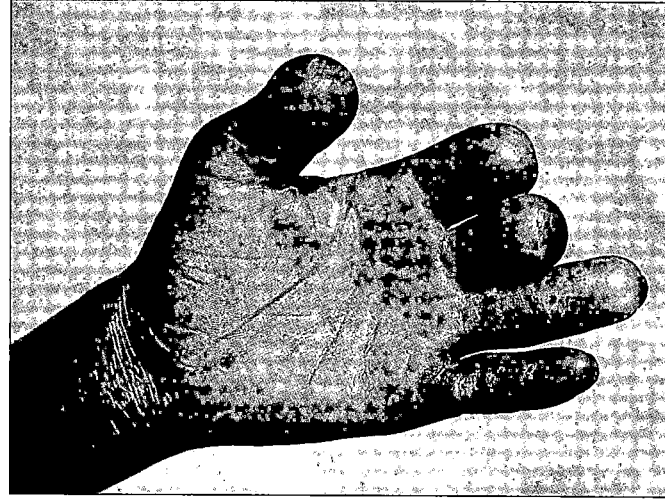

(b) Lepromatous (progressive) leprosy

FIGURE 22.8 Leprosy lesions. (a) The depigmented area of skin surrounded by a border of nodules is typical of tuberculoid (neural) leprosy. **(b)** If the immune system fails to control the disease, the result is lepromatous (progressive) leprosy. This severely deformed hand shows the progressive tissue damage to the cooler parts of the body typical of this later stage.

■ **What are the differences between tuberculoid leprosy and lepromatous leprosy?**

the toxin is type A. The toxin in food can similarly be identified by mouse inoculation.

The botulism pathogen can also grow in wounds in a manner similar to that of clostridia causing tetanus or gas gangrene (see Chapter 23). Such episodes of **wound botulism** occur occasionally.

Remarkably, the deadly toxin of botulism has a commercial use as a cosmetic: Local injections of botulism toxin at intervals of a few months eliminate forehead wrinkles (worry lines). It has also been approved to treat several other more serious conditions caused by excessive muscle contractions, for example, strabismus (cross-eyes) and blepharospasm (inability to keep eyelids raised).

Leprosy

Learning Objective

■ *Discuss the epidemiology of leprosy including mode of transmission, etiology, disease symptoms, and preventive measures.*

Mycobacterium leprae is probably the only bacterium that grows in the peripheral nervous system, although it can also grow in skin cells. It is an acid-fast rod closely related to the tuberculosis pathogen, *Mycobacterium tuberculosis*. The organism was first isolated and identified around 1870 by Gerhard A. Hansen of Norway; his discovery was one of the first links ever made between a specific bacterium and a disease. **Hansen's disease** is the more for-

mal name for **leprosy;** it is sometimes used to avoid the dreaded name leprosy.

The organism has an optimum growth temperature of 30°C and shows a preference for the outer, cooler portions of the human body. A very long generation time of about 12 days has been estimated. *M. leprae* has never been grown on artificial media. Armadillos have a body temperature of only 30–35°C and are often infected in the wild. They have proven to be a valuable experimental animal and are used in studies of the disease itself and in evaluating the effectiveness of chemotherapeutic agents.

Leprosy occurs in two main forms (although borderline forms are also recognized) that apparently reflect the effectiveness of the host's cell-mediated immune system. The *tuberculoid (neural) form* is characterized by regions of the skin that have lost sensation and are surrounded by a border of nodules (Figure 22.8a). This disease form is roughly the same as *paucibacillary* in the WHO leprosy classification system. This form of the disease occurs in people with effective immune reactions. Recovery sometimes occurs spontaneously. The disease can be diagnosed by detecting acid-fast rods in the fluids from a slit cut in a cool site, such as an earlobe. The **lepromin test** uses an extract of lepromatous tissue injected into the skin. A visible skin reaction that develops at the injection site indicates that the body has developed an immune response to the leprosy bacillus. This test is negative during the later lepromatous stage of the disease.

In the *lepromatous (progressive) form* of leprosy, which is much the same as *multibacillary* in the WHO system, skin cells are infected, and disfiguring nodules form all over the body. Patients with this type of leprosy have the least effective cell-mediated immune response, and the disease has progressed from the tuberculoid stage. Mucous membranes of the nose tend to become affected, and a lion-faced appearance is associated with this type of leprosy. Deformation of the hand into a clawed form and considerable necrosis of tissue can also occur (Figure 22.8b). The progression of the disease is unpredictable, and remissions may alternate with rapid deterioration.

The exact means of transfer of the leprosy bacillus is uncertain, but patients with lepromatous leprosy shed large numbers in their nasal secretions and in exudates (oozing matter) of their lesions. Most people probably acquire the infection when secretions containing the pathogen contact their nasal mucosa. However, leprosy is not very contagious, and transmission usually occurs only between people in fairly intimate and prolonged contact. The time from infection to the appearance of symptoms is usually measured in years, although children can have a much shorter incubation period. Death is not usually a result of the leprosy itself but of complications, such as tuberculosis.

Much of the public's fear of leprosy can probably be attributed to biblical and historical references to the disease. In the Middle Ages, people with leprosy were rigidly excluded from normal European society and sometimes even wore bells so that people could avoid them. This isolation might have contributed to the near disappearance of the disease in Europe. But patients with leprosy are no longer kept in isolation, because they can be made non-contagious within a few days by the administration of sulfone drugs. The National Leprosy Hospital in Carville, Louisiana, once housed several hundred patients but now houses only a few elderly ones and will eventually be closed. Most patients today are treated at centers on an outpatient basis.

The number of leprosy cases in the United States is gradually increasing. Currently, 100 to 150 cases are reported each year. Most are imported; the disease is usually found in tropical climates. Millions of people, most of them in Asia and Africa, suffer from leprosy today, and over half a million new cases are reported each year.

Dapsone (a sulfone drug), rifampin, and clofazimine, a fat-soluble dye, are the principal drugs used for treatment, usually in combination. A vaccine became commercially available in India in 1998. It is used as an adjunct to chemotherapy. Other vaccines that might be useful in prevention are being tested. One encouraging development is that the BCG vaccine for tuberculosis (also caused by a *Mycobacterium* species) has been found to be somewhat protective against leprosy.

Viral Diseases of the Nervous System

Learning Objective

- Discuss the epidemiology of poliomyelitis, rabies, and arboviral encephalitis, including mode of transmission, etiology, and disease symptoms.

Most viruses affecting the nervous system enter it by circulation in the blood or lymph. However, some viruses can enter peripheral nerve axons and move along them toward the CNS.

Poliomyelitis

Learning Objective

- Compare the Salk and Sabin polio vaccines.

Poliomyelitis (polio) is best known as a cause of paralysis. However, the paralytic form of poliomyelitis probably affects fewer than 1% of those infected with the poliovirus. The great majority of cases are asymptomatic or exhibit only mild symptoms, such as headache, sore throat, fever, and nausea. Asymptomatic or mild cases of polio are most common in the very young. In some parts of the world, because of poor sanitary conditions and the absence of vaccination, many infants contract asymptomatic poliomyelitis (while they are still protected by maternal antibodies) and thus develop immunity. When infection occurs in adolescence or early adulthood, the paralytic form of the disease occurs more frequently.

Polioviruses are more stable than most other viruses and can remain infectious for relatively long periods in water and food. The primary mode of transmission is ingestion of water contaminated with feces containing the virus.

Because the infection is initiated by ingestion of the virus, its primary areas of multiplication are the throat and small intestine. This accounts for the initial sore throat and nausea. Next, the virus invades the tonsils and the lymph nodes of the neck and ileum (the terminal portion of the small intestine). From the lymph nodes, the virus enters the blood, resulting in *viremia*. In most cases the viremia is only transient, the infection does not progress past the lymphatic stage, and clinical disease does not result. If the viremia is persistent, however, the virus penetrates the capillary walls and enters the central nervous system. Once in the CNS, the virus displays a high affinity for nerve cells, particularly motor nerve cells in the upper spinal cord. The virus does not infect the peripheral nerves or the muscles. As the virus multiplies within the cytoplasm of the motor nerve cells, the cells

FIGURE 22.9 Iron lungs in the polio ward of a Boston hospital in the 1950s. Many polio patients were able to breathe only with these mechanical aids. A few survivors from these polio epidemics still use these machines, at least part of the time. Others are able to use portable respiratory aids.

die and paralysis results. Death can result from respiratory failure (Figure 22.9).

Diagnosis of polio is usually based on isolation of the virus from feces and throat secretions. Cell cultures can be inoculated, and cytopathic effects on the cells can be observed (see Table 15.4).

The incidence of polio in the United States has decreased markedly since the availability of the polio vaccines (Figure 22.10). The last cases attributed to a wild-type virus were reported in 1979. The development of the first polio vaccine was made possible by the introduction of practical techniques for cell culture, because

the virus does not grow in any common laboratory animal. The polio vaccine was the prototype for the vaccines for mumps, measles, and rubella.

There are three different serotypes of the poliovirus, and immunity must be provided for all three. Two vaccines are available. The *Salk vaccine*, developed in 1954, uses viruses that have been inactivated by treatment with formalin. Vaccines of this type, called inactivated polio vaccines (IPV), require a series of injections. Their effectiveness rate may be as high as 90% against paralytic polio. The antibody levels decline with time, and booster shots are needed every few years to maintain full immunity. Using only IPV, several European countries have almost eliminated polio from their populations. A newer IPV has been introduced that is produced on human diploid cells. Known as *enhanced inactivated polio vaccine (E-IPV)*, it has replaced the original IPV in the United States.

The *Sabin vaccine*, introduced in 1963, contains three living, attenuated strains of the virus and has been more popular in the United States than the Salk vaccine. It is less expensive to administer, and most people prefer taking a sip of orange-flavored drink containing the virus to having a series of injections. The Sabin vaccine is also called oral polio vaccine, or OPV. The immunity achieved with the OPV resembles that acquired by natural infection, and the virus is excreted.

One disadvantage is that, on rare occasions—one in 750,000 first doses, one in about 2.4 million on subsequent doses—one of the attenuated strains of the excreted virus (type 3) may revert to virulence and transmit the disease. These cases often occur in secondary contacts, not in the person who received the vaccine. This has caused 8 or 9 cases a year. This illustrates that recipients of

FIGURE 22.10 Reported cases of poliomyelitis in the United States, 1950–1999. Notice the decrease after introduction of the Salk and Sabin vaccines. The inset detail shows the annual cases from 1980 to 1999; notice that most are associated with the Sabin vaccine. [SOURCES: CDC *Summary of Notifiable Diseases 1995*, and *Summary of Notifiable Diseases 1998*, MMWR 48 (51 and 52) (1/7/00).]

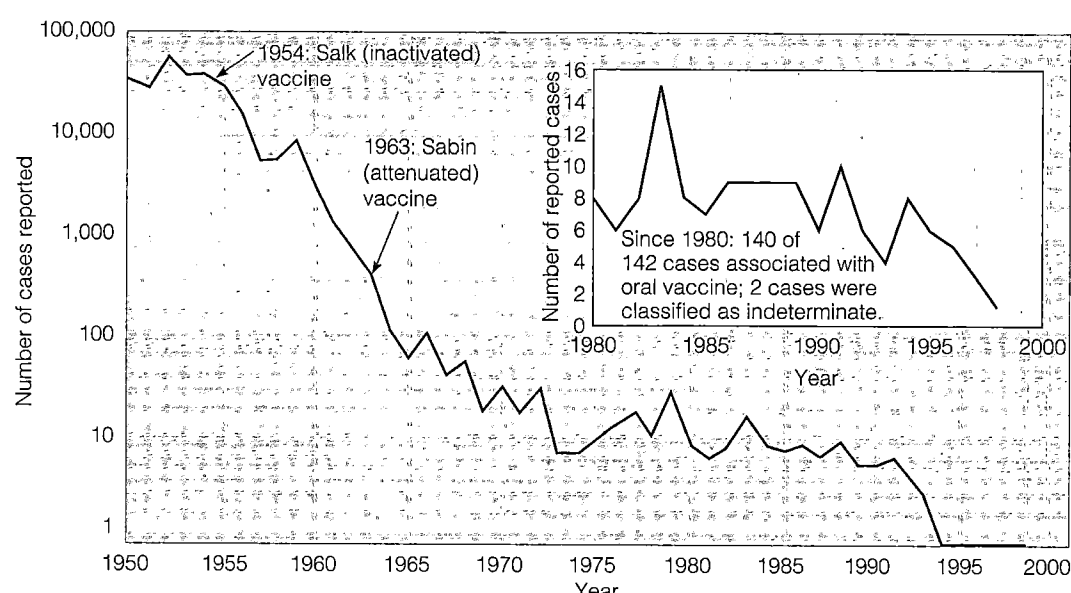

the Sabin vaccine can infect contacts, leading, in most cases, to immunization.

In 2000 the CDC decided that the advantages of the OPV no longer outweighed the risks and their recommendations are now to use only IPV for the routine immunization of children. The OPV should be used only to control widespread outbreaks, for protection of children traveling to areas of high risk, or for children who do not receive all four shots of IPV on schedule.

Immunosuppressed individuals should receive E-IPV to reduce the risk of vaccine-related polio from a live virus.

During the 1980s, many middle-aged adults who had had polio as children began showing a muscle weakness now called *post-polio syndrome*. It may be that nerve cells that had survived polio originally are now beginning to die. Fortunately, progress of the disease is extremely slow.

The disease appears to have been eliminated in North and South America, Europe, and the Western Pacific where cases caused by a wild-type virus have not been reported for several years. Humans are the only host for polio virus. China and India have carried out campaigns in which more than 100 million people were immunized in a single day. But eliminating polio will be more difficult than eradicating smallpox. Whereas smallpox expressed obvious symptoms, many cases of polio are asymptomatic, and the fecal-oral route of transmission makes containment more difficult. Eradication in parts of Africa, the Middle East, and south Asia will be the most difficult.

Rabies

Learning Objective

■ *Compare the preexposure and postexposure treatments for rabies.*

Rabies (the word is from the Latin for rage or madness), is a disease that almost always results in fatal encephalitis. The causative agent is the rabies virus, a rhabdovirus having a characteristic bullet shape. Humans usually acquire the rabies virus from the bite of an infected animal. Even the lick of such an animal is dangerous because rabies viruses in its saliva can enter minute scratches. The virus can cross the intact mucous membrane lining the eyes or body openings. Scientists investigating bat behavior in caves have contracted rabies, presumably from rabies viruses present in aerosols of bat secretions in the cave atmosphere. Humans have also acquired rabies from aerosols of the virus when macerating infected tissue in a laboratory blender. Several cases have been traced to corneal transplants.

Rabies is unique in that the incubation period is usually long enough to allow immunity to develop from post-

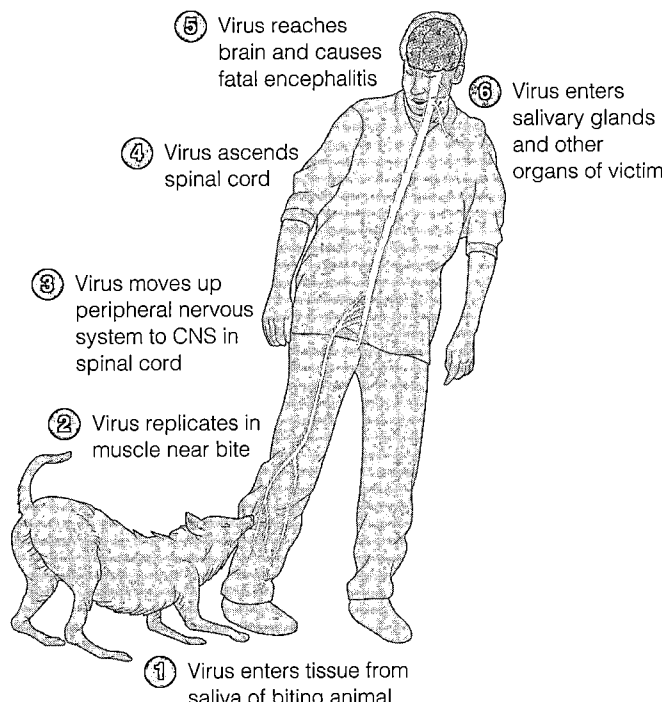

FIGURE 22.11   Pathology of rabies infection.

exposure vaccination. (The amount of virus introduced into the wound is usually too small to provoke adequate natural immunity in time.) Initially, the virus multiplies in skeletal muscle and connective tissue, where it remains localized for periods ranging from days to months. Then it enters and travels, at the rate of 15–100 mm per day, along peripheral nerves to the CNS, where it causes encephalitis (Figure 22.11). In some extreme cases, incubation periods of as long as 6 years have been reported, but the average is 30–50 days. Bites in areas rich in nerve fibers, such as the hands and face, are especially dangerous and the resulting incubation period tends to be short.

Once the virus enters the peripheral nerves, it is not accessible to the immune system until cells of the CNS begin to be destroyed, which triggers a belated immune response.

Preliminary symptoms are mild and varied, resembling several common infections. When the CNS becomes involved, the patient tends to alternate between periods of agitation and intervals of calm. At this time, a frequent symptom is spasms of the muscles of the mouth and pharynx that occur when the patient feels air drafts or swallows liquids. In fact, even the mere sight or thought of water can set off the spasms—thus the common name *hydrophobia* (fear of water). The final stages of the disease result from extensive damage to the nerve cells of the brain and the spinal cord.

Animals with **furious rabies** are at first restless, then become highly excitable and snap at anything within reach. The biting behavior is essential to maintaining the virus in the animal population. Humans also exhibit similar symptoms of rabies, even biting others. When paralysis sets in, the flow of saliva increases as swallowing becomes difficult, and nervous control is progressively lost. The disease is almost always fatal within a few days.

Some animals suffer from **paralytic rabies,** in which there is only minimal excitability. This form is especially common in cats. The animal remains relatively quiet and even unaware of its surroundings, but it might snap irritably if handled. A similar manifestation of rabies occurs in humans and is often misdiagnosed as Guillain-Barré syndrome, a form of paralysis that is usually transient but sometimes fatal, or other conditions. There is some speculation that the two forms of the disease may be caused by slightly different forms of the virus.

Laboratory diagnosis of rabies in humans and animals is based on several findings. When the patient or animal is alive, a diagnosis can sometimes be confirmed by immunofluorescence studies, in which viral antigens are detected in saliva, serum, or cerebrospinal fluid. After death, diagnosis is confirmed by a fluorescent-antibody test performed on brain tissue.

Rabies Treatments

Any person bitten by an animal that is positive for rabies must undergo *postexposure prophylaxis*—meaning a series of antirabies vaccine and immune globulin injections. Another indication for antirabies treatment is any unprovoked bite by a skunk, bat, fox, coyote, bobcat, or raccoon not available for examination. Treatment after a dog or cat bite, if the animal cannot be found, is determined by the prevalence of rabies in the area. The bite of a bat may be no more perceptible than a hypodermic needle mark; therefore, any encounter with a bat, unless a bite can be excluded, is considered by the CDC as a recommendation for vaccination—unless the bat can be tested and shown to be negative for rabies. Furthermore, it may be impossible to exclude a bite in cases where the bat had access to sleeping persons or small children.

The original Pasteur treatment, in which the virus was attenuated by drying in the dissected spinal cords of rabies-infected rabbits, has been replaced in the United States by human diploid cell vaccine (HDCV), or chick embryo–grown vaccines. These vaccines are administered in a series of 5–6 injections at intervals during a 28-day period. Passive immunization is provided simultaneously by injecting human rabies immune globulin (RIG) that has been harvested from people who are immunized against rabies, such as laboratory workers.

Rabies occurs all over the world. In the United States, 20,000–30,000 people get the rabies vaccine each year. Nearly twice that many *die* of rabies each year in India,

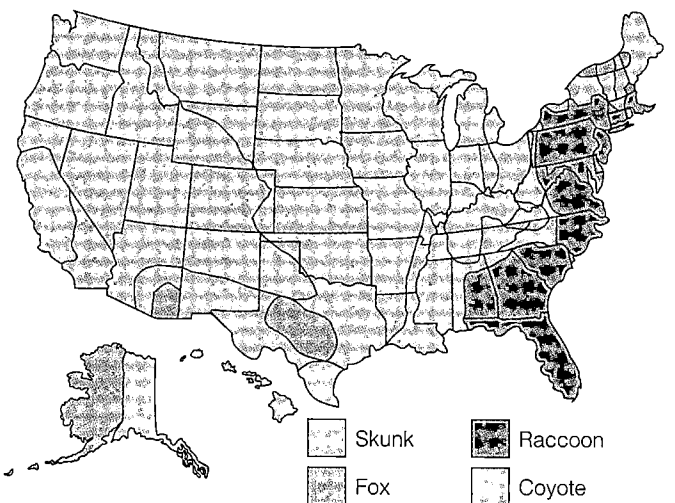

Skunk	Raccoon
Fox	Coyote

Predominant wildlife species affected by rabies in the United States. Several antigenically different strains of the virus are associated with different animals. This is often useful in tracing the possible origin of cases when it is otherwise unknown (as described in the box in this chapter).

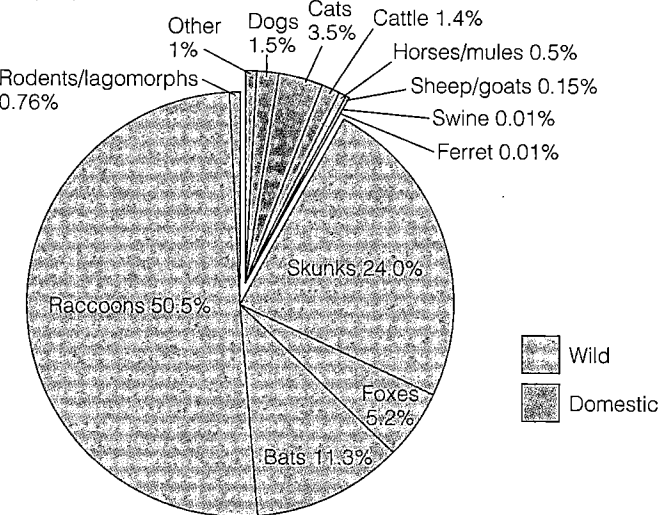

Rabies cases in various wild and domestic animals in the U.S.A. in 1998.

FIGURE 22.12 Reported cases of rabies in animals. All of the 48 contiguous states have reported rabies in bats. Rabies in foxes include different species in different geographical areas. [SOURCE: J. W. Krebs, J. S. Smith, C. E. Rupprecht, J. E. Childs. "Rabies surveillance in the United States during 1997." *Journal of the American Veterinary Medical Association* 213(12)(Dec. 15, 1998).]

■ What is the primary reservoir for the rabiesvirus in your area?

where the vaccine is given to about 3 million people. Australia, Great Britain, New Zealand, and Hawaii are free of the disease, a condition maintained by rigid quarantine. In North America, rabies is widespread among wildlife, predominantly skunks, foxes, and raccoons, although it is also found in domesticated animals (Figure 22.12). In fact, worldwide, dogs are the most common reservoirs of rabies. Rabies is almost never found in squirrels, rabbits, rats, or mice. The disease has long been endemic in vampire bats of South America. In Europe and North America, there are ongoing experiments to immunize wild animals with live rabies vaccine produced in genetically engineered vaccinia viruses that are added to food dropped for the animals to find. It has been found that gray foxes prefer dog-food bait flavored with vanilla. Coyotes are not at all choosy. These experiments have been highly successful in Europe.

In the United States 7000–8000 cases of rabies are diagnosed in animals each year, but in recent years, only 1–6 cases have been diagnosed in humans annually (see the box on page 617).

Arboviral Encephalitis

Learning Objective

■ *Explain how arboviral encephalitis can be prevented.*

Encephalitis caused by mosquito-borne viruses (called arboviruses) is rather common in the United States. (Arbovirus is short for *arthropod-borne virus.*) Figure 22.13 shows the incidence over a sequence of years. The increase in the summer months coincides with the proliferation of adult mosquitoes during these months. *Sentinel animals,* such as caged rabbits or chickens, are tested periodically for antibodies to arboviruses. This gives health officials information on the incidence and types of virus in their area.

A number of clinical types of arboviral encephalitis have been identified; all can cause symptoms ranging from subclinical to severe, and even rapid death. Active cases of these diseases are characterized by chills, headache, and fever. As the disease progresses, mental confusion and coma occur. Survivors may suffer from permanent neurological problems.

Horses as well as humans are frequently affected by these viruses; thus, there are strains causing *Eastern equine encephalitis (EEE)* and *Western equine encephalitis (WEE).* These two viruses are the most likely to cause severe disease in humans. EEE is the most severe. The mortality rate is 35% or more, and survivors experience a high incidence of brain damage, deafness, and other neurological problems. EEE is uncommon (its main mosquito vector prefers to feed on birds); only about 100 cases a year are reported. WEE has been only rarely reported in recent years.

St. Louis encephalitis (SLE) acquired its name from the location of an early major outbreak (in which it was discovered that mosquitoes are involved in the transmission of these diseases). SLE is the most common form of arboviral encephalitis, even though fewer than 1% of infec-

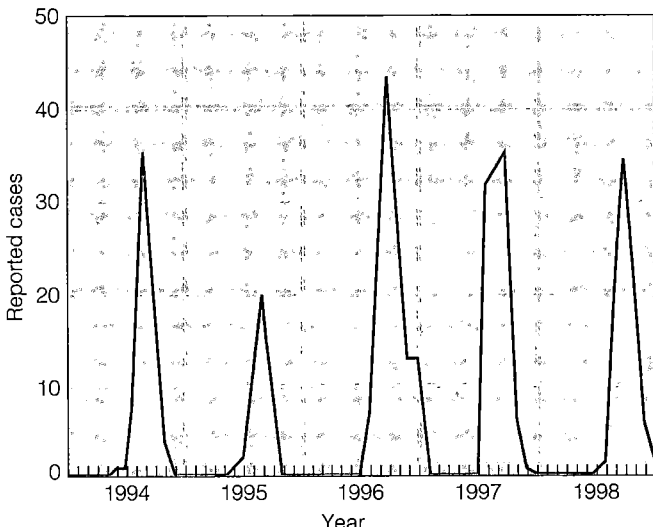

FIGURE 22.13 Arboviral infections of the central nervous system. Cases per month caused by California encephalitis viruses in the United States, 1993–1997. Notice the seasonal occurrence of cases.
[SOURCE: Centers for Disease Control and Prevention, *"Summary of Notifiable Diseases, 1997." MMWR* 46(54), 44(53) (1/3/97).]

■ **Why do arboviral infections occur during the summer months?**

tions exhibit symptoms. *California encephalitis (CE)* was first identified in that state, but most cases occur elsewhere. The La Crosse strain of CE is the most commonly encountered arbovirus. Some important characteristics of arboviral encephalitis types are summarized in Table 22.2.

In the New York City area in the fall of 1999 a veterinary pathologist at a zoo reported deaths among certain of their exotic birds, and local health authorities observed increased deaths in wild birds, especially crows. The cause was identified as a mosquito-borne virus not previously reported in the United States, the *West Nile virus.* The outbreak also caused cases in humans and resulted in several deaths. It is suspected that the virus was introduced from the Middle East, although there have also been outbreaks in Eastern Europe. If infected migratory birds spread the disease to areas of the country where mosquitoes survive year round, the disease may become endemic in the United States.

The Far East also has endemic encephalitis. **Japanese B encephalitis** is the best known; it is a serious public health problem, especially in Japan, Korea, and China. Vaccines are used to control the disease in these countries and are often recommended for visitors.

Diagnosis of arboviral encephalitis is made by serological tests, usually ELISA tests to identify IgM antibodies. The most effective preventive measure is local control of the mosquitoes.

table 22.2	Types of Arboviral Encephalitis			
Disease	Mosquito Vector	Host Animals	U.S. Distribution	Comment
Eastern equine encephalitis (EEE)	Aedes, Culiseta	Birds, horses	East Coast	More severe even than WEE; affects mostly young children and younger adults; relatively uncommon in humans.
Western equine encephalitis (WEE)	Culex	Birds, horses	Western United States	Severe disease; frequent neurological damage, especially in infants.
St. Louis encephalitis (SLE)	Culex	Birds	Throughout country	Mostly urban outbreaks; affects mainly adults over 40.
California encephalitis (CE)	Aedes	Small mammals	North–central states, New York State	Affects mostly 5–18 year age group in rural or suburban areas; La Crosse strain medically most important. Rarely fatal, about 10% have neurological damage.

Fungal Disease of the Nervous System

The central nervous system is seldom invaded by fungi. However, one pathogenic fungus in the genus *Cryptococcus* is well adapted to growth in CNS fluids.

Cryptococcus neoformans Meningitis (Cryptococcosis)

Learning Objective

- *Identify the causative agent, vector, symptoms, and treatment for cryptococcosis.*

Fungi of the genus *Cryptococcus* are spherical cells resembling yeasts; they reproduce by budding and produce polysaccharide capsules, some much thicker than the cells themselves (Figure 22.14). Only one species, *Cryptococcus neoformans*, is pathogenic for humans, causing the disease called **cryptococcosis.** The organism is widely distributed in soil, especially soil contaminated with pigeon droppings. It is also found in pigeon roosts and nests on the window ledges of urban buildings. Most cases of cryptococcosis occur in urban areas, and it is thought to be transmitted by the inhalation of dried infected pigeon droppings. The disease was relatively rare before the advent of AIDS.

Inhalation of *C. neoformans* initially causes infection of the lungs, frequently subclinical, and often the disease does not proceed beyond this stage. However, it can spread through the bloodstream to other parts of the body, including the brain and meninges, especially in immunosuppressed individuals and those receiving steroid treatments for major illnesses. The disease is usually expressed as chronic meningitis, which is often progressive and fatal if untreated.

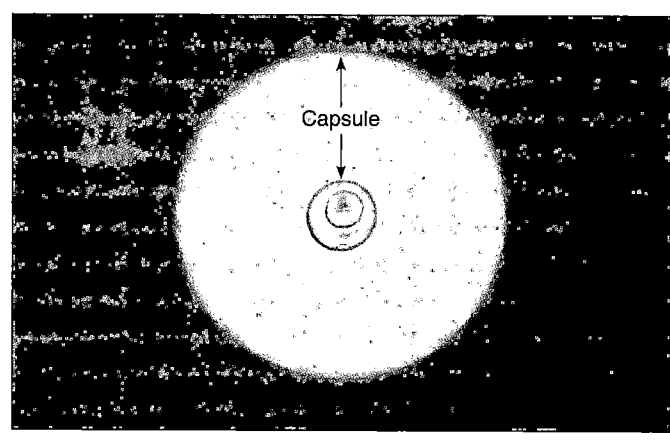

Capsule

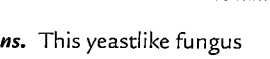

LM   10 mm

FIGURE 22.14 *Cryptococcus neoformans.* This yeastlike fungus has an unusually thick capsule. In this photomicrograph, the capsule is made visible by suspending the cells in dilute India ink.

■ **What disease does *C. neoformans* cause?**

The best serological diagnostic test is a latex agglutination test to detect cryptococcal antigens in serum or cerebrospinal fluid. The drugs of choice for treatment are amphotericin B and flucytosine in combination. Even so, the mortality rate may approach 30%.

Protozoan Diseases of the Nervous System

Learning Objective

- *Identify the causative agent, vector, symptoms, and treatment for African trypanosomiasis and* Naegleria *meningoencephalitis.*

CLINICAL PROBLEM SOLVING

A Neurological Disease

You will see questions as you read through this problem. The questions are those that clinicians ask themselves as they proceed through a diagnosis and treatment. Try to answer each question before going on to the next one.

1. On December 14, a 29-year-old Virginia man with muscle pains, vomiting, and abdominal cramps was treated with the pain reliever acetaminophen at a clinic. After he returned home, his signs progressed to include persistent right wrist pain, muscle tremors in his right arm, and difficulty walking. On December 18, he was transported to an emergency department, where his temperature was found to be 39.4°C. Physical examination revealed increased tone in his right forearm, unequal pupil sizes, and abnormally high sensitivity to stimuli. He initially was alert and oriented but had visual hallucinations. *What diseases are possible? What additional information do you need?*

2. Toxicology tests of his blood were negative for pesticides and Jimson weed, and tests of CSF were negative for human herpesvirus and bacteria. The patient's condition worsened, with wide fluctuations in body temperature and blood pressure. He was intubated and heavily sedated on December 20. *What diseases are possible? What additional information do you need?*

3. A skin biopsy was positive for rabies virus by direct fluorescent anti-

body test on December 22, and saliva and skin samples were positive by reverse transcriptase-PCR assay on December 23. On December 28, serum samples collected on December 21 and December 28 had rabies virus neutralizing antibody titers of 1:50 and 1:1200, respectively. *What does the change in antibody titer indicate? What is the disease? What treatment do you recommend?*

4. The presence of the virus and the rising antibody titer indicate rabies. There is no treatment after the disease process begins. By December 28, the man had loss of brain stem reflexes, and he died on December 31. *How would you treat people who had contact with the patient between December 14 and his death?*

5. Postexposure prophylaxis (PEP) was administered to 48 people, including 15 health care providers and the pathologist who conducted the autopsy. *Did the delay in testing the December 21 serum sample affect the outcome of the disease? Sedating the patient may have masked the classic symptoms of hydrophobia and hypersalivation. What problem did this cause?*

6. Early diagnosis cannot save a patient, however, it may help minimize the number of potential exposures and the need for PEP. *What else must be determined about this case?*

7. The nucleotide sequence of the PCR product was used to identify a rabies virus variant associated with eastern pipistrelle bats and silver-haired bats. The patient had no known animals bites, and no evidence of bats was found near his home. *Why is rabies surveillance and case reporting important in the United States?*

8. Bat rabies is endemic in wild animals in the United States and has been documented in all 48 contiguous states. Bats have accounted for an increasing percentage of rabies cases transmitted from wildlife reservoirs to humans. Since 1990, 27 human rabies cases have occurred in the United States. Although 20 have been attributed to bat-associated variants of the rabies virus, a definitive history of a bat bite was established for only one of these cases. Of the 20 attributed to bat-associated variants, 15 have been caused by the pipistrelle/silver-haired bat variant.

    The eastern pipistrelle and silver-haired bats are insectivorous bats with small teeth that may not cause an obvious wound in human skin. Accordingly, it is important to treat persons for rabies exposure when the possibility of a bat bite cannot be reasonably excluded.

SOURCE: Adapted from *MMWR* 48(5):95–97 (2/12/99).

Protozoa capable of invading the central nervous system are rare. However, those that can reach it cause devastating effects.

African Trypanosomiasis

African trypanosomiasis, or sleeping sickness, is a protozoan disease that affects the nervous system. In 1907, Winston Churchill described Uganda during an epidemic of sleeping sickness as a "beautiful garden of

death." Even today, the disease affects about 1 million people in Central and East Africa, and about 20,000 new cases are reported each year.

The disease is caused by *Trypanosoma brucei gambiense* and *Trypanosoma brucei rhodesiense,* flagellates that are injected by the bite of a tsetse fly. Animal reservoirs for the two trypanosomes are similar, but they occur in different habitats and are spread by different species of tsetse fly vector. Human-to-human transmission by bites of the insect vector is more likely with infections by *T. b. gambiense,*

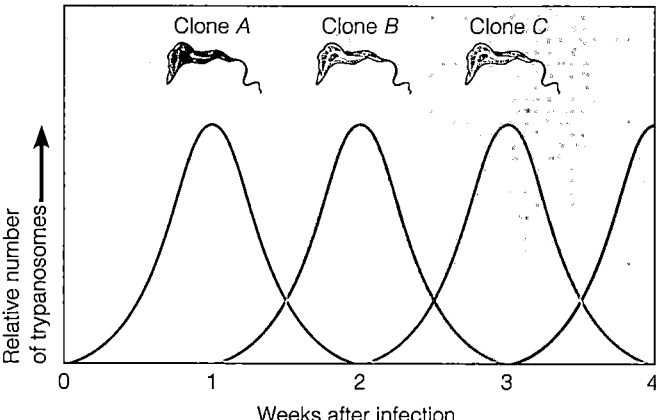

FIGURE 22.15 How trypanosomes evade the immune system.
The population of each trypanosome clone drops nearly to zero as the immune system suppresses its members, but a new clone with a different antigenic surface then replaces the previous clone. The black line represents the population of clone D.

■ Different trypanosomal cells can express different genes for their protein coat, and the immune system will not be able to respond to all the different antigens at once.

which circulates in the patient's blood during the 2- to 4-year course of the disease. *T. b. rhodesiense* causes a disease that is more acute, running its course in only a few months.

During the early stages of either form of the disease, a few trypanosomes can be found in the blood. During later stages, the pathogens move into the CSF. Symptoms include decreases in physical activity and mental acuity. Untreated, the host enters a coma, and death is almost inevitable.

There are some moderately effective chemotherapeutic agents, such as suramin and pentamidine isethionate. A new drug, eflornithine, which blocks an enzyme required for proliferation of the parasite, has come into use. It is so dramatically effective against *T. b. gambiense* that it has been called the resurrection drug. However, its effectiveness against *T. b. rhodesiense* is variable.

A vaccine is being developed, but a major obstacle is that the trypanosome is able to change protein coats at least 100 times and can thus evade antibodies aimed at only one or a few of the proteins. Each time the body's immune system is successful in suppressing the trypanosome, a new clone of parasites appears with a different antigenic coat (Figure 22.15).

Naegleria Meningoencephalitis

Naegleria fowleri (nī-glē'rē-ä fou'ler-ē) is a protozoan (amoeba) that causes the neurological disease *Naegleria* **meningoencephalitis** (Figure 22.16). Although cases of it are reported in most parts of the world, only a few annual cases are reported in the United States. The most

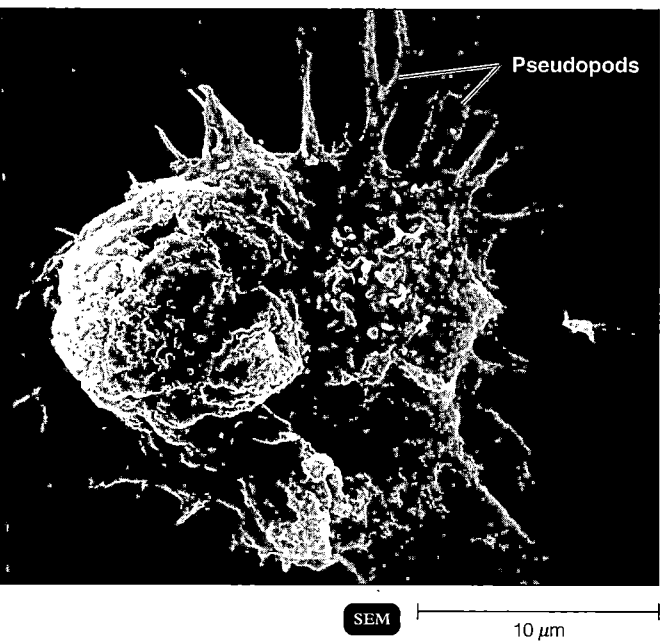

SEM ├──────── 10 μm ────────┤

FIGURE 22.16 *Naegleria fowleri,* **the cause of** *Naegleria* **meningoencephalitis.** This micrograph shows the numerous pseudopods of this deadly parasite. Shown is the trophozoite form. The infectious stage is a pear-shaped, flagellated form that is motile in its water habitat.

■ How is *N. fowleri* transmitted?

common victims are children who swim in ponds or streams. The organism initially infects the nasal mucosa and later proliferates in the brain. The fatality rate is nearly 100%; diagnosis is typically made at autopsy.

Nervous System Diseases Caused by Prions

Learning Objective

■ List the characteristics of diseases caused by prions.

A number of fatal diseases of the human central nervous system are caused by prions, self-replicating proteins with no detectable nucleic acids. (See the discussion of prions in Chapter 13, page 398.) These diseases have long incubation times, measured in years. CNS damage is insidious and slowly progressive, without the fever and inflammation seen in encephalitis. Autopsies show a characteristic spongiform (porous, like a sponge) degeneration of the brain (Figure 22.17a). Also present in brain tissue are characteristic fibrils (Figure 22.17b). In recent years, the study of these diseases has been one of the most interesting areas of medical microbiology.

A typical prion disease is **sheep scrapie.** The diseased animal scrapes itself against fences and walls until areas on

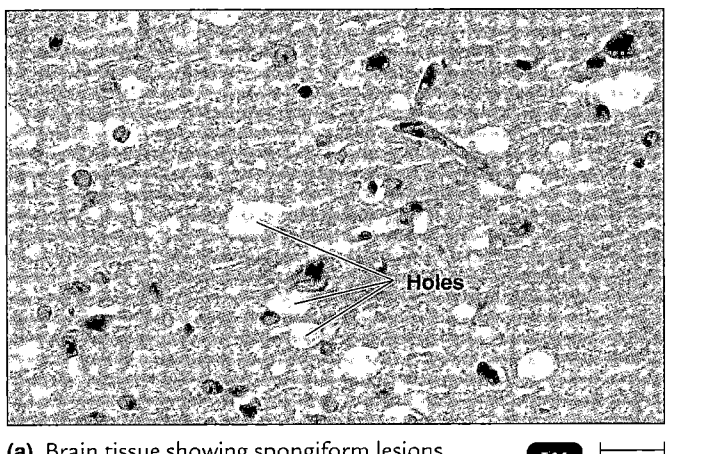

(a) Brain tissue showing spongiform lesions   **LM** ⊢——⊣ 25 μm

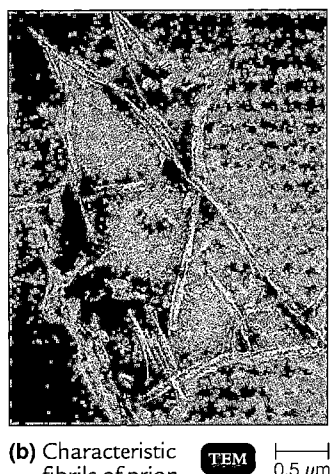

(b) Characteristic **TEM** ⊢——⊣ 0.5 μm
fibrils of prion-
caused diseases

FIGURE 22.17 Spongiform encephalopathies. These diseases, presumed to be caused by prions, include bovine spongiform encephalopathy, scrapie in sheep, and Creutzfeld-Jakob disease in humans. All are similar in their pathology. **(a)** Note the clear holes that give this slide of brain tissue a spongy appearance. This pathology is responsible for the term *spongiform*. **(b)** Brain tissue showing characteristic fibrils produced by prion diseases. Prions themselves are not visible by any known technology.

■ **What are prions?**

its body are raw. During a period of several weeks or months, the animal gradually loses motor control and dies. The prion can be passed to other animals by the injection of brain tissue from one animal to the next.

Humans suffer from neurological diseases similar to scrapie. **Creutzfeldt-Jakob disease (CJD)** is an example. CJD is rare (about 200 cases per year in the United States), and it often occurs in families. There is no doubt that an infective agent is involved because transmission via corneal transplants and accidental scalpel nicks of a surgeon during autopsy have been reported. Several cases have been traced to the injection of a growth hormone derived from human tissue. Incineration is apparently the only certain way of disinfecting prion-contaminated tissues or materials (Figure 22.18). Boiling and irradiation have no effect, and even autoclaving is not completely reliable. A temperature of about 132°C applied for an hour appears to be required.

Some tribes in New Guinea have suffered from a prion disease called **kuru.** Kuru is apparently related to the practice of smearing brain tissue onto the body during cannibalistic rituals. The infection results when the agent is introduced into open sores or cuts. The disease is disappearing as cannibalism dies out.

A current international public health concern is **bovine spongiform encephalopathy (BSE).** Since 1986, thousands of cattle in Great Britain have contracted this condition, known locally as *mad cow disease.* BSE causes the animals to become unmanageable, and they must be destroyed. The current theory is that the disease

is probably due to prions transmitted in feed supplements derived from animals such as sheep. Although this theory is considered plausible, the definitive experiments to prove it have not been done. It is hoped that rigid quarantine measures, including the slaughter of any herds suspected of infection, will soon extinguish BSE. Several tests have been developed in Europe that have the potential to reliably detect BSE in animals.

Of more immediate concern is the fact that in Britain, several cases of CJD have appeared in relatively young humans. CJD rarely occurs in this age group.

FIGURE 22.18 Disposal of animals infected with bovine spongiform encephalopathy. Herds in Great Britain are being culled to arrest the spread of this disease. Incineration is the only reliable method for disposing of contaminated animal carcasses.

There is great concern that CJD may have been acquired by the ingestion of beef. However, the apparent causative agent, called **bovine variant,** of the new cases of CJD shares many characteristics with the agent of BSE. Experts at the WHO concluded that there was no definite link between BSE and the British new-variant CJD (nv CJD) cases, but that circumstantial evidence suggested that exposure to BSE was the most likely explanation. The primary concern now is whether large numbers of humans have been infected, and the fact that it will be years before symptoms appear. In 1997, the FDA banned all feeding of mammal-derived products to cattle or other ruminants. Also, as a precaution, restrictions have been placed on blood donations by persons who spent significant time (six months) in Britain during the height of the BSE outbreak. The U.S. Department of Agriculture conducts systematic surveys to detect BSE in the United States. No cases have been found.

Table 22.3 summarizes the main microbial diseases associated with the nervous system.

table 22.3	Microbial Diseases of the Nervous System	
Disease	**Pathogen**	**Comments**
Bacterial Diseases		
Haemophilus influenzae meningitis	H. influenzae	Occurs primarily in children under age 4; childhood vaccine is decreasing incidence of disease.
Meningococcal meningitis	Neisseria meningitidis	Affects mostly children under age 2. Group A causes widespread epidemics in Africa; group C in United States causes local outbreaks.
Pneumococcal meningitis	Streptococcus pneumoniae	Occurs in children under age 4 and hospitalized elderly. Highest mortality rate of bacterial meningitis.
Listeriosis	Listeria monocytogenes	Usually transmitted by contaminated food. Main danger is to fetus.
Tetanus	Clostridium tetani	Toxin formed in contaminated wound causes uncontrolled muscle contractions, eventual respiratory failure.
Botulism	Clostridium botulinum	Toxin preformed in food is ingested; causes paralysis and respiratory failure.
Leprosy	Mycobacterium leprae	Bacteria grow in PNS; eventually cause extensive tissue damage.
Viral Diseases		
Poliomyelitis	Poliovirus	Transmitted mainly with ingested water. About 1% of cases result in at least partial paralysis.
Rabies	Rabies virus	Usually transmitted by animal bites; death from respiratory failure.
Arboviral encephalitis	Arboviruses	Mosquito-borne; seriousness varies with the viral species.
Fungal Disease		
Cryptococcosis	Cryptococcus neoformans	A severe meningitis transmitted by inhalation of fungus, often from bird droppings that support its growth.
Protozoan Diseases		
African trypanosomiasis	Trypanosoma brucei rhodesiense, T. b. gambiense	Transmitted by tsetse fly bites in central Africa. Eventually affects CNS, leading to coma and death.
Naegleria meningoencephalitis	Naegleria fowleri	Rare but deadly; transmitted by amoeba in fresh water during activities such as swimming.
Prion Diseases		
Creutzfeldt-Jakob disease	Prion	A rare but fatal neurological disorder apparently transmitted from infected blood or tissue.
Kuru	Prion	A fatal neurological disorder in isolated New Guinea tribes. Transmitted by contact with brain and tissue of dead victims, now declining as practices that led to this contact decline.

Study Outline

STRUCTURE AND FUNCTION OF THE NERVOUS SYSTEM (pp. 602–603)

1. The central nervous system (CNS) consists of the brain, which is protected by the skull bones, and the spinal cord, which is protected by the backbone.

2. The peripheral nervous system (PNS) consists of the nerves that branch from the CNS.

3. The CNS is covered by three layers of membranes called meninges: the dura mater, arachnoid, and pia mater. Cerebrospinal fluid (CSF) circulates between the arachnoid and the pia mater in the subarachnoid space.

4. The blood-brain barrier normally prevents many substances, including antibiotics, from entering the brain.

5. Microorganisms can enter the CNS through trauma, along peripheral nerves, and through the bloodstream and lymphatic system.

6. An infection of the meninges is called meningitis. An infection of the brain is called encephalitis.

BACTERIAL DISEASES OF THE NERVOUS SYSTEM (pp. 603–611)

Bacterial Meningitis (pp. 603–607)

1. Meningitis can be caused by viruses, bacteria, fungi, and protozoa.

2. The three major causes of bacterial meningitis are *Haemophilus influenzae, Streptococcus pneumoniae,* and *Neisseria meningitidis.*

3. Nearly 50 species of opportunistic bacteria can cause meningitis.

Haemophilus influenzae *Meningitis* (pp. 604–605)

1. *H. influenzae* is part of the normal throat microbiota.

2. *H. influenzae* requires blood factors for growth; there are six types of *H. influenzae* based on capsule differences.

3. *H. influenzae* type b is the most common cause of meningitis in children under 4 years old.

4. A conjugated vaccine directed against the capsular polysaccharide antigen is available.

Neisseria *Meningitis (Meningococcal Meningitis)* (p. 605)

1. *N. meningitidis* causes meningococcal meningitis. This bacterium is found in the throats of healthy carriers.

2. The bacteria probably gain access to the meninges through the bloodstream. The bacteria may be found in leukocytes in CSF.

3. Symptoms are due to endotoxin. The disease occurs most often in young children.

4. Military recruits are vaccinated with purified capsular polysaccharide to prevent epidemics in training camps.

Streptococcus pneumoniae *Meningitis (Pneumococcal Meningitis)* (pp. 605–606)

1. *S. pneumoniae* is commonly found in the nasopharynx.

2. Hospitalized patients and young children are most susceptible to *S. pneumoniae* meningitis. It is rare but has a high mortality rate.

3. The vaccine for pneumococcal pneumonia may provide some protection against pneumococcal meningitis.

Diagnosis and Treatment of the Most Common Types of Bacterial Meningitis (p. 606)

1. Cephalosporins may be administered before identification of the pathogen.

2. Diagnosis is based on Gram stain and serological tests of the bacteria in CSF.

3. Cultures are usually made on blood agar and incubated in an atmosphere containing reduced oxygen levels.

Listeriosis (pp. 606–607)

1. *Listeria monocytogenes* causes meningitis in newborns, the immunosuppressed, pregnant women, and cancer patients.

2. Acquired by ingestion of contaminated food, it may be asymptomatic in healthy adults.

3. *L. monocytogenes* can cross the placenta and cause spontaneous abortion and stillbirth.

Tetanus (pp. 607–608)

1. Tetanus is caused by a localized infection of a wound by *Clostridium tetani.*

2. *C. tetani* produces the neurotoxin tetanospasmin, which causes the symptoms of tetanus: spasms, contraction of muscles controlling the jaw, and death resulting from spasms of respiratory muscles.

3. *C. tetani* is an anaerobe that will grow in deep, unclean wounds and wounds with little bleeding.

4. Acquired immunity results from DPT immunization that includes tetanus toxoid.

5. Following an injury, an immunized person may receive a booster of tetanus toxoid. An unimmunized person may receive (human) tetanus immune globulin.

6. Debridement (removal of tissue) and antibiotics may be used to control the infection.

Botulism (pp. 608–610)

1. Botulism is caused by an exotoxin produced by *C. botulinum* growing in foods.

2. Serological types of botulinum toxin vary in virulence, with type A being the most virulent.

3. The toxin is a neurotoxin that inhibits the transmission of nerve impulses.

4. Blurred vision occurs in 1–2 days; progressive flaccid paralysis follows for 1–10 days, possibly resulting in death from respiratory and cardiac failure.

5. *C. botulinum* will not grow in acidic foods or in an aerobic environment.

6. Endospores are killed by proper canning. The addition of nitrites to foods inhibits growth after endospore germination.

7. The toxin is heat labile and is destroyed by boiling (100°C) for 5 minutes.

8. Infant botulism results from the growth of *C. botulinum* in an infant's intestines.

9. Wound botulism occurs when *C. botulinum* grows in anaerobic wounds.

10. For diagnosis, mice protected with antitoxin are inoculated with toxin from the patient or foods.

Leprosy (pp. 610–611)

1. *Mycobacterium leprae* causes leprosy, or Hansen's disease.

2. *M. leprae* has never been cultured on artificial media. It can be cultured in armadillos.

3. The tuberculoid form of the disease is characterized by loss of sensation in the skin surrounded by nodules. The lepromin test is positive.

4. Laboratory diagnosis is based on observations of acid-fast rods in lesions or fluids and the lepromin test.

5. In the lepromatous form, disseminated nodules and tissue necrosis occur. The lepromin test is negative.

6. Leprosy is not highly contagious and is spread by prolonged contact with exudates.

7. Untreated individuals often die of secondary bacterial complications, such as tuberculosis.

8. Patients with leprosy are made noncontagious within 4–5 days with sulfone drugs and then treated as outpatients.

9. Leprosy occurs primarily in the tropics.

VIRAL DISEASES OF THE NERVOUS SYSTEM (pp. 611–615)

Poliomyelitis (pp. 611–613)

1. The symptoms of poliomyelitis are usually headache, sore throat, fever, stiffness of the back and neck, and occasionally paralysis (fewer than 1% of cases).

2. Poliovirus is transmitted by the ingestion of water contaminated with feces.

3. Poliovirus first invades lymph nodes of the neck and small intestine. Viremia and spinal cord involvement may follow.

4. Diagnosis is based on isolation of the virus from feces and throat secretions.

5. The Salk vaccine (an inactivated polio vaccine, or IPV) involves the injection of formalin-inactivated viruses and boosters every few years. The Sabin vaccine (an oral polio

vaccine, or OPV) contains three live, attenuated strains of poliovirus and is administered orally.

6. Polio will be eliminated through vaccination.

Rabies (pp. 613–615)

1. Rabies virus (a rhabdovirus) causes an acute, usually fatal, encephalitis called rabies.

2. Rabies may be contracted through the bite of a rabid animal, by inhalation of aerosols, or invasion through minute skin abrasions. The virus multiplies in skeletal muscle and connective tissue.

3. Encephalitis occurs when the virus moves along peripheral nerves to the CNS.

4. Symptoms of rabies include spasms of mouth and throat muscles followed by extensive brain and spinal cord damage and death.

5. Laboratory diagnosis may be made by direct FA tests of saliva, serum, and CSF or brain smears.

6. Reservoirs for rabies in the U.S. include skunks, bats, foxes, and raccoons. Domestic cattle, dogs, and cats may get rabies. Rodents and rabbits seldom get rabies.

7. Current postexposure treatment includes administration of human rabies immune globulin (RIG) along with multiple intramuscular injections of vaccine.

8. Preexposure treatment consists of vaccination.

Arboviral Encephalitis (p. 615)

1. Symptoms of encephalitis are chills, headache, fever, and eventually coma.

2. Many types of viruses (called arboviruses) transmitted by mosquitoes cause encephalitis.

3. The incidence of arboviral encephalitis increases in the summer months when mosquitoes are most numerous.

4. Horses are frequently infected by EEE and WEE viruses.

5. Diagnosis is based on serological tests.

6. Control of the mosquito vector is the most effective way to control encephalitis.

FUNGAL DISEASE OF THE NERVOUS SYSTEM (p. 616)

Cryptococcus neoformans Meningitis (Cryptococcosis) (p. 616)

1. *Cryptococcus neoformans* is an encapsulated yeastlike fungus that causes cryptococcosis.

2. The disease may be contracted by inhalation of dried infected pigeon droppings.

3. The disease begins as a lung infection and may spread to the brain and meninges.

4. Immunosuppressed individuals are most susceptible to *Cryptococcus neoformans* meningitis.

5. Diagnosis is based on latex agglutination tests for cryptococcal antigens in serum or CSF.

PROTOZOAN DISEASES OF THE NERVOUS SYSTEM (pp. 616–618)

African Trypanosomiasis (pp. 617–618)

1. African trypanosomiasis is caused by the protozoa *Trypanosoma brucei gambiense* and *T. b. rhodesiense* and transmitted by the bite of the tsetse fly.

2. The disease affects the nervous system of the human host, causing lethargy and eventually coma. It is commonly called sleeping sickness.

3. Vaccine development is hindered by the protozoan's ability to change its surface antigens.

Naegleria Meningoencephalitis (p. 618)

1. Encephalitis caused by the protozoan *N. fowleri* is almost always fatal.

2. The protozoan invades the brain from the nasal mucosa.

NERVOUS SYSTEM DISEASES CAUSED BY PRIONS (pp. 618–620)

1. Diseases of the CNS that progress slowly and cause spongiform degeneration are caused by prions.

2. Sheep scrapie and bovine spongiform encephalopathy (BSE) are examples of diseases caused by prions that are transferable from one animal to another.

3. Creutzfeldt-Jakob disease and kuru are human diseases similar to scrapie. They are transmitted between humans.

4. Prions are self-replicating proteins with no detectable nucleic acid.

Study Questions

REVIEW

1. Differentiate meningitis from encephalitis.

2. Fill in the following table:

Causative Agent of Meningitis	Susceptible Population	Mode of Transmission	Treatment
N. meningitidis			
H. influenzae			
S. pneumoniae			
L. monocytogenes			
C. neoformans			

3. Briefly explain the derivation of the name *Haemophilus influenzae*.

4. If *Clostridium tetani* is relatively sensitive to penicillin, why doesn't penicillin cure tetanus?

5. Compare and contrast the Salk and Sabin vaccines with respect to composition, advantages, and disadvantages.

6. What treatment is used against tetanus under the following conditions?
   (a) before a person suffers a deep puncture wound
   (b) after a person suffers a deep puncture wound

7. Why is the following description used for wounds that are susceptible to *C. tetani* infection: ". . . Improperly cleaned deep puncture wounds . . . ones with little or no bleeding . . ."?

8. List the following information for botulism: etiologic agent, suspect foods, symptoms, treatment, conditions necessary for microbial growth, basis for diagnosis, prevention.

9. Provide the following information on leprosy: etiology, method of transmission, symptoms, treatment, prevention, and susceptible population.

10. Provide the following information on poliomyelitis: etiology, method of transmission, symptoms, prevention. Why aren't the Salk and Sabin vaccines considered treatments for poliomyelitis?

11. Provide the etiology, method of transmission, reservoirs, and symptoms for rabies.

12. Outline the procedures for treating rabies after exposure. Outline the procedures for preventing rabies prior to exposure. What is the reason for the differences in the procedures?

13. Fill in the following table.

Disease	Etiology	Vector	Symptoms	Treatment
Arboviral encephalitis				
African trypanosomiasis				

14. Why are meningitis and encephalitis generally difficult to treat?

15. Provide evidence that Creutzfeldt-Jakob disease is caused by a transmissible agent.

MULTIPLE CHOICE

1. Which of the following is *not* true?
   a. Only puncture wounds by rusty nails result in tetanus.
   b. Rabies is seldom found in rodents (e. g., rats, mice).
   c. Polio is transmitted by the fecal-oral route.
   d. Arboviral encephalitis is rather common in the United States.
   e. none of the above

2. Which of the following does *not* have an animal reservoir or vector?

 a. listeriosis
 b. cryptococcosis
 c. *Naegleria* meningoencephalitis
 d. rabies
 e. African trypanosomiasis

3. A 12-year-old girl hospitalized for Guillain-Barré syndrome had a 4-day history of headache, dizziness, fever, sore throat, and weakness of legs. Seizures began 2 weeks later. Bacterial cultures were negative. She died 3 weeks after hospitalization. An autopsy revealed inclusions in brain cells that tested positive in an immunofluoresence test. She probably had

 a. rabies.
 b. Creutzfeldt-Jakob disease.
 c. botulism.
 d. tetanus.
 e. leprosy.

4. After receiving a corneal transplant, a woman developed dementia and loss of motor function; she then became comatose and died. Cultures were negative. Serological tests were negative. Autopsy revealed spongiform degeneration of her brain. She most likely had

 a. rabies.
 b. Creutzfeldt-Jakob disease.
 c. botulism.
 d. tetanus.
 e. leprosy.

5. The endotoxin is responsible for symptoms caused by which of the following organisms?

 a. *N. meningitidis*
 b. *S. pyogenes*
 c. *L. monocytogenes*
 d. *C. tetani*
 e. *C. botulinum*

6. The increased incidence of encephalitis in the summer months is due to

 a. maturation of the viruses.
 b. increased temperature.
 c. the presence of adult mosquitoes.
 d. an increased population of birds.
 e. an increased population of horses.

Match the following choices to the statements in questions 7 and 8:
 a. antirabies antibodies
 b. HDVC

7. Produces the highest antibody titer.

8. Used for passive immunization.

Use the following choices to answer questions 9 and 10:
 a. *Cryptococcus*
 b. *Haemophilus*
 c. *Listeria*
 d. *Naegleria*
 e. *Neisseria*

9. Microscopic examination of cerebrospinal fluid reveals gram-positive rods.

10. Microscopic examination of cerebrospinal fluid from a person who washes windows reveals ovoid cells.

CRITICAL THINKING

1. Most of us have been told that a rusty nail causes tetanus. What do you suppose is the origin of this adage?

2. A BCG vaccination will result in positive lepromin and tuberculin tests. What is the relationship between leprosy and tuberculosis?

3. Some health care professionals feel that the OPV should no longer be used. Provide an argument to support this view.

CLINICAL APPLICATIONS

1. A 1-year-old infant was lethargic and had a fever. When admitted to the hospital, he had multiple brain abscesses with gram-negative coccobacillary rods. Identify the disease, etiology, and treatment.

2. A 40-year-old bird handler was admitted to the hospital with soreness over his upper jaw, progressive vision loss, and bladder dysfunction. He had been well 2 months earlier. Within weeks he lost reflexes in his lower extremities and subsequently died. Examination of CSF showed lymphocytes. What etiology do you suspect? What further information do you need?

3. One week after bathing in a hot spring, a 9-year-old girl was hospitalized after a 3-day history of progressive, severe headaches, nausea, lethargy, and stupor. Examination of CSF revealed amoeboid organisms. Identify the disease, etiology, and treatment.

Learning with Technology

[MP] = **The Microbiology Place website**   [ST] = **Student Tutorial CD-ROM**   [VU] = **VirtualUnknown CD-ROM**

[MP]   Don't forget to go to The Microbiology Place website (http://www.microbiologyplace.com) to take the practice tests, explore the interactive activity and case study, and check out the news articles and web links for this chapter.

[ST]   Remember there is also a quiz for this chapter on the Microbiology Interactive Student Tutorial CD-ROM.

[VU]   Enter the Virtual Lab, click the arrow next to the Session field, click Textbook Exercises, and select Chapter 22. Read the Case Study carefully, identify the unknown, and use what you learn to answer the following questions (consult your textbook for additional information).

1. Based on the symptoms observed, what would be the logical diagnosis?

2. The organism responsible for this infection is not one routinely encountered. What organisms are more likely to be recovered from patients with this disease? Under what circumstances is each one most likely to be found?

3. Provide a scenario to explain why this unlikely organism was able to cause this infection.

chapter 23

Microbial Diseases of the Cardiovascular and Lymphatic Systems

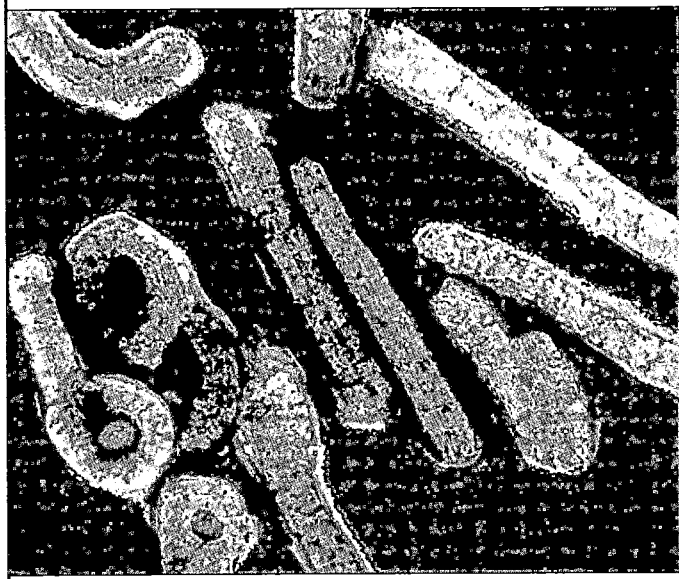

Ebola virus. This is one of several emerging hemorrhagic fever viruses. A human first contracts the virus from an unknown source, then the virus can be spread to other people via blood and secretions from the infected person.

The **cardiovascular system** consists of the heart, blood, and blood vessels. The **lymphatic system** consists of the lymph, lymph vessels, lymph nodes, and the lymphoid or- gans (tonsils, appendix, spleen, and thymus gland). Because both systems circulate various substances throughout the body, they can serve as vehicles for the spread of infection.

Structure and Function of the Cardiovascular and Lymphatic Systems

Learning Objective

- *Identify the role of the cardiovascular and lymphatic systems in spreading and eliminating infections.*

The center of the cardiovascular system is the heart (Figure 23.1). The function of the cardiovascular system is to circulate blood through the body's tissues so it can deliver certain substances to cells and remove other substances from them.

Blood is a mixture of formed elements and a liquid called plasma (see page 458). The *plasma* transports dissolved nutrients to body cells and carries wastes away from the cells. The formed elements in blood include red blood cells, white blood cells, and platelets. Red blood cells, or *erythrocytes*, carry oxygen and some carbon dioxide, although most of the carbon dioxide in blood is dissolved in the plasma. White blood cells, or *leukocytes*, play several roles in defending the body against infection, as discussed in Part Three.

The lymphatic system is an essential part of the circulation of blood (Figure 23.2). As the blood circulates, some plasma filters out of the blood capillaries into spaces between tissue cells called *interstitial spaces*. The circulating fluid is called *interstitial fluid*. Microscopic lymphatic vessels that surround tissue cells are called *lymph capillaries;* they are larger and more permeable than blood capillaries. As the interstitial fluid moves around the tissue cells, it is picked up by the lymph capillaries; the fluid is then called *lymph*.

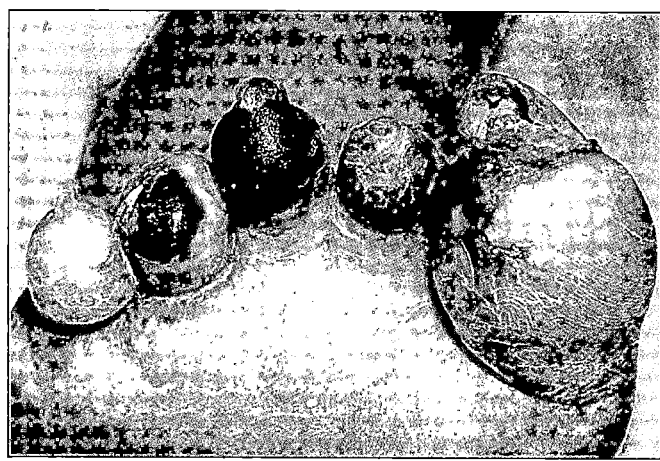

FIGURE 23.8 The toes of a patient with gangrene. This disease is caused by *Clostridium perfringens* and other clostridia. The black, necrotic tissue, resulting from poor circulation or injury, furnishes anaerobic growth conditions for the bacteria, which then progressively destroy adjoining tissue.

■ How can gangrene be prevented?

FIGURE 23.9 Hyperbaric chambers used to treat gas gangrene. A multiplace hyperbaric chamber that can accommodate several patients is shown. Such chambers are usually available at major medical centers. They are also used to treat victims of carbon monoxide poisoning.

■ Saturating the tissues with oxygen, in a hyperbaric chamber, prevents the growth of clostridia.

Substances released from dying and dead cells provide nutrients for many bacteria. Various species of the genus *Clostridium,* which are gram-positive, endospore-forming anaerobes widely found in soil and in the intestinal tracts of humans and domesticated animals, grow readily in such conditions. *C. perfringens* is the species most commonly involved in gangrene, but other clostridia and several other bacteria can also grow in such wounds.

Once ischemia and the subsequent necrosis have developed, **gas gangrene** can develop, especially in muscle tissue. As the *C. perfringens* microorganisms grow, they ferment carbohydrates in the tissue and produce gases (carbon dioxide and hydrogen) that swell the tissue. The bacteria produce toxins that move along muscle bundles, killing cells and producing necrotic tissue that is favorable for further growth. Eventually, these toxins and bacteria enter the bloodstream and cause systemic illness. Enzymes produced by the bacteria degrade collagen and proteinaceous tissue, facilitating the spread of the disease. Without treatment, the condition is fatal.

One complication of improperly performed abortions is the invasion of the uterine wall by *C. perfringens,* which resides in the genital tract of about 5% of all women. This infection can lead to gas gangrene and result in a life-threatening invasion of the bloodstream.

The surgical removal of necrotic tissue and amputation are the most common medical treatments for gas gangrene. When gas gangrene occurs in such regions as the abdominal cavity, the patient can be treated in a **hyperbaric chamber,** which contains a pressurized oxy-

gen-rich atmosphere (Figure 23.9). The oxygen saturates the infected tissues and thereby prevents the growth of the obligately anaerobic clostridia. Small chambers are available that can accommodate a gangrenous limb. The prompt cleaning of serious wounds and precautionary antibiotic treatment are the most effective steps in the prevention of gas gangrene. Penicillin is effective against *C. perfringens.*

Systemic Diseases Caused by Bites and Scratches

Learning Objective

■ List three pathogens that are transmitted by animal bites and scratches.

Animal bites, which account for about 1% of emergency room visits in U.S. hospitals, can result in serious infections. Most bites are by domestic animals, such as dogs and cats, because they live in close contact with humans. It has been reported that 69% of dog bites become infected unless effectively cleansed. Domestic animals often harbor *Pasteurella multocida* (pas-tyĕr-el'lä mul-tō'si-dä), a gram-negative rod similar to the *Yersinia* bacterium that causes plague (discussed on page 633). *P. multocida* is primarily a pathogen of animals and it causes septicemia (hence the name *multocida,* meaning many-killing).

Humans infected with *P. multocida* have varied responses. For example, local infections with severe swelling and pain can develop at the site of the wound. Development of forms of pneumonia and septicemia is possible and life-threatening. Penicillin and tetracycline are usually effective in treating these infections.

Clostridium species and other anaerobes, such as species of *Bacteroides* and *Fusobacterium*, can also infect deep animal bites. Bites by humans are also prone to infection and often cause serious, mixed bacterial wounds.

Cat-Scratch Disease

Cat-scratch disease, although it receives little attention, is surprisingly common. An estimate is that more than 40,000 cases occur annually in the United States, many more than the well-known Lyme disease. People who own or are closely exposed to cats are at risk. The pathogen is an aerobic, gram-negative bacterium, *Bartonella henselae,* which is commonly found on cats. A bite or scratch may not be necessary to transmit the disease. It is hypothesized that the bacteria in the cat's saliva can be deposited on the fur and transferred to the pet's owner when the person touches the cat and then rubs his or her eyes. In these instances, conjunctivitis and other symptoms of an eye infection sometimes occur. Fleas may transmit the disease between cats and possibly to humans. However, it is transmitted to humans mostly by a scratch or bite. The initial sign is a papule at the infection site, which appears 3–10 days after exposure. Swelling of the lymph nodes and usually malaise and fever follow in a couple of weeks. Cat-scratch disease is ordinarily self-limiting, with a duration of a few weeks, but in severe cases antibiotic therapy may be effective.

Vector-Transmitted Diseases

Learning Objective

- *Compare and contrast the causative agents, vectors, reservoirs, symptoms, treatments, and preventive measures for plague, relapsing fever, and Lyme disease.*

Plague

Few diseases have affected human history more dramatically than **plague,** known in the Middle Ages as the Black Death. This term comes from one of its characteristics, the dark blue areas of skin caused by hemorrhages.

The disease is caused by a gram-negative, rod-shaped bacterium, *Yersinia pestis.* Normally a disease of rats, plague is transmitted from one rat to another by the rat flea, *Xenopsylla cheopis* (ze-nop-sil′lä ke̅-o̅′pis) (see Figure 12.30b). In the far West and Southwest, the disease is endemic in wild rodents, especially ground squirrels and prairie dogs.

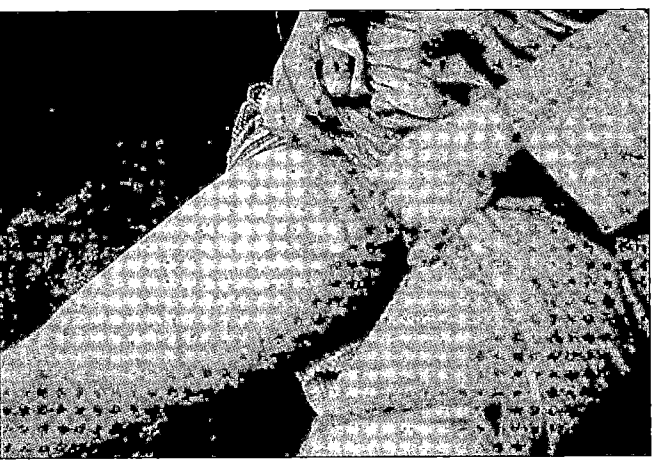

FIGURE 23.10 A case of bubonic plague. Bubonic plague is caused by infection with *Yersinia pestis.* This photograph shows a bubo (swollen lymph node) on the upper leg of a patient. Swollen lymph nodes are a common indication of systemic infection.

■ **In what two ways is plague transmitted?**

If its host dies, the flea seeks a replacement host, which may be another rodent or a human. It can jump about 3½ inches. A plague-infected flea is hungry for a meal because the growth of the bacteria blocks the flea's digestive tract, and the blood the flea ingests is quickly regurgitated. An arthropod vector is not always necessary for plague transmission. Contact from the skinning of infected animals, scratches of domestic cats, and similar incidents have been reported to cause infection.

In the United States, exposure to plague is increasing, as residential areas encroach on areas with infected animals. In parts of the world where human proximity to rats is common, infection from this source still prevails.

From the flea bite, bacteria enter the human's bloodstream and proliferate in the lymph and blood. One factor in the virulence of the plague bacterium is its ability to survive and proliferate inside phagocytic cells rather than being destroyed by them. An increased number of highly virulent organisms eventually emerges, and an overwhelming infection results. The lymph nodes in the groin and armpit become enlarged, and fever develops as the body's defenses react to the infection. Such swellings, called buboes, account for the name **bubonic plague** (Figure 23.10). The mortality rate of untreated bubonic plague is 50–75%. Death, if it occurs, is usually within less than a week after the appearance of symptoms.

A particularly dangerous condition called **septicemic plague** arises when the bacteria enter the blood and proliferate, causing septic shock. Eventually, the bacteria are carried by the blood to the lungs, and a form of the disease called **pneumonic plague** results. The mortality

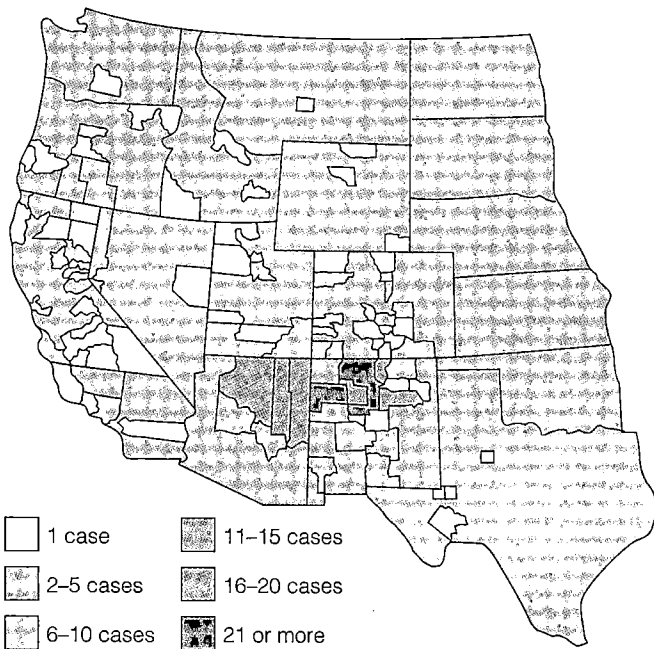

☐ 1 case	▓ 11–15 cases
▓ 2–5 cases	▓ 16–20 cases
▓ 6–10 cases	▓ 21 or more

FIGURE 23.11 The U.S. geographic distribution of human plague, 1970–1997.
[SOURCE: CDC, 1999.]

■ What area reporting plague is closest to you?

rate for this type of plague is nearly 100%. Even today, this disease can rarely be controlled if it is not recognized within 12–15 hours of the onset of fever. People with pneumonic plague usually die within 3 days.

Like influenza, pneumonic plague is easily spread by airborne droplets from humans or animals. Great care must be taken to prevent airborne infection of people in contact with patients.

Europe was ravaged by repeated pandemics of the plague; from the years 542 to 767, outbreaks occurred repeatedly in cycles of a few years. After a lapse of centuries, the disease reappeared in devastating form in the fourteenth and fifteenth centuries. It is estimated to have killed more than 25% of the population, which had lasting effects on the social and economic structure of Europe. A nineteenth-century pandemic primarily affected Asiatic countries; 12 million are estimated to have died in India. The last major rat-associated urban outbreak in the United States occurred in Los Angeles in 1924 and 1925. Following this, the disease became a rarity until it reappeared in 1965 on the Navajo reservation in the Southwest. Plague, once established in the ground squirrel and prairie dog communities in this area, has gradually spread over much of the western states (Figure 23.11). A peak incidence of 40 cases occurred in 1983. A few cases have

also arisen from cats, a new animal reservoir, and one from urban tree squirrels. Awareness of the disease in endemic areas and control measures have again reduced the number of cases—only 8 were reported in 1999.

Diagnosis of plague is most commonly done by isolating the bacterium and then sending it to a laboratory for identification. People exposed to infection can be given prophylactic antibiotic protection. A number of antibiotics, including streptomycin and tetracycline, are effective. Recovery from the disease confers reliable immunity. A vaccine is available for people likely to come into contact with infected fleas during field operations or for laboratory workers exposed to the pathogen.

Relapsing Fever

Learning Objective

■ *List four diseases transmitted by ticks.*

Except for the species that causes Lyme disease (discussed below), all members of the spirochete genus *Borrelia* cause **relapsing fever.** In the United States, the disease is transmitted by soft ticks that feed on rodents. The incidence of relapsing fever increases during the summer months, when the activity of rodents and arthropods increases.

The disease is characterized by fever, sometimes in excess of 40.5°C (105°F), jaundice, and rose-colored skin spots. After 3–5 days, the fever subsides. Three or four relapses may occur, each shorter and less severe than the initial fever. Each recurrence is caused by a different antigenic type of the spirochete, which evades existing immunity. Diagnosis is made by observing the bacteria in the patient's blood, unusual for a spirochete disease.

Lyme Disease (Lyme Borreliosis)

In 1975, a cluster of disease cases in young people that was first diagnosed as rheumatoid arthritis was reported near the city of Lyme, Connecticut. The seasonal occurrence (summer months), lack of contagiousness among family members, and descriptions of an unusual skin rash that appeared several weeks before the first symptoms suggested a tickborne disease. The fact that penicillin alleviated the progression of symptoms suggested a bacterial pathogen. In 1983, a spirochete that was later named *Borrelia burgdorferi* was identified as the cause. **Lyme disease** may now be the most common tickborne disease in the United States. More than 10,000 cases are reported annually (Figure 14.11a). It is also found in Europe, China, Japan, and Australia. In the United States, Lyme disease is most prevalent on the Atlantic Coast (Figure 23.12). Field mice are the most important animal reservoir that maintains the spiro-

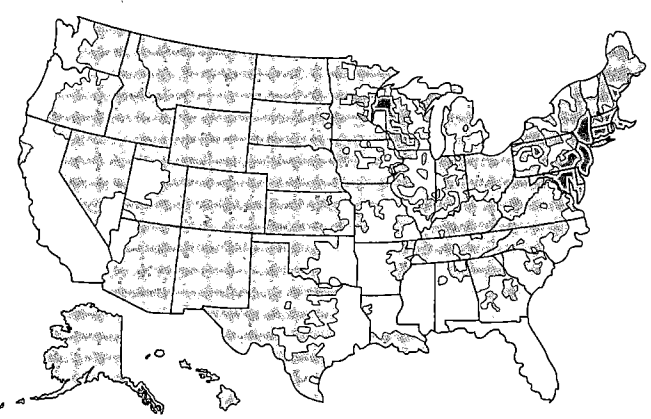

Areas of predicted Lyme disease transmission

High risk   Low risk

Moderate risk   Minimal or no risk

FIGURE 23.12 The U.S. geographic distribution of Lyme disease.
[SOURCE: CDC, 1999.]

■ What factors affect the geographic distribution of Lyme disease?

chete in the tick population. Deer do not support the bacteria, although the ticks feed and mate on them.

The tick (one of two *Ixodes* species) feeds three times during its life cycle (Figure 23.13a). The first and second feedings, as a larva and then as a nymph, are usually on a field mouse. The third feeding, as an adult, is usually on a deer. These feedings are separated by several months, and the ability of the spirochete to remain viable in the disease-tolerant field mice is crucial to maintaining the disease in the wild.

On the Pacific Coast, the tick is the western black-legged tick *Ixodes pacificus* (iks-ō'dēs pas-i'fi-kus) (see also Figure 12.30). In the rest of the country, it is mostly *Ixodes scapularis* (scap-ū-lãr'is). This latter tick is so small that it is often missed (Figure 23.13b). On the Atlantic Coast, almost all *Ixodes* ticks carry the spirochete (Figure 23.13c); on the Pacific Coast few are infected, because that tick feeds on lizards that do not carry the spirochete effectively. The spirochete is maintained by a higher infection rate in a second tick that does not bite humans. However, it does infect a wood rat upon which both ticks feed.

The first symptom of Lyme disease is usually a rash that appears at the bite site. It is a red area that clears in the center as it expands to a final diameter of about 15 cm (Figure 23.14 on page 637). This distinctive rash occurs in about 75% of cases. Flulike symptoms appear in a couple

of weeks as the rash fades. Antibiotics taken during this interval are very effective in limiting the disease.

During a second phase (when it occurs), there is often evidence that the heart is affected. The heartbeat may become so irregular that a pacemaker is required. Neurological symptoms, such as facial paralysis, meningitis, and encephalitis, may be seen. Months or years later, some patients develop arthritis that may affect them for years. Immune responses to the presence of the bacteria are probably the cause of this joint damage. Many of the symptoms of long-term Lyme disease resemble those of the later stages of syphilis, also caused by a spirochete.

Diagnosis of Lyme disease depends partly on the symptoms and an index of suspicion based on the prevalence in the geographic area. The most current procedure is the use of an ELISA screening test, followed by a more specific test similar to the Western blot. Physicians are cautioned that serological tests must be interpreted in conjunction with clinical symptoms and the likelihood of exposure to infection. Several antibiotics are effective in treatment of the disease, although in the later stages, large amounts may be needed. Vaccines are available for people at high risk of tick exposure.

Other Tickborne Diseases

The deer tick, *Ixodes scapularis*, is also the vector of other dangerous diseases. **Ehrlichiosis,** a flulike disease caused by *Ehrlichia* bacteria (which multiply in various leukocytes), was until 1986 considered only a veterinary concern. Now there are increasing reports of human ehrlichiosis, occasionally fatal. *Ehrlichia* are rickettsialike, resembling the pathogens causing tickborne typhus, a similar disease that is also sometimes fatal. **Human granulocytic ehrlichiosis** is found in northern states, occasionally as a coinfection with Lyme disease. **Human monocytic ehrlichiosis** is reported mostly in southern states. It is caused by a different species of *Ehrlichia* that is closely related to known animal pathogens and is carried by a different tick, familiarly known as the Lone Star tick. Doxycycline is the recommended antibiotic.

Typhus

Learning Objective
■ *Describe the epidemiologies of epidemic typhus, endemic murine typhus, and spotted fevers.*

The various typhus diseases are caused by rickettsias, bacteria that are obligate intracellular parasites of eukaryotes. Rickettsias, which are spread by arthropod vectors, infect mostly the endothelial cells of the vascular system and

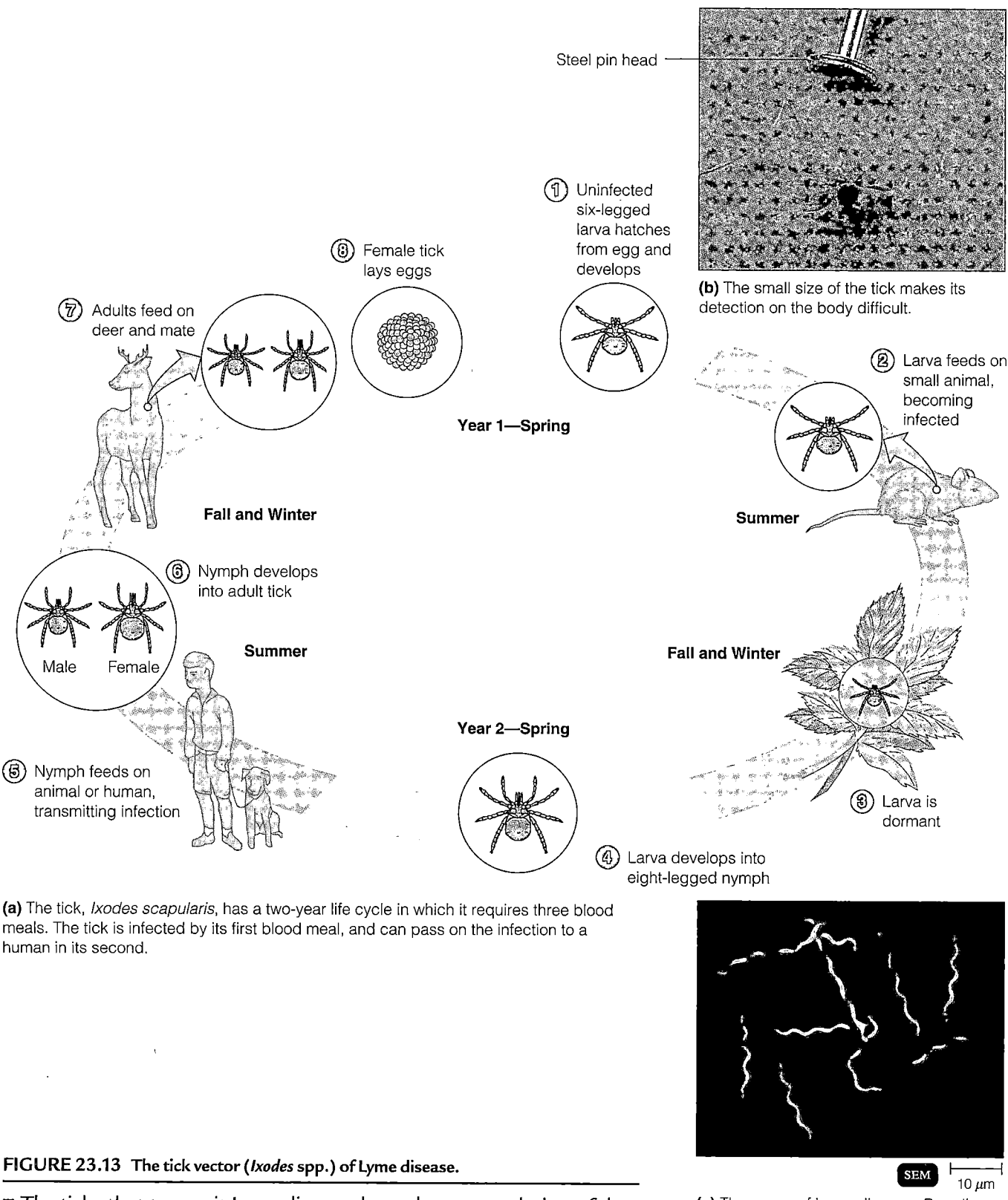

Steel pin head

(①) Uninfected six-legged larva hatches from egg and develops

(b) The small size of the tick makes its detection on the body difficult.

(⑧) Female tick lays eggs

(⑦) Adults feed on deer and mate

Year 1—Spring

Fall and Winter

(②) Larva feeds on small animal, becoming infected

Summer

(⑥) Nymph develops into adult tick

Male   Female

Summer

Fall and Winter

Year 2—Spring

(⑤) Nymph feeds on animal or human, transmitting infection

(③) Larva is dormant

(④) Larva develops into eight-legged nymph

(a) The tick, *Ixodes scapularis*, has a two-year life cycle in which it requires three blood meals. The tick is infected by its first blood meal, and can pass on the infection to a human in its second.

FIGURE 23.13 The tick vector (*Ixodes* spp.) of Lyme disease.

■ The ticks that transmit Lyme disease depend on a population of deer for completion of their life cycle.

SEM ⊢———⊣ 10 μm

(c) The cause of Lyme disease, *Borrelia burgdorferi*.

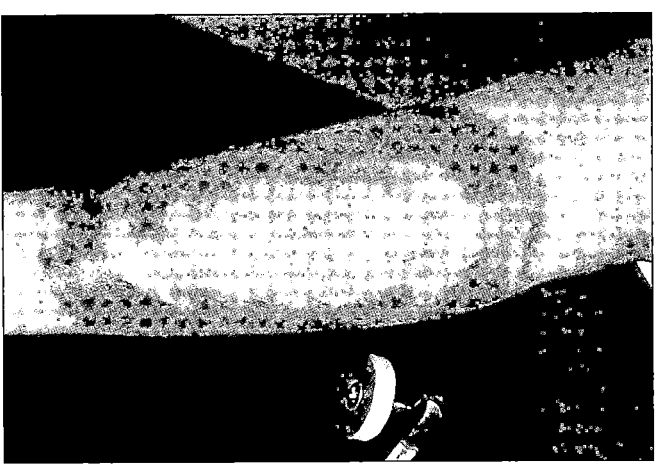

FIGURE 23.14 The common bull's-eye rash of Lyme disease.
The rash is not always this obvious.

■ **What symptoms occur once the rash fades?**

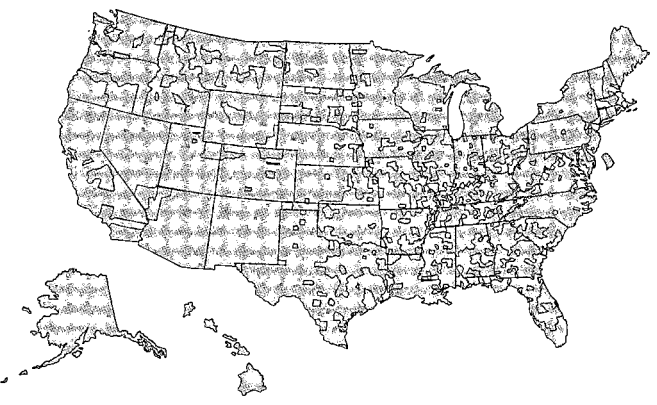

▨ Counties reporting cases of Rocky Mountain spotted fever

FIGURE 23.15 The U.S. geographic distribution of Rocky Mountain spotted fever (tickborne typhus) by county, 1984–1995.
SOURCE: CDC, 1995.

■ **What is the nearest area to you that has reported Rocky Mountain spotted fever?**

multiply within them. The resulting inflammation causes local blockage and rupture of the small blood vessels.

Epidemic Typhus   Epidemic typhus (louseborne typhus) is caused by *Rickettsia prowazekii* and carried by the human body louse *Pediculus humanus corporis* (ped-ik'ū-lus hü'ma-nus kôr'pô-ris) (see Figure 12.31a). The pathogen grows in the gastrointestinal tract of the louse and is excreted by it. It is not transmitted directly by the bite of an infected louse; rather, it is transmitted when the feces of the louse are rubbed into the wound when the bitten host scratches the bite. The disease can flourish only in crowded and unsanitary surroundings, when lice can transfer readily from an infected host to a new host. Anne Frank died of typhus contracted in concentration camp conditions.

Epidemic typhus disease produces a high and prolonged fever for at least 2 weeks. Stupor and a rash of small red spots caused by subcutaneous hemorrhaging are characteristic, as the rickettsias invade blood vessel linings. Mortality rates are very high when untreated.

Tetracycline and chloramphenicol are usually effective against epidemic typhus, but elimination of conditions in which the disease flourishes is more important. The microbe is considered especially hazardous, and attempts to culture it require extreme care. Vaccines are available for military populations, which historically have been highly susceptible to the disease.

Endemic Murine Typhus   Endemic murine typhus occurs sporadically rather than in epidemics. The term

murine (derived from Latin for mouse) refers to the fact that rodents, such as rats and squirrels, are the common hosts for this type of typhus. Endemic murine typhus is transmitted by the rat flea *Xenopsylla cheopis* (see Figure 12.31b), and the pathogen responsible for the disease is *Rickettsia typhi,* a common inhabitant of rats. Texas has had a number of outbreaks of murine typhus in recent years, often associated with campaigns to eliminate rodents, which caused the rat fleas to seek new hosts. With a mortality rate of less than 5%, the disease is considerably less severe than the epidemic form of typhus. Except for the reduced severity of the disease, endemic murine typhus is clinically indistinguishable from epidemic typhus.

Tetracycline and chloramphenicol are effective treatments for endemic murine typhus, and rat control is the best preventive measure.

Spotted Fevers   Tickborne typhus, or Rocky Mountain spotted fever, is probably the best-known rickettsial disease in the United States. It is caused by *Rickettsia rickettsii.* Despite its name (it was first recognized in the Rocky Mountain area), it is most common in the southeastern states and Appalachia (Figure 23.15). This rickettsia is a parasite of ticks and is usually passed from one generation of ticks to another through their eggs, a mechanism called *transovarian passage* (Figure 23.16). Surveys show that in endemic areas, perhaps 1 out of

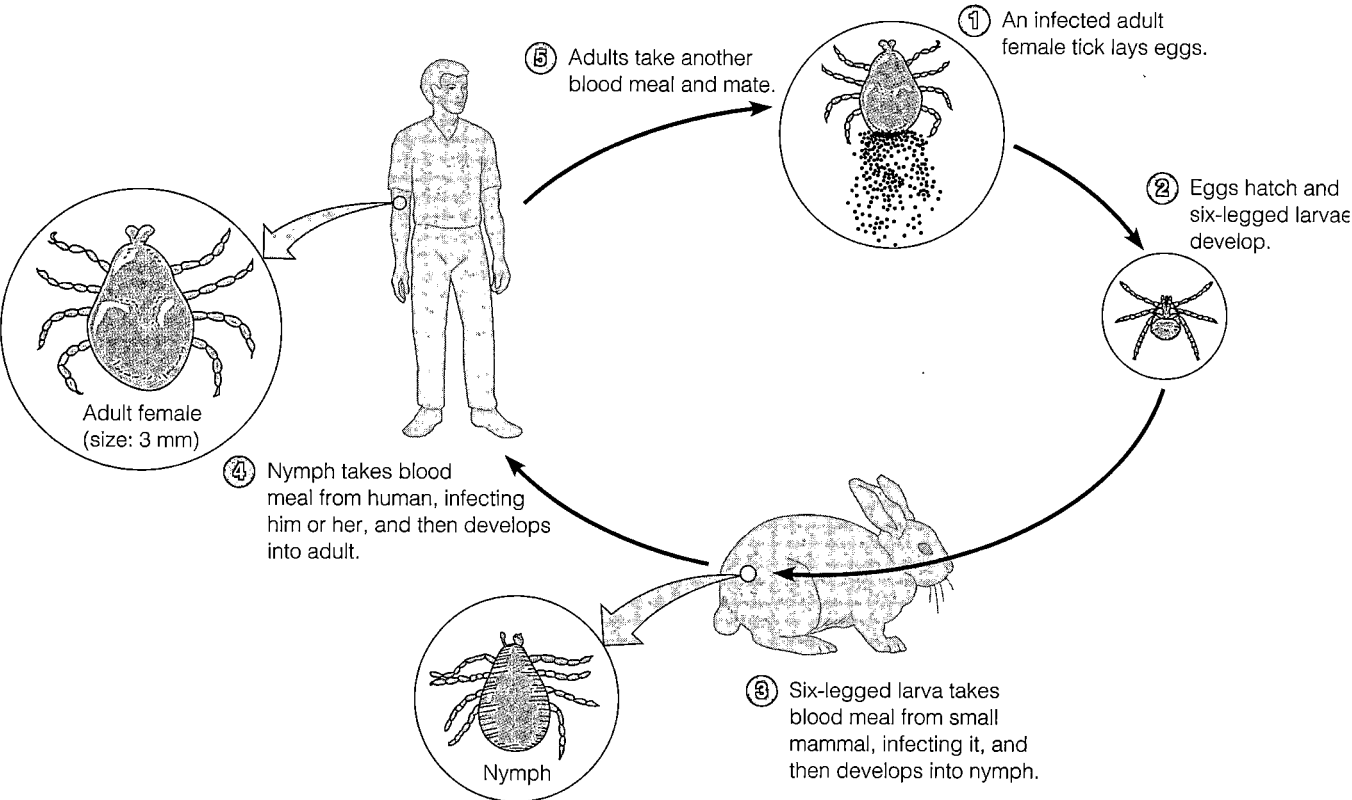

(5) Adults take another blood meal and mate.

(1) An infected adult female tick lays eggs.

(2) Eggs hatch and six-legged larvae develop.

Adult female (size: 3 mm)

(4) Nymph takes blood meal from human, infecting him or her, and then develops into adult.

(3) Six-legged larva takes blood meal from small mammal, infecting it, and then develops into nymph.

Nymph

FIGURE 23.16 **The life cycle of the tick vector (*Dermacentor* spp.) of Rocky Mountain spotted fever.** Mammals are not essential to survival of the pathogen, *Rickettsia rickettsii*, in the tick population; the bacteria may be passed by transovarian passage, so new ticks are infected upon hatching. A blood meal is required for ticks to advance to the next stage in the life cycle.

■ What is meant by transovarian passage?

FIGURE 23.17 **The rash caused by Rocky Mountain spotted fever.** This rash is often mistaken for measles. People with dark skin have a higher mortality rate because the rash is often not recognized early enough for effective treatment.

■ How can Rocky Mountain Spotted Fever be prevented?

every 1000 ticks is infected. In different parts of the United States, different ticks are involved—in the west, the wood tick *Dermacentor andersoni* (dėr-mä-sen'tôr an-dėr-sōn'ē); in the east, the dog tick *Dermacentor variabilis* (vār-ē-a'bil-is).

About a week after the tick bites, a rash develops that is sometimes mistaken for measles (Figure 23.17); however, it often appears on palms and soles, which does not occur with viral rashes. The rash is accompanied by fever and headache. Death, which occurs in about 3% of the approximately 800 cases reported each year, is usually caused by kidney and heart failure. Antibiotics such as tetracycline and chloramphenicol are very effective if administered early enough. No vaccine is available.

Serological tests do not become positive until late in the illness. Diagnosis before the typical rash appears is difficult; symptoms vary widely. Also, on dark-skinned individuals, the rash is difficult to see. A misdiagnosis can be costly; if treatment is not prompt and correct, the mortality rate is about 20%.

Viral Diseases of the Cardiovascular and Lymphatic Systems

Viruses are the cause of a number of cardiovascular and lymphatic diseases, mostly prevalent in tropical areas. However, one viral disease of this type, infectious mononucleosis, is an especially familiar infectious disease among American college-age individuals.

Burkitt's Lymphoma

Learning Objective

- *Describe the epidemiologies of Burkitt's lymphoma and infectious mononucleosis.*

In the 1950s, Denis Burkitt, an Irish physician working in eastern Africa, noticed the frequent occurrence in children of a fast-growing tumor of the jaw (Figure 23.18). Known as **Burkitt's lymphoma,** this is the most common childhood cancer in Africa. It has a limited geographic distribution similar to that of malaria in central Africa.

Burkitt suspected a viral cause of the tumor and a mosquito vector. At that time, there was no known virus that caused human cancer, although several viruses were clearly associated with animal cancers. Intrigued by this possibility, the British virologist Tony Epstein and his student, Yvonne Barr, performed biopsies on the tumors. A virus was cultured from this material, and the electron microscope showed a herpeslike virus in the culture cells; it was named the Epstein-Barr (EB) virus. The official name of this virus is human herpesvirus 4.

EB virus is clearly associated with Burkitt's lymphoma, but the mechanism by which it causes the tumor is not understood. Research eventually showed, however, that mosquitoes do not transmit the virus or the disease. Instead, mosquito-borne malarial infections apparently foster the development of Burkitt's lymphoma by impairing the immune response to the EB virus, which is almost universally present in human adults worldwide.

In areas without endemic malaria, such as the United States, Burkitt's lymphoma is rare. The appearance of the lymphoma in AIDS patients is an indication of the importance of immune surveillance in preventing expression of the disease. Another medical problem that has been associated with the EB virus is **nasopharyngeal** (of the nose and pharynx) **carcinoma.** This disease is a major cause of death in southeast Asia, where it has an incidence of 10–20 cases per 100,000 population.

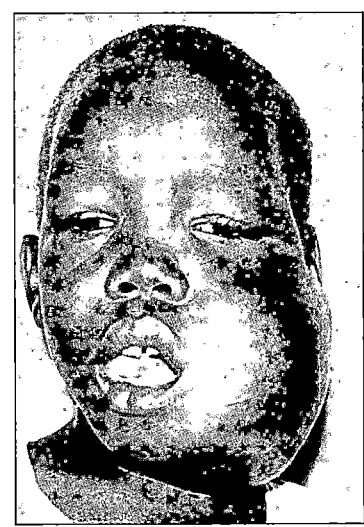

FIGURE 23.18 A child with Burkitt's lymphoma. Cancerous tumors of the jaw caused by Epstein-Barr virus are seen mainly in children. This child was successfully treated.

- **EB virus causes infectious mononucleosis and Burkitt's lymphoma.**

Infectious Mononucleosis

The identification of the EB virus as the cause of **infectious mononucleosis,** or *mono,* was the result of one of the accidental discoveries that often advance science. A technician in a laboratory investigating the EB virus served as a negative control for the virus. While on vacation, she contracted an infection characterized by fever, sore throat, swollen lymph nodes in the neck, and general weakness. The most interesting aspect of the technician's disease was that she now tested serologically positive for the EB virus. It was soon confirmed that the same virus that is associated with Burkitt's lymphoma also causes almost all cases of infectious mononucleosis.

In developing parts of the world, infection with the EB virus occurs in early childhood, and 90% of the children over age 4 have acquired antibodies. Childhood EB viral infections are usually asymptomatic, but if infection is delayed until young adulthood, as is often the case in the United States, the result is more symptomatic. The peak U.S. incidence of the disease occurs at about age 15–25 (see Figure 25.14). College populations, particularly those in upper socioeconomic levels, have a high incidence of the disease. Most college students have no immunity, and about 15% of them can expect to contract the disease. The disease is generally self-limiting and seldom fatal. A principal cause of the rare deaths is rupture of the enlarged spleen (a common response to a systemic infection) during vigorous activity. Recovery is usually complete in a few weeks, and immunity is permanent.

The usual route of infection is by the transfer of saliva by kissing or, for example, by sharing drinking vessels. Reproduction of the virus appears to occur mainly in the parotid glands, which accounts for its presence in saliva. It does not spread among casual household contacts, so aerosol transmission is unlikely. The incubation period before appearance of symptoms is from 4–7 weeks.

Although reproduction is considered to occur in epithelial cells of the parotid glands, the infection is almost exclusively selective for B cells. The virus does not reproduce significantly in B cells, but the infection causes B cells to transform into plasma cells, which reproduce rapidly. These are then attacked by cytotoxic T cells of the cell-mediated immune system. The symptoms of infectious mononucleosis are associated with B-cell proliferation and the immune response. After recovery, the virus remains latent in a small number of B cells.

The disease name *mononucleosis* refers to lymphocytes with unusual lobed nuclei that proliferate in the blood during the acute infection. The infected B cells produce nonspecific antibodies called heterophil antibodies. If this test is negative, the symptoms may be caused by cytomegalovirus (see page 700). A fluorescent-antibody test that detects IgM antibodies against the EB virus is the most specific diagnostic method.

Classic Viral Hemorrhagic Fevers

Learning Objective

- *Compare and contrast the causative agents, vectors, reservoirs, and symptoms of yellow fever, dengue, and dengue hemorrhagic fever.*

Most hemorrhagic fevers are zoonotic diseases; they appear in humans only from infectious contact with their normal animal hosts. Some of them have been medically familiar for so long that they are considered "classic" hemorrhagic fevers. First among these is **yellow fever.** The yellow fever virus is injected into the skin by a mosquito, *Aedes aegypti.*

In the early stages of severe cases of the disease, the person experiences fever, chills, and headache, followed by nausea and vomiting. This stage is followed by jaundice, a yellowing of the skin that gave the disease its name. This coloration reflects liver damage, which results in the deposit of bile pigments in the skin and mucous membranes. The mortality rate for yellow fever is high, about 20%.

Yellow fever is still endemic in many tropical areas, such as Central America, tropical South America, and Africa. At one time, the disease was endemic in the United States and occurred as far north as Philadelphia. The last U.S. case of yellow fever occurred in Louisiana in 1905 during an outbreak resulting in about 1000 deaths. Mosquito eradication campaigns initiated by the U.S. Army surgeon Walter Reed were effective in eliminating yellow fever in the United States.

Monkeys are a natural reservoir for the virus, but human-to-human transmission can maintain the disease. Local control of mosquitoes and immunization of the exposed population are effective controls in urban areas.

Diagnosis is usually by clinical signs, but it can be confirmed by a rise in antibody titer or isolation of the virus from the blood. There is no specific treatment for yellow fever. The vaccine used is an attenuated live viral strain and yields a very effective immunity with few adverse effects.

Dengue is a similar but milder viral disease also transmitted by the *Aedes aegypti* mosquito. This disease is endemic in the Caribbean and other tropical environments, where an estimated 100 million cases occur each year. It is characterized by fever, severe muscle and joint pain, and rash. Except for the painful symptoms, which have led to the name **breakbone fever,** classic dengue fever is a relatively mild disease and is rarely fatal.

The countries surrounding the Caribbean are reporting an increasing number of cases of dengue. In most years, more than 100 cases are imported into the United States, mostly by travelers from the Caribbean and South America. The disease does not appear to have an animal reservoir. The mosquito vector for dengue is common in the Gulf states, and there is some worry that the virus will sooner or later be introduced into this region and become endemic. Health officials are concerned about the American introduction of an Asian mosquito, *Aedes albopictus* (al-bō-pik'tus), an efficient carrier of the virus and an aggressive biter. It transmits the virus by transovarian passage and from person to person. The range of this mosquito can potentially cover much of the country. Control measures are directed at eliminating *Aedes* mosquitoes. These are urban mosquitoes that proliferate in sites such as tree holes and discarded plasticware.

Dengue also occurs in a second form, **dengue hemorrhagic fever (DHF),** primarily in southeast Asia. Recent outbreaks have also occurred in Mexico, South America, and the Caribbean. DHF can induce shock in the victim (usually a child) and kill in a few hours; it is a leading cause of death among southeast Asian children.

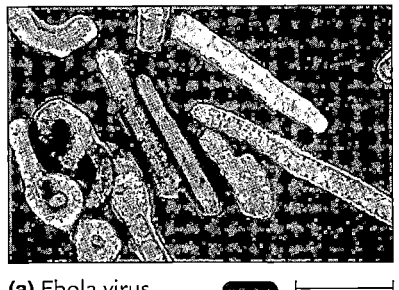

(a) Ebola virus   TEM   ⊢————⊣ 1 nm

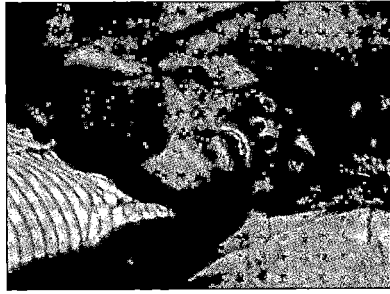

(b) A patient suffering from Ebola hemorrhagic fever

FIGURE 23.19   Ebola hemorrhagic fever.

▨ What are three hemorrhagic fevers and their causative agents?

Emerging Viral Hemorrhagic Fevers

Learning Objective

- *Compare and contrast the causative agents, reservoirs, and symptoms of Ebola hemorrhagic fever and hantavirus pulmonary syndrome.*

Certain other hemorrhagic diseases are considered new or "emerging" hemorrhagic fevers. In 1967, 31 people became ill and 7 died when some African monkeys were imported into Europe. The virus was strangely shaped, in the form of a filament (filoviruses) and was named for the site of the outbreak, **Marburg,** Germany. Nine years later, outbreaks in Africa of another highly lethal hemorrhagic fever, caused by a similar filovirus, caused a disease with a mortality rate approaching 90%. Named **Ebola,** for a regional river, this is now a well-publicized disease, the subject of films and books (Figure 23.19). The natural host is unknown, but most cases have been transmitted by contact with blood especially through unsterilized needles.

Lassa fever appeared in Africa in 1969, from a rodent reservoir. Like Ebola, it is mostly spread among humans by contact with body fluids. Outbreaks occur regularly and kill thousands. South America has several hemorrhagic fevers caused by Lassa-like viruses (arenaviruses) that are maintained in the rodent population. **Argentine** and **Bolivian hemorrhagic fevers** are transmitted in rural areas by contact with rodent excretions.

Hantavirus pulmonary syndrome (HPS), which has recently become well known in the United States, is actually a disease with a long history, especially in Asia and Europe. It is best known as **hemorrhagic fever with renal syndrome.** The U.S. disease manifests itself as a frequently fatal pulmonary infection, in which the lungs fill with fluids. These related diseases are transmitted by the inhalation of hantaviruses in dried urine from infected rodents.

Protozoan Diseases of the Cardiovascular and Lymphatic Systems

Learning Objectives

- *Compare and contrast the causative agents, modes of transmission, reservoirs, symptoms, and treatments for American trypanosomiasis, toxoplasmosis, malaria, leishmaniasis, and babesiosis.*
- *Discuss the worldwide effects of these diseases on human health.*

Protozoa that cause diseases of the cardiovascular and lymphatic systems often have complex life cycles, and their presence may affect human hosts seriously.

American Trypanosomiasis (Chagas' Disease)

American trypanosomiasis, also known as **Chagas' disease,** is a protozoan disease of the cardiovascular system. The causative agent is *Trypanosoma cruzi* (tri-pa-nō-sō'mä kruz'ē), a flagellated protozoan (Figure 23.20). The protozoan was discovered in its insect vector by the Brazilian microbiologist Carlos Chagas in 1910. He named it for the Brazilian epidemiologist Oswaldo Cruz. The disease occurs in southern Texas, Mexico, Central America, and parts of South America. The disease infects 40–50% of the population in some rural areas of South America. An estimated 100,000 infected immigrants carry the disease in the United States.

The reservoir for *T. cruzi* is a wide variety of wild animals, including rodents, opossums, and armadillos. The arthropod vector is the reduviid bug, called the "kissing bug" because it often bites people near the lips (see Figure 12.31d). The insects live in the cracks and crevices of mud or stone huts with thatched roofs. The trypanosomes, which grow in the gut of the bug, are passed on if the bug

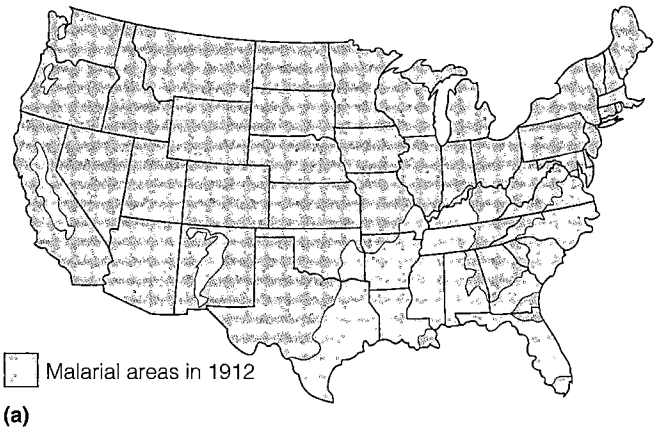

Malarial areas in 1912

(a)

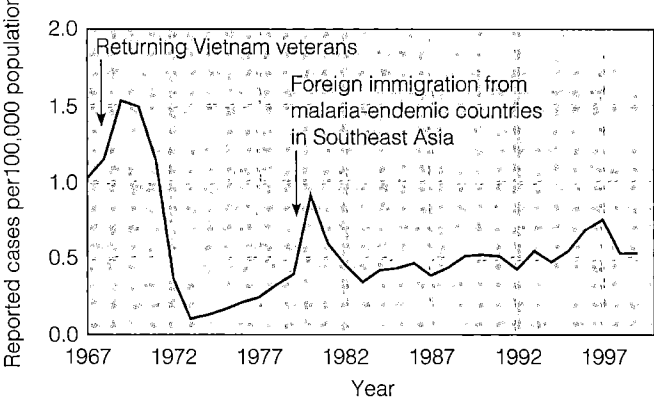

(b)

FIGURE 23.22 Malaria in the United States. (a) Areas where malaria was endemic as recently as 1912. **(b)** Graph showing reported cases of malaria in the U.S. 1967–1998.
SOURCES: CDC, *Summary of Notifiable Diseases 1998, MMWR* 47(53)(12/31/99); *MMWR* 48(51)(1/7/00).

■ **How can malaria be prevented?**

cause relatively benign forms of malaria, but even so, they are debilitating, and the victims lack energy. These two malarial types are relatively lower in incidence and rather restricted geographically.

The most dangerous malaria is that caused by *P. falciparum*. It is believed that humans have been exposed to this parasite (through contact with birds) only in relatively recent history. Perhaps one reason for the virulence of this type of malaria is that humans and the parasite have had less time to become adapted to each other. Referred to as "malignant" malaria it is the most likely to be fatal—untreated, it eventually kills about half of those infected. *P. falciparum* suppresses the ability of dendritic cells (Chapter 17) to initiate an immune response. More red blood cells (RBCs) are infected and destroyed than in other forms of malaria. The resulting anemia weakens the victim generally. Furthermore, the RBCs tend to stick to the walls of the capillary vessels, which become clogged.

This prevents the infected RBCs from reaching the spleen, where phagocytic cells would eliminate them. The blocked capillaries and subsequent loss of blood supply leads to death of the tissues. Kidney and liver damage is caused in this fashion. Because the brain is frequently affected, *P. falciparum* is the usual cause of cerebral malaria.

The mosquito carries the *sporozoite* form of the *Plasmodium* protozoa in its saliva (see Figure 12.18). The sporozoite enters the bloodstream of the bitten human and within about 30 minutes enters the liver cells. The sporozoites in the liver cells undergo reproductive schizogony by a series of stages that finally results in the release of large numbers of *merozoite* forms into the bloodstream. These merozoites infect RBCs and undergo reproductive schizogony again. Laboratory diagnosis of malaria is made by examining a blood smear for infected RBCs (Figure 23.23a). Eventually, the RBCs rupture almost simultaneously and release large numbers of merozoites (Figure 23.23b). There is also a nearly simultaneous release of toxic compounds, which is the cause of the paroxysms (recurrent intensifications of symptoms) of chills and fever characteristic of malaria. The fever reaches 40°C (104°F), and a sweating stage begins as the fever subsides. Between paroxysms, the patient feels normal. Anemia results from the loss of RBCs, and excessive enlargement of the liver and spleen is an added complication.

When the RBCs rupture, many of the released merozoites infect other RBCs within a few seconds to renew the cycle in the bloodstream. If only 1% of the RBCs contain parasites, an estimated 100,000,000,000 parasites will be in circulation at one time in a typical malaria patient! Some of the merozoites develop into male or female *gametocytes*. When these enter the digestive tract of a feeding mosquito, they pass through a sexual cycle that produces new infective sporozoites. It took the combined labors of several generations of scientists to discover this complex life cycle of the malaria parasite.

The highest mortality rates from malaria occur in young children. People who survive malaria acquire a limited immunity. Although they can be reinfected, they tend to have a less severe form of the disease. This relative immunity almost disappears if the person leaves an endemic area with its periodic reinfections. People who have the genetic sickle-cell trait, common in many areas where malaria is endemic, are relatively resistant to malaria.

Much effort is being expended on the search for an effective vaccine. Several are being field tested. The sporozoite stage is of primary interest as a target for a vaccine because neutralizing it would prevent the initial infection from becoming well established. However, some believe that any effective vaccine would have to protect against all three stages.

As we mentioned earlier, the most common diagnostic test for malaria is the blood smear. This requires equipment,

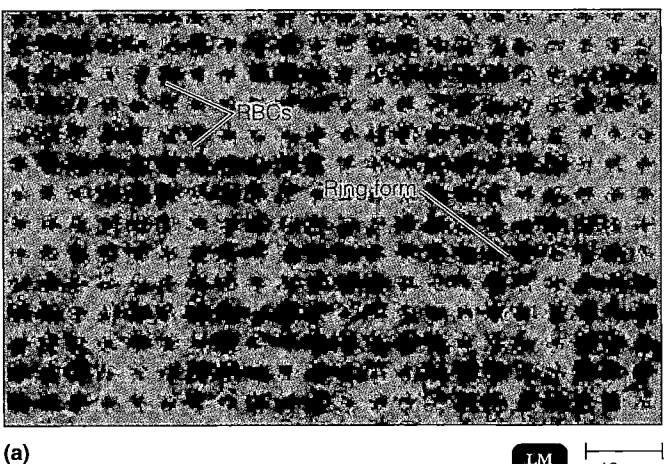

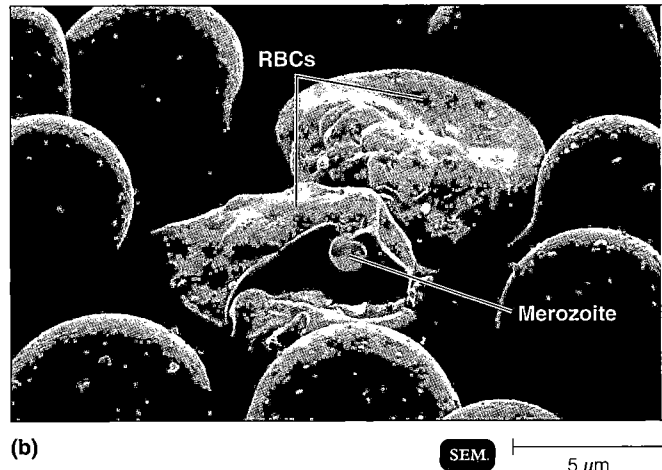

(a)   LM ├─────┤ 10 μm   (b)   SEM ├─────┤ 5 μm

FIGURE 23.23 Malaria. (a) Blood smears are used in the diagnosis of malaria; the protozoa can be detected growing in the RBCs. The feeding protozoan appears in early stages to resemble a ring within RBCs. The light central area within the circular ring is the food vacuole of the protozoan, and the dark spot of the ring is its nucleus. **(b)** Some of the RBCs are lysing and releasing merozoites that will infect new RBCs.

■ Plasmodia cause red blood cells to lyse to release gametocytes at the time of day that mosquitoes are feeding.

such as a microscope, and skill in interpretation. Also, inspection of more than 300 microscopic fields on a slide is recommended, which is time-consuming. In recent years, however, several serological tests have been developed that can yield results in a few minutes which are comparable to those obtained by microscopic inspection of blood smears.

Malaria was once fairly effectively treated with quinine, but newer quinine derivatives, such as primaquine and especially chloroquine, have been the mainstays for prevention and treatment for many years. Resistance to these drugs is spreading rapidly, and current chemical alternatives are more expensive, prohibiting widespread use in the parts of the world most troubled by malaria. The latest drug of choice for many applications is mefloquine, a quinine derivative. Another approach to chemotherapy has been to use drug combinations such as Fansidar®, a mixture of pyrimethamine and sulfadoxine.

Effective control of malaria is not in sight. It will probably require a combination of chemotherapeutic and immunological approaches. Currently, the most promising control method is the use of insecticide-treated bed nets, because the *Anopheles* mosquito is a night feeder. The expense and the need for an effective political organization in malarial areas are probably going to be as important in controlling the disease as advances in medical research.

Leishmaniasis

Leishmaniasis is a widespread and complex disease that exhibits several clinical forms. The protozoan pathogens are of about 20 different species, often categorized into three groups for reasons of simplicity. One group is *Leishmania donovani* (lĭsh′mă-nē-ä don.-ō-van′ē) which causes a visceral leishmaniasis where parasites invade the internal organs. The *L. tropica* (lĭsh′mă-nē-ä trop′i-kä) and *L. braziliensis* (brä-sil′ē-en-sis) groups grow preferentially at cooler temperatures and cause lesions of the skin or mucous membranes. Leishmaniasis is transmitted by the bite of female sandflies, species of which are found in much of the tropical world and around the Mediterranean. These insects are smaller than mosquitoes and often penetrate the mesh of standard netting. Small mammals are an unaffected reservoir of the protozoans. The infective form, the *promastigote,* is in the saliva of the insect. It loses its flagellum when it penetrates the skin of the mammalian victim, becoming an *amastigote* that proliferates in phagocytic cells, mostly in fixed locations in tissue. These amastigotes are then ingested by feeding sandflies, renewing the cycle.

Leishmania donovani Infection

Leishmania donovani infection occurs in may parts of Asia, Africa, and southeast Asia. Often known as *kala azar* or *dum dum fever,* it is often fatal. Early symptoms, following infection by as long as a year, resemble the chills and sweating of malaria. As the protozoa proliferate in the liver and spleen, these organs enlarge greatly. Eventually, kidney function is also lost as these organs are invaded. This is a debilitating disease that, if untreated, will lead to death within a year or two.

CLINICAL PROBLEM SOLVING

A Vectorborne Disease—Or Is It?

You will see questions as you read through this problem. The questions are those that clinicians ask themselves as they solve a clinical problem. Try to answer each question as a clinician.

1. On February 19, a 49-year-old man presented with a 5-day history of fever and hypotension. Physical examination revealed kidney failure, and the patient was hospitalized.
What disease do you suspect? What additional information do you need?

2. The patient had no history of travel outside the United States or injected drug use. He had received 4 units of packed RBCs during surgery for hip replacement on January 15.
Will you modify your hypothesis based on this information?

3. Three days later, examination of a peripheral blood smear showed that 12% of his RBCs contained intracellular parasites.
What diseases are possible? What treatment would you recommend?

4. A diagnosis of *Plasmodium falciparum* infection was confirmed by PCR. The patient responded to treatment with quinine and exchange blood transfusion.
What is the disease, and what was the source of his infection? Can his infection be transmitted to another person?

5. The man acquired malaria from the RBC transfusion. His infection can be transmitted because competent vectors for malaria are present in the United States. Between 1958 and 1998, about 100 cases of locally acquired malaria were reported.

Stored serum samples from donors used for the transfusion he received in January were tested for antibodies by immunofluoresence antibody testing. One donor had a titer of 16,384 against *P. falciparum*.
What does the presence of antibodies in the donor's serum indicate?

6. Antibodies could indicate a current or past infection.
What other information do you need?

7. PCR on the stored serum detected *P. falciparum* DNA. The donor reported no fever at the time of donation. Blood smears obtained from this donor in March demonstrated a ring-shaped parasite in the RBCs. He was treated with quinine and doxycycline.
Why was the donor treated?

8. The presence of the ring stage of *Plasmodium* verifies that the donor has parasitemia (parasites in the blood). He was treated to kill the parasites to protect his health as well as to prevent transmission of the disease.
What information do you need from the donor?

9. The donor was born in west Africa, had lived in Europe, then had returned to west Africa, where he lived for 20 years before immigrating to the United States 2 years before donating blood.
Transfusion-transmitted malaria is a rare but serious complication of blood transfusion. From 1958 to 1998, 103 cases of transfusion-transmitted malaria occurred in the United States. At present, it is diffi-

cult to screen blood for malaria because commercial test kits are not available.
What type of test kit would you design: immunofluorescence antibody testing, direct microscopic exam, or PCR?

10. Immunity results from recovery from malaria, therefore, antibody detection to screen blood donations would result in exclusion of otherwise healthy donors. If only a few sporozoites are present, they could be missed in a microscopic examination. PCR is a more sensitive direct test. Malaria continues to be a leading cause of morbidity and mortality worldwide, particularly because of the development of drug-resistant strains.
Until a screening test is available, what recommendations do you have for blood banks to prevent transfusion-transmitted malaria?

11. Persons who emigrate from highly malarious areas and have acquired immunity may have asymptomatic parasitemia. In the case cited here, a history of having been in a malarious area within the previous 3 years was elicited only during questioning after the transfusion. The American Association of Blood Banks is recommending use of uniform donor-history questions that ask about travel to specific countries and regions.

SOURCE: Adapted from *MMWR* 48(12):253–256 (4/2/99).

Leishmania tropica Infection

Leishmania tropica infection is a cutaneous form of leishmaniasis sometimes called *Oriental sore*. A papule appears at the bite site after a few weeks of incubation (Figure 23.24). The papule ulcerates and, after healing, leaves a prominent scar. This form of the disease is found in much of Asia, Africa, and the Mediterranean region. It has been reported in Mexico, Central America, and the northern part of South America.

Leishmania braziliensis Infection

Leishmania braziliensis infection is known as mucocutaneous leishmaniasis because it affects mucous membranes as well as skin. It causes disfiguring destruction of the tis-

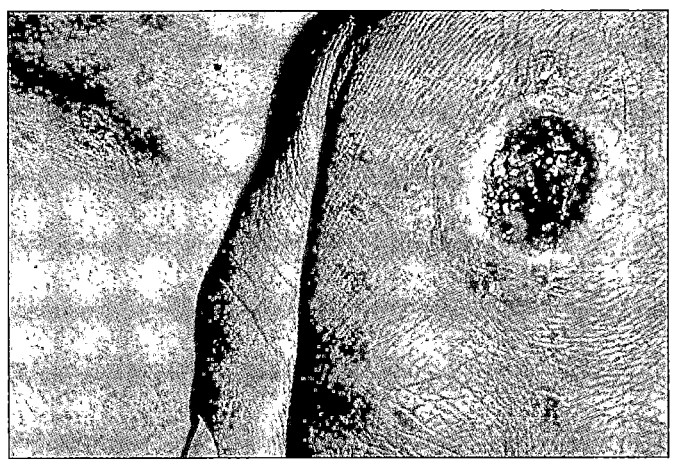

FIGURE 23.24 Cutaneous leishmaniasis. Lesion on the back of the hand of a patient.

sues of the nose, mouth, and upper throat. This form of leishmaniasis is most commonly found in the Yucatan Peninsula of Mexico and in the rain forest areas of Central and South America; it often affects workers harvesting the chicle sap used for making chewing gum. This disease is often referred to as American leishmaniasis.

Current treatments for leishmaniasis are 4 weeks of administration of drugs containing the toxic metal antimony, such as sodium stibogluconate, or a few expensive alternatives, such as amphotericin B. Very recently, an experimental, orally administered drug, miltefosine, has shown promise as an effective treatment.

A number of cases of leishmaniasis occurred among troops fighting the Persian Gulf War. It was once endemic in countries of southern Europe, such as Spain, Italy, Portugal, and the Balkan peninsula. Occasional cases of leishmaniasis as an opportunistic disease of HIV-infected persons are beginning to reappear in these areas.

Babesiosis

There have been increased reports of a tickborne protozoan disease called **babesiosis** in some parts of the United States. Endemic areas for the disease are the Great Lakes and northeastern states. The tick vector may also carry the pathogens causing Lyme disease and human granulocytic ehrlichiosis. Coinfections with these diseases often occur and may confuse the diagnosis.

Subclinical cases in endemic areas are common; it is much more serious in immunocompromised patients. Symptoms usually include chills and fever resembling malaria; the microbes replicate within RBCs. Anemia from the hemolysis of RBCs occurs and is difficult to treat.

Helminthic Diseases of the Cardiovascular and Lymphatic Systems

Learning Objective

■ *Diagram the life cycle of* Schistosoma, *and show where the cycle can be interrupted to prevent human disease.*

Many helminths use the cardiovascular system for part of their life cycle. Schistosomes find a home there, shedding eggs that are distributed in the bloodstream.

Schistosomiasis

Schistosomiasis is a debilitating disease caused by a small fluke. The life cycle of *Schistosoma* is depicted in Figure 23.25b. The disease is spread by human feces or urine carrying eggs of the schistosome that enter water supplies with which humans come into contact. In the developed world, sewage and water treatment minimizes the contamination of the water supply. Also, snails of certain species are essential for one stage of the life cycle of the schistosomes. They produce the cercariae that penetrate the skin of a human entering contaminated water. In most areas of the United States, a suitable host snail is not present. Therefore, even though it is estimated that schistosome eggs are being shed by approximately 400,000 immigrants, the disease is not being propagated.

The symptoms of the disease result from eggs shed by adult schistosomes in the human host. These adult helminths are 15–20 mm long, and the slender female lives permanently in a groove in the body of the male, from which is derived the name: *schisto-some*, or split-body (Figure 23.25a). The union between the male and female produces a continuing supply of new eggs. Some of these eggs lodge in tissues. Defensive reactions of the human host to these foreign bodies cause local tissue damage called **granulomas** (Figure 23.26). Other eggs are excreted and enter the water to continue the cycle.

There are three primary types of schistosomiasis. The disease caused by *Schistosoma haematobium,* sometimes called urinary schistosomiasis, results in inflammation of the urinary bladder wall. Similarly, *S. japonicum* and *S. mansoni* cause intestinal inflammation. Depending on the species, schistosomiasis can cause damage to many different organs when eggs migrate in the bloodstream to different areas. For example, damage to the liver or lungs, urinary bladder cancer, or, when eggs lodge in the brain, neurological symptoms. Geographically, *S. japonicum* is found in east Asia. *S. haematobium* infects many people throughout Africa and the Middle East, most particularly

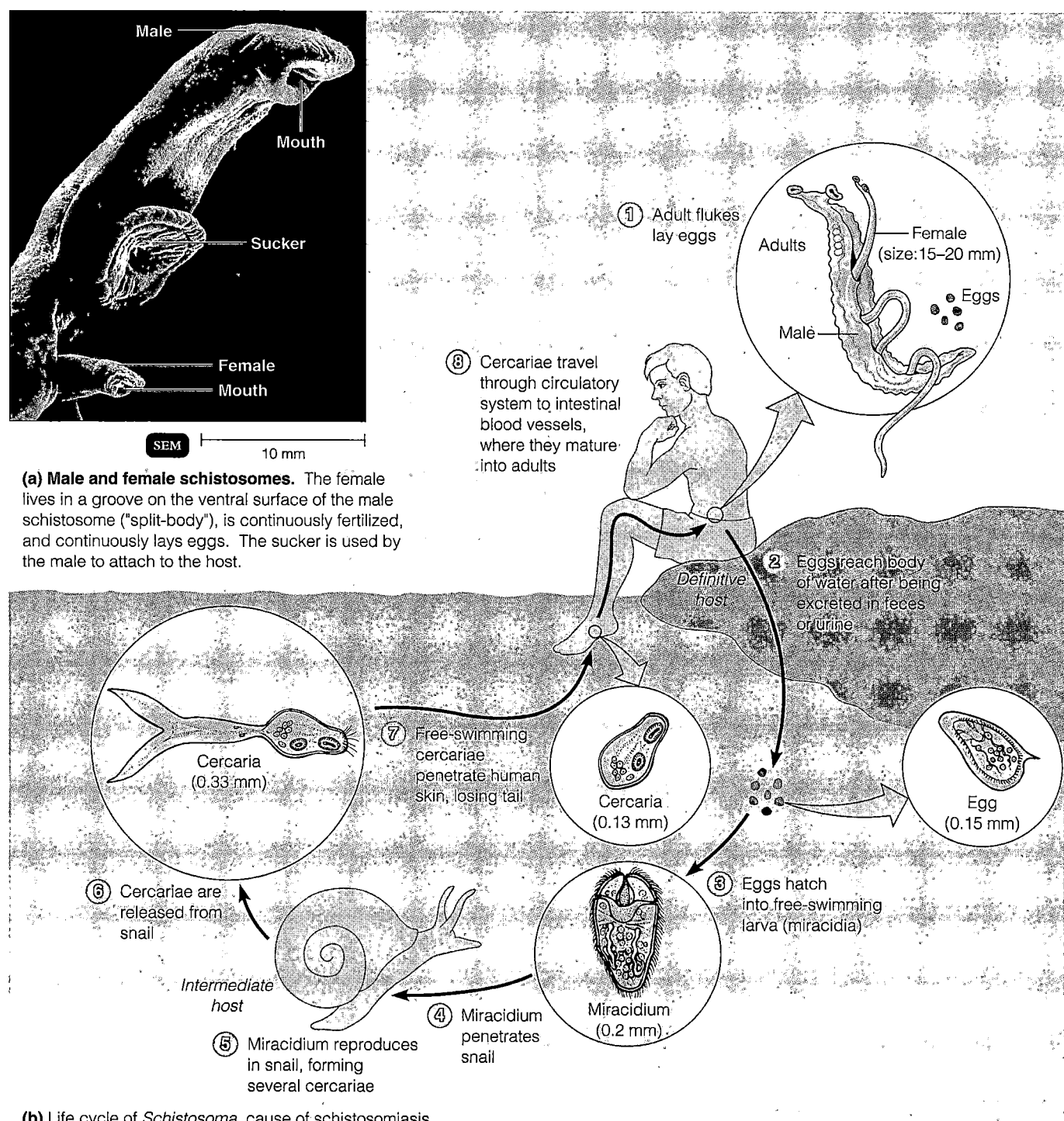

Male — **Mouth** — **Sucker** — **Female** — **Mouth**

SEM |——————| 10 mm

(a) Male and female schistosomes. The female lives in a groove on the ventral surface of the male schistosome ("split-body"), is continuously fertilized, and continuously lays eggs. The sucker is used by the male to attach to the host.

① Adult flukes lay eggs

Adults — Female (size:15–20 mm) — Eggs — Male

⑧ Cercariae travel through circulatory system to intestinal blood vessels, where they mature into adults

Definitive host

② Eggs reach body of water after being excreted in feces or urine

Cercaria (0.33 mm)

⑦ Free-swimming cercariae penetrate human skin, losing tail

Cercaria (0.13 mm)

Egg (0.15 mm)

⑥ Cercariae are released from snail

Intermediate host

⑤ Miracidium reproduces in snail, forming several cercariae

④ Miracidium penetrates snail

Miracidium (0.2 mm)

③ Eggs hatch into free-swimming larva (miracidia)

(b) Life cycle of *Schistosoma*, cause of schistosomiasis.

FIGURE 23.25 Schistosomiasis.

Egypt. *S. mansoni* has a similar distribution but also is endemic in South America and the Caribbean, including Puerto Rico. It is estimated that more than 250 million of the world's population are affected.

The adult worms appear to be unaffected by the host's immune system. Apparently, they quickly coat themselves with a layer that mimics the host's tissues.

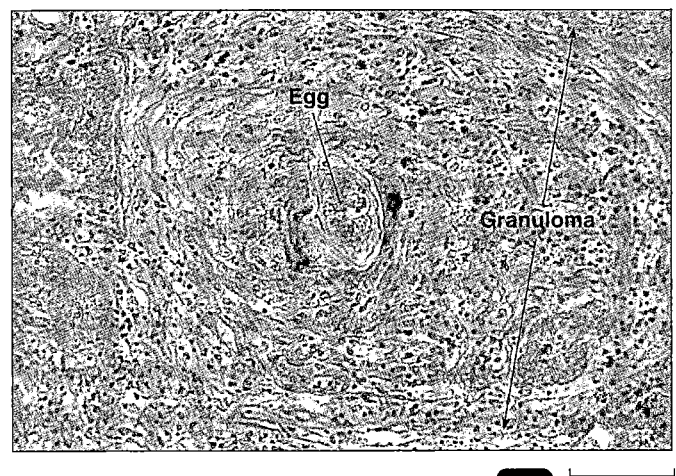

LM ⊢ 150 μm ⊣

FIGURE 23.26 A granuloma from a patient with schistosomes. Some of the eggs laid by the adult schistosomes lodge in the tissue, and the body responds to the irritant by surrounding it with scarlike tissue, forming a granuloma.

■ Why is the immune system ineffective against adult schistosomes?

Laboratory diagnosis consists of microscopic identification of the flukes or their eggs in fecal and urine specimens, intradermal tests, and serological tests such as complement-fixation and precipitin tests.

Praziquantel and oxamniquine are approved for use against schistosomes in the United States. Sanitation and elimination of the host snail are also useful forms of control.

Swimmer's Itch

Swimmers in lakes in the northern United States are sometimes troubled by **swimmer's itch.** This is a cutaneous allergic reaction to cercariae, similar to that of schistosomiasis. However, these parasites mature only in wildfowl and not in humans, so infection does not progress beyond penetration of the skin and a local inflammatory response.

★ ★ ★

The diseases covered in this chapter are summarized in taxonomic order in Table 23.1.

table 23.1	Microbial Diseases of the Cardiovascular and Lymphatic Systems	
Disease	**Pathogen**	**Comments**
Bacterial Diseases		
Septic shock, puerperal sepsis, and infections related to abortions	Gram-negatives and *Streptococcus pyogenes*	Life-threatening complications of septicemia; gram-negatives, which release endotoxins, are especially hard to treat.
Endocarditis		
Subacute bacterial	Mostly α-hemolytic streptococci	Bacteria lodge in heart valves and may cause fatal damage.
Acute bacterial	*Staphylococcus aureus*	More rapidly progressive damage to heart valves.
Pericarditis	*Streptococcus pyogenes*	Affects sac surrounding heart; rapidly progressive.
Rheumatic fever	Group A β-hemolytic streptococci	Probably an autoimmune condition; repeated streptococcal infections result in antibodies that damage heart tissue.
Tularemia	*Francisella tularensis*	Results from infection from handling small animals such as rabbits; enters by skin abrasions, ingestion, inhalation, bites.
Brucellosis	*Brucella* spp.	Formerly mostly acquired by ingesting cow or other milk; now mostly by contact with animal carcasses. Pathogen grows within phagocytic cells.
Anthrax	*Bacillus anthracis*	Endospores occur in soils and infect grazing animals, proliferating in blood; inhalation of endospores causes especially dangerous type.
Gangrene	*Clostridium perfringens*	Caused by contamination of an open wound by clostridial endospores; toxins destroy adjacent tissue.
Cat-scratch disease	*Bartonella henselae*	Systematic infection, prolonged fever; may be fatal.
Plague	*Yersinia pestis*	Classically transmitted by fleas from a reservoir in rats; in western U.S., endemic in rodents.

➤

table 23.1 *Microbial Diseases of the Cardiovascular and Lymphatic Systems (continued)*

Disease	Pathogen	Comments
Relapsing fever	*Borrelia* spp.	Symptoms include a series of fever peaks as new populations of bacteria evade the host's developing immunity.
Lyme disease	*Borrelia burgdorferi*	A tickborne disease in areas of high deer populations; complications may include cardiac and neurological problems.
Ehrlichiosis	*Ehrlichia* spp.	A tickborne flulike disease caused by rickettsias; human granulocytic ehrlichiosis can be fatal.
Epidemic typhus	*Rickettsia prowazekii*	A louseborne rickettsial disease, characterized by high fever; high mortality rate.
Endemic murine typhus	*Rickettsia typhi*	A fleaborne rickettsial disease; rodents are the reservoir. Resembles epidemic typhus but mortality rate is low.
Rocky Mountain spotted fever	*Rickettsia rickettsii*	A tickborne rickettsial disease, characterized by rash, fever, headache; high mortality rate.
Viral Diseases		
Burkitt's lymphoma	EB virus	A tumor endemic to central Africa.
Infectious mononucleosis	EB virus	A mild disease common in young people; transmitted by oral secretions.
Yellow fever	Arbovirus (yellow fever virus)	A mosquito-borne disease in Central and South America; high mortality rate.
Dengue	Arbovirus (dengue fever virus)	A mosquito-borne disease, rarely fatal but painful symptoms. Dengue hemorrhagic fever (DHF), common in Southeast Asia, is more serious.
Viral hemorrhagic fevers (Marburg, Ebola, Lassa)	Filovirus, Arenavirus	Highly fatal viral hemorrhagic fevers found in tropical Africa; spread by contact with contaminated blood.
Hantavirus pulmonary syndrome	Sin Nombre hantavirus	A highly fatal disease that causes the lungs to flood with fluid; spread by aerosols from field mouse excretions.
Protozoan Diseases		
American trypanosomiasis (Chagas' disease)	*Trypanosoma cruzi*	Common in Central and South America, transmitted by bite of reduviid bug; damages heart muscle or peristaltic movement in esophagus and colon.
Toxoplasmosis	*Toxoplasma gondii*	A mild disease in immunocompetent adults; may cause severe fetal damage if initial infection occurs during pregnancy. Reactivation in AIDS patients causes serious illness.
Malaria	*Plasmodium* spp.	A mosquito-borne disease common in hot climates, characterized by fever and chills at intervals. *P. falciparum* causes most serious type, often fatal in small children.
Leishmaniasis	As many as 20 species of *Leishmania,* primarily *L. donovani, L. tropica,* and *L. braziliensis* groups	*L. donovani* causes systemic disease of deep body organs; *L. tropica* causes skin sores; *L. braziliensis* causes skin sores and disfiguring damage to mucous membranes of nose, mouth, and so on. Insect vector is the sandfly.
Babesiosis	*Babesia microti*	A tickborne disease that resembles malaria, though mostly subclinical; serious in immunocompromised patients.
Helminthic Disease		
Schistosomiasis	*Schistosoma* spp.	Eggs produced by schistosomes lodge in tissue and induce damaging inflammation.
Swimmer's itch	Larvae of schistosomes of nonhuman animals.	An allergic reaction to parasite in skin.

Study Outline

INTRODUCTION (p. 625)

1. The heart, blood, and blood vessels make up the cardiovascular system.

2. Lymph, lymph vessels, lymph nodes, and lymphoid organs constitute the lymphatic system.

STRUCTURE AND FUNCTION OF THE CARDIOVASCULAR AND LYMPHATIC SYSTEMS (pp. 625–626)

1. The heart circulates substances to and from tissue cells.

2. Blood is a mixture of plasma and cells.

3. Plasma transports dissolved substances. Red blood cells carry oxygen. White blood cells are involved in the body's defense against infection.

4. Fluid that filters out of capillaries into spaces between tissue cells is called interstitial fluid.

5. Interstitial fluid enters lymph capillaries and is called lymph; vessels called lymphatics return lymph to the blood.

6. Lymph nodes contain fixed macrophages, B cells, and T cells.

BACTERIAL DISEASES OF THE CARDIOVASCULAR AND LYMPHATIC SYSTEMS (pp. 626–638)

Septicemia, Sepsis, and Septic Shock (pp. 626–627)

1. The growth of microorganisms in blood is called septicemia. Signs include lymphangitis (inflamed lymph vessels).

2. Septicemia usually results from a focus of infection in the body.

3. Gram-negative septicemia can lead to septic shock, characterized by decreased blood pressure. Endotoxin causes the symptoms.

Puerperal Sepsis (pp. 627–628)

1. Puerperal sepsis begins as an infection of the uterus following childbirth or abortion; it can progress to peritonitis or septicemia.

2. *Streptococcus pyogenes* is the most frequent cause.

3. Oliver Wendell Holmes and Ignaz Semmelweiss demonstrated that puerperal sepsis was transmitted by the hands and instruments of midwives and physicians.

4. Puerperal sepsis is now uncommon because of modern hygienic techniques and antibiotics.

Bacterial Infections of the Heart (p. 628)

1. The inner layer of the heart is the endocardium.

2. Subacute bacterial endocarditis is usually caused by α-hemolytic streptococci, staphylococci, or enterococci.

3. The infection arises from a focus of infection, such as a tooth extraction.

4. Preexisting heart abnormalities are predisposing factors.

5. Signs include fever, anemia, and heart murmur.

6. Acute bacterial endocarditis is usually caused by *Staphylococcus aureus*.

7. The bacteria cause rapid destruction of heart valves.

Rheumatic Fever (p. 629)

1. Rheumatic fever is an autoimmune complication of streptococcal infections.

2. Rheumatic fever is expressed as arthritis or inflammation of the heart. It can result in permanent heart damage.

3. Antibodies against group A β-hemolytic streptococci react with streptococcal antigens deposited in joints or heart valves or cross-react with the heart muscle.

4. Rheumatic fever can follow a streptococcal infection, such as streptococcal sore throat. Streptococci might not be present at the time of rheumatic fever.

5. Prompt treatment of streptococcal infections can reduce the incidence of rheumatic fever.

6. Penicillin is administered as a preventive measure against subsequent streptococcal infections.

Tularemia (pp. 629–630)

1. Tularemia is caused by *Francisella tularensis*. The reservoir is small wild mammals, especially rabbits.

2. Signs include ulceration at the site of entry, followed by septicemia and pneumonia.

3. Humans contract tularemia by handling diseased carcasses, eating undercooked meat of diseased animals, and being bitten by certain vectors (such as deer flies).

4. *F. tularensis* is resistant to phagocytosis.

5. Laboratory diagnosis is based on an agglutination test on isolated bacteria.

Brucellosis (Undulant Fever) (pp. 630–631)

1. Brucellosis can be caused by *Brucella abortus, B. melitensis,* and *B. suis.*

2. In the United States, elk and bison constitute the reservoir for *B. abortus.*

3. The bacteria enter through minute breaks in the mucosa or skin, reproduce in macrophages, and spread via lymphatics to liver, spleen, or bone marrow.

4. Signs include malaise and fever that spikes each evening (undulant fever).

5. Diagnosis is based on serological tests.

Anthrax (p. 631)

1. *Bacillus anthracis* causes anthrax. In soil, endospores can survive for up to 60 years.

2. Grazing animals acquire an infection after ingesting the endospores.

3. Humans contract anthrax by handling hides from infected animals. The bacteria enter through cuts in the skin or through the respiratory tract.

4. Entry through the skin results in a pustule that can progress to septicemia. Entry through the respiratory tract can result in pneumonia.

5. Diagnosis is based on isolation and identification of the bacteria.

Gangrene (pp. 631–632)

1. Soft tissue death from ischemia (loss of blood supply) is called gangrene.

2. Microorganisms grow on nutrients released from gangrenous cells.

3. Gangrene is especially susceptible to the growth of anaerobic bacteria such as *Clostridium perfringens,* the causative agent of gas gangrene.

4. *C. perfringens* can invade the wall of the uterus during improperly performed abortions.

5. Surgical removal of necrotic tissue, hyperbaric chambers, and amputation are used to treat gas gangrene.

Systemic Diseases Caused by Bites and Scratches (pp. 632–633)

1. *Pasteurella multocida,* introduced by the bite of a dog or cat, can cause septicemia.

2. Anaerobic bacteria such as *Clostridium, Bacteroides,* and *Fusobacterium* infect deep animal bites.

3. Cat-scratch disease is caused by *Bartonella henselae.*

Vector-Transmitted Diseases (pp. 633–638)

Plague (pp. 633–634)

1. Plague is caused by *Yersinia pestis.* The vector is usually the rat flea (*Xenopsylla cheopis*).

2. Reservoirs for plague include European rats and North American rodents.

3. Signs of bubonic plague include bruises on the skin and enlarged lymph nodes (buboes).

4. The bacteria can enter the lungs and cause pneumonic plague.

5. Laboratory diagnosis is based on isolation and identification of the bacteria.

6. Antibiotics are effective in treating plague, but they must be administered promptly after exposure to the disease.

Relapsing Fever (p. 634)

1. Relapsing fever is caused by *Borrelia* species and transmitted by soft ticks.

2. The reservoir for the disease is rodents.

3. Signs include fever, jaundice, and rose-colored spots. Signs recur three or four times after apparent recovery.

4. Laboratory diagnosis is based on the presence of spirochetes in the patient's blood.

Lyme Disease (Lyme Borreliosis) (pp. 634–635)

1. Lyme disease is caused by *Borrelia burgdorferi* and is transmitted by a tick *(Ixodes).*

2. Lyme disease is prevalent on the U.S. Atlantic Coast.

3. Field mice provide the animal reservoir.

4. Diagnosis is based on serological tests and clinical symptoms.

Other Tickborne Diseases (p. 635)

1. Ehrlichiosis is caused by *Ehrlichia* species.

Typhus (pp. 635–637)

1. Typhus is caused by rickettsias, obligate intracellular parasites of eukaryotic cells.

Epidemic Typhus (p. 637)

1. The human body louse *Pediculus humanus corporis* transmits *Rickettsia prowazekii* in its feces, which are deposited while the louse is feeding.

2. Epidemic typhus is prevalent in crowded and unsanitary living conditions that allow the proliferation of lice.

3. The signs of typhus are rash, prolonged high fever, and stupor.

4. Tetracyclines and chloramphenicol are used in treatment.

Endemic Murine Typhus (p. 637)

1. Endemic murine typhus is a less severe disease caused by *Rickettsia typhi* and transmitted from rodents to humans by the rat flea.

Spotted Fevers (pp. 637–638)

1. *Rickettsia rickettsii* is a parasite of ticks (*Dermacentor* spp.) in the southeastern U.S., Appalachia, and the Rocky Mountain states.

2. The rickettsia may be transmitted to humans, in whom it causes tickborne typhus fever.

3. Chloramphenicol and tetracyclines effectively treat Rocky Mountain spotted fever, or tickborne typhus.

4. Serological tests are used for laboratory diagnosis.

VIRAL DISEASES OF THE CARDIOVASCULAR AND LYMPHATIC SYSTEMS (pp. 639–641)

Burkitt's Lymphoma (p. 639)

1. Epstein-Barr (EB) virus causes Burkitt's lymphoma and nasopharyngeal carcinoma.

2. Burkitt's lymphoma tends to occur in patients whose immune system has been weakened, for example, by malaria or AIDS.

Infectious Mononucleosis (pp. 639–640)

1. Infectious mononucleosis is caused by the EB virus.

2. The virus multiplies in the parotid glands and is present in saliva. It causes the proliferation of atypical lymphocytes.

3. The disease is transmitted by the ingestion of saliva from infected individuals.

4. Diagnosis is made by an indirect fluorescent-antibody technique.

Classic Viral Hemorrhagic Fevers (p. 640)

1. Yellow fever is caused by a virus (yellow fever virus). The vector is the mosquito *Aedes aegypti*.

2. Signs and symptoms include fever, chills, headache, nausea, and jaundice.

3. Diagnosis is based on the presence of virus-neutralizing antibodies in the host.

4. No treatment is available, but there is an attenuated, live viral vaccine.

5. Dengue is caused by a virus (dengue fever virus) and is transmitted by the mosquito *Aedes aegypti*.

6. Signs are fever, muscle and joint pain, and rash.

7. Mosquito abatement is necessary to control the disease.

8. Dengue hemorrhagic fever (DHF) occurs when a person with antibodies is reinfected with the same dengue virus.

Emerging Viral Hemorrhagic Fevers (p. 641)

1. Human diseases caused by Marburg, Ebola, and Lassa fever viruses were first noticed in the late 1960s.

2. Marburg virus is found in nonhuman primates; Lassa fever viruses are found in rodents.

3. Rodents are the reservoirs for Argentine and Bolivian hemorrhagic fevers.

4. *Hantavirus* pulmonary syndrome is caused by hantavirus. The virus is contracted by inhalation of dried rodent urine.

PROTOZOAN DISEASES OF THE CARDIOVASCULAR AND LYMPHATIC SYSTEMS (pp. 641–647)

American Trypanosomiasis (Chagas' Disease) (pp. 641–642)

1. *Trypanosoma cruzi* causes Chagas' disease. The reservoir includes many wild animals. The vector is a reduviid, the "kissing bug."

2. Xenodiagnosis allows for the identification of trypanosomes in the intestinal tract of the reduviid bug, which confirms the diagnosis.

Toxoplasmosis (p. 642)

1. Toxoplasmosis is caused by the sporozoan *Toxoplasma gondii*.

2. *T. gondii* undergoes sexual reproduction in the intestinal tract of domestic cats, and oocysts are eliminated in cat feces.

3. In the host cell, sporozoites reproduce to form either tissue-invading tachyzoites or bradyzoites.

4. Humans contract the infection by ingesting tachyzoites or tissue cysts in undercooked meat from an infected animal or contact with cat feces.

5. Congenital infections can occur. Signs and symptoms include severe brain damage or vision problems.

6. Toxoplasmosis can be identified by serological tests, but interpretation of the results is uncertain.

Malaria (pp. 642–645)

1. The signs and symptoms of malaria are chills, fever, vomiting, and headache, which occur at intervals of 2–3 days.

2. Malaria is transmitted by *Anopheles* mosquitoes. The causative agent is any one of four species of *Plasmodium*.

3. Sporozoites reproduce in the liver and release merozoites into the bloodstream, where they infect red blood cells and produce more merozoites.

4. Laboratory diagnosis is based on microscopic observation of merozoites in red blood cells.

5. New drugs are being developed as the protozoa develop resistance to drugs such as chloroquine.

Leishmaniasis (pp. 645–647)

1. *Leishmania spp.*, which are transmitted by sandflies, cause leishmaniasis.

2. The protozoa reproduce in the liver, spleen, and kidneys.

3. Antimony compounds are used for treatment.

Babesiosis (p. 647)

1. Babesiosis is caused by the protozoan *Babesia microti* and transmitted to humans by ticks.

HELMINTHIC DISEASES OF THE CARDIOVASCULAR AND LYMPHATIC SYSTEMS (pp. 647–649)

Schistosomiasis (pp. 647–649)

1. Species of the blood fluke *Schistosoma* cause schistosomiasis.

2. Eggs eliminated with feces hatch into larvae that infect the intermediate host, a snail. Free-swimming cercariae are released from the snail and penetrate the skin of a human.

3. The adult flukes live in the veins of the liver or urinary bladder in humans.

4. Granulomas are from the host's defense to eggs that remain in the body.

5. Observation of eggs or flukes in feces, skin tests, or indirect serological tests may be used for diagnosis.

6. Chemotherapy is used to treat the disease; sanitation and snail eradication are used to prevent it.

Swimmer's Itch (p. 649)

1. Swimmer's itch is a cutaneous allergic reaction to cercariae that penetrate the skin. The definitive hosts for this fluke are wildfowl.

Study Questions

REVIEW

1. What are the signs of septicemia?

2. How can septicemia result from a single focus of infection, such as an abscess?

3. Complete the following table:

Disease	Frequent Causative Agent	Predisposing Condition(s)
Puerperal sepsis		
Subacute bacterial endocarditis		
Acute bacterial endocarditis		

4. Describe the probable cause of rheumatic fever. How is rheumatic fever treated? How is it prevented?

5. Compare and contrast epidemic typhus, endemic murine typhus, and tickborne typhus.

6. Complete the following table:

Disease	Causative Agent	Vector	Signs/ Symptoms	Treatment
Malaria				
Yellow fever				
Dengue				
Relapsing fever				
Leishmaniasis				

7. Complete the following table:

Disease	Causative Agent	Method of Transmission	Reservoir	Signs/ Symptoms	Prevention
Tularemia					
Brucellosis					
Anthrax					
Lyme disease					

8. Provide the following information on plague: causative agent, vector, U.S. reservoir, control, treatment, prognosis (probable outcome).

9. List the causative agent, method of transmission, and reservoir for schistosomiasis, toxoplasmosis, and American trypanosomiasis. Which disease are you most likely to get in the United States? Where are the other diseases endemic?

10. Compare and contrast cat-scratch disease and toxoplasmosis.

11. Why is *Clostridium perfringens* likely to grow in gangrenous wounds?

12. List the causative agents and methods of transmission of infectious mononucleosis.

13. Differentiate between the transmission and symptoms of bubonic plague and pneumonic plague.

MULTIPLE CHOICE

Use the following choices to answer questions 1–4:
- **a.** ehrlichiosis
- **b.** Lyme disease
- **c.** septic shock
- **d.** toxoplasmosis
- **e.** viral hemorrhagic fever

1. A patient presents with vomiting and diarrhea and a history of fever and headache. Bacterial cultures of blood, CSF, and stool are negative. What is your diagnosis?

2. A patient was hospitalized because of continuing fever and progression of symptoms including headache, fatigue, and back pain. Tests for antibodies to *Borrelia burgdorferi* were negative. What is your diagnosis?

3. A patient complained of headache. A CT (computed tomography) scan revealed cysts of varying size in her brain. What is your diagnosis?

4. A patient presents with mental confusion, rapid breathing and heartbeat, and low blood pressure. What is your diagnosis?

5. A patient has a red circular rash on his arm and fever, malaise, and joint pain. The most appropriate treatment is
- **a.** penicillin.
- **b.** chloroquine.
- **c.** anti-inflammatory drugs.
- **d.** rifampin.
- **e.** no treatment.

6. Which of the following is not a tickborne disease?
- **a.** babesiosis
- **b.** ehrlichiosis
- **c.** Lyme disease
- **d.** relapsing fever
- **e.** tularemia

Use the following choices to answer questions 7 and 8:
- **a.** brucellosis
- **b.** malaria
- **c.** relapsing fever
- **d.** Rocky Mountain spotted fever
- **e.** Ebola hemorrhagic fever

7. The patient's fever spikes each evening. Oxidase-positive, gram-negative cocci were isolated from a lesion on his arm. What is your diagnosis?

8. The patient was hospitalized with fever and headache. Spirochetes were observed in her blood. What is your diagnosis?

9. Which of the following diseases has the highest incidence in the United States?
- **a.** brucellosis
- **b.** Ebola hemorrhagic fever
- **c.** malaria
- **d.** plague
- **e.** Rocky Mountain spotted fever

10. Nineteen workers in a slaughterhouse developed fever and chills, with the fever spiking to 40°C each evening. The most likely method of transmission of this disease is
- **a.** a vector.
- **b.** the respiratory route.
- **c.** a puncture wound.
- **d.** an animal bite.
- **e.** water.

CRITICAL THINKING

1. Indirect FA tests on the serum of three 25-year-old women, each of whom is considering pregnancy, provided the information below. Which of these women may have toxoplasmosis? What advice might be given to each woman with regard to toxoplasmosis?

Patient	Antibody Titer		
	Day 1	Day 5	Day 12
Patient A	1024	1024	1024
Patient B	1024	2048	3072
Patient C	0	0	0

2. What is the most effective way to control malaria and dengue?

3. In adults, the second dengue virus infection results in dengue hemorrhagic fever (DHF), which is characterized by bleeding from the skin and mucosa. DHF can be fatal. In infants under 1 year old, the first dengue virus infection results in DHF. Offer an explanation for this.

CLINICAL APPLICATIONS

1. A 19-year-old man went deer hunting. While on the trail, he found a partially dismembered dead rabbit. The hunter picked up the front paws for good luck charms and gave them to another hunter in the party. The rabbit had been handled with bare hands that were bruised and scratched from the hunter's work as an automobile mechanic. Festering sores on his hands, legs, and knees were noted two days later. What infectious disease do you suspect the hunter has? How would you proceed to prove it?

2. On March 30, a 35-year-old veterinarian experienced fever, chills, and vomiting. On March 31, he was hospitalized with diarrhea, left armpit bubo, and secondary bilateral pneumonia. On March 27, he had treated a cat that had labored respiration; an X ray revealed pulmonary infiltrates. The cat died on March 28 and was disposed of. Chloramphenicol was administered to the veterinarian. On April 10, his temperature returned to normal, and on April 20, he was released from the hospital. Sixty human contacts were given tetracycline. Identify the incubation and prodromal periods for this case. Explain why the 60 contacts were treated. What was the etiologic agent? How would you identify the agent?

3. Three of five patients who underwent heart valve replacement surgery developed bacteremia. The causative agent was *Enterobacter cloacae*. What were the patients' signs and symptoms? How would you identify this bacterium? A manometer used in the operations was culture-positive for *E. cloacae*. What is the most likely source of this contaminant? Suggest a way of preventing such occurrences.

4. In August and September, six people who each at different times spent a night in the same cabin developed the symptoms shown in the following graph. Three recovered after

tetracycline therapy, two recovered without therapy, and one was hospitalized with septic shock. What is the disease? What is the incubation period of this disease? How do you account for the periodic temperature changes? What caused septic shock in the sixth patient?

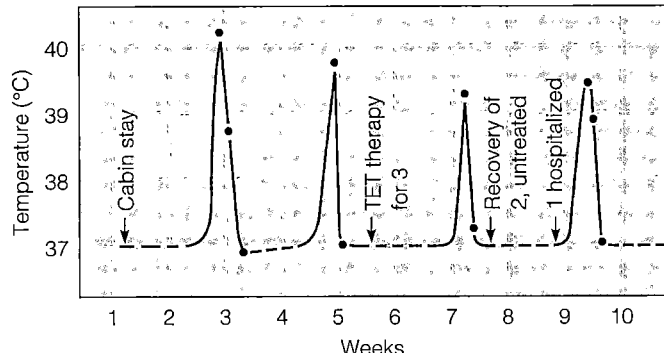

5. A 67-year-old man worked in a textile mill that processed imported goat hair into fabrics. He noticed a painless, slightly swollen pimple on his chin. Two days later he developed a 1-cm ulcer at the pimple site and a temperature of 37.6°C. He was treated with tetracycline. What is the etiology of this disease? Suggest ways to prevent it.

Learning with Technology

MP = The Microbiology Place website   **ST** = Student Tutorial CD-ROM   **VU** = VirtualUnknown CD-ROM

MP   Don't forget to go to The Microbiology Place website (http://www.microbiologyplace.com) to take the practice tests, explore the interactive activity and case study, and check out the news articles and web links for this chapter.

ST   Remember there is also a quiz for this chapter on the Microbiology Interactive Student Tutorial CD-ROM.

VU   Enter the Virtual Lab, click the arrow next to the Session field, click Textbook Exercises, and select Chapter 23. Read the Case Study carefully, identify the unknown, and use what you learn to answer the following questions (consult your textbook for additional information).

1. Based on the signs and symptoms observed and the agent identified, what would be the diagnosis for this patient?

2. Provide a likely scenario by which the victim would have been exposed to this organism.

3. What is the mode of transmission for this organism in this case?

4. How does the growth of the organism in the host lead to the signs and symptoms associated with this disease?

chapter 24

Microbial Diseases of the Respiratory System

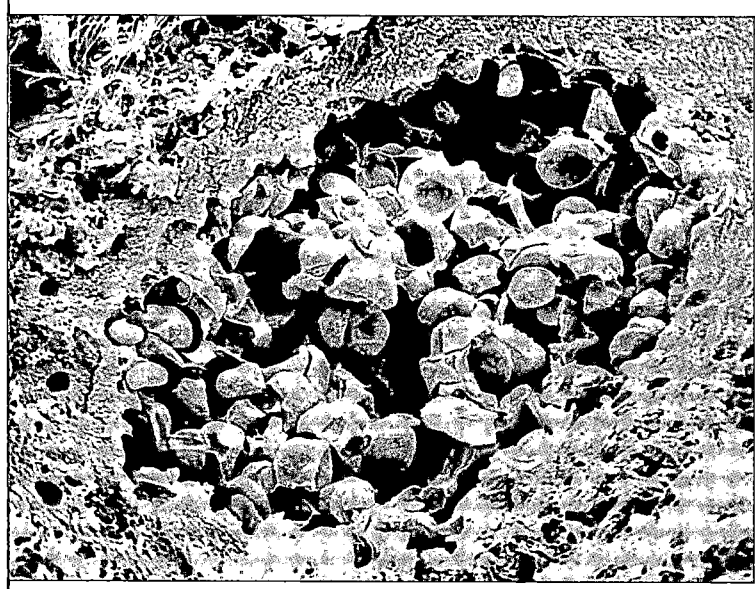

Pneumocystis carinii. *This fungus causes pneumonia in immunosuppressed patients.*

With every breath, we inhale several microorganisms; therefore, the upper respiratory system is a major portal of entry for pathogens. In fact, respiratory system infections are the most common type of infection—and among the most damaging. Some pathogens that enter via the respiratory route can infect other parts of the body, such as those that cause measles, mumps, and rubella.

Structure and Function of the Respiratory System

Learning Objective

- *Describe how microorganisms are prevented from entering the respiratory system.*

It is convenient to think of the respiratory system as being composed of two divisions: the upper respiratory system and the lower respiratory system. The **upper respiratory system** consists of the nose, the pharynx (throat), and the structures associated with them, including the middle ear and the auditory (eustachian) tubes (Figure 24.1). Ducts from the sinuses, and nasolacrimal ducts from the lacrimal (tear-forming) apparatus, empty into the nasal cavity (see Figure 16.3). The auditory tubes from the middle ear empty into the upper portion of the throat.

The upper respiratory system has several anatomical defenses against airborne pathogens. Coarse hairs in the nose filter large dust particles from the air. The nose is lined with a mucous membrane that contains numerous mucus-secreting cells and cilia. The upper portion of the throat also contains a ciliated mucous membrane. The mucus moistens inhaled air and traps dust and microorganisms, especially particles larger than 4–5 μm. The cilia help remove these particles by moving them toward the mouth for elimination.

At the junction of the nose and throat are masses of lymphoid tissue, the tonsils, which contribute immunity to certain infections. Occasionally, however, these tissues become infected and help spread infection to the ears via the auditory tubes. Because the nose and throat are connected to the sinuses, nasolacrimal apparatus, and middle ear, infections commonly spread from one region to another.

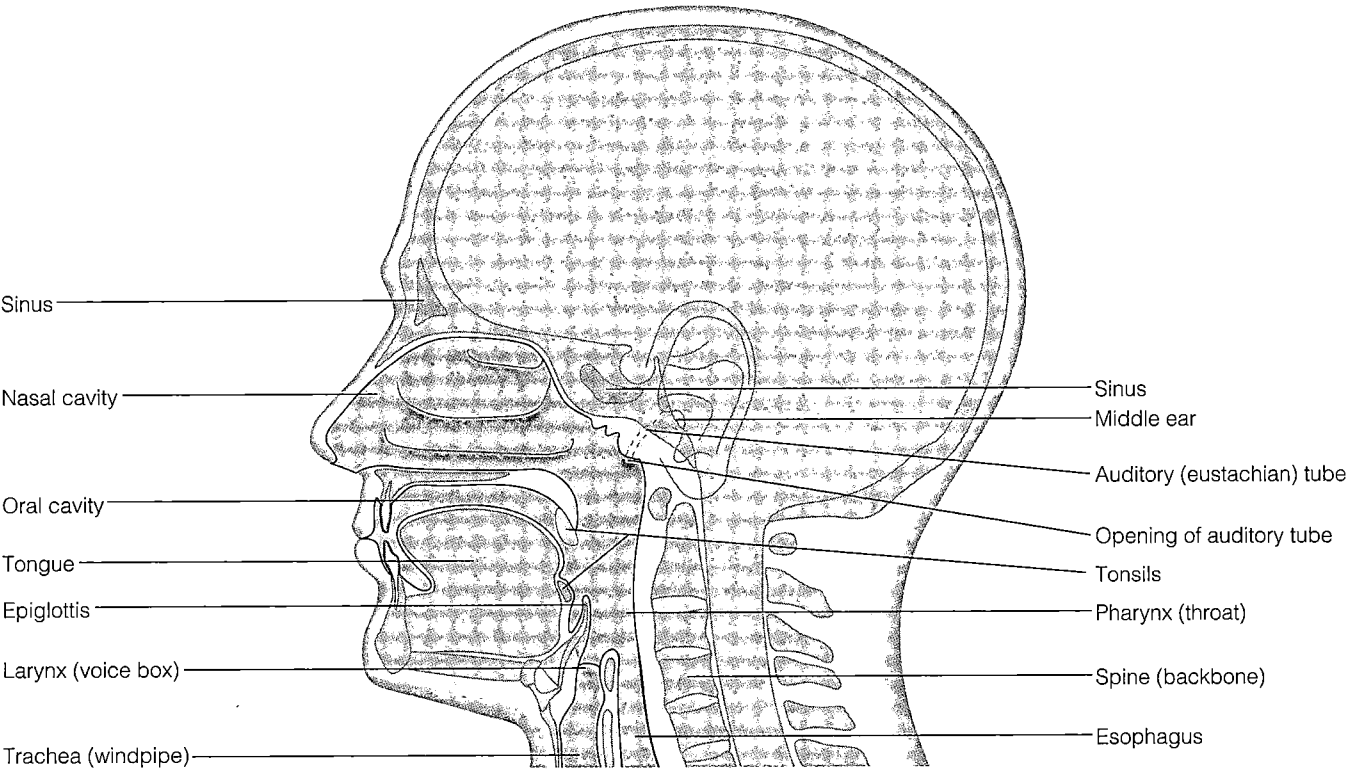

Sinus

Nasal cavity

Oral cavity

Tongue

Epiglottis

Larynx (voice box)

Trachea (windpipe)

Sinus
Middle ear
Auditory (eustachian) tube
Opening of auditory tube
Tonsils
Pharynx (throat)
Spine (backbone)
Esophagus

FIGURE 24.1 Structures of the upper respiratory system.

■ Name the upper respiratory system's defenses against disease.

The **lower respiratory system** consists of the larynx (voice box), trachea (windpipe), bronchial tubes, and *alveoli* (Figure 24.2). Alveoli are air sacs that make up the lung tissue; within them, oxygen and carbon dioxide are exchanged between the lungs and blood. Our lungs contain more than 300 million alveoli, with an area for gas exchange of 70 or more square meters. The double-layered membrane enclosing the lungs is the *pleura,* or pleural membranes. A ciliated mucous membrane lines the lower respiratory system down to the smaller bronchial tubes and helps prevent microorganisms from reaching the lungs.

As discussed in Chapter 16, particles trapped in the larynx, trachea, and larger bronchial tubes are moved up toward the throat by a ciliary action called the *ciliary escalator* (see Figure 16.4 on page 457). If microorganisms actually reach the lungs, phagocytic cells called *alveolar macrophages* usually locate, ingest, and destroy most of them. IgA antibodies in such secretions as respiratory mucus, saliva, and tears also help protect mucosal surfaces of the respiratory system from many pathogens. Thus, the

body has several mechanisms for removing the pathogens that cause airborne infections.

Normal Microbiota of the Respiratory System

Learning Objective

■ *Characterize the normal microbiota of the upper and lower respiratory systems.*

A number of potentially pathogenic microorganisms are part of the normal microbiota in the upper respiratory system. However, they usually do not cause illness because the predominant microorganisms of the normal microbiota suppress their growth by competing with them for nutrients and producing inhibitory substances.

By contrast, the lower respiratory tract is nearly sterile—although the trachea may contain a few bacteria—because of the normally efficient functioning of the ciliary escalator in the bronchial tubes.

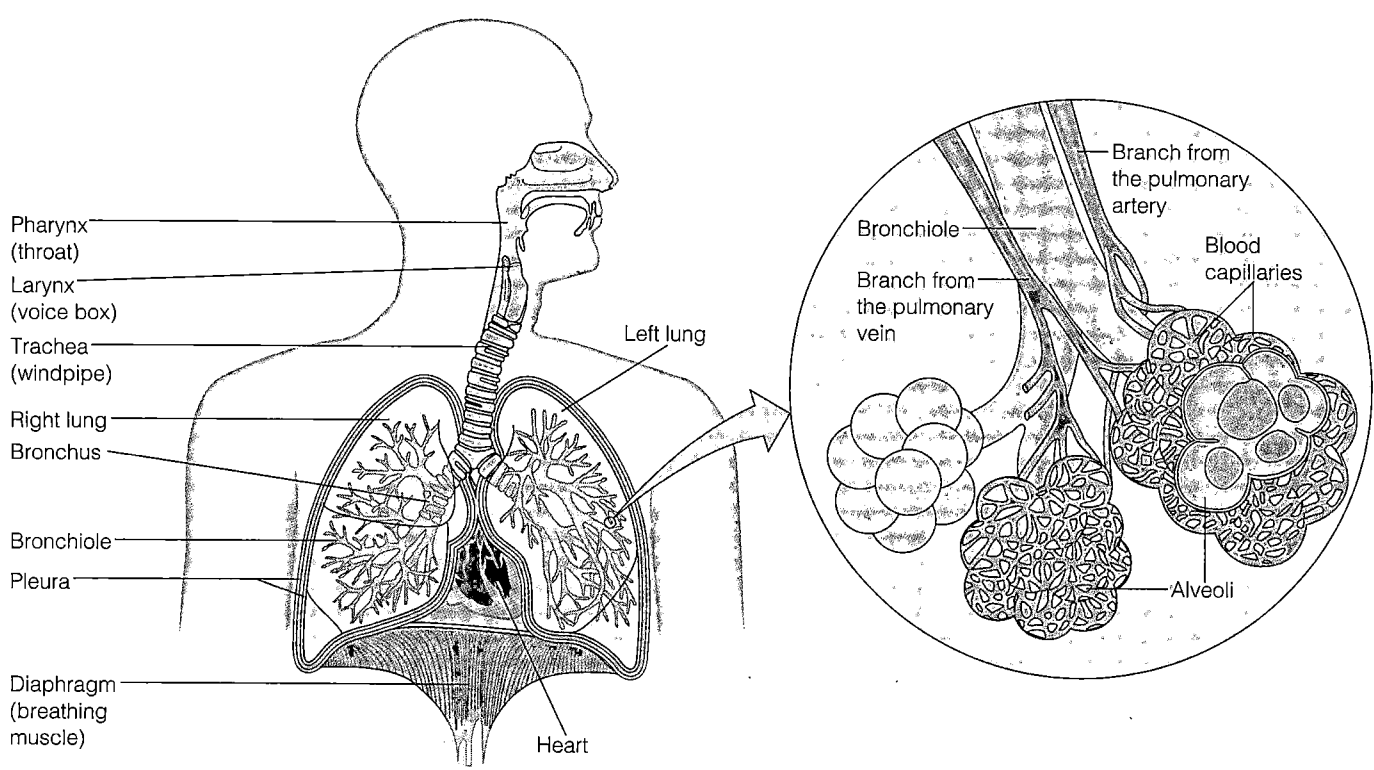

FIGURE 24.2   Structures of the lower respiratory system.

■ Name the lower respiratory system's defenses against disease.

MICROBIAL DISEASES OF THE UPPER RESPIRATORY SYSTEM

Learning Objective

■ *Differentiate among pharyngitis, laryngitis, tonsillitis, sinusitis, and epiglottitis.*

As most of us know from personal experience, the respiratory system is the site of many common infections. We will soon discuss **pharyngitis,** inflammation of the mucous membranes of the throat, or sore throat. When the larynx is the site of infection, we suffer from **laryngitis,** which affects our ability to speak. This infection is caused by bacteria, such as *Streptococcus pneumoniae* or *S. pyogenes,* or viruses, often in combination. The microbes that cause pharyngitis also can cause inflamed tonsils, or **tonsillitis.**

The nasal sinuses are cavities in certain cranial bones that open into the nasal cavity. They have a mucous membrane lining that is continuous with that of the nasal cavity. When a sinus becomes infected with such organisms as *S. pneumoniae* or *Haemophilus influenzae,* the mucous membranes become inflamed, and there is a heavy nasal discharge of mucus. This condition is called **sinusitis.** If the opening by which the mucus leaves the sinus becomes blocked, internal pressure can cause pain or a sinus headache. These diseases are almost always *self-limiting,* meaning that recovery will usually occur even without medical intervention.

Probably the most threatening infectious disease of the upper respiratory system is **epiglottitis,** inflammation of the epiglottis. The epiglottis is a flaplike structure of cartilage that prevents ingested material from entering the larynx (see Figure 24.1). Epiglottitis is a rapidly developing disease that can result in death within a few hours. It is caused by opportunistic pathogens, usually *H. influenzae* type b. The newly introduced Hib vaccine, although directed primarily at meningitis (see Figure 22.3 on page 605), has significantly reduced the incidence of epiglottitis in the vaccinated population.

Bacterial Diseases of the Upper Respiratory System

Learning Objective

■ *List the causative agent, symptoms, prevention, preferred treatment, and laboratory identification tests for streptococcal pharyngitis, scarlet fever, diphtheria, cutaneous diphtheria, and otitis media.*

Airborne pathogens make their first contact with the body's mucous membranes as they enter the upper respiratory system. Many respiratory or systemic diseases initiate infections here.

Streptococcal Pharyngitis (Strep Throat)

Streptococcal pharyngitis (strep throat) is an upper respiratory infection caused by group A β-hemolytic streptococci. This gram-positive bacterial group consists solely of *Streptococcus pyogenes,* the same bacterium responsible for many skin and soft tissue infections, such as impetigo, erysipelas, and acute bacterial endocarditis.

The pathogenicity of group A streptococci is enhanced by their resistance to phagocytosis. They are also able to produce special enzymes, called *streptokinases,* that lyse fibrin clots and *streptolysins* that are cytotoxic to tissue cells, red blood cells, and protective leukocytes.

Without some analysis, streptococcal pharyngitis cannot be distinguished from pharyngitis caused by other bacteria or viruses. Probably no more than half of so-called strep throats are actually streptococcal in origin. At one time, the diagnosis of strep throat was based on culturing bacteria from a throat swab. Results took overnight or longer, and this test has been almost universally replaced by new indirect agglutination diagnostic tests for streptococcal pharyngitis using microscopic latex particles coated with antibodies against group A streptococci. Many of these tests take as little as 10 minutes to perform and are highly specific in detecting the presence or absence of group A streptococci. In fact, *S. pyogenes* can be detected in the throats of many people who are only asymptomatic carriers. Therefore, a positive result is not a certain indication that the symptoms are caused by the detected streptococci. A rise in IgM antibody titer is the best indication of a sore throat of genuinely streptococcal origin, but, for reasons of time and expense, this is not a practical option. A negative result should be checked by culturing a throat swab.

Streptococcal pharyngitis is characterized by local inflammation and a fever (Figure 24.3). Frequently, tonsilli-

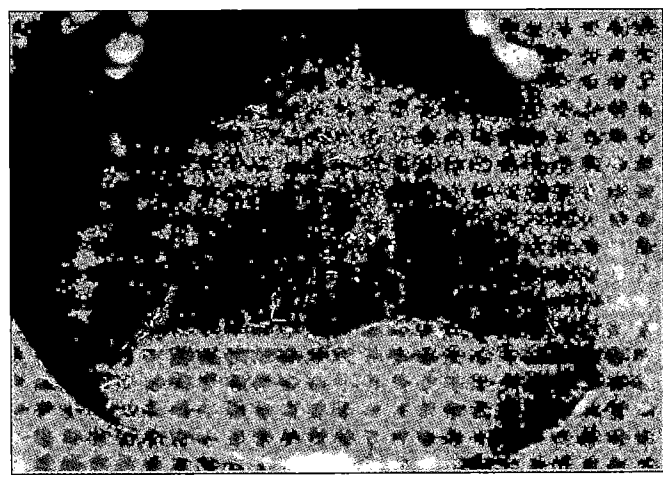

FIGURE 24.3 Streptococcal pharyngitis.

■ What are some complications of streptococcal pharyngitis?

tis occurs, and the lymph nodes in the neck become enlarged and tender. Another frequent complication is otitis media (infection of the middle ear). Penicillin remains the drug of choice for treatment of group A streptococcal infections.

More than 80 serological types of group A streptococci are responsible for at least a dozen different illnesses. Because resistance to streptococcal diseases is type-specific, a person who has recovered from infection by one type is not necessarily immune to infection by another type.

Strep throat is now most commonly transmitted by respiratory secretions, but epidemics spread by unpasteurized milk were once frequent.

Scarlet Fever

When the *S. pyogenes* strain causing streptococcal pharyngitis produces an *erythrogenic* (reddening) *toxin,* the resulting infection is called **scarlet fever.** When the strain produces this toxin, it has been lysogenized by a bacteriophage (see Figure 13.12 on page 384). Recall that this means the genetic information of a bacteriophage (bacterial virus) has been incorporated into the chromosome of the bacterium, so the characteristics of the bacterium have been altered. The toxin causes a pinkish-red skin rash, which is probably the skin's hypersensitivity reaction to the circulating toxin, and a high fever (Figure 24.4). The tongue has a spotted, strawberrylike appearance and then, as it loses its upper membrane, becomes very red

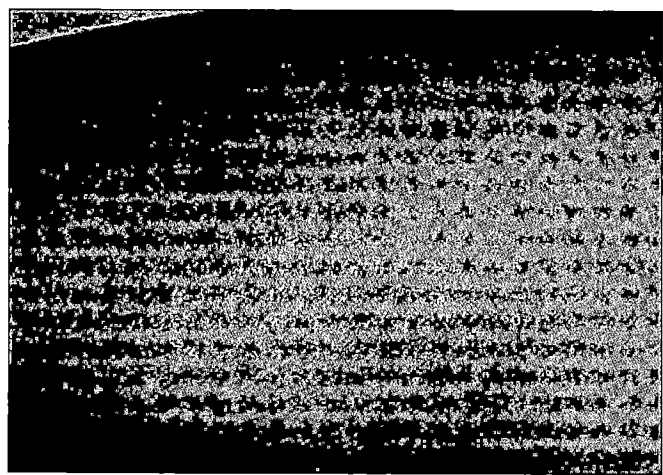

FIGURE 24.4 The typical red skin rash caused by scarlet fever.
It spreads to cover most of the body, except the palms and soles.

■ A lysogenic phage in *S. pyogenes* carries the gene for the erythrogenic toxin that causes scarlet fever.

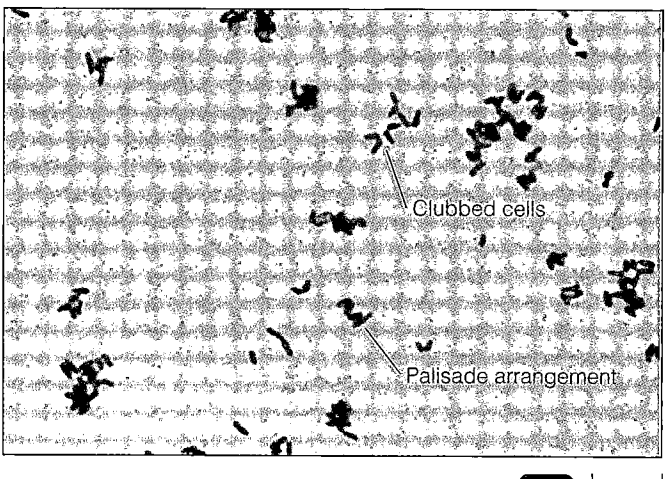

FIGURE 24.5 *Corynebacterium diphtheriae,* the cause of diphtheria. This Gram stain shows the club-shaped morphology; the dividing cells are often observed to fold together to form V- and Y-shaped figures. Also notice the side-by-side palisade arrangement.

■ A lysogenic phage in *C. diphtheriae* carries the gene for diphtheria toxin.

and enlarged. As the disease runs its course, the affected skin frequently peels off, as if sunburned, very similar to the scalded skin syndrome caused by *Staphylococcus aureus* (see Figure 21.4 on page 583).

Scarlet fever seems to vary in severity and frequency at different times and in different locations but has been declining in recent years. It is a communicable disease spread mainly by inhalation of infective droplets from an infected person. Classically, scarlet fever has been considered to be associated with streptococcal pharyngitis, but it might accompany a streptococcal skin infection.

Diphtheria

Another bacterial infection of the upper respiratory system is **diphtheria.** Until 1935, it was the leading infectious killer of children in the United States. The disease begins with a sore throat and fever, followed by general malaise and swelling of the neck. The organism responsible is *Corynebacterium diphtheriae,* a gram-positive, non–endospore-forming rod. Its morphology is pleomorphic, frequently club-shaped, and it stains unevenly (Figure 24.5).

Part of the normal immunization program for American children is the **DTP vaccine.** The D stands for diphtheria toxoid, an inactivated toxin that causes the body to produce antibodies against the diphtheria toxin.

C. diphtheriae has adapted to a generally immunized population, and relatively nonvirulent strains are found in the throats of many symptomless carriers. The bacterium

is well suited to airborne transmission and is very resistant to drying.

Characteristic of diphtheria (from the Greek word for leather) is a tough grayish membrane that forms in the throat in response to the infection (Figure 24.6). It contains fibrin, dead tissue, and bacterial cells and can totally block the passage of air to the lungs.

Although the bacteria do not invade tissues, those that have been lysogenized by a phage can produce a powerful exotoxin. Historically, it was the first disease for which a toxic cause was identified. Circulating in the bloodstream, the toxin interferes with protein synthesis. Only 0.01 mg of this highly virulent toxin is enough to kill a 91-kg (200-lb) person. Thus, if antitoxin therapy is to be effective, it must be administered before the toxin enters the tissue cells. When such organs as the heart and kidneys are affected by the toxin, the disease can rapidly be fatal. In other cases the nerves can be involved, and partial paralysis results.

Laboratory diagnosis by bacterial identification is difficult, requiring several selective and differential media. Identification is complicated by the need to differentiate between toxin-forming isolates and strains that are not toxigenic; both may be found in the same patient.

Even though antibiotics such as penicillin and erythromycin control the growth of the bacteria, they do not neutralize the diphtheria toxin. Thus antibiotics should be used only in conjunction with antitoxin.

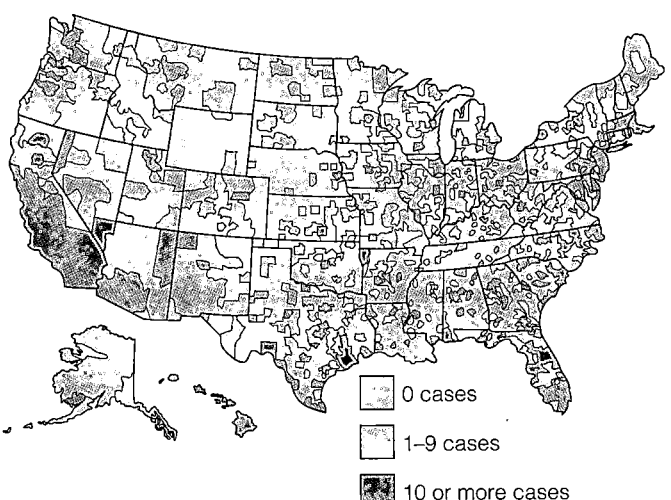

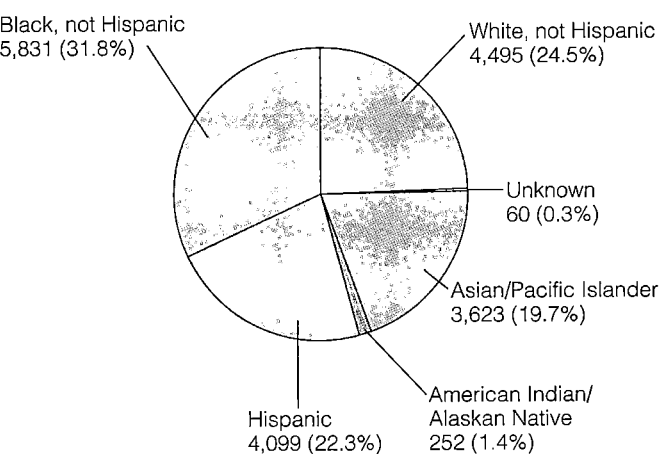

(a) Tuberculosis cases reported by county in the U.S.A.
Source: CDC, 1999.

Black, not Hispanic 5,831 (31.8%)

White, not Hispanic 4,495 (24.5%)

Unknown 60 (0.3%)

Asian/Pacific Islander 3,623 (19.7%)

American Indian/ Alaskan Native 252 (1.4%)

Hispanic 4,099 (22.3%)

(b) Percentage of cases by race or ethnic origin in the U.S.A.
Source: *MMWR* 47(53) (12/31/99).

FIGURE 24.12 U.S. distribution of tuberculosis.
[SOURCE: CDC, 1999.]

■ **Patients with tuberculosis must complete a long regimen of antibiotic treatment.**

States, certain ethnic groups tend to have much higher rates of TB. For example, African Americans, Native Americans, Asians, and Hispanics account for about two-thirds of cases (Figure 24.12). Most cases in the white population occur among the very elderly.

In the United States, 10–12 million people are estimated to be infected by the tubercle bacillus, usually with only latent infections. Worldwide, an estimated one-third of the total population is infected; at least 3 million die of the disease each year. Tuberculosis remains the leading killer among the world's infectious diseases.

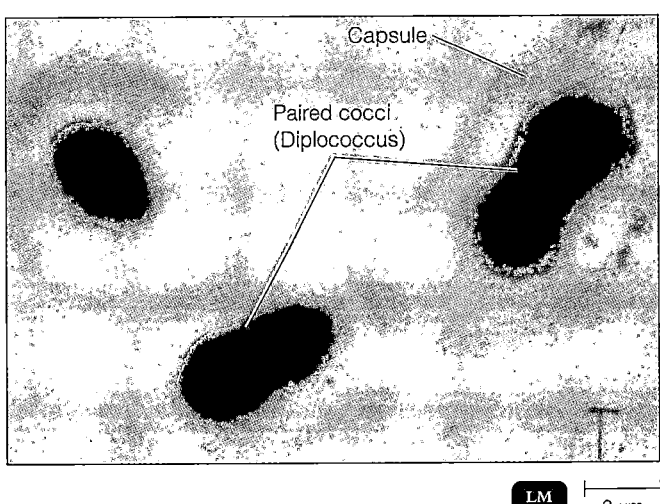

FIGURE 24.13 *Streptococcus pneumoniae,* **the most common cause of pneumococcal pneumonia.** Notice the paired arrangement of the cells. The capsule has been made more apparent here by reaction with a specific pneumococcal antiserum that makes it appear to swell.

■ *S. pneumoniae's* **capsule prevents phagocytosis.**

Bacterial Pneumonias

Learning Objective

■ *Compare and contrast the seven bacterial pneumonias discussed in this chapter.*

The term *pneumonia* is applied to many pulmonary infections, most of which are caused by bacteria. Pneumonia caused by *Streptococcus pneumoniae* is the most common and is therefore referred to as *typical pneumonia* (see the box on page 668). Pneumonias caused by other microorganisms, which can include fungi, protozoa, viruses, as well as other bacteria, are termed *atypical pneumonias.* This distinction is becoming increasingly blurred in practice.

Pneumonias also are named after the portions of the lower respiratory tract they affect. For example, if the lobes of the lungs are infected, it is called *lobar pneumonia;* pneumonias caused by *S. pneumoniae* are usually of this type. *Bronchopneumonia* indicates that the alveoli of the lungs adjacent to the bronchi are infected. *Pleurisy* is often a complication of various pneumonias, in which the pleural membranes become painfully inflamed.

Pneumococcal Pneumonia

Pneumonia caused by *S. pneumoniae* is called **pneumococcal pneumonia.** *S. pneumoniae* is a gram-positive, ovoid bacterium (Figure 24.13). This microbe is also a

CLINICAL PROBLEM SOLVING

A Respiratory Infection

You will see questions as you read through this problem. The questions are those that the primary care physician and public health epidemiologist ask themselves and each other as they solve a clinical problem. Try to answer each question as you read through the problem.

1. A 39-year-old woman presented with a 4-day history of fever, chills, fatigue, cough, and pain in the back upper left quadrant. Physical examination revealed diminished breath sounds, a temperature of 39°C, and increased heart rate. The patient had chronic sinus infections for which she received three courses of antibiotic therapy during the previous year. A chest X-ray film showed the left lobe infiltrates.
   What additional information do you need?

2. Results of Gram stains of sputum and a washing of the bronchial area are shown in Figure A. A blood sample was taken for culturing.
   What is the disease? What treatment is indicated?

3. The patient's symptoms and the presence of gram-positive cocci are compatible with a diagnosis of pneumonia, for which a 7-day course of penicillin is indicated. The patient's symptoms continued over the next 2 days.
   What other lab work should be done?

4. The blood sample was inoculated onto blood agar; the results are shown in Figure B. Antibiotic-sensitivity tests on the isolate are shown in Figure C.
   What do these results indicate?

5. The α-hemolytic colonies were identified as *Streptococcus pneumoniae* by antibody agglutination. The prescribed penicillin may not have cleared the infection because the *S. pneumoniae* isolate is resistant to penicillin, _____.
   What drug(s) is (are) effective against this strain?

6. Sensitivity to the following antibiotics was tested: penicillin (P), vancomycin (VA), erythromycin (E),

tetracycline (TE), and streptomycin (S). This strain is sensitive to vancomycin and erythromycin.

Strains of drug-resistant *S. pneumoniae* have become increasingly common in the United States and in other parts of the world. In some areas, as many as 35% of pneumococcal isolates have been reported to have intermediate resistance to penicillin. Many penicillin-resistant pneumococci are also resistant to other antibiotic drugs. Penicillin resistance and multidrug resistance often complicate the management of pneumococcal infection. Treating patients infected with nonsusceptible organisms may require expensive alternative antimicrobial agents and may result in prolonged hospitalization and increased medical costs. Emerging antimicrobial resistance further emphasizes the need for preventing pneumococcal infections by vaccination.

SOURCE: Multidrug data from *MMWR* 46(RR-08)(4/4/97).

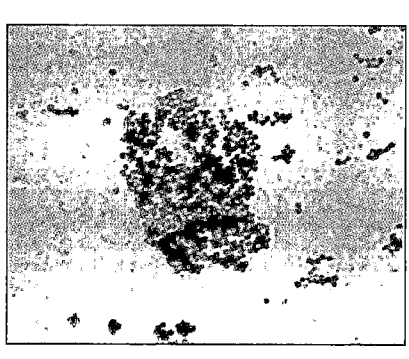

(a)

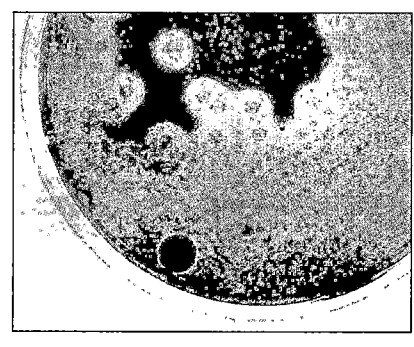
(b)

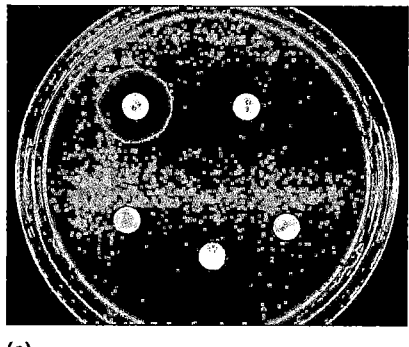
(c)

common cause of otitis media, meningitis, and septicemia. Because it usually forms cell pairs, the genus was formerly named Diplococcus pneumoniae. The cell pairs are surrounded by a dense capsule that makes the pathogen resistant to phagocytosis. These capsules are also the basis of serological differentiation of pneumococci into some 83 serotypes. Before antibiotic therapy became available, antisera directed at these capsular antigens were used to treat the disease.

Pneumococcal pneumonia involves both the bronchi and the alveoli (see Figure 24.2). Symptoms include high fever, breathing difficulty, and chest pain. (Atypical pneumonias usually have a slower onset and less fever and chest pain.) The lungs have a reddish appearance because blood vessels are dilated. In response to the infection, alveoli fill with some red blood cells, neutrophils (see Table 16.1 on page 458), and fluid from surrounding tissues. The sputum is often rust-colored from blood coughed up from the lungs. Pneumococci can invade the bloodstream, the pleural cavity surrounding the lung, and occasionally the meninges. No bacterial toxin has been clearly related to pathogenicity.

A presumptive diagnosis can be made by isolating the pneumococci from the throat, sputum, and other fluids. Pneumococci can be distinguished from other α-hemolytic streptococci by observing the inhibition of growth next to a disk of optochin (ethylhydrocupreine hydrochloride) or by performing a bile solubility test. They can also be serologically typed.

There are many healthy carriers of the pneumococcus. Virulence of the bacteria seems to be based mainly on the carrier's resistance, which can be lowered by stress. Many illnesses of older adults terminate in pneumococcal pneumonia.

A recurrence of pneumococcal pneumonia is not uncommon, but the second serological type is usually different. Before chemotherapy was available, the mortality rate was as high as 25%. This has now been lowered to less than 1% for younger patients treated early in the course of their disease. For elderly patients admitted to a hospital, mortality can approach 20%. Penicillin is the drug of choice, but strains of penicillin-resistant pneumococci have been reported. This is an increasing problem, in some areas involving 25% or more of isolates. A vaccine has been developed from the purified capsular material of the 23 types of pneumococci that cause at least 90% of the pneumococcal pneumonias in the United States. This vaccine is used for the groups most susceptible to infection: the elderly and debilitated individuals. A conjugated pneumococcal vaccine has recently been introduced (see page 606).

Haemophilus influenzae Pneumonia

Haemophilus influenzae is a gram-negative coccobacillus, and a Gram stain of sputum will differentiate this type of pneumonia from pneumococcal pneumonia. Patients with such conditions as alcoholism, poor nutrition, cancer, or diabetes are especially susceptible. Second-generation cephalosporins are resistant to the β-lactamases produced by many *H. influenzae* strains and are therefore usually the drugs of choice.

Mycoplasmal Pneumonia

The mycoplasmas, which do not have cell walls, do not grow under the conditions normally used to recover most bacterial pathogens. Because of this characteristic, pneumonias caused by mycoplasmas are often confused with viral pneumonias.

The bacterium *Mycoplasma pneumoniae* is the causative agent of **mycoplasmal pneumonia.** This type of pneumonia was first discovered when such atypical infections

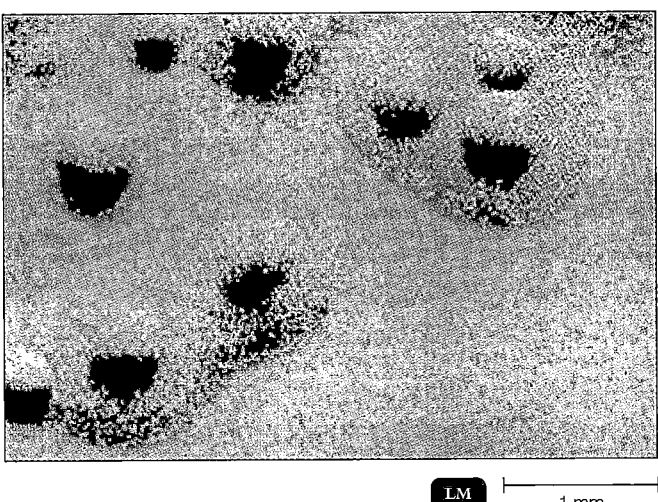

FIGURE 24.14 Colonies of *Mycoplasma pneumoniae,* the cause of mycoplasmal pneumonia.

■ These colonies are so tiny that they must be viewed under magnification of 50× or 100×.

responded to tetracyclines, indicating that the pathogen was nonviral. Mycoplasmal pneumonia is a common type of pneumonia in young adults and children. It may account for as much as 20% of pneumonias, although it is not a reportable disease. The symptoms, which persist for 3 weeks or longer, are low-grade fever, cough, and headache. Occasionally, they are severe enough to lead to hospitalization. Other terms for the disease are *primary atypical* (that is, the most common), and *walking pneumonia.*

When isolates from throat swabs and sputum grow on a medium containing horse serum and yeast extract, they form distinctive colonies with a "fried-egg" appearance (Figure 24.14). The colonies are so small that they must be observed with magnification. The mycoplasmas are highly varied in appearance because they lack cell walls (see Figure 11.17). Their flexibility allows them to pass through filters with pores as small as 0.2 μm in diameter, which is small enough to retain most other bacteria.

Diagnosis based on recovering the pathogens might not be useful in treatment because as long as 2 weeks may be required for the slow-growing organisms to develop. Diagnostic tests have improved greatly in recent years, however. They include PCR and serological tests that detect IgM antibodies against *M. pneumoniae.*

Legionellosis

Legionellosis, or **Legionnaires' disease,** first received public attention in 1976, when a series of deaths occurred among members of the American Legion who had attended a meeting in Philadelphia. A total of 182 people contracted pulmonary disease, apparently at this meeting,

29 of whom died. Because no obvious bacterial cause could be found, the deaths were attributed to viral pneumonia. Close investigation, mostly with techniques directed at locating a suspected rickettsial agent, eventually identified a previously unknown bacterium, an aerobic gram-negative rod now known as *Legionella pneumophila*. At least 39 species of *Legionella* have now been identified; not all of them cause disease.

The disease is characterized by a high fever of 40.5°C (105°F), cough, and general symptoms of pneumonia. No person-to-person transmission seems to be involved. Recent studies have shown that the bacterium can be readily isolated from natural waters. In addition, the microbes can grow in the water of air-conditioning cooling towers, which might mean that some epidemics in hotels, urban business districts, and hospitals were caused by airborne transmission. Recent outbreaks have been traced to whirlpool spas, humidifiers, showers, and decorative fountains.

The organism has also been found to inhabit the water lines of many hospitals. Most hospitals keep the temperature of hot water lines relatively low (43–55°C) as a safety measure, and in cooler parts of the system this inadvertently maintains a good growth temperature for *Legionella*. This bacterium is considerably more resistant to chlorine than most other bacteria and can survive for long periods in water with a low level of chlorine. One possible explanation for the resistance of *Legionella* to chlorine and heat is an interrelationship with waterborne amoebae. The bacteria are ingested by the amoebae but continue to proliferate and may even survive within encysted amoebae. They are also resistant to destruction by phagocytes.

The disease now seems to have always been fairly common, even if unrecognized. More than 1000 cases are reported each year, but the actual incidence is estimated at over 25,000 annually. Men over age 50 are the most likely to contract legionellosis, especially heavy smokers, alcohol abusers, or the chronically ill.

L. pneumophila is also responsible for *Pontiac fever;* its symptoms include fever, muscular aches, and usually a cough. The condition is mild and self-limiting.

The best diagnostic method is culture on a selective charcoal-yeast extract medium. Examination of respiratory specimens can be done by fluorescent antibody methods, and a DNA probe test is available. Erythromycin and other macrolide antibiotics, such as azithromycin, are the drugs of choice for treatment.

Psittacosis (Ornithosis)

The term **psittacosis** is derived from the disease's association with psittacine birds, such as parakeets and other parrots. It was later found that the disease can also be contracted from many other birds, such as pigeons, chickens, ducks, and turkeys. Therefore, the more general term **ornithosis** has come into use.

The causative agent is *Chlamydia psittaci* (sit'tä-sē), a gram-negative, obligate intracellular bacterium. One way chlamydias differ from rickettsias, which are also obligate intracellular bacteria, is that chlamydias form tiny **elementary bodies** as one part of their life cycle (see Figure 11.14 on page 319). Unlike most rickettsias, elementary bodies are resistant to environmental stress; therefore, they can be transmitted through air and do not require a bite to transfer the infective agent directly from one host to another.

Psittacosis is a form of pneumonia that usually causes fever, headache, and chills. Subclinical infections are very common, and stress appears to enhance susceptibility to the disease. Disorientation or even delirium, in some cases, indicates that the nervous system can be involved.

The disease is seldom transmitted from one human to another but is usually spread by contact with the droppings and other exudates of fowl. One of the most common modes of transmission is inhalation of dried particles from droppings. The birds themselves usually have diarrhea, ruffled feathers, respiratory illness, and a generally droopy appearance. Parakeets and other parrots sold commercially are usually (but not always) free of the disease. Many birds carry the pathogen in their spleen without symptoms, becoming ill only when stressed. Pet store employees and people involved in raising turkeys are at greatest risk of contracting the disease.

Diagnosis is made by isolating the bacterium in embryonated eggs or by cell culture. Serological tests can be used to identify the isolated organism. No vaccine is available, but tetracyclines are effective antibiotics in treating humans and animals. Effective immunity does not result from recovery, even when high titers of antibody are present in the serum.

Most years, fewer than 100 cases and very few deaths are reported in the United States. The main danger is late diagnosis. Before antibiotic therapy was available, the mortality rate was about 20%.

Chlamydial Pneumonia

Outbreaks of a respiratory illness in populations of college students were found to be caused by a chlamydial organism. Originally the pathogen was considered a strain of *C. psittaci,* but it has been assigned the species name *Chlamydia pneumoniae,* and the disease is known as **chlamydial pneumonia.** Clinically, it resembles mycoplasmal pneumonia. (There is also a suspected association between *C. pneumoniae* and atherosclerosis, the deposition of fatty deposits that block arteries.)

Coxiella burnetii in vacuole

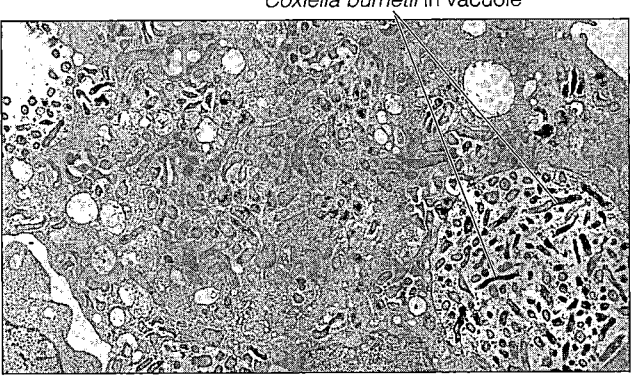

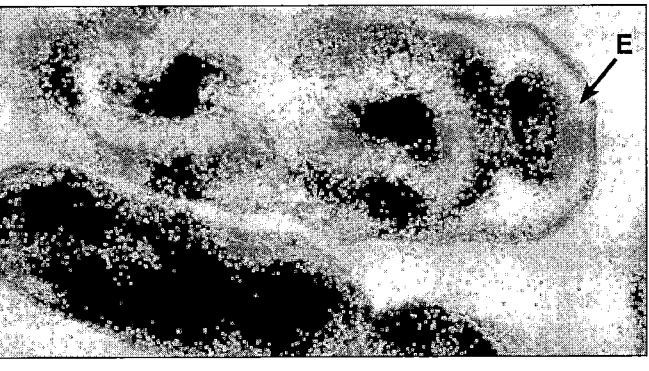

(a) Notice the intracellular growth of *Coxiella burnetii* in vacuoles of the host cell.   **TEM** ⊢————⊣ 5 μm

(b) This cell has just divided; notice the endospore-like body (E), which is probably responsible for the relative resistance of the organism.   **TEM** ⊢————⊣ 0.5 μm

FIGURE 24.15 *Coxiella burnetii,* **the cause of Q fever.**

■ By what two methods is Q fever transmitted?

The disease is apparently transmitted from person to person, probably by the respiratory route, but not as readily as infections such as influenza. Nearly half the U.S. population has antibodies against the organism, an indication that this is a common illness. Several serological tests are useful in diagnosis, but results are complicated by antigenic variation. The most effective antibiotic is tetracycline.

Q Fever

In Australia during the mid-1930s, a disease occurred that was characterized by a fever lasting 1–2 weeks, chills, chest pain, severe headache, and other evidence of a pneumonia-type infection. The disease was rarely fatal. In the absence of an obvious cause, the affliction was labeled **Q fever** (for *query*), much as one might say "X fever." The causative agent was subsequently identified as the obligately parasitic, intracellular bacterium *Coxiella burnetii* (käks′ē-el-lä bér-ne′tē-ē) (Figure 24.15a). Most intracellular bacteria, such as rickettsia, are not resistant enough to survive airborne transmission, but this microorganism is an exception.

The most serious complication of Q fever is endocarditis, which occurs in about 10% of cases. Some 5–10 years might elapse between the initial infection and the appearance of endocarditis. The pathogens apparently reside in the liver during this interval, where it may cause a form of hepatitis.

C. burnetii is a parasite of several arthropods, especially cattle ticks, and it is transmitted among animals by tick bites. The infection in animals is usually subclinical. The cattle ticks first spread the disease among dairy herds, and the microbes are shed in the feces, milk, and urine of in-

fected cattle. Once the disease is established in a herd, it is maintained by aerosol transmission. The disease is spread to humans by the ingestion of unpasteurized milk and by inhaling aerosols of microbes generated in dairy barns, especially from placental material at calving time. Inhaling a single pathogen is enough to cause infection, and many dairy workers have acquired at least subclinical infections. Workers in meat- and hide-processing plants are also at risk. The pasteurization temperature of milk, which was originally aimed at eliminating TB bacilli, was raised slightly in 1956 to ensure the killing of *C. burnetii*. In 1981, an endosporelike body was discovered, which may account for this heat resistance (Figure 24.15b).

The pathogen can be identified by isolation and growth in chick embryos in eggs or in cell culture. Laboratory workers testing for *Coxiella*-specific antibodies in a patient's serum can use serological tests.

Most cases of Q fever in the United States occur in the western states. The disease is endemic to California, Arizona, Oregon, and Washington. A vaccine for laboratory workers and other high-risk personnel is available. Tetracyclines are very effective in treatment.

Other Bacterial Pneumonias

An expanding number of bacteria have been implicated as causes of pneumonia. Important among them are *Staphylococcus aureus, M. catarrhalis, Streptococcus pyogenes,* and certain anaerobes found in the oral cavity. Gram-negative bacilli such as *Pseudomonas* species and *Klebsiella pneumoniae* occasionally cause bacterial pneumonia. *Klebsiella* pneumonia tends to affect older people with low resistance, such as those hospitalized or suffering from alcoholism.

Viral Diseases of the Lower Respiratory System

Learning Objective

■ *List the causative agent, symptoms, prevention, and preferred treatment for viral pneumonia, RSV, and influenza.*

For a virus to reach the lower respiratory system and initiate disease, it must pass numerous host defenses designed to trap and destroy it.

Viral Pneumonia

Viral pneumonia can occur as a complication of influenza, measles, or even chickenpox. A number of enteric and other viruses have been shown to cause viral pneumonia, but viruses are isolated and identified in fewer than 1% of pneumonia-type infections because few laboratories are equipped to properly test clinical samples for viruses. In those cases of pneumonia for which no cause is determined, viral etiology is often assumed if mycoplasmal pneumonia has been ruled out.

Respiratory Syncytial Virus (RSV)

Respiratory syncytial virus (RSV) is probably the most common cause of viral respiratory disease in infants; nearly 100,000 are hospitalized. There are about 4500 deaths each year in the United States, mostly in infants 2–6 months old. It can also cause a life-threatening pneumonia in the elderly. Epidemics occur during the winter and early spring. Virtually all children become infected by age 2. We have previously mentioned that RSV is sometimes implicated in cases of otitis media. The name of the virus is derived from its characteristic of causing cell fusion (syncytium formation, Figure 15.6b on page 447) when grown in cell culture. The symptoms are coughing and wheezing that last for more than a week. Fever occurs only when there are bacterial complications. Several rapid serological tests are now available that use samples of respiratory secretions to detect both the virus and its antibodies.

Naturally acquired immunity is very poor. An immune globulin product has been approved to protect infants with lung problems that put them at high risk. Protective vaccines are being clinically tested. For chemotherapy of severely ill patients, the antiviral drug ribavirin is administered by aerosol. It significantly diminishes the severity of symptoms.

Influenza (Flu)

The developed countries of the world are probably more aware of **influenza (flu)** than of any other disease, except for the common cold. The flu is characterized by chills,

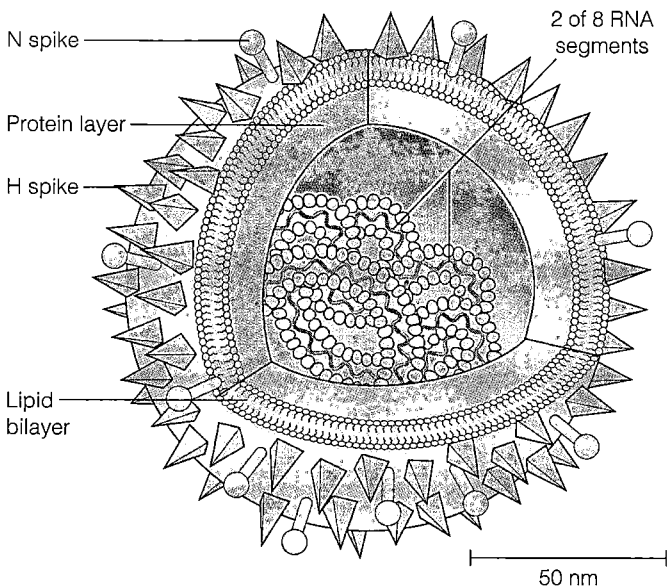

FIGURE 24.16 Detailed structure of the influenza virus. The envelope is composed of a protein layer, a lipid bilayer, and two types of spikes. Inside the envelope are eight segments of RNA. Morphologically, under certain environmental conditions the influenza virus assumes a filamentous form.

fever, headache, and general muscular aches. Recovery normally occurs in a few days, and coldlike symptoms appear as the fever subsides. Nonetheless, an estimated 10,000 to 20,000 Americans die annually of the flu, even in nonepidemic years. Diarrhea is not a normal symptom of the disease, and the intestinal discomforts attributed to "stomach flu" are probably from some other cause.

The Influenza Virus

Viruses in the genus *Influenzavirus* consist of eight separate RNA segments of differing lengths enclosed by an inner layer of protein and an outer lipid bilayer (Figures 13.3b and 24.16). Embedded in the lipid bilayer are numerous projections that characterize the virus. There are two types of projections: *hemagglutinin (H) spikes* and *neuraminidase (N) spikes*.

The H spikes, of which there are about 500 on each virus, allow the virus to recognize and attach to body cells before infecting them. Antibodies against the influenza virus are directed mainly at these spikes. The term *hemagglutinin* refers to the agglutination of red blood cells (hemagglutination) that occurs when the viruses are mixed with them. This reaction is important in serological tests, such as the hemagglutination inhibition test often used to identify influenza and some other viruses.

The N spikes, of which there are about 100 per virus, differ from the H spikes in appearance and function. Apparently they enzymatically help the virus separate from the infected cell as the virus exits after intracellular reproduction. N spikes also stimulate the formation of antibodies, but these are less important in the body's resistance to the disease than those produced in response to the H spikes.

Viral strains are identified by variation in the H and N antigens. The different forms of the antigens are assigned numbers—for example, H_1, H_2, H_3, N_1, and N_2. Each number change represents a substantial alteration in the protein makeup of the spike. These changes are called **antigenic shifts,** and they are great enough to evade most of the immunity developed in the human population. This ability is responsible for the outbreaks, including the pandemics of 1918, 1957, and 1968, that are summarized in Table 24.1. Incidentally, influenza viruses were not isolated before 1933, and the antigenic makeup of viruses causing outbreaks before about this time depended on analysis of antibodies taken from persons who had been infected.

Antigenic shifts are probably caused by a major genetic recombination. Because influenza viral RNA occurs as eight segments, recombination is likely in infections caused by more than one strain. Recombination between the RNA of animal viral strains (found in swine, horses, and birds, for example) and the RNA of human strains might be involved. Swine, ducks, and chickens (especially swine, which can be infected by both human and fowl influenza virus strains) in southern Chinese farming communities have come under suspicion as being the animals most likely to be involved in genetic shifts (and are thus called "mixing vessels"). Wild ducks and other migratory birds then become symptomless carriers that spread the virus over large geographic areas.

Epidemiology of Influenza

Between episodes of such major antigenic shifts, there are minor annual variations in the antigenic makeup called **antigenic drift.** The virus might still be designated as H_3N_2, for example, but viral strains arise reflecting minor antigenic changes within the antigenic group. These strains are sometimes assigned names related to the locality in which they were first identified. They usually reflect an alteration of only a single amino acid in the protein makeup of the H or N spike. Such a minor, one-step mutation is probably a response to selective pressure by antibodies (usually IgA in the mucous membranes) that neutralizes all viruses except for the new mutations. Such mutations can be expected about once in each million multiplications of the virus. High mutation rates are a characteristic of RNA viruses, which lack much of the "proofreading" ability of DNA viruses.

The usual result of antigenic drift is that a vaccine effective against H_3, for example, will be less effective against H_3 isolates circulating 10 years after the event.

table 24.1	Human Influenza Viruses*		
Type	**Antigenic Subtype**	**Year**	**Disease Severity**
A	H_3N_2 (the first "modern" pandemic; originated in southern China)	1889	Moderate
	H_1N_1 (Spanish)	1918	Severe
	H_2N_2 (Asian)	1957	Severe
	H_3N_2 (Hong Kong)	1968	Moderate
B	None	1940	Moderate
C	None	1947	Very mild

*The conventional wisdom is that H_1, H_2, and H_3 are human-infecting strains; H_4 and H_5 infect animals; both H_4 and H_5, however, are found in swine. In 1997, an outbreak of H_5N_1 in chickens occurred in Hong Kong. Only a handful of fatal cases occurred in humans, but the entire chicken population in Hong Kong was slaughtered as a precaution.
SOURCE: Adapted from C. Mims, J. Playfair, I. Roitt, D. Wakelin, and R. Williams, *Medical Microbiology,* 2nd ed. London: Mosby International, 1998.

There will have been enough drift in that time that the virus can largely evade the antibodies originally stimulated by the earlier strain.

Influenza viruses are also classified into major groups according to the antigens of their protein coats. These groups are A, B, and (rarely) C. The A-type viruses are responsible for the major pandemics. The B-type virus also circulates and mutates, but it is usually responsible for more geographically limited and milder infections.

Thus far, it has not been possible to make a vaccine for influenza that gives long-term immunity to the general population. Although it is not difficult to make a vaccine for a particular antigenic strain of virus, each new strain of circulating virus must be identified in time, usually about February, for the useful development and distribution of a new vaccine later that year. Representatives from the CDC have trained medical workers in mainland China in the techniques needed to detect antigenic changes in the influenza virus. By combining information from China, Japan, and Taiwan, we now get earlier warning of new influenza virus types. This usually allows greater opportunity to develop annual vaccines directed at the likely antigenic types. Unless the strain appears to be unusually virulent, vaccine administration is usually limited to older adults, hospital workers, and other high-risk groups. The vaccines are often *multivalent*—that is, directed at several strains in circulation at the time. At present, influenza viruses for manufacturing vaccines are grown in embryonated egg cultures. The vaccines are usually 70–90% effective, but the duration of protection is probably no more than 3 years for that strain. New types of influenza vaccine are under development. A vaccine that is administered as a nasal spray is likely to be approved shortly. Vaccines in which the viruses are grown in cell culture are currently undergoing trials. Should they be approved, these vaccines would eliminate the production bottleneck and allergic reactions associated with egg-grown viruses.

Almost every year, epidemics of the flu spread rapidly through large populations. The disease is so readily transmissible that epidemics are quickly propagated through populations susceptible to the newly changed strain of virus. The mortality rate from the disease is not high, usually less than 1%, and these deaths are mainly among the very young and the very old. However, so many people are infected in a major epidemic that the total number of deaths is often large. The cause of death is frequently not the influenza virus but secondary bacterial infections. The bacterium *H. influenzae* was named under the mistaken belief that it was the principal pathogen causing influenza rather than being a secondary invader. *S. aureus* and *S. pneumoniae* are other prominent secondary bacterial invaders.

In any discussion of influenza, the great pandemic of 1918–1919 must be mentioned.* Worldwide, more than 20 million people died. No one is sure why it was so unusually lethal. Today, the very young and very old are the principal victims, but in 1918–1919, young adults had the highest mortality rate, often dying within a few hours. The infection is usually restricted to the upper respiratory system, but some change in virulence allowed the virus to invade the lungs and cause viral pneumonia.

Evidence also suggests that the virus was able to infect cells in many organs of the body. Efforts are still underway to determine the genetic differences in this virus that account for its pathogenicity. These efforts have included analysis by PCR technology of preserved tissue from 1918 victims and even the exhumation of bodies from permanently frozen graves in Arctic areas.

Bacterial complications also frequently accompanied the infection and, in those preantibiotic days, were often fatal. The 1918 viral strain apparently became endemic in the U.S. swine population and may have originated there.

*Although there will always be uncertainty concerning the origin of this most famous pandemic, the best reliable reports place the first well-documented cases among U.S. Army recruits at Camp Funston, Kansas, in March of 1918. From there it spread rapidly among the military, quickly reaching the Western Front in France as reinforcements were dispatched overseas. Military censorship concealed the incapacitation of troops on both sides of the front, and the first newspaper descriptions were published when the outbreak reached the population of Spain, hence the name frequently assigned to the pandemic: the "Spanish flu."

Occasionally, influenza is still spread to humans from this reservoir, but the disease has not propagated like the virulent disease of 1918 did.

The antiviral drugs amantadine and rimantadine significantly reduce the symptoms of A-type influenza if administered promptly. More recently, two drugs for treatment of influenza have been introduced. They are inhibitors of neuraminidase, which the virus uses to separate itself from the host cell after it replicates. These are zanamivir (Relenza™), which is inhaled, and oseltamivir phosphate (Tamiflu™), which is administered orally. If taken within 30 hours of onset of influenza, these drugs can shorten the duration of symptoms. Neither drug should be considered a substitute for vaccination. The bacterial complications of influenza are amenable to treatment with antibiotics.

A complete serological identification of influenza virus isolates is generally conducted in a large, central laboratory. Diagnostic tests that will yield results within an hour in a physician's office are under development.

Fungal Diseases of the Lower Respiratory System

Learning Objective

■ List the causative agent, mode of transmission, preferred treatment, and laboratory identification tests for four fungal diseases of the respiratory system.

Fungi often produce spores that are disseminated through the air. It is therefore not surprising that several serious fungal diseases affect the lower respiratory system. The rate of fungal infections has been increasing in recent years. Opportunistic fungi are able to grow in immunosuppressed patients, and AIDS, transplant drugs, and anticancer drugs have created more immunosuppressed people than ever before.

Histoplasmosis

Histoplasmosis superficially resembles tuberculosis. In fact, it was first recognized as a widespread disease in the United States when X-ray surveys showed lung lesions in many people who were tuberculin-test negative. Although the lungs are most likely to be initially infected, the pathogens may spread in the blood and lymph, causing lesions in almost all organs of the body.

Symptoms are usually poorly defined and mostly subclinical, and the disease passes for a minor respiratory infection. In a few cases, perhaps fewer than 0.1%, histoplasmosis becomes progressive and is a severe, generalized disease. This occurs with an unusually heavy inoculum or

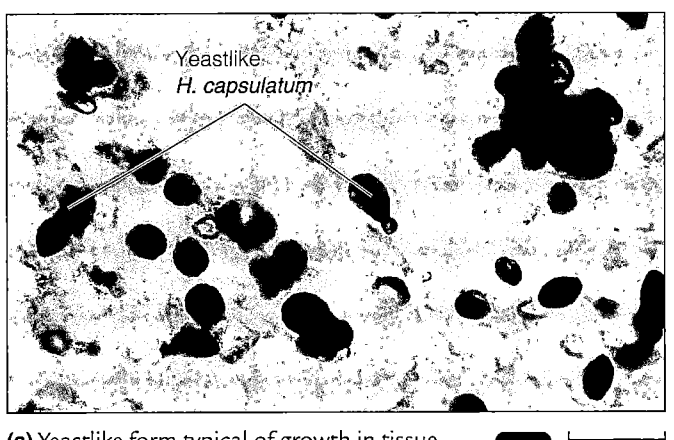

(a) Yeastlike form typical of growth in tissue at 37°C. Notice that one cell near the center is budding.   LM   5 μm

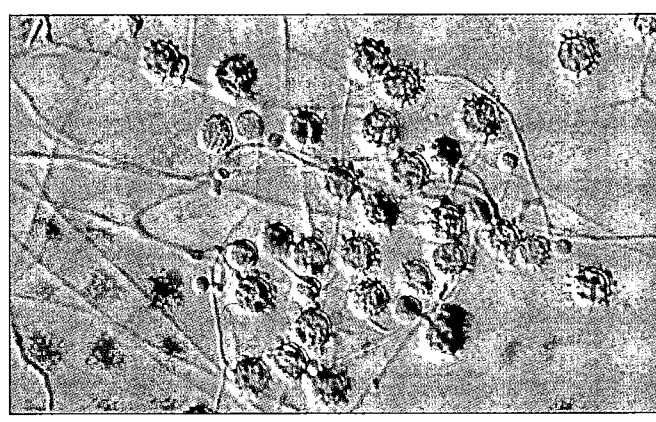

(b) Filamentous, spore-forming phase found in soil or at temperatures below 35°C; the spores are usually the infectious particle.   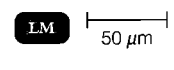 LM   50 μm

FIGURE 24.17 *Histoplasma capsulatum,* a dimorphic fungus that causes histoplasmosis.

▨ What does the term *dimorphic* mean?

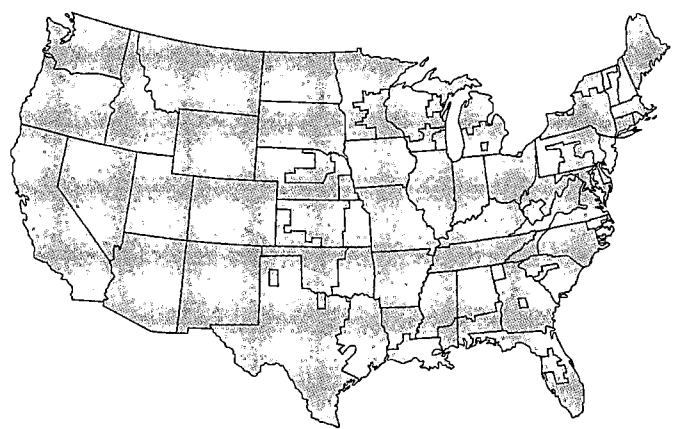

FIGURE 24.18 Histoplasmosis distribution. Gold indicates the U.S. geographic distribution.

■ Histoplasmosis is a systemic mycosis.

upon reactivation, when the infected person's immune system is compromised.

The causative organism, *Histoplasma capsulatum* (his-tō-plaz'mä kap-su-lä'tum), is a dimorphic fungus; that is, it has a yeastlike morphology in tissue growth (Figure 24.17a), and, in soil or artificial media, it forms a filamentous mycelium carrying reproductive conidia (Figure 24.17b). In the body, the yeastlike form is found intracellularly in macrophages, where it survives and multiplies.

Although histoplasmosis is rather widespread throughout the world, it has a limited geographic range in the United States (Figure 24.18). In general, the disease is found in the states adjoining the Mississippi and Ohio rivers. More than 75% of the population in some of these states have antibodies against the infection. In other states—Maine, for example—a positive test is a rare

event. Approximately 50 deaths are reported in the United States each year from histoplasmosis.

Humans acquire the disease from airborne conidia produced under conditions of appropriate moisture and pH levels. These conditions occur especially where droppings from birds and bats have accumulated. Birds themselves, because of their high body temperature, do not carry the disease, but their droppings provide nutrients, particularly a source of nitrogen, for the fungus. Bats, which have a lower body temperature than birds, carry the fungus, shed it in their feces, and infect new soil sites.

Clinical signs and history, serological tests, DNA probes, and, most importantly, isolation of the pathogen or its identification in tissue specimens are necessary for proper diagnosis. Currently, the most effective chemotherapy is with amphotericin B or itraconazole.

Coccidioidomycosis

Another fungal pulmonary disease, also rather restricted geographically, is **coccidioidomycosis.** The causative agent is *Coccidioides immitis,* a dimorphic fungus. The spores are found in dry, alkaline soils of the American Southwest and in similar soils of South America and northern Mexico. Because of its frequent occurrence in the San Joaquin Valley of California, it is sometimes known as *valley fever* or *San Joaquin fever.* In tissues, the organism forms a thick-walled body filled with spores called *spherules* (Figure 24.19). In soil, it forms filaments that reproduce by the formation of *arthrospores.* The wind carries the arthrospores to transmit the infection. Arthrospores are often so abundant that simply driving through an endemic area can result in infection, especially during a dust storm. An estimated 100,000 infections occur each year.

FIGURE 24.19 The life cycle of *Coccidioides immitis*, the cause of coccidioidomycosis.

■ Arthrospores are one type of conidiospore.

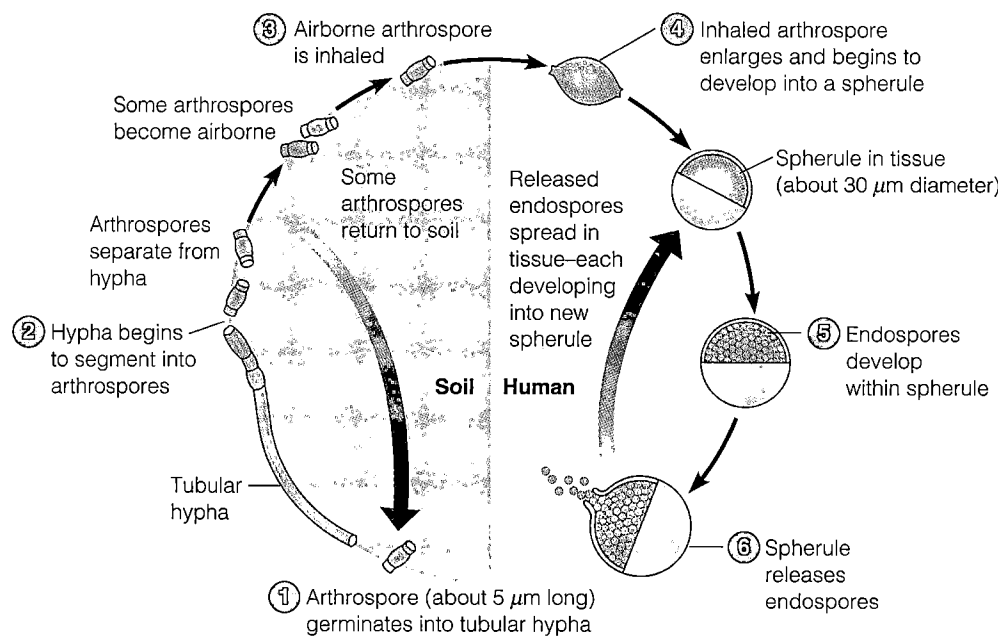

③ Airborne arthrospore is inhaled

Some arthrospores become airborne

Arthrospores separate from hypha

② Hypha begins to segment into arthrospores

Tubular hypha

④ Inhaled arthrospore enlarges and begins to develop into a spherule

Some arthrospores return to soil

Released endospores spread in tissue–each developing into new spherule

Soil | Human

Spherule in tissue (about 30 μm diameter)

⑤ Endospores develop within spherule

⑥ Spherule releases endospores

① Arthrospore (about 5 μm long) germinates into tubular hypha

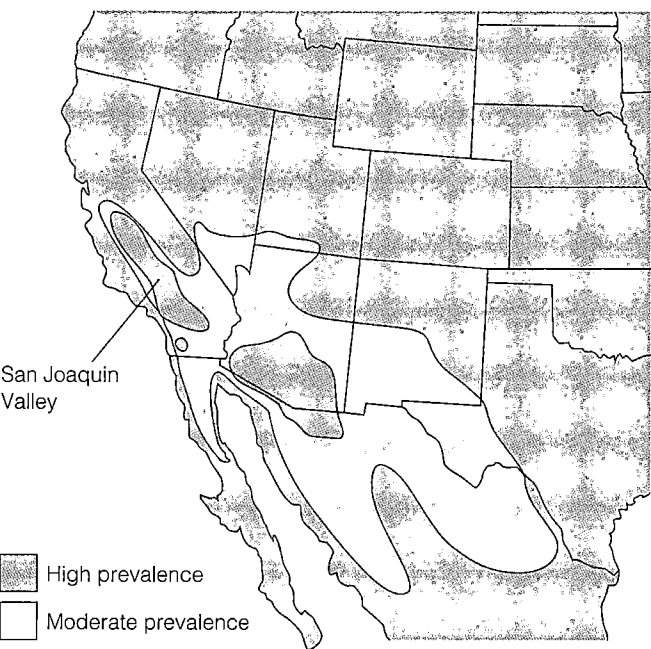

San Joaquin Valley

High prevalence

Moderate prevalence

FIGURE 24.20 The U.S. endemic area for coccidioidomycosis. High-altitude and nonarid regions within the indicated area are associated with only a slightly increased risk of exposure. The dark gold area in California is the San Joaquin Valley. Because of very high incidence there, the disease is also sometimes called Valley fever.

■ Why does the incidence of coccidioidomycosis increase after ecological disturbances, such as earthquakes and construction?

Most infections are not apparent, and almost all patients recover in a few weeks, even without treatment. The symptoms of coccidioidomycosis include chest pain and perhaps fever, coughing, and weight loss. In less than 1% of cases, a progressive disease resembling tuberculosis disseminates throughout the body. A substantial proportion of adults who are long-time residents of areas where the disease is endemic have evidence of prior infection with *C. immitis* by the skin test.

The resemblance to TB is so close that isolation of the pathogen is necessary to properly diagnose coccidioidomycosis. Diagnosis is most reliably made by identifying the spherules in tissue or fluids. The organism can be cultured from fluids or lesions, but laboratory workers must use great care because of the possibility of infectious aerosols. Several serological tests and DNA probes are available for identifying isolates. A tuberculin-like skin test is used in screening.

The incidence of coccidiodomycosis has been increasing recently in California and Arizona (Figure 24.20). Contributing factors include an increased number of older residents, an increased prevalence of HIV/AIDS, and a severe drought in California that facilitated dust-borne transmission. About 50–100 deaths occur annually from this disease in the United States.

Amphotericin B has been used to treat serious cases. However, less toxic imidazole drugs, such as ketoconazole and itraconazole, are useful alternatives.

Pneumocystis Pneumonia

***Pneumocystis* pneumonia** is caused by *Pneumocystis carinii* (Figure 24.21). The taxonomic position of this microbe has been uncertain ever since its discovery in 1909, when it was thought to be a developmental stage of a try-

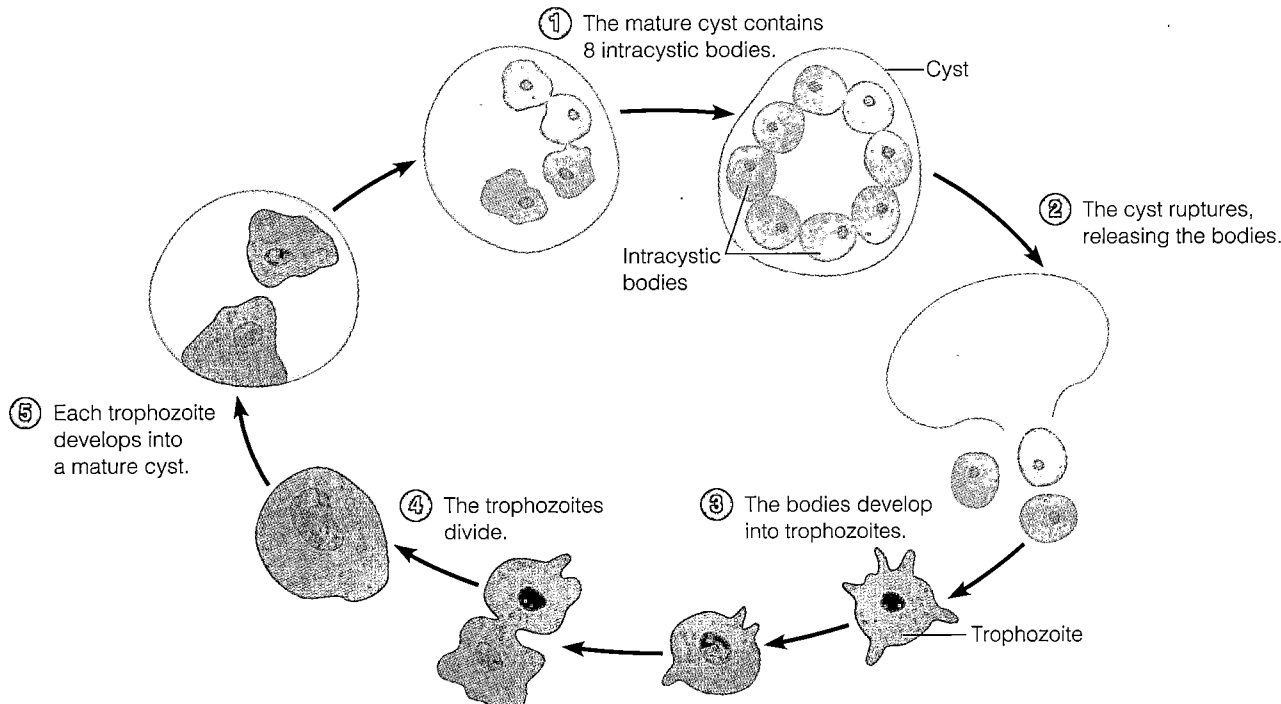

① The mature cyst contains 8 intracystic bodies.

Cyst

② The cyst ruptures, releasing the bodies.

Intracystic bodies

⑤ Each trophozoite develops into a mature cyst.

④ The trophozoites divide.

③ The bodies develop into trophozoites.

Trophozoite

FIGURE 24.21 The life cycle of *Pneumocystis carinii*, the cause of *Pneumocystis* pneumonia. Long classified as a protozoan, this organism is now usually considered to be a fungus, but it has characteristics of both groups.

Ribosomal RNA analyses show that *Pneumocystis* is related to fungi.

panosome. Since that time, there has been no universal agreement about whether it is a protozoan or a fungus. It has some characteristics of both groups. Recent analysis of RNA and certain other structural characteristics indicate that it is closely related to certain yeasts.

The disease occurs throughout the world and can be endemic in hospitals. The pathogen is found in healthy human lungs but causes disease among immunosuppressed patients. This group includes people receiving immunosuppressive drugs to minimize the rejection of transplanted tissue and those whose immunity is depressed because of cancer. People with AIDS are also very susceptible to to *P. carinii*, probably by the reactivation of an asymptomatic infection.

Before the AIDS epidemic, *Pneumocystis* pneumonia was an uncommon disease; perhaps 100 cases occurred each year. By 1993, it was the indicator of AIDS in more than 20,000 cases.

In the human lung, the microbes are found mostly in the lining of the alveoli. There, they form a thick-walled cyst in which spherical intracystic bodies successively divide as part of a sexual cycle. The mature cyst contains eight such bodies (see Figure 24.21). Eventually the cyst ruptures and releases them, and each body develops into a trophozoite. The trophozoite cells can reproduce asexually by fission, but they may also enter the encysted sexual stage (Figure 24.22).

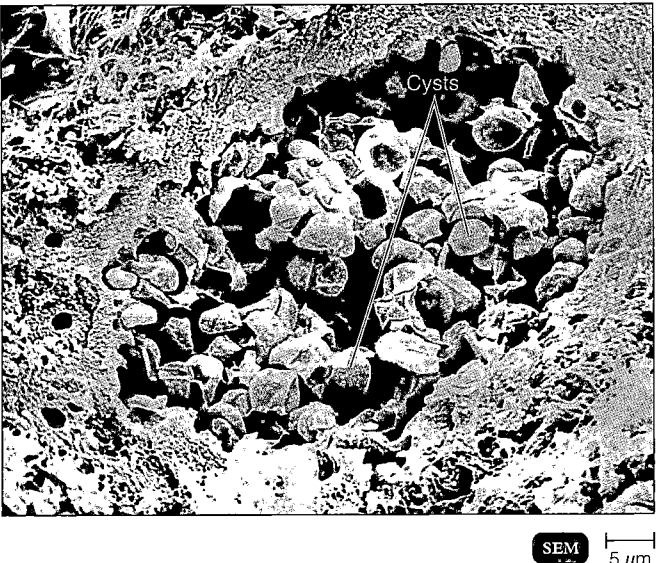

SEM ⊢——⊣ 5 μm

FIGURE 24.22 *Pneumocystis carinii.* Cyst stage of *P. carinii* in alveolus of a monkey lung.

The drug of choice for treatment is currently trimethoprim-sulfamethoxazole, but pentamidine isethionate is a frequent alternative.

table 24.2   Microbial Diseases of the Respiratory System

Disease	Pathogen	Characteristics
Bacterial Diseases of the Upper RespiratorySystem		
Streptococcal pharyngitis (strep throat)	Streptococci, especially *Streptococcus pyogenes*	Inflamed mucous membranes of the throat.
Scarlet fever	Erythrogenic toxin-producing strains of *Streptococcus pyogenes*	Streptococcal exotoxin causes skin and tongue reddening, peeling of affected skin.
Diphtheria	*Corynebacterium diphtheriae*	Bacterial exotoxin interferes with protein synthesis; damages heart, kidneys, and other organs. Membrane forms in throat. Cutaneous form also occurs.
Otitis media	Several agents, especially *Staphylococcus aureus, Streptococcus pneumoniae,* and *Haemophilus influenzae*	Accumulations of pus in middle ear build up painful pressure on eardrum.
Viral Disease of the Upper Respiratory System		
Common cold	Coronaviruses, rhinoviruses	Familiar symptoms of coughing, sneezing, runny nose.
Bacterial Diseases of the Lower Respiratory System		
Pertussis (whooping cough)	*Bordetella pertussis*	Cilia in upper respiratory tract inactivated, mucus accumulates, spasms of intense coughing to clear mucus.
Tuberculosis	*Mycobacterium tuberculosis Mycobacterium bovis*	Tubercle bacilli entering lungs survive phagocytosis, reproduce in macrophages. Tubercles formed to isolate pathogen. Defenses eventually fail and infection becomes systemic.
Pneumococcal pneumonia	*Streptococcus pneumoniae*	Infected alveoli of lung fill with fluids; interferes with oxygen uptake.
Haemophilus influenzae pneumonia	*Haemophilus influenzae*	Symptoms resemble pneumococcal pneumonia.
Mycoplasmal pneumonia	*Mycoplasma pneumoniae*	Mild but persistent respiratory symptoms; low fever, cough, headache.

Blastomycosis (North American Blastomycosis)

Blastomycosis is usually called **North American blastomycosis** to differentiate it from a similar South American blastomycosis. It is caused by the fungus *Blastomyces dermatitidis* (blas-tō-mī'sēz dėr-mä-tit'i-dis), a dimorphic fungus found most often in the Mississippi valley, where it probably grows in soil. Approximately 30–60 deaths are reported each year, although most infections are asymptomatic.

The infection begins in the lungs and can spread rapidly. Cutaneous ulcers commonly appear, and there is extensive abscess formation and tissue destruction. The pathogen can be isolated from pus and biopsy specimens. Amphotericin B is usually an effective treatment.

Other Fungi Involved in Respiratory Disease

Many other opportunistic fungi may cause respiratory disease, particularly in immunosuppressed hosts or when there is exposure to massive numbers of spores. **Aspergillosis** is an important example; it is transmitted by the spores of *Aspergillus fumigatus* (fü-mi-gä'tus) and other species of *Aspergillus,* which are widespread in decaying vegetation. Compost piles are ideal sites for growth, and farmers and gardeners are most often exposed to infective amounts of such spores.

Similar pulmonary infections sometimes result when individuals are exposed to spores of other mold genera, such as *Rhizopus* and *Mucor.* Such diseases can be very dangerous, particularly invasive infections of pulmonary aspergillosis. Predisposing factors include an impaired immune system, cancer, and diabetes. As with most systemic fungal infections, there is only a limited arsenal of antifungal agents available; amphotericin B has proved the most useful.

★ ★ ★

Table 24.2 summarizes the microbial respiratory diseases discussed in this chapter.

table 24.2 (continued)		
Disease	**Pathogen**	**Characteristics**
Legionellosis	*Legionella pneumophila*	Potentially fatal pneumonia that tends to affect older males who drink or smoke heavily. Pathogen grows in water such as air-conditioning towers.
Psittacosis (ornithosis)	*Chlamydia psittaci*	Symptoms, if any, are fever, headache, chills. Most commonly spread by contact with exudates of fowl.
Chlamydial pneumonia	*Chlamydia pneumoniae*	Mild respiratory illness common in young people; resembles mycoplasmal pneumonia.
Q fever	*Coxiella burnetii*	Mild respiratory disease lasting 1–2 weeks. Occasional complications such as endocarditis occur.
Viral Diseases of the Lower Respiratory System		
Respiratory syncytial virus (RSV) disease	Respiratory syncytial virus	A serious respiratory disease of infants.
Influenza	*Influenzavirus;* several serotypes	An illness characterized by chills, fever, headache, and muscular aches. Virus changes antigenic character rapidly, so there is limited immunity following recovery.
Fungal Diseases of the Lower Respiratory System		
Histoplasmosis	*Histoplasma capsulatum*	Fungal pathogen grows in soil, especially if contaminated with bird droppings. Widespread in Ohio and Mississippi river valleys; occasionally fatal.
Coccidioidomycosis	*Coccidioides immitis*	Fungal pathogen grows in desert soils of U.S. Southwest. Widespread infections in that area, occasionally fatal.
Pneumocystis pneumonia	*Pneumocystis carinii*	A common, serious complication of AIDS; no symptoms in people with healthy immune system.
Blastomycosis	*Blastomyces dermatitidis*	Rare but extremely serious fungal infection in Mississippi Valley area where pathogen grows in soil.

Study Outline

INTRODUCTION (p. 656)

1. Infections of the upper respiratory system are the most common type of infection.

2. Pathogens that enter the respiratory system can infect other parts of the body.

STRUCTURE AND FUNCTION OF THE RESPIRATORY SYSTEM (pp. 656–657)

1. The upper respiratory system consists of the nose, pharynx, and associated structures, such as the middle ear and auditory tubes.

2. Coarse hairs in the nose filter large particles from air entering the respiratory tract.

3. The ciliated mucous membranes of the nose and throat trap airborne particles and remove them from the body.

4. Lymphoid tissue, tonsils, and adenoids provide immunity to certain infections.

5. The lower respiratory system consists of the larynx, trachea, bronchial tubes, and alveoli.

6. The ciliary escalator of the lower respiratory system helps prevent microorganisms from reaching the lungs.

7. Microbes in the lungs can be phagocytized by alveolar macrophages.

8. Respiratory mucus contains IgA antibodies.

NORMAL MICROBIOTA OF THE RESPIRATORY SYSTEM (p. 658)

1. The normal microbiota of the nasal cavity and throat can include pathogenic microorganisms.

2. The lower respiratory system is usually sterile because of the action of the ciliary escalator.

MICROBIAL DISEASES OF THE UPPER RESPIRATORY SYSTEM (pp. 658–662)

1. Specific areas of the upper respiratory system can become infected to produce pharyngitis, laryngitis, tonsillitis, sinusitis, and epiglottitis.

2. These infections may be caused by several bacteria and viruses, often in combination.

3. Most respiratory tract infections are self-limiting.

4. *H. influenzae* type b can cause epiglottitis.

BACTERIAL DISEASES OF THE UPPER RESPIRATORY SYSTEM (pp. 659–661)

Streptococcal Pharyngitis (Strep Throat) (p. 659)

1. This infection is caused by group A β-hemolytic streptococci, the group that consists of *Streptococcus pyogenes.*

2. Symptoms of this infection are inflammation of the mucous membrane and fever; tonsillitis and otitis media may also occur.

3. Preliminary rapid diagnosis is made by indirect agglutination tests. Definitive diagnosis is based on a rise in IgM antibodies.

4. Penicillin is used to treat streptococcal pharyngitis.

5. Immunity to streptococcal infections is type-specific.

6. Strep throat is usually transmitted by droplets but at one time was commonly associated with unpasteurized milk.

Scarlet Fever (pp. 659–660)

1. Strep throat, caused by an erythrogenic toxin-producing *S. pyogenes,* results in scarlet fever.

2. *S. pyogenes* produces erythrogenic toxin when lysogenized by a phage.

3. Symptoms include a red rash, high fever, and a red, enlarged tongue.

Diphtheria (pp. 660–661)

1. Diphtheria is caused by exotoxin-producing *Corynebacterium diphtheriae.*

2. Exotoxin is produced when the bacteria are lysogenized by a phage.

3. A membrane, containing fibrin and dead human and bacterial cells, forms in the throat and can block the passage of air.

4. The exotoxin inhibits protein synthesis, and heart, kidney, or nerve damage may result.

5. Laboratory diagnosis is based on isolation of the bacteria and the appearance of growth on differential media.

6. Antitoxin must be administered to neutralize the toxin, and antibiotics can stop growth of the bacteria.

7. Routine immunization in the U.S. includes diphtheria toxoid in the DTP vaccine.

8. Slow-healing skin ulcerations are characteristic of cutaneous diphtheria.

9. There is minimal dissemination of the exotoxin in the bloodstream.

Otitis Media (p. 661)

1. Earache, or otitis media, can occur as a complication of nose and throat infections.

2. Pus accumulation causes pressure on the eardrum.

3. Bacterial causes include *Streptococcus pneumoniae,* nonencapsulated *Haemophilus influenzae, Moraxella catarrhalis, Streptococcus pyogenes,* and *Staphylococcus aureus.*

VIRAL DISEASE OF THE UPPER RESPIRATORY SYSTEM (p. 662)

The Common Cold (p. 662)

1. Any one of approximately 200 different viruses can cause the common cold; rhinoviruses cause about 50% of all colds.

2. Symptoms include sneezing, nasal secretions, and congestion.

3. Sinus infections, lower respiratory tract infections, laryngitis, and otitis media can occur as complications of a cold.

4. Colds are most often transmitted by indirect contact.

5. Rhinoviruses prefer temperatures slightly lower than body temperature.

6. The incidence of colds increases during cold weather, possibly because of increased interpersonal indoor contact or physiological changes.

7. Antibodies are produced against the specific viruses.

MICROBIAL DISEASES OF THE LOWER RESPIRATORY SYSTEM (p. 662)

1. Many of the same microorganisms that infect the upper respiratory system also infect the lower respiratory system.

2. Diseases of the lower respiratory system include bronchitis and pneumonia.

BACTERIAL DISEASES OF THE LOWER RESPIRATORY SYSTEM (pp. 663–671)

Pertussis (Whooping Cough) (p. 663)

1. Pertussis is caused by *Bordetella pertussis.*

2. The initial stage of pertussis resembles a cold and is called the catarrhal stage.

3. The accumulation of mucus in the trachea and bronchi causes deep coughs characteristic of the paroxysmal (second) stage.

4. The convalescence (third) stage can last for months.

5. Laboratory diagnosis is based on isolation of the bacteria on enrichment and selective media, followed by serological tests.

6. Regular immunization for children has decreased the incidence of pertussis.

Tuberculosis (pp. 663–667)

1. Tuberculosis is caused by *Mycobacterium tuberculosis.*

2. Large amounts of lipids in the cell wall account for the bacterium's acid-fast characteristic as well as its resistance to drying and disinfectants.

3. *M. tuberculosis* may be ingested by alveolar macrophages; if not killed, the bacteria reproduce in the macrophages.

4. Lesions formed by *M. tuberculosis* are called tubercles; dead macrophages and bacteria form the caseous lesion that might calcify and appear in an X ray as a Ghon complex.

5. Liquefaction of the caseous lesion results in a tuberculous cavity in which *M. tuberculosis* can grow.

6. New foci of infection can develop when a caseous lesion ruptures and releases bacteria into blood or lymph vessels; this is called miliary tuberculosis.

7. Miliary tuberculosis is characterized by weight loss, coughing, and loss of vigor.

8. Chemotherapy usually involves two drugs taken for 1–2 years; multidrug-resistant *M. tuberculosis* is becoming prevalent.

9. A positive tuberculin skin test can indicate either an active case of TB, or prior infection, or vaccination and immunity to the disease.

10. Laboratory diagnosis is based on the presence of acid-fast bacilli and isolation of the bacteria, which requires incubation of up to 8 weeks.

11. *Mycobacterium bovis* causes bovine tuberculosis and can be transmitted to humans by unpasteurized milk.

12. *M. bovis* infections usually affect the bones or lymphatic system.

13. BCG vaccine for tuberculosis consists of a live, avirulent culture of *M. bovis.*

14. *M. avium-intracellulare* complex infects patients in the late stages of HIV infection.

Bacterial Pneumonias (pp. 667–671)

1. Typical pneumonia is caused by *S. pneumoniae.*

2. Atypical pneumonias are caused by other microorganisms.

Pneumococcal Pneumonia (pp. 667–669)

1. Pneumococcal pneumonia is caused by encapsulated *Streptococcus pneumoniae.*

2. Symptoms are fever, breathing difficulty, chest pain, and rust-colored sputum.

3. The bacteria can be identified by the production of α-hemolysins, inhibition by optochin, bile solubility, and through serological tests.

4. A vaccine consists of purified capsular material from 23 serotypes of *S. pneumoniae.*

Haemophilus influenzae *Pneumonia* (p. 669)

1. Alcoholism, poor nutrition, cancer, and diabetes are predisposing factors for *H. influenzae* pneumonia.

2. *H. influenzae* is a gram-negative coccobacillus.

Mycoplasmal Pneumonia (p. 669)

1. *Mycoplasma pneumoniae* causes mycoplasmal pneumonia; it is an endemic disease.

2. *M. pneumoniae* produces small "fried-egg" colonies after 2 weeks' incubation on enriched media containing horse serum and yeast extract.

3. A complement-fixation test, used to diagnose the disease, is based on rising antibody titer.

Legionellosis (pp. 669–670)

1. The disease is caused by the aerobic gram-negative rod *Legionella pneumophila.*

2. The bacterium can grow in water, such as air-conditioning cooling towers, and then be disseminated in the air.

3. This pneumonia does not appear to be transmitted from person to person.

4. Bacterial culture, FA tests, and DNA probes are used for laboratory diagnosis.

Psittacosis (Ornithosis) (p. 670)

1. *Chlamydia psittaci* is transmitted by contact with contaminated droppings and exudates of fowl.

2. Elementary bodies allow the bacteria to survive outside a host.

3. Commercial bird handlers are most susceptible to this disease.

4. The bacteria are isolated in embryonated eggs, mice, or cell culture; identification is based on FA staining.

Chlamydial Pneumonia (pp. 670–671)

1. *Chlamydia pneumoniae* causes pneumonia; it is transmitted from person to person.

2. A fluorescent-antibody test is used for diagnosis.

Q Fever (p. 671)

1. Obligately parasitic, intracellular *Coxiella burnetii* causes Q fever.

2. The disease is usually transmitted to humans through unpasteurized milk or inhalation of aerosols in dairy barns.

3. Laboratory diagnosis is made with the culture of bacteria in embryonated eggs or cell culture.

Other Bacterial Pneumonias (p. 671)

1. Gram-positive bacteria that cause pneumonia include *S. aureus* and *S. pyogenes.*

2. Gram-negative bacteria that cause pneumonia include *M. catarrhalis, K. pneumoniae,* and *Pseudomonas* species.

VIRAL DISEASES OF THE LOWER RESPIRATORY SYSTEM (pp. 672–674)

Viral Pneumonia (p. 672)

1. A number of viruses can cause pneumonia as a complication of infections such as influenza.

2. The etiologies are not usually identified in a clinical laboratory because of the difficulty in isolating and identifying viruses.

Respiratory Syncytial Virus (RSV) (p. 672)

1. RSV is the most common cause of pneumonia in infants.

Influenza (Flu) (pp. 672–674)

1. Influenza is caused by *Influenzavirus* and is characterized by chills, fever, headache, and general muscular aches.

2. Hemagglutinin (H) and neuraminidase (N) spikes project from the outer lipid bilayer of the virus.

3. Viral strains are identified by antigenic differences in the H and N spikes; they are also divided by antigenic differences in their protein coats (A, B, and C).

4. Viral isolates are identified by hemagglutination-inhibition tests and immunofluorescence testing with monoclonal antibodies.

5. Antigenic shifts that alter the antigenic nature of the H and N spikes make natural immunity and vaccination of questionable value. Minor antigenic changes are caused by antigenic drift.

6. Deaths during an influenza epidemic are usually from secondary bacterial infections.

7. Multivalent vaccines are available for the elderly and other high-risk groups.

8. Amantadine and rimantadine are effective prophylactic and curative drugs against A-type *Influenzavirus.*

FUNGAL DISEASES OF THE LOWER RESPIRATORY SYSTEM (pp. 674–678)

1. Fungal spores are easily inhaled; they may germinate in the lower respiratory tract.

2. The incidence of fungal diseases has been increasing in recent years.

3. The mycoses below can be treated with amphotericin B.

Histoplasmosis (pp. 674–675)

1. *Histoplasma capsulatum* causes a subclinical respiratory infection that only occasionally progresses to a severe, generalized disease.

2. The disease is acquired by inhalation of airborne conidia.

3. Isolation of the fungus or identification of the fungus in tissue samples is necessary for diagnosis.

Coccidioidomycosis (pp. 675–676)

1. Inhalation of the airborne arthrospores of *Coccidioides immitis* can result in coccidioidomycosis.

2. Most cases are subclinical, but when there are predisposing factors such as fatigue and poor nutrition, a progressive disease resembling tuberculosis can result.

Pneumocystis Pneumonia (pp. 676–677)

1. *Pneumocystis carinii* is found in healthy human lungs.

2. *Pneumocystis carinii* causes disease in immunosuppressed patients.

3. *Pneumocystis* pneumonia is currently being treated with trimethoprim or pentamidine.

Blastomycosis (North American Blastomycosis) (p. 678)

1. *Blastomyces dermatitidis* is the causative agent of blastomycosis.

2. The infection begins in the lungs and can spread to cause extensive abscesses.

Other Fungi Involved in Respiratory Disease (p. 678)

1. Opportunistic fungi can cause respiratory disease in immunosuppressed hosts, especially when large numbers of spores are inhaled.

2. Among these fungi are *Aspergillus, Rhizopus,* and *Mucor.*

Study Questions

REVIEW

1. Respiratory diseases are usually transmitted by _____.

2. Describe how microorganisms are prevented from entering the upper respiratory system. How are they prevented from causing infections in the lower respiratory system?

3. How do the normal microbiota of the respiratory system illustrate microbial antagonism?

4. Compare and contrast mycoplasmal pneumonia and viral pneumonia.

5. How is otitis media contracted? What causes otitis media? Why was otitis media included in a chapter on diseases of the respiratory system?

6. List the causative agent, symptoms, and treatment for four viral diseases of the respiratory system. Separate the diseases according to whether they infect the upper or lower respiratory system.

7. Complete the following table:

Disease	Causative Agent	Symptoms	Treatment
Streptococcal pharyngitis			
Scarlet fever			
Diphtheria			
Whooping cough			
Tuberculosis			
Pneumococcal pneumonia			
H. influenzae pneumonia			
Chlamydial pneumonia			
Legionellosis			
Psittacosis			
Q fever			
Epiglottitis			

8. Under what conditions can the saprophytes *Aspergillus* and *Rhizopus* cause infections?

9. A patient has been diagnosed as having pneumonia. Is this sufficient information to begin treatment with antimicrobial agents? Briefly discuss why or why not.

10. List the causative agent, mode of transmission, and endemic area for the diseases histoplasmosis, coccidioidomycosis, blastomycosis, and *Pneumocystis* pneumonia.

11. Briefly describe the procedures and positive results of the tuberculin test and what is indicated by a positive test.

12. Discuss reasons for the increased incidences of colds and pneumonias during cold weather.

13. Match the bacteria involved in respiratory infections with the following laboratory test results:
Gram-positive cocci
  Catalase-positive _____
  Catalase-negative
    β-Hemolytic, bacitracin inhibition _____
    α-Hemolytic, optochin inhibition _____
Gram-positive rods
  Not acid-fast _____
  Acid-fast _____
Gram-negative cocci _____
Gram-negative rods
  Aerobes
    Coccobacilli _____
    Rods _____
  Facultative anaerobes
    Coccobacilli _____
    Rods _____
Intracellular parasites
  Form elementary bodies _____
  Do not form elementary bodies _____
Wall-less _____

MULTIPLE CHOICE

1. A patient has fever, difficulty breathing, chest pains, fluid in the alveoli, and a positive tuberculin skin test. Gram-positive cocci are isolated from the sputum. The recommended treatment is
  a. penicillin.   **d.** tetracyclines.
  b. antitoxin.   **e.** none of the above
  c. isoniazid.

2. No bacterial pathogen can be isolated from the sputum of a patient with pneumonia. Antibiotic therapy has not been successful. The next step should be
  a. culturing for *Mycobacterium tuberculosis.*
  b. culturing for *Mycoplasma pneumoniae.*
  c. culturing for fungi.
  d. a change in antibiotics.
  e. none; nothing more can be done

Match the following choices to the culture descriptions in questions 3–6:
  a. *Chlamydia*   **d.** *Mycobacterium*
  b. *Coccidioides*   **e.** *Mycoplasma*
  c. *Histoplasma*

3. Your culture from a pneumonia patient appears not to have grown. You do see colonies, however, when the plate is viewed at 100×.

4. This pneumonia etiology requires cell culture.

5. Microscopic examination of a lung biopsy shows ovoid cells in macrophages. You suspect these are the cause of the patient's symptoms, but your culture grows a filamentous organism.

6. Microscopic examination of a lung biopsy shows spherules.

7. In San Francisco, ten animal-health care technicians developed pneumonia 2 weeks after 130 goats were moved to the animal shelter where they worked. Which of the following is *not* true?
  a. Diagnosis is made by a blood agar culture of sputum.
  b. The cause is *Coxiella burnetii.*
  c. The cause is rickettsia.
  d. The disease was transmitted by aerosols.
  e. Diagnosis is made by complement-fixation tests for antibodies.

8. Which of the following leads to all the rest?
  a. catarrhal stage   **d.** mucus accumulation
  b. cough   **e.** trachaeal cytotoxin
  c. loss of cilia

Match the following choices to the statements in questions 9 and 10:
  a. *Bordetella pertussis*
  b. *Corynebacterium diphtheriae*
  c. *Legionella pneumophila*
  d. *Mycobacterium tuberculosis*
  e. none of the above

9. Causes the formation of a membrane across the throat.

10. Resistant to destruction by phagocytes.

CRITICAL THINKING

1. Differentiate between the following:
   a. *S. pyogenes* causing strep throat and *S. pyogenes* causing scarlet fever.
   b. Diphtheroids and *C. diphtheriae.*

2. Why might the influenza vaccine be less effective than other vaccines?

3. Explain why it would be impractical to include cold and influenza vaccinations in the required childhood vaccinations.

CLINICAL APPLICATIONS

1. In August, a 24-year-old man from Virginia developed difficulty breathing and bilateral lobe infiltrates 2 months after driving through California. During initial evaluation, typical pneumonia was suspected, and he was treated with antibiotics. Efforts to diagnose the pneumonia were unsuccessful. In October, a laryngeal mass was detected and laryngeal cancer was suspected; treatment with steroids and bronchodilators did not result in improvement. Lung biopsy and laryngoscopy detected diffuse granular tissue. He was treated with amphotericin B and discharged after 5 days. What was the disease? What might have been done differently to decrease the patient's recovery time from 3 months to 1 week?

2. During a 6-month period, 72 clinic staff members became tuberculin-positive. A case-control study was undertaken to determine the most likely source of *M. tuberculosis* infection among the staff. A total of 16 cases and 34 tuberculin-negative controls were compared. Pentamidine isethionate is not used for TB treatment. What disease was probably being treated with this drug? What is the most likely source of infection?

	Cases	Control
Works ≥ 40 hr/week	100	62
In room during aerosolized pentamidine isethionate therapy in TB patients	31	3
Patient contact	94	94
Lunch eaten in staff lounge	38	35
Resident of western Palm Beach	75	65
Female	81	77
Cigarette smoker	6	15
Contact with nurse diagnosed with TB	15	12
In unventilated room during collection of TB-positive sputum samples	13	8

3. In March, six members of a family had fever, anorexia, sore throat, cough, headache, vomiting, and muscle pain. Two people were hospitalized. All six improved after therapy with doxycycline. Convalescent serum samples gave titers of 64 and 32. The family had purchased a cockatiel in mid-February and noticed that the bird was irritable. The bird was euthanized in April. A fluorescent-antibody test for antigens was diagnostic. What is the disease? How was it transmitted?

4. A 27-year-old male with a history of asthma was hospitalized with a 4-day history of progressive cough and 2 days of spiking fevers. Gram-positive cocci in pairs were cultured from a blood sample. What is the diagnosis?

5. Three weeks after working on the demolition of an abandoned building in Kentucky, a worker was hospitalized for acute respiratory illness. At the time of demolition, a colony of bats inhabited the building. An X-ray examination revealed a lung mass. A purified protein derivative test was negative; a cytolological examination for cancer was also negative. The mass was surgically removed. Microscopic examination of the mass revealed ovoid yeast cells. What is the disease? How was it contracted?

Learning with Technology

MP = The Microbiology Place website   **ST** = Student Tutorial CD-ROM   **VU** = VirtualUnknown CD-ROM

MP   Don't forget to go to The Microbiology Place website (http://www.microbiologyplace.com) to take the practice tests, explore the interactive activity and case study, and check out the news articles and web links for this chapter.

ST   Remember there is also a quiz for this chapter on the Microbiology Interactive Student Tutorial CD-ROM.

VU   Enter the Virtual Lab, click the arrow next to the Session field, click Textbook Exercises, and select Chapter 24. Read the Case Study carefully, identify the unknown, and use what you learn to answer the following questions (consult your textbook for additional information):

1. Compare your identification with the organisms listed in Table 24.2 on pages 678–679. Would this organism be considered a common causative agent for respiratory diseases?

2. What traits of this organism equip it to cause a variety of infections?

3. What predisposing factors could have led to this patient's disease?

chapter 25

Microbial Diseases of the Digestive System

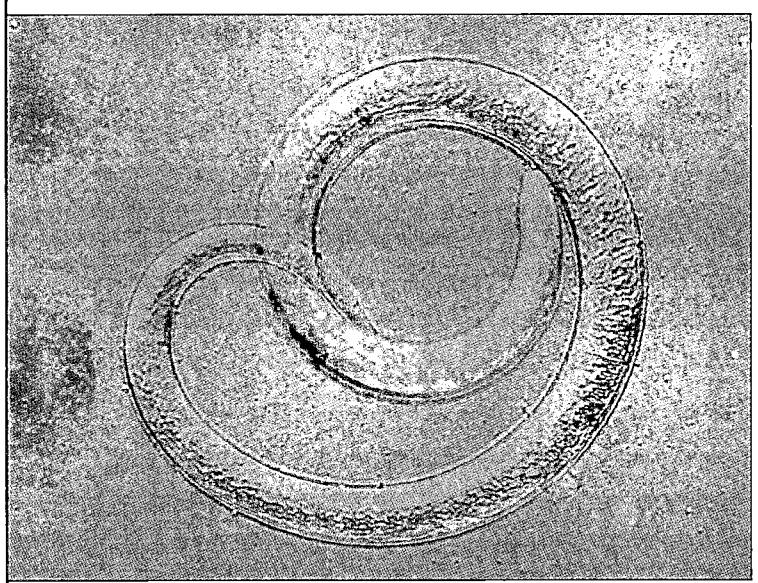

Trichinella spiralis *adult worm. These small worms, about 1 mm in length, produce larvae that become encysted in muscle. Large numbers of these larvae in muscle cause the disease trichinosis.*

Microbial diseases of the digestive system are second only to respiratory diseases as causes of illness in the United States. Most such diseases result from the ingestion of food or water contaminated with pathogenic microorganisms or their toxins. These pathogens usually enter the food or water supply after being shed in the feces of people or animals infected with them. Therefore, microbial diseases of the digestive system are typically transmitted by a *fecal-oral cycle*. This cycle is interrupted by effective, responsible sanitation practices in food handling and by modern methods of sewage treatment and disinfection of drinking water. However, as more of our food—especially fruits and vegetables—is being grown in countries with poor sanitation, outbreaks of foodborne disease from imported pathogens will increase.

Structure and Function of the Digestive System

Learning Objective

- *Name the structures of the digestive system that contact food.*

The **digestive system** may be divided into two principal groups of organs (Figure 25.1). One group is the *gastrointestinal (GI) tract* or *alimentary canal,* essentially a tubelike structure that includes the mouth, pharynx (throat), esophagus (food tube), stomach, small intestine, and large intestine. The other group of organs, the *accessory structures,* consists of the teeth, tongue, salivary glands, liver, gallbladder, and pancreas. Except for the teeth and tongue, the accessory structures lie outside the GI tract and produce secretions that are conveyed by ducts into it.

The purpose of the digestive system is to digest foods—that is, to break them down into small molecules that can be taken up and used by body cells. In a process called *absorption,* these end-products of digestion pass from the small intestine into the blood or lymph for distribution to body cells. Then the food moves through the large intestine where water, vitamins, and nutrients are absorbed from it. Over the course of an average life span, about 25 tons of food pass through the digestive system. The resulting undigested solids, called *feces,* are eliminated

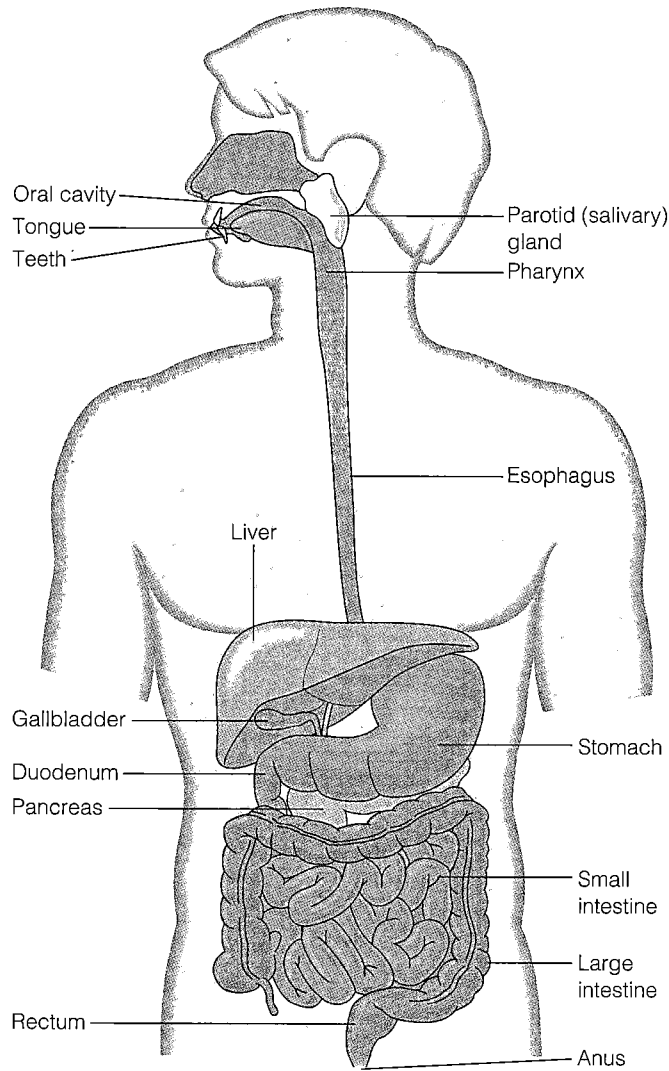

FIGURE 25.1 The human digestive system.

▨ How would you distinguish the gastrointestinal tract from the accessory organs of digestion?

from the body through the anus. Intestinal gas, or *flatus*, is a mixture of nitrogen from swallowed air and microbially produced carbon dioxide, hydrogen, and methane. On average, we produce 0.5–2.0 liters of flatus every day.

Normal Microbiota of the Digestive System

Learning Objective

■ List examples of the microbiota for each part of the gastrointestinal tract.

Bacteria heavily populate most of the digestive system. In the mouth, each milliliter of saliva can contain millions of

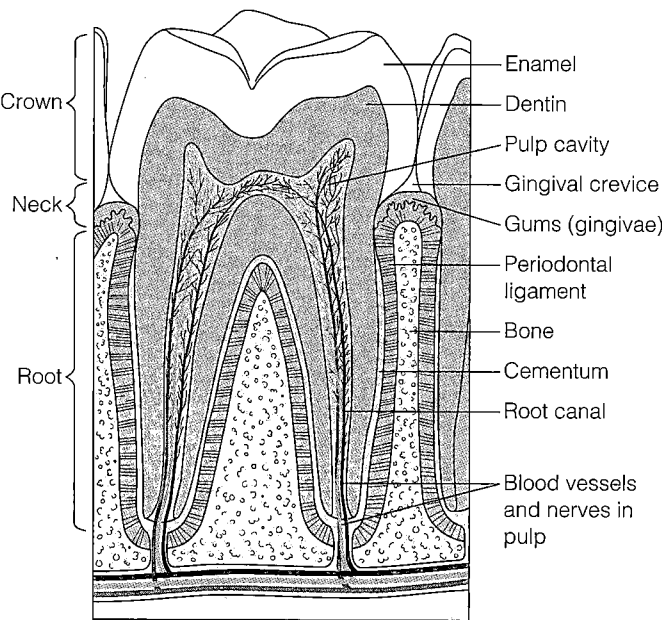

FIGURE 25.2 A healthy human tooth.

bacteria. The stomach and small intestine have relatively few microorganisms because of the hydrochloric acid produced by the stomach and the rapid movement of food through the small intestine. By contrast, the large intestine has enormous microbial populations, exceeding 100 billion bacteria per gram of feces. (Up to 40% of fecal mass is microbial cell material.) The population of the large intestine is composed mostly of anaerobes of the genera *Lactobacillus* and *Bacteroides,* and facultative anaerobes, such as *E. coli,* and *Enterobacter, Klebsiella,* and *Proteus spp.* Most of these bacteria assist in the enzymatic breakdown of foods, and some of them synthesize useful vitamins.

Bacterial Diseases of the Mouth

Learning Objective

■ Describe the events that lead to dental caries and periodontal disease.

The mouth, which is the entrance to the digestive system, provides an environment that supports a large and varied microbial population.

Dental Caries (Tooth Decay)

The teeth are unlike any other exterior surface of the body. They are hard, and they do not shed surface cells (Figure 25.2). This allows the accumulation of masses of microorganisms and their products. These accumulations, called **dental plaque,** are intimately involved in the formation of **dental caries,** or tooth decay.

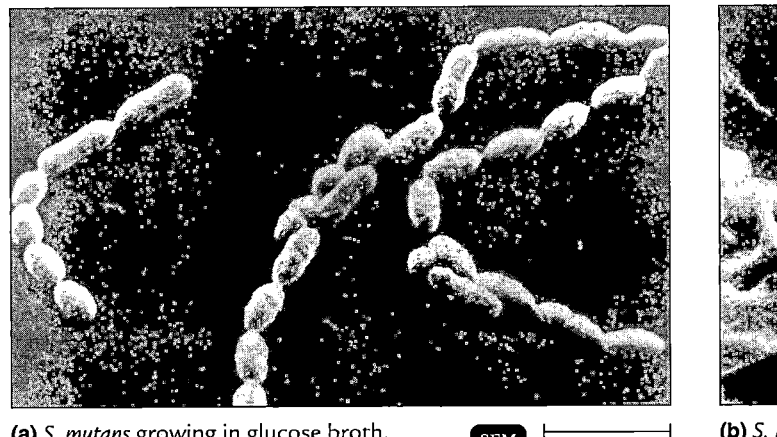

(a) *S. mutans* growing in glucose broth.   SEM |———————| 5 μm

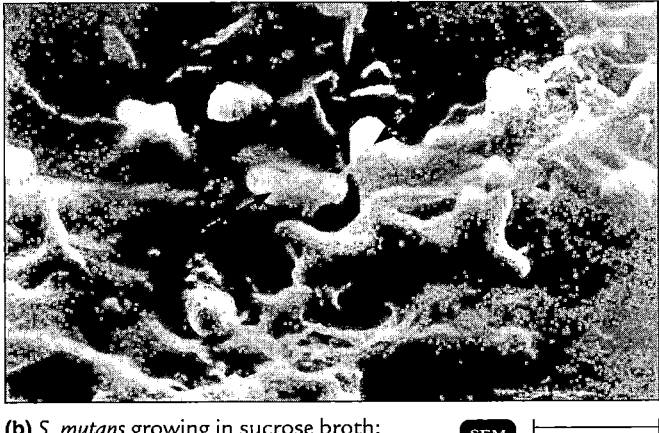

(b) *S. mutans* growing in sucrose broth; note the accumulations of dextran. Arrows point to S. mutans cells.   SEM |———————| 5 μm

FIGURE 25.3 The role of *Streptococcus mutans* and sucrose in dental caries.

■ What is dental plaque?

Oral bacteria convert sucrose and other carbohydrates into lactic acid, which in turn attacks the tooth enamel. The microbial population on and around the teeth is very complex; more than 300 species have been identified. Probably the most important *cariogenic* (caries-causing) bacterium is *Streptococcus mutans*, a gram-positive coccus. Some other species of streptococci are also cariogenic but play a lesser role in initiating caries.

The initiation of caries depends on the attachment of *S. mutans*, or other streptococci, to the tooth (Figure 25.3). These bacteria do not adhere to a clean tooth, but within minutes a freshly brushed tooth will become coated with a pellicle (thin film) of proteins from saliva. Within a couple of hours, cariogenic bacteria become established on this pellicle and begin to produce a gummy polysaccharide of glucose molecules called *dextran* (Figure 25.3b). In the production of dextran, the bacteria first hydrolyze sucrose into its component monosaccharides, fructose and glucose. The enzyme glucosyltransferase then assembles the glucose molecules into dextran. The residual fructose is the primary sugar fermented into lactic acid. Accumulations of bacteria and dextran adhering to the teeth make up dental plaque.

The bacterial population of plaque is predominantly streptococci and filamentous members of the genus *Actinomyces*. (Older, calcified deposits of plaque are called *dental calculus* or *tartar*.) *S. mutans* especially favors crevices or other sites on the teeth protected from the shearing action of chewing, or the flushing action of the liter or so of saliva produced in the mouth each day. On protected areas of the teeth, plaque accumulations can be several hundred cells thick. Because plaque is not very permeable to saliva, the lactic acid produced by bacteria is not di-

luted or neutralized, and it breaks down the enamel of the teeth to which the plaque adheres.

Although saliva contains nutrients that encourage the growth of bacteria, it also contains antimicrobial substances, such as lysozyme, that help protect exposed tooth surfaces. Some protection is also provided by *crevicular fluid*, a tissue exudate that flows into the gingival crevice (see Figure 25.2) and is closer in composition to serum than to saliva. It protects teeth by virtue of both its flushing action and its phagocytic cells and immunoglobulin content.

Localized acid production within deposits of dental plaque results in a gradual softening of the external *enamel*. Enamel low in fluoride is more susceptible to the effects of the acid. This is the reason for fluoridation of water and toothpastes, which has been a significant factor in the decline in tooth decay in the United States.

Figure 25.4 shows the stages of tooth decay. If the initial penetration of the enamel by caries remains untreated, bacteria can penetrate into the interior of the tooth. The composition of the bacterial population involved in spreading the decayed area from the enamel into the *dentin* is entirely different from that of the population initiating the decay. The dominant microorganisms are gram-positive rods and filamentous bacteria; *S. mutans* is present in small numbers only. Although once considered the cause of dental caries, *Lactobacillus* spp. actually play no role in initiating the process. However, these very prolific lactic acid producers are important in advancing the front of the decay once it has become established.

The decayed area eventually advances to the *pulp* (see Figure 25.4), which connects with the tissues of the jaw and contains the blood supply and the nerve cells. Almost

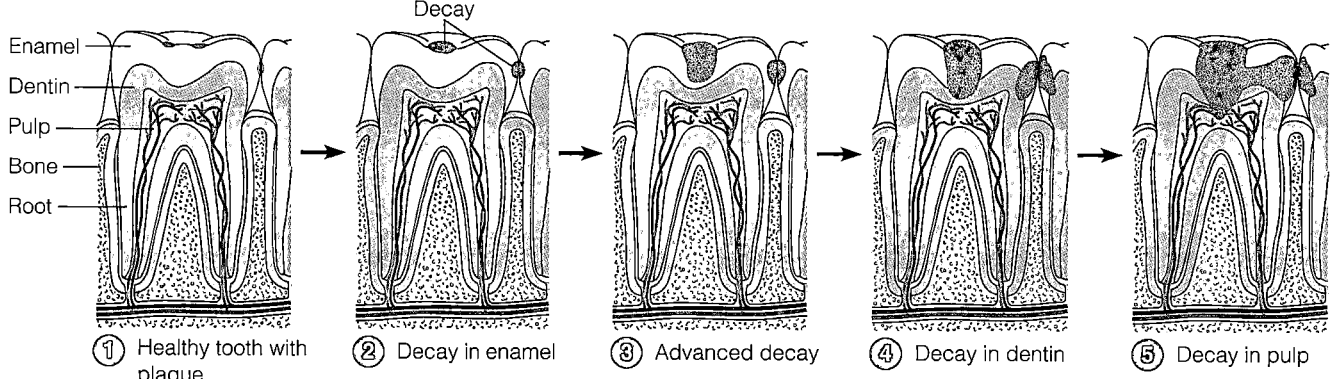

Enamel
Dentin
Pulp
Bone
Root

Decay

① Healthy tooth with plaque
② Decay in enamel
③ Advanced decay
④ Decay in dentin
⑤ Decay in pulp

FIGURE 25.4 **The stages of tooth decay.** ① A tooth with plaque accumulation in difficult-to-clean areas. ② Decay begins as enamel is attacked by acids formed by bacteria. ③ Decay advances through the enamel. ④ Decay advances into the dentin. ⑤ Decay enters the pulp and may form abscesses in the tissues surrounding the root.

any member of the normal microbiota of the mouth can be isolated from the infected pulp and roots. Once this stage is reached, root canal therapy is required to remove the infected and dead tissue and to provide access for antimicrobial drugs that suppress renewed infection. If untreated, the infection may advance from the tooth to the soft tissues, producing dental abscesses caused by mixed bacterial populations that contain many anaerobes. These, and most other dental soft tissue infections, can be treated by penicillin and its derivatives.

Although dental caries are probably one of the more common infectious diseases in humans today, they were scarce in the Western world until about the seventeenth century. In human remains from older times, only about 10% of the teeth contain caries. The introduction of table sugar, or sucrose, into the diet is highly correlated with our present level of caries in the Western world. Studies have shown that sucrose, a disaccharide composed of glucose and fructose, is much more cariogenic than either glucose or fructose individually (see Figure 25.3). People living on high-starch diets (starch is a polysaccharide of glucose) have a low incidence of tooth decay, unless sucrose is also a significant part of their diet. The contribution of bacteria to tooth decay has been shown by experiments with germ-free animals. Such animals do not develop caries even when fed a sucrose-rich diet designed to encourage their formation.

Sucrose is pervasive in the modern Western diet. However, if sucrose is ingested only at regular mealtimes, the protective and repair mechanisms of the body are usually not overwhelmed. It is the sucrose that is ingested between meals that is most damaging to teeth. Sugar alcohols, such as mannitol, sorbitol, and xylitol, are not cariogenic; xylitol appears to inhibit carbohydrate metab-

olism in *S. mutaus*. This is why they are used to sweeten "sugarless" candies and chewing gum.

The best strategies for preventing dental caries are a minimal ingestion of sucrose; brushing, flossing, and professional cleaning to remove plaque; and the use of fluoride. As for mouthwashes that claim to prevent or reduce plaque, chlorhexidine is probably the most effective. However, proper brushing and flossing are more important. The ancient Chinese used human urine as a mouthwash to improve oral health. Although urine does tend to lower acidity, this preventive measure is not recommended.

Periodontal Disease

Even people who avoid tooth decay might, in later years, lose their teeth to **periodontal disease,** a term for a number of conditions characterized by inflammation and degeneration of structures that support the teeth (Figure 25.5). The roots of the tooth are protected by a covering of specialized connective tissue called *cementum.* As the gums recede with age, the formation of caries on the cementum becomes more common.

Gingivitis

In many cases of periodontal disease, the infection is restricted to the gums, or *gingivae.* This resulting inflammation, called **gingivitis,** is characterized by bleeding of the gums while the teeth are being brushed (see Figure 25.5). This is a condition experienced by almost everyone. It has been shown experimentally that gingivitis will appear in a few weeks if brushing is discontinued and plaque is allowed to accumulate. An assortment of streptococci, actinomycetes, and anaerobic gram-negative bacteria predominate in these infections.

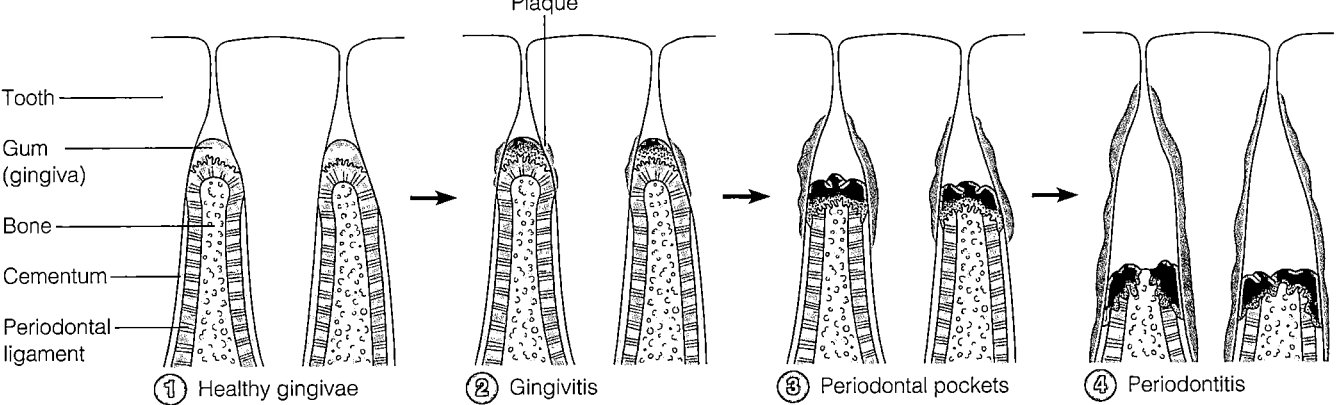

FIGURE 25.5 **The stages of periodontal disease.** ① Teeth firmly anchored by healthy bone and gum tissue (gingiva). ② Toxins in plaque irritate gums, causing gingivitis. ③ Periodontal pockets form as the tooth separates from the gingiva. ④ Gingivitis progresses to periodontitis. Toxins destroy the gingiva and bone that support the tooth and the cementum that protects the root.

■ Periodontal disease refers to a number of conditions that are characterized by inflammation and degeneration of supporting structures of the teeth.

Periodontitis

Gingivitis can progress to a chronic condition called **periodontitis,** which is responsible for nearly 10% of tooth loss in older adults. This is an insidious condition that generally causes little discomfort. The gums are inflamed and bleed easily. Sometimes pus forms in pockets surrounding the teeth (periodontal pockets; see Figure 25.5). As the infection continues, it progresses toward the root tips. The bone and tissue that support the teeth are destroyed, leading eventually to loosening and loss of the teeth. Numerous bacteria of many different types, primarily *Porphyromonas* species, are found in these infections; the damage to tissue is done by an inflammatory response to the presence of these bacteria. Periodontitis is treated either by surgically eliminating the periodontal pockets or by using specialized cleaning techniques on the tooth surfaces normally protected by the gums.

Acute **necrotizing ulcerative gingivitis,** also termed **Vincent's disease** or **trench mouth,** is one of the more common serious mouth infections. The disease causes enough pain that normal chewing is difficult. Foul breath (halitosis) also accompanies the infection. Among the bacteria usually associated with this condition is *Prevotella intermedia,* averaging up to 24% of the isolates. Because these pathogens are usually anaerobic, treatment with oxidizing agents, debridement, and the administration of metronidazole may be temporarily effective.

Bacterial Diseases of the Lower Digestive System

Learning Objective

■ *List the causative agents, suspect foods, signs and symptoms, and treatments for staphylococcal food poisoning, shigellosis, salmonellosis, typhoid fever, cholera, gastroenteritis, and peptic ulcer disease.*

Diseases of the digestive system are essentially of two types: infections and intoxications.

An **infection** occurs when a pathogen enters the GI tract and multiplies. Microorganisms can penetrate into the intestinal mucosa and grow there, or they can pass through to other systemic organs. Infections are characterized by a delay in the appearance of gastrointestinal disturbance while the pathogen increases in numbers or affects invaded tissue. There is also usually a fever, one of the body's general responses to an infective organism.

Some pathogens cause disease by forming toxins that affect the GI tract. An **intoxication** is caused by the ingestion of such a preformed toxin. Most intoxications, such as that caused by *Staphylococcus aureus,* are characterized by a very sudden appearance (usually in only a few hours) of symptoms of a GI disturbance. Fever is less often one of the symptoms.

Both infections and intoxications often cause *diarrhea,* which most of us have experienced. Severe diarrhea accompanied by blood or mucus is called **dysentery.** Both types of digestive system diseases are also frequently accompanied by *abdominal cramps, nausea,* and *vomiting.* Diarrhea and vomiting are both defensive mechanisms designed to rid the body of harmful material.

The general term **gastroenteritis** is applied to diseases causing inflammation of the stomach and intestinal mucosa. Botulism is a special case of intoxication because the ingestion of the preformed toxin affects the nervous system rather than the GI tract (see Chapter 22 on page 608).

In developing countries, diarrhea is a major factor in infant mortality. Approximately one child in every ten dies of it before the age of 5. Diarrhea also affects the absorption of nutrients from food and adversely affects the growth of the survivors.

The cause of diarrhea may be any of several pathogens. Most commonly, rotaviruses are identified as responsible, but enterotoxigenic *E. coli* and *Shigella* species are also frequent isolates. It is estimated that mortality from childhood diarrhea could be halved by *oral rehydration therapy.* This is usually a solution of sodium chloride, potassium chloride, glucose, and sodium bicarbonate to replace lost fluids and electrolytes. These solutions are sold in the infant supply department of many stores.

Diseases of the digestive system are often related to food ingestion. Estimates of the incidence of so-called food poisoning in the United States range from 6 to 80 million illnesses and up to 9000 deaths annually.

Staphylococcal Food Poisoning (Staphylococcal Enterotoxicosis)

A leading cause of gastroenteritis is **staphylococcal food poisoning,** an intoxication caused by ingesting an enterotoxin produced by *S. aureus.* Staphylococci are comparatively resistant to environmental stresses, as discussed on pages 323–324. They also have a fairly high resistance to heat; vegetative cells can tolerate 60°C for half an hour. Their resistance to drying and radiation helps them survive on skin surfaces. Resistance to high osmotic pressures helps them grow in foods, such as cured ham, in which the high osmotic pressure of salts inhibits the growth of competitors.

S. aureus is often an inhabitant of the nasal passages, from which it contaminates the hands. It is also a frequent cause of skin lesions on the hands. From these sources, it can readily enter food. If the microbes are allowed to incubate in the food, a situation called **temperature abuse,** they reproduce and release enterotoxin into the food. These events, which lead to outbreaks of staphylococcal intoxication, are illustrated in Figure 25.6.

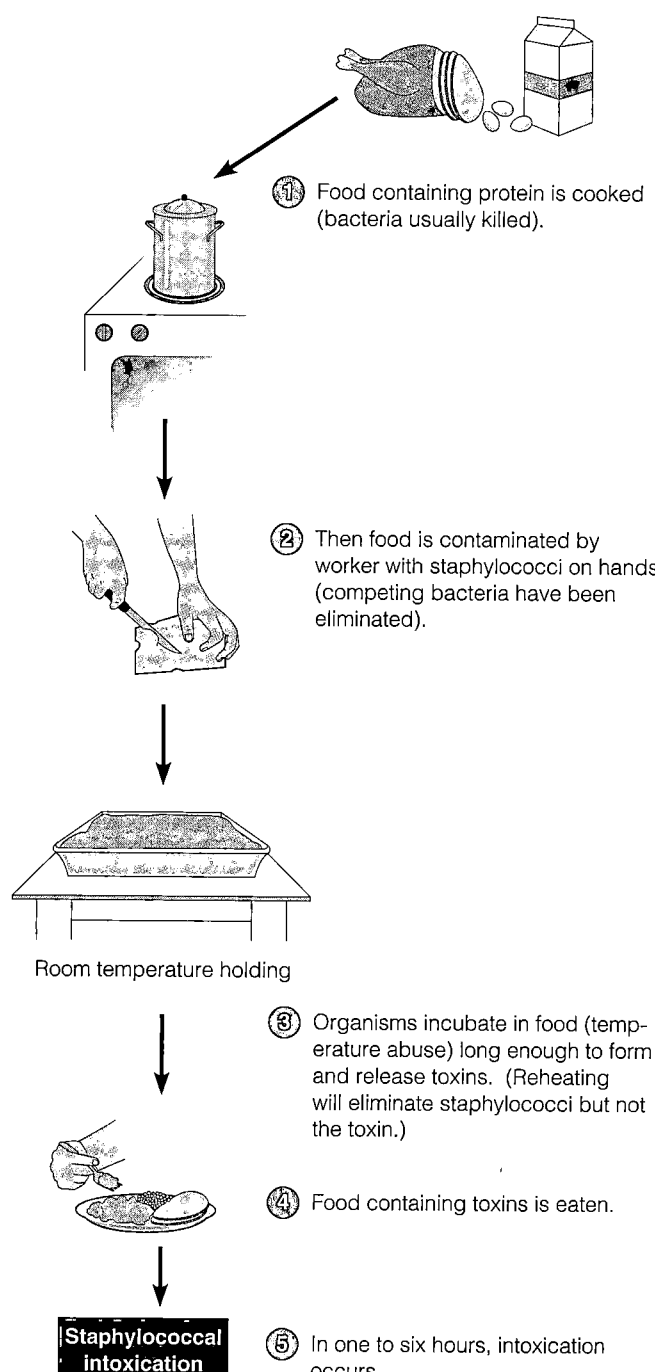

① Food containing protein is cooked (bacteria usually killed).

② Then food is contaminated by worker with staphylococci on hands (competing bacteria have been eliminated).

Room temperature holding

③ Organisms incubate in food (temperature abuse) long enough to form and release toxins. (Reheating will eliminate staphylococci but not the toxin.)

④ Food containing toxins is eaten.

Staphylococcal intoxication ⑤ In one to six hours, intoxication occurs.

FIGURE 25.6 The sequence of events in a typical outbreak of staphylococcal food poisoning.

■ Staphylococcal food poisoning is an intoxication caused by ingesting enterotoxins produced by *S. aureus.*

S. aureus produces several toxins that damage tissues or increase the microorganism's virulence. The production of the toxin of serological type A (which is responsible for most cases) is often correlated with the production of an enzyme that coagulates blood plasma. Such bacteria are described as *coagulase-positive*. No direct pathogenic effect can be attributed to the enzyme, but it is useful in the tentative identification of types that are likely to be virulent.

Generally, a population of about 1 million bacteria per gram of food will produce enough enterotoxin to cause illness. The growth of the microbe is facilitated if the competing microorganisms in the food have been eliminated—by cooking, for example. It is also more likely to grow if competing bacteria are inhibited by a higher-than-normal osmotic pressure or by a relatively low moisture level. *S. aureus* tends to outgrow most competing bacteria under these conditions.

Custards, cream pies, and ham are examples of high-risk foods. Competing microbes are minimized in custards by the high osmotic pressure of sugar and by cooking. In ham they are inhibited by curing agents, such as salts and preservatives. Poultry products can also harbor staphylococci if they are handled and allowed to stand at room temperatures. Because staphylococci do not compete well with the large number of microorganisms hamburger contains, it is rarely a factor in this type of food poisoning. Any foods prepared in advance and not kept chilled are a potential source of staphylococcal food poisoning. Because food contamination by human handlers cannot be avoided completely, the most reliable method of preventing staphylococcal food poisoning is adequate refrigeration during storage to prevent toxin formation.

The toxin itself is heat stable and can survive up to 30 minutes of boiling. Therefore, once the toxin is formed, it is not destroyed when the food is reheated, although the bacteria will be killed.

The toxin quickly triggers the brain's vomiting reflex center; abdominal cramps and usually diarrhea then ensue. This reaction is essentially immunological in character; the staphylococcal enterotoxin is a model example of a superantigen (see page 520). Recovery is usually complete within 24 hours.

The mortality rate of staphylococcal food poisoning is almost zero among otherwise healthy people, but it can be significant in weakened individuals, such as residents of nursing homes. No reliable immunity results from recovery. However, there is a great deal of variation in individual susceptibility to the toxin, and it is suspected that immunity acquired from a previous exposure might account for some of this variation.

The diagnosis of staphylococcal food poisoning is usually based on the symptoms, particularly the short incubation time characteristic of intoxication. If the food has not been reheated so that the bacteria are not killed, the pathogen can be recovered and grown. *S. aureus* isolates can be tested by *phage-typing,* a method used in tracing the source of the contamination (see Figure 10.13 on page 294). These bacteria grow well in 7.5% sodium chloride, so this concentration is often used in media for their selective isolation. Pathogenic staphylococci usually ferment mannitol, produce hemolysins and coagulase, and form golden-yellow colonies. They cause no obvious spoilage when growing in foods. Detecting the toxin in food samples has always been a problem; there may be only 1–2 nanograms in 100 g of food. Reliable serological methods have become commercially available only recently.

Shigellosis (Bacillary Dysentery)

In bacterial infections, disease is a result of microbial growth in body tissues rather than of the ingestion of food and drink already contaminated by preformed toxins resulting from microbial growth. Bacterial infections, such as salmonellosis and shigellosis, usually have longer incubation periods (12 hours to 2 weeks) than bacterial intoxications, reflecting the time needed for the microorganism to grow in the host. Bacterial infections are often characterized by some fever, indicating the host's response to the infection.

Shigellosis, also known as **bacillary dysentery,** is a severe form of diarrhea caused by a group of facultatively anaerobic gram-negative rods of the genus *Shigella.* (The genus was named for the Japanese microbiologist Kiyoshi Shiga.) There are four species of pathogenic *Shigella: S. sonnei* (sōn′ne-ē), *S. dysenteriae* (dis-en-te′rē-ī), *S. flexneri* (fleks′ner-ē), and *S. boydii* (boi′dē-ē). These bacteria are residents only of the intestinal tract of humans, apes, and monkeys. They are closely related to the pathogenic *E. coli.*

The most common species in the United States is *S. sonnei;* it causes a relatively mild dysentery. Many cases of so-called traveler's diarrhea might be mild forms of shigellosis. At the other extreme, infection with *S. dysenteriae* often results in a severe dysentery and prostration. The toxin responsible is unusually virulent and is known as the **Shiga toxin** (see enterohemorrhagic *E. coli,* on page 697). *S. dysenteriae* is the least common species in the United States.

The infective dose required to cause disease is small; the bacteria are not much affected by stomach acidity. They proliferate to immense numbers in the small intestine, but cause damage to the large intestine. There, they destroy tissue in the intestinal mucosa, causing severe diarrhea with blood and mucus in the stool. Virulent shigellae produce exotoxins that inhibit protein synthesis,

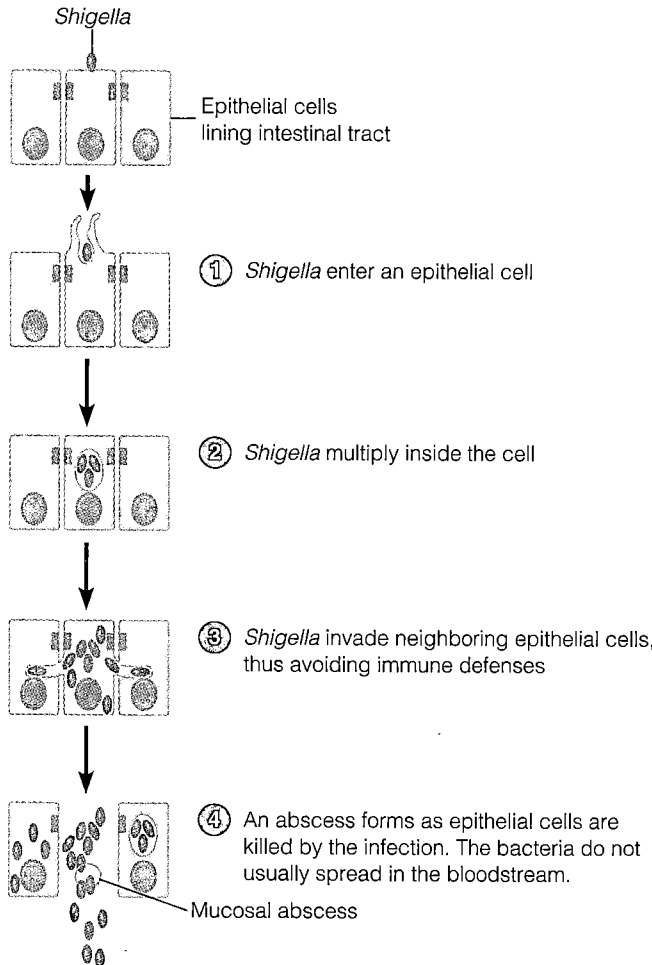

Shigella

Epithelial cells
lining intestinal tract

① Shigella enter an epithelial cell

② Shigella multiply inside the cell

③ Shigella invade neighboring epithelial cells,
thus avoiding immune defenses

④ An abscess forms as epithelial cells are
killed by the infection. The bacteria do not
usually spread in the bloodstream.

Mucosal abscess

FIGURE 25.7 Shigellosis. This sequence shows how the infection of epithelial cells in the intestinal wall can cause intestinal damage.

■ **What are the four species of Shigella?**

thereby killing cells (Figure 25.7). Infected people may have as many as 20 bowel movements in one day.

Additional symptoms of infection are abdominal cramps and fever. Except for *S. dysenteriae*, *Shigella* rarely invades the bloodstream. Diagnosis is usually based on recovery of the microbes from rectal swabs.

In recent years, the number of cases reported in the United States has been about 20,000–30,000, with 5–15 deaths. (See the box on page 694). *S. dysenteriae* has a significant mortality rate, however, and the death rate in tropical areas where it is prevalent can be as high as 20%. Some immunity seems to result from recovery, but a satisfactory vaccine has not yet been developed.

In severe cases of shigellosis, antibiotic therapy and oral rehydration are indicated. At present, fluoroquinolones are the antibiotics of choice.

Salmonellosis (*Salmonella* Gastroenteritis)

The *Salmonella* bacteria (named for their discoverer, Daniel Salmon) are gram-negative, facultatively anaerobic, non–endospore-forming rods. Their normal habitat is the intestinal tracts of humans and many animals. All salmonellae are considered pathogenic to some degree, causing **salmonellosis,** or *Salmonella* **gastroenteritis.**

The nomenclature of the *Salmonella* microbes is confusing. Rather than recognized species, there are more than 2000 serotypes (or serovars), only about 50 of which are isolated with any frequency in the United States. For a discussion of the nomenclature of the salmonellae, see page 312. Briefly, many consider them to belong to only two species, primarily *Salmonella enterica*. Therefore, you might encounter nomenclature such as *S. enterica* serotype Typhimurium, instead of the species-like name *S. typhimurium*.

Salmonellosis has an incubation time of about 12–36 hours. The salmonellae first invade the intestinal mucosa and multiply there. Sometimes they manage to pass through the intestinal mucosa to enter the lymphatic and cardiovascular systems, and from there they may spread to eventually affect many organs (Figure 25.8). The fever associated with *Salmonella* infections might be from endotoxins released by lysed cells, but this relationship is not certain. There is usually a moderate fever accompanied by nausea, abdominal pain and cramps, and diarrhea. As many as 1 billion salmonellae per gram can be found in an infected person's feces during the acute phase of the illness.

The mortality rate is overall very low, probably less than 1%. However, the death rate is higher in infants and among the very old; death is usually from septicemia. The severity and incubation time can depend on the number of *Salmonella* ingested. Normally, recovery will be complete in a few days, but many patients will continue to shed the organisms in their feces for up to 6 months. Antibiotic therapy is not useful in treating salmonellosis or, indeed, many diarrheal diseases; treatment consists of oral rehydration therapy.

Salmonellosis is probably greatly underreported. About 40,000–50,000 cases are reported each year, but an estimated 2–4 million occur, with 500–2000 deaths (Figure 25.9).

Meat products are particularly susceptible to contamination by *Salmonella*. The sources of the bacteria are the intestinal tracts of many animals, and meats can be contaminated readily in processing plants. Poultry, eggs, and egg products are especially often contaminated by *Salmonella*. Considerable experimental work has been done to

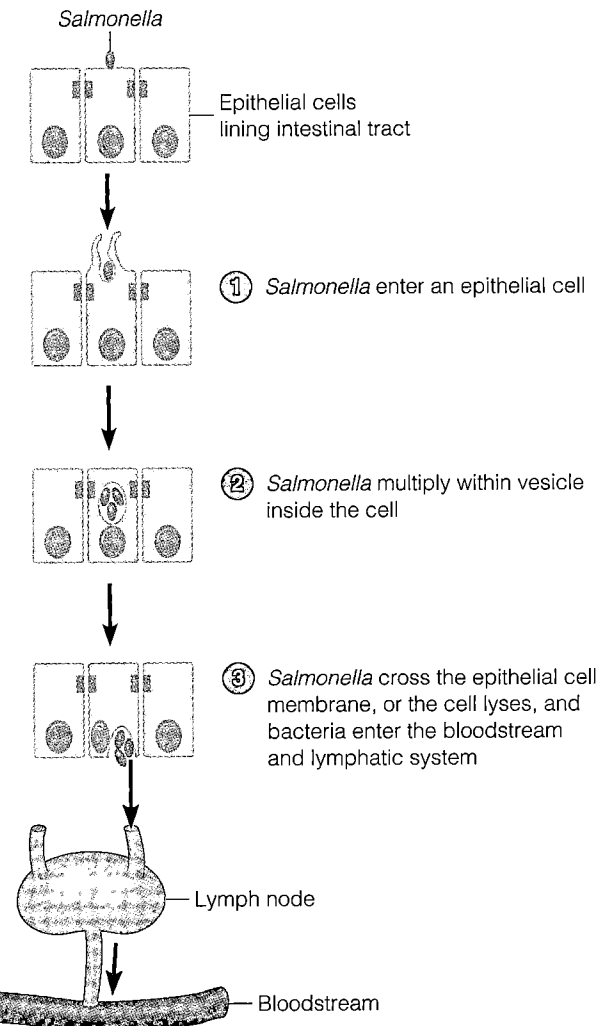

FIGURE 25.8 **Salmonellosis.** This sequence shows how the infection of epithelial cells in the intestinal wall spreads to other parts of the body.

■ Bacterial infections such as salmonellosis usually have longer incubation periods than bacterial intoxications.

suppress *Salmonella* populations in chicks by colonizing their intestinal tracts with competing, harmless, bacteria. Pet reptiles are also a source; their carriage rate is as high as 90%.

Outbreaks of salmonellosis have been traced to contaminated whole eggs. A reported 0.01% contain salmonellae. Apparently the bacteria are transmitted to the eggs before they are laid, although the chickens may be asymptomatic. Health authorities caution the public to eat only well-cooked eggs. Eggs fried "sunny side up" or boiled for only 4 minutes are not well enough cooked to reliably kill *Salmonella*. An often unsuspected factor is the

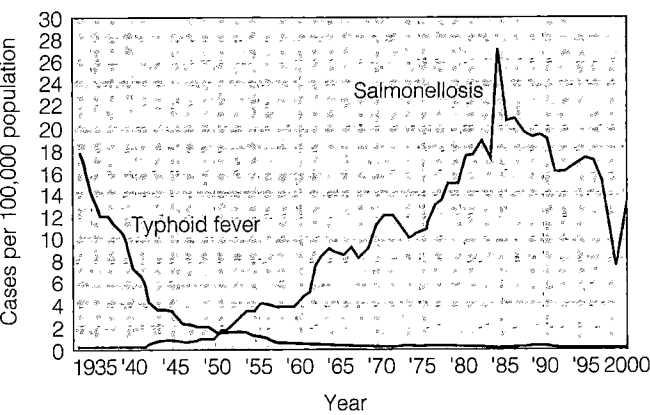

FIGURE 25.9 **The incidence of salmonellosis and typhoid fever.** An important factor in comparing the two diseases is that typhoid transmission is almost entirely human, and salmonellosis transmission is primarily between animal products and humans. SOURCES: CDC, *Summary of Notifiable Diseases 1998, MMWR* 47(53)(12/31/99); *MMWR* 48(51)(1/17/00).

■ The incidence of typhoid fever has been declining in the United States, while that of salmonellosis has been increasing.

presence of inadequately cooked or raw eggs in foods such as hollandaise sauce, cookie batter, and caesar salad.

Prevention also depends on good sanitation practices to deter contamination and on proper refrigeration to prevent increases in bacterial numbers. Recently, it has been possible to offer eggs in which any salmonella have been killed by a special hot water pasteurization procedure that does not cook the eggs. However, these eggs are more expensive. The microbes are generally destroyed by normal cooking that heats the food to an internal temperature of at least 60°C (140°F). However, contaminated food can contaminate a surface, such as a cutting board. Although the food first prepared on the board might later be cooked and its bacteria killed, another food subsequently prepared on the board might not be cooked.

Diagnosis usually depends on isolating the pathogen from the patient's stool or from leftover food. Isolation requires specialized selective and differential media; these methods are relatively slow. Also, the small numbers of *Salmonella* generally found in foods present a special problem in detection. Because this disease is important, much effort has been applied to improving detection and identification methods. What is needed are tests that will detect, within minutes or a few hours, small numbers of *Salmonella* directly in foods without any need for culturing. The infective dose (ID_{50}) may be as small as 1000 bacteria. Currently, one-step PCR-based tests are the most promising.

CLINICAL PROBLEM SOLVING

A Foodborne Epidemic

You will see questions as you read through this problem. The questions are those that public health epidemiologists ask themselves and each other as they solve a clinical problem. Try to answer each question as you read through the problem.

1. On July 24, a 24-year-old woman in Minnesota went to the emergency room with a 3-day history of nausea, vomiting, and diarrhea. She had a temperature of 39.5°C, and she was dehydrated. Her abdomen was tender on physical examination, and examination showed that she had blood in a stool sample.
   What sample is needed from the patient to determine the cause of her signs and symptoms?

2. Microscopic example of a stool sample revealed leukocytes and gram-negative rods.
   What selective medium would you use to culture the bacteria?

3. Colorless colonies grew from a culture of the stool sample on MacConkey agar. The bacteria were H_2S negative, urease negative, and nonmotile.
   Can you identify the bacteria? How would you treat this patient?

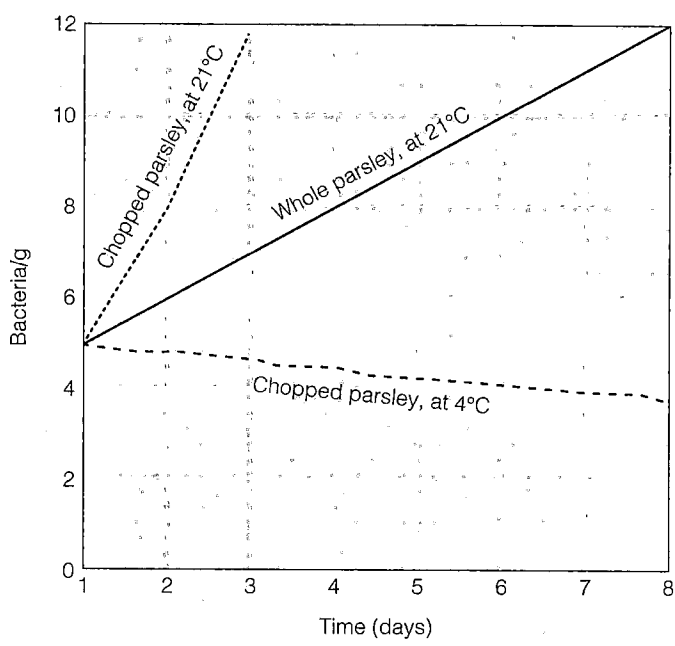

Time (days)

4. The bacteria are non–lactose-fermenting, and the other biochemical tests indicate that it is *Shigella sonnei*. Fluids and electrolytes are needed to replace those lost by vomiting and diarrhea. Fluoroquinolones may be administered. On August 17, the Minnesota Department of Health was notified of *S. sonnei* in 167 people with onset between July 24 and August 17.
   What information would you try to obtain in interviews with these people?

5. All of the ill people had eaten in the same restaurant, where eight of the 44 restaurant workers had a similar illness.
   What information do you need to obtain next?

6. In a case-control study of 172 ill and 95 well restaurant patrons, five items were associated with illness: water, ice, potatoes, uncooked parsley, and raw tomatoes. The CDC received reports of seven additional outbreaks of shigellosis in the United States and Canada.
   The occurrence of eight outbreaks in different geographic locations indicates that the Minnesota restaurant workers were

not the source of infection. How should you proceed?

7. The *Shigella* from seven of the outbreaks were resistant to ampicillin, TMP-SMZ, tetracycline, and streptomycin. DNA tests of *Shigella* from seven of the outbreaks had the same restriction fragment length polymorphism (RFLP) pattern.
   What additional information do you need?

8. In the Canadian outbreak, 20 of the 35 ill people reported eating a smoked salmon and pasta dish made with fresh chopped parsley. Stool samples from four restaurant employees who handled the parsley were negative for *Shigella*.
   What would you do next to try to identify the source of infection?

9. Parsley preparation was studied because parsley was implicated in all of the outbreaks. Food handlers reported washing parsley before chopping it. Usually parsley was chopped in the morning and left at room temperature, sometimes until the end of the day, before it was served to customers.
   What other information about the parsley would you like?

10. All of the parsley came from one farm. The well water that supplied the farm's packing shed was unchlorinated. This water was used for chilling the parsley immediately after harvest and for making ice with which the parsley was packaged for transport.
    Why didn't the farm workers get sick?

11. Farm workers reported drinking bottled water only. The results obtained in a laboratory study of parsley handling are in the figure.
    What do the findings indicate? What are your recommendations?

The results indicate that the number of bacteria present on the parsley increases with incubation at room temperature. Therefore, the risk of gastroenteritis can be reduced by storing chopped parsley for shorter times, keeping it refrigerated, and chopping smaller batches. Changes in parsley production on the farm should include use of adequately chlorinated water for chilling and icing the parsley and possibly the use of postharvest control measures such as irradiation.

SOURCE: Adapted from *MMWR* 48(14):285–289 (4/16/99).

Typhoid Fever

A few serotypes of *Salmonella* are much more virulent than others. The most virulent, *S. typhi,* causes the bacterial infection **typhoid fever.** This pathogen is not found in animals; it is spread only in the feces of other humans.

The incubation period is much longer than that of salmonellosis, normally about 2 weeks. The patient first suffers from a high fever of 40°C (104°F) and continual headache. Diarrhea appears only during the second or third week, and the fever tends to decline. Unlike in salmonellosis, the bacteria do not multiply in the intestinal epithelial cells. Rather, they differ by continuing to multiply in phagocytic cells. The microbe becomes disseminated in the body and can be isolated from the blood, urine, and feces. In severe cases, perforation of the intestinal wall can occur.

The mortality rate is now about 1–2%; at one time, it was at least 10%. Before the days of proper sewage disposal, water treatment, and food sanitation, typhoid was an extremely common disease. Its incidence has been declining in the United States, whereas that of salmonellosis has been increasing (see Figure 25.9). Typhoid fever is still a frequent cause of death in parts of the world with poor sanitation.

A substantial number of recovered patients, about 1–3%, become chronic carriers. They harbor the pathogen in the gallbladder and continue to shed bacteria for several months. A number of such carriers continue to shed the organism indefinitely. The classic example of a typhoid carrier was Typhoid Mary. Her name was Mary Mallon; she worked as a cook in New York state in the early part of the twentieth century and was responsible for several outbreaks of typhoid and three deaths. Her case became well known through the attempts of the state to restrain her from working at her chosen trade.

Recently, there have been about 350–500 annual cases of typhoid fever in the United States, of which 70% were acquired during foreign travel. Normally, there are fewer than three deaths each year.

The third-generation cephalosporins, such as ceftriaxone, are the usual drugs of choice for the treatment of severe typhoid fever. The carrier state can be successfully treated with several weeks of antibiotics therapy. Recovery from typhoid confers lifelong immunity.

Immunization is normally not done in developed countries, except for high-risk laboratory and military personnel. The vaccine that has long been in use is a killed-organism type, which must be injected. Live, orally ingested vaccines have become available. Their effectiveness is not much greater than older, killed vaccines (con-

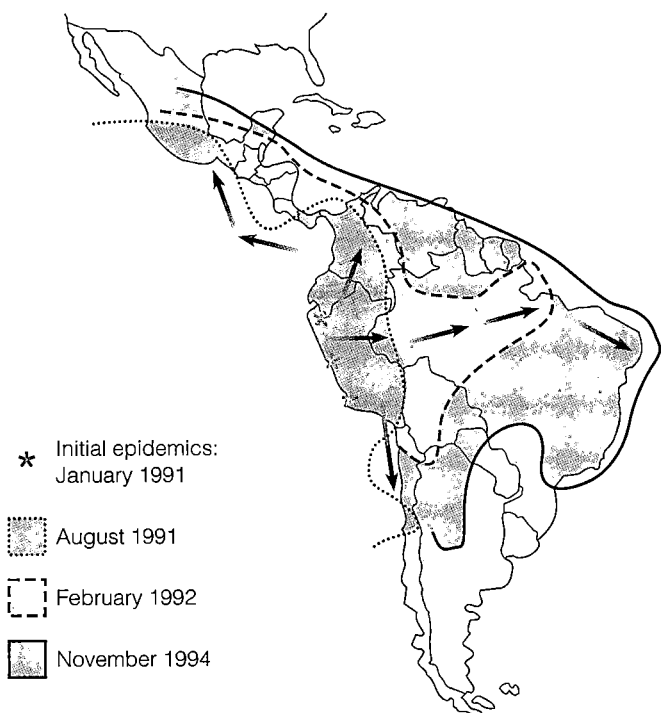

* Initial epidemics: January 1991

[shaded] August 1991

[dashed box] February 1992

[filled box] November 1994

FIGURE 25.10 The extent and progress of a cholera epidemic in Latin America, 1991–1994.
SOURCE: *MMWR* 44(11) (3/24/95).

■ **What is the main cause of cholera outbreaks?**

ferring immunity lasting only a few years on 65% of recipients), but they have fewer adverse effects.

Cholera

During the 1800s, the bacterial infection called Asiatic cholera crossed Europe and North America in repeated epidemics. Today, **cholera** is endemic in Asia, particularly India, and has only occasional outbreaks in Western countries. These outbreaks are caused by temporary lapses in sanitation practices. The 1991–1994 epidemic in Latin America resulted in over 1 million cases and 9600 deaths (Figure 25.10). The origin of the outbreak was probably seafood contaminated by the ballast water (taken aboard to stabilize the ship) that was picked up in Asia and emptied into harbors in Peru.

Cholera bacteria are strongly associated with brackish (salty) waters characteristics of estuaries, although they are also readily spread in contaminated fresh water. The organism can survive indefinitely, even without fresh fecal contamination. Under unfavorable conditions, the bacterial cell shrinks drastically into a nonculturable, spherical, dormant state. This has been described as a sporelike state without formation of a true spore coat. A change in the environment causes them to revert rapidly

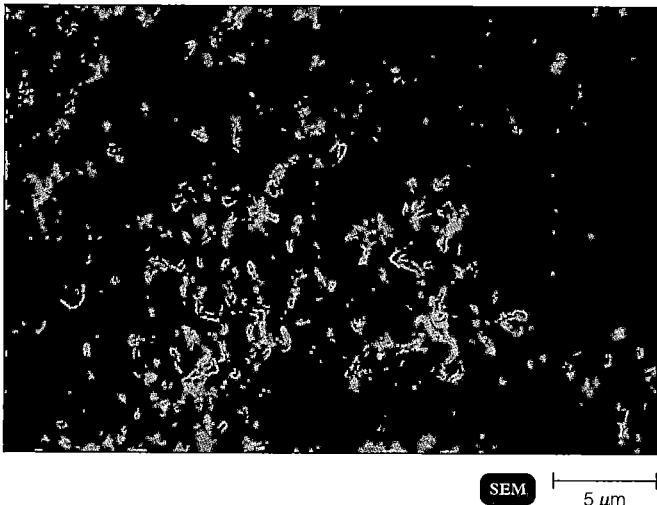

SEM ├─────────┤
                5 μm

FIGURE 25.11 *Vibrio cholerae,* **the cause of cholera.** Notice the slightly curved morphology.

■ **What are the effects of the sudden loss of fluid and electrolytes during infection with *V. cholerae?***

to the culturable form. Both forms are infectious. Tiny crustaceans, called copepods, are colonized by the bacteria, as are algae and other aquatic plants and plankton.

The causative agent is *Vibrio cholerae,* a slightly curved, gram-negative rod with a single polar flagellum (Figure 25.11). The serogroup O:1 of *V. cholerae* causes the classically recognized epidemic form of the disease. It can be subdivided into two *biotypes* (differentiated other than by serology; also called biovars): classical or *eltor* (named for the original culture isolated at a quarantine camp, El Tor, for pilgrims to Mecca). At present, the most widespread cholera pathogen is serogroup O:1, biotype eltor. A variant of this group, O:139, has recently caused wide epidemics in India and Bangladesh. Cholera bacilli grow in the small intestine and produce an enterotoxin that results in the secretion of chlorides, bicarbonates, and water. The excess water and these mineral electrolytes are excreted, taking on the typical appearance of "rice water stools" from the masses of intestinal mucus, epithelial cells, and bacteria. The sudden loss of these fluids and electrolytes causes shock, collapse, and often death (12–20 liters, or 3–5 gallons, of fluid might be lost in a day). Because of the loss of fluid, the blood becomes so viscous that vital organs are unable to function properly. Violent vomiting sometimes occurs. The microbes are not invasive, and a fever is usually not present. The severity of the disease varies considerably, and the number of subclinical cases might be several times the number reported.

In the United States, there have been occasional cases of cholera caused by the O:1 serogroup. These have all been in the Gulf Coast area, and the pathogen may be en-

demic in coastal waters. Most gastroenteritis due to *V. cholerae* in America has been caused by the non-O:1 serogroup, usually ingested in contaminated seafood. The waters of the Gulf and Pacific coasts support an indigenous population of these organisms. They differ from the O:1 serotypes in several ways, and they are more likely to invade the intestinal mucosa, causing bloody stools and fever.

Cholera bacteria can be readily isolated from the feces, partly because the microorganisms can grow in media that are alkaline enough to suppress the growth of many other organisms. The non-O:1 serogroup is occasionally isolated from blood and wounds.

Recovery from the disease results in an effective immunity based on the antigenic activity of both the cells and the enterotoxin. However, because of antigenic differences among bacterial strains, the same person can have cholera more than once. Most victims in endemic areas are children. The available vaccines provide immunity of relatively short duration and moderate effectiveness compared with immunity conferred by a natural infection.

Tetracycline is used for treatment, but chemotherapy is not as effective as replacement of the lost fluids and electrolytes. Untreated cases of cholera may have a 50% mortality rate, whereas the rate for cases having proper supportive care can be less than 1%.

Vibrio Gastroenteritis

Vibrio parahaemolyticus is found in salt water estuaries in many parts of the world. It is morphologically similar to *V. cholerae,* but it is halophilic and requires 2% or more sodium chloride for optimum growth. It is the most common cause of gastroenteritis in Japan, with thousands of cases reported annually. The bacterium is present in coastal waters of the continental United States and Hawaii. Raw oysters and crustaceans, such as shrimp and crabs, have been associated with several outbreaks of ***Vibrio* gastroenteritis** in the United States in recent years. A few large restaurant chains pasteurize raw oysters to avoid this mode of transmission.

Symptoms include abdominal pain, vomiting, a burning sensation in the stomach, and watery stools, resembling those of cholera. The incubation time is normally less than 24 hours. The generation time under optimum conditions is less than 10 minutes, and large numbers are reached quickly. Recovery usually follows in a few days.

Because *V. parahaemolyticus* has a requirement for sodium, as well as for a high osmotic pressure, isolation media containing 2–4% sodium chloride are used in diagnosing the disease.

Another important vibrio is *Vibrio vulnificus,* which is also found in estuaries. People with compromised immune systems are at special risk. In those suffering from liver disease, the mortality rate from septicemia may exceed 50%. Isolation media for *V. vulnificus* contain 1% sodium chloride.

Escherichia coli Gastroenteritis

One of the most prolific microorganisms in the human intestinal tract is *E. coli.* Because it is so common and so easily cultivated, microbiologists often regard it as something of a laboratory pet. Such *E. coli* are normally harmless, but certain strains can be pathogenic. All pathogenic strains have specialized fimbriae that allow them to bind to certain intestinal epithelial cells. They also produce toxins that cause gastrointestinal disturbances, collectively termed **E. coli** **gastroenteritis.**

There are several distinct pathogenic groups of *E. coli.* The *enterotoxigenic* strain is not invasive but forms an enterotoxin that produces a watery diarrhea that resembles a mild type of cholera. This group of microbes is a primary cause of **traveler's diarrhea** and, in developing countries, much infant diarrhea. The causative agent is not always identified, but 50–65% of traveler's diarrhea is probably caused by enterotoxigenic *E. coli.*

Less common are traveler's diarrhea cases caused by *enteroinvasive* strains of *E. coli.* These microorganisms invade the intestinal wall, resulting in inflammation, fever, and sometimes *Shigella*-like dysentery. Of the remaining cases, many are probably caused by *Shigella.* Other enteric bacteria, such as *Salmonella* and *Campylobacter,* as well as various unidentified bacterial pathogens, viruses, and protozoan parasites, may also be involved.

In adults, the disease is usually self-limiting, and chemotherapy is not attempted. Once contracted, the best treatment is the usual oral rehydration recommended for all diarrhea. In severe cases, antimicrobial drugs may be necessary.

In recent years, *enterohemorrhagic* strains of *E. coli* have become well known in the United States as the cause of several outbreaks of disease. The best known pathogen in this group is *E. coli* serotype O157:H7 (for the source of this nomenclature, see the discussion of *Salmonella* on page 312). It is an occasional inhabitant of animal intestinal tracts, especially cattle, where it has no pathogenic effect. These microorganisms produce Shiga-like toxins, which are responsible for *hemorrhagic colitis,* an inflammation of the colon with bleeding. (The colon is essentially the large intestine above the rectum.) Most persons infected suffer only a self-limiting diarrhea, but about 6% have stools that are combined with copious amounts of

blood. Another dangerous complication is *hemolytic uremic syndrome* (blood in the urine leading to kidney failure), which occurs when the kidneys are affected by the toxin. Some 5–10% of small children infected progress to this stage, which has a mortality rate of about 5%. Some children may require kidney dialysis or even transplants. An estimated 200–500 deaths occur annually.

Because of the attention this pathogen has attracted, researchers have been working, with some success, to develop rapid methods of detecting its presence in food without the need for time-consuming culturing methods. For isolation and identification, clinical laboratories use media that differentiate *E. coli* O157:H7 by its inability to ferment sorbitol.

Tests have shown that it is present in 1% or more of meat samples taken from cattle. Poultry and other meats may also be contaminated, but raw alfalfa sprouts have actually been responsible for most cases reported to date. The infective dose is estimated to be fewer than 10 bacteria.

Campylobacter Gastroenteritis

Campylobacter are gram-negative, microaerophilic, spirally curved bacteria that have emerged as the leading cause of foodborne illness in the United States. They are well-adapted to the intestinal environment of animal hosts, especially poultry. Culturing *Campylobacter* requires conditions of low oxygen and high carbon dioxide developed in special apparatus (see Figure 6.7b on page 168). Before their special cultural requirements were recognized, they were not considered an important pathogenic group. Their optimum growth temperature of about 42°C approximates that of their animal hosts. Almost all retail chicken is contaminated with *Campylobacter.* Also, nearly 60% of cattle excrete the organism in feces and milk, but red meats are less likely to be contaminated.

There are more than an estimated 2 million cases of **Campylobacter gastroenteritis** in the United States annually, usually caused by *C. jejuni.* Clinically, it is characterized by fever, cramping abdominal pain, and diarrhea or dysentery. Normally, recovery follows within a week.

An unusual complication of campylobacterial infection is that it is linked, in about 1 in 1000 cases, to the neurological disease Guillain-Barré syndrome, a temporary paralysis. Apparently, a surface molecule of the bacteria resembles a lipid component of nervous tissue and provokes an autoimmune attack.

Helicobacter Peptic Ulcer Disease

In 1982, a physician in Australia cultured a spiral-shaped, microaerophilic bacterium observed in the biopsied tissue of stomach ulcer patients. Now named *Helicobacter*

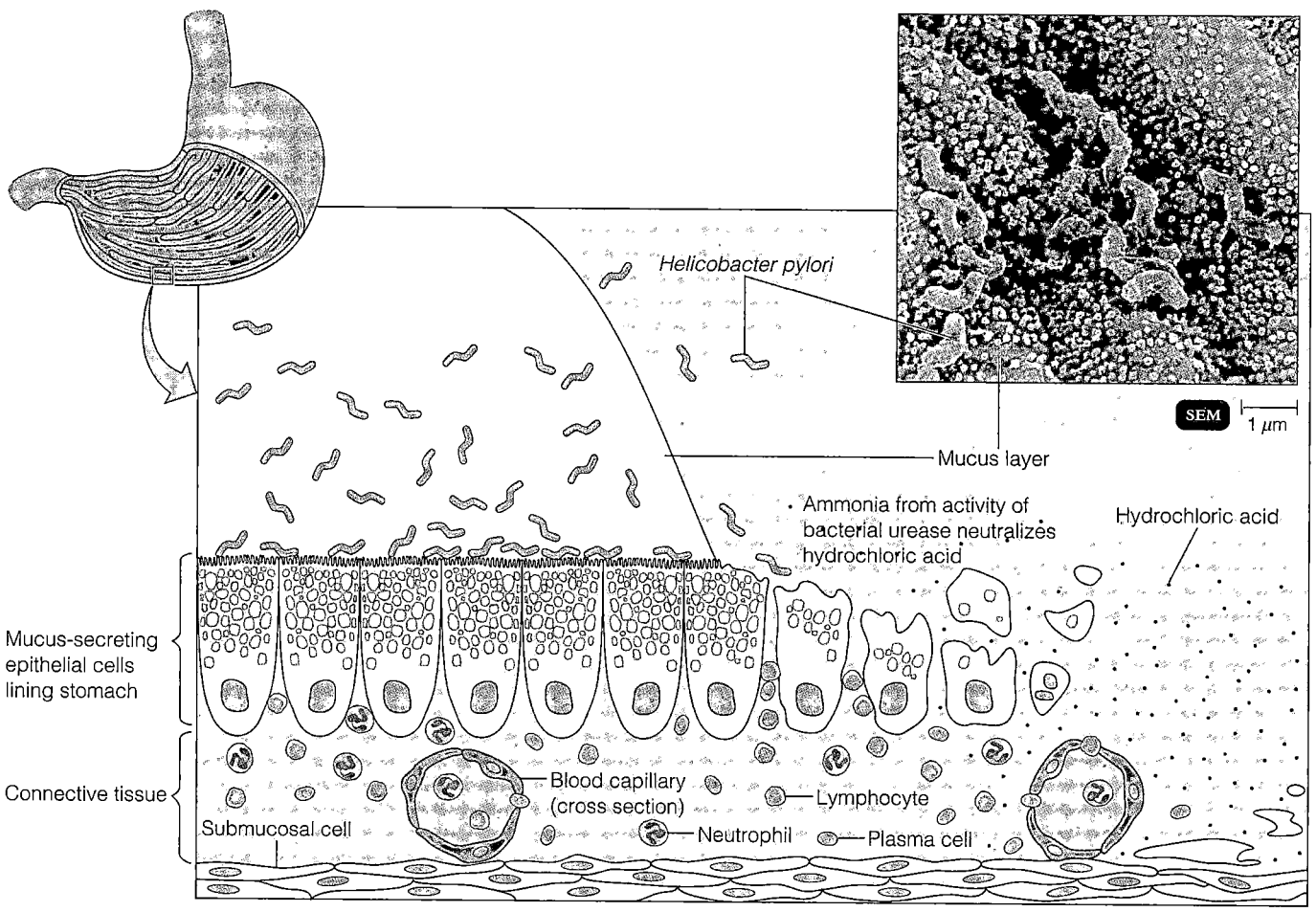

FIGURE 25.12 *Helicobacter pylori* infection, leading to ulceration of the stomach wall.

■ *H. pylori* can grow and cause damage in the acidic environment of the stomach by producing urease, which neutralizes the acidity; such destruction of the protective mucus layer leads to erosion of the wall.

pylori, it is widely accepted that this microbe is responsible for most cases of peptic ulcer disease. This syndrome includes gastric and duodenal ulcers. (The duodenum is the first few inches of the small intestine.) About 30–50% of the population in the developed world become infected; the infection rate is higher elsewhere. Only about 15% of those infected develop ulcers, so certain host factors are probably involved. For example, people with type O blood are more susceptible.

The stomach mucosa contains cells that secrete gastric juice containing proteolytic enzymes and hydrochloric acid that activates these enzymes. Other specialized cells produce a layer of mucus that protects the stomach itself from digestion. If this defense is disrupted, an inflammation of the stomach (gastritis) results. This inflammation can then progress to an ulcerated area (Figure 25.12).

Acid has been thought to be the primary agent of this inflammation, and therapy has been directed at drugs that inhibit acid release. Current thinking is that the immune system reacts to the presence of *H. pylori* and initiates the inflammation. Through an interesting adaptation, *H. pylori* can grow in the highly acidic environment of the stomach, which is lethal for most microorganisms. *H. pylori* produces large amounts of an especially efficient urease, an enzyme that converts urea to the alkaline compound ammonia, resulting in a locally high pH in the area of growth. This bacterium is recognized as a cause of stomach cancer (see the box in Chapter 8, page 234).

The eradication of *H. pylori* with antimicrobial drugs usually leads to the disappearance of peptic ulcers, and their recurrence rate is low. Several antibiotics, usually administered in combination, have proven effective. Bismuth subsalicylate (Pepto-Bismol®), which has antibacterial properties, is also effective and is often part of the drug regimen.

The most reliable diagnostic test requires a biopsy of tissue and culture of the organism. Several serological tests are available; they are simple and inexpensive, but only about 80% specific. An interesting diagnostic approach is the urea breath test. The patient swallows radioactivity-labeled urea, and, if the test is positive, within about 30 minutes CO_2 labeled with radioactivity can be detected in the breath. This test is most useful for determining the effectiveness of chemotherapy because a positive test is an indication of live *H. pylori*.

Yersinia Gastroenteritis

Other enteric pathogens being identified with increasing frequency are *Yersinia enterocolitica* (en-tér-ō-kōl-it'ik-ä) and *Y. pseudotuberculosis* (sū-dō-tü-bér-kü-lō'sis). These gram-negative bacteria are intestinal inhabitants of many domestic animals and are often transmitted in meat and milk. Both microbes are distinctive in their ability to grow at refrigerator temperatures of 4°C (39°F). *Yersinia* have occasionally been the cause of severe reactions when they contaminate transfused blood. Their ability to grow at low temperatures increases their numbers in stored refrigerated blood until their endotoxins can result in shock to the blood recipient.

These pathogens cause **Yersinia gastroenteritis**, or **yersiniosis.** The symptoms are diarrhea, fever, headache, and abdominal pain. The pain is often severe enough to cause a misdiagnosis of appendicitis. Diagnosis requires culturing the organism, which can then be evaluated by serological tests.

Clostridium perfringens Gastroenteritis

One of the more common, if underrecognized, forms of food poisoning in the United States is caused by *Clostridium perfringens,* a large, gram-positive, endospore-forming, obligately anaerobic rod. This bacterium is also responsible for human gas gangrene (see Chapter 23 on page 632).

Most outbreaks of *C. perfringens* **gastroenteritis** are associated with meats or meat stews contaminated with intestinal contents of the animal during slaughter. The pathogen's nutritional requirement for amino acids is met by such foods, and when the meats are cooked, the oxygen level is lowered enough for clostridial growth. The endospores survive most routine heatings, and the generation time of the vegetative bacterium is less than 20 minutes under ideal conditions. Large populations can therefore build up rapidly when foods are being held for serving, or when inadequate refrigeration leads to slow cooling.

The microbe grows in the intestinal tract and produces an exotoxin that causes the typical symptoms of abdominal pain and diarrhea. Most cases are mild and self-limiting and probably are never clinically diagnosed. The symptoms usually appear 8–12 hours after ingestion. Diagnosis is usually based on isolation and identification of the pathogen in stool samples.

Bacillus cereus Gastroenteritis

Bacillus cereus (se're-us) is a large, gram-positive, endospore-forming bacterium that is very common in soil and vegetation and is generally considered harmless. It has, however, been identified as the cause of outbreaks of foodborne illness. Heating the food does not always kill the spores, which germinate as the food cools. Because competing microbes have been eliminated in the cooked food, *B. cereus* grows rapidly and produces toxins. Rice dishes served in Asian restaurants seem especially susceptible.

Some cases of *B. cereus* **gastroenteritis** resemble *C. perfringens* infections and are almost entirely diarrheal in nature (usually 8–16 hours after ingestion). Other episodes involve nausea and vomiting (usually 2–5 hours after ingestion). It is suspected that different toxins are involved in producing the differing symptoms. Both forms of the disease are self-limiting.

Viral Diseases of the Digestive System

Learning Objective

■ *List the causative agents, modes of transmission, sites of infection, and symptoms for mumps and CMV inclusion disease.*

Although viruses do not reproduce within the contents of the digestive system like bacteria, they invade many organs associated with the system.

Mumps

The targets of the mumps virus, the parotid glands, are located just below and in front of the ears (see Figure 25.1). Because the parotids are one of the three pairs of salivary glands of the digestive system, it is appropriate to include a discussion of mumps in this chapter.

Mumps typically begins with painful swelling of one or both parotid glands 16–18 days after exposure to the virus (Figure 25.13). The virus is transmitted in saliva and respiratory secretions, and its portal of entry is the respiratory tract. An infected person is most infective to others

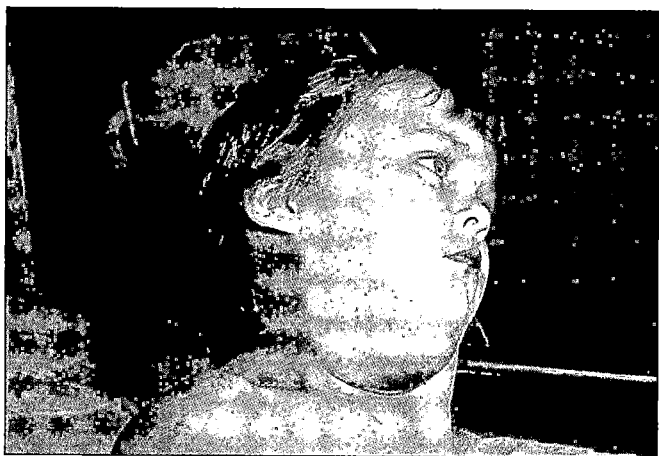

FIGURE 25.13 A case of mumps. This patient shows the typical swelling of mumps.

■ **How is the mumps virus transmitted?**

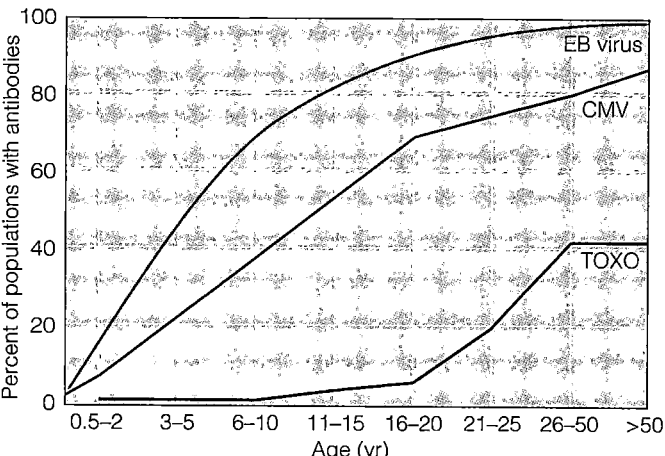

FIGURE 25.14 The typical U.S. prevalence of antibodies against Epstein-Barr (EB) virus, cytomegalovirus (CMV), and *Toxoplasma gondii* (TOXO) by age.
SOURCE: *Laboratory Management,* June 1987, p. 23ff.

■ **It is estimated that about 80% of the population eventually carries the cytomegalovirus.**

during the first 48 hours before clinical symptoms appear. Once the viruses have begun to multiply in the respiratory tract and local lymph nodes in the neck, they reach the salivary glands via the blood. Viremia (the presence of virus in the blood) begins several days before the onset of mumps symptoms and before the virus appears in saliva. The virus is present in the blood and saliva for 3–5 days after the onset of the disease and in the urine after about 10 days.

Mumps is characterized by inflammation and swelling of the parotid glands, fever, and pain during swallowing. About 4–7 days after the onset of symptoms, the testes can become inflamed, a condition called *orchitis*. This happens in about 20–35% of males past puberty. Sterility is a possible but rare consequence. Other possible complications include meningitis, inflammation of the ovaries, and pancreatitis.

An effective attenuated live vaccine is available and is often administered as part of the trivalent measles, mumps, rubella (MMR) vaccine. The number of cases of mumps has dropped sharply since the introduction of the vaccine in 1968. Second attacks are rare, and cases involving only one parotid gland, or subclinical cases (about 30% of those infected), are as effective as bilateral mumps in conferring immunity.

Serological diagnosis is not usually necessary. If confirmation of a diagnosis based only on symptoms is desired, the virus can be isolated by embryonated egg or cell culture techniques and identified by ELISA tests.

Cytomegalovirus (CMV) Inclusion Disease

Cytomegalovirus inclusion disease is caused by a cytomegalovirus (CMV), a very large herpesvirus that

causes nuclear inclusions and swelling, or *cytomegaly*, of host cells. The official name is human herpesvirus 5. These swollen cells are known as "owl's eyes" because of their distinctive appearance. Once a person is infected with CMV, the infection persists for life. Antibodies are formed but fail to clear the virus. Blood macrophages and T lymphocytes are the probable sites of latency. The virus is shed at intervals in such body secretions as saliva, urine, semen, cervical secretions, and breast milk.

Some estimate that 80% of the population of the United States may carry the virus. It has been said that if CMV were accompanied by a skin rash, it would be one of the better-known childhood diseases. Figure 25.14 plots the prevalence of antibodies against CMV, Epstein-Barr (EB) virus, and *Toxoplasma gondii*. In developing parts of the world, infection may approach 100%.

The disease can be spread by kissing and other personal contacts, especially among day-care workers. It can also be transmitted sexually, by transfused blood, and by transplanted tissue. In adults and older children, most infections are subclinical or, at worst, much like a mild case of infectious mononucleosis. However, immunosuppressed individuals may develop life-threatening pneumonia. Commercial products are now available that contain antibodies that passively neutralize CMV present in donated kidneys.

A primary CMV infection acquired during pregnancy by a nonimmune mother can seriously harm the fetus. At the time of her first pregnancy, a woman of affluent background has about a 50% chance of being naturally immune to CMV infection; a woman of disadvantaged background, who has probably experienced poor sanitation and more crowding, has about an 80% chance. Viral transmission to the fetus by an immune mother will sometimes occur, but almost no damage ensues. Tests for immune status are readily available, and physicians should determine the immune status of female patients of childbearing age. All nonimmune women should be informed of the risks of pregnancy; each year, nearly 8000 American infants are born with symptomatic disease; many die, and the survivors are severely damaged. The main source of infection is young children.

A reliable vaccine is not yet available. The virus is susceptible to the antiviral drug ganciclovir. It has beneficial effects when used on immunocompromised adult patients. One of the leading microbial complications for AIDS patients is reactivated CMV infection, especially affecting the eyesight; this is called *CMV retinitis.* One of the first antisense drugs (see Chapter 20 on page 571), fomivirsen, is nearing approval for treatment of CMV retinitis.

The most reliable diagnostic method is isolation of the virus from body fluids during the first 2 weeks of life; the sample is usually shipped to a central reference laboratory.

Hepatitis

Learning Objective

- *Differentiate among hepatitis A, hepatitis B, hepatitis C, hepatitis D, and hepatitis E.*

Hepatitis is an inflammation of the liver. Viral hepatitis is now the second most frequently reported infectious disease in the United States. At least five different viruses cause hepatitis, and probably more remain to be discovered or become better known. Hepatitis is also an occasional result of infections by other viruses such as EB virus or CMV. Drug and chemical toxicity can also cause acute hepatitis that is clinically identical to viral hepatitis. The characteristics of the various forms of viral hepatitis are summarized in Table 25.1.

table 25.1	**Characteristics of Viral Hepatitis**				
Characteristic	**Hepatitis A**	**Hepatitis B**	**Hepatitis C**	**Hepatitis D**	**Hepatitis E**
Route of transmission	Fecal-oral (ingestion of contaminated food and water)	Parenteral (injection of contaminated blood or other body fluids), including sexual contact	Parenteral	Parenteral (host must be coinfected with hepatitis B)	Fecal-oral
Causative agent	Hepatitis A virus (HAV); single-stranded RNA; no envelope	Hepatitis B virus (HBV); double-stranded DNA; envelope	Hepatitis C virus (HCV); single-stranded RNA; envelope	Hepatitis D virus (HDV); single-stranded RNA; envelope from HBV	Hepatitis E virus (HEV); single-stranded RNA; no envelope
Incubation period	2–6 weeks	4–26 weeks	2–22 weeks	6–26 weeks	2–6 weeks
Manifestations or symptoms	Mostly subclinical; severe cases: fever, headache, malaise, jaundice	Frequently subclinical; similar to HAV, but fever, no headache, and more likely to progress to severe liver damage	Similar to HBV, but more likely to become chronic	Severe liver damage; high mortality rate	Similar to HAV, but pregnant women may have high mortality rate
Antibody prevalence in U.S.	33%	5–10%	1.8%	Unknown	0.5%
Chronic liver disease	No	Yes	Yes	Yes	No
Vaccine	Inactivated vaccines; immune globulin provides temporary protection	Genetically engineered vaccine produced in yeasts	None	HBV vaccine is protective because coinfection required	Under development

Hepatitis A

The *hepatitis A virus (HAV)* is the causative agent of **hepatitis A.** HAV contains single-stranded RNA and lacks an envelope. It can be grown in cell culture.

After a typical entrance via the oral route, HAV multiplies in the epithelial lining of the intestinal tract. Viremia eventually occurs, and the virus spreads to the liver, kidneys, and spleen. The virus is shed in the feces and can also be detected in the blood and urine. The amount of virus excreted is greatest before symptoms appear and then declines rapidly. Therefore, a food handler responsible for spreading the virus might not appear to be ill at the time. The virus can probably survive for several days on such surfaces as cutting boards. Contamination of food or drink by feces is aided by the resistance of HAV to chlorine disinfectants at concentrations ordinarily used in water. Mollusks, such as oysters, that live in contaminated waters are also a source of infection.

At least 50% of infections with HAV are subclinical, especially in children. In clinical cases, the initial symptoms are anorexia (loss of appetite), malaise, nausea, diarrhea, abdominal discomfort, fever, and chills. These symptoms are more likely to appear in adults; they last 2–21 days, and the mortality rate is low. Nationwide epidemics occur about every 10 years, mostly in people under 14. In some cases, there is also jaundice, with yellowing of the skin and the whites of the eyes, and the dark urine typical of liver infections. In these cases the liver becomes tender and enlarged.

There is no chronic form of hepatitis A, and the virus is usually shed only during the acute stage of disease, although it is difficult to detect. The incubation time averages 4 weeks and ranges from 2–6 weeks, which makes epidemiological studies for the source of infections difficult. There are no animal reservoirs.

In the United States, the percentage of the population with HAV is much higher among lower socioeconomic groups (72–88%) than among middle and upper socioeconomic groups (18–30%). The 30,000 or more cases reported in the United States each year represent only a fraction of the actual number.

Acute disease is diagnosed by the detection of IgM anti-HAV, because these antibodies appear about 4 weeks after infection and disappear about 3–4 months after infection. Recovery results in lifelong immunity.

No specific treatment for the disease exists, but people at risk for exposure to hepatitis A can be given immune globulin, which provides protection for several months. Inactivated vaccines are now available and are recommended for travelers to areas of endemic disease. In 2000 the CDC suggested HAV vaccination for children in certain western states with historically high rates of the disease. Vaccination is also recommended for carriers of the more serious disease hepatitis C. Protection is desirable because the HAV infections superimposed upon the HCV infection may cause life-threatening liver damage.

Hepatitis B

Hepatitis B is caused by the *hepatitis B virus (HBV).* HBV and HAV are completely different viruses: HBV is larger, its genome is double-stranded DNA, and it is enveloped. HBV is a unique DNA virus; instead of replicating its DNA directly, it passes through an intermediate RNA stage resembling a retrovirus. Because HBV has often been transmitted by blood transfusions, this virus has been intensely studied in order to determine how to identify contaminated blood.

The serum from patients with hepatitis B contains three distinct particles. The largest, the *Dane particle,* is the complete virion; it is infectious and capable of replicating. There are also smaller *spherical particles,* about half the size of a Dane particle, and *filamentous particles,* which are tubular particles similar in diameter to the spherical particles but about ten times as long (Figure 25.15). The spherical and filamentous particles are unassembled components of Dane particles without nucleic acids; assembly is evidently not very efficient, and large numbers of these unassembled components accumulate. Fortunately, these numerous unassembled particles contain *hepatitis B surface antigen (HB$_s$Ag),* which can be detected with antibodies to them. Such antibody tests make convenient screening of blood for HBV possible.

Physicians, nurses, dentists, medical technologists, and others who are in daily contact with blood have a considerably higher incidence of hepatitis B than members of the general population. It is estimated that as many as 10,000 health care workers become infected each year in the United States. Federal regulations require that employees exposed to blood be offered free vaccinations by their employers. There have also been instances of transmission to patients by surgeons and dentists. Injecting drug users (IDU) often share needles and fail to sterilize them properly; as a consequence, they also have a high incidence of hepatitis B. Blood may contain up to a billion viruses per milliliter. Therefore, it is not surprising that it is also present in many body fluids, such as saliva, breast milk, and semen, but not in blood-free feces or urine. Transmission by semen donated for artificial insemination has been documented, and semen has been implicated in transmission between heterosexuals with multiple partners and in male homosexuals. Precautions taken to prevent HIV transmission have also had an effect on the incidence of HBV infections. A mother who is positive for HB$_s$Ag, especially if she is a chronic carrier,

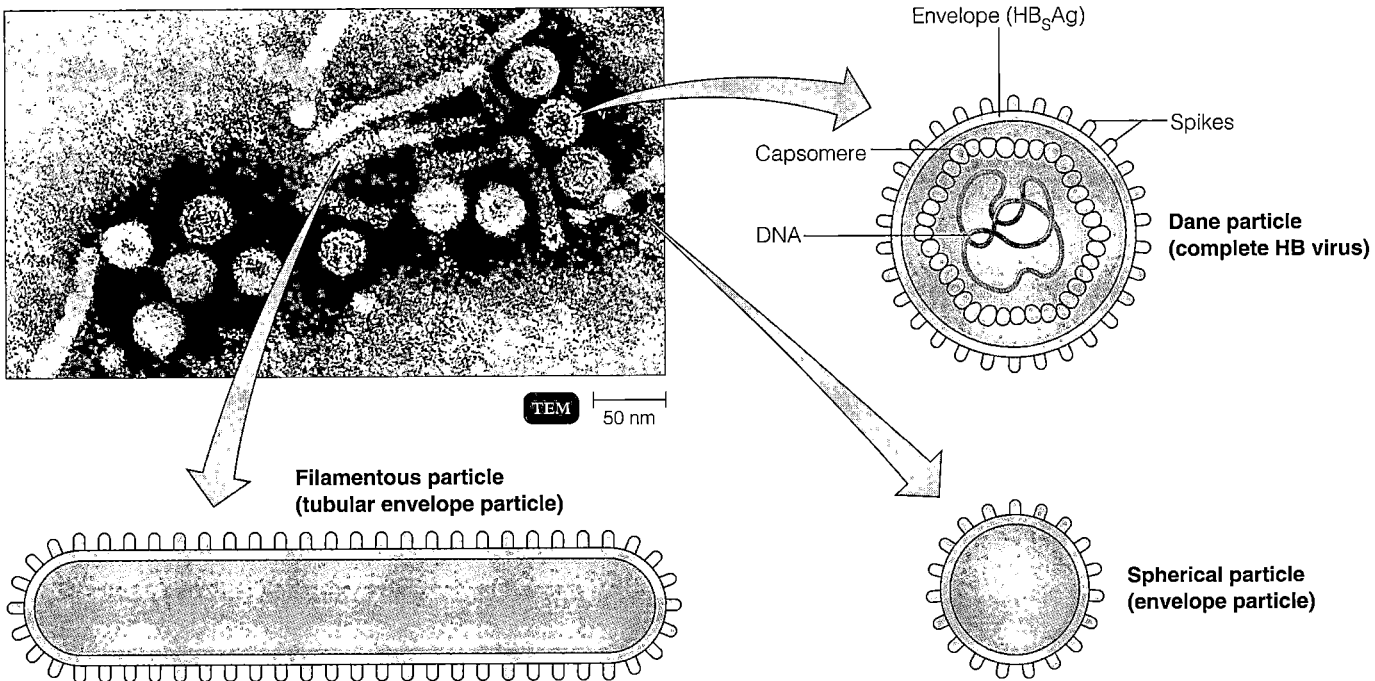

FIGURE 25.15 Hepatitis B virus (HBV). The micrograph and illustrations depict the distinct types of HBV particles discussed in the text.

■ What are some of the causes of hepatitis?

may transmit the disease to her infant, usually at birth. In most cases, this type of transmission can be prevented by administering hepatitis B immune globulin (HBIG) to the newborn immediately after birth. These babies should also be vaccinated.

The incubation period averages about 12 weeks; the range is 4–26 weeks. Because this incubation period is lengthy and of uneven duration, determining the origin of an infection can be difficult.

An estimated 130,000 Americans, mostly young adults, are infected with hepatitis B each year. Only about 10,000 cases are actually reported. About 5000 people die each year of HBV-related liver disease, ranging from cirrhosis (hardening and degeneration) to cancer. Clinical signs of hepatitis B vary widely. Probably about half the cases are entirely asymptomatic. Symptoms are highly variable and include, in the early stages, loss of appetite, low-grade fever, and joint pains. Later, jaundice usually appears. These clinical manifestations are more likely to appear in adults. It is difficult to distinguish between hepatitis A and B solely on clinical grounds.

At least 90% of acute hepatitis B infections end in complete recovery. The mortality rate is higher than that of hepatitis A, but it is probably less than 1% of the infected population. It might be as high as 2–3% among hospitalized patients, of which there are about 10,000 U.S. cases each year.

Overall, up to 10% of patients become chronic carriers. This is age-related; as many as 90% of infected newborns become carriers, but only 3–5% of adults. These carriers are reservoirs for transmission of the virus, and they also have a high rate of liver disease. There are an estimated 1.25 million HBV carriers in the United States. A special concern is the strong correlation between the occurrence of liver cancer and the incidence of chronic hepatitis B infections. Chronic carriers are about 200 times more likely to get liver cancer than the general population. Liver cancer is the most prevalent form of cancer in sub-Saharan Africa and the Far East, areas where hepatitis B is extremely common. Worldwide, there are an estimated 400 million HBV carriers.

Researchers have been unable to cultivate HBV in cell culture, a step that was necessary for the development of vaccines for polio, mumps, measles, and rubella. The available HBV vaccines use HB_sAg produced by a genetically engineered yeast. Vaccination is recommended for high-risk groups; a partial listing would include health care workers exposed to blood and blood products, people undergoing hemodialysis, patients and staff at mental health care institutions, and homosexually active males. In the United States, vaccination for HBV is now part of the childhood immunization schedule.

Treatments for HBV infection are currently limited. Persons with chronic HBV infections receive α-interferon

(α-IFN), which is expensive but results in improvement in a significant number of recipients. Several nucleoside analog drugs such as lamivudine are nearing approval for this application. Tests have shown that lamivudine often arrests progress of chronic HBV—but does not eradicate the virus.

Hepatitis C

In the 1960s, a previously unsuspected form of transfusion-transmitted hepatitis, now called **hepatitis C,** appeared. As we have mentioned, a reliable test for detection of HBV in blood has been in regular use and transfusions as a source of HBV have become rare. As a consequence, however, the new form of hepatitis soon constituted almost all transfusion-transmitted hepatitis. Eventually, serological tests to detect hepatitis C virus (HCV) antibodies were developed that similarly reduced the transmission of HCV to very low levels. However, there is a delay of about 70 to 80 days between infection and the appearance of detectable HCV antibodies. The presence of HCV in contaminated blood cannot be detected during this interval and about 1 in 100,000 transfusions can still result in infection. Blood-collecting facilities in the United States are introducing new tests, and eventually HCV-contaminated blood should be detectable within only 10 to 30 days of infection. If the virus can be detected directly without the need for development of antibodies, no delay would exist. A PCR test that can detect viral RNA is being introduced.

HCV has a single strand of RNA and is enveloped. It is capable of rapid genetic variation to evade the immune system. This characteristic, along with the fact that currently it cannot be cultured in vitro, will probably complicate the search for an effective vaccine.

Hepatitis C has been described as a silent epidemic, killing more people than AIDS in the United States. It is often clinically inapparent—few people have recognizable symptoms until about 20 years have elapsed. Even today, probably only a minority of infections have been diagnosed. Often, hepatitis C is detected only during some routine testing, such as for insurance or blood donation. A majority of cases, perhaps as high as 85%, progress to chronic hepatitis, a much higher rate than with HBV. Surveys indicated an estimated 1.8%, or nearly 4 million people, of the U. S. population is infected. More than 100,000 people are newly infected each year, and more than 8000 die. About 20% of chronically infected patients develop liver cirrhosis or liver cancer. Hepatitis C is probably the major reason for liver transplantation. This situation has led to concern that a "time bomb" of life-threatening liver disease may go off during the next couple of decades as HCV infections progress.

A common source of infection is the sharing of injection equipment among IDU. At least 80% of this group is infected with HCV. In one exceptional case the disease was transmitted by means of a straw shared for inhaling cocaine. Interestingly, in more than one-third of the cases, a mode of transmission—by contaminated blood, sexual contact, or other means—cannot be identified.

Some chronic cases of hepatitis C respond to treatment with α-IFN, but relapses are frequent. One new commercial drug combines interferon with the antiviral ribavirin. This treatment is expensive, and only about 30% of cases have a favorable sustained response.

Hepatitis D (Delta Hepatitis)

In 1977, a new hepatitis virus, now known as *hepatitis D virus (HDV),* was discovered in carriers of HBV in Italy. People who carried this so-called *delta antigen* and were also infected with HBV had a much higher incidence of severe liver damage and a much higher mortality rate than people who had antibodies against HBV alone. With time, it became clearer that **hepatitis D** can occur as either acute (*coinfection form*) or chronic (*superinfection form*) hepatitis. In people with a case of self-limiting acute hepatitis B, coinfection with HDV disappeared as the HBV was cleared from the system, and the condition resembled a typical case of acute hepatitis B. However, if the HBV infection progressed to the chronic stage, superinfection with HDV was often accompanied by progressive liver damage and a fatality rate several times that of people infected with HBV alone.

Epidemiologically, hepatitis D is linked to the epidemiology of hepatitis B. In the United States and northern Europe, the disease occurs predominantly in high-risk groups such as injecting drug users.

Structurally, the HDV antigen is a single strand of RNA, which is shorter than in any other animal-infecting virus. This particle is not capable of causing an infection. It becomes infectious when an external envelope of HB_sAg, whose formation is controlled by the genome of HBV, covers the HDV protein core (the delta antigen) (see Figure 25.15).

Hepatitis E

Hepatitis E is spread by fecal-oral transmission, much like hepatitis A, which it clinically resembles. The pathogen, known as *hepatitis E virus (HEV),* is endemic in areas of the world with poor sanitation, especially India and southeast Asia. It resembles HAV in being a nonenveloped virus with a single strand of RNA but is not related serologically to it. Like hepatitis A virus, HEV does not cause chronic liver disease, but for some unexplained reason it is responsible for a mortality rate in excess of 20% in pregnant women.

Other Types of Hepatitis

New techniques in molecular biology and serology have provided evidence of blood-transmitted viruses known as hepatitis F and hepatitis G. The virus of hepatitis G is related to HCV and is similarly widespread, but it apparently causes no significant disease condition. The presence of even other hepatitis viruses is suspected.

Viral Gastroenteritis

Learning Objective

- List the causative agents, mode of transmission, and symptoms of viral gastroenteritis.

A number of viruses, such as polioviruses, echoviruses, and coxsackieviruses, are transmitted by the fecal-oral route. However, despite the name *enteroviruses,* these viruses generally do not directly affect the digestive system. About 90% of cases of viral gastroenteritis are caused by rotavirus or by the Norwalk agent.

Rotavirus (Figure 25.16) is the most common cause of viral gastroenteritis. It is estimated to cause about 3 million cases, but fewer than 100 deaths, every year in the United States, mostly among small children. Worldwide, almost all children become infected by their first birthday. The mortality is much higher in less developed areas because rehydration therapy is not as available. In most cases, following an incubation period of 2–3 days, the patient suffers from low-grade fever, diarrhea, and vomiting, which persists for about a week. There is more than one serotype of rotovirus, and immunity upon recovery is only partly effective. An ELISA and several other serological tests can be used to detect the virus in feces.

Major epidemics of viral gastroenteritis have been caused by a virus known as the **Norwalk agent** (named after an outbreak in Norwalk, Ohio, in 1968). Following a 2-day incubation period, infected people typically suffer from nausea, abdominal cramps, diarrhea, and vomiting for 1–3 days. About 50% of middle-aged Americans show evidence, by serum antibodies, that they have been infected. One estimate is that two-thirds of foodborne illness, or more than 9,000,000 cases, in the United States is caused by viruses categorized as *Norwalk-like.* Susceptibility to the virus varies considerably; some people are not affected and do not even respond with an immune reaction. Others develop gastroenteritis and an immunity that might be only short-term.

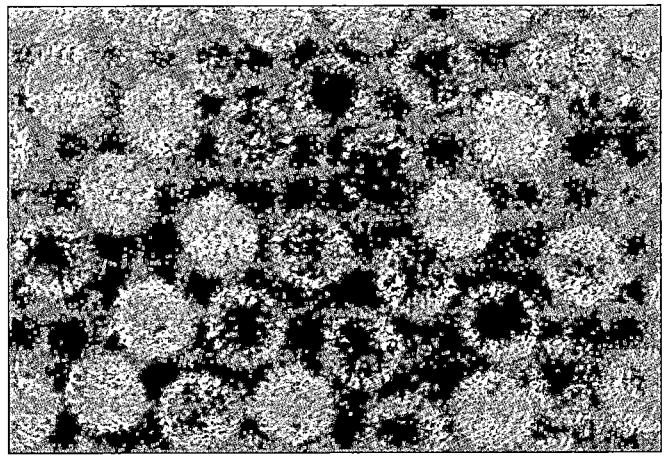

FIGURE 25.16 Rotavirus. This electron micrograph shows the morphology of the rotavirus (*rota* = wheel) that gives it its name.

The only treatment for viral gastroenteritis is oral rehydration or, in exceptional cases, intravenous rehydration.

Fungal Diseases of the Digestive System

Learning Objective

- Identify the causes of ergot poisoning and aflatoxin poisoning.

Some fungi produce toxins called *mycotoxins.* These toxins cause blood diseases, nervous system disorders, kidney damage, liver damage, and even cancer.

Ergot Poisoning

Some mycotoxins are produced by *Claviceps purpurea,* a fungus causing smut infections on grain crops. The mycotoxins produced by *C. purpurea* cause **ergot poisoning,** or *ergotism,* which results from the ingestion of rye or other cereal grains contaminated with the fungus. Ergot poisoning was very widespread during the Middle Ages. The toxin can restrict blood flow in the limbs, and gangrene results. It may also cause hallucinogenic symptoms, producing bizarre behavior similar to that caused by LSD.

Aflatoxin Poisoning

Aflatoxin is a mycotoxin produced by the fungus *Aspergillus flavus,* a common mold. Aflatoxin has been found in many foods but is particularly likely to be found on peanuts. **Aflatoxin poisoning** can cause serious damage

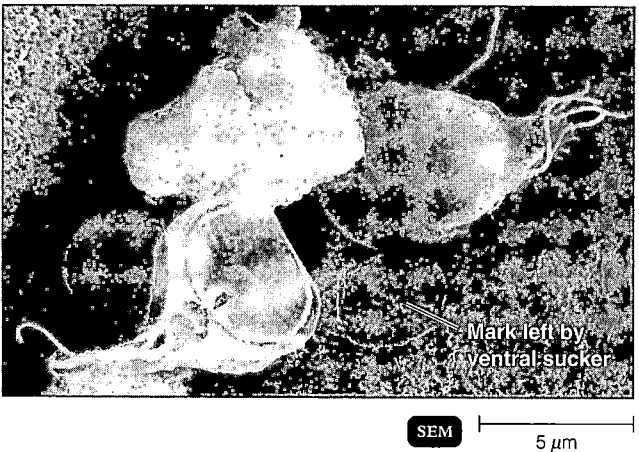

SEM ├─────────┤
                5 μm

FIGURE 25.17 The trophozoite form of *Giardia lamblia*, the flagellated protozoan that causes giardiasis. Notice the circular mark left on the intestinal wall by the ventral sucker disc the parasite uses to attach itself. The dorsal side is smoothly streamlined, and the intestinal contents move easily around the attached microorganism.

▨ What is the string test for giardiasis?

to livestock when their feed is contaminated with *A. flavus*. Although the risk to humans is unknown, there is strong evidence that aflatoxin contributes to cirrhosis of the liver and cancer of the liver in parts of the world, such as India and Africa, where food is subject to aflatoxin contamination.

Protozoan Diseases of the Digestive System

Learning Objective

■ List the causative agents, modes of transmission, symptoms, and treatments for giardiasis, amoebic dysentery, cryptosporidiosis, and Cyclospora diarrheal infection.

Several pathogenic protozoa complete their life cycles in the human digestive system. Usually they are ingested as resistant, infective cysts and are shed in greatly increased numbers as newly produced cysts.

Giardiasis

Giardia lamblia is a flagellated protozoan that is able to attach firmly to a human's intestinal wall (Figure 25.17). In 1681, van Leeuwenhoek described them as having "bodies...somewhat longer than broad and their belly, which was flatlike, furnished with sundry little paws." Taxonomically, there is some confusion and *G. lamblia* may also be referred to as *G. duodenalis* or *G. intestinalis*.

G. lamblia is the cause of **giardiasis,** a prolonged diarrheal disease. Sometimes persisting for weeks, giardiasis

is characterized by malaise, nausea, flatulence (intestinal gas), weakness, weight loss, and abdominal cramps. The distinctive odor of hydrogen sulfide can often be detected in the breath or stools. The protozoa, reproducing by binary fission, sometimes occupy so much of the intestinal wall that they interfere with food absorption.

Outbreaks of giardiasis in the United States occur often. About 7% of the population are healthy carriers and shed the cysts in their feces. The pathogen is also shed by a number of wild mammals, especially beavers, and the disease occurs in backpackers who drink from wilderness waters.

Most outbreaks are transmitted by contaminated water supplies. In a recent national survey of surface waters serving as sources for U.S. municipalities, the protozoan was detected in 18% of the samples. Because the cyst stage is relatively insensitive to chlorine, filtration or boiling of water supplies is usually necessary to eliminate the cysts from water.

G. lamblia is not reliably found in stools by microscopic examination, the *string test* is sometimes used for diagnosis. In this test, a gelatin capsule packed with about 140 cm of fine string is swallowed by the patient. One end of the string is taped to the cheek. The gelatin capsule dissolves in the stomach, and an enclosed weighted rubber bag attached to the other end of the string enters the upper bowel. After a few hours, the string is drawn up through the mouth and examined for trophozoite forms of *G. lamblia*. Several commercially available ELISA tests detect both ova and the parasite in stool specimens. These tests are especially useful for epidemiological screening. Testing of drinking water for *Giardia* is difficult, but often necessary to prevent or trace disease outbreaks. These tests are frequently combined with tests for *Cryptosporidium* protozoa, discussed in the next section.

Treatment with metronidazole or quinacrine hydrochloride is usually effective within a week.

Cryptosporidiosis

Cryptosporidiosis is caused by the protozoan *Cryptosporidium parvum,* which was not recognized as a pathogen of humans until 1976. Infection occurs when humans ingest the cryptosporidian oocysts (Figure 25.18). The oocysts eventually release sporozoites in the small intestine. The motile sporozoites invade the epithelial cells of the intestine and undergo a cycle that eventually releases oocysts to be excreted in the feces. (Compare with the similar life cycle of toxoplasmosis in Figure 23.21 on page 643). The disease is a choleralike diarrhea of 10–14 days duration. In immunodeficient individuals, including AIDS patients, the diarrhea becomes progressively worse and is life-threatening. Other than oral rehydration, there is no satisfactory treatment.

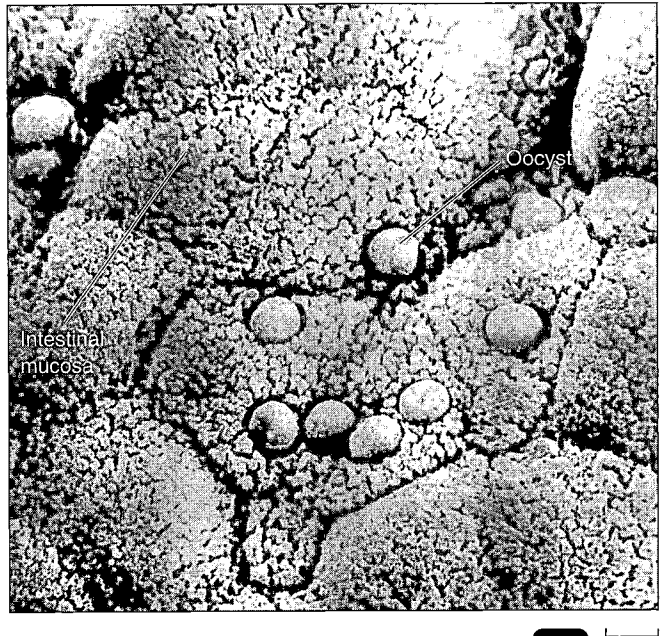

SEM |———|
          5 µm

FIGURE 25.18 Cryptosporidiosis. Oocysts of *Cryptosporidium parvum* are shown here embedded in the intestinal mucosa.

▨ How is cryptosporidiosis transmitted?

The infection is transmitted to humans largely through recreational and drinking water systems contaminated with oocysts of *Cryptosporidium,* mostly from animal wastes, especially cattle. Studies in the United States show that many, if not most, lakes, streams, and even wells are contaminated. The oocysts, like the cysts of *G. lamblia,* are resistant to chlorination and must be removed from water by filtration. Even filtration sometimes fails. This is especially true of swimming pools, where both chlorination and filtration systems are ineffective in removing oocysts. An infective dose may be as low as ten oocysts. Fecal-oral transmission resulting from poor sanitation also occurs; many outbreaks have occurred in day-care settings.

Because of the low density of the parasite in most samples, concentration by some method is required. The parasite is then identified by a modified acid-fast staining procedure, or some serological method such as fluorescent-antibody (FA) or ELISA assays. These diagnostic tests have a low sensitivity.

Testing of water is important, but the currently available methods have been described as being cumbersome, time-consuming, and inefficient. Most widely used is an FA test that can simultaneously detect both *G. lamblia* cysts and *C. parvum* oocysts. Regular water testing will probably become mandatory, and research for simpler and more reliable methods has a high priority in public health science.

Cyclospora Diarrheal Infection

A series of diarrheal disease outbreaks caused by relatively unknown protozoa has occurred in recent years. The pathogen has since been named *Cyclospora cayetanensis.*

The symptoms of ***Cyclospora* diarrheal infection** are a few days of watery diarrhea, but in some cases it may persist for weeks. The disease is especially debilitating for immunosuppressed people, such as AIDS patients. Whether humans are the only host for the protozoan is uncertain. Most outbreaks have been associated with the ingestion of oocysts in water, on contaminated berries, or similar uncooked foods. The foods are presumed to have been contaminated by oocysts shed in human feces, or possibly from birds in the field.

Microscopic examination can identify the oocysts, which are about twice the diameter of those of *Cryptosporidium.* There is really no satisfactory test for contamination of foods. The antibiotic combination of trimethoprim and sulfamethoxazole is used for treatment.

Amoebic Dysentery (Amoebiasis)

Amoebic dysentery, or **amoebiasis,** is spread mostly by food or water contaminated by cysts of the protozoan amoeba *Entamoeba histolytica* (see Figure 12.17b on page 351). Although stomach acid can destroy trophozoites, it does not affect the cysts. In the intestinal tract the cyst wall is digested away, and the trophozoites are released. They then multiply in the epithelial cells of the wall of the large intestine. A severe dysentery results, and the feces characteristically contain blood and mucus. The trophozoites feed on tissue in the gastrointestinal tract (Figure 25.19).

Severe bacterial infections result if the intestinal wall is perforated. Abscesses might have to be treated surgically, and the invasion of other organs, particularly the liver, is not uncommon. Perhaps 5% of the U.S. population are asymptomatic carriers of *E. histolytica.* Worldwide, one person in ten is estimated to be infected, mostly asymptomatically, and about 10% of these infections progress to the more serious stages.

Diagnosis largely depends on recovering and identifying the pathogens in feces. (Red blood cells, ingested as the parasite feeds on intestinal tissue and observed within the trophozoite stage of an amoeba, help identify *E. histolytica.*) Several serological tests can also be used for diagnosis, including latex agglutination and fluorescent-antibody tests. Such tests are especially useful when the affected areas are outside the intestinal tract and the patient is not passing amoebae.

Metronidazole plus iodoquinol are the drugs of choice in treatment.

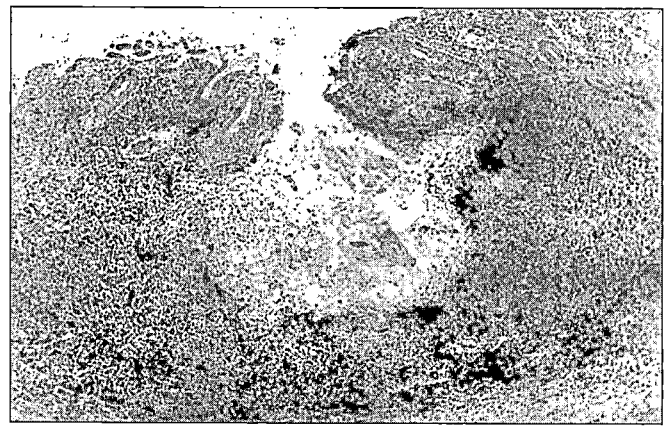

FIGURE 25.19 Section of intestinal wall showing a typical flask-shaped ulcer caused by *Entamoeba histolytica*.

■ Hydrochloric acid in gastric juice does not destroy the cysts of *E. histolytica*.

Helminthic Diseases of the Digestive System

Learning Objective

■ List the causative agents, modes of transmission, symptoms, and treatments for tapeworms, hydatid disease, pinworms, hookworms, ascariasis, and trichinosis.

Helminthic parasites are very common in the human intestinal tract, especially under conditions of poor sanitation. Figure 25.20 shows the worldwide incidence of infection with some intestinal helminths. In spite of their size and formidable appearance, they often produce few symptoms. They have become so well adapted to their human hosts, and vice versa, that when their presence is revealed, it is often a surprise.

Tapeworms

The life cycle of a typical **tapeworm** extends through three stages (see Figure 12.25 on page 360). The adult worm lives in the intestine of a human host, where it produces eggs that are excreted in the feces. The eggs are ingested by animals such as grazing cattle, where the egg hatches into a larval form called a *cysticercus* (plural cysticerci) that lodges in the animal's muscles. Human infections by tapeworms begin with the consumption of undercooked beef, pork, or fish containing cysticerci. The cysticerci develop into adult tapeworms that attach to the intestinal wall by suckers on the scolex (see the photo in Figure 12.25). The adult beef tapeworm, *T. saginata,* can

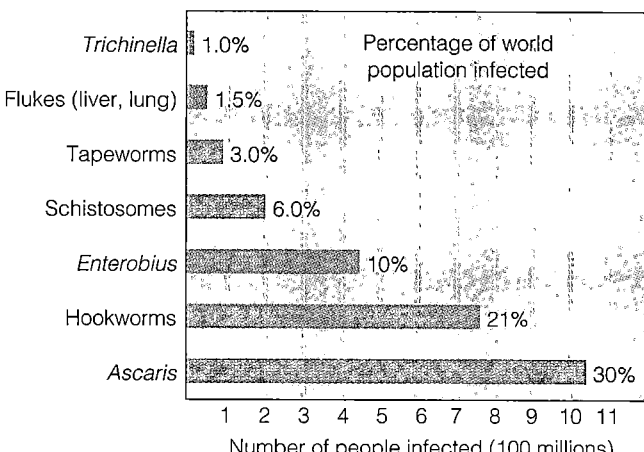

FIGURE 25.20 The worldwide prevalence of human infections with selected intestinal helminths.
SOURCE: World Health Organization (WHO)

live in the human intestine for 25 years and reaches a length of 6 meters (18 feet) or longer. Even a worm of this size seldom causes significant symptoms beyond a vague abdominal discomfort. There is, however, psychological distress when a meter or more of detached segments (proglottids) break loose and unexpectedly slip out of the anus, which happens occasionally.

Taenia solium (sō'lē-um), the pork tapeworm, has a life cycle similar to the beef tapeworm. An important difference is that *T. solium* may produce the larval stage in the human host, a disease condition called **neurocysticercosis**. In this disease, the infection can arise from the tapeworm egg rather than only from eating undercooked meat containing cysticerci. The eggs can be ingested as a result of poor sanitary practices or possibly by autoinfection, when eggs produced by the adult tapeworm in the intestinal tract somehow enter the stomach.

Cysticerci can develop in many human organs. When this occurs in muscle tissue, the symptoms are seldom severe. Cysticerci that develop in the eyes and in the brain, however, may be more serious (Figure 25.21). Neurocysticercosis, which is endemic in Mexico and Central America, has become a fairly common condition in parts of the United States with large Mexican and Central American immigrant populations.

The symptoms often mimic those of a brain tumor. The number of cases reported reflects, in part, the use of CT (computed tomography) scanning in diagnosis. These expensive machines X-ray the body in continuous "slices." In endemic areas, neurological patients can be screened with serological tests for antibodies to *T. solium*.

The fish tapeworm *Diphyllobothrium latum* (dī-fil-lō-bo'thrē-um lā'tum) is found in pike, trout, perch, and salmon. The CDC has issued warnings about the risks of fish tapeworm infection from the increasingly popular

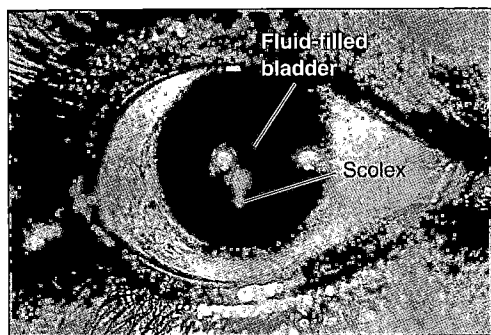

FIGURE 25.21 Neurocysticercosis in humans. Cysticerci can occur in many tissues; this one has formed in an eye.

■ How is human infestation with tapeworms diagnosed?

sashimi (Japanese dishes prepared from raw fish). To relate a vivid example, about 10 days after eating, one person developed symptoms of abdominal distention, flatulence, belching, intermittent abdominal cramping, and diarrhea. Eight days later, the patient passed a tapeworm 1.2 m (4 ft) long, identified as a species of *Diphyllobothrium*.

Laboratory diagnosis of tapeworms consists of identifying the tapeworm eggs or segments in feces. The drug of choice for eliminating tapeworms is niclosamide.

Hydatid Disease

Not all tapeworms are large. One of the most dangerous is *Echinococcus granulosus*, which is only a few millimeters in length (see Figure 12.26a on page 361). Humans are not the definitive hosts. The adult form lives in the intestinal tract of carnivorous animals, such as dogs and wolves. Typically, humans become infected from the feces of a dog that has become infected by eating the flesh of a sheep or deer containing the cyst form of the tapeworm. Unfortunately, humans can be an intermediate host, and cysts can develop in the body. The disease occurs most frequently in people who raise sheep or hunt or trap wild animals.

Once ingested by a human, the eggs of *E. granulosus* may migrate to various tissues of the body. The liver and lungs are the most common sites, but the brain and numerous other sites also may be infected. Once in place, the egg develops into a **hydatid cyst** that can grow to a diameter of 1 cm in a few months (Figure 25.22). In some locations, cysts may not be apparent for many years. Some, where they are free to expand, become enormous, containing up to 15 liters (4 gallons) of fluid.

Damage may arise from the size of the cyst in such areas as the brain or the interior of bones. If the cyst ruptures in the host, it can lead to the development of a great many daughter cysts. Another factor in the pathogenicity

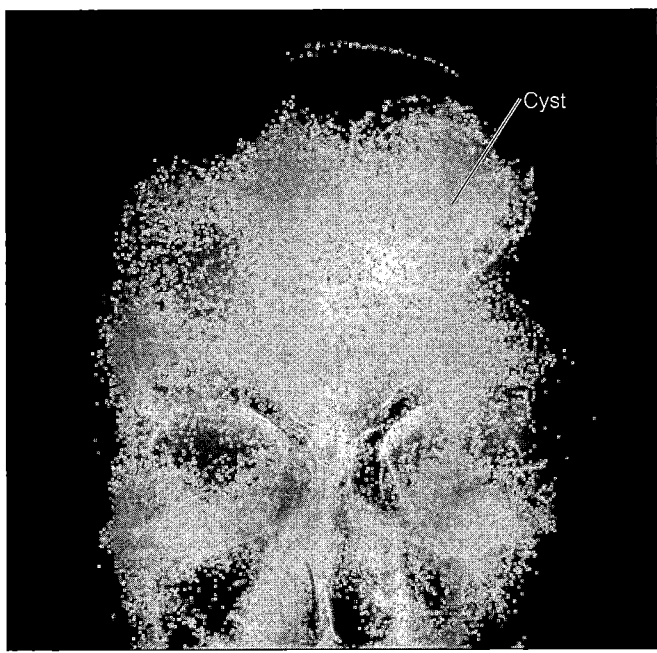

FIGURE 25.22 A hydatid cyst formed by *Echinococcus granulosus*. A large cyst can be seen in this X ray of the brain of an infected individual.

■ How do hydatid cysts affect the body?

of such cysts is that the fluid contains proteinaceous material, to which the host becomes sensitized. If the cyst suddenly ruptures, the result can be life-threatening anaphylactic shock.

Treatment is usually surgical removal, but care must be taken to avoid release of the fluid and the potential spread of infection or anaphylactic shock. If removal is not feasible, the drug albendazole can kill the cysts.

Nematodes

Pinworms

Most of us are familiar with the **pinworm,** *Enterobius vermicularis* (see Figure 12.27 on page 362). This tiny worm migrates out of the anus of the human host to lay its eggs, causing local itching. Whole households may become infected. Such drugs as pyrantel pamoate (often available without a prescription) and mebendazole are usually effective in treatment.

Hookworms

Hookworm infections were once a very common parasitic disease in the southeastern states. In the United States, the species most often seen is *Necator americanus* (see Figure 12.28 on page 363). Another species, *Acyclostoma duodenale*, is widely distributed around the world.

The hookworm attaches to the intestinal wall and feeds on blood and tissue rather than on partially digested

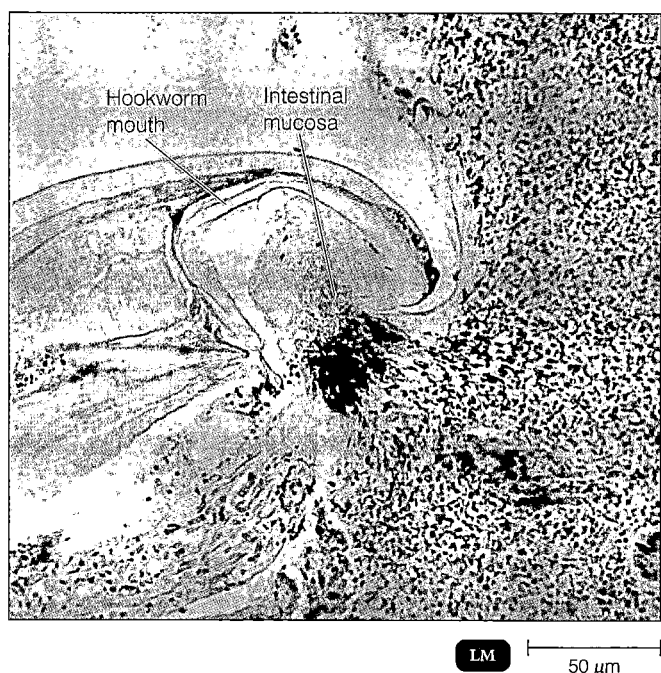

LM |————————|
50 μm

FIGURE 25.23 An *Ancylostoma* hookworm attached to intestinal mucosa. Notice how the mouth of the worm is adapted to feeding on the tissue.

■ **Why can a hookworm infection lead to anemia?**

food (Figure 25.23), so the presence of large numbers of worms can lead to anemia and lethargic behavior. Heavy infections can also lead to a bizarre symptom known as *pica,* a craving for peculiar foods, such as laundry starch or soil containing a certain type of clay. Pica is a symptom of iron deficiency anemia.

Because the life cycle of the hookworm requires human feces to enter the soil and bare skin to contact contaminated soil, the incidence of the disease has declined greatly with improved sanitation and the practice of wearing shoes. Hookworm infections can be treated effectively with mebendazole.

Ascariasis

One of the most widespread helminthic infections is **ascariasis,** caused by *Ascaris lumbricoides.* This condition is familiar to many American physicians. As described in Chapter 12 on page 362, diagnosis is often made when an adult worm emerges from the anus, mouth, or nose. These worms can be quite large, up to 30 cm (about 1 ft) in length (Figure 25.24). In the intestinal tract, they live on partially digested food and cause few symptoms.

The worm's life cycle begins when eggs are shed in a person's feces and, under poor sanitary conditions, are ingested by another person. In the upper intestine, the eggs

FIGURE 25.24 *Ascaris lumbricoides,* the cause of ascariasis. This photograph shows a male (the smaller worm with the coiled end) and a female. These worms are up to 30 cm (more than 1 foot) in length.

▨ **What are the principal features of the life cycle of *A. lumbricoides*?**

hatch into small wormlike larvae that pass into the bloodstream and then into the lungs. There they migrate into the throat and are swallowed. The larvae develop into egg-laying adults in the intestines. (All this migration just to return to the place where they started!)

In the lungs, the tiny larvae may cause some pulmonary symptoms. Extremely large numbers may block the intestine, bile duct, or pancreatic duct. The worms do not usually cause severe symptoms, but their presence can be manifested in distressing ways. The most dramatic consequences of infection with *A. lumbricoides* are from the migrations of adult worms. Worms have been known to leave the body of small children through the umbilicus (navel) and to escape through the nostrils of a sleeping person. Once ascariasis is diagnosed, it can be effectively treated with mebendazole.

Trichinosis

Most infections by the small roundworm *Trichinella spiralis,* called **trichinosis,** are insignificant. The larvae, in encysted form, are located in muscles of the host. In 1970, routine autopsies of human diaphragm muscles showed that about 4% of those tested carried this parasite.

The severity of the disease is generally proportional to the number of larvae ingested. Ingesting undercooked pork is probably the most common mode of infection (Figure 25.25), but eating the flesh of animals that feed on garbage (bears, for example) is increasingly causing outbreaks. Quite a few human cases of trichinosis have occurred in France from horsemeat infected in the United

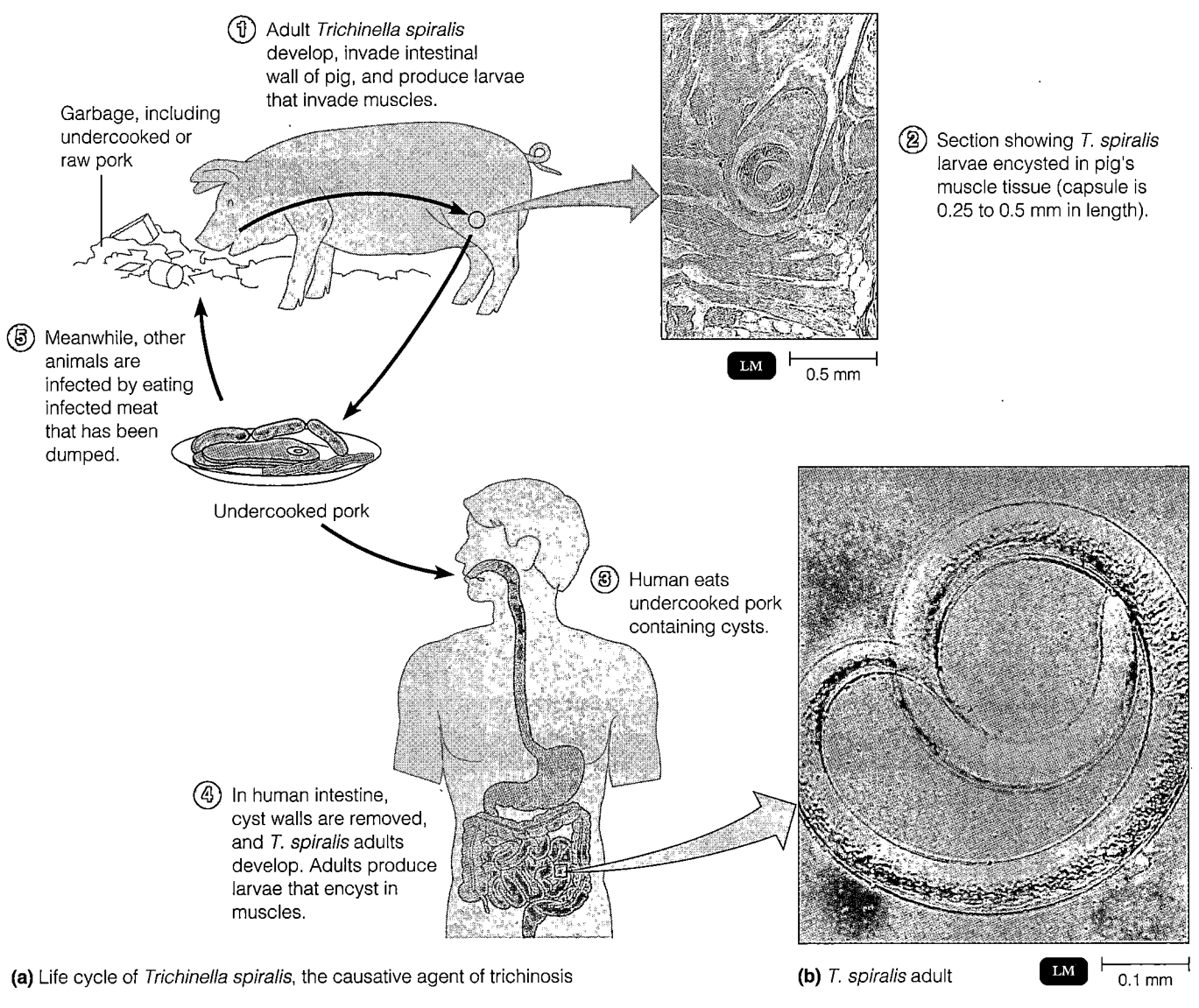

① Adult *Trichinella spiralis* develop, invade intestinal wall of pig, and produce larvae that invade muscles.

Garbage, including undercooked or raw pork

② Section showing *T. spiralis* larvae encysted in pig's muscle tissue (capsule is 0.25 to 0.5 mm in length).

LM   0.5 mm

⑤ Meanwhile, other animals are infected by eating infected meat that has been dumped.

Undercooked pork

③ Human eats undercooked pork containing cysts.

④ In human intestine, cyst walls are removed, and *T. spiralis* adults develop. Adults produce larvae that encyst in muscles.

(a) Life cycle of *Trichinella spiralis*, the causative agent of trichinosis

(b) *T. spiralis* adult   LM   0.1 mm

FIGURE 25.25   Life cycle of *Trichinella spiralis*.

■ What is the most common vehicle of infection of *T. spiralis*?

States and exported to restaurants. Severe cases can be fatal—sometimes in only a few days.

Any ground meat can be contaminated from machinery previously used to grind contaminated meats. Eating raw sausage or hamburger is a risky habit. One person acquired trichinosis by chewing the fingernails after handling infected pork. Prolonged freezing of meats containing *Trichinella* tends to eliminate the worms, but freezing should not be considered a substitute for thorough cooking.

In the muscles of hosts such as pigs, the *T. spiralis* larvae are encysted in the form of short worms about 1 mm in length. When the flesh of an infected animal is ingested by humans, the cyst wall is removed by digestive action in

the intestine. The organism then matures into the adult form. The adult worms spend only about a week in the intestinal mucosa and produce larvae that invade tissue. Eventually, the encysted larvae localize in muscle (common sites include the diaphragm and eye muscles), where they are barely visible in biopsied specimens.

Symptoms of trichinosis include fever, swelling around the eyes, and gastrointestinal upset. Small hemorrhages under the fingernails are often observed. Biopsy specimens, as well as a number of serological tests, can be used in diagnosis. Recently, a serological ELISA test that detects the parasite in meats has been developed. Treatment consists of administering mebendazole to kill intestinal worms and corticosteroids to reduce inflammation.

table 25.2	Microbial Diseases of the Digestive System	
Disease	Pathogen	Comments
Bacterial Diseases of the Mouth		
Dental caries	Primarily *Streptococcus mutans*	Accumulations of plaque allow localized acid production by bacteria, forming hole in tooth.
Periodontal disease	Various; primarily *Porphyromonas* spp.	Presence of bacterial plaque initiates inflammatory response that destroys bone and tissue.
Bacterial Diseases of the Lower Digestive System		
Staphylococcal food poisoning	*Staphylococcus aureus*	An exotoxin in food causes rapid onset of nausea, vomiting, and diarrhea.
Shigellosis (bacillary dysentery)	*Shigella* spp.	Bacteria are shed in human feces; ingested, they invade and multiply in intestinal epithelial cells. Infection spreads to neighboring cells, causing tissue damage and dysentery.
Salmonellosis	*Salmonella enterica*	Bacterial inhabitants of animal intestinal tracts contaminate foods; when ingested, they invade and multiply in intestinal epithelial cells. They do not invade neighboring cells but can enter the bloodstream, causing nausea and diarrhea.
Typhoid fever	*Salmonella typhi*	Typhoid pathogen is shed in human feces; incubation period about 2 weeks. Symptoms include high fever, disseminated infection, significant mortality rate.
Cholera	*Vibrio cholerae*	Exotoxin causes diarrhea with large loss of water and electrolytes; no invasion of tissue.
Vibrio parahaemolyticus gastroenteritis	*V. parahaemolyticus*	Exotoxin causes choleralike diarrhea, but generally milder.
V. vulnificus gastroenteritis	*V. vulnificus*	Very dangerous for people suffering from liver disease.
Enterotoxigenic *E. coli* gastroenteritis	*Escherichia coli*	Watery diarrhea that resembles mild form of cholera; typical traveler's diarrhea.
Enteroinvasive *E. coli* gastroenteritis	*E. coli*	Enterotoxin causes *Shigella*-like dysentery.
Enterohemorrhagic *E. coli* gastroenteritis	*E. coli* O157:H7	Causes hemorrhagic colitis (very bloody stools) and hemolytic uremic syndrome (blood in urine, possible kidney failure).
Campylobacter gastroenteritis	*Campylobacter jejuni*	Microaerophilic pathogen found in animal intestinal tracts; very common cause of gastroenteritis.
Helicobacter peptic ulcer disease	*Helicobacter pylori*	Pathogen is adapted to survive in stomach; presence leads to peptic ulcers.
Yersinia gastroenteritis	*Yersinia enterocolitica*	Pathogen is inhabitant of intestinal tract of animals; grows slowly at refrigerator temperatures. Symptoms are abdominal pain and diarrhea, usually mild. May be confused with appendicitis.
Clostridium perfringens gastroenteritis	*Clostridium perfringens*	Usually limited to diarrhea.
Bacillus cereus gastroenteritis	*Bacillus cereus*	May take form of diarrhea or nausea and vomiting; probably caused by different toxins.

In the past 10 years, the number of cases reported annually in the United States has varied from 16 to 129. Deaths are rare, and in most years there are none.

Table 25.2 summarizes the microbial diseases associated with the digestive system.

table 25.2   *(continued)*

Disease	Pathogen	Comments
Viral Diseases of the Digestive System		
Mumps	Mumps virus	Painful swelling of parotid glands.
Cytomegalovirus (CMV) inclusion disease	Cytomegalovirus (CMV)	Very common, mostly asymptomatic; if initial infection is acquired during pregnancy, can be damaging to fetus.
Hepatitis A	Hepatitis A virus (HAV)	Mild disease, mostly malaise; often subclinical. Fecal-oral transmission; low mortality rate.
Hepatitis B	Hepatitis B virus (HBV)	Transmitted by blood and other body fluids, including sexual activity. Severe disease likely to cause liver damage; about 10% of cases become chronic.
Hepatitis C	Hepatitis C virus (HCV)	Similar to hepatitis B but much more likely to become chronic.
Hepatitis D	Hepatitis D virus (HDV)	Very severe liver damage with high mortality rate. Must be coinfected with HBV.
Hepatitis E	Hepatitis E virus (HEV)	Similar to hepatitis A; fecal-oral transmission. Pregnant women may have high mortality rate.
Viral gastroenteritis	Rotavirus or Norwalk agent	Self-limiting.
Fungal Diseases of the Digestive System		
Ergot poisoning	Mycotoxin produced by *Claviceps purpurea*	Ingestion causes neurological or circulatory problems.
Aflatoxin poisoning	Mycotoxin produced by *Aspergillus flavus*	Mycotoxin probably contributes to liver cancer.
Protozoan Diseases of the Digestive System		
Giardiasis	*Giardia lamblia*	Protozoan adheres to intestinal wall, may inhibit nutritional absorption. Causes diarrhea.
Cryptosporidiosis	*Cryptosporidium parvum*	Shed in animal feces, protozoan enters water supply; causes self-limiting diarrhea but may be life-threatening if immunosuppressed.
Cyclospora diarrheal infection	*Cyclospora cayetanensis*	Usually ingested with fruits and vegetables; causes watery diarrhea.
Amoebic dysentery (amoebiasis)	*Entamoeba histolytica*	Amoeba lyses epithelial cells of intestine, causes abscesses; significant mortality rate.
Helminthic Diseases of the Digestive System		
Tapeworms	*Taenia saginata* (beef tapeworm); *T. solium* (pork tapeworm); *Diphyllobothrium latum* (fish tapeworm)	Helminth lives off undigested intestinal contents with few symptoms; pork tapeworm may cause larvae to form in many organs (neurocycticercosis) and cause damage; in this case, eggs are infectious. Usually transmitted by ingesting larvae in meats.
Hydatid disease	*Echinococcus granulosus*	Larvae form in body; may be very large and cause damage. Transmitted by ingesting tapeworm eggs.
Pinworms	*Enterobius vermicularis*	Itching around anus.
Hookworms	*Necator americanus, Ancyclostoma duodenale*	Larvae enter through skin. Large infections may result in anemia.
Ascariasis	*Ascaris lumbricoides*	Helminths live off undigested intestinal contents. Transmitted by ingesting eggs from feces. Usually few symptoms.
Trichinosis	*Trichinella spiralis*	Larvae encyst in striated muscle. Transmitted by ingestion of larvae in meats. Usually few symptoms, but large infections may be fatal.

4. A measles, mumps, rubella (MMR) vaccine is available.

5. Diagnosis is based on symptoms, or an ELISA test is performed on viruses cultured in embryonated eggs or cell culture.

Cytomegalovirus (CMV) Inclusion Disease (pp. 700–701)

1. CMV (a herpesvirus) causes intranuclear inclusion bodies and cytomegaly of host cells.

2. CMV is transmitted by saliva, urine, semen, cervical secretions, and human milk.

3. CMV inclusion disease can be asymptomatic, a mild disease, or progressive and fatal. Immunosuppressed patients may develop pneumonia.

4. If the virus crosses the placenta, it can cause congenital infection of the fetus, resulting in impaired mental development, neurological damage, and stillbirth.

5. Diagnosis is based on isolation of the virus or detection of IgG and IgM antibodies.

Hepatitis (p. 701)

1. Inflammation of the liver is called hepatitis. Symptoms include loss of appetite, malaise, fever, and jaundice.

2. Viral causes of hepatitis include hepatitis viruses, Epstein-Barr (EB) virus, and CMV.

Hepatitis A (p. 702)

1. Hepatitis A virus (HAV) causes hepatitis A; at least 50% of all cases are subclinical.

2. HAV is ingested in contaminated food or water, grows in the cells of the intestinal mucosa, and spreads to the liver, kidneys, and spleen in the blood.

3. The virus is eliminated with feces.

4. The incubation period is 2–6 weeks; the period of disease is 2–21 days, and recovery is complete in 4–6 weeks.

5. Diagnosis is based on tests for IgM antibodies.

6. Passive immunization can provide temporary protection. A vaccine is available.

Hepatitis B (pp. 702–704)

1. Hepatitis B virus (HBV) causes hepatitis B, which is frequently serious.

2. HBV is transmitted by blood transfusions, contaminated syringes, saliva, sweat, breast milk, and semen.

3. Blood is tested for HB_sAg before being used in transfusions.

4. The average incubation period is 3 months; recovery is usually complete, but some patients develop a chronic infection or become carriers.

5. A vaccine against HB_sAg is available.

Hepatitis C (p. 704)

1. Hepatitis C virus (HCV) is transmitted via blood.

2. The incubation period is 2–22 weeks; the disease is usually mild, but some patients develop chronic hepatitis.

3. Blood is tested for HCV antibodies before being used in transfusions.

Hepatitis D (Delta Hepatitis) (p. 704)

1. Hepatitis D virus (HDV) has a circular strand of RNA and uses HB_sAg as a coat.

Hepatitis E (pp. 704–705)

1. Hepatitis E virus (HEV) is spread by the fecal-oral route.

Other Types of Hepatitis (p. 705)

1. There is evidence of the existence of hepatitis types F and G.

Viral Gastroenteritis (p. 705)

1. Viral gastroenteritis is most often caused by a rotavirus or the Norwalk agent.

FUNGAL DISEASES OF THE DIGESTIVE SYSTEM (pp. 705–706)

1. Mycotoxins are toxins produced by some fungi.

2. Mycotoxins affect the blood, nervous system, kidneys, or liver.

Ergot Poisoning (p. 705)

1. Ergot poisoning, or ergotism, is caused by the mycotoxin produced by Claviceps purpurea.

2. Cereal grains are the crop most often contaminated with the Claviceps mycotoxin.

Aflatoxin Poisoning (pp. 705–706)

1. Aflatoxin is a mycotoxin produced by Aspergillus flavus.

2. Peanuts are the crop most often contaminated with aflatoxin.

PROTOZOAN DISEASES OF THE DIGESTIVE SYSTEM (pp. 706–707)

Giardiasis (p. 706)

1. Giardia lamblia grows in the intestines of humans and wild animals and is transmitted in contaminated water.

2. Symptoms of giardiasis are malaise, nausea, flatulence, weakness, and abdominal cramps that persist for weeks.

3. Diagnosis is based on identification of the protozoa in the small intestine.

Cryptosporidiosis (pp. 706–707)

1. Crytosporidium parvum causes diarrhea; in immunosuppressed patients, the disease is prolonged for months.

2. The pathogen is transmitted in contaminated water.

3. Diagnosis is based on the identification of oocysts in feces.

Cyclospora Diarrheal Infection (p. 707)

1. C. cayetanensis causes diarrhea; the protozoan was first identified in 1993.

2. It is transmitted in contaminated produce.

3. Diagnosis is based on the identification of oocysts in feces.

Amoebic Dysentery (Amoebiasis) (p. 707)

1. Amoebic dysentery is caused by *Entamoeba histolytica* growing in the large intestine.

2. The amoeba feeds on red blood cells and GI tract tissues. Severe infections result in abscesses.

3. Diagnosis is confirmed by observing trophozoites in feces and by several serological tests.

HELMINTHIC DISEASES OF THE DIGESTIVE SYSTEM (pp. 708–713)

Tapeworms (pp. 708–709)

1. Tapeworms are contracted by the consumption of undercooked beef, pork, or fish containing encysted larvae (cysticerci).

2. The scolex attaches to the intestinal mucosa of humans (the definitive host) and matures into an adult tapeworm.

3. Eggs are shed in the feces and must be ingested by an intermediate host.

4. Adult tapeworms can be undiagnosed in a human.

5. Diagnosis is based on the observation of proglottids and eggs in feces.

6. Neurocysticercosis in humans occurs when the pork tapeworm larvae encyst in humans.

Hydatid Disease (p. 709)

1. Humans infected with the tapeworm *Echinococcus granulosus* might have hydatid cysts in their lungs or other organs.

2. Dogs and wolves are usually the definitive hosts, and sheep or deer are the intermediate hosts for *E. granulosus*.

Nematodes (pp. 709–713)

Pinworms (p. 709)

1. Humans are the definitive host for pinworms, *Enterobius vermicularis*.

2. The disease is acquired by ingesting *Enterobius* eggs.

Hookworms (pp. 709–710)

1. Hookworm larvae bore through skin and migrate to the intestine to mature into adults.

2. In the soil, hookworm larvae hatch from eggs shed in feces.

Ascariasis (p. 710)

1. *Ascaris lumbricoides* adults live in human intestines.

2. The disease is acquired by ingesting *Ascaris* eggs.

Trichinosis (pp. 710–712)

1. *Trichinella spiralis* larvae encyst in muscles of humans and other mammals to cause trichinosis.

2. The roundworm is contracted by ingesting undercooked meat containing larvae.

3. Adult females mature in the intestine and lay eggs; the new larvae migrate to invade muscles.

4. Symptoms include fever, swelling around the eyes, and gastrointestinal upset.

5. Biopsy specimens and serological tests are used for diagnosis.

Study Questions

REVIEW

1. Name examples of the representative normal microbiota, if any, in the mouth, pharynx, esophagus, stomach, small intestine, large intestine, and rectum.

2. What properties of *S. mutans* implicate this bacterium in the formation of dental caries? Why is sucrose, more than any other carbohydrate, responsible for the formation of dental caries?

3. Complete the following table:

Disease	Causative Agent	Suspect Foods	Symptoms	Treatment
Staphylococcal food poisoning				
Shigellosis				
Salmonellosis				
Cholera				
Gastroenteritis				
Traveler's diarrhea				

4. Differentiate salmonellosis from typhoid fever.

5. You probably listed *E. coli* in answer to questions 1 and 3. Explain why this one bacterial species is both beneficial and harmful.

6. What preventive treatments are currently used for hepatitis A? For hepatitis B? For hepatitis C?

7. How is blood that is to be used for transfusions tested for HBV? For HCV?

8. Define mycotoxin. Give an example of a mycotoxin.

9. Explain how the following diseases differ and how they are similar: giardiasis, amoebic dysentery, *Cyclospora* diarrheal infection, and cryptosporidiosis.

10. Differentiate between amoebic dysentery and bacillary dysentery (shigellosis).

11. List the general symptoms of gastroenteritis. Because there are many etiologies, what is the laboratory diagnosis usually based on?

12. Differentiate among the following factors of bacterial intoxication and bacterial infection: prerequisite conditions, causative agents, onset, duration of symptoms, and treatment.

13. Complete the following table:

Disease	Causative Agent	Mode of Transmission	Site of Infection	Symptoms	Prevention
Mumps					
CMV inclusion disease					
Hepatitis A					
Hepatitis B					
Viral gastroenteritis					

14. Diagram the life cycle of a human tapeworm.

15. Diagram the life cycle of *Trichinella*, and include humans in the cycle.

16. How can bacterial and protozoan infections of the GI tract be prevented?

17. Look at your diagrams for questions 14 and 15. Indicate sequences in the life cycles that could be easily broken to prevent these diseases.

MULTIPLE CHOICE

1. All of the following can be transmitted by recreational (i.e., swimming) water sources *except*
   a. amoebic dysentery.
   b. cholera.
   c. giardiasis.
   d. hepatitis B.
   e. salmonellosis.

2. A patient with nausea, vomiting, and diarrhea within 5 hours after eating most likely has
   a. shigellosis.
   b. cholera.
   c. *E. coli* gastroenteritis.
   d. salmonellosis.
   e. staphylococcal food poisoning.

3. Isolation of *E. coli* from a stool sample is diagnostic proof that the patient has
   a. cholera.
   b. *E. coli* gastroenteritis.
   c. salmonellosis.
   d. typhoid fever.
   e. none of the above

4. Gastric ulcers are caused by
   a. stomach acid.
   b. *Helicobacter pylori.*
   c. spicy food.
   d. acidic food.
   e. stress.

5. Microscopic examination of a patient's fecal culture shows comma-shaped bacteria. These bacteria require 2–4% NaCl to grow. The bacteria probably belong to the genus
   a. *Campylobacter.*
   b. *Escherichia.*
   c. *Salmonella.*
   d. *Shigella.*
   e. *Vibrio.*

6. A recent cholera epidemic in Peru had all of the following characteristics. Which one *led* to the others?
   a. eating raw fish
   b. sewage contamination of water
   c. catching fish in contaminated water
   d. *Vibrio* in fish intestine
   e. including fish intestines with edibles

Use the following choices to answer questions 7–10:
   a. *Campylobacter*
   b. *Cryptosporidium*
   c. *Escherichia*
   d. *Salmonella*
   e. *Trichinella*

7. Identification is based on the observation of oocysts in feces.

8. A characteristic disease symptom caused by this microorganism is swelling around the eyes.

9. Microscopic observation of a stool sample reveals gram-negative helical cells.

10. This microbe is frequently transmitted to humans via raw eggs.

CRITICAL THINKING

1. Why is a human infection of trichinosis considered a dead-end for the parasite?

2. Complete the following table:

Disease	Conditions Necessary for Microbial Growth	Basis for Diagnosis	Prevention
Staphylococcal food poisoning			
Salmonellosis			

3. Match the following foods with the genus of microorganism most likely to contaminate each:
   Beef _____   a. *Vibrio*
   Delicatessen meats _____   b. *Campylobacter*
   Chicken _____   c. *E. coli* O157:H7
   Milk _____   d. *Listeria*
   Oysters _____   e. *Salmonella*
   Pork _____   f. *Trichinella*

   What disease does each microbe cause? How can these diseases be prevented?

4. Which diseases of the gastrointestinal tract can be acquired by swimming in a pool or lake? Why are these diseases not likely to be acquired while swimming in the ocean?

CLINICAL APPLICATIONS

1. In New York on April 26, patient A was hospitalized with a 2-day history of diarrhea. An investigation revealed that patient B had onset of watery diarrhea on April 22. On April 24, three other people (patients C, D, and E) had onset of diarrhea. All three had vibriocidal antibody titers ≥ 640. In Ecuador on April 20, B bought crabs that were boiled and shelled. He shared crabmeat with two people (F and G), then froze the remaining crab in a bag. Patient A returned to New York on April 21 with the bag of crabmeat in his suitcase. The bag was placed in a freezer overnight and thawed on April 22 in a double-boiler for 20 minutes. The crab was served 2 hours later in a crab salad. The crab was consumed during a 6-hour period by A, C, D, and E. F and G did not become ill. What is the etiology of this disease? How was it transmitted, and how could it have been prevented?

2. The 2130 students and employees of a public school system developed diarrheal illness on April 2. The cafeteria served chicken that day. On April 1, part of the chicken was placed in water-filled pans and cooked in an oven for 2 hours at a dial setting of 177°C. The oven was turned off, and the chicken was left overnight in the warm oven. The remainder of the chicken was cooked for 2 hours in a steam cooker and then left in the device overnight at the lowest possible setting (43°C). Two serotypes of a gram-negative, cytochrome oxidase-negative, lactose-negative rod were isolated from 32 patients. What is the pathogen? How could this outbreak have been prevented?

3. Staff members of one hospital ward noted an increase in the number of cases of HBV. Fifty cases occurred during a 6-month period compared with four cases during the previous 6 months. Between January 1 and 15, all 50 patients had multiple invasive procedures as shown below:

   - Transfusion, fingerstick, IV catheter, heparin injection: 78%
   - Transfusion, insulin injection, surgery, fingerstick: 64%
   - Fingerstick, IV catheter, insulin injection, heparin injection: 80%
   - Transfusion, heparin injection, surgery, IV catheter: 2%
   - Heparin injection, IV catheter, insulin injection, surgery: 0%

   How did the patients acquire HBV? Provide an explanation for the 2% and 0%.

4. A 31-year-old male became feverish 4 days after arriving at a vacation resort in Idaho. During his stay, he ate at two restaurants that were not associated with the resort. At the resort, he drank soft drinks with ice, used the hot tub, and went fishing. The resort is supplied by a well that was dug 3 years ago. He went to the hospital when he developed vomiting and bloody diarrhea. Gram-negative, lactose-negative bacteria were cultured from his stool. The patient recovered after receiving intravenous fluids. What microorganism most likely caused his symptoms? How is this disease transmitted? What is the most likely source of his infection, and how would you verify the source?

5. Three to five days after eating Thanksgiving dinner at a restaurant, 112 people developed fever and gastrointeritis. All the food had been consumed except for five "doggie" bags. Bacterial analysis of the mixed contents of the bags (containing roast turkey, giblet gravy, and mashed potatoes) showed the same bacterium that was isolated from the patients. The gravy had been prepared from giblets of 43 turkeys that had been refrigerated for three days prior to preparation. The uncooked giblets were ground in a blender and added to a thickened hot stock mixture. The gravy was not reboiled and was stored at room temperature throughout Thanksgiving Day. What was the source of the illness? What was the most likely etiologic agent? Was this an infection or an intoxication?

6. An outbreak of typhoid fever occurred following a family gathering of 293 people. Cultures from 17 of these people yielded *Salmonella typhi*. Nine foods prepared by three food-handlers were available at the event.

Foods Consumed	Percent Ill
Green salad and roast beef	60
Noodles and baked beans	42
Noodles and egg salad	12
Egg salad and roast beef	0
Baked beans and fruit	0

What is the most likely source of the Salmonella? Which tests would verify this?

Learning with Technology

MP = The Microbiology Place website **ST** = Student Tutorial CD-ROM **VU** = VirtualUnknown CD-ROM

MP Don't forget to go to The Microbiology Place website (http://www.microbiologyplace.com) to take the practice tests, explore the interactive activity and case study, and check out the news articles and web links for this chapter.

ST Remember there is also a quiz for this chapter on the Microbiology Interactive Student Tutorial CD-ROM.

VU Enter the Virtual Lab, click the arrow next to the Session field, click Textbook Exercises, and select Chapter 25. Read the Case Study carefully. Consult Table 25.2 on pages 712–713, and make a list of the possible causes of this infection.

1. From the information provided, does it appear that this disease is due to infection or to intoxication? Explain.

2. Complete the identification of this pathogen using VirtualUnknown™ Microbiology. How does the occupation of the victim complicate the investigation of his disease?

3. Based on the identification and the appropriate information provided by the textbook, predict the source of his infection.

chapter 26

Microbial Diseases

of the Urinary and

Reproductive Systems

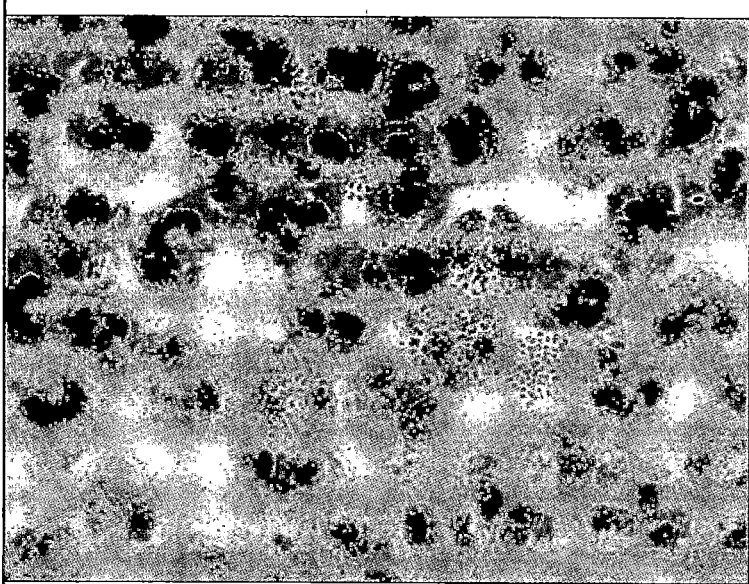

Neisseria gonorrhoeae. *The gram-negative cocci are seen inside phagocytes in this Gram stain. Antibiotic resistance of this bacterium has contributed to increased incidence of gonorrhea in urban areas.*

The **urinary system** is composed of organs that regulate the chemical composition and volume of the blood and excrete the waste products of metabolism. The **reproductive system** is composed of organs that produce gametes to propagate the species and, in the female, organs that support and nourish the developing embryo and fetus. Both systems open to the external environment and thus share portals of entry for microorganisms that can cause disease.

Structure and Function of the Urinary System

Learning Objective

■ *List the antimicrobial features of the urinary system.*

The **urinary system** consists of two *kidneys*, two *ureters*, a single *urinary bladder*, and a single *urethra* (Figure 26.1). Certain wastes, collectively called *urine*, are removed from the blood as it circulates through the kidneys. The urine passes through the ureters into the urinary bladder, where it is stored prior to elimination from the body through the urethra. In the female, the urethra conveys only urine to the exterior. In the male, the urethra is a common tube for both urine and seminal fluid.

Where the ureters enter the urinary bladder, valves prevent the backflow of urine to the kidneys. This mech-

anism helps shield the kidneys from lower urinary tract infections. In addition, the acidity of normal urine has some antimicrobial properties. The flushing action of urine during urination tends to remove potentially infectious microbes.

Structure and Function of the Reproductive System

Learning Objective

■ *Identify the portals of entry for microbes into the reproductive system.*

The **female reproductive system** consists of two *ovaries*, two *uterine (fallopian) tubes*, the *uterus*, including the *cervix*, the *vagina*, and *external genitals* (Figure 26.2). The ovaries produce female sex hormones and ova

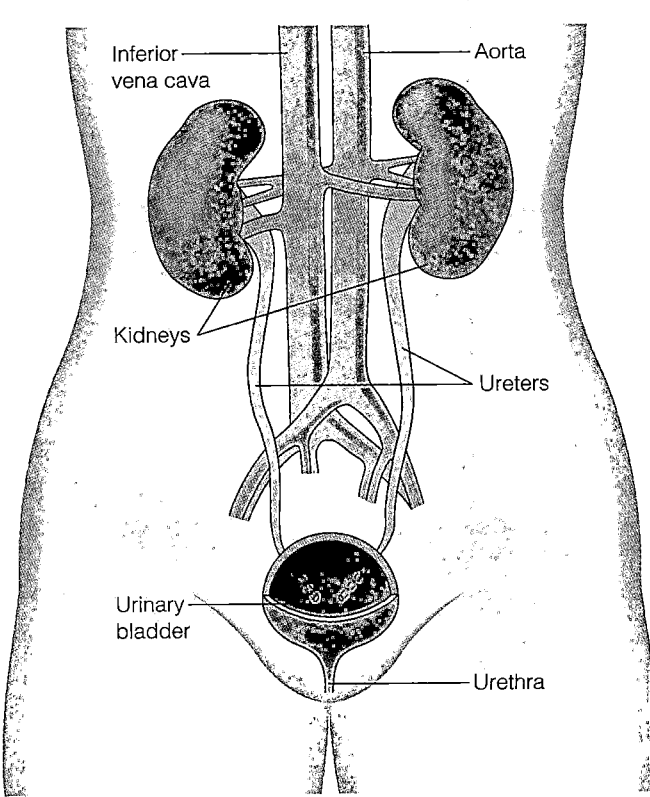

FIGURE 26.1   Organs of the human urinary system, shown here in the female.

■ **What anatomical features of the urinary system help prevent colonization by microbes?**

(eggs). When an ovum is released during the process of ovulation, it enters a uterine tube, where fertilization may occur if viable sperm are present. The fertilized ovum (zygote) descends the tube and enters the uterus. It implants in the inner wall and remains there while it develops into an embryo and, later, a fetus. The external genitals (*vulva*) include the clitoris, labia, and glands that produce a lubricating secretion during copulation.

The **male reproductive system** consists of two *testes,* a system of *ducts, accessory glands,* and the *penis* (Figure 26.3). The testes produce male sex hormones and sperm. To exit the body, sperm cells pass through a series of ducts: the epididymis, ductus (vas) deferens, ejaculatory duct, and urethra.

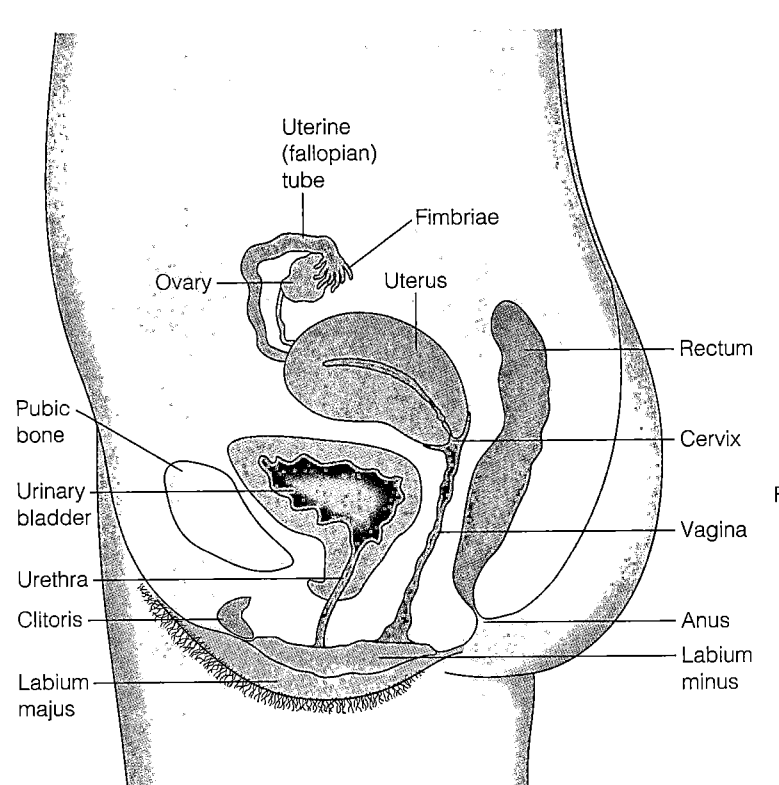

(a) Side view section of female pelvis showing reproductive organs.

(b) Front view of female reproductive organs, with the uterine tube and ovary to the left in the drawing sectioned. The fimbriae move to create fluid movement that moves the egg into the uterine tube.

FIGURE 26.2   Female reproductive organs.

■ **What are the major functions of the female reproductive organs?**

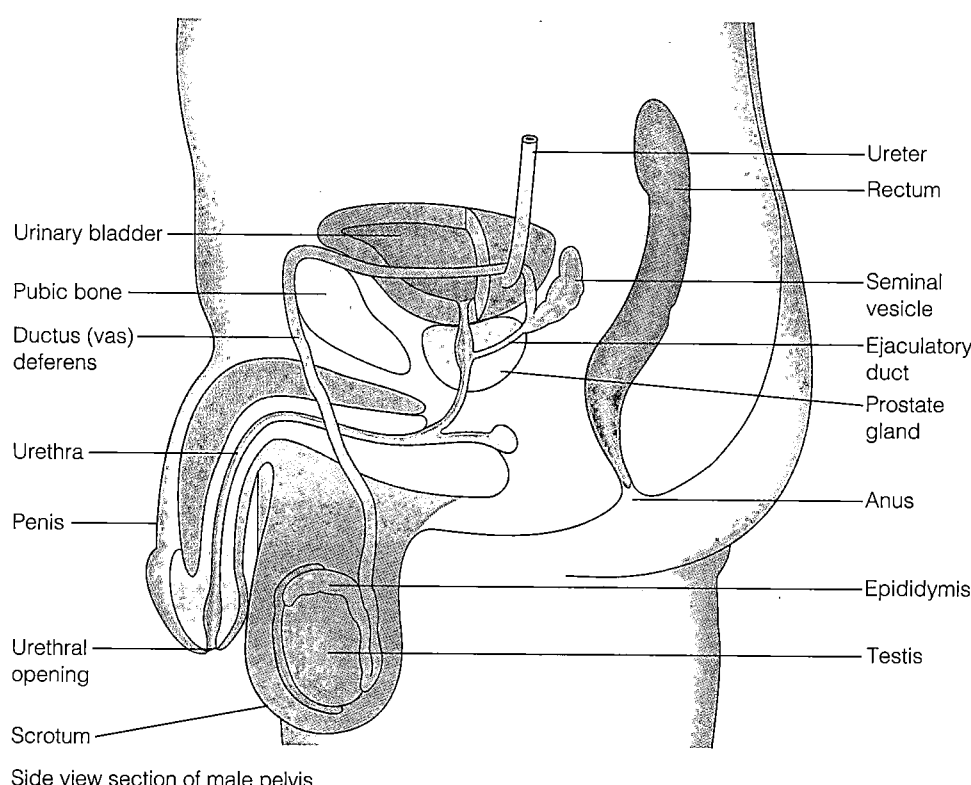

Side view section of male pelvis.

FIGURE 26.3 Male reproductive organs. A side view section of a male pelvis.

■ What are the major functions of the male reproductive organs?

Normal Microbiota of the Urinary and Reproductive Systems

Learning Objective

■ *Describe the normal microbiota of the upper urinary tract, the male urethra, and the female urethra and vagina.*

Normal urine is sterile, but it may become contaminated with members of the skin microbiota near the end of its passage through the urethra. Therefore, urine collected directly from the urinary bladder has fewer microbial contaminants than voided urine.

In the female genital system, the normal microbiota of the vagina are greatly influenced by sex hormones. For example, within a few weeks after birth, the female infant's vagina is populated by lactobacilli. This population grows because estrogens are transferred from maternal to fetal blood and cause glycogen to accumulate in the cells lining the vagina. Lactobacilli convert the glycogen to lactic acid, and the pH of the vagina becomes acidic. This glycogen–lactic acid sequence provides the conditions

under which acid-tolerant normal microbiota grow in the vagina.

The physiological effects of estrogens diminish several weeks after birth, and other bacteria, including corynebacteria and a variety of cocci and bacilli, become established and dominate the microbiota. As a result, the pH of the vagina becomes more neutral until puberty. At puberty, estrogen levels increase, lactobacilli again dominate, and the vagina again becomes acidic. In the adult, a disturbance of this ecosystem by an increase in glycogen (caused by oral contraceptives or pregnancy, for example), or elimination of the normal microbiota by antibiotics, can lead to *vaginitis*, an inflammation of the vagina. When the female reaches menopause, estrogen levels again decrease, the composition of the microbiota returns to that of childhood, and the pH again becomes neutral. Pregnancy and menopause are factors that increase the risk of urinary tract infections, which are probably related to lowered acidity. The use of spermicides, which may inhibit lactobacilli, is also associated with urinary tract infections.

The male urethra is usually sterile, except for a few contaminating microbes near the external opening.

DISEASES OF THE URINARY SYSTEM

The urinary system normally contains few microbes, but it is subject to opportunistic infections that can be quite troublesome. Almost all such infections are bacterial, although occasional infections by pathogens such as schistosome parasites, protozoa, and fungi occur. In addition, as we will see in this chapter, sexually transmitted diseases often affect the urinary system as well as the reproductive system.

Bacterial Diseases of the Urinary System

Learning Objectives

- *Describe the modes of transmission for urinary and reproductive system infections.*
- *List the microorganisms that cause cystitis, pyelonephritis, and leptospirosis, and name the predisposing factors for these diseases.*

Urinary system infections are most frequently initiated by an inflammation of the urethra, or *urethritis*. Infection of the urinary bladder is called *cystitis,* and infection of the ureters is *ureteritis.* The most significant danger from lower urinary tract infections is that they may move up the ureters and affect the kidneys, causing *pyelonephritis.* Occasionally the kidneys are affected by systemic bacterial diseases, such as *leptospirosis.* The pathogens causing these diseases are found in excreted urine.

Bacterial infections of the urinary system are usually caused by microbes that enter the system from external sources. In the United States, about 900,000 annual cases are of nosocomial origin. Probably 90% of these are associated with urinary catheters. Because of the proximity of the anus to the urinary opening, intestinal bacteria predominate in urinary tract infections. More than half of nosocomial infections of the urinary tract are caused by *E. coli.* Also common are infections by *Proteus, Klebsiella,* and *Enterococcus.* Infections by *Pseudomonas,* because of their natural resistance to antibiotics, are especially troublesome.

Diagnosis of urinary tract infections is usually based on symptoms such as painful urination or a sensation that the urinary bladder does not feel empty even after urination. Urine may be cloudy or have a light bloody tinge. The traditional guideline—that urine containing more than 100,000 bacteria per milliliter is an indication of urinary tract infection—has been modified. Counts as low as 1000/ml of any single bacterial type, or as few as 100/ml of coliforms (intestinal bacteria such as *E. coli*), are now considered an indication of significant infection. Before initiating therapy, urine bacteria are usually cultured to determine antibiotic sensitivity.

Cystitis

Cystitis is a common inflammation of the urinary bladder in females. Symptoms often include *dysuria* (difficult, painful, urgent urination) and *pyuria* (the presence of leukocytes in the urine).

The female urethra is less than 2 inches long, and microorganisms traverse it readily. It is also closer than the male urethra to the anal opening and its contaminating intestinal bacteria. These considerations are reflected in the fact that the rate of urinary tract infections in women is eight times that of men. In either gender, most cases are due to infection by *E. coli.* The second most common bacterial cause is the coagulase-negative *Staphylococcus saprophyticus* (sap-rō-fit'i-kus).

Trimethoprim-sulfamethoxazole usually clears cases of cystitis quickly. Quinolone antibiotics or ampicillin are often successful if drug resistance is encountered.

Pyelonephritis

In 25% of untreated cases, cystitis may progress to **pyelonephritis,** an inflammation of one or both kidneys. Symptoms are fever and flank or back pain. In females, it is often a complication of lower urinary tract infections. The causative agent in about 75% of the cases is *E. coli.* If pyelonephritis becomes chronic, scar tissue forms in the kidneys and severely impairs their function. Because pyelonephritis is a potentially life-threatening condition, treatment usually begins with intravenous, extended-term administration of a broad-spectrum antibiotic, such as a second- or third-generation cephalosporin.

Leptospirosis

Leptospirosis is primarily a disease of domestic or wild animals, but it can be passed to humans and sometimes causes severe kidney or liver disease. The causative agent is the spirochete *Leptospira interrogans* (in-tèr'rä-ganz), shown in Figure 26.4. *Leptospira* has a characteristic shape: an exceedingly fine spiral, only about 0.1 μm in diameter, wound so tightly that it is barely discernible under a darkfield microscope. Like other spirochetes, *L. interrogans* (so named because the hooked ends suggest a question mark) stains poorly and is difficult to see under

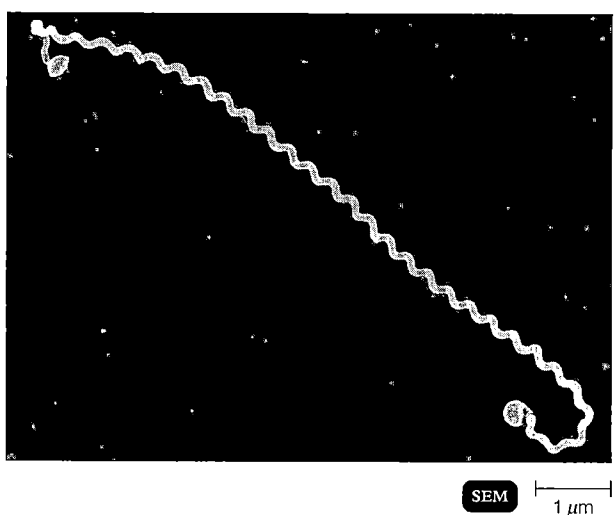

SEM |———————| 1 μm

FIGURE 26.4 *Leptospira interrogans*, the cause of leptospirosis. The hooked ends are often seen in preparations of this spirochete.

■ On what basis is *L. interrogans* named?

a normal light microscope. It is an obligate aerobe that can be grown in a variety of artificial media supplemented with rabbit serum.

Animals infected with the spirochete shed the bacteria in their urine for extended periods. Humans become infected by contact with urine-contaminated water or soil, or sometimes with animal tissue. People whose occupations expose them to animals or animal products are most at risk. Usually the pathogen enters through minor abrasions in the skin or mucous membranes. When ingested, it enters through the mucosa of the upper digestive system. In the United States, dogs and rats are the most common sources. Domestic dogs have a sizable rate of infection; even when immunized, they may continue to shed leptospira. A recent outbreak of leptospirosis is described in the box below.

CLINICAL PROBLEM SOLVING

An Unexplained Acute Fever

You will see questions as you read though this problem. The questions are those that the primary care physician and public health epidemiologist ask themselves and each other as they solve a clinical problem. Try to answer each question as you read through the problem.

1. On July 6, an athlete sought medical care for fever, muscle aches, and headache. The next day, a second athlete went to the emergency room with fever and kidney failure, and on July 10, a third athlete saw his physician because of fever and jaundice.
   What samples do you need from the patient?

2. Blood cultures were negative. However, darkfield microscopy of blood revealed spirochetes.
   What do you suspect? How will you verify your diagnosis?

3. ELISA tests for IgM antibodies tested positive for leptospirosis. The presence of IgM antibodies is used to indicate a current infection.

What is the normal habitat of Leptospira?

4. Leptospirosis is a widespread zoonosis that is endemic in most temperate and tropical climates. Leptospires infect various animals that then excrete the bacteria in their urine; the bacteria persist in fresh water, damp soil, vegetation, and mud.
   How is this disease transmitted to humans?

5. Human infection occurs through exposure to water or soil that has been contaminated by infected animal urine. It has been associated with wading, swimming, and whitewater rafting in contaminated lakes and rivers. Leptospires may enter the body through cut or abraded skin, mucous membranes, and conjunctivae.
   What additional information do you need?

6. The three athletes participated in a triathlon held in Springfield, Illinois, on June 21. A triathlon race consists of swimming, biking, and running

competitions. The swimming leg of the Springfield triathlon took place in Lake Springfield.
   What other data do you need?

7. To identify additional cases of illness, lists of the triathlon participants were obtained, and a questionnaire was administered to 1194 people who participated in the two events. Of these, 110 described having an illness meeting the case definition. Seventy-three of these people sought medical care, and 23 of those were hospitalized.
   As a public health officer, what would you do next?

On July 24, the Illinois Department of Health issued an adivsory not to swim or water skin in Lake Springfield. Investigators from the CDC are trying to develop prevention and control measures for this outbreak.

SOURCE: *MMWR* 47(28):585–588 (7/24/98) and *MMWR* 47(32):673–676 (8/21/98).

After an incubation period of 1–2 weeks, headaches, muscular aches, chills, and fever abruptly appear. Several days later, the acute symptoms disappear and the temperature returns to normal. A few days later, however, a second episode of fever may occur. In a small number of cases the kidneys and liver become seriously infected (*Weil's disease*); kidney failure is the most common cause of death. Recovery results in a solid immunity. There are usually about 50 cases reported each year in the United States, but because the clinical symptoms are not distinc-

tive, many cases are probably not diagnosed correctly. A recent study in a clinic serving the urban poor in a large eastern U.S. city found that up to 16% of the patients tested positive for exposure.

Serological testing is usually done by central reference laboratories. Most diagnoses are made by isolating the pathogen from the blood or cerebrospinal fluid. Treatment of leptospirosis with antibiotics in later stages is often unsatisfactory. That immune reactions are responsible for pathogenesis in this stage may be an explanation.

DISEASES OF THE REPRODUCTIVE SYSTEM

Microbes causing infections of the reproductive system are usually very sensitive to environmental stresses and require intimate contact for transmission.

Bacterial Diseases of the Reproductive System

Learning Objectives

- List the causative agents, symptoms, methods of diagnosis, and treatments for gonorrhea, NGU, PID, syphilis, LGV, chancroid, and bacterial vaginosis.
- List reproductive system diseases that can cause congenital and neonatal infections, and explain how these infections can be prevented.

Most diseases of the reproductive system are transmitted by sexual activity and are called **sexually transmitted diseases (STDs)**. More than 30 bacterial, viral, or parasitic diseases have been identified as sexually transmitted. In the United States, it is estimated that over 15 million new cases of STDs occur annually. Many of these diseases can be successfully treated with antibiotics and can be largely prevented by the use of condoms. However, over 60 million Americans have STDs, mostly viral, for which there is no effective cure.

Gonorrhea

One of the most common reportable, or notifiable, communicable diseases in the United States is **gonorrhea,** an STD caused by the gram-negative diplococcus *Neisseria gonorrhoeae*. An ancient disease, gonorrhea was described and given its present name by the Greek physician Galen in A.D. 150. (*gon* = semen + *rhea* = flow; a flow of semen. Apparently, he confused pus with semen). The incidence of gonorrhea has tended to de-

crease in recent years (Figure 26.5a), but more than 300,000 cases are still reported in the United States each year. The true number of cases is probably much larger, probably two or three times those reported. More than 60% of patients with gonorrhea are age 15–24.

In order to infect, the gonococcus must attach to the mucosal cells of the epithelial wall by means of fimbriae. The pathogen invades the spaces separating columnar epithelial cells, which are found in the oral-pharyngeal area, the eyes, rectum, urethra, opening of the cervix, and the external genitals of prepubertal females. The invasion sets up an inflammation, and, when leukocytes move into the inflamed area, the characteristic pus formation results. In men, a single unprotected exposure results in infection with gonorrhea 20–35% of the time. Women become infected 60–90% of the time from a single exposure.

Males become aware of a gonorrheal infection by painful urination and a discharge of pus-containing material from the urethra (Figure 26.6). About 80% of infected males show these obvious symptoms after an incubation period of only a few days; most others show symptoms in less than a week. In the days before antibiotic therapy, symptoms persisted for weeks. In untreated cases, recovery may eventually occur without complication, but, when complications do occur, they can be serious. In some instances, the urethra is scarred and partially blocked. Sterility can result when the testes become infected or when the ductus (vas) deferens, the tube carrying sperm from the testes, becomes blocked by scar tissue.

In females, the disease is more insidious. Only the cervix, which contains columnar epithelial cells, is infected. The vaginal walls are composed of stratified squamous epithelial cells, which are not colonized. Very few women are aware of the infection. Later in the course of the disease, there might be abdominal pain from complications such as pelvic inflammatory disease (discussed shortly).

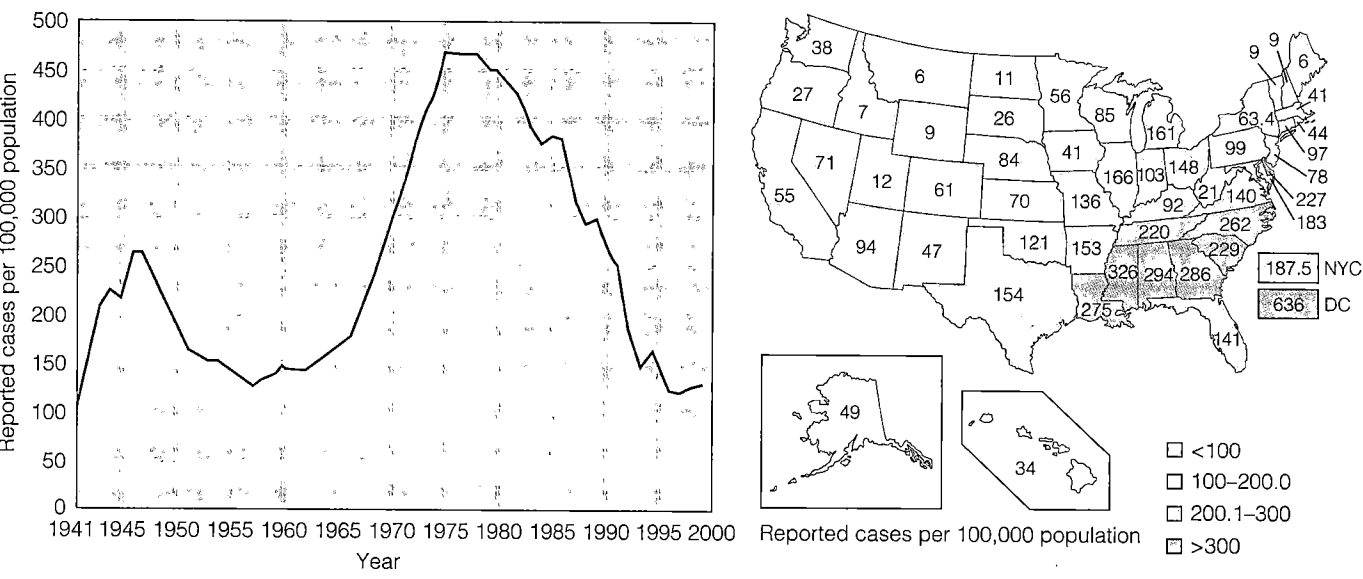

(a) Incidence of gonorrhea in the United States from 1941 to 2000

(b) Geographical distribution of cases in 1999

FIGURE 26.5 The U.S. incidence and distribution of gonorrhea.
[SOURCES: CDC, *Summary of Notifiable Diseases 1998, MMWR* 47(53)(12/31/99); *MMWR* 48(51) (1/7/00).]

■ How does a gonococcus attach to mucosal epithelial cells?

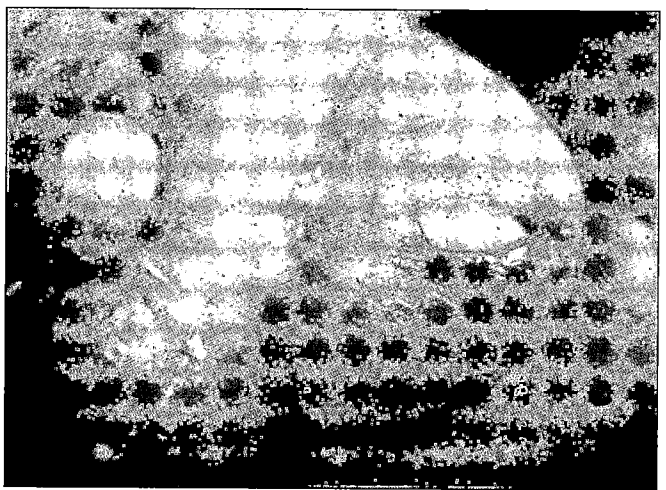

FIGURE 26.6 Pus-containing discharge from the urethra of a male with an acute case of gonorrhea.

■ What causes pus formation in gonorrhea?

In both males and females, untreated gonorrhea can disseminate and become a serious, systemic infection. Complications of gonorrhea can involve the joints, heart (**gonorrheal endocarditis**), meninges (**gonorrheal meningitis**), eyes, pharynx, or other parts of the body. **Gonorrheal arthritis,** which is caused by the growth of the gonococcus in fluids in joints, occurs in about 1% of gonorrhea cases. Joints commonly affected include the wrist, knee, and ankle.

Gonorrheal eye infections occur most often in newborns. If the mother is infected, the eyes of the infant can become infected as it passes through the birth canal. This condition, **ophthalmia neonatorum,** can result in blindness. Because of the seriousness of this condition and the difficulty of being sure the mother is free of gonorrhea, antibiotics are placed in the eyes of all newborn infants. If the mother is known to be infected, an intramuscular injection of antibiotic is also administered to the infant. Some sort of prophylaxis is required by law in most states. Gonorrheal infections can also be transferred by hand contact from infected sites to the eyes of adults.

Gonorrheal infections can be acquired at any point of sexual contact; pharyngeal and anal gonorrhea are not uncommon. The symptoms of **pharyngeal gonorrhea** often resemble those of the usual septic sore throat. **Anal gonorrhea** can be painful and accompanied by discharges of pus. In some cases, however, the symptoms are limited to itching.

Increased sexual activity with a series of partners, and the fact that the disease in the female may go unrecognized, contributed considerably to the increased incidence of gonorrhea and other STDs during the 1960s and 1970s. The widespread use of oral contraceptives also

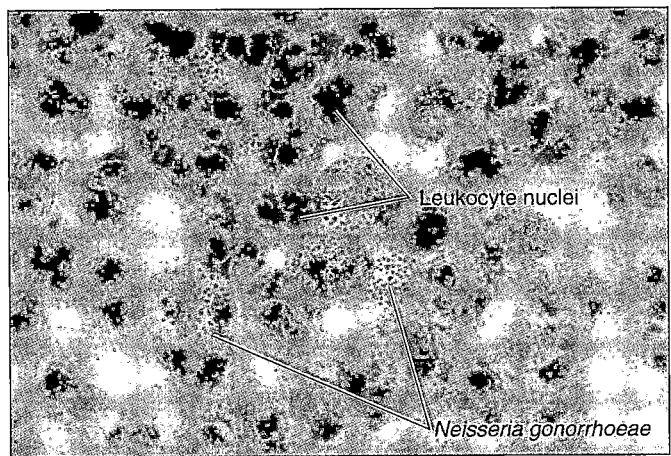

LM 10 μm

FIGURE 26.7 A smear of pus from a patient with gonorrhea.
The *Neisseria gonorrhoeae* bacteria are contained within phagocytic leukocytes. These gram-negative bacteria are visible here as pairs of cocci. The large stained bodies are the nuclei of the leukocytes.

■ How is gonorrhea diagnosed?

contributed to the increase. Oral contraceptives often replaced condoms and spermicides, which help prevent disease transmission. Recovery does not lead to immunity to reinfection. One reason is probably rapid changes in the antigenic makeup of the gonococcus.

Antibiotic resistance to the gonococcus has recently accelerated at an alarming rate. Although penicillin has been an effective treatment for gonorrhea for many years, the dosages have had to be increased substantially because of penicillin-resistant bacteria. Ceftriaxone, a third-generation cephalosporin, is currently the drug of choice, but this is subject to change. Because of frequent coinfection with chlamydia, antibiotic treatment should include an effective antichlamydial antibiotic, usually of the tetracycline group.

Gonorrhea in men is diagnosed by finding gonococci in a stained smear of pus from the urethra. The typical gram-negative diplococci within the phagocytic leukocytes are readily identified (Figure 26.7). Gram staining of exudates is not as reliable with women. Usually, a culture is taken from within the cervix and grown on special media. Cultivation of the nutritionally fastidious bacterium requires an atmosphere enriched in carbon dioxide. The gonococcus is very sensitive to adverse environmental influences (desiccation and temperature) and survives poorly outside the body. It even requires special transporting media to keep it viable for short intervals before the cultivation is under way. Cultivation has the advantage of allowing determination of antibiotic sensitivity.

Diagnosis of gonorrhea has been aided by the development of an ELISA that detects *N. gonorrhoeae* in urethral pus or on cervical swabs within about 3 hours with high accuracy. Other rapid tests now available use monoclonal antibodies against antigens on the surface of the gonococcus. DNA probe tests are very accurate for identifying clinical isolates from suspected cases.

Nongonococcal Urethritis (NGU)

Nongonococcal urethritis (NGU), also known as **nonspecific urethritis (NSU)**, refers to any inflammation of the urethra not caused by *Neisseria gonorrhoeae*. Symptoms include painful urination and a watery discharge.

The most common pathogen associated with NGU is *Chlamydia trachomatis*. Many people suffering from gonorrhea are coinfected with *C. trachomatis*, which infects the same columnar epithelial cells as the gonococcus. *C. trachomatis* is also responsible for the STD lymphogranuloma venereum (discussed on page 731) and trachoma (see page 596). This is the most common sexually transmitted pathogen in the United States, responsible for an estimated 4 million cases of NGU annually. Of special importance is the fact that five times as many cases are reported in females than males. In women, it is responsible for many cases of pelvic inflammatory disease (discussed on page 728), plus eye infections and pneumonia in infants born to infected mothers.

Because the symptoms are often mild in males, and females are usually asymptomatic, many cases of NGU go untreated. Although complications are not common, they can be serious. Males may develop inflammation of the epididymis. In females, inflammation of the uterine tubes may cause sterility by scarring the tubes. As many as 60% of such cases may be from chlamydial rather than gonococcal infection.

There is still a need for completely reliable diagnostic tests for NGU infections, especially *C. trachomatis*. Culturing is the most reliable method but requires cell culture techniques that are lengthy, expensive, and not always conveniently available. Currently, there are numerous rapid tests that yield results in a matter of hours. They include ELISA polymerase chain reaction (PCR), nucleic acid probes, and fluorescent-antibody (FA) methods. They can detect chlamydia in urine samples of either males or females or from cervical swabs of females. These tests are generally less sensitive than cell culture but are especially useful for screening, because two-thirds of infected women and one-fourth of infected men have no symptoms.

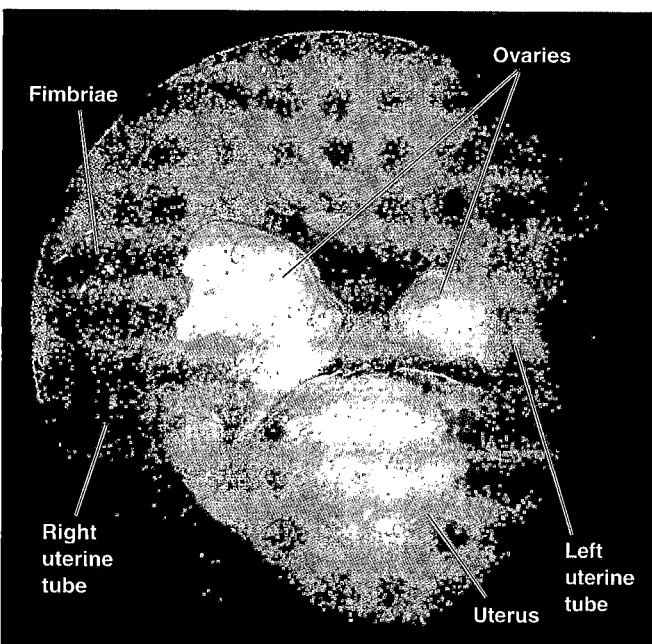

FIGURE 26.8 Salpingitis. This photograph, taken through a laparoscope (a specialized endoscope), shows an acutely inflamed right uterine tube and inflamed, swollen fimbriae and ovary, caused by salpingitis. The left tube is only mildly inflamed. (See Figure 26.2.) The use of a laparoscope is the most reliable diagnostic method for PID.

■ What is pelvic inflammatory disease?

Bacteria other than *C. trachomatis* can also be implicated in NGU. Another cause of urethritis and infertility is *Ureaplasma urealyticum* (ū-rē-ä-lit'i-kum). This pathogen is a member of the mycoplasma (bacteria without a cell wall). Another mycoplasma, *Mycoplasma hominis* (ho'minis), commonly inhabits the normal vagina but can opportunistically cause uterine tube infection.

Both chlamydia and mycoplasma are sensitive to tetracycline-type antibiotics such as doxycycline, or macrolide-type antibiotics such as azithromycin.

Pelvic Inflammatory Disease (PID)

Pelvic inflammatory disease (PID) is a collective term for any extensive bacterial infection of the female pelvic organs, particularly the uterus, cervix, uterine tubes, or ovaries. During their reproductive years, one in ten women suffers from PID, and one in four of these will have serious complications such as infertility or chronic pain.

PID is often caused by *N. gonorrhoeae*. About 20–30% of untreated gonorrhea cases progress to PID. However, coinfection with chlamydial bacteria is frequent, and they are also an important agent of this type of infection. In later stages of PID, anaerobes and various facultative anaerobes predominate.

The bacteria may attach to sperm cells and be transported by them from the cervical region to the uterine tubes. Women who use barrier contraceptives, especially with spermicides, have a significantly lower rate of PID.

Infection of the uterine tubes, or **salpingitis,** is the most serious form of PID (Figure 26.8). Salpingitis can result in scarring that blocks the passage of ova from the ovary to the uterus, possibly causing sterility. One episode of salpingitis causes infertility in 10–15% of women; 50–75% become infertile after three or more such infections.

A blocked uterine tube may cause a fertilized ovum to be implanted in the tube rather than the uterus. This is called an *ectopic* (or *tubal*) *pregnancy,* and it is life-threatening because of the possibility of rupture of the tube and resulting hemorrhage. The reported cases of ectopic pregnancies have been increasing steadily, corresponding to the increasing occurrence of PID.

The recommended treatment for PID is the simultaneous administration of doxycycline and cefoxitin (a cephalosporin). This combination is active against both the gonococcus and chlamydia. Such recommendations are constantly being reviewed.

Syphilis

The earliest reports of **syphilis** date back to the end of the fifteenth century in Europe, when the return of Columbus from the New World gave rise to a hypothesis that syphilis was introduced to Europe by his men. Another hypothesis is that syphilis existed in Europe and Asia before the fifteenth century but became widespread as urban living became more common. One English description of the "Morbus Gallicus" (French disease) seems clearly to describe syphilis as early as 1547 and ascribes its transmission in these terms: ". . . It is taken when one pocky person doth synne in lechery one with another."

The number of new syphilis cases in the United States has remained fairly stable (Fig 26.9). Compared with gonorrhea (see Figure 26.5), this relative stability is remarkable because the epidemiology of the two diseases is quite similar, and concurrent infections are not uncommon.

Many states discontinued the requirements for premarital syphilis tests because so few cases were detected. At present, the population most at risk is economically disadvantaged inner city residents, especially drug-using prostitutes of both sexes.

The causative agent of syphilis is a weak-staining, gram-negative spirochete, *Treponema pallidum*. The spiro-

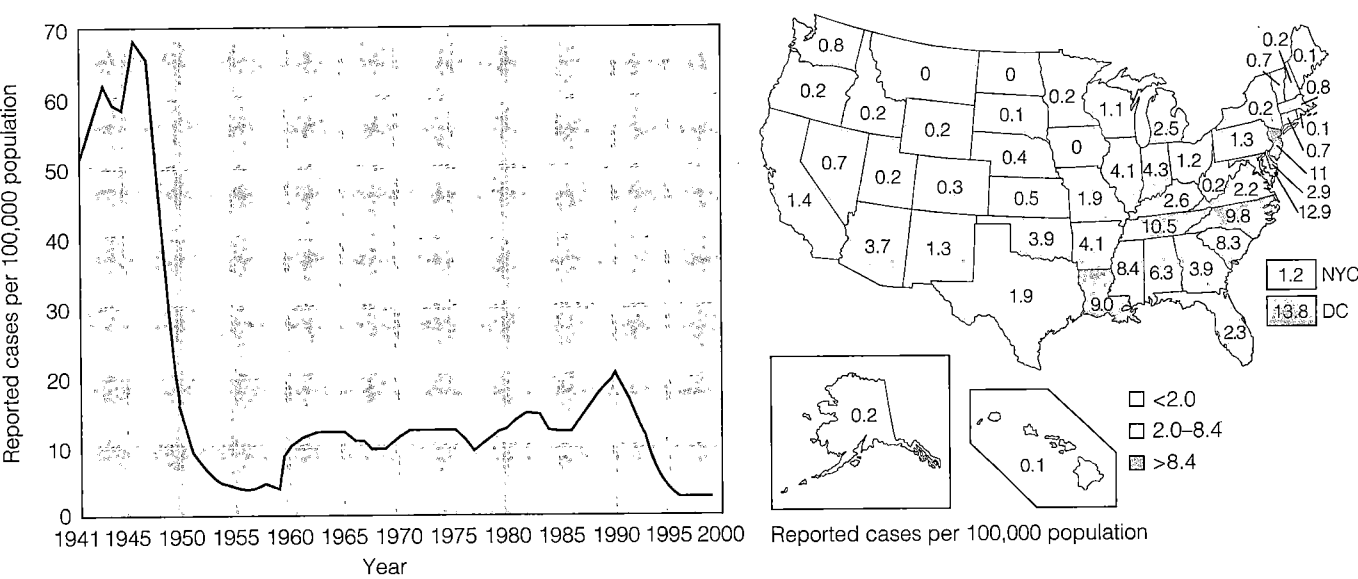

(a) Incidence of syphilis in the United States from 1941 to 1999

(b) Geographical distribution of cases in 1999

FIGURE 26.9 The U.S. incidence and distribution of primary and secondary syphilis.
[SOURCES: CDC, *Summary of Notifiable Diseases 1998, MMWR* 47(53)(12/31/99); *MMWR* 48(51)(1/7/00).]

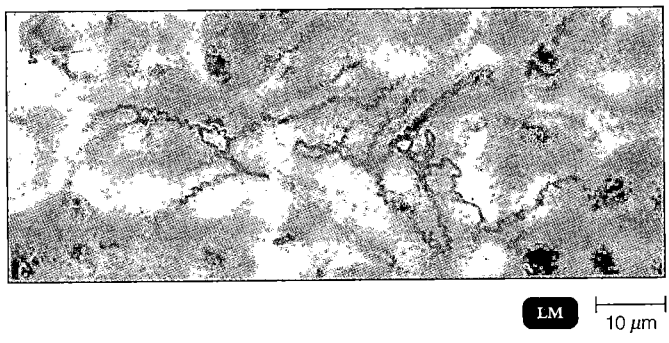

FIGURE 26.10 *Treponema pallidum,* **the cause of syphilis.** The microbes are made more visible in this brightfield micrograph by the use of a special silver stain.

■ How is syphilis transmitted?

chete is a thin, tightly coiled helix no more than 20 μm long (Figure 26.10) that lacks the enzymes necessary to build many complex molecules. Therefore, it relies on the host for many of the compounds necessary for life. The virulent strains have been successfully cultured only in cell cultures, which is not very useful for routine clinical diagnosis. Separate strains of *T. pallidum* are responsible for certain tropically endemic skin diseases such as **yaws.** These diseases are not sexually transmitted. Recently, investigators have been attempting to confirm speculation that syphilis may have arisen by mutation of *T. pallidum* in

New World cases of yaws after introduction into the new environment of Europe.

Syphilis is transmitted by sexual contact of all kinds, via syphilitic infections of the genitals or other body parts. The incubation period averages 3 weeks but can range from 2 weeks to several months. The disease progresses through several recognized stages.

In the *primary stage* of the disease, the initial sign is a small, hard-based **chancre,** or sore, which usually appears at the site of infection (Figure 26.11a). The chancre is painless, and an exudate of serum forms in the center. This fluid is highly infectious, and examination with a darkfield microscope shows many spirochetes. In a few weeks, this lesion disappears. None of these symptoms causes any distress. In fact, many women are entirely unaware of the chancre, which is often on the cervix. In males, the chancre sometimes forms in the urethra and is not visible. During this stage, bacteria enter the bloodstream and lymphatic system, which distribute them widely in the body.

Several weeks after the primary stage (the exact length of time varies), the disease enters the *secondary stage,* characterized mainly by skin rashes of varying appearance (Figure 26.11b). Other symptoms often observed are the loss of patches of hair, malaise, and mild fever. The rash is widely distributed on the skin and is also found in the mucous membranes of the mouth, throat, and cervix. At this stage, the lesions of the rash contain

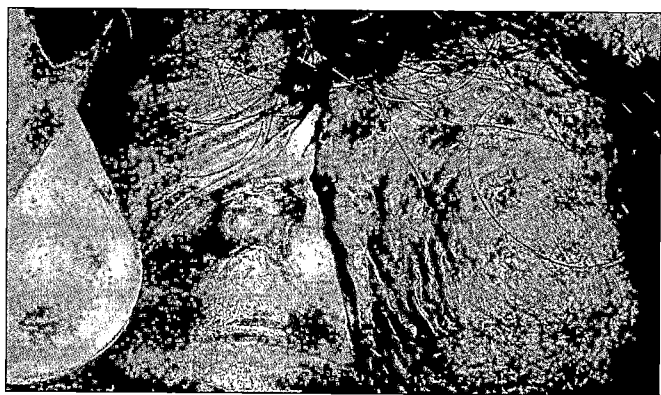

(a) Chancre of primary stage on a male in genital area.

FIGURE 26.11 **Characteristic lesions associated with various stages of syphilis.**

■ **How are the primary, secondary, and tertiary stages of syphilis distinguished?**

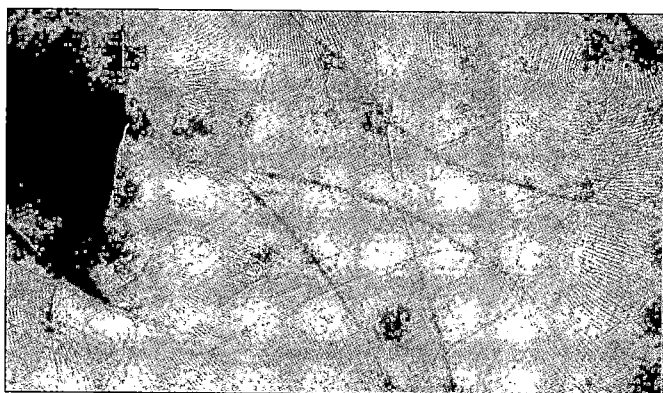

(b) Lesions of secondary syphilis rash on a palm; any surface area of the body may be afflicted with such lesions.

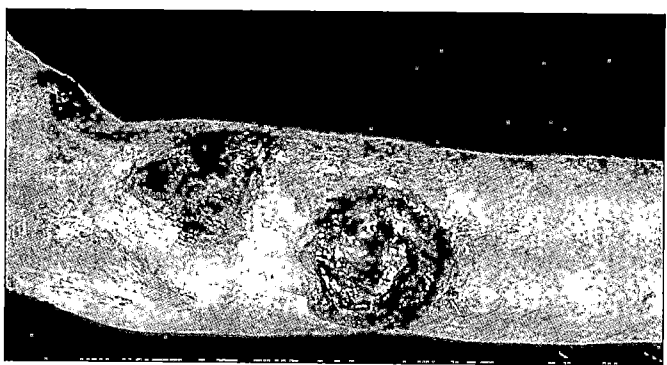

(c) Gummas of tertiary stage on the back of an arm; gummas such as these are rarely seen today in the era of antibiotics.

many spirochetes and are very infectious. Dentists and other health care workers coming into contact with fluid from these lesions can become infected by the spirochete entering through minute breaks in the skin. Such non-sexual transmission is possible, but the microbes do not survive long on environmental surfaces and are very un-likely to be transmitted via such objects as toilet seats.

The symptoms of secondary syphilis usually subside after a few weeks, and the disease enters a *latent period*. During this period, there are no symptoms. After 2–4 years of latency, the disease is not normally infectious, ex-cept for transmission from mother to fetus. The majority of cases do not progress beyond the latent stage, even without treatment.

Because the symptoms of primary and secondary syphilis are not disabling, people may enter the latent period without having received medical attention. In less than half of untreated cases, the disease reappears in a *tertiary stage*. This stage occurs only after an interval of many years, after the onset of the latent phase.

T. pallidum has an outer layer of lipids that stimulates remarkably little effective immune response, especially from cell-destroying complement reactions. It has been described as a Teflon® pathogen. Nonetheless, most of the symptoms of tertiary syphilis are probably due to the body's immune reactions, of a cell-mediated nature, to surviving spirochetes. Inflammatory responses by phago-cytic cells such as neutrophils and macrophages also con-tribute. This causes lesions called **gummas,** rubbery masses of tissue that appear in many organs and some-times on the external skin (Figure 26.11c). Although many of these lesions are not very harmful, some ulcerate

and cause extensive tissue damage, such as perforation of the palate (roof of the mouth), which interferes with speech. The cardiovascular system can be affected, most seriously by a weakening of the aorta. If the central ner-vous system is damaged, a loss of motor control occurs (*tabes dorsalis*); the brain may be affected (*paresis*), with personality changes, blindness, and seizures. Few, if any, pathogens are found in the lesions of the tertiary stage, and they are not considered very infectious. Today, cases of syphilis allowed to progress to this stage are rare.

One of the most distressing and dangerous forms of syphilis, called **congenital syphilis,** is transmitted across the placenta to the unborn fetus. Damage to mental de-velopment and other neurological symptoms are among the more serious consequences. This type of infection is most common when pregnancy occurs during the latent period of the disease. A pregnancy during the primary or secondary stage is likely to produce a stillbirth.

Diagnosis of syphilis is complex because each stage of the disease has unique requirements. Tests fall into three general groups: visual microscopic inspection, nontrep-onemal serological tests, and treponemal serological tests.

For preliminary screening laboratories use either non-treponemal serological tests, or microscopic examination of exudates from lesions when these are present. If a screening test is positive, the results are confirmed by treponemal serological tests.

Microscopic tests are important for screening for primary syphilis because serological tests for this stage are not reliable; antibodies take 1–4 weeks to form. The spirochetes can be detected in exudates of lesions by microscopic examination with a darkfield microscope (see Figure 3.4b on page 61). A darkfield microscope is necessary because the bacteria stain poorly and are only about 0.2 μm in diameter, near the lower limit of resolution for a brightfield microscope. Similarly, a direct fluorescent-antibody test (DFA-TP) using monoclonal antibodies (see Figure 18.10a on page 512) will both show and identify the spirochete. Figure 26.10 shows *T. pallidum* under brightfield illumination made possible by a special silver-impregnated stain.

At the secondary stage, when the spirochete has invaded almost all body organs, serological tests are reactive. Nontreponemal serological tests are so called because they are nonspecific; they do not detect antibodies produced against the spirochete itself but detect *reagin-type antibodies*. Reagin-type antibodies are apparently a response to lipid materials the body forms as an indirect reaction to infection by the spirochete. The antigen used in such tests is thus not the syphilis spirochete but an extract of beef heart (cardiolipin) that seems to contain lipids similar to those that stimulated the reagin-type antibody production. These tests will detect only about 70–80% of primary syphilis cases, but they will detect 99% of secondary syphilis cases. An example of nontreponemal tests is the slide agglutination **VDRL test** (for Venereal Disease Research Laboratory). Also used are modifications of the **rapid plasma reagin (RPR)** test, which is similar. The newest nontreponemal test is an ELISA test that uses the VDRL antigen.

Treponemal-type tests that react directly with the spirochete are used to check for false-positive nontreponemal tests and to diagnose late-stage syphilis. Almost 30% of patients fail at this stage to respond to nontreponemal tests. The treponemal tests in use detect antibodies in the patient with the use of antigens from *T. pallidum*. An example is the fluorescent treponemal antibody absorption test; **FTA-ABS** (an indirect fluorescent-antibody test, (see Figure 18.10b). Treponemal tests are not used for screening because about 1% of the results will be false-positives, but a positive test with both treponemal and nontreponemal types is highly specific.

Benzathine penicillin, a long-acting formulation that remains effective in the body for about 2 weeks, is the usual antibiotic treatment of syphilis. The serum concentrations achieved by this formulation are low, but the spirochete has remained very sensitive to this treatment.

For penicillin-sensitive people, several other antibiotics, such as doxycycline and tetracycline, have also proven effective. Antibiotic therapy to treat gonorrhea and other infections will not likely eliminate syphilis as well because such therapy is usually administered for too short a period to affect the slow-growing spirochete.

Lymphogranuloma Venereum (LGV)

A number of STDs that are uncommon in the United States occur frequently in the tropical areas of the world. For example, *Chlamydia trachomatis*, the cause of the eye infection trachoma and a major cause of NGU, is also responsible for **lymphogranuloma venereum (LGV),** a disease found in tropical and near-tropical regions. It is apparently caused by a strain of *C. trachomatis* that is invasive and tends to infect lymphoid tissue. In the United States, there are usually 200–400 cases each year, mostly in the southeast.

The microorganisms invade the lymphatic system, and the regional lymph nodes become enlarged and tender. Suppuration (a discharge of pus) may also occur. Inflammation of the lymph nodes results in scarring, which occasionally obstructs the lymph vessels. This blockage sometimes leads to massive enlargement of the external genitals in males. In females, rectal narrowing results from involvement of the lymph nodes in the rectal region. These conditions may eventually require surgery.

For diagnosis, pus can be aspirated from infected lymph nodes. When infected cells are properly stained with an iodine preparation, the clumped, intracellular chlamydias can be seen as inclusions. The isolated organisms can also be grown in cell culture or in embryonated eggs. The drug of choice for treatment is doxycycline.

Chancroid (Soft Chancre)

The STD known as **chancroid (soft chancre)** occurs most frequently in tropical areas, where it is seen more often than syphilis. The number of reported cases in the United States has been declining from a peak of 5000 cases in 1988. Almost all occur in New York, Texas, California, Florida, and Georgia. Like syphilis, its incidence is strongly associated with drug use. Because chancroid is so seldom seen by some physicians, and difficult to diagnose, it is probably underreported. It is very common in Africa, Asia, and Latin America.

In chancroid, a swollen, painful ulcer that forms on the genitals involves an infection of the adjacent lymph

table 26.1	Characteristics of the Most Common Types of Vaginitis and Vaginosis		
	Candidiasis	**Bacterial vaginosis**	**Trichomoniasis**
Causative agent	Fungus, *Candida albicans*	Bacterium, *Gardnerella vaginalis*	Protozoan, *Trichomonas vaginalis*
Odor	Yeasty or none	Fishy	Foul
Color of discharge	White	Gray-white	Greenish-yellow
Consistency of discharge	Curdy	Thin, frothy	Frothy
Amount of discharge	Varies	Copious	Copious
Appearance of vaginal mucosa	Dry, red	Pink	Tender, red
pH (normal pH is 4.0–4.5)	Below 4	Above 4.5	5–6

nodes. Infected lymph nodes in the groin area sometimes even break through and discharge pus to the surface. Such lesions are an important factor in the sexual transmission of HIV, especially in Africa. Lesions might also occur on such diverse areas as the tongue and lips. The causative agent is *Haemophilus ducreyi* (dü-krā'ē), a small gram-negative rod that can be isolated from exudates of lesions. The recommended antibiotics include erythromycin and ceftriaxone.

Bacterial Vaginosis

Inflammation of the vagina due to infection, or **vaginitis,** is most commonly caused by one of three organisms: the fungus *Candida albicans* (kan'did-ä al'bi-kans), the protozoan *Trichomonas vaginalis* (trik-ō-mōn'as va-jin-al'is), or the bacterium *Gardnerella vaginalis,* a small, pleomorphic gram-variable rod (Table 26.1). Most of these cases are attributed to the presence of *G. vaginalis* and are termed **bacterial vaginosis.** There is no sign of inflammation (and therefore the term *vaginosis* is preferred to vaginitis).

The condition is something of an ecological mystery. The population of *G. vaginalis* is many times higher than in uninfected women, the population of *Lactobacillus* bacteria is correspondingly decreased, and there is an overgrowth by anaerobic bacteria such as *Bacteroides* species. It is not exactly known whether the decrease in acid-producing lactobacilli, and the higher pH characteristic of the condition, is a cause or an effect of the overall microbial changes. Treatment with acetic acid gels or cultured lactic acid–producing bacteria (including yogurt!) has not been shown conclusively to be effective. The condition may be sexually transmitted, but also occurs in

women who are not sexually active. The microbe is a common inhabitant of the vaginas of asymptomatic women. Studies have identified the condition in 17–19% of women in family planning and student health clinics. There is no corresponding disease condition in males, but *Gardnerella* is often present in their urethras.

Bacterial vaginosis occurs when the vaginal pH is above 4.5. The condition is characterized by a pronounced fishy odor of a frothy vaginal discharge, which is usually copious. Diagnosis is based on the fishy odor, the vaginal pH, and the microscopic observation of *clue cells* in the discharge. These clue cells are sloughed-off vaginal epithelial cells covered with bacteria, mostly *G. vaginalis* (Figure 26.12). The disease has been considered more of a nuisance than a serious infection, but it is now seen as a factor in many premature births and low-birth-weight infants.

Treatment is primarily by metronidazole, a drug that eradicates the anaerobes essential to continuation of the disease but allows the normal lactobacilli to repopulate the vagina.

Viral Diseases of the Reproductive System

Learning Objective

■ *Discuss the epidemiology of genital herpes and genital warts.*

Viral diseases of the reproductive system are difficult to treat effectively, and so they represent an increasing health problem.

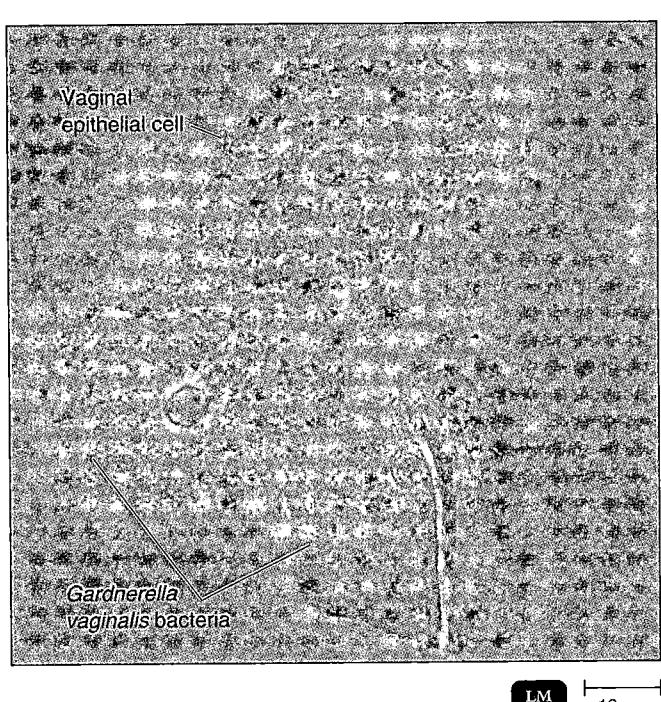

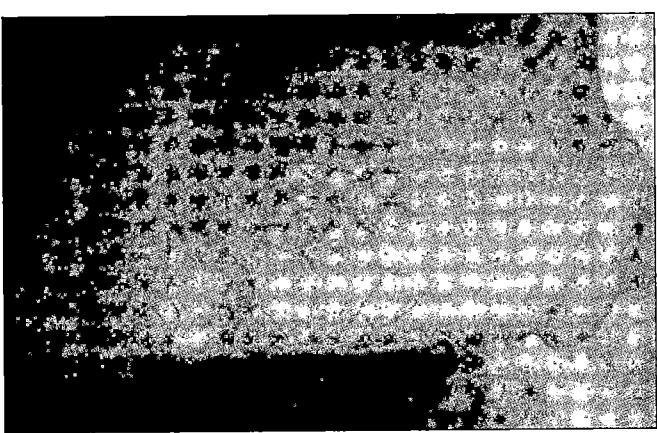

FIGURE 26.13 Vesicles of genital herpes on a penis.

■ What microbes causes genital herpes?

LM |————| 10 μm

FIGURE 26.12 Clue cells.

■ In the diagnosis of bacterial vaginosis, it is essential to find clue cells, vaginal epithelial cells coated with bacilli, mostly *Gardnerella vaginalis.*

Genital Herpes

A much publicized STD is **genital herpes,** usually caused by *herpes simplex virus type 2 (HSV-2).* (The herpes simplex virus occurs as either type 1 or type 2.) Herpes simplex virus type 1 (HSV-1) is primarily responsible for cold sores or fever blisters (see page 500), but it can also cause genital herpes. The official names are human herpesvirus 1 and 2.

In the United States, there are probably about 44 million people who are infected with HSV-2, and 4 million infected with HSV-1.

Genital herpes lesions appear after an incubation period of within 1 week and cause a burning sensation. After this, vesicles appear (Figure 26.13). In both males and females, urination can be painful, and walking is quite uncomfortable; the patient is even irritated by clothing. Usually, the vesicles heal in a couple of weeks.

The vesicles contain infectious fluid, but many times the disease is transmitted when no lesions or symptoms are apparent. Semen may contain the virus. Condoms may not provide protection because in women, the vesicles are usually on the external genitals (seldom on the

cervix or within the vagina), and in men, the vesicles may be on the base of the penis. Oral-genital contact may be a factor in transmission of HSV-1.

One of the most distressing characteristics of genital herpes is the possibility of recurrences. There is an element of truth in the medical adage that, unlike love, herpes is forever. As in other herpes infections, such as cold sores or chickenpox-shingles, the virus enters a lifelong latent state in nerve cells. Some people have several recurrences a year; for others, recurrence is rare. Men are more likely to experience recurrences than women. Reactivation appears to be triggered by several factors, including menstruation, emotional stress or illness (especially if accompanied by fever, a factor that is also involved in the appearance of cold sores), and perhaps just scratching the affected area. About 88% of patients with HSV-2 and about 50% of those with HSV-1 will have recurrences. If a person is going to experience recurrences, the first will usually appear within about 6 months after the initial infection.

Neonatal herpes is a serious consideration for women of childbearing age. The virus can cross the placental barrier and affect the fetus. The result can be spontaneous abortion or serious fetal damage, such as mental retardation and defective vision and hearing. Herpes infection of the newborn is most likely to have serious consequences when the mother acquires the initial herpes infection during the pregnancy. Therefore, any pregnant woman without a history of genital herpes should avoid sexual contact with anyone who might be carrying the virus. Damage to the fetus or newborn is much less likely, by a factor of about ten, from exposure to recurrent or asymptomatic

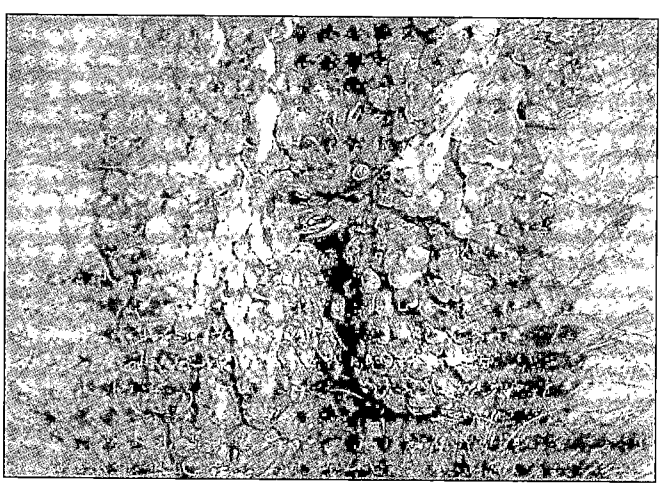

FIGURE 26.14 Genital warts on a vulva.

■ What is the relationship between genital warts and cervical cancer?

herpes. Maternal antibodies are apparently somewhat protective.

For practical purposes, the fetus is considered infected if the virus can be grown from amniotic fluid. If the fetus is free of the virus, it still needs to be protected from infection during passage through the birth canal. Clinical signs of infection cannot always be seen, and tests to detect asymptomatic shedding of the viruses are not very reliable. Therefore, screening for them is of very limited use. If there are obvious viral lesions at delivery time, a cesarean section is probably wise. The operation should be done before the fetal membrane ruptures and the viruses spread to the uterus.

There is no cure for genital herpes, although research on its prevention and treatment is intensive. Discussions of chemotherapy use terms such as *suppression* or *management* rather than *cure*. Acyclovir and other antiviral drugs are reasonably effective in alleviating the symptoms of a primary outbreak. There is some relief of pain and other symptoms and slightly faster healing. Continuous administration of antiviral drugs for several months has been found to lower the chances of recurrence during that time.

Genital Warts

Warts are an infectious disease; since 1907 it has been known that they are caused by viruses known as papillomaviruses. It is probably less well known that warts can be

transmitted sexually, and that this is an increasing problem. Nearly a million new cases of **genital warts** are estimated to occur in the United States each year.

There are more than 60 serotypes of papillomaviruses, and certain serotypes tend to be linked with certain forms of genital warts (technically, *condyloma acuminata*). Morphologically, some warts are extremely large and "warty" in appearance, with multiple fingerlike projections resembling cauliflower; others are relatively smooth or flat (Figure 26.14). The incubation period is usually a matter of a few weeks or months.

The greatest danger from genital warts is their connection to cancer. A few of the many serotypes are associated with a progression to cancer. In women this is usually cervical cancer, and in men it is usually cancer of the penis. Genital warts in women are much more likely to be precancerous than those in men. It is hoped that serological typing of warts will become a routine aid in determining which are the most dangerous, but at present this is expensive and limited to relatively few laboratories. Treatments of warts were discussed on page 587. Genital warts, because of the possibility of cancer, are more likely to be treated. Two patient-applied gels, podofilox and imiquimod, are often effective. Imiquimod (Aldara) stimulates the body to produce interferon (page 469), which appears to account for its antiviral activity.

AIDS

AIDS is a viral disease that is frequently transmitted by sexual contact. However, it affects the immune system, so it was discussed on pages 535–544 and the boxes in Chapter 13, page 395, and Chapter 19, page 543. It is important to remember that the lesions resulting from many of the diseases of bacterial and viral origin facilitate the transmission of HIV.

Fungal Disease of the Reproductive System

Learning Objective
■ Discuss the epidemiology of candidiasis.

The fungal disease described here is the well-known *yeast infection* for which nonprescription treatments are advertised.

Candidiasis

Candida albicans is a yeastlike fungus that often grows on mucous membranes of the mouth, intestinal tract, and genitourinary tract (Table 26.1; see Figure 21.16 on page 594). Infections are usually a result of opportunistic overgrowth when the competing microbiota are suppressed by antibiotics or other factors. As discussed in Chapter 21, *C. albicans* is the cause of **oral candidiasis,** or thrush. It is also responsible for occasional cases of NGU in males and for **vulvovaginal candidiasis,** which is the most common cause of vaginitis. About 75% of all women experience at least one episode.

The lesions of vulvovaginal candidiasis resemble those of thrush but produce more irritation; severe itching; a thick, yellow, cheesy discharge; and yeasty or no odor. *C. albicans,* the *Candida* species responsible for most cases, is an opportunistic pathogen. Predisposing conditions include the use of oral contraceptives and pregnancy, which cause an increase of glycogen in the vagina (see the discussion of the normal vaginal microbiota earlier in this chapter). Yeast infections are a frequent symptom in women suffering from uncontrolled diabetes; also, the use of broad-spectrum antibiotics suppresses the normal, competing bacterial microbiota, which leads to opportunistic fungal infections. Thus, diabetes and antibiotic therapy are predisposing factors to *C. albicans* vaginitis.

A yeast infection is diagnosed by microscopic identification of the fungus in scrapings of lesions and by isolation of the fungus in culture. Treatment consists of topical application of nonprescription antifungal drugs such as clotrimazole and miconazole. A single dose of oral fluconazole is an alternative form of treatment.

Protozoan Disease of the Reproductive System

Learning Objective

■ *Discuss the epidemiology of trichomoniasis.*

The only STD caused by a protozoan affects only women. Although common, it is not widely known.

Trichomoniasis

The anaerobic protozoan *Trichomonas vaginalis* is frequently a normal inhabitant of the vagina in females and of the urethra in many males (see Table 26.1 and Figure 26.15). If the normal acidity of the vagina is disturbed, the protozoan may overgrow the normal microbial population of the genital mucosa and cause **trichomoniasis.** (Males rarely have any symptoms as a result of the presence of the protozoan.) It is often accompanied by a

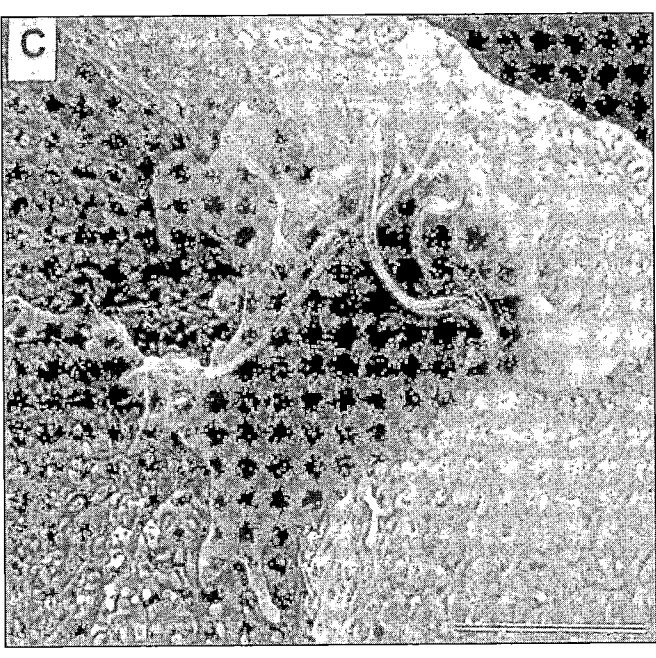

SEM ⊢————⊣ 5 μm

FIGURE 26.15 *Trichomonas vaginalis* **adhering to the surface of an epithelial cell in a cell culture preparation.** The flagella are clearly visible.
[SOURCE: D. Petrin et al., "Clinical and Microbiological Aspects of *Trichomonas vaginalis.*" *ASM Clinical Microbiology Reviews* 11:300–317 (1998).]

coinfection with gonorrhea. In response to the protozoan infection, the body accumulates leukocytes at the infection site. The resulting discharge is profuse, greenish-yellow, and characterized by a foul odor. Up to half the cases, however, are asymptomatic. This discharge is accompanied by irritation and itching. Trichomoniasis is usually sexually transmitted.

Diagnosis is usually made by microscopic examination and identification of the organisms in the discharge. They can also be isolated and grown on laboratory media. The pathogen can be found in semen or urine of male carriers. New rapid tests making use of DNA probes and monoclonal antibodies are now available. Treatment is by oral metronidazole, administered to both sex partners, which readily clears the infection.

★ ★ ★

The major microbial diseases of the urinary and reproductive systems are summarized in Table 26.2.

table 26.2 Microbial Diseases of the Urinary and Reproductive Systems

Disease	Pathogen	Comments
Bacterial Diseases of the Urinary System		
Cystitis (urinary bladder infection)	Escherichia coli, Staphylococcus saprophyticus	Difficulty or pain in urination.
Pyelonephritis (kidney infection)	Primarily E. coli	Fever; back or flank pain.
Leptospirosis (kidney infection)	Leptospira interrogans	Headaches, muscular aches, fever; kidney failure a possible complication.
Bacterial Diseases of the Reproductive System		
Gonorrhea	Neisseria gonorrhoeae	Males: painful urination and discharge of pus. Females: few symptoms but possible complications such as PID.
Nongonococcal urethritis (NGU)	Chlamydia or other bacteria, including Mycoplasma hominis and Ureaplasma urealyticum	Painful urination and watery discharge. In females, possible complications such as PID.
Pelvic inflammatory disease (PID)	N. gonorrhoeae, Chlamydia trachomatis	Chronic abdominal pain; possible infertility.
Syphilis	Treponema pallidum	Initial sore at site of infection, later skin rashes and mild fever. Final stages may be severe lesions, damage to cardiovascular and nervous systems. Today, few progress to tertiary stage.
Lymphogranuloma venereum (LGV)	Chlamydia trachomatis	Swelling in lymph nodes in groin.
Chancroid (soft chancre)	Haemophilus ducreyi	Painful ulcers of genitals, swollen lymph nodes in groin.
Bacterial vaginosis	Gardnerella vaginalis	Fishy odor, frothy vaginal discharge.
Viral Diseases of the Reproductive System		
Genital herpes	Herpes simplex virus type 2; HSV type 1	Painful vesicles in genital area.
Genital warts	Papillomavirus	Warts in genital area.
Fungal Disease of the Reproductive System		
Candidiasis	Candida albicans	Severe vaginal itching, yeasty odor, yellow discharge.
Protozoan Disease of the Reproductive System		
Trichomoniasis	Trichomonas vaginalis	Vaginal itching, greenish-yellow discharge.

Study Outline

INTRODUCTION (p. 720)

1. The urinary system regulates the chemical composition of the blood and excretes nitrogenous waste.

2. The reproductive system produces gametes for reproduction and, in the female, supports the growing embryo.

3. Microbial diseases of these systems can result from infection from an outside source or from opportunistic infection by members of the normal microbiota.

STRUCTURE AND FUNCTION OF THE URINARY SYSTEM (p. 720)

1. Urine is transported from the kidneys through ureters to the urinary bladder and is eliminated through the urethra.

2. Valves prevent urine from flowing back to the urinary bladder and kidneys.

3. The flushing action of urine and the acidity of normal urine have some antimicrobial value.

STRUCTURE AND FUNCTION OF THE REPRODUCTIVE SYSTEM (pp. 720–721)

1. The female reproductive system consists of two ovaries, two uterine tubes, the uterus, the cervix, the vagina, and the external genitals.

2. The male reproductive system consists of two testes, ducts, accessory glands, and the penis; seminal fluid leaves the male body through the urethra.

NORMAL MICROBIOTA OF THE URINARY AND REPRODUCTIVE SYSTEMS (p. 722)

1. The urinary bladder and upper urinary tract are sterile under normal conditions.

2. Lactobacilli dominate the vaginal microbiota during the reproductive years.

3. The male urethra is normally sterile.

DISEASES OF THE URINARY SYSTEM (pp. 723–725)

BACTERIAL DISEASES OF THE URINARY SYSTEM (pp. 723–725)

1. Urethritis, cystitis, and ureteritis are terms describing inflammations of tissues of the lower urinary tract.

2. Pyelonephritis can result from lower urinary tract infections or from systemic bacterial infections.

3. Opportunistic gram-negative bacteria from the intestines often cause urinary tract infections.

4. Nosocomial infections following catheterization occur in the urinary system. *E. coli* causes more than half of these infections.

5. More than 1000 bacteria of one species per milliliter of urine, or 100 coliforms per milliliter of urine, indicates an infection.

6. Treatment of urinary tract infections depends on the isolation and antibiotic sensitivity testing of the causative agents.

Cystitis (p. 723)

1. Inflammation of the urinary bladder, or cystitis, is common in females.

2. Microorganisms at the opening of the urethra and along the length of the urethra, careless personal hygiene, and sexual intercourse contribute to the high incidence of cystitis in females.

3. The most common etiologies are *E. coli* and *S. saprophyticus.*

Pyelonephritis (p. 723)

1. Inflammation of the kidneys, or pyelonephritis, is usually a complication of lower urinary tract infections.

2. About 75% of pyelonephritis cases are caused by *E. coli.*

Leptospirosis (pp. 723–725)

1. The spirochete *Leptospira interrogans* is the cause of leptospirosis.

2. The disease is transmitted to humans by urine-contaminated water.

3. Leptospirosis is characterized by chills, fever, headache, and muscle aches.

4. Diagnosis is based on isolation of the bacteria and serological identification.

DISEASES OF THE REPRODUCTIVE SYSTEM (pp. 725–735)

BACTERIAL DISEASES OF THE REPRODUCTIVE SYSTEM (pp. 725–732)

1. Most diseases of the reproductive system are sexually transmitted diseases (STDs).

2. Most STDs can be prevented by the use of condoms and are treated with antibiotics.

Gonorrhea (pp. 725–727)

1. *Neisseria gonorrhoeae* causes gonorrhea.

2. Gonorrhea is a common reportable communicable disease in the United States.

3. *N. gonorrhoeae* attaches to mucosal cells of the oral-pharyngeal area, genitals, eyes, and rectum by means of fimbriae.

4. Symptoms in males are painful urination and pus discharge. Blockage of the urethra and sterility are complications of untreated cases.

5. Females might be asymptomatic unless the infection spreads to the uterus and uterine tubes (see pelvic inflammatory disease).

6. Gonorrheal endocarditis, gonorrheal meningitis, and gonorrheal arthritis are complications that can affect both sexes if gonorrheal infections are untreated.

7. Ophthalmia neonatorum is an eye infection acquired by infants during passage through the birth canal of an infected mother.

8. Gonorrhea is diagnosed by Gram stain, ELISA, or DNA probe.

Nongonococcal Urethritis (NGU) (pp. 727–728)

1. Nongonococcal urethritis (NGU), or nonspecific urethritis (NSU), is any inflammation of the urethra not caused by *N. gonorrhoeae*.

2. Most cases of NGU are caused by *Chlamydia trachomatis*.

3. *C. trachomatis* infection is the most common STD.

4. Symptoms of NGU are often mild or lacking, although uterine tube inflammation and sterility may occur.

5. *C. trachomatis* can be transmitted to infants' eyes at birth.

6. Diagnosis is based on the detection of chlamydial DNA in urine.

7. *Ureaplasma urealyticum* and *Mycoplasma hominis* also cause NGU.

Pelvic Inflammatory Disease (PID) (p. 728)

1. Extensive bacterial infection of the female pelvic organs, especially of the reproductive system, is called pelvic inflammatory disease (PID).

2. PID is caused by *N. gonorrhoeae, Chlamydia trachomatis,* and other bacteria that gain access to the uterine tubes. Infection of the uterine tubes is called salpingitis.

3. PID can result in blockage of the uterine tubes and sterility.

Syphilis (pp. 728–731)

1. Syphilis is caused by *Treponema pallidum,* a spirochete that has not been cultured in vitro. Laboratory cultures are grown in cell cultures.

2. *T. pallidum* is transmitted by direct contact and can invade intact mucous membranes or penetrate through breaks in the skin.

3. The primary lesion is a small, hard-based chancre at the site of infection. The bacteria then invade the blood and lymphatic system, and the chancre spontaneously heals.

4. The appearance of a widely disseminated rash on the skin and mucous membranes marks the secondary stage. Spirochetes are present in the lesions of the rash.

5. The patient enters a latent period after the secondary lesions spontaneously heal.

6. At least 10 years after the secondary lesion, tertiary lesions called gummas can appear on many organs.

7. Congenital syphilis, resulting from *T. pallidum* crossing the placenta during the latent period, can cause neurological damage in the newborn.

8. *T. pallidum* is identifiable through darkfield microscopy of fluid from primary and secondary lesions.

9. Many serological tests, such as VDRL, RPR, and FTA-ABS, can be used to detect the presence of antibodies against *T. pallidum* during any stage of the disease.

Lymphogranuloma Venereum (LGV) (p. 731)

1. *C. trachomatis* causes lymphogranuloma venereum (LGV), which is primarily a disease of tropical and subtropical regions.

2. The initial lesion appears on the genitals and heals without scarring.

3. The bacteria are spread in the lymph system and cause enlargement of the lymph nodes, obstruction of lymph vessels, and swelling of the external genitals.

4. The bacteria are isolated and identified from pus taken from infected lymph nodes.

Chancroid (Soft Chancre) (pp. 731–732)

1. Chancroid, a swollen, painful ulcer on the mucous membranes of the genitals or mouth, is caused by *Haemophilus ducreyi*.

Bacterial Vaginosis (p. 732)

1. Vaginosis is an infection without inflammation caused by *Gardnerella vaginalis*.

2. Diagnosis of *G. vaginalis* is based on increased vaginal pH, fishy odor, and the presence of clue cells.

VIRAL DISEASES OF THE REPRODUCTIVE SYSTEM (pp. 732–734)

Genital Herpes (pp. 733–734)

1. Herpes simplex virus type 2 (HSV-2) causes genital herpes.

2. Symptoms of the infection are painful urination, genital irritation, and fluid-filled vesicles.

3. Neonatal herpes is contracted during fetal development or birth. It can result in neurological damage or infant fatalities.

4. The virus might enter a latent stage in nerve cells. Vesicles reappear following trauma and hormonal changes.

Genital Warts (p. 734)

1. Papillomaviruses cause warts.

2. The papillomaviruses that cause genital warts have been associated with cancer of the cervix or penis.

AIDS (p. 734)

1. AIDS is a sexually transmitted disease of the immune system (see Chapter 19, pages 535–544).

FUNGAL DISEASE OF THE REPRODUCTIVE SYSTEM (pp. 734–735)

Candidiasis (pp. 734–735)

1. *Candida albicans* causes NGU in males and vulvovaginal candidiasis, or yeast infection, in females.

2. Vulvovaginal candidiasis is characterized by lesions that produce itching and irritation.

3. Predisposing factors are pregnancy, diabetes, tumors, and broad-spectrum antibacterial chemotherapy.

4. Diagnosis is based on observation of the fungus and its isolation from lesions.

PROTOZOAN DISEASE OF THE REPRODUCTIVE SYSTEM (p. 735)

Trichomoniasis (p. 735)

1. *Trichomonas vaginalis* causes trichomoniasis when the pH of the vagina increases.

2. Diagnosis is based on observation of the protozoa in purulent discharges from the site of infection.

Study Questions

REVIEW

1. List the members of the normal microbiota of the urinary system, and show their habitats in Figure 26.1.

2. List the normal microbiota of the genital system, and show their habitats in Figures 26.2 and 26.3.

3. How are urinary tract infections transmitted?

4. Explain why *E. coli* is frequently implicated in cystitis in females. List some predisposing factors for cystitis.

5. Name one organism that causes pyelonephritis. What are the portals of entry for microbes that cause pyelonephritis?

6. Complete the following table:

Disease	Causative Agent	Symptoms	Method of Diagnosis	Treatment
Bacterial vaginosis				
Gonorrhea				
Syphilis				
PID				
NGU				
LGV				
Chancroid				

7. Leptospirosis is a kidney infection of humans and other animals. How is this disease transmitted? What types of activities would increase one's exposure to this disease? What is the etiology?

8. Describe the symptoms of genital herpes. What is the causative agent? When is this infection least likely to be transmitted?

9. Name one fungus and one protozoan that can cause genital system infections. What symptoms would lead you to suspect these infections?

10. List the genital infections that cause congenital and neonatal infections. How can transmission to a fetus or newborn be prevented?

MULTIPLE CHOICE

1. Which of the following is usually transmitted by contaminated water?
   a. *Chlamydia*
   b. leptospirosis
   c. syphilis
   d. trichomoniasis
   e. none of the above

Use the following choices to answer questions 2–5:
   a. *Candida*
   b. *Chlamydia*
   c. *Gardnerella*
   d. *Neisseria*
   e. *Trichomonas*

2. Microscopic examination of vaginal smear shows flagellated eukaryotes.

3. Microscopic examination of vaginal smear shows ovoid eukaryotic cell.

4. Microscopic examination of vaginal smear shows epithelial cells covered with bacteria.

5. Microscopic examination of vaginal smear shows gram-negative cocci in phagocytes.

Use the following choices to answer questions 6–8:
   a. candidiasis
   b. bacterial vaginosis
   c. genital herpes
   d. lymphogranuloma venereum
   e. trichomoniasis

6. Difficult to treat with chemotherapy

7. Fluid-filled vesicles

8. Frothy, fishy discharge

Use the following choices to answer questions 9 and 10:
   a. *Chlamydia trachomatis*
   b. *Escherichia coli*
   c. *Mycobacterium hominis*
   d. *Staphylococcus saprophyticus*

9. The most common cause of cystitis.

10. In cases of NGU, diagnosis is made using PCR to detect microbial DNA.

CRITICAL THINKING

1. The tropical skin disease called yaws is transmitted by direct contact. Its causative agent, *Treponema pallidum pertenue,* is indistinguishable from *T. pallidum.* The appearance of syphilis in Europe coincides with the return of Columbus from the New World. How might *T. pallidum pertenue* have evolved into *T. pallidum* in the temperate climate of Europe?

2. Why can frequent douching be a predisposing factor to bacterial vaginosis, vulvovaginal candidiasis, or trichomoniasis?

3. *Neisseria* is cultured on Thayer-Martin media, consisting of chocolate agar and nystatin, incubated in a 5% CO_2 environment. How is this selective for *Neisseria?*

4. The list below is a key to selected microorganisms that cause genitourinary infections. Complete this key by listing genera discussed in this chapter in the blanks that correspond to their respective characteristics.
   Gram-negative bacteria
      Spirochete
         Aerobic   _____
         Anaerobic   _____
      Coccus
         Oxidase-positive   _____
      Bacillus, nonmotile
         Requires X factor   _____
         Gram-positive wall   _____
      Obligate intracellular parasite   _____
      Lacking cell wall
         Urease-positive   _____
         Urease-negative   _____
   Fungus
      Pseudohyphae   _____
   Protozoa
      Flagella   _____
   No organism observed/cultured from
   patient   _____

CLINICAL APPLICATIONS

1. A previously healthy 19-year-old female was admitted to a hospital after 2 days of nausea, vomiting, headache, and neck stiffness. Cerebrospinal fluid and cervical cultures showed gram-negative diplococci in leukocytes; a blood culture was negative. What disease did she have? How was it probably acquired?

2. A 28-year-old woman was admitted to a Wisconsin hospital with a 1-week history of arthritis of the left knee. Four days later, a 32-year-old man was examined for a 2-week history of urethritis and a swollen, painful left wrist. A 20-year-old woman seen in a Philadelphia hospital had pain in the right knee, left ankle, and left wrist for 3 days. Pathogens cultured from synovial fluid or urethral culture were gram-negative diplococci that required proline to grow. Antibiotic sensitivity tests gave the following results:

Antibiotic	MIC Tested (μg/ml)	Susceptible MIC (μg/ml)
Cefoxitin sodium	0.5	$\leq$2
Penicillin	8	$\leq$0.06
Spectinomycin	64	$\leq$32
Tetracycline	4	$\leq$0.25

What is the pathogen, and how is this disease transmitted? Which of the antibiotics should be used for treatment? What is the evidence that these cases are related?

3. Using the following information, determine what the disease is and how the infant's illness might have been prevented:
   May 11:   A 23-year-old woman has her first prenatal examination. She is 4½ months pregnant. Her VDRL results are negative.
   June 6:   The woman returns to her physician complaining of a labial lesion of a few days' duration. A biopsy is negative for malignancy, and herpes test results are negative.
   July 1:   The woman returns to her physician because the labial lesion continues to cause some discomfort.
   Sept. 15: The baby's father has multiple penile lesions and a generalized body rash.
   Sept. 25: The woman delivers her baby. Her RPR is 1:32 and the infant's is 1:128.
   Oct. 1:   The woman takes her infant to a pediatrician because the baby is lethargic. She is told the infant is healthy and not to worry.
   Oct. 2:   The baby's father has a persistent body rash and plantar and palmar rashes.
   Nov. 8:   The infant becomes acutely ill with pneumonia and is hospitalized. The admitting physician finds signs of osteochondritis.

Learning with Technology

VU = **VirtualUnknown CD-ROM**

VU   Enter the Virtual Lab, click the arrow next to the Session field, click Textbook Exercises, and select Chapter 26. Read the Case Study carefully. Consult with Table 26.2 on page 736 of your textbook, and make a list of the possible causes of this infection.

1. Is this infection most likely cystitis, pyelonephritis, urethritis, or ureteritis? Explain your reasoning.

2. Complete the identification of this pathogen using VirtualUnknown™ Microbiology. Is this one of the more common causative agents of the disease you indicated in your response to question 1?

3. Does the occupation of the victim complicate the investigation of this disease? What role could it play in the development of this disease?

Environmental and Applied Microbiology

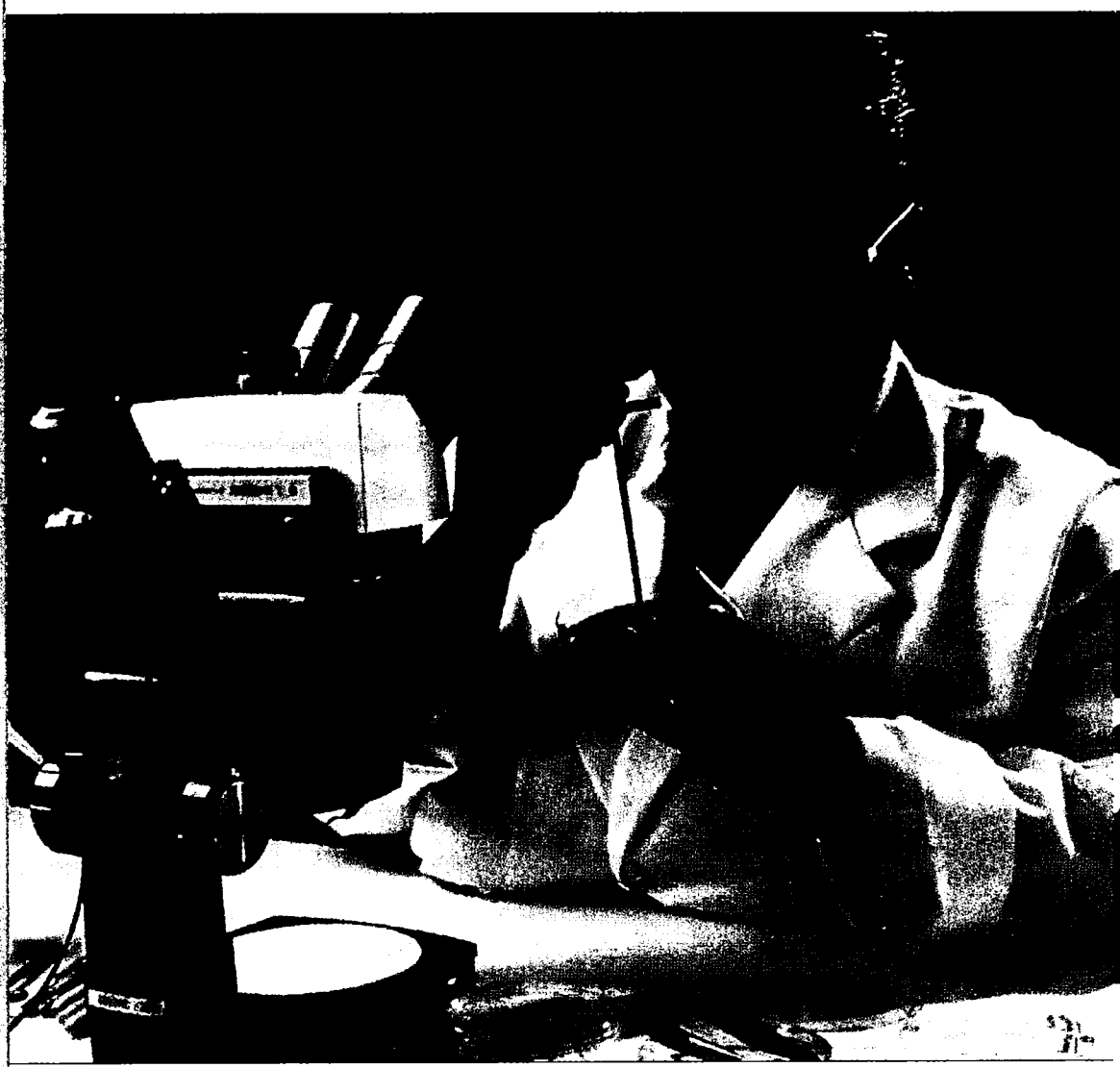

The use of microbial diseases of pests can provide a safe and environmentally sound alternative to chemical pesticides. When the disease-causing microbe has killed all of the pests, the microbe dies and is decomposed. Here, a microbiologist is studying the effects of infection on an insect that destroys agricultural crops.

Microbes often stick to surfaces in masses called biofilms.

n previous chapters, we focused primarily on the disease-causing capabilities of microorganisms. In this chapter, you will learn about many of the positive functions microbes perform in the environment. Bacteria and other microorganisms are, in fact, essential to the maintenance of life on Earth.

Metabolic Diversity

Microbes, especially Bacteria and Archaea, live in the most widely varied habitats on Earth. They are found in frozen antarctic regions as well as in boiling hot springs. Microbes have been recovered from rock a kilometer or more down and from the thin atmosphere thousands of meters above the ground. Explorations of the deepest ocean have revealed large numbers of microbes living there, in eternal darkness and subjected to incredible pressures (see the box in Chapter 6, page 160). Microbes are also found in clear mountain streams flowing from a melting glacier and in waters nearly saturated with salts, such as those of the Dead Sea (see the box in Chapter 5, page 144). The ability of microorganisms to live in so many habitats is due to their *metabolic diversity;* that is, different microbes can use a variety of carbon and energy sources and can grow under many different physical conditions.

Habitat Variety

Learning Objective

■ *Define extremophile, and identify two "extreme" habitats.*

The diversity of microbial populations indicates that they take advantage of any niches found in their environment. Different amounts of oxygen, light, or nutrients may exist within a few millimeters in the soil. As a population of aerobic organisms uses up the available oxygen, anaerobes are able to grow. If the soil is disturbed by plowing, earthworms, or other activity the aerobes will again be able to grow to repeat this succession.

Microbes that live in extreme conditions of temperature, acidity, alkalinity, or salinity are called **extremophiles.** Many are members of the Archaea. The enzymes that make growth possible under these conditions have been of great interest to industries, because they can tolerate extremes of temperature and pH that would inactivate other enzymes. The heat-resistant *Taq* polymerase enzyme from the source organism *Thermus aquaticus,* discussed in the context of PCR on page 295, is an example.

Microorganisms live in an intensely competitive environment and must exploit any advantage they can. They may metabolize common nutrients more rapidly or use nutrients that competing organisms cannot metabolize. Some, such as the lactic acid bacteria that are so useful in making dairy products, are able to make an environmental niche inhospitable to competing organisms. The lactic acid bacteria are unable to use oxygen as an electron acceptor and are able to ferment sugars only to lactic acid, leaving most of the energy unused. However, the acidity inhibits the growth of more efficient, competing microbes.

Symbiosis

Learning Objectives

- *Define symbiosis, differentiate parasitism from mutualism, and give an example of each.*
- *Define mycorrhiza.*

Recall from Chapter 14 that **symbiosis** is the interaction between coexisting organisms or populations. **Parasitism** is a type of symbiosis in which one microbe derives its nutrients and reproductive capability from another organism. *Bdellovibrio* bacteria, for example, prey on other bacteria (see the box in Chapter 3, page 60). **Mutualism,** another type of symbiosis, is an association between two organisms or populations that benefits both partners. Lichens are an example of a mutually advantageous association between a fungus and an alga or cyanobacterium.

Economically, the most important example of an animal-microbe symbiosis is that of the ruminants, animals that have a tanklike digestive organ called a *rumen.* Ruminants, such as cattle and sheep, graze on cellulose-rich plants. Bacteria in the rumen ferment the cellulose into compounds that are absorbed into the animal's blood, to be used for carbon and energy. Rumen fungi probably hydrolyze other plant components such as wood. Rumen protozoa keep the bacterial population under control by eating bacteria. In addition, many of the microbes in the rumen are digested for protein. (Another example of an animal-microbe symbiosis is described in the box in Chapter 6, page 160.)

A very important contribution to plant growth is made by **mycorrhizae,** or mycorrhizal symbionts (*myco* = fungus; *rhiza* = root). There are two primary types of these fungi: *endomycorrhizae,* also known as *vesicular-arbuscular mycorrhizae;* and *ectomycorrhizae.* Both types

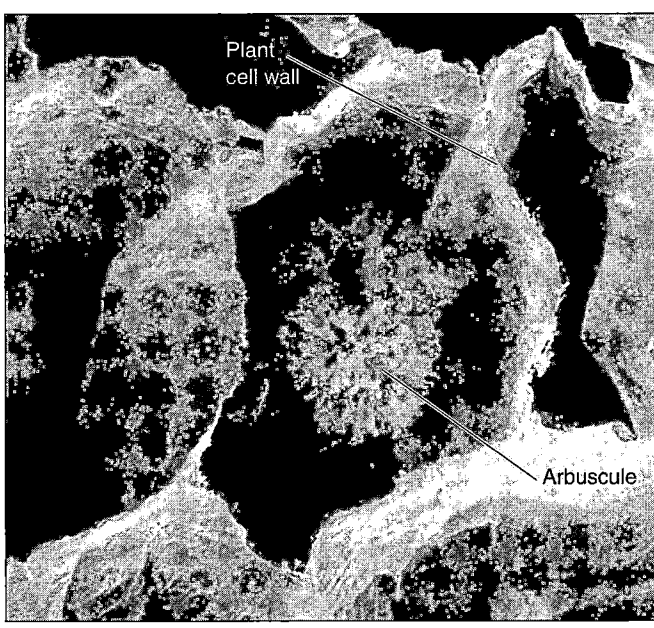

(a) Endomycorrhiza (vesicular-arbuscular mycorrhiza) A fully developed arbuscule of an endomycorrhiza in a plant cell. (The term arbuscule means little bush). As the arbuscule decomposes, it releases nutrients for the plant.

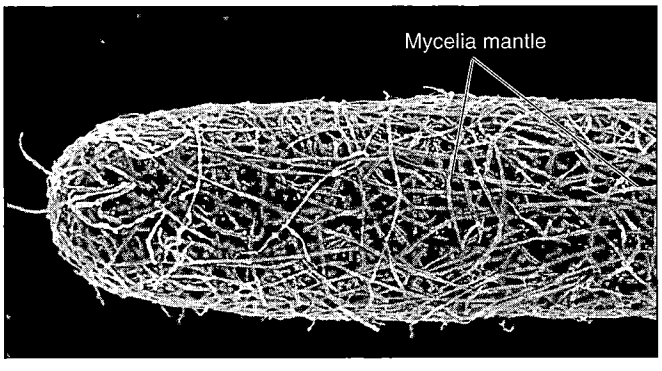

(b) Ectomycorrhiza The mycelial mantle of a typical ectomycorrhizal fungus surrounding a eucalyptus tree root.

FIGURE 27.1 Mycorrhizae.

■ **Of what value is a mycorrhiza to a plant?**

function as do root hairs on plants; that is, they extend the surface area through which the plant can absorb nutrients, especially phosphorus, which is not very mobile in soil.

Vesicular-arbuscular mycorrhizae form large spores that can be isolated easily from soil by sieving. The hyphae from these germinating spores penetrate into the plant root and form two types of structures: vesicles and arbuscules. **Vesicles** are smooth oval bodies that probably function as storage structures. **Arbuscules,** tiny bushlike structures, are formed inside plant cells (Figure 27.1a). Nutrients travel from the soil through fungal hyphae to these arbuscules, which gradually break down and release the nutrients to the plants. Most grasses and other plants are surprisingly dependent on these fungi for proper growth, and their presence is nearly universal in the plant kingdom.

Ectomycorrhizae mainly infect trees such as pine and oak. The fungus forms a mycelial *mantle* over the smaller roots of the tree (Figure 27.1b). Ectomycorrhizae do not form vesicles or arbuscules. Managers of commercial pine tree farms must ensure that seedlings are inoculated with soil containing effective mycorrhizae. Truffles, known as a food delicacy, are ectomycorrhizae. In Europe, trained pigs or dogs are often used to find them by smell and root them up. To the female pig, the odor is that of a potential mate. In nature, proliferation of the fungus depends on ingestion by an animal, which distributes the undigested spores into new locations.

Soil Microbiology and Biogeochemical Cycles

Learning Objective

■ *Define biogeochemical cycle.*

Billions of organisms, microscopic as well as comparatively huge insects and earthworms, form a vibrant living community in the soil. Typical soil has millions of bacteria in each gram. As Table 27.1 shows, the population is largest in the top few centimeters of soil and declines rapidly with depth. The most numerous organisms in soil are bacteria. Although actinomycetes are bacteria, they are usually considered separately. Many important antibiotics, such as streptomycin and tetracycline, were discovered by microbiologists investigating actinomycetes in soil.

Bacterial soil populations are usually estimated using plate counts on nutrient media, and the actual numbers are probably greatly underestimated by this method. No single nutrient medium or growth condition can possibly meet all the nutrient and other requirements of soil microorganisms.

We can think of soil as a "biological fire." A leaf falling from a tree is consumed by this "fire" as its organic matter is metabolized by microbes in the soil. Elements in the leaf enter the **biogeochemical cycles** for carbon, nitrogen, and sulfur that we will discuss in this chapter. In biogeochemical cycles, elements are oxidized and reduced by microorganisms to meet their metabolic needs. (See the discussion of oxidation-reduction in Chapter 5, page 121.) Without biogeochemical cycles, life on Earth would cease to exist.

The Carbon Cycle

Learning Objective

■ *Outline the carbon cycle, and explain the roles of microorganisms in this cycle.*

The primary biogeochemical cycle is the **carbon cycle** (Figure 27.2). All organisms, including plants, microbes, and animals, contain large amounts of carbon in the form of organic compounds such as cellulose, starches, fats, and proteins. Let's take a closer look at how these organic compounds are formed.

Recall from Chapter 5 that autotrophs, an essential component of all life on Earth, reduce carbon dioxide to form organic matter. When you look at a tree, you might think that its mass is from the soil where it grows. In fact, its great mass of cellulose is derived from the 0.03% of carbon dioxide in the atmosphere. This occurs as a result of photosynthesis, the first step of the carbon cycle in

table 27.1	Microorganisms per Gram of Typical Garden Soil at Various Depths			
Depth (cm)	Bacteria	Actinomycetes*	Fungi	Algae
3–8	9,750,000	2,080,000	119,000	25,000
20–25	2,179,000	245,000	50,000	5000
35–40	570,000	49,000	14,000	500
65–75	11,000	5000	6000	100
135–145	1400	—	3000	—

*Filamentous bacteria

SOURCE: Adapted from M. Alexander, *Introduction to Soil Microbiology,* 2nd ed. New York: Wiley, 1991.

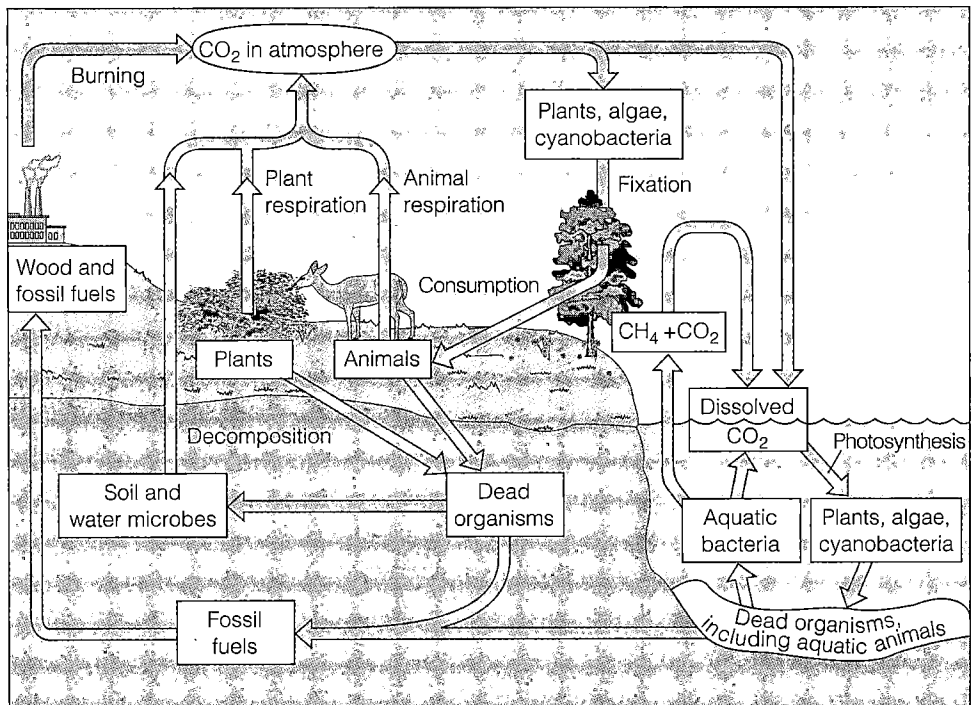

FIGURE 27.2 The carbon cycle. On a global scale, the return of CO_2 to the atmosphere by respiration closely balances its removal by fixation. However, the burning of wood and fossil fuels adds more CO_2 to the atmosphere; as a result, the amount of atmospheric CO_2 is steadily increasing.

■ Carbon-containing compounds are oxidized and reduced in the carbon cycle, making carbon available in the correct form for each organism.

which photoautotrophs such as cyanobacteria, green plants, algae, and green and purple sulfur bacteria *fix,* or incorporate carbon dioxide into organic matter using energy from sunlight. Chemoautotrophs such as *Thiobacillus* and *Beggiatoa* also fix carbon dioxide into organic matter, while metabolizing compounds such as hydrogen sulfide for energy.

In the next step of the cycle, chemoheterotrophs such as animals and protozoa eat autotrophs, and may in turn be eaten by other animals. Thus, as the organic compounds of the autotrophs are digested and resynthesized, the carbon atoms of carbon dioxide are transferred from organism to organism up the food chain.

Some of the organic molecules are used by chemoheterotrophs, including animals, to satisfy their energy requirements. When this energy is released through respiration, carbon dioxide immediately becomes available to start the cycle over again. Much of the carbon remains within the organisms until they excrete it as wastes or die. When plants and animals die, these organic compounds are decomposed by bacteria and fungi. During decomposition, the organic compounds are oxidized, and CO_2 is returned to the cycle.

Carbon is stored in rocks, such as limestone ($CaCO_3$), and is dissolved as carbonate ions (CO_3^{2-}) in oceans. Vast deposits of fossil organic matter exist in the form of fossil fuels, such as coal and petroleum. Burning these fossil fuels releases CO_2, resulting in an increased amount of CO_2

in the atmosphere. Many scientists believe the increased atmospheric carbon dioxide may be causing a **global warming** of the Earth (see the box on page 747).

The Nitrogen Cycle

Learning Objectives

■ Outline the nitrogen cycle, and explain the roles of microorganisms in this cycle.

■ Define ammonification, nitrification, denitrification, and nitrogen fixation.

The **nitrogen cycle** is shown in Figure 27.3. Nitrogen is needed by all organisms for the synthesis of protein, nucleic acids, and other nitrogen-containing compounds. Molecular nitrogen (N_2) makes up almost 80% of the Earth's atmosphere. For assimilation and use by plants, nitrogen must be *fixed,* that is, taken up and combined into organic compounds. The activities of specific microorganisms are important to the conversion of nitrogen to usable forms.

Ammonification

Almost all the nitrogen in the soil exists in organic molecules, primarily in proteins. When an organism dies, the process of microbial decomposition results in the hydrolytic breakdown of proteins into amino acids. In a process called **deamination,** the amino groups of amino

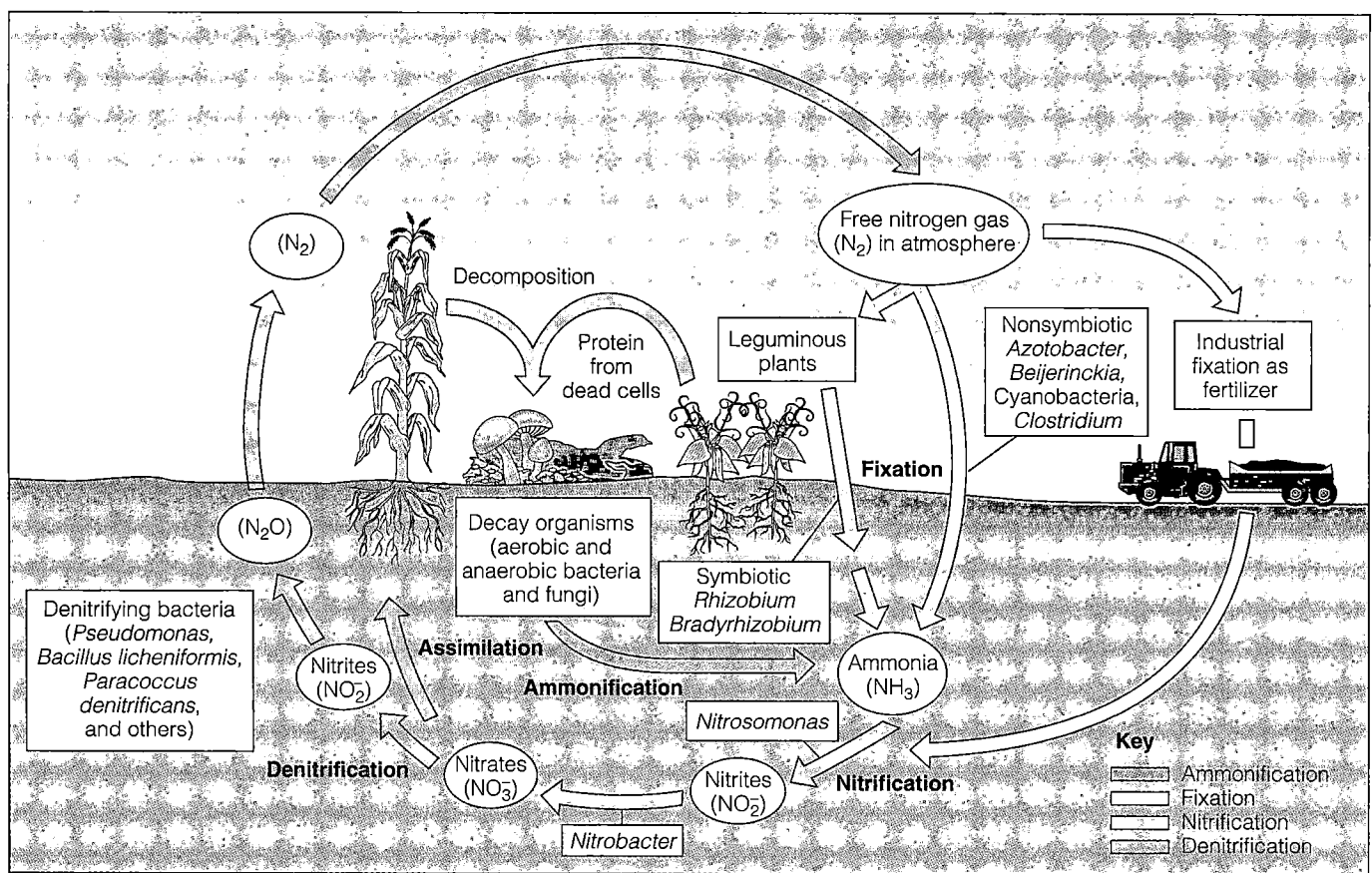

FIGURE 27.3 The nitrogen cycle. In general, nitrogen in the atmosphere goes through fixation, nitrification, and denitrification. Nitrates assimilated into plants and animals after nitrification go through decomposition, ammonification, and then nitrification again.

■ Which processes are performed exclusively by bacteria?

---

acids are removed and converted into ammonia (NH_3) (see Figure 27.3). This release of ammonia is called **ammonification.** Ammonification, brought about by numerous bacteria and fungi, can be represented as follows:

Proteins from dead cells and waste products $\xrightarrow[\text{decomposition}]{\text{Microbial}}$ Amino acids

Amino acids $\xrightarrow[\text{ammonification}]{\text{Microbial}}$ Ammonia (NH_3)

Microbial growth releases extracellular proteolytic enzymes that decompose proteins. The resulting amino acids are transported into the microbial cells, where ammonification occurs. The fate of the ammonia produced by ammonification depends on soil conditions (see the discussion of denitrification, which follows). Because ammonia is a gas, it rapidly disappears from dry soil, but in

moist soil it becomes solubilized in water, and ammonium ions (NH_4^+) are formed:

$$NH_3 + H_2O \longrightarrow NH_4^+ \; OH \longrightarrow NH_4^+ + OH^-$$

Ammonium ions from this sequence of reactions are used by bacteria and plants for amino acid synthesis.

Nitrification

The next sequence of reactions in the nitrogen cycle involves the oxidation of the nitrogen in the ammonium ion to produce nitrate, a process called **nitrification.** Living in the soil are autotrophic nitrifying bacteria, such as those of the genera *Nitrosomonas* and *Nitrobacter.* These microbes obtain energy by oxidizing ammonia or nitrite. In the first stage, *Nitrosomonas* oxidizes ammonium to nitrites:

$$\underset{\text{Ammonium ion}}{NH_4^+} \xrightarrow{\textit{Nitrosomonas}} \underset{\text{Nitrite ion}}{NO_2^-}$$

Water, methane (CH_4), and carbon dioxide (CO_2) in Earth's atmosphere absorb infrared radiation that would otherwise radiate out into space and reradiate it back to Earth. This phenomenon is important for maintaining Earth's temperature, because without these so-called "greenhouse gases," the average temperature would be about $-18°C$.

However, CH_4 levels have more than doubled since 1860, and CH_4 is 21 times more effective as a greenhouse gas than CO_2. The rising CH_4 content could eventually warm the atmosphere enough to melt the polar ice caps and flood major coastal cities. In addition, the more pessimistic scientists have warned that the warmer air could cause chronic droughts in major crop-growing areas, resulting in worldwide famine.

Atmospheric CH_4 comes from leaks in municipal gas pipes, coal mining, and microbial metabolism in rice paddies, livestock, and landfills. Dr. Boyd Strain, a botanist at Duke University, suggests that much of the increase in agricultural productivity after the Industrial Revolution came not from the use of fertilizers and pesticides but from the increasing abundance of CO_2 in the atmosphere.

Additional plant growth also results in a corresponding increase in the amount of decaying plant matter. The dangerous increase in the CH_4 content of the air may be largely a result of the

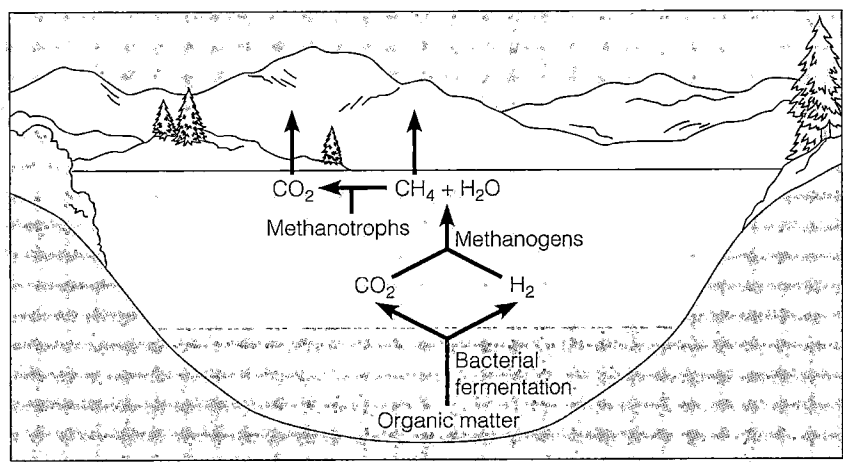

Methane is generated in the sediments of lakes and bays by the action of methanogens. Methanotrophs oxidize methane at the air-water interface.

decay of the extra plant matter whose growth was stimulated by the increased amount of CO_2. When dead plants decay in anaerobic pockets of organic debris in forests or grasslands, in the sediment of lakes and bays, or in rice paddies, bacterial fermentation breaks down the organic matter, releasing CO_2 and hydrogen (H_2). Anaerobic archaea known as *methanogens* then combine these into CH_4 and water (see the figure).

There are aerobic methaneoxidizing bacteria (methanotrophs) at the air-water interface and in topsoil. Methanotrophs rapidly oxidize methane to CO_2. Although it is not known what organisms consume *atmospheric* methane,

the actions of methanotrophs and methanogens are expected to be in balance. Why, then, is CH_4 increasing at such an alarming rate? One problem is that methane transport to cells is inhibited in water and in waterlogged soil. Are the activities of humans, such as pollution and the use of fertilizers, inhibiting the growth of methanotrophs? While some scientists are measuring the increasing levels of CH_4 in the atmosphere, others are isolating methanotrophs and determining their numbers and optimum growth conditions. With more information, we may be able to use methanotrophs to remove greenhouse gases.

In the second stage, such organisms as *Nitrobacter* oxidize nitrites to nitrates:

$$NO_2^- \xrightarrow{\text{Nitrobacter}} NO_3^-$$

Nitrite ion   Nitrate ion

Plants tend to use nitrate as their source of nitrogen for protein synthesis because nitrate is highly mobile in soil and is more likely to encounter a plant root than ammonium. Ammonium ions would actually make a more efficient source of nitrogen because they require less en-

ergy to incorporate into protein, but these positively charged ions are usually bound to negatively charged clays in the soil, whereas the negatively charged nitrate ions are not bound.

Denitrification

The form of nitrogen resulting from nitrification is fully oxidized and no longer contains any biologically usable energy. However, it can be used as an electron acceptor by microbes metabolizing other organic energy sources in the absence of atmospheric oxygen (see *anaerobic respiration* in Chapter 5). This process, called **denitrification,** can lead to a loss of nitrogen to the atmosphere, especially as nitrogen gas. Denitrification can be represented as follows:

$$NO_3^- \longrightarrow NO_2^- \longrightarrow N_2O \longrightarrow N_2$$

Nitrate ion   Nitrite ion   Nitrous   Nitrogen
                               oxide   gas

Pseudomonas species appear to be the most important group of bacteria in denitrification in soils. Denitrification occurs in waterlogged soils where little oxygen is available. In the absence of oxygen as an electron acceptor, denitrifying bacteria substitute the nitrates of agricultural fertilizer. This converts much of the valuable nitrate into gaseous nitrogen that enters the atmosphere and represents a considerable economic loss.

Nitrogen Fixation

We live at the bottom of an ocean of nitrogen gas. The air we breathe is about 79% nitrogen, and above every acre of soil (the area of an American football field from the goal line to the opposite 10-yard line) stands a column of nitrogen weighing about 32,000 tons. But the only creatures on Earth that can use it directly as a nitrogen source are a few species of bacteria, including cyanobacteria. The process by which they convert nitrogen gas to ammonia is known as **nitrogen fixation.**

The nitrogenase enzyme responsible for nitrogen fixation is anaerobic. In fact it is inactivated by oxygen. Therefore, it probably evolved early in the history of the planet, before the atmosphere contained much molecular oxygen and before nitrogen-containing compounds were available from decaying organic matter. Nitrogen fixation is brought about by two types of microorganisms: free-living and symbiotic. (Agricultural fertilizers are made up of nitrogen that has been fixed by industrial physical-chemical processes.)

Free-Living Nitrogen-Fixing Bacteria These bacteria are found in particularly high concentrations in the *rhizosphere,* the region where the soil and roots make contact, especially in grasslands. Among these bacteria that can fix nitrogen are aerobic species such as *Azotobacter.* These aerobic organisms apparently shield the anaerobic nitrogenase enzyme from oxygen by, among other things, having a very high rate of oxygen use that minimizes the diffusion of oxygen into the interior of the cell, where the enzyme is located.

Another free-living obligate aerobe that fixes nitrogen is *Beijerinckia* (bī-yĕ-rink'ē-ä). Some anaerobic bacteria, such as certain species of *Clostridium,* also fix nitrogen. The bacterium *C. pasteurianum* (pas-tyĕr-ē-a'num), an obligately anaerobic, nitrogen-fixing microorganism, is a prominent example.

There are many species of aerobic, photosynthesizing cyanobacteria that fix nitrogen. Because their energy supply is independent of carbohydrates in soil or water, they are especially useful suppliers of nitrogen to the environment. Cyanobacteria usually carry their nitrogenase enzymes in specialized structures called **heterocysts** that provide anaerobic conditions for fixation (see Figure 11.13a on page 317).

Most of the free-living nitrogen-fixing bacteria are capable of fixing large amounts of nitrogen under laboratory conditions. However, in the soil there is usually a shortage of usable carbohydrates to supply the energy needed for the reduction of nitrogen to ammonia, which is then incorporated into protein. Nevertheless, these nitrogen-fixing bacteria make important contributions to the nitrogen economy of such areas as grasslands, forests, and the arctic tundra.

Symbiotic Nitrogen-Fixing Bacteria These bacteria play an even more important role in plant growth for crop production. Members of the genera *Rhizobium, Bradyrhizobium,* and others infect the roots of leguminous plants, such as soybeans, beans, peas, peanuts, alfalfa, and clover. (These agriculturally important plants are only a few of the thousands of known leguminous species, many of which are bushy plants or small trees found in poor soils in many parts of the world.) Rhizobia are specially adapted to particular leguminous plant species, on which they form **root nodules** (Figure 27.4). Nitrogen is then fixed by a symbiotic process of the plant and the bacteria. The plant furnishes anaerobic conditions and growth nutrients for the bacteria, and the bacteria fix nitrogen to be incorporated into plant protein.

There are similar examples of symbiotic nitrogen fixation in nonleguminous plants, such as alder trees. These trees are among the first to appear in forests after fires or glaciation. The alder tree is symbiotically infected with an actinomycete (*Frankia*) and forms nitrogen-fixing root nodules. About 50 kg of nitrogen can be fixed each year by the growth of 1 acre of alder trees; the trees thus make a valuable addition to the forest economy.

Another important contribution to the nitrogen economy of forests is made by **lichens,** which are a combination fungus and an alga or a cyanobacterium in a mutualistic relationship (see Figure 12.10 on page 344). When one symbiont is a nitrogen-fixing cyanobacterium, the product is fixed nitrogen that eventually enriches the forest soil. Free-living cyanobacteria can fix significant amounts of nitrogen in desert soils after rains and on the surface of arctic tundra soils. Rice paddies can accumulate heavy growths of such nitrogen-fixing organisms. The cyanobacteria also form a symbiosis with a small floating fern, *Azolla,* which grows thickly in rice

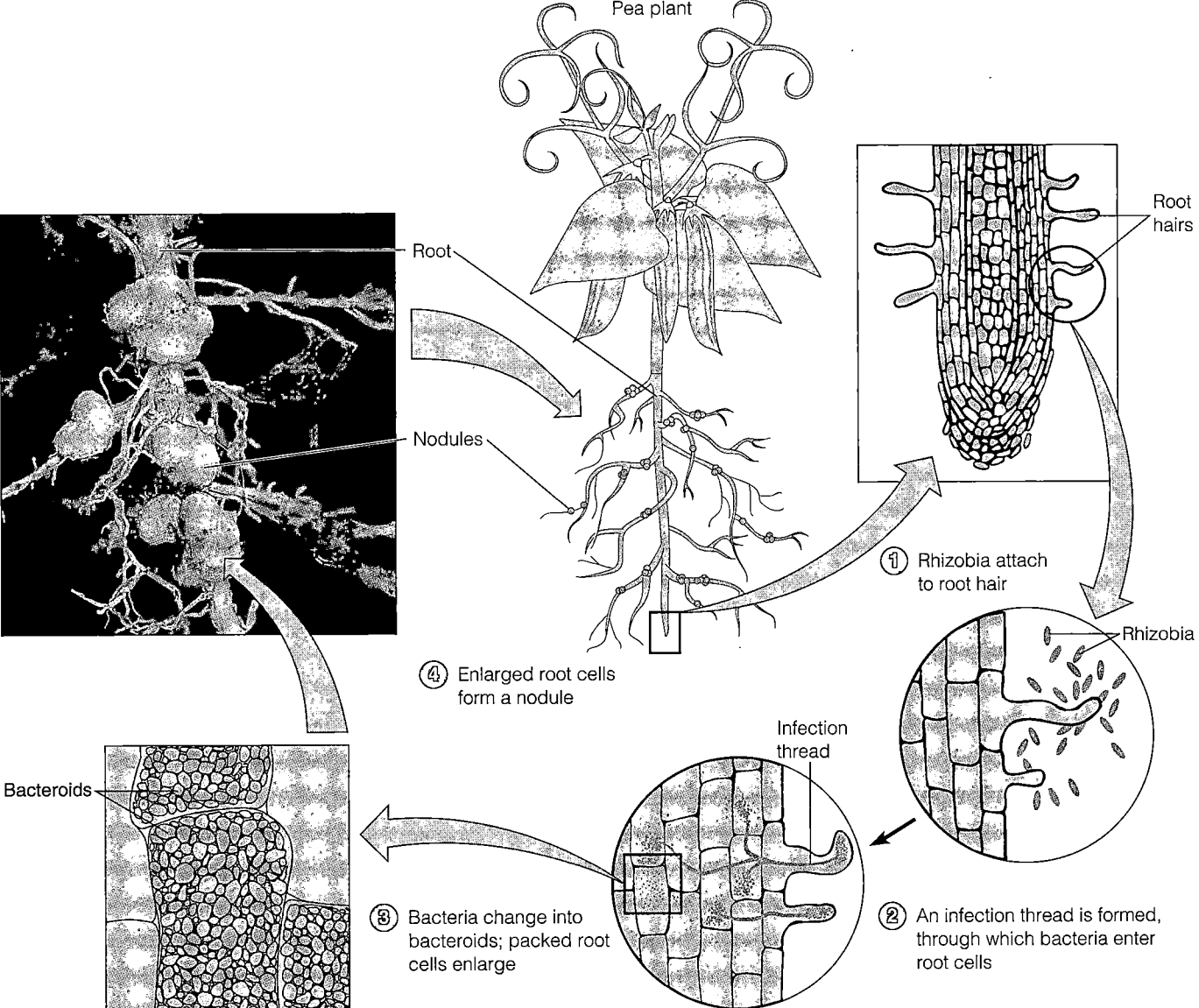

FIGURE 27.4 The formation of a root nodule. Members of the nitrogen-fixing genera *Rhizobium* and *Bradyrhizobium* form these nodules on legumes. This mutualistic association is beneficial to both the plant and the bacteria.

■ Symbiotic bacteria fix nitrogen in the roots of legumes.

paddy waters (Figure 27.5). So much nitrogen is fixed by these microbes that other nitrogenous fertilizers are often unnecessary for rice cultivation.

The Sulfur Cycle

Learning Objective

■ *Outline the sulfur cycle, and explain the roles of microorganisms in this cycle.*

The **sulfur cycle** (Figure 27.6) and nitrogen cycle resemble each other in the sense that they represent numerous oxidation states of these elements. The most reduced forms of sulfur are the sulfides, such as the odorous gas hydrogen sulfide (H_2S). Like the ammonium ion of the nitrogen cycle, this is a reduced compound that generally forms under anaerobic conditions. It represents a source of energy for autotrophic bacteria. These bacteria convert the reduced sulfur in H_2S into elemental sulfur granules and fully oxidized sulfates (SO_4^{2-}).

An interesting aspect of the sulfur cycle is the work of Winogradsky in first discovering chemoautotrophy, by which microbes obtain energy from inorganic chemicals. In Paris in the early 1900s, he was studying an exceptionally large, filamentous bacterium called *Beggiatoa*, which grew on the surface of stagnant ponds and in the slime of

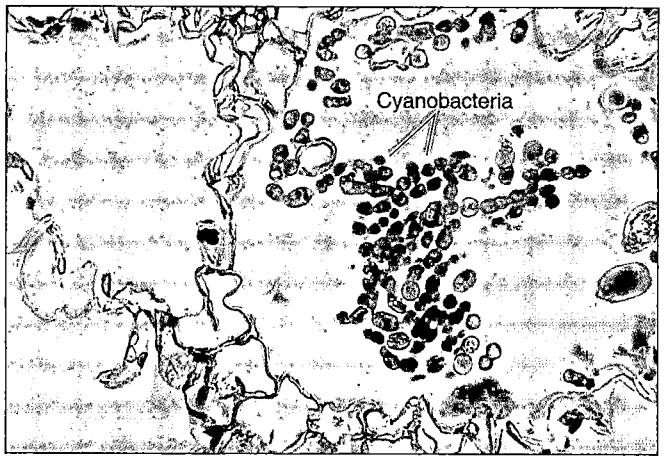

FIGURE 27.5 The *Azolla*-cyanobacteria symbiosis. A section through the leaf of an *Azolla* freshwater fern. The cyanobacteria *Anabaena azollae* is visible as chains of cells within the leaf cavity.

▨ What is the major contribution of cyanobacteria as a symbiont?

sulfurous springs (see Chapter 11, page 310). Winogradsky allowed water containing H_2S to flow past the microbes, which formed numerous granules of sulfur within the cell. This was an oxidative change that may have indicated a reaction with the oxygen in the air. When water without H_2S was then allowed to flow past the cultures, the sulfur granules slowly disappeared but the bacteria continued to thrive. *Beggiatoa* was obviously obtaining energy from the reaction that converted H_2S into elemental sulfur. The elemental sulfur, he later determined, disappeared because it

was being converted into fully oxidized sulfate ions that were rinsed away in the passing water. Therefore, both H_2S and elemental sulfur were being used as inorganic energy sources by *Beggiatoa* (Figure 27.6). This discovery of the autotrophic physiology of bacteria is central to an understanding of biogeochemical cycles.

Frequently, elemental sulfur is released from decaying microbes. Elemental sulfur is essentially insoluble in temperate waters, and microbes have difficulty absorbing it. This is probably the origin of huge, prehistoric underground accumulations of sulfur.

Several phototrophic bacteria, such as the green and purple sulfur bacteria, also oxidize H_2S, forming colorful internal sulfur granules (see Figure 11.10 on page 315). Like *Beggiatoa*, they can further oxidize the sulfur to sulfate ions. It is important to recognize that these organisms are using light for energy; the hydrogen sulfide is used to reduce CO_2 (see Chapter 5, page 139).

Hydrogen sulfide can be used as an energy source by *Thiobacillus* to produce sulfate ions and sulfuric acid. *Thiobacillus* can grow well at a pH as low as 2 and have practical uses in mining. (See Figure 28.13 on page 785). Sulfates are incorporated by plants and bacteria to become part of sulfur-containing amino acids for humans and other animals. There, they form disulfide links that give structure to proteins. As proteins are decomposed, in a process called **dissimilation**, the sulfur is released as hydrogen sulfide to reenter the cycle.

Acid Deposition

Sulfur is an important component of a major environmental problem that especially affects the developed

FIGURE 27.6 The sulfur cycle. Note the importance of aerobic and anaerobic conditions.

■ Why is a source of sulfur necessary for all organisms?

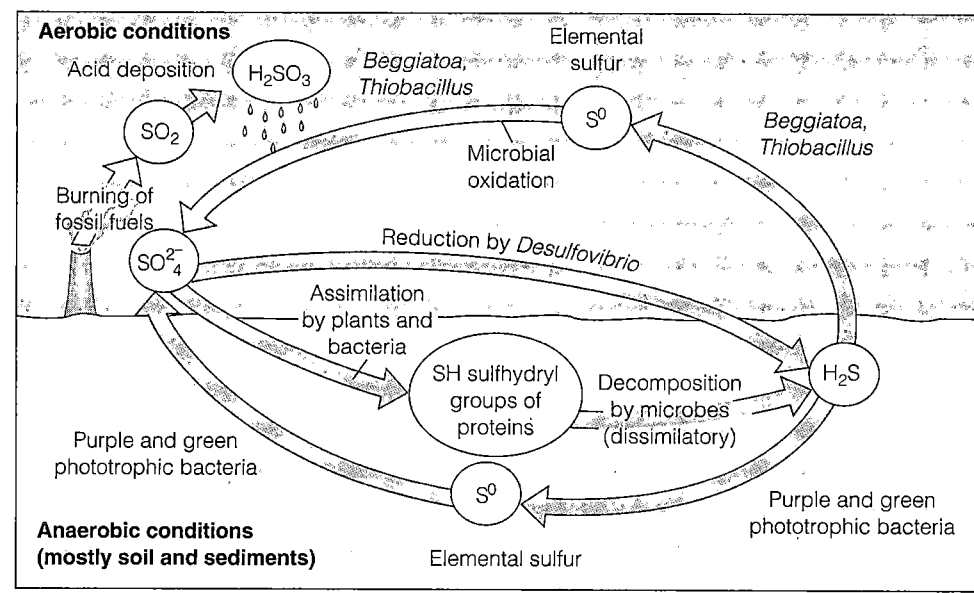

world. The burning of fossil fuels, which contain sulfur from the remains of dead organisms, releases sulfur in the form of sulfur dioxide (SO_2). Natural phenomena, such as volcanic eruptions, also release large amounts. This SO_2 reacts with rainwater to form sulfurous acid (H_2SO_3), which is familiarly called *acid rain*. (See Figure 27.6.) **Acid deposition** is probably a better term because many of the acid-forming substances that fall to earth are not dissolved in rainwater, but fall as dry particles of sulfate or nitrate salts. Acid deposition often impairs the growth of trees, reacts with and dissolves marble and limestone, and corrodes metal structures. The acidification of lakes and streams is probably the best-known effect. Sulfates deposited in lakes can change the pH to levels that are too acidic to permit the growth of fish or essential units of their food chain. Nitrogen oxides from motor vehicles, a large component of urban smog, enters waters as acidic forms of nitrogen and can contribute to eutrophication, which is discussed on page 758.

Life Without Sunshine

Learning Objective

- *Describe how an ecological community can exist without light energy.*

Interestingly, it is possible for entire biological communities to exist without photosynthesis by exploiting the energy in H_2S. In Chapter 11, page 314, equations are presented to show that photosynthesis and chemoautotrophic use of H_2S are similar in certain respects. An example of such a community around deep-sea vents is described in the box in Chapter 6, page 160. Deep caves, totally isolated from sunlight, have been discovered that also support an entire biological community in a similar manner. The **primary producers** in these systems are chemoautotrophic bacteria rather than photoautotrophic plants or microbes.

Recently, another microbial ecosystem operating far from sunlight has been discovered over 1 km deep within rocks, including shales, granites, and basalts. Such bacteria are called **endoliths** (inside rocks), which must grow in the near absence of oxygen and with minimal nutrient supplies. Sedimentary shales often contain trapped organic nutrients or sulfates that can support limited life. Rocks such as granite or basalt are minutely porous or have fractures that contain some water. Sulfate reduction is often a possible energy source. Also, in these rocks, chemical reactions produce hydrogen that can be used for energy by autotropic endolithic bacteria. Carbon dioxide dissolved in the water serves as a carbon source, and cellular organic matter is produced. Some is excreted, or is released upon the death and lysis of the microbe, and becomes available for the growth of other microbes. Nutrient inputs, especially of nitrogen, are very small in this environment, and generation times may be measured in many years. Various survival strategies have developed for life with minimal nutrition. For example, suspended in a state between life and death, certain of these organisms become dramatically smaller. Ecologists speculating upon forms of life that might be found in the harsh environment of Mars are very interested in endoliths.

The Phosphorus Cycle

Learning Objective

- *Compare and contrast the carbon cycle and the phosphorus cycle.*

Another important nutritional element that is part of a biogeochemical cycle is phosphorus. The availability of phosphorus may determine whether plants and other organisms can grow in an area. The problems associated with excess phosphorus are described later in the chapter.

Phosphorus exists primarily as phosphate ions (PO_4^{3-}) and undergoes very little change in its oxidation state. The **phosphorus cycle** involves changes from soluble to insoluble forms and from organic to inorganic phosphate, often in relation to pH. For example, phosphate in rocks can be solubilized by the acid produced by bacteria such as *Thiobacillus*. Unlike the other cycles, there is no volatile phosphorus-containing product to return phosphorus to the atmosphere in the way carbon dioxide, nitrogen gas, and sulfur dioxide are returned. Therefore, phosphorus tends to accumulate in the seas. It can be retrieved by mining the above-ground sediments of ancient seas, mostly as deposits of calcium phosphate. Seabirds also mine phosphorus from the sea by eating phosphorus-containing fish and depositing it as guano (bird droppings). Certain small islands inhabited by such birds have long been mined for these deposits as a source of phosphorus for fertilizers.

The Degradation of Synthetic Chemicals in Soil and Water

Learning Objective

- *Give two examples of the use of bacteria to remove pollutants.*

We seem to take for granted that soil microorganisms will degrade materials entering the soil. Natural organic matter, such as falling leaves or animal residues, are in fact readily degraded. However, in this industrial age many

FIGURE 27.7 2,4-D (black) and 2,4,5-T (magenta). This graph shows the structures and rates of microbial decomposition of the herbicides 2,4-D (black) and 2,4,5-T (magenta).

■ Slight structural differences in synthetic chemicals can result in great differences in their biodegradability.

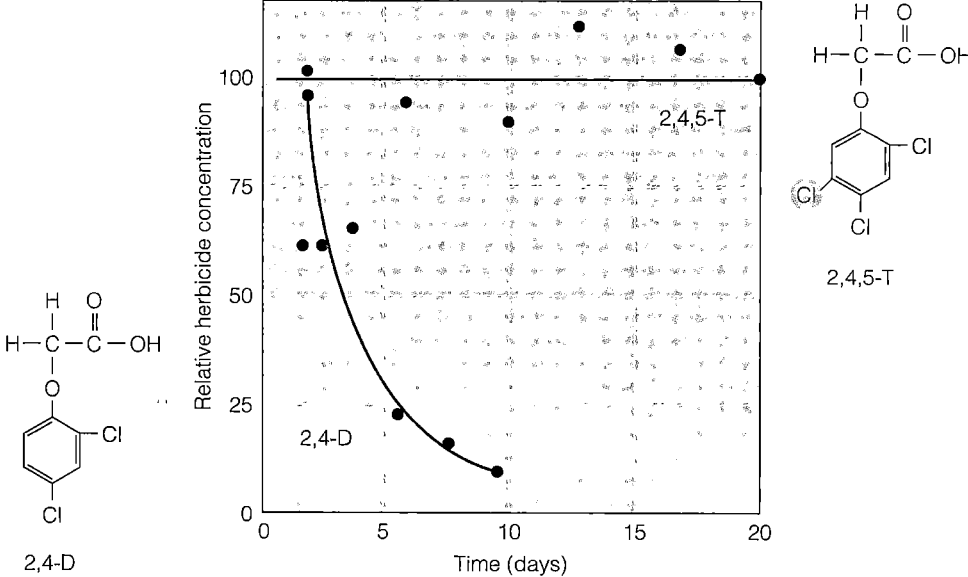

2,4-D

chemicals that do not occur in nature, such as plastics, enter the soil in large amounts. A new development is the production of plastic from natural chemicals of plants. The starchy sugars from plants is fed to lactic acid bacteria. The lactic acid is then reacted with a catalyst to make a biodegradable plastic. Many synthetic chemicals, such as pesticides, are highly resistant to degradation by microbial attack. A well-known example is the insecticide DDT, which proved so resistant that it accumulated to damaging levels in the environment. Polychlorinated biphenyls (PCBs), at one time much used as a heat-transfer fluid in electrical transformers and a component in some plastics, resemble DDT in many ways. Both are resistant to biological degradation, and both tend to become concentrated in the final predators of the food chain—for example, fish-eating birds. Microbial ecologists are currently investigating the use of microbes to eliminate PCBs from contaminated soils and waters. (Other examples are described in the box in Chapter 2, page 35.)

Some synthetic chemicals are made up of bonds and subunits that are subject to attack by bacterial enzymes. Small differences in chemical structure can make large differences in biodegradability. The classic example is that of two herbicides: 2,4-D, the common chemical used to kill lawn weeds, and 2,4,5-T (the notorious Agent Orange), which is used to kill shrubs. The addition of a single chlorine atom to the structure of 2,4-D extends its life in soil from a few days to an indefinite period (Figure 27.7).

A growing problem is the leaching into groundwaters of toxic materials that are not biodegradable or that degrade very slowly. The sources of these materials may include landfills, illegal industrial dumps, or pesticides applied to agricultural crops. Once groundwater becomes contam-

inated, the environmental and economic damage can be devastating. Researchers are developing processes and isolating bacteria that promote degradation and detoxification.

Bioremediation

Learning Objective

■ Define bioremediation.

The use of microbes to detoxify or degrade pollutants is called **bioremediaton.** Oil spills from wrecked tankers represent some of the most dramatic examples of chemical pollution. The economic losses from contaminated fisheries and beaches can be enormous. To some degree, bioremediation occurs naturally, as microbes attack the petroleum if conditions are aerobic. However, microbes usually obtain their nutrients in aqueous solution, and oil-based products are relatively nonsoluble. Also, petroleum hydrocarbons are deficient in essential elements, such as nitrogen and phosphorus. Bioremediation of oil spills is greatly enhanced if the resident bacteria are provided with "fertilizer" containing nitrogen and phosphorus (Figure 27.8). Bioremediaton may also make use of microbes that have been selected for growth on a certain pollutant or genetically modified bacteria that are specially adapted to metabolize petroleum products. The addition of specialized microbes is called **bioaugmentation.**

Subsurface petroleum spills from leaking gasoline storage tanks, for example, can be removed from ground water by pumping the water into aerating tanks, providing "fertilizer" nutrients, and returning the water after the gasoline has been decomposed. Similar principles are used to clean up pesticide and other chemical spills. Radiation-resistant bacteria have been genetically altered to

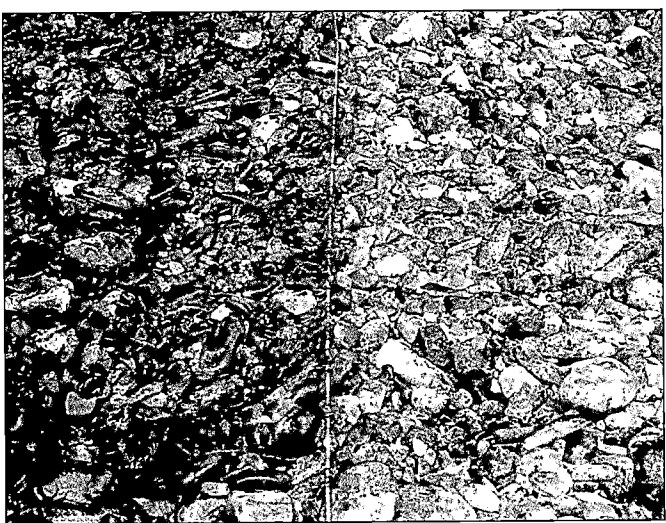

FIGURE 27.8 Bioremediation at an oil spill in Alaska. The portion of beach on the left is uncleaned, the beach on the right has been treated with applications of carbon free nutrients (fertilizer). However, below the surface layers, where conditions are anaerobic, oil often remains for much longer periods.

make them more useful in cleaning up sites contaminated with radioactive solvents. Bioremediation may also be used to remove certain natural pollutants. Selenium, for example, is an essential nutrient required in very small amounts by animals and bacteria; in high concentration, it is toxic. Bacteria are being studied as a method of neutralizing excessive selenium accumulations in California irrigation water. They convert the selenium to less toxic forms.

Solid Municipal Waste

Solid municipal waste (garbage) is most frequently placed into large compacted landfills. Conditions are largely anaerobic, and even presumably biodegradable materials such as paper are not very effectively attacked by microorganisms. In fact, recovering a 20-year-old newspaper in readable condition is not at all unusual. But such anaerobic conditions do promote the activity of the same methanogens that will be discussed with the operation of anaerobic sludge digesters in sewage treatment. The methane they produce can be tapped with drill holes and burned to generate electricity, or purified and introduced into natural gas pipeline systems (see Figure 28.14 on page 785). Such systems are part of the design of more than 100 large landfills in the United States, some of which provide energy for several thousand homes.

The amount of organic matter entering landfills can be considerably reduced if it is first separated from material that is not biodegradable and composted. **Composting** is a process used by gardeners to convert plant remains into the equivalent of natural humus (Figure 27.9). A pile of leaves or grass clippings will undergo microbial degradation. Under favorable conditions, thermophilic bacteria will raise the temperature of the compost to 55–60°C in a couple of days. After the temperature declines, the pile can be turned to renew the oxygen supply, and a second temperature rise will occur. Over time, the thermophilic microbial populations are replaced by mesophilic populations that slowly continue the conversion to a stable material similar to humus. Where space is available, municipal wastes are composted in windrows (long, low piles) that

(a) Solid municipal wastes being turned by a specially designed machine.

(b) Compost made from municipal wastes awaiting trucks to spread on agricultural fields.

FIGURE 27.9 Composting municipal wastes.

■ What occurs in a compost heap?

are distributed and periodically turned over by specialized machinery. Municipal waste disposal now also makes increasing use of composting methods.

Aquatic Microbiology and Sewage Treatment

Aquatic microbiology refers to the study of microorganisms and their activities in natural waters, such as lakes, ponds, streams, rivers, estuaries, and oceans. Domestic and industrial wastewater enters lakes and streams, and its degradation and effects on the microbial life is an important factor in aquatic microbiology. We will also see that the method of treating wastewater by municipalities mimics a natural filtering process.

Biofilms

Learning Objective

- *Describe the importance of biofilms.*

In nature, microorganisms seldom live in the isolated, single-species colonies we see on laboratory plates. They more typically live in slime communities, called **biofilms,** where they can share nutrients. Formation of a biofilm begins when a free-swimming bacterium attaches to a surface. If these bacteria grew in a thick layer, they would become overcrowded, nutrients would not be available in lower depths, and toxic wastes could accumulate. Biofilms, to avoid these problems, form into pillarlike structures (Figure 27.10a) with channels between them through which water can carry incoming nutrients and

outgoing wastes. This primitive circulatory system is constructed in response to chemical communication signals among the bacteria. Individual bacteria and clumps of slime occasionally leave the established biofilm and move to new locations where the biofilm becomes extended. A biofilm is generally composed of a surface layer about 10 μm thick, with pillars that extend up to 200 μm above it.

Like the cells that make up animal tissue, biofilms can form specialized groups in specialized environments and work cooperatively to carry out complex tasks. For example, ruminant digestive systems, in which the microbes are mostly located within biofilm communities, require at least five different species to break down cellulose. Biofilms are an essential element of the functioning of sewage treatment systems, which we will be discussing in this chapter.

Biofilms are an important factor in human health. For example, microbes in biofilms are probably 1000 times more resistant to microbicides. Experts at the CDC estimate that 65% of human bacterial infections involve biofilms.

Biofilms can be found just about any place where water and solid supports meet: teeth, medical catheters, contact lenses, and the interior of water pipes.

Aquatic Microorganisms

Learning Objective

- *Describe the freshwater and seawater habitats of microorganisms.*

Large numbers of microorganisms in a body of water generally indicate high nutrient levels in the water. Water

(a) Water currents move, as shown by the blue arrow, among pillars of slime formed by the growth of bacteria attached to solid surfaces. This allows efficient access to nutrients and removal of bacterial waste products. Individual slime-forming bacteria or bacteria in clumps of slime detach and move to new locations. 100 μm

(b) Bacterial biofilm growing in the pipes of a cooling system.  1 μm

FIGURE 27.10 Biofilms.

▣ **Microbial populations accumulate on surfaces in biofilms.**

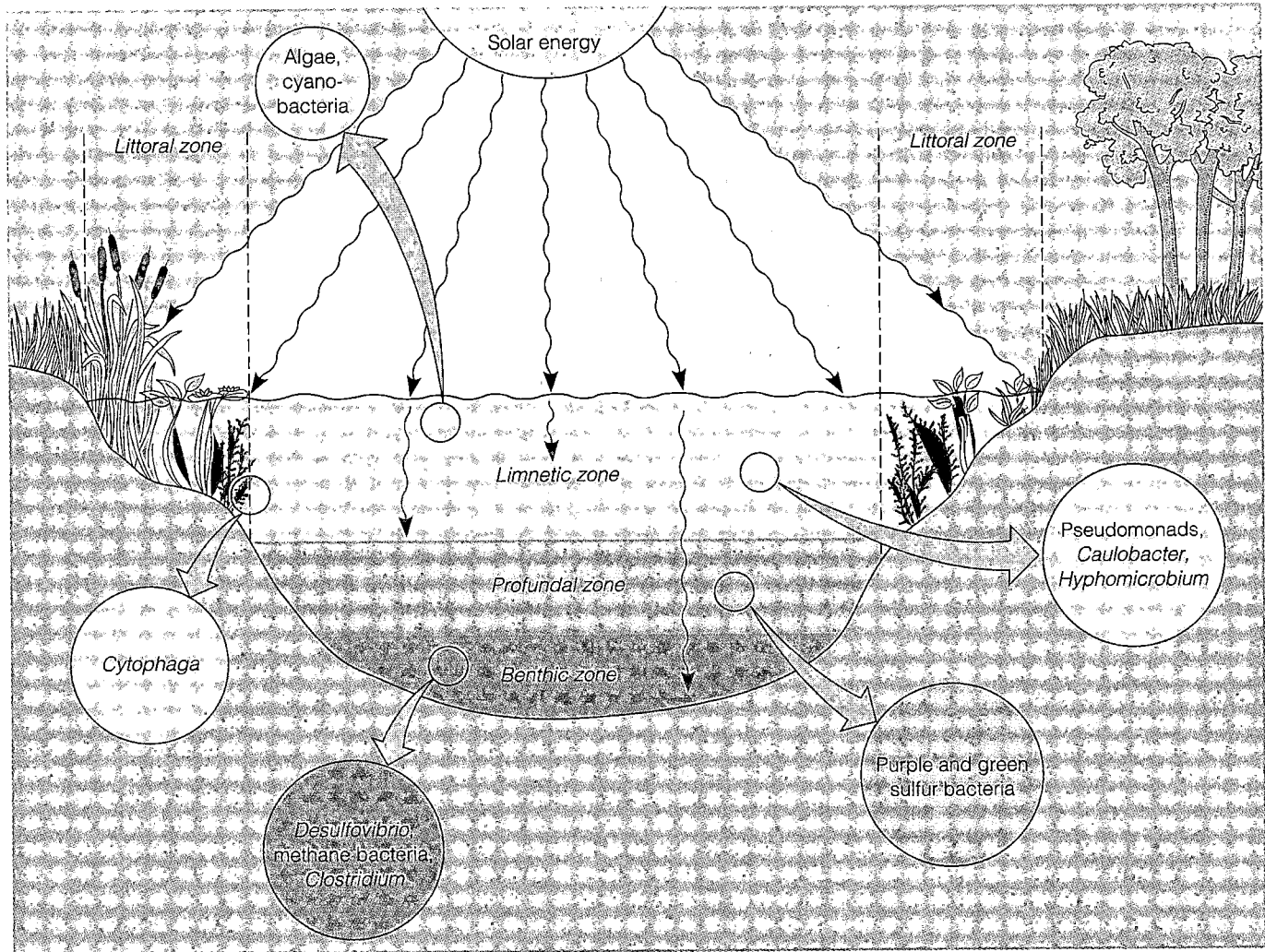

FIGURE 27.11 The zones of a typical lake or pond and some representative microorganisms of each zone. The microbes fill niches that vary in light, nutrients, and oxygen availability.

■ Which of the microorganisms listed are most likely aerobes?

contaminated by inflows from sewage systems or from biodegradable industrial organic wastes is relatively high in bacterial counts. Similarly, ocean estuaries (fed by rivers) have higher nutrient levels and therefore larger microbial populations than other shoreline waters.

In water, particularly water with low nutrient concentrations, microorganisms tend to grow on stationary surfaces and on particulate matter. In this way, a microorganism has contact with more nutrients than if it were randomly suspended and floating freely with the current. Many bacteria whose main habitat is water often have appendages and holdfasts that attach to various surfaces. One example is *Caulobacter* (see Figure 11.2 on page 308).

Freshwater Microbiota

Figure 27.11 shows a typical lake or pond that serves as an example to represent the various zones and the kinds of microbiota found in a body of fresh water. The **littoral zone** along the shore has considerable rooted vegetation, and light penetrates throughout it. The **limnetic zone** consists of the surface of the open water area away from the shore. The **profundal zone** is the deeper water under the limnetic zone. The **benthic zone** contains the sediment at the bottom.

Microbial populations of freshwater bodies tend to be affected mainly by the availability of oxygen and light. In many ways, light is the more important resource because

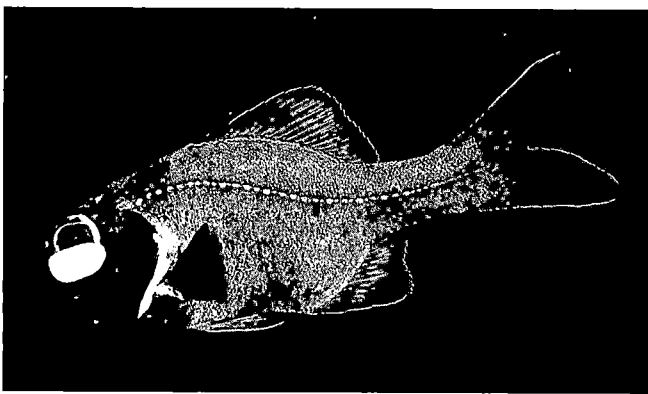

FIGURE 27.12 Bioluminescent bacteria as light organs in fish.
This is a deep-sea flashlight fish (*Photoblepharon palpebratus*).

■ **What enzyme is responsible for bioluminescence?**

photosynthetic algae are the main source of organic matter, and hence of energy, for the lake. These organisms are the primary producers of a lake that supports a population of bacteria, protozoa, fish, and other aquatic life. Photosynthetic algae are located in the limnetic zone.

Areas of the limnetic zone with sufficient oxygen contain pseudomonads and species of *Cytophaga, Caulobacter,* and *Hyphomicrobium*. Oxygen does not diffuse into water very well, as any aquarium owner knows. Microorganisms growing on nutrients in stagnant water quickly use up the dissolved oxygen in the water. In the oxygenless water, fish die, and odors (from hydrogen sulfide and organic acids, for example) are produced from anaerobic activity. Wave action in shallow layers, or water movement in rivers, tends to increase the amount of oxygen throughout the water and aid in the growth of aerobic populations of bacteria. Movement thus improves the quality of water and aids in the degradation of polluting nutrients.

Deeper waters of the profundal and benthic zones have low oxygen concentrations and less light. Algal growth near the surface often filters the light, and it is not unusual for photosynthetic microbes in deeper zones to use different wavelengths of light from those used by surface-layer photosynthesizers (see Figure 12.11 on page 345).

Purple and green sulfur bacteria are found in the profundal zone. These bacteria are anaerobic photosynthetic organisms that metabolize H_2S to sulfur and sulfate in the bottom sediments of the benthic zone.

The sediment in the benthic zone includes bacteria such as *Desulfovibrio* that use sulfate (SO_4^{2-}) as an electron acceptor and reduce it to H_2S. Methane-producing bacteria are also part of these anaerobic benthic populations. In swamps, marshes, or bottom sediments, they pro-

duce methane gas. *Clostridium* species are common in bottom sediments and may include botulism organisms, particularly those causing outbreaks of botulism in waterfowl.

Seawater Microbiota

Much of the microscopic life of the ocean is composed of photosynthetic diatoms and other algae. Largely independent of preformed organic nutrient sources, these microbes use energy from photosynthesis and atmospheric carbon dioxide for carbon. They constitute the marine **phytoplankton** community, the basis of the oceanic food chain. Krill (shrimplike crustaceans) feed on the phytoplankton and in turn are an important food supply of larger sea life. Many fish and whales are also able to feed directly on the phytoplankton.

Microbial **bioluminescence,** or light emission, is an interesting aspect of deep-sea life. Many bacteria are luminescent, and some have established symbiotic relationships with benthic-dwelling fish. These fish sometimes use the glow of their resident bacteria as an aid in attracting and capturing prey in the complete darkness of the ocean depths (Figure 27.12). These bioluminescent organisms have an enzyme called luciferase that picks up electrons from flavoproteins in the electron transport chain and then emits some of the electron's energy as a photon of light. (An industrial application of bioluminescence is discussed in the box in Chapter 28, page 786).

The Role of Microorganisms in Water Quality

Learning Objectives

■ *Explain how wastewater pollution is a public health problem and an ecological problem.*

■ *Discuss the causes and effects of eutrophication.*

Water in nature is seldom totally pure. Even rainfall is contaminated as it falls to Earth.

Water Pollution

The form of water pollution that is our primary interest is microbial pollution, especially by pathogenic organisms.

The Transmission of Infectious Diseases Water that moves below the ground's surface undergoes a filtering that removes most microorganisms. For this reason, water from springs and deep wells is generally of good quality. The most dangerous form of water pollution occurs when feces enter the water supply. Many diseases are perpetuated by the fecal-oral route of transmission, in which a pathogen is shed in human or animal feces, contami-

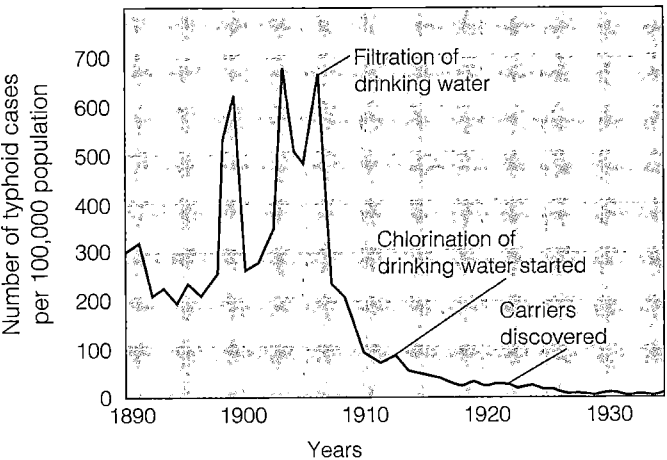

FIGURE 27.13 The incidence of typhoid fever in Philadelphia, 1890–1935. This graph clearly shows the effect of water treatments on incidence of typhoid.
[SOURCE: E. Steel, *Water Supply and Sewerage*, New York: McGraw-Hill, 1953.]

■ Why did the incidence of typhoid fever decrease?

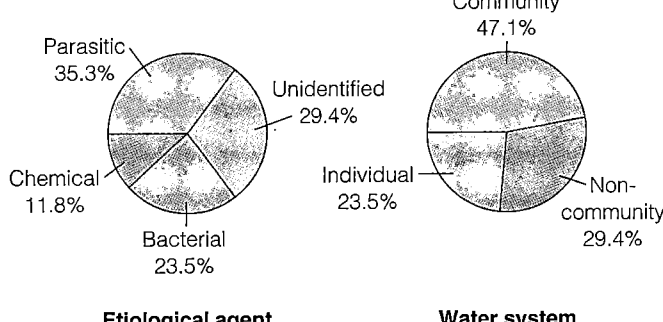

Etiological agent   Water system

(a) Waterborne-disease outbreaks associated with drinking water.

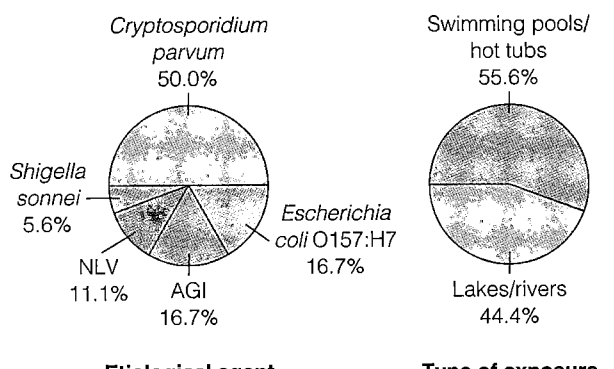

Etiological agent   Type of exposure

(b) Waterborne-disease outbreaks of gastroenteritis associated with recreational water.

FIGURE 27.14 U.S. outbreaks of waterborne diseases associated with drinking water 1997–1998.
[SOURCE: *MMWR* 49(SS-4)(5/26/00).]

■ What microorganisms are most often associated with drinking water and recreational water?

nates water, and is ingested (see Chapter 25). The CDC estimates that in the United States 900,000 people become ill each year from waterborne infections.

Examples of such diseases are typhoid fever and cholera, caused by bacteria that are shed only in human feces. About 100 years ago, the *Journal of the American Medical Association* reported that the typhoid fever mortality rate in Chicago had declined from 159.7 per 100,000 people in 1891 to 31.4 per 100,000 in 1894. This advance in public health had been accomplished by extending the city water supply intake pipes in Lake Michigan to a distance of 4 miles from shore. The medical journal commented that this diluted the sewage contaminating the water supply, which at that time was not treated further. This same article speculated on the need to remove microorganisms that caused specific diseases. They suggested the use of sand filter beds, already widely used in Europe at the time. Sand filtration mimics the natural purification of spring water. Figure 27.13 illustrates the effect of the introduction of such filtration of water supplies on the incidence of typhoid fever in Philadelphia.

As good sanitation practices have nearly eliminated diseases such as typhoid fever and cholera in the United States, attention has turned to other waterborne diseases. Figure 27.14 shows reported cases of illnesses associated with water intended for drinking.

Chemical Pollution Preventing chemical contamination of water is a difficult problem. Industrial and agricultural chemicals leached from the land enter water in great amounts and in forms that are resistant to biodegradation. Rural waters often have excessive amounts of nitrite from agricultural fertilizers. When ingested, the nitrate is converted to nitrite by bacteria in the gastrointestinal tract. Nitrite competes for oxygen in the blood and is especially likely to harm infants. Pesticides frequently contaminate water, and even fluorides, intended to prevent dental caries, may be accidentally added in excessive amounts.

A striking example of industrial water pollution involved mercury in wastewater from paper manufacturing. The metallic mercury was allowed to flow into waterways as waste. It was assumed that the mercury was inert and would remain segregated in the sediments. However, bacteria in the sediments converted the mercury into a soluble chemical compound, methyl mercury, which was then taken up by fish and invertebrates in the

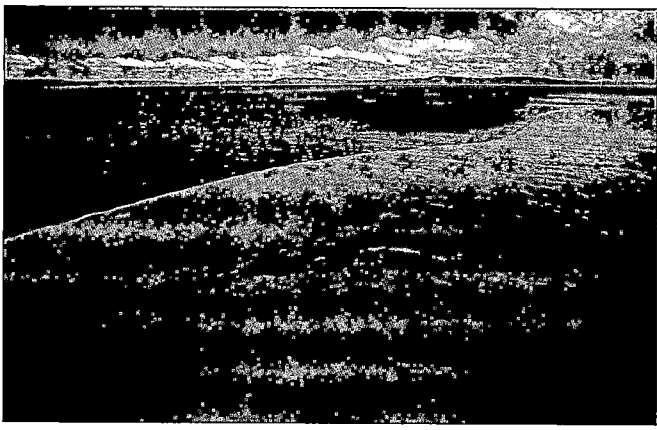

FIGURE 27.15 A red tide. These blooms of aquatic growth are caused by excess nutrients in water. The color is from the pigmentation of the dinoflagellates.

waters. When such seafood is a substantial part of the human diet, the mercury concentrations can accumulate with devastating effects on the nervous system. At present, 33 American states have released warnings about consuming fish from mercury-contaminated waters. Bioremediation efforts using bacteria to detoxify mercury in one wildlife refuge are discussed in the box in Chapter 2, page 35.

Another example of chemical pollution is the synthetic detergents developed immediately after World War II. These rapidly replaced many of the soaps then in use. Because these new detergents were not biodegradable, they rapidly accumulated in the waterways. In some rivers, large rafts of detergent suds could be seen traveling downstream. These detergents were replaced in 1964 by new biodegradable synthetic formulations.

Biogradable detergents, however, still present a major environmental problem because they often contain phosphates. Unfortunately, phosphates pass almost unchanged through sewage systems and can cause **eutrophication,** which is caused by an overabundance of nutrients in lakes and streams.

To understand the concept of eutrophication, recall that algae and cyanobacteria get their energy from sunlight and their carbon from carbon dioxide dissolved in water. In most waters only nitrogen and phosphorus supplies are inadequate for algal growth. Both of these nutrients can enter water from domestic, farm, and industrial wastes when waste treatment is absent or inefficient. These additional nutrients cause dense aquatic growths called **algal blooms.** Because many cyanobacteria can fix nitrogen from the atmosphere, these photosynthesizing organisms require only traces of phosphorus to initiate blooms. Once

eutrophication results in blooms of algae or cyanobacteria, the eventual effect is the same as the addition of biodegradable organic matter. In the short run, these algae and cyanobacteria produce oxygen. However, they eventually die and are degraded by bacteria. During the degradation process, the oxygen in the water is used up, which may kill the fish. Undegraded remnants of organic matter settle to the bottom and hasten the filling of the lake.

Red tides of toxin-producing phytoplankton (Figure 27.15), which were mentioned in Chapter 12, are probably caused by excessive nutrients from oceanic upwellings or terrestrial wastes. In addition to eutrophication effects, this type of biological bloom can affect human health. Seafood, especially clams or similar mollusks, that ingest these plankton become toxic to humans. (See the box in Chapter 12, page 348.)

Municipal waste containing detergents is likely to be the main source of phosphates in lakes and streams. As a result, phosphate-containing detergents are banned in many places.

Coal-mining wastes, particularly in the eastern United States, are very high in sulfur content, mostly iron sulfide (FeS_2). In the process of obtaining energy from the oxidation of the ferrous ion (Fe^{2+}), bacteria such as *Thiobacillus ferrooxidans* convert the sulfide into sulfate. The sulfate enters streams as sulfuric acid, which lowers the pH of the water and damages aquatic life. The low pH also promotes the formation of insoluble iron hydroxides, which form the yellow precipitates often seen clouding such polluted waters.

Water Purity Tests

Learning Objective

■ *Explain how water is tested for bacteriological quality.*

Historically, most of our concern about water purity has been related to the transmission of disease. Therefore, tests have been developed to determine the safety of water; many of these tests are also applicable to foods.

It is not practical, however, to look only for pathogens in water supplies. For one thing, if we were to find the pathogen causing typhoid or cholera in the water system, the discovery would already be too late to prevent an outbreak of the disease. Moreover, such pathogens would probably be present only in small numbers and might not be included in tested samples.

The tests for water purity in use today are aimed instead at detecting particular **indicator organisms.** There are several criteria for an indicator organism, the most important being that the microbe is consistently present in human feces in substantial numbers so that its detection is a good indication that human wastes are entering the water. The indicator organisms should also survive in the water at least as well as the pathogens would. The indicator organisms must also be detectable by simple tests that can be carried out by people with relatively little training in microbiology.

In the United States, the usual indicator organisms are the *coliform bacteria.* **Coliforms** are defined as aerobic or facultatively anaerobic, gram-negative, non–endospore-forming, rod-shaped bacteria that ferment lactose to form gas within 48 hours of being placed in lactose broth at 35°C. Because some coliforms are not solely enteric bacteria but are more commonly found in plant and soil samples, many standards for food and water specify the identification of *fecal coliforms.* The predominant fecal coliform is *E. coli,* which constitutes a large proportion of the human intestinal population. There are specialized tests to distinguish between fecal coliforms and nonfecal coliforms. Note that coliforms are not themselves pathogenic under normal conditions, although certain strains can cause diarrhea (see Chapter 25, page 697) and opportunistic urinary tract infections (see Chapter 26, page 723).

The methods for determining the presence of coliforms in water are largely based on the lactose-fermenting ability of coliform bacteria. The multiple-tube method can be used to estimate coliform numbers by the most probable number (MPN) method (see Figure 6.18 on page 177). The membrane filtration method is a more direct method of determining the presence and numbers of coliforms. This is possibly the most widely used method in North America and Europe. It makes use of a filtration apparatus similar to that shown in Figure 7.4 on page 191. In this application, though, the bacteria collected on the surface of a removable membrane filter are placed on an appropriate medium and incubated. Coliform colonies have a distinctive appearance, Figure 6.9b and 6.9c on page 169, and are counted. This method is suitable for low turbidity waters that do not clog the filter and have relatively few noncoliform bacteria that would mask the results.

A newer and more convenient method of detecting coliforms, specifically the fecal coliform *E. coli,* makes use of media containing the two substrates *o*-nitrophenyl-β-D-galactopyranoside (ONPG) and 4-methylumbelliferyl-β-D-glucuronide (MUG). Coliforms produce the enzyme β-galactosidase, which acts on ONPG and forms a yellow color, indicating their presence in the sample. *E. coli* is unique among coliforms in almost always producing the enzyme β-glucuronidase, which acts on MUG to form a fluorescent compound that glows blue when illuminated by long-wave ultraviolet light (Figure 27.16). These simple tests, or variants of them, can detect

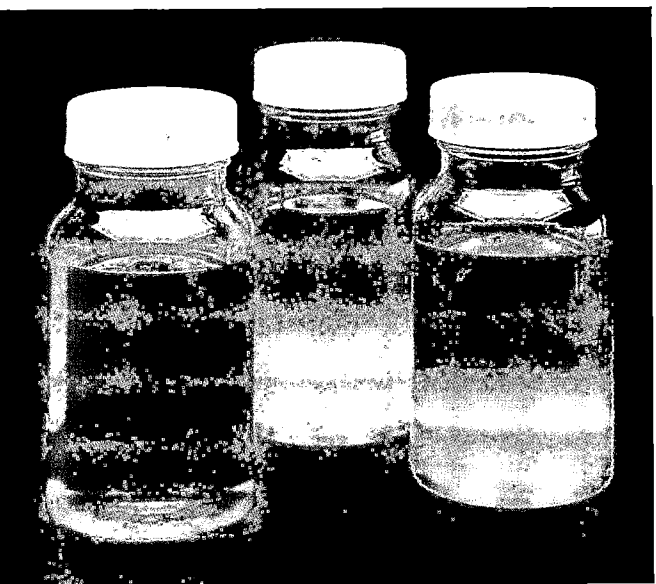

FIGURE 27.16 The ONPG and MUG coliform test. A yellow color (positive ONPG) indicates the presence of coliforms. Blue fluorescence (positive MUG) indicates the presence of the fecal coliform *E. coli.* The clear medium indicates an uncontaminated sample.

■ **What causes the formation of the compound in a positive MUG test?**

the presence or absence of coliforms or *E. coli* and can be combined with the multiple-tube method to enumerate them. It can also be applied to solid media, such as in the membrane filtration method. The colonies fluoresce under UV light.

Coliforms have been very useful as indicator organisms in water sanitation, but they have limitations. One problem is the growth of coliform bacteria embedded in biofilms on the inner surfaces of water pipes. These coliforms do not, then, represent external fecal contamination of the water, and they are not considered a threat to public health. Standards governing the presence of coliforms in drinking water require that any positive water sample be reported, and occasionally these indigenous coliforms have been detected. This has led to unnecessary community orders to boil water.

A more serious problem is that some pathogens, especially viruses and protozoan cysts and oocysts, are more resistant than coliforms to chemical disinfection. Through the use of sophisticated methods of detecting viruses, it has been found that chemically disinfected water samples that are free of coliforms are often still contaminated with enteric viruses. The cysts of *Giardia lamblia* and oocysts of *Cryptosporidium* are so resistant to chlorination that completely eliminating them by this method is probably impractical; mechanical methods such as filtration are necessary.

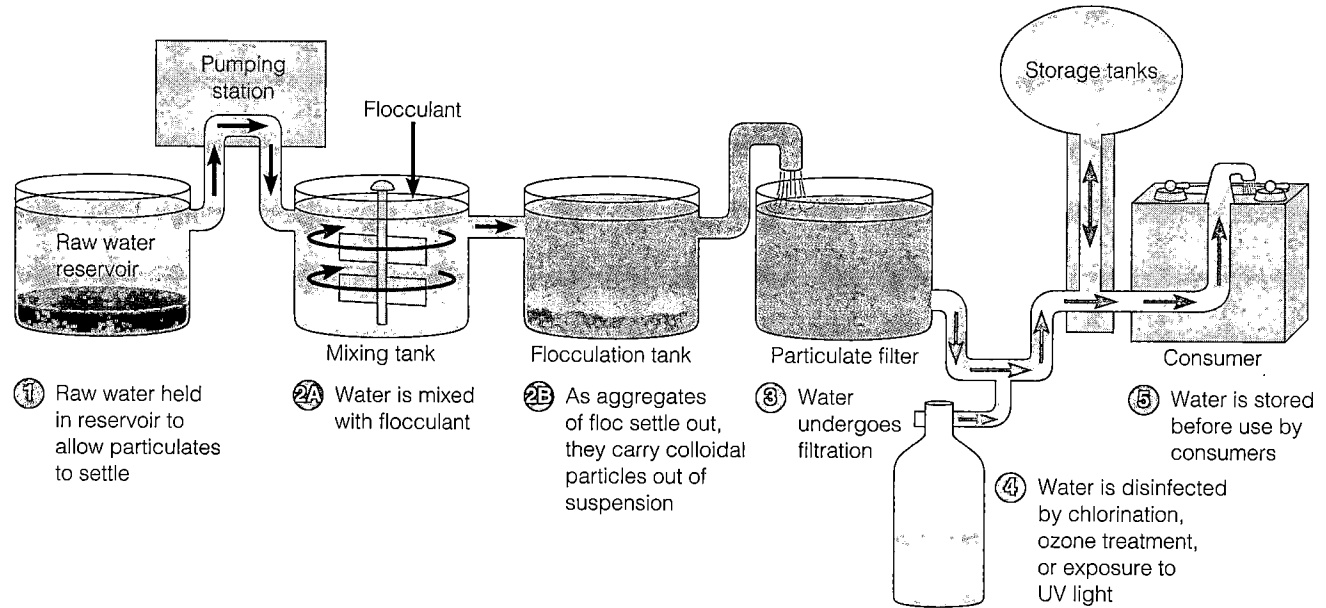

FIGURE 27.17 The steps involved in water treatment in a typical municipal water purification plant.

■ Municipal drinking water is usually filtered and disinfected.

Water Treatment

Learning Objective

■ *Describe how pathogens are removed from drinking water.*

When water is obtained from uncontaminated reservoirs fed by clear mountain streams or from deep wells, it requires minimal treatment to make it safe to drink. Many cities, however, obtain their water from badly polluted sources, such as rivers that have received municipal and industrial wastes upstream. The steps used to purify this water are shown in Figure 27.17.

Very turbid (cloudy) water is allowed to stand in a holding reservoir for a time to allow as much particulate suspended matter as possible to settle out. The water then undergoes **flocculation,** the removal of colloidal materials such as clay, which is so small that it would otherwise remain in suspension indefinitely. A flocculant chemical, such as aluminum potassium sulfate (alum), forms aggregations of fine suspended particles called *floc.* As these aggregations slowly settle out, they entrap colloidal material and carry it to the bottom. Large numbers of viruses and bacteria are also removed this way. Alum was used to clear muddy river water during the first half of the nineteenth century in the military forts of the American West, long before the germ theory of disease was developed.

After flocculation, the water is treated by **filtration—** that is, passing it through beds of 2–4 feet of fine sand or diatomaceous earth. As mentioned previously, some pro-

tozoan cysts and oocysts are removed from water only by such filtration treatment. The microorganisms are trapped mostly by surface adsorption onto the sand particles. They do not penetrate the tortuous routing between the particles, even though the openings might be larger than the microbes that are filtered out. These filters are periodically backflushed to clear them of accumulations. Water systems of cities that have an exceptional concern for toxic chemicals supplement sand filtration with filters of activated charcoal (carbon). Charcoal removes not only particulate matter but also most dissolved organic chemical pollutants. Low-pressure membrane filtration systems are now coming into use. These systems have pore openings as small as 0.2 μm and are more reliable for removal of *Giardia* and *Cryptosporidium.*

Before entering the municipal distribution system, the filtered water is chlorinated. Because organic matter neutralizes chlorine, the plant operators must pay constant attention to maintaining effective levels of chlorine. There has been some concern that chlorine itself might be a health hazard because it could react with organic contaminants of the water to form carcinogenic compounds. At present, this possibility is considered an acceptable risk when compared with the proven usefulness of chlorinating of water.

As noted in Chapter 7 on page 202, another disinfectant for water is ozone treatment. Ozone (O_3) is a highly reactive form of oxygen that is formed by electrical spark discharges and UV light. (The fresh odor of air following an electrical storm or around a UV light bulb is from

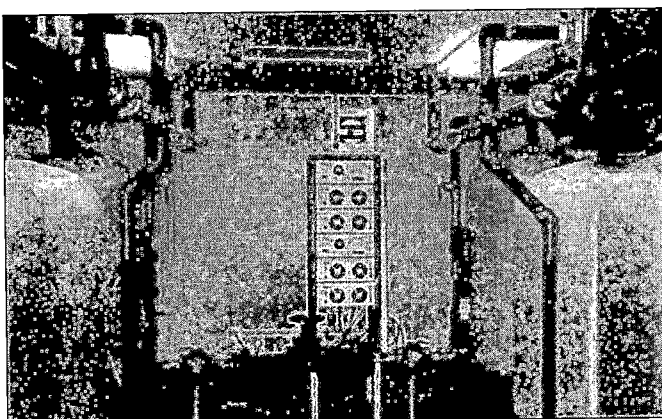

FIGURE 27.18 Ozone generation. Water treatment plants produce ozone by passing dry air between high-voltage electrodes in tanks called ozonators such as those shown here.

■ Ozone can be used to disinfect water.

ozone.) Ozone for water treatment is generated electrically at the site of treatment (Figure 27.18). Ozone treatment is also valued because it leaves no taste or odor. Because it has little residual effect, ozone is usually used as a primary disinfectant treatment and is followed by chlorination. The use of UV light is also a supplement or alternative to chemical disinfection. UV tube lamps are arranged so that water flows close to them. This is necessary because of the low penetrating power of UV radiation.

Sewage (Wastewater) Treatment

Learning Objectives

■ *Compare primary, secondary, and tertiary sewage treatment.*
■ *List some of the biochemical activities that take place in an anaerobic sludge digester.*
■ *Define BOD, activated sludge system, trickling filter, septic tank, and oxidation pond.*

Sewage or wastewater, includes all the water from a household that is used for washing and toilet wastes. Rainwater flowing into street drains, and some industrial wastes, enter the sewage system in some cities. Sewage is mostly water and contains little particulate matter, perhaps only 0.03%. Even so, in large cities this portion of sewage can total more than 1000 tons of solid material per day.

Until environmental awareness intensified, a surprising number of large American cities had only a rudimentary sewage-treatment system or no system at all. Raw sewage, untreated or nearly so, was simply discharged into rivers or oceans. A flowing, well-aerated stream is capable of considerable self-purification. Therefore, until expanding popula-

tions and their wastes exceeded this capability, this casual treatment of municipal wastes did not cause problems. In the United States, most cases of simple discharge have been improved. But this is not true in much of the world. Some 70% of the communities bordering the Mediterranean dump their unprocessed sewage into the sea.

Primary Sewage Treatment

The usual first step in sewage treatment is called **primary sewage treatment** (Figure 27.19a). In this process, large floating materials in incoming wastewater are screened out, the sewage is allowed to flow through settling chambers to remove sand and similar gritty material, skimmers remove floating oil and grease, and floating debris is shredded and ground. After this step, the sewage passes through sedimentation tanks, where more solid matter settles out. Sewage solids collecting on the bottom are called **sludge;** at this stage, *primary sludge.* About 40–60% of suspended solids are removed from sewage by this settling treatment, and flocculating chemicals that increase the removal of solids are sometimes added at this stage. Biological activity is not particularly important in primary treatment, although some digestion of sludge and dissolved organic matter can occur during long holding times. The sludge is removed on either a continuous or an intermittent basis, and the effluent (the liquid flowing out) then undergoes secondary treatment.

Biochemical Oxygen Demand

An important concept in sewage treatment and in the general ecology of waste management, **biochemical oxygen demand (BOD)** is a measure of the biologically degradable organic matter in water. Primary treatment removes about 25–35% of the BOD of sewage.

BOD is determined by the amount of oxygen required by bacteria to metabolize the organic matter. The classic method of measurement is the use of special bottles with airtight stoppers. Each bottle is first filled with test water or dilutions. The water is initially aerated to provide a relatively high level of dissolved oxygen and is seeded with bacteria if necessary. The filled bottles are incubated in the dark for 5 days at 20°C, and the decrease in dissolved oxygen is determined by a chemical or electronic testing method. The more oxygen that is used up as the bacteria degrade the organic matter in the sample, the greater the BOD, which is usually expressed in milligrams of oxygen per liter of water. The amount of oxygen that normally can be dissolved in water is only about 10 mg/liter; typical BOD values of wastewater may be twenty times this amount. If this wastewater enters a lake, for example, bacteria in the lake begin to consume the

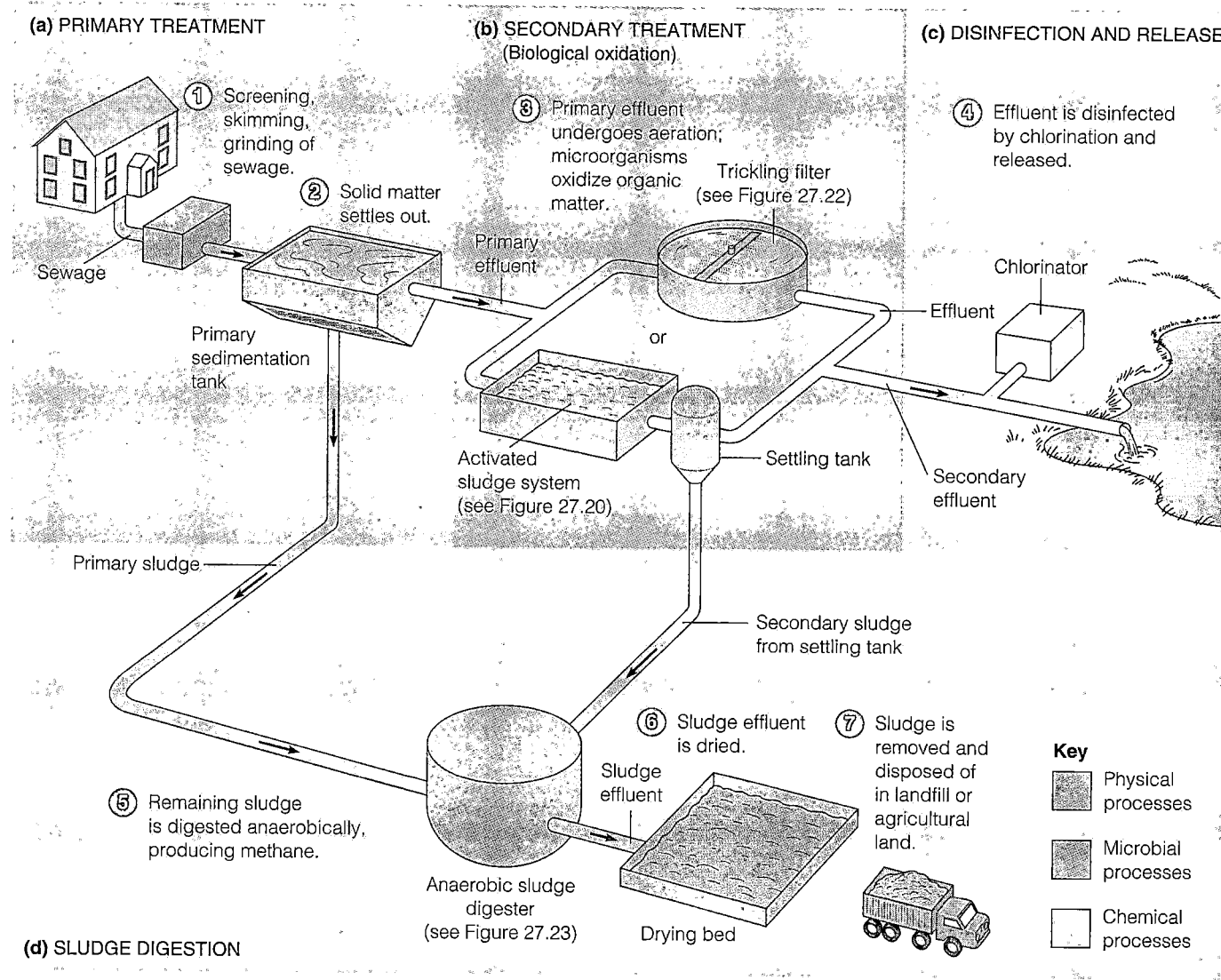

(a) PRIMARY TREATMENT

① Screening, skimming, grinding of sewage.

Sewage

Primary sedimentation tank

② Solid matter settles out.

Primary effluent

Primary sludge

(b) SECONDARY TREATMENT (Biological oxidation)

③ Primary effluent undergoes aeration; microorganisms oxidize organic matter.

Trickling filter (see Figure 27.22)

or

Activated sludge system (see Figure 27.20)

Settling tank

Secondary effluent

Secondary sludge from settling tank

(c) DISINFECTION AND RELEASE

④ Effluent is disinfected by chlorination and released.

Chlorinator

Effluent

⑥ Sludge effluent is dried.

Sludge effluent

⑦ Sludge is removed and disposed of in landfill or agricultural land.

⑤ Remaining sludge is digested anaerobically, producing methane.

Anaerobic sludge digester (see Figure 27.23)

Drying bed

(d) SLUDGE DIGESTION

Key

Physical processes

Microbial processes

Chemical processes

FIGURE 27.19 The stages in typical sewage treatment. Microbial activity occurs aerobically in trickling filters or activated sludge aeration tanks and anaerobically in the anaerobic sludge digester. A particular system would use either activated sludge aeration tanks or trickling filters, not both as shown in this figure. Methane produced by sludge digestion is burned off or used to power heaters or pump motors.

■ Which processes require oxygen?

organic matter responsible for the high BOD, rapidly depleting the oxygen in the lake water. (See the discussion of eutrophication earlier in the chapter, page 758.)

Secondary Sewage Treatment

After primary treatment, the greater part of the BOD remaining in the sewage is in the form of dissolved organic matter. **Secondary sewage treatment,** which is predominantly biological, is designed to remove most of this organic matter and reduce the BOD (Figure 27.19b). In this process, the sewage undergoes strong aeration to encourage the growth of aerobic bacteria and other microorganisms that oxidize the dissolved organic matter to carbon dioxide and water. Two commonly used methods of secondary treatment are activated sludge systems and trickling filters.

In the aeration tanks of an **activated sludge system,** air or pure oxygen is passed through the effluent from primary treatment (Figure 27.20). The name is derived from the practice of adding to the incoming sewage some of the sludge from a previous batch. This inoculum is termed *activated sludge* because it contains large numbers of sewage-metabolizing microbes. The activity of these aerobic microorganisms oxidizes much of the sewage or-

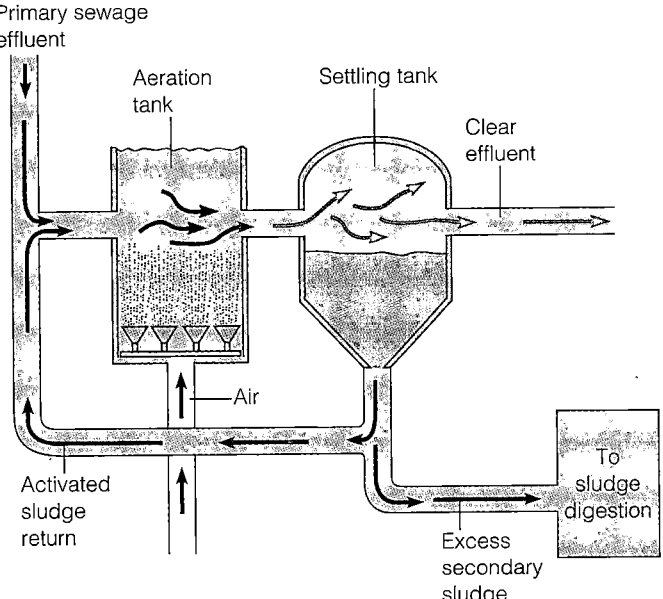

Primary sewage effluent

(a) Diagram of an activated sludge system

(b) An aeration tank. Note surface is frothing from aeration.

FIGURE 27.20 An activated sludge system of secondary sewage treatment.

■ What is occurring during the activated sludge process?

ganic matter into carbon dioxide and water. Especially important members of this microbial community are species of *Zoogloea* bacteria, which form flocculant masses in the aeration tanks (Figure 27.21). Soluble organic matter in the sewage is incorporated into the floc and its microorganisms. Aeration is discontinued after 4–8 hours, and the contents of the tank are transferred to a settling tank, where the floc settles out, removing much of the or-

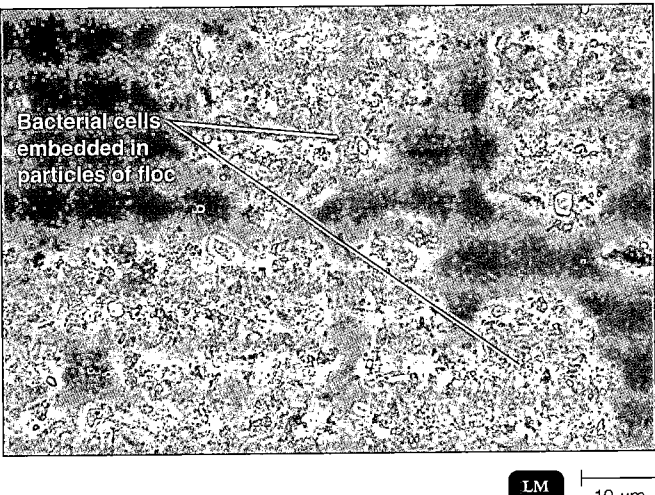

FIGURE 27.21 Floc formed by an activated sludge system.
Gelatinous masses of floc are formed by a species of *Zoogloea* bacteria. Notice the bacterial cells embedded in particles of floc.

■ Contact with microbes forming floc is responsible for much of the activity of activated sludge systems.

ganic matter. These solids are subsequently treated in an anaerobic sludge digester, which will be described shortly. Probably more organic matter is removed by this settling-out process than by the relatively short-term aerobic oxidation by microbes. The clear effluent is disinfected and discharged.

Occasionally, the sludge will float rather than settle out; this phenomenon is called **bulking.** When this happens, the organic matter in the floc flows out with the discharge effluent, resulting in local pollution. Bulking is caused by the growth of filamentous bacteria of various types; *Sphaerotilus natans* and *Nocardia* species are frequent offenders. Activated sludge systems are quite efficient: they remove 75–95% of the BOD from sewage.

Trickling filters are the other commonly used method of secondary treatment. In this method, the sewage is sprayed over a bed of rocks or molded plastic (Figure 27.22a). The components of the bed must be large enough so that air penetrates to the bottom but small enough to maximize the surface area available for microbial activity. A biofilm of aerobic microbes grows on the rock or plastic surfaces (Figure 27.22b). Because air circulates throughout the rock bed, these aerobic microorganisms in the slime layer can oxidize much of the organic matter trickling over the surfaces into carbon dioxide and water. Trickling filters remove 80–85% of the BOD, so they are generally less efficient than activated sludge systems. However, they are usually less troublesome to operate and have fewer problems from overloads or toxic sewage. Note that sludge is also a product of trickling filter systems.

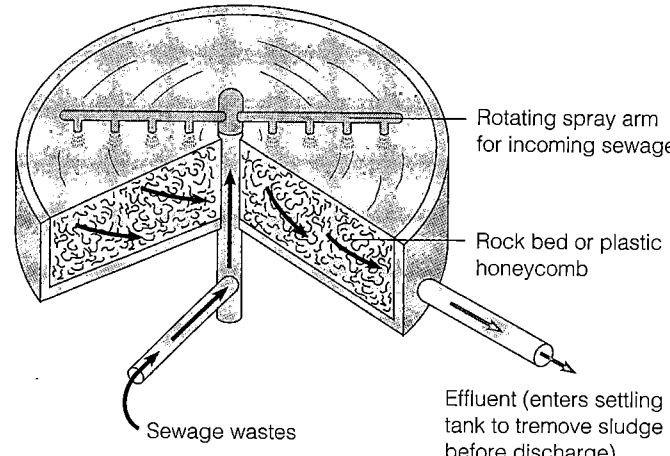

Rotating spray arm for incoming sewage

Rock bed or plastic honeycomb

Effluent (enters settling tank to tremove sludge before discharge)

Sewage wastes

(a) Rotating spray arm of a trickling filter system

(b) A cutaway view of a trickling filter system

FIGURE 27.22 A trickling filter of secondary sewage treatment. The sewage is sprayed from the system of rotating pipes onto a bed of rocks or plastic honeycomb designed to have a maximum surface area and to allow oxygen to penetrate deeply into the bed.

■ Microorganisms grow on the enormous surface area, forming a biofilm that aerobically metabolizes the organic matter in the sewage trickling down through the bed.

Another biofilm-based design for secondary sewage treatment is the **rotating biological contactor** system. This is a series of disks several feet in diameter, mounted on a shaft. The disks rotate slowly with their lower 40% submerged in wastewater. Rotation provides aeration and contact between the biofilm on the disks and the wastewater. The rotation also tends to cause the accumulated biofilm to slough off when it becomes too thick. This is about the equivalent of floc accumulation in activated sludge systems.

Disinfection and Release

Treated sewage is disinfected, usually by chlorination, before being discharged (Figure 27.19c). The discharge is usually into an ocean or into flowing streams, although spray irrigation fields are sometimes used to avoid phosphorus and heavy metal contamination of waterways. After disinfection, the water may be dechlorinated so the chlorine will not kill desirable aquatic organisms when the water is discharged. Chlorine is removed in an exchange reaction by the addition of sulfur dioxide.

In at least one U.S. city with limited freshwater supplies, some treated sewage is introduced into a water-treatment system just like any other incoming water and recycled into drinking water. This practice will probably be expanded; use of treated sewage as irrigation water is already widespread. Scientists are very interested in sewage recycling during long-term stays in planned space

stations. It costs thousands of dollars to send a gallon of drinking water into orbit.

Sludge Digestion

Primary sludge accumulates in primary sedimentation tanks; sludge also accumulates in activated sludge and in trickling filter secondary treatments. For further treatment, these sludges are often pumped to **anaerobic sludge digesters** (Figures 27.19 and 27.23). The process of sludge digestion is carried out in large tanks from which oxygen is almost completely excluded.

In secondary treatment, emphasis is placed on the maintenance of aerobic conditions so that organic matter is converted to carbon dioxide, water, and solids that can settle out. An anaerobic sludge digester, however, is designed to encourage the growth of anaerobic bacteria, especially methane-producing bacteria that decrease these organic solids by degrading them to soluble substances and gases, mostly methane (60–70%) and carbon dioxide (20–30%). Methane and carbon dioxide are relatively innocuous end-products, comparable to the carbon dioxide and water from aerobic treatment. The methane is routinely used as a fuel for heating the digester and is also frequently used to run power equipment in the plant.

There are essentially three stages in the activity of an anaerobic sludge digester. The first stage is the production of carbon dioxide and organic acids from anaerobic fermentation of the sludge by various anaerobic and faculta-

An anaerobic sludge digester at a California sewage-treatment plant. Much or all of a typical digester is below ground level, especially in cold climates. Methane from such a digester is often used to run pumps or heaters in the treatment plant. Excess methane is being burned off in the flame shown at the top of the digester.

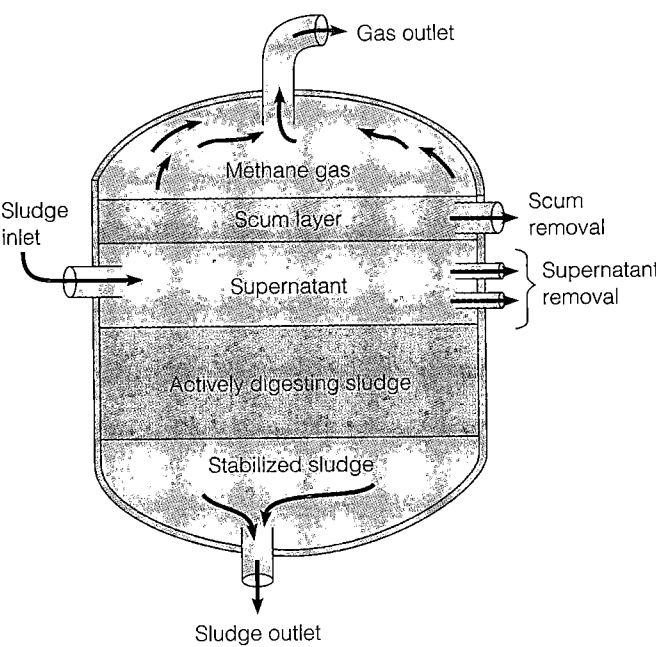

Section of a sludge digester. The scum and supernatant layers are low in solids and are recirculated through secondary treatment.

FIGURE 27.23 Sludge digestion.

■ What is more efficient, an activated sludge system or an anaerobic digester?

tively anaerobic microorganisms. In the second stage, the organic acids are metabolized to form hydrogen and carbon dioxide as well as organic acids, such as acetic acid. These products are the raw materials for a third stage, in which the methane-producing bacteria produce methane

(CH_4). Most of the methane is derived from the energy-yielding reduction of carbon dioxide by hydrogen gas:

$$CO_2 + 4\,H_2 \longrightarrow CH_4 + 2\,H_2O$$

Other methane-producing microbes split acetic acid (CH_3COOH) to yield methane and carbon dioxide:

$$CH_3COOH \longrightarrow CH_4 + CO_2$$

After anaerobic digestion is completed, large amounts of undigested sludge still remain, although it is relatively stable and inert. To reduce its volume, this sludge is pumped to shallow drying beds or water-extracting filters. Following this step, the sludge can be used for land-fill or as a soil conditioner. Heavy metals—especially cadmium, which can accumulate in plants, and copper, which is toxic to plants—are a potential problem with land disposal. Sludge has about one-fifth the growth-enhancing value of normal commercial lawn fertilizers but has desirable soil-conditioning qualities, much as do humus and mulch.

Septic Tanks

Homes and businesses in areas of low population density that are not connected to municipal sewage systems often use a **septic tank,** a device whose operation is similar in principle to primary treatment (Figure 27.24). Sewage enters a holding tank, and suspended solids settle out. The sludge in the tank must be pumped out periodically and disposed of. The effluent flows through a system of perforated piping into a leaching (soil drainage) field. The effluent entering the soil is decomposed by soil microorganisms.

These systems work well when not overloaded and when the drainage system is properly sized to the load and soil type. Heavy clay soils require extensive drainage systems because of the soil's poor permeability. The high porosity of sandy soils can cause chemical or bacterial pollution of nearby water supplies.

Oxidation Ponds

Many small communities and many industries use **oxidation ponds,** also called *lagoons* or *stabilization ponds,* for water treatment. These are inexpensive to build and operate but require large areas of land. Designs vary, but most incorporate two stages. The first stage is analogous to primary treatment; the sewage pond is deep enough that conditions are almost entirely anaerobic. Sludge settles out in this stage. In the second stage, which roughly corresponds to secondary treatment, effluent is pumped into an adjoining pond or system of ponds shallow enough to be aerated by wave action. Because it is difficult to maintain aerobic conditions for bacterial growth in ponds

FIGURE 27.24 A septic tank system.

■ Solids settle out in the septic tank as sludge, which is pumped out periodically through the access manhole.

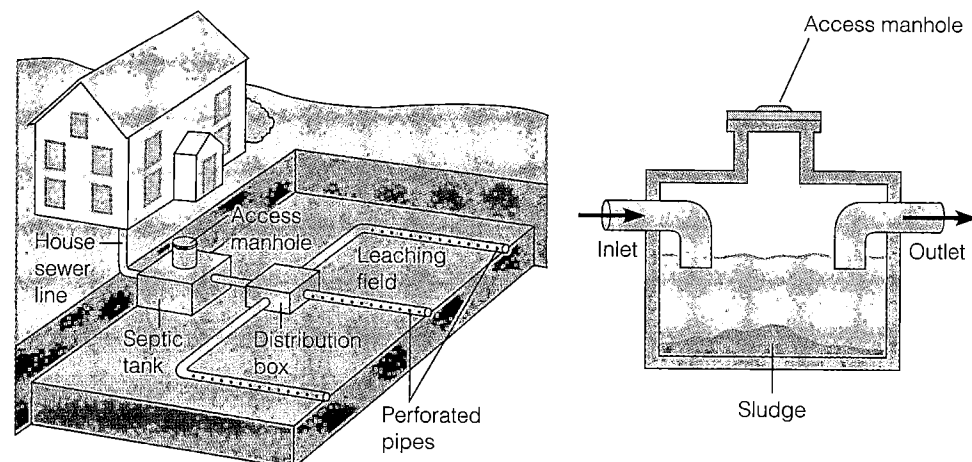

(a) Overall plan. Most soluble organic matter is disposed of by percolation into the soil.

(b) A section of a septic tank

with so much organic n couraged to produce o\ posing the organic mat; dioxide. Algae, which us synthetic metabolism, g' in turn encourages th sewage. Large amou··· · algae accumulate, but on oxidation pond, unli·· a ·al· load.

Some small sewa,c-p·uds ·ng ···· ·ations, such as isolated campgrounds a 1 hi·· .., ·' ··p areas, use an *oxidation ditch* for sewage w··m ···, in this method, a small oval channel in the sh·p· ·a·eec·ck is filled with sewage water. A p·d·l·· wh·· similar to that on a Mississippi steamboat, but in a fix·· l·········, propels the water in a self-contained fl··· ing ··· ··· ···ed enough to oxidize the wastes.

Tertiary Sewage Treatment

As we have seen, primary and secondary treatments of sewage do not remove all the biologically degradable organic matter. Amounts of organic matter that are not excessive can be released into a flowing stream without causing a serious problem. In many; however, the pressures of increased population might increase wastes beyond a body of water's carrying capacity, and additional treatments might be required. Even now, primary and secondary treatments are inadequate in certain situations, such as when the effluent is discharged into small streams or recreational lakes. Some communities have therefore developed **tertiary sewage treatment** plants. Lake Tahoe in the Sierra Nevada Mountains, surrounded by extensive development, is the site of one of the best-known tertiary sewage-treatment systems.

'e growth of algae is encerial action in decom-··astes generates carbon dioxide in their photo-·. duce oxygen, which ··obic microbes in the ·c ··tter in the form of ·· · problem because the ····· has a large nutrient

The effluent from secondary treatment plants contains some residual BOD. It also contains about 50% of the original nitrogen and 70% of the original phosphorus, which can greatly affect a lake's ecosystem. Tertiary treatment is designed to remove essentially all the BOD, nitrogen, and phosphorus. Tertiary treatment depends less on biological treatment than on physical and chemical treatments. Phosphorus is precipitated out by combining with such chemicals as lime, alum, and ferric chloride. Filters of fine sands and activated charcoal remove small particulate matter and dissolved chemicals. Nitrogen is converted to ammonia and discharged into the air in stripping towers. Some systems encourage denitrifying bacteria to form volatile nitrogen gas. Finally, the purified water is chlorinated.

Tertiary treatment provides water that is suitable for drinking, but the process is extremely costly. Secondary treatment is less costly, but water that has undergone only secondary treatment still contains many water pollutants. Much work is being done to design secondary treatment plants in which the effluent can be used for irrigation. This design would eliminate a source of water pollution, provide nutrients for plant growth, and reduce the demand on already scarce water supplies. The soil to which this water is applied would act as a trickling filter to remove chemicals and microorganisms before the water reaches groundwater and surface water supplies.

★ ★ ★

We hope this chapter on environmental microbiology, as well as previous chapters in the book, has left you with a greater appreciation of the microbial influences around us. Without the natural and human-directed applications of microbes, life would be very different, and perhaps could not sustain itself at all.

Study Outline ![ST] Student Tutorial CD-ROM

METABOLIC DIVERSITY (pp. 742–744)

1. Microorganisms live in a wide variety of habitats because of their metabolic diversity, their ability to use a variety of carbon and energy sources and grow under different physical conditions.

Habitat Variety (pp. 742–743)

1. Extremophiles live in extreme conditions of temperature, acidity, alkalinity, or salinity.

Symbiosis (pp. 743–744)

1. Symbiosis is a relationship between two different organisms or populations.
2. Parasitism is a type of symbiosis in which one organism gets its nutrients and reproductive capability from another organism.
3. Mutualism is a type of symbiosis in which both partners benefit.
4. Symbiotic fungi called mycorrhizae live in and on plant roots; they increase the surface area and nutrient absorption of the plant.

SOIL MICROBIOLOGY AND BIOGEOCHEMICAL CYCLES (pp. 744–754)

1. In biogeochemical cycles, certain chemical elements are recycled.
2. Microorganisms in the soil decompose organic matter and transform carbon-, nitrogen-, and sulfur-containing compounds into usable forms.
3. Microbes are essential to the continuation of biogeochemical cycles.
4. Elements are oxidized and reduced by microorganisms during these cycles.

The Carbon Cycle (pp. 744–745)

1. Carbon dioxide is incorporated, or fixed, into organic compounds by photoautotrophs and chemoautotrophs.
2. These organic compounds provide nutrients for chemoheterotrophs.
3. Chemoheterotrophs release CO_2 that is then used by photoautotrophs.
4. Carbon is removed from the cycle when it is in $CaCO_3$ and fossil fuels.

The Nitrogen Cycle (pp. 745–749)

![ST] *To review, go to Biodiversity: Biochemical cycles: Nitrogen cycle*

1. Microorganisms decompose proteins from dead cells and release amino acids.
2. Ammonia is liberated by microbial ammonification of the amino acids.

3. The nitrogen in ammonia is oxidized to produce nitrates for energy by nitrifying bacteria.
4. Denitrifying bacteria reduce the nitrogen in nitrates to molecular nitrogen (N_2).
5. N_2 is converted into ammonia by nitrogen-fixing bacteria.
6. Nitrogen-fixing bacteria include free-living genera such as *Azotobacter*, cyanobacteria, and the symbiotic bacteria *Rhizobium* and *Frankia*.
7. Ammonium and nitrate are used by bacteria and plants to synthesize amino acids that are assembled into proteins.

The Sulfur Cycle (pp. 749–751)

1. Hydrogen sulfide (H_2S) is used by autotrophic bacteria; the sulfur is oxidized to form S^0 or SO_4^{2-}.
2. Winogradsky discovered that *Beggiatoa* bacteria oxidize sulfur (H_2S and S^0) for energy.
3. Plants and other microorganisms can reduce SO_4^{2-} to make certain amino acids. These amino acids are in turn used by animals.
4. H_2S is released by decay or dissimilation of these amino acids.
5. Sulfur dioxide produced by combustion of fossil fuels combines with water to form sulfurous acid.

The Phosphorus Cycle (p. 751)

1. Phosphorus (as PO_4^{3-}) is found in rocks and bird guano.
2. When solubilized by microbial acids, the PO_4^{3-} is available for plants and microorganisms.
3. Endolithic bacteria live in solid rock; these autotrophic bacteria use hydrogen as an energy source.

The Degradation of Synthetic Chemicals in Soil and Water (pp. 751–754)

1. Many synthetic chemicals, such as pesticides, are resistant to degradation by microbes.
2. Ecologists are trying to use bacteria to degrade PCBs.

Bioremediation (pp. 752–753)

1. The use of microorganisms to remove pollutants is called bioremediation.
2. The growth of oil-degrading bacteria can be enhanced by the addition of nitrogen and phosphorus fertilizer.

Solid Municipal Waste (pp. 753–754)

1. Municipal landfills prevent decomposition of solid wastes because they are dry and anaerobic.
2. In some landfills, methane produced by methanogens can be recovered for an energy source.
3. Composting can be used to promote biodegradation of organic matter.

AQUATIC MICROBIOLOGY AND SEWAGE TREATMENT (pp. 754–766)

Biofilms (p. 754)

1. Microbes adhere to surfaces and accumulate as biofilms on solid surfaces in contact with water.

Aquatic Microorganisms (pp. 754–756)

1. The study of microorganisms and their activities in natural waters is called aquatic microbiology.

2. Natural waters include lakes, ponds, streams, rivers, estuaries, and the oceans.

3. The concentration of bacteria in water is proportional to the amount of organic material in the water.

4. Most aquatic bacteria tend to grow on surfaces rather than in a free-floating state.

Freshwater Microbiota (pp. 755–756)

1. The number and location of freshwater microbiota depend on the availability of oxygen and light.

2. Photosynthetic algae are the primary producers of a lake; they are found in the limnetic zone.

3. Pseudomonads, *Cytophaga, Caulobacter,* and *Hyphomicrobium* are found in the limnetic zone, where oxygen is abundant.

4. Microbes in stagnant water use available oxygen and can cause odors and the death of fish.

5. The amount of dissolved oxygen is increased by wave action.

6. Purple and green sulfur bacteria are found in the profundal zone, which contains light and H_2S but no oxygen.

7. *Desulfovibrio* reduces SO_4^{2-} to H_2S in benthic mud.

8. Methane-producing bacteria are also found in the benthic zone.

Seawater Microbiota (p. 756)

1. Phytoplankton, consisting mainly of diatoms, are the primary producers of the open ocean.

2. Some algae and bacteria are bioluminescent. They possess the enzyme luciferase, which can emit light.

The Role of Microorganisms in Water Quality (pp. 756–759)

Water Pollution (pp. 756–758)

1. Microorganisms are filtered from water that percolates into groundwater supplies.

2. Some pathogenic microorganisms are transmitted to humans in drinking and recreational waters.

3. Resistant chemical pollutants may be concentrated in animals in an aquatic food chain.

4. Mercury is metabolized by certain bacteria into a soluble compound that is concentrated in animals.

5. Nutrients such as phosphates cause algal blooms, which can lead to eutrophication of aquatic ecosystems.

6. Eutrophication, meaning well nourished, is the result of the addition of pollutants or natural nutrients.

7. *Thiobacillus ferrooxidans* produces sulfuric acid at coal-mining sites.

Water Purity Tests (pp. 758–759)

1. Tests for the bacteriological quality of water are based on the presence of indicator organisms, the most common of which are coliforms.

2. Coliforms are aerobic or facultatively anaerobic, gram-negative, non–endospore-forming rods that ferment lactose with the production of acid and gas within 48 hours of being placed in a medium at 35°C.

3. Fecal coliforms, predominantly *E. coli,* are used to indicate the presence of human feces.

Water Treatment (pp. 760–761)

1. Drinking water is held in a holding reservoir long enough that suspended matter settles.

2. Flocculation treatment uses a chemical such as alum to coalesce and then settle colloidal material.

3. Filtration removes protozoan cysts and other microorganisms.

4. Drinking water is disinfected with chlorine to kill remaining pathogenic bacteria.

Sewage (Wastewater) Treatment (pp. 761–766)

1. Domestic wastewater is called sewage; it includes household water, toilet wastes, industrial wastes, and rainwater.

Primary Sewage Treatment (p. 761)

1. Primary sewage treatment is the removal of solid matter called sludge.

2. Biological activity is not very important in primary treatment.

Biochemical Oxygen Demand (pp. 761–762)

STI *Biotechnology: Bioremediation: Sewage Treatment*

1. Biochemical oxygen demand (BOD) is a measure of the biologically degradable organic matter in water.

2. Primary treatment removes about 25–35% of the BOD of sewage.

3. BOD is determined by measuring the amount of oxygen bacteria required to degrade the organic matter.

Secondary Sewage Treatment (pp. 762–764)

1. Secondary sewage treatment is the biological degradation of organic matter after primary treatment.

2. Activated sludge systems, trickling filters, and rotating biological contactors are methods of secondary treatment.

3. Microorganisms degrade the organic matter aerobically.

4. Secondary treatment removes up to 95% of the BOD.

Disinfection and Release (p. 764)

1. Treated sewage is disinfected, usually by chlorination, before discharge onto land or into water.

Sludge Digestion (pp. 764–765)

1. Sludge is placed in an anaerobic sludge digester; bacteria degrade organic matter and produce simpler organic compounds, methane, and CO_2.

2. The methane produced in the digester is used to heat the digester and operate other equipment.

3. Excess sludge is periodically removed from the digester, dried, and disposed of (as landfill or as soil conditioner) or incinerated.

Septic Tanks (p. 765)

1. Septic tanks can be used in rural areas to provide primary treatment of sewage.

2. They require a large leaching field for the effluent.

Oxidation Ponds (pp. 765–766)

1. Small communities can use oxidation ponds for secondary treatment.

2. These require a large area in which to build an artificial lake.

Tertiary Sewage Treatment (p. 766)

1. Tertiary sewage treatment uses physical filtration and chemical precipitation to remove all the BOD, nitrogen, and phosphorus from water.

2. Tertiary treatment provides drinkable water, whereas secondary treatment provides water usable only for irrigation.

Study Questions

REVIEW

1. Give two examples of extremophiles.

2. The koala is a leaf-eating animal. What prediction can you make about its digestive system?

3. Give one possible explanation of why *Penicillium* would make penicillin, since the fungus does not get bacterial infections.

4. Diagram the carbon cycle in the presence and absence of oxygen. Name at least one microorganism that is involved at each step.

5. In the sulfur cycle, microbes degrade organic sulfur compounds, such as _____ to release H_2S, which can be oxidized by *Thiobacillus* to _____. This ion can be assimilated into amino acids by _____ or reduced by *Desulfovibrio* to _____. H_2S is used by photoautotrophic bacteria as an electron donor to synthesize _____. The sulfur-containing by-product of this metabolism is _____.

6. Why is the phosphorus cycle important?

7. Fill in the following table:

Process	Chemical Reactions	Microorganisms
Ammonification		
Nitrification		
Denitrification		
Nitrogen fixation		

8. The following organisms have important roles as symbionts with plants and fungi; describe the symbiotic relationship of each organism with its host: cyanobacteria, mycorrhizae, *Rhizobium*, *Frankia*.

9. Outline the treatment process for drinking water.

10. What is the purpose of a coliform count on water?

11. The following processes are used in wastewater treatment. Match the stage of treatment with the processes. Each choice can be used once, more than once, or not at all.

Processes	Treatment Stage
_____ Leaching field	a. Primary
_____ Removal of solids	b. Secondary
_____ Biological degradation	c. Tertiary
_____ Activated sludge system	
_____ Chemical precipitation of phosphorus	
_____ Trickling filter	
_____ Results in drinking water	

12. Why is activated sludge a more efficient means of removing BOD than a sludge digester?

13. Why are septic tanks and oxidation ponds not feasible for large municipalities?

14. Explain the effect of dumping untreated sewage into a pond on the eutrophication of the pond. The effect of sewage that has primary treatment? The effect of sewage that has secondary treatment? Contrast your previous answers with the effect of each type of sewage on a fast-moving river.

15. Bioremediation refers to the use of living organisms to remove pollutants. Describe three examples of bioremediation.

MULTIPLE CHOICE

For questions 1–4, answer whether
    a. the process takes place under aerobic conditions.
    b. the process takes place under anaerobic conditions.
    c. the amount of oxygen doesn't make any difference.

1. Activated sludge system.

2. Denitrification.

3. Nitrogen fixation.

4. Methane production.

5. The water used to prepare intravenous solutions in a hospital contained endotoxins. Infection control personnel performed plate counts to find the source of the bacteria. Their results:

	Bacteria/100 ml
Municipal water pipes	0
Boiler	0
Hot water line	300

All of the following conclusions about the bacteria can be drawn *except* which one?
a. It was present as a biofilm in the pipes.
b. It is gram-negative.
c. It comes from fecal contamination.
d. It comes from the city water supply.
e. none of the above

Use the following choices to answer questions 8–10:
a. aerobic respiration
b. anaerobic respiration
c. anoxygenic photoautotroph
d. oxygenic photoautotroph

6. $CO_2 + H_2S \xrightarrow{\text{Light}} C_6H_{12}O_6 + S^0$

7. $SO_4^{2-} + 10\ H^+ + 10\ e^- \rightarrow H_2S + 4\ H_2O$

8. $CO_2 + 8\ H^+ + 8\ e^- \rightarrow CH_4 + 2\ H_2O$

9. All of the following are effects of water pollution *except*
a. the spread of infectious diseases.
b. increased eutrophication.
c. increased BOD.
d. increased growth of algae.
e. none of the above

10. Coliforms are used as indicator organisms of sewage pollution because
a. they are pathogens.
b. they ferment lactose.
c. they are abundant in human intestines.
d. they grow within 48 hours.
e. all of the above

CRITICAL THINKING

1. Here are the formulas of two detergents that have been manufactured:

$$C-C-C-C-C-C-C-C-C-C...$$

```
              C
              |
              C     C
              |     |
      C—C—C—C—C—C...
          |
          C
```

Which of these would be resistant, and which would be readily degraded by microorganisms? (*Hint:* Refer to the degradation of fatty acids in Chapter 5.)

2. Complete the following graph to show the effect of dumping phosphates into a body of water at time x.

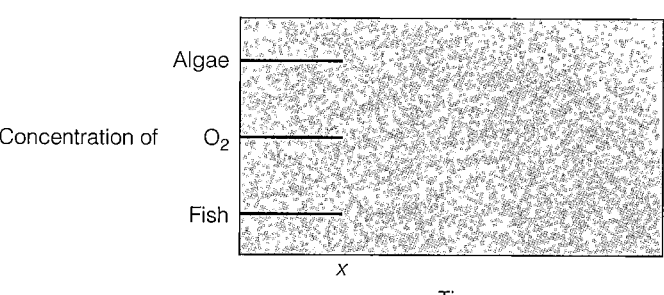

Concentration of — Algae — O_2 — Fish

x

Time

CLINICAL APPLICATIONS

1. A patient with a heart pacemaker received antibiotic therapy for streptococcal bacteremia. One month later, he was treated for recurrence of the bacteremia. When he returned 6 weeks later, again with bacteremia, the physician recommended replacing the pacemaker. Why did this cure his condition?

2. The bioremediation process shown in the photograph is used to remove benzene and other hydrocarbons from soil contaminated by petroleum. The pipes are used to add nitrates, phosphates, oxygen, or water. Why are each of these added? Why is it not always necessary to add bacteria?

Learning with Technology

VU Enter the Virtual Lab, click the arrow next to the Session field, click Textbook Exercises, and select Chapter 27. Read the Case Study carefully, identify the unknown, and use what you learn to answer the following questions (consult your textbook for additional information).

1. Based on your test results, does this organism fit the definition for a coliform? Would it be a satisfactory indicator organism for fecal contamination of the soil?

2. Which test in VirtualUnknown™ Microbiology would be useful in determining this microbe's role in the nitrogen cycle?

3. Would this microbe be more likely classified as a nitrifying bacterium, a denitrifying bacterium, a nitrogen-fixing bacterium, or an ammonifying bacterium?

chapter 28

Applied and Industrial Microbiology

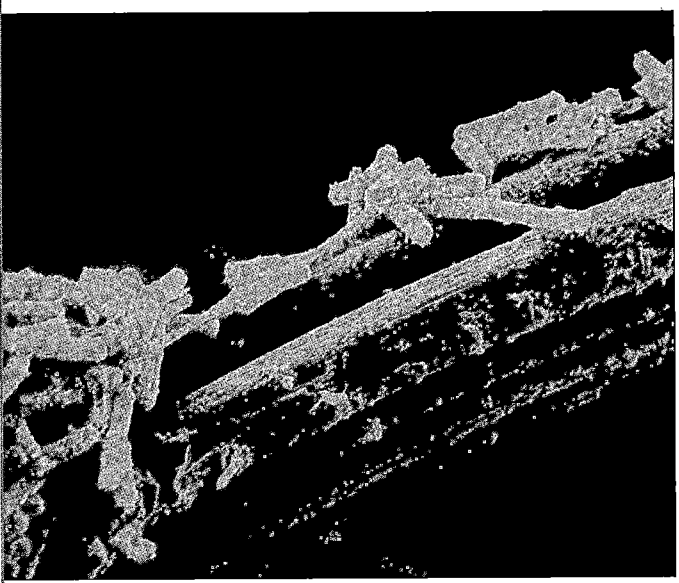

Bacteria that have been immobilized on silk fibers can promote a continuous conversion of substrate to product.

In the previous chapter on environmental microbiology, we saw that microbes are an essential factor in many natural phenomena that make life possible on Earth. In this chapter we will look at how microorganisms are harnessed in such useful applications as the making of food and industrial products. Many of these processes—especially baking, wine-making, brewing, and cheese-making—have origins long lost in history (see the box in Chapter 1, page 5). In Chapter 9, we discussed industrial applications of genetically engineered microorganisms that are at the cutting edge of our knowledge of molecular biology. Many of these applications are now essential to modern industry.

Food Microbiology

Modern civilization, with its large population, could not be supported without methods of preserving food. In fact, civilization arose only after agriculture produced a year-round stable food supply in a single location, so that people were able to give up a nomadic hunting-and-gathering way of life.

Many of the methods of food preservation used today were probably discovered by chance in centuries past. People in early cultures observed that dried meat and salted fish resisted decay. Nomads must have noticed that soured animal milk resisted further decomposition and was still palatable. Moreover, if the curd of the soured milk was pressed to remove moisture and allowed to ripen (in effect, cheese-making), it was even more effec-tively preserved and tasted better. Farmers soon learned that if grains were kept dry, they did not become moldy.

Industrial Food Canning

Learning Objective

■ Describe thermophilic anaerobic spoilage and flat sour spoilage by mesophilic bacteria.

In Chapter 7, you learned that preserving foods by heating a properly sealed container, as in home canning, is not difficult. The challenge in commercial canning is to use the right amount of heat necessary to kill spoilage organisms and dangerous microbes, such as the endospore-forming *Clostridium botulinum*, without degrading the appearance and palatability of food. Thus, much research

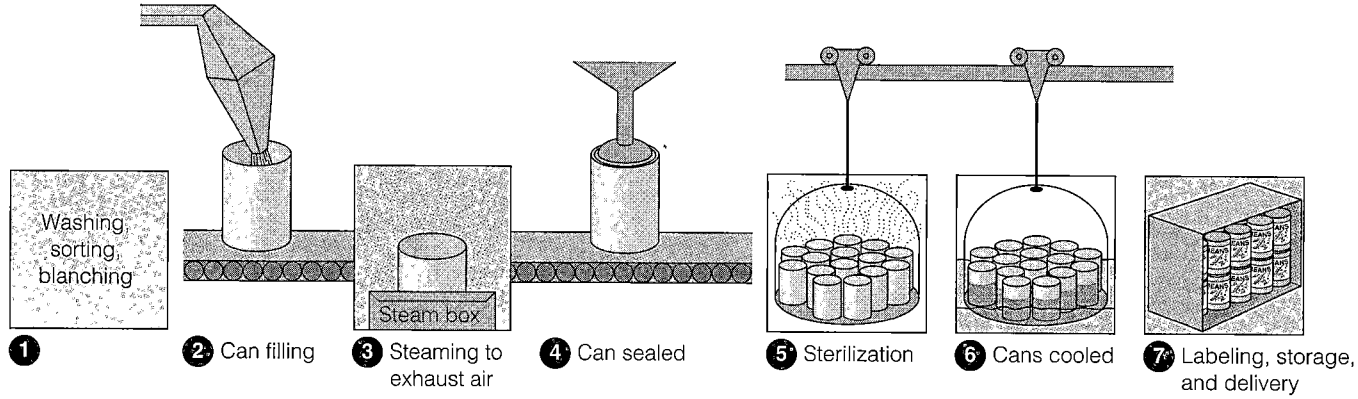

① Washing, sorting, blanching   ② Can filling   ③ Steaming to exhaust air   ④ Can sealed   ⑤ Sterilization   ⑥ Cans cooled   ⑦ Labeling, storage, and delivery

Retort (see Figure 28.2)

FIGURE 28.1   The commercial sterilization process in industrial canning. ① Blanching is a treatment with hot water or steam intended to soften the product so the can will fill better. It also destroys enzymes that might alter the color, flavor, or texture of the product and lower the microbial population. ② Cans are filled to capacity, leaving as little dead space as possible. ③ To exhaust (drive out) most dissolved air, cans are heated in a steam box. ④ The cans are sealed. ⑤ Cans are sterilized by steam under pressure. ⑥ Cans are cooled by submerging them or spraying them with water. ⑦ Cans are labeled for sale.

▨ How does commercial sterilization differ from complete sterilization?

FIGURE 28.2   Three commercial canning retorts, note size compared to worker at extreme left.

▨ **A retort uses steam under pressure to achieve sterilization temperatures.**

is applied to determining the exact minimum heat treatment that will accomplish both these goals.

Industrial food canning is much more technically sophisticated than home canning (Figure 28.1). Industrially canned goods undergo what is called **commercial sterilization** by steam under pressure in a large **retort** (Figure 28.2) which operates on the same principle as an autoclave (see Figure 7.2 on page 188). Commercial sterilization is intended to destroy *C. botulinum* endospores

and is not as rigorous as complete sterilization. The reasoning is that if *C. botulinum* endospores are destroyed, then any other significant spoilage or pathogenic bacteria will also be destroyed.

To ensure commercial sterilization, enough heat is applied for the **12D treatment** (12-decimal reductions), by which a theoretical population of *C. botulinum* endospores would be decreased by 12 logarithmic cycles. What this means is that if there were 10^{12} (1,000,000,000,000) endospores in a can, after treatment there would be only one survivor. Because 10^{12} is an improbably large population, this treatment is considered quite safe. Certain thermophilic endospore-forming bacteria have endospores that are more resistant to heat treatment than those of *C. botulinum*. However, these bacteria are obligate thermophiles and generally remain dormant at temperatures lower than about 45°C (113°F). Therefore, they are not a spoilage problem at normal storage temperatures.

Spoilage of Canned Food

If canned foods are incubated at high temperatures, such as in a truck in the hot sun or next to a steam radiator, the thermophilic bacteria that often survive commercial sterilization can germinate and grow. **Thermophilic anaerobic spoilage** is therefore a fairly common cause of spoilage in low-acid canned foods. The can usually swells from gas, and the contents have a lowered pH and a sour

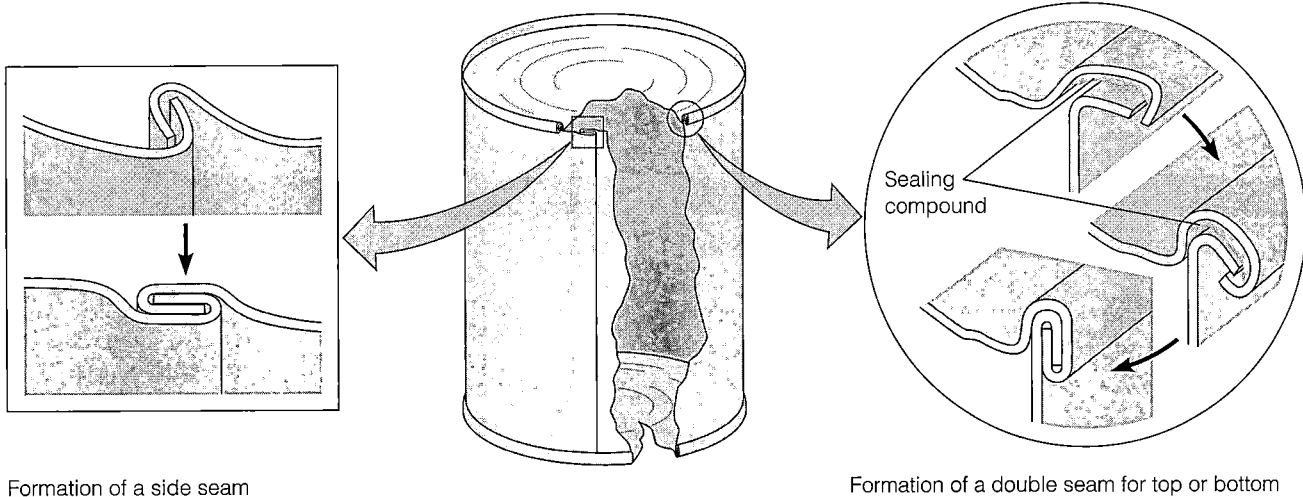

Formation of a side seam

Formation of a double seam for top or bottom

Sealing
compound

FIGURE 28.3 The construction of a metal can. Notice the seam construction. During cooling after sterilization (see Figure 28.1, step 6), the vacuum formed in the can may actually force contaminating organisms into the can along with water.

■ Why isn't the can sealed before putting it in the steam box?

odor. A number of thermophilic species of *Clostridium* can cause this type of spoilage. When thermophilic spoilage occurs but the can is not swollen by gas production, the spoilage is termed **flat sour spoilage.** This type of spoilage is caused by thermophilic organisms such as *Bacillus stearothermophilus* (ste-rō-thér-mä'fil-us), which is found in the starch and sugars used in food preparation. Many industries have standards for the numbers of such thermophilic bacteria permitted in raw materials. Both types of spoilage occur only when the cans are stored at higher than normal temperatures, which permits the growth of bacteria whose endospores are not destroyed by normal processing.

Mesophilic bacteria can spoil canned foods if the food is underprocessed or if the can leaks. Underprocessing is more likely to result in spoilage by endospore formers; the presence of non–endospore-forming bacteria strongly suggests that the can leaks. Leaking cans are often contaminated during the cooling of cans after processing by heat. The hot cans are sprayed with cooling water or passed through a trough filled with water. As the can cools, a vacuum is formed inside, and external water can be sucked through a leak past the heat-softened sealant in the crimped lid (Figure 28.3). Contaminating bacteria in the cooling water are drawn into the can with the water. Spoilage from underprocessing or can leakage is likely to produce odors of putrefaction, at least in high-protein foods, and occurs at normal storage temperatures. In such types of spoilage, there is always the potential that botulinal bacteria will be present.

Some acidic foods, such as tomatoes or preserved fruits, are preserved by processing temperatures of 100°C or below. The reasoning is that the only spoilage organisms that will grow in such acidic foods are easily killed by even 100°C temperatures. Primarily, these would be molds, yeasts, and certain vegetative bacteria.

Occasional problems in acidic foods develop from a few microorganisms that are both heat-resistant and acid-tolerant. Examples of heat-resistant fungi are the mold *Byssochlamys fulva* (bis-sō-klam'is fül'vä), which produces a *heat-resistant ascospore,* and a few molds, especially species of *Aspergillus,* that sometimes produce specialized resistant bodies called *sclerotia.* A spore-forming bacterium, *Bacillus coagulans* (kō-ag'ū-lanz), is unusual in that it is capable of growth at a pH of almost 4.0.

Table 28.1 lists types of spoilage in low- and medium-acid foods.

Aseptic Packaging

Learning Objective

■ Compare and contrast food preservation by industrial food canning, aseptic packaging, and radiation.

The use of **aseptic packaging** to preserve food has been increasing recently. Packages are usually made of some material that cannot tolerate conventional heat treatment, such as laminated paper or plastic. The packaging materials come in continuous rolls that are fed into a machine that sterilizes the material with a hot hydrogen peroxide

table 28.1	Common Types of Spoilage in Low-Acid and Medium-Acid Canned Foods (pH above 4.5)		
		Indications of Spoilage	
Type of Spoilage		Appearance of Can	Contents of Can
Flat sour (*Bacillus stearothermophilus*)		Can not swollen	Appearance not usually altered; pH markedly lowered; sour; may have slightly abnormal odor; sometimes cloudy liquid
Thermophilic anaerobic (*Clostridium thermosaccharolyticum*)		Swollen	Fermented, sour, cheesy, or butyric acid odor
Putrefactive anaerobic (*Clostridium sporogenes;* possibly *C. botulinum*)		Swollen	May be partially digested; pH slightly above normal; typical putrid odor

FIGURE 28.4 Aseptic packaging. Rolls of packaging material in foreground, filled packages at right center.

■ Why is the use of this procedure increasing in recent years?

solution, sometimes aided by ultraviolet light (Figure 28.4). Metal containers can be sterilized with superheated steam or other high-temperature methods. High-energy electron beams can also be used to sterilize the packaging materials. While still in the sterile environment, the material is formed into packages, which are then filled with liquid foods that have been conventionally sterilized by heat. The filled package is not sterilized after it is sealed.

Radiation and Industrial Food Preservation

There has been considerable research into the use of ionizing radiation for large-scale food preservation, espe-

cially for military applications. It is possible to sterilize food completely by radiation, but this method is unlikely to replace heat sterilization for most purposes. Although this treatment does not render the food harmful or radioactive, it does change the tastes of certain foods. The use of irradiation in foods is more likely to resemble pasteurization by heat treatment. Table 28.2 lists foods that have been approved by the U.S. Food and Drug Administration (FDA) to receive ionizing radiation. Only one specialized application for space flight programs involves sterilization. Chilled or frozen meats, especially poultry, are often contaminated by pathogenic bacteria such as *Salmonella* and *Listeria* that are a frequent cause of human illness. Beef, especially ground beef, occasionally contains pathogenic *E. coli* O157:H7. It is bacteria of these types that are the intended target of irradiation treatment. Handling meats that contain them often results in the contamination of other foods; undercooking may worsen the problem by allowing the pathogens to survive. Foods such as herbs, spices, and seasonings are also irradiated to lower the microbial counts, although the organisms in these foods are seldom a threat to health. Irradiation has been permitted since 1963 to kill insects in wheat products, and potatoes are often irradiated to inhibit sprouting that limits their long-term storage. Similarly, irradiation can delay the ripening of fruits during storage. At least 28 countries, developing as well as developed, have at least one irradiated food commercially available, most commonly spices. Irradiated meats are sold in 18 countries, including France and the Netherlands. Irradiated food is marked in the United States with a radura symbol (Figure 28.5), and a printed notice. Unfortunately, this symbol has often been interpreted as a warning rather than the description of an approved processing treatment or preservative. In fact, irradiated foods are not radioactive; consider that the X-ray table in a hospital does not become radioactive from repeated daily exposure to ionizing radiation.

table 28.2	**Applications of Ionizing Radiation Accepted in the United States by the Food and Drug Administration**	
Product	**Purpose**
Wheat, wheat flour	Insect disinfestation
White potatoes	Sprout inhibition
Pork	*Trichinella spiralis* control
Enzymes (dehydrated)	Microbial control
Fruit	Disinfestation, ripening delay
Vegetables, fresh	Disinfestation
Herbs	Microbial control
Spices	Microbial control
Vegetable seasonings	Microbial control
Poultry, fresh or frozen	Microbial control
Meat, frozen, packaged*	Sterilization
Animal feed and pet food	*Salmonella* control
Meat, uncooked, chilled	Microbial control
Meat, uncooked, frozen	Microbial control

*For meats used solely in the National Aeronautics and Space Administration space flight programs.

The preferred method for irradiation, when deep penetration is a requirement, is gamma rays produced by cobalt-60. However, this type of treatment requires several hours of exposure in isolation behind protective walls (Figure 28.6). High-energy electron accelerators are much faster and sterilize in a few seconds, but this treatment has low penetrating power and is suitable only for sliced meats, bacon, or similar thin products. Also, plasticware used in microbiology is usually sterilized in this way.

The Role of Microorganisms in Food Production

Learning Objective

- *Name four beneficial activities of microorganisms in food production.*

In the latter part of the nineteenth century, microbes used in food production were grown in pure culture for the first time. This development quickly led to an improved understanding of the relationships between specific microbes and their products and activities. This period can be considered the beginning of industrial food microbiology. For example, once it was understood that a certain yeast grown under certain conditions produced beer and that certain bacteria could spoil the beer, brewers were better able to control the quality of their products. Specific industries became active in microbiological research and selected certain microbes for their special qualities. For example, the brewing industry extensively investigated the isolation and identification of yeasts and selected those that could produce more alcohol. In this section, we will discuss the role of microorganisms in the production of several common foods.

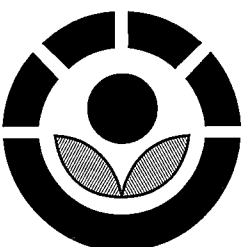

FIGURE 28.5 Irradiation logo. This logo, the international radura symbol, indicates that a food has received irradiation treatment.

■ Why would food be irradiated?

Cheese

The United States leads the world in the making of cheese, producing millions of tons every year. Although there are many types of cheeses, all require the formation of a **curd,** which can be separated from the main liquid fraction, or **whey** (Figure 28.7). The curd is made up of a protein, **casein,** and is usually formed by the action of an enzyme, **rennin** (or chymosin), which is aided by acidic conditions provided by certain lactic acid–producing bacteria. These inoculated lactic acid bacteria also provide the characteristic flavors and aromas of fermented dairy products during the ripening process. The curd undergoes a microbial ripening process, except for a few unripened cheeses, such as ricotta and cottage cheese.

Cheeses are generally classified by their hardness, which is produced in the ripening process. The more moisture lost from the curd and the more the curd is compressed, the harder the cheese. Romano and Parmesan cheeses, for example, are classified as very hard cheeses; Cheddar and Swiss are hard cheeses. Limburger, blue, and Roquefort cheeses are classified as semisoft; Camembert is an example of a soft cheese.

The hard Cheddar and Swiss cheeses are ripened by lactic acid bacteria growing anaerobically in the interior.

Irradiation sources
lifted from storage
pool for processing
period

Shielding

Shielding

Material to be
irradiated

Conveyors to
move material
in and out of
processing
position

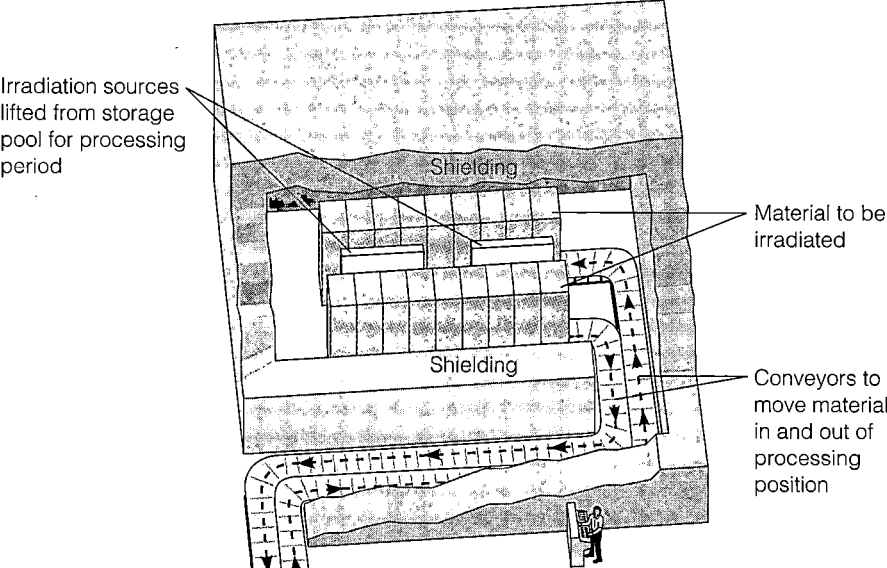

(a) An irradiation facility showing path of material to be irradiated.

FIGURE 28.6   A gamma-ray irradiation facility.

(b) The irradiation source is in the lowered position in the storage pool. The blue glow is Cerenkov radiation caused by charged particles exceeding the speed of light in water.

■ Can microwaves be used to sterilize foods?

(a) The milk has been coagulated by the action of rennin (forming curd) and is inoculated with ripening bacteria for flavor and acidity. Here the workers are cutting the curd into slabs.

(b) The curd is chopped into small cubes to facilitate efficient draining of whey.

(c) The curd is milled to allow even more drainage of whey and is compressed into blocks for extended ripening. The longer the ripening period, the more acidic (sharper) the cheese.

FIGURE 28.7   Making cheddar cheese.

■ What accounts for the characteristic flavor and aromas of cheeses?

Such hard, interior-ripened cheeses can be quite large. The longer the incubation time, the higher the acidity and the sharper the taste of the cheese. A *Propionibacterium* (prō-pē-on-ē-bak-ti'rē-um) species produces carbon dioxide, which forms the holes in Swiss cheese. Semisoft cheeses, such as Limburger, are ripened by bacteria and other contaminating organisms growing on the surface. Blue and Roquefort cheeses are ripened by *Penicillium* molds inoculated into the cheese. The texture of the cheese is loose enough that adequate oxygen can reach the aerobic molds. The growth of the *Penicillium* molds is visible as blue-green clumps in the cheese. Camembert cheese is ripened in small packets so that the enzymes of *Penicillium* mold growing aerobically on the surface will diffuse into the cheese for ripening.

Other Dairy Products

Butter is made by churning cream until the fatty globules of butter separate from the liquid *buttermilk*. The typical flavor and aroma of butter and buttermilk are from diacetyls, a combination of two acetic acid molecules that is a metabolic end-product of fermentation by some lactic acid bacteria. Today, buttermilk is usually not a by-product of butter-making but is made by inoculating skim milk with bacteria that form lactic acid and the diacetyls. *Cultured sour cream* is made from cream inoculated with microorganisms similar to those used to make buttermilk.

A wide variety of slightly acidic dairy products— probably a heritage of a nomadic past—are found around the world. Many of them are part of the daily diet in the Balkans, eastern Europe, and Russia. One such product is *yogurt*, which is also popular in the United States. Commercial yogurt is made from low-fat milk, from which much of the water has been evaporated in a vacuum pan. The resulting thickened milk is inoculated with a mixed culture of *Streptococcus thermophilus*, primarily for acid production, and *Lactobacillus bulgaricus* (bul-gā'ri-kus), to contribute flavor and aroma. The temperature of the fermentation is about 45°C for several hours, during which time *S. thermophilus* outgrows *L. bulgaricus*. Maintaining the proper balance between the flavor-producing and the acid-producing microbes is the secret of making yogurt.

Kefir and *kumiss* are fermented milk beverages that are popular in eastern Europe. The usual lactic acid–producing bacteria are supplemented with a lactose-fermenting yeast to give these drinks an alcohol content of 1–2%.

Nondairy Fermentations

Historically, milk fermentation allowed dairy products to be stored and then consumed much later. Other microbial fermentations were used to make certain plants edible. For example, pre-Columbian people in Central and South America learned to ferment chocolate seeds before consumption. It is the microbial products released during fermentation that produce the chocolate flavor.

Microorganisms are also used in baking, especially for *bread*. The sugars in bread dough are fermented by yeasts, much as they are in the fermentation of alcoholic beverages (discussed below). Anaerobic conditions for the production of ethanol by the yeasts are mandatory for producing alcoholic beverages. In baking, carbon dioxide forms the typical bubbles of leavened bread. Aerobic conditions favor carbon dioxide production and are encouraged as much as possible. This is the reason the bread dough is kneaded repeatedly. Whatever ethanol is produced evaporates during baking. In some breads, such as rye or sourdough, the growth of lactic acid bacteria produce the typical tart flavor (see the box in Chapter 1, page 5).

Fermentation is also used in the production of such foods as *sauerkraut, pickles,* and *olives*. In Asia, extremely large amounts of *soy sauce* are produced by molds that form starch-degrading enzymes to produce fermentable sugars. This principle is used in making other Asian fermented foods, such as *miso*. In soy sauce production, molds such as *Aspergillus oryzae* (a-spėr-jil'lus ô'ri-zī) are grown on wheat bran and then are allowed to act with lactic acid bacteria on cooked soybean and crushed wheat mixtures. After this process has produced fermentable carbohydrates, a prolonged fermentation results in soy sauce.

Table 28.3 lists many fermented foods.

Alcoholic Beverages and Vinegar

Microorganisms are involved in the production of almost all alcoholic beverages. *Beer* and *ale* are products of grain starches fermented by yeast (Table 28.4). Because yeasts are unable to use starch directly, the starch from grain must be converted to glucose and maltose, which the yeasts can ferment into ethanol and carbon dioxide. In this conversion, called **malting,** starch-containing grains, such as malting barley, are allowed to sprout and then are dried and ground. This product, called **malt,** contains starch-degrading enzymes (amylases) that convert cereal starches into carbohydrates that can be fermented by yeasts. Light beers use amylases or selected strains of yeast to convert more of the starch to fermentable glucose and maltose, resulting in fewer carbohydrates and more alcohol. The beer is then diluted to arrive at an alcohol percentage in the usual range. **Sake,** the Japanese rice wine, is made from rice without malting because the mold *Aspergillus* is first used to convert the rice's starch to sugars that can be fermented. For *distilled spirits,* such as *whiskey,*

table 28.3 Fermented Foods and Related Products

Foods and Products	Raw Ingredients	Fermenting Microorganism(s)	Location Produced
Dairy Products			
Cheeses (ripened)	Milk curd	*Streptococcus* spp., *Leuconostoc* spp.	Worldwide
Kefir	Milk	*Streptococcus lactis, Lactobacillus bulgaricus, Candida* spp.	Primarily southwestern Asia
Kumiss	Mare's milk	*Lactobacillus bulgaricus, L. leichmannii, Candida* spp.	Russia
Yogurt	Milk, milk solids	*S. thermophilus, L. bulgaricus*	Worldwide
Meat and Fish Products			
Country-cured hams	Pork hams	*Aspergillus, Penicillium* spp.	Southern United States
Dry sausages	Pork, beef	*Pediococcus cerevisiae*	Europe, United States
Fish sauces	Small fish	Halophilic *Bacillus* spp.	Southeast Asia
Nonbeverage Plant Products			
Cocoa beans (chocolate)	Cacao fruits (pods)	*Candida krusei, Geotrichum* spp.	Africa, South America
Coffee beans	Coffee cherries	*Erwinia dissolvens, Saccharomyces* spp.	Brazil, Congo, Hawaii, India
Kimchi	Cabbage and other vegetables	Lactic acid bacteria	Korea
Miso	Soybeans	*Aspergillus oryzae, Saccharomyces rouxii*	Primarily Japan
Olives	Green olives	*Leuconostoc mesenteroides, Lactobacillus plantarum*	Worldwide
Poi	Taro roots	Lactic acid bacteria	Hawaii
Sauerkraut	Cabbage	*Leuconostoc mesenteroides, Lactobacillus plantarum*	Worldwide
Soy sauce	Soybeans	*A. oryzae* or *A. soyae, S. rouxii, Lactobacillus delbrueckii*	Japan, China, United States
Breads			
Rolls, cakes, breads, and so on	Wheat flours	*Saccharomyces cerevisiae*	Worldwide
San Francisco sourdough bread	Wheat flour	*S. exiguus, Lactobacillus sanfrancisco*	Northern California

vodka, and *rum,* carbohydrates from cereal grains, potatoes, and molasses are fermented to alcohol. The alcohol is then distilled to make a concentrated alcoholic beverage.

Wines are made from fruits, typically grapes, that contain sugars that can be used directly by yeasts for fermentation; malting is unnecessary in wine-making. Grapes usually need no additional sugars, but other fruits might be supplemented with sugars to ensure enough alcohol production. The steps of wine-making are shown in Figure 28.8. Lactic acid bacteria are important when wine is made from grapes that are especially acidic from high concentrations of malic acid. These bacteria convert the

malic acid to the weaker lactic acid in a process called **malolactic fermentation.** The result is a less acidic, better-tasting wine than would otherwise be produced.

Wine producers who allowed wine to be exposed to air found that it soured from the growth of aerobic bacteria that converted the ethanol in the wine to acetic acid. The result was *vinegar* (*vin* = wine; *aigre* = sour). The process is now used deliberately to make vinegar. Ethanol is first produced by anaerobic fermentation of carbohydrates by yeasts. The ethanol is then aerobically oxidized to acetic acid by acetic acid–producing bacteria of the genera *Acetobacter* and *Gluconobacter.*

table 28.4 The Production of Alcoholic Beverages by Yeasts

Beverage	Yeast	Method of Preparation	Function of Yeast
Beer and Wine			
Beer, lager	S. uvarum (bottom yeast)	Germinated barley releases starches and amylase enzymes (malting). Enzymes in malt hydrolyze starch to fermentable sugars (mashing). Liquid (wort) sterilized. Hops added for flavor. Yeast added, incubated at 3–10°C.	Converts sugar into alcohol and carbon dioxide; > 6% alcohol. Yeast grows on bottom of fermenting vessel.
Beer, ale	S. cerevisiae (top yeast)	As in lager; incubated at 10–21°C.	Converts sugar into alcohol; and CO_2;< 4% alcohol. Yeast grows at top of fermentation vessel.
Sake	S. cerevisiae	Aspergillus oryzae converts starch in steamed rice into sugar; yeast added; incubated at 20°C.	Converts sugar into alcohol; 14–16% alcohol.
Wine, natural	S. cerevisiae	Strain of grape provides various flavors and sugar concentrations. Grapes crushed into must; sulfur dioxide added to inhibit wild yeast; yeast added. Red wines: incubated at 25°C. Aged in oak 3–5 years and in bottle 5–15 years. White wines: incubated at 10–15°C. Aged 2–3 years in bottle.	Converts grape sugar into alcohol; 14% or less alcohol.
Wine, sparkling (champagne)	S. cerevisiae	As natural wine, with secondary fermentation in bottle. 2.5% sugar and yeast added to bottled wine; incubated at 15°C; bottle inverted to collect yeast in neck.	In secondary fermentation, produces carbon dioxide; yeast settles quickly.
Distilled Beverages			
Rum	Wild yeast	Cane molasses inoculated from previous fermentation. Oak aging adds color. Distilled to concentrate.	Converts sugar to alcohol; 50–95% alcohol.
Brandy	S. cerevisiae	Fruits pressed; yeast added. Distilled to concentrate alcohol, blended with other brandies.	Converts sugar into alcohol; 40–43% alcohol.
Whiskey	S. cerevisiae	Wort (see beer) is fermented by yeast. Distilled to concentrate alcohol; aged in charred oak barrels.	Converts sugar to alcohol; 50–95% alcohol.

Industrial Microbiology

The industrial uses of microbiology had their beginnings in large-scale food fermentations that produced lactic acid from dairy products and ethanol from brewing. These two chemicals also proved to have many industrial uses unrelated to foods. During World Wars I and II, microbial fermentation and similar technologies were used in the production of armament-related chemical compounds such as glycerol and acetone. Present industrial microbiology largely dates from the technology developed to produce antibiotics following World War II. There is now renewed interest in some of the classic microbial fermentations, especially if they can use as feedstocks renewable or waste resources.

In recent years, industrial microbiology has been revolutionized by the application of genetically engineered organisms. An example of a genetically engineered *biosensor* to detect pollution is explained in the box on page 786. In Chapter 9, we discussed the methods for making these modified organisms using recombinant DNA technology and described some of the products derived from them; this technology is now known as **biotechnology.**

Fermentation Technology

Learning Objectives

- *Define industrial fermentation and bioreactor.*
- *Differentiate primary from secondary metabolites.*

The industrial production of microbial products usually involves fermentation. *Industrial fermentation* is the large-scale cultivation of microbes or other single cells to produce a commercially valuable substance. (See the box in

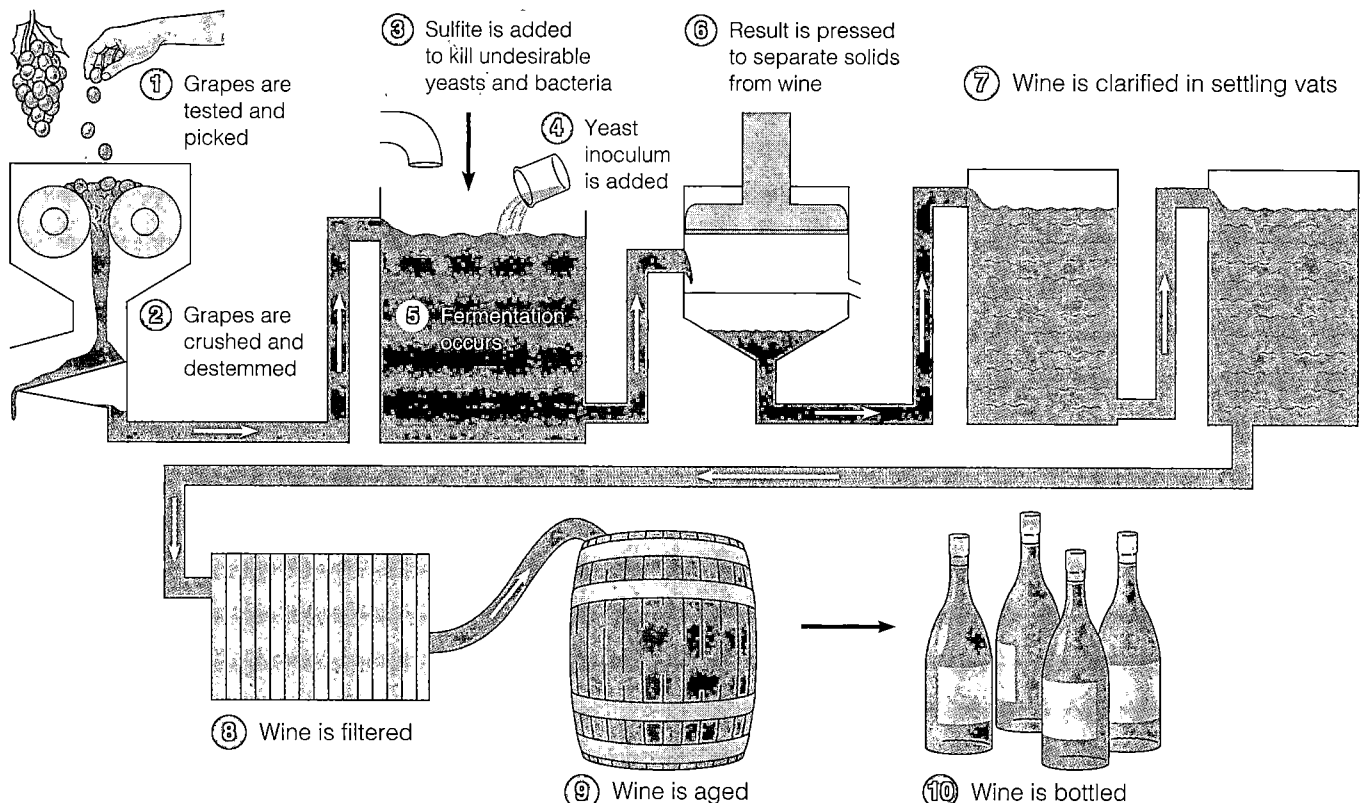

FIGURE 28.8 The basic steps in making red wine. For white wines, the pressing precedes fermentation so that the color is not extracted from the solid matter.

■ What is the purpose of adding yeast in step 4?

Chapter 5, page 133, for other definitions of fermentation). We have just discussed the most familiar examples: the anaerobic food fermentations used in the dairy, brewing, and wine-making industries. Much of the same technology, with the frequent addition of aeration, has been adapted to make other industrial products, such as insulin and human growth hormone, from genetically engineered microorganisms. Industrial fermentation is also used in biotechnology to obtain useful products from genetically engineered plant and animal cells (see Chapter 9). For example, animal cells are used to make monoclonal antibodies (see Chapter 17, page 486).

Vessels for industrial fermentation are called **bioreactors;** they are designed with close attention to aeration, pH control, and temperature control. There are many different designs, but the most widely used bioreactors are of the continuously stirred type (Figure 28.9). The air is introduced through a diffuser at the bottom (which breaks up the incoming airstream to maximize aeration), and a series of impeller paddles and stationary wall baffles keep the microbial suspension agitated. Oxygen is not very soluble in water, and keeping the heavy microbial suspension well aerated is difficult. Highly sophisticated designs

have been developed to achieve maximum efficiency in aeration and other growth requirements, including medium formulation. The high value of the products of genetically engineered microorganisms and eukaryotic cells has stimulated the development of newer types of bioreactors and computerized controls for them.

Bioreactors are sometimes very large, holding as much as 500,000 liters. When the product is harvested at the completion of the fermentation, this is known as *batch production.* There are other designs of fermentors, for *continuous flow production,* in which the substrates, usually a carbon source, are fed continuously past immobilized enzymes or into a culture of growing cells; spent medium and desired product are continuously removed.

Generally speaking, the microbes in industrial fermentation produce either primary metabolites, such as ethanol, or secondary metabolites, such as penicillin. A **primary metabolite** is formed essentially at the same time as the new cells, and the production curve follows the cell population curve almost in parallel, with only minimal lag (Figure 28.10a). **Secondary metabolites** are not produced until the microbe has largely completed its logarithmic growth phase, known as the **trophophase,**

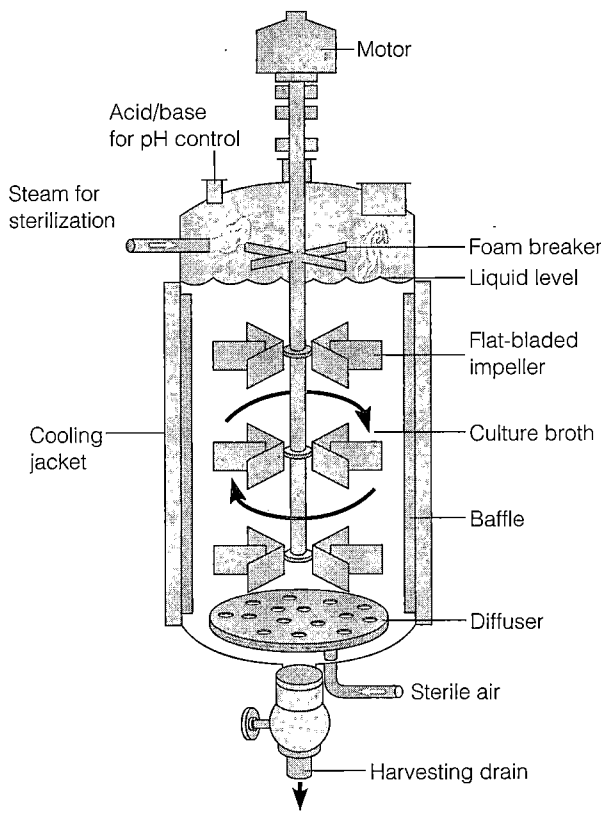

(a) Section of a continuously stirred bioreactor

(b) A bioreactor tank is at the left.

FIGURE 28.9 Bioreactors for industrial fermentations.

and has entered the stationary phase of the growth cycle (Figure 28.10b). The following period, during which most of the secondary metabolite is produced, is known as the **idiophase.** The secondary metabolite may be a microbial conversion of a primary metabolite. Alternatively, it may be a metabolic product of the original growth medium that the microbe makes only after considerable numbers of cells and a primary metabolite have accumulated.

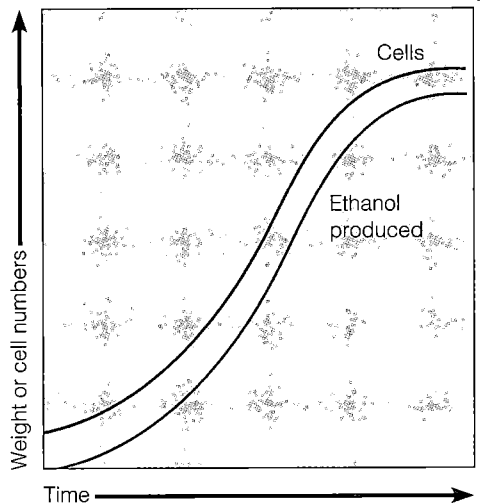

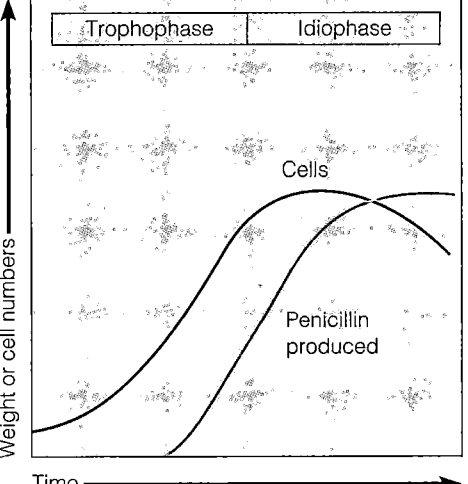

FIGURE 28.10 Primary and secondary fermentation.

(a) A primary metabolite, such as ethanol from yeast, has a production curve that lags only slightly behind the line showing cell growth.

(b) A secondary metabolite, such as penicillin from mold, begins to be produced only after the logarithmic growth phase of the cell (trophophase) is completed. The main production of the secondary metabolite occurs during the stationary phase of cell growth (idiophase).

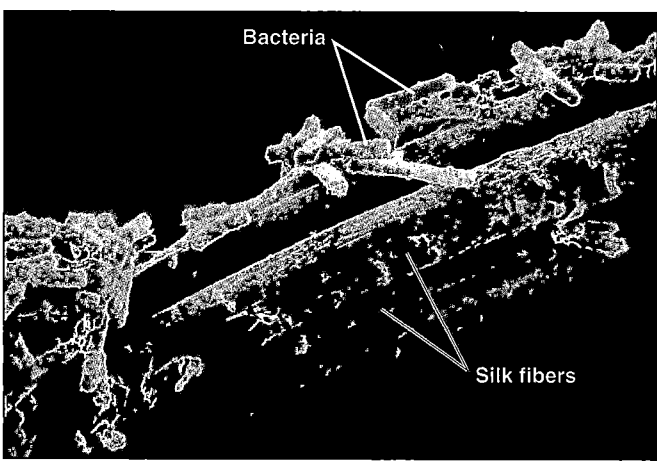

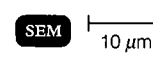
10 μm

FIGURE 28.11 Immobilized cells. In some industrial processes, the cells are immobilized on surfaces such as the silk fibers shown here. The substrate flows past the immobilized cells.

■ **Bacteria can be used to make chemicals quickly and without producing toxic by-products.**

Strain improvement is also an ongoing activity in industrial microbiology. (A microbial strain differs physiologically in some significant way. For example, it has an enzyme to carry out some additional activity, or the lack of such an ability, but this is not enough difference to change its species identity). A well-known example is that of the mold used for penicillin production. The original culture of *Penicillium* did not produce penicillin in large enough quantities for commercial use. A more efficient culture was isolated from a moldy cantaloupe from a Peoria, Illinois, supermarket. This strain was treated variously with UV light, X rays, and nitrogen mustard (a chemical mutagen). Selections of mutants, including some that arose spontaneously, quickly increased the production rates by a factor of more than 100. Today, the original penicillin-producing molds produce, not the original 5 mg/L, but 60,000 mg/L. Improvements in fermentation techniques have nearly tripled even this yield. An example of a strain that was developed by enrichment and selection is described in the box in Chapter 11 (page 328).

Immobilized Enzymes and Microorganisms

In many ways, microbes are packages of enzymes. Industries are increasing their use of free enzymes isolated from microbes to manufacture many products such as high-fructose syrups, paper, and textiles. The demand for such enzymes is high because they are specific and do not produce costly or toxic waste products. And, unlike traditional chemical processes that require heat or acids, enzymes work under moderate conditions and are safe and

biodegradable. For most industrial purposes, the enzyme must be immobilized on the surface of some solid support or otherwise manipulated so that it can convert a continuous flow of substrate to product without being lost.

Continuous-flow techniques have also been adapted to live whole cells, and sometimes even to dead cells (Figure 28.11). Whole-cell systems are difficult to aerate, and they lack the single-enzyme specificity of immobilized enzymes. However, whole cells are advantageous if the process requires a series of steps that can be carried out by one microbe's enzymes. They also have the advantage of allowing continuous flow processes with large cell populations operating at high reaction rates. Immobilized cells, which are usually anchored to microscopically small spheres or fibers, are currently used to make high-fructose syrup, aspartic acid, and numerous other products of biotechnology.

Industrial Products

Learning Objective

■ Describe the role of microorganisms in the production of industrial chemicals and pharmaceuticals.

As mentioned earlier, cheese-making produces an organic waste called whey. The whey must be disposed of as sewage or dried and burned as solid waste. Both of these processes are costly and ecologically problematic. However, microbiologists have discovered an alternative use for whey, as discussed in the box in Chapter 11 (page 328). In this way, microbiologists are devising uses for old products and creating new ones. In this section, we will discuss some of the more important commercial microbial products and the growing alternative energy industry.

Amino Acids

Amino acids have become a major industrial product from microorganisms. For example, over 600,000 tons of *glutamic acid* (L-glutamate), used to make the flavor enhancer monosodium glutamate, are produced every year. Certain amino acids, such as *lysine* and *methionine,* cannot be synthesized by animals and are present only at low levels in the normal diet. Therefore, the commercial synthesis of lysine and some of the other essential amino acids as cereal food supplements is an important industry. More than 70,000 tons of lysine or methionine are produced every year.

Two microbially synthesized amino acids, *phenylalanine* and *aspartic acid* (L-aspartate), have become important as ingredients in the sugar-free sweetener aspartame (NutraSweet®). Some 3000–4000 tons of each of these amino acids are produced annually in the United States.

table 28.5 Microbial Enzymes Produced Commercially

Enzyme	Microorganism	Use of Enzyme
α-Amylase	Aspergillus spp.	Laundry detergent
β-Amylase	Bacillus subtilis	Brewing
Cellulase	Trichoderma viride	Fruit juices, coffee, paper
Invertase	Saccharomyces cerevisiae	Candy manufacture
Lactase	S. fragilis	Digestive aid, candy manufacture
Lipase	A. niger	Laundry detergent, tanning leather, cheese production
Oxidases	A. niger	Paper bleaching and fabrics, glucose test papers
Pectinase	A. niger	Fruit juice
Proteases	A. oryzae	Meat tenderizer, digestive aid, tanning leather
Rennin (chymosin)	Mucor, Escherichia coli	Cheese production
Streptokinase	Group C β-hemolytic Streptococcus	Lysis of blood clots

In nature, microbes rarely produce amino acids in excess of their own needs because feedback inhibition prevents wasteful production of primary metabolites (see Chapter 5, page 120). Commercial microbial production of amino acids depends on specially selected mutants and sometimes on ingenious manipulations of metabolic pathways. For example, in applications in which only the L-isomer of an amino acid is desired, microbial production, which forms only the L-isomer, has an advantage over chemical production, which forms both the **D-isomer** and the **L-isomer** (see Figure 2.14, page 47).

Citric Acid

Citric acid is a constituent of citrus fruits, such as oranges and lemons, and at one time these were its only industrial source. However, over 100 years ago, citric acid was identified as a product of mold metabolism. Citric acid has an extraordinary range of uses beyond the obvious ones of giving tartness and flavor to foods. It is an antioxidant and pH adjuster in many foods, and in dairy products it often serves as an emulsifier. Well over 300,000 tons of citric acid are produced every year in the United States. Much of it is produced by a mold, *Aspergillus niger* (nī′jer), using molasses as a substrate.

Enzymes

Enzymes are widely used in different industries. For example, *amylases* are used in the production of syrups from corn starch, in the production of paper sizing (a coating for smoothness, as on this page), and in the production of glucose from starch. *Glucose isomerase* is an important enzyme; it converts the glucose that amylases form from

starches into fructose, which is used in place of sucrose as a sweetener in many foods. Probably half of the bread baked in this country is made with the aid of *proteases,* which adjust the amount of glutens (protein) in wheat so that baked goods are improved or made uniform. Other proteolytic enzymes are used as meat tenderizers or in detergents as an additive to remove proteinaceous stains. *Rennin,* an enzyme used to form curds in milk, is usually produced commercially by fungi but more recently by genetically engineered bacteria. Several enzymes produced commercially by microorganisms are listed in Table 28.5.

Vitamins

Vitamins are sold in large quantities combined in tablet form and are used as individual food supplements. Microbes can provide an inexpensive source of some vitamins. *Vitamin B_{12}* is produced by *Pseudomonas* and *Propionibacterium* species. *Riboflavin* is another vitamin produced by fermentation, mostly by fungi such as *Ashbya gossypii* (ash′bē-ä gos-sip′ē-ē). *Vitamin C* (ascorbic acid) is produced by a complicated modification of glucose by *Acetobacter* species.

Pharmaceuticals

Modern pharmaceutical microbiology developed after World War II, with the introduction of the production of antibiotics.

All antibiotics were originally the products of microbial metabolism. Many are still produced by microbial fermentations, and work continues on the selection of more productive mutants by nutritional and genetic manipulations. At least 6000 antibiotics have been described.

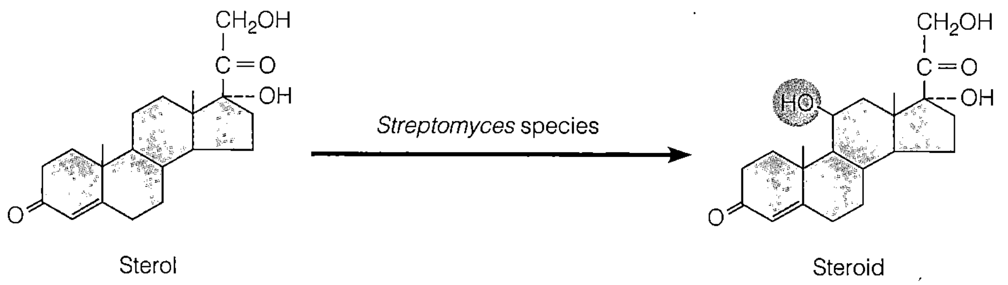

FIGURE 28.12 **The production of steroids.** Shown here is the conversion of a precursor compound such as a sterol into a steroid by *Streptomyces*. The addition of a hydroxyl group to carbon number II (highlighted in purple on the steroid) might require more than 30 steps by chemical means, but the microorganism can add it in only one step.

■ **Bacteria can often perform chemical syntheses more easily than a chemist.**

One organism, *Streptomyces hydroscopius*, has different strains that make almost 200 different antibiotics. Antibiotics are typically made industrially by inoculating a solution of growth medium with spores of the appropriate mold or streptomycete and vigorously aerating it. After the antibiotic reaches a satisfactory concentration, it is extracted by solution, precipitation, and other standard industrial procedures.

Vaccines are a product of industrial microbiology. Many antiviral vaccines are mass-produced in chicken eggs or cell cultures. The production of vaccines against bacterial diseases usually requires the growth of large amounts of the bacteria. Genetic engineering techniques are increasingly important in the development and production of subunit vaccines (see Chapter 18, page 503).

Steroids are a very important group of chemicals that include *cortisone,* which is used as an anti-inflammatory drug, and *estrogens* and *progesterone,* which are used in oral contraceptives. Recovering steroids from animal sources, or chemically synthesizing them, is difficult, but microorganisms can synthesize steroids from sterols or from related, easily obtained compounds. For example, Figure 28.12 illustrates the conversion of a sterol into a valuable steroid.

Copper Extraction by Leaching

Thiobacillus ferrooxidans is used in the recovery of otherwise unprofitable grades of copper ore, which sometimes contain as little as 0.1% copper. *Thiobacillus* bacteria get their energy from the oxidation of a reduced form of iron, ferrous sulfide (Fe^{2+}), to an oxidized form, ferric sulfate (Fe^{3+}). Sulfuric acid (H_2SO_4) is also a product of the reaction. The acidic solution of Fe^{3+}-containing water is applied by sprinklers and allowed to percolate down slope through the ore body (Figure 28.13). The ferrous

iron, Fe^{2+} and *T. ferrooxidans* are normally present in the ore and continue to contribute to the reactions. The Fe^{3+} in the sprinkling water reacts with insoluble *copper sulfides* (Cu^+) in the ore to form soluble *copper sulfates* (Cu^{2+}). In order to maintain a low enough pH, more sulfuric acid can be added. The soluble copper sulfate moves down slope to collection tanks where it contacts metallic scrap iron. The copper sulfates react chemically with the iron and precipitate out as metallic copper (Cu^0). In this reaction the metallic iron (Fe^0) is converted into ferrous iron (Fe^{2+}) that is recycled to an aerated oxidation pond where *Thiobacillus* bacteria use it as energy to renew the cycle. This process, although very time consuming, is economical and can recover as much as 70% of the copper in the ore. The entire arrangement resembles a continuous-flow bioreactor.

Microorganisms as Industrial Products

Microorganisms themselves sometimes constitute an industrial product. *Baker's yeast* is produced in large aerated fermentation tanks. At the end of the fermentation, the contents of the tank are about 4% yeast solids. The cells are harvested by continuous centrifuges and are pressed into the familiar yeast cakes or packets sold in the supermarket for home baking. Wholesale bakers purchase yeast in 50-lb boxes.

Other important microbes that are sold industrially are the symbiotic nitrogen-fixing bacteria *Rhizobium* and *Bradyrhizobium*. These organisms are usually mixed with peat moss to preserve moisture; the farmer mixes the peat moss and bacterial inoculum with the seeds of legumes to ensure infection of the plants with efficient nitrogen-fixing strains (see Chapter 27). For many years, gardeners have used the insect pathogen *Bacillus thuringiensis* (Bt) to control leaf-eating insect larvae. *Bt* subspecies *israelensis* is

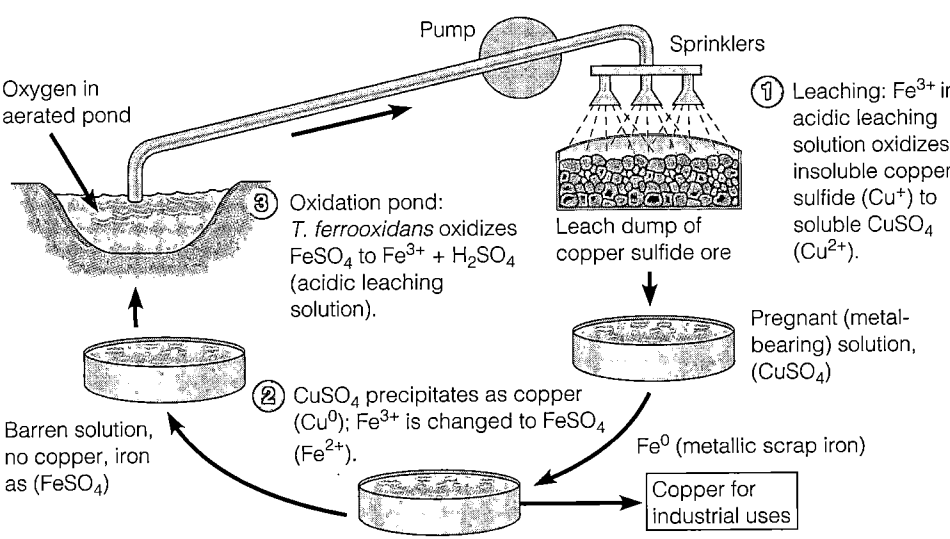

Pump

Sprinklers

Oxygen in aerated pond

③ Oxidation pond: *T. ferrooxidans* oxidizes $FeSO_4$ to $Fe^{3+} + H_2SO_4$ (acidic leaching solution).

Leach dump of copper sulfide ore

① Leaching: Fe^{3+} in acidic leaching solution oxidizes insoluble copper sulfide (Cu^+) to soluble $CuSO_4$ (Cu^{2+}).

Pregnant (metal-bearing) solution, ($CuSO_4$)

② $CuSO_4$ precipitates as copper (Cu^0); Fe^{3+} is changed to $FeSO_4$ (Fe^{2+}).

Barren solution, no copper, iron as ($FeSO_4$)

Fe^0 (metallic scrap iron)

Copper for industrial uses

(a) Simplified copper ore leaching process

(b) Leaching solution being sprayed at top of ore dump.

FIGURE 28.13 Biological leaching of copper ores. The chemistry of the process is much more complicated than shown here. Essentially, *Thiobacillus ferrooxidans* bacteria are used in a biological/chemical process that changes insoluble copper in the ore into soluble copper that leaches out and is precipitated as metallic copper. The solutions are continuously recirculated.

especially active against mosquito larvae and is widely used in municipal control programs. Commercial preparations containing toxic crystals and endospores of this organism are available at almost any gardening supply store. For an example of microbes being developed to detect chemicals, see the box on page 786.

Alternative Energy Sources Using Microorganisms

Learning Objective

■ *Define bioconversion, and list its advantages.*

As our supplies of fossil fuels diminish or become more expensive, interest in the use of renewable energy resources will increase. Prominent among these is **biomass,** the collective organic matter produced by living organisms, including crops, trees, and municipal wastes. Microbes can be used for **bioconversion,** the process of converting biomass into alternative energy sources. Bioconversion can also decrease the amount of waste materials requiring disposal.

Methane is one of the most convenient energy sources produced from bioconversion. Methane was discussed in the box in Chapter 27 (page 747) as a product of the anaerobic treatment of sewage sludge. Many communities produce useful amounts of methane from wastes in landfill sites (Figure 28.14). Large cattle-feeding lots

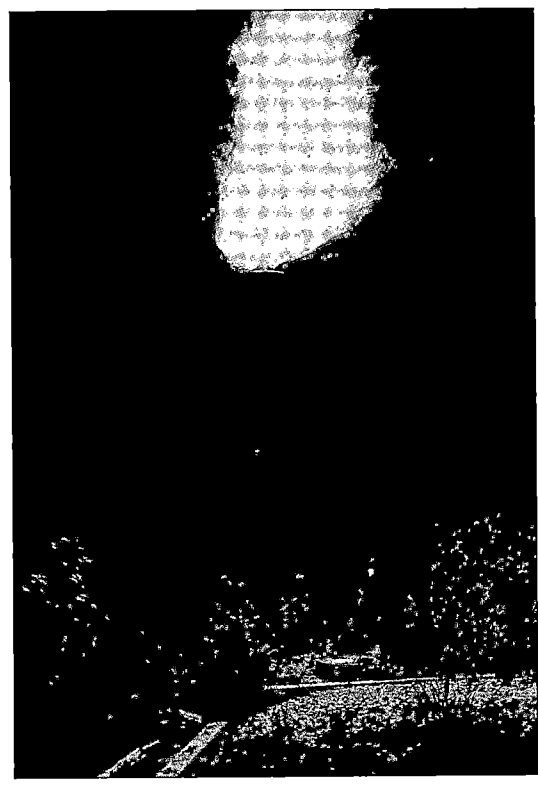

FIGURE 28.14 Methane production from solid waste. Methane begins to accumulate a few months after a landfill is completed and sealed. It remains productive for 5–10 years. This photograph shows a methane flare from such a landfill.

■ **How is methane produced in a landfill?**

MICROBIOLOGY IN THE NEWS

Biosensors: Bacteria That Detect Pollutants and Pathogens

Each year in the United States, industrial plants generate 265 million metric tons of hazardous waste, 80% of which make their way into landfills. Burying these chemicals does not remove them from the ecosystem, however; it just moves them to other places, where they may still find their way into bodies of water. Traditional chemical analyses to locate these chemicals are expensive and cannot distinguish between chemicals that affect biological systems from those that lie inert in the environment.

In response to this problem, scientists are developing biosensors, bacteria that can locate biologically active pollutants. Biosensors do not require costly chemicals or equipment, and they work quickly—within minutes.

In order to work, bacterial biosensors require both a receptor that is activated in the presence of pollutants and a reporter that will make such a change apparent. Biosensors use the *lux* operon from *Vibrio* or *Photobacterium* as a reporter. This operon contains inducer and structural genes for the enzyme luciferase. In the presence of a coenzyme called FMNH$_2$, luciferase reacts with the molecule in such a way that the enzyme–substrate complex emits blue-green light, which then oxidizes the FMNH$_2$ to produce FMN. Therefore, a bacterium containing the *lux* gene will emit visible light when the receptor is activated (see the photographs).

The *lux* operon is readily transferred to many bacteria. Scientists in South Korea are using *E. coli* containing the *lux* operon as an early-warning to detect treatment failures from wastewater treatment plants. Before being discharged into the environment, water from treatment plants is continuously run into a bioreactor containing the *E. coli* bacteria. The bacteria will emit light as long as they are healthy but will stop emitting light if they have been killed by toxic pollutants.

In another application, *Lactococcus* bacteria containing the *lux* operon are being used to detect the presence of antibiotics in milk that is to be used for cheese production. (If milk contains antibiotics, the cheese starter cultures will not grow.) Because the emission of light requires a living cell, the presence of antibiotics is measured as a decrease in light output by such recombinant *Lactococcus* bacteria.

The Microtox System developed in Great Britain uses the marine bacterium *Photobacterium* directly to detect toxic pollutants. These bacteria cannot emit light if they are killed by pollutants.

Other biosensors use recombinant bacteriophages that contain the *lux* genes. These phages are being used to detect *Listeria* and *E. coli* in foods and to detect drug-resistant mycobacteria.

Detecting harmful pollutants and pathogens in soil and water is necessary to protect humans and animals. However, after detection, bioremediation processes are still required to remove the pollutants.

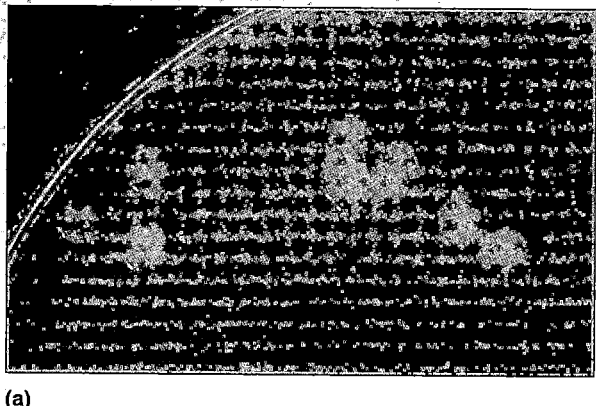

(a)

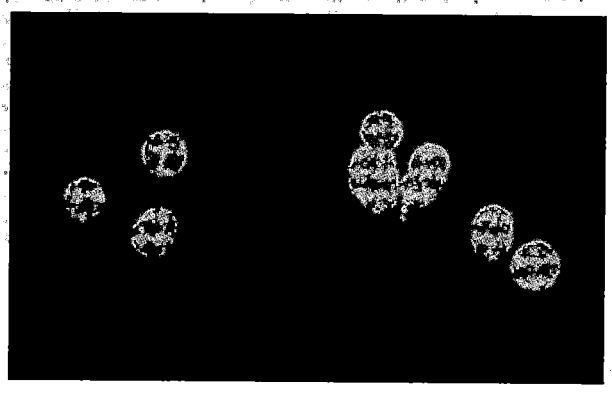

(b)

Vibrio fischeri *emits light when energy is released by the transport of electrons to luciferase. Shown here are colonies of* V. fischeri *photographed (a) in daylight and (b) in the dark, illuminated by their own light.*

must dispose of immense amounts of animal manure, and much effort has been devoted to devising practical methods for producing methane from these wastes. A major problem with any scheme for large-scale methane production is the need to economically concentrate the widespread biomass material. If it could be economically concentrated, the animal and human wastes in the United States could provide much of our energy now supplied by fossil fuels and natural gas.

The agricultural industry has encouraged the production of **ethanol** from agricultural products. Gasoline containing ethanol (90% gasoline + 10% ethanol) is available in many parts of the United States and is used in automobiles worldwide. Corn is currently the most frequently used substrate, but eventually agricultural waste products may be used.

Industrial Microbiology and the Future

Microbes have always been exceedingly useful to humankind, even when their existence was unknown. They will remain an essential part of many basic food-processing technologies. The development of genetic engineering has further intensified interest in industrial microbiology by expanding the potential for new products and applications (see the box in Chapter 9, page 251). As new biotechnology applications and products enter the marketplace, they will affect our lives and well-being in ways that we can only speculate about today.

Study Outline 🖳 Student Tutorial CD-ROM

FOOD MICROBIOLOGY (pp. 771–778)

1. The earliest methods of preserving foods were drying, the addition of salt or sugar, and fermentation.

Industrial Food Canning (pp. 771–773)

1. Commercial sterilization of food is accomplished by steam under pressure in a retort.

2. Commercial sterilization heats canned foods to the minimum temperature necessary to destroy *Clostridium botulinum* endospores while minimizing alteration of the food.

3. The commercial sterilization process uses sufficient heat to reduce a population of *C. botulinum* by 12 logarithmic cycles (12D treatment).

4. Endospores of thermophiles can survive commercial sterilization.

5. Canned foods stored above 45°C can be spoiled by thermophilic anaerobes.

6. Thermophilic anaerobic spoilage is sometimes accompanied by gas production; if no gas is formed, the spoilage is called flat sour spoilage.

7. Spoilage by mesophilic bacteria is usually from improper heating procedures or leakage.

8. Acidic foods can be preserved by heat of 100°C because microorganisms that survive are not capable of growth in a low pH.

9. *Byssochlamys, Aspergillus,* and *Bacillus coagulans* are acid-tolerant and heat-resistant microbes that can spoil acidic foods.

Aseptic Packaging (pp. 773–774)

1. Presterilized materials are assembled into packages and aseptically filled with heat-sterilized liquid foods.

Radiation and Industrial Food Preservation (pp. 774–775)

1. Gamma radiation can be used to sterilize food, kill insects and parasitic worms, and prevent the sprouting of fruits and vegetables.

The Role of Microorganisms in Food Production (pp. 775–779)

Cheese (pp. 775–777)

1. The milk protein casein curdles because of the action by lactic acid bacteria or the enzyme rennin.

2. Cheese is the curd separated from the liquid portion of milk, called whey.

3. Hard cheeses are produced by lactic acid bacteria growing in the interior of the curd.

4. The growth of microbes in cheese is called ripening.

5. Semisoft cheeses are ripened by bacteria growing on the surface; soft cheeses are ripened by *Penicillium* growing on the surface.

Other Dairy Products (p. 777)

1. Old-fashioned buttermilk was produced by lactic acid bacteria growing during the butter-making process.

2. Commercial buttermilk is made by letting lactic acid bacteria grow in skim milk for 12 hours.

3. Sour cream, yogurt, kefir, and kumiss are produced by lactobacilli, streptococci, or yeasts growing in low-fat milk.

Nondairy Fermentations (p. 777)

1. Sugars in bread dough are fermented by yeast to ethanol and CO_2; the CO_2 causes the bread to rise.

2. Sauerkraut, pickles, olives, and soy sauce are products of microbial fermentations.

Alcoholic Beverages and Vinegar (pp. 777–779)

1. Carbohydrates obtained from grains, potatoes, or molasses are fermented by yeasts to produce ethanol in the production of beer, ale, sake, and distilled spirits.

2. The sugars in fruits such as grapes are fermented by yeasts to produce wines.

3. In wine-making, lactic acid bacteria convert malic acid into lactic acid in malolactic fermentation.

4. *Acetobacter* and *Gluconobacter* oxidize ethanol in wine to acetic acid (vinegar).

INDUSTRIAL MICROBIOLOGY (pp. 779–787)

1. Microorganisms produce alcohols and acetone that are used in industrial processes.

2. Industrial microbiology has been revolutionized by the ability of genetically engineered cells to make many new products.

3. Biotechnology is a way of making commercial products by using living organisms.

Fermentation Technology (pp. 779–782)

     To review, go to Biotechnology: Bioreactor

1. The growth of cells on a large scale is called industrial fermentation.

2. Industrial fermentation is carried on in bioreactors, which control aeration, pH, and temperature.

3. Primary metabolites such as ethanol are formed as the cells grow (during the trophophase).

4. Secondary metabolites such as penicillin are produced during the stationary phase (idiophase).

5. Mutant strains that produce a desired product can be selected.

Immobilized Enzymes and Microorganisms (p. 782)

1. Enzymes or whole cells can be bound to solid spheres or fibers. When substrate passes over the surface, enzymatic reactions change the substrate to the desired product.

2. They are used to make paper, textiles, and leather and are environmentally safe.

Industrial Products (pp. 782–785)

1. Most amino acids used in foods and medicine are produced by bacteria.

2. Microbial production of amino acids can be used to produce L-isomers; chemical production results in both D- and L-isomers.

3. Lysine and glutamic acid are produced by *Corynebacterium glutamicum*.

4. Citric acid, used in foods, is produced by *Aspergillus niger.*

5. Enzymes used in manufacturing foods, medicines, and other goods are produced by microbes.

6. Some vitamins used as food supplements are made by microorganisms.

7. Vaccines, antibiotics, and steroids are products of microbial growth.

8. The metabolic activities of *Thiobacillus ferrooxidans* can be used to recover uranium and copper ores.

9. Yeasts are grown for wine- and bread-making; other microbes (*Rhizobium, Bradyrhizobium,* and *Bacillus thuringiensis*) are grown for agricultural use.

Alternative Energy Sources Using Microorganisms (pp. 785–787)

1. Organic waste, called biomass, can be converted by microorganisms into alternative fuels, a process called bioconversion.

2. Fuels produced by microbial fermentation are methane and ethanol.

Industrial Microbiology and the Future (p. 787)

1. Genetic engineering will continue to enhance the ability of industrial microbiology to produce medicines and other useful products.

Study Questions

REVIEW

1. What is industrial microbiology? Why is it important?

2. How does commercial sterilization differ from sterilization procedures used in a hospital or laboratory?

3. Why is a can of blackberries preserved by commercial sterilization typically heated to 100°C instead of at least 116°C?

4. Describe aseptic packaging.

5. Outline the steps in the production of cheese, and compare the production of hard and soft cheeses.

6. Outline the steps in the process of wine production.

7. Beer is made with water, malt, and yeast; hops are added for flavor. What is the purpose of the water, malt, and yeast? What is malt?

8. Label the trophophase and idophase in this graph. Indicate when primary and secondary metabolites are formed.

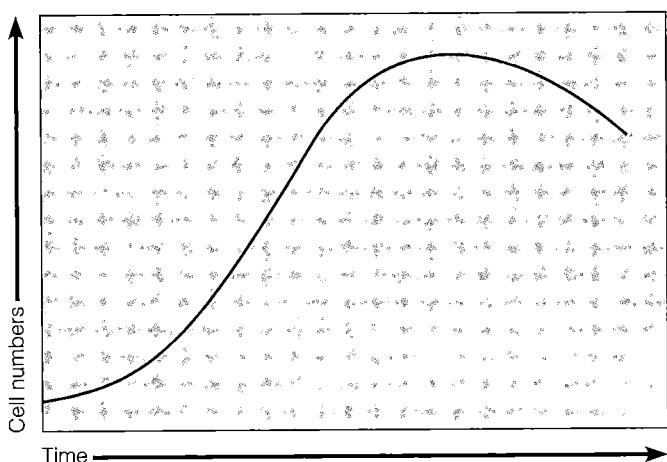

Cell numbers (y-axis)

Time ⟶ (x-axis)

9. Why is a bioreactor better than a large flask for industrial production of an antibiotic?

10. The manufacture of paper includes the use of bleach and formaldehyde-based glue. The microbial enzyme xylanase whitens paper by digesting dark lignins. Oxidase causes the fibers to stick together, and cellulase will remove ink. List three advantages of using these microbial enzymes over traditional chemical methods for making paper.

11. Describe an example of bioconversion. What metabolic processes can result in fuels?

MULTIPLE CHOICE

1. Foods packed in plastic for microwaving are
  a. dehydrated.
  b. freeze-dried.
  c. packaged aseptically.
  d. commercially sterilized.
  e. autoclaved.

2. *Acetobacter* is necessary for only one of the steps of vitamin C manufacture. The easiest way to accomplish this step would be to
  a. add substrate and *Acetobacter* to a test tube.
  b. affix *Acetobacter* to a surface and run substrate over it.
  c. add substrate and *Acetobacter* to a bioreactor.
  d. find an alternative to this step.
  e. none of the above

Use the following choices to answer questions 3–5:
  a. *Bacillus coagulans*
  b. *Byssochlamys*
  c. flat sour spoilage
  d. *Lactobacillus*
  e. thermophilic anaerobic spoilage

3. The spoilage of canned foods due to inadequate processing, accompanied by gas production.

4. The spoilage of canned foods caused by *Bacillus stearothermophilus.*

5. A heat-resistant fungus that causes spoilage in acidic foods.

6. The term 12D treatment refers to
  a. heat treatment sufficient to kill 12 bacteria.
  b. the use of·12 different treatments to preserve food.
  c. a 10^{12} reduction in *C. botulinum* endospores.
  d. any process that destroys thermophilic bacteria.

7. Microorganisms themselves are industrial products. Which of the following pairs is mismatched?
  a. *Penicillium*—treatment of disease
  b. *S. cerevisiae*—for fermentation
  c. *Rhizobium*—increases nitrogen in the soil
  d. *B. thuringiensis*—insecticide

8. Which type of radiation is used to preserve foods?
  a. ionizing
  b. nonionizing
  c. radiowaves
  d. microwaves
  e. all of the above

9. Which of the following reactions is undesirable in winemaking?
  a. Sucrose → ethanol
  b. Ethanol → acetic acid
  c. Malic acid → lactic acid
  d. Glucose → pyruvic acid

10. Which of the following reactions is an oxidation carried out by *Thiobacillus ferrooxidans*?
  a. $Fe^{2+} \rightarrow Fe^{3+}$
  b. $Fe^{3+} \rightarrow Fe^{2+}$
  c. $CuS \rightarrow CuSO_4$
  d. $Fe^0 \rightarrow Cu^0$
  e. none of the above

CRITICAL THINKING

1. Which bacteria seem to be most frequently used in the production of food? Propose an explanation for this.

2. *Methylophilus methylotrophus* can convert methane (CH_4) into proteins. Amino acids are represented by this structure:

$$H_2N-\overset{\overset{\displaystyle H}{|}}{\underset{\underset{\displaystyle R}{|}}{C}}-C\overset{\displaystyle O}{\underset{\displaystyle OH}{}}$$

Diagram a pathway illustrating the production of at least one amino acid.

3. "Stone-washed" denim is produced with cellulase. How does cellulase accomplish the look and feel of stone-washing? What is the source of the cellulase?

APPLICATIONS

1. Suppose you are culturing a microorganism that produces enough lactic acid to kill itself in a few days.

    a. How can the use of a bioreactor help you maintain the culture for weeks or months? The graph below shows conditions in the bioreactor:

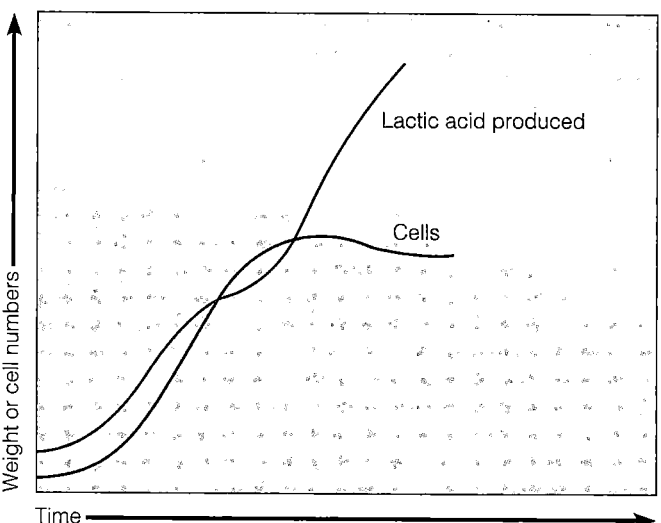

    b. If your desired product is a secondary metabolite, when can you begin collecting it?

    c. If your desired product is the cells themselves and you want to maintain a continuous culture, when can you begin harvesting?

2. Researchers at the CDC inoculated apple cider with 10^5 *E. coli* O157:H7 cells/ml to determine the fate of the bacteria in apple cider (pH 3.7). They obtained the following results:

	Number of *E. coli* O157:H7 cells/ml after 25 days
Apple cider at 25 °C	10^4 (mold growth evident by 10 days)
Apple cider with potassium sorbate at 25 °C	10^3
Apple cider at 8 °C	10^2

    What conclusions can you reach from these data? What disease is caused by *E. coli* O157:H7? (*Hint:* See Chapter 25.)

3. The antibiotic efrotomycin is produced by *Nocardia lactamdurans. N. lactamdurans* was grown in 40,000 liters of medium. The medium consisted of glucose, maltose, soybean oil, $(NH_4)_2SO_4$, NaCl, KH_2PO_4, and Na_2HPO_4. The culture was aerated and maintained at 28°C. The following results were obtained from analyses of the culture medium during cell growth:

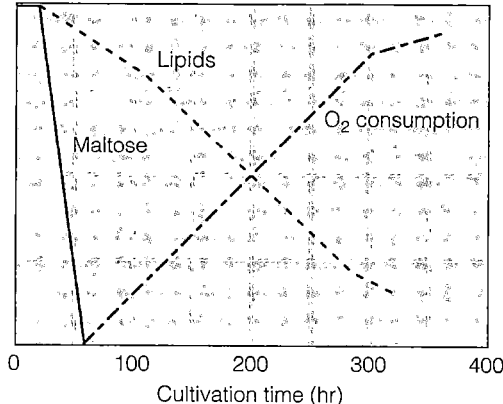

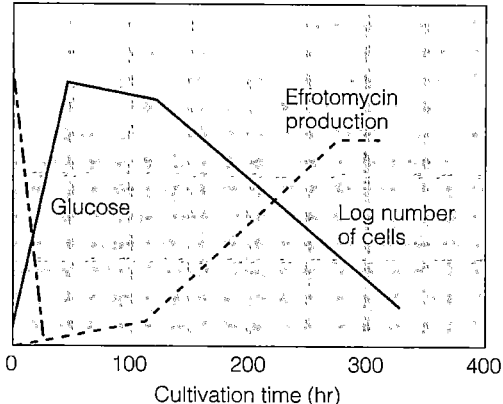

    a. Under what conditions is the most efrotomycin produced? Is it a primary or secondary metabolite?

    b. Which is used first, maltose or glucose? Suggest a reason for this.

    c. What is the purpose of each ingredient in the grow medium? (*Hint:* See Chapter 6.)

    d. What is *Nocardia?* (*Hint:* See Chapter 11.)

Learning with Technology

MP = The Microbiology Place website   **ST** = Student Tutorial CD-ROM   **VU** = VirtualUnknown CD-ROM

MP   Don't forget to go to The Microbiology Place website (http://www.microbiologyplace.com) to take the practice tests, explore the interactive activity and case study, and check out the news articles and web links for this chapter.

ST   Remember there is also a quiz for this chapter on the Microbiology Interactive Student Tutorial CD-ROM.

VU   Enter the Virtual Lab, click the arrow next to the Session field, click Textbook Exercises, and select Chapter 28. View the list of tests supported by Virtual Unknown™ Microbiology by selecting View Identification Matrix from the Main Menu bar. Use the information you find to answer the following questions:

1. Which enzymes listed in Table 28.5 on page 783 are expressed by bacteria included in VirtualUnknown™ Microbiology?

2. Which bacterium listed in the Identification Matrix possesses the greatest number of these commercially valuable enzymes?

Appendix A

*Classification of Bacteria According to Bergey's Manual**

Domain: Archaea
  Phylum I: Crenarchaeota
      Class I: 'Thermoprotei'
         Order I: Thermoproteales
            Family I: Thermoproteaceae
              Caldivirga
              Pyrobaculum 3 spp.
              Thermocladium 1 sp.
              Thermoproteus 2 spp.
            Family II: Thermofilaceae
              Thermofilum 1sp.
         Order II: Desulfurococcales
            Family I: Desulfurococcaceae
              Aeropyrum 1 sp.
              Desulfurococcus 2 spp.
              Igniococcus 1 sp.
              Staphylothermus 1 sp.
              Stetteria 1 sp.
              Thermodiscus 1 sp.
              Thermosphaera 1 sp.
            Family II: Pyrodictiaceae
              Hyperthermus 1 sp.
              Pyrodictium 3 spp.
              Pyrolobus 1 sp.
         Order III: Sulfolobales
            Family I: Sulfolobaceae
              Acidianus 3 spp.
              Metallosphaera 2 spp.
              Stygiolobus 1 sp.
              Sulfolobus 6 spp.
              Sulfurisphaera 1 sp.
              Sulfurococcus 2 spp.
  Phylum II: Euryarchaeota
      Class I: Methanobacteria
         Order I: Methanobacteriales
            Family I: Methanobacteriaceae
              Methanobacterium 19 spp.
              Methanobrevibacter 7 spp.

              Methanosphaera 2 spp.
              Methanothermobacter 5 spp.
            Family II: Methanothermaceae
              Methanothermus 2 spp.
      Class II: Methanococci
         Order I: Methanococcales
            Family I: Methanococcaceae
              Methanococcus 11 spp.
              Methanothermococcus 1 sp.
            Family II: Methanocaldococcaceae
              Methanocaldococcus 4 sp.
              Methanotorris 1 sp.
         Order II: Methanomicrobiales
            Family I: Methanomicrobiaceae
              Methanoculleus 6 spp.
              Methanogenium 11 spp.
              Methanolacinia 1 sp.
              Methanomicrobium 2 spp.
              Methanoplanus 3 spp.
              Methanofollis 2 spp.
            Family II: Methanocorpusculaceae
              Methanocorpusculum 5 spp.
            Family III: Methanospirillaceae
              Methanospirillum 1 sp.
         Order III: Methanosarcinales
            Family I: Methanosarcinaceae
              Methanococcoides 2 spp.
              Methanohalobium 1 sp.
              Methanohalophilus 5 spp.
              Methanolobus 5 spp.
              Methanosarcina 8 spp.
              Methanosalsum 1 sp.
            Family II: Methanosaetaceae
              Methanosaeta 2 spp.
      Class III: Halobacteria
         Order I: Halobacteriales
            Family I: Halobacteriaceae
              Haloarcula 6 spp.
              Halobacterium 13 spp.
              Halobaculum 1 sp.
              Halococcus 4 spp.
              Halogeometricum 1 sp.
              Halorubrum 7 spp.
              Haloterrigena 1 sp.*Natrialba* 2 spp.
              Natrinema 2 spp.
              Natronobacterium 4 spp.
              Natronococcus 2 spp.
              Natronomonas 1 sp.
              Natronorubrum 2 spp.

---

*Bergey's Manual of Systematic Bacteriology, 2nd ed., 5 vols. (2000), is the reference for classification. *Bergey's Manual of Determinative Bacteriology,* 9th ed. (1994), should be used for identification of culturable bacteria and archaea.

Note: Names in single quotation marks have not been validated by publication in the *International Journal of Systematic Bacteriology* as of March 1999.

Class IV: Thermoplasmata
    Order I: Thermoplasmatales
        Family I: Thermoplasmataceae
            Thermoplasma 2 spp.
        Family II: Picrophilaceae
            Picrophilus 2 spp.
Class V: Thermococci
    Order II: Thermococcales
        Family I: Thermococcaceae
            Pyrococcus 2 spp.
            Thermococcus 12 spp.
Class VI: Archaeoglobi
    Order I: Archaeoglobales
        Family I: Archaeoglobaceae
            Archaeoglobus 3 spp.
            Ferroglobus 1 sp.
Class VII: Methanopyri
    Order I: Methanopyrales
        Family I: Methanopyraceae
            Methanopyrus 1 sp.

Domain: Bacteria
  Phylum I: Aquificae
    Class I: Aquificae
        Order I: Aquificales
            Family I: Aquificaceae
                Aquifex 1 sp.
                Calderobacterium 1 sp.
                Hydrogenobacter 2 spp.
  Phylum II: Thermotogae
    Class I: Thermotogae
        Order I: Thermotogales
            Family I: Thermotogaceae
                Fervidobacterium 4 spp.
                Geotoga 2 spp.
                Petrotoga 2 spp.
                Thermosipho 2 spp.
                Thermotoga 5 spp.
  Phylum III: Thermodesulfobacteria
        Order I: Thermodesulfobacteriales
            Family I: 'Thermodesulfobacteriaceae'
                Thermodesulfobacterium 2 spp.
  Phylum IV: 'Deinococcus-Thermus'
    Class I: Deinococci
        Order I: Deinococcales
            Family I: Deinococcaceae
                Deinococcus 8 spp.
        Order II: Thermales
            Family I: Thermaceae
                Meiothermus 4 spp.
                Thermus 9 spp.
  Phylum V: Chrysiogenetes
    Class I: Chrysiogenetes
        Order I: Chrysiogenales
            Family I: Chrysiogenaceae
                Chrysiogenes 1 sp.

Phylum VI: Chloroflexi
  Class I: 'Chloroflexi'
    Order I: 'Chloroflexales'
        Family I: 'Chloroflexaceae'
            Chloroflexus 2 spp.
            Chloronema 1 sp.
            Heliothrix 1 sp.
            Oscillochloris 2 spp.
    Order II: 'Herpetosiphonales'
        Family I: 'Herpetosiphonaceae'
            Herpetosiphon 5 spp.
Phylum VII: Thermomicrobia
  Class I: Thermomicrobia
    Order I: Thermomicrobiales
        Family I: Thermomicrobiaceae
            Thermomicrobium 2 spp.
Phylum VIII: Nitrospira
  Class I: 'Nitrospira'
    Order I: 'Nitrospirales'
        Family I: 'Nitrospiraceae'
            Leptospirillum 1 sp.
Phylum IX: Deferribacteres
  Class I: Deferribacteres
    Order I: Deferribacterales
        Family I: Deferrivacter
            Deferribacter 1 sp.
Phylum X: 'Cyanobacteria'
  Class I: Cyanobacteria
    SubSection I
        Chamaesiphon
        Chroococcus
        Cyanobacterium
        Cyanothece
        Dactylococcopsis (Myxobaktron)
        Gloeobacter
        Gloeocapsa
        Gloeothece
        Microcystis
        Prochlorococcus
        Prochloron
        Synechococcus
        Synechocystis
    SubSection II
        Chroococcidiopsis
        Cyanocystis
        Dermocarpella
        Stanieria
        Xenococcus
    SubSection III
        Arthrospira
        Borzia
        Crinalium
        Geitlerinema
        Leptolyngbia
        Limnothrix

Lyngbya
Microcoleus
Oscillatoria
Planktothrix
Prochlorothrix
Pseudoanabaena
Spirulina
Starria
Symploca
Trichodesmium
Tychonema
Subsection IV
 Anabaena
 Anabaenopsis
 Aphanizomenon
 Calothrix
 Cyanospira
 Cylindrospermum
 Microchaete
 Nodularia
 Nostoc
 Rivularia
 Scytonema
 Tolypothrix
SubSection V
 Chlorogloeopsis
 Fischerella
 Geitleria
 Iyengariella
 Nostochopsis
 Stigonema

Phylum XI: Chlorobi
Class I: Chlorobia
 Order I: Chlorobiales
  Family I: Chlorobiaceae
   Ancalochloris 1 sp.
   Chlorobium 6 spp.
   Chloroherpeton 1 sp.
   Pelodictyon 4 spp.
   Prosthecochloris 1 sp.

Phylum IX: Proteobacteria
Class I: Alpha Proteobacteria
 Order I: Rhodospirillales
  Family I: Rhodosprillaceae
   Azospirillum 7 spp.
   Magnetospirillum 2 spp.
   Phaeospirillum 2 spp.
   Rhodocista 1 sp.
   Rhodospira 1 sp.
   Rhodospirillum 9 spp.
   Rhodothalassium 1 sp.
   Rhodovibrio 2 spp.
   Roseospira 1 sp.
   Shermanella

  Family II: Acetobacteraceae
   Acetobacter 20 spp.
   Acidiphilium 8 spp.
   Acidocella 2 spp.
   Acidomonas 1 sp.
   Craurococcus 1 sp.
   Gluconacetobacter 6 spp.
   Gluconobacter 8 spp.
   Paracraurococcus 1 sp.
   Rhodopila 1 sp.
   Roseococcus 1 sp.
   Stella 2 spp.
   Zavarzinia 1 sp.
 Order II: Rickettsiales
  Family I: Rickettsiaceae
   Orientia 1 sp.
   Rickettsia 22 spp.
   Wolbachia
  Family II: Ehrlichiaceae
   Aegyptianella 1 sp.
   Anaplasma 4 spp.
   Cowdria 1 sp.
   Ehrlichia 8 spp.
   Neorickettsia 1 sp.
  Family III: 'Holosporaceae'
   Caedibacter 5 spp.
   Holospora 4 spp.
   Lyticum 2 spp.
   Polynucleobacter 1 sp.
   Pseudocaedibacter 3 spp.
   Symbiotes 1 sp.
   Tectibacter 1 sp.
 Order III: 'Rhodobacterales'
  Family I: 'Rhodobacteraceae'
   Ahrensia 1 sp.
   Amaricoccus 4 spp.
   Antarctobacter 1 sp.
   Gemmobacter 1 sp.
   Hirschia 1 sp.
   Hyphomonas 5 spp.
   Octadecabacter 2 spp.
   Paracoccus 13 spp.
   Rhodobacter 8 spp.
   Rhodovulum 4 spp.
   Roseobacter 4 spp.
   Roseovarius 1 sp.
   Rubrimonas 1 sp.
   Ruegeria 3 spp.
   Sagittula 1 sp.
   Stappia 2 spp.
   Staleya
   Sulfitobacter 1 sp.
   Roseovarius 1 sp.

Order IV: 'Sphingomonadales'
  Family I: 'Sphingomonodaceae'
    Blastomonas 1 sp.
    Erythrobacter 2 spp.
    Erythromicrobium 1 sp.
    Erythromonas 1 sp.
    Porphyrobacter 2 spp.
    Rhizomonas 1 sp.
    Sandaracinobacter 1 sp.
    Sphingomonas 19 spp.
    Zymomonas 2 spp.
Order V: Caulobacterales
  Family I: Caulobacteraceae
    Asticcacaulis 2 spp.
    Brevundimonas 2 spp.
    Caulobacter 11 spp.
    Phenylobacterium 1 sp.
Order VI: 'Rhizobiales'
  Family I: Rhizobiaceae
    Agrobacterium 10 spp.
    Carbophilus 1 sp.
    Chelatobacter 1 sp.
    Ensifer 1 sp.
    Rhizobium 20 spp.
    Sinorhizobium 6 spp.
  Family II: Bartonellaceae
    Bartonella 14 spp.
  Family III: Brucellaceae
    Brucella 6 spp.
    Mycoplana 4 spp.
    Ochrobactrum 2 spp.
  Family IV: 'Phyllobacteriaceae'
    Mesorhizobium 7 spp.
    Phyllobacterium 2 spp.
  Family V: 'Methylocystaceae'
    Methylocystis 2 spp.
    Methylosinus 2 spp.
  Family VI: 'Beijerinckiaceae'
    Beijerinckia 6 spp.
    Chelatococcus 1 sp.
    Derxia 1 sp.
  Family VII: 'Bradyrhizobiaceae'
    Afipia 3 spp.
    Agromonas 1 sp.
    Blastobacter 5 spp.
    Bosea 1 sp.
    Bradyrhizobium 3 spp.
    Nitrobacter 1 sp.
    Oligotropha 1 sp.
    Rhodopseudomonas 15 spp
  Family VIII: Hyphomicrobiaceae
    Ancalomicrobium 1 sp.
    Ancylobacter 1 sp.
    Angulomicrobium 1 sp.
    Aquabacter 1 sp.
    Azorhizobium 1 sp.

Blastochloris 2 spp.
Devosia 1 ap.
Dichotomicrobium 1 sp.
Filomicrobium 1 sp.
Gemmiger 1 sp.
Hyphomicrobium 12 spp.
Labrys 1 sp.
Nevskia 1 sp.
Methylorhabdus 1 sp.
Pedomicrobium 4 spp.
Prosthecomicrobium 4 spp.
Rhodomicrobium 1 sp.
Rhodoplanes 2 spp.
Seliberia 1 sp.
Xanthobacter 4 spp.
  Family IX: 'Methylobacteriaceae'
    Methylobacterium 1 sp.
    Protomonas 1 sp.
    Roseomonas 3 spp.
  Family X: 'Rhodobiaceae'
    Rhodobium 2 spp.
Class II: 'Beta Proteobacteria '
  Order I: 'Burkholderiales'
    Family I: 'Burkholderiaceae'
      Burkholderia 20 spp.
      Cupriavidus 1 sp.
      Lautropia 1 sp.
      Thermothrix 2 spp.
    Family II: 'Ralstoniaceae'
      Ralstonia 3 spp.
    Family III: 'Oxalobacteraceae'
      Duganella 1 sp.
      Herbaspirillum 2 spp.
      Janthinobacterium 1 sp.
      Oxalogacter 2 spp.
      Telluria 2 spp.
    Family IV: Alcaligenaceae
      Achromobacter 4 spp.
      Alcaligenes 16 spp.
      Bordetella 7 spp.
      Pelistega 1 sp.
      Sutterella 1 sp.
      Taylorella 1 sp.
    Family V: Comamonadaceae
      Acidovorax 7 spp.
      Brachymonas 1 sp.
      Comamonas 3 spp.
      Ideonella 1 sp.
      Leptothrix 5 spp.
      Polaromonas 1 sp.
      Rhodoferax 1 sp.
      Rubrivivax 1 sp.
      Sphaerotilus 1 sp.
      Thiomonas 4 spp.
      Variovorax 1 sp.

Order II: 'Hydrogenophilales'
   Family I: 'Hydrogenophilus'
      Hydrogenophaga 4 spp.
      Thiobacillus 21 spp.
Order III: 'Methylophilales'
   Family I: 'Methylophilaceae'
      Methylobacillus 2 spp.
      Methylophilus 1 sp.
      Methylovorus 1 sp.
Order IV: 'Neisseriales'
   Family I: Neisseriaceae
      Alysiella 1 sp.
      Aquaspirillum 21 spp.
      Catenococcus 1 sp.
      Chromobacterium 2 spp.
      Eikenella 1 sp.
      Iodobacter 1 sp.
      Kingella 4 spp.
      Microvirgula 1 sp.
      Neisseria 24 spp.
      Prolinoborus 1 sp.
      Simonsiella 3 spp.
      Vogesella 1 sp.
Order V: 'Nitrosomonadales'
   Family I: 'Nitrosomonadaceae'
      Nitrosomonas 1 sp.
      Nitrosospira 3 spp.
   Family II: Spirillaceae
      Spirillum 1 sp.
   Family III: Gallionellacea
      Gallionella 1 sp.
Order VI: 'Rhodocyclales'
   Family I: 'Rhodocyclaceae'
      Azoarcus 5 spp.
      Propionibacter
      Rhodocyclus 3 spp.
      Thauera 4 spp.
      Zoogloea 1 sp.
Class III: 'Gamma Proteobacteria'
   Order I: 'Chromatiales'
      Family I: Chromatiaceae
         Allochromatium 3 spp.
         Amoebobacter 4 spp.
         Chromatium 13 spp.
         Halochromatium 2 spp.
         Isochromatium 1 sp.
         Lamprobacter 1 sp.
         Lamprocystis 1 sp.
         Marichromatium 2 spp.
         Nitrosococcus 2 spp.
         Pfennigia
         Rhabdochromatium 1 sp.
         Thermochromatium 1 sp.
         Thiocapsa 5 spp.
         Thiococcus 1 sp.
         Thiocystis 4 spp.

         Thiodictyon 2 spp.
         Thiohalocapsa 1 sp.
         Thiolamprovum 1 sp.
         Thiopedia 1 sp.
         Thiorhodococcus 1 sp.
         Thiorhodovibrio 1 sp.
         Thiospirillum 1 sp.
      Family II: Ectothiorhodospiraceae
         Arhodomonas 1 sp.
         Ectothiorhodospira 9 spp.
         Halorhodospira 3 spp.
         Nitrococcus 1 sp.
Order II: 'Xanthomonadales'
   Family I: 'Xanthomonadaceae'
      Lyssobacter 5 spp.
      Stenotrophomonas 2 spp.
      Xanthomonas 24 spp.
      Xylella 1 sp.
Order III: 'Cardiobacteriales'
   Family I: Cardiobacteriaceae
      Cardiobacterium 1 sp.
      Dichelobacter 1 sp.
      Suttonella 1 sp.
Order IV: 'Thiotrichales'
   Family I: 'Thiotrichaceae'
      Achromatium 1 sp.
      Beggiatoa 1 sp.
      Leucothrix 1 sp.
      Macromonas 2 spp.
      Thiobacterium 1 sp.
      Thiomargarita
      Thioploca 4 spp.
      Thiospira 1 sp.
      Thiothrix 1 sp.
   Family II: 'Piscirickettsiaceae'
      Cycloclasticus 1 sp.
      Hydrogenovibrio 1 sp.
      Piscirickettsia 1 sp.
      Thiomicrospira 4 spp.
   Family III: 'Francisellaceae'
      Francisella 5 spp.
Order V: 'Legionellales'
   Family I: Legionellaceae
      Legionella 44 spp.
   Family II: 'Coxiellaceae'
      Coxiella 1 sp.
      Rickettsiella 4 spp.
Order VI: 'Methylococcales'
   Family I: Methylococcaceae
      Methylobacter 6 spp.
      Methylocaldum 3 spp.
      Methylococcus 8 spp.
      Methylomicrobium 3 spp.
      Methylomonas 4 spp.
      Methylosphaera 1 sp.

Order VII: 'Oceanospirillales'
    Family I: 'Oceanospirillaceae'
        Balneatrix 1 sp.
        Fundibacter
        Marinomonas 2 spp.
        Marinospirillum 2 spp.
        Neptunomonas 1 sp.
        Oceanospirillum 16 spp.
    Family II: Halomonadaceae
        Alcanivorax 1 sp.
        Carnimonas 1 sp.
        Chromohalobacter 1 sp.
        Deleya 8 spp.
        Halomonas 19 spp.
        Zymobacter 1 sp.
Order VIII: Pseudomonadales
    Family I: Pseudomonadaceae
        Azomonas 3 spp.
        Azotobacter 9 spp.
        Cellvibrio 2 spp.
        Chryseomonas 2 spp.
        Flavimonas 1 sp.
        Lampropedia 1 sp.
        Mesophilobacter 1 sp.
        Morococcus 1 sp.
        Oligella 2 spp.
        Pseudomonas 117 spp.
        Rhizobacter 1 sp.
        Rugamonas 1 sp.
        Serpens 1 sp.
        Thermoleophilum 2 spp.
        Xylophilus 1 sp.
    Family II: Moraxellaceae
        Acinetobacter 7 spp.
        Moraxella 8 spp.
        Psychrobacter 5 spp.
Order IX: 'Alteromonadales'
    Family I: 'Alteromonadaceae'
        Alteromonas 21 spp.
        Colwellia 7 spp.
        Ferrimonas 1 sp.
        Marinobacter 1 sp.
        Marinobacterium 1 sp.
        Microbulbifer 1 sp.
        Pseudoalteromonas 17 spp.
        Shewanella 10 spp.
Order X: 'Vibrionales'
    Family I: Vibrionaceae
        Allomonas 1 sp.
        Enhydrobacter 1 sp.
        Listonella 3 spp.
        Photobacterium 10 spp.
        Salvinivibrio 1 sp.
        Vibrio 46 spp.

Order XI: 'Aeromonadales'
    Family I: Aeromonadaceae
        Aeromonas 23 spp.
        Tolumonas 1 sp.
    Family II: Succinivibrionaceae
        Anaerobiospirillumm 2 spp.
        Ruminobacter 1 sp.
        Succinomonas 1 sp.
        Succinivibrio 1 sp.
Order XII: 'Enterobacteriales'
    Family I: Enterobacteriaceae
        Arsenophonus 1 sp.
        Brenneria 6 spp.
        Buchnera 1 sp.
        Budvicia 1 sp.
        Buttiauxella 7 spp.
        Calymmatobacterium 1 sp.
        Cedecea 3 spp.
        Citrobacter 10 spp.
        Edwardsiella 4 spp.
        Enterobacter 15 spp.
        Erwinia 30 spp.
        Escherichia 6 spp.
        Ewingella 1 sp.
        Hafnia 1 sp.
        Klebsiella 11 spp.
        Kluyvera 4 spp.
        Leclercia 1 sp.
        Leminorella 2 spp.
        Moellerella 1 sp.
        Morganella 2 spp.
        Obesumbacterium 1 sp.
        Pantoea 8 spp.
        Pectobacterium 11 spp.
        Photorhabdus 1 sp.
        Plesiomonas 1 sp.
        Pragia 1 sp.
        Proteus 1 sp.
        Providencia 6 spp.
        Rahnella 1 sp.
        Saccharobacter 1 sp.
        Salmonella 12 spp.
        Serratia 12 spp.
        Shigella 4 spp.
        Sodalis 1 sp.
        Tatumella 1 sp.
        Trabulsiella 1 sp.
        Wigglesworthia 1 sp.
        Xenorhabdus 9 spp.
        Yersinia 12 spp.
        Yokenella 1 sp.

Order XIII: 'Pasteurellales'
 Family I: Pasteurellaceae
  Actinobacillus 17 spp.
  Haemophilus 20 spp.
  Lonepinella 1 sp.
  Pasteurella 23 spp.
  Mannheimia 5 spp.
Class VI: 'Delta Proteobacteria'
 Order I: 'Desulfurales''Desulfovibrionales'
  Family I: 'Desulfurellaceae'
   Desulfurella 4 spp.
   Hippea
 Order II: 'Desulfovibrionales'
  Family I: 'Desulfovibrionaceae'
   Bilophila 1 sp.
   Desulfovibrio 29 spp.
   Lawsonia 1 sp.
  Family II: 'Desulfomicrobiaceae'
   Demicrobium 1 sp.
 Order III: 'Desulfobacterales'
  Family I: 'Desulfobacteraceae'
   Desulfobacter 6 spp.
   Desulfobacterium 7 spp.
   Desulfococcus 2 spp.
   Desulfosarcina 1 sp.
   Desulfospira 1 sp.
   Desulfocella 1 sp.
  Family II: 'Desulfobulbaceae'
   Desulfobulbus 3 spp.
   Desulfocapsa 1 sp.
   Desulfofustis 1 sp.
  Family III: 'Desulfoarculaceae'
 Order IV: 'Desulfuromonadales'
  Family I: 'Desulfomonadaceae'
   Desulfuromonas 4 sp.
  Family II: 'Geobacteraceae'
   Geobacter 2 spp.
  Family III: 'Pelobacteriaceae'
   Pelobacter 6 spp.
 Order V: 'Syntrophobacterales'
  Family I: 'Syntrophobacteraceae
   Desulfacinum 1 sp.
   Syntrophobacter 3 spp.
   Desulforhabdus 1 sp.
   Thermodesulforhabdus 1 sp.
  Family II: 'Syntrophaceae
 Order VI: 'Bdellovibrionales'
  Family VI: 'Bdellovibrionaceae'
   Bdellovibrio 3 spp.
   Micavibrio 1 sp.
   Vampirovibrio 1 sp.

Order VII: Myxococcales
 Family I: Myxococcaceae
  Angiococcus 1 sp.
  Myxococcus 8 spp.
 Family II: Archangiaceae
  Archangium 1 sp.
 Family III: Cystobacteraceae
  Cystobacter 3 spp.
  Melittangium 3 spp.
  Stigmatella 2 spp.
 Family IV: Polyangiaceae
  Chondromyces 5 spp.
  Nannocystis 1 sp.
  Polyangium 10 spp.
Class V: 'Epsilon Proteobacteria'
 Order I: 'Campylobacterales'
  Family I: Campylobacteraceae
   Arcobacter 4 spp.
   Campylobacter 28 spp.
   Sulfurospirillum 2 spp.
   Thiovulum 1 sp.
  Family II: 'Helicobacteraceae'
   Helicobacter 18 spp.
   Wolinella 3 spp.
Phylum XIII: 'Firmicutes'
 Class I: 'Clostridia'
  Order I: Clostridiales
   Family I: Clostridiaceae
    Anaerobacter 1 sp.
    Caloramator 3 spp.
    Clostridium 146 spp.
    Oxobacter 1 sp.
    Sarcina 2 spp.
    Sporobacter 1 sp.
    Thermobrachium 1 sp.
   Family II: 'Lachnospiraceae'
    Acetitomaculum 1 sp.
    Anaerofilum 2 spp.
    Butyrivibrio 2 spp.
    Catonella 1 sp.
    Coprococcus 3 spp.
    Johnsonella 1 sp.
    Lachnospira 2 spp.
    Pseudobutyrivibrio 1 sp.
    Roseburia 1 sp.
    Ruminococcus 14 spp.
    Sporobacterium
   Family III: 'Peptostreptococcace'
    Filifactor 1 sp.
    Fusibacter
    Helcoccus 1 sp.
    Peptostreptococcus 18 spp.
    Tissierella 3 spp.

Family IV: 'Eubacteriaceae'
Eubacterium 52 spp.
Pseudoramibacter 1 sp.
Family V: Peptococcaceae
Desulfitobacterium 3 spp.
Desulfosporosinus 1 sp.
Desulfotomaculum 18 spp.
Mitsuokella 2 spp.
Peptococcus 8 spp.
Propionispira 1 sp.
Succinispira
Syntrophobotulus 1 sp.
Thermoterrabacterium 1 sp.
Family VI: 'Heliobacteriaceae'
Heliobacterium 3 spp.
Heliobacillus 1 sp.
Heliophilum 1 sp.
Family VII: Acidaminocaccaceae
Acetonema 1 sp.
Acidaminococcus 1 sp.
Dialister 1 sp.
Megasphaera 2 spp.
Pectinatus 2 spp.
Phascolarctobacterium 1 sp.
Quinella 1 sp.
Schwartzia 1 sp.
Selenomonas 11 spp.
Sporomusa 7 spp.
Succiniclasticum 1 sp.
Veillonella 14 spp.
Zymophilus 2 spp.
Family VII: Syntrophomonadaceae
Acetogenium 1 sp.
Anaerobaculum 1 sp.
Anaerobranca 1 sp.
Caldicellulosiruptor 3 spp.
Dethiosulfovibrio 1 sp.
Synthrophomonas 3 spp.
Syntrophospora 1 sp.
Thermohydrogenium 1 sp.
Thermosyntropha 1 sp.
Order II: 'Thermoanaerobactriales'
Family I: 'Thermoanaerobacteriaceae'
Thermoanaerobacter 13 spp.
Thermoanaerobacterium 5 spp.
Thermoanaerobium 2 spp.
Ammonifex 1 sp.
Moorella 3 spp.
Sporotomaculum 1 sp.
Order III: Haloanaerobiales
Family I: Haloanaerobiaceae
Haloanaerobium 8 spp.
Halocella 1 sp.
Halothermothrix 1 sp.
Natroniella 1 sp.

Family II: Halobacteroidaceae
Acetohalobium 1 sp.
Haloanaerobacter 3 spp.
Halobacteroides 4 spp.
Orenia 1 sp.
Sporohalobacter 2 spp.
Class II: Mollicutes
Order I: Mycoplasmatales
Family I: Mycoplasmataceae
Mycoplasma 110 spp.
Ureaplasma 6 spp.
Order II: Entoplasmatales
Family I: Entoplasmatales
Entomoplasma 6 spp.
Mesoplasma 12 spp.
Family II: Spiroplasmataceae
Spiroplasma 33 spp.
Order III: Acholeplasmataceae
Family I: Anaeroplasmataceae
Anaeroplasma 4 spp.
Acholeplasma 16 spp.
Order IV: Anaeroplasmtales
Anaeroplasma 4 spp.
Asteroleplasma 1 sp.
Order V: Incertae sedis
Family V: 'Erysipelothrichaeceae'
Erysipelothrix 2 spp.
Holdemania 1 sp.
Class III: 'Bacilli'
Order I: Bacillales
Family I: Bacillaceae
Amphibacillus 1 sp.
Bacillus 114 spp.
Exiguobacterium 2 spp.
Halobacillus 3 spp.
Saccharococus 1 sp.
Virgibacillus 1 sp.
Family II: Planococcaceae
Filibacter 1 sp.
Kurthia 3 spp.
Planococcus 5 spp.
Sporosarcina 2 spp.
Family III: Caryophanaceae
Caryophanon 2 spp.
Family IV: 'Listeriaceae'
Brochothrix 2 spp.
Listeria 9 spp.
Family V: 'Staphylococcaceae'
Gemella 4 spp.
Macrococcus 4 spp.
Salinicoccus 2 spp.
Staphylococcus 47 spp.
Family VI: 'Sporolactobacillaceae'
Marinococcus 3 spp.
Sporolactobacillus 6 spp.

Family VII: 'Paenibacillaceae'
   Ammoniphilus 2 spp.
   Aneurinibacillus 3 spp.
   Brevibacillus 10 spp.
   Oxalophagus 1 sp.
   Paenibacillus 27 spp.
Family VIII: 'Alicyclobacillaceae'
   Alicyclobacillus 3 spp.
   Pasteuria 4 spp
   Sulfobacillus 3 spp.
Family VII: 'Thermoactinomycetaceae'
   Thermoactinomyces 8 spp.
Order II: 'Lactobacillales'
  Family I: Lactobacillaceae
   Lactobacillus 100 spp.
   Pediococcus 8 spp.
  Family II: 'Aerococcaceae'
   Abiotrophia 3 spp.
   Aerococcus 2 spp.
   Facklamia 2 spp.
   Globicatella 1 sp.
   Tetragenococcus 2 spp.
   Ignavigranum 1 sp.
  Family III: 'Carnobacteriaceae'
   Agitococcus 1 sp.
   Alloiococcus 1 sp.
   Carnobacterium 6 spp.
   Desemzia 1 sp.
   Dolosigranulum 1 sp.
   Lactosphaera 1 sp.
   Trichococcus 1 sp.
  Family IV: 'Enterococcaceae'
   Enterococcus 20 spp.
   Melissococcus 1 sp.
   Vagococcus 2 spp.
  Family V: 'Leuconostocaceae'
   Leuconostoc 15 spp.
   Oenococcus 1 sp.
   Weissela 7 spp.
  Family VI: Streptococcaceae
   Lactococcus 7 spp.
   Streptococcus 68 spp.

Phylum XIV: The Actinobacteria
Class I: Actinobacteria
  Order I: Acidimicrobiales
   Family I: Acidimicrobiaceae
    Acidimicrobium 1 sp.
  Order II: Rubrobacterales
   Family I: Rubrobacteraceae
    Rubrobacter 2 spp.
  Order III: Coriobacteridae
   Family I: Coriobacteriaceae
    Atophobium 3 spp.
    Coriobacterium 1 sp.

Order IV: 'Sphaerobacterales'
  Family I: Sphaerobacteraceae
   Sphaerobacter 1 sp.
Order V: Actinomycetales
  Suborder: Actinomycineae
   Family I: Actinomycetaceae
    Actinobaculum 2 spp.
    Actinomyces 23 spp.
    Arcanobacterium 4 spp.
    Mobiluncus 3 spp.
  Suborder: Micrococcineae
   Family I: Micrococcaceae
    Arthrobacter 32 spp.
    Bogoriella 1 sp.
    Demetria 1 sp.
    Kocuria 6 spp.
    Leucobacter 1 sp.
    Micrococcus 9 spp.
    Nesterenkonia 1 sp.
    Renibacterium 1 sp.
    Rothia 1 sp.
    Stomatococcus 1 sp.
    Terracoccus 1 sp.
   Family II: 'Brevibacteriaceae'
    Brevibacterium 26 spp.
   Family III: Cellulomonadaceae
    Cellulomonas 1 sp.
    Oerskovia 2 spp.
    Rarobacter 2 spp.
   Family IV: Dermabacteraceae
    Brachybacterium 7 spp.
    Dermabacter 1 sp.
   Family V: Dermatophilaceae
    Dermacoccus 1 sp.
    Dermatophilus 2 spp.
    Kytococcus 1 sp.
   Family VI: Intrasporangiaceae
    Intrasporangium 1 sp.
    Janibacter 1 sp.
    Sanguibacer 3 spp.
    Terrabacter 1 sp.
   Family VII: Jonesiaceae
    Jonesia 1 sp.
   Family VIII: Microbacteriaceae
    Agrococcus 1 sp.
    Agromyces 6 spp.
    Aureobacterium 14 spp.
    Clavibacter 11 spp.
    Cryobacterium 1 sp.
    Curtobacterium 8 spp.
    Microbacterium 27 spp.
    Rathayibacter 4 spp.
   Family IX: 'Beutenbergiaceae'
    Beutenbergia

Family X: Promicromonosporaceae
  Promicromonospora 3 spp.
Suborder: Corynebacterineae
  Family I: Corynebacteriaceae
    Corynebacterium 67 spp.
  Family II: Dietziaceae
    Dietzia 2 spp.
  Family III: Gordoniaceae
    Gordonia 9 spp.
    Skermania 1 sp.
  Family IV: Mycobacteriaceae
    Mycobacterium 85 spp.
  Family V: Nocardiaceae
    Nocardia 29 spp.
    Rhodococus 25 spp.
  Family VI: Tsukamurellaceae
    Tsukamurella 5 spp.
  Family VII: 'Williamsiaceae'
    Williamsia ??
Suborder: Micromonosporaceae
  Family I: Micromonosporaceae
    Actinoplanes 23 spp.
    Catellatospora 5 spp.
    Catenuloplanes 6 spp.
    Couchioplanes 2 spp.
    Dactylosporangium 6 spp.
    Micromonospora 19 spp.
    Pilimelia 4 spp.
    Spirilliplanes 1 sp.
    Verrucosispora 1 sp.
Suborder: Propionibacterineae
  Family I: Propionibacteriaceae
    Luteococcus 1 sp.
    Microlunatus 1 sp.
    Propionibacterium 12 spp.
    Propioniferax 1 sp.
  Family II: Nocardioidaceae
    Aeromicrobium 2 spp.
    Friedmanniella 1 sp.
    Nocardiodes 7 spp.
Suborder: Pseudonocardineae
  Family I: Pseudonocardiaceae
    Actinopolyspora 3 spp.
    Amycolatopsis 11 spp.
    Kibdelosporangium 4 spp.
    Kutzneria 3 spp.
    Pseudonocardia 12 spp.
    Saccharomonospora 5 spp.
    Saccharopolyspora 10 spp.
    Streptoalloteichus 1 sp.
    Thermobispora 1 sp.
    Thermocrispum 2 spp.
  Family II: Actinosynnemataceae
    Actinosynnema 3 spp.
    Lentzea 1 sp.
    Saccharothrix 15 spp.

Suborder: Streptomycineae
  Family I: Streptomycetaceae
    Streptomyces 509 spp.
    Streptoverticillium 48 spp.
Suborder: Streptosporangineae
  Family I: Streptosporangiaceae
    Herbidospora 1 sp.
    Microbispora 15 spp.
    Micropolyspora 5 spp.
    Microtetraspora 20 spp.
    Nonomuria 15 spp.
    Planobispora 2 spp.
    Planomonospora 5 spp.
    Planopolyspora 1 sp.
    Planotetraspora 1 sp.
    Streptosporangium 19 spp.
  Family II: Nocardiopsaceae
    Nocardiopsis 17 spp.
    Thermobifida 2 spp.
  Family III: Thermomonosporaceae
    Actinomadura 51 spp.
    Spirillospora 2 spp.
    Thermomonospora 7 spp.
Suborder: Frankineae
  Family I: Frankiaceae
    Frankia 1 sp.
  Family II: Geodermatophilaceae
    Blastococcus 1 sp.
    Geodermatophilus 1 sp.
  Family III: Microsphaeraceae
    Microsphaera 1 sp.
  Family IV: Sporichthyaceae
    Sporichthya 1 sp.
  Family V: Acidothermaceae
    Acidothermus 1 sp.
  Family VI: 'Kineosporaceae'
    Cryptosporangium 2 spp.
    Kineococcus 1 sp.
    Kineosporia 5 spp.
Suborder XI: Glycomycineae
  Family I: Glycomycetaceae
    Glycomyces 3 spp.
Order VI: Bifidobacteriales
  Family I: Bifidobacteriaceae
    Bifidobacterium 33 spp.
    Falcivibrio 2 spp.
    Gardnerella 1 sp.
  Family II: 'Unknown Affiliation'
    Actinobispora 1 sp.
    Actinocorallia 1 sp.
    Excellospora 1 sp.
    Pelczaria 1 sp.
    Turicella 1 sp.

Phylum XV: Planctomycetes

Order I: Planctomycetales

Family I: Planctomycetaceae

Gemmata 1 sp.

Isosphaera 1 sp.

Pirellula 2 spp.

Planctomyces 6 spp.

Phylum XVI: Chlamydiae

Order I: Chlamydiales

Family I: Chlamydiaceae

Chlamydia 4 spp.

Family II: Parachlamydiaceae

Family III: Simkaniaceae

Family VI: Waddliaceae

Phylum XII: Spirochetes

Class I: 'Spirochaetes'

Order I: Spirochaetales

Family I: Spirochaetaceae

Borrelia 30 spp.

Brevinema 1 sp.

Clevelandina 1 sp.

Cristispira 1 sp.

Diplocalyx 1 sp.

Hollandina 1 sp.

Pillotina 1 sp.

Spirochaeta 14 spp.

Treponema 18 spp.

Family II: Serpulinaceae

Brachyspira 5 spp.

Serpulina 6 spp.

Family III: Leptospiraceae

Leptonema 1 sp.

Leptospira 12 spp.

Phylum XVIII: Fibrobacteres

Class I: 'Fibrobacteres'

Family I: 'Fibrobacteraceae'

Fibrobacter 3 spp.

Phylum IXX: Actinobacter

Family I: 'Acidobacteriaceae'

Acidobacterium 1 sp.

Holophaga 1 sp.

Phylum XX: Bacteroidetes

Class I: 'Bacteroides'

Order I: 'Bacteroidales'

Family I: Bacteroidaceae

Acetofilamentum 1 sp.

Acetomicrobium 2 spp.

Acetothermus 1 sp.

Anaerorhabdus 1 sp.

Bacteroides 65 spp.

Megamonas 1 sp.

Family II: 'Rikenellaceae'

Marinilabilia 2 spp.

Rikenella 1 sp.

Family III: 'Porphyromonadaceae'

Porphyromonas 13 spp.

Family IV: 'Prevotellaceae'

Prevotella 26 spp.

Class II: 'Flavobacteria'

Order I: 'Flavobacteriales'

Family I: Flavobacteriaceae

Bergeyella 1 sp.

Capnocytophaga 7 spp.

Cellulophaga

Chryseobacterium 6 spp.

Coenonia

Empedobacter 1 sp.

Flavobacterium 40 spp.

Gelidibacter 1 sp.

Ornithobacterium 1 sp.

Polaribacter 4 spp.

Psychroflexus 2 spp.

Psychroserpens 1 sp.

Riemerella 2 spp.

Weeksella 2 spp.

Family II: 'Myroideaceae'

Myroides 2 spp.

Psychromonas 1 sp.

Family III: 'Blattabacteriaceae'

Blattabacterium 1 sp.

Class III: 'Sphingobacteria'

Order I: 'Sphingobacteriales'

Family I: Sphingobacteriaceae

Pedobacter 4 spp.

Sphingobacterium 8 spp.

Family II: 'Saprospiraceae'

Haliscomenobacter 1 sp.

Lewinella 3 spp.

Saprospira 1 sp.

Family III: 'Flexibacteraceae'

Cyclobacterium 1 sp.

Cytophaga 23 spp.

Flectobacillus 3 spp.

Flexibacter 17 spp.

Meniscus 1 sp.

Microscilla 1 sp.

Runella 1 sp.

Spirosoma 1 sp.

Sporocytophaga 1 sp.

Family IV: 'Flammeovirgaceae'

Flammeovirga 1 sp.

Flexithrix 1 sp.

Persicobacter 1 sp.

Thermonema 2 spp.

Family V: 'Crenotrichaceae'

Chitinophaga 1 sp.

Crenothrix 1 sp.

Rhodothermus 2 spp.

Toxothrix 1 sp.

Phylum XXI: Fusobacteria
   Class I: 'Fusobacteria'
      Order I: 'Fusobacteriales'
         Family I: 'Fusobacteriaceae'
            Fusobacterium 23 spp.
            Ilyobacter 3 spp.
            Leptotrichia 1 sp.
            Propionigenium 2 spp.
            Sebaldella 1 sp.
            Streptobacillus 1 sp.
         Genera incertae sedis
            Cetobacterium 1 sp.

Phylum XXII: Verrucomicrobia
   Class I: Verrucomicrobiae
      Order I: Verrucomicrobiales
         Family I: Verrucomicrobiaceae
            Prosthecobacter 4 spp.
            Verrucomicrobium 1 sp.

Appendix B

Methods for Taking Clinical Samples

To diagnose a disease, it is often necessary to obtain a sample of material that may contain the pathogenic microorganism. Samples must be taken aseptically. The sample container should be labeled with the patient's name, room number (if hospitalized), date, time, and medications being taken. Samples must be transported to the laboratory immediately for culture. Delay in transport may result in the growth of some organisms, and their toxic products may kill other organisms. Pathogens tend to be fastidious and die if not kept in optimum environmental conditions.

In the laboratory, samples from infected tissues are cultured on differential and selective media in an attempt to isolate and identify any pathogens or organisms that are not normally found in association with that tissue.

Universal Precautions

The following procedures should be used by all health care workers, including students, whose activities involve contact with patients or with blood or other body fluids. These procedures were developed to minimize the risk of transmitting HIV or AIDS in a health care environment, but adherence to these guidelines will minimize the transmission of *all* nosocomial infections.

1. Gloves should be worn when touching blood and body fluids, mucous membranes, and nonintact skin and when handling items or surfaces soiled with blood or body fluids. Gloves should be changed after contact with each patient.

2. Hands and other skin surfaces should be washed immediately and thoroughly if contaminated with blood or other body fluids. Hands should be washed immediately after gloves are removed.

3. Masks and protective eyewear or face shields should be worn during procedures that are likely to generate droplets of blood or other body fluids.

4. Gowns or aprons should be worn during procedures that are likely to generate splashes of blood or other body fluids.

5. To prevent needlestick injuries, needles should not be recapped, purposely bent or broken, or otherwise manipulated by hand. After disposable syringes and needles, scalpel blades, and other sharp items are used, they should be placed in puncture-resistant containers for disposal.

6. Although saliva has not been implicated in HIV transmission, mouthpieces, resuscitation bags, and other ventilation devices should be available for use in areas in which the need for resuscitation is predictable. Emergency mouth-to-mouth resuscitation should be minimized.

7. Health care workers who have exudative lesions or weeping dermatitis should refrain from all direct patient care and from handling patient-care equipment.

8. Pregnant health care workers are not known to have a greater risk of contracting HIV infection than health care workers who are not pregnant; however, if a health care worker develops HIV infection during pregnancy, the infant is at risk of infection. Because of this risk, pregnant health care workers should be especially familiar with, and strictly adhere to, precautions to minimize the risk of HIV transmission.

Source: Centers for Disease Control and Prevention and National Institutes of Health. *Biosafety in Microbiological and Biomedical Laboratories,* 2nd ed. Washington, D. C.: U. S. Government Printing Office, 1988.

Instructions for Specific Procedures

Wound or Abscess Culture

1. Cleanse the area with a sterile swab moistened in sterile saline.
2. Disinfect the area with 70% ethanol or iodine solution.
3. If the abscess has not ruptured spontaneously, a physician will open it with a sterile scalpel.
4. Wipe the first pus away.
5. Touch a sterile swab to the pus, taking care not to contaminate the surrounding tissue.
6. Replace the swab in its container, and properly label the container.

Ear Culture

1. Clean the skin and auditory canal with 1% tincture of iodine.
2. Touch the infected area with a sterile cotton swab.
3. Replace the swab in its container.

Eye Culture

This procedure is often performed by an ophthalmologist.

1. Anesthetize the eye with topical application of a sterile anesthetic solution.
2. Wash the eye with sterile saline solution.
3. Collect material from the infected area with a sterile cotton swab. Return the swab to its container.

Blood Culture

1. Close the room's windows to avoid contamination.
2. Clean the skin around the selected vein with 2% tincture of iodine on a cotton swab.
3. Remove dried iodine with gauze moistened with 80% isopropyl alcohol.
4. Draw a few milliliters of venous blood.
5. Aseptically bandage the puncture.

Urine Culture

1. Provide the patient with a sterile container.
2. Instruct the patient to collect a midstream sample by first voiding a small volume from the urinary bladder before collection. This washes away extraneous bacteria of the skin microbiota.
3. A urine sample may be stored under refrigeration ($4°$–$6°C$) for up to 24 hours.

Fecal Culture

For bacteriological examination, only a small sample is needed. This may be obtained by inserting a sterile swab into the rectum or feces. The swab is then placed in a tube of sterile enrichment broth for transport to the laboratory. For examination for parasites, a small sample may be taken from a morning stool. The sample is placed in a preservative (polyvinyl alcohol, buffered glycerol, saline, or formalin) for microscopic examination for eggs and adult parasites.

Sputum Culture

1. A morning sample is best because microorganisms will have accumulated while the patient is sleeping.
2. The patient should rinse his or her mouth thoroughly to remove food and normal microbiota.
3. The patient should cough deeply from the lungs and expectorate into a sterile glass wide-mouth jar.
4. Care should be taken to avoid contaminating health care workers.
5. In cases such as tuberculosis in which there is little sputum, stomach aspiration may be necessary.
6. Infants and children tend to swallow sputum. A fecal sample may be of some value in these cases.

Appendix C
Metabolic Pathways

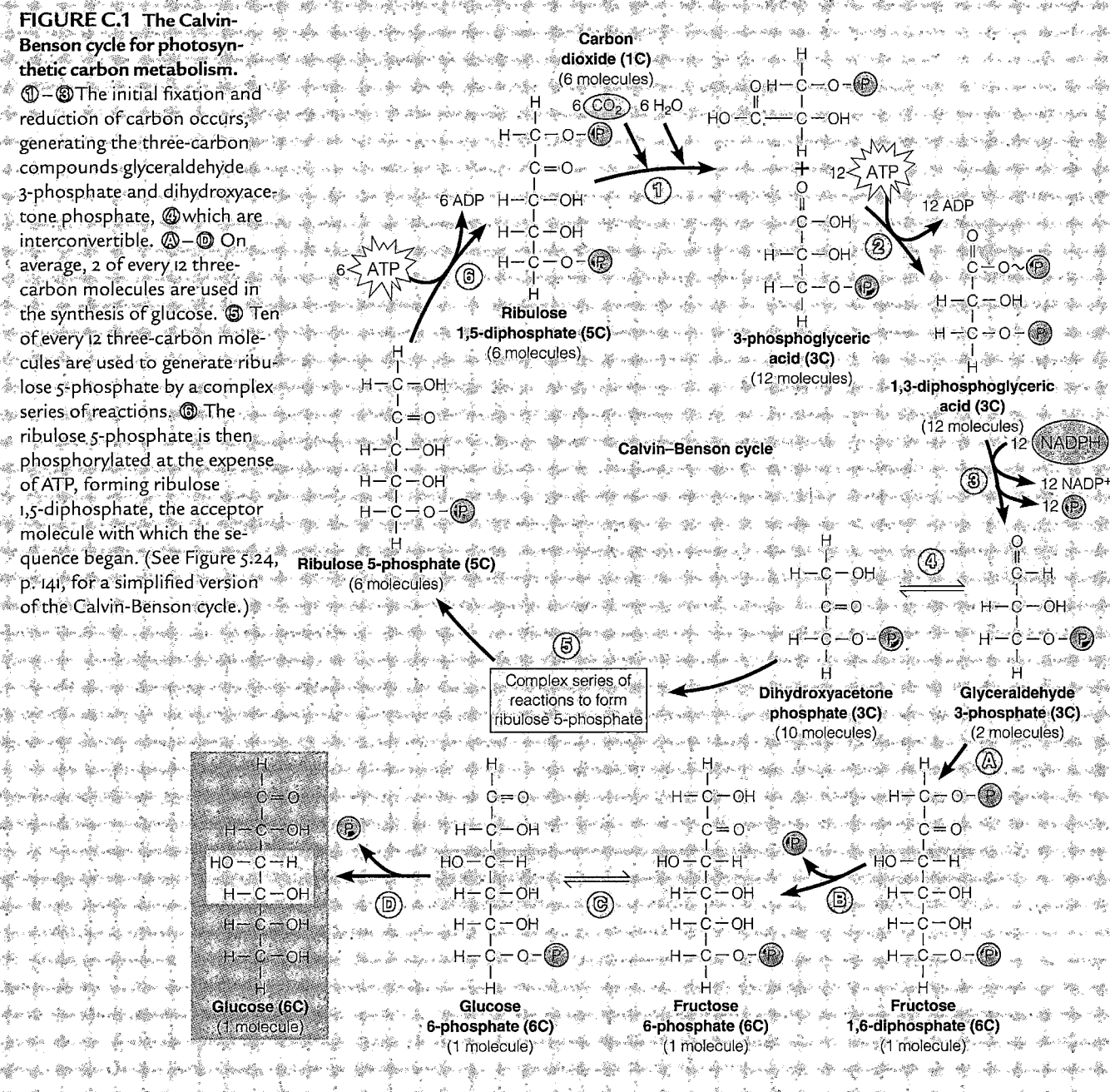

FIGURE C.1 The Calvin-Benson cycle for photosynthetic carbon metabolism.
①–⑤ The initial fixation and reduction of carbon occurs, generating the three-carbon compounds glyceraldehyde 3-phosphate and dihydroxyacetone phosphate, ④ which are interconvertible. Ⓐ–Ⓓ On average, 2 of every 12 three-carbon molecules are used in the synthesis of glucose. ⑤ Ten of every 12 three-carbon molecules are used to generate ribulose 5-phosphate by a complex series of reactions. ⑥ The ribulose 5-phosphate is then phosphorylated at the expense of ATP, forming ribulose 1,5-diphosphate, the acceptor molecule with which the sequence began. (See Figure 5.24, p. 141, for a simplified version of the Calvin-Benson cycle.)

Carbon dioxide (1C)
(6 molecules)

$6\ CO_2$ $6\ H_2O$

Ribulose 1,5-diphosphate (5C)
(6 molecules)

3-phosphoglyceric acid (3C)
(12 molecules)

12 ATP
12 ADP

1,3-diphosphoglyceric acid (3C)
(12 molecules)

Calvin–Benson cycle

Ribulose 5-phosphate (5C)
(6 molecules)

12 NADPH
12 NADP+
12 P

Complex series of reactions to form ribulose 5-phosphate

Dihydroxyacetone phosphate (3C)
(10 molecules)

Glyceraldehyde 3-phosphate (3C)
(2 molecules)

Glucose (6C)
(1 molecule)

Glucose 6-phosphate (6C)
(1 molecule)

Fructose 6-phosphate (6C)
(1 molecule)

Fructose 1,6-diphosphate (6C)
(1 molecule)

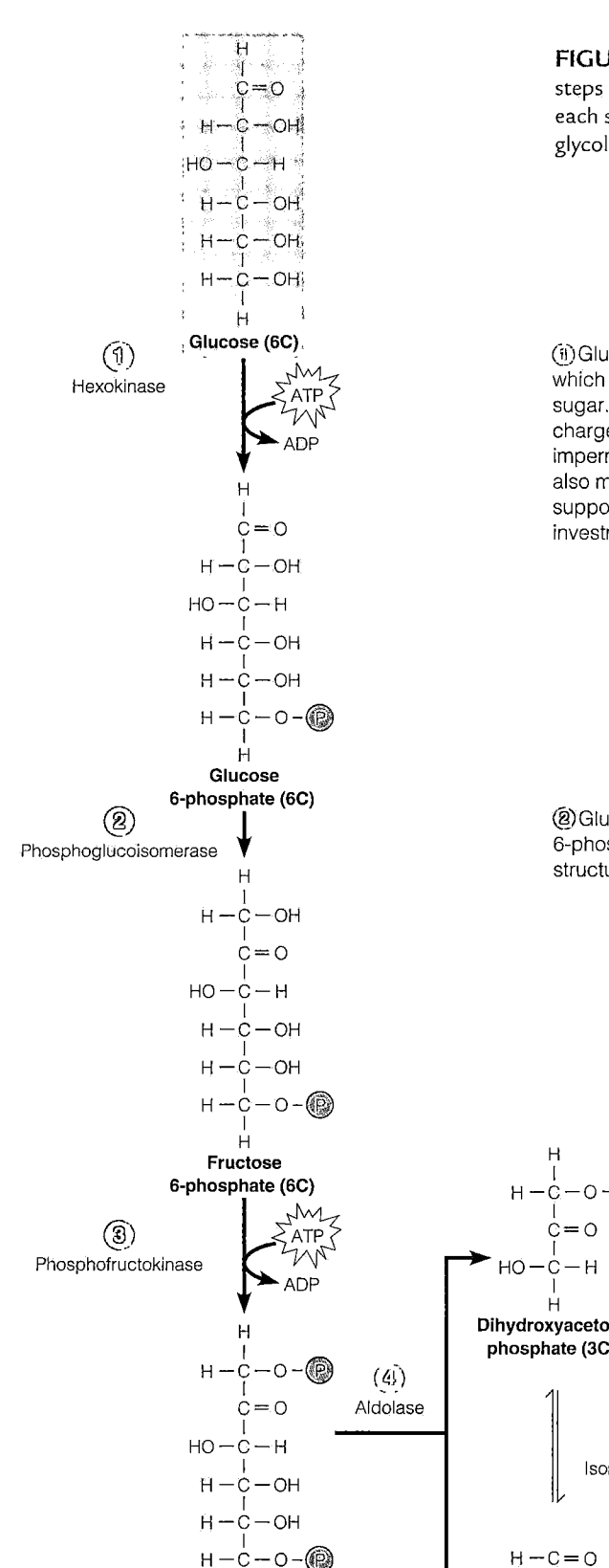

FIGURE C.2 Glycolysis (Embden-Meyerhof pathway). Each of the ten steps of glycolysis is catalyzed by a specific enzyme, which is named under each step number. (See Figure 5.11, p. 126, for a simplified version of glycolysis.)

(1) Glucose enters the cell and is phosphorylated by the enzyme hexokinase, which transfers a phosphate group from ATP to the number 6 carbon of the sugar. The product of the reaction is glucose 6-phosphate. The electrical charge of the phosphate group traps the sugar in the cell because of the impermeability of the plasma membrane to ions. Phosphorylation of glucose also makes the molecule more chemically reactive. Although glycolysis is supposed to *produce* ATP, in step (1), ATP is actually consumed—an energy investment that will be repaid with dividends later in glycolysis.

(2) Glucose 6-phosphate is rearranged to convert it to its isomer, fructose 6-phosphate. Isomers have the same number and types of atoms but in different structural arrangements.

(3) In this step, still another molecule of ATP is invested in glycolysis. An enzyme transfers a phosphate group from ATP to the sugar, producing fructose 1,6-diphosphate.

(4) This is the reaction from which glycolysis gets its name ("sugar splitting"). An enzyme cleaves fructose 1,6-diphosphate into two different three-carbon sugars: glyceraldehyde 3-phosphate and dihydroxyacetone phosphate. These two sugars are isomers.

(5) The enzyme isomerase interconverts the three-carbon sugars. The next enzyme in glycolysis uses only glyceraldehyde 3-phosphate as its substrate. This pulls the equilibrium between the two three-carbon sugars in the direction of glyceraldehyde 3-phosphate, which is removed as fast as it forms.

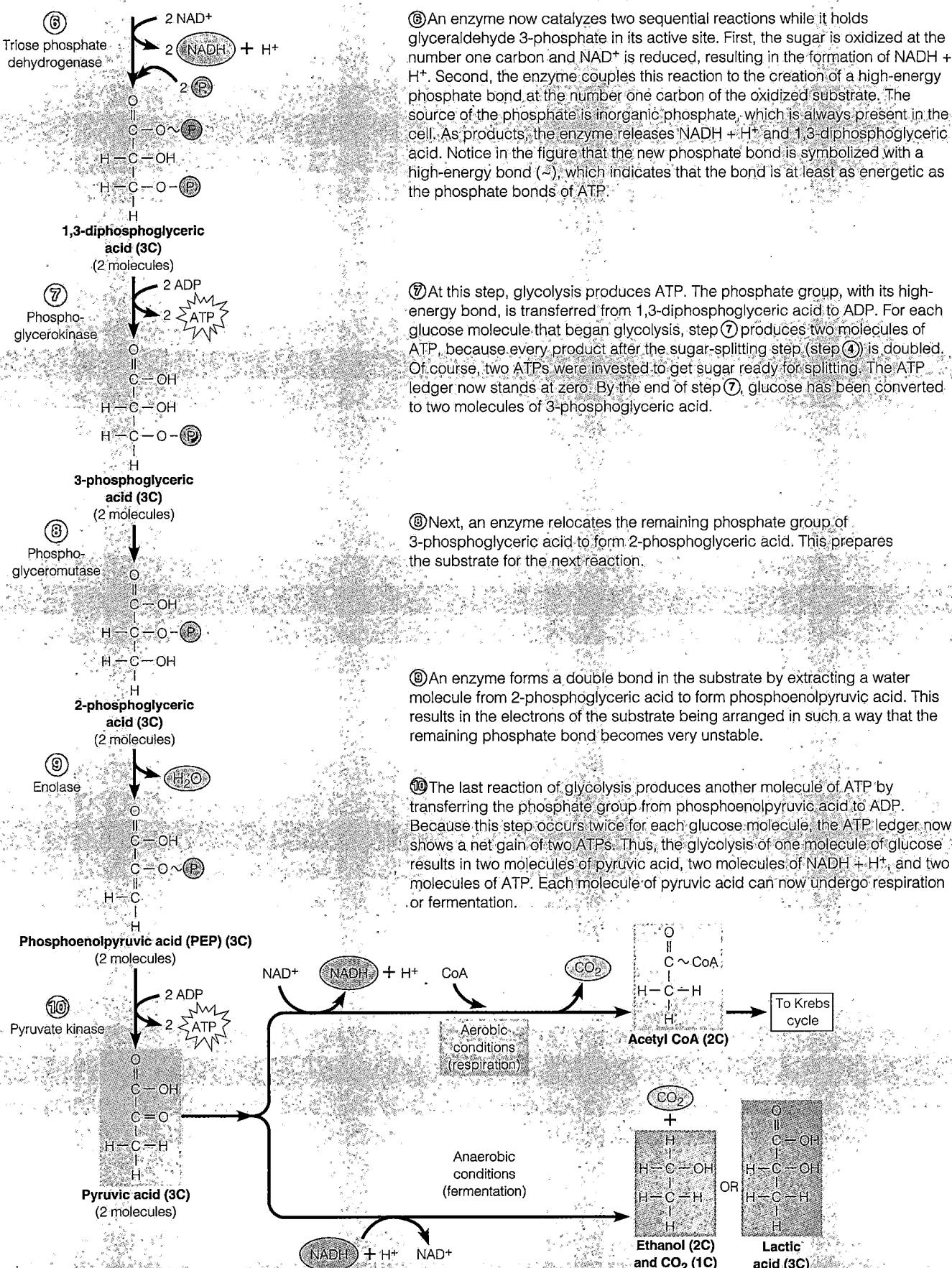

⑥ Triose phosphate dehydrogenase

2 NAD⁺

2 NADH + H⁺

2 P

1,3-diphosphoglyceric acid (3C)
(2 molecules)

⑥ An enzyme now catalyzes two sequential reactions while it holds glyceraldehyde 3-phosphate in its active site. First, the sugar is oxidized at the number one carbon and NAD⁺ is reduced, resulting in the formation of NADH + H⁺. Second, the enzyme couples this reaction to the creation of a high-energy phosphate bond at the number one carbon of the oxidized substrate. The source of the phosphate is inorganic phosphate, which is always present in the cell. As products, the enzyme releases NADH + H⁺ and 1,3-diphosphoglyceric acid. Notice in the figure that the new phosphate bond is symbolized with a high-energy bond (~), which indicates that the bond is at least as energetic as the phosphate bonds of ATP.

⑦ Phospho-glycerokinase

2 ADP

2 ATP

3-phosphoglyceric acid (3C)
(2 molecules)

⑦ At this step, glycolysis produces ATP. The phosphate group, with its high-energy bond, is transferred from 1,3-diphosphoglyceric acid to ADP. For each glucose molecule that began glycolysis, step ⑦ produces two molecules of ATP, because every product after the sugar-splitting step (step ④) is doubled. Of course, two ATPs were invested to get sugar ready for splitting. The ATP ledger now stands at zero. By the end of step ⑦, glucose has been converted to two molecules of 3-phosphoglyceric acid.

⑧ Phospho-glyceromutase

2-phosphoglyceric acid (3C)
(2 molecules)

⑧ Next, an enzyme relocates the remaining phosphate group of 3-phosphoglyceric acid to form 2-phosphoglyceric acid. This prepares the substrate for the next reaction.

⑨ Enolase

H₂O

Phosphoenolpyruvic acid (PEP) (3C)
(2 molecules)

⑨ An enzyme forms a double bond in the substrate by extracting a water molecule from 2-phosphoglyceric acid to form phosphoenolpyruvic acid. This results in the electrons of the substrate being arranged in such a way that the remaining phosphate bond becomes very unstable.

⑩ Pyruvate kinase

2 ADP

2 ATP

Pyruvic acid (3C)
(2 molecules)

⑩ The last reaction of glycolysis produces another molecule of ATP by transferring the phosphate group from phosphoenolpyruvic acid to ADP. Because this step occurs twice for each glucose molecule, the ATP ledger now shows a net gain of two ATPs. Thus, the glycolysis of one molecule of glucose results in two molecules of pyruvic acid, two molecules of NADH + H⁺, and two molecules of ATP. Each molecule of pyruvic acid can now undergo respiration or fermentation.

NAD⁺   NADH + H⁺   CoA   CO₂

Aerobic conditions (respiration)

Acetyl CoA (2C)   To Krebs cycle

Anaerobic conditions (fermentation)

CO₂ +

Ethanol (2C) and CO₂ (1C)   OR   **Lactic acid (3C)**

NADH + H⁺   NAD⁺

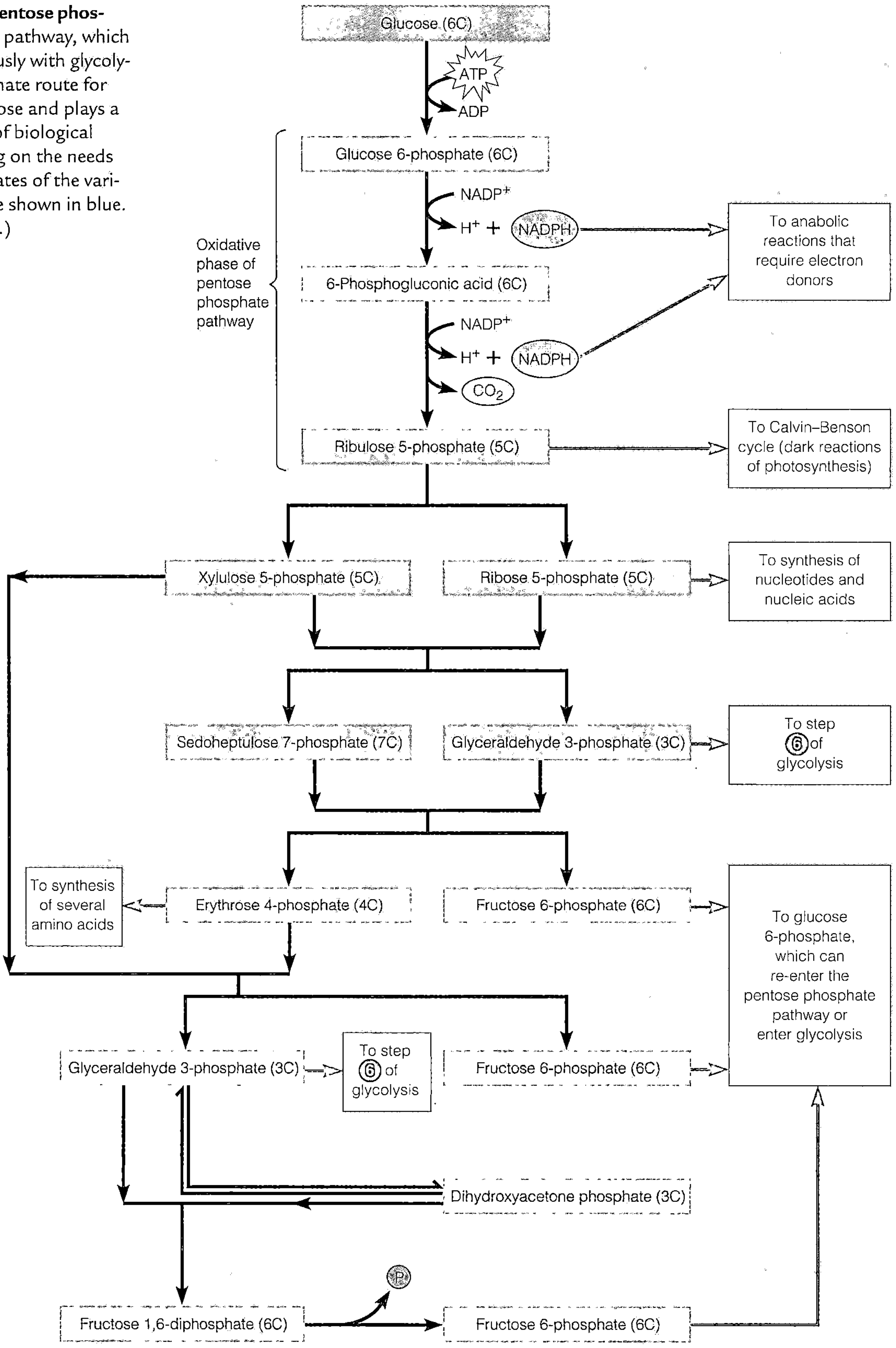

FIGURE C.3 The pentose phosphate pathway. This pathway, which operates simultaneously with glycolysis, provides an alternate route for the oxidation of glucose and plays a role in the synthesis of biological molecules, depending on the needs of the cell. Possible fates of the various intermediates are shown in blue. (See Chapter 5, p. 125.)

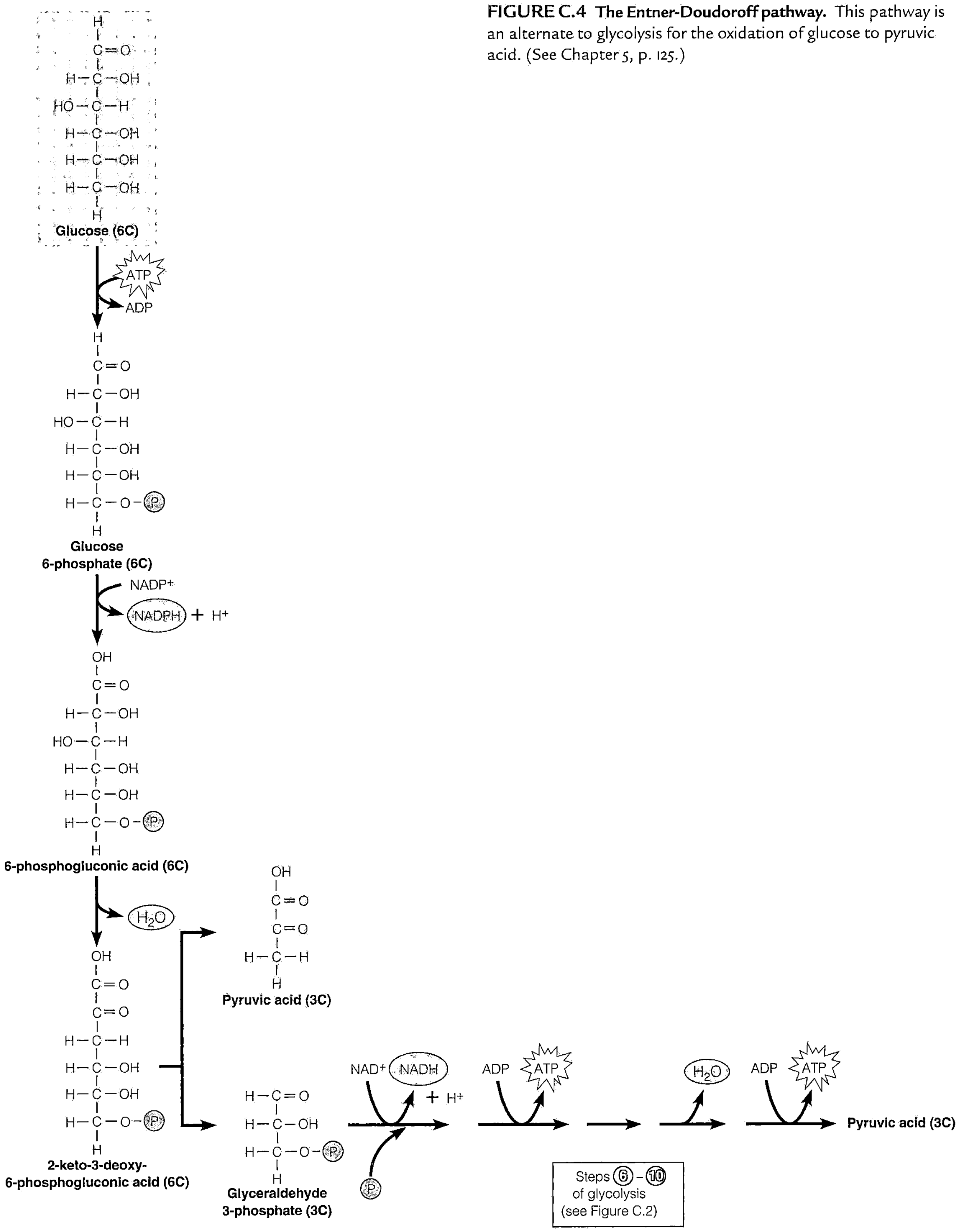

FIGURE C.4 The Entner-Doudoroff pathway. This pathway is an alternate to glycolysis for the oxidation of glucose to pyruvic acid. (See Chapter 5, p. 125.)

FIGURE C.5 The Krebs cycle.

(See Figure 5.12, p. 128, for a simplified version.)

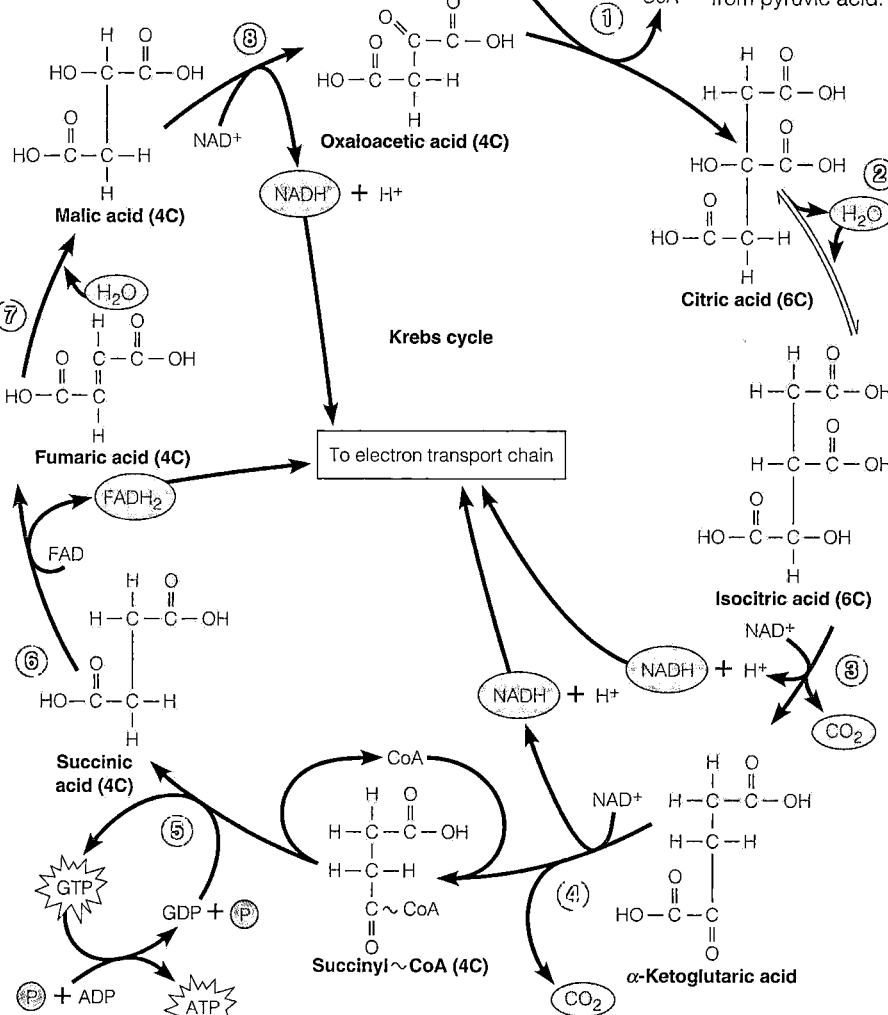

① Acetyl CoA adds its two-carbon acetyl fragment (*pink*) to oxaloacetic acid, a four-carbon compound. The unstable bond of acetyl CoA is broken as oxaloacetic acid displaces the coenzyme and attaches to the acetyl group. The product is the six-carbon citric acid. CoA is then free to prime another two-carbon fragment derived from pyruvic acid.

⑧ The last oxidative step produces another molecule of NADH + H$^+$ and regenerates oxaloacetic acid, which accepts a two-carbon fragment from acetyl CoA for another turn of the cycle.

② A molecule of water is removed, and another is added back. The net result is the conversion of citric acid to its isomer, isocitric acid.

⑦ Bonds in the substrate are rearranged in this step by the addition of a water molecule.

⑥ In another oxidative step, two hydrogens are transferred to FAD to form FADH$_2$. The function of this coenzyme is similar to that of NADH + H$^+$, but FADH$_2$ stores less energy.

③ The substrate loses a CO$_2$ molecule (*gray*), and the remaining five-carbon compound is oxidized, reducing NAD$^+$ to NADH + H$^+$.

⑤ Substrate-level phosphorylation occurs in this step. CoA is displaced by a phosphate group, which is then transferred to GDP to form guanosine triphosphate (GTP). GTP is similar to ATP, which is formed when GTP donates a phosphate group to ADP.

④ CO$_2$ (*gray*) is lost; the remaining four-carbon compound is oxidized by the transfer of electrons to NAD$^+$ to form NADH + H$^+$ and is then attached to CoA by an unstable bond.

Appendix D

Exponents, Exponential Logarithms, and Generation Time

Exponents and Exponential Notation

Very large and very small numbers, such as 4,650,000,000 and 0.00000032, are cumbersome to work with. It is more convenient to express such numbers in exponential notation—that is, as a power of 10. For example, 4.65×10^9 is in standard exponential notation, or **scientific notation:** 4.65 is the *coefficient,* and 9 is the power or *exponent.* In standard exponential notation, the coefficient is always a number between 1 and 10, and the exponent can be positive or negative.

To change a number into exponential notation, follow two steps. First, determine the coefficient by moving the decimal point so there is only one nonzero digit to the left of it. For example,

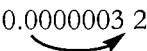

$$0.0000003\ 2$$

The coefficient is 3.2. Second, determine the exponent by counting the number of places you moved the decimal point. If you moved it to the left, the exponent is positive. If you moved it to the right, the exponent is negative. In the example, you moved the decimal point seven places to the right, so the exponent is −7. Thus

$$0.00000032 = 3.2 \times 10^{-7}$$

Now suppose you are working with a larger number instead of a very small number. The same rules apply, but the exponential value will be positive rather than negative. For example,

$$4,650,000,000 = 4.65 \times 10^{+9}$$
$$= 4.65 \times 10^9$$

To multiply numbers written in exponential notation, multiply the coefficients and *add* the exponents. For example,

$$(3 \times 10^4) \times (2 \times 10^3) =$$
$$(3 \times 2) \times (10^{4+3}) = 6 \times 10^7$$

To divide, divide the coefficient and *subtract* the exponents. For example,

$$\frac{3 \times 10^4}{2 \times 10^3} = \frac{3}{2} \times 10^{4-3} = 1.5 \times 10^1$$

Microbiologists use exponential notation in many situations. For instance, exponential notation is used to describe the number of microorganisms in a population. Such numbers are often very large (see Chapter 6). Another application of exponential notation is to express concentrations of chemicals in a solution—chemicals such as media components (Chapter 6), disinfectants (Chapter 7), or antibiotics (Chapter 20). Such numbers are often very small. Converting from one unit of measurement to another in the metric system requires multiplying or dividing by a power of 10, which is easiest to carry out in exponential notation.

Logarithms

A **logarithm (log)** is the power to which a base number is raised to produce a given number. Usually we work with logarithms to the base 10, abbreviated $\log_{10}$. The first step in finding the $\log_{10}$ of a number is to write the number in standard exponential notation. If the coefficient is exactly 1, the $\log_{10}$ is simply equal to the exponent. For example

$$\log_{10} 0.00001 = \log_{10}(1 \times 10^{-5})$$
$$= -5$$

If the coefficient is not 1, as is often the case, the logarithm function on a calculator must be used to determine the logarithm.

Microbiologists use logs for calculating pH levels and for graphing the growth of microbial populations in culture (see Chapter 6).

Calculating Generation Time

As a cell divides, the population increases exponentially. Numerically this is equal to 2 (because one cell divides into two) raised to the number of times the cell divided (generations);

$$2^{\text{number of generations}}$$

To calculate the final concentration of cells:

Initial number of cells $\times\ 2^{\text{number of generations}}\ =$

number of cells

For example if 5 cells were allowed to divide 9 times, this would result in

$$5 \times 2^9 = 2560 \text{ cells}$$

To calculate the number of generations a culture has undergone, cell numbers must be converted to logarithms. Standard logarithm values are based on 10. The log of 2 (0.301) is used because one cell divides into two.

$$\text{No. of generations} = \frac{\log \text{number of cells (end)} - \log \text{number of cells (beginning)}}{0.301}$$

To calculate the generation of time for a population:

$$\frac{60 \text{ min} \times \text{hours}}{\text{number of generations}} = \text{minutes/generation}$$

As an example, we will calculate the generation time if 100 bacterial cells growing for 5 hours produced 1,720,320 cells:

$$\frac{\log 1,720,320 - \log 100}{0.301} = 14 \text{ generations}$$

$$\frac{60 \text{ min} \times 5 \text{ hr}}{14 \text{ generation}} = 21 \text{ minutes/generation}$$

A practical application of the calculation is determining the effect of a newly developed food preservative on the culture. Suppose 900 of the same species were grown under the same conditions as the previous example, except that the preservative was added to the culture medium. After 15 hours, there were 3,276,800 cells. Calculate the generation time, and decide whether the preservative inhibited growth.

Answer: 75 min/generation. The preservative did inhibit growth.

Appendix E

Pronunciation of Scientific Names

Rules of Pronunciation

The easiest way to learn new material is to talk about it, and that requires saying scientific names. Scientific names may look difficult at first glance, but keep in mind that generally every *syllable* is pronounced. The primary requirement in saying a scientific name is to communicate it.

The rules for the pronunciation of scientific names depend, in part, on the derivation of the root word and its vowel sounds. We have provided some general guidelines here. Pronunciations frequently do not follow the rules because a common usage has become "accepted," or the derivation of the name cannot be determined. And for many scientific names there are alternative correct pronunciations.

Vowels

Pronounce all the vowels in scientific names. Two vowels written together and pronounced as one sound are called a *diphthong* (for example, the *ou* in *"sound"*). A special comment is needed about the pronunciation of the vowel endings -*i* and -*ae:* There are two alternative ways to pronounce each of these. In this book, we usually give the pronunciation of a long *e* (*ē*) to the -*i* ending and a long *i* (*ī*) to the -*ae* ending. However, the reverse pronunciations are also correct and in some cases are preferred. For example, *coli* is usually pronounced kō′lī.

Consonants

When *c* or *g* is followed by *ae, e, oe, i,* or *y,* it has a soft sound. When *c* or *g* is followed by *a, o, oi,* or *u,* it has a hard sound. When a double *c* is followed by *e, i,* or *y,* it is pronounced as *ks* (e.g., cocci).

Accent

The accented syllable is usually the next-to-last or third-to-last syllable.

1. The accent is on the next-to-last syllable:
   a. When the name contains only two syllables. Example: pes′tis.
   b. When the next-to-last syllable is a diphthong. Example: a-kan-thä-mē′bä.
   c. When the vowel of the next-to-last syllable is long. Example: tre-pō-nē′mä. The vowel in the next-to-last syllable is long in words ending in the following suffixes:

Suffix	Example
-ales	Orders such as Eubacteriales
-ina	Sarcina
-anus, anum	pasteurianum
-uta	diminuta

   d. When the word ends in one of the following suffixes:

Suffix	Example
-atus, atum	caudatum
-ella	Salmonella

2. The accent is on the third-to-last syllable in family names. Families end in -*aceae,* which is always pronounced -ā′sē-ē.

Pronunciation of Microorganisms in This Text

Pronunciation key:

a	hat	ē	see	o	hot	th	thin
ā	age	è	term	ō	go	u	cup
ã	care	g	go	ô	order	ù	put
ä	father	i	sit	oi	oil	ü	rule
ch	child	ī	ice	ou	out	ū	use
e	let	ng	long	sh	she	zh	seizure

Acanthamoeba polyphaga a-kan-thä-mē′bä pol′if-ä-gä
Acetobacter a-sē′tō-bak-tėr
Achinanthes minutissima ā′kin-an-thēs min′ü-tis-sē-mä
Acinetobacter a-si-ne′tō-bak-tėr
Actinomyces israelii ak-tin-ō-mī′sēs is-rā′lē-ē
Aedes aegypti ā′ē-dēz ē-jip′tē
A. albopictus al-bō-pik′tus
Aeromonas hydrophilia ār′ō-mō-nas hī′dro-fil-ē-ä
Afipia felis ä-fi′pē-ä fē′lis
Agrobacterium tumefaciens ag′rō-bak-ti′rē-um tü′me-fāsh-enz
Ajellomyces ä-jel-lō-mī′sēs
Alcaligenes al′kä-li-gen-ēs
Alexandrium äl-eg-zan′drē-um
Amanita phalloides am-an-ī′ta fal-loi′dēz
Anabaena an-ä-bē′nä
Ancylostoma an-sil-ō′stō-mä
Anopheles an-of′e-l-ēz
Aquaspirillum bengal ä-kwä-spī-ril′lum ben′gal
A. graniferum gra-ni′fėr-um
A. magnetotacticum mag-ne-tō-tak′ti-kum
Arthroderma är-thrō-dėr′mä
Ascaris lumbricoides as′kar-is lum-bri-koi′dēz
Ashbya gossypii ash′bē-ä gos-sip′ē-ē
Aspergillus flavus a-spėr-jil′lus flā′vus
A. fumigatus fü-mi-gä′tus
A. niger nī′jėr
A. oryzae ô′ri-zī
A. soyae soi′ī
Azolla ā-zō′lä
Azomonas ā-zō-mō′nas
Azospirillum ā-zō-spī′ril-lum
Azotobacter ä-zo′tō-bak-tėr
Babesia microti ba-bē′sē-ä mī-krō′tē
Bacillus anthracis bä-sil′lus an-thrā′sis
B. brevis brev′is
B. cereus se′rē-us
B. circulans sėr′ku-lans
B. coagulans kō-ag′ū-lanz
B. licheniformis lī-ken-i-fôr′mis
B. megaterium meg-ä-tėr′ē-um
B. polymyxa po-lē-miks′ä
B. sphaericus sfe′ri-kus
B. stearothermophilus ste-rō-thėr-mä′fil-us
B. subtilis su′til-us
B. thuringiensis thúr-in-jē-en′sis
Bacteroides hypermegas bak-tė-roi′dēz hī-pėr′meg-äs
Balantidium coli bal-an-tid′ē-um kō′lī (or kō′lē)
Bartonella henselae bär′tō-nel-lä hen′sel-ī
Bdellovibrio bacteriovorus del-lō-vib′rē-ō bak-tė-rē-o′vô-rus
Beauveria bō-vär′ē-a
Beggiatoa bej′jē-ä-tō-ä
Beijerinckia bī-yė-rink′ē-ä
Bifidobacterium globosum bī-fi-dō-bak-ti′rē-um glob-ō′sum
B. pseudolongum sū-dō-lông′um
Blastomyces dermatitidis blas-tō-mī′sēz dėr-mä-tit′i-dis
Blattabacterium blat-tä-bak-ti′rē-um
Boletus edulis bō-lē′tus e′dū-lis
Bordetella pertussis bòr-de-tel′lä pėr-tus′sis
Borrelia burgdorferi bòr-rel′ē-ä burg-dôr′fėr-ē
Bradyrhizobium brad-ē-rī-zō′bē-um
Brucella abortus brü′sel-lä ä-bôr′tus
B. melitensis me-li-ten′sis
B. suis sü′is
Burkholderia berk′hōld-ér-ē-ä
Byssochlamys fulva bis-sō-klam′is fül′vä
Campylobacter fetus kam′pi-lō-bak-tėr fē′tus

C. jejuni jē-jū′nē
Candida albicans kan′did-ä al′bi-kanz
C. krusei krūs′ä-ē
C. oleophila ō-lē-of′il-ä
C. utilis ū′til-is
Canis familiaris kānis fa-mil′ē-är-is
Carpenteles kär-pen′tel-ēz
Caulobacter kô-lō-bak′tėr
Cephalosporium sef-ä-lō-spô′rē-um
Ceratocystis ulmi sē-rä-tō-sis′tis ul′mē
Chilomastix kē′lō-ma-sticks
Chlamydia pneumoniae kla-mi′dē-ä nü-mō′nē-ī
C. psittaci sit′tä-sē
C. trachomatis trä-kō′mä-tis
Chlamydomonas klam-i-dō-mō′näs
Chlorobium klô-rō′bē-um
Chloroflexus klô-rō-flex′us
Chromatium vinosum krō-mā′tē-um vi-nō′sum
Chroococcus turgidus krō-ō-kok′kus tėr′ji-dus
Chrysops krī′sops
Citrobacter sit′rō-bak-tėr
Claviceps purpurea kla′vi-seps púr-pú-rē′ä
Clonorchis sinensis klo-nôr′kis si-nen′sis
Clostridium acetobutylicum klôs-tri′dē-um a-sē-tō-bū-til′li-kum
C. botulinum bo-tū-lī′num
C. butyricum bü-ti′ri-kum
C. difficile dif′fi-sil-ē
C. pasteurianum pas-tyėr-ē-ā′num
C. perfringens pėr-frin′jens
C. sporogenes spô-rä′jen-ēz
C. tetani te′tan-ē
C. thermosaccharolyticum thėr-mō-sak-kär-ō-li′ti-kum
Coccidioides immitis kok-sid-ē-oi′dēz im′mi-tis
Corynebacterium diphtheriae kôr′ī-nē-bak-ti-rē-um dif-thi′rē-ī
C. glutamicum glü-tam′i-kum
C. xerosis ze-rō′sis
Coxiella burnetii käks′ē-el-lä bér-ne′tē-ē
Cristispira pectinis kris-tē-spī′rä pek′tin-is
Crucibulum vulgare krü-si-bū′lum vul′gär-ē
Cryphonectria parasitica kri-fō-nek′trē-ä par-ä-si′ti-kä
Cryptococcus neoformans krip′tō-kok-kus nē-ō-fôr′manz
Cryptosporidium parvum krip′tō-spô-ri-dē-um pär′vum
Culex kū′leks
Culiseta kū-li′se-tä
Cyclospora cayetanensis sī′klō-spô-rä kī′ē-tan-en-sis
Cytophaga sī-täf′äg-ä
Daptobacter dap′to-bak-tėr
Dermacentor andersoni dėr-mä-sen′tôr an-dėr-sōn′ē
D. variabilis vär-ē-a′bil-is
Desulfotomaculum nigrificans dē′sul-fō-to-ma-kū-lum nī′gri-fi kans
Desulfovibrio desulfuricans dē′sul-fō-vib-rē-ō dē-sul-fėr′i-kans
Didinium nasutum dī-di′nē-um nä-sūt′um
Diphyllobothrium latum dī-fil-lō-bo′thrē-um lā′tum
Dracunculus medininsis dra-kun′ku-lus med-in′in-sis
Echinococcus granulosus ē-kīn-ō-kok′kus gra-nū-lō′sus
Ectothiorhodospira mobilis ek′tō-thī-ō-rō-dō-spī-rä mō′bil-is
Emericella em′ér-ē-sel-lä
Emmonsiella em′mon-sē-el-lä
Entamoeba histolytica en-tä-mē′bä his-tō-li′ti-kä
Enterobacter aerogenes en-te-rō-bak′tėr ā-rä′jen-ēz
E. cloacae klō-ā′kī
Enterobius vermicularis en-te-rō′bē-us ver-mi-kū-lar′is
Enterococcus faecalis en-te-rō-kok′kus fē-kä′lis
Entomorphaga en-tō-môr′fag-ä
Epidermophyton ep-i-dėr-mō-fī′ton
Epulopiscium ep′ū-lō-pis-sē-um

Ehrlichia èr'lik-ē-ä
Erwinia dissolvens èr-wi'nē-ä dis-solv'ens
Erysipelothrix rhusiopathiae är-i-si-pel'ō-thrix rus-ē-ō-path'ē-ī
Erysiphe grammis är'i-sīf gram'mis
Escherichia coli esh-èr-i'kē-ä kō'lī (or kō'lē)
Euglena ū-glē'nä
Eupenicillium ū-pen-i-sil'lē-um
Eurotium yèr-ō'tē-um
Filobasidiella fi-lō-ba-si-dē-el'lä
Fonsecaea pedrosoi fon-se'kē-ä pe-drō'sō-ē
Francisella tularensis fran'sis-el-lä tü'lä-ren-sis
Frankia frank'ē-ä
Fusobacterium fü-sō-bak-ti'rē-um
Gambierdiscus toxicus gam'bē-èr-dis-kus toks'i-kus
Gardnerella vaginalis gärd-nè-rel'lä va-jin-al'is
Gelidium jel-id'ē-um
Geotrichum jē-ō-trik'um
Giardia lamblia jē-är'dē-ä lam'lē-ä
Gloeocapsa glē-ō-kap'sä
Glomus fasciculatus glo'mus fa-sik-ū-lä'tus
Glossina gläs-sē'nä
Gluconobacter glü'kon-ō-bak-tèr
Gymnoascus jim-nō-as'kus
Gymnodynium breve jim-nō-din'ē-um brev'ē
Haemophilus ducreyi hē-mä'fil-us dü-krā'ē
Haloarcula hā'lō-är-kū-lä
Halobacterium halobium ha-lō-bak-ti'rē-um hal-ō'bē-um
Halococcus hā'lō kok-kus
Helicobacter pylori hē'lik-ō-bak-tèr pī'lô-rē
H. influenzae in-flü-en'zī
H. vaginalis va-jin-al'is
Histoplasma capsulatum his-tō-plaz'mä kap-su-lä'tum
Holospora hō-lo'spô-rä
Homo sapiens hō'mō sā'pē-ens
Hydrogenomonas hī-drō-je-nō-mō'näs
Hyphomicrobium hī-fō-mī-krō'bē-um
Isospora ī-so'spô-rä
Ixodes scapularis iks-ō'dēs scap-ū-lär'is
I. pacificus pas-i'fi-kus
Klebsiella pneumoniae kleb-sē-el'lä nü-mō'nē-ī
Lactobacillus bulgaricus lak-tō-bä-sil'lus bul-gā'ri-kus
L. delbrueckii del-brük'ē-ē
L. leichmannii līk-man'nē-ē
L. plantarum plan-tä'rum
L. sanfrancisco sän-fran-sis'kō
Lactococcus lactis lak-tō-kok'kus lak'tis
Laminaria japonica lam'i-när-e-ä ja-pon'i-kä
Legionella pneumophila lē-jä-nel'lä nü-mō'fi-lä
Leishmania braziliensis lish'mä-nē-ä brä-sil'ē-en-sis
L. donovani don'ō-van-ē
L. tropica trop'i-kä
Leptospira interrogans lep-tō-spī'rä in-tèr'rä-ganz
Leuconostoc mesenteroides lü-kō-nos'tok mes-en-ter-oi'dēz
Licmorphora lik-môr'fôr-ä
Listeria monocytogenes lis-te'rē-ä mo-nō-sī-tô'je-nēz
Meniscus mē-nis'kus
Methanobacterium meth-a-nō-bak-tèr'ē-um
Methanosarcina meth-a-nō-sär'sī-nä
Methylophilus methylotrophus meth-i-lo'fi-lus meth-i-lō-trōf'us
Microcladia mī-krō-klād'ē-ä
Micrococcus luteus mī-krō-kok'kus lū'tē-us
Micromonospora purpureae mī-krō-mo-nä'spô-rä pùr-pú-rē'ī
Microsporum gypseum mī-krō-spô'rum jip'sē-um
Mixotricha mix-ō-trik'ä
Moraxella catarrhalis mô-raks-el'lä ka-tär'al-is
M. lacunata la-kü-nä'tä

M. osloensis os'lō-en-sis
Mucor rouxii mū-kôr rō'ē-ē
Mycobacterium avium-intracellulare mī-kō-bak-ti'rē-um ā'vē-um-in'trä-cel-ū-lä-rē
M. bovis bō'vis
M. leprae lep'rī
M. smegmatis smeg-ma'tis
M. tuberculosis tü-bér-kū-lō'sis
M. ulcerans ul'sèr-anz
Mycoplasma hominis mī-kō-plaz'mä ho'mi-nis
M. pneumoniae nu-mō'nē-ī
Myxococcus fulvus micks-ō-kok'kus ful'vus
Naegleria fowleri nī-gle'rē-ä fou'lèr-ē
Nannizia nan'nē-zē-ä
Necator americanus ne-kā'tôr ä-me-ri-ka'nus
Neisseria gonorrhoeae nī-se'rē-ä go-nôr-rē'ī
N. meningitidis me-nin-ji'ti-dis
Nereocystis nē-rē-ō-sis'tis
Nitrobacter nī-trō-bak'tèr
Nitrosomonas nī-trō-sō-mō'näs
Nocardia asteroides nō-kär'dē-ä as-tèr-oi'dēz
Oocystis ō-ō-sis'tis
Opisthorchis sinensis [opisthorchis] si-nen'sis
Ornithodorus ôr-nith-ō'dô-rus
Paracoccus denitrificans pär-ä-kok'kus dē-nī-tri'fi-kanz
Paragonimus westermani pär-ä-gōn'e-mus we-stèr-ma'nē
Paramecium multimicronucleatum pär-ä-mē'sē-um mul-tē-mī-krō-nü-klē-ä'tum
Pasteurella haemolytica pas-tyèr-el'lä hēm-ō-lit'i-kä
P. multocida mul-tō'si-dä
Pediculus humanus corporis ped-ik'ū-lus hü'ma-nus kôr'pô-ris
Pediococcus cerevisiae pe-dē-ō-kok'kus se-ri-vis'ē-ī
Penicillium chrysogenum pen-i-sil'lē-um krī-so'gen-um
P. griseofulvum gri-sē-ō-fül'vum
P. notatum nō-tä'tum
Peridinium per-i-din'ē-um
Petriellidium pet-rē-el-li'dē-um
Pfiesteria fes-tèr'ē-ä
Phormidium luridum fôr-mi'dē-um ler'i-dum
Phytophthora infestans fī-tof'thô-rä in-fes'tans
Pichia pik'ē-ä
Pityrosporum ovale pit-i-ros'pô-rum ovale
Plasmodium falciparum plaz-mō'dē-um fal-sip'är-um
P. malariae mä-lä'rē-ī
P. ovale ō-vä'lē
P. vivax vī'vaks
Plesiomonas shigelloides ple-sē-ō-mō'nas shi-gel-loi'des
Pneumocystis carinii nü-mō-sis'tis kär-i'nē-ī (or kär-i'nē-ē)
Porphyromonas pôr'fī-rō-mō-nas
Prevotella intermedia prev'ō-tel-la in'tér-mē-dē-ä
Propionibacterium acnes prō-pē-on'ē-bak-ti-rē-um ak'nēz
P. freudenreichii froi-den-rīk'ē-ē
Proteus mirabilis prō'tē-us mi-ra'bi-lis
P. vulgaris vul-ga'ris
Prototheca prō-tō-thā'kä
Pseudomonas aeruginosa sū-dō-mō'nas ā-rü-ji-nō'sä
P. andropogonis än'dro-po-gō-nis
P. cepacia se-pā'sē-ä
P. fluorescens flôr-es'ens
P. haemolytica hē-mō-lit'i-kä
P. multocida mul'tō-sid-ä
P. syringae sèr-in'jī
Pyridictum abyssi pir-i'dik-tum a-bis'sē
Quercus kwer'kus
Rhizobium japonicum rī-zō'bē-um jap-on'i-kum
R. meliloti mel-i-lot'ē

Rhizopus nigricans   rī'zō-pùs nī'gri-kans
R. oryzae   ô'rī-zī
Rhodococcus bronchialis   rō-dō-kok'kus bron-kē'al-is
R. erythropolis   er-i-throp'ō-lis
Rhodopseudomonas   rō-dō-su-dō-mō'nas
Rhodospirillum rubrum   rō-dō-spī-ril'um rūb'rum
Rickettsia prowazekii   ri-ket'sē-ä prou-wä-ze'kē-ē
R. rickettsii   ri-ket'sē-ē
R. typhi   tī'fē
Riftia   rift'ē-ä
Rosa multiflora   rō-sä mul-ti-flô'rä
Saccharomyces beticus   sak-ä-rō-mī'sēs bet'i-kus
S. bayanus   bā'an-us
S. cerevisiae   se-ri-vis'ē-ī
S. ellipsoideus   ē-lip-soi'dē-us
S. exiguus   egz-ij'ū-us
S. fragilis   fra'jil-is
S. rouxii   rō-ē'ē
S. uvarum   ü-var'um
Salmonella bongori   sal-mōn-el'lä bon'gôr-ē
S. choleraesuis   kol-ėr-ä-sü'is
S. enterica   en-ter'i-kä
S. enteritidis   en-tėr-ī'ti-dis
S. typhi   tī'fē
S. typhimurium   tī-fi-mùr'ē-um
Sarcoptes   sär-kop'tēs
Sargassum   sär-gas'sum
Sartorya   sär-tô'rē-ä
Schistosoma   shis-tō-sō'mä or skis-tō-sō'mä
Schizosaccharomyces   skiz-ō-sak-ä-rō-mī'sēs
Serratia marcescens   ser-rä'tē-ä mär-ses'sens
Shigella boydii   shi-gel'lä boi'dē-ē
S. dysenteriae   dis-en-te'rē-ī
S. flexneri   fleks'nėr-ē
S. sonnei   sōn'ne-ē
Sphaerotilus natans   sfe-rä'ti-lus nä'tans
Spirillum ostreae   spī'ril-lum ost'rä-ī
S. volutans   vō'lū-tans
Spiroplasma   spī-rō-plaz'mä
Spirosoma   spī-rō-sō'mä
Spirulina   spī-rü-lī'nä
Spongomorphora   spon'jō-môr-fô-rä
Sporothrix schenkii   spô-rō'thriks shen'kē-ē
Stachybotrys   stak'ē-bo-tris
Staphylococcus aureus   staf-i-lō-kok'kus ô'rē-us
S. epidermidis   e-pi-der'mi-dis
S. saprophyticus   sap-rō-fit'i-kus
Stella   stel'lä
Stigmatella aurantiaca   stig-mä'tel-lä ô-rän-tē'ä-kä
Streptobacillus   strep-tō-bä-sil'lus
Streptococcus iniae   strep-tō-kok'kus in'ē-ī
S. mutans   mū'tans

S. pneumoniae   nü-mō'nē-ī
S. pyogenes   pī-äj'en-ēz
S. thermophilus   thėr-mo'fil-us
Streptomyces aureofaciens   strep-tō-mī'sēs ô-rē-ō-fa'si-ens
S. erythraeus   ā-rith'rē-us
S. fradiae   frā'dē-ī
S. griseus   gri'sē-us
S. nodosus   nō-dō'sus
S. noursei   nôr'sē-ī
S. olivoreticuli   ō-liv-ō-re-tik'ū-lē
S. venezuelae   ve-ne-zü-e'lī
Sulfolobus   sul'fō-lō-bus
Taenia saginata   te'nē-ä sa-ji-nä'tä
T. solium   sō'lē-um
Talaromyces   ta-lä-rō-mī'sēs
Tetrahymena   tet-rä-hī'me-nä
Thermoactinomyces vulgaris   thėr-mō-ak-tin-ō-mī'sēs vul-ga'ris
Thermoplasma   thėr-mō-plaz'mä
Thermotoga maritima   thėr'mō-tō-gä mar-it'ē-mä
Thiobacillus ferrooxidans   thī-ō-bä-sil'lus fer-rō-oks'i-danz
T. thiooxidans   thī-ō-oks'i-danz
Thiocapsa floridana   thī-ō-kap'sä flôr'i-dä-nä
Thiomargarita namibiensis   thī'ō-mär-gär-ē-tä na'mi-bē-n-sis
Thunnus   tün'nus
Toxoplasma gondii   toks-ō-plaz'mä gon'dē-ē
Trachelomonas   trä-kel-ō-mōn'as
Treponema pallidum   tre-pō-nē'mä pal'li-dum
Triatoma   trī-ä-tō'ma·
Tribonema vulgare   trī'bō-nē-mä vul'gär-ē
Trichinella spiralis   trik-in-el'lä spī-ra'lis
Trichoderma   trik'ō-dėr-mä
Trichomonas vaginalis   trik-ō-mōn'as va-jin-al'is
Trichonympha sphaerica   trik-ō-nimf'ä sfe'ri-kä
Trichophyton   trik-ō-fī'ton
Trichosporon   trik-ō-spôr'on
Trichuris trichiura   trik'ėr-is trik-ē-yėr'a
Tridacna   trī-dak'nä
Tropheryma   trō-fer-ē'mä
Trypanosoma brucei gambiense   tri-pa'nō-sō-mä brüs'ē gam-bē-ens'
T. brucei rhodesiense   rō-dē-sē-ens'
T. cruzi   kruz'ē
Ureaplasma urealyticum   ū-rē-ä-plaz'mä ū-rē-ä-lit'i-kum
Usnea   üs'nē-ä
Veillonella   vī-lo-nel'lä
Vibrio cholerae   vib'rē-ō kol'ėr-ī
V. parahaemolyticus   pa-rä-hē-mō-li'ti-kus
Wuchereria bancrofti   vū-kėr-ār'ē-ä ban-krof'tē
Xenopsylla cheopis   ze-nop'sil-lä kē-ō'pis
Yersinia enterocolitica   yėr-sin'ē-ä en'tėr-ō-kōl-it-ik-ä
Y. pestis   pes'tis
Y. pseudotuberculosis   sü-dō-tü-bėr-kū-lō'sis
Zoogloea   zō'ō-glē-ä

Appendix F

Taxonomic Guide to Diseases

Bacteria and the Diseases They Cause

Proteobacteria

α-Proteobacteria

Cat scratch fever	*Bartonella henselae*	p. 632
Ehrlichiosis	*Ehrlichia* sp.	p. 635
Endemic murine typhus	*Rickettsia typhi*	p. 637
Epidemic typhus	*R. prowazekii*	p. 637
Rocky Mountain spotted fever	*R. rickettsii*	pp. 637–638
Brucellosis	*Brucella* spp.	pp. 630–631

β-Proteobacteria

Gonorrhea	*Neisseria gonorrhoeae*	pp. 725–727
Meningitis	*N. meningitidis*	p. 605
Neonatal gonorrheal ophthalmia	*N. gonorrhoeae*	p. 596
Pelvic inflammatory disease	*N. gonorrhoeae*	p. 728
Nosocomial infections	*Burkholderia* spp.	p. 201
Whooping cough	*Bordetella pertussis*	p. 663

γ-Proteobacteria

Animal bites	*Pasteurella multocida*	p. 632
Bacillary dysentery	*Shigella* spp.	pp. 691–692
Epiglottis	*Haemophilus influenzae*	p. 658
Meningitis	*H. influenzae*	pp. 604–605
Otitis media	*H. influenzae*	p. 661
Pneumonia	*H. influenzae*	p. 669
Chancroid	*H. ducreyi*	pp. 731–732
Cholera	*Vibrio cholerae*	pp. 695–696
Gastroenteritis	*V. parahaemolyticus*	p. 696
Gastroenteritis	*V. vulnificus*	p. 697
Cystitis	*Escherichia coli*	p. 723
Gastroenteritis	*E. coli*	p. 697
Pyelonephritis	*E. coli*	p. 723
Pneumonia	*Klebsiella pneumoniae*	p. 671
Dermatitis	*Pseudomonas aeruginosa*	p. 586
Otits externa	*P. aeruginosa*	p. 586
Legionellosis	*Legionella pneumophila*	pp. 669–670
Plague	*Yersinia pestis*	pp. 633–634
Gastroenteritis	*Y. enterocolitica*	p. 699
Pneumonia	*Moraxella catarrhalis*	p. 671
Q-fever	*Coxiella burnetti*	p. 671
Salmonellosis	*Salmonella enterica*	pp. 692–694
Typhoid fever	*S. enterica typhi*	p. 695
Tularemia	*Francisella tularensis*	pp. 629–630

ε-Proteobacteria

Gastroenteritis	*Campylobacter jejuni*	p. 697
Gastritis	*Helicobacter pylori*	pp. 697–699
Peptic ulcers	*H. pylori*	pp. 697–699

Clostridia

Tetanus	*Clostridium tetani*	pp. 607–608
Gangrene	*C. perfringens*	pp. 631–632
Gastroenteritis	*C. perfringens*	p. 699
Botulism	*C. botulinum*	pp. 608–610

Mollicutes

Pneumonia	*Mycoplasma pneumoniae*	p. 669
Urethritis	*M. hominis*	p. 728
Urethritis	*Ureaplasma ureolyticum*	p. 728

Bacilli

Anthrax	*Bacillus anthracis*	p. 631
Gastroenteritis	*B. cereus*	p. 699
Listeriosis	*Listeria monocytogenes*	p. 606
Acute bacterial endocarditis	*Staphylococcus aureus*	p. 628
Follicultis	*S. aureus*	p. 583
Food poisoning	*S. aureus*	pp. 690–691
Impetigo	*S. aureus*	p. 583
Otitis media	*S. aureus*	p. 661
Scalded skin syndrome	*S. aureus*	p. 583
Toxic shock syndrome	*S. aureus*	p. 584
Cystitis	*S. saprophyticus*	p. 723
Erysipelas	*Streptococcus pyogenes*	p. 585
Impetigo	*S. pyogenes*	p. 585
Meningitis	*S. pyogenes*	p. 603
Necrotizing fasciitis	*S. pyogenes*	p. 586
Puerpual sepsis	*S. pyogenes*	pp. 627–628
Rheumatic fever	*S. pyogenes*	p. 629
Scarlet fever	*S. pyogenes*	pp. 659–660
Strep throat	*S. pyogenes*	p. 659
Streptococcal toxic shock syndrome	*S. pyogenes*	p. 423
Meningitis	*S. pneumoniae*	pp. 605–606
Otitis media	*S. pneumoniae*	p. 661
Pneumonia	*S. pneumoniae*	pp. 667–669
Dental caries	*S. mutans*	pp. 686–688
Subacute bacterial endocarditis	*α*-hemolytic streptococci	p. 628

Actinobacteria

Acne	*Propionibacterium acnes*	p. 587
Buruli Ulcer	*Mycobacterium ulcerans*	p. 584
Diphtheria	*Corynebacterium diphtheriae*	pp. 660–661
Leprosy	*Myobacterium leprae*	pp. 610–611
Tuberculosis	*M. tuberculosis*	pp. 663–667
Mycetoma	*Nocardia asteroides*	p. 325
Vaginosis	*Gardnerella vaginalis*	p. 732

Chlamydiae

Inclusion conjunctivitis	*Chlamydia trachomatis*	p. 596
Lymphogranuloma venereum	*C. trachomatis*	p. 731
Pelvic inflammatory disease	*C. trachomatis*	p. 728
Trachoma	*C. trachomatis*	p. 596
Urethritis	*C. trachomatis*	p. 727
Pneumonia	*C. pneumoniae*	pp. 670–671
Psittacosis	*C. psittaci*	p. 670

Spirochetes

Leptospirosis	*Leptospira interrogans*	pp. 723–725
Relapsing fever	*Borrelia* spp.	p. 634
Lyme Disease	*B. burgdorferi*	pp. 634–635
Syphilis	*Treponema pallidum*	pp. 728–731

Bacteroidetes

Periodontal disease	*Porphyromonas* spp.	p. 689
Acute necrotizing gingivitis	*Prevotella intermedia*	p. 689

Fungi and the Diseases They Cause

Ascomycetes

Aspergillosis	*Aspergillus fumigatus*	p. 678
Blastomycosis	*Blastomyces dermatidis*	p. 678
Histoplasmosis	*Histoplasma capsulatum*	pp. 674–675
Ringworm	*Microsporum, Trichophyton*	pp. 593–594
Mycotoxins		pp. 449, 705

Anamorphs

Candidiasis	*Candida albicans*	pp. 594–595, 734–735
Coccidioidomycosis	*Coccidoides immitis*	pp. 675–676
Pneumonia	*Pneumocystis carinii*	pp. 676–677
Sporotrichosis	*Sporothrix schenckii*	p. 594

Basidiomycetes

Meningitis	*Cryptococcus neoformans*	p. 616
Mycotoxins		p. 449

Protozoa and the Diseases They Cause

Archaezoa

Giardiasis	*Giardia lamblia*	p. 706
Trichomoniasis	*Trichomonas vaginalis*	p. 735

Apicomplexa

Babesiosis	*Babesia microti*	p. 647
Cryptosporidiosis	*Cryptosporidium parvum*	pp. 706–707
Cyclospora infection	*Cyclospora cayetanensis*	p. 707
Malaria	*Plasmodium* spp.	pp. 642–645
Toxoplasmosis	*Taxoplasma gondii*	p. 642

Rhizopoda

Amoebic dysentery	*Entamoeba histolytica*	p. 707
Keratitis	*Acanthamoeba* spp.	p. 597

Dinoflagellates

Paralytic shellfish poisoning	*Alexandrium* spp.	p. 347
Possible estuary-associated syndrome	*Pfiesteria* sp.	p. 348

Euglenzoa

African trypanosomiasis	*Trypanosoma brucei*	pp. 617–618
American trypanosomiasis	*T. cruzi*	pp. 641–642
Leishmaniasis	*Leishmania* spp.	pp. 645–647
Meningoencephalitis	*Naegleria fowleri*	p. 618

Helminths and Diseases They Cause

Platyhelminths

Hydatid disease	*Echinococcus granulosis*	p. 709
Schistosomiasis	*Schistosoma* spp.	pp. 647–649
Swimmer's itch	Schistosomes	p. 649
Tapeworm infections	*Taenia* spp.	p. 708

Nematodes

Ascariasis	*Ascaris lumbricoides*	p. 710
Hookworm disease	*Necator americanus*	pp. 709–710
Pinworm disease	*Enterobius vermicularis*	p. 709
Trichinosis	*Trichinella spiralis*	pp. 710–712

Algae and the Diseases They Cause

Red algae, diatoms, and dinoflagellates   pp. 346–349, 449

Arthropods and the Diseases They Cause

Scabies   *Sarcoptes scabiei*   pp. 595

Viruses and the Diseases They Cause

DNA Viruses

Genital warts	Papovavirus	p. 734
Warts	Papovavirus	p. 587
Smallpox	Poxvirus	pp. 587–588
Burkitt's lymphoma	Herpesvirus	p. 639
Chickenpox	Herpesvirus	pp. 588–590
Cold sores	Herpesvirus	p. 590
Cytomegalovirus inclusion disease	Herpesvirus	pp. 700–701
Genital Herpes	Herpesvirus	pp. 733–734
Infectious mononucleosis	Herpesvirus	pp. 639–640
Keratitis	Herpesvirus	p. 596
Roseola	Herpesvirus	p. 593
Shingles	Herpesvirus	p. 589
Hepatitis B	Hepadnavirus	pp. 702–704

RNA Viruses

Encephalitis	Bunyavirus	p. 615
Hantavirus pulmonary syndrome	Bunyavirus	p. 641
Gastroenteritis	Calcivirus	p. 705
Hepatitis E	Calcivirus	p. 704
Common Cold	Coronavirus	p. 662
Hepatitis D	Deltavirus	p. 704
Encephalitis	Flavivirus	p. 615
Hepatitis C	Flavivirus	p. 704
Yellow Fever	Flavivirus	p. 640
Hemorrhagic fever	Filovirus, Arenavirus	p. 640
Influenza	Orthomyxovirus	pp. 672–674
Fifth disease	Parvovirus	p. 592
Common Cold	Picornavirus	p. 662
Hepatitis A	Picornavirus	p. 702
Poliomyelitis	Picornavirus	pp. 611–613
Measles	Paramyxovirus	pp. 590–591
Mumps	Paramyxovirus	pp. 699–700
Pneumonia	Paramyxovirus	p. 672
RSV infection	Paramyxovirus	p. 672
AIDS	Retroviruses	pp. 395, 535–544
Rabies	Rhabdovirus	pp. 613–615, 617
Dengue	Togavirus	p. 640
Encephalitis	Togavirus	p. 615
Rubella	Togavirus	pp. 591–592
Gastroenteritis	Reovirus	p. 705

Prions and the Diseases They Cause

Creutzfeldt–Jakob Disease	pp. 398, 619–620
Kuru	pp. 398, 619–620

Appendix G

Instructions for Using VirtualUnknown™ Microbiology

VirtualUnknown™ Microbiology is a laboratory-simulation program that lets you practice bacterial identification experiments outside of a wet lab. At the beginning of each lab session, you are assigned an unknown microorganism and presented with a **Case Study** involving that unknown. It's then up to you to decide what tests to perform in order to identify your unknown! Numerous resources are available to help you along the way. **References** provide detailed information on the bacteria, media, tests, and reagents you will be working with. A **Virtual Laboratory Report** provides a chronology of your progress, giving you feedback on your understanding of laboratory procedures. An **Elimination Hierarchy** and an **Identification Matrix** let you view which microbes you have eliminated with each test that you perform. And an extensive **Tutorial** walks you through the laboratory process, from turning on the

Bunsen burner, to performing aseptic transfers, to analyzing your experimental results. Over 70 bacterial species and more than 40 laboratory tests are included on **VirtualUnknown™ Microbiology!**

You may have the program randomly generate unknowns for you, or you may select a chapter-specific session to assign an unknown that is associated with a particular chapter you are studying in *Microbiology: An Introduction*. You will find chapter-specific exercises in the **Learning with Technology** sections found at the end of most of the chapters in your textbook. Although many of these exercises emphasize bacterial identification, not all of them do—some of them will ask you to use the resources on **VirtualUnknown™ Microbiology** to answer questions related to topics and concepts covered in your textbook.

To begin, follow the instructions below.

Minimum System Requirements

- Microsoft® Windows® Operating System (95, 98, NT 4, 2000, or ME)
- Pentium-class processor (233 MHz recommended)
- 64MB RAM; 15MB free disk space (30+ recommended)
- 800 × 600 screen resolution
- 16-bit color graphics (65K colors)
- CD-ROM drive
- Printer

Installing the Software

Starting the Setup Utility

Insert the CD into the CD-ROM drive and follow the installation instructions.

Note: If installer does not automatically start, select **Start → Run** and type D:\setup.exe in the command prompt box (where D: is the drive letter of the CD-ROM drive.) You will be taken through several screens; click **Next** or **Yes** until installation is complete.

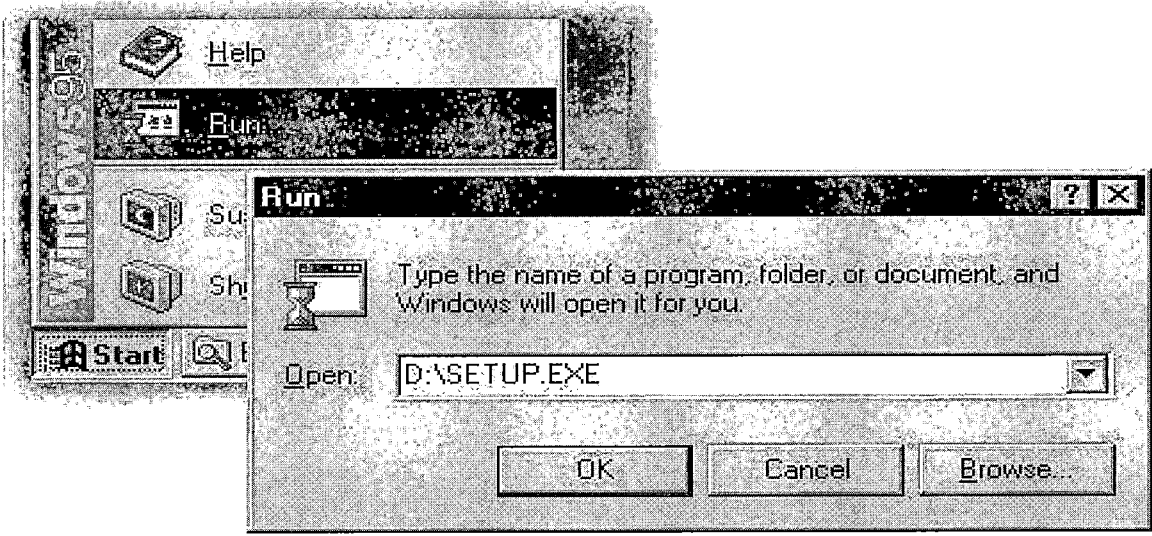

Getting Started

Launch the Software

Click **Start** → **Programs** → **VirtualUnknown** → **Microbiology Student Laboratory** to launch the software.

Work through the Tutorial

Select the **Tutorial** option and walk through the wizard-based tutorial. This feature provides an overview of the software design and hands-on practice with the skills needed to work in the virtual laboratory. Once you are familiar with the software, enter Virtual Lab (from the tutorial), or select **Enter Lab** from the welcome screen to proceed directly to the laboratory.

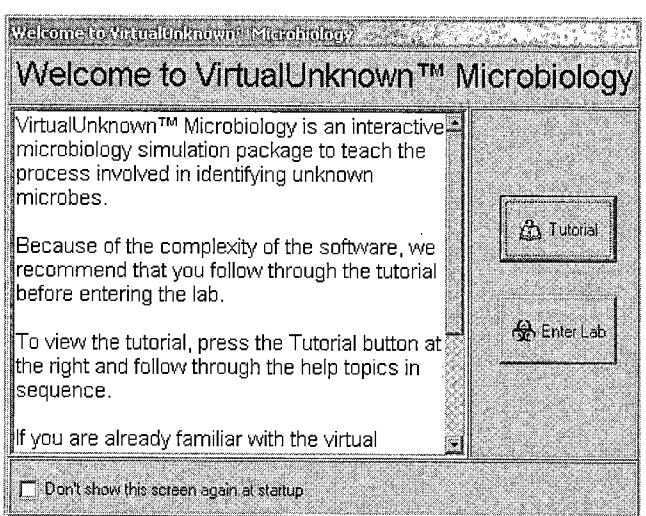
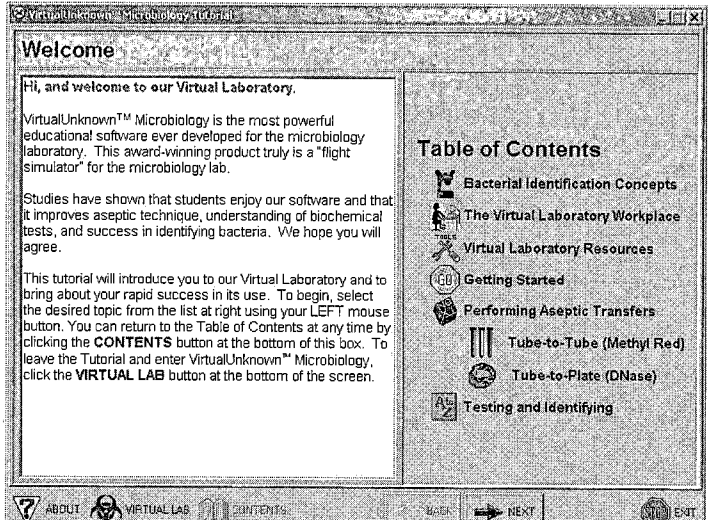

Open an Identification Session (Lab)

You must create an identification session or open an existing session to activate the lab. Click **File** → **New Session** to have the program randomly generate an unknown. To have the program generate a chapter-specific unknown, click on the arrow next to the **Session** field, click **Textbook Exercises**, then select the corresponding chapter to generate an unknown that is specific to that chapter.

Analyze the Case Study

Once a session has been activated, a Case Study will pop up on the screen. This provides valuable information about the source for the microbe you will be working with in this session. Case Studies can always be viewed later by selecting **Case Study** from the **View** menu.

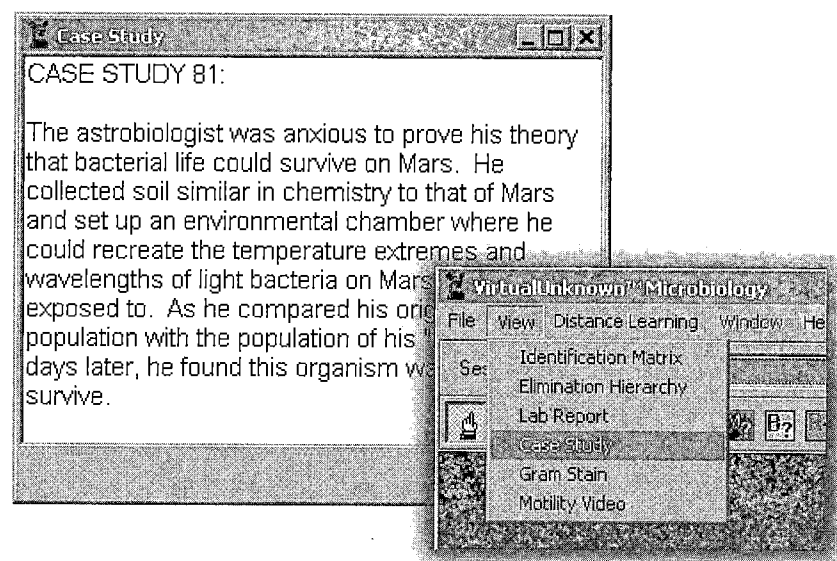

Identification Strategy

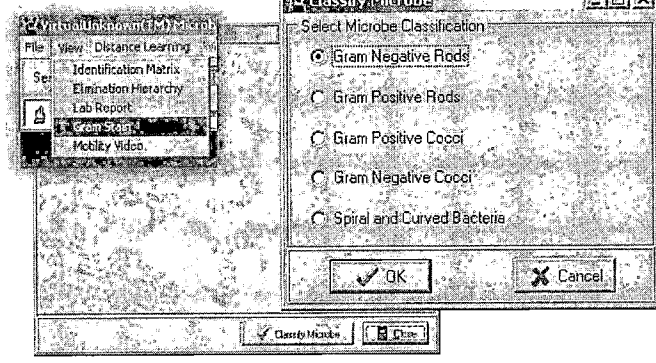

1. **VirtualUnknown™ Microbiology** contains a large database of microbes and their results for a variety of laboratory tests. It automatically eliminates from the database all bacteria with results that disagree with those you record. The first test you should perform is a **Gram Stain** to determine Gram reaction.

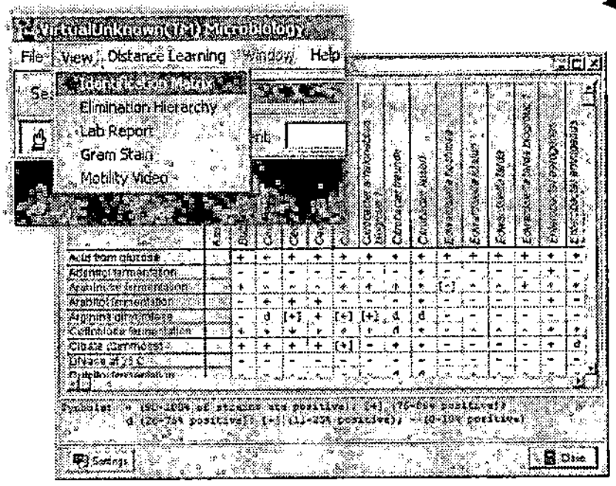

2. View the list of bacteria still in the database.

NO

3. Select a medium and perform a test to reduce the number of possible choices.

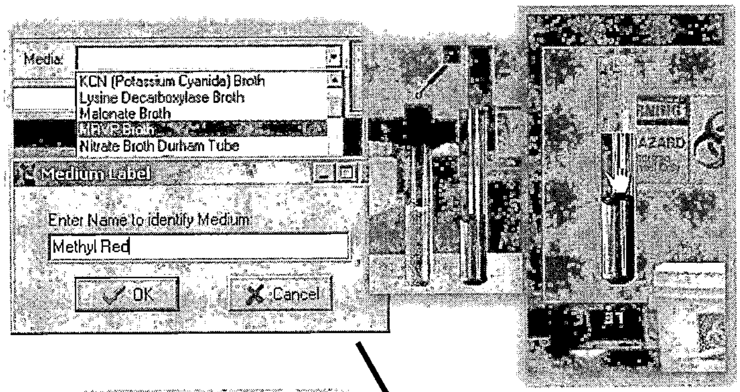

4. Record the results observed.

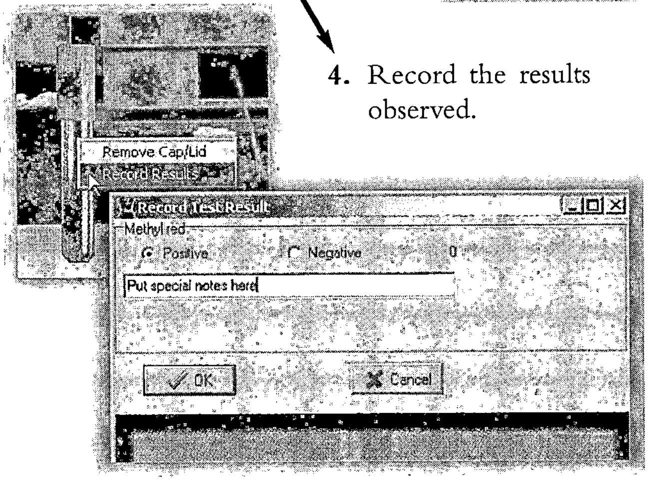

5. Do you know the identity of the unknown?

Not sure what to do? See the **Tutorial** in **VirtualUnknown™ Microbiology!**

YES

6. Identify the Unknown.

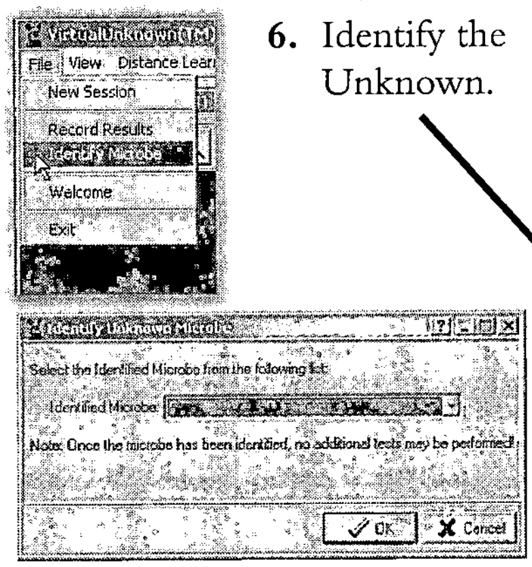

7. Print **Lab Report** for Instructor.

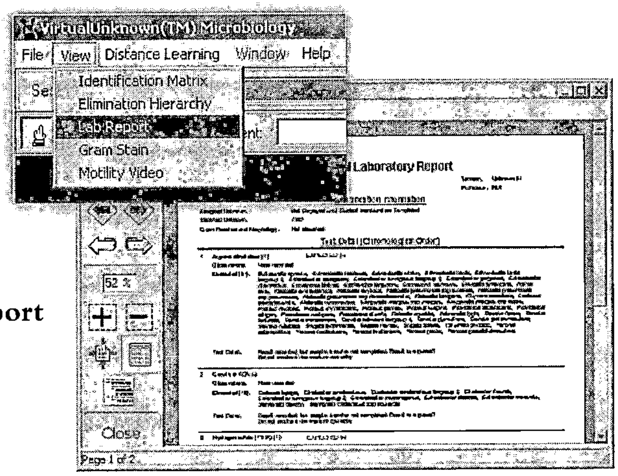

Help Files and Reference Resources

Under **Help** in the Main Menu Bar, a variety of options are available. These include, among other things:

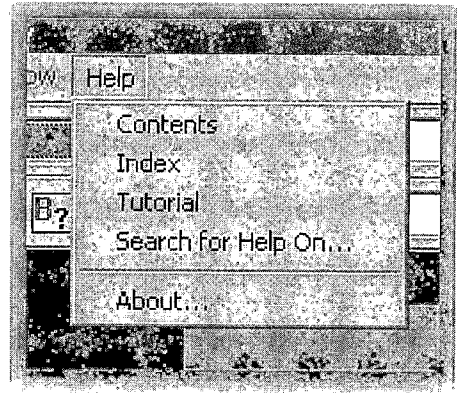

- **Contents**—Several books of information to orient you to the layout and use of the software
- **Index**—A list of all topics covered in the software and Help files
- **Tutorial**—Walks you through the software and its use
- **Search**—A search engine for locating topics in the software

Reference Buttons

Also, there are buttons located on the Menu Bar above the virtual lab that provide instant access to the reference files, your virtual library for working with the bacteria, tests, media, and reagents found in **VirtualUnknown™ Microbiology**. For all of the online references, a two-paned window appears containing a topic list in the left pane and corresponding topic details in the right pane.

 Biochemical Test Reference

This "book" provides an alphabetical list of the various test methods found in the software. The list is linked to descriptions of the tests—the chemistry involved, media and reagents used, incubation conditions required, and the method of interpretation. All test methods are provided in clear terms with hyperlinks that connect this reference with important information in the other reference books.

 Bacteria Reference

This "book" provides an alphabetical list of the bacteria found in the software. The list is linked to descriptions of each bacterium and their normal distribution. If a bacterium is associated with a disease, information about this relationship is provided. If the organism has undergone a recent renaming or repositioning in the taxonomic scheme of bacteria, this situation is discussed in this book.

 Test Media Reference

This "book" provides an alphabetical list of the various media found in the software. The list is linked to descriptions of the media—the content, uses, incubation conditions, and the method of interpretation. A link is also provided to an image of the sterile medium. All media are provided in clear terms with hyperlinks that connect this reference with important information in the other reference books.

 Reagent Reference

This "book" provides an alphabetical list of the reagents used to complete some laboratory tests. The list is linked to descriptions of the reagents—their content, uses, chemistry involved, and the method of interpretation. All reagents are provided in clear terms with hyperlinks that connect this reference with important information in the other reference books.

Other Helpful Resources

Identification Matrix

Included in the software are the results of over 40 tests conducted on over 70 species of bacteria. These results are presented in a matrix that is automatically reduced in size each time you record a test result. By consulting the matrix, smart decisions on further testing can be made. The **Identification Matrix** is accessed under the **View** menu.

Your "Assigned" Unknown

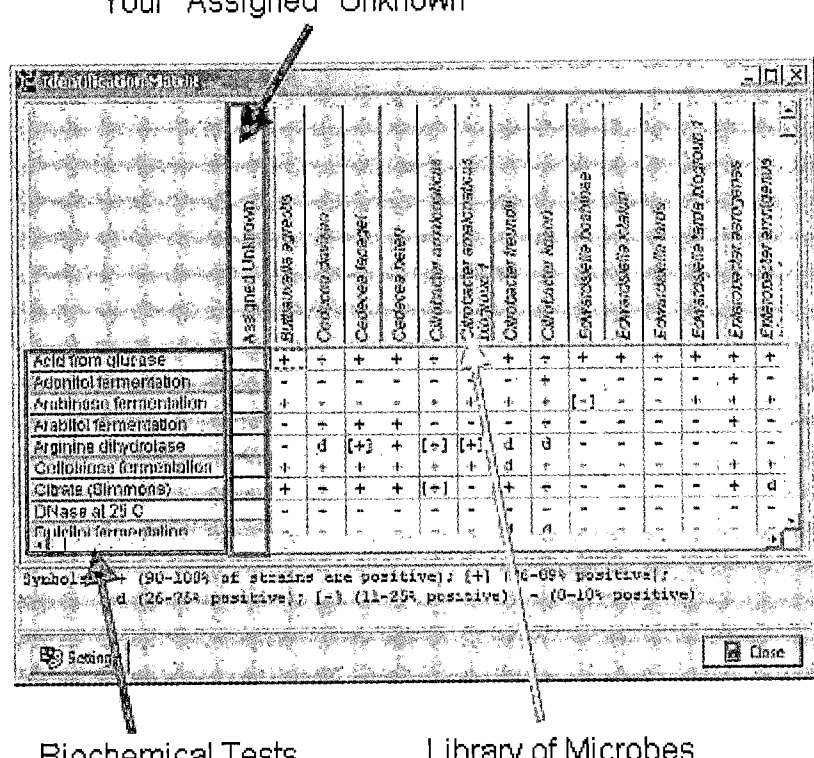

Biochemical Tests   Library of Microbes

Elimination Hierarchy

VirtualUnknown™ Microbiology maintains a dichotomous tree of eliminated and remaining microbes as biochemical test results are recorded. The **Elimination Hierarchy** is accessed from the **View** menu.

Virtual Laboratory Report

The **Virtual Laboratory Report** provides a chronology of the tests you have performed during a particular identification session. The software automatically inserts comments into the report that provide important feedback about how well you understand aseptic technique and the laboratory tests you have performed.

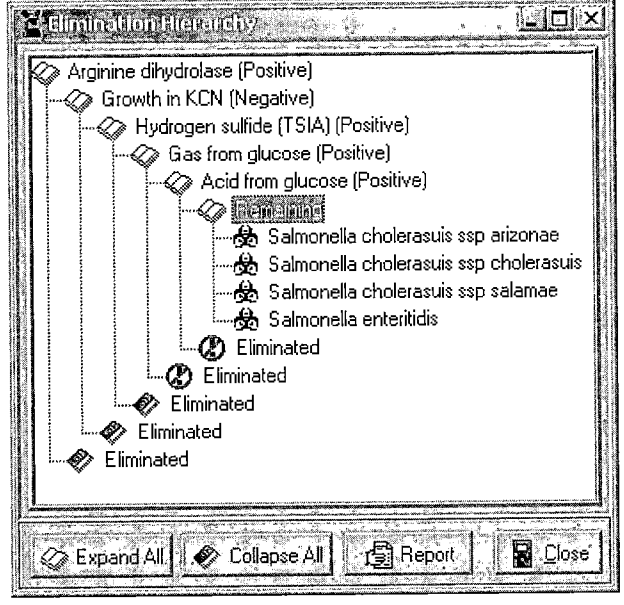

Technical Support: Contact Media Support at Pearson Education at 800-677-6337; Email: media.support@pearsoned.com; Hours: 8am to 5pm CST, Monday–Friday

For additional resources supporting **VirtualUnknown™ Microbiology** (including free downloads of instructional materials, as well as information about using this product for distance learning), **visit http://www.virtualunknown.com.**

Glossary

ABO blood group system   The classification of red blood cells based on the presence or absence of A and B carbohydrate antigens.

abscess   A localized accumulation of pus.

acellular vaccine   A vaccine consisting of antigenic parts of cells.

acetyl group
$$H_3C-\overset{\overset{\displaystyle O}{\|}}{C}-$$

acid   A substance that dissociates into one or more hydrogen ions (H^+) and one or more negative ions.

acid deposition   Rain, snow, sleet, hail, or drizzle containing acidic oxides of sulfur, nitrogen, etc. that are released into the atmosphere, usually from the combustion of fossil fuels, and return to Earth dissolved in precipitation.

acid-fast stain   A differential stain used to identify bacteria that are not decolorized by acid-alcohol.

acidic dye   A salt in which the color is in the negative ion; used for negative staining.

acidophile   A bacterium that grows below pH 4.

acquired immune deficiency   The inability, obtained during the life of an individual, to produce specific antibodies or T cells, due to drugs or disease.

acquired immunity   The ability, obtained during the life of the individual, to produce specific antibodies.

activated macrophage   A macrophage that has increased phagocytic ability and other functions after exposure to mediators released by T cells after stimulation by antigens.

activated sludge system   A process used in secondary sewage treatment in which batches of sewage are held in highly aerated tanks; to ensure the presence of microbes efficient in degrading sewage, each batch is inoculated with portions of sludge from a precious batch.

activation energy   The minimum collision energy required for a chemical reaction to occur.

active site   A region on an enzyme that interacts with the substrate.

active transport   Net movement of a substance across a membrane against a concentration gradient; requires the cell to expend energy.

acute disease   A disease in which symptoms develop rapidly but last for only a short time.

acute-phase proteins   Serum proteins whose concentration changes by at least 25% during inflammation.

adenocarcinoma   Cancer of glandular epithelial tissue.

adenosine diphosphate (ADP)   The substance formed when ATP is hydrolyzed and energy is released.

adenosine triphosphate (ATP)   An important intracellular energy source.

adherence   Attachment of a microbe or phagocyte to another's plasma membrane or other surface.

adhesin   A carbohydrate-specific binding protein that projects from prokaryotic cells; used for adherence, also called a ligand.

adjuvant   A substance added to a vaccine to increase its effectiveness.

aerobe   An organism requiring molecular oxygen (O_2) for growth.

aerobic respiration   Respiration in which the final electron acceptor in the electron transport chain is molecular oxygen (O_2).

aerotolerant anaerobe   An organism that does not use molecular oxygen (O_2) but is not affected by its presence.

aflatoxin   A carcinogenic toxin produced by *Aspergillus flavus*.

agar   A complex polysaccharide derived from a marine alga and used as a solidifying agent in culture media.

agglutination   A joining together or clumping of cells.

agranulocyte   A leukocyte without visible granules in the cytoplasm; includes monocytes and lymphocytes.

AIDS (acquired immunodeficiency syndrome)   An infectious disease caused by the human immunodeficiency virus (HIV) in which the virus infects CD4 cells.

airborne transmission   The spread of pathogens farther than 1 meter in air from reservoir to susceptible host.

alcohol   An organic molecule with the functional group —OH.

alcohol fermentation   A catabolic process, beginning with glycolysis, that produces ethyl alcohol to reoxidize NADH.

aldehyde   An organic molecule with the functional group

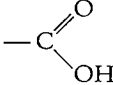

alga (plural: **algae**)   A photosynthetic eukaryote; may be unicellular, filamentous, or multicellular but lack the tissues found in plants.

algal bloom   An abundant growth of microscopic algae producing visible colonies in nature.

algin   A sodium salt of mannuronic acid ($C_6H_8O_6$); found in brown algae.

allergen   An antigen that evokes a hypersensitivity response.

allergy   *See* hypersensitivity.

allograft   A tissue graft that is not from a genetically identical donor (i.e., not from self or an identical twin).

allosteric inhibition   The process in which an enzyme's activity is changed because of binding to the allosteric site.

allosteric site   The site on an enzyme at which a noncompetitive inhibitor binds.

amanitin   A polypeptide mushroom toxin that causes liver and nerve damage.

Ames test   A procedure using bacteria to identify potential carcinogens.

amination   The addition of an amino group.

amino acid   An organic acid containing an amino group and a carboxyl group.

aminoglycoside   An antibiotic consisting of amino sugars and an aminocyclitol ring; for example, streptomycin.

amino group   —NH_2.

ammonification   The release of ammonia from nitrogen-containing organic matter by the action of microorganisms.

amoeba   A unicellular eukaryote that moves by means of pseudopods.

amphibolic pathway   A pathway that is both anabolic and catabolic.

amphitrichous   Having tufts of flagella at both ends of a cell.

anabolism   All synthesis reactions in a living organism; the building of complex organic molecules from simpler ones.

anaerobe   An organism that does not require molecular oxygen (O_2) for growth.

anaerobic respiration   Respiration in which the final electron acceptor in the electron transport chain is an inorganic molecule other than molecular oxygen (O_2); for example, a nitrate ion or CO_2.

anaerobic sludge digester   Anaerobic digestion used in secondary sewage treatment.

anal pore   A site in certain protozoa for elimination of waste.

analytic epidemiology   Comparison of a diseased group and a healthy group to determine the cause of the disease.

anamnestic response   *See* memory response.

anamorph   Ascomycete fungi that have lost the ability to reproduce sexually; the asexual stage of a fungus.

anaphylaxis   A hypersensitivity reaction involving IgE antibodies, mast cells, and basophils.

animalia   The kingdom composed of multicellular eukaryotes lacking cell walls.

anion   An ion with a negative charge.

anoxygenic   Not producing molecular oxygen; typical of cyclic photophosphorylation.

antagonism   Active opposition; for example, beyond two drugs or two microbes.

antibiotic   An antimicrobial agent, usually produced naturally by a bacterium or fungus.

antibody   A protein produced by the body in response to an antigen, and capable of combining specifically with that antigen.

antibody-dependent cell-mediated cytotoxicity (ADCC)   The killing of antibody-coated cells by natural killer cells and phagocytes.

antibody titer   The amount of antibody in serum.

anticodon   The three nucleotides by which a tRNA recognizes an mRNA codon.

antigen   Any substance that causes antibody formation and reacts only with its specific antibody; also called immunogen.

antigen-antibody complex   The combination of an antigen with the antibody that is specific for it; the basis of immune protection and many diagnostic tests.

antigen-binding sites   A site on an antibody that binds to an antigenic determinant.

antigenic determinant   A specific region on the surface of an antigen against which antibodies are formed; also called epitope.

antigenic drift   A minor variation in the antigenic makeup of influenza viruses that occurs with time.

antigenic shift   A major genetic change in influenza viruses causing changes in H and N antigens.

antigen-presenting cell (APC)   A macrophage or dendritic cell that engulfs an antigen and presents fragments to T cells.

antigen receptors   An antibody-like molecule on B and T cells that enables them to recognize and bind to their specific antigen.

antihuman immune serum globulin (anti-HISG)   An antibody that reacts specifically with human antibodies.

antimicrobial drug   A chemical that destroys pathogens without damaging body tissues.

antimicrobial peptide   An antibiotic that is bactericidal and has a broad spectrum of activity; *see* bacteriocin.

antisense DNA   DNA that is complementary to the DNA encoding a protein; the antisense RNA transcript will hybridize with the mRNA encoding the protein and inhibit synthesis of the protein.

antisense strand (− strand)   Viral RNA that cannot act as mRNA.

antisepsis   A chemical method for disinfection of the skin or mucous membranes; the chemical is called an antiseptic.

antiserum   A blood-derived fluid containing antibodies.

antitoxin   A specific antibody produced by the body in response to a bacterial exotoxin or its toxoid.

antiviral protein (AVP)   A protein made in response to interferon that blocks viral multiplication.

apicomplexa   Unicellular eukaryotes that are obligate intracellular parasites; possess a special organelle at the tip of the cell.

apoenzyme   The protein portion of an enzyme, which requires activation by a coenzyme.

apoptosis   The natural programmed death of a cell; the residual fragments are disposed of by phagocytosis.

aquatic microbiology   The study of microorganisms and their activities in natural waters.

arbuscule   Fungal mycelia in plant root cells.

archaea   Prokaryotic cells lacking peptidoglycan; one of the three domains.

archaezoa   Primitive, unicellular eukaryotes that lack mitochondria, such as *Giardia*.

arthrospore   An asexual fungal spore formed by fragmentation of a septate hypha.

Arthus reaction   Inflammation and necrosis at the site of injection of foreign serum, due to immune complex formation.

artificially acquired active immunity   The production of antibodies by the body in response to a vaccination.

artificially acquired passive immunity   The transfer of humoral antibodies formed by one individual to a susceptible individual, accomplished by the injection of antiserum.

artificial selection   Choosing one organism from a population to grow because of its desirable traits.

ascospore   A sexual fungal spore produced in an ascus, formed by the ascomycetes.

ascus   A saclike structure containing ascospores; found in the ascomycetes.

asepsis   The absence of contamination by unwanted organisms.

aseptic packaging   Commercial food preservation by filling sterile containers with sterile food.

aseptic surgery   Techniques used in surgery to prevent microbial contamination of the patient.

aseptic techniques   Laboratory techniques used to minimize contamination.

asexual spore   A reproductive cell produced by mitosis and cell division (eukaryotes) or binary fission (actinomycetes).

atom   The smallest unit of matter that can enter into a chemical reaction.

atomic number   The number of protons in the nucleus of an atom.

atomic weight   The total number of protons and neutrons in the nucleus of an atom.

attenuated whole-agent vaccine   A vaccine containing live, attenuated (weakened) microorganisms.

atomic force microscopy   *See* scanned-probe microscopy.

autoclave   Equipment for sterilization by steam under pressure, usually operated at 15 psi and 121°C.

autograph   A tissue graft from one's self.

autoimmune disease   Damage to one's own organs due to action of the immune system.

autotroph   An organism that uses carbon dioxide (CO_2) as its principal carbon source. *See also* chemoautotroph, photoautotroph.

auxotroph   A mutant microorganism with a nutritional requirement that is absent in the parent.

axial filament   The structure for motility found in spirochetes; also called endoflagellum.

azoles   Antifungal agents that interfere with sterol synthesis.

bacillus (plural: **bacilli**)   (1) Any rod-shaped bacterium. (2) When written as a genus *(Bacillus)* refers to rod-shaped, endospore-forming, facultatively anaerobic, gram-positive bacteria.

bacteremia   A condition in which there are bacteria in the blood.

bacteria   Kingdom of prokaryotic organisms, characterized by peptidoglycan cell walls; **bacterium** (singular) when referring to a single organism.

bacteriology   The scientific study of prokaryotes, including bacteria and archaea.

bacterial growth curve   A graph indicating the growth of a bacterial population over time.

bactericide   A substance capable of killing bacteria.

bacteriocin   An antimicrobial peptide produced by bacteria that kills other bacteria.

bacteriophage (phage)   A virus that infects bacterial cells.

bacteriostasis   A treatment capable of inhibiting bacterial growth.

base   A substance that dissociates into one or more hydroxide ions (OH^-) and one or more positive ions.

base pairs   The arrangement of nitrogenous bases in nucleic acids based on hydrogen bonding; in DNA, base pairs are A-T and G-C; in RNA, base pairs are A-U and G-C.

base substitution   The replacement of a single base in DNA by another base, causing a mutation; also called point mutation.

basic dye   A salt in which the color is in the positive ion; used for bacterial stains.

basidiospore   A sexual fungal spore produced in a basidium, characteristic of the basidiomycetes.

basidium   A pedestal that produces basidiospores; found in the basidiomycetes.

basophil   A granulocyte (leukocyte) that readily takes up basic dye and is not phagocytic; has receptors for IgE Fc regions.

batch flow   An industrial process in which cells are grown for a period of time after which the product is collected.

B cell   A type of lymphocyte; differentiates into antibody-secreting plasma cells and memory cells.

BCG vaccine   A live, attenuated strain of *Mycobacterium bovis* used to provide immunity to tuberculosis.

benthic zone   The sediment at the bottom of a body of water.

Bergey's Manual   *Bergey's Manual of Systematic Bacteriology,* the standard taxonomic reference on bacteria; also refers to *Bergey's Manual of Determinative Bacteriology.*

beta oxidation   The removal of two carbon units from a fatty acid to form acetyl CoA.

biguanide   A group of antimicrobial chemicals, including chlorhexidine, especially useful on skin and mucous membranes.

binary fission   Prokaryotic cell reproduction by division into two daughter cells.

binomial nomenclature   The system of having two names (genus and specific epithet) for each organism; also called scientific nomenclature.

bioaugmentation   The use of pollutant-acclimated microbes or genetically engineered microbes for bioremediation.

biochemical oxygen demand (BOD)   A measure of the biologically degradable organic matter in water.

biocide   A substance capable of killing microorganisms.

bioconversion   Changes in organic matter brought about by the growth of microorganisms.

biofilm   A microbial community that usually forms as a slimy layer on a surface.

biogenesis   The theory that living cells arise only from preexisting cells.

biogeochemical cycle   The recycling of chemical elements by microorganisms for use by other organisms.

biological transmission   The transmission of a pathogen from one host to another when the pathogen reproduces in the vector.

bioluminescence   The emission of light from the electron transport chain; requires the enzyme luciferase.

biomass   Organic matter produced by living organisms and measured by weight.

bioreactor   A fermentation vessel with controls for environmental conditions, e.g., temperature and pH.

bioremediation   The use of microbes to remove an environmental pollutant.

biotechnology   The industrial application of microorganisms, cells, or cell components to make a useful product.

biotype   *See* biovar.

biovar   A subgroup of a serovar based on biochemical or physiological properties; also called biotype.

bisphenol   A group of antimicrobial chemicals composed of two phenolic groups; includes triclosan.

blade   A flat leaflike structure of multicellular algae.

blastoconidium   An asexual fungal spore produced by budding from the parent cell.

blood-brain barrier   Cell membranes that allow some substances to pass from the blood to the brain but restrict others.

brightfield microscope   A microscope that uses visible light for illumination; the specimens are viewed against a white background.

broad-spectrum antibiotic   An antibiotic that is effective against a wide range of both gram-positive and gram-negative bacteria.

broth dilution test   A method of determining the MIC by using serial dilutions of an antimicrobial drug.

bubo   An enlarged lymph node caused by inflammation.

budding   (1) Asexual reproduction beginning as a protuberance from the parent cell that grows to become a daughter cell. (2) Release of an enveloped virus through the plasma membrane of an animal cell.

budding yeast   Following mitosis, a yeast cell that divides unevenly to produce a small cell (bud) from the parent cell.

buffer   A substance that tends to stabilize the pH of a solution.

bulking   A condition arising when sludge floats rather than settles in secondary sewage treatment.

bullae (singular: **bulla**)   Large serum-filled vesicles in the skin.

burst size   The number of newly synthesized bacteriophage particles released from a single cell.

burst time   The time required from bacteriophage attachment to release.

Calvin-Benson cycle   The fixation of CO_2 into reduced organic compounds; used by autotrophs.

capnophile   A microorganism that grows best at relatively high CO_2 concentrations.

capsid   The protein coat of a virus that surrounds the nucleic acid.

capsomere   A protein subunit of a viral capsid.

capsule   An outer, viscous covering on some bacteria composed of a polysaccharide or polypeptide.

carbapenems   Antibiotics that contain a β-lactam antibiotic and cilastatin.

carbohydrate   An organic compound composed of carbon, hydrogen, and oxygen, with the hydrogen and oxygen present in a 2:1 ratio; carbohydrates include starches, sugars, and cellulose.

carbon cycle   The series of processes that converts CO_2 to organic substances and back to CO_2 in nature.

carbon fixation The synthesis of sugars by using carbons from CO_2. *See also* Calvin–Benson cycle.

carbon skeleton The basic chain or ring of carbon atoms in a molecule; for example,

$$-\overset{|}{\underset{|}{C}}-\overset{|}{\underset{|}{C}}-\overset{|}{\underset{|}{C}}-$$

carboxyl group

$$-C\overset{\displaystyle O}{\underset{\displaystyle OH}{}}$$

carboxysome A prokaryotic inclusion containing ribulose 1,5-diphosphate carboxylase.

carcinogen Any cancer-causing substance.

casein Milk protein.

catabolic repression Inhibition of the metabolism of alternate carbon sources by glucose.

catabolism All decomposition reactions in a living organism; the breakdown of complex organic compounds into simpler ones.

catalase An enzyme that catalyzes the breakdown of hydrogen peroxide to water and oxygen.

catalyst A substance that increases the rate of a chemical reaction but is not altered itself.

cation A positively charged ion.

CD (cluster of determination) Number assigned to an epitope on a single antigen, for example, CD4 protein, which is found on helper T cells.

cDNA (complementary DNA) DNA made in vitro from an mRNA template.

cell-mediated immunity An immune response that involves T cells binding to antigens presented on infected cells; T cells then differentiate into several types of effector T cells, including helper and cytotoxic.

cellular respiration *See* respiration.

cell wall The outer covering of most bacterial, fungal, algal, and plant cells; in bacteria, it consists of peptidoglycan.

Centers for Disease Control and Prevention (CDC) A branch of the U.S. Public Health Service that serves a central source of epidemiological information.

central nervous system (CNS) The brain and the spinal cord. *See also* peripheral nervous system.

centriole A structure consisting of nine microtubule triplets, found in eukaryotic cells.

centrosome Region in a eukaryotic cell consisting of a pericentriolar area (protein fibers) and a pair of centrioles; involved in formation of the mitotic spindle.

cephalosporin An antibiotic produced by the fungus *Cephalosporium* that inhibits the synthesis of gram-positive bacterial cell walls.

cercaria A free-swimming larva of trematodes.

chancre A hard sore, the center of which ulcerates.

chemical bond An attractive force between atoms forming a molecule.

chemical element A fundamental substance composed of atoms that have the same atomic number and behave the same way chemically.

chemical energy The energy of a chemical reaction.

chemically defined medium A culture medium in which the exact chemical composition is known.

chemical reaction The process of making or breaking bonds between atoms.

chemiosmosis A mechanism that uses a proton gradient across a cytoplasmic membrane to generate ATP.

chemistry The science of the interactions between atoms and molecules.

chemoautotroph An organism that uses an inorganic chemical as an energy source and CO_2 as a carbon source.

chemoheterotroph An organism that uses organic molecules as a source of carbon and energy.

chemokine A cytokine that induces, by chemotaxis, the migration of leukocytes into infected areas.

chemotaxis Movement in response to the presence of a chemical.

chemotherapeutic agent A chemical (drug) used in the treatment of disease. *See also* chemotherapy.

chemotherapy Treatment with chemical substances.

chemotroph An organism that uses oxidation-reduction reactions as its primary energy source.

chimeric monoclonal antibody A genetically engineered antibody made of human constant regions and mouse variable regions.

chlamydospore An asexual fungal spore formed within a hypha.

chloramphenicol A broad-spectrum bacteriostatic chemical.

chloroplast The organelle that performs photosynthesis in photoautotrophic eukaryotes.

chlorosome Plasma membrane folds in green sulfur bacteria containing bacteriochlorophylls.

chromatin Threadlike, uncondensed DNA in an interphase eukaryotic cell.

chromatophore An infolding in the plasma membrane where bacteriochlorophyll is located in photoautotrophic bacteria; also known as thylakoids.

chromosome The structure that carries hereditary information, chromosomes contain genes.

chronic disease An illness that develops slowly and is likely to continue or recur for long periods.

ciliary escalator Ciliated mucosal cells of the lower respiratory tract that move inhaled particulates away from the lungs.

cilium (plural: cilia) A relatively short cellular projection from some eukaryotic cells, composed of nine pairs plus two microtubules. *See* flagellum.

ciliate A member of the protozoan phylum Ciliophora that uses cilia for locomotion.

cistern A flattened membranous sac in endoplasmic reticulum and the Golgi complex.

cladogram A dichotomous phylogenetic tree that branches repeatedly, suggesting the classification of organisms based on the time sequence in which evolutionary branches arose.

class A taxonomic group between phylum and order.

clonal deletion The elimination of B and T cells that react with self.

clonal selection The development of clones of B and T cells against a specific antigen.

clone A population of cells that are identical to the parent cell.

clue cells Sloughed-off vaginal cells covered with *Gardnerella vaginalis*.

coagulase A bacterial enzyme that causes blood plasma to clot.

coccobacillus (plural: coccobacilli) A bacterium that is an oval rod.

coccus (plural: cocci) A spherical or ovoid bacterium.

codon A sequence of three nucleotides in mRNA that specifies the insertion of an amino acid into a polypeptide.

coenocytic hypha A fungal filament that is not divided into uninucleate cell-like units because it lacks septa.

coenzyme A nonprotein substance that is associated with and that activates an enzyme.

coenzyme A (CoA) A coenzyme that functions in decarboxylation.

coenzyme Q   *See* ubiquinone.

cofactor   (1) The nonprotein component of an enzyme. (2) A microorganism or molecule that acts with others to synergistically enhance or cause disease.

coliforms   Aerobic or facultatively anaerobic, gram-negative, nonendospore-forming, rod-shaped bacteria that ferment lactose with acid and gas formation within 48 hours at 35°C.

collagenase   An enzyme that hydrolyzes collagen.

collision theory   The principle that chemical reactions occur because energy is gained as particles collide.

colony   A visible mass of microbial cells arising from one cell or from a group of the same microbes.

colony hybridization   The identification of a colony containing a desired gene by using a DNA probe that is complementary to that gene.

colony-stimulating factor (CSF)   A substance that induces certain cells to proliferate or differentiate.

commensalism   A symbiotic relationship in which two organisms live in association and one is benefited while the other is neither benefited nor harmed.

commercial sterilization   A process of treating canned goods aimed at destroying the endospores of *Clostridium botulinum.*

communicable disease   Any disease that can be spread from one host to another.

competence   The physiological state in which a recipient cell can take and incorporate a large piece of donor DNA.

competitive inhibitor   A chemical that competes with the normal substrate for the active site of an enzyme. *See also* noncompetitive inhibitor.

complement   A group of serum proteins involved in phagocytosis and lysis of bacteria.

complement fixation   The process in which complement combines with an antigen–antibody complex.

complex medium   A culture medium in which the exact chemical composition is not known.

complex virus   A virus with a complicated structure, such as a bacteriophage.

composting   A method of solid waste disposal, usually plant material, by encouraging its decomposition by microbes.

compound   A substance composed of two or more different chemical elements.

compound light microscope (LM)   An instrument with two sets of lenses that uses visible light as the source of illumination.

compromised host   A host whose resistance to infection is impaired.

condensation reaction   A chemical reaction in which a molecule of water is released; also called dehydration synthesis.

condenser   A lens system located below the microscope stage that directs light rays through the specimen.

confocal microscopy   A light microscope that uses fluorescent stains and laser to make two- and three-dimensional images.

congenital   Refers to a condition existing at birth; may be inherited or acquired in utero.

congenital immune deficiency   The inability, due to an individual's genotype, to produce specific antibodies or T cells.

conidiophore   An aerial hypha bearing conidiospores.

conidiospore   An asexual spore produced in a chain from a conidiophore.

conjugated vaccine   A vaccine consisting of the designed antigen and other proteins.

conjugation   The transfer of genetic material from one cell to another involving cell-to-cell contact.

conjugative plasmid   A prokaryotic plasmid that carries genes for sex pili and for transfer of the plasmid to another cell.

constitutive enzyme   An enzyme that is produced continuously.

contact inhibition   The cessation of animal cell movement and division as a result of contact with other cells.

contact transmission   The spread of disease by direct or indirect contact or via droplets.

contagious disease   A disease that is easily spread from one person to another.

continuous cell line   Animal cells that can be maintained through an indefinite number of generations in vitro.

continuous flow   An industrial fermentation in which cells are grown indefinitely with continual addition of nutrients and removal of waste and products.

corepressor   A molecule that binds to a repressor protein, enabling the repressor to bind to an operator.

cortex   The protective fungal covering of a lichen.

counterstain   A stain used to give contrast in a differential stain.

covalent bond   A chemical bond in which the electrons of one atom are shared with another atom.

crisis   The phase of a fever characterized by vasodilation and sweating.

crista (plural: cristae)   Folding of the inner membrane of a mitochondrion.

crossing over   The process by which a portion of one chromosome is exchanged with a portion of another chromosome.

culture   Microorganisms that grow and multiply in a container of culture medium.

culture medium   The nutrient material prepared for growth of microorganisms in a laboratory.

curd   The solid part of milk that separates from the liquid (whey) in the making of cheese, for example.

cutaneous mycosis   A fungal infection of the epidermis, nails, or hair.

cuticle   The outer covering of helminths.

cyanobacteria   Oxygen-producing photoautotrophic prokaryotes.

cyclic AMP (cAMP)   A molecule derived from ATP, in which the phosphate group has a cyclic structure; acts as a cellular messenger.

cyclic photophosphorylation   The movement of an electron from chlorophyll through a series of electron acceptors and back to chlorophyll; anoxygenic; purple and green bacterial photophosphorylation.

cyst   A sac with a distinct wall containing fluid or other material; also, a protective capsule of some protozoa.

cysticercus   An encysted tapeworm larva.

cystitis   Inflammation of the urinary bladder.

cytochrome oxidase   An enzyme that oxidizes cytochrome *c.*

cytochrome   A protein that functions as an electron carrier in cellular respiration and photosynthesis.

cytokine   A small protein released from human cells in response to bacterial infection; directly or indirectly may induce fever, pain, or T-cell proliferation.

cytolysis   The destruction of cells, resulting from damage to their cell membrane, that causes cellular contents to leak out.

cytopathic effect (CPE)   A visible effect on a host cell, caused by a virus, that may result in host cell damage or death.

cytoplasm   In a prokaryotic cell, everything inside the plasma membrane; in a eukaryotic cell, everything inside the plasma membrane and external to the nucleus.

cytoplasmic streaming The movement of cytoplasm in a eukaryotic cell.

cytoskeleton Microfilaments, intermediate filaments, and microtubules that provide support and movement for eukaryotic cytoplasm.

cytosol The fluid portion of cytoplasm.

cytostome The mouthlike opening in some protozoa.

cytotoxic T (T_C) cells A specialized T cell that destroys infected cells presenting antigens.

cytotoxin A bacterial toxin that kills host cells or alters their functions.

darkfield microscope A microscope that has a device to scatter light from the illuminator so that the specimen appears white against a black background.

dark (light-independent) reaction The process by which electrons and energy from ATP are used to reduce CO_2 to sugar. *See also* Calvin-Benson cycle.

deamination The removal of an amino group from an amino acid to form ammonia. *See also* ammonification.

death phase The period of logarithmic decrease in a bacterial population; also called logarithmic decline phase.

debridement Surgical removal of necrotic tissue.

decarboxylation The removal of CO_2 from an amino acid.

decimal reduction time (DRT) The time (in minutes) required to kill 90% of a bacterial population at a given temperature; also called D value.

decolorizing agent A solution used in the process of removing a stain.

decomposition reaction A chemical reaction in which bonds are broken to produce smaller parts from a large molecule.

deep-freezing Preservation of bacterial cultures at $-50°C$ to $-95°C$.

definitive host An organism that harbors the adult, sexually mature form of a parasite.

degeneracy Redundancy of the genetic code; that is, most amino acids are encoded by several codons.

degerming The removal of microorganisms in an area; also called degermation.

degranulation The release of contents of secretory granules from mast cells or basophils during anaphylaxis.

dehydration synthesis *See* condensation reaction.

dehydrogenation The loss of hydrogen atoms from a substrate.

delayed-type hypersensitivity Cell-mediated hypersensitivity.

delayed hypersensitivity T (T_D) cells A specialized T cell that produces lymphokines in type IV hypersensitivities.

denaturation A change in the molecular structure of a protein, usually making it nonfunctional.

dendritic cell A type of antigen-presenting cell characterized by long fingerlike extensions; found in lymphatic tissue and skin.

denitrification The reduction of nitrates to nitrites or nitrogen gas.

dental plaque A combination of bacterial cells, dextran, and debris adhering to the teeth.

deoxyribonucleic acid *See* DNA.

deoxyribose A five-carbon sugar contained in DNA nucleotides.

dermatomycosis A fungal infection of the skin; also known as tinea or ringworm.

dermatophyte A fungus that causes a cutaneous mycosis.

dermis The inner portion of the skin.

descriptive epidemiology The collection and analysis of all data regarding the occurrence of a disease to determine its cause.

desensitization The prevention of allergic inflammatory responses.

desiccation The removal of water.

diapedesis *See* emigration.

dichotomous key An identification scheme based on successive paired questions; answering one question leads to another pair of questions, until an organism is identified.

differential interference contrast (DIC) microscope An instrument that provides a three-dimensional, magnified image.

differential medium A solid culture medium that makes it easier to distinguish colonies of the desired organism.

differential stain A stain that distinguishes objects on the basis of reactions to the staining procedure.

differential white blood cell count The number of each kind of leukocyte in a sample of 100 leukocytes.

dimorphism The property of having two forms of growth.

dioecious Referring to organisms in which organs of different sexes are located in different individuals.

diplobacilli (singular: diplobacillus) Rods that divide and remain attached in pairs.

diplococci (singular: diplococcus) Cocci that divide and remain attached in pairs.

diploid cell A cell having two sets of chromosomes; diploid is the normal state of a eukaryotic cell.

diploid cell line Eukaryotic cells grown in vitro.

direct agglutination test The use of known antibodies to identify an unknown cell-bound antigen.

direct contact transmission A method of spreading infection from one host to another through some kind of close association between the hosts.

direct FA test A fluorescent-antibody test to detect the presence of an antigen.

direct microscopic count Enumeration of cells by observation through a microscope.

disaccharide A sugar consisting of two simple sugars, or monosaccharides.

disease An abnormal state in which part or all of the body is not properly adjusted or is incapable of performing normal functions; any change from a state of health.

disinfection Any treatment used on inanimate objects to kill or inhibit the growth of microorganisms; a chemical used is called a disinfectant.

disk–diffusion method An agar-diffusion test to determine microbial susceptibility to chemotherapeutic agents; also called Kirby-Bauer test.

D-isomer A stereoisomer.

dissimilation A metabolic process in which nutrients are not assimilated but are excreted as ammonia, hydrogen sulfide, and so on.

dissimilation plasmid A plasmid containing genes encoding production of enzymes that trigger the catabolism of certain unusual sugars and hydrocarbons.

dissociation The separation of a compound into positive and negative ions in solution. *See also* ionization.

disulfide bond A covalent bond that holds together two atoms of sulfur.

division A phylum; used in botany.

DNA The nucleotide of genetic material in all cells and some viruses.

DNA base composition   The moles-percentage of guanine plus cytosine in an organism's DNA.

DNA fingerprinting   Analysis of DNA by electrophoresis of restriction enzyme fragments of the DNA.

DNA ligase   An enzyme that covalently bonds a carbon atom of one nucleotide with the phosphate of another nucleotide.

DNA probe   A short, labeled, single strand of DNA or RNA used to locate its complementary strand in a quantity of DNA.

DNA sequencing   A process by which the nucleotide sequence of DNA is determined.

domain   A taxonomic classification based on rRNA sequences; above the kingdom level.

domoic acid intoxication   Diarrhea and memory loss caused by domoic acid, produced by diatoms.

donor cell   A cell that gives DNA to a recipient cell during genetic recombination.

DPT vaccine   A combined vaccine used to provide active immunity, containing diphtheria and tetanus toxoids and killed *Bordetella pertussis* cells or cell fragments.

droplet transmission   The transmission of infection by small liquid droplets carrying microorganisms.

D value   *See* decimal reduction time.

dysentery   A disease characterized by frequent, watery stools containing blood and mucus.

eclipse period   The time during viral multiplication when complete, infective virions are not present.

edema   An abnormal accumulation of interstitial fluid in body parts or tissues, causing swelling.

electron   A negatively charged particle in motion around the nucleus of an atom.

electron acceptor   An ion that picks up an electron that has been lost from another atom.

electron donor   An ion that gives up an electron to another atom.

electronic configuration   The arrangement of electrons in shells or energy levels in an atom.

electron microscope   A microscope that uses electrons instead of light to produce an image.

electron shell   A region of an atom where electrons orbit the nucleus, corresponding to an energy level.

electron transport chain, electron transport system   A series of compounds that transfer electrons from one compound to another, generating ATP by oxidative phosphorylation.

electroporation   A technique by which DNA is inserted into a cell using an electrical current.

elementary body   An infectious form of *Chlamydia.*

ELISA (enzyme-linked immunosorbent assay)   A group of serological tests that use enzyme reactions as indicators.

emerging infectious disease (EID)   A new or changing disease that is increasing or has the potential to increase in incidence in the near future.

emigration   The process by which phagocytes move out of blood vessels; also called diapedesis.

Embden-Meyerhof pathway   *See* glycolysis.

emulsification   The process of mixing two liquids that do not dissolve in each other.

encephalitis   Infection of the brain.

encystment   Formation of a cyst.

endemic disease   A disease that is constantly present in a certain population.

endergonic reaction   A chemical reaction that requires energy.

endocarditis   Infection of the lining of the heart (endocardium).

endocytosis   The process by which material is moved into a eukaryotic cell.

endoflagellum   *See* axial filament.

endolith   An organism that lives inside rock.

endoplasmic reticulum (ER)   A membranous network in eukaryotic cells connecting the plasma membrane with the nuclear membrane.

endospore   A resting structure formed inside some bacteria.

endosymbiotic theory   A model for the evolution of eukaryotes which states that organelles arose from prokaryotic cells living inside a host prokaryote.

endotoxic shock   Loss of blood pressure caused by the endotoxin component of a gram-negative cell wall. *See* septic shock.

endotoxin   Part of the outer portion of the cell wall (lipid A) of most gram-negative bacteria; released on destruction of the cell.

end-product inhibition   *See* feedback inhibition.

energy level   Potential energy of an electron in an atom. *See also* electron shell.

enrichment culture   A culture medium used for preliminary isolation that favors the growth of a particular microorganism.

enteric   The common name for a bacterium in the family Enterobacteriaceae.

enterotoxin   An exotoxin that causes gastroenteritis, such as those produced by *Staphylococcus, Vibrio,* and *Escherichia.*

Entner-Doudoroff pathway   An alternate pathway for the oxidation of glucose to pyruvic acid.

envelope   An outer covering surrounding the capsid of some viruses.

enzyme   A molecule that catalyzes biochemical reactions in a living organism, usually a protein. *See also* ribozyme.

enzyme induction   The process by which a substance can cause the synthesis of an enzyme.

enzyme-linked immunosorbent assay   *See* ELISA.

enzyme–substrate complex   A temporary union of an enzyme and its substrate.

eosinophil   A granulocyte whose granules take up the stain eosin.

epidemic disease   A disease acquired by many hosts in a given area in a short time.

epidemiology   The science that studies when and where diseases occur and how they are transmitted.

epidermis   The outer portion of the skin.

epitope   *See* antigenic determinant.

equilibrium   The point of even distribution.

equivalent treatments   Different methods that have the same effect on controlling microbial growth.

ergot   A toxin produced in sclerotia by the fungus *Claviceps purpurea* that causes ergotism.

E test   An agar diffusion test to determine antibiotic sensitivity using a plastic strip impregnated with varying concentrations of an antibiotic.

ethambutol   A synthetic antimicrobial agent that interferes with the synthesis of RNA.

ethanol

$$\begin{array}{c} \quad H \quad H \\ \quad | \quad | \\ H-C-C-OH \\ \quad | \quad | \\ \quad H \quad H \end{array}$$

etiology   The study of the cause of a disease.

eukarya   All eukaryotes (animals, plants, fungi, and protists); members of the Domain Eukarya.

eukaryote A cell having DNA inside a distinct membrane-enclosed nucleus.

eukaryotic species A group of closely related organisms that can interbreed.

eutrophication The addition of organic matter and subsequent removal of oxygen from a body of water.

exchange reaction A chemical reaction that has both synthesis and decomposition components.

exergonic reaction A chemical reaction that releases energy.

exon A region of a eukaryotic chromosome that encodes a protein.

exotoxin A protein toxin released from living, mostly gram-positive bacterial cells.

experimental epidemiology The study of a disease using controlled experiments.

exponential growth phase *See* log phase.

extracellular polysaccharide (EPS) A glycocalyx, composed of sugars, that permits bacteria to attach to various surfaces.

extreme halophile An organism that requires a high salt concentration for growth.

extreme thermophile *See* hyperthermophile.

extremophile A microorganism that lives in environmental extremes of temperature, acidity, alkalinity, salinity, or pressure.

eyespot A pigmented area in a cell, capable of detecting the presence of light.

facilitated diffusion The movement of a substance across a plasma membrane from an area of higher concentration to an area of lower concentration, mediated by transporter proteins.

facultative anaerobe An organism that can grow with or without molecular oxygen (O_2).

facultative halophile An organism capable of growth in, but not requiring, 1–2% salt.

FAD Flavin adenine dinucleotide; a coenzyme that functions in the removal and transfer of hydrogen ions (H^+) and electrons from substrate molecules.

family A taxonomic group between order and genus.

feedback inhibition Inhibition of an enzyme in a particular pathway by the accumulation of the end-product of the pathway; also called end-product inhibition.

fermentation The enzymatic degradation of carbohydrates in which the final electron acceptor is an organic molecule, ATP is synthesized by substrate-level phosphorylation, and O_2 is not required.

fermentation test Method used to determine whether a bacterium or yeast ferments a specific carbohydrate; usually performed in a peptone broth containing the carbohydrate, a pH indicator, and an inverted tube to trap gas.

fever An abnormally high body temperature.

F factor (fertility factor) A plasmid found in the donor cell in bacterial conjugation.

fibrinolysin A kinase produced by streptococci.

filtration The passage of a liquid or gas through a screenlike material; a 0.45-μm filter removes most bacteria.

fimbria (plural: fimbriae) An appendage on a bacterial cell used for attachment.

fission yeast Following mitosis, a yeast cell that divides evenly to produce two new cells.

fixed macrophage A macrophage that is located in a certain organ or tissue (e.g., liver, lungs, spleen, or lymph nodes); also called a histiocyte.

fixing (1) In slide preparation, the process of attaching a specimen to a slide. (2) Regarding chemical elements, combining elements so that a critical element can enter the food chain. *See also* Calvin-Benson cycle; nitrogen fixation.

flagellum (plural: flagella) A thin appendage from the surface of a cell; used for cellular locomotion; composed of flagellin in prokaryotic cells, composed of nine pairs plus two microtubules in eukaryotic cells.

flaming The process of sterilizing an inoculating loop by holding it in an open flame.

flat sour spoilage Thermophilic spoilage of canned goods not accompanied by gas production.

flatworm An animal belonging to the phylum Platyhelminthes.

flavin adenine dinucleotide *See* FAD.

flavin mononucleotide *See* FMN.

flavoprotein A protein containing the coenzyme flavin; functions as an electron carrier in electron transport chains.

flocculation The removal of colloidal material during water purification by adding a chemical that causes colloidal particles to coalesce.

flow cytometry A method of counting cells using a flow cytometer, which detects cells by the presence of a fluorescent tag on the cell surface.

fluid mosaic model A way of describing the dynamic arrangement of phospholipids and proteins comprising the plasma membrane.

fluke A flatworm belonging to the class Trematoda.

fluorescence The ability of a substance to give off light of one color when exposed to light of another color.

fluorescence-activated cell sorter (FACS) A modification of a flow cytometer that counts and sorts cells labeled with fluorescent antibodies.

fluorescence microscope A microscope that uses an ultraviolet light source to illuminate specimens that will fluoresce.

fluorescent-antibody (FA) technique A diagnostic tool using antibodies labeled with fluorochromes and viewed through a fluorescence microscope; also called immunofluorescence.

fluoroquinolone A synthetic antibacterial agent that inhibits DNA synthesis.

FMN Flavin mononucleotide; a coenzyme that functions in the transfer of electrons in the electron transport chain.

focal infection A systemic infection that began as an infection in one place.

folliculitis An infection of hair follicles, often occurring as pimples.

fomite A nonliving object that can spread infection.

forespore A structure consisting of chromosome, cytoplasm, and endospore membrane inside a bacterial cell.

frameshift mutation A mutation caused by the addition or deletion of one or more bases in DNA.

freeze-drying *See* lyophilization.

FTA-ABS test An indirect fluorescent-antibody test used to detect syphilis.

fulminating A condition that develops quickly and rapidly increases in severity.

functional group An arrangement of atoms in an organic molecule that is responsible for most of the chemical properties of that molecule.

fungus (plural: fungi) An organism that belongs to the Kingdom Fungi; a eukaryotic absorptive chemoheterotroph.

fusion The merging of plasma membranes of two different cells, resulting in one cell containing cytoplasm from both original cells.

gamete   A male or female reproductive cell.

gametocyte   A male or female protozoan cell.

gamma globulin   *See* immune serum globulin.

gastroenteritis   Inflammation of the stomach and intestine.

gas vacuole   A prokaryotic inclusion for buoyancy compensation.

gel electrophoresis   The separation of substances (such as serum proteins or DNA) by their rate of movement through an electrical field.

gene   A segment of DNA (a sequence of nucleotides in DNA) encoding a functional product.

gene library   A collection of cloned DNA fragments created by inserting restriction enzyme fragments in a bacterium, yeast, or phage.

gene therapy   Treating a disease by replacing abnormal genes.

generalized transduction   The transfer of bacterial chromosome fragments from one cell to another by a bacteriophage.

generation time   The time required for a cell or population to double in number.

genetic code   The mRNA codons and the amino acids they encode.

genetic engineering   Manufacturing and manipulating genetic material in vitro; also called recombinant DNA technology.

genetic recombination   The process of joining pieces of DNA from different sources.

genetics   The science of heredity.

genetic screening   Techniques for determining which genes are in a cell's genome.

genome   One complete copy of the genetic information in a cell.

genomics   The study of genes and their function.

genotype   The genetic makeup of an organism.

genus (plural: **genera**)   The first name of the scientific name (binomial); the taxon between family and species.

germicide   *See* biocide.

germination   The process of starting to grow from a spore or endospore.

germ theory of disease   The principle that microorganisms cause disease.

global warming   Retention of solar heat by gases in the atmosphere.

globulin   The class of proteins that includes antibodies. *See also* immunoglobulin.

glycocalyx   A gelatinous polymer surrounding a cell.

glycolysis   The main pathway for the oxidation of glucose to pyruvic acid; also called Embden-Meyerhof pathway.

Golgi complex   An organelle involved in the secretion of certain proteins.

graft-versus-host (GVH) disease   A condition that occurs when a transplanted tissue has an immune response to the tissue recipient.

gram-negative bacteria   Bacteria that lose the crystal violet color after decolorizing by alcohol; they stain red after treatment with safranin.

gram-negative cell wall   The cell wall of gram-negative bacteria, a peptidoglycan layer surrounded by a lipopolysaccharide outer membrane.

gram-positive bacteria   Bacteria that retain the crystal violet color after decolorizing by alcohol; they stain dark purple.

gram-positive cell wall   The cell wall of most gram-positive bacteria, consisting of peptidoglycan and teichoic acids.

Gram stain   A differential stain that classifies bacteria into two groups, gram-positive and gram-negative.

granulocyte   A leukocyte with visible granules in the cytoplasm; includes neutrophils, basophils, and eosinophils.

green nonsulfur bacteria   Gram-negative, nonproteobacteria; anaerobic and phototrophic; use reduced organic compounds as electron donors for CO_2 fixation.

green sulfur bacteria   Gram-negative, nonproteobacteria; strictly anaerobic and phototrophic; no growth in dark; use reduced sulfur compounds as electron donors for CO_2 fixation.

griseofulvin   A fungistatic antibiotic.

group translocation   In prokaryotes, active transport in which a substance is chemically altered during transport across the plasma membrane.

gumma   A rubbery mass of tissue characteristic of tertiary syphilis.

halogen   One of the following elements: fluorine, chlorine, bromine, iodine, or astatine.

haploid cell   A eukaryotic cell or organism with one of each type of chromosome.

hapten   A substance of low molecular weight that does not cause the formation of antibodies by itself but does so when combined with a carrier molecule.

helminth   A parasitic roundworm or flatworm.

helper T (T_H) cell   A specialized T cell that often interacts with an antigen before B cells interact with the antigen.

hemagglutination   The clumping of red blood cells.

hemoflagellate   A parasitic flagellate found in the circulatory system of its host.

hemolysin   An enzyme that lyses red blood cells.

herd immunity   The presence of immunity in most of a population.

hermaphroditic   Having both male and female reproductive capacities.

heterocyst   A large cell in certain cyanobacteria; the site of nitrogen fixation.

heterolactic   Describing an organism that produces lactic acid and other acids or alcohols as end-products of fermentation; e.g., *Escherichia*.

heterotroph   An organism that requires an organic carbon source; also called organotroph.

Hfr cell   A bacterial cell in which the F factor has become integrated into the chromosome; Hfr stands for high frequency of recombination.

high-efficiency particulate air (HEPA) filter   A screenlike material that removes particles larger than 0.3 μm from air.

high-temperature short-time (HTST) pasteurization   Pasteurizing at 72°C for 15 seconds.

histamine   A substance released by tissue cells that causes vasodilation, capillary permeability, and smooth muscle contraction.

histocompatibility antigen   An antigen on the surface of human cells.

histone   A protein associated with DNA in eukaryotic chromosomes.

holdfast   The branched base of an algal stipe.

haloenzyme   An enzyme consisting of an apoenzyme and a cofactor.

homolactic   Describing an organism that produces only lactic acid from fermentation; e.g., *Streptococcus*.

horizontal gene transfer   Transfer of genes between two organisms in the same generation. *See also* vertical gene transfer.

host   An organism infected by a pathogen. *See also* definitive host; intermediate host.

host range   The spectrum of species, strains, or cell types that a pathogen can infect.

hot–air sterilization Sterilization by the use of an oven at 170°C for approximately 2 hours.

human leukocyte antigen (HLA) complex Human cell surface antigens. *See also* major histocompatibility complex.

humoral immunity Immunity produced by antibodies dissolved in body fluids, mediated by B cells; also called antibody-mediated immunity.

hyaluronidase An enzyme secreted by certain bacteria that hydrolyzes hyaluronic acid and helps spread microorganisms from their initial site of infection.

hybridoma A cell made by fusing an antibody-producing B cell with a cancer cell.

hydrogen bond A bond between a hydrogen atom covalently bonded to oxygen or nitrogen and another covalently bonded oxygen or nitrogen atom.

hydrolysis A decomposition reaction in which chemicals react with the H^+ and OH^- of a water molecule.

hydroxyl radical A toxic form of oxygen (OH•) formed in cytoplasm by ionizing radiation and aerobic respiration.

hyperacute rejection Very rapid rejection of transplanted tissue, usually in the case of tissue from nonhuman sources.

hyperbaric chamber An apparatus to hold materials at pressures greater than 1 atmosphere.

hypersensitivity An altered, enhanced immune reaction leading to pathological changes; also called allergy.

hyperthermophile An organism whose optimum growth temperature is at least 80°C; also called extreme thermophile.

hypertonic (hyperosmotic) solution A solution that has a higher concentration of solutes than an isotonic solution.

hypha A long filament of cells in fungi or actinomycetes.

hypotonic (hypoosmotic) solution A solution that has a lower concentration of solutes than an isotonic solution.

ID$_{50}$ The number of microorganisms required to produce a demonstrable infection in 50% of the test host population.

idiophase The period in the production curve of an industrial cell population in which secondary metabolites are produced; a period of stationary growth following the phase of rapid growth. *See also* trophophase.

IgA The class of antibodies found in secretions.

IgD The class of antibodies found on B cells.

IgE The class of antibodies involved in hypersensitivities.

IgG The most abundant class of antibodies in serum.

IgM The first class of antibodies to appear after exposure to an antigen.

immune complex A circulating antigen-antibody aggregate capable of fixing complement.

immune deficiency The absence of an adequate immune response; may be congenital or acquired.

immune serum globulin The serum fraction containing immunoglobulins (antibodies); also called gamma globulin.

immunity The body's defense against particular pathogenic microorganisms; also called specific resistance.

immunization *See* vaccination.

immunodiffusion test A test consisting of precipitation reactions carried out in an agar gel medium.

immunoelectrophoresis The identification of proteins by electrophoretic separation followed by serological testing.

immunofluorescence *See* fluorescent-antibody technique.

immunogen *See* antigen.

immunoglobulin (Ig) A protein (antibody) formed in response to an antigen and can react with that antigen. *See also* globulin.

immunological escape Resistance of cancer cells to the immune response.

immunological surveillance The body's immune response to cancer.

immunology The study of a host's specific defenses to a pathogen.

immunosuppression Inhibition of the immune response.

immunotherapy Making use of the immune system to attack tumor cells, either by enhancing the normal immune response or by using toxin-bearing specific antibodies. *See also* immunotoxin.

immunotoxin An immunotherapeutic agent consisting of a poison bound to a monoclonal antibody.

inapparent infection *See* subclinical infection.

incidence The fraction of the population that contracts a disease during a particular period of time.

inclusion Material held inside a cell, often consisting of reserve deposits.

inclusion body A granule or viral particle in the cytoplasm or nucleus of some infected cells; important in the identification of viruses that cause infection.

incubation period The time interval between the actual infection and first appearance of any signs or symptoms of disease.

indicator organism A microorganism, such as a coliform, whose presence indicates conditions such as fecal contamination of food or water.

indirect (passive) agglutination test An agglutination test using soluble antigens attached to latex or other small particles.

indirect contact transmission The spread of pathogens by fomites (nonliving objects).

indirect FA test A fluorescent-antibody test to detect the presence of specific antibodies.

inducer A substance that initiates transcription of a gene.

induction The process that turns on the transcription of a gene.

infection The invasion or growth of microorganisms in the body.

infectious disease A disease in which pathogens invade a susceptible host and carry out at least part of their life cycle in the host.

inflammation A host response to tissue damage characterized by redness, pain, heat, and swelling; and sometimes loss of function.

innate resistance The resistance of an individual to diseases that affect other species and other individuals of the same species.

inoculum A culture medium in which microorganisms are implanted.

inorganic compound A small molecule that does not contain carbon and hydrogen.

insertion sequence (IS) The simplest kind of transposon.

interferon (IFN) An antiviral protein produced by certain animal cells in response to a viral infection.

interleukin A chemical that causes T-cell proliferation. *See also* cytokine.

intermediate host An organism that harbors the larval or asexual stage of a helminth or protozoan.

intoxication A condition resulting from the ingestion of a microbially produced toxin.

intron A region in a eukaryotic gene that does not code for a protein or mRNA.

invasin A surface protein produced by *Salmonella typhimurium* and *Escherichia coli* that rearranges nearby actin filaments in the cytoskeleton of a host cell.

iodophor A complex of iodine and a detergent.

ion A negatively or positively charged atom or group of atoms.

ionic bond A chemical bond formed when atoms gain or lose electrons in the outer energy levels.

ionization The separation (dissociation) of a molecule into ions.

ionizing radiation High-energy radiation with a wavelength less than 1nm; causes ionization. X rays and gamma rays are examples.

isograft A tissue graft from a genetically identical source (i.e., from an identical twin).

isomer One or two molecules with the same chemical formula but different structures.

isoniazid (INH) A bacteriostatic agent used to treat tuberculosis.

isotonic (isosmotic) solution A solution in which, after immersion of a cell, osmotic pressure is equal across the cell's membrane.

isotope A form of a chemical element in which the number of neutrons in the nucleus is different from the other forms of that element.

karyogamy Fusion of the nuclei of two cells; occurs in the sexual stage of a fungal life cycle.

kelp A multicellular brown alga.

keratin A protein found in epidermis, hair, and nails.

kinase (1) An enzyme that removes a ℗ from ATP and attaches it to another molecule. (2) A bacterial enzyme that breaks down fibrin (blood clots).

kingdom A taxonomic classification between domain and phylum.

kinin A substance released from tissue cells that causes vasodilation.

Kirby–Bauer test *See* disk-diffusion method.

Koch's postulates Criteria used to determine the causative agent of infectious diseases.

Krebs cycle A pathway that converts two-carbon compounds to CO_2, transferring electrons to NAD^+ and other carriers; also called tricarboxylic acid (TCA) cycle or critic acid cycle.

lactic acid fermentation A catabolic process, beginning with glycolysis, that produces lactic acid to reoxidize NADH.

lagging strand During DNA replication, the daughter strand that is synthesized discontinuously.

lag phase The time interval in a bacterial growth curve during which there is no growth.

larva The sexually immature stage of a helminth or arthropod.

latent disease A disease characterized by a period of no symptoms when the pathogen is inactive.

latent infection A condition in which a pathogen remains in the host for long periods without producing disease.

LD_{50} The lethal dose for 50% of the inoculated hosts within a given period.

leading strand During DNA replication, the daughter strand that is synthesized continuously.

lepromin test A skin test to determine the presence of antibodies to *Mycobacterium leprae*, the cause of leprosy.

leukocidins Substances produced by some bacteria that can destroy neutrophils and macrophages.

leukocyte A white blood cell.

leukotriene A substance produced by mast cells and basophils that causes increased permeability of blood vessels and helps phagocytes attach to pathogens.

lichen A mutualistic relationship between a fungus and an alga or a cyanobacterium.

ligand *See* adhesin.

light (light-dependent) reactions The process by which light energy is used to convert ADP and phosphate to ATP. *See also* photophosphorylation.

light-repair enzyme An enzyme that splits thymine dimers in the presence of visible light.

limnetic zone The surface zone of an inland body of water away from the shore.

***Limulus* amoebocyte lysate (LAL) assay** A test to detect the presence of bacterial endotoxins.

lipase An enzyme that breaks down triglycerides into their component glycerol and fatty acids.

lipid A non–water soluble organic molecule, including triglycerides, phospholipids, and sterols.

lipid A A component of the gram-negative outer membrane; endotoxin.

lipid inclusion *See* inclusion.

lipopolysaccharide (LPS) A molecule consisting of a lipid and a polysaccharide, forming the outer membrane of gram–negative cell walls.

L-isomer A stereoisomer.

lithotroph *See* autotroph.

littoral zone The region along the shore of the ocean or a large lake where there is considerable vegetation and where light penetrates to the bottom.

local infection An infection in which pathogens are limited to a small area of the body.

localized anaphylaxis An immediate hypersensitivity reaction that is restricted to a limited area of skin or mucous membrane; for example, hayfever, a skin rash, or asthma. *See also* systemic anaphylaxis.

logarithm decline phase *See* death phase.

log phase The period of bacterial growth or logarithmic increase in cell numbers; also called exponential growth phase.

lophotrichous Having two or more flagella at one end of a cell.

luciferase An enzyme that accepts electrons from flavoproteins and emits a photon of light in bioluminescence.

lymphangitis Inflammation of lymph vessels.

lymphocyte A leukocyte involved in specific immune responses.

lyophilization Freezing a substance and sublimating the ice in a vacuum; also called freeze-drying.

lysogenic conversion The acquisition of new properties by a host cell infected by a lysogenic phage.

lysogenic cycle Stages in viral development that result in the incorporation of viral DNA into host DNA.

lysis (1) Destruction of a cell by the rupture of the plasma membrane, resulting in a loss of cytoplasm. (2) In disease, a gradual period of decline.

lysogeny A state in which phage DNA is incorporated into the host cell without lysis.

lysosome An organelle containing digestive enzymes.

lysozyme An enzyme capable of hydrolyzing bacterial cell walls.

lytic cycle A mechanism of phage multiplication that results in host cell lysis.

macrolide An antibiotic that inhibits protein synthesis; for example, erythromycin.

macromolecule A large organic molecule.

macrophage A phagocytic cell; a mature monocyte.

macrophage activation factor Increases macrophages' efficiency at destroying ingested cells.

macrophage chemotactic factor Attracts macrophages to infection site.

macrophage migration-inhibiting factor Prevents macrophages from leaving infection site.

macule A flat, reddened skin lesion.

magnetosome An iron oxide inclusion, produced by some gram-negative bacteria, that acts like a magnet.

major histocompatibility complex (MHC) The genes that code for histocompatibility antigens; also known as human leukocyte antigen (HLA) complex.

malolactic fermentation The conversion of malic acid to lactic acid by lactic acid bacteria.

malt Germinated barley grains containing maltose, glucose, and amylase.

malting The germination of starchy grains resulting in glucose and maltose production.

margination The process by which phagocytes stick to the lining of blood vessels.

mast cell A type of cell found throughout the body that contains histamine and other substances that stimulate vasodilation.

maximum growth temperature The highest temperature at which a species can grow.

mechanical transmission The process by which arthropods transmit infections by carrying pathogens on their feet and other body parts.

medulla A lichen body consisting of algae (or cyanobacteria) and fungi.

membrane attack complex Complement proteins C5–C9, which together make lesions in cell membranes that lead to cell death.

membrane filter A screenlike material with pores small enough to retain microorganisms; a 0.45-μm filter retains most bacteria.

memory cells A long-lived B or T cell responsible for the memory, or secondary, response.

memory response A rapid rise in antibody titer following exposure to an antigen after the primary response to that antigen; also called anamestic response or secondary response.

meningitis Inflammation of the meninges, the three membranes covering the brain and spinal cord.

merozoite A trophozoite of *Plasmodium* found in red blood cells or liver cells.

mesophile An organism that grows between about 10°C and 50°C; a moderate-temperature–loving microbe.

mesosome An irregular fold in the plasma membrane of a prokaryotic cell that is an artifact of preparation for microscopy.

messenger RNA (mRNA) The type of RNA molecule that directs the incorporation of amino acids into proteins.

metabolic pathway A sequence of enzymatically catalyzed reactions occurring in a cell.

metabolism The sum of all the chemical reactions that occur in a living cell.

metacercaria The encysted stage of a fluke in its final intermediate host.

metachromatic granule A granule that stores inorganic phosphate and stains red with certain blue dyes; characteristic of *Corynebacterium diphtheriae*. Collectively known as volutin.

methane The hydrocarbon CH_4, a flammable gas formed by the microbial decomposition of organic matter; natural gas.

methylate Addition of a methyl group ($-CH_3$) to a molecule; methylated cytosine is protected from digestion by restriction enzymes.

microaerophile An organism that grows best in an environment with less molecular oxygen (O_2) than is normally found in air.

microbial antagonism *See* antagonism.

microinjection The introduction of material, such as DNA, directly into a cell, usually with a glass pipette.

micrometer (μm) A unit of measurement equal to 10^{-6} m.

microorganism A living organism too small to be seen with the naked eye; includes bacteria, fungi, protozoa, and microscopic algae; also includes viruses.

microsporidia Eukaryotic organisms that lack mitochondria and microtubules; obligate intracellular parasites.

microtubule A hollow tube made of the protein tubulin; the structural unit of eukaryotic flagella and centrioles.

microwave Electromagnetic radiation with wavelength between 10^{-1} and 10^{-3} m.

minimal bactericidal concentration (MBC) The lowest concentration of chemotherapeutic agent that will kill test microorganisms.

minimal inhibitory concentration (MIC) The lowest concentration of a chemotherapeutic agent that will prevent growth of the test microorganisms.

minimum growth temperature The lowest temperature at which a species will grow.

miracidium The free-swimming, ciliated larva of a fluke that hatches from the egg.

missense mutation A mutation that results in the substitution of an amino acid in a protein.

mitochondrion (plural: mitochondria) An organelle containing Krebs cycle enzymes and the electron transport chain.

mitosis A eukaryotic cell replication process in which the chromosomes are duplicated; followed by division of the cytoplasm of the cell.

MMWR *Morbidity and Mortality Weekly Report;* a CDC publication containing data on notifiable diseases and topics of special interest.

mole An amount of a chemical equal to the atomic weights of all the atoms in a molecule of the chemical.

molecular biology The science dealing with DNA and protein synthesis of living organisms.

molecular weight The sum of the atomic weights of all atoms making up a molecule.

molecule A combination of atoms forming a specific chemical compound.

monobactam A synthetic antibiotic with a β-lactam ring that is monocyclic in structure, in contrast to the bicyclic β-lactam structure of the penicillins and cephalosporins.

monoclonal antibody A specific antibody produced in vitro by a clone of B cells hybridized with cancerous cells.

monocyte A leukocyte that is the precursor of a macrophage.

monoecious Having both male and female reproductive capacities.

monomer A small molecule that collectively combines to form polymers.

mononuclear phagocytic system A system of fixed macrophages located in the spleen, liver, lymph nodes, and red bone marrow.

monosaccharide A simple sugar consisting of 3–7 carbon atoms.

monotrichous Having a single flagellum.

morbidity (1) The incidence of a specific disease. (2) The condition of being diseased.

morbidity rate   The number of people affected by a disease in a given period of time in relation to the total population.

mordant   A substance added to a staining solution to make it stain more intensely.

mortality   The number of deaths from a specific notifiable disease.

mortality rate   The number of deaths resulting from a disease in a given period of time in relation to the total population.

most probable number (MPN) method   A statistical determination of the number of coliforms per 100 ml of water or 100 g of food.

motility   The ability of an organism to move by itself.

M protein   A heat- and acid-resistant protein of streptococcal cell walls and fibrils.

mucous membranes   Membranes that line body openings, including the intestinal tract, open to the exterior; also called mucosa.

mutagen   An agent in the environment that brings about mutations.

mutation   Any change in the nitrogenous base sequence of DNA.

mutation rate   The probability that a gene will mutate each time a cell divides.

mutualism   A type of symbiosis in which both organisms or populations are benefited.

mycelium   A mass of long filaments of cells that branch and intertwine, typically found in molds.

mycology   The scientific study of fungi.

mycorrhiza   A fungus growing in symbiosis with plant roots.

mycosis   A fungal infection.

mycotoxin   A toxin produced by a fungus.

NAD$^+$ A coenzyme that functions in the removal and transfer of hydrogen ion (H$^+$) and electrons from substrate molecules.

NADP$^+$ A coenzyme similar to NAD$^+$

nanometer (nm)   A unit of measurement equal to 10^{-9} m, 10^{-3} μm.

natural killer (NK) cell   A lymphoid cell that destroys tumor cells and virus-infected cells.

natural penicillins   Penicillin molecules made by *Penicillium* spp.; penicillins G and V are examples. *See also* semisynthetic penicillins.

naturally acquired active immunity   Antibody production in response to an infectious disease.

naturally acquired passive immunity   The natural transfer of humoral antibodies, for example, transplacental transfer.

necrosis   Tissue death.

negative (indirect) selection   The process of identifying mutations by selecting cells that do not grow using replica plating.

negative staining   A procedure that results in colorless bacteria against a stained background.

neurotoxin   An exotoxin that interferes with normal nerve impulse conduction.

neutralization   An antigen–antibody reaction that inactivates a bacterial exotoxin or virus.

neutron   An uncharged particle in the nucleus of an atom.

neutrophil   A highly phagocytic granulocyte; also called polymorphonuclear leukocyte (PMN) or polymorph.

nicotinamide adenine dinucleotide   *See* NAD$^+$.

nicotinamide adenine dinucleotide phosphate   *See* NADP$^+$.

nitrogen cycle   The series of processes that converts nitrogen (N$_2$) to organic substances and back to nitrogen in nature.

nitrogen fixation   The conversion of nitrogen (N$_2$) into ammonia.

nitrosamine   A carcinogen formed by the combination of nitrite and amino acids.

nomenclature   The system of naming things.

noncommunicable disease   A disease that is not transmitted from one person to another.

noncompetitive inhibitor   An inhibitory chemical that does not compete with the substrate for an enzyme's active site. *See also* allosteric inhibition; competitive inhibitor.

noncyclic photophosphorylation   The movement of an electron from chlorophyll to NAD$^+$; plant and cyanobacterial photophosphorylation.

nonionizing radiation   Short-wavelength radiation that does not cause ionization; ultraviolet (UV) radiation is an example.

nonsense codon   A codon that does not encode any amino acid.

nonsense mutation   A base substitution in DNA that results in a nonsense codon.

nonspecific resistance   Host defenses that afford protection against any kind of pathogen. *See also* specific resistance.

normal microbiota   The microorganisms that colonize a host without causing disease; also called normal flora.

nosocomial infection   An infection that develops during the course of a hospital stay and was not present at the time the patient was admitted.

notifiable disease   A disease that physicians must report to the U.S. Public Health Service; also called reportable disease.

N (neuraminidase) spikes   Antigenic projections from the outer lipid bilayer of *Influenzavirus*.

nuclear envelope   The double membrane that separates the nucleus from the cytoplasm in a eukaryotic cell.

nuclear pore   An opening in the nuclear envelope through which materials enter and exit the nucleus.

nucleic acid   A macromolecule consisting of nucleotides; DNA and RNA are nucleic acids.

nucleic acid hybridization   The process of combining single complementary strands of DNA.

nucleic acid vaccine   A vaccine made up of DNA, usually in the form of a plasmid; also called DNA vaccine.

nucleoid   The region in a bacterial cell containing the chromosome.

nucleolus (plural: nucleoli)   An area in a eukaryotic nucleus where rRNA is synthesized.

nucleoside   A compound consisting of a purine or pyrimidine base and a pentose sugar.

nucleoside analog   A chemical that is structurally similar to the normal nucleosides in nucleic acids but with altered base-pairing properties.

nucleotide   A compound consisting of a purine or pyrimidine base, a five-carbon sugar, and a phosphate.

nucleotide excision repair (NER)   The repair of DNA involving removal of defective nucleotides and replacement with functional ones.

nucleus   (1) The part of an atom consisting of the protons and neutrons. (2) The part of a eukaryotic cell that contains the genetic material.

nutrient agar   Nutrient broth containing agar.

nutrient broth   A complex medium made of beef extract and peptone.

objective lenses   In a compound light microscope, the lenses closest to the specimen.

obligate aerobe   An organism that requires molecular oxygen (O$_2$) to live.

obligate anaerobe   An organism that does not use molecular oxygen (O$_2$) and is killed in the presence of O$_2$.

obligate halophile   An organism that requires high osmotic pressures such as high concentrations of NaCl.

ocular lens   In a compound light microscope, the lens closest to the viewer; also called the eyepiece.

oligodynamic action   The ability of small amounts of a heavy metal compound to exert antimicrobial activity.

oligoadenylate synthetase   A eukaryotic antiviral protein that degrades viral RNA.

oligosaccharide   A carbohydrate consisting of 2 to approximately 20 monosaccharides.

oncogene   A gene that can bring about malignant transformation.

oncogenic virus   A virus that is capable of producing tumors; also called oncovirus.

oocyst   An encysted apicomplexan zygote in which cell division occurs to form the next infectious stage.

operator   The region of DNA adjacent to structural genes that controls their transcription.

operon   The operator and promoter sites and structural genes they control.

opportunistic pathogen   A microorganism that does not ordinarily cause a disease but can become pathogenic under certain circumstances.

opsonization   The enhancement of phagocytosis by coating microorganisms with certain serum proteins (opsonins); also called immune adherence.

optimum growth temperature   The temperature at which a species grows best.

order   A taxonomic classification between class and family.

organelle   A membrane-enclosed structure within eukaryotic cells.

organic compound   A molecule that contains carbon and hydrogen.

organic growth factor   An essential organic compound that an organism is unable to synthesize.

osmosis   The net movement of solvent molecules across a selectively permeable membrane from an area of lower solute concentration to an area of higher solute concentration.

osmotic lysis   Rupture of the plasma membrane resulting from movement of water into the cell.

osmotic pressure   The force with which a solvent moves from a solution of lower solute concentration to a solution of higher solute concentration.

oxidation   The removal of electrons from a molecule.

oxidation pond   A method of secondary sewage treatment by microbial activity in a shallow standing pond of water.

oxidation-reduction   A coupled reaction in which one substance is oxidized and one is reduced; also called redox reaction.

oxidative phosphorylation   The synthesis of ATP coupled with electron transport.

oxygenic   Producing oxygen, as in plant and cyanobacterial photosynthesis.

PABA   Para-aminobenzoic acid; a precursor for folic acid synthesis.

pandemic disease   An epidemic that occurs worldwide.

parasite   An organism that derives nutrients from a living host.

parasitism   A symbiotic relationship in which one organism (the parasite) exploits another (the host) without providing any benefit in return.

parasitology   The scientific study of parasites (protozoa and parasitic worms).

parenteral route   A portal of entry for pathogens by deposition directly into tissues beneath the skin and mucous membranes.

pasteurization   The process of mild heating to kill particular spoilage microorganisms or pathogens.

pathogen   A disease-causing organism.

pathogenesis   The manner in which a disease develops.

pathogenicity   The ability of a microorganism to cause disease by overcoming the defenses of a host.

pathology   The scientific study of disease.

pellicle   (1) The flexible covering of some protozoa. (2) Scum on the surface of a liquid medium.

penicillins   A group of antibiotics produced either by *Penicillium* (natural penicillins) or by adding side chains to the β-lactam ring (semisynthetic penicillins).

pentose phosphate pathway   A metabolic pathway that can occur simultaneously with glycolysis to produce pentoses and NADH without ATP production; also called hexose monophosphate shunt.

peptide bond   A bond joining the amino group of one amino acid to the carboxyl group of a second amino acid with the loss of a water molecule.

peptidoglycan   The structural molecule of bacterial cell walls consisting of the molecules N-acetylglucosamine, N-acetylmuramic acid, tetrapeptide side chain, and peptide side chain.

perforin   Protein that makes a pore in a target cell membrane, released by T_C cells.

pericarditis   Inflammation of the pericardium, the sac around the heart.

period of convalescence   The recovery period, when the body returns to its predisease state.

peripheral nervous system (PNS)   The nerves that connect the outlying parts of the body with the central nervous system.

periplasm   The region of a gram-negative cell wall between the outer membrane and the cytoplasmic membrane.

peritrichous   Having flagella distributed over the entire cell.

peroxidase   An enzyme that breaks down hydrogen peroxide.

peroxide anion   An oxygen anion consisting of two atoms of oxygen.

peroxisome   Organelle that oxidizes amino acids, fatty acids, and alcohol.

peroxygen   A class of oxidizing-type sterilizing disinfectants.

persistent viral infection   *See* slow viral infection.

pH   The symbol for hydrogen ion (H^+) concentration; a measure of the relative acidity or alkalinity of a solution.

phage   *See* bacteriophage.

phage typing   A method of identifying bacteria using specific strains of bacteriophages.

phagocyte   A cell capable of engulfing and digesting particles that are harmful to the body.

phagocytosis   The ingestion of solids by eukaryotic cells.

phagolysosome   A digestive vacuole.

phagosome   A food vacuole of a phagocyte; also called a phagocytic vesicle.

phase-contrast microscope   A compound light microscope that allows examination of structures inside cells through the use of a special condenser.

phenol   Also called carbolic acid.

phenolic   A synthetic derivative of phenol used as a disinfectant.

phenotype   The external manifestations of an organism's genotype, or genetic makeup.

phosphate group   A portion of a phosphoric acid molecule attached to some other molecule, ⓟ,

$$PO_4^{3-}, \quad {}^-O-\overset{\overset{\displaystyle O}{\|}}{\underset{\underset{\displaystyle O^-}{|}}{P}}-O^-$$

phospholipid   A complex lipid composed of glycerol, two fatty acids, and a phosphate group.

phosphorous cycle   The various solubility stages of phosphorus in the environment.

phosphorylation   The addition of a phosphate group to an organic molecule.

photoautotroph   An organism that uses light as its energy source and carbon dioxide (CO_2) as its carbon source.

photoheterotroph   An organism that uses light as its energy source and an organic carbon source.

photophosphorylation   The production of ATP in a series of redox reactions; electrons from chlorophyll initiate the reactions.

photosynthesis   The conversion of light energy from the sun into chemical energy; the light-fueled synthesis of carbohydrate from carbon dioxide (CO_2).

phototaxis   Movement in response to the presence of light.

phototroph   An organism that uses light at its primary energy source.

phylogeny   The evolutionary history of a group of organisms; phylogenetic relationships are evolutionary relationships.

phylum   A taxonomic classification between kingdom and class.

phytoplankton   Free-floating photoautotrophs.

pilus (plural: **pili**)   An appendage on a bacterial cell used for the transfer of genetic material during conjugation.

pinocytosis   The engulfing of fluid by infolding of the plasma membrane, in eukaryotes.

plankton   Free-floating aquatic organisms.

plantae   The kingdom composed of multicellular eukaryotes with cellulose cell walls.

plaque   A clearing in a bacterial lawn resulting from lysis by phages. *See also* dental plaque.

plaque-forming units (pfu)   Visible viral plaques counted.

plasma   The liquid portion of blood in which the formed elements are suspended.

plasma cell   A cell that an activated B cell differentiates into; plasma cells manufacture specific antibodies.

plasma (cytoplasmic) membrane   The selectively permeable membrane enclosing the cytoplasm of a cell; the outer layer in animal cells, internal to the cell wall in other organisms.

plasmid   A small circular DNA molecule that replicates independently of the chromosome.

plasmodium   (1) A multinucleated mass of protoplasm, as in plasmodial slime molds. (2) When written as a genus, refers to the causative agent of malaria.

plasmogamy   Fusion of the cytoplasm of two cells; occurs in the sexual stage of a fungal life cycle.

plasmolysis   Loss of water from a cell in a hypertonic environment.

plate count   A method of determining the number of bacteria in a sample by counting the number of colony-forming units on a solid culture medium.

pleomorphic   Having many shapes, characteristic of certain bacteria.

pneumonia   Inflammation of the lungs.

point mutation   *See* base substitution.

polar molecule   A molecule with an unequal distribution of charges.

polyene antibiotic   An antimicrobial agent that alters sterols in eukaryotic plasma membranes and contains more than four carbon atoms and at least two double bonds.

polymer   A molecule consisting of a sequence of similar molecules, or monomers.

polymerase chain reaction (PCR)   A technique using DNA polymerase to make multiple copies of a DNA template in vitro. *See also* cDNA.

polymorphonuclear leukocyte (PMN)   *See* neutrophil.

polypeptide   (1) A chain of amino acids. (2) A group of antibiotics.

polysaccaride   A carbohydrate consisting of 8 or more monosaccharides joined through dehydration synthesis.

porins   A type of protein in the outer membrane of gram-negative cell walls that permits the passage of small molecules.

portal of entry   The avenue by which a pathogen gains access to the body.

portal of exit   The route by which a pathogen leaves the body.

positive (direct) selection   A procedure for picking out mutant cells by growing them.

pour plate method   A method of inoculating a solid nutrient medium by mixing bacteria in the melted medium and pouring the medium into a Petri dish to solidify.

precipitation reaction   A reaction between soluble antigens and multivalent antibodies to form visible aggregates.

precipitin ring test   A precipitation test performed in a capillary tube.

predisposing factor   Anything that makes the body more susceptible to a disease or alters the course of a disease.

prevalence   The fraction of a population having a specific disease at a given time.

primary cell line   Human tissue cells that grow for only a few generations in vitro.

primary infection   An acute infection that causes the initial illness.

primary metabolite   A product of an industrial cell population produced during the time of rapid logarithmic growth. *See also* secondary metabolite.

primary producer   An autotrophic organism, either chemotroph or phototroph, that converts carbon dioxide into organic compounds.

primary response   Antibody production in response to the first contact with an antigen. *See also* memory response.

primary sewage treatment   The removal of solids from sewage by allowing them to settle out and be held temporarily in tanks or ponds.

prion   An infectious agent consisting of a self-replicating protein, with no detectable nucleic acids.

privileged site (tissue)   An area of the body (or a tissue) that does not elicit an immune response.

prokaryote   A cell whose genetic material is not enclosed in a nuclear envelope.

prokaryotic species   A population of cells that share certain rRNA sequences; in conventional biochemical testing, it is a population of cells with similar characteristics.

prodromal period   The time following the incubation period when the first symptoms of illness appear.

profundal zone   The deeper water under the limnetic zone in an inland body of water.

proglottid   A body segment of a tapeworm containing both male and female organs.

promoter   The starting site on a DNA strand for transcription of RNA by RNA polymerase.

prophage   Phage DNA inserted into the host cell's DNA.

prostaglandin   A hormonelike substance that is released by damaged cells; intensifies inflammation.

prostheca   A stalk or bud protruding from a prokaryotic cell.

protease inhibitor   A chemical that inhibits proteases (proteolytic enzymes).

protein   A large molecule containing carbon, hydrogen, oxygen, and nitrogen (and sulfur); some proteins have a helical structure and others are pleated sheets.

protein kinase   An enzyme that activates another protein by adding a ⓟ from ATP.

proteobacteria   Gram-negative, chemoheterotrophic bacteria that possess a signature rRNA sequence.

protist   Term used for unicellular and simple multicellular eukaryotes; usually protozoa and algae.

proton   A positively charged particle in the nucleus of an atom.

protoplast   A gram-positive bacterium or plant cell treated to remove the cell wall.

protoplast fusion   A method of joining two cells by first removing their cell walls; used in genetic engineering.

protozoan (plural: protozoa)   Unicellular eukaryotic organisms; usually chemoheterotrophic.

provirus   Viral DNA that is integrated into the host cell's DNA.

pseudohypha   A short chain of fungal cells that results from the lack of separation of daughter cells after budding.

pseudopod   An extension of a eukaryotic cell that aids in locomotion and feeding.

psychrophile   An organism that grows best at about 15°C and does not grow above 20°C; a cold-loving microbe.

pscyhrotroph   An organism that is capable of growth between about 0°C and 30°C.

purines   The class of nucleic acid bases that includes adenine and guanine.

purple nonsulfur bacteria   Alpha-proteobacteria; strictly anaerobic and phototrophic; grow on yeast extract in dark; use reduced organic compounds as electron donors for CO_2 fixation.

purple sulfur bacteria   Gamma-proteobacteria; strictly anaerobic and phototrophic; use reduced sulfur compounds as electron donors for CO_2 fixation.

pus   An accumulation of dead phagocytes, dead bacterial cells, and fluid.

pyocyanin   A blue-green pigment produced by *Pseudomonas aeruginosa*.

pyrimidines   The class of nucleic acid bases that includes uracil, thymine, and cytosine.

quaternary ammonium compound (quat)   A cationic detergent with four organic groups attached to a central nitrogen atom; used as a disinfectant.

quinine   An antimalarial drug derived from the cinchona tree and effective against schizogomy stage in red blood cells.

quinolone   An antibiotic whose mode of action is to inhibit DNA replication by interfering with the enzyme DNA gyrase.

R   Used to represent nonfunctional groups of a molecule. *See also* resistance factor.

radioimmunoassay (RIA)   A method of measuring the amount of a compound using radioactively labeled antibodies.

rapid plasma reagin (RPR) test   A serological test for syphilis.

r-determinant   A group of genes for antibiotic resistance carried on R factors.

receptor   An attachment for a pathogen on a host cell.

recipient cell   A cell that receives DNA from a donor cell during genetic recombination.

recombinant DNA   A DNA molecule produced by recombination.

recombinant DNA technology   *See* genetic engineering.

recombinant vaccine   A vaccine made by recombinant DNA techniques.

redia   A trematode larval stage that reproduces asexually to produce cercariae.

redox reaction   *See* oxidation-reduction.

red tide   A bloom of planktonic dinoflagellates.

reducing medium   A culture medium containing ingredients that will remove dissolved oxygen from the medium to allow the growth of anaerobes.

reduction   The addition of electrons to a molecule.

refractive index   The relative velocity with which light passes through a substance.

rennin   An enzyme that forms curds as part of any dairy fermentation product; originally from calves' stomachs, now produced by molds and bacteria.

replica plating   A method of inoculating a number of solid minimal culture media from an original plate to produce the same pattern of colonies on each plate.

replication fork   The point where DNA strands separate and new strands will be synthesized.

repression   The process by which a repressor protein can stop the synthesis of a protein.

repressor   A protein that binds to the operator site to prevent transcription.

reservoir of infection   A continual source of infection.

resistance   The ability to ward off diseases through nonspecific and specific defenses.

resistance (R) factor   A bacterial plasmid carrying genes that determine resistance to antibiotics.

resistance transfer factor   A group of genes for replication and conjugation on the R factor.

resolution   The ability to distinguish fine detail with a magnifying instrument; also called resolving power.

respiration   A series of redox reactions in a membrane that generates ATP; the final electron acceptor is usually an inorganic molecule.

restriction enzyme   An enzyme that cuts double-stranded DNA at specific sites between nucleotides.

retort   A device for commercially sterilizing canned food by using steam under pressure; operates on the same principle as an autoclave but is much larger.

reverse transcriptase   An RNA-dependent DNA polymerase; an enzyme that synthesizes a complementary DNA from an RNA template.

reversible reaction   A chemical reaction in which the end-products can readily revert to the original molecules.

RFLP   Restriction fragment length polymorphism; a fragment resulting from restriction-enzyme digestion of DNA.

Rh factor   An antigen on red blood cells of rhesus monkeys and most humans; possession makes the cells Rh^+.

Rh blood group system   The classification of red blood cells based on the presence or absence of Rh antigens.

rhizine   A rootlike hypha that anchors a fungus to a surface.

rhizopoda   Eukaryotic organisms that move using pseudopods; amoebas.

ribonucleic acid (RNA)   The class of nucleic acids that comprises messenger RNA, ribosomal RNA, and transfer RNA.

ribose   A five-carbon sugar that is part of ribonucleotide molecules and RNA.

ribosomal RNA (rRNA)   The type of RNA molecule that forms ribosomes.

ribosomal RNA (rRNA) sequencing   Determination of the order of nucleotide bases in rRNA.

ribosome   The site of protein synthesis in a cell, composed of RNA and protein.

ribozyme   An enzyme consisting of RNA that specifically acts on strands of RNA to remove introns and splice together the remaining exons.

rifamycin   An antibiotic that inhibits bacterial RNA synthesis.

ring stage   A young *Plasmodium* trophozoite that looks like a ring in a red blood cell.

RNA primer   A short strand of RNA used to start synthesis of the lagging strand of DNA, and to start the polymerase chain reaction.

root nodule   A tumorlike growth on the roots of certain plants containing symbiotic nitrogen-fixing bacteria.

rotating biological contactor   A method of secondary sewage treatment in which large disks are rotated while partially submerged in a sewage tank exposing sewage to microorganisms and aerobic conditions.

rough ER   Endoplasmic reticulum with ribosomes on its surface.

roundworm   An animal belonging to the phylum Nematoda.

S (Svedberg unit)   Notes the relative rate of sedimentation during ultra-high-speed centrifugation.

salt   A substance that dissolves in water to cations and anions, neither of which is H^+ or OH^-.

sanitization   The removal of microbes from eating utensils and food preparation areas.

saprophyte   An organism that obtains its nutrients from dead organic matter.

sarcina (plural: sarcinae)   (1) A group of eight bacteria that remain in a packet after dividing. (2) When written as a genus, refers to gram-positive, anaerobic cocci.

sarcoma   A cancer of fleshy, nonepithelial tissue or connective tissue.

saturation   The condition in which the active site on an enzyme is occupied by the substrate or product at all times.

saxitoxin   A neurotoxin produced by some dinoflagellates.

scanned probe microscopy   Microscopic technique used to obtain images of molecular shapes, to characterize chemical properties, and to determine temperature variations within a specimen.

scanning electron microscope (SEM)   An electron microscope that provides three-dimensional views of the specimen magnified 1000–10,000×.

scanning tunneling microscopy   *See* scanned-probe microscopy.

schizogony   The process of multiple fission, in which one organism divides to produce many daughter cells.

scientific nomenclature   *See* binomial nomenclature.

sclerotia   The compact mass of hardened mycelia of the fungus *Claviceps purpurea* that fills infected rye flowers; produces the toxin ergot.

scolex   The head of a tapeworm, containing suckers and possibly hooks.

secondary infection   An infection caused by an opportunistic microbe after a primary infection has weakened the host's defenses.

secondary metabolite   A product of an industrial cell population produced after the microorganism has largely completed its period of rapid growth and is in a stationary phase of the growth cycle. *See also* primary metabolite.

secondary response   *See* memory response.

secondary sewage treatment   Biological degradation of the organic matter in wastewater following primary treatment.

secretory vesicle   A membrane-enclosed sac produced by the ER; transports synthesized material into cytoplasm.

selective medium   A culture medium designed to suppress the growth of unwanted microorganisms and encourage the growth of desired ones.

selective permeability   The property of a plasma membrane to allow certain molecules and ions to move through the membrane while restricting others.

selective toxicity   The property of some antimicrobial agents to be toxic for a microorganism and nontoxic for the host.

self-tolerance   The ability of an organism to recognize and not make antibodies against self.

semiconservative replication   The process of DNA replication in which each double-stranded DNA molecule contains one original strand and one new strand.

semisynthetic penicillins   Modifications of natural penicillins by introducing different side chains that extend the spectrum of antimicrobial activity and avoid microbial resistance.

sense codon   A codon that codes for an amino acid.

sense strand (+ strand)   Viral RNA that can act as mRNA.

sepsis   A toxic condition resulting from the growth and spread of bacteria in blood and tissue.

septa   A cross-wall in a fungal hypha.

septate hypha   A hypha consisting of uninucleate cell-like units.

septicemia   The proliferation of bacteria in the blood, accompanied by fever; sometimes causes organ damage.

septic shock   A sudden drop in blood pressure induced by bacterial toxins, usually endotoxins from gram-negative bacteria.

serial dilution   The process of diluting a sample several times.

seroconversion   A change in a person's response to an antigen in a serological test.

serological testing   Techniques for identifying a microorganism based on its reaction with antibodies.

serology   The branch of immunology that studies blood serum and antigen–antibody reactions in vitro.

serotype   *See* serovar.

serovar   A variation within a species; also called serotype.

serum   The liquid remaining after blood plasma is clotted; contains antibodies (immunoglobulins).

sexual dimorphism   The distinctly different appearance of adult male and female organisms.

sexual spore   A spore formed by sexual reproduction.

Shiga toxin   An exotoxin produced by *Shigella dysenteriae* and enterohemorrhagic *E. coli.*

shock   Any life-threatening loss of blood pressure. *See also* septic shock.

shuttle vector   A plasmid that can exist in several different species; used in genetic engineering.

sign   A change due to a disease that a person can observe and measure.

simple diffusion   The net movement of molecules or ions from an area of higher concentration to an area of lower concentration.

simple stain   A method of staining microorganisms with a single basic dye.

singlet oxygen   Highly reactive molecular oxygen (O_2).

site-directed mutagenesis   Techniques used to modify a gene in a specific location to produce the desired polypeptide.

slide agglutination test   A method of identifying an antigen by combining it with a specific antibody on a slide.

slime layer   A glycocalyx that is unorganized and loosely attached to the cell wall.

slime mold   A funguslike protist.

slow viral infection   A disease process that occurs gradually over a long period; also called persistent viral infection.

sludge   Solid matter obtained from sewage.

smear   A thin film of material containing microorganisms, spread over the surface of a slide.

smooth ER   Endoplasmic reticulum without ribosomes.

solute   A substance dissolved in another substance.

solvent   A dissolving medium.

Southern blotting   A technique that uses DNA probes to detect the presence of specific DNA in restriction fragments separated by electrophoresis.

specialized transduction   The process of transferring a piece of cell DNA adjacent to a prophage to another cell.

species   The most specific level in the taxonomic hierarchy. *See also* bacterial species; eukaryotic species; viral species.

specific epithet   The second or species name in a scientific binomial. *See also* species.

specific resistance   *See* immunity.

spectrum of microbial activity   The range of distinctly different types of microorganisms affected by an antimicrobial drug; a wide range is referred to as a broad spectrum of activity.

spheroplast   A gram-negative bacterium treated to damage the cell wall, resulting in a spherical cell.

spicule   One of two external structures on the male roundworm used to guide sperm.

spike   A carbohydrate-protein complex that projects from the surface of certain viruses.

spiral   *See* spirillum and spirochete.

spirillum (plural: spirilla)   (1) A helical or corkscrew-shaped bacterium. (2) When written as a genus, refers to aerobic, helical bacteria with clumps of polar flagella.

spirochete   A corkscrew-shaped bacterium with axial filaments.

splicing   A process by which an RNA molecule cuts out introns and rejoins exons to produce a molecule of mRNA.

spontaneous generation   The idea that life could arise spontaneously from nonliving matter.

spontaneous mutation   A mutation that occurs without a mutagen.

sporadic disease   A disease that occurs occasionally in a population.

sporangiophore   An aerial hypha supporting a sporangium.

sporangiospore   An asexual fungal spore formed within a sporangium.

sporangium   A sac containing one or more spores.

spore   A reproductive structure formed by fungi and actinomycetes.

sporogenesis   *See* sporulation.

sporozoite   A trophozoite of *Plasmodium* found in mosquitoes, infective for humans.

sporulation   The process of spore and endospore formation; also called sporogenesis.

spread plate method   A plate count method in which inoculum is spread over the surface of a solid culture medium.

staining   Colorizing a sample with a dye to view through a microscope or to visualize specific structures.

staphylococci (singular: staphylococcus)   Cocci in a grapelike cluster or broad sheet.

stationary phase   The period in a bacterial growth curve when the number of cells dividing equals the number dying.

stem cell   A fetal cell that gives rise to red bone marrow, blood cells, and B and T cells.

stereoisomers   Two molecules consisting of the same atoms, arranged in the same manner but differing in their relative positions; mirror images; also called D-isomer and L-isomer.

sterile   Free of microorganisms.

sterilization   The killing of all microorganisms, including endospores.

steroid   A specific group of lipids, including cholesterol and hormones.

stipe   A stemlike supporting structure of multicellular algae and basidiomycetes.

storage vesicle   Organelles that form from the Golgi complex; contain proteins made in the rough ER and processed in the Golgi complex.

strain   A genetically identical group of cells derived from a single cell. *See* serovar.

streak plate method   A method of isolating a culture by spreading microorganisms over the surface of a solid culture medium.

streptobacilli (singular: streptobacillus)   Rods that remain attached in chains after cell division.

streptococci (singular: streptococcus)   (1) Cocci that remain attached in chains after cell division. (2) When written as a genus, refers to gram-positive, catalase-negative bacteria.

structural gene   A gene that determines the amino acid sequence of a protein.

subacute disease   A disease with symptoms that are intermediate between acute and chronic.

subclinical infection   An infection that does not cause a noticeable illness; also called inapparent infection.

subcutaneous mycosis   A fungal infection of tissue beneath the skin.

substrate   Any compound with which an enzyme reacts.

substrate-level phosphorylation   The synthesis of ATP by direct transfer of a high-energy phosphate group from an intermediate metabolic compound to ADP.

subunit vaccine   A vaccine consisting of an antigenetic fragment.

sulfhydryl group   —SH.

sulfonamide   A bacteriostatic compound that interferes with folic acid synthesis by competitive inhibition; also called a sulfa drug.

sulfur cycle   The various oxidation and reduction stages of sulfur in the environment, mostly due to the action of microorganisms.

sulfur granule   *See* inclusion.

superantigen   An antigen that activates many different T cells, thereby eliciting a large immune response.

superficial mycosis   A fungal infection localized in surface epidermal cells and along hair shafts.

superinfection   The growth of a pathogen that has developed resistance to an antimicrobial drug being used; the growth of an opportunistic pathogen.

superoxide dismutase (SOD)   An enzyme that destroys superoxide free radicals.

superoxide free radical   A toxic form of oxygen ($O_2^{-\bullet}$) formed during aerobic respiration.

suppressor T cell (T_S)   A T cell that is thought to end an immune response after an antigen is no longer present.

surface–active agent   Any compound that decreases the tension between molecules lying on the surface of a liquid; also called surfactant.

susceptibility   The lack of resistance to a disease.

symbiosis   The living together of two different organisms or populations.

symptom   A change in body function that is felt by a patient as a result of a disease.

syncytium   A multinucleated giant cell resulting from certain viral infections.

syndrome   A specific group of signs or symptoms that accompany a disease.

synergism   (1) The effect of two microbes working together that is greater than the effect of either acting alone. (2) The principle whereby the effectiveness of two drugs used simultaneously is greater than that of either drug used alone.

synthesis reaction   A chemical reaction in which two or more atoms combine to form a new, larger molecule.

synthetic drug   A chemotherapeutic agent that is prepared from chemicals in a laboratory.

systemic anaphylaxis   A hypersensitivity reaction causing vasodilation and resulting in shock; also called anaphylactic shock.

systemic (generalized) infection   An infection throughout the body.

systemic mycosis   A fungal infection in deep tissues.

tachyzoite   A rapidly growing trophozoite form of a protozoan.

T antigen   An antigen in the nucleus of a tumor cell.

tapeworm   A flatworm belonging to the class Cestoda.

target cell   An infected body cell to which defensive cells of the immune system bind.

taxis   Movement in response to an environmental stimulus.

taxonomy   The science of the classification of organisms.

T cell   A type of lymphocyte, which develops from a stem cell processed in the thymus gland, that is responsible for cell-mediated immunity.

T-dependent antigen   An antigen that will stimulate the formation of antibodies only with the assistance of helper T cells. *See also* T-independent antigen.

teichoic acid   A polysaccharide found in gram-positive cell walls.

teleomorph   The sexual stage in the life cycle of a fungus; also refers to a fungus that produces both sexual and asexual spores.

temperature abuse   Improper food storage at a temperature that allows bacteria to grow.

terminator   The site on a DNA strand at which transcription ends.

tertiary sewage treatment   A method of waste treatment that follows conventional secondary sewage treatment; nonbiodegradable pollutants and mineral nutrients are removed, usually by chemical or physical means.

tetracycline   Broad-spectrum antibiotics that interfere with protein synthesis.

tetrad   A group of four cocci.

thallus   The entire vegetative structure or body of a fungus, lichen, or alga.

thermal death point (TDP)   The temperature required to kill all the bacteria in a liquid culture in 10 minutes.

thermal death time (TDT)   The length of time required to kill all bacteria in a liquid culture at a given temperature.

thermoduric   Heat resistant.

thermophile   An organism whose optimum growth temperature is between 50°C and 60°C; a heat loving microbe.

thermophilic anaerobic spoilage   Spoilage of canned foods due to the growth of thermophilic bacteria.

thylakoid   A chlorophyll-containing membrane in a chloroplast. A bacterial thylakoid is also known as a chromatophore.

tincture   A solution in aqueous alcohol.

T-independent antigen   An antigen that will stimulate the formation of antibodies without the assistance of helper T cells. *See also* T-dependent antigen.

Ti plasmid   An *Agrobacterium* plasmid carrying genes for tumor induction in plants.

titer   An estimate of the amount of antibodies or viruses in a solution; determined by serial dilution and expressed as the reciprocal of the dilution.

total magnification   The magnification of a microscopic specimen, determined by multiplying the ocular lens magnification by the objective lens magnification.

toxemia   The presence of toxins in the blood.

toxigenicity   The capacity of a microorganism to produce a toxin.

toxin   Any poisonous substance produced by a microorganism.

toxoid   An inactivated toxin.

trace element   A chemical element required in small amounts for growth.

transamination   The transfer of an amino group from an amino acid to another organic acid.

transcription   The process of synthesizing RNA from a DNA template.

transduction   The transfer of DNA from one cell to another by a bacteriophage. *See also* generalized transduction; specialized transduction.

transferrin   A human iron-binding protein that reduces iron available to a pathogen.

transfer RNA (tRNA)   The type of RNA molecule that brings amino acids to the ribosomal site where they are incorporated into proteins.

transfer vesicle   Membrane-bound sacs that move proteins from the Golgi complex to specific areas in the cell.

transformation   (1) the process in which genes are transferred from one bacterium to another as "naked" DNA in solution. (2) The changing of a normal cell into a cancerous cell.

transient microbiota   The microorganisms that are present in an animal for a short time without causing a disease.

translation   The use of mRNA as a template in the synthesis of protein.

transmembrane channel   A pore in a target cell membrane produced by complement.

transmission electron microscope (TEM)   An electron microscope that provides high magnifications (10,000– 100,000$\times$) of thin sections of a specimen.

transport vesicle   Membrane-bound sacs that move proteins from the rough ER to the Golgi complex.

transporter protein   A carrier protein in the plasma membrane.

transposon   A small piece of DNA that can move from one DNA molecule to another.

trickling filter   A method of secondary sewage treatment in which sewage is sprayed out of rotating arms onto a bed of rocks or similar materials, exposing the sewage to highly aerobic conditions and microorganisms.

triglyceride   A simple lipid consisting of glycerol and three fatty acids.

triplex agent   A short segment of DNA that binds to a target area on a double strand of DNA blocking transcription.

trophophase   The period in the production curve of an industrial cell population in which the primary metabolites are formed; a period of rapid, logarithmic growth. *See also* idiophase.

trophozoite   The vegetative form of a protozoan.

tuberculin skin test   A skin test used to detect the presence of antibodies to *Mycobacterium tuberculosis.*

tumor necrosis factor (TNF)   A polypeptide released by phagocytes in response to bacterial endotoxins; induces shock; also called cachectin.

tumor-specific transplantation antigen (TSTA)   A viral antigen on the surface of a transformed cell.

turbidity   The cloudiness of a suspension.

turnover number   The number of substrate molecules acted on per enzyme molecule per second.

12D treatment   A sterilization process that would result in a decrease of the number of *Clostridium botulinum* endospores by 12 logarithmic cycles.

ubiquinone   A low–molecular weight, nonprotein carrier in an electron transport chain; also called coenzyme Q.

UDP-N-acetylglucosamine (UDPNAc)   A compound necessary for the biosynthesis of peptidoglycan.

ultra-high-temperature (UHT) treatment   A method of treating food with high temperatures (140–150°C) for very short times to make the food sterile so that it can be stored at room temperature.

uncoating   The separation of viral nucleic acid from its protein coat.

undulating membrane   A highly modified flagellum on some protozoa.

use-dilution test   A method of determining the effectiveness of a disinfectant using serial dilutions.

vaccination   The process of conferring immunity by administering a vaccine; also called immunization.

vaccine   A preparation of killed, inactivated, or attenuated microorganisms or toxoids to induce artificially acquired active immunity.

vacuole   An intracellular inclusion, in eukaryotic cells, surrounded by a plasma membrane; in prokaryotic cells, surrounded by a proteinaceous membrane.

valence   The combining capacity of an atom or a molecule.

vancomycin   An antibiotic that inhibits cell wall synthesis.

variolation   An early method of vaccination using infected material from a patient.

vasodilation   Dilation or enlargement of blood vessels.

VDRL test   A rapid screening test to detect the presence of antibodies against *Treponema pallidum*. (VDRL stands for Venereal Disease Research Laboratory.)

vector   (1) A plasmid or virus used in genetic engineering to insert genes into a cell. (2) An arthropod that carries disease-causing organisms from one host to another.

vegetative   Referring to cells involved with obtaining nutrients, as opposed to reproduction.

vehicle transmission   The transmission of a pathogen by an inanimate reservoir.

V factor NAD^+ or $NADP^+$.

vertical gene transfer   Transfer of genes from an organism or cell to its offspring.

vesicle   (1) A small serum-filled elevation of the skin. (2) Smooth oval bodies formed in plant roots by mycorrhizae.

vibrio   (1) A curved or comma-shaped bacterium. (2) When written as a genus, a gram-negative, motile, facultatively anaerobic curved rod.

vibroid   A curved, helical bacterium that does not make a complete turn or twist.

viral hemagglutination   The ability of certain viruses to cause the clumping of red blood cells in vitro.

viral hemagglutination inhibition test   A neutralization test in which antibodies against particular viruses prevent the viruses from clumping red blood cells in vitro.

viral species   A group of viruses sharing the same genetic information and ecological niche.

viremia   The presence of viruses in the blood.

virion   A complete, fully developed viral particle.

viroid   Infectious RNA.

virology   The scientific study of viruses.

virulence   The degree of pathogenicity of a microorganism.

virus   A submicroscopic, parasitic, filterable agent consisting of a nucleic acid surrounded by a protein coat.

volutin   Stored inorganic phosphate in a prokaryotic cell. *See also* metachromatic granule.

wandering macrophage   A macrophage that leaves the blood and migrates to infected tissue.

Western blotting   A technique that uses antibodies to detect the presence of specific proteins separated by electrophoresis.

wheal   A small, burning or itching swelling of the skin, such as that resulting from a skin test or a mosquito bite.

whey   The fluid portion of milk that separates from curd.

xenodiagnosis   A method of diagnosis based on exposing a parasite-free normal host to the parasite and then examining the host for parasites.

xenograft   A tissue graft from another species; also called xenotransplant.

X factor   Substances from the heme fraction of blood hemoglobin.

yeast   Nonfilamentous, unicellular fungi.

yeast infection   Disease caused by growth of certain yeasts in a susceptible host.

zone of inhibition   The area of no bacterial growth around an antimicrobial agent in the disk-diffusion method.

zoonosis   A disease that occurs primarily in wild and domestic animals but can be transmitted to humans.

zygospore   A sexual fungal spore characteristic of the zygomycetes.

zygote   A diploid cell produced by the fusion of two haploid gametes.

Credits

Text and Illustration Credits

2.1, 2.3, 2.6, 2.10, 2.13, 2.15, 2.16: G.J. Tortora and S.R. Grabowski, *Principles of Anatomy and Physiology*, 8th Ed. Figures 2.1, 2.4, 2.6, 2.9, 2.12, 2.13, 2.14, pp. 28, 31, 36, 41, 44, 45. Copyright ©1996 by Biological Sciences Textbooks, Inc., A & P Textbooks, Inc., and Sandra Reynolds Grabowski. Reprinted by permission of Addison Wesley Longman, Inc. Illustrations by Jared Schneidman Design.

T2.4: Adapted from N. Campbell, *Biology*, 4th Ed. (Menlo Park, CA: Benjamin/Cummings 1996), Figure 5.15, p. 75. © 1996 by The Benjamin/Cummings Publishing Company, Inc.

Chapter 3: Text excerpt from *USA Today*, August 6, 1996, p. 2A.

4.21, 4.23, 4.24, 4.25: Adapted from G.J. Tortora and S.R. Grabowski, *Principles of Anatomy and Physiology*, 9th Ed. Figures 3.1, 3.25, 3.20, 3.21, pp. 61, 83, 80, 81. Copyright © 2000 by Biological Sciences Textbooks, Inc., A & P Textbooks, Inc. and Sandra Reynolds Grabowski. Reprinted by permission of John Wiley.

5.1 G.J. Tortora and S.R. Grabowski, *Principles of Anatomy and Physiology*, 8th Ed. Figure 25.1, p. 808. Copyright ©1996 by Biological Sciences Textbooks, Inc., A & P Textbooks, Inc., and Sandra Reynolds Grabowski. Reprinted by permission of Addison Wesley Longman, Inc. Illustration by Page Two Associates.

T7.2: O. Rahns, *Physiology of Bacteria*. New York: McGraw-Hill, 1932.

T7.6: Adapted from S.S. Block, *Disinfection, Sterilization, and Preservation*, 4th Ed., Philadelphia: Lea and Febiger, 1951, p. 194.

8.3: E.N. Marieb, *Human Anatomy and Physiology*, 3rd Ed. (Menlo Park, CA: Benjamin/Cummings, 1995), Figure 3.27, p. 98. © 1995 The Benjamin/Cummings Publishing Company, Inc.

8.4, 8.5: Adapted from N. Campbell *Biology*, 5th Edition (San Francisco, CA: Benjamin/Cummings, 1999), Figure 16.12 and 16.11, p. 287. The Benjamin/Cummings Publishing Company, Inc.

8.7b: Adapted from N. Campbell *Biology*, 4th Ed. (Menlo Park, CA: Benjamin/Cummings, 1996), Figure 17.9, p. 336. © 1996 The Benjamin/Cummings Publishing Company, Inc.

8.8: P. Berg and M. Singer, *Dealing With Genes: The Language of Heredity* (Mill Valley, CA: University Science Books, 1992). Figure 13.1, p. 61. © 1992 University Science Books.

8.17: Adapted from N. Campbell *Biology*, 4th Ed. (Menlo Park, CA: Benjamin/Cummings, 1996), Figure 16.23, p. 318. © 1996 The Benjamin/Cummings Publishing Company, Inc.

9.12: Adapted from N. Campbell *Biology*, 4th Ed. (Menlo Park, CA: Benjamin/Cummings, 1996), Figures 19.1, 19.6, pp. 371, 376. © 1996 The Benjamin/Cummings Publishing Company, Inc.

9.14: From *Recombinant DNA*, 2nd Ed. © 1992 by James D. Watson, Michael Gilman, Jan Witkowski, and Mark Zoller. Used with permission of W.H. Freeman and Company.

9.4: M. Bloom, G., and D. Micklos, *Laboratory DNA Science,* (Menlo Park, CA: Benjamin/Cummings, 1996). p. 283. ©1996 by The Benjamin/Cummings Publishing Company, Inc.

Chapter 11: Text excerpt from *JAMA*, November 18, 1988, Vol. 260, no. 19.

T12.2: After B.D. Davis et al., *Microbiology*, 4th ed. Philadelphia: J.B. Lippincott, 1990, p. 746.

T13.2: Art from R.I.B. Francki, C.M. Fauquet, D.L. Knudson, F. Brown, eds, "Classification and Nomenclature of Viruses. Fifth Report of the International Committee on Taxonomy of Viruses." *Archives of Virology: Supplementum 2* (Wien, New York: Springer-Verlag, 1991). Copyright ©1991 by Springer-Verlag.

14.4: Source: CDC

16.1: Adapted from N. Campbell *Biology*, 4th Ed. (Menlo Park, CA: Benjamin/Cummings, 1996), Figure 39.1, p. 853. © 1996 by The Benjamin/Cummings Publishing Company, Inc.

16.6, 16.8a: G.J. Tortora and S.R. Grabowski, *Principles of Anatomy and Physiology*, 7th Ed. Figures 22.1, 22.10, pp. 684, 696. Copyright © 1993 by Biological Sciences Textbooks, A & P Textbooks, Inc., and Sandra Reynolds Grabowski. Reprinted by permission of Addison Wesley Longman, Inc.

17.1, 17.Box1: E.N. Marieb, *Human Anatomy and Physiology*, 3rd Ed. (Menlo Park, CA: Benjamin/Cummings, 1995), Figures 22.10, 22.12, pp. 722, 725. © 1995 by The Benjamin/Cummings Publishing Company, Inc.

17.6, 17.12, 17.13: Adapted from N. Campbell, *Biology*, 4th Ed. (Menlo Park, CA: Benjamin/Cummings, 1996), Figures 39.6, 39.13, 39.14, pp. 858, 867, 868. © 1996 by The Benjamin/Cummings Publishing Company, Inc.

T19.2: E.N. Marieb, *Human Anatomy and Physiology*, 3rd Ed. (Menlo Park, CA: Benjamin/Cummings, 1995), Table 18.4, p. 605. © 1995 by The Benjamin/Cummings Publishing Company, Inc.

19.11: Hoth, Jr., Meyers, Stein, "Current Status of HIV Therapy," *Hospital Practice*, Vol. 27, No. 9, p. 145. Illustration Alan D. Iselin. Reprinted with permission.

19.12, 19.13: Greenberg, "Immunopathogenesis of HIV Infection," *Hospital Practice*, Vol. 27, No. 2, p. 109. Illustrations Ilil Arbel. Reprinted with permission.

19.15: "Estimated number of adults with HIV/AIDS, by region, 1997" from UNAIDS data by the Map Design Unit of the World Bank. Reprinted by permission of Confronting AIDS: Public Priorities in a Global Epidemic.

20.21: *Biology of Microorganisms*, 7th Ed. by T.D. Brock, M.T. Madigan, Martinko, and Parker p. 410. © 1993. Adapted by permission of Prentice-Hall Inc., Upper Saddle River, New Jersey.

21.12: Adapted from *Medical Microbiology*, 3rd Edition by Patrick R. Murray et al. Copyright © 1993 by Williams & Wilkins. Reprinted by permission.

22.3: Source: CDC

22.11: Adapted from *Medical Microbiology*, 3rd Edition by Patrick R. Murray et al. Copyright © 1993 by Williams & Wilkins.

23.1: Adapted from N. Campbell, *Biology*, 3rd Ed. (Redwood City, CA: Benjamin/Cummings, 1993), Figure 38.5a, p. 823 © 1993 by The Benjamin/Cummings Publishing Company, Inc.

23.13a: Steere, "Current Understanding of Lyme Disease," *Hospital Practice*, Vol. 28, No. 4, p. 37. Illustration Nancy Lou Makris Riccio. Reprinted by permission.

24.2: Adapted from N. Campbell, *Biology*, 3rd Ed. (Redwood City, CA: Benjamin/Cummings, 1993), Figure 38.22, p. 840 © 1993 by The Benjamin/Cummings Publishing Company, Inc.

24.10: Dannenberg, Jr., "Pulmonary Tuberculosis," *Hospital Practice*, Vol. 28, No. 1, p. 51 illustrations by Seward Hung, and by Laura Pardi Duprey. Reprinted by permission.

24.20: From "Coccidiomycosis: An Update" by Austin Vaz et al., *Hospital Practice*, September 15, 1998, p. 106. Reprinted by permission.

24.21: Adapted from J.W. Smith and M.S. Bartlett, *Laboratory Medicine* 10:430–435, 1979; illustration by Gwen Gloege. Copyright © 1979 by the American Society of Clinical Pathologists. Reprinted with permission.

25.7, 25.8: Schaechter et al., *Mechanisms of Microbial Disease*, 2nd Ed. (Baltimore, MD: Williams & Wilkins, 1993), Figure 18.1, p. 269. © 1993 Williams & Wilkins.

25.12: From "Helicobacterpylori Infection," *Hospital Practice*, Vol. 26, No. 2, Figure 1, p. 45. Illustration by Laura Pardi Duprey. Reprinted by permission.

25.14: "Percent of populations with antibodies" from *Laboratory Management*, June 1987, p. 23. Reprinted by permission.

T26.1: Adapted from P.A. Gilly, "Vaginal discharge: Its causes and cures" *Postgraduate Medicine* 80(8):213, December 1986.

T27.1: Adapted from M. Alexander, *Introduction to Soil Microbiology*, 2nd Ed. New York: Wiley, 1991.

27.13: From *Water Supply and Sewerage*, 3rd Ed. Copyright © 1953. Reproduced with permission of The McGraw-Hill Companies.

T28.4: J.M. Jay, *Modern Food Microbiology*, 5th Ed. New York: Chapman and Hall, 1996.

Photo Credits

PO 1   ©Agricultural Research Service, USDA
CO 1   ©Meckes/Ottawa/Photo Researchers, Inc.
1.1a   ©CNRI/Photo Science Library/Photo Researchers, Inc.
1.1b   ©G.L. Barron, Univ. of Guelph/Biological Photo Service
1.1c   ©K.W. Jean/Visuals Unlimited
1.1d   ©Cabisco/Visuals Unlimited
1.1e   ©Lee D. Simon/Photo Researchers, Inc.
1.2a   Courtesy of Christine Case
1.3   ©Charles O'Rear/Westlight
1.4 (top)   ©Corbis-Bettmann Archive
1.4 (middle)   ©Corbis-Bettmann Archive
1.4 (bottom)   Courtesy of the Rockefeller Archive Center (on page)
1.05b   ©Michael M. Kliks, PhD., CTS Foundation
1.6   Science Source/Photo Researchers, Inc.
1.7a   ©Institut Pasteur/CNRI/Phototake, NYC
1.7b   Hank Morgan/Photo Researchers, Inc.
1.B1   ©Ken Karp

CO 2   Photo by Dennis Strete
2.Box1   ©1997 Ken Graham/Ken Graham Agency

CO 3   ©Karl Aufderheide/Visuals Unlimited
3.1a   Courtesy of Leica
3.4a   Photo by Dennis Strete
3.4b   ©David M. Phillips/Visuals Unlimited
3.4c   ©David M. Phillips/Visuals Unlimited
3.5   ©David M. Phillips/Visuals Unlimited
3.6b   Courtesy of the CDC
3.7   Courtesy of Technical Instrument San Francisco
3.8a   ©Phototake Electra/Phototake
3.8b   ©Karl Aufderheide/Visuals Unlimited
3.9a   ©1996 Digital Instruments
3.9b   ©1996 Digital Instruments
3.10b   ©Jack Bostrack
3.11   ©John Cunningham/Visuals Unlimited
3.12a   ©M. Abbey/Photo Researchers, Inc.
3.12b   ©G.W. Willis/Biological Photo Service
3.12c   ©Eric Graves/Photo Researchers, Inc.
3.Box1   Jeffrey Burnham
3.T2.1   Photo by Dennis Strete
3.T2.2   ©David M. Phillips/Visuals Unlimited
3.T2.3   ©David M. Phillips/Visuals Unlimited
3.T2.4   ©David M. Phillips/Visuals Unlimited
3.T2.5   Courtesy of the CDC
3.T2.6   Courtesy of Technical Instrument San Francisco
3.T2.7   ©Phototake Electra/Phototake
3.T2.8   ©Karl Aufderheide/Visuals Unlimited
3.T2.9   ©Digital Instruments
3.T2.10   ©Digital Instruments
3 End of chapter   ©Biophoto Associates/Science Source/Photo Researchers, Inc.

CO4   ©Dr. Dennis Kunkel/Phototake, NYC
4.1a.1   ©David M. Phillips/Visuals Unlimited
4.1a.2   Meckes/Ottawa/Photo Researchers, Inc.
4.1b   G. Shih-R. Kessel/Visuals Unlimited
4.1c   ©R. Kessel & G. Shih/Visuals Unlimited
4.1d   ©David Scharf/Peter Arnold, Inc.

4.2ab   ©Manfred Kage/Peter Arnold, Inc.
4.2c   ©Dr. Dennis Kunkel/Phototake, NYC
4.2d   ©CNRI/Phototake, NYC
4.3a   ©Veronika Burmeister/Visuals Unlimited
4.3b   ©Stanley Flegler/Visuals Unlimited
4.3c   ©Charles Stratton/Visuals Unlimited
4.4a   Horst Volker and Heinz Schlesner, Institute fur Allgemeine Mikrobiologie, Kiel
4.4b   Dr. H.W. Jannasch, Woods Hole Oceanographic Institute
4.5b   ©Ralph A. Slepecky/Visuals Unlimited
4.6a   ©Cabisco/Visuals Unlimited
4.6b   ©Ed Reschke/Peter Arnold, Inc.
4.6c   ©E.C.S. Chan/Visuals Unlimited
4.6d   Science Source/Photo Researchers, Inc.
4.8b   peritrichous flagella of picteus Lee D. Simon/Photo Researchers, Inc.
4.9a   ©SPL/Custom Medical Stock Photography
4.10   ©CNRI/SPL/Photo Researchers, Inc.
4.13a   ©Biological Photo Service
4.14   ©H.D. Pankratz and R.L. Uffin MSU/Biological Photo Service
4.15b   Courtesy of Christine Case
4.19   ©D. Balkwill, D. Maratea
4.20b   ©Dr. Kari Lounatmaa/Photo Researchers, Inc.
4.21b (left)   ©Biophoto Associates/Photo Researchers, Inc.
4.21b (right)   ©D.W. Fawcett/Photo Researchers, Inc.
4.22a   D.M. Phillips/Visuals Unlimited
4.22b   D.M. Phillips/Visuals Unlimited
4.23c   ©CNRI/SPL/Photo Researchers, Inc.
4.24b   ©R. Bolender, D. Fawcett/Photo Researchers, Inc.
4.25b   M. Powell/Visuals Unlimited
4.26b   ©Keith Porter/Photo Researchers, Inc.
4.27b   ©E.H. Newcomb and T.D. Pugh, University of Wisconsin, Biological Photo Service
4.Box a   N.H. Mendelson and J.J. Thwaites, ASM News 59 (1):27 (1993). Reprinted by permission
4.Box b   Courtesy of Christine Case
4.Box c   ©Patricia L. Grilione/Phototake NYC
4.Table 1   Courtesy of Christine Case
4.Table 2   Courtesy of Christine Case

CO5   Courtesy of Christine Case
5.3b   ©Science Visuals Unlimited-IBMUKOU
5.21   Courtesy of Christine Case
5.22   Courtesy of Christine Case
5.Box 1   ©Helen E. Carr/Biological Photo Service

CO6   Courtesy of Christine Case
6.6   ©Hank Morgan/Photo Researchers, Inc.
6.7   Photo by Dennis Strete
6.9a   Courtesy of Christine Case
6.9b   Courtesy of Christine Case
6.9c   Courtesy of Christine Case
6.9d   Courtesy of Christine Case
6.10b   Courtesy of Christine Case
6.11   Lee D. Simon/Photo Researchers, Inc.
6.17b   ©K. Taiaro/Visuals Unlimited
6.17a   Pall/Visuals Unlimited
6.Box 1   ©D. Foster/WHO1/Visuals Unlimited

CO 7   Pall/Visuals Unlimited
7.3   Courtesy of Christine Case
7.6   Courtesy of Christine Case
7.8   Courtesy of the CDC

CO 8   Dr. Gopal Murti/Science Photo Library/Photo Researchers, Inc.
8.1a   Dr. Gopal Murti/Science Photo Library/Photo Researchers, Inc.
8.7a   J. Cairns, Imperial Cancer Research Fund Laboratory, Mill Hill, London
8.11   Reprinted with permission from O.L. Miller Jr., B.A. Hamkalo, and C.A. Thomas Jr., *Science* 169 (24 July 1970):392 ©1970 by the American Association for the Advancement of Science
8.26   Dr. Dennis Kunkel/Phototake
8.29a   ©Gopal Murti/Photo Researchers, Inc.

CO 9   Secchi-Lecaque/Roussel-UCLAF/SPL/Photo Researchers, Inc.
9.5b   ©Dr. Jeremy Burgess/Science Photo Library/Photo Researchers, Inc.
9.6   ©Matt Meadows/Peter Arnold, Inc.
9.7   ©Institut Pasteur/Phototake, NYC
9.10   ©Matt Meadows/Peter Arnold, Inc.
9.13   Secchi-Lecaque/Roussel-UCLAF/SPL/Photo Researchers, Inc.
9.15   Courtesy of Cellmark Diagnostics
9.16   ©Jack M. Bostrack/Visuals Unlimited
9. End of chapter   Courtesy of Christine Case

PO 2   ©Agricultural Research Service, USDA
CO 10   ©A.B. Dowsett/SPL/Photo Researchers, Inc.
10.3   ©J. Pickett-Heaps/Photo Researchers, Inc.
10.4a   ©M.D. Maset/Visuals Unlimited
10.4b   ©S.M. Awramik, University of California/Biological Photo Service
10.10   Courtesy of Christine Case
10.11a   Courtesy of the CDC
10.11b   Courtesy of the CDC
10.12   Courtesy of the CDC
10.13   Courtesy of the Microbial Diseases Laboratory, Berkeley, CA
10.14   Courtesy of Christine Case
10.B01   Courtesy of Marine World/Africa USA, Vallejo, CA. Photo by Darryl W. Bush
10.T01.01   ©Ralph Robinson/Visuals Unlimited
10.T01.02   ©A.B. Dowsett/Science Source/Photo Researchers, Inc.
10.T01.03   ©Eugene McArdle/Custom Medical Stock Photo

CO 11   ©Frederick P. Mertz
11.1b   ©Dr. Bruce Paster/Forsythe Dental Center
11.1a   Bergey's Manual Trust, Michigan State University
11.1b   Bergey's Manual Trust, Michigan State University
11.2   ©Biology Media/Science Source/Photo Researchers, Inc.
11.3a   ©R. Calentine/Visuals Unlimited
11.3   ©Biological Photo Service
11.4   ©John D. Cunningham/Visuals Unlimited
11.5   ©Chris Bjornberg/Photo Researchers, Inc.

Index

NOTE: A *t* following a page number indicates tabular material, an *f* following a page number indicates a figure, and a *b* following a page number indicates a boxed feature. Drugs are listed under their generic names. When a trade name is listed, the reader is referred to the generic name.

Dear Student:

As the authors of *Microbiology: An Introduction,* **Seventh Edition Update,** we set out to write a book that explains core microbiology topics in a way that is easy for students to understand and remember. We would appreciate hearing about your experiences with this textbook and its multimedia support, and we invite your suggestions for improvements!

Many thanks,

Gerard J. Tortora   Bert Funke   Christine Case

1. Did you use the Student Tutorial CD-ROM? ☐ Yes ☐ No
If so, which topics were most useful to you? Are there topics that should be added? How can we improve the Student Tutorial CD-ROM? _____

_____

_____

_____

_____

2. Did you use The Microbiology Place? ☐ Yes ☐ No
How was your user experience? How can we improve The Microbiology Place?

_____

_____

_____

_____

3. Did you use the VirtualUnknown CD-ROM? ☐ Yes ☐ No
How was your user experience? How can we improve the VirtualUnknown CD-ROM?

_____

_____

_____

_____

4. Which study tool(s) – e.g. Learning Objectives, Study Outlines, Study Questions – did you find most useful?

_____

_____

_____

_____

5. What did you like most about ***Microbiology: An Introduction,*** **Seventh Edition Update?** Please provide three examples, in order of priority. ______________________________

____

____

____

____

6. Do you have any ideas about how this book could be improved? Please write your specific suggestions, citing page numbers if appropriate. ______________________________

____

____

____

____

NO POSTAGE NECESSARY IF MAILED IN THE UNITED STATES

BUSINESS REPLY MAIL

FIRST-CLASS MAIL PERMIT NO. 275 SAN FRANCISCO CA

POSTAGE WILL BE PAID BY ADDRESSEE

BENJAMIN CUMMINGS
PEARSON EDUCATION
1301 SANSOME STREET
SAN FRANCISCO CA 94111-9328

School: ______________________________

Instructor's Name: ______________________________

Optional:

Your name: ______________________________

Email: ______________________________

Date: ______________________________

May Benjamin Cummings have permission to quote your comments in promotions for ***Microbiology: An Introduction?*** ☐ Yes ☐ No

meta– beyond, between, transition. Example: metabolism, chemical changes occurring within a living organism.

micro– smallness. Example: microscope, an instrument used to make small objects appear larger.

–mnesia memory. Examples: amnesia, loss of memory; anamnesia, return of memory.

molli– soft. Example: Mollicutes, a class of wall-less eubacteria.

–monas a unit. Example: *Methylomonas,* a unit (bacterium) that utilizes methane as its carbon source.

mono– singleness. Example: monotrichous, having one flagellum.

morpho– form. Example: morphology, the study of the form and structure of organisms.

multi– many. Example: multinuclear, having several nuclei.

mur– wall. Example: murein, a component of bacterial cell walls.

mus-, muri– mouse. Example: murine typhus, a form of typhus endemic in mice.

mut– to change. Example: mutation, a sudden change in characteristics.

myco-, –mycetoma, –myces a fungus. Example: *Saccharomyces,* sugar fungus, a genus of yeast.

myxo– slime, mucus. Example: Myxobacteriales, an order of slime-producing bacteria.

necro– a corpse. Example: necrosis, cell death or death of a portion of tissue.

–nema a thread. Example: *Treponema* has long, threadlike cells.

nigr– black. Example: *Aspergillus niger,* a fungus that produces black conidia.

ob– before, against. Example: obstruction, impeding or blocking up.

oculo– eye. Example: monocular, pertaining to one eye.

–oecium, –ecium a house. Examples: perithecium, an ascus with an opening that encloses spores; ecology, the study of the relationships among organisms and between an organism and its environment (household).

–oid like, resembling. Example: coccoid, resembling a coccus.

oligo– small, few. Example: oligiosaccharide, a carbohydrate composed of a few (7–10) monosaccharides.

–oma tumor. Example: lymphoma, a tumor of the lymphatic tissues.

–ont being, existing. Example: schizont, a cell existing as a result of schizogony.

ortho– straight, direct. Example: orthomyxovirus, a virus with a straight, tubular capsid.

–osis, –sis condition of. Examples: lysis, the condition of loosening; symbiosis, the condition of living together.

pan– all, universal. Example: pandemic, an epidemic affecting a large region.

para– beside, near. Example: parasite, an organism that "feeds beside" another.

peri– around. Example: peritrichous, projections from all sides.

phaeo– brown. Example: Phaeophyta, brown algae.

phago– eat. Example: phagocyte, a cell that engulfs and digests particles or cells.

philo-, –phil liking, preferring. Example: thermophile, an organism that prefers high temperatures.

–phore bears, carries. Example: conidiophore, a hypha that bears conidia.

–phyll leaf. Example: chlorophyll, the green pigment in leaves.

–phyte plant. Example: saprophyte, a plant that obtains nutrients from decomposing organic matter.

pil– a hair. Example: pilus, a hairlike projection from a cell.

plankto– wandering, roaming. Example: plankton, organisms drifting or wandering in water.

plast– formed. Example: plastid, a formed body within a cell.

–pnoea, –pnea breathing. Example: dyspnea, difficulty in breathing.

pod– foot. Example: pseudopod, a footlike structure.

poly– many. Example: polymorphism, many forms.

post– after, behind. Example: posterior, a place behind a (specific) part.

pre-, pro– before, ahead of. Examples: prokaryote, a cell with the first nucleus; pregnant, before birth.

pseudo– false. Example: pseudopod, false foot.

psychro– cold. Example: psychrophile, an organism that grows best at low temperatures.

–ptera wing. Example: Diptera, the order of true flies, insects with two wings.

pyo– pus. Example: pyogenic, pus-forming.

rhabdo– stick, rod. Example: rhabdovirus, an elongated, bullet-shaped virus.

rhin– nose. Example: rhinitis, inflammation of mucous membranes in the nose.

rhizo– root. Examples: *Rhizobium,* a bacterium that grows in plant roots; mycorrhiza, a fungus that grows in or on plant roots.

rhodo– red. Example: *Rhodospirillum,* a red-pigmented, spiral-shaped bacterium.

rod– gnaws. Example: rodents, the class of mammals with gnawing teeth.

rubri– red. Example: *Clostridiium rubrum,* red-pigmented colonies.

rumin– throat. Example: *Ruminococcus,* a bacterium associated with a rumen (modified esophagus).

saccharo– sugar. Example: disaccharide, a sugar consisting of two simple sugars.

sapr– rotten. Example: *Saprolegnia,* a fungus that lives on dead animals.

sarco– flesh. Example: sarcoma, a tumor of muscle or connective tissues.

schizo– split. Example: schizomycetes, organisms that reproduce by splitting and an early name for bacteria.

scolec– worm. Example: scolex, the head of a tapeworm.

–scope, –scopic watcher. Example: microscope, an instrument used to watch small things.

semi– half. Example: semicircular, having the form of half a circle.

sept– rotting. Example: septic, presence of bacteria that could cause decomposition.

septo– partition. Example: septum, a cross-wall in a fungal hypha.

serr– notched. Example: serrate, with a notched edge.

sidero– iron. Example: *Siderococcus,* a bacterium capable of oxidizing iron.

siphon– tube. Example: Siphonaptera, the order of fleas, insects with tubular mouths.

soma– body. Example: somatic cells, cells of the body other than gametes.

speci– particular things. Examples: species, the smallest group of organisms with similar properties; specify, to indicate exactly.

spiro– coil. Example: spirochete, a bacterium with a coiled cell.

sporo– spore. Example: sporangium, a structure that holds spores.

staphylo– grapelike cluster. Example: *Staphylococcus,* a bacterium that forms clusters of cells.

–stasis arrest, fixation. Example: bacteriostasis, cessation of bacterial growth.

strepto– twisted. Example: *Streptococcus,* a bacterium that forms twisted chains of cells.

sub– beneath, under. Example: subcutaneous, just under the skin.

super– above, upon. Example: superior, the quality or state of being above others.

sym–, syn– together, with. Examples: synapse, the region of communication between two neurons; synthesis, putting together.

–taxi to touch. Example: chemotaxis, response to the presence (touch) of chemicals.

taxis– orderly arrangement. Example: taxonomy, the science dealing with arranging organisms into groups.

tener– tender. Example: Tenericutes, the phylum containing wall-less eubacteria.

thallo– plant body. Example: thallus, an entire macroscopic fungus.

therm– heat. Example: *Thermus,* a bacterium that grows in hot springs (to 75°C).

thio– sulfur. Example: *Thiobacillus,* a bacterium capable of oxidizing sulfur-containing compounds.

–thrix See **trich–.**

–tome, –tomy to cut. Example: appendectomy, surgical removal of the appendix.

–tone, –tonic strength. Example: hypotonic, having less strength (osmotic pressure).

tox– poison. Example: antitoxin, effective against poison.

trans– across, through. Example: transport, movement of substances.

tri– three. Example: trimester, three-month period.

trich– a hair. Example: peritrichous, hairlike projections from cells.

–trope turning. Example: geotropic, turning toward the Earth (pull of gravity).

–troph food, nourishment. Example: trophic, pertaining to nutrition.

–ty condition of, state. Example: immunity, the condition of being resistant to disease or infection.

undul– wavy. Example: undulating, rising and falling, presenting a wavy appearance.

uni– one. Example: unicellular, pertaining to one cell.

vaccin– cow. Example: vaccination, injection of a vaccine (originally pertained to cows).

vacu– empty. Example: vacuoles, an intracellular space that appears empty.

vesic– bladder. Example: vesicle, a bubble.

vitr– glass. Example: in vitro, in culture media in a glass (or plastic) container.

–vorous eat. Example: carnivore, an animal that eats other animals.

xantho– yellow. Example: *Xanthomonas,* produces yellow colonies.

xeno– strange. Example: axenic, sterile, free of strange organisms.

xero– dry. Example: xerophyte, any plant that tolerates dry conditions.

xylo– wood. Example: xylose, a sugar obtained from wood.

zoo– animal. Example: zoology, the study of animals.

zygo– yoke, joining. Example: zygospore, a spore formed from the fusion of two cells.

–zyme ferment. Example: enzyme, any protein in living cells that catalyzes chemical reactions.